Primer on the Metabolic Bone Diseases and Disorders of Mineral Metabolism

Eighth Edition

An Official Publication of the American Society for Bone and Mineral Research

Primer on the Metabolic Bone Diseases and Disorders of Mineral Metabolism

Eighth Edition

Editor-in-Chief

Clifford J. Rosen, MD

Senior Associate Editors
Roger Bouillon, MD, PhD, FRCP
Juliet E. Compston, MD, FRCP, FRCPath, FMedSci
Vicki Rosen, PhD

Associate Editors
Douglas C. Bauer, MD
Marie Demay, MD
Theresa A. Guise, MD
Suzanne M. Jan de Beur, MD
Richard W. Keen, MD, PhD
Karen M. Lyons, PhD
Laurie K. McCauley, DDS, MS, PhD
Paul D. Miller, MD, FACP
Socrates E. Papapoulos, MD, PhD
Ego Seeman, BSc, MBBS, FRACP, MD
Rajesh V. Thakker, MD, ScD, FRCP, FRCPath, FMedSci
Mone Zaidi, MBBS, PhD

A John Wiley & Sons, Inc., Publication

This edition first published 2013 by John Wiley & Sons, Inc. © 2013 by American Society for Bone and Mineral Research
First Edition published by American Society for Bone and Mineral Research © 1990 American Society for Bone and Mineral Research
Second Edition published by Raven Press, Ltd. © 1993 American Society for Bone and Mineral Research
Third Edition published by Lippincott-Raven Publishers © 1996 American Society for Bone and Mineral Research
Fourth Edition published by Lippincott Williams & Wilkins © 1999 American Society for Bone and Mineral Research
Fifth Edition published by American Society for Bone and Mineral Research © 2003 American Society for Bone and Mineral Research
Sixth Edition published by American Society for Bone and Mineral Research © 2006 American Society for Bone and Mineral Research
Seventh Edition published by American Society for Bone and Mineral Research © 2008 American Society for Bone and Mineral Research

Editorial offices: 1606 Golden Aspen Drive, Suites 103 and 104, Ames, Iowa 50010, USA
The Atrium, Southern Gate, Chichester, West Sussex, PO19 8SQ, UK
9600 Garsington Road, Oxford, OX4 2DQ, UK

For details of our global editorial offices, for customer services, and for information about how to apply for permission to reuse the copyright material in this book please see our website at www.wiley.com/wiley-blackwell.

Authorization to photocopy items for internal or personal use, or the internal or personal use of specific clients, is granted by Blackwell Publishing, provided that the base fee is paid directly to the Copyright Clearance Center, 222 Rosewood Drive, Danvers, MA 01923. For those organizations that have been granted a photocopy license by CCC, a separate system of payments has been arranged. The fee codes for users of the Transactional Reporting Service are ISBN-13: 978-1-1184-5388-9/2013.

Designations used by companies to distinguish their products are often claimed as trademarks. All brand names and product names used in this book are trade names, service marks, trademarks or registered trademarks of their respective owners. The publisher is not associated with any product or vendor mentioned in this book.

Limit of Liability/Disclaimer of Warranty: While the publisher and author(s) have used their best efforts in preparing this book, they make no representations or warranties with respect to the accuracy or completeness of the contents of this book and specifically disclaim any implied warranties of merchantability or fitness for a particular purpose. It is sold on the understanding that the publisher is not engaged in rendering professional services and neither the publisher nor the author shall be liable for damages arising herefrom. If professional advice or other expert assistance is required, the services of a competent professional should be sought.

The American Society for Bone and Mineral Research, the Primer on the Metabolic Bone Diseases and Disorders of Mineral Metabolism, the author, editors, and John Wiley & Sons, Inc. make no representations or warranties with respect to the accuracy or completeness of the contents of this work and specifically disclaim all warranties, including without limitation any implied warranties of fitness for a particular purpose. The advice and strategies contained herein may not be suitable for every situation. In view of ongoing research, equipment modifications, changes in governmental regulations, and the constant flow of information relating to the use of medicines, equipment, and devices, the reader is urged to review and evaluate the information provided in the package insert or instructions for each medicine, equipment, or device for, among other things, any changes in the instructions or indication of usage and for added warnings and precautions. Readers should consult with a specialist where appropriate. No warranty may be created or extended by any promotional statements for this work. Neither the American Society for Bone and Mineral Research, the Primer on the Metabolic Bone Diseases and Disorders of Mineral Metabolism, the author, editors, nor John Wiley & Sons, Inc. shall be liable for any damages arising herefrom.

Library of Congress Cataloging-in-Publication Data

Primer on the metabolic bone diseases and disorders of mineral metabolism. – 8th ed. / editor-in-chief, Clifford J. Rosen; senior associate editors, Roger Bouillon, Juliet E. Compston, Vicki Rosen; associate editors, Douglas C. Bauer ... [et al.].
   p. ; cm.
   Previous editions published by American Society for Bone and Mineral Research.
   Includes bibliographical references and index.
   ISBN 978-1-118-45388-9 (softback : alk. paper) – ISBN 978-1-118-45389-6 (eMobi) – ISBN 978-1-118-45390-2 (ePDF) – ISBN 978-1-118-45391-9 (ePub) – ISBN 978-1-118-45392-6
   I. Rosen, Clifford J. II. American Society for Bone and Mineral Research.
   [DNLM: 1. Bone Diseases, Metabolic. 2. Bone and Bones–metabolism. 3. Minerals–metabolism. WE 250]
   RC931.M45
   616.7'16–dc23
                                                        2013002831

A catalogue record for this book is available from the British Library.

Wiley also publishes its books in a variety of electronic formats. Some content that appears in print may not be available in electronic books.

Cover image: Copyright Tobi Kahn, NEBILA, 1990, Acrylic on canvas over wood, 70 × 24 inches.
Cover design by Nicole Teut

Set in 9.5/11 pt Trump Mediaeval by Toppan Best-set Premedia Limited
Printed and bound in Singapore by Markono Print Media Pte Ltd

1 2013

Contents

Contributors **x**
Primer Corporate Sponsors **xxii**
Preface to the Eighth Edition of the *Primer*:
Clifford J. Rosen **xxiii**
About ASBMR **xxiv**
President's Preface: *Lynda F. Bonewald* **xxv**
About the Companion Website **xxvi**

Section I: Molecular, Cellular, and Genetic Determinants of Bone Structure and Formation **1**
Section Editor Karen M. Lyons

1. Skeletal Morphogenesis and Embryonic Development **3**
   Yingzi Yang

2. Signal Transduction Cascades Controlling Osteoblast Differentiation **15**
   David J.J. de Gorter and Peter ten Dijke

3. Osteoclast Biology and Bone Resorption **25**
   F. Patrick Ross

4. Osteocytes **34**
   Lynda F. Bonewald

5. Connective Tissue Pathways That Regulate Growth Factors **42**
   Gerhard Sengle and Lynn Y. Sakai

6. The Composition of Bone **49**
   Adele L. Boskey and Pamela Gehron Robey

7. Assessment of Bone Mass and Microarchitecture in Rodents **59**
   Blaine A. Christiansen

8. Animal Models: Genetic Manipulation **69**
   Karen M. Lyons

9. Animal Models: Allelic Determinants for BMD **76**
   Robert D. Blank

10. Neuronal Regulation of Bone Remodeling **82**
    Florent Elefteriou and Gerard Karsenty

11. Skeletal Healing **90**
    Michael J. Zuscik

12. Biomechanics of Fracture Healing **99**
    Elise F. Morgan and Thomas A. Einhorn

13. Human Genome-Wide Association (GWA) Studies **106**
    Douglas P. Kiel

14. Circulating Osteogenic Cells **111**
    Robert J. Pignolo and Moustapha Kassem

Section II: Skeletal Physiology **119**
Section Editor Ego Seeman

15. Human Fetal and Neonatal Bone Development **121**
    Tao Yang, Monica Grover, Kyu Sang Joeng, and Brendan Lee

16. Skeletal Growth and Peak Bone Strength **127**
    Qingju Wang and Ego Seeman

17. Ethnic Differences in Bone Acquisition **135**
    Shane A. Norris, Lisa K. Micklesfield, and John M. Pettifor

18. Calcium and Other Nutrients During Growth **142**
    Tania Winzenberg and Graeme Jones

19. Growing a Healthy Skeleton: The Importance of Mechanical Loading **149**
    Mark R. Forwood

20. Pregnancy and Lactation **156**
    Christopher S. Kovacs and Henry M. Kronenberg

21 Menopause 165
Ian R. Reid

Section III: Mineral Homeostasis 171
Section Editor Ego Seeman

22 Regulation of Calcium and Magnesium 173
Murray J. Favus and David Goltzman

23 Fetal Calcium Metabolism 180
Christopher S. Kovacs

24 Fibroblast Growth Factor-23 (FGF23) 188
Kenneth E. White and Michael J. Econs

25 Gonadal Steroids 195
Stavros C. Manolagas, Maria Almeida, and Robert L. Jilka

26 Parathyroid Hormone 208
Robert A. Nissenson and Harald Jüppner

27 Parathyroid Hormone-Related Protein 215
John J. Wysolmerski

28 Ca^{2+}-Sensing Receptor 224
Edward M. Brown

29 Vitamin D: Production, Metabolism, Mechanism of Action, and Clinical Requirements 235
Daniel Bikle, John S. Adams, and Sylvia Christakos

Section IV: Investigation of Metabolic Bone Diseases 249
Section Editor Douglas C. Bauer

30 DXA in Adults and Children 251
Glen Blake, Judith E. Adams, and Nick Bishop

31 Quantitative Computed Tomography in Children and Adults 264
C.C. Glüer

32 Magnetic Resonance Imaging of Bone 277
Sharmila Majumdar

33 Radionuclide Scintigraphy in Metabolic Bone Disease 283
Gary J.R. Cook, Gopinath Gnanasegaran, and Ignac Fogelman

34 FRAX®: Assessment of Fracture Risk 289
John A Kanis

35 Biochemical Markers of Bone Turnover in Osteoporosis 297
Pawel Szulc, Douglas C. Bauer, and Richard Eastell

36 Bone Biopsy and Histomorphometry in Clinical Practice 307
Robert R. Recker

37 Diagnosis and Classification of Vertebral Fracture 317
James F. Griffith, Judith E. Adams, and Harry K. Genant

38 Approaches to Genetic Testing 336
Christina Jacobsen, Yiping Shen, and Ingrid A. Holm

Section V: Osteoporosis 343
Section Editors Paul D. Miller and Socrates E. Papapoulos

39 Osteoporosis Overview 345
Michael Kleerekoper

40 The Epidemiology of Osteoporotic Fractures 348
Nicholas Harvey, Elaine Dennison, and Cyrus Cooper

41 Overview of Pathogenesis 357
Ian R. Reid

42 Nutrition and Osteoporosis 361
Connie M. Weaver and Robert P. Heaney

43 The Role of Sex Steroids in the Pathogenesis of Osteoporosis 367
Matthew T. Drake and Sundeep Khosla

44 Translational Genetics of Osteoporosis: From Population Association to Individualized Prognosis 376
Bich H. Tran, Jacqueline R. Center, and Tuan V. Nguyen

45 Prevention of Falls 389
Heike A. Bischoff-Ferrari

46 Exercise and the Prevention of Osteoporosis 396
Clinton T. Rubin, Janet Rubin, and Stefan Judex

47 Calcium and Vitamin D 403
    Bess Dawson-Hughes

48 Estrogens, Estrogen Agonists/Antagonists,
    and Calcitonin 408
    Nelson B. Watts

49 Bisphosphonates for Postmenopausal
    Osteoporosis 412
    Socrates E. Papapoulos

50 Denosumab 420
    Michael R. McClung

51 Parathyroid Hormone Treatment
    for Osteoporosis 428
    Felicia Cosman and Susan L. Greenspan

52 Strontium Ranelate in the Prevention
    of Osteoporotic Fractures 437
    René Rizzoli

53 Combination Anabolic and Antiresorptive
    Therapy for Osteoporosis 444
    John P. Bilezikian and Natalie E. Cusano

54 Compliance and Persistence with Osteoporosis
    Medications 448
    Deborah T. Gold

55 Cost-Effectiveness of Osteoporosis
    Treatment 455
    Anna N.A. Tosteson

56 Future Therapies of Osteoporosis 461
    Kong Wah Ng and T. John Martin

57 Juvenile Osteoporosis 468
    Nick Bishop and Francis H. Glorieux

58 Glucocorticoid-Induced Bone Disease 473
    Robert S. Weinstein

59 Inflammation-Induced Bone Loss
    in the Rheumatic Diseases 482
    Steven R. Goldring

60 Secondary Osteoporosis: Other Causes 489
    Neveen A.T. Hamdy

61 Transplantation Osteoporosis 495
    Peter R. Ebeling

62 Osteoporosis in Men 508
    Eric S. Orwoll

63 Premenopausal Osteoporosis 514
    Adi Cohen and Elizabeth Shane

64 Skeletal Effects of Drugs 520
    Juliet Compston

65 Orthopedic Surgical Principles
    of Fracture Management 527
    Manoj Ramachandran and David G. Little

66 Abnormalities in Bone and Calcium Metabolism
    After Burns 531
    Gordon L. Klein

Section VI: Disorders of Mineral Homeostasis 535
Section Editors Marie Demay and Suzanne M. Jan de Beur

67 Approach to Parathyroid Disorders 537
    John P. Bilezikian

68 Primary Hyperparathyroidism 543
    Shonni J. Silverberg

69 Familial Primary Hyperparathyroidism
    (Including MEN, FHH, and HPT-JT) 553
    Andrew Arnold and Stephen J. Marx

70 Non-Parathyroid Hypercalcemia 562
    Mara J. Horwitz, Steven P. Hodak,
    and Andrew F. Stewart

71 Hypocalcemia: Definition, Etiology, Pathogenesis,
    Diagnosis, and Management 572
    Anne L. Schafer and Dolores Shoback

72 Hypoparathyroidism and
    Pseudohypoparathyroidism 579
    Mishaela R. Rubin and Michael A. Levine

73 Pseudohypoparathyroidism 590
    Harald Jüppner and Murat Bastepe

74 Disorders of Phosphate Homeostasis 601
    Mary D. Ruppe and Suzanne M. Jan de Beur

75 Vitamin D–Related Disorders 613
    Paul Lips, Natasja M. van Schoor,
    and Nathalie Bravenboer

76 Vitamin D Insufficiency and Deficiency **624**
J. Christopher Gallagher

77 Pathophysiology of Chronic Kidney Disease Mineral Bone Disorder (CKD–MBD) **632**
Keith A. Hruska and Michael Seifert

78 Treatment of Chronic Kidney Disease Mineral Bone Disorder (CKD–MBD) **640**
Hala M. Alshayeb and L. Darryl Quarles

79 Disorders of Mineral Metabolism in Childhood **651**
Thomas O. Carpenter

80 Paget's Disease of Bone **659**
Ethel S. Siris and G. David Roodman

Section VII: Cancer and Bone 669
Section Editor Theresa A. Guise

81 Overview of Mechanisms in Cancer Metastases to Bone **671**
Gregory A. Clines

82 Clinical and Preclinical Imaging in Osseous Metastatic Disease **677**
Geertje van der Horst and Gabri van der Pluijm

83 Metastatic Solid Tumors to Bone **686**
Rachelle W. Johnson and Julie A. Sterling

84 Hematologic Malignancies and Bone **694**
Rebecca Silbermann and G. David Roodman

85 Osteogenic Osteosarcoma **702**
Jianning Tao, Yangjin Bae, Lisa L. Wang, and Brendan Lee

86 Skeletal Complications of Breast and Prostate Cancer Therapies **711**
Catherine Van Poznak and Pamela Taxel

87 Bone Cancer and Pain **720**
Patrick W. O'Donnell and Denis R. Clohisy

88 Radiation Therapy-Induced Osteoporosis **728**
Jeffrey S. Willey, Shane A.J. Lloyd, and Ted A. Bateman

89 Skeletal Complications of Childhood Cancer **734**
Ingrid A. Holm

90 Treatment and Prevention of Bone Metastases and Myeloma Bone Disease **741**
Jean-Jacques Body

91 Radiotherapy of Skeletal Metastases **754**
Edward Chow, Luluel M. Khan, and Øyvind S. Bruland

92 Concepts and Surgical Treatment of Metastatic Bone Disease **760**
Kristy Weber and Scott L. Kominsky

Section VIII: Sclerosing and Dysplastic Bone Diseases 767
Section Editor Richard W. Keen

93 Sclerosing Bone Disorders **769**
Michael P. Whyte

94 Fibrous Dysplasia **786**
Michael T. Collins, Mara Riminucci, and Paolo Bianco

95 Osteochondrodysplasias **794**
Yasemin Alanay and David L. Rimoin

96 Ischemic and Infiltrative Disorders **805**
Richard W. Keen

97 Tumoral Calcinosis—Dermatomyositis **810**
Nicholas Shaw

98 Fibrodysplasia Ossificans Progressiva **815**
Frederick S. Kaplan, Robert J. Pignolo, and Eileen M. Shore

99 Osteogenesis Imperfecta **822**
Joan C. Marini

100 Skeletal Manifestations in Marfan Syndrome and Related Disorders of the Connective Tissue **830**
Emilio Arteaga-Solis and Francesco Ramirez

101 Enzyme Defects and the Skeleton **838**
Michael P. Whyte

Section IX: Approach to Nephrolithiasis 843
Section Editor Rajesh V. Thakker

102 Renal Tubular Physiology of Calcium Excretion **845**
Peter A. Friedman and David A. Bushinsky

103 Epidemiology of Nephrolithiasis 856
Murray J. Favus

104 Diagnosis and Evaluation of Nephrolithiasis 860
Stephen J. Knohl and Steven J. Scheinman

105 Kidney Stones in the Pediatric Patient 869
Amy E. Bobrowski and Craig B. Langman

106 Treatment of Renal Stones 878
John R. Asplin

107 Genetic Basis of Renal Stones 884
Rajesh V. Thakker

Section X: Oral and Maxillofacial Biology and Pathology 893
Section Editor Laurie K. McCauley

108 Development of the Craniofacial Skeleton 895
Maiko Matsui and John Klingensmith

109 Development and Structure of Teeth and Periodontal Tissues 904
Petros Papagerakis and Thimios Mitsiadis

110 Craniofacial Disorders Affecting the Dentition: Genetic 914
Yong-Hee P. Chun, Paul H. Krebsbach, and James P. Simmer

111 Pathology of the Hard Tissues of the Jaws 922
Paul C. Edwards

112 Bisphosphonate-Associated Osteonecrosis of the Jaws 929
Hani H. Mawardi, Nathaniel S. Treister, and Sook-Bin Woo

113 Periodontal Diseases and Oral Bone Loss 941
Mary G. Lee and Keith L. Kirkwood

114 Oral Manifestations of Metabolic Bone Diseases 948
Roberto Civitelli and Charles Hildebolt

Section XI: The Skeleton and Its Integration with Other Tissues 959
Section Editor Mone Zaidi

115 Central Neuronal Control of Bone Remodeling 961
Shu Takeda and Paul Baldock

116 The Pituitary–Bone Connection 969
Mone Zaidi, Tony Yuen, Li Sun, Terry F. Davies, Alberta Zallone, and Harry C. Blair

117 Skeletal Muscle Effects on the Skeleton 978
William J. Evans

118 Glucose Control and Integration by the Skeleton 986
Patricia Ducy and Gerard Karsenty

119 Obesity and Skeletal Mass 993
Sue Shapses and Deeptha Sukumar

120 Neuropsychiatric Disorders and the Skeleton 1002
Itai Bab and Raz Yirmiya

121 Vascular Disease and the Skeleton 1012
Dwight A. Towler

122 Spinal Cord Injury: Skeletal Pathophysiology and Clinical Issues 1018
William A. Bauman and Christopher P. Cardozo

123 Hematopoiesis and Bone 1028
Benjamin J. Frisch and Laura M. Calvi

124 Bone and Immune Cell Interactions 1036
Brendan F. Boyce

Index 1043

Contributors

John S. Adams, MD
Departments of Orthopaedic Surgery, Medicine and Molecular Cell and Developmental Biology
University of California-Los Angeles
Los Angeles, California, USA

Judith E. Adams, MD
Department of Clinical Radiology
Manchester Academic Health Science Centre
Central Manchester University Hospitals NHS Foundation Trust
The Royal Infirmary
Manchester, UK

Yasemin Alanay, MD, PhD
Pediatric Genetics Unit
Department of Pediatrics
Acibadem University School of Medicine
Istanbul, Turkey

Maria Almeida
Division of Endocrinology and Metabolism
Center for Osteoporosis and Metabolic Bone Diseases
University of Arkansas for Medical Sciences
and the Central Arkansas Veterans Healthcare System
Little Rock, Arkansas, USA

Hala M. Alshayeb, MD
Division of Nephrology
Department of Medicine
Hashemite University
Zarqa, Jordan

Andrew Arnold, MD
Center for Molecular Medicine and Division of Endocrinology and Metabolism
University of Connecticut School of Medicine
Farmington, Connecticut, USA

Emilio Arteaga-Solis, MD, PhD
Pediatrics Pulmonary Division
Department of Pediatrics
Columbia University College of Physician and Surgeons
New York, New York, USA

John R. Asplin, MD, FASN
Litholink Corporation
and Department of Medicine
University of Chicago Pritzker School of Medicine
Chicago, Illinois, USA

Itai Bab
Bone Laboratory
The Hebrew University of Jerusalem
Jerusalem, Israel

Yangjin Bae
Departments of Molecular and Human Genetics
Baylor College of Medicine
Texas Children's Hospital
Houston, Texas, USA

Paul Baldock, PhD
Bone and Mineral Research Program
Garvan Institute of Medical Research
St. Vincent's Hospital
Sydney, Australia

Murat Bastepe, MD, PhD
Endocrine Unit
Department of Medicine
Massachusetts General Hospital
Harvard Medical School
Boston, Massachusetts, USA

Ted A. Bateman
Departments of Biomedical Engineering and Radiation Oncology
University of North Carolina
Chapel Hill, North Carolina, USA

Douglas C. Bauer, MD
Departments of Medicine and Epidemiology and Biostatistics
University of California, San Francisco
San Francisco, California, USA

William A. Bauman, MD
Department of Veterans Affairs Rehabilitation Research and Development Service
National Center of Excellence for the Medical Consequences of Spinal Cord Injury
and Medical Service
James J. Peters Veterans Affairs Medical Center
Bronx, New York, USA
Departments of Medicine and Rehabilitation Medicine
The Mount Sinai School of Medicine
New York, New York, USA

Paolo Bianco
Dipartimento di Medicina Molecolare
Universita' La Sapienza
Rome, Italy

Daniel Bikle, MD, PhD
Department of Medicine
Division of Endocrinology
VA Medical Center and
University of California, San Francisco
San Francisco, California, USA

John P. Bilezikian
Division of Endocrinology
Department of Medicine
College of Physicians and Surgeons
Columbia University
New York, New York, USA

Heike A. Bischoff-Ferrari, MD, DrPH
Centre on Aging and Mobility
University of Zurich, Switzerland
Jean Mayer USDA Human Nutrition
Research Center on Aging
Tufts University
Boston, Massachusetts, USA

Nick Bishop, MB, ChB, MRCP, MD, FRCPCH
Academic Unit of Child Health
Department of Human Metabolism
University of Sheffield
Sheffield Children's Hospital
Sheffield, UK

Harry C. Blair
The Pittsburgh VA Medical Center
and Departments of Pathology and of Cell Biology
University of Pittsburgh School of Medicine
Pittsburgh, Pennsylvania, USA

Glen Blake, PhD
Osteoporosis Research Unit
King's College London
Guy's Campus
London, UK

Robert D. Blank, MD, PhD
Division of Endocrinology
Department of Medicine
Medical College of Wisconsin
Milwaukee, Wisconsin, USA

Amy E. Bobrowski, MD, MSCI
Feinberg School of Medicine
Northwestern University
Chicago, Illinois, USA

Jean-Jacques Body, MD, PhD
University Hospital Brugmann
Université Libre de Bruxelles (U.L.B.) Internal Medicine
Brussels, Belgium

Lynda F. Bonewald, PhD
Department of Oral Biology
University of Missouri at Kansas City School of Dentistry
Kansas City, Missouri, USA

Adele L. Boskey, PhD
Research Division
Hospital for Special Surgery
and Department of Biochemistry and Graduate Field of Physiology, Biophysics, and Systems Biology
Cornell University Medical and Graduate Medical Schools
New York, New York, USA

Roger Bouillon, MD, PhD, FRCP
Department of Endocrinology
KU Leuven
Gasthuisberg, Belgium

Brendan F. Boyce, MBChB
Department of Pathology and Laboratory Medicine and
The Center for Musculoskeletal Research
University of Rochester Medical Center
Rochester, New York, USA

Nathalie Bravenboer, PhD
Department of Clinical Chemistry
VU University Medical Center
Amsterdam, The Netherlands

Edward M. Brown, MD
Division of Endocrinology, Diabetes and Hypertension
Brigham and Women's Hospital and Harvard Medical School
Boston, Massachusetts, USA

Øyvind S. Bruland, MD, PhD
Institute of Clinical Medicine, University of Oslo
Department of Oncology, The Norwegian Radium Hospital
Oslo University Hospital
Oslo, Norway

David A. Bushinsky, MD
Nephrology Division
University of Rochester School of Medicine and Dentistry
Rochester, New York, USA

Laura M. Calvi, MD
Endocrine Metabolism Division
Department of Medicine
University of Rochester School of Medicine and Dentistry
Rochester, New York, USA

Christopher P. Cardozo, MD
Department of Veterans Affairs Rehabilitation Research and Development Service
National Center of Excellence for the Medical Consequences of Spinal Cord Injury,
and Medical Service
James J. Peters Veterans Affairs Medical Center
Bronx, New York, USA
Departments of Medicine and Rehabilitation Medicine
The Mount Sinai School of Medicine
New York, New York, USA

Thomas O. Carpenter, MD
Yale University School of Medicine
New Haven, Connecticut, USA

Jacqueline R. Center, MBBS, MS (epi), PhD, FRACP
Osteoporosis and Bone Biology
Garvan Institute of Medical Research
St. Vincent's Hospital Clinical School
Department of Medicine
University of NSW
Sydney, Australia

Edward Chow, MBBS, MSc, PhD, FRCPC
Department of Radiation Oncology
University of Toronto
Sunnybrook Research Institute
Odette Cancer Centre
Sunnybrook Health Sciences Centre
Toronto, Ontario, Canada

Sylvia Christakos, PhD
Department of Biochemistry and Molecular Biology
UMDNJ–New Jersey Medical School
Newark, New Jersey, USA

Blaine A. Christiansen, PhD
Department of Orthopaedic Surgery
University of California
Davis Medical Center
Sacramento, California, USA

Yong-Hee P. Chun, DDS, MS, PhD
Department of Periodontics
University of Texas Health Science Center at San Antonio
San Antonio, Texas, USA

Roberto Civitelli, MD
Division of Bone and Mineral Diseases
Department of Internal Medicine
Musculoskeletal Research Center
Washington University in St. Louis
St. Louis, Missouri, USA

Gregory A. Clines, MD, PhD
Department of Medicine
Division of Endocrinology, Diabetes, and Metabolism
University of Alabama at Birmingham
and Veterans Affairs Medical Center
Birmingham, Alabama, USA

Denis R. Clohisy
Department of Orthopaedic Surgery and Masonic Cancer Center
University of Minnesota School of Medicine
Minneapolis, Minnesota, USA

Adi Cohen, MD, MHSc
Division of Endocrinology
Department of Medicine
College of Physicians and Surgeons
Columbia University
New York, New York, USA

Michael T. Collins, MD
Skeletal Clinical Studies Unit
Craniofacial and Skeletal Diseases Branch
National Institute of Dental and Craniofacial Research
National Institutes of Health
Department of Health and Human Services
Bethesda, Maryland, USA

Juliet E. Compston, MD, FRCP, FRCPath, FMedSci
Department of Medicine
University of Cambridge
Cambridge, UK

Gary J.R. Cook, MBBS, MSc, MD, FRCR, FRCP
Division of Imaging Sciences and Biomedical Engineering
Kings College London
London, UK

Cyrus Cooper, MA, DM, FRCP, FFPH, FMedSci
The MRC Lifecourse Epidemiology Unit
University of Southampton
Southampton General Hospital
Southampton, UK

Felicia Cosman
Columbia College of Physicians and Surgeons
Columbia University
Clinical Research Center
Helen Hayes Hospital
West Haverstraw
New York, New York, USA

Natalie E. Cusano
Division of Endocrinology
Department of Medicine
College of Physicians and Surgeons
Columbia University
New York, New York, USA

Terry F. Davies, MBBS, MD, FRCP, FACE
The Mount Sinai Bone Program
Department of Medicine
Mount Sinai School of Medicine
New York, New York, USA

Bess Dawson-Hughes, MD
Jean Mayer USDA Human Nutrition
 Research Center on Aging
Tufts University
Boston, Massachusetts, USA

David J.J. de Gorter, PhD
Institute for Molecular Cell Biology
University of Münster
Münster, Germany

Marie Demay, MD
Endocrine Unit
Massachusetts General Hospital and Harvard Medical
 School
Boston, Massachusetts, USA

Elaine Dennison, MA, MB, BChir, MSc, PhD
The MRC Lifecourse Epidemiology Unit
University of Southampton
Southampton General Hospital
Southampton, UK

Matthew T. Drake, MD, PhD
Department of Internal Medicine
Division of Endocrinology
College of Medicine
Mayo Clinic
Rochester, Minnesota, USA

Patricia Ducy, PhD
Department of Pathology and Cell Biology
Columbia University Medical Center
New York, New York, USA

Richard Eastell, MD, FRCP, FRCPath, FMedSci
Department of Human Metabolism
University of Sheffield
Sheffield, UK

Peter R. Ebeling, MD, FRACP
Australian Institute for Musculoskeletal Science
NorthWest Academic Centre
The University of Melbourne
Western Health
St. Albans, Victoria, Australia

Michael J. Econs, MD, FACP, FACE
Endocrinology and Metabolism
Medicine and Medical and Molecular Genetics
Indiana University School of Medicine
Indianapolis, Indiana, USA

Paul C. Edwards, MSc, DDS, FRCD(C)
Department of Periodontics and Oral Medicine
School of Dentistry
University of Michigan
Ann Arbor, Michigan, USA

Thomas A. Einhorn, MD
Department of Orthopaedic Surgery
Boston University Medical Center
Boston, Massachusetts, USA

Florent Elefteriou, PhD
Vanderbilt Center for Bone Biology
Vanderbilt University Medical Center
Nashville, Tennessee, USA

William J. Evans, PhD
Muscle Metabolism DPU
GlaxoSmithKline
Research Triangle Park, North Carolina, USA

Murray J. Favus, MD
Section of Endocrinology, Diabetes, and Metabolism
The University of Chicago
Chicago, Illinois, USA

Ignac Fogelman
Division of Imaging Sciences and Biomedical
 Engineering
Kings College London
London, UK

Mark R. Forwood, PhD
School of Medical Science and Griffith Health Institute
Griffith University
Gold Coast, Australia

Peter A. Friedman
University of Pittsburgh
Pittsburgh, Pennsylvania, USA

Benjamin J. Frisch, PhD
Wilmot Cancer Center
University of Rochester School of Medicine and
 Dentistry
Rochester, New York, USA

J. Christopher Gallagher, MD, MRCP
Creighton University Medical School
Omaha, Nebraska, USA

Harry K. Genant, MD, FACR, FRCR
Radiology, Medicine and Orthopedic Surgery
University of California
San Francisco, California, USA

Francis H. Glorieux, OC, MD, PhD
Genetics Unit
Shriners Hospital for Children–Canada and McGill University
Montreal, Quebec, Canada

C.C. Glüer
Sektion Biomedizinische Bildgebung
Klinik für Diagnostische Radiologie
Universitätsklinikum Schleswig-Holstein
Campus Kiel
Kiel, Germany

Gopinath Gnanasegaran
Department of Nuclear Medicine
Guy's and St. Thomas' NHS Foundation Trust
London, UK

Deborah T. Gold, PhD
Departments of Psychiatry and Behavioral Sciences, Sociology, and Psychology and Neuroscience
Duke University Medical Center
Durham, North Carolina, USA

Steven R. Goldring, MD
Hospital for Special Surgery
Weill Medical College of Cornell University
New York, New York, USA

David Goltzman, MD
Departments of Medicine and Physiology
McGill University
Montreal, Canada

Susan L. Greenspan, MD, FACP
Divisions of Geriatrics, Endocrinology and Metabolism
Department of Medicine
University of Pittsburgh School of Medicine
Pittsburgh, Pennsylvania, USA

James F. Griffith, MB, BCh, BAO, MD, MRCP (UK), FRCR
Department of Imaging and Interventional Radiology
The Chinese University of Hong Kong
Hong Kong, China

Monica Grover, MD
Department of Molecular and Human Genetics
Baylor College of Medicine
Houston, Texas, USA

Theresa A. Guise, MD
Department of Medicine
Division of Endocrinology
Indiana University School of Medicine
Indianapolis, Indiana, USA

Neveen A.T. Hamdy
Department of Endocrinology and Metabolic Diseases
Leiden University Medical Center
Leiden, The Netherlands

Nicholas Harvey, MA, MB, BChir, MRCP, PhD
The MRC Lifecourse Epidemiology Unit
University of Southampton
Southampton General Hospital
Southampton, UK

Robert P. Heaney, MD
Creighton University
Omaha, Nebraska, USA

Charles Hildebolt, DDS, PhD
Mallinckrodt Institute of Radiology
Washington University in St. Louis
St. Louis, Missouri, USA

Steven P. Hodak, MD
Division of Endocrinology and Metabolism
The University of Pittsburgh School of Medicine
Pittsburgh, Pennsylvania, USA

Ingrid A. Holm, MD, MPH
Children's Hospital Boston
Manton Center for Orphan Disease Research
Harvard Medical School
Boston, Massachusetts, USA

Mara J. Horwitz, MD
Division of Endocrinology and Metabolism
The University of Pittsburgh School of Medicine
Pittsburgh, Pennsylvania, USA

Keith A. Hruska, MD
Division of Pediatric Nephrology
Department of Pediatrics, Medicine and Cell Biology
Washington University
St. Louis, Missouri, USA

Christina Jacobsen, MD, PhD
Divisions of Endocrinology and Genetics
Boston Children's Hospital
Harvard Medical School
Boston, Massachusetts, USA

Suzanne M. Jan de Beur, MD
Division of Endocrinology and Metabolism
Department of Medicine
The Johns Hopkins School of Medicine
Baltimore, Maryland, USA

Robert L. Jilka, PhD
Division of Endocrinology and Metabolism
Center for Osteoporosis and Metabolic Bone Diseases
University of Arkansas for Medical Sciences
Central Arkansas Veterans Healthcare System
Little Rock, Arkansas, USA

Rachelle W. Johnson
Department of Veterans Affairs
Tennessee Valley Healthcare System
and Vanderbilt Center for Bone Biology
and Department of Cancer Biology
Vanderbilt University
Nashville, Tennessee, USA

Graeme Jones, MBBS, FRACP, MD
Menzies Research Institute Tasmania
University of Tasmania
Hobart, Tasmania, Australia

Stefan Judex, PhD
Department of Biomedical Engineering
Stony Brook University
Stony Brook, New York, USA

Harald Jüppner, MD
Endocrine Unit and Pediatric Nephrology Unit
Departments of Medicine and Pediatrics
Harvard Medical School
Massachusetts General Hospital
Boston, Massachusetts, USA

John A Kanis
WHO Collaborating Centre for Metabolic Bone
   Diseases
University of Sheffield Medical School
Sheffield, UK

Frederick S. Kaplan, MD
Departments of Orthopaedic Surgery and Medicine
Center for Research in FOP and Related Disorders
Perelman School of Medicine at The University of
   Pennsylvania
Philadelphia, Pennsylvania, USA

Gerard Karsenty, MD, PhD
Department of Genetics and Development
Columbia University
New York, New York, USA

Moustapha Kassem, MD, PhD, DSc
Department of Endocrinology
University Hospital of Odense
Odense, Denmark

Richard W. Keen, MD, PhD
Institute of Orthopaedics and Musculoskeletal Sciences
Royal National Orthopaedic Hospital
Brockley Hill
Stanmore
Middlesex, UK

Luluel M. Khan
University of Toronto
Department of Radiation Oncology
Odette Cancer Centre
Sunnybrook Health Sciences Centre
Toronto, Ontario, Canada

Sundeep Khosla, MD
Department of Internal Medicine
Division of Endocrinology
College of Medicine
Mayo Clinic
Rochester, Minnesota, USA

Douglas P. Kiel, MD, MPH
Institute for Aging Research
Hebrew SeniorLife
Department of Medicine
Harvard Medical School
Boston, Massachusetts, USA

Keith L. Kirkwood, DDS, PhD
Department of Craniofacial Biology
The Center for Oral Health Research
Medical University of South Carolina
Charleston, South Carolina, USA

Michael Kleerekoper, MBBS, FACB, FACP, MACE
Department of Medicine, Division of Endocrinology
University of Toledo Medical College
Toledo, Ohio, USA

Gordon L. Klein, MD, MPH
Department of Orthopaedic Surgery
University of Texas Medical Branch
Galveston, Texas, USA

John Klingensmith, PhD
Department of Cell Biology
Duke University Medical Center
Durham, North Carolina, USA

Stephen J. Knohl, MD
Department of Medicine Division of Nephrology
State University of New York Upstate Medical
   University
Syracuse, New York, USA

Scott L. Kominsky, PhD
Departments of Orthopaedic Surgery and Oncology
Johns Hopkins University School of Medicine
Baltimore, Maryland, USA

Christopher S. Kovacs, MD, FRCPC, FACP, FACE
Faculty of Medicine–Endocrinology
Health Sciences Centre
Memorial University of Newfoundland
St. John's, Newfoundland, Canada

Paul H. Krebsbach, DDS, PhD
Department of Biologic and Materials Sciences
School of Dentistry
The University of Michigan
Ann Arbor, Michigan, USA

Henry M. Kronenberg, MD
Endocrine Unit
Massachusetts General Hospital
and Harvard Medical School
Boston, Massachusetts, USA

Craig B. Langman, MD
Feinberg School of Medicine
Northwestern University
Chicago, Illinois, USA

Brendan Lee, MD, PhD
Howard Hughes Medical Institute
Department of Molecular and Human Genetics
Baylor College of Medicine
Houston, Texas, USA

Mary G. Lee, DMD, MSD
Department of Craniofacial Biology
and the Center for Oral Health Research
Medical University of South Carolina
Charleston, South Carolina, USA

Michael A. Levine, MD, FAAP, FACP, FACE
Division of Endocrinology and Diabetes
The Children's Hospital of Philadelphia and
  Department of Pediatrics
University of Pennsylvania Perelman School of
  Medicine
Philadelphia, Pennsylvania, USA

Paul Lips, MD, PhD
Department of Internal Medicine/Endocrinology
VU University Medical Center
Amsterdam, The Netherlands

David G. Little, MBBS, FRACS(Orth), PhD
Paediatrics and Child Health
University of Sydney
Orthopaedic Research and Biotechnology
The Children's Hospital at Westmead
Westmead, Australia

Shane A.J. Lloyd
Department of Orthopaedics and Rehabilitation
Division of Musculoskeletal Sciences
The Pennsylvania State University College of Medicine
Hershey, Pennsylvania, USA

Karen M. Lyons, PhD
UCLA Orthopaedic Hospital
Department of Orthopaedic Surgery
Department of Molecular, Cell & Developmental
  Biology
University of California
Los Angeles, California, USA

Sharmila Majumdar, PhD
Department of Radiology and Biomedical Imaging
and Orthopedic Surgery UCSF
and Department of Bioengineering UC Berkeley
San Francisco, California, USA

Stavros C. Manolagas, MD, PhD
Division of Endocrinology and Metabolism
Center for Osteoporosis and Metabolic Bone Diseases
University of Arkansas for Medical Sciences
and the Central Arkansas Veterans Healthcare System
Little Rock, Arkansas, USA

Joan C. Marini, MD, PhD
National Institute of Child Health and Human
  Development
Bone and Extracellular Matrix Branch
National Institutes of Health
Bethesda, Maryland, USA

T. John Martin, MD, DSc
St. Vincent's Institute of Medical Research
University of Melbourne Department of Medicine
Melbourne, Australia

Stephen J. Marx, MD
Genetics and Endocrinology Section
Metabolic Diseases Branch
National Institute of Diabetes and Digestive and
  Kidney Diseases
National Institutes of Health
Bethesda, Maryland, USA

Maiko Matsui, BSc
Department of Cell Biology
Duke University Medical Center
Durham, North Carolina, USA

Hani H. Mawardi, BDS, DMSc
Division of Oral Medicine
King Abdulaziz University
Faculty of Dentistry
Jeddah, Saudi Arabia

Laurie K. McCauley, DDS, MS, PhD
University of Michigan
School of Dentistry
Ann Arbor, Michigan, USA

Michael R. McClung
Oregon Osteoporosis Center
Portland, Oregon, USA

Lisa K. Micklesfield, PhD
MRC/Wits Developmental Pathways for Health
  Research Unit
Department of Paediatrics
Faculty of Health Sciences
University of the Witwatersrand
Johannesburg, South Africa

Paul D. Miller, MD, FACP
University of Colorado Health Sciences Center
Colorado Center for Bone Research
Lakewood, Colorado, USA

Thimios Mitsiadis, DDS, MS, PhD
Institute of Oral Biology
Department of Orofacial Development & Regeneration,
  ZZM
Department of Medicine
University of Zurich
Zurich, Switzerland

Elise F. Morgan, PhD
Department of Mechanical Engineering
Boston University
Boston, Massachusetts, USA

Tuan V. Nguyen
Osteoporosis and Bone Biology Research Program
Garvan Institute of Medical Research
Sydney, Australia

Robert A. Nissenson, PhD
Endocrine Research Unit
VA Medical Center
Departments of Medicine and Physiology
University of California
San Francisco, California, USA

Shane A. Norris, PhD
MRC/Wits Developmental Pathways for Health
  Research Unit
Department of Paediatrics
Faculty of Health Sciences
University of the Witwatersrand
Johannesburg, South Africa

Patrick W. O'Donnell, MD, PhD
Center for Musculoskeletal Oncology
Department of Orthopaedic Surgery
Markey Cancer Center
University of Kentucky
Lexington, Kentucky, USA

Eric S. Orwoll
Oregon Health and Science University
Portland, Oregon, USA

Petros Papagerakis, DDS, MS, PhD
Department of Orthodontics and Pediatric Dentistry
Center for Organogenesis
Center for Computational Medicine and Bioinformatics
Schools of Dentistry and Medicine
University of Michigan
Ann Arbor, Michigan, USA

Socrates E. Papapoulos, MD, PhD
Department of Endocrinology and Metabolic Diseases
Leiden University Medical Center
Leiden, The Netherlands

John M. Pettifor, MBBCh, PhD
MRC/Wits Developmental Pathways for Health
  Research Unit
Department of Paediatrics
Faculty of Health Sciences
University of the Witwatersrand
Johannesburg, South Africa

Robert J. Pignolo, MD, PhD
Center for Research in FOP and Related Disorders
Department of Orthopaedic Surgery
The University of Pennsylvania School of Medicine
Philadelphia, Pennsylvania, USA

L. Darryl Quarles
Division of Nephrology
Department of Medicine
University of Tennessee Health Science Center
Memphis, Tennessee, USA

Manoj Ramachandran, BSc, MBBS, MRCS, FRCS
The Royal London and St. Bartholomew's Hospitals
Barts Health NHS Trust
Barts and The London School of Medicine and
  Dentistry
Queen Mary
University of London
London, England

Francesco Ramirez, PhD
Department of Pharmacology and Systems
  Therapeutics
Mount Sinai School of Medicine
New York, New York, USA

Robert R. Recker, MD, MACP, FACE
Department of Medicine
Section of Endocrinology
Osteoporosis Research Center
Creighton University Medical Center
Omaha, Nebraska, USA

Ian R. Reid, MBChB, MD, FRACP
Department of Medicine
University of Auckland
Auckland, New Zealand

Mara Riminucci
Dipartimento di Medicina Molecolare
Universita' La Sapienza
Rome, Italy

David L. Rimoin, MD, PhD*
Medical Genetics Institute
Cedars-Sinai Medical Center
Los Angeles, California, USA

René Rizzoli, MD
Division of Bone Diseases
Department of Internal Medicine Specialties
Geneva University Hospital and Faculty of Medicine
Geneva, Switzerland

Pamela Gehron Robey
Craniofacial and Skeletal Diseases Branch
National Institute of Dental and Craniofacial Research
National Institutes of Health
Department of Health and Human Services
Bethesda, Maryland, USA

G. David Roodman
Division of Hematology/Oncology
Indiana University School of Medicine
Indianapolis, Indiana, USA

Clifford J. Rosen, MD
Tufts University School of Medicine
Maine Medical Center Research Institute
Scarborough, Maine, USA

Vicki Rosen, PhD
Department of Developmental Biology
Harvard School of Dental Medicine
Boston, Massachusetts, USA

F. Patrick Ross
Department of Pathology and Immunology
Washington University School of Medicine
St. Louis, Missouri, USA

Clinton T. Rubin, PhD
Department of Biomedical Engineering
Stony Brook University
Stony Brook, New York, USA

Janet Rubin, MD
Department of Medicine
University of North Carolina School of Medicine
Chapel Hill, North Carolina, USA

Mishaela R. Rubin, MD
Division of Endocrinology
Department of Medicine
Columbia University College of Physicians
New York, New York, USA

Mary D. Ruppe, MD
Department of Medicine
Division of Endocrinology
The Methodist Hospital
Houston, Texas, USA

Lynn Y. Sakai
Department of Biochemistry and Molecular Biology
Oregon Health and Science University
Shriners Hospital for Children
Portland, Oregon, USA

Kyu Sang Joeng, PhD
Department of Molecular and Human Genetics
Baylor College of Medicine
Houston, Texas, USA

Anne L. Schafer, MD
Department of Medicine
University of California, San Francisco
San Francisco, California, USA

Steven J. Scheinman, MD
The Commonwealth Medical College
Scranton, Pennsylvania, USA

Ego Seeman, BSc, MBBS, FRACP, MD
Department of Endocrinology and Medicine
Repatriation Campus
Austin Health
University of Melbourne
Melbourne, Australia

Michael Seifert, MD
Division of Pediatric Nephrology
Department of Pediatrics
Washington University and Southern Illinois University
St. Louis, Missouri, USA

*Deceased.

Gerhard Sengle
Department of Biochemistry and Molecular Biology
Oregon Health and Science University
and Shriners Hospital for Children
Portland, Oregon, USA

Elizabeth Shane
Division of Endocrinology
Department of Medicine
College of Physicians and Surgeons
Columbia University
New York, New York, USA

Sue Shapses, PhD
Department of Nutritional Sciences
Rutgers University
Newark, New Jersey, USA

Nicholas Shaw, MBChB, FRCPCH
Department of Endocrinology
Birmingham Children's Hospital
Birmingham, West Midlands, UK

Yiping Shen, PhD
Boston Children's Hospital
Harvard Medical School
Boston, Massachusetts, USA
Shanghai Jiaotong University School of Medicine
Shanghai, China

Dolores Shoback, MD
University of California, San Francisco
Endocrine Research Unit
San Francisco VA Medical Center
San Francisco, California, USA

Eileen M. Shore, PhD
Departments of Orthopaedic Surgery and Genetics
Center for Research in FOP and Related Disorders
Perelman School of Medicine at the University of Pennsylvania
Philadelphia, Pennsylvania, USA

Rebecca Silbermann, MD
Indiana University School of Medicine
Indianapolis, Indiana, USA

Shonni J. Silverberg, MD
Division of Endocrinology
Department of Medicine
College of Physicians and Surgeons
Columbia University
New York, New York, USA

James P. Simmer, DDS, PhD
Department of Biological and Materials Sciences
University of Michigan School of Dentistry
Ann Arbor, Michigan, USA

Ethel S. Siris, MD
Columbia University Medical Center
New York Presbyterian Hospital
New York, New York, USA

Julie A. Sterling, PhD
Department of Veterans Affairs
Tennessee Valley Healthcare System
and Vanderbilt Center for Bone Biology
Department of Medicine/Clinical Pharmacology
Vanderbilt University
Nashville, Tennessee, USA

Andrew F. Stewart, MD
Diabetes Obesity and Metabolism Institute
Icahn School of Medicine at Mount Sinai
New York, New York, USA

Deeptha Sukumar, PhD
Department of Nutritional Sciences
Rutgers University
Newark, New Jersey, USA

Li Sun
The Mount Sinai Bone Program
Department of Medicine
Mount Sinai School of Medicine
New York, New York, USA

Pawel Szulc, MD, PhD
INSERM
Université de Lyon
Lyon, France

Shu Takeda, MD, PhD
Section of Nephrology, Endocrinology and Metabolism
Department of Internal Medicine
School of Medicine
Keio University
Tokyo, Japan

Jianning Tao, PhD
Departments of Molecular and Human Genetics
Baylor College of Medicine
Texas Children's Hospital
Houston, Texas, USA

Pamela Taxel, MD
Division of Endocrinology and Metabolism
University of Connecticut Health Center
Farmington, Connecticut, USA

Peter ten Dijke, PhD
Department of Molecular Cell Biology
and Centre for Biomedical Genetics
Leiden University Medical Centre
Leiden, The Netherlands

Rajesh V. Thakker, MD, ScD, FRCP, FRCPath, FMedSci
Academic Endocrine Unit
Nuffield Department of Clinical Medicine
Oxford Centre for Diabetes
Endocrinology and Metabolism (OCDEM)
Churchill Hospital
Headington, Oxford, UK

Anna N.A. Tosteson, ScD
Multidisciplinary Clinical Research Center in Musculoskeletal Diseases
Geisel School of Medicine at Dartmouth
Lebanon, New Hampshire, USA

Dwight A. Towler, MD, PhD
Diabetes and Obesity Research Center
Sanford-Burnham Medical Research Institute at Lake Nona
Orlando, Florida, USA

Bich H. Tran
Osteoporosis and Bone Biology Research Program
Garvan Institute of Medical Research
Sydney, Australia

Nathaniel S. Treister, DMD, DMSc
Division of Oral Medicine and Dentistry
Brigham and Women's Hospital
Boston, Massachusetts, USA

Geertje van der Horst, PhD
Department of Urology
Leiden University Medical Centre
Leiden, The Netherlands

Gabri van der Pluijm, PhD
Department of Urology
Leiden University Medical Centre
Leiden, The Netherlands

Catherine Van Poznak, MD
Department of Internal Medicine
University of Michigan
Ann Arbor, Michigan, USA

Natasja M. van Schoor, PhD
Department of Epidemiology and Biostatistics
EMGO Institute for Health and Care Research
VU University Medical Center
Amsterdam, The Netherlands

Kong Wah Ng, MBBS, MD, FRACP, FRCP
Department of Endocrinology and Diabetes
St. Vincent's Health
St. Vincent's Institute
University of Melbourne Department of Medicine
Fitzroy, Victoria, Australia

Lisa L. Wang, MD
Texas Children's Cancer Center
Baylor College of Medicine
Houston, Texas, USA

Qingju Wang
Department Endocrinology and Medicine
Repatriation Campus
Austin Health
University of Melbourne
Melbourne, Australia

Nelson B. Watts, MD
Mercy Health Osteoporosis and Bone Health Services
Cincinnati, Ohio, USA

Connie M. Weaver, PhD
Purdue University
Nutrition Science
West Lafayette, Indiana, USA

Kristy Weber, MD
Departments of Orthopaedic Surgery and Oncology
Johns Hopkins School of Medicine
Baltimore, Maryland, USA

Robert S. Weinstein, MD
Division of Endocrinology and Metabolism
Center for Osteoporosis and Metabolic Bone Diseases
Department of Internal Medicine
and the Central Arkansas Veterans Healthcare System
University of Arkansas for Medical Sciences
Little Rock, Arkansas, USA

Kenneth E. White, PhD
Medical and Molecular Genetics
Indiana University School of Medicine
Indianapolis, Indiana, USA

Michael P. Whyte, MD
Division of Bone and Mineral Diseases
Washington University School of Medicine
Barnes-Jewish Hospital
and Center for Metabolic Bone Disease and Molecular Research
Shriners Hospital for Children
St. Louis, Missouri, USA

Jeffrey S. Willey, PhD
Department of Radiation Oncology
Wake Forest School of Medicine
Winston-Salem, North Carolina, USA

Tania Winzenberg, MBBS, FRACGP, MMedSc, PhD
Menzies Research Institute Tasmania
University of Tasmania
Hobart, Tasmania, Australia

Sook-Bin Woo, DMD, MMSc
Division of Oral Medicine and Dentistry
Brigham and Women's Hospital
Boston, Massachusetts, USA

John J. Wysolmerski
Section of Endocrinology and Metabolism
Department of Internal Medicine
Yale University School of Medicine
New Haven, Connecticut, USA

Tao Yang, PhD
Center for Skeletal Disease Research
Laboratory of Skeletal Biology
Van Andel Research Institute,
Grand Rapids, Michigan, USA

Yingzi Yang
Developmental Genetics Section
Genetic Disease Research Branch
National Human Genome Research Institute
Bethesda, Maryland, USA

Raz Yirmiya
Department of Psychology
The Hebrew University of Jerusalem
Jerusalem, Israel

Tony Yuen, PhD
The Mount Sinai Bone Program
Department of Medicine
Mount Sinai School of Medicine
New York, New York, USA

Mone Zaidi, MBBS, PhD
The Mount Sinai Bone Program
Department of Medicine
Mount Sinai School of Medicine
New York, New York, USA

Alberta Zallone
Department of Histology
University of Bari
Bari, Italy

Michael J. Zuscik, PhD
Department of Orthopaedics & Rehabilitation
Center for Musculoskeletal Research
University of Rochester Medical Center
Rochester, New York, USA

Primer Corporate Sponsors

Distribution to the ASBMR membership of the *Primer on the Metabolic Bone Diseases and Disorders of Mineral Metabolism, Eighth Edition*, is supported by unrestricted educational grants from the following companies:

Silver: Warner Chilcott

Bronze: Alexion

Partner: OsteoMetrics

Preface to the Eighth Edition of the *Primer*

Welcome to the eighth edition of the *Primer on the Metabolic Bone Diseases and Disorders of Mineral Metabolism*! Loyal readers of previous editions will find the same up-to-date summaries of clinical diseases of the skeleton as well as overviews of skeletal growth, remodeling, and pathophysiological disorders that they have seen in the past. As always, calcium, phosphorus, and vitamin D are reviewed extensively and updated. New visitors to the *Primer* will find a potpourri of information that can assist readers at the bench and at the bedside. The many tables, references, figures, and illustrations are certain to be helpful. In fact, many figures and illustrations appear in full color for the first time in this edition.

Importantly, the eighth edition has expanded by one-third, adding new chapters principally related to how the skeleton interacts with other systems and tissues, including hematopoietic, muscular, metabolic, and neuropsychiatric. A basic, but comprehensive, overview on bone as an organ system is updated. As with previous editions, it is the contributing authors, who are at the top in their respective fields, who make the *Primer* so valuable.

And as a real treat, our cover is an artistic image of bone painted by Mr. Toby Kahn. His creation reminds us that, since the last edition, the osteocyte has emerged as a major regulator of skeletal remodeling as well as mineral homeostasis.

The eighth edition is dedicated to the memory of Dr. Larry Raisz, a giant in our field who led and inspired all of us, and who particularly found the *Primer* to be the premier textbook of bone.

Clifford J. Rosen, MD

We Make the Discoveries That Keep Bones Healthy for a Lifetime

The American Society for Bone and Mineral Research (ASBMR) is home to the world's foremost bone and mineral researchers and clinicians. Our 4,000 members work in more than 60 countries and are experts in a diverse range of specialties, including endocrinology, cell biology, orthopedics, rheumatology, internal medicine, physiology, epidemiology, biomechanics and more.

ASBMR's mission is to promote excellence in bone and mineral research, to foster integration of basic and clinical science and to facilitate the translation of that science into clinical practice. For more than 30 years, our Society has remained true to its innovative and inclusive founding principles, fostering an encouraging environment for researchers and clinicians to develop and discuss the latest findings.

ASBMR's work is guided by the following key strategic objectives:

- **GENERATING AWARENESS** about bone diseases and the importance of bone research by serving as the leading authority on bone health and disease for health professionals and the public

- **PROMOTING CUTTING-EDGE RESEARCH** by fostering multidisciplinary connections among our investigators, awarding grants and advocating for funding for bone and mineral research

- **COMMUNICATING WITH THE NATIONAL INSTITUTES OF HEALTH** to both discover and influence the latest directions in basic, clinical and translational research

- **SUPPORTING YOUNG INVESTIGATORS** through continuing education, grant funding, professional networking and mentoring opportunities

Learn more about ASBMR and the vast learning and networking opportunities it offers by visiting www.asbmr.org today.

President's Preface

The American Society for Bone and Mineral Research is pleased and excited to present this fully revised, updated and expanded eighth edition of the *Primer on the Metabolic Bone Diseases and Disorders of Mineral Metabolism*. Dr. Clifford Rosen returns as Editor-in-Chief, overseeing even greater coverage of the field to include the addition of a completely new section on "The Skeleton and Its Integration with Other Tissues." In addition to new content, many new reader-friendly features have been incorporated into the design of the book.

This eighth edition presents numerous full-color figures for the first time, and the layout has been updated to make accessing information easier. Dr. Rosen and the editorial team have maintained the rigorous high standards of scientific content that have long been the trademark feature of the book, ensuring the *Primer* maintains its reputation as an essential reference in the field of bone health.

Along with these changes, a new companion site has been developed at www.asbmrprimer.com to provide researchers, instructors, clinicians, and students with valuable supplementary materials that support and expand the content of the book.

The eighth edition of our *Primer* maintains the strong tradition developed by its predecessors while continuing to evolve to serve the changing needs of the communities it serves. We thank and commend all those involved in making this new edition a reality.

Lynda F. Bonewald, PhD
President, ASBMR

About the Companion Website

This book is accompanied by a companion website:
  www.asbmrprimer.com

The website includes:

- Videos
- Editors' biographies
- PowerPoints of all figures from the book for downloading
- Useful website links

Section I
Molecular, Cellular, and Genetic Determinants of Bone Structure and Formation

Section Editor Karen M. Lyons

Chapter 1. Skeletal Morphogenesis and Embryonic Development 3
Yingzi Yang

Chapter 2. Signal Transduction Cascades Controlling Osteoblast Differentiation 15
David J.J. de Gorter and Peter ten Dijke

Chapter 3. Osteoclast Biology and Bone Resorption 25
F. Patrick Ross

Chapter 4. Osteocytes 34
Lynda F. Bonewald

Chapter 5. Connective Tissue Pathways That Regulate Growth Factors 42
Gerhard Sengle and Lynn Y. Sakai

Chapter 6. The Composition of Bone 49
Adele L. Boskey and Pamela Gehron Robey

Chapter 7. Assessment of Bone Mass and Microarchitecture in Rodents 59
Blaine A. Christiansen

Chapter 8. Animal Models: Genetic Manipulation 69
Karen M. Lyons

Chapter 9. Animal Models: Allelic Determinants for BMD 76
Robert D. Blank

Chapter 10. Neuronal Regulation of Bone Remodeling 82
Florent Elefteriou and Gerard Karsenty

Chapter 11. Skeletal Healing 90
Michael J. Zuscik

Chapter 12. Biomechanics of Fracture Healing 99
Elise F. Morgan and Thomas A. Einhorn

Chapter 13. Human Genome-Wide Association (GWA) Studies 106
Douglas P. Kiel

Chapter 14. Circulating Osteogenic Cells 111
Robert J. Pignolo and Moustapha Kassem

Primer on the Metabolic Bone Diseases and Disorders of Mineral Metabolism, Eighth Edition. Edited by Clifford J. Rosen.
© 2013 American Society for Bone and Mineral Research. Published 2013 by John Wiley & Sons, Inc.

1
Skeletal Morphogenesis and Embryonic Development

Yingzi Yang

Early Skeletal Patterning 3
Embryonic Cartilage and Bone Formation 6
Chondrocyte Proliferation and Differentiation in the Developing Cartilage 7

Regulation of Chondrocyte Survival 9
Conclusions 10
References 10

Formation of the skeletal system is one of the hallmarks that distinguish vertebrates from invertebrates. In higher vertebrates (i.e., birds and mammals), the skeletal system contains mainly cartilage and bone, which are mesoderm-derived tissues formed by chondrocytes and osteoblasts, respectively, during embryogenesis. A common mesenchymal progenitor cell, also referred as the osteochondral progenitor, gives rise to both chondrocytes and osteoblasts. The first overt sign of skeletal development is the formation of mesenchymal condensations, in which mesenchymal progenitor cells aggregate at future skeletal locations. Mesenchymal cells in different parts of the embryo come from different cell lineages. Neural crest cells give rise to craniofacial bones, the sclerotome compartment of the somites gives rise to most axial skeletal elements, and lateral plate mesoderm forms the limb mesenchyme, from which limb skeletons are derived (Fig. 1.1). Skeletal formation proceeds through two major mechanisms: intramembranous and endochondral ossification. In intramembranous ossification, osteochondral progenitors differentiate directly into osteoblasts to form membranous bone; during endochondral ossification, osteochondral progenitors differentiate into chondrocytes to form a cartilage template of the future bone. The location of each skeletal element determines its ossification mechanism and anatomic properties such as shape and size. This positional identity is acquired early in embryonic development, before mesenchymal condensation, through a process called pattern formation.

Cell–cell communication plays a critical role in pattern formation, and is mediated by several major signaling pathways. These include Wnts, Hedgehogs (Hhs), bone morphogenetic proteins (Bmps), fibroblast growth factors (Fgfs), and Notch/Delta. These pathways are also used later in skeletal development to control cell fate determination, proliferation, maturation, and polarity.

EARLY SKELETAL PATTERNING

Craniofacial patterning

Neural crest cells are the major source of cells establishing the craniofacial skeleton [1]. Reciprocal signaling between and among neural crest cells and epithelial cells (surface ectoderm, neural ectoderm or endodermal cells) ultimately establishes the identities of craniofacial skeletal elements [2].

Axial patterning

The most striking feature of axial skeletal patterning is the periodic organization of the vertebral column into multiple vertebrae along the anterior–posterior (A–P) axis. This pattern is established when somites, which are segmented mesodermal structures located on either side

---

Primer on the Metabolic Bone Diseases and Disorders of Mineral Metabolism, Eighth Edition. Edited by Clifford J. Rosen.
© 2013 American Society for Bone and Mineral Research. Published 2013 by John Wiley & Sons, Inc.

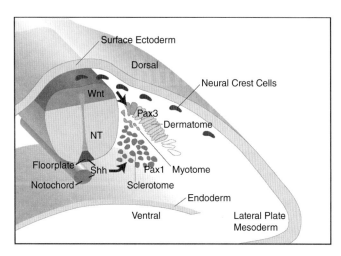

Fig. 1.1. Cell lineage contribution of chondrocytes and osteoblasts. Neural crest cells are born at the junction of dorsal neural tube and surface ectoderm. In the craniofacial region, neural crest cells from the branchial arches differentiate into chondrocytes and osteoblasts. In the trunk, axial skeletal cells are derived from the ventral somite compartment, sclerotome. Shh secreted from the notochord and floor plate of the neural tube induces the formation of sclerotome, which expresses Pax1. Wnts produced in the dorsal neural tube inhibit sclerotome formation and induce the dermomyotome, which expresses Pax3. Cells from the lateral plate mesoderm will form the limb mesenchyme, from which limb skeletons are derived.

of the neural tube, bud off at a defined pace from the anterior tip of the presomitic mesoderm (PSM) [3]. Somites give rise to the axial skeleton, striated muscle, and dorsal dermis [4–7]. The patterning of the axial skeleton is controlled by a molecular oscillator, or segmentation clock, that acts in the PSM [Fig. 1.2(A)]. The segmentation clock is operated by a traveling wave of gene expression (or cyclic gene expression) along the embryonic A–P axis, which is generated by an integrated network of the Notch, Wnt/β-catenin and fibroblast growth factor (FGF) signaling pathways [Fig. 1.2(B)] [8, 9].

The Notch signaling pathway mediates short-range communication between contacting cells [10]. The majority of cyclically expressed genes in the segmentation clock are targets of the Notch signaling pathway. The Wnt/β-catenin and FGF pathways mediate long-range signaling across several cell diameters. Upon activation of the Wnt pathway, β-catenin is stabilized and translocates to the nucleus where it activates the expression of downstream genes that are rhythmically expressed in the PSM [9, 11–13]. FGF signaling is also activated periodically in the posterior PSM [14, 15]. There is extensive cross-talk among these major oscillating signaling pathways; it is likely that each of the three pathways has the capacity to generate its own oscillations, while interactions among them allow efficient coupling and entrainment [16, 17]. Retinoic acid (RA) signaling controls somitogenesis by regulating the competence of PSM cells to undergo segmentation via antagonizing FGF signaling [Fig. 1.2(A)] [18, 19].

The functional significance of the segmentation clock in human skeletal development is highlighted by congenital axial skeletal diseases. For instance, mutations in Notch signaling components cause at least two human disorders, spondylocostal dysostosis (SCD, #277300, #608681, and #609813) and Alagille syndrome (AGS, OMIM# 118450 and #610205), both of which include vertebral column segmentation defects.

Once formed by the segmentation mechanism described above, somites are patterned along the dorsal–ventral axis by secreted signals derived from the surface ectoderm, neural tube and notochord (Fig. 1.1). The sclerotome forms from the ventral region of the somite, and gives rise to the axial skeleton and the ribs. Sonic hedgehog (Shh) from the notochord and ventral neural tube is required to induce sclerotome formation [20, 21] (Fig. 1.1) [22, 23]. In mice that lack *Shh*, the vertebral column and posterior ribs fail to form [24].

Limb patterning

Limb skeletons are patterned along the proximal–distal (P–D, shoulder to digit tip), anterior–posterior (A–P, thumb to little finger), and dorsal–ventral (D–V, back of the hand to palm) axes (Fig. 1.3). Along the P–D axis, the limb skeletons form three major segments: humerus or femur at the proximal end, radius and ulna or tibia and fibula in the middle, and carpal/tarsal, metacarpal/metatarsal, and digits in the distal end. Along the A–P axis, the radius and ulna have distinct morphological features; so do each of the five digits. Skeletal elements are also patterned along the D–V limb axis. For instance, the sesamoid processes are located ventrally whereas the patella forms on the dorsal side of the knee. Limb patterning events are regulated by three signaling centers in the early limb primordium, known as the limb bud, that act prior to mesenchymal condensation.

The apical ectoderm ridge (AER), a thickened epithelial structure formed at the distal tip of the limb bud, is the signaling center that directs P–D limb outgrowth (Fig. 1.3). Canonical Wnt signaling activated by Wnt3 induces AER formation [25], whereas BMP signaling leads to AER regression to halt limb extension [26]. Multiple FGF family members are expressed in the AER, but Fgf8 alone is sufficient to mediate the function of AER [27–29]. Fgf10 is expressed in the presumptive limb mesoderm and is required for initiation of limb bud formation; it subsequently controls limb outgrowth by maintaining *Fgf8* expression in the AER [30–32].

The second signaling center is the zone of polarizing activity (ZPA), a group of mesenchymal cells located at the posterior distal margin of the limb bud, immediately adjacent to the AER [Fig. 1.3(B)]. The ZPA patterns digit identity along the A–P axis. When ZPA tissue is grafted to a host limb bud on the anterior side under the AER, it leads to digit duplications in a mirror image of the

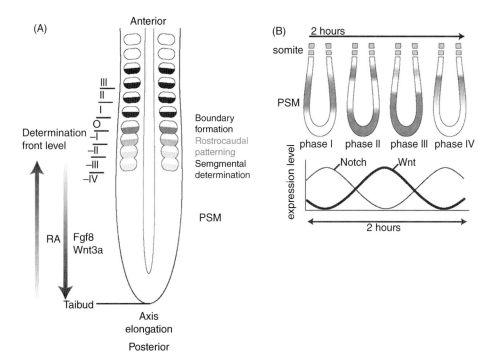

Fig. 1.2. Periodic and left–right symmetrical somite formation is controlled by signaling gradients and oscillations. (A) Somites form from the presomitic mesoderm (PSM) on either side of the neural tube in an anterior to posterior (A–P) wave. Each segment of the somite is also patterned along the A–P axis. Retinoic acid signaling controls the synchronization of somite formation on the left and right side of the neural tube. The most recent visible somite is marked by "0," whereas the region in the anterior PSM that is already determined to form somites is marked by a determination front that is determined by Fgf8 and Wnt3a gradients. This FGF signaling gradient is antagonized by an opposing gradient of retinoic acid. (B) Periodic somite formation (one pair of somite/2 hours) is controlled by a segmentation clock, the molecular nature of which is oscillated expression of signaling components in the Notch and Wnt pathways. Notch signaling oscillates out of phase with Wnt signaling.

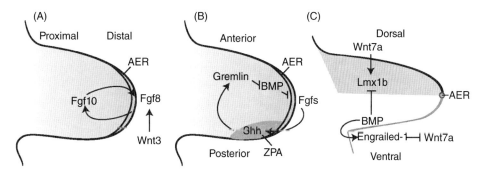

Fig. 1.3. Limb patterning and growth along the proximal–distal (P–D), anterior–posterior (A–P) and dorsal–ventral (D–V) axes are controlled by signaling interactions and feedback loops. (A) A signaling feedback loop between Fgf10 in the limb mesoderm and Fgf8 in the AER is required to direct P–D limb outgrowth. Wnt3 is required for AER formation. (B) Shh in the ZPA controls A–P limb patterning. A–P and P–D limb patterning and growth are also coordinated through a feedback loop between Shh and Fgfs expressed in the AER. Fgf signaling from the AER is required for Shh expression. Shh also maintains AER integrity by regulating Gremlin expression. Gremlin is a secreted antagonist of BMP signaling that promotes AER degeneration. (C) D–V patterning of the limb is determined by Wnt7a and BMP signaling through regulating the expression of Lmx1b in the limb mesenchyme.

endogenous ones [33]. *Shh* is expressed in the ZPA and is necessary and sufficient to mediate ZPA activity [34, 35]. However, the A–P axis of the limb is established prior to Shh signaling. This pre-Shh A–P limb patterning is controlled by combined activities of multiple transcription factors, including Gli3, Alx4, and the basic helix-loop-helix (bHLH) transcription factors dHand and Twist1. Mutations in the human *TWIST1* gene cause Saethre-Chotzen syndrome (SCS, OMIM#101400). The hallmarks of this syndrome are premature fusion of the calvarial bones and limb abnormalities. Mutations in the *GLI3* gene cause Greig cephalopolysyndactyly syndrome (GCPS, OMIM#175700) and Pallister-Hall syndrome (PHS, OMIM#146510), which are characterized by limb malformations.

The third signaling center is the non-AER limb ectoderm that covers the limb bud. This tissue controls D–V polarity of the ectoderm itself and also of the underlying mesoderm [Fig. 1.3(C)] (reviewed in Refs. 36 and 37). Wnt and BMP signaling control D–V limb polarity. *Wnt7a* is expressed in the dorsal limb ectoderm and activates the expression of *Lmx1b*, which encodes a dorsal-specific LIM homeobox transcription factor that determines the dorsal identity [38, 39]. *Wnt7a* expression is suppressed by the transcription factor En-1 in the ventral ectoderm [40]. The BMP signaling pathway is also ventralizing in the early limb [Fig. 1.3(C)]. The effects of BMP signaling in this ventralization are mediated by the transcription factors Msx1 and Msx2. The function of BMP signaling in the early limb ectoderm is upstream of En-1 in controlling D–V limb polarity [41]. However, BMPs also have En-1-independent ventralization activity by directly signaling to the limb mesenchyme to inhibit *Lmx1b* expression [42].

Limb development is a coordinated three-dimensional event. Indeed, the three signaling centers interact with each other through interactions of the mediating signaling molecules. First, there is a positive feedback loop between Shh expressed in the ZPA to maintain expression of FGFs in the AER, which connects A–P limb patterning with P–D limb outgrowth [Fig. 1.3(B)] [43–45]. This positive feedback look is antagonized by an FGF/Grem1 inhibitory loop that attenuates FGF signaling and thereby terminates limb outgrowth in order to maintain a proper limb size [46]. Second, the dorsalizing signal Wnt7a is also required for maintaining the expression of Shh that patterns the A–P axis [47, 48]. Third, Wnt/β-catenin signaling is both distalizing and dorsalizing [49–51].

EMBRYONIC CARTILAGE AND BONE FORMATION

The early patterning events described above determine where and when mesenchymal cells condense. Subsequently, the osteochondral progenitors in these condensations must form either chondrocytes or osteoblasts. Sox9 and Runx2 are master transcription factors that are required for the determination of chondrocyte and osteoblast cell fates, respectively [52–55]. Both are expressed in the osteochondral progenitor cells that constitute the mesenchymal condensations in the limb. *Sox9* expression precedes that of *Runx2* [56]. Coexpression of Sox9 and Runx2 in osteochondral progenitors is terminated when Sox9 and Runx2 expression is segregated into differentiated chondrocytes and osteoblasts, respectively [57]. The requirement for Runx2 in bone formation was demonstrated by the finding that *Runx2$^{-/-}$* mice have no differentiated osteoblasts [52, 53]. Humans carrying heterozygous null mutations of the *RUNX2* gene have cleidocranial dysplasia (CCD, OMIM#119600), an autosomal-dominant condition characterized by hypoplasia/aplasia of clavicles, patent fontanelles, supernumerary teeth, short stature, and other changes in skeletal patterning and growth [53].

A number of transcriptional regulators that interact with Runx2 to control osteoblast differentiation have been identified. Zfp521 regulates osteoblast differentiation by HDAC3-dependent attenuation of Runx2 activity [58]. In addition, Runx2 mediates the function of Notch signaling in regulating osteoblast differentiation [59, 60].

Signaling through the Wnt and Indian hedgehog (Ihh) pathways is required for cell fate determination of osteoprogenitors into chondrocytes or osteoblasts by controlling the expression of *Sox9* and *Runx2*. Enhanced Wnt/β-catenin signaling increased bone formation and *Runx2* expression, but inhibited chondrocyte differentiation and *Sox9* expression [61–63]. Conversely, blocking Wnt/β-catenin signaling by removing β-*catenin* or *Lrp5* and *Lrp6* in osteochondral progenitor cells resulted in ectopic chondrocyte differentiation at the expense of osteoblasts [63–66]. Therefore, Wnt/β-catenin signaling levels in the condensation determine the outcome of bone formation. Relatively high Wnt/β-catenin signaling in intramembranous ossification allows direct osteoblast differentiation in the condensation, whereas during endochondral ossification, Wnt/β-catenin signaling in the condensation is initially lower, such that only chondrocytes differentiate. At later stages of endochondral ossification, Wnt/β-catenin signaling is upregulated at the periphery of the cartilage, driving osteoblast differentiation.

Ihh signaling is required for osteoblast differentiation only during endochondral bone formation by activating Runx2 expression [67, 68]. When Ihh signaling is inactivated in perichondrial cells, they ectopically form chondrocytes that express Sox9 at the expense of Runx2. Genetic epistatic tests showed that that β-*catenin* is required downstream of *Ihh* to promote osteoblast maturation [69]. In accordance, Ihh signaling is not (required once osteoblasts express osterix *Osx*) [70], a maker for cells committed to the osteoblast fate [71].

BMPs are transforming growth factor β (TGFβ) superfamily members that were identified as secreted proteins able to promote ectopic cartilage and bone formation [72]. Unlike Ihh and Wnt signaling, BMP signaling promotes the differentiation of both osteoblasts and chondrocytes from mesenchymal progenitors. Reducing BMP signaling by removing BMP receptors leads to impaired chondrocyte and osteoblast differentiation and maturation [73]. The mechanisms underlying this unique property of BMPs have been under intense investigation for the past two decades. Our understanding of BMP action in chondrogenesis and osteogenesis has benefited greatly from molecular studies of BMP signal transduction [74].

The functions of FGF pathways in mesenchymal condensation and osteochondral progenitor differentiation remain to be elucidated, as complete genetic inactivation of FGF signaling in mesenchymal condensations has not been achieved. Nevertheless, it is clear that FGFs act in mesenchymal condensations to control intramembranous bone formation. FGF signaling can promote or inhibit osteoblast proliferation and differentiation depending on the cell context. Mutations in the genes encoding FGFR 1, 2, and 3 cause craniosynostosis (premature fusion of the cranial sutures). All of these mutations are autosomal dominant and many of them are activating mutations. The craniosynostosis syndromes involving FGFR1, 2, 3 include Apert syndrome (AS, OMIM# 101200), Beare-Stevenson cutis gyrata (OMIM#123790), Crouzon syndrome (CS, OMIM#123500), Pfeiffer syndrome (PS, OMIM#101600), Jackson-Weiss syndrome (JWS, OMIM#123150), Muenke syndrome (MS, OMIM#602849), crouzonodermoskeletal syndrome (OMIM#134934), and osteoglophonic dysplasia (OGD, OMIM#166250).

CHONDROCYTE PROLIFERATION AND DIFFERENTIATION IN THE DEVELOPING CARTILAGE

During endochondral bone formation, chondrocytes differentiate from osteochondral progenitor cells to form cartilage, which provides a growth template for the future bone. Chondrocytes undergo a tightly controlled program of progressive proliferation and hypertrophy, which is required for endochondral bone formation. In the developing cartilage of the long bone, chondrocytes at different stages of differentiation are located in distinct zones along the longitudinal axis and such organization is required for long bone elongation [Fig. 1.4(A)]. Proliferating chondrocytes express *Col2a1* (ColII), whereas hypertrophic chondrocytes express *Col10a1* (ColX). The chondrocytes that have exited the cell cycle, but have not yet become hypertrophic, are known as prehypertrophic chondrocytes. Chondrocytes either remain in one zone (i.e., those in the permanent cartilage) or transit to other zones (i.e., those in the growth plate) during development or homeostasis. This progression is precisely regulated by multiple signaling pathways.

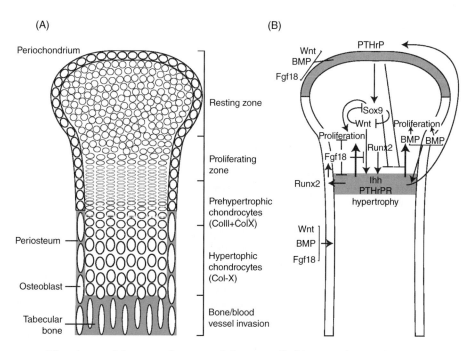

Fig. 1.4. Chondrocyte proliferation and hypertrophy are tightly controlled by signaling pathways and transcription factors. (A) Schematic drawing of a developing long bone cartilage. Chondrocytes with different properties of proliferation have different morphologies and are located in distinct locations along the longitudinal axis. See text for details. (B) Molecular regulation of chondrocyte proliferation and hypertrophy. Ihh, PTHrP, Wnt, FGF, and BMP are major signaling pathways that control chondrocyte proliferation and hypertrophy. A negative feedback loop between Ihh and PTHrP is fundamental in regulating the pace of chondrocyte hypertrophy. Transcription factors Sox9 and Runx2 act inside the cell to integrate signals from different pathways. See text for details.

Ihh is expressed in prehypertrophic and early hypertrophic chondrocytes and acts as a master regulator of endochondral bone development by promoting chondrocyte proliferation, controlling the pace of chondrocyte hypertrophy and coupling cartilage development with bone formation by inducing osteoblast differentiation in the adjacent perichondrium [67].

$Ihh^{-/-}$ mice have striking skeletal defects, including a lack of endochondral bone formation and smaller cartilage elements due to a marked decrease in chondrocyte proliferation and acceleration of hypertrophy [67, 75]. Ihh controls the pace of chondrocyte hypertrophy by activating the expression of parathyroid hormone related peptide (*PTHrP*) in articular cartilage and periarticular cells [67, 76]. PTHrP acts on the same G-protein-coupled receptors used by parathyroid hormone (PTH). These PTH/PTHrP receptors (*PPRs*) are expressed at high levels by prehypertrophic and early hypertrophic chondrocytes. PTHrP signaling is required to inhibit precocious chondrocyte hypertrophy primarily by keeping proliferating chondrocytes in the proliferating pool [77, 78]. Ihh and PTHrP form a negative feedback loop to control the chondrocyte's decision whether or not to leave the proliferating pool and become hypertrophic [Fig. 1.4(B)]. In this model, PTHrP, secreted from cells at the ends of cartilage, acts on proliferating chondrocytes to keep them proliferating. When chondrocytes displaced far enough from the source of PTHrP that the PPRs are no longer activated, they exit the cell cycle and become Ihh-producing prehypertrophic chondrocytes. Ihh diffuses through the growth plate to stimulate PTHrP expression at the ends of cartilage as way to slow down hypertrophy. This model is supported by experiments using chimeric mouse embryos [79]. Clones of $PPR^{-/-}$ chondrocytes differentiate into hypertrophic chondrocytes and produce Ihh within the wild type proliferating chondrocyte domain. This ectopic Ihh expression leads to ectopic osteoblast differentiation in the perichondrium, upregulation of PTHrP expression, and a consequent lengthening of the columns of wild type proliferating chondrocytes. These studies demonstrate that the lengths of proliferating columns, and hence the elongation potential of cartilages, are critically determined by the Ihh–PTHrP negative feedback loop. Indeed, mutations in *IHH* in humans cause brachydactyly Type A1 (OMIM#112500), which is characterized by shortened digit phalanges and short body statue [80].

Several Wnt ligands are expressed in the cartilage and perichondrium of mouse embryos [62, 81]. Some activate canonical (β-catenin-dependent) and others activate non-canonical (β-catenin-independent) pathways to regulate chondrocyte proliferation and hypertrophy. In the absence of either canonical or noncanonical Wnt signaling, chondrocyte proliferation is altered and hypertrophy is delayed [63, 81, 82]. Furthermore, both canonical and noncanonical Wnt pathways act in parallel with Ihh signaling to regulate chondrocyte proliferation and differentiation [69, 81]. Wnt and Ihh signaling may regulate common downstream targets such as Sox9 (see below) [81, 82].

Many FGF ligands and receptors (FGFRs) are expressed in the developing cartilage. The significant role of FGF signaling in skeletal development was first realized by the discovery that achondroplasia (ACH, OMIM#100800), the most common form of skeletal dwarfism in humans, was caused by a missense mutation in *FGFR3*. Later, hypochondroplasia (HCH, OMIM#146000), a milder form of dwarfism, and thanatophoric dysplasia (TD, OMIM# 187600 & 187601), a more severe form of dwarfism, were also found to result from mutations in *FGFR3*. Signaling through FGFR3 negatively regulates chondrocyte proliferation and hypertrophy [83–90], in part by direct signaling in chondrocytes [83, 84] to activate Janus kinase–signal transducer and activator of transcription-1 (Jak–Stat1) and the MAPK pathways [85]. FGFR3 signaling also interacts with the Ihh/PTHrP/BMP signaling pathways [86, 87].

Since $Fgf18^{-/-}$ mice exhibit a phenotype including increased chondrocyte proliferation that closely resembles the cartilage phenotypes of $Fgfr3^{-/-}$ mice, Fgf18 is likely a physiological ligand for FGFR3 in the mouse. However, the phenotype of the $Fgf18^{-/-}$ mouse is more severe than that of the $Fgfr3^{-/-}$ mice, suggesting that Fgf18 signals through FGFR1 in hypertrophic chondrocytes and through FGFR2 and -1 in the perichondrium. Mice conditionally lacking FGFR2 develop skeletal dwarfism with decreased bone mineral density [88, 89]. Osteoblasts also express FGFR3, and mice lacking *Fgfr3* are osteopenic [90, 91]. Thus in osteoblasts, FGF signaling positively regulates bone growth by promoting osteoblast proliferation. Interestingly, mice lacking *Fgf2* also show osteopenia, though much later in development than in $Fgfr2$-deficient mice [92], suggesting that Fgf2 may be a homeostatic factor that replaces the developmental growth factor, Fgf18, in adult bones. It is still not clear which FGFR responds to Fgf2/18 in osteoblasts.

Like the other major signaling pathways mentioned above, BMP signaling also acts during later stages of cartilage development. Both *in vitro* explant experiments and *in vivo* genetic studies showed that BMP signaling promotes chondrocyte proliferation and *Ihh* expression. The addition of BMPs to limb explants increases proliferation of chondrocytes whereas Noggin blocks chondrocyte proliferation [86, 93]. In addition, conditional removal of both *BmpRIA* and *BmpRIB* in differentiated chondrocytes leads to reduced chondrocyte proliferation and Ihh expression. BMP signaling also regulates chondrocyte hypertrophy, as removal of *BmpRIA* in chondrocytes leads to an expanded hypertrophic zone due to accelerated chondrocyte hypertrophy and delayed terminal maturation of hypertrophic chondrocytes [94]. BMP signaling regulates chondrocyte proliferation and hypertrophy at least in part through regulating Ihh expression.

BMP and FGF signaling pathways are mutually antagonistic in cartilage [86]. Comparison of cartilage phenotypes of BMP and FGF signaling mutants indicate that these two signaling pathways antagonize each other in regulating chondrocyte proliferation and hypertrophy [94].

The above signaling pathways mediate the majority of their effects on cell proliferation, differentiation, and survival by regulating the expression of key transcription factors. Sox9 and Runx2 are two critical transcription factors that integrate inputs from these signaling pathways. When *Sox9* was removed from differentiated chondrocytes, chondrocyte proliferation and the expression of matrix genes and the Ihh–PTHrP signaling components were reduced in the cartilage [56]. This phenotype is very similar to that of mice lacking both *Sox5* and *Sox6*, two other Sox-family members that require Sox9 for expression. Sox5 and Sox6 cooperate with Sox9 to maintain the chondrocyte phenotype to regulate chondrocyte specific gene expression [95]. Haploinsufficiency for *SOX9* in humans causes campomelic dysplasia (CD, OMIM# 114290), a condition that is recapitulated in $Sox9^{+/-}$ mice, and which includes cartilage hypoplasia and a perinatal lethal osteochondrodysplasia [96]. Chondrocyte hypertrophy is accelerated in the $Sox9^{+/-}$ cartilage, but delayed in Sox9-overexpressing cartilage [82, 96]. Sox9 acts in both the PTHrP and Wnt signaling pathways to control chondrocyte proliferation. PTHrP signaling in chondrocytes activates PKA, which promotes Sox9 transcriptional activity by phosphorylating it [97]. In addition, Sox9 inhibits Wnt/β-catenin signaling activity by promoting β-catenin degradation [82, 98]. Thus, Sox9 is a master transcription factor that acts in many critical stages of chondrocyte proliferation and differentiation as a central node inside prechondrocytes and chondrocytes to receive and integrate multiple signaling inputs.

In addition to its role in early osteoblast differentiation, Runx2 is expressed in prehypertrophic and hypertrophic chondrocytes and controls chondrocyte proliferation and hypertrophy. Chondrocyte hypertrophy is significantly delayed and Ihh expression is reduced in $Runx2^{-/-}$ mice, whereas *Runx2* overexpression in the cartilage results in accelerated chondrocyte hypertrophy [99, 100]. Furthermore, removing both *Runx2* and *Runx3* completely blocks chondrocyte hypertrophy and Ihh expression in mice, suggesting that Runx transcription factors control Ihh expression [101]. Thus, as with Sox9, Runx2 can be viewed as a master controlling transcription factor and a central node through which other signaling pathways are integrated in coordinate chondrocyte proliferation and hypertrophy. In chondrocytes, Runx2 acts in the Ihh-PThrP pathway to regulate cartilage growth by controlling the expression of Ihh. However, this cannot be its only function, as *Runx2* upregulation leads to accelerated chondrocyte hypertrophy, whereas *Ihh* upregulation leads to delayed chondrocyte hypertrophy. One of Runx2's Ihh-independent activities is to act in the perichondrium to inhibit chondrocyte proliferation and hypertrophy by regulating Fgf18 expression [102]. Interestingly, this role of Runx2 in the perichondrium is antagonistic to its role in chondrocytes. Recent studies have shown that histone deacetylase 4 (HDAC4), which governs chromatin structure and represses the activity of specific transcription factors, regulates chondrocyte hypertrophy and endochondral bone formation by inhibiting the activity of Runx2 [103]. Runx2 interacts with the Gli3 repressor form Gli3rep, which inhibits DNA binding by Runx2 [104]. Therefore, one mechanism whereby Hh signaling promotes osteoblast differentiation may be through enhancing Runx2 DNA binding by reducing the generation of Gli3rep.

Developing skeletal elements have distinct morphologies, which are required for their function. For example, the limb and the long bones preferentially elongate along the P–D axis. Although the molecular mechanism underlying such directional morphogenesis was poorly understood in the past, there is evidence that alignment of columnar chondrocytes in the growth plate is regulated by planar cell polarity (PCP) pathways during directional elongation of the cartilage [105, 106]. PCP is an evolutionarily conserved pathway that is required in many directional morphogenetic processes including left–right asymmetry, neural tube closure, body axis elongation and brain wiring [107–109]. Recently, a major breakthrough has been made by demonstrating that newly differentiated chondrocytes in developing long bones are polarized along the P–D axis. Vangl2 protein, a core regulatory component in the PCP pathway, is asymmetrically localized on the proximal side of chondrocytes [110]. The asymmetrical localization of Vangl2 requires a Wnt5a signaling gradient. In the $Wnt5a^{-/-}$ mutant limb, the cartilage forms a ball-like structure, and Vangl2 is symmetrically distributed on the cell membrane [110] (Fig. 1.5). Mutations in genes encoding PCP pathway components, such as *WNT5a* and *ROR2*, have been found in skeletal malformations such as the Robinow Syndrome and brachydactyly type B1, both of which are short-limb dwarfisms [111–115].

REGULATION OF CHONDROCYTE SURVIVAL

Apart from its proliferation, differentiation and polarity, chondrocyte survival is also highly regulated. The Wnt/β-catenin, Hh, and BMP pathways signaling are all important in chondrocyte survival. Chondrocyte cell death is significantly increased when β-catenin is removed [69]. Cartilage is also special as it is an avascular tissue that develops under hypoxia because chondrocytes, particularly those in the middle of the cartilage, do not have access to vascular oxygen delivery [116]. As in other hypoxic conditions, the transcription factor hypoxia-inducible factor 1 (Hif-1), and its oxygen-sensitive component Hif-1α, is the major mediator of the hypoxic response in developing cartilage. Removal of *Hif-1α* in cartilage results in chondrocyte cell death in the interior of the growth plate. A downstream target of Hif-1 in regulating the hypoxic response of chondrocytes is VEGF [117]. The extensive cell death seen in the cartilage of mice lacking *Vegfa* has a striking similarity to that observed in mice in which *Hif-1α* is removed in cartilage [116]. The Wnt/β-catenin, Hh, and BMP pathways signaling are all important in chondrocyte survival.

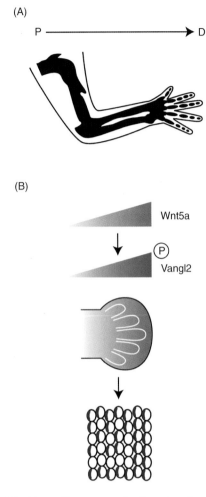

Fig. 1.5. Wnt5a gradient controls directional morphogenesis by regulating Vangl2 phosphorylation and asymmetrical localization. (A) Schematics of skeletons in a human limb that preferentially elongates along the proximal–distal axis. (B) A model of a Wnt5a gradient controlling P–D limb elongation by providing a global directional cue. Wnt5a is expressed in a gradient (orange) in the developing limb bud, and this Wnt5a gradient is translated into an activity gradient of Vangl2 by inducing different levels of Vangl2 phosphorylation (blue). In the distal limb bud of an E12.5 mouse embryo showing the forming digit cartilage, the Vangl2 activity gradient then induces asymmetrical Vangl2 localization (blue) and downstream polarized events.

Chondrocyte cell death is significantly increased when β-catenin is removed [69].

CONCLUSIONS

Skeletal formation is a process that has been perfected and highly conserved during vertebrate evolution. Understanding the molecular mechanisms regulating cartilage and bone formation during development will allow us to redeploy these pathways to promote skeletal tissue repair using endogenous cells, autologous cells and tissues, or iPS (induced pleuripotent stem) cells. Understanding skeletal development is also indispensable for understanding pathological mechanisms in skeletal diseases, finding therapeutic targets, promoting consistent cartilage or bone repair *in vivo* and eventually growing functional cartilage or bone *in vitro*.

REFERENCES

1. Santagati F, Rijli FM. 2003. Cranial neural crest and the building of the vertebrate head. *Nat Rev Neurosci* 4(10): 806–818.
2. Helms JA, Cordero D, Tapadia MD. 2005. New insights into craniofacial morphogenesis. *Development* 132(5): 851–861.
3. Pourquie O. 2011. Vertebrate segmentation: From cyclic gene networks to scoliosis. *Cell* 145(5): 650–663.
4. Christ B, Huang R, Scaal M. 2004. Formation and differentiation of the avian sclerotome. *Anat Embryol (Berl)* 208(5): 333–350.
5. Gossler A, Hrabe de Angelis M. 1998. Somitogenesis. *Curr Top Dev Biol* 38: 225–287.
6. Hirsinger E, Jouve C, Dubrulle J, Pourquie O. 2000. Somite formation and patterning. *Int Rev Cytol* 198: 1–65.
7. Scaal M, Christ B. 2004. Formation and differentiation of the avian dermomyotome. *Anat Embryol (Berl)* 208(6): 411–424.
8. Aulehla A, Pourquie O. 2006. On periodicity and directionality of somitogenesis. *Anat Embryol (Berl)* 211(Suppl 1): 3–8.
9. Dequeant ML, Glynn E, Gaudenz K, Wahl M, Chen J, Mushegian A, Pourquie O. 2006. A complex oscillating network of signaling genes underlies the mouse segmentation clock. *Science* 314(5805): 1595–1598.
10. Ilagan MX, Kopan R. 2007. SnapShot: Notch signaling pathway. *Cell* 128(6): 1246.
11. Aulehla A, Wehrle C, Brand-Saberi B, Kemler R, Gossler A, Kanzler B, Herrmann BG. 2003. Wnt3a plays a major role in the segmentation clock controlling somitogenesis. *Dev Cell* 4(3): 395–406.
12. Suriben R, Fisher DA, Cheyette BN. 2006. Dact1 presomitic mesoderm expression oscillates in phase with Axin2 in the somitogenesis clock of mice. *Dev Dyn* 235(11): 3177–3183.
13. Ishikawa A, Kitajima S, Takahashi Y, Kokubo H, Kanno J, Inoue T, Saga Y. 2004. Mouse Nkd1, a Wnt antagonist, exhibits oscillatory gene expression in the PSM under the control of Notch signaling. *Mech Dev* 121(12): 1443–1453.
14. Niwa Y, Masamizu Y, Liu T, Nakayama R, Deng CX, Kageyama R. 2007. The initiation and propagation of Hes7 oscillation are cooperatively regulated by Fgf and notch signaling in the somite segmentation clock. *Dev cell* 13(2): 298–304.

15. Hayashi S, Shimoda T, Nakajima M, Tsukada Y, Sakumura Y, Dale JK, Maroto M, Kohno K, Matsui T, Bessho Y. 2009. Sprouty4, an FGF inhibitor, displays cyclic gene expression under the control of the notch segmentation clock in the mouse PSM. *PloS One* 4(5): e5603.
16. Goldbeter A, Pourquie O. 2008. Modeling the segmentation clock as a network of coupled oscillations in the Notch, Wnt and FGF signaling pathways. *J Theor Biol* 252(3): 574–585.
17. Ozbudak EM, Lewis J. 2008. Notch signalling synchronizes the zebrafish segmentation clock but is not needed to create somite boundaries. *PLoS Genetics* 4(2): e15.
18. Moreno TA, Kintner C. 2004. Regulation of segmental patterning by retinoic acid signaling during Xenopus somitogenesis. *Dev Cell* 6(2): 205–218.
19. Diez del Corral R, Olivera-Martinez I, Goriely A, Gale E, Maden M, Storey K. 2003. Opposing FGF and retinoid pathways control ventral neural pattern, neuronal differentiation, and segmentation during body axis extension. *Neuron* 40(1): 65–79.
20. Fan CM, Tessier-Lavigne M. 1994. Patterning of mammalian somites by surface ectoderm and notochord: Evidence for sclerotome induction by a hedgehog homolog. *Cell* 79(7): 1175–1186.
21. Johnson RL, Laufer E, Riddle RD, Tabin C. 1994. Ectopic expression of Sonic hedgehog alters dorsal-ventral patterning of somites. *Cell* 79(7): 1165–1173.
22. Fan CM, Lee CS, Tessier-Lavigne M. 1997. A role for WNT proteins in induction of dermomyotome. *Dev Biol* 191(1): 160–165.
23. Capdevila J, Tabin C, Johnson RL. 1998. Control of dorsoventral somite patterning by Wnt-1 and beta-catenin. *Dev Biol* 193(2): 182–194.
24. Chiang C, Litingtung Y, Lee E, Young KE, Corden JL, Westphal H, Beachy PA. 1996. Cyclopia and defective axial patterning in mice lacking Sonic hedgehog gene function. *Nature* 383(6599): 407–413.
25. Barrow JR, Thomas KR, Boussadia-Zahui O, Moore R, Kemler R, Capecchi MR, McMahon AP. 2003. Ectodermal Wnt3/beta-catenin signaling is required for the establishment and maintenance of the apical ectodermal ridge. *Genes Dev* 17(3): 394–409.
26. Pizette S, Abate-Shen C, Niswander L. 2001. BMP controls proximodistal outgrowth, via induction of the apical ectodermal ridge, and dorsoventral patterning in the vertebrate limb. *Development* 128(22): 4463–4474.
27. Niswander L, Tickle C, Vogel A, Booth I, Martin GR. 1993. FGF-4 replaces the apical ectodermal ridge and directs outgrowth and patterning of the limb. *Cell* 75(3): 579–587.
28. Crossley PH, Minowada G, MacArthur CA, Martin GR. 1996. Roles for FGF8 in the induction, initiation, and maintenance of chick limb development. *Cell* 84(1): 127–136.
29. Sun X, Mariani FV, Martin GR. 2002. Functions of FGF signalling from the apical ectodermal ridge in limb development. *Nature* 418(6897): 501–508.
30. Ohuchi H, Nakagawa T, Yamamoto A, Araga A, Ohata T, Ishimaru Y, Yoshioka H, Kuwana T, Nohno T, Yamasaki M, Itoh N, Noji S. 1997. The mesenchymal factor, FGF10, initiates and maintains the outgrowth of the chick limb bud through interaction with FGF8, an apical ectodermal factor. *Development* 124(11): 2235–2244.
31. Sekine K, Ohuchi H, Fujiwara M, Yamasaki M, Yoshizawa T, Sato T, Yagishita N, Matsui D, Koga Y, Itoh N, Kato S. 1999. Fgf10 is essential for limb and lung formation. *Nat Genet* 21(1): 138–141.
32. Min H, Danilenko DM, Scully SA, Bolon B, Ring BD, Tarpley JE, DeRose M, Simonet WS. 1998. Fgf-10 is required for both limb and lung development and exhibits striking functional similarity to Drosophila branchless. *Genes Dev* 12(20): 3156–3161.
33. Saunders JWJ, Gasseling MT. 1968. Ectoderm-mesenchymal interaction in the origin of wing symmetry. In: Fleischmajer R, Billingham RE (eds.) *Epithelia-Mesenchymal Interactions*. Baltimore: Williams and Wilkins. pp. 78–97.
34. Riddle RD, Johnson RL, Laufer E, Tabin C. 1993. Sonic hedgehog mediates the polarizing activity of the ZPA. *Cell* 75(7): 1401–1416.
35. Chan DC, Laufer E, Tabin C, Leder P. 1995. Polydactylous limbs in Strong's Luxoid mice result from ectopic polarizing activity. *Development* 121(7): 1971–1978.
36. Tickle C. 2003. Patterning systems—from one end of the limb to the other. *Dev Cell* 4(4): 449–458.
37. Niswander L. 2002. Interplay between the molecular signals that control vertebrate limb development. *Int J Dev Biol* 46(7): 877–881.
38. Riddle RD, Ensini M, Nelson C, Tsuchida T, Jessell TM, Tabin C. 1995. Induction of the LIM homeobox gene Lmx1 by WNT7a establishes dorsoventral pattern in the vertebrate limb. *Cell* 83(4): 631–640.
39. Parr BA, Shea MJ, Vassileva G, McMahon AP. 1993. Mouse Wnt genes exhibit discrete domains of expression in the early embryonic CNS and limb buds. *Development* 119(1): 247–261.
40. Loomis CA, Harris E, Michaud J, Wurst W, Hanks M, Joyner AL. 1996. The mouse Engrailed-1 gene and ventral limb patterning. *Nature* 382(6589): 360–363.
41. Lallemand Y, Nicola MA, Ramos C, Bach A, Cloment CS, Robert B. 2005. Analysis of Msx1; Msx2 double mutants reveals multiple roles for Msx genes in limb development. *Development* 132(13): 3003–3014.
42. Ovchinnikov DA, Selever J, Wang Y, Chen YT, Mishina Y, Martin JF, Behringer RR. 2006. BMP receptor type IA in limb bud mesenchyme regulates distal outgrowth and patterning. *Dev Biol* 295(1): 103–115.
43. Khokha MK, Hsu D, Brunet LJ, Dionne MS, Harland RM. 2003. Gremlin is the BMP antagonist required for maintenance of Shh and Fgf signals during limb patterning. *Nat Genet* 34(3): 303–307.
44. Niswander L, Jeffrey S, Martin GR, Tickle C. 1994. A positive feedback loop coordinates growth and patterning in the vertebrate limb. *Nature* 371(6498): 609–612.

45. Laufer E, Nelson CE, Johnson RL, Morgan BA, Tabin C. 1994. Sonic hedgehog and Fgf-4 act through a signaling cascade and feedback loop to integrate growth and patterning of the developing limb bud. *Cell* 79(6): 993–1003.
46. Verheyden JM, Sun X. 2008. An Fgf/Gremlin inhibitory feedback loop triggers termination of limb bud outgrowth. *Nature* 454(7204): 638–641.
47. Parr BA, McMahon AP. 1995. Dorsalizing signal Wnt-7a required for normal polarity of D-V and A-P axes of mouse limb. *Nature* 374(6520): 350–353.
48. Yang Y, Niswander L. 1995. Interaction between the signaling molecules WNT7a and SHH during vertebrate limb development: Dorsal signals regulate anteroposterior patterning. *Cell* 80(6): 939–947.
49. Ten Berge D, Brugmann SA, Helms JA, Nusse R. 2008. Wnt and FGF signals interact to coordinate growth with cell fate specification during limb development. *Development* 135(19): 3247–3257.
50. Hill TP, Taketo MM, Birchmeier W, Hartmann C. 2006. Multiple roles of mesenchymal beta-catenin during murine limb patterning. *Development* 133(7): 1219–1229.
51. Cooper KL, Hu JK, ten Berge D, Fernandez-Teran M, Ros MA, Tabin CJ. 2011. Initiation of proximal–distal patterning in the vertebrate limb by signals and growth. *Science* 332(6033): 1083–1086.
52. Komori T, Yagi H, Nomura S, Yamaguchi A, Sasaki K, Deguchi K, Shimizu Y, Bronson RT, Gao YH, Inada M, Sato M, Okamoto R, Kitamura Y, Yoshiki S, Kishimoto T. 1997. Targeted disruption of Cbfa1 results in a complete lack of bone formation owing to maturational arrest of osteoblasts. *Cell* 89(5): 755–764.
53. Otto F, Thornell AP, Crompton T, Denzel A, Gilmour KC, Rosewell IR, Stamp GW, Beddington RS, Mundlos S, Olsen BR, Selby PB, Owen MJ. 1997. Cbfa1, a candidate gene for cleidocranial dysplasia syndrome, is essential for osteoblast differentiation and bone development. *Cell* 89(5): 765–771.
54. Ducy P, Zhang R, Geoffroy V, Ridall AL, Karsenty G. 1997. Osf2/Cbfa1: A transcriptional activator of osteoblast differentiation. *Cell* 89(5): 747–754.
55. Bi W, Deng JM, Zhang Z, Behringer RR, de Crombrugghe B. 1999. Sox9 is required for cartilage formation. *Nat Genet* 22(1): 85–89.
56. Akiyama H, Chaboissier MC, Martin JF, Schedl A, deCrombrugghe B. 2002. The transcription factor Sox9 has essential roles in successive steps of the chondrocyte differentiation pathway and is required for expression of Sox5 and Sox6. *Genes Dev* 16(21): 2813–2828.
57. Akiyama H, Kim JE, Nakashima K, Balmes G, Iwai N, Deng JM, Zhang Z, Martin JF, Behringer RR, Nakamura T, de Crombrugghe B. 2005. Osteo-chondroprogenitor cells are derived from Sox9 expressing precursors. *Proc Natl Acad Sci U S A* 102(41): 14665–14670.
58. Hesse E, Saito H, Kiviranta R, Correa D, Yamana K, Neff L, Toben D, Duda G, Atfi A, Geoffroy V, Horne WC, Baron R. 2010. Zfp521 controls bone mass by HDAC3-dependent attenuation of Runx2 activity. *J Cell Biol* 191(7): 1271–1283.
59. Engin F, Yao Z, Yang T, Zhou G, Bertin T, Jiang MM, Chen Y, Wang L, Zheng H, Sutton RE, Boyce BF, Lee B. 2008. Dimorphic effects of Notch signaling in bone homeostasis. *Nat Med* 14(3): 299–305.
60. Hilton MJ, Tu X, Wu X, Bai S, Zhao H, Kobayashi T, Kronenberg HM, Teitelbaum SL, Ross FP, Kopan R, Long F. 2008. Notch signaling maintains bone marrow mesenchymal progenitors bysuppressing osteoblast differentiation. *Nat Med* 14(3): 306–314.
61. Hartmann C, Tabin CJ. 2000. Dual roles of Wnt signaling during chondrogenesis in the chicken limb. *Development* 127(14): 3141–3159.
62. Guo X, Day TF, Jiang X, Garrett-Beal L, Topol L, Yang Y. 2004. Wnt/beta-catenin signaling is sufficient and necessary for synovial joint formation. *Genes Dev* 18(19): 2404–2417.
63. Day TF, Guo X, Garrett-Beal L, Yang Y. 2005. Wnt/beta-catenin signaling in mesenchymal progenitors controls osteoblast and chondrocyte differentiation during vertebrate skeletogenesis. *Dev Cell* 8(5): 739–750.
64. Hill TP, Spater D, Taketo MM, Birchmeier W, Hartmann C. 2005. Canonical Wnt/beta-catenin signaling prevents osteoblasts from differentiating into chondrocytes. *Dev Cell* 8(5): 727–738.
65. Hu H, Hilton MJ, Tu X, Yu K, Ornitz DM, Long F. 2005. Sequential roles of Hedgehog and Wnt signaling in osteoblast development. *Development* 132(1): 49–60.
66. Joeng KS, Schumacher CA, Zylstra-Diegel CR, Long F, Williams BO. 2011. Lrp5 and Lrp6 redundantly control skeletal development in the mouse embryo. *Devel Biol* 359: 222–229.
67. St-Jacques B, Hammerschmidt M, McMahon AP. 1999. Indian hedgehog signaling regulates proliferation and differentiation of chondrocytes and is essential for bone formation. *Genes Dev* 13(16): 2072–2086.
68. Long F, Chung UI, Ohba S, McMahon J, Kronenberg HM, McMahon AP. 2004. Ihh signaling is directly required for the osteoblast lineage in the endochondral skeleton. *Development* 131(6): 1309–1318.
69. Mak KK, Chen MH, Day TF, Chuang PT, Yang Y. 2006. Wnt/beta-catenin signaling interacts differentially with Ihh signaling in controlling endochondral bone and synovial joint formation. *Development* 133(18): 3695–3707.
70. Rodda SJ, McMahon AP. 2006. Distinct roles for Hedgehog and canonical Wnt signaling in specification, differentiation and maintenance of osteoblast progenitors. *Development* 133(16): 3231–3244.
71. Nakashima K, Zhou X, Kunkel G, Zhang Z, Deng JM, Behringer RR, de Crombrugghe B. 2002. The novel zinc finger-containing transcription factor osterix is required for osteoblast differentiation and bone formation. *Cell* 108(1): 17–29.
72. Wozney JM. 1989. Bone morphogenetic proteins. *Prog Growth Factor Res* 1(4): 267–280.
73. Yoon BS, Ovchinnikov DA, Yoshii I, Mishina Y, Behringer RR, Lyons KM. 2005. Bmpr1a and Bmpr1b have overlapping functions and are essential for chondrogen-

esis in vivo. *Proc Natl Acad Sci U S A* 102(14): 5062–5067.
74. Derynck R, Zhang YE. 2003. Smad-dependent and Smad-independent pathways in TGF-beta family signalling. *Nature* 425(6958): 577–584.
75. Long F, Zhang XM, Karp S, Yang Y, McMahon AP. 2001. Genetic manipulation of hedgehog signaling in the endochondral skeleton reveals a direct role in the regulation of chondrocyte proliferation. *Development* 128(24): 5099–5108.
76. Vortkamp A, Lee K, Lanske B, Segre GV, Kronenberg HM, Tabin CJ. 1996. Regulation of rate of cartilage differentiation by Indian hedgehog and PTH-related protein. *Science* 273(5275): 613–622.
77. Karaplis AC, Luz A, Glowacki J, Bronson RT, Tybulewicz VL, Kronenberg HM, Mulligan RC. 1994. Lethal skeletal dysplasia from targeted disruption of the parathyroid hormone-related peptide gene. *Genes Dev* 8(3): 277–289.
78. Lanske B, Karaplis AC, Lee K, Luz A, Vortkamp A, Pirro A, Karperien M, Defize LH, Ho C, Mulligan RC, Abou-Samra AB, Juppner H, Segre GV, Kronenberg HM. 1996. PTH/PTHrP receptor in early development and Indian hedgehog-regulated bone growth. *Science* 273(5275): 663–666.
79. Chung UI, Schipani E, McMahon AP, Kronenberg HM. 2001. Indian hedgehog couples chondrogenesis to osteogenesis in endochondral bone development. *J Clin Invest* 107(3): 295–304.
80. Gao B, Guo J, She C, Shu A, Yang M, Tan Z, Yang X, Guo S, Feng G, He L. 2001. Mutations in IHH, encoding Indian hedgehog, cause brachydactyly type A-1. *Nat Genet* 28(4): 386–388.
81. Yang Y, Topol L, Lee H, Wu J. 2003. Wnt5a and Wnt5b exhibit distinct activities in coordinating chondrocyte proliferation and differentiation. *Development* 130(5): 1003–1015.
82. Akiyama H, Lyons JP, Mori-Akiyama Y, Yang X, Zhang R, Zhang Z, Deng JM, Taketo MM, Nakamura T, Behringer RR, McCrea PD, De Crombrugghe B. 2004. Interactions between Sox9 and β-catenin control chondrocyte differentiation. *Genes Dev* 18(9): 1072–1087.
83. Dailey L, Laplantine E, Priore R, Basilico C. 2003. A network of transcriptional and signaling events is activated by FGF to induce chondrocyte growth arrest and differentiation. *J Cell Biol* 161(6): 1053–1066.
84. Henderson JE, Naski MC, Aarts MM, Wang D, Cheng L, Goltzman D, Ornitz DM. 2000. Expression of FGFR3 with the G380R achondroplasia mutation inhibits proliferation and maturation of CFK2 chondrocytic cells. *J Bone Miner Res* 15(1): 155–165.
85. Raucci A, Laplantine E, Mansukhani A, Basilico C. 2004. Activation of the ERK1/2 and p38 mitogen-activated protein kinase pathways mediates fibroblast growth factor-induced growth arrest of chondrocytes. *J Biol Chem* 279(3): 1747–1756.
86. Minina E, Kreschel C, Naski MC, Ornitz DM, Vortkamp A. 2002. Interaction of FGF, Ihh/Pthlh, and BMP signaling integrates chondrocyte proliferation and hypertrophic differentiation. *Dev Cell* 3(3): 439–449.
87. Naski MC, Colvin JS, Coffin JD, Ornitz DM. 1998. Repression of hedgehog signaling and BMP4 expression in growth plate cartilage by fibroblast growth factor receptor 3. *Development* 125(24): 4977–4988.
88. Eswarakumar VP, Monsonego-Ornan E, Pines M, Antonopoulou I, Morriss-Kay GM, Lonai P. 2002. The IIIc alternative of Fgfr2 is a positive regulator of bone formation. *Development* 129(16): 3783–3793.
89. Yu K, Xu J, Liu Z, Sosic D, Shao J, Olson EN, Towler DA, Ornitz DM. 2003. Conditional inactivation of FGF receptor 2 reveals an essential role for FGF signaling in the regulation of osteoblast function and bone growth. *Development* 130(13): 3063–3074.
90. Valverde-Franco G, Liu H, Davidson D, Chai S, Valderrama-Carvajal H, Goltzman D, Ornitz DM, Henderson JE. 2004. Defective bone mineralization and osteopenia in young adult FGFR3−/− mice. *Hum Mol Genet* 13(3): 271–284.
91. Xiao L, Naganawa T, Obugunde E, Gronowicz G, Ornitz DM, Coffin JD, Hurley MM. 2004. Stat1 controls postnatal bone formation by regulating fibroblast growth factor signaling in osteoblasts. *J Biol Chem* 279(26): 27743–27752.
92. Montero A, Okada Y, Tomita M, Ito M, Tsurukami H, Nakamura T, Doetschman T, Coffin JD, Hurley MM. 2000. Disruption of the fibroblast growth factor-2 gene results in decreased bone mass and bone formation. *J Clin Invest* 105(8): 1085–1093.
93. Minina E, Wenzel HM, Kreschel C, Karp S, Gaffield W, McMahon AP, Vortkamp A. 2001. BMP and Ihh/PTHrP signaling interact to coordinate chondrocyte proliferation and differentiation. *Development* 128(22): 4523–4534.
94. Yoon BS, Pogue R, Ovchinnikov DA, Yoshii I, Mishina Y, Behringer RR, Lyons KM. 2006. BMPs regulate multiple aspects of growth-plate chondrogenesis through opposing actions on FGF pathways. *Development* 133(23): 4667–4678.
95. Smits P, Li P, Mandel J, Zhang Z, Deng JM, Behringer RR, de Crombrugghe B, Lefebvre V. 2001. The transcription factors L-Sox5 and Sox6 are essential for cartilage formation. *Dev Cell* 1(2): 277–290.
96. Bi W, Huang W, Whitworth DJ, Deng JM, Zhang Z, Behringer RR, de Crombrugghe B. 2001. Haploinsufficiency of Sox9 results in defective cartilage primordia and premature skeletal mineralization. *Proc Natl Acad Sci U S A* 98(12): 6698–6703.
97. Huang W, Chung UI, Kronenberg HM, de Crombrugghe B. 2001. The chondrogenic transcription factor Sox9 is a target of signaling by the parathyroid hormone-related peptide in the growth plate of endochondral bones. *Proc Natl Acad Sci U S A* 98(1): 160–165.
98. Topol L, Chen W, Song H, Day TF, Yang Y. 2009. Sox9 inhibits Wnt signaling by promoting beta-catenin phosphorylation in the nucleus. *J Biol Chem* 284(5): 3323–3333.

99. Kim IS, Otto F, Zabel B, Mundlos S. 1999. Regulation of chondrocyte differentiation by Cbfa1. *Mech Dev* 80(2): 159–170.
100. Takeda S, Bonnamy JP, Owen MJ, Ducy P, Karsenty G. 2001. Continuous expression of Cbfa1 in nonhypertrophic chondrocytes uncovers its ability to induce hypertrophic chondrocyte differentiation and partially rescues Cbfa1-deficient mice. *Genes Dev* 15(4): 467–481.
101. Yoshida CA, Yamamoto H, Fujita T, Furuichi T, Ito K, Inoue K, Yamana K, Zanma A, Takada K, Ito Y, Komori T. 2004. Runx2 and Runx3 are essential for chondrocyte maturation, and Runx2 regulates limb growth through induction of Indian hedgehog. *Genes Dev* 18(8): 952–963.
102. Hinoi E, Bialek P, Chen YT, Rached MT, Groner Y, Behringer RR, Ornitz DM, Karsenty G. 2006. Runx2 inhibits chondrocyte proliferation and hypertrophy through its expression in the perichondrium. *Genes Dev* 20(21): 2937–2942.
103. Vega RB, Matsuda K, Oh J, Barbosa AC, Yang X, Meadows E, McAnally J, Pomajzl C, Shelton JM, Richardson JA, Karsenty G, Olson EN. 2004. Histone deacetylase 4 controls chondrocyte hypertrophy during skeletogenesis. *Cell* 119(4): 555–566.
104. Ohba S, Kawaguchi H, Kugimiya F, Ogasawara T, Kawamura N, Saito T, Ikeda T, Fujii K, Miyajima T, Kuramochi A, Miyashita T, Oda H, Nakamura K, Takato T, Chung UI. 2008. Patched1 haploinsufficiency increases adult bone mass and modulates Gli3 repressor activity. *Devel Cell* 14(5): 689–699.
105. Ahrens MJ, Li Y, Jiang H, Dudley AT. 2009. Convergent extension movements in growth plate chondrocytes require gpi-anchored cell surface proteins. *Development* 136(20): 3463–3474.
106. Li Y, Dudley AT. 2009. Noncanonical frizzled signaling regulates cell polarity of growth plate chondrocytes. *Development* 136(7): 1083–1092.
107. Gray RS, Roszko I, Solnica-Krezel L. 2011. Planar cell polarity: coordinating morphogenetic cell behaviors with embryonic polarity. *Devel Cell* 21(1): 120–133.
108. Goodrich LV, Strutt D. 2011. Principles of planar polarity in animal development. *Development* 138(10): 1877–1892.
109. Bayly R, Axelrod JD. 2011. Pointing in the right direction: new developments in the field of planar cell polarity. *Nat Rev Genet* 12(6): 385–391.
110. Gao B, Song H, Bishop K, Elliot G, Garrett L, English MA, Andre P, Robinson J, Sood R, Minami Y, Economides AN, Yang Y. 2011. Wnt signaling gradients establish planar cell polarity by inducing Vangl2 phosphorylation through Ror2. *Devel Cell* 20(2): 163–176.
111. Minami Y, Oishi I, Endo M, Nishita M. 2010. Ror-family receptor tyrosine kinases in noncanonical Wnt signaling: their implications in developmental morphogenesis and human diseases. *Dev Dyn* 239(1): 1–15.
112. Person AD, Beiraghi S, Sieben CM, Hermanson S, Neumann AN, Robu ME, Schleiffarth JR, Billington CJ, Jr, van Bokhoven H, Hoogeboom JM, Mazzeu JF, Petryk A, Schimmenti LA, Brunner HG, Ekker SC, Lohr JL. 2010. WNT5A mutations in patients with autosomal dominant Robinow syndrome. *Dev Dyn* 239(1): 327–337.
113. van Bokhoven H, Celli J, Kayserili H, van Beusekom E, Balci S, Brussel W, Skovby F, Kerr B, Percin EF, Akarsu N, Brunner HG. 2000. Mutation of the gene encoding the ROR2 tyrosine kinase causes autosomal recessive Robinow syndrome. *Nat Genet* 25(4): 423–426.
114. Schwabe GC, Tinschert S, Buschow C, Meinecke P, Wolff G, Gillessen-Kaesbach G, Oldridge M, Wilkie AO, Komec R, Mundlos S. 2000. Distinct mutations in the receptor tyrosine kinase gene ROR2 cause brachydactyly type B. *Am J Hum Genet* 67(4): 822–831.
115. DeChiara TM, Kimble RB, Poueymirou WT, Rojas J, Masiakowski P, Valenzuela DM, Yancopoulos GD. 2000. Ror2, encoding a receptor-like tyrosine kinase, is required for cartilage and growth plate development. *Nat Genet* 24(3): 271–274.
116. Schipani E, Ryan HE, Didrickson S, Kobayashi T, Knight M, Johnson RS. 2001. Hypoxia in cartilage: HIF-1alpha is essential for chondrocyte growth arrest and survival. *Genes Dev* 15(21): 2865–2876.
117. Zelzer E, Mamluk R, Ferrara N, Johnson RS, Schipani E, Olsen BR. 2004. VEGFA is necessary for chondrocyte survival during bone development. *Development* 131(9): 2161–2171.

2

Signal Transduction Cascades Controlling Osteoblast Differentiation

David J.J. de Gorter and Peter ten Dijke

Introduction 15
Runx2 and Osterix Transcription Factors 15
BMP Signaling 16
TGF-β Signaling 17
WNT Signaling 17
Hedgehog Signaling 19
PTH Signaling 19

IGF-1 Signaling 19
FGF Signaling 19
Notch Signaling 20
Concluding Remarks 20
Acknowledgments 21
Abbreviations 21
References 21

INTRODUCTION

Mesenchymal stem cells (MSCs) are pluripotent cells located in bone marrow, muscles, and fat that can differentiate into a variety of tissues, including bone, cartilage, muscle and fat [1, 2]. Differentiation toward these lineages is controlled by a multitude of cytokines, which regulate the expression of cell-lineage specific sets of transcription factors. Among the cytokines involved in osteoblast differentiation are the Hedgehogs, BMPs, TGF-β, PTH, and WNTs. The signal transduction cascades initiated by these cytokines and their effect on osteoblast differentiation will be discussed in this chapter. Osteoblasts and chondrocytes are thought to differentiate from a common mesenchymal precursor, the osteochondrogenic precursor (Fig. 2.1). The osteoblastic differentiation process can be divided into several stages, including proliferation, extracellular matrix deposition, matrix maturation, and mineralization [3]. To investigate osteoblast differentiation, the expression levels of distinct differentiation markers are used, including alkaline phosphatase (ALP), type I collagen (Col1), bone sialoprotein (BSP), osteopontin (OPN), and osteocalcin (OC). Whereas ALP is used as an early marker, OC is considered to be a late marker for osteoblast differentiation.

RUNX2 AND OSTERIX TRANSCRIPTION FACTORS

An essential event in osteoblast differentiation, and a point of convergence of many signal transduction pathways involved, is the activation of the transcription factor Runx2 [also known as core-binding factor α subunit (Cbfa1) or AML3]. Runx2 is a master switch for osteoblast differentiation. This is demonstrated by the fact that *Runx2*-deficient mice completely lack osteoblasts and fail to form hypertrophic chondrocytes, and they produce a cartilaginous skeleton that is completely devoid of mineralized matrix [4]. In humans, heterozygous insertions, deletions, and nonsense mutations leading to translational stop codons in the DNA-binding domain or in the C-terminal transactivating region of the *Runx2* gene underlie the rare skeletal disorder cleidocranial dysplasia (CCD). CCD is characterized by defective development of the cranial bones and the complete or partial absence of the collar bones, emphasizing the importance of Runx2 in bone formation [5]. By interacting with many transcriptional activators and repressors and other coregulatory proteins, Runx2 can either positively or negatively regulate expression of a variety of osteoblast-specific genes including Col1, ALP, OPN, osteonectin (ON) and OC [6, 7].

Primer on the Metabolic Bone Diseases and Disorders of Mineral Metabolism, Eighth Edition. Edited by Clifford J. Rosen.
© 2013 American Society for Bone and Mineral Research. Published 2013 by John Wiley & Sons, Inc.

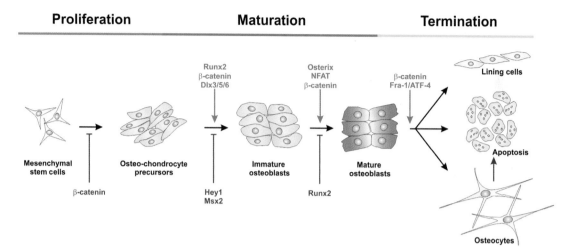

Fig. 2.1. Schematic model of MSC differentiation toward the osteoblastic linage and the impact of transcriptional regulators in this process. ATF4: Activating transcription factor-4; Dlx: Distalless homeobox; FRA: Fos-related antigen; Osx: Osterix; Runx2: Runt-related transcription factor2.

In addition, Runx2 regulates expression of the zinc-finger-containing transcription factor Osterix. The promoter of the *Osx (Sp7)* gene (which encodes Osterix) contains a consensus Runx2-binding sequence, suggesting that *Osx* is a direct Runx2 target gene [8]. Whereas Runx2 expression is not affected in $Osx^{-/-}$ mice, Osterix expression is lost in *Runx2*-deficient mice [9]. Similar to mice deficient in *Runx2*, $Osx^{-/-}$ mice lack osteoblasts, showing the requirement of this transcription factor in bone formation. Osterix can interact with the nuclear factor for activated T cells 2 (NFAT2), which cooperates with Osterix in controlling the transcription of target genes such as osteocalcin, osteopontin, osteonectin, and collagen1 [9, 10]. Since nuclear localization of NFAT transcription factors is regulated by the Ca^{2+}-calcineurin pathway, signaling pathways that modulate intracellular Ca^{2+} levels can potentially control Osterix-mediated osteoblast differentiation via NFAT activation.

Other transcription factors that are involved in osteoblast differentiation are homeobox proteins such as Msx2, Distalless homeobox(Dlx)-3, Dlx-5, Dlx-6, and members of the activator protein-1 (AP-1) family such as Fos, Fos-related antigen (Fra), and activating transcription factor-4 (ATF4) (Fig. 2.1). However, deficiency of these genes does not result in complete loss of osteoblasts like in $Runx2^{-/-}$ mice and $Osx^{-/-}$ mice, pointing at a facilitating role in osteoblastogenesis.

BMP SIGNALING

BMPs belong to the TGF-β superfamily and were originally identified as the active components in bone extracts capable of inducing bone formation at ectopic sites [11]. BMPs are expressed in skeletal tissue and are required for skeletal development and maintenance of adult bone homeostasis, and they play an important role in fracture healing [12, 13]. Conditional knockout mice deficient in specific BMP members in bone display skeletal defects, and several naturally occurring mutations in BMPs or their receptors underlie inherited skeletal disorders [12, 13]. For example, fibrodysplasia ossificans progressiva (FOP), in which bone is progressively formed at ectopic sites, has been linked to a heterozygous activating mutation in the BMP type I receptor activin receptor-like kinase (ALK)2 [14, 15]. Several preclinical studies in primates and other mammals have proven the effectiveness of using BMPs in restoring large segmental bone defects, and BMPs are used therapeutically in open fractures of long bones, non-unions, and spinal fusion [16].

BMPs bind as dimers to type-I and type-II serine/threonine receptor kinases, forming an oligomeric complex (Fig. 2.2). Upon oligomerization, the constitutively active type II receptors phosphorylate and consequently activate the type I receptors. Subsequently, the activated type I receptors phosphorylate the intracellular signaling mediators BMP receptor-regulated Smads, Smad1, -5 and -8, at their extreme C-termini. The receptor-regulated Smads then associate with the Co-Smad, Smad4, and translocate into the nucleus, where they together with other transcription factors bind promoters of target genes and control their expression (Fig. 2.2) [17, 18]. For example, Runx2 interacts with Smad1 and -5, and cooperates in controlling BMP-induced osteoblast-specific gene expression and osteogenic differentiation [19, 20]. Interestingly, a nonsense mutation found in a CCD patient results in the expression of a truncated Runx2 mutant that displayed impaired Smad1 interaction and inhibited BMP-induced ALP activity [21]. Moreover, BMP signaling induces expression of both BMPs and Runx2 thereby generating a positive feedback loop [19, 22]. In

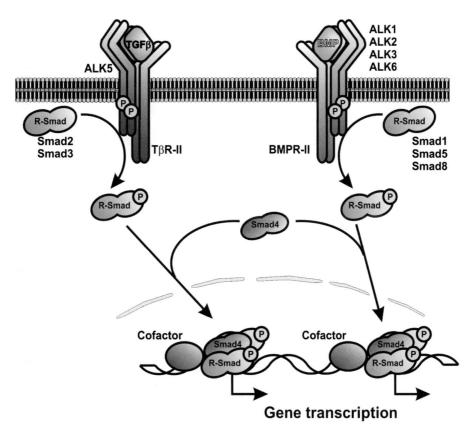

Fig. 2.2. Schematic model illustrating the TGF-β and BMP signaling pathway in osteoblasts. TGF-β and BMP signal via heteromeric complexes of type I and type II transmembrane serine/threonine kinase receptors and specific intracellular Smad effector proteins. Activated heteromeric complexes between R-Smad and Smad4 can act as transcription factors and regulate gene transcriptional responses, ultimately stimulating alkaline phosphatase activity and bone formation.

addition, BMPs control the expression of Id (inhibitor of differentiation or inhibitor of DNA binding) proteins [23, 24]. Id proteins are dominant negative inhibitors of basic helix–loop–helix proteins, which inhibit osteoblast differentiation. Indeed, BMP-induced bone formation *in vivo* was found to be suppressed in Id1/Id3 heterozygous knockout mice [25]. Furthermore, BMP-2 was found to induce expression of Osterix, which besides Runx2 also seems to be mediated by the p38 and JNK MAP kinases [9, 26, 27].

TGF-β SIGNALING

TGF-β is implicated in the control of proliferation, migration, differentiation, and survival of many different cell types. TGF-β is one of the most abundant cytokines in bone matrix and plays a major role in development and maintenance of the skeleton, affecting both cartilage and bone metabolism [28]. Interestingly, TGF-β can have both positive and negative effects on bone formation depending on the context and concentration [28-30].

TGF-β signals via a similar mechanism as the related BMPs. However, upon binding to its specific receptors, TGF-β induces activation of Smad2 and -3 (Fig. 2.2) [17, 18]. Smad3 overexpression in mouse osteoblastic MC3T3-E1 cells enhanced the levels of bone matrix proteins, ALP activity, and mineralization [31]. As is the case for Smad1 and -5, Runx2 also interacts with Smad3 and cooperates in regulating TGF-β-induced transcription [29]. This interaction requires a functional Runx2 C-terminal domain, since the truncated Runx2 mutant derived from a CCD patient is unable to interact with Smad3 [21, 29, 32]. The effects of TGF-β/Smad3 signaling on the function of Runx2 depend on the cell type and promoter context [29].

WNT SIGNALING

WNTs are secreted glycoproteins that transduce their signals via 7-transmembrane spanning receptors of the frizzled family, and co-receptors low-density lipoprotein receptor-related protein (LRP)-5 and -6, to β-catenin

18 *Molecular, Cellular, and Genetic Determinants of Bone Structure and Formation*

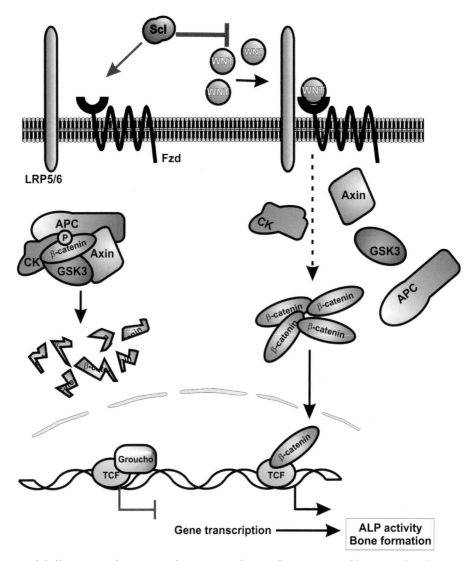

Fig. 2.3. Schematic model illustrating the canonical WNT signaling pathway in osteoblasts. In the absence of WNT, β-catenin is forming a complex with APC, Axin, GSK3, and CK1, and becomes phosphorylated, ubiquitinated and targeted for proteosomal degradation. WNT binding to frizzled (Fzd) receptors and LRP5/6 coreceptors prevents formation of the complex, leading to stabilization of β-catenin and regulation of gene expression through LEF/TCF transcription factors, ultimately stimulating alkaline phosphatase activity and bone formation. Sclerostin (Scl) proteins antagonize Wnt responses by binding to LRP5/6.

(Fig. 2.3) [33]. In the absence of a WNT ligand, β-catenin forms a complex with adenomatous polyposis coli (APC), Axin, glycogen synthase kinase 3 (GSK3), and casein kinase I (CK1). This complex facilitates phosphorylation and proteosomal degradation of β-catenin. In the presence of the WNT ligand, this complex dissociates, leading to accumulation of cytoplasmic β-catenin and its translocation into the nucleus, where it initiates the transcription of target genes via complex formation with T-cell factor (TCF)/lymphoid enhancing factor (Lef1) transcription factors [33]. Conditional deletion of β-catenin led osteochondroprogenitor cells to differentiate into chondrocytes instead of osteoblasts during both intramembranous and endochondral ossification, whereas ectopic WNT signaling enhanced osteoblast differentiation [34–36], indicating that WNT signaling drives differentiation of osteochondroprogenitor cells towards the osteoblast lineage (Fig. 2.1). Moreover, WNT signaling in osteocytes plays an essential role in controlling normal bone homeostasis, while mice deficient in β-catenin expression in osteocytes display progressive bone loss [37]. Furthermore, loss or gain of function mutations in *LRP5* have been associated with bone diseases characterized by low or high bone mass due to decreased or increased osteoblast activity, respectively [38–40]. In addition, mutations in the coding region or in regulatory elements of *SOST*, the gene encoding the osteocyte-derived WNT antagonist sclerostin, underlie the rare high bone mass disorders sclerosteosis and Van Buchem disease, respectively (Fig. 2.3) [41]. All these findings demonstrate the importance of WNT signaling in controlling bone formation.

HEDGEHOG SIGNALING

There are three Hedgehog proteins in mammals (Sonic, Indian, and Desert hedgehog), and they are critically important for development. In the endochondral skeleton, Indian hedgehog (Ihh) was found to be indispensable for osteoblast development, since mice deficient in Ihh completely lack osteoblasts in bones formed by endochondral ossification [42, 43].

Cellular responses to the Hh signal are controlled by two transmembrane proteins: the tumor suppressor 12-transmembrane protein Patched-1 (Ptch) and the 7-transmembrane receptor and oncoprotein Smoothened (Smo). The latter has homology to G protein–coupled receptors and transduces the Hh signal. In the absence of Hh, Ptch maintains Smo in an inactive state. With the binding of Hh, Ptch inhibition of Smo is released, and intracellular signaling is initiated [44]. The transcriptional response to Hh signaling is mediated by three closely related zinc finger transcription factors termed Gli proteins: Gli1, Gli2, and Gli3, each with different roles and a distinct set of target genes. Gli2 functions mainly as a transcriptional activator. In the absence of Hh, Gli3 is processed into a transcription repressor. In the presence of Hh, however, full-length Gli3 translocates into the nucleus, which has transcriptional activation properties. Gli1 acts only as transcriptional activator and is induced by Hh signals [44]. Gli3 has a pivotal role in limb bud development and regulation of digit number and identity [45, 46]. Ihh regulates osteoblast differentiation via a Gli2-mediated increase in expression and function of Runx2 [47].

PTH SIGNALING

Parathyroid hormone (PTH) and its related peptide PTHrP can have both anabolic and catabolic effects on bone. Whereas intermittent PTH administration induces bone formation, continuous treatment of PTH leads to bone loss [48]. The critical role of PTH(rP) in bone development is evident from loss of function of these genes and their receptors in mice and humans. Mice lacking PTHrP die perinatally, likely from respiratory failure due to abnormalities in endochondral bone development [49]. A less severe phenotype is found for mice deficient in PTH, which are viable and display a slightly expanded hypertrophic zone [50]. Transgenic mice that overexpress PTHrP under control of the collagen type II promoter develop shortened limbs due to delayed mineralization and chondrocyte maturation in the growth plate [51]. Mice lacking the PTH receptor PTHR1 demonstrate growth plate abnormalities due to premature chondrocyte maturation [52]. In humans, loss of function in PTH1R has been linked to Blomstrand lethal osteochondrodysplasia [53]. These patients suffer also from advanced skeletal maturation and premature ossification of the skeleton.

PTH(rP) signals via the 7-transmembrane G protein coupled receptor PTHR1; and, upon ligand binding, several intracellular signaling pathways can be activated, including the cAMP/protein kinase A (PKA) and protein kinase C (PKC) pathways. Different mechanisms have been suggested to explain the anabolic and catabolic effects of PTH; PTH may have diverse effects on the proliferation, commitment, differentiation or apoptosis of the osteoblasts. Effects of PTH and PTHrP appear highly context dependent; their action varies with cell type, stage of cell differentiation, dosage, and exposure time. For example, the bone anabolic effects of PTH involve an increase in expression and PKA-dependent activation of Runx2 (Fig. 2.4) [54]. On the other hand, PTH has also shown to repress Runx2 and Osterix expression in osteoblastic cells [55].

IGF-1 SIGNALING

Insulin-like growth factor-1 (IGF-1) is secreted by skeletal cells and is considered to be an auto- or paracrine regulator of osteoblastic cell function [56]. IGF-1-deficient mice were found to develop a smaller, but more compact bone structure [57]. IGF-1 signals via the IGF1 receptor, which like many other tyrosine kinase receptors activates the phosphatidylinositol 3-kinase (PI3K)-Akt and Ras-ERK MAP kinase pathways (Fig. 2.4). Interestingly, Akt1/Akt2 double-knockout mice show a phenotype that resembles that of mice deficient in the IGF-1 receptor, which includes impaired bone development [58]. Moreover, osteoblastic differentiation by forced expression of Runx2 was found to be inhibited by the co-expression of dominant negative Akt, or by treatment with IGF-1 antibodies or the pharmacological PI3K-inhibitor LY294002 [59]. IGF-1-induced Akt activation results in phosphorylation and nuclear exclusion of forkhead transcription factor Foxo1, thereby relieving the inhibitory effect of Foxo1 on the binding of Runx2 to the promoters of its target genes, and thus promoting osteoblast differentiation [60]. In addition, IGF-1 was found to induce expression of Osterix, which occurs in an ERK-, p38-, and c-Jun N-terminal kinase (JNK)-dependent manner [26]. JNK activity was also found to be required for late-stage osteoblast differentiation by controlling ATF-4 expression (Fig. 2.1) [61]. Via these mechanisms, IGF-1 can stimulate osteoblast differentiation.

FGF SIGNALING

Fibroblast growth factors (FGFs) are important regulators of endochondral and intramembranous bone formation,

20 Molecular, Cellular, and Genetic Determinants of Bone Structure and Formation

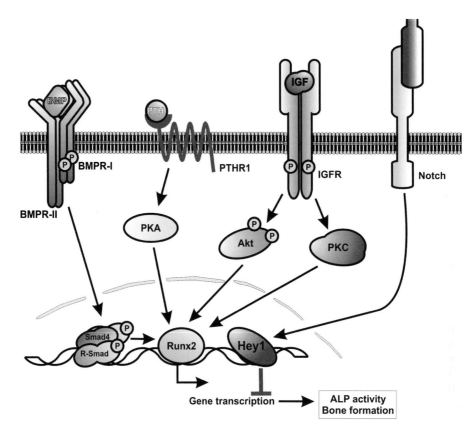

Fig. 2.4. Schematic model illustrating signaling pathways controlling Runx2-mediated osteoblast differentiation.

development and apoptosis, affecting both chondrogenesis and osteogenesis [62]. Many human craniosynostosis disorders have been linked to activating mutations in FGF receptors (FGFRs) [62]. Disruption of FGFR2 signaling in skeletal tissues results in skeletal dwarfism and decreased bone density due to disruption of the proliferation of osteoprogenitors and the anabolic function of mature osteoblasts while osteoblast differentiation was found not to be affected [63, 64]. FGFs are known to induce proliferation of immature osteoblasts via the Ras/ERK MAPK pathway. In addition, FGF signaling was demonstrated to stimulate expression, DNA-binding and transcriptional activities of Runx2, mainly in a PKC-dependent manner [65].

NOTCH SIGNALING

Notch proteins are transmembrane receptors that control cell fate decisions and inhibit osteoblastic differentiation [66–69]. The binding of the transmembrane Notch ligands Delta, Serrate, and Lag2 to Notch receptors induces cleavage of the Notch extracellular domain near the transmembrane region [66]. The resulting membrane-associated Notch is then cleaved by presenilin, generating the Notch intracellular domain (NICD), which then translocates into the nucleus. Here the NCID forms a complex with members of the CSL family (C promoter-binding factor 1 (CBF1), Suppressor of Hairless [Su(H)], and longevity assurance gene-1 (LAG-1)) of DNA binding proteins, which recruits coactivators to drive transcription of target genes [66]. In osteoblast precursor cells, Notch stimulates the expression of Hey1, which physically interacts with Runx2 and thereby inhibits its transcriptional activity and consequently the differentiation of MSCs into osteoblasts (Figs 2.1, 2.4) [68, 69]. Mice in which Notch signaling was disrupted in the limb skeletogenic mesenchyme displayed reduced numbers of mesenchymal stem cells and increased trabecular bone mass, and they developed severe osteopenia [69].

CONCLUDING REMARKS

The effect certain signaling events evoke during the osteoblast differentiation process is highly dependent on which signaling molecules are activated or inhibited, the magnitude and the duration of the response, and the differentiation stage of the responding cell. Given that regulation of Runx2 activity is a point of convergence of many of the signal transduction routes discussed in this chapter, there is also a high degree of cross-talk between these pathways. For example, apart from its C-terminal phosphorylation by BMP type I receptors, Smad1 can be phosphorylated by the ERK, p38, and JNK MAP kinases and subsequently by GSK3, which results in cytoplasmic retention and increased proteosomal degradation of Smad1 [70, 71]. In this manner, FGF and WNT signaling can control the duration of BMP signaling [70–72]. On the other hand, β-catenin-TCF/Lef1 can interact with Smad1 and -3 proteins to cooperate in transcription of target genes [73, 74]. TGF-β can inhibit BMP2-induced transcription and osteoblast differentiation [30, 75]; however, these inhibitory effects depend on the timing and environmental conditions of the co-stimulation, and under specific conditions TGF-β can actually promote BMP-induced differentiation toward the osteoblast lineage [76, 77]. Notch-1 overexpression inhibits osteoblast differentiation by suppressing WNT signaling [67], Gli2 mediates BMP2 expression in osteoblasts in response to Hh signaling [78], and Hh signaling is required for accurate β-catenin-mediated WNT signaling in osteoblasts [36]. tumor necrosis factor-α (TNF-α) inhibits BMP2-induced Smad activation through induction of NF κB [79] and in part through JNK signaling [80]. The TGF-β type II receptor associates with and phosphorylates the PTHR1, leading to internalization of both receptors and attenuation of both TGF-β and PTH signaling [81]. Thus, the combined action of the signal transduction pathways induced by bone-promoting cytokines determines the commitment of MSCs toward the osteoblast lineage and the efficiency of bone formation.

ACKNOWLEDGMENTS

Research on signaling pathways controlling osteoblast differentiation in the laboratory of Prof. Dr. Peter ten Dijke was supported by a grant of the Dutch Organization for Scientific Research (918.66.606).

This chapter is an updated version of *Signal transduction cascades controlling osteoblast differentiation* by C. Krause, D.J.J. de Gorter, M. Karperien, and P. ten Dijke, in: *Primer on the Metabolic Bone Diseases and Disorders of Mineral Metabolism, 7th Ed.* Section Editor V. Rosen. *Journal of Bone and Mineral Research*, ASBMR Publications, Primer, 2008: 7: 10–16, ISBN 978-0-9778882-1-4).

ABBREVIATIONS

ALK	activin receptor-like kinase
ALP	alkaline phosphatase
AP-1	activator protein-1
APC	adenomatous polyposis coli
ATF4	activating transcription factor-4
BMP	bone morphogenetic protein
BSP	bone sialoprotein
CCD	cleidocranial dysplasia
CK1	casein kinase I
Col1	type I collagen
Dlx	Distalless homeobox
ERK	extracellular signal-regulated kinase
FGF	fibroblast growth factor
FOP	fibrodysplasia ossificans progressiva
Fra	Fos-related antigen
Fzd	frizzled
GSK3	glycogen synthase kinase 3
Id	inhibitor of differentiation/inhibitor of DNA binding
IGF-1	insulin-like growth factor-1
Ihh	Indian hedgehog
JNK	c-Jun N-terminal kinase
LAG-1	longevity assurance gene-1
Lef1	lymphoid enhancing factor
LRP	low-density lipoprotein receptor-related protein
MAPK	mitogen-activated protein kinase
MSC	mesenchymal stem cell
NFAT	nuclear factor for activated T cells
NF κB	nuclear factor-κB
NICD	Notch intracellular domain
OC	osteocalcin
ON	osteonectin
OPN	osteopontin
PI3K	phosphatidylinositol 3-kinase
PKA	protein kinase A
PKC	protein kinase C
Ptch	patched
PTH	parathyroid hormone
Runx2	Runt-related transcription factor2
Smo	Smoothened
TCF	T-cell factor
TGF-β	transforming growth factor-β
TNF-α	tumor necrosis factor-α

REFERENCES

1. Caplan AI, Bruder SP. 2001. Mesenchymal stem cells: Building blocks for molecular medicine in the 21st century. *Trends MolMed* 7(6): 259–64.
2. Jiang Y, Jahagirdar BN, Reinhardt RL, Schwartz RE, Keene CD, Ortiz-Gonzalez XR, et al. 2002. Pluripotency of mesenchymal stem cells derived from adult marrow. *Nature* 418(6893): 41–9.
3. Stein GS, Lian JB. 1993. Molecular mechanisms mediating proliferation/differentiation interrelationships during progressive development of the osteoblast phenotype. *EndocrRev* 14(4): 424–42.
4. Otto F, Thornell AP, Crompton T, Denzel A, Gilmour KC, Rosewell IR, et al. 1997. Cbfa1, a candidate gene for cleidocranial dysplasia syndrome, is essential for osteoblast differentiation and bone development. *Cell* 89(5): 765–71.

5. Mundlos S. 1999. Cleidocranial dysplasia: Clinical and molecular genetics. *J Med Genet* 36(3): 177–82.
6. Harada H, Tagashira S, Fujiwara M, Ogawa S, Katsumata T, Yamaguchi A, et al. 1999. Cbfa1 isoforms exert functional differences in osteoblast differentiation. *J Biol Chem* 274(11): 6972–8.
7. Kern B, Shen J, Starbuck M, Karsenty G. 2001. Cbfa1 contributes to the osteoblast-specific expression of type I collagen genes. *J Biol Chem* 276(10): 7101–7.
8. Nishio Y, Dong Y, Paris M, O'Keefe RJ, Schwarz EM, Drissi H. 2006. Runx2-mediated regulation of the zinc finger Osterix/Sp7 gene. *Gene* 372: 62–70.
9. Nakashima K, Zhou X, Kunkel G, Zhang Z, Deng JM, Behringer RR, et al. 2002. The novel zinc finger-containing transcription factor osterix is required for osteoblast differentiation and bone formation. *Cell* 108(1): 17–29.
10. Koga T, Matsui Y, Asagiri M, Kodama T, de Crombrugghe B, Nakashima K, et al. 2005. NFAT and Osterix cooperatively regulate bone formation. *Nat Med* 11(8): 880–5.
11. Urist MR. 1965. Bone: formation by autoinduction. *Science* 150(698): 893–9.
12. Chen D, Zhao M, Mundy GR. 2004. Bone morphogenetic proteins. *Growth Factors* 22(4): 233–41.
13. Gazzerro E, Canalis E. 2006. Bone morphogenetic proteins and their antagonists. *Rev Endocr Metab Disord* 7(1–2): 51–65.
14. Shore EM, Xu M, Feldman GJ, Fenstermacher DA, Cho TJ, Choi IH, et al. 2006. A recurrent mutation in the BMP type I receptor ACVR1 causes inherited and sporadic fibrodysplasia ossificans progressiva. *Nat Genet* 38(5): 525–7.
15. van Dinther M, Visser N, de Gorter DJJ, Doorn J, Goumans MJ, de Boer J, et al. 2010. ALK2 R206H mutation linked to fibrodysplasia ossificans progressiva confers constitutive activity to the BMP type I receptor and sensitizes mesenchymal cells to BMP-induced osteoblast differentiation and bone formation. *J Bone Miner Res* 25(6): 1208–15.
16. Gautschi OP, Frey SP, Zellweger R. 2007. Bone morphogenetic proteins in clinical applications. *ANZ J Surg* 77(8): 626–31.
17. Feng XH, Derynck R. 2005. Specificity and versatility in tgf-β signaling through Smads. *Annu Rev Cell Dev Biol* 21: 659–93.
18. Massague J, Seoane J, Wotton D. 2005. Smad transcription factors. *Genes Dev* 19(23): 2783–810.
19. Lee KS, Kim HJ, Li QL, Chi XZ, Ueta C, Komori T, et al. 2000. Runx2 is a common target of transforming growth factor β1 and bone morphogenetic protein 2, and cooperation between Runx2 and Smad5 induces osteoblast-specific gene expression in the pluripotent mesenchymal precursor cell line C2C12. *Mol Cell Biol* 20(23): 8783–92.
20. Javed A, Bae JS, Afzal F, Gutierrez S, Pratap J, Zaidi SK, et al. 2008. Structural coupling of Smad and Runx2 for execution of the BMP2 osteogenic signal. *J Biol Chem* 283(13): 8412–22.
21. Zhang YW, Yasui N, Ito K, Huang G, Fujii M, Hanai J, et al. 2000. A RUNX2/PEBP2α A/CBFA1 mutation displaying impaired transactivation and Smad interaction in cleidocranial dysplasia. *Proc Natl Acad Sci U S A* 97(19): 10549–54.
22. Pereira RC, Rydziel S, Canalis E. 2000. Bone morphogenetic protein-4 regulates its own expression in cultured osteoblasts. *J Cell Physiol* 182(2): 239–46.
23. Korchynskyi O, ten Dijke P. 2002. Identification and functional characterization of distinct critically important bone morphogenetic protein-specific response elements in the Id1 promoter. *J Biol Chem* 277(7): 4883–91.
24. Ogata T, Wozney JM, Benezra R, Noda M. 1993. Bone morphogenetic protein 2 transiently enhances expression of a gene, Id (inhibitor of differentiation), encoding a helix-loop-helix molecule in osteoblast-like cells. *Proc Natl Acad Sci U S A* 90(19): 9219–22.
25. Maeda Y, Tsuji K, Nifuji A, Noda M. 2004. Inhibitory helix-loop-helix transcription factors Id1/Id3 promote bone formation in vivo. *J Cell Biochem* 93(2): 337–44.
26. Celil AB, Campbell PG. 2005. BMP-2 and insulin-like growth factor-I mediate Osterix (Osx) expression in human mesenchymal stem cells via the MAPK and protein kinase D signaling pathways. *J Biol Chem* 280(36): 31353–9.
27. Celil AB, Hollinger JO, Campbell PG. 2005. Osx transcriptional regulation is mediated by additional pathways to BMP2/Smad signaling. *J Cell Biochem* 95(3): 518–28.
28. Janssens K, ten Dijke P, Janssens S, Van Hul W. 2005. Transforming growth factor-β1 to the bone. *Endocr Rev* 26(6): 743–74.
29. Alliston T, Choy L, Ducy P, Karsenty G, Derynck R. 2001. TGF-β-induced repression of CBFA1 by Smad3 decreases cbfa1 and osteocalcin expression and inhibits osteoblast differentiation. *EMBO J* 20(9): 2254–72.
30. Maeda S, Hayashi M, Komiya S, Imamura T, Miyazono K. 2004. Endogenous TGF-β signaling suppresses maturation of osteoblastic mesenchymal cells. *EMBO J* 23(3): 552–63.
31. Sowa H, Kaji H, Yamaguchi T, Sugimoto T, Chihara K. 2002. Smad3 promotes alkaline phosphatase activity and mineralization of osteoblastic MC3T3-E1 cells. *J Bone Miner Res* 17(7): 1190–9.
32. Hanai J, Chen LF, Kanno T, Ohtani-Fujita N, Kim WY, Guo WH, et al. 1999. Interaction and functional cooperation of PEBP2/CBF with Smads. Synergistic induction of the immunoglobulin germline Cα promoter. *J Biol Chem* 274(44): 31577–82.
33. Clevers H. 2006. Wnt/β-catenin signaling in development and disease. *Cell* 127(3): 469–80.
34. Day TF, Guo X, Garrett-Beal L, Yang Y. 2005. Wnt/β-catenin signaling in mesenchymal progenitors controls osteoblast and chondrocyte differentiation during vertebrate skeletogenesis. *Dev Cell* 8(5): 739–50.
35. Hill TP, Spater D, Taketo MM, Birchmeier W, Hartmann C. 2005. Canonical Wnt/β-catenin signaling prevents osteoblasts from differentiating into chondrocytes. *Dev Cell* 8(5): 727–38.

36. Hu H, Hilton MJ, Tu X, Yu K, Ornitz DM, Long F. 2005. Sequential roles of Hedgehog and Wnt signaling in osteoblast development. *Development* 132(1): 49–60.
37. Kramer I, Halleux C, Keller H, Pegurri M, Gooi JH, Weber PB, et al. 2010. Osteocyte Wnt/β-catenin signaling is required for normal bone homeostasis. *Mol Cell Biol* 30(12): 3071–85.
38. Boyden LM, Mao J, Belsky J, Mitzner L, Farhi A, Mitnick MA, et al. 2002. High bone density due to a mutation in LDL-receptor-related protein 5. *N Engl J Med* 346(20): 1513–21.
39. Little RD, Carulli JP, Del Mastro RG, Dupuis J, Osborne M, Folz C, et al. 2002. A mutation in the LDL receptor-related protein 5 gene results in the autosomal dominant high-bone-mass trait. *Am J Hum Genet* 70(1): 11–9.
40. Van Wesenbeeck L, Cleiren E, Gram J, Beals RK, Benichou O, Scopelliti D, et al. 2003. Six novel missense mutations in the LDL receptor-related protein 5 (LRP5) gene in different conditions with an increased bone density. *Am J Hum Genet* 72(3): 763–71.
41. ten Dijke P, Krause C, de Gorter DJJ, Lowik CW, van Bezooijen RL. 2008. Osteocyte-derived sclerostin inhibits bone formation: Its role in bone morphogenetic protein and Wnt signaling. *J Bone Joint Surg Am* 90(Suppl 1): 31–5.
42. St Jacques B, Hammerschmidt M, McMahon AP. 1999. Indian hedgehog signaling regulates proliferation and differentiation of chondrocytes and is essential for bone formation. *Genes Dev* 13(16): 2072–86.
43. Long F, Chung UI, Ohba S, McMahon J, Kronenberg HM, McMahon AP. 2004. Ihh signaling is directly required for the osteoblast lineage in the endochondral skeleton. *Development* 131(6): 1309–18.
44. Hooper JE, Scott MP. 2005. Communicating with Hedgehogs. *Nat Rev Mol Cell Biol* 6(4): 306–17.
45. Hui CC, Joyner AL. 1993. A mouse model of greig cephalopolysyndactyly syndrome: The extra-toesJ mutation contains an intragenic deletion of the Gli3 gene. *Nat Genet* 3(3): 241–6.
46. Litingtung Y, Dahn RD, Li Y, Fallon JF, Chiang C. 2002. Shh and Gli3 are dispensable for limb skeleton formation but regulate digit number and identity. *Nature* 418(6901): 979–83.
47. Shimoyama A, Wada M, Ikeda F, Hata K, Matsubara T, Nifuji A, et al. 2007. Ihh/Gli2 signaling promotes osteoblast differentiation by regulating Runx2 expression and function. *Mol Biol Cell* 18(7): 2411–8.
48. Rubin MR, Bilezikian JP. 2003. New anabolic therapies in osteoporosis. *Endocrinol Metab Clin North Am* 32(1): 285–307.
49. Karaplis AC, Luz A, Glowacki J, Bronson RT, Tybulewicz VL, Kronenberg HM, et al. 1994. Lethal skeletal dysplasia from targeted disruption of the parathyroid hormone-related peptide gene. *Genes Dev* 8(3): 277–89.
50. Miao D, He B, Karaplis AC, Goltzman D. 2002. Parathyroid hormone is essential for normal fetal bone formation. *J Clin Invest* 109(9): 1173–82.
51. Weir EC, Philbrick WM, Amling M, Neff LA, Baron R, Broadus AE. 1996. Targeted overexpression of parathyroid hormone-related peptide in chondrocytes causes chondrodysplasia and delayed endochondral bone formation. *Proc Natl Acad Sci U S A* 93(19): 10240–5.
52. Lanske B, Karaplis AC, Lee K, Luz A, Vortkamp A, Pirro A, et al. 1996. PTH/PTHrP receptor in early development and Indian hedgehog-regulated bone growth. *Science* 273(5275): 663–6.
53. Zhang P, Jobert AS, Couvineau A, Silve C. 1998. A homozygous inactivating mutation in the parathyroid hormone/parathyroid hormone-related peptide receptor causing Blomstrand chondrodysplasia. *J Clin Endocrinol Metab* 83(9): 3365–8.
54. Krishnan V, Moore TL, Ma YL, Helvering LM, Frolik CA, Valasek KM, et al. 2003. Parathyroid hormone bone anabolic action requires Cbfa1/Runx2-dependent signaling. *Mol Endocrinol* 17(3): 423–35.
55. van der Horst G, Farih-Sips H, Lowik CW, Karperien M. 2005. Multiple mechanisms are involved in inhibition of osteoblast differentiation by PTHrP and PTH in KS483 Cells. *J Bone Miner Res* 20(12): 2233–44.
56. Canalis E. 2009. Growth factor control of bone mass. *J Cell Biochem* 108(4): 769–77.
57. Bikle D, Majumdar S, Laib A, Powell-Braxton L, Rosen C, Beamer W, et al. 2001. The skeletal structure of insulin-like growth factor I-deficient mice. *J Bone Miner Res* 16(12): 2320–9.
58. Peng XD, Xu PZ, Chen ML, Hahn-Windgassen A, Skeen J, Jacobs J, et al. 2003. Dwarfism, impaired skin development, skeletal muscle atrophy, delayed bone development, and impeded adipogenesis in mice lacking Akt1 and Akt2. *Genes Dev* 17(11): 1352–65.
59. Fujita T, Azuma Y, Fukuyama R, Hattori Y, Yoshida C, Koida M, et al. 2004. Runx2 induces osteoblast and chondrocyte differentiation and enhances their migration by coupling with PI3K-Akt signaling. *J Cell Biol* 166(1): 85–95.
60. Yang S, Xu H, Yu S, Cao H, Fan J, Ge C, et al. 2011. Foxo1 mediates insulin-like growth factor 1 (IGF1)/insulin regulation of osteocalcin expression by antagonizing Runx2 in osteoblasts. *J Biol Chem* 286(21): 19149–58.
61. Matsuguchi T, Chiba N, Bandow K, Kakimoto K, Masuda A, Ohnishi T. 2009. JNK activity is essential for Atf4 expression and late-stage osteoblast differentiation. *J Bone Miner Res* 24(3): 398–410.
62. Ornitz DM. 2005. FGF signaling in the developing endochondral skeleton. *Cytokine Growth Factor Rev* 16(2): 205–13.
63. Eswarakumar VP, Monsonego-Ornan E, Pines M, Antonopoulou I, Morriss-Kay GM, Lonai P. 2002. The IIIc alternative of Fgfr2 is a positive regulator of bone formation. *Development* 129(16): 3783–93.
64. Yu K, Xu J, Liu Z, Sosic D, Shao J, Olson EN, et al. 2003. Conditional inactivation of FGF receptor 2 reveals an essential role for FGF signaling in the regulation of osteoblast function and bone growth. *Development* 130(13): 3063–74.

65. Kim HJ, Kim JH, Bae SC, Choi JY, Ryoo HM. 2003. The protein kinase C pathway plays a central role in the fibroblast growth factor-stimulated expression and transactivation activity of Runx2. *J Biol Chem* 278(1): 319–26.
66. Ehebauer M, Hayward P, Martinez-Arias A. 2006. Notch signaling pathway. *Sci STKE* 2006(364): cm7.
67. Deregowski V, Gazzerro E, Priest L, Rydziel S, Canalis E. 2006. Notch 1 overexpression inhibits osteoblastogenesis by suppressing Wnt/β-catenin but not bone morphogenetic protein signaling. *J Biol Chem* 281(10): 6203–10.
68. Zamurovic N, Cappellen D, Rohner D, Susa M. 2004. Coordinated activation of notch, Wnt, and transforming growth factor-β signaling pathways in bone morphogenic protein 2-induced osteogenesis. Notch target gene Hey1 inhibits mineralization and Runx2 transcriptional activity. *J Biol Chem* 279(36): 37704–15.
69. Hilton MJ, Tu X, Wu X, Bai S, Zhao H, Kobayashi T, et al. 2008. Notch signaling maintains bone marrow mesenchymal progenitors by suppressing osteoblast differentiation. *Nat Med* 14(3): 306–14.
70. Fuentealba LC, Eivers E, Ikeda A, Hurtado C, Kuroda H, Pera EM, et al. 2007. Integrating patterning signals: Wnt/GSK3 regulates the duration of the BMP/Smad1 signal. *Cell* 131(5): 980–93.
71. Sapkota G, Alarcon C, Spagnoli FM, Brivanlou AH, Massague J. 2007. Balancing BMP signaling through integrated inputs into the Smad1 linker. *Mol Cell* 25(3): 441–54.
72. Nakayama K, Tamura Y, Suzawa M, Harada S, Fukumoto S, Kato M, et al. 2003. Receptor tyrosine kinases inhibit bone morphogenetic protein-Smad responsive promoter activity and differentiation of murine MC3T3-E1 osteoblast-like cells. *J Bone Miner Res* 18(5): 827–35.
73. Hu MC, Rosenblum ND. 2005. Smad1, β-catenin and Tcf4 associate in a molecular complex with the Myc promoter in dysplastic renal tissue and cooperate to control Myc transcription. *Development* 132(1): 215–25.
74. Labbe E, Letamendia A, Attisano L. 2000. Association of Smads with lymphoid enhancer binding factor 1/T cell-specific factor mediates cooperative signaling by the transforming growth factor-β and wnt pathways. *Proc Natl Acad Sci U S A* 97(15): 8358–63.
75. Spinella-Jaegle S, Roman-Roman S, Faucheu C, Dunn FW, Kawai S, Gallea S, et al. 2001. Opposite effects of bone morphogenetic protein-2 and transforming growth factor-β1 on osteoblast differentiation. *Bone* 29(4): 323–30.
76. de Gorter DJJ, van Dinther M, Korchynskyi O, ten Dijke P. 2011. Biphasic effects of transforming growth factor β on bone morphogenetic protein-induced osteoblast differentiation. *J Bone Miner Res* 26(6): 1178–87.
77. Matsaba T, Ramoshebi LN, Crooks J, Ripamonti U. 2001. Transforming growth factor-β1 supports the rapid morphogenesis of heterotopic endochondral bone initiated by human osteogenic protein-1 via the synergistic upregulation of molecular markers. *Growth Factors* 19(2): 73–86.
78. Zhao M, Qiao M, Harris SE, Chen D, Oyajobi BO, Mundy GR. 2006. The zinc finger transcription factor Gli2 mediates bone morphogenetic protein 2 expression in osteoblasts in response to hedgehog signaling. *Mol Cell Biol* 26(16): 6197–208.
79. Li Y, Li A, Strait K, Zhang H, Nanes MS, Weitzmann MN. 2007. Endogenous TNFα lowers maximum peak bone mass and inhibits osteoblastic Smad activation through NF-κB. *J Bone Miner Res* 22(5): 646–55.
80. Mukai T, Otsuka F, Otani H, Yamashita M, Takasugi K, Inagaki K, et al. 2007. TNF-α inhibits BMP-induced osteoblast differentiation through activating SAPK/JNK signaling. *Biochem Biophys Res Commun* 356(4): 1004–10.
81. Qiu T, Wu X, Zhang F, Clemens TL, Wan M, Cao X. 2010. TGF-β type II receptor phosphorylates PTH receptor to integrate bone remodelling signalling. *Nat Cell Biol* 12(3): 224–34.

3
Osteoclast Biology and Bone Resorption

F. Patrick Ross

Cell Biology of the Osteoclast 25
Integrin Signaling 28
Small GTPases 28
Factors Regulating Osteoclast Formation and/or Function 28

Cell–Cell Interactions in Bone Marrow 30
Intracellular Signaling Pathways 31
Human Genetics 31
References 31

CELL BIOLOGY OF THE OSTEOCLAST

Pathological bone loss, regardless of etiology, invariably represents an increase in the rate at which the skeleton is degraded by osteoclasts relative to its formation by osteoblasts. Thus, the prevention of conditions such as osteoporosis requires an understanding of the molecular mechanisms of bone resorption.

The osteoclast, the exclusive bone resorptive cell (Fig. 3.1), is a member of the monocyte-macrophage family and a polykaryon that can be generated *in vitro* from mononuclear phagocyte precursors resident in a number of tissues [1]. There is, however, general agreement that the principal physiological osteoclast precursor is the bone marrow macrophage. Two cytokines are essential and sufficient for basal osteoclastogenesis, the first being receptor activator of nuclear factor-κB ligand (RANKL) [1, 2] and the second being macrophage-colony stimulating factor (M-CSF), also designated CSF-1 [3]. These two proteins, which exist as both membrane-bound and soluble forms (the former is secreted by activated T cells) [4], are produced by marrow stromal cells and their derivative osteoblasts. Thus, physiological recruitment of osteoclasts from their mononuclear precursors requires the presence of these nonhematopoietic, bone-residing cells [1]. RANKL, a member of the tumor necrosis factor (TNF) superfamily, is the key osteoclastogenic cytokine,

because osteoclast formation requires its presence or its priming of precursor cells. M-CSF contributes to the proliferation, survival, and differentiation of osteoclast precursors, as well as the survival and cytoskeletal rearrangement required for efficient bone resorption (Fig. 3.2); a brief summary of the integrated signaling pathways for each osteoclastic regulator discussed in this review is provided in Fig. 3.3.

The discovery of RANKL was preceded by identification of its physiological inhibitor osteoprotegerin (OPG), to which it binds with high affinity [5]. In contrast, M-CSF is a moiety long known to regulate the broader biology of myeloid cells, including osteoclasts [3] (see Fig. 3.2).

Our understanding of how osteoclasts resorb bone derives from two major sources: biochemistry and genetics [2]. The unique osteoclastogenic properties of RANKL permit generation of pure populations of osteoclasts in culture and hence the performance of meaningful biochemical and molecular experiments that provide insights into the molecular mechanisms by which osteoclasts resorb bone. Further evidence has come from our capacity to generate mice that lack specific genes, plus the positional cloning of genetic abnormalities in people with abnormal osteoclast function. Key to the resorptive event is the capacity of the osteoclast to form a microenvironment between itself and the underlying bone matrix [Fig. 3.4(A)]. This compartment, which is isolated

---

Primer on the Metabolic Bone Diseases and Disorders of Mineral Metabolism, Eighth Edition. Edited by Clifford J. Rosen.
© 2013 American Society for Bone and Mineral Research. Published 2013 by John Wiley & Sons, Inc.

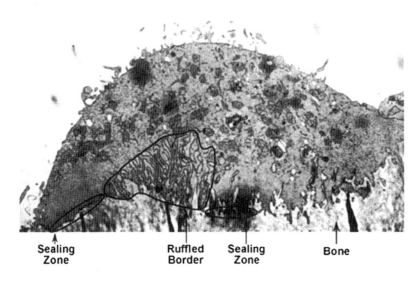

Fig. 3.1. The osteoclast as a resorptive cell. Transmission electron microscopy of a multinucleated primary rat osteoclast on bone. Note the extensive ruffled border, close apposition of the cell to bone, and the partially degraded matrix between the sealing zones. Courtesy of H. Zhao.

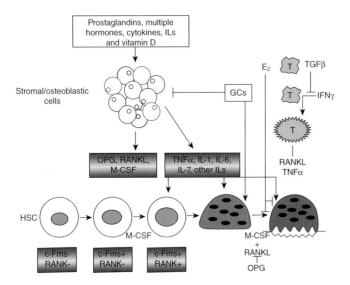

Fig. 3.2. Role of cytokines, hormones, steroids, and prostaglandins in osteoclast formation. Under the influence of other cytokines (data not shown), hematopoietic stem cells (HSCs) commit to the myeloid lineage, express c-Fms and RANK, the receptors for M-CSF and RANKL, respectively, and differentiate into osteoclasts. Mesenchymal cells in the marrow respond to a range of stimuli, secreting a mixture of pro- and anti-osteoclastogenic proteins, the latter primarily OPG. Glucocorticoids suppress bone resorption indirectly but possibly also target osteoclasts and/or their precursors. Estrogen (E2), by a complex mechanism, inhibits activation of T cells, decreasing their secretion of RANKL and TNF-α; the sex steroid also inhibits osteoblast and osteoclast differentiation and lifespan. A key factor regulating bone resorption is the RANKL/OPG ratio.

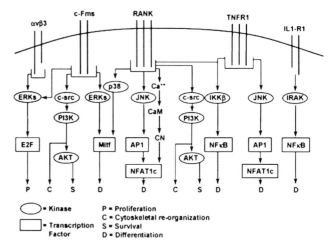

Fig. 3.3. Osteoclast signaling pathways. Summary of the major receptors, downstream kinases, and effector transcription factors that regulate osteoclast formation and function. Proliferation (P) of precursors is driven chiefly through ERKs and their downstream cyclin targets and E2F. Maximal activation of this pathway requires combined signals from c-Fms and the integrin αvβ3. As expected, the cyloskeleton (C) is independent of nuclear control but depends on a series of kinases and their cytoskeletal-regulating targets, whereas differentiation (D) is regulated largely by controlling gene expression. The calcium/calmodulin (CaM)/calcineurin (CN) axis enhances nuclear translocation of NFAT1c, the most distal transcription factor characterized to date. See Refs. 2, 3, 10, 28, 40, and 53–56 for details.

from the general extracellular space, is acidified by an electrogenic proton pump (H+- ATPase) and a Cl- channel to a pH of 4.5 [6]. The acidified milieu mobilizes the mineralized component of bone, exposing its organic matrix, consisting largely of type I collagen that is subsequently degraded by the lysosomal enzyme cathepsin K. The critical roles that the proton pump, Cl- channel, and cathepsin K play in osteoclast action is underscored by the fact that diminished function of each results in a human disease of excess bone mass, namely osteopetrosis or pyknodysostosis [2, 6]. Degraded protein fragments are endocytosed and transported in undefined vesicles to the basolateral surface of the cell, where they are discharged into the surrounding intracellular fluid [7, 8]. It

Osteoclast Biology and Bone Resorption 27

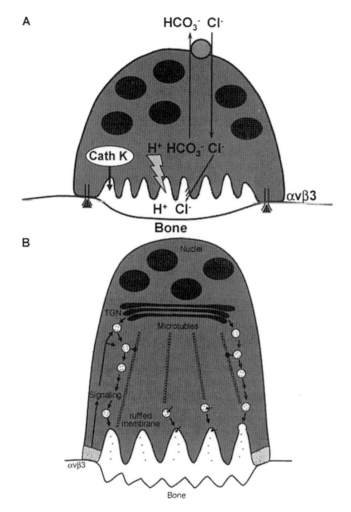

Fig. 3.4. Mechanism of osteoclastic bone resorption. (A) The osteoclast adheres to bone through the integrin αvβ3, creating a sealing zone, into which is secreted hydrochloric acid and acidic proteases such as cathepsin K, MMP9, and MMP13. The acid is generated by the combined actions of a vacuolar H^+ ATPase; it coupled chloride channel and a basolateral chloride–bicarbonate exchanger. Carbonic anhydrase converts CO_2 into H^+ and HCO^+ (data not shown). (B) Integrin engagement results in signals that target acidifying vesicles (+ = proton pump complex) containing specific cargo (black dots) to the bone-apposed face of the cell. Fusion of these vesicles with the plasma membrane generates a polarized cell capable of secreting the acid and proteases required for bone resorption.

is also likely that retraction of an osteoclast from the resorptive pit results in release of products of digestion.

The above model of bone degradation clearly depends on physical intimacy between the osteoclast and bone matrix, a role provided by integrins. Integrins are *alpha beta* heterodimers with long extracellular and single transmembrane domains [9]. In most instances, the integrin cytoplasmic region is relatively short, consisting of 40–70 amino acids. Integrins are the principal cell matrix attachment molecules and they mediate osteoclastic bone recognition. Members of the β1 family of integrins, which recognize collagen, fibronectin, and laminin, are present on osteoclasts, but αvβ3 is the principal integrin mediating bone resorption [10]. This heterodimer, like all members of the alpha V integrin family, recognizes the amino acid motif Arg- Gly-Asp (RGD), which is present in a variety of bone-residing proteins such as osteopontin and bone sialoprotein. Thus, osteoclasts attach to and spread on these substrates in an RGD-dependent manner, and, most importantly, competitive ligands arrest bone resorption *in vivo*. Proof of the pivotal role that αvβ3 has in the resorptive process came with the generation of the β3 integrin knockout mouse, which develops a progressive increase in bone mass because of osteoclast dysfunction. Based on a combination of these *in vitro* and *in vivo* observations, small molecule inhibitors of osteoclast function that target αvβ3 have been developed [11].

Bone resorption also requires a polarization event in which the osteoclast delivers effector molecules like HC1 and cathepsin K into the resorptive microenvironment. Osteoclasts are characterized by a unique cytoskeleton, which mediates the resorptive process. Specifically, when the cell contacts bone, it generates two polarized structures, which enable it to degrade skeletal tissue. In the first instance, a subset of acidified vesicles containing specific cargo, including cathepsin K and other matrix metalloproteases (MMPs), are transported, probably through microtubules and actin, to the bone-apposed plasma membrane [12] to which they fuse in a manner not currently understood, but which may involve PLEKHM1, a nonsecretory adaptor protein found in endosomal vesicles [13]. Insertion of these vesicles into the plasmalemma results in formation of a villous structure, unique to the osteoclast, called the ruffled membrane. This resorptive organelle contains the abundant H^+ transporting machinery to create the acidified microenvironment, whereas the accompanying exocytosis serves as the means by which cathepsin K is secreted [Fig. 3.4(B)].

In addition to inducing ruffled membrane formation, contact with bone also prompts the osteoclast to polarize its fibrillar actin into a circular structure known as the "actin ring." A separate "sealing zone" surrounds and isolates the acidified resorptive microenvironment in the active cell, but its composition is almost completely unknown. The actin ring, like the ruffled membrane, is a hallmark of the degradative capacity of the osteoclast, because structural abnormalities of either occur in conditions of arrested resorption [14]. In most cells, such as fibroblasts, matrix attachment prompts formation of stable structures known as "focal adhesions" that contain both integrins and a host of signaling and cytoskeletal molecules, which mediate contact and formation of actin stress fibers. In keeping with the substitution of the actin ring for stress fibers in osteoclasts, these cells form podosomes instead of focal adhesions. Podosomes, which in resorbing osteoclasts are present in the actin ring, consist of an actin core surrounded by αvβ3 and associated cytoskeletal proteins.

The integrin β3 subunit knockout mouse serves as an important tool for determining the role of αvβ3 in the capacity of the osteoclast to resorb bone. Failure to express αvβ3 results in a dramatic osteoclast phenotype, particularly regarding the actin cytoskeleton. The β3$^{-/-}$ osteoclast forms abnormal ruffled membranes *in vivo* and, whether generated *in vitro* or directly isolated from bone, the mutant cells fail to spread when plated on immobilized RGD ligand or mineralized matrix in physiological amounts of RANKL and M-CSF. Confirming their attenuated resorptive activity, β3$^{-/-}$ osteoclasts generate fewer and shallower resorptive lacunae on dentin slices than do their wild type counterparts. In keeping with attenuated bone resorption *in vivo*, β3$^{-/-}$ mice are substantially hypocalcemic [6].

INTEGRIN SIGNALING

Whereas integrins were viewed initially as merely cell attachment molecules, it is now apparent that their capacity to transmit signals to and from the cell interior is equally important, an event that requires that the integrin convert from a default low affinity state to one in which its capacity to bind matrix is significantly enhanced. The process, termed "activation," arises from either integrin ligation of their multivalent ligands or indirectly by growth factor signaling [15].

Integrin αvβ3 is absent from osteoclast precursors, but their differentiation under the action of RANKL results in marked upregulation of this heterodimer. The capacity of integrins to transmit intracellular signals to the cytoskeleton heightened interest in the cytoplasmic molecules mediating these events in osteoclasts, and αvβ3 signaling in this context is reasonably well understood. The initial signaling event involves the proto-oncogene c-src, which, acting as a kinase and an adaptor protein, regulates formation of lamellipodia and disassembly of podosomes, indicating that c-src controls formation of resorptive organelles of the cell, such as the ruffled membrane, and also arrests migration on the bone surface. There is continuing debate surrounding the molecules that link c-src to the cytoskeleton, one proposal being that the focal adhesion kinase family member Pyk2, acting in concert with c-Cbl, a proto-oncogene and ubiquitin ligase [16]. A second strong candidate is Syk, a nonreceptor tyrosine kinase that is recruited to the active conformation of αvβ3 in osteoclasts in a c-src-dependent manner [17], where it targets Vav3 [10], a member of the large family of guanine nucleotide exchange factors (GEFs) that convert Rho GTPases from their inactive GDP to their active GTP conformation.

SMALL GTPases

The Rho family of GTPases is central to remodeling of the actin cytoskeleton in many cell types [18] and as such

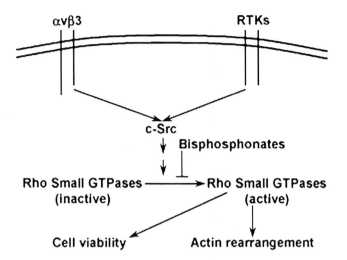

Fig. 3.5. Regulation and role of small GTPases in osteoclasts. Signals from αvβ3 and/or receptor tyrosine kinases (RTKs) activate small GTPases of the Rho family in a c-src-dependent manner. Bisphosphonates, the potent antiresorptive drugs, block addition of hydrophobic moieties onto the GTPases, preventing their membrane targeting and activation. The active GTPases also regulate cell viability and thus bisphosphonates induce osteoclast death.

plays a central role in osteoclastic bone resorption (Fig. 3.5). On attachment to bone, Rho and Rac bind GTP and translocate to the cytoskeleton. Whereas both small GTPases impact the actin cytoskeleton, Rac and Rho exert distinctive effects. Rho signaling mediates formation of the actin ring and a constitutively active form of the GTPase stimulates podosome formation, osteoclast motility, and bone resorption, whereas dominant negative Rho arrests these events [19]. Rac stimulation in osteoclast precursors prompts appearance of lamellipodia, thus forming the migratory front of the cell to which αvβ3 moves when activated [20]. In sum, it is likely that Rho's effect is principally on cell adhesion, whereas Rac mediates the cytoskeleton's migratory machinery. Importantly, absence of Vav3 blunts Rac but not Rho activity in the osteoclast [21].

FACTORS REGULATING OSTEOCLAST FORMATION AND/OR FUNCTION

Proteins

In addition to the two key osteoclastogenic cytokines MCSF and RANKL, a number of other proteins play important roles in osteoclast biology, either in physiological and/or pathophysiological circumstances.

As discussed previously, OPG, a high-affinity ligand for RANKL that acts as a soluble inhibitor of RANKL, is secreted by cells of mesenchymal origin, both basally and in response to other regulatory signals, including cytokines and bone-targeting steroids [5]. Pro-inflammatory cytokines suppress OPG expression while simultaneously enhancing that of RANKL, with the net effect being a marked increase in osteoclast formation and function. Genetic deletion of OPG in both mice and humans leads to profound osteoporosis [22], whereas overexpression of the molecule under the control of a hepatic promoter results in severe osteopetrosis [23]. Together, these observations indicate that skeletal, and perhaps circulating, OPG modulates the bone resorptive activity of RANKL and helps to explain the increased bone loss in clinical situations accompanied by increased levels of TNF-α, interleukin (IL)-1, parathyroid hormone (PTH), or PTH-related protein (PTHrP). Serum PTH levels are increased in hyperparathyroidism of whatever etiology, whereas PTHrP is secreted by metastatic lung and breast carcinoma [24, 25]. F antibodies or a soluble TNF receptor-IgG fusion protein potently suppress the bone loss in disorders of inflammatory osteolysis such as rheumatoid arthritis [26]. The molecular basis of this observation seems to be that the inflammatory cytokine synergizes with RANKL in a unique manner, most likely because RANKL and TNF-α each activate a number of key downstream effector pathways, leading to nuclear localization of a range of osteoclastogenic transcription factors (see Fig. 3.3). Recent evidence suggests a new paradigm linking TNF-α, IL-1, and the natural secreted inhibitor for the latter cytokine, IL-1 receptor antagonist, which blocks IL-1 function. Specifically, it seems that, at least in murine osteoclasts and their precursors, many of the effects of TNF are mediated through TNF's stimulation of IL-1, which in turn increases expression and secretion of IL-1 receptor antagonist (IL-1ra), a set of events that represent a complex control pathway. The significance of the IL receptor antagonist is shown by the fact an IgG fusion protein containing the active component of this molecule has been developed and enhances the ability of anti-TNF antibodies to decrease bone loss in rheumatoid arthritis [27].

Elegant studies suggest that interferon γ (IFNγ) is an important suppressor of osteoclast formation and function [28]. Nevertheless, these findings seem to be in conflict with other *in vivo* observations, including the report that IFNγ treatment of children with osteopetrosis ameliorates the disease [29] and the fact that a number of *in vivo* studies indicate that IFNγ stimulates bone resorption [30, 31]. This conundrum highlights the importance of discriminating between *in vitro* culture experiments using single cytokines and results *in vivo*. Many additional studies have implicated a range of other cytokines in the regulation of the osteoclast. These include a range of interleukins, GM-CSF, IFNβ, stromal cell-derived factor 1 (SDF-1), macrophage inflammatory protein 1 (MIP-1), and monocyte chemoattractant protein 1 (MCP-1), but at this time the results are either contradictory, as for GM-CSF in the murine vs human systems, or lack direct proof in humans. Future studies are likely to clarify the currently confusing data set. Finally, interactions between immune receptors such as DNAX activating protein of 12 kDa (DAP12) and FC receptor γ (FcRγ), present on osteoclasts and their precursors, and their ligands on cells of the stromal and myeloid/lymphoid lineages are important for transmission of RANK-derived signals [28]. IL-17 is a product of Th17 cells, a recently identified T-cell subset that is generated from uncommitted precursors under the influence of TGF-β, IL-23, and IL-6 [32, 33].

Small molecules

The active form of vitamin D, 1,25-dihydroxyvitamin D, has all the characteristics of a steroid hormone, including a high-affinity nuclear receptor that binds as a heterodimer with the retinoid X receptor to regulate transcription of a set of specific target genes. Generated by successive hydroxylation in the liver and kidney, this active form of vitamin D is a well-established stimulator of bone resorption when present at supraphysiological levels. Studies over many years have indicated that this steroid hormone increases mesenchymal cell transcription of the *RANKL* gene, while diminishing that of OPG [5]. Separately, 1,25-dihydroxyvitamin D suppresses synthesis of the pro-osteoclastogenic hormone PTH [34] and enhances calcium uptake from the gut. Taken together, the two latter effects would seem to be antiresorptive, but many studies in humans indicate the net osteolytic action resulting from high levels of this steroid hormone, suggesting that its ability to stimulate osteoclast function overrides any bone anabolic actions.

Loss of estrogen (E_2), most often seen in the context of menopause, is a major reason for the development of significant bone loss in aging. Interestingly, it is now clear that estrogen is the main sex steroid regulating bone mass in both men and women [35]. The mechanisms by which estrogen mediates its osteolytic effects are still not completely understood, but significant advances have been made over the past decade. The original hypothesis, now considered to be only part of the explanation, is that decreased serum E2 led to increased production, by circulating macrophages, of osteoclastogenic cytokines such as IL-6, TNF, and IL-1. These molecules act on stromal cells and osteoclast precursors to enhance bone resorption by regulating expression of pro- (RANKL, M-CSF) and anti- (OPG) osteoclastogenic cytokines (in the case of mesenchymal cells) and by synergizing with RANKL itself (in the case of myeloid osteoclast precursors; see Fig. 3.2). However, the understanding that lymphocytes play a key role in mediating several aspects of bone biology has led to a growing realization that the cellular and molecular targets for E2 are more widespread than previously believed. A model proposes that E2 impacts the resorptive component of bone turnover (the steroid has separate effects on osteoblasts), at least in part by modulating production by T cells of RANKL and TNF [30]. This effect is itself indirect, with E2 suppressing antigen presentation by dendritic cells and macrophages by enhancing expression by the same cells of TGFβ. Antigen presentation activates T cells, thereby enhancing their production of RANKL and TNF. As discussed previously, the first molecule is the key osteoclastogenic cytokine, whereas the second potentiates RANKL action and stimulates production by stromal cells of M-CSF and RANKL. This newly discovered interface between T cells and bone resorption also clarifies aspects of inflammatory osteolysis. Finally, some studies indicate that E2 modulates signaling in preosteoclasts and that, acting through reactive oxygen species, it increases the lifespan and/or function of mature osteoclasts [36].

Both endogenous glucocorticoids and their synthetic analogs, which have been and continue to be a major mainstay of immunosuppressive therapy, are members of a third steroid hormone family having a major impact on bone biology [37].

One consequence of their chronic mode of administration is severe osteoporosis arising from decreased bone formation and resorption with the latter absolutely decreased (low turnover osteoporosis). The majority of the evidence focuses on the osteoblast as the prime target with the steroid increasing apoptosis of these bone-forming cells. However, numerous human studies document a rapid initial decrease in bone resorption, suggesting that the osteoclast and/or its precursors may also be targets. The molecular basis for this latter finding is unclear. However, because osteoblasts are a requisite part of the resorptive cycle, one consequence of their long-term diminution could be decreased osteoclast formation and/or function secondary to lower levels of RANKL and/or M-CSF production. Alternatively, glucocorticoids have been shown to decrease osteoclast apoptosis [38].

A wide range of clinical information shows that excess prostaglandins stimulate bone loss, but once again, the cellular basis has not been established. Prostaglandins target stromal and osteoblastic cells, stimulating expression of RANKL and suppressing that of OPG [30]. This increase in the RANKL/OPG ratio, seen in a variety of human studies, is sufficient of itself to explain the clinical findings of increased osteoclastic activity. However, highlighting again the dilemma of interpreting *in vitro* studies, there have been a number of studies in which prostaglandins regulate osteoclastogenesis *per se* in murine cell culture.

Phosphoinositides play distinct and important roles in organization of the osteoclast cytoskeleton [39, 40]. Binding of M-CSF or RANKL to their cognate receptors, c-Fms and RANK, or activation of αvβ3, recruits phosphoinositol-3-kinase (PI3K) to the plasma membrane, where it converts membrane-bound phosphatidylinositol 4,5-bisphosphate into phosphatidylinositol 3,4,5-trisphosphate (Fig. 3.3). The latter compound is recognized by specific motifs in a wide range of cytoskeletally active proteins [41], and thus PI3K plays a central role in organizing the cytoskeleton of the osteoclast, including its ruffled membrane. Akt is a downstream target of PI3K and plays an important role in osteoclast function, particularly by mediating RANKL and/or M-CSF-stimulated proliferation and/or survival [40].

CELL–CELL INTERACTIONS IN BONE MARROW

Recent evidence has indicated that a number of additional cell types are important for osteoclast biology in a variety of situations (Fig. 3.6). First, as discussed previously, T cells play a key role in estrogen deficiency bone loss but also are important in a range of inflammatory diseases, most notably rheumatoid arthritis [42] and periodontal disease [43]; here the Th17 subset likely secretes TNF and IL-17, a newly described osteoclastogenic cytokine. Given that both osteoclast precursors and the various lymphocyte subsets, such as T, B, and NK cells, arise from the same stem cell, it is not surprising that some of the same receptors and ligands that mediate the immune process also govern the maturation of osteoclast precursors and the capacity of the mature cell to degrade bone. This interface has given rise to the new discipline of osteoimmunology, which promises to provide important and exciting findings in the future.

Second, whereas it is well established that mesenchymal cells are major mediators of cytokine and prostaglandin action on osteoclasts, it has become clear recently that cells of the same lineage, residing on cortical and trabecular bone, are the site of a hematopoietic stem cell (HSC) niche [44]. Specifically, HSCs reside close to osteoblasts as a result of multiple interactions involving recep-

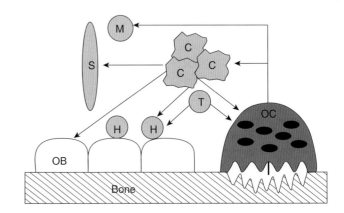

Fig. 3.6. Cell–cell interactions in bone marrow. Hematopoietic stem cells (H), the precursors of both T cells (T) and osteoclasts (OC), reside in a stem cell niche provided by osteoblasts (OB), which, together with stromal cells (S), derive from mesenchymal stem cells (M). Bone degradation results in release of matrix-associated growth factors (thick vertical line), which stimulate mesenchymal cells and thus bone formation. This "coupling" is an essential consequence of osteoclast activity. After activation, T cells secrete molecules that stimulate osteoclastogenesis and function. Cancer cells (C) release cytokines that activate bone resorption; in turn, matrix-derived factors stimulate cancer cell proliferation, the so-called vicious cycle.

tors and ligands on both cells types [45]. Furthermore, the mesenchymally derived cells secrete both membrane-bound and soluble factors that contribute to survival and proliferation of multipotent osteoclast precursors, as well as molecules that influence osteoclast formation and function. Both committed osteoblasts and the numerous stromal cells in bone marrow produce a range of proteins both basally and in response to hormones and growth factors, resulting in modulation of the capacity of HSCs to become functional osteoclasts.

Third, cancer cells facilitate their infiltration into the marrow cavity by stimulating osteoclast formation and function. An initial stimulus is PTHrP generation by lung and breast cancer cells [24, 25, 36, 46], thus enhancing mesenchymal production of RANKL and M-CSF, whereas decreasing that of OPG and possibly chemotactic factors. The resulting increase in matrix dissolution releases bone-residing cytokines and growth factors that, feeding back on the cancer cells, increase their growth and/or survival. This loop has been termed "the vicious cycle" [24]. Multiple myeloma seems to use a different but related strategy, namely secretion of MIPa and MCP-1, both of which are chemotactic and proliferative for osteoclast precursors [47, 48]. The latter compound has been reported to be secreted by osteoclasts in response to RANKL and enhances osteoclast formation [2]. It seems likely further future studies experiments will uncover additional molecules mediating bone loss in metastatic disease.

INTRACELLULAR SIGNALING PATHWAYS

The discussions above have not described in detail the intracellular signals by which osteoclasts are formed or those by which they degrade bone. The final major section of this review lays out the important pathways involved. Briefly, three major protein classes are involved: adaptors, kinases, and transcription factors (Fig. 3.3), with one significant exception, RANKL-induced release of Ca^{++}, a pathway that activates the calmodulin-dependent phosphatase calcineurin. NFATlc is a major substrate for this enzyme, resulting in its nuclear translocation and subsequent activation of osteoclast-specific genes. Importantly, the potent immunosuppressive drugs FK506 and cyclosporine inhibit calcineurin activity and therefore may target the osteoclast [49].

The multiplicity of adaptors that link the various receptors to downstream signals precludes providing a meaningful summary, and so we summarize only the modulatory effects of kinases and transcription factors, which together regulate receptor-driven proliferation and/or survival of precursors. Thus, proliferation is mediated by $\alpha v \beta 3$ and c-Fms [10, 50]; reorganization of the cytoskeleton by $\alpha v \beta 3$, c-Fms, and RANK [2, 10]; differentiation of mature osteoclasts from myeloid progenitors by c-Fms, RANK, TNFRl, and IL-lRl [2, 50, 51], and their function by RANK, TNFRl, and IL-lRl [52, 53]. Not shown is the fact that multiple other cytokines and growth factors, targeting the same or other less prominent pathways, or acting indirectly probably contribute to overall control of bone resorption.

HUMAN GENETICS

The text above might suggest that numerous mutations in many genes linked to the osteoclast are likely to have been discovered in humans. In fact, few such genetic changes have been defined, with more than 50% of those reported being in patients with osteopetrosis caused by defects in the chloride channel that modulates osteoclast acid secretion (Fig. 3.4). Rare reports link deficiencies in RANK, the proton pump, or carbonic anhydrase II to osteopetrosis, whereas decreased cathepsin K function leads to pyknodysostosis. In contrast, RANK activation manifests as osteolytic bone disease, whereas OPG deficiency leads to a severe form of high turnover osteoporosis.

REFERENCES

1. Suda T, Takahashi N, Udagawa N, Jimi E, Gillespie MT, Martin TJ. 1999. Modulation of osteoclast differentiation and function by the new members of the tumor necrosis factor receptor and ligand families. *Endocr Rev* 20: 345–57.
2. Boyle WJ, Simonet WS, Lacey DL. 2003. Osteoclast differentiation and activation. *Nature* 423: 337–42.
3. Pixley FJ, Stanley ER. 2004. CSF-1 regulation of the wandering macrophage: Complexity in action. *Trends Cell Biol* 14: 628–38.
4. Weitzmann MN, Cenci S, Rifas L, Brown C, Pacifici R. 2000. Interluekin-7 stimulates osteoclast formation by upregulating the T-cell production of soluble osteoclastogenic cytokines. *Blood* 96: 1873–78.
5. Kostenuik PJ, Shalhoub V. 2001. Osteoprotegerin: A physiological and pharmacological inhibitor of bone resorption. *Curr Pharm Des* 7: 613–35.
6. Teitelbaum SL, Ross FP. 2003. Genetic regulation of osteoclast development and function. *Nat Rev Genet* 4: 638–49.
7. Salo J, Lehenkari P, Mulari M, Metsikko K, Vaananen HK. 1997. Removal of osteoclast bone resorption products by transcytosis. *Science* 276: 270–73.
8. Stenbeck G, Horton MA. 2004. Endocytic trafficking in actively resorbing osteoclasts. *J Cell Sci* 117: 827–36.
9. Hynes RO. 2002. Integrins: Bidirectional, allosteric signaling machines. *Cell* 110: 673–87.
10. Ross FP, Teitelbaum SL. 2005. alphavbeta3 and macrophage colony-stimulating factor: Partners in osteoclast biology. *Immunol Rev* 208: 88–105.
11. Teitelbaum SL. 2005. Osteoporosis and integrins. *J Clin Endocrinol Metab* 90: 2466–68.

12. Teitelbaum SL, Abu-Amer Y, Ross FP. 1995. Molecular mechanisms of bone resorption. *J Cell Biochem* 59: 1–10.
13. Van Wesenbeeck L, Odgren PR, Coxon FP, Frattini A, Moens P, Perdu B, MacKay CA, Van Hul E, Timmermans JP, Vanhoenacker F, Jacobs R, Peruzzi B, Teti A, Helfrich MH, Rogers MJ, Villa A, Van Hul W. 2007. Involvement of PLEKHM1 in osteoclastic vesicular transport and osteopetrosis in incisors absent rats and humans. *J Clin Invest* 117: 919–30.
14. Vaananen HK, Zhao H, Mulari M, Halleen JM. 2000. The cell biology of osteoclast function. *J Cell Sci* 113: 377–81.
15. Schwartz MA, Ginsberg MH. 2002. Networks and crosstalk: Integrin signalling spreads. *Nat Cell Biol* 4: E65–8.
16. Horne WC, Sanjay A, Bruzzaniti A, Baron R. 2005. The role(s) of Src kinase and Cbl proteins in the regulation of osteoclast differentiation and function. *Immunol Rev* 208: 106–125.
17. Zou W, Kitaura H, Reeve J, Long F, Tybulewicz VLJ, Shattil SJ, Ginsberg MH, Ross FP, Teitelbaum SL. 2007. Syk, c-Src, the avp3 integrin, and ITAM immunoreceptors, in concert, regulate osteoclastic bone resorption. *J Cell Biol* 877–88.
18. Jaffe AB, Hall A. 2005. Rho GTPases: Biochemistry and biology. *Annu Rev Cell Dev Biol* 21: 247–69.
19. Chellaiah MA. 2005. Regulation of actin ring formation by rho GTPases in osteoclasts. *J Biol Chem* 280: 32930–43.
20. Fukuda A, Hikita A, Wakeyama H, Akiyama T, Oda H, Nakamura K, Tanaka S. 2005. Regulation of osteoclast apoptosis and motility by small GTPase binding protein Rac1. *J Bone Miner Res* 20: 2245–53.
21. Faccio R, Teitelbaum SL, Fujikawa K, Chappel JC, Zallone A, Tybulewicz VL, Ross FP, Swat W. 2005. Vav3 regulates osteoclast function and bone mass. *Nat Med* 11: 284–90.
22. Whyte MP, Obrecht SE, Finnegan PM, Jones JL, Podgornik MN, McAlister WH, Mumm S. 2002. Osteoprotegerin deficiency and juvenile Paget's disease. *N Engl J Med* 347: 175–84.
23. Simonet WS, Lacey DL, Dunstan CR, Kelley M, Chang MS, Luthy R, Nguyen HQ, Wooden S, Bennett L, Boone T, Shimamoto G, DeRose M, Elliott R, Colombero A, Tan HL, Trail G, Sullivan J, Davy E, Bucay N, Renshaw-Gegg L, Hughes TM, Hill D, Pattison W, Campbell P, Sander S, Van G, Tarpley J, Derby J, Lee R, Boyle WJ. 1997. Osteoprotegerin: A novel secreted protein involved in the regulation of bone density. *Cell* 89: 309–19.
24. Clines GA, Guise TA. 2005. Hypercalcaemia of malignancy and basic research on mechanisms responsible for osteolytic and osteoblastic metastasis to bone. *Endocr Relat Cancer* 12: 549–83.
25. Martin TJ. 2002. Manipulating the environment of cancer cells in bone: A novel therapeutic approach. *J Clin Invest* 110: 1399–401.
26. Zwerina J, Redlich K, Schett G, Smolen JS. 2005. Pathogenesis of rheumatoid arthritis: Targeting cytokines. *Ann NY Acad Sci* 105: 716–29.
27. Zwerina J, Hayer S, Tohidast-Akrad M, Bergmeister H, Redlich K, Feige U, Dunstan C, Kollias G, Steiner G, Smolen J, Schett G. 2004. Single and combined inhibition of tumor necrosis factor, interleukin-1, and RANKL pathways in tumor necrosis factor-induced arthritis: Effects on synovial inflammation, bone erosion, and cartilage destruction. *Arthritis Rheum* 50: 277–90.
28. Takayanagi H. 2005. Mechanistic insight into osteoclast differentiation in osteoimmunology. *J Mol Med* 83: 170–79.
29. Key LL, Rodriguiz RM, Willi SM, Wright NM, Hatcher HC, Eyre DR, Cure JK, Griffin PP, Ries WL. 1995. Long-term treatment of osteopetrosis with recombinant human interferon gamma. *N Engl J Med* 332: 1594–9.
30. Cenci S, Toraldo G, Weitzmann MN, Roggia C, Gao Y, Qian WP, Sierra O, Pacifici R. 2003. Estrogen deficiency induces bone loss by increasing T cell proliferation and lifespan through IFN-gamma-induced class II transactivator. *Proc Natl Acad Sci U S A* 100: 10405–10.
31. Kim MS, Day CJ, Selinger CI, Magno CL, Stephens SRJ, Morrison NA. 2006. MCP-1 -induced human osteoclast-like cells are tartrate- resistant acid phosphatase, NFATc1, and calcitonin receptor-positive but require receptor activator of NFkappaB ligand for bone resorption. *J Biol Chem* 281: 1274–85.
32. Stockinger B, Veldhoen M. 2007. Differentiation and function of Th17 T cells. *Curr Opin Immunol* 19: 281–6.
33. Udagawa N. 2003. The mechanism of osteoclast differentiation from macrophages: Possible roles of T lymphocytes in osteoclastogenesis. *J Bone Miner Metab* 21: 337–43.
34. Goltzman D, Miao D, Panda DK, Hendy GN. 2004. Effects of calcium and of the Vitamin D system on skeletal and calcium homeostasis: Lessons from genetic models. *J Steroid Biochem Mol Biol* 89–90: 485–89.
35. Syed F, Khosla S. 2005. Mechanisms of sex steroid effects on bone. *Biochem Biophys Res Commun* 32: 688–96.
36. Eastell R. 2005. Role of oestrogen in the regulation of bone turnover at the menarche. *J Endocrinol* 185: 223–34.
37. Canalis E, Bilezikian JP, Angeli A, Giustina A. 2004. Perspectives on glucocorticoid-induced osteoporosis. *Bone* 34: 593–98.
38. Weinstein RS, Chen J-R, Powers CC, Stewart SA, Landes RD, Bellido T, Jilka RL, Parfitt AM, Manolagas SC. 2002. Promotion of osteoclast survival and antagonism of bisphosphonate-induced osteoclast apoptosis by glucocorticoids. *J Clin Invest* 109: 1041–8.
39. Kobayashi T, Narumiya S. 2002. Function of prostanoid receptors: Studies on knockout mice. *Prostaglandins Other Lipids Mediat* 68–69: 557–3.
40. Golden LH, Insogna KL. 2004. The expanding role of PI3-kinase in bone. *Bone* 34: 3–12.
41. DiNitto JP, Cronin TC, Lambright DG. 2003. Membrane recognition and targeting by lipid-binding domains. *Sci STKE* 2003: re16.

42. Nakashima T, Wada T, Penninger JM. 2003. RANKL and RANK as novel therapeutic targets for arthritis. *Curr Opin Rheumatol* 15: 280–7.
43. Taubman MA, Valverde P, Han X, Kawai T. 2005. Immune response: The key to bone resorption in periodontal disease. *J Periodontol* 76: 2033–41.
44. Suda T, Arai F, Hirao A. 2005. Hematopoictic stem cells and their niche. *Trends Immunol* 26: 426–33.
45. Taichman RS. 2005. Blood and bone: Two tissues whose fates are intertwined to create the hematopoietic stem-cell niche. *Blood* 105: 2631–9.
46. Bendre M, Gaddy D, Nicholas RW, Suva LJ. 2003. Breast cancer metastasis to bone: It is not all about PTHrP. *Clin Orthop* (415 Suppl): S39–45.
47. Hata H. 2005. Bone lesions and macrophage inflammatory protein-1 alpha (MIP-la) in human multiple myeloma. *Leuk Lymphoma* 46: 967–72.
48. Kim MS, Day CJ, Morrison NA. 2005. MCP-1 is induced by receptor activator of nuclear factor κB ligand, promotes human osteoclast fusion, and rescues granulocyte macrophage colony stimulating factor suppression of osteoclast formation. *J Biol Chem* 280: 16163–9.
49. Seales EC, Micoli KJ, McDonald JM. 2006. Calmodulin is a critical regulator of osteoclastic differentiation, function, and survival. *J Cell Biochem* 97: 45–55.
50. Ross FP. 2006. M-CSF, c-Fms and signaling in osteoclasts and their precursors. *Ann NY Acad Sci* 1068: 110–6.
51. Rogers MJ. 2004. From molds and macrophages to mevalonate: A decade of progress in understanding the molecular mode of action of bisphosphonates. *Calcif Tissuc Int* 75: 451–61.
52. Blair HC, Robinson LJ, Zaidi M. 2005. Osteoclast signalling pathways. *Biochem Biophys Res Commun* 328: 728–38.
53. Feng X. 2005. Regulatory roles and molecular signaling of TNF family members in osteoclasts. *Gene* 350: 1–13.
54. Hershey CL, Fisher DE. 2004. Mitf and Tfe3: Members of a b-HLH-ZIP transcription factor family essential for osteoclast development and function. *Bone* 34: 689–96.
55. Lee ZH, Kim H-H. 2003. Signal transduction by receptor activator of nuclear factor kappa B in osteoclasts. *Biochem Biophys Res Commun* 305: 211–4.
56. Wagner EF, Eferl R. 2005. Fos/AP-1 proteins in bone and the immune system. *Immunol Rev* 208: 126–40.

4
Osteocytes

Lynda F. Bonewald

Introduction 34
Osteocyte Ontogeny 34
Osteocytes as Orchestrators of Bone (Re)Modeling 36
Osteocyte Cell Death and Apoptosis 37
Osteocyte Modification of Their Microenvironment 37
Mechanosensation and Transduction 38

Role of Gap Junctions and Hemichannels in Osteocyte Communication 38
The Potential Role of Osteocytes in Bone Disease 39
Acknowledgment 39
References 39

INTRODUCTION

In the adult skeleton, osteocytes make up over 90–95% of all bone cells, compared to 4–6% osteoblasts and approximately 1–2% osteoclasts. These cells are regularly dispersed throughout the mineralized matrix, connected to each other and cells on the bone surface through dendritic processes generally radiating towards the bone surface and the blood supply. The dendritic processes travel through the bone in tiny canals called canaliculi (250–300 nm) while the cell body is encased in a lacuna (15–20 um) (See Figs. 4.1, 4.2, and 4.3). Osteocytes are thought to function as a network of sensory cells mediating the effects of mechanical loading through this extensive lacunocanalicular network. Not only do these cells communicate with each other and with cells on the bone surface, but their dendritic processes extend past the bone surface into the bone marrow. Osteocytes have long been thought to respond to mechanical strain to send signals of resorption or formation, and evidence is accumulating to show that this is a major function of these cells. Recently, it has been shown that osteocytes have another important function: to regulate phosphate homeostasis; therefore, the osteocyte network may also function as an endocrine gland. Defective osteocyte function may play a role in a number of bone diseases, especially glucocorticoid induced bone fragility and osteoporosis in the adult, aging skeleton (for review see Ref. 1).

OSTEOCYTE ONTOGENY

Osteoprogenitor cells reside in the bone marrow before differentiating into plump, polygonal osteoblasts on the bone surface. (For a review of the osteoblast to osteocyte transformation, see Refs. 2 and 3.) By an unknown mechanism, some of these cells are destined to become osteocytes, while some become lining cells and some undergo programmed cell death known as apoptosis [4]. Osteoblasts, osteoid-osteocytes, and osteocytes may play distinct roles in the initiation and regulation of mineralization of bone matrix, but Bordier first proposed that osteoid-osteocytes are major regulators of the mineralization process [5]. Osteoid-osteocytes actively make a matrix while simultaneously calcifying this matrix. The osteoblast cell body reduces in size about 30% at the osteoid-osteocyte stage while cytoplasmic processes are forming and about 70% with complete maturation of the osteocyte (See Figs. 4.1 and 4.2).

Whereas numerous markers for osteoblasts have been identified, such as cbfa1, osterix, alkaline phosphatase, collagen type 1, etc., few markers have been available for osteocytes until recently (for a review, see Ref. 3). In 1996, the markers described for osteocytes were limited to low- or no-alkaline phosphatase, high casein kinase II, high osteocalcin protein expression, and high CD44 as compared to osteoblasts. Osteocyte markers such as E11/gp38, phosphate-regulating neutral endopeptidase

---

Primer on the Metabolic Bone Diseases and Disorders of Mineral Metabolism, Eighth Edition. Edited by Clifford J. Rosen.
© 2013 American Society for Bone and Mineral Research. Published 2013 by John Wiley & Sons, Inc.

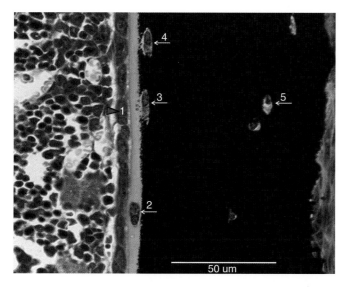

Fig. 4.1. Histological section of tetrachrome stained murine cortical bone showing osteoblast to osteocyte differentiation. 1 = matrix producing osteoblast; 2 = osteoid osteocyte; 3 = embedding osteocyte; 4 = newly embedded osteocyte; 5 = mature osteocyte. From this histological section, one would assume only the lacunae are the porosities within bone. However, as can be seen in Figs. 4.2 and 4.3, the osteocyte canaliculi provide extensive porosities within the mineralized bone matrix.

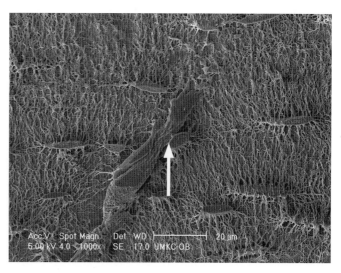

Fig. 4.3. The osteocyte lacunocanalicular network is intimately associated with the blood vessel network in the bone matrix. The white marker points to an osteocyte lacunae intimately associated with the blood vessel.

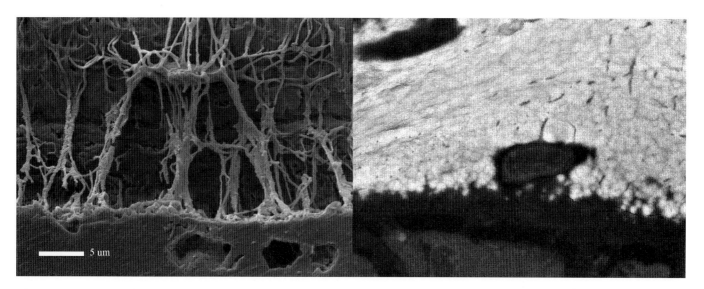

Fig. 4.2. The embedding osteocyte retains its connectivity with cells on the bone surface. The image on the right is of acid-etched plastic embedded murine cortical bone. With this technique, resin fills the lacunocanalicular system, osteoid, and marrow, but cannot penetrate mineral. Mild acid is used to remove the mineral, leaving behind a resin cast relief. Note the canaliculi connecting the lacunae with the bone surface at the bottom of the image. The image on the right is from transmission electron microscopy showing a fully embedded osteocyte and an osteoid-osteocyte becoming surrounded by mineral (white). The osteoid is black and the osteoblasts are at the bottom of the image.

Table 4.1. Osteocyte Markers

Marker	Expression	Function
E11/gp38	Early, embedding cell (Ref. 45)	Dendrite formation (Ref. 45)
CD44	More highly expressed in osteocytes compared to osteoblasts (Ref. 46)	Hyaluronic acid receptor associated with E11 and linked to cytoskeleton (Ref. 47)
Fimbrin	All osteocytes (Ref. 6)	Dendrite branching?
Phex	Early and late osteocytes (Refs. 48, 49)	Phosphate metabolism (Ref. 50)
OF45/MEPE	Late osteoblast through osteocytes (Ref. 51)	Inhibitor of bone formation (Ref. 51)/regulator of phosphate metabolism (Ref. 52)
DMP1	Early and mature osteocytes (Ref. 53) osteocytes (Ref. 53)	Phosphate metabolism and mineralization (Ref. 42)
Sclerostin	Late embedded osteocyte (Ref. 54)	Inhibitor of bone formation (Ref. 55)
FGF23	Early and mature osteocytes (Ref. 43)	Induces hypophosphatemia (Ref. 50)
ORP150	Mature osteocytes (Ref. 9)	Protection from hypoxia (Ref. 9)

on the chromosome X (Phex), dentin matrix protein 1 (DMP1), sclerostin, FGF23, and ORP150 have since then been identified (see Table 4.1). Some of these markers overlap in expression with osteoblasts, but some have been identified for specific stages of differentiation. The identification of these markers has led to identification of osteocyte functions (Table 4.1). The actin-bundling proteins—villin, alpha-actinin, and fimbrin—were shown to be markers for osteocytes with strong signals of fimbrin at branching points in dendrites [6]. It is likely that these actin-reorganizing proteins play a role in osteocyte cell body movement within its lacuna and the retraction and extension of dendritic processes [7, 8]. CapG and destrin have been shown to be more highly expressed in the embedding osteocyte [9].

Promoters for specific markers have been used to drive green fluorescent protein (GFP) to follow osteoblast to osteocyte differentiation in vivo. Collagen type 1-GFP is strongly expressed in both osteoblasts and osteocytes, osteocalcin-GFP is expressed in a few osteoblastic cells lining the endosteal bone surface and in scattered osteocytes, and the osteocyte-selective promoter, the 8kb DMP1 driving GFP showed selective expression in osteocytes [10]. A marker for the late, embedded osteocyte is sclerostin, which is coded by the gene Sost. This protein is a negative regulator of bone formation, targeting the osteoblast. Inhibition of sclerostin activity leads to increased bone mass; therefore, therapeutics such as neutralizing antibodies or small molecules are now the focus of much research and are being tested clinically (for a review, see Ref. 11).

OSTEOCYTES AS ORCHESTRATORS OF BONE (RE)MODELING

Considerable evidence is mounting that osteocytes can conduct and control both bone resorption and bone formation. Some of the earliest supporting data for the theory that osteocytes can send signals to initiate bone resorption were observations that isolated avian osteocytes can support osteoclast formation and activation in the absence of any osteotropic factors [12], as can the osteocyte-like cell line, MLO-Y4 [13]. It was suggested that expression of the receptor activator of NF-kappaB ligand (RANKL) along exposed osteocyte dendritic processes provides a potential means for osteocytes within bone to stimulate osteoclast precursors at the bone surface.

One of the major means by which osteocytes may support osteoclast activation and formation is through their death. Osteocyte apoptosis, also known as programmed cell death, is an orderly process that can occur at sites of microdamage; it is proposed that dying osteocytes are targeted for removal by osteoclasts. The expression of anti-apoptotic and pro-apoptotic molecules in osteocytes surrounding microcracks was mapped, and it was found that pro-apoptotic molecules are elevated in osteocytes immediately at the microcrack locus, whereas anti-apoptotic molecules are expressed 1–2mm from the microcrack [14]. Therefore, those osteocytes that do not undergo apoptosis are prevented from doing so by protective mechanisms, while those destined for removal by osteoclasts undergo apoptosis. Targeted ablation of osteocytes by necrosis was performed using the 10kb Dmp1 promoter to drive expression of the diptheria toxin receptor in mice [15]. Injection of a single dose of diphtheria toxin eliminated approximately 70% of osteocytes in cortical bone in these mice leading to dramatic osteoclast activation. Therefore, viable osteocytes are necessary to prevent osteoclast activation and maintain bone mass (see Fig. 4.4).

The osteocyte-like cell line MLO-Y4 supports osteoclast formation and osteoblast differentiation [16], and surprisingly, mesenchymal stem cell differentiation [17]. It is most likely (but remains to be proven) that primary osteocytes can perform all three functions, therefore possessing the unique capacity to regulate all phases of bone remodeling.

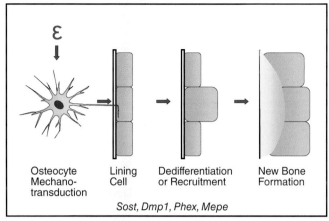

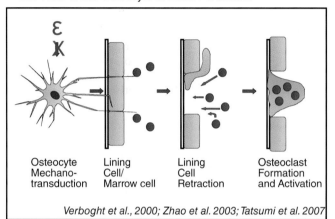

Fig. 4.4. Osteocytes as orchestrators of bone (re)modeling. Osteocytes play a role in bone formation and mineralization as promoters of mineralization through Dmp1 and Phex and inhibitors of mineralization and bone formation, such as Sost/sclerostin and MEPE/OF45, which are highly expressed in osteocytes (top figure). These supporters and inhibitors of bone formation and mineralization are most likely exquisitely balanced to maintain equilibrium in order to maintain bone mass.

Osteocytes also appear to play a major role in the regulation of osteoclasts, by both inhibiting and activating osteoclastic resorption. It has recently been shown that with loading, the osteocytes send signals inhibiting osteoclast activation (bottom figure) (Ref. 15). In contrast, compromised, hypoxic, apoptotic, or dying osteocytes, especially with unloading, appear to send unknown signals to osteoclasts/preosteoclasts on the bone surface to initiate resorption. Therefore, osteocytes within the bone regulate bone formation and mineralization and inhibit osteoclastic resorption, while also having the capacity to send signals of osteoclast activation under specific conditions.

OSTEOCYTE CELL DEATH AND APOPTOSIS

It has been proposed that the purpose and function of osteocytes is to die, thereby releasing signals of resorption. Osteocyte cell death can occur in association with pathological conditions, such as osteoporosis and osteoarthritis, leading to increased skeletal fragility [18]. Such fragility is considered to be due to loss of the ability to sense microdamage and/or to initiate repair. Several conditions have been shown to result in osteocyte cell death, such as oxygen deprivation as occurs during immobilization, withdrawal of estrogen, and glucocorticoid treatment [18]. TNFα and interleukin-1 (IL-1) have been reported to increase with estrogen deficiency and also induce osteocyte apoptosis (for a review, see Ref. 3). Voluntary exercise appears to preserve osteocyte viability and bone strength with ovariectomy [19].

Several agents have been found to reduce or inhibit osteoblast and osteocyte apoptosis. These include estrogen, selective estrogen receptor modulators, bisphosphonates, calcitonin, CD40 ligand, calbindin-D28k, monocyte chemotactic proteins MCP1 and 3, and recently mechanical loading through the release of prostaglandin (for a review, see Ref. 3). Osteocyte viability clearly plays a significant role in the maintenance of bone homeostasis and integrity. However, whereas blocking osteocyte apoptosis may improve diseases such as bone loss due to aging or to glucocorticoid therapy, osteocyte apoptosis may be essential for normal damage repair and skeletal replacement. Any agents that block this process may exacerbate conditions in which repair is required. The death processes, and consequently the resorption signals sent by dying osteocytes in an aging or glucocorticoid treated skeleton may be distinct from those in a normal, healthy skeleton in response to microdamage. It will be important to identify and characterize these differences.

Alternatively, instead of the osteocyte undergoing apoptosis, it can enter a state of self-preservation called autophagy where the cell "eats itself," i.e. "auto-phagy" to maintain viability until a favorable environment returns. Glucocorticoids can induce this state in osteocytes [20], and dose determines whether the osteocyte will undergo apoptosis or autophagy with high or low dose, respectively [21].

OSTEOCYTE MODIFICATION OF THEIR MICROENVIRONMENT

Almost 100 years ago, it was proposed that osteocytes may resorb their lacunar wall under particular conditions [22]. The term "osteolytic osteolysis" was initially used to describe the enlarged lacunae in patients with hyperparathyroidism and later in immobilized rats. "Osteolytic osteolysis" has a negative connotation as it was confused with osteoclastic bone resorption. When resorption "pits" similar to those observed with osteoclasts were not observed with primary avian osteocytes seeded onto dentin slices, it was concluded that osteocytes cannot remove the mineralized matrix. Removal of mineral by osteocytes would not be detectable using this approach as these cells are within a lacuna and do not form the characteristic sealed osteoclast resorption lacuna that rapidly decalcifies bone. In contrast to lacunar enlargement by the embedded, mature osteocyte, enlarged lacunae in bone from patients with renal osteodystrophy may be due to defective mineralization during embedding of the osteoid-osteocyte during bone formation (for reviews, see Refs. 3, 23–25).

In addition to enlargement of the lacunae, changes can take place in the perilacunar matrix. The term "osteocyte halos" was used to describe perilacunar demineralization in rickets [26] and later to describe periosteocytic lesions in X-linked hypophosphatemic rickets [27]. Glucocorticoids, in addition to having effects on osteocyte apoptosis, also appear to cause osteocytes to not only enlarge their lacunae but also to remove mineral from the perilacunar space, thereby generating "halos" of hypomineralized bone [28]. Glucocorticoid may therefore alter or compromise the metabolism and function of the osteocyte, not just induce cell death.

Over 3 decades ago, it was suggested that the osteocyte not only has matrix-destroying capability but also can form a new matrix [29]. Osteocyte lacunae were shown to uptake tetracycline, called "periosteocytic perilacunar tetracycline labeling," indicating the ability to calcify or form bone. Therefore, the osteocyte may be capable of both adding and removing mineral from its surroundings. The capacity to deposit or remove mineral from lacunae and canaliculi has important implications with regards to (1) mineral homeostasis, (2) magnitude of fluid shear stress applied to the cell, and (3) mechanical properties of bone. The surface area of the osteocyte lacunocanalicular system is several orders of magnitudes greater than the bone surface area; therefore, removal of only a few angstroms of mineral would have significant effects on circulating, systemic ion levels. Enlargement of the lacunae and canaliculi would reduce bone fluid flow shear stress, thereby reducing mechanical loading on the osteocyte. As holes in a material act as stress concentrators, enlargement of lacunae would enhance this effect in bone. Therefore, changes in lacunar size and matrix properties could have dramatic effects on bone properties and quality in addition to osteocyte function.

MECHANOSENSATION AND TRANSDUCTION

Mechanical strain is required for postnatal, but not for prenatal skeletal development and maintenance. The postnatal and adult skeletons are able to continually adapt to mechanical loading by the process of adaptive remodeling, where new bone is added to withstand increased amounts of loading, and bone is removed in response to unloading or disuse. The parameters for inducing bone formation or bone resorption in vivo are fairly well known and well characterized. Frequency, intensity, and timing of loading are all important parameters. Bone mass is influenced by peak applied strain [30], and bone formation rate is related to loading rate [31]. When rest periods are inserted, the loaded bone shows increased bone formation rates, compared to bone subjected to a single bout of mechanical loading, and improved bone structure and strength is greatest if loading is applied in shorter vs longer increments [32].

The major challenge in the field of mechanotransduction has been to translate these well-characterized in vivo parameters of mechanical loading to in vitro cell culture models. Theoretical models and experimental studies suggest that flow of bone fluid is driven by extravascular pressure as well as applied cyclic mechanical loading of osteocytes [33]. Mechanical forces applied to bone cause fluid flow through the canaliculi surrounding the osteocyte inducing shear stress and deformation of the cell membrane. It has also been proposed that mechanical information is relayed by the primary cilium, a flagellar-like structure found on every cell [34, 35]. Osteocytes may use a combination of means to sense mechanical strain [36]. Theoretical modeling predicts osteocyte wall and membrane shear stresses resulting from peak physiologic loads in vivo to be in the range of 8 to 30 dynes/cm^3 [33]. Recently in vivo experiments have been performed showing an estimated 5 Pa, which is within the 8 to 30 dynes/cm^3 peak shear stress along the osteocyte membrane [37], validating the use of this magnitude of strain in vitro.

ROLE OF GAP JUNCTIONS AND HEMICHANNELS IN OSTEOCYTE COMMUNICATION

A means by which osteocytes communicate intracellularly is through gap junctions, transmembrane channels that connect the cytoplasm of two adjacent cells, through which molecules with molecular weights less than 1 kD$_a$ can pass. Gap junction channels are formed by members of a family of proteins known as connexins, and Cx43 is the major connexin in bone cells. Much of mechanotransduction in bone is thought to be mediated through gap junctions.

Primary osteocytes and MLO-Y4 osteocyte-like cells [38] express large amounts of Cx43, suggesting that Cx43 has another function in addition to being a component of gap junctions. Recently, it has been shown that connexins can form and function as unapposed halves of gap junction channels called hemichannels. Hemichannels directly serve as the pathway for the exit of intracellular PGE$_2$ in osteocytes induced by fluid flow shear stress [39] and function as essential transducers of the anti-apoptotic effects of bisphosphonates [40]. Hemichannels are now one of several types of openings or channels to the extracellular bone fluid that also includes channels such as calcium, ion, voltage, stretch activated channels, and others (for a review, see Ref. 41). Therefore, gap junctions at the connecting tips of dendrites appear to mediate a form of intracellular communication, and hemichannels along the dendrite (and perhaps the cell body) appear to mediate a form of extracellular communication between osteocytes.

THE POTENTIAL ROLE OF OSTEOCYTES IN BONE DISEASE

Osteoid osteocytes play a role in phosphate homeostasis. Once the osteoblast begins to embed in osteoid, molecules such as Dmp1, Phex, and Mepe are elevated (See Table 4.1). Autosomal recessive hypophosphatemic rickets in patients is due to mutations in *Dmp1* [42]. *Dmp1* null mice have a similar phenotype to Hyp mice carrying a Pex mutation, that of osteomalacia and rickets due to elevated FGF23 levels in osteocytes [42, 43]. The osteocyte lacunocanalicular system should be viewed as an endocrine organ regulating phosphate metabolism. The unraveling of the interactions of these molecules should lead to insight into diseases of hyper and hypophosphatemia (See the chapter on Phex/FGF23).

The connectivity and structure of the osteocyte lacunocanalicular system most likely play a role in bone disease. Osteocyte dendricity may change with static and dynamic bone formation and has been shown to be disrupted in bone disease [44]. In osteoporotic bone there is disorientation of the canaliculi, as well as a marked decrease in connectivity, which increases in severity of disease. In contrast, in osteoarthritic bone, a decrease in connectivity is observed, but orientation is intact. In osteomalacic bone, the osteocytes appear viable with high connectivity, but the processes are distorted and the network chaotic [44]. Variability in complexity and number of dendrites and canaliculi could have a dramatic effect on osteocyte function and viability and on the mechanical properties of bone.

Osteocyte cell death may be responsible for some forms of osteonecrosis. Osteonecrosis is "dead" bone containing empty osteocyte lacunae that does not remodel but can remain in the bone for years. As reviewed above, viable osteocytes are necessary to send signals of (re)modeling. Early proposed mechanisms responsible for osteonecrosis include: the "mechanical theory," where osteoporosis and the accumulation of unhealed trabecular microcracks result in fatigue fractures; the "vascular theory" where ischemia is caused by microscopic fat emboli; and a newer theory, where agents induce osteocyte cell death, which results in dead bone that does not remodel [18]. Osteocyte health, compromised status, and viability and capacity to regulate its own death most likely play a highly significant role in the maintenance and integrity of bone. Bone loss in osteoporosis may be due in part to pathological and not physiological osteocyte cell death [4]. It will be important to develop therapeutics that maintain both osteocyte viability and physiological osteocyte cell death that leads to normal bone repair.

In conclusion, it is most likely that osteocytes utilize undiscovered specific molecules to regulate bone (re)modeling. With the dramatic increases or maintenance of bone mass being observed with neutralizing antibody to sclerostin, an osteocyte selective marker [11], greater effort is being made to identify additional markers and to unravel the mysteries surrounding osteocyte function. It is also likely that new functions will be discovered for these cells, making them a target of investigation, not only to understand basic bone physiology but also to understand and treat bone disease.

ACKNOWLEDGMENT

The author's work in osteocyte biology is supported by the National Institutes of Health AR-46798.

REFERENCES

1. Bonewald LF. 2011. The amazing osteocyte. *J Bone Miner Res* 26(2): 229–38.
2. Franz-Odendaal TA, Hall BK, Witten PE. 2006. Buried alive: How osteoblasts become osteocytes. *Dev Dyn* 235(1): 176–90.
3. Bonewald L. 2007. Osteocytes. In:,Marcus R, Feldman D, Nelson, D, Rosen, C (eds.) *Osteoporosis, 3rd Ed.* Boston: Elsevier. pp. 169–90.
4. Manolagas SC. 2000. Birth and death of bone cells: Basic regulatory mechanisms and implications for the pathogenesis and treatment of osteoporosis. *Endocr Rev* 21(2): 115–37.
5. Bordier PJ, Miravet L, Ryckerwaert A, Rasmussen H. 1976. Morphological and morphometrical characteristics of the mineralization front. A vitamin D regulated sequence of bone remodeling. In: Bordier PJ (ed.) *Bone Histomorphometry*, Paris: Armour Montagu. pp. 335–54.
6. Tanaka-Kamioka K, Kamioka H, Ris H, Lim SS. 1998. Osteocyte shape is dependent on actin filaments and osteocyte processes are unique actin-rich projections. *J Bone Miner Res* 13(10): 1555–68.
7. Dallas SL, Bonewald LF. 2010. Dynamics of the transition from osteoblast to osteocyte. *Ann NY Acad Sci* 1192(1): 437–43.
8. Veno P, Nicolella DP, Sivakumar P, Kalajzic I, Rowe D, Harris SE, Bonewald L, Dalls SL. 2006. Live imaging of osteocytes within their lacunae reveals cell body and dendrite motions. *J Bone Min Res* 21(Suppl 1): S38–S39.
9. Guo D, Keightley A, Guthrie J, Veno PA, Harris SE, Bonewald LF. 2010. Identification of osteocyte-selective proteins. *Proteomics* 10(20): 3688–98.
10. Kalajzic I, Braut A, Guo D, Jiang X, Kronenberg MS, Mina M, et al. 2004. Dentin matrix protein 1 expression during osteoblastic differentiation, generation of an osteocyte GFP-transgene. *Bone* 35(1): 74–82.
11. Paszty C, Turner CH, Robinson MK. 2010. Sclerostin: A gem from the genome leads to bone-building antibodies. *J Bone Miner Res* 25(9): 1897–904.

12. Tanaka K, Yamaguchi, Y, Hakeda, Y. 1995. Isolated chick osteocytes stimulate formation and bone-resorbing activity of osteoclast-like cells. *J Bone Miner Metab* 13: 61–70.
13. Zhao S, Zhang YK, Harris S, Ahuja SS, Bonewald LF. 2002. MLO-Y4 osteocyte-like cells support osteoclast formation and activation. *J Bone Miner Res* 17(11): 2068–79.
14. Verborgt O, Tatton NA, Majeska RJ, Schaffler MB. 2002. Spatial distribution of Bax and Bcl-2 in osteocytes after bone fatigue: complementary roles in bone remodeling regulation? *J Bone Miner Res* 17(5): 907–14.
15. Tatsumi S, Ishii K, Amizuka N, Li M, Kobayashi T, Kohno K, et al. 2007. Targeted ablation of osteocytes induces osteoporosis with defective mechanotransduction. *Cell Metab* 5(6): 464–75.
16. Heino TJ, Hentunen TA, Vaananen HK. 2002. Osteocytes inhibit osteoclastic bone resorption through transforming growth factor-beta: Enhancement by estrogen. *J Cell Biochem* 85(1): 185–97.
17. Heino TJ, Hentunen TA, Vaananen HK. 2004. Conditioned medium from osteocytes stimulates the proliferation of bone marrow mesenchymal stem cells and their differentiation into osteoblasts. *Exp Cell Res* 294(2): 458–68.
18. Weinstein RS, Nicholas RW, Manolagas SC. 2000. Apoptosis of osteocytes in glucocorticoid-induced osteonecrosis of the hip. *J Clin Endocrinol Metab* 85(8): 2907–12.
19. Fonseca H, Moreira-Goncalves D, Esteves JL, Viriato N, Vaz M, Mota MP, et al. 2011. Voluntary exercise has long-term in vivo protective effects on osteocyte viability and bone strength following ovariectomy. *Calcif Tissue Int* 88(6): 443–54.
20. Xia X, Kar R, Gluhak-Heinrich J, Yao W, Lane NE, Bonewald LF, et al. 2010. Glucocorticoid-induced autophagy in osteocytes. *J Bone Miner Res* 25(11): 2479–88.
21. Jia J, Yao W, Guan M, Dai W, Shahnazari M, Kar R, et al. 2011. Glucocorticoid dose determines osteocyte cell fate. *FASEB J* 25(10): 3366–76.
22. Recklinghausen FV (ed.) 1910. *Untersuchungen uber rachitis and osteomalacia*. Fischer, Jena, Germany.
23. Belanger LF. 1969. Osteocytic osteolysis. *Calcif Tissue Res* 4(1): 1–12.
24. Kremlien B, Manegold C, Ritz E, Bommer J. 1976. The influence of immobilization on osteocyte morphology: Osteocyte differential count and electron microscopic studies. *Virchows Arch A Pathol Anat Histol* 370(1): 55–68.
25. Qing H, Bonewald LF. 2009. Osteocyte remodeling of the perilacunar and pericanalicular matrix. *Int J Oral Sci* 1(2): 59–65.
26. Heuck F. 1970. Comparative investigations of the function of osteocytes in bone resorption. *Calcif Tissue Res* Suppl: 148–9.
27. Marie PJ, Glorieux FH. 1983. Relation between hypomineralized periosteocytic lesions and bone mineralization in vitamin D-resistant rickets. *Calcif Tissue Int* 35(4–5): 443–8.
28. Lane NE, Yao W, Balooch M, Nalla RK, Balooch G, Habelitz S, et al. 2006. Glucocorticoid-treated mice have localized changes in trabecular bone material properties and osteocyte lacunar size that are not observed in placebo-treated or estrogen-deficient mice. *J Bone Miner Res* 21(3): 466–76.
29. Baud CA, Dupont DH. 1962. The fine structure of the osteocyte in the adult compact bone. In: Breese, SS (ed.) *Electron Microscopy, Vol. 2*. New York: Academic Press. pp. QQ–10.
30. Rubin C. 1984. Skeletal strain and the functional significance of bone architecture. *Calcif Tissue Int* 36: S11–S8.
31. Turner CH, Forwood MR, Otter MW. 1994. Mechanotransduction in bone: Do bone cells act as sensors of fluid flow? *Faseb J* 8(11): 875–8.
32. Robling AG, Hinant FM, Burr DB, Turner CH. 2002. Shorter, more frequent mechanical loading sessions enhance bone mass. *Med Sci Sports Exerc* 34(2): 196–202.
33. Weinbaum S, Cowin SC, Zeng Y. 1994. A model for the excitation of osteocytes by mechanical loading-induced bone fluid shear stresses. *J Biomech* 27(3): 339–60.
34. Xiao Z, Zhang S, Mahlios J, Zhou G, Magenheimer BS, Guo D, et al. 2006. Cilia-like structures and polycystin-1 in osteoblasts/osteocytes and associated abnormalities in skeletogenesis and Runx2 expression. *J Biol Chem* 281(41): 30884–95.
35. Malone AM, Anderson CT, Tummala P, Kwon RY, Johnston TR, Stearns T, et al. 2007. Primary cilia mediate mechanosensing in bone cells by a calcium-independent mechanism. *Proc Natl Acad Sci U S A*. 104(33): 13325–30.
36. Bonewald LF. 2006. Mechanosensation and transduction in osteocytes. *Bonekey osteovision*. 3(10): 7–15.
37. Price C, Zhou X, Li W, Wang L. 2011. Real-time measurement of solute transport within the lacunar-canalicular system of mechanically loaded bone: direct evidence for load-induced fluid flow. *J Bone Miner Res* 26(2): 277–85.
38. Kato Y, Windle JJ, Koop BA, Mundy GR, Bonewald LF. 1997. Establishment of an osteocyte-like cell line, MLO-Y4. *J Bone Miner Res* 12(12): 2014–23.
39. Cherian PP, Siller-Jackson AJ, Gu S, Wang X, Bonewald LF, Sprague E, et al. 2005. Mechanical strain opens connexin 43 hemichannels in osteocytes: A novel mechanism for the release of prostaglandin. *Mol Biol Cell* 16(7): 3100–6.
40. Plotkin LI, Manolagas SC, Bellido T. 2002. Transduction of cell survival signals by connexin-43 hemichannels. *J Biol Chem* 277(10): 8648–57.
41. Klein-Nulend J, Bonewald, LF. 2008. The osteocyte. In: Bilezikian JP, Raisz, LG (eds.) *Principles of Bone Biology, Vol 2*. San Diego: Academic Press. pp. QQ–10.
42. Feng JQ, Ward LM, Liu S, Lu Y, Xie Y, Yuan B, et al. 2006. Loss of DMP1 causes rickets and osteomalacia and identifies a role for osteocytes in mineral metabolism. *Nat Genet* 38(11): 1310–5.
43. Liu S, Lu Y, Xie Y, Zhou J, Quarles LD, Bonewald L, et al. 2006. Elevated levels of FGF23 in dentin matrix

protein 1 (DMP1) null mice potentially explain phenotypic similarities to hyp mice. *J Bone Min Res* 21: S51.
44. Knothe Tate ML, Adamson JR, Tami AE, Bauer TW. 2004. The osteocyte. *Int J Biochem Cell Biol* 36(1): 1–8.
45. Zhang K, Barragan-Adjemian C, Ye L, Kotha S, Dallas M, Lu Y, et al. 2006. E11/gp38 selective expression in osteocytes: regulation by mechanical strain and role in dendrite elongation. *Mol Cell Biol* 26(12): 4539–52.
46. Hughes DE, Salter DM, Simpson R. 1994. CD44 expression in human bone: A novel marker of osteocytic differentiation. *J Bone Miner Res* 9(1): 39–44.
47. Ohizumi I, Harada N, Taniguchi K, Tsutsumi Y, Nakagawa S, Kaiho S, et al. 2000. Association of CD44 with OTS-8 in tumor vascular endothelial cells. *Biochim Biophys Acta* 1497(2): 197–203.
48. Westbroek I, De Rooij KE, Nijweide PJ. 2002. Osteocyte-specific monoclonal antibody MAb OB7.3 is directed against Phex protein. *J Bone Miner Res* 17(5): 845–53.
49. Ruchon AF, Tenenhouse HS, Marcinkiewicz M, Siegfried G, Aubin JE, DesGroseillers L, et al. 2000. Developmental expression and tissue distribution of Phex protein: Effect of the Hyp mutation and relationship to bone markers. *J Bone Miner Res* 15(8): 1440–50.
50. A gene (PEX) with homologies to endopeptidases is mutated in patients with X-linked hypophosphatemic rickets. The HYP Consortium. 1995. *Nat Genet* 11(2): 130–6.
51. Gowen LC, Petersen DN, Mansolf AL, Qi H, Stock JL, Tkalcevic GT, et al. 2003. Targeted disruption of the osteoblast/osteocyte factor 45 gene (OF45) results in increased bone formation and bone mass. *J Biol Chem* 278(3): 1998–2007.
52. Rowe PS, Kumagai Y, Gutierrez G, Garrett IR, Blacher R, Rosen D, et al. 2004. MEPE has the properties of an osteoblastic phosphatonin and minhibin. *Bone* 34(2): 303–19.
53. Toyosawa S, Shintani S, Fujiwara T, Ooshima T, Sato A, Ijuhin N, et al. 2001. Dentin matrix protein 1 is predominantly expressed in chicken and rat osteocytes but not in osteoblasts. *J Bone Miner Res* 16(11): 2017–26.
54. Poole KE, van Bezooijen RL, Loveridge N, Hamersma H, Papapoulos SE, Lowik CW, et al. 2005. Sclerostin is a delayed secreted product of osteocytes that inhibits bone formation. *Faseb J* 19(13): 1842–4.
55. Balemans W, Ebeling M, Patel N, Van Hul E, Olson P, Dioszegi M, et al. 2001. Increased bone density in sclerosteosis is due to the deficiency of a novel secreted protein (SOST). *Hum Mol Genet* 10(5): 537–43.

5
Connective Tissue Pathways That Regulate Growth Factors

Gerhard Sengle and Lynn Y. Sakai

Introduction 42
The Fibrillinopathies 42
Fibrillin Microfibrils 43
Regulation of Growth Factors by Fibrillin Microfibrils 43

Molecular Mechanisms Orchestrated on a Microfibril Scaffold 44
Summary 45
Abbreviations 46
References 46

INTRODUCTION

Different connective tissues perform different physiological functions. In order to perform these different functions, connective tissue cells secrete distinct sets of extracellular matrix (ECM) proteins that are arranged within the individual connective tissue in discrete patterns. The relative abundance of ECM proteins within a tissue and the histological patterns in which these proteins are organized endow connective tissues with their specific developmental, physiological, and homeostatic properties. For example, in bone, type I collagen is the most abundant ECM protein constituent, and type I collagen fibers are organized in long, thick bundles that are consistent with the mechanical properties required of bone.

The contributions of ECM proteins such as collagens and proteoglycans to the mechanical properties of cartilage and bone are better understood today than the contributions of many of the minor constituents of the ECM. However, recent and emerging knowledge of the function of a fibril-forming ECM protein named "fibrillin" [1] indicates important roles for fibrillin microfibrils in the development, growth, and maintenance of skeletal elements. This chapter summarizes current knowledge about fibrillin microfibrils, their molecular partners in the connective tissue, and their relevance to skeletal biology.

THE FIBRILLINOPATHIES

The importance of fibrillin to the skeleton was first appreciated when a mutation in the gene for fibrillin-1 (*FBN1*) was identified as the cause of Marfan syndrome [2]. Individuals with Marfan syndrome (OMIM #154700) display major disease features in the skeleton: tall stature and arachnodactyly, scoliosis and chest deformities, joint hypermobility and muscle wasting, pes planus, and craniofacial abnormalities, including a highly arched palate. Skeletal features of Marfan syndrome are thought to be largely due to the overgrowth of long bones. Multiple features in other organs (cardiovascular, ocular, skin, lung, and central nervous system) also distinguish Marfan syndrome, reflecting the ubiquitous tissue distribution of fibrillin-1 as well as the importance of fibrillin-1 to the affected tissues.

Mutations in the gene for fibrillin-2 (*FBN2*) cause Beals syndrome or congenital contractural arachnodactyly (CCA) [3]. Features of CCA (OMIM #121050) include contractures of the small and large joints, crumpled ears, and arachnodactyly. The more limited nature of disease features caused by mutations in *FBN2* is thought to reflect the low to null expression levels of *FBN2* mRNA in postnatal tissues and the compensatory high levels of *FBN1* mRNA in postnatal tissues.

Mutations in *FBN1* also cause autosomal dominant Weill-Marchesani syndrome [4, 5]. The skeletal features

---

Primer on the Metabolic Bone Diseases and Disorders of Mineral Metabolism, Eighth Edition. Edited by Clifford J. Rosen.
© 2013 American Society for Bone and Mineral Research. Published 2013 by John Wiley & Sons, Inc.

of Weill-Marchesani syndrome are the opposite of those found in the Marfan syndrome. Individuals with Weill-Marchesani syndrome (OMIM #608328) display short stature, brachydactyly, hypermuscularity, and joint stiffness. While these skeletal features are the opposite of Marfan, ectopia lentis (resulting from a weakness in the suspensory ligament of the lens) is typical of both syndromes. Weill-Marchesani syndrome is one of several others grouped together as the acromelic dysplasias. Mutations in *FBN1* have been recently identified in individuals with geleophysic and acromicric dysplasias, two additional syndromes in the acromelic group [6].

These genetic findings demonstrate the pivotal role of fibrillins in controlling bone growth. In particular, it is now clear that disturbances in the fibrillin-1 structure or function can lead to either tall or short bones. Further study of these genetic disorders will reveal the mechanisms by which fibrillin-1 influences bone growth.

FIBRILLIN MICROFIBRILS

Fibrillin was first identified as the major protein component of small (10 nm) diameter microfibrils that are ubiquitous in the extracellular space [1]. At the ultrastructural level, fibrillin microfibrils can be distinguished from collagen fibers by their uniform small diameter and a characteristic beaded or hollow appearance with no periodic banding pattern. Fibrillin microfibrils are often found as bundles of microfibrils and are always present in elastic fibers.

In developing cartilages, fibrillin microfibrils in the perichondrium ring the cartilage matrix and also extend into the cartilage matrix in a chicken-wire pattern. In aging cartilages, fibrillin microfibrils close to the chondrocyte aggregate laterally to form banded thick fibers that can be distinguished from amianthoid fibers [7]. In demineralized bone, fibrillin microfibrils are found in the cement lines, Haversian canals, osteocyte lacunae, canaliculi, and within fibrous structures containing type III collagen [8].

The protein components of microfibrils include fibrillin-1 [1], fibrillin-2 [9], and fibrillin-3 [10]. Interestingly, the expression levels of *FBN2* and *FBN3* mRNA are highly restricted to fetal development, while fibrillin-1 can be found from the time of gastrulation [11] through postnatal life. Microfibrils can be heteropolymers or homopolymers of fibrillins [12]. Gene knockout experiments in mice suggest that fibrillin-2 function is especially important in the interdigital space, since mice null for fibrillin-2 develop syndactyly [13]. Mice null for fibrillin-1 develop postnatal aortic aneurysms and dissection, indicating that fibrillin-1 function is required in the aorta after the second week of life [14].

The fibrillins and the LTBPs (latent TGFβ binding proteins) form a family of structurally related proteins. Each of the three fibrillins is a modular protein composed of tandemly repeated epidermal growth factor (EGF)-like domains of the calcium-binding type (cbEGF). These stretches of cbEGF domains are interspersed by domains that contain 8 cysteines (8-cys) and "hybrid" domains that appear to be similar to both the 8-cys and the cbEGF domains [15]. In addition, each of the fibrillins is flanked by N-terminal and C-terminal domains, and each contains a special region that is either proline-rich, glycine-rich, or proline- and glycine-rich [10]. While the fibrillins are homologous in primary structure, overall domain structure, and size, LTBPs vary in size and in primary structure but retain an overall similarity in domain organization. LTBPs are composed of the same types of modular domains as the fibrillins. One of the three 8-cys domains present in LTBP-1 was shown to bind covalently to LAP, the latency associated propeptide of TGFβ-1 and to facilitate the secretion of latent TGFβ complexes [16]. Hence, 8-cys domains are also referred to as TB (TGFβ-binding) domains. LTBP-1 was immunolocalized to fibrillin microfibrils in perichondrium and osteoblast cultures [17, 18], and direct binding of the C-terminal regions of LTBP-1 and -4 to fibrillin-1 was demonstrated [19].

Other proteins associated with fibrillin microfibrils include elastin, the fibulins [20, 21], MAGPs [22], perlecan [23], versican [24], decorin [25], and biglycan [26]. These associated proteins establish connective tissue pathways that extend from the fibrillin microfibril networks to basement membranes (through perlecan interactions) and from proteoglycan shells around the microfibrils to hyaluronan and to collagen. These connective tissue pathways integrate fibrillin microfibril networks into the histological patterns of specific organs (cartilage and bone compared with muscle and skin) and into the mechanical and physiological properties of specific organs. In addition, these connective tissue pathways serve to integrate growth factor signaling into the mechanical and physiological functions of specific organs.

REGULATION OF GROWTH FACTORS BY FIBRILLIN MICROFIBRILS

A working model for the extracellular regulation of the large latent TGFβ complex by fibrillin microfibrils was proposed, based on the demonstration that LTBPs interact directly with fibrillin-1 [19]. According to this model, LTBPs target latent TGFβ complexes to the extracellular matrix, where the large LTBP/TGFβ complex is stabilized through C-terminal interactions of LTBPs with fibrillin [19] and N-terminal interactions with other matrix components, possibly fibronectin [16]. Mutant fibrillin-1 (for example, in Marfan syndrome) or absent fibrillin-1 (in fibrillin-1 deficient mice) was expected to destabilize LTBP/TGFβ complexes. This hypothesis was tested in mouse models of deficient and mutant fibrillin-1, and activation of TGFβ signaling was found in the lung [27], mitral valve [28], and aorta [29].

Additional support that activation of TGFβ signaling underlies aortic dissection and aneurysm, the major life-threatening feature of Marfan syndrome, comes from genetic evidence in humans. Mutations in the receptors for TGFβ cause a Marfan-related disorder [30] that has been named Loeys-Dietz syndrome [31, 32]. The major phenotypic features of Loeys-Dietz syndrome (OMIM #609192) include aortic aneurysm and dissection, hypertelorism, bifid uvula and/or cleft palate, arterial tortuosity, and sometimes craniosynostosis and mental retardation. Dolichostenomelia, a feature shared in common with Marfan syndrome, was reported in less than 20% of individuals with Loeys-Dietz [32].

All together, the genetic evidence in humans and mice demonstrates that TGFβ signaling and fibrillin-1 share common pathways, at least in the aorta. However, skeletal features in Loeys-Dietz syndrome are not major, compared with those found in Marfan syndrome, suggesting that additional signaling mechanisms may be modulating skeletal phenotypes in these two genetic disorders. In *Fbn1* mutant mice, analyses of skeletal phenotypes may be precluded by early postnatal death in homozygotes [14, 33, 34]. Skeletal features (mainly kyphosis and overgrowth of the ribs) were noted in heterozygous C1039G *Fbn1* mutant mice [34] and in a hypomorphic *Fbn1* deficient mouse [35], but it has not been reported whether activation of TGFβ signaling is associated with these phenotypes. The tight skin (Tsk) mouse *(Fbn1Tsk)*, which harbors a large in-frame duplication within the *Fbn1* gene [36], demonstrates skin fibrosis as well as excessive long bone growth, but it is unknown whether TGFβ signaling is associated with Tsk skeletal phenotypes.

Another related mechanism that may be involved in the pathogenesis of phenotypic features in Marfan syndrome is the interaction between fibrillin and bone morphogenetic proteins (BMPs). Fibrillin-1 binds to the propeptide of BMP-7, and antibodies specific for BMP-7 propeptide and growth factor can be immunolocalized to fibrillin microfibrils [37]. BMP-7 propeptide and the BMP-7 propeptide/growth factor complex bind to fibrillin microfibrils isolated from tissues [38]. Studies of additional members of the TGFβ superfamily indicate that the propeptides of BMP-2, -4, -5, and -10, and GDF-5 also interact with fibrillin and that these growth factor complexes may utilize this interaction as a mechanism for proper and effective positioning of growth factors in the extracellular space [39]. The specificity of these interactions is supported by the findings that propeptides of TGFβs bind to LTBPs but not to fibrillins [40], that the propeptide of BMP-7 binds to fibrillin but not to LTBP-1 [37], and that the propeptide of myostatin (GDF-8) does not bind to fibrillin [39] but may interact with LTBP-3 [41].

In vivo evidence that fibrillin-2 affects BMP-7 signaling and limb patterning was demonstrated when fibrillin-2 was knocked out in mice [13]. *Fbn2* null mice were born with bilateral syndactyly of the forelimbs and hindlimbs. Complete fusion or close apposition of the cartilage elements of digits 2 and 3 or 2, 3, and 4 were observed, along with fewer apoptotic signals in the interdigital spaces. Expression of BMP-4 and responsiveness to BMP-loaded beads were retained in the *Fbn2* null autopods, demonstrating that loss of fibrillin-2 in the interdigital space was sufficient to dysregulate the BMP signaling required for proper formation of the cartilage elements and regression of interdigital tissue. Furthermore, compound heterozygous mice *(Fbn2+/-; Bmp-7+/-)* were both polydactylous and syndactylous (features of the single nulls, but not of the single heterozygous mice), indicating that both genes are in the same pathway that controls digit formation.

Fibrillin-1 and fibrillin-2 protein localization is non-overlapping in the interdigital space of the developing autopod: Fibrillin-1 is located mostly at the boundaries or edges of the cartilage elements whereas fibrillin-2 forms abundant rays extending through the interdigital space to connect the cartilage elements to the ectoderm [42]. These tissue-specific differences in the fibrillins may explain why BMP signaling is dysregulated in the developing autopod but apparently not in other tissues, where it is presumed that fibrillin-1 compensates for the loss of fibrillin-2.

MOLECULAR MECHANISMS ORCHESTRATED ON A MICROFIBRIL SCAFFOLD

In the TGFβ complex, the propeptides of TGFβs confer latency to the growth factor dimer, and the propeptides are disulfide-linked to LTBPs. Therefore, in order to initiate TGFβ signaling, the TGFβ growth factor dimer must first be released from this large latent complex. BMPs, on the other hand, are not believed to require activation. BMP-9 propeptide/growth factor complex has been shown to be as active as BMP-9 growth factor, using *in vitro* cell-based assays [43]. Similarly, BMP-7 propeptide/growth factor complex is as active as BMP-7 growth factor [44]. In solution, type II BMP receptors can outcompete BMP-7 propeptides for interaction with the BMP-7 growth factor dimer and displace the propeptides from the propeptide/growth factor complex, indicating that unlike TGFβ propeptides BMP-7 propeptide does not block receptor binding and does not confer latency to the growth factor [44].

Although these growth factors are similar in structure, extracellular mechanisms that control signaling are clearly different. Whereas TGFβ requires activation by proteolytic or dissociative mechanisms and also requires cofactors like thrombospondin, endoglin, or betaglycan to facilitate signaling, BMPs can activate signaling simply by binding to its receptors, and instead of activators, BMPs require inhibitors like noggin (and several others) to control the amount of extracellular presentation. The role of the matrix as a physical scaffold important in regulating extracellular signaling is only now becoming appreciated. However, the mechanisms involved in matrix regulation of extracellular signaling are largely unknown.

Direct interactions of fibrillins with BMPs and LTBPs with TGFβs have been documented. Nevertheless, it is not clearly understood whether interactions between these proteins, and with the microfibril scaffold composed by these proteins, sequester and store growth factors in the connective tissue or target and concentrate growth factors for immediate use. It is not known how other interactions (e.g., fibrillin/fibulin interactions) mediated by the microfibril scaffold affect these matrix/growth factor interactions. However, since both TGFβ and BMP signals are associated with the same microfibril scaffold, it may be predicted that the matrix scaffold physically integrates these disparate, and sometimes opposing, signals. Since perlecan binds to fibrillin [23], it is also possible that the microfibril scaffold helps to integrate FGF signals with BMP and TGFβ signals. The roles of heparan sulfate proteoglycans in the regulation of fibroblast growth factor (FGF) signaling and skeletal biology have been extensively reviewed [45, 46].

Knockouts of the genes encoding LTBP-2, LTBP-3, and LTBP-1L, and an LTBP-4 hypomorphic mouse have been described. Only *Ltbp3* null mice demonstrated skeletal phenotypes (craniofacial abnormalities, kyphosis, osteosclerosis of the long bones and vertebrae, and osteopetrosis). Many of the phenotypes found in these mutant mice were reported to be consistent with loss of TGFβ activity, indicating that LTBPs are required in specific tissues for appropriate TGFβ function. However, none of these mouse models phenocopies *Tgfb1*-null mice, suggesting functional redundancies among the four LTBPs and/or additional tissue-specific functions of the individual LTBPs [16].

An interesting study of LTBP-4 deficient lung fibroblasts revealed that BMP-4 signaling is enhanced and that expression of gremlin, an inhibitor of BMP signaling, was decreased [47]. These findings, which were confirmed in lung tissue, were within the context of loss of LTBP-4 and consequent impaired activation of TGFβ. This study directs attention to the unknown mechanisms by which BMP and TGFβ signals may be integrated across the microfibril scaffold.

Many genes that encode other proteins that interact with fibrillin microfibrils have also been knocked out in mice. Mice null for each of the five fibulin genes have been reported. Skeletal phenotypes have not been reported. However, disturbed TGFβ signaling was found associated with aneurysm formation in a fibulin-4 hypomorphic mouse [48]. In humans, mutations in *FBN4* can result in skeletal features (arachnodactyly) as well as aneurysm and cutis laxa [49], suggesting that defects in fibulin function may perturb pathways common to fibrillin microfibrils.

Biochemical results have also identified proteoglycans as potential interactors on the microfibril scaffold. Versican [24], perlecan [23], and decorin [25] bind to fibrillin, and biglycan binds to elastic fiber components [26]. Perlecan null *(Hspg2-/-)* mice have defective cartilage formation, possibly due to both structural effects and dysregulated FGF signaling, and die in the perinatal period [50]. Decorin binds to the TGFβ growth factor [51], and biglycan modulates BMP-4 signaling to osteoblasts [52]. More investigation is required in order to understand how these associated proteoglycans may cooperate with the microfibrillar scaffold in the extracellular regulation of growth factor signaling.

SUMMARY

Human genetic disorders have pointed to important functional roles of fibrillins in skeletal biology. Fibrillin-1 controls long bone growth, and mutations in *FBN1* can lead either to tall stature and arachnodactyly (in Marfan syndrome) or to short stature and brachydactyly (in Weill-Marchesani syndrome). Fibrillins also control joint function, since mutations in *FBN1* or *FBN2* can result in either joint hypermobility, joint contractures, or stiff joints. Recent advances have demonstrated that mutations in fibrillin genes result in dysregulated TGFβ and BMP signaling, and have led to the concept that the microfibrillar scaffold, together with its extended partners in specific connective tissue pathways, regulate TGFβ and BMP signaling.

According to this concept of extracellular regulation by the microfibrillar scaffold, growth factor signaling is "fine-tuned" [42], and resulting perturbations in signaling can lead to a broad range of pathologies. These pathologies may bear only limited resemblance to phenotypes that result from the complete loss of function mouse models of growth factor signaling. More sophisticated mouse models are required in order to dissect the mechanisms by which the microfibril scaffold regulates growth factor signaling and by which human pathologies are produced.

A model for the extracellular regulation of TGFβ and BMP signals is depicted in Fig. 5.1(A). Cells secrete the large latent TGFβ complex and BMP complexes. Through interactions with LTBPs, which target the latent TGFβ complex to the matrix, and with the propeptides of BMP complexes, fibrillins position and concentrate growth factor signals. Cells may activate or inhibit growth factor signals as they come into contact with the microfibril scaffold. These choices may be regulated by positional information contained in the microfibrils themselves or in associated proteins, some of which are shown in Fig. 5.1(B). In addition, the associated proteins may themselves regulate growth factor signaling by competitive mechanisms (for example, LTBPs and fibulins may compete for the same binding site on fibrillin) or by differential or sequential interactions with the growth factors themselves. In this manner, the microfibril scaffold may work as a well-oiled machine [53] to regulate growth factor signaling.

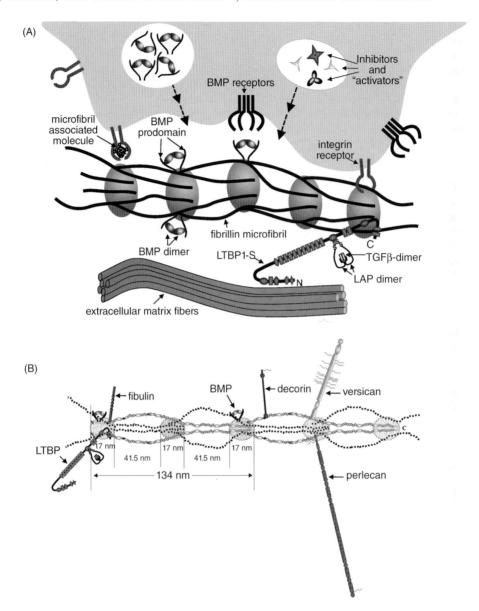

Fig. 5.1. Microfibril regulation of TGFβ and BMP signals.

ABBREVIATIONS

8-cys:	8-cysteine domain
BMP:	bone morphogenetic protein
cbEGF:	calcium-binding epidermal growth factor like
CCA:	congenital contractural arachnodactyly
ECM:	extracellular matrix
FBN:	fibrillin gene
GDF:	growth and differentiation factor
LTBP:	latent TGFβ binding protein
MAGP:	microfibril associated glycoprotein
TGF:	transforming growth factor
Tsk:	tight skin

REFERENCES

1. Sakai LY, Keene DR, Engvall E. 1986. Fibrillin, a new 350kD glycoprotein, is a component of extracellular microfibrils. *J Cell Biol* 103: 2499–2509.
2. Dietz HC, Cutting GR, Pyeritz RE, Maslen CL, Sakai LY, Corson GM, Puffenberger EG, Hamosh A, Nanthakumar EJ, Curristin SM, Stetten G, Meyers DA, Francomano CA. 1991. Marfan syndrome caused by a recurrent de novo missense mutation in the fibrillin gene. *Nature* 352: 337–339.
3. Gupta PA, Putnam EA, Carmical SG, Kaitila I, Steinmann B, Child A, Danesino C, Metcalfe K, Berry SA,

Chen E, Delorme CV, Thong MK, Ades LC, Milewicz DM. 2002. Ten novel FBN2 mutations in congenital contractural arachnodactyly: Delineation of the molecular pathogenesis and clinical phenotype. *Hum Mutat* 19: 39–48.
4. Faivre L, Gorlin RJ, Wirtz MK, Godfrey M, Dagoneau N, Samples JR, Le Merrer M, Collod-Beroud G, Boileau C, Munnich A, Cormier-Daire V. 2003. In frame fibrillin-1 gene deletion in autosomal dominant Weill-Marchesani syndrome. *J Med Genet* 40: 34–36.
5. Sengle G, Tsutsui K, Keene DR, Tufa SF, Carlson EJ, Charbonneau NL, Ono RN, Sasaki T, Wirtz MK, Samples JR, Fessler LI, Fessler JH, Sekiguchi K, Hayflick SJ, Sakai LY. 2012. Microenvironmental regulation by fibrillin-1. *Plos Genet* 8(1): e1002425.
6. LeGoff C, Mahaut C, Wang LW, Allali S, Abhyankar A, Jensen S, Zylberberg L, Collod-Beroud G, Bonnet D, Alanay Y, Brady AF, Cordier M-P, Devriendt K, Genevieve D, Kiper POS, Kitoh H, Krakow D, Lynch SA, LeMerrer M, Megarbane A, Mortier G, Odent S, Polak M, Rohrbach M, Sillence D, Stolte-Dijkstra I, Superti-Furga A, Rimoin DL, Topouchian V, Unger S, Zabel B, Bole-Feysot C, Nitschke P, Handford P, Casanova J-L, Boileau C, Apte SS, Munnich A, Cormier-Daire V. 2011. Mutations in the TGFβ binding-protein-like domain 5 of FBN1 are responsible for acromicric and geleophysic dysplasias. *Am J Hum Genet* 89: 7–14.
7. Keene DR, Jordan CD, Reinhardt DP, Ridgway CC, Ono RN, Corson GM, Fairhurst M, Sussman MD, Memoli VA, Sakai LY. 1997. Fibrillin-1 in human cartilage: Developmental expression and formation of special banded fibers. *J Histochem Cytochem* 45: 1069–1082.
8. Keene DR, Sakai LY, Burgeson RE. 1991. Human bone contains type III collagen, type VI collagen, and fibrillin. *J Histochem Cytochem* 39: 59–69.
9. Zhang H, Hu W, Ramirez F. 1995. Developmental expression of fibrillin genes suggests heterogeneity of extracellular microfibrils. *J Cell Biol* 129: 1165–1176.
10. Corson GM, Charbonneau NL, Keene DR, Sakai LY. 2004. Differential expression of fibrillin-3 adds to microfibril variety in human and avian, but not rodent, connective tissues. *Genomics* 83: 461–472.
11. Gallagher BC, Sakai LY, Little CD. 1993. Fibrillin delineates the primary axis of the early avian embryo. *Dev Dyn* 196: 70–78.
12. Charbonneau NL, Dzamba BJ, Ono RN, Keene DR, Corson GM, Reinhardt DP, Sakai LY. 2003. Fibrillins can co-assemble in fibrils, but fibrillin fibril composition displays cell-specific differences. *J Biol Chem* 278: 2740–2749.
13. Arteaga-Solis E, Gayraud B, Lee SY, Shum L, Sakai L, Ramirez F. 2001. Regulation of limb patterning by extracellular microfibrils. *J Cell Biol* 154: 275–281.
14. Carta L, Pereira L, Arteaga-Solis E, Lee-Arteaga SY, Lenart B, Starcher B, Merkel CA, Sukoyan M, Kerkis A, Hazeki N, Keene DR, Sakai LY, Ramirez F. 2006. Fibrillins 1 and 2 perform partially overlapping functions during aortic development. *J Biol Chem* 281: 8016–8023.
15. Corson GM, Chalberg SC, Dietz H C, Charbonneau NL, Sakai LY. 1993. Fibrillin binds calcium and is coded by cDNAs that reveal a multidomain structure and alternatively spliced exons at the 5′ end. *Genomics* 17: 476–484.
16. Rifkin DB. 2005. Latent transforming growth factor-β (TGF-β) binding proteins: Orchestrators of TGF-β availability. *J Biol Chem* 280: 7409–7412.
17. Dallas SL, Miyazono K, Skerry TM, Mundy GR, Bonewald LF. 1995. Dual role for the latent transforming growth factor-beta binding protein in storage of latent TGF-beta in the extracellular matrix and as a structural matrix protein. *J Cell Biol* 131: 539–549.
18. Dallas SL, Keene DR, Bruder SP, Saharinen J, Sakai LY, Mundy GR, Bonewald LF. 2000. Role of the latent transforming growth factor beta binding protein 1 in fibrillin-containing microfibrils in bone cells in vitro and in vivo. *J Bone Miner Res* 15: 68–81.
19. Isogai Z, Ono RN, Ushiro S, Keene DR, Chen Y, Mazzieri R, Charbonneau NL, Reinhardt DP, Rifkin DB, Sakai LY. 2003. Latent transforming growth factor β-binding protein 1 interacts with fibrillin and is a microfibril-associated protein. *J Biol Chem* 278: 2750–2757.
20. Reinhardt DP, Sasaki T, Dzamba BJ, Keene DR, Chu M-L, Göhring W, Timpl R, Sakai LY. 1996. Fibrillin-1 and fibulin-2 interact and are colocalized in some tissues. *J Biol Chem* 271: 19489–19496.
21. El Hallous E, Sasaki T, Hubmacher D, Getie M, Tiedemann K, Brinckmann J, Bätge B, Davis EC, Reinhardt DP. 2007. Fibrillin-1 interactions with fibulins depend on the first hybrid domain and provide an adaptor function to tropoelastin. *J Biol Chem* 282: 8935–8946.
22. Gibson MA, Hughes JL, Fanning JC, Cleary EG. 1986. The major antigen of elastin-associated microfibrils is a 31-kDa glycoprotein. *J Biol Chem* 261: 11429–11436.
23. Tiedemann K, Sasaki T, Gustafsson E, Göhring W, Bätge B, Notbohm H, Timpl R, Wedel T, Schlötzer-Schrehardt U, Reinhardt DP. 2005. Microfibrils at basement membrane zones interact with perlecan via fibrillin-1. *J Biol Chem* 280: 11404–11412.
24. Isogai Z, Aspberg A, Keene DR, Ono RN, Reinhardt DP, Sakai LY. 2002. Versican interacts with fibrillin-1 and links extracellular microfibrils to other connective tissue networks. *J Biol Chem* 277: 4565–4572.
25. Trask BC, Trask TM, Brockelmann T, Mecham RP. 2000. The microfibrillar proteins MAGP-1 and fibrillin-1 form a ternary complex with the chondroitin sulfate proteoglycan decorin. *Mol Biol Cell* 11: 1499–1507.
26. Reinboth B, Hanssen E, Cleary EG, Gibson MA. 2002. Molecular interactions of biglycan and decorin with elastic fiber components: Biglycan forms a ternary complex with tropoelastin and microfibril-associated glycoprotein 1. *J Biol Chem* 277: 3950–3957.
27. Neptune ER, Frischmeyer PA, Arking DE, Myers L, Bunton TE, Gayraud B, Ramirez F, Sakai LY, Dietz HC. 2003. Dysregulation of TGFβ activation contributes to pathogenesis in Marfan syndrome. *Nat Genet* 33: 407–411.
28. Ng CM, Cheng A, Myers LA, Martinez-Murillo F, Jie C, Bedja D, Gabrielson KL, Hausladen JM, Mecham RP,

Judge DP, Dietz HC. 2004. TGF-beta-dependent pathogenesis of mitral valve prolapse in a mouse model of Marfan syndrome. *J Clin Invest* 114: 1586–1592.

29. Habashi JP, Judge DP, Holm TM, Cohn RD, Loeys BL, Cooper TK, Myers L, Klein EC, Liu G, Calvi C, Podowski M, Neptune ER, Halushka MK, Bedja D, Gabrielson K, Rifkin DB, Carta L, Ramirez F, Huso DL, Dietz HC. 2006. Losartan, an AT1 antagonist, prevents aortic aneurysm in a mouse model of Marfan syndrome. *Science* 312: 117–121.

30. Mizuguchi T, Collod-Beroud G, Akiyama T, Abifadel M, Harada N, Morisaki T, Allard D, Varret M, Claustres M, Morisaki H, Ihara M, Kinoshita A, Yoshiura K, Junien C, Kajii T, Jondeau G, Ohta T, Kishino T, Furukawa Y, Nakamura Y, Niikawa N, Boileau C, Matsumoto N. 2004. Heterozygous TGFBR2 mutations in Marfan syndrome. *Nat Genet* 36: 855–860.

31. Loeys BL, Chen J, Neptune ER, Judge DP, Podowski M, Holm T, Meyers L, Leitch CC, Katsanis N, Sharifi N, Xu FL, Myers LA, Spevak PJ, Cameron DE, De Backer J, Hellemans J, Chen Y, Davis EC, Webb CL, Kress W, Coucke P, Rifkin DB, De Paepe AM, Dietz HC. 2005. A syndrome of altered cardiovascular, craniofacial, neurocognitive and skeletal development caused by mutations in TGFBR1 or TGFBR2. *Nat Genet* 37: 275–281.

32. Loeys BL, Schwarze U, Holm T, Callewaert BL, Thomas GH, Pannu H, De Backer JF, Oswald GL, Symoens S, Manouvrier S, Roberts AE, Faravelli F, Greco MA, Pyeritz RE, Milewicz DM, Coucke PJ, Cameron DE, Braverman AC, Byers PH, De Paepe AM, Dietz HC. 2006. Aneurysm syndromes caused by mutations in the TGF-beta receptor. *N Engl J Med* 355: 788–798.

33. Pereira L, Andrikopoulos K, Tian J, Lee SY, Keene DR, Ono R, Reinhardt DP, Sakai LY, Jensen-Biery N, Bunton T, Dietz HC, Ramirez F. 1997. Targeting of the gene encoding fibrillin-1 recapitulates the vascular aspect of Marfan syndrome. *Nat Gen* 17: 218–222.

34. Judge DP, Biery NJ, Keene DR, Geubtner J, Myers L, Huso DL, Sakai LY, Dietz HC. 2004. Evidence for a critical contribution of haploinsufficiency in the complex pathogenesis of Marfan syndrome. *J Clin Invest* 114: 172–181.

35. Pereira L, Lee SY, Gayraud B, Andrikopoulos K, Shapiro SD, Bunton T, Biery NJ, Dietz HC, Sakai LY, Ramirez F. 1999. Pathogenetic sequence for aneurysm revealed in mice underexpressing fibrillin-1. *Proc Natl Acad Sci U S A* 96: 3819–23.

36. Siracusa LD, McGrath R, Ma Q, Moskow JJ, Manne J, Christner PJ, Buchberg AM, Jimenez SA. 1996. A tandem duplication within the fibrillin 1 gene is associated with the mouse tight skin mutation. *Genome Res* 6: 300–313.

37. Gregory KE, Ono RN, Charbonneau NL, Kuo C-L, Keene DR, Bächinger HP, Sakai LY. 2005. The prodomain of BMP-7 targets the BMP-7 complex to the extracellular matrix. *J Biol Chem* 280: 27970–27980.

38. Kuo CL, Isogai Z, Keene DR, Hazeki N, Ono RN, Sengle G, Bächinger HP, Sakai LY. 2007. Effects of fibrillin-1 degradation on microfibril ultrastructure. *J Biol Chem* 282: 4007–4020.

39. Sengle G, Charbonneau NL, Ono RN, Sasaki T, Alvarez J, Keene DR, Bächinger HP, Sakai LY. 2008. Targeting of BMP growth factor complexes to fibrillin. *J Biol Chem* 283: 13874–13888.

40. Saharinen J, Keski-Oja J. 2000. Specific sequence motif of 8-cys repeats of TGF-beta binding proteins, LTBPs, repeats a hydrophobic interaction site for binding of small latent TGF-beta. *Mol Biol Cell* 11: 2691–2704.

41. Anderson SB, Goldberg AL, Whitman M. 2008. Identification of a novel pool of extracellular pro-myostatin in skeletal muscle. *J Biol Chem* 283: 7027–7035.

42. Charbonneau NL, Ono RN, Corson GM, Keene DR, Sakai, LY. 2004. Fine tuning of growth factor signals depends on fibrillin microfibril networks. *Birth Defects Res C Embryo Today* 72: 37–50.

43. Brown MA, Zhao Q, Baker KA, Naik C, Chen C, Pukac L, Singh M, Tsareva T, Parice Y, Mahoney A, Roschke V, Sanyal I, Choe S. 2005. Crystal structure of BMP-9 and functional interactions with pro-region and receptors. *J Biol Chem* 280: 25111–25118.

44. Sengle G, Ono RN, Lyons KM, Bächinger HP, Sakai LY. 2008. A new model for growth factor activation: Type II receptors compete with the prodomain for BMP-7. *J Mol Biol* 381: 1025–1039.

45. Jackson RA, Nurcombe V, Cool SM. 2006. Coordinated fibroblast growth factor and heparan sulfate regulation of osteogenesis. *Gene* 379: 79–91.

46. DeCarlo AA, Whitelock JM. 2006. The role of heparan sulfate and perlecan in bone-regenerative procedures. *J Dent Res* 85: 122–132.

47. Koli K, Wempe F, Sterner-Kock A, Kantola A, Komor M, Hofmann WK, von Melchner H, Keski-Oja J. 2004. Disruption of LTBP-4 function reduces TGF-beta activation and enhances BMP-4 signaling in the lung. *J Cell Biol* 167: 123–133.

48. Hanada K, Vermeij M, Garinis GA, de Waard MC, Kunen MG, Myers L, Maas A, Duncker DJ, Meijers C, Dietz HC, Kanaar R, Essers J. 2007. Perturbations of vascular homeostasis and aortic valve abnormalities in fibulin-4 deficient mice. *Circ Res* 100: 738–746.

49. Dasouki M, Markova D, Garola R, Sasaki T. 2007. Charbonneau NL, Sakai LY, Chu ML. 2007. Compound heterozygous mutations in fibulin-4 causing neonatal lethal pulmonary artery occlusion, aortic aneurysm, arachnodactyly, and mild cutis laxa. *Am J Med Genet A* 143: 2635–2641.

50. Arikawa-Hirasawa E, Watanabe H, Takami H, Hassell JR, Yamada Y. 1999. Perlecan is essential for cartilage and cephalic development. *Nat Genet* 23: 354–358.

51. Yamaguchi Y, Mann DM, Ruoslahti E. 1990. Negative regulation of transforming growth factor-beta by the proteoglycan decorin. *Nature* 346: 281–284.

52. Chen X-D, Fisher LW, Gehron-Robey P, Young MF. 2004. The small leucine-rich proteoglycan biglycan modulates BMP-4 induced osteoblast differentiation. *FASEB J* 18: 948–958.

53. Engel J. 2006. Molecular machines in the matrix? *Matrix Biol* 25: 200–201.

6

The Composition of Bone

Adele L. Boskey and Pamela Gehron Robey

Introduction 49
The Composite 49
Noncollagenous Proteins 50
Other Components 56
References 56

INTRODUCTION

Bone comprises the largest proportion of the body's connective tissue mass. Unlike most other connective tissue matrices, bone matrix is physiologically mineralized and is unique in that it is constantly regenerated throughout life as a consequence of bone turnover. Bone as an organ is made up of the cartilaginous joints, the calcified cartilage in the growth plate (in developing individuals), the marrow space, and the cortical and cancellous mineralized structures. Bone as a tissue consists of the mineralized and nonmineralized (osteoid) components of the cortical and cancellous regions of long and flat bones. There are three cell types on (and in) bone tissue: (1) the bone-forming osteoblasts, which when engulfed in mineral become (2) osteocytes, and (3) the bone-destroying osteoclasts. Each of these cells communicates with one another by either direct cell contact or through signaling molecules, and they respond to each other. The detailed properties of these cells have been discussed in numerous publications; see for example the excellent chapter by Lian and Stein [1]. This chapter focuses on the extracellular matrix, which is synthesized primarily by osteoblasts but also contains proteins adsorbed from the circulation. The preponderance of bone is extracellular matrix (ECM). Information on the gene and protein structure and potential function of bone ECM constituents has exploded during the past two decades. This information has been described in great detail in several recent reviews [2,3], to which the reader is referred for specific references; these are too numerous to be listed adequately here. Below is a summary of the composition of bone and the salient features of the classes of bone matrix proteins. The tables list specific details for the individual ECM components.

THE COMPOSITE

Bone is a composite material whose extracellular matrix consists of mineral, collagen, water, noncollagenous proteins, and lipids in decreasing proportion (depending on age, species, and site). These components have both mechanical and metabolic functions. Understanding of some of the biological functions of these components has come from mouse models and analyses of healthy and diseased human tissues, and from cell culture studies.

The mineral

The mineral phase of bone is a nanocrystalline, highly substituted analog of the naturally occurring mineral, hydroxylapatite $[Ca_{10}(PO_4)_6(OH)_2]$. The major substitutions are with carbonate, magnesium, and acid phosphate, along with other trace elements, the content of which depends on diet and environment. Although the precise chemical nature of the initial mineral formed has

---

Primer on the Metabolic Bone Diseases and Disorders of Mineral Metabolism, Eighth Edition. Edited by Clifford J. Rosen.
© 2013 American Society for Bone and Mineral Research. Published 2013 by John Wiley & Sons, Inc.

Table 6.1. Characteristics of Collagen-Related Genes and Proteins Found in Bone Matrix

Protein/Gene	Function	Disease/Animal Model/Phenotype
Type I—17q21.23, 7q21.3-22 [α1(I)$_2$α2(I)] α1(I)$_3$	Serves as scaffolding, binds and orients other proteins that nucleate hydroxyapatite deposition	Human mutations: Osteogenesis imperfecta (OMIM 166210; 166200; 610854; 259420; 166220) Mouse models: oim mouse; mov 14 mouse; brittle mouse, mechanically weak, mineral crystals small, some mineral outside collagen
Type X—6q21-22.3 [α1(X)]$_3$	Present in hypertrophic cartilage of the growth plate but does not appear to regulate matrix mineralization	Human mutations: Schmid metaphyseal chondrodysplasia (OMIM 120110); knockout mouse: no apparent skeletal phenotype, may have hematopoietic phenotype
Type III—2q24.3-31 [α1(III)]$_3$	Present in bone in trace amounts, may regulate collagen fibril diameter, paucity in bone may explain the large diameter size of bone collagen fibrils	Human mutations in type III: different forms of Ehlers-Danlos syndrome (OMIM 130050)
Type V—9q34.2-34.3;2q24.3-31, 9q34.2-34.3 [α1(V)$_2$α2(V)] [α1(V) α2(V) α3(V)]	Present in bone in trace amounts, may regulate collagen fibril diameter, paucity in bone may explain the large diameter size of bone collagen fibrils	Mutations in type V α1 and α2 (OMIM 120215; 120190) Mouse model: disrupted fibril arrangement

been debated [4, 5], it is well accepted that the majority of the "biomineral" present in the bones during development is apatitic. The physical and chemical properties of this mineral have been determined by a variety of techniques including chemical analyses, X-ray diffraction, vibrational spectroscopy, energy dispersive electron analysis, nuclear magnetic resonance, small angle scattering, and transmission and atomic force microscopy [6].

The functions of the mineral are to strengthen the collagen composite, providing more mechanical resistance to the tissue, and also to serve as a source of calcium, phosphate, and magnesium ions for mineral homeostasis. For physicochemical reasons, usually it is the smallest mineral crystals that are lost during remodeling; thus, in osteoporosis, it is not surprising that the larger more perfect crystals persist within the matrix [7] contributing to the brittle nature of osteoporotic bone. When remodeling is impaired, as in osteopetrosis, the mineral crystals remain small relative to age-matched controls [7].

Collagen

The basic building block of the bone matrix fiber network is type I collagen, which is a triple-helical molecule containing two identical α1(I) chains and a structurally similar, but genetically different α2(I) chain [8]. Collagen α chains are characterized by a Gly-X-Y repeating triplet (where X is usually proline, and Y is often hydroxyproline) and by several post-translational modifications including: (1) hydroxylation of certain lysyl or prolyl residues; (2) glycosylation of the hydroxylysine with glucose or galactose residues, or both; (3) addition of mannose at the propeptide termini; and (4) formation of intra- and intermolecular covalent cross-links that differ from those found in soft connective tissues. Measurement of these bone-derived collagen cross-links in urine has proven to be good measures of bone resorption [9]. Bone matrix proper consists predominantly of type I collagen; however, trace amounts of type III, V, and fibril-associated collagens (Table 6.1) may be present during certain stages of bone formation and may regulate collagen fibril diameter.

NONCOLLAGENOUS PROTEINS

Noncollagenous proteins (NCPs) comprise 10–15% of the total bone protein content. NCPs are multifunctional, having roles in organizing the extracellular matrix, coordinating cell-matrix and mineral-matrix interactions, and regulating the mineralization process. Knowledge of these functions have come from studies of the isolated proteins in solution, from analyses of mice in which the proteins are ablated (knocked out), or overexpressed, characterization of human diseases in which these proteins have mutations, and studies using appropriate cell cultures. The tables summarize the gene and protein structures, and the functions of these proteins are associated with each discussion of the protein families.

Serum-derived proteins

Approximately one-fourth of the total NCP content is exogenously derived (Table 6.2). This fraction is largely composed of serum-derived proteins, such as albumin

Table 6.2. Gene and Protein Characteristics of Serum Proteins found in Bone Matrix

Protein/Gene	Function	Disease/Animal Model/Phenotype
Albumin—2q11-13 69 kd, non-glycosylated, one sulfhydryl, 17 disulfide bonds, high affinity hydrophobic binding pocket	Inhibits hydroxyapatite crystal growth	
α2HS glycoprotein—3q27-29 Precursor protein of fetuin, cleaved to form A and B chains that are disulfide linked, Ala-Ala and Pro-Pro repeat sequences, N-linked oligosaccharides, cystatin-like domains	Promotes endocytosis, has opsonic properties, chemoattractant for monocytic cells, bovine analog (fetuin) is a growth factor; inhibits calcification	Knockout mouse: adult ectopic calcification

and α_2-HS-glycoprotein, which are acidic in character and bind to bone matrix because of their affinity for hydroxyapatite. Although these proteins are not endogenously synthesized, they may exert effects on matrix mineralization and bone cell proliferation. For example, α_2-HS-glycoprotein, the human analog of fetuin, when ablated in mice causes ectopic calcification [10] suggesting that the protein is a mineralization inhibitor. The remainder of the exogenous fraction is composed of growth factors and a large variety of other molecules present in trace amounts, which influence local bone cell activity [1, 2].

On a mole-to-mole basis, bone-forming cells synthesize and secrete as many molecules of NCPs as of collagen. These molecules can be classified into four general (and sometimes overlapping) groups: (1) proteoglycans, (2) glycosylated proteins, (3) glycosylated proteins with potential cell attachment activities, and (4) γ-carboxylated (gla) proteins. The physiological roles for individual bone protein constituents are not well defined; however, they may participate not only in regulating the deposition of mineral but also in the control of osteoblastic and osteoclastic metabolism.

Proteoglycans

Proteoglycans are macromolecules that contain acidic polysaccharide side chains (glycosaminoglycans) attached to a central core protein, and bone matrix contains several members of this family [11] (Table 6.3).

During the initial stages of bone formation, the large chondroitin sulfate proteoglycan, versican, and the glycosaminoglycan, hyaluronan (which is not attached to a protein core), are highly expressed and may delineate areas that will become bone. With continued osteogenesis, versican is replaced by two small chondroitin sulfate proteoglycans—decorin and biglycan—composed of tandem repeats of a leucine-rich repeat (LRR) sequence. Decorin has been implicated in the regulation of collagen fibrillogenesis and is distributed predominantly in the ECM space of connective tissues and in bone, whereas biglycan tends to be found in pericellular locales. A heparan sulfate proteoglycan, perlecan is involved in limb patterning and is found surrounding chondrocytes in the growth plate, whereas the glypican family of cell surface associated heparan sulfate proteoglycans also affect skeletal growth. In addition, there are other small leucine-rich proteoglycans (SLRPs) in bone, including osteoglycin (mimecan), keratocan [12], osteoadherin, lumican, asporin, and fibromodulin [13]. Although their exact physiological functions are not known, these proteoglycans are assumed to be important for the integrity of most connective tissue matrices. Deletion of the *biglycan* gene, for example, leads to a significant decrease in the development of trabecular bone, indicating that it is a positive regulator of bone formation [2]. Deletion of the *epiphican* gene, or the *epiphican* and *biglycan* genes together causes shortening of the femur during growth, and early onset osteoarthritis [14]. Other functions might arise from the ability of these proteoglycans to bind and modulate the activity of the growth factors in the extracellular space, thereby influencing cell proliferation and differentiation [1].

Glycosylated proteins

Glycosylated proteins with diverse functions abound in bone. One of the hallmarks of bone formation is the synthesis of high levels of alkaline phosphatase (Table 6.4).

Alkaline phosphatase, a glycoprotein enzyme, is primarily bound to the cell surface through a phosphoinositol linkage, but is cleaved from the cell surface and found within a mineralized matrix. The function of alkaline phosphatase in bone cell biology has been the subject of much speculation and remains undefined. Mice lacking tissue nonspecific alkaline phosphatase have impaired mineralization, suggesting the importance of this enzyme for mineral deposition [15].

Table 6.3. Gene and Protein Characteristics: Glycosaminoglycan-Containing Molecules in Bone

Protein/Gene	Function	Disease/Animal Model/Phenotype
Aggrecan—15q26.1 ~2.5×10^6 kd intact protein, ~180–370,000 core, ~100 CS chains of 25 kd, and some KS chains of similar size, G1, G2, and G3 globular domains with hyaluronan binding sites, EGF and CRP-like sequences	Matrix organization, retention of water and ions, resilience to mechanical forces	Human mutation: Spondyloepiphyseal dysplasia (OMIM 155760; 608361) Mouse models: brachymorphic mouse, accelerated growth plate calcification; cartilage matrix deficiency mouse- shortened stature Nanomelic chick (mutation)—abnormal bone shape
Versican (PG-100)—5q12-14 ~1×10^6 kd intact protein, ~360 kd core, ~12 CS chains of 45 kd, G1 and G3 globular domains with hyaluronan binding sites, EGF and CRP-like sequences	Regulates chondrogenesis; may "capture" space that is destined to become bone	Human mutation: Wagner syndrome (an ocular disorder) (OMIM 143200)
Decorin (Class 1 SLRP)—12q13.2 ~130 kd intact protein, ~38–45 kd core with 10 leucine rich repeat sequences, 1 CS chain of 40 kd	Binds to collagen and may regulate fibril diameter, binds to TGF-β and may modulate activity, inhibits cell attachment to fibronectin	Knockout mouse: no apparent skeletal phenotype although collagen fibrils are abnormal, DCN/BGN double knockout— progeroid form of Ehler's–Danlos syndrome
Biglycan (Class 1 SLRP)—Xq27 ~270 kd intact protein, ~38–45 kd core protein with 12 leucine rich repeat sequences, 2 CS chains of 40 kd	Binds to collagen, TGF-β, and other growth factors; pericellular environment, a genetic determinant of peak bone mass	Knockout mouse: osteopenia; thin bones, decreased mineral content, increased crystal size; short stature
Asporin (Class 1 SLRP)—9q21.3 67 kd, most likely no GAG chains	Regulates collagen structure	Human polymorphism associated with osteoarthritis (OMIM 608135)
Fibromodulin (Class 2 SLRP)—1q32 59 kd intact protein, 42 kd core protein, one N-linked KS chain	Binds to collagen, may regulate fibril formation, binds to TGF-β	Fibromodulin/biglycan double knockout mice: joint laxity and formation of supernumery sesmoid bones
Osteoadherin (Class 2 SLRP) 85 kd intact protein, 47 kd core protein, RGD sequence	May mediate cell attachment	
Lumican (Class 2 SLRP)—12q21.3-q22 70-80 kd intact protein, 37 kd core protein	Binds to collagen, may regulate fibril formation	Lumican/fibromodulin double knockout mouse: has ectopic calcification and a variant of Ehler's—Danlos syndrome (OMIM 130000)
Perlecan- 1p36.1 Five domain heparan sulfate proteoglycan, core protein 400 kd	Interacts with matrix components to regulate cell signaling; cephalic development	Transgenic mice with mutated perlecan: Schwartz-Jampel syndrome (OMIM 142461): impaired mineralization and mis-shapen skeletons and joint abnormalities Knockout mice: phenotype resembling thanatophoric dysplasia (TD) type I
Glypican-3 –Xq26 Lipid-linked heparan sulfate proteoglycan, 65 kd core protein, 14 conserved cysteine residues	Regulates BMP-SMAD signaling Regulates cell development	Human mutation: Simpson-Golabi-Behmel Syndrome (OMIM 300037) Knockout mouse: delayed endochondral ossification and impaired osteoclast development
Keratocan (Class 2 SLRP)—12q22 60–200 kDa intact protein, 52 kDa core protein with 6–10 LRRs, multiple keratan sulfate side chains	Regulates osteoblast differentiation	Knockout cells show delayed mineralization
Osteoglycin/Mimecan (Class 3 SLRP)—9q22 299 aa precursor, 105 aa mature protein, no GAG in bone, keratan sulfate in other tissues	Binds to TGF-β Regulates collagen fibrillogenesis	
Hyaluronan—Multi-gene complex Multiple proteins associated outside of the cell, structure unknown	May work with versican molecule to capture space destined to become bone	

Table 6.4. Gene and Protein Characteristics of Glycoproteins in Bone Matrix

Protein/Gene	Function	Disease/Animal Models/Phenotype
Alkaline Phosphatase (bone-liver-kidney isozyme)—1p34-36.1 Two identical subunits of ~80 kd, disulfide bonded, tissue specific post translational modifications	Potential Ca^{++} carrier, hydrolyzes inhibitors of mineral deposition such as pyrophosphates, increases local phosphate concentration	Human mutations: hypophosphatasia (OMIM 171760) (decreased activity) TNAP knockout mouse: growth impaired; decreased mineralization
Osteonectin—5q31.3-q32 ~35–45 kd, intramolecular disulfide bonds, α helical amino terminus with multiple low affinity Ca^{++} binding sites, two EF hand high affinity Ca^{++} sites, ovomucoid homology, glycosylated, phosphorylated, tissue specific modifications	Regulates collagen organization, may mediate deposition of hydroxyapatite, binds to growth factors, may influence cell cycle, positive regulator of bone formation	Knockout mouse: severe osteopenia, decreased trabecular connectivity; decreased mineral content; increased crystal size
Periostin—13q13.3 90 kd disulfide linked protein, homology to BIGH3, promotes cell attachment	Regulates collagen organization and response to mechanical signals	Knockout mouse: periodontal and vascular calcification
Tetranectin—3p22-p21.3 21 kd protein composed of four identical subunits of 5.8 kd, sequence homologies with asialoprotein receptor and G3 domain of aggrecan	Binds to plasminogen, may regulate matrix mineralization	Knockout mouse: no long bone phenotype, spinal deformity, increased mineralization in implant model
Tenascin-C—9q33 Hexameric structure, six identical chains of 320 kd, Cys rich, EGF-like repeats, FN type III repeats	Interferes with cell-FN interactions	Knockout mouse: no overt bone phenotype
Tenascin-X—6p21.3 Hexameric with 5 N-linked glycosylation sites and multiple EFG and 40 fibronectin type III repeats	Regulates cell-matrix interactions	Human mutation: Ehlers-Danlos II phenotype (OMIM 600985)
Secreted phosphoprotein 24—2q37 24 kd secreted phosphoprotein, shares sequence homology with members of the cystatin family of thiol protease inhibitors	Associates with regulators of mineralization in serum, regulates BMP mediated bone turnover	

The most abundant NCP produced by bone cells is osteonectin, a phosphorylated glycoprotein accounting for approximately 2% of the total protein of developing bone in most animal species. Osteonectin is transiently produced in non-bone tissues that are rapidly proliferating, remodeling, or undergoing profound changes in tissue architecture, and is also found constitutively expressed in certain types of epithelial cells, cells associated with the skeleton, and in platelets. Osteonectin, along with thrombospondin-2 (TSP-2) and periostin are members of the class of "matricellular proteins," each of which has a role in bone cell proliferation and differentiation, with some role in regulating mineralization [16]. Tetranectin (which is important for wound healing) [2], tenascin (which regulates the organization of the extracellular matrix) [17], and secreted phosphoprotein 24 (which regulates bone morphogenetic protein expression) along with periostin [18] are other glycoproteins found in the bone matrix.

Small integrin-binding ligand, N-glycosylated protein, and other glycoproteins with cell attachment activity

All connective tissue cells interact with their extracellular environment in response to stimuli that direct or coordinate (or both) specific cell functions, such as migration, proliferation, and differentiation (Tables 6.5 and 6.6). These particular interactions involve cell attachment through transient or stable focal adhesions to extracellular macromolecules, which are mediated by cell surface receptors that subsequently transduce intracellular signals. Bone cells synthesize at least 12 proteins that may mediate cell attachment: members of the small integrin-binding ligand, N-glycosylated protein (SIBLING) family (osteopontin, bone sialoprotein, dentin matrix protein-1, dentin sialophosphoprotein, and matrix extracellular phosphoprotein), type 1 collagen, fibronectin,

Table 6.5. Gene and Protein Characteristics of SIBLINGs (Small Integrin-Binding Ligands, N-Glycosylated Proteins)

Protein/Gene	Function	Disease/Animal Models/Phenotype
Osteopontin—4q21 ~44-75 kd, polyaspartyl stretches, no disulfide bonds, glycosylated, phosphorylated, RGD located 2/3 from the N-terminal	Binds to cells and collagen, may regulate mineralization, may regulate cell proliferation, inhibits nitric oxide synthase, may regulate resistance to viral infection	Knockout mouse: decreased crystal size; increased mineral content; not subject to osteoclastic remodeling
Bone Sialoprotein—4q21 ~46-75 kd, polyglutamyl stretches, no disulfide bonds, 50% carbohydrate, tyrosine-sulfated, RGD near the C-terminus	Binds to cells, may regulate bone remodeling, may initiate mineralization	Knockout mouse: impaired osteoblast and osteoclast function
DMP-1—4q21 513 amino acids predicted; serine-rich, acidic, RGD 2/3 from N-terminus	Regulator of biomineralization; regulates osteocyte function	Human mutation: dentinogenesis imperfecta and hypophosphatemia (OMIM 600980) Knockout mouse: undermineralized with craniofacial and growth plate abnormalities and defective osteocyte function
Dentin sialophosphoprotein—4q21.3 Gene produces three proteins: dentin sialoprotein, dentin phosphophoryn, and dentin glycoprotein. All have RGD sites; dentin phosphophoryn is highly phosphorylated	Regulation of biomineralization	Human mutations: dentinal dysplasias and dentinogenesis imperfecta; no bone disease (OMIM 125485) Knockout mouse: has thinner bones at 9 months, no significant other bone phenotype, and severe dentin abnormalities
MEPE—4q21.1 525 amino acids, 2 N-glycosylation motifs, a glycosaminoglycan-attachment site, an RGD cell-attachment motif, and phosphorylation motifs	Regulation of biomineralization; regulation of PHEX (phosphaturic hormone) activity	Humans: association with oncogenic osteomalacia Knockout mouse: increased bone mass and resistance to ovariectomy-induced bone loss
Other Siblings: Enamelin—4q21	Regulates enamel mineralization	Human mutations: amelogenesis imperfecta (OMIM 104500, 204650)

Table 6.6. Gene and Protein Characteristics of Other RGD-Containing Glycoproteins

Protein/Gene	Function	Disease/Animal Models
Thrombospondins (1-4, COMP)—15Q-1, 6q27, 1q21-24, 5q13, 19p13.1 ~450 kd molecules, three identical disulfide linked subunits of ~150–180 kd, homologies to fibrinogen, properdin, EGF, collagen, von Willebrand, *P. falciparum* and calmodulin, RGD at the C terminal globular domain	Cell attachment (but usually not spreading), binds to heparin, platelets, types I and V collagens, thrombin, fibrinogen, laminin, plasminogen and plasminogen activator inhibitor, histidine rich glycoprotein	Human mutation in COMP: pseudoachondroplasia (OMIM 600310) TSP-2 knockout mouse: large collagen fibrils, thickened bones; spinal deformities
Fibronectin—2q34 ~400 kd with 2 nonidentical subunits of ~200 kd, composed of type I, II, and III repeats, RGD in the 11th type III repeat 2/3 from N-terminus	Binds to cells, fibrin, heparin, gelatin, collagen	Knockout mouse: lethal prior to skeletal development
Vitronectin—17q11 ~70 kd, RGD close to N-terminus, homology to somatomedin B, rich in cysteines, sulfated, phosphorylated	Cell attachment protein, binds to collagen, plasminogen and plasminogen activator inhibitor, and to heparin	
Fibrillin 1 and 2—15q21.1, 5q23-q31 350 kd, EGF-like domains, RGD, cysteine motifs	May regulate elastic fiber formation	Human fibrillin 1 mutations: Marfan's syndrome (OMIM 134797) Human fibrillin 2 mutations: congenital contractural arachnodactyly (OMIM 121050)—.

thrombospondin(s) (predominantly TSP-2 with lower levels of TSP1, 3, and 4, and COMP), vitronectin, fibrillin, BAG-75, and osteoadherin (which is also a proteoglycan). Many of these proteins are phosphorylated and/or sulfated, and all contain RGD (Arg-Gly-Asn), the cell attachment consensus sequence that binds to the integrin class of cell surface molecules. However, in some cases, cell attachment seems to be independent of RGD, indicating the presence of other sequences or mechanisms of cell attachment [2]. Thrombospondin(s), fibronectin, vitronectin, fibrillin, and osteopontin are expressed in many tissues. Whereas certain types of epithelial cells synthesize bone sialoprotein, it is highly enriched in bone and is expressed by hypertrophic chondrocytes, osteoblasts, osteocytes, and osteoclasts. In bone, the expression of bone sialoprotein correlates with the appearance of mineral [19]. The bone sialoprotein knockout (KO) has impaired osteoblast and osteoclast function, although the precise mechanism of action of bone sialoprotein is not known [20]. In solution, bone sialoprotein can function as an hydroxyapatite nucleator [2], it is found in association with bone acidic glycoprotein-75 in mineralization foci [21], and it is upregulated during mineralization in culture [22].

Both osteopontin and bone sialoprotein are known to anchor osteoclasts to bone, and in addition to supporting cell attachment, they bind Ca^{2+} with extremely high affinity through polyacidic amino acid sequences. Each SIBLING protein regulates hydroxyapatite formation in solution, and their KOs have phenotypes that can be correlated with these *in vitro* functions [2]. It is not immediately clear why there are such a plethora of RGD-containing proteins in bone; however, the pattern of expression varies from one RGD protein to another, as does the pattern of the different integrins that bind to these proteins. This variability indicates that cell-matrix interactions change as a function of maturational stage, suggesting that they also may play a role in osteoblastic maturation [2]. Their post-translational modifications also vary, suggesting that these modifications may also determine their *in situ* functions [17, 23, 24].

Gla-containing proteins

Four bone-matrix NCPs, matrix gla protein (MGP), osteocalcin [bone gla protein (BGP)], and periostin (also a bone matrix glycoprotein), all of which are made endogenously, and protein S (made primarily in the liver but also made by osteogenic cells) are post-translationally modified by the action of vitamin K-dependent γ-carboxylases (Table 6.7) [2]. The dicarboxylic glutamyl (gla) residues enhance calcium binding. MGP is found in many connective tissues, whereas osteocalcin is more bone specific [2] and periostin is made in all connective tissues that respond to load [25]. The physiological roles of these proteins are still under investigation; MGP, osteocalcin and periostin may function in the control of mineral deposition and remodeling. MGP-deficient mice develop calcification in extraskeletal sites such as the aorta [26], implying that it is an inhibitor of mineralization. Expression of MGP in blood vessels prevents calcification, whereas expression in osteoblasts prevents mineralization [27]. Osteocalcin seems to be involved in regulating bone turnover. Osteocalcin-deficient mice are reported to have increased BMD compared with normal [28] but with age, the mineral properties did not show the changes that occurred in age-matched controls [29], which suggests a role for osteocalcin in osteoclast recruitment. In human bone, osteocalcin is concentrated in osteocytes, and its release may be a signal in the bone turnover cascade. Osteocalcin measurements in serum have proved valuable as a marker of bone turnover in metabolic disease states [9]. In contrast, uncarboxylated osteocalcin was reported to be a hormone involved in the regulation of energy and glucose metabolism in mice [30] but was not found to have this

Table 6.7. Gene and Protein Characteristics of Gamma-Carboxy Glutamic Acid-Containing Proteins in Bone Matrix

Protein/Gene	Function	Disease/Animal Model/Phenotype
Matrix Gla Protein—12p13.1 ~15 kd, five gla residues, one disulfide bridge, phosphoserine residues	May function in cartilage metabolism, a negative regulator of mineralization	Human mutations: Keutel syndrome (OMIM 245150), excessive cartilage calcification Knockout mouse: excessive cartilage calcification
Osteocalcin—1q25-31 ~5 kd, one disulfide bridge, gla residues located in α helical region	May regulate activity of osteoclasts and their precursors, may mark the turning point between bone formation and resorption, suggested to be a hormone	Knockout mouse: osteopetrotic, thickened bones, decreased crystal size, increased mineral content
Periostin—13q13.3	Regulates response to load	Knockout mouse: vascular calcification and periodontal calcification
Protein S—3p11-q11.2 ~72 kd	Primarily a liver product, but may be made by osteogenic cells	Human mutations: protein deficiency with osteopenia (OMIM 076080)

Table 6.8. Effects of Bone Matrix Molecules on Mineralization *In Vitro*

Promote or Support Apatite Formation	Inhibit Mineralization	Dual Function (Nucleate and Inhibit)	No Effect	Data not available
Type I collagen	Aggrecan	Biglycan	Decorin	Thrombospondin
Proteolipid (matrix vesicle nucleational core)	α2-HS glycoprotein	Osteonectin		Osteoadherin
	Matrix gla protein (MGP)	Fibronectin		Lumican
BAG-75	Osteocalcin	Bone sialoprotein		Mimecan
Alkaline phosphatase		Osteopontin		Tetranectin
		MEPE		Periostin
				Keratocan

function in human studies [31]. Periostin senses load and regulates periodontal and vascular calcification [32]. Periostin-deficient mice also have increased vascular calcifications.

OTHER COMPONENTS

The sections above summarize the major components of bone ECM, but there are other minor components that affect the properties of the tissue. For example, there are numerous enzymes that are important for processing the extracellular matrix components. Some of these are cell associated, and some are found in the ECM. Readers are referred to other reviews [2, 23, 33–35] for more details. Growth factors sequestered in bone regulate cell-matrix interactions and cell function [1]. Water accounts for approximately 10% of the weight of bone, depending on species and bone age. Water is important for cell and matrix nutrition, for control of ion flux, and for maintenance of the collagen structure because type I collagen contains the bulk of the tissue water.

Lipids make up less than 2% of the dry weight of bone; however, they have some significant effects on bone properties [36]. This is illustrated by a few recent examples including the neutral sphingomyelinase-deficient mouse that has a dwarfed phenotype [37], the fro/fro mouse that mimics severe osteogenesis imperfecta and has a chemically induced mutation in sphingomyelinase [38], the caveolin-deficient mouse that has altered mechanical properties [39], and the report that phospholipase D is involved in the initial formation of bone during embryogenesis [40].

Each of the components in the organic matrix of bone influences the mechanism of mineral deposition. Some promote mineralization; some inhibit the formation and/or growth of mineral crystals; and some are multifunctional, promoting in some cases and inhibiting in others. The known effects on hydroxyapatite formation in solution for each of the components discussed in this chapter are summarized in Table 6.8.

REFERENCES

1. Lian JB, Stein GS. 2006 The cells of bone. In: Seibel MJ, Robins SJ, Bilezikian JP (eds.) *Dynamics of Bone and Cartilage Metabolism*. San Diego: Academic Press. pp. 221–58.
2. Zhu W, Robey PG, Boskey AL. 2007. The regulatory role of matrix protines in mineralization of bone. In: Marcus R, Feldman D, Nelson DA, Rosen CJ (eds.) *Osteoporosis, Vol. 1*. San Diego: Academic Press. pp. 191–240.
3. Shekaran A, Garcia AJ. 2011. Extracellular matrix-mimetic adhesive biomaterials for bone repair. *J Biomed Mater Res A* 96(1): 261–72.
4. Grynpas MD, Omelon S. 2007. Transient precursor strategy or very small biological apatite crystals? *Bone* 41(2): 162–4.
5. Dorozhkin SV. 2010. Amorphous calcium (ortho)phosphates. *Acta Biomater* 6(12): 4457–75.
6. Boskey AL. 2006. Organic and inorganic matrices. In: Wnek G, Bowlin GL (eds.) *Encyclopedia of Biomaterials and Biomecial Engineering*. London, UK: Dekker Encyclopedias, Taylor & Francis Books. pp. 1–15.
7. Boskey AL. 2007. Osteoporosis and osteopetrosis. In: Baeuerlein E, Bchrens P, Epple M (eds.) *Biomineralization in Medicine, Vol. 7*. New York: Wiley. pp. 59–75.
8. Rossert J, de Crombrugghe B. 2002. Type I collagen: Structure, synthesis and regulation. In: Bilezikian JP, Raisz LA, Rodan GA (eds.) *Principles of Bone Biology, 2nd Ed., Vol 1*. San Diego: Academic Press. pp. 189–210.
9. Pagani F, Francucci CM, Moro L. 2005. Markers of bone turnover: Biochemical and clinical perspectives. *J Endocrinol Invest* 28(10 Suppl): 8–13.
10. Schafer C, Heiss A, Schwarz A, Westenfeld R, Ketteler M, Floege J, Muller-Esterl W, Schinke T, Jahnen-Dechent W. 2003. The serum protein alpha 2-Heremans-Schmid glycoprotein/fetuin-A is a systemically acting inhibitor of ectopic calcification. *J Clin Invest* 112(3): 357–66.
11. Robey PG. 2008. Non-collagenous bone matrix proteins. In: Bilezikian JP, Raisz LA, Rodan GA (eds.) *Principles of Bone Biology, Vol. 1*. San Diego: Academic Press. pp. 335–50.

12. Igwe JC, Gao Q, Kizivat T, Kao WW, Kalajzic I. 2011. Keratocan is expressed by osteoblasts and can modulate osteogenic differentiation. *Connect Tissue Res* 52(5): 401–7.
13. Schaefer L, Iozzo RV. 2008. Biological functions of the small leucine-rich proteoglycans: From genetics to signal transduction. *J Biol Chem* 283(31): 21305–9.
14. Nuka S, Zhou W, Henry SP, Gendron CM, Schultz JB, Shinomura T, Johnson J, Wang Y, Keene DR, Ramirez-Solis R, Behringer RR, Young MF, Hook M. 2010. Phenotypic characterization of epiphycan-deficient and epiphycan/biglycan double-deficient mice. *Osteoarthritis Cartilage* 18(1): 88–96.
15. Anderson HC, Sipe JB, Hessle L, Dhanyamraju R, Atti E, Camacho NP, Millan JL. 2004. Impaired calcification around matrix vesicles of growth plate and bone in alkaline phosphatase-deficient mice. *Am J Pathol* 164(3): 841–7.
16. Delany AM, Hankenson KD. 2009. Thrombospondin-2 and SPARC/osteonectin are critical regulators of bone remodeling. *J Cell Commun Signal* 3(3–4): 227–38.
17. Kimura H, Akiyama H, Nakamura T, de Crombrugghe B. 2007. Tenascin-W inhibits proliferation and differentiation of preosteoblasts during endochondral bone formation. *Biochem Biophys Res Commun* 356(4): 935–41.
18. Kii I, Nishiyama T, Li M, Matsumoto K, Saito M, Amizuka N, Kudo A. 2010. Incorporation of tenascin-C into the extracellular matrix by periostin underlies an extracellular meshwork architecture. *J Biol Chem* 285(3): 2028–39.
19. Paz J, Wade K, Kiyoshima T, Sodek J, Tang J, Tu Q, Yamauchi M, Chen J. 2005. Tissue- and bone cell-specific expression of bone sialoprotein is directed by a 9.0 kb promoter in transgenic mice. *Matrix Biol* 24(5): 341–52.
20. Wade-Gueye NM, Boudiffa M, Laroche N, Vanden-Bossche A, Fournier C, Aubin JE, Vico L, Lafage-Proust MH, Malaval L. 2010. Mice lacking bone sialoprotein (BSP) lose bone after ovariectomy and display skeletal site-specific response to intermittent PTH treatment. *Endocrinology* 151(11): 5103–13.
21. Huffman NT, Keightley JA, Chaoying C, Midura RJ, Lovitch D, Veno PA, Dallas SL, Gorski JP. 2007. Association of specific proteolytic processing of bone sialoprotein and bone acidic glycoprotein-75 with mineralization within biomineralization foci. *J Biol Chem* 282(36): 26002–13.
22. Gordon JA, Tye CE, Sampaio AV, Underhill TM, Hunter GK, Goldberg HA. 2007. Bone sialoprotein expression enhances osteoblast differentiation and matrix mineralization in vitro. *Bone* 41(3): 462–73.
23. Qin C, Baba O, Butler WT. 2004. Post-translational modifications of sibling proteins and their roles in osteogenesis and dentinogenesis. *Crit Rev Oral Biol Med* 15(3): 126–36.
24. Prasad M, Butler WT, Qin C. 2010. Dentin sialophosphoprein in biomineralization. *Connect Tissue Res* 51: 404–17.
25. Coutu DL, Wu JH, Monette A, Rivard GE, Blostein MD, Galipeau J. 2008. Periostin, a member of a novel family of vitamin K-dependent proteins, is expressed by mesenchymal stromal cells. *J Biol Chem* 283(26): 17991–8001.
26. Luo G, Ducy P, McKee MD, Pinero GJ, Loyer E, Behringer RR, Karsenty G. 1997. Spontaneous calcification of arteries and cartilage in mice lacking matrix GLA protein. *Nature* 386(6620): 78–81.
27. Murshed M, Schinke T, McKee MD, Karsenty G. 2004. Extracellular matrix mineralization is regulated locally; different roles of two gla-containing proteins. *J Cell Biol* 165(5): 625–30.
28. Ducy P, Desbois C, Boyce B, Pinero G, Story B, Dunstan C, Smith E, Bonadio J, Goldstein S, Gundberg C, Bradley A, Karsenty G. 1996. Increased bone formation in osteocalcin-deficient mice. *Nature* 382(6590): 448–52.
29. Boskey AL, Gadaleta S, Gundberg C, Doty SB, Ducy P, Karsenty G. 1998. Fourier transform infrared microspectroscopic analysis of bones of osteocalcin-deficient mice provides insight into the function of osteocalcin. *Bone* 23(3): 187–96.
30. Lee NK, Sowa H, Hinoi E, Ferron M, Ahn JD, Confavreux C, Dacquin R, Mee PJ, McKee MD, Jung DY, Zhang Z, Kim JK, Mauvais-Jarvis F, Ducy P, Karsenty G. 2007. Endocrine regulation of energy metabolism by the skeleton. *Cell* 130(3): 456–69.
31. Kumar R, Vella A. 2011. Carbohydrate metabolism and the skeleton: Picking a bone with the beta-cell. *J Clin Endocrinol Metab* 96(5): 1269–71.
32. Bonnet N, Standley KN, Bianchi EN, Stadelmann V, Foti M, Conway SJ, Ferrari SL. 2009. The matricellular protein periostin is required for sost inhibition and the anabolic response to mechanical loading and physical activity. *J Biol Chem* 284(51): 35939–50.
33. Trackman PC. 2005. Diverse biological functions of extracellular collagen processing enzymes. *J Cell Biochem* 96(5): 927–37.
34. Ge G, Greenspan DS. 2006. Developmental roles of the BMP1/TLD metalloproteinases. *Birth Defects Res C Embryo Today* 78(1): 47–68.
35. Yadav MC, Simao AM, Narisawa S, Huesa C, McKee MD, Farquharson C, Millan JL. 2011. Loss of skeletal mineralization by the simultaneous ablation of PHOSPHO1 and alkaline phosphatase function: A unified model of the mechanisms of initiation of skeletal calcification. *J Bone Miner Res* 26(2): 286–97.
36. Goldberg M, Boskey AL. 1996. Lipids and biomineralizations. *Prog Histochem Cytochem* 31(2): 1–187.
37. Stoffel W, Jenke B, Block B, Zumbansen M, Koebke J. 2005. Neutral sphingomyelinase 2 (smpd3) in the control of postnatal growth and development. *Proc Natl Acad Sci U S A* 102(12): 4554–9.
38. Aubin I, Adams CP, Opsahl S, Septier D, Bishop CE, Auge N, Salvayre R, Negre-Salvayre A, Goldberg M, Guenet JL, Poirier C. 2005. A deletion in the gene encoding sphingomyelin phosphodiesterase 3 (Smpd3) results

in osteogenesis and dentinogenesis imperfecta in the mouse. *Nat Genet* 37(8): 803–5.

39. Rubin J, Schwartz Z, Boyan BD, Fan X, Case N, Sen B, Drab M, Smith D, Aleman M, Wong KL, Yao H, Jo H, Gross TS. 2007. Caveolin-1 knockout mice have increased bone size and stiffness. *J Bone Miner Res* 22(9): 1408–18.

40. Gregory P, Kraemer E, Zurcher G, Gentinetta R, Rohrbach V, Brodbeck U, Andres AC, Ziemiecki A, Butikofer P. 2005. GPI-specific phospholipase D (GPI-PLD) is expressed during mouse development and is localized to the extracellular matrix of the developing mouse skeleton. *Bone* 37(2): 139–47.

7

Assessment of Bone Mass and Microarchitecture in Rodents

Blaine A. Christiansen

Introduction 59
Radiographs 59
Peripheral Dual-Energy X-Ray Absorptiometry 60
Peripheral Quantitative Computed Tomography 61
Magnetic Resonance Imaging 61

Microcomputed Tomography 62
Nanocomputed Tomography 63
Imaging Considerations 63
Conclusions 66
References 66

INTRODUCTION

Rodent models are crucially important research tools for investigating the musculoskeletal system. Analysis methods that are able to accurately distinguish bone mass and structure can help to determine the effects of aging or disease, as well as changes in skeletal morphology due to dietary, genetic, pharmacologic, or mechanical interventions. Until recently, quantitative histological techniques were the standard for assessing trabecular and cortical bone architecture. However, these techniques are limited with respect to assessment of trabecular microarchitecture since structural parameters are derived from analysis of a few two-dimensional (2D) sections, interpolating three-dimensional (3D) structure based on stereologic models [1]. Although histological analyses continue to provide unique information on cellularity and dynamic indices of bone remodeling, the use of 2D histological analysis to determine trabecular bone structure is now an antiquated technique that has largely been replaced by imaging techniques that are able to directly measure 3D bone structure.

Several imaging modalities are currently available for the assessment of skeletal morphology in animal models, offering a variety of costs and benefits for every application. Some imaging techniques, such as radiographs and peripheral dual-energy X-ray absorptiometry (pDXA) provide relatively inexpensive and fast assessments of bone mass and gross morphology *in vivo*, but have poor spatial resolution and are limited to planar images. In comparison, 3D imaging techniques such as microcomputed tomography (µCT) are able to directly measure bone microarchitecture without relying on stereologic models, but at a higher financial cost and greater scan time. Cutting-edge 3D imaging techniques are able to distinguish bone structure at extremely high resolutions (less than 1 micron), allowing for the imaging of bone nanostructure such as osteonal networks and cortical porosity, but with costly and difficult preparation techniques, and imaging systems that are not widely available.

In this chapter, we review the imaging techniques commonly used to assess bone mass and microarchitecture in rodents, paying particular attention to their advantages and disadvantages, as well as technical challenges associated with each technique.

RADIOGRAPHS

Though they are most likely underutilized, 2D radiographs can be useful tools for evaluating gross skeletal morphology both *in vivo* and *ex vivo*. For example, radiographs are commonly used to assess fracture healing [Fig. 7.1(a)] and have been used to assess limb development and range of motion in mouse models of injury [Fig. 7.1(b)] [2]. Radiographs are produced by the summation of attenuation along a single scan direction. The advantage to planar

---

Primer on the Metabolic Bone Diseases and Disorders of Mineral Metabolism, Eighth Edition. Edited by Clifford J. Rosen.
© 2013 American Society for Bone and Mineral Research. Published 2013 by John Wiley & Sons, Inc.

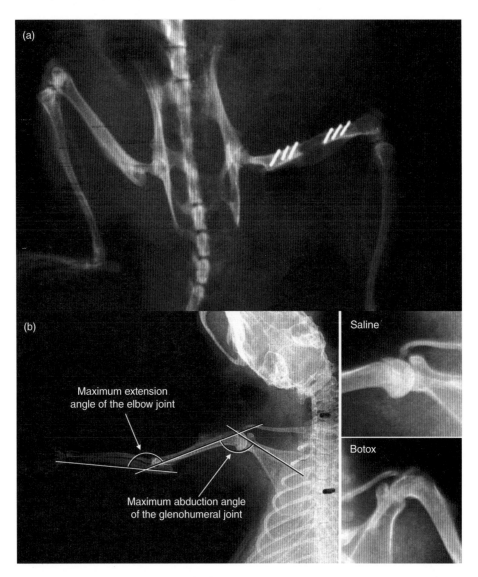

Fig. 7.1. (a) Radiograph image of a rat skeleton, with a healed segmental defect in the femur. Radiographs can be used for qualitative assessment of fracture healing and gross morphology of bone. (Image courtesy of J.C. Williams and M.A. Lee, University of California, Davis Medical Center.) (b) Radiographic images used to determine range of motion and limb development in growing mice injected with either botulinum toxin or saline. [Images courtesy of S. Thomopoulos, Washington University in St. Louis, and reproduced with permission from Kim, Galatz, Patel et al. (Ref. 2).]

radiographs is the rapid, relatively inexpensive visualization of skeletal morphology; however, they are limited to qualitative or semi-quantitative 2D evaluations.

PERIPHERAL DUAL-ENERGY X-RAY ABSORPTIOMETRY

Peripheral dual-energy X-ray absorptiometry (pDXA) is commonly used to measure bone mineral content (BMC, g), areal bone mineral density (BMD, g/cm^2), and body composition (percentage fat, lean tissue mass) of small animals both *in vivo* and *ex vivo*. DXA imaging uses a compact X-ray source to expose an animal to two X-ray beams with different energy levels. The ratio of attenuation of the high- and low-energy beams allows the separation of bone from soft tissue, as well as lean tissue from fat. A typical pixel size for pDXA measurement in mice and rats is 0.18×0.18 mm. *In vivo* pDXA measurements have been widely used to demonstrate the skeletal changes following estrogen deficiency [3], dietary and pharmacologic interventions [4, 5], as well as mouse strain-related differences in bone mass and body composition [6, 7].

Generally, pDXA provides highly reproducible measures of whole-body bone mass and body composition, with precision errors for whole body measurements less than 2% [8, 9] with relatively short scan times (less than 5 min). Due to the relatively large pixel size of DXA imaging, the precision of bone measurements at individual skeletal sites, such as the lumbar spine or distal femur, is much worse due to the challenges of consistently identifying the region of interest. Additionally, measurements of body composition suffer from poor accuracy, with general underestimation of lean tissue mass and overestimation of fat mass [8, 10, 11]. These errors can largely be eliminated with careful calibration of the DXA system to body fat content measured by carcass analysis [10, 11], though few investigators take the time to do this. DXA has relatively limited spatial resolution and assesses areal (i.e., 2D) images; therefore, it is not an appropriate imaging method for distinguishing cortical and trabecular bone compartments or for the assessment of trabecular microarchitecture.

PERIPHERAL QUANTITATIVE COMPUTED TOMOGRAPHY

Peripheral quantitative computed tomography (pQCT) is used for 3D assessment of bone geometry, bone mineral content (BMC), and volumetric bone mineral density (vBMD) in small animal studies, both *in vivo* and *ex vivo*. Commercially available pQCT scanners achieve a nominal voxel size of approximately 70 μm. The ability to measure bone mass and morphology *in vivo* makes pQCT useful for longitudinal assessment of the skeletal response to aging, disease, and/or interventions. The resolution of commercially available pQCT scanners does not allow for effective imaging of trabecular architecture, but they can be used to analyze compartment-specific changes in vBMD.

Many studies have used pQCT in rats to monitor changes in cortical bone geometry along with cortical and trabecular vBMD following pharmacologic [12, 13] or mechanical [14] interventions, or to monitor fracture healing [15]. Peripheral QCT has also been used to quantify bone density and geometry in mice [16, 17]. It has been suggested that this imaging method yields satisfactory *in vivo* precision and accuracy in skeletal characterization of mice [18]. However, the voxel size of pQCT (approximately 70 μm) is relatively large compared to the cortical thickness of mouse long bones (100–300 μm); therefore, this imaging method may have significant errors due to partial volume averaging (described below in the section "Voxel Size and Image Resolution"). Because of this, the accuracy of pQCT for assessment of compartment-specific bone parameters in mice is questionable.

MAGNETIC RESONANCE IMAGING

Magnetic resonance imaging (MRI) uses a powerful magnetic field and radio frequency fields to image the soft tissues of the body. A few studies have used high-resolution MRI to assess trabecular and cortical bone in small animals (Fig. 7.2) [19–21], although this technique

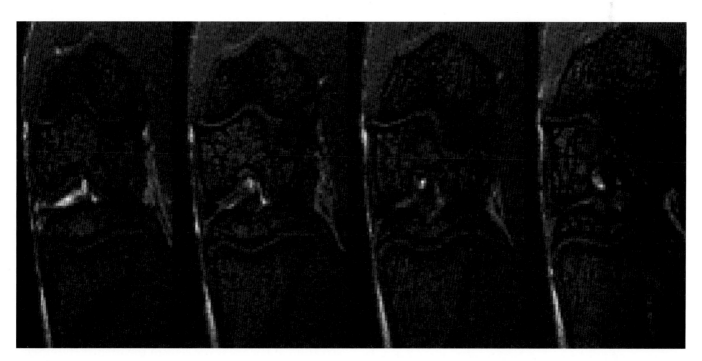

Fig. 7.2. Magnetic resonance imaging (MRI) of mouse knees. In MRI images, bone appears black because bone mineral lacks free protons and generates no MR signal, whereas soft tissues contain abundant free protons and give a strong signal. Typical voxel size for MRI images is 39–137 μm, making this imaging technique more suitable for use in humans or large animal models.

is more widely used for assessment of soft tissues. This is because bone mineral lacks free protons and generates no MR signal, whereas soft tissues contain abundant free protons and give a strong signal. The contrast between bone and the surrounding soft tissue allows for segmentation and quantification of bone architecture, which can yield results that are highly correlated with those obtained from 2D histology [19]. However, no information about the mineralization of bone tissue (BMD) can be obtained from MRI scans. Additionally, MRI systems with sufficiently powerful magnets to distinguish trabecular microstructure in small animal models are not widely available (typical voxel size is 39–137 µm), making this imaging technique more suitable for use in humans or large animal models [22].

MICROCOMPUTED TOMOGRAPHY

Microcomputed tomography (µCT) has become the gold standard for evaluation of bone morphology and microarchitecture of animal models. Microcomputed tomography uses X-ray attenuation data taken from multiple viewing angles to reconstruct a 3D representation of the specimen characterizing the spatial distribution of material density (Fig. 7.3). Current µCT scanners can achieve a nominal voxel size of 5 µm or less, which is sufficient for investigating structures such as mouse trabeculae that have widths of approximately 20 to 50 µm [23]. Microcomputed tomography allows for rapid and nondestructive 3D measurement of trabecular morphology, allowing samples to be used for subsequent analyses such as histology or mechanical testing. µCT scans can also provide an estimate of bone tissue mineralization on a voxel-by-voxel basis by comparing material attenuation to that of known standards, though this must be done with care given the constraints of the scan resolution and the polychromatic X-ray source [24].

Microcomputed tomography has been used for a wide range of studies of bone mass and bone morphology, including the analysis of growth and development [25], skeletal phenotypes in genetically altered mice, and animal models of disease states such as postmenopausal osteoporosis and renal osteodystrophy. Additionally, µCT has been used to assess the effects of pharmacologic interventions [26], as well as mechanical loading [27] and unloading [28]. Furthermore, µCT has been used to image macrocracks in cortical bone [29], evaluate fracture healing [30], and has also been used in combination with a perfused contrast agent to assess 3D vascular architecture [31].

All current µCT systems allow for *ex vivo* analysis of bone samples, with some systems providing the additional functionality of *in vivo* scanning. *In vivo* µCT provides the high resolution of µCT while allowing for longitudinal studies of morphological changes (Fig. 7.4). *In vivo* µCT is an ideal strategy for tracking bone changes that occur on a time scale of weeks or months, such as bone loss associated with disuse or ovariectomy [32, 33], or increased bone mass due to pharmacological or mechanical intervention [34–36]. By registering 3D images against images from previous time points, it is possible to determine the precise locations of bone formation or resorption [32, 37]. Altogether, the ability to perform longitudinal assessments of bone microstructure has the potential to reduce the number of animals needed in a given study and provide novel information about skeletal development, adaptation, repair, and response to disease or therapeutic interventions.

Despite the clear advantages of *in vivo* µCT, there are also several concerns about this method of imaging. First, there are concerns about the amount of ionizing radiation delivered during the *in vivo* µCT scan, particularly when animals are scanned multiple times throughout an experimental period. This radiation may introduce unwanted effects on the tissues or processes of interest, or on the animals in general. Young, growing animals and proliferative biologic processes, such as fracture healing or tumor growth, may be particularly susceptible to radiation exposure. It has previously been reported that weekly scans (8 weeks) have no effect on trabecular bone structure or bone marrow cells in the irradiated limbs of rats [38], while others have reported 8–20% decreases in trabecular bone volume (BV/TV) in irradiated limbs compared to nonirradiated limbs following weekly scans (5

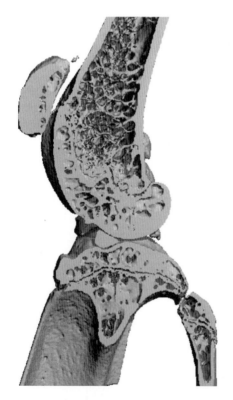

Fig. 7.3. Three-dimensional reconstruction of a mouse proximal tibia and distal femur scanned by µCT. Typical skeletal sites of interest for trabecular bone analysis in mice and rats are the proximal tibia epiphysis or metaphysis, distal femur epiphysis or metaphysis, and vertebral bodies of lumbar vertebrae.

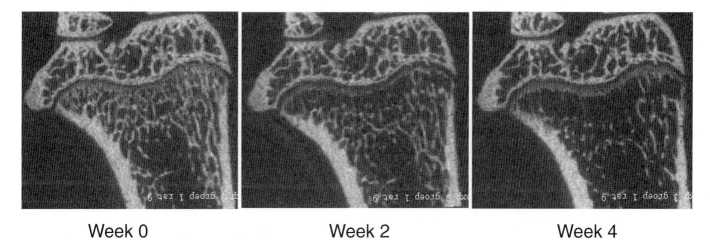

Fig. 7.4. *In vivo* longitudinal μCT images of the proximal tibia of a 30-week-old Wistar rat at baseline and then 2 and 4 weeks after ovariectomy. Note the marked deterioration of trabecular bone in the metaphysis, particularly adjacent to the growth plate, in comparison with little change in the epiphysis. (Image courtesy of J.E.M. Brouwers, Eindhoven University of Technology.)

weeks) in mice and rats [39]. Additional studies are needed to determine the potential effects of repeated *in vivo* μCT scans.

NANOCOMPUTED TOMOGRAPHY

Nanocomputed tomography (nCT) refers to the imaging of bone tissue at resolutions below 1 micron. This was first achieved by systems utilizing synchrotron radiation. Synchrotron radiation systems utilize a high photon flux monochromatic X-ray beam that is extracted from a synchrotron source, rather than the polychromatic X-ray source used for standard desktop μCT imaging systems. The use of a monochromatic X-ray beam eliminates beam-hardening artifacts and allows for accurate assessment of tissue mineral density [40]. Several commercially available imaging systems are also able to achieve submicron nominal voxel sizes without the use of a synchrotron radiation source.

The high spatial resolution associated with nCT systems allows for extremely precise assessment of trabecular bone architecture [23] and may be particularly useful for assessment of small-scale bone structures in young animals [41, 42], or ultrastructural properties of cortical bone (i.e., vascular canals and osteocyte lacunae; Fig. 7.5) [43, 44]. Although nCT imaging is typically performed on excised specimens, Kinney and colleagues [45] have used synchrotron radiation imaging *in vivo* in the rat proximal tibia to show the early deterioration in trabecular architecture following estrogen deficiency.

Altogether, nCT offers extremely high resolution imaging of microstructure and mineral density in excised bone specimens. The disadvantages of the technique are its high cost and limited availability, relatively small volume of tissue examined, and technical expertise needed to acquire and analyze the measurements.

IMAGING CONSIDERATIONS

There are several common issues associated with imaging of bone that must be considered in order to ensure the accuracy and reproducibility of measurements. Structural measurements obtained from μCT or other 3D imaging techniques are strongly dependent on a number of technical issues associated with the analysis, including (1) the scan resolution (voxel size), (2) the segmentation algorithm and threshold used to delineate soft tissue from bone, (3) the skeletal site(s) and volumes of interest, and (4) calibration of the system using density phantoms. These considerations are presented here in the context of μCT imaging but are also applicable to other 3D imaging techniques.

Voxel size and image resolution

A voxel, or volumetric pixel, is a 3D volume representing two dimensions within a reconstructed image and the slice thickness. Typically, voxels from μCT images have all three dimensions equal and therefore are described as isotropic voxels. Ideally, the smallest voxel size (i.e., highest scan resolution) available should be used for all scans; however, higher-resolution scans are more expensive with respect to scan time and memory requirements; therefore, the trade-off between image resolution and scan time/file size should be considered carefully. Differences in voxel size have little effect on the evaluation of structures with high relative thickness (i.e., 100–200 μm),

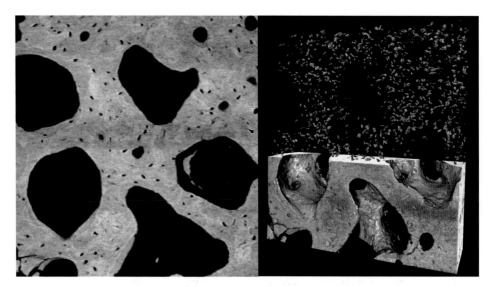

Fig. 7.5. Nanocomputed tomography allows for high-resolution imaging of cortical or trabecular bone with nominal voxel sizes below 1 micron. This enables research into the complex hierarchical structures of bone, including microstructure (e.g., 50–200μm osteons and trabeculae) and fine structure (e.g., 10–25-μm canal networks, and 3–15-μm osteocyte lacunae). (Images courtesy of Xradia and T.J. Wronski, University of Florida.)

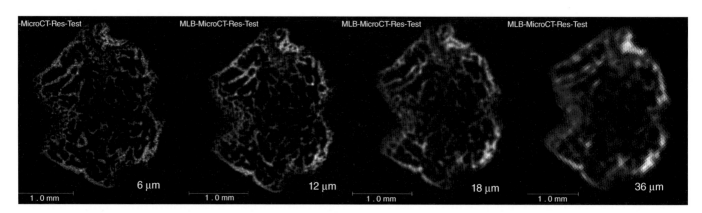

Fig. 7.6. Effect of voxel size on image quality. 2D gray-scale images of the distal femur of an adult mouse scanned with nominal voxel size of 6μm, 12μm, 18μm, and 36μm. Images acquired at 70kVp, 114mA, and 200-ms integration time. [Images courtesy of Rajaram Manoharan, Beth Israeal Deaconess Medical Center, and reprinted with permission from Bouxsein, Boyd, Christiansen et al. (Ref. 57).]

such as cortical bone or trabeculae in humans or large animal models. However, when analyzing smaller structures such as mouse or rat trabeculae with approximate dimensions of 20–60μm, voxel size can have significant effects on the results (Fig. 7.6) [46]. Scanning with low resolution (large voxel size) relative to the size of the structure of interest can cause an underestimation of bone mineral density owing to partial-volume averaging effects, in which a CT voxel samples a region containing two materials with different densities (bone and soft tissue). Depending on segmentation methods, partial volume averaging errors typically cause overestimation of cortical or trabecular thickness and underestimation of BMD. This has been confirmed in mouse femora as well as similarly sized aluminum tubes, where an object thickness to voxel size ratio of 9:1 or less was associated with errors of at least 15% in density measurements [47]. Generally, as the ratio of voxel size to object size decreases, so does the measurement of error regardless of scale (i.e., from mouse to human). The minimum ratio of voxels to object size is 2, but this is associated with substantial local errors (albeit smaller when averaged over an entire structure). Ideally, the ratio should be much higher for accurate morphological measurements.

Segmentation

The segmentation process involves separating the mineralized and nonmineralized structures (bone and non-bone) for subsequent quantitative analysis. The goal of selecting a threshold is to obtain results that are physiologically accurate (i.e., similar to those that would be obtained using 2D histology) [48]. However, this "physiologically accurate" threshold value depends on many factors, including the scan resolution, the bone volume fraction (BV/TV), and the mineral density of the bone in the volume of interest. The simplest approach to segmentation is to use a global threshold that extracts all voxels from the μCT data exceeding a given CT value (density). The advantage of using a global threshold is that it is efficient and requires setting only one parameter. In most studies, using a single global threshold for all scans is possible, ensuring that differences between study groups are due to experimental effects rather than image-processing effects. Yet there is no consensus on a threshold that should be used for all studies, and extreme care must be taken when selecting a threshold in studies where bone mineralization may not be constant for all groups (i.e., during growth and development or fracture healing) or when there are extreme ranges of bone volume fraction among groups [49, 50], in which case a single global threshold may not be adequate. Thus, in some cases, it is necessary to use more sophisticated segmentation tools, including specimen-specific thresholds [51, 52], and/or local segmentation methods, where the inclusion of each voxel is based on its local neighborhood. Examples include using the local neighborhood histogram to account for image inhomogeneities owing to beam hardening [53] or using the magnitude of the local image gradient (Sobel operator) to identify bone marrow edges for limited resolution images [54]. No matter what segmentation routine is applied, slice-wise 2D comparisons between the original and segmented scan must be performed to ensure accuracy of the segmentation (Fig 7.7).

Skeletal site and volume of interest

Skeletal sites of interest must be carefully selected depending on the specific research question. For trabecular

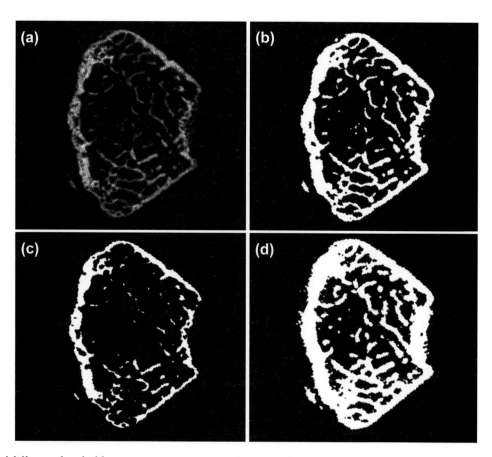

Fig. 7.7. Effect of different thresholds on image segmentation. (a) Original, unsegmented image of mouse distal femur. (b) Correctly segmented image showing reasonable binarization of bone structure. (c) Image segmented with too high a threshold such that key bone structures are missing and/or thinned relative to the original, unsegmented image. (d) Image segmented with too low a threshold such that bone structures appear too thick relative to the original, unsegmented image. Proper segmentation requires visual inspection and comparison of 2D and 3D binarized images with original gray-scale image. [Images courtesy of Rajaram Manoharan, Beth Israeal Deaconess Medical Center, and reprinted with permission from Bouxsein, Boyd, Christiansen et al. (Ref. 57).]

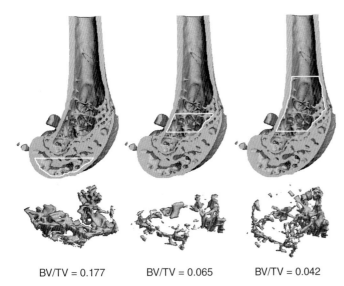

Fig. 7.8. Influence of the volume of interest on trabecular BV/TV at the distal femur of a 24-week-old female C57BL/6 mouse. Evaluation of epiphyseal trabecular bone (left) typically will yield a much higher BV/TV than evaluation of metaphyseal trabecular bone (center, right). Likewise, a volume of interest limited to the metaphyseal trabecular bone close to the growth plate (center) will yield a higher BV/TV than a larger volume of interest that includes a large amount of "empty space" in the diaphysis that contains few trabeculae (right).

bone analysis of mice or rats, the sites most often investigated are the proximal tibia, distal femur, and vertebral bodies. It is also often desirable to analyze more than one skeletal site, since heterogeneity of skeletal sites has been reported [28, 49]. For samples that will be mechanically tested, quantifying cortical bone (of the diaphysis for long bones) is critical, since the cortical bone bears the majority of the load during typical mechanical testing.

Similarly, selecting volumes of interest (VOIs) at a particular skeletal site is an important issue. Assessing trabecular bone requires a suitable VOI, and particular consideration should be paid to the distance that the VOI extends into the diaphysis of long bones. Extension of the VOI too far into the diaphysis will decrease the mean bone volume fraction relative to a VOI that is contained in the metaphyseal region (Fig. 7.8). To represent trabecular bone architecture accurately, the VOI should contain at least three to five intertrabecular lengths [55]. At the mid-diaphysis, 3D cortical thickness measurements must be based on a VOI that is longer than the cortex is thick; otherwise, the thickness will be underestimated.

Calibration

A calibration phantom is necessary to related CT values to a mineral-equivalent value, normally expressed in milligrams per cubic centimeter (mg/cm^3) of calcium hydroxyapatite (HA) from a solid-state phantom. Owing to the excellent linearity of modern μCT systems, calibration is possible with only two points, although some manufacturers use up to five points, covering a range from 0 up to 1,000 mg/cm^3 HA [56]. Many factors (e.g., beam-hardening correction, nominal voxel size, etc.) can influence density results, and absolute differences between scanners cannot be eliminated; however, with care, good relative measures can be made within any given study. Calibration should be performed routinely according to the manufacturer's recommendations.

Other considerations

Several other factors relating to image acquisition, image processing, and analysis can affect the ultimate outcomes of a scan. These factors include sample preparation and positioning, scanning medium, scan energy and intensity, beam-hardening adjustments, and image filtration. These factors have been reviewed previously, along with guidelines for reporting scan parameters and results from μCT analysis [57].

CONCLUSIONS

Assessment of skeletal mass and morphology in rodents via nondestructive imaging is an important component of current investigations aimed at improving our understanding of musculoskeletal development, growth, adaptation, and disease. Currently, several different imaging modalities aimed at whole animal, organ, tissue, and cellular levels are available, and will be increasingly used to study bone structure and adaptation.

REFERENCES

1. Parfitt AM, Drezner MK, Glorieux FH, Kanis JA, Malluche H, Meunier PJ, Ott SM, Recker RR. 1987. Bone histomorphometry: Standardization of nomenclature, symbols, and units. Report of the ASBMR Histomorphometry Nomenclature Committee. *J Bone Miner Res* 2(6): 595–610.
2. Kim HM, Galatz LM, Patel N, Das R, Thomopoulos S. 2009. Recovery potential after postnatal shoulder paralysis. An animal model of neonatal brachial plexus palsy. *J Bone Joint Surg Am* 91(4): 879–91.
3. Binkley N, Dahl DB, Engelke J, Kawahara-Baccus T, Krueger D, Colman RJ. 2003. Bone loss detection in rats using a mouse densitometer. *J Bone Miner Res* 18(2): 370–5.
4. Brochmann EJ, Duarte ME, Zaidi HA, Murray SS. 2003. Effects of dietary restriction on total body, femoral, and vertebral bone in SENCAR, C57BL/6, and DBA/2 mice. *Metabolism* 52(10): 1265–73.

5. Iida-Klein A, Hughes C, Lu SS, Moreno A, Shen V, Dempster DW, Cosman F, Lindsay R. 2006. Effects of cyclic versus daily hPTH(1–34) regimens on bone strength in association with BMD, biochemical markers, and bone structure in mice. *J Bone Miner Res* 21(2): 274–82.
6. Masinde GL, Li X, Gu W, Wergedal J, Mohan S, Baylink DJ 2002 Quantitative trait loci for bone density in mice: The genes determining total skeletal density and femur density show little overlap in F2 mice. *Calcif Tissue Int* 71(5): 421–8.
7. Reed DR, Bachmanov AA, Tordoff MG. 2007. Forty mouse strain survey of body composition. *Physiol Behav* 91(5): 593–600.
8. Nagy TR, Clair AL. 2000. Precision and accuracy of dual-energy X-ray absorptiometry for determining in vivo body composition of mice. *Obes Res* 8(5): 392–8.
9. Kolta S, De Vernejoul MC, Meneton P, Fechtenbaum J, Roux C. 2003. Bone mineral measurements in mice: Comparison of two devices. *J Clin Densitom* 6(3): 251–8.
10. Brommage R. 2003. Validation and calibration of DEXA body composition in mice. *Am J Physiol Endocrinol Metab* 285(3): E454–9.
11. Johnston SL, Peacock WL, Bell LM, Lonchampt M, Speakman JR. 2005. PIXImus DXA with different software needs individual calibration to accurately predict fat mass. *Obes Res* 13(9): 1558–65.
12. Gasser JA, Ingold P, Venturiere A, Shen V, Green JR. 2008. Long-term protective effects of zoledronic acid on cancellous and cortical bone in the ovariectomized rat. *J Bone Miner Res* 23(4): 544–51.
13. Armamento-Villareal R, Sheikh S, Nawaz A, Napoli N, Mueller C, Halstead LR, Brodt MD, Silva MJ, Galbiati E, Caruso PL, Civelli M, Civitelli R. 2005. A new selective estrogen receptor modulator, CHF 4227.01, preserves bone mass and microarchitecture in ovariectomized rats. *J Bone Miner Res* 20(12): 2178–88.
14. Silva MJ, Touhey DC. 2007. Bone formation after damaging in vivo fatigue loading results in recovery of whole-bone monotonic strength and increased fatigue life. *J Orthop Res* 25(2): 252–61.
15. McCann RM, Colleary G, Geddis C, Clarke SA, Jordan GR, Dickson GR, Marsh D. 2008. Effect of osteoporosis on bone mineral density and fracture repair in a rat femoral fracture model. *J Orthop Res* 26(3): 384–93.
16. Beamer WG, Donahue LR, Rosen CJ, Baylink DJ. 1996. Genetic variability in adult bone density among inbred strains of mice. *Bone* 18(5): 397–403.
17. Breen SA, Loveday BE, Millest AJ, Waterton JC. 1998. Stimulation and inhibition of bone formation: Use of peripheral quantitative computed tomography in the mouse in vivo. *Lab Anim* 32(4): 467–76.
18. Schmidt C, Priemel M, Kohler T, Weusten A, Muller R, Amling M, Eckstein F. 2003. Precision and accuracy of peripheral quantitative computed tomography (pQCT) in the mouse skeleton compared with histology and microcomputed tomography (microCT). *J Bone Miner Res* 18(8): 1486–96.
19. Weber MH, Sharp JC, Latta P, Sramek M, Hassard HT, Orr FW. 2005. Magnetic resonance imaging of trabecular and cortical bone in mice: Comparison of high resolution in vivo and ex vivo MR images with corresponding histology. *Eur J Radiol* 53(1): 96–102.
20. Gardner JR, Hess CP, Webb AG, Tsika RW, Dawson MJ, Gulani V. 2001. Magnetic resonance microscopy of morphological alterations in mouse trabecular bone structure under conditions of simulated microgravity. *Magn Reson Med* 45(6): 1122–5.
21. Kapadia RD, Stroup GB, Badger AM, Koller B, Levin JM, Coatney RW, Dodds RA, Liang X, Lark MW, Gowen M. 1998. Applications of micro-CT and MR microscopy to study pre-clinical models of osteoporosis and osteoarthritis. *Technol Health Care* 6(5–6): 361–72.
22. Jiang Y, Zhao J, White DL, Genant HK. 2000. Micro CT and micro MR imaging of 3D architecture of animal skeleton. *J Musculoskel Neuron Interact* 1: 45–51.
23. Martin-Badosa E, Amblard D, Nuzzo S, Elmoutaouakkil A, Vico L, Peyrin F. 2003. Excised bone structures in mice: Imaging at three-dimensional synchrotron radiation micro CT. *Radiology* 229(3): 921–8.
24. Fajardo R, Cory E, Patel N, Nazarian A, Snyder B, Bouxsein M. 2009. Specimen size and porosity can introduce error into µCT-based tissue mineral density measurements. *Bone* 44(1): 176–84.
25. Hankenson KD, Hormuzdi SG, Meganck JA, Bornstein P. 2005. Mice with a disruption of the thrombospondin 3 gene differ in geometric and biomechanical properties of bone and have accelerated development of the femoral head. *Mol Cell Biol* 25(13): 5599–606.
26. von Stechow D, Zurakowski D, Pettit AR, Muller R, Gronowicz G, Chorev M, Otu H, Libermann T, Alexander JM. 2004. Differential transcriptional effects of PTH and estrogen during anabolic bone formation. *J Cell Biochem* 93(3): 476–90.
27. Christiansen BA, Silva MJ. 2006. The effect of varying magnitudes of whole-body vibration on several skeletal sites in mice. *Ann Biomed Eng* 34(7): 1149–56.
28. Squire M, Donahue LR, Rubin C, Judex S. 2004. Genetic variations that regulate bone morphology in the male mouse skeleton do not define its susceptibility to mechanical unloading. *Bone* 35(6): 1353–60.
29. Uthgenannt BA, Silva MJ. 2007. Use of the rat forelimb compression model to create discrete levels of bone damage in vivo. *J Biomech* 40(2): 317–24.
30. Gardner MJ, Ricciardi BF, Wright TM, Bostrom MP, van der Meulen MC. 2008. Pause insertions during cyclic in vivo loading affect bone healing. *Clin Orthop Relat Res* 466(5): 1232–8.
31. Bolland BJ, Kanczler JM, Dunlop DG, Oreffo RO. 2008. Development of in vivo muCT evaluation of neovascularisation in tissue engineered bone constructs. *Bone* 43(1): 195–202.
32. Boyd SK, Davison P, Muller R, Gasser JA. 2006. Monitoring individual morphological changes over time in ovariectomized rats by in vivo micro-computed tomography. *Bone* 39(4): 854–62.

33. Campbell GM, Buie HR, Boyd SK. 2008. Signs of irreversible architectural changes occur early in the development of experimental osteoporosis as assessed by in vivo micro-CT. *Osteoporos Int* 19(10): 1409–19.
34. Brouwers JE, Lambers FM, Gasser JA, van Rietbergen B, Huiskes R. 2008. Bone degeneration and recovery after early and late bisphosphonate treatment of ovariectomized wistar rats assessed by in vivo micro-computed tomography. *Calcif Tissue Int* 82(3): 202–11.
35. McErlain DD, Appleton CT, Litchfield RB, Pitelka V, Henry JL, Bernier SM, Beier F, Holdsworth DW. 2008. Study of subchondral bone adaptations in a rodent surgical model of OA using in vivo micro-computed tomography. *Osteoarthritis Cartilage* 16(4): 458–69.
36. Morenko BJ, Bove SE, Chen L, Guzman RE, Juneau P, Bocan TM, Peter GK, Arora R, Kilgore KS. 2004. In vivo micro computed tomography of subchondral bone in the rat after intra-articular administration of monosodium iodoacetate. *Contemp Top Lab Anim Sci* 43(1): 39–43.
37. Waarsing JH, Day JS, van der Linden JC, Ederveen AG, Spanjers C, De Clerck N, Sasov A, Verhaar JA, Weinans H. 2004. Detecting and tracking local changes in the tibiae of individual rats: A novel method to analyse longitudinal in vivo micro-CT data. *Bone* 34(1): 163–9.
38. Brouwers JE, van Rietbergen B, Huiskes R. 2007. No effects of in vivo micro-CT radiation on structural parameters and bone marrow cells in proximal tibia of wistar rats detected after eight weekly scans. *J Orthop Res* 25(10): 1325–32.
39. Klinck RJ, Campbell GM, Boyd SK. 2008. Radiation effects on bone architecture in mice and rats resulting from in vivo micro-computed tomography scanning. *Med Eng Phys* 30(7): 888–95.
40. Nuzzo S, Lafage-Proust MH, Martin-Badosa E, Boivin G, Thomas T, Alexandre C, Peyrin F. 2002. Synchrotron radiation microtomography allows the analysis of three-dimensional microarchitecture and degree of mineralization of human iliac crest biopsy specimens: Effects of etidronate treatment. *J Bone Miner Res* 17(8): 1372–82.
41. Burghardt AJ, Wang Y, Elalieh H, Thibault X, Bikle D, Peyrin F, Majumdar S. 2007. Evaluation of fetal bone structure and mineralization in IGF-I deficient mice using synchrotron radiation microtomography and Fourier transform infrared spectroscopy. *Bone* 40(1): 160–8.
42. Matsumoto T, Yoshino M, Asano T, Uesugi K, Todoh M, Tanaka M. 2006. Monochromatic synchrotron radiation muCT reveals disuse-mediated canal network rarefaction in cortical bone of growing rat tibiae. *J Appl Physiol* 100(1): 274–80.
43. Raum K, Hofmann T, Leguerney I, Saied A, Peyrin F, Vico L, Laugier P. 2007. Variations of microstructure, mineral density and tissue elasticity in B6/C3H mice. *Bone* 41(6): 1017–24.
44. Schneider P, Stauber M, Voide R, Stampanoni M, Donahue LR, Muller R. 2007. Ultrastructural properties in cortical bone vary greatly in two inbred strains of mice as assessed by synchrotron light based micro- and nano-CT. *J Bone Miner Res* 22(10): 1557–70.
45. Kinney JH, Ryaby JT, Haupt DL, Lane NE. 1998. Three-dimensional in vivo morphometry of trabecular bone in the OVX rat model of osteoporosis. *Technol Health Care* 6(5–6): 339–50.
46. Muller R, Koller B, Hildebrand T, Laib A, Gianolini S, Ruegsegger P. 1996. Resolution dependency of microstructural properties of cancellous bone based on three-dimensional mu-tomography. *Technol Health Care* 4(1): 113–9.
47. Brodt MD, Pelz GB, Taniguchi J, Silva MJ. 2003. Accuracy of peripheral quantitative computed tomography (pQCT) for assessing area and density of mouse cortical bone. *Calcif Tissue Int* 73(4): 411–8.
48. Rajagopalan S, Lu L, Yaszemski MJ, Robb RA. 2005. Optimal segmentation of microcomputed tomographic images of porous tissue-engineering scaffolds. *J Biomed Mater Res A* 75(4): 877–87.
49. Glatt V, Canalis E, Stadmeyer L, Bouxsein ML. 2007. Age-related changes in trabecular architecture differ in female and male C57BL/6J mice. *J Bone Miner Res* 22(8): 1197–207.
50. Ominsky MS, Stolina M, Li X, Corbin TJ, Asuncion FJ, Barrero M, Niu QT, Dwyer D, Adamu S, Warmington KS, Grisanti M, Tan HL, Ke HZ, Simonet WS, Kostenuik PJ. 2009. One year of transgenic overexpression of osteoprotegerin in rats suppressed bone resorption and increased vertebral bone volume, density, and strength. *J Bone Miner Res* 24(7): 1234–46.
51. Meinel L, Fajardo R, Hofmann S, Langer R, Chen J, Snyder B, Vunjak-Novakovic G, Kaplan D. 2005. Silk implants for the healing of critical size bone defects. *Bone* 37(5): 688–98.
52. Ridler T, Calvard S. 1978. Picture thresholding using an iterative selection method. *IEEE Trans on Systems, Man and Cybernetics SMC* 8: 630–2.
53. Dufresne T. 1998. Segmentation techniques for analysis of bone by three-dimensional computed tomographic imaging. *Technol Health Care* 6(5–6): 351–9.
54. Waarsing JH, Day JS, Weinans H. 2004. An improved segmentation method for in vivo microCT imaging. *J Bone Miner Res* 19(10): 1640–50.
55. Harrigan TP, Jasty M, Mann RW, Harris WH. 1988. Limitations of the continuum assumption in cancellous bone. *J Biomech* 21(4): 269–75.
56. Ruegsegger P, Muller R. 1997. Quantitative computed tomography techniques in the determination of bone density and bone architecture. In: Leondes C (ed.) *Medical Imaging Systems Techniques and Applications—Brain and Skeletal Systems*. Singapore: Gordon and Breach Science Publishers. pp. 169–220.
57. Bouxsein ML, Boyd SK, Christiansen BA, Guldberg RE, Jepsen KJ, Muller R. 2010. Guidelines for assessment of bone microstructure in rodents using micro-computed tomography. *J Bone Miner Res* 25(7): 1468–86.

8
Animal Models: Genetic Manipulation

Karen M. Lyons

Introduction 69
Overexpression of Target Genes 69
Gene Targeting 70
Tissue-Specific and Inducible Knockout and Overexpression 71

Lineage Tracing and Activity Reporters 73
Functional Genomics 73
General Considerations 73
Acknowledgments 73
References 73

INTRODUCTION

Genetically manipulated mice have contributed enormously to our identification of genes controlling skeletal development and to the clarification of their mechanisms of action. Techniques are available to examine the effects of loss-of-function, gain-of-function, and altered structure of gene products. The ability to introduce defined mutations has facilitated the production of animal models of human diseases, cell lineage studies, examination of tissue-specific functions, and dissection of distinct gene functions at specific stages of differentiation within a single cell lineage.

OVEREXPRESSION OF TARGET GENES

The first widely used approach to study gene function *in vivo* was to produce transgenic mice that overexpress target genes. This requires the full-length coding sequence (cDNA) of a gene to be cloned downstream of a promoter. There are several promoters that have been well characterized and used to drive gene expression in skeletal tissues. In the following sections, promoters that have been used to drive transgene expression directly are described. The use of these and other promoters to enable overexpression or underexpression of target genes using Cre recombinase is discussed in a separate section.

Chondrocytes

The most widely used cartilage-specific promoter is derived from the mouse pro aI(II) collagen gene (*Col2a1*). This promoter drives high levels of expression beginning after the condensation stage in appendicular elements, and prior to condensation in axial elements in the sclerotomal compartment [1]. The *Col11a2* promoter has also been used to overexpress genes in chondrocytes, although some of these promoters also drive expression in perichondrium and osteoblasts [2, 3]. Overexpression in prechondrogenic limb mesenchyme has been achieved using the *Prx1* promoter [4]. Promoters that drive high levels of expression in hypertrophic chondrocytes have not been described. Chicken and mouse *Col10a1* promoters allow transgene expression in hypertrophic chondrocytes at low to moderate levels; however, expression is not seen in all hypertrophic chondrocytes and is weak [5]. The recent development of *Col10a1-Cre* knock-in mice, when used in conjunction with Cre-inducible transgenic lines (see below) [6] may permit high levels of gene expression in these cells.

Primer on the Metabolic Bone Diseases and Disorders of Mineral Metabolism, Eighth Edition. Edited by Clifford J. Rosen.
© 2013 American Society for Bone and Mineral Research. Published 2013 by John Wiley & Sons, Inc.

Osteoblasts

Several promoters allow overexpression of genes in osteoblasts. The most frequently used is a 2.3-kb fragment from the rat or mouse *Col1a1* proximal promoter *(2.3Col1a1)*. Strong and specific activity is seen in fetal and adult mature osteoblasts and osteocytes [7]. A second *2.3Col1a1* promoter has been described [8]; this promoter shows similar activity in bone, but expression in brain has recently been noted [9]. The 3.6-kb proximal *Col1a1* promoter drives strong gene expression at an earlier stage of differentiation (preosteoblasts), but it is also expressed in nonosseous tissues, including tendon, skin, muscle, and brain [8, 9]. The 1.7-kb mouse *Osteocalcin (OC; Bglap)* promoter has been used to express genes in mature osteoblasts. However, this promoter is expressed in a low percentage of osteoblasts and at a relatively low level. Consistent with this, the 1.7-kb *OC* promoter is unable to drive Cre recombinase expression at effective levels [7]. On the other hand, 3.5- to 3.9-kb human *OC* promoter fragments drive osteoblast-specific expression in a large proportion of mature osteoblasts and osteocytes [10]. Osteocyte-specific promoters have not been widely used, although osteocyte-specific transgene overexpression can potentially be achieved using *Dmp1*-Cre mice (see below) [11].

Tendon and ligament

Tendon patterning and differentiation are seldom studied genetically because of the lack of tissue-specific markers. *Scleraxis (Scx)* encodes a transcription factor expressed in developing tendons and ligaments and their progenitors. The development of *Scx*-Cre mice [12] provides a potential strategy for inducing expression of Cre-inducible transgenes (see below).

Osteoclasts

There are a variety of promoters that drive high levels of gene expression in osteoclasts and their progenitors. These include *CD11b*, expressed in monocytes, macrophages, and along the osteoclast differentiation pathway from monocuclated progenitor cells and into mature osteoclasts [13], and *TRACP*, expressed in mature osteoclasts and their precursors [14].

Advantages and disadvantages of overexpression approaches

The major advantages of the transgenic approach are that it is straightforward and inexpensive, high levels of gene expression can potentially be achieved, and transgenic mice often show obvious phenotypes. Furthermore, transgenic strains in which marker genes such as *LacZ*, *GFP*, and/or *ALP* are expressed under the control of tissue-specific promoters allow easy visualization of specific cell types *in vivo* and may permit their isolation with a resolution not possible using other methods. Other methods for detecting specific lineages based on site-specific recombination systems are discussed in a separate section.

A major caveat of the transgenic approach is that constitutive overexpression models often yield nonphysiological levels of protein expression, and this will confound interpretation of the role of the normal gene. Modified transgenic approaches can overcome some of this uncertainty. These include the use of transgenes encoding dominant negative variants or natural antagonists. These approaches lead to loss of function and thus target pathways in their normal physiological context.

The site of transgene integration can have major consequences on tissue specificity and levels of expression. This can be exploited to examine dose-dependent effects, but care must be taken to assess not only levels of expression but also sites of transgene expression, making comparisons of multiple transgenic lines potentially difficult to interpret.

Finally, overexpression of genes that have profound effects in skeletal tissues may confer embryonic lethality, precluding the establishment of stable transgenic lines. By the same token, transgenic lines that can be established overexpress genes only to an extent compatible with survival to sexual maturity. Several bigenic systems address this issue to permit establishment of lines that overexpress genes at levels that confer lethality. One uses the tetracycline (tet) responsive transactivator [15]. This system permits tissue-specific tet-responsive gene expression. A transactivator (tTA) whose activity is modified by tet or the tet analog doxycline (dox) is expressed under the control of a tissue-specific or ubiquitous promoter. The second component is a strain expressing the gene of interest under the control of the operator sequences of the tet operon (tetO). Depending on whether the transactivator is induced (tTA: tet-on) or repressed (rtTA: tet-off) by dox, expression of target genes can be induced or repressed. A second bigenic strategy uses the GAL4/UAS system. This uses one transgenic strain expressing the GAL4 transactivator under a tissue-specific or inducible promoter and a second transgenic strain expressing the gene of interest under the control of the UAS sequence, which requires GAL4 binding for activity. These systems have been used to develop transgenic lines that permit activation of genes in cartilage and bone [16, 17].

GENE TARGETING

The most widely used technique for genetic manipulation is gene targeting in mouse embryonic stem (ES) cells. Briefly, a targeting construct contains a portion of the gene of interest, along with a modification that renders the gene product inactive or modifies its activity.

Gene targeting in ES cells can also be used to generate knock-in models. In these, a locus of interest is modified such that it encodes a gene product with altered activity. A major application of this technology has been to generate mouse models of human genetic diseases.

The ability to eliminate or alter the structure of defined genes is possible in the mouse because of the unique properties of ES cells. These can be genetically manipulated in culture yet retain the ability to colonize the germline when injected into a mouse blastocyst. Once incorporated, they can give rise to germ cells, permitting the establishment of mouse strains carrying the modified gene. ES cells from the 129 strain were the first to be derived and are the most frequently used. However, 129 mice are poor breeders and exhibit abnormal immunological characteristics [18]. Germline competent ES cell lines from 129, C57Bl/6, and C3H are available commercially from multiple sources. These strains have different bone mineral density (BMD) profiles [19] that must be taken into consideration when interpreting skeletal phenotypes.

Germline-competent ES cells carrying defined mutations in many genes are now readily available. These can be found by searching the Mouse Genome Informatics website (www.informatics.jax.org). The most time-consuming step in gene targeting is the introduction of modified ES cells into the germline. This is most commonly done by blastocyst injection, resulting in an F_0 animal that is partially derived from the modified ES cells. These chimeric mice are bred to obtain F_1 mice that are heterozygous for the defined mutation. There are many university core facilities and commercial groups that routinely perform blastocyst injections.

Advantages and disadvantages of gene targeting

Phenotypes caused by loss of function provide direct insight into the physiological roles of the ablated gene product. Moreover, novel actions of targeted genes can emerge because, unlike transgenic models, global knockout models are not limited to a particular tissue or system.

A complication of the global knockout approach is that the deletion of genes essential for early development may preclude analysis of the roles of these genes in skeletal tissues. On the other hand, many knockout strains do not exhibit obvious phenotypes because of functional redundancy; the creation of double or even triple knockouts may be necessary. Another consideration is that global knockout strains usually contain a modified allele in which the selectable cassette used to screen the ES colonies is retained in the locus of interest. On occasion, this leads to effects on neighboring genes. These effects can be shown by comparing phenotypes of mice carrying null alleles in which the selection cassette is left in place with those in which it has been removed.

This is discussed below in the context of tissue-specific knockouts.

TISSUE-SPECIFIC AND INDUCIBLE KNOCKOUT AND OVEREXPRESSION

The ability to achieve site-specific recombination has revolutionized analysis of gene action in skeletal cells. Tissue-specific recombination circumvents the early lethality associated with global knockout or overexpression. Tools are available that allow researchers to ablate or express genes in specific types of skeletal cells at specific stages of commitment and differentiation.

Several methods can be used to achieve tissue-specific gene knockout or activation; these rely on site-specific recombinases derived from bacteriophage (Cre) or yeast (Flp) [20]. Cre and Flp recombine DNA at specific target sites. Depending on the orientation of the sites, the recombinase catalyzes excision or inversion of DNA flanked by the sites. Two mouse lines are required. For the Cre-*loxP* system, these are the "floxed" strain, in which the region of the gene targeted for deletion is flanked by *loxP* sites, and a second transgenic mouse line in which Cre recombinase is expressed under the control of an inducible and/or tissue-specific promoter. In mice carrying both the floxed gene and the Cre transgene, Cre deletes the sequence flanked by *loxP* sites. The *loxP* sites are usually placed in introns and generally do not interfere with the normal function of the gene. Hence, the floxed target gene usually functions normally except in tissues where Cre is expressed.

This technology is most commonly used to inactivate genes in which a critical exon is flanked by loxP sites. However, it has also been used to achieve site-specific inducible overexpression. In this case, a transgene is often generated under the control of a strong ubiquitious promoter such as *CAG* [21]. The expression of the transgene is prevented by placing a strong transcriptional stop signal, flanked by *loxP* sites. The gene is activated when Cre catalyzes excision of the stop signal [22]. Engin et al [23] provide an example of this approach in osteoblasts.

Most studies have used constitutively active forms of Cre. However, ligand-regulated forms enable temporal control of gene activity. The most popular strategy uses fusions of Cre to a ligand binding domain from a mutant estrogen receptor (ER) [24]. The ER domain recognizes the synthetic estrogen antagonist 4-OH tamoxifen (T), but is insensitive to endogenous β-estradiol. In the absence of T, the Cre-ER(T) fusion protein is retained in the cytoplasm. Binding of T to the ER domain induces a conformational change that permits the fusion protein to enter the nucleus and catalyze recombination. Many Cre strains have been developed that are suitable for analysis of skeletal tissues. A complete list of published Cre lines can be found on the Mouse Genome Informatics site (www.informatics.jax.org/recombinases). A few of the most widely used, and some promising newcomers, are discussed below.

Uncondensed mesenchyme, mesenchymal condensations, and neural crest

Prx1-Cre drives expression in early uncondensed limb and head mesenchyme [4]. *Dermo1-Cre* expresses Cre in mesenchymal condensations [25]. Sox9 is also expressed in mesenchymal condensations. A *Sox9-Cre* knock-in strain drives Cre-mediated excision in precursors of osteoblasts and chondrocytes in these condensations [26]. The *Wnt1-Cre* transgene drives expression in the migrating neural crest and can be used to ablate genes in all chondrocytes and osteoblasts derived from this source [27].

Cartilage

The most widely used Cre strain for cartilage is *Col2a1-Cre* [1]. The activity of this promoter seems to be restricted to chondrocytes in the majority of studies, but there appears to be a brief window during chondrogenesis when *Col2a1-Cre* is expressed in the perichondrium [28]. Hence, controls should be performed to determine the extent to which Cre is expressed in the perichondrium in specific experiments. To date, no published Cre lines exhibit sufficient robustness to permit gene deletion in hypertrophic chondrocytes, but several groups are working on this.

The development of tools that permit inducible recombination in postnatal cartilage has been a challenge. Several groups have generated transgenic lines using CreER(T) fusion proteins under the control of the *Col2a1* promoter. These strains permit ablation in articular chondrocytes if T is administered within 2 weeks after birth [28–30]. However, expression of *Col2a1* is known to be low in adults, and although not all of these lines have been tested for in adults, in the one study examining this, the efficiency of recombination was very low [28]. A major breakthrough has been the development of *Aggrecan-CreERT2* mice, in which CreERT2 has been knocked into the *Aggrecan (Acan)* locus [31]. These mice exhibit robust Cre expression in the adult growth plate and articular cartilage, as well as in fibrocartilage.

Osteoblasts

The transcription factor *Osterix (Osx1)* is expressed in osteoblast precursors. A Cre-GFP fusion protein has been inserted into the *Osx1* locus on a BAC transgene [32]. This construct allows the targeting of osteoblast precursors. Several transgenic lines in which Cre is expressed under the control of the *Col1a1* promoter permit excision of genes in osteoblasts. A 3.6-kb *Col1a1* promoter drives high levels of Cre expression in osteoblasts but, as discussed above, also targets tendon and fibrous cells types in the suture, skin, and several organs [8, 9]. *Col1a1-Cre* lines (2.3 kb) show more restricted expression to mature osteoblasts, but ectopic expression in the brain is noted in some lines [9]. *Osteocalcin-Cre* drives excision in mature osteoblasts but is not activated until just before birth [10]. Few inducible Cre transgenic strains have been developed for bone. An inducible 2.3 *Col1a1-ERT* strain has been described [33].

Osteoclasts

Several Cre strains permit ablation in myeloid cells. These include *LysMcre* mice, in which Cre has been introduced into the *M Lysozyme* locus [34], and a strain in which the *CD11h* promoter drives Cre expression in macrophages and osteoclasts [35]. Strains permitting Cre-mediated recombination in mature osteoclasts include a Cre knock-in into the *Cathepsin K (Ctsk)* locus [35] and transgenic lines expressing Cre under the control of the *TRAPC* and *Ctsk* promoters [36].

Considerations when using inducible knockouts

The most significant advantage of the Cre-*loxP* system is its flexibility, permitting exploration of gene function in multiple tissues at multiple time points. However, there are some caveats. It may be difficult to find a promoter that drives Cre expression with sufficient activity to result in complete excision of the target gene; this is highly dependent on the floxed allele. Moreover, Cre transgenic strains based on identical promoters but generated in different laboratories can exhibit different specificities and efficiencies. For this reason, every study should include controls to verify the extent of Cre-mediated recombination of the floxed line of interest. Floxed genes also vary with respect to the kinetics of Cre-mediated recombination. This must be borne in mind when attempting to compare phenotypes caused by excision of different genes using the same Cre transgenic line.

The presence of the drug selection cassette in a floxed gene can have a major effect on expression levels of the targeted allele even in the absence of the Cre transgene. Prominent examples include a floxed *Fgf8* strain in which retention of the neo cassette leads to a hypomorphic (reduced function) phenotype [37], and *Scleraxis* knockouts, in which the presence of the drug selection cassette led to embryonic lethality by day 9.5 of gestation [38]. In striking contrast, mice homozygous for a *Scleraxis* null allele in which the drug selection cassette was removed are viable as adults [39]. Another important consideration is that in some cases, Cre itself may confer a phenotype due to toxicity. This is especially true if the Cre is expressed as a fusion with GFP [40].

With respect to inducible Cre models, the inducer may have a significant impact on the phenotype. Both doxycycline and tamoxifen can have profound effects on cartilage, bone, and osteoclasts independently of target gene deletion. Even the low doses of tamoxifen used to catalyze Cre-ER(T)-mediated excision may have effects on bone [41]. Thus, a control group of Cre-negative mice treated with the inducer may need to be included to examine the impact of this variable on the mutant phenotype under study.

LINEAGE TRACING AND ACTIVITY REPORTERS

Genetically modified mice have permitted the determination of cell lineage relationships and the relative contributions of cells from various sources to a given organ with unprecedented resolution. These studies rely on strains that carry a floxed reporter gene, such as *LacZ* or *GFP*. For example, the *R26R* strain carries a floxed *LacZ* cassette introduced into the *ROSA26* locus [42]. When bred to a Cre-expressing strain, all cells in which Cre is expressed, and all of their descendants, express LacZ. *R26R* mice have been used to test the specificity and efficiency of Cre-expressing transgenic lines, although use of this strain in bone is limited by the fact that osteoblasts express endogenous LacZ.

Major insights have been made using *R26R* to study osteoprogenitors. For example, *Wnt1-Cre;R26R* mice revealed that frontal bones are derived from the neural crest, but parietal bones are derived from mesoderm [43]. A second example is the demonstration that immature osteoblasts move into developing bone along with invading blood vessels [44]. Most recently, lines have been generated that permit live imaging of the expression of fluorescent proteins in specific organelles (cell membrane, nucleus, etc.) in a tissue-specific and inducible manner; this system was used to mark sites of *Sox9-Cre* expression in skeletal tissues [45].

Transgenic reporter lines can also be used to monitor signaling pathway activity. For example, Wnt pathway activity has been monitored *in vivo* using TOPGAL mice to track β-catenin activity during endochondral bone formation [46]. Tools are also available to monitor canonical bone morphogenetic protein (BMP) pathway activity *in vivo* [47, 48].

FUNCTIONAL GENOMICS

Functional data are available for nearly 14,000 of the estimated 25,000 mouse genes. This information can be accessed through the Phenotype/Alleles project in Mouse Genome Informatics (www.informatics.jax.org/phenotypes). However, fewer than 4,000 of the annotations relate to gene function in skeletal tissue. Thus, there is much work remaining to be done. The International Mouse Knockout Consortium (IMKC) was established in 2007 to coordinate international efforts to generate conditional alleles for all mouse genes. As of 2011, ES cells carrying conditional alleles in nearly 8,000 genes have been generated. Mutant Mouse Regional Resource Centers (http://www.mmrrc.org) have permitted the acquisition and storage of another 320 strains developed in individual laboratories. A second major effort is the use of gene trapping, a high-throughput mutagenesis strategy. Progress on these and other mutagenesis efforts can be found on the Mouse Genome Informatics and IMKC websites [49, 50].

GENERAL CONSIDERATIONS

With all of the successes in genetic manipulation, the real bottleneck is phenotyping. Phenotype is dependent on genetic and environmental factors. As discussed, inbred strains vary considerably in their peak BMD. Moreover, housing conditions and food intake can have a significant impact on metabolic parameters and BMD [51, 52]. Hence, even genetically identical mice can have different phenotypes in different facilities. It is important to assess phenotypes at different ages because effects present at early stages may be compensated for later on. An example of this can be seen in matrix metalloproteinase (MMP)-9-deficient mice [53]. These exhibit prenatal expansion of the hypertrophic zone but are normal within a few weeks of birth. In contrast, other phenotypes manifest only in late stages or when a metabolic stress is applied.

Caution must be taken when extrapolating findings in mouse models to functions in humans. Biomechanical loading and hormonal effects on bones are clearly different in mice and humans. Moreover, linear growth in humans ceases after epiphyseal closure, whereas in mice the growth plate does not fuse. Nonetheless, similarities outweigh the differences by far, and genetic models are likely to play an increasingly prominent role in every aspect of research in skeletal biology.

ACKNOWLEDGMENTS

The author acknowledges funding from NIH (AR052686 and AR044528).

REFERENCES

1. Ovchinnikov DA, Deng JM, Ogunrinu G, Behringer RR. 2000. Col2a1-directed expression of Cre recombinase in differentiating chondrocytes in transgenic mice. *Genesis* 26: 145–6.
2. Horiki M, Imamura T, Okamoto M, Hayashi M, Murai J, Myoui A, Ochi T, Miyazono K, Yoshikawa H, Tsumaki

N. 2004. Smad6/Smurf1 overexpression in cartilage delays chondrocyte hypertrophy and causes dwarfism with osteopenia. *J Cell Biol* 165: 433–445.
3. Li SW, Arita M, Kopen GC, Phinney DG, Prockop DJ. 1998. A 1,064 bp fragment from the promoter region of the Col11a2 gene drives lacZ expression not only in cartilage but also in osteoblasts adjacent to regions undergoing both endochondral and intramembranous ossification in mouse embryos. *Matrix Biol* 17: 213–221.
4. Logan M, Martin JF, Nagy A, Lobe C, Olson EN, Tabin CJ. 2002. Expression of Cre Recombinase in the developing mouse limb bud driven by a Prx1 enhancer. *Genesis* 33: 77–80.
5. Campbell MR, Cress CJ, Appleman EH, Jacenko O. 2004. Chicken collagen X regulatory sequences restrict transgene expression to hypertrophic cartilage in mice. *Am J Pathol* 164: 487–499.
6. Kim Y, Murao, H, Yamamoto K, Deng JM, Behringer RR, Nakamura T, Akiyama H. 2011. Generation of transgenic mice for conditional overexpression of Sox9. *J Bone Miner Metab* 29: 123–129.
7. Dacquin R, Starbuck M, Schinke T, Karsenty G. 2002. Mouse alpha(1)-collagen promoter is the best known promoter to drive efficient Cre recombinase expression in osteoblasts. *Dev Dyn* 224: 245–251.
8. Liu F, Woitge HW, Braut A, Kronenberg MS, Lichtler AC, Mina M, Kream BE. 2004. Expression and activity of osteoblast-targeted Cre recombinase transgenes in murine skeletal tissues. *Int J Dev Biol* 48: 645–653.
9. Scheller EL, Leinninger GM, Hankenson KD, Myers MG Jr, Krebsbach PH. 2011. Ectopic expression of Col2.3 and Col3.6 promoters in the brain and association with leptin signaling. *Cells Tissues Organs* 194: 268–273.
10. Zhang M, Xuan S, Bouxsein ML, von Stechow D, Akeno N, Faugere MC, Malluche H, Zhao G, Rosen CJ, Efstratiadis A, Clemens TL. 2002. Osteoblast-specific knockout of the insulin-like growth factor (IGF) receptor gene reveals an essential role of IGF signaling in bone matrix mineralization. *J Biol Chem* 277: 44005–44012.
11. Lu Y, Xie Y, Zhang S, Dusevich V, Bonewald LF, Feng JQ. 2007. DMP1-targeted Cre expression in odontoblasts and osteocytes. *J Dent Res* 86: 320–325.
12. Blitz E, Viukov S, Sharir A, Shwartz Y, Galloway JL, Pryce BA, Johnson RL, Tabin CJ. 2009. Bone ridge patterning during musculoskeletal assembly is mediated through SCX regulation of Bmp4 at the tendon-skeleton junction. *Dev Cell* 17: 861–873.
13. Ferron M, Vacher J. 2005. Targeted expression of Cre recombinase in macrophages and osteoclasts in transgenic mice. *Genesis* 41: 138–145.
14. Reddy SV, Hundley JE, Windle JJ, Alcantara O, Linn R, Leach RJ, Boldt DH, Roodman GD. 1995. Characterization of the mouse tartrate-resistant acid phosphatase (TRAP) gene promoter. *J Bone Miner Res* 10: 601–606.
15. Branda CS, Dymecki SM. 2004. Talking about a revolution: The impact of site-specific recombinases on genetic analyses in mice. *Dev Cell* 6: 7–28.
16. Liu Z, Shi W, Ji X, Sun C, Jee WS, Wu Y, Mao Z, Nagy TR, Li Q, Cao X. 2004. Molecules mimicking Smad1 interacting with Hox stimulate bone formation. *J Biol Chem* 279: 11313–11319.
17. Kobayashi T, Lyons KM, McMahon AP, Kronenberg HM. 2005. BMP signaling stimulates cellular differentiation at multiple steps during cartilage development. *Proc Natl Acad Sci U S A* 102: 18023–18027.
18. McVicar DW, Winkler-Pickett R, Taylor LS, Makrigiannis A, Bennett M, Anderson SK, Ortaldo JR. 2002. Aberrant DAP12 signaling in the 129 strain of mice: Implications for the analysis of gene-targeted mice. *J Immunol* 169: 1721–1728.
19. Rosen CJ, Beamer WG, Donahue LR. 2001. Defining the genetics of osteoporosis: Using the mouse to understand man. *Osteoporosis Int* 12: 803–810.
20. Birling MC, Gofflot F, Warot X. 2009. Site-specific recombinases for manipulation of the mouse genome. *Methods Mol Biol* 561: 245–263.
21. Niwa H, Yamamura K, Miyazaki J. 1991. Efficient selection for high-expression transfectants with a novel eukaryotic vector. *Gene* 108: 193–199.
22. Saunders TL. 2011. Inducible transgenic mouse models. *Methods Mol Biol* 693: 103–115.
23. Engin F, Yao Z, Yang T, Zhou G, Bertin T, Jiang MM, Chen Y, Wang L, Zheng H, Sutton RE, Boyce BF, Lee B. 2008. Dimorphic effects of Notch signaling in bone homeostasis. *Nat Med* 14: 299–305.
24. Feil R, Brocard J, Mascrez B, LeMeur M, Metzger D, Chambon P. 1996. Ligand-activated site-specific recombination in mice. *Proc Natl Acad Sci U S A* 93: 10887–10890.
25. Yu K, Xu J, Liu Z, Sosic D, Shao J, Olson EN, Towler DA, Ornitz DM. 2003. Conditional inactivation of FGF receptor 2 reveals an essential role for FGF signaling in the regulation of osteoblast function and bone growth. *Development* 130: 3063–3074.
26. Akiyama H, Kim JE, Nakashima K, Balmes G, Iwai N, Deng JM, Zhang X, Martin JF, Behringer RR, Nakamura T, de Crombrugghe B. 2005. Osteo-chondroprogenitor cells are derived from Sox9 expressing precursors. *Proc Natl Acad Sci U S A* 102: 14665–14670.
27. Chai Y, Jiang X, Ito Y, Bringas P Jr, Han J, Rowitch DH, Soriano P, McMahon AP, Sucov HM. 2000. Fate of the mammalian cranial neural crest during tooth and mandibular morphogenesis. *Development* 127: 1671–1679.
28. Nakamura E, Nguyen MT, Mackem S. 2006. Kinetics of tamoxifen-regulated Cre activity in mice using a cartilage-specific CreER(T) to assay temporal activity windows along the proximodistal limb skeleton. *Dev Dyn* 235: 2603–26012.
29. Grover J, Roughley PJ. 2006. Generation of a transgenic mouse in which Cre recombinase is expressed under control of the type II collagen promoter and doxycycline administration. *Matrix Biol* 25: 158–65.
30. Chen M, Lichtler AC, Sheu TJ, Xie C, Zhang X, O'Keefe RJ, Chen D. 2007. Generation of a transgenic mouse model with chondrocyte-specific and tamoxifen-

inducible expression of Cre recombinase. *Genesis* 45: 44–50.
31. Henry SP, Jang CW, Deng JM, Zhang Z, Behringer RR, de Crombrugghe B. 2009. Generation of aggrecan-CreERT2 knockin mice for inducible Cre activity in adult cartilage. *Genesis* 47: 805–814.
32. Rodda SJ, McMahon AP. 2006. Distinct roles for Hedgehog and canonical Wnt signaling in specification, differentiation and maintenance of osteoblast progenitors. *Development* 133: 3231–3244.
33. Kim JE, Nakashima K, de Crombrugghe B. 2004. Transgenic mice expressing a ligand-inducible cre recombinase in osteoblasts and odontoblasts: A new tool to examine physiology and disease of postnatal bone and tooth. *Am J Pathol* 165: 1875–1882.
34. Clausen BE, Burkhardt C, Reith W, Renkawitz R, Forster I. 1999. Conditional gene targeting in macrophages and granulocytes using LysMcre mice. *Transgenic Res* 8: 265–277.
35. Nakamura T, Imai Y, Matsumoto T, Sato S, Takeuchi K, Igarashi K, Harada Y, Azuma Y, Krust A, Yamamoto Y, Nishina H, Takeda S, Takayanagi H, Metzger D, Kanno J, Takaoka K, Martin TJ, Chambon P, Kato S. 2007. Estrogen prevents bone loss via estrogen receptor alpha and induction of Fas ligand in osteoclasts. *Cell* 130: 811–823.
36. Chiu WS, McManus JF, Notini AJ, Cassady AI, Zajac JD, Davey RA. 2004. Transgenic mice that express Cre recombinase in osteoclasts. *Genesis* 39: 178–1785.
37. Meyers EN, Lewandoski M, Martin GR. 1998. An Fgf8 mutant allelic series generated by Cre- and Flp-mediated recombination. *Nat Genet* 18: 136–141.
38. Brown D, Wagner D, Li X, Richardson JA, Olson EN. 1999. Dual role of the basic helix-loop-helix transcription factor scleraxis in mesoderm formation and chondrogenesis during mouse embryogenesis. *Development* 126: 4317–4329.
39. Murchison ND, Price BA, Conner DA, Keene DR, Olson EN, Tabin CJ, Schweitzer R. 2007. Regulation of tendon differentiation by scleraxis distinguishes force-transmitting tendons from muscle-anchoring tendons. *Development* 134: 2697–2708.
40. Huang WY, Aramburu J, Douglas PS, Izumo S. 2000. Transgenic expression of green fluorescence protein can cause dilated cardiomyopathy. *Nat Med* 6: 482–483.
41. Starnes LM, Downey CM, Boyd SK, Jirik FR. 2007. Increased bone mass in male and female mice following tamoxifen administration. *Genesis* 45: 229–35.
42. Soriano P. 1999. Generalized lacZ expression with the ROSA26 Cre reporter strain. *Nat Genet* 21: 70–71.
43. Jiang X, Iseki S, Maxson RE, Sucov HM, Morriss-Kay GM. 2002. Tissue origins and interactions in the mammalian skull vault. *Dev Biol* 241: 106–116.
44. Maes C, Kobayashi T, Selig MK, Torrekens S, Roth SI, Mackem S, Carmeliet G, Kronenberg HM. 2010. Osteoblast precursors, but not mature osteoblasts, move into developing and fractured bones along with invading blood vessels. *Dev Cell* 19: 329–344.
45. Shioi G, Kiyonari H, Abe T, Nakao K, Fujimori T, Jang CW, Huang CC, Akiyama H, Behringer RR, Aizawa S. 2011. A mouse reporter line to conditionally mark nuclei and cell membranes for in vivo live-imaging. *Genesis* 49(7): 570–578.
46. Day TF, Guo X, Garrett-Beal L, Yang Y. 2005. Wnt/beta-catenin signaling in mesenchymal progenitors controls osteoblast and chondrocyte differentiation during vertebrate skeletogenesis. *Dev Cell* 8: 739–750.
47. Monteiro RM, de Sousa Lopes SM, Korchynskyi O, ten Dijke P, Mummery CL. 2004. Spatio-temporal activation of Smad1 and Smad5 in vivo: Monitoring transcriptional activity of Smad proteins. *J Cell Sci* 117: 4653–63.
48. Blank U, Seto ML, Adams DC, Wojchowski DM, Karolak MJ, Oxburgh L. 2008. An in vivo reporter of BMP signaling in organogenesis reveals targets in the developing kidney. *BMC Dev Biol* 8: 86.
49. Blake JA, Bult CJ, Kadin JA, Richardson JE, Eppig JT. 2011. The Mouse Genome Database (MGD): Premier model organism resource for mammalian genomics and genetics. *Nucleic Acids Res* 39: D842–D848.
50. Ringwald M, Iyer V, Mason JC, Stone KR, Tadepally HD, Kadin JA, Bult CJ, Eppig JT, Oakley DJ, Briois S, Stupka E, Maselli V, Smedley D, Liu S, Hansen J, Baldock R, Hicks GG, Skarnes WC. 2011. The IKMC web portal: A central point of entry to data and resources from the International Knockout Mouse Consortium. *Nucleic Acids Res* 39: D849–855.
51. Nagy TR, Krzywanski D, Li J, Meleth S, Desmond R. 2002. Effect of group vs. single housing on phenotypic variance in C57BL/6J mice. *Obes Res* 10: 412–415.
52. Champy MF, Selloum M, Piard L, Zeitler V, Caradec C, Chambon P, Auwerx J. 2004. Mouse functional genomics requires standardization of mouse handling and housing conditions. *Mamm Genome* 15: 768–783.
53. Vu TH, Shipley JM, Bergers G, Berger JE, Helms JA, Hanahan D, Shapiro SD, Senior RM, Werb Z. 1998. MMP-9/gelatinase B is a key regulator of growth plate angiogenesis and apoptosis of hypertrophic chondrocytes. *Cell* 93: 411–422.

9

Animal Models: Allelic Determinants for BMD

Robert D. Blank

Introduction 76
Phenotypes 76
Themes of Existing Data 77

Future Directions 79
Acknowledgments 79
References 79

INTRODUCTION

There is a large body of work studying the impact of allelic variation on skeletal phenotypes in mice, but it is beyond the scope of this brief review to provide encyclopedic coverage of this literature. Rather, examples will be used to illustrate specific aspects of the investigations. The review will encompass two broad areas. First, the phenotypes studied in mice will be described. Second, general themes emerging from the data will be summarized. Finally, expected directions for future work will be explored.

It is important for readers who are not familiar with laboratory mice to recognize that there are many specialized genetic resources available for mouse work. The material below assumes cursory familiarity with specially bred mouse strains. I have previously reviewed these and genetically engineered mice for the nonspecialist [1]. It is also important to note at the outset that the emphasis here is on the variations that exist within established mouse strains and stocks, not the rapidly growing collection of knockout and knock-in mice with interesting skeletal phenotypes. These are obviously of great importance, but beyond the scope of this review.

PHENOTYPES

Most phenotypes of interest to bone biologists are "complex traits." By complex traits, we mean properties that are determined by the combined influence of multiple genes and environmental conditions. Often, but not invariably, complex traits are also quantitative, meaning that there is a continuum of possible trait values rather than a categorical classification of the trait. Genetic loci contributing to quantitative traits are called quantitative trait loci (QTL). Two properties of traits make them attractive objects of genetic study. The first is that they be highly heritable, or that much of the variation in the trait can be attributed to genetics. In human studies, this is determined by twin studies and recurrence rates. In mouse studies, this is determined in backcrosses, intercrosses, or analysis of specially bred mice such as recombinant inbred lines. The second criterion is that the assay used for phenotyping is precise, i.e., highly reproducible. Less precise assays, although not precluding study, make it more difficult because larger sample sizes are needed to distinguish the genetic signal from the noise attending a less precise, "sloppy" phenotyping assay. Another desirable, but not essential, property is that phenotyping be quick and technically easy to perform, as genetic mapping studies require large samples. In addition to the genetic desiderata for phenotypes, it is also important to consider their potential utility in guiding interpretation of the biology. Thus, phenotypes that are difficult to measure reliably may still be worth pursuing if they provide insight that can't be gained otherwise.

Bone Mineral Density (BMD): The contemporary era of mouse bone genetics began with the demonstration that inbred mouse strains differ in apparent volumetric BMD, using peripheral quantitative computed tomography (pQCT) scanning [2]. Subsequent studies [3, 4] have

---

Primer on the Metabolic Bone Diseases and Disorders of Mineral Metabolism, Eighth Edition. Edited by Clifford J. Rosen.
© 2013 American Society for Bone and Mineral Research. Published 2013 by John Wiley & Sons, Inc.

used both pQCT and dual energy X-ray absorptiometry (DXA) technologies to map quantitative trait loci for volumetric and areal BMD in mice, respectively.

Trabecular Structure: Improvements in microcomputed tomography (micro-CT) technology have allowed trabecular structure, a phenotype previously assayable only by histomorphometry, to be studied with sufficiently high throughput for genetic analyses [5]. While micro-CT makes it feasible to analyze hundreds of specimens for trabecular structure, in practice these phenotypes have been studied primarily to characterize congenic mice bred on the basis of another phenotype, as these investigations require far smaller sample sizes.

Dimensions: Both lengths and cross-sectional dimensions of long bones have been widely studied in mice [6–9], using a variety of methods. Measurement of bone geometry greatly enhances interpretation of biomechanical tests, and therefore is included in virtually all studies that include biomechanical phenotypes. It is worth noting that the robustness of the genetic signals obtained for bone geometry typically exceeds those for any other class of traits, manifested by higher linkage statistics for bone dimensions than for bone density or mechanical performance.

Mechanical Performance: The ability to perform mechanical tests represents one of the great advantages of the mouse model system. Various aspects of mechanical performance have been widely studied [7, 9–12]. There are several distinct aspects of mechanical performance, and while each has been mapped successfully, the robustness of these for genetic analysis varies. In general, both strength (yield or maximum load or stress) performs best among these, reflecting its better reproducibility relative to displacement (or strain) or energy (or toughness) [13]. It is important to recall that the domains of mechanical performance are distinct and that each is important to function *in vivo*.

Gene Expression: Relatively inexpensive microarrays allow genome-wide measurement of message abundance. The abundance of a specific message may itself be considered a phenotype, and mapped genetically [14]. When considered in conjunction with pleiotropic traditional phenotypes, expression QTLs (eQTLs) offer the promise of improved identification of the gene underlying the QTL [15]. Expression QTLs typically explain a larger fraction of the genetic variance in gene expression than phenotypic QTLs do, in some cases approaching 50%. Chief among the possible reasons for eQTLs impressive performance is that the phenotype being assessed—gene expression—is far less removed physiologically from the causative variant than is the case for a "clinical" or physiological phenotype, thus the influences of adaptation and feedback are restricted.

Dynamic Phenotypes: All the phenotypes considered above are "snapshots" obtained at a single time. It is also possible to map genes for change in a trait, either as a consequence of time or in response to an intervention. In mice, this has been done for postmaturity change in BMD [16] and modeling in response to mechanical loading [17].

Principal Components and Other Composite Phenotypes: The phenotypes considered above are not independent, as each has some biological overlap with at least one other trait, and it is useful to attempt to extract the unique information from each. One approach to this challenge is to apply principal component (PC) analysis to the data [18]. PC analysis transforms the original phenotypes into an equal number of orthogonal PCs, each defined as a specific linear combination of the original phenotypes. While PCs have been used to study bone phenotypes in mice [7, 19], they suffer from two important limitations. First, as they are linear combinations of directly measured phenotypes chosen algorithmically, they defy intuitive biological interpretation. Second, because the PCs are dependent on the specific phenotypes contributing to them, PCs studied by different investigative teams can't readily be compared.

The genetics of other composite phenotypes have been studied as well. A particularly interesting example is mandibular shape [20]. The method used in this study employs software that converts the positions of multiple anatomical landmarks to normalized distances from the corresponding mean landmark positions. This approach allows a useful mathematical framework for comparing morphological differences independently of size.

THEMES OF EXISTING DATA

Heritability: It has long been recognized that inbred mouse strains have distinctive, reproducible phenotypes. This is true for volumetric BMD [2], areal BMD [4], architectural features [21], various aspects of biomechanical performance [21, 22], and responsiveness to mechanical loading and unloading [23, 24]. In mice, all of these traits are highly heritable, a prerequisite for successfully mapping the responsible genes.

Covariation: Bone properties are interrelated. Larger bones are stronger bones, more heavily mineralized bones are stiffer bones, and areal BMD is dependent on both bone size and mineralization. It is natural to ask which traits used in mapping are more "important" or "informative." There is no simple answer to the question, and several research groups have studied the interrelationship among bone phenotypes in detail [12, 25, 26]. The simplest way to gauge the interdependence of phenotypes is to construct a correlation matrix, in which the correlation of each trait with every other trait is tabulated. These uniformly reveal that some traits are largely redundant with each other, e.g., stiffness and maximum load, and therefore provide largely overlapping information. This redundancy is a motivation for performing PC analysis, as discussed above. More interesting from a biological perspective is the observed negative correlation between mineralization and long bone cross-sectional size [12, 26]. The relationship provides powerful evidence supporting the mechanostat model of the skeleton [27, 28].

Pleiotropy: Pleiotropy is the property of a single genetic locus affecting multiple traits. Unsurprisingly, many bone genes and loci identified in mouse experiments display pleiotropy. Pleiotropy is best appreciated in experiments that have studied multiple traits in the same population. Pleiotropy is commonly observed among mechanical and geometric phenotypes at a single site [6, 29], as well as between measures of the same phenotype at different anatomical sites [8, 30].

One possible interpretation of the data is that the observed pleiotropy reflects covariation among the traits, as discussed in the previous section. The extreme version of this view implies that the traits being studied are only approximations of an underlying, fundamental set of traits that can't be measured directly with current assessment methods, but can be approximated by existing phenotyping assays.

An alternative interpretation instead focuses on the large mechanistic gap that exists between most phenotypes and the proteins encoded by the QTLs. According to this formulation, a primary task for bone biologists is to fill in how differences in the level of expression or activity of specific protein cascades through integrative physiology to impact the skeleton. This is a daunting task, which entails detailed studies of protein function in multiple tissues in isolation, and, additionally, of dissecting the cross-talk among tissues to understand the whole-organism physiology.

The fraction of variance explained by allelic variation serves as a rough surrogate for the extent to which the physiologic impact of the polymorphism can be buffered by compensatory mechanisms. As a trivial example, consider a polymorphism that affects the activity of albumin transcription. Such a polymorphism would be expected to have a large impact on the abundance of albumin mRNA, a lesser effect on serum albumin concentration, even lesser impact on serum calcium, and negligible impact on serum phosphate. At each step in this hierarchy, additional regulatory feedback loops contribute to the ultimate phenotype.

Clustering: QTLs identified in intercrosses of inbred mouse strains, while accounting for much phenotypic variability, are only imprecisely located. The next step in identifying the responsible gene(s) is often construction of nested congenic strains, in which a donor chromosome segment is crossed into a recipient strain. In this way, the phenotypic consequence of substituting only a short genomic region can be assessed independently of the genetic contributions of other QTLs. Recombination, or crossing over, within the donor segment allows more precise chromosomal localization of the QTL. The result of several such experiments [30–32] has been that the donor segment contained not one, but multiple, closely linked QTLs for bone phenotypes. Physical linkage of genes contributing to a common set of functions is an evolutionary mechanism for keeping compatible alleles together (for a review, see Ref. 33). This mechanism is believed to underlie the emergence of sex chromosomes and may be operating with regard to the skeleton as well.

Sex Limitation: Multiple research groups have reported QTLs that affect only males, only females, or have significantly greater effects in one sex than the other [9, 32, 34]. The most important implication of these findings is that the genes underlying such QTLs are involved in pathways that are include sex hormone signaling or that display cross-talk with sex hormone signaling pathways. The additional mechanistic insight arising from sex specificity is a powerful tool for identifying the causative gene and an equally powerful tool for investigating the more general question of how sex hormones act on the skeleton.

Intersite Discordance: One of the most important insights gained from mice is that the genetic bases of cortical and trabecular bone properties are distinct [5], even though some QTLs affect both the cortical and trabecular compartments. In contrast, long bone length QTLs tend to affect multiple sites similarly [8]. This finding demonstrates that cortical bone and trabecular bone are subject to different physiological feedback. A common interpretation is that cortical bone is more responsive to regulation related to mechanical loading, while trabecular bone is more responsive to metabolic signals.

Concordance with Human Data: Not only are the sequences of genes conserved across species, but so are linkage relationships. Therefore, it has been possible to develop a detailed comparative genetic map of the human and the mouse. If one knows the location of a gene in one organism, then its position is also known in the other. Armed with this knowledge, it is possible to ask whether the genes that contribute to bone properties in one species also have an impact in the other. There is substantial overlap between mouse and human genes for BMD [35]. This analysis did not include any of the BMD-related phenotypes that have also been studied in mice, as most of these have not been amenable to measurement in humans.

The mouse data and the human data are complementary in important ways. The human data allow identification of specific DNA sequence variants that are associated with BMD, either because they have a functional significance themselves or because they are in linkage disequilibrium with functional variants. However, the analysis is limited to common sequence variants, so that the contribution of rare variants to BMD is not addressed. Consequently, human studies account for only a very small fraction of the phenotypic variability. In mice, the fraction of the variance explained is about five- to tenfold greater than in humans, but the localization of the responsible loci is much less precise than in humans. Moreover, the genetic linkage studies in mice are performed in experimental crosses in which only a small number of alleles are considered. The greater fraction of the variance explained in the mouse studies is likely due in part to the capturing of all the contributions of genetic differences to the phenotype, unlike the situation in the human studies. In addition, since the mouse linkage peaks include a greater fraction of the genome, they may also contain multiple genes, as discussed above. Nevertheless, the high concordance between the human and mouse QTL for BMD is a critical validation of the mouse as a model organism for studying bone genetics.

The above notwithstanding, it is important to recognize that mice are not a perfect model, as differences of body size and other features result in important differences between humans and mice. One obvious consequence of the body size difference is the absence of Haversian systems in mouse cortical bone. Thus, there are aspects of human bone architecture that can't be studied in mice.

FUTURE DIRECTIONS

In addition to the recognized limitations of mice in reproducing human bone biology, it is also important to recognize that there are important genetic limitations that attend the use of mouse models. These include the extent to which studies have been performed in highly inbred animals, the limited genetic variation being studied, and the poor genetic resolving power achievable in most experimental crosses, which only extend two or three generations.

In order to overcome these limitations, the collaborative cross is an ongoing effort produce a mouse resource for genetic mapping that overcomes the genetic limitations of currently available mouse resources [36]. The objective is to generate a large series of eight progenitor recombinant inbred strains. The progenitors are chosen to capture more than 80% of the known interstrain allelic diversity among present inbred mouse strains. F1 animals produced from such strains are isogenic, outbred, and possess haplotype blocks whose length approaches those of natural outbred populations. Combinational mating among these inbred strains will allow enough distinct genotypes to apply genome-wide association methods as are used in human studies. The short haplotype blocks will allow localization of functionally important genetic variation to short genomic segments. Thus, the advantages of isogeneity will be maintained, including the need to perform genotyping only once and the ability to estimate phenotype from the pooling of multiple animals sharing a common genotype. The first collaborative cross RI strains have been bred, and hundreds more are approaching completion. These strains will provide a powerful resource for future mouse studies of bone.

Another approach to the limited resolving power of short mouse breeding experiments is to study advanced intercross mice [37]. With every generation of breeding, additional recombination, or crossing over, occurs. This shortens the lengths of chromosome segments that have been inherited from a specific ancestor. For this reason, advanced intercross lines have improved genetic resolving power relative to F2 mice, with genetic resolving power increasing as a function of the number of generations of breeding. The challenge of using advanced intercross mice, however, is that statistical analysis of the data requires accounting for family structure, and is therefore more computationally difficult than for F2 or backcross experiments. Some work using advanced intercross populations to study skeletal phenotypes have already been published [30]. These will no doubt increase in frequency in the years ahead.

Of course, the usefulness of mouse models in advancing the study of bone genetics in the future will depend most on the ability to integrate functional, structural, and mechanical elements in innovative experiments. The genetic tools available in the mouse will allow resourceful investigators to continue to learn new biology that can be applied to improving the human condition.

ACKNOWLEDGMENTS

RDB gratefully acknowledges support from NIH grant AR54753.

REFERENCES

1. Blank RD. 2010. Mouse genetics: Breeding strategies and genetic engineering. In: *Up To Date*.Waltham: Wolters Kluwer.
2. Beamer WG, Donahue LR, Rosen CJ, Baylink, DJ. 1996. Genetic variability in adult bone density among inbred strains of mice. *Bone* 18(5): 397–403.
3. Beamer WG, Shultz KL, Churchill GA, Frankel WN, Baylink DJ, Rosen CJ, and Donahue LR. 1999. Quantitative trait loci for bone density in C57BL/6J and CAST/EiJ inbred mice. *Mamm Genome* 10(11): 1043–9.
4. Klein RF, Mitchell SR, Phillips TJ, Belknap JK, Orwoll ES. 1998. Quantitative trait loci affecting peak bone mineral density in mice. *J Bone Miner Res* 13(11): 1648–56.
5. Bouxsein ML, Uchiyama T, Rosen CJ, Shultz KL, Donahue LR, Turner CH, Sen S, Churchill GA, Muller R, Beamer WG. 2004. Mapping quantitative trait loci for vertebral trabecular bone volume fraction and microarchitecture in mice. *J Bone Miner Res* 19(4): 587–99.
6. Volkman SK, Galecki AT, Burke DT, Miller RA, Goldstein SA. 2004. Quantitative trait loci that modulate femoral mechanical properties in a genetically heterogeneous mouse population. *J Bone Miner Res* 19(9): 1497–505.
7. Koller DL, Schriefer J, Sun Q, Shultz KL, Donahue LR, Rosen CJ, Foroud T, Beamer WG, Turner CH. 2003. Genetic effects for femoral biomechanics, structure, and density in C57BL/6J and C3H/HeJ inbred mouse strains. *J Bone Miner Res* 18(10): 1758–65.
8. Kenney-Hunt JP, Wang B, Norgard EA, Fawcett G, Falk D, Pletscher LS, Jarvis JP, Roseman C, Wolf J, Cheverud JM. 2008. Pleiotropic patterns of quantitative trait loci for 70 murine skeletal traits. *Genetics* 178(4): 2275–88.

9. Saless N, Litscher SJ, Lopez Franco GE, Houlihan MJ, Sudhakaran S, Raheem KA, O'Neil TK, Vanderby R, Demant P, Blank RD. 2009. Quantitative trait loci for biomechanical performance and femoral geometry in an intercross of recombinant congenic mice: Restriction of the Bmd7 candidate interval. *FASEB J* 23(7): 2142–54.

10. Li X, Masinde G, Gu W, Wergedal J, Hamilton-Ulland M, Xu S, Mohan S, Baylink DJ. 2002. Chromosomal regions harboring genes for the work to femur failure in mice. *Funct Integr Genomics* 1(6): 367–74.

11. Li X, Masinde G, Gu W, Wergedal J, Mohan S, Baylink DJ. 2002. Genetic dissection of femur breaking strength in a large population (MRL/MpJ x SJL/J) of F2 Mice: Single QTL effects, epistasis, and pleiotropy. *Genomics* 79(5): 734–40.

12. Saless N, Lopez Franco GE, Litscher S, Kattappuram RS, Houlihan MJ, Vanderby R, Demant P, Blank RD. 2010. Linkage mapping of femoral material properties in a reciprocal intercross of HcB-8 and HcB-23 recombinant mouse strains. *Bone* 46(5): 1251–9.

13. Leppanen OV, Sievanen H, Jarvinen TL. 2008. Biomechanical testing in experimental bone interventions—May the power be with you. *J Biomech* 41(8): 1623–31.

14. Cookson W, Liang L, Abecasis G, Moffatt M, Lathrop M. 2009. Mapping complex disease traits with global gene expression. *Nat Rev Genet* 10(3): 184–94.

15. Farber CR, van Nas A, Ghazalpour A, Aten JE, Doss S, Sos B, Schadt EE, Ingram-Drake L, Davis RC, Horvath S, Smith DJ, Drake TA, Lusis AJ. 2009. An integrative genetics approach to identify candidate genes regulating BMD: Combining linkage, gene expression, and association. *J Bone Miner Res* 24(1): 105–16.

16. Szumska D, Benes H, Kang P, Weinstein RS, Jilka RL, Manolagas SC, Shmookler Reis RJ. 2007. A novel locus on the X chromosome regulates post-maturity bone density changes in mice. *Bone* 40(3): 758–66.

17. Kesavan C, Mohan S, Srivastava AK, Kapoor S, Wergedal JE, Yu H, Baylink DJ. 2006. Identification of genetic loci that regulate bone adaptive response to mechanical loading in C57BL/6J and C3H/HeJ mice intercross. *Bone* 39(3): 634–43.

18. Pearson K. On lines and planes of closest fit to systems of points in space. 1901. *Philosophical Magazine* 2: 559–572.

19. Saless N, Litscher SJ, Vanderby R, Demant P, Blank RD. 2011. Linkage mapping of principal components for femoral biomechanical performance in a reciprocal HCB-8 x HCB-23 intercross. *Bone* 48(3): 647–53.

20. Klingenberg CP, Leamy LJ, Cheverud JM. 2004. Integration and modularity of quantitative trait locus effects on geometric shape in the mouse mandible. *Genetics* 166(4): 1909–21.

21. Turner CH, Hsieh YF, Muller R, Bouxsein ML, Baylink DJ, Rosen CJ, Grynpas MD, Donahue LR, Beamer WG. 2000. Genetic regulation of cortical and trabecular bone strength and microstructure in inbred strains of mice. *J Bone Miner Res* 15(6): 1126–31.

22. Jepsen KJ, Pennington DE, Lee YL, Warman M, Nadeau J. 2001. Bone brittleness varies with genetic background in A/J and C57BL/6J inbred mice. *J Bone Miner Res* 16(10): 1854–62.

23. Akhter MP, Cullen DM, Pedersen EA, Kimmel DB, Recker RR. 1998. Bone response to in vivo mechanical loading in two breeds of mice. *Calcif Tissue Int* 63(5): 442–9.

24. Judex S, Donahue LR, Rubin C. 2002. Genetic predisposition to low bone mass is paralleled by an enhanced sensitivity to signals anabolic to the skeleton. *FASEB J* 16(10): 1280–2.

25. Jepsen KJ, Akkus OJ, Majeska RJ, Nadeau JH. 2003. Hierarchical relationship between bone traits and mechanical properties in inbred mice. *Mamm Genome* 14(2): 97–104.

26. Jepsen KJ, Hu B, Tommasini SM, Courtland HW, Price C, Terranova CJ, Nadeau JH. 2007. Genetic randomization reveals functional relationships among morphologic and tissue-quality traits that contribute to bone strength and fragility. *Mamm Genome* 18(6–7): 492–507.

27. Frost HM. 2000. The Utah paradigm of skeletal physiology: An overview of its insights for bone, cartilage and collagenous tissue organs. *J Bone Miner Metab* 18(6): 305–16.

28. Frost HM. 2001. From Wolff's law to the Utah paradigm: Insights about bone physiology and its clinical applications. *Anat Rec* 262(4): 398–419.

29. Volkman SK, Galecki AT, Burke DT, Paczas MR, Moalli MR, Miller RA, Goldstein SA. 2003. Quantitative trait loci for femoral size and shape in a genetically heterogeneous mouse population. *J Bone Miner Res* 18(8): 1497–505.

30. Norgard EA, Jarvis JP, Roseman CC, Maxwell TJ, Kenney-Hunt JP, Samocha KE, Pletscher LS, Wang B, Fawcett GL, Leatherwood CJ, Wolf JB, Cheverud JM. 2009. Replication of long-bone length QTL in the F9-F10 LG,SM advanced intercross. *Mamm Genome* 20(4): 224–35.

31. Beamer WG, Shultz KL, Ackert-Bicknell CL, Horton LG, Delahunty KM, Coombs HF, 3rd, Donahue LR, Canalis E, Rosen CJ. 2007. Genetic dissection of mouse distal chromosome 1 reveals three linked BMD QTLs with sex-dependent regulation of bone phenotypes. *J Bone Miner Res* 22(8): 1187–96.

32. Edderkaoui B, Baylink DJ, Beamer WG, Shultz KL, Wergedal JE, Mohan S. 2007. Genetic regulation of femoral bone mineral density: Complexity of sex effect in chromosome 1 revealed by congenic sublines of mice. *Bone* 41(3): 340–5.

33. Charlesworth B. 2002. The evolution of chromosomal sex determination. *Novartis Found Symp* 244: 207–19; discussion 220–4, 253–7.

34. Orwoll ES, Belknap JK, Klein RF. 2001. Gender specificity in the genetic determinants of peak bone mass. *J Bone Miner Res* 16(11): 1962–71.

35. Ackert-Bicknell CL, Karasik D, Li Q, Smith RV, Hsu YH, Churchill GA, Paigen BJ, Tsaih SW. 2010. Mouse BMD quantitative trait loci show improved concordance with human genome-wide association loci when recal-

culated on a new, common mouse genetic map. *J Bone Miner Res* 25(8): 1808–20.

36. Churchill GA, Airey DC, Allayee H, Angel JM, Attie AD, Beatty J, Beavis WD, Belknap JK, Bennett B, Berrettini W, Bleich A, Bogue M, Broman KW, Buck KJ, Buckler E, Burmeister M, Chesler EJ, Cheverud JM, Clapcote S, Cook MN, Cox RD, Crabbe JC, Crusio WE, Darvasi A, Deschepper CF, Doerge RW, Farber CR, Forejt J, Gaile D, Garlow SJ, Geiger H, Gershenfeld H, Gordon T, Gu J, Gu W, de Haan G, Hayes NL, Heller C, Himmelbauer H, Hitzemann R, Hunter K, Hsu HC, Iraqi FA, Ivandic B, Jacob HJ, Jansen RC, Jepsen KJ, Johnson DK, Johnson TE, Kempermann G, Kendziorski C, Kotb M, Kooy RF, Llamas B, Lammert F, Lassalle JM, Lowenstein PR, Lu L, Lusis A, Manly KF, Marcucio R, Matthews D, Medrano JF, Miller DR, Mittleman G, Mock BA, Mogil JS, Montagutelli X, Morahan G, Morris DG, Mott R, Nadeau JH, Nagase H, Nowakowski RS, O'Hara BF, Osadchuk AV, Page GP, Paigen B, Paigen K, Palmer AA, Pan HJ, Peltonen-Palotie L, Peirce J, Pomp D, Pravenec M, Prows DR, Qi Z, Reeves RH, Roder J, Rosen GD, Schadt EE, Schalkwyk LC, Seltzer Z, Shimomura K, Shou S, Sillanpaa MJ, Siracusa LD, Snoeck HW, Spearow JL, Svenson K, Tarantino LM, Threadgill D, Toth LA, Valdar W, de Villena FP, Warden C, Whatley S, Williams RW, Wiltshire T, Yi N, Zhang D, Zhang M, Zou F. 2004. The Collaborative Cross, a community resource for the genetic analysis of complex traits. *Nat Genet* 36(11): 1133–7.

37. Darvasi A, Soller M. 1995. Advanced intercross lines, an experimental population for fine genetic mapping. *Genetics* 141(3): 1199–207.

10

Neuronal Regulation of Bone Remodeling

Florent Elefteriou and Gerard Karsenty

Introduction 82
An Adipocyte-Driven Central Control of Bone Mass 82
Anatomical, Cellular, and Molecular Bases of the Central Control of Bone Mass 83
Brainstem-Derived Serotonin Controls Bone Formation 83
The Sympathetic Nervous System Is a Peripheral Mediator of Leptin Signaling in the Brain 83

Cart Is a Mediator of Leptin Regulation of Bone Mass 85
Both Arms of the Autonomic Nervous System Control Bone Remodeling 85
The Growing List of Skeletogenic Neuropeptides 85
What about Sensory Nerves? 86
Looking Forward 86
References 86

INTRODUCTION

Bone remodeling, the process whereby bone is constantly renewed during adulthood, is a prototypical homeostatic process in which the destruction of bone by osteoclasts is followed by *de novo* bone formation by osteoblasts to maintain bone mass. Given its importance, it is no surprise that this physiological function is regulated by local, endocrine, and neuronal factors. This review will detail the mechanisms underlying the central control of bone mass.

AN ADIPOCYTE-DRIVEN CENTRAL CONTROL OF BONE MASS

Many observations suggest a link between food, i.e., energy intake, and bone mass, as well as between reproduction and bone mass. For instance, obese patients are protected from osteoporosis, and osteoporosis typically follows gonadal deficiency. Faced with these and other observations, we hypothesized that there is a coordinated regulation, endocrine in nature, of bone mass, energy metabolism, and reproduction [1]. Since the regulation of reproduction and body weight include a central component, this hypothesis implied from its inception the existence of a central control of bone remodeling. Again, the homeostatic nature of bone remodeling adds conceptual credence to this hypothesis since the regulation of most homeostatic functions includes a central component.

Our working hypothesis was verified by showing that mice lacking leptin (*ob/ob* mice), an adipocyte-derived hormone that inhibits appetite, exhibited increased energy expenditure and bone mass along with enhanced reproduction. [1] Moreover, intracerebroventricular (ICV) infusion of leptin into the third ventricle of *ob/ob* mice corrected fully their bone phenotype and established the existence of a central control of bone mass. This experiment is particularly telling because of its genetic nature. In effect, leptin was reintroduced, in the brain only, into animals that were lacking it. At no point could leptin be detected in the blood of these animals, yet their high bone mass was completely corrected. Because this ICV infusion corrected fully the high bone mass phenotype of the *ob/ob* mice, this experiment not only established the existence of a leptin-dependent central regulation of bone formation but also showed that leptin had no other mechanism of regulating bone formation *in vivo*; otherwise, the rescue would never have been complete. This latter point was also established genetically. Indeed,

---

Primer on the Metabolic Bone Diseases and Disorders of Mineral Metabolism, Eighth Edition. Edited by Clifford J. Rosen.
© 2013 American Society for Bone and Mineral Research. Published 2013 by John Wiley & Sons, Inc.

mice lacking the signaling form of the leptin receptor in osteoblasts have a normal bone mass, whereas mice lacking this receptor in neurons only have a high bone mass [2]. Soon after this initial discovery was made in mice, the existence of a central control of bone mass was extended to rats, sheep, and humans [1, 3, 4]. In line with this work, the increased bone density and decreased fracture risk associated with obese patients is explained by the fact that obesity is a state of leptin resistance.

ANATOMICAL, CELLULAR, AND MOLECULAR BASES OF THE CENTRAL CONTROL OF BONE MASS

It is fair to say that the notion that bone mass was regulated centrally was unexpected. As a result, it was important to establish firm molecular bases for it. The questions to address here were: Where does leptin act in the brain? What is (are) the mediator(s) linking leptin signaling in the brain to bone cells?

Studies performed in wild-type (WT) and ob/ob mice, in which specific neuronal populations were lesioned and followed by leptin ICV infusions, identified the ventromedial hypothalamic (VMH) neurons as constituting a major leptin-sensitive central center regulating bone mass [5]. Destruction of VMH neurons (that highly express *ObRb*) indeed recapitulated the high bone formation/high bone mass phenotype of the ob/ob mice, and leptin ICV infusion could no longer rescue the high bone mass phenotype of VMH-lesioned ob/ob mice. Remarkably, genetic cell-specific ablation of *ObRb* in VMH neurons did not affect bone mass nor did it affect appetite or reproduction [6]. These observations suggested that leptin requires the integrity of VMH neurons to achieve its functions on bone but needs not to bind to them. In other words, leptin must act elsewhere in the brain to regulate the synthesis of neuromediator(s) that will then act in the hypothalamus.

BRAINSTEM-DERIVED SEROTONIN CONTROLS BONE FORMATION

Serotonin, an indoleamine produced by tryptophan hydroxylase 1 (*Tph1*) in enterochromaffin cells of the duodenum and by tryptophan hydroxylase 2 (*Tph2*) in serotonergic neurons of the brainstem, does not cross the blood–brain barrier. Hence, it is *de facto* a molecule with two distinct functional identities depending on its site of synthesis: a hormone when made in the gut and a neurotransmitter when made in the brain. Besides its well-known role in influencing cognitive functions, brainstem-derived serotonin emerged recently as a regulator of three homeostatic functions: bone remodeling, appetite, and energy expenditure. Lack of brain serotonin in $Tph2^{-/-}$ mice induced a low bone mass phenotype without any change in serum serotonin level, indicating that brainstem-derived serotonin is a major central stimulator of bone formation. Brain-derived serotonin is the first, and so far the only, neuropeptide regulating bone mass; indeed all other molecules incriminated (see below) can cross the blood–brain barrier.

Further studies indicated that *Tph2*-expressing neurons express *ObRb* and that leptin significantly decreases *Tph2* expression and action potential frequency in serotonergic neurons of WT mice, but not in serotonergic neurons of mice lacking *ObRb* in *Tph2*-expressing neurons ($ObRb_{SERT}^{-/-}$ mice). Accordingly, $ObRb_{SERT}^{-/-}$ mice lacking *ObRb* in serotonergic neurons of the brainstem developed a high bone mass phenotype. Axon guidance experiments showed the existence of a connection between VMH and brainstem neurons and the analysis of genetic mouse models in which *Htr2c*, the gene encoding the most highly expressed serotonin receptor in VMH neurons, was deleted, showed that brainstem-derived serotonin acts on VMH neurons via *Htr2c*, to favor bone mass accrual [7]. In VMH neurons, brain-derived serotonin uses a calmodulin kinase (CaMK) cascade involving CaMKKb and CaMKIV to phosphorylate the transcription factor CREB (cAMP response element binding protein) [8]. Together, these studies demonstrate that in order to regulate bone mass accrual, appetite, and energy expenditure, leptin needs to inhibit the synthesis and release of serotonin by *Tph2*-expressing neurons of the brainstem. They also establish that the serotonergic neuronal circuitry exerts a more fundamental influence on several homeostatic functions than previously thought.

THE SYMPATHETIC NERVOUS SYSTEM IS A PERIPHERAL MEDIATOR OF LEPTIN SIGNALING IN THE BRAIN

The nature of the downstream mechanism whereby brainstem and hypothalamic neurons regulate the activity of distant cells, such as osteoblasts and osteoclasts, has been determined through genetic and pharmacological approaches. Early on, a key experiment for the field pointed toward a neuronal mediation of leptin central regulation of bone mass. This experiment is a parabiosis, which consists of surgically connecting the blood circulations of two ob/ob mice. In one of them, leptin was then infused centrally at a dose known to not cross the blood–brain barrier. This mouse became the control of the experiment and the question was: Does the contralateral mouse lose bone? It actually did not, thus pointing toward a neural relay of leptin action. This experiment—and the fact that patients with reflex sympathetic dystrophy, a disease characterized by focal high sympathetic tone, associated with bone loss—led us to hypothesize that the sympathetic nervous system, whose activity is regulated by leptin, could be this neural mediator. Analysis of mice with autonomic dysfunction confirmed the existence of a sympathetic link between the brain and osteoblasts. For instance, mice lacking dopamine beta-hydroxylase (*Dbh*), the enzyme generating norepinephrine (NE), display a late onset increase in bone mass, indicating that the sympathetic nervous system inhibits bone formation [5]. Supporting this result, mice and rats treated with the nonselective beta-adrenergic receptor (AR) blocker propranolol exhibit a high bone mass, whereas mice treated with the nonselective beta-agonist isoproterenol or the beta2AR-selective agonists clenbuterol or salbutamol exhibit a low bone mass [5, 9–12]. More to the point, mice lacking *Adrβ2* (the gene encoding the beta2AR) globally or selectively in mature osteoblasts, display an increase in bone formation and a decrease in bone resorption, leading to a high bone mass phenotype [13, 14]. Similarly, mice lacking adenylyl cyclase 5, a downstream mediator of beta2AR signaling, display positive changes in bone mass and biomechanical properties [15]. Importantly, leptin ICV infusion failed to decrease bone mass in beta2AR-deficient mice, demonstrating not only that the sympathetic nervous system (SNS), via the beta2AR, mediates leptin regulation of bone formation, but also that there is no other mediator of this function [13, 14].

The increased number of osteoblasts and increased bone formation rate observed in beta2AR-deficient mice suggested that the autonomic nervous system inhibits osteoblast proliferation and function. Surprisingly, studies of mice lacking, in osteoblasts only, critical determinants of the clock machinery, including *Per1* and *-2* or *Cry1* and *-2-*, revealed that peripheral clock genes mediate the effect of the sympathetic nervous system on osteoblast proliferation [16]. These studies indicated that osteoblasts express peripheral subordinate clock genes that mediate leptin-dependent sympathetic inhibition of bone formation by suppressing the expression of *G1 cyclins* and their proliferation. Further studies revealed that osteoblastic clock genes are under the control of beta2AR signaling and uncovered that leptin and sympathetic signaling exert a countervailing and stimulatory effect on osteoblast proliferation through the AP-1 family of transcription factors [16]. These results are in agreement with the known daily variation in bone marrow cell proliferation, collagen synthesis, and turnover markers [17, 18]. More recently, the analysis of osteoblast-specific mutant mice indicated that CREB and cMyc are two major transcription factors mediating the effect of beta2AR stimulation on osteoblast proliferation [14].

Although osteoclasts express the beta2AR [19], the effect of the sympathetic nervous system on osteoclast differentiation is indirect and mediated by osteoblasts, via beta2AR and stimulation of the expression of the osteoclastogenic factor receptor activator of nuclear factor-κB ligand (RANKL) [13]. The transcription factor ATF4, involved in osteoblast differentiation [20], was identified as a target of beta2AR signaling. ATF4 becomes phosphorylated by PKA following beta2AR stimulation

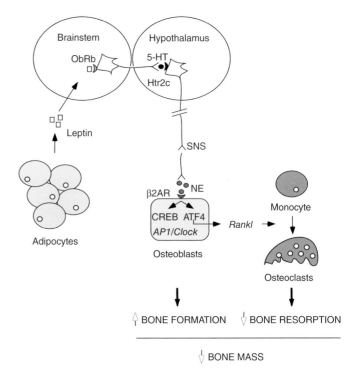

Fig. 10.1. Bone remodeling is under the control of brainstem and hypothalamic serotonergic neurons. Leptin-responsive (ObRb+) neurons in the brainstem, via serotonin signaling in hypothalamic centers, modulate sympathetic outflow to the skeletal system. Norepinephrine (NE) released by sympathetic nerve signals, via the beta2AR in osteoblasts, stimulates bone formation and indirectly inhibits bone resorption. This occurs by controlling AP1/Clock and Rankl expression in a CREB and ATF4-dependent manner, respectively.

and directly binds to the *Rankl* promoter to activate *Rankl* transcription [13, 14] (Fig. 10.1).

Is the leptin-dependent anti-osteogenic function of the sympathetic nervous system on bone remodeling conserved between mice and humans? To put this question in an evolutionary perspective, it is important to state that no molecule identified as a hormone in mice has lost this characteristic in humans. What has been established is that leptin is a negative independent predictor of bone mineral density (BMD) in adults [21]. Likewise, the majority of the available retrospective studies suggests a protective effect of beta-blockers on fracture risk, in agreement with the mouse and rat data [22–26]. Some reports, however, did not find any significant relationship between beta-blocker users and fracture risk [27, 28]. Therefore, long-term prospective randomized studies will be required to demonstrate unequivocally a potential protective effect of beta-blockers on fracture risk in humans. Another question is whether regulatory endogenous mechanisms that involve sympathetic signaling are in play in known pathological bone diseases. Sympathetic activation during aging or chronic stress and severe depression, for instance, might contribute to changes in bone mass observed under these conditions [29–34]. The response of osteoblasts to sympathetic signals might also be relevant to such conditions, as suggested by the stimulatory effect of glucocorticoids on beta2AR signaling [35].

CART IS A MEDIATOR OF LEPTIN REGULATION OF BONE MASS

An approach based on a screening for genes whose expression is regulated by leptin but that do not regulate appetite or reproduction led to the discovery that *Cart* (cocaine and amphetamine-regulated transcript) mediates leptin inhibition of bone resorption [13]. CART is a neuropeptide broadly expressed in the central nervous system (CNS), including hypothalamic neurons, as well as in peripheral tissues such as the pancreas [36]. The importance of CART in regulating bone remodeling and the hypothalamic nature of this regulation is supported by animal models characterized by low or high hypothalamic *Cart* expression and significant bone phenotypes. Low *Cart* expression in *ob/ob* mice accompanies the increased bone resorption observed in these mice, whereas increased hypothalamic *Cart* expression in obese and hyperleptinemic *Mc4r*-deficient mice correlates with their low bone resorption and high bone mass [13]. Furthermore, lack of one *Cart* copy in *Mc4r*-deficient mice rescues their resorption phenotype, indicating that an increased level of *Cart* in *Mc4r*-deficient mice causes their high bone mass phenotype [37]. The observation that an increase of CART serum concentration causes a high bone mass and rescues the low bone mass of $Cart^{-/-}$ mice suggests that CART regulates bone remodeling as a circulating molecule rather than a central neuropeptide. Regardless of its mode of action, this CART-mediated regulatory loop of bone resorption is conserved in humans, as individuals lacking *MC4R* have increased CART serum levels and increased BMD associated with decreased bone resorption [13, 37, 38]. The molecular mode of action of CART on bone resorption is not yet defined.

BOTH ARMS OF THE AUTONOMIC NERVOUS SYSTEM CONTROL BONE REMODELING

The crucial role of the sympathetic nervous system (SNS) in the regulation of bone remodeling leads to the question of the possible role that the other arm of the autonomic nervous system, the parasympathetic nervous system (PNS), may have in this context. The main neurotransmitter used by PNS endings is acetylcholine, which binds to five types of muscarinic acetylcholine receptors. Among four receptors tested, the M3 receptor (M3R) was shown to be the only muscarinic receptor subtype influencing bone remodeling [39]. *M3R-/-* mice indeed displayed a low bone mass phenotype with cellular changes similar to those observed in mice with increased SNS activity. The weak expression of this receptor in bone cells, as compared to central neurons, led to the hypothesis that M3R signaling in the CNS regulates bone remodeling, a hypothesis that was confirmed using neuronal and osteoblast-specific inactivation of *M3R-/-*. Furthermore, studies of double mutant mouse strains lacking one copy of *Adrβ2* and *M3R* supported the model whereby M3R fulfills this function by decreasing sympathetic activity. These observations broadened the importance of the autonomic nervous system in the regulation of bone remodeling and suggest that the PNS acts as a break on the SNS to dampen its anti-osteogenic influence on bone remodeling, in a similar fashion to other organs.

THE GROWING LIST OF SKELETOGENIC NEUROPEPTIDES

Neuropeptide Y (NPY) is a well-known target of leptin in the hypothalamus, with the potential to act through at least five Y-receptors (Y1, Y2, Y4, Y5, and Y6) that differ in their distribution in the CNS and the periphery. All of these receptors are expressed in the hypothalamus and several respond to other ligands, including peptide YY and pancreatic polypeptide (PP). Strong expression of NPY is found in the hypothalamic area, where NPY fibers project from the arcuate nucleus. NPY expression is inhibited by leptin, as shown by the high NPY content in the hypothalamus of *ob/ob* mice. Exogenous NPY ICV administration, however, induces bone loss, indicating that leptin and NPY do not antagonize each other's function in the control of bone formation, as they appear to do in the control of body weight [40, 41].

In support of a role of NPY receptor signaling in the regulation of bone formation, mice lacking the Y2 receptor ($Y2^{-/-}$ mice) display an increase in trabecular bone mass that can be phenocopied by hypothalamic-specific deletion of *Y2*, indicating that Y2 signaling in the hypothalamus inhibits bone formation [42, 43]. Interestingly, inactivation of both *Y2* and *Y4* receptors led to a further increase in bone mass compared to *Y2* alone, which was accompanied by reduced serum leptin level, suggesting a *Y4*-mediated additional effect of leptin deficiency on the $Y2^{-/-}$ bone phenotype [44]. Thus, Y2 receptor signaling clearly regulates, via a hypothalamic relay, the bone formation arm of the bone remodeling process. NPY and the Y1 receptor are expressed by bone cells, and lack of Y1R promotes the differentiation of mesenchymal progenitor cells as well as the activity of mature osteoblasts, constituting a likely mechanism for the high-bone-mass phenotype evident in $Y1R^{-/-}$ mice [41, 45, 46]. NPY thus may function as both a central neuromediator and a peripheral autocrine/paracrine factor in bone.

Neuropeptide U (NMU) is another neuropeptide whose expression is regulated by leptin and that may be involved in the regulation of bone remodeling. NMU is expressed in hypothalamic neurons as well as in the small intestine, and its functions include the regulation of appetite and sympathetic activation [47], as demonstrated by the obesity of NMU-deficient mice [48]. These mice display a high-bone-mass phenotype caused by an increase in bone formation [49]. The fact that the receptors for NMU, NMU1R and NMU2R, are not detectable in bone, the absence of effect of NMU treatment on osteoblast differentiation in vitro, the expression of NMU2R in hypothalamic neurons, and, most importantly, the rescue of the high bone mass of NMU-deficient mice by ICV infusion of NMU, collectively demonstrate that NMU acts via a central relay to regulate bone remodeling. That NMU ICV infusion could decrease the high bone mass of leptin-deficient mice indicates that NMU acts downstream of leptin to regulate bone formation. Most interestingly, NMU-deficient mice are resistant to the anti-osteogenic effect of leptin and adrenergic agonists. Furthermore, osteoblast number paradoxically increased in NMU-deficient mice treated by leptin ICV, as observed in *Clock*-deficient mice, suggesting that NMU regulates the clock gene's function in osteoblasts (see below). In support of this hypothesis, expression of *Per* genes was downregulated in NMU-deficient bones compared to WT bones.

The endocannabinoid system is involved in the regulation of bone remodeling in a central and peripheral manner as well. The cannabinoid type 1 (CB1) receptor, mostly known for its involvement in psychotropic, analgesic, and orexigenic processes, is expressed in the CNS and PNS, whereas the CB2 receptor is expressed mostly peripherally. The main CB1 and CB2 endogenous ligands are N-arachidonoylethanolamine (AEA or anandamide) and 2-arachidonoylglycerol (2-AG). Interestingly, leptin negatively regulates both 2-AG levels and bone mass, and traumatic brain injury increases both bone formation and central 2-AG production. These observations and the fact that CB1 signaling in peripheral neurons inhibits NE release [50] led several groups to assess the contribution of the cannabinoid system to bone remodeling and to analyze genetic mouse mutants lacking *Cb1* or *Cb2*. Mice lacking *Cb1* exhibit a bone phenotype that is dependent on strain background differences [51–53], indicating that indeed the endocannabinoid system contributes to the regulation of bone remodeling. Prejunctional CB1 stimulation by 2-AG and tonic inhibition of NE release by sympathetic nerves are proposed to suppress the anti-osteogenic effect of the sympathetic nervous system and to contribute to the osteoinductive effect of traumatic brain injury [54]. On the other hand, the CB2 receptor and endogenous cannabinoids produced by bone cells, including osteoblasts, have a stimulatory effect on bone formation [55–57]. Importantly, polymorphisms in *CNR2* are important genetic risk factors for osteoporosis [58], and the peripheral expression of CB2 as well as the availability of nonpsychoactive CB2 agonists position this receptor as an anti-osteoporosis target [59].

In contrast to the endothelial (eNOS) and inducible (iNOS) isoforms of NO synthase that are expressed in bone cells (see Ref. 60 for a review), the neuronal form of NO synthase (nNOS) is not expressed in bone cells under normal conditions [61], but is highly expressed in the CNS. This observation and the high-bone-mass phenotype of nNOS-deficient mice suggest that central NO signaling might also regulate bone mass by a central mechanism [62], although direct evidence for this is still lacking.

WHAT ABOUT SENSORY NERVES?

Although sensory neuropeptides are richly expressed in bones [63–66], the relevance of these nerves in bone remodeling is less well defined. Several studies, however, support the contribution of sensory neurons to this process. First, familial dysautomia is an autosomal recessive disease in which patients suffer from unmyelinated sensory neuron loss, reduced BMD, and frequent fractures [67]; second, destruction of unmyelinated sensory axons by capsaicin induces bone loss in rats, accompanied by reduced expression of substance P and cGRP (a calcitonin gene-related peptide) [68, 69]. Lastly, both *in vitro* and *in vivo* studies have led to the conclusion that cGRP is an anabolic factor that acts on osteoblasts and stimulates their proliferation and function [70, 71]. *In vitro* data also suggest that cGRP may be involved in modulating the proresorptive effect of the sympathetic nervous system by interfering with the action of RANKL [19].

LOOKING FORWARD

What are the implications of this growing body of work? The first and most obvious, given the powerful influence exerted by leptin and serotonin, is that there may still be novel systemic regulators of bone mass to be discovered. The second is that the notion that bone formation and resorption are always regulated in the same direction is erroneous since the sympathetic tone exerts opposite influence on each arm of bone remodeling. A third implication of therapeutic nature is that it may be possible to harness the control of bone mass by the sympathetic tone for clinical purposes.

REFERENCES

1. Ducy P, Amling M, Takeda S, Priemel M, Schilling AF, Beil FT, Shen J, Vinson C, Rueger JM, Karsenty G. 2000.

Leptin inhibits bone formation through a hypothalamic relay: A central control of bone mass. *Cell* 100(2): 197–207.

2. Shi Y, Yadav VK, Suda N, Liu XS, Guo XE, Myers MG Jr, Karsenty G. 2008. Dissociation of the neuronal regulation of bone mass and energy metabolism by leptin in vivo. *Proc Natl Acad Sci U S A* 105(51): 20529–33.

3. Guidobono F, Pagani F, Sibilia V, Netti C, Lattuada N, Rapetti D, Mrak E, Villa I, Cavani F, Bertoni L, Palumbo C, Ferretti M, Marotti G, Rubinacci A. 2006. Different skeletal regional response to continuous brain infusion of leptin in the rat. *Peptides* 27(6): 1426–33.

4. Pogoda P, Egermann M, Schnell JC, Priemel M, Schilling AF, Alini M, Schinke T, Rueger JM, Schneider E, Clarke I, Amling M. 2006. Leptin inhibits bone formation not only in rodents, but also in sheep. *J Bone Miner Res* 21(10): 1591–9.

5. Takeda S, Elefteriou F, Levasseur R, Liu X, Zhao L, Parker KL, Armstrong D, Ducy P, Karsenty G. 2002. Leptin regulates bone formation via the sympathetic nervous system. *Cell* 111(3): 305–17.

6. Balthasar N, Coppari R, McMinn J, Liu SM, Lee CE, Tang V, Kenny CD, McGovern RA, Chua SC Jr, Elmquist JK, Lowell BB. 2004. Leptin receptor signaling in POMC neurons is required for normal body weight homeostasis. *Neuron* 42(6): 983–91.

7. Yadav VK, Oury F, Suda N, Liu ZW, Gao XB, Confavreux C, Klemenhagen KC, Tanaka KF, Gingrich JA, Guo XE, Tecott LH, Mann JJ, Hen R, Horvath TL, Karsenty G. 2009. A serotonin-dependent mechanism explains the leptin regulation of bone mass, appetite, and energy expenditure. *Cell* 138(5): 976–89.

8. Oury F, Yadav VK, Wang Y, Zhou B, Liu XS, Guo XE, Tecott LH, Schutz G, Means AR, Karsenty G. 2010. CREB mediates brain serotonin regulation of bone mass through its expression in ventromedial hypothalamic neurons. *Genes Dev* 24(20): 2330–42.

9. Bonnet N, Brunet-Imbault B, Arlettaz A, Horcajada MN, Collomp K, Benhamou CL, Courteix D. 2005. Alteration of trabecular bone under chronic beta2 agonists treatment. *Med Sci Sports Exerc* 37(9): 1493–501.

10. Bonnet N, Laroche N, Vico L, Dolleans E, Benhamou CL, Courteix D. 2006. Dose effects of propranolol on cancellous and cortical bone in ovariectomized adult rats. *J Pharmacol Exp Ther* 318(3): 1118–27.

11. Bonnet N, Benhamou CL, Malaval L, Goncalves C, Vico L, Eder V, Pichon C, Courteix D. 2008. Low dose beta-blocker prevents ovariectomy-induced bone loss in rats without affecting heart functions. *J Cell Physiol* 217(3): 819–27.

12. Sato T, Arai M, Goto S, Togari A. 2010. Effects of propranolol on bone metabolism in spontaneously hypertensive rats. *J Pharmacol Exp Ther* 334(1): 99–105.

13. Elefteriou F, Ahn JD, Takeda S, Starbuck M, Yang X, Liu X, Kondo H, Richards WG, Bannon TW, Noda M, Clement K, Vaisse C, Karsenty G. 2005. Leptin regulation of bone resorption by the sympathetic nervous system and CART. *Nature* 434(7032): 514–20.

14. Kajimura D, Hinoi E, Ferron M, Kode A, Riley KJ, Zhou B, Guo XE, Karsenty G. 2011. Genetic determination of the cellular basis of the sympathetic regulation of bone mass accrual. *J Exp Med* 208(4): 841–51.

15. Yan L, Vatner DE, O'Connor JP, Ivessa A, Ge H, Chen W, Hirotani S, Ishikawa Y, Sadoshima J, Vatner SF. 2007. Type 5 adenylyl cyclase disruption increases longevity and protects against stress. *Cell* 130(2): 247–58.

16. Fu L, Patel MS, Bradley A, Wagner EF, Karsenty G. 2005. The molecular clock mediates leptin-regulated bone formation. *Cell* 122(5): 803–15.

17. Simmons DJ, Nichols G Jr. 1966. Diurnal periodicity in the metabolic activity of bone tissue. *Am J Physiol* 210(2): 411–8.

18. Gundberg CM, Markowitz ME, Mizruchi M, Rosen JF. 1985. Osteocalcin in human serum: A circadian rhythm. *J Clin Endocrinol Metab* 60(4): 736–9.

19. Arai M, Nagasawa T, Koshihara Y, Yamamoto S, Togari A. 2003. Effects of beta-adrenergic agonists on bone-resorbing activity in human osteoclast-like cells. *Biochim Biophys Acta* 1640(2–3): 137–42.

20. Yang X, Matsuda K, Bialek P, Jacquot S, Masuoka HC, Schinke T, Li L, Brancorsini S, Sassone-Corsi P, Townes TM, Hanauer A, Karsenty G. 2004. ATF4 is a substrate of RSK2 and an essential regulator of osteoblast biology; implication for Coffin-Lowry Syndrome. *Cell* 117(3): 387–98.

21. Lorentzon M, Landin K, Mellstrom D, Ohlsson C. 2006. Leptin is a negative independent predictor of areal BMD and cortical bone size in young adult Swedish men. *J Bone Miner Res* 21(12): 1871–8.

22. Pasco JA, Henry MJ, Sanders KM, Kotowicz MA, Seeman E, Nicholson GC. 2004. Beta-adrenergic blockers reduce the risk of fracture partly by increasing bone mineral density: Geelong Osteoporosis Study. *J Bone Miner Res* 19(1): 19–24.

23. Schlienger RG, Kraenzlin ME, Jick SS, Meier CR. 2004. Use of beta-blockers and risk of fractures. *JAMA* 292(11): 1326–32.

24. Rejnmark L, Vestergaard P, Kassem M, Christoffersen BR, Kolthoff N, Brixen K, Mosekilde L. 2004. Fracture risk in perimenopausal women treated with beta-blockers. *Calcif Tissue Int* 75(5): 365–72.

25. Wiens M, Etminan M, Gill SS, Takkouche B. 2006. Effects of antihypertensive drug treatments on fracture outcomes: A meta-analysis of observational studies. *J Intern Med* 260(4): 350–62.

26. Rejnmark L, Vestergaard P, Mosekilde L. 2006. Treatment with beta-blockers, ACE inhibitors, and calcium-channel blockers is associated with a reduced fracture risk: A nationwide case-control study. *J Hypertens* 24(3): 581–9.

27. Reid IR, Gamble GD, Grey AB, Black DM, Ensrud KE, Browner WS, Bauer DC. beta-Blocker use, BMD, and fractures in the study of osteoporotic fractures. 2005. *J Bone Miner Res* 20(4): 613–8.

28. Levasseur R, Marcelli C, Sabatier JP, Dargent-Molina P, Breart G. 2005. Beta-blocker use, bone mineral density, and fracture risk in older women: Results from the

Epidemiologie de l'Osteoporose prospective study. *J Am Geriatr Soc* 53(3): 550–2.
29. Schweiger U, Deuschle M, Korner A, Lammers CH, Schmider J, Gotthardt U, Holsboer F, Heuser I. 1994. Low lumbar bone mineral density in patients with major depression. *Am J Psychiatry* 151(11): 1691–3.
30. Michelson D, Stratakis C, Hill L, Reynolds J, Galliven E, Chrousos G, Gold P. 1996. Bone mineral density in women with depression. *N Engl J Med* 335(16): 1176–81.
31. Cizza G, Ravn P, Chrousos GP, Gold PW. 2001. Depression: A major, unrecognized risk factor for osteoporosis? *Trends Endocrinol Metab* 12(5): 198–203.
32. Yirmiya R, Goshen I, Bajayo A, Kreisel T, Feldman S, Tam J, Trembovler V, Csernus V, Shohami E, Bab I. 2006. Depression induces bone loss through stimulation of the sympathetic nervous system. *Proc Natl Acad Sci U S A* 103(45): 16876–81.
33. Diem SJ, Blackwell TL, Stone KL, Yaffe K, Haney EM, Bliziotes MM, Ensrud KE. 2007. Use of antidepressants and rates of hip bone loss in older women: The study of osteoporotic fractures. *Arch Intern Med* 167(12): 1240–5.
34. Eskandari F, Martinez PE, Torvik S, Phillips TM, Sternberg EM, Mistry S, Ronsaville D, Wesley R, Toomey C, Sebring NG, Reynolds JC, Blackman MR, Calis KA, Gold PW, Cizza G. 2007. Low bone mass in premenopausal women with depression. *Arch Intern Med* 167(21): 2329–36.
35. Ma Y, Nyman JS, Tao H, Moss HH, Yang X, Elefteriou F. 2011. β2-Adrenergic receptor signaling in osteoblasts contributes to the catabolic effect of glucocorticoids on bone. *J Biol Chem* 152(4): 1412–22.
36. Kristensen P, Judge ME, Thim L, Ribel U, Christjansen KN, Wulff BS, Clausen JT, Jensen PB, Madsen OD, Vrang N, Larsen PJ, Hastrup S. 1998. Hypothalamic CART is a new anorectic peptide regulated by leptin. *Nature* 393(6680): 72–6.
37. Ahn JD, Dubern B, Lubrano-Berthelier C, Clement K, Karsenty G. 2006. Cart overexpression is the only identifiable cause of high bone mass in melanocortin 4 receptor deficiency. *Endocrinology* 147(7): 3196–202.
38. Orwoll B, Bouxsein ML, Marks DL, Cone RD, Klein RF, editors. Increased bone mass and strength in melanocortin-4 receptor-deficient mice. 2004. ORS/AAOS Poster Presentation. 71st Annual Meeting of the AAOS, March 2004, San Francisco, CA.
39. Shi Y, Oury F, Yadav VK, Wess J, Liu XS, Guo XE, Murshed M, Karsenty G. 2010. Signaling through the M(3) muscarinic receptor favors bone mass accrual by decreasing sympathetic activity. *Cell Metab* 11(3): 231–8.
40. Elefteriou F, Takeda S, Liu X, Armstrong D, Karsenty G. 2003. Monosodium glutamate-sensitive hypothalamic neurons contribute to the control of bone mass. *Endocrinology* 144(9): 3842–7.
41. Baldock PA, Lee NJ, Driessler F, Lin S, Allison S, Stehrer B, Lin EJ, Zhang L, Enriquez RF, Wong IP, McDonald MM, During M, Pierroz DD, Slack K, Shi YC, Yulyaningsih E, Aljanova A, Little DG, Ferrari SL, Sainsbury A, Eisman JA, Herzog H. 2009. Neuropeptide Y knockout mice reveal a central role of NPY in the coordination of bone mass to body weight. *PLoS One* 4(12): e8415.
42. Baldock PA, Sainsbury A, Couzens M, Enriquez RF, Thomas GP, Gardiner EM, Herzog H. 2002. Hypothalamic Y2 receptors regulate bone formation. *J Clin Invest* 109(7): 915–21.
43. Shi YC, Lin S, Wong IP, Baldock PA, Aljanova A, Enriquez RF, Castillo L, Mitchell NF, Ye JM, Zhang L, Macia L, Yulyaningsih E, Nguyen AD, Riepler SJ, Herzog H, Sainsbury A. 2010. NPY neuron-specific Y2 receptors regulate adipose tissue and trabecular bone but not cortical bone homeostasis in mice. *PLoS One* 5(6): e11361.
44. Sainsbury A, Baldock PA, Schwarzer C, Ueno N, Enriquez RF, Couzens M, Inui A, Herzog H, Gardiner EM. 2003. Synergistic effects of Y2 and Y4 receptors on adiposity and bone mass revealed in double knockout mice. *Mol Cell Biol* 23(15): 5225–33.
45. Baldock PA, Allison SJ, Lundberg P, Lee NJ, Slack K, Lin EJ, Enriquez RF, McDonald MM, Zhang L, During MJ, Little DG, Eisman JA, Gardiner EM, Yulyaningsih E, Lin S, Sainsbury A, Herzog H. 2007. Novel role of Y1 receptors in the coordinated regulation of bone and energy homeostasis. *J Biol Chem* 282(26): 19092–102.
46. Lee NJ, Doyle KL, Sainsbury A, Enriquez RF, Hort YJ, Riepler SJ, Baldock PA, Herzog H. 2010. Critical role for Y1 receptors in mesenchymal progenitor cell differentiation and osteoblast activity. *J Bone Miner Res* 25(8): 1736–47.
47. Brighton PJ, Szekeres PG, Willars GB. 2004. Neuromedin U and its receptors: Structure, function, and physiological roles. *Pharmacol Rev* 56(2): 231–48.
48. Hanada R, Teranishi H, Pearson JT, Kurokawa M, Hosoda H, Fukushima N, Fukue Y, Serino R, Fujihara H, Ueta Y, Ikawa M, Okabe M, Murakami N, Shirai M, Yoshimatsu H, Kangawa K, Kojima M. 2004. Neuromedin U has a novel anorexigenic effect independent of the leptin signaling pathway. *Nat Med* 10(10): 1067–73.
49. Sato S, Hanada R, Kimura A, Abe T, Matsumoto T, Iwasaki M, Inose H, Ida T, Mieda M, Takeuchi Y, Fukumoto S, Fujita T, Kato S, Kangawa K, Kojima M, Shinomiya K, Takeda S. 2007. Central control of bone remodeling by neuromedin U. *Nat Med* 13(10): 1234–40.
50. Ishac EJ, Jiang L, Lake KD, Varga K, Abood ME, Kunos G. 1996. Inhibition of exocytotic noradrenaline release by presynaptic cannabinoid CB1 receptors on peripheral sympathetic nerves. *Br J Pharmacol* 118(8): 2023–8.
51. Tam J, Ofek O, Fride E, Ledent C, Gabet Y, Muller R, Zimmer A, Mackie K, Mechoulam R, Shohami E, Bab I. 2006. Involvement of neuronal cannabinoid receptor CB1 in regulation of bone mass and bone remodeling. *Mol Pharmacol* 70(3): 786–92.
52. Idris AI, van't Hof RJ, Greig IR, Ridge SA, Baker D, Ross RA, Ralston SH. 2005. Regulation of bone mass, bone loss and osteoclast activity by cannabinoid receptors. *Nat Med* 11(7): 774–9.
53. Idris AI, Sophocleous A, Landao-Bassonga E, Canals M, Milligan G, Baker D, van't Hof RJ, Ralston SH. 2009.

Cannabinoid receptor type 1 protects against age-related osteoporosis by regulating osteoblast and adipocyte differentiation in marrow stromal cells. *Cell Metab* 10(2): 139–47.

54. Tam J, Trembovler V, Di Marzo V, Petrosino S, Leo G, Alexandrovich A, Regev E, Casap N, Shteyer A, Ledent C, Karsak M, Zimmer A, Mechoulam R, Yirmiya R, Shohami E, Bab I. 2008. The cannabinoid CB1 receptor regulates bone formation by modulating adrenergic signaling. *FASEB J* 22(1): 285–94.

55. Ofek O, Karsak M, Leclerc N, Fogel M, Frenkel B, Wright K, Tam J, Attar-Namdar M, Kram V, Shohami E, Mechoulam R, Zimmer A, Bab I. 2006. Peripheral cannabinoid receptor, CB2, regulates bone mass. *Proc Natl Acad Sci U S A* 103(3): 696–701.

56. Sophocleous A, Landao-Bassonga E, Van't Hof RJ, Idris AI, Ralston SH. 2011. The type 2 cannabinoid receptor regulates bone mass and ovariectomy-induced bone loss by affecting osteoblast differentiation and bone formation. *Endocrinology* 152(6): 2141–9.

57. Ofek O, Attar-Namdar M, Kram V, Dvir-Ginzberg M, Mechoulam R, Zimmer A, Frenkel B, Shohami E, Bab I. 2011. CB2 cannabinoid receptor targets mitogenic Gi protein-cyclin D1 axis in osteoblasts. *J Bone Miner Res* 26(2): 308–16.

58. Karsak M, Cohen-Solal M, Freudenberg J, Ostertag A, Morieux C, Kornak U, Essig J, Erxlebe E, Bab I, Kubisch C, de Vernejoul MC, Zimmer A. 2005. Cannabinoid receptor type 2 gene is associated with human osteoporosis. *Hum Mol Genet* 14(22): 3389–96.

59. Bab I, Zimmer A, Melamed E. 2009. Cannabinoids and the skeleton: From marijuana to reversal of bone loss. *Ann Med* 41(8): 560–7.

60. van't Hof RJ, Ralston SH. 2001. Nitric oxide and bone. *Immunology* 103(3): 255–61.

61. Helfrich MH, Evans DE, Grabowski PS, Pollock JS, Ohshima H, Ralston SH. 1997. Expression of nitric oxide synthase isoforms in bone and bone cell cultures. *J Bone Miner Res* 12(7): 1108–15.

62. van't Hof RJ, Macphee J, Libouban H, Helfrich MH, Ralston SH. 2004. Regulation of bone mass and bone turnover by neuronal nitric oxide synthase. *Endocrinology* 145(11): 5068–74.

63. Bjurholm A, Kreicbergs A, Brodin E, Schultzberg M. 1988. Substance P- and CGRP-immunoreactive nerves in bone. *Peptides* 9(1): 165–71.

64. Bjurholm A. 1991. Neuroendocrine peptides in bone. *Int Orthop* 15(4): 325–9.

65. Hill EL, Elde R. 1991. Distribution of CGRP-, VIP-, D beta H-, SP-, and NPY-immunoreactive nerves in the periosteum of the rat. *Cell Tissue Res* 264(3): 469–80.

66. Hukkanen M, Konttinen YT, Rees RG, Gibson SJ, Santavirta S, Polak JM. 1992. Innervation of bone from healthy and arthritic rats by substance P and calcitonin gene related peptide containing sensory fibers. *J Rheumatol* 19(8): 1252–9.

67. Maayan C, Bar-On E, Foldes AJ, Gesundheit B, Pollak RD. 2002. Bone mineral density and metabolism in familial dysautonomia. *Osteoporos Int* 13(5): 429–33.

68. Offley SC, Guo TZ, Wei T, Clark JD, Vogel H, Lindsey DP, Jacobs CR, Yao W, Lane NE, Kingery WS. 2005. Capsaicin-sensitive sensory neurons contribute to the maintenance of trabecular bone integrity. *J Bone Miner Res* 20(2): 257–67.

69. Ding Y, Arai M, Kondo H, Togari A. 2010. Effects of capsaicin-induced sensory denervation on bone metabolism in adult rats. *Bone* 46(6): 1591–6.

70. Ballica R, Valentijn K, Khachatryan A, Guerder S, Kapadia S, Gundberg C, Gilligan J, Flavell RA, Vignery A. 1999. Targeted expression of calcitonin gene-related peptide to osteoblasts increases bone density in mice. *J Bone Miner Res* 14(7): 1067–74.

71. Schinke T, Liese S, Priemel M, Haberland M, Schilling AF, Catala-Lehnen P, Blicharski D, Rueger JM, Gagel RF, Emeson RB, Amling M. 2004. Decreased bone formation and osteopenia in mice lacking alpha-calcitonin gene-related peptide. *J Bone Miner Res* 19(12): 2049–56.

11
Skeletal Healing

Michael J. Zuscik

- The Process of Skeletal Healing 90
- Conditions That Impair Fracture Healing and Therapeutic Modalities 93
- Molecular Therapies to Enhance Bone Healing 94
- References 95

The process of skeletal repair is essential for resolution of (1) orthopedic trauma that has caused bony disjunction or (2) surgical interventions that are intended to create bony injury with the aim of inducing a repair response or a therapy. Understanding the cellular and molecular basis of this healing process has been the focus of intense research both in humans and animal models over the past two decades, with this work largely driven by the need to develop therapeutic strategies to enable or enhance healing of fibrous nonunions, critically sized defects, or other situations of impaired healing. Combined, failed, or delayed healing impacts up to 10% of all fracture patients seen clinically [1, 2] and can result from a series of common situations including comminution, inadequate fixation, infection, tumor, hypoxia/poor blood supply, metabolic dysfunction, and other chronic disease [3]. Overall, research efforts have led to a general understanding of the molecular and genetic control over the inflammatory, cellular, and tissue processes that are required for healing, which are generally conserved across species and are not different in structurally distinct skeletal elements. This chapter provides a concise and up-to-date overview of our understanding of the skeletal healing process at the cellular and molecular level, a discussion of a few key situations that complicate healing, and a summary of therapeutic modalities that are either in development or employed clinically to enhance repair or facilitate healing in nonunion situations.

THE PROCESS OF SKELETAL HEALING

The cellular contribution to healing tissues: The fracture healing process requires the coordinated activity of several different cell types, including inflammatory cells, chondro- and osteoprogenitors, chondrocytes, osteoblasts, and osteoclasts. Healing events that occur in various vertebrate species are similar, except that relative to humans the pace of repair is generally accelerated in smaller animals/rodents. Thus, the schematic presented in Fig. 11.1, which presents a pictorial representation of the unique morphogenesis of reparative tissue during the phases of bone fracture healing, provides a benchmark for the description of the healing process in general. This process begins immediately after fracture and usually involves both intramembranous and endochondral ossification (comprehensively reviewed in Refs. 4–6). The trauma that induces the fracture initially results in bleeding and formation of a hematoma at the injury site. Hematoma-associated cytokines including tumor necrosis factor-α (TNF-α) and interleukins (IL) -1, -6, -11 and -18 [5] lead to recruitment and infiltration of inflammatory cells to the fracture site, which themselves potentiate the inflammatory environment and induce the secondary recruitment of key mesenchymal stem cell (MSC) populations. These MSCs may derive from a number of niches, including bone marrow [7, 8], muscle

Primer on the Metabolic Bone Diseases and Disorders of Mineral Metabolism, Eighth Edition. Edited by Clifford J. Rosen.
© 2013 American Society for Bone and Mineral Research. Published 2013 by John Wiley & Sons, Inc.

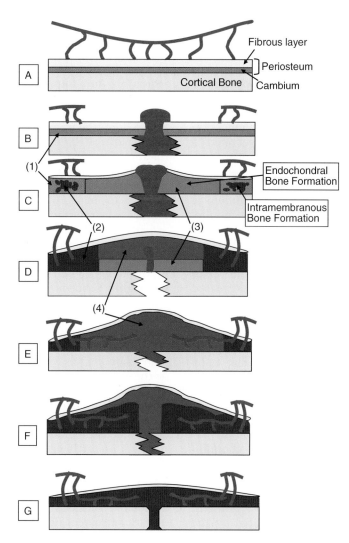

Fig. 11.1. Tissue morphogenesis during bone repair. (A) Periosteum is a well-microvascularized tissue (vessels in red) consisting of an outer fibrous layer and an inner cambium layer. The cambium layer contains abundant stem/progenitor cells that can differentiate into bone and cartilage. (B) Following fracture or osteotomy, blood supply is disrupted at the defect, and a blood clot (hematoma) forms near the disjunction. (C) Progenitor cells residing in the periosteum are recruited to differentiate into osteoblasts to facilitate intramembranous bone formation where intact blood supply is preserved, and into chondrocytes to facilitate endochondral bone formation adjacent to the fracture where the tissue is hypoxic. In this panel, osteogenic tissue is labeled (1), with newly mineralized tissue labeled (2). Tissues supporting chondrogenesis are labeled (3). (D) Intramembranous bone formation proceeds with robust matrix mineralization (1) where blood supply is present distal to the fracture site. Endochondral bone formation proceeds simultaneously with chondrogenic tissue supporting a growing population of chondrocytes that comprise the hypertrophic cartilage, which is labeled (4). (E) Cartilage tissue continues to mature, ultimately encompassing the callus nearest the fracture site. Revascularization of the callus also ensues. (F) Chondrocytes in the hypertrophic cartilage undergo terminal differentiation and the matrix is progressively mineralized expanding the portion of the callus that is comprised of woven bone (brown). (G) The remodeling process proceeds with osteoclasts and osteoblasts facilitating the conversion of woven bone into lamellar bone to ultimately support reestablishment of the appropriate anatomic shape.

[9, 10], periosteum [11, 12], and possibly the general circulation [13, 14]. While there is much debate about which of these populations are the most critical for initiating repair, data suggest that a key participant is the periosteal progenitor cell that responds to the inflammation by entering the osteogenic or chondrogenic lineage [12]. Endochondral bone formation always takes place closest to the fracture site where the oxygen tension is low and vascularity is disrupted. Intramembranous bone formation, on the other hand, always occurs distal to the disjunction where intact vasculature remains present. The mechanical stability of the fractured bone markedly affects the fate of the progenitor cells, with stabilized fractures healing with virtually no evidence of cartilage whereas non-stabilized fractures produce abundant cartilage at the fracture site [15].

Given that the periosteum represents a primary source of MSCs that contribute to bone repair, understanding its structure/function is critical for dissecting the tissue and cellular dynamics of the healing process. Overall, periosteum is a vascularized connective tissue that covers the outer surface of cortical bone. It can be separated into two distinct layers: an outer layer that contains fibroblasts and Sharpey's fibers (which facilitate connection to the underlying cortical bone) and an inner layer referred to as the cambium, which contains multipotent mesenchymal stem cells and osteoprogenitor cells that contribute to normal bone growth, healing, and regeneration [16, 17]. It is known that the cambium layer in children is much thicker and better vascularized than in adults, facilitating faster healing.

Once periosteal MSCs have committed to the chondrogenic or osteogenic lineage, chondrocyte and osteoblast differentiation takes place. Directly overlying the site of the fracture, the ends of the original bone have decreased perfusion due to disrupted vascularity, and necrosis occurs. In this central hypoxic region, MSCs differentiate into chondrocytes, and endochondral bone formation is initiated. This is consistent with the concept that hypoxia is a critical inducer of chondrogenesis [18, 19]. The tissue that forms as the cell population expands is referred to as the callus, and differentiation of MSCs into chondrocytes occurs directionally within the callus, with the process initiating in the most central avascular region. While these centrally positioned MSCs persist in the callus area directly overlying the fracture site, chondrocytes that differentiate radially recapitulate the maturation process that occurs in the growth plate, including phases of proliferation, hypertrophy, and terminal differentiation [5]. The calcified cartilage, which acts a template for primary bone formation, is populated by the most terminally differentiated hypertrophic chondrocytes that are contributing to the mineralization of the tissue. Ultimately, terminally differentiated chondrocytes residing in the calcified cartilage matrix undergo apoptosis, and an initial phase of osteoclast-driven remodeling occurs to remove the remaining cartilage in response to macrophage colony-stimulating factor (M-CSF), receptor activator of nuclear factor-κB (RANK) ligand (RANKL) and osteoprotegerin (OPG) [20]. Maximal resorption occurs when the ratio of OGP to RANKL expression is at its lowest. Distal to the fracture site and flanking the chondrocytes undergoing endochondral ossification is the location where intramembranous ossification occurs. This process, which proceeds in the zone of injury where blood supply has been better preserved, is characterized by the differentiation of periosteal cells into osteoblasts, which directly lay down new mineral without a cartilage intermediate. As mentioned, the better the fracture is fixed (minimizing instability), the greater the ratio of intramembranous to endochondral ossification in the overall healing process.

Fractures are considered healed when bone stability has been restored by the formation of new bone that bridges the area of fracture. However, this initial woven bone matrix is replaced by organized lamellar bone through a second remodeling process that is the critical final step in achieving an anatomically correct skeletal element. Again, this process is governed by osteoclasts, which become dominant in this final stage due to the induction of IL-1 and TNF-α and the subsequent expansion of the functional osteoclast population [21, 22] via RANKL in the remodeling callus [23]. Similarly, the initial cortical bone is remodeled and replaced at the fracture site, where necrosis occurred secondary to loss of vascularization due to the injury. The completion of this final remodeling phase results in an anatomically correct element with biomechanical stability that matches the pre-fracture state.

Gene expression profile during fracture healing: Given that the bone repair process is dependent on a combination of endochondral and intramembranous ossification followed by osteoclast remodeling, the genetic profile of the healing tissue is stage dependent and reflective of the differentiation of these cells. Since the endochondral healing process recapitulates the events that occur during skeletal development, it is no surprise that the genetic profile partially reflects the profile seen in the growth plate chondrocyte hypertrophic program. Overlying this is the genetic profile of osteoblast differentiation that occurs during intramembranous bone formation and in the process of cartilaginous callus conversion into woven bone. Regarding the endochondral process, mesenchymal cell condensation coincides with the expression of early markers of cartilage formation that include Sox-9 and type II collagen [24]. As chondrocyte differentiation ensues, there is a significant increase in cell volume that is associated with the expression of hypertrophy-associated genes that include type X collagen, MMP9 and 13, and osteocalcin [4, 5, 25], as well Indian hedgehog (Ihh) [26]. Partially in response to the upregulation of hypoxia-inducible factor-1α (HIF-1α) [18], terminally mature chondrocytes contribute to revascularization via expression of vascular endothelial growth factor (VEGF) [27] and also may initiate the remodeling process via the induction of osteoclast formation/activity via expression of RANKL [28]. Regarding the intramembranous process, markers of osteoblast differentiation are detected including type I collagen, osteopontin, and osteocalcin. Osteoblasts also contribute to callus revascularization by producing VEGF [29]. The differentiation process in these cells is driven by the expression of Runx2 [30], a transcription factor required for mineralization. By establishing the temporal and regional gene expression pattern during the healing process, a benchmark has emerged that facilitates the monitoring of healing rate that may be delayed or accelerated depending on comorbidities such as aging [31] or diabetes [32], or during therapeutic intervention such as BMP-2 treatment [33, 34].

Molecular control of fracture healing: The molecular signaling pathways involved in the initiation and tissue morphogenesis that occurs during fracture repair are only superficially understood. Although animals and humans have only very limited capacity to regenerate damaged tissues, it has long been suspected that postnatal bone repair involving endochondral ossification recapitulates some of the essential pathways/factors in limb development [6, 35]. Regarding the developmental process, the most notable regulators are the bone morphogenetic proteins (BMPs) that belong to the transforming growth factor beta (TGF-β) superfamily, Ihh, the mammalian homologs of wingless in *Drosophila* (Wnt) proteins, fibroblast growth factors (FGFs) and insulin-like growth factors (IGFs). Relative to fracture repair, BMP-2 expression has been observed in early periosteal callus just a few days following cortical bone fracture. Most recently, Tsuji et al. demonstrated that elimination of BMP-2 in the limb disrupts the initiation of postnatal fracture healing [33], establishing the essential role of BMP-2 in bone repair. In fact, BMP use has been approved in a number of bone healing situations (reviewed in Ref. 36 and discussed below). Evidence has also emerged indicating that Wnt/β-catenin signaling is required to drive osteoblast differentiation in the fracture callus [37–39], implicating this pathway as an important participant in the callus mineralization process. Hedgehogs are likely important during the cartilage differentiation phase of healing, where Ihh expression is the highest in a rat model of healing [26]. Similarly, FGFs [40] and IGFs [41] have also been implicated to play a role during skeletal healing. Overall, studies are ongoing in the fracture healing field with the aim of fully characterizing the role of these pathways and factors in the adult fracture healing process.

In addition to cell differentiation processes that recapitulate limb development, genes that are involved in injury and inflammatory responses during bone repair have been shown to play key roles in endochondral bone repair [42, 43]. For example, during the inflammatory phase, a constellation of cytokines drive MSC commitment to the chondrogenic and osteogenic lineages [5] as described above. Later, during the endochondral healing phase, a turnover of mineralized cartilage occurs that sets the stage for primary woven bone formation. As mentioned, this initial remodeling process coincides with the upregulation of M-CSF, RANKL, OPG, and TNF-α, implicating these factors as critical for the transition from cartilage to bone [20, 44]. During the second remodeling phase when woven bone is converted into lamellar bone, TNF-α and IL-1 and -6 expression is upregulated, implicating these factors in the recruitment of osteoclasts that are critical for this final remodeling step [21–23, 45]. Supporting this idea, IL-6 knockout mice display delayed callus mineralization and maturation during repair of femoral osteotomy [46]. Thus, there is a clear involvement of pro-inflammatory mediators in the bone repair process.

Several studies have also shown that cyclooxygenase activity is involved in normal bone metabolism and suggest that non-steroidal anti-inflammatory drugs (NSAIDs) have a negative impact on bone repair [47, 48]. The action of aspirin and other NSAIDs is via inhibition of cyclooxygenase (COX), an enzyme that catalyzes the formation of prostaglandins and thromboxanes from arachidonic acid [49]. The most compelling data implicating COX function during skeletal healing comes from genetic models that demonstrate a critical role for COX-2 in the process. While COX-2$^{-/-}$ mice develop normally, bone repair is impaired in adult knockout mice following fracture [50]. Defective healing in this model occurs at the early inflammatory phase and persists into the reparative phase of healing. Histology of fracture callus derived from COX-2$^{-/-}$ mice shows delayed chondrogenesis and persistent mesenchyme at the fracture site. The defective bone healing phenotype coincides with the early induction of COX-2 expression, further demonstrating the requirement of this enzyme in early chondrogenesis in skeletal repair.

Critical to successful bone repair is the revascularization of injured tissues to provide oxygen, facilitate nutrient/metabolic waste management, and deliver a population of precursor cells of hematopoietic origin that may have the potential to contribute to healing. As mentioned previously, support for angiogenesis during the repair process is thought to be modulated by VEGFs and their cognate receptors VEGFR1 and VEGFR2. It has been demonstrated that exogenous administration of VEGF during mouse femur fracture healing enhances vascular ingrowth into the callus and accelerates repair by promoting bony bridging [51]. This has been borne out in allograft bone healing, where VEGF gene therapy likewise enhanced the healing process [52]. While it has been suggested that osteoblasts are the primary regulators of angiogenesis during healing due to their production of VEGF in response to BMPs [53], a contribution of VEGF from hypertrophic cartilage in the callus may also occur [28].

CONDITIONS THAT IMPAIR FRACTURE HEALING AND THERAPEUTIC MODALITIES

The normal progression of the fracture healing process can be significantly compromised by a number of physiologic, pathologic, and environmental factors including aging, diabetes, and cigarette smoking. Clinical data provide evidence for this, and basic research has begun to reveal the details of the underlying biological basis in some cases. Below is a brief discussion of three of the most important conditions that are documented to impair the process of skeletal healing.

Aging: While it has been known for more than 30 years that the rate of fracture healing is reduced with aging, minimal progress has been made toward understanding the mechanisms involved. Studies have suggested that the rate of bone repair is progressively reduced with aging in the pediatric population [54]. Furthermore, numerous studies document that development of nonunion in the aging population is a significant clinical problem [55–57]. Several mechanisms have been proposed to explain reduced/delayed fracture healing in the elderly. Regarding the potential molecular basis of impaired healing in the elderly, recent work has suggested that reduced expression of COX-2 during the inflammatory phase leads to reduced chondrogenic potential and a truncation of the cartilage phase of healing [31]. Additionally, the normal upregulation of BMP-2 [58], Ihh [59], and various Wnts [60] during chondrocyte maturation and osteoblast differentiation is reduced in aged mice, further impairing the progression of healing in senescence. A reduced number/responsiveness of progenitors [61] enhanced adipogenic potential at the expense of chondrogenesis, and osteogenesis [62] or altered competency to support osteoclastogenesis at various stages of healing [63] may also be involved. There has also been an association between aging and a decrement in endothelial cells and the factors/pathways that modulate them [64, 65], suggesting that impaired blood vessel formation in elderly patients could also affect healing. Ongoing effort aims to address the relative mechanistic contribution of these and other processes to impaired skeletal healing in the elderly.

Diabetes: Documented clinical findings establish that fracture healing is impaired in patients with type 1 diabetes [66, 67]. Consistent with this, animal models of streptozotocin-induced type 1 diabetes show impaired healing evidenced by reduced mesenchymal cell proliferation in the early callus, reduced matrix deposition (collagen), and reduced biomechanical properties in the healed fracture [68, 69]. Additionally, diabetes-related overexpression of TNF-α leads to accelerated loss of cartilage in the callus [70] due to increased chondrocyte apoptosis [71]. While it is not known if the impaired healing is the result of hypoinsulinemia or hyperglycemia/formation of advanced glycation endproducts (AGE), insulin treatment to normalize blood glucose in a diabetic rat bone explant model [72] and a murine femur fracture model [73] has been shown to reverse the deficit in healing. Interestingly, during diabetic bone healing, osteoblast expression of RAGE (cell surface receptors for AGE) is enhanced, possibly facilitating the effects of AGE on reduced bone repair [74]. Lastly, in a diabetic rat fracture model, local intramedullary delivery of insulin to the fracture site, which does not provide systemic management of glucose, reversed the healing deficit at both early (mesenchymal cell proliferation and chondrogenesis) and late (mineralization and biomechanical strength) time points [75]. This supports a novel hypothesis predicting that there is a direct anabolic effect of insulin on cells at the fracture site. It should be noted that in the case of obesity-related type 2 diabetes, clinical data clearly suggest an impairment of fracture healing [76], but the molecular and metabolic basis of the impairment is not currently known. This is likely to be an active area of research given the growing size of the obese and type 2 diabetic population worldwide [77].

Cigarette smoking: Clinically, smoking has been shown to have a negative impact on skeletal healing following long bone fracture [78, 79] and spinal fusion surgery [80]. Little is known about the mechanism underlying these deficits in skeletal repair, with mesenchymal cell condensation and the process of chondrogenesis hypothesized to be important targets of cigarette smoke. This hypothesis is supported by studies demonstrating inhibited cartilage formation during distraction osteogenesis in rabbits [81] and delayed chondrogenesis in a murine tibial fracture model [82]. The most widely studied molecule in the context of smoking and bone healing is nicotine, which has been shown to inhibit distraction osteogenesis [83], spinal fusion [84, 85], and fracture healing in rabbits [86]. Conversely, cigarette smoke, but not nicotine, has been shown to inhibit bone healing around a titanium implant in rabbits [87–89], nor was nicotine found to affect mechanical strength in a rat fracture model [90]. Recently, another class of molecules in cigarette smoke, polycyclic aromatic hydrocarbons, has been implicated in impaired tibial fracture healing in the mouse via activation of the aryl hydrocarbon receptor [91]. In general, however, a full characterization of the healing process in smokers or animal models of smoke exposure is necessary, and work to identify which components of cigarette smoke are responsible for their effects is important if the underlying molecular mechanisms are to be fully understood.

MOLECULAR THERAPIES TO ENHANCE BONE HEALING

Currently, the only FDA-approved molecular therapy for bone healing is BMP-2. As mentioned previously, the underlying rationale for its therapeutic potential is based on the finding that elimination of BMP-2 in the limb disrupts the initiation of postnatal fracture healing [33], establishing its essential role in the process. Thus, it is no surprise that a number of animal studies have identified a positive effect of BMP-2 or activation of its signaling pathway on various bone healing situations. As its use has gained attention, clinical data have emerged that supports the use of BMP-2 in clinical situations. For example, recombinant BMP-2 delivered in a collagen sponge with a cancellous autograft aids in the healing of tibial diaphyseal fractures [92, 93]. Spine fusion patients were also found to have better neck disability and arm pain scores 24 months postoperatively when the INFUSE® bone graft (collagen sponge impregnated with BMP-2) was used [94]. Despite the positive outcomes of these and other studies, it should be noted that the clinical and cost effectiveness of BMP-2 in skeletal repair situations remains an open debate [3, 95].

Recent work has been focused on development of molecular therapies to enhance healing or induce repair in nonunion situations using agents known for their bone-anabolic capability: parathyroid hormone (PTH) and activators of the Wnt/β-catenin pathway. Since PTH is an FDA-approved treatment for enhancing bone mass in osteoporosis patients [96], its repurposing as a candidate therapy for fracture healing in patients with nonunion has been proposed. In addition to several case reports, recent studies in humans [97] and animals [98–102] show compelling evidence of positive actions of PTH on bone healing. While the mechanism underlying the ability of PTH to enhance healing remains to be determined, recent findings suggest that PTH increases mesenchymal cell proliferation and accelerates chondrogenesis via enhanced upregulation of SOX-9 in rat femur fractures [103]. Regarding the modulation of Wnt signaling, either genetic or molecular (via Wnt3a) enhancement of β-catenin signaling in osteo- and chondroprogenitors has been recently shown to accelerate fracture healing in the mouse [104], suggesting a potential therapeutic strategy involving agonism of this pathway in the treatment of delayed healing [105]. Overall, these recent and novel advances set PTH and Wnt3a treatment on the horizon as important emerging candidate molecular therapies for accelerating healing or alleviating/reversing the development of fracture nonunion.

REFERENCES

1. Calori GM, Albisetti W, Agus A, Iori S, Tagliabue L. 2007. Risk factors contributing to fracture non-unions. *Injury* 38(Suppl 2): S11–18.
2. Tzioupis C, Giannoudis PV. 2007. Prevalence of long-bone non-unions. *Injury* 38(Suppl 2): S3–9.
3. Garrison KR, Shemilt I, Donell S, Ryder JJ, Mugford M, Harvey I, Song F, Alt V. 2010. Bone morphogenetic protein (BMP) for fracture healing in adults. *Cochrane Database Syst Rev* CD006950.
4. Einhorn TA. 1998. The cell and molecular biology of fracture healing. *Clin Orthop* 355(Suppl): S7–21.
5. Gerstenfeld LC, Cullinane DM, Barnes GL, Graves DT, Einhorn TA. 2003. Fracture healing as a post-natal developmental process: Molecular, spatial, and temporal aspects of its regulation *J Cell Biochem* 88: 873–884.
6. Marsell R, Einhorn TA. 2011. The biology of fracture healing. *Injury* 42: 551–555.
7. Matsumoto T, Mifune Y, Kawamoto A, Kuroda R, Shoji T, Iwasaki H, Suzuki T, Oyamada A, Horii M, Yokoyama A, Nishimura H, Lee SY, Miwa M, Doita M, Kurosaka M, Asahara T. 2008. Fracture induced mobilization and incorporation of bone marrow-derived endothelial progenitor cells for bone healing. *J Cell Physiol* 215: 234–242.
8. Ueno M, Uchida K, Takaso M, Minehara H, Suto K, Takahira N, Steck R, Schuetz MA, Itoman M. 2011. Distribution of bone marrow-derived cells in the fracture callus during plate fixation in a green fluorescent protein-chimeric mouse model. *Exp Anim* 60: 455–462.
9. Henrotin Y. 2011. Muscle: A source of progenitor cells for bone fracture healing. *BMC Medicine* 9: 136.
10. Glass GE, Chan JK, Freidin A, Feldmann M, Horwood NJ, Nanchahal J. 2011. TNF-alpha promotes fracture repair by augmenting the recruitment and differentiation of muscle-derived stromal cells. *Proc Natl Acad Sci U S A* 108: 1585–1590.
11. Zhang X, Naik A, Xie C, Reynolds D, Palmer J, Lin A, Awad H, Guldberg R, Schwarz E, O'Keefe R. 2005. Periosteal stem cells are essential for bone revitalization and repair. *J Musculoskelet Neuronal Interact* 5: 360–362.
12. Ushiku C, Adams DJ, Jiang X, Wang L, Rowe DW. 2010. Long bone fracture repair in mice harboring GFP reporters for cells within the osteoblastic lineage. *J Orthop Res* 28: 1338–1347.
13. Granero-Molto F, Weis JA, Miga MI, Landis B, Myers TJ, O'Rear L, Longobardi L, Jansen ED, Mortlock DP, Spagnoli A. 2009. Regenerative effects of transplanted mesenchymal stem cells in fracture healing. *Stem Cells* 27: 1887–1898.
14. Kitaori T, Ito H, Schwarz EM, Tsutsumi R, Yoshitomi H, Oishi S, Nakano M, Fujii N, Nagasawa T, Nakamura T. 2009. Stromal cell-derived factor 1/CXCR4 signaling is critical for the recruitment of mesenchymal stem cells to the fracture site during skeletal repair in a mouse model. *Arthritis Rheum* 60: 813–823.
15. Thompson Z, Miclau T, Hu D, Helms JA. 2002. A model for intramembranous ossification during fracture healing. *J Orthop Res* 20: 1091–1098.
16. Augustin G, Antabak A, Davila S. 2007. The periosteum Part 1: Anatomy, histology and molecular biology. *Injury* 38: 1115–1130.
17. Orwoll ES. 2003. Toward an expanded understanding of the role of the periosteum in skeletal health. *J Bone Miner Res* 18: 949–954.
18. Komatsu DE, Hadjiargyrou M. 2004. Activation of the transcription factor HIF-1 and its target genes, VEGF, HO-1, iNOS, during fracture repair. *Bone* 34: 680–688.
19. Schipani E. 2005. Hypoxia and HIF-1 alpha in chondrogenesis. *Semin Cell Dev Biol* 16: 539–546.
20. Kon T, Cho TJ, Aizawa T, Yamazaki M, Nooh N, Graves D, Gerstenfeld LC, Einhorn TA. 2001. Expression of osteoprotegerin, receptor activator of NF-kappaB ligand (osteoprotegerin ligand) and related proinflammatory cytokines during fracture healing. *J Bone Miner Res* 16: 1004–1014.
21. Mountziaris PM, Mikos AG. 2008. Modulation of the inflammatory response for enhanced bone tissue regeneration. *Tissue Eng Part B Rev* 14: 179–186.
22. ZS Ai-Aql, AS Alagl, DT Graves, LC Gerstenfeld, TA Einhorn. 2008. Molecular mechanisms controlling bone formation during fracture healing and distraction osteogenesis. *J Dent Res* 87: 107–118.
23. Gerstenfeld LC, Sacks DJ, Pelis M, Mason ZD, Graves DT, Barrero M, Ominsky MS, Kostenuik PJ, Morgan EF, Einhorn TA. 2009. Comparison of effects of the bisphosphonate alendronate versus the RANKL inhibitor denosumab on murine fracture healing. *J Bone Miner Res* 24: 196–208.
24. Uusitalo H, Salminen H, Vuorio E. 2001. Activation of chondrogenesis in response to injury in normal and transgenic mice with cartilage collagen mutations. *Osteoarthritis Cartilage* 9(Suppl A): S174–S179.

25. Le AX, Miclau T, Hu D, Helms JA. 2001. Molecular aspects of healing in stabilized and non-stabilized fractures. *J Orthop Research* 19: 78–84.
26. Murakami S, Noda M. 2000. Expression of Indian hedgehog during fracture healing in adult rat femora. *Calcif Tissue Int* 66: 272–276.
27. Pufe T, Wildemann B, Petersen W, Mentlein R, Raschke M, Schmidmaier G. 2002. Quantitative measurement of the splice variants 120 and 164 of the angiogenic peptide vascular endothelial growth factor in the time flow of fracture healing: A study in the rat. *Cell Tissue Res* 309: 387–392.
28. Gerber HP, Vu TH, Ryan AM, Kowalski J, Werb Z, Ferrara N. 1999. VEGF couples hypertrophic cartilage remodeling, ossification and angiogenesis during endochondral bone formation. *Nat Med*, 5: 623–628.
29. Athanasopoulos AN, Schneider D, Keiper T, Alt V, Pendurthi UR, Liegibel UM, Sommer U, Nawroth PP, Kasperk C, Chavakis T. 2007. Vascular endothelial growth factor (VEGF)-induced up-regulation of CCN1 in osteoblasts mediates proangiogenic activities in endothelial cells and promotes fracture healing. *J Biol Chem* 282: 26746–26753.
30. Kawahata H, Kikkawa T, Higashibata Y, Sakuma T, Huening M, Sato M, Sugimoto M, Kuriyama K, Terai K, Kitamura Y, Nomura S. 2003. Enhanced expression of Runx2/PEBP2alphaA/CBFA1/AML3 during fracture healing. *J Orthop Sci* 8: 102–108.
31. Naik AA, Xie C, Zuscik MJ, Kingsley P, Schwarz EM, Awad H, Guldberg R, Drissi H, Puzas JE, Boyce B, Zhang X, O'Keefe RJ. 2009. Reduced COX-2 expression in aged mice is associated with impaired fracture healing. *J Bone Miner Res* 24: 251–264.
32. Lu H, Kraut D, Gerstenfeld LC, Graves DT. 2003. Diabetes interferes with the bone formation by affecting the expression of transcription factors that regulate osteoblast differentiation. *Endocrinology* 144: 346–352.
33. Tsuji K, Bandyopadhyay A, Harfe BD, Cox K, Kakar S, Gerstenfeld L, Einhorn T, Tabin CJ, Rosen V. 2006. BMP2 activity, although dispensable for bone formation, is required for the initiation of fracture healing. *Nat Genet* 38: 1424–1429.
34. Betz OB, Betz VM, Nazarian A, Egermann M, Gerstenfeld LC, Einhorn TA, Vrahas MS, Bouxsein ML, Evans CH. 2007. Delayed administration of adenoviral BMP-2 vector improves the formation of bone in osseous defects. *Gene Ther* 14: 1039–1044.
35. Vortkamp A, Pathi S, Peretti GM, Caruso EM, Zaleske DJ, Tabin C. 1998. Recapitulation of signals regulating embryonic bone formation during postnatal growth and in fracture repair. *Mech Dev* 71: 65–76.
36. De Biase P, Capanna R. 2005. Clinical applications of BMPs. *Injury* 36 Suppl 3: S43–S46.
37. Kakar S, Einhorn TA, Vora S, Miara LJ, Hon G, Wigner NA, Toben D, Jacobsen KA, Al-Sebaei MO, Song M, Trackman PC, Morgan EF, Gerstenfeld LC, Barnes GL. 2007. Enhanced chondrogenesis and Wnt signaling in PTH-treated fractures. *J Bone Miner Res* 22: 1903–1912.
38. Chen Y, Whetstone HC, Lin AC, Nadesan P, Wei Q, Poon R, Alman BA. 2007. Beta-catenin signaling plays a disparate role in different phases of fracture repair: Implications for therapy to improve bone healing. *PLoS Med* 4: e249.
39. Huang Y, Zhang X, Du K, Yang F, Shi Y, Huang J, Tang T, Chen D, Dai K. 2012. Inhibition of beta-catenin signaling in chondrocytes induces delayed fracture healing in mice. *J Orthop Res* 30: 304–310.
40. Szczesny G. 2002. Molecular aspects of bone healing and remodeling. *Pol J Pathol* 53: 145–153.
41. Weiss S, Henle P, Bidlingmaier M, Moghaddam A, Kasten P, Zimmermann G. 2007. Systemic response of the GH/IGF-I axis in timely versus delayed fracture healing. *Growth Horm IGF Res* 18(3): 205–12.
42. Lehmann W, Edgar CM, Wang K, Cho TJ, Barnes GL, Kakar S, Graves DT, Rueger JM, Gerstenfeld LC, Einhorn TA. 2005. Tumor necrosis factor alpha (TNF-alpha) coordinately regulates the expression of specific matrix metalloproteinases (MMPS) and angiogenic factors during fracture healing. *Bone* 36: 300–310.
43. Baldik Y, Diwan AD, Appleyard RC, Fang ZM, Wang Y, Murrell GA. 2005. Deletion of iNOS gene impairs mouse fracture healing *Bone* 37: 32–36.
44. Kimble RB, Bain S, Pacifici R. 1997. The functional block of TNF but not of IL-6 prevents bone loss in ovariectomized mice. *J Bone Miner Res* 12: 935–941.
45. Gerstenfeld LC, Shapiro FD. 1996. Expression of bone-specific genes by hypertrophic chondrocytes: implication of the complex functions of the hypertrophic chondrocyte during endochondral bone development. *J Cell Biochem* 62: 1–9.
46. Yang X, Ricciardi BF, Hernandez-Soria A, Shi Y, Pleshko CN, Bostrom MP. 2007. Callus mineralization and maturation are delayed during fracture healing in interleukin-6 knockout mice. *Bone* 41: 928–936.
47. Sudmann E, Hagen T. 1976. Indomethacin-induced delayed fracture healing. *Arch Orthop Unfallchir* 85: 151–154.
48. Ho ML, Chang JK, Wang GJ. 1998. Effects of ketorolac on bone repair: A radiographic study in modeled demineralized bone matrix grafted rabbits. *Pharmacology* 57: 148–159.
49. Vane JR. 1971. Inhibition of prostaglandin synthesis as a mechanism of action for aspirin-like drugs. *Nat New Biol* 231: 232–235.
50. Zhang X, Schwarz EM, Young DA, Puzas JE, Rosier RN, O'Keefe RJ. 2002. Cyclooxygenase-2 regulates mesenchymal cell differentiation into the osteoblast lineage and is critically involved in bone repair. *J Clin Invest* 109: 1405–1415.
51. Street J, Bao M, deGuzman L, Bunting S, Peale FV Jr, Ferrara N, Steinmetz H, Hoeffel J, Cleland JL, Daugherty A, van Bruggen N, Redmond HP, Carano RA, Filvaroff EH. 2002. Vascular endothelial growth factor stimulates bone repair by promoting angiogenesis and bone turnover. *Proc Natl Acad Sci U S A* 99: 9656–9661.
52. Ito H, Koefoed M, Tiyapatanaputi P, Gromov K, Goater JJ, Carmouche J, Zhang X, Rubery PT, Rabinowitz J,

Samulski RJ, Nakamura T, Soballe K, O'Keefe RJ, Boyce BF, Schwarz EM. 2005. Remodeling of cortical bone allografts mediated by adherent rAAV-RANKL and VEGF gene therapy. *Nat Med* 11: 291–297.
53. Deckers MM, van Bezooijen RL, van der Horst G, Hoogendam J, van Der Bent C, Papapoulos SE, Löwik CW. 2002. Bone morphogenetic proteins stimulate angiogenesis through osteoblast-derived vascular endothelial growth factor A. *Endocrinology* 143: 1545–1553.
54. Skak SV, Jensen TT. 1988. Femoral shaft fracture in 265 children. Log-normal correlation with age of speed of healing. *Acta Orthop Scand* 59: 704–707.
55. Nieminen S, Nurmi M, Satokari K. 1981. Healing of femoral neck fractures; influence of fracture reduction and age. *Ann Chir Gynaecol* 70: 26–31.
56. Nilsson BE, Edwards P. 1969. Age and fracture healing: A statistical analysis of 418 cases of tibial shaft fractures. *Geriatrics* 24: 112–117.
57. Hee HT, Wong HP, Low YP, Myers L. 2001. Predictors of outcome of floating knee injuries in adults: 89 patients followed for 2–12 years. *Acta Orthop Scand* 72: 385–394.
58. Meyer RA Jr, Desai BR, Heiner DE, Fiechtl J, Porter S, Meyer MH. 2006. Young, adult, and old rats have similar changes in mRNA expression of many skeletal genes after fracture despite delayed healing with age. *J Orthop Res* 24: 1933–1944.
59. Meyer RA Jr, Meyer MH, Tenholder M, Wondracek S, Wasserman R, Garges P. 2003. Gene expression in older rats with delayed union of femoral fractures, *J Bone Joint Surg Am* 85-A: 1243–1254.
60. Bajada S, Marshall MJ, Wright KT, Richardson JB, Johnson WE. 2009. Decreased osteogenesis, increased cell senescence and elevated Dickkopf-1 secretion in human fracture non union stromal cells. *Bone* 45: 726–735.
61. Gruber R, Koch H, Doll BA, Tegtmeier F, Einhorn TA, Hollinger JO. 2006. Fracture healing in the elderly patient. *Exp Gerontol* 41: 1080–1093.
62. Akune T, Ohba S, Kamekura S, Yamaguchi M, Chung UI, Kubota N, Terauchi Y, Harada Y, Azuma Y, Nakamura K, Kadowaki T, Kawaguchi H. 2004. PPARgamma insufficiency enhances osteogenesis through osteoblast formation from bone marrow progenitors *J Clin Invest* 113: 846–855
63. Cao JJ, Wronski TJ, Iwaniec U, Phleger L, Kurimoto P, Boudignon B, Halloran BP. 2005. Aging increases stromal/osteoblastic cell-induced osteoclastogenesis and alters the osteoclast precursor pool in the mouse. *J Bone Miner Res* 20: 1659–1668.
64. Brandes RP, Fleming I, Busse R. 2005. Endothelial aging. *Cardiovascular Research* 66: 286–294.
65. Edelberg JM, Reed MJ. 2003. Aging and angiogenesis. *Front Biosci* 8: s1199–s1209.
66. Loder RT. 1988. The influence of diabetes mellitus on the healing of closed fractures. *Clin Orthop Relat Res* 232: 210–216.
67. Blakytny R, Spraul M, Jude EB. 2011. Review: The diabetic bone: A cellular and molecular perspective. *Int J Low Extrem Wounds* 10: 16–32.
68. Beam HA, Parsons JR, Lin SS. 2002. The effects of blood glucose control upon fracture healing in the BB Wistar rat with diabetes mellitus. *J Orthop Res* 20: 1210–1216.
69. Funk JR, Hale JE, Carmines D, Gooch HL, Hurwitz SR. 2000. Biomechanical evaluation of early fracture healing in normal and diabetic rats. *J Orthop Res*, 18: 126–132.
70. Alblowi J, Kayal RA, Siqueira M, McKenzie E, Krothapalli N, McLean J, Conn J, Nikolajczyk B, Einhorn TA, Gerstenfeld L, Graves DT. 2009. High levels of tumor necrosis factor-alpha contribute to accelerated loss of cartilage in diabetic fracture healing. *Am J Pathol* 175: 1574–1585.
71. Kayal RA, Siqueira M, Alblowi J, McLean J, Krothapalli N, Faibish D, Einhorn TA, Gerstenfeld LC, Graves DT. 2010. TNF-alpha mediates diabetes-enhanced chondrocyte apoptosis during fracture healing and stimulates chondrocyte apoptosis through FOXO1. *J Bone Miner Res* 25: 1604–1615.
72. Hough S, Avioli LV, Bergfeld MA, Fallon MD, Slatopolsky E, Teitelbaum SL. 1981. Correction of abnormal bone and mineral metabolism in chronic streptozotocin-induced diabetes mellitus in the rat by insulin therapy. *Endocrinology* 108: 2228–2234.
73. Kayal RA, Alblowi J, McKenzie E, Krothapalli N, Silkman L, Gerstenfeld L, Einhorn TA, Graves DT. 2009. Diabetes causes the accelerated loss of cartilage during fracture repair which is reversed by insulin treatment. *Bone* 44: 357–363.
74. Santana RB, Xu L, Chase HB, Amar S, Graves DT, Trackman PC. 2003. A role for advanced glycation end products in diminished bone healing in type 1 diabetes. *Diabetes* 52: 1502–1510.
75. Gandhi A, Beam HA, O'Connor JP, Parsons JR, Lin SS. 2005. The effects of local insulin delivery on diabetic fracture healing. *Bone* 37: 482–490.
76. Khazai NB, Beck GR Jr, Umpierrez GE. 2009. Diabetes and fractures: An overshadowed association. *Curr Opin Endocrinol Diabetes Obes* 16: 435–445.
77. Shamseddeen H, Getty JZ, Hamdallah IN, Ali MR. 2011. Epidemiology and economic impact of obesity and type 2 diabetes. *Surg Clin North Am* 91: 1163–1172, vii.
78. Schmitz MA, Finnegan M, Natarajan R, Champine J. 1999. Effect of smoking on tibial shaft fracture healing. *Clin Orthop Relat Res* 365: 184–200.
79. Sloan A, Hussain I, Maqsood M, Eremin O, El-Sheemy M. 2010. The effects of smoking on fracture healing. *The Surgeon* 8: 111–116.
80. Hadley MN, Reddy SV. 1997. Smoking and the human vertebral column: A review of the impact of cigarette use on vertebral bone metabolism and spinal fusion. *Neurosurgery* 41: 116–124.
81. Ueng SW, Lee MY, Li AF, Lin SS, Tai CL, Shih CH. 1997. Effect of intermittent cigarette smoke inhalation on tibial lengthening: Experimental study on rabbits. *J Trauma* 42: 231–238.
82. El-Zawawy HB, Gill CS, Wright RW, Sandell LJ. 2006. Smoking delays chondrogenesis in a mouse model of

83. Ma L, Zheng LW, Cheung LK. 2007. Inhibitory effect of nicotine on bone regeneration in mandibular distraction osteogenesis. *Front Biosci* 12: 3256–3262.
84. Silcox DH III, Daftari T, Boden SD, Schimandle JH, Hutton WC, Whitesides TE Jr. 1995. The effect of nicotine on spinal fusion. *Spine* 20: 1549–1553.
85. Silcox DH III, Boden SD, Schimandle JH, Johnson P, Whitesides TE, Hutton WC. 1998. Reversing the inhibitory effect of nicotine on spinal fusion using an osteoinductive protein extract. *Spine* 23: 291–296.
86. Raikin SM, Landsman JC, Alexander VA, Froimson MI, Plaxton NA. 1998. Effect of nicotine on the rate and strength of long bone fracture healing. *Clin Orthop Relat Res* 353: 231–237.
87. Balatsouka D, Gotfredsen K, Lindh CH, Berglundh T. 2005. The impact of nicotine on osseointegration. An experimental study in the femur and tibia of rabbits. *Clin Oral Implants Res* 16: 389–395.
88. Balatsouka D, Gotfredsen K, Lindh CH, Berglundh T. 2005. The impact of nicotine on bone healing and osseointegration. *Clin Oral Implants Res* 16: 268–276.
89. Cesar-Neto JB, Duarte PM, Sallum EA, Barbieri D, Moreno H Jr, Nociti FH Jr. 2003. A comparative study on the effect of nicotine administration and cigarette smoke inhalation on bone healing around titanium implants. *J Periodontol* 74: 1454–1459.
90. Skott M, Andreassen TT, Ulrich-Vinther M, Chen X, Keyler DE, LeSage MG, Pentel PR, Bechtold JE, Soballe K. 2006. Tobacco extract but not nicotine impairs the mechanical strength of fracture healing in rats. *J Orthop Res* 24: 1472–1479.
91. Kung MH, Yukata K, O'Keefe RJ, Zuscik MJ. 2011. Aryl hydrocarbon receptor-mediated impairment of chondrogenesis and fracture healing by cigarette smoke and benzo(a)pyrene. *J Cell Physiol* 227(3): 1062–1070.
92. Jones AL, Bucholz RW, Bosse MJ, Mirza SK, Lyon TR, Webb LX, Pollak AN, Golden JD, Valentin-Opran A. 2006. Recombinant human BMP-2 and allograft compared with autogenous bone graft for reconstruction of diaphyseal tibial fractures with cortical defects. A randomized, controlled trial. *J Bone Joint Surg Am* 88: 1431–1441.
93. Swiontkowski MF, Aro HT, Donell S, Esterhai JL, Goulet J, Jones A, Kregor PJ, Nordsletten L, Paiement G, Patel A. 2006. Recombinant human bone morphogenetic protein-2 in open tibial fractures. A subgroup analysis of data combined from two prospective randomized studies. *J Bone Joint Surg Am* 88: 1258–1265.
94. Baskin DS, Ryan P, Sonntag V, Westmark R, Widmayer MA. 2003. A prospective, randomized, controlled cervical fusion study using recombinant human bone morphogenetic protein-2 with the CORNERSTONE-SR allograft ring and the ATLANTIS anterior cervical plate. *Spine* 28: 1219–1224.
95. Garrison KR, Donell S, Ryder J, Shemilt I, Mugford M, Harvey I, Song F. 2007. Clinical effectiveness and cost-effectiveness of bone morphogenetic proteins in the non-healing of fractures and spinal fusion: A systematic review. *Health Technol Assess* 11: 1–iv.
96. FDA. 2003. Forteo approved for osteoporosis treatment. *FDA Consum* 37: 4.
97. Aspenberg P, Genant HK, Johansson T, Nino AJ, See K, Krohn K, Garcia-Hernandez PA, Recknor CP, Einhorn TA, Dalsky GP, Mitlak BH, Fierlinger A, Lakshmanan MC. 2010. Teriparatide for acceleration of fracture repair in humans: A prospective, randomized, double-blind study of 102 postmenopausal women with distal radial fractures. *J Bone Miner Res* 25: 404–414.
98. Andreassen TT, Willick GE, Morley P, Whitfield JF. 2004. Treatment with parathyroid hormone hPTH(1-34), hPTH(1-31), and monocyclic hPTH(1-31) enhances fracture strength and callus amount after withdrawal fracture strength and callus mechanical quality continue to increase. *Calcif Tissue Int* 74: 351–356.
99. Alkhiary YM, Gerstenfeld LC, Krall E, Westmore M, Sato M, Mitlak BH, Einhorn TA. 2005. Enhancement of experimental fracture-healing by systemic administration of recombinant human parathyroid hormone (PTH 1-34). *J Bone Joint Surg Am* 87: 731–741.
100. Komatsubara S, Mori S, Mashiba T, Nonaka K, Seki A, Akiyama T, Miyamoto K, Cao Y, Manabe T, Norimatsu H. 2005. Human parathyroid hormone (1-34) accelerates the fracture healing process of woven to lamellar bone replacement and new cortical shell formation in rat femora. *Bone* 36: 678–687.
101. Kaback LA, Soung DY, Naik A, Geneau G, Schwarz EM, Rosier RN, O'Keefe RJ, Drissi H. Teriparatide (1-34 human PTH) regulation of osterix during fracture repair. *J Cell Biochem* 105: 219–226.
102. Reynolds DG, Takahata M, Lerner AL, O'Keefe RJ, Schwarz EM, Awad HA. 2011. Teriparatide therapy enhances devitalized femoral allograft osseointegration and biomechanics in a murine model. *Bone* 48: 562–570.
103. Nakazawa T, Nakajima A, Shiomi K, Moriya H, Einhorn TA, Yamazaki M. 2005. Effects of low-dose, intermittent treatment with recombinant human parathyroid hormone (1-34) on chondrogenesis in a model of experimental fracture healing. *Bone* 37: 711–719.
104. Minear S, Leucht P, Jiang J, Liu B, Zeng A, Fuerer C, Nusse R, Helms JA. 2010. Wnt proteins promote bone regeneration. *Sci Transl Med* 2: 29ra30.
105. Einhorn TA. 2010. The Wnt signaling pathway as a potential target for therapies to enhance bone repair. *Sci Transl Med* 2: 42ps36.

12
Biomechanics of Fracture Healing
Elise F. Morgan and Thomas A. Einhorn

Introduction 99
Biomechanical Assessment of Fracture Healing 99
Biomechanical Stages of Fracture Healing 100
Noninvasive Assessment of Fracture Healing 101

Mechanobiology of Fracture Healing 102
Summary 102
References 103

INTRODUCTION

Fracture healing involves a dynamic interplay of biological processes that when properly executed restore form and function to the injured bone. This chapter presents a biomechanical description of fracture healing, with emphasis on methods of assessing the extent of healing—as defined principally by the extent of the regaining of mechanical function—and on the role of the local mechanical environment. Fracture healing is often classified as either primary or secondary fracture healing, where the former is characterized by direct cortical reconstruction and the latter involves substantial periosteal callus formation. The techniques for assessing healing that are presented in this chapter apply equally well to primary and secondary healing; however, the overviews of the biomechanical stages of fracture healing and the mechanobiology of fracture healing are largely specific to secondary healing. We also note that this chapter does not include a discussion of the biomechanics of fracture fixation, as this topic has been extensively reviewed elsewhere [1–4].

BIOMECHANICAL ASSESSMENT OF FRACTURE HEALING

In the laboratory setting, the mechanical properties of a healing bone are commonly assessed by mechanical tests that load the bone in torsion or in three-point bending. Tension and compression tests are less common. The choice of the type of test is dictated by technical as well as physiological considerations. For example, bending and torsion are logical choices when studying fracture healing in long bones, because these bones experience bending and torsional moments *in vivo*. However, whereas torsion tests subject every cross-section of the callus to the same torque, three-point bending creates a nonuniform bending moment throughout the callus. As a result, failure of the callus during a three-point bend test does not necessarily occur at the weakest cross-section of the callus.

Regardless of the type of mechanical test, the outcome measures that can be obtained are the strength, stiffness, rigidity, and toughness of the healing bone (Fig. 12.1). For torsion tests, an additional parameter, twist to failure, can be used as a measure of the ductility of the callus. Although strength, a measure of the force or moment that causes failure, can only be measured once for a given callus, it is possible to obtain more than one measure of stiffness and rigidity. Multistage testing protocols have been reported that apply nondestructive loads to the callus in planes or in loading modes that are different from those used for the stage of the test in which the callus is loaded to failure. With these protocols, it is possible to quantify the bending stiffness in multiple planes [5] or the torsional as well as compressive stiffness [6].

The mechanical properties illustrated in Fig. 12.1 are structural, rather than material, properties. *Material properties* describe the intrinsic mechanical behavior of

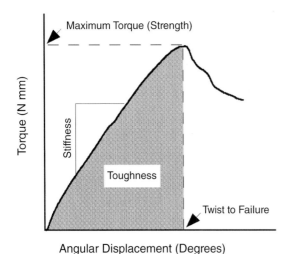

Fig. 12.1. Representative torque-twist curve for a mouse tibia 21 days post fracture. The curve is annotated to show definitions of basic biomechanical parameters. Torsional rigidity is computed by multiplying the torsional stiffness by the gage length. Analogous definitions hold for bending tests.

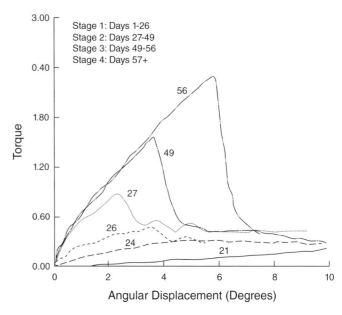

Fig. 12.2. Torque-twist curves for healing rabbit tibiae at various time points (in days) post fracture. [After White et al. (Ref. 10).]

a particular type of material (tissue), such as woven bone, fibrocartilage, or granulation tissue. The *structural properties* of a fracture callus depend on the material properties of the individual callus tissues as well as the spatial arrangement of the tissues and the overall geometry of the callus. While it is possible to use measurements of callus geometry together with those of structural properties to gain some insight into callus tissue material properties [7], true measurement of these material properties requires direct testing of individual callus tissues [8, 9].

BIOMECHANICAL STAGES OF FRACTURE HEALING

White et al. [10] used the results of torsion tests performed on healing rabbit tibiae at multiple time-points (Fig. 12.2) to define four biomechanical stages of secondary fracture healing. *Stage 1* is characterized by extremely low callus stiffness and strength, and failure during the torsion test occurs at the original fracture line. *Stage 2* corresponds to a notable increase in callus stiffness and, to a lesser extent, strength. However, it is not until *Stage 3* that failure during the torsion test occurs at least partly outside of the original fracture line. This stage is also characterized by an increase in callus strength from Stage 2. Finally, in *Stage 4*, failure during the torsion test occurs in the intact bone rather than through the original fracture line. Although fracture healing is commonly described in terms of four biological phases (inflammation, soft callus formation, bony callus formation, and remodeling), these phases do not map onto the four biomechanical stages in a one-to-one manner. Stage 1 does correspond to the inflammatory phase, yet Stage 2 encompasses the soft callus phase as well as the first part of the bony callus phase. It is the occurrence of bony bridging of the fracture line that is responsible for the increase in stiffness observed in Stage 2. The transition from Stage 3 to Stage 4 roughly corresponds to the start of the remodeling phase.

If the bony callus is sufficiently large, the rigidity and strength of the callus during Stage 3 can exceed that of the intact bone. Even though the callus tissues at this stage are not as rigid or as strong as those of well-mineralized lamellar bone, the larger cross-sectional area and moments of inertia of the callus as compared to the intact bone can overcompensate for the inferior material properties. However, though robust, the callus at this point in the healing progression is also mechanically inefficient. Through remodeling, the callus is able to retain sufficient mechanical integrity with less mass.

Results of several recent studies further illustrate the biomechanical consequences of individual biological phases of healing. For example, intermittent parathyroid hormone (PTH) (1-34) treatment has been shown to increase callus strength [11, 12] primarily as a result of enhanced chondrogenesis [13]. However, while PTH treatment leads to an increase in callus size, a slight decrease in the fraction of the callus that is comprised of mineralized tissue was observed [13], suggesting that the mechanical enhancement results purely from modulation of callus geometry [Fig. 12.3(A)]. The biomechanical importance of the extent of bony bridging (Stage 2), and

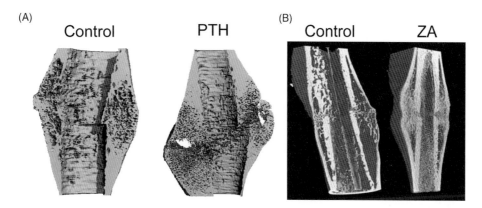

Fig. 12.3. (A) Longitudinal cut-away views of 3D microcomputed tomography reconstructions of representative saline- (control) and PTH-treated murine fracture calluses at 14 days post fracture. (Data from Ref. 13.) (B) Longitudinal cross-sections of the fracture callus and cortex at 6 weeks post fracture in rats treated with saline (control) and zolendronic acid (ZA) beginning 2 weeks after fracture. (From Ref. 10; images not to scale.)

in particular the extent of outer cortical bridging was demonstrated in a study of the effect of lovastatin treatment on fracture healing [14]. With respect to the later stages of healing, bisphosphonate treatment has been shown to enhance callus strength through inhibition of callus remodeling, resulting in a larger callus and larger proportion of mineralized tissue [Fig. 12.3(B)] [15, 16].

NONINVASIVE ASSESSMENT OF FRACTURE HEALING

While mechanical tests provide the gold-standard measures of healing in laboratory studies of fracture healing, clinical assessment of healing requires noninvasive methods. Multiple noninvasive approaches to measuring callus stiffness, whether in axial loading or bending, have been reported, and the clinical feasibility of several has been demonstrated. Typically, these measurements rely on measuring the displacement across the fracture gap or the pin-to-pin displacement under a known force or bending moment [17–19]. If an external fixator is present, it is necessary to consider only the fraction of the applied load that is borne by the callus as opposed to the fixator. From these approaches, quantitative criteria for healing have been put forth. For example, it has been proposed that a fracture can be considered healed when the bending stiffness (the ratio of the applied bending moment to the angular displacement) exceeds a certain threshold (15 Nm/deg in the case of human tibia fractures) [17], that "healing time" can be defined as the time required to achieve bony bridging of the callus (though assessment of bridging by radiographs is subjective) [20], and that in distraction osteogenesis, external fixation can be removed when the fraction of the axial force borne by the fixator is less than 10% [21].

Other noninvasive methods of assessing healing provide surrogate, rather than direct, measures of callus mechanical properties and include acoustic emission [22], ultrasound [23–28], and computed tomography (CT) imaging. Direct comparisons of CT and standard radiographic analyses have indicated that the former can yield comparable or better predictions of callus compressive strength[29], bending strength[14], and torsional strength and stiffness [30–32], and more definitive diagnoses of healing progression[33] and of nonunions [34]. No consensus currently exists, however, as to which CT-derived measures, or combinations of measures, best predict callus strength and stiffness for a range of types of fractures and/or bony defects.

Importantly, the vast majority of noninvasive approaches to monitoring healing focus on callus stiffness and not callus strength. While noninvasive measures of stiffness may provide valuable information about the healing process, a method to evaluate strength would be more clinically meaningful as it would theoretically provide information regarding the ability to bear weight and carry loads. In this respect, acoustic methods may pose a considerable advantage, as analysis of ultrasonic wave propagation across the fracture gap can be used to detect bony bridging of the gap. Another viable approach is CT-based finite element analysis, in which CT images are used to construct a finite element model of the callus. This approach was demonstrated for estimating callus stiffness [35]; however, this approach requires two key types of input for accurate estimates: (1) the elastic and failure properties of the callus tissues, and (2) the types of loads and/or displacements that the callus is subjected to *in vivo*. As mentioned earlier, direct measurements of the material properties of callus tissues have been reported [9, 36]. Recent studies have also made substantial progress in using techniques such as inverse dynamics analysis to

MECHANOBIOLOGY OF FRACTURE HEALING

Fracture healing is one of the most frequently employed scenarios for studies of the effects the local mechanical environment on skeletal tissue differentiation. Mechanical loading of a fracture callus occurs most commonly as a consequence of weight bearing; however, dynamization, or applied micromotion, of the fracture gap has also been investigated. Results of these studies have shown that the effects of loading depend heavily on the mode [39–42], rate [43, 44], and magnitude of loading [45, 46], as well as gap size [45] and the time during healing at which the dynamization is enacted [47–50]. Application of cyclic compressive displacements can enhance healing through increased callus formation and more rapid ossification and bridging [51, 52]. This effect was found to be greatest for an intermediate strain rate (40 mm/s) as opposed to fast (400 mm/s) or slow (2 mm/s) rates [53]. However, the benefits of applied cyclic compressive displacements appear to be limited to displacements that induce an interfragmentary strain (defined as the ratio of applied displacement to the gap size) of 7% or less [39, 45, 54, 55]. Moreover, dynamization of the fracture gap appears to be detrimental in the very early stage of healing [47] yet beneficial during later stages [48–50].

As evidenced by the success of distraction osteogenesis in both experimental and clinical settings, application of successive tensile displacements across an osteotomy gap can also promote bone formation. In contrast to the effects of cyclic compressive loading, however, bone formation in distraction osteogenesis occurs primarily via intramembranous ossification. These characteristics of distraction also appear to hold when the tensile displacements are applied for only 2 days at a time, followed by shortening of the osteotomy gap to its original length [56], but not when the tensile displacements are applied in a true oscillatory manner (e.g., 1–10 Hz frequency) [39]. The effects of shear or transverse movement at the fracture site are controversial, with some studies reporting enhanced bone healing [41, 42] and others reporting increased fibrous tissue formation and delayed bony healing [40, 57]. A series of studies investigating the use of shear movement and also a bending motion to an osteotomy gap reported that these two types of applied motion result in the formation of cartilage rather than bone within the gap [58, 59].

In parallel with some of the earlier experimental investigations summarized above, Perren [60, 61] and Perren and Cordey [43] proposed the interfragmentary strain theory, which states that only tissue that is capable of withstanding the present value of interfragmentary strain can form in the fracture gap. This theory is consistent with observations that granulation tissue forms initially in the gap, followed by cartilage and then bone. The successive formation of each type of tissue further reduces the interfragmentary strain that occurs as a result of the applied load and allows a stiffer tissue to form next.

The interfragmentary strain theory presents an oversimplified description of the mechanical environment within the fracture gap in that it uses one scalar (interfragmentary strain) to describe a multiaxial strain field that varies as a function of position within the gap. More recent models of the mechanobiology of skeletal tissue differentiation have sought to account for this complexity by considering the distributions of local mechanical stimuli present throughout the fracture gap [Fig. 12.4(A)–(C)] [62–64] and the interplay between osteogenesis and angiogenesis [Fig. 12.4(D)] [65, 66]. Carter and colleagues have proposed that different combinations of hydrostatic pressure and tensile strain promote formation of different skeletal tissues [62], while Claes and Heigle have postulated that these two stimuli regulate intramembranous versus endochondral ossification [63]. Prendergast and colleagues have instead proposed that the two key stimuli are shear strain and fluid flow [64, 67]. Direct comparison of these models' predictions to histological analyses of bone healing, and also experimental measurement of local mechanical stimuli such as shear strain within a bone defect, suggests that the most accurate predictions are those based on shear strain and fluid flow [68, 69]. However, each of these theories is unable to predict certain histological features of the fracture healing process [64, 68], indicating that the definitive role of the local mechanical environment in modulating healing has yet to be elucidated fully.

SUMMARY

An essential outcome in fracture healing is restoration of sufficient mechanical integrity to allow weight bearing and activities of daily living. Thus, biomechanical analyses of fracture healing are critical for thorough assessment of the repair process. At present, the biomechanical progression of secondary fracture healing is well characterized, and standardized *in vitro* methods of quantifying the extent of healing have been established. Noninvasive methods of measuring the regain of bone stiffness have also been reported; however, noninvasive methods of measuring the regain of strength are still under development. Studies to date on the effects of mechanical factors indicate that it is possible to augment healing via mechanical loading, and the growing body of literature in this area suggests that further enhancements in healing may be possible. Thus, an understanding of the biomechanics of fracture healing can be applied not only to the assessment of healing but also to development of new repair strategies.

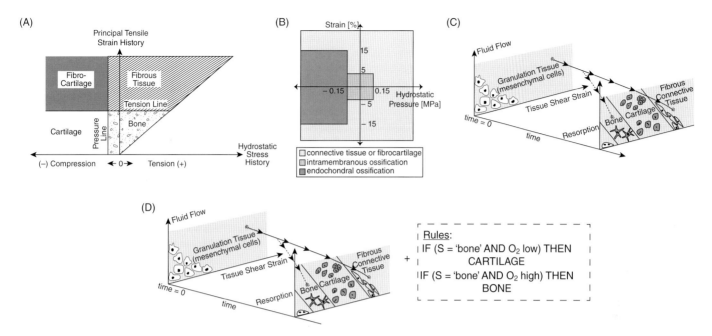

Fig. 12.4. Models of the mechanobiology of skeletal tissue differentiation by (A) Carter et al. (Ref. 62); (B) Claes and Heigle (Ref. 63); (C) Lacroix and Prendergast (Ref. 64); (D) Checa and Prendergast (Ref. 65).

REFERENCES

1. Chao EYS, Aro HT. 1997. Biomechanics of fracture fixation. In: Mow VC, Hayes WC (eds.) *Basic Orthopaedic Biomechanics*. Philadelphia: Lippincott-Raven. pp. 317–352.
2. Bottlang M. 2011. Biomechanics of far cortical locking. *J Orthop Trauma* 25(6): e60.
3. Moss DP, Tejwani NC. 2007. Biomechanics of external fixation: A review of the literature. *Bull NYU Hosp Jt Dis* 65(4): 294–9.
4. Bong MR, Kummer FJ, Koval KJ, Egol KA. 2007. Intramedullary nailing of the lower extremity: Biomechanics and biology. *J Am Acad Orthop Surg* 15(2): 97–106.
5. Toux A, Black RC, Uhthoff HK. 1990. Quantitative measures for fracture healing: An in-vitro biomechanical study. *J Biomech Eng* 112(4): 401–6.
6. Tsiridis E, Morgan EF, Bancroft JM, Song M, Kain M, Gerstenfeld L, Einhorn TA, Bouxsein ML, Tornetta P 3rd. 2007. Effects of OP-1 and PTH in a new experimental model for the study of metaphyseal bone healing. *J Orthop Res* 25(9): 1193–203.
7. Ulrich-Vinther M, Andreassen TT. 2005. Osteoprotegerin treatment impairs remodeling and apparent material properties of callus tissue without influencing structural fracture strength. *Calcif Tissue Int* 76(4): 280–6.
8. Leong PL, Morgan EF. 2008. Measurement of fracture callus material properties via nanoindentation. *Acta Biomater* 4(5): 1569–75.
9. Manjubala I, Liu Y, Epari DR, Roschger P, Schell H, Fratzl P, Duda GN. 2009. Spatial and temporal variations of mechanical properties and mineral content of the external callus during bone healing. *Bone* 45(2): 185–92.
10. White AA 3rd, Panjabi MM, Southwick WO. 1977. The four biomechanical stages of fracture repair. *J Bone Joint Surg Am* 59(2): 188–92.
11. Alkhiary YM, Gerstenfeld LC, Krall E, Sato M, Westmore M, Mitlak B, Einhorn TA. 2004. Parathyroid hormone (1-24; teriparitide) enhances experimental fracture healing. *Transactions of the Annual Meeting of the Orthopaedic Research Society, Vol 29*. San Francisco: Orthopaedic Research Society. p. 328.
12. Andreassen TT, Ejersted C, Oxlund H. 1999. Intermittent parathyroid hormone (1-34) treatment increases callus formation and mechanical strength of healing rat fractures. *J Bone Miner Res* 14(6): 960–8.
13. Kakar S, Einhorn TA, Vora S, Miara LJ, Hon G, Wigner NA, Toben D, Jacobsen KA, Al-Sebaei MO, Song M, Trackman PC, Morgan EF, Gerstenfeld LC, Barnes GL. 2007. Enhanced chondrogenesis and Wnt-signaling in parathyroid hormone treated fractures. *J Bone Miner Res* 22(12): 1903–12.
14. Nyman JS, Munoz S, Jadhav S, Mansour A, Yoshii T, Mundy GR, Gutierrez GE. 2009. Quantitative measures of femoral fracture repair in rats derived by microcomputed tomography. *J Biomech* 42(7): 891–7.
15. Amanat N, McDonald M, Godfrey C, Bilston L, Little D. 2007. Optimal timing of a single dose of zoledronic

acid to increase strength in rat fracture repair. *J Bone Miner Res* 22(6): 867–76.
16. Little DG, McDonald M, Bransford R, Godfrey CB, Amanat N. 2005. Manipulation of the anabolic and catabolic responses with OP-1 and zoledronic acid in a rat critical defect model. *J Bone Miner Res* 20(11): 2044–52.
17. Richardson JB, Cunningham JL, Goodship AE, O'Connor BT, Kenwright J. 1994. Measuring stiffness can define healing of tibial fractures. *J Bone Joint Surg Br* 76(3): 389–94.
18. Hente R, Cordey J, Perren SM. 2003. In vivo measurement of bending stiffness in fracture healing. *Biomed Eng Online* 2: 8.
19. Ogrodnik PJ, Moorcroft CI, Thomas PB. 2001. A fracture movement monitoring system to aid in the assessment of fracture healing in humans. *Proc Inst Mech Eng H* 215(4): 405–14.
20. Claes LE, Cunningham JL. 2009. Monitoring the mechanical properties of healing bone. *Clin Orthop Relat Res* 467(8): 1964–71.
21. Aarnes GT, Steen H, Ludvigsen P, Waanders NA, Huiskes R, Goldstein SA. 2005. In vivo assessment of regenerate axial stiffness in distraction osteogenesis. *J Orthop Res* 23(2): 494–8.
22. Watanabe Y, Takai S, Arai Y, Yoshino N, Hirasawa Y. 2001. Prediction of mechanical properties of healing fractures using acoustic emission. *J Orthop Res* 19(4): 548–53.
23. Gerlanc M, Haddad D, Hyatt GW, Langloh JT, St Hilaire P. 1975. Ultrasonic study of normal and fractured bone. *Clin Orthop Relat Res* (111): 175–80.
24. Glinkowski W, Gorecki A. 2006. Clinical experiences with ultrasonometric measurement of fracture healing. *Technol Health Care* 14(4-5): 321–33.
25. Brown SA, Mayor MB. 1976. Ultrasonic assessment of early callus formation. *Biomed Eng* 11(4): 124–7, 136.
26. Dodd SP, Cunningham JL, Miles AW, Gheduzzi S, Humphrey VF. 2008. Ultrasound transmission loss across transverse and oblique bone fractures: An in vitro study. *Ultrasound Med Biol* 34(3): 454–62.
27. Dodd SP, Miles AW, Gheduzzi S, Humphrey VF, Cunningham JL. 2007. Modelling the effects of different fracture geometries and healing stages on ultrasound signal loss across a long bone fracture. *Comput Methods Biomech Biomed Engin* 10(5): 371–5.
28. Saulgozis J, Pontaga I, Lowet G, Van der Perre G. 1996. The effect of fracture and fracture fixation on ultrasonic velocity and attenuation. *Physiol Meas* 17(3): 201–11.
29. Jamsa T, Koivukangas A, Kippo K, Hannuniemi R, Jalovaara P, Tuukkanen J. 2000. Comparison of radiographic and pQCT analyses of healing rat tibial fractures. *Calcif Tissue Int* 66(4): 288–91.
30. Augat P, Merk J, Genant HK, Claes L. 1997. Quantitative assessment of experimental fracture repair by peripheral computed tomography. *Calcif Tissue Int* 60(2): 194–9.
31. den Boer FC, Bramer JA, Patka P, Bakker FC, Barentsen RH, Feilzer AJ, de Lange ES, Haarman HJ. 1998. Quantification of fracture healing with three-dimensional computed tomography. *Arch Orthop Trauma Surg* 117(6-7): 345–50.
32. Nazarian A, Pezzella L, Tseng A, Baldassarri S, Zurakowski D, Evans CH, Snyder BD. 2010. Application of structural rigidity analysis to assess fidelity of healed fractures in rat femurs with critical defects. *Calcif Tissue Int* 86(5): 397–403.
33. Grigoryan M, Lynch JA, Fierlinger AL, Guermazi A, Fan B, MacLean DB, MacLean A, Genant HK. 2003. Quantitative and qualitative assessment of closed fracture healing using computed tomography and conventional radiography. *Acad Radiol* 10(11): 1267–73.
34. Kuhlman JE, Fishman EK, Magid D, Scott WW Jr, Brooker AF, Siegelman SS. 1988. Fracture nonunion: CT assessment with multiplanar reconstruction. *Radiology* 167(2): 483–8.
35. Shefelbine SJ, Simon U, Claes L, Gold A, Gabet Y, Bab I, Muller R, Augat P. 2005. Prediction of fracture callus mechanical properties using micro-CT images and voxel-based finite element analysis. *Bone* 36(3): 480–8.
36. Leong PL, Morgan EF. 2008. Measurement of fracture callus material properties via nanoindentation. *Acta Biomaterialia* 4(5): 1569–75.
37. Prasad J, Wiater BP, Nork SE, Bain SD, Gross TS. 2010. Characterizing gait induced normal strains in a murine tibia cortical bone defect model. *J Biomech* 43(14): 2765–70.
38. Histing T, Kristen A, Roth C, Holstein JH, Garcia P, Matthys R, Menger MD, Pohlemann T. 2010. In vivo gait analysis in a mouse femur fracture model. *J Biomech* 43(16): 3240–3.
39. Augat P, Merk J, Wolf S, Claes L. 2001. Mechanical stimulation by external application of cyclic tensile strains does not effectively enhance bone healing. *J Orthop Trauma* 15(1): 54–60.
40. Schell H, Epari DR, Kassi JP, Bragulla H, Bail HJ, Duda GN. 2005. The course of bone healing is influenced by the initial shear fixation stability. *J Orthop Res* 23(5): 1022–8.
41. Bishop NE, van Rhijn M, Tami I, Corveleijn R, Schneider E, Ito K. 2006. Shear does not necessarily inhibit bone healing. *Clin Orthop Relat Res* 443: 307–14.
42. Park SH, O'Connor K, McKellop H, Sarmiento A. 1998. The influence of active shear or compressive motion on fracture-healing. *J Bone Joint Surg Am* 80(6): 868–78.
43. Wolf S, Augat P, Eckert-Hubner K, Laule A, Krischak GD, Claes LE. 2001. Effects of high-frequency, low-magnitude mechanical stimulus on bone healing. *Clin Orthop* (385): 192–8.
44. Goodship AE, Cunningham JL, Kenwright J. 1998. Strain rate and timing of stimulation in mechanical modulation of fracture healing. *Clin Orthop* 355 Suppl: S105–15.
45. Claes L, Augat P, Suger G, Wilke HJ. 1997. Influence of size and stability of the osteotomy gap on the success of fracture healing. *J Orthop Res* 15(4): 577–84.
46. Claes L, Eckert-Hubner K, Augat P. 2002. The effect of mechanical stability on local vascularization and tissue

differentiation in callus healing. *J Orthop Res* 20(5): 1099–105.
47. Claes L, Blakytny R, Gockelmann M, Schoen M, Ignatius A, Willie B. 2009. Early dynamization by reduced fixation stiffness does not improve fracture healing in a rat femoral osteotomy model. *J Orthop Res* 27(1): 22–7.
48. Weaver AS, Su YP, Begun DL, Miller JD, Alford AI, Goldstein SA. 2010. The effects of axial displacement on fracture callus morphology and MSC homing depend on the timing of application. *Bone* 47(1): 41–8.
49. Claes L, Blakytny R, Besse J, Bausewein C, Ignatius A, Willie B. 2011. Late dynamization by reduced fixation stiffness enhances fracture healing in a rat femoral osteotomy model. *J Orthop Trauma* 25(3): 169–74.
50. Willie BM, Blakytny R, Glockelmann M, Ignatius A, Claes L. 2011. Temporal variation in fixation stiffness affects healing by differential cartilage formation in a rat osteotomy model. *Clin Orthop Relat Res* 469(11): 3094–101.
51. Goodship AE, Kenwright J. 1985. The influence of induced micromovement upon the healing of experimental tibial fractures. *J Bone Joint Surg Br* 67(4): 650–5.
52. Claes LE, Wilke HJ, Augat P, Rubenacker S, Margevicius KJ. 1995. Effect of dynamization on gap healing of diaphyseal fractures under external fixation. *Clin Biomech (Bristol, Avon)* 10(5): 227–34.
53. Goodship AE, Watkins PE, Rigby HS, Kenwright J. 1993. The role of fixator frame stiffness in the control of fracture healing. An experimental study. *J Biomech* 26(9): 1027–35.
54. Augat P, Margevicius K, Simon J, Wolf S, Suger G, Claes L. 1998. Local tissue properties in bone healing: influence of size and stability of the osteotomy gap. *J Orthop Res* 16(4): 475–81.
55. Claes LE, Heigele CA, Neidlinger-Wilke C, Kaspar D, Seidl W, Margevicius KJ, Augat P. 1998. Effects of mechanical factors on the fracture healing process. *Clin Orthop* 355 Suppl: S132–47.
56. Claes L, Augat P, Schorlemmer S, Konrads C, Ignatius A, Ehrnthaller C. 2008. Temporary distraction and compression of a diaphyseal osteotomy accelerates bone healing. *J Orthop Res* 26(6): 772–7.
57. Augat P, Burger J, Schorlemmer S, Henke T, Peraus M, Claes L. 2003. Shear movement at the fracture site delays healing in a diaphyseal fracture model. *J Orthop Res* 21(6): 1011–7.
58. Cullinane DM, Fredrick A, Eisenberg SR, Pacicca D, Elman MV, Lee C, Salisbury K, Gerstenfeld LC, Einhorn TA. 2002. Induction of a neoarthrosis by precisely controlled motion in an experimental mid-femoral defect. *J Orthop Res* 20(3): 579–86.
59. Cullinane DM, Salisbury KT, Alkhiary Y, Eisenberg S, Gerstenfeld L, Einhorn TA. 2003. Effects of the local mechanical environment on vertebrate tissue differentiation during repair: Does repair recapitulate development? *J Exp Biol* 206(Pt 14): 2459–71.
60. Perren SM. 1979. Physical and biological aspects of fracture healing with special reference to internal fixation. *Clin Orthop Relat Res* 138: 175–180.
61. Perren SM, Cordey J. 1980. The concept of interfragmentary strain. In: Uhthoff HK (ed.) *Current Concepts of Internal Fixation of Fractures*. Berlin: Springer. pp. 63–77.
62. Carter DR, Beaupre GS, Giori NJ, Helms JA. 1998. Mechanobiology of skeletal regeneration. *Clin Orthop* 355 Suppl: S41–55.
63. Claes LE, Heigele CA. 1999. Magnitudes of local stress and strain along bony surfaces predict the course and type of fracture healing. *J Biomech* 32(3): 255–66.
64. Lacroix D, Prendergast PJ. 2002. A mechano-regulation model for tissue differentiation during fracture healing: Analysis of gap size and loading. *J Biomech* 35(9): 1163–71.
65. Checa S, Prendergast PJ. 2009. A mechanobiological model for tissue differentiation that includes angiogenesis: A lattice-based modeling approach. *Ann Biomed Eng* 37(1): 129–45.
66. Simon U, Augat P, Utz M, Claes L. 2011. A numerical model of the fracture healing process that describes tissue development and revascularisation. *Comput Methods Biomech Biomed Engin* 14(1): 79–93.
67. Prendergast PJ, Huiskes R, Soballe K. 1997. ESB Research Award 1996. Biophysical stimuli on cells during tissue differentiation at implant interfaces. *J Biomech* 30(6): 539–48.
68. Isaksson H, Wilson W, van Donkelaar CC, Huiskes R, Ito K. 2006. Comparison of biophysical stimuli for mechano-regulation of tissue differentiation during fracture healing. *J Biomech* 39(8): 1507–16.
69. Morgan EF, Salisbury Palomares KT, Gleason RE, Bellin DL, Chien KB, Unnikrishnan GU, Leong PL. 2010. Correlations between local strains and tissue phenotypes in an experimental model of skeletal healing. *J Biomech* 43(12): 2418–24.

13
Human Genome-Wide Association (GWA) Studies

Douglas P. Kiel

References 109

Twin and family studies have demonstrated that bone mineral density (BMD), one of the most commonly studied bone phenotypes, is highly heritable, as is bone geometry and bone ultrasound measures [1–3]. Significant heritability of fractures has also been demonstrated for hip fracture occurring at younger ages and most recently for vertebral fractures [4, 5]. The search for genes underlying these skeletal phenotypes has matured over the past decade, beginning with linkage analyses in family-based cohorts [6–10] and candidate gene studies [11–14]. Linkage analyses have not been successful in narrowing quantitative trait loci to specific genes responsible for transmission of the trait through families, and they have relatively low power for complex disorders such as osteoporosis. Candidate gene studies, while meeting with some success in identifying genes related to bone density and geometry, are limited to the study of a set of genes known to be related to bone biology based on existing knowledge. Furthermore, many of the early positive candidate gene studies were not replicated by subsequent work. These early successes and failures set the stage for a more "agnostic" genome-wide search for osteoporosis genes with the advent of improvements in genotyping technology and with the development of computer storage platforms and statistical methods to analyze large amounts of data. Thus within the past five years, the field of skeletal genetics has moved into the era of genome-wide association (GWA) studies [15–17].

Genome-wide association studies use high throughput genotyping of hundreds of thousands and even millions of the most common form of genetic variant, the single nucleotide polymorphism (SNP), and relate these SNPs to various phenotypes. This approach is powerful in that it interrogates the entire genome for associations between the variants and the phenotypes. It has also introduced a considerable number of challenges, especially from a statistical point of view, as the number of associations tested across the genome requires an adjustment for multiple testing. The resulting stringent levels of statistical tests needed to achieve a "significant" association require large sample sizes. Thus, the GWA studies era has spawned a dramatic change in the collaborative efforts across the scientific community. On their own, the individual studies are not sufficiently powered to interrogate so many SNPs across the genome. To achieve the large samples required to perform GWA studies, cohorts across the world have developed consortia that cultivate the necessary trust and collegiality required to undertake large meta-analyses of the GWA studies from individual cohorts.

One of the challenges of these collaborative GWA study efforts includes proper harmonization of phenotypes to minimize measurement heterogeneity. There have been efforts to collect the best phenotypic measures as part of the PhenX project (https://www.phenx.org/). Further efforts to minimize heterogeneity require the standardization of data analysis to account for potential confounding, methods of minimizing the effects of possible population stratification, and appropriate quality metrics for the handling of genotyping data that come from a variety of genotyping platforms.

To combine data across cohorts, studies typically "impute" millions of SNPs that are not actually genotyped on the platforms. Robust methods have emerged

---

Primer on the Metabolic Bone Diseases and Disorders of Mineral Metabolism, Eighth Edition. Edited by Clifford J. Rosen.
© 2013 American Society for Bone and Mineral Research. Published 2013 by John Wiley & Sons, Inc.

allowing for the imputation of genotypes based on work from the HapMap project, which developed a human haplotype map that facilitated the imputation process [18, 19]. This project genotyped millions of SNPs in 270 samples from four populations with diverse geographic ancestry in two phases [20]. Most recently, the 1,000 Genomes Project (http://www.1000genomes.org/) is characterizing over 95% of the variants that are in genomic regions accessible to current high-throughput sequencing technologies and that have allele frequency of 1% or higher in each of five major population groups (populations in or with ancestry from Europe, East Asia, South Asia, West Africa, and the Americas). The 1,000 Genomes Project will also catalog lower frequency alleles in the range of 0.1% frequency because these low frequency alleles are often found in coding regions [21].

All publications of GWA study results are recorded on a website maintained by the National Genome Research Institute of all GWA studies (http://www.ebi.ac.uk/fgpt/gwas/). The field of skeletal genetics has been dominated by a focus on the BMD phenotype, although there has been some evolution from the use of areal BMD using dual energy X-ray absorptiometry (DXA) to volumetric measures of bone density using peripheral quantitative computed tomography (pQCT) and also to the study of fractures. The first GWA study for DXA-derived BMD and hip geometry traits was performed in the Framingham Study using a genotyping platform with only 100,000 SNPs in a sample of only 1,141 men and women, which yielded no genome-wide significant findings using the strict p-value criterion of 5×10^{-8} [22].

The largest meta-analyses to date have been conducted by the Genetic Factors for Osteoporosis (GEFOS) Consortium. The first meta-analysis from the GEFOS consortium included five GWA studies of femoral neck and lumbar spine BMD in 19,195 subjects of Northern European descent [23].

Using DXA-derived BMD phenotypes of the spine and hip, 20 loci reached genome-wide significance (GWS; $P < 5 \times 10^{-8}$), of which 13 mapped to regions not previously associated with BMD. These included 1p31.3 (*GPR177*), 2p21 (*SPTBN1*), 3p22 (*CTNNB1*), 4q21.1 (*MEPE*), 5q14 (*MEF2C*), 7p14 (*STARD3NL*), 7q21.3 (*FLJ42280*), 11p11.2 (*LRP4, ARHGAP1, F2*), 11p14.1 (*DCDC5*), 11p15 (*SOX6*), 16q24 (*FOXL1*), 17q21 (*HDAC5*), and 17q12 (*CRHR1*).

The other seven loci mapped to genes known to be involved in skeletal metabolism such as 1p36 (*ZBTB40*), 6q25 (*ESR1*), 8q24 (*TNFRSF11B* also known as osteoprotegerin), 11q13.4 (*LRP5*), 12q13 (*SP7* also known as osterix), 13q14 (*TNFSF11* also known as RANKL), and 18q21 (*TNFRSF11A* also known as RANK).

In a second study from the GEFOS consortium using DXA-derived BMD of the hip and spine from 17 GWA studies and 32,961 individuals of European and East Asian ancestry, the top-associated SNPs with BMD at either the femoral neck or lumbar spine were then tested for replication in 50,933 independent subjects [24]. These independent samples for replication came from a previously funded consortium called Genetic Markers for Osteoporosis (GENOMOS), which had produced several of the largest candidate gene meta-analyses prior to the advent of GWA study efforts. The availability of fracture phenotypes in the replication cohorts also permitted the GEFOS meta-analysis to include the testing of the top BMD associations for their association with the for low-trauma fracture in 31,016 cases and 102,444 controls. This meta-analysis identified 56 loci (32 novel) associated with BMD at a genome-wide significant level ($P < 5 \times 10^{-8}$). The overall results of this meta-analysis are displayed in Fig. 13.1 referred to as a "Manhattan" plot, as each tower of points indicate $-\log_{10}$ p-values. Each point represents a p-value for a SNP-phenotype association meta-analysis result. The regional plot of one of these loci, *MEF2C*, is shown in Fig. 13.2. The signal for association is located in the 5′ flanking region of the *MEF2C* gene. This gene codes for a protein that may play a role in osteoblasts and in maintaining the differentiated state of mature muscle cells is an essential transcriptional activator of the bone formation inhibitor, SOST; and this activation is controlled by parathyroid hormone [25].

With so many novel findings as well as confirmation of previous findings, the study also used a formal text mining algorithm called "GRAIL" [26] to determine the degree of relatedness of implicated genes using PubMed abstracts. This analysis found that the implicated genes clustered within the RANK-RANKL-OPG, mesenchymal-stem-cell differentiation, endochondral ossification, and the Wnt signaling pathways. In addition, the meta-analysis discovered loci containing genes not known to play a role in bone biology. When the loci associated with BMD were then tested for association with fracture risk, it was found that 14 BMD loci were also associated with fracture risk ($P < 5 \times 10^{-4}$), of which six reached $P < 5 \times 10^{-8}$, including: 18p11.21 (*C18orf19*), 7q21.3 (*SLC25A13*), 11q13.2 (*LRP5*), 4q22.1 (*MEPE*), 2p16.2 (*SPTBN1*) and 10q21.1 (*DKK1*).

The results from GWA study meta-analyses of DXA-derived BMD traits in individuals of European descent have been followed by a moderately large GWA study with close to 9,000 participants in the Korean Genome Epidemiology Study. However, in that study, bone density was measured at the radius, tibia, and calcaneus using quantitative ultrasound (speed of sound or "SOS") [27]. Perhaps not surprisingly, the results did not confirm many of the loci identified in the large GEFOS efforts, as the phenotype measure in the Asian samples was made at different skeletal sites and using a different technology that is not highly correlated with BMD by DXA. A small GWA study in premenopausal women from Indiana (n = 1,524) failed to identify any genome-wide significant associations; however, the top 50 findings were taken forward for replication in an African-American sample (n = 669) of premenopausal women, and one locus on chromosome 14 near the *CATSPERB* gene showed some evidence of replication. Only two other loci identified in larger GWA study efforts were suggestively associated

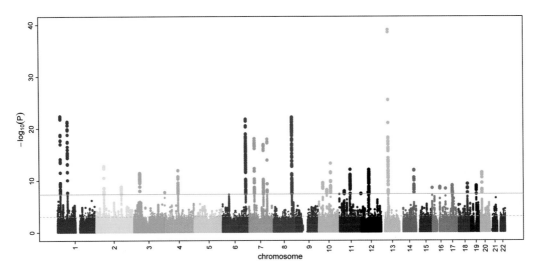

Fig. 13.1. Manhattan plots from the most recent GWA study meta-analysis of the GEFOS Consortium (Ref. 23). The plot displays the genome-wide significant associations between all of the 2,543,686 SNPs that were imputed using the HapMap data from the Centre d'Etude du Polymorphisme Humain (CEPH) samples consisting of 30 U.S. trios who provided samples, which were collected in 1980 from U.S. residents with northern and western European ancestry. Each point on the plot represents a p-value for association with lumbar spine BMD from the meta-analysis using a fixed-effects model. (Reproduced with permission from Estrada K, Styrkarsdottir U, Evangelou E, et al. 2012. Genome-wide meta-analysis identifies 56 bone mineral density loci and reveals 14 loci associated with risk of fracture. *Nat Genet* 44: 491–501.)

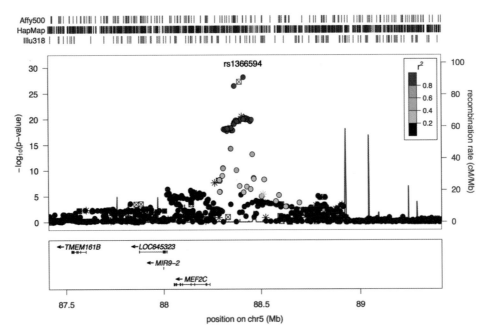

Fig. 13.2. This is a regional plot of p-values for association between SNPs in the 5' flanking region of the *MEF2C* gene and BMD from the GEFOS meta-analysis results (Ref. 23). (Reproduced with permission from Estrada K, Styrkarsdottir U, Evangelou E, et al. 2012. Genome-wide meta-analysis identifies 56 bone mineral density loci and reveals 14 loci associated with risk of fracture. *Nat Genet* 44: 491–501.)

with BMD of the spine or hip in the premenopausal sample [28]. These two additional GWA study efforts in sample sizes that were considerably smaller than the larger GEFOS meta-analyses highlight the consistent findings from GWA studies in general; namely, that the combined effects of many relatively common genetic variants on a complex trait like BMD require large sample sizes. There remains a considerable unexplained proportion of the high heritability observed for BMD, as is true of other complex phenotypes [29].

Most recently, a group of investigators from Europe used pQCT phenotypes in a GWA study meta-analysis in just over 1,500 teenagers and found an SNP in the receptor activator of nuclear factor-κB ligand (RANKL) gene *(TNFSF11)* to be associated with cortical BMD, with some evidence of a stronger effect and more significant association in the young male participants than in the young women [30]. The significance of this study lies in the use of a phenotype that provides skeletal compartment specific information. If cortical versus trabecular bone modeling and remodeling are influenced by different genes, it will be important to refine the skeletal phenotypes to better understand the underlying genetic architecture.

The unexplained heritability of bone phenotypes using GWA studies suggests that there may be larger numbers of variants with small effects that have not yet been identified, or rare variants with larger effects. Also, the possibility exists that there are structural variants, such as copy number variants, that have not yet been captured on some of the genotyping arrays or have not been adequately investigated. Very few GWA study efforts have investigated the X-chromosome due to early challenges in imputation and analytic approaches [31–33]. There may also be gene by gene interactions that the current studies have not been powered to detect.

The future of GWA studies for the field of skeletal genetics remains promising. As large consortia are able to identify novel loci, the promise of new biologic discoveries lies ahead. The next steps will involve a more thorough testing of the identified loci for discovery of the functional variants, finer mapping and sequencing of promising loci, and studies of gene and environment interactions. There are a number of obvious interactions that might be explored, including differences in the loading on the skeleton according to genetic makeup, or pharmacogenetic studies to identify the variants predicting responses to common medications used to treat osteoporosis, or even nutrigenomic studies to identify variations in response to nutritional interventions used to treat osteoporosis such as vitamin D and calcium.

REFERENCES

1. Karasik D, Myers RH, Cupples LA, et al. 2002. Genome screen for quantitative trait loci contributing to normal variation in bone mineral density: The Framingham Study. *J Bone Miner Res* 17: 1718–27.
2. Karasik DE, Myers RH, Hannan MT, et al. 2001. Mapping of quantitative ultrasound of the calcaneus to chromosomes 1 and 5 by genome-wide linkage analysis. *J Bone Min Res* 16(Suppl 1): S167.
3. Karasik D, Dupuis J, Cupples LA, et al. 2007. Bivariate linkage study of proximal hip geometry and body size indices: The Framingham study. *Calcif Tissue Int* 81: 162–73.
4. Michaelsson K, Melhus H, Ferm H, Ahlbom A, Pedersen NL. 2005. Genetic liability to fractures in the elderly. *Arch Intern Med* 165: 1825–30.
5. Liu CT, Karasik D, Zhou Y, Hsu YH, Genant HK, Broe KE, Lang TF, Samelson EJ, Demissie S, Bouxsein ML, Cupples LA, Kiel DP. 2012. Heritability of prevalent vertebral fracture and volumetric bone mineral density and geometry at the lumbar spine in three generations of the framingham study. *J Bone Miner Res* 27: 954–8.
6. Johnson ML, Gong G, Kimberling W, Recker SM, Kimmel DB, Recker RB. 1997. Linkage of a gene causing high bone mass to human chromosome 11 (11q12-13). *Am J Hum Genet* 60: 1326–32.
7. Koller DL, Rodriguez LA, Christian JC, et al. 1998. Linkage of a QTL contributing to normal variation in bone mineral density to chromosome 11q12-13. *J Bone Miner Res* 13: 1903–8.
8. Styrkarsdottir U, Cazier JB, Kong A, et al. 2003. Linkage of osteoporosis to chromosome 20p12 and association to BMP2. *PLoS Biol* 1: E69.
9. Deng HW, Deng XT, Conway T, Xu FH, Heaney R, Recker RR. 2002. Determination of bone size of hip, spine, and wrist in human pedigrees by genetic and lifestyle factors. *J Clin Densitom* 5: 45–56.
10. Deng HW, Livshits G, Yakovenko K, et al. 2002. Evidence for a major gene for bone mineral density/content in human pedigrees identified via probands with extreme bone mineral density. *Ann Hum Genet* 66: 61–74.
11. Langdahl BL, Uitterlinden AG, Ralston SH. 2008. Large-scale analysis of association between polymorphisms in the transforming growth factor beta 1 gene (TGFB1) and osteoporosis: The GENOMOS Study. *Bone* 42(5): 969–81
12. Ralston SH, Uitterlinden AG, Brandi ML, et al. 2006. Large-scale evidence for the effect of the COLIA1 Sp1 polymorphism on osteoporosis outcomes: The GENOMOS study. *PLoS Med* 3: e90.
13. van Meurs JB, Trikalinos TA, Ralston SH, et al. 2008. Large-scale analysis of association between LRP5 and LRP6 variants and osteoporosis. *JAMA* 299: 1277–90.
14. Ioannidis JP, Ralston SH, Bennett ST, et al. 2004. Differential genetic effects of ESR1 gene polymorphisms on osteoporosis outcomes. *JAMA* 292: 2105–14.
15. Altshuler D, Daly MJ, Lander ES. 2008. Genetic mapping in human disease. *Science* 322: 881–8.
16. Hindorff LA, Sethupathy P, Junkins HA, et al. 2009. Potential etiologic and functional implications of genome-wide association loci for human diseases and traits. *Proc Natl Acad Sci U S A* 106: 9362–7.
17. Manolio TA. 2009. Cohort studies and the genetics of complex disease. *Nat Genet* 41: 5–6.

18. de Bakker PI, Ferreira MA, Jia X, Neale BM, Raychaudhuri S, Voight BF. 2008. Practical aspects of imputation-driven meta-analysis of genome-wide association studies. *Hum Mol Genet* 17: R122–8.
19. Marchini J, Howie B, Myers S, McVean G, Donnelly P. 2007. A new multipoint method for genome-wide association studies by imputation of genotypes. *Nat Genet* 39: 906–13.
20. Manolio TA, Brooks LD, Collins FS. 2008. A HapMap harvest of insights into the genetics of common disease. *J Clin Invest* 118: 1590–605.
21. 1000 Genomes Project Consortium. 2010. A map of human genome variation from population-scale sequencing. *Nature* 467: 1061–73.
22. Kiel DP, Demissie S, Dupuis J, Lunetta KL, Murabito JM, Karasik D. 2007. Genome-wide association with bone mass and geometry in the Framingham Heart Study. *BMC Med Genet* 8(Suppl 1): S14.
23. Rivadeneira F, Styrkarsdottir U, Estrada K, et al. 2009. Twenty bone-mineral-density loci identified by large-scale meta-analysis of genome-wide association studies. *Nat Genet* 41(11): 1199–206.
24. Estrada K, Styrkarsdottir U, Evangelou E, et al. 2012. Genome-wide meta-analysis identifies 56 bone mineral density loci and reveals 14 loci associated with risk of fracture. *Nat Genet* 44: 491–501.
25. Leupin O, Kramer I, Collette NM, et al. 2007. Control of the SOST bone enhancer by PTH using MEF2 transcription factors. *J Bone Miner Res* 22: 1957–67.
26. Raychaudhuri S, Plenge RM, Rossin EJ, et al. 2009. Identifying relationships among genomic disease regions: Predicting genes at pathogenic SNP associations and rare deletions. *PLoS Genet* 5: e1000534.
27. Cho YS, Go MJ, Kim YJ, et al. 2009. A large-scale genome-wide association study of Asian populations uncovers genetic factors influencing eight quantitative traits. *Nat Genet* 41: 527–34.
28. Koller DL, Ichikawa S, Lai D, et al. 2010. Genome-wide association study of bone mineral density in premenopausal European-American women and replication in African-American women. *J Clin Endocrinol Metab* 95: 1802–9.
29. Manolio TA, Collins FS, Cox NJ, et al. 2009. Finding the missing heritability of complex diseases. *Nature* 461: 747–53.
30. Paternoster L, Lorentzon M, Vandenput L, et al. 2010. Genome-wide association meta-analysis of cortical bone mineral density unravels allelic heterogeneity at the RANKL locus and potential pleiotropic effects on bone. *PLoS Genet* 6: e1001217.
31. Clayton D. 2008. Testing for association on the X chromosome. *Biostatistics* 9: 593–600.
32. Hickey PF, Bahlo M. 2011. X chromosome association testing in genome wide association studies. *Genet Epidemiol* 35: 664–70.
33. Loley C, Ziegler A, Konig IR. 2011. Association tests for X-chromosomal markers—A comparison of different test statistics. *Hum Hered* 71: 23–36.

14
Circulating Osteogenic Cells

Robert J. Pignolo and Moustapha Kassem

Introduction 111
Parabiosis Experiments 111
Bone Marrow Transplantation Experiments 112
Ectopic Bone Formation Experiments 112
COP Cells in Human Diseases 112
MSC Mobilization Studies Using Pharmacological Agents 112
Approaches for Isolation and Characterization of COP Cells 112

What Are the Possible Physiologic and Pathologic Functions of COP Cells? 114
COP Cells: A Synthesis 114
Concluding Remarks 116
Acknowledgments 116
References 116

INTRODUCTION

Bone remodeling presupposes a continuous recruitment of osteoclastic and osteoblastic cells to newly established bone remodeling sites where bone resorption and bone formation are taking place, coupled in space and time. While osteoclastic cells originate from hematopoietic precursors that reach bone surfaces through the circulation, the nature and route by which the osteoblastic cells are recruited to bone formation surfaces are not completely understood. The classical description of osteoblastic cells is that they are fibroblast-like, derived from stem cells present primarily in bone marrow stroma and thus termed "marrow stromal stem cells" (MSCs) (also known as skeletal, or mesenchymal stem cells) and recent evidence suggests their presence in the perivascular niche on abluminal surfaces of blood vessels [1].

Traditionally, MSC are isolated from low-density mononuclear cells through plastic adherence and described as CD105+, CD73+, CD90+, CD14–, CD34–, CD45–, CD79–, CD19– cells [2]. In addition, they have been characterized as CD44+, CD63+, CD146+, Stro-1+ cells [3]. Some recent histomorphometric studies suggest that osteoblastic cells are recruited to bone formation surfaces through blood vessels since bone resorption and bone formation are visualized to take place in specialized "bone remodeling compartments" that are lined by a layer of flat cells connected to a capillary, which provides a conduit for both the osteoclastic and osteoblastic cells to reach bone surfaces [4]. However, in systemic infusion studies, plastic adherent MSCs exhibit poor homing to noninjured skeletal tissues [5], suggesting that MSCs are not circulating (i.e., they are solid-phase cells in contrast to hematopoietic stem cells, which are fluid-phase cells).

On the other hand, a number of experimental studies have tested for the existence of a circulating osteoblastic population, termed "circulating osteogenic precursor (COP) cells," which have the ability to access bone formation sites from the circulation. The biological characteristics and possible roles of COP cells have been studied using a variety of experimental approaches that are summarized here.

PARABIOSIS EXPERIMENTS

Parabiosis experiments are based on creating a conjoint pair of mice that share a common circulatory system. Using this model, Kumagai et al. created a conjoint pair of a mouse constitutively overexpressing green fluorescence protein (GFP) and a wild-type (WT) syngeneic mouse [6]. Following fibular fracture in the conjoined WT partner, GFP+ alkaline phosphatase (ALP)+ cells were found localized to the fracture callus, suggesting recruitment through the circulation. However, Boban et al. examined for the existence of COP cells in transgenic collagen I 2.3-GFP (Col.I2.3-GFP) or osteocalcin (OCN)-GFP mice that were joined to transgenic parabionts overexpressing thymidine kinase (TK) under the control of the Col.I2.3 promoter (Col2.3ΔTK) [7]. When Col2.3ΔTK transgenic mice are given ganciclovir (GCV), the osteoblastic cells are destroyed, but upon removal of the drug the bone recovers uneventfully. Following GCV-induced osteoblast ablation, the authors found no evidence for the presence of GFP marked cells in the Col2.3ΔTK mice, suggesting that matrix-forming osteoblastic cells and osteocytes expressing Col.I or OCN do not circulate.

BONE MARROW TRANSPLANTATION EXPERIMENTS

Olmstead-Davis et al. provided evidence for the presence of COP cells as belonging to the "side population" (SP) of bone marrow cells. SP cells are identified by their ability to expel a DNA binding dye. When transplanted in marrow-ablated mice, SP cells regenerated not only osteoblastic cells but also hematopoietic cells, suggesting that SP cells contain a common precursor for osteoblastic and hematopoietic lineages [8]. Similar results were obtained by Dominici et al., where GFP-marked, plastic nonadherent bone marrow cells generated osteoblasts, osteocytes, and hematopoietic cells following transplantation into lethally irradiated mice [9]. In addition, Hayakawa et al. reported that hematopoietic cells and MSCs can be reconstituted in lethally irradiated WT mice with GFP+ mononuclear bone marrow cells obtained from GFP-transgenic mice [10].

ECTOPIC BONE FORMATION EXPERIMENTS

Otsuru et al. demonstrated the presence of COP cells that contribute to bone morphogenetic protein (BMP)2-induced ectopic bone formation in mice [11]. Following lethal irradiation and subsequent GFP+-transgenic bone marrow transplantation, GFP+ osteoblastic cells expressing osteocalcin were found in the newly formed ectopic bone. Also, transplantation of GFP+ peripheral blood mononuclear cells isolated from BMP2-implanted GFP mice to BMP2-implanted wild-type nude mice led to accumulation of GFP+ osteocalcin+ cells in the ectopic bone [11]. The authors characterized COP cells as CD45– CD44+ CXCR4+, and capable of in vitro and in vivo differentiation into osteoblasts [12].

COP CELLS IN HUMAN DISEASES

Few studies have examined the presence of donor MSC and osteoblastic cells in bones of patients that have received successful bone marrow transplantation. Koc et al. examined if donor-derived MSCs were transferred to allogenic hematopoietic stem cell transplant recipients for treatment of lysosomal or peroxisomal storage diseases [13]. Bone marrow MSCs were cultured from 13 patients 1 to 14 years after transplantation. Despite successful donor-type hematopoietic engraftment, there was no evidence for the presence of donor MSCs based on fluorescent *in situ* hybridization (FISH) analysis using probes for X- and Y-chromosomes in gender-mismatched transplantations or radiolabeled PCR amplification of polymorphic simple sequence repeats [13]. In contrast, bone marrow transplantation was shown to improve lethal osteogenesis imperfecta (OI), and transplantation of plastic-adherent, bone marrow MSCs resulted in engraftment and clinical improvement in five out of six patients with severe OI [14]. In addition, Suda et al. reported the presence of donor-derived COP cells in patients that received gender-mismatched hematopoietic stem cell transplantation [15].

MSC MOBILIZATION STUDIES USING PHARMACOLOGICAL AGENTS

Hong et al. were able to isolate COP cells following their mobilization with substance P in mice and rats from the peripheral blood mononuclear cell fraction as CD45–, CD29+, and capable of mesoderm-type cell differentiation (osteoblast, adipocyte, and chondrocyte) [16]. Pitchford et al. [17] reported peripheral blood enrichment of MSC-like cells (plastic adherent, CD29+, CD105+, CD34–, CD45–, VE-Cadherin–, vWF-) following treatment with VEGF and the CXCR4 antagonist AMD3100, suggesting mobilization from bone marrow to peripheral blood. Also, Tondreau et al. have demonstrated that G-CSF-mobilized CD133+ cells in peripheral blood contain plastic-adherent MSC-like cells capable of osteoblast differentiation [18].

APPROACHES FOR ISOLATION AND CHARACTERIZATION OF COP CELLS

Different approaches have been utilized to obtain COP from peripheral blood with variable success.

Plastic adherence

Kuznetsov et al. identified circulating MSCs (termed "circulating connective tissue precursors" [19] or "circulating skeletal stem cells" [20]) from human and experimental animal peripheral blood through plastic-adherence, and the cells were described as osteonectin+ osteopontin+ collagen type I (Col. I)+, alpha smooth muscle (ASM)+ CD45− and endothelial marker negative cells. The frequency of these cells was extremely low in humans but more abundant in experimental animals, e.g., in the guinea pig [19, 20]. Interestingly, Zvaifler et al. modified this method and reported a more successful isolation of MSCs from peripheral blood of healthy volunteers with a similar phenotype [21]. Rochefort et al. reported that the frequency of plastic-adherent MSCs present in peripheral blood increased in a rat model for chronic hypoxia, suggesting that pathophysiological conditions including hypoxia can mobilize MSCs from bone marrow to peripheral blood [22]. Also, Otsuru et al. isolated plastic-adherent CD45− CD44+ CXCR4+ cells from murine peripheral blood and demonstrated their homing to sites of BMP2-induced ectopic bone formation [11,12]. Circulating MSC-like cells have also been isolated from umbilical cord blood [23, 24].

Plastic-nonadherent cells

Based on the notion that fluid-phase cells are poor at plastic adherence, Long et al. have reported the presence of osteocalcin (OCN)+ or osteopontin+ bone marrow plastic-nonadherent cells [25, 26]. Eghbali-Fatourechi et al. extended these studies to isolate a plastic-nonadherent cell population from peripheral blood [27], demonstrating that OCN+ cells exhibit mononuclear cell morphology and that up to apprxoimately 50% of these cells express CD34 [28]. However, these cells seem to exhibit under standard *in vitro* culture conditions low growth rates and less robust osteoblast differentiation capacity compared with bone marrow MSCs [27].

COP cells isolated from vascular lineage cells

A number of investigators have described the osteoblastic differentiation potential of vascular cells. For example, a common progenitor (termed "mesangioblasts") that can differentiate into endothelial and mesodermal cells (osteoblasts, adipocytes, myocytes) has been isolated in cultures of embryonic dorsal aorta [29, 30] as Flk1+ CD34+ VE-cadherin+ alpha smooth muscle+ cells [29, 30]. Some studies suggest the presence of a counterpart for this cell population in the bone marrow of postnatal organisms [31, 32].

Also, circulating endothelial cells have been described that acquire an osteoblast-like phenotype *in vitro* and enhance fracture healing when applied locally [33]. In a lineage tracing study to identify cells recruited to sites of ectopic bone formation in mice, Lounev et al. demonstrated that Tie2+ cells (as a putative marker for endothelial cells), but not MyoD+ (muscle cells) or smooth muscle myosin heavy chain+ cells, made significant contributions to the newly formed bone [34]. Furthermore, vascular endothelial cells may transform into mesodermal multipotent stem-like cells by an activin-like kinase-2 (ALK2) receptor–dependent mechanism [35]. Interestingly, endothelial markers are expressed in chondrocytes and osteoblasts in lesions from individuals with fibrodysplasia ossificans progressiva (FOP) [35]. Thus, expression of constitutively active ALK2 in endothelial cells may cause an endothelial-to-mesenchymal transition and acquisition of a stem cell–like phenotype. However, multipotent mesenchymal cells that reside in the skeletal muscle interstitium may also be a source of progenitors for heterotopic ossification and, at least in BMP2-induced skeletogenesis in the mouse, the endothelium may not exclusively supply these precursor cells [36].

Isolation of COP cells from fibrocytes, monocytes, and other cells from the hematopoietic lineage

Several hematopoietic-derived cell populations present in peripheral blood have been characterized as having osteogenic potential. Matsumoto et al. demonstrated that the human peripheral blood CD34+ cell fraction contains a minor population of OCN+ cells that upon intravenous infusion were identified at bone formation sites in a rat femoral fracture model [37]. Several studies have examined the biological characteristics of circulating connective tissue cells called "fibrocytes." The cells were first identified in *in vivo* assays of wound repair models [38]. Fibrocytes are cultured from the mononuclear cell fraction of peripheral blood through adherence to fibronectin-coated plastic, and characterized by fibroblast-like morphology and a combined hematopoietic (CD34+, CD45+, CD13+) and stromal mesodermal phenotype typified by collagen type I. Fibrocytes exhibit variable levels of CXCR4 expression and differentiate into mesoderm-type cells (e.g., osteoblast, adipocyte, and chondrocyte), but the efficiency of differentiation is low and no evidence for *in vivo* bone formation has been demonstrated [39].

Kuwana et al. identified a related population called "monocyte-derived mesenchymal progenitors" (MOMPs) in the peripheral blood mononuclear cell fraction based on adherence to fibronectin-coated plastic plates [40]. The cells initially exhibited monocyte-like morphology and then acquired fibroblast-like morphology in culture. The cells are CD14+ CD45+ CD34+ Col I+ and can differentiate into osteoblastic, adipocytic, and myocytic cells using standard differentiation protocols [39]. Suda et al. corroborated these results and described the cells as a CD14+ CD34+ CD45+ CXCR4+ col I+ ALP+ Tie2+ population that formed heterotopic bone *in vivo* upon implantation in immune-deficient mice [15]. Similar approaches have been reported to isolate stem cells with wide differentiation potential from peripheral blood monocytic cells [41, 42]. Thus, MOMPs and fibrocytes seem to be identical cell populations.

WHAT ARE THE POSSIBLE PHYSIOLOGIC AND PATHOLOGIC FUNCTIONS OF COP CELLS?

Several studies have suggested that COP cells participate in a number of physiological processes including long bone development and fracture healing. The number of OCN+ COP (plastic nonadherent) cells in the circulation of adolescent boys is more than five times as many as found in adults, and their number correlated with serum levels of insulin-like growth factor (IGF) I and IGF-binding protein 3 [28]. Also, Eghbali-Fatourechi et al. reported an increase in the number of circulating OCN+ COP cells in three men following recent fractures [28]. Similarly, circulating plastic-adherent COP cells (MSC-like, CD105+, CD73+, CD90+, CD45−, CD14−) were detected in peripheral blood from 22% of patients with hip fractures, 46% of younger patients with fractures, but not in an age- and sex-matched group of women with hip osteoarthritis [43].

Animal studies have provided supportive data for these findings in humans. In the above-mentioned studies of Kumagai et al., where parabiotic animals were formed by surgically conjoining transgenic mice constitutively expressing GFP and syngeneic WT mice, a transverse fibular fracture was created in the contralateral hind limb of the conjoined WT partner and assessed for the contribution of circulating cells to the fracture callus [6]. Based on analysis of GFP+ cells and colocalization of ALP staining, histomorphometric analysis of the fracture callus revealed a significant increase of GFP+ ALP+ cells after 2 and 3 weeks post fracture compared to nonfractured controls. Interestingly, bone healing assessed by biomechanical, radiological, and histological criteria was significantly enhanced by human CD34 cell transplantation in a nonhealing femoral fracture model in nude rats [37] and by plastic-adherent murine MSC-like and CXCR4+ cells injected in a tibia fracture mouse model [44].

It has also been suggested that COP cells participate in the pathogenesis of heterotopic ossification (HO), which can take place in pulmonary, vascular, cardiac, and periarticular soft tissue sites. Strong evidence exists that suggests that circulating fibrocytes are mediators of pulmonary fibrosis [45], so it is also plausible that fibrocytes can mediate both fibrosis and ossification in the lung. There is also increasing evidence that COP cells participate in HO following hip arthroplasty, in end-stage aortic valvular disease, and in a genetic syndrome of extraskeletal bone formation [15, 46, 47]. Animal models of ectopic bone formation also support the notion of COP cell involvement in lesion formation [11, 12, 15]. In FOP, patients with active episodes of extraskeletal bone formation have higher numbers of clonally derived COP cells than patients with stable disease or unaffected individuals, and these COP cells are present in early fibroproliferative lesions [15].

The process of vascular calcification was formerly considered the result of passive calcium deposition but is now recognized to be an active pathophysiologic process resulting in *de novo* bone formation in the late stages. Ossification is present in about 13–15% of carotid endarterectomy and stenotic aortic valve specimens [48,49]. The detection of COP cells in patients with end-stage aortic valve disease suggests that these cells are involved in the process of vascular ossification where there is preexisting calcification [46]. This observation is also supported by reports showing that levels of OCN-positive cells are elevated in patients with peripheral arterial disease and that expression of osteocalcin in endothelial progenitor cells is increased in patients with coronary artery disease [50, 51]. Ossified lesions in vascular disease develop in the setting of injury and inflammation, suggesting that ectopic ossification, regardless of location, shares similar initiating events.

Recent data also suggest that COP cells may reflect changes in bone remodeling due to metabolic bone disease. OCN+ COP cells were lower in patients with hypoparathyroidism compared to controls, and PTH (1-84) administration increased the number of COP cells threefold [52]. COP cells in patients with osteoporosis have been reported to increase, but their osteogenic differentiation is defective [53].

COP CELLS: A SYNTHESIS

The above experimental evidence suggests that COP cells are heterogenous populations of cells that span a continuum of phenotypes from hematopoietic cells to plastic-adherent stromal cells with several intermediate forms (Fig. 14.1) and reflecting functional heterogeneity. In both physiological and pathophysiological processes where COP cells may play etiological or ancillary roles, a common mechanistic link may be homing to sites of injury, inflammation, or relative hypoxia. In individuals undergoing normal physiologic growth or repair of bone tissue, COP cells may be recruited to target tissues by inflammatory signals, such as during fracture repair, or by signals released from a hypoxic microenvironment, such as the growth plate during long bone development. For example, a possible explanation for elevated levels of COP cells during pubertal growth may be formation of an oxygen gradient associated with intense remodeling, and as suggested by Canalis [54], the possibility that circulating osteogenic cells then return to the skeleton where they may function as mature osteoblasts.

The basis for the homing of COP cells to areas of injury and inflammation is best described for heterotopic ossification. Extraskeletal bone formation can be precipitated by soft tissue injury in skeletal muscle, causing the presumptive release of inflammatory cytokines and

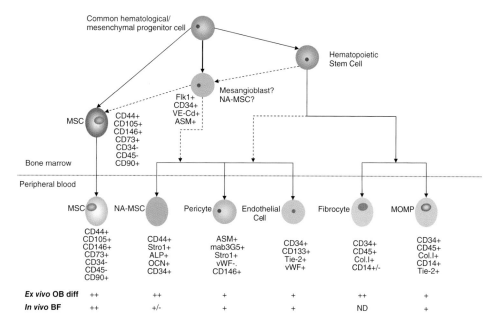

Fig. 14.1. Characterization of circulating osteogenic precursor cells and their possible lineage derivations. ---, presumptive relationships; BF, bone formation; diff, differentiation; MOMP, monocyte-derived mesenchymal progenitors; MSC, mesenchymal stem cells; NA, nonadherent; OB, osteoblast. Reprinted with permission from Pignolo RJ, Kassem M. 2011. Circulating osteogenic cells: Implications for injury, repair, and regeneration. *J Bone Miner Res* 26: 1–9.

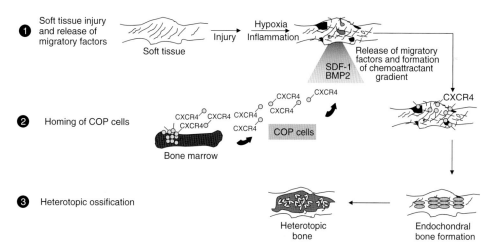

Fig. 14.2. Putative mechanism for COP cell homing in heterotopic bone formation. BMP, bone morphogenetic protein; SDF, stromal cell-derived factor. Reprinted with permission from Pignolo RJ, Kassem M. 2011. Circulating osteogenic cells: Implications for injury, repair, and regeneration. *J Bone Miner Res* 26: 1–9.

migratory factors (Fig. 14.2). Inflammatory signals appear to be necessary for BMP-induced HO, and cells of the monocyte lineage appear to be necessary for triggering ectopic bone formation following injury [55].

In an inflammatory milieu, stromal cell-derived factor-1 (SDF-1) and BMP may serve as important chemoattractant molecules [45, 56, 57]. It is well documented that SDF-1 (CXCL12) is induced by hypoxic tissue injury and forms a gradient that attracts cells expressing its cognate receptor CXCR4. Release of COP cells from bone marrow and their homing to sites of injury may thus occur through CXCR4 (Fig. 14.2). This homing mechanism is established in fibrocyte localization to lesions of pulmonary fibrosis [45]. It is also established in a mouse model of BMP2-induced HO, where osteoprogenitor cells expressing CXCR4 migrate from the bone marrow to regions of ectopic bone formation by SDF-1 chemoattraction [11, 12]. BMP may play roles in both bone formation as well as in attraction of inflammatory cells [58].

The SDF-1/CXCR4 axis may be the final common pathway for mobilization of bone marrow-derived progenitor cells by hypoxia, angiogenic peptides, inflammatory cytokines, and injury [59–61]. This axis has been implicated in processes as diverse as development, regeneration, and tumorgenesis/metastasis. It is not so surprising then that the same homing mechanism may be involved in COP cell mobilization and targeting.

CONCLUDING REMARKS

Several key concepts appear to be generally true regarding the physiological functions of COP cells. The first is that the bone-forming function of COP cells may not be their primary role but is an adaptive response in conditions of injury, repair, or abnormal cytokine signaling. The ultimate fate of COP cells may be to participate in tissue regeneration that under certain circumstances dictates *de novo* bone formation. A corollary of this hypothesis is that local MSCs serve as the primary osteochondro progenitors, whereas COP cells likely play a role in bone formation at nonskeletal sites and during tissue injury, e.g., fracture healing.

Second, COP cell homing may be mediated by the CXCR4/SDF-1 axis that is shared by multiple processes requiring the migration of stem cells. COP cells may share a common mechanism for progenitor cell migration, especially given that their putative roles are likely precipitated by injury and inflammation.

Finally, the burgeoning area of COP cell biology holds promise for development of gene and cell therapy protocols to enhance bone formation as well as diagnostic tests based on COP cell levels as a biomarker for disease states.

ACKNOWLEDGMENTS

This work was supported by the National Institutes of Health grants R01AG028873 (Robert J. Pignolo) and AG025929 (Robert J. Pignolo), The Ian Cali Endowment/University of Pennsylvania Center for Research in FOP and Related Disorders Developmental Grant Award (Robert J. Pignolo), a grant from the Novo Nordisk Foundation (Moustapha Kassem), the Lundbeck Foundation (Moustapha Kassem), and a grant from the region of Southern Denmark (Moustapha Kassem).

REFERENCES

1. Sacchetti B, Funari A, Michienzi S, Di Cesare S, Piersanti S, Saggio I, Tagliafico E, Ferrari S, Robey PG, Riminucci M, Bianco P. 2007. Self-renewing osteoprogenitors in bone marrow sinusoids can organize a hematopoietic microenvironment. *Cell* 131: 324–336.
2. Dominici M, Le Blanc K, Mueller I, laper-Cortenbach I, Marini F, Krause D, Deans R, Keating A, Prockop Dj, Horwitz E. 2006. Minimal criteria for defining multipotent mesenchymal stromal cells. The International Society for Cellular Therapy position statement. *Cytotherapy* 8: 315–317.
3. Gronthos S, Graves SE, Ohta S, Simmons PJ. 1994. The STRO-1+ fraction of adult human bone marrow contains the osteogenic precursors. *Blood* 84: 4164–4173.
4. Hauge EM, Qvesel D, Eriksen EF, Mosekilde L, Melsen F. 2001. Cancellous bone remodeling occurs in specialized compartments lined by cells expressing osteoblastic markers. *J Bone Miner Res* 16: 1575–1582.
5. Bentzon JF, Stenderup K, Hansen FD, Schroder HD, Abdallah BM, Jensen TG, Kassem M. 2005. Tissue distribution and engraftment of human mesenchymal stem cells immortalized by human telomerase reverse transcriptase gene. *Biochem Biophys Res Commun* 330: 633–640.
6. Kumagai K, Vasanji A, Drazba JA, Butler RS, Muschler GF. 2008. Circulating cells with osteogenic potential are physiologically mobilized into the fracture healing site in the parabiotic mice model. *J Orthop Res* 26: 165–175.
7. Boban I, Barisic-Dujmovic T, Clark SH. Parabiosis model does not show presence of circulating osteoprogenitor cells. *Genesis* 48: 171–182.
8. Olmsted-Davis EA, Gugala Z, Camargo F, Gannon FH, Jackson K, Kienstra KA, Shine HD, Lindsey RW, Hirschi KK, Goodell MA, Brenner MK, Davis AR. 2003. Primitive adult hematopoietic stem cells can function as osteoblast precursors. *Proc Natl Acad Sci U S A* 100: 15877–15882.
9. Dominici M, Pritchard C, Garlits JE, Hofmann TJ, Persons DA, Horwitz EM. 2004. Hematopoietic cells and osteoblasts are derived from a common marrow progenitor after bone marrow transplantation. *Proc Natl Acad Sci U S A* 101: 11761–11766.
10. Hayakawa J, Migita M, Ueda T, Shimada T, Fukunaga Y. 2003. Generation of a chimeric mouse reconstituted with green fluorescent protein-positive bone marrow cells: A useful model for studying the behavior of bone marrow cells in regeneration in vivo. *Int J Hematol* 77: 456–462.
11. Otsuru S, Tamai K, Yamazaki T, Yoshikawa H, Kaneda Y. 2007. Bone marrow-derived osteoblast progenitor cells in circulating blood contribute to ectopic bone formation in mice. *Biochem Biophys Res Commun* 354: 453–458.
12. Otsuru S, Tamai K, Yamazaki T, Yoshikawa H, Kaneda Y. 2008. Circulating bone marrow-derived osteoblast progenitor cells are recruited to the bone-forming site by CXCR4/SDF-1 pathway. *Stem Cells* 26: 223–234.
13. Koc ON, Peters C, Aubourg P, Raghavan S, Dyhouse S, DeGasperi R, Kolodny EH, Yoseph YB, Gerson SL, Lazarus HM, Caplan AI, Watkins PA, Krivit W. 1999. Bone marrow-derived mesenchymal stem cells remain host-derived despite successful hematopoietic engraftment after allogeneic transplantation in patients with

lysosomal and peroxisomal storage diseases. *Exp Hematol* 27: 1675–1681.
14. Horwitz EM, Gordon PL, Koo WK, Marx JC, Neel MD, McNall RY, Muul L, Hofmann T. 2002. Isolated allogeneic bone marrow-derived mesenchymal cells engraft and stimulate growth in children with osteogenesis imperfecta: Implications for cell therapy of bone. *Proc Natl Acad Sci U S A* 99: 8932–8927.
15. Suda RK, Billings PC, Egan KP, Kim JH, McCarrick-Walmsley R, Glaser DL, Porter DL, Shore EM, Pignolo RJ. 2009. Circulating osteogenic precursor cells in heterotopic bone formation. *Stem Cells* 27: 2209–2219.
16. Hong HS, Lee J, Lee E, Kwon YS, Lee E, Ahn W, Jiang MH, Kim JC, Son Y. 2009. A new role of substance P as an injury-inducible messenger for mobilization of CD29(+) stromal-like cells. *Nat Med* 15: 425–435.
17. Pitchford SC, Furze RC, Jones CP, Wengner AM, Rankin SM. 2009. Differential mobilization of subsets of progenitor cells from the bone marrow. *Cell Stem Cell* 4: 62–72.
18. Tondreau T, Meuleman N, Delforge A, Dejeneffe M, Leroy R, Massy M, Mortier C, Bron D, Lagneaux L. 2005. Mesenchymal stem cells derived from CD133-positive cells in mobilized peripheral blood and cord blood: Proliferation, Oct4 expression, and plasticity. *Stem Cells* 23: 1105–1112.
19. Kuznetsov SA, Mankani MH, Leet AI, Ziran N, Gronthos S, Robey PG. 2007. Circulating connective tissue precursors: Extreme rarity in humans and chondrogenic potential in guinea pigs. *Stem Cells* 25: 1830–1839.
20. Kuznetsov SA, Mankani MH, Gronthos S, Satomura K, Bianco P, Robey PG. 2001. Circulating skeletal stem cells. *J Cell Biol* 153: 1133–1140.
21. Zvaifler NJ, Marinova-Mutafchieva L, Adams G, Edwards CJ, Moss J, Burger JA, Maini RN. 2000. Mesenchymal precursor cells in the blood of normal individuals. *Arthritis Research* 2: 477–488.
22. Rochefort GY, Delorme B, Lopez A, Herault O, Bonnet P, Charbord P, Eder V, Domenech J. 2006. Multipotential mesenchymal stem cells are mobilized into peripheral blood by hypoxia. *Stem Cells* 24: 2202–2208.
23. Rosada C, Justesen J, Melsvik D, Ebbesen P, Kassem M. 2003. The human umbilical cord blood: A potential source for osteoblast progenitor cells. *Calcif Tissue Int* 72: 135–142.
24. Kern S, Eichler H, Stoeve J, Kluter H, Bieback K. 2006. Comparative analysis of mesenchymal stem cells from bone marrow, umbilical cord blood, or adipose tissue. *Stem Cells* 24: 1294–1301.
25. Long MW, Williams JL, Mann KG. 1990. Expression of human bone-related proteins in the hematopoietic microenvironment. *J Clin Invest* 86: 1387–1395.
26. Long MW, Robinson JA, Ashcraft EA, Mann KG. 1995. Regulation of human bone marrow-derived osteoprogenitor cells by osteogenic growth factors. *J Clin Invest* 95: 881–887.
27. Eghbali-Fatourechi GZ, Modder UI, Charatcharoenwitthaya N, Sanyal A, Undale AH, Clowes JA, Tarara JE, Khosla S. 2007. Characterization of circulating osteoblast lineage cells in humans. *Bone* 40: 1370–1307.
28. Eghbali-Fatourechi GZ, Lamsam J, Fraser D, Nagel D, Riggs BL, Khosla S. 2005. Circulating osteoblast-lineage cells in humans. *N Engl J Med* 352: 1959–1966.
29. Minasi MG, Riminucci M, De Angelis L, Borello U, Berarducci B, Innocenzi A, Caprioli A, Sirabella D, Baiocchi M, De Maria R, Boratto R, Jaffredo T, Broccoli V, Bianco P, Cossu G. 2002. The meso-angioblast: A multipotent, self-renewing cell that originates from the dorsal aorta and differentiates into most mesodermal tissues. *Development* 129: 2773–2783.
30. Cossu G, Bianco P. 2003. Mesoangioblasts—Vascular progenitors for extravascular mesodermal tissues. *Curr Opin Genet Dev* 13: 537–542.
31. Reyes M, Dudek A, Jahagirdar B, Koodie L, Marker PH, Verfaillie CM. 2002. Origin of endothelial progenitors in human postnatal bone marrow. *J Clin Invest* 109: 337–346.
32. Qi H, Aguiar DJ, Williams SM, La Pean A, Pan W, Verfaillie CM. 2003. Identification of genes responsible for osteoblast differentiation from human mesodermal progenitor cells. *Proc Natl Acad Sci U S A* 100: 3305–3310.
33. Rozen N, Bick T, Bajayo A, Shamian B, Schrift-Tzadok M, Gabet Y, Yayon A, Bab I, Soudry M, Lewinson D. 2009. Transplanted blood-derived endothelial progenitor cells (EPC) enhance bridging of sheep tibia critical size defects. *Bone* 45: 918–924.
34. Lounev VY, Ramachandran R, Wosczyna MN, Yamamoto M, Maidment AD, Shore EM, Glaser DL, Goldhamer DJ, Kaplan FS. 2009. Identification of progenitor cells that contribute to heterotopic skeletogenesis. *J Bone Joint Surg Am* 91: 652–663.
35. Medici D, Shore EM, Lounev VY, Kaplan FS, Kalluri R, Olsen BR. 2010. Conversion of vascular endothelial cells into multipotent stem-like cells. *Nat Med* 16: 1400–1406.
36. Wosczyna, MN, Biswas, AA, Cogswell, CA, Goldhamer, DJ. 2012. Multipotent progenitors resident in the skeletal muscle interstitium exhibit robust BMP-dependent osteogenic activity and mediate heterotopic ossification. *J Bone Miner Res* 27: 1004–1017.
37. Matsumoto T, Kawamoto A, Kuroda R, Ishikawa M, Mifune Y, Iwasaki H, Miwa M, Horii M, Hayashi S, Oyamada A, Nishimura H, Murasawa S, Doita M, Kurosaka M, Asahara T. 2006. Therapeutic potential of vasculogenesis and osteogenesis promoted by peripheral blood CD34-positive cells for functional bone healing. *Am J Pathol* 169: 1440–1457.
38. Bucala R, Spiegel LA, Chesney J, Hogan M, Cerami A. 1994. Circulating fibrocytes define a new leukocyte subpopulation that mediates tissue repair. *Molecular Medicine* 1: 71–81.
39. Choi YH, Burdick MD, Strieter RM. 2010. Human circulating fibrocytes have the capacity to differentiate osteoblasts and chondrocytes. *Int J Biochem Cell Biol* 42: 662–671.
40. Kuwana M, Okazaki Y, Kodama H, Izumi K, Yasuoka H, Ogawa Y, Kawakami Y, Ikeda Y. 2003. Human circulating CD14+ monocytes as a source of progenitors

that exhibit mesenchymal cell differentiation. *J Leukoc Biol* 74: 833–845.
41. Zhao Y, Glesne D, Huberman E. 2003. A human peripheral blood monocyte-derived subset acts as pluripotent stem cells. *Proc Nat Acad Sci U S A* 100: 2426–2431.
42. Ratajczak MZ, Kucia M, Reca R, Majka M, Janowska-Wieczorek A, Ratajczak J. 2004. Stem cell plasticity revisited: CXCR4-positive cells expressing mRNA for early muscle, liver and neural cells "hide out" in the bone marrow. *Leukemia* 18: 29–40.
43. Alm JJ, Koivu HM, Heino TJ, Hentunen TA, Laitinen S, Aro HT. 2010. Circulating plastic adherent mesenchymal stem cells in aged hip fracture patients. *J Orthop Res* 28: 1634–1642.
44. Granero-Molto F, Weis JA, Miga MI, Landis B, Myers TJ, O'Rear L, Longobardi L, Jansen ED, Mortlock DP, Spagnoli A. 2009. Regenerative effects of transplanted mesenchymal stem cells in fracture healing. *Stem Cells* 27: 1887–1898.
45. Phillips RJ, Burdick MD, Hong K, Lutz MA, Murray LA, Xue YY, Belperio JA, Keane MP, Strieter RM. 2004. Circulating fibrocytes traffic to the lungs in response to CXCL12 and mediate fibrosis. *J Clin Invest* 114: 438–446.
46. Egan KP, Kim J-H, Mohler ER 3rd, Pignolo RJ. 2011. Role for circulating osteogenic precursor (COP) cells in aortic valvular disease. *Arterioscler Thromb Vasc Biol* 31: 2965–2971.
47. Egan KP, Pignolo RJ. 2010. COP cells in periarticular non-hereditary heterotopic ossification. *J Bone Miner Res* 25: S1.
48. Mohler ER 3rd, Gannon F, Reynolds C, Zimmerman R, Keane MG, Kaplan FS. 2001. Bone formation and inflammation in cardiac valves. *Circulation* 103: 1522–1528.
49. Hunt JL, Fairman R, Mitchell ME, Carpenter JP, Golden M, Khalapyan T, Wolfe M, Neschis D, Milner R, Scoll B, Cusack A, Mohler ER 3rd. 2002. Bone formation in carotid plaques: A clinicopathological study. *Stroke* 33: 1214–1219.
50. Gossl M, Modder UI, Atkinson EJ, Lerman A, Khosla S. 2008. Osteocalcin expression by circulating endothelial progenitor cells in patients with coronary atherosclerosis. *J Am Coll Cardiol* 52: 1314–1325.
51. Pal SN, Rush C, Parr A, Van Campenhout A, Golledge J. 2010. Osteocalcin positive mononuclear cells are associated with the severity of aortic calcification. *Atherosclerosis* 210: 88–93.
52. Rubin MR, Manavalan JS, Dempster DW, Shah J, Cremers S, Kousteni S, Zhou H, McMahon DJ, Kode A, Sliney J, Shane E, Silverberg SJ, Bilezikian JP. 2011. Parathyroid hormone stimulates circulating osteogenic cells in hypoparathyroidism. *J Clin Endocrinol Metab* 96: 176–186.
53. Dalle Carbonare L, Valenti MT, Zanatta M, Donatelli L, Lo Cascio V. 2009. Circulating mesenchymal stem cells with abnormal osteogenic differentiation in patients with osteoporosis. *Arthritis Rheum* 60: 3356–3365.
54. Canalis E. 2005. The fate of circulating osteoblasts. *N Engl J Med* 352: 2014–2016.
55. Kan L, Liu Y, McGuire TL, Palila Berger DM, Awatramani RB, Dymecki SM, Kessler JA. 2009. Dysregulation of local stem/progenitor cells as a common cellular mechanism for heterotopic ossification. *Stem Cells* 27: 150–156.
56. Ceradini DJ, Kulkarni AR, Callaghan MJ, Tepper OM, Bastidas N, Kleinman ME, Capla JM, Galiano RD, Levine JP, Gurtner GC. 2004. Progenitor cell trafficking is regulated by hypoxic gradients through HIF-1 induction of SDF-1. *Nat Med* 10: 858–864.
57. Du R, Lu KV, Petritsch C, Liu P, Ganss R, Passegue E, Song H, Vandenberg S, Johnson RS, Werb Z, Bergers G. 2008. HIF1alpha induces the recruitment of bone marrow-derived vascular modulatory cells to regulate tumor angiogenesis and invasion. *Cancer Cell* 13: 206–220.
58. Cunningham NS, Paralkar V, Reddi AH. 1992. Osteogenin and recombinant bone morphogenetic protein 2B are chemotactic for human monocytes and stimulate transforming growth factor beta 1 mRNA expression. *Proc Natl Acad Sci U S A* 89: 11740–11744.
59. Schober A. 2008. Chemokines in vascular dysfunction and remodeling. *Arterioscler Thromb Vasc Biol* 28: 1950–1959.
60. Hoenig MR, Bianchi C, Sellke FW. 2008. Hypoxia inducible factor-1 alpha, endothelial progenitor cells, monocytes, cardiovascular risk, wound healing, cobalt and hydralazine: A unifying hypothesis. *Curr Drug Targets* 9: 422–435.
61. Dar A, Kollet O, Lapidot T. 2006. Mutual, reciprocal SDF-1/CXCR4 interactions between hematopoietic and bone marrow stromal cells regulate human stem cell migration and development in NOD/SCID chimeric mice. *Exp Hematol* 34: 967–975.

Section II
Skeletal Physiology

Section Editor Ego Seeman

Chapter 15. Human Fetal and Neonatal Bone Development   121
Tao Yang, Monica Grover, Kyu Sang Joeng, and Brendan Lee

Chapter 16. Skeletal Growth and Peak Bone Strength   127
Qingju Wang and Ego Seeman

Chapter 17. Ethnic Differences in Bone Acquisition   135
Shane A. Norris, Lisa K. Micklesfield, and John M. Pettifor

Chapter 18. Calcium and Other Nutrients During Growth   142
Tania Winzenberg and Graeme Jones

Chapter 19. Growing a Healthy Skeleton: The Importance of Mechanical Loading   149
Mark R. Forwood

Chapter 20. Pregnancy and Lactation   156
Christopher S. Kovacs and Henry M. Kronenberg

Chapter 21. Menopause   165
Ian R. Reid

Primer on the Metabolic Bone Diseases and Disorders of Mineral Metabolism, Eighth Edition. Edited by Clifford J. Rosen.
© 2013 American Society for Bone and Mineral Research. Published 2013 by John Wiley & Sons, Inc.

15

Human Fetal and Neonatal Bone Development

Tao Yang, Monica Grover, Kyu Sang Joeng, and Brendan Lee

Introduction 121
Physiology of Fetal and Neonatal Bone Development 121
Extrinsic Factors that Affect Fetal/Neonatal Bone Development 122

Inherited Fetal/Neonatal Bone Disorders 123
References 124

INTRODUCTION

Our understanding of human bone development, especially that occurring *in utero*, has been greatly accelerated through the analysis of animal models. However, direct studies of human bone development are still invaluable because pathological and genetic findings from human bone disorders have been extremely important for generating novel hypotheses, validating model organism studies, and uncovering new mechanisms for bone development. Moreover, not all human conditions can be recapitulated by animal models. In this chapter, we will focus on human data related to the physiology of fetal and neonatal bone development and the intrinsic and extrinsic factors that lead to fetal and neonatal bone disorders.

PHYSIOLOGY OF FETAL AND NEONATAL BONE DEVELOPMENT

At the beginning of human fetal development (8 weeks post fertilization), the patterning of skeleton has been largely determined. Compared to the earliest fetus, newborns are approximately 12 times longer in body length (30 mm vs 360 mm in crown-rump length). Hence, the major theme of bone development during the fetal stage is very rapid growth. For example, the rate of femur elongation during gestation between 16 and 41 weeks is 0.35 mm per day on average [1]. Ossification is an important component of bone development and growth, and it involves the coordination of osteoblast differentiation, matrix production, mineralization, and vasculogenesis. Studies have shown that the majority of bones commence ossification during the first several weeks of the fetal stage and that there is a sequential appearance of ossification centers in each individual bone. For example, the ossification of clavicle, humerus, and mandible occur during the embryonic stage (6 or 7 weeks). In contrast, the ossification of the talus or cuboid starts late at 28 weeks or after birth [2]. In order to maintain bone shape while accommodating rapid growth, the progression of ossification must be tightly coupled with bone resorption, and this is mediated by osteoclasts acting both inside and outside the bone. This bone remodeling begins during the fetal period and becomes prominent by the fourth and fifth months of gestation[2].

To adapt to the rapid growth and ossification of the fetal skeleton, a fetus requires a large quantity of building blocks, including proteins and minerals. These substances are transported against a concentration gradient across the placenta from the mother. More than 150 g of calcium and 70 g of phosphorus per kilogram of fetal body weight are transferred via active transport during the third trimester [3]. The actual steps of mineral transport are not completely understood. It has been proposed that calcium is transported across the placenta via a three-step model. TRPV6, a voltage-dependent calcium channel, is present on the maternal side of the placenta.

Primer on the Metabolic Bone Diseases and Disorders of Mineral Metabolism, Eighth Edition. Edited by Clifford J. Rosen.
© 2013 American Society for Bone and Mineral Research. Published 2013 by John Wiley & Sons, Inc.

Some studies have suggested that this channel transfers calcium to calbindin D9K, which is an intracellular binding protein in trophoblastic cells. Finally, calcium is transferred via PMCA3, a plasma membrane calcium–ATPase protein on the basolateral membrane to the fetal bloodstream [4–6]. Transport of phosphorus across the placenta is less well understood, but NaPi-IIb, a sodium-dependent inorganic phosphorus transporter, is believed to play an important role in transplacental phosphorus transport [7].

The primary hormone responsible for the active transport of minerals across the placenta to the fetus is parathyroid hormone related peptide (PTHrP) [8]. The fetus, placenta, umbilical cord, and breast tissue produce this hormone. Mice lacking PTHrP exhibit lethal skeletal dysplasia characterized by premature mineralization of all bones that are formed through an endochondral process. In the placenta, it is known to act through a receptor distinct from PTH1R [9]. On the other hand, parathyroid hormone (PTH) and vitamin D, the two critical hormones for maintaining calcium and phosphorus homeostasis in adults, are present at low levels in fetal serum, which may be a response to high serum calcium levels [10]. PTH is important in the mineralization of fetal bones but not in active transport of calcium across the placenta. Similarly, vitamin D does not have a major role in mineral transport, although maternal vitamin D deficiency has been associated with congenital rickets [11]. Calcitonin, on the other hand, may not play a major role in fetal bone development as seen in calcitonin or calcitonin gene related peptide ablated mice [12]. Other hormones that play a role in skeletal health during adulthood, such as growth hormone and cortisol, have been shown to influence birth weight and weight gain during infancy. Furthermore, levels of growth hormone and cortisol were found to be determinants of prospective bone loss rate. This is compatible with the hypothesis that environmental influences during intrauterine life may alter sensitivity of the skeleton to growth hormone and cortisol [13].

After birth, the skeleton maintains a fast rate of growth and requires substantial mineral input to support bone development. Different from the fetus whose calcium level is higher than that in its mother's serum, the newborn exhibits a reduced calcium level, rapidly reaching a base level because the placental source is removed; concomitantly, the PTH level rapidly rises. The neonate becomes dependent on intestinal calcium absorption and the primary hormones responsible for maintaining serum calcium levels are PTH and vitamin D. Skeletal calcium is stored, and renal calcium is reabsorbed to maintain these serum levels. Prematurity, small size for gestational age status, maternal vitamin D deficiency, and maternal diabetes make the calcium nadir more dramatic. During infancy, a phase of rapid bone mineralization, vitamin D deficiency can lead to rickets and hypocalcemia. Hence the American Academy of Pediatrics recommends supplementing all infants with 600 IU of vitamin D daily.

EXTRINSIC FACTORS THAT AFFECT FETAL/NEONATAL BONE DEVELOPMENT

Nutritional influences

Maternal nutrition during pregnancy influences fetal nutrition. Animal models show that a low protein diet during pregnancy leads to low bone area and bone mineral content with altered growth plate morphology in adult rats [14]. In addition, neonatal bone mass was found to be strongly and positively associated with birth weight, birth length, and placental weight, after adjusting for sex and gestational age, indicating the importance of maternal nutrition during pregnancy [15]. There are also studies suggesting that genetic influences on bone mineral density (BMD) and adult bone size may be modified by undernutrition *in utero* [16].

Mechanical influences

Fetal movement *in utero* is a form of mechanical stimulation against resistance, which leads to mineral accretion. The importance of muscle–bone interaction (probably regulated by a network of osteocytes) *in utero* is evident in newborns with muscular disease or hypotonia as they have a lower BMD [17]. Physiological osteoporosis of infancy, a condition of decreased cortical density, presents within 6 months following birth. Although it is mainly attributed to expansion of the bone marrow cavity size [18], lack of resistance in movement after delivery may also be a contributing factor [17]. Whether this is of clinical significance is controversial.

Environmental influences

Approximately 1 in 1000 live births is affected with axial skeletal defects. Many toxins and drugs have been implicated in its etiology, including retinoic acid, valproic acid, arsenic, and carbon monoxide. These can lead to vertebral body defects such as block vertebra and nonsegmented hemivertebra. Uncontrolled maternal diabetes mellitus can cause fetal skeletal defects, specifically caudal dysgenesis along with neonatal hypocalcemia by mechanisms not completely understood [19]. Rats with uncontrolled diabetes have decreased calbindin mRNA in the placenta; this could explain decreased calcium transport across the placenta [20]. In addition, maternal smoking has been associated with decreased numbers of ossification centers, and maternal alcohol consumption affects calciotropic hormones and thus can cause fetal bone defects [19].

Other influences

Seasonal variation has been shown to influence newborn bone mineral content (BMC), possibly due to the effect on maternal vitamin D levels. Prematurity and small gestational age status are also associated with increased risk of rickets and osteoporosis due to multiple factors including hypoxia, immobility, and decreased mineral supply/intake. In addition, gender and race appear to play a role. In some studies, BMC for male newborns was higher than females and higher for African-American newborns than for Caucasian newborns [21].

INHERITED FETAL/NEONATAL BONE DISORDERS

Multiple signaling and metabolic pathways are involved in fetal bone development, and the identification of human mutations has served as a major guide for uncovering these signaling pathways and mechanisms. Although genetically related dysregulation of these pathways can eventually lead to human skeletal diseases, many of them are difficult to diagnose in newborns. This is because the milder spectrum of diseases may not cause pronounced deformity of the skeleton, and the clinical consequences of abnormal bone mass, such as fracture, may not be evident given the relatively mild mechanical loading in the fetal or neonatal stages. Here we have selected several examples of severe bone diseases that underscore key developmental processes affecting fetal and neonatal bone development to review.

Defects in bone matrix production

Osteogenesis imperfecta (OI) is a group of inborn bone diseases in humans characterized by brittle bones. The most severe forms of OI can lead to bone fractures and lethality in fetuses and neonates. Etiologies of these severe OIs are related to abnormal production, post-translational modification, or metabolism of fibrillar collagens, especially type I collagen, which are the major substrate of bone matrix. For example, dominantly inherited point mutations in COL1A1 and COL1A2, encoding the proα1(I) and proα2(I) chains of type I collagen, lead to post-translational overmodification of collagen chains and severe forms of osteogenesis imperfecta (types II and III) [22]. More recently described recessive mutations in genes that are important for modification or trafficking of type I collagen also cause OI. These genes (and the corresponding gene products) include CRTAP (cartilage associated protein) [23], LEPRE1 (prolyl 3 hydroxylase 1) [24], PPIB (cyclophilin B) [25], FKBP10 (FK506 binding protein 10) [26], SERPINH1 (heat shock protein 47) [27], and SERPINF1 (pigment epithelial derived factor) [28].

Defects in mineral homeostasis

Recessive inactivating mutations of the calcium sensing receptor gene (CASR) are the cause of neonatal severe primary hyperparathyroidism (NSHPT) [29, 30]. This disease is characterized by extreme hypercalcemia and severe neonatal hyperparathyroidism, including demineralization of the skeleton, respiratory distress and parathyroid hyperplasia. Without prompt parathyroidectomy of the affected infants, NSHPT is usually lethal. In contrast, familial hypocalciuric hypercalcemia (FHH), caused by haploinsufficiency of CASR, affords a much milder hypocalcaemia and does not exhibit the complexity of hyperparathyroidism.

Defects in mineral deposition

Perinatal and infantile hypophosphatasia is a pernicious inborn metabolic disease manifesting in utero with profound hypomineralization that results in caput membraneceum, deformed or shortened limbs, and rapid death due to respiratory failure. Infantile hypophosphatasia is caused by recessive mutations in the gene encoding tissue-nonspecific iso-enzyme of alkaline phosphatase (TNSALP), a glycoprotein localized to the plasma membranes of osteoblasts and chondrocytes that hydrolyzes monophosphate esters at an alkaline pH optimum [31]. The deficiency of TSNALP activity leads to extracellular accumulation of inorganic pyrophosphate (PPi), which potently inhibits growth of hydroxyapatite crystal and causes severe hypomineralization in the infant's bone [32]. Haploinsufficiency of TNSALP also causes hypophosphatasia, but in a milder manner, usually diagnosed later in life.

Defects in osteoclastic function

Infantile malignant osteopetrosis (IMO) is a group of severe autosomal recessive osteopetrosis. The affected bones become very brittle, although bone mass is markedly higher than normal. IMO arises in the fetal stage; thus fractures of the clavicle can be found during delivery and frequent bone fractures occur during infancy. The affected infants suffer from hypocalcemia. Moreover, due to defective osteoclastic function, the bone marrow space that accommodates hematopoiesis is gradually diminished. Hence, if not properly treated in the first year, most of affected infants develop anemia and thrombocytopenia because of encroachment of bone on marrow [33]. Genetically, IMO is caused by mutations in the genes important for osteoclast activity. The bone resorption of osteoclasts primarily relies on the acidification of bone resorption lacuna. Hence, defects in the machinery for acid secretion, such as that caused by mutations in either CLCN7 or OSTM1 (CLCN7 encodes the chloride channel 7, which complexes with and is stabilized by the OSTM1 gene product, the osteopetrosis-associated transmembrane protein 1) or in TRCIRG1 (encoding T cell immune regulator 1, a subunit of a vacuolar proton pump) have been identified in the osteoclast-rich IMO patients [34–37].

Defects in cranial suture closure and osteogenesis

The skull of neonates is composed of separate cranial bones connected by fibrous cranial sutures (fontanels). These sutures provide flexibility for the skull to facilitate its passage through the birth canal without damaging the infant's brain. Moreover, cranial sutures contain osteogenic mesenchymal cells, serving as important sites for cranial bone growth to adapt to the rapid brain growth of infancy [38, 39]. The fusion of cranial bones normally starts after infancy and completes by adulthood. Disorders characterized by delayed or premature closure of cranial sutures are not rare in newborns. Cleidocranial dysplasia (CCD) patients have persistently open and unossified skull sutures. This is caused by a haploinsufficiency of Runx2, a master gene that regulates multiple steps of osteoblast differentiation [40–42]. In contrast, premature suture closure leads to craniosynostosis, which can severely restrain growth of the skull, thus leading to increased intracranial pressure that can severely impair neural development [38]. The etiologies of craniosynostosis include dominant activating mutations in the FGF receptors (*FGFR1*, *2*, and *3*) [43–46] or haploinsufficiency of *TWIST1* [47, 48]. Mutations in *MSX2* [49], EFNB1 [50], Gli3 [51], *RAB23* [52], *POR* [53], and *RECQL4* [54] have also been identified in some rare types of craniosystosis.

REFERENCES

1. Salle BL, Rauch F, Travers R, Bouvier R, Glorieux FH. 2002. Human fetal bone development: Histomorphometric evaluation of the proximal femoral metaphysis. *Bone* 30: 823–828.
2. Gardner E. 1971. Osteogenesis in the human embryo and fetus. In: *The Biochemistry and Physiology of Bone, 2nd Ed., Vol.3, Chapter 2.* London: Academic. pp. 77–118.
3. Neer R, Berman M, Fisher L, Rosenberg LE. 1967. Multicompartmental analysis of calcium kinetics in normal adult males. *J Clin Invest* 46: 1364–1379.
4. Belkacemi L, Bedard I, Simoneau L, Lafond J. 2005. Calcium channels, transporters and exchangers in placenta: A review. *Cell Calcium* 37: 1–8.
5. Bianco SD, Peng JB, Takanaga H, Suzuki Y, Crescenzi A, Kos CH, Zhuang L, Freeman MR, Gouveia CH, Wu J, Luo H, Mauro T, Brown EM, Hediger MA. 2007. Marked disturbance of calcium homeostasis in mice with targeted disruption of the Trpv6 calcium channel gene. *J Bone Miner Res* 22: 274–285.
6. Suzuki Y, Kovacs CS, Takanaga H, Peng JB, Landowski CP, Hediger MA. 2008. Calcium channel TRPV6 is involved in murine maternal-fetal calcium transport. *J Bone Miner Res* 23: 1249–1256.
7. Shibasaki Y, Etoh N, Hayasaka M, Takahashi MO, Kakitani M, Yamashita T, Tomizuka K, Hanaoka K. 2009. Targeted deletion of the tybe IIb Na(+)-dependent Pi-cotransporter, NaPi-IIb, results in early embryonic lethality. *Biochem Biophys Res Commun* 381: 482–486.
8. Kovacs CS, Kronenberg HM. 1997. Maternal-fetal calcium and bone metabolism during pregnancy, puerperium, and lactation. *Endocr Rev* 18: 832–872.
9. Kovacs CS, Lanske B, Hunzelman JL, Guo J, Karaplis AC, Kronenberg HM. 1996. Parathyroid hormone-related peptide (PTHrP) regulates fetal-placental calcium transport through a receptor distinct from the PTH/PTHrP receptor. *Proc Natl Acad Sci U S A* 93: 15233–15238.
10. Salle BL, Glorieux FH, Delvin EE. 1988. Perinatal vitamin D metabolism. *Biol Neonate* 54: 181–187.
11. Mahon P, Harvey N, Crozier S, Inskip H, Robinson S, Arden N, Swaminathan R, Cooper C, Godfrey K. 2010. Low maternal vitamin D status and fetal bone development: Cohort study. *J Bone Miner Res* 25: 14–19.
12. McDonald KR, Fudge NJ, Woodrow JP, Friel JK, Hoff AO, Gagel RF, Kovacs CS. 2004. Ablation of calcitonin/calcitonin gene-related peptide-alpha impairs fetal magnesium but not calcium homeostasis. *Am J Physiol Endocrinol Metab* 287: E218–E226.
13. Dennison EM, Syddall HE, Rodriguez S, Voropanov A, Day IN, Cooper C. 2004. Polymorphism in the growth hormone gene, weight in infancy, and adult bone mass. *J Clin Endocrinol Metab* 89: 4898–4903.
14. Mehta G, Roach HI, Langley-Evans S, Taylor P, Reading I, Oreffo RO, ihie-Sayer A, Clarke NM, Cooper C. 2002. Intrauterine exposure to a maternal low protein diet reduces adult bone mass and alters growth plate morphology in rats. *Calcif Tissue Int* 71: 493–498.
15. Godfrey K, Walker-Bone K, Robinson S, Taylor P, Shore S, Wheeler T, Cooper C. 2001. Neonatal bone mass: Influence of parental birthweight, maternal smoking, body composition, and activity during pregnancy. *J Bone Miner Res* 16: 1694–1703.
16. Dennison EM, Arden NK, Keen RW, Syddall H, Day IN, Spector TD, Cooper C. 2001. Birthweight, vitamin D receptor genotype and the programming of osteoporosis. *Paediatr Perinat Epidemiol* 15: 211–219.
17. Land C, Schoenau E. 2008. Fetal and postnatal bone development: Reviewing the role of mechanical stimuli and nutrition. *Best Pract Res Clin Endocrinol Metab* 22: 107–118.
18. Rauch F, Schoenau E. 2001. Changes in bone density during childhood and adolescence: An approach based on bone's biological organization. *J Bone Miner Res* 16: 597–604.
19. Alexander PG, Tuan RS. 2010. Role of environmental factors in axial skeletal dysmorphogenesis. *Birth Defects Res C Embryo Today* 90: 118–132.
20. Husain SM, Birdsey TJ, Glazier JD, Mughal MZ, Garland HO, Sibley CP. 1994. Effect of diabetes mellitus on maternofetal flux of calcium and magnesium and calbindin9K mRNA expression in rat placenta. *Pediatr Res* 35: 376–381.
21. Namgung R, Tsang RC. 2000. Factors affecting newborn bone mineral content: In utero effects on newborn bone mineralization. *Proc Nutr Soc* 59: 55–63.

22. Marini JC, Forlino A, Cabral WA, Barnes AM, San Antonio JD, Milgrom S, Hyland JC, Korkko J, Prockop DJ, De PA, Coucke P, Symoens S, Glorieux FH, Roughley PJ, Lund AM, Kuurila-Svahn K, Hartikka H, Cohn DH, Krakow D, Mottes M, Schwarze U, Chen D, Yang K, Kuslich C, Troendle J, Dalgleish R, Byers PH. 2007. Consortium for osteogenesis imperfecta mutations in the helical domain of type I collagen: Regions rich in lethal mutations align with collagen binding sites for integrins and proteoglycans. *Hum Mutat* 28: 209–221.

23. Morello R, Bertin TK, Chen Y, Hicks J, Tonachini L, Monticone M, Castagnola P, Rauch F, Glorieux FH, Vranka J, Bachinger HP, Pace JM, Schwarze U, Byers PH, Weis M, Fernandes RJ, Eyre DR, Yao Z, Boyce BF, Lee B. 2006. CRTAP is required for prolyl 3-hydroxylation and mutations cause recessive osteogenesis imperfecta. *Cell* 127: 291–304.

24. Cabral WA, Chang W, Barnes AM, Weis M, Scott MA, Leikin S, Makareeva E, Kuznetsova NV, Rosenbaum KN, Tifft CJ, Bulas DI, Kozma C, Smith PA, Eyre DR, Marini JC. 2007. Prolyl 3-hydroxylase 1 deficiency causes a recessive metabolic bone disorder resembling lethal/severe osteogenesis imperfecta. *Nat Genet* 39: 359–365.

25. van Dijk FS, Nesbitt IM, Zwikstra EH, Nikkels PG, Piersma SR, Fratantoni SA, Jimenez CR, Huizer M, Morsman AC, Cobben JM, van Roij MH, Elting MW, Verbeke JI, Wijnaendts LC, Shaw NJ, Hogler W, McKeown C, Sistermans EA, Dalton A, Meijers-Heijboer H, Pals G. 2009. PPIB mutations cause severe osteogenesis imperfecta. *Am J Hum Genet* 85: 521–527.

26. Alanay Y, Avaygan H, Camacho N, Utine GE, Boduroglu K, Aktas D, Alikasifoglu M, Tuncbilek E, Orhan D, Bakar FT, Zabel B, Superti-Furga A, Bruckner-Tuderman L, Curry CJ, Pyott S, Byers PH, Eyre DR, Baldridge D, Lee B, Merrill AE, Davis EC, Cohn DH, Akarsu N, Krakow D. 2010. Mutations in the gene encoding the RER protein FKBP65 cause autosomal-recessive osteogenesis imperfecta. *Am J Hum Genet* 86: 551–559.

27. Christiansen HE, Schwarze U, Pyott SM, AlSwaid A, Al Balwi M, Alrasheed S, Pepin MG, Weis MA, Eyre DR, Byers PH. 2010. Homozygosity for a missense mutation in SERPINH1, which encodes the collagen chaperone protein HSP47, results in severe recessive osteogenesis imperfecta. *Am J Hum Genet* 86: 389–398.

28. Becker J, Semler O, Gilissen C, Li Y, Bolz HJ, Giunta C, Bergmann C, Rohrbach M, Koerber F, Zimmermann K, de Vries P, Wirth B, Schoenau E, Wollnik B, Veltman JA, Hoischen A, Netzer C. 2011. Exome sequencing identifies truncating mutations in human SERPINF1 in autosomal-recessive osteogenesis imperfecta. *Am J Hum Genet* 88: 362–371.

29. Pollak MR, Brown EM, Chou YH, Hebert SC, Marx SJ, Steinmann B, Levi T, Seidman CE, Seidman JG. 1993. Mutations in the human Ca(2+)-sensing receptor gene cause familial hypocalciuric hypercalcemia and neonatal severe hyperparathyroidism. *Cell* 75: 1297–1303.

30. Bai M, Pearce SH, Kifor O, Trivedi S, Stauffer UG, Thakker RV, Brown EM, Steinmann B. 1997. In vivo and in vitro characterization of neonatal hyperparathyroidism resulting from a de novo, heterozygous mutation in the Ca2+-sensing receptor gene: Normal maternal calcium homeostasis as a cause of secondary hyperparathyroidism in familial benign hypocalciuric hypercalcemia. *J Clin Invest* 99: 88–96.

31. Weiss MJ, Cole DE, Ray K, Whyte MP, Lafferty MA, Mulivor RA, Harris H. 1988. A missense mutation in the human liver/bone/kidney alkaline phosphatase gene causing a lethal form of hypophosphatasia. *Proc Natl Acad Sci U S A* 85: 7666–7669.

32. Whyte MP. 2010. Physiological role of alkaline phosphatase explored in hypophosphatasia. *Ann N Y Acad Sci* 1192: 190–200.

33. Stark Z, Savarirayan R. 2009. Osteopetrosis. *Orphanet J Rare Dis* 4: 5.

34. Kornak U, Kasper D, Bosl MR, Kaiser E, Schweizer M, Schulz A, Friedrich W, Delling G, Jentsch TJ. 2001. Loss of the ClC-7 chloride channel leads to osteopetrosis in mice and man. *Cell* 104: 205–215.

35. Frattini A, Orchard PJ, Sobacchi C, Giliani S, Abinun M, Mattsson JP, Keeling DJ, Andersson AK, Wallbrandt P, Zecca L, Notarangelo LD, Vezzoni P, Villa A. 2000. Defects in TCIRG1 subunit of the vacuolar proton pump are responsible for a subset of human autosomal recessive osteopetrosis. *Nat Genet* 25: 343–346.

36. Ramirez A, Faupel J, Gocbel I, Stiller A, Beyer S, Stockle C, Hasan C, Bode U, Kornak U, Kubisch C. 2004. Identification of a novel mutation in the coding region of the grey-lethal gene OSTM1 in human malignant infantile osteopetrosis. *Hum Mutat* 23: 471–476.

37. Lange PF, Wartosch L, Jentsch TJ, Fuhrmann JC. 2006. ClC-7 requires Ostm1 as a beta-subunit to support bone resorption and lysosomal function. *Nature* 440: 220–223.

38. Morriss-Kay GM, Wilkie AO. 2005. Growth of the normal skull vault and its alteration in craniosynostosis: Insights from human genetics and experimental studies. *J Anat* 207: 637–653.

39. Opperman LA. 2000. Cranial sutures as intramembranous bone growth sites. *Dev Dyn* 219: 472–485.

40. Lee B, Thirunavukkarasu K, Zhou L, Pastore L, Baldini A, Hecht J, Geoffroy V, Ducy P, Karsenty G. 1997. Missense mutations abolishing DNA binding of the osteoblast-specific transcription factor OSF2/CBFA1 in cleidocranial dysplasia. *Nat Genet* 16: 307–310.

41. Mundlos S, Otto F, Mundlos C, Mulliken JB, Aylsworth AS, Albright S, Lindhout D, Cole WG, Henn W, Knoll JH, Owen MJ, Mertelsmann R, Zabel BU, Olsen BR. 1997. Mutations involving the transcription factor CBFA1 cause cleidocranial dysplasia. *Cell* 89: 773–779.

42. Otto F, Thornell AP, Crompton T, Denzel A, Gilmour KC, Rosewell IR, Stamp GW, Beddington RS, Mundlos S, Olsen BR, Selby PB, Owen MJ. 1997. Cbfa1, a candidate gene for cleidocranial dysplasia syndrome, is essential for osteoblast differentiation and bone development. *Cell* 89: 765–771.

43. Bellus GA, Gaudenz K, Zackai EH, Clarke LA, Szabo J, Francomano CA, Muenke M. 1996. Identical mutations

in three different fibroblast growth factor receptor genes in autosomal dominant craniosynostosis syndromes. *Nat Genet* 14: 174–176.

44. Meyers GA, Orlow SJ, Munro IR, Przylepa KA, Jabs EW. 1995. Fibroblast growth factor receptor 3 (FGFR3) transmembrane mutation in Crouzon syndrome with acanthosis nigricans. *Nat Genet* 11: 462–464.

45. Rutland P, Pulleyn LJ, Reardon W, Baraitser M, Hayward R, Jones B, Malcolm S, Winter RM, Oldridge M, Slaney SF, et al. 1995. Identical mutations in the FGFR2 gene cause both Pfeiffer and Crouzon syndrome phenotypes. *Nat Genet* 9: 173–176.

46. Muenke M, Schell U, Hehr A, Robin NH, Losken HW, Schinzel A, Pulleyn LJ, Rutland P, Reardon W, Malcolm S, et al. 1994. A common mutation in the fibroblast growth factor receptor 1 gene in Pfeiffer syndrome. *Nat Genet* 8: 269–274.

47. el Ghouzzi V, Le Merrer M, Perrin-Schmitt F, Lajeunie E, Benit P, Renier D, Bourgeois P, Bolcato-Bellemin AL, Munnich A, Bonaventure J. 1997. Mutations of the TWIST gene in the Saethre-Chotzen syndrome. *Nat Genet* 15: 42–46.

48. Howard TD, Paznekas WA, Green ED, Chiang LC, Ma N, Ortiz de Luna RI, Garcia DC, Gonzalez-Ramos M, Kline AD, Jabs EW. 1997. Mutations in TWIST, a basic helix-loop-helix transcription factor, in Saethre-Chotzen syndrome. *Nat Genet* 15: 36–41.

49. Jabs EW, Muller U, Li X, Ma L, Luo W, Haworth IS, Klisak I, Sparkes R, Warman ML, Mulliken JB, et al. 1993. A mutation in the homeodomain of the human MSX2 gene in a family affected with autosomal dominant craniosynostosis. *Cell* 75: 443–450.

50. Wieland I, Jakubiczka S, Muschke P, Cohen M, Thiele H, Gerlach KL, Adams RH, Wieacker P. 2004. Mutations of the ephrin-B1 gene cause craniofrontonasal syndrome. *Am J Hum Genet* 74: 1209–1215.

51. Vortkamp A, Gessler M, Grzeschik KH. 1991. GLI3 zinc-finger gene interrupted by translocations in Greig syndrome families. *Nature* 352: 539–540.

52. Jenkins D, Seelow D, Jehee FS, Perlyn CA, Alonso LG, Bueno DF, Donnai D, Josifova D, Mathijssen IM, Morton JE, Orstavik KH, Sweeney E, Wall SA, Marsh JL, Nurnberg P, Passos-Bueno MR, Wilkie AO. 2007. RAB23 mutations in Carpenter syndrome imply an unexpected role for hedgehog signaling in cranial-suture development and obesity. *Am J Hum Genet* 80: 1162–1170.

53. Fluck CE, Tajima T, Pandey AV, Arlt W, Okuhara K, Verge CF, Jabs EW, Mendonca BB, Fujieda K, Miller WL. 2004. Mutant P450 oxidoreductase causes disordered steroidogenesis with and without Antley-Bixler syndrome. *Nat Genet* 36: 228–230.

54. Mendoza-Londono R, Lammer E, Watson R, Harper J, Hatamochi A, Hatamochi-Hayashi S, Napierala D, Hermanns P, Collins S, Roa BB, Hedge MR, Wakui K, Nguyen D, Stockton DW, Lee B. 2005. Characterization of a new syndrome that associates craniosynostosis, delayed fontanel closure, parietal foramina, imperforate anus, and skin eruption: CDAGS. *Am J Hum Genet* 77: 161–168.

16
Skeletal Growth and Peak Bone Strength

Qingju Wang and Ego Seeman

Introduction 127
Modeling and Remodeling and Bone's External Size, Shape, and Internal Architecture 128
Sex Differences in Bone Morphology 130
Growth of Metaphyses and Fractures in Childhood 130

Effect of Illnesses on Bone Morphology Is Maturational Stage–Specific 131
Summary 131
References 131

INTRODUCTION

The magnitude of the variance of bone traits such as bone mass and size around their age-specific mean is large; 1 standard deviation (SD) is about 10–15% of the mean. Thus, individuals at the 95th and 5th percentiles for bone size differ by approximately 50%. The variance in the rate of bone loss is about an order of magnitude less (1 SD = 1% of the mean), so differences in the percentile location of bone traits established at the completion of growth are likely to be more important determinants of fracture risk in adulthood than differences in the rates of bone loss over many years [1].

Whether trait differences appear during intrauterine life or postnatally is uncertain, but whatever the timing, trait variances are established well before puberty and probably within the first 2 years of life [2, 3]. For example, Wang et al. reported that the magnitude of variance in tibial cross-sectional area (CSA) and cortical volumetric bone mineral density (vBMD) in prepubertal children was no different from their premenopausal mothers [3]. However, variance in cortical vBMD, a function of the degree of bone matrix mineralization and intracortical porosity, decreased in the adolescent years, suggesting that bone's material composition becomes more similar among individuals as maturity is reached [2, 4]. If so, then variance in bone strength is more likely to be due to differences in bone structure than material composition.

Even though trait variance are largely genetically determined, we found no evidence that ranking in femur length occurred *in utero*; only 7% of the variance in the percentile location of femur length at birth was predicted by the percentile location in earlier gestation [5]. Studies by Wang et al. suggest that morphology at puberty and adulthood is predicted by morphology after 6 months of age, not at birth [6]. Tracking appeared to become established during the first year of life [7].

For example, Wang et al. report crown-heel length (CHL) in infants at 6 months and thereafter, but not at birth, predicted body height, bone size, mass, and strength almost 2 decades later and predicted these traits in their parents [6]; CHL or height then tracked in a trajectory of growth from 6 months of age through adolescence to adulthood. This also applied to bone traits such as total and regional bone mass and size, tibial and radial cross-sectional area, and indices of bending and compressive strength first measured at 11.5 years of age and then during 7 years to maturity at 18 years of age. These observations suggest that the location of an individual's skeletal trait relative to others, and the familial resemblance of bone traits at maturity, appear during the first year of postnatal life.

Loro et al. report that the percentile ranking of traits at Tanner stage 2 was unchanged during 3 years and 60–90% of the variance at maturity was accounted for by the variance before puberty [2]. Thus, an individual with a larger vertebral or femoral shaft cross section, or higher

---

Primer on the Metabolic Bone Diseases and Disorders of Mineral Metabolism, Eighth Edition. Edited by Clifford J. Rosen.
© 2013 American Society for Bone and Mineral Research. Published 2013 by John Wiley & Sons, Inc.

vertebral vBMD, or femoral cortical area than their peers of the same age before puberty retained this relative position to maturity.

Cheng et al. reported that ranking of total body bone mineral content (BMC) was maintained during 7 years follow-up in adolescence [8]. Emaus et al. reported distal and ultra distal radial bone mineral density measured by single-energy X-ray absorptiometry tracked during 7 years follow-up in 5,366 women and men aged 45–84 years [9]. Healthy premenopausal daughters of women with spine or hip fracture have structural abnormalities at the corresponding site [10]. Girls with forearm fractures during puberty have reduced distal radius vBMD before puberty, and this deficit in vBMD persisted into young adulthood [11].

MODELING AND REMODELING AND BONE'S EXTERNAL SIZE, SHAPE, AND INTERNAL ARCHITECTURE

Bone modeling assembles bone size and shape according to a genetic program; fetal limb buds grown *in vitro* develop the shape of the proximal femur so the variance in bone shape is largely determined by genetic factors. Bone modeling also adapts a bone's size, shape, and the spatial distribution of its mineralized mass to the prevailing loads [12–14]. Differences in bone size in the playing arm and nonplaying arm of tennis players attest to the ability of periosteal apposition to adapt a structure to its loading circumstances during growth) [15, 16]. Comparable effects of exercise begun in adulthood have never been reported.

In prepubertal girls, tibial cross-sectional shape is already elliptical at 10 years of age [3]. Within 2 years, periosteal apposition increased ellipticity by adding twice the amount of bone anteriorly and posteriorly rather than adding it medially and laterally (Fig. 16.1). Consequently, estimates of bending strength increased more in the antero–posterior (Imax) than medio–lateral direction (Imin). Resistance to bending increased by 44% along the principal axis (Imax) with a 22% increase in bone mass.[3]

Greater periosteal apposition on the anterior and posterior surfaces than medial and lateral surfaces creates the elliptical shape of the tibia and demonstrates how strength is optimized and mass minimized by modifying the spatial distribution of the material rather than using more material. If cortical thickness increased by the same amount of periosteal apposition at each point around the perimeter of a cross section, the amount of bone producing the same increase in bending resistance would be fourfold more than what is observed [3].

This is further illustrated by the heterogeneity in femoral neck shape (Fig. 16.2) [17]. At the junction with the shaft, the size and ellipticity of the femoral neck (FN) cross section is greatest and lessens moving proximally toward the midpoint. Then, the size increases but the

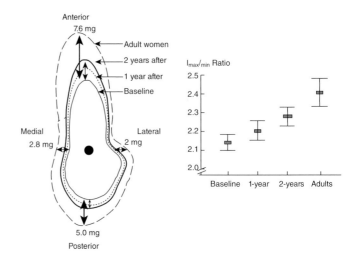

Fig. 16.1. Bone mass distribution around the bone center (the black dot in the center of the bone). Focal periosteal apposition varied from site to site around the bone perimeter. More bone was deposited at the anterio–posterior (AP) regions than at the medio–lateral (ML) regions, increasing the ellipticity and bending strength more along the AP axis (Imax) than the ML axis (Imin) (Ref. 3; reproduced with permission from The Endocrine Society).

shape becomes increasingly circular. This diversity in total cross-sectional size and shape, from cross section to cross section, is achieved using the *same* amount of material distributed differently in space at each cross section. The same amount of bone is assembled as a larger cross section, with mainly cortical bone adjacent to the femoral shaft and most of this cortical mass is distributed inferiorly. More proximally, the proportion of bone that is cortical decreases while the proportion that is trabecular increases. Cortices become thinner, and at the femoral head junction most of the bone is trabecular and the cortex is thin and evenly distributed around the perimeter. These structural features are likely to be partly adaptations to the differing loading patterns throughout the femoral neck.

Differing periosteal apposition at each point around the bone perimeter is accompanied by concurrent resorption on the endosteal or inner surface of the bone. In tubular bones lightness is achieved by endocortical resorption, which excavates the marrow cavity, shifting the thickening cortex outward; distance from the neutral axis increases the bone's resistance to bending [12]. Wang et al. reported that the amount of bone deposited in 2 years on the periosteal surface in prepubertal children with larger tibial cross sections was no different to the amount deposited on the periosteal surface of smaller cross sections [3]. Thus, larger bones deposited less bone *relative* to their starting cross-sectional size than did smaller bones. Larger bones also excavated a larger medullary canal by higher rates of endocortical resorption as reflected in those with the higher bone resorption marker tartrate-resistant acid phosphatase isoform 5b (TRACP).

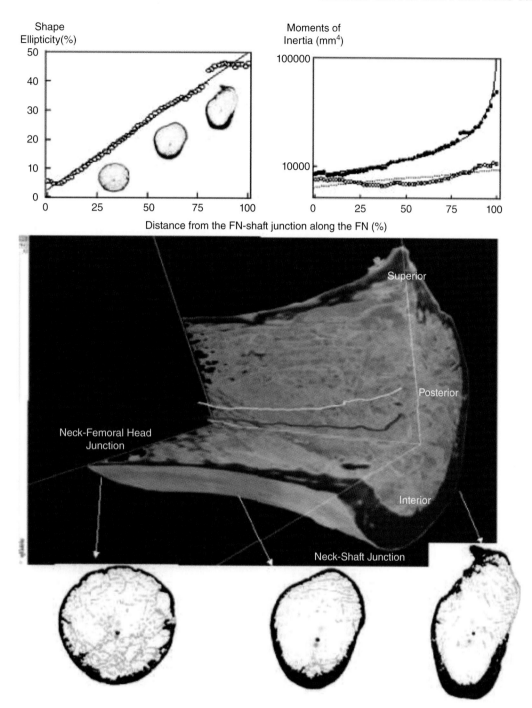

Fig. 16.2. Cross-sectional shape, size, and spatial distribution of the mineralized bone tissue mass, and geometric indices of strength of the femoral neck. From the femoral neck–head junction, there is a gradual shift of the external shape toward increasing ellipticity (upper left panel). Alongside this increasing ellipticity, there is an exponential increase in Imax (closed circles), the geometric index of strength orientated in the supero–inferior direction, while the changes in Imin (open circles), the strength in the antero–posterior direction, is minimal (upper right panel) (Ref. 17; reproduced with permission).

These individuals had a thinner cortex so larger bones were relatively lighter; they had a lower apparent vBMD [3]. Individuals with smaller tubular bone cross sections assemble them with relatively more mass forming a relatively denser bone. In bones with a smaller cross section, the vulnerability to fracture due to slenderness is offset by more periosteal apposition relative to their starting cross-sectional size and less excavation of the marrow cavity so that vBMD is greater [3]. A high peak vBMD is not the result of increased bone formation (which is costly), it is the result of reduced bone resorption. (The resorption is not followed by formation and is a modeling, not remodeling.) Similarly, a lower vBMD is the result of more bone resorption not less bone formation.

SEX DIFFERENCES IN BONE MORPHOLOGY

Growth in stature is the result of differing contributions of appendicular and axial growth. Appendicular growth continues at a greater rate (approximately 2 times) than axial growth until the onset of puberty at which time appendicular growth decelerates while axial growth accelerates [18–21]. At the first 2 years of puberty (11–13 years of age in girls and 13–15 in boys), the contribution of axial and appendicular growth to the standing height is similar (7.7 vs 7.4 cm in girls and 8.5 vs 8.0 cm in boys), while in the late pubertal stage, the increase in standing height is more derived from axial than appendicular growth (4.5 vs 1.5 cm in both sexes) [18–21]. Males may have a larger skeleton at birth and have a 1–2 year longer prepubertal growth than females because puberty occurs later in males.

Before puberty, bones do not differ in length between the sexes but are wider in males than females in some studies, but not in others [22, 23]. This difference in bone widths are likely to originate *in utero* or the first half-year of life, perhaps because of exposure to sex hormones [24]. Sex differences in bone length, width, mass, and strength emerge largely during puberty.

During puberty, periosteal apposition increases bone width while endosteal resorption enlarges the medullary cavity in boys [25, 26]. Cortical thickness increases because periosteal apposition is greater than endocortical resorption. In girls, periosteal apposition decelerates earlier, and there is no change in medullary size in girls at some sites, but there is medullary contraction at other sites [4, 25–28]. The net effect of cessation of periosteal apposition and medullary contraction in girls is the construction of a bone with a smaller total and medullary size than in boys, but with a similar cortical thickness [26–30]. Estrogen at high levels in postpubertal girls likely promotes the endosteal apposition [31, 32].

At the metaphyses of long bones, vBMD of the trabecular compartment remains constant from 5 years of age to young adulthood in both sexes [33, 34]. At this region, sex differences in BMC, vBMD, and cross-sectional size emerge after puberty; males are reported to have thicker trabeculae and higher bone volume per unit total volume (BV/TV) [35, 36]. Further investigations are needed to confirm these observations.

As the size of the vertebral body increases during growth and the amount of bone within it increases, there is no increase in vBMD before puberty [37]. At puberty, trabecular vBMD increases in both sexes due to an increase in trabecular thickness, not number, but the increase is no different by sex [38]. The vertebral body cross section is approximately 15% larger in boys than in girls before puberty and approximately 25% greater at maturity, but there is no sex difference in trabecular number or thickness [39–42]. That is, the sex differences in morphology is in size, not density; vertebral total CSA, but neither vertebral height nor trabecular density, differs before puberty.

GROWTH OF METAPHYSES AND FRACTURES IN CHILDHOOD

Incidence of fracture, especially at the distal metaphysis of radius, peaks at 10–12 years of age in girls and 12–14 years in boy, coinciding with the pubertal growth spurt [43]. The rate of linear growth peaks earlier than that of bone mass. In particular, growth of the distal radius is more rapid at the distal metaphysis, and at this site, the longitudinal growth outpaces bone formation upon the surfaces of trabeculae emerging from the growth plate that thereby coalesce, forming the cortical bone of the metaphyseal region [44]. This is "corticalization" of the trabeculae, and the relative delay results in a transitory phase of porosity [28, 45]. The vBMD of the metaphysis decrease during puberty due to this transitory phase of porosity, which predisposes to fracture [4, 33, 46]. With increase in weight and bone length and decrease in vBMD, the safety factor or ratio of bone strength index (SSI) to the product of weight and bone length, decreases until later puberty and increases thereafter [47]. In contrast, vBMD of the diaphysis of long bone continues to increase during puberty, although slower in early than in later puberty [4].

The porosity and reduced cortical density in children is transitory. In late puberty, cortical porosity decreased and cortical vBMD increased at the end of growth [48]. This is the result of the slowing of longitudinal bone growth while the trabeculae coalescence continues as bone formation proceeds on trabecular surfaces (Fig. 16.3). Earlier exposure to sex steroids in peri- and postpubertal girls may enhance the consolidation of the metaphyseal cortex at the endocortical surface, decreasing the residual cortical porosity. This partly explains the higher cortical density in girls than boys [49, 50].

The wider metaphysis must be modeled to fit the relative slender diaphysis during longitudinal growth. Unlike the diaphysis where bone diameter increases by periosteal apposition, the metaphyseal cortex is resorbed on periosteal surface while bone formation occurs at the endosteal surface by trabecular coalescence [51]. This modeling by resorption on the periosteum and formation on the endocortical surface with transitory porosity as well as decreased cortical thickness during pubertal growth spurt, makes this region more susceptible to fracture due to stress concentration and buckling during a compressive stress in a fall onto the outstretched hand.

Trabecular structure dominates the volume of metaphyses. It is derived from the cartilage of growth plate (primary trabeculae). Following vascular invasion, osteo-

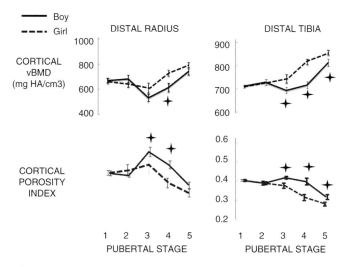

Fig. 16.3. Cortical vBMD and cortical porosity at the distal metaphyses of radius and tibia in girls (dashed line) and boys (solid line). Cortical vBMD is the lowest at Tanner stage III than earlier and later stages at the distal radius due to the rapid growth at this region (Ref. 44; reproduced with permission).

blasts deposit woven bone on the surface of calcified cartilage septa forming primary trabeculae. It is remodeled into thicker but more separated secondary trabeculae. There is little variance in the primary trabecular structural traits between children [52]. However, the subsequent remodeling of the primary trabeculae results in diverse trabecular architecture from individual to individual. Thicker and densely connected trabeculae are more easily coalesced. High trabecular BV/TV in girls predicted thicker cortices in their premenopausal mothers.[53]

The trabecular density at metaphyseal region in females is independent of age in childhood and adolescence. Using high-resolution peripheral quantitative computed tomography, we observed no increase in trabecular number, thickness, or separation at the distal metaphyses of radius or tibia in girls aged 5 to 18 years [44]. In contrast, trabecular vBMD in males increased during puberty, due to increased trabecular thickness without change in trabecular number. Thus, in young adulthood, males have thicker trabeculae but similar trabecular numbers relative to females. This sex difference may have important implication in later life when bone resorption occurs. In females, thin trabeculae are more readily perforated.

EFFECT OF ILLNESSES ON BONE MORPHOLOGY IS MATURATIONAL STAGE-SPECIFIC

Appendicular and axial growth differ, with the former preceding the later; the tempo of growth in bone size and mass also differs. Therefore, the effects of illness during growth depend on the maturational stage at the time of disease exposure, not just the severity of the illness. As longitudinal growth is more rapid in the appendicular than the axial skeleton before and during early puberty, illness may produce greater deficits in appendicular morphology. For example, disease affecting radial growth before puberty and especially during early puberty, compromises the gain in bending strength [54, 55]. Illness during late puberty may produce greater deficits in the axial than in the appendicular morphology, while in postpuberty is unlikely to produce deficits in external morphology [28]. This regional specificity in growth and the effects of illness are obscured by the study of standing height or BMD alone.

Diseases leading to sex hormone deficiency in females during puberty produce loss of sexual dimorphism in leg length as estrogen deficiency allows continued growth in females so the epiphyses do not fuse, and periosteal apposition continues. Periosteal apposition continues increasing bone width while endocortical apposition fails to occur. Cortical thickness is reduced, but only modestly, and the bone is wider. In contrast, the pubertal growth spurt in trunk length may be affected, creating a shorter but wider vertebral body and a shorter sitting height [56–58]. Delayed puberty in males reduces periosteal apposition, producing a narrower bone with a thinner cortex, while appendicular growth in length continues and produces a longer, more slender bone with a thinner cortex predisposed to greater fragility in males than in females. The greater bone diameter in females with delayed puberty produces a biomechanical advantage and less cortical thinning because the lack of endocortical apposition is offset by continued periosteal apposition [59–62].

SUMMARY

Skeletal fragility has its antecedence during growth because the variance in bone traits achieved during growth is an order of magnitude greater than the rates of loss during aging. Thus, factors that modify skeletal morphology, such as exercise and nutrition, are likely to be best instituted during growth. Metaphyseal trabecular morphology predicts both trabecular and cortical morphology and remains largely unchanged from early in life to young adulthood. Advances in imaging allow us to quantify the material composition and microstructure of bone, opening the door to quantifying the determinants of bone strength so that persons at risk for fracture can be identified and offered treatment.

REFERENCES

1. Hui SL, Slemenda CW, Johnston CC. 1990. The contribution of bone loss to postmenopausal osteoporosisi. *Osteoporos Int* 1(1): 30–4.

2. Loro ML, Sayre J, Roe TF, Goran MI, Kaufman FR, Gilsanz V. 2000. Early identification of children predisposed to low peak bone mass and osteoporosis later in life. *J Clin Endocrinol Metab* 85(10): 3908–18.
3. Wang Q, Cheng S, Alén M, Seeman E; Finnish Calex Study Group. 2009. Bone's structural diversity in adult females is established before puberty. *J Clin Endocrinol Metab* 94(5): 1555–61.
4. Wang Q, Alen M, Nicholson P, Lyytikainen A, Suuriniemi M, Helkala E, Suominen H, Cheng S. 2005. Growth patterns at distal radius and tibial shaft in pubertal girls: A 2-year longitudinal study. *J Bone Miner Res* 20(6): 954–61.
5. Bjornerem A, Johnsen SL, Nguyen TV, Kiserud T, Seeman E. 2010. The shifting trajectory of growth in femur length during gestation. *J Bone Miner Res* 25(5): 1029–33.
6. Wang Q, Alen M, Lyytikainen A, Xu L, Tylavsky FA, Kujala UM, Kroger H, Seeman E, Cheng S. 2010. Familial resemblance and diversity in bone mass and strength in the population are established during the first year of postnatal life. *J Bone Miner Res* 25(7): 1512–20.
7. Pietilainen KH, Kaprio J, Rasanen M, Winter T, Rissanen A, Rose RJ. 2001. Tracking of body size from birth to late adolescence: contribution of birth length, birth weight, duration of gestation, parents' body size, and twinship. *Am J Epidemiol* 154(1): 21–9.
8. Cheng S, Volgyi E, Tylavsky FA, Lyytikainen A, Tormakangas T, Xu L, Cheng SM, Kroger H, Alen M, Kujala UM. 2009. Trait-specific tracking and determinants of body composition: A 7-year follow-up study of pubertal growth in girls. *BMC Med* 7: 5.
9. Emaus N, Berntsen GK, Joakimsen R, Fonnebo V. 2006. Longitudinal changes in forearm bone mineral density in women and men aged 45–84 years: The Tromso Study, a population-based study. *Am J Epidemiol* 163(5): 441–449.
10. Seeman E, Tsalamandris C, Formica C, Hopper JL, McKay J. 1994. Reduced femoral neck bone density in the daughters of women with hip fractures: The role of low peak bone density in the pathogenesis of osteoporosis. *J Bone Miner Res* 9(5): 739–743.
11. Cheng S, Xu L, Nicholson PH, Tylavsky F, Lyytikainen A, Wang Q, Suominen H, Kujala UM, Kroger H, Alen M. 2009. Low volumetric BMD is linked to upper-limb fracture in pubertal girls and persists into adulthood: A seven-year cohort study. *Bone* 45: 480–6.
12. Ruff CB, Hayes WC. 1982. Subperiosteal expansion and cortical remodeling of the human femur and tibia with aging. *Science* 217(4563): 945–948.
13. Ruff C. 2003. Growth in bone strength, body size, and muscle size in a juvenile longitudinal sample. *Bone* 33(3): 317–329.
14. Lanyon LE. 1992. Control of bone architecture by functional load bearing. *J Bone Miner Res* 7 Suppl 2: S369–375.
15. Bass SL, Saxon L, Daly RM, Turner CH, Robling AG, Seeman E, Stuckey S. 2002. The effect of mechanical loading on the size and shape of bone in pre-, peri-, and postpubertal girls: A study in tennis players. *J Bone Miner Res* 17(12): 2274–2280.
16. Kontulainen S, Sievanen H, Kannus P, Pasanen M, Vuori I. 2002. Effect of long-term impact-loading on mass, size, and estimated strength of humerus and radius of female racquet-sports players: A peripheral quantitative computed tomography study between young and old starters and controls. *J Bone Miner Res* 17(12): 2281–2289.
17. Zebaze RM, Jones A, Welsh F, Knackstedt M, Seeman E. 2005. Femoral neck shape and the spatial distribution of its mineral mass varies with its size: Clinical and biomechanical implications. *Bone* 37(2): 243–252.
18. Maresh MM. 1955. Linear growth of long bones of extremities from infancy through adolescence; continuing studies. *AMA Am J Dis Child* (1960) 89: 725–42.
19. Tanner JM, Whitehouse RH. 1976. Clinical longitudinal standards for height, weight, height velocity, weight velocity and stages of puberty. *Arch Dis Child* 51: 170–9.
20. Hensinger RN. 1986. *Standards in Pediatric Orthopedics*. New York: Raven Press.
21. Karlberg J. 1990. The infancy-childhood growth spurt. *Act Paediatr Scand* 367: 111–8.
22. Clark EM, Ness AR, Tobias JH. 2007. Gender differences in the ratio between humerus width and length are established prior to puberty. *Osteoporos Int* 18(4): 463–70.
23. Hogler W, Blimkie CJ, Cowell CT, Kemp AF, Briody J, Wiebe P, Farpour-Lambert N, Duncan CS, Woodhead HJ. 2003. A comparison of bone geometry and cortical density at the mid-femur between prepuberty and young adulthood using magnetic resonance imaging. *Bone* 33(5): 771–8.
24. Bolton NJ, Tapanainen J, Koivisto M, Vihko R. 1989. Circulating sex hormone-binding globulin and testosterone in newborns and infants. *Clin Endocr* 31(2): 201–7.
25. Neu C, Rauch F, Manz F, Schoenau E. 2001. Modeling of cross-sectional bone size, mass and geometry at the proximal radius: A study of normal bone development using peripheral quantitative computed tomography. *Osteoporos Int* 12: 538–47.
26. Tanner JM, Hughes PC, Whitehouse RH. 1981. Radiographically determined widths of bone muscle and fat in the upper arm and calf from age 3–18 years. *Ann Hum Biol* 8(6): 495–517.
27. Garn SM. 1972. The course of bone gain and the phases of bone loss. *Orth Clin North Am* 3(3): 503–20.
28. Bass S, Delmas PD, Pearce G, Hendrich E, Tabensky A, Seeman E. 1999. The differing tempo of growth in bone size, mass, and density in girls is region-specific. *J Clin Invest* 104(6): 795–804.
29. Garn SM, Miller RL, Larsen KE. 1976. *Metacarpal lengths, cortical diameters and areas from the 10-state nutrition survey*. Ann Arbor, MI: University of Michigan, Center for Human Growth and Development.
30. Neu CM, Rauch F, Manz F, Schoenau E. 2001. Modeling of cross-sectional bone size, mass and geometry at the proximal radius: A study of normal bone development using peripheral quantitative computed tomography. *Osteoporos Int* 12(7): 538–47.

31. Wang Q, Alen M, Nicholson PH, Halleen JM, Alatalo SL, Ohlsson C, Suominen H, Cheng S. 2006. Differential effects of sex hormones on peri- and endocortical bone surfaces in pubertal girls. *J Clin Endocinol Metab* 91(1): 277–82.
32. Wang Q, Nicholson PH, Suuriniemi M, Lyytikainen A, Helkala E, Alen M, Suominen H, Cheng S. 2004. Relationship of sex hormones to bone geometric properties and mineral density in early pubertal girls. *J Clin Endocrinol Metab* 89(4): 1698–703.
33. Rauch F, Schoenau E. 2005. Peripheral quantitative computed tomography of the distal radius in young subjects—New reference data and interpretation of results. *J Musculoskelet Neuronal Interact* 5(2): 119–26.
34. Moyer-Mileur LJ, Quick JL, Murray MA. 2007. Peripheral quantitative computed tomography of the tibia: Pediatric reference values. *J Clin Densitom* 11: 283–94.
35. Khosla S, Riggs BL, Atkinson EJ, Oberg AL, McDaniel LJ, Holets M, Peterson JM, Melton LJ 3rd. 2006. Effects of sex and age on bone microstructure at the ultradistal radius: a population-based noninvasive in vivo assessment. *J Bone Miner Res* 21(1): 124–31.
36. Seeman E, Delmas PD. 2006. Bone quality—The material and structural basis of bone strength and fragility. *N Engl J Med* 354(21): 2250–61.
37. Gilsanz V, Roe TF, Mora S, Costin G, Goodman WG. 1991. Changes in vertebral bone density in black girls and white girls during childhood and puberty. *N Engl J Med* 325(23): 1597–600.
38. Parfitt AM, Travers R, Rauch F, Glorieux FH. 2000. Structural and cellular changes during bone growth in healthy children. *Bone* 27(4): 487–94.
39. Gilsanz V, Boechat MI, Roe TF, Loro ML, Sayre JW, Goodman WG. 1994. Gender differences in vertebral body sizes in children and adolescents. *Radiology* 190(3): 673–7.
40. Ebbesen EN, Thomsen JS, Beck-Nielsen H, Nepper-Rasmussen HJ, Mosekilde L. 1999. Age- and gender-related differences in vertebral bone mass, density, and strength. *J Bone Miner Res* 14(8): 1394–403.
41. Schultz AB, Sorensen SE, Andersson GB. 1984. Measurement of spine morphology in children, ages 10–16. *Spine* 9(1): 70–3.
42. Veldhuizen AG, Baas P, Webb PJ. 1986. Observations on the growth of the adolescent spine. *J Bone Joint Surg Br* 68(5): 724–8.
43. Cooper C, Dennison E, Leufkins H, Bishop N, van Staa T. 2004. Epidemiology of childhood fractures in britain: A study using the general practice research database. *J Bone Miner Res* 19(12): 1976–81.
44. Wang Q, Wang XF, Iuliano-Burns S, Ghasem-Zadeh A, Zebaze R, Seeman E. 2010. Rapid growth produces transient cortical weakness: A risk factor for metaphyseal fractures during puberty. *J Bone Miner Res* 25(7): 1521–6.
45. Bailey DA, McKay H, Mirwald RL, Crocker P, Faulkner DL. 1999. A six-year longitudinal study of the relationship of physical activity to bone mineral accrual in growing children: The Univeriusity of Saskatchewan bone mineral accrual study. *J Bone Miner Res* 14(10): 1672–9.
46. Kirmani S, McCready L, Holets M, Fischer PR, Riggs BL, Melton LJ, Khosla S. 2007. Decreases in cortical thickness, and not changes in trabecular microstructure, are associated with the pubertal increase in forearm fractures in girls. *J Bone Miner Res* 22(Suppl): Abstract No.1193.
47. Rauch F, Neu C, Manz F, Schoenau E. 2001. The development of metaphyseal cortex—Implications for distal radius fractures during growth. *J Bone Miner Res* 16(8): 1547–55.
48. Kirmani S, Christen D, van Lenthe GH, Fischer PR, Bouxsein ML, McCready LK, Melton LJ, Riggs BL, Amin S, Muller R, Khosla S. 2008. Bone structure at the distal radius during adolescent growth. *J Bone Miner Res* 24: 1033–42.
49. Havill LM, Mahaney MC, L Binkley T, Specker BL. 2007. Effects of genes, sex, age, and activity on BMC, bone size, and areal and volumetric BMD. *J Bone Miner Res* 22(5): 737–46.
50. Schoenau E, Neu CM, Rauch F, Manz F. 2002. Gender-specific pubertal changes in volumetric cortical bone mineral density at the proximal radius. *Bone* 31(1): 110–3.
51. Cadet ER, Gafni RI, McCarthy EF, McCray DR, Bacher JD, Barnes KM, Baron J. 2003. Mechanisms responsible for longitudinal growth of the cortex: Coalescence of trabecular bone into cortical bone. *J Bone Joint Surg* 85-A(9): 1739–48.
52. Fazzalari NL, Moore AJ, Byers S, Byard RW. 1997. Quantitative analysis of trabecular morphogenesis in the human costochondral junction during the postnatal period in normal subjects. *Anat Rec* 248(1): 1–12.
53. Wang Q, Ghasem-Zadeh A, Wang XF, Iuliano-Burns S, Seeman E. 2011. Trabecular bone of growth plate origin influences both trabecular and cortical morphology in adulthood. *J Bone Miner Res* 26(7): 1577–83.
54. Macdonald H, Kontulainen S, Petit M, Janssen P, McKay H. 2006. Bone strength and its determinants in pre- and early pubertal boys and girls. *Bone* 39(3): 598–608.
55. Seeman E, Karlsson MK, Duan Y. 2000. On exposure to anorexia nervosa, the temporal variation in axial and appendicular skeletal development predisposes to site-specific deficits in bone size and density: A cross-sectional study. *J Bone Miner Res* 15(11): 2259–65.
56. Poyrazoglu S, Gunoz H, Darendeliler F, Saka N, Bundak R, Bas F. 2005. Constitutional delay of growth and puberty: From presentation to final height. *J Pediatr Endocrinol Metab* 18(2): 171–9.
57. Morishima A, Grumbach MM, Simpson ER, Fisher C, Qin K. 1995. Aromatase deficiency in male and female siblings caused by a novel mutation and the physiological role of estrogens. *J Clin Endocrinol Metab* 80(12): 3689–98.
58. Conte FA, Grumbach MM, Ito Y, Fisher CR, Simpson ER. 1994. A syndrome of female pseudohermaphrodism, hypergonadotropic hypogonadism, and multicystic ovaries associated with missense mutations in the gene

encoding aromatase (P450arom). *J Clin Endocrinol Metab* 78(6): 1287–92.
59. Yap F, Hogler W, Briody J, Moore B, Howman-Giles R, Cowell CT. 2004. The skeletal phenotype of men with previous constitutional delay of puberty. *J Clin Endocrinol Metab* 89(9): 4306–11.
60. Finkelstein JS, Klibanski A, Neer RM. 1996. A longitudinal evaluation of bone mineral density in adult men with histories of delayed puberty. *J Clin Endocrinol Metab* 81(3): 1152–5.
61. Finkelstein JS, Neer RM, Biller BM, Crawford JD, Klibanski A. 1992. Osteopenia in men with a history of delayed puberty. *New Eng J Med* 326(9): 600–4.
62. Zhang XZ, Kalu DN, Erbas B, Hopper JL, Seeman E. 1999. The effects of gonadectomy on bone size, mass, and volumetric density in growing rats are gender-, site-, and growth hormone-specific. *J Bone Miner Res* 14(5): 802–9.

17
Ethnic Differences in Bone Acquisition

Shane A. Norris, Lisa K. Micklesfield, and John M. Pettifor

Ethnic Differences in Bone and Mineral Metabolism   135
Ethnic Differences in Bone Acquisition, Mass, and Geometry   136
Ethnic Differences in Dietary Calcium and Bone Mass   138
South Africa: A Case Study   138
Summary   138
References   139

Bone acquisition during childhood and adolescence is a critical time to maximize peak bone mass, which has been shown to occur by the end of the second or early in the third decade of life [1]. Although it is widely accepted that peak bone mass is an important determinant of future bone health and fracture risk, there is also evidence that bone mass achieved during childhood influences fracture rates during adolescence [2, 3]. A number of genetic and environmental factors play important roles in determining the rate and magnitude of bone acquisition during the early years. These factors include nutritional status, such as dietary calcium and protein intakes [4], growth rates [5], and the rate and timing of pubertal development [6]. Ethnicity is also an important factor determining not only peak bone mass, but also fracture rates during childhood and adolescence [7] and in later life [8].

A significant variation in the incidence of hip fractures exists among different communities across the world and among different ethnic groups. The highest incidence of hip fractures has been reported in Scandinavia, with a reported age standardized hip fracture rate of 920 per 100,000 per annum in Norwegian women, and the lowest in Africa, at 4.1 per 100,000 persons over 35 years of age in women living in Cameroon [9, 10], and 4.5 and 4.2 fractures per 100,000 per annum for black South African men and women, respectively [8]. Similarly in the United States of America, osteoporotic fractures are less common in African-Americans and occur later in African-American, Asian, and Hispanic populations than in non-Hispanic white populations [11]. Recently, studies conducted in South Africa have shown that fracture rates in blacks during childhood and adolescence are less than half those found in whites of the same age [7].

Understanding the role that genetic and environmental factors play in determining bone mass within ethnic groups in different geographical locations is key in identifying risk factors for fracture and osteoporosis in different populations. Some evidence suggests that the region of residence early in life is more closely associated with the variation in hip fracture incidence than the current region of residency [9, 12]. Within some countries, such as South Africa, it has been reported that the contribution of lifestyle and socioeconomic factors to bone mineral density differs between ethnic groups [13], but despite these differences the overriding contribution of genetics cannot be underestimated. In order to identify the predictors of skeletal health in adults, in particular peak bone mass, understanding ethnic differences in mineral metabolism and bone mass acquisition is an important research area.

ETHNIC DIFFERENCES IN BONE AND MINERAL METABOLISM

Bone turnover rates and urinary calcium excretion have been measured in various ethnic groups in an attempt

---

Primer on the Metabolic Bone Diseases and Disorders of Mineral Metabolism, Eighth Edition. Edited by Clifford J. Rosen.
© 2013 American Society for Bone and Mineral Research. Published 2013 by John Wiley & Sons, Inc.

to understand differences in bone mineral density and fracture incidence. The majority of studies have been conducted comparing black and white ethnic groups. Adult studies in the U.S. show equivocal results, with some studies showing lower bone turnover in African-Americans compared to Caucasians [14, 15], while others have shown no difference [16]. In South Africa, using iliac crest bone histomorphometry, it has been suggested that black adults have a higher bone turnover than white adults [17]. Differences in calcium absorption, serum 25-hydoxyvitamin D and 1,25-dihydroxyvitamin D levels, and urinary calcium excretion between ethnic groups have been explored in children and adolescents [18–23]. Some studies have shown greater fractional absorption of calcium, lower urinary calcium excretion, and therefore higher calcium retention, in African-American compared to Caucasian girls. A similar study in Caucasian and Mexican-American children, none of whom were vitamin D deficient, found higher parathyroid hormone (PTH) and lower 25(OH)D concentrations in the Mexican-American girls, although these differences did not significantly affect calcium absorption, excretion, or bone calcium kinetics [19]. Many of these comparative studies are difficult to interpret because of differences in habitual calcium intakes and vitamin D status between the groups, which themselves influence intestinal calcium absorption, bone turnover, and urinary calcium excretion.

ETHNIC DIFFERENCES IN BONE ACQUISITION, MASS, AND GEOMETRY

Ethnic differences in bone mass have been well described in the literature and clearly show that axial and appendicular bone mass is greater in African-Americans compared to Caucasians [24–27] (see Table 17.1). These findings have led to the generalization that African population groups have greater bone mass and strength compared to Caucasian population groups, and these data have been used as an explanation for the very low fracture rates observed in Africa. Although there are few studies comparing bone mass among children and adolescents of different ethnic groups in other countries, it is clear that black children and adolescents in South Africa and the Gambia do not show the consistently greater bone mass at all sites that is seen in countries such as the U.S. In South Africa, black children have higher bone mass at the hip but not at other sites, while in the Gambia, black children had lower radial bone mineral content than U.K. white children [28–32]. These data illustrate the difficulty in generalizing about an ethnic group, when obvious ethnic gradations in bone mass exist. Similarly, when comparing 9-year-old black South African children to U.S. children of a similar age, after adjusting for age, sex, and body size, whole body bone mineral content (BMC) was significantly higher in the South African group compared to U.S. children [29]. These differences between the same "ethnic" group living in different geographical areas possibly provides evidence that environmental factors have a significant influence on bone mass.

Pediatric studies using computed tomography indicate that, regardless of sex, ethnicity has significant and differential effects on the density and the size of the bones in the axial and appendicular skeletons [33]. In the axial skeleton, the density of cancellous bone in the vertebral bodies is greater in African-American than in U.S. Caucasian adolescents, with the magnitude of the increase from prepubertal to postpubertal values being substantially greater in African-American than in Caucasian subjects (34% vs 11%, respectively). This difference, which first becomes apparent during late stages of puberty then persists throughout life [34]. The cross-sectional areas of the vertebral bodies, however, do not differ between African-American and Caucasian children [33]. Thus, theoretically, the structural basis for the lower vertebral bone strength and the greater incidence of fractures in the axial skeleton of Caucasian subjects resides in their lower cancellous bone density, which probably reflects a smaller number or thickness of cancellous trabeculae. In contrast, in the appendicular skeleton, ethnicity influences the cross-sectional areas of the femora, but not the cortical bone area nor the material density of cortical bone [33]. Although values for femoral cross-sectional area increase with height, weight, and other anthropometric parameters in all children, this measurement is substantially greater in African-American children [33]. As the same amount of cortical bone placed further from the center of the bone results in greater bone strength, the skeletal advantage for African-Americans in the appendicular skeleton is likely the consequence of the greater cross-sectional size of the bones [35]. Data from Asian and Hispanic youth suggest that their bone mass is similar to that of Caucasians, but lower than that of African-American children [36]. Differences in bone and body size account for much of the observed ethnic differences in bone mineral density (BMD) among non-Hispanic, Hispanic, and Asian children [36], and similarly in South African children, site-specific bone length predicts bone mass better than height, with black children having greater regional segment lengths than white children [5].

More recently peripheral quantitative computed tomography (pQCT) has been used to compare measures of volumetric density and bone geometry at appendicular bone sites between ethnic groups in children [37–40] and adults [41–43]. A comparison between 10-year-old African-American, Hispanic, and Caucasian children revealed greater bone strength in the African-American and Hispanic children compared to Caucasian children, as a result of greater volumetric density and cortical area at the 50% tibia and radius sites, after adjusting for differences in age, sex, tibial length, and tibia muscle cross-sectional area [38]. Leonard et al., in a recent cross-sectional study [40], suggest that ethnic differences may

Table 17.1. Comparison of Ethnic Differences in South African (SA) and United States (US) Body Composition and Bone Data in Children. [From Micklesfield et al. (Ref. 57); used with permission.]

	SA	US	References
Growth			
Weight	*B < W (boys only)*	*B > W*	25, 28
Height	*B < W*	*B = W; B > W*	5, 25, 28, 37, 38, 40
Sitting height	*B < W (boys only)*	*B = W; B < W*	5, 33, 40
Body Composition			
Fat mass	B = W	B = W	29
Lean mass	*B < W*	*B > W*	25, 28, 29, 37
DXA Measures			
Whole body BMC	B > W; B = W	B > W	25–29, 31
Lumbar spine BMC	B > W (girls only); B = W	B > W*; B = W; B > W	24–26, 28, 30, 31, 58
Femoral neck BMC	B > W	B > W	24, 30, 31
Mid-radius BMC	B > W	B > W* (girls only)	24, 31
pQCT Measures			
Leg muscle CSA	*B < W*	*B > W*	37, 38
Arm muscle CSA	*B < W (boys only)*	*B = W*	37, 38
Leg fat CSA	B = W	B = W	37, 38
Tibial metaphysis [trabecular BMD]	*B = W***	*B > W***	37, 38
Tibial metaphysis [total BMD]	*B = W***	*B > W***	37, 38
Tibial metaphysis [total area]	B = W**	B = W**	37, 38
Tibial metaphysis [BSI]	*B = W***	*B > W***	37, 38
Radial metaphysis [trabecular BMD]	B > W (girls only)**	B > W**	37, 38
Radial metaphysis [total BMD]	B = W**	B = W**	37, 38
Radial metaphysis [total area]	B > W (boys only)**	*B = W***	37, 38
Radial metaphysis [BSI]	B = W**	B = W**	37, 38
Tibial diaphysis [total area]	B > W**	B > W**	37, 38
Tibial diaphysis [cortical area]	*B = W***	*B > W***	37, 38
Tibial diaphysis [cortical density]	B > W (boys only)**	B > W**	37, 38
Tibial diaphysis [cortical thickness]	*B < W***	Data not available	37, 38
Tibial diaphysis [endosteal diameter]	B > W**	Data not available	37, 38
Tibial diaphysis [tibial diameter]	B > W**	Data not available	37, 38
Tibial diaphysis [periosteal circumference]	B > W**	Data not available	37, 38
Tibial diaphysis [polar strength-strain index]	B > W**	B > W**	37, 38

*not adjusted for differences in body size.
**SA data: adjusted for bone age; US data: adjusted for age, sex, tibial length, and tibial muscle CSA. Italicized and bold data are different between SA and US populations. B = black children; W = white children.

be maturation-dependent as they showed that after adjusting for a number of covariates, polar section modulus was 13.4% higher in the African-American group compared to the Caucasian group in Tanner stage 1, but decreased to only 2.5% higher in Tanner stage 5. Recent South African data have shown ethnic differences in bone size and strength in 13-year-old black children compared to white children at the 38% tibia, a predominantly cortical site, but not at the metaphyseal sites that are predominantly composed of trabecular bone [37]. These findings are consistent with the ethnic differences that have been found using dual-energy X-ray absorptiometry (DXA) when comparing South African black and white adults and children in that differences have been shown at the hip, a predominantly cortical site, but not at the lumbar spine, a predominantly trabecular site.

Data comparing pre- and postmenopausal Chinese-American and white women using high-resolution pQCT has provided additional information on skeletal microarchitecture to help understand why the fracture rate in Chinese women is lower than white women despite lower BMD as measured by DXA [41, 42]. Trabecular and cortical bone density as well as trabecular and cortical thickness were greater in the premenopausal Chinese women compared to their white peers before and after adjusting for covariates [41], and these and other microarchitectural advantages appear to be maintained with aging in Chinese women [42].

ETHNIC DIFFERENCES IN DIETARY CALCIUM AND BONE MASS

Data from Croatia suggest that dietary calcium intake does influence peak bone mass, and that these differences are present at 30 years of age, indicating that the effects of dietary calcium probably occurred during growth rather than in adulthood [44]. Moreover, some epidemiological studies have shown an increased prevalence of osteoporosis in regions where dietary calcium intake is extremely low [45]. Yet the converse is also true with the prevalence of osteoporosis and hip fracture being reported to be highest in those countries (Scandinavia) whose national average dietary calcium intakes are among the highest in the world [46].

There is evidence that a plateau threshold for calcium intake exists above which additional calcium does not increase bone mass. This threshold for adolescents is approximately 1,600 mg/day, but calcium retention is influenced by body size [47], which has been shown to differ between ethnic groups in South Africa and internationally (see Table 17.1).

The most convincing evidence that calcium consumption influences rates of bone mineral accrual comes from controlled supplementation trials in young healthy subjects. A 7-year randomized controlled trial of calcium supplementation, extending through adolescence, showed that calcium supplementation of approximately 670 mg/day on top of habitual dietary calcium intake was associated with greater increases in distal and proximal radius BMD, whole body BMD, and metacarpal cortical indexes, compared to the placebo group [4]. The differences between the treated and control groups subsequently diminished for reasons that are unclear. Whether these benefits are maintained once calcium supplementation is withdrawn was investigated by Winzenberg and colleagues through a meta-analysis of randomized controlled trials. They showed that after the cessation of calcium supplementation, the modest effect on bone mineral density disappeared, except at the upper limb where it was considered unlikely to reduce the risk of fracture in later life [48].

Few studies have addressed the effect of calcium supplementation in children of different ethnic groups. A study conducted on 10-year-old Chinese girls found that a 2-year fortified school milk supplementation trial improved bone mineral acquisition; however, 3 years after supplement withdrawal there were no long-lasting effects on bone mass [49]. Dibba and colleagues have studied the effect of calcium supplements in a group of prepubertal black children living in the Gambia, whose calcium intakes averaged approximately 300 mg/d prior to supplementation [50]. During supplementation, BMC and BMD at the midshaft and distal radius increased, and osteocalcin levels fell in the treatment group, but no effect on radial size or height was detected. At follow-up 12–24 months later, the calcium supplemented group still showed a 5% and 5.1% greater size-adjusted BMC and BMD, respectively, compared to the placebo group at the midshaft of the radius [51].

SOUTH AFRICA: A CASE STUDY

The role of environmental factors, including lifestyle and socioeconomic status (SES), in determining bone mass and strength is very clearly illustrated in the South African population. SES measured using an indicator of social support and disposable income has been positively associated with whole body bone area and bone mineral content in pre- and early pubertal South African children [52]. The very low calcium intake of black South African children has been well characterized and is estimated to be below half the recommended dietary allowance (RDA) for Caucasian children in the United States [53, 54]. This low intake may explain the absence of a higher lumbar spine bone mass in South African black children, which has been observed in the U.S. black population, compared to the respective white populations. The lumbar spine consists of more than 66% trabecular bone, the turnover of which is influenced by the hormonal and metabolic milieu; therefore the suboptimal nutritional status associated with a low calcium intake in black South African children [55] may explain these findings. Similarly, although white South African women participate in more vigorous leisure-time activity, black South African women use walking as a means of transport, and therefore accumulate more moderate-intensity activity [56]. It appears, however, that the response of the bone, particularly at the hip, to physical activity is ethnic-dependent. Recent research has shown that leisure time activity contributes significantly to hip and lumbar spine BMD in white women; however, physical activity does not make a significant contribution at any of the bone sites in black women [13]. Similarly, in children, greater mechanical loading was associated with BMD in white children, but not in black children [28].

SUMMARY

The literature on ethnic variation in hip fracture incidence confirms that many factors contribute to skeletal health. Skeletal mass is accrued throughout childhood and adolescence, and many ethnic differences appear to be present throughout growth. Interethnic studies in bone mass acquisition are important to help elucidate possible etiologies in the early pathogenesis of osteoporosis and the development of interventions that may ameliorate the manifestations of the disease in later life. In particular, exploring the environmental and cultural milieu during bone mass acquisition and its impact on dietary exposures, calcium metabolism, bone mass, and geometry is needed to understand the broad range of diversity in different ethnic groups and differing risk profiles.

REFERENCES

1. Baxter-Jones AD, Faulkner RA, Forwood MR, Mirwald RL, Bailey DA. 2011. Bone mineral accrual from 8 to 30 years of age: An estimation of peak bone mass. *J Bone Miner Res* 26(8): 1729–1739.
2. Clark EM, Ness AR, Bishop NJ, Tobias JH. 2006. Association between bone mass and fractures in children: A prospective cohort study. *J Bone Miner Res* 21(9): 1489–1495.
3. Chevalley T, Bonjour JP, van RB, Ferrari S, Rizzoli R. 2011. Fractures during childhood and adolescence in healthy boys: Relation with bone mass, microstructure, and strength. *J Clin Endocrinol Metab* 96(10): 3134–3142.
4. Matkovic V, Goel PK, Badenhop-Stevens NE, et al. 2005. Calcium supplementation and bone mineral density in females from childhood to young adulthood: A randomized controlled trial. *Am J Clin Nutr* 81(1): 175–188.
5. Nyati LH, Norris SA, Cameron N, Pettifor JM. 2006. Effect of ethnicity and sex on the growth of the axial and appendicular skeleton of children living in a developing country. *Am J Phys Anthropol* 130(1): 135–141.
6. Iuliano-Burns S, Hopper J, Seeman E. 2009. The age of puberty determines sexual dimorphism in bone structure: A male/female co-twin control study. *J Clin Endocrinol Metab* 94(5): 1638–1643.
7. Thandrayen K, Norris SA, Pettifor JM. 2009. Fracture rates in urban South African children of different ethnic origins: The Birth to Twenty cohort. *Osteoporos Int* 20(1): 47–52.
8. Solomon L. 1968. Osteoporosis and fracture of the femoral neck in the South African Bantu. *J Bone Joint Surg Br* 50(1): 2–13.
9. Dhanwal DK, Cooper C, Dennison EM. 2010. Geographic variation in osteoporotic hip fracture incidence: The growing importance of Asian influences in coming decades. *J Osteoporos* 2010: 757102.
10. Cheng SY, Levy AR, Lefaivre KA, Guy P, Kuramoto L, Sobolev B. 2011. Geographic trends in incidence of hip fractures: A comprehensive literature review. *Osteoporos Int* 22(10): 2575–2586.
11. Maggi S, Kelsey JL, Litvak J, Heyse SP. 1991. Incidence of hip fractures in the elderly: A cross-national analysis. *Osteoporos Int* 1(4): 232–241.
12. Lauderdale DS, Thisted RA, Goldberg J. 1998. Is geographic variation in hip fracture rates related to current or former region of residence? *Epidemiology* 9(5): 574–577.
13. Chantler S, Dickie K, Goedecke JH, et al. 2011. Site-specific differences in bone mineral density in black and white premenopausal South African women. *Osteoporos Int* 9(5): 574–577.
14. Kleerekoper M, Nelson DA, Peterson EL, et al. 1994. Reference data for bone mass, calciotropic hormones, and biochemical markers of bone remodeling in older (55–75) postmenopausal white and black women. *J Bone Miner Res* 9(8): 1267–1276.
15. Finkelstein JS, Sowers M, Greendale GA, et al. 2002. Ethnic variation in bone turnover in pre- and early perimenopausal women: Effects of anthropometric and lifestyle factors. *J Clin Endocrinol Metab* 87(7): 3051–3056.
16. Perry HM 3rd, Horowitz M, Morley JE, et al. 1996. Aging and bone metabolism in African American and Caucasian women. *J Clin Endocrinol Metab* 81(3): 1108–1117.
17. Schnitzler CM, Pettifor JM, Mesquita JM, Bird MD, Schnaid E, Smyth AE. 1990. Histomorphometry of iliac crest bone in 346 normal black and white South African adults. *Bone Miner* 10(3): 183–199.
18. Abrams SA, O'Brien KO, Liang LK, Stuff JE. 1995. Differences in calcium absorption and kinetics between black and white girls aged 5–16 years. *J Bone Miner Res* 10(5): 829–833.
19. Abrams SA, Copeland KC, Gunn SK, Stuff JE, Clarke LL, Ellis KJ. 1999. Calcium absorption and kinetics are similar in 7- and 8-year-old Mexican-American and Caucasian girls despite hormonal differences. *J Nutr* 129: 666–671.
20. Braun M, Palacios C, Wigertz K, et al. 2007. Racial differences in skeletal calcium retention in adolescent girls with varied controlled calcium intakes. *Am J Clin Nutr* 85(6): 1657–1663.
21. Bryant RJ, Wastney ME, Martin BR, et al. 2003. Racial differences in bone turnover and calcium metabolism in adolescent females. *J Clin Endocrinol Metab* 88(3): 1043–1047.
22. Poopedi MA, Norris SA, Pettifor JM. 2011. Factors influencing the vitamin D status of 10-year-old urban South African children. *Public Health Nutr* 14(2): 334–339.
23. Weaver CM, McCabe LD, McCabe GP, et al. 2008. Vitamin D status and calcium metabolism in adolescent black and white girls on a range of controlled calcium intakes. *J Clin Endocrinol Metab* 93(10): 3907–3914.
24. Bell NH, Shary J, Stevens J, Garza M, Gordon L, Edwards J. 1991. Demonstration that bone mass is greater in black than in white children. *J Bone Miner Res* 6(7): 719–723.
25. Nelson DA, Simpson PM, Johnson CC, Barondess DA, Kleerekoper M. 1997. The accumulation of whole body skeletal mass in third- and fourth-grade children: Effects of age, gender, ethnicity, and body composition. *Bone* 20(1): 73–78.
26. Hui SL, Dimeglio LA, Longcope C, et al. 2003. Difference in bone mass between black and white American children: Attributable to body build, sex hormone levels, or bone turnover? *J Clin Endocrinol Metab* 88(2): 642–649.
27. Horlick M, Thornton J, Wang J, Levine LS, Fedun B, Pierson RN Jr. 2000. Bone mineral in prepubertal

children: Gender and ethnicity. *J Bone Miner Res* 15(7): 1393–1397.
28. McVeigh JA, Norris SA, Cameron N, Pettifor JM. 2004. Associations between physical activity and bone mass in black and white South African children at age 9 yr. *J Appl Physiol* 97(3): 1006–1012.
29. Micklesfield LK, Norris SA, Nelson DA, Lambert EV, van der Merwe L, Pettifor JM. 2007. Comparisons of body size, composition, and whole body bone mass between North American and South African children. *J Bone Miner Res* 22(12): 1869–1877.
30. Micklesfield LK, Norris SA, van der Merwe L, Lambert EV, Beck T, Pettifor JM. 2009. Comparison of site-specific bone mass indices in South African children of different ethnic groups. *Calcif Tissue Int* 85(4): 317–325.
31. Vidulich L, Norris SA, Cameron N, Pettifor JM. 2006. Differences in bone size and bone mass between black and white 10-year-old South African children. *Osteoporos Int* 17(3): 433–440.
32. Prentice A, Laskey MA, Shaw J, Cole TJ, Fraser DR. 1990. Bone mineral content of Gambian and British children aged 0–36 months. *Bone Miner* 10(3): 211–224.
33. Gilsanz V, Skaggs DL, Kovanlikaya A, et al. 1998. Differential effect of race on the axial and appendicular skeletons of children. *J Clin Endocrinol Metab* 83(5): 1420–1427.
34. Kleerekoper M, Nelson DA, Flynn MJ, Pawluszka AS, Jacobsen G, Peterson EL. 1994. Comparison of radiographic absorptiometry with dual-energy x-ray absorptiometry and quantitative computed tomography in normal older white and black women. *J Bone Miner Res* 9(11): 1745–1749.
35. van der Meulen MC, Beaupre GS, Carter DR. 1993. Mechanobiologic influences in long bone cross-sectional growth. *Bone* 14(4): 635–642.
36. Bachrach LK, Hastie T, Wang MC, Narasimhan B, Marcus R. 1999. Bone mineral acquisition in healthy Asian, Hispanic, black, and Caucasian youth: A longitudinal study. *J Clin Endocrinol Metab* 84(12): 4702–4712.
37. Micklesfield LK, Norris SA, Pettifor JM. 2011. Determinants of bone size and strength in 13-year-old South African children: The influence of ethnicity, sex and pubertal maturation. *Bone* 48(4): 777–785.
38. Wetzsteon RJ, Hughes JM, Kaufman BC, et al. 2009. Ethnic differences in bone geometry and strength are apparent in childhood. *Bone* 44(5): 970–975.
39. Pollock NK, Laing EM, Taylor RG, et al. 2011. Comparisons of trabecular and cortical bone in late adolescent black and white females. *J Bone Miner Metab* 29(1): 44–53.
40. Leonard MB, Elmi A, Mostoufi-Moab S, et al. 2010. Effects of sex, race, and puberty on cortical bone and the functional muscle bone unit in children, adolescents, and young adults. *J Clin Endocrinol Metab* 95(4): 1681–1689.
41. Walker MD, McMahon DJ, Udesky J, Liu G, Bilezikian JP. 2009. Application of high-resolution skeletal imaging to measurements of volumetric BMD and skeletal microarchitecture in Chinese-American and white women: Explanation of a paradox. *J Bone Miner Res* 24(12): 1953–1959.
42. Walker MD, Liu XS, Stein E, et al. 2011. Differences in bone microarchitecture between postmenopausal Chinese-American and white women. *J Bone Miner Res* 26(7): 1392–1398.
43. Wang XF, Wang Q, Ghasem-Zadeh A, et al. 2009. Differences in macro- and microarchitecture of the appendicular skeleton in young Chinese and white women. *J Bone Miner Res* 24(12): 1946–1952.
44. Matkovic V, Kostial K, Simonovic I, Buzina R, Brodarec A, Nordin BE. 1979. Bone status and fracture rates in two regions of Yugoslavia. *Am J Clin Nutr* 32(3): 540–549.
45. Heaney RP. 1992. Calcium in the prevention and treatment of osteoporosis. *J Intern Med* 231(2): 169–180.
46. Hjartaker A, Lagiou A, Slimani N, et al. 2002. Consumption of dairy products in the European Prospective Investigation into Cancer and Nutrition (EPIC) cohort: Data from 35 955 24-hour dietary recalls in 10 European countries. *Public Health Nutr* 5(6B): 1259–1271.
47. Hill KM, Braun MM, Egan KA, et al. 2011. Obesity augments calcium-induced increases in skeletal calcium retention in adolescents. *J Clin Endocrinol Metab* 96(7): 2171–2177.
48. Winzenberg T, Shaw K, Fryer J, Jones G. 2006. Effects of calcium supplementation on bone density in healthy children: Meta-analysis of randomised controlled trials. *BMJ* 333(7572): 775.
49. Zhu K, Zhang Q, Foo LH, et al. 2006. Growth, bone mass, and vitamin D status of Chinese adolescent girls 3 y after withdrawal of milk supplementation. *Am J Clin Nutr* 83(3): 714–721.
50. Dibba B, Prentice A, Ceesay M, Stirling DM, Cole TJ, Poskitt EM. 2000. Effect of calcium supplementation on bone mineral accretion in Gambian children accustomed to a low-calcium diet. *Am J Clin Nutr* 71(2): 544–549.
51. Dibba B, Prentice A, Ceesay M, et al. 2002. Bone mineral contents and plasma osteocalcin concentrations of Gambian children 12 and 24 mo after the withdrawal of a calcium supplement. *Am J Clin Nutr* 76(3): 681–686.
52. Norris SA, Sheppard ZA, Griffiths PL, Cameron N, Pettifor JM. 2008. Current socio-economic measures, and not those measured during infancy, affect bone mass in poor urban South african children. *J Bone Miner Res* 23(9): 1409–1416.
53. Labadarios D, Steyn NP, Maunder E, et al. 2005. The National Food Consumption Survey (NFCS): South Africa, 1999. *Public Health Nutr* 8(5): 533–543.
54. MacKeown JM, Pedro TM, Norris SA. 2007. Energy, macro- and micronutrient intake among a true longitudinal group of South African adolescents at two interceptions (2000 and 2003): The Birth-to-Twenty (Bt20) Study. *Public Health Nutr* 10(6): 635–643.
55. McVeigh JA, Norris SA, Pettifor JM. 2007. Bone mass accretion rates in pre- and early-pubertal South African

black and white children in relation to habitual physical activity and dietary calcium intakes. *Acta Paediatr* 96(6): 874–880.
56. Goedecke JH, Levitt NS, Lambert EV, et al. 2009. Differential effects of abdominal adipose tissue distribution on insulin sensitivity in black and white South African women. *Obesity (Silver Spring)* 17(8): 1506–1512.
57. Micklesfield LK, Norris SA, Pettifor JM. 2011. Ethnicity and bone: A South African perspective. *J Bone Miner Metab* 29(3): 257–267.
58. McCormick DP, Ponder SW, Fawcett HD, Palmer JL. 1991. Spinal bone mineral density in 335 normal and obese children and adolescents: Evidence for ethnic and sex differences. *J Bone Miner Res* 6(5): 507–513.

18

Calcium and Other Nutrients During Growth

Tania Winzenberg and Graeme Jones

Introduction 142
Calcium 142
Vitamin D 143
Fruit and Vegetables 143
Diet in Pregnancy 143

Breastfeeding 144
Salt 144
Soft Drinks and Milk Avoidance 145
Acknowledgments 145
References 145

INTRODUCTION

Bone mineral density (BMD) in later life is a function of peak bone mass and the rate of subsequent bone loss [1]. Childhood is potentially an important time to intervene, as modeling suggests that a 10% increase in peak bone mass will delay the onset of osteoporosis by 13 years [2]. In addition, low BMD in childhood is a risk factor for childhood fractures [3], suggesting that optimizing age-appropriate bone mass could also have a more immediate benefit on childhood fracture rates. This chapter reviews key nutritional influences on childhood bone development.

CALCIUM

It is widely accepted that an adequate calcium intake in childhood is important for bone development, though the results of observational and intervention studies are mixed [4]. In case-control studies, low calcium/dairy intake has been found to be associated with increased fracture risk in 11- to 13-year-old boys, but this has not been confirmed in other groups [5–7]. Low calcium/dairy intake has been found to be associated with recurrent fracture in both sexes [8, 9].

High levels of calcium intake for children are recommended in many developed countries. Current World Health Organization recommendations based on North American and Western European data are from 300 to 400 mg/day for infants, 400 to 700 mg/day for children, and 1,300 mg/day for adolescents) [10]. Modeling of data from calcium balance studies in 348 children [11] suggest that there is a calcium threshold below which skeletal calcium accumulation was related to intake, but above which skeletal accumulation remained constant. This varied with age to up to 1,730 mg in 9- to 17-year-olds. A similar threshold at about 1300 mg was described in girls aged 12–15 years [12]. However, the relationship between short-term calcium balance studies and achieving bone outcome improvements from longer-term calcium supplementation is open to question. In a meta-analysis of randomized controlled trials (RCTs) [13, 14], bone outcomes were no different above or below calcium intakes of 1,400 mg/day, casting doubt on the clinical relevance of the balance studies' results.

This meta-analysis [13, 14] also found that calcium supplementation had no effect on BMD at the femoral neck (FN) or lumbar spine (LS). Supplementation had a small effect on total body (TB) bone mineral content (BMC) but this did not persist once supplementation ceased. There was a small persistent effect on upper limb BMD, equivalent to a 1.7 percentage point greater increase in BMD in the supplemented group compared to the

---

Primer on the Metabolic Bone Diseases and Disorders of Mineral Metabolism, Eighth Edition. Edited by Clifford J. Rosen.
© 2013 American Society for Bone and Mineral Research. Published 2013 by John Wiley & Sons, Inc.

control group, which might reduce the absolute risk of fracture at the peak childhood fracture incidence by at most 0.2% per annum (p.a.). Thus, the small increase in bone density at the upper limb from increasing intake from an average 700 mg/day to 1,200 mg/day is unlikely to result in a clinically significant decrease in fracture risk. Furthermore the evidence did not suggest that increasing the duration of supplementation led to increasing effects. In the meta-analysis, the effect size did not vary with baseline calcium intakes, down to a level of less than 600 mg/day. A subsequent RCT targeting children (mean age 12 years) with an habitual calcium intake less than 650 mg/day resulted in greater increases in TB BMC (2.3%) and total hip (TH) and LS BMD (2.5 and 2.2%, respectively) in children supplemented with an average of 555 mg calcium/day after 18 months, but as in the meta-analysis, the effects did not persist once supplements ceased [15]. As the meta-analysis only included placebo-controlled trials, some RCTs of dairy products were not included [16–22]. Qualitatively, the results of these studies were similar, mainly demonstrating no effect [16] or only small to moderate short-term effects [18–21], which did not persist after supplementation ceased [19, 22]. In the only study with a larger effect, the intervention group had substantially higher levels of vitamin D intake [17], making it unclear how much of the effect was due to calcium and how much was due to vitamin D.

VITAMIN D

Vitamin D deficiency is common in children, especially in late adolescence [23–26]. There is observational evidence that mild vitamin D deficiency (less than 50 nmol/l) may affect bone turnover. The effectiveness of vitamin D supplementation for improving bone density in children was examined in a recent meta-analysis of six RCTs [27, 28]. In all children, vitamin D supplementation had no statistically significant effects on TB BMC, hip BMD, or forearm BMD, and effect sizes were small [standardized mean difference (SMD) 0.10 or less at all 3 sites]. There was a trend to a small effect on LS BMD [SMD +0.15, (95% CI −0.01 to +0.31), p = 0.07]. However, in studies in which the mean baseline serum vitamin D level of the children was low (less than 35 nmol/L), there were significant effects on TB BMC and LS BMD. These were approximately equivalent to a 2.6% and 1.7% percentage point greater increase from the baseline in the supplemented group. No studies were powered for fracture, and it is not known if effects accumulate with ongoing supplementation. Nonetheless, the data suggest that vitamin D supplementation of deficient children could result in clinically useful improvements particularly if in future trials it is demonstrated that effects accumulate with ongoing supplementation.

FRUIT AND VEGETABLES

Fruit and vegetable intake is postulated to have effects on bone through a number of mechanisms. These include the induction of a mild metabolic alkalosis, vitamin K, vitamin C, antioxidants, and phytoestrogens, though phytoestrogens alone have little effect on bone turnover in children [29]. Observational data support a positive relationship between fruit and vegetable intake and bone outcomes in children. Cross sectionally, in 8-year-old children [30], urinary potassium was positively associated with both fruit and vegetable intake and BMD. In addition, girls at Tanner stage 2 [31] who consumed three or more servings of fruit and vegetables daily had higher bone area, lower urinary calcium excretion, and lower parathyroid hormone levels than those consuming fewer than three servings daily, though there were no differences in BMD or bone turnover markers. In other cross-sectional studies, 12-year-old girls who consumed high amounts of fruit had higher heel BMD than moderate fruit consumers [32], and in adolescent boys and girls [33], fruit intake was positively associated with spine size-adjusted BMC (SA-BMC) and, in boys, with FN SA-BMC. Longitudinal data also suggest benefits. Over 7 years, fruit and vegetable intake was an independent predictor of TB BMC in boys, but not girls [34]. In children aged 10–15 years [35], over 1 year, girls who increased their fruit intake had a 4.7% greater increase in stiffness index (SI) (measured by quantitative ultrasound) than those who did not, girls who increased vegetable intake had a 3.6% greater increase, and boys who increased vegetable intake had a 2.4% greater increase. In children who were followed from age 3.8 to 7.8 years of age, a dietary pattern characterized by a high intake of dark-green and deep-yellow vegetables was associated with high bone mass [36]. Fruit and vegetable intake in children can be increased by 0.3 to 0.99 servings per day by dietary intervention [37]. Further research is needed to confirm if bone health is changed by clinically significant amounts by such increases.

DIET IN PREGNANCY

Nutritional influences on childhood bone development may begin *in utero*, and because of *in utero* programming, such influences may affect both early skeletal development and the acquisition of bone mass throughout childhood. RCTs of calcium supplementation during pregnancy have given inconsistent results so it remains unclear whether improving maternal calcium intake in pregnancy is beneficial for *in utero* bone development. One trial demonstrated that either 600 mg or 300 mg of calcium daily resulted in higher neonatal bone density of the ulna, radius, fibula, and tibia (measured by X-ray) [38]. Two grams of calcium in 256 women did not demonstrate any effect on neonatal TB or LS BMC overall, though in mothers with a baseline intake of less than

600mg/day, TB BMC was higher with supplementation [39]. Conversely, calcium carbonate (1,500mg/day) given to pregnant Gambian women with a calcium intake of less than 400mg/day was not beneficial for TB or radial BMC or BMD [40]. Lastly, an RCT in pregnant adolescents compared 1,200mg/day of calcium from either calcium-supplemented orange juice/calcium carbonate supplements or dairy foods with no-intervention controls. The dairy—but not the calcium supplementation—group had higher total body calcium than the controls, which may have been due to the higher vitamin D content of the dairy foods [41].

RCTs of other supplements in pregnancy with childhood bone outcomes are lacking, and evidence comes from observational data. Maternal serum 25-hydroxy vitamin D (25-OH D) in late pregnancy was positively associated with whole body and LS BMC in the children at age 9 [42]. In another cohort, maternal serum 25-OH D was associated with alterations in femoral development (determined by high resolution 3-dimensional ultrasound) in the fetus as early as 19 weeks gestation [43]. Viljakainen et al. reported that tibial BMC and cross-sectional area (CSA) were both higher in neonates whose mother's first trimester vitamin D was above 42.6nmol/l compared to those whose mother's vitamin D was below this level [44]. By 14 months, serum vitamin D levels in the children were similar at 64nmol/l in both groups [45]; the difference in tibial BMC was no longer apparent, but that in CSA persisted, suggesting only partial correction of the original bone deficits. Zinc supplementation in pregnancy in a disadvantaged area in a developing country resulted in increased fetal femur diaphysis length [46]. Maternal folate intake at 32 weeks was positively associated with spinal SA BMC after adjusting for children's weight and height [47] in 9-year-olds, and maternal red blood cell folate at 28 weeks gestation with spine BMD in 6-year-olds [48].

Other nutritional factors have been investigated with observational studies. Maternal dietary intake of magnesium, phosphorus, potassium, and protein during the third trimester of pregnancy has been shown to be positively associated, and maternal fat intake negatively associated, with bone density in their children at age 8 [49]. In the same children at age 16, FN and LS BMD associations with magnesium density and fat density in the maternal diet in pregnancy persisted. LS BMD was also positively associated with maternal milk intake, calcium, and phosphorus density. With all significant nutrients in the same model, fat density remained negatively associated for the FN and LS, whereas magnesium density remained positively associated for the FN [50]. In another cohort [47], maternal magnesium intake at 32 weeks gestation was positively associated with TB BMC and BMD at age 9 years until adjusted for the child's height. Maternal intake of potassium was positively associated with spinal BMC and BMD, until adjusted for the child's weight. A principal component analysis of the maternal diet in the same study identified a pattern of a high intake of fruits, vegetables, and wholemeal bread, pasta, and rice, and low intake of processed foods, which was quantified by a "prudent diet score." A high score was associated with higher TB and lumbar BMC and BMD [51]. Similarly, in a cohort of rural Indian mother–child pairs, intake of milk products, pulses, and fruits were all positively associated with spine BMD in children at age 6 years [48].

Though limited, these data support the need for further research into nutritional interventions in pregnancy.

BREASTFEEDING

Generally, studies show that infants fed human milk have lower bone accretion compared to formula-fed infants, possibly due to low vitamin D content and the decreasing phosphorus content of human milk with continued lactation [52]. However, data on the long-term effects of breastfeeding on bone health in children born at term suggest that this initial lower bone accretion is temporary, and catch-up growth occurs later in childhood. This includes data from an RCT of infant feeding that compared two different formulas and breastfeeding, in which initial differences in BMC accretion did not persist past 12 months of age [53], as well as longitudinal observational data. In 8-year-old children [54], breastfed children had higher FN, LS, and TB BMD compared with bottle-fed children, and the effect was most marked in children breastfed for more than 3 months. In 7- to 9-year-old children, being breastfed was not associated with broadband ultrasound (BUA) or speed of sound (SOS), but in breastfed children, duration of breastfeeding was positively associated with metacarpal diameter [55]. The effects of prolonged breastfeeding may differ from shorter exposures and between sexes, though this is not certain. In one observational study, breastfeeding for more than 7 months resulted in lower TB BMD, LS bone area, and LS BMC at age 32 years in males, but not females [56]. However, breastfeeding duration was not associated with bone density outcomes at age 4 in a different cohort [57].

Other observational studies with bone measures at younger ages [58, 59] did not demonstrate associations between breastfeeding and bone density. However, in a retrospective study, premenopausal women who had been breastfed for more than 3 months had greater cortical thickness at the radius and a trend toward greater cortical area and cortical BMC at the radius, but not at other sites [60]. Importantly, breastfeeding was protective for childhood fractures in a longitudinal study of prepubertal children [61] and in a case-control study of children aged 4–15 years [9], though this was not observed in a longitudinal study of fracture risk from birth to 18 years [62].

SALT

Urinary sodium excretion has been shown to be associated with urinary calcium excretion in girls [63–65], though not with an acute sodium chloride load [65]. Despite this, in the few studies assessing bone outcomes in children, urinary sodium excretion has not in turn been shown to be associated with bone density [30, 64], though dietary sodium intake was associated with size-adjusted bone area (but not BMC) in a cross-sectional study of 10-year-old girls [66]. Urinary sodium has also been shown to be associated with a high bone turnover state in adolescent boys [67]. Whether high dietary sodium intake adversely affects other bone outcomes in children is uncertain. Initially, more longitudinal studies are needed to determine if sodium intake does in fact have a clinically important effect on bone in children.

SOFT DRINKS AND MILK AVOIDANCE

Carbonated beverage consumption has been linked with decreased BMD in girls but not boys [68, 69] and with increased fracture risk in both sexes. Low milk intake and a higher consumption of carbonated beverages were independent fracture risk factors in children with recurrent fractures [9]. Other studies have reported increased fracture risk with higher cola intake but not non-cola carbonated beverage intake [70, 71]. It is unclear if this effect is due to milk replacement. Two studies have demonstrated that associations between fracture risk [7], peripheral quantitative computed tomography (pQCT) measures [72], and cola drinks persist after adjustment for milk intake, which suggests independent effects. Milk avoidance also appears to have deleterious effects on children's bones. Prepubertal children who avoid milk have lower TB BMC and areal BMD [73], as well as an increased risk of childhood fracture [74]. The effects of low milk consumption in childhood may extend into adult life, with associations seen with lower BMD [75] and higher risk of fracture in adult women [76].

In conclusion, there is increasing evidence linking a number of nutritional factors with children's bone development. Calcium supplementation has been investigated to the greatest extent, but its effects are of limited public health significance. This makes the exploration of other nutritional approaches of key importance.

ACKNOWLEDGMENTS

Graeme Jones receives a NHMRC Practitioner Fellowship, which supported this work.

REFERENCES

1. Hansen MA, Overgaard K, Riis BJ, Christiansen C. 1991. Role of peak bone mass and bone loss in postmenopausal osteoporosis: 12 year study. *BMJ* 303(6808): 961–4.
2. Hernandez CJ, Beaupre GS, Carter DR. 2003. A theoretical analysis of the relative influences of peak BMD, age-related bone loss and menopause on the development of osteoporosis. *Osteoporos Int* 14(10): 843–7.
3. Clark EM, Tobias JH, Ness AR. 2006. Association between bone density and fractures in children: A systematic review and meta-analysis. *Pediatrics* 117(2): e291–7.
4. Lanou AJ, Berkow SE, Barnard ND. 2005. Calcium, dairy products, and bone health in children and young adults: A reevaluation of the evidence. *Pediatrics* 115(3): 736–43.
5. Goulding A, Jones IE, Taylor RW, Williams SM, Manning PJ. 2001. Bone mineral density and body composition in boys with distal forearm fractures: A dual-energy x-ray absorptiometry study. *J Pediatr* 139(4): 509–15.
6. Goulding A, Cannan R, Williams SM, Gold EJ, Taylor RW, Lewis-Barned NJ. 1998. Bone mineral density in girls with forearm fractures. *J Bone Miner Res* 13(1): 143–8.
7. Ma D, Jones G. 2004. Soft drink and milk consumption, physical activity, bone mass, and upper limb fractures in children: A population-based case-control study. *Calcif Tissue Int* 75(4): 286–91.
8. Goulding A, Grant AM, Williams SM. 2005. Bone and body composition of children and adolescents with repeated forearm fractures. *J Bone Miner Res* 20(12): 2090–6.
9. Manias K, McCabe D, Bishop N. 2006. Fractures and recurrent fractures in children; varying effects of environmental factors as well as bone size and mass. *Bone* 39(3): 652–7.
10. World Health Organization, Food and Argiculture Organization of the United Nations. 2004. *Vitamin and Mineral Requirements in Human Nutrition, 2nd Ed.*
11. Matkovic V, Heaney RP. 1992. Calcium balance during human growth: evidence for threshold behavior. *Am J Clin Nutr* 55(5): 992–6.
12. Jackman LA, Millane SS, Martin BR, Wood OB, McCabe GP, Peacock M, Weaver CM. 1997. Calcium retention in relation to calcium intake and postmenarcheal age in adolescent females. *Am J Clin Nutr* 66(2): 327–33.
13. Winzenberg TM, Shaw K, Fryer J, Jones G. 2006. Calcium supplementation for improving bone mineral density in children. *Cochrane Database Syst Rev* CD005119. doi: 10.1002/14651858.CD005119.pub2.
14. Winzenberg T, Shaw K, Fryer J, Jones G. 2006. Effects of calcium supplementation on bone density in healthy children: Meta-analysis of randomised controlled trials. *BMJ* 333(7572): 775.
15. Lambert HL, Eastell R, Karnik K, Russell JM, Barker ME. 2008. Calcium supplementation and bone mineral accretion in adolescent girls: An 18-mo randomized controlled trial with 2-y follow-up. *Am J Clin Nutr* 87(2): 455–62.
16. Lau EMC, Lee WTK, Leung S, Cheng J. 1992. Milk supplementation—A feasible and effective way to enhance bone gain for Chinese adolescents in Hong Kong? *J Appl Nutr* 44(3–4): 16–21.

17. Chan GM, Hoffman K, McMurry M. 1995. Effects of dairy products on bone and body composition in pubertal girls. *J Pediatr* 126(4): 551–6.
18. Cadogan J, Eastell R, Jones N, Barker ME. 1997. Milk intake and bone mineral acquisition in adolescent girls: Randomised, controlled intervention trial. *BMJ* 315(7118): 1255–60.
19. Merrilees MJ, Smart EJ, Gilchrist NL, Frampton C, Turner JG, Hooke E, March RL, Maguire P. 2000. Effects of diary food supplements on bone mineral density in teenage girls. *Eur J Nutr* 39(6): 256–62.
20. Lau EM, Lynn H, Chan YH, Lau W, Woo J. 2004. Benefits of milk powder supplementation on bone accretion in Chinese children. *Osteoporos Int* 15(8): 654–8.
21. Du X, Zhu K, Trube A, Zhang Q, Ma G, Hu X, Fraser DR, Greenfield, H. 2004. School-milk intervention trial enhances growth and bone mineral accretion in Chinese girls aged 10–12 years in Beijing. *Br J Nutr* 92(1): 159–68.
22. Zhu K, Zhang Q, Foo LH, Trube A, Ma G, Hu X, Du X, Cowell CT, Fraser DR, Greenfield H. 2006. Growth, bone mass, and vitamin D status of Chinese adolescent girls 3 y after withdrawal of milk supplementation. *Am J Clin Nutr* 83(3): 714–21.
23. Looker AC, Dawson-Hughes B, Calvo MS, Gunter EW, Sahyoun NR. 2002. Serum 25-hydroxyvitamin D status of adolescents and adults in two seasonal subpopulations from NHANES III. *Bone* 30(5): 771–7.
24. Jones G, Blizzard C, Riley MD, Parameswaran V, Greenaway TM, Dwyer T. 1999. Vitamin D levels in prepubertal children in Southern Tasmania: Prevalence and determinants. *Eur J Clin Nutr* 53(10): 824–9.
25. Jones G, Dwyer T, Hynes KL, Parameswaran V, Greenaway TM. 2005. Vitamin D insufficiency in adolescent males in Southern Tasmania: Prevalence, determinants, and relationship to bone turnover markers. *Osteoporos Int* 16(6): 636–41.
26. Rockell JE, Skeaff CM, Williams SM, Green TJ. 2006. Serum 25-hydroxyvitamin D concentrations of New Zealanders aged 15 years and older. *Osteoporos Int* 17(9): 1382–9.
27. Winzenberg T, Powell S, Shaw KA, Jones G. 2011. Effects of vitamin D supplementation on bone density in healthy children: Systematic review and meta-analysis. *BMJ* 342: c7254.
28. Winzenberg TM, Powell S, Shaw KA, Jones G. 2010. Vitamin D supplementation for improving bone mineral density in children. *Cochrane Database Syst Rev* 10: CD006944.
29. Jones G, Dwyer T, Hynes K, Dalais FS, Parameswaran V, Greenaway TM. 2003. A randomized controlled trial of phytoestrogen supplementation, growth and bone turnover in adolescent males. *Eur J Clin Nutr* 57(2): 324–7.
30. Jones G, Riley MD, Whiting S. 2001. Association between urinary potassium, urinary sodium, current diet, and bone density in prepubertal children. *Am J Clin Nutr* 73(4): 839–44.
31. Tylavsky FA, Holliday K, Danish R, Womack C, Norwood J, Carbone L. 2004. Fruit and vegetable intakes are an independent predictor of bone size in early pubertal children. *Am J Clin Nutr* 79(2): 311–7.
32. McGartland CP, Robson PJ, Murray LJ, Cran GW, Savage MJ, Watkins DC, Rooney MM, Boreham CA. 2004. Fruit and vegetable consumption and bone mineral density: The Northern Ireland Young Hearts Project. *Am J Clin Nutr* 80(4): 1019–23.
33. Prynne CJ, Mishra GD, O'Connell MA, Muniz G, Laskey MA, Yan L, Prentice A, Ginty F. 2006. Fruit and vegetable intakes and bone mineral status: A cross sectional study in 5 age and sex cohorts. *Am J Clin Nutr* 83(6): 1420–8.
34. Vatanparast H, Baxter-Jones A, Faulkner RA, Bailey DA, Whiting SJ. 2005. Positive effects of vegetable and fruit consumption and calcium intake on bone mineral accrual in boys during growth from childhood to adolescence: The University of Saskatchewan Pediatric Bone Mineral Accrual Study. *Am J Clin Nutr* 82(3): 700–6.
35. Hirota T, Kusu T, Hirota K. 2005. Improvement of nutrition stimulates bone mineral gain in Japanese school children and adolescents. *Osteoporos Int* 16(9): 1057–64.
36. Wosje KS, Khoury PR, Claytor RP, Copeland KA, Hornung RW, Daniels SR, Kalkwarf HJ. 2010. Dietary patterns associated with fat and bone mass in young children. *Am J Clin Nutr* 92(2): 294–303.
37. Knai C, Pomerleau J, Lock K, McKee M. 2006. Getting children to eat more fruit and vegetables: A systematic review. *Prev Med* 42(2): 85–95.
38. Raman L, Rajalakshmi K, Krishnamachari KA, Sastry JG. 1978. Effect of calcium supplementation to undernourished mothers during pregnancy on the bone density of the bone density of the neonates. *Am J Clin Nutr* 31(3): 466–9.
39. Koo WW, Walters JC, Esterlitz J, Levine RJ, Bush AJ, Sibai B. 1999. Maternal calcium supplementation and fetal bone mineralization. *Obstet Gynecol* 94(4): 577–82.
40. Jarjou LM, Prentice A, Sawo Y, Laskey MA, Bennett J, Goldberg GR, Cole TJ. 2006. Randomized, placebo-controlled, calcium supplementation study in pregnant Gambian women: Effects on breast-milk calcium concentrations and infant birth weight, growth, and bone mineral accretion in the first year of life. *Am J Clin Nutr* 83(3): 657–66.
41. Chan GM, McElligott K, McNaught T, Gill G. 2006. Effects of dietary calcium intervention on adolescent mothers and newborns: A randomized controlled trial. *Obstet Gynecol* 108(3 Pt 1): 565–71.
42. Javaid MK, Crozier SR, Harvey NC, Gale CR, Dennison EM, Boucher BJ, Arden NK, Godfrey KM, Cooper C. 2006. Maternal vitamin D status during pregnancy and childhood bone mass at age 9 years: A longitudinal study. *Lancet* 367(9504): 36–43.
43. Mahon P, Harvey N, Crozier S, Inskip H, Robinson S, Arden N, Swaminathan R, Cooper C, Godfrey K. 2009.

Low maternal vitamin D status and fetal bone development: Cohort study. *J Bone Miner Res* 25(1): 14–9.
44. Viljakainen HT, Saarnio E, Hytinantti T, Miettinen M, Surcel H, Makitie O, Andersson S, Laitinen K, Lamberg-Allardt, C. 2010. Maternal vitamin D status determines bone variables in the newborn. *J Clin Endocrinol Metab* 95(4): 1749–57.
45. Viljakainen HT, Korhonen T, Hytinantti T, Laitinen EK, Andersson S, Makitie O, Lamberg-Allardt C. 2011. Maternal vitamin D status affects bone growth in early childhood—A prospective cohort study. *Osteoporos Int* 22(3): 883–91.
46. Merialdi M, Caulfield LE, Zavaleta N, Figueroa A, Costigan KA, Dominici F, Dipietro JA. 2004. Randomized controlled trial of prenatal zinc supplementation and fetal bone growth. *Am J Clin Nutr* 79(5): 826–30.
47. Tobias JH, Steer CD, Emmett PM, Tonkin RJ, Cooper C, Ness AR. 2005. Bone mass in childhood is related to maternal diet in pregnancy. *Osteoporos Int* 16(12): 1731–41.
48. Ganpule A, Yajnik CS, Fall CH, Rao S, Fisher DJ, Kanade A, Cooper C, Naik S, Joshi N, Lubree H, Deshpande V, Joglekar C. 2006. Bone mass in Indian children—Relationships to maternal nutritional status and diet during pregnancy: The Pune Maternal Nutrition Study. *J Clin Endocrinol Metab* 91(8): 2994–3001.
49. Jones G, Riley MD, Dwyer T. 2000. Maternal diet during pregnancy is associated with bone mineral density in children: A longitudinal study. *Eur J Clin Nutr* 54(10): 749–56.
50. Yin J, Dwyer T, Riley M, Cochrane J, Jones G. 2010. The association between maternal diet during pregnancy and bone mass of the children at age 16. *Eur J Clin Nutr* 64(2): 131–7.
51. Cole ZA, Gale CR, Javaid MK, Robinson SM, Law C, Boucher BJ, Crozier SR, Godfrey KM, Dennison EM, Cooper C. 2009. Maternal dietary patterns during pregnancy and childhood bone mass: A longitudinal study. *J Bone Miner Res* 24(4): 663–8.
52. Specker B. 2004. Nutrition influences bone development from infancy through toddler years. *J Nutr* 134(3): 691S–5S.
53. Specker BL, Beck A, Kalkwarf H, Ho M. 1997. Randomized trial of varying mineral intake on total body bone mineral accretion during the first year of life. *Pediatrics.* 99(6): E12.
54. Jones G, Riley M, Dwyer T. 2000. Breastfeeding in early life and bone mass in prepubertal children: A longitudinal study. *Osteoporos Int* 11(2): 146–52.
55. Micklesfield L, Levitt N, Dhansay M, Norris S, van der Merwe L, Lambert E. 2006. Maternal and early life influences on calcaneal ultrasound parameters and metacarpal morphometry in 7- to 9-year-old children. *J Bone Miner Metab* 24(3): 235–42.
56. Pirila S, Taskinen M, Viljakainen H, Kajosaari M, Turanlahti M, Saarinen-Pihkala UM, Makitie O. 2011. Infant milk feeding influences adult bone health: A prospective study from birth to 32 years. *PLoS One* 6(4): e19068.
57. Harvey NC, Robinson SM, Crozier SR, Marriott LD, Gale CR, Cole ZA, Inskip HM, Godfrey KM, Cooper C. 2009. Breast-feeding and adherence to infant feeding guidelines do not influence bone mass at age 4 years. *Br J Nutr* 102(6): 915–20.
58. Kurl S, Heinonen K, Jurvelin JS, Lansimies E. 2002. Lumbar bone mineral content and density measured using a Lunar DPX densitometer in healthy full-term infants during the first year of life. *Clin Physiol Funct Imaging* 22(3): 222–5.
59. Young RJ, Antonson DL, Ferguson PW, Murray ND, Merkel K, Moore TE. 2005. Neonatal and infant feeding: Effect on bone density at 4 years. *J Pediatr Gastroenterol Nutr* 41(1): 88–93.
60. Laskey MA, de Bono S, Smith EC, Prentice A. 2007. Influence of birth weight and early diet on peripheral bone in premenopausal Cambridge women: A pQCT study. *J Musculoskelet Neuronal Interact* 7(1): 83.
61. Ma DQ, Jones G. 2002. Clinical risk factors but not bone density are associated with prevalent fractures in prepubertal children. *J Paediatr Child Health* 38(5): 497–500.
62. Jones IE, Williams SM, Goulding A. 2004. Associations of birth weight and length, childhood size, and smoking with bone fractures during growth: Evidence from a birth cohort study. *Am J Epidemiol* 159(4): 343–50.
63. O'Brien KO, Abrams SA, Stuff JE, Liang LK, Welch TR. 1996. Variables related to urinary calcium excretion in young girls. *J Pediatr Gastroenterol Nutr* 23(1): 8–12.
64. Matkovic V, Ilich JZ, Andon MB, Hsieh LC, Tzagournis MA, Lagger BJ, Goel PK. 1995. Urinary calcium, sodium, and bone mass of young females. *Am J Clin Nutr* 62(2): 417–25.
65. Duff TL, Whiting SJ. 1998. Calciuric effects of short-term dietary loading of protein, sodium chloride and potassium citrate in prepubescent girls. *J Am Coll Nutr* 17(2): 148–54.
66. Hoppe C, Molgaard C, Michaelsen KF. 2000. Bone size and bone mass in 10-year-old Danish children: Effect of current diet. *Osteoporos Int* 11(12): 1024–30.
67. Jones G, Dwyer T, Hynes KL, Parameswaran V, Udayan R, Greenaway TM. 2007. A prospective study of urinary electrolytes and bone turnover in adolescent males. *Clin Nutr* 26(5): 619–23.
68. Whiting S, Heaky A, Psiuk S, Mirwald R, Kowalski K, Bailey DA. 2001. Relationship between carbonated and other low nutrient dense beverages and bone mineral content of adolescents. *Nutr Res* 21: 1107–15.
69. McGartland C, Robson PJ, Murray L, Cran G, Savage MJ, Watkins D, Rooney M, Boreham C. 2003. Carbonated soft drink consumption and bone mineral density in adolescence: The Northern Ireland Young Hearts project. *J Bone Miner Res* 18(9): 1563–9.
70. Wyshak G, Frisch RE. 1994. Carbonated beverages, dietary calcium, the dietary calcium/phosphorus ratio, and bone fractures in girls and boys. *J Adolesc Health* 15(3): 210–5.
71. Petridou E, Karpathios T, Dessypris N, Simou E, Trichopoulos D. 1997. The role of dairy products and non

alcoholic beverages in bone fractures among schoolage children. *Scand J Soc Med* 25(2): 119–25.
72. Libuda L, Alexy U, Remer T, Stehle P, Schoenau E, Kersting M. 2008. Association between long-term consumption of soft drinks and variables of bone modeling and remodeling in a sample of healthy German children and adolescents. *Am J Clin Nutr* 88(6): 1670–7.
73. Black RE, Williams SM, Jones IE, Goulding A. 2002. Children who avoid drinking cow milk have low dietary calcium intakes and poor bone health. *Am J Clin Nutr* 76(3): 675–80.
74. Goulding A, Rockell JE, Black RE, Grant AM, Jones IE, Williams SM. 2004. Children who avoid drinking cow's milk are at increased risk for prepubertal bone fractures. *J Am Diet Assoc* 104(2): 250–3.
75. Vatanparast H, Whiting SJ. 2004. Early milk intake, later bone health: Results from using the milk history questionnaire. *Nutr Rev* 62(6 Pt 1): 256–60.
76. Kalkwarf HJ, Khoury JC, Lanphear BP. 2003. Milk intake during childhood and adolescence, adult bone density, and osteoporotic fractures in US women. *Am J Clin Nutr* 77(1): 257–65.

19
Growing a Healthy Skeleton: The Importance of Mechanical Loading

Mark R. Forwood

Introduction 149
What Is the Window of Opportunity? 149
Characteristics of an Effective Loading Prescription 150
Peak Bone Mass or Peak Bone Strength? 151
Persistence of Childhood Bone Adaptation 152
Conclusions 153
References 153

INTRODUCTION

The basic morphology of the skeleton is determined genetically, but its final mass and architecture are modulated by adaptive mechanisms sensitive to mechanical factors. When subjected to loading, the ability of bones to resist fracture depends on their mass, material properties, geometry, and tissue quality [1]. Knowing whether an intervention during childhood optimizes skeletal mechanics as an adult requires understanding of the effects on each mechanical determinant. Unfortunately, it is only recently that studies have moved beyond areal bone mineral density (aBMD) as an outcome measure to include volumetric density and geometry [2]. None are yet able to determine if childhood adaptations translate into anti-fracture efficacy in adults, but at least the latter variables provide surrogates of bone strength. There is convincing evidence that growing bone has greater capacity to respond to increased mechanical loading than the adult skeleton [2–9]. The question is what to prescribe, what is the window of opportunity, and does adherence to the prescription optimize bone strength so that the risk of fracture is reduced in adulthood and old age.

WHAT IS THE WINDOW OF OPPORTUNITY?

Osteoporotic fractures occur because reduced bone mass decreases the safety factor for skeletal loading. This can result from age-related bone loss and/or failure to achieve optimal peak bone mass [10, 11]. During childhood and adolescence the skeleton undergoes rapid change due to growth, modeling and remodeling, the processes that maximize bone accrual. It is generally believed that bone mass increases substantially during adolescence, reaching a plateau [peak bone mass (PBM)] in the late teen or young adult years [12]. Until recently, the timing of this event was still disputed. Some studies suggested that it was reached by 20 years of age [13, 14], but others concluded that PBM was attained in the third decade of life [15]. Longitudinal bone mineral accrual data measured from childhood (age 8) to young adulthood (up to age 30), demonstrate that PBM occurs by the end of the second, or very early in the third, decade of life [16] (Fig. 19.1). On a regional basis, the lower limb achieves a plateau earliest, within 1 year of peak height velocity (PHV), and the lumbar spine the latest, some 4 years after PHV. Importantly, depending on the skeletal site, 33–46% of the adult BMC was accrued over a 4-year period of adolescent growth surrounding PHV. In females, this total accrual represents double the amount of bone mineral that will subsequently be lost during the postmenopausal years from age 50 to 80 years [17].

Although the variance in peak bone mass is genetically determined [10], it is influenced by mechanical factors such as exercise. The evidence, while not universal [18, 19], is mounting that the prepubertal and early pubertal periods are more advantageous than postpuberty to elicit an adaptive response in growing bones [8, 9, 20–30]. This

Primer on the Metabolic Bone Diseases and Disorders of Mineral Metabolism, Eighth Edition. Edited by Clifford J. Rosen.
© 2013 American Society for Bone and Mineral Research. Published 2013 by John Wiley & Sons, Inc.

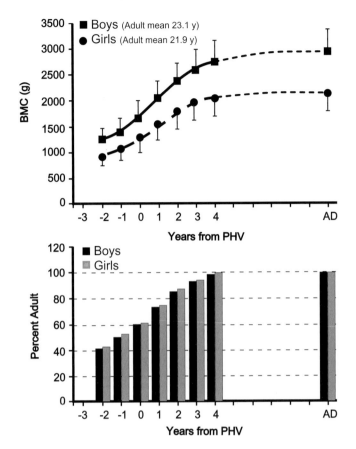

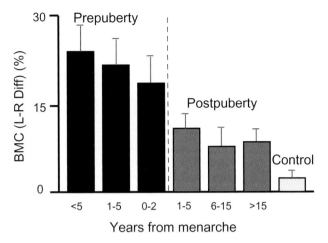

Fig. 19.1. Bone mineral accrual (total body BMC) from preadolescence [2 years before peak height velocity (PHV)] to adulthood (AD). When controlled for maturity, total body BMC reached a plateau by 7 years post-PHV, representing about 18.8 and 20.5 years of age in girls and boys, respectively. Data adapted from Baxter-Jones et al. (Ref. 16).

Fig. 19.2. The differences in BMC between the playing and nonplaying arms of women in racquet sports were greatest in those whose training started before puberty, compared to those who started during adulthood (postpuberty). Adapted from Kannus et al. (Ref. 8).

CHARACTERISTICS OF AN EFFECTIVE LOADING PRESCRIPTION

If fracture prevention is a consequence of an activity program, then we need to know the characteristics of mechanical stimuli that achieve adaptive morphological changes. We know that static, or isometric, loading provides minimal adaptive stimulus to bone [34–37], and can even inhibit normal appositional growth [38]. Activation of new bone formation also requires that a threshold magnitude of loading is exceeded [39], but that there is an interaction between strain rate and amplitude of loading that modulates this threshold [37, 39–42]. The moderating effect of strain rate occurs because bone tissue is viscoelastic, and interstitial fluid mechanics underlie transduction of applied dynamic loads into bone cell responses [43]. For external loads, such as those experienced during exercise, these responses are optimal in a range of loading frequencies up to about 2.0 Hz [42]. In terms of physical activity, exercises that create relatively high strain rates will be more adaptive than loads applied gently, or in which the load is held constant for a period of time. That is, jumping exercises will create a greater osteogenic effect than simply walking or doing isometric strength exercises.

A key characteristic of loading is that very few loading cycles are required to elicit adaptive responses [36, 43, 44]. The loading effect saturates relatively quickly so that increasing the duration of loading beyond about 40 cycles per day has little additional effect [36]. Cardiovascular health notwithstanding, long exercise sessions will therefore have diminishing returns in terms of growing strong bones. Children do not need to participate in long

observation is exemplified in unilateral racquet sports in which the differences between the bones of dominant and nondominant arms are greatest in players who started training before puberty, compared to those who started playing during adulthood [8, 9] (Fig. 19.2).

The consequence of physical activity on the skeleton of adults is conservation, not acquisition [5], so factors that optimize PBM must be implemented during growth, when the stimulus of mechanical loading can augment modeling processes that are already active, and the surfaces have a greater proportion of active bone cells. There is also evidence that the increase in estrogen levels in males and females during adolescence augments the amount of functional ERα available to facilitate strain-related responses in bone [31, 32], explaining enhanced sensitivity to physical activity during early puberty [19, 21, 24–26, 28, 33].

periods of exercise, nor disrupt their normal schedule of activities (or inactivity), to provide an adequate adaptive stimulus to their skeleton. Moreover, a given physical activity will be more osteogenic if it is divided into shorter bouts with rest periods in between [45–48]. This is because the sensitivity of the bone cells to the loading stimulus returns after a period of rest. For example, new bone formation was 80% greater in tibiae subjected to four bouts of 90 cycles per day for 2 weeks, compared to one bout of 360 cycles [45]. When a similar program was extended to 16 weeks, groups that received four bouts of 90 cycles per day had significantly greater bending strength in the loaded ulnae compared to those who received a single bout of 360 cycles [47]. In ulnae loaded four times per day, the increased strength was attributed to greater geometric adaptations that resist axial bending (Fig. 19.3) [47].

These mechanical loading principles have been adopted increasingly in physical activity interventions [18, 24–26, 28, 49–53]. But stronger bones can also be achieved in children who undertake greater levels of normal physical activity than sedentary children [22], or compete in high-impact sports such as gymnastics [54, 55]. Although numerous controlled trials have applied some of these principles to maximize modeling in growing bone [2, 21, 27, 49, 52, 56], the Healthy Bones II [25, 26] and "Action Schools! BC" programs [18, 24, 50] in Vancouver, British Columbia, were specifically designed around principles to optimize the osteogenic index [57] of a practical and sustainable activity intervention. In Action Schools! BC, the bone-loading component of the program included an extra 15 minutes of simple activities for 5 days per week, and "bounce at the bell" [50], in which just 3 minutes of variegated jumping activities were implemented 3 times per day (at each school bell) for 4 days per week. During initial trials, the program induced an increase in bone mass [bone mineral content (BMC)] at the lumbar spine and femoral neck of about 2% in both boys and girls. Simple additions to other physical education programs in schools have also improved the indices of bone strength in children quite effectively [51].

Controlled trials, using programs like Action Schools! BC, have increased bone strength, estimated using pQCT, in the distal tibia [18], and 2–4% increases in spine and total body BMC [24] of prepubertal boys, and BMC and section modulus of the femoral neck (an index of bending strength) in peripubertal girls [24]. By optimizing the osteogenic index, this program is able to achieve modest, but significant, increases in parameters of bone strength that are similar to other studies that involve more intensive bone-loading programs [2, 21, 49, 52, 58].

PEAK BONE MASS OR PEAK BONE STRENGTH?

Skeletal adaptation to mechanical loading must increase bone strength without unduly increasing the metabolic cost of locomotion. This design gives rise to a trade-off between the goals of strength and lightness. Efficient adaptation, therefore, cannot simply increase bone mass but must effect efficient increases in the geometry, and hence the structural properties, of bone (Fig. 19.4). These

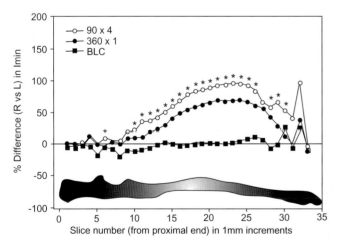

Fig. 19.3. When daily loading was divided into 4 bouts of 90 cycles per day, instead of a single bout of 360 cycles, ulnae showed significantly greater increases in the geometric property responsible for resistance to bending (cross-section moment of inertia—CSMI). Graph illustrates the difference between loaded and unloaded limbs (%) for the CSMI (Imin) of loaded animals, and baseline controls (BLC). Adapted from Robling et al. (47).

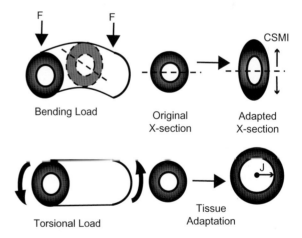

Fig. 19.4. The aim of skeletal adaptation is not to increase bone mass per se, but to improve resistance to increased loads efficiently. Thus, increased strength is achieved, without jeopardizing lightness. Modeling on appropriate bone surfaces improves the biomechanical determinants for enhanced bending or torsional strength [cross-sectional moment of inertia (CSMI) or polar moment of inertia (J), respectively]. These changes improve bending and torsional strength, disproportionately to the change in BMC or BMD, and hence can be misinterpreted by DXA. Adapted from Forwood (Ref. 1).

adaptations are often independent of the material properties. The contribution of altered bone geometry to fracture risk is unappreciated by clinical assessment using dual-energy X-ray absorptiometry (DXA), because it cannot distinguish geometry and density, nor cortical and cancellous bone [1, 59]. The resolution of DXA is also too low to detect small changes in bone dimensions that can provide substantial increases in bone strength. An excellent example of this characteristic is the modest 5% increase in the aBMD of rat ulnae after 16 weeks of axial loading three times per day [47]. This is starkly contrasted by an incredible 64% increase in ultimate breaking strength. Such a marked discrepancy occurs because the new bone formation occurs at the periosteal surface where a relatively small increase in bone apposition provides a disproportionate mechanical advantage at the locations of greatest strain. That is, the small amount of bone is strategically placed away from the axis of bending where it has an exponential effect to resist bending loads (Figs. 19.3 and 19.4). Nonetheless, the vast majority of physical activity studies have relied upon measures of BMD or BMC, derived from DXA, to assess the adaptive response. The use of hip structural analysis, derived from DXA, has provided some mechanical indices to assess maturation of bone strength [22, 29]. But the increasing use of microcomputed tomography (micro-CT) in animal studies, and peripheral quantitative CT (pQCT) in children [2, 9, 18, 60, 61, 62], specifically allows adaptations in density and geometry to be distinguished and proves that increased bone strength, not mass, is the goal of adaptation [9, 62]. The important question to address is whether these changes are maintained into adulthood and old age.

PERSISTENCE OF CHILDHOOD BONE ADAPTATION

During childhood and adolescence it is clear that the mechanical component of physical activity effects adaptations in the growing skeleton. The increases in bone mass [bone mineral density (BMD)] are relatively modest, between 1% and 5%, but may be manifested by geometric adaptations that augment bone strength by a significantly greater degree. It is practically impossible to prove that the adaptations of childhood prevent osteoporotic fracture in old age. Confounding variables associated with retrospective studies, such as self-selection for physique, reduce the certainty that a given activity created long-lasting skeletal protection. But a small number of studies have now traversed childhood and followed adolescents into adulthood. Early studies of this phenomenon suggested that the gains in bone mass from childhood training were lost in adulthood [63–65]. Some of these studies were cross sectional, or started relatively late in adolescence, when it is harder to control for confounding variables such as maturity. Longitudinal studies from childhood through adolescence report a sustained effect

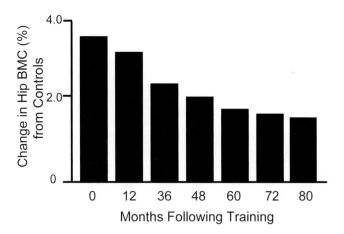

Fig. 19.5. Results of a school-based jumping intervention for 7 months in children aged approximately 8 years [49]. At the end of the 7-month program (0 months), hip BMC was significantly greater than controls (3.6%). When followed for up to 8 years of detraining, the exercise group retained greater hip BMC (1.4%) after controlling for baseline age, change in height, weight and sports participation (Refs. 66, 67). Adapted from Gunter et al. (Ref. 67).

on BMC accrual up to 8 years after training, or cessation of a physical activity intervention [49, 66–68] (Fig. 19.5). In addition, physically active children achieve greater bone mass (BMC) during adolescence than their sedentary peers [20, 22], and their higher BMC is maintained into adulthood [69].

Retention of childhood bone mass into adulthood may be modest, but preservation of skeletal architecture and geometry, underlying bone strength, could be more significant. The distinction between preservation of bone mass or architecture is elegantly illustrated by lifelong preservation of adapted bone structure after short-term exercise in rapidly growing rodents [70]. Forearm training started at 5 weeks of age, with a short 7-week exercise program, following which animals were limited to cage activity for up to 92 weeks (2 years of age, equivalent to senescence for rodents). Increases in bone mass induced by exercise (aBMD and BMC) were not retained into adulthood. There was, however, long-term preservation of bone structural changes, which was manifested in superior strength and fatigue life in the trained animals [70]. Similar retention of exercise-induced structural adaptation from childhood has been observed following cessation of racquet sports in male and female tennis players who had started training during childhood (about 10 years of age) [71, 72]. Up to 3 years after retirement a greater bone mass was retained at the age of 30, modeled into an architecture that provided greater indices of bone strength, such as cortical area and cross-sectional moment of inertia. Taken together, these data support the argument that physical activity during childhood can effect structural adaptations in the skeleton that persist well in to adulthood. The greater strength afforded by these structural changes could reduce the risk of fracture in adults, more than that predicted by bone mass alone.

CONCLUSIONS

Compared to adults, the skeleton of children and adolescents is capable of greater structural adaptations in response to the mechanical stimulus of physical activity. The evidence is persuasive that the prepubertal and early pubertal periods are more advantageous to effect adaptive responses. To grow healthy bones, physical activity should not consist of static or isometric exercises but should incorporate repetitive cyclical loads that include a range of strain magnitudes and directions, such as running and jumping. Since only a few cycles of loading are required to elicit an adaptive response, distributed bouts of loading that incorporate rest periods are more osteogenic than single sessions of long duration. These parameters of loading have been translated into feasible public health interventions that have achieved improved bone mass and strength in children and adolescents. Those architectural adaptations can persist into adulthood and be translated into a lower risk of fracture.

REFERENCES

1. Forwood MR. 2001. Mechanical effects on the skeleton: Are there clinical implications? *Osteoporos Int* 12: 77–83.
2. Heinonen A, Sievänen H, Kannus P, Oja P, Pasanen M, Vuori I. 2000. High-impact exercise and bones of growing girls: A 9-month controlled trial. *Osteoporos Int* 11: 1010–1017.
3. Forwood MR, Burr DB. 1993. Physical activity and bone mass: Exercises in futility? *Bone Miner* 21: 89–112.
4. Jarvinen TL, Pajamaki I, Sievanen H, Vuohelainen T, Tuukkanen J, Jarvinen M, Kannus P. 2003. Femoral neck response to exercise and subsequent deconditioning in young and adult rats. *J Bone Miner Res* 18: 1292–1299.
5. Parfitt AM. 1994. The two faces of growth: Benefits and risks to bone integrity. *Osteoporos Int* 4: 382–398.
6. Rubin CT, Bain SD, McLeod KJ. 1992. Suppression of the osteogenic response in the aging skeleton. *Calcif Tissue Int* 50: 306–313.
7. Turner CH, Takano Y, Owan I. 1995. Aging changes mechanical loading thresholds for bone formation in rats. *J Bone Miner Res* 10: 1544–1549.
8. Kannus P, Haapasalo H, Sankelo M, Sievanen H, Pasanen M, Heinonen A, Oja P, Vuori I. 1995. Effect of starting age of physical activity on bone mass in the dominant arm of tennis and squash players. *Ann Intern Med* 123: 27–31.
9. Kontulainen S, Sievanen H, Kannus P, Pasanen M, Vuori I. 2002. Effect of long-term impact-loading on mass, size, and estimated strength of humerus and radius of female racquet-sports players: A peripheral quantitative computed tomography study between young and old starters and controls. *J Bone Miner Res* 17: 2281–2289.
10. Ferrari S, Rizzoli R, Slosman D, Bonjour JP. 1998. Familial resemblance for bone mineral mass is expressed before puberty. *J Clin Endocrino Metab* 83: 358–361.
11. Hui SL, Slemenda CW, Johnston CC. 1990. The contribution of bone loss to post menopausal osteoporosis. *Osteoporos Int* 1: 30–34.
12. Faulkner RA, Bailey DA. 2007. Osteoporosis: A pediatric concern? *Med Sport Sci* 51: 1–12.
13. Bachrach LK, Hastie T, Wang M-C, Narasimhan B, Marcus R. 1999. Bone mineral acquisition in healthy Asian, Hispanic, Black and Caucasian youth: A longitudinal study. *J Clin Endocrinol Metab* 84: 4702–4712.
14. Faulkner RA, Bailey DA, Drinkwater DT, McKay HA, Arnold C, Wilkinson AA. 1996. Bone densitometry in Canadian children 8–17 years of age. *Calcif Tissue Int* 59: 344–351.
15. Recker EE, Davies KM, Hinders SM, Heaney RP, Stegman MR, Kimmel DB. 1992. Bone gain in young adult women. *JAMA* 268: 2403–2408.
16. Baxter-Jones ADG, Faulkner RA, Forwood MR, Mirwald RL, Bailey DA. 2011. Bone mineral accrual from 8 to 30 years of age: An estimation of peak bone mass. *J Bone Miner Res* 26: 1729–1739.
17. Arlot M, Sornay-Rendu E, Garnero P, VeyMarty B, Delmas PD. 1997. Apparent pre- and postmenopausal bone loss evaluated by DXA at different skeletal sites in women: The OFELY cohort. *J Bone Miner Res* 12: 683–690.
18. Macdonald HM, Kontulainen SA, Khan KM, McKay HA. 2007. Is a school-based physical activity intervention effective for increasing tibial bone strength in boys and girls? *J Bone Miner Res* 22: 434–446.
19. Sundberg M, Gardsell P, Johnell O, Karlsson MK, Ornstein E, Sandstedt B, Sernbo I. 2001. Peripubertal moderate exercise increases bone mass in boys but not in girls: A population-based intervention study. *Osteoporos Int* 12: 230–238.
20. Bailey DA, McKay HA, Mirwald RL, Crocker PR, Faulkner RA. 1999. A six-year longitudinal study of the relationship of physical activity to bone mineral accrual in growing children: The University of Saskatchewan bone mineral accrual study. *J Bone Miner Res* 14: 1672–1679.
21. Bradney M, Pearce G, Naughton G, Sullivan C, Bass S, Beck T, Carlson J, and Seeman E. 1998. Moderate exercise during growth in prepubertal boys: Changes in bone mass, size, volumetric density, and bone strength: A controlled prospective study. *J Bone Miner Res* 13: 1814–1821.
22. Forwood MR, Baxter-Jones AD, Beck TJ, Mirwald RL, Howard A, Bailey DA. 2006. Physical activity and strength of the proximal femur during the adolescent growth spurt: A longitudinal analysis. *Bone* 38: 576–583.
23. Kannus P, Haapasalo H, Sankelo M, Sievanen H, Pasanen M, Heinonen A, Oja P, Vuori I. 1995. Effect of starting

age of physical activity on bone mass in the dominant arm of tennis and squash players. *Ann Intern Med* 123: 27–31.
24. MacDonald HM, Kontulainen S, Petit M, Khan K, McKay HA. 2008. Does a novel school-based physical activity model benefit femoral neck bone strength in pre- and early pubertal children? *Osteoporos Int* 9: 1445–1456.
25. Mackelvie KJ, McKay HA, Khan KM, Crocker PR. 2001. A school-based exercise intervention augments bone mineral accrual in early pubertal girls. *J Pediatr* 139: 501–508.
26. MacKelvie KJ, Petit MA, Khan KM, Beck TJ, McKay HA. 2004. Bone mass and structure are enhanced following a 2-year randomized controlled trial of exercise in prepubertal boys. *Bone* 34: 755–764.
27. Morris FL, Naughton GA, Gibbs JL, Carlson JS, Wark JD. 1997. Prospective ten-month exercise intervention in premenarcheal girls: Positive effects on bone and lean mass. *J Bone Miner Res* 12: 1453–1462.
28. Petit MA, McKay HA, MacKelvie KJ, Heinonen A, Khan KM, Beck TJ. 2002. A randomized school-based jumping intervention confers site and maturity-specific benefits on bone structural properties in girls: A hip structural analysis study. *J Bone Miner Res* 17: 363–372.
29. Sundberg M, Gardsell P, Johnell O, Karlsson MK, Ornstein E, Sandstedt B, Sernbo I. 2002. Physical activity increases bone size in prepubertal boys and bone mass in prepubertal girls: A combined cross-sectional and 3-year longitudinal study. *Calcif Tissue Int* 71: 406–415.
30. Zouch M, Jaffré C, Thomas T, Frère D, Courteix D, Vico L, Alexandre C. 2008. Long-term soccer practice increases bone mineral content gain in prepubescent boys. *J Bone Spine* 75: 41–49.
31. Damien E, Price JS, Lanyon LE. 2000. Mechanical strain stimulates osteoblast proliferation through the estrogen receptor in males as well as females. *J Bone Miner Res* 15: 2169–2177.
32. Zaman G, Jessop HL, Muzylak M, De Souza RL, Pitsillides AA, Price JS, Lanyon LE. 2006. Osteocytes use estrogen receptor alpha to respond to strain but their ERalpha content is regulated by estrogen. *J Bone Miner Res* 21: 1297–1306.
33. Hind K, Burrows M. 2007. Weight bearing exercise and bone mineral accrual in children and adolescents: A review of controlled trials. *Bone* 40: 14–27.
34. Hert J, Liskova M, Landa J. 1971. Reaction of bone to mechanical stimuli. 1. Continuous and intermittent loading of tibia in rabbit. *Folia Morphol (Praha)* 19: 290–300.
35. Hert J, Liskova M, Landrgot B. 1969. Influence of the long-term, continuous bending on the bone. An experimental study on the tibia of the rabbit. *Folia Morphol (Praha)* 17: 389–399.
36. Rubin CT, Lanyon LE. 1984. Regulation of bone formation by applied dynamic loads. *J Bone Joint Surg Am* 66: 397–402.
37. Turner CH, Owan I, Takano Y. 1995. Mechanotransduction in bone: Role of strain rate. *Am J Physiol* 269: E438–442.

38. Robling AG, Duijvelaar KM, Geevers JV, Ohashi N, Turner CH. 2001. Modulation of appositional and longitudinal bone growth in the rat ulna by applied static and dynamic force. *Bone* 29: 105–113.
39. Turner C, Forwood M, Rho J, Yoshikawa T. 1994. Mechanical loading thresholds for lamellar and woven bone formation. *J Bone Miner Res* 9: 87–97.
40. Mosley JR, Lanyon LE. 1998. Strain rate as a controlling influence on adaptive modeling in response to dynamic loading of the ulna in growing male rats. *Bone* 23: 313–318.
41. O'Connor JA, Lanyon LE, MacFie H. 1982. The influence of strain rate on adaptive bone remodelling. *J Biomech* 15: 767–781.
42. Turner CH, Forwood MR, Otter MW. 1994. Mechanotransduction in bone: Do bone cells act as sensors of fluid flow? *Faseb J* 8: 875–878.
43. Rubin CT, Lanyon LE. 1987. Kappa Delta Award paper. Osteoregulatory nature of mechanical stimuli: Function as a determinant for adaptive remodeling in bone. *J Orthop Res* 5: 300–310.
44. Rubin CT, Lanyon LE. 1985. Regulation of bone mass by mechanical strain magnitude. *Calcif Tissue Int* 37: 411–417.
45. Robling AG, Burr DB, Turner CH. 2000. Partitioning a daily mechanical stimulus into discrete loading bouts improves the osteogenic response to loading. *J Bone Miner Res* 15: 1596–1602.
46. Robling AG, Burr DB, Turner CH. 2001. Recovery periods restore mechanosensitivity to dynamically loaded bone. *J Exp Biol* 204: 3389–3399.
47. Robling AG, Hinant FM, Burr DB, Turner CH. 2002. Improved bone structure and strength after long-term mechanical loading is greatest if loading is separated into short bouts. *J Bone Miner Res* 17: 1545–1554.
48. Robling AG, Hinant FM, Burr DB, Turner CH. 2002. Shorter, more frequent mechanical loading sessions enhance bone mass. *Med Sci Sports Exerc* 34: 196–202.
49. Fuchs RK, Bauer JJ, Snow CM. 2001. Jumping improves hip and lumbar spine bone mass in prepubescent children: A randomized controlled trial. *J Bone Miner Res* 16: 148–156.
50. McKay HA, MacLean L, Petit M, MacKelvie-O'Brien K, Janssen P, Beck T, Khan KM. 2005. "Bounce at the Bell": A novel program of short bouts of exercise improves proximal femur bone mass in early pubertal children. *Br J Sports Med* 39: 521–526.
51. Löfgren B, Detter F, Dencker M, Stenevi-Lundgren S, Nilsson JÅ, Karlsson MK. 2011. Influence of a 3-year exercise intervention program on fracture risk, bone mass, and bone size in prepubertal children. *J Bone Miner Res* 26: 1740–1747.
52. Linden C, Ahlborg HG, Besjakov J, Gardsell P, Karlsson MK. 2006. A school curriculum-based exercise program increases bone mineral accrual and bone size in prepubertal girls: Two-year data from the pediatric osteoporosis prevention (POP) study. *J Bone Miner Res* 21: 829–835.

53. Weeks BK, Young CM, Beck BR. 2008. Eight months of regular in-school jumping improves indices of bone strength in adolescent boys and Girls: The POWER PE study. *J Bone Miner Res* 23: 1002–1011.
54. Faulkner RA, Forwood MR, Beck TJ, Mafukidze JC, Russell K, Wallace W. 2003. Strength indices of the proximal femur and shaft in prepubertal female gymnasts. *Med Sci Sports Exerc* 35: 513–518.
55. Erlandson MC, Kontulainen SA, Chilibeck PD, Arnold CM, Baxter-Jones AD. 2011. Bone mineral accrual in 4- to 10-year-old precompetitive, recreational gymnasts: A 4-year longitudinal study. *J Bone Miner Res* 26: 1313–1320.
56. Specker B, Binkley T. 2003. Randomized trial of physical activity and calcium supplementation on bone mineral content in 3- to 5-year-old children. *J Bone Miner Res* 18: 885–892.
57. Turner CH, Robling AG. 2003. Designing exercise regimens to increase bone strength. *Exerc Sport Sci Rev* 31: 45–50.
58. MacKelvie KJ, McKay HA, Petit MA, Moran O, Khan KM. 2002. Bone mineral response to a 7-month randomized controlled, school-based jumping intervention in 121 prepubertal boys: Associations with ethnicity and body mass index. *J Bone Miner Res* 17: 834–844.
59. Prentice A, Parsons TJ, Cole TJ. 1994. Uncritical use of bone mineral density in absorptiometry may lead to size-related artifacts in the identification of bone mineral determinants. *Am J Clin Nutr* 60: 837–842.
60. Daly RM, Saxon L, Turner CH, Robling AG, Bass SL. 2004. The relationship between muscle size and bone geometry during growth and in response to exercise. *Bone* 34: 281–287.
61. Macdonald H, Kontulainen S, Petit M, Janssen P, McKay H. 2006. Bone strength and its determinants in pre- and early pubertal boys and girls. *Bone* 39: 598–608.
62. Macdonald HM, Cooper DM, McKay HA. 2009. Anterior–posterior bending strength at the tibial shaft increases with physical activity in boys: Evidence for non-uniform geometric adaptation. *Osteoporos Int* 20: 61–70.
63. Gustavsson A, Olsson T, Nordström P. 2003. Rapid loss of bone mineral density of the femoral neck after cessation of ice hockey training: A 6-year longitudinal study in males. *J Bone Miner Res* 18: 1964–1969.
64. Karlsson MK, Linden C, Karlsson C, Johnell O, Obrant K, Seeman E. 2000. Exercise during growth and bone mineral density and fractures in old age. *Lancet* 355: 469–470.
65. Nordström A, Olsson T, Nordström P. 2004. Bone gained from physical activity and lost through detraining: A longitudinal study in young males. *Osteopor Int* 16: 835–841.
66. Gunter K, Baxter-Jones AD, Mirwald RL, Almstedt H, Fuller A, Durski S, Snow C. 2008. Jump starting skeletal health: A 4-year longitudinal study assessing the effects of jumping on skeletal development in pre and circum pubertal children. *Bone* 42: 710–718.
67. Gunter K, Baxter-Jones AD, Mirwald RL, Almstedt H, Fuchs RK, Durski S, Snow C. 2008. Impact exercise increases BMC during growth: An 8-year longitudinal study. *J Bone Miner Res* 23: 986–993.
68. Scerpella TA, Dowthwaite JN, Rosenbaum PF. 2011. Sustained skeletal benefit from childhood mechanical loading. *Osteoporos Int* 22: 2205–2210.
69. Baxter-Jones AD, Kontulainen SA, Faulkner RA, Bailey DA. 2008. A longitudinal study of the relationship of physical activity to bone mineral accrual from adolescence to young adulthood. *Bone* 43: 1101–1109.
70. Warden SJ, Fuchs RK, Castillo AB, Nelson IR, Turner CH. 2007. Exercise when young provides lifelong benefits to bone structure and strength. *J Bone Miner Res* 22: 251–259.
71. Haapasalo H, Kontulainen S, Sievanen H, Kannus P, Jarvinen M, Vuori I. 2000. Exercise-induced bone gain is due to enlargement in bone size without a change in volumetric bone density: A peripheral quantitative computed tomography study of the upper arms of male tennis players. *Bone* 27: 351–357.
72. Kontulainen S, Kannus P, Haapasalo H, Sievänen H, Pasanen M, Heinonen A, Oja P, Vuori I. 2001. Good maintenance of exercise-induced bone gain with decreased training of female tennis and squash players: A prospective 5-year follow-up study of young and old starters and controls. *J Bone Miner Res* 16: 195–201.

20

Pregnancy and Lactation

Christopher S. Kovacs and Henry M. Kronenberg

Pregnancy 156
Lactation 159
Adolescent Pregnancy and Lactation 162
Implications 162
References 162

Normal pregnancy places a demand on the calcium homeostatic mechanisms of women, as the fetus and placenta draw calcium from the maternal circulation in order to mineralize the fetal skeleton. Similar demands are placed on lactating women, in order to supply sufficient calcium to breast milk and enable continued skeletal growth in a nursing infant. Despite a similar magnitude of calcium demand presented to pregnant and lactating women, the adjustments made in each of these reproductive periods differ significantly. These hormone-mediated adjustments normally satisfy the daily calcium needs of the fetus and infant without long-term consequences to the maternal skeleton. Detailed references on this subject are available in several comprehensive reviews [1–3] (see Fig. 20.1).

PREGNANCY

The developing fetal skeleton accretes about 30 g of calcium by term, and about 80% of it during the third trimester when the fetal skeleton is rapidly mineralizing [4]. This calcium demand appears to be largely met by a doubling of maternal intestinal calcium absorption, mediated by 1,25-dihydroxyvitamin D (calcitriol or 1,25-D) and other factors.

Mineral ions and calcitropic hormones

Normal pregnancy results in characteristic alterations in serum chemistries and calciotropic hormones [1]. The total serum calcium falls early in pregnancy due to a fall in the serum albumin, but the ionized calcium (the physiologically important fraction) remains constant. Serum phosphate and magnesium levels are normal.

In studies of women from North America and Europe, the serum parathyroid hormone (PTH) level, when measured with two-site "intact" assays, falls to the low–normal range (i.e., 10–30% of the mean nonpregnant value) during the first trimester, but increases steadily to the mid–normal range by term [5–9]. In contrast, the PTH value did not suppress in studies of women from Asia and Gambia and may reflect the lower calcium and vitamin D intakes in those populations [10]. Total 1,25-D levels double early in pregnancy and maintain this increase until term; free 1,25-D levels are increased from the third trimester [1, 11]. The rise in 1,25-D does not appear to be driven by PTH, since PTH levels are falling while 1,25-D is increasing. Maternal kidneys, and not the placenta, likely account for most, if not all, of the rise in 1,25-D during pregnancy, as shown by an anephric woman who had very low 1,25-D levels before and during pregnancy [12]. The renal 1α-hydroxylase is upregulated in response to factors such as PTH-related protein (PTHrP), estradiol, prolactin and placental lactogen. Serum calcitonin levels are also increased during pregnancy.

PTHrP levels increase during pregnancy and may derive from multiple tissues in the mother and fetus. PTHrP may contribute to the elevation in 1,25-D and suppression of PTH during pregnancy. PTHrP may have other roles such as regulating placental calcium transport in the fetus [1, 13]. Also, PTHrP may have a role in protecting the maternal skeleton during pregnancy, since the

---

Primer on the Metabolic Bone Diseases and Disorders of Mineral Metabolism, Eighth Edition. Edited by Clifford J. Rosen.
© 2013 American Society for Bone and Mineral Research. Published 2013 by John Wiley & Sons, Inc.

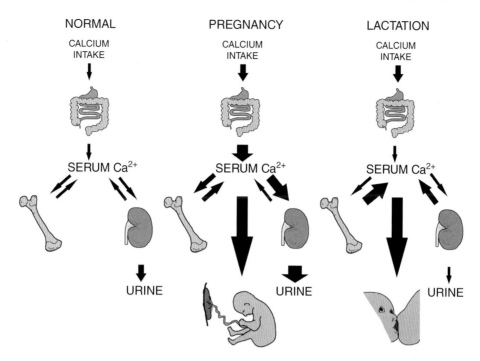

Fig. 20.1. Schematic illustration contrasting calcium homeostasis in human pregnancy and lactation, as compared to normal. The thickness of arrows indicates a relative increase or decrease with respect to the normal and nonpregnant state. Although not illustrated, the serum (total) calcium is decreased during pregnancy while the ionized calcium remains normal during both pregnancy and lactation. Adapted from Ref. 1, © 1997, The Endocrine Society.

carboxyl-terminal portion of PTHrP ("osteostatin") has been shown to inhibit osteoclastic bone resorption [14].

Pregnancy induces significant changes in other hormones, including sex steroids, prolactin, placental lactogen, and insulin-like growth factor-1 (IGF-1). Each of these may have direct or indirect effects on calcium and bone metabolism during pregnancy, but these issues have been largely unexplored.

Intestinal absorption of calcium

Intestinal absorption of calcium is doubled during pregnancy from as early as 12 weeks of gestation; this appears to be a major maternal adaptation to meet the fetal need for calcium. The increased intestinal calcium absorption may be largely the result of a 1,25-D-mediated action on intestinal cells, although it is notable that intestinal calcium absorption doubles months before an increase in free 1,25-D levels occur. Also, studies in rodents indicate that a pregnancy-induced increase in intestinal calcium absorption occurs despite severe vitamin D deficiency or absence of the vitamin D receptor [15–17]. Prolactin, placental lactogen, and other factors also stimulate intestinal calcium absorption, as demonstrated in rodents [18, 19]. Enhancing calcium absorption early in pregnancy causes a positive calcium balance in women [20], and this may enable the maternal skeleton to store calcium prior to the peak fetal demand during the third trimester.

Renal handling of calcium

The 24-hour urine calcium excretion increases as early as the 12th week of gestation and often exceeds the normal range [1]. Since fasting urine calcium values are normal or low, the increased 24-hour urine calcium reflects the increased intestinal absorption of calcium (absorptive hypercalciuria). The elevated calcitonin levels of pregnancy might also promote renal calcium excretion.

Skeletal calcium metabolism

Animal models indicate that histomorphometric parameters of bone turnover are increased during pregnancy and that bone mineral content of normal mice may increase or decrease depending upon the genetic background [1, 21, 22]. Comparable histomorphometric data are not available for human pregnancy, but one small study indicated that bone resorption parameters were increased at 8–10 weeks in 15 women who electively terminated pregnancy as compared to nonpregnant controls and pregnant women at term [23].

Most human studies of skeletal calcium metabolism in pregnancy have examined changes in serum markers of bone formation, and urine markers of bone resorption. These studies are fraught with a number of confounding variables, including lack of prepregnancy baseline values;

effects of hemodilution in pregnancy on serum markers; increased glomerular filtration rate (GFR); altered creatinine excretion; placental, uterine, and fetal contribution to the levels of markers in blood; degradation and clearance by the placenta; and lack of diurnally timed or fasted specimens. Given these limitations, many studies have reported that urinary markers of bone resorption (24-hr collection) are increased from early to midpregnancy (including deoxypyridinoline, pyridinoline, and hydroxyproline). Conversely, serum markers of bone formation (generally not corrected for hemodilution or increased GFR) are often decreased from prepregnancy or nonpregnant values in early or mid-pregnancy, rising to normal or above before term (including osteocalcin, procollagen I carboxypeptides, and bone specific alkaline phosphatase). Total alkaline phosphatase rises early in pregnancy due largely to contributions from the placental fraction; it is not a useful marker of bone formation in pregnancy.

Based on the scant bone biopsy data, and the measurements of bone markers (with aforementioned confounding factors), one may cautiously conclude that bone resorption is increased in pregnancy, from as early as the 10th week of gestation. There is comparatively little maternal–fetal calcium transfer occurring at this stage of pregnancy, as compared to the peak rate of calcium transfer in the third trimester. One might have anticipated that markers of bone resorption would increase particularly in the third trimester, but no marked increase is seen at that time.

Changes in skeletal calcium content have been assessed through the use of sequential areal bone mineral density (aBMD) studies during pregnancy. Due to concerns about fetal radiation exposure, few such studies have been done. Such studies are confounded by the changes in body composition, weight and skeletal volumes during normal pregnancy that can lead to artifactual changes in the aBMD reading obtained. Using single and/or dualphoton absorptiometry, several prospective studies did not find a significant change in cortical or trabecular aBMD during pregnancy [1]. Several recent studies have utilized dual energy X-ray absorptiometry (DXA) before conception (range 1–8 months prior, but not always stated) and after delivery (range 1–6 weeks postpartum) [studies reviewed in detail in Ref. 24]. Most studies involved 16 or fewer subjects. One study found no change in lumbar spine aBMD measurements obtained preconception and within 1–2 weeks postdelivery, whereas the other studies reported decreases of 4–5% in lumbar aBMD with the postpartum measurement taken between 1 and 6 weeks postdelivery. Since the puerperium is associated with bone density losses of 1–3% per month (see lactation section, below), it is possible that obtaining the second measurement 2 to 6 weeks after delivery contributed to the bone loss documented in many of the studies. Other longitudinal studies have found a progressive decrease during pregnancy in indices thought to correlate with volumetric BMD, as determined by ultrasonographic measurements at another peripheral site, the os calcis. None of all the aforementioned studies can address the question as to whether skeletal calcium content is increased early in pregnancy in advance of the third trimester. Further studies, with larger numbers of patients, will be needed to clarify the extent of bone loss during pregnancy.

It seems certain that any acute changes in bone metabolism during pregnancy do not cause long-term changes in skeletal calcium content or strength. Numerous studies of osteoporotic or osteopenic women have failed to find a significant association of parity with bone density or fracture risk [1, 25].

Osteoporosis in pregnancy

Women occasionally present with fragility fractures and low bone mineral density (BMD) during or shortly after pregnancy; in most cases, the possibility that low bone density was present prior to pregnancy cannot be excluded. Some women may experience excessive resorption of calcium from the skeleton due to changes in mineral metabolism induced by pregnancy and other factors, such as low dietary calcium intake and vitamin D insufficiency. The apparently increased rate of bone resorption in pregnancy may contribute to fracture risk, because a high rate of bone turnover is an independent risk factor for fragility fractures outside of pregnancy. Therefore, fragility fractures in pregnancy or the puerperium may be a consequence of preexisting low bone density and increased bone resorption, among other possible factors. During lactation, additional changes in mineral metabolism occur that may further increase fracture risk in some women (see below).

Focal, transient osteoporosis of the hip is a rare, self-limited form of pregnancy-associated osteoporosis. It is probably not a manifestation of altered calciotropic hormone levels or mineral balance during pregnancy, but rather might be a consequence of local factors. These patients present with unilateral or bilateral hip pain, limp, and/or hip fracture in the third trimester. There is objective evidence of reduced bone density of the symptomatic femoral head and neck that has been shown by magnetic resonance imaging (MRI) to be the consequence of increased water content of the femoral head and the marrow cavity; a joint effusion may also be present. The symptoms and the radiological appearance usually resolve within 2 to 6 months postpartum.

Primary hyperparathyroidism

Although probably a rare condition (there are no data available on its prevalence), primary hyperparathyroidism in pregnancy has been associated in the literature with an alarming rate of adverse outcomes in the fetus and neonate, including a 30% rate of spontaneous abortion or stillbirth. The adverse postnatal outcomes are thought to result from suppression of the fetal and neonatal parathyroid glands; this suppression may occasionally be prolonged after birth for months. To prevent these adverse outcomes, surgical correction of primary hyperparathyroidism during the second trimester has been almost universally recommended. Several case series have found elective surgery to be well tolerated and to dramatically reduce the rate of adverse events when compared to the earlier cases reported in the literature. Many of the women in those early cases had a relatively severe form of primary hyperparathyroidism that is not often seen today (symptomatic, with nephrocalcinosis and renal insufficiency). While mild, asymptomatic primary hyperparathyroidism during pregnancy has been followed conservatively with successful outcomes, complications continue to occur so that in the absence of definitive data, surgery during the second trimester remains the most common recommendation [26].

Familial hypocalciuric hypercalcemia

Although familial hypocalciuric hypercalcemia (FHH) has not been reported to adversely affect the mother during pregnancy, the maternal hypercalcemia has caused fetal and neonatal parathyroid suppression with subsequent tetany.

Hypoparathyroidism and pseudohypoparathyroidism

Early in pregnancy, some hypoparathyroid women have fewer hypocalcemic symptoms and require less supplemental calcium. This is consistent with a limited role for PTH in the pregnant woman and suggests that an increase in 1,25-D and/or increased intestinal calcium absorption will occur in the absence of PTH. However, it is clear from other case reports that some pregnant hypoparathyroid women require increased calcitriol replacement in order to avoid worsening hypocalcemia. It is important to maintain a normal ionized calcium level in pregnant women since maternal hypocalcemia due to hypoparathyroidism has been associated with the development of intrauterine fetal hyperparathyroidism and fetal death. The ionized calcium, rather than the total calcium, should be followed, because of the fall of serum albumin during pregnancy. Late in pregnancy, hypercalcemia may occur in hypoparathyroid women unless the calcitriol dosage is substantially reduced or discontinued. This effect may be mediated by the increasing levels of PTHrP in the maternal circulation in late pregnancy.

In limited case reports of pseudohypoparathyroidism, pregnancy has been noted to normalize the serum calcium level, reduce the PTH level by half, and increase the 1,25-D level two- to threefold [27]. The mechanism by which these changes occur despite pseudohypoparathyroidism remains unclear.

Vitamin D deficiency and insufficiency

Maternal 25-hydroxyvitamin D levels do not change significantly during pregnancy [28–31], even in women starting with the extremely low level of 20 nmol/l (8 ng/ml) [31], and so it does not appear that pregnant women require higher intakes of vitamin D to maintain a set 25-hydroxyvitamin D level. There are as yet no large randomized trials that have examined the effects of vitamin D deficiency or insufficiency on human pregnancy. However, available data from small clinical trials of vitamin D supplementation, observational studies, and case reports suggest that, consistent with animal studies, vitamin D deficiency is not associated with any worsening of maternal calcium homeostasis and that the fetus will have a normal serum calcium and fully mineralized skeleton at term (this topic is reviewed in detail in Refs. 32 and 33). The only consistent benefit seen in randomized trials is that vitamin D supplementation during pregnancy increases maternal and cord blood 25(OH)D levels without altering cord blood calcium or anthropometric parameters in the baby.

Low calcium intake

Since absorptive hypercalciuria typically occurs during pregnancy, this may be viewed as evidence that calcium intake normally exceeds maternal requirements. Calcium supplementation did not improve maternal or neonatal bone density in one study [34]. By contrast, in a randomized trial of 2 g of calcium supplementation versus placebo found there was an increase in neonatal BMD in the women with the lowest quintile of calcium intake (less than 600 mg per day) [35]. This suggests that calcium supplementation may benefit only those women (and their babies) with very low dietary intakes of calcium.

Low calcium intake has been associated with increased risk of preeclampsia. Calcium supplementation appears to reduce the risk of preeclampsia when the dietary calcium intake is low, whereas there is no effect when dietary calcium intake is adequate [36–39].

LACTATION

The average daily loss of calcium in breast milk is 210 mg, with losses as high as 1,000 mg calcium reported for women nursing twins. A temporary demineralization of the skeleton appears to be the main mechanism by which lactating humans meet these calcium requirements. This demineralization does not appear to be mediated by PTH or 1,25-D but may be mediated by PTHrP in the setting of a fall in estrogen levels.

Mineral ions and calciotropic hormones

The mean ionized calcium level of exclusively lactating women is increased, although it remains within the normal range. Serum phosphate levels are also higher during lactation, and the level may exceed the normal range. Since reabsorption of phosphate by the kidneys appears to be increased, the increased serum phosphate levels may, therefore, reflect the combined effects of increased flux of phosphate into the blood from diet and from skeletal resorption in the setting of decreased renal phosphate excretion.

Intact PTH, as determined by a two-site immunoradioactive monoclondal antibody (IRMA) assay, has been found to be reduced 50% or more during the first several months of lactation. It rises to normal at weaning but may rise above normal postweaning. In contrast to the high 1,25-D levels of pregnancy, maternal free and bound 1,25-D levels fall to normal within days of parturition and remain there throughout lactation. Calcitonin levels fall to normal after the first 6 weeks postpartum. Mice lacking the calcitonin gene lose twice the normal amount of bone mineral content during lactation, which indicates that physiological levels of calcitonin may protect the maternal skeleton from excessive resorption during this time period [21]. Whether calcitonin plays a similar role in human physiology is unknown. Thyroidectomized women have not been studied during pregnancy and lactation, though studies in such women might not be definitive because of extrathyroidal production of calcitonin in placenta, pituitary, and lactating breast.

PTHrP levels, as measured by two-site IRMA assays, are significantly higher in lactating women than in nonpregnant controls. The source of PTHrP may be the breast, since PTHrP has been detected in breast milk at concentrations exceeding 10,000 times the level found in the blood of patients with hypercalcemia of malignancy or normal human controls. Further, lactating mice with the PTHrP gene ablated only from mammary tissue have lower blood levels of PTHrP than control lactating mice [40]. Studies in animals suggest that PTHrP may regulate mammary development and blood flow, and the calcium content of milk. In addition, PTHrP reaching the maternal circulation from the lactating breast may cause resorption of calcium from the maternal skeleton, renal tubular reabsorption of calcium, and (indirectly) suppression of PTH. In support of this hypothesis, deletion of the PTHrP gene from mammary tissue at the onset of lactation resulted in more modest losses of bone mineral content during lactation in mice [40]. In humans, PTHrP levels correlate with the amount of BMD lost: negatively with PTH levels and positively with the ionized calcium levels of lactating women [41–43]. Furthermore, observations in aparathyroid women provide evidence of the impact of PTHrP in calcium homeostasis during lactation (see below).

Intestinal absorption of calcium

Intestinal calcium absorption decreases to the nonpregnant rate from the increased rate of pregnancy. This corresponds to the fall in 1,25-D levels to normal.

Renal handling of calcium

In humans, the GFR falls during lactation, and the renal excretion of calcium is typically reduced to levels as low as 50 mg per 24 hours. This suggests that the tubular reabsorption of calcium must be increased, perhaps by the actions of PTHrP.

Skeletal calcium metabolism

Histomorphometric data from animals consistently show increased bone turnover during lactation, and losses of 35% or more of bone mineral are achieved during 2–3 weeks of normal lactation in the rat [reviewed in Ref. 1]. Comparative histomorphometric data are lacking for humans, and in place of that, serum markers of bone formation, and urinary markers of bone resorption have been assessed in numerous cross-sectional and prospective studies of lactation. Some of the confounding factors discussed with respect to pregnancy apply to the use of these markers in lactating women. In this instance, the GFR is reduced and the intravascular volume is contracted. Urinary markers of bone resorption (24-hr collection) have been reported to be elevated two- to threefold during lactation and are higher than the levels attained in the third trimester. Serum markers of bone formation (not adjusted for hemoconcentration or reduced GFR) are generally high during lactation and increased over the levels attained during the third trimester. Total alkaline phosphatase falls immediately postpartum due to loss of the placental fraction, but may still remain above normal due to the elevation in the bone-specific fraction. Despite the confounding variables, these findings suggest that bone turnover is significantly increased during lactation.

Serial measurements of aBMD during lactation [by single photon absorptionmetry (SPA), dual photon absorptionmetry (DPA) or DXA] have shown a fall of 3.0 to 10.0% in bone mineral content after 2 to 6 months of lactation at trabecular sites (lumbar spine, hip, femur, and distal radius), with smaller losses at cortical sites and whole body [1, 25]. These aBMD changes are in accord with studies in rats, mice, and primates in which the skeletal resorption has been shown to occur largely at trabecular and to a lesser degree at endocortical surfaces. The loss occurs at a peak rate of 1–3% *per month*, far exceeding the rate of 1–3% per year that can occur in women after menopause who are considered to be losing bone rapidly. Loss of mineral from the maternal skeleton appears to be a normal consequence of lactation and may not be preventable by raising the calcium intake above the recommended dietary allowance. Several studies have demonstrated that calcium supplementation does not significantly reduce the amount of bone lost during lactation [44–47]. Not surprisingly, the lactational decrease in BMD correlates with the amount of calcium lost in the breast milk [48].

The mechanisms controlling the rapid loss of skeletal calcium content are not well understood. The reduced estrogen levels of lactation are clearly important, but are unlikely to be the sole explanation. To estimate the effects of estrogen deficiency during lactation, it is worth noting the alterations in calcium and bone metabolism that occur in reproductive-age women who have estrogen deficiency induced by GnRH agonist therapy for endometriosis and other conditions. Six months of acute estrogen deficiency induced by GnRH agonist therapy leads to 1–4% losses in trabecular (but not cortical) aBMD, increased urinary calcium excretion, and suppression of 1,25-D and PTH [reviewed in Ref. 1]. In lactation, women are not as estrogen deficient but lose more aBMD (at both trabecular and cortical sites), have normal (as opposed to low) 1,25-D levels, and have reduced (as opposed to increased) urinary calcium excretion. The difference between isolated estrogen deficiency and lactation is due to the effects of PTHrP, which complements the effects of estrogen withdrawal in lactation. Stimulated in part by suckling and high prolactin levels, the effects of PTHrP and estrogen deficiency combine to coordinate the marked skeletal resorption that occurs during lactation (see Fig. 20.2).

The bone density losses of lactation appear to be substantially reversed during weaning [1, 25, 45], although the speed and completeness of recovery may differ by skeletal site and technique used. This corresponds to a gain in bone density of 0.5 to 2% per month in the woman who has weaned her infant. The mechanism for this restoration of bone density is uncertain. In the long-term, the consequences of lactation-induced depletion of bone mineral appear clinically unimportant. The vast majority of epidemiologic studies of pre- and postmenopausal women have found no adverse effect of a history of lactation on peak bone mass, bone density, or hip fracture risk.

Osteoporosis of lactation

Rarely, women suffer fragility fractures during lactation, and osteoporotic readings are confirmed by DXA. Like osteoporosis in pregnancy, this may represent a coincidental, unrelated disease; the woman may have had low bone density prior to conception. Alternatively, some cases might represent an exacerbation of the normal degree of skeletal demineralization that occurs during lactation, and a continuum from changes in bone density and bone turnover that may have occurred during pregnancy. For example, excessive PTHrP release from the lactating breast into the maternal circulation, combined with estrogen deficiency, could conceivably cause excessive bone resorption, osteoporosis, and fractures. Consistent with this, PTHrP levels were high in one case of lactational osteoporosis and were found to remain elevated for months after weaning [49]. Also, in most cases where vertebral compression fractures occurred during lactation, bone density spontaneously increases after weaning, consistent with lactation-induced loss of bone causing fragility [50].

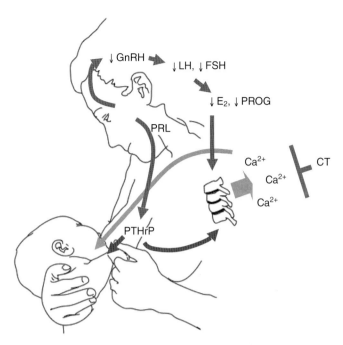

Fig. 20.2. The breast is a central regulator of skeletal demineralization during lactation. Suckling and prolactin both inhibit the hypothalamic gonadotropin-releasing hormone (GnRH) pulse center, which in turn suppresses the gonadotropins [luteinizing hormone (LH) and follicle-stimulating hormone (FSH)], leading to low levels of the ovarian sex steroids (estradiol and progesterone). PTHrP production and release from the breast is controlled by several factors including suckling, prolactin, and the calcium receptor. PTHrP enters the bloodstream and combines with systemically low estradiol levels to markedly upregulate bone resorption. Increased bone resorption releases calcium and phosphate into the blood stream, which then reaches the breast ducts and is actively pumped into the breast milk. PTHrP also passes into milk at high concentrations, but whether swallowed PTHrP plays a role in regulating calcium physiology of the neonate is unknown. Calcitonin (CT) may inhibit skeletal responsiveness to PTHrP and low estradiol. Reprinted from Ref. 64, ©2005, with kind permission of Springer Science and Business Media B.V.

Hypoparathyroidism and pseudohypoparathyroidism

Levels of calcitriol and calcium supplementation required for treatment of hypoparathyroid women fall early in the postpartum period, especially if the woman breastfeeds, and hypercalcemia may occur if the calcitriol dosage is not substantially reduced [51]. As observed in one case, this is consistent with PTHrP reaching the maternal circulation in amounts sufficient to allow stimulation of 1,25-D synthesis and maintenance of normal (or slightly increased) maternal serum calcium [52].

The management of pseudohypoparathyroidism has been less well documented. Since these patients are likely resistant to the renal actions of PTHrP, and the placental sources of 1,25-D are lost at parturition, the calcitriol requirements might well increase and may require further adjustments during lactation.

Vitamin D deficiency and insufficiency

During lactation, maternal 25-hydroxyvitamin D levels are unchanged [53, 54] because very little vitamin D or 25-hydroxyvitamin D passes into breast milk. The available data from small clinical trials, observational studies, and case reports indicate that lactation proceeds normally regardless of vitamin D status, and breast milk calcium content is unaffected by vitamin D deficiency or by supplementation with doses as high as 6,400 IU per day (topic reviewed in detail in Refs. 32 and 33). This is likely because maternal calcium homeostasis is dominated by skeletal resorption induced by estrogen deficiency and PTHrP. It is the neonate who will suffer the consequences of being born of a vitamin D deficient mother and especially if exclusively breastfed without supplements or sunlight exposure. A small pilot study indicated that 6,400 IU daily of vitamin D raised the vitamin D and 25(OH)D content of breast milk enough for the suckling neonate to attain a 25(OH)D level over 75 nmol/l (30 ng/ml); the same level was achieved by giving 300 IU directly to the baby [55].

Low calcium intake

The calcium content of milk appears to be largely derived from skeletal resorption; consequently, low calcium intake does not alter breast milk calcium content nor does it accentuate maternal bone loss during lactation [56–59]. Instead, the calcium receptor, expressed by lactating mammary epithelial cells, regulates the calcium content of milk by upregulating expression of the plasma membrane calcium ATPase isoform 2 (PMCA2) [60, 61]. The calcium receptor also regulates the fluid content of milk and mammary production of PTHrP [60].

ADOLESCENT PREGNANCY AND LACTATION

Adolescent pregnancy and lactation do not reduce peak bone mass, as previously feared [62]. In a NHANES III analysis of 819 women aged 20–25, women who had been pregnant as adolescents had the same BMD as nulliparous women and women who had been pregnant as adults [63]. Women who breastfed as adolescents had higher BMD than women who had not breastfed, as well as nulliparous women [63]. Thus it appears that, as with adults, adolescents lose bone during lactation and recover fully afterwards without any long-term adverse effects on the skeleton.

IMPLICATIONS

In both pregnancy and lactation, novel regulatory systems specific to these settings complement the usual regulators of calcium homeostasis. The fetal calcium demand is met in large part by intestinal calcium absorption that more than doubles from early in pregnancy, an adaptation that may not be fully explained by the increase in 1,25-D levels. In comparison, skeletal calcium resorption is a dominant mechanism by which calcium is supplied to the breast milk, while renal calcium conservation is also apparent. These changes during lactation appear to be driven by PTHrP in association with estrogen deficiency rather than PTH and vitamin D. Consistent with this, treatment of lactating women with calcium supplements has little or no impact on bone loss.

These observations indicate that the maternal adaptations to pregnancy and lactation have evolved differently over time, such that dietary calcium absorption dominates in pregnancy, whereas lactation programs an obligatory but temporary skeletal calcium loss that is completely restored after weaning. The rapidity of calcium regain by the skeleton of the lactating woman occurs through a mechanism that is not understood. A full elucidation of the mechanism of bone restoration after lactation might lead to the development of novel approaches to the treatment of osteoporosis and other metabolic bone diseases. Finally, while it is apparent that some women will experience fragility fractures as a consequence of pregnancy or lactation, the vast majority of women can be assured that the changes in calcium and bone metabolism during pregnancy and lactation are normal, healthy, and without adverse consequences in the long-term.

REFERENCES

1. Kovacs CS, Kronenberg HM. 1997. Maternal-fetal calcium and bone metabolism during pregnancy, puerperium and lactation. *Endocr Rev* 18: 832–72.
2. Wysolmerski JJ. 2007. Conversations between breast and bone: Physiological bone loss during lactation as evolutionary template for osteolysis in breast cancer and pathological bone loss after menopause. *BoneKEy* 4(8): 209–25.
3. Kovacs CS. 2011. Calcium and bone metabolism disorders during pregnancy and lactation. *Endocrinol Metab Clin N America* 40(4): 795–826.
4. Trotter M, Hixon BB. 1974. Sequential changes in weight, density, and percentage ash weight of human skeletons from an early fetal period through old age. *Anat Rec* 179: 1–18.
5. Dahlman T, Sjoberg HE, Bucht E. 1994. Calcium homeostasis in normal pregnancy and puerperium. A longitudinal study. *Acta Obstet Gynecol Scand* 73: 393–8.
6. Gallacher SJ, Fraser WD, Owens OJ, Dryburgh FJ, Logue FC, Jenkins A, Kennedy J, Boyle IT. 1994. Changes in

calciotrophic hormones and biochemical markers of bone turnover in normal human pregnancy. *Eur J Endocrinol* 131: 369–74.

7. Cross NA, Hillman LS, Allen SH, Krause GF, Vieira NE. 1995. Calcium homeostasis and bone metabolism during pregnancy, lactation, and postweaning: A longitudinal study. *Am J Clin Nutr* 61: 514–23.

8. Rasmussen N, Frolich A, Hornnes PJ, Hegedus L. 1990. Serum ionized calcium and intact parathyroid hormone levels during pregnancy and postpartum. *Br J Obstet Gynaecol* 97: 857–9.

9. Seki K, Makimura N, Mitsui C, Hirata J, Nagata I. 1991. Calcium-regulating hormones and osteocalcin levels during pregnancy: A longitudinal study. *Am J Obstet Gynecol* 164: 1248–52.

10. Singh HJ, Mohammad NH, Nila A. 1999. Serum calcium and parathormone during normal pregnancy in Malay women. *J Matern Fetal Med* 8(3): 95–100.

11. Kovacs CS. 2001. Calcium and bone metabolism in pregnancy and lactation. *J Clin Endocrinol Metab* 86(6): 2344–8.

12. Turner M, Barre PE, Benjamin A, Goltzman D, Gascon-Barre M. 1988. Does the maternal kidney contribute to the increased circulating 1,25-dihydroxyvitamin D concentrations during pregnancy? *Miner Electrolyte Metab* 14: 246–52.

13. Kovacs CS, Lanske B, Hunzelman JL, Guo J, Karaplis AC, Kronenberg HM. 1996. Parathyroid hormone-related peptide (PTHrP) regulates fetal-placental calcium transport through a receptor distinct from the PTH/PTHrP receptor. *Proc Natl Acad Sci U S A* 93: 15233–8.

14. Cornish J, Callon KE, Nicholson GC, Reid IR. 1997. Parathyroid hormone-related protein-(107-139) inhibits bone resorption in vivo. *Endocrinology* 138: 1299–304.

15. Halloran BP, DeLuca HF. 1980. Calcium transport in small intestine during pregnancy and lactation. *Am J Physiol* 239: E64–8.

16. Brommage R, Baxter DC, Gierke LW. 1990. Vitamin D-independent intestinal calcium and phosphorus absorption during reproduction. *Am J Physiol* 259: G631–G8.

17. Fudge NJ, Kovacs CS. 2010. Pregnancy up-regulates intestinal calcium absorption and skeletal mineralization independently of the vitamin D receptor. *Endocrinology* 151(3): 886–95.

18. Pahuja DN, DeLuca HF. 1981. Stimulation of intestinal calcium transport and bone calcium mobilization by prolactin in vitamin D-deficient rats. *Science* 214: 1038–9.

19. Mainoya JR. 1975. Effects of bovine growth hormone, human placental lactogen and ovine prolactin on intestinal fluid and ion transport in the rat. *Endocrinology* 96: 1165–70.

20. Heaney RP, Skillman TG. 1971. Calcium metabolism in normal human pregnancy. *J Clin Endocrinol Metab* 33: 661–70.

21. Woodrow JP, Sharpe CJ, Fudge NJ, Hoff AO, Gagel RF, Kovacs CS. 2006. Calcitonin plays a critical role in regulating skeletal mineral metabolism during lactation. *Endocrinology* 147: 4010–21.

22. Kirby BJ, Ardeshirpour L, Woodrow JP, Wysolmerski JJ, Sims NA, Karaplis AC, Kovacs CS. 2011. Skeletal recovery after weaning does not require PTHrP. *J Bone Miner Res* 26(6): 1242–51.

23. Purdie DW, Aaron JE, Selby PL. 1988. Bone histology and mineral homeostasis in human pregnancy. *Br J Obstet Gynaecol* 95(9): 849–54.

24. Kovacs CS, El-Hajj Fuleihan G. 2006. Calcium and bone disorders during pregnancy and lactation. *Endocrinol Metab Clin N America* 35(1): 21–51.

25. Sowers M. 1996. Pregnancy and lactation as risk factors for subsequent bone loss and osteoporosis. *J Bone Miner Res* 11: 1052–60.

26. Schnatz PF, Curry SL. 2002. Primary hyperparathyroidism in pregnancy: Evidence-based management. *Obstet Gynecol Surv* 57(6): 365–76.

27. Breslau NA, Zerwekh JE. 1986. Relationship of estrogen and pregnancy to calcium homeostasis in pseudohypoparathyroidism. *J Clin Endocrinol Metab* 62: 45–51.

28. Hillman LS, Slatopolsky E, Haddad JG. 1978. Perinatal vitamin D metabolism. IV. Maternal and cord serum 24,25-dihydroxyvitamin D concentrations. *J Clin Endocrinol Metab* 47: 1073–7.

29. Morley R, Carlin JB, Pasco JA, Wark JD. 2006. Maternal 25-hydroxyvitamin D and parathyroid hormone concentrations and offspring birth size. *J Clin Endocrinol Metab* 91(3): 906–12.

30. Ardawi MS, Nasrat HA, BA'Aqueel HS. 1997. Calcium-regulating hormones and parathyroid hormone-related peptide in normal human pregnancy and postpartum: A longitudinal study. *Eur J Endocrinol* 137(4): 402–9.

31. Brooke OG, Brown IR, Bone CD, Carter ND, Cleeve HJ, Maxwell JD, Robinson VP, Winder SM. 1980. Vitamin D supplements in pregnant Asian women: Effects on calcium status and fetal growth. *Br Med J* 280: 751–4.

32. Kovacs CS. 2008. Vitamin D in pregnancy and lactation: Maternal, fetal, and neonatal outcomes from human and animal studies. *Am J Clin Nutr* 88(2): 520S–8S.

33. Kovacs CS. 2011. Fetus, Neonate and Infant. In: Feldman D, Pike WJ, Adams JS (eds.). *Vitamin D: Third Edition*. New York: Academic Press. pp. 625–46.

34. Prentice A. 2003. Pregnancy and Lactation. In: Glorieux FH, Petifor JM, Jüppner H (eds.) *Pediatric Bone: Biology & Diseases*. New York: Academic Press. pp. 249–69.

35. Koo WW, Walters JC, Esterlitz J, Levine RJ, Bush AJ, Sibai B. 1999. Maternal calcium supplementation and fetal bone mineralization. *Obstet Gynecol* 94(4): 577–82.

36. Hofmeyr GJ, Lawrie TA, Atallah AN, Duley L. 2010. Calcium supplementation during pregnancy for preventing hypertensive disorders and related problems. *Cochrane Database Syst Rev* (8): CD001059.

37. Kumar A, Devi SG, Batra S, Singh C, Shukla DK. 2009. Calcium supplementation for the prevention of preeclampsia. *Int J Gynaecol Obstet* 104(1): 32–6.

38. Hiller JE, Crowther CA, Moore VA, Willson K, Robinson JS. 2007. Calcium supplementation in pregnancy and its impact on blood pressure in children and women:

Follow up of a randomised controlled trial. *Aust N Z J Obstet Gynaecol* 47(2): 115–21.
39. Villar J, Abdel-Aleem H, Merialdi M, Mathai M, Ali MM, Zavaleta N, Purwar M, Hofmeyr J, Nguyen TN, Campodonico L, Landoulsi S, Carroli G, Lindheimer M. 2006. World Health Organization randomized trial of calcium supplementation among low calcium intake pregnant women. *Am J Obstet Gynecol* 194(3): 639–49.
40. VanHouten JN, Dann P, Stewart AF, Watson CJ, Pollak M, Karaplis AC, Wysolmerski JJ. 2003. Mammary-specific deletion of parathyroid hormone-related protein preserves bone mass during lactation. *J Clin Invest* 112(9): 1429–36.
41. Kovacs CS, Chik CL. 1995. Hyperprolactinemia caused by lactation and pituitary adenomas is associated with altered serum calcium, phosphate, parathyroid hormone (PTH), and PTH-related peptide levels. *J Clin Endocrinol Metab* 80: 3036–42.
42. Dobnig H, Kainer F, Stepan V, Winter R, Lipp R, Schaffer M, Kahr A, Nocnik S, Patterer G, Leb G. 1995. Elevated parathyroid hormone-related peptide levels after human gestation: Relationship to changes in bone and mineral metabolism. *J Clin Endocrinol Metab* 80: 3699–707.
43. Sowers MF, Hollis BW, Shapiro B, Randolph J, Janney CA, Zhang D, Schork A, Crutchfield M, Stanczyk F, Russell-Aulet M. 1996. Elevated parathyroid hormone-related peptide associated with lactation and bone density loss. *J Am Med Assoc* 276: 549–54.
44. Kolthoff N, Eiken P, Kristensen B, Nielsen SP. 1998. Bone mineral changes during pregnancy and lactation: A longitudinal cohort study. *Clin Sci (Colch)* 94(4): 405–12.
45. Polatti F, Capuzzo E, Viazzo F, Colleoni R, Klersy C. 1999. Bone mineral changes during and after lactation. *Obstet Gynecol* 94(1): 52–6.
46. Kalkwarf HJ, Specker BL, Bianchi DC, Ranz J, Ho M. 1997. The effect of calcium supplementation on bone density during lactation and after weaning. *N Engl J Med* 337(8): 523–8.
47. Cross NA, Hillman LS, Allen SH, Krause GF. 1995. Changes in bone mineral density and markers of bone remodeling during lactation and postweaning in women consuming high amounts of calcium. *J Bone Miner Res* 10: 1312–20.
48. Laskey MA, Prentice A, Hanratty LA, Jarjou LM, Dibba B, Beavan SR, Cole TJ. 1998. Bone changes after 3 mo of lactation: Influence of calcium intake, breast-milk output, and vitamin D-receptor genotype. *Am J Clin Nutr* 67(4): 685–92.
49. Reid IR, Wattie DJ, Evans MC, Budayr AA. 1992. Post-pregnancy osteoporosis associated with hypercalcaemia. *Clin Endocrinol (Oxf)* 37: 298–303.
50. Phillips AJ, Ostlere SJ, Smith R. 2000. Pregnancy-associated osteoporosis: Does the skeleton recover? *Osteoporos Int* 11(5): 449–54.
51. Caplan RH, Beguin EA. 1990. Hypercalcemia in a calcitriol-treated hypoparathyroid woman during lactation. *Obstet Gynecol* 76: 485–9.
52. Mather KJ, Chik CL, Corenblum B. 1999. Maintenance of serum calcium by parathyroid hormone-related peptide during lactation in a hypoparathyroid patient. *J Clin Endocrinol Metab* 84(2): 424–7.
53. Kent GN, Price RI, Gutteridge DH, Smith M, Allen JR, Bhagat CI, Barnes MP, Hickling CJ, Retallack RW, Wilson SG. 1990. Human lactation: Forearm trabecular bone loss, increased bone turnover, and renal conservation of calcium and inorganic phosphate with recovery of bone mass following weaning. *J Bone Miner Res* 5: 361–9.
54. Sowers M, Zhang D, Hollis BW, Shapiro B, Janney CA, Crutchfield M, Schork MA, Stanczyk F, Randolph J. 1998. Role of calciotrophic hormones in calcium mobilization of lactation. *Am J Clin Nutr* 67(2): 284–91.
55. Wagner CL, Hulsey TC, Fanning D, Ebeling M, Hollis BW. 2006. High-dose vitamin D3 supplementation in a cohort of breastfeeding mothers and their infants: A 6-month follow-up pilot study. *Breastfeed Med* 1(2): 59–70.
56. Prentice A. 2000. Calcium in pregnancy and lactation. *Annu Rev Nutr* 20: 249–72.
57. Prentice A, Jarjou LM, Cole TJ, Stirling DM, Dibba B, Fairweather-Tait S. 1995. Calcium requirements of lactating Gambian mothers: Effects of a calcium supplement on breast-milk calcium concentration, maternal bone mineral content, and urinary calcium excretion. *Am J Clin Nutr* 62: 58–67.
58. Prentice A, Jarjou LM, Stirling DM, Buffenstein R, Fairweather-Tait S. 1998. Biochemical markers of calcium and bone metabolism during 18 months of lactation in Gambian women accustomed to a low calcium intake and in those consuming a calcium supplement. *J Clin Endocrinol Metab* 83(4): 1059–66.
59. Prentice A, Yan L, Jarjou LM, Dibba B, Laskey MA, Stirling DM, Fairweather-Tait S. 1997. Vitamin D status does not influence the breast-milk calcium concentration of lactating mothers accustomed to a low calcium intake. *Acta Paediatr* 86(9): 1006–8.
60. VanHouten J, Dann P, McGeoch G, Brown EM, Krapcho K, Neville M, Wysolmerski JJ. 2004. The calcium-sensing receptor regulates mammary gland parathyroid hormone-related protein production and calcium transport. *J Clin Invest* 113(4): 598–608.
61. VanHouten JN, Neville MC, Wysolmerski JJ. 2007. The calcium-sensing receptor regulates plasma membrane calcium adenosine triphosphatase isoform 2 activity in mammary epithelial cells: A mechanism for calcium-regulated calcium transport into milk. *Endocrinology* 148(12): 5943–54.
62. Bezerra FF, Mendonca LM, Lobato EC, O'Brien KO, Donangelo CM. 2004. Bone mass is recovered from lactation to postweaning in adolescent mothers with low calcium intakes. *Am J Clin Nutr* 80(5): 1322–6.
63. Chantry CJ, Auinger P, Byrd RS. 2004. Lactation among adolescent mothers and subsequent bone mineral density. *Arch Pediatr Adolesc Med* 158(7): 650–6.
64. Kovacs CS. 2005. Calcium and bone metabolism during pregnancy and lactation. *J Mammary Gland Biol Neoplasia* 10(2): 105–18.

21
Menopause
Ian R. Reid

Introduction 165
Effects on Bone 165
Effects on Calcium Metabolism 166

Summary 167
References 167

INTRODUCTION

Menopause refers to the cessation of menstruation, which occurs at about 48–50 years of age in healthy women. The decline in ovarian hormone production is gradual and starts several years before a woman's last period. Changes in bone mass and calcium metabolism are evident during this perimenopausal transition. Estrogen is the ovarian product that has the greatest impact on mineral metabolism, though both progesterone and ovarian androgens may have some influence. Menopause ushers in a period of bone loss that extends until the end of life, and which is the central contributor to the development of osteoporotic fractures in older women.

EFFECTS ON BONE

Before the menopause, there is virtually no bone loss in most regions of the skeleton [1, 2], and fracture rates are stable. The most obvious effect of menopause on bone is an increase in the incidence of fractures; in the forearms and vertebrae this is clearly apparent within the first postmenopausal decade. It is attributable to the rapid decline in bone mass that occurs in the perimenopausal years. Bone loss is more marked in trabecular than in cortical bone because the former has a greater surface area over which bone resorption can take place (e.g., a loss of 1.4% per year at the hip compared with 1.6% per year at the spine, in a recent study [3]). Thus, the fractures that occur early in the menopause are in trabecular-rich regions of the skeleton such as the distal forearm and vertebrae. The loss of bone and increase in fracture rates are preventable with estrogen replacement.

The perimenopausal increase in bone loss is driven by increased bone resorption [4]. Bone biopsies in normal postmenopausal women show an increase in the proportion of bone surfaces at which resorption is taking place and an increase in the depth of resorption pits. These changes follow from an increase in the activation frequency of remodeling units and a prolongation of their resorptive phase. Indices of bone resorption are twice the levels found in premenopausal women whereas markers of bone formation are only about 50% above premenopausal levels [5], leading to negative bone balance. The resulting loss of bone leads to the perforation and loss of trabeculae [6], and endosteal resorption, cortical thinning, and increased porosity in cortical bone [7]. The changes in histomorphometric indices and biochemical markers can be returned to premenopausal levels with estrogen replacement therapy.

The changes in bone turnover that accompany menopause are in part accounted for by the direct actions of estrogen on bone cells. Estrogen receptors are present in both osteoblasts and osteoclasts. Estrogen promotes the development of osteoblasts in preference to adipocytes from their common precursor cell [8], increases osteoblast proliferation [9], and increases production of a

Primer on the Metabolic Bone Diseases and Disorders of Mineral Metabolism, Eighth Edition. Edited by Clifford J. Rosen.
© 2013 American Society for Bone and Mineral Research. Published 2013 by John Wiley & Sons, Inc.

number of osteoblast proteins [e.g., insulin-like growth factor-1, type I procollagen, transforming growth factor-β (TGFβ), and bone morphogenetic protein-6]. Thus, estrogen tends to have an anabolic effect on the isolated osteoblast, which is complemented by its inhibition of apoptosis in osteocytes [10] and osteoblasts [11]. *In vivo*, however, the initiation of estrogen replacement therapy is usually associated with a reduction in osteoblast numbers and activity [12]. This is accounted for by the tight coupling of osteoblast activity to that of osteoclasts, and the overriding effect of estrogen to reduce osteoclastic bone resorption. However, there is now evidence that high concentrations of estrogen increase some histomorphometric indices of osteoblast activity (e.g., mean wall thickness) in humans, possibly by increasing osteoblast synthesis of growth factors [13].

Estrogen's suppression of osteoclast activity is contributed to by increased osteoclast apoptosis [14], by reduced osteoblast/stromal cell production of receptor activator of nuclear facgtor-κB ligand (RANKL), and by increased production of osteoprotegerin [15]. These direct effects are buttressed by estrogen action on bone marrow stromal and mononuclear cells, which produce cytokines, such as interleukin-1 (IL-1), interleukin-6 (IL-6) and tumor necrosis factor-α (TNF-α), potent stimulators of osteoclast recruitment and/or activity [16]. Estrogen also regulates T-cell production of TNF-α, via changes in IL-7 [17] and interferon-γ [18]. Estrogen decreases production of each of these cytokines [19], and modulates levels of IL-1 receptors [20]. Bone loss following ovariectomy is reduced by blockers of these cytokines, and blockade of TNF-α and IL-1 has been shown to reduce bone resorption in postmenopausal women [21]. IL-1 and TNF-α may act in part by regulating stromal cell production of IL-6 and macrophage colony stimulating factor [22]. The CD40 ligand on the surface of T cells binds to CD40 on stromal cells and osteoblasts and is pivotal to estrogen actions on these cells and, through regulating their production of macrophage colony stimulating factor (M-CSF), RANKL, and osteoprotegerin, to estrogen effects on bone resorption [23]. These cytokine interactions have recently been reviewed in detail [24]. Estrogen also has antioxidant effects, decreasing reactive oxygen species in mouse bone [25], thereby reducing osteocyte apoptosis [26] and contributing to reduced bone resorption via increased levels of nitric oxide [27], and by decreasing T-cell proliferation and resultant TNF-α production [24]. Estrogen's effects on osteoclasts and on bone marrow cytokines may also be supported by its regulation of other systemic bone-active factors, such as growth hormone [28] and follicle-stimulating hormone (FSH) [29].

EFFECTS ON CALCIUM METABOLISM

The bone loss that follows menopause is accompanied by negative changes in external calcium balance. Decreases in intestinal calcium absorption and increases in urinary calcium loss contribute approximately equally to these negative changes [30]. Menopause is associated with reduced circulating concentrations of total, but not free, 1,25-dihydroxyvitamin D (1,25(OH)$_2$D), implying that its main effect is on vitamin D binding protein. However, intestinal mucosal cells contain estrogen receptors and respond directly to 17β-estradiol with enhanced calcium transport [31], probably through regulation of the epithelial calcium channel CaT1 [32], suggesting estrogen's effects are independent of vitamin D.

In the kidney, it is clear that tubular reabsorption of calcium is higher in the presence of estrogen [33, 34]. One study [34] found higher parathyroid hormone (PTH) concentrations in the presence of estrogen and inferred that this was the mechanism of the renal calcium conservation. However, higher PTH levels have not been the finding in a number of other studies. Thus, it is likely that estrogen directly modulates renal tubular calcium absorption via its own receptor in the kidney, as suggested by *in vitro* studies of renal tubule cells, which show a stimulatory effect of 17β-estradiol on calcium membrane transport [35].

The changes in the handling of calcium by the gut and kidney could each be a *cause* of postmenopausal bone loss, or they could represent homeostatic *responses* to it. If the former were the case, then PTH concentrations would be elevated in postmenopausal women to maintain plasma calcium concentrations in the face of intestinal and renal losses. This, in turn, would cause bone loss. If, on the other hand, bone loss were the primary event, then suppression of PTH would be expected, leading to secondary declines in intestinal and renal calcium absorption. The effect of menopause on PTH concentrations has been addressed many times without any consistent pattern emerging. This suggests that estrogen has direct effects on bone, kidney, and gut, and the opposing effects of these actions on PTH secretion leads to inconsistent changes in PTH concentrations. Further, estrogen itself may directly modulate PTH secretion [36, 37].

There are small but consistently demonstrable effects of menopause on circulating concentrations of calcium. Total calcium is 0.05 mmol/L higher after the menopause [33, 38]. This is partly attributable to a contraction of the plasma volume and resulting increase in albumin concentrations that occur in the absence of estrogen [39, 40], and partly to an increase in plasma bicarbonate, which leads to an increase in the complexed fraction of plasma calcium. The higher bicarbonate levels of postmenopausal women are attributable to a respiratory acidosis that results from the loss of the respiratory stimulatory effects of progesterone on the central nervous system, an action that is potentiated by estrogen [41, 42]. Despite changes in protein-bound and complexed calcium fractions, ionized calcium concentrations are usually found to be the same in pre- and postmenopausal women.

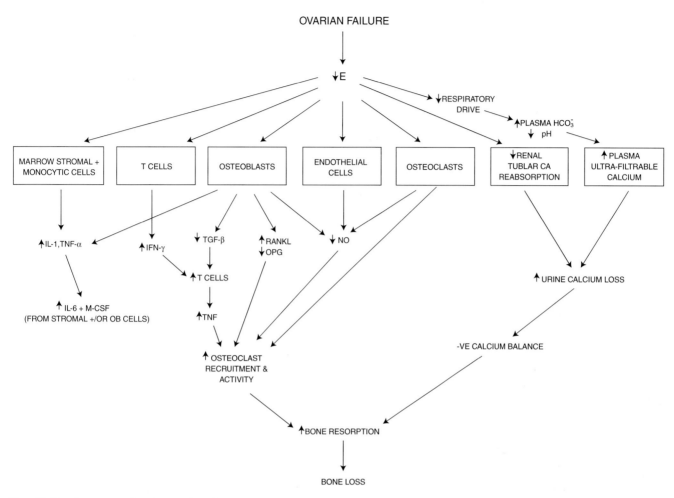

Fig. 21.1. The potential pathways by which menopause leads to bone loss. For simplicity, the figure does not show a contribution from loss of any anabolic effect of estrogen on the osteoblast. The fall in ovarian production of androgens and progesterone also contributes to some of these changes.

SUMMARY

The effects of menopause on skeletal physiology are summarized in Fig. 21.1. The major effect is an increase in bone turnover, which is dominantly an increase in bone resorption. This results in bone loss that may be contributed to by reductions in both intestinal and renal tubular absorption of calcium. Bone loss persists throughout the entire postmenopausal period and results in a high risk of fractures in those women whose peak bone mass was in the lower part of the normal range.

REFERENCES

1. Chapurlat RD, Garnero P, Sornay-Rendu E, Arlot ME, Claustrat B, Delmas PD. 2000. Longitudinal study of bone loss in pre- and perimenopausal women: Evidence for bone loss in perimenopausal women. *Osteoporos Int* 11(6): 493–8.
2. Sowers MR, Jannausch M, McConnell D, Little R, Greendale GA, Finkelstein JS, Neer RM, Johnston J, Ettinger B. 2006. Hormone predictors of bone mineral density changes during the menopausal transition. *J Clin Endocrinol Metab* 91(4): 1261–7.
3. Macdonald HM, New SA, Campbell MK, Reid DM. 2005. Influence of weight and weight change on bone loss in perimenopausal and early postmenopausal Scottish women. *Osteoporos Int* 16(2): 163–71.
4. Heaney RP, Recker RR, Saville PD. 1978. Menopausal changes in bone remodeling. *J Lab Clin Med* 92: 964–70.
5. Garnero P, Sornayrendu E, Chapuy MC, Delmas PD. 1996. Increased bone turnover in late postmenopausal women is a major determinant of osteoporosis. *J Bone Miner Res* 11(3): 337–49.
6. Akhter MP, Lappe JM, Davies KM, Recker RR. 2007. Transmenopausal changes in the trabecular bone structure. *Bone* 41: 111–6.

7. Cooper DM, Thomas CD, Clement JG, Turinsky AL, Sensen CW, Hallgrímsson B. 2007. Age-dependent change in the 3D structure of cortical porosity at the human femoral midshaft. *Bone* 40(4): 957–65.
8. Okazaki R, Inoue D, Shibata M, Saika M, Kido S, Ooka H, Tomiyama H, Sakamoto Y, Matsumoto T. 2002. Estrogen promotes early osteoblast differentiation and inhibits adipocyte differentiation in mouse bone marrow stromal cell lines that express estrogen receptor (ER) alpha or beta. *Endocrinology* 143(6): 2349–56.
9. Fujita M, Urano T, Horie K, Ikeda K, Tsukui T, Fukuoka H, Tsutsumi O, Ouchi Y, Inoue S. 2002. Estrogen activates cyclin-dependent kinases 4 and 6 through induction of cyclin D in rat primary osteoblasts. *Biochem Biophys Res Commun* 299(2): 222–8.
10. Tomkinson A, Reeve J, Shaw RW, Noble BS. 1997. The death of osteocytes via apoptosis accompanies estrogen withdrawal in human bone. *J Clin Endocrinol Metab* 82(9): 3128–35.
11. Gohel A, McCarthy MB, Gronowicz G. 1999. Estrogen prevents glucocorticoid-induced apoptosis in osteoblasts in vivo and in vitro. *Endocrinology* 140(11): 5339–47.
12. Vedi S, Compston JE. 1996. The effects of long-term hormone replacement therapy on bone remodeling in postmenopausal women. *Bone* 19(5): 535–9.
13. Bord S, Beavan S, Ireland D, Horner A, Compston JE. 2001. Mechanisms by which high-dose estrogen therapy produces anabolic skeletal effects in postmenopausal women: Role of locally produced growth factors. *Bone* 29(3): 216–22.
14. Kameda T, Mano H, Yuasa T, Mori Y, Miyazawa K, Shiokawa M, Nakamaru Y, Hiroi E, Hiura K, Kameda A, Yang NN, Hakeda Y, Kumegawa M. 1997. Estrogen inhibits bone resorption by directly inducing apoptosis of the bone-resorbing osteoclasts. *J Exp Med* 186(4): 489–95.
15. Syed F, Khosla S. 2005. Mechanisms of sex steroid effects on bone. *Biochem Biophys Res Comm* 328(3): 688–96.
16. Manolagas SC, Jilka RL. 1995. Mechanisms of disease: Bone marrow, cytokines, and bone remodeling. Emerging insights into the pathophysiology of osteoporosis. *N Engl J Med* 332(5): 305–11.
17. Ryan MR, Shepherd R, Leavey JK, Gao YH, Grassi F, Schnell FJ, Qian WP, Kersh GJ, Weitzmann MN, Pacifici R. 2005. An IL-7-dependent rebound in thymic T cell output contributes to the bone loss induced by estrogen deficiency. *Proc Natl Acad Sci U S A* 102(46): 16735–40.
18. Cenci S, Toraldo G, Weitzmann MN, Roggia C, Gao YH, Qian WP, Sierra O, Pacifici R. 2003. Estrogen deficiency induces bone loss by increasing T cell proliferation and lifespan through IFN-gamma-induced class II transactivator. *Proc Natl Acad Sci U S A* 100(18): 10405–10.
19. Rogers A, Eastell R. 1998. Effects of estrogen therapy of postmenopausal women on cytokines measured in peripheral blood. *J Bone Miner Res* 13(10): 1577–86.
20. Sunyer T, Lewis J, Collin-Osdoby P, Osdoby P. 1999. Estrogen's bone-protective effects may involve differential IL-1 receptor regulation in human osteoclast-like cells. *J Clin Invest* 103(10): 1409–18.
21. Charatcharoenwitthaya N, Khosla S, Atkinson EJ, McCready LK, Riggs BL. 2007. Effect of blockade of TNF-alpha and interleukin-1 action on bone resorption in early postmenopausal women. *J Bone Miner Res* 22(5): 724–9.
22. Kimble RB, Srivastava S, Ross FP, Matayoshi A, Pacifici R. 1996. Estrogen deficiency increases the ability of stromal cells to support murine osteoclastogenesis via an interleukin-1-and tumor necrosis factor-mediated stimulation of macrophage colony-stimulating factor production. *J Biol Chem* 271(46): 28890–7.
23. Li JY, Tawfeek H, Bedi B, Yang XY, Adams J, Gao KY, Zayzafoon M, Weitzmann MN, Pacifici R. 2011. Ovariectomy disregulates osteoblast and osteoclast formation through the T-cell receptor CD40 ligand. *Proc Natl Acad Sci U S A* 108(2): 768–73.
24. Pacifici R. 2010. T cells: Critical bone regulators in health and disease. *Bone* 47(3): 461–71.
25. Almeida M, Han L, Martin-Millan M, Plotkin LI, Stewart SA, Roberson PK, Kousteni S, O'Brien CA, Bellido T, Parfitt AM, Weinstein RS, Jilka RL, Manolagas SC. 2007. Skeletal involution by age-associated oxidative stress and its acceleration by loss of sex steroids. *J Biol Chem* 282: 27285–97.
26. Mann V, Huber C, Kogianni G, Collins F, Noble B. 2007. The antioxidant effect of estrogen and Selective Estrogen Receptor Modulators in the inhibition of osteocyte apoptosis in vitro. *Bone* 40(3): 674–84.
27. Ralston SH. 1997. The Michael-Mason-Prize essay 1997. Nitric oxide and bone: What a gas! *Br J Rheumatol* 36(8): 831–8.
28. Friend KE, Hartman ML, Pezzoli SS, Clasey JL, Thorner MO. 1996. Both oral and transdermal estrogen increase growth hormone release in postmenopausal women—A clinical research center study. *J Clin Endocrinol Metab* 81(6): 2250–6.
29. Sun L, Peng Y, Sharrow AC, Iqbal J, Zhang Z, Papachristou DJ, Zaidi S, Zhu LL, Yaroslavskiy BB, Zhou H, Zallone A, Sairam MR, Kumar TR, Bo W, Braun J, Cardoso-Landa L, Schaffler MB, Moonga BS, Blair HC, Zaidi M. 2006. FSH directly regulates bone mass. *Cell* 125(2): 247–60.
30. Heaney RP, Recker RR, Saville PD. 1978. Menopausal changes in calcium balance performance. *J Lab Clin Med* 92: 953–63.
31. Arjandi BH, Salih MA, Herbert DC, Sims SH, Kalu DN. 1993. Evidence for estrogen receptor-linked calcium transport in the intestine. *Bone Miner* 21(1): 63–74.
32. Van Cromphaut SJ, Rummens K, Stockmans I, Van Herck E, Dijcks FA, Ederveen A, Carmeliet P, Verhaeghe J, Bouillon R, Carmeliet G. 2003. Intestinal calcium transporter genes are upregulated by estrogens and the reproductive cycle through vitamin D receptor-independent mechanisms. *J Bone Miner Res* 18(10): 1725–36.
33. Nordin BEC, Wlshart JM, Clifton PM, McArthur R, Scopacasa F, Need AG, Morris HA, O'Loughlin PD, Horow-

itz M. 2004. A longitudinal study of bone-related biochemical changes at the menopause. *Clin Endocrinol* 61(1): 123–30.
34. McKane WR, Khosla S, Burritt MF, Kao PC, Wilson DM, Ory SJ, Riggs BL. 1995. Mechanism of renal calcium conservation with estrogen replacement therapy in women in early postmenopause—A clinical research center study. *J Clin Endocrinol Metab* 80(12): 3458–64.
35. Dick IM, Liu J, Glendenning P, Prince RL. 2003. Estrogen and androgen regulation of plasma membrane calcium pump activity in immortalized distal tubule kidney cells. *Mol Cell Endocrinol* 212(1–2): 11–8.
36. Duarte B, Hargis GK, Kukreja SC. 1988. Effects of estradiol and progesterone on parathyroid hormone secretion from human parathyroid tissue. *J Clin Endocrinol Metab* 66(3): 584–7.
37. Greenberg C, Kukreja SC, Bowser EN, Hargis GK, Henderson WJ, Williams GA. 1987. Parathyroid hormone secretion: Effect of estradiol and progesterone. *Metabolism* 36(2): 151–4.
38. Sokoll LJ, Dawson-Hughes B. 1989. Effect of menopause and aging on serum total and ionized calcium and protein concentrations. *Calcif Tissue Int* 44: 181–5.
39. Aitken JM, Lindsay R, Hart DM. 1974. The redistribution of body sodium in women on long-term estrogen therapy. *Clin Sci Mol Med* 47: 179–87.
40. Minkoff JR, Young G, Grant B, Marcus R. 1986. Interactions of medroxyprogesterone acetate with estrogen on the calcium-parathyroid axis in post-menopausal women. *Maturitas* 8: 35–45.
41. Bayliss DA, Millhorn DE. 1992. Central neural mechanisms of progesterone action: Application to the respiratory system. *J Appl Physiol* 73: 393–404.
42. Orr-Walker BJ, Horne AM, Evans MC, Grey AB, Murray MAF, McNeil AR, Reid IR. 1999. Hormone replacement therapy causes a respiratory alkalosis in normal postmenopausal women. *J Clin Endocrinol Metab* 84(6): 1997–2001.

Section III
Mineral Homeostasis

Section Editor Ego Seeman

Chapter 22. Regulation of Calcium and Magnesium   173
Murray J. Favus and David Goltzman

Chapter 23. Fetal Calcium Metabolism   180
Christopher S. Kovacs

Chapter 24. Fibroblast Growth Factor-23 (FGF23)   188
Kenneth E. White and Michael J. Econs

Chapter 25. Gonadal Steroids   195
Stavros C. Manolagas, Maria Almeida, and Robert L. Jilka

Chapter 26. Parathyroid Hormone   208
Robert A. Nissenson and Harald Jüppner

Chapter 27. Parathyroid Hormone-Related Protein   215
John J. Wysolmerski

Chapter 28. Ca^{2+}-Sensing Receptor   224
Edward M. Brown

Chapter 29. Vitamin D: Production, Metabolism, Mechanism of Action, and Clinical Requirements   235
Daniel Bikle, John S. Adams, and Sylvia Christakos

Primer on the Metabolic Bone Diseases and Disorders of Mineral Metabolism, Eighth Edition. Edited by Clifford J. Rosen.
© 2013 American Society for Bone and Mineral Research. Published 2013 by John Wiley & Sons, Inc.

22

Regulation of Calcium and Magnesium

Murray J. Favus and David Goltzman

Calcium (Ca) 173	References 177
Magnesium 177	

CALCIUM (CA)

Distribution

Total body distribution

In adults, the body contains about 1,000 g of calcium, of which 99% is located in the mineral phase of bone as the hydroxyapatite crystal [$Ca_{10}(PO4)_6(OH)_2$]. The crystal plays a key role in the mechanical weight-bearing properties of bone and serves as a ready source of calcium to support a number of calcium-dependent biological systems and to maintain blood ionized calcium within the normal range. The remaining 1% of total body calcium is located in the blood, extracellular fluid, and soft tissues. In serum, total calcium is 10^{-3} M and is the most frequent measurement of serum calcium levels. Of the total calcium, the ionized fraction (50%) is the biologically functional portion of total calcium and can be measured clinically; 40% of the total is bound to albumin in a pH-dependent manner; and the remaining 10% exists as a complex of either citrate or PO_4 ions.

Cell levels

Cytosol calcium is about 10^{-6} M, which creates a 1,000-fold gradient across the plasma membrane [extracellular fluid (ECF) calcium is 10^{-3} M] that favors calcium entry into the cell. There is also an electrical charge across the plasma membrane of about 50 mV with the cell interior negative. Thus, the chemical and electrical gradients across the plasma membrane favor calcium entry, which the cell must defend against to preserve cell viability. Calcium-induced cell death is largely prevented by several mechanisms including extrusion of calcium from the cell by adenosine triphosphate (ATP)-dependent energy driven calcium pumps and calcium channels; Na–Ca exchangers; and the binding of intracellular calcium by proteins located in the cytosol, endoplasmic reticulum (ER), and mitochondria. Calcium binding to ER and mitochondrial sites buffer intracellular calcium and can be mobilized to maintain cytosol calcium levels and to create pulsatile peaks of calcium to mediate membrane receptor signaling that regulate a variety of biologic systems.

Blood levels

Calcium in the blood is normally transported partly bound to plasma proteins (about 45%), notably albumin; partly bound to small anions such as phosphate and citrate (about 10%); and partly in the free or ionized state (about 45%) [1]. Although only the ionized calcium is available to move into cells and activate cellular processes, most clinical laboratories report total serum calcium concentrations. Concentrations of total calcium in normal serum generally range between 8.5 and 10.5 mg/dL (2.12 to 2.62 mM); levels above this are considered to be hypercalcemic. The normal range of ionized calcium is 4.65–5.25 mg/dL (1.16–1.31 mM). When protein concentrations, and especially albumin concentrations, fluctuate, total calcium levels may vary, whereas the ionized calcium may remain relatively stable. Dehydration or hemoconcentration during venipuncture may elevate serum albumin and falsely elevate total serum calcium. Such elevations in total calcium, when albumin levels

---

Primer on the Metabolic Bone Diseases and Disorders of Mineral Metabolism, Eighth Edition. Edited by Clifford J. Rosen.
© 2013 American Society for Bone and Mineral Research. Published 2013 by John Wiley & Sons, Inc.

are increased, can be "corrected" by subtracting 0.8 mg/dL from the total calcium for every 1.0 g/dL by which the serum albumin concentration is greater than 4 g/dL. Conversely, when albumin levels are low, total calcium can be corrected by adding 0.8 mg/dL for every 1.0 g/dL by which the albumin is less than 4 g/dL. Even in the presence of a normal serum albumin, changes in blood pH can alter the equilibrium constant of the albumin–Ca^{2+} complex, with acidosis reducing the binding and alkalosis enhancing it. Consequently, major shifts in serum protein or pH requires direct measurement of the ionized calcium level to determine the physiologic serum calcium level.

Mineral homeostasis

The ECF concentration of calcium is tightly maintained within a rather narrow range because of the importance of the calcium ion to numerous cellular functions including cell division, cell adhesion and plasma membrane integrity, protein secretion, muscle contraction, neuronal excitability, glycogen metabolism, and coagulation.

The skeleton, the gut, and the kidney each play a major role in assuring calcium homeostasis. Overall, in a typical individual, if 1,000 mg of calcium are ingested in the diet per day, approximately 200 mg will be absorbed. Approximately 10 g of calcium will be filtered daily through the kidney and most will be reabsorbed, with about 200 mg being excreted in the urine. The normal 24-hour excretion of calcium may, however, vary between 100 and 300 mg per day (2.5 to 7.5 mmoles per day). The skeleton, a storage site of about 1 kg of calcium, is the major calcium reservoir in the body. Ordinarily, as a result of normal bone turnover, approximately 500 mg of calcium is released from bone per day and the equivalent amount is accreted per day (Fig. 22.1).

Tight regulation of the ECF calcium concentration is maintained through the action of calcium-sensitive cells that modulate the production of hormones [2–5]. These hormones act on specific cells in bone, gut, and kidney, which can respond by altering fluxes of calcium to maintain ECF calcium. Thus a reduction in ECF calcium stimulates release of parathyroid hormone (PTH) from the parathyroid glands in the neck. This hormone can then act to enhance bone resorption and liberate both calcium and phosphate from the skeleton. PTH can also enhance calcium reabsorption in the kidney while at the same time inhibit phosphate reabsorption producing phosphaturia. Hypocalcemia and PTH itself can both stimulate the conversion of the inert metabolite of vitamin D, 25-hydroxyvitamin D_3 [25(OH)D_3], to the active moiety 1,25-dihydroxyvitamin D_3 [1,25(OH)$_2D_3$] [6], which in turn will enhance intestinal calcium absorption, and to a lesser extent renal phosphate reabsorption. The net effect of the mobilization of calcium from bone, the increased absorption of calcium from the gut, and the increased reabsorption of filtered calcium along the nephron is to restore the ECF calcium to normal and to

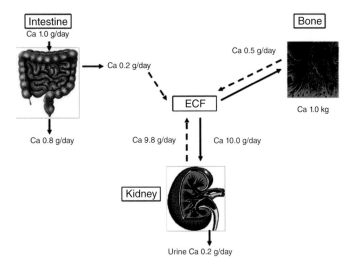

Fig. 22.1. Calcium balance. On average, in a typical adult approximately 1 g of elemental calcium (Ca^{+2}) is ingested per day. Of this, about 200 mg/day will be absorbed and 800 mg/day excreted. Approximately 1 kg of calcium is stored in bone, and about 500 mg/day is released by resorption or deposited during bone formation. Of the 10 g of calcium filtered through the kidney per day, only about 200 mg or less appears in the urine, the remainder being reabsorbed.

inhibit further production of PTH and 1,25(OH)$_2D_3$. The opposite sequence of events, i.e., diminished PTH and 1,25(OH)$_2D_3$ secretion, along with stimulation of renal calcium-sensing receptor (CaSR) occurs when the ECF calcium is raised above the normal range. The effect of suppressing the release of PTH and 1,25(OH)$_2D_3$ and stimulating CaSR diminishes skeletal calcium release, decreases intestinal calcium absorption and renal calcium reabsorption, and restores the elevated ECF calcium to normal.

PTH and 1,25(OH)$_2D_3$ actions on target tissues

Intestinal calcium transport

Net intestinal calcium absorption can be determined by the external balance technique in which a diet of known composition with a known amount of calcium is ingested, and urine calcium excretion and fecal calcium loss are measured. Negative absorption occurs when net absorption declines to about 200 mg calcium per day (5.0 mmol). The portion of dietary calcium absorbed varies with age and amount of calcium ingested and may vary from 20% to 60%. Rates of net calcium absorption are high in growing children, during growth spurts in adolescence, and during pregnancy and lactation. The efficiency of calcium absorption increases during prolonged dietary calcium restriction to absorb the greatest portion of that ingested. Net absorption declines with age in men and women, and so increased calcium intake is required to compensate for the lower absorption rate. Fecal calcium losses vary between 100 and 200 mg per day (2.5 to 5.0 mmol). Fecal calcium is composed of unabsorbed dietary calcium and calcium contained in intestinal, pancreatic, and biliary secretions. Secreted calcium is not regulated by hormones or serum calcium.

About 90% of absorbed calcium occurs in the large surface area of the duodenum and jejunum. Increased calcium requirements stimulate expression of the epithelial calcium active transport system in the duodenum, ileum, and throughout the colon sufficient to increase fractional calcium absorption from 20% to 45% in older men and women to 55% to 70% in children and young adults. $1,25(OH)_2D_3$ increases the efficiency of the small intestine and colon to absorb dietary calcium. Active calcium absorption accounts for absorption of 10–15% of a dietary load [7]. Active transcellular intestinal absorption involves three sequential cellular steps; a rate-limiting step involving transfer of luminal calcium into the intestinal cell via the epithelial calcium channel; expression of TRPV6, a channel-associated protein; annexin2 calbindin-D9K; and to a lesser extent, the basolateral extrusion system PMCA1b [8, 9]. Reductions in dietary calcium intake can increase PTH secretion and $1,25(OH)_2D_3$ production. Increased $1,25(OH)_2D_3$ can then increase expression of TRPV6, resulting in enhanced fractional calcium absorption and compensation for the dietary reduction [10].

Intestinal epithelial calcium transport includes both an energy-dependent, cell-mediated saturable active process that is largely regulated by $1,25(OH)_2D_3$, and a passive, diffusional paracellular path of absorption that is driven by transepithelial electrochemical gradients. The cell-mediated pathway involving the TRPV6 calcium channel is saturable with a Kt (1/2 maximal transport) of 1.0 mM. Passive diffusion increases linearly with luminal calcium concentration. In adults fed a diet low in calcium, enhanced $1,25(OH)_2D_3$ production increases the efficiency of absorption through an increase in saturable calcium transport. During high dietary calcium intake absorption $1,25(OH)_2D_3$ is suppressed, and passive paracellular transport accounts for most all absorption. Causes of increased and decreased intestinal calcium absorption are listed in Table 22.1.

Renal calcium handling

The kidney plays a central role in ensuring calcium balance, and PTH has a major role in fine-tuning this renal function [11–13] by stimulating both renal calcium reabsorption (proximal tubule) and excretion (distal nephron). Multiple influences on calcium handling are listed in Table 22.2 (Chapters 26, 28, and 73 contain descriptions of the molecular actions of PTH on the kidney.) PTH has little effect on modulating calcium fluxes in the proximal tubule where 65% of the filtered calcium is reabsorbed, coupled to the bulk transport of solutes such as sodium and water [12]. In this nephron region, PTH can also stimulate the $25(OH)D_3$-1α hydrox-

Table 22.1. Conditions That Increase or Decrease Intestinal Calcium Absorption

Increased Calcium Absorption	Decreased Calcium Absorption
Increased renal $1,25(OH)_2D_3$ production	Decreased renal $1,25(OH)_2D_3$ production
Growth	Vitamin D deficiency
Pregnancy	Chronic renal insufficiency
Lactation	Hypoparathyroidism
Primary hyperparathyroidism	Aging
Vitamin D-dependent rickets type 1	
Idiopathic hypercalciuria	
Increased extra-renal $1,25(OH)_2D_3$ production	Normal $1,25(OH)_2D_3$ production
Sarcoid and other granulomatous diseases	Glucocorticoid excess
B-cell lymphoma	

ylase [$1\alpha(OH)$ase], leading to increased synthesis of $1,25(OH)_2D_3$ [14]. A reduction in ECF calcium can itself stimulate $1,25(OH)_2D_3$ production, but whether this occurs via the CaSR is presently unknown. Finally, PTH can also inhibit Na and HCO_3^- reabsorption in the proximal tubule by inhibiting the apical type 3 Na^+/H^+ exchanger [15], and the basolateral Na^+/K^+-ATPase [16] by inhibiting apical Na+/Pi– cotransport.

About 20% of filtered calcium is reabsorbed in the cortical thick ascending limb of the loop of Henle (CTAL), and 15% is reabsorbed in the distal convoluted tubule (DCT). At both sites PTH binds to the PTH receptor (PTHR) [17, 18] and enhances calcium reabsorption. In the CTAL, at least, this appears to occur by increasing the activity of the Na/K/2Cl cotransporter that drives NaCl reabsorption and stimulates paracellular calcium and magnesium reabsorption [19]. The CaSR is also resident in the CTAL [20], where increased ECF calcium activates phospholipase A2, thereby reducing the activity of the Na/K/2Cl cotransporter and of an apical K channel, and diminishing paracellular calcium reabsorption. Consequently, a raised ECF calcium antagonizes the effect of PTH in this nephron segment, and ECF calcium can, in fact, participate in this way in the regulation of its own homeostasis. Inhibition of NaCl reabsorption and loss of NaCl in the urine may contribute to the volume depletion observed in severe hypercalcemia. ECF calcium may therefore act in a manner analogous to "loop" diuretics such as furosemide.

In the DCT, PTH can influence [8] luminal calcium transfer into the renal tubule cell via the transient receptor potential channel (TRPV5). It can also influence translocation of calcium across the cell from apical to

Table 22.2. Hormones and Conditions That Regulate Urine Ca and Mg Excretion

Hormones/Conditions	Calcium	Magnesium
Glomerular filtration		
Hypercalcemia	I	D
Hypocalcemia	D	I
Hypermagnesemia	-	I
Hypomagnesmia	D	D
Renal insufficiency	D	D
Tubular reabsorption		
Increased		
ECF volume contraction	I	I
Hypocalcemia	I	I
Thiazide diuretics	I	-
Phosphate administration	I	I
Metabolic alkalosis	I	I
Parathyroid hormone	I	I
Parathyroid hormone related peptide	I	I
Familial hypocalciuric hypercalcemia	I	-
Decreased		
ECF volume expansion	D	D
Hypercalcemia	D	D
Phosphate deprivation	D	D
Metabolic acidosis	D	-
Loop diuretics	D	D
Cyclosporin A	D	D
Autosomal dominant hypocalcemia	D	-
Dent's disease	D	-
Bartter's syndrome	D	-
Gittelman's syndrome	-	D

D = decreased GFR or tubule reabsorption
I = increased GFR or tubule reabsorption
"-" indicates either modest effects are present or that no specific information is available

basolateral surface involving proteins such as calbindin-D28K, and active extrusion of calcium from the cell into the blood via a Na+/Ca exchanger, designated NCX1. PTH markedly stimulates calcium reabsorption in the DCT primarily by augmenting NCX1 activity via a cyclic AMP-mediated mechanism.

Bone resorption and calcium release

In bone, the PTHR is localized on cells of the osteoblast phenotype, which are of mesenchymal origin [21] but not on osteoclasts, which are of hematogenous origin. A major physiologic role of PTH appears to be to maintain normal calcium homeostasis by enhancing osteoclastic bone resorption and liberating calcium into the ECF. Bone formation and resorption are discussed in detail in Section I and Chapters 25–29.

Mediators of bone remodeling

Calciotropic hormones PTH, PTHrP, and $1,25(OH)_2D_3$ initiate osteoclastic bone resorption and increase the activation frequency of bone remodeling. Physiologic control of bone turnover can be disrupted by an excess of each of these calciotropic hormones, resulting in altered ECF calcium homeostasis and hypercalcemia. The molecular basis for physiologic and pathologic states of bone turnover is detailed in Chapters 2–4, 25, 26, 29, and in Section VII.

Regulation of hormone production and actions on calcium homeostasis

PTH production

A major regulator of parathyroid gland secretion of PTH is ECF calcium. The relationship between ECF calcium and PTH secretion is governed by a steep inverse sigmoidal curve that is characterized by a maximal secretory rate at low ECF calcium, a midpoint or "set point," which is the level of ECF calcium that half-maximally suppresses PTH, and a minimal secretory rate at high ECF calcium [22, 23]. The parathyroid glands detect ECF calcium via a CaSR [24]. Sustained hypocalcemia can eventually lead to parathyroid cell proliferation [25] and an increased total secretory capacity of the parathyroid gland. $1,25(OH)_2D_3$ reduces PTH synthesis and parathyroid cell proliferation [26]. Molecular events in PTH secretion and CaSR function are found in Chapters 26 and 28.

Vitamin D production and actions

The renal production of $1,25(OH)_2D_3$ is stimulated by hypocalcemia, hypophosphatemia, and elevated PTH levels. The renal $1\alpha(OH)$ase is also potently inhibited by $1,25(OH)_2D_3$ as part of a negative feedback loop. The molecular details of the vitamin D metabolic pathway are described in Chapter 29.

Vitamin D is essential for normal mineralization of bone that may be due to an indirect effect by enhancing intestinal calcium and phosphate absorption and maintaining these ions within a range that facilitates hydroxyapatite deposition in bone matrix. A major indirect function of $1,25(OH)_2D_3$ on bone appears to be to enhance mobilization of calcium stores when dietary calcium is insufficient to maintain a normal ECF calcium [27]. As with PTH [28], $1,25(OH)_2D_3$ enhances osteoclastic bone resorption by binding to receptors in the preosteoblastic stromal cell and stimulating the RANK/RANK system to enhance the proliferation, differentiation, and activation of the osteoclastic system from its monocytic precursors [29]. Endogenous and exogenous $1,25(OH)_2D_3$ have also been reported to have an anabolic role *in vivo* [30,31]. $1,25(OH)_2D_3$ has a direct effect on renal calcium handling through stimulation of CaSR. It remains controversial whether $1,25(OH)_2D_3$ plays a direct role in enhancing tubular calcium reabsorption.

MAGNESIUM

Total body distribution

There is about 1.04 mol (25 g) of magnesium in the adult, of which about 66% is within the skeleton, 33% is intracellular, and 1% is in the ECF, including blood [1, 2]. Magnesium content of the hydroxyapatite crystal in bone varies widely and is mainly on the surface of bone where a portion is in equilibrium with ECF magnesium. Magnesium is the most abundant divalent cation within cells where it is found at a concentration of about 5×10^{-4} M in the cytosol. In the cells it serves as a cofactor and regulates a number of essential biological systems [1]. The concentration of magnesium in the ECF approaches that of the intracellular environment. Both intracellular and ECF are tightly regulated by factors that are poorly understood.

Cellular content

Ionic cytosolic magnesium accounts for 5% to 10% of total cellular magnesium. Cytosol magnesium is regulated by binding to intracellular organelles, of which 60% is within mitochondria where it participates in phosphate transport and ATP utilization. Control of intracellular magnesium is poorly understood.

Homeostasis

Of the total serum magnesium, 70% is either ionic or complexed, and the remaining 30% is protein bound [1, 2]. Blood levels are not as tightly regulated as calcium but fluctuate with influx and efflux across the ECF with changes in intestinal magnesium absorption, net renal magnesium reabsorption, and influx and efflux across bone. Blood ionic magnesium regulates PTH secretion but the potency is less than that of calcium.

Intestinal absorption

Magnesium is a requirement for bone health; however, unlike calcium, magnesium is found in all food groups and is especially rich in foods of cellular origin. Therefore, magnesium deficiency due to inadequate intake does not occur in the absence of severe defects in intestinal or renal function. Net intestinal magnesium absorption is in direct proportion to dietary magnesium ingestion. Under conditions of stable magnesium intake, external magnesium balance studies show that when magnesium intake is greater than 28 mg (2 mmol), magnesium absorption exceeds magnesium secretion, and magnesium balance becomes positive. The efficiency of magnesium absorption is 35% to 40% over the range of usual intakes (168 to 720 mg/day; or 7 to 30 mmol). Net magnesium absorption also varies with dietary constituents such as phosphate, which forms insoluble complexes with magnesium and thereby reduces magnesium absorption. In contrast to its actions on calcium and phosphorus absorption, $1,25(OH)_2D_3$ does not stimulate magnesium absorption.

There is no correlation between serum $1,25(OH)_2D_3$ levels and net magnesium absorption [32].

Absorptive and secretory magnesium fluxes across both small intestine and colon are largely voltage dependent, indicating the presence of a large paracellular pathway of magnesium transport that is driven by luminal magnesium concentrations. The magnesium ion channel TRPM6 has been identified in the apical membrane of intestinal brush border epithelial cells that appears to play an important role in magnesium homeostasis [32]. Whether TRPM6 is regulated by PTH or $1,25(OH)2D3$ has yet to be determined.

Renal handling

Ultrafilterable magnesium is 70% of the total serum magnesium (ionized plus complexed). Based on the urine magnesium excretion (about 24 mmol per 24 hours), about 95% of the filtered load of magnesium undergoes tubular reabsorption before the final urine is formed. A small fraction of reabsorbed magnesium (15%) occurs along the proximal tubule, whereas about 70% of filtered magnesium is reabsorbed along the cortical TALH [19, 33]. The magnesium ion may also stimulate basolateral membrane CaR, which decreases renal magnesium reabsorption. DCT magnesium reabsorption is via a transcellular transport process and accounts for about 10% of magnesium reabsorption.

Magnesium reabsorption is highly regulated, with a number of factors that may increase or decrease renal tubule magnesium reabsorption (Table 22.2). As there is little distal tubule magnesium reabsorption, ECF volume expansion decreases magnesium reabsorption and increases urine magnesium excretion. Hypermagnesemia increases urine magnesium excretion, at least in part through an activation of CaR [20]. In contrast, hypomagnesemia increases TALH magnesium reabsorption and decreases urine magnesium excretion. Loop diuretics increase urine magnesium excretion, and thiazide diuretic agents have a minimal effect of magnesium transport (Table 22.2). The magnesium ion channel TRPM6 is found in the apical membrane of the renal distal convoluted tubule and may be involved in magnesium homeostasis in both the kidney and intestine.

REFERENCES

1. Walser M. 1961. Ion association: VI. Interactions between calcium, magnesium, inorganic phosphate, citrate, and protein in normal human plasma. *J Clin Invest* 40: 723–30.
2. Parfitt AM, Kleerekoper M. 1980. Clinical disorders of calcium, phosphorus and magnesium metabolism. In: Maxwell MH, Kleeman CR (eds.) *Clinical Disorders of Fluid and Electrolyte Metabolism, 3rd Ed.* New York: McGraw-Hill. pp. 947.
3. Stewart AF, Broadus AE. 1987. Mineral metabolism. In: Felig P, Baxter JD, Broadus AE, Frohman LA (eds.)

Endocrinology and Metabolism, 2nd Ed. New York: McGraw-Hill. p. 1317.
4. Bringhurst FR, Demay MB, Kronenberg HM. 1998. Hormones and disorders of mineral metabolism. In: Wilson JD, Foster DW, Kronenberg HM, Larsen PR (eds.) Williams Textbook of Endocrinology, 9th Ed. Philadelphia: Saunders. p. 1155.
5. Brown EM. 2001. Physiology of calcium homeostasis. In: Bilezikian JP, Marcus R, Levine MA (eds.) The Parathyroids: Basic and Clinical Concepts, 2nd Ed. San Diego: Academic Press. p. 167.
6. Fraser DR, Kodicek E. 1973. Regulation of 25-hydroxycholecalciferol-1-hydroxylase activity in kidney by parathyroid hormone. Nat New Biol 241: 163–6.
7. Favus MF. 1992. Intestinal absorption of calcium, magnesium and phosphorus. In: Coe FL, Favus MJ (eds.) Disorders Of Bone and Mineral Metabolism. New York: Raven. p. 57.
8. Hoenderop JGJ, Nilius B, Bindels RJM. 2005. Calcium absorption across epithelia. Physiol Rev 85: 373–422.
9. Van de Graaf SFJ, Boullart I, Hoenderop JGJ, Bindels RJM. 2004. Regulation of the epithelial Ca^{2+} channels TRPV5 and TRPV6 by 1α,25-dihydroxy Vitamin D3 and dietary Ca^{2+}. J Steroid Biochem Molec Biol 89–90: 303–8.
10. Lieben L, Benn BS, Ajibade D, Stockmans I, Moermans K, Hediger MA, Peng JB, Christakos S, Bouillon R, Carmeliet G. 2010. Trpv6 mediates intestinal calcium absorption during calcium restriction and contributes to bone homeostasis. Bone 47: 301–8.
11. Friedman PA, Gesek FA. 1995. Cellular calcium transport in renal epithelia: Measurement, mechanisms, and regulation. Physiol Rev 75: 429–71.
12. Nordin BE, Peacock M. 1969. Role of kidney in regulation of plasma-calcium. Lancet 2: 1280–3.
13. Rouse D, Suki WN. 1995. Renal control of extracellular calcium. Kidney Int 38: 700–8.
14. Brenza HL, Kimmel-Jehan C, Jehan F, et al. 1998. Parathyroid hormone activation of the 25-hydroxyvitamin D3-1a-hydroxylase gene promoter. Proc Natl Acad Sci U S A 95: 1387–91.
15. Azarani A, Goltzman D, Orlowski J. 1995. Parathyroid hormone and parathyroid hormone-related peptide inhibit the apical Na+/H+ exchanger NHE-3 isoform in renal cells (OK) via a dual signaling cascade involving protein kinase A and C. J Biol Chem 270: 20004–10.
16. Derrickson BH, Mandel LJ. 1997. Parathyroid hormone inhibits Na(+)-K(+)-ATPase through Gq/G11 and the calcium-independent phospholipase A2. Am J Physiol 272: F781–8.
17. Juppner H, Abou-Samra AB, Freeman MW, et al. 1991. A G protein-linked receptor for parathyroid hormone and parathyroid hormone-related peptide. Science 254: 1024–6.
18. Abou-Samra AB, Juppner H, Force T, et al. 1992. Expression cloning of a common receptor for parathyroid hormone and parathyroid hormone-related peptide from rat osteoblast-like cells: A single receptor stimulates intracellular accumulation of both cAMP and inositol triphosphates and increases intracellular free calcium. Proc Natl Acad Sci U S A 89: 2732–6.
19. De Rouffignac C, Quamme GA. 1994. Renal magnesium handling and its hormonal control. Physiol Rev 74: 305–22.
20. Hebert SC. 1996. Extracellular calcium-sensing receptor: Implications for calcium and magnesium handling in the kidney. Kidney Int 50: 2129–39.
21. Rouleau MF, Mitchell J, Goltzman D. 1988. In vivo distribution of parathyroid hormone receptors in bone: Evidence that a predominant osseous target cell is not the mature osteoblast. Endocrinology 123: 187–91.
22. Potts JT Jr, Juppner H. 1997. Parathyroid hormone and parathyroid hormone-related peptide in calcium homeostasis, bone metabolism, and bone development: The proteins, their genes, and receptors. In: Avioli LV, Krane SM (eds.) Metabolic Bone Disease, 3rd Ed. New York: Academic Press. p. 51.
23. Grant FD, Conlin PR, Brown EM. 1990. Rate and concentration dependence of parathyroid hormone dynamics during stepwise changes in serum ionized calcium in normal humans. J Clin Endocrinol Metab 71: 370–8.
24. Brown EM, Gamba G, Riccardi D, et al. 1993. Cloning and characterization of an extracellular Ca(2+)-sensing receptor from bovine parathyroid. Nature 366: 575–80.
25. Kremer R, Bolivar I, Goltzman D, et al. 1989. Influence of calcium and 1,25-dihydroxycholecalciferol on proliferation and proto-oncogene expression in primary cultures of bovine parathyroid cells. Endocrinology 125: 935–41.
26. Goltzman D, Miao D, Panda DK, Hendy GN. 2004. Effects of calcium and of the vitamin D system on skeletal and calcium homeostasis: lessons from genetic models. J Steroid Biochem Mol Biol 89–90: 485–9.
27. Li YC, Pirro, AE, Amling M, et al. 1997. Targeted ablation of the vitamin D receptor: An animal model of vitamin D-dependent rickets type II with alopecia. Proc Natl Acad Sci U S A 94: 9831–5.
28. Lee SK, Lorenzo JA. 1999. Parathyroid hormone stimulates TRANCE and inhibits osteoprotegerin messenger ribonucleic acid expression in murine bone marrow cultures: Correlation with osteoclast-like cell formation. Endocrinology 140: 3552–61.
29. Takahashi N, Udagawa N, Takami M, et al. 2002. Cells of bone: Osteoclast generation. In: Bilezikian JP, Raisz LG, Rodan GA (eds.) Principles of Bone Biology, 2nd Ed. San Diego: Academic Press. p. 109.
30. Panda DK, Miao D, Bolivar I, Li J, Huo R, Hendy GN, Goltzman D. 2004. Inactivation of the 25-dihydroxyvitamin D-1alpha-hydroxylase and vitamin D receptor demonstrates independent effects of calcium and vitamin D on skeletal and mineral homeostasis. J Biol Chem 279: 16754–66.
31. Xue Y, Karaplis AC, Hendy GN, Goltzman D, Miao D. 2006. Exogenous 1,25-dihydroxyvitamin D3 exerts a skeletal anabolic effect and improves mineral ion homeostasis in mice which are homozygous for both the

1alpha hydroxylase and parathyroid hormone null alleles. *Endocrinology* 147: 4801–10.
32. Schmulen AC, Leman M, Pak CY, Zerwekh J, Morawski S, Fordtran JS, Vergne-Marini P. 1980. Effect of 1,25(OH)2D3 on jejunal absorption of magnesium in patients with chronic renal disease. *Am J Physiol* 238: G349–52.
33. Yu ASL. 2004. Renal transport of calcium, magnesium, and phosphate. In: Brenner BM (ed.) *The Kidney, 7th Ed.* Philadelphia: Saunders. pp. 535–72.

23
Fetal Calcium Metabolism

Christopher S. Kovacs

Mineral Ions and Calciotropic Hormones 180
Fetal Parathyroids 181
Calcium Sensing Receptor (CaSR) 181
Fetal Kidneys and Amniotic Fluid 181
Placental Mineral Ion Transport 181
Fetal Skeleton 182

Fetal Response to Maternal Hyperparathyroidism 182
Fetal Response to Maternal Hypoparathyroidism 183
Fetal Response to Maternal Vitamin D Deficiency 183
Integrated Fetal Calcium Homeostasis 184
References 185

Much of normal mineral and bone homeostasis in the adult can be explained by the interactions of parathyroid hormone (PTH), 1,25-dihydroxyvitamin D or calcitriol (1,25-D), calcitonin, and the sex steroids. In contrast, less is known about how mineral and bone homeostasis is regulated in the fetus. Due to obvious limitations in studying human fetuses, human regulation of fetal mineral homeostasis must be largely inferred from studies in animals, and some observations in animals may not apply to humans. This chapter briefly reviews existing human and animal data; two comprehensive reviews on the subject contain more detailed references [1, 2].

Fetal mineral metabolism has adapted to maintain a high extracellular level of calcium (and other minerals) that is physiologically appropriate for fetal tissues and to provide sufficient calcium (and other minerals) to fully mineralize the skeleton before birth. Mineralization occurs rapidly in late gestation, such that a human accretes 80% of the required 20–30g of calcium in the third trimester [3, 4], while a rat accretes 95% of the required 12.5mg of calcium in the last 5 days of its 3-week gestation [5].

MINERAL IONS AND CALCIOTROPIC HORMONES

Human and other mammalian fetuses consistently maintain the serum calcium (both total and ionized) at significantly higher values than in the mother during late gestation. Similarly, serum phosphate is significantly elevated while serum magnesium is minimally elevated above the maternal concentration.

These increased serum mineral concentrations have physiological importance. Maintaining the fetal calcium concentration above the maternal level is required to achieve normal mineralization of the fetal skeleton, as discussed below. Survival to the end of gestation is unaffected by significant hypocalcemia in *Pthrp* null, *Pthr1* null, *Hoxa3* null, *Trpv6* null, and *Hoxa3/Pthrp* double mutant fetuses [6–9]. But survival after birth may be aided by a high blood calcium level *in utero*. The serum calcium declines 20–30% after birth in humans [10–12] and 40% in rodents [13, 14] before increasing to adult values over the succeeding 24 hours. A lower fetal blood calcium may predispose to an even lower trough level being reached after birth, thereby increasing the risk of tetany and death. The early postnatal mortality of *Pthrp* null, *Pthr1* null, and *Hoxa3* null fetuses is consistent with this possibility [8, 15, 16], whereas hypocalcemic *Tpv6* null fetuses do not show perinatal mortality [9, 17].

The high level of calcium in the fetal circulation is robustly maintained despite maternal hypocalcemia from a variety of causes. For example, the serum calcium is normal in fetal mice lacking the vitamin D receptor (*Vdr* null fetuses) and also in pups born of severely vitamin D deficient rodents [18–22]. Similarly, the cord blood calcium is normal in human babies with severe vitamin D deficiency (25-hydroxyvitamin D levels of 10 nmol/l versus 138 nmol/l in offspring of vitamin

---

Primer on the Metabolic Bone Diseases and Disorders of Mineral Metabolism, Eighth Edition. Edited by Clifford J. Rosen.
© 2013 American Society for Bone and Mineral Research. Published 2013 by John Wiley & Sons, Inc.

D-treated mothers) [23], while children with naturally occurring deletions in the VDR do not present with hypocalcemia or rickets until their second year of life [2].

Calciotropic hormone levels are maintained at levels that differ from the adult. These differences appear to reflect the relatively different roles that these hormones play in the fetus and are not an artifact of altered metabolism or clearance of hormones. Intact PTH levels are much lower than maternal PTH levels near the end of gestation, but PTH is important for fetal development since fetal mice lacking parathyroids or PTH are hypocalcemic and have undermineralized skeletons [7, 8, 24, 25]. Circulating 1,25-D levels are also low in late gestation, owing to suppression of the fetal 1α-hydroxylase by high serum calcium and phosphate, and low PTH. 1,25-D may also be relatively unimportant for fetal mineral homeostasis, since several vitamin D deficiency models, 1α-hydroxylase null pigs, and Vdr null mice, all have normal serum mineral concentrations and fully mineralized skeletons at term (reviewed in detail in Ref. 2). Fetal calcitonin levels are higher than maternal levels and may be responding to the higher serum calcium in the fetus, but calcitonin does not play an essential role in fetal calcium homeostasis [26].

At term, cord blood levels of parathyroid hormone-related protein (PTHrP) are up to 15-fold higher than simultaneous PTH levels. PTHrP is produced in many tissues and plays multiple roles during embryonic and fetal development (see the chapter on parathyroid hormone-related protein). Absence of PTHrP (in Pthrp null fetuses) leads to abnormal endochondral bone development, modest hypocalcemia [15], and reduced placental calcium transfer [15, 27]. These Pthrp null fetuses have secondary hyperparathyroidism [7], but the blood calcium is reduced to the maternal level, confirming that PTH does not make up for lack of PTHrP in maintaining a normal fetal calcium concentration. Conversely, PTHrP does not make up for absence of PTH, given that fetuses lacking either parathyroids or PTH are hypocalcemic and have no compensatory increase in circulating PTHrP levels [7, 8, 24].

The role (if any) of the sex steroids in fetal skeletal development and mineral accretion is uncertain. Mice lacking estrogen receptor alpha and beta or the aromatase appear normal at birth and develop altered skeletal metabolism postnatally, but the fetal skeleton has not been examined [28–32]. Similarly, while postnatal skeletal roles of receptor activator of nuclear factor-κB (RANK), RANK ligand (RANKL), and osteoprotegerin are well known in humans and various knockout mouse models, the role that this system plays in fetal mineral metabolism is not yet known. Notably, mice lacking RANK or RANKL appear normal at birth and maintain normal serum calcium and phosphorus until weaning [33, 34].

FETAL PARATHYROIDS

Intact parathyroid glands are required for maintenance of normal fetal calcium, magnesium, and phosphate levels. Lack of parathyroids causes the fetal blood calcium to fall below the maternal level in mice [7, 8], whereas lack of either PTH or PTHrP causes the fetal calcium to fall to the maternal level [24]. Fetal parathyroids and PTH are also required for normal accretion of mineral by the skeleton and may be required for regulation of placental mineral transfer, as discussed below. Studies in fetal lambs have indicated that the fetal parathyroids may contribute to mineral homeostasis by producing both PTH and PTHrP, while detailed study of rats indicated that the fetal parathyroids produce only PTH (reviewed in Ref. 35). Whether human fetal parathyroids produce PTH alone, or PTH and PTHrP together, is unclear.

CALCIUM SENSING RECEPTOR (CaSR)

The parathyroid CaSR regulates the serum calcium level in adults by inhibiting PTH, but it does not appear to set the high serum calcium level of fetuses. Instead, the CaSR likely suppresses fetal PTH in response to the elevated fetal blood calcium [36]. On the other hand, inactivating mutations of the CaSR (Casr null fetuses) disrupt fetal homeostasis by inducing hyperparathyroidism with increased serum calcium, 1,25-D, and bone turnover, and this results in a lower skeletal calcium content by term [36]. The CaSR is also expressed within human and murine placentas [37], and may play some role in regulating placental mineral transfer. Casr null fetuses have a reduced rate of placental calcium transfer but whether this is a direct consequence of the loss of placental CaSR is not known [36].

FETAL KIDNEYS AND AMNIOTIC FLUID

Fetal kidneys partly regulate calcium homeostasis by adjusting the relative reabsorption and excretion of calcium, magnesium and phosphate in response to the filtered load and other regulatory factors, such as PTHrP and PTH. The fetal kidneys also synthesize 1,25-D, but since Vdr null fetal mice and severely vitamin D deficient rodent fetuses show no alteration in serum minerals or skeletal mineral content [2], it appears likely that production of 1,25-D by the fetal kidneys is relatively unimportant.

Renal calcium handling during fetal life may have minimal importance because calcium excreted by the kidneys is not permanently lost. Fetal urine is the major source of fluid and solute in amniotic fluid, and with swallowing the excreted calcium is made available again to the fetus.

PLACENTAL MINERAL ION TRANSPORT

As noted above, the bulk of placental calcium and other mineral transfer occurs late in gestation at a rapid rate.

Active transport of calcium, magnesium, and phosphate across the placenta is necessary for the fetal requirement to be met; only placental calcium transfer has been studied in detail. Analogous to calcium transfer across the intestinal mucosa, it has been theorized that calcium enters calcium-transporting cells through channels in maternal-facing basement membranes, is carried across these cells by calcium binding proteins, and is then actively extruded at the fetal-facing basement membranes by Ca^{2+}-ATPase.

Data from animal models indicate that a normal rate of maternal-to-fetal calcium transfer can usually be maintained despite the presence of maternal hypocalcemia or maternal hormone deficiencies such as aparathyroidism, vitamin D deficiency, and absence of the VDR [1, 6]. Whether the same is true for human pregnancies is less certain (see below, fetal response to maternal hypoparathyroidism). A "normal" rate of maternal–fetal calcium transfer does not necessarily imply that the fetus is unaffected by maternal hypocalcemia. Instead, it is an indication of the resilience of the fetal–placental unit to be able to extract the required amount of calcium from a hypocalcemic maternal circulation.

Fetal regulation of placental calcium transfer has been studied in a number of different animal models. Thyroparathyroidectomy in fetal lambs results in a reduced rate of calcium transfer across isolated, perfused placentas, suggesting that the parathyroids regulate placental calcium transfer [38]. In contrast, mice lacking parathyroids as a consequence of ablation of the *Hoxa3* gene have a normal rate of placental calcium transfer [8]. The discrepancy between these findings in lambs and mice may be due to whether or not the parathyroids are an important source of PTHrP in the circulation, as discussed above. Studies in fetal lambs and in *Pthrp* null fetal mice are in agreement that PTHrP, and in particular mid-molecular forms of PTHrP, stimulate placental calcium transfer [27, 39, 40]. There is also evidence that PTH is expressed within murine placenta from where it may stimulate placental transfer of calcium and other cations [24]. Conversely, calcitonin and 1,25-D are not required for placental calcium transport [18, 26].

FETAL SKELETON

A complete cartilaginous skeleton with digits and intact joints is present by the 8th week of gestation in humans. Primary ossification centers form in the vertebrae and long bones between the 8th to 12th weeks, but it is not until the third trimester that the bulk of mineralization occurs. At the 34th week of gestation, secondary ossification centers form in the femurs, but otherwise most epiphyses are cartilaginous at birth, with secondary ossification centers appearing in other bones in the neonate and child [41].

The skeleton must undergo substantial growth and be sufficiently mineralized by the end of gestation to support the organism, but as in the adult, the fetal skeleton participates in the regulation of mineral homeostasis. Calcium accreted by the fetal skeleton can be subsequently resorbed to help maintain the concentration of calcium in the blood, and this resorption can become more pronounced with severe maternal hypocalcemia [6] or in *Casr* null fetal mice [36]. Functioning fetal parathyroid glands are needed for normal skeletal mineral accretion, and both hypoparathyroidism (thyroparathyroidectomized fetal lambs, aparathyroid fetal mice, *Pth* null fetuses) and hyperparathyroidism (including *Casr* null fetal mice) reduce the net amount of skeletal mineral accreted by term.

Further comparative study of fetal mice lacking parathyroids, PTH, or PTHrP has demonstrated the interlocking roles of PTH and PTHrP in regulating the development and mineralization of the fetal skeleton [35]. PTHrP is produced locally in proliferating chondrocytes and perichondrium from where it directs the development of the cartilaginous scaffold that is later broken down and transformed into bone [42]. PTHrP is also expressed in preosteoblasts and osteoblasts from where it acts in an autocrine and paracrine manner to stimulate osteoblast function [43]. PTH acts from the systemic circulation to stimulate osteoblasts and contribute to maintaining the fetal blood calcium and magnesium at levels that facilitate mineralization. In the absence of PTHrP, a severe chondrodysplasia results from rapid differentiation and early apoptosis of chondrocytes [15]; accelerated and abnormal calcification also occurs, resulting in an apparently normal mineral content [7, 15]. Secondary hyperparathyroidism enables the *Pthrp* null growth plates to maintain normal expression of collagen α1(I) and collagenase-3, upregulated expression of osteocalcin and osteopontin, and increased mineralization [15, 44]. But when the PTH/PTHrP receptor (PTH1R) is ablated (*Pthr1* null fetuses), the resultant phenotype combines the *Pthrp* null chondrodysplasia with decreased osteoblast function, as evidenced by reduced expression of collagenase-3, osteocalcin, and osteopontin, and reduced mineralization [16, 44].

In the absence of parathyroids or PTH, the chondrocytic aspects of endochondral bone formation proceed normally but the bone compartment is significantly undermineralized at term [7, 24, 25]. Whether defects in osteoblast function are present when PTH is absent is unclear, since the expression of osteoblast-specific genes is normal in aparathyroid *Hoxa3* null fetuses but inconsistently reduced in *Pth* null fetuses [7, 24, 25]. The blood calcium and magnesium are significantly reduced in fetuses lacking parathyroids or PTH, and so the lack of PTH may impair mineralization simply by reducing the amount of mineral presented to the skeletal surface and to osteoblasts.

FETAL RESPONSE TO MATERNAL HYPERPARATHYROIDISM

In humans, maternal primary hyperparathyroidism has been associated with adverse fetal outcomes, including spontaneous abortion and stillbirth, which are thought to result from suppression of the fetal parathyroid glands [45]. Since PTH cannot cross the placenta [46, 47], fetal parathyroid suppression may result from increased calcium flux across the placenta to the fetus, facilitated by maternal hypercalcemia. The suppressed parathyroid function may persist for months in the neonate and has been permanent in some cases [48, 49]. Similar suppression of fetal parathyroids occurs when the mother has hypercalcemia due to familial hypocalciuric hypercalcemia [50, 51]. Chronic elevation of the maternal serum calcium in mice results in suppression of the fetal PTH level [36], but fetal survival is not notably affected by this.

FETAL RESPONSE TO MATERNAL HYPOPARATHYROIDISM

Maternal hypoparathyroidism during human pregnancy can cause fetal hyperparathyroidism. This is characterized by fetal parathyroid gland hyperplasia, generalized skeletal demineralization, subperiosteal bone resorption, bowing of the long bones, osteitis fibrosa cystica, rib and limb fractures, low birth weight, spontaneous abortion, stillbirth, and neonatal death [6]. Similar skeletal findings have been reported in the fetuses and neonates of women with pseudohypoparathyroidism, renal tubular acidosis, and chronic renal failure [6]. These changes in human skeletons differ from what has been found in animal models of maternal hypoparathyroidism, in which the fetal skeleton and the blood calcium are generally normal.

FETAL RESPONSE TO MATERNAL VITAMIN D DEFICIENCY

Cord blood levels of 25OHD are usually within 80–100% of the maternal value at term [2] because 25-hydroxyvitamin D (25OHD) readily crosses the placenta. This means that if the mother is vitamin D deficient or insufficient, then the fetus will be as well.

As mentioned earlier, fetal calcium metabolism and skeletal mineral content are normal in animal models of severe vitamin D deficiency, and in *Vdr* null mice [2]. The available but limited human data indicate that calcium homeostasis and skeletal mineral content in human fetuses may also be unaffected by severe vitamin D deficiency and absence of VDR.

These data include the finding of normal ash weight, skeletal mineral content (by atomic absorption spectroscopy), and absence of radiological signs of rickets, in severely vitamin D deficient infants who died of obstetrical accidents [52]. Observational data of infants with severe vitamin D deficiency or genetic absence of 1α-hydroxylase or VDR show that hypocalcemia and rickets do not normally develop or become recognized until at least several months after birth, with the peak incidence occurring in the second year [2]. Studies of bone mineral density in newborns have not shown any association with vitamin D sufficiency [2]. Recently, two large randomized trials examined the use of up to 4,000 IU of vitamin D per day during pregnancy and found no effect of vitamin D use or 25-hydroxyvitamin D levels on cord blood calcium or skeletal parameters in the newborns [53].

The available animal and human data predict that human fetuses will have a normal skeleton and serum calcium despite a severe vitamin D deficiency. It is the vitamin D-deficient neonates and infants who are at risk for postnatal hypocalcemia followed by the later development of rickets [2]. The reason maternal vitamin D deficiency may have little or no effect on fetal calcium homeostasis is that secondary hyperparathyroidism in the mother minimizes the hypocalcemia as compared to what occurs with maternal hypoparathyroidism, and because placental transport of calcium (and relevant genes involved in cation transport) does not require calcitriol or VDR.

However, recent clinical studies have questioned whether the fetal skeleton is truly normal with even maternal insufficiency of vitamin D. These studies examined associations between single measurements of maternal serum 25OHD during pregnancy and various skeletal outcomes in the fetus, neonate, or child. In none of these was an association found with birth weight, cord blood calcium, skeletal lengths, and bone mineral density [54–57]. In one study, a 25OHD level below 28 nmol/L was reportedly associated with a slightly shorter knee–heel length, but the difference was not statistically significant after correcting for gestational age [54]. A second study found that maternal 25OHD levels *below* 50 nmol/L were associated with greater distal metaphyseal cross-sectional area in the femur and concluded that this was evidence of early rickets [55]. But a third study found that maternal 25OHD levels *above* 42.6 nmol/L were associated with greater metaphyseal cross-sectional area in the tibia, and concluded that this meant stronger bone [56]. The latter pair of studies exemplify how subjective the interpretations can be, with greater metaphyscal cross-sectional area considered an adverse effect in one study and a benefit in another.

A fourth well-publicized study by Javaid found no associations between maternal serum 25OHD and birth weight, length, placental weight, abdominal circumference, head circumference, or cord blood calcium [57]. At the 9-*month* follow-up there was still no association of maternal 25OHD with skeletal and anthropometric parameters in the infants. However, a maternal serum 25OHD level below 27.5 nmol/L during pregnancy was associated with a modestly lower bone mineral content in offspring at 9 *years of age* compared to offspring of mothers whose 25OHD levels had been 50 nmol/L or higher. These findings have been used to promote the theory that vitamin D exposure during fetal development programs childhood peak bone mass [58].

There are several caveats to keep in mind about these associational studies. Unstated are which radiological parameters were prespecified and how many were examined before a single statistically significant finding was reported in each study. Why the tibia in one study and the femur in another, and not both or more skeletal sites? The possibility of chance and clinically insignificant findings cannot be excluded. The studies are confounded by factors that predispose to low maternal 25OHD levels, including obesity, lower socioeconomic status, poorer nutrition and lack of exercise, and prenatal care or vitamin supplementation, etc. Therefore, is lower 25OHD simply a marker of a less healthy pregnant woman? In the Javaid study, much time elapsed between birth (when no effect was seen) and 9 years of age (when lower bone mineral content was found). Does low 25OHD *in utero* program lower bone mineral content at 9 years of age? Or does lower maternal 25OHD in late pregnancy signal that lower socioeconomic status, poorer nutrition, and other factors in the mother will remain unchanged and continue to be shared with the child? Ultimately, associational studies are hypothesis generating and do not prove causality, thus clinical trials of vitamin D supplementation are needed to control for confounding and definitively determine if vitamin D supplementation during pregnancy confers any skeletal benefit on the fetus, neonate, infant, or child. The two largest clinical trials described above did not find any skeletal benefit [53]; unfortunately, bone mineral density measurements in the newborns were apparently not done as originally planned.

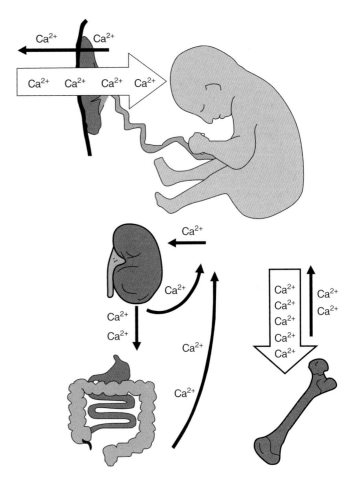

Fig. 23.1. Calcium sources in fetal life. Reproduced with permission from Ref. 59. ©2003 Academic Press.

INTEGRATED FETAL CALCIUM HOMEOSTASIS

The evidence discussed in the preceding sections suggests the following summary models.

Calcium sources

The main flux of calcium and other minerals is across the placenta and into fetal bone, but calcium is also made available to the fetal circulation through several routes (see Fig. 23.1). The kidneys reabsorb calcium; calcium excreted by the kidneys into the urine and amniotic fluid may be swallowed and reabsorbed, and calcium is also resorbed from the developing skeleton. Some calcium returns to the maternal circulation (backflux). The maternal skeleton is a potential source of mineral, and it may be compromised in mineral deficiency states in order to provide to the fetus.

Blood calcium regulation

The fetal blood calcium is set at a level higher than the maternal value through the actions of PTHrP and PTH acting in concert (among other potential factors). The CaSR suppresses PTH in response to the high calcium level, but the low level of PTH is required for maintaining the blood calcium and facilitating mineral accretion by the skeleton. The synthesis and secretion of 1,25-D are, in turn, suppressed by low PTH, and high blood calcium and phosphate. The parathyroids may play a central role by producing PTH and PTHrP, or may produce PTH alone. PTHrP and PTH are produced by the placenta, and PTHrP is also produced by many other fetal tissues.

PTH and PTHrP, both present in the fetal circulation, independently and additively regulate the fetal blood calcium. Neither hormone can make up for absence of the other: if one is missing, the blood calcium is reduced, and if both are missing, the blood calcium is reduced even further. PTH upregulates in the absence of PTHrP, but PTHrP does not upregulate when parathyroids or PTH are absent. PTH may contribute to the blood calcium through actions on the PTH1R in several target tissues (kidney, bone, placenta), while PTHrP may act through the PTH1R and other novel receptors that respond to the mid or carboxy-terminal portions of the molecule.

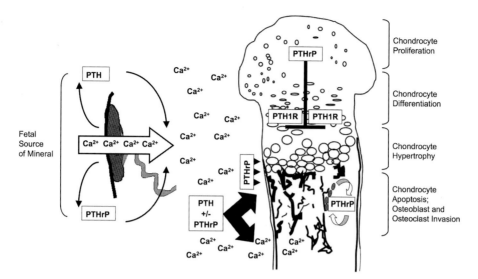

Fig. 23.2. Relative roles of PTH, PTHrP, and 1,25-D during fetal life. The placenta is the main source of minerals. PTH and PTHrP are expressed within the placenta but may also act on it from systemic sources. PTHrP stimulates calcium and possibly magnesium transfer; PTH also regulates calcium transfer and the expression of cation transporters. The regulators of placental phosphorus transfer are unknown. Within the endochondral skeleton, PTHrP is produced by proliferating chondrocytes and perichondrial cells (arrowheads) from which sites it acts on prehypertrophic chondrocytes (where the PTH1R is expressed) in order to delay their differentiation into hypertrophic chondrocytes. Hypertrophic chondrocytes undergo apoptosis, vascular invasion occurs, chondroclasts and osteoclasts resorb the cartilaginous matrix, and osteoblasts lay down primary spongiosa. PTHrP is also produced within pre-osteoblasts and osteoblasts from where it acts in a paracrine and autocrine fashion to stimulate bone formation (semicircular arrows). During fetal life PTHrP and PTH both regulate the fetal blood calcium, magnesium, and phosphorus, the concentrations of which are maintained above ambient maternal levels in order to facilitate mineralization. The regulation of blood calcium, placental calcium transfer, endochondral bone formation, and skeletal mineralization do not require 1,25-D or VDR during fetal life. Adapted with permission from Ref. 60. 2012 Elsevier.

The increased fetal blood calcium level is not a simple consequence of active placental calcium transfer because placental calcium transfer is normal in aparathyroid and *Pth* null mice and relatively increased in mice lacking the PTH1R, and yet each model is hypocalcemic [8, 24, 27]. Also, *Casr* null fetuses have reduced placental calcium transfer but markedly increased blood calcium levels that appear to result from fetal bone resorption [36].

Placental calcium transfer

Placental calcium transfer is regulated by both PTHrP and PTH, and the placenta (and possibly the parathyroids) is likely an important source of both hormones. The CaSR may also be involved in some aspect of calcium sensing within the calcium-transporting cells of the placenta.

Skeletal mineralization

PTH and PTHrP have separate roles with respect to skeletal development and mineralization (Fig. 23.2). PTH normally acts systemically to direct the mineralization of the bone matrix by maintaining the blood calcium at the adult level, and by direct actions on osteoblasts within the bone matrix. In contrast, PTHrP acts locally to direct endochondral bone development and osteoblast function, and acts from outside of bone to affect skeletal development and mineralization by contributing to the regulation of the blood calcium and placental calcium transfer. PTH may have the more dominant effect to maintain skeletal mineral accretion since the absence of PTH leads to an undermineralized skeleton while absence of PTHrP does not.

REFERENCES

1. Kovacs CS. Fetal mineral homeostasis. 2011. In: Glorieux FH, Pettifor JM, Jüppner H (eds.) *Pediatric Bone: Biology and Diseases*, 2nd Ed. San Diego: Elsevier/Academic Press. pp. 247–75.
2. Kovacs CS. 2011. Fetus, neonate and infant. In: Feldman D, Pike WJ, Adams JS (eds.) *Vitamin D: Third Edition*. New York: Academic Press. pp. 625–46.
3. Givens MH, Macy IC. 1933. The chemical composition of the human fetus. *J Biol Chem* 102: 7–17.
4. Widdowson EM, Dickerson JW. 1964. Chemical composition of the body. In: Comar CL, Bronner F (eds.) *Mineral Metabolism: An Advanced Treatise, Volume II*,

The Elements, Part A. New York: Academic Press. pp. 1–247.
5. Comar CL. 1956. Radiocalcium studies in pregnancy. *Ann N Y Acad Sci* 64: 281–98.
6. Kovacs CS, Kronenberg HM. 1997. Maternal–fetal calcium and bone metabolism during pregnancy, puerperium and lactation. *Endocr Rev* 18: 832–72.
7. Kovacs CS, Chafe LL, Fudge NJ, Friel JK, Manley NR. 2001. PTH regulates fetal blood calcium and skeletal mineralization independently of PTHrP. *Endocrinology* 142(11): 4983–93.
8. Kovacs CS, Manley NR, Moseley JM, Martin TJ, Kronenberg HM. 2001. Fetal parathyroids are not required to maintain placental calcium transport. *J Clin Invest* 107(8): 1007–15.
9. Suzuki Y, Kovacs CS, Takanaga H, Peng JB, Landowski CP, Hediger MA. 2008. Calcium TRPV6 is involved in murine maternal-fetal calcium transport. *J Bone Miner Res* 23(8): 1249–56.
10. Loughead JL, Mimouni F, Tsang RC. 1988. Serum ionized calcium concentrations in normal neonates. *Am J Dis Child* 142: 516–8.
11. David L, Anast CS. 1974. Calcium metabolism in newborn infants. The interrelationship of parathyroid function and calcium, magnesium, and phosphorus metabolism in normal, sick, and hypocalcemic newborns. *J Clin Invest* 54: 287–96.
12. Schauberger CW, Pitkin RM. 1979. Maternal-perinatal calcium relationships. *Obstet Gynecol* 53: 74–6.
13. Garel JM, Barlet JP. 1976. Calcium metabolism in newborn animals: The interrelationship of calcium, magnesium, and inorganic phosphorus in newborn rats, foals, lambs, and calves. *Pediatr Res* 10: 749–54.
14. Krukowski M, Smith JJ. 1976. pH and the level of calcium in the blood of fetal and neonatal albino rats. *Biol Neonate* 29: 148–61.
15. Karaplis AC, Luz A, Glowacki J, Bronson RT, Tybulewicz VL, Kronenberg HM, Mulligan RC. 1994. Lethal skeletal dysplasia from targeted disruption of the parathyroid hormone-related peptide gene. *Genes Dev* 8: 277–89.
16. Lanske B, Karaplis AC, Lee K, Luz A, Vortkamp A, Pirro A, Karperien M, Defize L, Ho C, Abou-Samra AB, Jüppner H, Segre GV, Kronenberg HM. 1996. PTH/PTHrP receptor in early development and Indian hedgehog-regulated bone growth. *Science* 273: 663–6.
17. Bianco SD, Peng JB, Takanaga H, Suzuki Y, Crescenzi A, Kos CH, Zhuang L, Freeman MR, Gouveia CH, Wu J, Luo H, Mauro T, Brown EM, Hediger MA. 2007. Marked disturbance of calcium homeostasis in mice with targeted disruption of the Trpv6 calcium channel gene. *J Bone Miner Res* 22(2): 274–85.
18. Kovacs CS, Woodland ML, Fudge NJ, Friel JK. 2005. The vitamin D receptor is not required for fetal mineral homeostasis or for the regulation of placental calcium transfer. *Am J Physiol Endocrinol Metab* 289(1): E133–E44.
19. Halloran BP, De Luca HF. 1981. Effect of vitamin D deficiency on skeletal development during early growth in the rat. *Arch Biochem Biophys* 209: 7–14.
20. Halloran BP, DeLuca HF. 1979. Vitamin D deficiency and reproduction in rats. *Science* 204: 73–4.
21. Miller SC, Halloran BP, DeLuca HF, Jee WS. 1983. Studies on the role of vitamin D in early skeletal development, mineralization, and growth in rats. *Calcif Tissue Int* 35: 455–60.
22. Brommage R, DeLuca HF. 1984. Placental transport of calcium and phosphorus is not regulated by vitamin D. *Am J Physiol* 246: F526–F9.
23. Brooke OG, Brown IR, Bone CD, Carter ND, Cleeve HJ, Maxwell JD, Robinson VP, Winder SM. 1980. Vitamin D supplements in pregnant Asian women: Effects on calcium status and fetal growth. *Br Med J* 280: 751–4.
24. Simmonds CS, Karsenty G, Karaplis AC, Kovacs CS. 2010. Parathyroid hormone regulates fetal-placental mineral homeostasis. *J Bone Miner Res* 25(3): 594–605.
25. Miao D, He B, Karaplis AC, Goltzman D. 2002. Parathyroid hormone is essential for normal fetal bone formation. *J Clin Invest* 109(9): 1173–82.
26. McDonald KR, Fudge NJ, Woodrow JP, Friel JK, Hoff AO, Gagel RF, Kovacs CS. 2004. Ablation of calcitonin/calcitonin gene related peptide-a impairs fetal magnesium but not calcium homeostasis. *Am J Physiol Endocrinol Metab* 287(2): E218–26.
27. Kovacs CS, Lanske B, Hunzelman JL, Guo J, Karaplis AC, Kronenberg HM. 1996. Parathyroid hormone-related peptide (PTHrP) regulates fetal-placental calcium transport through a receptor distinct from the PTH/PTHrP receptor. *Proc Natl Acad Sci U S A* 93: 15233–8.
28. Mueller SO, Korach KS. 2001. Estrogen receptors and endocrine diseases: Lessons from estrogen receptor knockout mice. *Curr Opin Pharmacol* 1(6): 613–9.
29. Vidal O, Lindberg MK, Hollberg K, Baylink DJ, Andersson G, Lubahn DB, Mohan S, Gustafsson JA, Ohlsson C. 2009. Estrogen receptor specificity in the regulation of skeletal growth and maturation in male mice. *Proc Natl Acad Sci U S A* 97(10): 5474–9.
30. Windahl SH, Andersson G, Gustafsson JA. 2002. Elucidation of estrogen receptor function in bone with the use of mouse models. *Trends Endocrinol Metab* 13(5): 195–200.
31. Lubahn DB, Moyer JS, Golding TS, Couse JF, Korach KS, Smithies O. 1993. Alteration of reproductive function but not prenatal sexual development after insertional disruption of the mouse estrogen receptor gene. *Proc Natl Acad Sci U S A* 90(23): 11162–6.
32. Couse JF, Curtis SW, Washburn TF, Lindzey J, Golding TS, Lubahn DB, Smithies O, Korach KS. 1995. Analysis of transcription and estrogen insensitivity in the female mouse after targeted disruption of the estrogen receptor gene. *Mol Endocrinol* 9(11): 1441–54.
33. Kong YY, Yoshida H, Sarosi I, Tan HL, Timms E, Capparelli C, Morony S, Oliveira-dos-Santos AJ, Van G, Itie A, Khoo W, Wakeham A, Dunstan CR, Lacey DL, Mak TW, Boyle WJ, Penninger JM. 1999. OPGL is a key regulator of osteoclastogenesis, lymphocyte development and lymphnode organogenesis. *Nature* 397(6717): 315–23.
34. Li J, Sarosi I, Yan XQ, Morony S, Capparelli C, Tan HL, McCabe S, Elliott R, Scully S, Van G, Kaufman S, Juan

SC, Sun Y, Tarpley J, Martin L, Christensen K, McCabe J, Kostenuik P, Hsu H, Fletcher F, Dunstan CR, Lacey DL, Boyle WJ. 2009. RANK is the intrinsic hematopoietic cell surface receptor that controls osteoclastogenesis and regulation of bone mass and calcium metabolism. *Proc Natl Acad Sci U S A* 97(4): 1566–71.
35. Simmonds CS, Kovacs CS. 2010. Role of parathyroid hormone (PTH) and PTH-related protein (PTHrP) in regulating mineral homeostasis during fetal development. *Crit Rev Eukaryot Gene Expr* 20(3): 235–73.
36. Kovacs CS, Ho-Pao CL, Hunzelman JL, Lanske B, Fox J, Seidman JG, Seidman CE, Kronenberg HM. 1998. Regulation of murine fetal-placental calcium metabolism by the calcium-sensing receptor. *J Clin Invest* 101: 2812–20.
37. Kovacs CS, Chafe LL, Woodland ML, McDonald KR, Fudge NJ, Wookey PJ. 2002. Calcitropic gene expression suggests a role for intraplacental yolk sac in maternal-fetal calcium exchange. *Am J Physiol Endocrinol Metab* 282(3): E721–32.
38. Care AD, Caple IW, Abbas SK, Pickard DW. 1986. The effect of fetal thyroparathyroidectomy on the transport of calcium across the ovine placenta to the fetus. *Placenta* 7: 417–24.
39. Care AD, Abbas SK, Pickard DW, Barri M, Drinkhill M, Findlay JB, White IR, Caple IW. 1990. Stimulation of ovine placental transport of calcium and magnesium by mid-molecule fragments of human parathyroid hormone-related protein. *Exp Physiol* 75: 605–8.
40. Rodda CP, Kubota M, Heath JA, Ebeling PR, Moseley JM, Care AD, Caple IW, Martin TJ. 1988. Evidence for a novel parathyroid hormone-related protein in fetal lamb parathyroid glands and sheep placenta: Comparisons with a similar protein implicated in humoral hypercalcaemia of malignancy. *J Endocrinol* 117: 261–71.
41. Moore KL, Persaud TVN. 1998. *The Developing Human, 6th Ed.* Philadelphia, PA: W. B. Saunders.
42. Karsenty G. 2001. *Chondrogenesis just ain't what it used to be. J Clin Invest* 107(4): 405–7.
43. Miao D, He B, Jiang Y, Kobayashi T, Soroceanu MA, Zhao J, Su H, Tong X, Amizuka N, Gupta A, Genant HK, Kronenberg HM, Goltzman D, Karaplis AC. 2005. Osteoblast-derived PTHrP is a potent endogenous bone anabolic agent that modifies the therapeutic efficacy of administered PTH 1-34. *J Clin Invest* 115(9): 2402–11.
44. Lanske B, Divieti P, Kovacs CS, Pirro A, Landis WJ, Krane SM, Bringhurst FR, Kronenberg HM. 1998. The parathyroid hormone/parathyroid hormone-related peptide receptor mediates actions of both ligands in murine bone. *Endocrinology* 139: 5192–204.
45. Schnatz PF, Curry SL. 2002. Primary hyperparathyroidism in pregnancy: Evidence-based management. *Obstet Gynecol Surv* 57(6): 365–76.
46. Northrop G, Misenhimer HR, Becker FO. 1977. Failure of parathyroid hormone to cross the nonhuman primate placenta. *Am J Obstet Gynecol* 129: 449–53.
47. Garel JM, Dumont C. 1972. Distribution and inactivation of labeled parathyroid hormone in rat fetus. *Horm Metab Res* 4: 217–21.
48. Bruce J, Strong JA. 1955. Maternal hyperparathyroidism and parathyroid deficiency in the child, with account of effect of parathyroidectomy on renal function, and of attempt to transplant part of tumor. *Q J Med* 24: 307–19.
49. Better OS, Levi J, Grief E, Tuma S, Gellei B, Erlik D. 1973. Prolonged neonatal parathyroid suppression. A sequel to asymptomatic maternal hyperparathyroidism. *Arch Surg* 106: 722–4.
50. Powell BR, Buist NR. 1990. Late presenting, prolonged hypocalcemia in an infant of a woman with hypocalciuric hypercalcemia. *Clin Pediatr (Phila)* 29: 241–3.
51. Thomas BR, Bennett JD. 1995. Symptomatic hypocalcemia and hypoparathyroidism in two infants of mothers with hyperparathyroidism and familial benign hypercalcemia. *J Perinatol* 15: 23–6.
52. Maxwell JP, Miles LM. 1925. Osteomalacia in China. *J Obstet Gynaecol Br Empire* 32(3): 433–73.
53. Wagner CL. Vitamin D supplementation during pregnancy: Impact on maternal outcomes. Presented at the Centers for Disease Control and Prevention Conference on Vitamin D Physiology in Pregnancy: Implications for Preterm Birth and Preeclampsia. April 26–27, 2011. Atlanta, Georgia: Centers for Disease Control and Prevention.
54. Morley R, Carlin JB, Pasco JA, Wark JD. 2006. Maternal 25-hydroxyvitamin D and parathyroid hormone concentrations and offspring birth size. *J Clin Endocrinol Metab* 91(3): 906–12.
55. Mahon P, Harvey N, Crozier S, Inskip H, Robinson S, Arden N, Swaminathan R, Cooper C, Godfrey K. 2010. Low maternal vitamin D status and fetal bone development: Cohort study. *J Bone Miner Res* 25(1): 14–9.
56. Viljakainen HT, Saarnio E, Hytinantti T, Miettinen M, Surcel H, Makitie O, Andersson S, Laitinen K, Lamberg-Allardt C. 2010. Maternal vitamin D status determines bone variables in the newborn. *J Clin Endocrinol Metab* 95(4): 1749–57.
57. Javaid MK, Crozier SR, Harvey NC, Gale CR, Dennison EM, Boucher BJ, Arden NK, Godfrey KM, Cooper C. 2006. Maternal vitamin D status during pregnancy and childhood bone mass at age 9 years: A longitudinal study. *Lancet* 367(9504): 36–43.
58. Cooper C, Westlake S, Harvey N, Javaid K, Dennison E, Hanson M. 2006. Review: Developmental origins of osteoporotic fracture. *Osteoporos Int* 17(3): 337–47.
59. Kovacs CS. 2003. Fetal mineral homeostasis. In: Glorieux FH, Pettifor JM, Jüppner H (eds.) *Pediatric Bone: Biology and Diseases*. San Diego: Academic Press. pp. 271–302.
60. Kovacs CS. 2012. Fetal control of calcium and phosphate homeostasis–Lessons from mouse models. In: Thakker RV, Whyte MP, Eisman JA, Igarashi T (eds). *Genetics of Bone Biology and Skeletal Disease*. San Diego: Academic Press/Elsevier. pp. 205–220.

24

Fibroblast Growth Factor-23 (FGF23)

Kenneth E. White and Michael J. Econs

Introduction 188
The FGF23 Gene and Protein 188
FGF23 Activity 188
Regulation of FGF23 In Vivo 189
FGF23 Receptors 189
Serum Assays 189
FGF23-Associated Syndromes 190
References 192

INTRODUCTION

Disorders of phosphate (Pi) homeostasis, in concert with powerful *in vitro* and *in vivo* studies have demonstrated that fibroblast growth factor-23 (FGF23) is central to the control of renal Pi and vitamin D homeostasis. Although the molecular mechanisms are unique to each disorder, elevated FGF23 is associated with syndromes manifested by hypophosphatemia with paradoxically low or normal 1,25(OH)$_2$ vitamin D (1,25(OH)$_2$D), and include: autosomal dominant hypophosphatemic rickets (ADHR), X-linked hypophosphatemic rickets (XLH), tumor-induced osteomalacia (TIO), and autosomal recessive hypophosphatemic rickets (ARHR1 and ARHR2). Heritable disorders of hyperphosphatemia and often elevated 1,25(OH)$_2$D, such as tumoral calcinosis (TC), are associated with reduced FGF23 activity. These collective findings have provided unique insight into the activity of FGF23 on renal Pi and vitamin D metabolism.

THE FGF23 GENE AND PROTEIN

The *FGF23* gene resides on human chromosome 12p13 (mouse chromosome 6). It comprises three coding exons and contains an open reading frame of 251 residues [1]. The tissue with the highest FGF23 expression is bone.

FGF23 mRNA is observed in osteoblasts, osteocytes, flattened bone-lining cells, and osteoprogenitor cells [2]. Quantitative PCR demonstrated that FGF23 mRNA was most highly expressed in long bone, followed by thymus, brain, and heart [3].

Western analyses revealed that wild type FGF23 is secreted as a full length 32 kD species, as well as cleavage products of 12 and 20 kD [3–5]. Cleavage of FGF23 occurs within a subtilisin-like proprotein convertase (SPC) proteolytic site ($_{176}$RXXR$_{179}$/S$_{180}$), which separates the conserved FGF-like N-terminal domain from the variable C-terminal tail.

FGF23 ACTIVITY

FGF23 has overlapping function with PTH to reduce renal Pi reabsorption, but it has the opposite effects on 1,25(OH)$_2$D. The two primary transport proteins responsible for Pi reabsorption in the kidney are the type II sodium–phosphate cotransporters NPT2a and NPT2c, expressed in the apical membrane of the proximal tubule. FGF23 delivery leads to renal Pi wasting through the downregulation of both Npt2a and Npt2c [6].

Normally, hypophosphatemia is a strong positive stimulator for increasing serum 1,25(OH)$_2$D. However, patients with TIO, ADHR, XLH, and ARHR manifest hypophosphatemia with paradoxically low or inappropri-

---

Primer on the Metabolic Bone Diseases and Disorders of Mineral Metabolism, Eighth Edition. Edited by Clifford J. Rosen.
© 2013 American Society for Bone and Mineral Research. Published 2013 by John Wiley & Sons, Inc.

ately normal 1,25(OH)$_2$D. In mice, the expression of the 1α(OH)ase enzyme and the catabolic 24(OH)ase are reduced, and elevated, respectively, when the animals are exposed to FGF23 [4]. Thus, the effects of FGF23 on the renal vitamin D metabolic enzymes is most likely responsible for the reductions in 1,25(OH)$_2$D in the setting of persistent hypophosphatemia in ADHR, XLH, TIO, and ARHR patients.

REGULATION OF FGF23 IN VIVO

In humans, dietary Pi supplementation increased FGF23 whereas Pi restriction and the addition of Pi binders suppressed serum FGF23 [7], indicating that FGF23 plays a role in maintenance of Pi homeostasis. In animal studies, the FGF23 response to serum Pi has been more dramatic than in the human studies. Mice given high and low Pi diets produce the expected correlations between FGF23 and dietary Pi intake [8].

Vitamin D has important regulatory effects on FGF23. In mice, injections of 20–200 ng 1,25(OH)$_2$D led to dose-dependent increases in serum FGF23 [9]. These changes in FGF23 occurred before changes in serum Pi, indicating that FGF23 is directly regulated by vitamin D. Physiologically, this would be consistent with results examining the role of FGF23 in vitamin D metabolism. FGF23 has been shown to downregulate the 1α(OH)ase mRNA [6, 9]; thus, as 1,25(OH)$_2$D rises in the blood as a product of 1α(OH)ase activity, vitamin D would then stimulate FGF23, which would complete the feedback loop and downregulate 1α(OH)ase expression.

FGF23 RECEPTORS

FGF23 is a member of unique class of FGFs, including FGF19 and FGF21, that are endocrine, as opposed to paracrine/autocrine, factors. FGF23 requires the co-receptor αKlotho (αKL) for bioactivity. αKL-null mice have severe calcifications as well as markedly elevated serum Pi [10], which parallels Fgf23-null mice [11, 12], and that of TC patients. However, both the αKL-null and Fgf23-null mice have more extreme phenotypes than that observed in patients. Importantly, these defects in the αKL-null and Fgf23-null mice can be ameliorated with a low Pi diet to reduce serum Pi [13]. In parallel with Fgf23-null mice, αKL-null mice have increased Npt2a in the proximal tubule [14], indicating that the hyperphosphatemia is secondary to increased renal reabsorption of Pi.

αKL is produced as several isoforms. Membrane bound KL (mKL) is a 130-kD single-pass transmembrane protein characterized by a large extracellular domain and a very short (10 residue) intracellular domain that does not possess signaling capabilities [15]. The mKL protein can also be cleaved extracellularly near the transmembrane domain, giving rise to a circulating form of αKL. A secreted isoform of KL (sKL) is approximately 80 kD and is spliced within exon 3 to result in an isoform that does not contain the transmembrane domain [15].

The most likely mechanism for FGF23 signaling through αKL is the recruitment of canonical FGF receptors (FGFRs) to form heteromeric complexes. One group has identified a specific complex between FGFR1c and αKL [16] and another found that FGFR3c and FGFR4 were also involved [17]. Signaling appears to be through mitogen activated protein kinase (MAPK) cascades [17]. More recently, animal studies showed that genetic deletion of FGFR3 and FGFR4 partially corrected the phenotype in the Hyp mouse, indicating that these receptors are physiologically relevant receptors for FGF23 [18]. FGFR1 may be more important for phosphate homeostasis, while FGFR3c and FGFR4 may be more relevant to vitamin D status [18]. Importantly, within the kidney, αKL localizes to the distal tubule [14]; however, FGF23 mediates its effects on NPT2a, NPT2c, and vitamin D within the proximal tubule [4, 6]. Acute delivery of FGF23 results in p-ERK1/2 signaling in the renal DCT [19]; therefore, the mechanisms underlying a local DCT-PT axis in the kidney following FGF23 delivery are unclear.

SERUM ASSAYS

FGF23 is measured in the circulation via several assays. One extensively used assay is a "C-terminal" FGF23 ELISA with both the capture and detection antibodies binding the C-terminal to the FGF23 $_{176}$RXXR$_{179}$/S cleavage site [20]. This assay thus recognizes full-length FGF23 as well as C-terminal proteolytic fragments. In a study with a large number of controls and TIO patients, this ELISA was used to test the levels of FGF23 in TIO and XLH [20], and showed that serum FGF23 is detectable in normal individuals. The mean FGF23 was greater than tenfold elevated in TIO patients and rapidly fell after tumor resection. Importantly, many XLH patients (13 out of 21) had elevated FGF23 compared to controls [20], and in those with "normal" FGF23, these levels may be "inappropriately normal" in the setting of hypophosphatemia.

An "intact" FGF23 ELISA assay has been developed that uses conformation-specific monoclonal antibodies that span the $_{176}$RXXR$_{179}$/S$_{180}$ SPC site and thus recognizes N- and C-terminal portions of FGF23 [21]. This assay detects a mean circulating concentration of 29 pg/ml in normal individuals [21]. The results of these two assays generally agree with regard to the relative ranges of FGF23 concentrations in XLH and TIO patients, and that FGF23 is elevated in most XLH patients. Based upon limited data from two TIO patients undergoing resection, the half-life of FGF23 is between 20 and 50 minutes [22, 23].

FGF23-ASSOCIATED SYNDROMES

Disorders associated with increased FGF23 bioactivity

Autosomal dominant hypophosphatemic rickets (ADHR; OMIM 193100)

Importantly, ADHR is distinguished from other hereditary hypophosphatemias by having either early or delayed onset with variable expressivity [24]. The ADHR mutations replace arginine (R) residues at positions 176 or 179 with glutamine (Q) or tryptophan (W) within the FGF23 subtilisin-like proprotein convertase (SPC) cleavage site, $_{176}RXXR_{179}/S_{180}$ [1, 4, 25] (Table 24.1). Following insertion of the ADHR mutations into wild type FGF23, FGF23 secreted from mammalian cells was primarily full length (32 kD), active polypeptide, as opposed to the 32 kD and cleavage products typically observed for wild type FGF23 expression [5].

Tumor-induced osteomalacia (TIO)

TIO is an acquired disorder of isolated renal Pi wasting that is associated with tumors. TIO patients present with similar biochemistries as patients with ADHR [26], and osteomalacia is seen upon bone biopsy. Clinical symptoms include muscle weakness, fatigue, and bone pain [26]. Insufficiency fractures are common and proximal muscle weakness can become severe [26]. FGF23 is elevated in patients with TIO [20, 21] and tumors that cause TIO have a dramatic over expression of FGF23 mRNA [25]. Surgical resection of the tumor results in rapid decreases in serum FGF23 [20].

X-linked hypophosphatemic rickets (XLH; OMIM 307800)

X-linked hypophosphatemic rickets is caused by inactivating mutations in *PHEX* (phosphate-regulating gene with homologies to endopeptidases on the X chromosome) [27]. PHEX is a member of the M13 family of membrane-bound metalloproteases and shows the highest expression in bone cells such as osteoblasts, osteocytes, and odontoblasts in teeth [28].

Reports have established that FGF23 is elevated in many XLH patients [20, 21]. Although it was initially thought that PHEX might cleave FGF23, this is not the case [3]. Instead, FGF23 mRNA expression is markedly increased in *Hyp* mice (mouse model of XLH) bone [3, 8]. The elevated FGF23 mRNA levels indicate that the increase in serum FGF23 in XLH is due to overproduction by skeletal cells, as opposed to a decreased rate of FGF23 degradation by cell surface proteases after secretion into the circulation. Although the interactions between FGF23 and PHEX are most likely indirect (see ARHR below), the encoded proteins have overlapping expression in bone [2, 3, 28]. At present, the PHEX substrate is unknown.

Table 24.1. Summary of Heritable and Acquired Disorders Involving FGF23

Disorder	Mutated Gene	Mutation Consequence	Relationship to FGF23	Effect on Serum Pi	Effect on Serum 1,25D	Intact FGF23 ELISA conc. (Kainos, Inc.)	C-terminal FGF23 ELISA conc. (Immutopics, Inc.)
ADHR	FGF23	Gain of function	Stabilize full-length, active FGF23	↓	↔	↔ or ↑	↔ or ↑
XLH	PHEX	Loss of function	Increased FGF23 production in osteocytes	↓	↔	↔ or ↑	↔ or ↑
ARHR1	DMP1	Loss of function	Increased FGF23 production in osteocytes	↓	↔	↔ or ↑	↔ or ↑
ARHR2	ENPP1	Loss of function	Increased FGF23 production	↓	↔	↔ or ↑	↔
TIO	–	–	FGF23 overproduced by tumor	↓	↔	↔ or ↑	↔ or ↑
TC/HHS	FGF23 or GALNT3	Loss of function	Destabilize full-length, active FGF23	↑	↔ or ↑	↓	↑
TC	KL	Loss of function	Decreased FGF23-dependent signaling	↑	↔ or ↑	↑	↑

Autosomal recessive hypophosphatemic rickets types 1 and 2

(ARHR1; OMIM #241520). Dentin matrix protein-1 (*DMP1*), a member of the small integrin-binding ligand, N-linked glycoprotein (SIBLING) family, is highly expressed in osteocytes. Both *Dmp1*-null mice and patients with ARHR manifest rickets and osteomalacia with isolated renal Pi wasting associated with elevated FGF23. Mutational analyses revealed that one ARHR family carried a mutation that ablated the DMP1 start codon, and a second family exhibited a deletion in the DMP1 C-terminus [29]. Mutations have also been identified in DMP1 splicing sites, which likely result in non-functional protein [30]. Mechanistic studies using the *Dmp1*-null mouse demonstrate that loss of DMP1 causes defective osteocyte maturation, leading to elevated FGF23 expression and pathological changes in bone mineralization [29]. Importantly, *Dmp1*-null mice are biochemical phenocopies of the *Hyp* mouse, and patients with ARHR and XLH (as well as the *Dmp1*-null and *Hyp* mice) share a unique bone histology characterized by distinctive periosteocytic lesions [29]. Thus, these findings suggest that PHEX may also have a role in osteocyte maturation in a parallel pathway to DMP1 that leads to over expression of FGF23.

ARHR2 (OMIM #613312). ARHR can also be caused by mutations in the ectonucleotide pyrophosphatase/phosphodiesterase-1 (*ENPP1*) gene that controls physiologic mineralization and pathologic chondrocalcinosis by generating inorganic pyrophosphate. Studies support that ENPP1 may regulate osteoblastic differentiation in an extracellular phosphate-independent manner [31]. Therefore, similar to DMP1, loss of function ENPP1 mutations could potentially result in an early-osteocyte differentiation defect and over expression of FGF23.

Other heritable disorders involving elevated FGF23

In addition to the disorders described above, FGF23 is also upregulated in several bone dysplasias that manifest documented isolated renal Pi wasting. These disorders include McCune Albright syndrome (OMIM 174800) [2] due to somatic activating mutations in G_s; opsismodysplasia (OMIM 258480) [32]; osteoglophonic dysplasia (OMIM 166250) [33] which is due to activating mutations in FGFR1; and epidermal nevus syndrome (ENS) [34].

Disorders associated with reduced FGF23 bioactivity

Familial tumoral calcinosis

Familial tumoral calcinosis (TC; OMIM 211900) is an autosomal recessive disorder characterized by dental abnormalities, as well as soft tissue periarticular and vascular calcification [35]. Biochemical abnormalities include hyperphosphatemia, increased percentage of TRP and inappropriately normal or elevated $1,25(OH)_2D$. Calcium and PTH are usually within the normal ranges, although PTH may be suppressed. Hyperostosis–hyperphosphatemia syndrome (HHS) is a rare metabolic disorder characterized by a biochemical profile that is identical to TC, with localized hyperostosis [36, 37].

TC/HHS due to GALNT3 mutations

The first gene identified for heritable TC was UDP-N-acetyl-alpha-D-galactosamine: polypeptide N-acetylgalactosaminyl transferase-3 (*GALNT3*) [38]. GALNT3 is expressed in the Golgi and initiates O-linked glycosylation of nascent proteins. These TC patients were originally reported to manifest serum FGF23 levels approximately 30-fold above the normal mean when assessed with the C-terminal FGF23 ELISA [38]. Importantly, it was subsequently demonstrated that the TC patients did indeed have elevated C-terminal FGF23, however the same individuals had low FGF23 when measured with the intact FGF23 ELISA (Table 24.1) [39]. These findings were then confirmed by demonstrating that loss of GALNT3 resulted in the production of nonfunctional FGF23 protein due to intracellular degradation [40]. FGF23 is O-glycosylated on specific residues within the $_{176}RH\underline{T}R_{179}/S_{180}$ site (at threonine 178), thus the lack of glycosylation at this residue is thought to destabilize intact active FGF23 [40].

HHS was also found to be due to inactivating mutations in *GALNT3* [36], and these patients also manifest inappropriate C-terminal to intact FGF23 ELISA values (Table 24.1) [36, 37]. Indeed, some of the HHS mutations are the same as those that result in TC [37], indicating that genetic background may influence disease phenotype and/or that TC and HHS may represent a spectrum of severity within the same disease.

TC due to FGF23 Mutations

TC can also be due to recessive, inactivating mutations in the *FGF23* gene [41–43]. These mutations have all been missense mutations (S71G, M96T, S129F) within the FGF23 N-terminal FGF-like domain. The TC alterations destabilize FGF23, as supported by the findings that the TC patients with *FGF23* mutations have the same FGF23 ELISA pattern as GALNT3-TC patients (i.e., markedly elevated C-terminal concentrations, in concert with low intact values [41, 42]) and the fact that these mutants are cleaved prior to cellular secretion [41–43]. Thus, the common denominator in GALNT3-TC and FGF23-TC is the lack of production of intact FGF23. This lack of intact FGF23 then results in elevation of serum Pi through increased renal reabsorption, which in turn results in elevated secretion of nonfunctional FGF23 fragments through a positive feedback cycle.

TC due to klotho mutations

αKlotho (αKL) is a co-receptor for FGF23, and was therefore tested as a candidate gene for TC in a 13-year old female with hypothesized end-organ defects in renal FGF23 bioactivity. This patient manifested hyperphosphatemia, hypercalcemia, elevated PTH, elevated intact and C-terminal FGF23 [44] (approximately 100- to 550-fold elevation of the normal means), as well as ectopic calcifications in the heel and brain. She had normal pubertal development, and her disease paralleled αKL-null mice with regard to ectopic calcifications, and dramatic elevation of circulating FGF23 [16]. This patient had a novel recessive mutation in a highly conserved residue (Histidine193 Arginine, or H193R) in the extracellular domain of αKL (KL1 domain). Mutant KL expression was markedly reduced compared to that of wild type αKL, which resulted in a striking reduction in the ability of αKL to mediate FGF23-dependent signaling [44]. Thus, an inactivating H193R αKL mutation results in a TC phenotype and demonstrates that αKL is required for FGF23 bioactivity.

Chronic kidney disease (CKD)

FGF23 is elevated in patients with CKD and recent studies indicate that this FGF23 is biologically active [45]. One report has shown that higher FGF23 levels are a predictor of increased progression of renal disease in patients with nondiabetic CKD [46]. Other reports demonstrate an association between high FGF23 concentrations and left ventricular hypertrophy in patients with CKD [47]. Furthermore, epidemiological studies demonstrate higher mortality in renal and nonrenal patients with higher FGF23 concentrations [48]. The pathophysiological significance of these findings remains to be elucidated, and it is unclear as to whether high concentrations of FGF23 are somehow toxic or if high FGF23 concentrations are simply a marker of more severe disease.

REFERENCES

1. ADHRConsortium. 2000. Autosomal dominant hypophosphataemic rickets is associated with mutations in FGF23. *Nat Genet* 26: 345–348
2. Riminucci M, Collins MT, Fedarko NS, Cherman N, Corsi A, White KE, Waguespack S, Gupta A, Hannon T, Econs MJ, Bianco P, Gehron Robey P. 2003. FGF-23 in fibrous dysplasia of bone and its relationship to renal phosphate wasting. *J Clin Invest* 112: 683–692.
3. Liu S, Guo R, Simpson LG, Xiao ZS, Burnham CE, Quarles LD. 2003. Regulation of fibroblastic growth factor 23 expression but not degradation by PHEX. *J Biol Chem* 278: 37419–37426
4. Shimada T, Mizutani S, Muto T, Yoneya T, Hino R, Takeda S, Takeuchi Y, Fujita T, Fukumoto S, Yamashita T. 2001. Cloning and characterization of FGF23 as a causative factor of tumor-induced osteomalacia. *Proc Natl Acad Sci U S A* 98: 6500–6505.
5. White KE, Carn G, Lorenz-Depiereux B, Benet-Pages A, Strom TM, Econs MJ. 2001. Autosomal-dominant hypophosphatemic rickets (ADHR) mutations stabilize FGF-23. *Kidney Int* 60: 2079–2086.
6. Larsson T, Marsell R, Schipani E, Ohlsson C, Ljunggren O, Tenenhouse HS, Juppner H, Jonsson KB. 2004. Transgenic mice expressing fibroblast growth factor 23 under the control of the alpha1(I) collagen promoter exhibit growth retardation, osteomalacia, and disturbed phosphate homeostasis. *Endocrinology* 145: 3087–3094.
7. Burnett SM, Gunawardene SC, Bringhurst FR, Juppner H, Lee H, Finkelstein JS. 2006. Regulation of C-terminal and intact FGF-23 by dietary phosphate in men and women. *J Bone Miner Res* 21: 1187–1196.
8. Perwad F, Azam N, Zhang MY, Yamashita T, Tenenhouse HS, Portale AA. 2005. Dietary and serum phosphorus regulate fibroblast growth factor 23 expression and 1,25-dihydroxyvitamin D metabolism in mice. *Endocrinology* 146: 5358–5364.
9. Shimada T, Hasegawa H, Yamazaki Y, Muto T, Hino R, Takeuchi Y, Fujita T, Nakahara K, Fukumoto S, Yamashita T. 2004. FGF-23 is a potent regulator of vitamin D metabolism and phosphate homeostasis. *J Bone Miner Res* 19: 429–435.
10. Tsujikawa H, Kurotaki Y, Fujimori T, Fukuda K, Nabeshima Y. 2003. Klotho, a gene related to a syndrome resembling human premature aging, functions in a negative regulatory circuit of vitamin D endocrine system. *Mol Endocrinol* 17: 2393–2403.
11. Shimada T, Kakitani M, Yamazaki Y, Hasegawa H, Takeuchi Y, Fujita T, Fukumoto S, Tomizuka K, Yamashita T. 2004. Targeted ablation of Fgf23 demonstrates an essential physiological role of FGF23 in phosphate and vitamin D metabolism. *J Clin Invest* 113: 561–568.
12. Sitara D, Razzaque MS, Hesse M, Yoganathan S, Taguchi T, Erben RG, Jüppner H, Lanske B. 2004. Homozygous ablation of fibroblast growth factor-23 results in hyperphosphatemia and impaired skeletogenesis, and reverses hypophosphatemia in Phex-deficient mice. *Matrix Biol* 23: 421–432.
13. Segawa H, Yamanaka S, Ohno Y, Onitsuka A, Shiozawa K, Aranami F, Furutani J, Tomoe Y, Ito M, Kuwahata M, Tatsumi S, Imura A, Nabeshima Y, Miyamoto KI. 2006. Correlation between hyperphosphatemia and type II Na/Pi cotransporter activity in klotho mice. *Am J Physiol Renal Physiol* 292: F769–779
14. Li SA, Watanabe M, Yamada H, Nagai A, Kinuta M, Takei K. 2004. Immunohistochemical localization of Klotho protein in brain, kidney, and reproductive organs of mice. *Cell Struct Funct* 29: 91–99.
15. Matsumura Y, Aizawa H, Shiraki-Iida T, Nagai R, Kuroo M, Nabeshima Y. 1998. Identification of the human klotho gene and its two transcripts encoding membrane and secreted klotho protein. *Biochem Biophys Res Commun* 242: 626–630.
16. Urakawa I, Yamazaki Y, Shimada T, Iijima K, Hasegawa H, Okawa K, Fujita T, Fukumoto S, Yamashita T. 2006. Klotho converts canonical FGF receptor into a specific receptor for FGF23. *Nature* 444(7120): 770–774.

17. Kurosu H, Ogawa Y, Miyoshi M, Yamamoto M, Nandi A, Rosenblatt KP, Baum MG, Schiavi S, Hu MC, Moe OW, Kuro-o M. 2006. Regulation of fibroblast growth factor-23 signaling by klotho. *J Biol Chem* 281: 6120–6123.
18. Li H, Martin A, David V, Quarles LD. 2011. Compound deletion of Fgfr3 and Fgfr4 partially rescues the Hyp mouse phenotype. *Am J Physiol Endocrinol Metab* 300: E508–517.
19. Farrow EG, Davis SI, Summers LJ, White KE. 2009. Initial FGF23-mediated signaling occurs in the distal convoluted tubule. *J Am Soc Nephrol* 20: 955–960.
20. Jonsson KB, Zahradnik R, Larsson T, White KE, Sugimoto T, Imanishi Y, Yamamoto T, Hampson G, Koshiyama H, Ljunggren O, Oba K, Yang IM, Miyauchi A, Econs MJ, Lavigne J, Juppner H. 2003. Fibroblast growth factor 23 in oncogenic osteomalacia and X-linked hypophosphatemia. *N Engl J Med* 348: 1656–1663.
21. Yamazaki Y, Okazaki R, Shibata M, Hasegawa Y, Satoh K, Tajima T, Takeuchi Y, Fujita T, Nakahara K, Yamashita T, Fukumoto S. 2002. Increased circulatory level of biologically active full-length FGF-23 in patients with hypophosphatemic rickets/osteomalacia. *J Clin Endocrinol Metab* 87: 4957–4960.
22. Khosravi A, Cutler CM, Kelly MH, Chang R, Royal RE, Sherry RM, Wodajo FM, Fedarko NS, Collins MT. 2007. Determination of the elimination half-life of fibroblast growth factor-23. *J Clin Endocrinol Metab* 92: 2374–2377.
23. Takeuchi Y, Suzuki H, Ogura S, Imai R, Yamazaki Y, Yamashita T, Miyamoto Y, Okazaki H, Nakamura K, Nakahara K, Fukumoto S, Fujita T. 2004. Venous sampling for fibroblast growth factor-23 confirms preoperative diagnosis of tumor-induced osteomalacia. *J Clin Endocrinol Metab* 89: 3979–3982.
24. Econs MJ, McEnery PT. 1997. Autosomal dominant hypophosphatemic rickets/osteomalacia: Clinical characterization of a novel renal phosphate-wasting disorder. *J Clin Endocrinol Metab* 82: 674–681.
25. White KE, Jonsson KB, Carn G, Hampson G, Spector TD, Mannstadt M, Lorenz-Depiereux B, Miyauchi A, Yang IM, Ljunggren O, Meitinger T, Strom TM, Jüppner H, Econs MJ. 2001. The autosomal dominant hypophosphatemic rickets (ADHR) gene is a secreted polypeptide overexpressed by tumors that cause phosphate wasting. *J Clin Endocrinol Metab* 86: 497–500.
26. Ryan EA, Reiss E. 1984. Oncogenous osteomalacia. Review of the world literature of 42 cases and report of two new cases. *Am J Med* 77: 501–512.
27. HypConsortium. 1995. A gene (PEX) with homologies to endopeptidases is mutated in patients with X-linked hypophosphatemic rickets. The HYP Consortium. *Nat Genet* 11: 130–136.
28. Beck L, Soumounou Y, Martel J, Krishnamurthy G, Gauthier C, Goodyer CG, Tenenhouse HS. 1997. Pex/PEX tissue distribution and evidence for a deletion in the 3′ region of the Pex gene in X-linked hypophosphatemic mice. *J Clin Invest* 99: 1200–1209.
29. Feng JQ, Ward LM, Liu S, Lu Y, Xie Y, Yuan B, Yu X, Rauch F, Davis SI, Zhang S, Rios H, Drezner MK, Quarles LD, Bonewald LF, White KE. 2006. Loss of DMP1 causes rickets and osteomalacia and identifies a role for osteocytes in mineral metabolism. *Nat Genet* 38: 1310–1315.
30. Lorenz-Depiereux B, Bastepe M, Benet-Pages A, Amyere M, Wagenstaller J, Müller-Barth U, Badenhoop K, Kaiser SM, Rittmaster RS, Shlossberg AH, Olivares JL, Loris C, Ramos FJ, Glorieux F, Vikkula M, Juppner H, Strom TM. 2006. DMP1 mutations in autosomal recessive hypophosphatemia implicate a bone matrix protein in the regulation of phosphate homeostasis. *Nat Genet* 38: 1248–1250.
31. Nam HK, Liu J, Li Y, Kragor A, Hatch NE. 2011. Ectonucleotide pyrophosphatase/phosphodiesterase-1 (Enpp1) regulates osteoblast differentiation. *J Biol Chem.* 286(45): 39059–71.
32. Zeger MD, Adkins D, Fordham LA, White KE, Schoenau E, Rauch F, Loechner KJ. 2007. Hypophosphatemic rickets in opsismodysplasia. *J Pediatr Endocrinol Metab* 20: 79–86.
33. White KE, Cabral JM, Davis SI, Fishburn T, Evans WE, Ichikawa S, Fields J, Yu X, Shaw NJ, McLellan NJ, McKeown C, Fitzpatrick D, Yu K, Ornitz DM, Econs MJ. 2005. Mutations that cause osteoglophonic dysplasia define novel roles for FGFR1 in bone elongation. *Am J Hum Genet* 76: 361–367.
34. Hoffman WH, Jueppner HW, Deyoung BR, O'Dorisio MS, Given KS. 2005. Elevated fibroblast growth factor-23 in hypophosphatemic linear nevus sebaceous syndrome. *Am J Med Genet A* 134: 233–236.
35. Prince MJ, Schaeffer PC, Goldsmith RS, Chausmer AB. 1982. Hyperphosphatemic tumoral calcinosis: Association with elevation of serum 1,25-dihydroxycholecalciferol concentrations. *Ann Intern Med* 96: 586–591.
36. Frishberg Y, Topaz O, Bergman R, Behar D, Fisher D, Gordon D, Richard G, Sprecher E. 2005. Identification of a recurrent mutation in GALNT3 demonstrates that hyperostosis-hyperphosphatemia syndrome and familial tumoral calcinosis are allelic disorders. *J Mol Med* 83: 33–38.
37. Ichikawa S, Guigonis V, Imel EA, Courouble M, Heissat S, Henley JD, Sorenson AH, Petit B, Lienhardt A, Econs MJ. 2007. Novel GALNT3 mutations causing hyperostosis-hyperphosphatemia syndrome result in low intact fibroblast growth factor 23 concentrations. *J Clin Endocrinol Metab* 92: 1943–1947.
38. Topaz O, Shurman DL, Bergman R, Indelman M, Ratajczak P, Mizrachi M, Khamaysi Z, Behar D, Petronius D, Friedman V, Zelikovic I, Raimer S, Metzker A, Richard G, Sprecher E. 2004. Mutations in GALNT3, encoding a protein involved in O-linked glycosylation, cause familial tumoral calcinosis. *Nat Genet* 36: 579–581.
39. Garringer HJ, Fisher C, Larsson TE, Davis SI, Koller DL, Cullen MJ, Draman MS, Conlon N, Jain A, Fedarko NS, Dasgupta B, White KE. 2006. The role of mutant UDP-N-acetyl-alpha-D-galactosamine-polypeptide

N-acetylgalactosaminyltransferase 3 in regulating serum intact fibroblast growth factor 23 and matrix extracellular phosphoglycoprotein in heritable tumoral calcinosis. *J Clin Endocrinol Metab* 91: 4037–4042.
40. Frishberg Y, Ito N, Rinat C, Yamazaki Y, Feinstein S, Urakawa I, Navon-Elkan P, Becker-Cohen R, Yamashita T, Araya K, Igarashi T, Fujita T, Fukumoto S. 2007. Hyperostosis-hyperphosphatemia syndrome: A congenital disorder of O-glycosylation associated with augmented processing of fibroblast growth factor 23. *J Bone Miner Res* 22: 235–242.
41. Benet-Pages A, Orlik P, Strom TM, Lorenz-Depiereux B. 2005. An FGF23 missense mutation causes familial tumoral calcinosis with hyperphosphatemia. *Hum Mol Genet* 14: 385–390.
42. Larsson T, Yu X, Davis SI, Draman MS, Mooney SD, Cullen MJ, White KE. 2005. A novel recessive mutation in fibroblast growth factor-23 causes familial tumoral calcinosis. *J Clin Endocrinol Metab* 90: 2424–2427.
43. Araya K, Fukumoto S, Backenroth R, Takeuchi Y, Nakayama K, Ito N, Yoshii N, Yamazaki Y, Yamashita T, Silver J, Igarashi T, Fujita T. 2005. A novel mutation in fibroblast growth factor 23 gene as a cause of tumoral calcinosis. *J Clin Endocrinol Metab* 90: 5523–5527.
44. Ichikawa S, Imel EA, Kreiter ML, Yu X, Mackenzie DS, Sorenson AH, Goetz R, Mohammadi M, White KE, Econs MJ. 2007. A homozygous missense mutation in human KLOTHO causes severe tumoral calcinosis. *J Clin Invest* 117: 2684–2691.
45. Shimada T, Urakawa I, Isakova T, Yamazaki Y, Epstein M, Wesseling-Perry K, Wolf M, Salusky IB, Juppner H. 2010. Circulating fibroblast growth factor 23 in patients with end-stage renal disease treated by peritoneal dialysis is intact and biologically active. *J Clin Endocrinol Metab* 95: 578–585.
46. Fliser D, Kollerits B, Neyer U, Ankerst DP, Lhotta K, Lingenhel A, Ritz E, Kronenberg F, Kuen E, Konig P, Kraatz G, Mann JF, Muller GA, Kohler H, Riegler P. 2007. Fibroblast growth factor 23 (FGF23) predicts progression of chronic kidney disease: The Mild to Moderate Kidney Disease (MMKD) Study. *J Am Soc Nephrol* 18: 2600–2608.
47. Mirza MA, Larsson A, Melhus H, Lind L, Larsson TE. 2009. Serum intact FGF23 associate with left ventricular mass, hypertrophy and geometry in an elderly population. *Atherosclerosis* 207: 546–551.
48. Gutiérrez OM, Mannstadt M, Isakova T, Rauh-Hain JA, Tamez H, Shah A, Smith K, Lee H, Thadhani R, Jüppner H, Wolf M. 2008. Fibroblast growth factor 23 and mortality among patients undergoing hemodialysis. *N Engl J Med* 359: 584–592.

25
Gonadal Steroids

Stavros C. Manolagas, Maria Almeida, and Robert L. Jilka

Introduction 195
Hormone Biosynthesis 195
Receptors and Molecular Mechanisms of Action 196
Effects of Sex Steroids on Skeletal Growth 197
Effects of Sex Steroids on Skeletal Maintenance 198

Loss of Sex Steroids and Aging 200
Sex Steroid Deficiency and the Development of Osteoporosis 201
Role of Sex Steroids in the Treatment of Osteoporosis 203
References 204

INTRODUCTION

Estrogens and androgens influence the development of the skeleton during growth and its maintenance during adulthood. Absence or dysfunction of the sex glands is associated with skeletal abnormalities. This relationship was known to Aristotle (384–322 BC), who emphasized the importance of age on the effects of sex steroids on bone as follows: "If castrated at an early age, animals become taller and more delicate than non-castrate; if they are, however, developed (at the time of castration) they do not increase their size." The interest in the role of sex steroids on bone was greatly intensified in the 1940s by the original suggestion of Fuller Albright that there was an association between the menopause and loss of bone mass [1]. Since that time, it has been extensively documented that a deficiency of estrogens in females or both androgens and estrogens in males adversely affects skeletal development during growth and homeostasis during adulthood, and contributes to the development of osteoporosis in either sex. However, the cellular and molecular mechanism(s) responsible for the adverse effects of the deficiency of either class of sex steroid on the male or female skeleton are still not well understood, nor is the molecular mechanism(s) by which estrogens contribute to the maintenance of the male skeleton. Moreover, until very recently, it has remained unknown whether and how sex steroid deficiency and old age may influence each other's negative impact on bone. In this chapter we will briefly review what is currently known about the biosynthesis of gonadal steroids and how androgens can be converted to estrogens, the role of the estrogen and androgen receptors in the effects of these two hormones on bone, the signaling pathways activated by these receptors, the biologic effects of estrogens or androgens on different bone cell types, their effects on skeletal growth, the interaction between sex steroid deficiency and aging on bone, the contribution of sex steroid deficiency to the development of osteoporosis, and finally the role of natural estrogens or related synthetic compounds in the treatment of osteoporosis. Besides estrogens and androgens, the gonads produce progesterone, but the role of this hormone in bone is minor, if any, and therefore it will not be discussed further in this chapter.

HORMONE BIOSYNTHESIS

Both estrogens and androgens are derived from C19 metabolites of cholesterol [2, 3]. Estrogens in vertebrates comprise a large number of molecules, the most abundant of which are 17β-estradiol (E_2), estrone (E_1), and estriol. E_2 is made in the ovaries and is the main circulating estrogen in premenopausal women (Fig. 25.1). The circulating concentration of E_2 in women is 20–200 pg/ml and fluctuates depending on the stage of the ovulatory cycle. The predominant form of estrogens in postmenopausal

---

Primer on the Metabolic Bone Diseases and Disorders of Mineral Metabolism, Eighth Edition. Edited by Clifford J. Rosen.
© 2013 American Society for Bone and Mineral Research. Published 2013 by John Wiley & Sons, Inc.

women is E_1, which is synthesized in several extra ovarian sites including bone.

The predominant sex steroid in men is testosterone (T), 95% of which is produced by the testes and reaches circulating levels of 3–10 ng/ml. The remaining 5% of T is made in the adrenals by the conversion of dehydroepiandrosterone (DHEA) by 3β-hydroxysteroid dehydrogenase (HSD) and 17β-HSD. Dihydrotestosterone (DHT) is the second most abundant circulating androgen and is made by the conversion of T by 5α-reductase in peripheral target tissues of androgen action. Even though DHT circulates at lower levels than T (0.25–0.75 ng/ml), it has considerably higher affinity for the androgen receptor (AR), and it is therefore more potent than T. E_2 is also present in the circulation of men (approximately 50 pg/ml), and it is formed by aromatization of testosterone. Of the total circulating level of E_2 in men, approximately 20% is derived from the testes via the action of P450 aromatase on T. The other 80% is derived from DHEA in peripheral tissues via aromatization and dehydrogenation to form the weak estrogen E_1, which is then converted to E_2 via 17β-HSD (Fig. 25.1).

In women, ovarian C19 steroid intermediates also give rise to T, most of which is then converted to E_2 by P450 aromatase, which is highly expressed in the ovary. The relatively low amounts of T and DHT present in women (less than 3 ng/ml) are synthesized in extragonadal tissues via the same pathways as in men.

Fifty to 60% of the total T, and 20% to 40% of the total E_2, present in the circulation are bound with high affinity to sex hormone binding globulin (SHBG) [4]. Most of the remainder is bound nonspecifically to albumin with lower affinity. Of the total circulating level of either estrogens or androgens only approximately 2% is free to enter the cells and exert biologic effects.

After menopause, the circulating levels of estrogens are generally lower than in men of the same age. The circulating levels of T decrease only marginally (–1%/year) during aging in men, whereas the total E_2 level remains constant. However, SHBG increases markedly in elderly men, resulting in a greater decrease of bioavailable T than E_2 [5].

RECEPTORS AND MOLECULAR MECHANISMS OF ACTION

Similar to other steroid hormones, estrogens or androgens exert their biologic effects by binding to specific receptor proteins: the estrogen receptor α (ERα) or β (ERβ) and the androgen receptor (AR), respectively [6]. These receptors are ligand-activated transcription factors, that homo- or heterodimerize upon ligand binding and directly attach to specific DNA sequences called hormone response elements (HREs) in regulatory regions of target genes (Fig. 25.2). DNA-bound receptors interact with

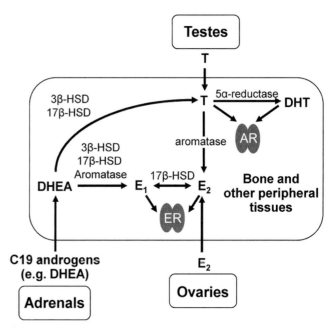

Fig. 25.1. Biosynthesis of estrogens and androgens in the gonads and peripheral tissues. T = testosterone; E_1 = estrone; E_2 = 17β-estradiol; DHT = dihydrotestosterone; DHEA = dehydroepiandrosterone; AR = androgen receptor; ER = estrogen receptor; HSD = hydroxysteroid dehydrogenase.

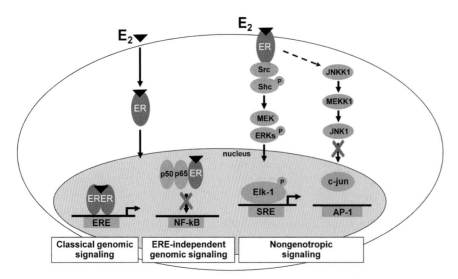

Fig. 25.2. Signaling pathways activated by the estrogen receptor. In the classical genomic signaling pathway 17β-estradiol (depicted by the triangle) binds to its receptor and translocates into the nucleus where it undergoes dimerization, attaches to estrogen response elements (ERE) on DNA, and activates or represses transcription. In the ERE-independent genomic signaling pathway, the ligand-activated receptor binds to other transcription factors (depicted by the p50 and p65 subunits of NF-κB), thereby preventing them from binding to their response elements on DNA. In the nongenotropic mode of action, the ligand-activated receptor (residing in the plasma membrane or the cytoplasm) interacts with cytoplasmic kinases and triggers cascades that positively or negatively regulate the activation of transcription factors such as Elk-1 and c-jun.

several coregulator proteins to form multiprotein complexes that activate or repress the transcriptional machinery [7]. Besides direct protein–DNA interactions, steroid receptors are able to interact indirectly with gene promoters through protein–protein interactions with other transcription factors, as is the case of ER binding to nuclear factor κB (NF–κB). This association inhibits NF–κB activation and the transcription of genes like interleukin 6 (IL-6) [8].

Estrogens and androgens are also able to evoke nongenomic or nongenotropic actions triggered by the activation of cytoplasmic signaling cascades, such as Src, Shc, mitogen activated protein (MAP) kinases including extracellular regulated kinases (ERKs), phosphatidyl inositol-3-kinase (PI3K), and c-jun N-terminal kinases (JNKs). Such actions result from ligand binding to ERs localized in the cytoplasm or the plasma membrane followed by downstream kinase-induced changes in the activation of transcription factors [9]. Via this latter mode of action ERs control an array of genes larger than the one regulated by their direct association with DNA [10].

ERα, ERβ, and the AR have been detected in all skeletal cell types including chondrocytes, bone marrow stromal cells, osteoblasts, osteocytes, and osteoclasts and their progenitors [11]. Nevertheless, compared to reproductive organs, the level of ER expression in bone cells is low in both males and females; and ERβ expression is two to three orders of magnitude lower than ERα [12]. Osteoblast number and bone mass are unaffected in mice lacking ERβ [13]. In addition, studies in mice with a genuine null mutation of ERβ indicate that with the exception of impaired ovarian function, this isotype of the ER is not required (at least in the mouse) for the development and homeostasis of the major body systems [14]. Hence, at this stage it seems that ERβ may not play a major role in bone, with the possible exception that it may exert an inhibitory effect on periosteal apposition [15].

EFFECTS OF SEX STEROIDS ON SKELETAL GROWTH

Sex is an important determinant of the size and shape of the skeleton. The difference in shape between sexes reflects in part the need of the female pelvis to accommodate gestation and delivery of the offspring. Bones grow in length at the growth plate of the epiphyses by the process of endochondral bone formation, whereby calcified cartilage made by chondrocytes is resorbed by osteoclasts and gradually replaced by mineralized bone made by osteoblasts. Simultaneously, bones expand radially, their cortices thicken, and the medullary cavities become larger as a result of bone formation at the periosteum and unbalanced remodeling at the endocortical surface—with resorption exceeding formation [16].

At the initiation of puberty, low levels of estrogens and perhaps androgens are responsible for a spurt in linear bone growth in both sexes, and this effect results from the stimulation of endochondral bone formation [16]. At the end of puberty, high levels of estrogens are essential for the closure of the epiphyses and the cessation of linear growth. In parallel to the acceleration of linear growth, pubertal boys and girls experience an accelerated enlargement of the outer perimeter and further widening of the medullary cavity of long bones. These changes lead to bigger bones in boys than in girls primarily due to a larger increase in periosteal bone formation. In contrast, girls exhibit more endocortical bone formation than boys. Puberty starts earlier in girls, but lasts longer in boys, without major differences in absolute growth rate between sexes. This may account for part of the differences in the size of the skeleton between the two sexes.

Effects on linear growth

Estrogens are essential for the pubertal bone changes in both girls and boys. Thus, aromatase-deficient females and males lacking estrogens do not exhibit a growth spurt and do not undergo closure of their epiphyses [17]. Estrogen replacement of patients with aromatase deficiency enhances bone growth. In addition, administration of E_2 to pre- or early-pubertal girls increases longitudinal growth [18]. Furthermore, overexpression of the aromatase gene in both sexes accelerates growth and leads to premature closure of the epiphyses [19]. The importance of estrogens for the male skeleton is further highlighted by the absence of a pubertal growth spurt in a man with a loss of function mutation of ERα [20]. Linear skeletal growth is dependent on the growth hormone (GH)/insulin-like growth factor (IGF) axis. The stimulatory effect of estrogens in the process is in part attributed to activation of the GH–IGF axis [16].

It is unclear whether androgens have a significant effect on linear growth. In support of the contention that androgens may not play a significant role in linear bone growth, serum testosterone was high normal in the man with the ERα mutation or aromatase-deficient men and women [17, 20]. On the other hand, men with androgen insensitivity syndrome due to AR mutations have intermediate height between normal males and females [21]. Moreover, administration of DHT to boys with delayed growth, or to growing rats and rabbits, stimulates longitudinal bone growth, supporting a role of androgens in this process [16]. To date, there are no consistent data in the murine model of the effects of androgens, estrogens, or their receptors, on pubertal bone growth spurt.

Effects on periosteal expansion

The greater periosteal expansion in boys as compared to girls during puberty has been ascribed to the higher levels of androgens in the male. Nevertheless, administration of E_2 to an aromatase-deficient young man increased bone size; this was presumed to be the result of increased periosteal apposition [22]. Consistent with this finding, studies in rodents have revealed that both estrogens and androgens, acting via their respective receptors, are involved in expansion of the periosteum in growing males. Thus, male mice with a mutation or deletion of the AR exhibit reduced periosteal expansion [23]. DHT may be the androgen responsible for periosteal expansion in the male. Thus, male mice lacking 5α-reductase type I exhibit reduced cortical thickness [24]. However, deletion of ERα, but not ERβ, also causes decreased periosteal bone formation and decreased femoral width in male mice [13]. Pharmacologic inhibition of E_2 synthesis by an aromatase inhibitor further reduces periosteal expansion in orchidectomized male mice, indicating that the estrogens needed for radial bone growth are derived from peripheral tissues [25]. In line with the need for both AR and ERα for optimal periosteal expansion, periosteal circumference is lower in mice that lack both receptors, as compared to mice lacking only one of the receptors [26].

Unlike the situation in males, estrogens restrain radial bone growth in females, as evidenced by the finding that ovariectomy of growing rats or mice increases periosteal expansion during early, but not late, puberty [25]. Moreover, female mice that lack ERβ exhibit increased periosteal circumference [15]. Whether the opposite effect of estrogens on periosteal apposition in males versus females also occurs in humans is unknown.

Cessation of longitudinal bone growth at the end of puberty is due to a decline in the replication of chondrocytes in the proliferative zone of the growth plate. This results in reduced cartilage synthesis at the distal end of the growth plate, as well as reduced replacement of the cartilage by bone at the proximal end. Consequently, the growth plate closes. The closure of the growth plate at the end of puberty is clearly mediated by E_2 in both men and women as indicated by failure of growth plate closure and continued longitudinal growth in the man who lacks ERα [20], and in aromatase-deficient men and women [17, 27]. Moreover, chondrocyte-specific deletion of ERα results in failure of growth plate closure, indicating a direct suppressive effect of E_2 on chondrocyte proliferation [28]. Consistent with this, ovariectomy increased the number of proliferating chondrocytes in the proliferative zone of growing rats [29].

EFFECTS OF SEX STEROIDS ON SKELETAL MAINTENANCE

Throughout life, older bone is periodically resorbed and replaced with new bone by teams of short-lived osteoclasts and osteoblasts comprising the basic multicellular unit. This process is known as remodeling. It is now abundantly clear that remodeling is orchestrated and targeted to a particular site that is in need for repair by osteocytes: long-lived former osteoblasts that are buried within mineralized bone and can sense and respond to changes in mechanical forces [30].

An oversupply of osteoclasts relative to the need for remodeling or an undersupply of osteoblasts relative to the need for cavity repair are critical pathophysiological changes for most acquired metabolic bone diseases, including osteoporosis [31, 32]. Estrogen or androgen deficiency causes loss of bone associated with an increase in the bone remodeling rate, increased osteoclast and osteoblast numbers, and increased resorption and formation, albeit unbalanced. Conversely, estrogens or androgens decrease bone resorption, restrain the rate of bone remodeling, and help to maintain a focal balance between bone formation and resorption. These effects are evidently the result of hormonal influences on the birth rate of osteoclast and osteoblast progenitors in the bone marrow as well as pro-apoptotic effects on osteoclasts and anti-apoptotic effects on mature osteoblasts and osteocytes.

Over the previous 30 years, a large number of studies with cell models, primary cell cultures, and gonadectomized rodents have suggested several mechanisms to account for the protective effects of sex steroids on skeletal homeostasis. The list includes direct effects of estrogens or androgens on osteoblast and osteoclast progenitor proliferation, differentiation, and lifespan, as well as indirect effects mediated via cytokines (including IL-1β, IL-6, IL-7, TNFα, M-CSF, RANKL, OPG, and prostaglandins) produced by bone marrow stromal cells, T and B lymphocytes, macrophages, and dendritic cells [32, 33]. A study of bone marrow cells from pre- and postmenopausal women suggested that estrogen deficiency may increase RANKL and decrease OPG production by osteoblast progenitors, and T and B cells [34]. Nonetheless, estrogens do not seem to suppress the transcription of the RANKL gene directly (C.A. O'Brien, unpublished observations); RANKL production by osteoblast progenitors or mature osteoblasts does not contribute to osteoclastogenesis during remodeling [30]; and the levels of circulating OPG in pre- and postmenopausal women are not different [35].

During the past decade the relative contribution of some of the aforementioned mechanisms to the overall effects of sex steroids on skeletal homeostasis has been investigated by more vigorous means involving genetic approaches; in particular, the generation of mice in which the ERs or the AR have been deleted or modified either globally (i.e., in all somatic cells) or selectively in different bone cell subtypes. Unfortunately, mice with global deletion of the ERα have not been very informative in regard to the physiologic effects of estrogens on bone either because the deletion was incomplete or because loss of the negative feedback control of sex steroids on gonadotrophins at the hypothalamic/pituitary level led to high serum levels of E_2 and T, thus confounding the interpretation of the bone phenotype. Recent attempts with cell type specific deletion of the ERα, however, have provided more meaningful insights.

Effects on osteoclasts

Selective deletion of the ERα from osteoclasts by two laboratories, one using the cathepsin K promoter

($ER\alpha^{\Delta Oc/\Delta Oc}$ mice) that is expressed in mature osteoclasts [36], and the other using the lysozyme M (LysM) gene promoter ($ER\alpha_{LysM}^{-/-}$ mice), which is expressed in cells of the entire monocyte/macrophage lineage and neutrophils [37], have independently demonstrated that cell autonomous effects of estrogens on osteoclasts mediated via the ERα can account for most, if not all, the antiresorptive properties of estrogens on the female skeleton.

Specifically, $ER\alpha^{\Delta Oc/\Delta Oc}$ female mice exhibited decreased bone volume due to an increase in the number of osteoclasts, resulting from the loss of a cell autonomous pro-apoptotic effect of estrogens on mature osteoclasts [36]. The pro-apoptotic effect of estrogens on osteoclasts, according to the authors of the study, was mediated by an increase in Fas ligand (FasL) production by osteoclasts. The $ER\alpha^{\Delta Oc/\Delta Oc}$ mice did not exhibit changes in osteoclast progenitors under basal conditions and did not undergo any bone loss two weeks following ovariectomy (OVX). Based on these results, Nakamura and colleagues concluded that the pro-apoptotic effect of estrogens on mature osteoclasts accounts for their osteoprotective properties. However, in contrast to the findings of Nakamura et al., others were unable to confirm that estrogens increase the transcription of the *FasL* gene in osteoclasts [37, 38].

In the $ER\alpha_{LysM}^{-/-}$ mice, the number of osteoclast progenitors was twofold higher, as was the number of osteoclasts in cancellous bone. In the estrogen replete state the $ER\alpha_{LysM}^{-/-}$ mice had lower cancellous bone mass, and they did not exhibit the expected decrease of cancellous bone in the estrogen-deficient state [37]. Yet, following loss of estrogens, they experienced a similar decrease in cortical bone as the littermate controls. The evidence that selective abrogation of the effects of estrogens on osteoclasts prevented loss of cancellous bone after estrogen withdrawal indicates that the cell autonomous effects of estrogens via ERα on osteoclast precursors and their progeny are sufficient for the protective effects of estrogens on this bone compartment.

The pro-apoptotic effect of estrogens on osteoclasts was completely abrogated in the $ER\alpha_{LysM}^{-/-}$ mice but was unaffected in mice bearing an ERα knock-in mutation that prevents binding to DNA ($ER\alpha^{NERKI/-}$) [37]. Moreover, a polymeric form of estrogen that is not capable of stimulating the nuclear-initiated actions of ERα was as effective as E_2 in inducing osteoclast apoptosis in cells with the wild-type ERα. These results demonstrate that the attenuation of osteoclast generation and lifespan cannot be exerted by the classical estrogen response element- (ERE) mediated mechanism of estrogen action. Deletion of steroid receptor coactivator-1 leads to skeletal resistance to E_2 in cancellous (but not cortical) bone, raising the possibility that a genotropic mechanism of action may be also required for the antiresorptive effects of estrogens [39] or that at least nongenotropic and genotropic pathways are interdependent.

The extent to which the effects of estrogens or androgens on T and B lymphocytes contribute to the overall antiresorptive effects of sex steroids and the role of the ERα in these cell types remains unclear [33, 40].

Effects on osteoblasts

The ERα deletion in osteoclasts did not affect cortical bone in females and had no effect in either cancellous or cortical bone in males [37]. These findings suggest that whereas elimination of the pro-apoptotic effect of estrogen on osteoclasts is sufficient for loss of bone in the cancellous compartment in which complete perforation of trabeculae by osteoclastic resorption precludes subsequent refilling of the cavities by the bone forming osteoblasts, the effects of estrogens on osteoblasts, and perhaps other cell types, are indispensable for their protective effects on the cortical compartment. ERα deletion from mesenchymal progenitors (Osx1 expressing cells) decreased cortical thickness, and this effect was manifested in both female and male mice, suggesting that ERα promotes proliferation/differentiation and/or survival of osteoblast progenitors in both sexes [41]. Collectively, these lines of evidence favor the view that the effects of estrogens on the birth and death of both osteoclasts and osteoblasts result from cell autonomous actions, mediated by ERα.

The significance of the anti-apoptotic effects of sex steroids on mature osteoblasts remains unclear. Likewise the significance of their pro- and antiproliferative or pro- and antidifferentiation effects on osteoblast progenitors and their mitotic progeny remain unknown. It is also unclear whether the increased bone formation seen in the sex steroid deficient state is the result of the loss of cell autonomous actions restraining the generation of osteoblasts as opposed to indirect effects secondary to increased resorption.

Effects on osteocytes

Osteocytes are by far the most abundant cells present in mammalian bone, and their lifespan corresponds to the age of the bone, as opposed to the short-lived osteoclasts and osteoblasts that are present only during remodeling [42]. Osteocytes are one of the cellular targets of estrogen and androgen action. Estrogens or androgens inhibit osteoblast and osteocyte apoptosis [9, 43], and this effect requires activation of the Src/Shc/ERK signaling pathway and downstream transcription factors such as Elk-1, C/EBPß and CREB [9]. Conversely, estrogen or androgen deficiency increases the prevalence of osteocyte apoptosis in humans [44], rats [45], and mice [43]. Frost had posited that mechanical stain, perceived by a hypothetical skeletal mechanostat, leads to changes in bone remodeling in order to adjust bone mass to a level that is appropriate for the current ambient mechanical forces. He also hypothesized that estrogen decreases the minimum effective strain necessary to initiate bone formation. In support of this hypothesis, estrogens and exercise may exert additive effect on bone mass in humans [46]. Moreover, Lanyon and colleagues have shown in mice that the increased bone formation that normally occurs in response to mechanical loading is diminished in the estrogen deficient state [47].

Reduced physical activity in old age, bed rest, or space flight invariably leads to bone loss [46]. Reduced mechanical forces in the murine model of unloading increase the prevalence of osteocyte apoptosis, followed by bone resorption and loss of bone mineral and strength [48]. On the other hand, mechanical strain is a requirement of osteocyte viability. Physiological levels of mechanical strain prevent apoptosis of cultured osteocytic cells [49]. The evidence for a role of the ER in the pro-survival effect of mechanical strain on osteocytes is consistent with the poor osteogenic response to loading exhibited by mice lacking the estrogen receptors α or β [47].

To conclude this section, the osteocyte ERα may play a role in the mechanosensing function of osteocytes and thereby affect bone strength. Given the fact that osteocytes produce molecules that control osteoclast or osteoblast formation, increased osteocyte apoptosis following sex steroid deficiency could also affect bone mass [30, 50, 51].

LOSS OF SEX STEROIDS AND AGING

For over 60 years, skeletal involution with advancing age has been attributed primarily to the decline of ovarian function at menopause and to a later and smaller decline of estrogens in elderly males [52, 53]. Nonetheless, recent epidemiological evidence in humans has made it clear that the balance between bone formation and resorption becomes progressively negative with advancing age in both women and men prior to, and independently from, any appreciable decline in sex steroids. Consistent with this evidence, most fractures in old age are nonvertebral and occur at predominantly cortical sites after age 65, and most bone loss occurs after this age [54]. Furthermore, extrapolation of the changes in bone mass found in perimenopausal women before the beginning of the menopausal estrogen drop and after its completion indicates that between menopause and the age of 75, women lose approximately 22% of their total body bone mineral; of this, 13.3% is due to aging and 7.75% to estrogen deprivation. In the femoral neck, 14% of the loss is "age-related" and only 5.3% because of estrogen deprivation [55]. In addition, a more recent analysis of cortical bone loss with high resolution peripheral CT of the radius and postmortem femurs of women between ages 50 and 80 years has revealed that most bone loss in old age is the result of increased intracortical porosity [54]. On the other hand, studies by Khosla and colleagues have suggested that while the onset of cortical bone loss in humans is closely tied to estrogen deficiency, a significant proportion of trabecular bone loss is estrogen-independent [53].

In agreement with the clinical evidence, studies in the murine model have revealed that age-related mechanisms intrinsic to bone such as oxidative stress (OS) are protagonists and age-related changes in other organs and tissues, such as ovaries, may be contributory [52].

Reactive oxygen species (ROS) influence the birth and death of bone cells

Oxidative stress (OS) has been thought for many years as a common mechanism in the pathogenesis of several degenerative disorders associated with aging. The same seems to be true for osteoporosis. Indeed, several lines of evidence implicate an increase in reactive oxygen species (ROS) in the decreased bone formation associated with advancing age, as well as the increased resorption associated with estrogen deficiency [12, 52, 56]. In line with this evidence, OS caused by increased ROS production in osteoblasts stimulates apoptosis and decreases bone formation [12, 57, 58]. On the other hand, ROS are a critical requirement for RANKL-induced osteoclast generation, activation, and survival (Fig. 25.3) [52]. In agreement with the results from the study of ROS in bone biology, it is now widely appreciated that ROS can be both: harmful byproducts of aerobic metabolism that damage proteins, lipids, and DNA leading to cell demise, as well as signalers deliberately produced by cells for the purpose of propagating intracellular signaling from cell surface receptors [52].

To protect against OS, organisms ranging from prokaryotes to mammals, scavenge ROS by a network of overlapping mechanisms including various forms of superoxide dismutases (SODs) and catalase as well as thiol-containing oligopeptides with redox-active sulfhydryl moieties, the most abundant of which are glutathione and thioredoxin. Reduction of these scavenging mechanisms along with increased mitochondrial respira-

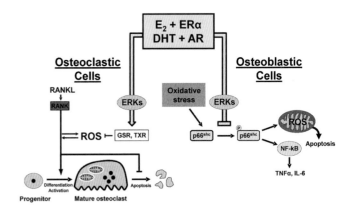

Fig. 25.3. Reactive oxygen species (ROS) influence the birth and death of osteoclasts and osteoblasts, and their effects are counterregulated by sex steroids. In osteoclasts, ROS are required for RANKL-induced differentiation, activation, and survival. On the other hand, ROS stimulate osteoblast apoptosis and attenuate osteoblastogenesis. Estrogens or androgens, acting via cytoplasmic kinases like ERKs, antagonize the production or the effects of ROS. These antioxidant actions are responsible for the ability of sex steroids to attenuate osteoclast generation and survival, promote osteoblast and osteocyte survival, and attenuate NF-κB and cytokine production. Reproduced with modifications and permission from *Endocrine Reviews* (Ref 52).

tory chain leakage and increased activity of oxidases in other cellular compartments are the three main mechanisms for OS [52]. Over the past few years, FoxO transcription factors have emerged as another important defense mechanism against OS, serving primarily to maintain the integrity of long-lived cells, including tissue stem cells. Consistent with this, global deletion of FoxOs recapitulates the effects of old age on the murine skeleton in young mice [57], suggesting that oxidative defense provides a mechanism to handle the oxygen free radicals constantly generated by the aerobic metabolism of osteoblasts and is thereby indispensable for bone mass homeostasis throughout life.

Estrogens or androgens attenuate oxidative stress in bone cells

Sex steroid deficiency, similar to aging, increases the generation of ROS and the activity of the tumor suppressor p53 and p66^{Shc} (a 66-Kd isoform of the growth factor adapter Shc with a pivotal role in OS, apoptosis, and aging) in the murine skeleton; the adverse effects of the acute loss of ovarian or testicular function on bone can be prevented by antioxidants (Fig. 25.4) [12, 56]. Thus, sex steroid deficiency accelerates the effect of aging [42, 52].

Conversely, estrogens or nonaromatizable androgens decrease OS and p66^{Shc} activation and antagonize ROS-induced osteoblast apoptosis, NF-κB activation, and cytokine production by attenuating a PKCβ/p66^{Shc} signaling cascade in osteoblastic cells [12, 59] (Fig. 25.3). They also decrease ROS production in bone marrow [12, 60]. On the other hand, estrogens or androgens attenuate the prosurvival effects of RANKL on osteoclasts also via antioxidant mechanisms. All the antioxidant effects of estrogens are preserved in the mice bearing the ERα knock-in mutation that prevents binding to DNA (ERα$^{NERKI/-}$) and are mimicked by an estradiol dendrimer conjugate (EDC) that is not capable of stimulating the nuclear-initiated actions of ERα [37, 61]. These observations suggest that the effects of estrogens on OS and the birth and death of osteoblasts and osteoclasts do not require the binding of ERα to DNA response elements, but instead they result from the activation of cytoplasmic kinases (Fig. 25.3).

A cell autonomous effect of estrogens that favors osteoblast formation by protecting osteoblast progenitors from the adverse effects of ROS (and perhaps the generation of ROS) is consistent with the finding that ERα deletion from mesenchymal progenitors (expressing Osx1) results in reduced cortical bone mass [41].

Based on these latest advances, it appears that estrogen deficiency may not only decrease mature osteoblast survival but also hinder osteoblast generation in such a way that the osteoblast number is insufficient relative to the demand caused by the increased resorption. In females, loss of estrogen actions in osteoclasts in and of itself is sufficient for loss of bone in the cancellous compartment in which complete perforation of trabeculae by osteoclastic resorption precludes subsequent refilling of the cavities by the bone-forming osteoblasts. On the other hand, loss of estrogen actions on osteoblast generation and lifespan may be responsible for the loss of cortical bone (Fig. 25.5).

SEX STEROID DEFICIENCY AND THE DEVELOPMENT OF OSTEOPOROSIS

Both women and men lose bone as a result of age, but men are less likely to develop osteoporosis than women for two reasons. First, men gain more bone during puberty, and second they lose less bone during aging because unlike women, men do not experience an abrupt loss of estrogens.

The accelerated cancellous bone loss caused by menopause results predominantly from trabecular perforation and loss of connectivity. This phase is followed a few years later by a phase of slower bone loss that primarily affects cortical sites. This later stage occurs in both women and men and is associated with a decrease in osteoblast number and bone formation rate and reduced number of trabeculae. In line with this, decreased wall width is the most consistent histologic findings in elderly women and men with osteoporosis. Moreover, bone loss in elderly men is associated with trabecular thinning rather than perforation [62]. The dissimilarity of the effects of acute estrogen deficiency (as in menopause) and prolonged estrogen deficiency (as in elderly osteoporotic male and females) suggests that the effects of the loss of sex steroids are drastically modified and perhaps some—for example the increased osteoclastogenesis—overridden by aging mechanisms intrinsic to bone. Besides aging *per se* and estrogen deficiency, two additional age-dependent pathogenetic mechanisms contribute to increased OS and loss of bone: lipid oxidation [63] and an increase in endogenous glucocorticoids [64]. It is therefore likely that these two mechanisms also participate in the later stage of bone loss by further exaggerating OS and restraining osteoblastogenesis. In support of this contention, the upregulation of osteoblast (and osteoclast) progenitors caused by gonadectomy in mice is prevented by simultaneous administration of glucocorticoids [65].

Estrogen versus androgen deficiency in men with osteoporosis

Estrogen deficiency may be an important mechanism of the development of osteoporosis not only in females but also in males. Support for this view is provided from three types of evidence: (1) the genetic evidence from men with ERα or aromatase mutations discussed earlier [17, 20]; (2) results of short-term clinical experimentation with administration of aromatase inhibitors [27]; and (3) cross-sectional correlations between free serum estradiol levels menopause, has nowadays significantly reduced the long-term use of estrogen replacement for the prevention or treatment of osteoporosis [76, 77]. The FDA-approved selective estrogen receptor modulator, raloxifene, obviates the adverse effects of estrogens on the uterus and breast and is effective in reducing fracture risk. However, its efficacy is lower than natural estrogens and even lower compared to FDA-approved alternative antiresorptive agents, such as the latest bisphosphonates and the anti-RANKL antibody denosumab. Currently, there is no clear understanding of how selective estrogen receptor modulators(SERMs) act as agonists in some tissues and antagonists in others. This situation hinders rational design of an ideal SERM that can be used long-term in the prevention and treatment of osteoporosis without any of the side effects associated with estrogens, including the adverse effects on clotting mechanisms and venous thromboembolic events [78, 79]. Selective AR modulators (SARMs) have been shown in preclinical studies to retain the anabolic efficacy of androgens on bone and muscle while acting as partial antagonists in the prostate [80]. However, the development of SARMs as therapeutics has been hindered by evidence that, as a class, they may have adverse effects on the heart. Therefore, the future of such compounds is at this stage uncertain.

REFERENCES

1. Reifenstein EC, Albright F. 1947. The metabolic effects of steroid hormones in osteoporosis. *J Clin Invest* 26(1): 24–56.
2. Longcope C. 1998. Metabolism of estrogens and progestins. In: Fraser IS, Jansen RP, Lobo RA (eds.) *Estrogens and Progestens in Clinical Practice*. New York: Churchill Livingstone. pp. 89–94.
3. Griffin JE, Wilson JD. 1998. Disorders of the testes and male reproductive tract. In: Wilson JD, Foster DW, Kronenberg HM, et al. (eds.) *Williams Textbook of Endocrinology*. Philadelphia: W.B. Saunders Co. pp. 819–75.
4. Siiteri PK, Murai JT, Hammond GL, Nisker JA, Raymoure WJ, Kuhn RW. 1982. The serum transport of steroid hormones. *Recent Prog Horm Res* 38: 457–510.
5. Khosla S, Melton LJ 3rd, Atkinson EJ, O'Fallon WM, Klee GG, Riggs BL. 1998. Relationship of serum sex steroid levels and bone turnover markers with bone mineral density in men and women: A key role for bioavailable estrogen. *J Clin Endocrinol Metab* 83(7): 2266–74.
6. Tsai MJ, O'Malley BW. 1994. Molecular mechanisms of action of steroid/thyroid receptor superfamily members. *Annu Rev Biochem* 63: 451–86.
7. McKenna NJ, Lanz RB, O'Malley BW. 1999. Nuclear receptor coregulators: Cellular and molecular biology. *Endocr Rev* 20(3): 321–44.
8. Stein B, Yang MX. 1995. Repression of the interleukin-6 promoter by estrogen receptor is mediated by NF-kappa B and C/EBP beta. *Mol Cell Biol* 15: 4971–9.
9. Kousteni S, Bellido T, Plotkin LI, O'Brien CA, Bodenner DL, Han L, Han K, DiGregorio GB, Katzenellenbogen JA, Katzenellenbogen BS, Roberson PK, Weinstein RS, Jilka RL, Manolagas SC. 2001. Nongenotropic, sex-nonspecific signaling through the estrogen or androgen receptors: Dissociation from transcriptional activity. *Cell* 104: 719–30.
10. Rai D, Frolova A, Frasor J, Carpenter AE, Katzenellenbogen BS. 2005. Distinctive actions of membrane-targeted versus nuclear localized estrogen receptors in breast cancer cells. *Mol Endocrinol* 19(6): 1606–17.
11. Vanderschueren D, Vandenput L, Boonen S, Lindberg MK, Bouillon R, Ohlsson C. 2004. Androgens and bone. *Endocr Rev* 25(3): 389–425.
12. Almeida M, Han L, Martin-Millan M, Plotkin LI, Stewart SA, Roberson PK, Kousteni S, O'Brien CA, Bellido T, Parfitt AM, Weinstein RS, Jilka RL, Manolagas SC. 2007. Skeletal involution by age-associated oxidative stress and its acceleration by loss of sex steroids. *J Biol Chem* 282(37): 27285–97.
13. Sims NA, Dupont S, Krust A, Clement-Lacroix P, Minet D, Resche-Rigon M, Gaillard-Kelly M, Baron R. 2002. Deletion of estrogen receptors reveals a regulatory role for estrogen receptors-beta in bone remodeling in females but not in males. *Bone* 30(1): 18–25.
14. Antal MC, Krust A, Chambon P, Mark M. 2008. Sterility and absence of histopathological defects in nonreproductive organs of a mouse ERbeta-null mutant. *Proc Natl Acad Sci U S A* 105(7): 2433–8.
15. Windahl SH, Vidal O, Andersson G, Gustafsson JA, Ohlsson C. 1999. Increased cortical bone mineral content but unchanged trabecular bone mineral density in female ERbeta(-/-) mice. *J Clin Invest* 104(7): 895–901.
16. van der Eerden BC, Karperien M, Wit JM. 2003. Systemic and local regulation of the growth plate. *Endocr Rev* 24(6): 782–801.
17. Jones ME, Boon WC, McInnes K, Maffei L, Carani C, Simpson ER. 2007. Recognizing rare disorders: Aromatase deficiency. *Nat Clin Pract Endocrinol Metab* 3(5): 414–21.
18. Caruso-Nicoletti M, Cassorla F, Skerda M, Ross JL, Loriaux DL, Cutler GB Jr. 1985. Short term, low dose estradiol accelerates ulnar growth in boys. *J Clin Endocrinol Metab* 61(5): 896–8.
19. Stratakis CA, Vottero A, Brodie A, Kirschner LS, DeAtkine D, Lu Q, Yue W, Mitsiades CS, Flor AW, Chrousos GP. 1998. The aromatase excess syndrome is associated with feminization of both sexes and autosomal dominant transmission of aberrant P450 aromatase gene transcription. *J Clin Endocrinol Metab* 83(4): 1348–57.
20. Smith EP, Boyd J, Frank GR, Takahashi H, Cohen RM, Specker B, Williams TC, Lubahn DB, Korach KS. 1994. Estrogen resistance caused by a mutation in the estrogen-receptor gene in a man. *N Engl J Med* 331: 1056–61.
21. Quigley CA, De Bellis A, Marschke KB, el Awady MK, Wilson EM, French FS. 1995. Androgen receptor defects: Historical, clinical, and molecular perspectives. *Endocr Rev* 16(3): 271–321.

tory chain leakage and increased activity of oxidases in other cellular compartments are the three main mechanisms for OS [52]. Over the past few years, FoxO transcription factors have emerged as another important defense mechanism against OS, serving primarily to maintain the integrity of long-lived cells, including tissue stem cells. Consistent with this, global deletion of FoxOs recapitulates the effects of old age on the murine skeleton in young mice [57], suggesting that oxidative defense provides a mechanism to handle the oxygen free radicals constantly generated by the aerobic metabolism of osteoblasts and is thereby indispensable for bone mass homeostasis throughout life.

Estrogens or androgens attenuate oxidative stress in bone cells

Sex steroid deficiency, similar to aging, increases the generation of ROS and the activity of the tumor suppressor p53 and p66^{Shc} (a 66-Kd isoform of the growth factor adapter Shc with a pivotal role in OS, apoptosis, and aging) in the murine skeleton; the adverse effects of the acute loss of ovarian or testicular function on bone can be prevented by antioxidants (Fig. 25.4) [12, 56]. Thus, sex steroid deficiency accelerates the effect of aging [42, 52].

Conversely, estrogens or nonaromatizable androgens decrease OS and p66^{Shc} activation and antagonize ROS-induced osteoblast apoptosis, NF-κB activation, and cytokine production by attenuating a PKCβ/p66^{Shc} signaling cascade in osteoblastic cells [12, 59] (Fig. 25.3). They also decrease ROS production in bone marrow [12, 60]. On the other hand, estrogens or androgens attenuate the pro-survival effects of RANKL on osteoclasts also via antioxidant mechanisms. All the antioxidant effects of estrogens are preserved in the mice bearing the ERα knock-in mutation that prevents binding to DNA (ERα$^{NERKI/-}$) and are mimicked by an estradiol dendrimer conjugate (EDC) that is not capable of stimulating the nuclear-initiated actions of ERα [37, 61]. These observations suggest that the effects of estrogens on OS and the birth and death of osteoblasts and osteoclasts do not require the binding of ERα to DNA response elements, but instead they result from the activation of cytoplasmic kinases (Fig. 25.3).

A cell autonomous effect of estrogens that favors osteoblast formation by protecting osteoblast progenitors from the adverse effects of ROS (and perhaps the generation of ROS) is consistent with the finding that ERα deletion from mesenchymal progenitors (expressing Osx1) results in reduced cortical bone mass [41].

Based on these latest advances, it appears that estrogen deficiency may not only decrease mature osteoblast survival but also hinder osteoblast generation in such a way that the osteoblast number is insufficient relative to the demand caused by the increased resorption. In females, loss of estrogen actions in osteoclasts in and of itself is sufficient for loss of bone in the cancellous compartment in which complete perforation of trabeculae by osteoclastic resorption precludes subsequent refilling of the cavities by the bone-forming osteoblasts. On the other hand, loss of estrogen actions on osteoblast generation and lifespan may be responsible for the loss of cortical bone (Fig. 25.5).

SEX STEROID DEFICIENCY AND THE DEVELOPMENT OF OSTEOPOROSIS

Both women and men lose bone as a result of age, but men are less likely to develop osteoporosis than women for two reasons. First, men gain more bone during puberty, and second they lose less bone during aging because unlike women, men do not experience an abrupt loss of estrogens.

The accelerated cancellous bone loss caused by menopause results predominantly from trabecular perforation and loss of connectivity. This phase is followed a few years later by a phase of slower bone loss that primarily affects cortical sites. This later stage occurs in both women and men and is associated with a decrease in osteoblast number and bone formation rate and reduced number of trabeculae. In line with this, decreased wall width is the most consistent histologic findings in elderly women and men with osteoporosis. Moreover, bone loss in elderly men is associated with trabecular thinning rather than perforation [62]. The dissimilarity of the effects of acute estrogen deficiency (as in menopause) and prolonged estrogen deficiency (as in elderly osteoporotic male and females) suggests that the effects of the loss of sex steroids are drastically modified and perhaps some—for example the increased osteoclastogenesis—overridden by aging mechanisms intrinsic to bone. Besides aging *per se* and estrogen deficiency, two additional age-dependent pathogenetic mechanisms contribute to increased OS and loss of bone: lipid oxidation [63] and an increase in endogenous glucocorticoids [64]. It is therefore likely that these two mechanisms also participate in the later stage of bone loss by further exaggerating OS and restraining osteoblastogenesis. In support of this contention, the upregulation of osteoblast (and osteoclast) progenitors caused by gonadectomy in mice is prevented by simultaneous administration of glucocorticoids [65].

Estrogen versus androgen deficiency in men with osteoporosis

Estrogen deficiency may be an important mechanism of the development of osteoporosis not only in females but also in males. Support for this view is provided from three types of evidence: (1) the genetic evidence from men with ERα or aromatase mutations discussed earlier [17, 20]; (2) results of short-term clinical experimentation with administration of aromatase inhibitors [27]; and (3) cross-sectional correlations between free serum estradiol levels

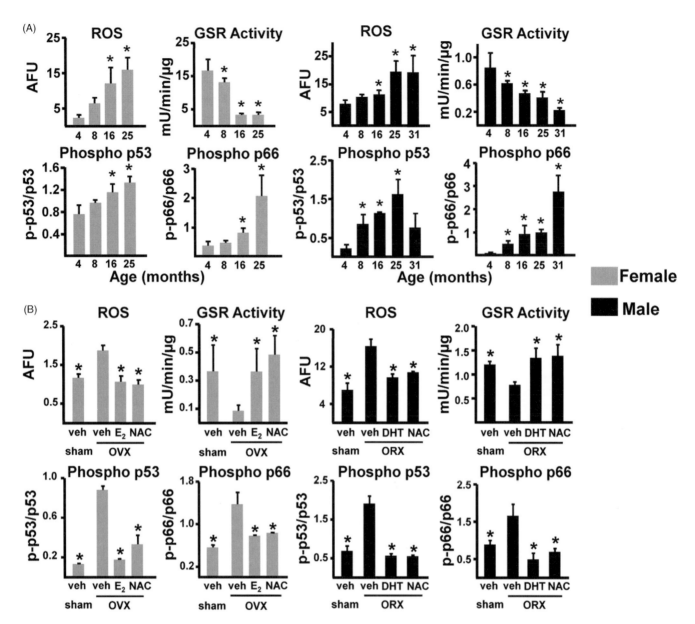

Fig. 25.4. Aging (A) and loss of estrogens or androgens (B) cause similar changes in oxidative stress in the skeleton of male and female mice. ROS and glutathione reductase (GSR) activity were measured in the bone marrow and the phosphorylation of p53 and p66 in vertebral lysates from C57BL6 mice. Ovariectomy (OVX) or orchidectomy (ORX) were performed at 5 months of age, and the measurements were made following euthanasia 6 weeks later. Reproduced with permission from *Endocrine Reviews* (Ref 52).

and bone mineral density (BMD) or bone remodeling markers [53]. However, longitudinal studies of the propositus of genetic ERα mutations indicate that these individuals experience a failure of normal bone development and growth, which is, of course, a different situation to that of the vast majority of patients with osteoporosis who have normal bone development and growth but start losing bone after the attainment of peak bone mass [66]. Furthermore, it remains unknown whether, and to what extent, small differences in estrogen or androgen levels in the circulation contribute to changes in bone markers or BMD in elderly males, especially as similar small differences do not seem to cause BMD differences in pre- or perimenopausal women [67]. More important, and as discussed above, whereas the accelerated cancellous bone loss caused by estrogen deficiency at menopause results predominantly from trabecular perforation and loss of connectivity, the later phase of slower bone loss that occurs in both elderly women and men primarily affects cortical sites [68] and is associated with a decrease in osteoblast number and bone formation rate and reduced thickness of trabeculae.

On the other hand, the important role of androgens and the AR in the homeostasis of the male skeleton in

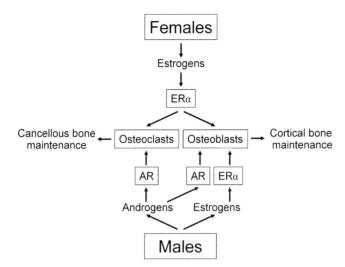

Fig. 25.5. Protective effects of estrogen and androgen on the female vs. the male skeleton. In females, estrogens protect against loss of cancellous and cortical bone mass by cell autonomous effects on osteoclasts and osteoblasts, respectively. In males, cell autonomous effects of androgens (acting via the AR) on osteoclasts are responsible for the protection of the cancellous compartment. However, cell autonomous effects of both androgens (acting via the AR) and aromatizable androgens converted to estrogens (acting via the ERα) are responsible for the protection of the cortical compartment.

humans is well manifested by the low bone mass of men with idiopathic hypogonadotropic hypogonadism or complete androgen insensitivity syndrome [69, 70]. In agreement with this, global AR deletion in male mice results in high bone turnover, increased resorption, and decreased trabecular and cortical bone volume [71]. Nonetheless, in sharp contrast to consistent findings that DHT suppresses osteoclastogenesis, stimulates mature osteoclast apoptosis, and fully protects against orchidectomy (ORX)-induced loss of bone [37, 43], osteoclastogenesis and osteoclast survival were unaffected in the systemic AR deletion model, leading the authors of that study to conclude that the increased resorption was attributed to increased RANKL production by osteoblasts. Importantly, and in contrast to the adverse effects of androgen deficiency on both trabecular and cortical bone mass, trabecular or cortical bone formation was increased in these mice. This makes it highly unlikely, if not implausible, that the bone loss was solely the result of the loss of cell autonomous effects on osteoblasts.

The uncertainty regarding the relative contribution of estrogens versus aging and OS to the development of osteoporosis in elderly men notwithstanding, ER-mediated actions of aromatizable androgens seem to play a minimal or no role in the protective effects of androgens on cancellous bone, but they may play a role in the protective effect of androgens on cortical bone. In support of the contention that the protective effect of androgens on trabecular bone cannot be the result of ERα-mediated action of aromatized androgens, deletion of the ERα selectively from mature osteoclasts or the entire osteoclast lineage in mice has no effect on trabecular bone in the male [36, 37]. On the other hand, nonaromatizable androgens prevent the ORX-induced loss of trabecular bone. Moreover, trabecular number is preserved in men with homozygous loss of function mutation of ERα, indicating that the AR can preserve the trabecular number in the absence of the ERα [66]. Consistent with an AR-mediated effect of androgens on the preservation of trabecular bone, deletion or overexpression of the AR in mature osteoblasts decreases and increases trabecular bone volume, respectively [72, 73]. Nonetheless, the effect of such genetic manipulation on trabecular bone is minimal, as compared to the global deletion of the AR, indicating that AR mediated effects of androgens on osteoclasts are the predominant mechanism of the protective effect of androgens on this site (Fig. 25.5).

In contrast to the global AR deletion, specific deletion of the AR from mature osteoblasts or osteoclasts has no effect on cortical thickness or periosteal diameter [74, 75]. The lack of an effect on cortical bone in mice that lack AR in mature osteoblasts, in the face of such an effect with global AR deletion, strongly suggests that the AR plays an important role in earlier stages of osteoblastogenesis. Support for this possibility has been provided by the selective ERα deletion in early osteoblast progenitors [41], indicating for the first time that sex steroids may positively regulate bone formation by cell autonomous actions on these early progenitors (Fig. 25.5).

To sum this section up, it remains currently unclear how much of the bone-sparing effect of androgens on the adult male skeleton results from effects of T or DHT acting via the AR as opposed to androgens converted to estrogens and acting via the ER [23]. It is also unclear to what extent the bone-sparing effect of androgens result from cell autonomous actions on osteoclasts, osteoblasts, or some other cell type, and whether their protective effects on the adult skeleton result from similar or different mechanisms to those responsible for their effects on skeletal growth. Thus, a considerable gap in the understanding of the basic and clinical aspects of androgen action on bone makes uncertain how much of the osteoporosis in old men results from sex steroid deficiency versus aging, and of the hormonal component, how much is due to true estrogen versus androgen deficiency and/or a decrease in bioavailable E_2 as opposed to T or DHT. However, several clinical studies show correlation between a decrease in bioavailable E_2, but not T, and bone mass in elderly men [53].

ROLE OF SEX STEROIDS IN THE TREATMENT OF OSTEOPOROSIS

The recognition of the serious side effects associated with natural estrogen-based therapies in the uterus, breast, and the cardiovascular system, as well as their decreased efficacy in older women the longer they are from the onset of menopause, has nowadays significantly reduced the long-term use of estrogen replacement for the prevention or treatment of osteoporosis [76, 77]. The FDA-approved selective estrogen receptor modulator, raloxifene, obviates the adverse effects of estrogens on the uterus and breast and is effective in reducing fracture risk. However, its efficacy is lower than natural estrogens and even lower compared to FDA-approved alternative antiresorptive agents, such as the latest bisphosphonates and the anti-RANKL antibody denosumab. Currently, there is no clear understanding of how selective estrogen receptor modulators(SERMs) act as agonists in some tissues and antagonists in others. This situation hinders rational design of an ideal SERM that can be used long-term in the prevention and treatment of osteoporosis without any of the side effects associated with estrogens, including the adverse effects on clotting mechanisms and venous thromboembolic events [78, 79]. Selective AR modulators (SARMs) have been shown in preclinical studies to retain the anabolic efficacy of androgens on bone and muscle while acting as partial antagonists in the prostate [80]. However, the development of SARMs as therapeutics has been hindered by evidence that, as a class, they may have adverse effects on the heart. Therefore, the future of such compounds is at this stage uncertain.

REFERENCES

1. Reifenstein EC, Albright F. 1947. The metabolic effects of steroid hormones in osteoporosis. *J Clin Invest* 26(1): 24–56.
2. Longcope C. 1998. Metabolism of estrogens and progestins. In: Fraser IS, Jansen RP, Lobo RA (eds.) *Estrogens and Progestens in Clinical Practice*. New York: Churchill Livingstone. pp. 89–94.
3. Griffin JE, Wilson JD. 1998. Disorders of the testes and male reproductive tract. In: Wilson JD, Foster DW, Kronenberg HM, et al. (eds.) *Williams Textbook of Endocrinology*. Philadelphia: W.B. Saunders Co. pp. 819–75.
4. Siiteri PK, Murai JT, Hammond GL, Nisker JA, Raymoure WJ, Kuhn RW. 1982. The serum transport of steroid hormones. *Recent Prog Horm Res* 38: 457–510.
5. Khosla S, Melton LJ 3rd, Atkinson EJ, O'Fallon WM, Klee GG, Riggs BL. 1998. Relationship of serum sex steroid levels and bone turnover markers with bone mineral density in men and women: A key role for bioavailable estrogen. *J Clin Endocrinol Metab* 83(7): 2266–74.
6. Tsai MJ, O'Malley BW. 1994. Molecular mechanisms of action of steroid/thyroid receptor superfamily members. *Annu Rev Biochem* 63: 451–86.
7. McKenna NJ, Lanz RB, O'Malley BW. 1999. Nuclear receptor coregulators: Cellular and molecular biology. *Endocr Rev* 20(3): 321–44.
8. Stein B, Yang MX. 1995. Repression of the interleukin-6 promoter by estrogen receptor is mediated by NF-kappa B and C/EBP beta. *Mol Cell Biol* 15: 4971–9.
9. Kousteni S, Bellido T, Plotkin LI, O'Brien CA, Bodenner DL, Han L, Han K, DiGregorio GB, Katzenellenbogen JA, Katzenellenbogen BS, Roberson PK, Weinstein RS, Jilka RL, Manolagas SC. 2001. Nongenotropic, sex-nonspecific signaling through the estrogen or androgen receptors: Dissociation from transcriptional activity. *Cell* 104: 719–30.
10. Rai D, Frolova A, Frasor J, Carpenter AE, Katzenellenbogen BS. 2005. Distinctive actions of membrane-targeted versus nuclear localized estrogen receptors in breast cancer cells. *Mol Endocrinol* 19(6): 1606–17.
11. Vanderschueren D, Vandenput L, Boonen S, Lindberg MK, Bouillon R, Ohlsson C. 2004. Androgens and bone. *Endocr Rev* 25(3): 389–425.
12. Almeida M, Han L, Martin-Millan M, Plotkin LI, Stewart SA, Roberson PK, Kousteni S, O'Brien CA, Bellido T, Parfitt AM, Weinstein RS, Jilka RL, Manolagas SC. 2007. Skeletal involution by age-associated oxidative stress and its acceleration by loss of sex steroids. *J Biol Chem* 282(37): 27285–97.
13. Sims NA, Dupont S, Krust A, Clement-Lacroix P, Minet D, Resche-Rigon M, Gaillard-Kelly M, Baron R. 2002. Deletion of estrogen receptors reveals a regulatory role for estrogen receptors-beta in bone remodeling in females but not in males. *Bone* 30(1): 18–25.
14. Antal MC, Krust A, Chambon P, Mark M. 2008. Sterility and absence of histopathological defects in nonreproductive organs of a mouse ERbeta-null mutant. *Proc Natl Acad Sci U S A* 105(7): 2433–8.
15. Windahl SH, Vidal O, Andersson G, Gustafsson JA, Ohlsson C. 1999. Increased cortical bone mineral content but unchanged trabecular bone mineral density in female ERbeta(-/-) mice. *J Clin Invest* 104(7): 895–901.
16. van der Eerden BC, Karperien M, Wit JM. 2003. Systemic and local regulation of the growth plate. *Endocr Rev* 24(6): 782–801.
17. Jones ME, Boon WC, McInnes K, Maffei L, Carani C, Simpson ER. 2007. Recognizing rare disorders: Aromatase deficiency. *Nat Clin Pract Endocrinol Metab* 3(5): 414–21.
18. Caruso-Nicoletti M, Cassorla F, Skerda M, Ross JL, Loriaux DL, Cutler GB Jr. 1985. Short term, low dose estradiol accelerates ulnar growth in boys. *J Clin Endocrinol Metab* 61(5): 896–8.
19. Stratakis CA, Vottero A, Brodie A, Kirschner LS, DeAtkine D, Lu Q, Yue W, Mitsiades CS, Flor AW, Chrousos GP. 1998. The aromatase excess syndrome is associated with feminization of both sexes and autosomal dominant transmission of aberrant P450 aromatase gene transcription. *J Clin Endocrinol Metab* 83(4): 1348–57.
20. Smith EP, Boyd J, Frank GR, Takahashi H, Cohen RM, Specker B, Williams TC, Lubahn DB, Korach KS. 1994. Estrogen resistance caused by a mutation in the estrogen-receptor gene in a man. *N Engl J Med* 331: 1056–61.
21. Quigley CA, De Bellis A, Marschke KB, el Awady MK, Wilson EM, French FS. 1995. Androgen receptor defects: Historical, clinical, and molecular perspectives. *Endocr Rev* 16(3): 271–321.

22. Bouillon R, Bex M, Vanderschueren D, Boonen S. 2004. Estrogens are essential for male pubertal periosteal bone expansion. *J Clin Endocrinol Metab* 89(12): 6025–9.
23. Callewaert F, Boonen S, Vanderschueren D. 2010. Sex steroids and the male skeleton: A tale of two hormones. *Trends Endocrinol Metab* 21(2): 89–95.
24. Windahl SH, Andersson N, Borjesson AE, Swanson C, Svensson J, Moverare-Skrtic S, Sjogren K, Shao R, Lagerquist MK, Ohlsson C. 2011. Reduced bone mass and muscle strength in male 5α-reductase type 1 inactivated mice. *PLoS ONE* 6(6): e21402.
25. Callewaert F, Venken K, Kopchick JJ, Torcasio A, van Lenthe GH, Boonen S, Vanderschueren D. 2010. Sexual dimorphism in cortical bone size and strength but not density is determined by independent and time-specific actions of sex steroids and IGF-1: Evidence from pubertal mouse models. *J Bone Miner Res* 25(3): 617–26.
26. Callewaert F, Venken K, Ophoff J, De Gendt K, Torcasio A, van Lenthe GH, Van Oosterwyck H, Boonen S, Bouillon R, Verhoeven G, Vanderschueren D. 2009. Differential regulation of bone and body composition in male mice with combined inactivation of androgen and estrogen receptor-alpha. *FASEB J* 23(1): 232–40.
27. Santen RJ, Brodie H, Simpson ER, Siiteri PK, Brodie A. 2009. History of aromatase: Saga of an important biological mediator and therapeutic target. *Endocr Rev* 30(4): 343–75.
28. Börjesson AE, Lagerquist MK, Liu C, Shao R, Windahl SH, Karlsson C, Sjögren K, Movérare-Skrtic S, Antal MC, Krust A, Mohan S, Chambon P, Sävendahl L, Ohlsson C. 2010. The role of estrogen receptor alpha in growth plate cartilage for longitudinal bone growth. *J Bone Miner Res* 25(12): 2690–700.
29. Tajima Y, Yokose S, Kawasaki M, Takuma T. 1998. Ovariectomy causes cell proliferation and matrix synthesis in the growth plate cartilage of the adult rat. *Histochem J* 30(7): 467–72.
30. Xiong J, Onal M, Jilka RL, Weinstein RS, O'Brien CA. 2011. Matrix-embedded cells control osteoclast formation. *Nat Med* 17(10): 1235–41.
31. Manolagas SC. 2000. Birth and death of bone cells: Basic regulatory mechanisms and implications for the pathogenesis and treatment of osteoporosis. *Endocr Rev* 21(2): 115–37.
32. Manolagas SC, Kousteni S, Jilka RL. 2002. Sex steroids and bone. *Recent Prog Horm Res* 57: 385–409.
33. Weitzmann MN, Pacifici R. 2006. Estrogen deficiency and bone loss: An inflammatory tale. *J Clin Invest* 116(5): 1186–94.
34. Eghbali-Fatourechi G, Khosla S, Sanyal A, Boyle WJ, Lacey DL, Riggs BL. 2003. Role of RANK ligand in mediating increased bone resorption in early postmenopausal women. *J Clin Invest* 111(8): 1221–30.
35. Clowes JA, Riggs BL, Khosla S. 2005. The role of the immune system in the pathophysiology of osteoporosis. *Immunol Rev* 208: 207–27.
36. Nakamura T, Imai Y, Matsumoto T, Sato S, Takeuchi K, Igarashi K, Harada Y, Azuma Y, Krust A, Yamamoto Y, Nishina H, Takeda S, Takayanagi H, Metzger D, Kanno J, Takaoka K, Martin TJ, Chambon P, Kato S. 2007. Estrogen prevents bone loss via estrogen receptor alpha and induction of Fas ligand in osteoclasts. *Cell* 130(5): 811–23.
37. Martin-Millan M, Almeida M, Ambrogini E, Han L, Zhao H, Weinstein RS, Jilka RL, O'Brien C, Manolagas SC. 2010. The estrogen receptor alpha in osteoclasts mediates the protective effects of estrogens on cancellous but not cortical bone. *Mol Endocrinol* 24(2): 323–34.
38. Krum SA, Miranda-Carboni GA, Hauschka PV, Carroll JS, Lane TF, Freedman LP, Brown M. 2008. Estrogen protects bone by inducing Fas ligand in osteoblasts to regulate osteoclast survival. *EMBO J* 27(3): 535–45.
39. Mödder UI, Sanyal A, Kearns AE, Sibonga JD, Nishihara E, Xu J, O'Malley BW, Ritman EL, Riggs BL, Spelsberg TC, Khosla S. 2004. Effects of loss of steroid receptor coactivator-1 on the skeletal response to estrogen in mice. *Endocrinology* 145(2): 913–21.
40. Lee SK, Kadono Y, Okada F, Jacquin C, Koczon-Jaremko B, Gronowicz G, Adams DJ, Aguila HL, Choi Y, Lorenzo JA. 2006. T lymphocyte-deficient mice lose trabecular bone mass with ovariectomy. *J Bone Miner Res* 21(11): 1704–12.
41. Almeida M, Iyer S, Martin-Millan M, Bartell SM, Han L, Ambrogini E, Onal M, Xiong J, Weinstein RS, Jilka RL, O'Brien CA, Manolagas SC. 2013. Estrogen receptor α signaling in osteoblast progenitors stimulates cortical bone accrual. *J Clin Invest* 123: 394–404.
42. Manolagas SC, Parfitt AM. 2010. What old means to bone. *Trends Endocrinol Metab* 21(6): 369–74.
43. Kousteni S, Chen JR, Bellido T, Han L, Ali AA, O'Brien CA, Plotkin L, Fu Q, Mancino AT, Wen Y, Vertino AM, Powers CC, Stewart SA, Ebert R, Parfitt AM, Weinstein RS, Jilka RL, Manolagas SC. 2002. Reversal of bone loss in mice by nongenotropic signaling of sex steroids. *Science* 298: 843–6.
44. Tomkinson A, Reeve J, Shaw RW, Noble BS. 1997. The death of osteocytes via apoptosis accompanies estrogen withdrawal in human bone. *J Clin Endocrinol Metab* 82(9): 3128–35.
45. Tomkinson A, Gevers EF, Wit JM, Reeve J, Noble BS. 1998. The role of estrogen in the control of rat osteocyte apoptosis. *J Bone Miner Res* 13(8): 1243–50.
46. Marcus R. 2002. Mechanisms of exercise effects on bone. In: Bilezikian JP, Raisz LG, Rodan GA (eds.) *Principles of Bone Biology*. San Diego: Academic Press. pp. 1477–88.
47. Lee K, Jessop H, Suswillo R, Zaman G, Lanyon L. 2003. Endocrinology: Bone adaptation requires oestrogen receptor-alpha. *Nature* 424(6947): 389.
48. Aguirre JI, Plotkin LI, Stewart SA, Weinstein RS, Parfitt AM, Manolagas SC, Bellido T. 2006. Osteocyte apoptosis is induced by weightlessness in mice and precedes osteoclast recruitment and bone loss. *J Bone Miner Res* 21(4): 605–15.
49. Plotkin LI, Mathov I, Aguirre JI, Parfitt AM, Manolagas SC, Bellido T. 2005. Mechanical stimulation prevents osteocyte apoptosis: Requirement of integrins, Src

50. O'Brien CA, Jia D, Plotkin LI, Bellido T, Powers CC, Stewart SA, Manolagas SC, Weinstein RS. 2004. Glucocorticoids act directly on osteoblasts and osteocytes to induce their apoptosis and reduce bone formation and strength. *Endocrinology* 145(4): 1835–41.
51. Winkler DG, Sutherland MK, Geoghegan JC, Yu C, Hayes T, Skonier JE, Shpektor D, Jonas M, Kovacevich BR, Staehling-Hampton K, Appleby M, Brunkow ME, Latham JA. 2003. Osteocyte control of bone formation via sclerostin, a novel BMP antagonist. *EMBO J* 22(23): 6267–76.
52. Manolagas SC. 2010. From estrogen-centric to aging and oxidative stress: A revised perspective of the pathogenesis of osteoporosis. *Endocr Rev* 31(3): 266–300.
53. Khosla S, Melton LJ 3rd, Riggs BL. 2011. The unitary model for estrogen deficiency and the pathogenesis of osteoporosis: Is a revision needed? *J Bone Miner Res* 26(3): 441–51.
54. Zebaze RM, Ghasem-Zadeh A, Bohte A, Iuliano-Burns S, Mirams M, Price RI, Mackie EJ, Seeman E. 2010. Intracortical remodelling and porosity in the distal radius and post-mortem femurs of women: A cross-sectional study. *Lancet* 375(9727): 1729–36.
55. Recker R, Lappe J, Davies K, Heaney R. 2000. Characterization of perimenopausal bone loss: A prospective study. *J Bone Miner Res* 15(10): 1965–73.
56. Lean JM, Davies JT, Fuller K, Jagger CJ, Kirstein B, Partington GA, Urry ZL, Chambers TJ. 2003. A crucial role for thiol antioxidants in estrogen-deficiency bone loss. *J Clin Invest* 112(6): 915–23.
57. Ambrogini E, Almeida M, Martin-Millan M, Paik JH, Depinho RA, Han L, Goellner J, Weinstein RS, Jilka RL, O'Brien CA, Manolagas SC. 2010. FoxO-mediated defense against oxidative stress in osteoblasts is indispensable for skeletal homeostasis in mice. *Cell Metab* 11(2): 136–46.
58. Jilka RL, Almeida M, Ambrogini E, Han L, Roberson PK, Weinstein RS, Manolagas SC. 2010. Decreased oxidative stress and greater bone anabolism in the aged, as compared to the young, murine skeleton by parathyroid hormone. *Aging Cell* 9(5): 851–67.
59. Almeida M, Han L, Ambrogini E, Bartell SM, Manolagas SC. 2010. Oxidative stress stimulates apoptosis and activates NF-kappaB in osteoblastic cells via a PKCbeta/p66shc signaling cascade: Counter regulation by estrogens or androgens. *Mol Endocrinol* 24(10): 2030–7.
60. Grassi F, Tell G, Robbie-Ryan M, Gao Y, Terauchi M, Yang X, Romanello M, Jones DP, Weitzmann MN, Pacifici R. 2007. Oxidative stress causes bone loss in estrogen-deficient mice through enhanced bone marrow dendritic cell activation. *Proc Natl Acad Sci U S A* 104(38): 15087–92.
61. Almeida M, Martin-Millan M, Ambrogini E, Bradsher R 3rd, Han L, Chen XD, Roberson PK, Weinstein RS, O'Brien CA, Jilka RL, Manolagas SC. 2010. Estrogens attenuate oxidative stress, osteoblast differentiation and apoptosis by DNA binding-independent actions of the ERalpha. *J Bone Mineral Res* 25(4): 769–81.
62. Ebeling PR. 2008. Clinical practice. Osteoporosis in men. *N Engl J Med* 358(14): 1474–82.
63. Almeida M, Ambrogini E, Han L, Manolagas SC, Jilka RL. 2009. Increased lipid oxidation causes oxidative stress, increased PPAR(gamma) expression and diminished pro-osteogenic Wnt signaling in the skeleton. *J Biol Chem* 284(40): 27438–48.
64. Weinstein RS, Wan C, Liu Q, Wang Y, Almeida M, O'Brien CA, Thostenson J, Roberson PK, Boskey AL, Clemens TL, Manolagas SC. 2010. Endogenous glucocorticoids decrease skeletal angiogenesis, vascularity, hydration, and strength in aged mice. *Aging Cell* 9: 147–61.
65. Weinstein RS, Jia D, Powers CC, Stewart SA, Jilka RL, Parfitt AM, Manolagas SC. 2004. The skeletal effects of glucocorticoid excess override those of orchidectomy in mice. *Endocrinology* 145(4): 1980–7.
66. Smith EP, Specker B, Bachrach BE, Kimbro KS, Li XJ, Young MF, Fedarko NS, Abuzzahab MJ, Frank GR, Cohen RM, Lubahn DB, Korach KS. 2008. Impact on bone of an estrogen receptor-alpha gene loss of function mutation. *J Clin Endocrinol Metab* 93(8): 3088–96.
67. Sowers MR, Greendale GA, Bondarenko I, Finkelstein JS, Cauley JA, Neer RM, Ettinger B. 2003. Endogenous hormones and bone turnover markers in pre- and perimenopausal women: SWAN. *Osteoporos Int* 14(3): 191–7.
68. Parfitt AM. 1992. The two-stage concept of bone loss revisited. *Triangle* 31: 99–110.
69. Finkelstein JS, Klibanski A, Neer RM, Greenspan SL, Rosenthal DI, Crowley WF Jr. 1987. Osteoporosis in men with idiopathic hypogonadotropic hypogonadism. *Ann Intern Med* 106: 354–61.
70. Marcus R, Leary D, Schneider DL, Shane E, Favus M, Quigley CA. 2000. The contribution of testosterone to skeletal development and maintenance: Lessons from the androgen insensitivity syndrome. *J Clin Endocrinol Metab* 85(3): 1032–7.
71. Kawano H, Sato T, Yamada T, Matsumoto T, Sekine K, Watanabe T, Nakamura T, Fukuda T, Yoshimura K, Yoshizawa T, Aihara K, Yamamoto Y, Nakamichi Y, Metzger D, Chambon P, Nakamura K, Kawaguchi H, Kato S. 2003. Suppressive function of androgen receptor in bone resorption. *Proc Natl Acad Sci U S A* 100(16): 9416–21.
72. Wiren KM, Semirale AA, Zhang XW, Woo A, Tommasini SM, Price C, Schaffler MB, Jepsen KJ. 2008. Targeting of androgen receptor in bone reveals a lack of androgen anabolic action and inhibition of osteogenesis: A model for compartment-specific androgen action in the skeleton. *Bone* 43(3): 440–51.
73. Chiang C, Chiu M, Moore AJ, Anderson PH, Ghasem-Zadeh A, McManus JF, Ma C, Seeman E, Clemens TL, Morris HA, Zajac JD, Davey RA. 2009. Mineralization and bone resorption are regulated by the androgen receptor in male mice. *J Bone Miner Res* 24(4): 621–31.

74. Nakamura T, Watanabe T, Nakamichi Y, Azuma Y, Yoshimura K, Matsumoto T, Fukuda T, Ochiai E, Metzger D, Chambon P, et al. 2005. Genetic evidence of androgen receptor function in osteoclasts. *J Bone Min Res* 20(Suppl 1): S104.
75. Notini AJ, McManus JF, Moore A, Bouxsein M, Jimenez M, Chiu WS, Glatt V, Kream BE, Handelsman DJ, Morris HA, Zajac JD, Davey RA. 2006. Osteoblast deletion of exon 3 of the androgen receptor gene results in trabecular bone loss in adult male mice. *J Bone Miner Res* 22(3): 347–56.
76. Rossouw JE, Anderson GL, Prentice RL, LaCroix AZ, Kooperberg C, Stefanick ML, Jackson RD, Beresford SA, Howard BV, Johnson KC, Kotchen JM, Ockene J; Writing Group for the Women's Health Initiative Investigators. 2002. Risks and benefits of estrogen plus progestin in healthy postmenopausal women: Principal results From the Women's Health Initiative randomized controlled trial. *JAMA* 288(3): 321–33.
77. North American Menopause Society. 2010. Estrogen and progestogen use in postmenopausal women: 2010 position statement of The North American Menopause Society. *Menopause* 17(2): 242–55.
78. Cummings SR, Ensrud K, Delmas PD, LaCroix AZ, Vukicevic S, Reid DM, Goldstein S, Sriram U, Lee A, Thompson J, Armstrong RA, Thompson DD, Powles T, Zanchetta J, Kendler D, Neven P, Eastell R; PEARL Study Investigators. 2010. Lasofoxifene in postmenopausal women with osteoporosis. *N Engl J Med* 362(8): 686–96.
79. Cummings SR, McClung M, Reginster JY, Cox D, Mitlak B, Stock J, Amewou-Atisso M, Powles T, Miller P, Zanchetta J, Christiansen C. 2011. Arzoxifene for prevention of fractures and invasive breast cancer in postmenopausal women. *J Bone Miner Res* 26(2): 397–404.
80. Rosen J, Negro-Vilar A. 2002. Novel, non-steroidal, selective androgen receptor modulators (SARMs) with anabolic activity in bone and muscle and improved safety profile. *J Musculoskelet Neuronal Interact* 2(3): 222–4.

26
Parathyroid Hormone
Robert A. Nissenson and Harald Jüppner

Introduction 208
Parathyroid Hormone and Parathyroid Hormone-Related Peptide 208
PTH Synthesis and Secretion 209
Regulation of PTH Secretion by Extracellular Calcium 210

Regulation of PTH Secretion by 1,25(OH)$_2$ Vitamin D 211
Regulation of PTH Secretion by Plasma Phosphate, α-Klotho, and FGF23 211
Mechanism of Action of PTH 212
References 212

INTRODUCTION

The parathyroid glands first appeared during evolution with the movement of animals from an aquatic environment to a terrestrial environment deficient in calcium. Maintenance of adequate levels of blood ionized calcium (1.1–1.3 mM) is required for normal neuromuscular function, bone mineralization, and many other physiological processes. Chief cells in the parathyroid gland secrete parathyroid hormone (PTH) in response to very small decrements in blood ionized calcium in order to maintain the normocalcemic state. As discussed later, PTH accomplishes this task by promoting bone resorption and releasing calcium from the skeletal reservoir; by reducing urinary calcium losses and increasing phosphate excretion; and by enhancing intestinal calcium absorption, indirectly through the renal production of the active vitamin D metabolite 1,25(OH)$_2$ vitamin D. Blood ionized calcium and 1,25(OH)$_2$ vitamin D contribute to the negative feedback inhibition of PTH secretion, whereas serum phosphate increases PTH secretion. Fibroblast growth factor 23 (FGF23) is the third hormone that contributes to the regulation of calcium and phosphate homeostasis; it promotes renal phosphate excretion and reduces the circulating levels of 1,25(OH)$_2$ vitamin D, thus diminishing intestinal calcium absorption. The interplay between serum calcium, PTH, FGF23, 1,25(OH)$_2$ vitamin D, and phosphate permit blood ionized calcium levels to be maintained within very narrow limits over a wide range of dietary calcium intake. The present chapter summarizes our current understanding of the biology of PTH secretion and action; for an historical perspective on the field, see Ref. 1.

PARATHYROID HORMONE AND PARATHYROID HORMONE-RELATED PEPTIDE

Mammalian PTH is synthesized as a pre-pro-peptide comprising 115 amino acids, but only the single chain 84 full-length polypeptide is secreted by the parathyroid glands; very limited expression has also been detected in the rodent hypothalamus and thymus. Normal development of these glands depends on the transcription factor GCM2 (Gcm2 in rodents) as well as several other upstream proteins, including SOX3, the transcription factor cascade Hoxa3-Pax1/9-Eya, GATA3, the transcription factor Tbx1, and the Shh-Bmp4 signaling network [2, 3]. Mutations in GATA3 and GCMB can be a cause of isolated hypoparathyroidism [4–7].

PTH displays considerable sequence homology with PTH-related peptide (PTHrP), which is, however, limited to the amino-terminal 1-34 region of both peptides [8]. PTH and PTHrP furthermore share some homology with TIP39 (tuberoinfundibular peptide of 39 amino acids) [9], a factor that is expressed in the brain and testes where it

---

Primer on the Metabolic Bone Diseases and Disorders of Mineral Metabolism, Eighth Edition. Edited by Clifford J. Rosen.
© 2013 American Society for Bone and Mineral Research. Published 2013 by John Wiley & Sons, Inc.

contributes to nociception and male fertility, respectively (Fig. 26.1). PTHrP, originally identified as the humoral mediator of hypercalcemia of malignancy [8, 10, 11], plays important physiological roles in the regulation of endochondral bone formation, smooth muscle function, and branching morphogenesis of the mammary gland [12]. The genes encoding PTH, PTHrP, and TIP39 have a similar structure and at least *PTH* and *PTHrP* are most likely derived from a common ancestral precursor gene (Fig. 26.2).

PTH SYNTHESIS AND SECRETION

In humans, there is a single mammalian PTH gene located on the short arm of chromosome 11, which encodes the precursor molecule preproPTH. After removal of the pre-sequence comprising 25 amino acid residues and of the pro-sequence comprising 6 amino acid residues, the mature PTH peptide of 84 amino acid residues is secreted [13]. The pre-sequence functions as a signal peptide that directs the nascent polypeptide to the machinery that transports it across the membrane of the endoplasmatic reticulum, where the pre-sequence is cleaved off. The role of the pro-sequence is not as clearly defined, but it appears to be required for efficient ER transport of the polypeptide and may play a role in subsequent events such as protein folding [14]. Once produced and packaged into secretory vesicles within the

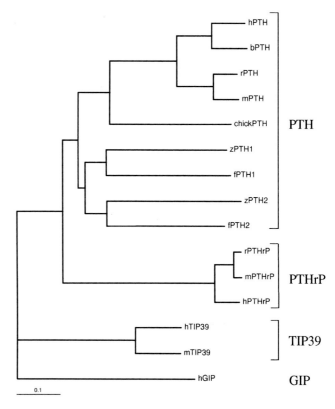

Fig. 26.1. Phylogenetic analysis of PTH and related polypeptides. Reproduced with permission from Gensure, Ponugoti, Gunes et al. (Ref. 58).

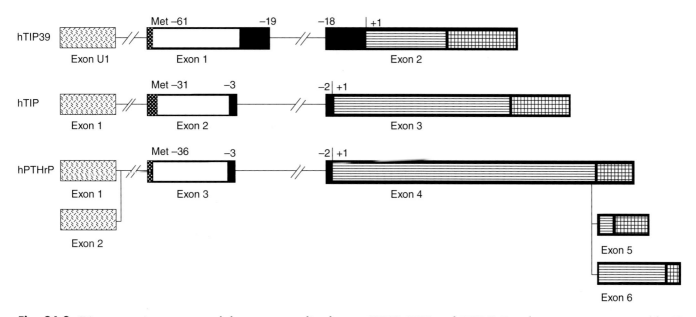

Fig. 26.2. Diagrammatic structures of the genes encoding human TIP39, PTH, and PTHrP. Boxed areas represent exons (the 5' end of exon U1 in the TIP39 gene is not known). White boxes denote pre-sequences, black boxes are pro-sequences, gray stippled boxes are mature protein sequences, and striped boxes are non-coding regions. The small striped boxes preceding the white boxes denote untranslated exonic sequences. The positions of the initiator methionines are also indicated. +1 represents the relative position of the beginning of the secreted protein. Reproduced with permission from John, Arai, Rubin et al. (Ref. 59).

parathyroid chief cells, PTH(1-84) is subject to alternative fates. The mature hormone can be secreted through a classical exocytotic mechanism or it may be cleaved by calcium-sensitive proteases present within secretory vesicles, resulting in the production and secretion of fragments of PTH(1-84) that lack portions of amino-terminal region and are thus inactive with respect to responses mediated by the PTH/PTHrP receptor [15]. Cleavage of circulating PTH(1-84) into carboxyl fragments can also occur in peripheral tissues such as the liver and kidney [16]. Historically, cleavage of PTH(1-84) has been viewed as a mechanism for biological inactivation of the hormone, but some evidence suggests that carboxyl-terminal fragments of PTH may display unique biological properties [17].

REGULATION OF PTH SECRETION BY EXTRACELLULAR CALCIUM

The major physiological function of the parathyroid glands is to act as a "calciostat," sensing the prevailing blood ionized calcium level and adjusting the secretion of PTH accordingly (Fig. 26.3). The relationship between ionized calcium and PTH secretion is a steep sigmoidal one, allowing significant changes in PTH secretion in response to very small changes in blood ionized calcium. The midpoint of this curve ("set point") is a reflection of the sensitivity of the parathyroid gland to suppression by extracellular calcium [18].

Alteration in blood ionized calcium affects the secretion of PTH(1-84) by multiple mechanisms. Short-term increases in extracellular ionized calcium produce increased levels of intracellular free calcium in the parathyroid cell, resulting in the activation of calcium-sensitive proteases in secretory vesicles. As a result, there is increased cleavage of PTH(1-84) into carboxyl-terminal fragments. Increased extracellular calcium also reduces the release of stored PTH from secretory vesicles, although the molecular details of this regulation are not well defined. Long-term changes in blood ionized calcium (e.g., chronic dietary calcium deficiency, resistance toward PTH) result in increased PTH mRNA expression and in the number of PTH-secreting parathyroid cells.

Extracellular calcium is "measured" at the surface of the parathyroid cell through a calcium-sensing receptor (CaSR) that is abundantly expressed at the plasma membrane of these cells [19]. Increased levels of extracellular calcium suppress PTH secretion, while diminished levels increase PTH secretion. Unlike intracellular calcium-binding proteins, which have an affinity for free calcium in the nanomolar range (consistent with intracellular levels of free calcium), CaSR is presumed to bind free calcium with an affinity in the millimolar range. The CaSR is a member of the G protein-coupled receptor superfamily, which contains calcium-binding elements in its extracellular domain and signaling determinants in its cytoplasmic regions. Binding of calcium (or calcimimetic agents such as Sensipar) to the CaSR triggers activation of heterotrimeric G proteins comprising the alpha-subunits Gq/11 and (to a lesser extent) Gi, which results in stimulation of phospholipase C and inhibition of adenylyl cyclase, respectively [20, 21]. This results in an increase in intracellular free calcium and a decrease in cyclic AMP levels in parathyroid cells. By mechanisms that are not fully defined, activation of these signaling pathways suppresses the synthesis and secretion of PTH. When blood ionized calcium falls, there is less intracellular signaling by the CaSR on the plasma membrane of parathyroid cells and PTH secretion consequently increases. The essential role of the CaSR can best be seen in humans bearing loss-of-function mutations in the CaSR gene. In the heterozygous state, such mutations cause familial hypocalciuric hypercalcemia (FHH), characterized by inappropriately elevated levels of PTH in the face of hypercalcemia [22, 23]. These individuals are quantitatively resistant to the suppressive effect of calcium on PTH secretion due to the reduced number of functional CaSRs; this disorder usually requires no surgical intervention. In the homozygous state, patients display a severe increase in PTH secretion with life-threatening hypercalcemia (neonatal severe hyperparathyroidism), which can be controlled with bisphosphonates

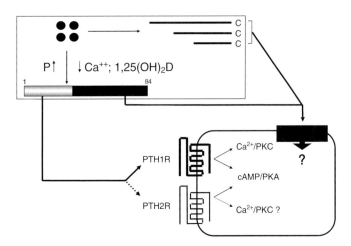

Fig. 26.3. PTH production and activation of different receptors. Intact PTH and different fragments are secreted from the parathyroid glands. Low ionized calcium and elevated phosphate increase PTH synthesis and secretion, while increased ionized calcium and 1,25(OH)$_2$D lead to a decrease; note that the regulatory actions of calcium are mediated through the calcium-sensing receptor. Different receptors interact with the amino- or carboxyl-terminal portion of intact PTH. Through its amino-terminal portion, PTH activates the PTH/PTHrP receptor (PTH1R), a G protein-coupled receptor that mediates its actions through at least two different signaling pathways: cAMP/KPA and Ca^{2+}/PKC. The closely related PTH2-receptor (PTH2R) is most likely the primary receptor for the tuberoinfundibular peptide of 39 residues (TIP39); however, at least the human PTH2R is also activated by amino-terminal PTH. Another receptor, which has not yet been cloned, interacts only with the carboxyl-terminal portion of PTH.

and possibly with calcimimetics, but usually require removal of all parathyroid glands during infancy. Mice with homozygous and heterozygous disruption of the CaSR gene display similar phenotypes [24]. Of interest, deletion in mice of the function of Gq/11, the two major G proteins associated with the CaSR, also results in neonatal severe hyperparathyroidism, confirming the role of Gq/11 in CaSR signaling [25]. Point mutations in the CaSR that induce constitutive signaling are the cause of autosomal dominant hypocalcemia [26, 27].

Limited information is available on the mechanisms by which CaSR signaling suppresses PTH gene expression. Some studies suggest that transcription of the PTH gene is negatively regulated by calcium [28], while others have provided evidence for a post-transcriptional effect of calcium that reduces PTH mRNA stability through binding of a factor(s) to the 3′ untranslated region [29]. This effect appears to be mediated by CaSR-dependent increases in the level of intracellular free calcium [30]. The identity of the relevant factor(s) remains to be established.

Under normal physiological conditions, there is minimal proliferation of parathyroid cells. However, chronic hypocalcemia elicits an increase in both the size and number of parathyroid cells [31].

It is of considerable interest that the CaSR is expressed in a number of tissues outside of the parathyroid gland including kidney, C cells of the thyroid gland, gut, bone, cartilage, and many others [26]. An important function of CaSR in the kidney is to contribute to the regulation of urinary calcium excretion independently of PTH; thus a reduction in signaling as observed in patients with FHH results in hypercalcemia and hypocalciuria due to increased calcium reabsorption in the distal tubules. Although the physiological role of the CaSR in other peripheral tissues is not well understood, recent studies with conditional knockout models suggests that the expression of CaSR in chondrocytes and osteoblasts is essential for normal endochondral bone development [32, 33].

Several pharmacological agents that interact with CaSR have been developed, and these are effective in altering second messenger signaling downstream of the CaSR [27, 34]. So-called calcimimetic drugs bind to the CaSR thereby increasing the receptor's sensitivity to extracellular calcium, increasing receptor signaling, and thus reducing PTH secretion. Calcimimetic drugs have a clinical utility in the medical management of primary or secondary hyperparathyroidism. Calcilytic drugs act as pharmacological antagonists at the CaSR, which reduce the receptor's sensitivity to extracellular calcium, thereby increasing PTH secretion.

REGULATION OF PTH SECRETION BY 1,25(OH)$_2$ VITAMIN D

For many years it has been known that vitamin D deficiency causes an increase in PTH production. This is due to diminished intestinal calcium absorption, which leads to an increase in PTH secretion due to insufficient 1,25(OH)$_2$ vitamin D levels, resulting hypocalcemia and thus diminished negative feedback regulation of PTH production by this biologically active vitamin D analog. 1,25(OH)$_2$ vitamin D production is frequently limited in the setting of chronic renal disease (CKD). Even in the early stages of CKD, levels of FGF23 are already increased to help promote the urinary excretion of phosphate. However, elevated FGF23 levels reduce the 1-alpha hydroxylase activity and enhance the 24-hydroxylase activity, thereby decreasing 1,25(OH)$_2$ vitamin D production and accelerating its metabolism into biologically inactive analogs. These changes in vitamin D metabolism contribute to the development of hypocalcemia and secondary hyperparathyroidism in CKD patients. Frank hyperphosphatemia, which usually does not develop until later in the course of CKD, contributes further to the increase in PTH secretion.

The suppression of PTH secretion by 1,25(OH)$_2$ vitamin D results from the inhibition of transcription of the PTH gene [35]. This appears to involve 1,25(OH)$_2$ vitamin D-induced binding of the vitamin D receptor to negative regulatory elements in the PTH gene promoter [36] and 1,25(OH)$_2$ vitamin D-induced association of the vitamin D receptor with a transcriptional repressor [37]. Calcium and 1,25(OH)$_2$ vitamin D act coordinately to suppress expression of the PTH gene and to inhibit parathyroid cell proliferation.

REGULATION OF PTH SECRETION BY PLASMA PHOSPHATE, α-KLOTHO, AND FGF23

It has long been known that hyperphosphatemia (as in the late stages of CKD) is associated with parathyroid hyperplasia and hyperparathyroidism. This effect of hyperphosphatemia is in part due to the binding of plasma phosphate to free calcium, which lowers blood ionized calcium thus stimulating PTH synthesis, secretion, and parathyroid cell number [38]. However, serum phosphate also directly affects the parathyroid gland, increasing PTH synthesis by promoting the stability of PTH mRNA [29].

Another regulator of parathyroid function is FGF23, which is secreted by osteocytes/osteoblasts in response to increased oral phosphate intake and possibly other factors. FGF23 acts on the kidney to reduce expression of NPT2a and NPT2c, two sodium-dependent phosphate transporters in the proximal renal tubules, thereby reducing phosphate reabsorption [39]. These actions require binding of FGF23 to its cognate FGF receptor and the co-receptor α-Klotho [40]. FGF23 appears to directly inhibit PTH secretion [41, 42] thus providing another level of complexity for the regulation of parathyroid gland activity.

MECHANISM OF ACTION OF PTH

Target cell actions of PTH are initiated by the PTH/PTHrP receptor, a member of the G protein-coupled receptor superfamily [43]. Binding of PTH to its receptor leads to at least three distinct signaling processes [12, 44]. The most important pathway is receptor-mediated activation of Gsα signaling, resulting in adenylyl cyclase activation and thus increased cellular levels of cyclic AMP, and activation of protein kinase A. The importance of this signaling pathway in the renal response to PTH is highlighted by the renal resistance associated with Gsα deficiency in patients with pseudohypoparathyroidism type Ia [45] and with a recurrent PRKAR1A mutation (the regulatory subunit of PKA) in patients with acrodysostosis [46]. The effects of PTH on the expression of key genes that regulate bone resorption and bone formation (e.g., RANKL, SOST) is mediated at least in part through the cyclic AMP pathway [47, 48]. The PTH/PTHrP receptor also couples to Gq, resulting in the activation of phospholipase C with consequent activation of protein kinase C and increased intracellular free calcium. This pathway appears to play a lesser, modulatory role in PTH action in kidney and bone [49, 50]. Binding of PTH to the PTH/PTHrP receptor also recruits the adaptor protein arrestin to the plasma membrane [51]. Arrestin has been shown to participate in the desensitization and downregulation of the PTH/PTHrP receptor [52], but arrestin also acts as a signaling molecule that may contribute to target cell responses to PTH [53]. Interestingly, analogs of PTH that recruit arrestin to the membrane without activating G protein signaling have been shown to promote bone formation *in vivo* [54].

Recent findings have uncovered new complexities in signaling by the PTH/PTHrP receptor. First, the PTH/PTHrP receptor has classically been thought to initiate signaling exclusively from its location in the plasma membrane. However, PTH(1-34) has recently been shown to remain bound to the PTH/PTHrP receptor during the process of receptor endocytosis, and this was associated with persistent Gsα activation/cyclic AMP signaling [55]. The temporal and spatial domains of PTH/PTHrP receptor signaling may be essential determinants of target cell responses. Secondly, PTH/PTHrP receptor activation may play a direct role modulating the activity of the canonical Wnt signaling pathway. The PTH/PTHrP receptor has been shown to physically interact with components of this pathway—LRP6 [56] and disheveled [57], and in both cases these interactions contributed to PTH-stimulated Wnt signaling. These provocative results may provide new insights into the mechanism of action of PTH on bone and kidney *in vivo*.

REFERENCES

1. Potts JT. 2005. Parathyroid hormone: Past and present. *J Endocrinol* 187(3): 311–25.
2. Zajac JD, Danks JA. 2008. The development of the parathyroid gland: From fish to human. *Curr Opin Nephrol Hypertens* 17(4): 353–6.
3. Gordon J, Patel SR, Mishina Y, Manley NR. 2010. Evidence for an early role for BMP4 signaling in thymus and parathyroid morphogenesis. *Dev Biol* 339(1): 141–54.
4. Adachi M, Tachibana K, Asakura Y, Tsuchiya T. 2006. A novel mutation in the GATA3 gene in a family with HDR syndrome (Hypoparathyroidism, sensorineural Deafness and Renal anomaly syndrome). *J Pediatr Endocrinol Metab* 19(1): 87–92.
5. Grigorieva IV, Mirczuk S, Gaynor KU, Nesbit MA, Grigorieva EF, Wei Q, Ali A, Fairclough RJ, Stacey JM, Stechman MJ, Mihai R, Kurek D, Fraser WD, Hough T, Condie BG, Manley N, Grosveld F, Thakker RV. 2010. Gata3-deficient mice develop parathyroid abnormalities due to dysregulation of the parathyroid-specific transcription factor Gcm2. *J Clin Invest* 120(6): 2144–55.
6. Ding C, Buckingham B, Levine MA. 2001. Familial isolated hypoparathyroidism caused by a mutation in the gene for the transcription factor GCMB. *J Clin Invest* 108(8): 1215–20.
7. Mannstadt M, Bertrand G, Muresan M, Weryha G, Leheup B, Pulusani SR, Grandchamp B, Jüppner H, Silve C. 2008. Dominant-negative GCMB mutations cause an autosomal dominant form of hypoparathyroidism. *J Clin Endocrinol Metab* 93(9): 3568–76.
8. Strewler GJ, Stern PH, Jacobs JW, Eveloff J, Klein RF, Leung SC, Rosenblatt M, Nissenson RA. 1987. Parathyroid hormonelike protein from human renal carcinoma cells. Structural and functional homology with parathyroid hormone. *J Clin Invest* 80(6): 1803–7.
9. Usdin TB, Hoare SR, Wang T, Mezey E, Kowalak JA. 1999. TIP39: A new neuropeptide and PTH2-receptor agonist from hypothalamus. *Nat Neurosci* 2(11): 941–3.
10. Suva LJ, Winslow GA, Wettenhall RE, Hammonds RG, Moseley JM, Diefenbach-Jagger H, Rodda CP, Kemp BE, Rodriguez H, Chen EY, et al. 1987. A parathyroid hormone-related protein implicated in malignant hypercalcemia: Cloning and expression. *Science* 237(4817): 893–6.
11. Broadus AE, Mangin M, Ikeda K, Insogna KL, Weir EC, Burtis WJ, Stewart AF. 1988. Humoral hypercalcemia of cancer. Identification of a novel parathyroid hormone-like peptide. *N Engl J Med* 319(9): 556–63.
12. Gensure RC, Gardella TJ, Jüppner H. 2005. Parathyroid hormone and parathyroid hormone-related peptide, and their receptors. *Biochem Biophys Res Commun* 328(3): 666–78.
13. Kemper B, Habener JF, Mulligan RC, Potts JT Jr, Rich A. 1974. Pre-proparathyroid hormone: A direct translation product of parathyroid messenger RNA. *Proc Natl Acad Sci U S A* 71(9): 3731–5.
14. Wiren KM, Ivashkiv L, Ma P, Freeman MW, Potts JT Jr, Kronenberg HM. 1989. Mutations in signal sequence cleavage domain of preproparathyroid hormone alter protein translocation, signal sequence cleavage, and

membrane-binding properties. *Mol Endocrinol* 3(2): 240–50.
15. Habener JF, Kemper B, Potts JT Jr. 1975. Calcium-dependent intracellular degradation of parathyroid hormone: A possible mechanism for the regulation of hormone stores. *Endocrinology* 97(2): 431–41.
16. D'Amour P. 2006. Circulating PTH molecular forms: What we know and what we don't. *Kidney Int Suppl* (102): S29–33.
17. Murray TM, Rao LG, Divieti P, Bringhurst FR. 2005. Parathyroid hormone secretion and action: Evidence for discrete receptors for the carboxyl-terminal region and related biological actions of carboxyl- terminal ligands. *Endocr Rev* 26(1): 78–113.
18. Brown EM. 1983. Four-parameter model of the sigmoidal relationship between parathyroid hormone release and extracellular calcium concentration in normal and abnormal parathyroid tissue. *J Clin Endocrinol Metab* 56(3): 572–81.
19. Brown EM, Gamba G, Riccardi D, Lombardi M, Butters R, Kifor O, Sun A, Hediger MA, Lytton J, Hebert SC. 1993. Cloning and characterization of an extracellular Ca(2+)-sensing receptor from bovine parathyroid. *Nature* 366(6455): 575–80.
20. Brown EM, MacLeod RJ. 2001. Extracellular calcium sensing and extracellular calcium signaling. *Physiol Rev* 81(1): 239–97.
21. Chang W, Chen TH, Pratt S, Shoback D. 2000. Amino acids in the second and third intracellular loops of the parathyroid Ca2+-sensing receptor mediate efficient coupling to phospholipase C. *J Biol Chem* 275(26): 19955–63.
22. Pearce SH, Williamson C, Kifor O, Bai M, Coulthard MG, Davies M, Lewis-Barned N, McCredie D, Powell H, Kendall-Taylor P, Brown EM, Thakker RV. 1996. A familial syndrome of hypocalcemia with hypercalciuria due to mutations in the calcium-sensing receptor. *N Engl J Med* 335(15): 1115–22.
23. Pollak MR, Seidman CE, Brown EM. 1996. Three inherited disorders of calcium sensing. *Medicine (Baltimore)* 75(3): 115–23.
24. Ho C, Conner DA, Pollak MR, Ladd DJ, Kifor O, Warren HB, Brown EM, Seidman JG, Seidman CE. 1995. A mouse model of human familial hypocalciuric hypercalcemia and neonatal severe hyperparathyroidism. *Nat Genet* 11(4): 389–94.
25. Wettschureck N, Lee E, Libutti SK, Offermanns S, Robey PG, Spiegel AM. 2007. Parathyroid-specific double knockout of Gq and G11 alpha-subunits leads to a phenotype resembling germline knockout of the extracellular Ca2+ -sensing receptor. *Mol Endocrinol* 21(1): 274–80.
26. Egbuna OI, Brown EM. 2008. Hypercalcaemic and hypocalcaemic conditions due to calcium-sensing receptor mutations. *Best Pract Res Clin Rheumatol* 22(1): 129–48.
27. Hu J, Spiegel AM. 2007. Structure and function of the human calcium-sensing receptor: Insights from natural and engineered mutations and allosteric modulators. *J Cell Mol Med* 11(5): 908–22.
28. Russell J, Sherwood LM. 1987. The effects of 1,25-dihydroxyvitamin D3 and high calcium on transcription of the pre-proparathyroid hormone gene are direct. *Trans Assoc Am Physicians* 100: 256–62.
29. Moallem E, Kilav R, Silver J, Naveh-Many T. 1998. RNA-Protein binding and post-transcriptional regulation of parathyroid hormone gene expression by calcium and phosphate. *J Biol Chem* 273(9): 5253–9.
30. Ritter CS, Pande S, Krits I, Slatopolsky E, Brown AJ. 2008. Destabilization of parathyroid hormone mRNA by extracellular Ca2+ and the calcimimetic R-568 in parathyroid cells: Role of cytosolic Ca and requirement for gene transcription. *J Mol Endocrinol* 40(1): 13–21.
31. Cozzolino M, Brancaccio D, Gallieni M, Galassi A, Slatopolsky E, Dusso A. 2005. Pathogenesis of parathyroid hyperplasia in renal failure. *J Nephrol* 18(1): 5–8.
32. Chang W, Tu C, Chen TH, Bikle D, Shoback D. 2008. The extracellular calcium-sensing receptor (CaSR) is a critical modulator of skeletal development. *Sci Signal* 1(35): ra1.
33. Dvorak-Ewell MM, Chen TH, Liang N, Garvey C, Liu B, Tu C, Chang W, Bikle DD, Shoback DM. 2011. Osteoblast extracellular Ca(2+) -sensing receptor regulates bone development, mineralization and turnover. *J Bone Miner Res* 26(12): 2935–47.
34. Steddon SJ, Cunningham J. 2005. Calcimimetics and calcilytics—Fooling the calcium receptor. *Lancet* 365(9478): 2237–9.
35. Silver J, Russell J, Sherwood LM. 1985. Regulation by vitamin D metabolites of messenger ribonucleic acid for preproparathyroid hormone in isolated bovine parathyroid cells. *Proc Natl Acad Sci U S A* 82(12): 4270–3.
36. Okazaki T, Igarashi T, Kronenberg HM. 1988l. 5'-flanking region of the parathyroid hormone gene mediates negative regulation by 1,25-(OH)2 vitamin D3. *J Biol Chem* 263(5): 2203–8.
37. Kim MS, Fujiki R, Murayama A, Kitagawa H, Yamaoka K, Yamamoto Y, Mihara M, Takeyama K, Kato S. 2007. 1Alpha,25(OH)2D3-induced transrepression by vitamin D receptor through E-box-type elements in the human parathyroid hormone gene promoter. *Mol Endocrinol* 21(2): 334–42.
38. Naveh-Many T, Rahamimov R, Livni N, Silver J. 1995. Parathyroid cell proliferation in normal and chronic renal failure rats. The effects of calcium, phosphate, and vitamin D. *J Clin Invest* 96(4): 1786–93.
39. Fukumoto S. 2008. Physiological regulation and disorders of phosphate metabolism–pivotal role of fibroblast growth factor 23. *Intern Med* 47(5): 337–43.
40. Urakawa I, Yamazaki Y, Shimada T, Iijima K, Hasegawa H, Okawa K, Fujita T, Fukumoto S, Yamashita T. 2006. Klotho converts canonical FGF receptor into a specific receptor for FGF23. *Nature* 444(7120): 770–4.
41. Ben-Dov IZ, Galitzer H, Lavi-Moshayoff V, Goetz R, Kuro-o M, Mohammadi M, Sirkis R, Naveh-Many T, Silver J. 2007. The parathyroid is a target organ for FGF23 in rats. *J Clin Invest* 117(12): 4003–8.
42. Krajisnik T, Bjorklund P, Marsell R, Ljunggren O, Akerström G, Jonsson KB, Westin G, Larsson TE. 2007.

Fibroblast growth factor-23 regulates parathyroid hormone and 1alpha-hydroxylase expression in cultured bovine parathyroid cells. *J Endocrinol* 195(1): 125–31.
43. Jüppner H, Abou-Samra AB, Freeman M, Kong XF, Schipani E, Richards J, Kolakowski LF Jr, Hock J, Potts JT Jr, Kronenberg HM, et al. 1991. A G protein-linked receptor for parathyroid hormone and parathyroid hormone-related peptide. *Science* 254(5034): 1024–6.
44. Datta NS, Abou-Samra AB. 2009. PTH and PTHrP signaling in osteoblasts. *Cell Signal* 21(8): 1245–54.
45. Weinstein LS, Liu J, Sakamoto A, Xie T, Chen M. 2004. Minireview: GNAS: Normal and abnormal functions. *Endocrinology* 145(12): 5459–64.
46. Linglart A, Menguy C, Couvineau A, Auzan C, Gunes Y, Cancel M, Motte E, Pinto G, Chanson P, Bougneres P, Clauser E, Silve C. 2011. Recurrent PRKAR1A mutation in acrodysostosis with hormone resistance. *N Engl J Med* 364(23): 2218–26.
47. Fu Q, Manolagas SC, O'Brien CA. 2006. Parathyroid hormone controls receptor activator of NF-kappaB ligand gene expression via a distant transcriptional enhancer. *Mol Cell Biol* 26(17): 6453–68.
48. Keller H, Kneissel M. 2005. SOST is a target gene for PTH in bone. *Bone* 37(2): 148–58.
49. Pfister MF, Forgo J, Ziegler U, Biber J, Murer H. 1999. cAMP-dependent and -independent downregulation of type II Na-Pi cotransporters by PTH. *Am J Physiol* 276(5 Pt 2): F720–5.
50. Guo J, Liu M, Yang D, Bouxsein ML, Thomas CC, Schipani E, Bringhurst FR, Kronenberg HM. 2010. Phospholipase C signaling via the parathyroid hormone (PTH)/PTH-related peptide receptor is essential for normal bone responses to PTH. *Endocrinology* 151(8): 3502–13.
51. Vilardaga JP, Frank M, Krasel C, Dees C, Nissenson RA, Lohse MJ. 2001. Differential conformational requirements for activation of G proteins and the regulatory proteins arrestin and G protein-coupled receptor kinase in the G protein-coupled receptor for parathyroid hormone (PTH)/PTH-related protein. *J Biol Chem* 276(36): 33435–43.
52. Bisello A, Manen D, Pierroz DD, Usdin TB, Rizzoli R, Ferrari SL. 2004. Agonist-specific regulation of parathyroid hormone (PTH) receptor type 2 activity: Structural and functional analysis of PTH- and tuberoinfundibular peptide (TIP) 39-stimulated desensitization and internalization. *Mol Endocrinol* 18(6): 1486–98.
53. Bianchi EN, Ferrari SL. 2009. Beta-arrestin2 regulates parathyroid hormone effects on a p38 MAPK and NFkappaB gene expression network in osteoblasts. *Bone* 45(4): 716–25.
54. Gesty-Palmer D, Flannery P, Yuan L, Corsino L, Spurney R, Lefkowitz RJ, Luttrell LM. 2009. A beta-arrestin-biased agonist of the parathyroid hormone receptor (PTH1R) promotes bone formation independent of G protein activation. *Sci Transl Med* 1(1): 1ra.
55. Ferrandon S, Feinstein TN, Castro M, Wang B, Bouley R, Potts JT, Gardella TJ, Vilardaga JP. 2009. Sustained cyclic AMP production by parathyroid hormone receptor endocytosis. *Nat Chem Biol* 5(10): 734–42.
56. Wan M, Yang C, Li J, Wu X, Yuan H, Ma H, He X, Nie S, Chang C, Cao X. 2008. Parathyroid hormone signaling through low-density lipoprotein-related protein 6. *Genes Dev* 22(21): 2968–79.
57. Romero G, Sneddon WB, Yang Y, Wheeler D, Blair HC, Friedman PA. 2010. Parathyroid hormone receptor directly interacts with dishevelled to regulate beta-Catenin signaling and osteoclastogenesis. *J Biol Chem* 285(19): 14756–63.
58. Gensure RC, Ponugoti B, Gunes Y, Papasani MR, Lanske B, Bastepe M, Rubin DA, Jüppner H. 2004. Identification and characterization of two parathyroid hormone-like molecules in zebrafish. *Endocrinology* 145(4): 1634–9.
59. John MR, Arai M, Rubin DA, Jonsson KB, Jüppner H. 2002. Identification and characterization of the murine and human gene encoding the tuberoinfundibular peptide of 39 residues. *Endocrinology* 143(3): 1047–57.

27

Parathyroid Hormone-Related Protein

John J. Wysolmerski

Introduction 215
The PTHrP Gene and the PTH/PTHrP Gene Family 215
PTHrP Is a Polyhormone 216
PTHrP Receptors 216
Nuclear PTHrP 216

Physiological Functions of PTHrP 217
Conclusions 220
Acknowledgments 220
References 220

INTRODUCTION

In a 1941 case report in the *New England Journal of Medicine*, Fuller Albright first postulated that tumors associated with hypercalcemia might elaborate a parathyroid hormone (PTH)-like humor [1]. Work in the 1980s and 1990s subsequently led to the biochemical characterization of humoral hypercalcemia of malignancy (HHM) and the fulfillment of Albright's predictions by the identification and characterization of parathyroid hormone-related protein (PTHrP) [2–6]. We now understand that PTHrP and PTH are related molecules that can both stimulate the same Type I PTH/PTHrP receptor (PTHR1) [7–9]. PTHrP usually serves a local autocrine, paracrine, or intracrine role, and does not circulate. However, in patients with HHM, PTHrP does reach the circulation and mimics the systemic actions of PTH. A later chapter will discuss malignancy-associated hypercalcemia in detail. This chapter will review the normal physiology of PTHrP.

THE PTHrP GENE AND THE PTH/PTHrP GENE FAMILY

In humans, PTHrP is encoded by a single-copy gene containing eight exons and at least three promoters located on the short arm of chromosome 12 [3, 6, 8]. Alternative splicing at the 3' end of the gene gives rise to three different classes of mRNAs coding for specific translation products of 139, 141, or 173 amino acids. The physiological significance of these different PTHrP transcripts remains unclear, and in rodents and lower vertebrates such as birds and fish, PTHrP 1-173 does not exist [7, 8, 10, 11]. PTHrP mRNA has been found in almost every organ at some time during its development or functioning. Many different hormones and growth factors regulate the transcription and/or stability of PTHrP mRNA. As with PTH, the calcium-sensing receptor (CaSR) has been found to regulate PTHrP gene expression in many cells [12, 13]. Another common theme is the observation that mRNA levels are induced by mechanical deformation [8]. The reader is referred to other reviews for a comprehensive discussion of the sites and regulation of PTHrP expression [7–9].

The PTHrP and PTH genes share structural elements and sequence homology, suggesting that they are related [3, 5, 6, 8]. The exon/intron organization of that portion of both genes encoding the pre-pro sequences and the initial portion of the mature peptides is identical. Furthermore, there is high sequence homology at the amino-terminal portion of both genes such that the peptides share 8 of the first 13 amino acids and a high degree of predicted secondary structure over the next 21 amino acids. These common sequences allow both peptides to bind and activate the same PTHR1, which ultimately

---

Primer on the Metabolic Bone Diseases and Disorders of Mineral Metabolism, Eighth Edition. Edited by Clifford J. Rosen.
© 2013 American Society for Bone and Mineral Research. Published 2013 by John Wiley & Sons, Inc.

explains the ability of PTHrP to cause hypercalcemia in HHM [4]. The above-mentioned structural similarities, together with the location of the two genes on related chromosomes in the human genome (short arm of chromosome 11 for PTH; short arm of chromosome 12 for PTHrP), indicate that the two genes likely arose from a common ancestor. The PTH family also contains the PTH-L gene and the more distantly related tuberoinfundibular peptide 39 (TIP 39) gene. All of these genes were derived from a common ancestor and emerged concurrent with the evolution of vertebrates [11]. Furthermore, fish have two PTH genes, two PTHrP genes, and one PTH-L gene; amphibians have one PTH gene, one PTHrP gene, and a PTH-L gene; and mammals only retain one PTH and one PTHrP gene [10, 11]. Thus, PTHrP is a member of an ancient family of PTH-like peptides that appears to be larger and more diverse in lower vertebrates than in mammals.

PTHrP IS A POLYHORMONE

Similar to the pro-opiomelanocortin (POMC) gene, the primary translation product of PTHrP can undergo a variety of post-translational processing events resulting in an overlapping series of biological peptides [8]. The details of cell-specific PTHrP processing and the biological significance of the different PTHrP peptides are not entirely clear, but several specific secreted forms of PTHrP have been defined. First, PTHrP 1-36 is secreted from several cell types [8, 14]. In addition, longer forms of amino-terminal-containing PTHrP are secreted from keratinocytes and mammary epithelial cells, and circulate in patients with cancer and during lactation [15–17]. The amino-terminus is necessary for interaction with the PTHR1. The secretion of mid-region peptides including amino acids 38-94, 38-95, and 38-101 has also been described [8, 18]. The biology of these specific secretory forms is unclear, but the mid-region of PTHrP stimulates placental calcium transport and modulates renal bicarbonate handling, and this portion of the molecule contains nuclear localization signals (see below) [19–21]. Finally, C-terminal fragments consisting of amino acids 107-138 and 109-138 have been described. These peptides have been suggested to inhibit osteoclast function and stimulate osteoblast proliferation [8, 20].

PTHrP RECEPTORS

The amino-terminus of PTHrP binds to and activates the PTHR1, a prototypical, 7-transmembrane-containing, G-protein coupled receptor (GPCR), which is a member of class B of the large family of GPCRs [7, 22]. Like the PTHrP gene, the PTHR1 is one of a family of several related PTH receptor genes. Although PTH can bind to other receptors in this family, PTHrP can only interact with the PTHR1. This receptor has been described to couple to both $G_{\alpha s}$ and $G_{\alpha q11}$ and signal via the cyclic adenine monophosphate (cAMP) and protein kinase A pathway as well as through the generation of inositol phosphates, diacylglycerol, and intracellular calcium transients [7, 22]. The vast majority of studies *in vitro* suggest that this receptor binds PTHrP and PTH with equal affinity and that both activate the receptor in an identical manner. This is also true when amino-terminal fragments of PTH and PTHrP are infused into animals [7, 8]. However, the human PTH1R may respond differently to PTH and PTHrP. Human subjects subjected to continuous infusion of the two peptides for 3–7 days were found to be become hypercalcemic with lower doses of PTH 1-34 than PTHrP 1-36 [23, 24]. PTHrP also is less efective than PTH at stimulating the renal 1-α–hydroxylase enzyme producing 1,25-dihydroxyvitamin D. This may be explained by physical differences in the binding of the two peptides to different conformational states of the receptor so that the duration of cAMP production is shorter for PTHrP 1-36 than for PTH 1-34 [25].

The existence of biological actions for mid-region and C-terminal peptides of PTHrP implies the possibility of additional receptors for these forms of PTHrP. However, no such receptors have been identified to date.

NUCLEAR PTHrP

Immunohistochemical studies have localized PTHrP to the nucleus of many different cell types [20, 26]. There are several potential mechanisms by which PTHrP can avoid secretion and remain in the cell [20, 26]. Once in the cytoplasm, PTHrP appears to shuttle into and out of the nucleus in a regulated fashion. This is dependent on a specific nuclear localization sequence (NLS) located between amino acids 84–93 and a specific shuttle protein known as importin β1, which displaces PTHrP from microtubules and allows it to transit the nuclear pore [20, 26, 27]. Nuclear export is facilitated by a related shuttle protein known as CRM1 and likely requires a different recognition sequence in the C-terminal region of the peptide [20]. The regulation of nuclear trafficking of PTHrP is not fully understood but phosphorylation at Thr^{85} by the cell-cycle-regulated, cyclin-dependent kinase, $p34^{cdc2}$ appears to regulate nuclear import in a cell-cycle-dependent fashion [20]. The function of nuclear PTHrP remains obscure, but it has been described to bind RNA and in some cells, PTHrP localizes to the nucleolus, suggesting it may be involved in regulating RNA trafficking, ribosomal dynamics, and/or protein translation [20, 26]. Whatever the exact functions of nuclear PTHrP, this pathway appears to be necessary for life. Two groups have replaced the endogenous mouse PTHrP gene with foreshortened versions of PTHrP that exclude the nuclear localization signals [28, 29]. In both cases, this results in failure of the animals to survive more than a couple of weeks due to a phenotype of widespread cellular senescence and growth arrest.

PHYSIOLOGICAL FUNCTIONS OF PTHrP

PTHrP has been found in at least some cells of almost all organs, and a variety of functions have been ascribed to PTHrP. The reader is referred to more comprehensive reviews for a complete discussion of these findings [8, 9]. What follows is a brief outline of areas where PTHrP has been rigorously documented to have physiological effects in intact organisms.

The skeleton

Disruption of the PTHrP gene disturbs chondrocyte differentiation in the growth plates of long bones and in costal cartilage leading to short-limbed dwarfism and a shield chest that interferes with breathing and causes perinatal death. Ablating the PTHR1 gene generates a similar phenotype and overexpressing PTHrP or a constitutively active PTHR1 within growth plate chondrocytes in transgenic mice produces the opposite effect [30–32]. These animal models have documented that amino-terminal PTHrP acts through the PTHR1 to coordinate the rate of chondrocyte differentiation in order to maintain the orderly growth of long bones during development [33]. As illustrated in Fig. 27.1, the growth plate consists of columns of proliferating and differentiating chondrocytes that progressively enlarge to prehypertrophic and then hypertrophic chondrocytes, which secrete matrix and undergo apoptosis in order to form a calcified scaffold that is remodeled into bone in the primary spongiosum. PTHrP is secreted primarily by immature chondrocytes at the top of the columns in response to another molecule known as Indian hedgehog (IHH) produced by differentiating hypertrophic chondrocytes. PTHrP, in turn, activates the PTHR1 located on proliferating and prehypertrophic cells to slow their rate of differentiation into hypertrophic cells. In this manner, IHH and PTHrP act in a local negative feedback loop to regulate the rate of chondrocyte differentiation (see Fig. 27.1) [33]. Emerging evidence suggests that PTHrP affects chondrocyte differentiation by regulating the movement of histone deacetylase 4 (HDAC4) into the nucleus in a PKA-dependent fashion, which, in turn, regulates the activity of a network of transcription factors such as Zfp521, MEF2, and Runx2 [34–36]. This may help to explain why mutations in the GNAS, HDAC4 and PTHrP genes all produce similar defects in bone development [37–39].

PTHrP is also produced in other cartilaginous sites such as the perichondrium that surrounds the costal cartilage and the subarticular chondrocyte population immediately subjacent to the hyaline cartilage lining the joint space [40, 41]. In both of these sites, PTHrP appears to prevent hypertrophic differentiation of chondrocytes

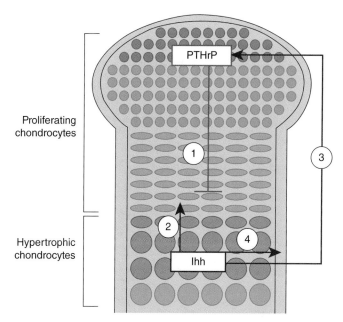

Fig. 27.1. PTHrP and Indian hedgehog (Ihh) act as part of a negative feedback loop regulating chondrocyte proliferation and differentiation. The chondrocyte differentiation program proceeds from undifferentiated chondrocytes at the end of the bone, to proliferative chondrocytes within the columns and then to prehypertrophic and terminally differentiated hypertrophic chondrocytes nearest the primary spongiosum. PTHrP is made by undifferentiated and proliferating chondrocytes at the ends of long bones. It acts through the PTH1R on proliferating and prehypertrophic chondrocytes to delay their differentiation, maintain their proliferation, and delay the production of Ihh, which is made by hypertrophic cells (1). Ihh, in turn, increases the rate of chondrocyte proliferation (2) and stimulates the production of PTHrP at the ends of the bone (3). Ihh also acts on perichondrial cells in order to generate osteoblasts of the bone collar (4). (Kronenberg HM. 2003. Developmental regulation of the growth plate. *Nature* 423: 332–336. Reprinted by permission from Macmillan Publishers Ltd. Copyright 2003.)

and the inappropriate encroachment of bone into these structures [40–42]. Similarly, recent data suggest that PTHrP is important for the maintenance of growth plate chondrocytes in mice. Disruption of the PTHrP gene in these cells after birth causes premature growth arrest due to fusion of the growth plates [43]. These experiments raise the intriguing possibility that PTHrP signaling might be involved in the fusion of growth plates during adolescence or in the loss of articular cartilage in osteoarthritis [42, 43].

In addition to cartilage, PTHrP has important anabolic functions in bone. Heterozygous PTHrP-null mice are normal at birth but develop trabecular osteopenia with age [44]. In addition, selective deletion of the PTHrP gene from osteoblasts results in decreased bone mass, reduced bone formation and mineral apposition, and a reduction in the formation and survival of osteoblasts [45]. Osteoblast cell lines in culture produce PTHrP, and its production *in vitro* can be stimulated by mechanical deformation,

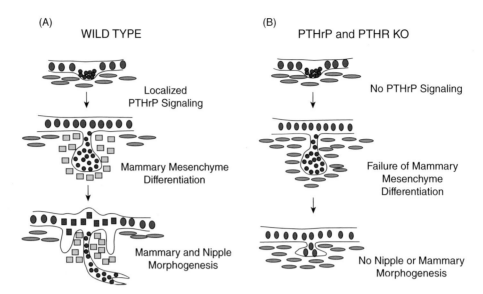

Fig. 27.2. PTHrP regulates mesenchymal cell fate during embryonic mammary development. (A) During normal mammary development, PTHrP is secreted by epithelial cells within the forming mammary bud (red circles) and interacts with the immature dermal mesenchyme (green ovals) to induce formation of the dense mammary mesenchyme (yellow squares). These cells, in response to PTHrP, maintain the fate of the mammary epithelial cells, initiate branching morphogenesis and induce the formation of the specialized nipple skin (purple squares). (B) In PTHrP- or PTHR1-knockout embryos, the mammary bud forms, but the mammary mesenchyme does not. As a result, the mammary epithelial cells revert to an epidermal fate (blue ovals), morphogenesis fails and the nipple never forms. (Adapted with permission from Foley J, Dann P, Hong J, Cosgrove J, Dreyer BE, Rimm D, Dunbar, ME, Philbrick WM, Wysolmerski JJ. 2001. Parathyroid hormone-related protein maintains mammary epithelial fate and triggers nipple skin differentiation during embryonic breast development. *Development* 128: 513–525 and the Company of Biologists Ltd.)

suggesting that it may be involved in mediating the anabolic response to skeletal loading. However, despite the clear phenotype in the osteoblast-specific PTHrP-knockout mice, there is disagreement over the PTHrP-expressing population(s) of osteoblasts in the skeleton and even if the gene is normally expressed within these cells [40, 41]. Nonetheless, the osteopenia in these animal models suggests that intermittent PTH treatment invokes an anabolic response in bone by mimicking the natural functions of local PTHrP in the skeleton.

Mammary gland

Not long after its discovery, PTHrP mRNA was found to be expressed in the lactating breast, and PTHrP protein was measured in high concentrations in milk [46, 47]. It is now known that PTHrP has important functions during breast development, is involved in regulating systemic calcium metabolism during lactation, and contributes to the pathophysiology of breast cancer.

Like other epidermal appendages, the mammary gland initially forms as a bud-like invagination of epidermal cells that grow down into a developing fatty stroma as a branching tube that becomes the mammary duct system. These processes are regulated by a series of sequential and reciprocal interactions between the epithelial cells in the bud and ducts and adjacent mesenchymal cells in the stroma [48]. In mice, as soon as the mammary bud begins to form, epithelial cells produce PTHrP, which interacts with the PTHR1 expressed on surrounding mesenchymal cells. This interaction is necessary for proper differentiation of the dense mammary mesenchyme that surrounds the embryonic mammary bud so that these mesenchymal cells can, in turn, support the proper development of the epithelial ducts [49]. PTHrP- or PTHR1-knockout mice lack mammary glands because loss of PTHrP signaling interrupts the vital crosstalk between epithelial and mesenchymal cells (Fig. 27.2). The formation of the breast in human fetuses is similar to the formation of the fetal mammary gland in mice and it also requires PTHrP [50].

PTHrP is made by breast epithelial cells during lactation and large quantities are secreted into milk [17, 46]. Although its function in milk is unclear, PTHrP is also secreted from the lactating breast into the circulation, where it participates in the regulation of systemic calcium metabolism. Milk production requires a great deal of calcium, an important source of which is the maternal skeleton. Elevated rates of bone resorption and rapid bone loss are well documented in both nursing women and rodents [51]. During lactation, elevated levels of PTHrP correlate with bone loss in humans, and circulating levels of PTHrP correlate directly with rates of bone resorption and inversely with bone mass in mice [16, 52]. In addition, disruption of the PTHrP gene in mammary glands during lactation reduces circulating PTHrP levels, lowers bone turnover, and preserves bone

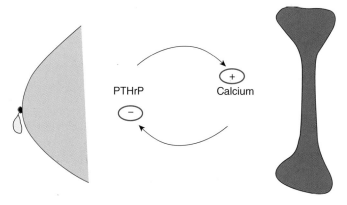

Fig. 27.3. The breast and the skeleton communicate during lactation in order to provide a steady supply of calcium for milk production. The lactating breast secretes PTHrP into the systemic circulation during lactation. PTHrP interacts with the PTH1R in bone cells in order to increase the rate of bone resorption and liberate skeletal calcium stores. Mammary epithelial cells in the lactating breast express the CaSR and suppress PTHrP production in response to increased delivery of calcium, defining a classical endocrine negative feedback loop between breast and bone.

mass, demonstrating that the lactating breast secretes PTHrP into the circulation to increase bone resorption [17]. The lactating breast also expresses the CaSR, which signals to suppress PTHrP secretion in response to increases in calcium delivery to the breast [13]. These interactions define a classical endocrine negative feedback loop, whereby mammary cells secrete PTHrP to mobilize calcium from the bone. Calcium, in turn, feeds back to inhibit further PTHrP secretion from the breast. Therefore, during lactation, the breast and bone engage in a conversation, which leads to the mobilization of skeletal calcium stores to ensure a steady supply of calcium for milk production (see Fig. 27.3). Interestingly, fish PTHrP fulfills a similar function in mobilizing calcium stored in scales to be used during egg production [10]. Thus, this reproductive function of PTHrP is ancient, and the actions of PTHrP during lactation may represent one of the principal evolutionary pressures that resulted in PTHrP and PTH retaining the use of the same PTHR1.

Placenta

During pregnancy, calcium must be actively transported across the placenta from mother to fetus. Furthermore, the responsible placental pump maintains a higher calcium concentration in the fetus as compared to the mother, so that calcium must be transported against a gradient [51]. In PTHrP$^{-/-}$ mice, this gradient is lost and PTHrP-deficient fetuses are relatively hypocalcemic, suggesting that fetal PTHrP is important in mediating placental calcium transport from the mother [21]. The source of the PTHrP is likely the placenta itself, and placental production of PTHrP has been shown to be regulated by the CaSR [53, 54]. Interestingly, experiments in sheep and mice have demonstrated that mid-region PTHrP, not amino-terminal PTHrP, is responsible for placental calcium transport [19, 21].

Smooth muscle and the cardiovascular system

PTHrP is expressed in many different smooth muscle cell beds in response to mechanical deformation and functions to relax the stretched muscle [8, 9, 55–57]. In the vasculature, PTHrP is induced by vasoconstrictive agents, as well as stretch itself, and acts as a vasodilator in resistance vessels. Given these actions, PTHrP may act as a local modulator of blood flow [58].

PTHrP regulates the proliferation of vascular smooth muscle cells. Secreted amino-terminal PTHrP can inhibit the proliferation of vascular smooth muscle cells by acting via the PTHR1 on the cell surface. However, the mid-region and C-terminal portions of PTHrP act in the nucleus to stimulate the proliferation of these cells by destabilizing a cell cycle regulatory protein known as p27^{kip1} and promoting progression through the G$_1$/S checkpoint [59, 60]. This pathway appears to be active during development, since the rate of proliferation of smooth muscle cells in the aorta of PTHrP$^{-/-}$ embryos is reduced [60]. Furthermore, PTHrP expression is upregulated by vascular damage following balloon angioplasty and in atherosclerotic lesions in rodents and in humans. Several studies have suggested that PTHrP plays an important role in the response of these cells to injury and may contribute to the pathophysiologic development of a neointima following angioplasty [61, 62].

Teeth

Developing teeth become surrounded by alveolar bone but must maintain a cavity or crypt that is free from bone to allow for proper morphogenesis. After teeth are formed, they must then erupt through the roof of the dental crypt in order to emerge into the oral cavity. The process of tooth eruption relies on geographically uncoupled bone turnover in which osteoclasts form over the crown of the tooth in order to resorb the overlying bone and osteoblasts at the base of the tooth propel it upward out of the crypt. In the absence of PTHrP, teeth develop but they do not erupt. PTHrP is normally produced by stellate reticulum cells and signals to dental follicle cells in order to drive the formation of osteoclasts above the crypt. In the absence of PTHrP, these osteoclasts do not appear, eruption fails to occur, and teeth become impacted [63–65].

Pancreatic islets

PTHrP is expressed by all four neuroendocrine cell types within the pancreatic islets [66]. In β cells, it is stored within secretory granules and is co-released with insulin [67]. Pancreatic islets express the PTHR1, and β cells respond to PTHrP by activating phospholipase C and increasing intracellular calcium [68]. Overexpression of PTHrP in β cells leads to an increased β-cell mass, hyperinsulinemia, and hypoglycemia, due to the combination of increased β-cell proliferation, increased insulin production, and inhibition of β-cell apoptosis [68, 69]. PTHrP also induces proliferation and improves glucose-stimulated insulin secretion in cultured human β cells. These actions of PTHrP are mediated by a pathway involving PKC-ζ, cyclin E, and cyclin-dependent kinase 2 (cdk2) [70].

Central and peripheral nervous systems

PTHrP and the PTH1R are both widely expressed within specific neurons within the brain, including regions of the cortex, the cerebellum, the hippocampus, hypothalamus, and pituitary [71, 72]. In cultured hippocampal neurons, PTHrP is secreted in response to calcium influx through L-type calcium channels. In turn, PTHrP can act on the PTHR1 on these same neurons to dampen L-type channel activity, giving rise to the idea that PTHrP acts in an autocrine/paracrine short feedback loop to protect neurons from damage due to prolonged or repeated depolarization, so-called excitotoxicity [73]. Consistent with this idea, disruption of the PTHrP gene sensitizes mice to kainite-induced seizures [73].

In addition to neurons, PTHrP has also been shown to be expressed in glia and astrocytes [74–76]. Interestingly, the PTHrP gene is expressed in fetal and malignant glial cells but not in mature glia in the adult brain [76]. However, its expression can be induced in reactive glia in an injury model in rats [75] and in dedifferentiated Schwann cells after crush injury in the peripheral nervous system [77]. PTHrP has been shown to inhibit the differentiation of these cells, suggesting that it contributes to maintaining the dedifferentiated state necessary for nerve regeneration.

CONCLUSIONS

PTHrP was discovered as the cause of the clinical syndrome of HHM. It is evolutionarily and functionally related to PTH and shares the same PTHR1. The common use of this receptor, in turn, allows PTHrP to act as a hormone mimicking the actions of PTH during reproduction, a function preserved from fish to mammals. Although the conservation of these relationships through evolution allows PTHrP to cause hypercalcemia when it is secreted into the circulation by tumors, we have come to understand that PTHrP is generally a locally produced and locally acting growth factor that participates in normal development and physiology at many diverse sites. The power of mouse genetics has provided insight into the functions of PTHrP. However, much still remains to be learned of its normal biology.

ACKNOWLEDGMENTS

This work was supported by grants DK077565, DK55501, and CA153702 from the NIH, and BC095546 from the DOD BCRP. The author would like to thank Drs. Arthur Broadus and Rupangi Vasavada for valuable conversations during the preparation of this chapter.

REFERENCES

1. Mallory TB. 1941. Case records of the Massachusetts General Hospital. Case #27461. *N Eng J Med* 225: 789–791.
2. Burtis WJ, Wu T, Bunch C, Wysolmerski JJ, Insogna KL, Weir EC, Broadus AE, Stewart AF. 1987. Identification of a novel 17,000-dalton parathyroid hormone-like adenylate cyclase-stimulating protein from a tumor associated with humoral hypercalcemia of malignancy. *J Biol Chem* 262: 7151–7156.
3. Mangin M, Webb AC, Dreyer BE, Posillico JT, Ikeda K, Weir EC, Stewart AF, Bander NH, Milstone LM, Barton DE, Francke U, Broadus AE. 1988. An identification of a cDNA encoding a parathyroid hormone-like peptide from a human tumor associated with humoral hypercalcemia of malignancy. *Proc Natl Acad Sci U S A* 85: 597–601.
4. Stewart AF, Horst RL, Deftos LJ, Cadman EC, Lang R, Broadus AE. 1980. Biochemical evaluation of patients with cancer-associated hypercalcemia: Evidence for humoral and nonhumoral groups. *N Eng J Med* 303: 1377–1383.
5. Strewler GJ, Stern PH, Jacobs JW, Evelott J, Klein RF, Leung SC, Rosenblatt M, Nissenson RA. 1987. Parathyroid hormone-like protein from human renal carcinoma cells. Structural and functional homology with parathyroid hormone. *J Clin Invest* 80: 1803–1807.
6. Suva LJ, Winslow GA, Wettenhall RE, Hammonds RG, Moseley JM, Diefenbach-Jagger H, Rodda CP, Kemp BE, Rodriguez H, Chen EY. 1987. A parathyroid hormone-related protein implicated in malignant hypercalcemia: Cloning and expression. *Science* 237: 893–896.
7. Gensure RC, Gardella TJ, Jüppner H. 2005. Parathyroid hormone and parathyroid hormone-related peptide, and their receptors. *Biochem Biophys Res Commun* 328: 666–678.
8. Philbrick WM, Wysolmerski JJ, Galbraith S, Holt EH, Orloff JJ, Yang KH, Vasavada R, Weir EC, Broadus AE, Stewart AF. 1996. Defining the roles of parathyroid hormone-related protein in normal physiology. *Physiol Rev* 76: 127–173.
9. Strewler GJ. 2000. The physiology of parathyroid hormone-related protein. *N Engl J Med* 342: 177–185.

10. Guerreiro PM, Renfro JL, Power DM, Canario AV. 2007. The parathyroid hormone family of peptides: Structure, tissue distribution, regulation, and potential functional roles in calcium and phosphate balance in fish. *Am J Physiol Regul Integr Comp Physiol* 292: R679–696.

11. Pinheiro PL, Cardoso JC, Gomes AS, Fuentes J, Power DM, Canario AV. 2010. Gene structure, transcripts and calciotropic effects of the PTH family of peptides in Xenopus and chicken. *BMC Evol Biol* 10: 373.

12. Chattopadhyay N. 2006. Effects of calcium-sensing receptor on the secretion of parathyroid hormone-related peptide and its impact on humoral hypercalcemia of malignancy. *Am J Physiol Endocrinol Metab* 290: E761–770.

13. VanHouten J, Dann P, McGeoch G, Brown EM, Krapcho K, Neville M, Wysolmerski JJ. 2004. The calcium-sensing receptor regulates mammary gland parathyroid hormone-related protein production and calcium transport. *J Clin Invest* 113: 598–608.

14. Orloff JJ, Reddy D, de Papp AE, Yang KH, Soifer NE, Stewart AF. 1994. Parathyroid hormone-related protein as a prohormone: Posttranslational processing and receptor interactions. *Endocr Rev* 15: 40–60.

15. Burtis WJ, Brady TG, Orloff JJ, Ersbak JB, Warrell RP Jr, Olson BR, Wu TL, Mitnick ME, Broadus AE, Stewart AF. 1990. Immunochemical characterization of circulating parathyroid hormone-related protein in patients with humoral hypercalcemia of cancer. *N Engl J Med* 322: 1106–1112.

16. Sowers MF, Hollis BW, Shapiro B, Randolph J, Janney CA, Zhang D, Schork A, Crutchfield M, Stanczyk F, Russell-Aulet M. 1996. Elevated parathyroid hormone-related peptide associated with lactation and bone density loss. *JAMA* 276: 549–554.

17. VanHouten JN, Dann P, Stewart AF, Watson CJ, Pollak M, Karaplis AC, Wysolmerski JJ. 2003. Mammary-specific deletion of parathyroid hormone-related protein preserves bone mass during lactation. *J Clin Invest* 112: 1429–1436.

18. Soifer NE, Dee KE, Insogna KL, Burtis WJ, Matovcik LM, Wu TL, Milstone LM, Broadus AE, Philbrick WM, Stewart AF. 1992. Parathyroid hormone-related protein. Evidence for secretion of a novel mid-region fragment by three different cell types. *J Biol Chem* 267: 18236–18243.

19. Care AD, Abbas SL, Pickard DW, Barri M, Drinkhill M, Findley JBC, White IR, Caple IW. 1990. Stimulation of ovine placental transport of calcium and magnesium by mid-molecule fragments of human parathyroid hormone-related protein. *Exp Physiol* 75: 605–608.

20. Jans DA, Thomas RJ, Gillespie MT. 2003. Parathyroid hormone-related protein (PTHrP): A nucleocytoplasmic shuttling protein with distinct paracrine and intracrine roles. *Vitam Horm* 66: 345–384.

21. Kovacs CS, Lanske B, Hunzelman JL, Guo J, Karaplis AC, Kronenberg HM. 1996. Parathyroid hormone-related peptide (PTHrP) regulates fetal-placental calcium transport through a receptor distinct from the PTH/PTHrP receptor. *Proc Natl Acad Sci U S A* 93: 15233–15238.

22. Juppner H, Abou-Samra AB, Freeman M, Kong XF, Schipani E, Richards J, Kolakowski LF Jr, Hock J, Potts JT Jr, Kronenberg HM, et al. 1991. A G protein-linked receptor for parathyroid hormone and parathyroid hormone-related peptide. *Science* 254: 1024–1026.

23. Horwitz MJ, Tedesco MB, Sereika SM, Syed MA, Garcia-Ocana A, Bisello A, Hollis BW, Rosen CJ, Wysolmerski JJ, Dann P, Gundberg C, Stewart AF. 2005. Continuous PTH and PTHrP infusion causes suppression of bone formation and discordant effects on 1,25(OH)2 vitamin D. *J Bone Miner Res* 20: 1792–1803.

24. Horwitz MJ, Tedesco MB, Sereika SM, Prebehala L, Gundberg CM, Hollis BW, Bisello A, Garcia-Ocana A, Carneiro RM, Stewart AF. 2011. A seven day continuous infusion of PTH or PTHrP suppresses bone formation and uncouples bone turnover. *J Bone Miner Res* 26: 2287–2297.

25. Dean T, Vilardaga JP, Potts JT Jr, Gardella TJ. 2008. Altered selectivity of parathyroid hormone (PTH) and PTH-related protein (PTHrP) for distinct conformations of the PTH/PTHrP receptor. *Mol Endocrinol* 22: 156–166.

26. Fiaschi-Taesch NM, Stewart AF. 2003. Minireview: Parathyroid hormone-related protein as an intracrine factor—Trafficking mechanisms and functional consequences. *Endocrinology* 144: 407–411.

27. Roth DM, Moseley GW, Pouton CW, Jans DA. 2011. Mechanism of microtubule-facilitated "fast track" nuclear import. *J Biol Chem* 286: 14335–14351.

28. Miao D, Su H, He B, Gao J, Xia Q, Goltzman D, Karaplis AC. 2005. Deletion of the mid- and carboxyl regions of PTHrP produces growth retardation and early senescence in mice. *J Bone Mineral Res* 20: S14.

29. Toribio RE, Brown HA, Novince CM, Marlow B, Hernon K, Lanigan LG, Hildreth BE 3rd, Werbeck JL, Shu ST, Lorch G, Carlton M, Foley J, Boyaka P, McCauley LK, Rosol TJ. 2010. The midregion, nuclear localization sequence, and C terminus of PTHrP regulate skeletal development, hematopoiesis, and survival in mice. *FASEB J* 24: 1947–1957.

30. Lanske B, Karaplis AC, Lee K, Luz A, Vortkamp A, Pirro A, Karperien M, Defize LH, Ho C, Mulligan RC, Abou-Samra A-B, Jueppner H, Segre GV, Kronenberg HM. 1996. PTH/PTHrP receptor in early development and Indian hedgehog-regulated bone growth. *Science* 273: 663–666.

31. Schipani E, Lanske B, Hunzelman JL, Luz A, Kovacs CS, Lee K, Pirro A, Kronenberg HM, Jueppner H. 1997. Targeted expression of constitutively active receptors for parathyroid hormone and parathyroid hormone-related peptide. *Proc Natl Acad Sci U S A* 94: 13689–13694.

32. Weir EC, Philbrick WM, Amling M, Niff LA, Baron R, Broadus AE. 1996. Targeted overexpression of parathyroid hormone-related peptide in chondrodysplasia and delayed endochondrial bone formation. *Proc Natl Acad Sci U S A* 93: 10240–10245.

33. Kronenberg HM. 2006. PTHrP and skeletal development. *Ann N Y Acad Sci* 1068: 1–13.

34. Correa D, Hesse E, Seriwatanachai D, Kiviranta R, Saito H, Yamana K, Neff L, Atfi A, Coillard L, Sitara D, Maeda Y, Warming S, Jenkins NA, Copeland NG, Horne WC, Lanske B, Baron R. 2010. Zfp521 is a target gene and key effector of parathyroid hormone-related peptide signaling in growth plate chondrocytes. *Dev Cell* 19: 533–546.
35. Kozhemyakina E, Cohen T, Yao TP, Lassar AB. 2009. Parathyroid hormone-related peptide represses chondrocyte hypertrophy through a protein phosphatase 2A/histone deacetylase 4/MEF2 pathway. *Mol Cell Biol* 29: 5751–5762.
36. Seriwatanachai D, Densmore MJ, Sato T, Correa D, Neff L, Baron R, Lanske B. 2011. Deletion of Zfp521 rescues the growth plate phenotype in a mouse model of Jansen metaphyseal chondrodysplasia. *FASEB J* 25: 3057–3067.
37. Klopocki E, Hennig BP, Dathe K, Koll R, de Ravel T, Baten E, Blom E, Gillerot Y, Weigel JF, Kruger G, Hiort O, Seemann P, Mundlos S. 2010. Deletion and point mutations of PTHLH cause brachydactyly type E. *Am J Hum Genet* 86: 434–439.
38. Maass PG, Wirth J, Aydin A, Rump A, Stricker S, Tinschert S, Otero M, Tsuchimochi K, Goldring MB, Luft FC, Bahring S. 2010. A cis-regulatory site downregulates PTHLH in translocation t(8;12)(q13;p11.2) and leads to Brachydactyly Type E. *Hum Mol Genet* 19: 848–860.
39. Williams SR, Aldred MA, Der Kaloustian VM, Halal F, Gowans G, McLeod DR, Zondag S, Toriello HV, Magenis RE, Elsea SH. 2010. Haploinsufficiency of HDAC4 causes brachydactyly mental retardation syndrome, with brachydactyly type E, developmental delays, and behavioral problems. *Am J Hum Genet* 87: 219–228.
40. Chen X, Macica C, Nasiri A, Judex S, Broadus AE. 2007. Mechanical regulation of PTHrP expression in entheses. *Bone* 41: 752–759.
41. Chen X, Macica CM, Dreyer BE, Hammond VE, Hens JR, Philbrick WM, Broadus AE. 2006. Initial characterization of PTH-related protein gene-driven lacZ expression in the mouse. *J Bone Miner Res* 21: 113–123.
42. Macica C, Liang G, Nasiri A, Broadus AE. 2011. Genetic evidence that parathyroid hormone-related protein regulates articular chondrocyte maintenance. *Arthritis Rheum* 63: 3333–3343.
43. Hirai T, Chagin AS, Kobayashi T, Mackem S, Kronenberg HM. 2011. Parathyroid hormone/parathyroid hormone-related protein receptor signaling is required for maintenance of the growth plate in postnatal life. *Proc Natl Acad Sci U S A* 108: 191–196.
44. Amizuka N, Karaplis AC, Henderson JE, Warshawsky H, Lipman ML, Matsuki Y, Ejiri S, Tanaka M, Izumi N, Ozawa H, Goltzman D. 1996. Haploinsufficiency of parathyroid hormone-related peptide (PTHrP) results in abnormal postnatal bone development. *Dev Biol* 175: 166–176.
45. Miao D, He B, Jiang Y, Kobayashi T, Soroceanu MA, Zhao J, Su H, Tong X, Amizuka N, Gupta A, Genant HK, Kronenberg HM, Goltzman D, Karaplis AC. 2005. Osteoblast-derived PTHrP is a potent endogenous bone anabolic agent that modifies the therapeutic efficacy of administered PTH 1-34. *J Clin Invest* 115: 2402–2411.
46. Budayr AA, Halloran BP, King JC, Diep D, Nissenson RA, Strewler GJ. 1989. High levels of a parathyroid hormone-like protein in milk. *Proc Natl Acad Sci U S A* 86: 7183–7185..
47. Thiede MA, Rodan GA. 1988. Expression of a calcium-mobilizing parathyroid hormone-like peptide in lactating mammary tissue. *Science* 242: 278–280.
48. Robinson GW. 2007. Cooperation of signalling pathways in embryonic mammary gland development. *Nat Rev Genet* 8: 963–972.
49. Hens JR, Wysolmerski JJ. 2005. Key stages of mammary gland development: Molecular mechanisms involved in the formation of the embryonic mammary gland. *Breast Cancer Res* 7: 220–224.
50. Wysolmerski JJ, Cormier S, Philbrick WM, Dann P, Zhang JP, Roume J, Delezoide AL, Silve C. 2001. Absence of functional type 1 parathyroid hormone (PTH)/PTH-related protein recpetors in humans is associated with abnormal breast development and tooth impaction. *J Clin Endocrinol Metab* 86: 1788–1794.
51. Kovacs CS. 2001. Calcium and bone metabolism in pregnancy and lactation. *J Clin Endocrinol Metab* 86: 2344–2348.
52. VanHouten JN, Wysolmerski JJ. 2003. Low estrogen and high parathyroid hormone-related peptide levels contribute to accelerated bone resorption and bone loss in lactating mice. *Endocrinology* 144: 5521–5529.
53. Hellman P, Ridefelt P, Juhlin C, Akerstrom G, Rastad J, Gylfe E. 1992. Parathyroid-like regulation of parathyroid-hormone-related protein release and cytoplasmic calcium in cytotrophoblast cells of human placenta. *Arch Biochem Biophys* 293: 174–180.
54. Kovacs CS, Ho C, Seidman CE, Seidman JG, Kronenberg HM. 1996. Parathyroid calcium sensing receptor regulates fetal blood calcium and fetal-maternal calcium gradient independently of the maternal calcium levels. *J Bone Miner Res* 22: S121.
55. Thiede MA, Daifotis AG, Weir EC, Brines ML, Burtis WJ, Ikeda K, Dreyer BE, Garfield RE, Broadus AE. 1990. Intrauterine occupancy controls expression of the parathyroid hormone-related peptide gene in pre-term rat myometrium. *Proc Natl Acad Sci U S A* 87: 6969–6973.
56. Thiede MA, Harm SC, McKee RL, Grasser WA, Duong LT, Leach RM Jr. 1991. Expression of the parathyroid hormone-related protein gene in the avian oviduct: Potential role as a local modulator of vascular smooth muscle tension and shell gland motility during the egg-laying cycle. *Endocrinology* 129: 1958–1966.
57. Yamamoto M, Harm SC, Grasser WA, Thiede MA. 1992. Parathyroid hormone-related protein in the rat urinary bladder: A smooth muscle relaxant produced locally in response to mechanical stretch. *Proc Natl Acad Sci U S A* 89: 5326–5330.
58. Massfelder T, Helwig JJ. 1999. Parathyroid hormone-related protein in cardiovascular development and blood pressure regulation. *Endocrinology* 140: 1507–1510.

59. Fiaschi-Taesch N, Sicari BM, Ubriani K, Bigatel T, Takane KK, Cozar-Castellano I, Bisello A, Law B, Stewart AF. 2006. Cellular mechanism through which parathyroid hormone-related protein induces proliferation in arterial smooth muscle cells: Definition of an arterial smooth muscle PTHrP/p27kip1 pathway. *Circ Res* 99: 933–942.
60. Massfelder T, Dann P, Wu TL, Vasavada R, Helwig JJ, Stewart AF. 1997. Opposing mitogenic and antimitogenic actions of parathyroid hormone-related protein in vascular smooth muscle cells: A critical role for nuclear targeting. *Proc Natl Acad Sci U S A* 94: 13630–13635.
61. Fiaschi-Taesch N, Takane KK, Masters S, Lopez-Talavera JC, Stewart AF. 2004. Parathyroid-hormone-related protein as a regulator of pRb and the cell cycle in arterial smooth muscle. *Circulation* 110: 177–185.
62. Ishikawa M, Akishita M, Kozaki K, Toba K, Namiki A, Yamaguchi T, Orimo H, Ouchi Y. 2000. Expression of parathyroid hormone-related protein in human and experimental atherosclerotic lesions: Functional role in arterial intimal thickening. *Atherosclerosis* 152: 97–105.
63. Boabaid F, Berry JE, Koh AJ, Somerman MJ, McCauley LK. 2004. The role of parathyroid hormone-related protein in the regulation of osteoclastogenesis by cementoblasts. *J Periodontol* 75: 1247–1254.
64. Calvi LM, Shin HI, Knight MC, Weber JM, Young MF, Giovannetti A, Schipani E. 2004. Constitutively active PTH/PTHrP receptor in odontoblasts alters odontoblast and ameloblast function and maturation. *Mech Dev* 121: 397–408.
65. Philbrick WM, Dreyer BE, Nakchbandi IA, Karaplis AC. 1998. Parathyroid hormone-related protein is required for tooth eruption. *Proc Natl Acad Sci U S A* 95: 11846–11851.
66. Asa SL, Henderson J, Goltzman D, Drucker DJ. 1990. Parathyroid hormone-like peptide in normal and neoplastic human endocrine tissues. *J Clin Endocrinol Metab* 71: 1112–1118.
67. Plawner LL, Philbrick WM, Burtis WJ, Broadus AE, Stewart AF. 1995. Cell type-specific secretion of parathyroid hormone-related protein via the regulated versus the constitutive secretory pathway. *J Biol Chem* 270: 14078–14084.
68. Vasavada RC, Wang L, Fujinaka Y, Takane KK, Rosa TC, Mellado-Gil JM, Friedman PA, Garcia-Ocana A. 2007. Protein kinase C-zeta activation markedly enhances beta-cell proliferation: An essential role in growth factor mediated beta-cell mitogenesis. *Diabetes* 56: 2732–2743.
69. Vasavada RC, Cavaliere C, D'Ercole AJ, Dann P, Burtis WJ, Madlener AL, Zawalich K, Zawalich W, Philbrick W, Stewart AF. 1996. Overexpression of parathyroid hormone-related protein in the pancreatic islets of transgenic mice causes islet hyperplasia, hyperinsulinemia, and hypoglycemia. *J Biol Chem* 271: 1200–1208.
70. Guthalu Kondegowda N, Joshi-Gokhale S, Harb G, Williams K, Zhang XY, Takane KK, Zhang P, Scott DK, Stewart AF, Garcia-Ocana A, Vasavada RC. 2010. Parathyroid hormone-related protein enhances human ss-cell proliferation and function with associated induction of cyclin-dependent kinase 2 and cyclin E expression. *Diabetes* 59: 3131–3138.
71. Weaver DR, Deeds JD, Lee K, Segre GV. 1995. Localization of parathyroid hormone-related peptide (PTHrP) and PTH/PTHrP receptor mRNAs in rat brain. *Brain Res Mol Brain Res* 28: 296–310.
72. Weir EC, Brines ML, Ikeda K, Burtis WJ, Broadus AE, Robbins RJ. 1990. Parathyroid hormone-related peptide gene is expressed in the mammalian central nervous system. *Proc Natl Acad Sci U S A* 87: 108–112.
73. Chatterjee O, Nakchbandi IA, Philbrick WM, Dreyer BE, Zhang JP, Kaczmarek LK, Brines ML, Broadus AE. 2002. Endogenous parathyroid hormone-related protein functions as a neuroprotective agent. *Brain Res* 930: 58–66.
74. Chattopadhyay N, Evliyaoglu C, Heese O, Carroll R, Sanders J, Black P, Brown EM. 2000. Regulation of secretion of PTHrP by Ca(2+)-sensing receptor in human astrocytes, astrocytomas, and meningiomas. *Am J Physiol Cell Physiol* 279: C691–699.
75. Funk JL, Trout CR, Wei H, Stafford G, Reichlin S. 2001. Parathyroid hormone-related protein (PTHrP) induction in reactive astrocytes following brain injury: A possible mediator of CNS inflammation. *Brain Res* 915: 195–209.
76. Shankar PP, Wei H, Davee SM, Funk JL. 2000. Parathyroid hormone-related protein is expressed by transformed and fetal human astrocytes and inhibits cell proliferation. *Brain Res* 868: 230–240.
77. Macica CM, Liang G, Lankford KL, Broadus AE. 2006. Induction of parathyroid hormone-related peptide following peripheral nerve injury: Role as a modulator of Schwann cell phenotype. *Glia* 53: 637–648.

28
Ca^{2+}-Sensing Receptor
Edward M. Brown

Introduction 224
Structure and Function of the CaSR 224

Roles of the CaSR in Tissues Maintaining Ca^{2+}_o Homeostasis 229
References 231

INTRODUCTION

Complex terrestrial organisms, including humans, maintain a virtually constant level of their extracellular ionized calcium (Ca^{2+}_o) concentration, with a normal range of 1.1–1.3 mM [1]. This provides a reliable supply of calcium ions for their numerous extracellular roles, i.e., acting as a cofactor for clotting factors, adhesion molecules, and numerous other proteins, modulating neuronal excitability and providing a source of calcium for its myriad intracellular functions [1]. In addition, salts of calcium and phosphate provide the mineral phase of the skeleton, which protects vital organs and facilitates locomotion and other movements. Bone is also a nearly inexhaustible supply of calcium and phosphate when dietary sources are insufficient for the body's requirements.

The resting level of the cytosolic calcium concentration (Ca^{2+}_i), in contrast, is about 100 nM, nearly 10,000-fold lower than that of Ca^{2+}_o [2]. Changes in Ca^{2+}_i function as a key intracellular second messenger, regulating diverse processes, such as cellular motility, differentiation, proliferation, and apoptosis as well as muscular contraction and hormonal secretion [2]. All intracellular Ca^{2+} ultimately derives from Ca^{2+} in the extracellular fluids (ECFs). Therefore, maintaining a nearly constant level of Ca^{2+}_o ensures that calcium is available for its host of intracellular roles.

The "guardian" of the near constancy of Ca^{2+}_o in mammals is a homeostatic system comprising the parathyroid glands, calcitonin (CT)-secreting C cells, kidney, bone, and intestines [1], as detailed elsewhere in this volume. Key components of this homeostatic mechanism are several types of cells that can "sense" small perturbations in Ca^{2+}_o from its normal value and respond so as to return Ca^{2+}_o to normal. The parathyroid glands play key roles in this process by secreting PTH in response to hypocalcemia, which then increases renal tubular reabsorption of Ca^{2+}, contributes to net release of Ca^{2+} from bone and enhances intestinal Ca^{2+} absorption by increasing renal synthesis of $1,25(OH)_2D_3$.

This chapter describes the properties and functions of the Ca^{2+}_o-sensing receptor (CaSR), a G-protein coupled receptor (GPCR) that plays a central role in Ca^{2+}_o homeostasis by virtue of its capacity to sense Ca^{2+}_o [1]. It is the principal mechanism in parathyroid cells, C cells, and several nephron segments in the kidney, as well as in bone and intestine for measuring the level of Ca^{2+}_o. As such, it can serve as the body's "thermostat for Ca^{2+}_o" or "calciostat" through its capacity to modulate the functions of those cell types just enumerated that participate in Ca^{2+}_o homeostasis.

STRUCTURE AND FUNCTION OF THE CaSR

The CaSR was originally cloned from bovine parathyroid gland [3] and then from several tissues of a variety of other species, including humans [4]. All exhibit highly similar amino acid sequences (≥84% identical) and pre-

Primer on the Metabolic Bone Diseases and Disorders of Mineral Metabolism, Eighth Edition. Edited by Clifford J. Rosen.
© 2013 American Society for Bone and Mineral Research. Published 2013 by John Wiley & Sons, Inc.

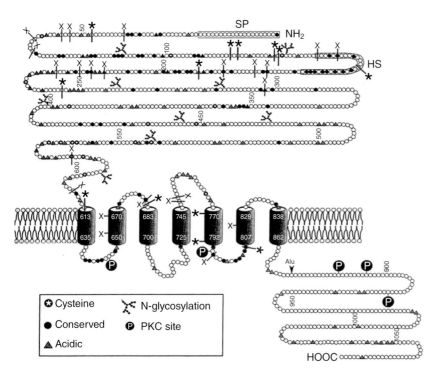

Fig. 28.1. Predicted structure of the human CaSR (see text for details). SP: signal peptide; HS: hydrophobic segment. The X's show examples of sites of naturally occurring inactivating mutations, and asterisks indicate locations of activating mutations. Reproduced in modified form with permission from Brown EM, Bai M, Pollak M. In: Avioli L, Krane SM (eds.) *Metabolic Bone Disease and Clinically Related Diseases*. San Diego: Academic Press. 1999, pp. 479–499.

dicted structures, and they represent tissue and species homologs of the same ancestral CaSR gene (so-called orthologs). Similar genes have been isolated from evolutionarily more distant species, such as salmon and dogfish shark [5], which exhibit approximately 60–70% amino acid identity to the sequence of the human CaSR. Thus the CaSR originated prior to the migration of vertebrates from the oceans onto dry land and is thought to regulate processes that maintain stability of Ca^{2+}_o in bony fish and elasmobranchs, e.g., sharks [5].

Figure 28.1 shows the predicted topology of the human CaSR protein. Its three principal structural domains include: (a) a large, 612 amino acid extracellular amino-terminal domain (ECD) containing a cysteine rich region followed by a peptide linker to the first transmembrane helix at its carboxy terminus; (b) a 250 amino acid, seven-transmembrane domain (TMD) motif characteristic of the superfamily of GPCRs; and (c) a carboxyl-terminal (C−) tail of 216 amino acids (for review, see Ref. 6).

The CaSR gene and properties of the CaSR

The human CaSR gene resides on the long arm of chromosome 3 in band 3q13.3–21. It has seven exons, the first of which encodes sequences upstream of the translational start site and contains two promoters. The remaining six exons contain the translational start site and the three domains of the CaSR just described. Exons 2–6 encode the ECD and exon 7 codes for the TMD and C-terminus [6]. Interleukins 1β [7] and 6 [8], 1,25(OH)$_2$D$_3$ [9] and activation of the CaSR itself [10] upregulate the receptor. The high Ca^{2+}_o-evoked increases and 1,25(OH)$_2$D$_3$- evoked increases in CaSR expression may contribute to the known inhibition of parathyroid function by these two factors. The chemokines, MCP-1 and SDF-1α, increase CaSR expression on the cell surface [11], probably by enhancing translocation of preformed receptor to the plasma membrane. The CaSR's expression is commonly downregulated in various forms of hyperparathyroidism [12].

Biogenesis and structure of the CaSR

Biogenesis of the CaSR in the endoplasmic reticulum (ER) is accompanied by cleavage of the 20 amino acid signal peptide that directs the nascent protein chain into the lumen of the ER. The receptor dimerizes in the ER through intermolecular disulfide bonds involving cysteines 129 and 131 [13]. Transport of the CaSR from the ER to the Golgi is followed by glycosylation of the receptor on a total of eight O-linked glycosylation sites, a process that is essential for surface expression of the receptor (for review, see Ref. 6). Several proteins have been identified that are important for the CaSR's efficient transport from the ER to the Golgi and, ultimately, the cell surface, including the low molecular weight, monomeric G protein, Rab1 [14], the "cargo receptor" p24a, and the so-called receptor activity modifying proteins (RAMP) 1 and 3 (for review, see Ref. 15). These may overcome the action of an ER retention signal present within the CaSR's C-terminus [6].

Following activation by its agonists, the CaSR can undergo desensitization by mechanisms involving phosphorylation of its C-terminus by G protein receptor kinases

(GRKs) and uncoupling from heterotrimeric G proteins that it activates, such as $G_{q/11}$ [16]. The receptor undergoes relatively little agonist-induced internalization [17], however, and the recycling of intracellular CaSR to the cell surface is facilitated by another monomeric G protein, Rab11a [15]. The continued availability of cell surface CaSR, resulting in part from agonist-induced insertion of intracellular CaSR into the plasma membrane (17a), is important to ensure continuous, efficient monitoring of Ca^{2+}_o by the CaSR in its role as "calciostat."

Binding of Ca^{2+}_o and activation of CaSR-mediated signaling

The ECD of the CaSR is thought to assume a bilobed configuration resembling a Venus flytrap (Fig. 28.2), based on homology modeling from the related, metabotropic glutamate receptors (mGluRs) [18]. Several of the mGluR ECDs have been crystallized and their structures solved by X-ray crystallography with and without their agonist, glutamate. Like the mGluRs, the CaSR has a binding site for its agonist, Ca^{2+}, in the crevice between the two lobes of each monomer (Fig. 28.2) [19, 20]. Binding of calcium to this site is thought to facilitate closure of the Venus fly trap (VFT) domain and initiation of intracellular signaling through additional conformational changes on the CaSR's TMD, intracellular loops and C-tail that activate G proteins (for review, see Ref. 6). It is likely that there are additional Ca^{2+}-binding sites elsewhere in the ECD (and possibly in the TMD), which are important for the receptor's high degree of positive cooperativity, i.e., the binding of Ca^{2+} to one site enhances binding of Ca^{2+} to other sites [20]. This generates the steep slope of the relationship between Ca^{2+}_o and CaSR-mediated biological responses that is essential for maintaining near constancy of Ca^{2+}_o [1]. Within its intracellular loops and C-tail, the human CaSR harbors five predicted protein kinase C (PKC) sites. CaSR-mediated activation of protein kinase C (PKC) diminishes stimulation of phospholipase C (PLC) by the receptor, primarily by phosphorylating a key PKC site in the CaSR's C-tail at threonine 888 (T888) [21]. This PKC-induced phosphorylation of the C-tail confers negative feedback regulation of CaSR-mediated stimulation of PLC.

On the plasma membrane, the CaSR in parathyroid resides in caveolae, which are flask-shaped invaginations of the plasma membrane [22]. In this locale, the CaSR binds to caveolin-1, a key cholesterol-binding, structural protein of caveolae that interacts with multiple other proteins, including signaling proteins (e.g., G proteins) [23]. The CaSR's C-tail interacts with filamin-A, an actin-binding protein, which, like caveolin-1, acts as a scaffold for multiple proteins [24]. The CaSR's ability to activate MAPKs, i.e., ERK1/2 (see below), is dependent, in part, on its binding to filamin-A [24]. The binding of filamin-A to the CaSR's C-tail also increases the receptor's resistance to degradation. In contrast, binding of the E3 ubiquitin ligase, dorfin, to the C-tail is thought to promote proteasomal degradation of the protein [25].

CaSR agonists and CaSR signaling

The CaSR is activated by a variety of agonists in addition to Ca^{2+}_o (Fig. 28.3) (for review, see Ref. 6), including other

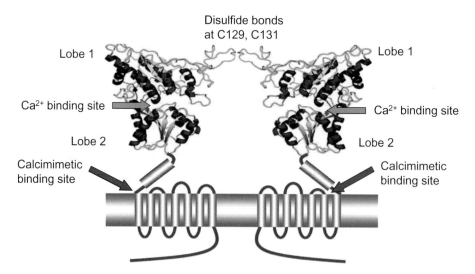

Fig. 28.2. Schematic illustration of the ECD of the CaSR based on the known structure of the ECDs of several metabotropic glutamate receptors. Note that the ECD is a dimer, with each monomer having a Venus flytrap-like conformation with a binding site for Ca^{2+} in the crevice between the two lobes of each monomer. Additional binding sites for Ca^{2+} are probably present elsewhere in the ECD. A binding site for amino acids (e.g., phenylalanine) is close to the binding site for Ca^{2+}_o depicted. Calcimimetics, such as Cinacalcet, in contrast, bind to a site in the TMD, with the amino group of the drug between the two hydrophobic ends anchored to Glu837. Reproduced in modified form from a figure originally published in the *Journal of Biological Chemistry*. Huang Y, Zhou Y, Yang W, Butters R, Lee HW, Li S, Castiblanco A, Brown EM, Yang JJ. 2007. Identification and dissection of Ca(2+)-binding sites in the extracellular domain of Ca(2+)-sensing receptor. *J Biol Chem* 282: 19000–10. Copyright The American Society for Biochemistry and Molecular Biology.

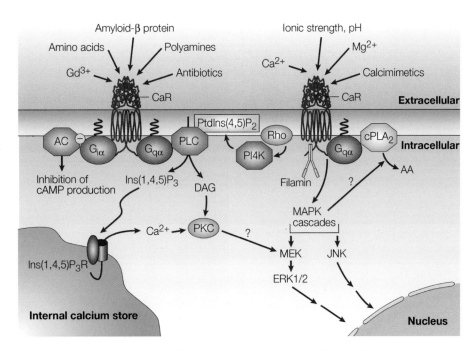

Fig. 28.3. Multiple agonists and other factors modulating the activity of the CaSR as well as the numerous intracellular signaling pathways through which the receptor can regulate cellular function. Ca^{2+}, Mg^{2+}, aminoglycoside antibiotics, spermine and other polyamines, and amyloid beta peptides are examples of polycationic agonists of the CaSR. Aromatic amino acids and calcimimetics—drugs that activate the receptor and are used to control hyperparathyroidism—are allosteric modulators of the CaSR. The former bind to a site near a putative calcium-binding site in the receptor's ECD, while the latter bind to the CaSR's transmembrane domains; both increase its apparent affinity for polycationic agonists. AA: arachidonic acid; AC: adenylate cyclase; cAMP: cyclic AMP; cPLA$_2$: cytosolic phospholipase A$_2$; DAG: diacylglycerol; ERK: extracellular signal-regulated kinase; Gα_i and Gα_q, α subunits of the i- and q-type heterotrimeric G-proteins, respectively, Ins(1,4,5)P$_3$: inositol 1,4,5-trisphosphate; Ins(1,4,5)P$_3$R: 1,4,5-trisphosphate receptor; JNK: Jun N-terminal kinase; MAPK: mitogen-activated protein kinase; MEK: MAPK kinase; PI4K: phosphatidylinositol 4-kinase; PKC: protein kinase C; PLC: phospholipase C; PtdIns(4,5)P$_2$: phoasphatidylinositol-4,5-bisphosphate. Reproduced from Hofer A, Brown EM. 2003. Extracellular calcium sensing and signaling. *Nat Rev Cell Mol Biol* 4: 530–538.

divalent (i.e., Mg^{2+} and Sr^{2+}) and trivalent cations (e.g., La^{3+} and Gd^{3+}), as well as organic polycations, such as polylysine, polyarginine, and neomycin. All of these are so-called type 1 agonists (i.e., they activate the receptor even in the absence of Ca^{2+}). Type II CaSR activators, in contrast, require the presence of extracellular Ca^{2+} for their activity [26]. These include amino acids, particularly aromatics, which bind to a site close to the Ca^{2+}-binding site between the two lobes of the ECD, and calcimimetics. The latter are drugs used to suppress parathyroid function in hyperparathyroid states [26]; they bind to a pocket within the TMD (Fig. 28.2). CaSR antagonists, so-called calcilytics [26], in contrast, inhibit the CaSR by binding to a site in the TMD overlapping, but not identical to, that binding calcimimetics [27].

Does the CaSR serve as an Mg^{2+}_o-sensor?

Some evidence supporting the CaSR's role in sensing and "setting" Mg^{2+}_o derives from experiments-in-nature that initially proved the CaSR's role as a central element in Ca^{2+}_o homeostasis. Namely, individuals with hypercalcemia owing to heterozygous inactivating CaSR mutations, viz., the syndrome familial hypocalciuric hypercalcemia (FHH), have serum Mg^{2+} levels in the high normal or mildly elevated range [28]. Conversely, persons who have activating CaSR mutations can manifest mild hypomagnesemia [1]. Therefore, inactivating and activating mutations of the CaSR reset not only Ca^{2+}_o but also Mg^{2+}_o. The regulation of Mg^{2+}_o homeostasis by the CaSR may occur, in part, in the parathyroid gland, where hypermagnesemia suppresses parathyroid hormone (PTH) release, and/or in the cortical thick ascending limb (cTAL) of the kidney, where elevated Mg^{2+}_o reduces the reabsorption of not only Mg^{2+} but also Ca^{2+} [29].

Intracellular signaling by the CaSR

Activation of the CaSR by its various agonists and activators modulates numerous signaling pathways (Table 28.1) (for reviews, see Ref. 6). These include activation of phospholipases C, A$_2$ (PLA$_2$), and D (PLD). The CaSR activates PLC via the G-proteins, G$_{q/11}$, PLA$_2$ through a mitogen-activated protein kinase (MAPK) (e.g., ERK1/2)- or

Table 28.1. Mechanisms of Intracellular Signaling by the CaSR*

1. Diverse extracellular ligands (di-, trivalent cations, organic polyvalent cations, amino acids, calcimimetics, calcilytics, etc.)
2. Changes in CaSR gene expression
3. Changes in CaSR desensitization, internalization, and degradation
4. Changes in forward trafficking of CaSR to the plasma membrane
5. Interaction of CaSR with scaffold proteins (caveolin-1, filamin) that bind signaling molecules (MAPK components, G proteins, etc.)
6. Activation of G proteins
   a. Heterotrimeric G proteins ($G_{i/o}$, $G_{q/11}$, $G_{12/13}$, G_s)
   b. Low molecular weight G proteins (Arf6, RhoA, Ras, Rab1, Rab11a)
7. Generation of second messengers
   a. Adenylate cyclase (forms cAMP)
   b. Phospholipase A_2 (forms arachidonic acid)
   c. Phospholipase C (forms IP_3, DAG)
   d. Phospholipase D (forms phosphatidic acid)
8. Activation of lipid kinases
   a. PI-3 kinase (forms PIP_3)
   b. PI-4 kinase (forms inositol 4-phosphate)
9. Activation of protein kinases
   a. Protein kinase A (PKA)
   b. Protein kinase B (Akt)
   c. Protein kinase C (PKC)
   d. Calmodulin-dependent kinases (CaMKII)
   e. Tyrosine kinases
   f. Mitogen-activated protein kinases (ERK1/2, P38 MAPK, JNK)

*See text for details and abbreviations.

Table 28.2. Functions of CaSR in Ca^{2+}_o Homeostatic Tissues

Tissue	Function
Parathyroid cells	Inhibits PTH secretion
	Reduces expression of prepoPTH mRNA
	Suppresses parathyroid cellular proliferation
	Inhibit PTH degradation
	Increase expression of CaSR and VDR
C cells	Stimulates calcitonin secretion
Kidney	Increases expression of VDR (PT)
	Inhibits Ca^{2+} and Mg^{2+} reabsorption (TAL)
	Activates TRPV5 (DCT)
	Stimulates acid secretion (CCD)
	Inhibits water reabsorption (IMCD)
Intestine	Upregulates genes involved in Ca^{2+} absorption
	Stimulates gastrin, gastric acid secretion
	Enhances cholecystokinin secretion
	Inhibits fluid secretion in colon
Osteoblast	Stimulates osteoblast proliferation, chemotaxis, differentiation, bone growth, mineralization
Osteoclast**	Promotes osteoclast differentiation
	Supports calcemic action of PTH
	Inhibits bone resorption
	Stimulates apoptosis Mentaverri
Lactating breast	Stimulates Ca^{2+} transport into milk
Placenta	Stimulates Ca^{2+} transport to fetus

**How the CaSR supports osteoclast differentiation, on the one hand (Refs. 68, 75), and inhibits osteoclast action and stimulates apoptosis (Ref. 68), on the other, has not been fully elucidated.

calmodulin-dependent mechanisms, and PLD via protein kinase C (PKC) or $G_{12/13}$. Arachidonic acid (AA) produced by PLA_2 can be further metabolized to produce biologically active products, including those arising from the P450 pathway [20-hydroxyeicosatetraenoic acid (i.e., 20-HETE)] [30], from the 12- and 15-lipoxygenases (e.g., 12- and 15-HETE) [31], and/or from cyclooxygenase (viz. PGE_2) [32]. The high Ca^{2+}_o-induced, transient elevation in Ca^{2+}_i in bovine parathyroid cells is the result of PLC activation and consequent IP_3-mediated release of intracellular Ca^{2+} stores [6]. High Ca^{2+}_o also produces sustained increases in Ca^{2+}_i via incompletely defined influx pathway(s) for Ca^{2+} that probably include activation of one or more Ca^{2+}-permeable, nonselective cation channels of the transient receptor potential (TRP) family (i.e., TRPC1)[33]. Not uncommonly, activation of the CaSR and PLC signaling initiates slow intracellular Ca^{2+}_i oscillations, which may encode information related to their amplitude and/or frequency [34]. High Ca^{2+}_o decreases agonist-evoked cAMP accumulation in parathyroid cells [6] by inhibiting adenylate cyclase via the inhibitory G protein, G_i [35]. High Ca^{2+}_o-elicited diminution of cAMP in other cells types, however, can result from inhibition of a Ca^{2+}-inhibitable isoform of adenylate cyclase [6] or increased degradation of cAMP by phosphodiesterase [36].

Another common form of CaSR signaling involves activation of MAPKs, including ERK1/2, p38 MAPK, and JNK through both PKC- and tyrosine kinase-dependent pathways [37]. Other signaling mechanisms/pathways activated by the CaSR include monomeric G proteins (e.g., Ras and RhoA), lipid kinases [phosphatidylinositol-3 kinase (PI3K) (producing PIP_3) and PI-4 kinase (PI4K)], and protein kinases [protein kinases A, B (also known as Akt), and C, as well as calmodulin kinase II (CaMKII)]. The interested reviewer is referred to recent comprehensive reviews on CaSR signaling [6, 37].

ROLES OF THE CaSR IN TISSUES MAINTAINING Ca^{2+}_o HOMEOSTASIS

Parathyroid

The parathyroid glands express arguably the highest levels of CaSR mRNA and protein in the body (Table 28.2) [38]. Abundant evidence supports the CaSR's key role as the mediator of the high Ca^{2+}_o-induced inhibition of PTH secretion. Through this action, i.e., CaR-mediated stimulation of PTH secretion in response to hypocalcemia, the CaSR in the parathyroid provides a "floor," which the Ca^{2+}_o homeostatic system utilizes to vigorously defend against hypocalcemia. Evidence supporting the CaSR's importance in Ca^{2+}_o-regulated PTH secretion is as follows: First, humans and mice homozygous for genetic inactivation ("knockout") of the CaSR exhibit severely impaired inhibition of PTH release by high Ca^{2+}_o [1, 39]. Second, calcimimetics acutely inhibit PTH secretion in vivo and in vitro [26], further documenting the CaSR's key role in regulating PTH secretion.

Despite several decades of study, the details of how the CaSR suppresses PTH release remain elusive. Recent evidence demonstrates that activating $G_{q/11}$ is essential, since mice with knockout of both of these G proteins have severe hyperparathyroidism similar to that present in mice homozygous for global knockout of the CaSR [40]. Downstream signaling pathways involved in Ca^{2+}_o-regulated PTH release may include products of the 12- and 15-lipoxygenase pathways of AA metabolism [41] and/or ERK1/2 [6, 37]. Through poorly defined mechanisms downstream of these signaling cascades, the CaSR eventually induces polymerization of the actin-based cytoskeleton, which may represent a physical barrier to the secretion of PTH-containing secretory vesicles [42]. The failure of the high Ca^{2+}_o-evoked, CaSR-mediated increase in Ca^{2+}_i to stimulate exocytosis, as it does in nearly all other secretory cells, may relate to the presence of a Ca^{2+}-insensitive isoform of synaptosomal-associated protein (SNAP-23) that participates in vesicle fusion [37].

Another CaSR-regulated process in parathyroid cells is PTH gene expression, since the calcimimetic CaSR activator, NPS R-568, decreases the elevated level of preproPTH mRNA [43]. The CaSR-mediated change in the level of PTH mRNA is the result of a change in preproPTH mRNA stability rather than in PTH gene transcription by a pathway involving stimulation of calmodulin (CaM) and protein phosphatase 2B (for review, see Ref. 44).

Finally, the CaSR suppresses parathyroid cellular proliferation, since individuals homozygous for inactivating CaSR mutations [1] or homozygous CaSR knockout mice [39] exhibit marked parathyroid cellular hyperplasia. Furthermore, treating rats with experimentally induced renal insufficiency with a calcimimetic mitigates the parathyroid hyperplasia that would otherwise occur in this setting [45]. The CaSR-mediated inhibition of parathyroid cellular proliferation is thought to be the result of induction of the cyclin-dependent kinase inhibitor, $p21^{WAF1}$, and downregulation of both the growth factor, TGF-α, and its receptor, the EGFR [46].

C cells

The cloning of the CaSR enabled direct documentation that C cells express the receptor [47]. Studies in CaSR knockout mice subsequently documented the mediatory role of the CaSR in high Ca^{2+}_o-stimulated CT secretion [48]. In contrast to the role of the CaSR in the parathyroid, the receptor in the C cell serves more as a "ceiling" to prevent hypercalcemia. A model for how the CaSR stimulates CT secretion [49] involves CaSR-induced activation of a nonselective cation channel, which causes cellular depolarization, thereby stimulating voltage-dependent calcium channels, elevating Ca^{2+}_i, and activating exocytosis.

Kidney

In the rat kidney, the CaSR resides along nearly the entire nephron [50], although some studies have reported a more limited distribution [51]. The highest levels of expression of the CaSR protein are at the basolateral surface of the cells of the cTAL [50]. This nephron segment plays a key role in PTH-regulated divalent cation reabsorption [52]. The CaSR also has a basolateral localization in the distal convoluted tubule (DCT), where PTH likewise stimulates Ca^{2+} reabsorption. The CaSR also resides at the base of the microvilli of the proximal tubular (PT) brush border, on the basolateral surface of the epithelial cells of the medullary thick ascending limb (mTAL) [50], on the type A (acid-secreting) intercalated cells of the cortical collecting duct (CCD) [50], and on the luminal surface of the inner medullary collecting duct (IMCD) [53]. These last four nephron segments do not participate directly in the regulation of renal Ca^{2+} handling but are sites where the CaSR could potentially modulate the handling of other solutes and/or water.

In the PT, the CaSR suppresses PTH-induced phosphaturia [54] and enhances vitamin D receptor expression [55]. The latter may participate in the direct, high Ca^{2+}_o-elicited lowering of circulating $1,25(OH)_2D_3$ levels, since activation of the vitamin D receptor (VDR) reduces production of $1,25(OH)_2D_3$ and increases production of $24,25(OH)_2D_3$.

The CaSR's location on the basolateral membrane in the cTAL supports its role as the mediator of the known inhibitory action of high peritubular but not luminal Ca^{2+}_o on Ca^{2+} and Mg^{2+} reabsorption in perfused tubular segments from this region of the nephron [54]. The CaSR does so by acting in a "lasix-like" manner to inhibit the overall activity of the Na–K–2Cl cotransporter and associated components of the mechanism by which Ca^{2+} is reabsorbed by the paracellular route in this nephron segment.

This cotransporter contributes to the generation of the lumen-positive, transepithelial potential gradient driving passive paracellular reabsorption of about 50% of NaCl, and most of the Ca^{2+} and Mg^{2+} in cTAL. Others, however, have reported discrepant findings, including high Ca^{2+}_o-induced inhibition of Ca^{2+} transport in CTAL without any concomitant decrease in NaCl or Mg^{2+} transport or reduced paracellular permeability to Ca^{2+} [54a]. Elevated levels of peritubular Ca^{2+}_o have also been reported to inhibit not only paracellular but also transepithelial Ca^{2+} transport (for review, see Ref. 54). Additional studies are needed to resolve these discrepancies. In any event, hypercalcemia-induced hypercalciuria has two distinct CaSR-mediated components: (1) inhibition of PTH release, which then reduces Ca^{2+} reabsorption, and (2) direct suppression of reabsorption of Ca^{2+} in cTAL. The direct inhibitory action of the CaSR on renal tubular Ca^{2+} reabsorption, like CaR-stimulated CT secretion, represents a "ceiling" that defends against hypercalcemia [48].

Recent data indicate that the CaSR enhances Ca^{2+} reabsorption in the DCT by stimulating the activity of the apical uptake channel, TRPV5, which is the influx mechanism for transcellular Ca^{2+} reabsorption in DCT [56]. In addition, extracellular Ca^{2+} increases the expression of key, vitamin D-inducible genes involved in transcellular Ca^{2+} transport in DCT. These include TRPV5, calbindin D_{28K}, the basolateral calcium pump, PMCA1b, and the sodium–calcium exchanger, NCX1 [57]. The contribution of these actions of Ca^{2+}_o to regulating Ca^{2+} transport in DCT under normal circumstances is not presently known. Recent studies have also shown that CaSR-induced stimulation of acid secretion by the type A intercalated cells of the CCD protects against the nephrolithiasis that occurs in hypercalciuric mice with knockout of TRPV5 [58]. The apical CaSR in the IMCD has also been suggested to defend against calcium stone formation by inhibiting vasopressin-stimulated reabsorption of water in IMCD when Ca^{2+}_o in the final urine is high, thereby diluting urinary Ca^{2+} [53].

Intestine

The CaSR is expressed throughout the rat intestine on the basal surface of the small intestinal epithelial cells, within the crypts of the large and small intestines, and in the enteric nervous system [59]. Does the intestinal CaSR contribute to systemic Ca^{2+}_o homeostasis? Ca^{2+}_o modulates several intestinal functions. Hypercalcemia reduces the absorption of dietary Ca^{2+} [60]. Recent studies have also documented apparently direct actions of dietary and/or blood Ca^{2+} on the expression of TRPV6 (the apical uptake channel in intestine that corresponds to TRPV5 in kidney), calbindin D_{9K} and PMCA1b in mice lacking the 1-hydroxylase gene and cannot, therefore, make $1,25(OH)_2D_3$. The increase in the levels of these genes in response to high dietary Ca^{2+} are of uncertain physiological relevance but suggest that the gastrointestinal (GI) tract per se has the ability to sense Ca^{2+}_o [57].

As noted earlier, in addition to responding to Ca^{2+}_o, the CaSR is also activated by amino acids [61]. Furthermore, it has recently been proven that the long-recognized but poorly understood stimulation of GI functions such as gastrin and gastric acid secretion [62] in the stomach and cholecystokinin in the small intestine are CaSR-mediated [63]. This likely enables the CaSR to serve as a GI "nutrient sensor," whereby it monitors levels of both minerals and amino acids in the luminal contents so as to make appropriate adjustments in the digestive process [61]. The CaSR in the enteric nervous system, which regulates secretomotor functions of the GI tract, could potentially contribute to the known capacity of hypo- and hypercalcemia to enhance and decrease, respectively, GI motility. Thus the CaSR could also participate as a nutrient sensor in the secretomotor control of GI function. Recent studies have shown that activation of the CaSR in the colon markedly reduces fluid secretion, potentially affording a novel approach to the treatment of diarrheal states, e.g., through the use of a calcimimetic agent [36].

Bone and cartilage

The level of Ca^{2+}_o within the microenvironment of bone fluctuates substantially during osteoclastic bone resorption and osteoblastic bone formation. In fact, Ca^{2+}_o underneath resorbing osteoclasts can be as high as 40 mM [64]. Moreover, Ca^{2+}_o has several actions on bone cells in vitro of potential physiological relevance. For instance, high Ca^{2+}_o stimulates the proliferation and chemotaxis of preosteoblasts [65], potentially enhancing their availability at sites of recent bone resorption, promotes their differentiation to mature osteoblasts, and increases their capacity to mineralize bone proteins in vitro [66, 67]. In addition, raising Ca^{2+}_o inhibits the activity of osteoclasts in vitro [68, 69]. If these actions of Ca^{2+}_o on bone cells take place in vivo, increases in Ca^{2+}_o could promote net transfer of Ca^{2+} into bone by enhancing bone formation and inhibiting bone resorption. Are these effects of Ca^{2+}_o mediated by the CaSR?

Some [68, 70, 71] but not all [72] investigators have detected CaSR expression in osteoblast-like and osteoclast-like cells in vitro and in bone sections in vivo. These CaSR-expressing cells include cells of both the osteoclast and osteoblast lineages. In addition, the use of dominant negative CaSR constructs or calcilytics has documented in vitro that the CaSR can mediate the stimulatory effects of high Ca^{2+}_o on important parameters of osteoblast function in vitro (i.e., proliferation [73], differentiation, and mineralization of bone [66]). In vivo, conditional knockout of the CaSR in cells of the osteoblast lineage produced mice with a runted phenotype that had small poorly mineralized skeletons and died after several weeks, strongly supporting a key biological role for the CaSR in osteoblasts [74].

With regard to cells of the osteoclast lineage, preosteoclast-like cells generated in vitro show CaSR expression, and osteoclasts derived from rabbit or mouse bone also express the receptor [68]. In sections from murine, rat, and bovine bones, however, only a minority of multinucleated osteoclasts expressed CaSR mRNA and protein [71]. *In vivo*, the CaSR seems to play a permissive/stimulatory role in the generation of osteoclasts [68, 75]. It is also necessary to achieve the full calcemic action of PTH *in vivo* in mice [75]. It also exerts inhibitory actions, however, on osteoclast function by mediating the suppressive effects of very high levels of Ca^{2+}_o (5–20 mM) on osteoclast activity and stimulating their apoptosis [68]. It is unclear how the CaSR exerts both stimulatory and inhibitory effects on osteoclast formation/function.

Some cartilage cells in intact bone express CaSR mRNA and protein, including the hypertrophic chondrocytes of the growth plate, which are key participants in the growth of long bones [71]. Raising Ca^{2+}_o dose dependently reduces the levels of the mRNAs that encode important cartilaginous proteins, including aggrecan, the α_1 chains of types II and X collagen, and alkaline phosphatase, in a chondrocytic cell line (RCJ3.1C5.18 cells) [71]. These actions are likely CaSR mediated, as the use of a CaSR antisense oligonucleotide to lower CaSR expression reversed the action of high Ca^{2+}_o on aggrecan mRNA expression [71]. Recent data have documented that conditional knockout of the CaSR in chondrocytes of mice results in an embryonic lethal phenotype with death before day 14 of embryonic life (E14), confirming an essential role of the CaSR in chondrogenesis [74].

The CaSR in breast and placenta

Expression of the CaSR in the breast increases greatly during lactation in the mouse [76]. The receptor has two functions, relevant to Ca^{2+}_o homeostasis: (1) it promotes transport of calcium from the blood to the milk of the mother, and (2) it inhibits secretion of parathyroid hormone related peptide (PTHrP) by breast epithelial cells into the milk and, probably, into the bloodstream [76]. Thus, when maternal calcium is restricted, PTHrP will be secreted, resulting in more efficient reabsorption of Ca^{2+} by the kidney and release of skeletal Ca^{2+}. The additional Ca^{2+} made available by kidney and bone can then be transported into the milk. Because of the inverse relationship just described between Ca^{2+}_o and PTHrP, the lactating breast can to some extent be viewed as an "accessory" parathyroid gland.

The placenta plays a key role in fetal development by providing adequate quantities of calcium for the developing fetal skeleton, particularly during the third trimester in humans. This takes place by "pumping" calcium transcellularly using the same machinery used in other Ca^{2+}-transporting epithelia, including TRPV6, calbindin D_{9K}, and PMCA. The CaSR in human placenta is expressed in trophoblasts, cytotrophoblasts, and syncytiotrophoblasts [77]. The role of the CaSR in regulating placental Ca^{2+} transport was explored by Kovacs, et al. [78] utilizing mice with the knockout of exon 5 of the CaSR. As expected CaSR–/– fetuses were severely hyperparathyroid, with increased bone resorption and urinary Ca^{2+} excretion. However, placental transport in the CaSR–/– fetuses was significantly less than in both CaSR+/+ and CaSR+/– fetuses [78]. Thus the CaSR normally enhances placental Ca^{2+} transport. This takes place, at least in part, through a PTHrP-dependent pathway, since knocking out PTHrP *in vivo* decreased Ca^{2+} transport to a level similar to that in CaSR–/– fetuses [78]. That is, the CaSR can only stimulate placental Ca^{2+} transport in the presence of PTHrP, which, therefore, is presumably "downstream" of the CaSR.

In addition to the roles that it plays in tissues participating in Ca^{2+}_o homeostasis, the CaSR is expressed in and modulates the functions of numerous other cells uninvolved in mineral ion homeostasis (http://biogps.gnf.org/#goto=genereport&id=846). Examples of the CaSR's roles in some of these tissues can be found in recent reviews [6, 37].

REFERENCES

1. Brown EM. 2007. Clinical lessons from the calcium-sensing receptor. *Nat Clin Pract Endocrinol Metab* 3: 122–33.
2. Berridge MJ, Bootman MD, Roderick HL. 2004. Calcium signalling: Dynamics, homeostasis and remodelling. *Nat Rev Mol Cell Biol* 4: 517–29.
3. Brown EM, Gamba G, Riccardi D, Lombardi M, Butters R, Kifor O, et al. 1993. Cloning and characterization of an extracellular Ca(2+)-sensing receptor from bovine parathyroid. *Nature* 366: 575–80.
4. Garrett JE, Capuano IV, Hammerland LG, Hung BC, Brown EM, Hebert SC, et al. 1995. Molecular cloning and functional expression of human parathyroid calcium receptor cDNAs. *J Biol Chem* 270: 12919–25.
5. Nearing J, Betka M, Quinn S, Hentschel H, Elger M, Baum M, et al. 2002. Polyvalent cation receptor proteins (CaRs) are salinity sensors in fish. *Proc Natl Acad Sci U S A* 99: 9231–6.
6. Magno AL, Ward BK, Ratajczak T. 2011. The calcium-sensing receptor: A molecular perspective. *Endocr Rev* 32: 3–30.
7. Nielsen PK, Rasmussen AK, Butters R, Feldt-Rasmussen U, Bendtzen K, Diaz R, et al. 1997. Inhibition of PTH secretion by interleukin-1 beta in bovine parathyroid glands in vitro is associated with an up-regulation of the calcium-sensing receptor mRNA. *Biochem Biophys Res Commun* 238: 880–5.
8. Canaff L, Zhou X, Hendy GN. 2008. The proinflammatory cytokine, interleukin-6, up-regulates calcium-sensing receptor gene transcription via Stat1/3 and Sp1/3. *J Biol Chem* 283: 13586–600.
9. Canaff L, Hendy GN. 2002. Human calcium-sensing receptor gene. Vitamin D response elements in promoters P1 and P2 confer transcriptional responsiveness

to 1,25-dihydroxyvitamin D. *J Biol Chem* 277: 30337–50.
10. Emanuel RL, Adler GK, Kifor O, Quinn SJ, Fuller F, Krapcho K, et al. 1996. Calcium-sensing receptor expression and regulation by extracellular calcium in the AtT-20 pituitary cell line. *Mol Endocrinol* 10: 555–65.
11. Olszak IT, Poznansky MC, Evans RH, Olson D, Kos C, Pollak MR, et al. 2000. Extracellular calcium elicits a chemokinetic response from monocytes in vitro and in vivo. *J Clin Invest* 105: 1299–305.
12. Goodman WG, Quarles LD. 2008. Development and progression of secondary hyperparathyroidism in chronic kidney disease: Lessons from molecular genetics. *Kidney Int* 74: 276–88.
13. Ray K, Hauschild BC, Steinbach PJ, Goldsmith PK, Hauache O, Spiegel AM. 1999. Identification of the cysteine residues in the amino-terminal extracellular domain of the human Ca(2+) receptor critical for dimerization. Implications for function of monomeric Ca(2+) receptor. *J Biol Chem* 274: 27642–50.
14. Zhuang X, Adipietro KA, Datta S, Northup JK, Ray K. 2010. Rab1 small GTP-binding protein regulates cell surface trafficking of the human calcium-sensing receptor. *Endocrinology* 151: 5114–23.
15. Huang C, Miller RT. 2007. The calcium-sensing receptor and its interacting proteins. *J Cell Mol Med* 11: 923–34.
16. Pi M, Oakley RH, Gesty-Palmer D, Cruickshank RD, Spurney RF, Luttrell LM, et al. 2005. Beta-arrestin- and G protein receptor kinase-mediated calcium-sensing receptor desensitization. *Mol Endocrinol* 19: 1078–87.
17. Lorenz S, Frenzel R, Paschke R, Breitwieser GE, Miedlich SU. 2007. Functional desensitization of the extracellular calcium-sensing receptor is regulated via distinct mechanisms: Role of G protein-coupled receptor kinases, protein kinase C and beta-arrestins. *Endocrinology* 148: 2398–404.
17a. Breitwieser GE. 2012. Minireview: The intimate link between calcium sensing receptor trafficking and signaling: Implications for disorders of calcium homeostasis. *Mol Endocrinol*. 26: 1482–95.
18. Hu J, Spiegel AM. 2003. Naturally occurring mutations in the extracellular Ca2+-sensing receptor: Implications for its structure and function. *Trends Endocrinol Metabol* 14: 282–8.
19. Silve C, Petrel C, Leroy C, Bruel H, Mallet E, Rognan D, et al. 2005. Delineating a Ca2+ binding pocket within the venus flytrap module of the human calcium-sensing receptor. *J Biol Chem* 280: 37917–23.
20. Huang Y, Zhou Y, Castiblanco A, Yang W, Brown EM, Yang JJ. 2009. Multiple Ca(2+)-binding sites in the extracellular domain of the Ca(2+)-sensing receptor corresponding to cooperative Ca(2+) response. *Biochemistry* 48: 388–98.
21. Davies SL, Ozawa A, McCormick WD, Dvorak MM, Ward DT. 2007. Protein kinase C-mediated phosphorylation of the calcium-sensing receptor is stimulated by receptor activation and attenuated by calyculin-sensitive phosphatase activity. *J Biol Chem* 282: 15048–56.
22. Kifor O, Diaz R, Butters R, Kifor I, Brown EM. 1998. The calcium-sensing receptor is localized in caveolin-rich plasma membrane domains of bovine parathyroid cells. *J Biol Chem* 273: 21708–13.
23. Williams TM, Lisanti MP. 2004. The Caveolin genes: From cell biology to medicine. *Ann Med* 36: 584–95.
24. Awata H, Huang C, Handlogten ME, Miller RT. 2001. Interaction of the calcium-sensing receptor and filamin, a potential scaffolding protein. *J Biol Chem* 276: 34871–9.
25. Huang Y, Niwa J, Sobue G, Breitwieser GE. 2006. Calcium-sensing receptor ubiquitination and degradation mediated by the E3 ubiquitin ligase dorfin. *J Biol Chem* 281: 11610–7.
26. Nemeth EF. 2004. Calcimimetic and calcilytic drugs: Just for parathyroid cells? *Cell Calcium* 35: 283–9.
27. Miedlich SU, Gama L, Seuwen K, Wolf RM, Breitwieser GE. 2004. Homology modeling of the transmembrane domain of the human calcium sensing receptor and localization of an allosteric binding site. *J Biol Chem* 279: 7254–63.
28. Strewler GJ. 1994. Familial benign hypocalciuric hypercalcemia—From the clinic to the calcium sensor [Editorial; comment]. *West J Med* 160: 579–80.
29. Quamme GA. 1997. Renal magnesium handling: New insights in understanding old problems. *Kidney Int* 52: 1180–95.
30. Wang WH, Lu M, Hebert SC. 1996. Cytochrome P-450 metabolites mediate extracellular Ca(2+)-induced inhibition of apical K+ channels in the TAL. *Am J Physiol* 271: C103–11.
31. Bourdeau A, Moutahir M, Souberbielle J, Bonnet P, Herviaux P, Sachs C, et al. 1994. Effects of lipoxygenase products of arachidonate metabolism on parathyroid hormone secretion. *Endocrinology* 135: 1109–12.
32. Wang D, An SJ, Wang WH, McGiff JC, Ferreri NR. 2001. CaR-mediated COX-2 expression in primary cultured mTAL cells. *Am J Physiol Renal Physiol* 281: F658–64.
33. Cai S, Fatherazi S, Presland RB, Belton CM, Roberts FA, Goodwin PC, et al. 2006. Evidence that TRPC1 contributes to calcium-induced differentiation of human keratinocytes. *Pflugers Arch* 452: 43–52.
34. Breitwieser GE, Gama L. 2001. Calcium-sensing receptor activation induces intracellular calcium oscillations. *Am J Physiol Cell Physiol* 280: C1412–21.
35. Gerbino A, Ruder WC, Curci S, Pozzan T, Zaccolo M, Hofer AM. 2005. Termination of cAMP signals by Ca2+ and G(alpha)i via extracellular Ca2+ sensors: A link to intracellular Ca2+ oscillations. *J Cell Biol* 171: 303–12.
36. Geibel J, Sritharan K, Geibel R, Geibel P, Persing JS, Seeger A, et al. 2006. Calcium-sensing receptor abrogates secretagogue- induced increases in intestinal net fluid secretion by enhancing cyclic nucleotide destruction. *Proc Natl Acad Sci U S A* 103: 9390–7.
37. Brennan SC, Conigrave AD. 2009. Regulation of cellular signal transduction pathways by the extracellular

calcium-sensing receptor. *Curr Pharm Biotechnol* 10: 270–81.
38. Kifor O, Moore FD Jr, Wang P, Goldstein M, Vassilev P, Kifor I, et al. 1996. Reduced immunostaining for the extracellular Ca2+-sensing receptor in primary and uremic secondary hyperparathyroidism [see Comments]. *J Clin Endocrinol Metab* 81: 1598–606.
39. Ho C, Conner DA, Pollak MR, Ladd DJ, Kifor O, Warren HB, et al. 1995. A mouse model of human familial hypocalciuric hypercalcemia and neonatal severe hyperparathyroidism [see Comments]. *Nat Genet* 11: 389–94.
40. Wettschureck N, Lee E, Libutti SK, Offermanns S, Robey PG, Spiegel AM. 2007. Parathyroid-specific double knockout of Gq and G11 alpha-subunits leads to a phenotype resembling germline knockout of the extracellular Ca2+ -sensing receptor. *Mol Endocrinol* 21: 274–80.
41. Bourdeau A, Moutahir M, Souberbielle JC, Bonnet P, Herviaux P, Sachs C, et al. 1994. Effects of lipoxygenase products of arachidonate metabolism on parathyroid hormone secretion. *Endocrinology* 135: 1109–12.
42. Quinn SJ, Kifor O, Kifor I, Butters RR Jr, Brown EM. 2007. Role of the cytoskeleton in extracellular calcium-regulated PTH release. *Biochem Biophys Res Commun* 354: 8–13.
43. Levi R, Ben-Dov IZ, Lavi-Moshayoff V, Dinur M, Martin D, Naveh-Many T, et al. 2006. Increased parathyroid hormone gene expression in secondary hyperparathyroidism of experimental uremia is reversed by calcimimetics: Correlation with posttranslational modification of the trans acting factor AUF1. *J Am Soc Nephrol* 17: 107–12.
44. Naveh-Many T, Nechama M. 2007. Regulation of parathyroid hormone mRNA stability by calcium, phosphate and uremia. *Curr Opin Nephrol Hypertens* 16: 305–10.
45. Colloton M, Shatzen E, Miller G, Stehman-Breen C, Wada M, Lacey D, et al. 2005. Cinacalcet HCl attenuates parathyroid hyperplasia in a rat model of secondary hyperparathyroidism. *Kidney Int* 67: 467–76.
46. Cozzolino M, Lu Y, Finch J, Slatopolsky E, Dusso AS. 2001. p21WAF1 and TGF-alpha mediate parathyroid growth arrest by vitamin D and high calcium. *Kidney Int* 60: 2109–17.
47. Freichel M, Zink-Lorenz A, Holloschi A, Hafner M, Flockerzi V, Raue F. 1996. Expression of a calcium-sensing receptor in a human medullary thyroid carcinoma cell line and its contribution to calcitonin secretion. *Endocrinology* 137: 3842–8.
48. Kantham L, Quinn SJ, Egbuna OI, Baxi K, Butters R, Pang JL, et al. 2009. The calcium-sensing receptor (CaSR) defends against hypercalcemia independently of its regulation of parathyroid hormone secretion. *Am J Physiol Endocrinol Metab* 297: E915–23.
49. McGehee DS, Aldersberg M, Liu KP, Hsuing S, Heath MJ, Tamir H. 1997. Mechanism of extracellular Ca2+ receptor-stimulated hormone release from sheep thyroid parafollicular cells. *J Physiol (Lond)* 502: 31–44.
50. Riccardi D, Hall AE, Chattopadhyay N, Xu JZ, Brown EM, Hebert SC. 1998. Localization of the extracellular Ca2+/polyvalent cation-sensing protein in rat kidney. *Am J Physiol* 274: F611–22.
51. Yang T, Hassan S, Huang YG, Smart AM, Briggs JP, Schnermann JB. 1997. Expression of PTHrP, PTH/PTHrP receptor, and Ca(2+)-sensing receptor mRNAs along the rat nephron. *Am J Physiol* 272: F751–8.
52. de Rouffignac C, Quamme G. 1994. Renal magnesium handling and its hormonal control. *Physiol Rev* 74: 305–22.
53. Sands JM, Naruse M, Baum M, Jo I, Hebert SC, Brown EM, et al. 1997. Apical extracellular calcium/polyvalent cation-sensing receptor regulates vasopressin-elicited water permeability in rat kidney inner medullary collecting duct. *J Clin Invest* 99: 1399–405.
54. Ba J, Friedman PA. 2004. Calcium-sensing receptor regulation of renal mineral ion transport. *Cell Calcium* 35: 229–37.
54a. Loupy A, Ramakrishnan SK, Wootla B, Chambrey R, de la Faille R, Bourgeois S, Bruneval P, Mandet C, Christensen EI, Faure H, Cheval L, Laghmani K, Collet C, Eladari D, Dodd RH, Ruat M, Houillier P. 2012. PTH-independent regulation of blood calcium concentration by the calcium-sensing receptor. *J Clin Invest*. 122: 3355–67.
55. Maiti A, Beckman MJ. 2007. Extracellular calcium is a direct effecter of VDR levels in proximal tubule epithelial cells that counter-balances effects of PTH on renal Vitamin D metabolism. *J Steroid Biochem Mol Biol* 122: 3355–67.
56. Topala CN, Schoeber JP, Searchfield LE, Riccardi D, Hoenderop JG, Bindels RJ. 2009. Activation of the Ca2+-sensing receptor stimulates the activity of the epithelial Ca2+ channel TRPV5. *Cell Calcium* 45: 331–9.
57. Thebault S, Hoenderop JG, Bindels RJ. 2006. Epithelial Ca2+ and Mg2+ channels in kidney disease. *Adv Chronic Kidney Dis* 13: 110–7.
58. Renkema KY, Velic A, Dijkman HB, Verkaart S, van der Kemp AW, Nowik M, et al. 2009. The calcium-sensing receptor promotes urinary acidification to prevent nephrolithiasis. *J Am Soc Nephrol* 20: 1705–13.
59. Chattopadhyay N, Cheng I, Rogers K, Riccardi D, Hall A, Diaz R, et al. 1998. Identification and localization of extracellular Ca(2+)-sensing receptor in rat intestine. *Am J Physiol* 274: G122–30.
60. Krishnamra N, Angkanaporn K, Deenoi T. 1994. Comparison of calcium absorptive and secretory capacities of segments of intact or functionally resected intestine during normo-, hypo-, and hyper-calcemia. *Can J Physiol Pharmacol* 72: 764–70.
61. Conigrave AD, Mun HC, Brennan SC. 2007. Physiological significance of L-amino acid sensing by extracellular Ca(2+)-sensing receptors. *Biochem Soc Trans* 35: 1195–8.
62. Geibel JP, Hebert SC. 2009. The functions and roles of the extracellular Ca2+-sensing receptor along the gastrointestinal tract. *Annu Rev Physiol* 71: 205–17.

63. Liou AP, Sei Y, Zhao X, Feng J, Lu X, Thomas C, et al. 2011. The extracellular calcium-sensing receptor is required for cholecystokinin secretion in response to L-phenylalanine in acutely isolated intestinal I cells. *Am J Physiol Gastrointest Liver Physiol* 300: G538–46.
64. Silver IA, Murrils RJ, Etherington DJ. 1988. Microlectrode studies on the acid microenvironment beneath adherent macrophages and osteoclasts. *Exp Cell Res* 175: 266–76.
65. Godwin SL, Soltoff SP. 1997. Extracellular calcium and platelet-derived growth factor promote receptor-mediated chemotaxis in osteoblasts through different signaling pathways. *J Biol Chem* 272: 11307–12.
66. Dvorak MM, Siddiqua A, Ward DT, Carter DH, Dallas SL, Nemeth EF, et al. 2004. Physiological changes in extracellular calcium concentration directly control osteoblast function in the absence of calciotropic hormones. *Proc Natl Acad Sci U S A* 101: 5140–5.
67. Quarles LD. 1997. Cation-sensing receptors in bone: A novel paradigm for regulating bone remodeling? *J Bone Miner Res* 12: 1971–4.
68. Mentaverri R, Yano S, Chattopadhyay N, Petit L, Kifor O, Kamel S, et al. 2006. The calcium sensing receptor is directly involved in both osteoclast differentiation and apoptosis. *FASEB J* 20: 2562–4.
69. Zaidi M, Adebanjo OA, Moonga BS, Sun L, Huang CL. 1999. Emerging insights into the role of calcium ions in osteoclast regulation. *J Bone Miner Res* 14: 669–74.
70. Yamaguchi T, Chattopadhyay N, Kifor O, Butters RR Jr, Sugimoto T, Brown EM. 1998. Mouse osteoblastic cell line (MC3T3-E1) expresses extracellular calcium (Ca2+o)-sensing receptor and its agonists stimulate chemotaxis and proliferation of MC3T3-E1 cells. *J Bone Miner Res* 13: 1530–8.
71. Chang W, Tu C, Chen T-H, Komuves L, Oda Y, Pratt S, et al. 1999. Expression and signal transduction of calcium-sensing receptors in cartilage and bone. *Endocrinology* 140: 5883–93.
72. Pi M, Hinson TK, Quarles L. 1999. Failure to detect the extracellular calcium-sensing receptor (CasR) in human osteoblast cell lines. *J Bone Miner Res* 14: 1310–9.
73. Chattopadhyay N, Yano S, Tfelt-Hansen J, Rooney P, Kanuparthi D, Bandyopadhyay S, et al. 2004. Mitogenic action of calcium-sensing receptor on rat calvarial osteoblasts. *Endocrinology* 145: 3451–62.
74. Chang W, Tu C, Chen TH, Bikle D, Shoback D. 2008. The extracellular calcium-sensing receptor (CaSR) is a critical modulator of skeletal development. *Sci Signal* 1: ra1.
75. Shu L, Ji J, Zhu Q, Cao G, Karaplis A, Pollak MR, et al. 2011. The calcium-sensing receptor mediates bone turnover induced by dietary calcium and parathyroid hormone in neonates. *J Bone Miner Res* 26: 1057–71.
76. VanHouten J, Dann P, McGeoch G, Brown EM, Krapcho K, Neville M, et al. 2004. The calcium-sensing receptor regulates mammary gland parathyroid hormone-related protein production and calcium transport. *J Clin Invest* 113: 598–608.
77. Bradbury RA, Sunn KL, Crossley M, Bai M, Brown EM, Delbridge L, et al. 1998. Expression of the parathyroid Ca(2+)-sensing receptor in cytotrophoblasts from human term placenta. *J Endocrinol* 156: 425–30.
78. Kovacs CS, Ho-Pao CL, Hunzelman JL, Lanske B, Fox J, Seidman JG, et al. 1998. Regulation of murine fetal-placental calcium metabolism by the calcium-sensing receptor. *J Clin Invest* 101: 2812–20.

29

Vitamin D: Production, Metabolism, Mechanism of Action, and Clinical Requirements

Daniel Bikle, John S. Adams, and Sylvia Christakos

Vitamin D_3 Production 235
Vitamin D Metabolism 235
Transport of Vitamin D in the Blood 237
Internalization of Vitamin D Metabolites 237
Molecular Mechanism of Action 238
Regulation of Calcium and Phosphate Metabolism 238
Classical Target Tissues 239

Nonclassical Target Tissues 240
Immunobiology of Vitamin D 241
Keratinocyte Function in Epidermis and Hair Follicles 242
Nutritional Considerations 242
Vitamin D Treatment Strategies 243
References 244

VITAMIN D_3 PRODUCTION

Vitamin D_3 is produced from 7-dehydrocholesterol (7-DHC) (Fig. 29.1). Although irradiation of 7-DHC was known to produce pre-vitamin D_3 (pre-D_3, which subsequently undergoes a temperature-dependent rearrangement of the triene structure to form D_3, lumisterol, and tachysterol), the physiologic regulation of this pathway was not well understood until the studies of Holick et al. [1–3]. They demonstrated that the formation of pre-D_3 under the influence of solar or UVB irradiation (maximal effective wavelength between 290 and 310 µm) is relatively rapid and reaches a maximum within hours. Both the degree of epidermal pigmentation and the intensity of exposure correlate with the time required to achieve this maximal concentration of pre-D_3 but do not alter the maximal level achieved. Although pre-D_3 levels reach a maximum level, the biologically inactive lumisterol and tachysterol accumulate with continued UV exposure. Thus, prolonged exposure to sunlight would not produce toxic amounts of D_3 because of the photoconversion of pre-D_3 to lumisterol and tachysterol. Melanin in the epidermis, by absorbing UV irradiation, can also reduce the effectiveness of sunlight in producing D_3 in the skin. Sunlight exposure increases melanin production and so provides another mechanism by which excess D_3 production can be prevented. As noted, the intensity of UV irradiation is also important for D_3 production and is dependent on latitude. In Edmonton, Canada (52° N) very little D_3 is produced in exposed skin from mid-October to mid-April, while in San Juan (18° N) the skin is able to produce D_3 all year long [4]. Clothing and sunscreen effectively prevent D_3 production in the covered areas.

VITAMIN D METABOLISM

Vitamin D_3 and its plant-derived counterpart vitamin D_2 (collectively referred to as "vitamin D") are by themselves biologically inert for most of their actions. After cutaneous synthesis and transport into the general circulation, vitamin D disappears from the serum within a week [5]. Vitamin D is bound to the serum vitamin D binding protein and ferried in the circulation to sites of storage (mainly fat and muscle) [6] and to the tissues, primarily the liver, where the first step in metabolism occurs, that of conversion to the prohormone 25-hydroxyvitamin D (25OHD; Fig. 29.2). There are a number of cytochrome P450 enzymes capable of converting vitamin D to 25OHD. These enzymes exhibit a high capacity for substrate vitamin D, and release product 25OHD back into the circulation and not into the bile. As such, the serum

Primer on the Metabolic Bone Diseases and Disorders of Mineral Metabolism, Eighth Edition. Edited by Clifford J. Rosen.
© 2013 American Society for Bone and Mineral Research. Published 2013 by John Wiley & Sons, Inc.

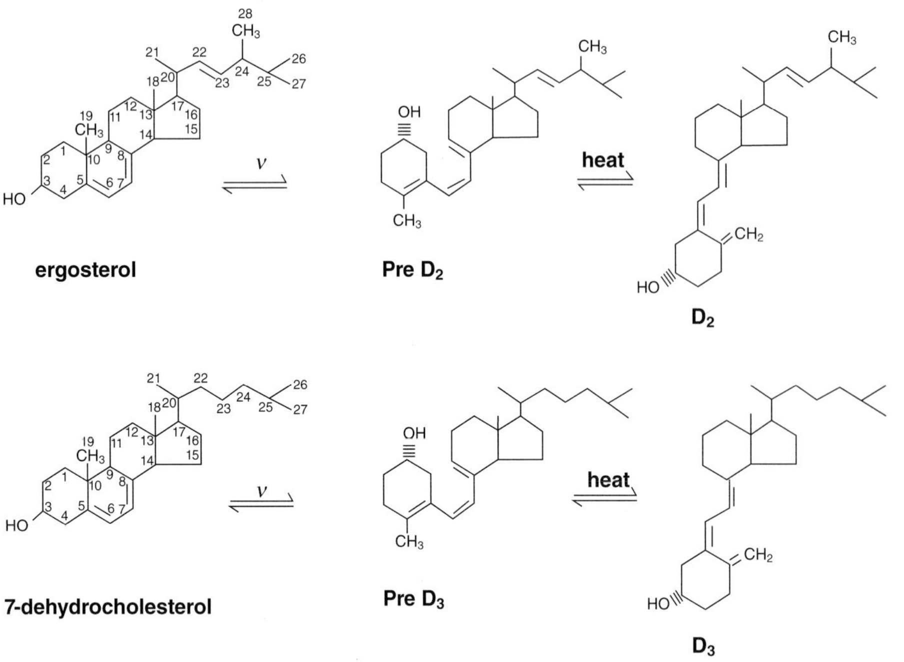

Fig. 29.1. The photolysis of ergosterol and 7-dehydrocholesterol to vitamin D_2 (ergocalciferol) and vitamin D_3 (cholecalciferol), respectively. An intermediate is formed after photolysis, which undergoes a thermal-activated isomerization to the final form of vitamin D. The rotation of the A-ring puts the 3β-hydroxyl group into a different orientation with respect to the plane of the A-ring during production of vitamin D.

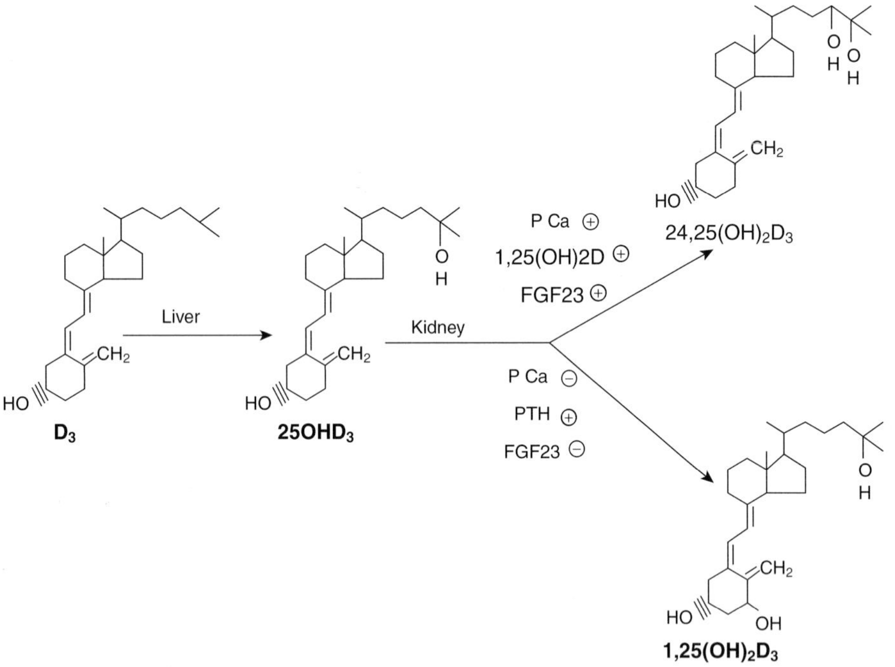

Fig. 29.2. The metabolism of vitamin D. The liver converts vitamin D to 25OHD. The kidney converts 25OHD to 1,25(OH)$_2$D$_3$ and 24,25(OH)$_2$D$_3$. Control of metabolism is exerted primarily at the level of the kidney, where low serum phosphorus, low serum calcium, low FGF23, and high parathyroid hormone (PTH) levels favor production of 1,25(OH)$_2$D$_3$, whereas high serum phosphorus, calcium, FGF23, and 1,25(OH)$_2$D$_3$ and low PTH favor 24,25(OH)$_2$D$_3$ production.

25OHD is the most reliable indicator of whether too little or too much vitamin D is entering the host [7]. 25OHD is biologically inert unless present in intoxicating concentrations in the blood owing to the ingestion of large amounts of vitamin D. Otherwise, it must be converted via the CYP27B1 1α-hydroxylase to 1,25-dihydroxyvitamin D (1,25(OH)$_2$D), the specific, naturally occurring ligand for the vitamin D receptor (VDR) (Fig. 29.2). CYP27B1 is a heme-containing, inner mitochondrial membrane-embedded, cytochrome P450 mixed function oxidase requiring molecular oxygen and a source of electrons for biological activity. Although the proximal renal tubular epithelial cell is the richest source of CYP27B1 and responsible for generating the relatively large amounts of 1,25(OH)$_2$D that are required to achieve the endocrine functions of the hormone in mineral ion homeostasis, this enzyme is also encountered in a number of extrarenal sites, including immune cells and a variety of normal and malignant epithelia [8], where it functions to provide 1,25(OH)$_2$D for intracrine or paracrine access to the VDR in these and neighboring cells. As discussed below, the VDR has an extraordinarily broad distribution among human tissues. There are four major recognized means of regulating CYP27B1: (1) controlling the availability of substrate 25OHD to the enzyme; (2) controlling the amount of CYP27B1; (3) altering the activity of the enzyme by cofactor availability; and (4) controlling the amount and activity of the alternatively-spliced, catabolic CYP24A, the 24-hydroxylase.

For the kidney, CYP27B1 substrate is provided by the endocytic internalization of filtered, megalin-bound vitamin D binding protein (DBP) carrying 25OHD into the proximal tubular cell from the urinary side of that cell. Regulation of the CYP27B1 in the proximal nephron is principally controlled at the level of transcription with circulating parathyroid hormone and FGF-23 being the major stimulator and inhibitor of CYP27B1 gene expression, respectively (see below). In the kidney, the CYP24A, also a mitochondrial P450, serves not only to limit the amount of 1,25(OH)$_2$D, leaving the kidney for distant target tissues by accelerating its catabolism to 1,24, 25(OH)$_3$D but also by shunting available substrate 25OHD away from CYP27B1. In both cases, the 24-hydroxylated products are degraded by the same enzyme to side chain-cleaved, water-soluble catabolites. The CYP24A gene is under stringent transcriptional control by 1,25(OH)$_2$D itself, providing a robust means of proximate, negative feedback regulation of the amount of 1,25(OH)$_2$D made in and released from the kidney [9]. By comparison, the activity of some of the extrarenal, intracrine/paracrine-acting CYP27B1, such as in keratinocytes and disease-activated macrophages, appears to be primarily governed by the availability of extracellular substrate 25OHD to the enzyme, cytokines such as tumor necrosis factor-α (TNF-α) and interferon-γ (IFN-γ), and toll-like receptor (TLR) activation. In macrophages, excess 1,25(OH)$_2$D may be produced, spilling into the circulation and causing hypercalcemia. It is postulated that this is due to the expression of an amino-terminally truncated splice variant product of the CYP24 gene that cannot be transported into mitochondria [10, 11]. The result is generation of a non-catalytically active enzyme, albeit one that can serve as a cytoplasmic reservoir for the CYP24 substrates, 1,25(OH)$_2$D and 25OHD. Also contrary to the renal CYP27B1, the extrarenal CYP27B1, at least in the macrophage, is (1) immune to control by either PTH or FGF-23 (receptors for these molecules are not expressed to any degree on inflammatory cells); (2) susceptible to induction by TLR ligands shed by microbial agents; and (3) can be upregulated by nontraditional electron donors such as nitric oxide [8]. The CYP27B1 in the keratinocyte, on the other hand, shares features of both the renal and macrophage CYP27B1 in that it is associated with a very active CYP24A, which limits the levels of 1,25(OH)$_2$D in the cell, is stimulated by cytokines such as TNF-α [12] and IFN-γ [13] but not by c-AMP, and is induced by TLR 2 activation [14].

TRANSPORT OF VITAMIN D IN THE BLOOD

In order for the hormone 1,25(OH)$_2$D to reach any of its target tissues, with the exception of the skin where it can be both produced and act locally as just described, vitamin D must be able to escape its synthetic site in the skin or its absorption site in the gut and be transported to tissues expressing one of the vitamin D-25-hydroxylase genes. From there 25OHD must travel to tissue sites expressing the CYP27B1 gene, and the synthesized 1,25(OH)$_2$D must be able to gain access to target tissues containing cells expressing the VDR in order for the genomic actions of the sterol hormone to be realized. The serum vitamin D binding protein (DBP), a member of the albumin family of proteins, is the specific chaperone for vitamin D and its metabolites in the serum [15]. It has a high capacity (less than 5% saturated with vitamin D metabolites in humans) and is bound with high affinity (nM range) by vitamin Ds, particularly the 25-hydroxylated metabolites 25OHD, 24,25(OH)$_2$D, and 1,25(OH)$_2$D [16]. DBP is produced mainly in the liver and is freely filterable across the glomerulus into the urine. DBP has a serum half-life of 2.5–3.0 days, indicating that it must be largely reclaimed from the urine once filtered. Reclamation is achieved by DBP being bound by the endocytic, LDL-like co-receptor molecules megalin and cubulin embedded in the plasma membrane of the proximal renal tubular epithelial cell, with eventual transcellular transport and return to the circulation via intracellular vitamin D binding protein (IDBP) chaperones in the heat protein-70 family [17]. Interestingly, 1,25(OH)$_2$D$_3$ infused into the DBP-null mouse is able to reach and act normally in the kidney, gut, and bone, despite failure to detect the infused hormone in the serum [18]. No human, DBP-null homozygote has yet been described suggesting that, unlike the DBP-null mouse, which is both viable and fertile [19], such a human genotype might be embryonically lethal.

INTERNALIZATION OF VITAMIN D METABOLITES

Once bound to DBP and shuttled to sites of metabolism, action, and/or catabolism, vitamin D metabolites must gain access to the interior of their target cell and arrive safely at their intracellular destination (i.e., nucleus for transaction via the VDR, inner mitochondrial membrane for access to the CYP27B1 and CYP24A hydroxylases). Although possible, it seems unlikely that simple diffusion of the sterol metabolite off the serum DBP in the extracellular space and subsequent diffusion through the plasma membrane to a specific intracellular destination, the so-called free hormone hypothesis, can account for the required specificity for targeted metabolite delivery to all cells. For example, as just described for the kidney, there likely exists a plasma-membrane anchored "acceptor" for DBP that acts to concentrate and/or endocytically internalize DBP and its cargo, with intracellular chaperones moving the cargo or metabolite(s) to specific intracellular destinations (e.g., the CYP27B1 and VDR) [20].

MOLECULAR MECHANISM OF ACTION

The mechanism of action of the active form of vitamin D, 1,25(OH)$_2$D, is similar to that of other steroid hormones. The intracellular mediator of the 1,25(OH)$_2$D$_3$ function is the VDR. 1,25(OH)$_2$D binds stereospecifically to VDR, which is a high affinity, low capacity intracellular receptor that has extensive homology with other members of the superfamily of nuclear receptors, including receptors for steroid and thyroid hormones. VDR functions as a heterodimer with the retinoid X receptor (RXR) for activation for vitamin D target genes. Once formed, the 1,25(OH)$_2$D$_3$–VDR–RXR heterodimeric complex interacts with specific DNA sequences [vitamin D response elements (VDREs)] in and around target genes resulting in either activation or repression of transcription [21–24]. In general, for activation of transcription, the VDRE consensus consists of two direct repeats of the hexanucleotide sequence GGGTGA separated by three nucleotide pairs. The mechanisms involved in VDR-mediated transcription following binding of the 1,25(OH)$_2$D–VDR–RXR heterodimeric complex to DNA are now beginning to be defined. TFIIB, several TATA binding protein associated factors (TAFs) as well as the p160 coactivators, steroid receptor activator-1, 2, and 3 (SRC-1, SRC-2, and SRC-3), that have histone acetylase (HAT) activity, have been reported to be involved in VDR-mediated transcription. In addition to acetylation, methylation also occurs on core histones. Recent studies have indicated that cooperativity between histone methyltransferases and p160 coactivators may also play a fundamental role in VDR-mediated transcriptional activation [25]. VDR-mediated transcription is also mediated by the coactivator complex DRIP (vitamin D receptor interacting protein more generally known as Mediator). This complex does not have HAT activity but rather functions, at least in part, through recruitment of RNA polymerase II [23, 26]. In addition, a number of specific transcription factors including YY1 and CCAAT, enhancer binding proteins β and δ have been reported to modulate VDR-mediated transcription [27–29]. SWI/SNF complexes that remodel chromatin, using the energy of ATP hydrolysis, are also involved in VDR-mediated transcription [30]. It has been suggested that cell- and gene-specific functions of VDR may be mediated through differential recruitment of coactivators. In addition, genome-wide studies of VDR binding sites have indicated that although many of the VDR regulatory regions are located in proximal promoters of target genes, many are also situated many kilobases upstream and downstream and in intronic and exonic sites. VDR binding to these sites has been reported to be largely but not exclusively dependent on activation by 1,25(OH)$_2$D [31]. These genome-wide studies have provided a new perspective on mechanisms involved in the regulation of gene expression by 1,25(OH)$_2$D.

REGULATION OF CALCIUM AND PHOSPHATE METABOLISM

The classical actions of 1,25(OH)$_2$D involve its regulation of calcium and phosphate flux across three target tissues: bone, gut, and kidney. The mechanisms by which 1,25(OH)$_2$D operates in these tissues will be described in more detail below. However, the receptor for 1,25(OH)$_2$D (VDR) is widespread and not limited to these classical target tissues. Indeed the list of tissues not containing the VDR is probably shorter than the list of tissues that contain the VDR. Furthermore, as discussed previously, a number of these tissues express CYP27B1 and so can make their own 1,25(OH)$_2$D. The biologic significance of these observations is found in the large number of non-classical actions of active vitamin D metabolites including effects on cellular proliferation and differentiation, regulation of hormone secretion, and immune modulation. Examples of these actions will be discussed below.

With regard to the classical endocrine actions of vitamin D, 1,25(OH)$_2$D acts in concert with two peptide hormones, parathyroid hormone (PTH) and FGF23 (Fig. 29.3). In each case, feed forward and feedback regulatory loops are operative. Parathyroid hormone (PTH) is the major stimulator of 1,25(OH)$_2$D$_3$ production in the kidney. 1,25(OH)$_2$D$_3$ in turn suppresses PTH production directly via a transcriptional mechanism and indirectly by increasing serum calcium levels. The parathyroid gland expresses both VDR and CYP27B1 so the regulation of PTH secretion may involve endogenous as well as exogenous 1,25(OH)$_2$D. Calcium acts via the calcium sensing receptor (CaSR) in the parathyroid gland to suppress PTH release. 1,25(OH)$_2$D increases the levels of CaSR in the parathyroid gland just as calcium increases the VDR in the parathyroid gland further enhancing the negative influence of calcium and 1,25(OH)$_2$D on PTH secretion. FGF23, on the other hand, inhibits 1,25(OH)$_2$D production by the kidney while increasing the expression of CYP24A, whereas 1,25(OH)$_2$D stimulates FGF23 production in bone. FGF23 requires klotho (a multifunctional protein involved in phosphate and calcium homeostasis) as a cofactor for FGF23 signaling. 1,25(OH)$_2$D also upregulates klotho and the loss of klotho results in induction of CYP27B1. Klotho-deficient mice and FGF23-deficient mice exhibit homologous phenotypes (including hyperphosphatemia and increased synthesis of 1,25(OH$_2$D), further indicating cooperation of klotho and

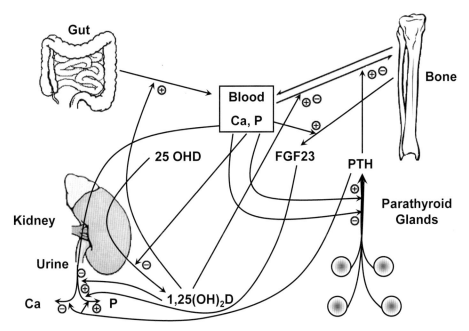

Fig. 29.3. Classical actions: Bone mineral homeostasis mineral feedback loops. 1,25(OH)$_2$D interacts with other hormones, in particular FGF23 and PTH, to regulate calcium and phosphate homeostasis. As noted in the legend to Fig. 29.2, FGF23 inhibits whereas PTH stimulates 1,25(OH)$_2$D production by the kidney. In turn, 1,25(OH)$_2$D inhibits PTH production but stimulates that of FGF23.

FGF23 [32]. Dietary phosphate also regulates FGF23 levels (high phosphate intake stimulates), an effect independent of 1,25(OH)$_2$D as demonstrated by phosphate regulation of FGF23 in the VDR null mouse. Similarly, high phosphate stimulates PTH secretion, an action independent of its inhibitory actions on 1,25(OH)$_2$D production. Whether phosphate has its own receptor, as calcium does, is unclear. FGF23 expression is found in a number of tissues including the parathyroid gland, but its greatest expression is in osteocytes, bone lining cells, and active osteoblasts. Thus, in considering the mechanisms of action of 1,25(OH)$_2$D in vivo, the roles of PTH and FGF23 must also be considered along with the minerals they regulate: calcium and phosphate.

CLASSICAL TARGET TISSUES

Bone

Whether 1,25(OH)$_2$D acts directly on bone or mediates its antirachitic effects indirectly by stimulation of intestinal calcium and phosphorus absorption is uncertain. VDR-ablated mice [VDR knockout (KO) mice] develop secondary hyperparathyroidism, hypocalcemia, and rickets after weaning [33, 34]. However, when VDR KO mice are fed a rescue diet containing high levels of calcium, phosphorus, and lactose, serum ionized calcium and PTH levels are normalized, and rickets and osteomalacia are prevented; these results suggest that the major effect of 1,25(OH)$_2$D$_3$ is the provision of calcium and phosphate to bone from the stimulated intestinal absorption of calcium and phosphorous rather than a direct action of 1,25(OH)$_2$D on bone [35]. In addition, transgenic expression of VDR in the intestine of VDR KO mice results in normalization of serum calcium, bone density, and bone volume [36]. However, in vivo studies using CYP27B1 KO mice and CYP27B1/VDR double KO mice showed that when hypocalcemia and secondary hyperparathyroidism are prevented by a rescue diet, not all changes in osteoblast number, mineral apposition rate, and bone volume are rescued, suggesting direct skeletal effects of the 1,25(OH)$_2$D-VDR system [37]. Studies performed in vitro also support a direct effect of 1,25(OH)$_2$D on bone [38]. Osteoblast differentiation and osteoclastogenesis can be stimulated by 1,25(OH)$_2$D [38]. The effect of 1,25(OH)$_2$D on osteoclastogenesis is indirect. VDR is not present in osteoclasts but rather in osteoprogenitor cells, osteoblast precursors, and mature osteoblasts. Stimulation of osteoclast formation by 1,25(OH)$_2$D involves upregulation of receptor activator of nuclear factor-κB ligand (RANKL) by 1,25(OH)$_2$D in osteoblastic cells and requires cell-to-cell contact between osteoblasts and osteoclast precursors [39]. The osteoclastogenesis inhibitory factor, osteoprotegerin (OPG), is a decoy receptor for RANKL that antagonizes RANKL function thus blocking osteoclastogenesis. OPG is downregulated by 1,25(OH)$_2$D [39]. 1,25(OH)$_2$D has also been reported to stimulate the production in osteoblasts of the calcium binding proteins osteocalcin and osteopontin [40, 41]. Runx2, a transcriptional regulator of osteoblast differentiation, is also regulated by 1,25(OH)$_2$D [42]. Transgenic mice overexpressing VDR in osteoblastic cells have increased bone formation, further indicating direct effects of 1,25(OH)$_2$D on bone [43]. Thus, the effects of 1,25(OH)$_2$D on bone are diverse and can affect formation or resorption.

Intestine

When the demand for calcium increases as a result of increased requirement associated with growth, pregnancy or lactation, 1,25(OH)$_2$D synthesis increases the efficiency of intestinal calcium absorption. In VDR KO mice, a major defect is in intestinal calcium absorption, suggesting that a principal action of 1,25(OH)$_2$D$_3$ to maintain calcium homeostasis is increased intestinal calcium absorption [35]. It is thought that intestinal calcium absorption comprises two different modes of calcium transport: the saturable process, which is mainly transcellular, and a diffusional mode, which is nonsaturating. The latter requires a lumenal free calcium concentration greater than 2–6 mM and is paracellular (i.e., between neighboring intestinal epithelial cells). In the paracellular pathway, the movement of calcium is across tight junctions and intercellular spaces and is directly related to the concentration of calcium in the intestinal lumen. The saturable component of intestinal calcium absorption is observed predominantly in the duodenum. The diffusional, nonsaturable process is observed all along the intestine (duodenum, jejunum, ileum, and colon). 1,25(OH)$_2$D has been reported to affect both the transcellular and the paracellular path [44–46]. The transcellular process comprises three 1,25(OH)$_2$D-regulated steps: the entry of calcium across the brush border membrane, intracellular diffusion, and the energy requiring extrusion of calcium across the basolateral membrane [44–46]. It is thought that the calcium binding protein, calbindin, which is induced by 1,25(OH)$_2$D in the intestine, acts to facilitate the diffusion of calcium through the cell interior toward the basolateral membrane. Interestingly, recent studies in calbindin-D$_{9k}$ null mutant mice showed no change in 1,25(OH)$_2$D$_3$-mediated intestinal calcium absorption and in serum calcium levels compared to wild-type mice [47, 48]. This observation provides evidence that calbindin alone is not responsible for 1,25(OH)$_2$D-mediated intestinal calcium absorption.

Calcium extrusion from the enterocyte is also affected by 1,25(OH)$_2$D. The plasma membrane calcium pump (PMCA) has been reported to be stimulated by 1,25(OH)$_2$D, suggesting that intestinal calcium absorption may involve a direct effect of 1,25(OH)$_2$D on calcium pump expression [45]. The rate of calcium entry into the enterocyte is also increased by 1,25(OH)$_2$D. Recently, a calcium selective channel, TRPV6, which is colocalized with calbindin and is induced by 1,25(OH)$_2$D, was cloned from rat duodenum [49, 50]. It has been suggested that TRPV6 plays a key role in vitamin D dependent calcium entry into the enterocyte. Since studies in calbindin-D$_{9k}$/TRPV6 double KO mice have shown that intestinal calcium absorption in response to low dietary calcium or to 1,25(OH)$_2$D is least efficient in the absence of both proteins, calbindin-D$_{9k}$ and TRPV6 may act together to affect calcium absorption [51]. It is possible that calbindin is not a facilitator of calcium diffusion (thus challenging the traditional model of vitamin D mediated transcellular calcium absorption) but acts rather as a modulator of calcium influx via TRPV6.

In addition to intestinal calcium absorption, 1,25(OH)$_2$D also acts to enhance intestinal phosphorus absorption. Although the mechanisms involved have been a matter of debate, it has been suggested that FGF23 and 1,25(OH)$_2$D stimulate the active transport of phosphorus [52, 53].

Kidney

A third target tissue involved in 1,25(OH)$_2$D-mediated mineral homeostasis is the kidney. It has been reported that 1,25(OH)$_2$D enhances the actions of PTH on calcium transport in the distal tubule, at least in part, by increasing the PTH receptor mRNA and binding activity in distal tubule cells [54]. 1,25(OH)$_2$D also induces the synthesis of the calbindins in the distal tubules [44]. It has been suggested that calbindin-D$_{28k}$ stimulates the high affinity calcium transport system in the distal luminal membrane, and calbindin-D$_{9k}$ enhances the adenosine triphosphate (ATP)-dependent calcium transport of the basolateral membrane [44]. Similar to studies in the intestine, an apical calcium channel, TRPV5, which is colocalized with the calbindins and induced by 1,25(OH)$_2$D, has been identified in the distal convoluted tubule and distal connecting tubules [49]. Calbindin-D$_{28k}$ was reported to associate directly with TRPV5 and to control TRPV5 mediated calcium influx [55]. Thus, 1,25(OH)$_2$D affects calcium transport in the distal tubule by enhancing the action of PTH and by inducing TRPV5 and the calbindins. Another important effect of 1,25(OH)$_2$D in the kidney is the inhibition of the CYP27B1 and the induction of the CYP24 [56]. Besides calcium transport in the distal nephron and modulation of CYP27B1 and CYP24A, effects of 1,25(OH)$_2$D on sodium phosphate transporter b (NPTB)-mediated phosphate reabsorption in the proximal tubule have been suggested [57]. Depending on the parathyroid status and on experimental conditions, 1,25(OH)$_2$D has been reported to increase or decrease renal phosphate reabsorption.

NONCLASSICAL TARGET TISSUES

Parathyroid glands

The parathyroid glands (PTGs) are an important target of 1,25(OH)$_2$D. As discussed previously, 1,25(OH)$_2$D inhibits the synthesis and secretion of PTH and prevents the proliferation of parathyroid-producing cells in the gland in order to maintain normal parathyroid status [58, 59]. It has also been shown that 1,25(OH)$_2$D upregulates calcium sensing receptor (CaSR) transcription [60], suggesting that 1,25(OH)$_2$D sensitizes the PTG to calcium inhibition. In addition to containing the VDR, the PTGs also express CYP27B1; as such, local production of

1,25(OH)$_2$D, as well as circulating 1,25(OH)$_2$D, may contribute to PTH production and secretion.

Pancreas

The pancreas was one of the first nonclassical target tissues in which receptors for 1,25(OH)$_2$D were identified [61]. Although 1,25(OH)$_2$D has been reported to play a role in insulin secretion, the exact mechanisms remain unclear. Autoradiographic data and immunocytochemical studies have localized the VDR and calbindin-D$_{28k}$, respectively, in pancreatic beta cells [62, 63]. Studies using calbindin-D$_{28k}$ null mutant mice have indicated that calbindin-D$_{28k}$, by regulating intracellular calcium, can modulate depolarization-stimulated insulin release [64]. In addition to modulating insulin release, calbindin-D$_{28k}$, by buffering calcium, can protect against cytokine-mediated destruction of beta cells [65]. These findings have important therapeutic implications for type 1 diabetes and the prevention of cytokine destruction of pancreatic beta cells as well as for type 2 diabetes and the potentiation of insulin secretion.

IMMUNOBIOLOGY OF VITAMIN D

Nonclassical regulation of immune responses by 1,25-(OH)$_2$D was first reported nearly 30 years ago with the discovery of the presence of the VDR in activated human inflammatory cells [66] and the ability of disease-activated macrophages to make 1,25-(OH)$_2$D [67]. Recent studies have shown that 1,25(OH)$_2$D regulates both innate and adaptive immunity, but in opposite directions, namely promoting the former while repressing the latter.

Vitamin D and innate immunity

Innate immunity encompasses the ability of the host immune system to recognize and respond to an offending antigen. In 1986, Rook and colleagues described studies using cultured human macrophages in which they showed that 1,25(OH)$_2$D can inhibit the growth of *Mycobacterium tuberculosis* (*M.tb.*) [68]. Although this seminal report was widely cited, it is only in the last few years that more comprehensive appraisals of the antimicrobial effects of vitamin D metabolites have been published. *In silico* screening of the human genome revealed the presence of a vitamin D response element in the promoter of the human gene for cathelicidin, whose product LL37 is an antimicrobial peptide capable of killing bacteria [69]. Subsequent investigations confirmed the ability of 1,25(OH)$_2$D [70] and its precursor 25OHD [71] to induce expression of cathelicidin in cells of the monocyte/macrophage and epidermal lineage, respectively, highlighting the potential for intracrine/autocrine induction of antimicrobial responses in cells that also express the 25OHD-activating enzyme, CYP27B1. Although detectable in many cell types, functionally significant expression and activity of CYP27B1 appears to be dependent on cell-specific stimulation of a broad spectrum of immune surveillance proteins, the Toll-like receptors (TLRs). The TLRs are an extended family of noncatalytic, host cell transmembrane pattern-recognition receptors (PRRs) that interact with specific pathogen-associated membrane patterns or PAMPs shed by infectious agents and trigger the innate immune response in the host [72].

In this regard, Liu and colleagues [73] recently used DNA array to characterize changes in gene expression following activation of the human macrophage TLR2/1 dimer by one of the PAMPs for *M. tb*. Macrophages, but not dendritic cells, so treated showed increased expression of both CYP27B1 and VDR genes and gene products, and demonstrated intracrine induction of the antimicrobial cathelicidin gene with subsequent mycobacterial killing in response to 25OHD as well as 1,25(OH)$_2$D. In fact, microbial killing was more efficiently achieved with the pro-hormone 25OHD than with 1,25(OH)$_2$D at similar extracellular concentrations, indicating that the robustness of the human innate response to microbial challenge is dependent upon the serum 25OHD status of the host. This concept was confirmed in these studies by the ability of 25OHD-sufficient serum to rescue a deficient, cathelicidin-driven antimicrobial response in human macrophages conditioned in vitamin D-deficient serum [74]. A similar vitamin D-directed antimicrobial generating capacity has been recently observed in wounded skin [14], suggesting that TLR-driven expression of cathelicidin, requiring the intracellular synthesis and genomic action of 1,25(OH)$_2$D, is a common response feature to infectious agent invasion. While clinical evidence in support of this notion is beginning to accumulate [75], the utility of vitamin D or 25OHD as adjuvant therapy to boost the human innate immune response in a therapeutic or preventive sense needs more in depth, prospective investigation in humans. Reinforcing these events is the ability of locally generated 1,25(OH)$_2$D to escape the confines of the cell in which it is made to act on neighboring, VDR-expressing monocytes to promote their maturation to mature macrophages [76], thus acting as a feed-forward signal to further enhance the innate immune response.

Vitamin D and the adaptive immune response

The adaptive immune response is generally defined by T and B lymphocytes and their ability to produce cytokines and immunoglobulins, respectively, to specifically combat the source of the antigen presented to them by cells (i.e., macrophages, dendritic cells, etc.) of the innate immune response. As previously noted [66], the presence of VDR in activated, but not resting, human T and B lymphocytes was the first observation implicating these cells as targets for the noncalciotropic responses to 1,25(OH)$_2$D. Contrary to the role of locally produced 1,25(OH)$_2$D to promote the innate immune response, the hormone exerts a generalized dampening effect on lymphocyte function. With respect to B cells, 1,25(OH)$_2$D suppresses proliferation and immunoglobulin production and retards the differentiation of B-lymphocyte precursors to mature plasma cells. With regard to T cells, 1,25(OH)$_2$D, acting through the VDR, inhibits the proliferation of pro-inflammatory interferon *gamma*-elaborating, macrophage-stimulating Th(helper)1 and chemokine-producing Th17 cells [77]. A much less robust antiproliferative effect of 1,25(OH)$_2$D is exerted on immunosuppressive Th2 and regulatory T cells (Tregs) [78], promoting their accumulation at sites of inflammation by stimulating expression of the dendritic cell-derived T cell homing molecule, CCL22 [79]. In fact, it is this generalized ability of 1,25(OH)$_2$D to quell the adaptive immune response that has prompted the use of the hormone and its analogues in the adjuvant treatment of inflammatory and neoplastic disorders.

In sum, the collective, concerted action of 1,25(OH)$_2$D is to promote the host's response to an invading pathogen while simultaneously acting to limit what might be an overzealous immune response to that pathogen, representative of the process of tolerance. Once again, a good example is that of infection with the intracellular pathogen *M. t.b.* In this case, the pathogen evokes an exceptionally robust innate immune response that is fueled by the endogenous production of 1,25(OH)$_2$D by macrophages at the site of host invasion. If substantial amounts of 1,25(OH)$_2$D escape the confines of the macrophage, then the immunostimulation of VDR-expressing, activated lymphocytes in that environment is quelled by 1,25(OH)$_2$D. If the innate immune response is extreme and enough 1,25(OH)$_2$D finds its way into the general circulation, as may occur in human granuloma-forming disorders like sarcoidosis and tuberculosis, an endocrine effect of the hormone, most notably hypercalciuria and hypercalcemia, can be observed.

KERATINOCYTE FUNCTION IN EPIDERMIS AND HAIR FOLLICLES

1,25(OH)$_2$D-regulated epidermal differentiation

Locally produced 1,25(OH)$_2$D is likely to be an autocrine or paracrine factor for epidermal differentiation since it is produced in the keratinocyte by the same enzyme, CYP27B1, as found in the kidney; under normal circumstances keratinocyte production of 1,25(OH)$_2$D does not appear to be substantial enough to contribute to circulating levels of the hormone [80]. The concentration of VDR and the production of 1,25(OH)$_2$D in the skin decrease with differentiation. Stimulation of differentiation is accompanied by the rise in mRNA and protein levels of involucrin and transglutaminase [81] as well as the late differentiation markers filaggrin and loricrin [82] with promotion of an intact barrier. The mechanisms by which 1,25(OH)$_2$D alters keratinocyte differentiation are multiple and include induction of the CaSR enhancing the effects of calcium on differentiation and induction of the phospholipase C family, which provides second messengers such as diacyl glycerol and inositol tris phosphate to the differentiation process. Although the most striking feature of the VDR-null mouse is the development of alopecia (also found in many but not all patients with mutations in the VDR), these mice also exhibit a defect in epidermal differentiation as shown by reduced levels of involucrin and loricrin, loss of keratohyalin granules, loss of the calcium gradient, and disruption of lamellar body production and secretion resulting in defective barrier function. Furthermore, both VDR and 1,25(OH)$_2$D production are required for normal antimicrobial peptide expression in response to epidermal injury and infections [14].

VDR regulation of hair follicle cycling

As noted above, alopecia is a well-known part of the phenotype of many patients with mutations in their VDR [83]. Vitamin D deficiency or lack of CYP27B1 *per se* is not associated with alopecia, and the alopecia can be rescued with mutants of VDR that fail to bind 1,25(OH)$_2$D or its coactivators [84]. Recent attention has been paid to both hairless (Hr), a putative transcription factor capable of binding the VDR and suppressing at least its ligand dependent transcriptional activity, and β-catenin, which, like Hr, binds to the VDR and may regulate its transcriptional activity (or vice versa). Hr mutations in both mice and humans and transcriptionally inactivating β-catenin mutations in mice result in phenocopies of the VDR-null animal with regard to the morphologic changes observed in hair cycling. In these models the abnormality leading to alopecia develops during catagen at the end of the developmental cycle, precluding the reinitiation of anagen. Both VDR-null mice and those with disrupted wnt signaling lose stem cells from the bulge, perhaps as a result of the loss of the interaction of the bulge with the dermal papilla [85, 86]. Thus, although the mechanism by which the VDR controls hair follicle cycling is not established, hair follicle cycling represents the best example by which the VDR regulates a physiologic process that is independent of its ligand, 1,25(OH)$_2$D and so points to a novel mechanism of action for this transcriptional regulator.

NUTRITIONAL CONSIDERATIONS

Defining vitamin D sufficiency

Serum 25OHD levels provide a useful surrogate for assessing vitamin D status, as the conversion of vitamin D to 25OHD is less well controlled (i.e., primarily substrate dependent) than the subsequent conversion of

25OHD to 1,25(OH)$_2$D. Levels of 1,25(OH)$_2$D, unlike 25OHD levels, are well maintained until vitamin D deficiency reaches extremes because of the secondary hyperparathyroidism and so do not provide a useful index for assessing vitamin D deficiency, at least in the initial stages. Historically, vitamin D sufficiency was defined as the level of 25OHD sufficient to prevent rickets in children and osteomalacia in adults. Levels of 25OHD below 5 ng/ml (or 12 nM) are associated with a high prevalence of rickets or osteomalacia. However, there is a growing consensus that these lower limits of normal are too low. Recently, an expert panel for the Institute of Medicine (IOM) recommended that a level of 20 ng/ml (50 nM) was sufficient for 97.5% of the population, although up to 50 ng/ml (125 nM) was safe [87]. For individuals between the ages of 1 and 70 years old, 600 IU vitamin D was thought to be sufficient to meet these goals, although up to 4,000 IU vitamin D was considered safe [87]. These recommendations are based primarily on data from randomized, placebo-controlled clinical trials (RCTs) that evaluated falls and fractures; data supporting the nonskeletal effects of vitamin D were considered too preliminary to be used in their recommendations, because of the lack of RCTs for these other actions. The lower end of these recommendations has been considered too low and the upper end too restrictive by a number of vitamin D experts, but the call for better clinical data especially for the nonskeletal actions is well taken. These guidelines, at least with respect to the lower recommended levels of vitamin D supplementation, are unlikely to correct vitamin D deficiency in individuals with obesity, dark complexions, limited capacity for sunlight exposure, or malabsorption. Furthermore, a large body of data from animal and cell studies as well and epidemiologic associations supports a large range of beneficial actions of vitamin D; as a consequence, adequate RCT data in these areas may eventually alter these IOM recommendations for nonskeletal benefits of vitamin D supplementation.

Impact on the musculoskeletal system

This rethinking of the definition of vitamin D sufficiency comes from the appreciation that vitamin D affects a large number of physiologic functions in addition to bone mineralization. The levels of 25OHD are inversely proportional to PTH levels such that PTH levels increase at levels of 25OHD below 20–30 ng/ml. Intestinal calcium transport has been reported to increase significantly when the 25OHD levels are increased from 20 to 32 ng/ml. Large epidemiologic surveys demonstrate a positive correlation between 25OHD levels and bone mineral density, with no evidence for a plateau below 30 ng/ml, and vitamin D (with calcium supplementation) showed improvement in bone mineral density (BMD) in older individuals. Similarly, a positive association between 25OHD levels and muscle function (e.g., walking speed, sit-to-stand) has been demonstrated, even over the interval of 20–38 ng/ml, although the correlation is strongest at lower levels. Vitamin D supplementation with at least 800 IU improved lower extremity function, decreased body sway, and reduced falls. Most importantly, adequate levels of vitamin D and calcium supplementation prevent fractures [88, 89].

Impact beyond the musculoskeletal system

The impact of vitamin D extends beyond the musculoskeletal system and the regulation of calcium homeostasis. Vitamin D deficiency is a well-known accompaniment of various infectious diseases such as tuberculosis, and 1,25(OH)$_2$D has long been recognized to potentiate the killing of mycobacteria by monocytes. The nutritional aspect of these observations has recently been illuminated by the observation that the monocyte, when activated by mycobacterial lipopeptides, expresses CYP27B1, producing 1,25(OH)$_2$D from circulating 25OHD, and in turn inducing cathelicidin, an antimicrobial peptide that enhances killing of the mycobacterium. Inadequate 25OHD levels abort this process [73]. Vitamin D deficiency and/or living at higher latitudes (with less sunlight) is associated with a number of autoimmune diseases including type 1 diabetes mellitus, multiple sclerosis, and Crohn's disease [90]. Levels of 25OHD are also inversely associated with type 2 diabetes mellitus and metabolic syndrome, and some studies have shown that vitamin D and calcium supplementation may prevent the progression to diabetes mellitus in individuals with glucose intolerance. Improvements in both insulin secretion and action have been observed with vitamin D supplementation. The potential to prevent certain cancers may be the most compelling reason for adequate vitamin D nutrition. A large body of epidemiologic data exists documenting the inverse correlation of 25OHD levels, latitude, and/or vitamin D intake with cancer incidence [7]. Although numerous types of cancers show reduction [91], most attention has been paid to breast, colon, and prostate. A relatively small prospective 4-year trial with 1,100 IU vitamin D and 1,400–1,500 mg calcium showed a 77% reduction in cancers after the first year of study [92], including a reduction in both breast and colon cancers. In this study, vitamin D supplementation raised the 25OHD levels from a mean of 28.8 ng/ml to 38.4 ng/ml with no changes in the placebo or calcium only arms of the study. However, other trials with vitamin D metabolites or analogs for cancer prevention or treatment have yielded less promising results.

VITAMIN D TREATMENT STRATEGIES

Adequate sunlight exposure is the most cost effective means of obtaining vitamin D. Whole body exposure to sunlight has been calculated to provide the equivalent of 10,000 IU vitamin D$_3$ [93]. A 0.5 minimal erythema dose of sunlight (i.e., half the dose required to produce a slight reddening of the skin) or UVB radiation to the arms and legs, which can be achieved in 5–10 min on a bright summer day in light-skinned individuals, has been calculated to be the equivalent of 3,000 IU vitamin D_3 [7]. However, concerns regarding the association between sunlight and skin cancer and/or solar aging of the skin have limited this approach, perhaps to the extreme, although it remains a viable option for those unable or unwilling to benefit from oral supplementation. Studies have demonstrated that on average for every 100 IU of a vitamin D_3 supplementation administered, the 25OHD levels rise by 0.5–1 ng/ml [93, 94]. For obese individuals or those with malabsorption (including after bariatric surgery) much higher doses are likely to be required. A number of studies suggest that 700–800 IU is the lower limit of vitamin D supplementation required to prevent fractures and falls, although as noted, the IOM has concluded that 600 IU suffices. Unfortified food contains little vitamin D with the exception of wild salmon and other fish products such as cod liver oil. Milk and other fortified beverages typically contain 100 IU/8 oz. serving. Vitamin D_2 does not appear to increase or sustain the 25OHD levels as well as vitamin D_3 in part because it is more rapidly cleared. Therefore, if vitamin D_2 is used, it needs to be given at least weekly. Toxicity due to vitamin D supplementation has not been observed at doses less than 10,000 IU per day in several studies, although the IOM report indicates that the upper limit should be restricted to 4000 IU at least in the general population. [95].

REFERENCES

1. Holick MF, McLaughlin JA, Clark MB, Doppelt SH. 1981. Factors that influence the cutaneous photosynthesis of previtamin D3. *Science* 211: 590–593.
2. Holick MF, MacLaughlin JA, Clark MB, Holick SA, Potts JT Jr, Anderson RR, Blank IH, Parrish JA, Elias P. 1980. Photosynthesis of previtamin D3 in human and the physiologic consequences. *Science* 210: 203–205.
3. Holick MF, Richtand NM, McNeill SC, Holick SA, Frommer JE, Henley JW, Potts JT Jr. 1979. Isolation and identification of previtamin D3 from the skin of exposed to ultraviolet irradiation. *Biochemistry* 18: 1003–1008.
4. Webb AR, Kline L, Holick MF. 1988. Influence of season and latitude on the cutaneous synthesis of vitamin D3: Exposure to winter sunlight in Boston and Edmonton will not promote vitamin D3 synthesis in human skin. *J Clin Endocrinol Metab* 67: 373–378.
5. Adams JS, Clemens TL, Parrish JA, Holick MF. 1982. Vitamin-D synthesis and metabolism after ultraviolet irradiation of normal and vitamin-D-deficient subjects. *N Engl J Med* 306: 722–725.
6. Heaney RP, Horst RL, Cullen DM, Armas LA. 2009. Vitamin D3 distribution and status in the body. *J Am Coll Nutr* 28: 252–256.
7. Holick MF. 2007. Vitamin D deficiency. *N Engl J Med* 357: 266–281.
8. Hewison M, Burke F, Evans KN, Lammas DA, Sansom DM, Liu P, Modlin RL, Adams JS. 2007. Extra-renal 25-hydroxyvitamin D3-1alpha-hydroxylase in human health and disease. *J Steroid Biochem Mol Biol* 103: 316–321.
9. Zierold C, Darwish HM, DeLuca HF. 1995. Two vitamin D response elements function in the rat 1,25-dihydroxyvitamin D 24-hydroxylase promoter. *J Biol Chem* 270: 1675–1678.
10. Ren S, Nguyen L, Wu S, Encinas C, Adams JS, Hewison M. 2005. Alternative splicing of vitamin D-24-hydroxylase: A novel mechanism for the regulation of extrarenal 1,25-dihydroxyvitamin D synthesis. *J Biol Chem* 280: 20604–20611.
11. Wu S, Ren S, Nguyen L, Adams JS, Hewison M. 2007. Splice variants of the CYP27b1 gene and the regulation of 1,25-dihydroxyvitamin D3 production. *Endocrinology* 148: 3410–3418.
12. Bikle DD, Pillai S, Gee E, Hincenbergs M. 1989. Regulation of 1,25-dihydroxyvitamin D production in human keratinocytes by interferon-gamma. *Endocrinology* 124: 655–660.
13. Bikle DD, Pillai S, Gee E, Hincenbergs M. 1991. Tumor necrosis factor-alpha regulation of 1,25-dihydroxyvitamin D production by human keratinocytes. *Endocrinology* 129: 33–38.
14. Schauber J, Dorschner RA, Coda AB, Buchau AS, Liu PT, Kiken D, Helfrich YR, Kang S, Elalieh HZ, Steinmeyer A, Zugel U, Bikle DD, Modlin RL, Gallo RL. 2007. Injury enhances TLR2 function and antimicrobial peptide expression through a vitamin D-dependent mechanism. *J Clin Invest* 117: 803–811.
15. Cooke NE, Haddad JG. 1989. Vitamin D binding protein (Gc-globulin). *Endocr Rev* 10: 294–307.
16. Liang C, Cooke N. 2005. Vitamin D-binding protein. In: Feldman D, Pike JW, Glorieux F (eds.) *Vitamin D, 2nd Ed.* San Diego: Elsevier Academic Press. pp. 117–134.
17. Willnow T, Nykjaer A. 2005. Endocytic pathways for 25-(OH) vitamin D3. In: Feldman D, Pike JW, Glorieux F (eds.) *Vitamin D, 2nd Ed.* San Diego: Elsevier Academic Press. pp. 153–163.
18. Zella LA, Shevde NK, Hollis BW, Cooke NE, Pike JW. 2008. Vitamin D-binding protein influences total circulating levels of 1,25-dihydroxyvitamin D3 but does not directly modulate the bioactive levels of the hormone in vivo. *Endocrinology* 149: 3656–3667.
19. Safadi FF, Thornton P, Magiera H, Hollis BW, Gentile M, Haddad JG, Liebhaber SA, Cooke NE. 1999. Osteopathy and resistance to vitamin D toxicity in mice null for vitamin D binding protein. *J Clin Invest* 103: 239–251.
20. Adams JS. 2005. "Bound" to work: The free hormone hypothesis revisited. *Cell* 122: 647–649.
21. Christakos S, Dhawan P, Liu Y, Peng X, Porta A. 2003. New insights into the mechanisms of vitamin D action. *J Cell Biochem* 88: 695–705.
22. DeLuca HF. 2004. Overview of general physiologic features and functions of vitamin D. *Am J Clin Nutr* 80: 1689S–1696S.
23. Rachez C, Freedman LP. 2000. Mechanisms of gene regulation by vitamin D(3) receptor: A network of coactivator interactions. *Gene* 246: 9–21.

24. Sutton AL, MacDonald PN. 2003. Vitamin D: More than a "bone-a-fide" hormone. *Mol Endocrinol* 17: 777–791.
25. Christakos S, Dhawan P, Benn BS, Porta A, Hediger M, Oh GT, Jeung EB, Zhong Y, Ajibade D, Dhawan K, Joshi S. 2007. Vitamin D: Molecular mechanism of action. *Ann NY Acad Sci* 1116: 340–348.
26. Christakos S, Dhawan P, Peng X, Obukhov AG, Nowycky MC, Benn BS, Zhong Y, Liu Y, Shen Q. 2007. New insights into the function and regulation of vitamin D target proteins. *J Steroid Biochem Mol Biol* 103: 405–410.
27. Dhawan P, Peng X, Sutton AL, MacDonald PN, Croniger CM, Trautwein C, Centrella M, McCarthy TL, Christakos S. 2005. Functional cooperation between CCAAT/enhancer-binding proteins and the vitamin D receptor in regulation of 25-hydroxyvitamin D3 24-hydroxylase. *Mol Cell Biol* 25: 472–487.
28. Guo B, Aslam F, van Wijnen AJ, Roberts SG, Frenkel B, Green MR, DeLuca H, Lian JB, Stein GS, Stein JL. 1997. YY1 regulates vitamin D receptor/retinoid X receptor mediated transactivation of the vitamin D responsive osteocalcin gene. *Proc Natl Acad Sci U S A* 94: 121–126.
29. Raval-Pandya M, Dhawan P, Barletta F, Christakos S. 2001. YY1 represses vitamin D receptor-mediated 25-hydroxyvitamin D(3)24-hydroxylase transcription: Relief of repression by CREB-binding protein. *Mol Endocrinol* 15: 1035–1046.
30. Christakos S, Dhawan P, Shen Q, Peng X, Benn B, Zhong Y. 2006. New insights into the mechanisms involved in the pleiotropic actions of 1,25dihydroxyvitamin D3. *Ann N Y Acad Sci* 1068: 194–203.
31. Pike JW, Meyer MB. 2010. The vitamin D receptor: New paradigms for the regulation of gene expression by 1,25-dihydroxyvitamin D(3). *Endocrinol Metab Clin North Am* 39: 255–269.
32. Imura A, Tsuji Y, Murata M, Maeda R, Kubota K, Iwano A, Obuse C, Togashi K, Tominaga M, Kita N, Tomiyama K, Iijima J, Nabeshima Y, Fujioka M, Asato R, Tanaka S, Kojima K, Ito J, Nozaki K, Hashimoto N, Ito T, Nishio T, Uchiyama T, Fujimori T, Nabeshima Y. 2007. alpha-Klotho as a regulator of calcium homeostasis. *Science* 316: 1615–1618.
33. Li YC, Pirro AE, Amling M, Delling G, Baron R, Bronson R, Demay MB. 1997. Targeted ablation of the vitamin D receptor: An animal model of vitamin D-dependent rickets type II with alopecia. *Proc Natl Acad Sci U S A* 94: 9831–9835.
34. Yoshizawa T, Handa Y, Uematsu Y, Takeda S, Sekine K, Yoshihara Y, Kawakami T, Arioka K, Sato H, Uchiyama Y, Masushige S, Fukamizu A, Matsumoto T, Kato S. 1997. Mice lacking the vitamin D receptor exhibit impaired bone formation, uterine hypoplasia and growth retardation after weaning. *Nat Genet* 16: 391–396.
35. Amling M, Priemel M, Holzmann T, Chapin K, Rueger JM, Baron R, Demay MB. 1999. Rescue of the skeletal phenotype of vitamin D receptor-ablated mice in the setting of normal mineral ion homeostasis: Formal histomorphometric and biomechanical analyses. *Endocrinology* 140: 4982–4987.
36. Xue Y, Fleet JC. 2009. Intestinal vitamin D receptor is required for normal calcium and bone metabolism in mice. *Gastroenterology* 136: 1317–1327, e1311–1312.
37. Panda DK, Miao D, Bolivar I, Li J, Huo R, Hendy GN, Goltzman D. 2004. Inactivation of the 25-hydroxyvitamin D 1alpha-hydroxylase and vitamin D receptor demonstrates independent and interdependent effects of calcium and vitamin D on skeletal and mineral homeostasis. *J Biol Chem* 279: 16754–16766.
38. Raisz LG, Trummel CL, Holick MF, DeLuca HF. 1972. 1,25-dihydroxycholecalciferol: A potent stimulator of bone resorption in tissue culture. *Science* 175: 768–769.
39. Yasuda H, Shima N, Nakagawa N, Yamaguchi K, Kinosaki M, Mochizuki S, Tomoyasu A, Yano K, Goto M, Murakami A, Tsuda E, Morinaga T, Higashio K, Udagawa N, Takahashi N, Suda T. 1998. Osteoclast differentiation factor is a ligand for osteoprotegerin/osteoclastogenesis-inhibitory factor and is identical to TRANCE/RANKL. *Proc Natl Acad Sci U S A* 95: 3597–3602.
40. Prince CW, Butler WT. 1987. 1,25-Dihydroxyvitamin D3 regulates the biosynthesis of osteopontin, a bone-derived cell attachment protein, in clonal osteoblast-like osteosarcoma cells. *Coll Relat Res* 7: 305–313.
41. Price PA, Baukol SA. 1980. 1,25-Dihydroxyvitamin D3 increases synthesis of the vitamin K-dependent bone protein by osteosarcoma cells. *J Biol Chem* 255: 11660–11663.
42. Drissi H, Pouliot A, Koolloos C, Stein JL, Lian JB, Stein GS, van Wijnen AJ. 2002. 1,25-(OH)2-vitamin D3 suppresses the bone-related Runx2/Cbfa1 gene promoter. *Exp Cell Res* 274: 323–333.
43. Gardiner EM, Baldock PA, Thomas GP, Sims NA, Henderson NK, Hollis B, White CP, Sunn KL, Morrison NA, Walsh WR, Eisman JA. 2000. Increased formation and decreased resorption of bone in mice with elevated vitamin D receptor in mature cells of the osteoblastic lineage. *FASEB J* 14: 1908–1916.
44. Raval-Pandya M, Porta A, Christakos S. 1998. Mechanism of action of 1,25 dihydroxyvitamin D3 on intestinal calcium absorption and renal calcium transportation. In: Holick MF (ed.) *Vitamin D Physiology, Molecular Biology and Clinical Applications*. Totowa, NJ: Humana Press. pp. 163–173.
45. Wasserman RH, Fullmer CS. 1995. Vitamin D and intestinal calcium transport: Facts, speculations and hypotheses. *J Nutr* 125: 1971S–1979S.
46. Fleet JC, Schoch RD. 2010. Molecular mechanisms for regulation of intestinal calcium absorption by vitamin D and other factors. *Crit Rev Clin Lab Sci* 47: 181–195.
47. Akhter S, Kutuzova GD, Christakos S, DeLuca HF. 2007. Calbindin D9k is not required for 1,25-dihydroxyvitamin D3-mediated Ca2+ absorption in small intestine. *Arch Biochem Biophys* 460: 227–232.
48. Kutuzova GD, Akhter S, Christakos S, Vanhooke J, Kimmel-Jehan C, Deluca HF. 2006. Calbindin D(9k) knockout mice are indistinguishable from wild-type

48. mice in phenotype and serum calcium level. *Proc Natl Acad Sci U S A* 103: 12377–12381.
49. Hoenderop JG, Nilius B, Bindels RJ. 2003. Epithelial calcium channels: From identification to function and regulation. *Pflugers Arch* 446: 304–308.
50. Peng JB, Chen XZ, Berger UV, Vassilev PM, Tsukaguchi H, Brown EM, Hediger MA. 1999. Molecular cloning and characterization of a channel-like transporter mediating intestinal calcium absorption. *J Biol Chem* 274: 22739–22746.
51. Benn BS, Ajibade D, Porta A, Dhawan P, Hediger M, Peng JB, Jiang Y, Oh GT, Jeung EB, Lieben L, Bouillon R, Carmeliet G, Christakos S. 2008. Active intestinal calcium transport in the absence of transient receptor potential vanilloid type 6 and calbindin-D9k. *Endocrinology* 149: 3196–3205.
52. Williams KB, DeLuca HF. 2007. Characterization of intestinal phosphate absorption using a novel in vivo method. *Am J Physiol Endocrinol Metab* 292: E1917–1921.
53. Sabbagh Y, O'Brien SP, Song W, Boulanger JH, Stockmann A, Arbeeny C, Schiavi SC. 2009. Intestinal npt2b plays a major role in phosphate absorption and homeostasis. *J Am Soc Nephrol* 20: 2348–2358.
54. Sneddon WB, Barry EL, Coutermarsh BA, Gesek FA, Liu F, Friedman PA. 1998. Regulation of renal parathyroid hormone receptor expression by 1, 25-dihydroxyvitamin D3 and retinoic acid. *Cell Physiol Biochem* 8: 261–277.
55. Lambers TT, Weidema AF, Nilius B, Hoenderop JG, Bindels RJ. 2004. Regulation of the mouse epithelial Ca2(+) channel TRPV6 by the Ca(2+)-sensor calmodulin. *J Biol Chem* 279: 28855–28861.
56. Omdahl JL, Bobrovnikova EA, Choe S, Dwivedi PP, May BK. 2001. Overview of regulatory cytochrome P450 enzymes of the vitamin D pathway. *Steroids* 66: 381–389.
57. Kaneko I, Segawa H, Furutani J, Kuwahara S, Aranami F, Hanabusa E, Tominaga R, Giral H, Caldas Y, Levi M, Kato S, Miyamoto K. 2011. Hypophosphatemia in vitamin D receptor null mice: Effect of rescue diet on the developmental changes in renal Na+ -dependent phosphate cotransporters. *Pflugers Arch J Physiol* 461: 77–90.
58. Demay MB, Kiernan MS, DeLuca HF, Kronenberg HM. 1992. Sequences in the human parathyroid hormone gene that bind the 1,25- dihydroxyvitamin D3 receptor and mediate transcriptional repression in response to 1,25-dihydroxyvitamin D3. *Proc Natl Acad Sci U S A* 89: 8097–8101.
59. Martin KJ, Gonzalez EA. 2004. Vitamin D analogs: Actions and role in the treatment of secondary hyperparathyroidism. *Semin Nephrol* 24: 456–459.
60. Canaff L, Hendy GN. 2002. Human calcium-sensing receptor gene. Vitamin D response elements in promoters P1 and P2 confer transcriptional responsiveness to 1,25-dihydroxyvitamin D. *J Biol Chem* 277: 30337–30350.
61. Christakos S, Norman AW. 1979. Studies on the mode of action of calciferol. XVIII. Evidence for a specific high affinity binding protein for 1,25 dihydroxyvitamin D3 in chick kidney and pancreas. *Biochem Biophys Res Commun* 89: 56–63.
62. Clark SA, Stumpf WE, Sar M, DeLuca HF, Tanaka Y. 1980. Target cells for 1,25 dihydroxyvitamin D3 in the pancreas. *Cell Tissue Res* 209: 515–520.
63. Morrissey RL, Bucci TJ, Richard B, Empson N, Lufkin EG. 1975. Calcium-binding protein: Its cellular localization in jejunum, kidney and pancreas. *Proc Soc Exp Biol Med* 149: 56–60.
64. Sooy K, Schermerhorn T, Noda M, Surana M, Rhoten WB, Meyer M, Fleischer N, Sharp GW, Christakos S. 1999. Calbindin-D(28k) controls [Ca(2+)](i) and insulin release. Evidence obtained from calbindin-d(28k) knockout mice and beta cell lines. *J Biol Chem* 274: 34343–34349.
65. Rabinovitch A, Suarez-Pinzon WL, Sooy K, Strynadka K, Christakos S. 2001. Expression of calbindin-D(28k) in a pancreatic islet beta-cell line protects against cytokine-induced apoptosis and necrosis. *Endocrinology* 142: 3649–3655.
66. Provvedini DM, Tsoukas CD, Deftos LJ, Manolagas SC. 1983. 1,25-dihydroxyvitamin D3 receptors in human leukocytes. *Science* 221: 1181–1183.
67. Adams JS, Sharma OP, Gacad MA, Singer FR. 1983. Metabolism of 25-hydroxyvitamin D3 by cultured pulmonary alveolar macrophages in sarcoidosis. *J Clin Invest* 72: 1856–1860.
68. Rook GA, Steele J, Fraher L, Barker S, Karmali R, O'Riordan J, Stanford J. 1986. Vitamin D3, gamma interferon, and control of proliferation of Mycobacterium tuberculosis by human monocytes. *Immunology* 57: 159–163.
69. Wang TT, Nestel FP, Bourdeau V, Nagai Y, Wang Q, Liao J, Tavera-Mendoza L, Lin R, Hanrahan JW, Mader S, White JH. 2004. Cutting edge: 1,25-dihydroxyvitamin D3 is a direct inducer of antimicrobial peptide gene expression. *J Immunol* 173: 2909–2912.
70. Gombart AF, Borregaard N, Koeffler HP. 2005. Human cathelicidin antimicrobial peptide (CAMP) gene is a direct target of the vitamin D receptor and is strongly up-regulated in myeloid cells by 1,25-dihydroxyvitamin D3. *FASEB J* 19: 1067–1077.
71. Weber G, Heilborn JD, Chamorro Jimenez CI, Hammarsjo A, Torma H, Stahle M. 2005. Vitamin D induces the antimicrobial protein hCAP18 in human skin. *J Invest Dermatol* 124: 1080–1082.
72. Medzhitov R. 2007. Recognition of microorganisms and activation of the immune response. *Nature* 449: 819–826.
73. Liu PT, Stenger S, Li H, Wenzel L, Tan BH, Krutzik SR, Ochoa MT, Schauber J, Wu K, Meinken C, Kamen DL, Wagner M, Bals R, Steinmeyer A, Zugel U, Gallo RL, Eisenberg D, Hewison M, Hollis BW, Adams JS, Bloom BR, Modlin RL. 2006. Toll-like receptor triggering of a vitamin D-mediated human antimicrobial response. *Science* 311: 1770–1773.
74. Adams JS, Ren S, Liu PT, Chun RF, Lagishetty V, Gombart AF, Borregaard N, Modlin RL, Hewison M.

2009. Vitamin D-directed rheostatic regulation of monocyte antibacterial responses. *J Immunol* 182: 4289–4295.
75. Hewison M. 2011. Antibacterial effects of vitamin D. *Nat Rev Endocrinol* 7: 337–345.
76. Kreutz M, Andreesen R, Krause SW, Szabo A, Ritz E, Reichel H. 1993. 1,25-dihydroxyvitamin D3 production and vitamin D3 receptor expression are developmentally regulated during differentiation of human monocytes into macrophages. *Blood* 82: 1300–1307.
77. Bruce D, Ooi JH, Yu S, Cantorna MT. 2010. Vitamin D and host resistance to infection? Putting the cart in front of the horse. *Exp Biol Med (Maywood)* 235: 921–927.
78. Penna G, Adorini L. 2000. 1 Alpha,25-dihydroxyvitamin D3 inhibits differentiation, maturation, activation, and survival of dendritic cells leading to impaired alloreactive T cell activation. *J Immunol* 164: 2405–2411.
79. Penna G, Amuchastegui S, Giarratana N, Daniel KC, Vulcano M, Sozzani S, Adorini L. 2007. 1,25-Dihydroxyvitamin D3 selectively modulates tolerogenic properties in myeloid but not plasmacytoid dendritic cells. *J Immunol* 178: 145–153.
80. Bikle DD, Nemanic MK, Whitney JO, Elias PW. 1986. Neonatal human foreskin keratinocytes produce 1,25-dihydroxyvitamin D3. *Biochemistry* 25: 1545–1548.
81. Su MJ, Bikle DD, Mancianti ML, Pillai S. 1994. 1,25-Dihydroxyvitamin D3 potentiates the keratinocyte response to calcium. *J Biol Chem* 269: 14723–14729.
82. Hawker NP, Pennypacker SD, Chang SM, Bikle DD. 2007. Regulation of human epidermal keratinocyte differentiation by the vitamin D receptor and its coactivators DRIP205, SRC2, and SRC3. *J Invest Dermatol* 127: 874.
83. Malloy PJ, Pike JW, Feldman D. 1999. The vitamin D receptor and the syndrome of hereditary 1,25-dihydroxyvitamin D-resistant rickets. *Endocr Rev* 20: 156–188.
84. Skorija K, Cox M, Sisk JM, Dowd DR, MacDonald PN, Thompson CC, Demay MB. 2005. Ligand-independent actions of the vitamin D receptor maintain hair follicle homeostasis. *Mol Endocrinol* 19: 855–862.
85. Bikle DD, Elalieh H, Chang S, Xie Z, Sundberg JP. 2006. Development and progression of alopecia in the vitamin D receptor null mouse. *J Cell Physiol* 207: 340–353.
86. Cianferotti L, Cox M, Skorija K, Demay MB. 2007. Vitamin D receptor is essential for normal keratinocyte stem cell function. *Proc Natl Acad Sci U S A* 104: 9428–9433.
87. Ross AC, Manson JE, Abrams SA, Aloia JF, Brannon PM, Clinton SK, Durazo-Arvizu RA, Gallagher JC, Gallo RL, Jones G, Kovacs CS, Mayne ST, Rosen CJ, Shapses SA. 2011. The 2011 report on dietary reference intakes for calcium and vitamin D from the Institute of Medicine: What clinicians need to know. *J Clin Endocrinol Metab* 96: 53–58.
88. Bischoff-Ferrari HA, Willett WC, Wong JB, Giovannucci E, Dietrich T, Dawson-Hughes B. 2005. Fracture prevention with vitamin D supplementation: A meta-analysis of randomized controlled trials. *JAMA* 293: 2257–2264.
89. Chapuy MC, Arlot ME, Duboeuf F, Brun J, Crouzet B, Arnaud S, Delmas PD, Meunier PJ. 1992. Vitamin D3 and calcium to prevent hip fractures in the elderly women. *N Engl J Med* 327: 1637–1642.
90. Ponsonby AL, McMichael A, van der Mei I. 2002. Ultraviolet radiation and autoimmune disease: Insights from epidemiological research. *Toxicology* 181–182: 71–78.
91. Boscoe FP, Schymura MJ. 2006. Solar ultraviolet-B exposure and cancer incidence and mortality in the United States, 1993–2002. *BMC Cancer* 6: 264.
92. Lappe JM, Travers-Gustafson D, Davies KM, Recker RR, Heaney RP. 2007. Vitamin D and calcium supplementation reduces cancer risk: Results of a randomized trial. *Am J Clin Nutr* 85: 1586–1591.
93. Vieth R. 1999. Vitamin D supplementation, 25-hydroxyvitamin D concentrations, and safety. *Am J Clin Nutr* 69: 842–856.
94. Heaney RP, Davies KM, Chen TC, Holick MF, Barger-Lux MJ. 2003. Human serum 25-hydroxycholecalciferol response to extended oral dosing with cholecalciferol. *Am J Clin Nutr* 77: 204–210.
95. Hathcock JN, Shao A, Vieth R, Heaney R. 2007. Risk assessment for vitamin D. *Am J Clin Nutr* 85: 6–18.

Section IV
Investigation of Metabolic Bone Diseases

Section Editor Douglas C. Bauer

Chapter 30. DXA in Adults and Children 251
Glen Blake, Judith E. Adams, and Nick Bishop

Chapter 31. Quantitative Computed Tomography in Children and Adults 264
C.C. Glüer

Chapter 32. Magnetic Resonance Imaging of Bone 277
Sharmila Majumdar

Chapter 33. Radionuclide Scintigraphy in Metabolic Bone Disease 283
Gary J.R. Cook, Gopinath Gnanasegaran, and Ignac Fogelman

Chapter 34. FRAX®: Assessment of Fracture Risk 289
John A Kanis

Chapter 35. Biochemical Markers of Bone Turnover in Osteoporosis 297
Pawel Szulc, Douglas C. Bauer, and Richard Eastell

Chapter 36. Bone Biopsy and Histomorphometry in Clinical Practice 307
Robert R. Recker

Chapter 37. Diagnosis and Classification of Vertebral Fracture 317
James F. Griffith, Judith E. Adams, and Harry K. Genant

Chapter 38. Approaches to Genetic Testing 336
Christina Jacobsen, Yiping Shen, and Ingrid A. Holm

30
DXA in Adults and Children

Glen Blake, Judith E. Adams, and Nick Bishop

Introduction 251
Technical Aspects 251
Scan Sites 252
Fracture Prediction 253
Scan Artifacts 256
Scan Interpretation Using T- and Z-Scores 256

Scan Interpretation Using FRAX® 257
Precision Errors and Monitoring BMD$_a$ Change 258
Vertebral Fracture Assessment (VFA) 258
DXA in Children 258
Conclusions 260
References 260

INTRODUCTION

Dual-energy X-ray absorptiometry (DXA) scanners were first introduced in the late 1980s, and DXA is now established as the most widely used method for measuring bone mineral density (BMD) [1]. An important reason for this choice is the consensus that DXA results in postmenopausal women and older men should be interpreted using the World Health Organization (WHO) T-score definition of osteoporosis of a spine, femoral neck, or total hip BMD measurement 2.5 standard deviations (SD) or more below the population mean for healthy young adults [2–5]. Other advantages of DXA include its excellent precision, low ionizing radiation dose and short scan times (Table 30.1) [6]. With the recent introduction of the WHO Fracture Risk Assessment Tool (FRAX), DXA measurements of femoral neck BMD have acquired an important role in estimating patients' 10-year risk of hip or major osteoporotic fracture [7]. DXA also has some important limitations (Table 30.1). Because DXA scans are two-dimensional (2D) projection images, they measure areal BMD (BMD$_a$, units g/cm^2) and the results are dependent on bone size. This is a particular problem for interpreting results in children [8] but may also explain some of the racial and gender differences in BMD$_a$ values between adults [9]. Another limitation is that the algorithms used to convert X-ray transmission measurements into BMD$_a$ assume that soft tissue is homogeneous in composition. In reality soft tissue is a heterogeneous mixture of lean and adipose tissue, and this is a source of measurement error [10]. As a result of these and other factors there is often significant discordance in scan findings when T-score results from different BMD$_a$ measurement sites are compared [11, 12].

TECHNICAL ASPECTS

DXA scanners measure the attenuation through the body of X-ray beams with two different photon energies [13]. They use the principle of basis set decomposition to resolve these measurements into the areal densities of two chosen reference materials, which, if present, would give the same transmission factors for the two beams [14]. One of these reference materials is bone mineral [hydroxyapatite, Ca$_5$(PO$_4$)$_3$OH], while the second is soft tissue with a composition defined by a reference area adjacent to the bone region of interest (ROI). A DXA scan is a set of point-by-point measurements of BMD$_a$ across the chosen anatomical site. Edge detection software is used to find the bone outline, and the pixels inside the

---

Primer on the Metabolic Bone Diseases and Disorders of Mineral Metabolism, Eighth Edition. Edited by Clifford J. Rosen.
© 2013 American Society for Bone and Mineral Research. Published 2013 by John Wiley & Sons, Inc.

Table 30.1. Advantages and Limitations of Central DXA

Advantages
Consensus that BMD$_a$ results can be interpreted using T-scores
Proven ability to predict fracture risk
Used in FRAX algorithm for predicting 10-year fracture risk
Proven for effective targeting of anti-fracture treatments
Good precision
Effective at monitoring response to treatment
Widely available
Stable calibration
Effective instrument quality control procedures
Short scan times
Rapid patient set up
Low ionizing radiation dose
Reliable reference ranges

Limitations
2D projection measurement affected by bone size and shape
Measurement errors caused by heterogeneity in soft tissue composition
Discordant findings between different measurement sites
Degenerative disease affects spine BMD$_a$ in older patients
Measurements for fracture and non-fracture populations overlap
Will not differentiate low bone calcium that is due to osteomalacia or osteoporosis

bone edges are summed to find the bone area (BA; cm^2). The BMD$_a$ result reported on the scan printout is the average BMD$_a$ measurement within the BA. Finally, BMD$_a$ is multiplied by BA to find the bone mineral content (BMC; g), equal to the total mass of hydroxyapatite within the bone ROI. The original DXA scanners used a pencil beam of X-rays with a single detector that scanned in a rectilinear fashion across the anatomical site with scan times of 5–10 minutes. Subsequently, fan-beam machines were introduced with an array of detectors, which resulted in shorter scan times of less than 1 minute per site and improved image quality. The radiation dose from DXA examinations is extremely low, and for a spine and hip examination is in the range of 1–10 µSv, a figure that is less than or comparable to the daily dose from natural background radiation (2.5 mSv/yr = 7 µSv/day) [15]. If adult scan modes are used for children the doses are slightly higher [16].

The accuracy error in a DXA measurement differs from patient to patient and reflects the heterogeneous distribution of adipose tissue including marrow fat at the scan site, which is different for every individual [10, 17, 18]. Precision errors (typically 1% to 2% [19]) are smaller than accuracy errors (typically 3% to 7% [10]) because when a patient is rescanned, the effect of adipose tissue on the measurement is the same. Accuracy errors are best determined by cadaver studies involving *in situ* scanning followed by ashing of bone samples to find the true mass of bone mineral [17]. Accuracy errors can also be estimated from computed tomography (CT) or magnetic resonance (MR) imaging of adipose tissue [18, 20]. It has been claimed that the accuracy errors of DXA are so large that they make the results too unreliable to be of any use [21, 22]. However, these studies were based on phantom measurements [21] that overestimated the true errors [20]. They also discount the overwhelming evidence from epidemiological studies that DXA predicts fracture risk [23–26]. Although DXA measurements are affected by accuracy errors, it should be remembered that similar questions have been raised about the accuracy of such basic clinical measurements as blood pressure and body temperature [27–30].

SCAN SITES

DXA is primarily used to scan the central skeleton to measure BMD$_a$ of the lumbar spine and hip (Fig. 30.1) [1]. Central DXA examinations have three main roles: (1) the diagnosis of osteoporosis, (2) assessment of patients' risk of fracture, and (3) monitoring response to treatment [6]. The reasons for choosing to measure the spine and hip include the fact that the hip is the most reliable measurement site for predicting hip fracture risk [23, 24], the use of femoral neck T-score in the FRAX scheme [7], the use of spine BMD$_a$ for monitoring treatment [31], and the consensus that osteoporosis and low bone mass (previously referred to as osteopenia) in postmenopausal women can be determined from spine and hip DXA measurements interpreted using the WHO T-score definitions of osteoporosis and low bone mass [3–5].

T-scores are calculated by taking the difference between a patient's measured BMD$_a$ and the mean BMD$_a$ in healthy young adults matched for gender and ethnicity, and expressing the difference relative to the young adult population SD:

$$\text{T-score} = \frac{\text{Measured BMDa} - \text{Young adult mean BMDa}}{\text{Young adult population SD}}$$

(Eq. 30.1)

A T-score measurement of less than or equal to –2.5 at the spine, femoral neck, or total hip sites is taken to indicate osteoporosis [3, 4]. Measurements between –2.5 and –1.0 are interpreted as showing low bone mass, while those greater than –1.0 at all three sites are classified as normal.

Manufacturers' scan printouts also report Z-scores, which are similar to T-scores except that instead of comparing the patient's BMD$_a$ with the young adult mean it is compared with the mean BMD$_a$ for a healthy subject matched for age, gender and ethnicity:

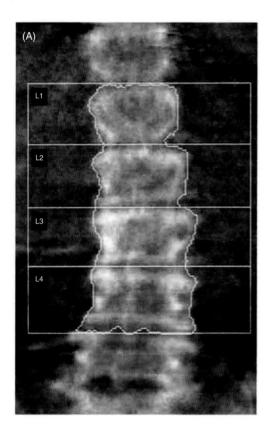
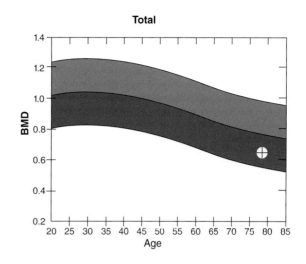

DXA Results Summary:

Region	Area (cm²)	BMC (g)	BMD (g/cm²)	T-score	PR (%)	Z-scroe	AM (%)
L1	12.57	7.24	0.576	−3.8	58	−1.5	78
L2	14.46	9.41	0.651	−3.4	63	−0.9	87
L3	14.78	10.30	0.697	−3.5	64	−0.8	89
L4	18.11	11.43	0.631	−3.9	59	−1.1	84
Total	59.91	38.38	0.641	−3.7	61	−1.1	84

Total BMD CV 1.0%, ACF = 1.025, BCF = 0.994, TH = 6.442
WHO Classsification: Osteoporosis

Fig. 30.1. (A) Scan printout of a spine DXA examination. The printout shows: (left) scan image of the lumbar spine; (top right) patient's age and BMD plotted with respect to the manufacturer's reference range; (bottom right) BMD figures for individual vertebrae and total spine (L1-L4), together with the interpretation in terms of T-scores and Z-scores. (*Continued on next page.*)

$$\text{Z-score} = \frac{\text{Measured BMDa} - \text{Age matched mean BMDa}}{\text{Age matched population SD}}$$

(Eq. 30.2)

Many DXA scanners can also measure forearm BMD_a. This is useful as an alternative site in circumstances when it is not possible to obtain a valid measurement at the spine or hip [3, 4]. This includes patients with bilateral hip replacement, elderly patients with severe degenerative disease in the spine, and patients whose weight exceeds the safety limit for the scanner. Forearm scans are also used to investigate patients with hyperparathyroidism by measuring cortical bone at the 33% radius site [32].

BMD_a measurements across the whole skeleton can be performed using total body scans [33]. These are usually interpreted after excluding the head from the scan analysis. They are useful in younger children with a growing skeleton in whom the adult hip scan may not be appropriate [3, 4, 34]. In adults, total body scans are most frequently used to study whole body and regional body composition by measuring BMC, and lean and fat mass [35].

A variety of peripheral DXA (pDXA) devices are available for measuring the forearm, heel, or hand. In principle, these are attractive because they offer a quick, inexpensive, and convenient method of investigating skeletal status. In practice, however, these alternative types of measurement correlate poorly with central DXA, with correlation coefficients in the range r = 0.5 to 0.65 [12]. The lack of agreement with central DXA has proved a barrier to reaching a consensus on how to interpret results from pDXA devices [36]. However, no type of DXA measurement is immune from the uncertainties caused by T-score discordance, since similar differences are observed when comparing spine and hip results [11, 12].

FRACTURE PREDICTION

The ability to identify patients at increased risk of fracture is the most important reason for performing DXA [37]. The earliest ideas for interpreting BMD_a measurements were based on the concept of a fracture threshold, a

(B)

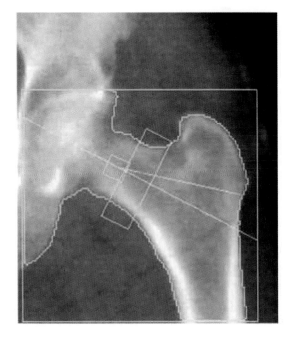
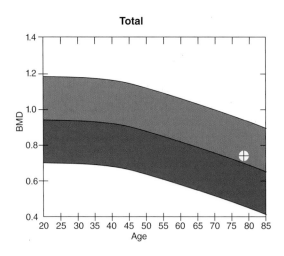

DXA Results Summary:

Region	Area (cm²)	BMC (g)	BMD (g/cm²)	T-score	PR (%)	Z-score	AM (%)
Neck	4.87	3.14	0.645	−1.8	76	0.4	107
Total	42.54	31.61	0.743	−1.6	79	0.4	106

Total BMD CV 1.0%, ACF = 1.025, BCF = 0.994, TH = 4.677
WHO Classification: Osteopenia

10-year Fracture Risk[1]

Risk Type	
Major Osteoporotic Fracture	10%
Hip Fracture	3.0%

Reported Risk Factors:
U.K., Neck BMD = 0.645, BMI = 20.2

[1]FRAX™ Version 1.00. Fracture probability calculated for an untreated patient. Fracture probability may be lower if the patient has received treatment.

Fig. 30.1. (*Continued from previous page.*) (B) Scan printout of a hip DXA examination. The printout shows: (left) scan image of the hip; (top right) patient's age and total hip BMD plotted with respect to the NHANES III) reference range (Ref. 43); (bottom right) BMD figures for the femoral neck and total hip regions of interest, together with the interpretation in terms of T-scores and Z-scores using the NHANES III reference range and the FRAX® evaluation of the 10-year risk of hip fracture or an osteoporotic fracture at a major site (hip, forearm, humerus, or clinical vertebral fracture).

BMD_a figure below which patients were at increased risk for suffering an insufficiency fracture [38]. However, once data from large epidemiological studies were available it became clear that the relationship between BMD_a and fracture incidence is best described by a gradient-of-risk model in which the risk increases progressively as BMD_a decreases following an approximately exponential relationship (Fig. 30.2) [23, 24]. The gradient-of-risk is described by the relative risk (RR), the increased risk of fracture for each SD decrease in BMD_a. Results from meta-analyses give RR figures between 1.5 and 2.5 (Fig. 30.3), and show that the ability of a BMD_a measurement to predict fracture is comparable to the use of blood pressure to predict stroke and serum cholesterol to predict myocardial infarction [24]. Not all types of fracture are equally well predicted by BMD_a measurements, which are best at predicting fractures at sites such as the hip, spine, forearm, humerus, and pelvis, and are relatively ineffective for fractures of the face, ankles, feet and toes [25].

The most effective BMD_a measurement site for predicting fractures is the one with the largest value of RR [6]. The reason is explained in Fig. 30.4. If we consider a specific group such as all postmenopausal women in a particular age range, the distribution of Z-scores follows

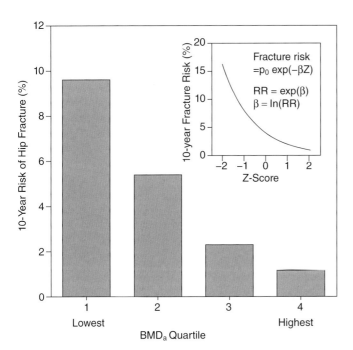

Fig. 30.2. Incidence of hip fracture risk by BMD$_a$ quartile for femoral neck BMD$_a$. Data are taken from the 2-year follow-up of the Study of Osteoporotic Fractures (Ref. 23). Inset diagram: Data from fracture studies are fitted using a gradient-of-risk model, in which the fracture risk varies exponentially with Z-score with gradient β. Results are expressed in terms of the RR, the increased risk of fracture for each unit decrease in Z-score. The value of RR is found from β using the relationship RR = exp(β). Alternatively, β is found by taking the natural logarithm of RR [β = ln(RR)].

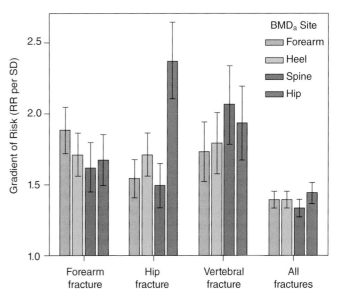

Fig. 30.3. Values of the RR (defined as the increased risk of fracture for a 1 SD decrease in BMD$_a$) for fractures at different skeletal sites (wrist, hip, spine, and any fracture) for BMD$_a$ measurements made at four different sites (forearm, heel, spine, and femoral neck). The error bars show the 95% confidence intervals. Data are taken from the 10-year follow-up of the Study of Osteoporotic Fractures (SOF) study population (Ref. 25). In the SOF data the largest value of RR is for the prediction of hip fracture risk from a hip BMD$_a$ measurement (RR = 2.4). This means that the clinically most effective type of DXA scan examination is to use hip BMD$_a$ to predict hip fracture risk.

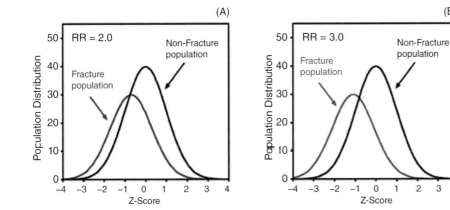
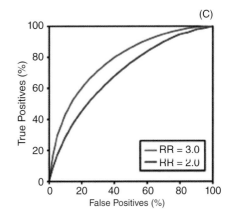

Fig. 30.4. (A) Distribution of Z-score values in a fracture population compared with an age-matched non-fracture population. The curve for the non-fracture population is a bell-shaped curve symmetrically distributed around its peak at Z = 0. The corresponding curve for the fracture population is a similar bell-shaped curve that is offset to lower (more negative) Z-scores by an amount that is greater for larger values of the RR. The curves shown are for RR = 2.0. (B) Similar to (A) but drawn for RR = 3.0. For a larger value of RR there is less overlap between the distribution curves for the fracture and non-fracture populations. (C) Plot of the ROC curves for the distribution curves shown in (A) and (B). The ROC curve shows the true positive fraction (those patients who sustain a fracture and were correctly identified as being at risk) against the false positive fraction (those patients identified as being at risk but who never actually have a fracture). The larger the value of RR the more effective BMD measurements are at discriminating the patients who will have a fracture as shown by the larger area under the ROC curve.

a bell-shaped curve. Multiplying the population distribution by the curve for increasing fracture risk leads to the conclusion that the corresponding fracture population has a similar bell-shaped curve with its peak shifted to lower Z-scores. This leads to one of the most important limitations of BMD_a measurements, namely that there is a wide overlap between the BMD_a distributions of the fracture and non-fracture populations, and that DXA is an imperfect tool for distinguishing between them [Fig. 30.4(A)]. However, the larger the value of RR the wider the separation between the two groups and the more effective BMD_a measurements are at distinguishing between patients who will have a fracture from those who will not [Fig. 30.4(B)] [6].

The clinical importance of choosing a measurement site with a large value of RR can be illustrated by using the distribution curves in Figs. 30.4(A) and (B) to draw a receiver operating characteristic (ROC) curve in which the percentage of true positive cases (patients who will suffer a fracture in the future and are correctly identified to be at risk) is plotted against the percentage of false positive cases (patients identified to be at risk but who will never have a fracture) [Fig. 30.4(C)]. Plotting ROC curves shows that the larger the RR value the greater the area under the curve (AUC) and the better the test will discriminate patients who are at the greatest risk of fracture.

SCAN ARTIFACTS

Careful visual inspection of the images included on the manufacturer's scan printout is important to ensure that BMD_a results are not affected by artifacts. For spine scans these include degenerative disc disease with osteophytes, spondylosis, osteoarthritis with hyperostosis of the facet joints, aortic calcification or vertebral fractures, which all cause BMD_a to be falsely elevated and particularly affect patients aged 65 years or more (Table 30.2) [39]. Technical artifacts include metal clips, navel rings, and buttons. The effect on scan interpretation can be assessed by noting the trend in T- or Z-score values at each vertebral level. Vertebrae significantly affected by artifacts should be excluded from the scan analysis, with the proviso that for clinical diagnosis a minimum of two unaffected vertebrae is recommended [3]. Laminectomy will cause underestimation of BMD_a at the affected level. For spine scans it is important to check that the correct vertebrae have been selected for analysis. Instead of L1-L4, scan analysis is sometimes erroneously performed for T12-L3 or L2-L5. One source of confusion is patients with abnormal segmentation [40]. In patients who have recently had an investigation with barium contrast medium it may be necessary to wait up to a month for this to clear. For patients who have recently had a radionuclide scan with technetium-99m, it may be necessary to wait 48 hours for this to decay [41, 42]. It is also important to recognize that DXA cannot differentiate between

Table 30.2. Causes of Artifacts Resulting in Errors in DXA BMD_a

Overestimation of BMD_a
Spinal degeneration and hyperostosis (osteophytes)
Vertebral fracture
Extraneous calcification (aortic calcification, lymph nodes)
Overlying metal (clips, coins, navel rings, surgical rods)
Other overlying objects (buttons, wallets)
Sclerotic metastases
Vertebral hemangioma
Ankylosing spondylitis with paravertebral ossification
Exposure to strontium-containing medications (strontium ranelate, health supplements)

Underestimation of BMD_a
Laminectomy
Lytic metastases
Barium contrast medium in bowel
Recent radionuclide examinations

a low BMD_a that is due to osteomalacia (e.g., in vitamin D deficiency) or osteoporosis.

In patients with extensive degenerative disease, the spine scan may be of little diagnostic value, and in such individuals a more reliable result is obtained from the hip. Hip scans also require careful scrutiny as there is a wide range of anatomical variations, some of which cause difficulties in correctly positioning the hip ROIs. Inspection of scan images is particularly important when interpreting follow-up scans, and a visual comparison should always be made with previous studies. For the spine a check should be made that the same vertebrae were used for the analysis. For hip scans it is important that the angles of rotation and abduction of the hip are the same and that the ROI boxes are placed in a consistent manner. Follow-up scan results showing unexpectedly large rates of change should be reviewed to exclude technical factors as a cause.

SCAN INTERPRETATION USING T- AND Z-SCORES

Once a BMD_a measurement has been made, a clinical report is issued interpreting the findings for the referring clinician. The International Society for Clinical Densitometry (ISCD) has issued guidance for the interpretation and reporting of DXA scans that is free to download from the ISCD website [3]; it is also available in published articles [4]. Results are interpreted using T- and Z-scores calculated using the appropriate age-, gender- and ethnically matched reference ranges (Eqs. 30.1 and 30.2). Osteoporosis may be diagnosed in postmenopausal women and men aged 50 or older if the T-score at the lumbar spine, femoral neck, or total hip site is −2.5 or lower [3, 4]. For the hip sites, the United States (U.S.) National Health and Nutrition Examination Survey (NHANES) III database should be used for calculating T-scores [43]. In women, the Caucasian female reference range is recommended for interpreting results for all ethnic groups, with a similar recommendation for men [3, 4]. In circumstances where it is necessary to perform a forearm scan the 33% radius (also called the 1/3 radius) site may be used [3, 4]. However, the WHO T-score definition of osteoporosis should only be applied to DXA measurements at the above-mentioned sites, since with other measurements [for example, quantitative computed tomography (QCT) or quantitative ultrasound (QUS)] it is likely to lead to the over or under diagnosis of osteoporosis [12, 44]. When applied appropriately, the WHO T-score threshold for defining osteoporosis is similar to the use of a blood pressure threshold to diagnose hypertension or serum cholesterol to diagnose hypercholesterolemia.

For premenopausal women and men younger than age 50, Z-scores rather than T-scores are preferred for scan interpretation [3, 4]. In particular, T-scores must never be used to interpret DXA results in children because of the effect of bone size on the measurements and the fact that peak bone mass has not yet been reached [8]. When Z-scores are used they should be based on reference data for the relevant population and ethnic group. A Z-score of −2.0 or lower is defined as below the expected range for age [3, 4].

SCAN INTERPRETATION USING FRAX®

A new approach to DXA scan interpretation based on using femoral neck BMD_a combined with information on clinical risk factors to estimate patients' 10-year probability of a fracture at the hip or one of the major osteoporotic sites (the latter defined as the hip, spine, forearm, or humerus) was introduced by the WHO and is designed for use in women and men aged 40 to 90 years [45]. The FRAX WHO fracture risk assessment tool is readily accessed using the website [7] and has also been incorporated into manufacturers' DXA scan reports [Fig. 30.1(B)]. A limitation of the T-score approach to making decisions about patient treatment is that factors such as age and a history of previous insufficiency fracture are independent risk factors that are as important as BMD_a in determining the 10-year risk of fracture. In the FRAX scheme a selected list of clinical risk factors (Table 30.3) is used with BMD_a to improve the ROC curve (Fig. 30.5).

The FRAX algorithm was developed based on a series of meta-analyses of data from nine different fracture studies from North America, Europe, and Asia. These enrolled a total of 46,000 men and women with 190,000 person-years of follow-up, and included 850 cases of hip fracture and 3,300 other osteoporotic fractures [45]. Once

Table 30.3. Clinical Risk Factors Included in FRAX® WHO Fracture Risk Algorithm*

Country or geographic region
Ethnic origin (U.S. only)
Age
Gender
Weight and height (BMI)
Previous history of low trauma fracture during adult life
Parental history of hip fracture
Current smoking habit
Current or past use of oral glucocorticoids
Rheumatoid arthritis
Secondary osteoporosis
Alcohol intake ≥ 3 units daily
Femoral neck BMD_a

*(Ref. 7).

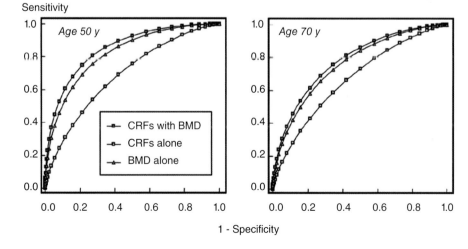

Fig. 30.5. ROC curves for the risk score for hip fracture prediction at the ages of 50 and 70 years. Results are from the FRAX® validation study (Ref. 45). Reproduced with permission. CRF = clinical risk factor.

Table 30.4. Treatment Criteria from the National Osteoporosis Foundation 2010 Guidelines*

Postmenopausal women and men age 50 and older presenting with the following should be considered for treatment:

A hip or vertebral (clinical or morphometric) fracture

T-score ≤−2.5 at the femoral neck or spine after appropriate evaluation to exclude secondary causes

Low bone mineral density (T-score between −1.0 and −2.5 at the femoral neck or spine) and a 10-year probability of a hip fracture ≥3% or a 10-year probability of a major osteoporosis related fracture ≥20% based on the U.S.-adapted WHO algorithm

*(Ref. 5).

the fracture risk algorithm had been constructed, a validation study was performed using data from 11 independent studies not used in the development of the original model involving a total of 230,000 subjects with over 1.2 million person-years of follow-up [45]. By reason of its large numbers, its international character, and the care taken in its construction and implementation, the FRAX algorithm is arguably the most important advance in bone densitometry since the introduction of DXA.

FRAX is gradually being incorporated into national guidelines on the treatment of osteoporosis. In the U.S. the National Osteoporosis Foundation (NOF) has added FRAX to the criteria for treating patients with a threshold of 3% for hip fracture and 20% for a major osteoporotic fracture (Table 30.4) [5]. In the U.K., a different treatment algorithm was issued by the National Osteoporosis Guidelines Group (NOGG), which recommended that FRAX be used in a triage scheme to select the patients who would benefit from having a DXA examination [46]. The introduction of FRAX has raised some interesting discussion and controversy [47–49]. As a result, the FRAX tool has been refined and a number of errors corrected, and at the time of writing, versions are available for use in 47 different countries. An up-to-date review of FRAX can be found in a recent position paper [50].

PRECISION ERRORS AND MONITORING BMD$_a$ CHANGE

The use of DXA scans to monitor patients undergoing treatment is more controversial than other applications [51, 52], and good precision is essential [53]. Precision errors measure the reproducibility of BMD$_a$ results in individual patients and can be measured by performing repeated scans on a representative group of subjects. Precision is usually expressed in terms of the percentage coefficient of variation (CV) and is typically around 1–1.5% for spine and total hip BMD$_a$ and 2–2.5% for femoral neck BMD$_a$ [19]. DXA scanners have excellent long-term precision because their calibration is very stable, and there are instrument quality control procedures provided by manufacturers to detect any long-term drifts should they occur. However, good precision depends on scanners being operated by skilled and appropriately trained staff and that rigorous quality assurance protocols are in place.

To reach a statistically significant conclusion when monitoring a patient, the BMD$_a$ change needs to equal or exceed the least significant change (LSC), which is defined as 2.8 times the precision error, i.e., a change of 3–4.5% for spine and total hip BMD$_a$ and 6–7.5% for femoral neck BMD$_a$ [53, 54]. In general the same DXA scanner should be used for follow-up examinations to avoid even larger errors. Typically for monitoring individual patients an adequate interval of time (18–24 months) is required between measurements unless particularly large changes in BMD$_a$ are expected, for example after large doses of oral glucocorticoids [3, 4].

VERTEBRAL FRACTURE ASSESSMENT (VFA)

Lateral views of the thoracic and lumbar spine (T4–L4) can be obtained with fan-beam DXA scanners, using dual- or single-energy scanning (dual-energy images are generally superior for visualizing vertebrae in the thoracic region), with the patient either in the supine ("C" arm scanners) or lateral decubitus position. From these images a visual assessment can be made as to whether or not vertebral fractures are present, and morphometric assessment of vertebral shape can be made [55, 56]. The method uses a lateral scan projection, with simultaneous movement of the X-ray source and detectors along the spine, so the X-ray beam is always parallel to the vertebral endplates. This avoids the "bean can" artifact of vertebral endplates due to the parallax effect of a conventional conical shaped X-ray beam, and the dose from ionising radiation is much lower (1/100th to 1/40th) than with spinal radiographs. Guidelines for indications for VFA and scan interpretation have been published by the ISCD [3, 4, 57]. The method has been shown to be satisfactory for excluding the presence of vertebral fractures [58]. However, more scientific studies are required to confirm the exact clinical role of this alternative technology to conventional spinal radiography for the identification of vertebral fractures.

DXA IN CHILDREN

The purpose of any assessment of bone size or mass in children is to provide data relevant to the current, or future, health of the skeleton. DXA is the most widely used quantitative bone imaging technique in pediatric practice, but many aspects of its use, and the interpretation of the data obtained, remain contentious [59–67].

DXA provides estimates of bone size in two dimensions and bone mass within that envelope, the value of bone mass adjusted for size being "areal" BMD_a (g/cm^2) [8]. The manufacturer-provided reference values for BMD_a, which increase in a similar way to height and weight during childhood and adolescence, clearly indicate that DXA is not measuring true volumetric bone density, but some composite measure of bone size and mass. This is not necessarily a disadvantage since bone size, especially in the tubular bones that children are most prone to fracture, is an important predictor of bone strength. Moreover, to some extent this limitation can be overcome by using an estimated volumetric BMD [bone mineral apparent density (BMAD, g/cm^3)] in the spine and femoral neck [68, 69].

The advantages of using DXA in children are the short scan time, low radiation dose, and general widespread availability. DXA measures of BMD_a in healthy children are predictive of fracture risk both at the measurement site (in the forearm) and elsewhere; total body BMD_a less head (Fig. 30.6), adjusted for weight, height, and bone area, is the measure that has been found in a prospective cohort study at age 9.9 years to be most strongly associated with fracture risk over the following 2 years [70]. However, there are no data for the predictive value of DXA at other ages in apparently healthy children and no similar data for children with bone disease.

Children are a difficult group to study, because as they grow so their bone mass should increase [64]. They also fracture frequently—by the end of teenage years, up to half of all boys and a third of girls will have sustained a fracture [71, 72]. A single fracture in an otherwise healthy child should not therefore require investigation of skeletal health.

Most requests for DXA scans in children will be in the groups thought to be at increased risk of fracture. These include children with primary bone diseases such as osteogenesis imperfecta and idiopathic juvenile osteoporosis; chronic immobilization (cerebral palsy and Duchenne muscular dystrophy); inflammatory conditions (Crohn's disease, cystic fibrosis, and juvenile idiopathic arthritis); conditions with endocrine disturbance such as anorexia nervosa and Cushing's syndrome (but not Turner's syndrome); following chemotherapy or organ transplantation; and in thalassemia major [73]. Other requests come when a child has recurrent fractures in the absence of an obvious underlying predisposition.

Measurement sites for DXA in children are typically the lumbar spine (L1-4) and total body less head, where precision is similar to that achieved in adults [74]. The forearm and proximal femur have been used in some studies, as has the lateral distal femur where deformity and contracture preclude the use of DXA in the normal measurement sites. Normative reference data are available for spine, femoral neck, total body, and lateral distal femur [34, 62, 75]. It should be noted that inconsistencies have been reported between different reference databases

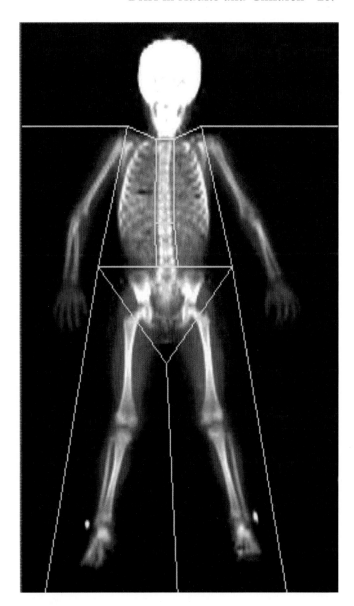

Fig. 30.6. Whole body DXA in a child is obtained in approximately 1 min on fan-beam scanners and provides total and regional information about the skeleton (less head) and body composition (lean muscle and fat mass).

that can lead to clinically relevant differences in BMD_a Z-scores [65].

Measurements are usually reported in relation to gender, age and race [62, 75]. Further adjustments have been made in research studies in apparently healthy children to account for the assumed shape of the vertebrae (cylindrical, cuboidal) to calculate BMAD [76, 77] or the expected relationship of bone size and mass with height or elements of body composition such as lean body mass [63, 67, 78–80]. Recent data suggest vertebral shape also influences BMD_a measurements [81]. Some of the current generation of scanners provide methods of adjusting for body size at both the spine and total body, but currently there is a lack of information as to which adjustment (if any) should be used in clinical practice. It is unclear from the published data whether any one of the proposed adjustments is optimal in terms of either fracture prediction or skeletal health assessment.

In diseases and disorders in childhood where fracture risk is increased, initiation of bone mass measurement by DXA should be at the discretion of the treating clinician, and the monitoring interval should reflect the severity of the disease. There is little evidence to support monitoring at intervals of less than 6 months in any setting [74]. Although there is a known increased risk of fracture for children receiving glucocorticoid therapy, it is difficult to disentangle the effects of these on bone from the effect of the underlying disease. It does not seem appropriate at this time to suggest that measurements should be any more frequent in children receiving such therapy [73].

Interpretation of the scan results depends on the clinical context. A diagnosis of osteoporosis should not be made on the basis of a BMD_a measurement in isolation [4]. Terminology such as "low bone density for chronological age" may be used if the Z-score is below –2.0 [3, 4, 8]. If a child presents with recurrent fractures but no clinically apparent underlying disease, and BMD_a is within the expected range, reassurance can be offered. If there is an underlying problem, further monitoring and additional imaging may be required. Such a clinical setting would be a child with apparently mild osteogenesis imperfecta with BMD_a in the normal range who may have occult fractures of thoracic vertebrae; such findings have also been reported in nephrotic syndrome [82]. Normal bone mass by DXA does not, therefore, preclude the presence of vertebral fractures that merit intervention. Low BMD_a in the presence of vertebral, or recurrent, fractures resulting in loss of independent mobility and chronic bone pain should prompt evaluation of the need for active intervention, as should falling bone mass in conditions that predispose to fracture.

CONCLUSIONS

DXA is the most widely used method of measuring BMD_a in adults and children and provides precise results with very low doses of ionizing radiation. It can be used in postmenopausal women and men aged 50 years and over to diagnose osteoporosis using the WHO definition (T-score of –2.5 or less at the lumbar spine, femoral neck, total hip, or 33% radius) and aid decisions on patient management. The basic rationale for DXA examinations is their ability to identify patients at increased risk of fracture, and the new WHO FRAX fracture risk assessment tool effectively exploits this ability and sets the clinical application of bone densitometry on a sound scientific basis. DXA has some important limitations. The discordant results between different sites reflect the relatively limited information that can be derived from a 2D projection scan as well as the errors caused by variations in soft tissue composition. Although DXA has excellent precision, it is easy to overlook the significance of this lack of concordance and to mistake good precision for good diagnostic accuracy. The degree of discordance between different sites indicates that DXA measurements should be viewed as a relatively rough and ready indicator of skeletal health. One merit of the FRAX scheme is that it treats BMD_a measurements as just another type of clinical risk factor rather than a uniquely special indicator of skeletal status. Another merit of FRAX is that it provides a focus for future research by directing attention to the importance of achieving the best ROC curve for fracture risk prediction. There are particular issues with interpreting DXA results in children in whom the dependency on bone size is a limitation, and to date there is no consensus on whether size correction should be applied and which method is optimum.

REFERENCES

1. Chun KJ. 2011. Bone densitometry. *Semin Nucl Med* 41(3): 220–8.
2. Kanis JA, Glüer CC. 2000. An update on the diagnosis and assessment of osteoporosis with densitometry. *Osteoporos Int* 11(3): 192–202.
3. International Society for Clinical Densitometry. 2007. *ISCD 2007 Official Positions & Pediatric Official Positions.* West Hartford (CT): International Society for Clinical Densitometry. Available from: http://www.iscd.org.
4. Lewiecki EM, Gordon CM, Baim S, Leonard MB, Bishop NJ, Bianchi ML, Kalkwarf HJ, Langman CB, Plotkin H, Rauch F, Zemel BS, Binkley N, Bilezikian JP, Kendler DL, Hans DB, Silverman S. 2008. International Society for Clinical Densitometry 2007 Adult and Pediatric Official Positions. *Bone* 43(6): 1115–21.
5. National Osteoporosis Foundation. 2010. *Clinician's Guide to Prevention and Treatment of Osteoporosis.* Washington, DC: National Osteoporosis Foundation. Available from: http://www.nof.org/professionals/clinical-guidelines.
6. Blake GM, Fogelman I. 2007. The role of DXA bone density scans in the diagnosis and treatment of osteoporosis. *Postgrad Med J* 83(982): 509–17.
7. World Health Organization Collaborating Centre for Metabolic Bone Diseases. FRAX® WHO Fracture Risk Assessment Tool Web Version 3.7 [Internet]. Sheffield, UK: University of Sheffield. September 14, 2012. Available from: http://www.shef.ac.uk/FRAX.
8. Adams JE, Ahmed SF, Alsop C, Bishop N, Crabtree N, Fewtrell M, Mughal MZ, Shaw NJ, Stevens MR, Ward K. 2004. *A Practical Guide to Bone Densitometry in Children.* Bath, UK: National Osteoporosis Society.
9. Seeman E. 1998. Growth in bone mass and size—Are racial and gender differences in bone mineral density

more apparent than real? *J Clin Endocrinol Metab* 83(5): 1414–9.
10. Blake GM, Fogelman I. 2008. How important are BMD accuracy errors for the clinical interpretation of DXA scans? *J Bone Miner Res* 23(4): 457–62.
11. Woodson G. 2000. Dual X-ray absorptiometry T-score concordance and discordance between the hip and spine measurement sites. *J Clin Densitom* 3(4): 319–24.
12. Lu Y, Genant HK, Shepherd J, Zhao S, Mathur A, Fuerst TP, Cummings SR. 2001. Classification of osteoporosis based on bone mineral densities. *J Bone Miner Res* 16(5): 901–10.
13. Blake GM, Fogelman I. 1997. Technical principles of dual-energy x-ray absorptiometry. *Semin Nucl Med* 27(3): 210–28.
14. Lehmann LA, Alvarez RE, Macovski A, Brody WR, Pelc NJ, Riederer SJ, Hall AL. 1981. Generalized image combinations in dual KVP digital radiography. *Med Phys* 8(5): 659–67.
15. Damilakis J, Adams JE, Guglielmi G, Link TM. 2010. Radiation exposure in X-ray-based imaging techniques used in osteoporosis. *Eur Radiol* 20(11): 2707–14.
16. Blake GM, Naeem M, Boutros M. 2006. Comparison of effective dose to children and adults from dual x-ray absorptiometry examinations. *Bone* 38(6): 935–42.
17. Svendsen OL, Hassager C, Skødt V, Christiansen C. 1995. Impact of soft tissue on in vivo accuracy of bone mineral measurements in the spine, hip and forearm: A human cadaver study. *J Bone Miner Res* 10(6): 868–73.
18. Tothill P, Pye DW. 1992. Errors due to non-uniform distribution of fat in dual x-ray absorptiometry of the lumbar spine. *Br J Radiol* 65(777): 807–13.
19. Shepherd JA, Fan B, Lu Y, Lewiecki EM, Miller P, Genant HK. 2006. Comparison of BMD precision for Prodigy and Delphi spine and femur scans. *Osteoporos Int* 17(9): 1303–8.
20. Blake GM, Griffith JF, Yeung DK, Leung PC, Fogelman I. 2009. Effect of increasing vertebral marrow fat content on BMD measurement, T-Score status and fracture risk prediction by DXA. *Bone* 44(3): 495–501.
21. Bolotin HH, Sievänen H, Grashuis JL, Kuiper JW, Järvinen TL. 2001. Inaccuracies inherent in patient-specific dual-energy X-ray absorptiometry bone mineral density measurements: Comprehensive phantom-based evaluation. *J Bone Miner Res* 16(2): 417–26.
22. Bolotin HH. 2007. DXA in vivo BMD methodology: An erroneous and misleading research and clinical gauge of bone mineral status, bone fragility, and bone remodelling. *Bone* 41(1): 138–54.
23. Cummings SR, Black DM, Nevitt MC, Browner W, Cauley J, Ensrud K, Genant HK, Palermo L, Scott J, Vogt TM. 1993. Bone density at various sites for prediction of hip fractures. *Lancet* 341(8837): 72–5.
24. Marshall D, Johnell O, Wedel H. 1996. Meta-analysis of how well measures of bone mineral density predict occurrence of osteoporotic fractures. *BMJ* 312(7041): 1254–9.
25. Stone KL, Seeley DG, Lui LY, Cauley JA, Ensrud K, Browner WS, Nevitt MC, Cummings SR. 2003. BMD at multiple sites and risk of fracture of multiple types: Long-term results from the Study of Osteoporotic Fractures. *J Bone Miner Res* 18(11): 1947–54.
26. Johnell O, Kanis JA, Oden A, Johansson H, De Laet C, Delmas P, Eisman JA, Fujiwara S, Kroger H, Mellstrom D, Meunier PJ, Melton LJ 3rd, O'Neill T, Pols H, Reeve J, Silman A, Tenenhouse A. 2005. Predictive value of BMD for hip and other fractures. *J Bone Miner Res* 20(7): 1185–94.
27. Campbell NR, Chockalingam A, Fodor JG, McKay DW. 1990. Accurate, reproducible measurement of blood pressure. *CMAJ* 143(1): 19–24.
28. Rotch AL, Dean JO, Kendrach MG, Wright SG, Woolley TW. 2001. Blood pressure monitoring with home monitors versus mercury sphygmomanometer. *Ann Pharmacother* 35(7–8): 817–2.
29. Modell JG, Katholi CR, Kumaramangalam SM, Hudson EC, Graham D. 1998. Unreliability of the infrared tympanic thermometer in clinical practice: A comparative study with oral mercury and oral electronic thermometers. *South Med J* 91(7): 649–54.
30. Rubia-Rubia J, Arias A, Sierra A, Aguirre-Jaime A. 2011. Measurement of body temperature in adult patients: Comparative study of accuracy, reliability and validity of different devices. *Int J Nurs Stud* 48(7): 872–80.
31. Faulkner KG. 1998. Bone densitometry: Choosing the proper site to measure. *J Clin Densitom* 1(3): 279–85.
32. Silverberg SJ, Lewiecki EM, Mosekilde L, Peacock M, Rubin MR. 2009. Presentation of asymptomatic primary hyperparathyroidism: Proceedings of the Third International Workshop. *J Clin Endocrinol Metab* 94(2): 351–65.
33. Nuti R, Martini G. 1992. Measurements of bone mineral density by DXA total body absorptiometry in different skeletal sites in postmenopausal osteoporosis. *Bone* 13(2): 173–8.
34. Ward KA, Ashby RL, Roberts SA, Adams JE, Zulf Mughal M. 2007. UK reference data for the Hologic QDR Discovery dual-energy x ray absorptiometry scanner in healthy children and young adults aged 6–17 years. *Arch Dis Child* 92(1): 53–9.
35. Schoeller DA, Tylavsky FA, Baer DJ, Chumlea WC, Earthman CP, Fuerst T, Harris TB, Heymsfield SB, Horlick M, Lohman TG, Lukaski HC, Shepherd J, Siervogel RM, Borrud LG. 2005. QDR 4500A dual energy X-ray absorptiometer underestimates fat mass in comparison with criterion methods in adults. *Am J Clin Nutr* 81(5): 1018–25.
36. Blake GM, Chinn DJ, Steel SA, Patel R, Panayiotou E, Thorpe J, Fordham JN. 2005. A list of device specific thresholds for the clinical interpretation of peripheral x-ray absorptiometry examinations. *Osteoporos Int* 16(12): 2149–56.
37. Kanis JA. 2002. Diagnosis of osteoporosis and assessment of fracture risk. *Lancet* 359(9321): 1929–36.
38. Nordin BE. 1987. The definition and diagnosis of osteoporosis. *Calcif Tissue Int* 40(2): 57–8.

39. Liu G, Peacock M, Eilam O, Dorulla G, Braunstein E, Johnston CC. 1997. Effect of osteoarthritis in the lumbar spine and hip on bone mineral density and diagnosis of osteoporosis in elderly men and women. *Osteoporos Int* 7(6): 564–9.
40. Peel NF, Johnson A, Barrington NA, Smith TW, Eastell R. 1993. Impact of anomalous vertebral segmentation on measurements of bone mineral density. *J Bone Miner Res* 8(6): 719–23.
41. McKiernan FE, Hocking J, Cournoyer S. 2006. Antecedent 99mTc-MDP and 99mTc-sestamibi administration corrupts bone mineral density measured by DXA. *J Clin Densitom* 9(2): 164–6.
42. Sala A, Webber C, Halton J, Morrison J, Beaumont L, Zietak A, Barr R. 2006. Effect of diagnostic radioisotopes and radiographic contrast media on measurements of lumbar spine bone mineral density and body composition by dual-energy x-ray absorptiometry. *J Clin Densitom* 9(1): 91–6.
43. Looker AC, Wahner HW, Dunn WL, Calvo MS, Harris TB, Heyse SP, Johnston CC Jr, Lindsay R. 1998. Updated data on proximal femur bone mineral levels of US adults. *Osteoporos Int* 8(5): 468–89.
44. Faulkner KG, Von Stetten E, Miller P. 1999. Discordance in patient classification using T-scores. *J Clin Densitom* 2(3): 343–50.
45. Kanis JA, Oden A, Johnell O, Johansson H, De Laet C, Brown J, Burckhardt P, Cooper C, Christiansen C, Cummings S, Eisman JA, Fujiwara S, Glüer C, Goltzman D, Hans D, Krieg MA, La Croix A, McCloskey E, Mellstrom D, Melton LJ 3rd, Pols H, Reeve J, Sanders K, Schott AM, Silman A, Torgerson D, van Staa T, Watts NB, Yoshimura N. 2007. The use of clinical risk factors enhances the performance of BMD in the prediction of osteoporotic fractures in men and women. *Osteoporos Int* 18(8): 1033–46.
46. Kanis JA, McCloskey EV, Johansson H, Strom O, Borgstrom F, Oden A. 2008. Case finding for the management of osteoporosis with FRAX-assessment and intervention thresholds for the UK. *Osteoporos Int* 19(10): 1395–408.
47. Binkley N, Lewiecki EM. 2010. The evolution of fracture risk estimation. *J Bone Miner Res* 25(10): 2098–100.
48. Watts NB, Diab DL. 2011. A backbone for FRAX? *J Bone Miner Res* 26(3): 458–9.
49. Leslie WD, Lix LM. 2011. Absolute fracture risk assessment using lumbar spine and femoral neck bone density measurements: Derivation and validation of a hybrid system. *J Bone Miner Res* 26(3): 460–7.
50. Kanis JA, Hans D, Cooper C, Baim S, Bilezikian JP, Binkley N, Cauley JA, Compston JE, Dawson-Hughes B, El-Hajj Fuleihan G, Johansson H, Leslie WD, Lewiecki EM, Luckey M, Oden A, Papapoulos SE, Poiana C, Rizzoli R, Wahl DA, McCloskey EV; Task Force of the FRAX Initiative. 2011. Interpretation and use of FRAX in clinical practice. *Osteoporos Int* 22(9): 2395–411.
51. Compston J. 2009. Monitoring bone mineral density during antiresorptive treatment for osteoporosis. *BMJ* 338: b1276.
52. Bell KJ, Hayen A, Macaskill P, Irwig L, Craig JC, Ensrud K, Bauer DC. 2009. Value of routine monitoring of bone mineral density after starting bisphosphonate treatment: Secondary analysis of trial data. *BMJ* 338: b2266.
53. Bonnick SL, Johnston CC Jr, Kleerekoper M, Lindsay R, Miller P, Sherwood L, Siris E. 2001. Importance of precision in bone density measurements. *J Clin Densitom* 4(2): 105–10.
54. Glüer CC. 1999. Monitoring skeletal changes by radiological techniques. *J Bone Miner Res* 14(11): 1952–62.
55. Link TM, Guglielmi G, van Kuijk C, Adams JE. 2005. Radiologic assessment of osteoporotic vertebral fractures: Diagnostic and prognostic implications. *Eur Radiol* 15(8): 1521–32.
56. Ferrar L, Jiang G, Adams J, Eastell R. 2005. Identification of vertebral fractures: An update. *Osteoporos Int* 16(7): 717–28.
57. Schousboe JT, Vokes T, Broy SB, Ferrar L, McKiernan F, Roux C, Binkley N. 2008. Vertebral Fracture Assessment: The 2007 ISCD Official Positions. *J Clin Densitom* 11(1): 92–108.
58. Rea JA, Li J, Blake GM, Steiger P, Genant HK, Fogelman I. 2000. Visual assessment of vertebral deformity by X-ray absorptiometry: A highly predictive method to exclude vertebral deformity. *Osteoporos Int* 11(8): 660–8.
59. van Rijn RR, van der Sluis IM, Link TM, Grampp S, Guglielmi G, Imhof H, Glüer C, Adams JE, van Kuijk C. 2003. Bone densitometry in children: A critical appraisal. *Eur Radiol* 13(4): 700–10.
60. Fewtrell MS; British Paediatric and Adolescent Bone Group. 2003. Bone densitometry in children assessed by dual X-ray absorptiometry: Uses and pitfalls. *Arch Dis Child* 88(9): 795–8.
61. Mughal M, Ward K Adams J. 2004. Assessment of bone status in children by densitometric and quantitative ultrasound techniques. In: Carty H, Brunelle F, Stringer DA, Kao SC (eds.) *Imaging in Children (Vol 1)*. Edinburgh: Elsevier Science. pp. 477–86.
62. Kalkwarf HJ, Zemel BS, Gilsanz V, Lappe JM, Horlick M, Oberfield S, Mahboubi S, Fan B, Frederick MM, Winer K, Shepherd JA. 2007. The bone mineral density in childhood study: Bone mineral content and density according to age, sex, and race. *J Clin Endocrinol Metab* 92(6): 2087–99.
63. Zemel BS, Leonard MB, Kelly A, Lappe JM, Gilsanz V, Oberfield S, Mahboubi S, Shepherd JA, Hangartner TN, Frederick MM, Winer KK, Kalkwarf HJ. 2010. Height adjustment in assessing dual energy x-ray absorptiometry measurements of bone mass and density in children. *J Clin Endocrinol Metab* 95(3): 1265–73.
64. Kalkwarf HJ, Gilsanz V, Lappe JM, Oberfield S, Shepherd JA, Hangartner TN, Huang X, Frederick MM, Winer KK, Zemel BS. 2010. Tracking of bone mass and density during childhood and adolescence. *J Clin Endocrinol Metab* 95(4): 1690–8.
65. Kocks J, Ward K, Mughal Z, Moncayo R, Adams J, Högler W. 2010. Z-score comparability of bone mineral density reference databases for children. *J Clin Endocrinol Metab* 95(10): 4652–9.

66. Crabtree NJ, Leonard MB, Zemel BS. 2010. Dual energy X-ray absorptiometry. In: Sawyer AJ, Bachrach LK, Fung EB (eds.) *Bone Densitometry in Growing Patients: Guidelines for Clinical Practice*. Totowa, NJ: Humana Press Inc. pp. 41–58.
67. Short DF, Zemel BS, Gilsanz V, Kalkwarf HJ, Lappe JM, Mahboubi S, Oberfield SE, Shepherd JA, Winer KK, Hangartner TN. 2011. Fitting of bone mineral density with consideration of anthropometric parameters. *Osteoporos Int* 22(4): 1047–57.
68. Katzman DK, Bachrach LK, Carter DR, Marcus R. 1991. Clinical and anthropometric correlates of bone mineral acquisition in healthy adolescent girls. *J Clin Endocrinol Metab* 73(6): 1332–9.
69. Lu PW, Cowell CT, Lloyd-Jones SA, Briody JN, Howman-Giles R. 1996. Volumetric bone mineral density in normal subjects, aged 5–27 years. *J Clin Endocrinol Metab* 81(4): 1586–90.
70. Clark EM, Ness AR, Bishop NJ, Tobias JH. 2006. Association between bone mass and fractures in children: A prospective cohort study. *J Bone Miner Res* 21(9): 1489–95.
71. Cooper C, Dennison EM, Leufkens HG, Bishop N, van Staa TP. 2004. Epidemiology of childhood fractures in Britain: A study using the general practice research database. *J Bone Miner Res* 19(12): 1976–81.
72. Jones IE, Williams SM, Dow N, Goulding A. 2002. How many children remain fracture-free during growth? A longitudinal study of children and adolescents participating in the Dunedin Multidisciplinary Health and Development Study. *Osteoporos Int* 13(12): 990–5.
73. Bishop N, Braillon P, Burnham J, Cimaz R, Davies J, Fewtrell M, Hogler W, Kennedy K, Mäkitie O, Mughal Z, Shaw N, Vogiatzi M, Ward K, Bianchi ML. 2008. Dual-energy X-ray absorptiometry assessment in children and adolescents with diseases that may affect the skeleton: The 2007 Pediatric Official Positions. *J Clin Densitom* 11(1): 29–42.
74. Shepherd JA, Wang L, Fan B, Gilsanz V, Kalkwarf HJ, Lappe J, Lu Y, Hangartner T, Zemel BS, Fredrick M, Oberfield S, Winer KK. 2011. Optimal monitoring time interval between DXA measures in children. *J Bone Miner Res* 26(11): 2745–52.
75. Zemel BS, Stallings VA, Leonard MB, Paulhamus DR, Kecskemethy HH, Harcke HT, Henderson RC. 2009. Revised pediatric reference data for the lateral distal femur measured by Hologic Discovery/Delphi dual-energy X-ray absorptiometry. *J Clin Densitom* 12(2): 207–18.
76. Carter DR, Bouxsein ML, Marcus R. 1992. New approaches for interpreting projected bone densitometry data. *J Bone Miner Res* 7(2): 137–45.
77. Kröger H, Kotaniemi A, Kröger L, Alhava E 1993. Development of bone mass and bone density of the spine and femoral neck: A prospective study of 65 children and adolescents. *Bone Miner* 23(3): 171–82.
78. Mølgaard C, Thomsen BL, Prentice A, Cole TJ, Michaelsen KF. 1997. Whole body bone mineral content in healthy children and adolescents. *Arch Dis Child* 76(1): 9–15.
79. Crabtree NJ, Kibirige MS, Fordham J, Banks LM, Muntoni F, Chinn D, Boivin CM, Shaw NJ. 2004. The relationship between lean body mass and bone mineral content in paediatric health and disease. *Bone* 35(4): 965–72.
80. Dimitri P, Wales JHK, Bishop N. 2010. Fat and bone in children: Differential effects of obesity on bone size and mass according to fracture history. *J Bone Miner Res* 25(3): 527–36.
81. Barlow T, Carlino W, Blades HZ, Crook J, Harrison R, Arundel P, Bishop NJ. 2011. The role of bone shape in determining gender differences in vertebral bone mass. *J Clin Densitom* 14(4): 440–6.
82. Sbrocchi AM, Rauch F, Matzinger M, Feber J, Ward LM. 2011. Vertebral fractures despite normal spine bone mineral density in a boy with nephrotic syndrome. *Pediatr Nephrol* 26(1): 139–42.

31

Quantitative Computed Tomography in Children and Adults

C.C. Glüer

Introduction 264
Methodology 264
Clinical Application 267
QCT-Based Finite Element Modeling 270

Summary and Perspectives 271
Acknowledgments 272
References 272

INTRODUCTION

Bone mineral density (BMD) is one of the strongest predictors of fracture risk, but today's standard densitometric method, dual X-ray absorptiometry (DXA)-based measurement of areal BMD (aBMD), has limitations, particularly for characterizing changes in bone status in subjects on treatment. Quantitative computed tomography (QCT) is a potential alternative for the future because in addition to volumetric density it permits assessment of bone microstructure and calculation of whole bone strength. Technical and clinical considerations are outlined in this chapter.

METHODOLOGY

X-ray-based computed tomography (CT) provides three-dimensional (3D) morphological and compositional information. On the reconstructed CT-images, tissue contrast is predominantly determined by X-ray absorption of its heavier elements (e.g., calcium). Image gray values are expressed as a CT number measured in Hounsfield units (HUs), but for QCT, a reference phantom of known composition is scanned together with the patient, which allows conversion of HUs into calcium-hydroxyapatite (Ca-HAP) equivalent BMD. For earlier measurements that were calibrated to K_2HPO_4, a correction factor needs to be applied to express results in Ca-HAP [1], an issue important, for example, for comparison with published reference ranges. Unlike DXA, BMD measured by QCT reflects volumetric BMD, not an areal projection.

Single-energy QCT of the skeleton was initially applied to the radius in 1976 by Rüegsegger et al. [2] and Isherwood et al. [3], followed by measurements at the spine by Genant and Boyd [4]. Measured BMD is decreased by marrow fat [5, 6]. Low tube voltage settings of about 80 kVp lead to smaller fat-related accuracy errors than measurements at 120 kVp. Dual-energy QCT approaches improve the accuracy of the measurement but at the expense of poorer precision and increased radiation exposure and these approaches have not achieved a significant clinical role. The recent advent of dual source CT scanners may facilitate dual-energy QCT in the future. For a QCT measurement, first a lateral planar overview measurement of the region to be scanned is generated [Fig. 31.1(A) and (B)], which allows placement of the measurement regions at the anatomically correct positions. Until the 1990s, the measurement region consisted of multiple individual and spatially separated CT slices, each about 5–10 mm thick [single-slice QCT; Fig. 31.1(A)]. Today, spiral CT approaches yield a complete 3D dataset of the volume of interest [multislice QCT; Fig. 31.1(B)], consisting of anywhere from tens of CT slices to hundreds. This permits automated and precise placement of volumes of

---

Primer on the Metabolic Bone Diseases and Disorders of Mineral Metabolism, Eighth Edition. Edited by Clifford J. Rosen.
© 2013 American Society for Bone and Mineral Research. Published 2013 by John Wiley & Sons, Inc.

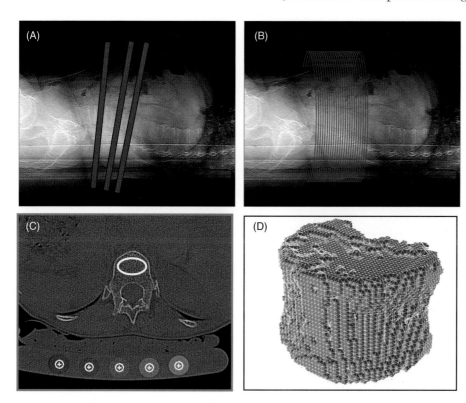

Fig. 31.1. Illustration of QCT procedure: (A) localizer image used for either tilted single slice placement at three adjacent vertebrae or (B) spiral volumetric multislice acquisition covering the same three vertebrae; (C) typical single slice technique image showing elliptical trabecular region of interest evaluated and calibration phantom underneath patient; (D) stack of images segmented along the cortical rim depicting the volume of interest evaluated in the multislice approach.

interest (VOIs) to study automatically well-defined compartments. Data acquisition is fast with scan times on the order of 1–10 s. The spatial resolution has improved substantially, approaching 100–200 μm for human studies *in vivo* at peripheral measurement sites [7]. At central measurement sites, in-plane resolution is poorer by a factor of 2, and slice thickness is greater than or equal to 300 μm.

Spinal QCT. For single-slice approaches, L1-L3 or T12-L3 are measured, and a minimum of two vertebrae should be evaluated. At 80 kVp, 140 mAs, and 8–10 mm slice thickness located at the center of the vertebrae [Fig. 31.1(C)], good precision at low radiation exposure and lower fat errors can be achieved [8]. With the more recent multislice approach [Fig. 31.1(D)] L1-L3 are measured, typically at 120 kVp, 50–120 mAs, slice thicknesses of 1–3 mm, pitch = 1. Trabecular bone is evaluated in an ellipse in the anterior part of the vertebral body or in larger regions encompassing most of the trabecular volume. The trabecular region shows high responsiveness early after menopause, and thus the assessment with QCT is superior to DXA. For accurate assessment of vertebral fragility the cortical rim should also be evaluated. It is very thin (300–600 μm) [9], but due to partial volume effects on CT images appears to be much thicker (more than 1 mm) [10]. Assessment of 3D vertebral cortical thickness and trabecular separation is substantially improved by high-resolution QCT (HRQCT) scan protocols (120 kVp, 350 mAs, a slice thickness of 0.3–0.5 mm, pitch = 1), particularly in nonaxial imaging planes (Fig. 31.2). Using gray-level based fuzzy distance transformation, the distance between trabeculae can be assessed with residual errors of less than 100 μm [11]. Differences in spatial resolution and reconstruction kernels of CT scanners represent a challenge for multicenter studies, and structural cross-calibration methods need to be developed.

Hip QCT. In recent years, QCT hip analysis software has been marketed by that is based on multislice data acquisition with 120 kVp, 100 mAs, and 3-mm slice thickness (Mindways Software Inc., Austin, TX, USA). The software allows both calculation of trabecular and cortical compartments in the femoral neck and trochanter. It has been used to study differences between cervical and trochanteric fractures [12] and for the assessment of treatment effects [13]. The software also provides fairly correct estimates of femoral aBMD results [computer tomography X-ray absorptiometry (CTXA)] calculated from their QCT data [14]. Accurate assessment of cortical thickness is important but challenging. Good results can be achieved by model-based methods [15]. This software is freely downloadable (http://mi.eng.cam.ac.uk/~rwp/stradwin/docs/thickness.htm). Other research software for flexible analysis of various VOIs within the proximal femur has been developed and used to assess changes in density and structure in disease or under treatment [16–18].

Peripheral QCT (pQCT) has mostly been carried out at the distal forearm and the distal tibia. Using a small-angle fan beam, typical single-slice pQCT X-ray tube settings are 45–60 kVp, 140–400 mAs, and at 1–3-mm slice thickness in plane pixel sizes of 100–300 μm can be

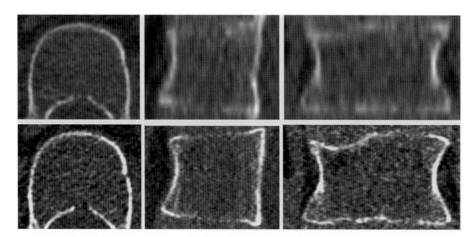

Fig. 31.2. Comparison of spatial resolution of QCT (top) and HRQCT (bottom) in the axial (left), sagittal (center), and coronal (right) planes. Protocols: QCT: 120 kVp, 100 mAs, $0.72 \times 0.72 \times 3$ mm^3, Kernel B, threshold 250 mg/ml. HRQCT: 120 kVp, 359 mAs, $0.16 \times 0.16 \times 0.33$ mm^3, Kernel D, threshold 250 mg/ml. Cortical thickness measured with QCT: 3.3 mm, for HRQCT: 1.9 mm; cortical density for QCT: 133 mg/ml, for HRQCT: 255 mg/ml. Microstructural assessment in 3D cannot be performed with QCT; the cortical shell can be displayed in 3D with HRQCT; for quantitative assessment, partial volume effects have to be considered.

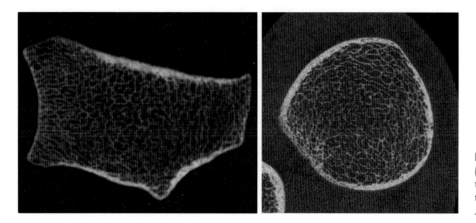

Fig. 31.3. HRpQCT of the distal forearm (left) or the tibia (right) allows visualization of trabecular microstructure and cortical porosity (courtesy Scanco Medical AG, Bruettisellen, Switzerland).

achieved. BMD is measured at the distal radius (4% or radius length, mostly trabecular) and at a more proximal location (15–65% radius length, cortical).

Recently, a **multislice high-resolution pQCT (HRpQCT)** device featuring an isotropic voxel size and slice thickness of 82 μm, sufficient to depict microstructural information, has been marketed (Fig 31.3). Measurements with a microfocus X-ray tube operated at 60 kVp, 0.9 mA are obtained at the distal radius and the distal tibia within a region of 9.02 mm length (110 slices). To calculate a variety of microstructural variables, the images are binarized. Variables include cortical thickness (derived from cortical area and perimeter) and trabecular number (Tb.N), directly determined from 3D distance transformation techniques. In its current implementation bone volume fraction (BV/TV) is NOT a microstructural variable but directly calculated from trabecular BMD, assuming a constant tissue mineral density of 1,200 mg Ca-HAP/ml. Since trabecular thickness and separation are derived from Tb.N and BV/TV, an impact of density should be noted. The precision of densitometric variables is high (less than 1%) and for most microstructural variables it is in the range of 4–5 % [19]. This method, for the first time, permits assessment of trabecular and cortical density and microstructure with high accuracy (modulation transfer function of 100 μm at 10% contrast) [20, 21]. In order to exploit this high level of spatial resolution, software for registration and microstructural characterization has been developed [22–24]; motion artifacts have to be avoided [25, 26].

Radiation Exposure. The level of desired image resolution, the location of the measurement region (proximity to radiation-sensitive organs), and the size of the measurement region determine the level of radiation exposure. By variation of the X-ray energy, radiation levels can be optimized, recognizing that the choice of the energy also affects accuracy. For earlier single-slice QCT approaches of T12-L2, low radiation levels down to 50–100 μSv have been reported [27]. For current multislice approaches the radiation exposure of QCT may vary substantially. Typical levels of effective dose are 1.5–2.3 mSv for QCT of L1-3 at 120 kVp, 100 mAs, and

3.3–3.7 mSv for HRQCT of T12 at 120 kVp, 360 mAs (values for men are slightly lower than those for women). For QCT of the hip at 120 kVp, 100 mAs the levels are 1.0–1.4 mSv for women and 2.1–3.0 mSv for men. The radiation exposure for pQCT measurements is small: about 3 µSv effective dose for radial HRpQCT [7]. Further improvements of spatial resolution are, however, restricted by high local skin radiation exposure. For comparison, typical levels of radiation exposure for DXA at the hip or spine are in the range of 0.5–5 µSv [28].

CLINICAL APPLICATION

Reference data

Pediatric spinal single-slice QCT reference data for L1-3 recently published [29] are presented in Fig. 31.4 (left). Adult reference ranges have been published for single-slice spinal QCT for a European [1] and a North American population [30] (and additional data have been published for a variety of other cohorts [31]). For these two studies QCT methods differed with regard to scanner, calibration standard, kVp, and slice thickness results depended on the model fit selected, but still BMD results deviated by less than one half of the population standard deviation (SD) [1]. Assuming that a dual-energy approach provides the most accurate data, volumetric BMD of a central slice remains fairly constant between ages 20 and 40 at about 140 mg/ml, and decreases to about 60 mg/ml at age 80, with a population SD of about 25 mg/ml [1]. Men have the same volumetric BMD at young ages, a similar population SD, and an only slightly slower decrease to a level of about 70 mg/ml at age 80 [1]. Single-energy approaches yield BMD values that are approximately 10–15% higher, depending on the kVp setting [Fig. 31.4 (right), adapted from Ref. 1]. Multislice spinal QCT BMD levels should be somewhat higher since the denser regions that are closer to the endplates are included in the volume evaluated. Some data on large cohorts have been published [32]. However, since there is no agreement on the placement of the VOIs, no standardized reference data are available yet.

For pQCT, reference ranges have been published for the single-slice approach [33]. For HRpQCT limited reference data have also been published [34]. In a multicenter setting, single and multislice approaches can be cross-calibrated using the European Forearm Phantom [35].

Diagnosis of BMD status in children

Z-scores should be used for diagnostic assessment of children. DXA as a projectional technique has many limitations for the assessment of bone status in children [36], most importantly the impact of bone size, but also variable geometry and changing tissue composition (trabecular vs cortical bone and red vs yellow marrow). QCT approaches may thus be of interest, but the radiation exposure is higher than for DXA. QCT is less affected by size, but it is more sensitive to bone marrow changes [5]. Unlike DXA, QCT allows differentiation between small bones and low BMD. If chronological age and bone age (e.g., according to Greulich-Peyl or Tanner-Whitehouse scores) differ, even QCT may not provide the complete answer, because it is not possible to distinguish low bone organ density from low tissue mineralization (e.g., in rickets). The stage of sexual development seems to be the strongest predictor of volumetric BMD [37]. Therefore, adjustment of the expected Z-score may have to be considered if pubertal status deviates from the norm.

Peripheral measurements have less of a radiation exposure problem. Moreover, pQCT of the forearm shows important relationships between bone growth and mechanical or hormonal factors impacting during adolescence. However, the many different variables generated from pQCT may confuse a clinical user. For single-slice pQCT, reference curves for trabecular or cortical bone at the 4% radius measurement site may be a good choice [38, 39]. Volumetric BMD does not change in girls and boys between ages 6 and 15 and later increases only for

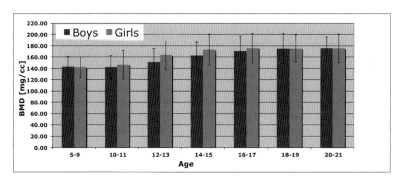 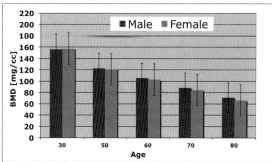

Fig. 31.4. Spinal trabecular BMD reference data measured by single-slice QCT as function of age. Published pediatric data (left, adapted from Ref. 29) were readjusted back to calibrated units of mg/ml K_2HPO_4 by applying a factor of .59, the ratio of bone density of 2.01 g/ml to mineral density of 1.183 mg/cm3 K_2HPO_4 (Ref. 109). Adult data (right) were adapted from the linear age-related fit from Ref. 1, which would closely match the pediatric data if extended to age 20). QCT was acquired in single energy mode at 80–85 kVp.

boys. However, pQCT results may also be biased: partial volume effects need to be taken in consideration, particularly at younger ages and for cortical measures. HRpQCT might be a better choice [40], but no pediatric reference data have been published to data. The techniques have recently be used to study a variety of disorders with detrimental effects on bone. For example, in children with chronic kidney disease, no significant difference compared to healthy controls was observed [41]. For both pQCT and HRpQCT, potential discordance of spinal, femoral, and peripheral BMD developments may have to be considered.

The indications for bone densitometry have been addressed in position papers by the International Society for Clinical Densitometry (ISCD) [42] and the British National Osteoporosis Society [36], but both are largely focused on DXA. QCT and pQCT can be used instead of DXA to examine with higher accuracy whether BMD deviates from age-specific reference data, but for spinal QCT, this comes at the expense of higher radiation exposure. To date, pediatric HRpQCT remains a tool for research.

Diagnosis of BMD status in adults

T-scores should be used for diagnostic assessment of adults. The diagnostic threshold given by the World Health Organization (WHO) of a T-score of −2.5 cannot be applied to QCT. Efforts have been made to define equivalent diagnostic thresholds for QCT, either based on equivalent fracture risk or equivalent fracture prevalence. However, QCT decreases faster with age than DXA. Therefore, diagnostic T-score thresholds for QCT would have to be lower than −2.5 and in order to be DXA equivalent they cannot be constant; they will decrease with age. For example, a level of 72 mg/ml for trabecular bone of the vertebrae [43], corresponding to a QCT T-score of approximately −3.2, may be DXA-T = −2.5-equivalent at age 50, but it is not low enough for older subjects. For QCT techniques using protocols that differ from those used for the above-mentioned published reference data, correction factors need to be applied. Additionally, for multislice approaches, the differences between the single and multislice regions need to be accounted for.

The correlations between spinal trabecular BMD by QCT (or spinal DXA-based BMD) and radial or tibial trabecular BMD by pQCT or HRpQCT are modest (r = 0.4 [44], r = 0.3 to r = 0.7 [45], respectively). Therefore, it is not possible to use pQCT or HRpQCT to accurately estimate BMD status at the spine. Because the diagnostic relevance of an abnormally low radial BMD is difficult to judge, low pQCT or HRpQCT results are not sufficient to diagnose osteoporosis.

Fracture risk assessment

QCT methods permit assessment of volumetric density and bone microstructure. Both are important determinants of bone strength [46, 47] and thus there is potential for improved fracture risk assessment [48, 49]. Bone density [44] and structure [50] vary considerably across different sites, and therefore the performance of imaging methods has to be assessed separately for each major fracture type. Results for standardized hazard ratios (sHR) and risk ratios (sRR) are expressed per standard deviation of the population variance.

Vertebral fractures

In men, spinal multislice QCT predicts incident clinical vertebral fractures significantly better than aBMD with areas under the curve (AUCs) for receiver operating characteristic (ROC) analyses of 0.83 vs. 0.76 (p < 0.05); age-adjusted sHRs were 5.7 (3.1, 10.3) compared to 3.2 (2.0, 5.2) [51]. For women, there is only one old prospective fracture study (in women) in which **spinal QCT** showed a predictive power similar to dual photon absorptiometry [52]. When compared to DXA in a various cross-sectional studies, single spinal QCT in general discriminated women with and without vertebral fractures at least as well as DXA [17, 44, 53–57], and in several of these studies it was significantly stronger. Recently, the stronger performance of QCT compared to DXA has also been reported for spinal multislice QCT evaluated cross-sectionally in women [58].

For **pQCT**, no prospective studies have been published on incidence of vertebral fractures. The cross-sectional studies (in women) for the single-slice approach are difficult to interpret because the authors usually tested a large number of pQCT variables, and thus it is unclear whether the observed association would hold in a retest in independent samples [44, 59–61]. For multislice pQCT, only data from one Japanese group on an older device are available, giving positive results for women [54].

For **HRpQCT**, two recent cross-sectional studies showed vertebral fracture discrimination (for densitometric and microstructural variables) in women independent of DXA for various structural variables [7, 62], but another study reported insignificant discrimination [58]. In men, a population-based study demonstrated significant age-adjusted discrimination of vertebral fractures for both density and microstructural variables, and cortical thickness contributed independently of aBMD. Two case control studies, one in men [63] and one in women [64], have reported significant association of HRpQCT variables with the *severity* of vertebral fractures; cortical thickness contributed independently of aBMD.

In summary, vertebral fracture risk prediction using QCT vBMD appears to be superior to DXA. HRpQCT measures at the radius or tibia also reflect fracture status, particularly in men and women with more severe fractures.

Hip fractures

The proximal femur is a site of substantial anatomic complexity and, therefore, 3D assessment of volumetric BMD and its distribution within the cancellous and cortical bone envelopes by **hip QCT** appears to be warranted. Beyond density, QCT permits evaluation of a variety of structural measures identified as potential predictors of hip fracture, including hip axis length [12, 65], acetabular bone width [66], cross-sectional area [12, 66], and neck-shaft angle [12, 67]. Density and structural measures can be combined to define biomechanically motivated strength indices [12, 68]; alternatively, the distribution of BMD can be used for finite element modeling (see below) or statistical parametric mapping, in which a model-based representation of the femur depicts fracture-critical subregions for cervical or trochanteric hip fractures [69]. Hip fracture prediction by hip QCT has been assessed prospectively in men [70]. Incident fractures were best predicted by integral BMD of the femoral neck (sHR = 3.6). Combining three independent predictors, i.e., trabecular BMD, percent cortical volume, and minimum cross-sectional area, resulted in the same predictive power as femoral neck aBMD measured with DXA (sHR = 4.1). Femoral neck structure was associated with fracture incidence independent of aBMD but QCT variables did not enhance predictive power above and beyond femoral neck aBMD alone. For QCT and DXA of the proximal femur hip fracture risk, increase was large and significant only in the lowest quartile of the respective index [70]. This pattern differs from that observed for DXA in women, which is more gradual [71, 72], and, if confirmed over longer follow-up time periods, perhaps permits a more selective case-finding strategy in men. The impact of muscle and fat tissue on hip fracture risk has also been studied prospectively. A decrease in thigh muscle density measured by QCT, a measure of fatty infiltration of muscle, was associated with an increase in hip fracture risk, even after adjusting for BMD (sRR = 1.5) [73]. For women, only cross-sectional hip QCT studies have been published indicating improved fracture discrimination for multivariate models that include variables reflecting both trabecular density and cortical thickness [74].

Spinal QCT discriminated trochanteric but not cervical hip fractures [75].

For **HRpQCT of the radius and tibia**, one cross-sectional study reported significant differences of three groups of women with hip fractures with wrist fractures and a control group without fractures. In the direct comparison of women with hip fractures and controls SD decrements of cortical HRpQCT measures looked favorable when compared with aBMD of the spine and the hip [45]. However, age adjustment was not performed and thus uncertainty remains.

In summary, aBMD of the proximal femur measured by DXA remains the most powerful and best tested predictor of hip fracture risk, but multivariate QCT measurements at this site provide similar predictive power. QCT-based evaluation of macrostructure and BMD distribution permits the determination of strength indices and finite element modeling, thereby providing more detailed insight into whole bone strength under varying loading conditions. To date HRpQCT data are too limited to judge the potential.

Mixed fracture groups

Fracture discrimination in general is similar for *peripheral, nonvertebral,* and *clinical* fractures. Among tomography-based approaches, prospective fracture risk for *nonvertebral* fractures to date has only been published for **pQCT** of the radius and the tibia [76]. For men, sHRs ranged from 1.4 to 1.6 at the tibia and from 1.6 to 2.2 at the distal third radius site, compared to 2.3 for aBMD of the femoral neck. The predictive power of aBMD of the femoral neck was significantly improved in a multivariate model by addition of pQCT-based strength indices (the best results: the AUCs of ROC analysis improved from 0.73 for aBMD to 0.80 when adding the cross-sectional moment of inertia of the radius).

For **HRpQCT**, trabecular density and microstructure of the radius and both trabecular and cortical density and microstructure of the tibia discriminated *peripheral* fractures in men, but there was no independent contribution of microstructural variables beyond density at these sites [63]. In women, *clinical* fractures could be significantly discriminated by a large range of density and microstructural variables despite lack of discrimination with aBMD of the spine or proximal femur [77]. For prediction of *all* fractures in women, very similar discrimination was reported for density and microstructural variables, and performance seemed to be as least as good as aBMD of the total hip; several density and microstructural variables contributed independently [62]. For men, two studies from the same cohort also reported discrimination of *all* fractures based on a large variety of density and structural variables but, unlike for women, only few remained significant after adjustment for aBMD [63,78].

Spinal QCT predicted non-spine fracture in black and white women and black men, but it was not a stronger predictor than total hip aBMD from DXA [79].

In summary, nonvertebral or clinical fractures risk can be assessed by pQCT and (at a lower level of evidence) also by HRpQCT, while prospective spine or hip QCT data are scant. At peripheral measurement sites the good performance is largely achieved by density or density distribution rather than microstructure. DXA-based aBMD remains the strongest and best tested predictor, but HRpQCT with stronger performance of trabecular measures at the radius and cortical measures at the tibia may add some independent information.

Clinical interpretation

For spinal single-slice QCT, the risk of fracture at a given T-score level is smaller than for DXA because a larger fraction of the T-score reflects age-associated risk, leaving less to BMD-associated risk. Therefore, for a given level of risk, the T-score of QCT has to be lower than the DXA T-score, and that difference increases with age. Given the limited performance of spinal QCT in predicting hip fracture and the increasing prevalence of hip fractures with advancing age, the differences between risk-equivalent T-scores of DXA and QCT increase further at older age. Evidence how to use QCT for treatment decisions is limited. A comparison of patient data with appropriate reference ranges (not yet available) would provide valuable insight. Most likely, intervention thresholds would have to be lower for QCT than for DXA if expressed as T-scores. From the perspective of evidence-based medicine, a valid DXA scan will in general still be preferable for making treatment decisions. For the future, the development of risk-based intervention criteria for QCT is indicated: Instead of using T-scores, a patient's risk could better be estimated by correcting the average age-associated risk by the QCT-associated risk derived from Z-scores and the technique's standardized gradient of risk [80]. However, this requires standardized methods, reference data, and more studies on QCT-associated fracture risk. These issues have been discussed in greater detail in the ISCD position statement on QCT [31].

Monitoring changes in bone status

A necessary requirement for monitoring is good longitudinal sensitivity (i.e., the ability of the technique to detect the changes early on). Longitudinal sensitivity is defined as the ratio of response rate divided by (long-term) precision errors [80]. Compared with DXA, precision errors of single-slice spinal CT approaches are somewhat higher, but for multislice approaches with automated placement of VOIs levels around 1%, i.e., comparable to spinal DXA, can be achieved [81]. For interpretation of changes in the individual patient, the least significant change [80], therefore, is similar to DXA. Response rates are about twice as large as for DXA for antiresorptive drugs [82] and three times as large during early osteoanabolic treatment [83, 84], and thus the monitoring time interval [80] is correspondingly shorter for QCT. As with DXA, it remains difficult to distinguish bone anabolic versus antiresorptive effects with QCT: increases in mineralization, reduction in porosity (number and size of Haversian canals), and apposition of low-mineral bone volume all result in increases in measured volumetric density. It is not clear yet to what extent QCT changes reflect the anti-fracture efficacy of drugs, but spinal BMD measured with QCT- or HRQCT-based improvements in finite element derived bone strength associated with treatment were larger than the DXA changes seen in the same patients, both for antiresorptive as well as for bone anabolic agents [84, 85].

QCT of the proximal femur has revealed differences in the specific effects bone anabolic and antiresorptive treatment on cortical and trabecular bone [86, 87]. Analysis of strength indices of specific compartments provided detailed insight into mechanism of bone anabolic action not achievable with DXA [13]. The impact of spaceflight on BMD has been investigated with QCT and a loss of 1–2.7% per month was reported [88].

The longitudinal sensitivity of **pQCT at the radius** for monitoring bisphosphonate effects is poorer than that of spinal DXA [89, 90] but treatment effects have been reported for trabecular BMD measured with pQCT [91]. Potentially, HRpQCT techniques may show improved performance. The good spatial resolution of HRpQCT permits differentiation of treatment-induced changes in the cortical and trabecular compartments. For example, in the comparison of denosumab and alendronate, comparable effects on trabecular bone but different effects on cortical bone, specifically cortical BMD and thickness, could be documented [92]. HRpQCT response to bisphosphonates in the tibia appear to be more pronounced than in the radius, driven by trabecular and endocortical changes [93]. Careful matching of the VOI is critical in longitudinal HRpQCT studies. The manufacturer-provided method, which is based on matching the outer perimeter of the bone, may yield misleading results, e.g., in the case of periosteal apposition or, more importantly, in pediatric studies. For such studies, the placement of the VOI at a fixed distance from the mid-joint line may also cause a problem since it disregards bone length. For the tibia, an 8% tibial length criterion might be a useful alternative [40].

There is growing interest in treatment effects on cortical thickness, porosity, and density. However, there are limitations to consider: The lumen of Haversian canals has an average smaller diameter of 70–80 μm [94]. Even if the Haversian cutting cone is larger with a diameter of 100–300 μm [95] and giant osteons with a diameter above 385 μm are more frequent in cases with hip fracture [96], given the 10% modulation transfer function (MTF) at 100-μm porosity, an HRpQCT-based measurement of porosity is substantially biased by partial volume effects. For the same reason, tissue mineral density (TMD) of thin cortices will be underestimated. The overall cortical treatment effects can be measured with HRpQCT but differentiation of effects on porosity, TMD, and cortical thickness is challenging. Assessment of other treatment-induced processes such as differentiation of endosteal apposition versus endosteal trabecularization, or endosteal versus periosteal apposition may be somewhat easier to achieve. For the assessment of treatment-induced microstructural changes in the trabecular compartment, directly measured Tb.N is a better measure than BV/TV since the latter, as currently implemented, only reflects BMD (see above). For HRpQCT-based assessment of the treatment effect of strontium ranelate [97], the incorporation of strontium into the bone has to be considered as an additional source of error. Beyond the overestimation of density, partial volume effects lead to biased structural measures, and thus treatment changes are difficult to interpret.

Finally, differences in the response profile of the measurement sites at the radius and tibia and the main fracture sites at the proximal femur and the vertebra need to be considered.

QCT-BASED FINITE ELEMENT MODELING

A major advantage of tomographic approaches is the feasibility to use the data as the basis for finite element (FE)

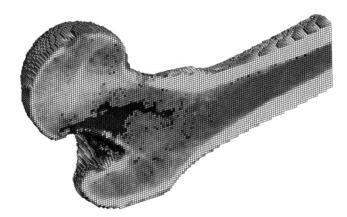

Fig. 31.5. Finite element modeling of the proximal femur (shown sectioned) under a sideways fall load. The areas of high plastic strain (red) are likely those where fractures occur first; grayscale values indicate bone mineral density. (Image provided courtesy of Tony Keaveny.)

analysis [98]. In voxel-based FE models, the bone to be modeled is segmented, and each voxel (or multi-voxel subvolume) of the segmented object is converted into a single element. These elements constitute the building blocks in a mesh that represents the entire the bone. The BMD of the mesh elements is converted into an elasticity tensor, assuming, e.g., orthotropic symmetry of bone tissue. Computer modeling then permits studying of the mechanical interaction of the elements under application of an external load. In this way the relatively simple mechanical properties of the elements can be computer-modeled to derive the complex mechanical properties of whole bones. The strains of the elements can be color-coded for visualization of the resulting strain distribution under the given load (Fig. 31.5). Linear modeling of the stress-strain relationship permits studying whole bone load-deformation processes to determine stiffness, while nonlinear modeling approaches permit to estimate whole bone strength, i.e., fracture loads.

In mechanical tests, the method has been shown to yield more accurate estimates of whole bone strength than DXA or QCT. FE analysis allows modeling of the mechanical competence of whole bones under specific loading conditions [48, 99–102]. It permits studying the respective contribution of cortical and cancellous bone compartments to whole bone strength [47], and it can also be used to identify weak regions that are most likely to fracture under loads [103]. FE analysis produces direct estimates of the forces that a specific bone can withstand. In men, spinal QCT-based FE-derived bone strength predicts incident vertebral fractures significantly better than aBMD with age-adjusted sHRs of 7.2 (3.6, 14.1) vs 3.2 (2.0, 5.2) [51] whereas for hip fractures, no improvement over DXA has yet been accomplished. In combination with estimates of the magnitude of impacting forces, it is possible to derive indices such as the "load-to-strength ratio" (also referred to as "factor or risk") [104];

at a value of greater than 1, the bone is expected to fracture. Whether this measure improves fracture risk assessment beyond FE-based estimates of bone strength remains controversial [51, 58]. Studies in recent years have shown that more detailed information about treatment effects can be obtained from FE modeling [84, 85, 105–107].

FE analysis has substantial potential and could perhaps be integrated as a surrogate marker of bone fragility. Patient-specific FE models could yield results on bone strength that are easier to interpret for clinicians and patients than T-scores. To date, however, only very few centers worldwide contribute to the development, and considerable efforts will be required before this tool will be applicable for clinical routine. Computing power is still a limiting factor, and a number of details of the modeling approach are controversial [98]. Moreover, accuracy errors affecting the underlying QCT data will also bias FE results, e.g., increases in measured BMD caused by strontium ranelate, changes in material properties in secondary osteoporosis; thus these results need to be interpreted with caution. For improved assessment of fracture risk and evaluation of treatment effects, FE models will benefit from further refinements:

- standardized assessment of different and more realistic loading configurations
- reduction of the size and adaptation of the shape of the mesh elements (current element sizes do not yet exploit the potential of HRQCT
- incorporation of the anisotropy of the trabecular network
- standardization of nonlinear modeling approaches
- benchmarking strength—density relationships
- treatment specific strength—density relationships
- addressing discrepancy of material properties across skeletal sites
- incorporation of the role of collagen and age

SUMMARY AND PERSPECTIVES

The renewed interest in QCT has two primary reasons: the deficiencies of DXA in monitoring treatment and the prospect that QCT may not only do better but also yield direct measures of strength (and, addressing the combination of these two aspects, variations in strength induced by treatment). Further refinement in imaging technology, i.e., improvement of spatial resolution (e.g., by use of flat-panel detectors [108]), improved image processing techniques, (e.g., automated placement of anatomically well-defined VOIs and gray-level-based parametric evaluation,) and progress on consensus about standardization (including imaging protocols, VOI definition, and microstructural cross-calibration procedures) may permit significant advances in the following areas:

- differentiation of changes in cortical porosity versus vs changes in TMD
- differentiation of bone anabolic versus antiresorptive effects
- differentiation of endosteal versus periosteal apposition

- improved predictive power, also for subjects already on treatment
- improved concordance of changes in BMD and treatment-induced reduction of fracture risk, i.e., better assessment of treatment efficacy in the individual patient
- incorporation of material properties

Already today, QCT is a valuable clinically applicable tool for the assessment of bone status [31]. However, prospective studies on risk assessment and clinical guidelines for appropriate use need to be expanded. In current clinical settings, a diagnostic QCT-based evaluation can only be carried out if the scanner, calibration method, and the protocol used are well characterized, and if results are compared with published reference data. Recent developments, however, document that the potential for further advances in QCT, specifically for clinical research, is very substantial.

ACKNOWLEDGMENTS

Valuable input by Judith Adams, Keenan Brown, Graeme Campbell, Christian Graeff, Tony Keaveny, Andres Laib, and Jaime Peña is gratefully acknowledged.

REFERENCES

1. Kalender WA, Felsenberg D, Louis O, Lopez P, Klotz E, Osteaux M, Fraga J. 1989. Reference values for trabecular and cortical vertebral bone density in single and dual-energy quantitative computed tomography. *Eur J Radiol* 9(2): 75–80.
2. Rüegsegger P, Elsasser U, Anliker M, Gnehm H, Kind H, Prader A. 1976. Quantification of bone mineralization using computed tomography. *Radiology* 121(1): 93–97.
3. Isherwood I, Rutherford RA, Pullan BR, Adams PH. 1976. Bone-mineral estimation by computer-assisted transverse axial tomography. *Lancet* 2(7988): 712–715.
4. Genant HK, Boyd D. 1977. Quantitative bone mineral analysis using dual energy computed tomography. *Invest Radiol* 12(6): 545–551.
5. Glüer CC, Genant HK. 1989. Impact of marrow fat on accuracy of quantitative CT. *J Comput Assist Tomogr* 13(6): 1023–1035.
6. Lang TF. 2010. Quantitative computed tomography. *Radiol Clin North Am* 48(3): 589–600.
7. Boutroy S, Bouxsein ML, Munoz F, Delmas PD. 2005. In vivo assessment of trabecular bone microarchitecture by high-resolution peripheral quantitative computed tomography. *J Clin Endocrinol Metab* 90(12): 6508–6515.
8. Cann CE. 1981. Low-dose CT scanning for quantitative spinal mineral analysis. *Radiology* 140(3): 813–815.
9. Vesterby A, Gundersen HJ, Melsen F, Mosekilde L. 1991. Marrow space star volume in the iliac crest decreases in osteoporotic patients after continuous treatment with fluoride, calcium, and vitamin D2 for five years. *Bone* 12(1): 33–37.
10. Silva MJ, Wang C, Keaveny TM, Hayes WC. 1994. Direct and computed tomography thickness measurements of the human, lumbar vertebral shell and endplate. *Bone* 15(4): 409–414.
11. Krebs A, Graeff C, Frieling I, Kurz B, Timm W, Engelke K, Glüer CC. 2009. High resolution computed tomography of the vertebrae yields accurate information on trabecular distances if processed by 3D fuzzy segmentation approaches. *Bone* 44(1): 145–152.
12. Ito M, Wakao N, Hida T, Matsui Y, Abe Y, Aoyagi K, Uetani M, Harada A. 2010. Analysis of hip geometry by clinical CT for the assessment of hip fracture risk in elderly Japanese women. *Bone* 46(2): 453–457.
13. Borggrefe J, Graeff C, Nickelsen TN, Marin F, Glüer CC. 2010. Quantitative computed tomographic assessment of the effects of 24 months of teriparatide treatment on 3D femoral neck bone distribution, geometry, and bone strength: Results from the eurofors study. *J Bone Miner Res* 25(3): 472–481.
14. Khoo BC, Brown K, Cann C, Zhu K, Henzell S, Low V, Gustafsson S, Price RI, Prince RL. 2009. Comparison of QCT-derived and DXA-derived areal bone mineral density and T scores. *Osteoporos Int* 20(9): 1539–1545.
15. Treece GM, Gee AH, Mayhew PM, Poole KE. 2010. High resolution cortical bone thickness measurement from clinical Ct data. *Med Image Anal* 14(3): 276–290.
16. Engelke K, Fuerst T, Dasic G, Davies RY, Genant HK. 2010. Regional distribution of spine and hip QCT BMD responses after one year of once-monthly ibandronate in postmenopausal osteoporosis. *Bone* 46(6): 1626–1632.
17. Lang TF, Guglielmi G, Van Kuijk C, De Serio A, Cammisa M, Genant HK. 2002. Measurement of bone mineral density at the spine and proximal femur by volumetric quantitative computed tomography and dual-energy X-ray absorptiometry in elderly women with and without vertebral fractures. *Bone* 30(1): 247–250.
18. Poole KE, Treece GM, Ridgway GR, Mayhew PM, Borggrefe J, Gee AH. 2011. Targeted regeneration of bone in the osteoporotic human femur. *Plos One* 6(1): E16190.
19. Krug R, Burghardt AJ, Majumdar S, Link TM. 2010. High-resolution imaging techniques for the assessment of osteoporosis. *Radiol Clin North Am* 48(3): 601–621.
20. Liu XS, Shane E, Mcmahon DJ, Guo XE. 2011. Individual trabecula segmentation (ITS)-based morphological analysis of microscale images of human tibial trabecular bone at limited spatial resolution. *J Bone Miner Res* 26(9): 2184–2193.
21. Sekhon K, Kazakia GJ, Burghardt AJ, Hermannsson B, Majumdar S. 2009. Accuracy of volumetric bone mineral density measurement in high-resolution peripheral quantitative computed tomography. *Bone* 45(3): 473–479.
22. Shi L, Wang D, Hung VW, Yeung BH, Griffith JF, Chu WC, Heng PA, Cheng JC, Qin L. 2010. Fast and accu-

23. Varga P, Zysset PK. 2009. Assessment of volume fraction and fabric in the distal radius using HR-pQCT. *Bone* 45(5): 909–917.
24. Varga P, Zysset PK. 2009. Sampling sphere orientation distribution: An efficient method to quantify trabecular bone fabric on grayscale images. *Med Image Anal* 13(3): 530–541.
25. Pialat JB, Burghardt AJ, Sode M, Link TM, Majumdar S. 2012. Visual grading of motion induced image degradation in high resolution peripheral computed tomography: Impact of image quality on measures of bone density and micro-architecture. *Bone* 50(1): 111–118.
26. Sode M, Burghardt AJ, Pialat JB, Link TM, Majumdar S. 2011. Quantitative characterization of subject motion in HR-pQCT images of the distal radius and tibia. *Bone* 48(6): 1291–1297.
27. Kalender WA. 1992. Effective dose values in bone mineral measurements by photon absorptiometry and computed tomography. *Osteoporos Int* 2(2): 82–87.
28. Blake GM, Wahner HW, Fogelman I. 1999. *The Evaluation of Osteoporosis*. London: Martin Dunitz.
29. Gilsanz V, Perez FJ, Campbell PP, Dorey FJ, Lee DC, Wren TA. 2009. Quantitative CT reference values for vertebral trabecular bone density in children and young adults. *Radiology* 250(1): 222–227.
30. Block JE, Smith R, Glueer CC, Steiger P, Ettinger B, Genant HK. 1989. Models of spinal trabecular bone loss as determined by quantitative computed tomography. *J Bone Miner Res* 4(2): 249–257.
31. Engelke K, Adams JE, Armbrecht G, Augat P, Bogado CE, Bouxsein ML, Felsenberg D, Ito M, Prevrhal S, Hans DB, Lewiecki EM. 2008. Clinical use of quantitative computed tomography and peripheral quantitative computed tomography in the management of osteoporosis in adults: the 2007 iscd official positions. *J Clin Densitom* 11(1): 123–162.
32. Sigurdsson G, Aspelund T, Chang M, Jonsdottir B, Sigurdsson S, Eiriksdottir G, Gudmundsson A, Harris TB, Gudnason V, Lang TF. 2006. Increasing sex difference in bone strength in old age: The Age, Gene/Environment Susceptibility-Reykjavik study (AGES-REYKJAVIK). *Bone* 39(3): 644–651.
33. Butz S, Wuster C, Scheidt-Nave C, Gotz M, Ziegler R. 1994. Forearm BMD as measured by peripheral quantitative computed tomography (pQCT) in a german reference population. *Osteoporos Int* 4(4): 179–184.
34. Dalzell N, Kaptoge S, Morris N, Berthier A, Koller B, Braak L, Van Rietbergen B, Reeve J. 2009. Bone microarchitecture and determinants of strength in the radius and tibia: age-related changes in a population-based study of normal adults measured with high-resolution pQCT. *Osteoporos Int* 20(10): 1683–1694.
35. Pearson J, Ruegsegger P, Dequeker J, Henley M, Bright J, Reeve J, Kalender W, Felsenberg D, Laval-Jeantet AM, Adams JE, et al. 1994. European semi-anthropomorphic phantom for the cross-calibration of peripheral bone densitometers: Assessment of precision accuracy and stability. *Bone Miner* 27(2): 109–120.
36. National Osteoporosis Society. 2004. *A Practical Guide to Bone Densitometry in Children*. Bath: National Osteoporosis Society.
37. Gilsanz V, Boechat MI, Roe TF, Loro ML, Sayre JW, Goodman WG. 1994. Gender differences in vertebral body sizes in children and adolescents. *Radiology* 190(3): 673–677.
38. Neu CM, Manz F, Rauch F, Merkel A, Schoenau E. 2001. Bone densities and bone size at the distal radius in healthy children and adolescents: A study using peripheral quantitative computed tomography. *Bone* 28(2): 227–232.
39. Rauch F, Schoenau E. 2005. Peripheral quantitative computed tomography of the distal radius in young subjects—New reference data and interpretation of results. *J Musculoskelet Neuronal Interact* 5(2): 119–126.
40. Burrows M, Liu D, Mckay H. 2010. High-resolution peripheral QCT imaging of bone micro-structure in adolescents. *Osteoporos Int* 21(3): 515–520.
41. Bacchetta J, Boutroy S, Vilayphiou N, Ranchin B, Fouque-Aubert A, Basmaison O, Cochat P. 2011. Bone assessment in children with chronic kidney disease: Data from two new bone imaging techniques in a single-center pilot study. *Pediatr Nephrol* 26(4): 587–595.
42. Rauch F, Plotkin H, Dimeglio L, Engelbert RH, Henderson RC, Munns C, Wenkert D, Zeitler P. 2008. Fracture prediction and the definition of osteoporosis in children and adolescents: The ISCD 2007 Pediatric Official Positions. *J Clin Densitom* 11(1): 22–28.
43. Lafferty FW, Rowland DY. 1996. Correlations of dual-energy X-ray absorptiometry, quantitative computed tomography, and single photon absorptiometry with spinal and non-spinal fractures. *Osteoporos Int* 6(5): 407–415.
44. Grampp S, Genant HK, Mathur A, Lang P, Jergas M, Takada M, Glüer CC, Lu Y, Chavez M. 1997. Comparisons of noninvasive bone mineral measurements in assessing age-related loss, fracture discrimination, and diagnostic classification. *J Bone Miner Res* 12(5): 697–711.
45. Vico L, Zouch M, Amirouche A, Frere D, Laroche N, Koller B, Laib A, Thomas T, Alexandre C. 2008. High-resolution pQCT analysis at the distal radius and tibia discriminates patients with recent wrist and femoral neck fractures. *J Bone Miner Res* 23(11): 1741–1750.
46. Wegrzyn J, Roux JP, Arlot ME, Boutroy S, Vilayphiou N, Guyen O, Delmas PD, Chapurlat R, Bouxsein ML. 2010. Role of trabecular microarchitecture and its heterogeneity parameters in the mechanical behavior of ex vivo human L3 vertebrae. *J Bone Miner Res* 25(11): 2324–2331.
47. Eswaran SK, Gupta A, Adams MF, Keaveny TM. 2006. Cortical and trabecular load sharing in the human vertebral body. *J Bone Miner Res* 21(2): 307–314.
48. Keaveny TM. 2010. Biomechanical computed tomography-noninvasive bone strength analysis using clinical computed tomography scans. *Ann N Y Acad Sci* 1192: 57–65.

49. Hansen S, Jensen JE, Ahrberg F, Hauge EM, Brixen K. 2011. The combination of structural parameters and areal bone mineral density improves relation to proximal femur strength: An in vitro study with high-resolution peripheral quantitative computed tomography. *Calcif Tissue Int* 89(4): 335–346.
50. Cohen A, Dempster DW, Müller R, Guo XE, Nickolas TL, Liu XS, Zhang XH, Wirth AJ, Van Lenthe GH, Kohler T, Mcmahon DJ, Zhou H, Rubin MR, Bilezikian JP, Lappe JM, Recker RR, Shane E. 2010. Assessment of trabecular and cortical architecture and mechanical competence of bone by high-resolution peripheral computed tomography: Comparison with transiliac bone biopsy. *Osteoporos Int* 21(2): 263–273.
51. Wang X, Sanyal A, Cawthon PM, Palermo L, Jekir M, Christensen J, Ensrud KE, Cummings SR, Orwoll E, Black DM, Keaveny TM. 2012. Prediction of new clinical vertebral fractures in elderly men using finite element analysis of CT scans. *J Bone Miner Res* 27(4): 808–816.
52. Ross PD, Genant HK, Davis JW, Miller PD, Wasnich RD. 1993. Predicting vertebral fracture incidence from prevalent fractures and bone density among non-black, osteoporotic women. *Osteoporos Int* 3(3): 120–126.
53. Yu W, Glüer CC, Grampp S, Jergas M, Fuerst T, Wu CY, Lu Y, Fan B, Genant HK. 1995. Spinal bone mineral assessment in postmenopausal women: A comparison between dual X-ray absorptiometry and quantitative computed tomography. *Osteoporos Int* 5(6): 433–439.
54. Tsurusaki K, Ito M, Hayashi K. 2000. Differential effects of menopause and metabolic disease on trabecular and cortical bone assessed by peripheral quantitative computed tomography (pQCT). *Br J Radiol* 73(865): 14–22.
55. Bergot C, Laval-Jeantet AM, Hutchinson K, Dautraix I, Caulin F, Genant HK. 2001. A comparison of spinal quantitative computed tomography with dual energy X-ray absorptiometry in european women with vertebral and nonvertebral fractures. *Calcif Tissue Int* 68(2): 74–82.
56. Guglielmi G, Cammisa M, De Serio A, Scillitani A, Chiodini I, Carnevale V, Fusilli S. 1999. Phalangeal US velocity discriminates between normal and vertebrally fractured subjects. *Eur Radiol* 9(8): 1632–1637.
57. Duboeuf F, Jergas M, Schott AM, Wu CY, Glüer CC, Genant HK. 1995. A comparison of bone densitometry measurements of the central skeleton in post-menopausal women with and without vertebral fracture. *Br J Radiol* 68(811): 747–753.
58. Melton LJ 3rd, Riggs BL, Keaveny TM, Achenbach SJ, Hoffmann PF, Camp JJ, Rouleau PA, Bouxsein ML, Amin S, Atkinson EJ, Robb RA, Khosla S. 2007. Structural determinants of vertebral fracture risk. *J Bone Miner Res* 22(12): 1885–1892.
59. Formica CA, Nieves JW, Cosman F, Garrett P, Lindsay R. 1998. Comparative assessment of bone mineral measurements using dual X-ray absorptiometry and peripheral quantitative computed tomography. *Osteoporos Int* 8(5): 460–467.
60. Clowes JA, Eastell R, Peel NF. 2005. The discriminative ability of peripheral and axial bone measurements to identify proximal femoral, vertebral, distal forearm and proximal humeral fractures: A case control study. *Osteoporos Int* 16(12): 1794–1802.
61. Grampp S, Lang P, Jergas M, Glüer CC, Mathur A, Engelke K, Genant HK. 1995. Assessment of the skeletal status by peripheral quantitative computed tomography of the forearm: Short-term precision in vivo and comparison to dual X-ray absorptiometry. *J Bone Miner Res* 10(10): 1566–1576.
62. Sornay-Rendu E, Boutroy S, Munoz F, Delmas PD. 2007. Alterations of cortical and trabecular architecture are associated with fractures in postmenopausal women, partially independent of decreased BMD measured by DXA: The OFELY study. *J Bone Miner Res* 22(3): 425–433.
63. Szulc P, Boutroy S, Vilayphiou N, Chaitou A, Delmas PD, Chapurlat R. 2011. Cross-sectional analysis of the association between fragility fractures and bone microarchitecture in older men: The STRAMBO study. *J Bone Miner Res* 26(6): 1358–1367.
64. Sornay-Rendu E, Cabrera-Bravo JL, Boutroy S, Munoz F, Delmas PD. 2009. Severity of vertebral fractures is associated with alterations of cortical architecture in postmenopausal women. *J Bone Miner Res* 24(4): 737–743.
65. Faulkner KG, Cummings SR, Black D, Palermo L, Glüer CC, Genant HK. 1993. Simple measurement of femoral geometry predicts hip fracture: The study of osteoporotic fractures. *J Bone Miner Res* 8(10): 1211–1217.
66. Glüer CC, Cummings SR, Pressman A, Li J, Glüer K, Faulkner KG, Grampp S, Genant HK. 1994. Prediction of hip fractures from pelvic radiographs: the study of osteoporotic fractures. The Study of Osteoporotic Fractures Research Group. *J Bone Miner Res* 9(5): 671–677.
67. Pulkkinen P, Partanen J, Jalovaara P, Jamsa T. 2004. Combination of bone mineral density and upper femur geometry improves the prediction of hip fracture. *Osteoporos Int* 15(4): 274–280.
68. Karlamangla AS, Barrett-Connor E, Young J, Greendale GA. 2004. Hip fracture risk assessment using composite indices of femoral neck strength: The Rancho Bernardo study. *Osteoporos Int* 15(1): 62–70.
69. Li W, Kornak J, Harris T, Keyak J, Li C, Lu Y, Cheng X, Lang T. 2009. Identify fracture-critical regions inside the proximal femur using statistical parametric mapping. *Bone* 44(4): 596–602.
70. Black DM, Bouxsein ML, Marshall LM, Cummings SR, Lang TF, Cauley JA, Ensrud KE, Nielson CM, Orwoll ES. 2008. Proximal femoral structure and the prediction of hip fracture in men: A large prospective study using QCT. *J Bone Miner Res* 23(8): 1326–1333.
71. Cummings SR, Black DM, Nevitt MC, Browner W, Cauley J, Ensrud K, Genant HK, Palermo L, Scott J, Vogt TM. 1993. Bone density at various sites for prediction of hip fractures. The Study of Osteoporotic Fractures Research Group. *Lancet* 341(8837): 72–75.

72. Schott AM, Cormier C, Hans D, Favier F, Hausherr E, Dargent-Molina P, Delmas PD, Ribot C, Sebert JL, Breart G, Meunier PJ. 1998. How hip and whole-body bone mineral density predict hip fracture in elderly women: The EPIDOS Prospective Study. *Osteoporos Int* 8(3): 247–254.
73. Lang T, Cauley JA, Tylavsky F, Bauer D, Cummings S, Harris TB. 2010. Computed tomographic measurements of thigh muscle cross-sectional area and attenuation coefficient predict hip fracture: The health, aging, and body composition study. *J Bone Miner Res* 25(3): 513–519.
74. Bousson VD, Adams J, Engelke K, Aout M, Cohen-Solal M, Bergot C, Haguenauer D, Goldberg D, Champion K, Aksouh R, Vicaut E, Laredo JD. 2011. In vivo discrimination of hip fracture with quantitative computed tomography: Results from the prospective European Femur Fracture Study (EFFECT). *J Bone Miner Res* 26(4): 881–893.
75. Lang TF, Augat P, Lane NE, Genant HK. 1998. Trochanteric hip fracture: strong association with spinal trabecular bone mineral density measured with quantitative CT. *Radiology* 209(2): 525–530.
76. Sheu Y, Zmuda JM, Boudreau RM, Petit MA, Ensrud KE, Bauer DC, Gordon CL, Orwoll ES, Cauley JA. 2011. Bone strength measured by peripheral quantitative computed tomography and the risk of nonvertebral fractures: the osteoporotic fractures in men (MrOS) study. *J Bone Miner Res* 26(1): 63–71.
77. Stein EM, Liu XS, Nickolas TL, Cohen A, Thomas V, Mcmahon DJ, Zhang C, Yin PT, Cosman F, Nieves J, Guo XE, Shane E. 2010. Abnormal microarchitecture and reduced stiffness at the radius and tibia in postmenopausal women with fractures. *J Bone Miner Res* 25(12): 2572–2581.
78. Vilayphiou N, Boutroy S, Szulc P, Van Rietbergen B, Munoz F, Delmas PD, Chapurlat R. 2011. Finite element analysis performed on radius and tibia HR-pQCT images and fragility fractures at all sites in men. *J Bone Miner Res* 26(5): 965–973.
79. Mackey DC, Eby JG, Harris F, Taaffe DR, Cauley JA, Tylavsky FA, Harris TB, Lang TF, Cummings SR; Health, Aging, And Body Composition Study Group. 2007. Prediction of clinical non-spine fractures in older black and white men and women with volumetric BMD of the spine and areal BMD of the hip: The Health, Aging, and Body Composition Study*. *J Bone Miner Res* 22(12): 1862–1868.
80. Glüer CC. 1999. Monitoring skeletal changes by radiological techniques. *J Bone Miner Res* 14(11): 1952–1962.
81. Engelke K, Mastmeyer A, Bousson V, Fuerst T, Laredo JD, Kalender WA. 2009. Reanalysis precision of 3D quantitative computed tomography (QCT) of the spine. *Bone* 44(4): 566–572.
82. Black DM, Greenspan SL, Ensrud KE, Palermo L, Mcgowan JA, Lang TF, Garnero P, Bouxsein ML, Bilezikian JP, Rosen CJ. 2003. The effects of parathyroid hormone and alendronate alone or in combination in postmenopausal osteoporosis. *N Engl J Med* 349(13): 1207–1215.
83. Graeff C, Timm W, Nickelsen TN, Farrerons J, Marin F, Barker C, Glüer CC. 2007. Monitoring teriparatide-associated changes in vertebral microstructure by high-resolution CT in vivo: Results from the EUROFORS study. *J Bone Miner Res* 22(9): 1426–1433.
84. Keaveny TM, Donley DW, Hoffmann PF, Mitlak BH, Glass EV, San Martin JA. 2007. Effects of teriparatide and alendronate on vertebral strength as assessed by finite element modeling of QCT scans in women with osteoporosis. *J Bone Miner Res* 22(1): 149–157.
85. Graeff C, Chevalier Y, Charlebois M, Varga P, Pahr D, Nickelsen TN, Morlock MM, Glüer CC, Zysset PK. 2009. Improvements in vertebral body strength under teriparatide treatment assessed in vivo by finite element analysis: Results from the EUROFORS study. *J Bone Miner Res* 24(10): 1672–1680.
86. Black DM, Greenspan SL, Ensrud KE, Palermo L, Mcgowan JA, Lang TF, Garnero P, Bouxsein ML, Bilezikian JP, Rosen CJ. 2003. The effects of parathyroid hormone and alendronate alone or in combination in postmenopausal osteoporosis. *N Engl J Med* 349(13): 1207–1215.
87. Mcclung MR, San Martin J, Miller PD, Civitelli R, Bandeira F, Omizo M, Donley DW, Dalsky GP, Eriksen EF. 2005. Opposite bone remodeling effects of teriparatide and alendronate in increasing bone mass. *Arch Intern Med* 165(15): 1762–1768.
88. Lang T, Leblanc A, Evans H, Lu Y, Genant H, Yu A. 2004. Cortical and trabecular bone mineral loss from the spine and hip in long-duration spaceflight. *J Bone Miner Res* 19(6): 1006–1012.
89. Qin L, Choy W, Au S, Fan M, Leung P. 2007. Alendronate increases BMD at appendicular and axial skeletons in patients with established osteoporosis. *J Orthop Surg Res* 2: 9.
90. Schneider PF, Fischer M, Allolio B, Felsenberg D, Schroder U, Semler J, Ittner JR. 1999. Alendronate increases bone density and bone strength at the distal radius in postmenopausal women. *J Bone Miner Res* 14(8): 1387–1393.
91. Sawada K, Morishige K, Nishio Y, Hayakawa J, Mabuchi S, Isobe A, Ogata S, Sakata M, Ohmichi M, Kimura T. 2009. Peripheral quantitative computed tomography is useful to monitor response to alendronate therapy in postmenopausal women. *J Bone Miner Metab* 27(2): 175–181.
92. Seeman E, Delmas PD, Hanley DA, Sellmeyer D, Cheung AM, Shane E, Kearns A, Thomas T, Boyd SK, Boutroy S, Bogado C, Majumdar S, Fan M, Libanati C, Zanchetta J. 2010. Microarchitectural deterioration of cortical and trabecular bone: Differing effects of denosumab and alendronate. *J Bone Miner Res* 25(8): 1886–1894.
93. Burghardt AJ, Kazakia GJ, Sode M, De Papp AE, Link TM, Majumdar S. 2010. A longitudinal HR-pQCT study of alendronate treatment in postmenopausal women with low bone density: Relations among

density, cortical and trabecular microarchitecture, biomechanics, and bone turnover. *J Bone Miner Res* 25(12): 2558–2571.
94. Currey JD. 1964. Some effects of ageing in human haversian systems. *J Anat* 98: 69–75.
95. Borah B, Dufresne T, Nurre J, Phipps R, Chmielewski P, Wagner L, Lundy M, Bouxsein M, Zebaze R, Seeman E. 2010. Risedronate reduces intracortical porosity in women with osteoporosis. *J Bone Miner Res* 25(1): 41–47.
96. Bell KL, Loveridge N, Power J, Garrahan N, Meggitt BF, Reeve J. 1999. Regional differences in cortical porosity in the fractured femoral neck. *Bone* 24(1): 57–64.
97. Rizzoli R, Chapurlat RD, Laroche JM, Krieg MA, Thomas T, Frieling I, Boutroy S, Laib A, Bock O, Felsenberg D. 2011. Effects of strontium ranelate and alendronate on bone microstructure in women with osteoporosis: Results of a 2-year study. *Osteoporos Int* 23(1): 305–315.
98. Christen D, Webster DJ, Müller R. 2010. Multiscale modelling and nonlinear finite element analysis as clinical tools for the assessment of fracture risk. *Philos Transact A Math Phys Eng Sci* 368(1920): 2653–2668.
99. Van Rietbergen B. 2001. Micro-FE analyses of bone: State of the art. *Adv Exp Med Biol* 496: 21–30.
100. Chevalier Y, Pahr D, Allmer H, Charlebois M, Zysset P. 2007. Validation of a voxel-based FE method for prediction of the uniaxial apparent modulus of human trabecular bone using macroscopic mechanical tests and nanoindentation. *J Biomech* 40(15): 3333–3340.
101. Crawford RP, Cann CE, Keaveny TM. 2003. Finite element models predict in vitro vertebral body compressive strength better than quantitative computed tomography. *Bone* 33(4): 744–750.
102. Keyak JH, Skinner HB, Fleming JA. 2001. Effect of force direction on femoral fracture load for two types of loading conditions. *J Orthop Res* 19(4): 539–544.
103. Keyak JH, Rossi SA, Jones KA, Les CM, Skinner HB. 2001. Prediction of fracture location in the proximal femur using finite element models. *Med Eng Phys* 23(9): 657–664.
104. Bouxsein ML, Melton LJ 3rd, Riggs BL, Muller J, Atkinson EJ, Oberg AL, Robb RA, Camp JJ, Rouleau PA, Mccollough CH, Khosla S. 2006. Age- and sex-specific differences in the factor of risk for vertebral fracture: A population-based study using qct. *J Bone Miner Res* 21(9): 1475–1482.
105. Keaveny TM, Hoffmann PF, Singh M, Palermo L, Bilezikian JP, Greenspan SL, Black DM. 2008. Femoral bone strength and its relation to cortical and trabecular changes after treatment with PTH, alendronate, and their combination as assessed by finite element analysis of quantitative CT scans. *J Bone Miner Res* 23(12): 1974–1982.
106. Chevalier Y, Quek E, Borah B, Gross G, Stewart J, Lang T, Zysset P. 2010. Biomechanical effects of teriparatide in women with osteoporosis treated previously with alendronate and risedronate: Results from quantitative computed tomography-based finite element analysis of the vertebral body. *Bone* 46(1): 41–48.
107. Keaveny TM, Mcclung MR, Wan X, Kopperdahl DL, Mitlak BH, Krohn K. 2011. Femoral strength in osteoporotic women treated with teriparatide or alendronate. *Bone*.
108. Gupta R, Cheung AC, Bartling SH, Lisauskas J, Grasruck M, Leidecker C, Schmidt B, Flohr T, Brady TJ. 2008. flat-panel volume CT: Fundamental principles, technology, and applications. *Radiographics* 28(7): 2009–2022.
109. Gilsanz V. 2011. Personal communication.

32

Magnetic Resonance Imaging of Bone

Sharmila Majumdar

Introduction 277
Magnetic Resonance Imaging of Trabecular Bone 277
Magnetic Resonance Imaging of Cortical Bone 279
Relationship of MR-Derived Structure Measures to Bone Strength, Fracture, Osteoporotic Status, and Response to Therapy 279

Applications Using MR-Derived Cortical Bone Measures 280
Summary 280
References 280

INTRODUCTION

Three-dimensional (3D) imaging techniques that reveal bone structure are emerging as important contenders for defining bone quality, at least partially. Techniques such as microcomputed tomography (μCT) have recently been developed and provide high-resolution images of the trabecular and cortical bone micro- and macroarchitecture. This method is routinely used in specimen evaluation and has recently been extended to *in vivo* animal and human extremity imaging. Magnetic resonance (MR) imaging has been used in addition to X-rays, computed tomography, and bone scintigraphy to detect insufficiency fractures, stress fractures, and metastatic fractures. The range of soft tissue contrast possible with different MR techniques has led to the development of techniques that differentiate osteoporotic fractures from metastatic disease, and others that identify stress fractures and related bone marrow changes. However, a more recent development in MR is in the assessment of trabecular bone and cortical bone structure that makes it possible to obtain noninvasive bone biopsies at multiple anatomic sites.

MAGNETIC RESONANCE IMAGING OF TRABECULAR BONE

Trabecular bone consists of a network of rod-like elements interconnected by plate-like elements, immersed in bone marrow composed partly of water and partly of fat. The magnetic susceptibility of trabecular bone is substantially different from that of bone marrow. This gives rise to susceptibility gradients at the bone–marrow interface. Magnetic inhomogeneity arising from these susceptibility gradients depends on the static magnetic field strength, number of bone–bone marrow interfaces, and the size of individual trabeculae [1–3]. These effects cause a rapid signal decay at a rate known as $T2^*$. In a voxel partly occupied by bone and partly by marrow, the static inhomogeneity induced effect leads to signal cancellation within the voxel. $T2^*$ methods have been used to quantify trabecular bone, and these measures have been related to bone strength, osteoporotic status, and therapeutic response [4].

Beside the tissue composition, the small dimensions of the trabecular elements (approximately 100 μm) require

Primer on the Metabolic Bone Diseases and Disorders of Mineral Metabolism, Eighth Edition. Edited by Clifford J. Rosen.
© 2013 American Society for Bone and Mineral Research. Published 2013 by John Wiley & Sons, Inc.

very high imaging resolutions. The suitability of an MR imaging method (acquisition and analysis) for depicting bone microstructure depends on its ability to yield images with a high enough signal in a reasonable acquisition time and its ability to derive trabecular structural measurements accurately and reproducibly from the images. The three competing factors to be considered in high-resolution MR imaging (HR-MRI) are signal-to-noise ratio, spatial resolution, and imaging time. Spatial resolution and signal-to-noise ratio (SNR) are both directly related to imaging time but are inversely related to each other. Recent technique developments in trabecular bone MRI technique reflect all these considerations, and the goal has been to increase SNR and accelerate total acquisition times.

By adjusting various times associated with switching on radio-frequency pulses, the duration of radio-frequency pulses that govern a parameter called flip angle in MR acquisitions, images with varying contrast and signal-to-noise features can be obtained. MR pulse sequences can be broadly classified into spin echo and gradient echo sequences. Ideally, 3D spin echo (SE) sequences are better suited for imaging of trabecular bone microarchitecture than gradient echo (GE)-based sequences because they are less sensitive to the thickening of trabeculae due to susceptibility differences. However, GE sequences can be employed with short repetition time (TR) because of their higher SNR efficiency and can thus acquire a 3D volume in a shorter scan time and avoid patient motion artifacts [5, 6]. 3D-SE type pulse sequences with a variable flip angle such as rapid SE excitation (RASEE) [7, 8], large-angle spin-echo imaging [9], and, subsequently, fast 3D large-angle spin-echo imaging (FLASE) [10], and a new fully balanced steady state 3D-spin-echo pulse sequence have also been developed [11]. The choice of pulse sequence for trabecular bone imaging is still a topic of active research. The availability of the sequences at multiple centers, their robustness, and total imaging time versus the anatomical coverage are typical considerations.

SNR is linearly proportional to the static magnetic field strength, perhaps making 3 T preferable over 1.5 T magnets. Phan et al. imaged the trabecular microarchitecture in 40 cadaveric calcaneus specimens at 1.5 T and 3 T, compared it with μCT as gold standard [12] and found that correlations between trabecular structural parameters derived from 3 T MR images and μCT were significantly higher ($p < 0.05$) than correlations between structural parameters obtained from 1.5 T MR imaging and μCT. Preliminary experiments conducted on a 7-T GE Signa scanner yielded a twofold increase in SNR for HR-MRI of trabecular bone [13, 14]. Figure 32.1 shows an example of an HR-MR image obtained at 3 T, in the radius, calcaneus, tibia, and femur, where the bone marrow is bright and trabecular bone is depicted as dark striations. Images such as these can be analyzed to derive structural measures of microarchitecture.

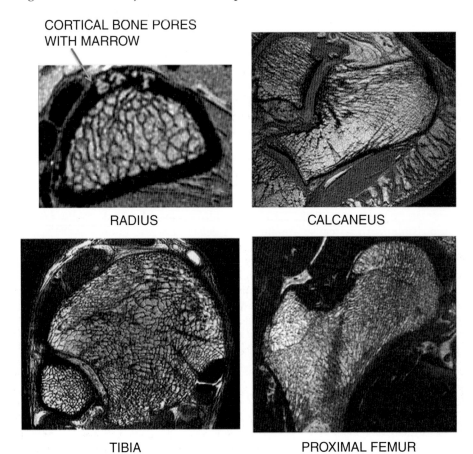

Fig. 32.1. Representative MR images showing trabecular bone in the radius, calcaneus, tibia, and femur. Cortical bone in the radius shows evidence of marrow-filled cortical porosity. Images were acquired at 3 T using a General Electric Signa scanner.

The most common structural measures analogous to quantitative histomorphometry derived from MR images include app.BV/TV, app.Tb.N, app.Tb.Sp, and app. Tb.Th [15, 16], and require the images to be subdivided into a bone and marrow component or binarized. Since the MR images are not acquired at true microscopic resolutions, Majumdar et al. [17] described these measures derived from MR images as "apparent" measurements, which, although obtained in the limited-resolution regime, is highly correlated to the "true" structure. Binarization of an MR image is not a trivial task, and mainly because of partial volume effects, multiple techniques have been developed that operate directly on the grayscale image. Recognizing the fuzzy nature of the images due to partial volume effects, Saha et al. [18] applied a fuzzy distance transform (FDT) technique, a method that obviates the need for a binary assignment of bone and marrow, for computing trabecular thickness and observed an improved robustness in the computation against loss of resolution. Mathematical formulations such as digital topological analysis techniques have also been applied to quantify the number of surface and curved edges, junctions and, interiors in the trabecular network [19].

MAGNETIC RESONANCE IMAGING OF CORTICAL BONE

MR imaging has been used to image cortical bone as well. Specifically for the proximal femur, the ability of MRI to align the image plane perpendicular to the femoral neck is a great advantage and enables more accurate acquisition of the cortical architecture [20]. MRI allows the visualization of soft tissues such as bone marrow and thus a quantification of the amount of cortical porosity that contains bone marrow.

Using advanced MRI methods with ultrashort echo times (UTE), the bone water content in the microscopic pores of the haversian and the lacuno-canalicular systems of cortical bone can be quantified. A smaller water fraction is also bound to collagen and the matrix substrate and imbedded in the crystal structure of the mineral [21]. These micropores usually have a very small size on the order of a few micrometers and are thus difficult to visualize, but the quantification of bone water using MRI could potentially provide a surrogate measure of bone porosity without resolving these individual small pores.

RELATIONSHIP OF MR-DERIVED STRUCTURE MEASURES TO BONE STRENGTH, FRACTURE, OSTEOPOROTIC STATUS, AND RESPONSE TO THERAPY

Several studies relating the measures of trabecular structure obtained using MR imaging to measures of bone strength *in vitro* have been conducted [17, 22–24]. The relationship between whole bone strength and bone structure measures have been demonstrated in radii and in the proximal femur [25, 26].

High-resolution MR images of the distal radius were obtained at 1.5 T in premenopausal normal, postmenopausal normal, and postmenopausal osteoporotic women [27]. Significant differences were evident in spinal bone mineral density (BMD), radial trabecular BMD, trabecular bone volume fraction, trabecular spacing, and trabecular number between the postmenopausal non-fracture and the postmenopausal osteoporotic subjects. Trabecular spacing and trabecular number showed moderate correlation with radial trabecular BMD but correlated poorly with radial cortical BMC.

Distance transformation techniques were applied to the 3D image of the distal radius of postmenopausal patients and structural indices such as app.Tb.N, app.Tb.Th, and app.Tb.Sp were determined without model assumptions [28]. A new metric index, the apparent intraindividual distribution of separations (app.Tb.Sp.SD), was introduced. It was found that app.Tb.Sp.SD discriminates fracture subjects from non-fracture patients as well as dual energy X-ray absorptiometry (DXA) measurements of the radius and the spine, but not as well as DXA of the hip. MR-derived measures of trabecular bone architecture in the distal radius [29] and calcaneus [30] were obtained in 20 subjects with hip fractures and 19 age-matched postmenopausal controls, in addition to BMD measures at the hip DXA and the distal radius [peripheral quantitative computed tomography (pQCT)]. Measures of app.Tb.Sp and app.Tb.N in the distal radius showed significant ($p < 0.05$) differences between the two groups, as did hip BMD measures. However, radial trabecular BMD measures showed only a marginal difference ($p = 0.05$). In the calcaneus, significant differences between both patient groups were obtained using morphological parameters, with odds ratios being slightly higher than BMD [30].

Sagittal MR images of the calcaneus were obtained in 50 men (26 patients with osteoporosis and 24 age-matched healthy control subjects) [31]. Structural parameters, especially connectivity parameters, showed significant differences between control subjects and patients ($p < .05$).

In addition, *in vivo* images have also been combined with micro-finite element analysis in a limited set of subjects. Newitt et al. [32] studied subjects in two groups: postmenopausal women with normal BMD (n = 22, mean age 58 ± 7 years) and postmenopausal women with spine or femur BMD −1 SD to −2.5 SD below young normal(n = 37, mean age 62 ± 11 years). Anisotropy of trabecular bone microarchitecture, as measured by the ratios of the mean intercept length (MIL) values (MIL1/MIL3, etc.), and the anisotropy in elastic modulus (E1/E3, etc.), were greater in the osteopenic group. To date, all studies exploring the role of MRI in predicting fracture risk are based on cross-sectional studies. To establish the role of MR measures for the prediction of fracture risk, it may be necessary to conduct prospective fracture trials.

The role of MR for assessing therapeutic effects has also been explored. Ninety-one postmenopausal osteoporotic women were followed for 2 years (n = 46 for nasal spray calcitonin; n = 45 for placebo) [33]. MRI measurements of trabecular structure were obtained at the distal radius and calcaneus in addition to DXA–BMD at the spine/hip/wrist/calcaneus (obtained yearly). MRI assessment of the trabecular microarchitecture at individual regions of the distal radius revealed preservation (no significant loss), in the treated group compared with significant deterioration in the placebo control group.

Trabecular bone structure of the tibia was studied in 10 men with severe, untreated hypogonadism and age- and race-matched eugonadal men. Two composite topological indices were determined: the ratio of surface voxels (representing plates) to curve voxels (representing rods), which is higher when architecture is more intact; and the erosion index, a ratio of parameters expected to increase upon architectural deterioration to those expected to decrease, which is higher when deterioration is greater. The surface/curve ratio was 36% lower (p = 0.004), and the erosion index was 36% higher (p = 0.003) in the hypogonadal men than in the eugonadal men [34]. In contrast, bone mineral density of the spine and hip were not significantly different between the two groups. After 24 months of testosterone treatment, BMD of the spine increased 7.4% (p < 0.001), and BMD of the total hip increased 3.8% (p = 0.008). Architectural parameters assessed by MRI also changed: The surface-to-curve ratio increased 11% (p = 0.004) and the topological erosion index decreased 7.5% (p = 0.004) [35].

Until recently, *in vivo* MRI of trabecular microarchitecture was limited to peripheral sites such as distal tibia and femur, radius, and calcaneus because of SNR limitations. However, the main sites of osteoporotic fractures are nonperipheral regions such as the vertebral bodies (spine) and the proximal femur (hip). HR-MRI has only recently been applied to the proximal femur [36] by using SNR efficient sequences, high magnetic field strength (3 T), and phased array coils.

APPLICATIONS USING MR-DERIVED CORTICAL BONE MEASURES

Although significant work has been done using MRI to measure trabecular bone structure, there is relatively little work on the macroarchitectural geometry of the cortex, which may play an equally important role for bone strength. Recently, Gomberg [20] investigated cortical shell geometry of the femur, further expanding the potential role of MR in characterizing bone. In a recent study, images of the distal radius and the distal tibia of 49 postmenopausal osteopenic women (age 56 ± 3.7) were acquired with both HR-pQCT and MRI [37]. It was found that the amount of cortical porosity did not vary greatly between subjects but the type of cortical pore containing marrow versus not containing marrow varied highly between subjects.

Bone water quantification measurements have been previously conducted in sheep and human cadaveric specimens, and the method's sensitivity to distinguish subjects of different ages and disease states has been evaluated [38]. The data were compared with areal and volumetric BMD from DXA and pQCT, respectively. The bone water content was calibrated with the aid of an external reference (10% H_2O in D_2O doped with 27 mmol/L $MnCl2$), which was attached anteriorly to the subject's tibial midshaft. Excellent agreement ($R^2 = 0.99$) was found in the specimen between the water displaced by using D_2O exchange and water measured with respect to the reference sample. Measurements *in vivo* revealed that the bone water content was increased 65% in the postmenopausal group compared to the premenopausal group [39]. Patients with renal osteodystrophy had 135% higher bone water content than the premenopausal group whereas conventional BMD measurements showed an opposite behavior, with much smaller group differences.

SUMMARY

Imaging trabecular and cortical microarchitecture and characterizing the features of trabecular and cortical bone has been an area of fertile and ongoing research. Beyond relating microarchitecture to the biomechanical properties of bone in specimens, advances have been made to extend these measures *in vivo* in human subjects. In this context, the relationship between, age, fracture status, and even post-therapeutic response has been studied. New advances in peripheral CT, MR (non-ionizing, peripheral sites, calcaneus, and femur) are ongoing and evolving at a rapid pace, and with the establishment of robust analysis methodologies and normative databases, they have the potential for further clinical utilization in the coming years.

REFERENCES

1. Majumdar S, Thomasson D, Shimakawa A, Genant HK. 1999. Quantitation of the susceptibility difference between trabecular bone and bone marrow: Experimental studies. *Magn Reson Med* 22(1): 111–27.
2. Weisskoff RM, Zuo CS, Boxerman JL, Rosen BR. 1994. Microscopic susceptibility variation and transverse relaxation: Theory and experiment. *Magn Reson Med* 31(6): 601–10.
3. Ford JC, Wehrli FW, Chung HW. 1993. Magnetic field distribution in models of trabecular bone. *Magn Reson Med* 30(3): 373–9.
4. Link TM, Majumdar S, Augat P, Lin JC, Newitt D, Lane NE, Genant HK. 1998. Proximal femur: Assessment for

osteoporosis with T2* decay characteristics at MR imaging. *Radiology* 209(2): 531–6.

5. Majumdar S, Link TM, Augat P, Lin JC, Newitt D, Lane NE, Genant HK. 1999. Trabecular bone architecture in the distal radius using magnetic resonance imaging in subjects with fractures of the proximal femur. Magnetic Resonance Science Center and Osteoporosis and Arthritis Research Group. *Osteoporos Int* 1999;10(3): 231–9.

6. Newitt DC, Van Rietbergen B, Majumdar S. 2002. Processing and analysis of in vivo high-resolution MR images of trabecular bone for longitudinal studies: Reproducibility of structural measures and micro-finite element analysis derived mechanical properties. *Osteoporos Int* 13: 278–87.

7. Jara H, Wehrli FW, Chung H, Ford JC. 1993. High-resolution variable flip angle 3D MR imaging of trabecular microstructure in vivo. *Magn Reson Med* 29(4): 528–39.

8. Bogdan AR, Joseph PM. 1990. RASEE: A rapid spin-echo pulse sequence. *Magn Reson Imaging* 8(1): 13–9.

9. DiIorio G, Brown JJ, Borrello JA, Perman WH, Shu HH. 1995. Large angle spin-echo imaging. *Magn Reson Imaging* 13(1): 39–44.

10. Ma J, Wehrli FW, Song HK. 1996. Fast 3D large-angle spin-echo imaging 3D FLASE. *Magn Reson Med* 35(6): 903–10.

11. Krug R, Han ET, Banerjee S, Majumdar S. 2006. Fully balanced steady-state 3D-spin-echo (bSSSE) imaging at 3 Tesla. *Magn Reson Med* 56(5): 1033–40.

12. Phan CM, Matsuura M, Bauer JS, Dunn TC, Newitt D, Lochmueller EM, Eckstein F, Majumdar S, Link TM. 2006. Trabecular bone structure of the calcaneus: Comparison of MR imaging at 3.0 and 1.5 T with micro-CT as the standard of reference. *Radiology* 239(2): 488–96.

13. Zuo J, Bolbos R, Hammond K, Li X, Majumdar S. 2008. Reproducibility of the quantitative assessment of cartilage morphology and trabecular bone structure with magnetic resonance imaging at 7 T. *Magn Reson Imaging* 26(4): 560–6.

14. Krug R, Carballido-Gamio J, Banerjee S, Stahl R, Carvajal L, Xu D, Vigneron D, Kelley DA, Link TM, Majumdar S. 2007. In vivo bone and cartilage MRI using fully-balanced steady-state free-precession at 7 tesla. *Magn Reson Med* 58(6): 1294–8.

15. Parfitt AM, Mathews CH, Villanueva AR, Kleerekoper M, Frame B, Rao DS. 1983. Relationships between surface, volume, and thickness of iliac trabecular bone in aging and in osteoporosis. Implications for the micro-anatomic and cellular mechanisms of bone loss. *J Clin Invest* 72(4): 1396–409.

16. Parfitt AM. Assessment of trabecular bone status. 1983. *Henry Ford Hosp Med J* 31(4): 196–8.

17. Majumdar S, Newitt D, Mathur A, Osman D, Gies A, Chiu E, Lotz J, Kinney J, Genant H. 1996. Magnetic resonance imaging of trabecular bone structure in the distal radius: Relationship with X-ray tomographic microscopy and biomechanics. *Osteoporos Int* 6(5): 376–85.

18. Saha PK, Wehrli FW. 2004. Measurement of trabecular bone thickness in the limited resolution regime of in vivo MRI by fuzzy distance transform. *IEEE Trans Med Imaging* 23(1): 53–62.

19. Gomberg BR, Saha PK, Song HK, Hwang SN, Wehrli FW. 2000. Topological analysis of trabecular bone MR images. *IEEE Trans Med Imaging* 19(3): 166–74.

20. Gomberg BR, Saha PK, Wehrli FW. 2005. Method for cortical bone structural analysis from magnetic resonance images. *Acad Radiol* 12(10): 1320–32.

21. Timmins PA, Wall JC. 1977. Bone water. *Calcif Tissue Res* 23(1): 1–5.

22. Hwang SN, Wehrli FW, Williams JL. 1997. Probability-based structural parameters from three-dimensional nuclear magnetic resonance images as predictors of trabecular bone strength. *Med Phys* 24(8): 1255–61.

23. Pothuaud L, Laib A, Levitz P, Benhamou CL, Majumdar S. 2002. Three-dimensional-line skeleton graph analysis of high-resolution magnetic resonance images: A validation study from 34-microm-resolution microcomputed tomography. *J Bone Miner Res* 17(10): 1883–95.

24. Majumdar S, Kothari M, Augat P, et al. 1998. High-resolution magnetic resonance imaging: Three-dimensional trabecular bone architecture and biomechanical properties. *Bone* 22: 445–54.

25. Ammann P, Rizzoli R. 2003. Bone strength and its determinants. *Osteoporos Int* 14 Suppl 3: S13–8.

26. Link TM, Bauer J, Kollstedt A, Stumpf I, Hudelmaier M, Settles M, Majumdar S, Lochmuller EM, Eckstein F. 2004. Trabecular bone structure of the distal radius, the calcaneus, and the spine: Which site predicts fracture status of the spine best? *Invest Radiol* 39(8): 487–97.

27. Majumdar S, Genant H, Grampp S, Newitt D, Truong V, Lin J, Mathur A. 1997. Correlation of trabecular bone structure with age, bone mineral density and osteoporotic status: In vivo studies in the distal radius using high resolution magnetic resonance imaging. *J Bone Miner Res* 12: 111–8.

28. Laib A, Newitt DC, Lu Y, Majumdar S. 2002. New model-independent measures of trabecular bone structure applied to in vivo high-resolution MR images. *Osteoporos Int* 13(2): 130–6.

29. Majumdar S, Link T, Augat P, et al. 1999. Trabecular bone architecture in the distal radius using MR imaging in subjects with fractures of the proximal femur. *Osteoporos Int* 10: 231–9.

30. Link TM, Majumdar S, Augat P, Lin JC, Newitt D, Lu Y, Lane NE, Genant HK. 1998. In vivo high resolution MRI of the calcaneus: Differences in trabecular structure in osteoporosis patients. *J Bone Miner Res* 13(7): 1175–82.

31. Boutry N, Cortet B, Dubois P, Marchandise X, Cotten A. 2003. Trabecular bone structure of the calcaneus: Preliminary in vivo MR imaging assessment in men with osteoporosis. *Radiology* 227(3): 708–17.

32. Newitt DC, Majumdar S, van Rietbergen B, von Ingersleben G, Harris ST, Genant HK, Chesnut C, Garnero P, MacDonald B. 2002. In vivo assessment of architecture and micro-finite element analysis derived indices of mechanical properties of trabecular bone in the radius. *Osteoporos Int* 13(1): 6–17.

33. Chesnut CH 3rd, Majumdar S, Newitt DC, Shields A, Van Pelt J, Laschansky E, Azria M, Kriegman A, Olson M, Eriksen EF, Mindeholm L. 2005. Effects of salmon calcitonin on trabecular microarchitecture as determined by magnetic resonance imaging: Results from the QUEST study. *J Bone Miner Res* 20(9): 1548–61.
34. Benito M, Gomberg B, Wehrli FW, Weening RH, Zemel B, Wright AC, Song HK, Cucchiara A, Snyder PJ. 2003. Deterioration of trabecular architecture in hypogonadal men. *J Clin Endocrinol Metab* 88(4): 1497–502.
35. Benito M, Vasilic B, Wehrli FW, Bunker B, Wald M, Gomberg B, Wright AC, Zemel B, Cucchiara A, Snyder PJ. 2005. Effect of testosterone replacement on trabecular architecture in hypogonadal men. *J Bone Miner Res* 20(10): 1785–91.
36. Krug R, Banerjee S, Han ET, Newitt DC, Link TM, Majumdar S. 2005. Feasibility of in vivo structural analysis of high-resolution magnetic resonance images of the proximal femur. *Osteoporos Int* 16(11): 1307–14.
37. Goldenstein J, Kazakia G, Majumdar S. 2010. In vivo evaluation of the presence of bone marrow in cortical porosity in postmenopausal osteopenic women. *Ann Biomed Eng* 38(2): 235–46.
38. Techawiboonwong A, Song HK, Wehrli FW. 2008. In vivo MRI of submillisecond T(2) species with two-dimensional and three-dimensional radial sequences and applications to the measurement of cortical bone water. *NMR Biomed* 21(1): 59–70.
39. Techawiboonwong A, Song HK, Leonard MB, Wehrli FW. 2008. Cortical bone water: In vivo quantification with ultrashort echo-time MR imaging. *Radiology* 248(3): 824–33.

33
Radionuclide Scintigraphy in Metabolic Bone Disease

Gary J.R. Cook, Gopinath Gnanasegaran, and Ignac Fogelman

Introduction 283
Bone Metastases 283
Osteoporosis 283
Paget's Disease 284
Hyperparathyroidism 285

Renal Osteodystrophy 286
Osteomalacia 287
Conclusion 287
References 287

INTRODUCTION

Radionuclide bone imaging remains the most widely used method for detection of benign and metastatic involvement of the skeleton. Current gamma cameras systems are able to perform high-resolution imaging in short scan times (whole body acquisitions or localized views of the skeleton). The radiation dose from bone scintigraphy is relatively low [0.005 mSv/MBq, giving a typical radiation dose of 3–5 mSv equivalent to 20–25% of that from a computed tomography (CT) scan from neck to pelvis]. More recently, single photon emission CT (SPECT) imaging has become widely available and is becoming routine in nuclear medicine, leading to improved sensitivity and specificity for lesion detection as a result of tomographic data. Currently, hybrid SPECT/CT and positron emission tomography (PET)/CT scanners are the newest additions to the diagnostic armamentarium.

Bisphosphonate compounds such as methylene diphosphonate (MDP), labeled with ^{99m}Tc, are the most commonly used radiopharmaceuticals for bone scintigraphy. Their exact mechanism of localization in bone is not fully understood. However, it is likely that they are adsorbed onto the surface of hydroxyapatite crystals at sites of active mineralization. The degree of accumulation in bone is dependent on local blood flow but is influenced more strongly by the degree of osteoblastic activity and hence bone formation. In general, pathologic processes that involve bone result in increased local bone turnover, with both osteoblast and osteoclast activity being increased. As osteoblastic and osteoclastic activity is coupled in most bone pathology, increased uptake of bone tracers is present in most bone pathology.

BONE METASTASES

Radionuclide bone imaging is routinely used for detecting bone metastases in high-risk or symptomatic patients with prostate, breast, and some other cancers. The success of imaging bone lesions depends on the osteoblastic response, which usually accompanies bone destruction by a metastasis. However, rarely with aggressive metastases where bone is unable to mount sufficient osteoblastic response there may be false negatives, and this is most commonly seen in multiple myeloma. The radionuclide imaging of bone metastases is extensively reviewed elsewhere [1–3] and is beyond the scope of this chapter.

OSTEOPOROSIS

The bone scan has no role in the diagnosis of osteoporosis *per se* but is most often used in established osteoporosis to aid with the diagnosis of vertebral or other fractures

---

Primer on the Metabolic Bone Diseases and Disorders of Mineral Metabolism, Eighth Edition. Edited by Clifford J. Rosen.
© 2013 American Society for Bone and Mineral Research. Published 2013 by John Wiley & Sons, Inc.

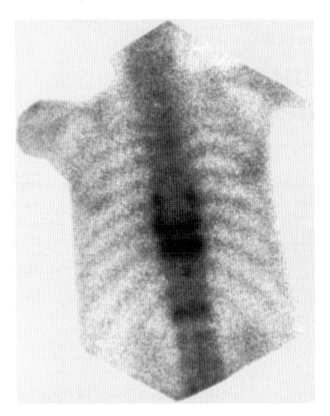

Fig. 33.1. A ^{99m}Tc-MDP bone scan showing typical linear uptake in vertebral fractures. The different intensities suggest they occurred at different times.

(e.g., sacrum, pelvis, or rib) and has a valuable role in the evaluation and management of patients with back pain [4]. The characteristic appearance of a vertebral fracture is of intense, linearly increased tracer uptake at the affected level (Fig. 33.1). Although the bone scan may become positive immediately after a fracture, this varies between skeletal sites, and in the spine it can take up to 2 weeks to become abnormal [5]. The level of uptake diminishes gradually over a period of time, with the scan normalizing between 3 and 18 months after the incident, the average being between 9 and 12 months. The degree of intensity is therefore also useful in assessing the age of fractures.

In general, if a patient is complaining of back pain with multiple vertebral fractures on radiographs but has a normal bone scan, this essentially excludes a recent fracture as a cause for the patient's symptoms. In these clinical scenarios, other causes of pain should be considered and may well be identified on the bone scan (e.g., facetal joint disease).

Vertebral fractures are defined on the basis of morphometry, but morphometric abnormalities are not specific to fracture and, for example, may be caused by congenital vertebral anomalies [6]. The bone scan may therefore have a role in deciding whether a morphometric abnormality is related to a fracture, provided that it is acquired within several months of the start of symptoms (it is generally positive for 6–18 months, with fading intensity). Similarly, edema on MRI scans indicates an acute fracture (often important to decide upon vertebroplasty or kyphoplasty), and other radiological features may be able to differentiate benign from malignant causes of vertebral collapse when this is considered to be a clinical issue [7, 8].

The radionuclide bone scan, being highly sensitive, is useful in identifying unsuspected osteoporotic fractures at other sites such as ribs, pelvis, and hip. It also has an important role in assessing suspected fractures where radiography is unhelpful, either because of poor sensitivity related to the anatomic site of the fracture (e.g., sacrum) or because adequate views are not obtainable because of the patient's discomfort or mobility.

A radionuclide bone scan also may be valuable in patients in whom back pain persists for longer than one would expect after a vertebral fracture (e.g., because of additional unsuspected vertebral fractures). In addition, we are increasingly becoming aware that osteoporotic patients with chronic back pain may have unsuspected abnormalities affecting the neighboring facet joints [9]. It is not known whether this is related to physical disruption of the joint at the time of vertebral collapse or is caused by subsequent secondary degenerative or inflammatory changes. To identify abnormalities in the facet joints, SPECT imaging is essential.

PAGET'S DISEASE

The radionuclide bone scan is invaluable in Paget's disease, both for diagnosis and to define the extent of skeletal involvement. It provides a simple way to evaluate the whole skeleton with greater sensitivity than radiographic skeletal surveys [10]. Characteristically, affected long bones show intensely increased activity, which starts at the end of a bone and spreads either proximally or distally, often showing a "V" or "flame-shaped" leading edge (Fig. 33.2). Another clue that a scintigraphic abnormality is caused by Paget's disease rather than other focal skeletal pathology is that a whole bone is often involved, and this is most often evident in the pelvis, scapula, and vertebrae. The common differential diagnosis is fibrous dysplasia. However, lack of preservation of bony outline in fibrous dysplasia can help differentiate it from Paget's disease.

The role of the bone scan with regard to treatment response assessment is not well defined. A radionuclide bone scan can be obtained 3 to 4 months after therapy, and it is important to note that pagetic lesions may often respond in a heterogeneous manner, even in individual patients. After intravenous bisphosphonate therapy, some bones may completely normalize, whereas the majority show some improvement, and a small proportion remain unchanged [11].

The bone scan may provide a sensitive means to assess persistent metabolically active disease or to identify reactivation of disease, influencing decisions on the timing of further treatment, although there is no evi-

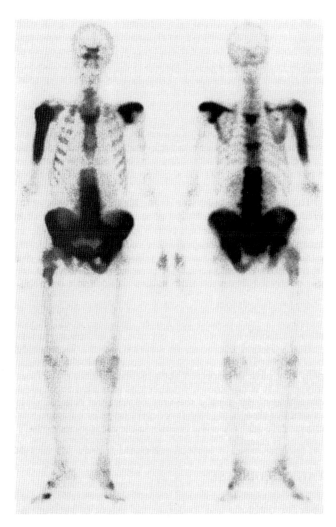

Fig. 33.2. A ^{99m}Tc-MDP bone scan (anterior and posterior images) showing intense activity in a number of skeletal sites typical of Paget's disease.

dence to support such an approach. However, it may be of particular use in patients with monostotic disease where biochemical markers are less sensitive [12]. It is important to be aware that the bone scan appearances can be unusual, and indeed bizarre, after successful bisphosphonate treatment, with resultant heterogeneous uptake sometimes mimicking metastatic disease, and hence knowledge of the relevant history and previous treatments is essential for correct interpretation. Several quantitative methods have been reported to evaluate the absolute skeletal uptake of the tracer, but none have been found to be of sufficient value to be used in routine clinical practice [13].

The radionuclide bone scan may occasionally identify complications of Paget's disease. Although osteosarcoma complicating Paget's disease is very rare (less than 1%), clues that sarcomatous change may have occurred include a change to heterogeneous and irregular uptake within an area of bone, perhaps with some photon-deficient areas corresponding to bone destruction. However, the bone scan may be misleading in the event of fracture or sarcomatous change or both. Thus, in general, the radionuclide bone scan is not reliable in the diagnosis of skeletal complications of Paget's disease, and complementary radiological correlation is necessary.

The radiopharmaceutical ^{18}F-fluorodeoxyglucose (FDG) PET evaluates tissue glucose metabolism and can differentiate benign from malignant tissue in many tumors. In principle, it could be a useful tool in distinguishing the benign changes of Paget's disease from associated osteosarcomas. However, it is important to be aware that some FDG uptake may be seen in patients with more active Paget's disease [14]. The bone-specific PET tracer ^{18}F-fluoride has a similar mechanism of uptake to ^{99m}Tc-diphosphonates. It is reported to be useful to measure the activity of Paget's disease of bone and as a promising noninvasive tool to monitor the therapeutic efficacy of bisphosphonate regimens in Paget's disease [15]. Although it essentially provides the same information as a conventional bone scan, quantitative accuracy is superior with PET.

HYPERPARATHYROIDISM

Most cases of primary hyperparathyroidism are asymptomatic and are unlikely to be associated with changes on bone scintigraphy. The diagnosis is made biochemically, and the bone scan therefore has no routine role in diagnosis. However, bone scans are often used to help differentiate the causes of hypercalcemia—in particular, hyperparathyroidism versus malignancy, and so typical features of metabolic bone disorders may be recognized. A bone scan may show a number of features in hyperparathyroidism, but the most important is the generalized increased uptake throughout the skeleton, identified as increased contrast between bone and soft tissues. Indeed, renal activity normally clearly seen on a bone scan may not be evident. This appearance has been termed a "superscan" because of the apparent high-quality images. Other typical features that have been described include a prominent calvarium and mandible, beading of the costochondral junctions (rosary beads), and prominent uptake in the sternum (the "tie" sign), sometimes with horizontal lines ("striped-tie" sign) [16]. Severe forms of hyperparathyroidism may be associated with ectopic calcification, which may lead to uptake of bone radiopharmaceuticals in soft tissues.

Nuclear medicine is the most frequently used modality for localizing abnormal parathyroid glands before surgery. Most commonly, it is performed using dual-phase ^{99m}Tc-sestamibi imaging (early and delayed imaging at 15 min and 2–3 hours, respectively) [17]. The current radionuclide tracer of choice for radionuclide localization of parathyroid adenomas is ^{99m}Tc-sestamibi. However, it is less useful in patients with secondary hyperparathyroidism, where the predominant pathology is often hyperplasia rather than adenoma.

The most frequent cause of false-positive results in ^{99m}Tc-sestamibi parathyroid imaging is the solid thyroid nodule (either solitary or multinodular gland), thyroid carcinoma, lymphoma, and other causes of lymphadenopathy. False-negative results may arise from the failure of ^{99m}Tc-sestamibi imaging to identify some smaller parathyroid lesions, which relates to system resolution and to the amount of tracer uptake by the parathyroid tissue. The addition of ^{99m}Tc-sestamibi SPECT to radionuclide scintigraphy is reported to increase the sensitivity and accuracy. More recently, ^{99m}Tc-sestamibi SPECT/CT has been used in the localization of adenomas. However, it has been reported to offer no significant additive value over conventional SPECT for identifying a normally located parathyroid adenoma, but it may assist in localizing ectopic parathyroid adenomas [18].

PET imaging using ^{11}C-methionine is reported to correctly localize abnormal parathyroid glands in patients with recurrent hyperparathyroidism for whom conventional nuclear medicine techniques have failed [19]. However, its role in routine practice is debatable because short-lived tracers that require a cyclotron for production will not be available in most centers.

RENAL OSTEODYSTROPHY

Renal osteodystrophy is caused by a combination of several metabolic bone disorders as a consequence of chronic renal dysfunction and often shows the most severe cases seen on the bone scan. It may be made up of osteoporosis, osteomalacia, adynamic bone, and secondary hyperparathyroidism in varying degrees. The most common bone scan appearance is a superscan reflecting hyperparathyroidism (Fig. 33.3). A clue in dif-

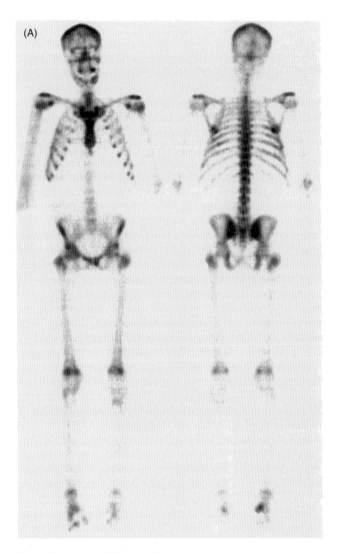
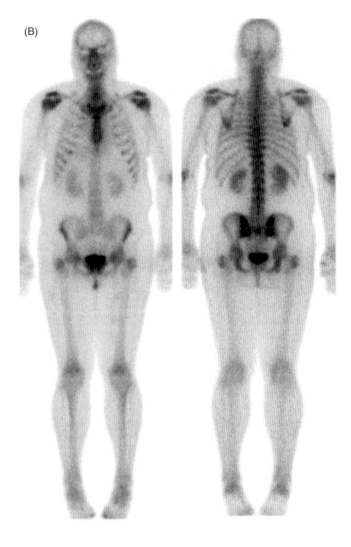

Fig. 33.3. (A) A ^{99m}Tc-MDP bone scan showing increased uptake throughout the skeleton (increased bone:soft tissue ratio) in a patient with chronic renal failure and biochemistry consistent with osteomalacia. (B) A normal bone scan is presented for comparison.

ferentiating this type of scintigraphic pattern from others is that there may be a lack of bladder activity. Although scintigraphic patterns may indicate the presence of renal osteodystrophy, the bone scan generally simply reflects the degree of hyperparathyroidism and cannot replace histologic analysis to assess the underlying determinants. Adynamic bone characteristically shows poor skeletal uptake of tracer, and the bone scan may therefore help differentiate this from the disorders with accelerated turnover.

OSTEOMALACIA

Patients with osteomalacia usually show similar features on a bone scan as described in hyperparathyroidism, although in the early stages of the disease, they may appear normal. Tracer avidity may reflect diffuse uptake in the osteoid, although it is more likely that it is caused by the secondary hyperparathyroidism that is present. The presence of focal lesions may represent pseudofractures or true fractures. Although osteomalacia may be suspected from the biochemistry, it is a histologic diagnosis. However, the typical bone scan features may be helpful in supporting the diagnosis. The detection of pseudofractures with this technique is more sensitive than that with radiography [20, 21].

CONCLUSION

The radioisotope bone scan remains a valuable method in the evaluation of malignant and metabolic bone diseases, providing functional and potentially quantitative information on bone metabolism. Recent advances in hybrid anatomical/functional imaging will consolidate the roles of radioisotope imaging in skeletal disease. In particular, hybrid SPECT/CT provides an opportunity to more accurately assess specific clinical problems, such as persistent back pain in osteoporotic patients.

REFERENCES

1. O'Sullivan JM, Cook GJ. 2002. A review of the efficacy of bone scanning in prostate and breast cancer. *Q J Nucl Med* 46: 152–9.
2. Gnanasegaran G, Cook GJ, Fogelman I. 2007. Musculoskeletal system. In: Biersack H, Freeman L (eds.) *Clinical Nuclear Medicine*. Berlin: Springer-Verlag. Chapter 10, pp. 241–62.
3. Van der Wall H, Clarke S. 2004. The evaluation of malignancy: Metastatic bone disease. In: Ell P, Gambhir, S (eds.) *Nuclear Medicine in Clinical Diagnosis and Treatment 3rd Ed.* Philadelphia: Churchill Livingstone Chapter 45, pp. 641–5.
4. Cook GJR, Hannaford E, Lee M, Clarke SEM, Fogelman I. 2002. The value of bone scintigraphy in the evaluation of osteoporotic patients with back pain. *Scand J Rheumatol* 31: 245–8.
5. Spitz J, Lauer I, Tittel K, Wiegand H. 1993. Scintimetric evaluation of remodeling after bone fractures in man. *J Nucl Med* 34: 1403–9.
6. Eastell R, Cedel SL, Wahner HW, Riggs BL, Melton LJ. 1991. Classification of vertebral fractures. *J Bone Miner Res* 6: 207–15.
7. Yamato M, Nishimura G, Kuramochi E, Saiki N, Fujioka M. 1998. MR appearance at different ages of osteoporotic compression fractures of the vertebrae. *Radiat Med* 16: 329–34.
8. Wang KC, Jeanmenne A, Weber GM, Thawait SK, Carrino JA. 2011. An online evidence-based decision support system for distinguishing benign from malignant vertebral compression fractures by magnetic resonance imaging feature analysis. *J Digit Imaging* 24: 507–15.
9. Ryan PJ, Evans PA, Gibson T, Fogelman I. 1992. Osteoporosis and chronic back pain: A study with single photon emission computed bone scintigraphy. *J Bone Miner Res* 7: 1455–1459.
10. Fogelman I, Carr D. 1980. A comparison of bone scanning and radiology in the assessment of patients with symptomatic Paget's disease. *Eur J Nucl Med* 5: 417–21.
11. Ryan PJ, Gibson T, Fogelman I. 1992. Bone scintigraphy following pamidronate therapy for Paget's disease of bone. *J Nucl Med* 33: 1589–93.
12. Patel S, Pearson D, Hosking DJ. 1995. Quantitative bone scintigraphy in the management of monostotic Paget's disease of bone. *Arthritis Rheum* 38: 1506–12.
13. Fogelman I, Bessent RG, Gordon D. 1981. A critical assessment of bone scan quantitation (bone to soft tissue ratios) in the diagnosis of metabolic bone disease. *Eur J Nucl Med* 6: 93–97.
14. Cook GJ, Maisey MN, Fogelman I. 1997. Fluorine-18-FDG PET in Paget's disease of bone. *J Nucl Med* 7: 1495–7.
15. Installe J, Nzeusseu A, Bol A, Depresseux G, Devogelaer JP, Lonneux M. 2005. 18F-fluoride PET for monitoring therapeutic response in Paget's disease of bone. *J Nucl Med* 6: 1650–8.
16. Fogelman I, Carr D. A comparison of bone scanning and radiology in the evaluation of patients with metabolic bone disease. *Clin Radiol* 31: 321–6.
17. Palestro CJ, Tomas MB, Tronco GG. 2005. Radionuclide imaging of the parathyroid glands. *Semin Nucl Med* 5: 262–6.
18. Gayed IW, Kim EE, Broussard WF, Evans D, Lee J, Broemeling LD, Ochoa BB, Moxley DM, Erwin WD, Podoloff DA. 2005. The value of 99mTc-sestamibi SPECT/CT over conventional SPECT in the evaluation

of parathyroid adenomas or hyperplasia. *J Nucl Med* 46: 248–52.
19. Cook GJ, Wong JC, Smellie WJ, Young AE, Maisey MN, Fogelman I. 1998. [11C]Methionine positron emission tomography for patients with persistent or recurrent hyperparathyroidism after surgery. *Eur J Endocrinol* 139: 195–7.
20. Fogelman I, McKillop JH, Bessent RG, Boyle IT, Turner JG, Greig WR. 1978. The role of bone scanning in osteomalacia. *J Nucl Med* 19: 245–8.
21. Fogelman I, McKillop JH, Greig WR, Boyle IT. 1977. Pseudofractures of the ribs detected by bone scanning. *J Nucl Med* 18: 1236–7.

34

FRAX®: Assessment of Fracture Risk

John A Kanis

Introduction 289
Input and Output 289
Performance Characteristics 290
Limitations 292
Intervention and Assessment Thresholds 292
Guidelines in the U.K. 293

Guidelines in North America 294
Guidelines without BMD Testing 294
Other Applications of FRAX 294
Conclusion 295
Competing Interests 295
References 295

INTRODUCTION

A major objective of fracture risk assessment is to enable the targeting of interventions to those at need and avoid unnecessary treatment in those at low risk of fracture. Historically, fracture risk assessment was largely based on the measurement of bone mineral density (BMD), since osteoporosis is defined operationally in terms of bone mass [1, 2]. Whereas BMD forms a central component in the assessment of risk, the accuracy of risk prediction is improved by taking into account other readily measured indices of fracture risk, particularly those that add information to that provided by BMD. Several risk prediction models have been developed, but the most widely used is FRAX®.

INPUT AND OUTPUT

FRAX® is a computer-based algorithm (http://www.shef.ac.uk/FRAX) developed by the World Health Organization Collaborating Centre for Metabolic Bone Diseases and first released in 2008 [3, 4]. The algorithm, intended for primary care, calculates fracture probability from easily obtained clinical risk factors (CRFs) in men aged 50 years or older, and postmenopausal women. The output of FRAX is the 10-year probability of a major osteoporotic fracture (hip, clinical spine, humerus, or wrist fracture) and the 10-year probability of hip fracture (Fig. 34.1).

Fracture probability is derived from the risk of fracture as well as the risk of death. Fracture risk is calculated from age, body mass index, and dichotomized risk factors comprising prior fragility fracture, parental history of hip fracture, current tobacco smoking, long-term oral glucocorticoid use, rheumatoid arthritis, excessive alcohol consumption, and other causes of secondary osteoporosis. Femoral neck BMD can be optionally input to enhance fracture risk prediction. Apart from rheumatoid arthritis and long-term use of glucocorticoids, the other secondary causes of osteoporosis considered (Table 34.1) are conservatively assumed to contribute to increased fracture risk because of low BMD.

The relationships between risk factors and fracture risk have been constructed using information derived from the primary data of population-based cohorts from around the world, including centers from North America, Europe, Asia, and Australia, based on a series of meta-analyses to identify clinical risk factors for fracture that provide independent information on fracture risk [3]. The use of primary data for the model construct permits the determination of the predictive importance in a multi-variable context of each of the risk factors, as well as interactions between risk factors, and thereby optimizes

---

Primer on the Metabolic Bone Diseases and Disorders of Mineral Metabolism, Eighth Edition. Edited by Clifford J. Rosen.
© 2013 American Society for Bone and Mineral Research. Published 2013 by John Wiley & Sons, Inc.

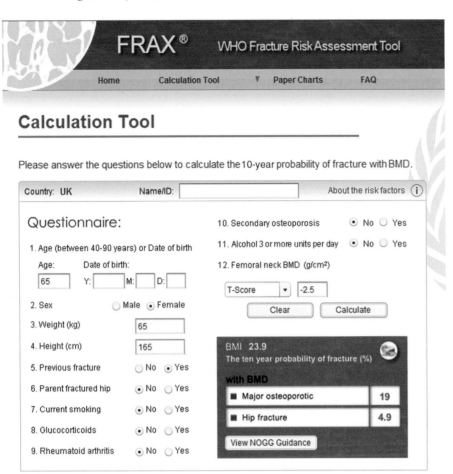

Fig. 34.1. Screen page for input of data and format of results in the U.K. version of the FRAX tool (U.K. model, version 3.1. http://www.shef.ac.uk/FRAX). (With permission of the World Health Organization Collaborating Centre for Metabolic Bone Diseases, University of Sheffield Medical School, U.K.).

the accuracy by which fracture risk can be computed [5, 6]. The fracture risk assessment with the combined use of these clinical risk factors with and without the use of BMD has been validated in independent cohorts with a similar geographic distribution with in excess of 1 million patient years [7–9].

Fracture probability is computed taking both the risk of fracture and the risk of death into account. The inclusion of the death hazard is important because those with a high immediate likelihood of death are less likely to fracture than individuals with longer life expectancy (Fig. 34.2). In addition, some of the risk factors affect the risk of death as well as the fracture risk. Examples include increasing age, low body mass index (BMI), low BMD, glucocorticoids, and smoking. Other risk tools calculate the risk of fracture without taking into account the possibility of death [9, 10]. Fracture probability varies markedly in different regions of the world [11]. Thus, the FRAX® models are calibrated to those countries where the epidemiology of fracture and death is known. Models are currently available for 52 countries and 24 languages. Ethnic-specific models are available in the U.S. and Singapore.

FRAX has been widely used for the assessment of patients since the launch of the website in 2008, and currently makes about 11,000 calculations per working day. Following regulatory review by the U.S. Food and Drug Administration, FRAX was incorporated into dual energy X-ray absorptiometry (DXA) scanners to provide FRAX probabilities at the time of DXA scanning. For those without Internet access, handheld calculators and an application for the iPhone and iPad have been developed by the International Osteoporosis Foundation (http://itunes.apple.com/us/app/frax/id370146412?mt=8).

PERFORMANCE CHARACTERISTICS

For the purpose of risk assessment, a characteristic of major importance is the ability of a technique to predict fractures. This is traditionally expressed as the increase in relative risk per standard deviation (SD) unit decrease in risk score—termed the "gradient of risk." The gradient of risk (GR) is shown in Table 34.2 for the use of the clinical risk factors alone, femoral neck BMD and the combination.

The use of clinical risk factors alone provides a gradient of risk that lies between 1.4 and 2.1, depending upon age and the type of fracture predicted. These gradients are comparable to the use of BMD alone to predict fractures [13, 14], indicating that clinical risk factors alone are of value in fracture risk prediction and might be used, therefore, in the many countries where DXA facilities are sparse [15]. Nevertheless, there are substantial gains in the use of the clinical risk factors in conjunction with BMD, particularly in the case of hip fracture prediction. At the age of 50, for example, the gradient of risk with BMD alone is 3.7/ SD, but with the addition of clinical risk factors is 4.2/SD. Although the improvement in GR with the addition of BMD appears relatively modest, particularly in the case of other osteoporotic fractures, it

Table 34.1. Secondary Causes of Osteoporosis Associated With an Increase in Fracture Risk*

Secondary Cause	Example
Glucocorticoids	Any dose, by mouth for 3 months or more
	High doses of inhaled glucocorticoids
	Cushing's disease
Rheumatoid arthritis	
Chronic liver disease	Alcoholism
Untreated hypogonadism	Bilateral oophorectomy or orchidectomy,
	Anorexia nervosa
	Chemotherapy for breast cancer
	Tamoxifen in premenopausal women
	Aromatase inhibitors
	GnRH inhibitors for prostate cancer
	Hypopituitarism
Prolonged immobility	Spinal cord injury
	Parkinson's disease
	Stroke
	Muscular dystrophy
	Ankylosing spondylitis
Organ transplantation	
Type 1 and 2 diabetes	
Thyroid disorders	Untreated hyperthyroidism
	Overtreated hypothyroidism
Gastrointestinal disease	Crohn's disease
	Ulcerative colitis
Chronic obstructive pulmonary disease	

*Adapted from Ref. 3 with permission of the World Health Organization Collaborating Centre for Metabolic Bone Diseases, University of Sheffield Medical School, U.K.

Table 34.2. Gradients of Risk (HR/per SD Change in Risk Score) (with 95% confidence intervals) With the Use of BMD at the Femoral Neck, Clinical Risk Factors, or the Combination*

	Gradient of Risk		
Age (years)	Clinical Risk Factors Alone	BMD Only	Clinical Risk Factors + BMD
(a) Hip fracture			
50	2.05 (1.58–2.65)	3.68 (2.61–5.19)	4.23 (3.12–5.73)
60	1.95 (1.63–2.33)	3.07 (2.42–3.89)	3.51 (2.85–4.33)
70	1.84 (1.65–2.05)	2.78 (2.39–3.23)	2.91 (2.56–3.31)
80	1.75 (1.62–1.90)	2.28 (2.09–2.50)	2.42 (2.18–2.69)
90	1.66 (1.47–1.87)	1.70 (1.50–1.93)	2.02 (1.71–2.38)
(b) Other osteoporotic fractures			
50	1.41 (1.28–1.56)	1.19 (1.05–1.34)	1.44 (1.30–1.59)
60	1.48 (1.39–1.58)	1.28 (1.18–1.39)	1.52 (1.42–1.62)
70	1.55 (1.48–1.62)	1.39 (1.30–1.48)	1.61 (1.54–1.68)
80	1.63 (1.54–1.72)	1.54 (1.44–1.65)	1.71 (1.62–1.80)
90	1.72 (1.58–1.88)	1.56 (1.40–1.75)	1.81 (1.67–1.97)

*(Ref. 7). Outcomes comprised (a) hip fracture and (b) other fractures at sites associated with osteoporosis.
With kind permission from Springer Science+Business Media B.V.

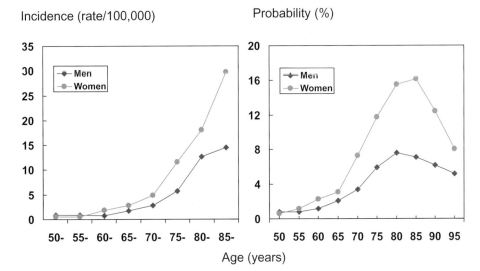

Fig. 34.2. Incidence of hip fracture by age in women from Sweden and the corresponding 10-year probability of hip fracture (data derived from Ref. 12).

Table 34.3. Percentage Adjustment of 10-year Probabilities of a Hip Fracture or a Major Osteoporotic Fracture By Age According to Doses of Glucocorticoids*

Dose	Prednisolone Equivalent (mg/day)	Age (years)						All Ages
		40	50	60	70	80	90	
Hip fracture								
Low	<2.5	−40	−40	−40	−40	−30	−30	−35
Medium[a]	2.5–7.5							
High	≥7.5	+25	+25	+25	+20	+10	+10	+20
Major osteoporotic fracture								
Low	<2.5	−20	−20	−15	−20	−20	−20	−20
Medium[a]	2.5–7.5							
High	≥ 7.5	+20	+20	+15	+15	+10	+10	+15

*Ref. 19
[a]No adjustment
With kind permission from Springer Science+Business Media B.V.

should be recognized that GRs are not multiplicative. For example, at the age of 70, BMD alone gave a GR of 2.8/SD for hip fracture. For the clinical risk factors the GR was 1.8/SD. If these two tests were totally independent, the combined GR would be $\sqrt{(2.08^2 + 1.8^2)} = 3.3$. The observed GR (2.9) falls short of the theoretical upper limit, since there is a significant correlation between the clinical risk factor score and BMD.

LIMITATIONS

FRAX should not be considered as a gold standard in patient assessment, but rather as a reference platform. The same argument applies to BMD testing. Thus, the result should not be uncritically used in the management of patients without an appreciation of its limitations as well as its strengths. In some instances, limitations (e.g., to experts in bone disease) are perceived as strengths to others (e.g., primary care physicians).

Assessment with FRAX takes no account of current or prior treatment, though the effect of treatment on the estimated fracture probability is modest, related in part to treatment-induced increases in BMD [16]. FRAX also takes no account of dose responses for several risk factors. For example, two or more prior vertebral fractures carry a much higher risk than a single prior fracture [17]. A prior clinical vertebral fracture carries an approximately twofold higher risk than other prior fractures. Dose responses are also evident for glucocorticoid use, smoking, and alcohol consumption [3, 18]. Since it is not possible to model all such scenarios with the FRAX algorithm, these limitations should temper clinical judgement.

In the case of glucocorticoids, simple arithmetic procedures have been formulated that can be applied to conventional FRAX estimates of probabilities of hip fracture and a major osteoporotic fracture to adjust the probability assessment with knowledge of the dose of glucocorticoids (Table 34.3). For example, an individual at the age of 60 years taking a high dose of glucocorticoids and with a probability of a major fracture of 18% would have the FRAX estimate uplifted by 15%, giving a revised probability of 21% (17 × 1.25). In contrast, if the patient were exposed to a low dose, the revised estimate would be 15%.

A further limitation is that the FRAX algorithm uses T-scores for femoral neck BMD from DXA and does not accommodate other sites or technologies. The lumbar spine is frequently measured by DXA for the assessment of patients and, indeed, is incorporated into many clinical guidelines [20–23]. It is also the site favored for monitoring treatment. There is, therefore, much interest in the incorporation into FRAX of measurements at the lumbar spine. Whereas this is not possible at present, some guidance is available in cases where there is a large discordance between the T-score at the femoral neck and the lumbar spine [24]. It has been proposed that the FRAX estimate for a major fracture is increased/decreased by one-tenth for each rounded T-score difference between the lumbar spine and femoral neck. An example is a case in which the T-score for femoral neck BMD is −2.2 SD with a FRAX-calculated fracture probability of 19% and a T-score of −3.5 SD at the lumbar spine. In this case, the T-score discordance is 1.3 SD (3.5 − 2.2). If the figure is rounded off (to 1.0 SD), the estimated probability with the inclusion of lumbar BMD is upward revised by 10% (19 + 1.9) to 21%.

INTERVENTION AND ASSESSMENT THRESHOLDS

The use of FRAX in clinical practice demands a consideration of the fracture probability at which to intervene,

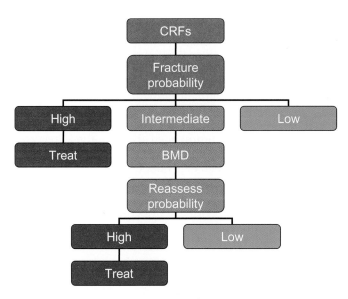

Fig. 34.3. Management algorithm for the assessment of individuals at risk of fracture (Ref. 25)]. With kind permission from Springer Science+Business Media B.V.

both for treatment (an intervention threshold) and for BMD testing (assessment thresholds). A general approach is shown in Fig. 34.3 [25]. The management process begins with the assessment of fracture probability and the categorization of fracture risk on the basis of age, sex, body mass index, and clinical risk factors. On this information alone, some patients at high risk may be offered treatment without recourse to BMD testing. Many guidelines recommend treatment in the absence of information on BMD in women with a previous fragility fracture (a prior vertebral or hip fracture in North America) [21, 23, 26]. Many physicians would also perform a BMD test, but frequently this is for reasons other than to decide on intervention, for example, as a baseline to monitor treatment. There will be other instances where the probability will be so low that a decision not to treat can be made without BMD. An example might be the woman at menopause in good health with no clinical risk factors. Thus, not all individuals require a BMD test. The size of the intermediate category in Fig. 34.3 will vary in different countries. In the U.S., this will be a large category, whereas in a large number of countries with limited or no access to densitometry [15], the size of the intermediate group will necessarily be small. In other countries (e.g. the U.K.), where provision for BMD testing is suboptimal, the intermediate category will lie between the two extremes.

FRAX has been accommodated to a greater or lesser extent in many assessment guidelines including those from Austria, Belgium, Canada, Europe, France, Greece, Hungary, Ireland, Italy, Japan, Netherlands, Poland, Singapore, Sri Lanka, Sweden, Switzerland, the U.K., and the U.S. The setting of intervention thresholds is problematic from an international perspective since the risk of fracture, the cost of fracture, the cost of treatment, reimbursement, and willingness to pay varies in different countries. Thus, probability-based guidelines differ in detail. Guidelines variously use an age-dependent fracture probability or a fixed probability threshold applied to all relevant ages. Examples of each are provided in the guidelines of the U.K. and North America, illustrated below in the case of postmenopausal women.

GUIDELINES IN THE U.K.

The U.K. guidance for the identification of individuals at high fracture risk, developed by the National Osteoporosis Guideline Group (NOGG) [25–27], recommends that postmenopausal women with a prior fragility fracture may be considered for intervention without the necessity for a BMD test. In women without a fragility fracture but with a FRAX risk factor, the intervention threshold set by NOGG is at the age-specific fracture probability equivalent to women with a prior fragility fracture. The same intervention threshold is applied to men, since the effectiveness and cost-effectiveness of intervention in men is broadly similar to that in women for equivalent risk [28, 29].

The NOGG management strategy considers two additional thresholds (Fig. 34.4):

- a threshold probability below which neither treatment nor a BMD test should be considered (lower assessment threshold)
- a threshold probability above which treatment may be recommended irrespective of BMD (upper assessment threshold)

In other words, some patients at high risk may be offered treatment without recourse to BMD testing. Conversely, some patients at low risk will be denied treatment without a BMD test. An attraction of the approach is that efficient use is made of BMD testing. For example, the NOGG strategy requires only 3.5 scans at the age of 50 years to identify one case of hip fracture, whereas the former guidelines of the Royal College of Physicians required 14. The lower number of BMD tests means that the acquisition costs for identifying a fracture case and the total costs (acquisition and treatment) per fracture averted are also lower [30].

The justification for this parsimonious approach derives from the correlation between fracture probability (without BMD) and BMD. It has been shown in several studies that the reclassification of high risk to low risk (and vice versa, from low risk to high risk) with the incorporation of BMD in the FRAX calculation is more or less confined to those close to an intervention threshold. Indeed, more than 80% of women who are reclassified have probabilities without BMD within 5% of the

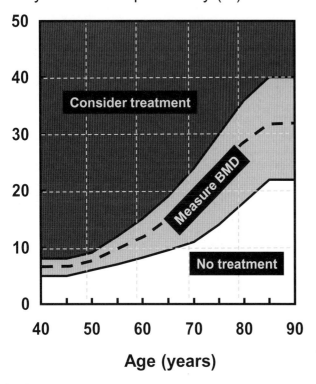

Fig. 34.4. Assessment guidelines of the National Osteoporosis Guideline Group based on the 10-year probability of a major fracture (%). The dotted line denotes the intervention threshold, which increases with age. Where assessment is made in the absence of BMD, a BMD test is recommended for individuals where the probability assessment lies in the lightly shaded region (adapted from Ref. 25).

treatment threshold. This figure rises to about 95% for probabilities without BMD within 10% of the treatment threshold. Thus, a BMD test is only required in individuals whose fracture probability lies within about 10% of the intervention threshold, thereby avoiding unnecessary testing in about 50% of eligible women without a prior fragility fracture [31].

GUIDELINES IN NORTH AMERICA

In the U.S. and Canada, the national guidelines recommend treatment for postmenopausal women who have had a prior spine or hip fracture and for women with a BMD at or below a T-score of -2.5 SD [20, 21, 23, 26]. Thus, as for the U.K., they recognize high-risk groups in whom treatment can be recommended without FRAX, but the high-risk groups differ (a prior fragility fracture in the U.K.). Conversely, the guidelines do not recommend treatment in postmenopausal women with a T-score of greater than −1.0 SD. In the U.K., the exclusion group is women without a clinical risk factor. Thus FRAX in North America becomes relevant only in women with a T-score between −1 and −2.5 SD. Treatment is recommended in such patients in whom the 10-year probability of a major fracture equals or exceeds 20% or (in the U.S. but not Canada) when the 10-year probability of a hip fracture equals or exceeds 3%.

Thus there are similarities and differences with the U.K. and North American guidelines. The similarities are that both guidelines select high-risk patients in whom treatment is recommended and low-risk groups in whom treatment is not recommended (though the categories differ somewhat). The differences lie in the intervention thresholds (fixed or age dependent) and the use of BMD in the intermediate group (all women in North America, assessment thresholds in the U.K.). Neither approach is right nor wrong, but both are imperfect and will remain so until fracture risk prediction can be improved still further.

GUIDELINES WITHOUT BMD TESTING

Many regions of the world have little or no access to BMD. In these circumstances, FRAX without BMD can be used. The clinical risk factors used in FRAX are not totally independent of BMD. Indeed there is a weak but significant correlation between the clinical risk factor score for hip fracture (assessed without BMD) and BMD at the femoral neck. This indicates that the selection of individuals with the use of FRAX, without knowledge of BMD, will preferentially select those with low BMD, and that the higher the fracture probability, the lower will be the BMD. When a fixed intervention threshold is chosen, the high-risk group has a BMD that is approximately 1SD lower than in the low-risk group [31].

As noted above, the prediction of fractures with the use of clinical risk factors alone in FRAX is comparable to the use of FRAX without BMD to predict fractures is suitable, therefore, in the many countries where DXA facilities are sparse. Notwithstanding, intervention thresholds still need to be determined.

OTHER APPLICATIONS OF FRAX

Several studies have examined the relationship between FRAX-based probability of fracture and efficacy of bone-active agents used in the treatment of osteoporosis. In some, but not all instances, efficacy (relative risk reduction) has been shown to be greater in patients with the higher baseline fracture probabilities [32]. This has implications for targeting treatments to high-risk patients in that the dividends in terms of fractures saved is amplified. This also has implications for health economic assessment and conventional meta-analyses of interventions used in osteoporosis [33].

CONCLUSION

FRAX represents a significant advance in the assessment of both women and men at risk for fracture and allows the tailoring of pharmacologic interventions to high-risk subjects. While FRAX does not define intervention thresholds, which depend on country-specific considerations, it provides a platform to assess fracture probability, which is needed to make rational treatment decisions by clinicians and public health agencies. The tool is, however, far from perfect, but better than BMD alone. The widespread use and interest in FRAX, and its adoption into management guidelines, has fueled interest as to how models can be improved, extended to other countries, and in particular, how the limitations of FRAX should temper clinical judgement.

COMPETING INTERESTS

The author has no competing interests with regard to the content of this chapter.

REFERENCES

1. World Health Organization. 1994. Assessment of fracture risk and its application to screening for postmenopausal osteoporosis. *Technical Report Series 843*. Geneva: WHO.
2. Kanis JA, McCloskey EV, Johansson H, Oden A, Melton LJ 3rd, Khaltaev N. 2008. A reference standard for the description of osteoporosis. *Bone* 42: 467–75.
3. Kanis JA, on behalf of the World Health Organization Scientific Group. 2008. Assessment of osteoporosis at the primary health-care level. *Technical Report*. WHO Collaborating Centre, University of Sheffield, UK. Available online at http://www.shef.ac.uk/FRAX/index.htm.
4. Kanis JA, Johnell O, Oden A, Johansson H, McCloskey EV. 2008. FRAX™ and the assessment of fracture probability in men and women from the UK. *Osteoporos Int* 19: 385–97.
5. De Laet C, Oden A, Johansson H, Johnell O, Jonsson B, Kanis JA. 2005. The impact of the use of multiple risk indicators for fracture on case-finding strategies: A mathematical approach. *Osteoporos Int* 16: 313–8.
6. Kanis JA, Johnell O, Oden A, De Laet C, Jonsson B, Dawson A. 2002. Ten-year risk of osteoporotic fracture and the effect of risk factors on screening strategies. *Bone* 30: 251–8.
7. Kanis JA, Oden A, Johnell O, et al. 2007. The use of clinical risk factors enhances the performance of BMD in the prediction of hip and osteoporotic fractures in men and women. *Osteoporos Int* 18: 1033–46.
8. Leslie WD, Lix LM, Johanansson H, Odén A, McCloskey E, Kanis JA. 2010. Independent clinical validation of a Canadian FRAX tool: Fracture prediction and model calibration. *J Bone Miner Res* 25: 2350–8.
9. Hippisley-Cox J, Coupland C. 2009. Predicting risk of osteoporotic fracture in men and women in England and Wales: Prospective derivation and validation of QFractures Scores. *Br Med J* 339: b4229.
10. Nguyen ND, Frost SA, Center JR, Eisman JA, Nguyen TV. 2008. Development of prognostic nomograms for individualizing 5-year and 10-year fracture risks. *Osteoporos Int* 19: 1431–44.
11. Kanis JA, Johnell O, De Laet C, Jonsson B, Oden A, Oglesby A. 2002. International variations in hip fracture probabilities: Implications for risk assessment. *J Bone Miner Res* 17: 1237–44.
12. Kanis JA, Johnell O, Oden A, et al, 2000, Long-term risk of osteoporotic fracture in Malmo. *Osteoporos Int* 11: 669–74.
13. Johnell O, Kanis JA, Oden A, et al, 2005, Predictive value of bone mineral density for hip and other fractures. *J Bone Miner Res* 20: 1185–1194; Erratum in 2007 *J Bone Miner Res* 22: 774.
14. Marshall D, Johnell O, Wedel H. 1996. Meta-analysis of how well measures of bone mineral density predict occurrence of osteoporotic fractures. *Br Med J* 312: 1254–59.
15. Kanis JA, Johnell O. 2004. Requirements for DXA for the management of osteoporosis in Europe. *Osteoporos Int* 16: 229–38.
16. Leslie WD, Lix LM, Johansson H, et al. 2012. Does osteoporosis therapy invalidate FRAX for fracture prediction? *J Bone Miner Res* 27: 1243–51.
17. Delmas PD, Genant HK, Crans GG, et al. 2003. Severity of prevalent vertebral fractures and the risk of subsequent vertebral and nonvertebral fractures: Results from the MORE trial. *Bone* 33: 522–32.
18. Van Staa TP, Leufkens HG, Abenhaim L, Zhang B, Cooper C. 2000. Use of oral corticosteroids and risk of fractures. *J Bone Miner Res* 15: 993–1000.
19. Kanis JA, Johansson H, Oden A, McCloskey EV. 2011. Guidance for the adjustment of FRAX according to the dose of glucocorticoids. *Osteoporos Int* 2: 809–16.
20. Baim S, Binkley N, Bilezikian JP, et al. 2008. Official Positions of the International Society for Clinical Densitometry and executive summary of the 2007 ISCD Position Development Conference. *J Clin Densitom* 11: 75–91.
21. Dawson-Hughes B. 2008. A revised clinician's guide to the prevention and treatment of osteoporosis. *J Clin Endocrinol Metab* 93: 2463–65.
22. National Osteoporosis Foundation. 2008. *Clinician's guide to prevention and treatment of osteoporosis.* Washington, DC: National Osteoporosis Foundation. www.nof.org.
23. Papaioannou A, Morin S, Cheung AM, et al. 2010. 2010 clinical practice guidelines for the diagnosis and management of osteoporosis in Canada: Summary. *CMAJ* 182: 1864–73.
24. Leslie WD, Lix LM, Johansson H, Oden A, McCloskey E, Kanis JA. 2011. Spine-hip discordance and fracture

risk assessment: A physician-friendly FRAX enhancement. *Osteoporos Int* 22: 839–47.
25. Kanis JA, McCloskey EV, Johansson H, Strom O, Borgstrom F, Oden A, and the National Osteoporosis Guideline Group. 2008. Case finding for the management of osteoporosis with FRAX®—Assessment and intervention thresholds for the UK. *Osteoporos Int* 19: 1395–1408; Erratum in 2009 *Osteoporos Int* 20: 499–502.
26. Grossman JM, Gordon R, Ranganath VK, et al. 2010. American College of Rheumatology 2010 recommendations for the prevention and treatment of glucocorticoid-induced osteoporosis. *Arthritis Care Res (Hoboken)* 62: 1515–26.
27. Compston J, Cooper A, Cooper C, Francis R, Kanis JA, Marsh D, McCloskey EV, Reid DM, Selby P, Wilkins M; National Osteoporosis Guideline Group (NOGG). 2009. Guidelines for the diagnosis and management of osteoporosis in postmenopausal women and men from the age of 50 years in the UK. *Maturitas* 62:105–8.
28. Kanis JA, Stevenson M, McCloskey EV, Davis S, Lloyd-Jones M. 2007. Glucocorticoid-induced osteoporosis: A systematic review and cost-utility analysis. *Health Technol Assess* 11: 1–256.
29. Tosteson AN, Melton LJ 3rd, Dawson-Hughes B, Baim S, Favus MJ, Khosla S, Lindsay RL; National Osteoporosis Foundation Guide Committee. 2008. Cost-effective osteoporosis treatment thresholds: The United States perspective. *Osteoporos Int* 19: 437–47.
30. Johansson H, Kanis JA, Oden A, Johnell O, Compston J, McCloskey EV. 2011. A comparison of case finding strategies in the UK for the management of hip fractures. *Osteoporos Int* 23: 907–15.
31. Johansson H, Oden A, Johnell O, Jonsson B, De Laet C, Oglesby A, et al. 2004. Optimisation of BMD measurements to identify high risk groups for treatment—A test analysis. *J Bone Miner Res* 19: 906–13.
32. Kanis JA, Oden A, Johansson H, Borgström F, Ström O, McCloskey E. 2009. FRAX® and its applications to clinical practice. *Bone* 44: 734–43.
33. Ström O, Borgström F, Kleman M, McCloskey E, Odén A, Johansson H, Kanis JA. 2010 FRAX and its applications in health economics—Cost-effectiveness and intervention thresholds using bazedoxifene in a Swedish setting as an example. *Bone* 47: 430–7.

35
Biochemical Markers of Bone Turnover in Osteoporosis

Pawel Szulc, Douglas C. Bauer, and Richard Eastell

Introduction 297
Analytical and Preanalytical Variability 297
BTMs and Reference Values 299
Bone Turnover Rate and Bone Loss 299
Bone Turnover Rate and Fracture Risk 299

Bone Turnover Rate and Monitoring 300
Bone Turnover Markers in Men 302
Conclusions 302
References 302

INTRODUCTION

Bone turnover is characterized by two opposite activities: bone formation and bone resorption [1]. During bone remodeling (after growth arrest and during aging), bone formation is preceded by bone resorption. Both activities are coupled at a basic multicellular unit (BMU) [also called bone remodeling unit (BRU)]. During bone resorption, dissolution of bone mineral and catabolism of bone matrix by osteoclasts results in the formation of resorptive cavity and the release of bone matrix components. Then, during bone formation, osteoblasts synthesize bone matrix, which fills in the resorption cavity and undergoes mineralization.

There are two groups of biochemical bone turnover markers (BTMs), markers of bone formation and markers of bone resorption (Table 35.1). Recently, an expert panel convened by the International Osteoporosis Foundation and International Federation of Clinical Chemistry and Laboratory Medicine proposed that PINP and serum CTX-I become the referent markers of bone formation and resorption, respectively [2]. PINP is derived from post-translational cleavage of type I procollagen molecules before assembly into fibrils. Circulating PINP originates primarily from bone, does not show circadian variation, and increases rapidly during bone formation-stimulating therapy. Serum CTX-I is a product of the breakdown of type I bone collagen. It is present predominantly in bone and rapidly decreases during antiresorptive treatment. However, it has a strong circadian rhythm, and blood for its measurement has to be collected in the fasting state in the morning. CTX-I occurs in its native (α) and β-isomerized forms, which may undergo racemization (D- and L-forms). Bone AP and TRACP5b are enzymes reflecting the metabolic activity, respectively, of osteoblasts and osteoclasts. Other BTMs are bone matrix components released into the blood during formation or resorption.

ANALYTICAL AND PREANALYTICAL VARIABILITY

The analytical variability (assessed by the intra-assay and inter-assay coefficients of variation) depends on the BTM, the measurement method, and the technician's expertise [3]. Several methods for measurements of BTMs are available (radioimmunoassay, immunoradiometric assay, enzymatic immunoassay, chemiluminescence). The use of monoclonal antibodies permits specific measurements of the molecules. Automated analyzers permit a rapid, convenient, fully automated, and precise measurement of BTM levels both in research projects and in clinical practice [4]. Moreover, miniaturized point-of-care (POC)

Primer on the Metabolic Bone Diseases and Disorders of Mineral Metabolism, Eighth Edition. Edited by Clifford J. Rosen.
© 2013 American Society for Bone and Mineral Research. Published 2013 by John Wiley & Sons, Inc.

Table 35.1. Biochemical Bone Turnover Markers

Bone formation
  Osteocalcin (**OC**)
  Bone alkaline phosphatase (**bone ALP**)
  N-terminal propeptide of type I procollagen (**PINP**)
  C-terminal propeptide of type I procollagen (**PICP**)

Bone resorption
  C-terminal cross-linking telopeptide of type I collagen (**CTX-I**)
  N-terminal cross-linking telopeptide of type I collagen (**NTX-I**)
  C-terminal cross-linking telopeptide of type I collagen generated by matrix metalloproteinases (**CTX-MMP, ICTP**)
  Helical peptide 620-633 of the α1 chain
  Deoxypyridinoline (**DPD**)
  Isoform 5b of tartrate resistant acid phosphatase (**TRACP5b**)

Table 35.2. Determinants of the Preanalytical Variability of Bone Turnover

Modifiable determinants
Circadian variation
Menstrual variation
Seasonal variation
Fasting and food intake (particularly serum CTX-I)
Exercise and physical activity

Determinants that cannot be easily modified
Age
Sex
Menopausal status
Vitamin D deficit and secondary hyperparathyroidism
Short- and long-term day-to-day variation
Diseases characterized by an acceleration of bone turnover
  Primary hyperparathyroidism
  Thyrotoxicosis
  Acromegaly
  Paget's disease
  Bone metastases

Diseases characterized by a dissociation of bone turnover
  Cushing's disease
  Multiple myeloma

Diseases characterized by a low bone turnover
  Hypothyroidism
  Hypoparathyroidism
  Hypopituitarism
  Growth hormone deficit

Renal impairment (depending on the stage)
Recent fracture
Depression
Chronic diseases associated with limited mobility
  Stroke
  Hemiplegia
  Dementia
  Alzheimer's disease
  Sarcopenia

Medications
  Oral corticosteroids
  Inhaled corticosteroids (only osteocalcin)
  Aromatase inhibitors (antiaromatases)
  Oral contraceptives
  Gonadoliberin agonists
  Anti-epileptic drugs
  Thiazolidinediones
  Heparin
  Vitamin K-antagonists

devices permit rapid measurement of urinary creatinine-corrected NTX-I [5].

The preanalytical variability has a strong effect on the BTM levels [6]. It comprises a large number of factors (Table 35.2), which may coexist in one person. Circadian rhythm has a strong impact on the BTM variability, especially serum bone resorption markers that peak in the second half of the night and have their nadir in the afternoon [7]. The amplitude is higher for CTX-I than for other BTMs. Food intake has a strong effect on bone resorption. This postprandial decrease in serum CTX-I is most probably stimulated by glucose, which induces the intestinal synthesis of glucagon-like peptide 2 [8]. Therefore, blood should be collected in standardized conditions, preferably in the fasting state in the morning. For urinary collection, the choice between a spot (preferably second morning void) and 24-hour collection is a trade-off between accuracy and feasibility. The spot sample may be easily collected with minimal patient burden. The 24-hour collection reflects the overall bone turnover. However, the 24-hour BTM excretion not corrected for creatinine may be artifically underestimated when renal function is not in a steady state. The excretion of BTM per mg of urinary creatinine may be artificially overestimated in the case of lower creatinine excretion due to sarcopenia. The BTM amount per glomerular filtrate volume assumes that the glomerular filtration is the same as that of creatinine and that there is no tubular reuptake of the BTM.

Bone metabolism is influenced by the vitamin D and calcium status, especially in the elderly. BTM levels are typically increased in the institutionalized and homebound vitamin D-deficient elderly who have lower 25-hydroxycholecalciferol (25OHD) and higher parathyroid hormone (PTH) concentrations compared to those who are ambulatory. As 25OHD is lowest during winter, especially in the elderly who rarely go out, the seasonal variation of the BTM levels is the most evident in this group. In contrast, the seasonal variability is minimal in young vitamin-D sufficient people.

BTM levels are usually increased in patients with bone metastases and their levels are correlated with the spread of bone metastases [9, 10]. ICTP and α-α-CTX-I (the native form of CTX-I) are considered the most sensitive markers of bone involvement [9–11].

BTM levels are influenced by a recent fracture [12, 13]. During the first hours after fracture, OC decreases because of stress-related cortisol secretion. Subsequently, bone formation and resorption increase reflecting the healing of the fracture. BTM levels are typically remain elevated for 4 months after fracture, and then decrease for up to 1 year.

Endogenous and exogenous corticosteroids inhibit bone formation [14]. The decrease in the OC level is most rapid and followed by a delayed and milder decrease in PICP and PINP. Bone resorption can increase; however, data are less consistent. Low-dose prednisone (5 mg/day) decreases bone formation but not bone resorption. Inhaled corticosteroids decrease OC levels in a dose- and drug-dependent manner without significant effect on other BTMs [15].

Aromatase inhibitors (used in the treatment of breast cancer) reduce the residual secretion of estrogens. GnRH agonists (used in the treatment of prostate cancer) inhibit the secretion of androgens and decrease the level of 17β-œstradiol. Consequently, both groups of drugs lead to an accelerated bone turnover and bone loss. In both cases, the expected increase in bone turnover is not observed with concomitant bisphosphonate treatment.

Premenopausal women taking oral contraceptives have lower BTMs; however, changes in BTM levels slightly depend on the composition of combined oral contraceptives [16].

Thiazolidinediones promote the differentiation of mesenchymal cells into adipocytes and may even suppress osteoblast formation. They decrease bone formation whereas bone resorption remains stable or slightly increases [17]. PINP and bone ALP, expressed by early stages of osteoblasts, decrease promptly, earlier than OC, expressed by mature osteoblasts.

BTMS AND REFERENCE VALUES

Reference intervals for BTMs have been reported from a number of studies with differing results. The reference values depend on the selection of subjects (age, sex, method of recruitment, inclusion and exclusion criteria), geographical region (including cultural habits), method of measurement, expertise of the laboratory, and the statistical approach. Given these concerns, reported reference values should interpreted with caution in clinical practice.

BONE TURNOVER RATE AND BONE LOSS

In young adults, the quantity of bone replaced at every BMU is approximately equal to the quantity of bone removed by resorption. After menopause and in diseases associated with accelerated bone loss, levels of bone resorption markers increase rapidly. Bone formation increases to fill in the higher number of resorption cavities, increasing serum levels of bone formation markers. As the quantity of bone formed is lower than the quantity of bone resorbed, there is a net bone loss at the BMU level. Therefore, the increased number of BMUs is the principal determinant of the postmenopausal BTM levels and bone loss.

In most studies, higher baseline BTM levels are associated with a faster subsequent bone loss. However, for a given BTM level, there is a large scatter of individual values of bone loss [18]. Thus, from a pathophysiological point of view, the rate of bone turnover seems to determine the subsequent bone loss. In contrast, from a clinical point of view, BTMs cannot be used for the prediction of the accelerated bone loss at the individual level.

BONE TURNOVER RATE AND FRACTURE RISK

Some, but not all, prospective cohort and case-control studies suggest that increased BTM levels predict fractures independently of age, bone mineral density (BMD), and prior fracture [19–21]. This association has been found in postmenopausal and elderly women, but not in men or in frail elderly where incident falls were the strongest predictor of fracture. BTM levels are predictive mainly of major osteoporotic fractures (vertebra, hip, multiple fractures), but not for minor peripheral osteoporotic fractures. BTMs predict fractures during short-term follow-up (less than 5 years) but not in the longer studies. Fracture risk is predicted mainly by urine bone resorption markers and, in some studies, bone ALP, but not by other bone formation markers (OC, PICP, PINP) or older nonspecific markers (total ALP, hydroxyproline).

High bone turnover is associated with lower BMD, faster bone loss, and poor bone microarchitecture in the trabecular compartment (trabecular perforations and loss, poor trabecular connectivity) and in the cortical compartment (cortical thinning, increased porosity) [22]. Consequently, remaining bone may sustain higher stress, leading to a more rapid fatigue of bone tissue and further deterioration of its material mechanical properties. Resorption cavities trigger stress risers, leading to a local weakening of a trabeculum [23]. High bone turnover is also associated with a higher fraction of recently formed partly mineralized bone, which may have suboptimal mechanical resistance. Shorter periods between remodeling cycles leave less time for the post-translational modifications of bone matrix proteins (such as cross-linking and β-isomerization of type I collagen). In one study of postmenopausal women, reduced isomerization of type I collagen, assessed by urinary α/β ratio of CTX-I, was associated with higher fracture risk independently of other predictors [24].

The potential clinical utility of BTMs is substantial; they may help identify women who will benefit the most from antiosteoporotic treatment and may improve the cost-effectiveness of treatment. However, both positive and negative data on BTMs and fracture risk should be interpreted cautiously. As mentioned above, BTM results may be inaccurate with fluctuation in renal function and improper collection. In particular, urinary bone resorption markers may be overestimated in patients with low muscle mass and high risk of falling.

Although promising, the clinical use of the BTMs for fracture prediction requires additional standardization concerning the time of collection of biological samples, choice of the BTM, expression of urinary markers, definition of the clinically valid thresholds, as well as choice of the type of fracture and of the duration of follow-up for which BTMs may be valid.

BONE TURNOVER RATE AND MONITORING

BTMs reflect the metabolic effect of drugs on bone turnover, help to establish the adequate dose, predict treatment-related increases in BMD and treatment-related reduction in fracture risk. Thus, BTMs have the potential to be helpful in the clinical management of osteoporosis.

Metabolic effect

Treatment-related changes in BTM levels depend on the mechanism of action of the drug. Antiresorptive drugs inhibit bone resorption and rapidly decrease levels of bone resorption markers. As bone formation continues in BMUs activated before treatment, bone formation markers are stable and decrease when osteoblasts fill in the lower number of BMUs formed during treatment. BMD increases rapidly during the early period, when bone resorption is reduced and bone formation is still at the baseline level. Changes in BTMs during antiresorptive therapy depend on the route of administration and dose of the drug, degree of inhibition of bone resorption, and the cellular mechanism of action of the drug. For instance, intravenous bisphosphonates or subcutaneous denosumab (monoclonal anti-RANKL antibody) decrease BTM levels faster than orally administered agents.

Inhibitors of cathepsin K (CatK), currently under development for clinical use, are an interesting group of antiresorptive drugs with regard to the effect on BTMs [25, 26]. CatK is a cysteine protease expressed by osteoclasts and degrades collagen under acidic conditions. CatK inhibitors decrease bone resorption by inhibiting the catabolism of type I collagen.

During bone resorption, collagen is degraded first by matrix metalloproteinases (MMPs), then by CatK. MMPs release CTX-MMP (C-telopeptide of type I collagen generated by MMPs, ICTP) [27]. Then, CatK degrades CTT-MMP to release CTX-I and generates NTX-I at the N-terminal end. Therefore, CatK inhibitors decrease the levels of CTX-I and NTX-I. In contrast, CTX-MMP is not further catabolized and its serum level increases. Thus, serum CTX-MMP does not reflect the suppression of bone turnover by CatK inhibitors.

Unlike bisphosphonates and denosumab, CatK inhibitors do not decrease the number of osteoclasts, only their activity. Therefore, TRACP5b concentration, which reflects the number of osteoclasts, decreases during the treatment with bisphosphonates or denosumab, but remains stable or even increases during the treatment with the CatK inhibitors [26-28]. In addition, the existing osteoclasts send signals to osteoprogenitor cells stimulating osteoblasts recruitment and differentiation [29]. Therefore, the decrease in bone formation is less prominent during the treatment with CatK inhibitors in comparison with the treatment by bisphosphonates or denosumab.

Potent bone formation-stimulating drugs, such as recombinant human PTH(1-34) (teriparatide) and PTH(1-84), induce a rapid increase in bone formation (especially PINP) followed by an increase in bone resorption [30, 31]. In the early phase of treatment, bone formation is increased, whereas bone resorption still remains low. It is during this phase, called the "anabolic window," that BMD increases most rapidly, mainly in trabecular bone. Recent data show that, in healthy subjects, the humanized sclerostin monoclonal antibody induces a rapid, dose-dependent increase in bone formation and a milder, transient decrease in serum CTX-I [32]. Strontium ranelate slightly increases bone ALP and slightly lowers serum CTX-I at the beginning of the therapy, and then both plateau throughout treatment [33].

BTMs and new treatments for osteoporosis

BTMs may be useful to establish the optimal dose of antiosteoporotic drugs because the treatment-related changes in BTMs are more rapid compared with BMD. In general, higher doses of antiresorptive agents as associated with greater reductions in BTM levels, and greater reductions in BTM levels are associated with greater increases in BMD. Transdermal 17β-estradiol, SERMs, and oral bisphosphonates induce a dose-dependent decrease in bone resorption (maximal after 3 months) and bone formation (maximal after 6 months), followed by an increase in BMD. A similar but more rapid trend was observed for an oral CatK inhibitor. The first dose of intermittent treatment with subcutaneous denosumab or intravenous bisphosphonates induces a very rapid dose-dependent decrease in bone resorption [34]. During treatment with bone formation-stimulating PTH(1-84), dose-dependent increases in BMD and BTM levels are found mainly for the lumbar spine and bone formation markers [35].

BTMs may also be useful for the assessment of the therapeutic equivalence of various doses of the same drug. Similar decreases in BTM levels in two groups of patients treated with different regimens of the same drug suggest that both regimens have similar efficacy [36, 37].

BTMs and therapeutic efficacy of antiosteoporotic treatment

Relative reduction of the risk of non-spine fracture induced by alendronate was greater in postmenopausal women with a higher baseline concentration of PINP (highest tertile) [38]. However, a similar trend was not found for spine fracture or for other BTM. Relative fracture risk reduction induced by risedronate or teraparatide did not depend on pretreatment BTM levels [39, 40]. However, in both studies, untreated women with higher BTM levels had a higher incidence of fractures. Therefore, the absolute fracture risk reduction (number of avoided fractures) was greatest for women with high pretreatment bone turnover.

The change in BMD induced by the antiresorptive treatment is a suboptimal surrogate measure of the anti-fracture efficacy. In contrast, the early decrease in BTM levels (6–12 months) is associated with the long-term increase in BMD and anti-fracture efficacy (2–3 years) of the antiresorptive therapy [41, 42]. For a given decrease in BTM levels and for a given BTM level during treatment, the incidence of vertebral fracture was similar in the active treatment and placebo groups [42, 43]. Although early teriparatide-induced increases in BTM levels correlated positively with the subsequent increase in BMD, especially with the increase in trabecular volumetric BMD [44], short-term changes in BTMs were not associated with fracture risk.

BTMs after discontinuation of antiosteoporotic treatments

Bisphosphonates are accumulated in bone and not metabolized. Therefore, the lower the cumulative dose of bisphosphonates, the sooner the BTMs return to the baseline. After the withdrawal of alendronate, which had administered for several years, BTMs increased and BMD decreased, but slowly and moderately [45]. In contrast, hormone replacement therapy, denosumab, and CatK inhibitors do not accumulate in bone. Therefore, their withdrawal is followed by a rapid increase in the BTM levels that may even attain levels exceeding pretreatment levels [46, 47]. This increase is followed by a decrease in BMD and potentially an increase in fracture risk. Withdrawal of PTH(1-84) after 1 year of treatment was followed by a return of BTM levels to baseline values and a decrease in trabecular volumetric BMD [48].

After discontinuation of denosumab followed by 12 months without treatment, re-treatment with the same agent rapidly decreased levels of BTMs to the values similar to those in patients who received continuous treatment [47]. After discontinuation of teriparatide followed by 12 months without treatment, re-treatment with the same agent induced an increase in BTM levels [49]. However, the increase was much lower compared with the initial treatment in the treatment-naïve subjects.

Additional studies are needed to determine if changes in BTMs after discontinuation of antiosteoporosis treatments are associated with subsequent fracture risk.

Combination therapy and BTMs

Several combinations of two different antiosteoporotic drugs have been studied in postmenopausal women. Alendronate and PTH(1-84) administered jointly rapidly decreased bone resorption (serum CTX-I) but less so than alendronate alone and temporarily increased bone formation (PINP, bone ALP) but less than PTH(1-84) alone [50]. Then, bone formation decreased and remained slightly below baseline levels. The time course of BTM levels during this therapy was more strongly determined by alendronate, which is consistent with the similar changes in BMD in the combination therapy group and in the group receiving alendronate alone.

The effect of PTH treatment on BTMs after antiresorptive treatment depends on the degree of inhibition of bone turnover, which itself depends on the antiresorptive drug and duration of therapy. Following prolonged treatment with alendronate, the increase in BTM levels induced by teriparatide was delayed and smaller than after raloxifene therapy [51]. In the women treated with risedronate, BTMs were higher, and teriparatide induced a greater increase in BTMs than in those treated with alendronate [52]. Alendronate administered after PTH(1-84) induces a marked decrease in BTM levels that are indistinguishable from those in women treated with alendronate alone [48, 53]. This strong inhibition of bone turnover may prevent the resorption of bone formed under PTH(1-84) treatment, which results in an additional increase in BMD.

In postmenopausal women who had been treated with alendronate for at least 6 months, administration of denosumab induced an additional decrease in BTM levels followed by an additional increase in BMD [54].

BTMs and treatment monitoring at the individual level

The goal of monitoring of antiresorptive therapy is to assess the degree of decrease in bone turnover. Changes in the BTM level that exceed the least significant change (2.8 × the inter-assay coefficient of variation of the BTM assay) exceed random BTM varibility and likely represent a true biologic effect of treatment. Ideally, the BTM level during the treatment should be below the mean for premenopausal women.

Poor adherence during antiosteoporotic treatment leads to a higher risk of fracture [55]. In postmenopausal women taking risedronate, NTX-I monitoring did not improve the persistence compared with usual care. The better the adherence the greater the average decrease in bone turnover, but it remains unclear if the association is sufficiently strong to be clinically useful [56]. Interestingly, persistence with the antiresorptive treatment was significantly better in women who received positive information corresponding to a substantial decrease in the NTX-I level [57, 58].

Available data are not sufficient to evaluate whether BTM measurement may be useful for the identification of patients at high risk of atypical subtrochanteric fracture or osteonecrosis of the jaw [59, 60]. Similarly, according to the available data, it is not possible to establish a threshold value of BTMs warranting discontinuation or reinitiation of treatment.

BONE TURNOVER MARKERS IN MEN

In boys, the growth spurt starts later and lasts longer than in girls. Therefore, young men enter the phase of consolidation (formation of peak BMD after growth arrest) later than women. Men have longer bones because they are taller than women. They also have wider bones even after adjustment for body size. At 20–25 years of age, men have BTM levels higher than women because they have more active bone turnover (later consolidation) in longer and wider bones. Then, BTMs decrease and attain their lowest levels between 50 and 60 years of age [61–63]. After the age of 60, bone formation markers remain stable or increase slightly. Bone resorption increases after the age of 60. In older men, urinary DPD and serum CTX-MMP increase with age, whereas serum CTX-I remains stable, which may reflect relative activities of various enzymes involved in the degradation of type I collagen [62, 64].

Men with high bone turnover have lower BMD as well as poor cortical microarchitecture [62, 65]. It indicates that age-related bone loss in men results at least in part from increased bone resorption. Elderly men with high BTM levels have a faster subsequent bone loss; however, this association is relatively weak [66–68]. In a nested case-control study, increased CTX-MMP level was associated with higher incidence of clinical fracture [69]. However, large prospective cohort studies showed that BTMs do not indepently predict osteoporotic fractures in elderly men [66, 67]. Interestingly, similarly to women, the higher α/β ratio of CTX-I was predictive of fragility fracture in older men independently of other predictors [70].

BTMs and antiosteoporotic treatment in men

Testosterone replacement therapy (TRT) inhibits bone turnover in overt hypogonadism (but not in men with borderline decreased testosterone concentration), if the normal concentration of bioavailable testosterone has been achieved [71, 72]. During TRT, bone resorption decreases promptly, but a decrease in urinary excretion per milligram creatinine may be partly related to the increase in muscle mass. Bone formation increases at the beginning of the TRT (direct stimulatory effect), levels off, and finally decreases, reflecting the general slowdown of bone turnover.

Oral (administered daily, weekly, or monthly) and intravenous bisphosphonates decrease BTM levels to a similar degree as in postmenopausal women [73–76]. These effects were observed in older osteoporotic men, in men with hypogonadism, in HIV-infected men, in men after cardiac transplantation, and in men after stroke. However, in men receiving androgen–deprivation therapy for prostate cancer, the decrease in BTM levels induced by denosumab is less than in postmenopausal women [77].

In men, teriparatide increased bone formation (PINP) after 1 month and bone resorption after 3 months of treatment [78]. In contrast, in eugonadal osteoporotic men, discontinuation of teriparatide was followed by a progressive decrease in the BTM levels [79]. In men treated with alendronate for 6 months, administration of teriparatide induced an increase in bone formation markers (but less than in men treated with teriparatide alone) and a slight increase in the serum NTX-I concentration which returned to baseline [80].

CONCLUSIONS

BTMs improve our understanding of the relationship between bone turnover, BMD, bone fragility, and the effect of antiosteoporotic treatment (biological mechanism, time course, anti-fracture efficacy). Data on BTMs show that the rate of bone turnover (spontaneous or modified by the therapy) is indepently associated with bone fragility in postmenopausal and elderly women (particularly urinary resorption markers). From a clinical point of view, there is optimism that measurement of BTMs may help to identify postmenopausal women at high risk of fracture and may improve persistence with antiresorptive treatment. It suggests that the use of BTMs may improve the cost-effectiveness of the antiosteoporotic treatment.

REFERENCES

1. Marti J, Seeman E. 2008. Bone remodelling: Its local regulation and the emergence of bone fragility. *Best Pract Res Clin Endocrinol Metab* 22: 701–722.
2. Vasikaran S, Cooper C, Eastell R, Griesmacher A, Morris HA, Trenti T, Kanis JA. 2011. International Osteoporosis Foundation and International Federation of Clinical Chemistry and Laboratory Medicine Position on bone

marker standards in osteoporosis. *Clin Chem Lab Med* 49(8): 1271–1274.

3. Schafer AL, Vittinghoff E, Ramachandran R, Mahmoudi N, Bauer DC. 2010. Laboratory reproducibility of biochemical markers of bone turnover in clinical practice. *Osteoporos Int* 21: 439–445.

4. Garnero P, Borel O, Delmas PD. 2001. Evaluation of a fully automated serum assay for C-terminal crosslinking telopeptide of type I collagen in osteoporosis. *Clin Chem* 47: 694–702.

5. Blatt JM, Allen MP, Baddam S, Chase CL, Dasu BN, Dickens DM, Hardt SJ, Hebert RT, Hsu YC, Kitazawa CT, Li SF, Mangan WM, Patel PJ, Pfeiffer JW, Quiwa NB, Scratch MA, Widunas JT. 1998. A miniaturized, self-contained, single-use, disposable assay device for the quantitative determination of the bone resorption marker, NTx, in urine. *Clin Chem* 44: 2051–2052.

6. Szulc P, Delmas PD. 2008. Biochemical markers of bone turnover: Potential use in the investigation and management of postmenopausal osteoporosis. *Osteoporos Int* 19: 1683–1704.

7. Qvist P, Christgau C, Pedersen BJ, Schlemmer A, Christiansen C. 2002. Circadian variation in the serum concentration of C-terminal telopeptide of type I collagen (serum CTx): Effects of gender, age, menopausal status, posture, daylight, serum cortisol, and fasting. *Bone* 31 57–61.

8. Yavropoulou MP, Tomos K, Tsekmekidou X, Anastasiou O, Zebekakis P, Karamouzis M, Gotzamani-Psarrakou A, Chassapopoulou E, Chalkia P, Yovos JG. 2011. Response of biochemical markers of bone turnover to oral glucose load in diseases that affect bone metabolism. *Eur J Endocrinol* 164: 1035–1041.

9. Voorzanger-Rousselot N, Juillet F, Mareau E, Zimmermann J, Kalebic T, Garnero P. 2006. Association of 12 serum biochemical markers of angiogenesis, tumour invasion and bone turnover with bone metastases from breast cancer: A crossectional and longitudinal evaluation. *Br J Cancer* 95: 506–514.

10. Leeming DJ, Koizumi M, Byrjalsen I, Li B, Qvist P, Tankó LB. 2006. The relative use of eight collagenous and noncollagenous markers for diagnosis of skeletal metastases in breast, prostate, or lung cancer patients. *Cancer Epidemiol Biomarkers Prev* 15: 32–38.

11. Leeming DJ, Delling G, Koizumi M, Henriksen K, Karsdal MA, Li B, Qvist P, Tankó LB, Byrjalsen I. 2006. Alpha CTX as a biomarker of skeletal invasion of breast cancer: Immunolocalization and the load dependency of urinary excretion. *Cancer Epidemiol Biomarkers Prev* 15: 1392–1395.

12. Ivaska KK, Gerdhem P, Akesson K, Garnero P, Obrant KJ. 2007. Effect of fracture on bone turnover markers: A longitudinal study comparing marker levels before and after injury in 113 elderly women. *J Bone Miner Res* 22: 1155–1164.

13. Stoffel K, Engler H, Kuster M, Riesen W. 2007. Changes in biochemical markers after lower limb fractures. *Clin Chem* 53: 131–134.

14. Dovio A, Perazzolo L, Osella G, Ventura M, Termine A, Milano E, Bertolotto A, Angeli A. 2004. Immediate fall of bone formation and transient increase of bone resorption in the course of high-dose, short-term glucocorticoid therapy in young patients with multiple sclerosis. *J Clin Endocrinol Metab* 89: 4923–4928.

15. Richy F, Bousquet J, Ehrlich GE, Meunier PJ, Israel E, Morii H, Devogelaer JP, Peel N, Haim M, Bruyere O, Reginster JY. 2003. Inhaled corticosteroids effects on bone in asthmatic and COPD patients: A quantitative systematic review. *Osteoporos Int* 14: 179–190.

16. Herrmann M, Seibel MJ. 2010. The effects of hormonal contraceptives on bone turnover markers and bone health. *Clin Endocrinol (Oxf)* 72: 571–583.

17. Grey A, Bolland M, Gamble G, Wattie D, Horne A, Davidson J, Reid IR. 2007. The peroxisome proliferator-activated receptor-gamma agonist rosiglitazone decreases bone formation and bone mineral density in healthy postmenopausal women: A randomized, controlled trial. *J Clin Endocrinol Metab* 92: 1305–1310.

18. Rogers A, Hannon RA, Eastell R. 2000. Biochemical markers as predictors of rates of bone loss after menopause. *J Bone Miner Res* 15: 1398–1404.

19. Vasikaran S, Eastell R, Bruyère O, Foldes AJ, Garnero P, Griesmacher A, McClung M, Morris HA, Silverman S, Trenti T, Wahl DA, Cooper C, Kanis JA. 2011. Markers of bone turnover for the prediction of fracture risk and monitoring of osteoporosis treatment: A need for international reference standards. *Osteoporos Int* 22: 391–420.

20. Daele PLA van, Seibel MJ, Burger H, Hofman A, Grobbee DE, van Leeuwen JPTM, Birkenhager JC, Pols HAP. 1996. Case-control analysis of bone resorption markers, disability, and hip fracture risk: The Rotterdam study. *Br Med J* 312: 482–483.

21. Garnero P, Hausher E, Chapuy MC, Marcelli C, Grandjean H, Muller C, Cormier C, Bréart G, Meunier PJ, Delmas PD. 1996. Markers of bone resorption predict hip fracture in elderly women: The Epidos prospective study. *J Bone Miner Res* 11: 1531–1538.

22. Bouxsein ML, Delmas PD. 2008. Considerations for development of surrogate endpoints for antifracture efficacy of new treatments in osteoporosis: A perspective. *J Bone Miner Res* 23: 1155–1167.

23. Dempster DW. 2000. The contribution of trabecular architecture to cancellous bone quality. *J Bone Miner Res* 15: 20–23.

24. Garnero P, Cloos P, Sornay-Rendu E, Qvist P, Delmas PD. 2002. Type I collagen racemization and isomerization and the risk of fracture in postmenopausal women: The OFELY prospective study. *J Bone Miner Res* 17: 826–833.

25. Bone HG, McClung MR, Roux C, Recker RR, Eisman JA, Verbruggen N, Hustad CM, DaSilva C, Santora AC, Ince BA. 2010. Odanacatib, a cathepsin-K inhibitor for osteoporosis: A two-year study in postmenopausal women with low bone density. *J Bone Miner Res* 25: 937–947.

26. Eastell R, Nagase S, Ohyama M, Small M, Sawyer J, Boonen S, Spector T, Kuwayama T, Deacon S. 2011.

26. Safety and efficacy of the cathepsin K inhibitor ONO-5334 in postmenopausal osteoporosis: The OCEAN study. *J Bone Miner Res* 26: 1303–1312.
27. Garnero P, Ferreras M, Karsdal MA, Nicamhlaoibh R, Risteli J, Borel O, Qvist P, Delmas PD, Foged NT, Delaissé JM. 2003. The type I collagen fragments ICTP and CTX reveal distinct enzymatic pathways of bone collagen degradation. *J Bone Miner Res* 18: 859–867.
28. Eastell R, Christiansen C, Grauer A, Kutilek S, Libanati C, McClung MR, Reid IR, Resch H, Siris E, Uebelhart D, Wang A, Weryha G, Cummings SR. 2011. Effects of denosumab on bone turnover markers in postmenopausal osteoporosis. *J Bone Miner Res* 26: 530–537.
29. Pederson L, Ruan M, Westendorf JJ, Khosla S, Oursler MJ. 2008. Regulation of bone formation by osteoclasts involves Wnt/BMP signaling and the chemokine sphingosine-1-phosphate. *Proc Natl Acad Sci U S A* 105: 20764–20769.
30. Glover SJ, Eastell R, McCloskey EV, Rogers A, Garnero P, Lowery J, Belleli R, Wright TM, John MR. 2009. Rapid and robust response of biochemical markers of bone formation to teriparatide therapy. *Bone* 45: 1053–1058.
31. Greenspan SL, Bone HG, Ettinger MP, Hanley DA, Lindsay R, Zanchetta JR, Blosch CM, Mathisen AL, Morris SA, Marriott TB. 2007. Effect of recombinant human parathyroid hormone (1-84) on vertebral fracture and bone mineral density in postmenopausal women with osteoporosis: A randomized trial. *Ann Intern Med* 146: 326–339.
32. Padhi D, Jang G, Stouch B, Fang L, Posvar E. 2011. Single-dose, placebo-controlled, randomized study of AMG 785, a sclerostin monoclonal antibody. *J Bone Miner Res* 26: 19–26.
33. Meunier PJ, Roux C, Seeman E, Ortolani S, Badurski JE, Spector TD, Cannata J, Balogh A, Lemmel EM, Pors-Nielsen S, Rizzoli R, Genant HK, Reginster JY. 2004. The effects of strontium ranelate on the risk of vertebral fracture in women with postmenopausal osteoporosis. *N Engl J Med* 350: 459–468.
34. McClung MR, Lewiecki EM, Cohen SB, Bolognese MA, Woodson GC, Moffett AH, Peacock M, Miller PD, Lederman SN, Chesnut CH, Lain D, Kivitz AJ, Holloway DL, Zhang C, Peterson MC, Bekker PJ. 2006. Denosumab in postmenopausal women with low bone mineral density. *N Engl J Med* 354: 821–831.
35. Hodsman AB, Hanley DA, Ettinger MP, Bolognese MA, Fox J, Metcalfe AJ, Lindsay R. 2003. Efficacy and safety of human parathyroid hormone-(1-84) in increasing bone mineral density in postmenopausal osteoporosis. *J Clin Endocrinol Metab* 88: 5212–5220.
36. Delmas PD, Benhamou CL, Man Z, Tlustochowicz W, Matzkin E, Eusebio R, Zanchetta J, Olszynski WP, Recker RR, McClung MR. 2008. Monthly dosing of 75 mg risedronate on 2 consecutive days a month: Efficacy and safety results. *Osteoporos Int* 19: 1039–1045.
37. Rizzoli R, Greenspan SL, Bone G 3rd, Schnitzer TJ, Watts NB, Adami S, Foldes AJ, Roux C, Levine MA, Uebelhart B, Santora AC 2nd, Kaur A, Peverly CA, Orloff JJ. 2002. Two-year results of once-weekly administration of alendronate 70 mg for the treatment of postmenopausal osteoporosis. *J Bone Miner Res* 17: 1988–1996.
38. Bauer DC, Garnero P, Hochberg MC, Santora A, Delmas P, Ewing SK, Black DM. 2006. Pretreatment levels of bone turnover and the antifracture efficacy of alendronate: The fracture intervention trial. *J Bone Miner Res* 21: 292–299.
39. Seibel MJ, Naganathan V, Barton I, Grauer A. 2004. Relationship between pretreatment bone resorption and vertebral fracture incidence in postmenopausal osteoporotic women treated with risedronate. *J Bone Miner Res* 19: 323–329.
40. Delmas PD, Licata AA, Reginster JY, Crans GG, Chen P, Misurski DA, Wagman RB, Mitlak BH. 2006. Fracture risk reduction during treatment with teriparatide is independent of pretreatment bone turnover. *Bone* 39: 237–243.
41. Bauer DC, Black DM, Garnero P, Hochberg M, Ott S, Orloff J, Thompson DE, Ewing SK, Delmas PD. 2004. Change in bone turnover and hip, non-spine, and vertebral fracture in alendronate-treated women: The fracture intervention trial. *J Bone Miner Res* 19: 1250–1258.
42. Eastell R, Hannon RA, Garnero P, Campbell MJ, Delmas PD. 2007. Relationship of early changes in bone resorption to the reduction in fracture risk with risedronate: Review of statistical analysis. *J Bone Miner Res* 22: 1656–1660.
43. Reginster JY, Sarkar S, Zegels B, Henrotin Y, Bruyere O, Agnusdei D, Collette J. 2004. Reduction in PINP, a marker of bone metabolism, with raloxifene treatment and its relationship with vertebral fracture risk. *Bone* 34: 344–351.
44. Chen P, Satterwhite JH, Licata AA, Lewiecki EM, Sipos AA, Misurski DM, Wagman RB. 2005. Early changes in biochemical markers of bone formation predict BMD response to teriparatide in postmenopausal women with osteoporosis. *J Bone Miner Res* 20: 962–970.
45. Black DM, Schwartz AV, Ensrud KE, Cauley JA, Levis S, Quandt SA, Satterfield S, Wallace RB, Bauer DC, Palermo L, Wehren LE, Lombardi A, Santora AC, Cummings SR. 2006. Effects of continuing or stopping alendronate after 5 years of treatment: The Fracture Intervention Trial Long-term Extension (FLEX): A randomized trial. *JAMA* 296: 2927–2938.
46. Sornay-Rendu E, Garnero P, Munoz F, Duboeuf F, Delmas PD. 2003. Effect of withdrawal of hormone replacement therapy on bone mass and bone turnover: The OFELY study. *Bone* 33: 159–166.
47. Miller PD, Bolognese MA, Lewiecki EM, McClung MR, Ding B, Austin M, Liu Y, San Martin J. 2008. Effect of denosumab on bone density and turnover in postmenopausal women with low bone mass after long-term continued, discontinued, and restarting of therapy: A randomized blinded phase 2 clinical trial. *Bone* 43: 222–229.
48. Black DM, Bilezikian JP, Ensrud KE, Greenspan SL, Palermo L, Hue T, Lang TF, McGowan JA, Rosen CJ.

2005. One year of alendronate after one year of parathyroid hormone (1-84) for osteoporosis. *N Engl J Med* 353: 555–565.

49. Finkelstein JS, Wyland JJ, Leder BZ, Burnett-Bowie SM, Lee H, Jüppner H, Neer RM. 2009. Effects of teriparatide retreatment in osteoporotic men and women. *J Clin Endocrinol Metab* 94: 2495–2501.

50. Black DM, Greenspan SL, Ensrud KE, Palermo L, McGowan JA, Lang TF, Garnero P, Bouxsein ML, Bilezikian JP, Rosen CJ. 2003. The effects of parathyroid hormone and alendronate alone or in combination in postmenopausal osteoporosis. *N Engl J Med* 349: 1207–1215.

51. Ettinger B, San Martin J, Crans G, Pavo I. 2004. Differential effects of teriparatide on BMD after treatment with raloxifene or alendronate. *J Bone Miner Res* 19: 745–751.

52. Miller PD, Delmas PD, Lindsay R, Watts NB, Luckey M, Adachi J, Saag K, Greenspan SL, Seeman E, Boonen S, Meeves S, Lang TF, Bilezikian JP. 2008. Early responsiveness of women with osteoporosis to teriparatide after therapy with alendronate or risedronate. *J Clin Endocrinol Metab* 93: 3785–3793.

53. Rittmaster RS, Bolognese M, Ettinger MP, Hanley DA, Hodsman AB, Kendler DL, Rosen CJ. 2000. Enhancement of bone mass in osteoporotic women with parathyroid hormone followed by alendronate. *J Clin Endocrinol Metab* 85: 2129–2134.

54. Kendler DL, Roux C, Benhamou CL, Brown JP, Lillestol M, Siddhanti S, Man HS, San Martin J, Bone HG. 2010. Effects of denosumab on bone mineral density and bone turnover in postmenopausal women transitioning from alendronate therapy. *J Bone Miner Res* 25: 72–81.

55. Siris ES, Harris ST, Rosen CJ, Barr CE, Arvesen JN, Abbott TA, Silverman S. 2006. Adherence to bisphosphonate therapy and fracture rates in osteoporotic women: Relationship to vertebral and nonvertebral fractures from 2 US claims databases. *Mayo Clin Proc* 81: 1013–1022.

56. Eastell R, Vrijens B, Cahall DL, Ringe JD, Garnero P, Watts NB. 2011. Bone turnover markers and bone mineral density response with risedronate therapy: Relationship with fracture risk and patient adherence. *J Bone Miner Res* 26: 1662–1669.

57. Clowes JA, Peel NF, Eastell R. 2004. The impact of monitoring on adherence and persistence with antiresorptive treatment for postmenopausal osteoporosis: A randomized controlled trial. *J Clin Endocrinol Metab* 89: 1117–1123.

58. Delmas PD, Vrijens B, Eastell R, Roux C, Pols HA, Ringe JD, Grauer A, Cahall D, Watts NB. 2007. Effect of monitoring bone turnover markers on persistence with risedronate treatment of postmenopausal osteoporosis. *J Clin Endocrinol Metab* 92: 1296–1304.

59. Baim S, Miller PD. 2009. Assessing the clinical utility of serum CTX in postmenopausal osteoporosis and its use in predicting risk of osteonecrosis of the jaw. *J Bone Miner Res* 24: 561–574.

60. Visekruna M, Wilson D, McKiernan FE. 2008. Severely suppressed bone turnover and atypical skeletal fragility. *J Clin Endocrinol Metab* 93: 2948–2952.

61. Fatayerji D, Eastell R. 1999. Age-related changes in bone turnover in men. *J Bone Miner Res* 14: 1203–1210.

62. Szulc P, Garnero P, Munoz F, Marchand F, Delmas PD. 2001. Cross-sectional evaluation of bone metabolism in men. *J Bone Miner Res* 16: 1642–1650.

63. Khosla S, Melton LJ 3rd, Atkinson EJ, O'Fallon WM, Klee GG, Riggs BL. 1998. Relationship of serum sex steroid levels and bone turnover markers with bone mineral density in men and women: a key role for bioavailable estrogen. *J Clin Endocrinol Metab* 83: 2266–2274.

64. Chandani AK, Scariano JK, Glew RH, Clemens JD, Garry PJ, Baumgartner RN. 2000. Bone mineral density and serum levels of aminoterminal propeptides and cross-linked N-telopeptides of type I collagen in elderly men. *Bone* 26: 513–518.

65. Chaitou A, Boutroy S, Vilayphiou N, Munoz F, Delmas PD, Chapurlat R, Szulc P. 2010. Association between bone turnover rate and bone microarchitecture in men: The STRAMBO study. *J Bone Miner Res* 25: 2313–2323.

66. Bauer DC, Garnero P, Harrison SL, Cauley JA, Eastell R, Ensrud KE, Orwoll E. 2009. Biochemical markers of bone turnover, hip bone loss, and fracture in older men: The MrOS study. *J Bone Miner Res* 24: 2032–2038.

67. Szulc P, Montella A, Delmas PD. 2008. High bone turnover is associated with accelerated bone loss but not with increased fracture risk in men aged 50 and over: The prospective MINOS study. *Ann Rheum Dis* 67: 1249–1255.

68. Dennison E, Eastell R, Fall CH, Kellingray S, Wood PJ, Cooper C. 1999. Determinants of bone loss in elderly men and women: A prospective population-based study. *Osteoporos Int* 10: 384–391.

69. Meier C, Nguyen TV, Center JR, Seibel MJ, Eisman JA. 2005. Bone resorption and osteoporotic fractures in elderly men: The dubbo osteoporosis epidemiology study. *J Bone Miner Res* 20: 579–587.

70. Bauer D, Garnero P, Litwack Harrison S, Cauley J, Ensrud K, Eastell R, Orwoll E. Type I Collagen Isomerization (Alpha/Beta CTX Ratio) and Risk of Clinical Vertebral Fracture in Men: A Prospective Study. http://www.asbmr.org/Meetings/AnnualMeeting/AbstractDetail.aspx?aid=7ed933e3-0487-4b5a-b2dd-747876d1ecde.

71. Amory JK, Watts NB, Easley KA, Sutton PR, Anawalt BD, Matsumoto AM, Bremner WJ, Tenover JL. 2004. Exogenous testosterone or testosterone with finasteride increases bone mineral density in older men with low serum testosterone. *J Clin Endocrinol Metab* 89: 503–510.

72. Wang C, Swerdloff RS, Iranmanesh A, Dobs A, Snyder PJ, Cunningham G, Matsumoto AM, Weber T, Berman N. 2001. Effects of transdermal testosterone gel on bone turnover markers and bone mineral density in hypogonadal men. *Clin Endocrinol (Oxf)* 54: 739–750.

73. Orwoll ES, Binkley NC, Lewiecki EM, Gruntmanis U, Fries MA, Dasic G. 2010. Efficacy and safety of monthly ibandronate in men with low bone density. *Bone* 46: 970–976.
74. Boonen S, Orwoll ES, Wenderoth D, Stoner KJ, Eusebio R, Delmas PD. 2009. Once-weekly risedronate in men with osteoporosis: Results of a 2-year, placebo-controlled, double-blind, multicenter study. *J Bone Miner Res* 24: 719–725.
75. Bolland MJ, Grey AB, Horne AM, Briggs SE, Thomas MG, Ellis-Pegler RB, Woodhouse AF, Gamble GD, Reid IR. 2007. Annual zoledronate increases bone density in highly active antiretroviral therapy-treated human immunodeficiency virus-infected men: A randomized controlled trial. *J Clin Endocrinol Metab* 92: 1283–1288.
76. Orwoll E, Ettinger M, Weiss S, Miller P, Kendler D, Graham J, Adami S, Weber K, Lorenc R, Pietschmann P, Vandormael K, Lombardi A. 2000. Alendronate for the treatment of osteoporosis in men. *N Engl J Med* 343: 604–610.
77. Smith MR, Egerdie B, Hernández Toriz N, Feldman R, Tammela TL, Saad F, Heracek J, Szwedowski M, Ke C, Kupic A, Leder BZ, Goessl C. 2009. Denosumab in men receiving androgen-deprivation therapy for prostate cancer. *N Engl J Med* 361: 745–755.
78. Orwoll ES, Scheele WH, Paul S, Adami S, Syversen U, Diez-Perez A, Kaufman JM, Clancy AD, Gaich GA. 2003. The effect of teriparatide [human parathyroid hormone (1-34)] therapy on bone density in men with osteoporosis. *J Bone Miner Res* 18: 9–17.
79. Leder BZ, Neer RM, Wyland JJ, Lee HW, Burnett-Bowie SM, Finkelstein JS. 2009. Effects of teriparatide treatment and discontinuation in postmenopausal women and eugonadal men with osteoporosis. *J Clin Endocrinol Metab* 94: 2915–2921.
80. Finkelstein JS, Leder BZ, Burnett SM, Wyland JJ, Lee H, de la Paz AV, Gibson K, Neer RM. 2006. Effects of teriparatide, alendronate, or both on bone turnover in osteoporotic men. *J Clin Endocrinol Metab* 91: 2882–2887.

36
Bone Biopsy and Histomorphometry in Clinical Practice

Robert R. Recker

Introduction 307
Organization and Function of Bone Cells 307
Basic Histomorphometric Variables 309
Interpretation of Findings 309
Findings in Metabolic Bone Disease 310
Obtaining the Specimen 313
Specimen Processing and Analysis 313
Indications for Bone Biopsy and Histomorphometry 314
Acknowledgments 314
References 314

INTRODUCTION

Histological examination of undecalcified transilial bone biopsy specimens is a valuable and well-established clinical and research tool for studying the etiology, pathogenesis, and treatment of metabolic bone diseases. In this chapter, we will review the underlying organization and function of bone cells; identify a set of basic structural and kinetic histomorphometric variables; outline an approach to interpretation of findings, with examples from a range of metabolic bone diseases; describe techniques for obtaining, processing, and analyzing transilial biopsy specimens; identify clinical situations in which bone histomorphometry can be useful; and relate histomorphometric measures to data from other methods for assessing bone properties and bone physiology.

ORGANIZATION AND FUNCTION OF BONE CELLS

Intermediary organization of the skeleton

In what he termed the intermediary organization (IO) of the skeleton, Frost [1] described four discrete functions of bone cells: growth, modeling, remodeling, and fracture repair. Although each involves osteoclasts and osteoblasts, the coordinated outcomes differ greatly. Growth elongates the skeleton; modeling shapes it during growth; remodeling removes and replaces bone tissue; and fracture repair heals sites of structural failure.

The remodeling IO, which predominates during adult life, is the focus of this chapter. Coordinated groups of bone cells (i.e., osteoclasts, osteoblasts, osteocytes, and lining cells) comprise the basic multicellular units (BMUs) that carry out bone remodeling. Basic structural units (BSUs) are the packets of new bone that BMUs form [2]. All adult-onset metabolic bone disease involves derangement of the remodeling IO.

Bone cells

Osteoclasts, large-to-giant cells that are typically multinucleated, resorb bone (both its matrix or osteoid, and mineral). They excavate shallow pits on the surface of cancellous bone, and they appear at the leading edge of tunnels ("cutting cones") in Haversian bone. Light microscopy discloses an irregular cell shape, foamy, acidophilic cytoplasm, a perimeter zone of attachment to the bone, a ruffled border that appears in the fluid cavity between the cell and the mineralized bone matrix, and positive staining for tartrate-resistant acid phosphatase (TRAP).

Primer on the Metabolic Bone Diseases and Disorders of Mineral Metabolism, Eighth Edition. Edited by Clifford J. Rosen.
© 2013 American Society for Bone and Mineral Research. Published 2013 by John Wiley & Sons, Inc.

Osteoblasts form new bone at sites of resorption. They produce the collagenous and noncollagenous constituents of bone matrix and participate in mineralization [3]. Under light microscopy, they appear as plump cells lined up at the surface of unmineralized osteoid. As the site matures, the cells lose their plump appearance.

Osteocytes, derived from osteoblasts, remain at the remodeling site, and are buried as bone formation advances at remodeling sites. They reside individually in small lacunae within the mineralized bone matrix. Their cytoplasmic processes extend through a fine network of narrow canaliculi to form an interconnected network that extends throughout living bone. This network is well situated to monitor the local strain environment and local microdamage, and to initiate organized bone cell work in response to changes in strain or microdamage.

Lining cells, also of osteoblast origin, cover cancellous and endocortical bone surfaces. By light microscopy, they appear as elongated, flattened, darkly stained nuclei. The localization and initiation of remodeling probably involves these cells.

Bone remodeling process

Remodeling occurs on cancellous and Haversian bone surfaces. The first step is activation of osteoclast precursors to form osteoclasts that then begin to excavate a cavity. After removal of about $0.05\,mm^3$ of bone tissue, the site remains quiescent for a short time. Then, activation of osteoblast precursors occurs at the site, and the excavation is refilled. The average length of time required to complete the remodeling cycle is approximately 6 months [4]: about 4 weeks for resorption and the rest for formation.

The healthy bone remodeling system accesses the required building materials within a favorable physiologic milieu to replace fully a packet of aged, microdamaged bone tissue with new, mechanically competent bone. However, overuse can overwhelm the capacity of the system to repair microdamage (the stress fractures that occur in military recruits are an example). The healthy bone remodeling system modifies bone architecture to meet changing mechanical needs. However, the system also promptly reduces the mass of underused bone (the bone loss of extended bed rest, paralysis, or space travel are examples). All bone loss occurs through bone remodeling. The bone remodeling system responds to nutritional and humoral as well as mechanical influences. Among the effects of vitamin D deficiency in adults, for example, is impaired mineralization of bone matrix. Finally, as other chapters describe, bone remodeling involves complex signaling processes between and within bone cells, and metabolic bone diseases of genetic origin involve defects at this level. Figures 36.1–36.3 present representative photomicrographs from human transilial biopsy specimens. An extensive atlas has also been published [5].

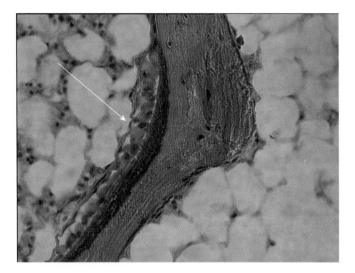

Fig. 36.1. A normal bone forming surface. Unmineralized osteoid is covered with plump osteoblasts, as identified by the arrow.

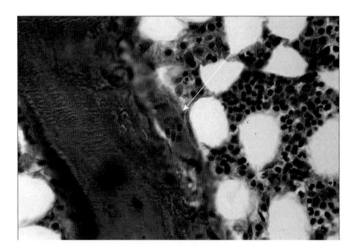

Fig. 36.2. A normal bone resorbing surface. The arrow locates a multinucleated osteoclast in a Howship's lacuna.

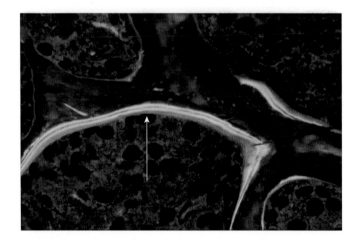

Fig. 36.3. The arrow identifies a mineralizing surface containing fluorescent double-labels.

BASIC HISTOMORPHOMETRIC VARIABLES

Bone biopsy specimens for histomorphometric examination are ordinarily obtained at the transilial site and shipped to specialized laboratories for processing and microscopic analysis. Later sections of this chapter outline these procedures. Of the dozens of measurements and calculations that have been devised, we provide here descriptions of several frequently used variables. Together they describe a basic set of structural and kinetic features. Nomenclature is as approved by a committee of the American Society of Bone and Mineral Research [6].

Structural features

Core width (C.Wi) represents the thickness of the ilium (i.e., distance between periosteal surfaces, in millimeters) at the point of biopsy. Cortical width (Ct.Wi) is the combined thickness, in millimeters, of both cortices. Cortical porosity (Ct.Po) is the area of intracortical holes as percent of total cortical area.

Cancellous bone volume (BV/TV) is the percent of total marrow area (including trabeculae) occupied by cancellous bone. Wall thickness (W.Th) is the mean distance in micrometers between resting cancellous surfaces (i.e., surfaces without osteoid or Howship's lacunae) and corresponding cement lines.

Trabecular thickness (Tb.Th) is the mean distance across individual trabeculae, in micrometers, and trabecular separation (Tb.Sp) is the mean distance, also in micrometers, between trabeculae. Trabecular number (Tb.N) per millimeter is calculated as (BV/TV)/Tb.Th. These variables can be used to evaluate trabecular connectivity [7]. Other measures of trabecular connectivity include the ratio of nodes to free ends [8], star volume [9, 10], and trabecular bone pattern factor (TBPf) [11].

Eroded surface (ES/BS) is the percent of cancellous surface occupied by Howship's lacunae, with and without osteoclasts. Osteoblast surface (Ob.S/BS) and osteoclast surface (Oc.S/BS) identify the percent of cancellous surface occupied by osteoblasts and osteoclasts, respectively. Osteoid surface (OS/BS) is the percent of cancellous surface with unmineralized osteoid, with and without osteoblasts. Osteoid thickness (O.Th) is the mean thickness, in micrometers, of the osteoid on cancellous surfaces.

Kinetic features

A fluorochrome labeling agent, taken orally on a strict schedule before biopsy, deposits a fluorescent double-label at sites of active mineralization and allows rates of change to be determined [12]. Mineralizing surface (MS/BS) is the percentage of cancellous surface that is mineralizing and thus labeled. The most accurate version of MS/BS includes surfaces with a double-label plus one-half of those with a single-label [13]. Clear definition of MS/BS is crucial, because it is used to calculate bone formation rates, bone formation periods, and mineralization lag time.

Mineral appositional rate (MAR), is the rate (in µm/day) at which new bone mineral is being added to cancellous surfaces. MAR represents distance between labels at doubly labeled surfaces divided by the marker interval (span in days between the midpoints of each labeling period). This and all measurements of thickness must be corrected for obliquity (i.e., the randomness of the angle between the plane of the section and the plane of the cancellous surface) by use of a scaling factor [7].

Activation frequency (Ac.f) is the probability that a new remodeling cycle will begin at any point on the cancellous bone surface. Bone formation rates (BFR/BV and BFR/BS) are estimates of cancellous bone volume (in $mm^3/mm^3/year$) and cancellous bone surface (in $mm^3mm^2/mm/year$), respectively, that are being replaced annually; BFR/BS = Ac.f × W.Th [14]. Formation period (FP) is the time in years required to complete a new cancellous BSU. Mineralization lag time (Mlt) is the interval in days between osteoid formation and mineralization. The most accurate version of Mlt is calculated as O.Th/MAR×MS/OS. Microcrack density (Cr.d.) is the number of microcracks per area if mineralized bone area ($\#/mm^2$), and microcrack length (Cr.L) is the average length (in millimeters) of visualized microcracks [15]. Apoptosis can be quantified as the percent of total osteocytes that are identified as apoptotic using special stains [16].

INTERPRETATION OF FINDINGS

Reference data

In 1988, Recker et al. [4] published the results of a study to establish reference values for histomorphometric variables in postmenopausal white women. The 34 healthy subjects were evenly distributed into three age groups: 45–54, 55–64, and 65–74 years. They ranged broadly in age at menopause and in years past menopause at the time of biopsy. A comparative study of 12 blacks and 13 whites, aged 19–46 years, has also been published [17].

In 2000, Glorieux et al. [18] reported histomorphometric data from 58 white subjects in each of five age groups: 1.5–6.9, 7.0–10.9, 11.0–13.9, 14.0–16.9, and 17.0–22.9 years. Biopsy specimens were obtained during corrective orthopedic surgeries, but the subjects had been ambulatory and otherwise healthy. The report includes within-subject coefficient of variation (CV) derived from analysis of adjacent duplicate biopsy specimens in eight subjects.

A recent paper from our center reported Ac.f in several sets of transilial biopsy specimens. In 50 paired transilial biopsy specimens taken during perimenopause and in early postmenopause, a year after last menses, median values for Ac.f increased from 0.13/year to 0.24/year, respectively (p < 0.001). Ac.f. was higher still (median, 0.37; p < 0.01) in another group of ostensibly normal women who were postmenopausal by an average of 13 years [19]. Others have published reference databases [20–23].

Replacement of normal marrow elements

A variety of hematopoietic cells and a varying proportion of fat cells normally occupy the marrow space at the transilial biopsy site. If these normal marrow elements have been displaced by fibrous tissue (osteitis fibrosa), clumps of tumor cells, or sheets of abnormal hematopoietic cells, this change will be obvious to the histomorphometrist. The biopsy preparations described here preserve cellular detail, spatial relationships, and architectural features. However, this approach is unsuitable for hematologic diagnosis because of the time that histomorphometry laboratories require to generate a report (typically, at least 4 weeks).

Cortical bone deficit

Both the angle of the biopsy and site-to-site variation in cortical thickness at the biopsy site influence Ct.Wi. Nevertheless, low bone density at the lumbar spine and/or proximal femur is often reflected in low values for Ct.Wi [24]. Evidence of trabeculation of the cortex (i.e., formation of a transitional zone with characteristic coarse trabeculae) indicates that cortical bone once present in the area adjacent to the marrow space, has been lost [25].

Cancellous bone deficit

Low BV/TV indicates a cancellous bone deficit. Generalized trabecular thinning (decreased Tb.Th) and/or complete loss of trabecular elements (poor trabecular connectivity) may contribute to this deficit. The latter finding (e.g., low Tb.N with high Tb.Sp) characterizes bone that is more fragile than its overall mass would suggest.

Altered bone remodeling

Ac.f is an indicator of the overall level of remodeling activity in cancellous bone. Values for Ac.f correlate with excretion of bone resorption markers (r = 0.71) (Recker, unpublished observations). In biopsy specimens from ostensibly healthy women, we have yet to see a case in which the subject had followed the fluorochrome labeling protocol but that label could not be found in cancellous areas. However, a recent paper from our laboratory, cited earlier, reports three cases of no label (i.e., zero Ac.f) among women with untreated postmenopausal osteoporosis [19]. Further, extensive attention has been directed at osteoporosis drugs that suppress remodeling, and whether they might sometimes suppress remodeling to the extent that microdamage repair is not adequate. This has led to questions concerning a histomorphometric definition of abnormally low remodeling rates and has been thoroughly discussed in a recent publication [26]. These authors concluded that absence of fluorochrome label in a human transiliac biopsy is evidence of abnormal reduction in remodeling.

Abnormal osteoid morphology

The characteristic arrangement of osteoid (collagen) fibers in lamellar and woven bone is readily apparent. Lamellar bone contains collagen fibers arranged in layers, and woven bone contains collagen fibers arranged in a random fashion. Woven bone in transilial specimens is generally associated with an intense stimulus to rapid bone formation, as in Paget's disease or renal osteodystrophy. It can also occur in osteitis fibrosa. In osteogenesis imperfecta, collagen abnormalities result in production of variable amounts of woven bone, but may be subtle enough to escape detection. The presence of woven bone can be suspected in stained sections by the presence of increased numbers of osteocytes, and by the hint of randomly arranged collagen. However, the best way to identify woven bone is by using polarizing lenses and light microscopy of unstained sections. One polarizing lens is placed under the substage of the microscope above the light source, and the other is placed in the light path above the section. When the polarizing lenses are rotated so that the polarity of one is arranged 90° relative to the polarity of the other (the views in the eyepieces are very dark) the differences between areas of woven and lamellar bone are readily apparent due to refraction of light in random directions by woven bone.

Accumulation of unmineralized osteoid

Parfitt has described the complex relationships between dynamic indices of bone formation and static indices of osteoid accumulation [14]. Increases in OS/BS, O.Th, and Mlt indicate failure of osteoid to mineralize normally. If mineralization is arrested completely, no double-label will be seen, and Mlt is unmeasurable [27].

FINDINGS IN METABOLIC BONE DISEASE

In Table 36.1, we identify key histomorphometric findings that characterize representative types of metabolic bone disease. For further information, we encourage the reader to consult disease-specific chapters in this volume and the current literature.

Table 36.1. Patterns of Key Histomorphometric Findings That Characterize Several Types of Metabolic Bone Disease

	Marrow Spaces	Cortical Bone	Cancellous Bone	Bone Remodeling	Osteoid Morphology	Osteoid Mineralization
Postmenopausal osteoporosis	—	Cortical bone deficit with endocortical trabeculation	Cancellous bone deficit with poor trabecular connectivity	Ac.f generally increased, but values vary widely	—	—
Glucocorticoid-induced osteoporosis	—	Cortical bone deficit	Cancellous bone deficit	Early, increased Ac.f; later, decreased Ac.f	—	—
Primary hyperparathyroidism	Peritrabecular fibrosis may be seen	Cortical bone deficit, incr. Ct.Po, endocortical trabeculation	Typically unremarkable	Increased Ac.f	Woven bone may be seen	—
Hypogonadism (males and females)	—	Cortical bone deficit	Cancellous bone deficit, sometimes with poor trabecular connectivity	Increased Ac.f	—	—
Hypovitaminosis D osteopathy	Fibrous tissue may be seen	—	—	Early, increased Ac.f	—	Early, increased OS/BS; later, increased MLT and O.Th, double-label may be absent
Hypophosphatemic osteopathy	Fibrous tissue may be seen	—	—	—	—	Increased MLT and O.Th; double label may be absent
Renal osteodystrophy (high turnover type)	Fibrous tissue may be seen	Endocortical trabeculation	Osteoblast, osteocyte, and trabecular abnormalities	Markedly increased remodeling activity	Woven bone may be seen	Increased OS/BS
Renal osteodystrophy (low turnover type)	—	—	—	Markedly decreased remodeling activity	—	Increased OS/BS (osteomalacic type); decreased OS/BS (adynamic type)
Renal osteodystrophy (mixed type)	Fibrous tissue may be seen	—	Variable BV/TV	Patchy remodeling activity	Irregular, woven bone and osteoid may be seen	Increased OS/BS and O.Th

Postmenopausal osteoporosis

Osteoporosis in postmenopausal women is characterized by a cortical bone deficit with trabeculation of endocortical bone and a cancellous bone deficit with poor trabecular connectivity. Decreases in Tb.Th are modest, and dynamic measures vary widely [28, 29]. Median Ac.f. remains high in specimens from women with postmenopausal osteoporosis, but values vary widely [19].

Glucocorticoid-induced osteoporosis

Early in treatment, Ac.f is increased; later, Ac.f, MAR, and MS/BS are all decreased. In femoral specimens from patients with glucocorticoid-induced osteonecrosis, abundant apoptotic osteocytes and lining cells have been reported [30]. This has led to questions as to how viable osteocytes function to maintain bone mechanical integrity independent of bone mass and/or bone remodeling. (See the chapter on glucocorticoid-induced osteoporosis and detection of apoptotic osteocytes.)

Primary hyperparathyroidism

Primary hyperparathyroidism leads to a cortical bone deficit, with increased Ct.Po and trabeculation of endocortical bone [31]. Ct.Po correlates positively with fasting serum PTH [32]. BV/TV is generally preserved, and normal cancellous bone architecture is maintained [33, 34]. Osteoid with a woven appearance and peritrabecular fibrosis is also seen [35].

Hypogonadism

Hypogonadism in both women and men increases Ac.f and leads to deficits of both cortical bone and trabecular bone. At low levels of BV/TV and/or Tb.Th, loss of trabecular connectivity occurs [36].

Hypovitaminosis D osteopathy

Vitamin D depletion of any etiology leads to hypovitaminosis D osteopathy (HVO). Parfitt [27] describes three stages. In HVOi ("pre-osteomalacia"), Ac.f and OS/BS are increased, but O.Th is not. Accumulation of unmineralized osteoid characterizes both HVOii and HVOiii (osteomalacia), with Mlt and O.Th clearly increased (i.e., Mlt >100 days and O.Th >12.5 μm after correction for obliquity) [27]. Some double-label can be seen in HVOii, but not in HVOiii. A cortical bone deficit also characterizes advanced HVO; secondary hyperparathyroidism in response to reduced serum ionized calcium is usual, and fibrous tissue in the marrow spaces is frequently seen. Lifelong subclinical vitamin D insufficiency may contribute to the development of osteoporosis later in life in both men and women.

Low bone mass and bone disease with osteomalacic features occurs among patients treated with antiepileptic drugs (AEDs) [37]. Hepatic enzyme-inducing AEDs have been most clearly associated with these problems, but the newer AEDs cannot be exonerated at this time [38].

Hypophosphatemic osteopathy

Phosphate depletion of any etiology also leads to osteomalacia, with histomorphometric findings similar to those of advanced HVO [27]. These cases involve defects in phosphorus metabolism that manifest as defects in renal tubular reabsorption of phosphorus. However, most cases are not due to a primary renal tubular abnormality, but instead, are due to an abnormality in plasma phosphorus homeostasis [39]. Secondary hyperparathyroidism occurs variably. Transilial biopsy can be quite useful to assess the efficacy of treatment.

Gastrointestinal bone disease

Evidence of HVO has been reported in a variety of absorptive and digestive disorders [40]. However, these conditions also may promote deficiency of calcium and other nutrients. Malabsorption is not the only issue. For example, a calcium balance study of asymptomatic patients with celiac disease showed increased endogenous fecal calcium; the gut appeared to "weep" calcium into its lumen [41]. Bone histomorphometry may also reflect the results of treatment (i.e., corticosteroids or surgery). Parfitt describes a histomorphometric profile of low bone turnover, often with evidence of HVO and secondary hyperparathyroidism, that represents the result of multiple insults to bone health in these patients [27].

Renal osteodystrophy

At least three patterns of histomorphometric findings have been described among patients with end-stage renal disease (ESRD): high bone turnover with osteitis fibrosa (hyperparathyroid bone disease); low bone turnover (including osteomalacic and adynamic subtypes); and mixed osteodystrophy with high bone turnover, altered bone formation, and accumulation of unmineralized osteoid [42–45]. (See the chapter on chronic kidney disease mineral bone disorder.)

Transilial bone biopsy remains a useful "gold standard" on which to base decisions about treatment of bone disease in ESRD [42]. A dramatic example is the evaluation of bone pain and fractures in a chronic dialysis patient with hypercalcemia. If the biopsy shows high bone turnover and osteitis fibrosa, partial parathyroidectomy may be indicated. However, if the biopsy shows little turnover (little or no fluorochrome label), with or without extensive aluminum deposits, then parathyroidectomy is contraindicated, and treatment with a chelating agent may be indicated. The same biopsy can also help determine the extent of vitamin D deprivation and indicate the adequacy of vitamin D treatment.

OBTAINING THE SPECIMEN

In this section, we outline the procedures for obtaining bone biopsy specimens, processing them, and carrying out histomorphometric analysis. For greater detail, we recommend another recent publication [46].

Fluorochrome labeling

In clinical settings, tetracycline antibiotics are the only suitable fluorochrome labeling agents [12]. Demeclocycline (150 mg, four times daily) or tetracycline hydrochloride (250 mg, four times daily) are commonly used. The double-labeling process involves two dosing periods, and close adherence to the dosing schedule is crucial. A schedule of 3 days on, 14 days off, 3 days on, and 5–14 days off before biopsy (abbreviated as 3-14-3-5) produces good results, with a marker interval of 17 days [13]. Tetracyclines must be taken on an empty stomach; Thus, oral intake must be avoided for at least 1 hour before and after each dose.

Biopsy procedure

Specimens for histomorphometric examination require use of a trephine with an inner diameter of 7.5 mm or less. The teeth should be sharpened (and reconditioned, if necessary) after every two to three procedures. Transilial bone biopsy is performed in outpatient minor surgery facilities, with the usual procedures (e.g., the surgeon scrubs and uses a cap, mask, gown, and gloves, and the site is prepared and draped) and precautions (e.g., pulse oximetry and blood pressure monitoring). Before the procedure, the patient should be off aspirin for at least 3 days and have nothing orally for 4 hours. If a second biopsy is done on another occasion, it should always be on the side opposite the first; there is thus a practical limit of two transilial biopsy specimens per patient. The gowned patient lies in the supine position on the surgical table, and midazolam (2.5–5 mg) is given through a forearm intravenous catheter.

The biopsy site is approximately 2 cm posterior to the anterior–superior spine, which is approximately 2 cm inferior to the iliac crest. The skin, subcutaneous tissues, and periosteum on both sides of the ilium are infiltrated with local anesthetic. The periosteum is accessed by a 2-cm skin incision and blunt dissection. The trephine is inserted and advanced with steady, gentle pressure and a deliberate pace. The specimen—an intact, unfractured core with both cortices and the intervening cancellous bone—is transferred into a 20-ml screw-cap vial containing 70% ethanol. (Note that certain special procedures, presently used in research settings, require unfixed specimens.)

The bony defect is then packed with Surgicel. After local pressure to facilitate hemostasis, the wound is closed with three to five stitches and covered by a pressure dressing. Follow-up care is specified clearly (i.e., dressing in place and absolutely dry for 48 hours; then a daily shower is allowed; no bathing or strenuous physical activity until suture removal, 1 week after the procedure). The procedure produces localized aching for about 2 days and a small scar at the site.

Patients typically describe feeling something "like a cramp" as the trephine advances, and the bone biopsy procedure described here rarely evokes more than mild discomfort. Although bleeding during the procedure is typically minimal, there is risk of bleeding in some situations (e.g., liver disease, hemodialysis, or medications that compromise hemostasis). Local bruising sometimes occurs, but hematoma is uncommon. In an early survey, physicians who were doing transilial biopsy specimens reported adverse events in 0.7% of 9,131 biopsy specimens, that is, 22 with hematomas, 17 with pain for more than 7 days, 11 with transient neuropathy, 6 with wound infection, 2 with fracture, and 1 with osteomyelitis. No cases of death or permanent disability were reported [47].

SPECIMEN PROCESSING AND ANALYSIS

Specimen handling and processing

For routine histomorphometry, the bone biopsy specimen should remain in 70% ethanol for at least 48 hours for proper fixation. This solution is suitable for shipping and long-term storage at room temperature. The specimen vials should be filled to capacity with 70% ethanol for shipping, handling, and storage.

Steps in laboratory processing include dehydrating, defatting, embedding, sectioning, mounting, deplasticizing, staining, and microscopic examination.

After proper trimming, the tissue block is sectioned parallel to the long axis of the biopsy core. Two or more sets of sections are obtained at an intervals of 400 µm, beginning 35–40% into the embedded specimen. Unstained sections 8–10-µm thick are used to examine osteoid morphology and to measure fluorochrome-labeled surfaces. Sections 5–7-µm thick stained with toluidine blue are used to measure wall thickness. Sections 5-µm thick with Goldner's stain [48] are used for other histomorphometric measurements.

Microscopy

The histomorphometric variables described earlier are derived from data gathered at the microscope. These data include the width of both cortices and—in defined sectors of cancellous bone—volumes of bone, osteoid, and marrow; total trabecular perimeter; perimeters with features of formation (see Fig. 36.1) or resorption (see Fig. 36.2); thickness of osteoid and osteon walls; and interlabel width. Methods have been described for unbiased sampling of microscopic features [49].

Our histomorphometry laboratory uses an interactive image analysis system (BIOQUANT OSTEO 2011 v11.2 Bone Biology Research System; BIOQUANT Image Analysis Corporation, Nashville, TN, USA). A digital camera mounted on the microscope presents the microscopic images on screen, and measurements are made using a mouse. Fluorescent light at a wavelength of 350 nm is used to examine fluorochrome labels (see Fig. 36.3).

INDICATIONS FOR BONE BIOPSY AND HISTOMORPHOMETRY

The purpose of bone histomorphometry in the clinical setting is to gather information (i.e., to establish a diagnosis, clarify a prognosis, or evaluate adherence or response to treatment) on which to base informed clinical decisions. As is the case for every invasive procedure, the risk, discomfort, and expense should be proportionate to the importance of the information to be gained. Given these caveats, the number of clinical indications for this procedure is limited. Clinicians can manage most metabolic bone diseases, including osteoporosis, without the aid of a bone biopsy. However, there are some situations in which bone biopsy after fluorochrome labeling is appropriate, as outlined in Table 36.2.

Bone histomorphometry has been, and remains, crucial for assessing the mechanism of action, safety, and efficacy of new bone-active agents. Preclinical animal work includes serial biopsy specimens at multiple skeletal sites, using different colored fluorochrome labels (e.g., calcein or xylenol orange). Testing of every new bone-active treatment should include bone biopsy in at least a subset of subjects.

Table 36.2. Examples of Clinical Situations in Which Bone Histomorphometry Can Provide Useful Information*

1. When there is excessive skeletal fragility in unusual circumstances
2. When a mineralizing defect is suspected
3. To evaluate adherence to treatment in a malabsorption syndrome
4. To characterize the bone lesion in renal osteodystrophy
5. To diagnose and assess response to treatment in vitamin D-resistant osteomalacia and similar disorders
6. When a rare metabolic bone disease is suspected

*Adapted with permission (Ref. 50).

Trabecular bone histomorphometry provides a method for examining both bone properties and bone physiology. Cortical bone histomorphometry is seldom used because it requires obtaining a rib biopsy, a procedure more risky, expensive, and painful than transilial biopsy. Because of the random orientation of the transilial specimen with regard to the long axis of haversian systems, minimal information on cortical bone remodeling can be obtained from transilial specimens.

ACKNOWLEDGMENTS

The author thanks Susan Bare, Kathy McCon, and Toni Howard for assistance in describing technical methods and preparing digital photomicrographs.

Dr. Recker has received research funding from Merck, Novartis, Procter & Gamble, Roche, and Wyeth.

REFERENCES

1. Frost HM. 1986. *Intermediary Organization of the Skeleton*. Boca Raton: CRC Press.
2. Frost HM. 1973. *Bone Remodeling and its Relationship to Metabolic Bone Diseases*. Springfield, IL: Charles C. Thomas.
3. Marotti G, Favia A, Zallone AZ. 1972. Quantitative analysis on the rate of secondary bone mineralization. *Calc Tiss Res* 10: 67–81.
4. Recker RR, Kimmel DB, Parfitt AM, Davies KM, Keshawarz N, Hinders S. 1988. Static and tetracycline-based bone histomorphometric data from 34 normal postmenopausal females. *J Bone Miner Res* 3: 133–144.
5. Malluche HH, Faugere MC. 1986. In: Malluche HH, Faugere MC (eds.) *Atlas of Mineralized Bone Histology*. New York: Karger.
6. Parfitt AM, Drezner MK, Glorieux FH, Kanis JA, Malluche H, Meunier PJ, Ott SM, Recker RR. 1987. Bone histomorphometry: Standardization of nomenclature, symbols, and units. *J Bone Miner Res* 2: 595–610.
7. Parfitt AM. 1983. The physiologic and clinical significance of bone histomorphometric data. In: Recker RR (ed.) *Bone Histomorphometry: Techniques and Interpretation*. Boca Raton: CRC Press. pp. 143–224.
8. Garrahan NJ, Mellish RWE, Compston JE. 1986. A new method for the two-dimensional analysis of bone structure in human iliac crest biopsies. *J Microsc* 142: 341–349.
9. Vesterby A, Gundersen HJG, Melsen F. 1989. Star volume of marrow space and trabeculae of the first lumbar vertebra: Sampling efficiency and biological variation. *Bone* 10: 7–13.
10. Vesterby A, Gundersen HJG, Melsen F, Mosekilde L. 1991. Marrow space star volume in the iliac crest decreases in osteoporotic patients after continuous

treatment with fluoride, calcium, and vitamin D2 for five years. *Bone* 12: 33–37.
11. Hahn M, Vogel M, Pompesius-Kempa M, Delling G. 1992. Trabecular bone pattern factor: A new parameter for simple quantification of bone microarchitecture. *Bone* 13: 327–330.
12. Frost HM. 1969. Measurement of human bone formation by means of tetracycline labelling. *Can J Biochem Physiol* 41: 331–342.
13. Schwartz MP, Recker RR. 1982. The label escape error: Determination of the active bone-forming surface in histologic sections of bone measured by tetracycline double labels. *Metab Bone Dis Relat Res* 4: 237–241.
14. Parfitt AM. 2002. Physiologic and pathogenetic significance of bone histomorphometric data. In: Coe FL, Favus M (eds.) *Disorders of Bone and Mineral Metabolism, 2nd Ed.* Philadelphia: Lippincott Williams & Wilkins. pp. 469–485.
15. Chapurlat RD, Arlot M, Burt-Pichat B, Chavassieux P, Roux JP, Portero-Muzy N, Delmas PD. 2007. Microcrack frequency and bone remodeling in postmenopausal osteoporotic women on long-term bisphosphonates: A bone biopsy study. *J Bone Miner Res* 22: 1502–1509.
16. Jilka RL, Weinstein RS, Parfitt AM, Manolagas SC. 2007. Perspective: Quantifying osteoblast and osteocyte apoptosis: Challenges and rewards. *J Bone Miner Res* 22: 1492–1505.
17. Weinstein RS, Bell NH. 1988. Diminished rates of bone formation in normal black adults. *N Engl J Med* 319: 1698–1701.
18. Glorieux FH, Travers R, Taylor A, Bowen JR, Rauch F, Norman M, Parfitt AM. 2000. Normative data for iliac bone histomorphometry in growing children. *Bone* 26: 103–109.
19. Recker R, Lappe J, Davies KM, Heaney R. 2004. Bone remodeling increases substantially in the years after menopause and remains increased in older osteoporosis patients. *J Bone Miner Res* 19: 1628–1633.
20. Parfitt AM, Travers R, Rauch F, Glorieux FH. 2000. Structural and cellular changes during bone growth in healthy children. *Bone* 27: 487–494.
21. Cosman F, Morgan D, Nieves J, Shen V, Luckey M, Dempster D, Lindsay R, Parisien M. 1997. Resistance to bone resorbing effects of PTH in black women. *J Bone Miner Res* 12: 958–966.
22. Han Z-H, Palnitkar S, Rao DS, Nelson D, Parfitt AM. 1997. Effects of ethnicity and age or menopause on the remodeling and turnover of iliac bone: Implications for mechanisms of bone loss. *J Bone Miner Res* 12: 498–508.
23. Dahl E, Nordal KP, Halse J, Attramadal A. 1988. Histomorphometric analysis of normal bone from the iliac crest of Norwegian subjects. *Bone Miner* 3: 369–377.
24. Cosman R, Schnitzer MB, McCann PD, Parisien MV, Dempster DW, Lindsay R. 1992. Relationships betwen quantitative histological measurements and noninvasive assessments of bone mass. *Bone* 13: 237–242.
25. Keshawarz NM, Recker RR. 1984. Expansion of the medullary cavity at the expense of cortex in postmenopausal osteoporosis. *Metab Bone Dis Relat Res* 5: 223–228.
26. Recker RR, Kimmel DB, Dempster D, Weinstein R, Wronski TJ, Burr DB. 2011. Issues in modern bone histomorphometry. *Bone* 49(5): 955–64..
27. Parfitt AM. 1998. Osteomalacia and related disorders. In: Avioli LV, Krane SM (eds.) *Metabolic Bone Disease and Clinically Related Disorders.* Boston: Academic Press. pp. 327–386.
28. Kimmel DB, Recker RR, Gallagher JC, Vaswani AS, Aloia JF. 1990. A comparison of iliac bone histomorphometric data in post-menopausal osteoporotic and normal subjects. *Bone Miner* 11: 217–235.
29. Recker RR, Barger-Lux MJ. 2001. Bone remodeling findings in osteoporosis. In: Marcus R, Feldman D, Kelsey J (eds.) *Osteoporosis, 2nd Ed.* San Diego: Academic Press. pp. 59–70.
30. Weinstein RS, Nicholas RW, Manolagas SC. Apoptosis of osteocytes in glucocorticoid-induced osteonecrosis of the hip. *J Clin Endocrinol Metab* 85: 2907–2912.
31. Ericksen E. 2002. Primary hyperparathyroidism: Lessons from bone histomorhometry. *J Bone Miner Res* 17 Suppl 2: N95–7.
32. van Doorn L, Lips P, Netelenbos JC, Hackeng WHL. 1993. Bone histomorphometry and serum concentrations of intact parathyroid hormone (PTH(1-84)) in patients with primary hyperparathyroidism. *Bone Miner* 23: 233–242.
33. Parisien M, Mellish RW, Silverberg SJ, Shane E, Lindsay R, Bilezikian JP, Dempster DW. 1992. Maintenance of cancellous bone connectivity in primary hyperparathyroidism: Trabecular strut analysis. *J Bone Miner Res* 7: 913–919.
34. Uchiyama T, Tanizawa T, Ito A, Endo N, Takahashi HE. 1999. Microstructure of the trabecula and cortex of iliac bone in primiary hyperparathyroidism patients determined using histomorphometry and node-strut analysis. *J Bone Miner Res* 17: 283–288.
35. Monier-Faugere M-C, Langub MC, Malluche HH. 1998. Bone biopsies: A modern approach. In: Avioli LV, Krane SM (eds.) *Metabolic Bone Disease and Clinically Related Disorders, 3rd Ed.* San Diego: Academic Press. pp. 237–273.
36. Audran M, Chappard D, Legrand E, Libouban H, Basle MF. 2001. Bone microarchitecture and bone fragility in men: DXA and histomorphometry in humans and in the orchidectomized rat model. *Calcif Tissue Int* 69: 214–217.
37. Pack AM, Morrell MJ. 2004. Epilepsy and bone health in adults. *Epilepsy Behav* 5: S24–S29.
38. Fitzpatrick LA. 2004. Pathophysiology of bone loss in patients receiving anticonvulsant therapy. *Epilepsy Behav* 5: S3–S15.
39. Antoniucci DM, Yamashita T, Portale AA. 2006. Dietary phosphorus regulates serum fibroblast growth factor-23 concentrations in healthy men. *J Clin Endocrinol Metab* 91: 3144–3149.
40. Arnala I, Kemppainen T, Kroger H, Janatuinen E, Alhava EM. 2001. Bone histomorphometry in celiac disease. *Ann Chir Gynaecol* 90: 100–104.

41. Ott SM, Tucci JR, Heaney RP, Marx SJ. 1997. Hypocalciuria and abnormalities in mineral and skeletal homeostasis in patients with celiac sprue without intestinal symptoms. *Endocrinol Metab* 4: 206.
42. Pecovnik BB, Bren A. 2000. Bone histomorphometry is still the golden standard for diagnosing renal osteodystrophy. *Clin Nephrol* 54: 463–469.
43. Parker CR, Blackwell PJ, Freemont AJ, Hosking DJ. 2002. Biochemical measurements in the prediction of histologic subtype of renal transplant bone disease in women. *Am J Kidney Dis* 40: 396.
44. Elder G. 2002. Pathophysiology and recent advances in the management of renal osteodystrophy. *J Bone Miner Res* 17: 2094–2105.
45. Malluche HH, Langub MC, Monier-Faugere MC. 1997. Pathogenisis and histology of renal osteodystrophy. *J Bone Miner Res* 7: S184–S187.
46. Recker RR, Barger-Lux MJ. 2001. Transilial bone biopsy. In: Bilezikian JP, Raisz L, Rodan GA (eds.) *Principles of Bone Biology. 2nd Ed.* San Diego: Academic Press. pp. 1625–1634.
47. Rao DS, Matkovic V, Duncan H. 1980. Transiliac bone biopsy: Complications and diagnostic value. *Henry Ford Hosp Med J* 28: 112–118.
48. Goldner J. 1938. A modification of the Masson trichrome technique for routine laboratory purposes. *Am J Pathol* 14: 237–243.
49. Kimmel DB, Jee WSW. 1983. Measurements of area, perimeter, and distance: Details of data collection in bone histomorphometry. In: Recker RR (ed.) *Bone Histomorphometry: Techniques and Interpretation*. Boca Raton: CRC Press. pp. 80–108.
50. Barger-Lux MJ, Recker RR. 2005. Towards understanding bone quality: Transilial bone biopsy and bone histomorphometry. *Clin Rev Bone Miner Metab* 4: 167–176.

37

Diagnosis and Classification of Vertebral Fracture

James F. Griffith, Judith E. Adams, and Harry K. Genant

Significance of Vertebral Fracture 317
Pathophysiology of Vertebral Fracture 318
Detection of Vertebral Fracture 319
Other Imaging Methods 327

Clinical Recommendations for Spine Imaging to Detect Vertebral Fracture 330
Conclusion 331
References 331

SIGNIFICANCE OF VERTEBRAL FRACTURE

Vertebral fracture is usually the first osteoporotic fracture to occur and is also the most common osteoporotic fracture, being present in 15% of women aged 50–59 years and in 50% of women aged more than 85 years [1, 2]. Accurate recognition of vertebral fracture is essential to comprehensive clinical evaluation, determination of population prevalence, fracture risk, and evaluating treatment efficacy [3–5]. Almost half of all vertebral fractures occur in patients with low bone mineral density (BMD) (T-score at or below −1.0; osteopenia) rather than osteoporosis (T-score at or below −2.5) as defined by dual energy X-ray absorptiometry (DXA), as low BMD is much more common than osteoporosis [6, 7]. The presence of a nontraumatic vertebral fracture in older subjects provides such indisputable evidence of reduced bone strength that these patients should be considered as having osteoporosis. For this reason the National Osteoporosis Foundation has recommended that patients aged over 50 years with nontraumatic new vertebral fracture receive appropriate bone protective/bone enhancing therapy, irrespective of DXA T-score [8].

The presence, severity, and number of osteoporotic vertebral fractures are important predictors of further vertebral fracture risk [9]. Over an 8-year period, subjects with prevalent (i.e., preexisting) vertebral fractures had a fivefold increased risk of further vertebral fracture and a threefold increased risk of proximal femoral fracture than those without an incident vertebral fracture [10]. Vertebral fracture risk is 20% in the year following the incident (i.e., new) vertebral fracture. However, this relative risk is fourfold greater in those with a severe, rather than a mild, fracture and threefold greater in those with multiple, rather than a single, vertebral fracture [11]. In addition to being an indicator of increased fracture risk, vertebral fracture is also associated with reduced quality of life and increased mortality, particularly from pulmonary disease and cancer [12]. If vertebral fractures are recognized early and treated with appropriate antifracture medication, there is a significant reduction in the occurrence of new vertebral and nonvertebral fractures [13].

Despite the clear clinical importance of vertebral fractures, they remain underdiagnosed in clinical practice [14, 15]. There are two main reasons for this underdiagnosis: firstly, the typical clinical symptoms of back pain and restricted movement are usually attributed to spondylosis rather than vertebral fracture so that most patients with vertebral fracture do not seek medical attention [15]. However, improving patient awareness may not improve diagnosis, as only about one-third of vertebral fractures diagnosed retrospectively have a symptomatic period [16]. Secondly, many vertebral fractures evident on imaging are not reported. For example, 17% of elderly subjects attending an accident and emergency center had a moderate or severe vertebral fracture on lateral chest

---

Primer on the Metabolic Bone Diseases and Disorders of Mineral Metabolism, Eighth Edition. Edited by Clifford J. Rosen.
© 2013 American Society for Bone and Mineral Research. Published 2013 by John Wiley & Sons, Inc.

radiographs, but only 50% of these fractures were included in the radiology reports, and even fewer affected patients received appropriate medical intervention [7]. Even this study underestimated the true fracture prevalence as mild fractures were not included and lumbar radiographs were not available [17].

As many vertebral fractures are clinically asymptomatic, radiologists and clinicians who review imaging studies are urged to look specifically for the presence of vertebral fractures [4, 5]. If a vertebral fracture is present, then it is imperative that the report states this clearly as a "vertebral fracture" and that ambiguous terminology (e.g., "vertebral collapse," "compressed vertebral body," "loss of vertebral height," "wedging of vertebral body," "wedge deformity," "biconcavity," or "codfish deformity") is avoided. The presence, location, and severity of a vertebral fracture should also be clearly stated.

The World Health Organization's 10-year fracture risk assessment tool (FRAX®) was released in 2008 to help individualize osteoporotic fracture risk. While DXA-measured BMD is predictive of overall relative vertebral fracture risk, this risk is difficult to apply to an individual patient in clinical practice. The FRAX Web-based prediction model integrates easily assessed clinical risk factors such as age, sex, ethnicity, weight, height, fracture history, smoking, alcohol, steroid use, and rheumatoid arthritis, and has been adapted for different populations worldwide (http://www.shef.ac.uk/FRAX). FRAX is an important initiative in allowing clinicians to individualize fracture risk. It helps identify patients who are at fracture risk and can also be applied without BMD measurements in patients with limited access to densitometry. FRAX is mainly applicable to osteopenic patients as the results do not influence clinical management in patients with normal BMD or those known to have osteoporosis [18, 19]. FRAX is being incorporated into national guidelines on the management and treatment of osteoporosis, though intervention algorithms vary between nations. In the U.S., the National Osteoporosis Foundation (NOF) has added FRAX to the criteria for treating patients with a threshold of 3% for hip fracture and 20% for a major osteoporotic fracture [20]. In the U.K., a different treatment algorithm was issued by the National Osteoporosis Guidelines Group (NOGG), which recommended that FRAX be used in a triage scheme to select patients who would benefit from having a DXA examination [20, 21].

PATHOPHYSIOLOGY OF VERTEBRAL FRACTURE

The vertebral body has a relatively large trabecular bone component and relies more on trabecular bone than cortical bone for its strength. As trabecular bone is thin with a large surface area, it is particularly responsive to changes in the microenvironment and is affected before cortical bone in the development of osteoporosis. As a result, the

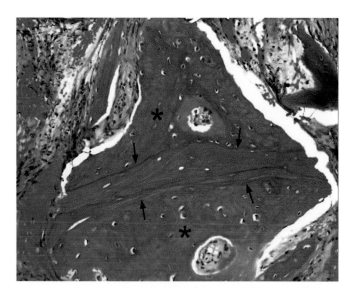

Fig. 37.1. Histological section from fractured human vertebral body showing a band of dead bone (arrows) containing osteocytes reduced to lacunae without viable nuclei. This band of dead bone is juxtaposed against less mature reparative new bone (*). Hematoxylin & Eosin x 180.

vertebral body is particularly prone to osteoporotic fracture. The weakest parts of the vertebral body are the central component of the endplates and anterosuperiorly where lower BMD is not compensated for by a higher trabecular strength [22]. As opposed to appendicular skeletal fractures, which occur as a definitive event, vertebral body fractures often progress incrementally over many months or years. This incremental progression of osteoporotic vertebral fracture is evident histologically with features of both recent fracture and bone repair occurring together [23, 24] (Fig 37.1).

Bone is architecturally an adaptive tissue that responds to mechanical load and use. Increase in vertebral body cross-sectional area will increase vertebral body strength. Males have larger bones than females, which are also substantially more heavily loaded. Periosteal bone apposition, particularly in males, may potentially compensate for the increased fragility caused by reduction in BMD, through an increase in vertebral cross-sectional area. A longitudinal volumetric quantitative CT (vQCT)-based study confirmed that females, in addition to having smaller vertebral bodies and losing BMD more rapidly, increase vertebral cross-sectional area more slowly than males, predisposing females to higher vertebral fracture risk [25].

A vertebral fracture occurs when the forces applied to a vertebral body exceed its strength. Functional loading of the spine is primarily compressive in nature. Micro-computed tomography, combined with finite element analysis (FEA), have shown that major load pathways through the vertebral body seem to be parallel to the columns of vertical trabeculae. The cortical shell and the horizontal trabeculae, although considered important in bracing vertical trabeculae against bending and buckling, appear to have a relatively modest role in resisting compressive forces [26]. While both horizontal and vertical trabeculae reduce in number with aging, only horizontal trabeculae reduce in thickness [27]. FEA suggests that this may be due to horizontal trabeculae being reabsorbed preferentially as the result of "adaptive strain resorption," while vertical trabeculae are reabsorbed as a result of microdamage [28]. This results in prominent vertical striated appearances of the vertebral bodies on imaging, which may be evident before vertebral fractures occur. When radiographic osteopenia is evident, osteoporosis should be suspected and DXA scanning would be appropriate. In addition to trabecular orientation being relevant to strength of the vertebral body, trabecular shape is also important with osteoporosis associated with a transition from plate-like to a more rod-like trabecular morphology [29].

Depending on the force applied and inherent vertebral body strength, fracture severity varies from a small peripheral fracture to a complete vertebral body fracture. Most vertebral fractures occur in the mid-thoracic and thoracolumbar regions [30]. Forces applied to the spine in these sites may exceed the inherent vertebral body strength. Compressive loading is accentuated in the mid-thoracic spine during flexion when there is a kyphosis present and also in the thoracolumbar region, which is the transition zone between the relatively fixed thoracic and more mobile lumbar segments. Osteoporotic vertebral fractures are very uncommon above the T4 level [31]. Loading on the spine is determined by gravitational forces and muscle contracture, which in turn are influenced by body weight, height, muscle action, coordination, and strength, as well as spinal curvature and intervertebral disks characteristics [32]. Fracture of a single vertebral body, particularly if of an anterior wedge type, shifts compressive forces toward the anterior aspects of the vertebral bodies to the extent that a "vertebral fracture cascade" may ensue, with fractures in adjacent vertebrae occurring in quite rapid succession [32].

DETECTION OF VERTEBRAL FRACTURE

Clinical detection

Increasing age, previous nonvertebral fracture, low BMD, low body mass index, current smoking, low milk consumption during pregnancy, low physical activity, having a fall, and regular use of aluminum-containing antacids all increase the likelihood of a first vertebral fracture [33]. Only about one in four vertebral fractures are recognized as a distinct clinical event as symptoms are nonspecific [4]. Probably the most effective clinical discriminator of incident vertebral fracture is a measured height loss of greater than 2 cm or a recalled height loss of greater than 4 cm. A measured height loss of greater than 2 cm had a 35% sensitivity and 94% specificity for incident vertebral fractures [34], while a recalled height loss of greater than 4 cm is associated with a threefold increased likelihood of identifying a vertebral fracture [35].

Spinal radiography

Radiography, as the most widely available and least expensive bone imaging modality, can reveal grades of bone density with normal bone being appreciably more dense than osteoporotic bone. On radiography, one can accurately distinguish between unequivocally normal bone and unequivocally osteoporotic bone. However, bone in most subjects is neither unequivocally normal nor unequivocally osteoporotic. In these subjects, the use of radiography to separate patients with mildly reduced bone density from those with osteoporosis is not very accurate. Osteoporotic vertebral fractures will only occur in the presence of reduced bone density. Radiographic detection of reduced bone density and osteoporosis is subjective and also dependent on radiographic technique, radiographic equipment, and patient body habitus. Additional radiographic signs of osteoporosis such as cortical thinning, trabecular rarefaction, cortical endosteal scalloping and intracortical tunneling can also be used to help recognize osteoporosis though appreciation of these radiographic signs is, similar to bone density, heavily dependent on the experience of the observer.

The majority of vertebral fractures are diagnosed using radiography of the thoracic and lumbar spine. Radiography is a good method for the diagnosis of vertebral fractures because it is easy and quick to perform, widely available, and of relatively low cost. For optimal radiographic assessment, a standardized high-quality image is essential. A radiographic protocol should visualize the C7-S1 vertebrae. Time and effort should be spent on obtaining finely collimated properly positioned anteroposterior (AP) and lateral views of the spine with the X-ray beam centered at T7 and L3 for the thoracic and lumbar spines respectively and a focus–film distance of 100 cm [Fig. 37.2(a), (b)]. The C7-T3 vertebral bodies are often not clearly seen on lateral thoracic spine radiographs due to superimposition of the scapulae and shoulder regions, but fortunately, isolated osteoporotic fractures are uncommon in this region. On the lateral projection, the spine must be parallel to the film so that the vertebral endplates at the level of the central X-ray beam are superimposed and seen as a single dense, well-defined cortical line. As a result of the divergent X-ray beam the endplates distant from the centering point will appear concave ("bean can" effect) and must not be mistaken for vertebral fractures. Usually a lateral spinal projection will suffice, although an AP projection is occasionally useful to (i) determine if scoliosis is present and (ii) determine the anatomical level of a vertebral fracture. The anteroposterior projection is particularly useful in assessment of the thoracic spine since the vertebral body outline is not so consistently seen as in the lumbar region. One should be aware, however, that mild

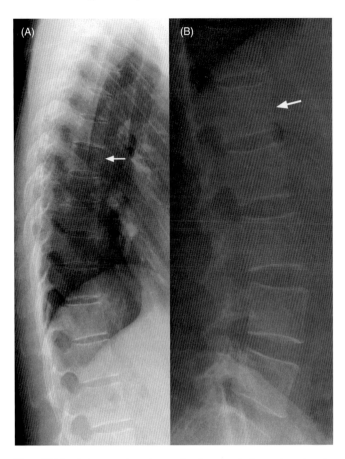

Fig. 37.2. (A) Lateral radiograph of normal thoracic spine in subject with low BMD. There is minor physiological anterior wedging of the T6 vertebral body (arrow). (B) Lateral radiograph of normal lumbar spine. There is slight physiological wedging of the L1 vertebral body (arrow).

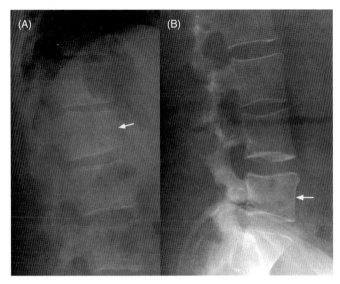

Fig. 37.3. (A) Anterior physiological wedging of the L1 vertebral body (arrow). (B) Posterior physiological wedging of the L5 vertebral body (arrow).

vertebral fractures may be overlooked on a frontal (anteroposterior) projection of the thoracic spine. The typical effective dose of ionizing radiation from a single lateral and AP projection of the thoracic spine are 0.3 mSv and 0.4 mSv, respectively, while for the lumbar spine they are 0.3 mSv and 0.7 mSv, respectively. For comparison, a 16-hour return transatlantic flight would amount to 0.07 mSv background radiation [36].

Problems in the diagnosis of vertebral fracture on radiographs

Although there is good agreement in the radiological diagnosis of moderate and severe vertebral fractures, there is considerable contention in the diagnosis of mild vertebral fractures. The difficulty in defining a mild vertebral fracture is a reflection of six pitfalls when assessing spinal radiographs for vertebral fracture:

(1) Mild physiological anterior or posterior wedging may be misinterpreted as a mild fracture but is a normal and entirely necessary feature of the thoracic and lumbar vertebral bodies as the spine changes from a thoracic kyphosis to a lumbar lordosis. Normal spinal curvature dictates that vertebral bodies are slightly anteriorly wedge shaped in the thoracic and upper lumbar spine, not wedged in the mid-lumbar region and slightly posteriorly wedge shaped in the lower lumbar region [Fig. 37.3(a), (b)] [37]. The degree of wedging is dependent on inherent sagittal spinal curvature, which varies between subjects.

(2) Short vertebral body height (SVH) is a common feature of this physiological wedging as well as increasing age and spondylosis in the absence of osteoporosis (Fig 37.4). In women, between about 30 and 70 years of age, the combined height of the anterior aspects of the vertebral bodies from T4-L5 decreases by about 1.5 mm per year while the combined middle and posterior heights decline by about 1.2 mm per year [38, 39]. Since true reduction in vertebral body height can only be ascertained on serial imaging studies, one cannot always be certain that a mild reduction in vertebral height evident on a baseline image is necessarily the result of a vertebral fracture. While SVH refers to reduction in vertebral height of up to about 20% expected height. Differentiating SVH from a mild vertebral fracture is probably the most contentious and difficult area in vertebral fracture diagnosis. The majority of evidence suggests that isolated SVH is not associated with low BMD or vertebral fracture [40]. In a DXA vertebral fracture assessment (VFA) study in which comparison was made between 250 premenopausal and 1,350 postmenopausal women, the prevalence of SVH was approximately 35%. The prevalence was similar in pre- and postmenopausal women and was not associated with low lumbar spine BMD. Premenopausal women with SVH tended to be older and heavier than those without SVH, while postmenopausal women with SVH tended to have higher spine BMD than those without SVH [41]. SVH, when isolated and in the absence of endplate changes, is most likely the result of physiological wedging exacerbated by

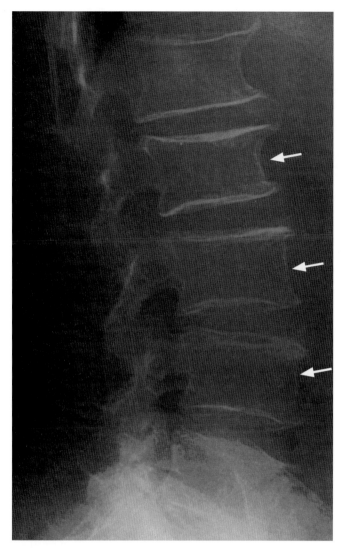

Fig. 37.4. Lateral radiograph of lumbar spine. There is "short vertebral height" with all of the lumbar vertebrae reduced in height by 15–20% compared to expected vertebral body height. Note associated spondylosis with marginal osteophytosis.

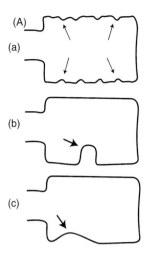

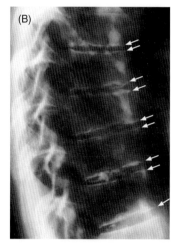

Fig. 37.5. (A) Schematic diagram showing endplate impressions caused by (a) Scheuermann's disease (b) Schmorl's node, and (c) Cupid's bow deformity. (B) Lateral radiograph midthoracic region showing diffuse endplate irregularity (arrows), narrowing of the intervertebral discs and elongated anteroposterior diameter of the vertebral bodies all consistent with Scheuermann's disease.

vertebral body remodeling due to increasing age or spondylosis.

(3) Vertebral deformity is a feature of an uncommon spinal disorder known as Scheuermann's disease [40]. Scheuermann's disease generally occurs during the teenage years and is characterized by endplate irregularity/indentation of the thoracic or lumbar vertebrae [Figs. 37.5A, 37.5B]. The disorder may involve only one or two vertebrae or longer segments of the spine. It is typically associated with reduced vertebral height, often in conjunction with an increased anteroposterior vertebral body diameter and usually associated with small discs and premature disc degeneration.

(4) Developmental, or more commonly degenerative, scoliosis may lead to obliquity of vertebral bodies and side-to side discrepancy in vertebral body height. On the lateral projection, this obliquity produces a spurious biconcave outline to the vertebral endplates that may be misinterpreted as a vertebral fracture. On the AP projection, the vertebral bodies, particularly at the apex of the curve, will be shortened on the concave side and of normal height or even elongated on the convex side. This scoliotic wedging, providing it is predominantly one-sided and commensurate with the severity of scoliosis, should not be misinterpreted as a vertebral fracture. Difficulty with interpretation of vertebral body height in moderate to severe scoliosis often leads to such patients being excluded from clinical trials, in part also due to the confounding effects on DXA BMD.

(5) Schmorl's nodes are discrete indentations of the endplates related to degenerative disc disease. Small Schmorl's nodes are a common finding, being present in 40–75% of imaging studies, particularly in degenerative disease of the lumbar spine. Medium-sized or large Schmorl's nodes occur much less frequently and may be misinterpreted as an endplate

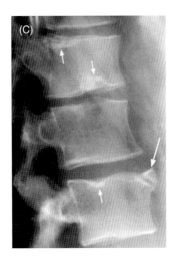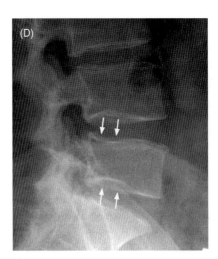

Fig. 37.5. (*Continued*) (C) Lateral radiograph of lumbar spine showing medium-sized Schmorl's nodes of the superior and inferior endplates (small arrows). This is also a limbus vertebra present (long arrow). (D) Lateral radiograph lower lumbar region showing smooth indentation of inferior and superior endplates of L5 (arrows) due to Cupid's bow deformity.

fracture. However, they can usually be readily distinguished from fractures by their well-defined rounded contour, intact sclerotic margin, and only focal involvement of the endplate [Figs. 37.5A, 37.5C].

(6) Cupid's bow deformity is a reasonably common developmental endplate contour abnormality most frequently affecting the inferior, and less frequently the superior, endplates of the fourth and fifth lumbar vertebral bodies [42] [Figs. 37.5A, 37.5D]. This contour deformity may also involve the endplates of the more cephalad lumbar vertebrae and those of the thoracolumbar vertebrae. Cupid's bow deformity most likely results from focal deficiency of the cartilage component of the endplate in the parasagittal regions of the vertebral body [42]. This absence of the cartilage component focally impairs endochondral growth of the vertebral body leading to the characteristic concave endplate depressions seen radiographically. The shape of the resulting deformity on the anteroposterior projection resembles "Cupid's bow." On the lateral projection, the posterior two-thirds of the inferior endplate are indented, simulating endplate fracture depression.

One can therefore appreciate that while all vertebral fractures result in vertebral deformity, not all vertebral deformities represent a vertebral fracture [31]. Vertebral fracture must be differentiated from these other etiologies, which may also change vertebral shape. The term "deformity" is appropriate when reporting such etiologies [43]. With careful scrutiny of imaging features, these various vertebral deformities can be differentiated from vertebral fractures.

Defining a vertebral fracture

Although there is no universally accepted definition of what constitutes a vertebral fracture, there is wide agreement that a vertebral fracture involves decrease in anterior, middle, or posterior vertebral height. Accurate identification of vertebral fracture on imaging studies is critically important not just for clinical management but also in research studies assessing the prevalence of vertebral fracture or efficacy of anti-fracture therapies. The over- or underreporting of vertebral fracture by an inexperienced reader can significantly skew research findings [44].

The approach to vertebral fracture diagnosis can vary considerably between a clinical setting and a research setting. In a clinical setting there is a specific clinical indication, only a single examination is being evaluated, previous radiographs or other imaging studies may be available for review, and a more experienced opinion or additional imaging can be sought to aid diagnosis if indicated. In the research setting, imaging is often performed without specific clinical need, evaluation is generally limited to lateral radiographs, a large number of subjects are involved requiring that high efficiency and assessment be performed by various readers with different levels of experience [31].

In an effort to produce definable, reproducible, and objective methods of detecting vertebral fracture, several methods have been developed and refined to diagnose and grade the severity of vertebral fracture on radiographs. These methods, which have also been applied to DXA VFA and computed tomography (CT) images, can be broadly considered as either qualitative, semi-quantitative (SQ), or quantitative morphometric (QM) in approach.

Semi-quantitative assessment

Semi-quantitative (SQ) analysis may involve either (a) measurement of vertebral height followed by evaluation of vertebrae reduced in height by an expert reader to determine the cause, or, more commonly, (b) evaluation of spinal radiographs by an experienced reader without prior measurement of vertebral heights. The most widely used SQ approach is that of Genant et al. [30, 45] (Fig. 37.6). Vertebral fractures are graded from 1 (mild) to 3 (severe), and incident fractures are defined as an increase of one grade or more on follow-up radiographs (Fig. 37.7). Grade 1 (mild) vertebral fracture corresponds to a ~20–25% reduction in anterior, middle, and/or posterior height compared to normal adjacent vertebrae or compared to expected height of the vertebral body based on experience (Fig. 37.6). Grade 2 (moderate) vertebral fracture is a ~25–40% reduction in vertebral height (Fig. 37.6), while grade 3 (severe) vertebral fracture is a ~>40% reduction in vertebral height (Fig. 37.6). The approximation symbol (~) is applied because the height reductions are generally visually estimated rather than measured directly. Additionally, other morphologic changes of endplate buckling or bowing and interruption of cortical margins are factored into the diagnosis of vertebral fracture. Applying these grades in research studies enables a spinal deformity index (SDI) to be assigned to each patient by summating the semi-quantitative scores for the T4 to L4 vertebrae [46].

Grading the severity of vertebral fracture recognizes the incremental nature of vertebral fracture and enables fracture progression from mild to moderate or moderate to severe to be meaningfully described on follow-up radiographs (Fig 37.7). The more severe the vertebral fracture, the greater the deterioration in bone architectural parameters as assessed by two-dimensional (2D) histomorphometry and three-dimensional (3D) microcomputed tomography [47], and this increases the risk of future fracture. It is therefore clearly important in both clinical practice and in a research setting to report and grade the severity of vertebral fracture in addition to simply recording the presence of a fracture.

The SQ method is a standardized method with excellent interreader reliability. For prevalent fractures, the agreement between each of three readers and a consensus reading yielded a kappa score of 0.84–0.87 for incident fractures and 0.86–0.96 for prevalent fractures [47, 48]. SQ analysis of spinal radiographs for vertebral fracture is quicker to perform than other methods of vertebral fracture assessment, is easy to implement in clinical practice, and suited to both epidemiological research studies and clinical therapeutic efficacy trials. Vertebral fractures detected by SQ analysis are associated with low BMD and are predictive of future fracture, independent of BMD [9, 14, 49, 50]. For follow-up studies, serial radiographs should be viewed together in a temporal order to fully appreciate changing vertebral morphology. Although visual assessment methods of vertebral fracture detection are more subjective than morphometric analysis, they do allow the experienced reader to address issues such as nonosteoporotic deformity. SQ analysis is also better suited to deal with errors introduced by radiographic technique such as varying magnification effect, which clearly would influence serial vertebral body measurements. The SQ method is a practical and reproducible method of vertebral fracture assessment when performed by trained and experienced readers [41, 51].

Vertebral quantitative morphometry

Vertebral quantitative morphometry (QM) is used in a research, but not in a clinical, setting [52]. The two main advantages of vertebral morphometry over other methods is that (i) it can be undertaken by relatively inexperienced or nonmedical research staff, and (ii) it provides an objective measure of loss of vertebral height on serial images. While the description and definition of the methodology is straightforward, the application in practice is often rather subjective.

The margins of each vertebral body from T4 to L4 are identified by six points on the upper and lower endplates—one for each corner and one for each of the endplate midpoints (Fig. 37.8). Marginal osteophytes, uncinate processes, and Schmorl's nodes are not included. For vertebrae off center from the central X-ray beam causing an oval shape to the endplates, the midpoint of the ovoid upper and lower endplate contour is chosen (Fig. 37.8).

Placement of the six points can be manual or automated. The automated method comprises computerized vertebral boundary recognition of a digitalized spinal radiograph. Automated point placement is checked and adjusted, if necessary, by a trained reader. The anterior (A_H), middle (M_H), and posterior (P_H) vertebral heights are immediately available for analysis and archiving.

Vertebral height ratios are used to define vertebral shape with A_H/P_H reflecting anterior wedging, M_H/P_H reflecting endplate concavity, and $P_H/P_{H'}$ of the adjacent normal vertebrae reflecting posterior compression [53]. Prevalent vertebral fracture is defined as a reduction in one or more of the three vertebral height ratios (A_H/P_H, M_H/P_H, or $P_H/P_{H'}$) of greater than 20% or 3 SD (standard deviations) from the mean of a reference population. Incident vertebral fracture is defined as a reduction in one of the three height ratios (A_H/P_H, M_H/P_H, or $P_H/P_{H'}$) greater than 15% or 3–4 mm compared to the baseline [1, 54]. While the reproducibility of QM is good in normal subjects (interobserver coefficient of variation less than 2%), it is lower in those with osteoporotic fractures [(interobserver and intraobserver coefficient of variation of 5% and 6.3% for M_H, respectively] [55].

Positioning QM reference points is partly subjective, especially for the vertebral midpoints where obliquity of the radiographic beam often produces a double endplate contour. Even with good radiographic technique, a mild degree of scoliosis will invariably lead to the endplate being visualized slightly en face. In such situations, observer experience will influence reference point placement for baseline and sequential imaging examinations. Small differences in reference point placement on follow-up radiographs can result in misdiagnosis of incident vertebral fracture by QM, though readily interpreted by the expert reader as simply due to a slight alteration in projection. QM also does not distinguish between

Grade 0: normal, unfractured vertebra.

Grade 0.5: uncertain or questionable fracture with borderline 20% reduction in anterior, middle or posterior heights relative to the same or adjacent vertebrae.

Grade 1: mid fracture with approximately 20-25% reduction in anterior, middle or posterior heights relative to the same or adjacent vertebrae.

Grade 2: moderate fracture with approximately 25-40% reduction in anterior, middle or posterior heights relative to the same or adjacent vertebrae.

Grade 3: severe fracture with approximately >40% reduction in anterior, middle or posterior relative to the same or adjacent vertebrae.

Fig. 37.6. Schematic diagram of Genant semi-quantitative analysis of vertebral fracture severity (adapted from Ref. 26).

vertebral fracture and non-fracture vertebral deformity (such as SVH or physiological wedging). While there is good concordance between QM and SQ methods in the detection of moderate or severe vertebral fractures, the concordance between both methods for detection of mild vertebral fractures is poor. In most instances, this is due to a false positive diagnosis of mild fractures by quantitative morphometry [15, 55]. Therefore, all fractures identified by vertebral morphometry should be confirmed by an expert reader before diagnosis of a vertebral fracture is made [55].

Defining normative databases for QM is fraught with difficulty, as is defining the threshold used for vertebral fracture diagnosis, as the quality of reference data, and, more particularly, the thresholds chosen will greatly influence fracture detection rate [2]. Vertebral morphometry is probably most suited to evaluating (a) large longitudinal studies and (b) individual vertebra in DXA VFA

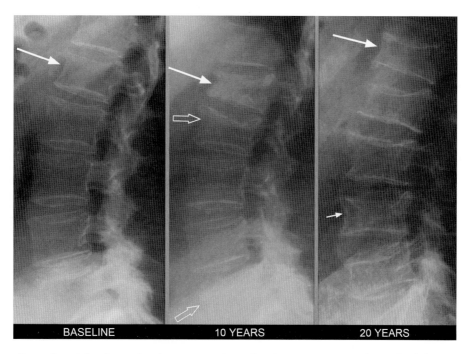

Fig. 37.7. Lateral radiographs of lumbar spine in same patient showing progression of vertebral fractures akin to "vertebral cascade." At baseline, there is a mild vertebral fracture of L1 (long arrow). Ten years later, this L1 vertebral fracture is now moderate in severity (long arrow) and there are new fractures of L2 and L3 (open arrows). Twenty years later, there is a new fracture of L4 (short arrow).

where large normative databases are available. Computerized vertebral morphometry is suited to analysis of a large number of spine radiographs and particularly for single center analysis in multicenter trials. More sophisticated automated models for vertebral morphometry have recently been developed using statistical-model-based and other approaches [2, 56].

Algorithm-based qualitative assessment

The algorithm-based qualitative (ABQ) method, as the name implies, emphasizes a qualitative assessment of vertebral fracture and relies more on detection of vertebral endplate abnormalities related to fracture than loss of vertebral body height. The ABQ method categories vertebrae as either (i) normal, (ii) osteoporotic fracture, or (iii) non-osteoporotic deformity or short vertebral height (SVH). The diagnosis of an osteoporotic vertebral fracture requires evidence of vertebral endplate fracture with loss of expected vertebral height but with no minimum threshold for apparent reduction in vertebral height [57]. If a fracture of the cortical margin in also visible radiographically, this provides clear-cut evidence that there is a fracture present and it is likely to be of recent origin. When one of more vertebral height (anterior, middle, or posterior) is shorter than expected but without specific endplate abnormalities of fracture (altered texture below endplate due to microfractures), this is designated as nonosteoporotic deformity.

Dual energy X-ray absorptiometry

Imaging vertebral fractures using dual energy X-ray absorptiometry (DXA) is known as vertebral fracture assessment (VFA) (Fig 37.9). VFA has several advantages over radiography, including greater convenience, as it can be performed at the same time as DXA and on the same equipment; lower radiation dose (less than 5% of spine radiography); and low cost [36, 58]. VFA allows a more ready detection of prevalent vertebral fractures [58]. Combining prevalent vertebral fracture status with BMD enhances fracture risk prediction of both vertebral and nonvertebral fractures [5, 9, 59].

VFA requires a fan-beam DXA scanner with appropriate software and can be performed either with the patient lying supine on a scanner with rotating "C" arm gantry or with the patient in a lateral decubitus position when performed in a scanner with a fixed gantry. Modern fan-beam DXA scanners can obtain single energy images of the spine from T4 to L4 in less than 10 seconds during suspended respiration. These low dose images can be used to assess for vertebral fracture at the time patients are undergoing DXA bone densitometry. The T4-T6 vertebral bodies can be adequately visualized in 40–70% of patients while vertebrae from T7 and below can be adequately identified in nearly all patients [60].

As VFA is a digital technique, it lends itself to computational analysis. Following image acquisition, manual or automated vertebral morphometry known as MXA is available [2]. The vertebral body is demarcated by four or

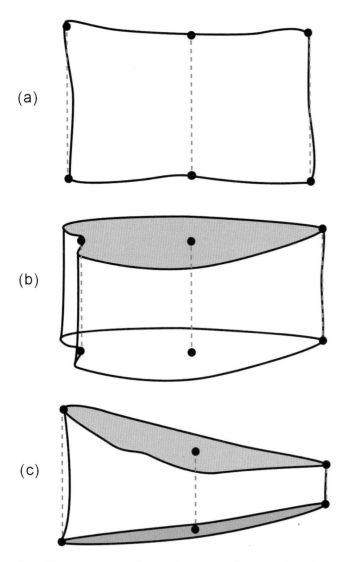

Fig. 37.8. Schematic diagram showing reference point placement for (a) normal vertebral body, (b) obliquely depicted vertebral body, and (c) anterior wedge fracture.

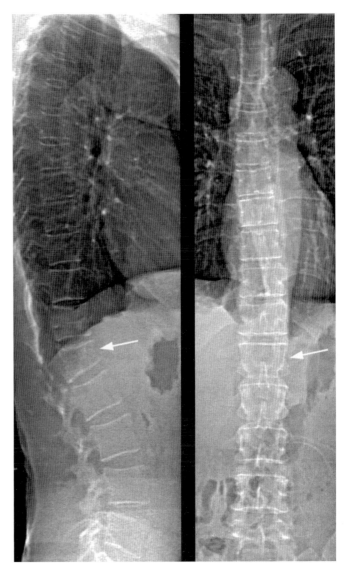

Fig. 37.9. Vertebral fracture assessment (VFA) by DXA showing moderate fracture of L1 vertebral body (arrow).

six reference points and vertebral body height, height ratios and average height calculated automatically (Fig. 37.10). An automated assessment of fracture status based on comparison with normative data is also available [2]. Superimposition of these baseline reference points on follow-up VFA spine images is also possible to allow more ready comparison of baseline and follow-up VFA [2]. VFA in children may be problematic as current software cannot consistently detect the vertebral outline in most children [61].

Although morphometric analysis is routinely undertaken on VFA, this alone is not recommended for fracture diagnosis. Visual inspection using the Genant semi-quantitative method is the International Society of Clinical Densitometry (ISCD) recommendation for diagnosing and grading severity of vertebral fracture on VFA. (http://www.iscd.org/Visitors/positions/OfficialPositionsText.

cfm). Semi-quantitative analysis of vertebral fracture on VFA compares well with radiography [45, 62], but as the images are of lower spatial resolution than radiography, VFA visualization of the upper thoracic spine is limited (although this is an uncommon site for fracture) but superior, and VFA has only moderate sensitivity for the diagnosis of mild grade 1 vertebral fracture (20–25% reduction in expected vertebral body height). However, the sensitivity/specificity for detecting moderate (more than 25–40% reduction in expected vertebral body height) or severe (greater than 40%) vertebral fractures is greater than 90% [63]. More sophisticated MXA analytical models, such as the active appearance model (AAS), may potentially help overcome this problem with mild vertebral fracture detection. AAS is a computational technique that analyzes statistical differences in vertebral morphometry and bone texture adjacent to the endplate, parallax effect on the endplates. However, VFA will have the same difficulties as radiography in assessment of vertebral fracture when scoliosis is present. Other advantages of VFA include convenience as both densitometry and vertebral fracture assessment can be performed at the same visit and on the same equipment, a low radiation dose [36], and relatively low cost. VFA will have an increasing important role in the diagnosis of vertebral fractures and in estimating fracture risk. In a DXA-based study of almost 1,000 patients (64% female, mean age 53 years), adding VFA to BMD analysis detected a high number of undiagnosed vertebral fractures and altered Canadian fracture risk classification in 20% of patients [65]. The added information provided by vertebral fracture by DXA assessment can be incorporated into the fracture risk assessment (FRAX) model along with DXA BMD of the femoral neck and clinical risk factors to improve individual prediction of 10-year probability (%) of developing a hip fracture or major osteoporotic fracture (distal radius, proximal femur, proximal humerus, or clinical vertebral fracture) [19] .

OTHER IMAGING METHODS

Computed tomography

The widespread availability of multidetector computed tomography (MDCT) and ease of midline sagittal reconstructions allows the thoracic or lumbar spine to be evaluated on all CT studies of the thorax or abdomen being performed for various clinical indications (Fig. 37.11). As with other image studies that may incidentally include the spine, CT enables fortuitous identification of vertebral fractures. It is important that such factures are always sought and, if present, clearly reported [4]. In a study of patients older than 55 years who had thoracic CT, one-fifth had a moderate or severe thoracic vertebral fracture but fewer than one-fifth of these fractures were reported [66]. The CT scout views should also be scrutinized routinely for vertebral fractures since these usually will include a greater length of the spine than that covered by axial sections [67, 68].

Although MDCT enables submillimeter imaging of vertebral structure (with a minimal slice thickness of 0.6 mm and minimal pixel size of 0.25 mm), this is still insufficient to image individual trabeculae. Nevertheless, textural analysis methods, rather than true structural analysis, can be applied to study vertebral architecture. Volumetric MDCT-derived structural parameters separated patients with and without vertebral fracture better than DXA-derived BMD [69], while the treatment effect of teriparatide was better evaluated by MDCT vertebral structural parameters than DXA-derived BMD [70].

Finite element analysis (FEA) is a well-established engineering computational technique that is used in virtual strength analysis of complex structures. FEA and

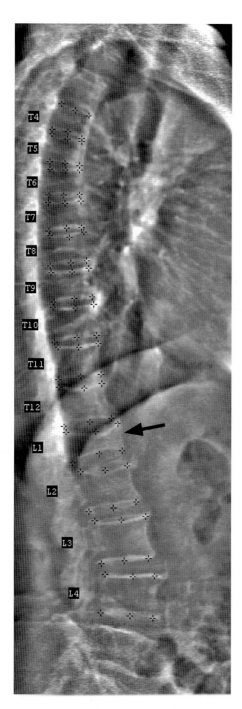

Fig. 37.10. Vertebral fracture assessment (VFA) by DXA with margins of vertebral bodies from T4 to L4 outlined by six reference points. There is a mild fracture of the L1 vertebral body (arrow).

which may improve discrimination of mild vertebral fractures from other non-fracture vertebral deformities [64].

One other advantage of VFA over radiography is that image distortion ("bean can" effect) is less as the scanning technique ensures that the X-ray beam is orthogonal, rather than divergent, to the spine, reducing the

328 *Investigation of Metabolic Bone Diseases*

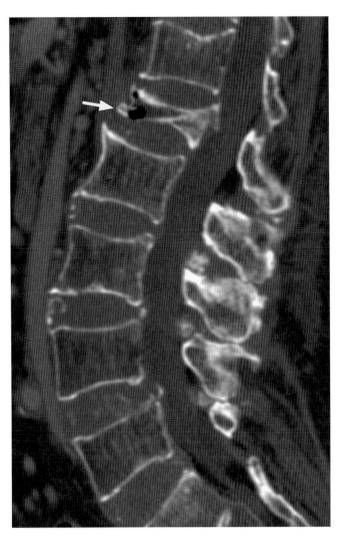

Fig. 37.11. Sagittal reconstruction of abdominal CT dataset (bone window) showing moderate generalized osteopenia and severe fracture of L1 (arrow) vertebral body with intravertebral gas.

other structural analytic methods can be applied to volumetric quantitative CT (vQCT) data to noninvasively estimate bone strength better than quantitative CT densitometry alone, with the potential to provide a more vigorous tailoring and monitoring of treatment protocols to prevent vertebral and other osteoporotic fractures [71]. FEA of microCT data from cadaveric vertebral bodies indicates how the integrated effects of trabecular and cortical bone on bone strength in different areas of the vertebral body can be assessed [72].

The major limitation to the more widespread primary, rather than fortuitous, use and evaluation of CT in vertebral fracture diagnosis is the cost and radiation dose involved. While the effective dose for DXA examination is 0.01–0.05 mSv, the dose for 2D QCT of the lumbar spine is 0.06–0.3 mSv, and the dose for examining vertebral microarchitecture by volumetric CT is 3 mSv (equivalent to 1.5 years of background radiation) [73].

Magnetic resonance imaging

Although spatial resolution issues currently limit the use of magnetic resonance imaging (MRI) in structurally evaluating the microarchitecture of vertebral bodies compared to CT, MRI does have distinct advantage as it does not use ionizing radiation, can measure physiological function such as molecular diffusion and perfusion, and can assess marrow changes better than any other imaging modality.

Because vertebral fractures are typically incremental in development, with occasional mild stepwise progression in severity, high sensitivity and specificity with radiographic assessment may not be achievable. Hence, a vertebral fracture is only diagnosed on radiography when there is more than 20% loss of vertebral height. This approach clearly overlooks a significant number of mild vertebral fractures that may not cause much vertebral deformity but can cause significant pain. MRI offers a solution to this problem by demonstrating marrow edema in even mild acute vertebral fractures. Marrow edema, in the absence of marrow infiltration, is a sensitive sign of acute or subacute vertebral fracture even when no fracture is visible radiographically. As MR scanners become more widely available, the use of MRI in the assessment of acute vertebral fractures may increase.

Determining the age of a vertebral fracture is usually difficult on radiographs in the absence of previous imaging to assess changes in appearance. Lack of cortical disruption, the presence of reparative sclerosis, and marginal hyperostosis or osteophytosis can help make the diagnosis of an old fracture. However, often these features may not be present. The best guide to the age of a fracture is the presence of marrow edema on fat-suppressed MR imaging. The presence and degree of edema on sagittal T2-weighted fat-suppressed MR images is a reliable guide as to the age and severity of a vertebral fracture. Conversely, vertebral fractures that lack marrow edema are not recent fractures, are not likely to respond to percutaneous vertebroplasty or balloon kyphoplasty and are less likely to be the cause of symptoms.

MRI studies have shown how perfusion is reduced in non-fractured osteoporotic vertebrae bodies compared to those of normal BMD [74, 75]. This reduced perfusion is most likely to be due to atherosclerosis, impaired endothelial function, or reduced demand for tissue oxygenation due to a relative decrease in the amount of hemopoetic marrow within osteoporotic vertebral bodies [76, 77]. MR-based perfusion parameters are reduced in osteoporotic vertebral fractures compared to adjacent non-fractured vertebrae [78]. Good perfusion is clearly a prerequisite for normal bone metabolism and fracture healing, including microdamage repair. The smaller the area of enhancing tissue within an acutely fractured vertebral body, the more likely that fractured vertebral body will reduce in height on subsequent follow up [79].

Nonunion affects about 10% of acute osteoporotic vertebral fractures. Nonunion is particularly prevalent in the T12 and L1 vertebrae and is evident radiographically as a vacuum cleft extending horizontally across the vertebral body. These nonunited vertebral fractures are associated with more severe back pain than united fractures [80]. Risk of nonunion is increased significantly if there is retropulsion of the posterior vertebral cortex with areas of localized high intensity on T2-weighted images or diffuse low intensity within the vertebral body on T2-weighted images [80]. MRI may therefore have the potential to distinguish those acute vertebral fractures particularly susceptible to progression or nonunion and consequently more likely to benefit from more aggressive fracture treatment such as vertebroplasty.

Differentiating osteoporotic fracture from neoplastic fracture

The spine is the most common site for skeletal metastases and osteoporotic fracture, with both being features of middle-aged and elderly subjects. Almost one-third of vertebral fractures in patients with known primary malignancy are due to osteoporosis and not neoplasia [81]. In the acute stage, the marrow cavity of the vertebral body in osteoporotic fracture is filled with blood and fluid. This is gradually reabsorbed and replaced by a variable amount of granulation and fibroblastic tissue. This reparative tissue is reabsorbed over time with restoration of normal fatty marrow. It is thus not difficult to differentiate a chronic osteoporotic vertebral fracture in which the marrow is filled with fat from a pathological fracture due to neoplasia (Fig. 37.12). Difficultly arises when differentiating acute/subacute osteoporotic fracture from neoplastic fracture.

Table 37.1 outlines the most helpful imaging signs used to discriminate between acute/subacute osteoporotic fracture and neoplastic fracture [82]. A combination of these signs should be applied when making the distinction, with suitable weighting being given to their relative discriminatory power. Residual marrow fat within the fractured vertebral body is a very helpful sign of osteoporotic fracture. Fluid in a cavity within the vertebral body is another helpful sign of osteoporotic fracture on MRI, being present in more (about 40%) osteoporotic fractures when compared to neoplastic fractures (about 6%) [83] (Fig. 37.12). Fluid accumulates in a fracture cavity adjacent to the endplates of a severely fractured vertebral body and is accounted for by the vacuum cleft on radiographs or CT. Gas within a vacuum cleft arises from compression/decompression forces leading to the release of nitrogen [84]. Intravertebral gas is thus generated by flexing or extending the spine, which will be replaced by fluid on resting supine for about 15 minutes (as is applicable to MRI scans) [84]. Hence, in the vertebral body fluid on MRI and gas on radiographs reflect similar pathology. Both indicate a medium to large cavity within the vertebral body and provide good evidence that a vertebral body is not filled with neoplastic or other soft tissue. Intravenous contrast medium enhancement is not a useful discriminatory criterion as

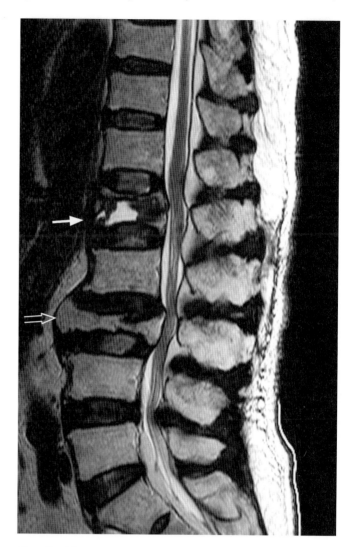

Fig. 37.12. T2-weighted sagittal MR image showing typical osteoporotic-type fracture of T12 vertebral body with preservation of some marrow fat and fluid-filled cavity within vertebral body (closed arrow). There is a chronic fracture of the L3 vertebral body with fat occupying the marrow cavity (open arrow).

acute/subacute vertebral fractures will enhance, as will most neoplastic related fractures [82]. However, occasionally postcontrast enhancement along fracture lines may be evident in osteoporotic vertebral fracture. This would not be an expected feature of neoplastic vertebral fracture.

The standard MR protocol for vertebral fracture should include (a) a T1-weighted sagittal sequence, mainly to assess fracture morphology and the presence of marrow fat; (b) a T2-weighted fat-suppressed sagittal sequence, mainly to assess marrow edema and fluid; and (c) a T2-weighed axial image to assess vertebral and paravertebral soft tissues. If doubt still exists as to the likely fracture etiology, standard MRI can be supplemented with functional imaging techniques such as diffusion

Table 37.1. Useful Imaging Signs That Help Distinguish Osteoporotic From Neoplastic Vertebral Fracture

1. Preservation of some marrow fat signal within marrow ***
2. No involvement of pedicles or posterior elements ***
3. Fluid or gas within vertebral body **
4. T1-hypointense fracture line within vertebral body (often near fractured endplate) **
5. Lack of discrete soft tissue mass ***
6. Only minimal or mild paravertebral soft tissue swelling **
7. Absence of epidural mass **
8. Fracture not occurring above T4 level ***
9. Posterior located triangular fracture fragment *
10. Non-convex posterior cortical margin *
11. Evidence of metastases elsewhere in spine *
12. Near complete fatty marrow of adjacent vertebrae *
13. Radiographic evidence of osteopenia *
14. Preservation of trabeculae within fractured vertebral body on CT *** (Fig 37.13)

The relative usefulness of these signs is indicated with *** = most useful, ** = less useful, and * = least useful signs. A known history of malignancy is moderately helpful. If doubt still exists after standard MR imaging, one can proceed to diffusion-weighted imaging or chemical shift imaging (Ref. 82).

raphy) imaging as part of clinical investigation, and this too can be helpful in differentiating osteoporotic from neoplastic vertebral fracture. As expected, a higher radioactive uptake is present in neoplastic vertebral fractures [standardized uptake value 2.2–7.1, mean 3.99±1.52 (SD)] than in osteoporotic fractures [standardized uptake value 0.7–4.9, mean 1.94±0.97]. A preliminary study indicates that the positive predictive value of FDG-PET for detecting malignant vertebral fracture is 71% while the negative predictive value is 91% [87].

CLINICAL RECOMMENDATIONS FOR SPINE IMAGING TO DETECT VERTEBRAL FRACTURE

In general, the indications for spine imaging for vertebral fracture detection are similar, irrespective of whether radiography or VFA is used. The current recommendations for using VFA by the ISCD are summarized in Table 37.2 [12]. Ultimately, the decision on whether to undertake spinal radiography or VFA by DXA depends on availability and local expertise.

If a patient has normal BMD, most clinicians would not undertake spine imaging unless there has been significant loss of height or back pain.

If the patient has osteoporosis on DXA examination, most clinicians would start treatment, and, in this instance, it can be argued that there is no added benefit from identifying a vertebral fracture [13] though the recognition of a vertebral fracture would influence the 10-year fracture prediction by FRAX [19]. Similarly, if moderate, severe, or multiple vertebral fractures are identified on imaging, this may alter the choice of treatment. For example, an anabolic drug (e.g., parathyroid hormone [PTH]) may be appropriate before considering long-term therapy with bone protective therapy (e.g., bisphosphonate).

If a patient has low BMD and an insufficiency fracture at any site (the most common of which is vertebral fracture), most clinicians would intervene with therapy. If a patient has low BMD and no previous history of insufficiency fractures, spinal radiographs or VFA are indicated as the presence of a vertebral fracture would influence management. Patients with low BMD and vertebral fracture do benefit from a range of antiresoptive therapies that reduce future fracture risk [12]. If a patient is taking long-term oral glucocorticoids and has a history of osteoporotic fracture at any site, most clinicians would consider antiosteoporotic therapy, irrespective of whether spine imaging shows a vertebral fracture. Therefore, additional imaging is probably not warranted unless a comprehensive fracture risk assessment by FRAX is required. Patients on long-term oral glucocorticoids with no history of osteoporotic fracture warrant spine imaging, particularly if there is a history of back pain, height loss, or they are over 65 years old.

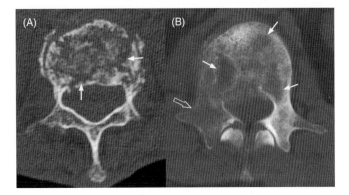

Fig. 37.13. Axial CT images of (A) osteoporotic fracture showing preservation of vertebral trabeculae with intervening fracture lines (arrows), (B) neoplastic fracture showing quite discrete lytic areas (arrows) due to metastatic infiltration and also expansion of right pedicle and adjacent neural arch (open arrow).

weighted imaging [85] or chemical shift imaging [86]. If doubt exists after MR imaging, CT is a useful additional investigation. In practice, one can usually distinguish accurately between osteoporotic and neoplastic vertebral fractures on imaging (radiography, CT, MRI) features alone, without the need for percutaneous vertebral biopsy.

Patients may, in certain circumstances, undergo FDG-PET (flurodeoxyglucose positron emission tomog-

Table 37.2. ISCD Recommendations for Screening for Vertebral Fractures Using VFA*

- Consider VFA when the results may influence clinical management
- Postmenopausal women with low bone mass (osteopenia) by BMD criteria, PLUS any one of the following:
  - Age greater than or equal to 70 years
  - Historical height loss greater than 4 cm (1.6 in)
  - Prospective height loss greater than 2 cm (0.8 in)
  - Self-reported vertebral fracture (not previously documented)
  - Two or more of the following:
    - Age 60 to 69 years
    - Self-reported prior nonvertebral fracture
    - Historical height loss of 2 to 4 cm
    - Chronic systemic diseases associated with increased risk of vertebral fractures (for example, moderate to severe COPD, seropositive rheumatoid arthritis, Crohn's disease)
- Men with low bone mass (osteopenia) by BMD criteria, PLUS any one of the following:
  - Age 80 years or older
  - Historical height loss greater than 6 cm (2.4 in)
  - Prospective height loss greater than 3 cm (1.2 in)
  - Self-reported vertebral fracture (not previously documented)
  - Two or more of the following:
    - Age 70 to 79 years
    - Self-reported prior nonvertebral fracture
    - Historical height loss of 3 to 6 cm
    - On pharmacologic androgen deprivation therapy or following orchiectomy
    - Chronic systemic diseases associated with increased risk of vertebral fractures (for example, moderate to severe COPD or COAD, seropositive rheumatoid arthritis, Crohn's disease)
- Women or men on chronic glucocorticoid therapy (equivalent to 5 mg or more of prednisone daily for 3 months or longer)
- Postmenopausal women or men with osteoporosis by BMD criteria, if documentation of one or more vertebral fractures will alter clinical management

*http://www.iscd.org/visitors/positions/OfficialPositionsText.cfm. (Ref. 48).

CONCLUSION

Vertebral fractures occurring in the presence of minor or no trauma provide evidence of reduced bone strength, irrespective of BMD. They are usually the first insufficiency fracture to occur. Recognition and appropriate treatment of vertebral fracture at this relatively early stage can reduce future fracture risk, patient pain, deformity, and suffering. Radiographic evaluation with qualitative, semi-quantitative, or quantitative assessment is the standard for detecting vertebral fracture. Good technique in performing spinal radiographs and a high level of observer experience in image interpretation are key to the reliable diagnosis of vertebral fractures. VFA is being utilized increasingly for vertebral fracture identification. Vertebral fracture diagnosis may be made fortuitously from any imaging method on which the spine is included. Density and structural parameters obtained by volumetric QCT can predict vertebral compressive strength *ex vivo*, and these parameters, together with nonlinear FEA, can also be applicable *in vivo*. In the future, virtual measurement of vertebral body strength with high-resolution imaging techniques may enable patients who are at risk of vertebral fracture to be identified more efficiently. MRI can detect minor acute or subacute vertebral fracture or re-fracture, determine fracture age, and distinguish between osteoporotic and neoplastic fracture with greater sensitivity than other imaging techniques.

REFERENCES

1. Melton LJ 3rd, Lane AW, Cooper C, Eastell R, O'Fallon WM, Riggs BL. 1993. Prevalence and incidence of vertebral deformities. *Osteoporos Int* 3: 113–119.
2. Diacinti D, Guglielmi G. 2010. Vertebral morphometry. *Radiol Clin North Am* 48: 561–575.
3. Szulc P, Bouxsein ML. 2010. *Overview of Osteoporosis. Epidemiology and Clinical Management*. International Osteoporosis Foundation Vertebral Fracture Initiative Resource Document Part I, pp. 1–64. http://www.iofbonehealth.org/health-professionals/educational-tools-and-slide-kits/vertebral-fracture-teaching-program.html.
4. Adams JE, Lenchik L, Roux C, Genant HK. 2010. *Radiological Assessment of Vertebral Fracture*. International Osteoporosis Foundation Vertebral Fracture Initiative Resource Document Part II, pp. 1–49. http://www.iofbonehealth.org/health-professionals/educational-

5. Lenchik L, Rogers LF, Delmas PD, Genant HK. 2004. Diagnosis of osteoporotic vertebral fractures: Importance of recognition and description by radiologists. *AJR Am J Roentgenol* 183: 949–958.
6. Siris ES, Miller PD, Barrett-Connor E, Faulkner KG, Wehren LE, Abbott TA, Berger ML, Santora AC, Sherwood LM. 2001. Identification and fracture outcomes of undiagnosed low bone mineral density in postmenopausal women: Results from the National Osteoporosis Risk Assessment. *JAMA* 286: 2815–2822.
7. Williams AL, Al-Busaidi A, Sparrow PJ, Adams JE, Whitehouse RW. 2009. Under-reporting of osteoporotic vertebral fractures on computed tomography. *Eur J Radiol* 69: 179–183.
8. National Osteoporosis Foundation. 2010. *The Clinician's Guide to Prevention and Treatment of Osteoporosis*. Washington, DC: National Osteoporosis Foundation. http://www.nof.org/files/nof/public/content/file/344/upload/159.pdf .
9. Siris ES, Genant HK, Laster AJ, Chen P, Misurski DA, Krege JH. 2007. Enhanced prediction of fracture risk combining vertebral fracture status and BMD. *Osteoporos Int* 18: 761–770.
10. Black DM, Arden NK, Palermo L, Pearson J, Cummings SR. 1999. Prevalent vertebral deformities predict hip fractures and new vertebral deformities but not wrist fractures. Study of Osteoporotic Fractures Research Group. *J Bone Miner Res* 14: 821–828.
11. Lindsay R, Silverman SL, Cooper C, Hanley DA, Barton I, Broy SB, Licata A, Benhamou L, Geusens P, Flowers K, Stracke H, Seeman E. 2001. Risk of new vertebral fracture in the year following a fracture. *JAMA* 285: 320–323.
12. Lips P, van Schoor NM. 2005. Quality of life in patients with osteoporosis. *Osteoporos Int* 16: 447–455.
13. Ensrud KE, Schousboe JT. 2011. Clinical practice. Vertebral fractures. *N Engl J Med* 364: 1634–1642.
14. Delmas PD, van de Langerijt L, Watts NB, Eastell R, Genant H, Grauer A, Cahall DL; IMPACT Study Group. 2005. Underdiagnosis of vertebral fractures is a worldwide problem: The IMPACT study. *J Bone Miner Res* 20: 557–563.
15. Grigoryan M, Guermazi A, Roemer FW, Delmas PD, Genant HK. 2003. Recognizing and reporting osteoporotic vertebral fractures. *Eur Spine J* 12 Suppl 2: S104–112.
16. Cooper C, Atkinson EJ, O'Fallon WM, Melton LJ 3rd. 1992. Incidence of clinically diagnosed vertebral fractures: A population-based study in Rochester, Minnesota, 1985–1989. *J Bone Miner Res* 7: 221–227.
17. Gehlbach SH, Bigelow C, Heimisdottir M, May S, Walker M, Kirkwood JR. 2000. Recognition of vertebral fracture in a clinical setting. *Osteoporos Int* 11: 577–582.
18. Kanis JA, Johnell O, Oden A, Johansson H, McCloskey E. 2008. FRAX and the assessment of fracture probability in men and women from the UK. *Osteoporos Int* 19: 385–397.
19. Kanis JA, Hans D, Cooper C, Baim S, Bilezikian JP, Binkley N, Cauley JA, Compston JE, Dawson-Hughes B, El-Hajj Fuleihan G, Johansson H, Leslie WD, Lewiecki EM, Luckey M, Oden A, Papapoulos SE, Poiana C, Rizzoli R, Wahl DA, McCloskey EV. 2011. Task Force of the FRAX Initiative. Interpretation and use of FRAX in clinical practice. *Osteoporos Int* 22: 2395–2411.
20. National Osteoporosis Guidelines Group (NOGG) guidelines. http://www.shef.ac.uk/NOGG/.
21. Kanis JA, McCloskey EV, Johansson H, Strom O, Borgstrom F, Oden A. 2008. Case finding for the management of osteoporosis with FRAX—Assessment and intervention thresholds for the UK. *Osteoporos Int* 19: 1395–1408.
22. Banse X, Devogelaer JP, Grynpas M. 2002. Patient-specific microarchitecture of vertebral cancellous bone: A peripheral quantitative computed tomographic and histological study. *Bone* 30: 829–835.
23. Vernon-Roberts B, Pirie CJ. 1973. Healing trabecular microfractures in the bodies of lumbar vertebrae. *Ann Rheum Dis* 32: 406–412.
24. Diamond TH, Clark WA, Kumar SV. 2007. Histomorphometric analysis of fracture healing cascade in acute osteoporotic vertebral body fractures. *Bone* 40: 775–780.
25. Riggs BL, Melton Iii LJ 3rd, Robb RA, Camp JJ, Atkinson EJ, Peterson JM, Rouleau PA, McCollough CH, Bouxsein ML, Khosla S. 2004. Population-based study of age and sex differences in bone volumetric density, size, geometry, and structure at different skeletal sites. *J Bone Miner Res* 19: 1945–1954.
26. Fields AJ, Lee GL, Liu XS, Jekir MG, Guo XE, Keaveny TM. 2011. Influence of vertical trabeculae on the compressive strength of the human vertebra. *J Bone Miner Res* 26: 263–269.
27. Thomsen JS, Ebbesen EN, Mosekilde LI. 2002. Age-related differences between thinning of horizontal and vertical trabeculae in human lumbar bone as assessed by a new computerized method. *Bone* 31: 136–142.
28. Mc Donnell P, Harrison N, Liebschner MA, Mc Hugh PE. 2009. Simulation of vertebral trabecular bone loss using voxel finite element analysis. *J Biomech* 42: 2789–2796.
29. Shi X, Liu XS, Wang X, Guo XE, Niebur GL. 2010. Effects of trabecular type and orientation on microdamage susceptibility in trabecular bone. *Bone* 46: 1260–1266.
30. Genant HK, Jergas M, Palermo L, Nevitt M, Valentin RS, Black D, Cummings SR. 1996. Comparison of semi-quantitative visual and quantitative morphornetric assessment of prevalent and incident vertebral fractures in osteoporosis. The Study of Osteoporotic Fractures Research Group. *J Bone Miner Res* 11: 984–996.
31. Genant HK, Jergas M. 2003. Assessment of prevalent and incident vertebral fractures in osteoporosis research. *Osteoporos Int* 14 Suppl 3: S43–55.
32. Christiansen BA, Bouxsein ML. 2010. Biomechanics of vertebral fractures and the vertebral fracture cascade. *Curr Osteoporos Rep* 8: 198–204.

33. Nevitt MC, Cummings SR, Stone KL, Palermo L, Black DM, Bauer DC, Genant HK, Hochberg MC, Ensrud KE, Hillier TA, Cauley JA. 2005. Risk factors for a first-incident radiographic vertebral fracture in women > or = 65 years of age: The study of osteoporotic fractures. *J Bone Miner Res* 20: 131–140.
34. Siminoski K, Jiang G, Adachi JD, Hanley DA, Cline G, Ioannidis G. Hodsman A, Josse RG, Kendler D, Olszynski WP, Ste Marie LG, Eastell R. 2005. Accuracy of height loss during prospective monitoring for detection of incident vertebral fractures. *Osteoporos Int* 16: 403–410.
35. Siminoski K, Warshawski RS, Jen H, Lee K. 2006. The accuracy of historical height loss for the detection of vertebral fractures in postmenopausal women. *Osteoporos Int* 17: 290–296.
36. Damilakis J, Adams JE, Guglielmi G, Link TM. 2010. Radiation exposure in X-ray-based imaging techniques used in osteoporosis. *Eur Radiol* 20: 2707–2714.
37. Masharawi Y, Salame K, Mirovsky Y, Peleg S, Dar G, Steinberg N, Hershkovitz I. 2008. Vertebral body shape variation in the thoracic and lumbar spine: Characterization of its asymmetry and wedging. *Clin Anat* 21: 46–54.
38. Diacinti D, Acca M, D'Erasmo E, Tomei E, Mazzuoli GF. 1995. Aging changes in vertebral morphometry. *Calcif Tissue Int* 57: 426–429.
39. Masunari N, Fujiwara S, Nakata Y, Nakashima E, Nakamura T. 2007. Historical height loss, vertebral deformity, and health-related quality of life in Hiroshima cohort study. *Osteoporos Int* 18: 1493–1499.
40. Ferrar L, Jiang G, Armbrecht G, Reid DM, Roux C, Glüer CC, Felsenberg D, Eastell R. 2007. Is short vertebral height always an osteoporotic fracture? The Osteoporosis and Ultrasound Study (OPUS). *Bone* 41: 5–12.
41. Ferrar L, Roux C, Reid DM, Felsenberg D, Glüer CC, Eastell R. 2011. Prevalence of non-fracture short vertebral height is similar in premenopausal and postmenopausal women: The osteoporosis and ultrasound study. *Osteoporos Int* 23: 1035–1040.
42. Chan KK, Sartoris DJ, Haghighi P, Sledge P, Barrett-Connor E, Trudell DT, Resnick D. 1997. Cupid's bow contour of the vertebral body: Evaluation of pathogenesis with bone densitometry and imaging-histopathologic correlation. *Radiology* 202: 253–256.
43. Link TM, Guglielmi G, van Kuijk C, Adams JE. 2005. Radiologic assessment of osteoporotic vertebral fractures: Diagnostic and prognostic implications. *Eur Radio* 15: 1521–1532.
44. Li EK, Tam LS, Griffith JF, Zhu TY, Li TK, Li M, Wong KC, Chan M, Lam CW, Chu FS, Wong KK, Leung PC, Kwok A. 2009. High prevalence of asymptomatic vertebral fractures in Chinese women with systemic lupus erythematosus. *J Rheumatol* 36: 1646–1652.
45. Genant HK, Wu CY, van Kuijk C, Nevitt MC. 1993. Vertebral fracture assessment using a semiquantitative technique. *J Bone Miner Res* 8: 1137–1148.
46. Genant HK, Siris E, Crans GG, Desaiah D, Krege JH. 2005. Reduction in vertebral fracture risk in teriparatide-treated postmenopausal women as assessed by spinal deformity index. *Bone* 37: 170–174.
47. Genant HK, Delmas PD, Chen P, Jiang Y, Eriksen EF, Dalsky GP, Marcus R, San Martin J. 2007. Severity of vertebral fracture reflects deterioration of bone microarchitecture. *Osteoporos Int* 18: 69–76.
48. Wu CY, Li J, Jergas M, Genant HK. 1995. Comparison of semiquantitative and quantitative techniques for the assessment of prevalent and incident vertebral fractures. *Osteoporos Int* 5: 354–370.
49. Siris E, Adachi JD, Lu Y, Fuerst T, Crans GG, Wong M, Harper KD, Genant HK. 2002. Effects of raloxifene on fracture severity in postmenopausal women with osteoporosis: Results from the MORE study. Multiple Outcomes of Raloxifene Evaluation. *Osteoporos Int* 13: 907–913.
50. Schousboe JT, Vokes T, Broy SB, Ferrar L, McKiernan F, Roux C, Binkley N. 2008. Vertebral Fracture Assessment: The 2007 ISCD Official Positions. *J Clin Densitom* 11: 92–108.
51. Buehring B, Krueger D, Checovich M, Gemar D, Vallarta-Ast N, Genant HK, Binkley N. 2010. Vertebral fracture assessment: Impact of instrument and reader. *Osteoporos Int* 21: 487–494.
52. Guglielmi G, Diacinti D, van Kuijk C, Aparisi F, Krestan C, Adams JE, Link TM. 2008. Vertebral morphometry: Current methods and recent advances. *Eur Radiol* 18: 1484–1496.
53. Grados F, Fechtenbaum J, Flipon E, Kolta S, Roux C, Fardellone P. 2009. Radiographic methods for evaluating osteoporotic vertebral fractures. *Joint Bone Spine* 76: 241–247.
54. Eastell R, Cedel SL, Wahner HW, Riggs BL, Melton LJ 3rd. 1991. Classification of vertebral fractures. *J Bone Miner Res* 6: 207–215.
55. Grados F, Roux C, de Vernejoul MC, Utard G, Sebert JL, Fardellone P. 2001. Comparison of four morphometric definitions and a semiquantitative consensus reading for assessing prevalent vertebral fractures. *Osteoporos Int* 12: 716–722.
56. Roberts MG, Pacheco EM, Mohankumar R, Cootes TF, Adams JE. 2010. Detection of vertebral fractures in DXA VFA images using statistical models of appearance and a semi-automatic segmentation. *Osteoporos Int* 21: 2037–2046.
57. Jiang G, Eastell R, Barrington NA, Ferrar L. 2004. Comparison of methods for the visual identification of prevalent vertebral fracture in osteoporosis. *Osteoporos Int* 15: 887–896.
58. Gallacher SJ, Gallagher AP, McQuillian C, Mitchell PJ, Dixon T. 2007. The prevalence of vertebral fracture amongst patients presenting with non-vertebral fractures. *Osteoporos Int* 18: 185–192.
59. Schousboe JT, Vokes T, Binkley N, Genant HK. 2010. *Densitometric Vertebral Fracture Assessment (VFA)*. International Osteoporosis Foundation Vertebral

Fracture Initiative Resource Document Part III, pp. 1–49. http://www.iofbonehealth.org/health-professionals/educational-tools-and-slide-kits/vertebral-fracture-teaching-program.html.
60. Ferrar L, Jiang G, Barrington NA, Eastell R. 2000. Identification of vertebral deformities in women: Comparison of radiological assessment and quantitative morphometry using morphometric radiography and morphometric X-ray absorptiometry. *J Bone Miner Res* 15: 575–585.
61. Mayranpaa MK, Helenius I, Valta H, Mayranpaa MI, Toiviainen-Salo S, Makitie O. 2007. Bone densitometry in the diagnosis of vertebral fractures in children: Accuracy of vertebral fracture assessment. *Bone* 41: 353–359.
62. Schousboe JT, Debold CR. 2006. Reliability and accuracy of vertebral fracture assessment with densitometry compared to radiography in clinical practice. *Osteoporos Int* 17: 281–289.
63. Fuerst T, Wu C, Genant HK, von Ingersleben G, Chen Y, Johnston C, Econs MJ, Binkley N, Vokes TJ, Crans G, Mitlak BH. 2009. Evaluation of vertebral fracture assessment by dual X-ray absorptiometry in a multicenter setting. *Osteoporos Int* 20: 1199–1205.
64. Roberts M, Cootes T, Pacheco E, Adams J. 2007. Quantitative vertebral fracture detection on DXA images using shape and appearance models. *Acad Radiol* 14: 1166–1178.
65. Jager PL, Slart RH, Webber CL, Adachi JD, Papaioannou AL, Gulenchyn KY. 2010. Combined vertebral fracture assessment and bone mineral density measurement: A patient-friendly new tool with an important impact on the Canadian Risk Fracture Classification. *Can Assoc Radiol J* 61: 194–200.
66. Williams AL, Al-Busaidi A, Sparrow PJ, Adams JE, Whitehouse RW. 2009. Under-reporting of osteoporotic vertebral fractures on computed tomography. *Eur J Radiol* 69: 179–183.
67. Takada M, Wu CY, Lang TF, Genant HK. 1998. Vertebral fracture assessment using the lateral scoutview of computed tomography in comparison with radiographs. *Osteoporos Int* 8: 197–203.
68. Samelson EJ, Christiansen BA, Demissie S, Broe KE, Zhou Y, Meng CA, Yu W, Cheng X, O'Donnell CJ, Hoffmann U, Genant HK, Kiel DP, Bouxsein ML. 2011. Reliability of vertebral fracture assessment using multidetector CT lateral scout views: The Framingham Osteoporosis Study. *Osteoporos Int* 22: 1123–1131.
69. Ito M, Ikeda K, Nishiguchi M, Shindo H, Uetani M, Hosoi T, Orimo H. 2005. Multi-detector row CT imaging of vertebral microstructure for evaluation of fracture risk. *J Bone Miner Res* 20: 1828–1836.
70. Graeff C, Timm W, Nickelsen TN, Farrerons J, Marín F, Barker C, Glüer CC; EUROFORS High Resolution Computed Tomography Substudy Group. 2007. Monitoring teriparatide-associated changes in vertebral microstructure by high-resolution CT in vivo: Results from the EUROFORS study. *J Bone Miner Res* 22: 1426–1433.
71. Keaveny TM, Donley DW, Hoffmann PF, Mitlak BH, Glass EV, San Martin JA. 2007. Effects of teriparatide and alendronate on vertebral strength as assessed by finite element modeling of QCT scans in women with osteoporosis. *J Bone Miner Res* 22: 149–157
72. Eswaran SK, Gupta A, Adams MF, Keaveny TM. 2006. Cortical and trabecular load sharing in the human vertebral body. *J Bone Miner Res* 21: 307–314.
73. Krug R, Burghardt AJ, Majumdar S, Link TM. 2010. High-resolution imaging techniques for the assessment of osteoporosis. *Radiol Clin North Am* 48: 601–621.
74. Griffith JF, Yeung DK, Antonio GE, Lee FK, Hong AW, Wong SY, Lau EM, Leung PC. 2005. Vertebral bone mineral density, marrow perfusion, and fat content in healthy men and men with osteoporosis: Dynamic contrast-enhanced MR imaging and MRspectroscopy. *Radiology* 236: 945–951.
75. Griffith JF, Yeung DK, Antonio GE, Wong SY, Kwok TC, Woo J, Leung PC. 2006. Vertebral marrow fat content and diffusion and perfusion indexes in women with varying bone density: MR evaluation. *Radiology* 241: 831–838.
76. Griffith JF, Wang YX, Zhou H, Kwong WH, Wong WT, Sun YL, Huang Y, Yeung DK, Qin L, Ahuja AT. 2010. Reduced bone perfusion in osteoporosis: Likely causes in an ovariectomy rat model. *Radiology* 254: 739–746.
77. Griffith JF, Kumta SM, Huang Y. 2011. Hard arteries, weak bones. *Skeletal Radiol* 40: 517–521.
78. Biffar A, Schmidt GP, Sourbron S, D'Anastasi M, Dietrich O, Notohamiprodjo M, Reiser MF, Baur-Melnyk A. 2011. Quantitative analysis of vertebral bone marrow perfusion using dynamic contrast-enhanced MRI: Initial results in osteoporotic patients with acute vertebral fracture. *J Magn Reson Imaging* 33: 676–683.
79. Kanchiku T, Taguchi T, Toyoda K, Fujii K, Kawai S. 2003. Dynamic contrast-enhanced magnetic resonance imaging of osteoporotic vertebral fracture. *Spine* 28: 2522–2526
80. Tsujio T, Nakamura H, Terai H, Hoshino M, Namikawa T, Matsumura A, Kato M, Suzuki A, Takayama K, Fukushima W, Kondo K, Hirota Y, Takaoka K. 2011. Characteristic radiographic or magnetic resonance images of fresh osteoporotic vertebral fractures predicting potential risk for nonunion: A prospective multicenter study. *Spine* 36: 1229–1235.
81. Fornasier VL, Czitrom AA. 1978. Collapsed vertebrae: A review of 659 autopsies. *Clin Orthop Relat Res* (131): 261–265.
82. Griffith JF, Guglielmi G. 2010. Vertebral fracture. *Radiol Clin North Am* 48: 519–529.
83. Baur A, Stäbler A, Arbogast S, Duerr HR, Bartl R, Reiser M. 2002. Acute osteoporotic and neoplastic vertebral compression fractures: Fluid sign at MR imaging. *Radiology* 225: 730–735.
84. Malghem J, Maldague B, Labaisse MA, Dooms G, Duprez T, Devogelaer JP, Vande Berg B. 1993. Intravertebral vacuum cleft: Changes in content after supine positioning. *Radiology* 187: 483–487.

85. Karchevsky M, Babb JS, Schweitzer ME. 2008. Can diffusion-weighted imaging be used to differentiate benign from pathologic fractures? A meta-analysis. *Skeletal Radiol* 37: 791–795.
86. Ragab Y, Emad Y, Gheita T, Mansour M, Abou-Zeid A, Ferrari S, Rasker JJ. 2009. Differentiation of osteoporotic and neoplastic vertebral fractures by chemical shift {in-phase and out-of phase} MR imaging. *Eur J Radiol* 72: 125–133.
87. Bredella MA, Essary B, Torriani M, Ouellette HA, Palmer WE. 2008. Use of FDG-PET in differentiating benign from malignant compression fractures. *Skeletal Radiol* 37: 405–413.

38
Approaches to Genetic Testing
Christina Jacobsen, Yiping Shen, and Ingrid A. Holm

Introduction 336
Overview of the Types of Genetic Testing Available 336
Genetic Tests Available for Skeletal Disorders 338
When to Order Genetic Testing 340
References 341

INTRODUCTION

Genetic testing is the analysis of human DNA, RNA, chromosomes, proteins, and certain metabolites in order to detect heritable disease-related genotypes, mutations, phenotypes, or karyotypes for clinical purposes [1, 2]. In practice, genetic testing mainly involves looking at an individual's DNA (genes and genome) for variants that may be the underlying cause of a clinical condition, and this chapter will focus on genetic testing of DNA. As knowledge about genetic disorders of bone has expanded greatly over the past few years, the number of disorders for which genetic testing is commercially available has increased. In this chapter we will discuss the types of genetic testing available, what testing is available for disorders of bone, and suggest an approach to genetic testing in individuals with a disorder of bone.

OVERVIEW OF THE TYPES OF GENETIC TESTING AVAILABLE

There are many ways in which an individual's genome can vary from "normal." Based on the impact on genome structure, variants can be classified as either small or large scale, and each category consists of different types of variants. Laboratory molecular techniques have been developed to specifically test the different types of variants. Here we briefly discuss the types of genetic testing available in molecular diagnostic laboratories to detect different types of variants associated with a variety of genetic disorders.

Types of variants

Small-scale variants

Base-pair substitutions
The replacement of one nucleotide base by another is the most abundant variant type in the human genome. The vast majority of single nucleotide variants are located in intergenic regions (stretches of DNA that contain few or no genes) and intronic regions (stretches of DNA within a gene that is removed by RNA splicing). Those located in the coding regions (exons) can be further classified based on the effect on the amino acid sequence of the protein.

Synonymous variants are also called "silent variants" and cause no change in the final protein product as they do not change the amino acid due to redundancy of the genetic code. In most cases, they are thought not to have clinical consequences, although theoretically a synonymous variant could be clinically significant if it affects coding usage. The consequences of synonymous variants can be hard to predict.

Non-synonymous variants are those that result in changes in the amino acid sequence. When the nucleotide change results in the replacement of one amino acid for another it is called a *missense variant*; when the

---

Primer on the Metabolic Bone Diseases and Disorders of Mineral Metabolism, Eighth Edition. Edited by Clifford J. Rosen.
© 2013 American Society for Bone and Mineral Research. Published 2013 by John Wiley & Sons, Inc.

nucleotide change results in the gain of a stop codon it is called a *nonsense mutation*. Sometimes a nucleotide change results in the loss of a stop codon (i.e., a stop codon changes to a codon that inserts an amino acid). A non-synonymous variant that causes a significant impact on protein structure may have clinical consequences.

Splicing variants are variants at the splicing junctions (the first two or last two nucleotides at the beginning or end of the exon, respectively), which results in an alteration in exon splicing. This can lead to exon skipping and the loss of an exon in the mRNA. If the number of nucleotides in the skipped (deleted) exon is not a multiple of 3, it results in a frameshift and a downstream change in the amino acid sequence usually leading to a new stop codon downstream, resulting in a truncated protein. This type of variant often has significant clinical consequences.

Indels

Indels are the insertion or deletion of one or several nucleotides. When the number of nucleotides is a multiple of 3, it usually results in an in-frame deletion or insertion of one amino acid. When the number of nucleotides is not a multiple of 3, it results in out-of-frame deletion or duplication, a frameshift, and an in the amino acid sequence and a new stop codon. Out-of-frame indels usually have larger impact on protein structure and function than in-frame indels and often have clinical consequences.

Repeat expansions

There are areas within the genome that contain repeating sequences, and an increase in the number of simple repeating sequences can occur in both coding and noncoding regions. This type of variant is known to be associated with a limited number of disease-causing genes, but constitutes an important category of mutation and disease mechanism.

Epigenetic variants

Some disorders are caused by changes in the epigenetic modification pattern such as methylation. Changes in the methylation pattern can affect gene expression and cause disease.

Methods to detect small scale changes in the DNA

Sanger sequencing using PCR (polymerase chain reaction) amplified product has been the most effective method for detecting most small scale variants. Genotyping methods are also useful in detecting targeted mutations but are less often used. For repeat expansions, PCR-based assays, and occasionally Southern-blot based assays, are necessary to assess the number of the repeats. Methylation specific PCR or MLPA (multiplex ligation dependent probe amplification) methods are often used to detect the methylation status of disease-related genes for epigenetic mutations. MPLA is a form of multiplex PCR commonly used not only to detect methylation status but also to detect mutations and gene deletions/duplications. Multiple genomic targets are amplified using a common primer pair located at the outside ends of the multiple oligonucleotide probe pairs specific to the genomic targets. Since PCR amplification of the genomic targets only occurs when the two probes hybridize to their target, probes that are unbound will not be amplified. Thus, the strength of amplification signal reflects the amount of genomic target (inferred as copy number) available for hybridization and ligation. MLPA is used to detect methylation status by using probes at the methylation loci and treating the genomic DNA with methylation sensitive restriction enzymes that disrupt the methylated DNA preventing the ligation, hence reducing the amplification signal. The strength of the amplification signal at the methylated sites compared to the signal generated from DNA that has not been treated with enzyme reflects the methylation level. MLPA is also used to complement Sanger sequencing, in which case Sanger sequencing is performed first, and if the result is normal, MLPA is then performed to detect possible exonic deletions; sometimes, the two tests are performed simultaneously. For disorders in which deletions or duplications are common, MLPA is performed first as it is much easier and fast than sequencing.

Large-scale variants

This type of variant affects at least one exon of a gene or a larger genomic segment. Large-scale variants cannot be reliably detected by conventional PCR and Sanger sequencing based assays. Large scale variants can be categorized into the following types.

Copy number variants (CNVs)

Copy number variants (CNVs) are imbalanced structure variants. Genomic DNA copy number gain or loss has been known for many years, but a renewed interest in CNVs is due to the finding that not only are CNVs much more abundant in the human genome than previously appreciated, they are now known to have a large impact on genome structure and function. Many CNVs, especially large ones, have significant clinical relevance. CNVs constitute the second most frequent mutation type associated with genetic disorders. Recently, microarray-based genomic profiling technologies have enabled the effective analysis of genomic imbalances in a genome-wide manner with much improved sensitivity, resolution, and reproducibility. The microarray-based genomic profiling techniques have been recommended as the first tier genetic test in many clinical scenarios.

There are two basic microarray-based genomic profiling types. One originated from comparative genomic hybridization (CGH) technology and is currently represented by the oligo-based CGH array. Recent addition of single nucleotide polymorphism (SNP) probes onto the CGH array has enabled detection of both CNV and LOH (loss of heterozygosity) by CGH array. Another array platform is derived from SNP genotyping arrays. The addition of CNV probes onto genotyping arrays significantly enhances the sensitivity of detecting CNVs by genotyping array. The two array types show a functional convergence. Next generation sequencing (NGS) may eventually replace array-based technologies for interrogating CNV, as well as for many other types of large-scale variants discussed below. Detecting CNV by NGS is an area of great bioinformatic research interest at this moment.

Balanced structure variants

Translocations and inversions represent another type of large-scale variant. Until very recently, balanced genomic variants could only be detected by conventional cytogenetic approaches. Microarray techniques are not able to detect balanced variants. Due to the limited resolution of cytogenetic approaches, the balanced genomic variants are believed to be underdetected. On the other hand, many apparently balanced rearrangements ascertained by cytogenetic approaches are later found to harbor cryptic imbalances when high-resolution microarray-based genomic profiling techniques are used. NGS-based approaches are poised to effectively solve the technical difficulty of detecting balanced rearrangements; it will become the method of choice for genome-wide detections of both balanced and imbalanced genomic variants when the informatic challenges are solved.

Loss of heterozygosity (LOH)

Loss of heterozygosity (LOH) results from deletion of one allele or from uniparental origin of both alleles, which results in copy neutral LOH (cn-LOH) and is often referred to as uniparental disomy (UPD). Within a family cn-LOH can be due to identity-by-descent (IBD), which indicates distant relatedness between two individuals, or to consanguinity, indicating recent relatedness. Traditionally, microsatellite marker-based genotyping methods have been used to detect UPD. Recent microarray platforms are able to detect large segments of LOH and UPD in a genome-wide manner. NGS data can also easily provide detailed genotyping information for the whole genome at base pair resolution, thus eventually providing a much more complete picture of cn-LOH in the whole genome.

Evolving approaches of DNA-based genetic testing

The nature of the variant dictates what method should be used for genetic testing. Technical innovations have been changing the way we perform genetic testing on patients. As described below, genetic testing is undergoing a significant paradigm shift with the advent of microarray-based and NGS-based technologies.

Conventional approaches deal with one mutation or one gene and one patient a time. Clinicians make a preliminary diagnosis and order the gene test that is most likely to explain a patient's clinical condition. Depending on the known mutation spectrum of the disease gene, a molecular diagnostic laboratory will either use Sanger sequencing to test for small-scale variants or targeted methods for CNV such as quantitative PCR or MLPA.

Whole-gene- or gene-panel-based tests are currently the most commonly used approaches in genetic testing. Since sequencing can only detect small-scale variants, it is often complemented by CNV testing using MLPA or quantitative PCR to detect potential exonic deletion or duplication.

Whole-genome chromosomal microarray (CMA) analysis for CNV and LOH has much enhanced clinical utility over conventional cytogenetic techniques, which justifies its use as the first tier test for patients with complex or unknown genetic conditions. CMA is a useful scanning tool for genome-wide detection of possible genomic defects associated with many genetic conditions.

NSG-based whole-exome or whole-genome testing generates sequence data by massive parallel sequencing of clonally amplified or single DNA molecules. Recent advances in NGS technologies are making the NGS-based diagnostic approach more cost-effective with a reasonable turnaround time. NGS is poised to bring a paradigm shift in genetic testing. The premise for using whole-exome, instead of whole-genome, sequencing for Mendalian disorders, both in research as well as in genetic testing, is that protein coding sequences constitute about 1% of the human genome but harbor about 85% of the known mutations that cause human diseases. However, given the fact that large-scale variants can be simultaneously detected by a whole-genome-based test, but not easily by whole-exome sequencing, it is predicted that when the cost for whole-genome sequencing becomes acceptable for routine testing, it will become the method of choice for both research and clinical diagnostics. Many challenges lie ahead, particularly in the area of data interpretation, which often involves database searching, segregation analysis, bioinformatic prediction, and functional demonstration. Even though NGS will eventually replace many current genetic testing methods, conventional sequencing, genotyping, and CNV detecting technologies are still useful for validating and confirming variants detected by NGS.

GENETIC TESTS AVAILABLE FOR SKELETAL DISORDERS

As knowledge about the genetic disorders of bone has expanded greatly over the past few years, the number of disorders for which genetic testing is commercially available has also increased. Unfortunately, there has been a lag between the discovery of causative genes on a research basis and the availability of commercial diagnostic testing [3]. However, genetic testing is available for a large number of skeletal diseases commonly seen in the clinic, both metabolic bone disorders and skeletal dysplasias.

Metabolic bone disease

Genetic testing for metabolic disorders of bone can be useful in the clinical setting for diagnostic purposes. There are a number of genetic causes of metabolic bone disease for which there is genetic testing available.

Familial hypophosphatemic rickets

Familial hypophosphatemic rickets (FHR) is the most common form of heritable rickets. Clinically, FHR presents in childhood with the typical signs of rickets, including bowing of the legs and growth delay as the child starts walking, and in the absence of a family history, FHR can be misidentified as nutritional rickets. The biochemical findings that distinguish FHR from other forms of rickets include a normal 25-hydroxyvitamin D, hypophosphatemia, and a low-normal 1,25-dihydroxyvitamin D level; in addition, calcium levels are normal and the PTH is normal or slightly elevated. Classic findings of rickets on radiographs exclude other possible diagnoses including physiologic bowing and a skeletal dysplasia. X-linked hypophosphatemic rickets (XLH) is by far the most common form of FHR and is due to mutations in the phosphate-regulating gene with homology to endopeptidases located on the X-chromosome (PHEX). XHR is inherited in an X-linked dominant manner and thus there are more affected females than males (2:1 ratio) and no male-to-male transmission. Other forms of hypophosphatemic rickets including autosomal dominant hypophosphatemic rickets (ADHR) and autosomal recessive hypophosphatemic rickets (ARHR). While XLH is by far the most common, genetic testing for all forms is available and can be used to differentiate between the forms when it is not clear from an inheritance pattern [4].

Vitamin D-related disorders

There are two genetic defects in the vitamin D pathway that lead to rickets. Vitamin D-dependent rickets type I (VDDR-I), also known as 1-alpha-hydroxylase deficiency. This disorder is characterized by low 1,25-dihydroxyvitamin D levels in the face of normal 25-hydroxyvitamin D levels, an elevated parathyroid hormone (PTH), and hypocalcemia. On the other hand, in vitamin D-dependent rickets type II (VDDR-II), also called vitamin D-resistant rickets, 1,25-dihydroxyvitamin D levels are markedly elevated due to the receptor resistance. VDDR-I is due to mutations in the 1-alpha-hydroxylase gene CYP27B, and VDDR-II is due to mutations in the vitamin D receptor, VDR; both genetic tests are available clinically.

Hypophosphatasia

Hypophosphatasia is an inherited bone disorder characterized by rickets in childhood and osteomalacia in adulthood due to a defect in mineralization of bone and teeth, and low serum and bone alkaline phosphatase activity. There are six clinical forms that span a wide spectrum of disease, from severe demineralization that leads to stillbirth, to rickets in childhood, to pathologic fractures in later adulthood, to odontohypophosphatasia (severe dental caries and/or premature exfoliation of primary teeth). Hypophosphatasia is due to mutations in ALPL, the gene encoding the alkaline phosphatase, tissue-nonspecific isozyme (TNSALP).

Skeletal dysplasias

Genetic testing for skeletal dysplasias can be used to confirm a clinical diagnosis, as well as for cases where a clinical diagnosis is not immediately clear but several different disorders are under consideration. All patients with a suspected skeletal dysplasia should have a through physical exam documenting any dysmorphic features, as well as a full radiographic skeletal survey in cases where the diagnosis is not suggested by history and physical exam alone. There are hundreds of skeletal dysplasias and many genes that are affected. Here we highlight some of the more common skeletal dysplasias for which there is genetic testing available.

Osteogenesis imperfecta

Osteogenesis imperfecta (OI) is characterized by low bone mass and increased fracture risk. Most patients with OI have mutations in the type I collagen genes, COL1A1 and COL1A2, or in genes encoding proteins that participate in the assembly, modification, and secretion of type I collagen. Dominant mutations in the COL1A1 and COL1A2 genes are responsible for OI types I through IV. Phenotypes range from the very severe type II form (perinatal lethal) to a "mild" type I form; type III OI is the most severe form that is not lethal, and type IV is moderate OI. The recessive forms of OI are much rarer in the general population (although they are more common in certain ethnic subgroups) and typically cause a severe phenotype.

Genetic testing in OI generally serves two purposes. In children with multiple fractures but who are otherwise normal without significant short stature or deformities, genetic testing can be used to make the diagnosis of type I OI. In children with the more severe forms of OI that are diagnosed on clinical exam, genetic testing can confirm the diagnosis of OI. In addition, genetic testing can be useful for prenatal counseling for patients and family members and help distinguish the more common COL1A1 and COL1A2 mutations in the dominant OI types I through IV from the recessive forms of OI. Genetic testing for recessive forms should be considered in patients with a severe OI phenotype who test negative for a COL1A1 and COL1A2 mutation. In addition, confirming the diagnosis of OI is helpful as there is available pharmacologic treatment of OI with bisphosphonates (typically pamidronate) in the pediatric population [5], although other treatments are being studied in the adult population. Bisphosphonates are used in many children with OI and are thought to act by preventing the high bone turnover seen in patients with OI due to their abnormal collagen [6]. Although bisphosphonate therapy does not correct the underlying genetic defect in patients with OI, the therapy has been reported to increase bone density and reduce fracture rates in treated patients [7].

Achondroplasia and other FGFR3-related disorders

Achondroplasia is a relatively common skeletal dysplasia. Affected patients have severe short stature, rhizomelic shortening of the limbs, and a relatively large head with frontal bossing. Although the diagnosis of achondroplasia is typically made based on the physical examination and radiographs, genetic testing for the disorder is useful to confirm the diagnosis, especially if the diagnosis is unclear. Nearly all cases of achondroplasia are caused by one of two mutations in the same nucleotide of the fibroblast growth factor receptor 3 (FGFR3) gene. Achondroplasia is inherited in an autosomal dominant manner, and approximately 80% of cases are new mutations [8, 9].

Mutations in FGFR3 are responsible for several other conditions, including hypochondroplasia, thanatophoric dysplasia types I and II, and severe achondroplasia with developmental delay and acanthosis nigricans (SADDAN). Of these, hypochondroplasia is the most common and presents as a milder form of achondroplasia, without the large head, and with milder short stature.

Multiple epiphyseal dysplasia

Multiple epiphyseal dysplasia (MED) is a skeletal dysplasia characterized by abnormal epiphyses of the long bones. Patients typically develop symptoms of joint pain in the hips and knees in childhood, leading to the development of early onset osteoarthritis. Mildly shortened stature is also common. There are two forms of MED: a dominantly inherited form as well as a recessive form. Although the symptoms of the two forms are similar, patients with the recessively inherited form may also have congenital defects such as cleft palate or clubfeet. While a clinical diagnosis of MED is typically made based on symptoms and abnormal epiphyses seen on radiographs, genetic testing can be very useful in confirming the diagnosis and differentiating between the dominant and recessive forms when the family history is not a suggestive. Dominant MED is a disorder with genetic heterogeneity, i.e., mutations in several different genes can cause the disorder. Mutations in five different genes—COMP, COL9A1, COL9A2, COL9A3, and MATN3—have all been shown to cause dominant MED. Given this scenario, a staged approach to genetic testing is recommended, where testing begins with the most commonly affected gene and proceeds sequentially through the next most affected until the gene defect is found [10, 11].

WHEN TO ORDER GENETIC TESTING

A metabolic bone disease may be suspected if there is bowing of the lower extremities, multiple fractures, a laboratory finding suggestive of metabolic bone disease (e.g., an elevated alkaline phosphatase, hypocalcemia, a low vitamin D level, or hypophosphatemia), or radiographs that suggest changes consistent with rickets. The initial evaluation of a suspected metabolic bone disorders should include X-rays and laboratory tests including: serum creatinine, calcium, magnesium, phosphorus, alkaline phosphatase, 1,25-dihydroxyvitamin D, 25-hydroxyvitamin D, PTH, and urine calcium, phosphorus, and creatinine. In addition, studies looking for systemic causes of metabolic bone disease, such as renal dysfunction, inflammatory bowel disease, and Celiac disease, should also be considered. In the case of a suspected skeletal dysplasia, the evaluation starts with a comprehensive physical examination and a skeletal radiographic series to characterize the skeletal findings.

The evaluation of a suspected metabolic bone disease may suggest a genetic condition. Examples include hypophosphatemic rickets (low phosphate, normal 1,25-dihydroxyvitamin D and 25-hydroxyvitamin D, and normal calcium) and pseudohypoparathyroidsim (low calcium, high PTH). Radiographs may confirm the clinical suspicion of skeletal dysplasia and often narrow the differential diagnosis to a limited group of disorders. Once a preliminary diagnosis of a genetic condition is made, the decision to order genetic testing is influenced by several factors, including the diagnostic and treatment necessity, the need for reproductive counseling, patient and family desire for testing, and cost and payment issues. Prior to ordering any genetic test, the provider needs to determine if the desired test is clinically available. In the United States, all genetic testing must be performed in a CLIA-certified clinical laboratory for both legal and reimbursement reasons [12]. The website www.genetests.org is an excellent source for determining which CLIA-certified clinical laboratories offer the desired testing. Any testing sent on a research basis must be done under the supervision of an institutional review board, and often the results cannot be legally revealed to the patient until the testing is repeated and the results verified in a CLIA-certified lab [13]. Fortunately, genetic testing tends to move rather quickly in most cases from the research setting to clinical testing.

The most common clinical situation where genetic testing is ordered is when the genetic testing is required to make a diagnosis or to confirm the diagnosis. Genetic testing is especially useful if a course of treatment is available and dependent on the diagnosis. In these cases, genetic testing will provide a clear benefit for the care of the affected patient. A clear example of this is OI, in which bisphosphonates may be beneficial.

Often if the diagnosis is unclear, a staged approach to testing, while perhaps prolonging the time to final diagnosis, can potentially conserve resources. This involves requesting one genetic test at a time and moving on to other genetic testing only if the initial tests are negative. Frequently, one blood sample can almost always provide enough DNA for a laboratory to carry out multiple sequencing analyses, thus preventing repetitive blood draws. This staged testing approach is also useful for disorders with genetic heterogeneity. Testing can begin with the most commonly affected gene and then proceed to subsequent genes only if the first test is negative. However, in cases where diagnosis will affect treatment and delaying therapy is not optimal, ordering multiple tests at once may benefit the patient.

Reproductive counseling is another reason that genetic testing may be useful in a clinical setting. It is not uncommon for skeletal disorders to be diagnosed clinically based on family history, physical exam, and radiographic evidence. While the diagnosis may not be in question, knowledge of the specific mutation causing the phenotype may be useful for future reproductive decisions. In the pediatric setting, parents of an affected child may desire prenatal testing with future pregnancies. In this situation, the patient's exact mutation needs to be known to determine if the parents are carriers (or affected themselves if the mutation is dominant) and to allow for prenatal testing with future pregnancies. In the case of recessive disorders, carrier testing is appropriate for adult family members but should not be performed on minor children until they reach maturity and can consent to their own testing [14].

Patients or family members may request or refuse genetic testing for various reasons. In cases where genetic testing is needed for initial diagnosis or to confirm a clinical diagnosis, a full discussion of the risks and benefits should be undertaken with the patient (or parents if the patient is a minor). This discussion should include the reasons why genetic testing is needed, how the results (whether a mutation is found or not) will affect the patient's immediate treatment, and how the results may affect the patient's care in the future. Discussion of these issues can help alleviate many misunderstandings and anxieties around genetic testing [15]. In the U.S., the 2008 Genetic Information Nondiscrimination Act (GINA) bars discrimination by health insurance providers and employers due to genetic testing results, although GINA does not protect from genetic discrimination in other areas such as long-term care or life insurance [16]. Finally, the discussion should include the limits of genetic testing, particularly when a negative test will not exclude a clinical diagnosis, which is often the case. Available genetic tests may only detect mutations in a small number of clinically diagnosed patients depending on the disorder. Patients and families may consider a negative test result "the final answer" and need to understand that a negative genetic test does not always exclude a given clinical diagnosis.

The cost of genetic testing can affect the ability to order genetic tests in different settings. Genetic tests can be among the most expensive laboratory-based tests available, ranging from hundreds to tens of thousands of dollars. Third-party payers are becoming increasingly reluctant to pay for the costs of genetic testing, particularly in the absence of documented clinical necessity. In the pediatric population, genetic testing for diagnostic purposes is typically covered by private insurance companies as long as the patient's pertinent history and physical findings are well documented. However, families with high-deductible plans should be warned that the cost of one genetic test may be greater than the entire deductible, requiring a single large payment. Carrier testing may not be covered, except when done for prenatal testing purposes. As genetic tests are ordered more frequently, institutions such as hospitals may limit the number of tests that can be ordered or require that tests be ordered only from laboratories that bill the patient's insurance company directly. This can effectively put testing for some disorders out of the reach of patients and providers. These increasing costs place a responsibility on providers to ensure that genetic tests are only ordered under appropriate circumstances where there is a clear benefit for the patient.

The increasing availability of genetic testing for all types of skeletal disorders can be of great benefit to patients and providers. Genetic testing can be useful for diagnosis and subsequent treatment as well as for reproductive counseling for patients and families. However, as with all medical testing, genetic testing has risks and benefits and may not be appropriate for every clinical situation.

REFERENCES

1. Holtzman NA, Watson MS (eds.) 1997. *Promoting Safe and Effective Genetic Testing in the United States*. Bethseda, MD: National Human Genome Research Institute. Available from: http://www.genome.gov/10001733.
2. Holtzman NA, Watson MS. 1999. Promoting safe and effective genetic testing in the United States. Final report of the Task Force on Genetic Testing. *J Child Fam Nurs* 2: 388–90.
3. Das S, Bale SJ, Ledbetter DH. 2008. Molecular genetic testing for ultra rare diseases: Models for translation from the research laboratory to the CLIA-certified diagnostic laboratory. *Genet Med* 10: 332–6.
4. Carpenter TO, Imel EA, Holm IA, Jan de Beur SM, Insogna KL. 2011. A clinician's guide to X-linked hypophosphatemia. *J Bone Miner Res* 26: 1381–8.
5. Glorieux FH, Bishop NJ, Plotkin H, Chabot G, Lanoue G, Travers R. 1998. Cyclic administration of pamidronate in children with severe osteogenesis imperfecta. *N Engl J Med* 339: 947–52.
6. Falk MJ, Heeger S, Lynch KA, DeCaro KR, Bohach D, Gibson KS, Warman ML. 2003. Intravenous bisphosphonate therapy in children with osteogenesis imperfecta. *Pediatrics* 111: 573–8.
7. Phillipi CA, Remmington T, Steiner RD. 2008. Bisphosphonate therapy for osteogenesis imperfecta. *Cochrane*

8. Carter EM, Davis JG, Raggio CL. 2007. Advances in understanding etiology of achondroplasia and review of management. *Curr Opin Pediatr* 19: 32–7.
9. Shirley ED, Ain MC. 2009. Achondroplasia: Manifestations and treatment. *J Am Acad Orthop Surg* 17: 231–41.
10. Briggs MD, Wright MJ, Mortier GR. 1993. Multiple Epiphyseal Dysplasia, Dominant. In: Pagon RA, Bird TD, Dolan CR, Stephens K, Adam MP (eds.) SourceGeneReviews™ [Internet]. Seattle, WA: University of Washington.
11. Li LY, Zhao Q, Ji SJ, Zhang LJ, Li QW. 2011. Clinical features and treatment of the hip in multiple epiphyseal dysplasia in childhood. *Orthopedics* 34: 352.
12. Rivers PA, Dobalian A, Germinario FA. 2005. A review and analysis of the clinical laboratory improvement amendment of 1988: Compliance plans and enforcement policy. *Health Care Manage Rev* 30: 93–102.
13. Hens K, Nys H, Cassiman JJ, Dierickx K. 2011. The return of individual research findings in paediatric genetic research. *J Med Ethics* 37: 179–83.
14. Borry P, Fryns JP, Schotsmans P, Dierickx K. 2006. Carrier testing in minors: A systematic review of guidelines and position papers. *Eur J Hum Genet* 14: 133–8.
15. Henneman L, Timmermans DR, Van Der Wal G. 2006. Public attitudes toward genetic testing: Perceived benefits and objections. *Genet Test* 10: 139–45.
16. Payne PW Jr, Goldstein MM, Jarawan H, Rosenbaum S. Health insurance and the Genetic Information Nondiscrimination Act of 2008: Implications for public health policy and practice. *Public Health Rep* 124: 328–31.

Section V
Osteoporosis

Section Editors Paul D. Miller and Socrates E. Papapoulos

Chapter 39. Osteoporosis Overview 345
Michael Kleerekoper

Chapter 40. The Epidemiology of
Osteoporotic Fractures 348
Nicholas Harvey, Elaine Dennison, and Cyrus Cooper

Chapter 41. Overview of Pathogenesis 357
Ian R. Reid

Chapter 42. Nutrition and Osteoporosis 361
Connie M. Weaver and Robert P. Heaney

Chapter 43. The Role of Sex Steroids in the Pathogenesis of Osteoporosis 367
Matthew T. Drake and Sundeep Khosla

Chapter 44. Translational Genetics of Osteoporosis: From Population Association to Individualized Prognosis 376
Bich H. Tran, Jacqueline R. Center, and Tuan V. Nguyen

Chapter 45. Prevention of Falls 389
Heike A. Bischoff-Ferrari

Chapter 46. Exercise and the Prevention of Osteoporosis 396
Clinton T. Rubin, Janet Rubin, and Stefan Judex

Chapter 47. Calcium and Vitamin D 403
Bess Dawson-Hughes

Chapter 48. Estrogens, Estrogen Agonists/Antagonists, and Calcitonin 408
Nelson B. Watts

Chapter 49. Bisphosphonates for Postmenopausal Osteoporosis 412
Socrates E. Papapoulos

Chapter 50. Denosumab 420
Michael R. McClung

Chapter 51. Parathyroid Hormone Treatment for Osteoporosis 428
Felicia Cosman and Susan L. Greenspan

Chapter 52. Strontium Ranelate in the Prevention of Osteoporotic Fractures 437
René Rizzoli

Chapter 53. Combination Anabolic and Antiresorptive Therapy for Osteoporosis 444
John P. Bilezikian and Natalie E. Cusano

Chapter 54. Compliance and Persistence with Osteoporosis Medications 448
Deborah T. Gold

Chapter 55. Cost-Effectiveness of Osteoporosis Treatment 455
Anna N.A. Tosteson

Chapter 56. Future Therapies of Osteoporosis 461
Kong Wah Ng and T. John Martin

Chapter 57. Juvenile Osteoporosis 468
Nick Bishop and Francis H. Glorieux

(Continued)

Primer on the Metabolic Bone Diseases and Disorders of Mineral Metabolism, Eighth Edition. Edited by Clifford J. Rosen.
© 2013 American Society for Bone and Mineral Research. Published 2013 by John Wiley & Sons, Inc.

Chapter 58. Glucocorticoid-Induced Bone Disease   473
Robert S. Weinstein

Chapter 59. Inflammation-Induced Bone Loss
in the Rheumatic Diseases   482
Steven R. Goldring

Chapter 60. Secondary Osteoporosis: Other Causes   489
Neveen A.T. Hamdy

Chapter 61. Transplantation Osteoporosis   495
Peter R. Ebeling

Chapter 62. Osteoporosis in Men   508
Eric S. Orwoll

Chapter 63. Premenopausal Osteoporosis   514
Adi Cohen and Elizabeth Shane

Chapter 64. Skeletal Effects of Drugs   520
Juliet Compston

Chapter 65. Orthopedic Surgical Principles
of Fracture Management   527
Manoj Ramachandran and David G. Little

Chapter 66. Abnormalities in Bone and Calcium Metabolism
After Burns   531
Gordon Klein

39
Osteoporosis Overview
Michael Kleerekoper

In March 2000, a National Institutes Consensus Development Conference redefined osteoporosis as "a skeletal disorder characterized by compromised bone strength predisposing to an increased risk of fracture. Bone strength reflects the integration of two main features: bone density and bone quality. Bone density is expressed as grams of mineral per area or volume and in any given individual is determined by peak bone mass and amount of bone loss. Bone quality refers to architecture, turnover, damage accumulation (e.g., microfractures), and mineralization. A fracture occurs when a failure-inducing force (e.g., trauma) is applied to osteoporotic bone. Thus, osteoporosis is a significant risk factor for fracture, and a distinction between risk factors that affect bone metabolism and risk factors for fracture must be made."

In the intervening decade much has been learned about this very common disease, and the next 27 chapters provide details of all aspects of osteoporosis—etiology, diagnosis, management. A synopsis of those chapters is provided here.

Understanding of the epidemiology and pathogenesis of osteoporosis as detailed in Chapters 40 and 41 have advanced beyond the age-related decline in bone mass assessed by dual energy X-ray densitometry (DXA) and an assessment of fracture risk determined by the WHO-developed FRAX model to include invasive (bone biopsy) and noninvasive (high-resolution computed tomography and magnetic resonance imaging) documentation of bone microarchitecture. It is now more clear that risk of fracture is heavily dependent on bone quality, possibly even more so than bone mass. Not to be forgotten in this emerging science is the contribution of falls to the likelihood of fracture occurrence. An osteoporosis-related fragility fracture is best defined as a fracture resulting from a fall from a standing height or less. The less bone mass, the more disrupted the bone microarchitecture, the greater the likelihood of sustaining a fracture (Chapter 40).

Not all fall-related fractures reflect osteoporosis, particularly in children. Figure 39.1 is a schematic of gain and loss of bone throughout life. Bone mass is low at birth and increases for the next 2 or 3 decades (it varies with the skeletal site being evaluated) before peak bone mass is acquired. Both bone formation and bone resorption are highest at birth, declining rapidly over the first few years of life. This is an age group where falls are not uncommon, either as the toddler is learning to stand and walk or as they grow and become involved in sporting activities. Advancing from a tricycle to a bicycle is a common cause of falls and fracture, but the high bone remodeling rate promotes rapid and complete healing in most children. The pubertal growth spurt results from a second surge of remodeling activity, but fractures are less common until engagement in more rigorous contact sports. In healthy people, peak bone mass and bone remodeling are stable unless a secondary cause of bone loss is present. In women, the decline in estrogen and/or the increase in gonadotropins at menopause (Chapters 43) result in a reversal of bone remodeling such that resorption now exceeds formation and bone mass decreases. Coincident with this is remodeling induced disruption of bone microarchitecture and most likely other as yet incompletely identified disturbances in skeletal integrity.

While osteoporosis is more prevalent after menopause, there are many factors that contribute to bone loss and osteoporosis with increased fracture risk in premenopausal women. In particular, anorexia, bulimia, and athletic amenorrhea interfere with the cyclical production of estrogen and progesterone (Chapter 43). Paradoxically, the alterations in menstrual function in patients with polycystic ovarian syndrome (PCOS) do not have an adverse effect on bone mass. Bone mass may even be higher in patients with PCOS than in age-matched controls with normal menses—presumably related to the increased levels of androgens.

Primer on the Metabolic Bone Diseases and Disorders of Mineral Metabolism, Eighth Edition. Edited by Clifford J. Rosen.
© 2013 American Society for Bone and Mineral Research. Published 2013 by John Wiley & Sons, Inc.

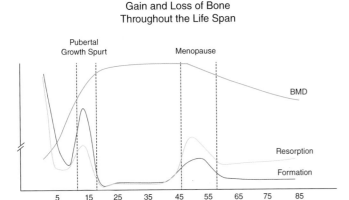

Fig. 39.1. Schematic representation of gain and loss of bone and changes in bone remodeling throughout life in women.

Age-related bone loss occurs in men but the mechanisms are not as well documented as they are in women. Age-related decline in testosterone and likely the change in diurnal excursion of testosterone as men age are important determinants of bone loss in men. There is also increasing data that links changes in fat mass (increase) and muscle mass (decrease) as major contributors to bone loss in women and men. The contributions of the sex hormones, sex-hormone binding proteins, gonadotrophins, and growth factors to gain and loss of bone in both men are elegantly described and detailed in Chapter 43.

Nutrition and lifestyle are critical factors in the development and maintenance of bone health (Chapters 42 and 47). Key nutritional factors are adequate intake of calcium and vitamin D. Guidelines for the optimal amount of calcium and vitamin D intake have recently been published by several major organizations but there is limited consistency in these guidelines. At issue is just "how much" or "how little" is regarded as adequate intake of calcium and vitamin D at different stages of life. It is possible to have inadequate intake of calcium, but dietary calcium overload is much less likely. Patients with calcium-containing kidney stones are often ill-advised to lower their calcium intake or do so of their own volition. Lowering dietary calcium is more likely to increase kidney stone risk as well as jeopardize skeletal health. Most recently, controversy has arisen about a possible link between excess calcium intake and the risk of coronary artery disease, but this issue remains unresolved. The issues regarding vitamin D are more complex. In particular there is a trend, at least in the United States, to measure 25- hydroxyvitamin D (25OHD) at annual physical examinations. Compared to the cost of the laboratory procedure, the cost of taking 1,000–2,000 units of vitamin D daily is substantially lower, with very limited likelihood of developing vitamin D toxicity. Patients with a history of malabsorption or overt malnutrition will benefit from regular monitoring of 25OHD, but this is a small fraction of the general population. As calcium intake has been linked to cardiovascular disease, there is a burgeoning literature linking 25OHD levels to diseases in seemingly every organ system and disease state. Hypothetically, this makes sense since the vitamin D receptor is present in many, perhaps most, tissues. However, to date the published data are inconsistent and inconclusive. In the 90 days before this chapter was written, there were 129 peer-reviewed publications listed in PubMed on the topic "diseases associated with vitamin D." The chapters "Premenopausal Osteoporosis" (Chapter 63) and "Prevention of Falls" (Chapter 45) provide additional critical information about maintenance of skeletal health and integrity throughout life.

The bulk of the synopsis presented above relates to "primary" osteoporosis in which bone loss can be attributed to aging *per se* or the known hormonal consequences of aging such as the decline in estrogen and testosterone. "Secondary" osteoporosis refers to those conditions that do not start as a skeletal issue but reflect adverse consequences of the primary disease itself (e.g., celiac disease with associated malabsorption) and/or as a result of therapy (e.g., glucocorticoid) for many diseases. Diseases and therapies that adversely affect bone remodeling and bone mass or are associated with an increased risk of falling are detailed in Chapter 45.

Much has already been learned about the cellular and molecular mechanisms underlying both primary and secondary osteoporosis and exciting new science is unraveling potential new factors in this scenario. This has resulted in the development of pharmacologic approaches to minimize further bone loss and to a great extent reverse the bone loss and decrease the likelihood of fracture. Chapters 48–53 describe the mechanism of action and the clinical use of therapies that have been documented to improve bone mineral density and reduce fracture risk and provide guidance into the appropriate use of these therapies for the several clinical scenarios described above. Each of these has demonstrated antifracture effectiveness for the clinical scenario for which a specific drug has received regulatory approval. Chapter 53 explores the potential use of combination or sequential antiresorptive and formation stimulation therapies. Since gain and loss of bone throughout the life span depends on the interaction between formation and resorption, it makes sense to determine the optimal combined use of drugs with different effects on bone remodeling. Bringing these forward into the clinical arena will take some time but must be pursued.

Chapter 56 describes future therapies for osteoporosis prevention and treatment that are founded on the rapidly increasing knowledge and pre-clinical studies of more intricate metabolic pathways affecting skeletal health. Denosumab (Chapter 50) is already approved as an effective therapy to reduce fracture risk with seemingly fewer adverse events than have been reported for other monoclonal antibodies used to treated diseases or conditions other than osteoporosis. Understanding of the mechanisms by which the Wnt signaling and LRP5, and molecules such as cathepsin K expressed in osteoclasts are linked will undoubtedly lead to development and marketing of other therapies.

While this is a very exciting step forward in understanding and maintaining skeletal health, it remains to be seen whether this will lead to medications more effective in reducing fracture risk than those drugs currently available—existing approved therapies all reduce fracture risk. Marketed therapies are not without potential adverse effects in a small number of patients, and the hope is that newer potential therapies under investigation will be associated with even fewer side effects. Some of these adverse effects (e.g., atypical femoral shaft fractures, osteonecrosis of the jaw) have a marked impact on the patient's well-being, and it is imperative that prescribing clinicians familiarize themselves with the potential side effects—as should be the case for any therapy for any disease. Guidelines for monitoring the effectiveness of therapy and minimizing the risk of an adverse event have been published, but there remain many unanswered questions. Pivotal clinical trials in osteoporosis leading to approval for marketing and dispensing have not resulted in any pre-marketing warning label with the exception of the formation stimulation agent teriparatide. The labeling was not based on any adverse events in patients, only on the occurrence of osteosarcoma in weanling rats. This has resulted in the duration of clinical use for teriparatide being limited to 2 years. The occurrence of osteosarcoma in patients with osteoporosis treated with teriparatide is rare! However, it is becoming apparent that overuse of some of the approved osteoporosis therapies does result in significant harm to some patients, but these occurrences cannot yet be predicted in individual patients. Carefully developed guidelines have provided directions by which clinicians can monitor the patient for adverse events and some less certain guidance about possible temporary interruption of therapy.

In sum, since the first edition of the *Primer* was published in 1990, our understanding of the pathogenesis of osteoporosis, both primary and secondary, has increased exponentially, as has the technology for assessing fracture risk and the availability of current therapies. The chapters in this section highlight our existing knowledge base and give insight into future therapeutic options.

40

The Epidemiology of Osteoporotic Fractures

Nicholas Harvey, Elaine Dennison, and Cyrus Cooper

Introduction 348
Fracture Epidemiology 348
Mortality and Morbidity 353
Conclusion 354
Acknowledgments 354
References 354

INTRODUCTION

Osteoporosis is a skeletal disease characterized by low bone mass and microarchitectural deterioration of bone tissue with a consequent increase in bone fragility and susceptibility to fracture [1]. The term "osteoporosis" was first introduced in France and Germany during the last century. It means "porous bone" and initially implied a histological diagnosis, but it was later refined to mean bone that was normally mineralized, but reduced in quantity. Clinically, osteoporosis has been difficult to define: A focus on bone mineral density (BMD) may not encompass all the risk factors for fracture, whereas a fracture-based definition will not enable identification of at risk populations. In 1994, the World Health Organization (WHO) [2] convened to resolve this issue, defining osteoporosis in terms of bone mineral density, and previous fracture. Thus the WHO definition does not take into account microarchitectural changes that may weaken bone independently of any effect on BMD. More recently, there has been a move toward assessment of individualized risk [3], with the development of the FRAX algorithm [4], a Web-based tool that uses clinical risk factors plus or minus BMD to calculate an individual's absolute risk of major osteoporotic or hip fracture over the next 10 years. This has the advantage of incorporating risk factors that are partly independent of BMD, such as age and previous fracture, and thus allows decisions regarding commencement of therapy to be made more readily. Osteoporosis-related fractures have a huge impact economically, in addition to their effect on health: The cost to the U.S. economy is around $17.9 billion per annum, with the burden in the U.K. being £1.7 billion (Table 40.1 summarizes fracture impact for a Western population) [5]. Hip fractures contribute most to these figures.

FRACTURE EPIDEMIOLOGY

Incidence and prevalence

Figure 40.1 shows the radiographic vertebral, hip and wrist fracture incidence by age and gender [6, 7].

The 2004 report from the U.S. Surgeon General highlighted the enormous burden of osteoporosis-related fractures [8]. An estimated 10 million Americans over 50 years old have osteoporosis, and there are about 1.5 million fragility fractures each year. Another 34 million Americans are at risk of the disease. A study of British fracture occurrence indicates that population risk is similar in the U.K. [6]. Thus, 1 in 2 women aged 50 years will have an osteoporosis-related fracture in their remaining lifetime; the figure for men is 1 in 5.

Fracture incidence in the community is bimodal, showing peaks in youth and in the very elderly. In young people, fractures of long bones predominate, usually after substantial trauma and are more frequent in males than females. In this group the question of bone strength rarely arises, although there are now data suggesting that this may not be entirely irrelevant as a risk factor [9]. Over

---

Primer on the Metabolic Bone Diseases and Disorders of Mineral Metabolism, Eighth Edition. Edited by Clifford J. Rosen.
© 2013 American Society for Bone and Mineral Research. Published 2013 by John Wiley & Sons, Inc.

the age of 35 years, fracture incidence in women climbs steeply, so that rates become twice those in men. Before studies that ascertained vertebral deformities radiographically, rather than by clinical presentation, this peak was thought to be mainly due to hip and distal forearm fracture, but as Fig. 40.1 shows, vertebral fracture can now be shown to make a significant contribution.

Hip fracture

In most populations hip fracture incidence increases exponentially with age (Fig. 40.1). Above 50 years of age, there is a female-to-male incidence ratio of around 2 to 1 [10]. Overall, about 98% of hip fractures occur among people aged 35 years and over, and 80% occur in women (because there are more elderly women than men). Worldwide there were an estimated 1.66 million hip fracture in 1990 [11]; about 1.19 million in women and 463,000 in men. The majority occur after a fall from standing height or less, and 90% occur in people over 50 years old [12]. Recent work has characterized the age and gender specific incidence in the U.K. population, using the General Practice Research Database (GPRD), which includes 6% of the U.K. population). Thus, the lifetime risk of hip fracture for 50-year-olds in the U.K. is 11.4% and 3.1% for women and men, respectively [6]. Most of this increased risk is accrued in old age, such that a 50-year-old woman's 10-year risk of hip fracture is 0.3%, rising to 8.7% when she is 80 years old [6]. The corresponding figures for men are 0.2% and 2.9%, respectively. Hip fractures are seasonal, with an increase in winter in temperate countries, but their occurrence mainly indoors would imply that this increase is not simply due to slipping on icy pavements: other possible causes may include slowed neuromuscular reflexes and lower light in winter weather. The direction of falling is important, and a fall directly onto the hip (sideways) is more likely to cause a fracture than falling forward [13].

Incidence rates vary substantially from one population to another, and incidence is usually greater in whites than nonwhites, although there are differences within populations of a given gender or race. In Europe, hip fracture rates vary sevenfold between countries [14]. These findings suggest an important role for environmental factors in the aetiology of hip fracture, but factors studied thus far, such as smoking, alcohol consumption, activity levels, obesity, and migration status, have not explained these trends.

Vertebral fracture

Recent data from the European Vertebral Osteoporosis Study (EVOS) have shown that the age-standardized population prevalence across Europe was 12.2% for men and 12.0% for women aged 50–79 years [15]. Historically, it

Table 40.1. Impact of Osteoporosis-Related Fractures (5)

	Hip	Spine	Wrist
Lifetime risk (%)			
Women	23	29	21
Men	11	14	5
Cases/year	620,000	810,000	574,000
Hospitalization (%)	100	2–10	5
Relative survival	0.83	0.82	1.00

Costs: All sites combined ~39 billion Euros.

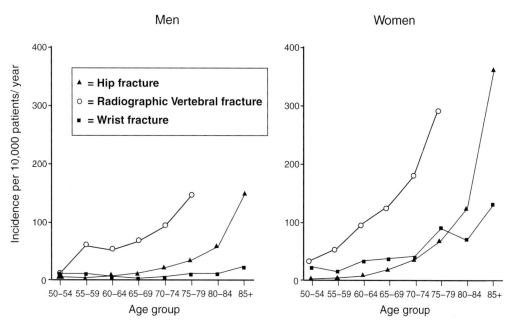

Fig. 40.1. Radiographic vertebral, hip, and wrist fracture incidence by age and gender (Refs. 6, 7).

was believed that vertebral fractures were more common in women than men, but the EVOS data suggest that this is not the case at younger ages: The prevalence of deformities in 50- to 60-year-olds is similar, if not higher in men, possibly because of a greater incidence of trauma [15]. The majority of vertebral fractures in elderly women occur through normal activities such as lifting, rather than through falling.

Many vertebral fractures are asymptomatic, and there is disagreement about the radiographic definition of deformities in those patients who do present. Thus, in studies using radiographic screening of populations, the incidence of all vertebral deformities has been estimated to be three times that of hip fracture, with only one-third of these coming to medical attention [16]. Data from EVOS have allowed accurate assessment of radiographically determined vertebral fractures in a large population. At age 75–79 years, the incidence of vertebral fractures so-defined was 13.6 per 1,000 person-years for men and 29.3 per 1,000 person-years for women [7]. This compares with 0.2 per 1,000 person-years for men and 9.8 per 1,000 person-years in 75–84 year olds where the fractures were defined by clinical presentation in an earlier study from Rochester, Minnesota [17]. The overall age-standardized incidence in EVOS was 10.7 per 1,000 person-years in women and 5.7 per 1,000 person-years in men.

Figure 40.2 shows the incidence rates of morphometrically (by radiograph) defined vertebral fracture in EVOS compared with rates derived from other population-based radiographic studies. It shows that the heterogeneity in vertebral fracture incidence is markedly lower than the geographic variation in age and gender-adjusted rates of hip fracture. The figure contrasts incidence rates for vertebral fractures defined radiographically with those identified by clinical diagnosis or hospitalization for the fracture. It clearly demonstrates the enormous shortfall in clinical recognition and hospital identification of this important osteoporotic fracture.

Distal forearm fracture

Wrist fractures show a different pattern of occurrence to hip and vertebral fractures. There is an increase in incidence in white women between the ages of 45 and 60 years, followed by a plateau [16]. This may relate to altered neuromuscular reflexes with aging, and as a result, a tendency to fall sideways or backward, and thus not to break the fall with an outstretched arm. Most wrist fractures occur in women, and 50% occur in women over 65 years old. Data from the GPRD show that a woman's lifetime risk of wrist fracture at 50 years old is 16.6%, falling to 10.4% at 70 years. The incidence in men is low and does not rise much with aging (lifetime risk 2.9% at age 50 years and 1.4% at age 70 years) [6].

Clustering of fractures in individuals

Epidemiological studies suggest that patients with different types of fragility fractures are at increased risk of developing other types of fracture. For example, the presence of a previous vertebral deformity leads to a seven- to tenfold increase in the risk of subsequent vertebral deformities [18]. This is a comparable level of increased risk to that seen for individuals who have sustained one hip fracture to then sustain a second. Furthermore, data from Rochester, Minnesota, suggest that the risk of a hip fracture is increased 1.4-fold in women and 2.7-fold in men after the occurrence of a distal forearm fracture [19]. The corresponding figures for subsequent vertebral fracture are 5.2 and 10.7. Data from EVOS demonstrate that prevalent vertebral deformity predicts incident hip fracture with a rate ratio of 2.8 to 4.5, and this increases with the number of vertebral deformities [20]. The number and morphometry of base line vertebral deformities also predict incident vertebral fractures [21]. The incidence of new vertebral fracture within a year of an index vertebral fracture is 19.2% [22], and in Rochester the cumulative incidence of any fracture 10 years after the baseline event was 70%. These data emphasize the importance of prompt therapeutic action upon discovering vertebral deformities. A recent study from the Netherlands, in which 4,140 postmenopausal women were followed, provides further support for this notion [23]. Within this cohort, 54% had re-fractured within 5 years, and 23% had re-fractured within 1 year of an initial fracture. The relative risk of subsequent fracture declined with time from initial fracture: Thus, the relative risk of subsequent fracture in the first year after the initial fracture was 5.3; within 2–5 years it was 2.8; within 6–10 years it was 1.4; and then it dropped to 0.41 for greater than 10 years post initial fracture. Finally, the Dubbo Osteoporosis Epidemiology Study demonstrated similar rates of re-fracture in men and women over a 10-year period; increased fracture rates were seen after most different types of baseline fractures, excluding rib (men) and ankle (women) [24].

Time trends and future projections

Life expectancy is increasing around the globe, and the number of elderly individuals is rising in every geographic region. The world population is expected to rise from the current 323 million individuals aged 65 years or over to 1.555 billion by the year 2050. These demographic changes alone can be expected to increase the number of hip fractures occurring among people aged 35 years and over worldwide: The incidence is estimated to rise from 1.66 million in 1990 to 6.26 million in 2050. Assuming a constant age-specific rate of fracture, as the number of over 65-year-olds increases from 32 million in 1990 to 69 million in 2050, the number of hip fractures in the U.S. will increase threefold [25]. In the U.K., the number of hip fractures may increase from 46,000 in 1985 to 117,000 in 2016 [26]. However, in the developed world, studies from Switzerland and Finland suggest that the age-adjusted incidence of hip fracture has declined over the past decade [2, 3, 27, 28]. The reason for these changes

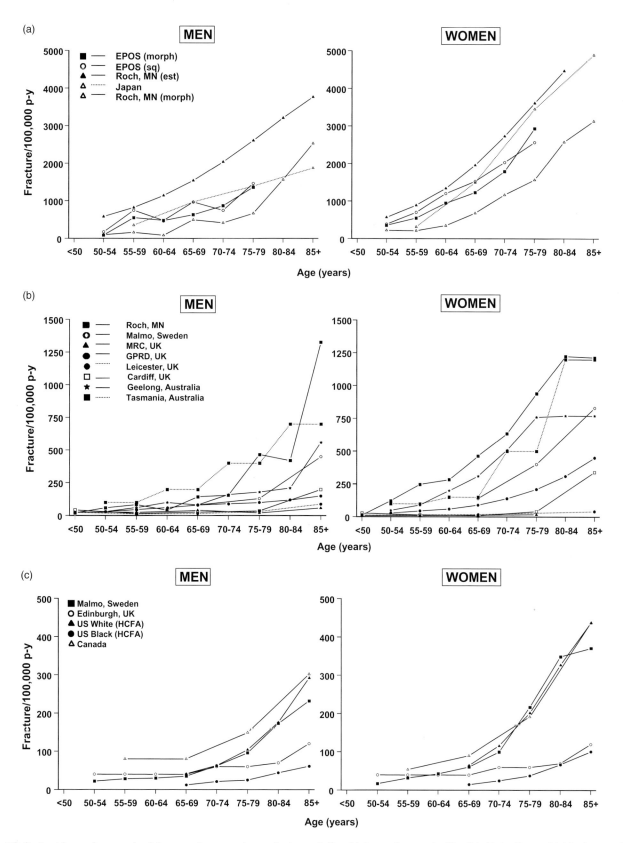

Fig. 40.2. Incidence for vertebral fracture in several populations, defined (a) morphometrically, (b) clinically, and (c) by hospitalization for fracture.

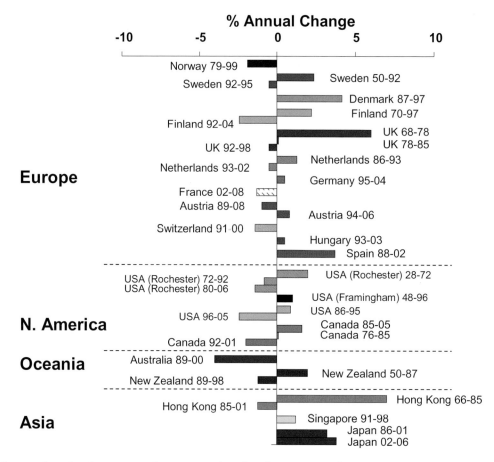

Fig. 40.3. Secular trends in hip fracture incidence. Reproduced with permission from Cooper et al. (Ref. 29).

may be a birth cohort effect, an increase in obesity or better screening and treatment for osteoporosis, and might potentially partly offset the impact of the projected increase in the elderly population. This pattern of declining age-adjusted incidence is supported by further studies reported in a recent systematic review [29]: Thus, in Western populations, whether in North America, Europe, or Oceania, while increases in age-adjusted hip fracture incidence have been reported through the second half of the past century, those studies that continue to follow trends over the past two decades have found that rates stabilize, with age-adjusted decreases being observed in some centers. In contrast, this decrease in age-adjusted incidence has not been recorded in the Third World, and thus the increase in the elderly population, together with the adoption of Westernized lifestyles, is likely to ensure an increase in the worldwide burden of osteoporotic fractures in future generations. Figure 40.3 summarizes these secular changes.

Thus the increase in fracture numbers is likely to be uneven across the globe, with the increase in the elderly population in Latin America and Asia potentially leading to a shift in the geographical distribution of hip fractures, with only a quarter occurring in Europe and North America [25].

Geography

There is variation in the incidence of hip fracture within populations of a given race and gender [14, 30]. Thus, age-adjusted hip fracture incidence rates are higher among white residents of Scandinavia than comparable subjects in the USA or Oceania. Within Europe the range of variation was approximately 11-fold [14, 30]. These differences were not explained by variation in activity levels, smoking, obesity, alcohol consumption, or migration status [14]. The EVOS study demonstrated a threefold difference in the prevalence of vertebral deformities between countries, with the highest rates in Scandinavia. The prevalence range between centers was 7.5–19.8% for men and 6.2–20.7% for women. The differences were not as great as those seen for hip fracture in Europe, and some of the differences could be explained by levels of physical activity and body mass index [15].

MORTALITY AND MORBIDITY

Mortality

Mortality patterns have been studied for the three most frequent osteoporotic fractures. In Rochester, Minnesota, survival rates 5 years after hip and vertebral fractures were found to be around 80% of those expected for men and women of similar age without fractures [31].

Hip fracture

Hip fracture mortality is higher in men than women, increases with age [31], and is greater for those with coexisting illnesses and poor pre-fracture functional status. There are around 31,000 excess deaths within 6 months of the approximately 300,000 hip fractures that occur annually in the U.S. About 8% of men and 3% of women older than 50 years die while hospitalized their fracture. In the U.K., the 12-month survival post hip fracture for men is 63.3% vs 90.0% expected, and for women, 74.9% vs 91.1% expected [6]. The risk of death is greatest immediately after the fracture and decreases gradually over time. The cause of death is not usually directly attributable to the fracture itself, but to other chronic diseases, which lead both to the fracture and to the reduced life expectancy.

However, recent data suggest that elevated mortality persists for up 10 years after hip fracture. In the Dubbo Epidemiology Study of community-dwelling women and men aged 60 years and older, risk of death was further elevated over another 5 years by a subsequent fracture [32]. Reduced survival was observed following all types of fracture except for minor fractures where mortality was increased only for those of 75 years old or greater. Only 25% of deaths from hip fracture are estimated to be directly due to the fracture or resulting complications such as infection, thrombo-embolism and subsequent surgery [33]. In the Dubbo study the main predictors of higher mortality were gender (male), increasing age, quadriceps weakness and subsequent fracture, but not comorbidities. Smoking, low BMD, and body sway were also predictors in women, as was a lower level of physical activity in men [32]. A recent Australian nested case-control study suggested that the pattern and causes of death may differ somewhat for those in institutional care. Thus, in 2,005 elderly people in residential care, survival rates for hip fracture and control patients became equal within a year post fracture repair [34]: After adjusting for age, gender, institution type, weight, immobility, cognitive function, comorbidities and number of medications, there was a threefold increased risk of death for cases compared with controls in the first three months following surgery for hip fracture; the hazard ratios decreased subsequently (1.99 for 3 to 9 months post fracture and 0.88 for greater than 9 months post fracture). Infections in females and cardiac disease in both genders were the main causes of death in the first 9 months, and bisphosphonate use was associated with a reduction in mortality post hip fracture repair. It seems likely that the swift equalization of mortality rates between cases and controls in this population reflects the high risk of death of institutionalized populations compared with those free-living individuals in the GPRD and Dubbo studies.

Vertebral fracture

Vertebral fractures are associated with increased mortality well beyond a year post fracture [31], with comorbid conditions contributing significantly to the decreased relative survival.

The impairment of survival following vertebral fracture also markedly worsens as time from diagnosis of the fracture increases. This is in contrast to the pattern of survival for hip fractures. In the U.K. GPRD study, the observed survival in women 12 months after vertebral fracture was 86.5% vs 93.6% expected. At 5 years, survival was 56.5% observed and 69.9% expected [6].

Morbidity

In the U.S., 7% of survivors of all types of fracture have some degree of permanent disability, and 8% require long-term nursing home care. Overall, a 50-year-old white American woman has a 13% chance of experiencing functional decline after any fracture [35].

Hip fracture

As with mortality, hip fractures contribute most to osteoporosis-associated disability. Patients are prone to developing acute complications such as pressure sores, bronchopneumonia, and urinary tract infections. Perhaps the most important long-term outcome is impairment of the ability to walk. Fifty percent of those ambulatory before the fracture are unable to walk independently afterwards. Age is an important determinant of outcome, with 14% of 50–55-year-old hip fracture victims being discharged to nursing homes, versus 55% of those older than 90 years [35].

Vertebral fracture

Despite only a minority of vertebral fractures coming to clinical attention, they account for 52,000 hospital admissions in the USA and 2,188 in England and Wales each year in patients aged 45 years and older. The major clinical consequences of vertebral fracture are back pain, kyphosis, and height loss. Quality of life (QUALEFFO) scores decrease as the number of vertebral fractures increases [36].

Distal forearm fracture

Wrist fractures do not appear to increase mortality [6]. Although wrist fractures may impact on some activities such as writing or meal preparation, overall few patients are severely disabled, although over half report only fair to poor function at 6 months post fracture [35].

Low bone mass in children

There has been considerably less investigation of the role of bone fragility in childhood fractures, probably because of the perception that the primary determinant of fracture in this age group is trauma. There are also considerable difficulties in reaching a definition of "osteoporosis" in a growing skeleton, when there is no straightforward relationship between BMD and fracture risk. Thus the consensus view is that the phrase "low bone mass for age (with or without fractures)" is used rather than "osteoporosis." Most evidence comes from two large European studies that describe the epidemiology of fractures in childhood [37–39]. In Malmo, Sweden, the overall incidence of fracture was 212 per 10,000 girls and 257 per 10,000 boys, with 27% of girls and 42% of boys sustaining a fracture between birth and 16 years of age. Fractures of the distal radius occurred most commonly, followed by fractures of the phalanges of the hand [38, 39]. A follow-up study in Malmo a decade later found the incidence of fracture had decreased by almost 10% since the original study [40].

A similar pattern was found in the U.K. GPRD [37]. The overall incidence of fracture was 133.1 per 10,000 children, with fractures being more common in boys than girls, with an incidence of 161.6 per 10,000 and 102.9 per 10,000, respectively. Again, the most common fracture site in both sexes was the radius/ulna with a total of 39.3 per 10,000 per year. Historically, most work has focused on the impact of trauma in the aetiology of childhood fractures, contrasting with the role of bone fragility in the elderly. However, several recent studies have documented lower areal and volumetric BMD in children with distal forearm fractures than age- and sex-matched controls [41, 42]. Consistent with these observations is the finding from GPRD that peak fracture incidence corresponds to age of entry to puberty, when the discordance between height gain and accrual of volumetric bone density is greatest [37]. Other studies have suggested that risk factors for childhood fracture include obesity and high levels of vigorous activity [42, 43]. As increased physical activity has been associated with increased bone mass in childhood [43, 44], this raises the possibility of there being different routes to increased fracture risk, relating separately to bone mass and risk of trauma.

Early life influences on adult fragility fracture

The importance of bone mineral accrual in childhood and achievement of adequate peak bone mass (PBM) in early adulthood has been emphasized in recent work, demonstrating that PBM is a major determinant of osteoporosis risk in later life [45]. Over the past twenty years, evidence has accrued that the early environment may have long-term influences on future bone health. This phenomenon of "developmental plasticity," whereby a single genotype may lead to different phenotypes, dependent upon the prevailing environmental milieu, is well established in the natural world. There is a growing body of epidemiological evidence that a poor intrauterine environment leads to lower bone mass in adult life, both in third and sixth/seventh decades [46–48]. Additionally work in Finland has demonstrated an association between poor infant and childhood growth and increased risk of hip fracture 7 decades later [49, 50]. Physiological studies have implicated the parathyroid hormone/vitamin D axis in mechanisms that underly this phenomenon, such that mothers who are deficient in vitamin D in late pregnancy have children with decreased bone mass in childhood [51–53]. This novel area of research may lead ultimately to innovative strategies to improve bone health in children, with a subsequent reduction in the burden of osteoporotic fracture in future generations.

CONCLUSION

Osteoporosis is a disease that has a huge effect on public health. The impact of osteoporotic fracture is massive, not just for individuals, but for the health service, economy, and population as a whole. Many risk factors for inadequate peak bone mass, excessive involutional loss and fracture have been characterized, and coupled with new pharmacologic therapies, we are now in a position to develop novel preventative and therapeutic strategies, both for the entire population and those at highest risk.

ACKNOWLEDGMENTS

We would like to thank Medical Research Council (UK), Arthritis Research UK, National Osteoporosis Society (UK), International Osteoporosis Foundation, and the European Union Network on Male Osteoporosis for funding this work.

REFERENCES

1. [No authors listed]. 1993. Consensus development conference: Diagnosis, prophylaxis and treatment of osteoporosis. *Am J Med* 941: 646–650.
2. World Health Organization Study Group. 1994. Assessment of fracture risk and its application to screening for postmenopausal osteoporosis. *World Health Organ Tech Rep Ser* 843: 1–129.
3. Kanis JA, Johnell O, Oden A, Dawson A, De Laet C, Jonsson B. 2001. Ten-year probabilities of osteoporotic fractures according to BMD and diagnostic thresholds. *Osteoporosis Int* 12: 989–995.
4. Kanis JA, McCloskey EV, Johansson H, Strom O, Borgstrom F, Oden A. 2008. Case finding for the management of osteoporosis with FRAX—Assessment and

intervention thresholds for the UK. *Osteoporos Int* 19: 1395–1408.
5. Ström O, Borgström F, Kanis JA, Compston J, Cooper C, McCloskey EV, Jönsson B. 2011. Osteoporosis: Burden, health care provision and opportunities in the EU: A report prepared in collaboration with the International Osteoporosis Foundation (IOF) and the European Federation of Pharmaceutical Industry Associations (EFPIA). *Arch Osteoporos.* 6(1–2): 59–155.
6. van Staa TP, Dennison EM, Leufkens HG, Cooper C. 2001. Epidemiology of fractures in England and Wales. *Bone* 29: 517–522.
7. The European Prospective Osteoporosis Study (EPOS) Group 2002 Incidence of vertebral fracture in Europe: Results from the European Prospective Osteoporosis Study (EPOS). *J Bone Miner Res* 17: 716–724.
8. U.S. Department of Health and Human Services. 2004. *Bone Health and Osteoporosis: A Report of the Surgeon General.* Rockville, MD.
9. Goulding A, Jones IE, Taylor RW, Manning PJ, Williams SM. 2000. More broken bones: A 4-year double cohort study of young girls with and without distal forearm fractures. *J Bone Miner Res* 15: 2011–2018.
10. Melton LJ. 1988. Epidemiology of fractures. In: Riggs BL, Melton LJ (eds). *Osteoporosis: Etiology, Diagnosis and Management.* New York: Raven Press. pp. 133–154.
11. Cooper C, Melton LJ. 1992. Epidemiology of osteoporosis. *Trends Endocrinol Metab* 314: 224–229.
12. Gallagher JC, Melton LJ, Riggs BL, Bergstrath E. 1980. Epidemiology of fractures of the proximal femur in Rochester, Minnesota. *Clin Orthop* 150: 163–171.
13. Nevitt MC, Cummings SR. 1993. Type of fall and risk of hip and wrist fractures: The study of osteoporotic fractures. The Study of Osteoporotic Fractures Research Group. *J Am Geriatr Soc* 41: 1226–1234.
14. Johnell O, Gullberg B, Allander E, Kanis JA. 1992. The apparent incidence of hip fracture in Europe: A study of national register sources. MEDOS Study Group. *Osteoporosis Int* 2: 298–302.
15. O'Neill TW, Felsenberg D, Varlow J, Cooper C, Kanis JA, Silman AJ. 1996. The prevalence of vertebral deformity in European men and women: The European Vertebral Osteoporosis Study. *J Bone Miner Res* 11: 1010–1018.
16. Melton LJ, Cooper C. 2001. Magnitude and impact of osteoporosis and fractures. In: Marcus R, Feldman D, Kelsey J (eds). *Osteoporosis, 2nd Ed., Vol 1.* San Diego: Academic Press. pp. 557–567.
17. Cooper C, Atkinson EJ, O'Fallon WM, Melton LJ. 1992. Incidence of clinically diagnosed vertebral fractures: A population-based study in Rochester, Minnesota, 1985–1989. *J Bone Miner Res* 7: 221–227.
18. Ross PD, Davis JW, Epstein RS, Wasnich RD. 1991. Pre-existing fractures and bone mass predict vertebral fracture incidence in women. *Ann Intern Med* 114: 919–923.
19. Cuddihy MT, Gabriel SE, Crowson CS, O'Fallon WM, Melton LJ. 1999. Forearm fractures as predictors of subsequent osteoporotic fractures. *Osteoporosis Int* 9: 469–475.
20. Ismail AA, Cockerill W, Cooper C, Finn JD, Abendroth K, Parisi G, et al. 2001. Prevalent vertebral deformity predicts incident hip though not distal forearm fracture: Results from the European Prospective Osteoporosis Study. *Osteoporosis Int* 12: 85–90.
21. Lunt M, O'Neill T, Armbrecht G, Reeve J, Felsenberg D, Cooper C, et al. 2002. Characteristics of prevalent vertebral deformity and the risk of incident vertebral fracture. *Rheumatology* 41 [Suppl 1]: 101–102. [Abstract]
22. Lindsay R, Silverman SL, Cooper C, Hanley DA, Barton I, Broy SB, et al. 2001. Risk of new vertebral fracture in the year following a fracture. *JAMA* 285: 320–323.
23. van Geel TA, van Helden S, Geusens PP, Winkens B, Dinant GJ. 2009. Clinical subsequent fractures cluster in time after first fractures. *Ann Rheum Dis* 68(1): 99–102.
24. Center JR, Bliuc D, Nguyen TV, Eisman JA. 2007. Risk of subsequent fracture after low-trauma fracture in men and women. *JAMA* 297(4): 387–394.
25. Cooper C, Campion G, Melton LJ. 1992. Hip fractures in the elderly: A world-wide projection. *Osteoporosis Int* 2: 285–289.
26. Royal College of Physicians. 1989. Fractured neck of femur: Prevention and management. Summary and report of the Royal College of Physicians. *J Roy Coll Physicians Lond* 23: 8–12.
27. Chevalley T, Guilley E, Herrmann FR, Hoffmeyer P, Rapin CH, Rizzoli R. 2007. Incidence of hip fracture over a 10-year period (1991–2000): Reversal of a secular trend. *Bone* 40: 1284–1289.
28. Kannus P, Niemi S, Parkkari J, Palvanen M, Vuori I, Jarvinen M. 2006. Nationwide decline in incidence of hip fracture. *J Bone Miner Res* 21: 1836–1838.
29. Cooper C, Cole ZA, Holroyd CR, Earl SC, Harvey NC, Dennison EM, Melton LJ, Cummings SR, Kanis JA. 2011. Secular trends in the incidence of hip and other osteoporotic fractures. *Osteoporos Int* 22: 1277–1288.
30. Elffors I, Allander E, Kanis JA, Gullberg B, Johnell O, Dequeker J, et al. 1994. The variable incidence of hip fracture in southern Europe: The MEDOS Study. *Osteoporosis Int* 4: 253–263.
31. Cooper C, Atkinson EJ, Jacobsen SJ, O'Fallon WM, Melton LJ. 1993. Population-based study of survival after osteoporotic fractures. *Am J Epidemiol* 137: 1001–1005.
32. Bliuc D, Nguyen ND, Milch VE, Nguyen TV, Eisman HA, Center JR. 2009. Mortality risk associated with low-trauma osteoporotic fracture and subsequent fracture in men and women. *JAMA* 310(5): 513–521.
33. Sernbo I, Johnell O. 1993. Consequences of a hip fracture: A prospective study over 1 year. *Ostoeporosis Int* 3: 148–153.
34. Cameron ID, Chen JS, March LM, Simpson JM, Cumming RG, Seibel MJ, Sambrook PN. 2009. Hip fracture causes excess mortality due to cardiovascular and infectious disease in institutionalized older people: A prospective five-year study. *J Bone Miner Res* 25: 866–872.
35. Chrischilles EA, Butler CD, Davis CS, Wallace RB. 1991. A model of lifetime osteoporosis impact. *Arch Intern Med* 151: 2026–2032.

36. Oleksik A, Lips P, Dawson A, Minshall ME, Shen W, Cooper C, et al. 2000. Health-related quality of life in postmenopausal women with low BMD with or without prevalent vertebral fractures. *J Bone Miner Res* 15: 1384–1392.
37. Cooper C, Dennison EM, Leufkens HG, Bishop N, van Staa TP. 2004. Epidemiology of childhood fractures in Britain: A study using the general practice research database. *J Bone Miner Res* 19: 1976–1981.
38. Landin LA. 1997. Epidemiology of children's fractures. *J Pediatr Orthop B* 6: 79–83.
39. Landin LA. 1983. Fracture patterns in children. Analysis of 8,682 fractures with special reference to incidence, etiology and secular changes in a Swedish urban population 1950–1979. *Acta Orthop Scand Suppl* 202: 1–109.
40. Tiderius CJ, Landin L, Duppe H. 1999. Decreasing incidence of fractures in children: An epidemiological analysis of 1,673 fractures in Malmo, Sweden, 1993–1994. *Acta Orthop Scand* 70: 622–626.
41. Goulding A, Jones IE, Taylor RW, Manning PJ, Williams SM. 2000 more broken bones: A 4-year double cohort study of young girls with and without distal forearm fractures. *J Bone Miner Res* 15: 2011–2018.
42. Clark EM, Ness AR, Bishop NJ, Tobias JH. 2006. Association between bone mass and fractures in children: A prospective cohort study. *J Bone Miner Res* 21: 1489–1495.
43. Clark EM, Ness AR, Tobias JH. 2008. Vigorous physical activity at age 9 increases the risk of childhood fractures, despite increasing bone mass. *J Bone Miner Res* 23(7). 1012–1022.
44. Harvey NC, Cole ZA, Crozier SR, Kim M, Ntani G, Goodfellow L, Robinson SM, Inskip HM, Godfrey KM, Dennison EM, Wareham N, Ekelund U, Cooper C. 2012. Physical activity, calcium intake and childhood bone mineral: A population-based cross-sectional study. *Osteoporos Int* 23(1): 121–130.
45. Hernandez CJ, Beaupre GS, Carter DR. 2003. A theoretical analysis of the relative influences of peak BMD, age-related bone loss and menopause on the development of osteoporosis. *Osteoporos Int* 14: 843–847.
46. Cooper C, Cawley M, Bhalla A, Egger P, Ring F, Morton L, et al. 1995. Childhood growth, physical activity, and peak bone mass in women. *J Bone Miner Res* 10: 940–947.
47. Gale CR, Martyn CN, Kellingray S, Eastell R, Cooper C. 2001. Intrauterine programming of adult body composition. *J Clin Endocrinol Metab* 86: 267–272.
48. Dennison EM, Syddall HE, Sayer AA, Gilbody HJ, Cooper C. 2005. Birth weight and weight at 1 year are independent determinants of bone mass in the seventh decade: The Hertfordshire cohort study. *Pediatr Res* 57: 582–586.
49. Cooper C, Eriksson JG, Forsen T, Osmond C, Tuomilehto J, Barker DJ. 2001. Maternal height, childhood growth and risk of hip fracture in later life: A longitudinal study. *Osteoporosis Int* 12: 623–629.
50. Javaid MK, Eriksson JG, Kajantie E, Forsen T, Osmond C, Barker DJ, Cooper C. 2011. Growth in childhood predicts hip fracture risk in later life. *Osteoporos Int* 22: 69–73.
51. Javaid MK, Crozier SR, Harvey NC, Gale CR, Dennison EM, Boucher BJ, Arden NK, Godfrey KM, Cooper C. 2006. Maternal vitamin D status during pregnancy and childhood bone mass at age 9 years: A longitudinal study. *Lancet* 367: 36–43.
52. Mahon P, Harvey N, Crozier S, Inskip H, Robinson S, Arden N, Swaminathan R, Cooper C, Godfrey K. 2010. Low maternal vitamin D status and fetal bone development: Cohort study. *J Bone Miner Res* 25: 14–19.
53. Harvey NC, Javaid MK, Poole JR, Taylor P, Robinson SM, Inskip HM, Godfrey KM, Cooper C, Dennison EM. 2008. Paternal skeletal size predicts intrauterine bone mineral accrual. *J Clin Endocrinol Metab* 93: 1676–1681.

41
Overview of Pathogenesis
Ian R. Reid

References 359

The pathogenesis of osteoporosis is as diverse as the influences that determine the strength of the skeleton and the trauma to which it is subjected. Skeletal trauma in the elderly is principally related to falls, and the propensity to fall is influenced by frailty, neuromuscular abnormalities, eyesight, sedative drugs, postural hypotension, and the safety of the home environment (presence of hazards such as loose carpets and electric cords, absence of safety devices such as handrails). Falls are dealt with elsewhere in this volume, so this chapter will focus on determinants of skeletal strength.

The elements that contribute to skeletal strength (and thus fragility) are listed in Table 41.1. In the past, the greatest emphasis has been on the mass of the skeleton, measured clinically as bone mineral density (BMD). Thus, factors that maximize bone mass during childhood and adolescence (e.g., global nutritional status, physical activity, genes) and delay or slow its loss during menopause and old age (fat mass, physical activity, calcium intake, vitamin D status) [1, 2] will reduce fracture risk. Throughout life, bone mass is influenced by intercurrent illnesses, body weight, lifestyle factors, and some medications, particularly glucocorticoids (Table 41.2). In women, bone mass is critically dependent on estrogen levels and can be maintained indefinitely with postmenopausal estrogen replacement [3].

One observation suggesting that bone mass is not the only determinant of fracture risk is the demonstration that age increases fracture risk independent of bone density. Thus, an 80-year-old woman with a femoral neck T-score of −3 is six times more likely to have a hip fracture than a 50-year-old woman with the same bone density [4]. A similar increase in fracture risk independent of bone density has been demonstrated in glucocorticoid users [5]. Accordingly, in recent years there has been an increasing focus on bone *quality*, as well as bone *quantity*. The components of bone quality are continuing to be defined, but include the material properties of both the proteinaceous matrix and the mineral phase of bone, as well as bone architecture (Table 41.1). Major perturbations of the matrix, as in osteogenesis imperfecta, clearly have a substantial effect on skeletal fragility, but it is now evident that collagen cross-linking, the state of cross-link isomerization, and levels of advanced glycation end products also have an impact on skeletal strength [6]. These aspects of bone matrix biology are thought to be influenced by rates of bone turnover, which is under the influence of sex hormones and cytokines, as well as genetic factors [7]. In addition, advanced glycation end products will be influenced by ambient glucose concentrations, and increased in the presence of diabetes mellitus. This probably contributes to the higher fracture risk of diabetics, independent of BMD [8]. Some studies have suggested that the presence of microcracks in bone might also compromise skeletal strength and that these are related to low turnover [9]. Adverse changes in these indices have been suggested to contribute to the deterioration in skeletal strength demonstrable in some animal studies of high-dose bisphosphonate use [6, 10]. Recent human studies, however, have suggested that these factors make only small contributions to bone strength, which is primarily determined by trabecular bone volume [11]. In pathological situations (e.g., Paget's disease) high turnover can result in disruption of the normal lamellar pattern of

---

Primer on the Metabolic Bone Diseases and Disorders of Mineral Metabolism, Eighth Edition. Edited by Clifford J. Rosen.
© 2013 American Society for Bone and Mineral Research. Published 2013 by John Wiley & Sons, Inc.

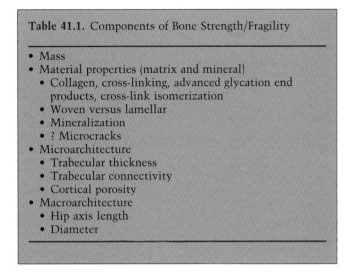

Table 41.1. Components of Bone Strength/Fragility

- Mass
- Material properties (matrix and mineral)
  - Collagen, cross-linking, advanced glycation end products, cross-link isomerization
  - Woven versus lamellar
  - Mineralization
  - ? Microcracks
- Microarchitecture
  - Trabecular thickness
  - Trabecular connectivity
  - Cortical porosity
- Macroarchitecture
  - Hip axis length
  - Diameter

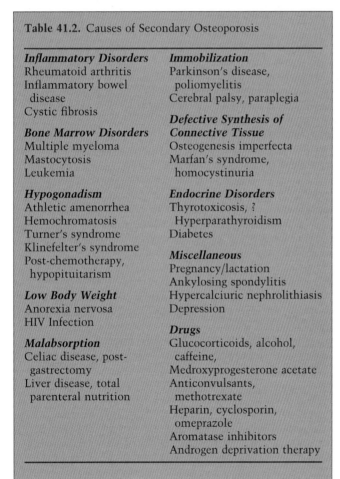

Table 41.2. Causes of Secondary Osteoporosis

Inflammatory Disorders
Rheumatoid arthritis
Inflammatory bowel disease
Cystic fibrosis

Bone Marrow Disorders
Multiple myeloma
Mastocytosis
Leukemia

Hypogonadism
Athletic amenorrhea
Hemochromatosis
Turner's syndrome
Klinefelter's syndrome
Post-chemotherapy, hypopituitarism

Low Body Weight
Anorexia nervosa
HIV Infection

Malabsorption
Celiac disease, post-gastrectomy
Liver disease, total parenteral nutrition

Immobilization
Parkinson's disease, poliomyelitis
Cerebral palsy, paraplegia

Defective Synthesis of Connective Tissue
Osteogenesis imperfecta
Marfan's syndrome, homocystinuria

Endocrine Disorders
Thyrotoxicosis, ? Hyperparathyroidism
Diabetes

Miscellaneous
Pregnancy/lactation
Ankylosing spondylitis
Hypercalciuric nephrolithiasis
Depression

Drugs
Glucocorticoids, alcohol, caffeine,
Medroxyprogesterone acetate
Anticonvulsants, methotrexate
Heparin, cyclosporin, omeprazole
Aromatase inhibitors
Androgen deprivation therapy

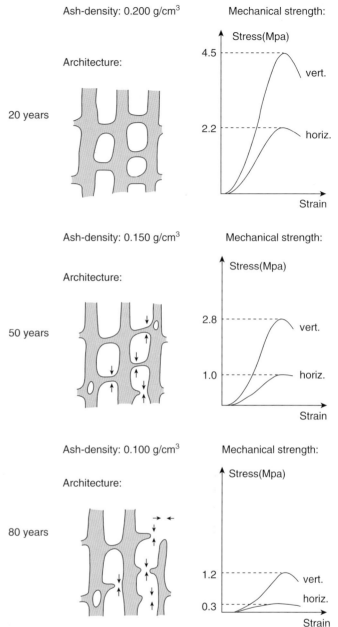

Fig. 41.1. Typical relationships among ash-density, architecture, and mechanical strength (vertical and horizontal) for vertebral trabecular bone obtained from three different age groups: 20 years, 50 years, and 80 years. From Mosekilde L. 1993. Vertebral structure and strength *in vivo* and *in vitro*. *Calcif Tissue Int* 53(Suppl 1): S121–S6. Used with permission.

collagen fibril assembly, leading to the deposition of woven bone, which has inferior mechanical properties. Hypocalcemia, hypophosphatemia, or the presence of interfering factors (such as fluoride or high-dose etidronate) can interfere with normal mineralization, with a resultant decrease in the compressive strength of bone.

The microarchitecture of bone impacts on its strength. As bone mass is lost, perforations occur in individual trabeculae, contributing to a much greater decrease in bone strength than the simple decrease in mass would predict (Fig. 41.1). Accordingly, loss of trabecular continuity is more frequently observed in patients with osteoporotic fractures than in controls [12]. High bone turnover contributes to trabecular perforation by increasing the likelihood of resorption lacunae coinciding on opposite sides of a trabecula. Even without leading to perforation, high turnover can increase trabecular fragility by causing critical trabecular narrowing where resorption lacunae are located close to one another on opposite sides of the trabeculae. This may explain the independent association of bone turnover with fracture risk [13], although this is probably also contributed to by the greater rate of bone loss observed in high turnover states [14]. In the cortex, increasing porosity resulting from increased osteoclast activity results in loss of strength and ultimately in the loss of cortical bone as a result of its "trabecularization" [15]. In the first 15 years after menopause, trabecular bone loss exceeds cortical, but subsequently this is reversed as the cortex becomes more porous and so has a greater area to be remodeled, whereas the loss of trabecular elements reduces the surface in that compartment [15].

Macroarchitecture also impacts on skeletal fragility. This has probably been most clearly demonstrated through the positive association between hip axis length and hip fracture risk [16]. Increases in hip axis length and height (also a risk factor for fracture) probably have contributed to the secular increases in hip fracture risk as global nutrition has improved over the past 60 years [17]. Macroarchitectural changes may counteract, to some extent, the loss of bone mass with age. Ahlborg [18] has elegantly demonstrated increasing medullary and periosteal diameters in the forearms of postmenopausal women, which substantially abrogate the loss of tissue-level strength caused by the sustained imbalance of osteoblast and osteoclast activity. The increase in diameter results in increased resistance to both bending and torsion.

The above paradigm dissects skeletal fragility in terms of the anatomical components that make up bone. Each of these components is under multiple influences including those of genes, nutrition (calcium, vitamin D, protein, fat mass, lean mass), and lifestyle (alcohol, smoking, physical activity) as well as the effects of pathologies and medications. It is equally valid to consider pathogenesis under these major headings, and that is the approach adopted in the following chapters. Thus, a matrix of multiple influences on multiple skeletal targets is established, and these interactions ultimately determine the fracture risk of a given individual. In the clinic, those elements of pathogenesis that are quantifiable can be combined to produce a numerical estimate of fracture risk, using one of the available indices [19, 20]. These analyses indicate that age, gender, body weight, bone density, and fracture history are the critical determinants of fracture risk, with further contributions from falls, smoking, alcohol intake, glucocorticoid use, height, family history of fracture, and the presence of other pathologies [21]. Again, this reflects the interaction of multiple factors on multiple end points in the determination of skeletal fragility.

REFERENCES

1. Reid IR, Ames R, Evans MC, Sharpe S, Gamble G, France JT, Lim TM, Cundy TF. 1992. Determinants of total body and regional bone mineral density in normal postmenopausal women—A key role for fat mass. *J Clin Endocrinol Metab* 75(1): 45–51.
2. Reid IR, Ames RW, Evans MC, Sharpe SJ, Gamble GD. Determinants of the rate of bone loss in normal postmenopausal women. *J Clin Endocrinol Metab* 1994;79(4): 950–4.
3. Lindsay R. 1988. Prevention and treatment of osteoporosis with ovarian hormones. *Ann Chir Gynaecol* 77 (5–6): 219–23.
4. Kanis JA, Johnell O, Oden A, Jonsson B, De Laet C, Dawson A. 2000. Risk of hip fracture according to the World Health Organization criteria for osteopenia and osteoporosis. *Bone* 27(5): 585–90.
5. Van Staa TP, Laan RF, Barton IP, Cohen S, Reid DA, Cooper C. 2003. Bone density threshold and other predictors of vertebral fracture in patients receiving oral glucocorticoid therapy. *Arthritis Rheum* 48(11): 3224–9.
6. Tang SY, Allen MR, Phipps R, Burr DB, Vashishth D. 2009. Changes in non-enzymatic glycation and its association with altered mechanical properties following 1-year treatment with risedronate or alendronate. *Osteoporos Int* 20(6): 887–94.
7. Kelly PJ, Hopper JL, Macaskill GT, Pocock NA, Sambrook PN, Eisman JA. 1991. Genetic factors in bone turnover. *J Clin Endocrinol Metab* 72(4): 808–13.
8. Schwartz AV, Vittinghoff E, Bauer DC, Hillier TA, Strotmeyer ES, Ensrud KE, Donaldson MG, Cauley JA, Harris TB, Koster A, Womack CR, Palermo L, Black DM. 2011. Association of BMD and FRAX score with risk of fracture in older adults with type 2 diabetes. *JAMA* 305: 2184–92.
9. Allen MR, Iwata K, Phipps R, Burr DB. 2006. Alterations in canine vertebral bone turnover, microdamage accumulation, and biomechanical properties following 1-year treatment with clinical treatment doses of risedronate or alendronate. *Bone* 39(4): 872–9.
10. Allen MR, Reinwald S, Burr DB. 2008. Alendronate reduces bone toughness of ribs without significantly increasing microdamage accumulation in dogs following 3 years of daily treatment. *Calcif Tissue Int* 82(5): 354–60.
11. Follet H, Viguet-Carrin S, Burt-Pichat B, Depalle B, Bala Y, Gineyts E, Munoz F, Arlot M, Boivin G, Chapurlat RD, Delmas PD, Bouxsein ML. 2011. Effects of preexisting microdamage, collagen cross-links, degree of mineralization, age, and architecture on compressive mechanical properties of elderly human vertebral trabecular bone. *J Orthop Res* 29: 481–8.
12. Parfitt AM, Mathews CHE, Villanueva AR, Kleerekoper M, Frame B, Rao DS. 1983. Relationships between surface, volume, and thickness of iliac trabecular bone in aging and osteoporosis. *J Clin Invest* 72: 1396–409.

13. Garnero P, Sornay-Rendu E, Claustrat B, Delmas PD. 2000. Biochemical markers of bone turnover, endogenous hormones and the risk of fractures in postmenopausal women: The OFELY study. *J Bone Miner Res* 15(8): 1526–36.
14. Rogers A, Hannon RA, Eastell R. 2000. Biochemical markers as predictors of rates of bone loss after menopause. *J Bone Miner Res* 15(7): 1398–404.
15. Zebaze RMD, Ghasem-Zadeh A, Bohte A, Iuliano-Burns S, Mirams M, Price RI, Mackie EJ, Seeman E. 2010. Intracortical remodelling and porosity in the distal radius and post-mortem femurs of women: A cross-sectional study. *Lancet* 375(9727): 1729–36.
16. Faulkner KG, Cummings SR, Black D, Palermo L, Gluer CC, Genant HK. 1993. Simple measurement of femoral geometry predicts hip fracture: The study of osteoporotic fractures. *J Bone Miner Res* 8(10): 1211–7.
17. Reid IR, Chin K, Evans MC, Jones JG. 1994. Relation between increase of hip axis in older women in 1950s and 1990s and increase in age specific rates of hip fracture. *BMJ* 309: 508–9.
18. Ahlborg HG, Johnell O, Turner CH, Rannevik G, Karlsson MK. 2003. Bone loss and bone size after menopause. *N Engl J Med* 349(4): 327–34.
19. Nguyen ND, Frost SA, Center JR, Eisman JA, Nguyen TV. 2007. Development of a nomogram for individualizing hip fracture risk in men and women. *Osteoporos Int* 18(8): 1109–17.
20. Kanis JA, Johnell O, Oden A, Johansson H, McCloskey E. 2008. FRAX and the assessment of fracture probability in men and women from the UK. *Osteoporos Int* 19(4): 385–97.
21. Bolland MJ, Siu ATY, Mason BH, Horne AM, Ames RW, Grey AB, Gamble GD, Reid IR. 2011. Evaluation of the FRAX and Garvan fracture risk calculators in older women. *J Bone Miner Res* 26(2): 420–7.

42

Nutrition and Osteoporosis

Connie M. Weaver and Robert P. Heaney

Introduction 361
Role of Diet In Building Peak Bone Mass 361
Role of Nutrition In Maintaining Bone Mass 362
Dietary Patterns 363

Dietary Bioactive Constituents 364
Conclusions 364
References 364

INTRODUCTION

Nutrients are essential to the viability of all cells including those in bone. However, it is the whole diet rather than individual nutrients that determines many factors that influence bone, including nutrition adequacy of all essential nutrients, the presence or absence of inhibitors to absorption and utilization of individual nutrients, the energy available for growth and maintenance of bone and adiposity, and the acid–base balance. Diet and other lifestyle choices around the world lead to shortages of some nutrients that are particularly important to bone (i.e., calcium, protein, and vitamin D). Our ability to accurately link and quantify the role of individual nutrients or whole diet in building and maintaining bone is handicapped by methodological limitations in assessing dietary intakes and the time lag for seeing consequences of diet on bone. Nutrition is an important component for treating those with osteoporosis, as addressed in another chapter. Diet, including dietary supplements, may be the most important complement to drugs for combination therapy. However, the more important role of diet is preventive. The cumulative effect of diet over the life span influences development of peak bone mass and its subsequent maintenance. Osteoporosis has been called a pediatric disorder because adult peak bone mass is largely determined during childhood.

ROLE OF DIET IN BUILDING PEAK BONE MASS

Rapid skeletal growth occurs in infancy and adolescence. During growth, there is a high demand for nutrients. For bone mineral matrix formation, calcium, phosphorus, and magnesium are particularly important. Vitamin D status is essential for active calcium absorption across the gut, and many nutrients are vital for collagen synthesis, including protein, copper, zinc, and iron. Meeting nutrient needs is easier during infancy through breastfeeding or carefully developed infant formulas. With the exception of its vitamin D content, the nutrient profile of breast milk is relatively constant and nearly independent of the diet of mothers [1]. In contrast, the pubertal growth spurt occurs at a life stage where diet becomes increasingly influenced by peers. This is an extremely important period for development of peak bone mass. During the 4 years surrounding peak bone mass accretion, approximately 40% of peak bone mass is acquired. Peak bone mass velocity determined from a longitudinal study in white boys and girls was 409 g/day in boys and 325 g/day in girls [2]. During peak bone mass accrual, controlled feeding studies on a range of calcium intakes in black and white girls showed that calcium intake explained 12.3% of the variance in skeletal calcium retention compared with the 13.7% explained by race,

---

Primer on the Metabolic Bone Diseases and Disorders of Mineral Metabolism, Eighth Edition. Edited by Clifford J. Rosen.
© 2013 American Society for Bone and Mineral Research. Published 2013 by John Wiley & Sons, Inc.

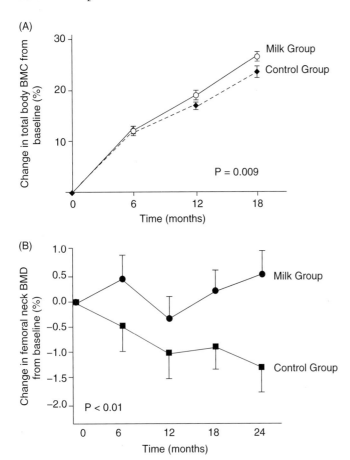

Fig. 42.1. (A) Milk supplementation in RCTs increases bone accrual in growing girls (Ref. 9), and (B) reduces bone loss in postmenopausal women (Ref. 25).

whereas a measure of sexual maturity explained an additional 4% [3]. The large contribution of just one nutrient [see Fig. 42.1(A)] emphasizes the importance of nutrition at this life stage. It also shows the large genetic influence on bone as race is a crude marker of genotype [4]. Interestingly, blacks do not require higher calcium intakes to achieve greater peak bone mass accrual than whites [3], and white boys do not need higher calcium intakes to achieve greater peak bone mass accrual than white girls [5], Chinese-American girls and boys have lower calcium requirements to maximize skeletal calcium accretion during puberty than do black and white adolescents [6]. Although, not many diet–gene interactions have been identified for bone accrual, calcium absorption efficiency has been associated with the Fok1 polymorphism of the Vitamin D receptor [7]. Genes associated with bone size are likely candidates for genetic differences in peak bone mass, as most racial variations in bone mineral content in early puberty disappear when adjusted for bone area [8]. Residual differences disappear when further adjusted for dairy calcium intake and physical activity [2].

In addition to providing raw materials for growth, diet can alter regulators of growth that affect bone accretion. In a randomized, controlled trial of a pint of milk a day in early pubertal girls, serum IGF-1 increased, which was thought to be partly causal for the increase in bone mineral density (BMD) in the intervention group relative to the control group [9]. Of the factors measured, serum IGF-1 was the greatest predictor of calcium retention after calcium intake in adolescent white boys; calcium intake predicted 21.7% and serum IGF-1 predicted 11.5% of calcium retention [10].

Diet can alter timing of menarche, perhaps through modulating growth hormones. This was shown in a study of girls 7.9 years of age who were randomized to products fortified with dairy minerals and followed for 16 years until approximate peak mass had been achieved [11]. Girls who had received the mineral complex, although only for 1 year, achieved menarche almost 5 months earlier on average than the control group. Earlier menarche with longer exposure to estrogen resulted in greater bone accretion at six skeletal sites.

The concern for low peak bone mass is risk of fracture, especially later in life. Fracture risk is also of concern in childhood, particularly during the period of relatively low BMD when bone consolidation lags behind growth [12]. Fracture incidence in childhood has dramatically increased over the past three decades [13], and the risk of fracture during childhood is compounded by the rapid increase in childhood obesity and suboptimal calcium intakes. In adolescence, as calcium intake increases, calcium retention increases to a plateau, but the plateau of calcium retention increases with increasing body mass index [14]. Thus, suboptimal calcium intakes may limit development of proportionally larger skeletons needed to reduce fracture risk in overweight and obese children.

The persistence of effects of short-term dietary interventions throughout life is controversial. The advantage in bone gain with supplementation of a calcium-rich source during randomized, controlled trials largely disappeared on follow-up after cessation of the intervention in some trials [15] but not others [16]. Additionally, the possibility of catch-up growth has been raised in a randomized controlled calcium supplementation trial from prepuberty to peak bone mass [17]. Although final BMD was not different on average for total body or radius, total hip BMD showed no catch-up growth, nor did taller women [18, 19]. Evidence of the lifelong consequences of diet early in life comes from those with eating disorders who fail to recover BMD [19].

ROLE OF NUTRITION IN MAINTAINING BONE MASS

Bone mass is ultimately determined by the genetic program as modified by current and past mechanical loading and limited or permitted by nutrition. The genetic potential cannot be reached or maintained if intake and absorption of essential nutrients are insufficient.

Calcium is the principal cation of bone mineral. Bone constitutes a very large nutrient reserve for calcium, which, over the course of evolution, acquired a secondary, structural function that is responsible for its importance for osteoporosis. Bone strength varies as the approximate second power of bone structural density. Accordingly, any decrease in bone mass produces a corresponding decrease in bone strength. The aggregate total of bone resorptive activity is controlled systemically by parathyroid hormone (PTH), which in turn responds to the demands of extracellular fluid calcium ion homeostasis and not to the structural need for bone mass. Whenever absorbed calcium intake is insufficient to meet the demands of growth and/or the drain of cutaneous and excretory losses, resorption will be stimulated and bone mass will be reduced as the body scavenges the calcium released in bone resorption. Whereas reserves are designed to be used in times of need, such use would normally be temporary. Sustained, unbalanced withdrawals deplete the reserves and thereby reduce bone strength. The principal skeletal role of calcium once peak bone mass has been achieved is to offset obligatory losses of calcium through sweat, desquamated skin, and excreta.

In addition to depleting or limiting bone mass, low calcium intakes directly cause fragility through the PTH-stimulated increase in bone remodeling. Resorption pits on trabeculae cause applied loads to shift to adjacent bone, leading to increased strain locally. In this way excessive remodeling is itself a fragility factor, altogether apart from its effect on bone mass. When adequate calcium is absorbed, PTH-stimulated remodeling decreases immediately [20] and with it, fragility. Fracture rate responses in the major treatment trials show the predicted prompt reduction in fracture risk [21, 22].

DIETARY PATTERNS

Recommended food patterns around the world attempt to meet national nutrient requirements and to promote health and reduce risk of disease. The food group most associated with bone health is the dairy group. This food group provides between 20% and 75% of recommended calcium, protein, phosphorus, magnesium, and potassium. Recommended intakes of dairy products around the world are two to three servings per day. Intakes on average are much less than the recommended levels for most populations. Dairy intakes in much of the world have been low persistently, possibly related to the high incidence of lactose maldigestion. Milk consumption has declined over the past half century in the United States, concurrent with increased consumption of soft drinks [23]. Adequacy of milk intake has been associated with adequacy of a number of nutrients in children including calcium, potassium, magnesium, zinc, iron, riboflavin, vitamin A, folate, and vitamin D [24]. Alternative sources to replace this whole package of nutrients are not typically consumed in sufficient amounts to replace milk [25]. Milk consumption within various cultures has been positively associated with bone health. High dairy-consuming regions have better bone measures than low dairy-consuming regions in Yugoslavia [26] and China [27]. A two-year randomized controlled trial (RCT) of milk (1,200 mg Ca/day) in 173 postmenopausal Chinese women reduced loss of femoral neck BMD relative to a control group [28] [See Fig. 42.1(B)]. Milk avoiders have higher risk of fracture than milk-drinking counterparts in children [29] and adults [30]. Retrospective studies show that milk drinking in childhood is inversely associated with risk of hip fracture later in life [31]. A European panel concluded that most individuals can tolerate 12 g of lactose in a single dose [32]. A National Institute of Health Consensus Conference concluded that the majority of people who self-identify as lactose malabsorbers do not have clinical lactose intolerance [33]. Unnecessary avoidance of dairy due to perceived lactose intolerance can predispose individuals to decrease bone accrual [34].

Two other food groups have also been associated with bone health by affecting the acid–base balance: fruits and vegetables (positively), and meats, fish, and poultry (negatively). Sulfur-rich amino acids from the meat groups favor an acidic ash that increases calciuria, whereas fruits and vegetables favor an alkaline ash largely because of potassium content, which has a protective effect against calcium loss in the urine. However, recent studies have challenged this paradigm. A meta-analysis of balance studies found no evidence for the hypothesis that dietary protein leads to negative calcium balance despite increased hypercalciuria [35]. The hypercalciuria is offset by increased calcium absorption or decreased endogenous secretion [36, 37]. Protein intake negatively predicts age-related bone loss, and protein supplements decrease fracture rates in the elderly [38]. A protein and calcium interaction has been identified [39] through retrospective analysis of an RCT. Subsequently, subjects in the Framingham Study with the highest tertile of animal protein intake who consumed less than 800 mg calcium/day had 2.8 times the risk of hip fracture as those in the lowest tertile of protein (p = 0.02), whereas higher protein intake was associated with decreased fracture incidence when calcium intakes were greater than 800 mg/day [40]. Similar to dietary proteins, potassium reduces calciuria, but it is offset by decreased calcium absorption so there is no change in calcium balance [41]. Some vegetables and herbs have the ability to decrease bone resorption, but the effect is independent of the alkaline load or potassium content [42].

Dietary salt is the largest predictor of urinary calcium excretion [43]. Sodium and calcium share transport proteins in the kidney. In adolescence, less sodium, and consequently less calcium, is excreted by black girls compared with white girls, presumably because of racial differences in renal transport [44].

DIETARY BIOACTIVE CONSTITUENTS

There is increasing interest in bioactive constituents that can replace or reduce the dose of drug therapies for ameliorating bone loss, or that can promote bone health in a preventative manner. Typically, these bioactive constituents must be added to the food supply or taken as dietary supplements, as effective doses may not be achievable in amounts naturally present in foods. There is much to learn about the nature and dose of bioactive constituents, their mechanisms of action, and under what conditions they may be effective.

One category of dietary bioactive constituents that has been studied more than other categories is the flavonoids that bind to estrogen receptors, notably soybean isoflavones [45]. Epidemiological evidence of soy isoflavone consumption in Asia is associated with reduced risk of hip fracture. But in Western countries, soy consumption is low so that the consumption of soy isoflavones is typically through dietary supplements. There is little support for soy isoflavone supplements as currently marketed to assess postmenopausal bone loss, although pure genistein aglycone [46] and higher doses have shown benefit [47].

Flavonoids from plant sources other than soy with promise in *in vitro* and in animal models are those in dried plums and blueberries. Osteoporosis and other chronic diseases are being considered as inflammatory disorders. To the extent that flavonoids can modify reactive oxygen species and redox status in bone cells involved in the regulation of bone turnover and survival of osteoblasts, osteoclasts, and osteocytes, they may have a protective role in the diet. Feeding blueberries activates the Wnt-β catenin signaling pathway to increase osteoblastogenesis and mineral apposition rate [48]. A mixture of phenolic acids could induce the same effects as whole blueberries [49]. Dried plums improved BMD of the ulna and spine more than dried apples at 100 g/day in a 1-year trial of postmenopausal women [50].

Certain carbohydrates and fibers that are fermented in the lower gut are of interest for their mineral-enhancing capacity with subsequent benefits to bone [51]. Upon fermentation of these prebiotics by gut microflora, short chain fatty acids are produced that may solubilize minerals and influence microbiota increasing the proportion of bifidobacteria.

CONCLUSIONS

Bone health rests on a combination of mechanical loading and adequate intakes of a broad array of macro- and micronutrients. Three important essential nutrients for bone health are calcium, vitamin D, and protein. Most diets that are inadequate in one key nutrient will be inadequate in several. Optimal protection of bone requires a diet rich in all the essential nutrients. Mononutrient supplementation regimens will often be inadequate to ensure optimal nutritional protection of bone health. Some bioactive ingredients may improve bone health by reducing chronic inflammation.

REFERENCES

1. Prentice A, Laskey A, Jarjon LMA. 1999. Lactation and bone development: Implications for calcium requirement of infants and lactating mothers. In: Bonjour JP, Tsang RC (eds). *Nutrition and Bone Development*. Philadelphia: Lippincott-Raven. pp. 127–145.
2. Bailey DA, McKay HA, Mirwald RL, Crocker PRE, Faulkner KA. 1999. A six-year longitudinal study of the relationship of physical activity to bone mineral accrual in growing children: The University of Saskatchewan bone mineral accrual study. *J Bone Miner Res* 14: 1672–1679.
3. Braun M, Palacios C, Wigertz K, Jackman LA, Bryant KJ, McCabe LD, Martin BR, McCabe GP, Peacock M, Weaver CM. 2007. Racial differences in skeletal calcium retention in adolescent girls on a range of controlled calcium intakes. *Am J Clin Nutr* 85: 1657–1663.
4. Walker, MD, Novotny R, Bilezikian JP, Weaver CM. 2008. Race and diet interactions in the acquisition, maintenance, and loss of bone. *J Nutr* 138: 1256S–1260S.
5. Braun MM, Martin BR, Kern M, McCabe GP, Peacock M, Jiang Z, Weaver CM. 2006. Calcium retention in adolescent boys on a range of controlled calcium intakes. *Am J Clin Nutr* 84: 414–418.
6. Wu L, Martin BR, Braun MM, Wastney ME, McCabe GP, McCabe LD, DiMeglio LA, Peacock M, Weaver CM. 2010. Calcium requirements and metabolism in Chinese American boys and girls. *J Bone Miner Res* 25(8): 1842–1849.
7. Ames SK, Ellis KJ, Gunn SR, Copeland KC, Abrams SA. 1999. Vitamin D receptor gene Fok1 polymorphism predicts calcium absorption and bone mineral density in children. *J Bone Miner Res* 14: 740–746.
8. Weaver CM, McCabe LD, McCabe GP, Novotny R, Van Loan M, Going S, Matkovic V, Boushey C, Savaiano DA; ACT Research Team. 2007. Bone mineral and predictors of bone mass in white, Hispanic, and Asian early pubertal girls. *Calcif Tissue Int* 81(5): 352–363.
9. Cadogan J, Eastell R, Jones N, Barker ME. 1997. Milk intake and bone mineral acquisition in adolescent girls: Randomized, controlled intervention trial. *BMJ* 315: 1255–1260.
10. Hill K, Braun MM, Kern M, Martin BR, Navalta J, Sedlock D, McCabe LD, McCabe GP, Peacock M, Weaver CM. Predictors of calcium retention in adolescent boys. *J Clin Endocrinol Metab* 93(12): 4743–4748.
11. Chevalley T, Rizzoli R, Hans D, Ferrari S, Bonjour J-P. 2005. Interaction between calcium intake and menarcheal age on bone mass gain: An eight-year follow-up study from prepuberty to postmenarche. *J Clin Endocrinol Metab* 90: 44–51.

12. Bailey DA, Martin AD, McKay AA, Whiting S, Mirwald R. 2000. Calcium accretion in girls and boys during puberty: A longitudinal analysis. *J Bone Miner Res* 15: 2245–2250.
13. Khosla S, Melton LJ 3rd, Dekutoski MB, Achenbach SJ, Oberg AL, Riggs BL. 2003. Incidence of childhood distal forearm fractures over 30 years. *JAMA* 290: 1479–1485.
14. Hill KM, Braun MM, Egan KA, Martin BR, McCabe LD, Peacock M, McCabe GP, Weaver CM. 2011. Obesity augments calcium-induced increases in skeletal calcium retention in adolescents. *J Clin Endocrinol Metab* 96: 2171–2177.
15. Lee WTK, Leung SSF, Leung DMY, Cheng JCY. 1996. A follow-up study on the effect of calcium-supplement withdrawal and puberty on bone acquisition of children. *Am J Clin Nutr* 64: 71–77.
16. Ghatge KD, Lambert HL, Barker ME, Eastell R. 2001. Bone mineral gain following calcium supplementation in teenage girls is preserved two years after withdrawal of the supplement. *J Bone Miner Res* 16: S173.
17. Matkovic V, Goel PK, Badenhop-Stevens NE, Landoll JD, Li B, Ilich JZ, Skugor M, Nagode L, Mobley SL, Ha EJ, Hangartner T, Clairmont A. 2005. Calcium supplementation and bone mineral density in females from childhood to young adulthood: A randomized controlled trial. *Am J Clin Nutr* 81: 175–188.
18. Matkovic V, Landoll JD, Badenhop-Stevens NE, Ha Y-Y, Crnevic-Orlic Z, Li B, Goel P. 2004. Nutrition influences skeletal development from childhood to adulthood: A study of hip, spine, and forearm in adolescent females. *J Nutr* 134: 701S–705S.
19. Biller BMK, Caughlin JF, Sake V, Schoenfeld D, Spratt DI, Klitanski A. 1991. Osteopenia in women with hypothalamic amenorrhea: A prospective study. *Obstet Gynecol* 78: 996–1001.
20. Wastney ME, Martin BR, Peacock M. Smith D, Jiang XY, Jackman LA, Weaver CM. 2000. Changes in calcium kinetics in adolescent girls induced by high calcium intake. *J Clin Endocrinol Metab* 85: 4470–4475.
21. Chapuy MC, Arlot ME, Duboeuf F, Brun J, Crouzet B, Arnaud S, Delmas PD, Meunier PJ. 1992. Vitamin D3 and calcium to prevent hip fractures in elderly women. *N Engl J Med* 327: 1637–1642.
22. Dawson-Hughes B, Harris SS, Krall EL, Dallal GE. 1997. Effect of calcium and vitamin D supplementation on bone density in men and women 65 years of age or older. *N Engl J Med* 337: 670–676.
23. U.S. Department of Health and Human Services and U.S. Department of Agriculture. 2005. *Dietary Guidelines for Americans 2005, 6th Ed.* Washington, DC: U.S. Government Printing Office.
24. Ballow C, Kuester S, Gillespie C. 2000. Beverage choices affect adequacy of children's nutrient intakes. *Arch Pediatr Adolesc Med* 154: 1148–1152.
25. Gao X, Wilde PE, Lichtenstein AH, Tucker KL. 2006. Meeting adequate intake for dietary calcium without dairy foods in adolescents aged 9 to 18 years (National Health and Nutrition Examination Survey 2001–2002). *J Am Diet Assoc* 106: 1759–1765.
26. Matkovic V, Kostial K, Siminovic I, Buzina R, Brodarec A, Nordin BEC. 1979. Bone status and fracture rates in two regions of Yugoslavia. *Am J Clin Nutr* 32: 540–549.
27. Hu J-F, Zbao X-H, Jia J-B, Parpia B, Campbell TC. 1993. Dietary calcium and bone density among middle-aged and elderly women in China. *Am J Clin Nutr* 58: 219–227.
28. Chee, WSS, Suriah, AR, Chan, SP, Zaitun, Y, Chang, YM. 2003. The effect of milk supplementation on bone mineral density in postmenopausal Chinese women living in Malaysia. *Osteoporos Int* 14: 828–834.
29. Goulding A, Rockell JE, Black RE, Grant AM, Jones IE, Williams SM. 2004. Children who avoid drinking cow's milk are at increased risk for prepubertal bone fractures. *J Am Diet Assoc* 104: 250–253.
30. Honkanen R, Kroger H, Alhava E, Turpeinen P, Tuppurainen M, Suarikoski S. 1997. Lactose intolerance associated with fractures in weight-bearing bones in Finnish women aged 38–57 years. *Bone* 21: 473–477.
31. Kalkwarf HJ, Khoury JC, Lanphear BP. 2003. Milk intake during childhood and adolescence, adult bone density, and osteoporotic fractures in US women. *Am J Clin Nutr* 77: 257–265.
32. European Food Safety Authority. 2010. Scientific opinion on lactose thresholds in lactose intolerance and galactosaemia. *EFSA Journal* 8(9): 1777.
33. Suchy FJ, Brannon PM, Carpenter TO, Fernandez JR, Gilsanz V, Gould JB, Hall K, Hui SL, Lupton J, Mennella J, Miller NJ, Osganian SK, Sellmeyer DE, Wolf MA. 2010. NIH Consensus Development Conference Statement: Lactose Intolerance and Health. *NIH Consens State Sci Statements* 27(2): 1–27.
34. Matlik L, Savaiano D, McCabe G, VanLoan M, Blue CL, Boushey CJ. 2007. Perceived milk intolerance is related to bone mineral content in 10- to 13-year-old female adolescents. *Pediatrics* 120: 3669.
35. Fenton TR, Eliasziu M, Lyon AN, Tough SC, Hanley DA. 2008. Meta-analysis of the quality of calcium excretion associated with the net acid excretion of the modern diet under the acid-ash diet hypothesis. *Am J Clin Nutr* 88: 1159–1166.
36. Kerstetter JE, O'Brien KO, Caseria DM, Wall DE, Insogna KL. 2005. The impact of dietary protein on calcium absorption and kinetic measures of bone turnover in women. *J Clin Endocrinol Metab* 90: 26 31.
37. Spence LA, Lipscomb ER, Cadogan J, Martin B, Wastney ME, Peacock M, Weaver CM. 2005. The effect of soy protein and soy isoflavones on calcium metabolism and renal handling in postmenopausal women: A randomized cross over study. *Am J Clin Nutr* 81: 916–922.
38. Bonjour JP. 2005. Dietary protein: An essential nutrient for bone health. *J Am Coll Nutr* 24: 526S–536S.
39. Dawson-Hughes B, Harris SS. 2002. Calcium intake influences the association of protein intake with rates of bone loss in elderly men and women. *Am J Clin Nutr* 75(4): 773–779.
40. Sahini S, Apples A, McLean RR, Tucker K, Broe KE, Kiel DP, Handon MT. 2010. Protective effect of high protein

40. and calcium intake on the risk of hip fracture in the Framingham offspring cohort. *J Bone Miner Res* 25: 2494–2500.
41. Rafferty K, Davies KM, Heaney RP. 2005. Potassium intake and the calcium economy. *J Am Coll Nutr* 24: 99–106.
42. Muhlbauer RC, Lozano A, Reiuli A. 2002. Onion and a mixture of vegetables, salads, and herbs affect bone resorption in the rat by a mechanism independent of their base exceeds. *J Bone Miner Res* 17: 1230–1236.
43. Nordin BE, Need AG, Morris HA, Horowitz M. 1993. The nature and significance of the relationship between urinary sodium and urinary calcium in women. *J Nutr* 123: 1615–1622.
44. Wigertz K, Palacios C, Jackman LA, Martin BR, McCabe LD, McCabe GP, Peacock M, Pratt JH, Weaver CM. 2005. Racial differences in calcium retention in response to dietary salt in adolescent girls. *Am J Clin Nutr* 81: 845–850.
45. North American Menopause Society. 2011. The role of soy isoflavones in menopausal health: report of The North American Menopause Society/Wulf H. Utian Translational Science Symposium in Chicago, IL (October 2010). *Menopause* 18(7): 732–753.
46. Marini H, Minutoli L, Polito F, Bitto A, Altavilla D, Ateritano M, Guadio A, Mazzaferro S, Frisina A, Lubrano C, Bonacluto M, D'Anna R, Cannata ML, Conado F, Adamo EB, Wilson S, Squadrito F. 2008. OPG and sRANKL serum concentrations in osteopenic, postmenopausal women after 2-year genistein administration. *J Bone Miner Res* 23: 715–720.
47. Weaver CM, Martin BR, Jackson GS, McCabe GP, Nolan JR, McCabe LD, Barnes S, Reinwald S, Boris ME, Peacock, M. 2009. Antiresorptive effects of phytoestrogen supplements compared to estradiol or risedronate in postmenopausal women Using 41Ca methodology. *J Clin Endocrin Metabol* 94(10): 3798–3805.
48. Chen JR, Lazarenko OP, Wu X, Kang J, Blackburn ML, Shankar K, Badyer TM, Ronis MJJ. 2010. Diet induced serum phenolic acids promote bone growth via p38 MAPK/β-catenin canonical Wnt signaling. *J Bone Miner Res* 25: 2399–2411.
49. Sellappan S, Akoh CC, Krewer G. 2002. Phenolic compounds and antioxidant capacity of Georgia-grown blueberries and blackberries. *J Agric Food Chem* 50: 2432–2438.
50. Hooshmand S, Chai SC, Saadat RL, Payton ME, Brummel-Smith K, Arjmandi BH. 2011. Comparative effects of dried plum and dried apple on bone in postmenopausal women. *Br J Nutr* 106(6): 923–30; doi:10.1017/S000711451100119X
51. Park, Clara Y, Weaver, Connie M. 2011. Calcium and bone health: Influence of prebiotics. *Functional Food Reviews* 3(2): 62–72.

43

The Role of Sex Steroids in the Pathogenesis of Osteoporosis

Matthew T. Drake and Sundeep Khosla

Introduction 367
Changes in Bone Mass and Structure with Aging 367
Sex Steroids and Bone Loss in Women 369
Sex Steroids and Bone Loss in Men 370
Osteoporosis and Non-Sex Steroid Hormone Changes with Aging 372
References 372

INTRODUCTION

Significant bone loss occurs with aging in both men and women, leading to alterations in skeletal microarchitecture and increased fracture incidence [1]. Much work has now significantly enhanced our understanding of the role that sex steroids (primarily estrogen and testosterone) play in the development and progression of osteoporosis in both men and women.

CHANGES IN BONE MASS AND STRUCTURE WITH AGING

The composite DXA data shown in Fig. 43.1 demonstrates that at the menopausal transition, women undergo rapid trabecular bone loss [2]. While somewhat variable in length, this period of accelerated bone loss extends for approximately 5–10 years, with loss of approximately 20–30% of trabecular bone, but only 5–10% of cortical bone. Following this initial phase of rapid bone loss, a second phase of slow and continuous bone loss becomes predominant. In this phase, which extends throughout the remaining female life span, cortical and trabecular bone loss occur at approximately equal rates. In comparison, men from middle life onward show slow progressive trabecular and cortical bone loss nearly equivalent to the latter phase seen in postmenopausal women. However, as men do not have a menopausal equivalent, the early accelerated phase of bone loss seen in women does not occur; thus overall loss of both trabecular and cortical bone is less in men compared to women.

Recent work has challenged the prevailing idea that there is relative preservation of skeletal integrity in both men and women from the end of puberty until middle age. Such work has been predicated upon studies performed using quantitative computed tomography (QCT), which permits measurement of volumetric bone mineral density (vBMD) and distinguishes trabecular and cortical components much more precisely than DXA imaging, which only provides areal BMD assessment. Thus, cross-sectional QCT studies show that in the spine (composed primarily of trabecular bone), large decreases (−55% in women and −45% in men) in vBMD occur beginning in the third decade (Fig. 43.2) [3]. In contrast, cortical vBMD (assessed at the distal radius) in the same cohort shows little change in either sex until midlife. Thereafter, roughly linear declines in cortical bone occur in both sexes, although cumulative decreases are greater in women than men (28% vs 18%, $p < 0.01$), reflecting the period of rapid bone loss seen in early menopausal women. Multivariate analysis in this same cohort showed that changes in trabecular microarchitecture at the wrist were most closely associated with reductions in sex steroid levels in subjects older than 60 years of age [4, 5]. Importantly, these cross-sectional findings have now been confirmed by longitudinal studies [6] in which vBMD was followed at both the distal radius and tibia,

---

Primer on the Metabolic Bone Diseases and Disorders of Mineral Metabolism, Eighth Edition. Edited by Clifford J. Rosen.
© 2013 American Society for Bone and Mineral Research. Published 2013 by John Wiley & Sons, Inc.

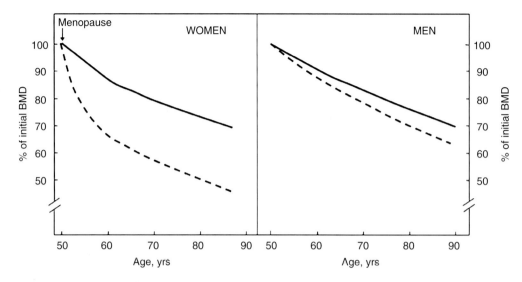

Fig. 43.1. Patterns of age-related bone loss in women and men. Dashed lines: trabecular bone; solid lines: cortical bone. The figure is based on multiple cross-sectional and longitudinal studies using DXA. (Reprinted with permission from Elsevier from Khosla S, Riggs BL. 2005. Pathophysiology of age-related bone loss and osteoporosis. *Endocrinol Metab Clin North Am* 34: 1015–1030. Copyright 2005.)

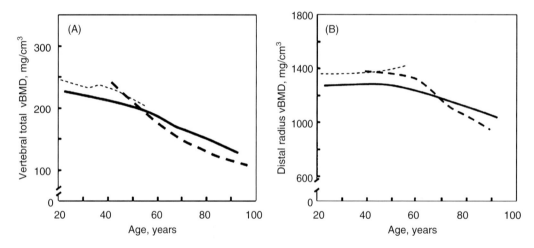

Fig. 43.2. (A) Values for vBMD (mg/cm^3) of the total vertebral body in a population sample of Rochester, MN, women and men between the ages of 20 and 97. Thin-dashed line: premenopausal women; thick-dashed line: postmenopausal women; solid line: men. (B) Values for cortical vBMD at the distal radius in the same cohort. Line coding as in (A). All changes with age were significant ($p < 0.05$). (Adapted with permission of the American Society for Bone and Mineral Research from Riggs BL, Melton LJ, Robb RA, Camp JJ, Atkinson EJ, Peterson JM, Rouleau PA, McCollough CH, Bouxsein ML, Khosla S. 2004. Population based study of age and sex differences in bone volumetric density, size, geometry, and structure at different skeletal sites. *J Bone Miner Res* 19: 1945–1954.)

with substantial loss of trabecular bone found to start shortly after completion of puberty in both men and women, an age range during which sex steroid levels are considered to be normal. The relative contribution of bone loss during these years to future skeletal fragility remains to be determined.

In addition to changes in bone mass that occur with aging, changes in bone cross-sectional area also occur in both sexes across skeletal sites. Notably, despite a net decrease in cortical area and thickness resulting from endocortical resorption, concomitant outward cortical displacement due to ongoing periosteal apposition has been demonstrated to occur in both men and women. This net outward displacement increases bone strength for bending stresses, and partially offsets the decrease in bone strength due to cortical thinning [7].

When taken together, these changes in both bone quantity and structure in aging women and men lead to changes in the annual incidence of osteoporotic fractures (Fig. 43.3). Thus, distal forearm fractures rise markedly

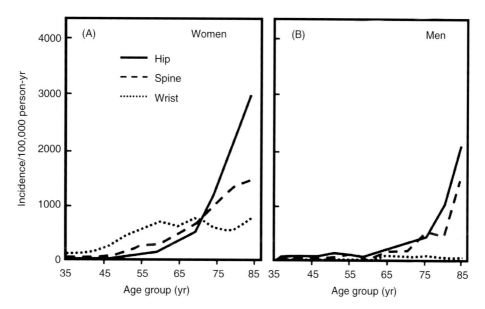

Fig. 43.3. Age specific incidence rates for proximal femur (hip), vertebral (spine), and distal forearm (wrist) fractures in Rochester, MN, women (A) and men (B). (Adapted with permission from Elsevier from Cooper C, Melton LJ. 1992. Epidemiology of osteoporosis. *Trends Endocrinol Metab* 3: 224–229. Copyright 1992.)

in women around the time of menopause and plateau approximately 15 years after menopause. Similarly, vertebral fracture incidence also begins to rise with the onset on menopause. In contrast to wrist fractures, however, vertebral fracture incidence continues to increase for the duration of the female life span. Hip fracture rates in women initially parallel those of vertebral fractures, but, as shown, rise markedly later in life.

In contrast, distal forearm fractures remain uncommon in men throughout life. Cross-sectional studies of the distal forearm in both sexes using peripheral quantitative computed tomography (pQCT) show that whereas women undergo both trabecular loss and increased trabecular spacing with aging, men begin young adult life with relatively thicker trabeculae and primarily sustain trabecular thinning rather than loss with aging [8]. Thus, in addition to having larger bones on average than women, elderly men also have comparatively more trabeculae at the distal forearm, likely contributing to the rarity of wrist fractures in aged men. Although delayed by about 1 decade relative to women, vertebral and hip fracture incidence in men is roughly equivalent to that in women, with the decade delay again likely reflecting the lack of a male menopause and the associated rapid skeletal loss seen during this period in women.

SEX STEROIDS AND BONE LOSS IN WOMEN

The relationship between diminishing estrogen levels in women caused by ovarian failure and the development of postmenopausal osteoporosis has been recognized for 7 decades [9]. Relative to premenopausal levels, serum estradiol (E2) levels decrease by 85–90%, and serum estrone (E1; a fourfold weaker estrogen) decline by 65–75% during the menopausal transition [10]. Temporally associated with this decline are changes in both bone formation and resorption rates. Whereas bone formation and resorption rates are approximately equivalent due to coupled bone remodeling prior to menopause, the onset of menopause heralds an increase in the basic multicellular unit (BMU) activation frequency rate, extension of the resorption period [11], and shortening of the formation period [12]. Accordingly, as assessed by biochemical markers, bone resorption at the menopause increases by 90%, whereas bone formation increases by only 45% [13]. This net imbalance leads to the accelerated phase of bone loss detailed above, and a net efflux of skeletal-derived calcium to the extracellular fluid. As a result, compensatory mechanisms to limit the development of hypercalcemia are employed, including increased renal calcium clearance [14], decreased intestinal calcium absorption [15], and partial suppression of parathyroid hormone (PTH) secretion [16]. Together, however, these compensatory mechanisms contribute to a net negative total body calcium balance from skeletal losses. Importantly, these compensatory effects appear directly related to estrogen deficiency, as estrogen repletion, at least in early menopause, leads to preservation of both renal calcium reclamation and intestinal calcium absorption [17].

The effects of estrogen on modulating skeletal metabolism at the cellular and molecular level remain the subject of active study. Estrogen plays a central role in osteoblast biology, where it promotes bone marrow stromal cell differentiation toward the osteoblast lineage, increases preosteoblast to osteoblast differentiation, and limits both osteoblast and osteocyte apoptosis [18–20]. In addition, estrogen increases the osteoblastic production of growth factors (IGF-1 and TGF-β) and procollagen synthesis [21, 22]. Recent work has also demonstrated that estrogen suppresses serum [23, 24] and bone marrow plasma [25] levels of the potent inhibitor of Wnt signaling, sclerostin, but further studies are needed to define the extent to which an estrogen-induced reduction in sclerostin production contributes to the effects of estrogen in maintaining bone formation. Impaired bone formation caused by estrogen deficiency becomes apparent soon after the onset of menopause. Direct evidence that pharmacologic dosages of estrogen have anabolic skeletal effects in postmenopausal women comes from histomorphometric studies of iliac crest bone biopsies from women given prolonged treatment with percutaneous estrogen, in whom a 61% increase in trabecular bone volume and a 12% increase in trabecular wall thickness relative to baseline was found following 6 years of continuous treatment [26].

In addition to its central role in regulating osteoblast biology, estrogen also plays a pivotal role in osteoclast biology. Both *in vitro* and *in vivo*, estrogen suppresses production of receptor activator of nuclear factor-κB ligand (RANKL), the central molecule in osteoclast development, from bone marrow stromal/osteoblast precursor cells, T cells, and B cells [27]. Furthermore, estrogen increases production of the soluble RANKL decoy receptor osteoprotegerin (OPG) by osteoblast lineage cells [28]. Thus, by modulating RANKL and OPG levels, estrogen limits exposure of osteoclast lineage cells (which express RANK) to RANKL, thereby effectively regulating osteoclast development. Declining estrogen levels as occur in menopausal women, however, alter the relative RANKL/OPG ratio, leading to increased osteoclast development and activity. Additional estrogen-suppressible cytokines produced by osteoblasts and bone marrow mononuclear cells, including interleukin-1 (IL-1), IL-6, tumor necrosis factor-α (TNF-α), macrophage colony stimulating factor (M-CSF), and prostaglandins also appear to play central roles in mediating bone resorption [29–32]. In support of this role for estrogen, a recent study showed that in early postmenopausal women induced to undergo acute estrogen withdrawal, pharmacologic blockade of either IL-1 or TNF-α activity was able to partially blunt a rise in bone resorption markers [33].

In addition to influencing osteoclast development, estrogen both directly and indirectly promotes the apoptosis of osteoclast lineage cells and mature osteoclasts. In osteoclast lineage cells, estrogen induces direct apoptosis through a decrease in c-jun activity, thereby limiting activator protein-1 (AP-1)-dependent transcription [34, 35]. In addition, estrogen can induce osteoblastic cell production of transforming growth factor-β (TGF-β), thereby indirectly leading to osteoclast apoptosis [11]. The importance of direct estrogen effects on osteoclast apoptosis was recently demonstrated in mice in which the estrogen receptor had been selectively deleted in osteoclasts so that the pro-apoptotic effects of estrogen deficiency on osteoclasts were lost [36, 37].

More recently, it has been noted that increases in bone resorption markers [38] and loss of spine and hip BMD [39] in perimenopausal women correlated more closely with the rise in follicle-stimulated hormone (FSH) levels than with E2 levels. Whether FSH has direct effects on bone has remained unclear, however, with evidence both for and against direct FSH effects in bone provided from rodent studies [40–42]. To address this issue directly in humans, Drake et al. recently performed a direct interventional study in women well past the menopausal transition, in whom hormonal levels other than FSH would be stable and low, to test whether FSH suppression in such postmenopausal women reduced bone resorption marker levels [43]. To suppress FSH levels, the experimental group received a gonadotropin releasing hormone (GnRH) agonist, while endogenous estrogen levels were controlled by aromatase inhibitor treatment of all subjects. In GnRH-treated subjects, FSH levels dropped rapidly into the premenopausal range, where they remained throughout the 4-month study period; as expected, FSH levels remained elevated in control subjects. Despite having markedly different circulating FSH levels, however, bone resorption [as assessed by serum C-terminal telopeptide of type I collagen (CTX) and tartrate-resistant acid phosphatase isoform type 5b (TRAP5b) levels] was not different between the groups when assessed at the study endpoint. In fact, both the control and GnRH-treated subjects had slightly increased bone resorption, likely due to the concomitant suppression of endogenous estrogen by aromatase inhibitor therapy in both groups. Together, these findings demonstrate that suppression of serum FSH levels in postmenopausal women into the premenopausal range does not reduce bone resorption.

Like estrogen, testosterone has a primary effect on bone to limit resorption, although at least part of this effect likely derives from aromatization of testosterone to estrogen [44]. *In vitro*, testosterone can both stimulate osteoblast proliferation [45], albeit weakly, and limit osteoblast apoptosis [12]. Whereas testosterone likely plays a role in increasing bone formation (and perhaps periosteal apposition) in women, there is at present little data suggesting a role of testosterone in maintaining postmenopausal skeletal integrity.

SEX STEROIDS AND BONE LOSS IN MEN

Relative to postmenopausal women, elderly men on average lose one-half as much bone and sustain one-third as many fragility fractures [16]. Although men do not undergo a hormonal menopausal equivalent, substantial changes in biologically available sex steroid levels occur over the male life span primarily as a result of the greater than twofold age-related increase in sex hormone binding globulin (SHBG) levels [46]. Whereas SHBG-bound circulating sex steroids have limited capacity to reach target tissues, free (1–3% of total) and albumin-associated (35–55% of total) sex steroids are biologically available. As a result of this rise in SHBG levels, bioavailable estrogen and testosterone levels decline an average of 47% and 64%, respectively, over the male life span, as demonstrated in a cross-sectional study of 346 men between the ages of 23 and 90 in Rochester, MN [46].

Although testosterone is the predominant sex steroid in men, evidence from both cross-sectional [47–52] and longitudinal [53] studies shows that male BMD at various skeletal sites is better correlated with circulating levels of bioavailable E2 than with testosterone. In a longitudinal study, young men (age 22–39 years) were compared with older men (age 60–90 years) over a 4-year interval to evaluate the effects of sex steroid levels on the final stages of skeletal maturation (the younger cohort) versus age-related bone loss (the older cohort) [53]. Whereas BMD in the distal forearm (which primarily reflects cortical bone changes) in the younger men increased by 0.42–0.43%/year, distal forearm BMD in the older men decreased by 0.49–0.66%/year. As shown in Table 43.1, both the increase in BMD seen in younger men and the decrease in BMD found in older men were more closely associated with bioavailable E2 levels than with testosterone levels. Perhaps most intriguingly, in older men, there appeared to be a threshold level of bioavailable E2 of approximately 40 pM (11 pg/mL) below which the rate of bone loss was most clearly associated with E2 levels. Interestingly, vBMD analysis showed this threshold effect to be more pronounced at cortical sites as compared to trabecular sites [4]. Above this level, however, there was no firm relationship between the rate of bone loss and bioavailable E2 levels. Similar findings by other investigators have been supportive of this threshold effect [54].

Although suggestive, the associations described above do not provide direct evidence for a causal role of estrogen in maintenance of the aging male skeleton. To address the relative roles of both estrogen and testosterone in skeletal maintenance in elderly men, Falahati-Nini et al. [44] used pharmacologic suppression of endogenous estrogen and testosterone production (through treatment with both a GnRH agonist and an aromatase inhibitor) in elderly men. Physiologic estrogen (E) and testosterone (T) levels were maintained by the placement of topical estrogen and testosterone patches, and baseline markers of bone resorption and formation were obtained. Subjects were then randomized to one of four groups: group A (–T, –E) discontinued both patches; group B (–T, +E) discontinued only the testosterone patch; group C (+T, –E) discontinued only the estrogen patch; and group D (+T, +E) continued both patches. As such, this direct human interventional study allowed for changes in bone metabolism resulting from either estrogen or testosterone to be determined, since suppression of endogenous sex steroid production was maintained for the entire intervention period.

As seen in Fig. 43.4(A), the significant increases in bone resorption seen in group A (–T, –E) were completely prevented by treatment with both testosterone and estrogen (group D). Whereas estrogen alone was almost completely able to prevent the rise in bone resorption (group B), however, testosterone alone was much less potent (group C). By comparison [Fig. 43.4(B)], the marked decreases in bone formation seen with dual sex steroid deficiency in group A were completely prevented by continuation of estrogen and testosterone (group D). Interestingly, serum osteocalcin levels were only slightly diminished with either estrogen or testosterone alone, whereas levels of serum amino-terminal propeptide of type I collagen (PINP) were sustained with estrogen (group B) but not testosterone. In summary, these results are consistent with a dominant role for estrogen in maintenance of skeletal integrity in aging men.

As noted above, bioavailable testosterone levels decrease to an even greater extent than bioavailable estrogen levels in aging men. Despite this decline, however, the role of testosterone in age-related bone loss in men is less clear than that of estrogen. As seen in Fig. 43.4, testosterone does have modest antiresorptive effects and also plays a role in mediating bone formation. As shown by Nair et al. [55], a statistically significant increase in BMD at the femoral neck was seen in elderly males provided low-dose testosterone replacement for 2 years. Interestingly, however, no increase in BMD was seen at any other site (anterior–posterior lumbar spine, total hip, or ultradistal radius) examined. In addition, whether the noted effects of testosterone replacement on BMD in this study were direct effects of testosterone, or were mediated through testosterone aromatization to estrogen, is unclear. Finally, testosterone likely also

Table 43.1. Spearman Correlation Coefficients Relating Rates of Change in BMD at the Radius and Ulna to Serum Sex Steroid Levels among a Sample of Rochester, MN, Men Stratified by Age

	Young		Elderly	
	Radius	Ulna	Radius	Ulna
T	–0.02	–0.19	0.13	0.14
E_2	0.33†	0.22*	0.21*	0.18*
E_1	0.35‡	0.34†	0.16	0.14
Bio T	0.13	–0.04	0.23†	0.27†
Bio E_2	0.30†	0.20	0.29†	0.33‡

*: $p < 0.05$;
†: $p < 0.01$;
‡: $p < 0.001$
T: testosterone; E2: estradiol; E1: estrone; Bio: bioavailable.
Adapted from Khosla S, Melton LJ, Atkinson EJ, O'Fallon WM. 2001. Relationship of serum sex steroid levels to longitudinal changes in bone density in young versus elderly men. *J Clin Endocrinol Metab* 86: 3555–3561. Copyright 2001, The Endocrine Society. Used with permission.

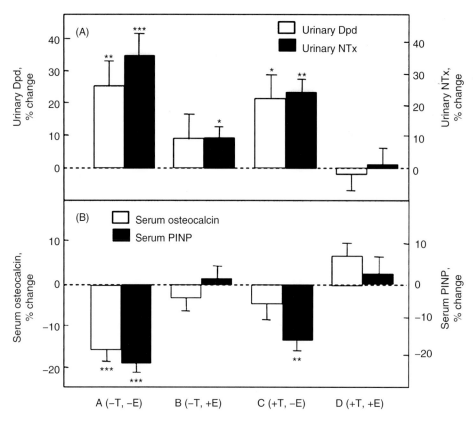

Fig. 43.4. Percent changes in (A) bone resorption markers (urinary deoxypyridinoline [Dpd] and N-telopeptide of type I collagen [NTX] and (B) bone formation markers (serum osteocalcin and N-terminal extension peptide of type I collagen [P1NP]) in a group of elderly men (mean age, 68 years) made acutely hypogonadal and treated with an aromatase inhibitor (group A), estrogen alone (group B), testosterone alone (group C), or both estrogen and testosterone (group D). Significance for change from baseline: *: $p < 0.05$; **: $p < 0.01$; ***: $p < 0.001$. (Adapted with permission from American Society for Clinical Investigation from Falahati-Nini A, Riggs BL, Atkinson EJ, O'Fallon WM, Eastell R, Khosla S. 2000. Relative contributions of testosterone and estrogen in regulating bone resorption and formation in normal elderly men. *J Clin Invest* 106: 1553–1560.)

plays a role in cortical appositional growth, although this has been most pronounced in studies of rodents [56] and may play a less important role in human biology.

OSTEOPOROSIS AND NON-SEX STEROID HORMONE CHANGES WITH AGING

Lastly, it is important to note that in addition to changes in sex steroid levels that occur with aging in both men and women, non-sex hormonal changes also occur. These include reductions in the production of growth factors important for osteoblast differentiation and function. Thus with aging, both the frequency and amplitude of growth hormone secretion is diminished [57], leading to decreased hepatic production of IGF-1 and IGF-2, an effect that may contribute to decreased bone formation with aging [58, 59]. Additionally, aging is associated with increased levels of the IGF inhibitory binding protein, IGFBP-2, which also correlates inversely with bone mass in the elderly [60]. Finally, it is likely that intrinsic changes occur in osteoblast and perhaps osteoclast lineage cells with aging [61]. These changes, which are likely independent of changes in sex steroids or other hormonal factors, are the focus of ongoing animal and human studies.

REFERENCES

1. Riggs BL, Khosla S, Melton LJ. 2002. Sex steroids and the construction and conservation of the adult skeleton. *Endocr Rev* 23: 279–302.
2. Khosla S, Riggs BL. 2005. Pathophysiology of age-related bone loss and osteoporosis. *Endocrinol Metab Clin North Am* 34: 1015–1030.
3. Riggs BL, Melton LJ, Robb RA, Camp JJ, Atkinson EJ, Peterson JM, Rouleau PA, McCollough CH, Bouxsein ML, Khosla S. 2004. Population-based study of age and sex differences in bone volumetric density, size, geometry, and structure at different skeletal sites. *J Bone Miner Res* 19: 1945–1954.

4. Khosla S, Riggs BL, Robb RA, Camp JJ, Achenbach SJ, Oberg AL, Rouleau PA, Melton LJ. 2005. Relationship of volumetric bone density and structural parameters at different skeletal sites to sex steroid levels in women. *J Clin Endocrinol Metab* 90: 5096–5103.
5. Khosla S, Melton LJ, Robb RA, Camp JJ, Atkinson EJ, Oberg AL, Rouleau PA, Riggs BL. 2005. Relationship of volumetric BMD and structural parameters at different skeletal sites to sex steroid levels in men. *J Bone Miner Res* 20: 730–740.
6. Riggs BL, Melton LJ, Robb RA, Camp JJ, Atkinson EJ, McDaniel L, Amin S, Rouleau PA, Khosla S. 2008. A population-based assessment of rates of bone loss at multiple skeletal sites: Evidence for substantial trabecular bone loss in young adult women and men. *J Bone Miner Res* 23: 205–214.
7. Seeman E. 1997. From density to structure: Growing up and growing old on the surfaces of bone. *J Bone Miner Res* 12: 509–521.
8. Khosla S, Riggs BL, Atkinson EJ, Oberg AL, McDaniel LJ, Holets M, Peterson JM, Melton LJ. 2006. Effects of sex and age on bone microstructure at the ultradistal radius: A population-based noninvasive in vivo assessment. *J Bone Miner Res* 21: 124–131.
9. Albright F, Smith PH, Richardson AM. 1941. Postmenopausal osteoporosis. *JAMA* 116: 2465–2474.
10. Khosla S, Atkinson EJ, Melton LJ, Riggs BL. 1997. Effects of age and estrogen status on serum parathyroid hormone levels and biochemical markers of bone turnover in women: A population-based study. *J Clin Endocrinol Metab* 82: 1522–1527.
11. Hughes DE, Dai A, Tiffee JC, Li HH, Mundy GR, Boyce BF. 1996. Estrogen promotes apoptosis of murine osteoclasts mediated by TGF-beta. *Nat Med* 2: 1132–1136.
12. Manolagas SC. 2000. Birth and death of bone cells: Basic regulatory mechanisms and implications for the pathogenesis and treatment of osteoporosis. *Endocr Rev* 21: 115–137.
13. Garnero P, Sornay-Rendu E, Chapuy M, Delmas PD. 1996. Increased bone turnover in late postmenopausal women is a major determinant of osteoporosis. *J Bone Miner Res* 11: 337–349.
14. Young MM, Nordin BEC. 1967. Effects of natural and artificial menopause on plasma and urinary calcium and phosphorus. *Lancet* 2: 118–120.
15. Gennari C, Agnusdei D, Nardi P, Civitelli R. 1990. Estrogen preserves a normal intestinal responsiveness to 1,25- dihydroxyvitamin D3 in oophorectomized women. *J Clin Endocrinol Metab* 71: 1288–1293.
16. Riggs BL. Khosla S, Melton LJ. 1998. A unitary model for involutional osteoporosis: Estrogen deficiency causes both type I and type II osteoporosis in postmenopausal women and contributes to bone loss in aging men. *J Bone Miner Res* 13: 763–773.
17. McKane WR, Khosla S, Burritt MF, Kao PC, Wilson DM, Ory SJ, Riggs BL. 1995. Mechanism of renal calcium conservation with estrogen replacement therapy in women in early postmenopause—A clinical research center study. *J Clin Endocrinol Metab* 80: 3458–3464.
18. Chow J, Tobias JH, Colston KW, Chambers TJ. 1992. Estrogen maintains trabecular bone volume in rats not only by suppression of bone resorption but also by stimulation of bone formation. *J Clin Invest* 89: 74–78.
19. Qu Q, Perala-Heape M, Kapanen A, Dahllund J, Salo J, Vaananen HK, Harkonen P. 1998. Estrogen enhances differentiation of osteoblasts in mouse bone marrow culture. *Bone* 22: 201–209.
20. Gohel A, McCarthy M-B, Gronowicz G. 1999. Estrogen prevents glucocorticoid-induced apoptosis in osteoblasts in vivo and in vitro. *Endocrinology* 140: 5339–5347.
21. Ernst M, Heath JK, Rodan GA. 1989. Estradiol effects on proliferation, messenger ribonucleic acid for collagen and insulin-like growth factor-I, and parathyroid hormone-stimulated adenylate cyclase activity in osteoblastic cells from calvariae and long bones. *Endocrinology* 125: 825–833.
22. Oursler MJ, Cortese C, Keeting PE, Anderson MA, Bonde SK, Riggs BL, Spelsberg TC. 1991. Modulation of transforming growth factor-beta production in normal human osteoblast-like cells by 17beta-estradiol and parathyroid hormone. *Endocrinology* 129: 3313–3320.
23. Mirza FS, Padhi ID, Raisz LG, Lorenzo JA. 2010. Serum sclerostin levels negatively correlate with parathyroid hormone levels and free estrogen index in postmenopausal women. *J Clin Endocrinol Metab* 95: 1991–1997.
24. Modder UIL, Clowes JA, Hoey K, Peterson JM, McCready L, Oursler MJ, Riggs BL, Khosla S. 2011. Regulation of circulating sclerostin levels by sex steroids in women and men. *J Bone Miner Res* 26: 27–34.
25. Modder UI, Roforth MM, Hoey K, McCready LK, Peterson JM, Monroe DG, Oursler MJ, Khosla S. 2011. Effects of estrogen on osteoprogenitor cells and cytokines/bone-regulatory factors in postmenopausal women. *Bone* 49: 202–207.
26. Khastgir G, Studd J, Holland N, Alaghband-Zadeh J, Fox S, Chow J. 2001. Anabolic effect of estrogen replacement on bone in postmenopausal women with osteoporosis: Histomorphometric evidence in a longitudinal study. *J Clin Endocrinol Metab* 86: 289–295.
27. Eghbali-Fatourechi C, Khosla S, Sanyal A, Boyle WJ, Lacey DL, Riggs BL. 2003. Role of RANK ligand in mediating increased bone resolution in early postmenopausal women. *J Clin Invest* 111: 1221–1230.
28. Hofbauer LC, Khosla S, Dunstan CR, Lacey DL, Spelsberg TC, Riggs BL. 1999. Estrogen stimulates gene expression and protein production of osteoprotegerin in human osteoblastic cells. *Endocrinology* 140: 4367–4370.
29. Jilka RL, Hangoc G, Girasole G, Passeri G, Williams DC, Abrams JS, Boyce B, Broxmeyer H, Manolagas SC. 1992. Increased osteoclast development after estrogen loss: Mediation by interleukin-6. *Science* 257: 88–91.
30. Ammann P, Rizzoli R, Bonjour J, Bourrin S, Meyer J, Vassalli P, Garcia I. 1997. Transgenic mice expressing soluble tumor necrosis factor-receptor are protected against bone loss caused by estrogen deficiency. *J Clin Invest* 99: 1699–1703.

31. Tanaka S, Takahashi N, Udagawa N, Tamura T, Akatsu T, Stanley ER, Kurokawa T, Suda T. 1993. Macrophage colony-stimulating factor is indispensable for both proliferation and differentiation of osteoclast progenitors. *J Clin Invest* 91: 257–263.
32. Kawaguchi H, Pilbeam CC, Vargas SJ, Morse EE, Lorenzo JA, Raisz LG. 1995. Ovariectomy enhances and estrogen replacement inhibits the activity of bone marrow factors that stimulate prostaglandin production in cultured mouse calvariae. *J Clin Invest* 96: 539–548.
33. Charatcharoenwitthaya N, Khosla S, Atkinson EJ, McCready LK, Riggs BL. 2007. Effect of blockade of TNF-α and interleukin-1 action on bone resorption in early postmenopausal women. *J Bone Miner Res* 22: 724–729.
34. Shevde NK, Bendixen AC, Dienger KM, Pike JW. 2000. Estrogens suppress RANK ligand-induced osteoclast differentiation via a stromal cell independent mechanism involving c-Jun repression. *Proc Natl Acad Sci U S A* 97: 7829–7834.
35. Srivastava S, Toraldo G, Weitzmann MN, Cenci S, Ross FP, Pacifici R. 2001. Estrogen decreases osteoclast formation by downregulating receptor activator of NF-kB ligand (RANKL)-induced JNK activation. *J Biol Chem* 276: 8836–8840.
36. Nakamura T, Imai Y, Matsumoto T, Sato S, Takeuchi K, Igarashi K, Harada Y, Azuma Y, Krust A, Yamamoto Y, Nishina H, Takeda S, Takayanagi H, Metzger D, Kanno J, Takaoka K, Martin TJ, Chambon P, Kato S. 2007. Estrogen prevents bone loss via estrogen receptor alpha and induction of fas ligand in osteoclasts. *Cell* 130: 811–823.
37. Martin-Millan M, Almeida M, Ambrogini E, Han L, Zhao H, Weinstein RS, Jilka RL, O'Brien CA, Manolagas SC. 2010. The estrogen receptor-alpha in osteoclasts mediates the protective effects of estrogens on cancellous but not cortical bone. *Mol Endocrinol* 24: 323–334.
38. Ebeling PR, Atley LM, Guthrie JR, Burger HG, Dennerstein L, Hopper JL, Wark JD. 1996. Bone turnover markers and bone density across the menopausal transition. *J Clin Endocrinol Metab* 81: 3366–3371.
39. Sowers MR, Jannausch M, McConnell D, Little R, Greendale GA, Finkelstein JS, Neer RM, Johnston J, Ettinger B. 2006. Hormone predictors of bone mineral density changes during the menopausal transition. *J Clin Endocrinol Metab* 91: 1261–1267.
40. Sun L, Peng Y, Sharrow AC, Iqbal J, Zhang Z, Papachristou DJ, Zaidi S, Zhu LL, Yaroslavskiy BB, Zhou H, Zallone A, Sairam MR, Kumar TR, Bo W, Braun JJ, Cardoso-Landa L, Schaffler MB, Moonga BS, Blair HC, Zaidi M. 2006. FSH directly regulates bone mass. *Cell* 125: 247–260.
41. Gao J, Tiwari-Pandey R, Samadfam R, Yang Y, Miao D, Karaplis AC, Sairam MR, Goltzman D. 2007. Altered ovarian function affects skeletal homeostasis independent of the action of follicle-stimulating hormone (FSH). *Endocrinology* 148: 2613–2621
42. Allan CM, Kalak R, Dunstan CR, McTavish KJ, Zhou H, Handelsman DJ, Seibel MJ. 2010. Follicle-stimulating hormone increases in female mice. *Proc Natl Acad Sci U S A* 107: 22629–22634.
43. Drake MT, McCready LK, Hoey KA, Atkinson EJ, Khosla S. 2010. Effects of suppression of follicle-stimulating hormone secretion on bone resorption markers in postmenopausal women. *J Clin Endocrinol Metab* 95: 5063–5068.
44. Falahati-Nini A, Riggs BL, Atkinson EJ, O'Fallon WM, Eastell R, Khosla S. 2000. Relative contributions of testosterone and estrogen in regulating bone resorption and formation in normal elderly men. *J Clin Invest* 106: 1553–1560.
45. Kasperk CH, Wergedal JE, Farley JR, Linkhart TA, Turner RT, Baylink DJ. 1989. Androgens directly stimulate proliferation of bone cells in vitro. *Endocrinology* 124: 1576–1578.
46. Khosla S, Melton LJ, Atkinson EJ, O'Fallon WM, Klee GG, Riggs BL. 1998. Relationship of serum sex steroid levels and bone turnover markers with bone mineral density in men and women: A key role for bioavailable estrogen. *J Clin Endocrinol Metab* 83: 2266–2274.
47. Slemenda CW, Longcope C, Zhou L, Hui SL, Peacock M, Johnston C. 1997. Sex steroids and bone mass in older men: Positive associations with serum estrogens and negative associations with androgens. *J Clin Invest* 100: 1755–1759.
48. Greendale GA, Edelstein S, Barrett-Connor E. 1997. Endogenous sex steroids and bone mineral density in older women and men: The Rancho Bernardo study. *J Bone Miner Res* 12: 1833–1843.
49. Center JR, Nguyen TV, Sambrook PN, Eisman JA. 1999. Hormonal and biochemical parameters in the determination of osteoporosis in elderly men. *J Clin Endocrinol Metab* 84: 3626–3635.
50. van den Beld AW, de Jong FH, Grobbee DE, Pols HAP, Lamberts SWJ. 2000. Measures of bioavailable serum testosterone and estradiol and their relationships with muscle strength, bone density, and body composition in elderly men. *J Clin Endocrinol Metab* 85: 3276–3282.
51. Amin S, Zhang Y, Sawin CT, Evans SR, Hannan MT, Kiel DP, Wilson PW, Felson DT. 2000. Association of hypogonadism and estradiol levels with bone mineral density in elderly men from the Framingham study. *Ann Intern Med* 133: 951–963.
52. Szulc P, Munoz F, Claustrat B, Garnero P, Marchand F, Duboeuf F, Delmas PD. 2001. Bioavailable estradiol may be an important determinant of osteoporosis in men: The MINOS study. *J Clin Endocrinol Metab* 86: 192–199.
53. Khosla S. Melton LJ, Atkinson EJ, O'Fallon WM. 2001. Relationship of serum sex steroid levels to longitudinal changes in bone density in young versus elderly men. *J Clin Endocrinol Metab* 86: 3555–3561.
54. Gennari L, Merlotti D, Martini G, Gonnelli S, Franci B, Campagna S, Lucani B, Canto ND, Valenti R, Gennari C, Nuti R. 2003. Longitudinal association between sex hormone levels, bone loss and bone turnover in elderly men. *J Clin Endocrinol Metab* 88: 5327–5333.
55. Nair KS, Rizza RA, O'Brien P, Dhatariya KR, Short KR, Nehra A, Vittone JL, Klee GG, Basu A, Basu R, Cobelli

C, Toffolo G, Dalla Man C, Tindall DJ, Melton LJ, Smith GE, Khosla S, Jensen MD. 2006. DHEA in elderly women and DHEA or testosterone in elderly men. *N Engl J Med* 355: 1647–1659.

56. Wakley GK, Shutte DE, Hannon KS, Turner RT. 1991. The effects of castration and androgen replacement therapy on bone: A histomorphometric study in the rat. *J Bone Miner Res* 6: 325–330.

57. Ho KY, Evans WS, Blizzard RM, Veldhuis JD, Merriam GR, Samojlik E, Furlanetto R, Rogol AD, Kaiser DL, Thorner MO. 1987. Effects of sex and age on the 24-hour profile of growth hormone secretion in man: Importance of endogenous estradiol concentrations. *J Clin Endocrinol Metab* 64: 51–58.

58. Bennett A, Wahner HW, Riggs BL, Hintz RL. 1984. Insulin-like growth factors I and II, aging and bone density in women. *J Clin Endocrinol Metab* 59: 701–704.

59. Boonen S, Mohan S, Dequeker J, Aerssens J, Vanderschueren D, Verbeke G, Broos P, Bouillon R, Baylink DJ. 1999. Downregulation of the serum stimulatory components of the insulin-like growth factor (IGF) system (IGF-I, IGF-II, IGF binding protein [BP]-3, and IGFBP-5) in age-related (type II) femoral neck osteoporosis. *J Bone Miner Res* 14: 2150–2158.

60. Amin S, Riggs BL, Atkinson EJ, Oberg AL, Melton LJ, Khosla S. 2004. A potentially deleterious role of IGFBP-2 on bone density in aging men and women. *J Bone Miner Res* 19: 1075–1083.

61. Moerman EJ, Teng K, Lipschitz DA, Lecka-Czernik B. 2004. Aging activates adipogenic and suppresses osteogenic programs in mesenchymal marrow stroma/stem cells: The role of PPAR-gamma2 transcription factor and TGF-beta/BMP signaling pathways. *Aging Cell* 3: 379–389.

44

Translational Genetics of Osteoporosis: From Population Association to Individualized Prognosis

Bich H. Tran, Jacqueline R. Center, and Tuan V. Nguyen

Introduction 376
Osteoporosis As a Complex Disease 377
Genetics of Bone Phenotypes 377
Candidate Gene Studies 378
Genome-Wide Studies 378

Clinical Application: Individualized Prognosis 381
Clinical Application: Pharmacogenetics 382
Conclusion 382
References 383

INTRODUCTION

Genetic factors have been found to play an important role in osteoporosis and fracture risk. Initial studies of monozygotic (identical) and dizygotic (nonidentical) twins and subsequently family studies demonstrated a high heritability, up to 60–80% of various measures of bone structure and a clear, albeit modest, heritability of fracture risk.

The demonstration of high heritability has led to a large body of studies aimed at identifying the genes responsible. The types of studies can be broadly divided into a candidate gene approach or genome-wide search strategy using either linkage or association studies. The candidate gene approach involves directly testing variation in genes known to be involved in bone biology for their role in osteoporosis and fracture risk. The **genome-wide search strategy** involves systematically screening all genes using DNA markers uniformly distributed throughout the entire genome. Localization is then progressively refined until a gene can be identified. **Association studies** are either population-based or case-control studies relating a polymorphism of a certain candidate gene to the desired phenotype. Association studies are also suitable for a genome-wide analysis where identified SNPs (single-nucleotide polymorphisms) are in linkage disequilibrium with the phenotype studied. **Linkage studies** relate the inheritance of genetic markers to the inheritance of phenotypes within families.

These studies have not always been very successful, for reasons related both to the complexity of the disease itself and methodological and statistical issues related to the types of studies. Osteoporosis is now widely accepted as being multifactorial with several genes involved, each having a small to moderate effect on various parameters affecting bone physiology and risk of fracture. Gene–gene interactions and gene–environment interactions potentially increase this complexity.

Linkage studies have been very successful in identifying rare monogenic diseases, and while they have been applied to complex diseases such as osteoporosis, they suffer from low statistical power when the gene effect is only modest and inherently are unable to determine the size of the genetic effect. In contrast, association studies are easier to perform, and the large numbers required for enough power to demonstrate a small effect may be easier to obtain; however, to date these studies in osteoporosis have often resulted in inconsistent results due to sample sizes that are too small, problems with population stratification, and variations in phenotype classification.

Nevertheless, the genetic studies performed over the past few decades have provided valuable insights into the pathophysiology of osteoporosis. With the decreasing costs of genome-wide scanning, consortia have been able to pool much larger populations in genome-wide association studies for complex diseases that overcome some of the statistical problems that have plagued reproducibility from earlier studies in osteoporosis. The challenge facing the researcher is whether genetic information obtained can be utilized in a clinically relevant manner that will enhance either the prediction of fracture or response to treatment after readily measurable clinical and historical factors have been taken into account.

OSTEOPOROSIS AS A COMPLEX DISEASE

While bone mineral density (BMD) is one of the best predictors of fracture risk, there is no threshold of BMD that discriminates absolutely between those who will or will not have a clinical event. Hence, a nonosteoporotic BMD is no guarantee that fracture will not occur, only the risk is relatively low. Conversely, if BMD is in the osteoporotic range, then fractures are more likely, but still may not occur. It should be noted that up to 50% of women and 70% of men with a fracture do not have osteoporotic BMD [1]. This suggests that factors other than BMD play important role in the determination of fracture liability.

The remaining contributory factors are a combination of both familial and nonfamilial factors. Indeed, compelling evidence of genetic determination of fracture risk and BMD has accumulated over the past 30 years or so. Women whose mothers have experienced a hip fracture exhibit a twofold increase in risk of hip fracture compared with controls [2], but the penetrance is not complete. Although fracture risk segregates within families, the segregation does not follow the genetic laws seen in single-gene Mendelian disorders [3].

The involvement of multiple familial and nonfamilial factors suggests that osteoporosis is a complex disease. Like many other multifactorial diseases, osteoporosis is determined by environmental factors, by genetic susceptibility, and likely by the interaction between these factors. Genetic variations do not necessarily cause osteoporosis or fracture, but they can influence an individual's susceptibility to specific environmental factors and so modify the disease risk. The familial effect on fracture risk can be characterized by the heritability of fracture risk and its risk factors.

GENETICS OF BONE PHENOTYPES

The assessment of the genetic contribution to osteoporosis has largely been based on biometric methods (e.g., complex segregation analysis), family studies, and associations with candidate genes. At the population level, the key measure of the genetic influence on a trait is the index of heritability, which is defined as the extent to which genetic individual differences contribute to individual differences in the observed trait. Because the difference between individuals is often quantified in terms of variance, the index of heritability can be operationally interpreted as the proportion of phenotypic variance attributable to genetic factors.

Using the biometric approach, it has been shown consistently that the liability to fracture is partly determined by genetic factors. In a Finnish twin study, approximately 35% of the variance in the liability to fracture (in both males and females) was attributable to genetic factors [4]. In a recent family study, approximately 25% of the liability to one fracture type, i.e., Colles' fracture of the wrist, was attributable to genetic factors [5]. Familial analysis within the Study of Osteoporotic Fracture [2] suggests that women whose mothers had had a hip fracture, had a twofold increase in the risk of hip fracture compared with controls. The risk of hip or other fractures was threefold higher with a paternal history of wrist fracture. In two small studies of osteoporotic women with vertebral or hip fractures, their daughters had bone density deficits intermediate between their mothers' and "expected" bone density at the site of their mothers' fracture, i.e., lumbar spine or proximal femur [6, 7]. Similar observations have been made in both elderly men and women [8].

Variation in BMD among individuals is also largely determined by genetic factors. It has been estimated from twin studies that 70% to 80% of variance of BMD measured at the lumbar spine and femoral neck is attributable to genetic factors [9–11]. The heritability of forearm BMD appears to be lower than that in either the femoral neck or lumbar spine [12, 13]. In these studies, there is evidence for pleiotropic effects, i.e., BMD in various skeletal sites being determined by both common and site-specific sets of genes [14, 15].

Change in BMD during adult life is the result of the net imbalance between bone formation and bone resorption. These are typically assessed by measurements (in blood or urinary excretion) of various products of osteoblast (bone formation) and osteoclasts (bone resorption) cell activity. Genetic factors have been shown to contribute significantly to the interindividual variance of bone formation markers (both osteocalcin and collagen C-terminal propeptide of type 1 collagen) in premenopausal twins [15–17].

Other markers of fracture risk have also been found to be heritable. For example, the genetic influence on different types of quantitative ultrasound (QUS) measurements, namely broadband ultrasound attenuation (BUA) and speed of sound (SOS), has been shown to be 0.53 to 0.82 [18]. BUA measurements have been reported to be more strongly correlated between mothers and their postmenopausal, rather than their premenopausal, daughters, i.e., the reverse of what has been reported for dual energy X-ray absorptiometry (DXA) measurements [19, 20]. These observations suggest that different genetic influences act on components of the bone phenotype as measured using

QUS and DXA. There are no data on heritability of SOS measures along cortical bone. Genetic correlations observed between transmission QUS and BMD measurements have been moderate, 0.32 to 0.59 [18]. Thus, genes that influence variation in BMD might, but not necessarily, influence variation in QUS, and vice versa. This is consistent with QUS measuring additional non-density characteristics of bone. In any case, a significant part of the variability of QUS and DXA BMD measurements appears unrelated, consistent with their assessment of some distinct bone phenotypic characteristics.

CANDIDATE GENE STUDIES

The recognition that fracture risk and various bone-related traits are largely determined by genetic factors has led to an intensive search for specific genes linked with either these traits or with fracture risk. Gene-search studies have been based on the two major approaches, namely, candidate gene and genome-wide studies [21]. The candidate gene approach has been used within the context of association studies, in which a specific DNA variant or variants are analyzed in cases and unrelated controls. This is a straightforward design that has been used extensively in the field of osteoporosis, but it also suffers from a number of shortcomings. The selection of appropriate controls can be a challenge, particularly for fracture that occurs mainly in later life. The arbitrary classification of BMD into a dichotomous variable (e.g., osteoporosis versus nonosteoporosis) does not exclude the possibility that the individual will develop "osteoporosis" in the future. Furthermore, any statistically significant association between a specific gene variant and fracture may not necessarily indicate a causative relationship, because such an association can have arisen from linkage disequilibrium and population stratification [22].

Based on this commonly used approach, several gene polymorphisms (including vitamin D receptor, collagen type Iα1, osteocalcin, IL-1 receptor antagonist, calcium-sensing receptor, α2HS glycoprotein, osteopontin, osteonectin, estrogen receptor α, interleukin-6, calcitonin receptor, collagen type Iα2, parathyroid hormone, and transforming growth factor α1 polymorphisms) have been proposed [23] (Table 44.1). However, the decade in which candidate gene association studies have blossomed has also been accompanied by increasing frustration with conflicting findings and a lack of independent replication, mainly due to a lack of statistical power [24] and to false positives [25].

GENOME-WIDE STUDIES

Instead of focusing on a biologically plausible candidate gene, genome-wide studies scan the entire genome to identify chromosomal regions harboring genes likely to influence a trait. There are two analytic strategies can be used in a genome-wide study: linkage analysis and association analysis. A genome-wide study, using either linkage or association analysis, is essentially is a hypothesis-free approach, because it makes no assumptions about the location and functional significance of associated loci or their products [26]. Genome-wide studies can examine the linkage or association between representative tagging-SNPs spreading throughout the genome and the susceptibility of diseases. As such, genome-wide studies can overcome weaknesses of the candidate gene design [27, 28] by providing a holistic picture of genes that are likely to contribute to the susceptibility of disease. However, genome-wide studies can have a major problem of multiplicity of hypothesis testing. Because genome-wide studies test hypotheses for hundreds of thousands of alleles, a statistically significant linkage or association is not necessarily a real phenomenon [27].

Genome-wide linkage studies aim at identifying anonymous markers that cosegregate in families with multiple affected members. Linkage analysis is based on the principle that alleles at loci close together will tend to be inherited jointly; therefore, if a marker allele and a locus for a specific trait are close together on the same chromosome they are unlikely to be separated by crossover during meiosis. A common metric of linkage is the log-of-the-odds (LOD) score, which measures the likelihood of linkage compared with no linkage. A linkage with a LOD score of greater than 3.6 is considered significant at the genome-wide level, while a LOD score greater than 2.2 is usually taken to indicate the presence of "suggestive linkage" [29].

Linkage analysis has been successfully used to map causative genes linked to BMD and fracture. Genome-wide linkage analysis has identified several regions that are linked to variation in BMD (Table 44.2). Furthermore, from linkage analysis of data from a family with osteoporosis-pseudoglioma syndrome (OPPG), a disorder characterized by severely low bone mass and eye abnormalities, investigators were able to localize the OPPG locus to chromosomal region 11q12-13 [30]. At the same time, a genome-wide linkage analysis of an extended family with 22 members, among whom 12 had very high bone mass (HBM), suggested that the HBM locus also located within the 30cM region of the same locus [31]. In follow-up studies using the positional candidate approach, both research groups found that a gene encoding the low-density lipoprotein receptor-related protein 5 (LRP5) was linked to both OPPG and high bone mass [32–34]. The finding that the LRP5 gene was linked to HBM was subsequently confirmed in a family study that included individuals with exceptionally high BMD but who were otherwise phenotypically normal [33]. This study showed that a missense mutation (G171V) was found in individuals with high BMD [34]. A recent family study further identified six novel mutations in the LRP5 gene among 13 confirmed polymorphisms that had been associated with different conditions characterized by

Table 44.1. Candidate Genes for Osteoporosis

Gene	Gene Name	Location	Reference
ARHGEF3	Rho guanine nucleotide exchange factor 3	3p14-p21	[91]
COL1A1	Collagen type I alpha 1	17q21.33	[92, 93]
CYP19A1	Cytochrome P450, family 19, subfamily A, polypeptide 1	15q21.1	[94]
DBP	Vitamin D binding protein	19q13.3	[95]
ESR1	Estrogen receptor 1	6q25.1	[96–98]
ESR2	Estrogen receptor 2	14q	[99]
FLNB	Filamin B, beta	3p14.3	[100]
FOXC2	Forkhead box C2	16q24.4	[101, 102]
ITGA1	Integrin, alpha 1	5q11.2	[48, 103]
LRP4	LDL receptor-related protein 4	11p11.2	[52]
LRP5	LDL receptor-related protein 5	11q13.4	[48, 104]
MHC	Major histocompatibility complex	6p21	[46, 52]
MTHFR	Methylenetetrahydrofolate reductase	1p36.3	[105]
PTH	Parathyroid hormone	11p15.3-p15.1	[106, 107]
RHOA	Ras homologue gene family, member A	3p21.3	[108]
SFRP1	Secreted frizzled-related protein 1	8p12-p11.1	[109]
SOST	Sclerosteosis	17q11.2	[48, 57, 110]
SPP1	Secreted phosphoprotein 1 (osteopontin)	4q21-q25	[48]
TNFSF11	Tumor necrosis factor ligand superfamily, member 11 (RANKL)	13q14	[48, 111]
TNFRSF11A	Tumor necrosis factor ligand superfamily, member 11a, NFκB activator (RANKL)	18q22.1	[48, 51]
TNFRSF11B	Tumor necrosis factor ligand superfamily, member 11b (OPG)	8q24	[48, 112]
VDR	Vitamin D receptor	12q13.11	[113–115]
WNT10B	Wingless-type MMTV integration site family, member 10B	12q13	[116]
ZBTB40	Zinc finger and BTB domain-containing protein 40	1p36	[48]

Table 44.2. Linkage Studies of the Quantitative Trait Loci (QTL) for BMD

Phenotype	Locus/Marker	LOD Score	Reference
Hip BMD	1p36 (D1S540)	3.51	[64]
Spinal BMD	2p23-24 (D2S149)	2.07	
Forearm BMD	2p21 (D2S2141, D2S1400, D2S405)	2.15	[117]
Distal forearm	13q34 (D12S788, D13S800)	1.67	
Hip BMD	6p21 (D6S2427)	2.93	[118]
Spinal BMD	12q24 (D12S395)	2.08	
Trochanteric BMD	21qter (D21S1446)	2.34	
Spinal BMD	3p21	2.1-2.7	[119]
Whole body BMD	1p36	2.4	
Spinal BMD	4q32 (D4S413)	2.12	[120]
	7p22 (D7S531)	2.28	
	12q24 (D12S1723)	2.17	
Wrist BMD	4q32 (D4S413)	2.53	
Spinal BMD	1q21-23 (D1S484)	3.11	[121]
	6p11-12 (D6S462)	1.94	
	11q12-13 (D11S987)	1.97	
	22q12-13 (D22S423)	2.13	
Hip BMD	5q33-35 (D5S422)	1.87	

increased BMD [35]. The conditions included endosteal hyperostosis, van Buchem disease, autosomal dominant osteosclerosis, and osteopetrosis type I. It is thus reasonable to state that the discovery of the LRP5 gene has opened up a new chapter of research in the genetics of osteoporosis.

Genome-wide association (GWA) studies examine hundreds of thousands of common SNPs (minor allele frequency greater than 5%) to identify chromosomal regions harboring genes likely to influence a trait. The basic idea is to test for differences in allelic frequency of anonymous genetic variants between cases and controls. In the presence of hundreds of thousands of tests, there is the real possibility of false positive findings. Just as with the evaluation of a diagnostic test, where one needs to know specificity, sensitivity, and positive predictive value (PPV), the reliability of a statistical association can also be evaluated by three analogous parameters: the observed P-value, the observed power (sensitivity) given an effect size, and the prior probability of a true association [36]. The P-value is equivalent to the false positive rate of a diagnostic test; it is the probability of observing the current data (or more extreme data) given that there is no true association. Power is the probability that a study will identify a true association if it exists. Prior probability is a subjective probability of a true association. Based on these three parameters and by using the Bayesian approach, it is possible to determine the probability of no true association given a statistically significant finding or the false positive report probability (FPRP) [37]. Of the three parameters for evaluating FPRP, the prior probability is the most difficult parameter to put a weight on. This probability is dependent on the number of gene variants that affect fracture susceptibility, which is unknown. Indeed, we do not know how many genes are involved in the regulation of, or are relevant to, the underlying susceptibility to osteoporotic fracture. However, we do know that in the human genome, there are about 3 billion base pairs [38], and that on average, more than 90% of the differences between any two individuals is due to common variants where both alleles are present in at least 1% of the population [39]. Therefore, it has been hypothesized that the susceptibility to common diseases such as osteoporosis is caused by a few common genetic variants with low effect size (i.e., the "common gene–common variant" hypothesis) [40]. Under this hypothesis, it has been estimated that the number of genetic variants that are associated with a common disease is about 100 or fewer [41]. It has also been estimated that the number of common variants in the human population is about 10 million [42]. Therefore, it may be reasonable to assume that the probability that a randomly selected common variant is associated with the risk of fracture is 1/100,000 or 0.000001. It has thus been suggested that a claim of association from a GWA study can be made if P-value is less than 5×10^{-5} [43] or 5×10^{-8} [44].

The earliest GWA study in osteoporosis examined the association between 71K genetic variants and BMD measured at different skeleton sites, and found evidence of association for 40 SNPs. Although the study was then considered to be underpowered, several SNPs identified in this study were located in potential osteoporosis-associated genes, such as *MTHFR*, *ESR1*, *LRP5*, *VDR*, and *COL1A1* genes [45]. Another GWA study screened 300K variants in an Icelandic population, and found that variants in the *ZBTB40*, *ESR1*, *OPG*, *RANKL* genes, and those in a novel region 6p21 were significantly associated with BMD at the genome-wide threshold ($P < 5 \times 10^{-8}$) [46]. This study also suggested some loci associated with fracture risk, including variants in the 1p36, 2p16, *OPG*, *MHC*, *LRP4*, and *RANK*. In the meantime, a GWA study in U.K. and Rotterdam cohorts found that variants in the *TNFRSF11B* and *LRP5* genes were associated with BMD, whereas the *LRP5* gene was also associated with fracture risk [47].

Two meta-analyses of GWA studies showed that variants in the *ZBTB40*, *ESR1*, *LRP4*, *LRP5*, *TNFSF11*, *SOST*, and *TNFRSF11A* genes were associated with BMD [48], and that variants in the *LRP5*, *SOST*, and *TNFRSF11A* were associated with fracture risk [48]. Overall, results from GWA studies and meta-analyses indicate that genes involved in the receptor activator of nuclear factor-κB– RANK ligand– osteoprotegerin (RANK–RANKL–OPG) pathway (*TNFRSF11B*, *TNFRSF11A*, and *TNFSF11* genes), the Wnt-β-catenin pathway (*LRP5*, *LRP4*, and *SOST* genes), the estrogen endocrine pathway (*ESR1* gene), and the 1p36 region (*ZBTB40* gene) are those strongly associated with osteoporosis (Table 44.3).

The RANK–RANKL–OPG pathway includes the *TNFRSF11A* and *TNFSF11* genes encoding the RANK and RANKL, respectively. The genes are members of the tumor necrosis factor (TNF) superfamily. Binding of the RANKL to RANK stimulates the formation and differentiation of osteoclasts, which in turn regulates bone resorption [49]. However, OPG, which is an inhibitor of this pathway by binding to RANKL through the TNF-receptor, may prevent the interaction between RANKL and RANK. Genetic variants within the *TNFRSF11A* gene [50] and *TNFRSF11* [50] and *OPG* [50, 51] have been found to be associated with BMD and fracture in both candidate gene association studies and GWA studies [46, 52].

The Wnt-β-catenin signaling pathway contributes to the process of bone formation by regulating the differentiation and proliferation of osteoblasts, and bone mineralization. Expression of the *LRP5* gene, which is an element of this pathway, transfers signal into the nucleus for activating the bone formation. The *LRP5* gene has been identified as a candidate gene for BMD [53] and fracture risk [47, 54]. Although several polymorphisms in the *LRP5* gene have been studied, variants in exon 9 (V776M) and exon 18 (A1330V) are the two most widely investigated [55]. An inhibitor of the Wnt-pathway, SOST, is also a candidate gene for osteoporosis. Expression of *SOST* has been found to inhibit bone formation by preventing the binding of Wnt to the LRP5 [56]. Polymorphisms of the *SOST* gene have been found to be associated with BMD variation in both a candidate gene study [57] and a GWA study [52].

Table 44.3. Genes Identified From Genome-Wide Association Studies

Gene	Gene name	Location and SNP	Reference
ADAMTS18	ADAM metallopeptidase with thrombospondin motif 18	16q23; Rs16945612	[122]
ALDH7A1	Aldehyde dehydrogenase 7 family member A1	5q31; Rs13182402	[123]
CTNNB1	Catenin (cadherin-associated protein), beta 1	3p22; Rs87939	[66]
CRHR1	Corticotropin releasing hormone receptor 1	17q12-q22; Rs9303521	[66]
DCDC5	Doublecortin domain containing 5	11p14; Rs16921914	[66]
FAM3C	Family with sequence similarity 3 member C	7q31; Rs7776725	[124]
FLJ42280	Putative uncharacterized protein FLJ42280	7q21; Rs4729260 Rs7781370	[66]
GPR177	G-protein coupled receptor 177	1p31; Rs1430742 Rs2566755	[66]
IL21R	Interleukin 21 receptor	16p11; Rs8057551 Rs8061992 Rs7199138	[125]
JAG1	Protein jagged-1	22p11-p12; Rs2273061	[126]
MARK3	Microtubule affinity-regulating kinase 3	14q32; Rs2010281	[46,66]
MEF2C	MADS box transcription enhancer factor 2C	5q14; Rs1366594	[66]
MHC	Major histocompatibility complex	6p21; Rs3130340	[46]
OSBPL1A	Oxysterol binding protein-like 1A	18q11; Rs7227401	[127]
PLCL1	Phospholipase C-like 1	2q33; Rs7595412	[128]
RAP1A	RAP1A, member of RAS oncogene family	1p13; Rs494453	[127]
RTP3	Receptor (chemosensory) transporting protein 3	3p21; Rs7430431	[129]
SFRP4	Secreted frizzled-related protein 4	7p14; Rs1721400	[124]
SOX6	SRY-box 6 gene	11p15; Rs297325, Rs4756846	[130]
SP7	Transcription factor 7	12q13; Rs10876432	[52]
STARD3NL	STARD3 N-terminal like	7p14-p13; Rs1524058	[66]
TBC1D8	TBC1 domain family, member 8	2q11; Rs2278729	[127]
TGFBR3	TGF-beta receptor type 3	1p33-p32; Rs7524102	[122]

The estrogen pathway. The estrogen endocrine pathway has long been known to play a key role in the formation and maintenance of bone mass [58]. The *ESR1* gene has been shown to be a candidate gene for the genetic effect on BMD [59] and fracture risk [59–61] in several populations. Moreover, polymorphisms of the *ESR1* gene have been found to be associated with bone loss [62]. Although the mechanism of effect of these genetic variants on BMD is unclear, intronic variants in this gene are thought to exert an effect on the efficiency of gene transcription [63].

The 1p36 region. The chromosomal region 1p36 has been implicated in the regulation of bone mass by linkage analysis [64, 65]. The *ZBTB40* gene situated in this region has been identified as a candidate gene for BMD variation in some recent GWA studies [46, 52] and in a meta-analysis [66]. However, the function of the *ZBTB40* in bone homeostasis was unknown.

CLINICAL APPLICATION: INDIVIDUALIZED PROGNOSIS

A major priority in osteoporosis research at present is to develop prognostic models for identifying individuals who have a high risk of fracture. Using established clinical risk factors, a number of prognostic models have recently been developed and implemented [67–69]. The predictive accuracy of these models has been less than perfect, with the area under the receiver operating characteristic curve ranging between 0.70 and 0.80 [68, 69]. Most prognostic models have low sensitivity and high specificity. Thus, there is room for further improvement of prognostic accuracy of the current models by incorporating genetic markers.

There are some major advantages for using genetic markers as a prognostic factor for fracture risk. First, since an individual's genotype is time invariant, it is easier to estimate its effect size and to incorporate its information in a prognostic model. Second, as the association between a genetic variant and fracture risk appears to be independent of clinical risk factors, the use of such a genetic marker can potentially improve the predictive value. Third, although there is no "genetic" therapy for individuals at high risk of fracture, the use of genetic markers could help segregate individuals at high risk from those at low risk of fracture, and help manage the burden of osteoporosis in the community.

It is clear from the above review that fracture risk is determined by several genes. This is perhaps not surprising given the number of complex phenotypes and the number of regulatory proteins involved in calcium, collagen, bone metabolism, bone strength and bone size. Two common features of these genes are that their allelic frequency in the general population is highly variable (ranging from 1% to 61%), and the effect size is very small, with relative risk ranging between 1.05 and 1.5. In the presence of such small effect size, it can be anticipated that the contribution of any single gene or SNP to fracture prognosis would be minimal. Even for a gene conferring a relative risk of 3, the area under the curve (AUC) attributable to this gene is barely 0.51 (in the absence of clinical risk factors). Even with five genes, each conferring a relative risk of between 2 and 3, the AUC is still 0.60, which is not useful for predicting fracture. This finding suggests that the contribution of any single gene to fracture prognosis, no matter how large the effect size is likely limited and would not be useful particularly in the clinical setting.

However, the integration of genetic profiling, either in the form of a genetic risk score or individual genes, into the current prognostic models could significantly improve the predictive accuracy of fracture risk for an individual. A recent simulation study suggested that a profile of up to 25 genes (each with relative risk of 1.1 to 1.35 and gene frequency ranging from 0.25 to 0.60) in the presence of clinical risk factors—with or without BMD—is required to achieve an AUC of 0.80, indicative of clinical usefulness [70]. Until now, very few genes have been implicated in the determination of fracture risk. A recent meta-analysis of 150 SNPs found that only five SNPs from four genes were consistently associated with fracture risk with relative risk (RR) ranging from 1.1 to 1.4 [48]. Thus, given the ongoing progress of finding new genes of osteoporosis, the prospect of using genetic profile in the prognosis of fracture is a real possibility.

The aim of individualized prognosis is to provide an accurate and reliable prognosis of fracture for an individual, and to help improve the management of the individual's predisposition to fracture. At present, individuals with low bone mineral density (i.e., T-scores being less than –2.5) or with a history of prior low trauma fracture are recommended for therapeutic intervention [71, 72]. This recommendation is logical and appropriate, since these individuals have higher risk of fracture [73, 74], and treatment can reduce this risk [75–77]. However, because fracture is a multifactorial event, there is more than one way that an individual can attain the risk conferred by either low BMD or a prior fracture [78]. Moreover, each individual is a unique case, because there is no "average individual" in the population. The uniqueness of an individual can be defined in terms of the individual's environmental and genetic profile. Thus, the knowledge of genetics, in combination with clinical risk factors, can shift our current risk stratification (i.e., "one-size-fits-all") approaches to a more individualized evaluation and treatment of osteoporosis.

CLINICAL APPLICATION: PHARMACOGENETICS

Research into the genetic background of osteoporosis and identification of osteoporosis-related genes can advance our understanding of pathophysiologic mechanisms of fracture. However, while the magnitude of association between genetic variants and osteoporosis is equivalent to what has been found for some clinical risk factors, there is currently no additional advice for people who carry a high-risk gene profile. However, there is an opportunity for using genetic information for therapeutic treatments of osteoporosis. Pharmacogenetics aims to select the optimal strategy for patients because different persons may respond differently to the same therapy [79].

There is evidence that genetic variants in the vitamin D receptor gene (*VDR*) influence the intestinal absorption of calcium leading an effect on bone mass [80, 81]. In a longitudinal study of calcium supplementation in the elderly, participants with *VDR BsmI* variant carrying the BB genotype had a higher proportion of bone loss at the lumbar spine than those with Bb and bb genotypes [82, 83]. Women with the BB genotype were also found to have reduced efficiency of calcium absorption on a regimen of low-calcium intake [84]. These results could imply that people with BB genotype should be considered for a calcium-rich diet.

Bisphosphonates are often prescribed to osteoporotic patients to decrease bone resorption for fracture prevention [85]. However, there is a high variability among people's responses to this treatment with about 15% of people on the treatment demonstrating continuing bone loss [86]. Osteoporotic women carrying bb genotype in the *VDR BsmI* showed more bone gain while on treatment with alendronate than those with BB genotype [87]. One of the extremely rare side effects of longer term bisphosphonate treatment is osteonecrosis of the jaw. Interestingly, results from a GWA study demonstrated that people with a genetic variant in the cytochrome P450-2C gene (*CYP2C8*) have a remarkable 12-fold increase in the disorder [88].

CONCLUSION

It is well documented that genetic factors play an important role in the regulation of BMD variation and fracture risk. While it is clear that fracture risk is determined by multiple genes, it is not known how many genes are involved. The mode of inheritance is also unknown. In fact, with current methodology, it is unlikely that we will completely understand the causes of fracture, and why some individuals fracture and others do not. However, it is possible to find risk factors that account for a substantial number of cases and that are amenable to intervention. Newly identified genetic variants in combination with clinical risk factors may help improve the accuracy of fracture prognosis for an individual, segregate individuals at high risk of fracture from those with lower risk, and hence lead to better management of the burden of osteoporosis in the general community.

With a rapid improvement in genotyping technology, the next generation of GWA studies will be adding more variants at a low frequency to cover as many SNPs as possible. This will not only increases the chance of detecting true associations, but also decreases the chance of false-positive findings [89]. Moreover, there is an emerging interest in studying RNA splicing in which synonymous variants could affect regulatory elements such as translational enhancers or RNA stability [90]. These molecular approaches could potentially have significant impact on the future direction of drug discovery and development, and open the possibility of personalized regimens for fracture risk prevention.

REFERENCES

1. Nguyen ND, Eisman JA, Center JR, Nguyen TV. 2007. Risk factors for fracture in nonosteoporotic men and women. *J Clin Endocrinol Metab* 92(3): 955–62.
2. Cummings SR, Nevitt MC, Browner WS, Stone K, Fox KM, Ensrud KE, Cauley J, Black D, Vogt TM. 1995. Risk factors for hip fracture in white women. Study of Osteoporotic Fractures Research Group. *N Engl J Med* 332(12): 767–73.
3. Nguyen TV, Eisman JA. 2000. Genetics of fracture: Challenges and opportunities. *J Bone Miner Res* 15(7): 1253–6.
4. Kannus P, Palvanen M, Kaprio J, Parkkari J, Koskenvuo M. 1999. Genetic factors and osteoporotic fractures in elderly people: Prospective 25 year follow up of a nationwide cohort of elderly Finnish twins. *BMJ* 319(7221): 1334–7.
5. Deng HW, Chen WM, Recker S, Stegman MR, Li JL, Davies KM, Zhou Y, Deng H, Heaney R, Recker RR. 2000. Genetic determination of Colles' fracture and differential bone mass in women with and without Colles' fracture. *J Bone Miner Res* 15(7): 1243–52.
6. Seeman E, Hopper JL, Bach LA, Cooper ME, Parkinson E, McKay J, Jerums G. 1989. Reduced bone mass in daughters of women with osteoporosis. *N Engl J Med* 320(9): 554–8.
7. Seeman E, Tsalamandris C, Formica C, Hopper JL, McKay J. 1994. Reduced femoral neck bone density in the daughters of women with hip fractures: The role of low peak bone density in the pathogenesis of osteoporosis. *J Bone Miner Res* 9(5): 739–43.
8. Evans RA, Marel GM, Lancaster EK, Kos S, Evans M, Wong SY. 1988. Bone mass is low in relatives of osteoporotic patients. *Ann Intern Med* 109(11): 870–3.
9. Nguyen TV. 1998. Contributions of Genetics and Environmental Factors to the Determinants of Osteoporosis Fractures, PhD Thesis. Garvan Institute of Medicine, Department of Community Medicine, Faculty of Medicine. The University of New South Wales (Australia), Sydney. p. 388.
10. Pocock NA, Eisman JA, Hopper JL, Yeates MG, Sambrook PN, Eberl S. 1987. Genetic determinants of bone mass in adults. A twin study. *J Clin Invest* 80(3): 706–10.
11. Young D, Hopper JL, Nowson CA, Green RM, Sherwin AJ, Kaymakci B, Smid M, Guest CS, Larkins RG, Wark JD. 1995. Determinants of bone mass in 10- to 26-year-old females: A twin study. *J Bone Miner Res* 10(4): 558–67.
12. Flicker L, Hopper JL, Rodgers L, Kaymakci B, Green RM, Wark JD. 1995. Bone density determinants in elderly women: A twin study. *J Bone Miner Res* 10(11): 1607–13.
13. Smith DM, Nance WE, Kang KW, Christian JC, Johnston CCJ. 1973. Genetic factors in determining bone mass. *J Clin Invest* 52(11): 2800–8.
14. Nguyen TV, Howard GM, Kelly PJ, Eisman JA. 1998. Bone mass, lean mass, and fat mass: Same genes or same environments? *Am J Epidemiol* 147(1): 3–16.
15. Garnero P, Arden NK, Griffiths G, Delmas PD, Spector TD. 1996. Genetic influence on bone turnover in postmenopausal twins. *J Clin Endocrinol Metab* 81(1): 140–6.
16. Tokita A, Kelly PJ, Nguyen TV, Qi JC, Morrison NA, Risteli L, Risteli J, Sambrook PN, Eisman JA. 1994. Genetic influences on type I collagen synthesis and degradation: Further evidence for genetic regulation of bone turnover. *J Clin Endocrinol Metab* 78(6): 1461–6.
17. Harris M, Nguyen TV, Howard GM, Kelly PJ, Eisman JA. 1998. Genetic and environmental correlations between bone formation and bone mineral density: A twin study. *Bone* 22(2): 141–5.
18. Howard GM, Nguyen TV, Harris M, Kelly PJ, Eisman JA. 1998. Genetic and environmental contributions to the association between quantitative ultrasound and bone mineral density measurements: A twin study. *J Bone Miner Res* 13(8): 1318–27.
19. Arden NK, Spector TD. 1997. Genetic influences on muscle strength, lean body mass, and bone mineral density: A twin study. *J Bone Miner Res* 12(12): 2076–81.
20. Danielson ME, Cauley JA, Baker CE, Newman AB, Dorman JS, Towers JD, Kuller LH. 1999. Familial resemblance of bone mineral density (BMD) and calcaneal ultrasound attenuation: The BMD in mothers and daughters study. *J Bone Miner Res* 14(1): 102–10.
21. Nguyen TV, Blangero J, Eisman JA. 2000. Genetic epidemiological approaches to the search for osteoporosis genes. *J Bone Miner Res* 15(3): 392–401.
22. Campbell H, Rudan I. 2002. Interpretation of genetic association studies in complex disease. *Pharmacogenomics J* 2(6): 349–60.
23. Ralston SH, de Crombrugghe B. 2006. Genetic regulation of bone mass and susceptibility to osteoporosis. *Genes Dev* 20(18): 2492–506.
24. Huang QY, Recker RR, Deng HW. 2003. Searching for osteoporosis genes in the post-genome era: Progress and challenges. *Osteoporos Int* 14(9): 701–15.
25. Ioannidis JP. 2005. Why most published research findings are false. *PLoS Med* 2(8): e124.
26. Hirschhorn JN, Daly MJ. 2005. Genome-wide association studies for common diseases and complex traits. *Nat Rev Genet* 6(2): 95–108.

27. Pearson TA, Manolio TA. 2008. How to interpret a genome-wide association study. *JAMA* 299(11): 1335–44.
28. Hunter DJ, Altshuler D, Rader DJ. 2008. From Darwin's finches to canaries in the coal mine—Mining the genome for new biology. *N Engl J Med* 358(26): 2760–3.
29. Lander E, Kruglyak L. 1995. Genetic dissection of complex traits: Guidelines for interpreting and reporting linkage results. *Nat Genet* 11(3): 241–7.
30. Gong Y, Vikkula M, Boon L, Liu J, Beighton P, Ramesar R, Peltonen L, Somer H, Hirose T, Dallapiccola B, De Paepe A, Swoboda W, Zabel B, Superti-Furga A, Steinmann B, Brunner HG, Jans A, Boles RG, Adkins W, van den Boogaard MJ, Olsen BR, Warman ML. 1996. Osteoporosis-pseudoglioma syndrome, a disorder affecting skeletal strength and vision, is assigned to chromosome region 11q12-13. *Am J Hum Genet* 59(1): 146–51.
31. Johnson ML, Gong G, Kimberling W, Recker SM, Kimmel DB, Recker RB. 1997. Linkage of a gene causing high bone mass to human chromosome 11 (11q12-13). *Am J Hum Genet* 60(6): 1326–32.
32. Gong Y, Slee RB, Fukai N, Rawadi G, Roman-Roman S, Reginato AM, Wang H, Cundy T, Glorieux FH, Lev D, Zacharin M, Oexle K, Marcelino J, Suwairi W, Heeger S, Sabatakos G, Apte S, Adkins WN, Allgrove J, Arslan-Kirchner M, Batch JA, Beighton P, Black GC, Boles RG, Boon LM, Borrone C, Brunner HG, Carle GF, Dallapiccola B, De Paepe A, Floege B, Halfhide ML, Hall B, Hennekam RC, Hirose T, Jans A, Juppner H, Kim CA, Keppler-Noreuil K, Kohlschuetter A, LaCombe D, Lambert M, Lemyre E, Letteboer T, Peltonen L, Ramesar RS, Romanengo M, Somer H, Steichen-Gersdorf E, Steinmann B, Sullivan B, Superti-Furga A, Swoboda W, van den Boogaard MJ, Van Hul W, Vikkula M, Votruba M, Zabel B, Garcia T, Baron R, Olsen BR, Warman ML. 2001. LDL receptor-related protein 5 (LRP5) affects bone accrual and eye development. *Cell* 107(4): 513–23.
33. Little RD, Carulli JP, Del Mastro RG, Dupuis J, Osborne M, Folz C, Manning SP, Swain PM, Zhao SC, Eustace B, Lappe MM, Spitzer L, Zweier S, Braunschweiger K, Benchekroun Y, Hu X, Adair R, Chee L, FitzGerald MG, Tulig C, Caruso A, Tzellas N, Bawa A, Franklin B, McGuire S, Nogues X, Gong G, Allen KM, Anisowicz A, Morales AJ, Lomedico PT, Recker SM, Van Eerdewegh P, Recker RR, Johnson ML. 2002. A mutation in the LDL receptor-related protein 5 gene results in the autosomal dominant high-bone-mass trait. *Am J Hum Genet* 70(1): 11–9.
34. Boyden LM, Mao J, Belsky J, Mitzner L, Farhi A, Mitnick MA, Wu D, Insogna K, Lifton RP. 2002. High bone density due to a mutation in LDL-receptor-related protein 5. *N Engl J Med* 346(20): 1513–21.
35. Van Wesenbeeck L, Cleiren E, Gram J, Beals RK, Benichou O, Scopelliti D, Key L, Renton T, Bartels C, Gong Y, Warman ML, De Vernejoul MC, Bollerslev J, Van Hul W. 2003. Six novel missense mutations in the LDL receptor-related protein 5 (LRP5) gene in different conditions with an increased bone density. *Am J Hum Genet* 72(3): 763–71.
36. Browner WS, Newman TB. 1987. Are all significant P values created equal? The analogy between diagnostic tests and clinical research. *JAMA* 257(18): 2459–63.
37. Wacholder S, Chanock S, Garcia-Closas M, El Ghormli L, Rothman N. 2004. Assessing the probability that a positive report is false: An approach for molecular epidemiology studies. *J Natl Cancer Inst* 96(6): 434–42.
38. International HapMap Consortium. 2003. The International HapMap Project. *Nature* 426(6968): 789–96.
39. Wang WY, Barratt BJ, Clayton DG, Todd JA. 2005. Genome-wide association studies: Theoretical and practical concerns. *Nat Rev Genet* 6(2): 109–18.
40. Reich DE, Lander ES. 2001. On the allelic spectrum of human disease. *Trends Genet* 17(9): 502–10.
41. Yang Q, Khoury MJ, Friedman J, Little J, Flanders WD. 2005. How many genes underlie the occurrence of common complex diseases in the population? *Int J Epidemiol* 34(5): 1129–37.
42. Kruglyak L, Nickerson DA. 2001. Variation is the spice of life. *Nat Genet* 27(3): 234–6.
43. Colhoun HM, McKeigue PM, Davey Smith G. 2003. Problems of reporting genetic associations with complex outcomes. *Lancet* 361(9360): 865–72.
44. Risch N, Merikangas K. 1996. The future of genetic studies of complex human diseases. *Science* 273(5281): 1516–7.
45. Kiel DP, Demissie S, Dupuis J, Lunetta KL, Murabito JM, Karasik D. 2007. Genome-wide association with bone mass and geometry in the Framingham Heart Study. *BMC Med Genet* 8 Suppl 1: S14.
46. Styrkarsdottir U, Halldorsson BV, Gretarsdottir S, Gudbjartsson DF, Walters GB, Ingvarsson T, Jonsdottir T, Saemundsdottir J, Center JR, Nguyen TV, Bagger Y, Gulcher JR, Eisman JA, Christiansen C, Sigurdsson G, Kong A, Thorsteinsdottir U, Stefansson K. 2008. Multiple genetic loci for bone mineral density and fractures. *N Engl J Med* 358(22): 2355–65.
47. Richards JB, Rivadeneira F, Inouye M, Pastinen TM, Soranzo N, Wilson SG, Andrew T, Falchi M, Gwilliam R, Ahmadi KR, Valdes AM, Arp P, Whittaker P, Verlaan DJ, Jhamai M, Kumanduri V, Moorhouse M, van Meurs JB, Hofman A, Pols HA, Hart D, Zhai G, Kato BS, Mullin BH, Zhang F, Deloukas P, Uitterlinden AG, Spector TD. 2008. Bone mineral density, osteoporosis, and osteoporotic fractures: A genome-wide association study. *Lancet* 371(9623): 1505–12.
48. Richards JB, Kavvoura FK, Rivadeneira F, Styrkarsdottir U, Estrada K, Halldorsson BV, Hsu YH, Zillikens MC, Wilson SG, Mullin BH, Amin N, Aulchenko YS, Cupples LA, Deloukas P, Demissie S, Hofman A, Kong A, Karasik D, van Meurs JB, Oostra BA, Pols HA, Sigurdsson G, Thorsteinsdottir U, Soranzo N, Williams FM, Zhou Y, Ralston SH, Thorleifsson G, van Duijn CM, Kiel DP, Stefansson K, Uitterlinden AG, Ioannidis JP, Spector TD. 2009. Collaborative meta-analysis: Associations of 150 candidate genes with osteoporosis

and osteoporotic fracture. *Ann Intern Med* 151(8): 528–37.
49. Boyce BF, Xing L. 2008. Functions of RANKL/RANK/OPG in bone modeling and remodeling. *Arch Biochem Biophys* 473(2): 139–46.
50. Hsu YH, Niu T, Terwedow HA, Xu X, Feng Y, Li Z, Brain JD, Rosen CJ, Laird N. 2006. Variation in genes involved in the RANKL/RANK/OPG bone remodeling pathway are associated with bone mineral density at different skeletal sites in men. *Hum Genet* 118(5): 568–77.
51. Choi JY, Shin A, Park SK, Chung HW, Cho SI, Shin CS, Kim H, Lee KM, Lee KH, Kang C, Cho DY, Kang D. 2005. Genetic polymorphisms of OPG, RANK, and ESR1 and bone mineral density in Korean postmenopausal women. *Calcif Tissue Int* 77(3): 152–9.
52. Styrkarsdottir U, Halldorsson BV, Gretarsdottir S, Gudbjartsson DF, Walters GB, Ingvarsson T, Jonsdottir T, Saemundsdottir J, Snorradottir S, Center JR, Nguyen TV, Alexandersen P, Gulcher JR, Eisman JA, Christiansen C, Sigurdsson G, Kong A, Thorsteinsdottir U, Stefansson K. 2009. New sequence variants associated with bone mineral density. *Nat Genet* 41(1): 15–7.
53. Tran BN, Nguyen ND, Eisman JA, Nguyen TV. 2008. Association between LRP5 polymorphism and bone mineral density: A Bayesian meta-analysis. *BMC Med Genet* 9: 55.
54. Bollerslev J, Wilson SG, Dick IM, Islam FM, Ueland T, Palmer L, Devine A, Prince RL. 2005. LRP5 gene polymorphisms predict bone mass and incident fractures in elderly Australian women. *Bone* 36(4): 599–606.
55. Kiel DP, Ferrari SL, Cupples LA, Karasik D, Manen D, Imamovic A, Herbert AG, Dupuis J. 2007. Genetic variation at the low-density lipoprotein receptor-related protein 5 (LRP5) locus modulates Wnt signaling and the relationship of physical activity with bone mineral density in men. *Bone* 40(3): 587–96.
56. Balemans W, Piters E, Cleiren E, Ai M, Van Wesenbeeck L, Warman ML, Van Hul W. 2008. The binding between sclerostin and LRP5 is altered by DKK1 and by high-bone mass LRP5 mutations. *Calcif Tissue Int* 82(6): 445–53.
57. Uitterlinden AG, Arp PP, Paeper BW, Charmley P, Proll S, Rivadeneira F, Fang Y, van Meurs JB, Britschgi TB, Latham JA, Schatzman RC, Pols HA, Brunkow ME. 2004. Polymorphisms in the sclerosteosis/van Buchem disease gene (SOST) region are associated with bone-mineral density in elderly whites. *Am J Hum Genet* 75(6): 1032–45.
58. Ralston SH. 2010. Genetics of osteoporosis. *Ann N Y Acad Sci* 1192: 181–9.
59. Ioannidis JP, Stavrou I, Trikalinos TA, Zois C, Brandi ML, Gennari L, Albagha O, Ralston SH, Tsatsoulis A. 2002. Association of polymorphisms of the estrogen receptor alpha gene with bone mineral density and fracture risk in women: A meta-analysis. *J Bone Miner Res* 17(11): 2048–60.
60. van Meurs JB, Schuit SC, Weel AE, van der Klift M, Bergink AP, Arp PP, Colin EM, Fang Y, Hofman A, van Duijn CM, van Leeuwen JP, Pols HA, Uitterlinden AG. 2003. Association of 5' estrogen receptor alpha gene polymorphisms with bone mineral density, vertebral bone area and fracture risk. *Hum Mol Genet* 12(14): 1745–54.
61. Wang JT, Guo Y, Yang TL, Xu XH, Dong SS, Li M, Li TQ, Chen Y, Deng HW. 2008. Polymorphisms in the estrogen receptor genes are associated with hip fractures in Chinese. *Bone* 43(5): 910–4.
62. Salmen T, Heikkinen AM, Mahonen A, Kroger H, Komulainen M, Saarikoski S, Honkanen R, Maenpaa PH. 2000. Early postmenopausal bone loss is associated with PvuII estrogen receptor gene polymorphism in Finnish women: Effect of hormone replacement therapy. *J Bone Miner Res* 15(2): 315–21.
63. Herrington DM, Howard TD, Brosnihan KB, McDonnell DP, Li X, Hawkins GA, Reboussin DM, Xu J, Zheng SL, Meyers DA, Bleecker ER. 2002. Common estrogen receptor polymorphism augments effects of hormone replacement therapy on E-selectin but not C-reactive protein. *Circulation* 105(16): 1879–82.
64. Devoto M, Shimoya K, Caminis J, Ott J, Tenenhouse A, Whyte MP, Sereda L, Hall S, Considine E, Williams CJ, Tromp G, Kuivaniemi H, Ala-Kokko L, Prockop DJ, Spotila LD. 1998. First-stage autosomal genome screen in extended pedigrees suggests genes predisposing to low bone mineral density on chromosomes 1p, 2p and 4q. *Eur J Hum Genet* 6(2): 151–7.
65. Devoto M, Specchia C, Li HH, Caminis J, Tenenhouse A, Rodriguez H, Spotila LD. 2001. Variance component linkage analysis indicates a QTL for femoral neck bone mineral density on chromosome 1p36. *Hum Mol Genet* 10(21): 2447–52.
66. Rivadeneira F, Styrkarsdottir U, Estrada K, Halldorsson BV, Hsu YH, Richards JB, Zillikens MC, Kavvoura FK, Amin N, Aulchenko YS, Cupples LA, Deloukas P, Demissie S, Grundberg E, Hofman A, Kong A, Karasik D, van Meurs JB, Oostra B, Pastinen T, Pols HA, Sigurdsson G, Soranzo N, Thorleifsson G, Thorsteinsdottir U, Williams FM, Wilson SG, Zhou Y, Ralston SH, van Duijn CM, Spector T, Kiel DP, Stefansson K, Ioannidis JP, Uitterlinden AG. 2009. Twenty bone-mineral-density loci identified by large-scale meta-analysis of genome-wide association studies. *Nat Genet* 41(11): 1199–206.
67. Kanis JA, Johnell O, Oden A, Johansson H, McCloskey E. 2008. FRAX and the assessment of fracture probability in men and women from the UK. *Osteoporos Int* 19(4): 385–97.
68. Nguyen ND, Frost SA, Center JR, Eisman JA, Nguyen TV. 2007. Development of a nomogram for individualizing hip fracture risk in men and women. *Osteoporos Int* 18(8): 1109–17.
69. Nguyen ND, Frost SA, Center JR, Eisman JA, Nguyen TV. 2008. Development of prognostic nomograms for individualizing 5-year and 10-year fracture risks. *Osteoporos Int* 19(10): 1431–44.
70. Tran BN, Nguyen ND, Nguyen VX, Center JR, Eisman JA, Nguyen TV. 2011. Genetic profiling and

individualized prognosis of fracture. *J Bone Miner Res* 26(2): 414–9.
71. Cummings SR. 2006. A 55-year-old woman with osteopenia. *JAMA* 296(21): 2601–10.
72. Sambrook PN, Eisman JA. 2000. Osteoporosis prevention and treatment. *Med J Aust* 172(5): 226–9.
73. Nguyen ND, Pongchaiyakul C, Center JR, Eisman JA, Nguyen TV. 2005. Identification of high-risk individuals for hip fracture: A 14-year prospective study. *J Bone Miner Res* 20(11): 1921–8.
74. Dargent-Molina P, Douchin MN, Cormier C, Meunier PJ, Breart G. 2002. Use of clinical risk factors in elderly women with low bone mineral density to identify women at higher risk of hip fracture: The EPIDOS prospective study. *Osteoporos Int* 13(7): 593–9.
75. Cranney A, Guyatt G, Griffith L, Wells G, Tugwell P, Rosen C. 2002. Meta-analyses of therapies for postmenopausal osteoporosis. IX: Summary of meta-analyses of therapies for postmenopausal osteoporosis. *Endocr Rev* 23(4): 570–8.
76. Nguyen ND, Eisman JA, Nguyen TV. 2006. Anti-hip fracture efficacy of bisphosphonates: A Bayesian analysis of clinical trials. *J Bone Miner Res* 21(1): 340–49.
77. Vestergaard P, Jorgensen NR, Mosekilde L, Schwarz P. 2007. Effects of parathyroid hormone alone or in combination with antiresorptive therapy on bone mineral density and fracture risk—A meta-analysis. *Osteoporos Int* 18(1): 45–57.
78. Nguyen TV. 2007. Individualization of osteoporosis risk. *Osteoporos Int* 18(9): 1153–6.
79. Nguyen TV, Eisman JA. 2006. Pharmacogenomics of osteoporosis: Opportunities and challenges. *J Musculoskelet Neuronal Interact* 6(1): 62–72.
80. Gennari L, Merlotti D, De Paola V, Martini G, Nuti R. 2009. Update on the pharmacogenetics of the vitamin D receptor and osteoporosis. *Pharmacogenomics* 10(3): 417–33.
81. Salamone LM, Glynn NW, Black DM, Ferrell RE, Palermo L, Epstein RS, Kuller LH, Cauley JA. 1996. Determinants of premenopausal bone mineral density: The interplay of genetic and lifestyle factors. *J Bone Miner Res* 11(10): 1557–65.
82. Ferrari S, Rizzoli R, Chevalley T, Slosman D, Eisman JA, Bonjour JP. 1995. Vitamin-D-receptor-gene polymorphisms and change in lumbar-spine bone mineral density. *Lancet* 345(8947): 423–4.
83. Kiel DP, Myers RH, Cupples LA, Kong XF, Zhu XH, Ordovas J, Schaefer EJ, Felson DT, Rush D, Wilson PW, Eisman JA, Holick MF. 1997. The BsmI vitamin D receptor restriction fragment length polymorphism (bb) influences the effect of calcium intake on bone mineral density. *J Bone Miner Res* 12(7): 1049–57.
84. Dawson-Hughes B, Harris SS, Finneran S. 1995. Calcium absorption on high and low calcium intakes in relation to vitamin D receptor genotype. *J Clin Endocrinol Metab* 80(12): 3657–61.
85. Nguyen ND, Eisman JA, Nguyen TV. 2006. Anti-hip fracture efficacy of biophosphonates: a Bayesian analysis of clinical trials. *J Bone Miner Res* 21(2): 340–9.
86. Francis RM. 2004. Non-response to osteoporosis treatment. *J Br Menopause Soc* 10(2): 76–80.
87. Palomba S, Numis FG, Mossetti G, Rendina D, Vuotto P, Russo T, Zullo F, Nappi C, Nunziata V. 2003. Effectiveness of alendronate treatment in postmenopausal women with osteoporosis: Relationship with BsmI vitamin D receptor genotypes. *Clin Endocrinol (Oxf)* 58(3): 365–71.
88. Sarasquete ME, Garcia-Sanz R, Marin L, Alcoceba M, Chillon MC, Balanzategui A, Santamaria C, Rosinol L, de la Rubia J, Hernandez MT, Garcia-Navarro I, Lahuerta JJ, Gonzalez M, San Miguel JF. 2008. Bisphosphonate-related osteonecrosis of the jaw is associated with polymorphisms of the cytochrome P450 CYP2C8 in multiple myeloma: A genome-wide single nucleotide polymorphism analysis. *Blood* 112(7): 2709–12.
89. Cordell HJ, Clayton DG. 2005. Genetic association studies. *Lancet* 366(9491): 1121–31.
90. Bracco L, Kearsey J. 2003. The relevance of alternative RNA splicing to pharmacogenomics. *Trends Biotechnol* 21(8): 346–53.
91. Mullin BH, Prince RL, Dick IM, Hart DJ, Spector TD, Dudbridge F, Wilson SG. 2008. Identification of a role for the ARHGEF3 gene in postmenopausal osteoporosis. *Am J Hum Genet* 82(6): 1262–9.
92. Yazdanpanah N, Rivadeneira F, van Meurs JB, Zillikens MC, Arp P, Hofman A, van Duijn CM, Pols HA, Uitterlinden AG. 2007. The -1997 G/T and Sp1 polymorphisms in the collagen type I alpha1 (COLIA1) gene in relation to changes in femoral neck bone mineral density and the risk of fracture in the elderly: The Rotterdam study. *Calcif Tissue Int* 81(1): 18–25.
93. Grant SF, Reid DM, Blake G, Herd R, Fogelman I, Ralston SH. 1996. Reduced bone density and osteoporosis associated with a polymorphic Sp1 binding site in the collagen type I alpha 1 gene. *Nat Genet* 14(2): 203–5.
94. Riancho JA, Sanudo C, Valero C, Pipaon C, Olmos JM, Mijares V, Fernandez-Luna JL, Zarrabeitia MT. 2009. Association of the aromatase gene alleles with BMD: Epidemiological and functional evidence. *J Bone Miner Res* 24(10): 1709–18.
95. Fang Y, van Meurs JB, Arp P, van Leeuwen JP, Hofman A, Pols HA, Uitterlinden AG. 2009. Vitamin D binding protein genotype and osteoporosis. *Calcif Tissue Int* 85(2): 85–93.
96. Wang CL, Tang XY, Chen WQ, Su YX, Zhang CX, Chen YM. 2007. Association of estrogen receptor alpha gene polymorphisms with bone mineral density in Chinese women: a meta-analysis. *Osteoporos Int* 18(3): 295–305.
97. Lai BM, Cheung CL, Luk KD, Kung AW. 2008. Estrogen receptor alpha CA dinucleotide repeat polymorphism is associated with rate of bone loss in perimenopausal women and bone mineral density and risk of osteoporotic fractures in postmenopausal women. *Osteoporos Int* 19(4): 571–9.
98. Sano M, Inoue S, Hosoi T, Ouchi Y, Emi M, Shiraki M, Orimo H. 1995. Association of estrogen receptor dinu-

cleotide repeat polymorphism with ostcoporosis. *Biochem Biophys Res Commun* 217(1): 378–83.
99. Rivadeneira F, van Meurs JB, Kant J, Zillikens MC, Stolk L, Beck TJ, Arp P, Schuit SC, Hofman A, Houwing-Duistermaat JJ, van Duijn CM, van Leeuwen JP, Pols HA, Uitterlinden AG. 2006. Estrogen receptor beta (ESR2) polymorphisms in interaction with estrogen receptor alpha (ESR1) and insulin-like growth factor I (IGF1) variants influence the risk of fracture in postmenopausal women. *J Bone Miner Res* 21(9): 1443–56.
100. Wilson SG, Jones MR, Mullin BH, Dick IM, Richards JB, Pastinen TM, Grundberg E, Ljunggren O, Surdulescu GL, Dudbridge F, Elliott KS, Cervino AC, Spector TD, Prince RL. 2009. Common sequence variation in FLNB regulates bone structure in women in the general population and FLNB mRNA expression in osteoblasts in vitro. *J Bone Miner Res* 24(12): 1989–97.
101. Yamada Y, Ando F, Shimokata H. 2006. Association of polymorphisms in forkhead box C2 and perilipin genes with bone mineral density in community-dwelling Japanese individuals. *Int J Mol Med* 18(1): 119–27.
102. Yerges LM, Klei L, Cauley JA, Roeder K, Kammerer CM, Moffett SP, Ensrud KE, Nestlerode CS, Marshall LM, Hoffman AR, Lewis C, Lang TF, Barrett-Connor E, Ferrell RE, Orwoll ES, Zmuda JM. 2009. High-density association study of 383 candidate genes for volumetric BMD at the femoral neck and lumbar spine among older men. *J Bone Miner Res* 24(12): 2039–49.
103. Lee HJ, Kim SY, Koh JM, Bok J, Kim KJ, Kim KS, Park MH, Shin HD, Park BL, Kim TH, Hong JM, Park EK, Kim DJ, Oh B, Kimm K, Kim GS, Lee JY. 2007. Polymorphisms and haplotypes of integrinalpha1 (ITGA1) are associated with bone mineral density and fracture risk in postmenopausal Koreans. *Bone* 41(6): 979–86.
104. Urano T, Shiraki M, Ezura Y, Fujita M, Sekine E, Hoshino S, Hosoi T, Orimo H, Emi M, Ouchi Y, Inoue S. 2004. Association of a single-nucleotide polymorphism in low-density lipoprotein receptor-related protein 5 gene with bone mineral density. *J Bone Miner Metab* 22(4): 341–5.
105. Riancho JA, Valero C, Zarrabeitia MT. 2006. MTHFR polymorphism and bone mineral density: meta-analysis of published studies. *Calcif Tissue Int* 79(5): 289–93.
106. Tenne M, McGuigan F, Jansson L, Gerdhem P, Obrant KJ, Luthman H, Akesson K. 2008. Genetic variation in the PTH pathway and bone phenotypes in elderly women: Evaluation of PTH, PTHLH, PTHR1 and PTHR2 genes. *Bone* 42(4): 719–27.
107. Guo Y, Zhang LS, Yang TL, Tian Q, Xiong DH, Pei YF, Deng HW. 2010. IL21R and PTH may underlie variation of femoral neck bone mineral density as revealed by a genome-wide association study. *J Bone Miner Res* 25(5): 1042–8.
108. Mullin BH, Prince RL, Mamotte C, Spector TD, Hart DJ, Dudbridge F, Wilson SG. 2009. Further genetic evidence suggesting a role for the RhoGTPase-RhoGEF pathway in osteoporosis. *Bone* 45(2): 387–91.
109. Sims AM, Shephard N, Carter K, Doan T, Dowling A, Duncan EL, Eisman J, Jones G, Nicholson G, Prince R, Seeman E, Thomas G, Wass JA, Brown MA. 2008. Genetic analyses in a sample of individuals with high or low BMD shows association with multiple Wnt pathway genes. *J Bone Miner Res* 23(4): 499–506.
110. Balemans W, Foernzler D, Parsons C, Ebeling M, Thompson A, Reid DM, Lindpaintner K, Ralston SH, Van Hul W. 2002. Lack of association between the SOST gene and bone mineral density in perimenopausal women: Analysis of five polymorphisms. *Bone* 31(4): 515–9.
111. Hsu YH, Venners SA, Terwedow HA, Feng Y, Niu T, Li Z, Laird N, Brain JD, Cummings SR, Bouxsein ML, Rosen CJ, Xu X. 2006. Relation of body composition, fat mass, and serum lipids to osteoporotic fractures and bone mineral density in Chinese men and women. *Am J Clin Nutr* 83(1): 146–54.
112. Langdahl BL, Carstens M, Stenkjaer L, Eriksen EF. 2002. Polymorphisms in the osteoprotegerin gene are associated with osteoporotic fractures. *J Bone Miner Res* 17(7): 1245–55.
113. Grundberg E, Lau EM, Pastinen T, Kindmark A, Nilsson O, Ljunggren O, Mellstrom D, Orwoll E, Redlund-Johnell I, Holmberg A, Gurd S, Leung PC, Kwok T, Ohlsson C, Mallmin H, Brandstrom H. 2007. Vitamin D receptor 3' haplotypes are unequally expressed in primary human bone cells and associated with increased fracture risk: The MrOS Study in Sweden and Hong Kong. *J Bone Miner Res* 22(6): 832–40.
114. Moffett SP, Zmuda JM, Cauley JA, Ensrud KE, Hillier TA, Hochberg MC, Li J, Cayabyab S, Lee JM, Peltz G, Cummings SR. 2007. Association of the VDR translation start site polymorphism and fracture risk in older women. *J Bone Miner Res* 22(5): 730–6.
115. Morrison NA, Yeoman R, Kelly PJ, Eisman JA. 1992. Contribution of trans-acting factor alleles to normal physiological variability: Vitamin D receptor gene polymorphism and circulating osteocalcin. *Proc Natl Acad Sci U S A* 89(15): 6665–9.
116. Zmuda JM, Yerges LM, Kammerer CM, Cauley JA, Wang X, Nestlerode CS, Wheeler VW, Patrick AL, Bunker CH, Moffett SP, Ferrell RE. 2009. Association analysis of WNT10B with bone mass and structure among individuals of African ancestry. *J Bone Miner Res* 24(3): 437–47.
117. Niu T, Chen C, Cordell H, Yang J, Wang B, Wang Z, Fang Z, Schork NJ, Rosen CJ, Xu X. 1999. A genome-wide scan for loci linked to forearm bone mineral density. *Hum Genet* 104(3): 226–33.
118. Karasik D, Myers RH, Cupples LA, Hannan MT, Gagnon DR, Herbert A, Kiel DP. 2002. Genome screen for quantitative trait loci contributing to normal variation in bone mineral density: The Framingham Study. *J Bone Miner Res* 17(9): 1718–27.
119. Wilson SG, Reed PW, Andrew T, Barber MJ, Lindersson M, Langdown M, Thompson D, Thompson E, Bailey M, Chiano M, Kleyn PW, Spector TD. 2004. A

genome-screen of a large twin cohort reveals linkage for quantitative ultrasound of the calcaneus to 2q33-37 and 4q12-21. *J Bone Miner Res* 19(2): 270–7.

120. Deng HW, Xu FH, Huang QY, Shen H, Deng H, Conway T, Liu YJ, Liu YZ, Li JL, Zhang HT, Davies KM, Recker RR. 2002. A whole-genome linkage scan suggests several genomic regions potentially containing quantitative trait Loci for osteoporosis. *J Clin Endocrinol Metab* 87(11): 5151–9.

121. Koller DL, Liu G, Econs MJ, Hui SL, Morin PA, Joslyn G, Rodriguez LA, Conneally PM, Christian JC, Johnston CC Jr, Foroud T, Peacock M. 2001. Genome screen for quantitative trait loci underlying normal variation in femoral structure. *J Bone Miner Res* 16(6): 985–91.

122. Xiong DH, Liu XG, Guo YF, Tan LJ, Wang L, Sha BY, Tang ZH, Pan F, Yang TL, Chen XD, Lei SF, Yerges LM, Zhu XZ, Wheeler VW, Patrick AL, Bunker CH, Guo Y, Yan H, Pei YF, Zhang YP, Levy S, Papasian CJ, Xiao P, Lundberg YW, Recker RR, Liu YZ, Liu YJ, Zmuda JM, Deng HW. 2009. Genome-wide association and follow-up replication studies identified ADAMTS18 and TGFBR3 as bone mass candidate genes in different ethnic groups. *Am J Hum Genet* 84(3): 388–98.

123. Guo Y, Tan LJ, Lei SF, Yang TL, Chen XD, Zhang F, Chen Y, Pan F, Yan H, Liu X, Tian Q, Zhang ZX, Zhou Q, Qiu C, Dong SS, Xu XH, Guo YF, Zhu XZ, Liu SL, Wang XL, Li X, Luo Y, Zhang LS, Li M, Wang JT, Wen T, Drees B, Hamilton J, Papasian CJ, Recker RR, Song XP, Cheng J, Deng HW. 2010. Genome-wide association study identifies ALDH7A1 as a novel susceptibility gene for osteoporosis. *PLoS Genet* 6(1): e1000806.

124. Cho YS, Go MJ, Kim YJ, Heo JY, Oh JH, Ban HJ, Yoon D, Lee MH, Kim DJ, Park M, Cha SH, Kim JW, Han BG, Min H, Ahn Y, Park MS, Han HR, Jang HY, Cho EY, Lee JE, Cho NH, Shin C, Park T, Park JW, Lee JK, Cardon L, Clarke G, McCarthy MI, Lee JY, Oh B, Kim HL. 2009. A large-scale genome-wide association study of Asian populations uncovers genetic factors influencing eight quantitative traits. *Nat Genet* 41(5): 527–34.

125. Guo Y, Zhang LS, Yang TL, Tian Q, Xiong DH, Pei YF, Deng HW. 2010. IL21R and PTH may underlie variation of femoral neck bone mineral density as revealed by a genome-wide association study. *J Bone Miner Res* 25(5): 1042–8.

126. Kung AW, Xiao SM, Cherny S, Li GH, Gao Y, Tso G, Lau KS, Luk KD, Liu JM, Cui B, Zhang MJ, Zhang ZL, He JW, Yue H, Xia WB, Luo LM, He SL, Kiel DP, Karasik D, Hsu YH, Cupples LA, Demissie S, Styrkarsdottir U, Halldorsson BV, Sigurdsson G, Thorsteinsdottir U, Stefansson K, Richards JB, Zhai G, Soranzo N, Valdes A, Spector TD, Sham PC. 2010. Association of JAG1 with bone mineral density and osteoporotic fractures: a genome-wide association study and follow-up replication studies. *Am J Hum Genet* 86(2): 229–39.

127. Hsu YH, Zillikens MC, Wilson SG, Farber CR, Demissie S, Soranzo N, Bianchi EN, Grundberg E, Liang L, Richards JB, Estrada K, Zhou Y, van Nas A, Moffatt MF, Zhai G, Hofman A, van Meurs JB, Pols HA, Price RI, Nilsson O, Pastinen T, Cupples LA, Lusis AJ, Schadt EE, Ferrari S, Uitterlinden AG, Rivadeneira F, Spector TD, Karasik D, Kiel DP. 2010. An integration of genome-wide association study and gene expression profiling to prioritize the discovery of novel susceptibility Loci for osteoporosis-related traits. *PLoS Genet* 6(6): e1000977.

128. Liu YZ, Wilson SG, Wang L, Liu XG, Guo YF, Li J, Yan H, Deloukas P, Soranzo N, Chinappen-Horsley U, Cervino A, Williams FM, Xiong DH, Zhang YP, Jin TB, Levy S, Papasian CJ, Drees BM, Hamilton JJ, Recker RR, Spector TD, Deng HW. 2008. Identification of PLCL1 gene for hip bone size variation in females in a genome-wide association study. *PLoS ONE* 3(9): e3160.

129. Zhao LJ, Liu XG, Liu YZ, Liu YJ, Papasian CJ, Sha BY, Pan F, Guo YF, Wang L, Yan H, Xiong DH, Tang ZH, Yang TL, Chen XD, Guo Y, Li J, Shen H, Zhang F, Lei SF, Recker RR, Deng HW. 2010. Genome-wide association study for femoral neck bone geometry. *J Bone Miner Res* 25(2): 320–9.

130. Liu YZ, Pei YF, Liu JF, Yang F, Guo Y, Zhang L, Liu XG, Yan H, Wang L, Zhang YP, Levy S, Recker RR, Deng HW. 2009. Powerful bivariate genome-wide association analyses suggest the SOX6 gene influencing both obesity and osteoporosis phenotypes in males. *PLoS ONE* 4(8): e6827.

45
Prevention of Falls

Heike A. Bischoff-Ferrari

Introduction 389
Epidemiology of Falls and Cost of Falls 389
Fall Definition and Inclusion of Fall Risk in Fracture Risk Prediction 389
Fall Mechanics and Risk of Fracture 390
Risk Factors for Falls 390
Fall Prevention Strategies 391
Fall Prevention Strategies with Evidence for Fracture Reduction 391
Conclusion 392
References 392

INTRODUCTION

Close to 75% of hip and non-hip fractures occur among seniors age 65 and older [1]. Notably, the primary risk factor for a hip fracture is a fall, and over 90% of all fractures occur after a fall [2]. Thus, the key to understanding and preventing fractures at a later age is based on their close relationship with muscle weakness [3] and falling [4, 5]. In fact, antiresorptive treatment alone may not reduce fractures among individuals 80 years and older in the presence of nonskeletal risk factors for fractures despite an improvement in bone metabolism [6]. This chapter will review the epidemiology of falls, and their importance in regard to fracture risk. Finally, fall prevention strategies and how these translate into fracture reduction are evaluated based on data from randomized controlled trials.

EPIDEMIOLOGY OF FALLS AND COST OF FALLS

Thirty percent of those 65 years or older, and 40–50% of those 80 years or older, report having had a fall over the past year [7, 8]. Serious injuries occur with 10–15% of falls, resulting in fractures in 5% and hip fracture in 1–2% [4]. As an independent determinant of functional decline [9], falls lead to 40% of all nursing home admissions [10]. The primary risk factor for a hip fracture is a fall, and over 90% of all fractures occur after a fall [3]. Recurrent fallers may have close to a fourfold increased odds of sustaining a fall-related fracture compared to individuals with a single fall [11]. As the number of seniors aged 65 and older is predicted to increase from 25% to 40% by 2030 [12–16], the number of fall-related fractures will increase substantially. Notably, even today 75% of fractures occur among seniors age 65 and older [1], and by 2050 the worldwide incidence in hip fractures is expected to increase by 240% among women and 310% among men [17]. Because of the increasing proportion of older individuals, annual costs from all fall-related injuries in the U.S. in people 65 years or older were projected to increase from $20.3 billion in 1994 to $32.4 billion in 2020, including medical, rehabilitation, hospital costs, and the costs of morbidity and mortality [18]. Thus, therapeutic interventions that are effective in fall prevention are urgently needed.

FALL DEFINITION AND INCLUSION OF FALL RISK IN FRACTURE RISK PREDICTION

Buchner and colleagues created a useful fall definition for the common data base of the FICSIT (Frailty and Injuries: Cooperative Studies of Intervention Techniques)

Primer on the Metabolic Bone Diseases and Disorders of Mineral Metabolism, Eighth Edition. Edited by Clifford J. Rosen.
© 2013 American Society for Bone and Mineral Research. Published 2013 by John Wiley & Sons, Inc.

trials [19]. Falls were defined as "unintentionally coming to rest on the ground, floor, or other lower level." Coming to rest against furniture or a wall was not counted as a fall [19]. A challenge for their assessment is that falls tend to be forgotten if not associated with significant injury [20], requiring short periods of follow-up. Thus, high-quality fall assessment in older persons requires a prospective ascertainment of falls and their circumstances, ideally in short periods of time periods (less than 3 months) [20].

Fall reports may be performed by postcards, phone calls, or diary/calendar, although the usefulness and comprehensiveness of different ascertainment methods have not been compared directly. Notably, fall assessment has not been standardized across randomized controlled trials or large epidemiologic data sets [21], which prevented falls from being included in the WHO FRAX tool (http://www.shef.ac.uk/FRAX/) that estimates the probability of a major osteoporotic fracture in the next 10 years [21]. Based on one Australian cohort study (Dubbo study), the Garvan nomogram has been developed as an alternative fracture prediction tool that includes falling as a risk factor of fracture (www.fractureriskcalculator.com). In a comparative assessment, however, the predictive accuracy of the two tools showed similar performance in postmenopausal women and a possible advantage of the Garvan nomogram over FRAX among men [22, 23]. One explanation for a similar predictive accuracy between FRAX and the Garvan nomogram may be the relatively long time interval of fall assessment in the Garvan nomogram (fall recall in the previous 12 months), which may lead to the underreporting of falls not associated with significant injury [20].

FALL MECHANICS AND RISK OF FRACTURE

Mechanistically, the circumstances [2] and the direction [24] of a fall determine the type of fracture, whereas bone density and factors that attenuate a fall, such as better strength or better padding, critically determine whether a fracture will take place when the faller lands on a certain bone [25]. Moreover, falling may affect bone density through increased immobility from self-restriction of activities [26]. It is well known that falls may lead to psychological trauma known as fear of falling [27]. After their first fall, about 30% of individuals develop a fear of falling [26], which results in self-restriction of activities and decreased quality of life [26]. Figure 45.1 illustrates the fall–fracture construct that describes the complexity of osteoporosis prevention introduced by nonskeletal risk factors for fractures among older individuals.

In support of the a concept that falling is a key determinant of fracture risk, antiresorptive treatment alone may not reduce fractures among individuals 80 years and older in the presence of nonskeletal risk factors for fractures, despite an improvement in bone metabolism [6].

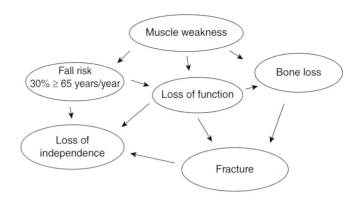

Fig. 45.1. Nonskeletal fall–fracture construct.

Further, consistent with the understanding that factors unrelated to bone are at play in fracture epidemiology, the circumstances of different fractures are strikingly different. Hip fractures tend to occur in less active individuals who fall indoors from a standing height with little forward momentum, and they tend to fall sideways or straight down on their hip [25]. On the other hand, other nonvertebral fractures, such as distal forearm or humerus fractures tend to occur among more active older individuals who are more likely to be outdoors and have a greater forward momentum when they fall [28].

Supporting the notion of bone not to be seen in isolation, fracture risk due to falling is increased among individuals with osteoarthritis of the weight-bearing joints despite having increased bone density compared to controls [29]. One prospective study found that prevalent knee pain due to osteoarthritis increased the risk of falling by 26% and the risk of hip fracture twofold [30]. A recent study applied a biomechanical risk measure, the factor-of-risk (ratio of force on the hip in a fall to femoral strength), for the prediction of hip fracture risk [31]; a 1 standard deviation increase in peak factor-of-risk was associated with a 1.88-fold increased risk of hip fracture in men and a 1.23-fold increased risk of hip fracture in women. Notably, examining the components of factor-of-risk, fall force, and soft tissue thickness were predictive of hip fracture risk independent of femoral strength estimated from bone mineral density (BMD) [31].

RISK FACTORS FOR FALLS

Falls are a hallmark of age and becoming frail, and falls are often heralded by the onset of gait instability, visual impairment or its correction by multifocal glasses, drug treatment with antidepressants, anticonvulsants/barbiturates, or benzodiazepines, weakness, cognitive impairment, vitamin D deficiency, poor mental health, home hazards, or often a combination of several risk factors [32, 33]. Some studies indicate that falls due to snow and ice may play an important role in seasonality of fractures [34, 35]. One cause of the increased fracture risk in winter compared to summer may be that older people are more likely to slip and fall during periods of snow and ice [36]. On the other hand, hip fractures, which mostly occur indoors [37], may be less affected by snow and ice, with a smaller seasonal swing compared to distal forearm, humerus, and ankle fractures [38].

The seemingly inseparable relationship of falls to worsening health status and the complexity of factors involved in falling has led to pessimism on the part of physicians when faced with falling, especially recurrent falling. However, there is a growing body of literature that should encourage the standardized assessment of falls and application of fall prevention strategies for fracture prevention.

FALL PREVENTION STRATEGIES

Fall prevention by risk factor reduction has been tested in a number of approaches. Multifactorial approaches, such as medical and occupational-therapy assessment or adjustment in medications, behavioral instructions, and exercise programs, as demonstrated in the PROFET-trial (Prevention of Falls in the Elderly Trial) [39] and FICSIT-trials [40]. Multifactorial approaches may be especially useful in high-risk populations for falls, such as older individuals in care institutions [41].

Many studies demonstrate that simple *weight-bearing exercise programs* improve gait speed, muscle strength, and balance in community-dwelling and frail seniors, which translates into fall reduction by 25% to 50% [41–43]. As falls are the primary risk factor for fractures, the rationale is that these interventions should also protect against fractures, although this needs confirmation in a large clinical study.

Significant limitations of exercise programs are their cost and high implementation time. To overcome this barrier, exercise as a strategy of fall prevention may be applied at a smaller expense when the program is instructed but unsupervised. Such a home exercise program reduced falls in a randomized trial by Campbell and colleagues among community dwelling elderly women age 80 and older [44]. Consistently, a simple unsupervised exercise home program, taught during acute care after hip fracture repair, reduced falls significantly by 25% over a 12-month follow-up among senior hip fracture patients with a mean age of 84 years [45]. Although not powered for a fracture end point, there was a suggestion that the unsupervised home exercise program contributed to reduction in repeat fractures among acute hip fracture patients (relative fracture rate difference was −56% for the exercise home program versus a control; 95% confidence interval; −82%, +9%; p-value = 0.08) [45].

As a precaution of exercise programs in frail seniors with poor balance, increased mobility may lead to an increased opportunity to fall and fracture. Therefore, exercise programs should be supervised and supported by strength and balance training.

Tai chi has been successful in reducing falls among healthy older individuals [42, 46], and physically inactive community-dwelling older individuals [47], while frail older individuals [48] and fallers [46] may not benefit as much. Furthermore, tai chi may not improve bone density [49], and fracture prevention has not been explored as an end point with tai chi intervention programs.

As an extension of exercise, programs that support dual tasking may be of great value for fall prevention. Earlier studies suggested that fall risk is increased in seniors unable to walk while talking ("stop walking when talking" [50, 51]). Thus, dual tasking assessments may best identify those at the greatest risk of falling, and programs that improve dual tasking may be useful in fall prevention at higher age. This concept was tested in a recently published trial that showed that a music-based multitask exercise program improved gait and balance and reduced fall risk significantly by 39% in community-dwelling seniors [43].

FALL PREVENTION STRATEGIES WITH EVIDENCE FOR FRACTURE REDUCTION

Two interventions among older individuals resulted in both fall and fracture reduction. One is **cataract surgery** with limited evidence from one trial, in which 306 women aged over 70, with cataracts, were randomized to expedited (approximately 4 weeks) or routine (12-month wait) surgery. Over a 12-month follow-up, the rate of falling was reduced by 34% in the expedited group (rate ratio 0.66, 95% confidence interval 0.45 to 0.96) accompanied by a significantly lower number of people with a new fracture (p = 0.04) [52].

With evidence from several double-blind randomized controlled trials (RCTs) (12 RCTs for fractures and 8 RCTs for falls) summarized in two 2009 meta-analyses, supplementation with **vitamin D** should reduce the risk of falls [53] and nonvertebral fractures, including those at the hip [54], by about 20%. Notably, however, for both end points, this benefit was dose dependent and only observed at an adherence-adjusted dose greater than 480 IU per day for fractures [54], and a treatment dose of at least 700 IU per day for falls [53].

Muscle weakness is an important risk factor for falls and is a prominent feature of the clinical syndrome of vitamin D deficiency [55]. Thus, muscle weakness due to vitamin D deficiency may plausibly mediate fracture risk through an increased susceptibility to falls. The vitamin D receptor (VDR) is expressed in human muscle tissue as suggested in several [56–59] but not all [60] studies. Vitamin D bound to its nuclear receptor in muscle tissue may lead to *de novo* protein synthesis [61, 62], followed by a relative increase in the diameter and number of fast type II muscle fibers [62]. Notably, fast type II muscle fibers decline with age relative to slow type I muscle fibers, resulting in an increased propensity to fall [63].

In 2010, the Institute of Medicine (IOM) did a thorough review on the effect of vitamin D on fall prevention [64]. Their synopsis was that the evidence of vitamin D on fall prevention is inconsistent, which is in contrast to the 2011 assessment of the Agency for Healthcare Research and Quality (AHRQ) for the U.S. Preventive Services Task Force [65], the 2010 American Geriatric Society/British Geriatric Society Clinical Practice Guideline [66], and the 2010 assessment by the IOF [67], all three of which reviewed the same evidence and identified vitamin D as an effective intervention to prevent falling in older adults. Notably, the main inconsistency raised by the IOM is based on four studies that may not be considered reliable indicators of true treatment efficacy, as those studies either used low dose vitamin D [68], had less than 50% adherence [69], had a low-quality fall assessment [70] or used one large bolus dose of vitamin D among seniors in unstable health [71]. Further, including these four studies in a pooled analysis of a total of 12 blinded and open-design trial evaluated by the IOM, there was a significant benefit overall (OR = 0.89; 95% CI 0.80–0.99), most pronounced in 6 of 12 studies that fulfilled the criteria for a high-quality fall ascertainment (OR = 0.79; 0.65–0.96) [64]. As falling is a challenging end point to assess, as discussed above [20], the latter analysis by the IOM restricted to the six trials with high-quality fall assessment may best reflect true treatment efficacy with a 21% fall reduction [64], confirming the earlier peer-reviewed meta-analysis that indentified eight double-blind RCTs with a high-quality fall assessment [53, 72].

Finally, it is important to note that vitamin D may address several components of the fall–fracture construct, including strength [73], balance [74], lower extremity function [75, 76], falling [77, 78], bone density [79, 80], the risk of hip and nonvertebral fractures [81], and the risk of nursing home admission [82]. In a most recent trial that compared 2,000 IU vitamin D per day with the current standard of care of 800 IU vitamin D per day in a double-blind RCT of 173 acute hip fracture patients, the higher dose of vitamin D did not reduce the rate of falls better than the 800-IU dose of vitamin D [45]. However, the higher dose compared to 800 IU vitamin D reduced the rate of hospital re-admission significantly by 39%, which was driven by a significant 60% reduction of fall-related injuries, primarily re-fracture [45].

CONCLUSION

Fall risk reduction is a significant component of fracture prevention at older age, and the public health impact of falls is significant. Falls can be reduced by a number of interventions with vitamin D offering efficacy established in several randomized controlled trials extending to fracture reduction in some of the same trials. In order to more effectively study falls and the fall–fracture risk profile from different interventions and cohort studies, fall definition and ascertainment need to be standardized.

REFERENCES

1. Melton LJ 3rd, Crowson CS, O'Fallon WM. 1999. Fracture incidence in Olmsted County, Minnesota: Comparison of urban with rural rates and changes in urban rates over time. *Osteoporos Int* 9(1): 29–37.
2. Cummings SR, Nevitt MC. 1994. Non-skeletal determinants of fractures: The potential importance of the mechanics of falls. Study of Osteoporotic Fractures Research Group. *Osteoporos Int* 4 Suppl 1: 67–70.
3. Cummings SR, Nevitt MC, Browner WS, Stone K, Fox KM, Ensrud KE, et al. 1995. Risk factors for hip fracture in white women. Study of Osteoporotic Fractures Research Group. *N Engl J Med* 332(12): 767–73.
4. Centers for Disease Control and Prevention (CDC). 2006. Fatalities and injuries from falls among older adults—United States, 1993–2003 and 2001–2005. *MMWR Morb Mortal Wkly Rep* 55(45): 1221–4.
5. Schwartz AV, Nevitt MC, Brown BW Jr, Kelsey JL. 2005. Increased falling as a risk factor for fracture among older women: The study of osteoporotic fractures. *Am J Epidemiol* 161(2): 180–5.
6. McClung MR, Geusens P, Miller PD, Zippel H, Bensen WG, Roux C, et al. 2001. Effect of risedronate on the risk of hip fracture in elderly women. Hip Intervention Program Study Group. *N Engl J Med* 344(5): 333–40.
7. Tinetti ME. 1988. Risk factors for falls among elderly persons living in the community. *N Engl J Med* 319: 1701–7.
8. Campbell AJ, Reinken J, Allan BC, Martinez GS. 1981. Falls in old age: A study of frequency and related clinical factors. *Age Ageing* 10(4): 264–70.
9. Tinetti ME, Williams CS. 1998. The effect of falls and fall injuries on functioning in community-dwelling older persons. *J Gerontol A Biol Sci Med Sci* 53(2): M112–9.
10. Tinetti ME, Williams CS. 1997. Falls, injuries due to falls, and the risk of admission to a nursing home. *N Engl J Med* 337(18): 1279–84.
11. Pluijm SM, Smit JH, Tromp EA, Stel VS, Deeg DJ, Bouter LM, et al. 2006. A risk profile for identifying community-dwelling elderly with a high risk of recurrent falling: Results of a 3-year prospective study. *Osteoporos Int* 17(3): 417–25.
12. Economic Policy Committee and the European Commission (DG ECFIN). *European Economy: Special Report No. 1/2006. The Impact of Ageing on Public Expenditure: Projections for the EU-25 Member States on Pensions, Healthcare, Long-Term Care, Education and Unemployment Transfers (2004–50)*. Brussels, Belgium: Office for the Official Publications of the European Community.
13. Eberstadt N, Groth, H. 2008. *Europe's Coming Demographic Challenge: Unlocking the Value of Health*. Washington, DC: American Enterprise Institute for Public Policy Research.
14. European Population Committee of the Council of Europe. 2006. *Recent Demographic Developments in Europe 2005*. Strasbourg: Council of Europe Publishing.

15. Eurostat. 2006. *Statistics in Focus: First Demographic Estimates for 2005.* http://epp.eurostat.ec.europa.eu/cache/ITY_OFFPUB/KS-NK-06-001/EN/KS-NK-06-001-EN.PDF.
16. Lee RD. 2007. *Global Population Aging and its Economic Consequences.* Washington, DC: AEI Press.
17. Gullberg B, Johnell O, Kanis JA. 1997. World-wide projections for hip fracture. *Osteoporos Int* 7(5): 407–13.
18. Englander F, Hodson TJ, Terregrossa RA. 1996. Economic dimensions of slip and fall injuries. *J Forensic Sci* 41(5): 733–46.
19. Buchner DM, Hornbrook MC, Kutner NG, Tinetti ME, Ory MG, Mulrow CD, et al. 1993. Development of the common data base for the FICSIT trials. *J Am Geriatr Soc* 41(3): 297–308.
20. Cummings SR, Nevitt MC, Kidd S. 1988. Forgetting falls. The limited accuracy of recall of falls in the elderly. *J Am Geriatr Soc* 36(7): 613–6.
21. Kanis JA, Borgstrom F, De Laet C, Johansson H, Johnell O, Jonsson B, et al. 2005. Assessment of fracture risk. *Osteoporos Int* 16(6): 581–9.
22. Sandhu SK, Nguyen ND, Center JR, Pocock NA, Eisman JA, Nguyen TV. 2010. Prognosis of fracture: Evaluation of predictive accuracy of the FRAX algorithm and Garvan nomogram. *Osteoporos Int* 21(5): 863–71.
23. van den Bergh JP, van Geel TA, Lems WF, Geusens PP. 2010. Assessment of individual fracture risk: FRAX and beyond. *Curr Osteoporos Rep* 8(3): 131–7.
24. Nguyen ND, Frost SA, Center JR, Eisman JA, Nguyen TV. 2007. Development of a nomogram for individualizing hip fracture risk in men and women. *Osteoporos Int* 17: 17.
25. Nevitt MC, Cummings SR. 1993. Type of fall and risk of hip and wrist fractures: The study of osteoporotic fractures. The Study of Osteoporotic Fractures Research Group. *J Am Geriatr Soc* 41(11): 1226–34.
26. Vellas BJ, Wayne SJ, Romero LJ, Baumgartner RN, Garry PJ. 1997. Fear of falling and restriction of mobility in elderly fallers. *Age Ageing* 26(3): 189–93.
27. Arfken CL, Lach HW, Birge SJ, Miller JP. 1994. The prevalence and correlates of fear of falling in elderly persons living in the community. *Am J Public Health* 84(4): 565–70.
28. Graafmans WC, Ooms ME, Hofstee HM, Bezemer PD, Bouter LM, Lips P. 1996. Falls in the elderly: A prospective study of risk factors and risk profiles. *Am J Epidemiol* 143(11): 1129–36.
29. Arden NK, Nevitt MC, Lane NE, Gore LR, Hochberg MC, Scott JC, et al. 1999. Osteoarthritis and risk of falls, rates of bone loss, and osteoporotic fractures. Study of Osteoporotic Fractures Research Group. *Arthritis Rheum* 42(7): 1378–85.
30. Arden NK, Crozier S, Smith H, Anderson F, Edwards C, Raphael H, et al. 2006. Knee pain, knee osteoarthritis, and the risk of fracture. *Arthritis Rheum* 55(4): 610–5.
31. Dufour AB, Roberts B, Broe KE, Kiel DP, Bouxsein ML, Hannan MT. 2011. The factor-of-risk biomechanical approach predicts hip fracture in men and women: The Framingham Study. *Osteoporos Int* 23(2): 513–2.
32. Tinetti ME, Inouye SK, Gill TM, Doucette JT. 1995. Shared risk factors for falls, incontinence, and functional dependence. Unifying the approach to geriatric syndromes. *JAMA* 273(17): 1348–53.
33. Mowe M, Haug E, Bohmer T. 1999. Low serum calcidiol concentration in older adults with reduced muscular function. *J Am Geriatr Soc* 47(2): 220–6.
34. Bulajic-Kopjar M. 2000. Seasonal variations in incidence of fractures among elderly people. *Inj Prev* 6(1): 16–9.
35. Hemenway D, Colditz GA. 1990. The effect of climate on fractures and deaths due to falls among white women. *Accid Anal Prev* 22(1): 59–65.
36. Ralis ZA. 1981. Epidemic of fractures during period of snow and ice. *Br Med J (Clin Res Ed)* 282(6264): 603–5.
37. Carter SE, Campbell EM, Sanson-Fisher RW, Gillespie WJ. 2000. Accidents in older people living at home: A community-based study assessing prevalence, type, location and injuries. *Aust N Z J Public Health* 24(6): 633–6.
38. Bischoff-Ferrari HA, Orav JE, Barrett JA, Baron JA. 2007. Effect of seasonality and weather on fracture risk in individuals 65 years and older. *Osteoporos Int* 24: 24.
39. Close J, Ellis M, Hooper R, Glucksman E, Jackson S, Swift C. 1999. Prevention of falls in the elderly trial (PROFET): a randomized controlled trial. *Lancet* 353: 93–7.
40. Province MA, Hadley EC, Hornbrook MC, Lipsitz LA, Miller JP, Mulrow CD, et al. 1995. The effects of exercise on falls in elderly patients. A preplanned meta-analysis of the FICSIT Trials. Frailty and Injuries: Cooperative Studies of Intervention Techniques. *JAMA* 273(17): 1341–7.
41. Oliver D, Connelly JB, Victor CR, Shaw FE, Whitehead A, Genc Y, et al. 2007. Strategies to prevent falls and fractures in hospitals and care homes and effect of cognitive impairment: Systematic review and meta-analyses. *BMJ* 334(7584): 82.
42. Wolf SL, Barnhart HX, Kutner NG, McNeely E, Coogler C, Xu T. 1996. Reducing frailty and falls in older persons: An investigation of Tai Chi and computerized balance training. Atlanta FICSIT Group. Frailty and Injuries: Cooperative Studies of Intervention Techniques. *J Am Geriatr Soc* 44(5): 489–97.
43. Trombetti A, Hars M, Herrmann FR, Kressig RW, Ferrari S, Rizzoli R. 2011. Effect of music-based multitask training on gait, balance, and fall risk in elderly people: A randomized controlled trial. *Arch Intern Med* 171(6): 525–33.
44. Campbell AJ, Robertson MC, Gardner MM, Norton RN, Tilyard MW, Buchner DM. Randomised controlled trial of a general practice programme of home based exercise to prevent falls in elderly women. *BMJ* 315(7115): 1065–9.
45. Bischoff-Ferrari HA, Dawson-Hughes B, Platz A, Orav EJ, Stahelin HB, Willett WC, et al. 2010. Effect of high-dosage cholecalciferol and extended physiotherapy on complications after hip fracture: A randomized controlled trial. *Arch Intern Med* 170(9): 813–20.

46. Voukelatos A, Cumming RG, Lord SR, Rissel C. 2007. A randomized, controlled trial of tai chi for the prevention of falls: The Central Sydney tai chi trial. *J Am Geriatr Soc* 2007;55(8): 1185–91.

47. Li F, Harmer P, Fisher KJ, McAuley E, Chaumeton N, Eckstrom E, et al. 2005. Tai Chi and fall reductions in older adults: A randomized controlled trial. *J Gerontol A Biol Sci Med Sci* 60(2): 187–94.

48. Wolf SL, Sattin RW, Kutner M, O'Grady M, Greenspan AI, Gregor RJ. 2003. Intense tai chi exercise training and fall occurrences in older, transitionally frail adults: A randomized, controlled trial. *J Am Geriatr Soc* 51(12): 1693–701.

49. Lee MS, Pittler MH, Shin BC, Ernst E. 2008. Tai chi for osteoporosis: A systematic review. *Osteoporos Int* 19(2): 139–46.

50. Lundin-Olsson L, Nyberg L, Gustafson Y. 1997. "Stops walking when talking" as a predictor of falls in elderly people. *Lancet* 349(9052): 617.

51. de Hoon EW, Allum JH, Carpenter MG, Salis C, Bloem BR, Conzelmann M, et al. 2003. Quantitative assessment of the stops walking while talking test in the elderly. *Arch Phys Med Rehabil* 84(6): 838–42.

52. Harwood RH, Foss AJ, Osborn F, Gregson RM, Zaman A, Masud T. 2005. Falls and health status in elderly women following first eye cataract surgery: A randomised controlled trial. *Br J Ophthalmol* 89(1): 53–9.

53. Bischoff-Ferrari HA, Dawson-Hughes B, Staehelin HB, Orav JE, Stuck AE, Theiler R, et al. 2009. Fall prevention with supplemental and active forms of vitamin D: A meta-analysis of randomised controlled trials. *BMJ* 339(1): 339: b3692.

54. Bischoff-Ferrari HA, Willett WC, Wong JB, Stuck AE, Staehelin HB, Orav EJ, et al. 2009. Prevention of nonvertebral fractures with oral vitamin D and dose dependency: A meta-analysis of randomized controlled trials. *Arch Intern Med* 169(6): 551–61.

55. Schott GD, Wills MR. 1976. Muscle weakness in osteomalacia. *Lancet* 1(7960): 626–9.

56. Endo I, Inoue D, Mitsui T, Umaki Y, Akaike M, Yoshizawa T, et al. 2003. Deletion of vitamin D receptor gene in mice results in abnormal skeletal muscle development with deregulated expression of myoregulatory transcription factors. *Endocrinology* 144(12): 5138–44.

57. Bischoff-Ferrari HA, Borchers M, Gudat F, Durmuller U, Stahelin HB, Dick W. 2004. Vitamin D receptor expression in human muscle tissue decreases with age. *J Bone Miner Res* 19(2): 265–9.

58. Ceglia L, da Silva Morais M, Park LK, Morris E, Harris SS, Bischoff-Ferrari HA, et al. 2010l. Multi-step immunofluorescent analysis of vitamin D receptor loci and myosin heavy chain isoforms in human skeletal muscle. *J Mol Histol* 41(2–3): 137–42.

59. Bischoff-Ferrari HA, Borchers M, Gudat F, Durmuller U, Stahelin HB, Dick W. Vitamin D receptor expression in human muscle tissue decreases with age. *J Bone Miner Res* 2004;19(2): 265–9.

60. Wang Y, DeLuca HF. Is the vitamin d receptor found in muscle? 2011. *Endocrinology* 152(2): 354–63.

61. Boland R. 1986. Role of vitamin D in skeletal muscle function. *Endocr Rev* 7: 434–47.

62. Sorensen OH, Lund B, Saltin B, Andersen RB, Hjorth L, Melsen F, et al. 1979. Myopathy in bone loss of ageing: Improvement by treatment with 1 alpha-hydroxycholecalciferol and calcium. *Clin Sci (Colch)* 56(2): 157–61.

63. von Haehling S, Morley JE, Anker SD. 2010. An overview of sarcopenia: Facts and numbers on prevalence and clinical impact. *J Cachex Sarcopenia Muscle* 1(2): 129–33.

64. Medicine Io. 2010. Dietary Reference Ranges for Calcium and Vitamin D. http://wwwiomedu/Reports/2010/Dietary-Reference-Intakes-for-Calcium-and-Vitamin-D/Report-Briefaspx.

65. Michael YL, Whitlock EP, Lin JS, Fu R, O'Connor EA, Gold R. 2011. Primary care-relevant interventions to prevent falling in older adults: A systematic evidence review for the U.S. Preventive Services Task Force. *Ann Intern Med* 153(12): 815–25.

66. AGS/BGS. 2010. AGS/BGS Guidelines on Fall Prevention in older Persons. http://wwwamericangeriatricsorg/files/documents/health_care_pros/FallsSummaryGuide pdf.

67. Dawson-Hughes B, Mithal A, Bonjour JP, Boonen S, Burckhardt P, Fuleihan GE, et al. 2010. IOF position statement: Vitamin D recommendations for older adults. *Osteoporos Int* 21(7): 1151–4.

68. Graafmans WC, Ooms ME, Hofstee HM, Bezemer PD, Bouter LM, Lips P. 1996. Falls in the elderly: A prospective study of risk factors and risk profiles. *Am J Epidemiol* 143(11): 1129–36.

69. Grant AM, Avenell A, Campbell MK, McDonald AM, MacLennan GS, McPherson GC, et al. 2005. Oral vitamin D3 and calcium for secondary prevention of low-trauma fractures in elderly people (Randomised Evaluation of Calcium Or vitamin D, RECORD): A randomised placebo-controlled trial. *Lancet* 365(9471): 1621–8.

70. Trivedi DP, Doll R, Khaw KT. 2003. Effect of four monthly oral vitamin D3 (cholecalciferol) supplementation on fractures and mortality in men and women living in the community: Randomised double blind controlled trial. *BMJ* 326(7387): 469.

71. Latham NK, Anderson CS, Lee A, Bennett DA, Moseley A, Cameron ID. 2003. A randomized, controlled trial of quadriceps resistance exercise and vitamin D in frail older people: The Frailty Interventions Trial in Elderly Subjects (FITNESS). *J Am Geriatr Soc* 51(3): 291–9.

72. Bischoff-Ferrari HA, Willett WC, Orav EJ, Kiel DP, Dawson-Hughes B. 2011. Re: Fall prevention with Vitamin D. Clarifications needed. http://www.bmj.com/content/339/bmj.b3692?tab=responses.

73. Bischoff HA, Stahelin HB, Dick W, Akos R, Knecht M, Salis C, et al. 2003. Effects of vitamin D and calcium supplementation on falls: A randomized controlled trial. *J Bone Miner Res* 18(2): 343–51.

74. Pfeifer M, Begerow B, Minne HW, Abrams C, Nachtigall D, Hansen C. 2000. Effects of a short-term vitamin D

and calcium supplementation on body sway and secondary hyperparathyroidism in elderly women. *J Bone Miner Res* 15(6): 1113–8.
75. Bischoff-Ferrari HA, Dietrich T, Orav EJ, Hu FB, Zhang Y, Karlson EW, et al. 2004. Higher 25-hydroxyvitamin D concentrations are associated with better lower-extremity function in both active and inactive persons aged >=60 y. *Am J Clin Nutr* 80(3): 752–8.
76. Wicherts IS, van Schoor NM, Boeke AJ, Visser M, Deeg DJ, Smit J, et al. 2007. Vitamin D status predicts physical performance and its decline in older persons. *J Clin Endocrinol Metab* 6: 6.
77. Broe KE, Chen TC, Weinberg J, Bischoff-Ferrari HA, Holick MF, Kiel DP. 2007. A higher dose of vitamin D reduces the risk of falls in nursing home residents: A randomized, multiple-dose study. *J Am Geriatr Soc* 55(2): 234–9.
78. Bischoff-Ferrari HA, Orav EJ, Dawson-Hughes B. 2006. Effect of cholecalciferol plus calcium on falling in ambulatory older men and women: A 3-year randomized controlled trial. *Arch Intern Med* 166(4): 424–30.
79. Dawson-Hughes B, Harris SS, Krall EA, Dallal GE. 1997. Effect of calcium and vitamin D supplementation on bone density in men and women 65 years of age or older. *N Engl J Med* 337(10): 670–6.
80. Bischoff-Ferrari HA, Dietrich T, Orav EJ, Dawson-Hughes B. Positive association between 25-hydroxy vitamin d levels and bone mineral density: A population-based study of younger and older adults. *Am J Med* 116(9): 634–9.
81. Bischoff-Ferrari HA, Willett WC, Wong JB, Giovannucci E, Dietrich T, Dawson-Hughes B. 2005. Fracture prevention with vitamin D supplementation: A meta-analysis of randomized controlled trials. *JAMA* 293(18): 2257–64.
82. Visser M, Deeg DJ, Puts MT, Seidell JC, Lips P. 2006. Low serum concentrations of 25-hydroxyvitamin D in older persons and the risk of nursing home admission. *Am J Clin Nutr* 84(3): 616–22; quiz 71–2.

46

Exercise and the Prevention of Osteoporosis

Clinton T. Rubin, Janet Rubin, and Stefan Judex

Mechanical Factors Regulating Bone Cell Response 396
Mechanically Responsive Bone Cells 397
Transducing Mechanical Signals into Cellular Response 398
Translating Mechanical Signals to the Clinic 398
Summary 400
References 400

Osteopenia, a condition of diminished bone mass, becomes osteoporosis when mechanical demands exceed the ability of the skeletal structure to support them. Consequences of bone loss are exacerbated by an age-related decrease in muscle strength [1] and postural stability [2], markedly increasing the risk of falls, fracture and—ultimately—mortality [3]. Mechanical signals generated by exercise can mitigate bone loss as well as help preserve the musculoskeletal "system." The physical and/or biologic basis of how mechanical signals are transformed into anabolic agents for bone and other tissues is called mechanotransduction and may represent the foundation for a nondrug approach to treat osteoporosis [4].

Bone quantity and quality is enhanced by exercise and compromised by inactivity, age, and disuse [5]. Cross-sectional studies illustrate the sensitivity of bone morphology to physical extremes. Astronauts subject to microgravity lose up to 2% of hip bone density each month [6], while professional tennis players possess 35% more bone in the dominant arm as compared to the arm that simply throws the ball into the air [7]. A broader benefit can be seen in professional athletes involved in a variety of demanding activities, including soccer players, weight lifters, speed skaters, and gymnasts [8, 9].

Several prospectively designed trials emphasize that functional loading result in increased bone mass. Intense exercise in young army recruits [10] stimulated large increases in bone mineral density (BMD), while a 10-month, high-impact strength building regimen in children significantly increased femoral neck BMD [11]. A number of longitudinal exercise studies, however, have reported only modest increases in bone mass [12]. For example, a 1-year high-resistance strength training study in young women significantly increased muscle strength but failed to influence bone mass [13]. And even if the goal of exercise is to slow the loss of bone in the elderly, there is increasing evidence that bone tissue is less responsive to mechanical stimuli as we age [14]. Before exercise in general, and mechanical signals in particular, can be effectively used to prevent the degradation of the musculoskeletal system with age, we must improve our understanding of both the mechanical milieu generated by exercise and the complex cellular machinery that perceive and respond to these key regulatory signals [15].

MECHANICAL FACTORS REGULATING BONE CELL RESPONSE

Skeletal loads and bending moments resolve into strain in the bone tissue, reaching up to 0.3% (3,000 microstrain) during strenuous activity, a level of matrix deformation common across a range of species [16]. The strain levels actually "experienced" by bone cells *in vivo* is unclear, but may be as much as 10 times that experienced by the matrix [17]. Bone cells also experience interstitial fluid flow dynamic pressure changes during mechanical loading [18]. Furthermore, functional loading also induces pressure in the intramedullary cavity [19], shear forces through canaliculi [20], and dynamic electric fields as interstitial fluid flows past charged bone crystals [21]. As

---

Primer on the Metabolic Bone Diseases and Disorders of Mineral Metabolism, Eighth Edition. Edited by Clifford J. Rosen.
© 2013 American Society for Bone and Mineral Research. Published 2013 by John Wiley & Sons, Inc.

such, the complex loading environment of the skeleton generates a diverse range of mechanical signals that are ultimately inseparable [22], but each of these biophysical factors may differentially target tissue, cell, and molecular activity.

Animal models demonstrate that bone remodeling is sensitive to changes in strain magnitude [23], the number of loading cycles [24], the distribution of the loading [25], and the rate of strain [26]. Importantly, the load signal must be dynamic (time varying), as static loads are ignored by the skeleton [27], while the anabolic potential increases when rest periods are inserted between the mechanical events [28]. Even extremely low-magnitude bone strains, three orders of magnitude below peak strains generated during strenuous activity, when induced at high frequencies similar to the spectral content of muscle contractibility [29], are anabolic to bone tissue [30], enhancing not only bone mass but bone quality and strength (Fig. 46.1). Together, these findings emphasize that the "key ingredient" to an osteogenic exercise regimen cannot simply be distilled to "bigger is better."

Physiologic levels of strain reduce osteocyte apoptosis, suggesting matrix deformation is critical to the survival of these cells [31, 32]. Too much strain, though, can cause matrix microdamage and exacerbate death of adjacent cells [33]. Factors other than matrix strain also cause an adaptive cellular response; e.g., acceleration, achieved independently of direct loading, can be anabolic to bone [34], suggesting that cells may act as "accelerometers" that respond to dynamic changes in force [35], emphasizing that *by-products* of the strain signal, such as shear stress or strain-generated potentials, may be critical to regulating the biology of the adaptive response. But *how* do the cells sense these mechanical signals?

MECHANICALLY RESPONSIVE BONE CELLS

The sensitivity of bone cells to mechanical signals, including stromal cells, osteoblasts, and osteocytes has been well documented [36], but it is difficult to designate a critically responsive cell. While the osteoblast is critical for the adaptive response, the osteocyte, representing 95% of adult skeletal cells, may prove key to bone tissue plasticity [37]. The antenna-like three-dimensional morphology of this osteocyte syncytium, interconnected by regulated gap-junctional connexins [38], are ideally configured to perceive and even amplify biophysical stimuli [20]. In particular, the removal of load is tied to osteocyte activity; loading regulates the release of sclerostin from osteocytes, an osteocyte product involved in Wnt signaling [39]. The osteoclastic bone resorption accompanying hind limb unloading in mice is ablated when osteocytes are absent [40], as the unloaded osteocytes release receptor activator of nuclear factor-κB ligand (RANKL) [41].

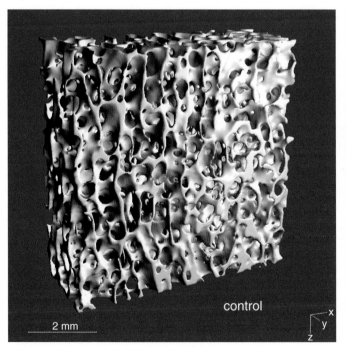
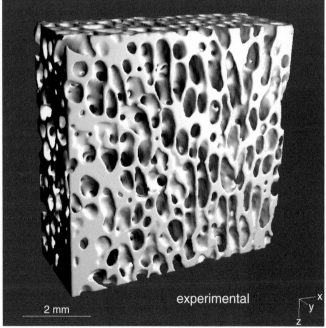

Fig. 46.1. Microcomputed tomography of the distal femur of adult (8-year-old) sheep, comparing a control animal (left) to an animal subject to 20 min/day of 30 Hz (cycles per second) of a low-level (0.3 g) mechanical vibration for 1 year (Ref. 76). The large increase in trabecular bone density results in enhanced bone strength, achieved with tissue strains three orders of magnitude below those that cause damage to the tissue. These data suggest that specific mechanical parameters may represent a nonpharmacologic basis for the treatment of osteoporosis (Ref. 71).

The connectivity of the osteocyte network deteriorates markedly with age and may contribute to the progressive loss of sensitivity of bone to chemical and physical signals [14].

Marrow stromal cells change proliferation and gene expression in response to mechanical stimulation [42] and through mechanical regulation of RANKL expression [43] also affect osteoclast number and function. The osteoclast also responds directly to mechanical signals limiting bone resorption [44]. Other cells present in bone, such as shear sensitive endothelial cells in the penetrating vasculature, likely contribute to the adaptive response by producing nitric oxide [45], which has pleiomorphic effects on the skeleton [46].

The role of mesenchymal stem cells (MSCs) in regulating the adaptive response to mechanical signals is also being investigated, with the demonstration that exercise can bias MSCs toward osteoblastogenesis [47], while disuse increases adipogenesis within the bone marrow [48]. Even low magnitude mechanical signals have been shown to influence the fate of MSCs, driving them toward a musculoskeletal lineage [49] while suppressing adipogenesis [50]. Considering the interdependence of fat and bone tissue [51], it may be that exercise-based prevention of obesity and osteoporosis could be achieved through regulation of MSC lineage rather than necessarily the resident cell population in fat (adipocytes) or bone (osteocytes) [52].

TRANSDUCING MECHANICAL SIGNALS INTO CELLULAR RESPONSE

When bone strains of 3,000 microstrain, realized during strenuous activity [16] are resolved to the level of the cell, these deformations are on the order of Angstroms, requiring an exquisitely sensitive receptor system. Further, cell mechanoreceptors must either be in contact with the outside, through the cell membrane and its attachment to substrate, or the mechanoreceptor must be able to sense by-products of load, such as fluid shear on the apical membrane. While in sensory organs there are examples of channels that are regulated by movement of mechanosensory bristles [53], or by tension waves [54], a unified model of proximal events inducing intracellular signal transduction in non-sensory tissues does not yet exist. There are, however, several components of the cell that could act as the mechanoreceptor, transducing a physical challenge into a cellular response.

Ion channel activity in osteoblasts stimulated by stretch/strain of the membrane [55] or by parathyroid hormone (PTH) [56] have been associated with bone cell activation. Patch-clamp techniques demonstrate at least three classes of mechanosensitive ion channels [57]. In limb bone cultures, gadolinium chloride, which blocks some stretch/shear-sensitive cation channels, blocked load-related increases in PGI2 and nitric oxide [58].

Membrane deformation and shear across the membrane, as well as pressure transients, are transmitted to the cytoskeleton and ultimately to the cell-matrix adhesion proteins that anchor the cell in place [59]. Membrane-spanning integrins, which couple the cell to its extracellular environment, and a large number of adhesion-associated linker proteins are potential molecular mechanotransducers. The architecture of the cytoskeleton with its microfilamentous and microtubular network linking adhesion receptors to the cell nucleus plays a role in perceiving small deformations and directly informing the nucleus [60]. Application of strain to MSCs induces focal adhesion assembly, which amplifies force generated signaling [61].

Cellular cytoskeletal adaptation to force is even more subtly reflected in a compartmentalization of signals within the several phases of liquid-ordered and liquid-disordered lipid making up the plasma membrane [62]. Organized lipid rafts are thought to act as mechanical sensors: in endothelial cells, shear stress causes signaling molecules to translocate to caveolar lipid rafts, and if the caveolae are disassembled, both proximal and downstream mechanical signals, including MAPK activation, are abrogated [63]. In bone cells, many of the outside-in signaling systems that respond to force are sequestered in lipid rafts.

With the multiplicity of mechanical signals presented to the cell, it is likely that no single mechanosensor or receptor mechanism is responsible for perceiving and responding to the mechanical environment (Fig. 46.2). At the very least, multiple mechanosensors are likely to interact to integrate both mechanical and chemical information from the microenvironment.

Since the distal responses to mechanical factors are similar to those elicited by ligand-receptor pairing and result in changes in gene expression, mechanotransduction must eventually end up utilizing similar intracellular signaling cascades. Mechanical forces have been shown to activate every type of signal transduction cascade, including β-catenin [64], mTORC2 [65], cAMP [66], IP3 and intracellular calcium [42, 67], guanine regulatory proteins [68], and MAPK [69], among others [70]. Given the pivotal role of mechanical stimuli in regulating anabolic and catabolic cell activity, perhaps it should not be so surprising that so many signaling pathways are involved.

TRANSLATING MECHANICAL SIGNALS TO THE CLINIC

There is growing clinical evidence that mechanical signals, even those that are extremely low magnitude, can be anabolic to bone [30], particularly in the young skeleton. In the first study, the ability of these low level signals to improve bone mass, delivered using low intensity vibration, was examined in children with disabling conditions [71]. Children were randomized to stand on an actively vibrating (0.3 g, 90 Hz) or placebo device for 10 min/day. Over a 6-month trial, proximal tibial volumetric trabecular BMD (vTBMD) in children on active devices increased by 6.3% while vTBMD decreased by 11.9% on placebo devices ($p < 0.01$). In the second study,

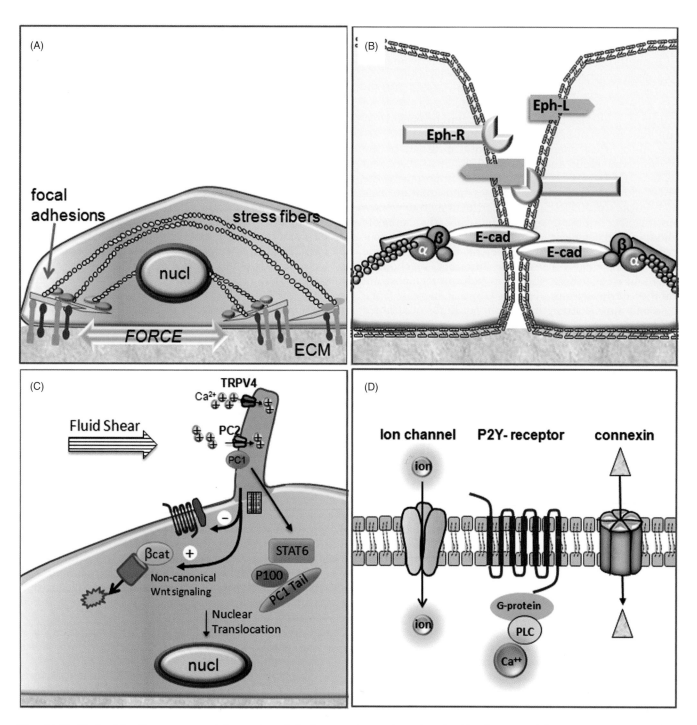

Fig. 46.2. Skeletal loading generates deformation of the bone matrix and acceleration/deceleration of the tissue system, resulting in strain across the cell and pressure within the marrow cavity and mineralized cortices. Functional load bearing also results in shear forces through the canaliculi, with the bulk fluid movement causing drag over cells and streaming potentials from charged interactions with the bone crystals. Some aggregate of these physical signals distorts the morphology of the cell. Some composite of these physical signals interacts with the morphology of the cell through interactions such as potentiated by the membrane–matrix or membrane–nucleus structures, or distortion of the membrane itself, to regulate transcriptional activity of the cell. Several candidate mechanotransducer systems are illustrated. (A) Cell cytoskeleton senses loading at the membrane through integrins that transmit force through focal adhesions and F-actin stress fibers. (B) Cadherins, which connect to the cytoskeleton, are examples of outside-in signaling modifiers. Ephrins exemplify an intercellular signaling system regulated by movement of components within the plasma membrane. (C) Primary cilia may sense flow, pressure, and strain, activating ion flux through PC1 and TRPV4, which can activate Stat signals. Cilia also modulate Wnt signaling via noncanonical antagonism that leads to β-cat degradation. (D) Membrane spanning proteins such as ion channels, purinergic receptors and connexins can be regulated through shear and strain. Modified from Thompson WR, Rubin CT, Rubin J. 2012. Mechanical regulation of signaling pathways in bone. *Gene* 503(2): 179–93.

a 12-month trial was conducted in 48 young women in the lowest quartile of bone density and at least one skeletal fracture [72]. Subjects were randomly assigned either into a daily, low-magnitude whole body vibration group (10 min/day, 30 Hz, 0.3 g) or control. Intention-to-treat data indicated that cancellous bone in the lumbar vertebrae and cortical bone in the femoral midshaft of the experimental group increased by 2.0% (p = 0.06) and 2.3% (p = 0.04), as compared to controls. Importantly, the cross-sectional area of paraspinous musculature was 4.9% greater (p = 0.002) in the experimental group versus controls. Per protocol analysis indicates that subjects who used the device at least two minutes each day realized a much greater musculoskeletal benefit from the mechanical signals. These low intensity mechanical signals have also been shown to help protect postural stability during chronic bedrest [73] and increase bone mass and muscle strength in the upper limbs of disabled children, thereby improving their autonomy [74].

SUMMARY

As bone geometry and material properties vary between individuals, there are also genome-specific sensitivities to mechanical loading which may help explain the variability in exercise based trials.[75] However, evidence in the animal and human indicates that exercise in general, and mechanical signals in particular, are both anabolic and anti-catabolic to the musculoskeletal system, and benefit both bone quantity and quality. The challenge remains to identify those parameters within the complex mechanical milieu induced by exercise that are critical to driving anabolism in bone, and that ultimately may represent the basis of a nondrug strategy to control musculoskeletal health. While bone-targeted pharmaceutical interventions represent effective means of curbing osteoporosis, exercise, in its many varied forms, is self-targeting, endogenous to the musculoskeletal system, and auto-regulated, causing site-specific positive adaptation in bone mass and structure. And as effective as drug treatments for osteoporosis may be, physical signals have had a 525 million year head start in terms of optimizing skeletal structure to withstand the loads placed upon them.

REFERENCES

1. Rosenberg IH. 1997. Sarcopenia: Origins and clinical relevance. *J Nutr* 127: 990S–991S.
2. Melzer I, Benjuya N, Kaplanski J. 2004. Postural stability in the elderly: A comparison between fallers and non-fallers. *Age and Ageing* 33: 602–607.
3. Spaniolas K, Cheng JD, Gestring ML, Sangosanya A, Stassen NA, Bankey PE. 2010. Ground level falls are associated with significant mortality in elderly patients. *J Trauma* 69: 821–825.
4. Rubin J, Rubin C, Jacobs CR. 2006. Molecular pathways mediating mechanical signaling in bone. *Gene* 367: 1–16.
5. Frost HM. 1987. Bone "mass" and the "mechanostat": A proposal. *Anat Rec* 219: 1–9.
6. Lang T, LeBlanc A, Evans H, Lu Y, Genant H, Yu A. 2004. Cortical and trabecular bone mineral loss from the spine and hip in long-duration spaceflight. *J Bone Miner Res* 19: 1006–1012.
7. Jones HH, Priest JD, Hayes WC, Tichenor CC, Nagel DA. 1977. Humeral hypertrophy in response to exercise. *J Bone Joint Surg Am* 59: 204–208.
8. Heinonen A, Oja P, Kannus P, Sievanen H, Haapasalo H, Manttari A, Vuori I. 1995. Bone mineral density in female athletes representing sports with different loading characteristics of the skeleton. *Bone* 17: 197–203.
9. Snow-Harter C, Whalen R, Myburgh K, Arnaud S, Marcus R. 1992. Bone mineral density, muscle strength, and recreational exercise in men. *J Bone Miner Res* 7: 1291–1296.
10. Leichter I, Simkin A, Margulies JY, Bivas A, Steinberg R, Giladi M, Milgrom C. 1989. Gain in mass density of bone following strenuous physical activity. *J Orthop Res* 7: 86–90.
11. McKay HA, MacLean L, Petit M, MacKelvie-O'Brien K, Janssen P, Beck T, Khan KM. 2005. "Bounce at the Bell": A novel program of short bouts of exercise improves proximal femur bone mass in early pubertal children. *Br J Sports Med* 39: 521–526.
12. Colberg SR, Sigal RJ, Fernhall B, Regensteiner JG, Blissmer BJ, Rubin RR, Chasan-Taber L, Albright AL, Braun B. 2010. Exercise and type 2 diabetes: The American College of Sports Medicine and the American Diabetes Association: Joint position statement executive summary. *Diabetes Care* 33: 2692–2696.
13. Heinonen A, Sievanen H, Kannus P, Oja P, Vuori I. 1996 Effects of unilateral strength training and detraining on bone mineral mass and estimated mechanical characteristics of the upper limb bones in young women. *J Bone Miner Res* 11: 490–501.
14. Rubin CT, Bain SD, McLeod KJ. 1992. Suppression of the osteogenic response in the aging skeleton. *Calcif Tissue Int* 50: 306–313.
15. Ozcivici E, Luu YK, Adler B, Qin YX, Rubin J, Judex S, Rubin CT. 2010. Mechanical signals as anabolic agents in bone. *Nat Rev Rheumatol* 6: 50–59.
16. Rubin CT, Lanyon LE. 1984. Dynamic strain similarity in vertebrates; an alternative to allometric limb bone scaling. *J Theor Biol* 107: 321–327.
17. Nicolella DP, Moravits DE, Gale AM, Bonewald LF, Lankford J. 2006. Osteocyte lacunae tissue strain in cortical bone. *J Biomech* 39(9): 1735–1743.
18. Piekarski K, Munro M. 1977. Transport mechanism operating between blood supply and osteocytes in long bones. *Nature* 269: 80–82.
19. Qin YX, Kaplan T, Saldanha A, Rubin C. 2003. Fluid pressure gradients, arising from oscillations in intramedullary pressure, is correlated with the formation of bone and inhibition of intracortical porosity. *J Biomech* 36: 1427–1437.

20. Han Y, Cowin SC, Schaffler MB, Weinbaum S. 2004. Mechanotransduction and strain amplification in osteocyte cell processes. *Proc Natl Acad Sci U S A* 101: 16689–16694.
21. Pollack SR, Salzstein R, Pienkowski D. 1984. The electric double layer in bone and its influence on stress-generated potentials. *Calcif Tissue Int* 36 Suppl 1: S77–S81.
22. Gross TS, McLeod KJ, Rubin CT. 1992. Characterizing bone strain distributions in vivo using three triple rosette strain gages. *J Biomech* 25: 1081–1087.
23. Rubin CT, Lanyon LE. 1985. Regulation of bone mass by mechanical strain magnitude. *Calcif Tissue Int* 37: 411–417.
24. Rubin CT, Lanyon LE. 1984. Regulation of bone formation by applied dynamic loads. *J Bone Joint Surg Am* 66: 397–402.
25. Lanyon LE, Goodship AE, Pye CJ, MacFie JH. 1982. Mechanically adaptive bone remodelling. *J Biomech* 15: 141–154.
26. O'Connor JA, Lanyon LE, MacFie H. 1982. The influence of strain rate on adaptive bone remodelling. *J Biomech* 15: 767–781.
27. Lanyon LE, Rubin CT. 1984. Static vs dynamic loads as an influence on bone remodelling. *J Biomech* 17: 897–905.
28. Srinivasan S, Weimer DA, Agans SC, Bain SD, Gross TS. 2002. Low-magnitude mechanical loading becomes osteogenic when rest is inserted between each load cycle. *J Bone Miner Res* 17: 1613–1620.
29. Huang RP, Rubin CT, McLeod KJ. 1999. Changes in postural muscle dynamics as a function of age. *J Gerontol A Biol Sci Med Sci* 54: B352–B357.
30. Rubin C, Turner AS, Bain S, Mallinckrodt C, McLeod K. 2001. Anabolism: Low mechanical signals strengthen long bones. *Nature* 412: 603–604.
31. Noble BS, Peet N, Stevens HY, Brabbs A, Mosley JR, Reilly GC, Reeve J, Skerry TM, Lanyon LE. 2003. Mechanical loading: Biphasic osteocyte survival and targeting of osteoclasts for bone destruction in rat cortical bone. *Am J Physiol Cell Physiol* 284: C934–C943.
32. Gross TS, Akeno N, Clemens TL, Komarova S, Srinivasan S, Weimer D A, Mayorov S. 2001. Selected contribution: Osteocytes upregulate HIF-1alpha in response to acute disuse and oxygen deprivation. *J Appl Physiol* 90: 2514–2519.
33. Verborgt O, Gibson GJ, Schaffler MB. 2000. Loss of osteocyte integrity in association with microdamage and bone remodeling after fatigue in vivo. *J Bone Miner Res* 15: 60–67.
34. Garman R, Gaudette G, Donahue LR, Rubin C, Judex S. 2007. Low-level accelerations applied in the absence of weight bearing can enhance trabecular bone formation. *J Orthop Res* 25: 732–740.
35. Hwang SJ, Lublinsky S, Seo YK, Kim IS, Judex S. 2009. Extremely small-magnitude accelerations enhance bone regeneration: A preliminary study. *Clin Orthop Relat Res* 467: 1083–1091.
36. Rubin J, Rubin C, Jacobs CR. 2005. Molecular pathways mediating mechanical signaling in bone. *Gene* 367: 1–16.
37. Cowin SC, Weinbaum S. 1998. Strain amplification in the bone mechanosensory system. *Am J Med Sci* 316: 184–188.
38. Yellowley CE, Li Z, Zhou Z, Jacobs CR, Donahue HJ. 2000. Functional gap junctions between osteocytic and osteoblastic cells. *J Bone Miner Res* 15: 209–217.
39. Gaudio A, Pennisi P, Bratengeier C, Torrisi V, Lindner B, Mangiafico RA, Pulvirenti I, Hawa G, Tringali G, Fiore CE. 2010. Increased sclerostin serum levels associated with bone formation and resorption markers in patients with immobilization-induced bone loss. *J Clin Endocrinol Metab.* 95: 2248–2253.
40. Tatsumi S, Ishii K, Amizuka N, Li MQ, Kobayashi T, Kohno K, Ito M, Takeshita S, Ikeda K. 2007. Targeted ablation of osteocytes induces osteoporosis with defective mechanotransduction. *Cell Metab* 5: 464–475.
41. Cui S, Xiong F, Hong Y, Jung JU, Li XS, Liu JZ, Yan RQ, Mei L, Feng X, Xiong WC. 2011. APPswe/A beta regulation of osteoclast activation and RAGE expression in an age-dependent manner. *J Bone Miner Res* 26: 1084–1098.
42. Li YJ, Batra NN, You L, Meier SC, Coe IA, Yellowley CE, Jacobs CR. 2004. Oscillatory fluid flow affects human marrow stromal cell proliferation and differentiation. *J Orthop Res* 22: 1283–1289.
43. Rubin J, Fan X, Biskobing DM, Taylor WR, Rubin CT. 1999. Osteoclastogenesis is repressed by mechanical strain in an in vitro model. *J Orthop Res* 17: 639–645.
44. Wiltink A, Nijweide PJ, Scheenen WJ, Ypey DL, Van Duijn B. 1995. Cell membrane stretch in osteoclasts triggers a self-reinforcing Ca2+ entry pathway. *Pflugers Arch* 429: 663–671.
45. Davis, ME, Cai H, Drummond GR, Harrison DG. 2001. Shear stress regulates endothelial nitric oxide synthase expression through c-Src by divergent signaling pathways. *Circ Res* 89: 1073–1080.
46. Fan X, Roy E, Zhu L, Murphy TC, Ackert-Bicknell C, Hart CM, Rosen C, Nanes MS, Rubin J. 2004. Nitric oxide regulates receptor activator of nuclear factor-kappaB ligand and osteoprotegerin expression in bone marrow stromal cells. *Endocrinology* 145: 751–759.
47. David V, Martin A, Lafage-Proust MH, Malaval L, Peyroche S, Jones DB, Vico L, Guignandon A. 2007. Mechanical loading down-regulates peroxisome proliferator-activated receptor gamma in bone marrow stromal cells and favors osteoblastogenesis at the expense of adipogenesis. *Endocrinology* 148: 2553–2562.
48. Zayzafoon M, Gathings WE, McDonald JM. 2004. Modeled microgravity inhibits osteogenic differentiation of human mesenchymal stem cells and increases adipogenesis. *Endocrinology* 145: 2421–2432.
49. Xie L, Rubin C, Judex S. 2008. Enhancement of the adolescent murine musculoskeletal system using low-level mechanical vibrations. *J Appl Physiol* 104: 1056–1062.
50. Rubin CT, Capilla E, Luu YK, Busa B, Crawford H, Nolan DJ, Mittal V, Rosen CJ, Pessin J E., Judex S. 2007. Adipogenesis is inhibited by brief, daily exposure to high-frequency, extremely low-magnitude mechanical signals. *Proc Natl Acad Sci U S A* 104: 17879–17884.

51. Rosen CJ, Bouxsein M L. 2006. Mechanisms of disease: is osteoporosis the obesity of bone? *Nat Clin Pract Rheumatol* 2: 35–43.
52. Ozcivici E, Luu YK, Rubin CT, Judex S. 2010. Low-level vibrations retain bone marrow's osteogenic potential and augment recovery of trabecular bone during reambulation. *PLoS ONE* 5: e11178.
53. Sukharev S, Corey DP. 2004. Mechanosensitive channels: Multiplicity of families and gating paradigms. *Sci STKE* 2004: re4–
54. Morris CE. 1990. Mechanosensitive ion channels. *J Membr Biol* 113: 93–107.
55. Duncan RL, Hruska KA, Misler S. 1992. Parathyroid hormone activation of stretch-activated cation channels in osteosarcoma cells (UMR-106.01). *FEBS Lett* 307: 219–223.
56. Ferrier J, Ward A, Kanehisa J, Heersche JN. 1986. Electrophysiological responses of osteoclasts to hormones. *J Cell Physiol* 128: 23–26.
57. Davidson RM, Tatakis DW, Auerbach AL. 1990. Multiple forms of mechanosensitive ion channels in osteoblast-like cells. *Pflugers Arch* 416: 646–651.
58. Rawlinson SC, Pitsillides AA, Lanyon LE. 1996. Involvement of different ion channels in osteoblasts' and osteocytes' early responses to mechanical strain. *Bone* 19: 609–614.
59. Katsumi A, Orr AW, Tzima E, Schwartz MA. 2004. Integrins in mechanotransduction. *J Biol Chem* 279: 12001–12004.
60. Ingber DE. 2005. Mechanical control of tissue growth: Function follows form. *Proc Natl Acad Sci U S A* 102: 11571–11572.
61. Sen B, Guilluy C, Xie Z, Case N, Syner M, Thomas J, Oguz I, Rubin CT, Burridge K, Rubin J. 2011. Mechanically induced focal adhesion assembly amplifies anti-adipogenic pathways in mesenchymal stem cells. *Stem Cells* 29(11): 1829–1836
62. Simons K, Toomre D. 2000. Lipid rafts and signal transduction. *Nat Rev Mol Cell Biol* 1: 31–39.
63. Rizzo V, Sung A, Oh P, Schnitzer JE. 1998. Rapid mechanotransduction in situ at the luminal cell surface of vascular endothelium and its caveolae. *J Biol Chem* 273: 26323–26329.
64. Armstrong VJ, Muzylak M, Sunters A, Zaman G, Saxon LK, Price JS, Lanyon LE. 2007. Wnt/beta-catenin signaling is a component of osteoblastic bone cell early responses to load-bearing and requires estrogen receptor alpha. *J Biol Chem* 282: 20715–20727.
65. Case N, Sen B, Thomas JA, Styner M, Xie Z, Jacobs CR, Rubin J. 2011. Steady and oscillatory fluid flows produce a similar osteogenic phenotype. *Calcif Tissue Int* 88: 189–197.
66. Lavandero S, Cartagena G, Guarda E, Corbalan R, Godoy I, Sapag-Hagar M, Jalil JE. 1993. Changes in cyclic AMP dependent protein kinase and active stiffness in the rat volume overload model of heart hypertrophy. *Cardiovasc Res* 27: 1634–1638.
67. Dassouli A, Sulpice JC, Roux S, Crozatier B. 1993. Stretch-induced inositol trisphosphate and tetrakisphosphate production in rat cardiomyocytes. *J Mol Cell Cardiol* 25: 973–982.
68. Gudi S, Huvar I, White CR, McKnight NL, Dusserre N, Boss GR, Frangos JA. 2003. Rapid activation of Ras by fluid flow is mediated by Galpha(q) and Gbetagamma subunits of heterotrimeric G proteins in human endothelial cells. *Arterioscler Thromb Vasc Biol* 23: 994–1000.
69. Rubin J, Murphy TC, Fan X, Goldschmidt M, Taylor WR. 2002. Activation of extracellular signal-regulated kinase is involved in mechanical strain inhibition of RANKL expression in bone stromal cells. *J Bone Miner Res* 17: 1452–1460.
70. Judex S, Zhong N, Squire ME, Ye K, Donahue LR, Hadjiargyrou M, Rubin CT. 2005. Mechanical modulation of molecular signals which regulate anabolic and catabolic activity in bone tissue. *J Cell Biochem* 94: 982–994.
71. Ward K, Alsop C, Caulton J, Rubin C, Adams J, Mughal Z. 2004. Low magnitude mechanical loading is osteogenic in children with disabling conditions. *J Bone Miner Res* 19: 360–369.
72. Gilsanz V, Wren TA, Sanchez M, Dorey F, Judex S, Rubin C. 2006. Low-level, high-frequency mechanical signals enhance musculoskeletal development of young women with low BMD. *J Bone Miner Res* 21: 1464–1474.
73. Muir J W, Judex S, Qin YX, Rubin C. 2011. Postural instability caused by extended bed rest is alleviated by brief daily exposure to low magnitude mechanical signals. *Gait Posture* 33: 429–435.
74. Reyes M L, Hernandez M, Holmgren LJ, Sanhueza E, Escobar RG. 2011. High-frequency, low-intensity vibrations increase bone mass and muscle strength in upper limbs, improving autonomy in disabled children. *J Bone Miner Res* 26: 1759–1766.
75. Judex S, Donahue LR, Rubin CT. 2002. Genetic predisposition to osteoporosis is paralleled by an enhanced sensitivity to signals anabolic to the skeleton. *FASEB J* 16(10): 1280–1282:
76. Rubin C, Turner AS, Muller R, Mittra E, McLeod K, Lin W, Qin YX. 2002. Quantity and quality of trabecular bone in the femur are enhanced by a strongly anabolic, noninvasive mechanical intervention. *J Bone Miner Res* 17: 349–357.

47
Calcium and Vitamin D

Bess Dawson-Hughes

- Impact on Bone Mineral Density 403
- Impact on Muscle Strength, Balance, and Falling 404
- Impact on Fracture Rates 404
- Role in Pharmacotherapy 404
- Intake Requirements 405
- Safety 405
- References 406

Calcium is required for the bone formation phase of bone remodeling. Typically, about 5 nmol (200 mg) of calcium is removed from the adult skeleton and replaced each day. To supply this amount, one would need to consume about 600 mg of calcium, since calcium is not very efficiently absorbed. Calcium also affects bone mass through its impact on the remodeling rate. An inadequate intake of calcium results in reduced calcium absorption, a lower circulating ionized calcium concentration, and an increased secretion of parathyroid hormone (PTH), a potent bone-resorbing agent. A high remodeling rate leads to bone loss; it is also an independent risk factor for fracture. Dietary calcium at sufficiently high levels, usually 1,000 mg per day or more, lowers the bone remodeling rate by about 10% to 20% in older men and women and the degree of suppression appears to be dose related [1]. The reduction in remodeling rate accounts for the increase in BMD that occurs in the first 12 to 18 months of treatment with calcium.

With aging there is a decline in calcium absorption efficiency in men and women. This may be related to loss of intestinal vitamin D receptors or resistance of these receptors to the action of 1,25-dihydroxyvitamin D. Diet composition, season, and race also influence calcium absorption efficiency.

Vitamin D is acquired from the diet and from skin synthesis, upon exposure to ultraviolet B rays. The best clinical indicator of vitamin D status is the serum 25-hydroxyvitamin D (25OHD) level. Serum 25OHD levels are lower in individuals using sunscreens and in those with more pigmented skin. Season is an important determinant of vitamin D levels. In much of the temperate zone, skin synthesis of vitamin D does not occur during the winter. Consequently, 25OHD levels fall in the winter and early spring. Serum PTH levels vary inversely with 25OHD levels. These cyclic changes are not benign. Bone loss is greater in the winter/spring when 25OHD levels are lowest (and PTH levels are highest) than in the summer/fall when 25OHD levels are highest (and PTH levels are lowest).

Serum 25OHD levels decline with aging for several reasons. There is less efficient skin synthesis of vitamin D with aging as a result of an age-related decline in the amount of 7-dehydrocholesterol, the precursor to vitamin D, in the epidermal layer of skin [2]. Also, older individuals as a group spend less time out-of-doors. There does not appear to be an impairment in the intestinal absorption of vitamin D with aging [3].

IMPACT ON BONE MINERAL DENSITY

Calcium and vitamin D support bone growth in children and adolescents and lower rates of bone loss in adults and the elderly. However, the association of calcium with bone mass appears to be influenced by vitamin D status. In men and women age 20 years and older in National Health and Nutrition Evaluation Survey III (NHANES III), higher calcium intake was associated with higher femoral neck bone mineral density (BMD) at 25OHD levels below 50 nmol/l (or 20 ng/ml) but not at higher

---

Primer on the Metabolic Bone Diseases and Disorders of Mineral Metabolism, Eighth Edition. Edited by Clifford J. Rosen.
© 2013 American Society for Bone and Mineral Research. Published 2013 by John Wiley & Sons, Inc.

25OHD levels [4]. A meta-analysis of 15 trials found that calcium alone in adults caused positive mean percentage BMD changes from a baseline of 1.7% at the lumbar spine, 1.6% at the hip, and 1.9% at the distal radius [5]. In one trial, the effects of calcium from food (milk powder) and supplement sources on changes in BMD in older postmenopausal women were compared and found to be similar [6].

Higher serum 25OHD levels have been associated with higher BMD of the hip in young and older adult men and women in NHANES III [4]. This association was present across the full range of 25OHD values. Supplementation with vitamin D also reduces rates of bone loss in older adults [7]. In order to sustain the reduced turnover rate and higher bone mass induced by increased calcium and vitamin D intakes, the higher intakes need to be maintained.

IMPACT ON MUSCLE STRENGTH, BALANCE, AND FALLING

In NHANES III women age 60 and older, higher 25OHD levels were associated with improved lower extremity function (faster walking and sit-to-stand speeds) [8]. Similarly, in 1,234 elderly men and women living in the Netherlands, concentrations of 25(OH)D below 50 nmol/L (or 20 ng/ml) were associated with reduced physical performance [9]. Results of vitamin D intervention studies have been more variable. In a meta-analysis of 17 trials, Stockton and colleagues found that supplemental vitamin D had no significant effect on lower extremity muscle strength except in individuals with low starting serum 25OHD levels, less than 25 nmol/L (or 10 ng/ml) [10]. The mechanism(s) by which vitamin D influences muscle performance and strength are not well established but are likely to involve the nuclear vitamin D receptors known to be present in muscle. Supplementation with 1,000 IU of vitamin D_2 daily over a 2-year period, when compared with placebo, significantly increased the diameter of fast twitch type II muscle fibers in older women who had had a recent stroke [11].

Vitamin D appears to have a favorable effect on balance in older adults. Sway, a measure of balance, is assessed with the subject standing on a force plate that measures the maximum displacement in the anteroposterior and medial-lateral directions, the average speed of displacement, and other parameters [12]. In adults age 60 years and older, the amplitude of sway in the medial–lateral direction was a strong predictor of falling more than once per year [13]. In two independent randomized controlled trials, 800 IU of vitamin D_3 plus 1,000 mg of calcium per day, compared with calcium alone, reduced sway by up to 28% over periods of 2 and 12 months [14, 15]. The mechanisms by which vitamin D affects balance have not been defined.

Vitamin D is recommended by several organizations to lower the risk of falling [16–18]. A meta-analysis of eight randomized placebo-controlled vitamin D intervention trials revealed that the effect of supplementation was dose dependent. Higher dose trials (700–1,000 IU/day) showed a risk reduction, whereas lower dose trials did not [19]. The magnitude of the risk reduction in the higher dose trials averaged 19%, or, when recalculated, 34% [20]. A higher dose of 2,000 IU per day, when compared with 800 IU per day, showed no fall risk reduction in elderly acute hip fracture patients [21], suggesting that 800 IU per day is adequate for this outcome. The serum 25OHD level needed to reduce falls was estimated to be at least 60 nmol/l (or 24 ng/ml) [19]. The impact of vitamin D on falls is likely to be mediated by its effect on muscle strength and balance.

IMPACT ON FRACTURE RATES

Several small studies have examined the impact of supplemental calcium on fracture rates. The Shea meta-analysis of these studies (13) found that calcium alone (versus placebo) tended to lower risk of vertebral fractures [RR 0.77 (CI 0.54–1.09)] but not nonvertebral fractures [RR 0.86 (CI 0.43–1.72)]. The studies in this analysis ranged from 18 months to 4 years in duration. A more recent meta-analysis reached similar conclusions, citing a trend toward a modest reduction in risk of nonvertebral fracture [RR 0.92 (95% CI: 0.81, 1.05)] and no significant effect on hip fracture risk [RR 1.64 (95% CI: 1.02, 2.64)] [22].

The effect of supplemental vitamin D on fracture incidence has been examined in several randomized controlled trials with varying results. A recent meta-analysis of these trials revealed that the response to supplementation was dose dependent. The trials using received doses >400 IU/day per day were positive for nonvertebral fractures [RR 0.80 (0.72–0.89)] and hip fractures [RR 0.82 (0.69–0.97)], whereas those using received doses of less than 400 IU per day were null [23]. Received dose is the product of administered dose and percentage adherence. Serum 25OHD levels were measured in most of these trials and among these, a group mean value of approximately 75 nmol/l (or 30 ng/ml) was needed to lower hip fracture risk [23]. The reduced risk of fracture likely results from effects of vitamin D on muscle, balance, fall risk, and bone metabolism. In many of the higher dose vitamin D trials, calcium was given along with vitamin D; based on this and other evidence, it is reasonable to replete both calcium and vitamin D. Several other meta-analyses have reached different conclusions, at least in part because different specific questions were posed and different selection criteria for study inclusion were applied [24, 25].

ROLE IN PHARMACOTHERAPY

In recent randomized controlled trials testing the antifracture efficacy of the antiresorptive therapies alendronate, risedronate, raloxifene, and calcitonin, and the anabolic drug, PTH 1-34, calcium and vitamin D have been given to both the control and intervention groups. This allows one to define the impact of these drugs in calcium- and vitamin D-replete patients and to conclude that any efficacy of the drugs is beyond that associated with calcium and vitamin D alone. However, one cannot conclude that these drugs would have the same efficacy in calcium- and vitamin D-deficient patients.

INTAKE REQUIREMENTS

Calcium intake recommendations vary enormously worldwide. Recommendations by the Institute of Medicine (IOM) of the U.S. National Academy of Sciences are among the highest. The IOM recommended intakes of calcium are: ages 1–3 years, 500 mg; 4–8 years, mg; 9–18 years, 1,100 mg; 19–50 years, 800 mg; 51–70 years, 800 mg for men and 1,000 mg for women; older than 70 years, 1,000 mg per day [26]. Lower calcium intakes would likely be adequate for populations with lower intakes of salt and protein, two diet components that promote calcium excretion in the urine.

Among females in the U.S., fewer than 1 in 4 meet the calcium requirement through their diets; when calcium from supplement and food sources is considered, about half of U.S. females meet the requirement [26]. Among males, calcium intake from food is somewhat higher and from supplements somewhat lower than among females, with the result that, considering combined calcium sources, about half of U.S. males meet the calcium requirement [26]. Calcium from calcium carbonate, the most commonly used supplement, is better absorbed when taken with a meal [27, 28]. Absorption from all supplements is more efficient in doses up to 500 mg than from higher doses [29]. Thus, individuals requiring more than 500 mg per day from supplements should take it in divided doses.

The vitamin D intake recommendations of the IOM are: for males and females age 1–70 years, 1.5 mcg (600 IU), and for ages older than 70 years, 2 mcg (800 IU) per day [26]. For older adults these recommendations are based on BMD and fracture risk. The IOM concluded that "a 25OHD level of 40 nmol/l was consistent with the intended nature of an average requirement, in that it reflects the desired level for a population median—it meets needs for approximately half of the population" [26]. The IOM recommends a level of 50 nmol/l to meet the need of 97.5% of the population [26]. There is lack of consensus in the professional community about whether the target group mean 25OHD level should be 40, 50, 75 nmol/l or another number. Several organizations recommend 75 nmol/l as the target [16–18].

A vitamin D intake of 800 IU (15 mcg) per day is not adequate to bring more than about half of the elderly population to 25OHD levels of 75 nmol/L (or 30 ng/ml). Older men and women who are at average risk for low 25OHD levels will need an intake of 20 to 25 mcg (800 to 1000 IU) per day to maintain a serum 25OHD level of 75 nmol/L (30 ng/ml). Individuals at increased risk for low serum 25OHD levels, such as those with limited regular skin production of vitamin D (related dark skin, little time out of doors, sunscreen use, protective clothing, high latitude), obesity, malabsorption, and other conditions that limit absorption or alter vitamin D metabolism will need more than 800 to 1,000 IU to maintain a 25OHD level of 75 nmol/l (or 30 ng/ml). The increase in 25OHD with supplementation is inversely related to the starting level. At low starting levels, 1 mcg (40 IU) of vitamin D will increase serum 25OHD by 1.2 nmol/L (or 0.48 ng/ml); at a higher starting level of 70 nmol/L (28 ng/ml), the increase from this dose would be only about 0.7 nmol/L (or 0.28 ng/ml) [30, 31]. Vitamin D is available in two forms: the plant-derived ergocalciferol (D_2) and animal-derived cholecalciferol (D_3). For years these forms were considered to be equipotent in humans but recent evidence indicates that vitamin D_3 increases serum 25OHD levels more efficiently than vitamin D_2 [32]. Moreover, vitamin D_2 is not accurately measured in all 25OHD assays [33]. For these reasons, vitamin D_3, when available, is the preferred form for clinical use.

SAFETY

Recent reports have raised the issue of potential risk associated with excessive calcium supplement use. Bolland and colleagues reported that calcium supplement use without coadministered vitamin D increased risk of myocardial infarction [34]. However, increased risk was not observed in another earlier meta-analysis [35]. A detailed report from the Women's Health Initiative revealed a 17% increase in renal stones in the women treated with calcium and vitamin D as compared to the women treated with placebo [36]. Individuals with high calcium intake from food sources do not share this risk and may in fact have reduced risk of nephrolithiasis [37]. All things considered, it would seem prudent to obtain calcium from food sources to the greatest extent possible and to use supplements only as needed to bring total intake up to recommended levels. The IOM and others have identified no risk associated with serum 25OHD levels up through 125 nmol/l (or 50 ng/ml) [26].

The safe upper limits for calcium set by the IOM are: ages 1–8 years, 2,500 mg; 9–18 years, 3,000 mg; 19–50 years, 2,500 mg; older than 50 years, 2,000 mg per day [26]. The IOM has placed the safe upper limit for vitamin D at: ages 1–3 years, 2,500 IU; ages 4–8 years, 3,000 IU; and ages older than 8 years, 4,000 IU per day [26].

In conclusion, adequate intakes of calcium and vitamin D are essential preventative measures and essential components of any therapeutic regimen for osteoporosis. Many men and women will need supplements to meet the intake requirements. There is no known advantage but there is potential risk in exceeding current calcium intake recommendations, particularly with use of supplements. Currently, the most common target serum 25OHD levels for musculoskeletal health are 50 nmol/l (20 ng/ml), recommended by the IOM, and 75 nmol/l (30 ng/ml), recommended by many specialty groups and others. There is a consensus that both of these target levels are safe. Additional research is needed to fully define the musculoskeletal and other health effects of different doses of vitamin D and serum 25OHD levels.

REFERENCES

1. Elders PJ, Lips P, Netelenbos JC, et al. 1994. Long-term effect of calcium supplementation on bone loss in perimenopausal women. *J Bone Miner Res* 9(7): 963–970.
2. MacLaughlin J, Holick MF. 1985. Aging decreases the capacity of human skin to produce vitamin D3. *J Clin Invest* 76(4): 1536–1538.
3. Harris SS, Dawson-Hughes B. 2002. Plasma vitamin D and 25OHD responses of young and old men to supplementation with vitamin D3. *J Am Coll Nutr* 21(4): 357–362.
4. Bischoff-Ferrari HA, Kiel DP, Dawson-Hughes B, et al. 2009. Dietary calcium and serum 25-hydroxyvitamin D status in relation to BMD among U.S. adults. *J Bone Miner Res* 24(5): 935–942.
5. Shea B, Wells G, Cranney A, et al. 2002. VII. Meta-analysis of calcium supplementation for the prevention of postmenopausal osteoporosis. *Endocrine Rev* 23(4): 552–559.
6. Prince R, Devine A, Dick I, et al. 1995. The effects of calcium supplementation (milk powder or tablets) and exercise on bone density in postmenopausal women. *J Bone Miner Res* 10(7): 1068–1075.
7. Ooms ME, Roos JC, Bezemer PD, van der Vijgh WJ, Bouter LM, Lips P. 1995. Prevention of bone loss by vitamin D supplementation in elderly women: A randomized double-blind trial. *J Clin Endocrin Metab* 80(4): 1052–1058.
8. Bischoff-Ferrari HA, Dietrich T, Orav EJ, et al. 2004. Higher 25-hydroxyvitamin D concentrations are associated with better lower-extremity function in both active and inactive persons aged > or =60 y. *Am J Clin Nutr* 80(3): 752–758.
9. Wicherts IS, van Schoor NM, Boeke AJ, et al. 2007. Vitamin D status predicts physical performance and its decline in older persons. *J Clin Endocrinol Metab* 92(6): 2058–2065.
10. Stockton KA, Mengersen K, Paratz JD, Kandiah D, Bennell KL. 2011. Effect of vitamin D supplementation on muscle strength: A systematic review and meta-analysis. *Osteoporos Int* 22(3): 859–871.
11. Sato Y, Iwamoto J, Kanoko T, Satoh K. 2005. Low-dose vitamin D prevents muscular atrophy and reduces falls and hip fractures in women after stroke: A randomized controlled trial. *Cerebrovasc Dis* 20(3): 187–192.
12. Swanenburg J, de Bruin ED, Favero K, Uebelhart D, Mulder T. 2008. The reliability of postural balance measures in single and dual tasking in elderly fallers and non-fallers. *BMC Musculoskelet Disord* 9: 162.
13. Swanenburg J, de Bruin ED, Uebelhart D, Mulder T. 2010. Falls prediction in elderly people: A 1-year prospective study. *Gait Posture* 31(3): 317–321.
14. Pfeifer M, Begerow B, Minne HW, Abrams C, Nachtigall D, Hansen C. 2000. Effects of a short-term vitamin D and calcium supplementation on body sway and secondary hyperparathyroidism in elderly women. *J Bone Miner Res* 15(6): 1113–1118.
15. Bischoff HA, Stahelin HB, Dick W, et al. 2003. Effects of vitamin D and calcium supplementation on falls: A randomized controlled trial. *J Bone Miner Res* 18(2): 343–351.
16. Michael YL, Whitlock EP, Lin JS, Fu R, O'Connor EA, Gold R. 2010. Primary care-relevant interventions to prevent falling in older adults: A systematic evidence review for the U.S. Preventive Services Task Force. *Ann Intern Med* 153(12): 87–825.
17. Dawson-Hughes B, Mithal A, Bonjour JP, et al. IOF position statement: Vitamin D recommendations for older adults. *Osteoporos Int* 24(4): 1151–1154.
18. Holick MF, Binkley NC, Bischoff-Ferrari HA, et al. 2011. Evaluation, treatment, and prevention of vitamin d deficiency: An Endocrine Society clinical practice guideline. *J Clin Endocrinol Metab* 96(7): 1911–1930.
19. Bischoff-Ferrari HA, Dawson-Hughes B, Staehelin HB, et al. 2009. Fall prevention with supplemental and active forms of vitamin D: A meta-analysis of randomised controlled trials. *Br Med J* 339: b3692.
20. Bischoff-Ferrari HA, Willett WC, Orav JE, Kiel DP, Dawson-Hughes B. 2011. Fall prevention with vitamin D: Author's reply. *Br Med J* 342: d2608
21. Bischoff-Ferrari HA, Dawson-Hughes B, Platz A, et al. 2010. Effect of high-dosage cholecalciferol and extended physiotherapy on complications after hip fracture: A randomized controlled trial. *Arch Intern Med* 170(9): 813–820.
22. Bischoff-Ferrari HA, Dawson-Hughes B, Baron JA, et al. 2007. Calcium intake and hip fracture risk in men and women: A meta-analysis of prospective cohort studies and randomized controlled trials. *Am J Clin Nutr* 86(6): 1780–1790.
23. Bischoff-Ferrari HA, Willett WC, Wong JB, et al. 2009. Prevention of nonvertebral fractures with oral vitamin D and dose dependency: A meta-analysis of randomized controlled trials. *Arch Intern Med* 169(6): 551–561.
24. Tang BM, Eslick GD, Nowson C, Smith C, Bensoussan A. 2007. Use of calcium or calcium in combination with vitamin D supplementation to prevent fractures and bone loss in people aged 50 years and older: A meta-analysis. [See comment]. *Lancet* 370(9588): 657–666.
25. Boonen S, Lips P, Bouillon R, Bischoff-Ferrari HA, Vanderschueren D, Haentjens P. 2007. Need for additional calcium to reduce the risk of hip fracture with vitamin d supplementation: evidence from a comparative meta-analysis of randomized controlled trials. *J Clin Endocrinol Metab* 92(4): 1415–1423.

26. IOM (Institute of Medicine). 2011. *Dietary Reference Intakes for Calcium and Vitamin D*. Washington, DC: The National Academies Press.
27. Heaney RP, Smith KT, Recker RR, Hinders SM. 1989. Meal effects on calcium absorption. *Am J Clin Nutr* 49(2): 372–376.
28. Recker RR. 1985. Calcium absorption and achlorhydria. *N Engl J Med* 313(2): 70–73.
29. Harvey JA, Zobitz MM, Pak CY. 1988. Dose dependency of calcium absorption: A comparison of calcium carbonate and calcium citrate. *J Bone Miner Res* 3(3): 253–258.
30. Vieth R, Ladak Y, Walfish PG. 2003. Age-related changes in the 25-hydroxyvitamin D versus parathyroid hormone relationship suggest a different reason why older adults require more vitamin D. *J Clin Endocrinol Metab* 88(1): 185–191.
31. Heaney RP, Davies KM, Chen TC, Holick MF, Barger-Lux MJ. 2003. Human serum 25-hydroxycholecalciferol response to extended oral dosing with cholecalciferol. *Am J Clin Nutr* 77(1): 204–210.
32. Heaney RP, Recker RR, Grote J, Horst RL, Armas LA. 2011. Vitamin D(3) is more potent than vitamin D(2) in humans. *J Clin Endocrinol Metab* 96(3): E447–452.
33. Binkley N, Krueger D, Cowgill CS, et al. 2004. Assay variation confounds the diagnosis of hypovitaminosis D: A call for standardization. *J Clin Endocrinol Metab* 89(7): 3152–3157.
34. Bolland MJ, Avenell A, Baron JA, et al. 2010. Effect of calcium supplements on risk of myocardial infarction and cardiovascular events: Meta-analysis. *Br Med J* 341: c3691.
35. Wang L, Manson JE, Song Y, Sesso HD. 2010. Systematic review: Vitamin D and calcium supplementation in prevention of cardiovascular events. *Ann Intern Med* 152(5): 315–323.
36. Wallace RB, Wactawski-Wende J, O'Sullivan MJ, et al. 2011. Urinary tract stone occurrence in the Women's Health Initiative (WHI) randomized clinical trial of calcium and vitamin D supplements. *Am J Clin Nutr* 94(1): 270–277.
37. Serio A, Fraioli A. 1999. Epidemiology of nephrolithiasis. *Nephron* 81 Suppl 1: 26–30.

48

Estrogens, Estrogen Agonists/Antagonists, and Calcitonin

Nelson B. Watts

Estrogen 408
Estrogen Agonists/Antagonists (Selective Estrogen Receptor Modulators or Serms) 409

Calcitonin 409
Summary 410
Selected Readings 410

ESTROGEN

It is well established that osteoporosis is more common in women than men and fracture risk increases dramatically after menopause. The relationship between estrogen deficiency and osteoporosis was first suggested in the mid-1900s when Albright showed that treatment with estrogen reversed the negative calcium balance in postmenopausal women. The positive effects of estrogen on peak bone mass and on bone loss have been demonstrated in both men and women. For years, estrogen was considered the treatment of choice for postmenopausal women and widely advocated for prevention of bone loss, with numerous trials showing prevention of bone loss in recently menopausal women and increased of bone mineral density (BMD) in women with postmenopausal osteoporosis. Part of the emphasis on estrogen therapy was the assumption of extraskeletal benefits (lower rates of cardiovascular events and dementia in women who chose to take estrogen compared with those who chose not to) and also that the benefits would be durable after estrogen was stopped.

The Women's Health Initiative (WHI) confirmed the value of estrogen administration (with or without a progestin) in the treatment of osteoporosis and reduction of risk for hip and other clinical fractures. However, the presumed extraskeletal benefits were not confirmed, and the effects on BMD and fracture reduction disappeared within a year of discontinuation of estrogen (loss of BMD of approximately 5% in the first year after stopping).

Thus, it appears that estrogen has protective effects on the skeleton only for as long as it is taken. However, women who take estrogen and then stop should be no worse off compared with women who have never taken estrogen at all.

Following results of the WHI, estrogen treatment is recommended for relief of menopausal symptoms, but in the lowest dose and for the shortest time necessary. Most women are not sufficiently symptomatic at menopause to require estrogen therapy, and most who require estrogen will be able to stop after a few years. However, a minority will require estrogen long-term. For those women, estrogen may be sufficient for prevention and/or treatment of osteoporosis.

Several different forms of estrogen are available for therapeutic use (e.g., estradiol, conjugated estrogen esterified estrogens, etc.) as well as several routes of administration (e.g., oral, transdermal). In addition to different compounds and routes of administration, the dose can vary as well. After hysterectomy, estrogen can be use "unopposed"; however, for women who still have their uterus, "opposing" estrogen with a progestin (of which there are several, e.g., progesterone, medroxyprogesterone acetate) is recommended for protection against endometrial hyperplasia and endometrial carcinoma. The specific therapy that was shown to reduce fracture risk in WHI was orally administered conjugated estrogen, 0.625 mg daily (also effective was the combination of 0.625 mg conjugated estrogen plus medroxyprogesterone acetate 5 mg daily; however, there is little or no evidence that progestins have any effect on bone). It is likely, but

Primer on the Metabolic Bone Diseases and Disorders of Mineral Metabolism, Eighth Edition. Edited by Clifford J. Rosen.
© 2013 American Society for Bone and Mineral Research. Published 2013 by John Wiley & Sons, Inc.

not proven, that other estrogen preparations, as well as lower doses, might have beneficial effects on maintenance of BMD and reducing fracture risk.

Estrogens are considered by many to be "antiresorptive agents"; however, they reduce the rates of both bone resorption and bone formation but reduce resorption more than formation.

In summary, despite its important role in physiology and pathophysiology, estrogen is not recommended for prevention or treatment of postmenopausal osteoporosis, but should provide protection against bone loss and fractures for women who require estrogen from relief of menopausal symptoms.

ESTROGEN AGONISTS/ANTAGONISTS (SELECTIVE ESTROGEN RECEPTOR MODULATORS OR SERMS)

Tamoxifen, widely used for treatment and prevention of breast cancer, was thought of as an antiestrogen, based on the concept that a compound that blocked estrogen receptors in breast tissue would reduce the risk of recurrence or spread of breast cancer. Clinical trials of tamoxifen in postmenopausal women that were expected to show acceleration of bone loss actually showed modest gains in BMD, as well as decreases in bone turnover markers, establishing that this compound blocks the effect of estrogen in some tissues but acts like estrogen in others.

Rather than being an antiestrogen, tamoxifen and functionally related compounds are classified as estrogen agonists/antagonists, also known as selective estrogen receptor modulators (SERMs) and tissue-selective estrogens. They bind with estrogen receptors, activating estrogen pathways in some tissues and blocking them in others. Effects appear to be similar and consistent in some tissues (e.g., antagonists to estrogen in breast tissue and partial agonists in bone) but compound-specific in others (e.g., some stimulate the endometrium, while others are neutral, and still others are antagonists). The ideal compound would relieve menopausal symptoms (including vasomotor symptoms and vaginal dryness), maintain or increase BMD and reduce the risks of fractures at all skeletal sites, as well as reducing the risks of cardiovascular disease, dementia, genitourinary problems, and breast, endometrial, and ovarian cancer without increasing the risk of venous thromboembolic events. This "holy grail" has not yet been attained. Several promising compounds failed or have been limited because of safety concerns (tamoxifen increases the risk of endometrial hyperplasia and carcinoma) or because the skeletal effects were limited to reducing fractures in the spine, with no effect on nonvertebral sites.

Raloxifene, the first of these agents specifically marketed for positive bone effects and given orally in a dose of 60 mg daily, was shown to prevent bone loss in recently menopausal women and increase BMD in postmenopausal women, although the gains in BMD appear to be about half of what is observed with estrogen. The risk of new and worsening vertebral fractures was significantly reduced, but raloxifene has not been shown to have an effect on hip and nonvertebral fractures. Confirming a reduction in breast cancer noted in the osteoporosis study, raloxifene was shown in the Study of Tamoxifen and Raloxifene (STAR) trial to reduce the risk of breast cancer in postmenopausal women at increased risk of breast cancer and in postmenopausal women with osteoporosis. Raloxifene does not stimulate the endometrium. Although raloxifene has effects on lipids that suggest it might be cardioprotective (decreased triglycerides, total, and LDL cholesterol), no benefit was shown in the large Raloxifene Use for the Heart (RUTH) trial. The occurrence of stroke was not different between raloxifene and placebo groups; however, there were more fatal strokes in the raloxifene group. Enthusiasm for the use of raloxifene has been limited because of the lack of evidence for protection against hip and other nonvertebral fractures, as well as side effects and safety concerns (increased menopausal symptoms, increased risk of venous thromboembolic events such a deep vein thrombosis, pulmonary embolus, and retinal vein thrombosis). The increase in venous thromboembolic events is similar to estrogen and is highest during the initial months of treatment.

Several estrogen agonists/antagonists have fallen by the wayside, while a few have made it into Phase 3 trials for osteoporosis. Commercialization of arzoxifene was abandoned because of lack of protection against hip and other nonvertebral fractures. Lasofoxifene and bazedoxifene are approved in Europe but not in the U.S. Bazedoxifene has been studied in combination with estrogen, with the estrogen improving vasomotor symptoms and vulvovaginal atrophy.

CALCITONIN

In humans, calcitonin is a 32-amino acid peptide produced by specialized C cells in the thyroid. Osteoclasts express receptors for calcitonin and respond to calcitonin with a rapid decrease in resorptive capacity. Unlike estrogen deficiency in adults, which has been shown to cause bone loss and increase fracture risk, there is no evidence for an important physiologic effect of calcitonin in adults, and patients with calcitonin deficiency (i.e., after total thyroidectomy) show no changes in skeletal status or mineral homeostasis. As a pharmacologic agent, human, eel, and salmon calcitonin have all been tried, but synthetic salmon calcitonin has been the major product in clinical use. Given by subcutaneous injection, salmon calcitonin has been use to treat Paget's disease, hypercalcemia, and osteoporosis but with significant side effects of injection site reactions, flushing, and nausea in up to 20% of patients. In women with osteoporosis who were at least 5 years postmenopausal, nasal spray salmon calcitonin in a dose of 200 IU daily was shown to reduce the risk of vertebral fracture and was well tolerated; however, lower (100 IU/day) and higher (400 IU/day) doses did not show an anti-fracture effect. Changes in BMD and bone turnover markers were minimal. There have been small trials of salmon calcitonin treatment in men with osteoporosis and men and women with glucocorticoid-induced osteoporosis.

Some patients appear to develop resistance to calcitonin, which could be due to the development of tachyphylaxis or the development of neutralizing antibodies. Treatment with calcitonin may lead to a reduction in plasma lithium due to increased renal clearance of lithium.

Although not included in the approved indications for calcitonin, studies suggest a possible analgesic effect, possibly mediated through an increase in circulating and central nervous system (CNS) endorphins, which may be clinically useful in patients with acute painful vertebral fractures.

An oral form of calcitonin is being studies as possible treatment for osteoarthritis and osteoporosis.

SUMMARY

Estrogen is indicated for relief of menopausal symptoms; for women who are candidates for estrogen either short term or long term, bone effects may be considered a "side benefit." Raloxifene has a limited role for prevention and treatment of osteoporosis and for reducing the risk of breast cancer. Nasal spray calcitonin has questionable evidence for beneficial effects on BMD, bone turnover markers, or fracture risk, but may have a role in reducing pain in patients with acute painful vertebral fractures.

SELECTED READINGS

Estrogen

1. Anderson GL, Limacher M, Assaf AR, Bassford T, Beresford SA, Black H, Bonds D, Brunner R, Brzyski R, Caan B, Chlebowski R, Curb D, Gass M, Hays J, Heiss G, Hendrix S, Howard BV, Hsia J, Hubbell A, Jackson R, Johnson KC, Judd H, Kotchen JM, Kuller L, LaCroix AZ, Lane D, Langer RD, Lasser N, Lewis CE, Manson J, Margolis K, Ockene J, O'Sullivan MJ, Phillips L, Prentice RL, Ritenbaugh C, Robbins J, Rossouw JE, Sarto G, Stefanick ML, Van Horn L, Wactawski-Wende J, Wallace R, Wassertheil-Smoller S; Women's Health Initiative Steering Committee. 2004. Effects of conjugated equine estrogen in postmenopausal women with hysterectomy: The Women's Health Initiative randomized controlled trial. *JAMA* 291: 1701–1712.
2. Ansbacher R. 2001. The pharmacokinetics and efficacy of different estrogens are not equivalent. *Am J Obstet Gynecol* 184: 255–263.
3. Cauley JA, Robbins J, Chen Z, Cummings SR, Jackson RD, LaCroix AZ, LeBoff M, Lewis CE, McGowan J, Neuner J, Pettinger M, Stefanick ML, Wactawski-Wende J, Watts NB, Women's Health Initiative Investigators. 2003. Effects of estrogen plus progestin on risk of fracture and bone mineral density: The Women's Health Initiative randomized trial. *JAMA* 290: 1729–1738.
4. Gallagher JC, Rapuri PB, Haynatzki G, Detter JR. 2002. Effect of discontinuation of estrogen, calcitriol, and the combination of both on bone density and bone markers. *J Clin Endocrinol Metab* 87: 4914–4923.
5. Greendale GA, Espeland M, Slone S, Marcus R, Barrett-Connor E. 2002. Bone mass response to long-term hormone discontinuation replacement therapy: Results from the Postmenopausal Estrogen/Progestin Interventions (PEPI) safety follow-up study. *Arch Intern Med* 162: 665–672.
6. Heiss G, Wallace R, Anderson GL, Aragaki A, Beresford SAA, Brzyski R, Chlebowski RT, Gass M, Lacroix A, Manson JE, Prentice RL, Rossouw J, Stefanick ML; WHI Investigators. 2008. Health risks and benefits 3 years after stopping randomized treatment with estrogen and progestin. *JAMA* 399: 1036–1045.
7. Lindsay R, Gallagher JC, Kleerekoper M, Pickar JH. 2002. Effect of lower doses of conjugated equine estrogens with and without medroxyprogesterone acetate on bone in early postmenopausal women. *JAMA* 287: 2668–2676.
8. MellstromD, Vandenput L, Mallmin H, Holmberg A, Lorentzon M, Oden A, Johansson H, Orwoll E, Labrie F, Karlsson M, Ljunggren Ö, Ohlsson C. 2008. Older men with low serum estradiol and high serum SHBG have an increased risk of fractures. *J Bone Miner Res* 23: 1552–1560.
9. Prestwood KM, Thompson DL, Kenny AM, Seibel MJ, Pilbeam CC, Raisz LG. 1999. Low dose estrogen and calcium have an additive effect on bone resorption in older women. *J Clin Endocrinol Metab* 84: 179–183.
10. ReckerR, Lappe J, Davies K, Heaney R. 2000. Characterization of perimenopausal bone loss: A prospective study. *J Bone Miner Res* 15: 1965–1973.
11. Rossouw, JE, Anderson GL, Prentice RL, LaCroix AZ, Kooperberg C, Stefanick ML, Jackson RD, Beresford SA, Howard BV, Johnson KC, Kotchen JM, Ockene J. 2002. Risks and benefits of estrogen plus progestin in healthy postmenopausal women. Results from the Women's Health Initiative randomized trial. *JAMA* 288: 321–333.
12. Watts NB, Nolan JC, Brennan JJ, Yang H-M; ESTRATAB/Osteoporosis Study Group. 2000. Esterified estrogen therapy in postmenopausal women. Relationships of bone marker changes and plasma estradiol to BMD changes: A two-year study. *Menopause* 7: 375–382.

Estrogen agonists/antagonists

1. Barrett-Connor E, Mosca L, Collins P, Geiger MJ, Grady D, Kornitzer M, McNabb MA, Wenger NK; Raloxifene

Use for The Heart (RUTH) Trial Investigators. 2006. Effects of raloxifene on cardiovascular events and breast cancer in postmenopausal women. *N Engl J Med* 355: 125–137.
2. Bolognese MA. 2010. SERMs and SERMs with estrogen for postmenopausal osteoporosis. *Rev Endo Metabol Disord* 11: 253–259.
3. Cranney A, Tugwell P, Zytaruk N, Robinson V, Weaver B, Adachi J, Wells G, Shea B, Guyatt G; Osteoporosis Methodology Group and The Osteoporosis Research Advisory Group. 2002. Meta-analysis of raloxifene for the prevention and treatment of postmenopausal osteoporosis. *Endocr Rev* 23: 524–528.
4. Cummings SR, Ensrud K, Delmas PD, LaCroix AZ, Vukicevic S, Reid DM, Goldstein S, Sriram U, Lee A, Thompson J, Armstrong RA, Thompson DD, Powles T, Zanchetta J, Kendler D, Neven P, Eastell R, for the PEARL Study Investigators. 2010. Lasofoxifene in postmenopausal women with osteoporosis. *N Engl J Med* 362: 868–896.
5. Ensrud K, Genazzani AR, Geiger MJ, McNabb M, Dowsett SA, Cox DA, Barrett-Connor E. 2006. Effect of *raloxifene* on cardiovascular adverse events in postmenopausal women with osteoporosis. *Am J Cardiol* 97: 520–527.
6. Ettinger B, Black DM, Mitlak BH, Knickerbocker RK, Nickelsen T, Genant HK, Christiansen C, Delmas PD, Zanchetta JR, Stakkestad J, Glüer CC, Krueger K, Cohen FJ, Eckert S, Ensrud KE, Avioli LV, Lips P, Cummings SR. 1999. Reduction of vertebral fracture risk in postmenopausal women with osteoporosis treated with raloxifene: Results from a 3-year randomized clinical trial. Multiple Outcomes of Raloxifene Evaluation (MORE) Investigators. *JAMA* 282: 637–645.
7. Goldstein SR, Neven P, Cummings S, Colgan T, Runowicz CD, Krpan D, Proulx J, Johnson M, Thompson D, Thompson J, Sriram U. 2010. Postmenopausal Evaluation and Risk Reduction with Lasofoxifene (PEARL) trial: 5-year gynecological outcomes. *Menopause* 18: 17–22.
8. Lobo RA, Pinkerton JV, Gass ML, Dorin MH, Ronkin S, Pickar JH, Constantine G. 2009. Evaluation of bazedoxifene/conjugated estrogens for the treatment of menopausal symptoms and effects on metabolic parameters and overall safety profile. *Fertil Steril* 92: 1025–1038.
9. Pickar JH, Mirkin S. 2010. Tissue-selective agents: Selective estrogen receptor modulators and the tissue-selective estrogen complex. *Menopause Int* 16: 121–128.
10. Silverman SL. 2010. New selective estrogen receptor modulators (SERMs) in development. *Curr Osteoporos Rep* 8: 151–153.
11. Silverman SL, Christiansen C, Genant HK, Vukicevic S, Zanchetta JR, de Villiers TJ, Constantine GD, Chines AA. 2008. Efficacy of bazedoxifene in reducing new vertebral fracture risk in postmenopausal women with osteoporosis: Results from a 3-year, randomized, placebo-, and active-controlled clinical trial. *J Bone Miner Res* 12: 1923–1934.
12. Stefanick ML. 2006. Risk-benefit profiles of raloxifene for women. *N Engl J Med* 355: 190–192.

Calcitonin

1. Azria M, Copp DH, Zanelli JM. 1995. 25 years of salmon calcitonin: From synthesis to therapeutic use. *Calcif Tissue Int* 57: 1–4.
2. Chesnut CH III, Azria M, Silverman S, Engelhardt M, Olson M, Mindeholm L. 2008. Salmon calcitonin: A review of current and future therapeutic indications. *Osteoporos Int* 19: 479–491.
3. Chesnut CH 3rd, Silverman S, Andriano K, Genant H, Gimona A, Harris S, Kiel D, LeBoff M, Maricic M, Miller P, Moniz C, Peacock M, Richardson P, Watts N, Baylink DJ, 2000. A randomized trial of nasal spray salmon calcitonin in postmenopausal women with established osteoporosis: The prevent recurrence of osteoporotic fractures study. PROOF Study Group *Am J Med* 102: 267–276.
4. Cranney A, Tugwell P, Zytaruk N, Robinson V, Weaver B, Shea B, Wells G, Adachi J, Waldegger L, Guyatt G; Osteoporosis Methodology Group and The Osteoporosis Research Advisory Group. 2002. Meta-analyses of therapies for postmenopausal osteoporosis. VI. Meta-analysis of calcitonin for the treatment of postmenopausal osteoporosis. *Endocr Rev* 23: 540–551.
5. Henriksen K, Bay-Jensen AC, Christiansen C, Karsdal MA. 2010. Oral salmon calcitonin—Pharmacology in osteoporosis. *Exp Opin Biol Therap* 10: 1617–1627.
6. Huang CL, Sun L, Moonga BS, Zaidi M 2006. Molecular physiology and pharmacology of calcitonin. *Cell Molec Biol* 52: 33–43.
7. Knopp JA, Diner BM, Blitz M, Lyritis GP, Rowe, BH. 2005. Calcitonin for treating acute pain of osteoporotic vertebral compression fractures: A systematic review of randomized controlled trials. *Osteoporos Int* 16: 1281–1290.
8. Tanko, LB, Bagger YZ, Alexandersen, P, Devogelaer JP, Reginster JY, Chick R, Olson M, Benmammar H, Mindeholm L, Azria M, Christiansen C. 2004. Safety and efficacy of a novel salmon calcitonin (sCT) technology-based oral formulation in healthy postmenopausal women: acute and 3-month effects on biomarkers of bone turnover. *J Bone Miner Res* 19: 1531–1538.

49
Bisphosphonates for Postmenopausal Osteoporosis

Socrates E. Papapoulos

Pharmacology 412
Antifracture Efficacy 413
Long-Term Effects on Bone Fragility 415
Special Issues Related to Treatment of Osteoporosis with Bisphosphonates 415
Conclusions 417
Acknowledgments 417
References 417

Bisphosphonates (BPs) are synthetic compounds that have high affinity for calcium crystals, concentrate selectively in the skeleton, and decrease bone resorption. The first BP was synthesized in the 19th century but their relevance to medicine was recognized in the 1960s, and they were first given to patients with osteoporosis in the early 1970s. Currently, alendronate, ibandronate, risedronate, and zoledronate are approved for the treatment of osteoporosis worldwide while other BPs are also available in some countries.

PHARMACOLOGY

BPs are synthetic analogs of inorganic pyrophosphate in which the oxygen atom that connects the two phosphates is replaced by a carbon (Fig. 49.1). This substitution renders BPs resistant to biological degradation and suitable for clinical use. BPs have two additional side chains (R1 and R2) that allow the synthesis of a large number of analogs with different pharmacological properties (Fig. 49.1). A hydroxyl substitution at R1 enhances the affinity of BPs for calcium crystals, while the presence of a nitrogen atom in R2 enhances their potency and determines their mechanism of action. The whole molecule is responsible for the action of BPs on bone resorption and probably also for their affinity for bone mineral [1, 2].

The intestinal absorption of BPs is poor (less than 1%) and decreases further in the presence of food, calcium, or other minerals that bind them. Oral BPs should be given in the fasting state 30 to 60 minutes before meals, with water. BPs are cleared rapidly from the circulation; about 50% of the administered dose concentrates in the skeleton, primarily at active remodeling sites, while the rest is excreted unmetabolized in urine. Skeletal uptake depends on the rate of bone turnover, renal function, as well as on the structure of BPs [3]. The capacity of the skeleton to retain BP is large, and saturation of binding sites with the doses used in the treatment of osteoporosis is unlikely even if these are given for a very long time. At the bone surface, BPs inhibit bone resorption and are subsequently embedded in bone where they remain for long and are pharmacologically inactive. The elimination of BPs from the body is multiexponential; the calculated terminal half-life of elimination from the skeleton can be as long as 10 years, and pamidronate has been detected in urines of patients for up to 8 years after discontinuation of treatment. This slow release of the BP from the skeleton is probably responsible for the slow speed of reversal of the effect of BPs on bone following cessation of treatment, which is different from that of all other antiosteoporotic treatments. The rate of reversal of the effect may be different among BPs depending on their pharmacological properties, particularly their affinity for bone mineral, but no head-to-head studies have addressed this issue in humans.

Primer on the Metabolic Bone Diseases and Disorders of Mineral Metabolism, Eighth Edition. Edited by Clifford J. Rosen.
© 2013 American Society for Bone and Mineral Research. Published 2013 by John Wiley & Sons, Inc.

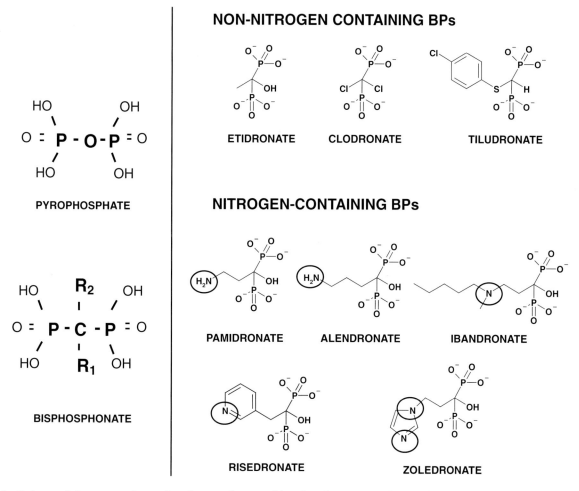

Fig. 49.1. *Left panel*: Structure of pyrophosphate and geminal bisphosphonates. *Right panel*: Structures of clinically used bisphosphonates (acid forms are depicted).

The decrease of bone resorption by BPs is followed by a slower decrease in the rate of bone formation, due to the coupling of the two processes, so that a new steady state at a lower rate of bone turnover is reached 3 to 6 months after the start of treatment. This level of bone turnover remains constant during the whole period of treatment, demonstrating that the accumulation of BP in the skeleton is not associated with a cumulative effect on bone turnover. In addition to decreasing the rate of bone turnover to premenopausal levels, BPs maintain or may improve trabecular and cortical architecture, improve the hypomineralization of osteoporotic bone, increase areal mineral density and may reduce the rate of osteocyte apoptosis. The relevant clinical outcome of these actions is the decrease of the risk of fractures (Fig. 49.2).

At the cellular level BPs inhibit the activity of osteoclasts [1, 4]. BPs bound to bone hydroxyapatite are released in the acidic environment of the resorption lacunae under the osteoclasts and are taken up by them. BPs without a nitrogen atom in their molecule (Fig. 49.1) incorporate into ATP and generate metabolites that induce osteoclast apoptosis. Nitrogen-containing BPs (N-BPs) induce changes in the cytoskeleton of osteoclasts leading to their inactivation and potentially apoptosis. This action is mainly the result of inhibition of farnesyl pyrophosphate synthase (FPPS), an enzyme of the mevalonate biosynthetic pathway. FPPS is responsible for the formation of isoprenoid metabolites required for the prenylation of small GTPases that are important for cytoskeletal integrity and function of osteoclasts. There is a close relation between the degree of inhibition of FPPS and the antiresorptive potencies of N-BPs. In addition, the inhibition of FPPS by N-BPs leads to accumulation of IPP, a metabolite immediately upstream of FPPS, which reacts with adenosine monophosphate (AMP) leading to the production of a new metabolite which induces osteoclast apoptosis.

ANTIFRACTURE EFFICACY

All BPs given daily in adequate doses reduce significantly the risk of vertebral fractures by 35 to 65 % (Fig. 49.3)

414 *Osteoporosis*

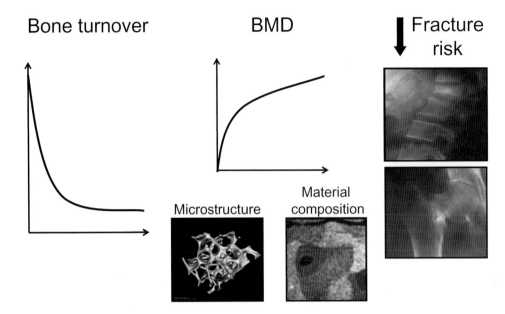

Fig. 49.2. Schematic presentation of effects of bisphosphonates on bone metabolism and strength in osteoporosis.

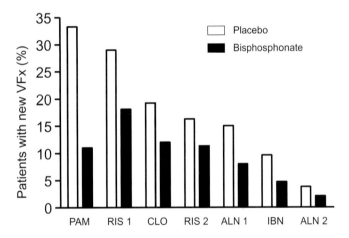

Fig. 49.3. Incidence of fractures in patients with osteoporosis treated with daily oral placebo (open bars) or bisphosphonate (closed bars) after 3 years. PAM = pamidronate (Ref. 5); RIS 1 = risedronate (VERT multinational study; Ref. 9); CLO = clodronate (Ref. 11); RIS 2 = risedronate (VERT North America study; Ref. 8); ALN 1 = alendronate (FIT 1 study; Ref. 6); IBN = ibandronate (BONE study; Ref. 10); ALN 2 = alendronate (FIT 2 study; Ref. 7).

[5–11]. Also illustrated in Fig. 49.3 are the large differences in the incidence of fractures among placebo-treated patients. Results, therefore, of different clinical trials should not be used to compare efficacy of individual BPs. For that, head-to-head studies are needed, but these are not available. The overall efficacy and consistency of daily BPs in reducing the risk of vertebral fractures has been demonstrated by meta-analyses of randomized controlled trials (RCTs) for alendronate and risedronate [12–14]. In studies in which radiographs were taken annually (e.g., the Vertebral Efficacy with Risedronate Therapy (VERT) study with risedronate), the effect of the BP in reducing the risk of vertebral fractures was already evident after 1 year, demonstrating rapid protection of skeletal integrity. This was also shown for moderate and severe vertebral fractures with ibandronate [15] and for clinical vertebral fractures with alendronate [16]. A *post hoc* analysis of risedronate trials reported a significant reduction in clinical vertebral fractures as early as 6 months after the start of treatment [17].

The efficacy of daily oral BPs in reducing the risk of nonvertebral fractures was explored in a number of RCTs. It should be noted that definitions and adjudication procedures of nonvertebral fractures were different among clinical trials. A meta-analysis of the Cochrane Collaboration reported an overall reduction of the risk of nonvertebral fractures in women with osteoporosis of 23% (RR 0.77; 95% CI 0.74–0.94) with alendronate and 20% (RR 0.80; 95% CI 0.72–0.90) with risedronate [13, 14]. The corresponding risk reductions for hip fractures were 53% (RR 0.47; 95% CI 0.26–0.85) with alendronate and 26% (RR 0.74; 95% CI 0.59–0.94%) with risedronate. These estimates are in agreement with earlier published meta-analyses [18, 19]. With daily ibandronate, a reduction (69%) in the risk of nonvertebral fractures was reported in a population at high risk [femoral neck bone mineral density (BMD) below –3.0] by *post hoc* analysis [10]. As with vertebral fractures, the effect of BPs on nonvertebral fractures occurs early after the start of treatment.

Daily administration of BPs, though highly efficacious, is inconvenient and may also be associated with gastrointestinal adverse effects. These reduce adherence to treatment and can diminish the therapeutic response [20,21]. To overcome both these problems, once-weekly formulations, the sum of seven daily doses, have been developed for alendronate and risedronate, and were shown to significantly improve patient adherence to treatment while sustaining the same pharmacodynamic response as the daily treatment [22, 23]. A once-weekly slow-release formulation of risedronate that can be administered after breakfast is also available. Daily and weekly BPs are pharmacologically equivalent and should be considered as continuous administration while the term "intermittent" or "cyclical administration" should be reserved for treatments with drug-free intervals longer than 2 weeks [3].

Intermittent administration of BPs

Results of early attempts to give bisphosphonates intermittently to patients with osteoporosis were equivocal but a meta-analysis of studies with cyclical etidronate showed a significant reduction in the risk of vertebral, but not nonvertebral, fractures [24]. The efficacy of intermittent administration of N-BPs was explored in studies with ibandronate, which indicated that dose and dosing intervals are important determinants of the response to intermittent BP therapy, which in turn depends on the safety and tolerability of the administered dose [25]. An oral ibandronate preparation given once monthly and an intravenous preparation given once every 3 months, providing higher cumulative doses than the daily regimen, were developed and were shown to significantly increase BMD and reduce the risk of nonvertebral fractures by 38% compared to the oral daily dose [26, 27]. A once-monthly oral preparation of risedronate is also available.

The efficacy of intermittent administration of zoledronate, the most potent N-BP, in reducing the risk of osteoporotic fractures was examined in the Health Outcomes and Reduced Incidence with Zoledronic Acid Once Yearly (HORIZON) trial, in which postmenopausal women with osteoporosis were randomized to receive 15-minute infusions of zoledronate 5 mg or placebo once-yearly [28]. Compared to placebo, zoledronate reduced the incidence of vertebral fractures by 70%, hip fractures by 41%, and nonvertebral fractures by 25% after 3 years. The effect of zoledronate on vertebral fractures was already significant at 1 year. In a second controlled study [29], zoledronate infusions given within 90 days after surgical repair of a hip fracture decreased significantly the rate of new clinical fractures by 35% and improved patient survival (28% reduction in all-cause mortality). Epidemiological studies reported also survival benefits in patients treated with oral BPs [30–32].

LONG-TERM EFFECTS ON BONE FRAGILITY

Skeletal fragility on long-term BP therapy has been examined in extensions of four clinical trials for 6 to 10 years [33–36]. None of these extension studies were specifically designed to assess antifracture efficacy; rather, safety and efficacy of surrogate end points, as well as the consistency of the effect of BPs over longer periods, were evaluated. In all four studies, the incidence of nonvertebral fractures was constant with time. In the extension of the FIT (Fracture Intervention Trial) (FLEX), patients who received on average alendronate for 5 years were randomized to placebo, alendronate 5 mg/day or alendronate 10 mg/day, and were followed for another 5 years [35]. Continuation of alendronate treatment led to further increases in BMD of the spine and stabilization of that of the hip, whereas there was a slow progressive decrease of the total hip BMD in patients who received a placebo during the extension. At the end of the 10-year observation period, the incidence of nonvertebral and hip fractures in the ALN/PBO group was similar to that of the ALN/ALN groups. In addition, the incidence of clinical vertebral fractures was lower in the ALN/ALN groups compared to the ALN/PBO group (2% vs 5 %). In a *post hoc* analysis, women who entered the extension with a femoral neck BMD T-score below −2.5 BMD and no vertebral fractures and continued treatment with alendronate showed a significant reduction in the risk of nonvertebral fractures during the 5-year extension. These results suggest that alendronate treatment should be continued in patients at high risk whereas discontinuation of treatment after 5 years may be considered in patients with lower fracture risk. Similar BMD and fracture data were recently reported in the extension of the HORIZON trial in which patients treated with zoledronate for 3 years were randomized to 3 additional years of zoledronate or placebo [36].

SPECIAL ISSUES RELATED TO TREATMENT OF OSTEOPOROSIS WITH BISPHOSPHONATES

Excessive suppression of bone remodeling

There have been concerns that the long-term decrease of bone remodeling by BPs may compromise bone integrity leading to increased bone fragility. Numerous studies in different animal models with N-BPs given at a wide range of doses and time intervals have consistently shown preservation or improvement of bone strength. In only one study with high doses of clodronate given to healthy dogs an increase in fracture incidence has been reported. Earlier reports of potential compromise of the biomechanical competence of bone due to increases in microdamage accumulation in bone biopsies of healthy dogs treated with high BP doses were not substantiated by later animal and human studies [37, 38]. In human controlled studies of osteoporosis, the incidence of nonvertebral fractures was not increased with long-term therapy, and bone turnover markers increased after cessation of treatment, which indicated metabolically active bone. In addition, an analysis of the FIT data showed that higher decreases of bone turnover were associated with larger decreases in the incidence of nonvertebral and hip fractures [39], a finding supported by the above-mentioned analysis of the ibandronate studies. Moreover, in studies of patients treated with BPs followed by treatment with teriparatide, early significant increases in bone markers have been reported, indicating that BP-treated bone can readily respond to stimuli [40]. This conclusion is further supported by a study of zoledronate treatment of patients previously treated with alendronate, which showed that alendronate-treated bone reacted normally to an acute BP load, as provided by zoledronate, indicating that metabolic activity was preserved [41].

Atypical fractures of the femur

In recent years there has been growing concern about the potential relationship between unusual low-energy subtrochanteric/diaphyseal femoral fractures, termed atypical, and long-term use of BPs. Atypical fractures of the femur are often preceded by prodromal pain, can be bilateral and healing may be delayed (Fig. 49.4). Criteria for the identification and diagnosis of atypical fractures of the femur have been proposed by a Task Force of the ASBMR [42]. These fractures are rare, about 1% of all femoral fractures, and occur more frequently in patients treated with BPs than in untreated patients [43, 44]. However, a causal association between BPs and atypical fractures has not been established but appears that the risk rises with increasing duration of exposure.

Osteonecrosis of the jaw

Osteonecrosis of the jaw (ONJ) is defined as exposed bone in the mandible, maxilla, or both that persists for at least 8 weeks in the absence of previous irradiation or metastases in the jaw. It has been reported mainly in patients with malignant diseases receiving high intravenous doses of BPs. The background incidence in the population and its pathogenesis are poorly defined, and a causal relation with BPs has not been established. In patients with osteoporosis treated with BPs, ONJ is rare; an incidence between 1:10,000 and <1:100,000 patient-years has been estimated, and appears to increase with duration of treatment [45–47]. In the two clinical trials of yearly infusions of zoledronate up to 3 years, two adjudicated cases of ONJ were reported among 9,892 patients with osteoporosis, one in the placebo-treated group and one in the zoledronate-treated group after 3 years [28, 29].

Adverse effects

BPs are relatively safe compounds, and their benefits outweigh their potential risks [48]. Specific adverse effects related to the use of BPs in osteoporosis include gastrointestinal toxicity associated with the oral, particularly daily, use of N-BPs and symptoms related to an acute phase reaction, mainly after first exposure to intravenous N-BPs. Gastrointestinal toxicity appears to be higher with generic preparations of oral BPs, of which many are currently available, resulting in significantly poorer adherence and effectiveness [49]. Case reports have suggested a relationship between oral N-BP treat-

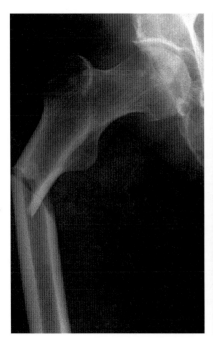
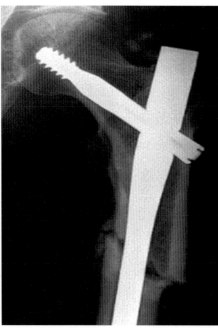

Fig. 49.4. Bilateral atypical fractures of the femur of a patient with rheumatoid arthritis on long-term treatment with alendronate and prednisone. Adapted from Somford MP, Draijer FW, Thomassen BJ, Chavassieux PM, Boivin G, Papapoulos SE. 2009. Bilateral fractures of the femur diaphysis in a patient with rheumatoid arthritis on long-term treatment with alendronate: clues to the mechanism of increased bone fragility. *J Bone Miner Res* 24: 1736–40.

ment and esophageal cancer. This was not confirmed in analyses of large databases, but a reduction in the incidence of gastric and colon cancer was recently reported in alendronate users [31, 50]. The kidney is the principal route of BP elimination, and BP use is contraindicated in patients with severely impaired renal function. Renal toxicity of intravenous BP is not a concern, provided that the indications for treatment and instructions for administration are closely followed. A significant increase in the incidence of atrial fibrillation, reported as a serious adverse event, was observed in one [28], but not in another, study [29] of patients receiving zoledronate compared to those receiving a placebo. A biological explanation for this effect is not apparent and further analyses of clinical trials with alendronate, ibandronate, and risedronate did not confirm such association.

CONCLUSIONS

BPs, because of their efficacy, safety, and ease of administration, are generally accepted as first-line therapy for osteoporosis. Selection of a BP for the treatment of an individual patient should be based on review of efficacy data, risk profile of the BP, and values and preferences of the patient. Despite progress in our understanding of the anti-fracture action of BPs and their long-term effects on bone, there are still questions that remain to be addressed. These include potential, clinically relevant, differences among BPs, optimal selection of patients for treatment and duration of use and their use in combination with bone-forming agents.

ACKNOWLEDGMENTS

S. Papapoulos has received research support and/or honoraria from bisphosphonate manufacturers: Merck & Co., Novartis, Procter & Gamble, and Roche/GSK.

REFERENCES

1. Russell RGG, Watts NB, Ebetino FH, Rogers MJ. 2008. Mechanisms of action of bisphosphonates: Similarities and differences and their potential influence on clinical efficacy. *Osteoporos Int* 19(6): 733–759.
2. Papapoulos SE. 2006. Bisphosphonate actions: Physical chemistry revisited. *Bone* 38: 613–616.
3. Cremers SC, Pillai G, Papapoulos SE. 2005. Pharmacokinetics/pharmacodynamics of bisphosphonates: Use for optimisation of intermittent therapy for osteoporosis. *Clin Pharmacokinet* 44: 551–570.
4. Rogers MJ, Crockett JC, Coxon FP, Monkkonen J. 2011. Biochemical and molecular mechanisms of action of bisphosphonates. *Bone* 49: 34–41.
5. Brumsen C, Papapoulos SE, Lips P, Geelhoed-Duijvestijn PHLM, Hamdy NAT, Landman JO, McCloskey EV, Netelenbos JC, Pauwels EKJ, Roos JC, Valentijn RM, Zwinderman AH. 2002. Daily oral pamidronate in women and men with osteoporosis: A 3-year randomized placebo-controlled clinical trial with a 2-year open extension. *J Bone Miner Res* 17: 1057–1064.
6. Black DM, Cummings SR, Karpf DB, Cauley JA, Thompson DE, Nevitt MC, Bauer DC, Genant HK, Haskell WL, Marcus R, Ott SM, Torner JC, Quandt SA, Reiss TF, Ensrud KE. 1996. Randomised trial of effect of alendronate on risk of fracture in women with existing vertebral fractures. Fracture Intervention Trial Research Group. *Lancet* 348: 1535–1541.
7. Cummings SR, Black DM, Thompson DE, Applegate WB, Barrett-Connor E, Musliner TA, Palermo L, Prineas R, Rubin SM, Scott JC, Vogt T, Wallace R, Yates AJ, LaCroix AZ 1998. Effect of alendronate on risk of fracture in women with low bone density but without vertebral fractures: Results from the Fracture Intervention Trial. *JAMA* 280: 2077–2082.
8. Harris ST, Watts NB, Genant HK, McKeever CD, Hangartner T, Keller M, Chesnut CH 3rd, Brown J, Eriksen EF, Hoseyni MS, Axelrod DW, Miller PD. 1999. Effects of risedronate treatment on vertebral and nonvertebral fractures in women with postmenopausal osteoporosis: A randomized controlled trial. Vertebral Efficacy With Risedronate Therapy (VERT) Study Group. *JAMA* 282: 1344–1352.
9. Reginster J, Minne HW, Sorensen OH, Hooper M, Roux C, Brandi ML, Lund B, Ethgen D, Pack S, Roumagnac I, Eastell R. 2000. Randomized trial of the effects of risedronate on vertebral fractures in women with established postmenopausal osteoporosis. Vertebral Efficacy with Risedronate Therapy (VERT) Study Group. *Osteoporos Int* 11: 83–91.
10. Chesnut CH, Ettinger MP, Miller PD, Baylink DJ, Emkey R, Harris ST, Wasnich RD, Watts NB, Schimmer RC, Recker RR. 2004. Effects of oral ibandronate administered daily or intermittently on fracture risk in postmenopausal osteoporosis. *J Bone Miner Res* 19: 1241–1249.
11. McCloskey E, Selby P, Davies M, Robinson J, Francis RM, Adams J, Kayan K, Beneton M, Jalava T, Pylkkänen L, Kenraali J, Aropuu S, Kanis JA. 2004. Clodronate reduces vertebral fracture risk in women with postmenopausal or secondary osteoporosis: Results of a double-blind, placebo-controlled 3-year study. *J Bone Miner Res* 19: 728–736.
12. Cranney A, Guyatt G, Griffith L, Wells G, Tugwell P, Rosen C; Osteoporosis Methodology Group and The Osteoporosis Research Advisory Group. 2002. Meta-analyses of therapies for postmenopausal osteoporosis. IX: Summary of meta-analyses of therapies for postmenopausal osteoporosis. *Endocr Rev* 23: 570–578.
13. Wells G, Cranney A, Peterson J, Boucher M, Shea B, Robinson V, Coyle D, Tugwell P. 2008. Risedronate for the primary and secondary prevention of osteoporotic fractures in postmenopausal women. *Cochrane Database Syst Rev* 23 (1): CD004523.

14. Wells GA, Cranney A, Peterson J, Boucher M, Shea B, Robinson V, Coyle D, Tugwell P. 2008. Alendronate for the primary and secondary prevention of osteoporotic fractures in postmenopausal women. *Cochrane Database Syst Rev* 23(1): CD001155.
15. Felsenberg D, Miller P, Armbrecht G, Wilson K, Schimmer RC, Papapoulos SE. 2005. Oral ibandronate significantly reduces the risk of vertebral fractures of greater severity after 1, 2, and 3 years in postmenopausal women with osteoporosis. *Bone* 37: 651–654.
16. Black DM, Thompson DE, Bauer DC, Ensrud K, Musliner T, Hochberg MC, Nevitt MC, Suryawanshi S, Cummings SR; Fracture Intervention Trial. 2000. Fracture risk reduction with alendronate in women with osteoporosis: The Fracture Intervention Trial. FIT Research Group. *J Clin Endocrinol Metab* 85: 4118–4124.
17. Roux C, Seeman E, Eastell R, Adachi J, Jackson RD, Felsenberg D, Songcharoen S, Rizzoli R, Di Munno O, Horlait S, Valent D, Watts NB. 2004. Efficacy of risedronate on clinical vertebral fractures within six months. *Curr Med Res Opin* 20: 433–439.
18. Papapoulos SE, Quandt SA, Liberman UA, Hochberg MC, Thompson DE. 2005. Meta-analysis of the efficacy of alendronate for the prevention of hip fractures in postmenopausal women. *Osteoporos Int* 16: 468–474.
19. Nguyen ND, Eisman JA, Nguyen TV. 2006. Anti-hip fracture efficacy of biophosphonates: A Bayesian analysis of clinical trials. *J Bone Miner Res* 21: 340–349.
20. Caro JJ, Ishak KJ, Huybrechts KF, Raggio G, Naujoks C. 2004. The impact of compliance with osteoporosis therapy on fracture rates in actual practice. *Osteoporos Int* 15: 1003–1008.
21. Siris ES, Harris ST, Rosen CJ, Barr CE, Arvesen JN, Abbott TA, Silverman S. 2006. Adherence to bisphosphonate therapy and fracture rates in osteoporotic women: Relationship to vertebral and nonvertebral fractures from 2 US claims databases. *Mayo Clin Proc* 81: 1013–1022.
22. Schnitzer T, Bone HG, Crepaldi G, Adami S, McClung M, Kiel D, Felsenberg D, Recker RR, Tonino RP, Roux C, Pinchera A, Foldes AJ, Greenspan SL, Levine MA, Emkey R, Santora AC 2nd, Kaur A, Thompson DE, Yates J, Orloff JJ. 2000. Therapeutic equivalence of alendronate 70 mg once-weekly and alendronate 10 mg daily in the treatment of osteoporosis. Alendronate Once-Weekly Study Group. *Aging (Milano)* 12: 1–12.
23. Brown JP, Kendler DL, McClung MR, Emkey RD, Adachi JD, Bolognese MA, Li Z, Balske A, Lindsay R. 2002. The efficacy and tolerability of risedronate once a week for the treatment of postmenopausal osteoporosis. *Calcif Tissue Int* 71: 103–111.
24. Cranney A, Guyatt G, Welch V, Griffith L, Adachi JD, Shea B, Tugwell P, Wells G. 2001. A meta-analysis of etidronate for the treatment of postmenopausal osteoporosis. *Osteoporos Int* 12: 140–151.
25. Papapoulos SE, Schimmer RC. 2007. Changes in bone remodelling and antifracture efficacy of intermittent bisphosphonate therapy: Implications from clinical studies with ibandronate. *Ann Rheum Dis* 66: 853–858.
26. Reginster JY, Adami S, Lakatos P, Greenwald M, Stepan JJ, Silverman SL, Christiansen C, Rowell L, Mairon N, Bonvoisin B, Drezner MK, Emkey R, Felsenberg D, Cooper C, Delmas PD, Miller PD. 2006. Efficacy and tolerability of once-monthly oral ibandronate in postmenopausal osteoporosis: 2 year results from the MOBILE study. *Ann Rheum Dis* 65: 654–661.
27. Cranney A, Wells GA, Yetisir E, Adami S, Cooper C, Delmas PD, Miller PD, Papapoulos S, Reginster JY, Sambrook PN, Silverman S, Siris E, Adachi JD. 2009. Ibandronate for the prevention of nonvertebral fractures: A pooled analysis of individual patient data. *Osteoporos Int* 20: 291–297.
28. Black DM, Delmas PD, Eastell R, Reid IR, Boonen S, Cauley JA, Cosman F, Lakatos P, Leung PC, Man Z, Mautalen C, Mesenbrink P, Hu H, Caminis J, Tong K, Rosario-Jansen T, Krasnow J, Hue TF, Sellmeyer D, Eriksen EF, Cummings SR; HORIZON Pivotal Fracture Trial. 2007. Once-yearly zoledronic acid for treatment of postmenopausal osteoporosis. *N Engl J Med* 356: 1809–1822.
29. Lyles KW, Colón-Emeric CS, Magaziner JS, Adachi JD, Pieper CF, Mautalen C, Hyldstrup L, Recknor C, Nordsletten L, Moore KA, Lavecchia C, Zhang J, Mesenbrink P, Hodgson PK, Abrams K, Orloff JJ, Horowitz Z, Eriksen EF, Boonen S; HORIZON Recurrent Fracture Trial. 2007. Zoledronic acid and clinical fractures and mortality after hip fracture. *N Engl J Med* 357: 1799–809.
30. Center JR, Bliuc D, Nguyen ND, Nguyen TV, Eisman JA. 2011. Osteoporosis medication and reduced mortality risk in elderly women and men. *J Clin Endocrinol Metab* 96: 1006–1014.
31. Pazianas M, Abrahamsen B, Eiken PA, Eastell R, Russell RGG. 2012. Reduced colon cancer incidence and mortality in postmenopausal women treated with an oral bisphosphonate-Danish National Register based Cohort Study. 2012. *J Bone Miner Res* 23(11): 2693–2701.
32. Sambrook PN, Cameron ID, Chen JS, March LM, Simpson JM, Cumming RG, Seibel MJ. 2011. Oral bisphosphonates are associated with reduced mortality in frail older people: A prospective five-year study. *Osteoporos Int* 22: 2551–2556.
33. Bone HG, Hosking D, Devogelaer JP, Tucci JR, Emkey RD, Tonino RP, Rodriguez-Portales JA, Downs RW, Gupta J, Santora AC, Liberman UA; Alendronate Phase III Osteoporosis Treatment Study Group. 2004. Ten years' experience with alendronate for osteoporosis in postmenopausal women. *N Engl J Med* 350: 1189–1199.
34. Mellström DD, Sörensen OH, Goemaere S, Roux C, Johnson TD, Chines AA. 2004. Seven years of treatment with risedronate in women with postmenopausal osteoporosis. *Calcif Tissue Int* 75: 462–468.
35. Black DM, Schwartz AV, Ensrud KE, Cauley JA, Levis S, Quandt SA, Satterfield S, Wallace RB, Bauer DC, Palermo L, Wehren LE, Lombardi A, Santora AC, Cummings SR; FLEX Research Group. 2006. Effects of continuing or stopping alendronate after 5 years of treatment: The Fracture Intervention Trial Long-term Extension (FLEX): A randomized trial. *JAMA* 296: 2927–2938.

36. Black DM, Reid IR, Boonen S, Bucci-Rechtweg C, Cauley JA, Cosman F, Cummings SR, Hue TF, Lippuner K, Lakatos P, Leung PC, Man Z, Martinez R, Tan M, Ruzycky ME, Eastell R. 2012. The effect of 3 versus 6 years of zoledronic acid treatment of osteoporosis: A randomized extension to the HORIZON-Pivotal Fracture Trial (PFT). *J Bone Miner Res* 27(2): 243–254.
37. Allen MR, Iwata K, Phipps R, Burr DB. 2006. Alterations in canine vertebral bone turnover, microdamage accumulation, and biomechanical properties following 1-year treatment with clinical treatment doses of risedronate or alendronate. *Bone* 39: 872–879.
38. Chapurlat RD, Arlot M, Burt-Pichat B, Chavassieux P, Roux JP, Portero-Muzy N, Delmas PD, Chapurlat RD, Arlot M, Burt-Pichat B, Chavassieux P, Roux J-P, Portero-Muzy N, Delmas PD. 2007. Microcrack frequency and bone remodeling in postmenopausal osteoporotic women on long-term bisphosphonates: A bone biopsy study. *J Bone Miner Res* 22: 1502–1509.
39. Bauer DC, Black DM, Garnero P, Hochberg M, Ott S, Orloff J, Thompson DE, Ewing SK, Delmas PD; Fracture Intervention Trial Study Group. 2004. Change in bone turnover and hip, non-spine, and vertebral fracture in alendronate-treated women: The fracture intervention trial. *J Bone Miner Res* 19: 1250–1258.
40. Miller PD, Delmas PD, Lindsay R, Watts NB, Luckey M, Adachi J, Greenspan SL, Seeman E, Boonen S, Meeves S, Lang TF, Bilezikian JP. 2008. Early responsiveness of women with osteoporosis to teriparatide after therapy with alendronate or risedronate. *J Clin Endocrinol Metab* 93: 3785–3793.
41. McClung M, Recker R, Miller P, Fiske D, Minkoff J, Kriegman A, Zhou W, Adera M, Davis J. 2007. Intravenous zoledronic acid 5 mg in the treatment of postmenopausal women with low bone density previously treated with alendronate. *Bone* 41: 122–128.
42. Shane E, Burr D, Ebeling PR, Abrahamsen B, Adler RA, Brown TD, Cheung AM, Cosman F, Curtis JR, Dell R, Dempster D, Einhorn TA, Genant HK, Geusens P, Klaushofer K, Koval K, Lane JM, McKiernan F, McKinney R, Ng A, Nieves J, O'Keefe R, Papapoulos S, Sen HT, van der Meulen MCH, Weinsten RS, Whyte M. 2010. Atypical subtrochanteric and diaphyseal femoral fractures: Report of a task force of the American Society for Bone and Mineral Research. *J Bone Miner Res* 25: 2267–2294.
43. Giusti A, Hamdy NAT, Dekkers OM, Ramautar SR, Dijkstra S, Papapoulos SE. 2011. Atypicalfractures and bisphosphonate therapy: A cohort study of patients with femoral fracture with radiographic adjudication of fracture site and features. *Bone* 48: 966–971.
44. Schilcher J, Michaelsson K, Aspenberg P. 2011. Bisphosphonate use and atypical fractures of the femoral shaft. *N Engl J Med* 364: 1728–1737.
45. Khosla S, Burr D, Cauley J, Dempster DW, Ebeling PR, Felsenberg D, Gagel RF, Gilsanz V, Guise T, Koka S, McCauley LK, McGowan J, McKee MD, Mohla S, Pendrys DG, Raisz LG, Ruggiero SL, Shafer DM, Shum L, Silverman SL, Van Poznak CH, Watts N, Woo SB, Shane E; American Society for Bone and Mineral Research. 2007. Bisphosphonate-associated osteonecrosis of the jaw: Report of a task force of the American Society for Bone and Mineral Research. *J Bone Miner Res* 22: 1479–1491.
46. Rizzoli R, Burlet N, Cahall D, Delmas PD, Eriksen EF, Felsenberg D, Grbic J, Jontell M, Landesberg R, Laslop A, Wollenhaupt M, Papapoulos S, Sezer O, Sprafka M, Reginster JY. 2008. Osteonecrosis of the jaw and bisphosphonate treatment for osteoporosis. *Bone* 42: 841–847.
47. Compston J. 2011. Pathophysiology of atypical femoral fractures and osteonecrosis of the jaw. *Osteoporos Int* 22: 2951–2961.
48. Pazianas M, Abrahamsen B. 2011. Safety of bisphosphonates. *Bone* 49: 103–110.
49. Kanis JA, Reginster J-Y, Kaufman J-M, Ringe JD, Adachi JD, Hiligsmann M, Rizzoli R, Cooper C. 2012. A reappraisal of generic bisphosphonates in osteoporosis. *Osteoporos Int* 23: 213–221.
50. Abrahamsen B, Pazianas M, Eiken P, Russell RG, Eastell R. 2012. Esophageal and gastric cancer incidence and mortality in alendronate users. *J Bone Miner Res* 27(3): 679–686.

50
Denosumab

Michael R. McClung

Introduction 420
Inhibition of Rankl: Early Studies 420
Pivotal Trials 421
Longer-Term Responses 422
Response to Discontinuing Denosumab 422

Safety and Tolerability 423
Unanswered Clinical Concerns about Rankl Inhibition 423
Denosumab for other Clinical Conditions 424
Summary 424
References 424

INTRODUCTION

Denosumab is the first inhibitor of receptor activator of nuclear factor κ-B ligand (RANKL) available for the treatment of osteoporosis and other bone diseases. The binding of RANKL to its receptor RANK on preosteoclasts is required for the proliferation, maturation, activation, and survival of osteoclasts [1]. Osteoprotogerin (OPG), a soluble form of RANK, is a decoy receptor for RANKL, interfering with the RANKL/RANK association, and inhibiting osteoclast activation. OPG administration decreased bone resorption and increased bone mass in rats and monkeys [2, 3]. Genetic animal models confirm the importance of this pathway in bone metabolism. Mice that lack OPG had osteopenia and skeletal fragility [4]. Mice deficient in RANKL or RANK and rats that overexpress OPG developed high bone mass of good quality [5, 6].

Parathyroid hormone, estrogen, glucocorticoids, vitamin D, transforming growth factor-β (TGF-β), inflammatory cytokines and other factors modulate the expression of RANKL and OPG [7–10]. Imbalances in the relative amounts of RANKL and OPG are implicated in clinical disorders characterized by increased bone resorption by osteoclasts such as osteoporosis, Paget's disease of bone, and skeletal metastases.

Genetic deficiency of RANKL, presenting as an osteopetrosis phenotype, provided evidence that the RANKL/RANK pathway is important in human skeletal health [11]. Rapidly progressive osteoporosis in a patient with autoantibodies that neutralized OPG activity confirmed the importance of OPG in humans, the delicacy of the balance between RANKL and OPG, and the potential magnitude of effect that occurs when that balance is shifted in favor of RANKL [12]. Inhibition of RANKL is an appealing strategy to treat osteoporosis [13].

INHIBITION OF RANKL: EARLY STUDIES

The administration of single doses (0.01 to 3 mg/kg) of an OPG-Fc fragment (of immunoglobulin G1) fusion molecule to healthy volunteers resulted in a dose-dependent reduction of urinary cross-linked N-telopeptides of type I Collagen (NTX) [14]. Higher doses induced a greater inhibition of bone resorption for a longer time. The highest dose given reduced NTX by 80% at 5 days; the treatment effect had resolved by 1 month. This study confirmed that inhibition of RANKL could reduce bone resorption in a clinical setting.

Denosumab is a fully human IgG_2 antibody that binds RANKL with very high affinity ($Kd=3\times10^{-12}$ M) and high specificity [15]. In a phase 1 study, nonlinear pharmacokinetics were observed after the administration of denosumab in single doses ranging from 0.01 to 3 mg/kg to healthy postmenopausal women; larger doses were

---

Primer on the Metabolic Bone Diseases and Disorders of Mineral Metabolism, Eighth Edition. Edited by Clifford J. Rosen.
© 2013 American Society for Bone and Mineral Research. Published 2013 by John Wiley & Sons, Inc.

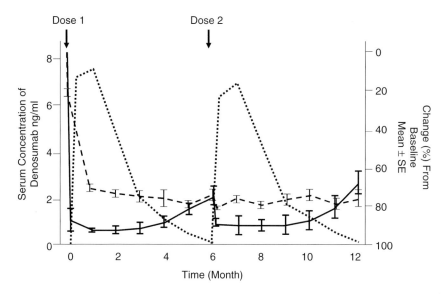

Fig. 50.1. Pharmacokinetics and pharmacodynamics of denosumab therapy 60 mg subcutaneously every 6 months. (.....) serum denosumab concentration; (——) serum CTX with denosumab; (-----) serum CTX with alendronate 70 mg po every week. Data modified from Ref. 18.

cleared more slowly than smaller doses [16] (Fig. 50.1). Clearance was modestly affected by body weight but not by age or renal function [17]. No dose adjustment is required as a function of age or weight or in patients with impaired renal function.

In the denosumab phase I study, subcutaneous doses ranging from 0.01 to 3 mg/kg caused very rapid reduction of urinary NTX by about 80% from baseline within 24 hours [16]. The duration of the inhibition of bone resorption was dose dependent with the effects of doses of 60 mg and higher persisting for at least 6 months. These data suggest that denosumab functions like OPG to reduce bone resorption by inhibiting the activity of RANKL. However, denosumab has these attributes that make it a more attractive therapy then OPG: (a) greater specificity to minimize the potential of off-target effects of other members of the tumor necrosis factor (TNF) family of cytokines such as TRAIL or CD40; (b) more favorable pharmacokinetics, allowing a more convenient dosing regimen; (c) avoidance of the theoretical possibility of inducing formation of anti-OPG antibodies that could have detrimental skeletal effects.

The development of denosumab proceeded with the phase II dose ranging study that evaluated the effects of therapy on bone mineral density (BMD), bone turnover markers, and tolerability in postmenopausal women with low bone density [18,19]. Subcutaneous doses of 6 mg once every 3 months to 210 mg once every 6 months significantly increased BMD at important skeletal sites. With all but the smallest doses, the BMD responses to denosumab at the lumbar spine, proximal femur and mid-radius were similar to or greater than the responses in women randomized to receive 70 mg of alendronate each week. All doses of denosumab resulted in a similar prompt and marked decrease in serum CTX, decreasing by 85% at 3 days after dosing. The duration of this effect was dose related. After dosing with 60 mg of denosumab, serum C-terminal cross-linking telopeptide of type-I collagen (CTX) reached a nadir at about 1 month, then gradually increased during the 6 months between doses, reaching a level similar to that observed in women receiving continuous alendronate therapy (Fig. 50.1). Markers of bone formation decreased after 2–3 months of treatment, and the response to denosumab paralleled that which occurred with alendronate therapy.

PIVOTAL TRIALS

Healthy postmenopausal women (7,808) with osteoporosis (BMD T-score below -2.5 at lumbar spine or total hip) were enrolled in FREEDOM, the pivotal phase III fracture end point study [20]. Subjects with T-score values of below -4 or with more than one mild vertebral deformity or with any moderate or severe vertebral fracture at baseline were excluded. Enrolled subjects were randomly assigned to receive denosumab 60 mg subcutaneously every 6 months or placebo injections; all received calcium and vitamin D. After 3 years of treatment, denosumab decreased the incidence of new morphometric vertebral fractures from 7.2% to 2.3% (68% relative reduction; CI 59%, 74%) (Fig. 50.2). A decrease of at least 60% was also seen at the 1- and 2-year time points. The incidence of hip and nonvertebral fracture was 1.2% and 8.0%, respectively, in subjects receiving placebo, and was 0.7% and 6.5%, respectively, in the denosumab group, resulting in a relative risk reduction of hip fracture of 40% (CI 3%, 63%) and 20% (CI 5%, 33%) for nonvertebral fracture. Because of the entry criteria (only 23% of subjects in FREEDOM had one mild vertebral fracture at baseline), the incidence of new morphometric vertebral and hip fractures in the placebo group of this study was lower than observed in pivotal trials of all other drugs approved for osteoporosis treatment. Thus, while the relative reduction in osteoporotic fracture risk was similar to or

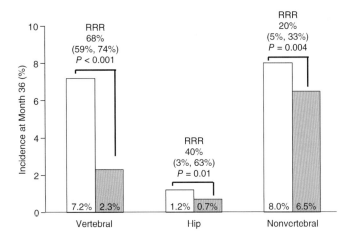

Fig. 50.2. Effect of denosumab therapy on fracture risk. Incidence of fracture after 36 months therapy with placebo (open bar) or denosumab 60mg subcutaneously every 6 months. Numbers in bars are incidence (%). RRR = relative risk reduction (95% confidence interval). Data taken from Ref. 20.

better than that observed with other antiresorptive agents, the absolute reduction in risk was smaller (and the number needed to treat (NNT) was greater) in FREEDOM than in the HORIZON, FIT, VERT, and TROPOS studies [21–24]. This demonstrates the fallacy of using NNT to compare effectiveness of the therapies in populations with differences in fracture risk [25].

A pre-planned subgroup analysis of FREEDOM demonstrated that denosumab reduced spine fractures irrespective of risk factors [26]. Nonvertebral fracture risk was decreased in patients with femoral neck BMD values consistent with osteoporosis but not in subjects with higher BMD. In a *post hoc* subgroup analysis of FREEDOM, denosumab reduced spine and hip fractures in patients at high risk due to advanced age, low BMD, or history of prior fracture [27]. Denosumab was effective in increasing bone density and decreasing the incidence of vertebral fracture in women of the FREEDOM trial across the spectrum of baseline renal function [28]. This included a total of 2,817 women with estimated glomerular filtration rate (GFR) between 30 and 59 cc per minute and 73 women with estimated GFR or 15–29 cc per minute. The efficacy and safety of denosumab therapy in patients with renal failure on dialysis has not been studied.

As was observed in the phase II study, significant and progressive increases in BMD were noted over the 3 years of denosumab therapy in FREEDOM [20]. Compared to placebo, increases were 9.2% at the lumbar spine and 6.0% at the total hip. The change in total hip BMD at time points up to 36 months explained a substantial proportion of the effect of denosumab in reducing risk of vertebral and nonvertebral fractures. All subjects in a subgroup receiving denosumab for whom serial measurements of biochemical markers were measured had significant and sustained reductions in serum CTX and P1NP levels [29].

Denosumab 60mg once every 6 months increased BMD in postmenopausal women with low bone mass [30]. Increases in BMD from baseline occurred in the lumbar spine (6.5%) and total hip (2.4%) over 2 years, while modest decreases occurred in the placebo control group. Similarly, denosumab increased BMD in women who were receiving aromatase inhibitor therapy for nonmetastatic breast cancer and in men treated with androgen deprivation therapy for nonmetastatic prostate cancer [31, 32]. After 3 years of denosumab therapy, vertebral fracture risk in the men on androgen deprivation therapy was reduced from 3.9% without treatment to 1.5% with denosumab therapy (relative risk reduction 62%; CI 22%, 81%) [32]. The effects of denosumab in men with low BMD or osteoporosis are being evaluated.

LONGER-TERM RESPONSES

In an extension of FREEDOM, about 4,500 women will take denosumab 60mg every 6 months for up to 10 years. In women who received denosumab during the first 3 years of the study, treatment during years 4 and 5 resulted in continued increase in BMD, resulting in 5-year gains of 13.7% and 7.0% in the lumbar spine and total hip, respectively [33]. Annualized incidences of vertebral and nonvertebral fracture were similar during years 4 and 5 compared to fracture rates during the first 3 years of therapy. The phase II study extension also provides information about the long-term effects of denosumab therapy. BMD of the spine and hip regions increased by 13.3% and 6.1% from baseline in the lumbar spine and total hip regions, respectively, in subjects who continued taking denosumab for 6 years [34]. Pretreatment reductions in bone turnover markers (BTMs) were maintained with the long-term group. No neutralizing antibodies and no evidence of resistance to therapy have been observed.

RESPONSE TO DISCONTINUING DENOSUMAB

Upon discontinuing denosumab after 2 years, indices of bone turnover returned to baseline within 3 months of not taking a dose (9 months from the last dose) [35, 36]. Resorption and formation markers rose above baseline values for the next several months, and then returned to baseline values within 24 months after stopping treatment (30 months after the last dose). Gains in BMD achieved on therapy were lost within the first year off treatment. By 2 years off therapy, BMD values had returned to the original baseline values. Re-treatment with denosumab after 1 year off therapy quickly reduced bone turnover, and BMD increased to levels almost as high in patients who had received denosumab continually for 4 years [35]. These effects are similar to the effects observed when estrogen therapy is discontinued

[37, 38]. In accord with the pharmacokinetics of denosumab, the return to baseline values occurs over a somewhat longer time interval upon stopping denosumab as compared to stopping estrogen. Because these studies were small and included women at low risk for fracture, the effect on fracture risk of the rebound in bone turnover and rapid, pronounced bone loss is not known.

SAFETY AND TOLERABILITY

The FREEDOM study provides the most robust evaluation of the safety of denosumab therapy [20]. The incidence of total or severe adverse events and frequency of infection, malignancies, or cardiovascular disease was similar between denosumab and placebo groups. Injection site reactions and post-dose symptoms were not observed. Numerically fewer deaths occurred with denosumab (70, 1.8%) compared to placebo (90, 2.3%) (P = 0.08).

Skin rash or eczema was more common with denosumab therapy (3%) compared to placebo (1%) (P < 0.001). Cellulitis associated with hospitalization, not associated with injection site or time of dosing, occurred in 12 denosumab subjects compared to 1 with placebo. The incidence of skin disorders did not increase with longer-term therapy in the extension of FREEDOM, nor in subjects switched from placebo to denosumab [33].

Serum calcium decreases transiently when denosumab therapy is begun. Subjects with vitamin D deficiency were excluded from the FREEDOM trial, and supplements of calcium and vitamin D were begun before denosumab was given. No cases of symptomatic hypocalcemia were observed in FREEDOM subjects who received denosumab.

UNANSWERED CLINICAL CONCERNS ABOUT RANKL INHIBITION

Oversuppression of bone turnover

Theoretical concern exists of "oversuppression" of bone turnover with long-term treatment of potent antiresorptive agents. In head-to-head comparisons with alendronate, denosumab is a more potent inhibitor of BTMs [19, 39]. In iliac crest bone biopsies obtained in FREEDOM after 24 and 36 months of denosumab therapy, median bone formation rate was decreased by 97% compared to placebo patients [40]. Double tetracycline labels were observed in only 19% of those receiving denosumab versus 94% with placebo. In patients previously treated with bisphosphonates, double tetracycline labels in trabecular bone were present in 20% of biopsies of patients who had taken denosumab for 12 months versus 94% in patients who continued taking alendronate [40]. Normal indices of bone remodeling and double tetracycline labels were observed in biopsies of all 15 subjects who had discontinued denosumab therapy for a mean of 25 months [41].

Osteonecrosis of the jaw (ONJ) was not observed in any of the main clinical osteoporosis trials. Two subjects, both of whom had received placebo during the first 3 years of FREEDOM, developed oral lesions that met predefined criteria for ONJ during the first 2 years of the FREEDOM extension [33]. The incidence of ONJ was similar between high-dose denosumab and zoledronic acid in the oncology studies [42–44]. No evidence of impaired fracture healing and no cases of atypical fractures have been described to date. While untoward skeletal effects with long-term treatment with denosumab would not be a surprise, it is not certain that they will occur. The pharmacodynamic response to denosumab (profound early reduction with modest attenuation between doses) differs compared to bisphosphonates (almost constant reduction in turnover). Some bisphosphonates bind avidly to bone and may concentrate at sites of stress reaction; denosumab does not. Distribution into compartments of cortical bone, including osteocyte lacunae and canalicular network, may be different. The rapid and compete reversal of the inhibition of bone turnover upon stopping denosumab therapy contrasts with the persistent effects on remodeling following withdrawal of long-acting bisphosphonates [35].

Immune compromise

The effects of RANKL inhibition on immune function is of interest because RANKL exists in T-helper and dendritic cells, and there is some evidence that RANKL is involved in the maturation of dendritic cells and in regulation of T-cell-dependent immune response [45]. Adults with RANKL deficiency may manifest hypoglobulinemia but no evidence of T-cell dysfunction [46]. No laboratory evidence of altered immune function was observed in thorough evaluations of subjects in the phase II study [18].

In all clinical trials with denosumab, the overall incidence of infection has not differed among treatment groups [47, 48]. In FREEDOM and some of the smaller studies, serious adverse events related to infection occurred more frequently in patients who received denosumab than placebo. The number of events was small, and in no study was the difference in frequency statistically significant. Diverse diagnoses were represented among those clinical events and included noninfectious inflammatory disorders (labyrinthitis, diverticulitis, appendicitis). No relationships between these events and the timing of dose or duration of denosumab treatment were observed [48]. The incidence of opportunistic infection did not differ between denosumab and placebo.

Vascular effects

OPG-deficient mice develop medial calcification of large arteries [5]. Observational studies have demonstrated a correlation between higher levels of serum OPG levels and vascular disease [49]. No evidence of cardiovascular consequences of denosumab therapy have been observed in clinical trials.

DENOSUMAB FOR OTHER CLINICAL CONDITIONS

In patients with rheumatoid arthritis who received methotrexate, denosumab increased BMD and prevented the progression of erosive bone lesions compared to placebo [50]. No unusual adverse events were noted in this unique population. The BMD response was not affected by glucocorticoid therapy, but denosumab has not been formally evaluated in patients receiving glucocorticoid therapy.

Denosumab would likely be effective in treating Paget's disease of bone and in controlling hypercalcemia related to malignancy. No studies have yet reported responses to denosumab therapy in such patients. The long-lasting effect of a single dose of zoledronic acid on bone metabolism in these conditions is likely an advantage compared to denosumab.

Larger doses of denosumab (120mg subcutaneously once every 4 weeks) are effective in management of patients with cancer-related bone diseases. Compared to zoledronic acid, denosumab delayed the occurrence of skeletal-related events in patients with metastatic breast or prostate cancer, while responses were similar in patients with other solid tumors metastatic to bone and in patients with lytic lesions due to multiple myeloma [42, 44]. Risk of infection did not increase with denosumab therapy in these immunocompromised patients.

SUMMARY

By inhibiting RANKL action on osteoclasts and bone resorption, denosumab is a novel strategy for treating osteoporosis. Unlike bisphosphonates that inhibit the activity of mature, active osteoclasts, denosumab prevents the proliferation and maturation of preosteoclasts, thereby limiting the number of active osteoclasts and reducing bone resorption. Both denosumab and bisphosphonates indirectly decrease bone formation, so both agents preserve but do not reconstruct the damaged trabecular architecture found in patients with severe osteoporosis. In postmenopausal women with osteoporosis, denosumab is at least as effective as other potent antiresorptive agents in reducing the risk of clinically important fractures. Denosumab also prevents bone loss in postmenopausal women without osteoporosis and in patients receiving hormone ablation therapy for nonmetastatic cancer, and it likely to be effective in treating men with low bone mass.

The pharmacokinetics of denosumab allow for a convenient dosing schedule, and the drug is easily administered by medical office personnel without requiring intravenous infusion. Parenteral dosing avoids the possibility of upper gastrointestinal intolerance, concerns about poor compliance, and suboptimal intestinal absorption. Persistence of treatment effect for at least 6 months is assured. Unlike intravenous bisphosphonates, denosumab can be used in patients with impaired renal function. There is no justification for the use of denosumab in combination with other antiresorptive agents. The use of denosumab in combination with parathyroid hormone (PTH) molecules has not yet been studied.

To date, no clear concerns about the safety and tolerability of denosumab have been observed. Both the large extension cohort of the FREEDOM study, some of whom will receive therapy for up to 10 years, and the experience being gleaned from the use of denosumab in clinical practice will provide important information about possible effects of denosumab on immune function, infection, skin lesions, and other off-target effects, and about the possible effects of possible oversuppression of bone remodeling. Denosumab is a more potent inhibitor of bone resorption than are bisphosphonates. Whether the persistent, marked inhibition of bone turnover with denosumab will have untoward skeletal consequences is not yet known.

REFERENCES

1. Boyce BF, Xing L. 2007. Biology of RANK, RANKL, and osteoprotegerin. *Arthritis Res Ther* 9 Suppl 1: S1.
2. Ominsky MS, Kostenuik PJ, Cranmer P, Smith SY, Atkinson JE. 2007. The RANKL inhibitor OPG-Fc increases cortical and trabecular bone mass in young gonad-intact cynomolgus monkeys. *Osteoporos Int* 18: 1073–1082.
3. Ominsky MS, Li X, Asuncion FJ, Barrero M, Warmington KS, Dwyer D, Stolina M, Geng Z, Grisanti M, Tan HL, Corbin T, McCabe J, Simonet WS, Ke HZ, Kostenuik PJ. 2008. RANKL inhibition with osteoprotegerin increases bone strength by improving cortical and trabecular bone architecture in ovariectomized rats. *J Bone Miner Res* 23: 672–682.
4. Iotsova V, Caamaño J, Loy J, Yang Y, Lewin A, Bravo R. 1997. Osteopetrosis in mice lacking NF-kappaB1 and NF-kappa B2. *Nat Med* 3: 1285–1289.
5. Bucay N, Sarosi I, Dunstan CR, Morony S, Tarpley J, Capparelli C, Scully S, Tan HL, Xu W, Lacey DL, Boyle WJ, Simonet WS. 1998. Osteoprotegerin-deficient mice develop early onset osteoporosis and arterial calcification. *Gene Dev* 12: 1260–1268.
6. Ominsky MS, Stolina M, Li X, Corbin TJ, Asuncion FJ, Barrero M, Niu QT, Dwyer D, Adamu S, Warmington KS, Grisanti M, Tan HL, Ke HZ, Simonet WS, Kostenuik PJ. 2009. One year of transgenic overexpression of osteoprotegerin in rats suppressed bone resorption and

increased vertebral bone volume, density, and strength. *J Bone Miner Res* 24: 1234–1246.

7. Huang JC, Sakata T, Pfleger LL, Bencsik M, Halloran BP, Bikle DD, Nissenson RA. 2004. PTH differentially regulates expression of RANKL and OPG. *J Bone Miner Res* 19: 235–244.

8. Hofbauer LC, Lacey DL, Dunstan CR, Spelsberg TC, Riggs BL, Khosla S. 1999. Interleukin-1beta and tumor necrosis factor-alpha, but not interleukin-6, stimulate osteoprotegerin ligand gene expression in human osteoblastic cells. *Bone* 25: 255–259.

9. Hofbauer LC, Gori F, Riggs BL, Lacey DL, Dunstan CR, Spelsberg TC, Khosla S. 1999. Stimulation of osteoprotegerin ligand and inhibition of ostcoprotegerin production by glucocorticoids in human osteoblastic lineage cells. *Endocrinology* 140: 4382–4389.

10. Hofbauer LC, Dunstan CR, Spelsberg TC, Riggs BL, Khosla S. 1998. Osteoprotegerin production by human osteoblast lineage cells is stimulated by vitamin D, bone morphogenetic protein-2, and cytokines. *Biochem Biophys Res Commun* 250: 776–781.

11. Sobacchi C, Frattini A, Guerrini MM, Abinun M, Pangrazio A, Susani L, Bredius R, Mancini G, Cant A, Bishop N, Grabowski P, Del Fattore A, Messina C, Errigo G, Coxon FP, Scott DI, Teti A, Rogers MJ, Vezzoni P, Villa A, Helfrich MH. 2007. Osteoclast-poor human osteopetrosis due to mutations in the gene encoding RANKL. *Nat Genet* 39: 960–962.

12. Riches PL, McRorie E, Fraser WD, Determann C, van't Hof R, Ralston SH. 2009. Osteoporosis associated with neutralizing autoantibodies against osteoprotegerin. *N Engl J Med* 361: 1459–1465.

13. Hofbauer LC, Schoppet M. 2004. Clinical implications of the osteoprotegerin/RANKL/RANK system for bone and vascular diseases. *JAMA* 292: 490–495.

14. Bekker PJ, Holloway D, Nakanishi A, Arrighi M, Leese PT, Dunstan CR. 2001. The effect of a single dose of osteoprotegerin in postmenopausal women. *J Bone Miner Res* 16: 348–360.

15. Kostenuik PJ, Smith SY, Jolette J, Schroeder J, Pyrah I, Ominsky MS. 2011. Decreased bone remodeling and porosity are associated with improved bone strength in ovariectomized cynomolgus monkeys treated with denosumab, a fully human RANKL antibody. *Bone* 49: 151–161.

16. Bekker PJ, Holloway D, Nakanishi A, Arrighi M, Leese PT, Dunstan CR. 2004. A single-dose placebo-controlled study of AMG 162, a fully human monoclonal antibody to RANKL, in postmenopausal women. *J Bone Miner Res* 19: 1059–1066.

17. Sutjandra L, Rodriguez RD, Doshi S, Ma M, Peterson MC, Jang GR, Chow AT, Pérez-Ruixo JJ. 2011. Population pharmacokinetic meta-analysis of denosumab in healthy subjects and postmenopausal women with osteopenia or osteoporosis. *Clin Pharmacokinet* 50: 793–807.

18. McClung MR, Lewiecki EM, Cohen SB, Bolognese MA, Woodson GC, Moffett AH, Peacock M, Miller PD, Lederman SN, Chesnut CH, Lain CD, Kivitz AJ, Holloway DL, Zhang C, Peterson MC, Bekker PJ. 2006. Denosumab in postmenopausal women with low bone mineral density. *N Engl J Med* 354: 821–831.

19. Lewiecki EM, Miller PD, McClung MR, Cohen SB, Bolognese MA, Liu Y, Wang A, Siddhanti S, Fitzpatrick LA; AMG 162 Bone Loss Study Group. 2007. Two-year treatment with denosumab (AMG 162) in a randomized phase 2 study of postmenopausal women with low bone mineral density. *J Bone Miner Res* 22: 1832–1841.

20. Cummings SR, San Martin J, McClung MR, Siris ES, Eastell R, Reid IR, Delmas P, Zoog HB, Austin M, Wang A, Kutilek S, Adami S, Zanchetta J, Libanati C, Siddhanti S, Christiansen C. 2009. Denosumab for prevention of fractures in postmenopausal women with osteoporosis. *N Engl J Med* 361: 756–765.

21. Black DM, Cummings SR, Karpf DB, Cauley JA, Thompson DE, Nevitt MC, Bauer DC, Genant HK, Haskell WL, Marcus R, Ott SM, Torner JC, Quandt SA, Reiss TF, Ensrud KE. 1996. Randomised trial of effect of alendronate on risk of fracture in women with existing vertebral fractures. Fracture Intervention Trial Research Group. *Lancet* 348: 1535–1541.

22. Reginster J, Minne HW, Sorensen OH, Hooper M, Roux C, Brandi ML, Lund B, Ethgen D, Pack S, Roumagnac I, Eastell R. 2000. Randomized trial of the effects of risedronate on vertebral fractures in women with established postmenopausal osteoporosis. Vertebral Efficacy with Risedronate Therapy (VERT) Study Group. *Osteoporos Int* 11: 83–91.

23. Reginster JY, Seeman E, De Vernejoul MC, Adami S, Compston J, Phenekos C, Devogelaer JP, Curiel MD, Sawicki A, Goemaere S, Sorensen OH, Felsenberg D, Meunier PJ. 2005. Strontium ranelate reduces the risk of nonvertebral fractures in postmenopausal women with osteoporosis: Treatment of Peripheral Osteoporosis (TROPOS) study. *J Clin Endocrinol Metab* 90: 2816–2822.

24. Black DM, Delmas PD, Eastell R, Reid IR, Boonen S, Cauley JA, Cosman F, Lakatos P, Leung PC, Man Z, Mautalen C, Mesenbrink P, Hu H, Caminis J, Tong K, Rosario-Jansen T, Krasnow J, Hue TF, Sellmeyer D, Eriksen EF, Cummings SR; HORIZON Pivotal Fracture Trial. 2007. Once-yearly zoledronic acid for treatment of postmenopausal osteoporosis. *N Engl J Med* 356: 1809–1822.

25. Ringe JD, Doherty JG. 2010. Absolute risk reduction in osteoporosis: Assessing treatment efficacy by number needed to treat. *Rheumatol Int* 30: 863–869.

26. McClung MR, Boonen S, Törring O, Roux C, Rizzoli R, Bone HG, Benhamou C-L, Lems WL, Minisola S, Halse J, Hoeck JC, Wang A, Siddhanti S, Cummings SR. 2012. Effect of denosumab treatment on the risk of fractures in subgroups of women with postmenopausal osteoporosis. *J Bone Miner Res* 27: 211–218.

27. Boonen S, Adachi JD, Man Z, Cummings SR, Lippuner K, Törring O, Gallagher JC, Farrerons J, Wang A, Franchimont N, San Martin J, Grauer A, McClung M. 2011. Treatment with denosumab reduces the incidence of new vertebral and hip fractures in postmenopausal

women at high risk. *J Clin Endocrinol Metab* 96: 1727–1736.
28. Jamal SA, Ljunggren O, Stehman-Breen C, Cummings SR, McClung MR, Goemaere S, Ebeling PR, Franek E, Yang YC, Egbuna OI, Boonen S, Miller PD. 2011. Effects of denosumab on fracture and bone mineral density by level of kidney function. *J Bone Miner Res* 26: 1829–1835.
29. Eastell R, Christiansen C, Grauer A, Kutilek S, Libanati C, McClung MR, Reid IR, Resch H, Siris E, Uebelhart D, Wang A, Weryha G, Cummings SR. 2011. Effects of denosumab on bone turnover markers in postmenopausal osteoporosis. *J Bone Miner Res* 26: 530–537.
30. Bone HG, Bolognese MA, Yuen CK, Kendler DL, Wang H, Liu Y, San Martin J. 2008. Effects of denosumab on bone mineral density and bone turnover in postmenopausal women. *J Clin Endocrinol Metab* 93: 2149–2157.
31. Ellis GK, Bone HG, Chlebowski R, Paul D, Spadafora S, Smith J, Fan M, Jun S. 2008. Randomized trial of denosumab in patients receiving adjuvant aromatase inhibitors for nonmetastatic breast cancer. *J Clin Oncol* 26: 4875–4882.
32. Smith MR, Egerdie B, Hernández Toriz N, Feldman R, Tammela TL, Saad F, Heracek J, Szwedowski M, Ke C, Kupic A, Leder BZ, Goessl C; Denosumab HALT Prostate Cancer Study Group. 2009. Denosumab in men receiving androgen-deprivation therapy for prostate cancer. *N Engl J Med* 361: 745–755.
33. Papapoulos S, Chapurlat R, Libanati C, Brandi M, Brown J, Czerwiski E, Krieg MA, Man Z, Mellström D, Radominski S, Reginster JY, Resch H, Román J, Roux C, Vittinghoff E, Austin M, Daizadeh N, Bradley M, Grauer A, Cummings S, Bone H. 2011. Five years of denosumab exposure in women with postmenopausal osteoporosis: Results from the first two years of the FREEDOM extension. *J Bone Miner Res* 27: 694–701.
34. Miller PD, Wagman RB, Peacock M, Lewiecki EM, Bolognese MA, Weinstein RL, Ding B, San Martin J, McClung MR. 2011. Effect of denosumab on bone mineral density and biochemical markers of bone turnover: Six-year results of a phase 2 clinical trial. *J Clin Endocrinol Metab* 96: 394–402.
35. Miller PD, Bolognese MA, Lewiecki EM, McClung MR, Ding B, Austin M, Liu Y, San Martin J. 2008. Effect of denosumab on bone density and turnover in postmenopausal women with low bone mass after long-term continued, discontinued, and restarting of therapy: A randomized blinded phase 2 clinical trial. *Bone* 43: 222–229.
36. Bone HG, Bolognese MA, Yuen CK, Kendler DL, Miller PD, Yang YC, Grazette L, San Martin J, Gallagher JC. 2011. Effects of denosumab treatment and discontinuation on bone mineral density and bone turnover markers in postmenopausal women with low bone mass. *J Clin Endocrinol Metab* 96: 972–980.
37. Greenspan SL, Emkey RD, Bone HG, Weiss SR, Bell NH, Downs RW, McKeever C, Miller SS, Davidson M, Bolognese MA, Mulloy AL, Heyden N, Wu M, Kaur A, Lombardi A. 2002. Significant differential effects of alendronate, estrogen, or combination therapy on the rate of bone loss after discontinuation of treatment of postmenopausal osteoporosis. A randomized, double-blind, placebo-controlled trial. *Ann Intern Med* 137: 875–883.
38. Wasnich RD, Bagger YZ, Hosking DJ, McClung MR, Wu M, Mantz AM, Yates JJ, Ross PD, Alexandersen P, Ravn P, Christiansen C, Santora AC 2nd; Early Postmenopausal Intervention Cohort Study Group. 2004. Changes in bone density and turnover after alendronate or estrogen withdrawal. *Menopause* 11: 622–630.
39. Brown JP, Prince RL, Deal C, Recker RR, Kiel DP, de Gregorio LH, Hadji P, Hofbauer LC, Alvaro-Gracia JM, Wang H, Austin M, Wagman RB, Newmark R, Libanati C, San Martin J, Bone HG. 2009. Comparison of the effect of denosumab and alendronate on BMD and biochemical markers of bone turnover in postmenopausal women with low bone mass: A randomized, blinded, phase 3 trial. *J Bone Miner Res* 24: 153–161.
40. Reid IR, Miller PD, Brown JP, Kendler DL, Fahrleitner-Pammer A, Valter I, Maasalu K, Bolognese MA, Woodson G, Bone H, Ding B, Wagman RB, San Martin J, Ominsky MS, Dempster DW; Denosumab Phase 3 Bone Histology Study Group. 2010. Effects of denosumab on bone histomorphometry: The FREEDOM and STAND studies. *J Bone Miner Res* 25: 2256–2265.
41. Brown JP, Dempster DW, Ding B, Dent-Acosta R, San Martin J, Grauer A, Wagman RB, Zanchetta J. 2011. Bone remodeling in postmenopausal women who discontinued denosumab treatment: Off-treatment biopsy study. *J Bone Miner Res* 26: 2737–2744.
42. Stopeck AT, Lipton A, Body JJ, Steger GG, Tonkin K, de Boer RH, Lichinitser M, Fujiwara Y, Yardley DA, Viniegra M, Fan M, Jiang Q, Dansey R, Jun S, Braun A. 2010. Denosumab compared with zoledronic acid for the treatment of bone metastases in patients with advanced breast cancer: A randomized, double-blind study. *J Clin Oncol* 28: 5132–5139.
43. Henry DH, Costa L, Goldwasser F, Hirsh V, Hungria V, Prausova J, Scagliotti GV, Sleeboom H, Spencer A, Vadhan-Raj S, von Moos R, Willenbacher W, Woll PJ, Wang J, Jiang Q, Jun S, Dansey R, Yeh H. 2011. Randomized, double-blind study of denosumab versus zoledronic acid in the treatment of bone metastases in patients with advanced cancer (excluding breast and prostate cancer) or multiple myeloma. *J Clin Oncol* 29: 1125–1132.
44. Fizazi K, Carducci M, Smith M, Damião R, Brown J, Karsh L, Milecki P, Shore N, Rader M, Wang H, Jiang Q, Tadros S, Dansey R, Goessl C. 2011. Denosumab versus zoledronic acid for treatment of bone metastases in men with castration-resistant prostate cancer: A randomised, double-blind study. *Lancet* 377(9768): 813–822.
45. Leibbrandt A, Penninger JM. 2011. TNF Conference 2009: Beyond bones - RANKL/RANK in the immune system. *Adv Exp Med Biol* 691: 5–22.
46. Guerrini MM, Sobacchi C, Cassani B, Abinun M, Kilic SS, Pangrazio A, Moratto D, Mazzolari E, Clayton-Smith J, Orchard P, Coxon FP, Helfrich MH, Crockett JC, Mellis D, Vellodi A, Tezcan I, Notarangelo LD, Rogers

MJ, Vezzoni P, Villa A, Frattini A. 2008. Human osteoclast-poor osteopetrosis with hypogammaglobulinemia due to TNFRSF11A (RANK) mutations. *Am J Hum Genet* 83: 64–76.

47. Ferrari-Lacraz S, Ferrari S. 2011. Do RANKL inhibitors (denosumab) affect inflammation and immunity? *Osteoporos Int* 22: 435–446.

48. Watts NB, Roux C, Modlin JF, Brown JP, Daniels A, Jackson S, Smith S, Zack DJ, Zhou L, Grauer A, Ferrari S. 2012. Infections in postmenopausal women with osteoporosis treated with denosumab or placebo: Coincidence or causal association? *Osteoporos Int* 23: 327–337.

49. Vik A, Mathiesen EB, Brox J, Wilsgaard T, Njølstad I, Jørgensen L, Hansen JB. 2011. Serum osteoprotegerin is a predictor for incident cardiovascular disease and mortality in a general population: The Tromsø Study. *J Thromb Haemost* 9: 638–644.

50. Cohen SB, Dore RK, Lane NE, Ory PA, Peterfy CG, Sharp JT, van der Heijde D, Zhou L, Tsuji W, Newmark R; Denosumab Rheumatoid Arthritis Study Group. 2008. Denosumab treatment effects on structural damage, bone mineral density, and bone turnover in rheumatoid arthritis: A twelve-month, multicenter, randomized, double-blind, placebo-controlled, phase II clinical trial. *Arthritis Rheum* 58: 1299–1309.

51

Parathyroid Hormone Treatment for Osteoporosis

Felicia Cosman and Susan L. Greenspan

Introduction 428
Candidates for Anabolic Therapy 428
Postmenopausal Osteoporosis 429
PTH Treatment of Men 432
PTH in Special Populations 433

Persistence of Effect 433
Rechallenge with PTH 433
Conclusion 433
References 434

INTRODUCTION

As a result of its unique mechanism of action, parathyroid hormone (PTH), the only approved anabolic therapy for bone, produces larger increments in bone mass (particularly in the spine), than those seen with antiresorptive therapies. PTH treatment first stimulates bone formation and subsequently stimulates both bone resorption and formation; the balance remains positive for formation, even in this latter phase of PTH activity [1–3]. The growth of new bone with PTH permits restoration of bone microarchitecture, including improved trabecular connectivity and enhanced cortical thickness [4, 5]. Bone formation may also be induced on the outer periosteal surface [6–8], possibly affecting bone size and geometry, with additional beneficial effects on bone strength [6–12], though this has not been conclusively proven.

This chapter reviews the clinical trial data using PTH as both monotherapy and in combination and sequential regimens with antiresorptive agents in women and men and briefly overviews trials in a few special populations [13]. PTH will be referred to as teriparatide when it is the recombinant or biochemically synthesized human PTH aminoterminal (1-34) fragment and PTH (1-84) as the intact human recombinant molecule. PTH without other designation denotes either of the compounds. Currently, PTH is routinely given as a daily subcutaneous injection.

CANDIDATES FOR ANABOLIC THERAPY

Good candidates for PTH are women and men who are at high risk of future osteoporosis-related fractures, including those with vertebral compression fractures (clinical or radiographic), other osteoporosis-related fractures with bone mineral density (BMD) in the osteoporosis range, or very low BMD even in the absence of fractures (T-score below –3). PTH should also be recommended for individuals who have been on prior antiresorptive agents, and who have had a suboptimal response to treatment, defined as incident fractures or active bone loss during therapy, or who have persistent osteoporosis despite therapy. Individuals who might be at elevated risk for osteosarcoma, such as those with a history of Paget's disease, bone irradiation, unexplained elevations in alkaline phosphatase, adults with open epiphyses, and children should not receive PTH treatment. Furthermore, people with metastatic bone cancer, primary bone cancer, myeloma, hyperparathyroidism and hypercalcemia should not receive PTH. The PTH treatment course is 18–24 months, a function of the duration of the pivotal fracture trial [14] as well as the finding that the effect appears to wane after this time.

POSTMENOPAUSAL OSTEOPOROSIS

Teriparatide (TPTD) as monotherapy

The largest study of TPTD action by Neer and colleagues included 1,637 postmenopausal women with prevalent vertebral fractures and an average age of 70 [14] who were randomized to receive TPTD (20 or 40 mcg), or placebo by daily subcutaneous injection. After a median treatment period of 19 months, TPTD increased spine BMD by 9.7% (20 mcg dose) and 13.7% (40 mcg dose), and hip and total body bone densities to a lesser extent. A small decline in radius BMD was seen (significant at the higher dose). Vertebral fracture risk reductions were 65% and 69%, respectively, with an absolute risk of 4% in the high dose group (19/434), and 5% in the low dose group (22/444), versus 14% in the placebo group (64/448). There was also a reduction in the incidence of new or worsening back pain in both TPTD groups. In patients with incident vertebral fractures, height loss was reduced (mean 0.21 cm lost in TPTD compared with 1.11 cm in placebo). Incident nonvertebral fractures were reduced by 40% (6% incidence in TPTD versus 10% in placebo) and by 50% for those defined as fragility fractures, as determined by individual investigators (no differences between TPTD groups). Despite the small decline in radius BMD, there was an apparent reduction in wrist fracture occurrence in TPTD-treated women (though too small a number to statistically evaluate). There were also numerically fewer hip fractures in TPTD-treated patients, though again too few to evaluate statistically.

Although transient increases in serum calcium were common when measured within 6 hours of the TPTD injection, sustained increases (confirmed with at least one subsequent measurement) were seen in only 3% of patients assigned to the 20-mcg group and 11% of those assigned to the 40-mcg group. There were no significant differences between TPTD or placebo groups with respect to deaths, hospitalizations, cardiovascular disorders, renal stones, or gout, despite an average increase in 24-hour urine calcium of 40 mg per day and an increase in serum uric acid of up to 25%. Animal studies have shown that administration of high-dose TPTD to rodents is associated with osteogenic sarcoma, dependent on dose and duration of administration [15, 16]. In patients with endogenous hyperparathyroidism or parathyroid cancer there is no evidence of an increased risk of osteogenic sarcoma. Furthermore, in over 9 years of postmarketing experience and about 3 million TPTD prescriptions, there is no evidence of an increased risk of osteosarcoma [17, 18]. Overall new cancer diagnoses occurred in fewer patients assigned to TPTD compared with placebo (2% vs 4%, p = 0.03 for the 20-mcg and 0.07 for the 40-mcg group) [14]. Possible side effects of TPTD are dizziness and leg cramps, redness and irritation at injection sites, headache, nausea, arthralgias, myalgias, lethargy, and weakness. The higher dose produced more side effects and withdrawals. TPTD-induced BMD changes in the Neer trial were not dependent on patient age, baseline BMD, or prior fracture history [19], but were related to baseline biochemical bone turnover indices [20]. Furthermore, early PTH-induced changes in bone turnover markers (at 1 and 3 months) were predictive of the ultimate change in spine BMD and bone structure [20, 21]. Finally, when women over age 75 years were examined relative to women younger than 75 years, age did not affect the safety or efficacy [22]. A longer duration of TPTD 14 months or more was associated with greater reduction in nonvertebral fracture incidence and reduced back pain, compared with shorter duration of therapy [23].

Two smaller studies have evaluated surrogate endpoints comparing TPTD to alendronate [1, 12]. In the first [12], where 146 women were randomized to receive TPTD (40 mcg/day) versus alendronate (10 mg/day), spine BMD increased 15% in the TPTD group versus 6% in the alendronate group after 1 year. Although there were fewer fractures in the TPTD than the alendronate group at the end of 14 months (3/73 vs 10/73), several of the fractures were minor (toe fractures).

McClung et al. studied 203 postmenopausal women with osteoporosis randomized to receive TPTD (20 mcg/day) or alendronate for 18 months [1]. Biochemical turnover increased substantially in the TPTD group (formation earlier than resorption) and declined substantially in the alendronate group (resorption earlier than formation). In TPTD-treated women, markers peaked within 6 months, suggesting developing resistance, as has been seen in other TPTD trials [14, 24, 25]. Spine BMD by dual energy X-ray absorptiometry (DXA) increased 10.3% in TPTD-treated women versus 5.5% in alendronate-treated women. Volumetric spine BMD by quantitative computed tomography (QCT) in a subset of women increased 19% in the TPTD versus 3.8% in the alendronate group. Femoral neck BMD by DXA increased similarly in both groups, though by QCT, cortical volumetric femoral neck BMD increased 7.7% in the alendronate group and declined 1.2% in the TPTD group. The spine BMD change correlated with the PINP increment in the TPTD group and with the PINP decrement in the alendronate group (r = 0.53 and −0.51, respectively). Clinical fracture incidence was similar in the TPTD (nine fractures) and alendronate (eight fractures) groups, but no radiographs were done to evaluate vertebral fractures. Moderate or severe back pain was reported significantly less often in women assigned to TPTD versus alendronate (15% vs 33%, p = 0.003).

TPTD delivered by a transdermal microneedle patch as a 30-minute daily wear in doses of 20, 30, or 40 mcg/day, versus placebo or subcutaneous injection of TPTD 20 mcg, over 6 months demonstrated that the 40-mcg patch had a BMD spine increase similar to the subcutaneous TPTD and the BMD response at the total hip was greater [26, 27]. In an investigation to determine the importance of TPTD with an escalating daily dose (20 to 30 to 40 mcg) versus constant daily dose (30 mcg) in patients with osteoporosis, no difference was found over 6 months [28].

PTH(1-84) as monotherapy

In a dose ranging study, 217 women were randomized to 50, 75, or 100 mcg of PTH(1-84) or placebo. There was a dose-dependent increase in spine BMD; however, no increase in hip or total body BMD. The Treatment of Osteoporosis (TOP) trial was an 18-month, randomized, double-blind, study of 2,532 postmenopausal women with osteoporosis randomized to 100 mg of recombinant PTH(1-84) or placebo by daily subcutaneous injection [27]. The mean age was 64, and 19% of subjects had a prevalent vertebral fracture. The average change in spine BMD was 7% in PTH(1-84) treated subjects compared to those on placebo. In the per protocol adherent population (n = 1870), new or worsened vertebral fracture incidence was 3.4% in placebo and 1.4% in PTH(1-84) (relative risk reduction 58%), with reductions in both those with and without prevalent vertebral fracture, but nonvertebral fracture incidence was not reduced. The incidence of hypercalcemia was significantly higher in PTH(1-84)-treated women (28.3% vs 4.5% in placebo) [27]. PTH(1-84) therapy is currently available in Europe. There have been no head to head trials comparing PTH(1-84) with TPTD.

PTH and antiresorptive combination/ sequential therapy

Although PTH and antiresorptive agents could theoretically produce additive or even synergistic effects on bone strength, studies on combination therapy have shown different outcomes based on skeletal site (spine vs hip), type of measurement (DXA vs QCT), specific antiresorptive therapy utilized, and whether patients are previously treatment naïve or treatment experienced. Furthermore, in treatment-experienced patients, there appear to be differences in outcome based on whether the prior antiresorptive agent is continued or stopped when TPTD is initiated.

Treatment naïve women: PTH and bisphosphonates

Black et al. randomized 238 previously treatment naïve women to PTH(1-84) with alendronate versus each agent alone in a blinded fashion [29]. The BMD of the anteroposterior (AP) spine by DXA increased similarly in the PTH(1-84) alone and combination groups (6.3% and 6.1%, respectively). Total hip BMD (by DXA) increased in the combination group (1.9%) but not with PTH(1-84) alone (0.3%). Radial BMD declined more with PTH(1-84) alone (-3.4%) than with combination therapy (-1.1%). Although QCT measured increases in the integral spine and total hip were similar between the PTH(1-84) alone and combination groups, trabecular spine BMD increased more with PTH(1-84) alone (25.5%) than with the combination (12.6%). In contrast, QCT-assessed cortical bone density declined in the hip (-1.7%) with PTH(1-84) alone but was unchanged in the combination group. The cortical volume of the femoral neck of the hip (but not the total hip) increased significantly in PTH(1-84)-treated versus combination-treated women, but this was not due to periosteal expansion. The DXA results demonstrated no clear evidence of the additive effect with combination therapy compared to PTH(1-84) alone in the spine. However, hip BMD increments were superior with the combination. Evidence of a blunted effect with combination treatment was apparent only by QCT. Since single-energy QCT-based BMD increments induced by PTH may be artifactually elevated by reductions in bone marrow fat [30, 31], it is unclear how important these findings are compared to DXA-based results. There were only a small number of fractures, with no group differences; incident morphometric vertebral fractures were not reported.

An unblinded study where 93 women were randomized to receive alendronate for 6 months prior to giving TPTD (versus either agent alone) suggested that BMD gains in both spine and hip by DXA were lower in those given TPTD after the brief course of alendronate (and with continued alendronate), compared to those given TPTD alone [32]. The difference in the DXA BMD of the spine was not significant if the groups were restricted to those who did not discontinue study medication prematurely. A larger difference in BMD between the TPTD alone and combination groups was seen by QCT. This is one of few PTH trials where treatment duration was a full 24 months; hip and femoral neck BMD levels increased most markedly during the latter year of treatment. However, radius BMD declined more in women who received TPTD alone compared to combination treatment, and the change in total body BMD did not differ between groups. Furthermore, the PTH dose used here was double the approved dose (40 mcg daily TPTD), and since PTH effects on BMD are clearly dose dependent, the clinical significance of these data is unclear.

Cosman et al. randomized 412 treatment naïve postmenopausal women (mean age 65, mean spine BMD T-score -2.9), to receive daily TPTD, an intravenous zoledronic acid (ZOL) infusion, or a combination of daily TPTD and an intravenous zoledronic acid infusion in a partial double-blind fashion (TPTD open label) [33]. With combination therapy, the bone resorption marker CTX declined similarly to that with ZOL alone, whereas the bone formation marker PINP declined only modestly compared to that seen with ZOL alone. Also with combination therapy, the spine BMD increase was similar to that seen with TPTD alone (7.5% vs 7.0% TPTD alone), whereas the hip BMD increase was larger (2.3% vs 1.1%) as was the femoral neck BMD increase (2.2% vs 0.1%). In the combination groups, peak BMD increments were reached the fastest at both sites. Although fractures were reported only as adverse events, the fractures were adjudicated. Clinical fractures occurred in 9.5% of patients in the ZOL alone group, 5.8% of the TPTD alone group and 2.9% of the combination group (p < 0.05 vs ZOL alone).

Treatment naïve women: Teriparatide and raloxifene

In a 6-month, double-blind, placebo-controlled trial, Deal et al. randomized 137 postmenopausal treatment-naïve women to receive TPTD or TPTD plus raloxifene. The bone formation marker PINP rose similarly in the two groups, while the bone resorption marker (CTX) increased more in the TPTD alone group. Spine BMD increments were similar in the two groups, while hip BMD increased more in the TPTD plus raloxifene group [34].

PTH therapy in women on established bisphosphonate or raloxifene treatment

Patients maintained and stabilized on long-term antiresorptive treatment are a distinct, but clinically very important population, since many of these patients have fractures or do not achieve a BMD above the osteoporotic range, and thus might benefit from anabolic therapy. At least 50% of all PTH treatment is initiated in patients who have received prior antiresorptive agents. Possible explanations for differences between treatment-naïve and treatment-experienced women include: reduced active bone surface in the treatment experienced, the increase in endogenous PTH seen for up to 12 months when potent antiresorptive agents are administered to treatment naïve individuals (which might produce a different response to exogenously administered PTH) and perhaps unique effects on osteoclast and/or osteoblast activation in treatment experienced individuals.

Prior studies evaluating TPTD treatment in treatment experienced women have followed two basic designs: antiresorptive agents are stopped when TPTD is started [35, 36, 37], or antiresorptive agents are continued when TPTD is started [38, 39]. Outcomes differ with these distinct study designs, particularly when the antiresorptive agents are oral bisphosphonates. In studies where bisphosphonates are discontinued, the spine BMD increment is of lesser magnitude, and hip BMD declines consistently over the first year, an effect not seen in protocols where TPTD is added to ongoing bisphosphonate.

Studies where antiresorptive therapy was stopped when TPTD was started

In an observational study where TPTD was given to women after cessation of long-term alendronate or raloxifene [35], bone turnover markers increased, as did spine BMD, but these increases were somewhat delayed and of lower magnitude in patients pretreated with alendronate compared to those in patients pretreated with raloxifene. A transient reduction in hip BMD was seen at 6 months in the group previously on alendronate but this reversed by the 18-month measurement.

Similarly, in an observational study of women previously treated with risedronate (n = 146) or alendronate (n = 146) biochemical responses showed increments in bone resorption already within 1 month in both groups of patients [36], an outcome not seen within the first month in treatment-naïve patients treated with TPTD [1, 14]. Furthermore, increases in bone resorption at 1 month are not seen in patients on prior antiresorptive therapy when the antiresorptive agent is continued during administration of TPTD [24, 38, 40]. In this trial [36], Miller found that spine BMD increases were not as great in patients receiving TPTD when the bisphosphonate was stopped, and hip BMD was below baseline for both the patients on prior risedronate as well as those on prior alendronate for the duration of the 1-year trial.

Finally, the average spine BMD in a cohort of women who had been on prior bisphosphonates and then switched to TPTD increased less than in a cohort of treatment-naïve women (9.8–10.2% for the bisphosphonate treated and 13.1% for treatment-naïve women). Furthermore, women on prior bisphosphonates who were switched to TPTD had a decline in hip BMD over the first year of treatment [41].

In contrast, in 126 women previously treated with long-term alendronate (mean age 68 years, mean alendronate duration 3.2 years), the subjects were randomized to continue alendronate and to receive daily TPTD, cyclic TPTD (given in a 3-month on/3-month off regimen), or alendronate alone [38]. In just over 15 months, spine BMD rose 6.1% in the daily TPTD group and 5.4% in the cyclic TPTD group (p < 0.001 for each TPTD group, no group difference), both higher increments than the average changes seen in the studies above when the underlying bisphosphonate was discontinued. Moreover, mean hip BMD did not decline at any time point during this study.

In a separate study, Cosman et al. evaluated postmenopausal women on raloxifene for at least 1 year (n = 42) with persistent osteoporosis and randomized them to stay on raloxifene alone or to receive raloxifene plus TPTD. The TPTD plus raloxifene group had an increment of about 10% in the lumbar spine and 3% in the total hip, whereas those randomized to the raloxifene alone group had no BMD change [39]. Increases in both biochemical turnover markers at 3 months correlated with increases in spine BMD at 1 year.

In order to formally compare the effect of continuing versus stopping the antiresorptive agent when TPTD is begun in a randomized trial, 198 women treated with prior antiresorptive agents for at least 1 year (102 women on alendronate and 96 on raloxifene) were studied [42]. Women within each antiresorptive category were randomized to continue or stop their antiresorptive when TPTD was initiated. Although an anabolic response was seen both biochemically and densitometrically in all groups of patients, biochemical turnover markers increased more in those randomized to the switch design. Of particular interest was the early increase in CTX, which was already significantly elevated at 1 month in the patients who were switched from alendronate to

TPTD, suggesting a truncation of the anabolic window in patients following this approach. As a result, BMD declined in the first 6 months in the hip (as seen in all switch studies above) and did not increase as much in the spine. The increases at both 6 and 18 months at both spine and hip were greater in those patients in whom TPTD was added to the ongoing alendronate, compared to those who switched to TPTD, and at no point in time did hip BMD decline in the combination group. Differences between combination and switch protocols were less marked with raloxifene pretreatment though hip BMD increased more in the combination than switch group.

These results suggest that there may be a role for combination therapy, particularly after prior bisphosphonate treatment. This might be particularly important in patients who begin with very low hip BMD and/or those in whom hip fractures have already occurred, where any early decline in BMD might be detrimental and a greater increase in BMD over 18 months a favorable outcome.

PTH and hormone therapy

In 52 women with osteoporosis (average age 60 years) treated with hormone therapy (HT) [24, 40], daily TPTD produced rapid increases in markers of bone formation, and delayed increases in markers of bone resorption [40]. This period of time, where augmentation of bone formation exceeds stimulation of bone resorption, has been referred to as the "anabolic window" and may represent the most efficient bone-building opportunity with TPTD. Furthermore, bone turnover levels remained elevated for only 18–24 months, after which marker levels declined [24]. The mechanism of this apparent resistance to TPTD has still not been determined. BMD increased by about 14% over 3 years in women receiving TPTD + HT, with evidence of the most rapid rise in BMD within the first 6 months. Total body and hip BMD increased by 4% in patients on TPTD + HT. Although the study was not powered to assess fracture occurrence, after 3 years of treatment, vertebral deformity occurrence was significantly reduced in patients receiving TPTD + HT compared to HT alone [24].

Another study of similar design performed in women who had previously been treated with HT showed BMD increments by DXA in the TPTD group of 30% in the lumbar spine and 12% in the femoral neck versus placebo [43]. No fracture data were presented from this trial, and the data have never been published in a peer-reviewed journal. A third study was performed in 247 women, where one subgroup had been on prior HT (as in the previously discussed two trials) and a second subgroup consisted of treatment-naïve women about to receive HT for the first time [44]. In the former group, there were BMD increments of approximately 11% in the spine and 3% in the total hip in women randomized to TPTD (40 mcg per day). In the women receiving *de novo* HT, there were increases due to HT itself (4% in the spine and 2% in the total hip), and larger increases in the group receiving HT with TPTD (16% in the spine and 6% in the hip). The increases from TPTD appeared additive to those of HT, although not synergistic.

PTH TREATMENT OF MEN

In a small study, men with idiopathic osteoporosis were randomized to receive TPTD or placebo [25]. Biochemical markers of bone turnover increased rapidly with TPTD administration, and spine BMD rose about 12%, with a plateau between 12 and 18 months. In the femoral neck and total hip, BMD increased 5% and 4%, respectively, and radius BMD did not change significantly.

A subsequent multicenter trial of TPTD [18] was performed in 437 men (mean age 49 years) with primary idiopathic or hypogonadal osteoporosis. Subjects were randomized to TPTD 20 or 40 mcg daily, or placebo. After approximately 1 year, spine BMD rose 5.4% and 8.5% in the 20 mcg and 40 mcg groups, respectively, with no change in the placebo group. There were also dose-dependent increases in BMD at hip sites and total body. Of the original enrollees, 355 men participated in an observational follow-up study. Lateral spine radiographs repeated after approximately 18 months of follow-up (including the use of antiresorptive therapy in a substantial proportion of the men) showed a 50% reduction in vertebral fracture risk in those men initially assigned to TPTD compared to those who had received placebo (p = 0.07) [45].

In a third study, 83 men with osteoporosis were assigned to TPTD at 40 mcg/day, alendronate alone, or TPTD after 6 months of alendronate pretreatment (with ongoing alendronate [26, 46]. A substantial proportion of men in both TPTD groups required dose adjustment (by 25–50%) due to hypercalcemia or side effects. After a total of 24 months of TPTD administration, spine BMD increased most in the TPTD alone group (18.1%), compared to that in the combination group (14.8%) or alendronate alone (7.9%). Similar trends were seen for the lateral spine and femoral neck, but for the total hip and total body, increases were similar in the three treatment groups. In contrast, in the radius, BMD declined in the TPTD alone group with slight increases in the other groups. Spine trabecular bone density on QCT increased 48% with TPTD alone, 17% with the combination, and 3% with alendronate alone.

Leder and colleagues [47] examined the BMD changes after discontinuation of TPTD. While the initial improvements in spine and hip BMD were similar in men and women, after TPTD was discontinued following 24 months of TPTD in eugonadal men and postmenopausal women, spine BMD decreased 4.1% in men compared to 7.1% in women over the next 12 months. Bone mass was stable at the hip in men, but decreased in women suggesting an even greater need for antiresorptive follow-up treatment in women.

PTH IN SPECIAL POPULATIONS

Glucocorticoid-Treated patients

PTH could conceivably be a preferred treatment for glucocorticoid osteoporosis, since some of the major pathophysiologic skeletal problems with glucocorticoid administration are reduced osteoblast function and lifespan, both of which might be counteracted by PTH. Women with a variety of rheumatologic conditions on glucocorticoids and treated with hormone therapy were randomized to TPTD + HT or continued HT alone [48]. TPTD resulted in a 12% increase in spine BMD by DXA and a smaller increase in femoral neck BMD. No fracture results were reported.

In an 18-month active comparator trial of TPTD versus alendronate for the treatment of glucocorticoid-induced osteoporosis, patients treated with TPTD had BMD increases of 7.2% at the spine and 3.8% at the total hip, both significantly greater than the changes of 3.4% at the spine and 2.4% at the hip seen with alendronate therapy [49]. Furthermore, fewer new vertebral fractures occurred in the TPTD group compared to the alendronate group (0.6% vs 6.1%, p = 0.004). At 36 months, TPTD compared to alendronate resulted in 11% vs 5.3% increases at the lumbar spine, 5.2% vs 2.7% at the total hip, 6.3% vs 3.4% at the femoral neck (all p = 0.001), in addition to fewer vertebral fractures (1.7% TPTD vs 7.72 alendronate) [50]. There were no differences in nonvertebral fractures between the groups.

Fracture repair

PTH could potentially accelerate fracture repair. In a study with postmenopausal women who sustained a radial fracture, TPTD 20 mcg shortened time to healing, though this effect was not seen with TPTD 40 mcg [51]. Further studies are needed.

PERSISTENCE OF EFFECT

A series of observational studies suggests that BMD is lost in individuals who do not take antiresorptive agents after cessation of TPTD or PTH(1-84), whereas antiresorptive therapy can maintain PTH-induced gains or even provide further increments in BMD after a course of PTH [24, 45, 48, 52–54]. Black et al. have now provided clinical trial confirmation of this observation [55]. Subjects originally randomized to 1 year of treatment with PTH(1-84) were subsequently randomized to receive alendronate or placebo for an additional year. Over 2 years, women who received alendronate following PTH(1-84) had significant increases in spine BMD of 12.1% compared to 4.1% in women in the PTH(1-84) followed by placebo. Trabecular bone at the spine assessed by QCT demonstrated a 31% increase in women on PTH(1-84) followed by alendronate, compared to 14% in those assigned to PTH(1-84) followed by placebo. BMD at the femoral neck and total hip was increased above baseline in all groups except those receiving PTH followed by placebo. These data suggest that after 1 year of PTH treatment, the gains in bone mass are preserved or further improved with alendronate but lost in patients not on antiresorptive therapy. In women enrolled in the Fracture Prevention Trial [14], a 30-month observational follow-up, after discontinuation of TPTD indicated that nonvertebral fracture reduction risk remained lower compared with the prior placebo group, but the difference was not significant for the 20-mcg dose. The *ad lib* use of antiresorptive treatments in both groups complicates the interpretation of these findings[56].

RECHALLENGE WITH PTH

Women originally randomized to daily or cyclic TPTD in addition to ongoing alendronate were followed for a year after TPTD was discontinued [57]. BMD remained stable in these women during this year. A second 15-month course of TPTD was given to those volunteers who still had osteoporosis. The rechallenge with TPTD produced similar biochemical and BMD changes to those seen during the first course of therapy [57]. In a study by Finkelstein and colleagues [58], men and women who completed a 30-month randomized trial that included TPTD, alendronate, or the combination were rechallenged with TPTD following a 12-month discontinuation phase of TPTD. After 12 months of TPTD rechallenge, BMD of the spine increased but the increment was attenuated compared to the initial TPTD treatment.

Cost-Effectiveness of TPTD

Liu and colleagues examined the cost-effectiveness of TPTD, alendronate, or usual care of calcium and vitamin D only [59]. The cost of 2 years of TPTD followed by alendronate was $156,500 per quality-adjusted life year (QALY) compared with alendronate. Alendronate alone was $11,600 per QALY compared to usual care, and TPTD alone was $172,300 per QALY compared to usual care. The cost-effectiveness of TPTD improved with increasing age and lower BMD, suggesting TPTD may be cost effective for high risk patients.

CONCLUSION

PTH is a unique approach to osteoporosis treatment. Because of the underlying effects it produces on the microarchitecture, macroarchitecture, and mass of bone, PTH may be able to ensure more long-term protection against fracture occurrence than antiresorptive agents alone; however, data proving this principle are lacking. Antiresorptive agents are clearly needed after PTH to maintain PTH-induced gains. There are still many unanswered questions concerning PTH therapy, including the optimal duration and regimen of therapy, and the mechanism underlying resistance to PTH effect after 18 months. Different PTH peptides and alternative forms of delivery (oral, nasal, inhaled, transdermal) are currently under study.

REFERENCES

1. McClung MR, San Martin J, Miller PD, Civitelli R, Bandeira F, Omizo M, et al. 2005. Opposite bone remodeling effects of teriparatide and alendronate in increasing bone mass. *Arch Intern Med* 165(15): 1762–8.
2. Arlot M, Meunier PJ, Boivin G, Haddock L, Tamayo J, Correa-Rotter R, et al. 2005. Differential effects of teriparatide and alendronate on bone remodeling in postmenopausal women assessed by histomorphometric parameters. *J Bone Miner Res* 20(7): 1244–53.
3. Lindsay R, Cosman F, Zhou H, Bostrom MP, Shen VW, Cruz JD, et al. 2006. A novel tetracycline labeling schedule for longitudinal evaluation of the short-term effects of anabolic therapy with a single iliac crest bone biopsy: Early actions of teriparatide. *J Bone Miner Res* 21(3): 366–73.
4. Jiang Y, Zhao JJ, Mitlak BH, Wang O, Genant HK, Eriksen EF. 2003. Recombinant human parathyroid hormone (1-34) [teriparatide] improves both cortical and cancellous bone structure. *J Bone Miner Res* 18(11): 1932–41.
5. Dempster DW, Cosman F, Kurland ES, Zhou H, Nieves J, Woelfert L, et al. 2001. Effects of daily treatment with parathyroid hormone on bone microarchitecture and turnover in patients with osteoporosis: A paired biopsy study. *J Bone Miner Res* 16(10): 1846–53.
6. Burr DB. 2005. Does early PTH treatment compromise bone strength? The balance between remodeling, porosity, bone mineral, and bone size. *Curr Osteoporos Rep* 3(1): 19–24.
7. Parfitt AM. 2002. Parathyroid hormone and periosteal bone expansion. *J Bone Miner Res* 17(10): 1741–3.
8. Lindsay R, Zhou H, Cosman F, Nieves J, Dempster DW, Hodsman AB. 2007. Effects of a one-month treatment with PTH(1-34) on bone formation on cancellous, endocortical, and periosteal surfaces of the human ilium. *J Bone Miner Res* 22(4): 495–502.
9. Rehman Q, Lang TF, Arnaud CD, Modin GW, Lane NE. 2003. Daily treatment with parathyroid hormone is associated with an increase in vertebral cross-sectional area in postmenopausal women with glucocorticoid-induced osteoporosis. *Osteoporos Int* 14(1): 77–81.
10. Zanchetta JR, Bogado CE, Ferretti JL, Wang O, Wilson MG, Sato M, et al. 2003. Effects of teriparatide [recombinant human parathyroid hormone (1-34)] on cortical bone in postmenopausal women with osteoporosis. *J Bone Miner Res* 18(3): 539–43.
11. Uusi-Rasi K, Semanick LM, Zanchetta JR, Bogado CE, Eriksen EF, Sato M, et al. 2005. Effects of teriparatide [rhPTH (1-34)] treatment on structural geometry of the proximal femur in elderly osteoporotic women. *Bone* 36(6): 948–58.
12. Body JJ, Gaich GA, Scheele WH, Kulkarni PM, Miller PD, Peretz A, et al. 2002. A randomized double-blind trial to compare the efficacy of teriparatide [recombinant human parathyroid hormone (1-34)] with alendronate in postmenopausal women with osteoporosis. *J Clin Endocrinol Metab* 87(10): 4528–35.
13. Marcus R. 2011. Present at the beginning: A personal reminiscence on the history of teriparatide. *Osteoporos Int* 22(8): 2241–8.
14. Neer RM, Arnaud CD, Zanchetta JR, Prince R, Gaich GA, Reginster JY, et al. 2001. Effect of parathyroid hormone (1-34) on fractures and bone mineral density in postmenopausal women with osteoporosis. *N Engl J Med* 344(19): 1434–41.
15. Vahle JL, Long GG, Sandusky G, Westmore M, Ma YL, Sato M. 2004. Bone neoplasms in F344 rats given teriparatide [rhPTH(1-34)] are dependent on duration of treatment and dose. *Toxicol Pathol* 32(4): 426–38.
16. Vahle JL, Sato M, Long GG, Young JK, Francis PC, Engelhardt JA, et al. 2002. Skeletal changes in rats given daily subcutaneous injections of recombinant human parathyroid hormone (1-34) for 2 years and relevance to human safety. *Toxicol Pathol* 30(3): 312–21.
17. Harper KD, Krege JH, Marcus R, Mitlak BH. 2007. Osteosarcoma and teriparatide? *J Bone Miner Res* 22(2): 334.
18. Subbiah V, Madsen VS, Raymond AK, Benjamin RS, Ludwig JA. 2010. Of mice and men: Divergent risks of teriparatide-induced osteosarcoma. *Osteoporos Int* 21(6): 1041–5.
19. Marcus R, Wang O, Satterwhite J, Mitlak B. 2003. The skeletal response to teriparatide is largely independent of age, initial bone mineral density, and prevalent vertebral fractures in postmenopausal women with osteoporosis. *J Bone Miner Res* 18(1): 18–23.
20. Chen P, Satterwhite JH, Licata AA, Lewiecki EM, Sipos AA, Misurski DM, et al. 2005. Early changes in biochemical markers of bone formation predict BMD response to teriparatide in postmenopausal women with osteoporosis. *J Bone Miner Res* 20(6): 962–70.
21. Dobnig H, Sipos A, Jiang Y, Fahrleitner-Pammer A, Ste-Marie LG, Gallagher JC, et al. 2005. Early changes in biochemical markers of bone formation correlate with improvements in bone structure during teriparatide therapy. *J Clin Endocrinol Metab* 90(7): 3970–7.
22. Boonen S, Marin F, Mellstrom D, Xie L, Desaiah D, Krege JH, et al. 2006. Safety and efficacy of teriparatide in elderly women with established osteoporosis: Bone anabolic therapy from a geriatric perspective. *J Am Geriatr Soc* 54(5): 782–9.
23. Lindsay R, Miller P, Pohl G, Glass EV, Chen P, Krege JH. 2009. Relationship between duration of teriparatide

therapy and clinical outcomes in postmenopausal women with osteoporosis. *Osteoporos Int* 20(6): 943–8.
24. Cosman F, Nieves J, Woelfert L, Formica C, Gordon S, Shen V, et al. 2001. Parathyroid hormone added to established hormone therapy: Effects on vertebral fracture and maintenance of bone mass after parathyroid hormone withdrawal. *J Bone Miner Res* 16(5): 925–31.
25. Kurland ES, Cosman F, McMahon DJ, Rosen CJ, Lindsay R, Bilezikian JP. 2000. Parathyroid hormone as a therapy for idiopathic osteoporosis in men: Effects on bone mineral density and bone markers. *J Clin Endocrinol Metab* 85(9): 3069–76.
26. Cosman F, Lane NE, Bolognese MA, Zanchetta JR, Garcia-Hernandez PA, Sees K, et al. 2010. Effect of transdermal teriparatide administration on bone mineral density in postmenopausal women. *J Clin Endocrinol Metab* 95(1): 151–8. PMCID: 2805490.
27. Greenspan SL, Bone HG, Ettinger MP, Hanley DA, Lindsay R, Zanchetta JR, et al. 2007. Effect of recombinant human parathyroid hormone (1-84) on vertebral fracture and bone mineral density in postmenopausal women with osteoporosis: A randomized trial. *Ann Intern Med* 146(5): 326–39.
28. Yu EW, Neer RM, Lee H, Wyland JJ, de la Paz AV, Davis MC, et al. 2011. Time-dependent changes in skeletal response to teriparatide: Escalating vs. constant dose teriparatide (PTH 1-34) in osteoporotic women. *Bone* 48(4): 713–9. PMCID: 3073572.
29. Black DM, Greenspan SL, Ensrud KE, Palermo L, McGowan JA, Lang TF, et al. 2003. The effects of parathyroid hormone and alendronate alone or in combination in postmenopausal osteoporosis. *N Engl J Med* 349(13): 1207–15.
30. Gluer CC, Genant HK. 1989. Impact of marrow fat on accuracy of quantitative CT. *J Comput Assist Tomogr* 13(6): 1023–35.
31. Kuiper JW, van Kuijk C, Grashuis JL, Ederveen AG, Schutte HE. 1996. Accuracy and the influence of marrow fat on quantitative CT and dual-energy X-ray absorptiometry measurements of the femoral neck in vitro. *Osteoporos Int* 6(1): 25–30.
32. Finkelstein JS, Wyland JJ, Lee H, Neer RM. 2010. Effects of teriparatide, alendronate, or both in women with postmenopausal osteoporosis. *J Clin Endocrinol Metab* 95(4): 1838–45. PMCID: 2853981.
33. Cosman F, Eriksen EF, Recknor C, Miller PD, Guanabens N, Kasperk C, et al. 2011. Effects of intravenous zoledronic acid plus subcutaneous teriparatide [rhPTH(1-34)] in postmenopausal osteoporosis. *J Bone Miner Res* 26(3): 503–11.
34. Deal C, Omizo M, Schwartz EN, Eriksen EF, Cantor P, Wang J, et al. 2005. Combination teriparatide and raloxifene therapy for postmenopausal osteoporosis: Results from a 6-month double-blind placebo-controlled trial. *J Bone Miner Res* 20(11): 1905–11.
35. Ettinger B, San Martin J, Crans G, Pavo I. 2004. Differential effects of teriparatide on BMD after treatment with raloxifene or alendronate. *J Bone Miner Res* 19(5): 745–51.
36. Miller PD, Delmas PD, Lindsay R, Watts NB, Luckey M, Adachi J, et al. 2008. Early responsiveness of women with osteoporosis to teriparatide after therapy with alendronate or risedronate. *J Clin Endocrinol Metab* 93(10): 3785–93.
37. Boonen S, Marin F, Obermayer-Pietsch B, Simoes ME, Barker C, Glass EV, et al. 2008. Effects of previous antiresorptive therapy on the bone mineral density response to two years of teriparatide treatment in postmenopausal women with osteoporosis. *J Clin Endocrinol Metab* 93(3): 852–60.
38. Cosman F, Nieves J, Zion M, Woelfert L, Luckey M, Lindsay R. 2005. Daily and cyclic parathyroid hormone in women receiving alendronate. *N Engl J Med* 353(6): 566–75.
39. Cosman F, Nieves JW, Zion M, Barbuto N, Lindsay R. 2008. Effect of prior and ongoing raloxifene therapy on response to PTH and maintenance of BMD after PTH therapy. *Osteoporos Int* 19(4): 529–35.
40. Lindsay R, Nieves J, Formica C, Henneman E, Woelfert L, Shen V, et al. 1997. Randomised controlled study of effect of parathyroid hormone on vertebral-bone mass and fracture incidence among postmenopausal women on oestrogen with osteoporosis. *Lancet* 350(9077): 550–5.
41. Obermayer-Pietsch BM, Marin F, McCloskey EV, Hadji P, Farrerons J, Boonen S, et al. 2008. Effects of two years of daily teriparatide treatment on BMD in postmenopausal women with severe osteoporosis with and without prior antiresorptive treatment. *J Bone Miner Res* 23(10): 1591–600.
42. Cosman F, Wermers RA, Recknor C, Mauck KF, Xie L, Glass EV, et al. 2009. Effects of teriparatide in postmenopausal women with osteoporosis on prior alendronate or raloxifene: Differences between stopping and continuing the antiresorptive agent. *J Clin Endocrinol Metab* 94(10): 3772–80.
43. Roe EB, Sanchez SD, del Puerto GA, Pierini E, Bacchetti P, Cann CE, et al. 1992. Parathyroid hormone 1–34 (hPTH 1–34) and estrogen produce dramatic bone density increases in postmenopausal osteoporosis- results from a placebo-controlled randomized trial. *J Bone Miner Res* 12(Suppl. 1): S137 [Abstract].
44. Ste-Marie LG, Schwartz SL, Hossain A, Desaiah D, Gaich GA. 2006. Effect of teriparatide [rhPTH(1-34)] on BMD when given to postmenopausal women receiving hormone replacement therapy. *J Bone Miner Res* 21(2): 283–91.
45. Kaufman JM, Orwoll E, Goemaere S, San Martin J, Hossain A, Dalsky GP, et al. 2005. Teriparatide effects on vertebral fractures and bone mineral density in men with osteoporosis: Treatment and discontinuation of therapy. *Osteoporos Int* 16(5): 510–6.
46. Finkelstein JS, Hayes A, Hunzelman JL, Wyland JJ, Lee H, Neer RM. 2003. The effects of parathyroid hormone, alendronate, or both in men with osteoporosis. *N Engl J Med* 349(13): 1216–26.
47. Leder BZ, Neer RM, Wyland JJ, Lee HW, Burnett-Bowie SM, Finkelstein JS. 2009. Effects of teriparatide

treatment and discontinuation in postmenopausal women and eugonadal men with osteoporosis. *J Clin Endocrinol Metab* 94(8): 2915–21.
48. Lane NE, Sanchez S, Modin GW, Genant HK, Pierini E, Arnaud CD. 2000. Bone mass continues to increase at the hip after parathyroid hormone treatment is discontinued in glucocorticoid-induced osteoporosis: Results of a randomized controlled clinical trial. *J Bone Miner Res* 15(5): 944–51.
49. Saag KG, Shane E, Boonen S, Marin F, Donley DW, Taylor KA, et al. 2007. Teriparatide or alendronate in glucocorticoid-induced osteoporosis. *N Engl J Med* 357(20): 2028–39.
50. Saag KG, Zanchetta JR, Devogelaer JP, Adler RA, Eastell R, See K, et al. 2009. Effects of teriparatide versus alendronate for treating glucocorticoid-induced osteoporosis: Thirty-six-month results of a randomized, double-blind, controlled trial. *Arthritis Rheum* 60(11): 3346–55.
51. Aspenberg P, Genant HK, Johansson T, Nino AJ, See K, Krohn K, et al. 2010. Teriparatide for acceleration of fracture repair in humans: A prospective, randomized, double-blind study of 102 postmenopausal women with distal radial fractures. *J Bone Miner Res* 25(2): 404–14.
52. Lindsay R, Scheele WH, Neer R, Pohl G, Adami S, Mautalen C, et al. 2004. Sustained vertebral fracture risk reduction after withdrawal of teriparatide in postmenopausal women with osteoporosis. *Arch Intern Med* 164(18): 2024–30.
53. Kurland ES, Heller SL, Diamond B, McMahon DJ, Cosman F, Bilezikian JP. 2004. The importance of bisphosphonate therapy in maintaining bone mass in men after therapy with teriparatide [human parathyroid hormone(1-34)]. *Osteoporos Int* 15(12): 992–7.
54. Rittmaster RS, Bolognese M, Ettinger MP, Hanley DA, Hodsman AB, Kendler DL, et al. 2000. Enhancement of bone mass in osteoporotic women with parathyroid hormone followed by alendronate. *J Clin Endocrinol Metab* 85(6): 2129–34.
55. Black DM, Bilezikian JP, Ensrud KE, Greenspan SL, Palermo L, Hue T, et al. 2005. One year of alendronate after one year of parathyroid hormone (1-84) for osteoporosis. *N Engl J Med* 353(6): 555–65.
56. Prince R, Sipos A, Hossain A, Syversen U, Ish-Shalom S, Marcinowska E, et al. 2005. Sustained nonvertebral fragility fracture risk reduction after discontinuation of teriparatide treatment. *J Bone Miner Res* 20(9): 1507–13.
57. Cosman F, Nieves JW, Zion M, Barbuto N, Lindsay R. 2009. Retreatment with teriparatide one year after the first teriparatide course in patients on continued long-term alendronate. *J Bone Miner Res* 24(6): 1110–5.
58. Finkelstein JS, Wyland JJ, Leder BZ, Burnett-Bowie SM, Lee H, Juppner H, et al. 2009. Effects of teriparatide retreatment in osteoporotic men and women. *J Clin Endocrinol Metab* 94(7): 2495–501.
59. Liu H, Michaud K, Nayak S, Karpf DB, Owens DK, Garber AM. 2006. The cost-effectiveness of therapy with teriparatide and alendronate in women with severe osteoporosis. *Arch Intern Med* 166(11): 1209–17.

52

Strontium Ranelate in the Prevention of Osteoporotic Fractures

René Rizzoli

Introduction 437
Strontium Ranelate Antifracture Efficacy 437
Subgroup Analysis 438
Safety of Strontium Ranelate 439
Reversibility of BMD and Effects of Previous Therapy 439
Quality of Life and Cost-Effectiveness 439

Recent Developments 440
Mechanisms of Anti-Fracture Efficacy 440
Preclinical Studies 440
Fracture Healing and Implant Osseointegration 440
References 441

INTRODUCTION

Strontium (Sr) is an alkaline earth divalent cation, which is a trace element in the human body, representing 0.00044% of the body mass [1]. Ninety-nine percent of body strontium is localized in bone. With an atomic weight of 87.6, it is more than twice heavier than calcium. A normal diet contains between 2 and 4 mg of strontium per day, mostly from vegetables and cereals. Strontium competes with calcium for intestinal absorption, which is about 20% of the intake, with a ratio of strontium to calcium intestinal absorption of approximately 0.6 to 0.7 [2]. Strontium is eliminated through the kidney and through gastrointestinal tract secretion. Though strontium and calcium share a common transport system in the renal tubule, the clearance of the former appears to be higher than that of the latter. In bone, strontium is mainly adsorbed onto the crystal surface with less than one atom of strontium replacing one calcium atom in the hydroxyapatite crystal [3], without altering the lattice structure. Strontium is primarily found in newly deposited bone tissue. After stopping treatment, more than 50% of bone strontium disappears within 10 weeks [3]. Strontium ranelate (5[bis(carboxymethyl)amino]-2-carboxy-4-cyano-3-thiophenacetic acid distrontium salt), which is approved for the treatment of osteoporosis in Europe and in many countries worldwide, comprises an organic anion (ranelate) and two stable strontium cations. Less than 3% of ranelate is absorbed in the intestine. Ranelate is not metabolized and is mainly excreted through the kidney. The therapeutic dose of 2 g per day of strontium ranelate provides 8 mmoles of strontium. For example, it is worth noting that 8 mmoles of calcium represents 320 mg.

STRONTIUM RANELATE ANTIFRACTURE EFFICACY (Fig. 52.1)

Strontium ranelate was investigated in a phase III program consisting of two studies. The Spinal Osteoporosis Therapeutic Intervention (SOTI) study aimed at assessing the effect of 2 g daily of strontium ranelate on the risk of vertebral fractures, and the Treatment of Peripheral Osteoporosis (TROPOS) trial aimed at evaluating the effect of strontium ranelate on nonvertebral fractures. All patients included in these two studies had previously participated in a run-in study, the Fracture International Run-in Strontium Ranelate Trials (FIRST), aimed at normalizing the calcium and vitamin D status of all patients prior to entry into either the SOTI or TROPOS trials.

Primer on the Metabolic Bone Diseases and Disorders of Mineral Metabolism, Eighth Edition. Edited by Clifford J. Rosen.
© 2013 American Society for Bone and Mineral Research. Published 2013 by John Wiley & Sons, Inc.

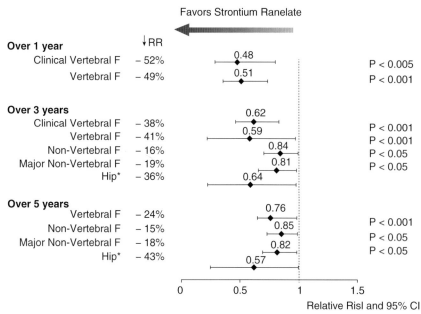

Fig. 52.1. Effects of strontium ranelate on vertebral and nonvertebral fracture risk (reprinted from Ref. 44).

Patients received a calcium/vitamin D supplement throughout the studies, which varied from 500 to 1,000 mg of calcium, and from 400 to 800 IU of vitamin D3. Among more than 9,000 postmenopausal women with osteoporosis who took part in the FIRST study, 1,649 patients, with a mean age of 70 years, were included in SOTI [4], and 5,091 patients, with a mean age of 77, were included in TROPOS [5].

In the SOTI study, treatment with strontium ranelate for 3 years was associated with a 41% reduction in the relative risk (RR) of experiencing a new vertebral fracture (RR = 0.59; 95% CI, 0.48–0.73), as assessed by semiquantitative evaluation. The relative risk of experiencing a new vertebral fracture was already significantly reduced by 49% (RR = 0.59; 95% CI, 0.36–0.74) in the strontium ranelate group by the end of the first year of treatment.

In the TROPOS study, which was designed to assess strontium ranelate efficacy in preventing nonvertebral fractures in postmenopausal women with osteoporosis, ambulatory postmenopausal women were eligible if they had femoral neck bone mineral density (BMD) corresponding to a T-score less than –2.5 and were older than 74 years, or aged between 70 and 74 years, but with one additional risk factor for fracture. In the 5,091 patients recruited, strontium ranelate was associated with a 16% (RR = 0.84; 95% CI, 0.70–0.99) relative risk reduction in all nonvertebral fractures over a 3-year follow-up period (p = 0.04), and with a 19% (RR = 0.81; 95% CI, 0.66–0.98) reduction in risk of major nonvertebral osteoporotic fractures (p = 0.031). In a high-risk fracture subgroup [women older than 74 years and with femoral neck BMD T-score less than –2.4 SD, using the Third National Health and Nutrition Examination Survey (NHANES III) reference range], treatment was associated with a 36% (RR = 0.64; 95% CI, 0.41–1.00) reduction in risk of hip fracture (p = 0.046). Yearly vertebral X-rays were performed in 3,640 patients of the TROPOS trial, in whom the relative risk of new vertebral fracture was reduced by 39% (RR = 0.61; 95% CI, 0.51–0.73) over 3 years in the strontium ranelate group (p < 0.001), and of 45% (RR = 0.55; 95% CI, 0.39–0.77) (p < 0.001) by the first year of treatment, indicating a highly consistent effect between both trials. In these 3,640 patients, 66.4% had no prevalent vertebral fracture at inclusion. The risk of experiencing a first vertebral fracture was reduced by 45% (RR = 0.55; 95% CI, 0.42–0.72) (p < 0.001). In the subgroup of patients with at least one prevalent fracture (n = 1,224), the risk of experiencing a first vertebral fracture was reduced by 32% (RR = 0.67; 95% CI, 0.53–0.85) (p < 0.001). An extension of the phase 3 trials up to 5 years for TROPOS has shown a sustained anti-fracture efficacy with –24% and –15% reduction in vertebral and nonvertebral fracture, respectively, after 5 years of placebo-controlled study.

SUBGROUP ANALYSIS

Data from SOTI and TROPOS were pooled to address several prespecified research questions. The efficacy of strontium ranelate was specifically demonstrated in elderly patients (over 80 years of age) (n = 1488, mean age: 84 ± 3 years) with a significant reduction in both vertebral (−32% and −31%) and nonvertebral (−31% and −26%) risks of fracture over 3 years [6] and 5 years [7], respectively. In the 353 patients of the SOTI trial aged 50 to 65 years, strontium ranelate decreased the risk of vertebral fracture by 43% (R = 0.57, 95% CI, 0.57–0.95) (p = 0.019) and by 35% (RR = 0.65; 95% CI, 0.42–0.99) (p = 0.049), by 3 and 4 years of treatment, respectively [8]. The response to strontium ranelate was evaluated in 1166 women with lumbar spine osteopenia after 3 years of treatment. Vertebral fracture risk was reduced by 41% (RR = 0.61; 95% CI, 0.43–0.83). A reduction of 59% (RR = 0.41; 95% CI, 0.17–0.99) and of 38% (RR = 0.62; 95% CI, 0.44–0.88) was recorded in the 447 and 719 women without and with prevalent fractures, respectively [9]. The anti-fracture efficacy of strontium ranelate could be demonstrated whatever the severity of osteoporosis or risk factors (age, baseline BMD, number of prevalent fracture, family history, body mass index or smoking) [10] or the level of bone remodelling [11]. In untreated patients, there was a worsening in spine kyphosis over 3 years. Kyphosis progression was lower under strontium ranelate therapy [12].

In patients in the pooled data from SOTI and TROPOS trials, identified as frail (n = 264), intermediate (n = 2472), or robust (n = 2346), using adapted Fried's criteria [13], vertebral fracture risk was reduced by 58% (RR = 0.42; 95% CI, 0.24–0.74), by 45% (RR = 0.55; 95% CI, 0.46–0.67), and by 30% (RR = 0.70; 95% CI, 0.57–0.86), respectively [14], by 3 years of treatment. In the frail group, the number needed to treat was as low as 5 over 3 years.

SAFETY OF STRONTIUM RANELATE

In general, strontium ranelate was well tolerated, without major adverse events. During clinical trials, the most common side effects were nausea and diarrhoea (about 7% vs 5% in the placebo group over 5 years), headache, and skin irritation. All were mild and transient and did not lead to withdrawal from the studies. By pooling both SOTI and TROPOS trials data, a significant increase in the risk of venous-thrombosis embolism event was found (RR: 1.42, p = 0.036). However, no coagulation abnormality has been detected in relation with strontium ranelate treatment. In a retrospective large cohort study, there was no difference in venous thromboembolism between strontium ranelate- or alendronate-treated patients, and osteoporotic untreated patients [15]. Recently a few cases of DRESS syndrome (Drug Rash with Eosinophilia and Systemic Symptoms) were recorded. This adverse event is very rare, and requires an immediate stop of the medication [16].

In two extension uncontrolled studies, of 8 years and 10 years with continuous treatment, the cumulative incidence of vertebral and nonvertebral fractures was quite similar to that observed in the equivalent first years of controlled treatment, despite participants being 5 and 7 years older, suggesting a sustained antifracture efficacy [17, 18]. There were no further significant adverse events.

REVERSIBILITY OF BMD AND EFFECTS OF PREVIOUS THERAPY

After 4 years of treatment in the SOTI trial, vertebral fracture risk was reduced by 33% (RR = 0.67; 95% CI, 0.55–0.81). The number needed to treat was 11. During the fifth year, the patients previously treated with strontium ranelate and switched to a placebo had a decrease in BMD of 3.2 and 2.5% at the spine and total hip, respectively, whereas the changes in those continuously treated were +1.2 and +0.4 (p = 0.001 for both comparisons) [19].

Paired transiliac bone biopsies performed before and after 12 months of strontium ranelate, in patients previously treated with bisphosphonates, showed an increase of trabecular thickness and of the number of remodeling sites [20]. In contrast, BMD increase in response to strontium ranelate appeared to be blunted in previously bisphosphonate treated as compared with naïve patients [21], particularly during the first 6 months.

QUALITY OF LIFE AND COST-EFFECTIVENESS

Using a validated instrument specifically developed to assess quality of life in patients with osteoporosis, strontium ranelate produced beneficial effects on both physical and emotional measures (p = 0.019 and 0.092 versus placebo, respectively), together with a significant increase in the number of patients free of back pain (+31%, p = 0.005) by the first year and over 3 years of treatment in the patients enrolled in the SOTI trial [22].

Adapting a Markov model to the conditions of SOTI and TROPOS trials, with Swedish costs and epidemiological data, strontium ranelate treatment was shown to be cost-effective, and even cost saving in patients older than 74 years and with a T-Score below −2.4, or in patients older than 80 years of age [23]. A similar assessment was undertaken for U.K. conditions. For a willingness-to-pay threshold of 30,000 pounds per quality-adjusted life year (QALY), strontium ranelate was cost-effective in women from age 65 years and older with a prior fracture, with and without information on BMD [24].

Using country-specific FRAX tool data applied to the combined SOTI and TROPOS cohorts, fracture risk reduction was evaluated in relation with the baseline 10-year fracture risk. The effects of strontium ranelate on clinical osteoporotic fracture and on morphometric fracture were not dependent on the level of baseline fracture risk assessed by FRAX [25].

RECENT DEVELOPMENTS

Based on preclinical data suggesting that strontium ranelate may influence chondrocytes metabolism, a subgroup of 2,617 postmenopausal women from the TROPOS study had measurement of urinary type II collagen degradation products. Strontium ranelate was associated with a 10–20% lower value as compared with placebo [26]. Spine osteoarthritis progression (osteophytes, disc space narrowing, and sclerosis) was evaluated in 1,105 osteoporotic women form SOTI and TROPOS. A 42% reduction (p = 0.0005) in the proportion of patients with radiological worsening was observed over 3 years [27a]. In a randomized placebo controlled trial, the effects of strontium ranelate on knee joint space narrowing were assessed in 1371 patients. There was a smaller degradation in joint space narrowing (0.27 ± 0.63 vs 0.37 ± 0.59 mm) with a between-group difference of 0.10 (95% CI: 0.02–0.19, p = 0.018) in the strontium ranelate treated group. This was associated with a larger reduction in the WOMAC score (p = 0.145) [27b].

MECHANISMS OF ANTI-FRACTURE EFFICACY

In the SOTI trial, lumbar spine BMD increased from baseline by 12.7% at the lumbar spine, 7.2% at the femoral neck, and 8.6% at the total hip [4] under strontium ranelate treatment. BMD adjusted for strontium content [28] increased by 6.8% from baseline at the lumbar spine after 3 years compared with a decrease of 1.3% in the placebo group [4]. The 8-year and 10-year extension trials showed a continuous increase in spine and proximal femur BMD, of more than 30% versus pretreatment values for the former, and more than 12% for the latter [17, 18]. In a pooled analysis of both trials, changes in femoral neck BMD were predictive of the reduction of vertebral fracture risk [29]. Each percent increase in femoral neck BMD at either 1 year or 3 years was associated with a 3% decrease in new vertebral fracture risk, with changes in femoral neck BMD explaining 76% of the variance of vertebral fracture risk reduction.

In both trials, bone-specific alkaline phosphatase increased in the strontium ranelate group by approximately 8%, while serum type I collagen C-telopeptide cross links decreased as well by 12%, indicating a mild decreased bone resorption and maintained, or even increased, bone formation. This pattern is quite different from what is observed with inhibitors of bone resorption, where markers of bone formation are decreased in a commensurate way to the changes in the markers of bone resorption. Changes in biochemical markers accounted for less than 8% of BMD change [30]. None of the 3-month changes in biochemical markers was predictive of fracture incidence.

Comparing 49 transiliac bone biopsies collected in strontium ranelate treated patients to 92 biopsies obtained either at baseline or in the placebo group, higher mineral apposition rate in cancellous bone (+9%, p = 0.019) and osteoblast perimeters (+38%, p = 0.047) were found [30]. Submitting 20 and 21 biopsies obtained at 3 years of strontium ranelate treatment and a placebo, respectively, to microcomputerized tomography, higher cortical thickness (+18%, p = 0.008) and trabecular number (+14%, p = 0.008), together with lower trabecular separation (−22%, p = 0.01), without any modification in cortical porosity were demonstrated [31]. In a head-to-head comparative exploratory trial, strontium ranelate appeared to influence more structural variables such as cortical thickness than alendronate, at the distal tibia, as assessed by high-resolution peripheral quantitative computed tomography [32, 33], with the limitations related to different X-ray attenuation by various minerals. In this trial, the load to failure estimated by finite element analysis was higher in the strontium ranelate treated group. There was no difference in cortical porosity [33].

PRECLINICAL STUDIES

A large series of preclinical studies have shown that strontium ranelate treatment increases ultimate bone strength, mainly through an increase in plastic energy [34], and improves material level properties, by increasing elastic modulus, hardness, and dissipated energy, as assessed by nanoindentation [35]. It reduces bone loss in various experimental models [36, 37], and even decreases spontaneous fracture in a mice model overexpressing Runx-2 transcription factor (−60%) [38]. In treated animals, higher cancellous bone volume and cortical thickness were also associated with increased force to failure (+120%). A direct inhibition of strontium on osteoclast differentiation and activity has been reported [39, 40]. Strontium stimulates proliferation and differentiation of cells of the osteoblast lineage, through mechanisms possibly involving a calcium-sensing receptor for the former [41] and prostaglandins for the latter [42–45]. Increased osteoprotegerin and decreased RANKL expressions have been reported in osteoblasts [45] incubated with strontium. The latter has also been shown to promote an osteocyte-like phenotype [46]. Fibroblast growth factor (FGF) receptors could also be involved in strontium-mediated osteoblastic cell growth [47].

FRACTURE HEALING AND IMPLANT OSSEOINTEGRATION

In the animal model of fracture healing, strontium treatment has been shown either to not impair the process [48, 49] or even to improve it [50, 51]. Systemic treatment with strontium ameliorated osseointegration of titanium implant or screw, as indicated by a higher strength required to remove the metallic implant from bone (Fig. 52.2) [52, 53].

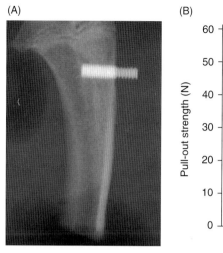
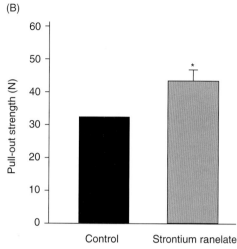

Fig. 52.2. (A) Implant in rats, proximal tibia. (B) Effects of strontium ranelate (325 mg/kg BW and 5 days/week) on pull out strength (p < 0.05 vs controls) (from Ref. 52).

REFERENCES

1. Pors Nielsen S. 2004. The biological role of strontium. *Bone* 35(3): 583–8.
2. Marcus CS, Lengemann FW. 1962. Absorption of Ca45 and Sr85 from solid and liquid food at various levels of the alimentary tract of the rat. *J Nutr* 77: 155–60.
3. Farlay D, Boivin G, Panczer G, Lalande A, Meunier PJ. 2005. Long-term strontium ranelate administration in monkeys preserves characteristics of bone mineral crystals and degree of mineralization of bone. *J Bone Miner Res* 20(9): 1569–78.
4. Meunier PJ, Roux C, Seeman E, Ortolani S, Badurski JE, Spector TD, Cannata J, Balogh A, Lemmel EM, Pors-Nielsen S, Rizzoli R, Genant HK, Reginster JY. 2004. The effects of strontium ranelate on the risk of vertebral fracture in women with postmenopausal osteoporosis. *N Engl J Med* 350(5): 459–68.
5. Reginster JY, Felsenberg D, Boonen S, Diez-Perez A, Rizzoli R, Brandi ML, Spector TD, Brixen K, Goemaere S, Cormier C, Balogh A, Delmas PD, Meunier PJ. 2008. Effects of long-term strontium ranelate treatment on the risk of nonvertebral and vertebral fractures in postmenopausal osteoporosis: Results of a five-year, randomized, placebo-controlled trial. *Arthritis Rheum* 58(6). 1687–95.
6. Seeman E, Vellas B, Benhamou C, Aquino JP, Semler J, Kaufman JM, Hoszowski K, Varela AR, Fiore C, Brixen K, Reginster JY, Boonen S. 2006. Strontium ranelate reduces the risk of vertebral and nonvertebral fractures in women eighty years of age and older. *J Bone Miner Res* 21(7): 1113–20.
7. Seeman E, Boonen S, Borgstrom F, Vellas B, Aquino JP, Semler J, Benhamou CL, Kaufman JM, Reginster JY. 2010. Five years treatment with strontium ranelate reduces vertebral and nonvertebral fractures and increases the number and quality of remaining life-years in women over 80 years of age. *Bone* 46(4): 1038–42.
8. Roux C, Fechtenbaum J, Kolta S, Isaia G, Andia JB, Devogelaer JP. 2008. Strontium ranelate reduces the risk of vertebral fracture in young postmenopausal women with severe osteoporosis. *Ann Rheum Dis* 67(12): 1736–8.
9. Seeman E, Devogelaer JP, Lorenc R, Spector T, Brixen K, Balogh A, Stucki G, Reginster JY. 2008. Strontium ranelate reduces the risk of vertebral fractures in patients with osteopenia. *J Bone Miner Res* 23(3): 433–8.
10. Roux C, Reginster JY, Fechtenbaum J, Kolta S, Sawicki A, Tulassay Z, Luisetto G, Padrino JM, Doyle D, Prince R, Fardellone P, Sorensen OH, Meunier PJ. 2006. Vertebral fracture risk reduction with strontium ranelate in women with postmenopausal osteoporosis is independent of baseline risk factors. *J Bone Miner Res* 21(4): 536–42.
11. Collette J, Bruyere O, Kaufman JM, Lorenc R, Felsenberg D, Spector TD, Diaz-Curiel M, Boonen S, Reginster JY. 2010. Vertebral anti-fracture efficacy of strontium ranelate according to pre-treatment bone turnover. *Osteoporos Int* 21(2): 233–41.
12. Roux C, Fechtenbaum J, Kolta S, Said-Nahal R, Briot K, Benhamou CL. 2010. Prospective assessment of thoracic kyphosis in postmenopausal women with osteoporosis. *J Bone Miner Res* 25(2): 362–8.
13. Fried LP, Tangen CM, Walston J, Newman AB, Hirsch C, Gottdiener J, Seeman T, Tracy R, Kop WJ, Burke G, McBurnie MA. 2001. Frailty in older adults: Evidence for a phenotype. *J Gerontol A Biol Sci Med Sci* 56(3): M146–56.
14. Rolland Y, Abellan Van Kan G, Gillette-Guyonnet S, Roux C, Boonen S, Vellas B. 2011. Strontium ranelate and risk of vertebral fractures in frail osteoporotic women. *Bone* 48(2): 332–8.
15. Breart G, Cooper C, Meyer O, Speirs C, Deltour N, Reginster JY. 2010. Osteoporosis and venous thromboembolism: A retrospective cohort study in the UK General Practice Research Database. *Osteoporos Int* 21(7): 1181–7.

16. Musette P, Brandi ML, Cacoub P, Kaufman JM, Rizzoli R, Reginster JY. 2010. Treatment of osteoporosis: Recognizing and managing cutaneous adverse reactions and drug-induced hypersensitivity. *Osteoporos Int* 21(5): 723–32.
17. Reginster JY, Bruyere O, Sawicki A, Roces-Varela A, Fardellone P, Roberts A, Devogelaer JP. 2009. Long-term treatment of postmenopausal osteoporosis with strontium ranelate: results at 8 years. *Bone* 45(6): 1059–64.
18. Reginster JY, Kaufman JM, Goemaere S, et al. 2012. Maintenance of antifracture efficacy over 10 years with strontium ranelate in postmenopausal osteoporosis. *Osteoporos Int* 23(3): 1115–22.
19. Meunier PJ, Roux C, Ortolani S, Diaz-Curiel M, Compston J, Marquis P, Cormier C, Isaia G, Badurski J, Wark JD, Collette J, Reginster JY. 2009. Effects of long-term strontium ranelate treatment on vertebral fracture risk in postmenopausal women with osteoporosis. *Osteoporos Int* 20(10): 1663–73.
20. Busse B, Jobke B, Hahn M, Priemel M, Niecke M, Seitz S, Zustin J, Semler J, Amling M. 2010. Effects of strontium ranelate administration on bisphosphonate-altered hydroxyapatite: Matrix incorporation of strontium is accompanied by changes in mineralization and microstructure. *Acta Biomater* 6(12): 4513–21.
21. Middleton ET, Steel SA, Aye M, Doherty SM. 2010. The effect of prior bisphosphonate therapy on the subsequent BMD and bone turnover response to strontium ranelate. *J Bone Miner Res* 25(3): 455–62.
22. Marquis P, Roux C, de la Loge C, Diaz-Curiel M, Cormier C, Isaia G, Badurski J, Wark J, Meunier PJ. 2008. Strontium ranelate prevents quality of life impairment in post-menopausal women with established vertebral osteoporosis. *Osteoporos Int* 19(4): 503–10.
23. Borgstrom F, Jonsson B, Strom O, Kanis JA. 2006. An economic evaluation of strontium ranelate in the treatment of osteoporosis in a Swedish setting: Based on the results of the SOTI and TROPOS trials. *Osteoporos Int* 17(12): 1781–93.
24. Borgstrom F, Strom O, Kleman M, McCloskey E, Johansson H, Oden A, Kanis JA. 2010. Cost-effectiveness of bazedoxifene incorporating the FRAX(R) algorithm in a European perspective. *Osteoporos Int* 22(3): 955–65.
25. Kanis JA, Johansson H, Oden A, McCloskey EV. 2011. A meta-analysis of the effect of strontium ranelate on the risk of vertebral and non-vertebral fracture in postmenopausal osteoporosis and the interaction with FRAX(®). *Osteoporos Int* 22(8): 2347–55.
26. Alexandersen P, Karsdal MA, Qvist P, Reginster JY, Christiansen C. 2007. Strontium ranelate reduces the urinary level of cartilage degradation biomarker CTX-II in postmenopausal women. *Bone* 40(1): 218–22.
27a. Bruyere O, Delferriere D, Roux C, Wark JD, Spector T, Devogelaer JP, Brixen K, Adami S, Fechtenbaum J, Kolta S, Reginster JY. 2008. Effects of strontium ranelate on spinal osteoarthritis progression. *Ann Rheum Dis* 67(3): 335–9.
27b. Reginster JY, Badurski J, Bellamy N, Bensen W, Chapurlat R, Chevalier X, Christiansen C, Genant H, Navarro F, Nasonov E, Sambrook PN, Spector TD & Cooper C. 2013. Efficacy and safety of strontium ranelate in the treatment of knee osteoarthritis: Results of a double-blind, randomised placebo-controlled trial. *Ann Rheum Dis* 72: 179–86.
28. Blake GM, Fogelman I. 2005. Long-term effect of strontium ranelate treatment on BMD. *J Bone Miner Res* 20(11): 1901–4.
29. Bruyere O, Roux C, Detilleux J, Slosman DO, Spector TD, Fardellone P, Brixen K, Devogelaer JP, Diaz-Curiel M, Albanese C, Kaufman JM, Pors-Nielsen S, Reginster JY. 2007. Relationship between bone mineral density changes and fracture risk reduction in patients treated with strontium ranelate. *J Clin Endocrinol Metab* 92(8): 3076–81.
30. Bruyere O, Collette J, Rizzoli R, Decock C, Ortolani S, Cormier C, Detilleux J, Reginster JY. 2010. Relationship between 3-month changes in biochemical markers of bone remodelling and changes in bone mineral density and fracture incidence in patients treated with strontium ranelate for 3 years. *Osteoporos Int* 21(6): 1031–6.
31. Arlot ME, Jiang Y, Genant HK, Zhao J, Burt-Pichat B, Roux JP, Delmas PD, Meunier PJ. 2008. Histomorphometric and microCT analysis of bone biopsies from postmenopausal osteoporotic women treated with strontium ranelate. *J Bone Miner Res* 23(2): 215–22.
32. Rizzoli R, Laroche M, Krieg MA, Frieling I, Thomas T, Delmas P, Felsenberg D. 2010. Strontium ranelate and alendronate have differing effects on distal tibia bone microstructure in women with osteoporosis. *Rheumatol Int* 30(10): 1341–8.
33. Rizzoli R, Chapurlat RD, Laroche J-M, Krieg MA, Thomas T, Frieling I, Boutroy S, Laib A, Bock O, Felsenberg D. 2012. Effects of strontium ranelate and alendronate on bone microstructure in women with osteoporosis: Results of a 2-year study. *Osteoporos Int* 23(1): 305–15
34. Ammann P, Shen V, Robin B, Mauras Y, Bonjour JP, Rizzoli R. 2004. Strontium ranelate improves bone resistance by increasing bone mass and improving architecture in intact female rats. *J Bone Miner Res* 19(12): 2012–20.
35. Ammann P, Badoud I, Barraud S, Dayer R, Rizzoli R. 2007. Strontium ranelate treatment improves trabecular and cortical intrinsic bone tissue quality, a determinant of bone strength. *J Bone Miner Res* 22(9): 1419–25.
36. Marie PJ. 2005. Strontium as therapy for osteoporosis. *Curr Opin Pharmacol* 5(6): 633–6.
37. Marie PJ, Hott M, Modrowski D, De Pollak C, Guillemain J, Deloffre P, Tsouderos Y. 1993. An uncoupling agent containing strontium prevents bone loss by depressing bone resorption and maintaining bone formation in estrogen-deficient rats. *J Bone Miner Res* 8(5): 607–15.

38. Geoffroy V, Chappard D, Marty C, Libouban H, Ostertag A, Lalande A, de Vernejoul MC. 2011. Strontium ranelate decreases the incidence of new caudal vertebral fractures in a growing mouse model with spontaneous fractures by improving bone microarchitecture. *Osteoporos Int* 22(1): 289–97.
39. Baron R, Tsouderos Y. 2002. In vitro effects of S12911-2 on osteoclast function and bone marrow macrophage differentiation. *Eur J Pharmacol* 450(1): 11–7.
40. Takahashi N, Sasaki T, Tsouderos Y, Suda T. 2003. S 12911-2 inhibits osteoclastic bone resorption in vitro. *J Bone Miner Res* 18(6): 1082–7.
41. Chattopadhyay N, Quinn SJ, Kifor O, Ye C, Brown EM. 2007. The calcium-sensing receptor (CaR) is involved in strontium ranelate-induced osteoblast proliferation. *Biochem Pharmacol* 74(3): 438–47.
42. Choudhary S, Halbout P, Alander C, Raisz L, Pilbeam C. 2007. Strontium ranelate promotes osteoblastic differentiation and mineralization of murine bone marrow stromal cells: Involvement of prostaglandins. *J Bone Miner Res* 22(7): 1002–10.
43. Pi M, Quarles LD. 2004. A novel cation-sensing mechanism in osteoblasts is a molecular target for strontium. *J Bone Miner Res* 19(5): 862–9.
44. Marie PJ, Felsenberg D, Brandi ML. 2011. How strontium ranelate, via opposite effects on bone resorption and formation, prevents osteoporosis. *Osteoporos Int* 22(6): 1659–67.
45. Brennan TC, Rybchyn MS, Green W, Atwa S, Conigrave AD, Mason RS. 2009. Osteoblasts play key roles in the mechanisms of action of strontium ranelate. *Br J Pharmacol* 157(7): 1291–300.
46. Atkins GJ, Welldon KJ, Halbout P, Findlay DM. 2009. Strontium ranelate treatment of human primary osteoblasts promotes an osteocyte-like phenotype while eliciting an osteoprotegerin response. *Osteoporos Int* 20(4): 653–64.
47. Caverzasio J, Thouverey C. 2011. Activation of FGF receptors is a new mechanism by which strontium ranelate induces osteoblastic cell growth. *Cell Physiol Biochem* 27(3–4): 243–50.
48. Cebesoy O, Tutar E, Kose KC, Baltaci Y, Bagci C. 2007. Effect of strontium ranelate on fracture healing in rat tibia. *Joint Bone Spine* 74(6): 590–3.
49. Bruel A, Olsen J, Birkedal H, Risager M, Andreassen TT, Raffalt AC, Andersen JE, Thomsen JS. 2011. Strontium is incorporated into the fracture callus but does not influence the mechanical strength of healing rat fractures. *Calcif Tissue Int* 88(2): 142–52.
50. Habermann B, Kafchitsas K, Olender G, Augat P, Kurth A. 2010. Strontium ranelate enhances callus strength more than PTH 1-34 in an osteoporotic rat model of fracture healing. *Calcif Tissue Int* 86(1): 82–9.
51. Li YF, Luo E, Feng G, Zhu SS, Li JH, Hu J. 2010. Systemic treatment with strontium ranelate promotes tibial fracture healing in ovariectomized rats. *Osteoporos Int* 21(11): 1889–97.
52. Maimoun L, Brennan TC, Badoud I, Dubois-Ferriere V, Rizzoli R, Ammann P. 2010. Strontium ranelate improves implant osseointegration. *Bone* 46(5): 1436–41.
53. Li Y, Feng G, Gao Y, Luo E, Liu X, Hu J. 2010. Strontium ranelate treatment enhances hydroxyapatite-coated titanium screws fixation in osteoporotic rats. *J Orthop Res* 28(5): 578–82.

53

Combination Anabolic and Antiresorptive Therapy for Osteoporosis

John P. Bilezikian and Natalie E. Cusano

Introduction 444
Antiresorptive Therapy Followed by Anabolic Therapy 444
Concurrent Use of Anabolic and Antiresorptive Therapy 445

Antiresorptives After Anabolic Therapy 446
Conclusions 446
References 446

INTRODUCTION

The advent of teriparatide [PTH(1-34)] and the full-length hormone [PTH(1-84)] provided an attractive therapeutic alternative to antiresorptive drugs like the bisphosphonates for individuals with advanced osteoporosis. The anabolic actions of teriparatide and PTH(1-84) are described by an initial effect to primarily stimulate bone formation, followed thereafter by an increase in bone resorption. The time between the initial stimulation of bone formation and the subsequent stimulation of bone resorption has been described as an anabolic window [1, 2]. The idea of the anabolic window includes the point that it eventually closes, i.e., there is a downturn in the actions of parathyroid hormone (PTH). Since most patients who receive teriparatide or PTH(1-84) have previously received an antiresorptive, these important questions have been addressed: is it better to *add* the osteoanabolic agent to the ongoing antiresorptive regimen or it is better to *switch* from the antiresorptive to the osteoanabolic alone? In addition, attempts have been made to add to the efficacy of osteoanabolic therapy by combining it simultaneously with an antiresorptive. Implicit in the attempts to improve osteoanabolic effects with an antiresorptive used simultaneously is the idea of expanding the anabolic window. Finally, since teriparatide and PTH(1-84) are approved for only an 18–24 month period of time, decisions about following up this period of osteoanabolic therapy are important. This chapter reviews these three key aspects of combination therapy.

ANTIRESORPTIVE THERAPY FOLLOWED BY ANABOLIC THERAPY

Adding osteoanabolic therapy to ongoing antiresorptive therapy

Postmenopausal women who had undergone antecedent estrogen therapy [3] and postmenopausal women with glucocorticoid-induced osteoporosis [4] showed rapid and sustained increases in vertebral bone mineral density (BMD) with the use of teriparatide.

When Cosman et al. [5] used the bisphosphonate alendronate as the antecedent antiresorptive, the addition of teriparatide was also associated with prompt increases in BMD. The same BMD gains were seen whether teriparatide was used in a 3-month cyclical fashion or continuously with a backdrop of continuous alendronate administration.

Switching from antiresorptive therapy to osteoanabolic therapy

After 28 months of raloxifene or alendronate, Ettinger et al. [6] showed rapid densitometric gains with teriparatide following raloxifene but a delayed response following alendronate. It appeared that alendronate, by virtue of its potent antiresorptive effects, was interfering with the rapid effects of teriparatide to stimulate bone forma-

---

Primer on the Metabolic Bone Diseases and Disorders of Mineral Metabolism, Eighth Edition. Edited by Clifford J. Rosen.
© 2013 American Society for Bone and Mineral Research. Published 2013 by John Wiley & Sons, Inc.

tion. In support of this idea, Kurland et al. have shown that the response to teriparatide is a function of the level of baseline bone turnover in subjects not previously treated with any therapy for osteoporosis: the lower the level of turnover, the more sluggish the initial densitometric response to teriparatide [7]. Additional support for this hypothesis comes from a study in which patients who had previously been treated with either risedronate or alendronate were then switched to teriparatide [8]. Individuals who were first treated with risedronate had bone turnover marker levels that were higher than those first treated with alendronate and responded more exuberantly to teriparatide, with more rapid increases in BMD. The results were consistent with the smaller effect of risedronate on bone turnover, thus permitting a more rapid response to teriparatide. Eventually, and within a rather short period of time, the delay, if present is overcome [9, 10].

A recent view of the switching versus adding conundrum

In a study that combined ideas of potency of antiresorptives as well as *switching* versus *adding*, the recent study of Cosman et al. is noteworthy [11]. The cohort studied by Cosman et al. had been treated previously for at least 18 months with either raloxifene or alendronate. For the next 18 months, subjects were treated either with teriparatide and antiresorptive (*add* regimen) or with teriparatide alone (*switch* regimen, antiresorptive discontinued). Bone turnover markers increased more in the *switch* group (from either alendronate or raloxifene to teriparatide) but the densitometric gains by DXA were greatest when teriparatide was *added* to the antiresorptive drug. The results, at first, are not readily explained, but the data show that the anabolic window (the difference between increases in bone formation over bone resorption) was greater with the *add* regimen [12]. The study had too few subjects to know whether the greater gains in BMD with the *add* regimen might have been associated with better fracture results. Another interesting aspect of this study is that raloxifene was associated with greater changes in bone turnover markers and bone density at the lumbar spine than was alendronate, a finding consistent with the study by Ettinger et al. [6].

CONCURRENT USE OF ANABOLIC AND ANTIRESORPTIVE THERAPY

It is attractive to consider simultaneous combination therapy with an antiresorptive and teriparatide or PTH(1-84) as potentially more beneficial than monotherapy with either class of therapeutic given that their mechanisms of action are quite different from each other. If bone resorption is being inhibited (antiresorptive) while bone formation is being stimulated (anabolic), combination therapy might give better results than therapy with either agent alone. This postulate would argue, under these conditions, for an expansion of the anabolic window. Despite the intuitive appeal of this reasoning, important data to the contrary have been provided by Black et al. [13] in a study called *PaTH* and by Finkelstein et al. [14]. These two investigators independently conducted trials using a form of PTH alone, alendronate alone, or the combination of PTH and alendronate. Black et al. studied postmenopausal women with 100 μg of PTH(1-84). The study of Finkelstein et al. involved men treated with 40 μg of teriparatide. Both studies used dual-energy X-ray absorptiometry (DXA) and quantitative computed tomography (QCT) to measure areal or volumetric BMD, respectively. With either DXA or QCT, monotherapy with PTH exceeded densitometric gains with combination therapy or alendronate alone at the lumbar spine. Measurement of the trabecular compartment by QCT showed that combination therapy was associated with substantially smaller increases in BMD than therapy with PTH alone, and equivalent to the changes seen with alendronate alone. Bone turnover markers followed the expected course for anabolic (increases) or antiresorptive (decreases) therapy alone. However, for combination therapy, bone markers followed the course of alendronate, not PTH therapy, with reductions in bone formation and bone resorption markers. This suggests that the impaired response to combination therapy, in contrast to PTH alone, might be due to the dominating effects of alendronate on bone remodeling dynamics when both drugs are used together. The results of these two combination therapy studies led to the concept that an antiresorptive agent that did not impair the anabolic actions of teriparatide to increase bone formation while mitigating its affects on bone resorption might be a more effective approach to combination therapy. Deal et al. [15] addressed this point by studying the effects of raloxifene, a less potent antiresorptive agent than the bisphosphonates, in combination with teriparatide. As hypothesized, the combination of teriparatide and raloxifene had greater densitometric effects than monotherapy with teriparatide in postmenopausal osteoporosis. Bone formation markers increased to the same extent in the teriparatide-only group and in the teriparatide plus raloxifene groups. In contrast, when raloxifene was present along with teriparatide, bone resorption markers were significantly lower than the teriparatide-only group. The change in total hip BMD was significantly greater in subjects treated with both teriparatide and raloxifene than with teriparatide alone. In the presence of raloxifene, teriparatide appears to stimulate bone formation, unimpeded, but its ability to stimulate bone resorption is reduced.

Another approach to combination therapy has employed zoledronic acid and teriparatide. The reasoning here is that a single does of zoledronic acid might lead to more evident anabolic actions of PTH when used in combination. Early 3–6-month data [16] did show densitometric advantages of the combination arm over either teriparatide or zoledronic acid alone but at the 12 month time point, the combination arm was not better uniformly over teriparatide (lumbar spine) or zoledronic acid (hip).

ANTIRESORPTIVES AFTER ANABOLIC THERAPY

Studies of Kurland et al. [17] with teriparatide were followed by the PaTH study to provide prospective data in a rigorously controlled, blinded fashion to address the question whether antiresorptive therapy is necessary after PTH(1-84) is discontinued [18]. Postmenopausal women who had received PTH(1-84) for 12 months were randomly assigned to an additional 12 months of therapy with 10 mg of alendronate daily or placebo. In subjects who received alendronate, there was a further 4.9% gain in lumbar spine BMD, whereas those who received placebo experienced a substantial decline. By QCT analysis, the net increase over 24 months in lumbar spine BMD among those treated with alendronate after PTH(1-84) was 30%. In those who received placebo after PTH(1-84), the net change in BMD was only 13%. There were similar, dramatic differences in hip BMD when those who followed PTH with alendronate were compared with those who were treated with placebo after PTH (13% vs 5%). The results of this study establish the importance of following PTH or teriparatide therapy with an antiresorptive with regard to BMD.

There are only observational data to examine the question of fracture protection after teriparatide or PTH(1-84) is stopped. In a 30-month observational cohort following the pivotal clinical trial of teriparatide [19], subjects were given the option of switching to a bisphosphonate or not taking any further medications after teriparatide [20]. A majority (60%) were treated with antiresorptive therapy after PTH discontinuation. As noted from the PaTH trial, gains in BMD were maintained in those who chose to begin antiresorptive therapy immediately after teriparatide. Reductions in BMD were progressive throughout the 30-month observational period in subjects who elected not to follow teriparatide with any therapy. In a group that did not begin antiresorptive therapy until 6 months after teriparatide discontinuation, major reductions in BMD were seen during these first 6 months but no further reductions were observed after antiresorptive therapy was instituted [21]. Despite these densitometric data, the effect of previous therapy with teriparatide and/or subsequent therapy with a bisphosphonate on fracture prevention persisted for as long as 31 months after teriparatide discontinuation. Nonvertebral fragility fractures were reported by proportionately fewer women previously treated with PTH (followed with or without a bisphosphonate) compared with those treated with placebo (with or without a bisphosphonate; $p < 0.03$). In a logistic regression model, bisphosphonate use for 12 months or longer was said to add little to overall risk reduction of new vertebral fractures in this post-treatment period. However, it is hard to be sure of this conclusion because the data were not actually separately analyzed into those who did or did not follow teriparatide treatment with an antiresorptive. One might anticipate a residual but transient protection against fracture after PTH treatment without follow-up antiresorptive therapy, which could wane over time. Additional studies are needed to address fracture outcomes specifically. However, based particularly on the PaTH trial, the importance of following PTH or teriparatide therapy with an antiresorptive agent to maintain increases in bone mass is clear.

CONCLUSIONS

With the availability of antiresorptives and anabolic therapy for the treatment of osteoporosis, we are still exploring ways in which combination and sequential therapy can be used to greatest advantage in osteoporosis. However, the information available at this time gives guidance into ways in which simultaneous or sequential therapy with antiresorptives can be used to maximal advantage.

REFERENCES

1. Bilezikian J. 2008. Combination anabolic and antiresorptive therapy for osteoporosis: Opening the anabolic window. *Curr Osteoporos Rep* 6(1): 24–30.
2. Rubin MR, Bilezikian JP. 2003. New anabolic therapies in osteoporosis. *Endocrinol Metab Clin North Am* 32(1): 285–307.
3. Lindsay R, Nieves J, Formica C, Henneman E, Woelfert L, Shen V, Dempster D, Cosman F. 1997. Randomised controlled study of effect of parathyroid hormone on vertebral-bone mass and fracture incidence among postmenopausal women on oestrogen with osteoporosis. *Lancet* 350(9077): 550–5.
4. Lane NE, Sanchez S, Modin GW, Genant HK, Pierini E, Arnaud CD. 1998. Parathyroid hormone treatment can reverse corticosteroid-induced osteoporosis. Results of a randomized controlled clinical trial. *J Clin Invest* 102(8): 1627–33.
5. Cosman FJ, Nieves M, Zion L, Woelfert M, Luckey M, Lindsay R. 2005. Daily and cyclic parathyroid hormone in women receiving alendronate. *N Engl J Med* 353(6): 566–75.
6. Ettinger BJ, San Martin G, Crans G, Pavo I. 2004. Differential effects of teriparatide on BMD after treatment with raloxifene or alendronate. *J Bone Miner Res* 19(5): 745–51.
7. Kurland ES, Cosman F, McMahon DJ, Rosen DJ, Lindsay R, Bilezikian JP. 2000. Parathyroid hormone as a therapy for idiopathic osteoporosis in men: Effects on bone

mineral density and bone markers. *J Clin Endocrinol Metab* 85(9): 3069–76.
8. Miller PD, Delmas PD, Lindsay R, Watts NB, Luckey M, Adachi J, Saag K, Greenspan SL, Seeman E, Boonen S, Meeves S, Lang TF, Bilezikian JP. 2008. Early responsiveness of women with osteoporosis to teriparatide after therapy with alendronate or risedronate. *J Clin Endocrinol Metab* 93(10): 3785–93.
9. Boonen S, Millisen K, Gielen E, Vanderschueren D. 2011. Sequential therapy in the treatment of osteoporosis. *Curr Med Res Opin* 27(6): 1149–55.
10. Cusano NE, Bilezikian JP. 2011. Combination antiresorptive and osteoanabolic therapy for osteoporosis: We are not there yet. *Curr Med Res Opin* 27(9): 1705–7.
11. Cosman C, Wermers RA, Recknor C, Mauck KF, Xie L, Glass EV, Krege JH. 2009. Effects of teriparatide in postmenopausal women with osteoporosis on prior alendronate or raloxifene: Differences between stopping and continuing the antiresorptive agent. *J Clin Endocrinol Metab* 94(10): 3772–80.
12. Cusano NE, Bilezikian JP. 2010. Teriparatide: Variations on the theme of a 2-year therapeutic course. *IBMS BoneKEy* 7: 84–7.
13. Black DM, Greenspan SL, Ensrud KE, Palermo L, McGowan JA, Lang TF, Garnero P, Bouxsein ML, Bilezikian JP, Rosen CJ. 2003. The effects of parathyroid hormone and alendronate alone or in combination in postmenopausal osteoporosis. *N Engl J Med* 349(13): 1207–15.
14. Finkelstein JS, Hayes A, Hunzelman JL, Wyland JJ, Lee H, Neer RM. 2003. The effects of parathyroid hormone, alendronate, or both in men with osteoporosis. *N Engl J Med* 349(13): 1216–26.
15. Deal CM, Omizo EN, Schwartz EF, Eriksen EF, Cantor P, Wang J, Glass EV, Myers SL, Krege JH. 2005. Combination teriparatide and raloxifene therapy for postmenopausal osteoporosis: Results from a 6-month double-blind placebo-controlled trial. *J Bone Miner Res* 20(11): 1905–11.
16. Cosman F, Eriksen EF, Recknor C, Miller PD, Guanabens N, Kasperk C, Papanastasio P, Readie A, Rao H, Gasser JA, Bucci-Rechtweg C, Boonen S. 2011. Effects of intravenous zoledronic acid plus subcutaneous teriparatide [rhPTH(1-34)] in postmenopausal osteoporosis. *J Bone Miner Res* 26(3): 503–11.
17. Kurland ES, Heller SL, Diamond B, McMahon D, Cosman F, Bilezikian JP. 2004. The importance of bisphosphonate therapy in maintaining bone mass in men after therapy with teriparatide [human parathyroid hormone (1-34)]. *Osteoporos Int* 15(12): 992–7.
18. Black DM, Bilezikian JP, Ensrud KE, Greenspan SL, Palermo L, Hue T, Lang TF, McGowan JA, Rosen CJ. 2005. One year of alendronate after one year of parathyroid hormone (1-84) for osteoporosis. *N Engl J Med* 353(6): 555–65.
19. Neer RM, Arnaud CD, Zanchetta JR, Prince R, Gaich GA, Reginster JY, Hodsman AB, Eriksen EF, Ish-Shalom S, Genant HK, Wang L, Mitlak BH. 2001. Effect of parathyroid hormone (1-34) on fractures and bone mineral density in postmenopausal women with osteoporosis. *N Engl J Med* 344(19): 1434–41.
20. Lindsay R, Scheele WH, Neer R, Pohl G, Adami S, Mautalen C, Reginster JY, Stepan JJ, Myers SL, Mitlak BH. 2004. Sustained vertebral fracture risk reduction after withdrawal of teriparatide in postmenopausal women with osteoporosis. *Arch Intern Med* 164(18): 2024–30.
21. Prince R, Sipos A, Hossain A, Syversen U, Ish-Shalom S, Marcinowska E, Hake J, Lindsay R, Dalsky GP, Mitlak BH. 2005. Sustained nonvertebral fragility fracture risk reduction after discontinuation of teriparatide treatment. *J Bone Miner Res* 20(9): 1507–13.

54
Compliance and Persistence with Osteoporosis Medications

Deborah T. Gold

No Controversy: Compliance and Persistence
with Osteoporosis Medications Are Poor 448
Controversy #1: What Should We Call It? Compliance?
Persistence? Adherence? 449
Controversy #2: How Should We Measure Compliance and
Persistence? 450
Controversy #3: Do Patients Forget to Take Their Medicine,
or Are Noncompliance and Nonpersistence More
Complex? 451

Controversy #4: Is There a Healthy Adherer Effect? 451
Controversy #5: Does Extending Time between Doses or
Changing the Delivery System for Osteoporosis Medications
Improve Compliance? 452
Summary 452
References 453

Peer-reviewed research articles on compliance and persistence with osteoporosis medication abound in the recent literature. It takes only a quick review of leading journals in the field to see how important this issue has become. In the 17 years between the launch of alendronate and today, the availability of prescription medications for this chronic metabolic bone disease has risen exponentially. From a short list prior to 1995, we now have nine FDA-approved drugs with which to prevent or treat osteoporosis. Why is it that osteoporosis remains as potent a public health problem today as it was a decade or two ago?

The answer is simple. We can provide all the medications, all the delivery methods, and the dosing options imaginable. But if patients do not take their medication as directed, health care professionals are only fooling themselves if we think we are adequately diagnosing and treating this disease. The bottom line is this: Why should we bother to research and develop new and expensive pharmaceutical agents for bone loss when patients are not taking the medications available now? If there is one issue that scientists can agree on about compliance and persistence with osteoporosis medications, it is this:

Patients are not now and have not been taking their treatments as directed, and nothing in the literature has suggested that these behaviors will change in the near future.

In this chapter, there is no need to prove that compliance and persistence are poor for those taking osteoporosis medicines. Multiple studies have already provided strong evidence on this issue [1–4]. Instead, I will review five key controversies about research into compliance and persistence with osteoporosis treatments and provide supporting evidence about those controversies.

NO CONTROVERSY: COMPLIANCE AND PERSISTENCE WITH OSTEOPOROSIS MEDICATIONS ARE POOR

All researchers who study compliance and persistence with osteoporosis medications agree on one issue: compliance and persistence with osteoporosis medications are dismal. This, in itself, is not unusual. Medication

---

Primer on the Metabolic Bone Diseases and Disorders of Mineral Metabolism, Eighth Edition. Edited by Clifford J. Rosen.
© 2013 American Society for Bone and Mineral Research. Published 2013 by John Wiley & Sons, Inc.

behaviors for other asymptomatic chronic diseases are much the same. For example, hypertension [5, 6] and hypercholesterolemia [7] are two examples of chronic asymptomatic diseases in which compliance and persistence with medication are poor. Even symptomatic chronic diseases such as diabetes mellitus [8] and Parkinson's disease [9] do not compel patients to comply or persist with appropriate medication regimens. Sometimes, patients simply do not take their medications. At other times, they take suboptimal amounts of the medication. Although the reasons are not entirely clear, an abundance of evidence proves that medications—and especially those prescribed for osteoporosis—are being taken incorrectly or not at all.

The long-term outcomes of poor compliance and persistence are legion. They include substantially increased severity of the disease state, significantly increased health care costs, more and more frequent hospitalizations, surgeries, other expensive procedures to name only a few. In the case of osteoporosis, increased fractures are the typical outcome. In addition, reasons for noncompliance and nonpersistence are somewhat predictable: medication side effects, cost, forgetfulness, misunderstood directions, health beliefs in medication effectiveness, and polypharmacy.

Despite general agreement on poor compliance and poor persistence with osteoporosis medications, several controversies still exist in this literature. Until those controversies are resolved, improved compliance and persistence are not likely. In the remaining pages of this chapter, I highlight some of the primary controversies in this field as well as make some suggestions about how to improve overall compliance and persistence. Although there is no "quick fix" for this problem, the elimination of controversies could help make significant progress toward solving the problems of noncompliance and nonpersistence in osteoporosis patients.

CONTROVERSY #1: WHAT SHOULD WE CALL IT? COMPLIANCE? PERSISTENCE? ADHERENCE?

The act of a patient taking medications as prescribed over an extended period of time has been poorly described in the past—poorly in the sense that different authors use different words to describe the same phenomenon. The reasons for this are not clear. Three words have been used interchangeably in the literature on medication taking: compliance, persistence, and adherence. Throughout the decades of this literature, we have made this situation more confusing by changing terminology to fit into current trends. In 2003, the World Health Organization (WHO) released its report titled, "Adherence to Long-Term Therapies: Evidence for Action" [10]. In this document, the authors explain why *adherence* was the term they chose: "The idea of compliance is associated too closely with blame, be it of providers or patients and the concept of adherence is a better way of capturing the dynamic and complex changes required of many players over long periods to maintain optimal health in people with chronic diseases" (p. v). Somewhat later in the document, the authors state that adherence suggests that patient agrees with the treatment alternative and is not simply following doctor's orders [10].

Unfortunately, this had little impact on terminology selected by researchers publishing in this area. A *Web of Science* search (2003–2008) resulted in the identification of articles using the three terms: *compliance* (113 hits), *persistence* (78 hits), and *adherence* (115 hits). All scientists publishing in this area are guilty of mixing terminology depending on journal preferences, study titles, and personal choices.

Another organization tried to solve the problem of confusing terminology as well. The International Society of Pharmacoeconomic and Outcomes Research (ISPOR) established a Medication Compliance and Persistence Special Interest Group which, for 3 years, explored the use and perceptions of these three terms by providers and patients. In a 2008 publication, Cramer et al. [11] found that the terms *compliance* and *persistence* were distinct from each other in concept and measurement and should be used in the literature; they also determined that *adherence* did not have a distinct definition and was, therefore, a synonym for *compliance* and not a separate concept. The authors did comment about earlier hypotheses that patients found the term *compliance* degrading: "We found no authoritative support for the assumption that 'adherence' is a less derogatory term or whether it is preferred by patients" (pp. 14–15). Unfortunately, the impact of these guidelines was no greater than that of the WHO document. Another search of *Web of Science*, this time including publications through 2011 (i.e., after the ISPOR guidelines had been published) found the following: *compliance* (194 hits), *persistence* (134 hits), and *adherence* (215 hits). Therefore we must conclude that there is no gold standard for the terminology used to describe medication-related behaviors.

It has been difficult enough to figure out what to call various medication-related behaviors. The use of adherence as a synonym for compliance is confusing but could be straightforward. Unfortunately, a number of researchers have chosen to operationalize adherence as the sum of compliance and persistence. For example, in a 2006 *Mayo Clinic Proceedings* editorial, Badamgarav and Fitzpatrick [12] say, "Adherence comprises compliance and persistence and refers to taking medications as instructed during a given period. Adherence is typically estimated by the medication possession ratio (MPR)" (p. 1009). This sentence actually contains an explicit contradiction. If compliance can be measured as MPR (see next section) and adherence equals compliance plus persistence, how can adherence be measured as MPR as well? As noted above, ISPOR suggests the use of adherence ONLY as a synonym of compliance and, therefore, measurable by MPR.

CONTROVERSY #2: HOW SHOULD WE MEASURE COMPLIANCE AND PERSISTENCE?

Measuring a patient's compliance and persistence with medications is extremely difficult, to say the least. Any prospective measurement that involves questioning a patient has the potential to influence that patient's behavior and discredit results.

Claims data

For this reason—and because retrospective data analyses avoid some of the pitfalls of ongoing data collection—many compliance and persistence studies have used retrospective analyses of claims data. For example, Siris et al. [13] used data from two claims databases over a 5-year period to assess medication behaviors and to determine the relationship between adherence and fracture prevention. They found that 43% of women with postmenopausal osteoporosis (PMO) were refill compliant (as determined by MPR) with their bisphosphonate therapy while only 20% were persistent across the entire study. Those who were refill compliant demonstrated significantly better fracture protection when compared to noncompliant patients ($p < 0.001$). An earlier study by Caro et al. [14] examined health services databases to assess compliance and persistence. Not surprisingly, Caro also found that compliant patients experienced a 25% ($p < 0.00001$) reduction in fractures. Compliance was defined as taking at least 80% of medication doses, and the better the compliance, the greater the fracture reduction. Finally, using administrative claims data from Medicare patients (n = 19,987), Patrick and colleagues [15] completed a cohort study in which adherence was assessed in sequential 60-day periods. Using various modeling techniques, they found that adherence was consistently associated with hip, vertebral, and overall osteoporotic fracture reduction.

McCombs et al. [3] looked at compliance in a slightly different way. Examining over 58,000 osteoporosis patients and their pattern of medication use, they examined compliance for different osteoporosis drugs (hormone replacement, bisphosphonates, and raloxifene were available at the time of this study). Sadly, 1-year compliance rates were below 25% for all osteoporosis medications. However, in this study—as in the two previously mentioned—investigators had no way to examine why noncompliance occurred. Given that noncompliance can often result from factors not related to osteoporosis, such information would be valuable in understanding reasons why this low level of compliance seems so widespread.

Prospective studies

In the early studies of compliance and persistence, there was virtually no use of prospective studies. One of the first was done by Papaioannou et al. [16]. In it, investigators measured adherence to three osteoporosis drugs [etidronate, alendronate and hormone replacement (HR)] using the Canadian Database of Osteoporosis and Osteopenia(CANDOO). Surprisingly, 1-year adherence rates were extremely high: 90% of etidronate users, 80% of HR users, and 78% of alendronate users. However, as time passed, persistence rates decreased; after 6 years, less than 50% of participants were still persistent with any medication. Unfortunately, the study did not assess reasons for discontinuation.

In 2007, Ringe et al. [17] conducted a 1-year prospective study of more than 5,000 patients on four osteoporosis medications: alendronate daily, alendronate weekly, raloxifene, and riscdronate daily. They reported moderate levels of compliance after 1 year (76–80%) and found that side effects were the primary reason for discontinuation.

Measuring persistence

Persistence is the duration of time from initiation to discontinuation of pharmaceutical therapy. For some time, researchers in medication behaviors believed that compliance plus persistence equaled adherence, but there has been little support for this and much opposition to it.

MPR versus other measures

In studies of compliance and persistence, measurement of the two primary outcomes continues to be a challenge. Using something such as electronic pill bottle caps reminds patients that they are participating in a study. Pill counts at follow-up clinic visits can easily be made inaccurate either accidentally or deliberately. Patient diaries are often inaccurate and may be influenced by social desirability. Questionnaires are now available that can assess compliance and persistence, but they are only as accurate as the patients completing them [18]. There is no question that, currently, medication possession ratio (MPR) is the most commonly used measure of compliance. MPR is calculated as the number of days of medication supplied within the refill interval/number of days in the refill interval. This information is easily available and calculated using claims data; it is the standard in virtually all claims data analyses [e.g., 3, 13, 19] as well as in prospective studies.

But as theoretically accurate and easy as MPR is, there is a substantial difference between being in possession of medication and actually taking medication. Thus, MPR is not the gold standard we have looked for, although it may be better than other options. Unfortunately, no one has come up with a way to measure compliance and persistence directly without having patient behavior influenced by that measurement. This is a challenge that remains unsolved.

CONTROVERSY #3: DO PATIENTS FORGET TO TAKE THEIR MEDICINE, OR ARE NONCOMPLIANCE AND NONPERSISTENCE MORE COMPLEX?

In the literally dozens of studies of compliance and persistence with osteoporosis medications, no strong answer emerges to the question, "Why don't people with osteoporosis take their medications as directed?" After all, we currently have nine FDA-approved medications to prevent or treat this disease. Within this family of medications, we have different delivery systems (oral, nasal spray, subcutaneous injection by self or healthcare professional, intravenous infusion) and different dosing intervals (daily, weekly, monthly, every 3 months, every 6 months, yearly). This is an impressive menu of medication options and appears to offer something for everyone. Yet even though we have options in osteoporosis medications, patients are still not compliant or persistent.

Researchers differ in the reasons they give for poor compliance and persistence with these medications. These behaviors are not easily identified or explained, but it does appear that there are five primary ways in which people can be noncompliant and nonpersistent: (1) they do not fill the prescription, (2) they do not take the full dose, (3) they take the medication at the wrong time, (4) they forget to take the medication, and (5) they stop taking the medication. Barriers to compliance and persistence are also heterogeneous and include, but are not limited to, side effects of medication, cost, complex dosing schedule (e.g., oral bisphosphonates), forgetting, inadequate understanding of disease and need for medication, doubt that medication can help, and so on [20].

Which of these barriers to compliance and persistence is most common and problematic is still up for debate. Some researchers believe that forgetting is the major cause of not taking medication [e.g., 21]. However, others would dispute that and say that more often, patients make a conscious choice not to take their medicines. Recent evidence from a Harris poll confirms that [22]. According to this report, only 24% of patients reported forgetting to take their medications. Twenty percent reported not wanting to experience side effects; 17% said that cost was a problem; 14% did not believe they really needed the drug. Further, one in three patients reported taking their medications less frequently than prescribed. Other medications behaviors included taking smaller than prescribed doses, stopping the prescription before they were supposed to, delaying in filling a prescription, and not filling a prescription at all. These data strongly support the concept that *forgetting medication* is, in some ways, the least of our problem with compliance and persistence.

The concept of *intentional nonadherence* appeared in the early 1990s in an article by Donvan and Blake [23]. They noted that patients often made reasoned decisions not to take medicine, but the lack of communication between physicians and patients led to healthcare professionals not understanding such decisions. In a more recent examination of this proposition, Lowry et al. [24] reported that respondents in the Veterans' Study to Improve the Control of Hypertension who were nonwhite, had better education, were able to pay bills and had more side effects were significantly more likely to decide intentionally *not* to take their medication.

Although no studies of which I am aware have looked at the intentionally of not taking osteoporosis medications, experts in this field confirm theoretically that intentional nonadherence by osteoporosis patients may well be a strong factor in medication behaviors [4].

CONTROVERSY #4: IS THERE A HEALTHY ADHERER EFFECT?

In 1985, the *New England Journal of Medicine* published a paper by Stamfer and colleagues titled, "A Prospective Study of Postmenopausal Estrogen Therapy and Coronary Heart Disease"[25] in which the authors suggested that there were multiple causes of bias operating in observational studies. The bias of interest here was the *healthy user effect*. This occurs when patients choosing one preventive service also choose others; therefore, they are healthy adherers to preventive efforts in general rather than just to a single medication or disease. Another article in the *New England Journal* also discussed the *healthy adherer effect* [26] in which they suggested that people who adhere to one healthy behavior (such as taking a medication as prescribed) are more likely to adhere to other healthy behaviors. In fact, in this study, patients who adhered well to the placebo were found to have lower mortality rates and fewer cardiovascular events.

How are these concepts relevant to compliance and persistence with osteoporosis medications? Several relatively recent publications have suggested that the *healthy adherer effect* has contaminated the results of observational studies designed to examine medication compliance and fracture reduction. First, Cadarette and colleagues [27] noted that multiple studies of health utilization data have found strong, positive correlations between adherence to osteoporosis medications and fracture reduction [e.g., 28, 29]. Cadarette et al. [27] suggest that there is a plausible alternative hypothesis to that which says good adherers have reduced fractures, and that is the *healthy adherer bias*. Patients who adhere to medication may also adhere to calcium and vitamin D supplementation, exercise, and may avoid behaviors that could increase fracture likelihood. Using claims data from the Pennsylvania Pharmaceutical Assistance Contract for the Elderly (PACE), they looked at the relationship between adherence with medications and fracture rates. Although they found that adherence to bisphosphonates does lower nonvertebral fracture rates on the whole, there was no other evidence of a healthy adherer effect.

In contradiction to this finding, Curtis et al. [30] examined compliance with placebo in the Fracture Intervention Trial (FIT) (n = 3169 in the placebo arm) to see if there were evidence of a healthy adherer effect. Eighty-two percent of the women in the placebo group had 80% or better compliance and, compared to the low compliers in that arm, did show less bone mineral density (BMD) loss in the total hip ($p < 0.04$). Although there were fewer hip fractures in this group than in the low-adherence placebo group, the difference was not statistically significant. Despite the lack of significant findings, the authors still concluded, ". . . our findings provide some support for the existence of the healthy adherer effect in this population" (p. 687).

This conclusion is somewhat surprising and appears not to be supported by the data. In the same issue of *Journal of Bone and Mineral Research*, Silverman and Gold [31] question the findings of Curtis et al. [30]. Previous *observational* studies have shown a healthy adherer effect, despite that fact that the Cadarette et al. study [27] did not support this. But randomized clinical trials have not supported it in the past. Given that the findings of Curtis et al. [30] show only a nonsignificant reduction in fractures—the truly critical variable upon which medication success rests—the assumption that a healthy adherer effect exists seems to have been theorized beyond the data. Future research in this area is essential as little congruence of findings occurs, especially in studies of osteoporosis and fractures.

CONTROVERSY #5: DOES EXTENDING TIME BETWEEN DOSES OR CHANGING THE DELIVERY SYSTEM FOR OSTEOPOROSIS MEDICATIONS IMPROVE COMPLIANCE?

Prior to 1995, only two FDA-approved medications for osteoporosis were available: estrogen/hormone therapy [32] and injectable calcitonin [33]. Estrogen was a daily therapy; calcitonin was given several times a week. Both were hormonal. Although patients had some degree of success with both medications, estrogen/hormone therapy was approved for both prevention and treatment of postmenopausal osteoporosis (PMO), while calcitonin was approved for treatment of PMO in women who were at least 5-year postmenopausal.

With the approval of alendronate in 1995, prevention and management of osteoporosis changed dramatically, and compliance with medications became significantly more important AND challenging [34]. Because of the burden associated with taking daily oral bisphosphonates [35], pharmaceutical companies began to look at medications that would have longer intervals between doses. Between 1995 and 2007, four bisphosphonates were approved by the FDA (alendronate, risedronate, ibandronate, zoledronic acid) with increasing dosing intervals (i.e., daily to weekly to monthly to yearly). However, despite many predictions to the contrary, these changes did not unequivocally lead to improved compliance. While Emkey and colleagues [36] reported that significantly more postmenopausal women preferred monthly dosing to weekly dosing, this study received considerable criticism. In it, women were not told that the efficacy of weekly and monthly bisphosphonates differed in terms of anti-fracture efficacy. Subsequent studies [e.g., 37] did not support this, especially those in which participants were told in advance of efficacy differences.

Several studies also examined differences in drug delivery as a factor in medication compliance and consistence. Kendler and colleagues [38] found, in two large double-blind, double-dummy trials of postmenopausal women with bone loss, that significantly more women preferred and were more satisfied with a 6-month injection rather than a weekly tablet. Additionally, investigators who compared persistence with teriparatide (daily injection) to that with oral anti-osteoporosis therapies found that persistence levels with teriparatide were very high and likely higher than that with oral therapies [39], especially when patients participated in a concurrent educational program [40]. Although these findings may seem counterintuitive (after all, who would choose an injection over an oral medication?), the evidence appears to be robust. Perhaps women who agree to begin an injectable therapy are more committed to it than women who start oral therapies.

SUMMARY

Significant empirical evidence illustrates the problems we face in helping patients remain compliant and persistent with osteoporosis medications regardless of delivery system or dose duration. As noted earlier, this problem is not unique to osteoporosis. However, given the increasing awareness of the prevalence of this disease and its long-term impact when untreated, health care professionals realize that changing patient behavior is of paramount importance in continuing to master this major public health problem. Further, the controversies in the research findings leave both patients and professionals searching for the best way to combat the negative forces that cause many postmenopausal women to stop their osteoporosis therapy. Several well-researched controversies were not addressed; for example, the role of side effects of the osteoporosis drugs is well known and not directly addressed here. As a result, the problems caused by patients not taking medications as directed in osteoporosis may be significantly worse than it appears.

No single answer will solve this problem. Limited success with educational programs and provider–patient interactions suggest that additional efforts in those areas may have a positive effect. It is helpful to remind ourselves that we have not yet found the key to this problem and that healthcare professionals and educators must continue to search for it.

REFERENCES

1. Cramer JA, Lynch NO, Gaudin AF, Walker M, Cowell W. 2006. The effect of dosing frequency on compliance and persistence with bisphosphonate therapy in postmenopausal women: A comparison of studies in the United States, the United Kingdom, and France. *Clin Ther* 28(10): 1686–94.
2. Huybrechts KF, Ishak KJ, Caro JJ. 2006. Assessment of compliance with osteoporosis treatment and its consequences in a managed care population. *Bone* 38(6): 922–8.
3. McCombs JS, Thiebaud P, McLaughlin-Miley C, Shi J. 2005. Compliance with drug therapies for the treatment and prevention of osteoporosis. *Maturitas* 48(3): 271–87.
4. Silverman SL, Schousboe JT, Gold DT. 2011. Oral bisphosphonate compliance and persistence: A matter of choice? *Osteoporos Int* 22(1): 21–6.
5. Hill MN, Miller NH, Degeest S, American Society of Hypertension Writing Group, Materson BJ, Black HR, Izzo JL Jr, Oparil S, Weber MA. 2011. Adherence and persistence with taking medication to control high blood pressure. *J Amer Soc Hypertens* 5(1): 56–63.
6. Vawter L, Tong X, Gemilyan M, Yoon PW. 2008. Barriers to antihypertensive medication adherence among adults—United States, 2005. *J Clin Hypertens* 10(12): 922–9.
7. Huser MA, Evans TS, Berger V. 2005. Medication adherence trends with statins. *Adv Ther* 22(2): 163–71.
8. Rubin RR. 2005. Adherence to pharmacologic therapy in patients with type 2 diabetes mellitus. *Amer J Med* 118 Suppl 5A: 27S–34S.
9. Kulkarni KS, Balkrishnan R, Anderson RT, Edin HM, Kirsch J, Stacy MA. 2008. Medication adherence and associated outcomes in Medicare health maintenance organization-enrolled older adults with Parkinson's disease. *Move Disord* 23(3): 359–365.
10. World Health Organization. 2003. *Adherence to Long-Term Therapies: Evidence for Action*. Geneva: World Health Organization.
11. Cramer JA, Roy A, Burrell A, Fairchild CJ, Fuldeore MJ, Ollendorf DA, Wong PK. 2008. Medication compliance and persistence: Terminology and definitions. *Value Health* 11(1): 44–7.
12. Badamgarav E, Fitzpatrick LA. 2006. A new look at osteoporosis outcomes: The influence of treatment, compliance, persistence, and adherence. *Mayo Clin Proc* 81(8): 1009–12.
13. Siris ES, Harris ST, Rosen CJ, Barr CE, Arvesen JN, Abbott TA, Silverman S. 2006. Adherence to bisphosphonate therapy and fracture rates in osteoporotic women: Relationship to vertebral and nonvertebral fractures from 2 US claims databases. *Mayo Clin Proc* 81(8): 1013–22
14. Caro JJ, Ishak KJ, Huybrechts KF, Raggio G, Naujoks C. 2004. The impact of compliance with osteoporosis therapy on fracture rates in actual practice. *Osteoporos Int* 15(12): 1003–8.
15. Patrick AR, Brookhart MA, Losina E, Schousboe JT, Cadarette SM, Mogun H, Solomon DH. 2010. The complex relation between bisphosphonate adherence and fracture reduction. *J Clin Endocrinol Metab* 95(7): 3251–9.
16. Papaioannou A, Ioannidis G, Adachi JD, Sebaldt RJ, Ferko N, Puglia M, Brown J, Tenenhouse A, Olszynski WP, Boulos P, Hanley DA, Josse R, Murray TM, Petrie A, Goldsmith CH. 2003. Adherence to bisphosphonates and hormone replacement therapy in a tertiary care setting of patients in the CANDOO database. *Osteoporos Int* 14(10): 808–13.
17. Ringe JD, Christodoulakos GE, Mellstrom D, Petto H, Nickelsen T, Marin F, Pavo I. 2007. Patient compliance with alendronate, risedronate and raloxifene for the treatment of osteoporosis in postmenopausal women. *Curr Med Res Opin* 23(11): 2677–87.
18. Farmer KC. 1999. Methods for measuring and monitoring medication regimen adherence in clinical trials and clinical practice. *Clin Ther* 21(6): 1074–90.
19. Blandford L, Dans PE, Ober JD, Wheelock C. 1999. Analyzing variations in medication compliance related to individual drug, drug class, and prescribing physician. *J Managed Care Pharm* 5(1): 47–51.
20. Silverman S. 2006. Adherence to medications for the treatment of osteoporosis. *Rheum Dis Clin North Am* 32(4): 721–31.
21. Doshi JA, Zuckerman IH, Picot SJ, Wright JT Jr, Hill-Westmoreland EE. 2003. Antihypertensive use and adherence and blood pressure stress response among black caregivers and non-caregivers. *Appl Nurs Res* 16(4): 266–77.
22. Boston Consulting Group. Analysis: Harris interactive 10,000 patient survey 2002. Available from: http://www.bcg.com/documents/file14265.pdf. Accessed: October 28, 2011.
23. Donovan JL, Blake DR. 1992. Patient non-compliance: Deviance or reasoned decision-making? *Soc Sci Med* 34(5): 507–13.
24. Lowry KP, Dudley TK, Oddone EZ, Bosworth HB. 2005. Intentional and unintentional nonadherence to antihypertensive medication. *Ann Pharmacother* 39(7–8): 1198–203.
25. Stampfer MJ, Willett WC, Colditz GA, Rosner B, Speizer FE, Hennekens CH. 1985. A prospective study of postmenopausal estrogen therapy and coronary heart disease. *N Engl J Med* 313(17): 1044–9.
26. Coronary Drug Project Research Group. 1980. Influence of adherence to treatment and response of cholesterol on mortality in the Coronary Drug Project. *N Engl J Med* 303(18): 1038–41.
27. Cadarette SM, Solomon DH, Katz JN, Patrick AR, Brookhart MA. 2011. Adherence to osteoporosis drugs and fracture prevention: No evidence of healthy adherer bias in a frail cohort of seniors. *Osteoporos Int* 22(3): 943–54.
28. Kothawala P, Badamgarav E, Ryu S, Miller RM, Halbert RJ. 2007. Systematic review and meta-analysis of real-world adherence to drug therapy for osteoporosis. *Mayo Clin Proc* 82(12): 1493–501.

29. Siris ES, Selby PL, Saag KG, Borgstrom F, Herings RM, Silverman SL. 2009. Impact of osteoporosis treatment adherence on fracture rates in North America and Europe. *Am J Med* 122(2 Suppl): S3–13.
30. Curtis JR, Delzell E, Chen L, Black D, Ensrud K, Judd S, Safford MM, Schwartz AV, Bauer DC. 2011. The relationship between bisphosphonate adherence and fracture: Is it the behavior or the medication? Results from the placebo arm of the fracture intervention trial. *J Bone Miner Res* 26(4): 683–8.
31. Silverman SL, Gold DT. 2011. Healthy users, healthy adherers, and healthy behaviors? *J Bone Miner Res* 26(4): 681–2.
32. Lindsay R. 1987. Estrogen therapy in the prevention and management of osteoporosis. *Am J Obstet Gynecol* 156(5): 1347–51.
33. Riggs BL. 1979. Postmenopausal and senile osteoporosis: Current concepts of etiology and treatment. *Endocrinol Jpn* 26(Suppl): 31–41.
34. Scoville EA, Ponce de Leon Lovaton P, Shah ND, Pencille LJ, Montori VM. 2011. Why do women reject bisphosphonates for osteoporosis? A videographic study. *PLoS ONE* 6(4): e18468.
35. Hadji P, Claus V, Ziller V, Intorcia M, Kostev K, Steinle T. 2012. GRAND: the German retrospective cohort analysis on compliance and persistence and the associated risk of fractures in osteoporotic women treated with oral bisphosphonates. *Osteoporos Int* 23(1): 223–31.
36. Emkey R, Koltun W, Beusterien K, Seidman L, Kivitz A, Devas V, Masanauskaite D. 2005. Patient preference for once-monthly ibandronate versus once-weekly alendronate in a randomized, open-label, cross-over trial: The Boniva Alendronate Trial in Osteoporosis (BALTO). *Curr Med Res Opin* 21(12): 1895–903.
37. Gold DT, Safi W, Trinh H. 2006. Patient preference and adherence: Comparative US studies between two bisphosphonates, weekly risedronate and monthly ibandronate. *Curr Med Res Opin* 22(12): 2383–91.
38. Kendler DL, Bessette L, Hill CD, Gold DT, Horne R, Varon SF, Borenstein J, Wang H, Man HS, Wagman RB, Siddhanti S, Macarios D, Bone HG. 2010. Preference and satisfaction with a 6-month subcutaneous injection versus a weekly tablet for treatment of low bone mass. *Osteoporos Int* 21(5): 837–46.
39. Arden NK, Earl S, Fisher DJ, Cooper C, Carruthers S, Goater M. 2006. Persistence with teriparatide in patients with osteoporosis: The UK experience. *Osteoporos Int* 17(11): 1626–9.
40. Briot K, Ravaud P, Dargent-Molina P, Zylberman M, Liu-Leage S, Roux C. 2009. Persistence with teriparatide in postmenopausal osteoporosis; impact of a patient education and follow-up program: The French experience. *Osteoporos Int* 20(4): 625–30.

55
Cost-Effectiveness of Osteoporosis Treatment

Anna N.A. Tosteson

Introduction 455
Overview of Methods for Cost-Effectiveness Analysis 455
The Cost-Effectiveness of Osteoporosis Treatment 457

Summary 458
Acknowledgment 458
References 458

INTRODUCTION

Osteoporosis affects a large proportion of the elderly population and is associated with fractures that are costly in both human and economic terms [1]. In 2005, the United States (U.S.) population sustained an estimated 2 million incident fractures at a cost of $16.9 billion [2]. With annual fracture-related expenditures projected to increase to $25.3 billion by the year 2025, there is widespread recognition that growing elderly populations and constrained health care budgets will continue to challenge health care systems to find cost-effective approaches to osteoporosis care. Cost-effectiveness analysis is a form of economic evaluation that estimates the value of an intervention by weighing the expected net increase in cost of an intervention against its expected net gain in health [3]. The rationale for cost-effectiveness analysis is that when health care resources are limited, expenditures should be planned to maximize health outcomes within available resources. The cost-effectiveness of new treatments relative to current care standards is one attribute that policy makers may consider when making formulary coverage decisions. In this chapter, the methodology of cost-effectiveness analysis is described, recent developments in the cost-effectiveness of osteoporosis care are discussed, and key findings are highlighted.

OVERVIEW OF METHODS FOR COST-EFFECTIVENESS ANALYSIS

Cost-effectiveness ratio

The incremental cost-effectiveness ratio (ICER), which estimates expected cost per unit of health gained, is the primary outcome measure used to characterize value in cost-effectiveness studies. Consider two alternative treatments, A and B, where the average cost of A is higher than the average cost of B. The ICER is defined as follows:

$$\text{ICER} = \frac{(\text{Cost}_A - \text{Cost}_B)}{(\text{Effectiveness}_A - \text{Effectiveness}_B)}$$

Using this definition, the value of each more costly intervention is judged relative to the improvement in health that it provides over and above health outcomes associated with the less costly alternative.

Choice of comparator

When assessing cost-effectiveness of a new osteoporosis intervention, the standard of care that is used as the basis for comparison (i.e., the comparator) may have a marked

Primer on the Metabolic Bone Diseases and Disorders of Mineral Metabolism, Eighth Edition. Edited by Clifford J. Rosen.
© 2013 American Society for Bone and Mineral Research. Published 2013 by John Wiley & Sons, Inc.

impact on the intervention's estimated value. Cost-effectiveness analyses of osteoporosis prevention conducted prior to 2002, when the Women's Health Initiative findings were published [4], typically included hormone therapy as a comparator. The choice of comparator today depends on whether treatment is being considered for a man or woman and whether or not the individual has established osteoporosis. Unless cost-effectiveness is measured relative to a reasonable alternative, the estimated ICER may not provide a meaningful estimate of an intervention's value. While ICERs computed relative to "no intervention" have meaning for the minority of patients who have no other viable treatment option, for the majority of patients in whom less costly treatments are possible, these ICERs can be potentially misleading estimates of value. In general, when new interventions are compared with "no intervention" (technically an average rather than an ICER), they will have more favorable values than when compared with active treatment comparators.

Model-based analyses

Estimating an ICER typically requires mathematical modeling to project expected health and cost implications of alternative treatments over a longer time horizon than can be observed in any clinical trial [5], and/or to expand the treatments and/or population subgroups considered. Most analyses utilize Markov state-transition models [6], which comprise a discrete number of health states, each with an associated cost and health state value (i.e., health utility), along with annual probabilities of transition among the health states. Other modeling methods that detail the biological processes related to bone health have also been proposed [7].

Estimating the cost of osteoporosis treatment

To estimate the net difference in cost of a new treatment relative to a comparator, several types of direct medical costs should be considered (Table 55.1). The cost of medical care in future years of life may also be included. Against these costs, potential savings that may accrue due to fracture prevention are considered and include the cost of acute fracture care, rehabilitation services (if required), and costs of ongoing fracture-related disability (if present). Differences in the cost of providing health care from country to country make the generalizability of cost-effectiveness findings across countries challenging.

Indirect costs of an illness are those that are associated with a loss in productivity due to morbidity and mortality. Such costs may be incurred by the individual who sustains a fracture and/or by their caregivers. However, there is scant evidence available to address the latter. The human capital approach, which values productivity changes based on lost earnings [8], has been applied to assess the cost of fractures in some U.S. cost-of-illness studies [9–11], but to date such costs have not been included in cost-effectiveness analyses of osteoporosis treatment. Some argue that productivity costs are adequately reflected in the denominator of the cost-effectiveness ratio when quality-adjusted life years (QALYs) are used to measure effectiveness [3].

Whether or not each of these potential costs/savings is included in the analysis depends on the perspective that is taken. For informing public policy decision makers, the societal perspective is generally most desirable. The marked impact that perspective may have on the cost-effectiveness of osteoporosis treatment under a health care system such as in the U.S., where different payers are responsible for health care at different ages, is shown by an example that underscores the disparity between who pays for the prevention and who realizes potential long-term savings. Consider 5-year treatment of high-risk 55-year-old women from two perspectives: (1) a private insurer who pays for health care services up until age 65, and (2) a government insurer who pays after age 65 (i.e., Medicare in the U.S.). For the private insurer who pays for the costs of the treatment and monitoring but will realize limited savings due to fractures averted, treatment does not appear cost-effective. In contrast, the government payer only benefits due to fractures averted and sees treatment as cost-saving. This simplistic example suggests that optimal decisions for public health require a broad perspective that considers the full time horizon of costs and benefits.

Estimating the effectiveness of osteoporosis treatment

Quality-adjusted life years

The recommended measure for assessing the effectiveness of health interventions is the quality-adjusted life year (QALY) [3], which takes both length of life and quality of life into account. The use of QALYs facilitates comparisons of economic value across disease areas (e.g., interventions to control diabetes can be compared with

---

Table 55.1. Components of Direct Medical Costs to Consider When Assessing the Cost-Effectiveness of Osteoporosis Treatment

Cost Component
   Medication
      Acquisition
      Health care services for routine monitoring
      Health care services for treatment of side effects/sequelae
   Fracture
      Acute care services
      Rehabilitation services
      Ongoing disability services
   Extended Life Years
      Health care services

osteoporosis treatments). Cost-effectiveness studies that report ICERs as cost per QALY gained are often referred to as cost-utility analyses, because to estimate QALYs, health state values or "utilities" that reflect preferences for various health states, are used.

While QALYs have the potential to incorporate the intangible fracture-related costs of pain and suffering, data on health state values for fracture-related health outcomes are required. Evidence on the impact of fractures on QALYs has been summarized in several reviews [12, 13]. The absolute QALY losses associated with fracture vary based on who is asked (e.g., a patient who sustained a vertebral fracture vs a patient imagining a vertebral fracture) and how they are asked (e.g., visual analog scale, time trade-off), yet published studies consistently report health state values for fracture-related outcomes that are significantly below ideal health. Although a growing literature addresses preference-based measures of health in osteoporosis, many cost-effectiveness studies continue to rely on expert opinion regarding the quality-of-life impact of fractures both initially and in the long term [14–16].

When evaluating the value of osteoporosis treatment, it is important to consider the potential adverse impact that treatment side effects may have on estimates of quality-adjusted life expectancy. The potential for side effects to offset quality-of-life gains due to fracture prevention was first highlighted in studies of the role of hormone therapy in osteoporosis prevention [17, 18].

Number of fractures prevented

The value of osteoporosis treatment is sometimes reported in disease-specific terms as number of fractures prevented, which is problematic for two reasons. First, some osteoporosis interventions, such as raloxifene, have extraskeletal health effects that go unaccounted for when value is reported in terms of cost per fracture prevented. Second, inherent differences in human and economic costs of different fracture types (e.g., wrist versus hip) make it challenging to interpret cost per fracture prevented. To address this, analysts sometimes report costs in terms of specific fracture types (e.g., cost per hip fracture prevented or per vertebral fracture prevented) or in "hip fracture equivalent units" [19].

THE COST-EFFECTIVENESS OF OSTEOPOROSIS TREATMENT

Cost-effectiveness analysis and clinical practice guidelines

As constraints are increasingly felt on health care budgets, guideline developers recognize that costs cannot be entirely ignored [20]. One approach to setting treatment thresholds that has seen growing application in the osteoporosis literature is to identify the absolute fracture risk at which the cost per QALY gained falls below a "willingness to pay" per QALY gained threshold [12, 14, 15, 19, 21, 22]. This approach was utilized by the National Osteoporosis Foundation (NOF) to identify a 10-year absolute hip fracture risk at which treatment cost $60,000 per QALY gained or lower for treatment relative to no intervention [22]. The World Health Organization (WHO) fracture risk assessment tool, FRAX®, facilitates such risk predictions for previously untreated populations [23] and a report from the NOF guide committee provided insight into specific clinical factors that meet the intervention thresholds (3% for 10-year hip fracture risk or 20% for hip, wrist, spine, and shoulder fracture risks combined) based on an adaptation of FRAX® for the U.S. population [24, 25].

Cost-effectiveness of osteoporosis treatment

Fracture risk, treatment cost, the impact that fractures have on health-related quality of life, treatment persistence, and the durability of treatment [26] all influence the value of osteoporosis treatment. A U.S. analysis of an unspecified treatment that reduces fracture incidence by 35% relative to no intervention [22] demonstrates the marked improvement in cost-effectiveness that average-risk women attain with advancing age due to their higher absolute fracture risk is shown in Fig. 55.1. For example, a treatment costing $900 per year costs in excess of $580,000 per QALY gained for a 50-year-old woman whose 10-year hip fracture risk is 2.5%, compared to only $4,000 per QALY gained for an 80-year-old white woman whose 10-year hip fracture risk is 4%.

Prior to 1993, most osteoporosis cost-effectiveness studies assessed the value of hormone therapy [27, 28]. More recently, several reviews and technology assessments have addressed the value of other osteoporosis treatments with evaluation of bisphosphonates being a frequent focus [28–32]. Studies also address the value of calcium and vitamin D [33], raloxifene [14, 34–38], teriparatide [39, 40], calcitonin [41], strontium ranelate [42, 43], vitamin K [44], and denosumab [45], as well as the cost-effectiveness of selective treatment strategies [46–48] and special populations [49].

Bisphosphonate treatment is generally cost-effective when used in moderately high-risk populations such as women over age 65 with osteoporosis [30]. While treatment of elderly populations with calcium and vitamin D is potentially cost-saving [33], studies of raloxifene differ in their findings depending on the comparators included, the side effects considered, and an individual woman's risks [34–37, 50]. Discrepant findings have also been reported for teriparitide use among women at high risk of fracture [39, 40]. Cost-effectiveness studies that distinguish between agents on the basis of treatment persistence are of relevance due to the advent of agents with differing modes of administration (e.g., weekly oral agent versus annual injection) [51]. Likewise, there is increasing interest in programs to support treatment adherence [52].

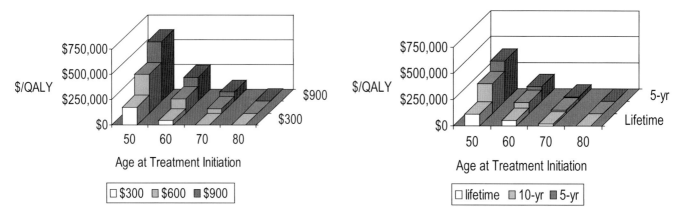

Fig. 55.1. The impact that annual treatment cost ($300, $600, or $900) has on cost per QALY gained at different ages of treatment initiation when treatment reduces fracture incidence by 35% and 5-year losses in health-related quality of life following fracture are modeled (Ref. 22).

SUMMARY

Due to the size of the elderly population that is at-risk for complications of osteoporosis-related fractures, it is imperative that cost-effective approaches to osteoporosis management be identified. While a growing literature addresses the value of specific treatments for various population subgroups, clinical practice guidelines identify cost-effective intervention thresholds on the basis of absolute 10-year fracture risk. For the U.S. population, cost-effective treatment intervention thresholds of 3% or greater for 10-year hip fracture risk, or 20% or greater for hip, wrist, clinical spine, and shoulder fracture risk combined, have been recommended [25]. Risk assessment tools for predicting 10-year fracture risk [53], provide a tool for efficiently targeting therapy to those individuals who stand to benefit most from osteoporosis treatment.

ACKNOWLEDGMENT

This work is supported by the National Institute of Arthritis and Musculoskeletal and Skin Diseases (P60-AR062799). The author thanks Ms. Loretta Pearson, MPhil, for research assistance and editorial support.

REFERENCES

1. Office of the Surgeon General. 2004. *Bone Health and Osteoporosis: A Report of the Surgeon General.* Rockville, MD: U.S. Department of Health and Human Services.
2. Burge R, Dawson-Hughes B, Solomon DH, Wong JB, King A, Tosteson A. 2007. Incidence and economic burden of osteoporosis-related fractures in the United States, 2005–2025. *J Bone Miner Res* 22(3): 465–75.
3. Gold M, Siegel J, Russell L, Weinstein M. 1996. *Cost-Effectiveness in Health and Medicine.* New York: Oxford University Press.
4. Rossouw JE, Anderson GL, Prentice RL, LaCroix AZ, Kooperberg C, Stefanick ML, Jackson RD, Beresford SA, Howard BV, Johnson KC, Kotchen JM, Ockene J. 2002. Risks and benefits of estrogen plus progestin in healthy postmenopausal women: Principal results from the Women's Health Initiative randomized controlled trial. *JAMA* 288(3): 321–33.
5. Tosteson AN, Jonsson B, Grima DT, O'Brien BJ, Black DM, Adachi JD. 2001. Challenges for model-based economic evaluations of postmenopausal osteoporosis interventions. *Osteoporos Int* 12: 849–57.
6. Sonnenberg FA, Beck JR. 1993. Markov models in medical decision making: A practical guide. *Med Decis Making* 13(4): 322–38.
7. Vanness DJ, Tosteson AN, Gabriel SE, Melton LJ 3rd. 2005. The need for microsimulation to evaluate osteoporosis interventions. *Osteoporos Int* 16(4): 353–8.
8. Hodgson TA, Meiners MR. 1982. Cost-of-illness methodology: A guide to current practices and procedures. *Milbank Mem Fund Q Health Soc* 60(3): 429–62.
9. Holbrook T, Grazier K, Kelsey J, Sauffer R. 1984. *The Frequency of Occurrence, Impact, and Cost Of Musculoskeletal Conditions in the United States.* Rosemont, IL: American Academy of Orthopaedic Surgeons.
10. Praemer A, Furner S, Rice D. 1992. *Musculoskeletal Conditions in the United States.* Rosemont, IL: American Academy of Orthopaedic Surgeons.
11. Praemer A, Furner S, Rice D. 199. *Musculoskeletal Conditions in the United States.* Rosemont, IL: American Academy of Orthopaedic Surgeons.
12. Brazier JE, Green C, Kanis JA. 2002. A systematic review of health state utility values for osteoporosis-related conditions. *Osteoporos Int* 13(10): 768–76.
13. Tosteson ANA, Hammond CS. 2002. Quality of life assessment in osteoporosis: Health status and preference-based instruments. *Pharmacoeconomics* 20(5): 289–303.

14. Kanis JA, Borgstrom F, Zethraeus N, Johnell O, Oden A, Jonsson B. 2005. Intervention thresholds for osteoporosis in the UK. *Bone* 36(1): 22–32.
15. Kanis JA, Johnell O, Oden A, Borgstrom F, Johansson H, De Laet C, Jonsson B. 2005. Intervention thresholds for osteoporosis in men and women: A study based on data from Sweden. *Osteoporos Int* 2005 16(1): 6–14.
16. Kanis JA, Johnell O, Oden A, De Laet C, Oglesby A, Jonsson B. 2002. Intervention thresholds for osteoporosis. *Bone* 31(1): 26–31.
17. Weinstein MC. 1980. Estrogen use in postmenopausal women—Costs, risks, and benefits. *N Engl J Med* 303(6): 308–16.
18. Weinstein MC, Schiff I. 1983. Cost-effectiveness of hormone replacement therapy in the menopause. *Obstet Gynecol Surv* 38(8): 445–55.
19. Kanis JA, Oden A, Johnell O, Jonsson B, de Laet C, Dawson A. 2001. The burden of osteoporotic fractures: A method for setting intervention thresholds. *Osteoporos Int* 12(5): 417–27.
20. Guyatt G, Baumann M, Pauker S, Halperin J, Maurer J, Owens DK, Tosteson AN, Carlin B, Gutterman D, Prins M, Lewis SZ, Schunemann H. 2006. Addressing resource allocation issues in recommendations from clinical practice guideline panels: Suggestions from an American College of Chest Physicians task force. *Chest* 129(1): 182–7.
21. Borgstrom F, Johnell O, Kanis JA, Jonsson B, Rehnberg C. 2006. At what hip fracture risk is it cost-effective to treat? International intervention thresholds for the treatment of osteoporosis. *Osteoporos Int* 17(10): 1459–71.
22. Tosteson AN, Melton LJ 3rd, Dawson-Hughes B, Baim S, Favus MJ, Khosla S, Lindsay RL. 2008. Cost-effective osteoporosis treatment thresholds: The United States perspective. *Osteoporos Int* 19(4): 437–47.
23. Kanis JA, McCloskey EV, Johansson H, Strom O, Borgstrom F, Oden A. 2008. Case finding for the management of osteoporosis with FRAX—Assessment and intervention thresholds for the UK. *Osteoporos Int* 19(10): 1395–408.
24. Dawson-Hughes B. 2008. A revised clinician's guide to the prevention and treatment of osteoporosis. *J Clin Endocrinol Metab* 93(7): 2463–5.
25. National Osteoprosis Foundation. 2008. *Clinician's Guide to Prevention and Treatment of Osteoporosis.* Washington, DC: National Osteoporosis Foundation.
26. Jonsson B, Kanis J, Dawson A, Oden A, Johnell O. 1999. Effect and offset of effect of treatments for hip fracture on health outcomes. *Osteoporos Int* 10(3): 193–9.
27. [No authors listed]. 1998. Osteoporosis: Review of the evidence for prevention, diagnosis and treatment and cost-effectiveness analysis. Introduction. *Osteoporos Int* 8(Suppl 4): S7–80.
28. Fleurence RL, Iglesias CP, Torgerson DJ. 2006. Economic evaluations of interventions for the prevention and treatment of osteoporosis: A structured review of the literature. *Osteoporos Int* 17(1): 29–40.
29. Zethraeus N, Borgstrom F, Strom O, Kanis JA, Jonsson B. 2007. Cost-effectiveness of the treatment and prevention of osteoporosis—A review of the literature and a reference model. *Osteoporos Int* 18(1): 9–23.
30. Fleurence RL, Iglesias CP, Johnson JM. 2007. The cost effectiveness of bisphosphonates for the prevention and treatment of osteoporosis: A structured review of the literature. *Pharmacoeconomics* 25(11): 913–33.
31. Urdahl H, Manca A, Sculpher MJ. 2006. Assessing generalisability in model-based economic evaluation studies: A structured review in osteoporosis. *Pharmacoeconomics* 24(12): 1181–97.
32. Stevenson M, Jones ML, De Nigris E, Brewer N, Davis S, Oakley J. 2005. A systematic review and economic evaluation of alendronate, etidronate, risedronate, raloxifene and teriparatide for the prevention and treatment of postmenopausal osteoporosis. *Health Technol Assess* 9(22): 1–160.
33. Torgerson D, Kanis J. 1995. Cost-effectiveness of preventing hip fracture in the elderly using vitamin D and calcium. *QJM* 88: 135–9.
34. Ivergard M, Strom O, Borgstrom F, Burge RT, Tosteson AN, Kanis J. 2010. Identifying cost-effective treatment with raloxifene in postmenopausal women using risk algorithms for fractures and invasive breast cancer. *Bone* 47(5): 966–74.
35. Goeree R, Blackhouse G, Adachi J. 2006. Cost-effectiveness of alternative treatments for women with osteoporosis in Canada. *Curr Med Res Opin* 22(7): 1425–36.
36. Mobley LR, Hoerger TJ, Wittenborn JS, Galuska DA, Rao JK. 2006. Cost-effectiveness of osteoporosis screening and treatment with hormone replacement therapy, raloxifene, or alendronate. *Med Decis Making* 26(2): 194–206.
37. Borgstrom F, Johnell O, Kanis JA, Oden A, Sykes D, Jonsson B. 2004. Cost effectiveness of raloxifene in the treatment of osteoporosis in Sweden: An economic evaluation based on the MORE study. *Pharmacoeconomics* 22(17): 1153–65.
38. Armstrong K, Chen TM, Albert D, Randall TC, Schwartz JS. 2001. Cost-effectiveness of raloxifene and hormone replacement therapy in postmenopausal women: Impact of breast cancer risk. *Obstet Gynecol* 98(6): 996–1003.
39. Lundkvist J, Johnell O, Cooper C, Sykes D. 2006. Economic evaluation of parathyroid hormone (PTH) in the treatment of osteoporosis in postmenopausal women. *Osteoporos Int* 17(2): 201–11.
40. Liu H, Michaud K, Nayak S, Karpf DB, Owens DK, Garber AM. 2006. The cost-effectiveness of therapy with teriparatide and alendronate in women with severe osteoporosis. *Arch Intern Med* 166(11): 1209–17.
41. Coyle D, Cranney A, Lee KM, Welch V, Tugwell P. 2001. Cost effectiveness of nasal calcitonin in postmenopausal women: Use of Cochrane Collaboration methods for meta-analysis within economic evaluation. *Pharmacoeconomics* 19(5 Pt 2): 565–75.
42. Borgstrom F, Jonsson B, Strom O, Kanis JA. 2007. An economic evaluation of strontium ranelate in the treatment of osteoporosis in a Swedish setting: Based on the results of the SOTI and TROPOS trials. *Osteoporos Int* 17(12): 1781–93.

43. Stevenson M, Davis S, Lloyd-Jones M, Beverley C. 2007. The clinical effectiveness and cost-effectiveness of strontium ranelate for the prevention of osteoporotic fragility fractures in postmenopausal women. *Health Technol Assess* 11(4): 1–134.
44. Stevenson M, Lloyd-Jones M, Papaioannou D. 2009. Vitamin K to prevent fractures in older women: Systematic review and economic evaluation. *Health Technol Assess* 13(45): iii–xi, 1–134.
45. Jonsson B, Strom O, Eisman JA, Papaioannou A, Siris ES, Tosteson A, Kanis JA. 2011. Cost-effectiveness of Denosumab for the treatment of postmenopausal osteoporosis. *Osteoporos Int* 22(3): 967–82.
46. Schousboe JT, Nyman JA, Kane RL, Ensrud KE. 2005. Cost-effectiveness of alendronate therapy for osteopenic postmenopausal women. *Ann Intern Med* 142(9): 734–41.
47. Schousboe JT, Bauer DC, Nyman JA, Kane RL, Melton LJ, Ensrud KE. 2007. Potential for bone turnover markers to cost-effectively identify and select post-menopausal osteopenic women at high risk of fracture for bisphosphonate therapy. *Osteoporos Int* 18(2): 201–10.
48. Schousboe JT, Taylor BC, Fink HA, Kane RL, Cummings SR, Orwoll ES, Melton LJ 3rd, Bauer DC, Ensrud KE. 2007. Cost-effectiveness of bone densitometry followed by treatment of osteoporosis in older men. *JAMA* 298(6): 629–37.
49. Kanis JA, Stevenson M, McCloskey EV, Davis S, Lloyd-Jones M. 2007. Glucocorticoid-induced osteoporosis: A systematic review and cost-utility analysis. *Health Technol Assess* 11(7): iii–iv, ix–xi, 1–231.
50. Kanis JA, Borgstrom F, De Laet C, Johansson H, Johnell O, Jonsson B, Oden A, Zethraeus N, Pfleger B, Khaltaev N. 2005. Assessment of fracture risk. *Osteoporos Int* 16(6): 581–9.
51. Kanis JA, Cooper C, Hiligsmann M, Rabenda V, Reginster JY, Rizzoli R. 2011. Partial adherence: A new perspective on health economic assessment in osteoporosis. *Osteoporos Int* 22(10): 2565–73.
52. Hiligsmann M, Rabenda V, Bruyere O, Reginster JY. 2010. The clinical and economic burden of non-adherence with oral bisphosphonates in osteoporotic patients. *Health Policy* 96(2): 170–7.
53. Kanis JA, McCloskey EV, Johansson H, Strom O, Borgstrom F, Oden A. 2008. Case finding for the management of osteoporosis with FRAX—Assessment and intervention thresholds for the UK. *Osteoporos Int* 19(10): 1395–408.

56

Future Therapies of Osteoporosis

Kong Wah Ng and T. John Martin

Antiresorptives 461
Anabolic Agents 463

Conclusion 464
References 465

Insights into the mechanisms that govern the formation and function of osteoblasts and osteoclasts and their communication processes have led to specific points of intervention that are guiding the development of current and next generation therapies for osteoporosis. The clinical outcomes of these new therapies may be predicted because the actions of the selected targets are known, and in some cases preclinical evidence is fulfilling those predictions. The aims are to develop therapies to improve fracture risk reduction if possible, to avoid the possibility of long-term effects on bone structure, to find drugs whose effects reverse with cessation of therapy, and drugs that stimulate bone formation or that inhibit resorption without inhibiting bone formation at the same time. This chapter will discuss several candidate molecules that have been identified with the potential to act as antiresorptive or anabolic agents. Studies with some of these molecules are still either in preclinical or in early investigational stages without fracture data. Nonetheless, preliminary results hold the promise that at least some of these new therapies may develop into effective means of treating and preventing osteoporosis.

ANTIRESORPTIVES

Several bisphosphonates and a selective estrogen receptor modulator (SERM) have been shown in careful, thorough clinical trials to reduce fracture incidence in osteoporosis by 30–50% [1]. Although it might be asked whether we need other antiresorptives, the real and potential limitations of existing therapies are sufficient to warrant the continued search for new approaches. Newer therapies may be better suited for particular indications or provide greater efficacy, safety, or convenience. There may be a limit to the safe reduction in fracture risk that can be achieved with resorption inhibitors, and it remains to be seen whether that limit can be reached simply by effecting more powerful inhibition of bone resorption [2]. The candidate resorption inhibitors to be considered below can be divided into those whose predominant action is to inhibit osteoclast formation, those that inhibit osteoclast activity, and a potentially interesting new class that might inhibit resorption without concomitant inhibition of bone formation.

Inhibition of osteoclast formation

Monoclonal antibody inhibiting RANKL action

Osteoblasts express a membrane protein, receptor activator of NF-κB ligand (RANKL) regulated by osteotropic hormones, parathyroid hormone (PTH) and calcitriol, as well as cytokines such as interleukin-6. RANKL plays a critical role in osteoclast differentiation, activation, and survival. The binding of RANKL to its receptor, RANK, expressed in mononuclear hematopoietic precursors, initiates the processes that ultimately leads to the formation of multinucleate osteoclasts. Osteoprotegerin (OPG) acts as a decoy receptor for RANKL to suppress osteoclast formation. Studies in genetically altered mice have clearly established the essential physiological roles of RANKL and OPG in controlling osteoclast formation and

---

Primer on the Metabolic Bone Diseases and Disorders of Mineral Metabolism, Eighth Edition. Edited by Clifford J. Rosen.
© 2013 American Society for Bone and Mineral Research. Published 2013 by John Wiley & Sons, Inc.

activity, and revealed a pathway obviously rich in targets for pharmaceutical development. Denosumab (Amgen) is a fully human monoclonal antibody that binds with high affinity and specificity to RANKL to inhibit its action. In the phase III study, subcutaneous injections of denosumab to postmenopausal women with low bone density (T-score below 2.5 at the spine) every 6 months for 36 months was associated with significant reduction in fractures of vertebrae (relative decrease 68%), hip (relative decrease 40%) and nonvertebral sites (relative decrease 20%) [3]. In this clinical trial with 7,868 women, no significant side effects involving the immune system were noted and no cases of osteonecrosis of the jaw. The earlier phase II study had shown that neutralizing RANKL with denosumab resulted in an exceptionally prolonged and powerful action to decrease bone resorption and increase bone mineral density (BMD) at the lumbar spine, hip, and distal third of radius [4]. Bone resorption marker indices decreased rapidly following denosumab injection, as did markers of bone formation, reflecting the coupling of formation to resorption. Extension of this study to 6 years resulted in progressive gain in BMD and sustained reduction in resorption markers [5].

Thus neutralization of RANKL presents as a resorption inhibitor at least as powerful as the most effective bisphosphonates, with a prolonged action, but one that is clearly different in its pharmacodynamics and more readily reversible than is the case with bisphosphonates [6]. Indeed, cessation of denosumab treatment results in rapid reversal of resorption markers to levels even above those of control subjects, and significant BMD decrease within 12 months [7].

Inhibition of osteoclast action

Can formation be uncoupled from resorption?

With existing marketed resorption inhibitory drugs, the coupling of bone formation to resorption is illustrated by the decrease in bone formation that accompanies resorption inhibition, e.g., with bisphosphonates, estrogens, and SERMs. This is the case also with the anti-RANKL, denosumab. Will it be possible to develop drugs that inhibit bone resorption without inhibiting formation—in other words, uncoupling bone formation from resorption? There are some indications that it might be possible.

An essential component of osteoclastic bone resorption is acidification of the resorption lacuna, which reduces the pH to about 4 in order to dissolve the bone mineral. Passive transport of chloride through chloride channels preserves electroneutrality in the course of the acidification process, and preventing chloride transport will lead to a rapid hyperpolarization of the membrane, preventing further secretion of protons, thus resulting in an inhibition of further bone resorption. A vacuolar H+ATPase in the osteoclast membrane plays a key role in this process by mediating the active transport of protons. Inhibitors of this enzyme, for example, bafilomycin, have been shown to inhibit osteoclastic bone resorption in vitro and in vivo. A relatively osteoclast-selective H+ATPase inhibitor has been shown to inhibit ovariectomy-induced bone loss in rats [8]. In mice deficient for either c-src [9] or the chloride-7 channel (CLCN7) [10] bone resorption is inhibited without any inhibition of the rate or extent of formation. In each of these mouse mutations, osteoclast numbers are maintained, but the osteoclasts are unable to resorb bone. This is the case also in human subjects with inactivating mutations either of CLCN7 (OMIM 166600 and OMIM 25900) [11] or the vacuolar H+ATPase [12].

One possibility is that osteoclasts are able to generate a factor (or factors) that can contribute to bone formation, but are not involved in bone resorption [13–15]. Early data with an orally delivered CLCN7 inhibitor showed that it inhibited bone loss in the ovariectomized rat without inhibiting bone formation [10]. A target in a related category is Src, a protein tyrosine kinase. $Src^{-/-}$ mice formed increased numbers of osteoclasts, but the cells failed to resorb bone because they did not form ruffled borders [16]. In a phase I study in healthy males, AZD0530, a highly selective, dual-specific, orally available inhibitor of Src and Abl kinases was shown to suppress bone resorption markers reversibly with variable changes in bone formation markers [17].

It is possible that such inhibitors of resorption, at the moment cathepsin K, CLCN7, vacuolar H+ATPase, and Src, might conceivably provide a new class of resorption inhibitory drug that does not inhibit bone formation. If they are safe and at least as effective in fracture reduction as other inhibitors, they could provide a real advance. For example they might more effectively be combined with anabolic therapy than those resorption inhibitors (e.g., bisphosphonates and, likely, anti-RANKL) that lead to inhibition of bone formation.

Cathepsin K is selectively expressed in osteoclasts and is the predominant cysteine protease in these cells. It accumulates in lysosomal vesicles and is localized at the ruffled border in actively resorbing osteoclasts, discharging into the acidified, sealed resorption space beneath the osteoclast when the lysosomal vesicles fuse with the cell membrane. Defects in the gene encoding cathepsin K are linked to the clinical condition pycnodysostosis (OMIM 265800), an autosomal recessive dysplasia characterized by skeletal defects including dense, brittle bones, short stature and poor bone remodeling [18]. Similarly, the deletion of the cathepsin K gene in mice results in osteopetrosis [19]. Preclinical data confirms that cathepsin K inhibition effectively inhibits bone resorption. In preclinical models in mouse and rabbit and in some monkey studies, cathepsin K inhibition reduced bone resorption without inhibiting bone formation [20].

Cathepsin K inhibitors; current studies with odanacatib

Peptide inhibitors designed to inhibit cysteine proteases by binding at the substrate site to mimic a cathepsin

K-substrate complex have been in clinical development [21, 22]. Preclinical studies showed that cathepsin K inhibitors act as antiresorptive agents that prevent bone loss while allowing bone formation to continue. A 12-month study with an orally bioavailable specific inhibitor of human cathepsin K, balicatib (compound AAE581, Novartis), in postmenopausal women showed an increase in lumbar spine and hip BMD associated with statistically significant decreases in markers of bone resorption but not in those of bone formation [23]. Other potent cathepsin K inhibitors investigated include SB-462795 (relacatib, GlaxoSmithKline) and CRA-013783 (Merck). The most advanced in development is odanacatib (Merck, MK-0822), which has completed phase II and is now in a phase III clinical trial.

Odanacatib is a potent, selective cathepsin K inhibitor with a long half-life (45–50 hours) that has allowed it to be used in weekly oral dosage in clinical study. In a phase II study in postmenopausal women with spinal BMD T-scores between −2.0 and −3.5, treatment for 3 years at the highest (50 mg) dose resulted in significant increases in BMD at the hip and spine, with a 50% reduction in a urinary resorption marker but no change in bone-specific alkaline phosphatase [24]. That treatment is reversible is shown by the rapid increase in a resorption marker and the loss of bone toward baseline values at all sites when drug treatment was stopped.

The phase III fracture study, due to be completed in the second half of 2012, is awaited with great interest. Assuming its positive outcome, among features of great interest with this new class of compound will be how well maintained bone formation is, what will this mean for effects on bone quality, and whether combination therapy with PTH will be more effective with a resorption inhibitor that does not inhibit bone formation.

ANABOLIC AGENTS

Antiresorptive agents do not reconstruct the skeleton, but until recently no therapeutic approach was available to restore bone once it had been lost. That situation has changed with the development of PTH as a highly effective anabolic therapy for the skeleton, despite its better-known action as a resorptive hormone. The anabolic effectiveness of PTH requires that it be administered intermittently, and this has been achieved with the use of daily injections that rapidly achieve a peak level in blood, which is not maintained [25].

Recent research has shed new light on the control of bone formation, and the effect of PTH treatment is such that it is important to learn from the action of PTH and to determine whether any of the newly recognized pathways play any part in PTH action. Such studies are at a very early stage, but some potential targets are emerging for development of anabolic agents. There is particular interest in the possibility of modulating the activity of components of the Wnt canonical signaling pathway to produce a net anabolic effect. The first link between Wnt signaling and human bone disease came from observations that inactivating mutations in the low density lipoprotein receptor-related protein 5 (LRP5) cause the osteoporosis-pseudoglioma syndrome (OPPG, OMIM 259770) characterized by severely decreased bone mass [26]. Conversely, a syndrome of high bone mass was found to be caused by a gain-of-function mutation of LRP5 (OMIM 601884) [27]. These genetic syndromes were reproduced with the appropriate genetic manipulations in mice [28, 29].

Wnt signaling targets

The Wnt/β-catenin signaling pathway offers several targets that may be suitable for pharmacological intervention at a number of specific points. These include extracellular agonists and the points of interaction of antagonists, especially the secreted frizzled-related proteins (SFRPs), dickkopf (DKK) proteins, and sclerostin, as well as regulation within the cell of glycogen synthase kinase-3β (GSK-3β), the enzyme that plays a crucial role in determining availability of β-catenin for the transcriptional effects that are essential for Wnt signaling (see http://www.stanford.edu/group/nusselab/cgi-bin/wnt/). The primary aim of these interventions is to increase Wnt/β-catenin canonical signaling in order to increase bone mass. Initial success in animal models has been reported with the inhibition of DKK-1, GSK-3, and sclerostin. Another potential target is SFRP-1. Deletion of *Sfrp-1* enhances osteoblast proliferation, differentiation and function, while it suppresses osteoblast and osteocyte apoptosis [30].

Inhibition of dickkopf 1 (DKK-1) action

LRP5 can form a ternary complex with DKK-1 and Kremen (a receptor for DKK), which triggers rapid internalization and depletion of LRP5, leading to inhibition of the canonical Wnt signaling pathway. Inhibition of the interactions between DKK-1 and LRP5 would release LRP5 to activate the Wnt pathway. Genetic studies with mice lacking a single allele of DKK-1 showed a markedly increased trabecular bone volume and elevated trabecular bone formation rate [31], while transgenic overexpression of DKK-1 under the control of the *Col1A1* promoter caused severe osteopenia [32]. The production of DKK-1 by multiple myeloma cells has been invoked as a contributing factor to the reduced bone formation in the lytic bone lesions of myeloma [33]. A study using antibodies raised against DKK-1 in the treatment of a mouse model of multiple myeloma showed increased numbers of osteoblasts, reduced number of osteoclasts, and reduced myeloma burden in the antibody-treated mice [34]. Initial drug development attempts in osteoporosis also use monoclonal anti-DKK-1 reagents, but the possibility of directing small molecules to prevent the DKK-1-LRP5 interaction is being explored.

Inhibition of GSK-3

Inhibition of GSK-3 would prevent the phosphorylation of β-catenin, leading to stabilization of β-catenin independently of Wnt interactions with the receptor complex. Mice treated with lithium chloride as a GSK-3 inhibitor showed increased bone formation and bone mass [35]. Treatment of ovariectomized rats with an orally active dual GSK α/β inhibitor, LY603281-31-8, for 2 months resulted in an increase in the number of trabeculae and connectivity as well as trabecular area and thickness. Bone mineral density at cancellous and cortical sites was increased, and this was associated with increased strength. The increased bone formation shown on histomorphometric analysis was associated with increased expression of mRNA markers of the osteoblast phenotype, such as bone sialoprotein, type 1 collagen, osteocalcin, alkaline phosphate, and runx-2 [36]. When the same drug was compared with PTH in a cDNA microarray study, ovariectomy in 6-month-old rats resulted in decreased markers of osteogenesis and chondrogenesis, as well as increased adipogenesis, both of which were reversed by PTH and the GSK-3 α/β inhibitor [37].

Inhibition of sclerostin

Sclerostin, the protein product of SOST, is produced in bone exclusively by osteocytes and is a circulating inhibitor of the Wnt-signaling pathway that achieves this by binding to LRP5 and LRP6 [38]. High bone density in sclerosteosis is caused by an inactivating mutation in SOST gene (OMIM 607363) [39]. Inhibition of production or action of sclerostin resulting in enhanced Wnt canonical signaling would be predicted to lead to increased bone mass. Indeed, in preclinical studies, monoclonal antibody against sclerostin has been shown to promote bone formation rapidly in monkeys and ovariectomized rats. Considerable increases in bone formation rates and in the amounts of trabecular bone took place rapidly without increases in resorption parameters [40].

At the time of writing, a phase I study of anti-sclerostin (AMG 785, Amgen) has been published [41], in which healthy men and women were treated for up to 85 days with escalating doses of AMG 785, resulting in dose-related increases in bone formation markers and a decrease in the resorption marker, serum CTx. The latter observation is interesting, with the authors suggesting that it contributes to a "large anabolic window." It may relate to changes in osteoblast differentiation, with less RANKL-producing cells of the osteoblast lineage available for presentation to osteoclast precursors. In this short study, BMD increased significantly at the spine (5.3%) and hip (2.8%), with five subjects at the highest dose developing detectable antibodies, two of which were neutralizing. The treatment with AMG 785 was generally well tolerated, and further clinical studies will be expected with what appears to be a most promising way of increasing the amount of bone.

Of particular interest is the fact that PTH rapidly reduces sclerostin mRNA and protein production by osteoblasts in vitro and in bone in vivo [42, 43], raising the possibility that transient reduction of sclerostin output by osteocytes in response to intermittent PTH could mediate enhanced osteoblast differentiation and bone formation [44] and reduced osteoblast apoptosis [45]. Such a mechanism would offer real possibilities as a drug target, and the mechanism of this inhibition is all the more interesting with the recent finding [46] that the cyclic AMP-mediated effect of PTH to diminish sclerostin production operates through a long range enhancer, MEF2, the discovery of which came from the pursuit of the nature of the sclerosteosis mutation. There may be small molecule approaches amenable to sclerostin regulation, in addition to antibody neutralization of its activity.

Safety and specificity

Any new therapy emerging from manipulation of the Wnt canonical signaling pathway will need to ensure firstly that it is safe, and secondly, that its action can be targeted specifically to bone. Wnt proteins are critical signaling proteins involved in developmental biology, with roles in early axis specification, brain patterning, intestinal development, and limb development. In adults, Wnt proteins play a vital role in tissue maintenance, with aberrations in Wnt signaling leading to diseases such as adenomatous polyposis [47]. Inhibition of GSK-3 results in increased cyclin D1, cyclin E, and c-Myc, and overexpression of these cell cycle regulators has been linked to tumour formation [48]. All relevant possibilities of side effects of enhanced Wnt signaling need to be kept in mind throughout preclinical studies.

CONCLUSION

The approach of targeting specific peptides that are known to play important roles in osteoblast formation or osteoclast action have yielded several promising candidates with the potential to become effective antiresorptive or anabolic agents to treat osteoporosis. Exciting possibilities of new anabolic therapies are evident, and the benefit of this approach is clear from the satisfactory increase in bone formation that is achieved with parathyroid hormone treatment. The early evidence that it might be possible to develop resorption inhibitors that can uncouple bone resorption from bone formation is exciting, potentially offering an advantage over currently available antiresorptive agents. Nonetheless, all these predictions are based largely on preclinical data, and it is hoped that properly conducted clinical trials in the coming years will see the emergence of new therapies that are effective, durable, and safe, at an affordable cost.

REFERENCES

1. Delmas PD. 2002. Treatment of postmenopausal osteoporosis. *Lancet* 359(9322): 2018–2026.
2. Martin TJ, Seeman E. 2007. New mechanisms and targets in the treatment of bone fragility. *Clin Sci (Lond)* 112(2): 77–91.
3. Cummings SR, San Martin J, McClung MR, Siris ES, Eastell R, Reid IR, Delmas P, Zoog HB, Austin M, Wang A, Kutilek S, Adami S, Zanchetta J, Libanati C, Siddhanti S, Christiansen C. 2009. Denosumab for prevention of fractures in postmenopausal women with osteoporosis. *N Engl J Med* 361(8): 756–765.
4. McClung MR, Lewiecki EM, Cohen SB, Bolognese MA, Woodson GC, Moffett AH, Peacock M, Miller PD, Lederman SN, Chesnut CH, Lain D, Kivitz AJ, Holloway DL, Zhang C, Peterson MC, Bekker PJ. 2006. Denosumab in postmenopausal women with low bone mineral density. *N Engl J Med* 354(8): 821–831.
5. Miller PD, Wagman RB, Peacock M, Lewiecki EM, Bolognese MA, Weinstein RL, Ding B, Martin JS, McClung MR. 2011. Effect of denosumab on bone mineral density and biochemical markers of bone turnover: Six-year results of a phase 2 clinical trial. *J Clin Endocrinol Metab* 96(2): 394–402.
6. Baron R, Ferrari S, Russell RG. 2011. Denosumab and bisphosphonates: Different mechanisms of action and effects. *Bone* 48(4): 677–692.
7. Miller PD, Bolognese MA, Lewiecki EM, McClung MR, Ding B, Austin M, Liu Y, San Martin J. 2008. Effect of denosumab on bone density and turnover in postmenopausal women with low bone mass after long-term continued, discontinued, and restarting of therapy: A randomized blinded phase 2 clinical trial. *Bone* 43(2): 222–229.
8. Visentin L, Dodds RA, Valente M, Misiano P, Bradbeer JN, Oneta S, Liang X, Gowen M, Farina C. 2000. A selective inhibitor of the osteoclastic V-H(+)-ATPase prevents bone loss in both thyroparathyroidectomized and ovariectomized rats. *J Clin Invest* 106(2): 309–318.
9. Marzia M, Sims NA, Voit S, Migliaccio S, Taranta A, Bernardini S, Faraggiana T, Yoneda T, Mundy GR, Boyce BF, Baron R, Teti A. 2000. Decreased c-Src expression enhances osteoblast differentiation and bone formation. *J Cell Biol* 151(2): 311–320.
10. Schaller S, Henriksen K, Sorensen MG, Karsdal MA. 2005. The role of chloride channels in osteoclasts: ClC-7 as a target for osteoporosis treatment. *Drug News Perspect* 18(8): 489–495.
11. Brockstedt H, Bollerslev J, Melsen F, Mosekilde L. 1996. Cortical bone remodeling in autosomal dominant osteopetrosis: A study of two different phenotypes. *Bone* 18(1): 67–72.
12. Del Fattore A, Peruzzi B, Rucci N, Recchia I, Cappariello A, Longo M, Fortunati D, Ballanti P, Iacobini M, Luciani M, Devito R, Pinto R, Caniglia M, Lanino E, Messina C, Cesaro S, Letizia C, Bianchini G, Fryssira H, Grabowski P, Shaw N, Bishop N, Hughes D, Kapur RP, Datta HK, Taranta A, Fornari R, Migliaccio S, Teti A. 2006. Clinical, genetic, and cellular analysis of 49 osteopetrotic patients: Implications for diagnosis and treatment. *J Med Genet* 43(4): 315–325.
13. Martin TJ, Sims NA. 2005. Osteoclast-derived activity in the coupling of bone formation to resorption. *Trends Mol Med* 11(2): 76–81.
14. Lee SH, Rho J, Jeong D, Sul JY, Kim T, Kim N, Kang JS, Miyamoto T, Suda T, Lee SK, Pignolo RJ, Koczon-Jaremko B, Lorenzo J, Choi Y. 2006. v-ATPase V0 subunit d2-deficient mice exhibit impaired osteoclast fusion and increased bone formation. *Nat Med* 12(12): 1403–1409.
15. Karsdal MA, Martin TJ, Bollerslev J, Christiansen C, Henriksen K. 2007. Are nonresorbing osteoclasts sources of bone anabolic activity? *J Bone Miner Res* 22(4): 487–494.
16. Boyce BF, Xing L, Yao Z, Shakespeare WC, Wang Y, Metcalf CA 3rd, Sundaramoorthi R, Dalgarno DC, Iuliucci JD, Sawyer TK. 2006. Future anti-catabolic therapeutic targets in bone disease. *Ann N Y Acad Sci* 1068 447–457.
17. Hannon RA, Clack G, Swisland A, Churchman C, Finkelman RD, Eastell R. 2005. The effect of AZD0530, a highly selective Src inhibitor, on bone turnover in healthy males. *J Bone Miner Res* 20 (Suppl 1): S372 (Abstract).
18. Gelb BD, Shi GP, Chapman HA, Desnick RJ. 1996. Pycnodysostosis, a lysosomal disease caused by cathepsin K deficiency. *Science* 273(5279): 1236–1238.
19. Saftig P, Hunziker E, Wehmeyer O, Jones S, Boyde A, Rommerskirch W, Moritz JD, Schu P, von Figura K. 1998. Impaired osteoclastic bone resorption leads to osteopetrosis in cathepsin-K-deficient mice. *Proc Natl Acad Sci U S A* 95(23): 13453–13458.
20. Pennypacker BL, Duong le T, Cusick TE, Masarachia PJ, Gentile MA, Gauthier JY, Black WC, Scott BB, Samadfam R, Smith SY, Kimmel DB. 2011. Cathepsin K inhibitors prevent bone loss in estrogen-deficient rabbits. *J Bone Miner Res* 26(2): 252–262.
21. Yamashita DS, Dodds RA. 2000. Cathepsin K and the design of inhibitors of cathepsin K. *Curr Pharm Des* 6(1): 1–24.
22. Tavares FX, Boncek V, Deaton DN, Hassell AM, Long ST, Miller AB, Payne AA, Miller LR, Shewchuk LM, Wells-Knecht K, Willard DH Jr, Wright LL, Zhou HQ. 2004. Design of potent, selective, and orally bioavailable inhibitors of cysteine protease cathepsin k. *J Med Chem* 47(3): 588–599.
23. Adami S, Supronik J, Hala T, Brown JP, Garnero P, Haemmerle S, Ortmann CE, Bouisset F, Trechsel U. 2006. Effect of one year treatment with the cathepsin-k inhibitor, balicatib, on bone mineral density (BMD) in postmenopausal women with osteopenia/osteoporosis. *J Bone Miner Res* 21(Suppl 1): S24 [Abstract].
24. Eisman JA, Bone HG, Hosking DJ, McClung MR, Reid IR, Rizzoli R, Resch H, Verbruggen N, Hustad CM,

DaSilva C, Petrovic R, Santora AC, Ince BA, Lombardi A. 2011. Odanacatib in the treatment of postmenopausal women with low bone mineral density: Three-year continued therapy and resolution of effect. *J Bone Miner Res* 26(2): 242–251.

25. Frolik CA, Black EC, Cain RL, Satterwhite JH, Brown-Augsburger PL, Sato M, Hock JM. 2003. Anabolic and catabolic bone effects of human parathyroid hormone (1-34) are predicted by duration of hormone exposure. *Bone* 33(3): 372–379.

26. Gong Y, Slee RB, Fukai N, Rawadi G, Roman-Roman S, Reginato AM, Wang H, Cundy T, Glorieux FH, Lev D, Zacharin M, Oexle K, Marcelino J, Suwairi W, Heeger S, Sabatakos G, Apte S, Adkins WN, Allgrove J, Arslan-Kirchner M, Batch JA, Beighton P, Black GC, Boles RG, Boon LM, Borrone C, Brunner HG, Carle GF, Dallapiccola B, De Paepe A, Floege B, Halfhide ML, Hall B, Hennekam RC, Hirose T, Jans A, Juppner H, Kim CA, Keppler-Noreuil K, Kohlschuetter A, LaCombe D, Lambert M, Lemyre E, Letteboer T, Peltonen L, Ramesar RS, Romanengo M, Somer H, Steichen-Gersdorf E, Steinmann B, Sullivan B, Superti-Furga A, Swoboda W, van den Boogaard MJ, Van Hul W, Vikkula M, Votruba M, Zabel B, Garcia T, Baron R, Olsen BR, Warman ML. 2001. LDL receptor-related protein 5 (LRP5) affects bone accrual and eye development. *Cell* 107(4): 513–523.

27. Boyden LM, Mao J, Belsky J, Mitzner L, Farhi A, Mitnick MA, Wu D, Insogna K, Lifton RP. 2002. High bone density due to a mutation in LDL-receptor-related protein 5. *N Engl J Med* 346(20): 1513–1521.

28. Kato M, Patel MS, Levasseur R, Lobov I, Chang BH, Glass DA 2nd, Hartmann C, Li L, Hwang TH, Brayton CF, Lang RA, Karsenty G, Chan L. 2002. Cbfa1-independent decrease in osteoblast proliferation, osteopenia, and persistent embryonic eye vascularization in mice deficient in Lrp5, a Wnt coreceptor. *J Cell Biol* 157(2): 303–314.

29. Babij P, Zhao W, Small C, Kharode Y, Yaworsky PJ, Bouxsein ML, Reddy PS, Bodine PV, Robinson JA, Bhat B, Marzolf J, Moran RA, Bex F. 2003. High bone mass in mice expressing a mutant LRP5 gene. *J Bone Miner Res* 18(6): 960–974.

30. Bodine PV, Zhao W, Kharode YP, Bex FJ, Lambert AJ, Goad MB, Gaur T, Stein GS, Lian JB, Komm BS. 2004. The Wnt antagonist secreted frizzled-related protein-1 is a negative regulator of trabecular bone formation in adult mice. *Mol Endocrinol* 18(5): 1222–1237.

31. Morvan F, Boulukos K, Clement-Lacroix P, Roman Roman S, Suc-Royer I, Vayssiere B, Ammann P, Martin P, Pinho S, Pognonec P, Mollat P, Niehrs C, Baron R, Rawadi G. 2006. Deletion of a single allele of the Dkk1 gene leads to an increase in bone formation and bone mass. *J Bone Miner Res* 21(6): 934–945.

32. Li J, Sarosi I, Cattley RC, Pretorius J, Asuncion F, Grisanti M, Morony S, Adamu S, Geng Z, Qiu W, Kostenuik P, Lacey DL, Simonet WS, Bolon B, Qian X, Shalhoub V, Ominsky MS, Zhu Ke H, Li X, Richards WG. 2006. Dkk1-mediated inhibition of Wnt signaling in bone results in osteopenia. *Bone* 39(4): 754–766.

33. Tian E, Zhan F, Walker R, Rasmussen E, Ma Y, Barlogie B, Shaughnessy JD Jr. 2003. The role of the Wnt-signaling antagonist DKK1 in the development of osteolytic lesions in multiple myeloma. *N Engl J Med* 349(26): 2483–2494.

34. Yaccoby S, Ling W, Zhan F, Walker R, Barlogie B, Shaughnessy JD Jr. 2007. Antibody-based inhibition of DKK1 suppresses tumor-induced bone resorption and multiple myeloma growth in vivo. *Blood* 109(5): 2106–2111.

35. Clement-Lacroix P, Ai M, Morvan F, Roman-Roman S, Vayssiere B, Belleville C, Estrera K, Warman ML, Baron R, Rawadi G. 2005. Lrp5-independent activation of Wnt signaling by lithium chloride increases bone formation and bone mass in mice. *Proc Natl Acad Sci U S A* 102(48): 17406–17411.

36. Kulkarni NH, Onyia JE, Zeng Q, Tian X, Liu M, Halladay DL, Frolik CA, Engler T, Wei T, Kriauciunas A, Martin TJ, Sato M, Bryant HU, Ma YL. 2006. Orally bioavailable GSK-3alpha/beta dual inhibitor increases markers of cellular differentiation in vitro and bone mass in vivo. *J Bone Miner Res* 21(6): 910–920.

37. Kulkarni NH, Wei T, Kumar A, Dow ER, Stewart TR, Shou J, N'Cho M, Sterchi DL, Gitter BD, Higgs RE, Halladay DL, Engler TA, Martin TJ, Bryant HU, Ma YL, Onyia JE. 2007. Changes in osteoblast, chondrocyte, and adipocyte lineages mediate the bone anabolic actions of PTH and small molecule GSK-3 inhibitor. *J Cell Biochem* 102(6): 1504–1518.

38. Li X, Zhang Y, Kang H, Liu W, Liu P, Zhang J, Harris SE, Wu D. 2005. Sclerostin binds to LRP5/6 and antagonizes canonical Wnt signaling. *J Biol Chem* 280(20): 19883–19887.

39. Ott SM. 2005. Sclerostin and Wnt signaling—The pathway to bone strength. *J Clin Endocrinol Metab* 90(12): 6741–6743.

40. Li X, Warmington KS, Niu QT, Asuncion FJ, Barrero M, Grisanti M, Dwyer D, Stouch B, Thway TM, Stolina M, Ominsky MS, Kostenuik PJ, Simonet WS, Paszty C, Ke HZ. 2010. Inhibition of sclerostin by monoclonal antibody increases bone formation, bone mass, and bone strength in aged male rats. *J Bone Miner Res* 25(12): 2647–2656.

41. Padhi D, Jang G, Stouch B, Fang L, Posvar E. 2011. Single-dose, placebo-controlled, randomized study of AMG 785, a sclerostin monoclonal antibody. *J Bone Miner Res* 26(1): 19–26.

42. Keller H, Kneissel M. 2005. SOST is a target gene for PTH in bone. *Bone* 37(2): 148–158.

43. Silvestrini G, Ballanti P, Leopizzi M, Sebastiani M, Berni S, Di Vito M, Bonucci E. 2007. Effects of intermittent parathyroid hormone (PTH) administration on SOST mRNA and protein in rat bone. *J Mol Histol* 38(4): 261–269.

44. van Bezooijen RL, ten Dijke P, Papapoulos SE, Lowik CW. 2005. SOST/sclerostin, an osteocyte-derived negative regulator of bone formation. *Cytokine Growth Factor Rev* 16(3): 319–327.

45. Sutherland MK, Geoghegan JC, Yu C, Winkler DG, Latham JA. 2004. Unique regulation of SOST, the sclerosteosis gene, by BMPs and steroid hormones in human osteoblasts. *Bone* 35(2): 448–454.

46. Leupin O, Kramer I, Collette NM, Loots GG, Natt F, Kneissel M, Keller H. 2007. Control of the SOST bone enhancer by PTH using MEF2 transcription factors. *J Bone Miner Res* 22(12): 1957–1967.
47. Krishnan V, Bryant HU, Macdougald OA. 2006. Regulation of bone mass by Wnt signaling. *J Clin Invest* 116(5): 1202–1209.
48. Dong J, Peng J, Zhang H, Mondesire WH, Jian W, Mills GB, Hung MC, Meric-Bernstam F. 2005. Role of glycogen synthase kinase 3beta in rapamycin-mediated cell cycle regulation and chemosensitivity. *Cancer Res* 65(5): 1961–1972.

57

Juvenile Osteoporosis

Nick Bishop and Francis H. Glorieux

Introduction 468
Pathophysiology 469
Clinical Features 469
Radiological Features 469
Biochemical Findings 470

Bone Biopsy 470
Differential Diagnosis 470
Treatment 471
Prognosis 471
References 471

INTRODUCTION

Taken literally, the phrase "juvenile osteoporosis" means "osteoporosis in children and adolescents," and thus does not refer to any particular form of osteoporosis in this age group. However, in the scientific literature and in clinical practice, the term "juvenile osteoporosis" is usually used to refer to idiopathic juvenile osteoporosis (IJO). This chapter therefore discusses IJO as a primary disease rather than the entirety of largely secondary osteoporotic conditions that may occur in young people.

Osteoporosis in childhood and adolescence may result from mutations in genes principally affecting the amount and quality of the fibrous component of bone, presenting clinically as osteogenesis imperfecta (OI), or secondary to a spectrum of diverse conditions, such as prolonged immobilization, and chronic inflammatory disease. Bone loss may be worsened by treatment with anticonvulsants and steroids, respectively, but may also improve as the underlying condition improves. Life-threatening diseases such as leukemia also may present with osteoporotic fractures, particularly of the vertebrae. It is clearly important to exclude such causes of osteoporosis. If no underlying cause can be detected, IJO is said to be present.

IJO was first described as a separate entity by Dent and Friedman [1] 4 decades ago. According to the classical description, IJO is a self-limiting disease that develops in a prepubertal, previously healthy child, leads to metaphyseal and vertebral compression fractures, and is characterized radiologically by radiolucent areas in the metaphyses of long bones, dubbed "neo-osseous osteoporosis" [2]. This implies a disorder of trabecular bone architecture and mass, possibly related to changes in the hormonal milieu around the time of the growth spurt.

It is clear, however, that there are many children and adolescents who have low bone mass (defined as a body size-adjusted bone mineral content (BMC) or bone mineral density (BMD) measured by DXA at the spine or total body that is more than 2 SD below the mean; i.e., a Z-score below −2) and who sustain recurrent fractures after minimal trauma, but whose clinical findings do not correspond to the classical description of Dent and Friedman.

In the Pediatric Position Statements of the International Society for Clinical Densitometry published in 2007, these patients would simply fulfill the diagnosis of "osteoporosis." Thus, it may be useful to distinguish "classical IJO" (for patients whose presentation is similar to the description of Dent and Friedman) from "osteoporosis in the wider sense" (for patients who do not match the description of Dent and Friedman but nevertheless have unexplained fractures with low bone mass).

Most reviews on the topic state that IJO is an extremely rare disease, because fewer than 200 patients have been published under that label. This is probably because few patients present with the classical picture. However, most clinicians who see children and adolescents with

---

Primer on the Metabolic Bone Diseases and Disorders of Mineral Metabolism, Eighth Edition. Edited by Clifford J. Rosen.
© 2013 American Society for Bone and Mineral Research. Published 2013 by John Wiley & Sons, Inc.

fractures could probably list a few of their patients who have osteoporosis without recognizable etiology. In our clinical settings, IJO is approximately 10 times less common than OI.

PATHOPHYSIOLOGY

The etiology of IJO is unknown. One study found a normal rise in serum osteocalcin in six IJO patients after calcitriol was administered orally, which was postulated to indicate "normal osteoblast function " [3]. However, the fact that osteoblasts in this test released normal amounts of osteocalcin into the circulation does not necessarily mean that they also deposited matrix on the bone surface in a normal fashion. Early histomorphometric reports on IJO were limited to static methods to quantify bone metabolism, described single cases, [4–8] or did not have adequate control data. No conclusive picture emerged from these reports.

More recent studies using dynamic histomorphometry showed that IJO was characterized by a markedly reduced activation frequency and, therefore, low remodeling activity [9]. In addition, the amount of bone formed at each remodeling site was abnormally low. No evidence was found for increased bone resorption. Interestingly, the bone formation defect was limited to bone surfaces that were exposed to the bone marrow environment; no abnormalities were detected in intracortical and periosteal surfaces [10].

These results suggested that, in IJO, impaired osteoblast performance decreases the ability of cancellous bone to adapt to the increasing mechanical needs during growth. This results in load failure at sites where cancellous bone is essential for stability. Nevertheless, the initial trigger of the decrease in osteoblast performance remains elusive. Two recent reports have indicated that heterozygous mutations in the low-density lipoprotein receptor-related protein 5 (LRP5) can result in low bone mass with fractures in some children [11, 12]. The frequency of such mutations in children with classical IJO, as opposed to childhood osteoporosis, generally remains to be determined.

CLINICAL FEATURES

Classical IJO typically develops in a prepubertal (mostly between 8 and 12 years of age), previously healthy child of either sex [13]. However, in one series, 21 children who were presented as having IJO were recorded as having a mean age at onset of 7 years with a range of 1–13 years [14].

Symptoms generally begin with an insidious onset of pain in the lower back, hips, and feet, and difficulty walking. Knee and ankle pain and fractures of the lower extremities may be present, as well as diffuse muscle

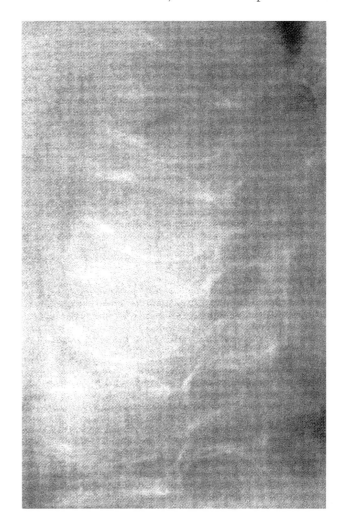

Fig. 57.1. Lateral lumbar spine radiograph of a 10-year-old girl with IJO. Compression fractures of all vertebral bodies and severe osteoporosis are evidence. At the time of this radiograph, lumbar spine areal BMD Z-score was −4.9 (at a height Z-score of −2.5).

weakness. Vertebral compression fractures are frequent, resulting in a short back (Fig. 57.1). Long bone fractures, mostly at metaphyseal sites, may occur. Physical examination may be entirely normal or show thoracolumbar kyphosis or kyphoscoliosis, pigeon chest deformity, loss of height, deformities of the long bones, and limp.

RADIOLOGICAL FEATURES

Children with fully expressed classical IJO present with generalized osteopenia and collapsed or biconcave vertebrae. Disc spaces may be widened asymmetrically because of wedging of the vertebral bodies. Long bones usually have normal diameter and cortical width, unlike the thin, gracile bones of children with OI. The typical radiographic finding in IJO is neoosseous osteoporosis, a radiolucent band at sites of newly formed metaphyseal bone. This localized metaphyseal weakness can give rise to fractures, often at the distal tibias and adjacent to the knee and hip joints. Nevertheless, "neoosseous osteoporosis" is not a prerequisite for diagnosing IJO.

BIOCHEMICAL FINDINGS

Biochemical studies of bone and mineral metabolism have not detected any consistent abnormality in children with IJO [15, 16]; bone resorption was increased in a young woman prior to pregnancy and increased further, with associated bone loss (25% decrease in spine BMD) and vertebral crush fractures [17].

BONE BIOPSY

Iliac bone biopsies show low trabecular bone volume but largely preserved core width (i.e., a normal outer size of the biopsy specimen) and cortical width [9, 10]. Tetracycline double labeling shows a low extent of mineralizing surface (a sign of decreased remodeling activity) and low mineral apposition rate (a sign of weakness of the individual osteoblast team at a remodeling site). There is no indication of a mineralization defect. Osteoclasts are normal in appearance and number.

DIFFERENTIAL DIAGNOSIS

The diagnosis of IJO is made by the exclusion of known etiologies for low bone mass and fractures. The list of conditions that may be associated with bone fragility in children and adolescents is shown in Table 57.1. The exclusion of most of these disorders is usually not difficult. The most frequent diagnostic problem facing a clinician is probably to separate IJO from OI type I.

Table 57.2 presents the typical distinguishing features between IJO and OI type I. Apart from bone fragility and low bone mass, most patients with OI type I have associated extraskeletal connective tissue signs, such as blue or grey scleral hue, dentinogenesis imperfecta, joint hyperlaxity, and Wormian bones (on skull X-rays). However, the extraskeletal involvement can be absent or too subtle to be clinically recognizable in some OI patients. In this situation, genetic analysis of the genes that code for the two collagen type I α-chains (*COL1A1* and *COL1A2*) can be helpful. Mutations affecting a glycine residue in either gene or those leading to a quantitative defect in *COL1A2* expression are diagnostic of OI. DNA-based collagen type I analysis now detects mutations in more than 99% of those individuals where mutations exist. LRP5 sequencing may be informative in 10–15% of cases [11, 12].

Table 57.1. Forms of Osteoporosis in Children, According to Current Literature

I. Primary
Osteogenesis imperfecta
Idiopathic juvenile osteoporosis

II. Secondary
Endocrine disorders
    Cushing syndrome
    Thyrotoxicosis
    Anorexia nervosa

Inflammatory disorders
    Juvenile arthritis
    Dermatomyositis
    Systemic lupus erythematosus
    Inflammatory bowel disease
    Cystic fibrosis
    Chronic hepatitis

Malabsorption syndromes
    Biliary atresia

Inborn errors of metabolism
    Homocystinuria
    Glycogen storage disease type 1

Immobilization
    Cerebral palsy
    Duchenne dystrophy

Hematology/oncology
    Acute lymphoblastic leukemia
    Thalassemia
    Severe congenital neutropenia

An iliac bone biopsy, preferably after tetracycline double labeling, may also contribute to clarifying the diagnosis. Microscopically, a "lack of activity" is usually noted in IJO, whereas there is "hypercellularity" in OI. In histomorphometric terms, this translates into low activation frequency and bone surface based remodeling parameters in IJO and an increase in these values in OI. Also, hyperosteocytosis is a common feature in OI, whereas the amount of osteocytes appears to be normal in IJO.

Fractures and low bone mass may also occur in healthy prepubertal children. Indeed, during late prepuberty and early puberty, fracture rates are almost as high as in postmenopausal women [18–20]. Similar to IJO, such fractures frequently involve metaphyseal bone sites, especially the distal radius. This may reflect problems in the adaptation of the skeleton, in particular the metaphyseal cortex, to the increasing mechanical needs during growth [21]. Growing children and adolescents who have had a few forearm fractures and who have borderline low areal BMD at the spine are frequently encountered in pediatric bone clinics. We propose that classical IJO should only be diagnosed when vertebral compression fractures are present (with or without extremity fractures).

Table 57.2. Differential Diagnosis Between Idiopathic Juvenile Osteoporosis (IJO) and Osteogenesis Imperfecta (OI) Type I

	IJO	OI type I
Family history	Negative	Often positive
Onset	Late prepubertal	Birth or soon after
Duration	1–5 yr	Lifelong
Clinical findings	Metaphyseal fractures	Long bone diaphyseal fractures
	No signs of connective tissue involvement	Blue sclerae, joint hyperlaxity, sometimes abnormal dentition
	Abnormal gait	
Growth rate	Normal	Normal or low
Radiologic findings	Vertebral compression fractures	Vertebral compression fractures
	Long bones: predominantly	"Narrow bones" (low diameter of metaphyseal involvement ["neodiaphyses osseous osteoporosis"])
	No Wormian bones	Wormian bones (skull)
Bone biopsy	Decreased bone turnover	Increased bone turnover
	Normal amount of osteocytes	Hyperosteocytosis
Genetic testing	Patients	Mutations affecting collagen type I in most patients

TREATMENT

There is no treatment with proven benefit to the patient. The effect of any kind of medical intervention is difficult to judge in IJO, because the disease is rare, has a variable course, and is said to resolve without treatment. Long-term outcome studies are lacking, however.

Given the current enthusiasm for pediatric bisphosphonate therapy, many IJO patients probably are receiving treatment with such drugs. We would normally restrict such an intervention to those children with multiple vertebral crush fractures, who may also experience debilitating chronic bone pain.

A number of case reports have described increasing BMD and clinical improvement after treatment with bisphosphonates was started [22–24]. Medical therapies should complement orthopedic and rehabilitative measures such as physiotherapy in all such cases. Review at 6-month intervals is also warranted in children who are not receiving bisphosphonates. Changes in the shape of the spine should be monitored carefully, and early referral to a specialist pediatric spine surgeon should be made in any progressive cases; it is unclear whether bisphosphonate therapy prevents scoliosis or its progression once established in these cases.

PROGNOSIS

The disease process appears to be active only in growing children, and spontaneous recovery is the rule after 3–5 years of evolution [14]. However, in some of the most severe cases reported to date, deformities and severe functional impairment persisted, leaving the children with cardiorespiratory abnormalities and confined to wheelchairs. Preventing such deformities with attendant loss of function should be the focus of attention during the active phase of the disease.

REFERENCES

1. Dent CE, Friedman M. 1965. Idiopathic juvenile osteoporosis. *Q J Med* 34: 177–210.
2. Dent CE. 1977. Osteoporosis in childhood. *Postgrad Med J* 53:450–457.
3. Bertelloni S, Baroncelli GI, Di Nero G, Saggese G. 1992. Idiopathic juvenile osteoporosis: Evidence of normal osteoblast function by 1,25-dihydroxyvitamin D3 stimulation test. *Calcif Tissue Int* 51: 20–23.
4. Cloutier MD, Hayles AB, Riggs BL, Jowsey J, Bickel WH. 1967. Juvenile osteoporosis: Report of a case including a description of some metabolic and microradiographic studies. *Pediatrics* 40: 649–655.
5. Gooding CA, Ball JH. 1969. Idiopathic juvenile osteoporosis. *Radiology* 93: 1349–1350.
6. Jowsey J, Johnson KA. 1972. Juvenile osteoporosis: Bone findings in seven patients. *J Pediatr* 81: 511–517.
7. Smith R. 1980. Idiopathic osteoporosis in the young. *J Bone Joint Surg Br* 62-B: 417–427.
8. Evans RA, Dunstan CR, Hills E. 1983. Bone metabolism in idiopathic juvenile osteoporosis: A case report. *Calcif Tissue Int* 35: 5–8.
9. Rauch F, Travers R, Norman ME, Taylor A, Parfitt AM, Glorieux FH. 2000. Deficient bone formation in idiopathic

juvenile osteoporosis: A histomorphometric study of cancellous iliac bone. *J Bone Miner Res* 15: 957–963.
10. Rauch F, Travers R, Norman ME, Taylor A, Parfitt AM, Glorieux FH. 2002. The bone formation defect in idiopathic juvenile osteoporosis is surface-specific. *Bone* 31: 85–89.
11. Toomes C, Bottomley HM, Jackson RM, Towns KV, Scott S, Mackey DA, Craig JE, Jiang L, Yang Z, Trembath R, Woodruff G, Gregory-Evans CY, Gregory-Evans K, Parker MJ, Black GC, Downey LM, Zhang K, Inglehearn CF. 2004. Mutations in LRP5 or FZD4 underlie the common familial exudative vitreoretinopathy locus on chromosome 11q. *Am J Hum Genet* 74: 721–730.
12. Hartikka H, Makitie O, Mannikko M, Doria AS, Daneman A, Cole WG, Ala-Kokko L, Sochett EB. 2005. Heterozygous mutations in the LDL receptor-related protein 5 (LRP5) gene are associated with primary osteoporosis in children. *J Bone Miner Res* 20: 783–789.
13. Teotia M, Teotia SP, Singh RK. 1979. Idiopathic juvenile osteoporosis. *Am J Dis Child* 133: 894–900.
14. Smith R. 1995. Idiopathic juvenile osteoporosis: Experience of twenty-one patients. *Br J Rheumatol* 34: 68–77.
15. Saggese G, Bertelloni S, Baroncelli GI, Di Nero G. 1992. Serum levels of carboxyterminal propeptide of type I procollagen in healthy children from 1st year of life to adulthood and in metabolic bone diseases. *Eur J Pediatr* 151: 764–768.
16. Saggese G, Bertelloni S, Baroncelli GI, Perri G, Calderazzi A. 1991. Mineral metabolism and calcitriol therapy in idiopathic juvenile osteoporosis. *Am J Dis Child* 145: 457–462.
17. Black AJ, Reid R, Reid DM, MacDonald AG, Fraser WD. 2003. Effect of pregnancy on bone mineral density and biochemical markers of bone turnover in a patient with juvenile idiopathic osteoporosis. *J Bone Miner Res* 18: 167–171.
18. Landin LA. 1997. Epidemiology of children's fractures. *J Pediatr Orthop B* 6: 79–83.
19. Cooper C, Dennison EM, Leufkens HG, Bishop N, van Staa TP. 2004. Epidemiology of childhood fractures in Britain: A study using the general practice research database. *J Bone Miner Res* 19: 1976–1981.
20. Khosla S. Melton W 3rd, Dekutoski MB, Achenbach SJ, Oberg AL, Riggs BL. 2003. Incidence of childhood distal forearm fractures over 30 years: A population-based study. *JAMA* 290: 1479–1485.
21. Rauch F, Neu C, Manz F, Schoenau E. 2001. The development of metaphyseal cortex-implications for distal radius fractures during growth. *J Bone Miner Res* 16: 1547–1555.
22. Hoekman K, Papapoulos SE, Peters AC, Bijvoet OL. 1985. Characteristics and bisphosphonate treatment of a patient with juvenile osteoporosis. *J Clin Endocrinol Metab* 61: 952–956.
23. Brumsen C, Hamdy NA, Papapoulos SE. 1997. Long-term effects of bisphosphonates on the growing skeleton. Studies of young patients with severe osteoporosis. *Medicine (Baltimore)* 76: 266–283.
24. Kauffman RP, Overton TH, Shiflett M, Jennings JC 2001 Osteoporosis in children and adolescent girls: Case report of idiopathic juvenile osteoporosis and review of the literature. *Obstet Gynecol Surv* 56: 492–504.

58

Glucocorticoid-Induced Bone Disease

Robert S. Weinstein

Introduction 473
Clinical Features 473
Basic Science 474
Bone Histomorphometry 474

Osteonecrosis 475
Management 475
References 478

INTRODUCTION

Glucocorticoid-induced osteoporosis (GIO) is the second most common form of osteoporosis and is the most common iatrogenic form of the disease [1, 2]. Fractures may occur in 30% to 50% of patients receiving chronic glucocorticoid therapy and many are asymptomatic, possibly due to glucocorticoid-induced analgesia [3, 4]. However, even asymptomatic prevalent vertebral fractures are important because they further reduce vital capacity in patients with chronic lung disease and increase the risk of subsequent fractures independently of bone mineral density (BMD) [1, 3]. Bone loss in GIO is biphasic, with a reduction in BMD of 6–12% within the first year, followed by a slower annual loss of about 3%; however, there is more to consider than just the reduction in BMD [5]. Bone quality is also an issue [6, 7]. The risk of fracture escalates by as much as 75% within the first 3 months of therapy, typically before a significant decline in BMD, suggesting a glucocorticoid-induced defect in bone strength not captured by bone densitometry [3, 6, 7]. GIO is diffuse, affecting both cortical and cancellous bone, but there is a distinct predilection for fractures in regions of the skeleton with abundant cancellous bone such as the lumbar spine. Several large case-controlled studies show strong associations between glucocorticoid exposure and fracture [8–10]. In a cohort study of patients receiving glucocorticoids for a variety of disorders, continuous treatment with prednisone, 10 mg/day, for more than 90 days was associated with a 7-fold increase in hip fractures and a 17-fold increase in vertebral fractures [10]. Furthermore, in addition to fractures, glucocorticoid administration is the most common cause of nontraumatic osteonecrosis [11].

CLINICAL FEATURES

The occasional patient who becomes Cushingoid when treated with relatively small amounts of glucocorticoids while other patients appear to be remarkably resistant has long perplexed physicians. Another puzzle is that glucocorticoid-induced fractures or osteonecrosis may occur without a Cushingoid habitus. The explanation may be that the sensitivity to exogenous glucocorticoids in different tissues is mediated by inherited or acquired gradations in the local activity of the 11β-hydroxysteroid dehydrogenase (11β-HSD) system. This remarkable system is a natural pre-receptor modulator of corticosteroid action. Two isoenzymes, 11β-HSD1 and 11β-HSD2, catalyze the interconversion of hormonally active glucocorticoids (such as cortisol or prednisolone) and inactive glucocorticoids (such as cortisone or prednisone). The 11β-HSD1 enzyme is an activator, and the 11β-HSD2 enzyme is an inactivator. The ability of any glucocorticoid to bind to the glucocorticoid receptor (GR) depends on the presence of a hydroxyl group at C-11. Therefore, any tissue expressing 11β-HSDs can regulate

---

Primer on the Metabolic Bone Diseases and Disorders of Mineral Metabolism, Eighth Edition. Edited by Clifford J. Rosen.
© 2013 American Society for Bone and Mineral Research. Published 2013 by John Wiley & Sons, Inc.

Table 58.1. Risk Factors for Glucocorticoid-Induced Osteoporosis

Advancing age
Low body mass index (<24 kg/m2)
Underlying diseases (e.g., rheumatoid arthritis, polymyalgia rheumatica, inflammatory bowel disease, chronic pulmonary disease, and transplantation)
Family history of hip fracture, prevalent fractures, smoking, excessive alcohol consumption, frequent falls
Polymorphisms in the glucocorticoid receptor gene
11β-hydroxysteroid dehydrogenase type 1 (11β-HSD1) expression increases with aging and glucocorticoid administration
Glucocorticoid dose (peak, current or cumulative; duration of therapy)
Glucocorticoid-induced fractures occur independently of a decline in bone mass but patients with very low bone density may be at higher risk

the exposure of the resident cells to active glucocorticoids. The increased risk of fractures due to GIO in the elderly may be due to increases in 11β-HSD1 that occur with aging and glucocorticoid exposure [12]. In addition, increased glucocorticoid sensitivity has also been postulated to result from polymorphisms in the GR gene [13]. An additional explanation for individual susceptibility to fractures is the underlying condition that requires glucocorticoid therapy [1, 2, 9]. Fractures may be more common with high dose glucocorticoids and other immunosuppressive drugs used in transplant recipients [14], inflammatory bowel disease and the accompanying malabsorption [15–17], severe rheumatoid arthritis and the relative immobilization in addition to the systemic inflammation [18], and chronic pulmonary disease and the rib fractures provoked by fits of coughing [19] than with myasthenia gravis [20] or multiple sclerosis [21]. Additional risk factors [22–24] are listed in Table 58.1.

However, the absence of severe underlying disease or the presence of youth does not convey protection from glucocorticoid-induced bone disease. Osteoporosis and osteonecrosis have been reported in young as well as older patients receiving depot, topical, or nasal glucocorticoids for dermatitis, rhinitis, or hay fever, and even with modest over-replacement for pituitary or adrenal insufficiency or congenital adrenal hyperplasia [25–34]. In addition, replacing oral with inhaled glucocorticoids, alternate day regimens and high-dose intermittent therapy have failed to prevent GIO [35–38].

BASIC SCIENCE

The adverse effects of glucocorticoids on the skeleton are exerted directly on the osteoblasts, osteocytes, and osteoclasts, decreasing the production of both osteoblasts and osteoclasts and increasing the apoptosis of osteoblasts and osteocytes while prolonging the lifespan of osteoclasts [39–43]. This is evident from a series of experiments in transgenic mice overexpressing the inactivating enzyme 11β-HSD2 [40]. Mice harboring the transgene in osteoblasts and osteocytes are protected from prednisolone-induced apoptosis and the resultant decrease in osteoblast number and bone formation, but still lose BMD because the osteoclasts remain exposed to the prednisolone. However, bone strength is preserved in these animals in spite of the loss of BMD, suggesting that osteocyte viability independently contributes to bone strength. Using the same approach, overexpression of 11β-HSD2 in osteoclasts preserves BMD, but does not prevent the prednisolone-induced decrease in osteoblast lifespan, osteoblast number, and bone formation [41]. Glucocorticoids also have indirect actions on the production of osteoblasts, enhancing the expression of Dickkopf-1, an antagonist of the wingless signaling pathway (Wnt) important for bone formation [2, 9] and suppressing bone morphogenetic proteins and runt-related transcription factor 2, factors required to induce osteoblast differentiation. Glucocorticoids increase production of peroxisome proliferator-activated receptor γ, a transcription factor that induces terminal adipocyte differentiation while suppressing osteoblast differentiation, contributing to an increase in marrow fat at the cost of osteoblasts and cancellous bone [2, 9].

BONE HISTOMORPHOMETRY

Histomorphometric studies in patients receiving long-term glucocorticoid treatment consistently show reduced numbers of osteoblasts on cancellous bone and diminished wall width, a measure of the work performed by these cells [39, 44, 45]. The decreased osteoblasts are due to glucocorticoid-induced reductions in the production of new osteoblast precursors as well as to premature apoptosis of mature, matrix-secreting osteoblasts [39]. Therefore, a decrease in the rate of bone formation is an expected histological finding in glucocorticoid-treated patients [1]. Inadequate numbers of osteoblasts are also an important cause of the reduction in cancellous bone area and decrease in trabecular width, a result of incomplete cavity repair during bone remodeling [1, 39, 44]. With glucocorticoid excess, cancellous bone area is often less than 12% (normal is 21.6% ± 4.5 SD) and correlates with the decreased wall width, thus supplying additional evidence of the glucocorticoid-induced adverse effects on osteoblasts [45]. Increased cortical and cancellous osteocyte apoptosis also occurs and is associated with decreases in vascular endothelial growth factor, skeletal angiogenesis, bone interstitial fluid, hydraulic support, and bone strength [46]. Therefore, glucocorticoid-induced osteocyte apoptosis could account for the loss of bone strength that occurs before the loss of BMD [2, 6, 7] and the mismatch between BMD and fracture risk in patients with GIO [3]. Cortical bone at the iliac crest demonstrates increased porosity while cortical width ranges from clearly subnormal to within normal limits [47].

Some clinical histomorphometric studies of GIO have reported moderate increases in the cancellous erosion perimeter (the crenated cancellous margin with or without the presence of osteoclasts), but others showed no significant change. However, increased erosion perimeter may occur merely because of a glucocorticoid-induced delay in the appearance of osteoblasts and start of bone formation [39]. When carefully measured, osteoclast numbers in patients receiving chronic glucocorticoid treatment are within the normal range or just slightly above normal [39, 44]. This occurs because glucocorticoid excess directly reduces osteoclast production but, in contrast to the glucocorticoid-induced increase in osteoblast apoptosis, the apoptosis of osteoclasts is decreased [41, 43]. The histological features of GIO are quite distinct from those found in other forms of osteoporosis. Loss of gonadal function and secondary hyperparathyroidism are characterized by marked increases in osteoclasts, osteoblasts, and the rate of bone formation. These different histological features, along with the evidence that the skeletal effects of glucocorticoid excess override those caused by sex steroid deficiency, that fracture prevalence is similar in amenorrheic and eumenorrheic women with Cushing's syndrome, and that parathyroid hormone levels are normal with glucocorticoid administration, clearly indicate that, in contrast to prior teaching, hypogonadism and secondary hyperparathyroidism are not central to the pathogenesis of GIO. [48–53].

OSTEONECROSIS

A devastating accompaniment of long-term glucocorticoid therapy is osteonecrosis (also known as aseptic necrosis, avascular necrosis, or ischemic necrosis). Osteonecrosis of the femoral neck, distal femur and proximal tibia, or proximal humerus occurs in as many as 40% of patients who receive high-dose or long-term therapy, although it may also occur with short-term exposure to high doses, after intra-articular injection, and without osteoporosis [11]. The name, "osteonecrosis," is misleading, since it has not been demonstrated that the bone cells die by necrosis with glucocorticoid excess. Indeed, the cell swelling and inflammatory responses that characterize necrosis in soft tissues usually do not occur in glucocorticoid-induced osteonecrosis. The disorder has been attributed to fat emboli, microvascular tamponade of the blood vessels of the femoral head by marrow fat or fluid retention and poorly mending fatigue fractures. However, abundant apoptotic osteocytes have been identified in sections of whole femoral heads obtained during total hip replacement for glucocorticoid-induced osteonecrosis, whereas apoptotic bone cells were absent from femoral specimens removed because of traumatic or sickle cell osteonecrosis, suggesting that the so-called glucocorticoid-induced osteonecrosis actually is osteocyte apoptosis [54, 55]. The apoptotic osteocytes were found juxtaposed to the subchondral fracture crescent in femurs from the patients with glucocorticoid excess. Apoptotic osteocytes persist because they are anatomically unavailable for phagocytosis, and, with glucocorticoid excess, decreased bone remodeling retards their replacement. Glucocorticoid-induced osteocyte apoptosis, a cumulative and unrepairable defect, could uniquely disrupt the mechanosensory function of the osteocyte–lacunar–canalicular system and thus start the inexorable sequence of events leading to collapse of the femoral head. Glucocorticoid-induced osteocyte apoptosis would also explain the correlation between total steroid dose and the incidence of osteonecrosis and its occurrence after glucocorticoid administration has ceased [11, 56, 57]. Persistent hip, knee, or shoulder pain, especially with joint movement, tenderness, or reduced range of motion, in patients receiving glucocorticoid therapy, warrants magnetic resonance imaging to exclude osteonecrosis [1].

MANAGEMENT

Responsibilities of glucocorticoid therapy

Under some circumstances (vasculitis, rheumatoid arthritis, lupus, status asthmaticus, inflammatory bowel disease), high-dose glucocorticoid therapy is an emergency; the clinician always hopes that the course of treatment will be brief, and, therefore, does not address the complications of these drugs. However, it is the responsibility of physicians who prescribe glucocorticoids to educate their patients about these side effects and complications including osteoporosis and osteonecrosis, cataract and glaucoma, hypokalemia, hyperglycemia, hypertension, hyperlipidemia, weight gain, fluid retention, easy bruisability, susceptibility to infection, impaired healing, myopathy, adrenal insufficiency, and the steroid withdrawal syndrome. All patients receiving long-term glucocorticoid therapy should carry a steroid therapy card or medication identification jewelry. Malpractice suits for failure to document disclosure of the skeletal complications to patients are not rare [11]. In spite of this, the bone complications are ignored by more than 50% of the specialists who prescribe glucocorticoids [58].

General measures

Laboratory testing should include measurement of serum 25-hydroxyvitamin D, creatinine, and calcium (in addition to glucose, potassium, and lipids). Bone turnover after long-term glucocorticoid therapy is low, so biochemical markers of bone metabolism are not usually helpful. If biomarkers have been already obtained and are elevated, another problem may be present. Adequate calcium (1,200–1,500 mg/day) and vitamin D (2,000 units/day) supplementation should be recommended, but the evidence is against any substantial effect of these measures to prevent fractures [1, 9]. Prevalent fractures, vertebral morphological assessment, spinal radiographs, or a loss of height as determined using a stadiometer may help identify patients at risk of additional fractures but the disparity between bone quantity and quality in GIO makes BMD or ultrasound measurements inadequate for identifying which patients are at risk [1, 3]. However, yearly BMD measurements may be useful after therapeutic intervention to prevent fractures. The World Health Organization fracture prevention algorithm (FRAX®) underestimates the risk of glucocorticoid-induced fractures as the current and cumulative glucocorticoid dose and duration of therapy are not considered. Furthermore, the algorithm uses femoral neck BMD, but vertebral fractures are more common than hip fractures in GIO, and inclusion of the common risk factors for postmenopausal osteoporosis in the algorithm may not be applicable to GIO [59]. More data are needed to establish the minimum dose and duration of treatment that requires protection from fractures and to develop strategies to ensure the incorporation of this knowledge into clinical practice. There is insufficient evidence to begin anti-fracture medication with sporadic dose-pack prescriptions, annual short-term (e.g., 7 to 10 days) intravenous treatment, or replacement therapy for patients with hypopituitarism, adrenal insufficiency, or congenital adrenal hyperplasia, provided the replacement doses and increases during stress are not excessive. However, the available evidence strongly suggests that any patient who receives or whose physician plans to administer 10 mg or more per day of prednisone for more than 3 months should be given specific treatment to prevent glucocorticoid-induced fractures [8–10].

Specific treatments

Bisphosphonates are considered first-line options for GIO even though they do not directly address the decreased bone formation characteristic of the disorder [1, 39] (Table 58.2). Randomized, placebo-controlled trials have shown that alendronate, risedronate, and zoledronic acid are effective in GIO [60–62]. Nitrogen-containing bisphosphonates induce apoptosis of osteoclasts and inhibit bone resorption [63], but glucocorticoids antagonize this effect [43] and this may account for the limited ability of these antiresorptive agents to protect BMD in GIO as compared to other forms of osteoporosis [60, 64, 65]. In addition, alendronate decreases glucocorticoid-induced osteocyte apoptosis [66], which may play a role in the preservation of bone strength [6, 7]. The evidence for the use of anti-fracture regimens in GIO is not as strong as that for postmenopausal osteoporosis because the primary end point in the glucocorticoid treatment trials was BMD rather than fracture, and glucocorticoids induce a susceptibility to fracture that is independent of BMD. In addition, most studies were only 12 to 18 months in duration, and insufficient numbers of patients were recruited to study hip fractures.

Oral bisphosphonates can be stopped if glucocorticoids are discontinued. However, compliance with oral therapy is poor. Yearly infusions of zoledronic acid solve this problem and provide rapid skeletal protection. If glucocorticoid administration has already been long-term, there may not be time to wait for the more gradual protective effects of oral bisphosphonates with their average oral absorption of 0.7%, weekly dosage, and lower molar potency as compared with intravenous zolendronate. For example, 63 mg of alendronate is absorbed after 90 days of therapy with the typical 70 mg/week dosage as compared to 5 mg of zoledronic acid after a 15-min infusion, but the potency of alendronate is about tenfold less than that of zoledronic acid [67]. Substantial BMD loss occurs in patients who discontinue bisphosphonates while receiving glucocorticoids, so the recommended duration of therapy is typically at least as long as the steroids are prescribed and "drug holidays" are not recommended [68]. About 15% of patients with GIO who are receiving their first infusion of zoledronic acid experience an acute phase reaction within 2 to 3 days and lasting 3 days or less [62]. The mild pyrexia, musculoskeletal pains, and flu-like symptoms are effectively managed with acetaminophen or ibuprofen and seldom occur with subsequent infusions.

In patients with glucocorticoid induced osteonecrosis of the femoral head, a 2-year, randomized, placebo-controlled, open-label trial showed decreased pain, delayed lesion expansion, and less need for surgery with alendronate [69]. An 8-year, prospective, observational study reported sustained improvement in pain and ambulation in patients with osteonecrosis within months of initiating alendronate and with courses lasting only 1–3 years [70]. Although bisphosphonates may be useful in the treatment of osteonecrosis of the hip, the drugs are associated with osteonecrosis of the jaw (ONJ) [71]. Distinctly different from osteonecrosis of the hip, ONJ is characterized by exposed maxillofacial bone for at least 8 weeks, poor dental hygiene, typical presentation after a dental extraction or other invasive procedure, and infection with Actinomyces. ONJ occurs mainly in patients with osteolytic breast cancer or multiple myeloma who receive frequent (often monthly) high-dose intravenous bisphosphonates. In patients with osteoporosis treated with bisphosphonates, the risk of ONJ is between 1 in 10,000 to 100,000 patient-years. Glucocorticoid use may slightly increase the risk for ONJ. Before prescribing bisphosphonates, the clinician should perform an oral examination and encourage patients to be seen by a dentist. Bisphosphonates may also be associated with atypical subtrochanteric femoral fractures, but if so, the risk is small [72].

An alternative treatment for GIO is teriparatide, recombinant human parathyroid hormone [rhPTH 1-34].

Table 58.2. Pharmaceutical Agents for Glucocorticoid-Induced Osteoporosis (GIO)

Intervention	Advantages	Disadvantages
Oral bisphosphonates *Alendronate 10 mg/day or 70 mg/week; *Risedronate 5 mg/day or 35 mg/week	Osteoclast inhibition reduces bone loss. Alendronate also prevents glucocorticoid-induced osteocyte apoptosis. If glucocorticoids are discontinued, these drugs may be stopped.	Antiresorptive agents do not directly address the decreased bone formation characteristic of GIO and have not been shown to reduce hip fractures. Additional problems include gastrointestinal side effects, rare uveitis, poor compliance with oral therapy, and the time required to obtain skeletal protection. Avoid in patients with a creatinine clearance less than 30 mL/min.
Intravenous bisphosphonate *Zoledronic acid (5 mg IV/year)	Osteoclast inhibition reduces bone loss. Increased compliance compared with oral treatment and rapid onset of skeletal effects. Gastrointestinal side effects are unlikely.	Does not address the reduced bone formation caused by glucocorticoid excess. Avoid in patients with a creatinine clearance less than 30 mL/min. Acute phase reaction (flu-like syndrome) may occur within 2 to 3 days and lasting ≤3 days, particularly with the first dose; effectively managed with acetaminophen or ibuprofen.
*Teriparatide rhPTH 1-34 (20 μg subcut/day)	Directly addresses the increase in osteoblast and osteocyte apoptosis and decrease in osteoblast number, bone formation, and bone strength characteristic of GIO. Rapid onset of skeletal effects. Reduces vertebral fractures.	Costs (greater than oral or IV bisphosphonates), daily injections are required, reduced response with high dose glucocorticoids. Not studied in patients with elevated parathyroid hormone levels. Adverse effects: mild hypercalcemia, headache, nausea, leg cramps, dizziness. Caution with preexisting nephrolithiasis. Check serum calcium at least once ≥16 hours after injection and adjust oral calcium intake as needed.
Denosumab (60 mg subcut every 6 months)	Potent inhibitor of osteoclasts with ease of administration. Can be stopped if glucocorticoids are discontinued. Can be used in patients with creatinine clearance ≤30 mL/min or with intolerance to other treatment options.	Does not address the reduced bone formation caused by glucocorticoid excess. Hypocalcemia and vitamin D deficiency must be treated prior to the use of denosumab.

*FDA approved for glucocorticoid-induced osteoporosis. In Europe, only the once-daily oral bisphosphonate regimens, zoledronic acid, and teriparatide are approved for GIO.

In an 18-month randomized, double-blind, controlled, head-to-head trial, teriparatide increased spinal BMD faster and to a greater extent than alendronate and also reduced vertebral fractures (0.6% vs 6.1%, P = 0.004) [73]. Teriparatide represents a particularly rational approach to GIO by counteracting several fundamental aspects of its pathophysiology. The expected glucocorticoid-induced increase in osteoblast and osteocyte apoptosis and decrease in osteoblast number, bone formation, and bone strength are prevented by teriparatide. Decreased osteoblast apoptosis is related to an increase in bone formation and decreased osteocyte apoptosis is associated with preservation of bone strength [74, 75]. Furthermore, teriparatide abrogates the negative impact of glucocorticoids on Akt (protein kinase B) phosphorylation, and Wnt signaling [75]. However, the effect of teriparatide is somewhat compromised by high-dose glucocorticoid therapy [75–77]. In addition, host factors such as severity of the underlying illness, weight loss, concurrent medications, renal function, and low insulin-like growth factor I levels may contribute to the diminished efficacy of teriparatide in GIO as compared with other forms of osteoporosis [1, 75]. Disadvantages of teriparatide include the need for daily subcutaneous injections, refrigeration, cost, side effects (headache, nausea, dizziness, leg cramps), occasional mild hypercalcemia, and the caution required in patients with nephrolithiasis or elevated levels of parathyroid hormone [78].

Another potential treatment option is denosumab, a humanized monoclonal antibody to the receptor activator of nuclear factor-κB ligand (RANKL), approved for the prevention of vertebral, nonvertebral, and hip fractures in women with postmenopausal osteoporosis, but not as yet for GIO [79]. In a randomized, double-blind, placebo-controlled trial of denosumab in 61 patients with rheumatoid arthritis who received concurrent prednisone (15 mg/day or less) and methotrexate, the BMD of the spine and hip increased with denosumab to the same extent as in 88 patients receiving methotrexate and denosumab alone [80]. Furthermore, there was no difference in adverse effects as compared with placebo and methotrexate. Denosumab may be considered for glucocorticoid-treated patients with renal insufficiency and stable serum calcium levels who are not candidates for bisphosphonates or teriparatide or who are intolerant of other treatment options. The ease of administration as a subcutaneous injection every 6 months may increase compliance.

In patients with GIO, incapacitating adjacent vertebral fractures have been reported days after kyphoplasty, suggesting that great caution should be exercised before recommending the procedure in patients receiving glucocorticoid therapy [81, 82].

REFERENCES

1. Weinstein RS. 2011. Clinical practice: Glucocorticoid-induced bone disease. *N Engl J Med* 364: 44–52.
2. Canalis E, Mazziotti G, Giustina A, Bilezikian JP. 2007. Glucocorticoid-induced osteoporosis: Pathophysiology and therapy. *Osteoporosis Int* 18: 1319–1328.
3. Angeli A, Guglielmi G, Dovio A, Capelli G, de Feo D, Giannini S, Giorgino R, Moro L, Giustina A. 2006. High prevalence of asymptomatic vertebral fractures in postmenopausal women receiving chronic glucocorticoid therapy: A cross-sectional outpatient study. *Bone* 39: 259–259.
4. Salerno A, Hermann R. Efficacy and safety of steroid use for postoperative pain relief. *J Bone Joint Surg* 88A: 1361–1372.
5. LoCascio V, Bonucci E, Imbimbo B, Ballanti P, Adami S, Milani S, Tartarotti D, DellaRocca C. 1990. Bone loss in response to long-term glucocorticoid therapy. *Bone Miner* 8: 39–51.
6. Weinstein RS. 2000. Perspective: True strength. *J Bone Miner Res* 15: 621–625.
7. Seeman E. Bone quality-the material and structural basis of bone strength and fragility. *N Engl J Med* 354: 2250–2261.
8. van Staa TP, Laan RF, Barton IP, Cohen S, Reid DM, Cooper C. 2003. Bone density threshold and other predictors of vertebral fracture in patients receiving oral glucocorticoid therapy. *Arth Rheum* 48: 3224–3229.
9. Adler RA, Curtis JR, Saag K, Weinstein RS. 2008. Glucocorticoid-induced osteoporosis. In: Marcus R, Feldman D, Nelsen DA, Rosen CJ (eds.) *Osteoporosis, 3rd Ed.* San Diego, CA: Elsevier-Academic Press. pp. 1135–1166.
10. Steinbuch M, Youket TE, Cohen S. 2004. Oral glucocorticoid use is associated with an increased risk of fracture. *Osteoporos Int* 15: 323–328.
11. Mankin HF. 1992. Nontraumatic necrosis of bone (osteonecrosis). *N Engl J Med* 326: 1473–1479.
12. Cooper MS, Rabbitt EH, Goddard PE, Bartlett WA, Hewison M, Stewart PM. 2002. Osteoblastic 11β-hydroxysteroid dehydrogenase type 1 activity increases with age and glucocorticoid exposure. *J Bone Miner Res* 17: 979–986.
13. Russcher H, Smit P, van den Akker EL, van Rossum EF, Brinkmann AO, de Jong FH, Lamberts SW, Koper JW. 2005. Two polymorphisms in the glucocorticoid receptor gene directly affect glucocorticoid-regulated gene expression. *J Clin Endocrinol Metab* 90: 5804–5810.
14. Maalouf NM, Shanc E. 2005. Osteoporosis after solid organ transplantation. *J Clin Endocrinol Metab* 90: 2456–2465.
15. Bernstein CN, Blanchard JF, Leslie W, Wajda A, Yu BN. 2000. The incidence of fractures among patients with inflammatory bowel disease. *Ann Intern Med* 133: 795–799.
16. Jahnsen J, Falch JA, Mowinckel P, Aadland E. 2002. Vitamin D status, parathyroid hormone and bone mineral density in patients with inflammatory bowel disease. *Scand J Gastroenterol* 37: 192–199.
17. Stockbrugger RW, Schoon EJ, Bollani S, Mills PR, Israeli E, Landgraf L, Felsenberg D, Ljunghall S, Nygard G, Persson T, Graffner H, Bianchi Porro G, Ferguson A. 2002. Discordance between the degree of osteopenia and the prevalence of spontaneous vertebral fractures in Crohn's disease. *Aliment Pharmacol Ther* 16: 1519–1527.
18. Saag KG, Koehnke R, Caldwell JR, Brasington R, Burmeister LF, Zimmerman B, Kohler JA, Furst DE. 1994. Low dose long-term corticosteroid therapy in rheumatoid arthritis: An analysis of serious adverse events. *Am J Med* 96: 115–123.
19. Jorgensen NR, Schwarz P, Holme I, Hendriksen BM, Petersen LJ, Becker V. 2007. The prevalence of osteoporosis in patients with chronic obstructive pulmonary disease: A cross sectional study. *Respir Med* 101: 177–185.
20. Wakata N, Nemoto H, Sugimoto H, Nomoto N, Konno S, Hayashi N, Arak Y, Nakazato A. 2004. Bone density in myasthenia gravis patients receiving long-term prednisolone therapy. *Clin Neurol Neurosurg* 106: 139–141.
21. Khachanova NV, Demina TL, Smirnov AV, Gusev EI. 2006. Risk factors of osteoporosis in women with multiple sclerosis. *Zh Nevrol Psikhiatr Im S S Korsakova* 3: 56–63.
22. Thompson JM, Modin GW, Arnaud CD, Lane NE. 1997. Not all postmenopausal women on chronic steroids and estrogen treatment are osteoporotic: Predictors of bone mineral density. *Calcif Tissue Int* 61: 377–381.
23. Tatsuno I, Sugiyama T, Suzuki S, Yoshida T, Tanaka T, Sueishi M, Saito Y. 2009. Age dependence of early symptomatic vertebral fracture with high-dose glucocorticoid treatment for collagen vascular diseases. *J Clin Endocrinol Metab* 94: 1671–1677.
24. van Staa TP, Leufkins H, Cooper C. 2001. Use of inhaled glucocorticoids and risk of fractures. *J Bone Miner Res* 16: 581–588.
25. Nathan AW, Rose GL. 1979. Fatal iatrogenic Cushing's syndrome. *Lancet* I: 207.
26. Nasser SMS, Ewan PW. 2001. Depot corticosteroid treatment for hay fever causing avascular necrosis of both hips. *Brit Med J* 322: 1589–1591.

27. McLean CJ, Lobo RFJ, Brazier DJ. 1995. Cataracts, glaucoma, and femoral avascular necrosis caused by topical corticosteroid ointment. *Lancet* 345: 330.
28. Champion PK. 1974. Cushing's syndrome secondary to abuse of dexamethasone nasal spray. *Arch Intern Med* 134: 750–751.
29. Licata AA. 2005. Systemic effects of fluticasone nasal spray: Report of two cases. *Endocr Pract* 11: 194–196.
30. Kubo T, Kojima A, Yamazoe S, Ueshima K, Yamamoto T, Hirasawa Y. 2001. Osteonecrosis of the femoral head that developed after long-term topical steroid application. *J Orthop Sci* 6: 92–94.
31. Williams PL, Corbett M. 1983. Avascular necrosis of bone complicating corticosteroid replacement therapy. *Ann Rheum Dis* 42: 276–279.
32. Snyder S. 1984. Avascular necrosis and corticosteroids. *Ann Intern Med* 100: 770.
33. Zelissen PMJ, Croughs RJM, van Rijk PP, Raymakers JA. 1994. Effect of glucocorticoid replacement therapy on bone mineral density in patients with Addison disease. *Ann Intern Med* 120: 207–210.
34. King JA, Wisniewski AB, Bankowski BJ, Carson KA, Zacur HA, Migeon CJ. 2006. Long-term corticosteroid replacement and bone mineral density in adult women with classical congenital adrenal hyperplasia. *J Clin Endocrinol Metab* 91: 865–869.
35. van Staa TP, Leufkins H, Cooper C. 2001. Use of inhaled glucocorticoids and risk of fractures. *J Bone Miner Res* 16: 581–588.
36. Gluck OS, Murphy WA Hahn TJ, Hahn B. 1981. Bone loss in adults receiving alternate day glucocorticoid therapy. A comparison with daily therapy. *Arth Rheum* 24: 892–898.
37. Samaras K, Pett S, Gowers A, McMurchie M, Cooper DA. 2005. Iatrogenic Cushing's syndrome with osteoporosis and secondary adrenal failure in human immunodeficiency virus-infected patients receiving inhaled corticosteroids and ritonavir-boosted protease inhibitors: Six cases. *J Clin Endocrinol Metab* 90: 4394–4398.
38. de Vries F, Bracke M, Leufkens HGM, Lammers JWJ, Cooper C, van Staa TP. 2007. Fracture risk with intermittent high-dose oral glucocorticoid therapy. *Arth Rheum* 56: 208–214.
39. Weinstein RS, Jilka RL, Parfitt AM, Manolagas SC. 1998. Inhibition of osteoblastogenesis and promotion of apoptosis of osteoblasts and osteocytes by glucocorticoids: Potential mechanisms of the deleterious effects on bone. *J Clin Invest* 102: 274–282.
40. O'Brien CA, Jia D, Plotkin LI, Bellido T, Powers CC, Stewart SA, Manolagas SC, Weinstein RS. 2004. Glucocorticoids act directly on osteoblasts and osteocytes to induce their apoptosis and reduce bone formation and strength. *Endocrinol* 145: 1835–1841.
41. Jia D, O'Brien CA, Stewart SA, Manolagas SC, Weinstein RS. 2006. Glucocorticoids act directly on osteoclasts to increase their lifespan and reduce bone density. *Endocrinol* 147: 5592–5599.
42. Sambrook PN, Hughes DR, Nelsen AE, Robinson BG, Mason RS. 2003. Osteocyte viability with glucocorticoid treatment: Relation to histomorphometry. *Ann Rheum Dis* 62: 1215–1217.
43. Weinstein RS, Chen JR, Powers CC, Stewart SA, Landes RD, Bellido T, Jilka RL, Parfitt AM, Manolagas SC. 2002. Promotion of osteoclast survival and antagonism of bisphosphonate-induced osteoclast apoptosis by glucocorticoids. *J Clin Invest* 109: 1041–1048.
44. Dempster DW. 1989. Bone histomorphometry in glucocorticoid-induced osteoporosis. *J Bone Miner Res* 4: 137–141.
45. Dempster DW, Arlot MA, Meunier PJ. 1983. Mean wall thickness and formation periods of trabecular bone packets in corticosteroid-induced osteoporosis. *Calcif Tissue Int* 35: 410–417.
46. Weinstein RS, Wan C, Liu Q, Wang Y, Almeida M, O'Brien CA, Thostenson J, Roberson PK, Boskey AL, Clemens TL, Manolagas SC. 2010. Endogenous glucocorticoids decrease angiogenesis, vascularity, hydration, and strength in aged mice. *Aging Cell* 9: 147–161.
47. Vedi S, Elkin SL, Compston JE. 2005. A histomorphometric study of cortical bone of the iliac crest in patients treated with glucocorticoids. *Calcif Tissue Int* 77: 79–83.
48. Weinstein RS, Jia D, Powers CC, Stewart SA, Jilka RL, Parfitt AM, Manolagas SC. 2004. The skeletal effects of glucocorticoid excess override those of orchidectomy in mice. *Endocrinol* 145: 1980–1987.
49. Pearce G, Tabensky DA, Delmas PD, Gordon Baker HW, Seeman E. 1998. Corticosteroid-induced bone loss in men. *J Clin Endocrinol Metab* 83: 801–806.
50. Tauchmanovà L, Pivonello R, Di Somma C, Rossi R, De Martino MC, Camera L, Klain M, Salvatore M, Lombardi G, Colao A. 2006. Bone demineralization and vertebral fractures in endogenous cortisol excess: Role of disease etiology and gonadal status. *J Clin Endocrinol Metab* 91: 1779–1784.
51. Rubin MA, Bilezikian JP. 2002. The role of parathyroid hormone in the pathogenesis of glucocorticoid-induced osteoporosis: A reexamination of the evidence. *J Clin Endocrinol Metab* 87: 4033–4041.
52. Hattersley AT, Meeran K, Burrin J, Hill P, Shiner R, Ibbertson HK. 1994. The effect of long-and short-term corticosteroids on plasma calcitonin and parathyroid hormone levels. *Calcif Tissue Int* 54: 198–202.
53. Pas-Pacheco E, Fuleihan GEH, LeBoff MS. 1995. Intact parathyroid hormone levels are not elevated in glucocorticoid-treated subjects. *J Bone Miner Res* 10: 1713–1718.
54. Weinstein RS, Nicholas RW, Manolagas SC. 2000. Apoptosis of osteocytes in glucocorticoid-induced osteonecrosis of the hip. *J Clin Endocrinol Metab* 85: 2907–2912.
55. Calder JDF, Buttery L, Revell PA, Pearse M, Polak JM. 2004. Apoptosis—A significant cause of bone cell death in osteonecrosis of the femoral head. *J Bone Joint Surg* 86B: 1209–1213.
56. Zizic TM, Marcoux C, Hungerford DS, Dansereau J-V, Stevens MB. 1985. Corticosteroid therapy associated with ischemic necrosis of bone in systemic lupus erythematosus. *Am J Med* 79: 596–604.

57. Felson DT, Anderson JJ. 1987. Across-study evaluation of association between steroid dose and bolus steroids and avascular necrosis of bone. *Lancet* I: 902–905.
58. Curtis JR, Westfall AO, Allison JJ, Becker A, Casebeer L, Freeman A, Spettell CM, Weissman NW, Wilke S, Saag KG. 2005. Longitudinal patterns in the prevention of osteoporosis in glucocorticoid-treated patients. *Arth Rheum* 52: 2485–2494.
59. Grossman JM, Gordon R, Ranganath VK, Deal C, Caplan L, Chen W, Curtis JR, Furst DE, McMahon M, Patkar NM, Volkmann E, Saag KG. 2010. American College of Rheumatology 2010 recommendations for the prevention and treatment of glucocorticoid-induced osteoporosis. *Arthritis Care Res (Hoboken)* 62: 1515–1526.
60. Adachi JD, Saag KG, Delmas PD, Liberman UA, Emkey RD, Seeman E, Lane NE, Kaufman JM, Poubelle PE, Hawkins F, Correa-Rotter R, Menkes CJ, Rodriguez-Portales JA, Schnitzer TJ, Block JA, Wing J, McIlwain HH, Westhovens R, Brown J, Melo-Gomes JA, Gruber BL, Yanover MJ, Leite MO, Siminoski KG, Nevitt MC, Sharp JT, Malice MP, Dumortier T, Czachur M, Carofano W, Daifotis A. 2001. Two-year effects of alendronate on bone mineral density and vertebral fracture in patients receiving glucocorticoids: A randomized, double-blind, placebo-controlled extension trial. *Arth Rheum* 44: 202–211.
61. Reid DM, Hughes RA, Laan RF, Sacco-Gibson NA, Wenderoth DH, Adami S, Eusebio RA, Devogelaer JP. 2000. Efficacy and safety of daily risedronate in the treatment of corticosteroid-induced osteoporosis in men and women: A randomized trial. *J Bone Miner Res* 15: 1006–1013.
62. Reid DM, Devogelaer J-P, Saag K, Roux C, Lau CS, Reginster JY, Papanastasiou P, Ferreira A, Hartl F, Fashola T, Mesenbrink P, Sambrook PN; HORIZON investigators. 2009. Zoledronic acid and risedronate in the prevention and treatment of glucocorticoid-induced osteoporosis (HORIZON): A multicenter, double-blind, double-dummy, randomized controlled trial. *Lancet* 373: 1253–1263.
63. Weinstein RS, Roberson PK, Manolagas SC. 2009. Giant osteoclast formation and long-term oral aminobisphosphonate therapy. *N Engl J Med* 360: 53–62.
64. Liberman UA, Weiss SR, Bröll J, Minne HW, Quan H, Bell NH, Rodriguez-Portales J, Downs RW, Dequeker J, Favus M. 1995. Effect of oral alendronate on bone mineral density and the incidence of fractures in postmenopausal osteoporosis. The Alendronate Phase III Osteoporosis Treatment Study Group. *N Engl J Med* 333: 1437–1443.
65. Orwoll E, Ettinger M, Weiss S, Miller P, Kendler D, Graham J, Adami S, Weber K, Lorenc R, Pietschmann P, Vandormael K, Lombardi A. 2000. Alendronate for the treatment of osteoporosis in men. *N Engl J Med* 343: 604–610.
66. Plotkin LI, Weinstein RS, Parfitt AM, Manolagas SC, Bellido T. 1999. Prevention of osteocyte and osteoblast apoptosis by bisphosphonates and calcitonin. *J Clin Invest* 104: 1363–1374.
67. Fleisch H. 2000. *Bisphosphonates in Bone Disease: From the Laboratory to the Patient*, 4th Ed. San Diego: Academic Press. p 42.
68. Emkey R, Delmas PD, Goemaere S, Liberman UA, Poubelle PE, Daifotis AG, Verbruggen N, Lombardi A, Czachur M. 2003. Changes in bone mineral density following discontinuation or continuation of alendronate therapy in glucocorticoid-treated patients: A retrospective, observational study. *Arth Rheum* 48: 1102–1108.
69. Lai K-A, Shen W-J, Yang C-Y, Shao C-J, Hsu J-T, Lin R-M. 2005. The use of alendronate to prevent early collapse of the femoral head in patients with nontraumatic osteonecrosis. *J Bone Joint Surg Am* 87: 2155–2159.
70. Agarwala S, Shah S, Joshi VR. 2009. The use of alendronate in the treatment of avascular necrosis of the femoral head: Follow-up to eight years. *J Bone Joint Surg Br* 91: 1013–1018.
71. Khosla S, Burr D, Cauley J, Dempster DW, Ebeling PR, Felsenberg D, Gagel RF, Gilsanz V, Guise T, Koka S, McCauley LK, McGowan J, McKee MD, Mohla S, Pendrys DG, Raisz LG, Ruggiero SL, Shafer DM, Shum L, Silverman SL, Van Poznak CH, Watts N, Woo SB, Shane E; American Society for Bone and Mineral Research. 2007. Bisphosphonate-associated osteonecrosis of the jaw: Report of a task force of the American Society for Bone and Mineral Research. *J Bone Miner Res* 22: 1479–1491.
72. Black DM, Kelly MP, Genant HK, Palermo L, Eastell R, Bucci-Rechtweg C, Cauley J, Leung PC, Boonen S, Santora A, de Papp A, Bauer DC; Fracture Intervention Trial Steering Committee; HORIZON Pivotal Fracture Trial Steering Committee. 2010. Bisphosphonates and fractures of the subtrochanteric or diaphyseal femur. *N Engl J Med* 362: 1761–1771.
73. Saag KG, Shane E, Boonen S, Martin F, Donley DW, Taylor KA, Dalsky GP Marcus R. 2007. Teriparatide or alendronate in glucocorticoid-induced osteoporosis. *N Engl J Med* 357: 2028–2039.
74. Jilka RL, Weinstein RS, Bellido T, Roberson P, Parfitt AM, Manolagas SC. 1999. Increased bone formation by prevention of osteoblast apoptosis with PTH. *J Clin Invest* 104: 439–446.
75. Weinstein RS, Jilka RJ, Roberson PK, Manolagas SC. 2010. Intermittent parathyroid hormone administration prevents glucocorticoid-induced osteoblast and osteocyte apoptosis, decreased bone formation, and reduced bone strength in mice. *Endocrinol* 151: 2641–2649.
76. Oxlund H, Ortoft G, Thomsen JS, Danielsen CC, Ejersted C, Andreassen TT. 2006. The anabolic effect of PTH on bone is attenuated by simultaneous glucocorticoid treatment. *Bone* 39: 244–252.
77. Devogelaer J-P, Adler RA, Recknor C, See K, Warner MR, Wong M, Krohn K. 2010. Baseline glucocorticoid dose and bone mineral density response with teriparatide or alendronate therapy in patients with glucocorticoid-induced osteoporosis. *J Rheumatol* 37: 141–148.
78. Miller PD. 2008. Safety of parathyroid hormone for the treatment of osteoporosis. *Curr Osteo Reports* 6: 12–16.

79. Cummings SR, San Martin J, McClung MR, Siris ES, Eastell R, Reid IR, Delmas P, Zoog HB, Austin M, Wang A, Kutilek S, Adami S, Zanchetta J, Libanati C, Siddhanti S, Christiansen C; FREEDOM Trial. 2009. Denosumab for prevention of fractures in postmenopausal women with osteoporosis. *N Engl J Med* 361: 756–765.
80. Dore RK, Cohen SB, Lane NE, Palmer W, Shergy W, Zhou L, Wang H, Tsuji W, Newmark R; Denosumab RA Study Group. 2010. Effects of denosumab on bone mineral density and bone turnover in patients with rheumatoid arthritis receiving concurrent glucocorticoids or bisphosphonates. *Ann Rheum Dis* 69: 872–875.
81. Donovan MA, Khandji AG, Siris E. 2004. Multiple adjacent vertebral fractures after kyphoplasty in a patient with steroid-induced osteoporosis. *J Bone Miner Res* 19: 712–713.
82. Syed MI, Patel NA, Jan S, Shaikh A, Grunden B, Morar K. 2006. Symptomatic refractures after vertebroplasty in patients with steroid-induced osteoporosis. *Am J Neuroradiol* 27: 1938–1943.

59

Inflammation-Induced Bone Loss in the Rheumatic Diseases

Steven R. Goldring

Rheumatoid Arthritis 483
Ankylosing Spondylitis 484
Systemic Lupus Erythematosus 485

Acknowledgment 485
References 486

The inflammatory joint diseases include a diverse group of disorders that share in common the presence of inflammatory and destructive changes that adversely affect the structure and function of articular and periarticular tissues. In many of these disorders, the inflammatory processes that target the joint tissues may affect extra-articular tissues and organs, and, in addition, there may be generalized effects on systemic bone remodeling. Attention will focus on rheumatoid arthritis (RA), systemic lupus erythematosus (SLE), and the seronegative spondyloarthropathies, which include ankylosing spondylitis, reactive arthritis (formerly designated as Reiter's syndrome), the arthritis of inflammatory bowel disease, juvenile-onset spondyloarthropathy, and psoriatic arthritis. The discussion will be limited to ankylosing spondylitis, which is the prototypical spondyloarthropathy.

In RA and SLE, the synovial lining is the initial site of the inflammatory process. Under physiological conditions, the synovium forms a thin membrane that lines the surface of the joint cavity and is responsible for generation of the synovial fluid that contributes to joint lubrication and nutrition for the chondrocytes that populate the articular cartilage. In patients with RA and SLE, the synovium becomes a site of an intense immune-mediated inflammatory process that results in synovial proliferation and production of potent inflammatory cytokines and soluble mediators that are responsible for the clinical signs of joint inflammation [1, 2]. In patients with RA, this inflammatory process ultimately leads to destruction of the joint tissues. Although the clinical signs of inflammation in SLE and RA are similar, the synovitis associated with SLE characteristically does not lead to direct destruction of articular cartilage and bone. Of interest, Toukap et al. [3] have reported that patients with SLE exhibit a synovial gene profile that is distinct from the patterns observed in RA and osteoarthritis, suggesting that differential inflammatory and immunologic process are involved in the pathogenesis and biologic activity of the synovium in these conditions. The potential mechanisms involved in the differential effects on bone resorption in SLE and RA will be reviewed in the following discussion. Despite the absence of destructive changes, joint deformities (referred to as Jaccoud's arthritis) do develop in patients with SLE, but these are attributable to alterations in the integrity of periarticular connective tissues rather than destruction of the articular cartilage and bone [4].

Synovial inflammation is also present in the seronegative spondyloarthropathies. Unlike the pattern of joint inflammation in RA and SLE, the joint inflammation is most often asymmetrical, involving distal as well as proximal joints, and, importantly, the axial skeleton is also affected. Anatomic and histopathological analyses have established that the entheses, which are the sites of tendon or ligament attachment to bone, are the initial sites of inflammation in the spondyloarthropathies [5]. Subsequently, the extension of the inflammatory process to the joint margins and the development of synovial pannus may be accompanied by the development of marginal joint erosions. In contrast to the findings in RA, the inflammatory process in the spondyloarthropathies may be associated with calcification and ossification at the

Primer on the Metabolic Bone Diseases and Disorders of Mineral Metabolism, Eighth Edition. Edited by Clifford J. Rosen.
© 2013 American Society for Bone and Mineral Research. Published 2013 by John Wiley & Sons, Inc.

enthesis and eventual bony ankylosis of the joint [6]. A similar inflammatory process may affect the axial skeleton leading to enhanced bone formation and fusion or ankylosis of adjacent vertebrae and the formation of so-called syndesmophytes [5–7].

RHEUMATOID ARTHRITIS

Rheumatoid arthritis (RA) is a systemic inflammatory disorder characterized by symmetrical polyarthritis. Four major forms of pathologic skeletal remodeling can be observed in this disorder, including focal marginal articular erosions, subchondral bone loss, periarticular osteopenia, and systemic osteoporosis. The focal marginal erosions are the radiologic hallmark of RA. Histopathologic examination of these sites of focal bone loss reveals the presence of inflamed synovial tissue that has attached to the bone surface forming a mantle or covering referred to as "pannus." The interface between the pannus and adjacent bone is frequently lined by resorption lacunae containing mono- and multinucleated cells with phenotypic features of authentic osteoclasts, thus implicating osteoclasts as the principal cell type responsible for the focal synovial resorptive process [8, 9]. Similar sites of focal bone loss are present on the endosteal surface of the subchondral bone, and cells with phenotypic features of ostcoclasts also are present on these bone surfaces. Erosion of the subchondral bone at these sites contributes to joint destruction by providing access to the deep zones of the articular cartilage, which is subject to degradation by the invading inflammatory tissue. These regions of subchondral bone erosion frequently conform to sites of so-called bone marrow edema visualized by magnetic resonance imaging. Histologic analysis of the bone marrow in these regions reveals that the bone marrow has been replaced by a fibrovascular stroma populated by inflammatory cells [10]. Importantly, the presence of bone marrow lesions is strongly predictive of the subsequent development of local bone erosions at these sites [11, 12].

More definitive evidence implicating osteoclasts in the pathogenesis of focal articular bone erosions has come from the use of genetic approaches in which investigators have induced inflammatory arthritis with features of RA in mice lacking the ability to form osteoclasts [13–15]. In these models, the inability to form osteoclasts results in protection from focal articular bone resorption despite the presence of extensive synovial inflammation.

The propensity of the synovial lesion in RA to induce osteoclast-mediated bone resorption can be attributed to the production by cells within the inflamed tissue of a wide variety of products with the capacity to recruit osteoclast precursors and induce their differentiation and activation. These include a spectrum of chemokines, as well as receptor activator of nuclear factor-κB ligand (RANKL), interleukin-1 (IL-1), IL-6, IL-11, IL-15, IL-17, monocyte colony-stimulating factor, tumor necrosis factor-α (TNF-α), prostaglandins, and parathyroid hormone-related peptide [1, 2]. Among these products, particular attention has focused on RANKL, which is produced by both synovial fibroblasts and T cells within the synovial tissue [16–18]. The critical role of this cytokine in the pathogenesis of focal bone erosions is suggested by the observations that blocking the activity of RANKL in animal models of RA with osteoprotegerin (OPG) results in marked attenuation of articular bone erosions [17, 19, 20]. The pivotal role of RANKL in the resorptive process is further supported by the results obtained in the genetic models in which deletion of RANKL [14] or disruption of its signaling pathway [15] protects animals from articular bone erosions in models of inflammatory arthritis. More recently, blockade of RANKL with denosumab, a monoclonal antibody that blocks RANKL activity, was shown to significantly reduce articular bone erosions in a group of patients with RA, providing further evidence that osteoclasts and osteoclast-mediated bone resorption represent a rational therapeutic target for preventing articular bone destruction in RA [21, 22]. Interestingly, bisphosphonates, which demonstrate beneficial effects in protecting from systemic bone loss in RA, have not been effective in reducing focal joint destruction [23], with the exception of a publication in which the investigators used a protocol involving the sequential administration of zoledronic acid [24]. Although there may be limitations with respect to the use of these agents to prevent joint destruction, as discussed below, there clearly is a role for bisphosphonates in treating and preventing systemic bone loss in RA [23]. In addition, several recent studies have also shown beneficial effects of targeting RANKL with denosumab in the prevention of systemic bone loss in patients with RA, including patients concomitantly treated with glucocorticoids [25, 26].

An additional striking feature of the focal marginal and subchondral bone loss in RA is the virtual absence of bone repair. The recent studies of Diarra et al. [27] have provided insights into the mechanism involved in the uncoupling of bone resorption and formation in this form of inflammatory arthritis. They demonstrated that cells in the inflamed RA synovial tissue produced dickkopf-1 (DKK-1), the inhibitor of the wingless (Wnt)-signaling pathway that plays a critical role in osteoblast-mediated bone formation. Studies by Walsh et al. have confirmed these observations and identified additional Wnt family antagonists, including members of the DKK and secreted Frizzled-related protein families in the RA synovium [28]. In the Diarra studies, synovial fibroblasts, endothelial cells, and chondrocytes were the principal sources of the DKK-1 [27]. They furthermore showed that TNF-α was a potent inducer of DKK-1, thus implicating this proinflammatory mediator in both bone formation and bone resorption. An additional, somewhat surprising observation was that inhibition of DKK-1 with a blocking antibody produced beneficial effects not only on bone formation, but also suppressed osteoclast-mediated bone resorption. The effects on suppression of bone resorption were attributed to downregulation of RANKL production by the inflamed synovium and upregulation of OPG. These observations have clear implications with respect to future therapeutic strategies to prevent bone loss in RA [29].

Focal marginal bone erosions are the radiographic hallmark of RA, but the earliest skeletal feature of RA is the development of periarticular osteopenia. Of importance, there is evidence that the juxta-articular bone loss has high predictive value with respect to the subsequent development of marginal joint erosions in the hand [30–32]. There are few studies examining the histopathologic changes associated with periarticular osteopenia. Shimizu et al. [33] examined the periarticular bone obtained from a series of patients with RA who underwent joint arthroplasty and observed evidence of both increased bone resorption and formation based on histomorphometric analysis. Examination of the bone marrow in the juxta-articular tissues frequently reveals the presence of focal accumulations of inflammatory cells, including lymphocytes and macrophages, and these cells are a likely source of cytokines and related proinflammatory mediators that could adversely affect bone remodeling [34]. Immobilization and reduced mechanical loading are additional factors that have been implicated in the pathogenesis of periarticular bone loss.

The final skeletal feature of RA is the presence of generalized osteoporosis. Numerous studies have documented that patients with RA have lower bone mineral density and an increased risk of fracture compared to disease controls [35–38]. In patients with RA, the presence of multiple confounding factors that influence bone remodeling has made it difficult to define in a given patient the underlying pathogenic mechanism responsible for the reduced bone mass. These include the effects of sex, age, nutritional state, level of physical activity, disease duration and severity, and the use of medications such as glucocorticoids that can adversely affect bone remodeling. Lodder et al. [37] evaluated the relationship between bone mass and disease activity in a cohort of patients with RA with low to moderate disease activity and observed that disease activity was a significant contributory factor to systemic bone loss, supporting the earlier observations made by several other investigators. Solomon and coworkers [39] recently examined the relationship between focal bone erosions and generalized osteoporosis in a cohort of postmenopausal women with RA. Although they observed an association between low hip bone mineral density (BMD) and joint erosions, the association disappeared after multivariable adjustment, suggesting that the relationship between erosions and BMD is complex and influenced by multiple disease- and treatment-related factors.

Several different approaches have been utilized to gain insights into the mechanism responsible for systemic bone loss in RA, including histomorphometric analysis of bone biopsies, measurement of urinary and serum biomarkers of bone remodeling and the assessment of serum cytokine levels. Earlier studies employing histomorphometric analysis suggested that the decrease in bone mass was attributable to depressed bone formation [40]. In contrast, Gough et al. [41], as well as several other investigators, observed the presence of increased bone resorption based on assessment of urinary markers. Of interest, Garnero et al. observed that a high urinary CTX-1 level (a marker of bone resorption) predicted risk of radiographic progression of joint damage independent of rheumatoid factor or erythrocyte sedimentation rate [42]. More recently, several groups of investigators have used the indices of bone remodeling and/or serum cytokine levels to assess the effects of treatment interventions on focal articular and systemic bone loss in RA patients [25, 43–46]. Results indicate that suppression of signs of inflammation and improved functional status are reflected in improvement in the level and pattern of bone remodeling indices.

The disturbance in systemic bone remodeling in RA has been attributed to the adverse effects of proinflammatory cytokines that are released into the circulation from sites of synovial inflammation and act in a manner similar to endocrine hormones to regulate systemic bone remodeling. Although the serum levels of multiple osteoclastogenic cytokines are elevated in RA patients, particular attention has focused on the levels of RANKL and OPG. Geusens and colleagues have recently reported results from a cohort of patients with RA and shown that circulating OPG/RANKL in early RA predicted subsequent bone destruction over an 11-year follow-up [47, 48]. In another study, Vis et al. [43] showed that anti-TNF therapy with infliximab was accompanied by decreased systemic bone loss, and these effects correlated with a fall in serum RANKL levels. There also is evidence that cytokines and mediators released from inflamed joints can adversely affect bone formation. This conclusion is supported by the observations of Diarra et al. [27], who detected elevated levels of DKK-1, an inhibitor of bone formation, in the sera of patients with RA. Similar findings have been reported by other authors [49, 50]. Of interest, in the Diarra study, the authors observed that levels of DKK-1 were not increased compared to controls in patients with AS, which is associated with focal increases in periarticular bone formation, as discussed in the following section.

These studies and the related investigations described in the preceding discussion highlight the importance of monitoring patients with RA for evidence of systemic bone loss and for the institution of early therapeutic interventions that have been shown to reduce the long-term risks of fracture and disability. Similar approaches should be considered in patients with SLE and related forms of inflammatory arthritis who also are at risk for the development of systemic osteoporosis and fracture.

ANKYLOSING SPONDYLITIS

As described above, ankylosing spondylitis (AS) is characterized by inflammation in the entheses, as well as the synovial lining of peripheral joints. Examination of the synovial lesion reveals many of the same features as the RA synovium, including synovial lining hyperplasia, lymphocytic infiltration and pannus formation. In contrast to the pattern of articular bone remodeling in RA, in patients with AS, the inflammatory process may be accompanied by evidence of increased bone formation. This is particularly the case at sites of entheseal inflammation, such as ligament and tendon insertion sites, especially in the spine. To investigate the mechanism responsible for the enhanced bone formation, Braun and coworkers [51] obtained biopsies from the sacroiliac joints of patients with AS. They noted the presence of dense infiltrates of lymphocytes, similar to the RA synovial lesion, but unlike the RA synovium, they also detected foci of endochondral ossification. Using in situ hybridization, they noted the presence of increased expression of TGF-β_2 mRNA in these regions and speculated that the upregulation of this growth factor could be responsible for the enhanced bone formation. Lories et al. [52] analyzed synovial tissues from a series of patients with AS or RA and noted the presence of elevated levels of bone morphogenic proteins–2 and –6 in tissues from both patient populations. They speculated that the differential pattern of new bone formation in AS could be related to the localization of the inflammatory process to the periosteal bone at the entheses. Lories and coworkers [53] have extended these observations in DBA/1 mice, which spontaneously develop an inflammatory arthritis that recapitulates the excessive bone formation characteristic of AS and showed that systemic delivery of noggin, a bone morphogenetic protein (BMP) antagonist, attenuated the new bone formation. Direct evidence supporting a role for BMPs in the new bone formation associated with AS was provided by immunohistochemical analysis of entheseal biopsies from patients with AS demonstrating the presence of phosphorylated smad 1/5, consistent with local activation of the BMP signaling pathway.

Maksymowych and coworkers [54] have examined the relationship between inflammation and new bone formation in the axial skeleton in patients with AS. MRI and radiographs of the spine revealed that new syndesmophytes developed more frequently at the vertebral body margins with inflammation than in sites without inflammation. Importantly, they noted that syndesmophytes continued to develop despite resolution of the inflammation with anti-TNF therapy. These observations are supported by results reported by van der Heijde and coworkers [55, 56] who showed that anti-TNF therapy, although producing significant benefit with respect to axial skeletal inflammation, did not prevent the development of syndesmophytes.

Despite the tendency of patients with AS to produce excessive bone formation at sites of inflammation, many individuals exhibit evidence of spinal osteopenia. This has been attributed to the adverse effects of immobilization that results from spinal ankylosis, although decreased bone density also has been detected in patients even in the absence of bony ankylosis [57–59]. These authors and others have suggested that, as in the other forms of inflammatory arthritis, the bone loss is related to the adverse effects of inflammation on systemic bone remodeling.

SYSTEMIC LUPUS ERYTHEMATOSUS

Systemic lupus erythematosus (SLE), similar to RA, is a systemic inflammatory disease, which, in addition to targeting joint structures, may be associated with widespread extra-articular organ damage. Although the pattern and distribution of joint inflammation in SLE and RA are similar, the joint inflammation in SLE most often does not result in extensive articular bone erosions or cartilage destruction. Joint deformity and subluxation do occur, but these have been attributed primarily to ligamentous laxity related to persistent periarticular soft tissue inflammation. This pattern of arthritis has been referred to as Jaccoud's arthritis and is characterized by the presence of "hook" erosions that occur on the radial aspect of metacarpal bones [60]. These local bone changes are distinct from the marginal erosions seen in RA. A similar pattern of joint deformity has been described in other inflammatory disorders, including rheumatic fever and sarcoid, and so the condition is not unique to SLE. As described above, analysis of the transcriptome in synovial tissue from patients with SLE reveals a gene profile that exhibits substantial differences from the patterns observed in RA [3]. The most prominent finding in the SLE synovial tissue is the upregulation of interferon inducible genes. A type I interferon signature has been reported in the peripheral blood cells from patients with SLE compared to disease control subjects [61, 62]. Of interest, both interferon-α and -β have been shown to inhibit osteoclastogenesis in vitro, and the upregulation of these genes and their products in the SLE synovium could contribute to the protection from the development of osteoclast-mediated bone erosions [63]. Studies by Mensah et al. [64] in the (NZB×NZW)F1 mouse model of SLE provide experimental support implicating synovial-derived interferon-α in the inhibition of osteoclast-mediated bone erosion in SLE. They showed that interferon-α shifted osteoclast precursors toward myeloid dendritic cell differentiation and away from osteoclast formation both in vivo and in vitro. Similar to patients with RA, SLE patients also are at risk for the development of systemic osteoporosis and associated fragility fractures [65]. In addition to the adverse effects of chronic inflammation on bone remodeling, additional factors, including use of glucocorticoid therapy, renal impairment, and vitamin D insufficiency likely contribute to the low bone mass.

ACKNOWLEDGMENT

Research grant from Boehringer Ingelheim.

REFERENCES

1. Schett G. 2011. Effects of inflammatory and anti-inflammatory cytokines on the bone. *Eur J Clin Invest* 10: 1–6.
2. Walsh NC, Crotti TN, Goldring SR, Gravallese EM. 2005. Rheumatic diseases: The effects of inflammation on bone. *Immunol Rev* 208: 228–51.
3. Nzeusseu Toukap A, Galant C, Theate I, Maudoux AL, Lories RJ, Houssiau FA, et al. 2007. Identification of distinct gene expression profiles in the synovium of patients with systemic lupus erythematosus. *Arthritis Rheum* 56(5): 1579–88.
4. Santiago MB, Galvao V. 2008. Jaccoud arthropathy in systemic lupus erythematosus: Analysis of clinical characteristics and review of the literature. *Medicine (Baltimore)* 87(1): 37–44.
5. Benjamin M, McGonagle D. 2009. The enthesis organ concept and its relevance to the spondyloarthropathies. *Adv Exp Med Biol* 649: 57–70.
6. Lories RJ, Luyten FP, de Vlam K. 2009. Progress in spondylarthritis. Mechanisms of new bone formation in spondyloarthritis. *Arthritis Res Ther* 11(2): 221.
7. Braun J, Baraliakos X, Golder W, Hermann KG, Listing J, Brandt J, et al. 2004. Analysing chronic spinal changes in ankylosing spondylitis: A systematic comparison of conventional x rays with magnetic resonance imaging using established and new scoring systems. *Ann Rheum Dis* 2004;63(9): 1046–55.
8. Gravallese EM, Harada Y, Wang JT, Gorn AH, Thornhill TS, Goldring SR. 1998. Identification of cell types responsible for bone resorption in rheumatoid arthritis and juvenile rheumatoid arthritis. *Am J Pathol* 152(4): 943–51.
9. Bromley M, Woolley DE. 1984. Chondroclasts and osteoclasts at subchondral sites of erosion in the rheumatoid joint. *Arthritis Rheum* 27(9): 968–75.
10. Jimenez-Boj E, Nobauer-Huhmann I, Hanslik-Schnabel B, Dorotka R, Wanivenhaus AH, Kainberger F, et al. 2007. Bone erosions and bone marrow edema as defined by magnetic resonance imaging reflect true bone marrow inflammation in rheumatoid arthritis. *Arthritis Rheum* 56(4): 1118–24.
11. Hetland ML, Ejbjerg B, Horslev-Petersen K, Jacobsen S, Vestergaard A, Jurik AG, et al. 2009. MRI bone oedema is the strongest predictor of subsequent radiographic progression in early rheumatoid arthritis. Results from a 2-year randomised controlled trial (CIMESTRA). *Ann Rheum Dis* 68(3): 384–90.
12. Boyesen P, Haavardsholm EA, van der Heijde D, Ostergaard M, Hammer HB, Sesseng S, et al. 2011. Prediction of MRI erosive progression: A comparison of modern imaging modalities in early rheumatoid arthritis patients. *Ann Rheum Dis* 70(1): 176–9.
13. Redlich K, Hayer S, Ricci R, David J, Tohidast-Akrad M, Kollias G, et al. 2002. Osteoclasts are essential for TNF-alpha-mediated joint destruction. *J Clin Invest* 110: 1419–27.
14. Pettit AR, Ji H, von Stechow D, Muller R, Goldring SR, Choi Y, et al. 2001. TRANCE/RANKL knockout mice are protected from bone erosion in a serum transfer model of arthritis. *Am J Pathol* 159(5): 1689–99.
15. Li P, Schwarz EM, O'Keefe RJ, Ma L, Boyce BF, Xing L. 2004. RANK signaling is not required for TNFalpha-mediated increase in CD11(hi) osteoclast precursors but is essential for mature osteoclast formation in TNFalpha-mediated inflammatory arthritis. *J Bone Miner Res* 19(2): 207–13.
16. Romas E, Bakharevski O, Hards DK, Kartsogiannis V, Quinn JM, Ryan PF, et al. 2000. Expression of osteoclast differentiation factor at sites of bone erosion in collagen-induced arthritis. *Arthritis Rheum* 43(4): 821–6.
17. Kong YY, Feige U, Sarosi I, Bolon B, Tafuri A, Morony S, et al. 1999. Activated T cells regulate bone loss and joint destruction in adjuvant arthritis through osteoprotegerin ligand. *Nature* 402(6759): 304–9.
18. Gravallese EM, Goldring SR. 2000. Cellular mechanisms and the role of cytokines in bone erosions in rheumatoid arthritis. *Arthritis Rheum* 43(10): 2143–51.
19. Romas E, Gillespie MT, Martin TJ. 2002. Involvement of receptor activator of NFkappaB ligand and tumor necrosis factor-alpha in bone destruction in rheumatoid arthritis. *Bone* 30(2): 340–6.
20. Redlich K, Hayer S, Maier A, Dunstan C, Tohidast-Akrad M, Lang S, et al. 2002. Tumor necrosis factor-a-mediated joint destruction is inhibited by targeting osteoclasts with osteoprotegerin. *Arthritis Rheum* 46: 785–92.
21. Cohen SB, Dore RK, Lane NE, Ory PA, Peterfy CG, Sharp JT, et al. 2008. Denosumab treatment effects on structural damage, bone mineral density, and bone turnover in rheumatoid arthritis: a twelve-month, multicenter, randomized, double-blind, placebo-controlled, phase II clinical trial. *Arthritis Rheum* 58(5): 1299–309.
22. Deodhar A, Dore RK, Mandel D, Schechtman J, Shergy W, Trapp R, et al. 2010. Denosumab-mediated increase in hand bone mineral density associated with decreased progression of bone erosion in rheumatoid arthritis patients. *Arthritis Care Res (Hoboken)* 62(4): 569–74.
23. Goldring SR, Gravallese EM. 2004. Bisphosphonates: Environmental protection for the joint? *Arthritis Rheum* 50(7): 2044–7.
24. Jarrett SJ, Conaghan PG, Sloan VS, Papanastasiou P, Ortmann CE, O'Connor PJ, et al. 2006. Preliminary evidence for a structural benefit of the new bisphosphonate zoledronic acid in early rheumatoid arthritis. *Arthritis Rheum* 54(5): 1410–4.
25. Miller PD, Wagman RB, Peacock M, Lewiecki EM, Bolognese MA, Weinstein RL, et al. 2011. Effect of denosumab on bone mineral density and biochemical markers of bone turnover: six-year results of a phase 2 clinical trial. *J Clin Endocrinol Metab* 96(2): 394–402.
26. Dore RK, Cohen SB, Lane NE, Palmer W, Shergy W, Zhou L, et al. 2010. Effects of denosumab on bone mineral density and bone turnover in patients with rheumatoid arthritis receiving concurrent glucocorti-

coids or bisphosphonates. *Ann Rheum Dis* 69(5): 872–5.
27. Diarra D, Stolina M, Polzer K, Zwerina J, Ominsky MS, Dwyer D, et al. 2007. Dickkopf-1 is a master regulator of joint remodeling. *Nat Med* 13(2): 156–63.
28. Walsh NC, Reinwald S, Manning CA, Condon KW, Iwata K, Burr DB, et al. 2009. Osteoblast function is compromised at sites of focal bone erosion in inflammatory arthritis. *J Bone Miner Res* 24(9): 1572–85.
29. Goldring SR, Goldring MB. 2007. Eating bone or adding it: The Wnt pathway decides. *Nat Med* 13(2): 133–4.
30. Hoff M, Haugeberg G, Odegard S, Syversen S, Landewe R, van der Heijde D, et al. 2009. Cortical hand bone loss after 1 year in early rheumatoid arthritis predicts radiographic hand joint damage at 5-year and 10-year follow-up. *Ann Rheum Dis* 68(3): 324–9.
31. Goldring SR. 2009. Periarticular bone changes in rheumatoid arthritis: pathophysiological implications and clinical utility. *Ann Rheum Dis* 68(3): 297–9.
32. Stewart A, Mackenzie LM, Black AJ, Reid DM. 2004. Predicting erosive disease in rheumatoid arthritis. A longitudinal study of changes in bone density using digital X-ray radiogrammetry: A pilot study. *Rheumatology (Oxford)* 43(12): 1561–4.
33. Shimizu S, Shiozawa S, Shiozawa K, Imura S, Fujita T. 1985. Quantitative histologic studies on the pathogenesis of periarticular osteoporosis in rheumatoid arthritis. *Arthritis Rheum* 28(1): 25–31.
34. Schett G. 2007. Joint remodelling in inflammatory disease. *Ann Rheum Dis* 66 Suppl 3: iii42–44.
35. Vis M, Haavardsholm EA, Boyesen P, Haugeberg G, Uhlig T, Hoff M, et al. 2011. High incidence of vertebral and non-vertebral fractures in the OSTRA cohort study: A 5-year follow-up study in postmenopausal women with rheumatoid arthritis. *Osteoporos Int* 22(9): 2413–9.
36. van Staa TP, Geusens P, Bijlsma JW, Leufkens HG, Cooper C. 2006. Clinical assessment of the long-term risk of fracture in patients with rheumatoid arthritis. *Arthritis Rheum* 54(10): 3104–12.
37. Lodder MC, de Jong Z, Kostense PJ, Molenaar ET, Staal K, Voskuyl AE, et al. 2004. Bone mineral density in patients with rheumatoid arthritis: Relation between disease severity and low bone mineral density. *Ann Rheum Dis* 63(12): 1576–80.
38. Haugeberg G, Orstavik RE, Kvien TK. 2003. Effects of rheumatoid arthritis on bone. *Curr Opin Rheumatol* 15(4): 469–75.
39. Solomon DH, Finkelstein JS, Shadick N, LeBoff MS, Winalski CS, Stedman M, et al. 2009. The relationship between focal erosions and generalized osteoporosis in postmenopausal women with rheumatoid arthritis. *Arthritis Rheum* 60(6): 1624–31.
40. Compston JE, Vedi S, Croucher PI, Garrahan NJ, O'Sullivan MM. 1994. Bone turnover in non-steroid treated rheumatoid arthritis. *Ann Rheum Dis* 53(3): 163–6.
41. Gough A, Sambrook P, Devlin J, Huissoon A, Njeh C, Robbins S, et al. 1998. Osteoclastic activation is the principal mechanism leading to secondary osteoporosis in rheumatoid arthritis. *J Rheumatol* 25(7): 1282–9.
42. Garnero P, Landewe R, Boers M, Verhoeven A, Van Der Linden S, Christgau S, et al. 2002. Association of baseline levels of markers of bone and cartilage degradation with long-term progression of joint damage in patients with early rheumatoid arthritis: The COBRA study. *Arthritis Rheum* 46(11): 2847–56.
43. Vis M, Havaardsholm EA, Haugeberg G, Uhlig T, Voskuyl AE, van de Stadt RJ, et al. 2006. Evaluation of bone mineral density, bone metabolism, osteoprotegerin and receptor activator of the NFkappaB ligand serum levels during treatment with infliximab in patients with rheumatoid arthritis. *Ann Rheum Dis* 65(11): 1495–9.
44. Seriolo B, Paolino S, Sulli A, Ferretti V, Cutolo M. 2006. Bone metabolism changes during anti-TNF-alpha therapy in patients with active rheumatoid arthritis. *Ann N Y Acad Sci* 1069: 420–7.
45. Barnabe C, Hanley DA. 2009. Effect of tumor necrosis factor alpha inhibition on bone density and turnover markers in patients with rheumatoid arthritis and spondyloarthropathy. *Semin Arthritis Rheum* 39(2): 116–22.
46. Syversen SW, Haavardsholm EA, Boyesen P, Goll GL, Okkenhaug C, Gaarder PI, et al. 2011. Biomarkers in early rheumatoid arthritis: Longitudinal associations with inflammation and joint destruction measured by magnetic resonance imaging and conventional radiographs. *Ann Rheum Dis* 69(5): 845–50.
47. Geusens PP, Landewe RB, Garnero P, Chen D, Dunstan CR, Lems WF, et al. 2006. The ratio of circulating osteoprotegerin to RANKL in early rheumatoid arthritis predicts later joint destruction. *Arthritis Rheum* 54(6): 1772–7.
48. van Tuyl LH, Voskuyl AE, Boers M, Geusens P, Landewe RB, Dijkmans BA, et al. 2011. Baseline RANKL:OPG ratio and markers of bone and cartilage degradation predict annual radiological progression over 11 years in rheumatoid arthritis. *Ann Rheum Dis* 69(9): 1623–8.
49. Garnero P, Tabassi NC, Voorzanger-Rousselot N. 2008. Circulating dickkopf-1 and radiological progression in patients with early rheumatoid arthritis treated with etanercept. *J Rheumatol* 35(12): 2313–5.
50. Wang SY, Liu YY, Ye H, Guo JP, Li R, Liu X, et al. 2011. Circulating Dickkopf-1 is correlated with bone erosion and inflammation in rheumatoid arthritis. *J Rheumatol* 38(5): 821–7.
51. Braun J, Bollow M, Neure L, Seipelt E, Seyrekbasan F, Herbst H, et al. 1995. Use of immunohistologic and in situ hybridization techniques in the examination of sacroiliac joint biopsy specimens from patients with ankylosing spondylitis. *Arthritis Rheum* 38(4): 499–505.
52. Lories RJ, Derese I, Ceuppens JL, Luyten FP. 2003. Bone morphogenetic proteins 2 and 6, expressed in arthritic synovium, are regulated by proinflammatory cytokines and differentially modulate fibroblast-like synoviocyte apoptosis. *Arthritis Rheum* 48(10): 2807–18.
53. Lories RJ, Derese I, Luyten FP. 2005. Modulation of bone morphogenetic protein signaling inhibits the onset and progression of ankylosing enthesitis. *J Clin Invest* 115(6): 1571–9.

54. Maksymowych WP, Chiowchanwisawakit P, Clare T, Pedersen SJ, Ostergaard M, Lambert RG. 2009. Inflammatory lesions of the spine on magnetic resonance imaging predict the development of new syndesmophytes in ankylosing spondylitis: Evidence of a relationship between inflammation and new bone formation. *Arthritis Rheum* 60(1): 93–102.
55. van der Heijde D, Landewe R, Baraliakos X, Houben H, van Tubergen A, Williamson P, et al. 2008. Radiographic findings following two years of infliximab therapy in patients with ankylosing spondylitis. *Arthritis Rheum* 58(10): 3063–70.
56. van der Heijde D, Landewe R, Einstein S, Ory P, Vosse D, Ni L, et al. 2008. Radiographic progression of ankylosing spondylitis after up to two years of treatment with etanercept. *Arthritis Rheum* 58(5): 1324–31.
57. Will R, Palmer R, Bhalla AK, Ring F, Calin A. 1989. Osteoporosis in early ankylosing spondylitis: A primary pathological event? *Lancet* 2(8678–8679): 1483–5.
58. Geusens P, Vosse D, van der Linden S. 2007. Osteoporosis and vertebral fractures in ankylosing spondylitis. *Curr Opin Rheumatol* 19(4): 335–9.
59. Ralston SH, Urquhart GD, Brzeski M, Sturrock RD. 1990. Prevalence of vertebral compression fractures due to osteoporosis in ankylosing spondylitis. *BMJ* 300(6724): 563–5.
60. Ostendorf B, Scherer A, Specker C, Modder U, Schneider M. 2003. Jaccoud's arthropathy in systemic lupus erythematosus: Differentiation of deforming and erosive patterns by magnetic resonance imaging. *Arthritis Rheum* 48(1): 157–65.
61. Crow MK. 2007. Type I interferon in systemic lupus erythematosus. *Curr Top Microbiol Immunol* 316: 359–86.
62. Bennett L, Palucka AK, Arce E, Cantrell V, Borvak J, Banchereau J, et al. 2003. Interferon and granulopoiesis signatures in systemic lupus erythematosus blood. *J Exp Med* 197(6): 711–23.
63. Coelho LF, Magno de Freitas Almeida G, Mennechet FJ, Blangy A, Uzé G. 2005. Interferon-alpha and -beta differentially regulate osteoclastogenesis: Role of differential induction of chemokine CXCL11 expression. *Proc Natl Acad Sci U S A* 102(33): 11917–22.
64. Mensah KA, Mathian A, Ma L, Xing L, Ritchlin CT, Schwarz EM. 2011. Mediation of nonerosive arthritis in a mouse model of lupus by interferon-alpha-stimulated monocyte differentiation that is nonpermissive of osteoclastogenesis. *Arthritis Rheum* 62(4): 1127–37.
65. Alele JD, Kamen DL. 2010. The importance of inflammation and vitamin D status in SLE-associated osteoporosis. *Autoimmun Rev* 9(3): 137–9.

60

Secondary Osteoporosis: Other Causes

Neveen A.T. Hamdy

Introduction 489
Osteoporosis Associated with Systemic Inflammatory Disorders 489
Osteoporosis Associated with Diabetes Mellitus 490
Osteoporosis Associated with Mastocytosis 491

Who Needs to Be Screened for Secondary Causes for Osteoporosis? 491
Conclusions 491
References 492

INTRODUCTION

Bone loss is an inevitable consequence of aging, starting some years before menopause, accelerating after its onset, and continuing throughout life in both men and women. A very large number of heterogeneous causes, collectively grouped as "secondary causes of osteoporosis" may also lead to bone loss through a number of mechanisms, independently of age or estrogen deficiency. A secondary cause for osteoporosis can be found in about two-thirds of men, in more than half of premenopausal women, and in about one-fifth of postmenopausal women [1, 2]. Secondary causes of osteoporosis are legion, ranging from easily identifiable specific disease states such as systemic inflammatory disorders, malignancy, endocrinopathies, and use of medication, particularly glucocorticoids, to more "occult" conditions such as vitamin D deficiency, hypercalciuria, and hyperparathyroidism. Although these latter secondary causes of osteoporosis are the most frequently encountered causes of unexpected bone loss, they can only be diagnosed by a high degree of suspicion, easily confirmed by undertaking the appropriate investigations [2–4]. A large number of these secondary osteoporoses are individually discussed elsewhere in the *Primer*. This chapter focuses on osteoporosis associated with systemic inflammatory diseases, diabetes mellitus, and mastocytosis.

OSTEOPOROSIS ASSOCIATED WITH SYSTEMIC INFLAMMATORY DISORDERS

The RANKL/OPG (receptor activator of nuclear factor-κB ligand/osteoprotegerin) ratio is the primary determinant of osteoclastogenesis and therefore of the maintenance of bone mass [5, 6]. In inflammatory disorders, T-cell activation leads to increased expression of T-cell-derived RANKL [7, 8]. Glucocorticoids, often used to control disease activity, decrease osteoblast number and function and inhibit OPG expression [9]. In these disorders, underlying disease activity alters the RANKL/OPG ratio, and this is further exacerbated by the use of glucocorticoids to control the inflammatory process, the combined effect of which potentially leads to significant bone loss.

Inflammatory arthritis

Rheumatoid arthritis, discussed elsewhere in the *Primer*, represents the prototype of a systemic inflammatory disorder, in which inflammation triggers the increased expression of RANKL from activated T cells and from synovial fibroblasts, which is not matched by an increase in OPG, resulting in local bone loss (joint erosions) and generalized bone loss (osteoporosis) [10].

Primer on the Metabolic Bone Diseases and Disorders of Mineral Metabolism, Eighth Edition. Edited by Clifford J. Rosen.
© 2013 American Society for Bone and Mineral Research. Published 2013 by John Wiley & Sons, Inc.

Inflammatory bowel diseases

In Crohn's disease, the pathophysiology of osteoporosis is multifactorial, including the effect of inflammatory cytokines mediating disease activity [interleukin-6 (IL-6), IL-1, tumor necrosis factor-α (TNF-α)], intestinal malabsorption due to disease activity or intestinal resection, the use of glucocorticoids, inability to achieve peak bone mass when the disease starts in childhood, malnutrition, immobilization, low body mass index (BMI), smoking, and hypogonadism [11, 12]. Ileum resection has been identified as the single most significant risk factor for osteoporosis, followed by age, which is of relevance in predicting overall lifetime risk, as Crohn's disease peaks in the second and third decade of life, with osteoporosis potentially becoming clinically significant only as patients grow older [13]. Patients with Crohn's disease are indeed relatively young, and the exact relationship between the host of factors potentially deleterious to the skeleton and the increased risk for osteoporosis and fractures remains unclear. Opinion is also divided on the prevalence of fractures [13–19], and on bone loss in the long term [13, 20–24]. Maintaining a vitamin D-replete status prevents bone loss, and the judicious use of corticosteroids may counteract the deleterious effects of cytokine-driven disease activity on the skeleton [13, 25]. Pharmacodynamic studies suggest that in patients with reasonably well-controlled disease activity, the nitrogen-containing bisphosphonate alendronate is adequately absorbed from the gut and retained in the skeleton, despite underlying chronic inflammatory gut changes and/or gut resection [26]. It is not clear whether this also applies during an exacerbation of Crohn's disease, as acute gut inflammation may be potentially associated with decreased or indeed increased absorption of an orally administered bisphosphonate. Whether all patients with Crohn's disease should be treated with bone protective agents, or whether these agents should be restricted to patients at increased risk for fracture is as yet to be established.

Chronic obstructive pulmonary disease

In chronic obstructive pulmonary disease (COPD), pro-inflammatory cytokines, particularly TNF-α, are the driving force behind the pathophysiology of the disease process [27]. Elevated inflammatory markers reflect not only the severity of lung disease but also the likelihood of increased risk for comorbidities, particularly cardiovascular disease, diabetes, and osteoporosis [28, 29]. A high prevalence of osteoporosis was thus observed within the first year after diagnosis among 2,699 COPD patients from the United Kingdom General Practice Research Database (GPRD) [30]. Data from over 9,500 subjects from the Third National Health and Nutrition Examination Survey conducted in the U.S. between 1988 and 1994 also showed that airflow obstruction was associated with increased odds of osteoporosis compared with no airflow obstruction [odds ratio (OR) 1.9; 95% CI 1.4 to 2.5] and that these odds increased with increased severity of airways obstruction (OR 2.4; 95% CI 1.3 to 4.4, $p < 0.005$) [31]. Loss of bone mass appears to be associated with increased excretion of bone collagen protein breakdown products, suggesting a protein catabolic state, which may not only lead to bone loss, but also to loss of skeletal muscle mass and function and progressive disability [32]. Continuous users of systemic glucocorticoids are more than twice as likely to have one or more vertebral fractures compared with nonusers [33]. In a large case-controlled study including more than 100,000 cases from the GPRD, an association between inhaled corticosteroids at daily doses equivalent to more than 1,600 μg beclomethasone and increased fracture risk disappeared after adjustment for disease severity, suggesting that in COPD, it is disease severity rather than inhaled corticosteroids that increase fracture risk [34]. Factors other than chronic inflammation and corticosteroid use also contribute to bone loss and increased fracture risk in patients with COPD. These include vitamin D deficiency or insufficiency, reduced skeletal muscle mass and strength, immobilization, low BMI and changes in body composition, hypogonadism, reduced levels of insulin-like growth factors (IGFs), smoking, increased alcohol intake, and genetic factors [35]. The morbidity associated with vertebral fractures is particularly high in patients with COPD as these are associated with restrictive changes in pulmonary function, significant decreases in forced expiratory volume, and up to 9% reduction of predicted lung vital capacity for each additional thoracic vertebral compression fracture [36, 37].

OSTEOPOROSIS ASSOCIATED WITH DIABETES MELLITUS

The deleterious effects of diabetes mellitus (DM) on the skeleton are multifactorial and both types 1 and 2 DM are associated with increased fracture risk [35–41]. Data from the Iowa Women's Health study suggest that women with type 1 DM are 12 times more likely to sustain hip fractures than women without DM and that women with type 2 DM have a 1.7-fold increased risk of sustaining hip fractures despite maintaining a normal bone mass [40]. The high prevalence of fractures in type 2 DM is likely to be influenced by long-term complications such as retinopathy-induced visual impairment and neuropathy-induced decreased balance, both increasing the risk for falls [41]. The osteoporosis of DM is one of low bone turnover, with the main mechanism of bone loss being decreased bone formation [42, 43]. Insulin and amylin have an anabolic effect on bone and their decrease in type 1 DM may lead to impaired bone formation, primarily because of a decrease in IGF-1 concentrations. *In vitro* studies also demonstrate that sustained exposure to high glucose concentrations results in osteoblast dysfunction, and poor metabolic control has a clear negative impact on bone mass. In type 1 DM, decreased peak bone mass also plays a role when the disease manifests itself before skeletal growth is complete. Microvascular complications and decreased mechanical stress due to neuropathy and/or myopathy contribute to the increased fracture risk at later stages of the disease [44]. In DM, there is increased bone marrow adiposity, which has also been linked with the osteoporosis of aging, glucocorticoid use, and immobility [45]. Several members of the nuclear hormone receptor family control the critical adipogenic and osteogenic steps, and evidence has been mounting for a clear interdependence of adipogenesis and osteogenesis [45, 46]. In the bone marrow microenvironment, the inverse relationship between adipogenic and osteogenic differentiation was shown to be mediated at least in part through cross talk between pathways activated by steroid receptors (estrogen, thyroid, corticosteroid and growth hormone receptors), the peroxisome proliferator activator receptors (PPARs), and other cytokine and paracrine factors. PPARs play a central role in initiating adipogenesis in bone marrow and other stromal-like cells in vitro and in vivo [47], and their ligands (rosiglitazone and pioglitazone) play a prominent role in the treatment of type 2 DM. These ligands induce adipogenesis and inhibit osteogenesis in vitro, which may explain the increased incidence of fractures reported in patients with DM using these agents [48, 49].

OSTEOPOROSIS ASSOCIATED WITH MASTOCYTOSIS

In all forms of mastocytosis, the proximity of the mast cell to bone remodeling surfaces and the production by this cell of a large number of chemical mediators and cytokines capable of modulating bone turnover translates into skeletal involvement, ranging from severe osteolysis to significant osteosclerosis, with osteoporosis being the most frequently observed pathology [50, 51]. Bone loss is also exacerbated by the use of glucocorticoids. An important clinical manifestation of skeletal involvement includes generalized bony pain, which may be incapacitating, and is often resistant to conventional analgesia, particularly in cases of extensive bone marrow involvement or rapidly progressive disease. Osteoporosis may be associated with systemic manifestations of enhanced mast cell activity such as flushes and gastrointestinal symptoms [51], or may also be the sole presentation of bone marrow mastocytosis [52–54], in which case the osteoporotic process may be severe and progressive. The diagnosis can only be confidently established by histologic examination of bone marrow biopsies that demonstrate the pathognomonic feature of bone marrow infiltration with a large number of morphologically abnormal mast cells, individually or in aggregates of more than 15 cells [51]. Bone marrow mastocytosis is an important "occult" cause of secondary osteoporosis, shown to be present in up to 9% of men with "idiopathic osteoporosis" [54]. Serum tryptase may be normal and the measurement of the 24-hour urine excretion of N-methyl histamine represents a valuable noninvasive surrogate to bone marrow biopsies in establishing the diagnosis and evaluating the degree of mast cell load [54, 55].

WHO NEEDS TO BE SCREENED FOR SECONDARY CAUSES FOR OSTEOPOROSIS?

The high prevalence of potentially reversible secondary causes for osteoporosis, which may be identified with a sensitivity of 92% by cost-effective laboratory investigations [56], dictates that the majority of patients with osteoporosis would require a basic battery of laboratory tests before the start of treatment, including a full blood count, serum biochemistry panel, 24-hour urine calcium excretion, and 25 hydroxy-vitamin D measurements [2, 57]. Secondary causes for osteoporosis should be particularly sought in young patients [58], premenopausal women [59], men under the age of 65 [60], in all patients with unexpected or severe osteoporosis, in those with accelerated bone loss, and in those experiencing bone loss under treatment with conventional osteoporosis therapy. Further laboratory tests should be requested to confirm or exclude hypogonadism, thyrotoxicosis, celiac disease, hypercortisolism, mastocytosis, and multiple myeloma. If suspicion remains high, or in the case of fragility fractures in the presence of a normal bone mineral density (BMD), a double tetracycline-labeled transiliac bone biopsy with bone marrow evaluation may be indicated to establish a mineralization defect, or a bone marrow disorder, particular a nonsecretory myeloma or mastocytosis.

CONCLUSIONS

Secondary causes of osteoporosis are very common, particularly in premenopausal women and in men with osteoporosis, while also being the cause of accelerated bone loss in postmenopausal and age-related osteoporosis. In addition to representing significant comorbidity in specific disease entities such as inflammatory disorders, malignant disease, bone marrow disorders, and endocrinopathies, secondary osteoporosis is also commonly associated with often silent disturbances in calcium homeostasis such as vitamin D deficiency, hypercalciuria, and hyperparathyroidism, all of which are easily detectable by standard laboratory testing. The ubiquitous nature of "secondary osteoporosis" suggests that diverse medical disciplines need to better interact to meet some of the challenges presented by osteoporosis as a chronic comorbidity of specific disease entities. Screening for secondary causes for osteoporosis should represent an intrinsic part of the optimal management of any patient with osteoporosis.

REFERENCES

1. Painter SE, Kleerekoper M, Camacho PM. 2006. Secondary osteoporosis: A review of the recent evidence. *Endocr Pract* 12: 436–445.
2. Gabaroi DC, Peris P, Monegal A, Albaladejo C, Martinez MA, Muxi A, Martinez de Osaba MJ, Suris X, Guanabens N. 2010. Search for secondary causes in postmenopausal women with osteoporosis. *Menopause* 17: 135–139.
3. Johnson BE, Lucasey B, Robinson RG, Lukert BP. 1989. Contributing diagnoses in osteoporosis. The value of a complete medical evaluation. *Arch Intern Med* 149: 1069–1072.
4. Deutschmann HA, Weger M, Weger W, Kotanko P, Deutschmann MJ, Skrabal F. 2002. Search for occult secondary osteoporosis: Impact of identified possible risk factors on bone mineral density. *J Intern Med* 252: 389–397.
5. Boyle WJ, Simonet WS, Lacey DL. 2003. Osteoclast differentiation and activation. *Nature* 423: 337–342.
6. Walsh MC, Kim N, Kadono Y, Rho J, Lee SY, Lorenzo J, Choi Y. 2006. Osteoimmunology: Interplay between the immune system and bone metabolism. *Annu Rev Immunol* 24: 33–36.
7. Teitelbaum SL. 2006. Osteoclasts: Culprits in inflammatory osteolysis. *Arthritis Res Ther* 8: 201.
8. Boyce BF, Schwartz EM, Xing L. 2006. Osteoclast precursors: Cytokine stimulated immunomodulators of inflammatory bone disease. *Curr Opin Rheumatol* 18: 427–432.
9. Hofbauer LC, Gori F, Riggs BL, Lacey DL, Dunstan CR, Spelsberg TC, Khosla S. 1999. Stimulation of osteoprotegerin ligand and inhibition of osteoprotegerin production by glucocorticoids in human osteoblastic lineage cells: Potential paracrine mechanisms of glucocorticoid-induced osteoporosis. *Endocrinology* 140: 4382–4389.
10. Kong YY, Feige U, Sarosi I, Bolon B, Tafuri A, Morony S, Caparelli C, Li J, Elliott R, McCabe S, Wong T, Campagnuolo G, Moran E, Bogoch ER, Van G, Nguyen LT, Ohashi PS, Lacey DL, Fish E, Boyle WJ, Penninger JM. 1999. Activated T cells regulate bone loss and joint destruction in adjuvant arthritis through osteoprotegerin ligand. *Nature* 402: 304–309.
11. Moschen AR, Kaser A, Enrich B, Ludwiczek O, Gabriel M, Obrist P, Wolf AM, Tilg H. 2005. The RANKL/OPG system is activated in inflammatory bowel disease and relates to the state of bone loss. *Gut* 54: 479–487.
12. Compston J. 2003. Osteoporosis in inflammatory bowel disease. *Gut* 52: 63–64
13. van Hogezand RA, Banffer D, Zwinderman AH, McCloskey EV, Griffoen G, Hamdy NA. 2006. Ileum resection is the most predictive factor for osteoporosis in patients with Crohn's disease. *Osteoporos Int* 17: 535–542.
14. Loftus EV, Crowson CS, Sandborn WJ, Tremaine WJ, O'Fallon WM, Melton LJ 3rd. 2002. Long-term fracture risk in patients with Crohn's disease: A population-based study in Olmsted County, Minnesota. *Gastroenterology* 123: 468–475.
15. Bernstein CN, Blanchard JF, Leslie W, Wajda A, Yu BN. 2000. The incidence of fracture among patients with inflammatory bowel disease. A population-based cohort study. *Ann Intern Med* 133: 795–799.
16. van Staa TP, Cooper C, Brosse LS, Leufkens H, Javaid MK, Arden NK. 2003. Inflammatory bowel disease and the risk of fracture. *Gastroenterology* 125: 1591–1597.
17. Card T, West J, Hubbard R, Logan F. 2004. Hip fractures in patients with inflammatory bowel disease and their relationship to corticosteroid use: A population-based study. *Gut* 53: 251–255.
18. Klaus J, Armbrecht G, Steinkamp M, Bruckel J, Rieber A, Adler G, Reinshagen M, Felsenberg D, von Tirpitz C. 2002. High prevalence of osteoporotic vertebral fractures in patients with Crohn's disease. *Gut* 51: 654–658.
19. Vestergaard P, Mosekilde L. 2000. Fracture risk is increased in Crohn's disease, but not ulcerative colitis. *Gut* 46: 176–181.
20. Vestergaard P, Mosekilde L. 2002. Fracture risk in patients with celiac disease, Crohn's disease, and ulcerative colitis: A nationwide follow-up study of 16,416 patients in Denmark. *Am J Epidemiol* 156: 1–10.
21. Schulte C, Dignass AU, Mann K, Goebell H. 1999. Bone loss in patients with inflammatory bowel disease is less than expected: A follow-up study. *Scand J Gastroenterol* 34: 696–702.
22. Clements D, Motley RJ, Evans WD, Harries AD, Rhodes J, Coles RJ, Compston JE. 1992. Longitudinal study of cortical bone loss in patients with inflammatory bowel disease. *Scand J Gastroenterol* 27: 1055–1060.
23. Roux C, Abitbol V, Chaussade S, Kolta S, Guillemant S, Dougados M, Amor B, Couturier D. 1995. Bone loss in patients with inflammatory bowel disease: A prospective study. *Osteoporos Int* 5: 156–160.
24. Jahnsen J, Falch JA, Mowinckel, Aadland E. 2004. Bone mineral density in patients with inflammatory bowel disease: A population-based prospective two-year follow-up study. *Scand J Gastroenterol* 39: 145–153.
25. Turk N, Cukovic-Cavka S, Korsic M, Turk Z, Vucelic B. 2009. Proinflammatory cytokines and receptor activator of nuclear factor κB-ligand/osteoprotegerin associated with bone deterioration in patients with Crohn's disease. *Eur J Gastroenterol Hepatol* 21: 159–166.
26. Cremers SC, van Hogezand R, Banffer D, den Hartigh J, Vermeij P, Papapoulos SE, Hamdy NA. 2005. Absorption of the oral bisphosphonate alendronate in osteoporotic patients with Crohn's disease. *Osteoporos Int* 16: 1727–1730.
27. Franciosi LG, Page CP, Celli BR, Cazzola M, Walker MJ, Danhof M, Rabe KF, Della Pasqua OE. 2006. Markers of disease severity in chronic obstructive pulmonary disease. *Pulm Pharmacol Ther* 19: 189–199.
28. Gan WQ, Man SF, Senthilselvan A, Sin DD. 2004. Association between chronic obstructive pulmonary disease

and systemic inflammation: A systematic review and a meta-analysis. *Thorax* 59: 574–580.
29. Sevenoaks MJ, Stockley RA. 2006. Chronic obstructive pulmonary disease, inflammation and co-morbidity: A common inflammatory phenotype? *Respiratory Res* 7: 70–78.
30. Soriano JB, Visick GT, Muellerova H, Payvandi N, Hansell AL. 2005. Patterns of comorbidities in newly diagnosed COPD and asthma in primary care. *Chest* 128: 2099–2107.
31. Sin DD, Man JP, Man SF. 2003. The risk of osteoporosis in Caucasian men and women with obstructive airways disease. *Am J Med* 114: 10–14.
32. Bolton CE, Ionexcu AA, Shiels KM, Pettit RJ, Edwards PH, Stone MD, Nixon LS, Evans WD, Griffiths TL, Shale DJ. 2004. Associated loss of fat-free mass and bone mineral density in chronic obstructive pulmonary disease. *Am J Respir Crit Care Med* 170: 1286–1293.
33. McEvoy CE, Ensrud KE, Bender E, Genant HK, Yu W, Griffith JM, Niewoehner DE. 1998. Association between corticosteroid use and vertebral fractures in older men with chronic obstructive pulmonary disease. *Am J Resp Crit Care Med* 157: 704–709.
34. de Vries F, van Staa TP, Bracke MS, Cooper C, Leufkens HG, Lammers JW. 2005. Severity of obstructive airway disease and risk of osteoporotic fracture. *Eur Respir J* 25: 879–884.
35. Ionescu AA, Schoon E. 2003. Osteoporosis in chronic obstructive pulmonary disease. *Eur Respir J* 22(Suppl46): S64–S75.
36. Schlaich C, Minne HW, Bruckner T, Wagner G, Gebest HJ, Grunze M, Ziegler R, Leidig-Bruckner G. 1998. Reduced pulmonary function in patients with spinal osteoporotic fractures. *Osteoporos Int* 8: 261–267.
37. Leech JA, Dulberg C, Kellie S, Pattee L, Gay J. 1990. Relationship of lung function to severity of osteoporosis in women. *Am Rev Respir Dis* 141: 68–71.
38. Hofbauer LC, Brueck CC, Singh SK, Dobnig H. 2007. Osteoporosis in patients with diabetes mellitus. *J Bone Miner Res* 22: 1317–1328.
39. Inzerillo AM, Epstein S. 2004. Osteoporosis and diabetes mellitus. *Rev Endocr Metab Disord* 5: 261–268.
40. Nicodemus KK, Folsom AR, Iowa Women's Health Study. 2001. Type 1 and type 2 diabetes and incident hip fractures in postmenopausal women. *Diabetes Care* 24: 1192–1197.
41. de Liefde II, van der Klift M, de Laet CE, van Daele PL, Hofman A, Pols HA. 2005. Bone mineral density and fracture risk in type-2 diabetes mellitus: The Rotterdam Study. *Osteoporos Int* 16: 1713–1720.
42. Bouillon R, Bex M, Van Herck E, Laureys J, Dooms L, Lesaffre E, Ravussin E. 1995. Influence of age, sex, and insulin on osteoblast function: Osteoblast dysfunction in diabetes mellitus. *J Clin Endocrinol Metab* 80: 1194–1202.
43. Goodman WG, Hori MT. 1984. Diminished bone formation in experimental diabetes. Relationship to osteoid maturation and mineralization. *Diabetes* 33: 825–831.
44. Kemink SA, Hermus AR, Swinkels LM, Lutterman JA, Smals AG. 2000. Osteopenia in insulin-dependent diabetes mellitus; prevalence and aspects of pathophysiology. *J Endocrinol Invest* 23: 295–303.
45. Rosen CJ, Bouxsein ML. 2006. Mechanisms of disease: Is osteoporosis the obesity of bone? *Nature Clin Prac Rheum* 2: 35–43.
46. Gimble JM, Zvonic S, Floyd ZE, Kassem M, Nuttall ME. 2006. Playing with fat and bone. *J Cell Biochem* 98: 251–266.
47. Botolin S, Faugere MC, Malluche H, Orth M, Meyer R, McCabe LR. 2005. Increased bone adiposity and peroxisomal proliferator-activated receptor-gamma2 expression in type I diabetic mice. *Endocrinology* 146: 3622–3631.
48. Kahn SE, Zinman B, Lachin JM, Haffner SM, Herman WH, Holman RR, Kravitz BG, Yu D, Heise MA, Aftring RP, Viberti G; Diabetes Outcome Progression Trial (ADOPT) Study Group. 2008. Rosiglitazone-associated fractures in type 2 diabetes: An Analysis from A Diabetes Outcome Progression Trial (ADOPT). *Diabetes Care* 31: 845–851.
49. Bilik D, McEwen LN, Brown MB, Pomeroy NE, Kim C, Asao K, Crosson JC, Duru OK, Ferrara A, Hsiao VC, Karter AJ, Lee PG, Marrero DG, Selby JV, Subramanian U, Herman WH. 2010. Thiazolidinediones and fractures: Evidence from translating research into action for diabetes. *J Clin Endocrinol Metab* 95: 4560–4565.
50. Barete S, Assous N, de Gennes C, Grandpeix C, Feger F, Palmerini F, Dubreuil P, Arock M, Roux C, Launay JM, Fraitag S, Canioni D, Billemont B, Suarez F, Lanternier F, Lortholary O, Hermine O, Frances C. 2010. Systemic mastocytosis and bone involvement in a cohort of 75 patients. *Ann Rheum Dis* 69: 1838–1841.
51. Valent P, Akin C, Escribano L, Fodinger M, Hartmann K, Brockow K, Castells M, Sperr WR, Kluin-Nelemans HC, Hamdy NA, Lortholary O, Robyn J, van Doormaal J, Sotlar K, Hauswirth AW, Arock M, Hermine O, Hellman A, Triggiani M, Niedoszytko M, Schwartz LB, Orfao A, Horny HP, Metcalfe DD. 2007. Standards and standardization in mastocytosis: Consensus statements on diagnostics, treatment recommendations and response criteria. *Eur J Clin Invest* 37: 435–453.
52. Lidor C, Frisch B, Gazit D, Gepstein R, Hallel T, Mekori YA. 1990. Osteoporosis as the sole presentation of bone marrow mastocytosis. *J Bone Miner Res* 5: 871–876.
53. De Gennes C, Kuntz D, de Vernejoul MC. 1992. Bone mastocytosis: A report of nine cases with a bone histomorphometric study. *Clin Orthop Rel Res* 279: 281–291.
54. Brumsen C, Papapoulos SE, Lentjes EG, Kluin PM, Hamdy NA. 2002. A potential role for the mast cell in the pathogenesis of idiopathic osteoporosis in men. *Bone* 31: 556–561.
55. Oranje AP, Mulder PG, Heide R, Tank B, Riezebos P, van Toorenenbergen AW. 2002. Urinary N-methylhistamine as an indicator of bone marrow involvement in mastocytosis. *Clin Exp Dermatol* 27: 502–506.

56. Tannenbaum C, Clark J, Schwartzman K, Wallenstein S, Lapinski R, Meier D, Luckey M. 2002. Yield of laboratory testing to identify secondary contributors to osteoporosis in otherwise healthy women. *J Clin Endocrinol Metab* 87: 4431–4437.
57. [No authors listed]. 2010. Management of osteoporosis in postmenopausal women: 2010 position statement of the North American Menopause Society. *Menopause* 17: 25–54.
58. Khosla S, Lufkin EG, Hodgson SF, Fitzpatrick LA, Melton LJ 3rd. 1994. Epidemiology and clinical features of osteoporosis in young individuals. *Bone* 15: 551–555.
59. Peris P, Guanabens N, Martinez de Osaba MJ, Monegal A, Alvarez L, Pons F, Ros I, Cerda D, Munoz-Gomez J. 2002. Clinical characteristics and etiologic factors of premenopausal osteoporosis in a group of Spanish women. *Semin Arthritis Rheum* 32: 64–70.
60. Ebeling PR. 1998. Osteoporosis in men. New insights into aetiology, pathogenesis, prevention and management. *Drugs Aging* 13: 421–434.

61

Transplantation Osteoporosis

Peter R. Ebeling

Abstract 495
Introduction 495
Preexisting Bone Disease 495
Skeletal Effects of Immunosuppressive Drugs 496

Management of Transplantation Osteoporosis 497
Summary and Conclusion 503
Acknowledgments 504
References 504

ABSTRACT

Transplantation is an established therapy for end-stage diseases of the kidney, endocrine pancreas, heart, liver and lung, intestines, and for many hematological disorders. Current immunosuppressive regimens with glucocorticoids and calcineurin inhibitors produce excellent patient and graft survival rates. This has resulted in both increases in transplant numbers and an increased recognition of previously neglected long-term complications of transplantation such as fractures and osteoporosis. Both pre-transplantation bone disease and immunosuppressive therapy result in high bone turnover, rapid bone loss, and increased fracture rates, particularly early after transplantation. The bone health of candidates for organ transplantation should be assessed with bone densitometry of the hip and spine. Spinal X-rays should be performed to diagnose prevalent fractures. Secondary causes of osteoporosis should be identified and treated. Vitamin D deficiency should be corrected to achieve a serum 25(OH)D concentration 30 ng/mL or higher. Patients with kidney failure should be evaluated and treated for chronic kidney disease–mineral and bone disorder (CKD-MBD), including renal osteodystrophy. Secondary hyperparathyroidism, in particular, should be treated.

Treatment is indicated in the immediate post-transplantation period irrespective of bone mineral density, since further rapid bone loss will occur in the first several months after transplantation. The duration of therapy will depend on the type of transplant. Long-term organ transplant recipients should also have bone mass measurement and treatment of osteoporosis.

Oral and intravenous bisphosphonates are the most promising approach for the management of transplantation osteoporosis and reduction of the number of patients with vertebral fractures following transplantation. Active vitamin D metabolites may have additional benefits in reducing hyperparathyroidism, particularly after kidney transplantation.

INTRODUCTION

Transplantation is an established therapy for end-stage diseases of the kidney, endocrine pancreas, heart, liver and lung, intestines, and for many hematological disorders. Improved survival rates, due to the addition of calcineurin inhibitors, cyclosporine A, and tacrolimus to immunosuppressive treatment, have been accompanied by a greater awareness of the long-term complications of transplantation such as fractures and osteoporosis [1, 2].

PREEXISTING BONE DISEASE

Chronic kidney disease

Renal osteodystrophy or chronic kidney disease–bone and mineral disorder (CKD-MBD) is the most complex

---

Primer on the Metabolic Bone Diseases and Disorders of Mineral Metabolism, Eighth Edition. Edited by Clifford J. Rosen.
© 2013 American Society for Bone and Mineral Research. Published 2013 by John Wiley & Sons, Inc.

form of pre-transplant bone disease. One or more types of bone disease may be present, including osteitis fibrosa cystica as a result of secondary hyperparathyroidism (SHPT), low turnover bone disease (osteomalacia, adynamic bone disease, or aluminum bone disease), osteoporosis, mixed bone disease, and β_2-microglobulin amyloidosis. In addition, hypogonadism, both in men and women, metabolic acidosis, and certain medications (loop diuretics, heparin, warfarin, glucocorticoids, or immunosuppressive agents) also adversely affect bone health. Patients with chronic kidney disease (CKD) who have low bone mineral density (BMD) and bone turnover markers in the upper half of the normal premenopausal range are at the highest risk of fracture [3].

Adynamic bone disease needs exclusion prior to treatment with bisphosphonates, which reduce bone turnover further. It is commonly associated with osteoporosis and occurs early in CKD. On bone histomorphometry, there is a scarcity of bone cells, reduced osteoid thickness, and a low bone formation rate [4]. The factors reducing bone turnover are a low vitamin D system, high phosphate, and FGF-23, which override the stimulatory effect of parathyroid hormone (PTH) in early CKD. The use of cinacalcet, calcium, and calcitriol may also reduce bone turnover. The perturbation in bone turnover in CKD therefore needs careful evaluation before treatment is initiated. Bone histomorphometry is the best method. The combination of a low bone formation marker and a slightly increased or normal PTH level is less specific.

In hemodialysis patients, the prevalence of both low BMD and fractures is increased. All skeletal sites are affected, including the spine, hip, and distal radius. Vertebral fracture prevalence is as high as 21%, and the relative risk of hip fracture is increased 2-fold to 14-fold. Fracture risk is increased with older age, female gender, Caucasian race [5], duration of hemodialysis [6], diabetic nephropathy, peripheral vascular disease [7], low spine BMD, and low bone turnover states.

Congestive heart failure

Osteoporotic BMD affects up to 40% of patients with congestive heart failure (CHF), with a 2.45-fold increase in fracture risk in one study [8]. In another study of patients awaiting heart transplantation, lumbar spine (LS) osteopenia was found in 43%, and osteoporosis in 7% [9]. Mild renal insufficiency, vitamin D deficiency, SHPT, and increased bone resorption markers, and use of loop diuretics may contribute.

End-stage liver disease

Osteoporosis and fractures commonly accompany chronic liver disease and low BMD is seen in the majority of patients undergoing liver transplantation. Osteoporosis at the spine or hip has been reported in 11–52% of patients awaiting liver transplantation [1, 10]. Low body mass index (BMI) prior to lung transplantation (LT), cholestatic liver disease, and older age are important risk factors [11, 12] for osteoporosis.

Chronic respiratory failure

Osteoporosis may be most common in patients awaiting lung transplantation. Hypoxia, hypercapnia, smoking, and glucocorticoids all contribute. Fragility fractures are extremely common in cystic fibrosis (CF) because additional risk factors (pancreatic insufficiency, vitamin D deficiency, calcium malabsorption, hypogonadism, genetic factors, and inactivity) exist. Up to 61% of patients with end-stage pulmonary disease have osteoporosis. Chronic glucocorticoid use, low BMI, and decreased pulmonary function are all associated with low BMD [13].

Candidates for bone marrow transplantation

Bone loss in bone marrow transplantation (BMT) recipients is related both to the underlying diseases and to chemotherapeutic drugs. These include glucocorticoid (GC)-induced decreases in bone formation and serum $1,25\text{-}(OH)_2D_3$, as well as hypogonadism secondary to the effects of high-dose chemotherapy, total body irradiation (TBI), and GCs. Women are particularly sensitive to the adverse effects of TBI and chemotherapy on gonadal function. Ovarian insufficiency occurs in the majority [14, 15], although some young, premenarchal women may recover ovarian function. Testosterone levels decline acutely after BMT related to a reduction in luteinizing hormone, then return to normal in most men [16–18]. There may be long-term impairment of spermatogenesis with elevated follicle stimulating hormone (FSH) occurring in 47% of men [14, 15]. In patients studied after chemotherapy but before BMT, osteopenia was present in 24% and osteoporosis in 4% [18].

Candidates for intestinal transplantation

Osteoporosis occurs in 36% of candidates for intestinal transplantation, with age and duration of parenteral nutrition being significant risk factors. Bone density at the spine and hip is reduced by about 1.5 standard deviations below normal age- and sex-matched mean levels [19].

SKELETAL EFFECTS OF IMMUNOSUPPRESSIVE DRUGS

Glucocorticoids

Exposure to glucocorticoids (GCs) varies with the organ transplanted and the number of rejection episodes. High doses are commonly prescribed immediately after transplantation and are weaned rapidly. Doses are increased at the time of rejection episodes. The highest GC-associated rates of bone loss are in the first 3 to 12 months post transplant. Trabecular sites are predominantly affected. The use of calcineurin inhibitors and more recent immunosuppressive regimens both have limited GC use. As a result, more recent studies show lower rates of post-transplant bone loss.

However, even small doses of glucocorticoids are associated with marked increases in fracture risk in epidemiological studies [20]. Glucocorticoids reduce bone formation by decreasing osteoblast replication and differentiation, and increasing apoptosis. Osteoblast genes including type I collagen, osteocalin, insulin-like growth factors, bone matrix proteins, transforming growth factor β (TGFβ) are downregulated, whereas receptor activator for nuclear factor-κB ligand (RANKL), is upregulated. Direct and indirect effects of GCs to increase bone resorption also contribute to the rapid increase in fracture risk post transplant. The immediate post-transplant period is characterized by high bone remodeling and increased bone resorption. Hyperparathyroidism may result from GC-induced reductions of intestinal and renal calcium absorption.

More recently, data from the United States Renal Data System have shown the use of early steroid withdrawal after kidney transplantation has been associated with a 31% fracture risk reduction [21]. Fractures associated with hospitalization were also significantly lower with regimens that withdraw corticosteroids. As this study likely underestimated overall fracture incidence, prospective studies are needed to determine the differences in overall fracture risk in patients managed with and without corticosteroids after kidney transplantation.

Calcineurin inhibitors

Cyclosporine (CsA) has independent adverse effects to increase bone turnover [22]. Although CsA treatment could result in high bone turnover after transplantation, it is reassuring that kidney transplant patients receiving CsA without GCs [23, 24] do not lose bone, and fractures are also reduced [21].

Tacrolimus (FK506), another calcineurin inhibitor (CI), also causes trabecular bone loss in the rat [22]. Both cardiac [25] and liver [26] transplant recipients sustained rapid bone loss with tacrolimus. However, tacrolimus may cause less bone loss in humans than CsA [27, 28] and may also protect the skeleton by reducing GC use.

Other immunosuppressive agents

Limited information is available regarding the effects of other immunosuppressive drugs on BMD and bone metabolism. However, azathioprine, sirolimus (rapamycin), mycophenelate mofetil, and daclizumab may also protect the skeleton by reducing GC use. *In vitro* studies suggest rapamycim inhibits osteoblast proliferation and differentiation [29], but more clinical data are required.

MANAGEMENT OF TRANSPLANTATION OSTEOPOROSIS

Diagnostic strategies

Before organ transplantation

All candidates going onto the waiting list for organ transplantation should have bone densitometry by dual energy X-ray absorptiometry (DXA) of the hip and spine. Spinal X-rays should be performed to diagnose prevalent fractures. Any secondary causes of osteoporosis should be identified and treated. Common secondary causes include hyperparathyroidism, hypogonadism, smoking, use of loop diuretics, low dietary calcium intake, and vitamin D deficiency (less than 20 ng/mL).

Vitamin D deficiency may be related to disease-related factors, decreased sunlight exposure and low dietary intakes [30]. Vitamin D deficiency should be corrected and all patients should receive adequate calcium and vitamin D (1,000–1,300 mg of calcium and at least 800 IU of vitamin D per day). Replacement doses of vitamin D may need to be higher (2,000 IU of vitamin D per day), but should be selected to achieve a 25(OH)D concentration of 20–30 ng/mL or greater. Patients with kidney failure should be evaluated and treated for renal osteodystrophy and secondary hyperparathyroidism, in particular.

Individuals with osteoporosis awaiting solid organ transplants should be evaluated and treated similarly to others with this condition, with the exception of patients awaiting kidney transplantation, in whom CKD–MBD is more complex. In this group, it is important to diagnose and treat secondary hyperparathyroidism and to exclude adynamic bone disease.

After organ transplantation

Risk factors for post-transplant bone loss and fractures are shown in Table 61.1. Bone loss is most rapid immediately after transplantation. Fractures often occur in the first year after transplantation and may affect patients with either low or normal pre-transplant BMD. Therefore, the majority of patients may benefit from treatment instituted immediately after transplantation, with exception of patients with CKD–MBD and adynamic bone disease. Patients who present after being transplanted months or years before should also be assessed for treatment.

Vitamin D deficiency is common post transplantation and in long-term graft recipients. Vitamin D status is partly determined by demographic and lifestyle factors and deficiency is associated with poorer general health, lower serum albumin levels, and even decreased survival in these groups [30].

Table 61.1. Risk Factors for Post-Transplant Bone Loss and Fractures

Contributing Factors	Mechanisms
Aging	**Low pre-transplant BMD**
Low body mass index	
Hypogonadism	
Calcium and vitamin D deficiency	
Tobacco	
Alcohol abuse	
Cholestasis (liver disease)	
Organ failure (heart, lung, liver, kidney)	
Pancreatic insufficiency (cystic fibrosis)	
Physical inactivity	
High dose prednisone	**Decreased bone formation**
	Direct effect
	Decreased gonadal function
	Reduced intestinal and renal calcium transport
Calcineurin inhibitors	**Increased bone resorption**
Cyclosporine or FK506	Decreased renal function and 1,25(OH)$_2$D
	Increased PTH secretion
	Possible direct effect
Calcineurin inhibitor	**Decreased bone formation**
Sirolimus	Possible direct effect

Most therapeutic trials have focused on the use of active vitamin D metabolites and antiresorptive drugs, particularly oral and intravenous bisphosphonates. Hormone therapy with estrogen ± progestin helps protect the skeleton in women receiving liver, lung, and bone marrow transplantation. Because amenorrhea is a common sequela of BMT in premenopausal women, they should receive hormone replacement therapy (HRT). However, it does not prevent bone loss after BMT. Hypogonadism is common in male cardiac and bone marrow transplant recipients, due to chronic illness and hypothalamic-pituitary-adrenal suppression by GCs and CsA. Testosterone levels fall immediately after transplantation and normalize 6 to 12 months later. However, testosterone treatment alone does not prevent bone loss after cardiac transplantation or BMT in men.

Recent studies examined prevention of bone loss after transplantation (Table 61.2).

Kidney transplantation

Renal osteodystrophy improves after transplantation; however, hyperparathyroidism (HPT) may persist. Bone resorption remains elevated in a substantial proportion of kidney transplant recipients, and there is GC-induced osteoblast dysfunction [31, 32]. Cross-sectional studies of patients evaluated several years after kidney transplantation have reported osteoporosis in 17–49% at the spine, 11–56% at the femoral neck, and 22–52% at the radius [1]. There is a correlation between cumulative GC dose and BMD. Rates of bone loss are greatest in the first 6–18 months after transplantation, and range from 4% to 9% at the spine and 5% to 8% at the hip. Bone loss has not been consistently related to gender, patient age, cumulative GC dose, rejection episodes, activity level, or PTH levels. Although increasing time since transplantation is considered to be a risk factor for low BMD, studies examining BMD after the first year or two do not consistently show ongoing bone loss; however, BMD remains low up to 20 years after transplantation. Secondary HPT and low 1,25(OH)$_2$D levels also often persist [3, 33].

Fractures affect appendicular sites (hips, long bones, ankles, feet) more commonly than axial sites (spine and ribs) [33]. Women and patients transplanted for diabetic nephropathy are at particularly increased risk of fractures. The majority of fractures occur within the first 3 years; however, fractures continue to increase the longer the post-transplant period [34].

Prevention and treatment

Calcium and vitamin D supplementation alone does not prevent bone loss in renal transplant patients [35]. Bisphosphonates reduce bone loss after kidney transplantation (KT) [36]. In a small study, the combination of alendronate (10 mg/day), calcium carbonate (2 g/day), and calcitriol (0.25 μg/day) resulted in a 6.3% increase in spinal BMD in the first 6 months after KT, compared with a 5.8% decrease with calcium and calcitriol alone [37]. Alendronate, calcitriol, and calcium treatment was also superior to calcitriol and calcium treatment alone, beginning 5 years post KT [38]. Intermittent calcitriol (0.5 μg/48 hours) during the first 3 months after renal transplantation, preserved total hip BMD more than calcium (500 mg/day) supplementation alone over 1 year [39].

Renal safety issues and dosing schedules are different for intravenous bisphosphonates. Zoledronic acid (ZA) may cause acute renal failure, with the induction of acute tubular necrosis related to the rate of its infusion rather than the dose. The risk of renal failure is also increased by preexisting renal impairment, so the infusion rate should be reduced to half the recommended rate in patients with a glomerular filtration rate (GFR) less than 30 mls/min or baseline serum creatinine concentration greater than 2 mg/dL [40]. The former measurement is the more accurate.

Intravenous ibandronate was effective at preventing spinal and hip bone loss post KT [41]. There were also fewer spinal deformities in the ibandronate group after 12 months. A small, randomized study compared the long-term effects of two infusions of 4 mg of ZA or placebo at 2 weeks and 3 months after KT. Although early bone loss was prevented by ZA, both treatment groups had significant and similar later increases in femoral neck (FN) BMD [42].

Table 61.2. Randomized Controlled Trials Using Vitamin D Analogues or Bisphosphonates for Prevention of Bone Loss After Heart, Lung, Liver, and Bone Marrow Transplantation

Transplant type	First author, yr	n	Duration	Treatment regimen	Control regimen	Findings/Summary
Heart and Lung	Sambrook (2000) [Ref. 51]	65	24 months	**Calcitriol** 0.5–0.75 μg for 12 months or 24 months Calcium 600 mg/day	Placebo Calcium 600 mg/day	BMD: FN (but not LS) bone loss was attenuated in the calcitriol groups at 12 months. LS bone loss was similar among all 3 groups. Fracture: Not powered.
Lung (CF)	Aris (2000) [Ref. 47]	37	24 months	**Pamidronate** 30 mg IV q 3 months Calcium 1000 mg/day Vitamin D 800 IU/day	Calcium 1,000 mg/day Vitamin D 800 IU/day	BMD: LS and TH BMD increased significantly more in the pamidronate group vs controls. Fracture: No difference.
Heart	Shane (2004) [Ref. 58]	149[a]	12 months	**Alendronate** 10 mg/day or **Calcitriol** 0.5 μg/day Calcium 945 mg/day Vitamin D 1,000 IU/day	Non-randomized reference group	BMD: Similar small losses at LS and TH in both groups. Significantly less bone loss at LS and TH than reference group. Fracture: No difference.
Heart	Gil-Fraguas (2005) [Ref. 56]	87	12 months	**Alendronate** 10 mg/day	Calcitonin 200 IU/day	BMD: Less bone loss from FN in the alendronate group. Fracture: Fewer vertebral fractures than in calcitonin group (6 vs 15).
Heart	Fahrleitner-Pammer (2009) [Ref. 57]	35	12 months	**Ibandronate** 2 mg IV q 3 months calcium 1000 mg/day Vitamin D 400 IU/day	Placebo calcium 1,000 mg/day Vitamin D 400 IU/day	BMD: Bone loss from LS and FN prevented in ibandronate group. Fracture: Fewer vertebral fractures than in control group (2 vs 17).
Liver and multivisceral	Hommann (2002) [Ref. 53]	36	12 months	**Ibandronate** 2 mg IV q 3 months Calcium 1,000 mg/day Vitamin D 1,000 IU/day	Calcium 1,000 mg/day Vitamin D 1,000 IU/day	BMD: LS, FN and forearm BMD decreased initially in both groups. Reversal of bone loss with ibandronate seen after 12 months.
Liver	Ninkovic (2000) [Ref. 62]	99	12 months	**Pamidronate** 60 mg IV given once prior to transplantation	No treatment	BMD: Significant, comparable bone loss at FN in pamidronate and control groups. Fracture: No difference.
Liver	Crawford (2006) [Ref. 64]	62	12 months	**Zoledronic acid** 4 mg IV administered within 7 days of transplantation and at 1, 3, 6, and 9 months after transplant Calcium 600 mg/day Vitamin D 1,000 IU/day	Placebo Calcium 600 mg/day Vitamin D 1,000 IU/day	BMD: At 3 months, difference in bone loss from baseline was decreased in ZA group vs placebo. At 12 months, the differences in % bone loss was less. Fracture: Not powered.
Liver	Bodingbauer (2007) [Ref. 55]	69	12 months	**Zoledronic acid** 4 mg IV 1–6, 9, and 12 months Calcium 600 mg/day Vitamin D 1,000 IU/day	Calcium 600 mg/day Vitamin D 1,000 IU/day	BMD: Less bone loss from LS (but not FN) in Zoledronic acid group. Fracture: Fewer vertebral facture than in control group (4 vs 11).

(Continued)

Table 61.2. (Continued) Randomized Controlled Trials Using Vitamin D Analogues or Bisphosphonates for Prevention of Bone Loss After Heart, Lung, Liver, and Bone Marrow Transplantation

Transplant type	First author, yr	n	Duration	Treatment regimen	Control regimen	Findings/Summary
Liver	Monegal (2009) [Ref. 66]	79	12 months	**Pamidronate** 90 mg IV 0 and 3 months Calcium 1,000 mg/day Vitamin D 16,000 IU q 15 days	Calcium 1,000 mg/day Vitamin D 16,000 IU q 15 days	BMD: Increase at LS in pamidronate group. Decrease at FN in both groups. Fracture: More fractures in pamidronate group (15 vs 3).
Liver	Kaemmerer (2010) [Ref. 67]	74	12 months	**Ibandronate** 2 mg IV q 3 months Calcium 1,000 mg/day Vitamin D 800–1,000 IU/day	Calcium 1,000 mg/day Vitamin D 800–1,000 IU/day	BMD: Increase at LS and less bone loss from FN in ibandronate group. Fracture: fewer fractures in ibandronate group (2 vs 8).
BMT	Tauchmanova (2003) [Ref. 78]	34	12 months	**Risedronate** 5 mg/day Calcium 1 g/day Vitamin D 800 IU/day	Calcium 1 g/day Vitamin D 800 IU/day	BMD: LS BMD significantly increased in risedronate group at 6 and 12 months and decreased in the control group at 6 months. FN BMD decreased significantly in control group at 6 months only.
BMT	Tauchmanova (2005) [Ref. 79]	32	12 months	**Zoledronic acid** 4 mg IV administered at 1, 2, and 3 months Calcium 500 mg/day Vitamin D 400 IU/day	Calcium 500 mg/day Vitamin D 400 IU/day	BMD: LS and FN BMD significantly increased in ZA group and did not change in the control group at 12 months.
BMT	Kananen (2005) [Ref. 17]	99	12 months	**Pamidronate** 60 mg IV administered prior to transplant and 1, 2, 3, 6, and 9 months after transplant Calcium 1,000 mg/day Vitamin D 800 IU/day estrogen – women testosterone – men	Calcium 1,000 mg/day Vitamin D 800 IU/day estrogen – women testosterone – men	BMD: At 12 months, difference in bone loss from baseline at LS and TH was decreased in pamidronate group vs no infusion. No difference in bone loss from baseline at the FN. Fracture: Not powered.
BMT	Grigg (2006) [Ref. 80]	116	24 months	**Pamidronate** 90 mg IV administered prior to transplant and every month after transplant for 12 months Calcium 1,000 mg/day Calcitriol 0.25 μg/day for 24 months	Calcium 1000 mg/day Calcitriol 0.25 μg/day	BMD: At 12 months, difference in bone loss from baseline was decreased at LS, FN and TH in pamidronate group vs no infusion. At 24 months, the difference in bone loss from baseline was only significant at the TH (3.9%). Fracture: Not powered.

[a] Number randomized to alendronate or calcitriol, 27 prospectively recruited non-randomized patients served as a reference group.

A large, recent systematic review of 24 trials with 1,299 patients showed any treatment for bone disease reduced the risk of fracture by 49% (95%; CI 0.27–0.99) compared with placebo [43]. Bisphosphonates and active vitamin D analogues had beneficial effects on the BMD at the spine and FN. Bisphosphonates were better at preventing bone loss compared with vitamin D analogues. An unexpected finding was a reduction in the risk of graft rejection associated with bisphosphonate therapy. A trial using fractures as a primary end point is now required to compare treatment with oral or parenteral bisphosphonates with calcitriol after KT.

Kidney-pancreas transplantation

Severe osteoporosis complicates kidney-pancreas transplants in recipients with type 1 diabetes, occurring in 23% and 58% at the LS and FN, respectively. Vertebral or nonvertebral fractures were documented in 45% [1]. Other retrospective studies have documented a fracture prevalence of 26–49% up to 8.3 years after transplantation [44].

A prospective study addressed osteoporosis and secondary hyperparathyroidism in simultaneous pancreas-kidney transplantation (SPK) recipients before transplant and 4 years after. Prior to transplantation, 68% had hyperparathyroidism. After 6 months, bone loss of 6.0% and 6.9% occurred at both LS and FN sites, respectively, and fractures were related to low pre-transplant FN BMD [45].

Lung transplantation

The prevalence of osteoporosis is as high as 73% in lung transplantation (LT) recipients. During the first year after lung transplantation, rates of bone loss at the LS and FN range from 2% to 5% [1]. Fracture rates are also high during the first year, ranging from 18% to 37%. Bone turnover is also increased [46]. Repeated doses of intravenous pamidronate prevented LS and FN bone loss in LT recipients [47, 48].

Cardiac transplantation

The most rapid rate of bone loss occurs in the first year post transplant. Spinal BMD declines by 6–10% during the first 6 months, while FN BMD falls by 6–11% in the first year and stabilizes thereafter in most cases. BMD declines at the largely cortical proximal radius site over the second and third years, perhaps reflecting post-transplant SHPT. Vitamin D deficiency and testosterone deficiency (in men) are associated with more severe bone loss. Testosterone levels fall immediately after cardiac transplantation (CT) and normalize after 6–12 months. Some studies have found correlations between GC dose and bone loss. Vertebral fracture incidence ranges from 33% to 36% during the first 1 to 3 years after CT [49, 50].

Prevention and treatment
Vitamin D and calcitriol. Calcium and vitamin D alone do not prevent bone loss after CT [1]. Early studies showed calcitriol was effective at reducing bone loss, particularly at the FN, after CT [51]. Another study compared rates of bone loss in patients randomized to receive calcitriol (0.5 µg/day) or two cycles of etidronate during the first 6 months after CT or LT [52]. Significant and similar bone loss (3–8%) occurred at the spine and FN in both treatment groups, but was less than in historical controls [51, 52].

Other studies observed that CT recipients randomized to either alphacalcidol or cyclic etidronate sustained considerable bone loss at the spine and FN during the first year after transplantation [1], while another study of calcitriol [53] found no protective benefit. Thus, data regarding calcitriol and prevention of post-CT bone loss are inconsistent. Monitoring of serum and urine calcium levels is also required.

Intranasal calcitonin. One small study showed spinal BMD was higher 1–3 years, but not 7 years after CT in those treated with intranasal salmon calcitonin [54].

Testosterone. Because low post-transplant testosterone concentrations are often transient, only hypogonadal men should receive testosterone therapy.

Bisphosphonates. An open-label study of a single intravenous dose of pamidronate (60 mg) followed by four cycles of etidronate (400 mg every 3 months) and daily low-dose calcitriol (0.25 µg), prevented spinal and FN bone loss and reduced fracture rates in CT recipients compared with historical controls [55]. Compared with calcitonin (200 IU/day), alendronate (10 mg/day) treatment reduced hip bone loss and resulted in fewer vertebral fractures [56]. In a small study of 35 men post CT, intravenous ibandronate (2 mg every 3 months) prevented spine and hip bone loss and resulted in fewer morphometric vertebral fractures [57].

In the largest study, where 149 patients were randomized immediately after CT to receive either alendronate (10 mg/day) or calcitriol (0.25 µg twice daily) for 1 year, bone loss at the spine and hip was prevented by both regimens compared with a prospectively recruited, non-randomized reference group who received only calcium and vitamin D [58]. After 1 year of treatment withdrawal, BMD did not change in either the former alendronate or calcitriol group, but bone resorption increased in the calcitriol group [59]. This suggests that antiresorptive therapy may be discontinued 1 year post transplant in CT recipients without inducing rapid bone loss. However, these patients still require observation to ensure that BMD remains stable in the long-term.

Exercise. Resistance exercise significantly improved lumbar spine BMD after lung [60] and heart [61] transplantation when used alone and in combination with alendronate. However, these small studies utilized highly variable lateral BMD measurements.

Liver transplantation

Bone loss and fracture rates after liver transplantation (LIT) are highest in the first 6–12 months. Spine BMD declines by 2–24% during the first year in earlier studies.

In more recent studies, rates of bone loss have been lower or absent. Fracture rates range from 24–65%, and the ribs and vertebrae are the most common sites. Women with primary biliary cirrhosis and the most severe preexisting bone disease appear to be at greatest risk. Older age and pre-LIT spinal and FN BMD predicted post-LIT fractures in one recent prospective study, while the presence of pre-LIT vertebral fractures also predicted post-LIT vertebral fractures [1, 62]. Post LIT, bone turnover is increased compared with low bone turnover pre-LIT.

Prevention and treatment
Both oral and intravenous bisphosphonates are effective in reducing post-LIT bone loss, although one early study of intravenous pamidronate was ineffective due to insufficient dosing [62]. A randomized trial of intravenous ibandronate in liver [63] transplant recipients found a significant protective effect on BMD at 1 year. In a randomized, double-blind trial, 62 LIT recipients received treatment with either infusions of 4 mg ZA or saline within 7 days of transplantation and again at 1, 3, 6, and 9 months post LIT [64]. All patients also received calcium and vitamin D. ZA significantly prevented bone loss from the LS, FN, and total hip by 3.8–4.7%, with differences being greatest 3 months post LIT. At 12 months post LIT, differences only remained significant at the total hip. Similar findings were identified in another study using 4 mg intravenous ZA at 1–6, 9, and 12 months post LIT. Bone loss from the spine (but not hip) was prevented and fewer vertebral fractures occurred in the ZA group [65]. Vitamin D deficiency should be corrected before giving bisphosphonates post LIT to prevent hypocalcemia.

One study using two intravenous doses of pamidronate (90 mg) at baseline and 3 months post LIT showed an increase in spinal BMD in the pamidronate group, but more fractures in the pamidronate group [66]. Intravenous ibandronate given every 3 months post LIT resulted in increases in spinal BMD and less bone loss from the FN, and fewer fractures in the ibandronate group [67].

Two studies examined effects of alendronate on bone after LIT. An uncontrolled, prospective study of 136 LIT patients showed alendronate prevented bone loss in patients with osteopenia and led to an increase in BMD at the spine and FN in patients with osteoporosis over 4 years [9]. Another study of 59 LIT patients used historical controls to examine effects of alendronate combined with calcium and calcitriol 0.5 μg daily [68]. Increases in spinal, FN, and total hip BMD at 12 months were higher than in historical controls.

Small bowel transplantation

Small bowel transplantation (SBT) is being increasingly used for severe inflammatory bowel disease. It may also include concomitant liver, pancreas, and stomach transplantation. In a cross-sectional study of 81 patients who had SBT 2.2 years previously, BMD at the spine, total hip, and FN was reduced by about 0.8 SD compared with age- and sex-matched controls with similar small bowel diseases. Long-term SBT recipients are at risk of both osteoporosis (44%) and fractures (20%) [19]. In a small longitudinal study of nine patients, significant bone loss occurred at both the spine (2.6%) and the total hip and FN (by about 15%) 1.3 years after SBT [69]. A larger longitudinal study (n = 24) documented acceleration (p = 0.025) of bone loss after SBT with a decline of 13.4% (FN), 12.7% (total hip), and 2.1% (spine) over 2.5 years. Alendronate reduced (p <0.05), but did not prevent bone loss [19].

Bone marrow transplantation

Bone marrow or stem cell transplantation (BMT) is the treatment of choice for patients with many hematological malignancies, the majority of whom will survive for many years. However, up to 29% and 52% of survivors have osteopenia at the spine or FN, respectively [1]. Osteoporosis is more common at proximal femur sites. The pathogenesis of post-BMT osteoporosis is complex, relating both to effects of treatment and effects on the bone marrow stromal cell (MSC) compartment [70, 71]. Bone resorption increases while bone formation decreases [1, 16], resulting in early, rapid bone loss. In addition to osteoporosis, osteomalacia, and avascular necrosis may occur.

Dramatic bone loss from the proximal femur occurs within the first 12 months of allogeneic BMT [1, 72, 73]. Spinal bone loss is less. Most studies suggest that little additional bone loss occurs after this time. Studies of long-term survivors of BMT have shown that losses from the proximal femur are not regained [74]. After autologous BMT, bone loss from the proximal femur is less (about 4%), occurs as early as 3 months and persists at 2 years, while spine BMD returns to baseline [75].

Bone loss after BMT is related to both cumulative GC exposure and duration of CsA exposure [72]. There may also be a direct effect of graft versus host disease (GVHD) itself on bone cells. Abnormal cellular or cytokine-mediated bone marrow function may affect bone turnover and BMD after BMT [1]. Both myeloablative treatment and BMT stimulate the early release of cytokines. BMT also has adverse effects on bone marrow osteoprogenitors. Osteocyte viability is decreased after BMT and bone marrow stromal cells are damaged by high-dose chemotherapy, TBI, GCs, and CsA, thereby reducing osteoblastic differentiation from osteoprogenitor cells [76]. In this regard, colony forming units-fibroblasts (CFU-f) are reduced for up to 12 years after BMT [1].

Avascular necrosis develops in 10% to 20% of allo-BMT survivors, a median of 12 months after BMT [71, 77]. Glucocorticoid treatment of chronic GVHD-inducing osteoblast apoptosis is the most important risk factor. Avascular necrosis appears to be related to decreased numbers of bone marrow CFU-f colonies *in vitro*, but not to BMD [77]. Avascular necrosis may thus be facilitated by a deficit in bone marrow stromal stem cell regeneration with low osteoblast numbers post BMT [76].

Prevention and treatment

Risedronate or intravenous ZA given 12 months after BMT prevents spinal and proximal femoral bone loss [78, 79]. ZA effects may be related to improved osteoblast recovery and increased osteoblast numbers post BMT as increases in *ex vivo* growth of bone marrow CFU-f have been shown.

Two randomized trials recently assessed the effectiveness of intravenous pamidronate in preventing bone loss after BMT. The first studied 99 allogeneic BMT recipients, who were randomized to receive calcium and vitamin D daily, hormone therapy with estrogen in females or testosterone in men, or the same treatments plus intravenous 60 mg pamidronate infusions before and 1, 2, 3, 6, and 9 months after BMT [16]. In the pamidronate group, LS BMD remained stable but decreased significantly in the control group. Total hip BMD and FN BMD decreased by 5.1% and 4.2%, respectively, in the pamidronate group, and by 7.8% and 6.2%, respectively, in the control group at 12 months. Thus, pamidronate reduced bone loss more than in those treated with calcium, vitamin D, and sex steroid replacement alone.

A larger randomized, multicenter open-label 12-month prospective study compared intravenous pamidronate (90 mg/month) beginning prior to conditioning versus no pamidronate [80]. All 116 patients also received calcitriol (0.25 μg/day) and calcium, which were continued for a further year. Pamidronate significantly reduced bone loss at the spine, FN, and total hip at 12 months. However, BMD of the FN and total hip was still 2.8% and 3.5% lower than baseline, respectively, following pamidronate. Only the BMD benefit at the total hip remained significant between the two groups at 24 months. This study also showed the benefits of pamidronate therapy were restricted to patients receiving an average daily prednisolone dose greater than 10 mg and cyclosporin therapy for more than 5 months within the first 6 months of alloSCT (stem cell transplantation). Importantly, most BMD benefits were lost 12 months after stopping pamidronate.

A small uncontrolled, prospective study of a single 4 mg ZA infusion in allogeneic BMT patients with either osteoporosis or rapid bone loss post-allogeneic BMT [81] showed reduced bone loss at the spine and FN.

SUMMARY AND CONCLUSION

Pre-transplantation bone disease and immunosuppressive therapy result in a severe form of osteoporosis characterized by rapid bone loss and increased fracture rates, early after transplantation. There is increased bone resorption and decreased bone formation, suggestive of uncoupling of bone turnover. In the late post-transplant period, with weaning of glucocorticoid doses, bone formation begins to increase and underlying high bone turnover resulting in osteoporosis. Although rates of bone loss and fractures reported in recent studies are lower than those of 10 years ago, they remain too high. Transplant candidates should be assessed, and pre-transplantation bone disease should be treated. Preventive therapy initiated in the immediate post-transplantation period is indicated in patients with osteopenia or osteoporosis, since further bone loss will occur immediately after transplantation. All organ transplant recipients should be considered at risk for post-transplantation bone loss and fractures, as it is impossible to identify patients with the highest fracture risk. Long-term organ transplant recipients should also have bone mass measurement and treatment of osteoporosis.

A recent meta-analysis showed treatment with a bisphosphonate or active vitamin D metabolite during the first year after solid organ transplantation is associated with a 50% reduction in the number of subjects with fractures and 76% fewer vertebral fractures [Fig. 61.1(A)] [82]. Bisphosphonate treatment was associated with a 47% reduction in the number of subjects with fractures [Fig. 61.1(B)], but no significant reduction in vertebral

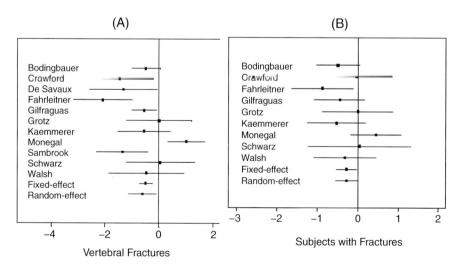

Fig. 61.1. (A) Log odds ratios (OR) and 95% confidence intervals for the effect of treatment with bisphosphonates or vitamin D analogues after organ transplantation on the number of vertebral fractures (OR 0.24, 95% CI 0.07, 0.78 by random effects model). (B) Log odds ratios (OR) and 95% confidence intervals for the effect of treatment with bisphosphonates after organ transplantation on the number of subjects with vertebral fractures (OR 0.53, 95% CI 0.30, 0.91 by fixed effect model). Reproduced with permission from Ref. 82.

fractures. Overall, bisphosphonates are the most promising approach for the prevention and treatment of transplantation osteoporosis. Active vitamin D metabolites may have additional benefits in reducing hyperparathyroidism, particularly after kidney transplantation. Potential new agents for transplantation osteoporosis include anabolic agents that stimulate bone formation, namely PTH(1-34) or teriparatide, and the anticatabolic drugs, human antibodies to RANKL (denosumab), and cathepsin K inhibitors. PTH(1-34) and other PTH1 receptor agonists may have a specific role after BMT in stimulating MSC cell differentiation into the osteoblast lineage and reducing adipogenesis [83, 84].

Several issues remain regarding the administration of bisphosphonates for transplantation bone disease, including the optimal route of administration and duration of therapy. Treatment may only need to be given for 1 year following cardiac transplantation, but its optimal duration is less clear following other transplants. It is also uncertain at what level of renal impairment oral bisphosphonates should be avoided and whether this level is the same for intravenous bisphosphonates. Another special consideration in using bisphosphonates in kidney transplant recipients is adynamic bone disease (see above). Large multicenter trials comparing treatment with oral or parenteral bisphosphonates and calcitriol, and commencing at the time of transplantation that are powered to detect differences in fracture rates are recommended here.

Despite some continuing uncertainties, much has been learned about transplantation osteoporosis. Armed with this information, it is now critical to act to prevent and treat this disabling disease.

ACKNOWLEDGMENTS

I thank Dr. Elizabeth Shane for her mentorship in this area.

REFERENCES

1. Cohen A, Sambrook P, Shane E. 2004. Management of bone loss after organ transplantation. *J Bone Miner Res* 19(12): 1919–32.
2. Cohen A, Shane E. 2003. Osteoporosis after solid organ and bone marrow transplantation. *Osteoporos Int* 14(8): 617–30.
3. Nickolas TL, Cremers S, Zhang A, Thomas V, Stein E, Cohen A, Chauncey R, Nikkel L, Yin MT, Liu XS, Boutroy S, Staron RB, Leonard MB, McMahon DJ, Dworakowski E, Shane E. 2011. Discriminants of prevalent fractures in chronic kidney disease. *J Am Soc Nephrol* 22(8): 1560–72.
4. Gal-Moscovici A, Sprague SM. 2007. Osteoporosis and chronic kidney disease. *Semin Dial* 20(5): 423–30.
5. Stehman-Breen CO, Sherrard DJ, Alem AM, Gillen DL, Heckbert SR, Wong CS, Ball A, Weiss NS. 2000. Risk factors for hip fracture among patients with end-stage renal disease. *Kidney Int* 58(5): 2200–5.
6. Alem AM, Sherrard DJ, Gillen DL, Weiss NS, Beresford SA, Heckbert SR, Wong C, Stehman-Breen C. 2000. Increased risk of hip fracture among patients with end-stage renal disease. *Kidney Int* 58(1): 396–9.
7. Ball AM, Gillen DL, Sherrard D, Weiss NS, Emerson SS, Seliger SL, Kestenbaum BR, Stehman-Breen C. 2002. Risk of hip fracture among dialysis and renal transplant recipients. *JAMA* 288(23): 3014–8.
8. Majumdar S, Ezekowitz JA, Lix LM, Leslie W. 2012. Heart failure is a clinically and densitometrically independent and novel risk factor for major osteoporotic fractures: Population-based cohort study of 45,509 subjects. *J Clin Endocrinol Metab* 97(4): 1179–86.
9. Shane E, Mancini D, Aaronson K, Silverberg SJ, Seibel MJ, Addesso V, McMahon DJ. 1997. Bone mass, vitamin D deficiency and hyperparathyroidism in congestive heart failure. *Am J Med* 103: 197–207.
10. Monegal A, Navasa M, Guanabens N, Peris P, Pons F, Martinez de Osaba MJ, Ordi J, Rimola A, Rodes J, Munoz-Gomez J. 2001. Bone disease after liver transplantation: A long-term prospective study of bone mass changes, hormonal status and histomorphometric characteristics. *Osteoporos Int* 12(6): 484–92.
11. Millonig G, Graziadei IW, Eichler D, Pfeiffer KP, Finkenstedt G, Muehllechner P, Koenigsrainer A, Margreiter R, Vogel W. 2005. Alendronate in combination with calcium and vitamin D prevents bone loss after orthotopic liver transplantation: A prospective single-center study. *Liver Transpl* 11: 960–6.
12. Ninkovic M, Love SA, Tom B, Alexander GJ, Compston JE. 2001. High prevalence of osteoporosis in patients with chronic liver disease prior to liver transplantation. *Calcif Tissue Int* 69(6): 321–6.
13. Tschopp O, Boehler A, Speich R, Weder W, Seifert B, Russi EW, Schmid C. 2002. Osteoporosis before lung transplantation: Association with low body mass index, but not with underlying disease. *Am J Transplant* 2(2): 167–72.
14. Keilholz U, Max R, Scheibenbogen C, Wuster C, Korbling M, Haas R. 1997. Endocrine function and bone metabolism 5 years after autologous bone marrow/blood-derived progenitor cell transplantation. *Cancer* 79(8): 1617–22.
15. Tauchmanova L, Selleri C, Rosa GD, Pagano L, Orio F, Lombardi G, Rotoli B, Colao A. 2002. High prevalence of endocrine dysfunction in long-term survivors after allogeneic bone marrow transplantation for hematologic diseases. *Cancer* 95(5): 1076–84.
16. Valimaki M, Kinnunen K, Volin L, Tahtela R, Loyttniemi E, Laitinen K, Makela P, Keto P, Ruutu T. 1999. A prospective study of bone loss and turnover after allogeneic bone marrow transplantation: Effect of calcium supplementation with or without calcitonin. *Bone Marrow Transplant* 23: 355–61.

17. Kananen K, Volin L, Laitinen K, Alfthan H, Ruutu T, Valimaki MJ. 2005. Prevention of bone loss after allogeneic stem cell transplantation by calcium, vitamin D, and sex hormone replacement with or without pamidronate. *J Clin Endocrinol Metab* 90: 3877–85.
18. Schulte C, Beelen D, Schaefer U, Mann K. 2000. Bone loss in long-term survivors after transplantation of hematopoietic stem cells: A prospective study. *Osteoporos Int* 11: 344–53.
19. Resnick J, Gupta N, Wagner J, Costa G, Cruz RJ Jr, Martin L, Koritsky DA, Perera S, Matarese L, Eid K, Schuster B, Roberts M, Greenspan S, Abu-Elmagd K. 2010. Skeletal integrity and visceral transplantation. *Am J Transplan* 10(10): 2331–40.
20. Van Staa TP, Leufkens HG, Abenhaim L, Zhang B, Cooper C. 2000. Use of oral corticosteroids and risk of fractures. *J Bone Miner Res* 15: 993–1000.
21. Nikkel LE, Mohan S, Zhang A, McMahon DJ, Boutroy S, Dube G, Tanriover B, Cohen D, Ratner L, Hollenbeak CS, Leonard MB, Shane E, Nickolas TL. 2012. Reduced fracture risk with early corticosteroid withdrawal after kidney transplant. *Am J Transplant* 12(3): 649–59.
22. Epstein S. 1996. Post-transplantation bone disease: The role of immunosuppressive agents on the skeleton. *J Bone Miner Res* 11: 1–7.
23. Ponticelli C, Aroldi A. 2001. Osteoporosis after organ transplantation. *Lancet* 357(9268): 1623.
24. McIntyre HD, Menzies B, Rigby R, Perry-Keene DA, Hawlcy CM, Hardie IR. 1995. Long-term bone loss after renal transplantation: Comparison of immunosuppressive regimens. *Clin Transplant* 9(1): 20–4.
25. Stempfle HU, Werner C, Echtler S, Assum T, Meiser B, Angermann CE, Theisen K, Gartner R. 1998. Rapid trabecular bone loss after cardiac transplantation using FK506 (tacrolimus)-based immunosuppression. *Transplant Proc* 30(4): 1132–3.
26. Park KM, Hay JE, Lee SG, Lee YJ, Wiesner RH, Porayko MK, Krom RA. 1996. Bone loss after orthotopic liver transplantation: FK 506 versus cyclosporine. *Transplant Proc* 28(3): 1738–40.
27. Goffin E, Devogelaer JP, Depresseux G, Squifflet JP, Pirson Y. 2001. Osteoporosis after organ transplantation. *Lancet* 357(9268): 1623.
28. Monegal A, Navasa M, Guanabens N, Peris P, Pons F, Martinez de Osaba MJ, Rimola A, Rodes J, Munoz-Gomez J. 2001. Bone mass and mineral metabolism in liver transplant patients treated with FK506 or cyclosporine A. *Calcif Tissue Int* 68: 83–6.
29. Singha UK, Jiang Y, Yu S, Luo M, Lu Y, Zhang J, Xiao G. 2008. Rapamycin inhibits osteoblast proliferation and differentiation in MC3T3-E1 cells and primary mouse bone marrow stromal cells. *J Cell Biochem* 103(2): 434–46.
30. Stein EM, Shane E. 2011. Vitamin D in organ transplantation. *Osteoporos Int* 22: 2107–18.
31. Julian BA, Laskow DA, Dubovsky J, Dubovsky EV, Curtis JJ, Quarrles LD. 1991. Rapid loss of vertebral bone density after renal transplantation. *N Engl J Med* 325: 544–50.
32. Monier-Faugere M, Mawad H, Qi Q, Friedler R, Malluche HH. 2000. High prevalence of low bone turnover and occurrence of osteomalacia after kidney transplantation. *J Am Soc Nephrol* 11: 1093–9.
33. Ramsey-Goldman R, Dunn JE, Dunlop DD, Stuart FP, Abecassis MM, Kaufman DB, Langman CB, Salinger MH, Sprague SM. 1999. Increased risk of fracture in patients receiving solid organ transplants. *J Bone Miner Res* 14(3): 456–63.
34. Sprague SM, Josephson MA. 2004. Bone disease after kidney transplantation. *Semin Nephrol* 24: 82–90.
35. Wissing KM, Broeders N, Moreno-Reyes R, Gervy C, Stallenberg B, Abramowicz D. 2005. A controlled study of vitamin D3 to prevent bone loss in renal-transplant patients receiving low doses of steroids. *Transplantation* 79: 108–15.
36. Grotz W, Nagel C, Poeschel D, Cybulla M, Petersen KG, Uhl M, Strey C, Kirste G, Olschewski M, Reichelt A, Rump LC. 2001. Effect of ibandronate on bone loss and renal function after kidney transplantation. *J Am Soc Nephrol* 12(7): 1530–7.
37. Kovac D, Lindic J, Kandus A, Bren AF. 2001. Prevention of bone loss in kidney graft recipients. *Transplant Proc* 33: 1144–5.
38. Giannini S, Dangel A, Carraro G, Nobile M, Rigotti P, Bonfante L, Marchini F, Zaninotto M, Dalle Carbonare L, Sartori L, Crepaldi G. 2001. Alendronate prevents further bone loss in renal transplant recipients. *J Bone Miner Res* 16(11): 2111–7.
39. Torres A, García S, Gómez A, González A, Barrios Y, Concepción MT, Hernández D, García JJ, Checa MD, Lorenzo V, Salido E. 2004. Treatment with intermittent calcitriol and calcium reduces bone loss after renal transplantation. *Kidney Int* 65(2): 705–12.
40. Miller PD. 2005. Treatment of osteoporosis in chronic kidney disease and end-stage renal disease. *Curr Osteoporos Rep* 3: 5–12.
41. Grotz W, Nagel C, Poeschel D, Cybulla M, Petersen KG, Uhl M, Strey C, Kirste G, Olschewski M, Reichelt A, Rump LC. 2001. Effect of ibandronate on bone loss and renal function after kidney transplantation. *J Am Soc Nephrol* 12(7): 1530–7.
42. Schwarz CL, Mitterbauer CL, Heinze G, Woloszczuk W, Haas M, Oberbauer R. 2004. Nonsustained effect of short-term bisphosphonate therapy on bone turnover three years after renal transplantation. *Kidney Int* 65(1): 304–9.
43. Palmer SC, McGregor DO, Strippoli GFM. 2007. Interventions for preventing bone disease in kidney transplant recipients. *Cochrane Database Syst Rev* (3): CD005015.
44. Chiu MY, Sprague SM, Bruce DS, Woodle ES, Thistlethwaite JR Jr, Josephson MA. 1998. Analysis of fracture prevalence in kidney-pancreas allograft recipients. *J Am Soc Nephrol* 9(4): 677–83.
45. Smets YF, de Fijter JW, Ringers J, Lemkes HH, Hamdy NA. 2004. Long-term follow-up study on bone mineral density and fractures after simultaneous pancreas-kidney transplantation. *Kidney Int* 66(5): 2070–6.

46. Shane E, Papadopoulos A, Staron RB, Addesso V, Donovan D, McGregor C, Schulman LL. 1999. Bone loss and fracture after lung transplantation. *Transplantation* 68: 220–7.
47. Aris RM, Lester GE, Renner JB, Winders A, Denene Blackwood A, Lark RK, Ontjes DA. 2000. Efficacy of pamidronate for osteoporosis in patients with cystic fibrosis following lung transplantation. *Am J Respir Crit Care Med* 162(3 Pt 1): 941–6.
48. Trombetti A, Gerbase MW, Spiliopoulos A, Slosman DO, Nicod LP, Rizzoli R. 2000. Bone mineral density in lung-transplant recipients before and after graft: Prevention of lumbar spine post-transplantation-accelerated bone loss by pamidronate. *J Heart Lung Transplant* 19(8): 736–43.
49. Shane E, Rivas M, Staron RB, Silverberg SJ, Seibel M, Kuiper J, Mancini D, Addesso V, Michler RE, Factor-Litvak P. 1996. Fracture after cardiac transplantation: A prospective longitudinal study. *J Clin Endocrinol Metab* 81: 1740–6.
50. Leidig-Bruckner G, Hosch S, Dodidou P, Ritchel D, Conradt C, Klose C, Otto G, Lange R, Theilmann L, Zimmerman R, Pritsch M, Zeigler R. 2001. Frequency and predictors of osteoporotic fractures after cardiac or liver transplantation: A follow-up study. *Lancet* 357: 342–7.
51. Sambrook P, Henderson NK, Keogh A, MacDonald P, Glanville A, Spratt P, Bergin P, Ebeling P, Eisman J. 2000. Effect of calcitriol on bone loss after cardiac or lung transplantation. *J Bone Miner Res* 15(9): 1818–24.
52. Henderson K, Eisman J, Keogh A, MacDonald P, Glanville A, Spratt P, Sambrook P. 2001. Protective effect of short-tem calcitriol or cyclical etidronate on bone loss after cardiac or lung transplantation. *J Bone Miner Res* 16(3): 565–71.
53. Stempfle HU, Werner C, Echtler S, Wehr U, Rambeck WA, Siebert U, Uberfuhr P, Angermann CE, Theisen K, Gartner R. 1999. Prevention of osteoporosis after cardiac transplantation: A prospective, longitudinal, randomized, double-blind trial with calcitriol. *Transplantation* 68(4): 523–30.
54. Kapetanakis EI, Antonopoulos AS, Antoniou TA, Theodoraki KA, Zarkalis DA, Sfirakis PD, Chilidou DA, Alivizatos PA. 2005. Effect of long-term calcitonin administration on steroid-induced osteoporosis after cardiac transplantation. *J Heart Lung Transplant* 24(5): 526–32.
55. Bianda T, Linka A, Junga G, Brunner H, Steinert H, Kiowski W, Schmid C. 2000. Prevention of osteoporosis in heart transplant recipients: A comparison of calcitriol with calcitonin and pamidronate. *Calcif Tissue Int* 67: 116–21.
56. Gil-Fraguas L, Jodar E, Martinez G, Escalona MA, Vara J, Robles E, Hawkins F. 2005. Evolution of bone density after heart transplantation: Influence of anti-resorptive therapy. *J Bone Miner Res* 20 (Suppl 1): S439–40.
57. Fahrleitner-Pammer A, Piswanger-Soelkner JC, Pieber TR, Obermayer-Pietsch BM, Pilz S, Dimai HP, Prenner G, Tscheliessnigg KH, Hauge E, Portugaller RH, Dobnig H. 2009. Ibandronate prevents bone loss and reduces vertebral fracture risk in male cardiac transplant patients: A randomized double-blind, placebo-controlled trial. *J Bone Miner Res.* 24: 1335–44.
58. Shane E, Addesso V, Namerow PB, McMahon, DJ, Lo SH, Staron RB, Zucker M, Pardi S, Maybaum S, Mancini D. 2004. Alendronate versus calcitriol for the prevention of bone loss after cardiac transplantation. *N Engl J Med* 350: 767–76.
59. Cohen A, Addesso V, McMahon DJ, Staron RB, Namerow P, Maybaum S, Mancini D, Shane E. 2006. Discontinuing antiresorptive therapy one year after cardiac transplantation: Effect on bone density and bone turnover. *Transplantation* 81: 686–91.
60. Mitchell MJ, Baz MA, Fulton MN, Lisor CF, Braith R. 2003. Resistance training prevents vertebral osteoporosis in lung transplant recipients. *Transplantation* 76: 557–62.
61. Braith RW, Magyari PM, Fulton MN, Lisor CF, Vogel SE, Hill JA, Aranda JM Jr. 2006. Comparison of calcitonin versus calcitonin and resistance exercise as prophylaxis for osteoporosis in heart transplant recipients. *Transplantation* 81: 1191–5.
62. Ninkovic M, Skingle SJ, Bearcroft PW, Bishop N, Alexander CJ, Compston JE. 2000. Incidence of vertebral fractures in the first three months after orthotopic liver transplantation. *Eur J Gastroenterol Hepatol* 12(8): 931–5.
63. Hommann M, Abendroth K, Lehmann G, Patzer N, Kornberg A, Voigt R, Seifert S, Hein G, Scheele J. 2002. Effect of transplantation on bone: Osteoporosis after liver and multivisceral transplantation. *Transplant Proc* 34(6): 2296–8.
64. Crawford BA, Kam C, Pavlovic J, Byth K, Handelsman DJ, Angus PW, McCaughan GW. 2006. Zoledronic acid prevents bone loss after liver transplantation: A randomized, double-blind, placebo-controlled trial. *Ann Intern Med* 144(4): 239–48.
65. Bodingbauer M, Wekerle T, Pakrah B, Roschger P, Peck-Radosavljevic M, Silberhumer G, Grampp S, Rockenschaub S, Berlakovich G, Steininger R, Klaushofer K, Oberbauer R, Muhlbacher F. 2007. Prophylactic bisphosphonate treatment prevents bone fractures after liver transplantation. *Am J Transplant* 7: 1763–69.
66. Monegal A, Guanabens N, Suarez MJ, Suarez F, Clemente G, Garcia-Gonzalez M, De la Mata M, Serrano T, Casafont F, Torne S, Barrios C, Navasa M. 2009. Pamidronate in the prevention of bone loss after liver transplantation: A randomized controlled trial. *Transpl Int* 22: 198–206.
67. Kaemmerer D, Lehmann G, Wolf G, Settmacher U, Hommann M. 2010. Treatment of osteoporosis after liver transplantation with ibandronate. *Transpl Int* 23: 753–9.
68. Karasu Z, Kilic M, Tokat Y. 2006. The prevention of bone fractures after liver transplantation: Experience with alendronate treatment. *Transplant Proc* 38: 1448–52.
69. Awan KS, Wagner JM, Martin D, Medich DL, Perera S, Abu Elmagd K, Greenspan SL. 2007. Bone loss following

69. small bowel transplantation. ASBMR 29th Annual Meeting. *J Bone Miner Res* 22: s352–s401. Abstract T497, s356.
70. Banfi A, Podesta M, Fazzuoli L, Sertoli MR, Venturini M, Santini G, Cancedda R, Quarto R. 2001. High-dose chemotherapy shows a dose-dependent toxicity to bone marrow osteoprogenitors: A mechanism for post-bone marrow transplantation osteopenia. *Cancer* 92(9): 2419–28.
71. Lee WY, Cho SW, Oh ES, Oh KW, Lee JM, Yoon KH, Kang MI, Cha BY, Lee KW, Son HY, Kang SK, Kim CC. 2002. The effect of bone marrow transplantation on the osteoblastic differentiation of human bone marrow stromal cells. *J Clin Endocrinol Metab* 87(1): 329–35.
72. Ebeling P, Thomas D, Erbas B, Hopper L, Szer J, Grigg A. 1999. Mechanism of bone loss following allogeneic and autologous hematopoeitic stem cell transplantation. *J Bone Miner Res* 14: 342–50.
73. Ebeling PR. 2005. Bone disease after bone marrow transplantation. In: Compston J, Shane E (eds.) *Bone Disease of Organ Transplantation*. San Diego: Elsevier Academic Press. Chapter 19. pp. 339–52.
74. Lee WY, Kang MI, Baek KH, Oh ES, Oh KW, Lee KW, Kim SW, Kim CC. 2002. The skeletal site-differential changes in bone mineral density following bone marrow transplantation: 3-year prospective study. *J Korean Med Sci* 17(6): 749–54.
75. Gandhi MK, Lekamwasam S, Inman I, Kaptoge S, Sizer L, Love S, Bearcroft PW, Milligan TP, Price CP, Marcus RE, Compston JE. 2003. Significant and persistent loss of bone mineral density in the femoral neck after haematopoietic stem cell transplantation: Long-term follow-up of a prospective study. *Br J Haematol* 121: 462–8.
76. Ebeling PR. 2005. Is defective osteoblast function responsible for bone loss from the proximal femur despite pamidronate therapy? *J Clin Endocrinol Metab* 90: 4414–6.
77. Tauchmanova L, De Rosa G, Serio B, Fazioli F, Mainolfi C, Lombardi G, Colao A, Salvatore M, Rotoli B, Selleri C. 2003. Avascular necrosis in long-term survivors after allogeneic or autologous stem cell transplantation: A single center experience and a review. *Cancer* 97: 2453–61.
78. Tauchmanova L, Selleri C, Esposito M, Di Somma C, Orio F Jr, Bifulco G, Palomba S, Lombardi G, Rotoli B, Colao A. 2003. Beneficial treatment with risedronate in long-term survivors after allogeneic stem cell transplantation for hematological malignancies. *Osteoporos Int* 14: 1013–9.
79. Tauchmanova L, Ricci P, Serio B, Lombardi G, Colao A, Rotoli B, Selleri C. 2005. Short-term zoledronic acid treatment increases bone mineral density and marrow clonogenic fibroblast progenitors after allogeneic stem cell transplantation. *J Clin Endocrinol Metab* 90: 627–34.
80. Grigg AP, Shuttleworth P, Reynolds J, Schwarer AP, Szer J, Bradstock K. 2006. Pamidronate reduces bone loss after allogeneic stem cell transplantation. *J Clin Endocrinol Metab* 91(10): 3835–43.
81. D'Souza AB, Grigg AP, Szer J, Ebeling PR. 2006. Zoledronic acid prevents bone loss after allogeneic haemopoietic stem cell transplantation. *Int Med J* 36: 600–3.
82. Stein EM, Ortiz D, Jin Z, McMahon DJ, Shane E. 2011. Prevention of fractures after solid organ transplantation: A meta-analysis. *J Clin Endocrinol Metab* 96: 3457–65.
83. Rickard DJ, Wang FL, Rodriguez-Rojas AM, Wu Z, Trice WJ, Hoffman SJ, Votta B, Stroup GB, Kumar S, Nuttall ME. 2006. Intermittent treatment with parathyroid hormone (PTH) as well as a non-peptide small molecule agonist of the PTH1 receptor inhibits adipocyte differentiation in human bone marrow stromal cells. *Bone* 39(6): 1361–72.
84. Chan GK, Miao D, Deckelbaum R, Bolivar I, Karaplis A, Goltzman D. 2003. Parathyroid hormone-related peptide interacts with bone morphogenetic protein 2 to increase osteoblastogenesis and decrease adipogenesis in pluripotent C3H10T mesenchymal cells. *Endocrinology* 144: 5511–20.

62
Osteoporosis in Men
Eric S. Orwoll

Introduction 508
Skeletal Development 508
Effects of Aging on the Skeleton in Men 508
Fracture Epidemiology 509
Causes of Osteoporosis in Men 509

Evaluation of Osteoporosis in Men 510
Osteoporosis Prevention in Men 511
Treatment of Osteoporosis in Men 511
References 512

INTRODUCTION

Osteoporosis in men is now recognized as an important public health problem, and there is a much greater understanding of the disorder. Effective diagnostic, preventive, and treatment strategies have been developed. Moreover, the study of osteoporosis in men has revealed male–female differences that in turn have fostered a greater understanding of bone biology in general. Nevertheless, there are important pathophysiological and clinical issues that remain unresolved and research continues to be very active.

SKELETAL DEVELOPMENT

Bone mass accumulation in males occurs gradually during childhood and accelerates dramatically during adolescence. Peak bone mass is closely tied to pubertal development, and male–female differences in the skeleton appear during adolescence [1]. The rapid increase in bone mass occurs somewhat later in boys than girls; the majority of the increase has occurred by an average age of 16 years in girls and age 18 years in boys. Moreover, whereas trabecular bone mineral density (BMD) accumulation is similar in boys and girls, boys generally develop thicker cortices and larger bones than do girls, even when adjusted for body size. These differences may provide important biomechanical advantages that could in part underlie the lower fracture risk observed in men later in life. The reasons for these sexual differences in skeletal development are unclear but could be related to differences in sex steroid action (androgens may stimulate periosteal bone formation and bone expansion), growth factor concentrations, and mechanical forces exerted on bone (e.g., by greater muscle action or activity). Sex-specific effects of a variety of genetic loci have been reported in animals and humans, suggesting that the origin of sex differences in skeletal phenotypes is complex. Finally, despite these average sex differences, there is wide variation in bone mass and structure in men after adjustment for body size, and considerable overlap with the range of similar measures in women.

EFFECTS OF AGING ON THE SKELETON IN MEN

As in women, aging is associated with large changes in bone mass and architecture in men [2]. Trabecular bone loss (e.g., in the vertebrae and proximal femur) occurs during midlife and accelerates in later life. The magnitude of these changes is similar, but probably slightly less, than those in women. In men there is more trabecu-

---

Primer on the Metabolic Bone Diseases and Disorders of Mineral Metabolism, Eighth Edition. Edited by Clifford J. Rosen.
© 2013 American Society for Bone and Mineral Research. Published 2013 by John Wiley & Sons, Inc.

lar thinning and less trabecular dropout than in women. Endocortical bone loss with resulting cortical thinning takes place in long bones, and cortical porosity increases with age, but these processes may be accompanied by a concomitant increase in periosteal bone expansion that tends to preserve the breaking strength of bone [3]. In general, the pattern of age-related bone loss in men is similar in men and women, but in men, there is no concomitant to the accelerated phase of loss associated with the menopause in women. In the elderly of both sexes, the rate of bone loss accelerates with increasing age.

FRACTURE EPIDEMIOLOGY

Fractures are common in men. Fracture data in men are derived primarily from the study of white populations. In them, the incidence of fracture is bimodal, with a peak incidence in adolescence and mid-adulthood, a lower incidence between ages 40 and 60, and a dramatic increase after the age of 70 (Fig 62.1) [4]. The types of fractures sustained in younger and older men are different, with long bone fractures being common in younger men, whereas vertebral and hip fractures predominate in the elderly. These differences suggest that the etiologies of fractures at these two periods of life are distinct. In younger men, trauma may play a larger role, whereas in older men skeletal fragility and fall propensity are likely to be major factors.

The exponential increase in fracture incidence as men age is as dramatic as the similar increase that occurs in women, but it begins 5–10 years later in life. This delay, combined with the longer life expectancy in women, underlies the greater burden of osteoporotic fractures in women. Nevertheless, the age-adjusted incidence of hip fracture in men is one-quarter to one-third that in women, and 20–25% of hip fractures occur in men [2]. The consequences of fracture in men are at least as great as in women, and in fact, elderly men seem to be more likely to die and to suffer disability than women after a hip fracture. Older men suffer lower rates of long bone fractures than do women [5]. There is less information concerning vertebral fracture epidemiology in men, but the age-adjusted incidence appears to be high (approximately 50% of that in women) [6]. In younger men, the prevalence of vertebral fracture is actually greater in men than in women, at least in part the result of higher rates of spinal trauma experienced by men. Although there are inadequate data, the epidemiology of fracture in men seems to be dramatically influenced by both race and geography [7, 8]. For instance, black men have a much lower likelihood of fractures than whites, and Asian men have a lower likelihood of suffering hip fracture than whites. Much more information is needed concerning these differences and their causation.

Over the past few decades, the population incidence of fractures has apparently been changing in both men and women. In Western societies the rate of hip fracture increased dramatically until about 10 years ago, but after that it began to decline, particularly in women but also in men [9, 10]. The reasons for the change are not clear but may be related to increased efforts in screening and treatment of osteoporosis, a greater prevalence of obesity, reduced smoking, etc. In contrast, recent data suggest that the rates of fracture are increasing quickly in Asian societies [11], potentially because of urbanization and other cultural changes. These divergent trends emphasize the importance of environmental influences on fracture causation. Finally, anticipated increases in the size of older populations may result in a greater number of fractures despite gradual reductions in fracture incidence.

CAUSES OF OSTEOPOROSIS IN MEN

Fractures in men are related to a variety of risk factors. Certainly, skeletal fragility makes fracture more likely. This trait is most commonly measured as reduced BMD, but almost certainly has other components (biomechanically important alterations in bone geometry, material properties, etc.). Aging and a previous history of fracture are independently associated with a higher probability of future fracture, and men of lower weight have a higher fracture risk [2, 12]. Finally, falling becomes much more common with increasing age in men, and falls are strongly associated with increased fracture risk [13].

The causation of osteoporosis in men is commonly heterogeneous, and most osteoporotic men have several factors that contribute to the disease. One-half to two-thirds of men with osteoporosis have multiple risk factors, including other medical conditions, medications, or lifestyle issues that result in bone loss and fragility (Table 62.1). [2, 7]. The most important include alcohol abuse, glucocorticoid excess, and hypogonadism. An

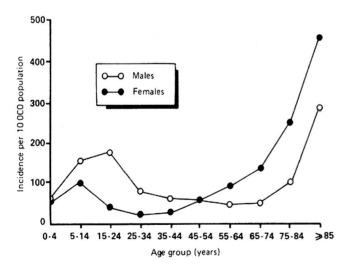

Fig. 62.1. Average annual fracture incidence rate per 10,000 population in Leicester, UK, by age group and by sex.

Table 62.1. Causes of Osteoporosis and Bone Loss in Men

Primary
    Aging
    Idiopathic

Secondary
    Hypogonadism
    Glucocorticoid excess
    Alcoholism, tobacco abuse
    Renal insufficiency
    Gastrointestinal, hepatic disorders; malabsorption
    Hyperparathyroidism
    Hypercalciuria
    Anticonvulsants
    Thyrotoxicosis
    Chronic respiratory disorders
    Anemias, hemoglobinopathies
    Immobilization
    Osteogenesis imperfecta
    Homocystinuria
    Systemic mastocytosis
    Neoplastic diseases and chemotherapy
    Rheumatoid arthritis

decline with age in men, and it has been postulated that the decline may be an important risk factor for age-related bone loss and fracture risk. The strength of this association remains somewhat uncertain.

Although both are important, the relative roles of estrogens and androgens in skeletal physiology in men are uncertain [15–17]. Estrogen is essential for normal bone development in young men, as evidenced by immature delay in development and low bone mass in men with aromatase deficiency and their reversal with estrogen therapy. Moreover, estrogen is correlated with bone remodeling, BMD, and rate of BMD loss in older men, apparently more strongly than is testosterone. However, testosterone is independently related to indices of bone resorption and formation and may stimulate periosteal bone [18–20]. Low levels of estradiol have been clearly linked to increased fracture risk in older men [21, 22]. Testosterone levels appear less strongly related to fracture but may have an effect, particularly at very low concentrations [23]. High sex hormone-binding globulin (SHBG) concentrations have also been linked to increased fracture propensity [22]. The relative roles of estrogen, androgen, and SHBG must be better defined, and how measures of their levels can be used in clinical situations must be clarified.

important fraction of osteoporotic men, however, have idiopathic disease.

Idiopathic osteoporosis

Osteoporosis of unknown etiology can present in men of any age [14], but its presentation is most dramatic in younger men who are otherwise unlikely to be affected by osteoporosis. A low bone formation rate has been a consistent finding in these patients. Several possible etiologies have been considered. Most prominent among them are genetic factors, because BMD and the risk of fracture are highly heritable. The specific genes that may be responsible are uncertain.

Hypogonadism

Sex steroids are clearly important for skeletal health in men, both during growth and the attainment of peak bone mass as well as in the maintenance of bone strength in adults [14]. Hypogonadism is associated with low BMD; the development of hypogonadism results in increased bone remodeling and rapid bone loss (at least in the early phases of hypogonadism), and testosterone replacement increases BMD in hypogonadal men. One of the most important causes of severe hypogonadism is androgen deprivation therapy for prostate cancer; in this situation, bone loss is rapid, and the risk of fractures is clearly increased. Gonadal function and sex steroid levels

EVALUATION OF OSTEOPOROSIS IN MEN

Guidelines for the evaluation of osteoporosis in men are not well validated, but there are several recommendations that can be made confidently.

BMD measurements

BMD measures are at least as effective in men as in women in predicting the risk of future fractures [24]. In light of the prevalence of osteoporosis and the high incidence of fractures in men, BMD measures are performed too infrequently. Two groups of men would benefit from BMD testing:

- Men over 50 years of age who have suffered a fracture, including those with vertebral deformity. Younger men who suffer low-trauma fractures should also be assessed.
- Men who have known secondary causes of bone loss should have their BMD determined. These include men treated with glucocorticoids or other medications associated with osteoporosis, men with hypogonadism of any cause, including those treated with androgen deprivation therapy, or men who have alcoholism. Many other risk factors may also prompt BMD measures (Table 62.1).

Screening BMD measures in older men have been recommended (e.g., older than 70 years of age) [25], and recent cost-effectiveness analysis has suggested that screening may be appropriate at that age, especially as treatment costs have declined [26, 27].

Currently, the presence of reduced BMD in men is commonly quantified with T-scores using a grading system parallel to that used in women (BMD T-score −1.0 to −2.5 = low bone mass; BMD T-score below −2.5 = osteoporosis). Whether BMD measurements in men should be interpreted using T-scores based on a male-specific reference range or using the same reference range used in women has been controversial. Analyses of large population level data suggest that the BMD-fracture risk association is the same in men and women. Some reports from carefully studied cohorts indicate there may be some sex differences [28, 29], but those differences appear to be relatively small. The use of FRAX or other fracture risk calculators may improve the estimation of fracture risk and the selection of men for therapy.

Men who have been selected for androgen deprivation therapy deserve special note because the risk of bone loss and fractures is clearly increased, especially in the first 5 years after sex steroid deficiency is induced [30, 31]. Men starting high-dose glucocorticoid therapy present the same challenges and should be similarly managed. When antiandrogen (or glucocorticoid) therapy is begun, a BMD assessment is appropriate. If it is normal, routine preventative measures are reasonable. A repeat BMD measurement should be done in 1–2 years. If BMD is reduced at the onset of therapy, more aggressive preventive measures should be considered (e.g., bisphosphonate therapy). Pharmacological approaches to prevent further bone loss or fractures are warranted in men with osteoporosis, even before antiandrogen (or glucocorticoid) therapy is begun.

Clinical evaluation

The clinical evaluation of men found to have low BMD should include a careful history and physical examination designed to identify any factors that may contribute to deficits in bone mass. Attention should be paid to lifestyle factors, nutrition (especially calcium, vitamin D, and protein nutrition), activity level, and family history. A history of previous fracture should be identified, and fall risk should be assessed. This information should be used to formulate recommendations for prevention and treatment.

Laboratory testing

In a man undergoing an evaluation for osteoporosis, laboratory testing is intended to identify correctable causes of bone loss. Appropriate tests are shown in Table 62.2.

OSTEOPOROSIS PREVENTION IN MEN

The essentials of fracture prevention in men are similar to those in women. In early life, excellent nutrition and exercise seem to have positive effects on bone mass. These principles and the avoidance of lifestyle factors known to be associated with bone loss (Table 62.1) remain important throughout life. Calcium and vitamin D probably provide beneficial effects on bone mass and fractures in men as in women. Recent Institute of Medicine recommendations for men include 1,000 mg of calcium for those 30–70 years of age and 1,200 mg for those over 70 years of age, with suggested vitamin D intakes of 600 IU/day until age 70 and 800 IU/day thereafter [32]. In those at risk for falls (e.g., with reduced strength, poor balance, previous falls), attempts to increase strength and balance may be beneficial.

Table 62.2. Evaluation of Osteoporosis in Men: Laboratory Tests

Serum calcium, phosphorus, creatinine, alkaline phosphatase, liver function tests
Complete blood count (protein electrophoresis in those over 50 years)
Serum 25(OH) vitamin D and parathyroid hormone
Serum testosterone and luteinizing hormone
24-hour urine calcium and creatinine
Targeted diagnostic testing in men with signs, symptoms, or other indications of secondary disorders
When an etiology is not apparent after the above, additional testing may be appropriate: thyroid function tests, 24-hour urine cortisol, biochemical indices of remodeling, immunological tests for sprue

TREATMENT OF OSTEOPOROSIS IN MEN

Ensuring adequate calcium and vitamin D intake and appropriate physical activity are essential foundations for preserving and enhancing bone mass in men who have osteoporosis. Secondary causes of osteoporosis should be identified and treated. In addition, there are pharmacological therapies that have been shown to enhance BMD, and in some cases, reduce fracture risk in men. Although the available data are not as extensive as in women, these therapies seem to be as effective in affecting BMD and in reducing fracture risk in men. The treatment indications for these drugs are similar in men and women.

Idiopathic and age-related osteoporosis

Alendronate, risedronate, ibandronate, zoledronate and teriparatide are effective in improving BMD [33, 34] regardless of age or gonadal function. Although the trials are relatively small, each is also apparently effective in reducing vertebral fracture risk. Zoledronate reduced the risk of recurrent fracture in men and women following hip fracture [35], and although the independent effects in men could not be reliably ascertained, the effects sizes on fracture risk were similar in men and women.

Glucocorticoid-induced osteoporosis

Bisphosphonate therapy (e.g., alendronate, risedronate) is effective in improving BMD, and, although the data are not extensive, also probably reduces fracture [36, 37].

Hypogonadal osteoporosis

Bisphosphonate, denosumab and parathyroid hormone (PTH) therapy are effective in increasing BMD in hypogonadal men. Moreover, bisphosphonate and denosumab treatment can prevent the bone loss that is common after androgen deprivation therapy for prostate cancer, and denosumab reduces vertebral fracture risk in those men [38]. Testosterone replacement therapy results in increases in serum levels of both estradiol and testosterone and improves BMD in men with established hypogonadism [2], but whether fracture risk is reduced is unknown. In older men with less severe, age-related reductions in gonadal function, the usefulness of testosterone is less certain. Relatively high doses (200 mg of intramuscular testosterone every 2 weeks) is associated with an increase in BMD and strength in older men with low testosterone levels [39, 40], but its impact on fracture risk has not been examined. Lower doses (e.g., dermal administration of testosterone) seem to have lesser effects [41]. Once again, the effects of testosterone replacement on fracture risk are uncertain. Moreover, the long-term risks of testosterone therapy in older men are unknown. Therefore, testosterone replacement therapy is appropriate for management of the hypogonadal symptoms, but the treatment of osteoporosis in a man with low testosterone levels is most confidently undertaken with an osteoporosis drug for which there are more data concerning fracture risk reduction (e.g. a bisphosphonate or teriparatide) [42]

REFERENCES

1. Seeman E. 2001. Sexual dimorphism in skeletal size, density, and strength. *J Clin Endocrinol Metab* 86(10): 4576–84.
2. Marcus R, Feldman D, Kelsey JL. 2001. *Osteoporosis, 2nd Ed.* San Diego: Academic Press.
3. Seeman E. 2002. Pathogenesis of bone fragility in women and men. *Lancet* 359(9320): 1841–50.
4. Donaldson LJ, Cook A, Thomson RG. 1990. Incidence of fractures in a geographically defined population. *J Epi Comm Health* 44: 241–5.
5. Ismail AA, Pye SR, Cockerill WC, Lunt M, Silman AJ, Reeve J, Banzer D, Benevolenskaya LI, Bhalla A, Armas JB, Cannata JB, Cooper C, Delmas PD, Dequeker J, Dilsen G, Falch JA, Felsch B, Felsenberg D, Finn JD, Gennari C, Hoszowski K, Jajic I, Janott J, Johnell O, Kanis JA, Kragl G, Vaz AL, Lorenc R, Lyritis G, Marchand F, Masaryk P, Matthis C, Miazgowski T, Naves-Diaz M, Pols HAP, Poor G, Rapido A, Raspe HH, Reid DM, Reisinger W, Scheidt-Nave C, Stepan J, Todd C, Weber K, Woolf AD, O'Neill TW. 2002. Incidence of limb fracture across Europe: Results from the European prospective osteoporosis study (EPOS). *Osteoporos Int* 13: 565–71.
6. Group EPOSE. 2002. Incidence of vertebral fracture in Europe: Results from the European prospective osteoporosis study (EPOS). *J Bone Miner Res* 17(4): 716–24.
7. Amin S, Felson DT. 2001. Osteoporosis in men. *Rheum Dis Clin North Am* 27(1): 19–47.
8. Schwartz AV, Kelsey JL, Maggi S, Tuttleman M, Ho SC, Jonsson PV, Poor G, Sisson de Castro JA, Xu L, Matkin CC, Nelson LM, Heyse SP. 1999. International variation in the incidence of hip fractures: Cross-national project on osteoporosis for the world health organization program for research on aging. *Osteoporos Int* 9: 242–53.
9. Cooper C, Cole ZA, Holroyd CR, Earl SC, Harvey NC, Dennison EM, Melton LJ, Cummings SR, Kanis JA. 2011. Secular trends in the incidence of hip and other osteoporotic fractures. *Osteoporos Int* 22(5): 1277–88.
10. Leslie WD, O'Donnell S, Jean S, Lagace C, Walsh P, Bancej C, Morin S, Hanley DA, Papaioannou A. 2009. Trends in hip fracture rates in Canada. *JAMA* 302(8): 883–9.
11. Xia WB, He SL, Xu L, Liu AM, Jiang Y, Li M, Wang O, Xing XP, Sun Y, Cummings SR. 2011. Rapidly increasing rates of hip fracture in Beijing, China. *J Bone Miner Res* 27(1): 125–9.
12. Nguyen TV, Eisman JA, Kelly PJ, Sambrook PN. 1996. Risk factors for osteoporotic fractures in elderly men. *Am J Epidemiol* 144(3): 258–61.
13. Chan BKS, Marshall LM, Lambert LC, Cauley JA, Ensrud KE, Orwoll ES, Cummings SR. 2005. The risk of non-vertebral and hip fracture and prevalent falls in older men: The MrOS Study. *J Bone Miner Res* 20(Suppl 1): S385.
14. Vanderschueren D, Boonen S, Bouillon R. 2000. Osteoporosis and osteoportic fractures in men: A clinical perspective. *Bailliers Best Pract Res Clin Endocrinol Metab* 14(2): 299–315.
15. Khosla S, Melton J 3rd. 2002. Estrogen and the male skeleton. *J Clin Endocrinol Metab* 87(4): 1443–50.
16. Vanderscheuren D, Boonen S, Bouillon R. 1998. Action of androgens versus estrogens in male skeletal homeostasis. *Bone* 23: 391–4.
17. Orwoll ES. 2003. Men, bone and estrogen: Unresolved issues. *Osteoporos Int* 14(2): 93–8.
18. Leder BZ, Le Blanc KM, Schoenfeld DA, Eastell R, Finkelstein J. 2003. Differential effects of androgens and estrogens on bone turnover in normal men. *J Clin Endocrinol Metab* 88(1): 204–10.
19. Falahati-Nini A, Riggs BL, Atkinson EJ, O'Fallon WM, Eastell R, Khosla S. 2000. Relative contributions of tes-

tosterone and estrogen in regulating bone resorption and formation in normal elderly men. *J Clin Invest* 106: 1553–60.
20. Orwoll ES. 2001. Androgens: Basic biology and clinical implication. *Calcif Tissue Int* 69: 185–8.
21. Mellström D, Vandenput L, Mallmin H, Holmberg AH, Lorentzon M, Odén A, Johansson H, Orwoll ES, Labrie F, Karlsson MK. 2008. Older men with low serum estradiol and high serum SHBG have an increased risk of fractures. *J Bone Miner Res* 23: 1552–60.
22. LeBlanc ES, Nielson CM, Marshall LM, Lapidus JA, Barrett-Connor E, Ensrud KE, Hoffman AR, Laughlin G, Ohlsson C, Orwoll ES. 2009. The effects of serum testosterone, estradiol, and sex hormone binding globulin levels on fracture risk in older men. *J Clin Endocrinol Metab* 94(9): 3337–46.
23. Fink HA, Ewing SK, Ensrud KE, Barrett-Connor E, Taylor BC, Cauley JA, Orwoll ES. 2006. Association of testosterone and estradiol deficiency with osteoporosis and rapid bone loss in older men. *J Clin Endocrinol Metab* 91(10): 3908–15.
24. Nguyen ND, Pongchaiyakul C, Center JR, Eisman JA, Nguyen TV. 2005. Identification of high-risk individuals for hip fracture: A 14-year prospective study. *J Bone Miner Res* 20(11): 1921–8.
25. Binkley NC, Schmeer P, Wasnich RD, Lenchik L. 2002. What are the criteria by which a densitometric diagnosis of osteoporosis can be made in males and non-caucasians? *J Clin Densitom* 5(Suppl): 19–27.
26. Schousboe JT, Taylor BC, Fink HA, Kane RL, Cummings SR, Orwoll ES, Melton LJ, 3rd, Bauer DC, Ensrud KE. 2007. Cost-effectiveness of bone densitometry followed by treatment of osteoporosis in older men. *JAMA* 298(6): 629–37.
27. Dawson-Hughes B, Tosteson ANA, Melton LJ, Baim S, Favus MJ, Khosla S, Lindsay RL. 2008. Implications of absolute fracture risk assessment for osteoporosis practice guidelines in the USA. *Osteoporos Int* 19(4): 449–58.
28. Johnell O, Kanis JA, Oden A, Johansson H, De Laet C, Delmas P, Eisman JA, Fujiwara S, Kroger H, Mellstrom D, Meunier PJ, Melton LJ 3rd, O'Neill T, Pols H, Reeve J, Silman A, Tenenhouse A. 2005. Predictive value of BMD for hip and other fractures. *J Bone Miner Res* 20(7): 1185–94.
29. Cummings SR, Cawthon PM, Ensrud KE, Cauley JA, Fink HA, Orwoll ES. 2006. BMD and risk of hip and nonvertebral fractures in older men: A prospective study and comparison with older women. *J Bone Miner Res* 21(10): 1550–56.
30. Shahinian VB, Kuo YF, Freeman JL, Goodwin JS. 2005. Risk of fracture after androgen deprivation for prostate cancer. *N Engl J Med* 352(2): 154–64.
31. Nielsen M, Brixen K, Walter S, Andersen J, Eskildsen P, Abrahamsen. 2007. Fracture risk is increased in danish men with prostate cancer: A nation-wide register study. Abstracts of the 29th Annual Meeting of the American Society for Bone and Mineral Research, Honolulu, HI.
32. Ross CA, Taylor CL, Yaktine AL, Del Valle HB (eds.); Committee to Review Dietary Reference Intakes for Vitamin D and Calcium; Institute of Medicine. 2011. *Dietary Reference Intakes for Calcium and Vitamin D*. Washington, DC: National Academies Press.
33. Orwoll E, Ettinger M, Weiss S, Miller P, Kendler D, Graham J, Adami S, Weber K, Lorenc R, Pietschmann P, Vandormael K, Lombardi A. 2000. Alendronate for the treatment of osteoporosis in men. *N Engl J Med* 343(9): 604–10.
34. Orwoll ES, Scheele WH, Paul S, Adami S, Syversen U, Diez-Perez A, Kaufman JM, Clancy AD, Gaich GA. 2003. The effect of teriparatide [human parathyroid hormone (1-34)] therapy on bone density in men with osteoporosis. *J Bone Miner Res* 18(1): 9–17.
35. Lyles KW, Colon-Emeric CS, Magaziner JS, Adachi JD, Pieper CF, Mautalen C, Hyldstrup L, Recknor C, Nordsletten L, Moore KA, Lavecchia C, Zhang J, Mesenbrink P, Hodgson PK, Abrams K, Orloff JJ, Horowitz Z, Eriksen EF, Boonen S. 2007. Zoledronic acid in reducing clinical fracture and mortality after hip fracture. *N Engl J Med* 357: nihpa40967.
36. Adachi JD, Bensen WG, Brown J, Hanley D, Hodsman A, Josse R, Kendler DL, Lentle B, Olszynski W, Ste-Marie LG, Tenenhouse A, Chines AA. 1997. Intermittent etidronate therapy to prevent corticosteroid-induced osteoporosis. *N Engl J Med* 337(6): 382–7.
37. Reid DM, Hughes RA, Laan RF, Sacco-Gibson NA, Wenderoth DH, Adami S, Eusebio RA, Devogelaer JP. 2000. Efficacy and safety of daily residronate in the treatment of corticosteroid-induced osteoporosis in men and women: A randomized trial. *J Bone Miner Res* 15: 1006–13.
38. Smith MR, Egerdie B, Hernandez Toriz N, Feldman R, Tammela TL, Saad F, Heracek J, Szwedowski M, Ke C, Kupic A, Leder BZ, Goessl C. 2009. Denosumab in men receiving androgen-deprivation therapy for prostate cancer. *N Engl J Med* 361(8): 745–55.
39. Page ST, Amory JK, Bowman FD, Anawalt BD, Matsumoto AM, Bremner WJ, Tenover JL. 2005. Exogenous testosterone (T) alone or with finasteride increases physical performance, grip strength, and lean body mass in older men with low serum T. *J Clin Endocrinol Metab* 90(3): 1502–10.
40. Amory JK, Watts NB, Easley KA, Sutton PR, Anawalt BD, Matsumoto AM, Bremner WJ, Tenover JL. 2004. Exogenous testosterone or testosterone with finasteride increases bone mineral density in older men with low serum testosterone. *J Clin Endocrinol Metab* 89(2): 503–10.
41. Snyder PJ, Peachey H, Berlin JA, Hannoush P, Haddad G, Dlewati A, Santanna J, Loh L, Lenrow DA, Holmes JH, Kapoor SC, Atkinson LE, Strom BL. 2000. Effects of testosterone replacement in hypogonadal men. *J Clin Endocrinol Metab* 85(8): 2670–7.
42. Watts NB, Adler RA, Bilezikian JP, Drake MT, Eastell R, Orwoll ES, Finkelstein JS. 2012. Osteoporosis in men: An Endocrine Society clinical practice guideline. *J Clin Endocrinol Metab* 97(6): 1802–22.

63
Premenopausal Osteoporosis
Adi Cohen and Elizabeth Shane

Introduction 514
Premenopausal Women with a History
of Low-Trauma Fracture 514
Premenopausal Women with Low Bone Mineral Density 514
Special Issues Related to Bone Mineral Density Interpretation
in Premenopausal Women 515
Secondary Causes of Osteoporosis
in Premenopausal Women 515

Evaluation of the Premenopausal Woman with
Osteoporosis 516
Management Issues 516
Summary and Conclusions 518
References 518

INTRODUCTION

In this chapter, we will discuss issues specific to the diagnosis, clinical evaluation, and management of premenopausal women who present with low trauma fracture and/or low bone mineral density (BMD).

PREMENOPAUSAL WOMEN WITH A HISTORY OF LOW-TRAUMA FRACTURE

The diagnosis of osteoporosis in premenopausal women is most secure when there is a history of low trauma fracture(s). A fracture (excluding fracture of the digits) that occurs with trauma equivalent to a fall from a standing height or less may be a sign of decreased bone strength, regardless of BMD.

Several studies have shown that fractures before menopause predict postmenopausal fractures [1–3]. In the Study of Osteoporotic Fractures (SOF), women with a history of premenopausal fracture were 35% more likely to fracture during the early postmenopausal years than women without a history of premenopausal fracture [1]. These findings suggest that certain lifelong traits, such as fall frequency, neuromuscular protective response to falls, bone mass, or various aspects of bone quality can affect lifelong fracture risk [2].

PREMENOPAUSAL WOMEN WITH LOW BONE MINERAL DENSITY

In postmenopausal women, osteoporosis may be diagnosed before a fracture has occurred by using BMD T-scores (see Chapter 49). However, in premenopausal women, the World Health Organization (WHO) criteria for diagnosis of osteoporosis and osteopenia do not apply to, and generally should not be used to categorize, BMD measurements. This is because in premenopausal women, the relationship between BMD and fracture risk is not clear. Since premenopausal women have a reported incidence and prevalence of fractures that is orders of magnitude lower than in postmenopausal women [1–4], the relationship between BMD and fracture risk is likely to be quite different in this younger age group.

The International Society for Clinical Densitometry (ISCD) recommends using Z-scores (comparison to an age-matched reference population), to categorize BMD measurements in premenopausal women. Young women with BMD Z-scores below −2.0 should be categorized as having BMD that is "below expected range for age" and those with Z-scores above −2.0 should be categorized as having BMD that is "within the expected range for age" [4]. Because Z-scores rather than T-scores are used, the diagnostic categories of "osteoporosis" and "osteopenia" based on T-scores should not be applied to premeno-

Primer on the Metabolic Bone Diseases and Disorders of Mineral Metabolism, Eighth Edition. Edited by Clifford J. Rosen.
© 2013 American Society for Bone and Mineral Research. Published 2013 by John Wiley & Sons, Inc.

pausal women. An exception to these recommendations occurs in perimenopausal women, in whom use of T-scores is appropriate.

According to current guidelines, the diagnosis of "osteoporosis" should not be based solely upon BMD measurements in premenopausal women. This is because the clinical significance of low BMD in a premenopausal woman remains unclear since there are no longitudinal data that allow us to evaluate the extent to which BMD predicts fracture incidence in this population. However, several studies have shown that young women with low BMD are at higher risk for fractures than young women with normal BMD [5, 6]. Premenopausal women with Colles fractures have been found to have significantly lower BMD at the non-fractured radius [7], lumbar spine, and femoral neck [8] than controls without fractures. Stress fractures in female military recruits and athletes are associated with lower BMD than controls [6]. In addition, high-resolution imaging and transiliac bone biopsy studies have found that healthy, normally menstruating, premenopausal women with unexplained low BMD and no fractures have similar microarchitectural disruption to a comparable cohort of premenopausal women with low-trauma fractures, suggesting that very low BMD may represent a presymptomatic phase of osteoporosis in this group [9, 10a].

Additionally, the International Osteoporosis Foundation has recommended maintaining the T score cutoff of –2.5 at the spine or hip for the diagnosis of osteoporosis in those young adults who are thought to have completed growth and who suffer from a chronic disorder known to affect bone mass or to have an ongoing secondary cause of bone loss or fragility [10b].

SPECIAL ISSUES RELATED TO BONE MINERAL DENSITY INTERPRETATION IN PREMENOPAUSAL WOMEN

Although the majority of bone mass acquisition occurs during adolescence, BMD may continue to increase slightly between ages 20 and 30 [11]. Thus, very young women with low BMD measurements may not have yet achieved peak bone mass.

There are expected changes in bone mass associated with both pregnancy and lactation. At the lumbar spine, longitudinal studies document losses of 3–5% over a pregnancy and 3–10% over a 6-month period of lactation [12], with recovery of bone mass expected over 6–12 months, thereafter. Therefore, when interpreting a low BMD measurement in a premenopausal woman, the clinician must take the timing of recent pregnancy and lactation into account. Note that a young woman with fractures or low BMD diagnosed in the setting of pregnancy or lactation still requires an evaluation for potential secondary causes of osteoporosis (see below).

SECONDARY CAUSES OF OSTEOPOROSIS IN PREMENOPAUSAL WOMEN

Most premenopausal women with low-trauma fractures or low BMD have an underlying disorder or medication exposure that has interfered with bone mass accrual during adolescence and/or has caused excessive bone loss after reaching peak bone mass. In a population study from Olmstead County, Minnesota, 90% of men and women aged 20–44 with osteoporotic fractures were found to have a secondary cause [13]. In contrast, several case series of young women with osteoporosis evaluated in tertiary centers report that only 50% have secondary causes [14, 15], likely reflecting referral bias of more obscure cases to specialists.

Potential secondary causes are listed in Table 63.1. Many of these are discussed elsewhere in this *Primer*. The main goal of the evaluation of a premenopausal woman with low-trauma fractures or low BMD is to identify any secondary cause, and to institute specific treatment for that cause if it is correctable. Often this can be accomplished by a detailed history and physical examination, though an exhaustive biochemical evaluation may be necessary.

Table 63.1. Secondary Causes of Osteoporosis in Premenopausal Women

Premenopausal amenorrhea (e.g., pituitary diseases, medications)
Anorexia nervosa
Cushing's syndrome
Hyperthyroidism
Primary hyperparathyroidism
Vitamin D, calcium, and/or other nutrient deficiency
Gastrointestinal malabsorption (celiac disease, inflammatory bowel disease, cystic fibrosis, postoperative states)
Rheumatoid arthritis, systemic lupus erythematosus (SLE)
SLE, other inflammatory conditions
Renal disease
Liver disease
Hypercalciuria
Alcoholism
Connective tissue diseases
  Osteogenesis imperfecta
  Marfan syndrome
  Ehlers-Danlos Syndrome
Medications
- Glucocorticoids
- Immunosuppressants (e.g., cyclosporine)
- Antiepileptic drugs (particularly cytochrome P450 inducers such as phenytoin, carbamazepine)
- Cancer chemotherapy
- GnRH agonists (when used to suppress ovulation)
- Heparin

Idiopathic osteoporosis

Idiopathic osteoporosis

Premenopausal women with osteoporosis, in whom no definable cause can be found after a detailed evaluation, are said to have idiopathic osteoporosis (IOP). IOP is predominantly reported in Caucasians, and family history of osteoporosis is common [13–15]. Mean age at diagnosis is 35 years. Multiple vertebral and/or nonvertebral fractures may occur over 5 to 15 years; alternatively, there may be a single, major osteoporotic fracture, such as a low-trauma spine, hip, or long bone fracture [13, 15, 16]. Studies of bone microarchitecture in premenopausal women with IOP have shown that women with unexplained low BMD and those with low-trauma fractures have comparably abnormal bone microstructure, with thinner cortices and thinner, more widely spaced and heterogeneously distributed trabeculae [9, 10a]. Bone turnover, as assessed by tetracycline-labeled transiliac bone biopsies, was heterogeneous, with very low, normal, and high remodeling observed, suggesting diverse pathogeneses could account for the microarchitectural deterioration. Women with high bone turnover had a biochemical pattern resembling idiopathic hypercalciuria (mildly increased 24-hour urinary calcium and higher serum $1,25(OH)_2D$ concentrations compared to controls). In women with low bone turnover, microstructural deficits were more profound, serum insulin-like growth factor-1 (IGF-1) concentrations were higher, and osteoblasts appeared to synthesize less bone matrix per remodeling site, suggesting osteoblast resistance to IGF-1 [9, 16].

EVALUATION OF THE PREMENOPAUSAL WOMAN WITH OSTEOPOROSIS

Premenopausal women with low BMD (Z-score below −2.0), and those with a low-trauma fracture regardless of whether their BMD is frankly low, should undergo a thorough evaluation for secondary causes of bone loss. Identification of a contributing condition often helps to guide management of the affected individual.

A careful medical history is essential, including information about family history, fractures, kidney stones, oligo- or amenorrhea, or other history consistent with premenopausal estrogen deficiency, timing of recent pregnancies and lactation, dieting and exercise behavior, subtle gastrointestinal symptoms and medications, including over-the-counter supplements. During the physical examination, look for signs of Cushing's syndrome, thyrotoxicosis, or connective tissue disorders (e.g., blue sclerae in some forms of osteogenesis imperfecta, joint hypermobility in Ehlers-Danlos syndrome).

The laboratory evaluation (Table 63.2) should be aimed at identifying secondary causes such as hyperthyroidism, hyperparathyroidism, Cushing's syndrome, early menopause, renal or liver disease, celiac disease, malabsorption, and idiopathic hypercalciuria. Bone turnover markers and follow-up bone density testing may help to distinguish those with stable low BMD from those with ongoing bone loss who may be at higher short-term risk of fracture. In premenopausal women with a history of low-trauma fracture(s) and no known secondary cause, bone biopsy may be indicated. In the authors' experience, a bone biopsy may occasionally identify unsuspected causes of bone fragility, such as Gaucher's disease or mastocytosis.

Table 63.2. Laboratory Evaluation

Initial Laboratory Evaluation
- Complete blood count
- Electrolytes, renal function
- Serum calcium, phosphate
- Serum albumin, transaminases, total alkaline phosphatase
- Serum TSH
- Serum 25-hydroxyvitamin D
- 24-hour urine for calcium and creatinine

Additional Laboratory Evaluation
- Estradiol, LH, FSH, prolactin
- PTH
- 1,25-dihydroxyvitamin D
- 24-hour urine for free cortisol
- Iron/TIBC, Ferritin
- Celiac screen
- Serum/urine protein electrophoresis
- ESR or CRP
- Bone turnover markers
- Transiliac crest bone biopsy

MANAGEMENT ISSUES

General measures

For all patients, one should recommend a set of general measures that benefit bone health: adequate weight-bearing exercise [17, 18], nutrition (protein, calories, calcium, vitamin D), and lifestyle modifications (smoking cessation, avoidance of excess alcohol).

In the authors' opinion, pharmacological therapy is rarely justified for premenopausal women with isolated low BMD and no history of fractures, in whom there is no identifiable secondary cause, particularly if the Z-score is above −3.0. Low BMD in such young woman may be due to genetic low peak bone mass, or to past insults to the growing or adult skeleton (nutritional deficiency, alcohol excess, medications, estrogen deficiency) that are no longer operative. Such young women usually have low short-term risk of fracture. Moreover, Peris et al. recently reported slight BMD improvement and no further fractures in women with unexplained osteoporosis managed with only calcium (total intake of 1,500 mg/day), vitamin D (400–800 IU/daily) and exercise [19]. Bone density should be remeasured after 1 or 2 years to confirm that it is stable and identify patients with ongoing bone loss.

In women with low BMD or low-trauma fractures and a known secondary cause, address the underlying cause if possible. Women with estrogen deficiency should receive estrogen (unless contraindicated), those with celiac disease should begin a gluten-free diet, those with primary hyperparathyroidism may benefit from parathyroidectomy (see Chapter 68), and those with idiopathic hypercalciuria may benefit from thiazide diuretics.

In some women, it is not possible to address or alleviate the secondary cause directly. Premenopausal women requiring long-term glucocorticoids and those being treated for breast cancer may require pharmacological therapy to prevent excessive bone loss or fractures. Treatment options include antiresorptive drugs such as estrogen, calcitonin, bisphosphonates, and denosumab, or anabolic agents such as teriparatide. Selective estrogen receptor modulators (SERMS), such as raloxifene, should not be used to treat bone loss in menstruating women since they block estrogen action on bone and lead to further bone loss [20].

Bisphosphonates

Bisphosphonates have been shown to prevent bone loss in premenopausal women with various conditions [21–26]. However, large randomized trials are scarce and the U.S. Food and Drug Administration has approved oral bisphosphonates only for premenopausal women on glucocorticoids. Because bisphosphonates accumulate in the maternal skeleton, cross the placenta, accumulate in the fetal skeleton [27], and cause toxic effects in pregnant rats [28], they should be used with caution in women who may become pregnant. While several reports document normal pregnancies and fetal outcomes in women receiving bisphosphonates [23, 29–31], the potential for fetal abnormalities should be considered when prescribing bisphosphonates for a premenopausal woman.

Because there are so few data regarding the long-term efficacy and safety of bisphosphonates in young women, the decision to initiate treatment must be made on a case-by-case basis with consideration of individual fracture risk and with a plan for the shortest possible duration of use. In general, bisphosphonates should be reserved for those with fragility fractures or ongoing bone loss.

Glucocorticoid-induced osteoporosis

Bisphosphonates are approved for prevention and treatment of glucocorticoid-induced osteoporosis. However, relatively few premenopausal women participated in the relevant large registration trials for bisphosphonates in glucocorticoid-induced osteoporosis and none of the premenopausal women in those trials fractured [32–34]. A few studies have demonstrated protective effects of intermittent cyclical etidronate and oral pamidronate in premenopausal women with autoimmune and connective tissue diseases [24, 25]. Guidelines from the American College of Rheumatology suggest that bisphosphonates be considered for prevention and treatment of glucocorticoid-induced osteoporosis in premenopausal women taking at least 7.5 mg of prednisone or equivalent per day for 3 months or more [35]. However, because of potential harm to the fetus in women who may become pregnant, they also urge great caution in the use of bisphosphonates in premenopausal women [35].

Bisphosphonate use for other secondary causes of osteoporosis

Intravenous and oral bisphosphonates prevent bone loss in premenopausal women experiencing ovarian failure in the setting of treatment for breast cancer [36–40]. Bisphosphonates may lower fracture risk in young individuals with osteogenesis imperfecta (see Chapter 99). Both alendronate and risedronate have been shown to significantly increase BMD in young women with anorexia [21, 22]. Bisphosphonates have also been associated with substantial increases of 11–23% in lumbar spine bone density [23] in women with pregnancy- and lactation-associated osteoporosis. Since bone density is expected to increase postpartum and after weaning in normal women, and there was no untreated control group, it is not clear to what extent bisphosphonate use provided an incremental benefit for these patients.

Human PTH(1-34)

There are even fewer data on the effects of teriparatide or PTH(1-34) in premenopausal women, but this medication has been studied in women with medication-induced amenorrhea, women with IOP, and those on glucocorticoids. In young women treated with the GnRH analog nafarelin for endometriosis, spine BMD declined by 4.9%, while those treated with PTH(1-34) 40 μg daily together with nafarelin had an increase of 2.1% ($p < 0.001$) [41]. It is not clear whether these results would apply to premenopausal women with normal gonadal status. A recent study comparing teriparatide and alendronate for glucocorticoid-induced osteoporosis included some premenopausal women. Overall, teriparatide was associated with significantly greater increases in lumbar spine and total hip BMD and resulted in significantly fewer incident vertebral fractures than alendronate [42]. The BMD responses were similar in premenopausal women as in men and postmenopausal women, but no fractures occurred in either premenopausal group.

In an observational study of teriparatide 20 μg daily in 21 premenopausal women with IOP, BMD increased by 12.2% at the lumbar spine and 6.4% at the total hip (both $p < 0.05$) after 12 months of treatment [26]. However, among this unique cohort, a small subset with very low baseline bone turnover had little or no increase in BMD on this medication [26]. Because the long-term effects of teriparatide in young women are not known, use of this medication should be reserved for those at highest risk for fracture or those who are experiencing recurrent fractures. In young women younger than 25 years of age, documentation of fused epiphyses is recommended prior to consideration of teriparatide treatment, since continued bone growth is considered a contraindication to use of this medication.

SUMMARY AND CONCLUSIONS

Premenopausal woman with low-trauma fracture(s) or low BMD (Z-score below −2.0) should have a thorough evaluation for secondary causes of osteoporosis and bone loss. In most, a secondary cause can be found, the most common being glucocorticoid excess, anorexia nervosa, premenopausal estrogen deficiency, and celiac disease. Where possible, identification and treatment of the underlying cause should be the focus of management. Although pharmacologic therapy is rarely justified in premenopausal women, those with an ongoing cause of bone loss and those who have had or continue to have low-trauma fractures may require pharmacological intervention, such as bisphosphonates or teriparatide. Few high-quality clinical trials exist to provide guidance, and there are no data that such intervention actually reduces the risk of future fractures.

REFERENCES

1. Hosmer WD, Genant HK, Browner WS. 2002. Fractures before menopause: A red flag for physicians. *Osteoporos Int* 13(4): 337–41.
2. Wu F, Mason B, Horne A, Ames R, Clearwater J, Liu M, Evans MC, Gamble GD, Reid IR. 2002. Fractures between the ages of 20 and 50 years increase women's risk of subsequent fractures. *Arch Intern Med* 162(1): 33–6.
3. Honkanen R, Tuppurainen M, Kroger H, Alhava E, Puntila E. 1997. Associations of early premenopausal fractures with subsequent fractures vary by sites and mechanisms of fractures. *Calcif Tissue Int* 60(4): 327–31.
4. Lewiecki EM, Gordon CM, Baim S, Leonard MB, Bishop NJ, Bianchi ML, et al. 2008. International Society for Clinical Densitometry 2007 adult and pediatric official positions. *Bone* 43(6): 1115–21.
5. Lauder TD, Dixit S, Pezzin LE, Williams MV, Campbell CS, Davis GD. 2000. The relation between stress fractures and bone mineral density: Evidence from active-duty Army women. *Arch Phys Med Rehabil* 81(1): 73–9.
6. Lappe J, Davies K, Recker R, Heaney R. 2005. Quantitative ultrasound: Use in screening for susceptibility to stress fractures in female army recruits. *J Bone Miner Res* 20(4): 571–8.
7. Wigderowitz CA, Cunningham T, Rowley DI, Mole PA, Paterson CR. 2003. Peripheral bone mineral density in patients with distal radial fractures. *J Bone Joint Surg Br* 85(3): 423–5.
8. Hung LK, Wu HT, Leung PC, Qin L. 2005. Low BMD is a risk factor for low-energy Colles' fractures in women before and after menopause. *Clin Orthop Relat Res* (435): 219–25.
9. Cohen A, Dempster D, Recker R, Stein EM, J L, Zhou H, Wirth AJ, van Lenthe GH, Kohler T, Zwahlen A, Muller R, Rosen CJ, Cremers S, Nickolas TL, DJ M, Rogers H, Staron RB, Lemaster J, Shane E. 2011. Abnormal bone microarchitecture and evidence of osteoblast dysfunction in premenopausal women with idiopathic osteoporosis. *J Clin Endocrinol Metab* 96(10): 3095–105.
10a. Cohen A, Liu XS, Stein EM, McMahon DJ, Rogers HF, Lemaster J, Recker RR, Lappe JM, Guo XE, Shane E. 2009. Bone microarchitecture and stiffness in premenopausal women with idiopathic osteoporosis. *J Clin Endocrinol Metab* 94(11): 4351–60.
10b. Ferrari S, Bianchi ML, Eisman JA, Foldes AJ, Adami S, Wahl DA, et al. 2012. Osteoporosis in young adults: Pathophysiology, diagnosis, and management. *Osteoporosis Int* 23(12): 2735–48.
11. Recker RR, Davies KM, Hinders SM, Heaney RP, Stegman MR, Kimmel DB. 1992. Bone gain in young adult women. *JAMA* 268(17): 2403–8.
12. Karlsson MK, Ahlborg HG, Karlsson C. 2005. Maternity and bone mineral density. *Acta Orthop* 76(1): 2–13.
13. Khosla S, Lufkin EG, Hodgson SF, Fitzpatrick LA, Melton LJ 3rd. 1994. Epidemiology and clinical features of osteoporosis in young individuals. *Bone* 15(5): 551–5.
14. Moreira Kulak CA, Schussheim DH, McMahon DJ, Kurland E, Silverberg SJ, Siris ES, Bilezikian JP, Shane E. 2000. Osteoporosis and low bone mass in premenopausal and perimenopausal women. *Endocr Pract* 6(4): 296–304.
15. Peris P, Guanabens N, Martinez de Osaba MJ, Monegal A, Alvarez L, Pons F, Ros I, Cerda D, Munoz-Gomez J. 2002. Clinical characteristics and etiologic factors of premenopausal osteoporosis in a group of Spanish women. *Semin Arthritis Rheum* 32(1): 64–70.
16. Cohen A, Recker RR, Lappe J, Dempster DW, Cremers S, McMahon DJ, Stein EM, Fleischer J, Rosen CJ, Rogers H, Staron RB, Lemaster J, Shane E. 2012. Premenopausal women with idiopathic low-trauma fractures and/or low bone mineral density. *Osteoporos Int* 23(1): 171–82.
17. Wallace BA, Cumming RG. 2000. Systematic review of randomized trials of the effect of exercise on bone mass in pre- and postmenopausal women. *Calcif Tissue Int* 67(1): 10–8.
18. Mein AL, Briffa NK, Dhaliwal SS, Price RI. 2004. Lifestyle influences on 9-year changes in BMD in young women. *J Bone Miner Res* 19(7): 1092–8.
19. Peris P, Monegal A, Martinez MA, Moll C, Pons F, Guanabens N. 2007. Bone mineral density evolution in young premenopausal women with idiopathic osteoporosis. *Clin Rheumatol* 26(6): 958–61.

20. Powles TJ, Hickish T, Kanis JA, Tidy A, Ashley S. 1996. Effect of tamoxifen on bone mineral density measured by dual-energy x-ray absorptiometry in healthy premenopausal and postmenopausal women. *J Clin Oncol* 14(1): 78–84.

21. Golden NH, Iglesias EA, Jacobson MS, Carey D, Meyer W, Schebendach J, Hertz S, Shenker IR. 2005. Alendronate for the treatment of osteopenia in anorexia nervosa: A randomized, double-blind, placebo-controlled trial. *J Clin Endocrinol Metab* 90(6): 3179–85.

22. Miller KK, Grieco KA, Mulder J, Grinspoon S, Mickley D, Yehezkel R, Herzog DB, Klibanski A. 2004. Effects of risedronate on bone density in anorexia nervosa. *J Clin Endocrinol Metab* 89(8): 3903–6.

23. O'Sullivan SM, Grey AB, Singh R, Reid IR. 2006. Bisphosphonates in pregnancy and lactation-associated osteoporosis. *Osteoporos Int* 17(7): 1008–12.

24. Nzeusseu Toukap A, Depresseux G, Devogelaer JP, Houssiau FA. 2005. Oral pamidronate prevents high-dose glucocorticoid-induced lumbar spine bone loss in premenopausal connective tissue disease (mainly lupus) patients. *Lupus* 14(7): 517–20.

25. Nakayamada S, Okada Y, Saito K, Tanaka Y, 2004, Etidronate prevents high dose glucocorticoid induced bone loss in premenopausal individuals with systemic autoimmune diseases. *J Rheumatol* 31(1): 163–6.

26. Cohen A, Stein EM, Recker RR, Lappe JM, Dempster DW, Zhou H, Cremers S, McMahon DJ, Nickolas TL, Müller R, Zwahlen A, Young P, Stubby J, Shane E. Teriparatide for idiopathic osteoporosis in premenopausal women: A Pilot Study. *J Clin Endocrinol Metab*. In Press.

27. Patlas N, Golomb G, Yaffe P, Pinto T, Breuer E, Ornoy A. 1999. Transplacental effects of bisphosphonates on fetal skeletal ossification and mineralization in rats. *Teratology* 60(2): 68–73.

28. Minsker DH, Manson JM, Peter CP. 1993. Effects of the bisphosphonate, alendronate, on parturition in the rat. *Toxicol Appl Pharmacol* 121(2): 217–23.

29. Biswas PN, Wilton LV, Shakir SA. 2003. Pharmacovigilance study of alendronate in England. *Osteoporos Int* 14(6): 507–14.

30. Chan B, Zacharin M. 2006. Maternal and infant outcome after pamidronate treatment of polyostotic fibrous dysplasia and osteogenesis imperfecta before conception: A report of four cases. *J Clin Endocrinol Metab* 91(6): 2017–20.

31. Levy S, Fayez I, Taguchi N, Han JY, Aiello J, Matsui D, Moretti M, Koren G, Ito S. 2009. Pregnancy outcome following in utero exposure to bisphosphonates. *Bone* 44(3): 428–30.

32. Adachi JD, Bensen WG, Brown J, Hanley D, Hodsman A, Josse R, Kendler DL, Lentle B, Olszynski W, Ste-Marie LG, Tenenhouse A, Chines AA. 1997. Intermittent etidronate therapy to prevent corticosteroid-induced osteoporosis. *N Engl J Med* 337(6): 382–7.

33. Saag KG, Emkey R, Schnitzer TJ, Brown JP, Hawkins F, Goemaere S, Thamsborg G, Liberman UA, Delmas PD, Malice MP, Czachur M, Daifotis AG. 1998. Alendronate for the prevention and treatment of glucocorticoid-induced osteoporosis. Glucocorticoid-Induced Osteoporosis Intervention Study Group. *N Engl J Med* 339(5): 292–9.

34. Wallach S, Cohen S, Reid DM, Hughes RA, Hosking DJ, Laan RF, Doherty SM, Maricic M, Rosen C, Brown J, Barton I, Chines AA. 2000. Effects of risedronate treatment on bone density and vertebral fracture in patients on corticosteroid therapy. *Calcif Tissue Int* 67(4): 277–85.

35. Grossman JM, Gordon R, Ranganath VK, Deal C, Caplan L, Chen W, Curtis JR, Furst DE, McMahon M, Patkar NM, Volkmann E, Saag KG. 2010. American College of Rheumatology 2010 recommendations for the prevention and treatment of glucocorticoid-induced osteoporosis. *Arthritis Care Res (Hoboken)* 62(11): 1515–26.

36. Hershman DL, McMahon DJ, Crew KD, Cremers S, Irani D, Cucchiara G, Brafman L, Shane E. 2008. Zoledronic acid prevents bone loss in premenopausal women undergoing adjuvant chemotherapy for early-stage breast cancer. *J Clin Oncol* 26(29): 4739–45.

37. Hershman DL, McMahon DJ, Crew KD, Shao T, Cremers S, Brafman L, Awad D, Shane E. 2010. Prevention of bone loss by zoledronic acid in premenopausal women undergoing adjuvant chemotherapy persist up to one year following discontinuing treatment. *J Clin Endocrinol Metab* 95(2): 559–66.

38. Fuleihan Gel H, Salamoun M, Mourad YA, Chehal A, Salem Z, Mahfoud Z, Shamseddine A. 2005. Pamidronate in the prevention of chemotherapy-induced bone loss in premenopausal women with breast cancer: A randomized controlled trial. *J Clin Endocrinol Metab* 90(6): 3209–14.

39. Delmas PD, Balena R, Confravreux E, Hardouin C, Hardy P, Bremond A. 1997. Bisphosphonate risedronate prevents bone loss in women with artificial menopause due to chemotherapy of breast cancer: A double-blind, placebo-controlled study. *J Clin Oncol* 15(3): 955–62.

40. Gnant MF, Mlineritsch B, Luschin-Ebengreuth G, Grampp S, Kaessmann H, Schmid M, Menzel C, Piswanger-Soelkner JC, Galid A, Mittlboeck M, Hausmaninger H, Jakesz R. 2007. Zoledronic acid prevents cancer treatment-induced bone loss in premenopausal women receiving adjuvant endocrine therapy for hormone-responsive breast cancer: A report from the Austrian Breast and Colorectal Cancer Study Group. *J Clin Oncol* 25(7): 820–8.

41. Finkelstein JS, Klibanski A, Arnold AL, Toth TL, Hornstein MD, Neer RM. 1998. Prevention of estrogen deficiency-related bone loss with human parathyroid hormone-(1-34): A randomized controlled trial. *JAMA* 280(12): 1067–73.

42. Langdahl BL, Marin F, Shane E, Dobnig H, Zanchetta JR, Maricic M, Krohn K, See K, Warner MR. 2009. Teriparatide versus alendronate for treating glucocorticoid-induced osteoporosis: An analysis by gender and menopausal status. *Osteoporos Int* 20(12): 2095–104.

64

Skeletal Effects of Drugs

Juliet Compston

Introduction 520
Antihormonal Drugs 520
Thiazolidinediones 521
Acid-Suppressive Medications 522
Antiepileptic Drugs 522

Selective Serotonin Receptor Uptake Inhibitors 523
Heparin 523
Drugs That May Protect Against Osteoporosis 523
References 523

INTRODUCTION

Iatrogenic bone loss, caused by therapies for nonskeletal diseases, is a growing and important cause of osteoporosis (Table 64.1). Drugs that have been implicated include aromatase inhibitors, androgen deprivation therapy, thiazolidenediones, and proton pump inhibitors. In contrast, some interventions for nonskeletal diseases may be protective to the skeleton. Osteoporosis associated with glucocorticoids, progestagens, excess thyroid hormone, chemotherapy, and calcineurin inhibitors is described in other sections; this chapter focuses on the remaining drugs listed in Table 64.1. Drugs that may have protective skeletal effects are also considered.

ANTIHORMONAL DRUGS

Androgen deprivation therapy

Androgen deprivation therapy (ADT) for carcinoma of the prostate encompasses a number of options, including bilateral orchidectomy, gonadotrophin-releasing hormone (GnRH) analog therapy, and antiandrogenic agents (cyproterone acetate, flutamide, and bicalutamide). Its use has increased considerably in recent years, and osteoporosis has emerged as a common complication. ADT is associated with increased rates of bone loss at multiple skeletal sites, and increased fracture rates have also been reported [1–4]. Examination of a large database containing medical records of more than 50,000 men revealed that, in those treated with ADT for prostate cancer and surviving at least 5 years after diagnosis, 19.4% had a fracture compared with 12.6% of men not receiving ADT. In men who received at least nine doses of GnRH analogs, the relative risk was 1.45 (95% CI, 1.36–1.56) [2]. In another retrospective study, the clinical fracture rate in men treated for prostate cancer with GnRH analogs was 7.91/100 person-years compared with 6.55/100 person years in men with early prostate cancer who were not treated with GnRH analogs, translating into a relative risk (RR) of 1.23 (95% CI, 1.09–1.34); the largest increase in risk was seen for hip fractures (RR, 1.76; 95% CI, 1.33–2.33) [3]. Finally, in a retrospective population-based cohort study in 742 men with prostate cancer, overall fracture risk was increased 1.9-fold; the increase in risk in men on ADT being mainly accounted for by pathological fractures [4]. In this study, a 1.7-fold increase in fracture risk was also seen in men with prostate cancer who were not receiving ADT.

A number of interventions have been shown to have beneficial effects on bone mineral density (BMD) in men treated with ADT. These include raloxifene, toremifene, risedronate, pamidronate, zoledronic acid, alendronate and denosumab [5–9]. Reduction in vertebral fracture risk has been reported with two of these treatments, namely

---

Primer on the Metabolic Bone Diseases and Disorders of Mineral Metabolism, Eighth Edition. Edited by Clifford J. Rosen.
© 2013 American Society for Bone and Mineral Research. Published 2013 by John Wiley & Sons, Inc.

Table 64.1. Drugs Associated with Osteoporosis

Glucocorticoids
Antihormonal drugs
Thiazolidinediones
Proton pump inhibitors
Heparin
Anticonvulsants
Calcineurin inhibitors
Thyroxine
Chemotherapy
Selective serotonin reuptake inhibitors

toremifene (80 mg/day) and denosumab (60 mg sc every 6 months) [8, 9]. All men treated with ADT should undergo assessment of fracture risk using BMD measurement and clinical risk factors at the start of treatment, with repeat BMD measurement at 1–2 years as clinically indicated. Bone protective therapy should be started in men with a history of hip or vertebral fracture and/or a T-score at or below −2.5. Additionally, drug therapy has been recommended in men with a low T-score (−1.0 to −2.5) and a 10-year risk of 3% or higher for hip fracture or 20% or lower for major osteoporotic fracture, as assessed by FRAX (fracture risk assessment tool) [10].

Aromatase inhibitors

Aromatase inhibitors (AIs) are now regarded as front-line adjuvant therapy in women with estrogen receptor-positive breast cancer. They reduce endogenous estrogen production by 80–90% by blocking the peripheral conversion of androgens to estrogen and have largely replaced the selective estrogen receptor modulator, tamoxifen, as the preferred treatment option for postmenopausal women. The most commonly used AIs in clinical practice are exemestane, anastrozole, and letrozole; the former is steroidal, whereas the latter two are nonsteroidal [11].

In contrast to tamoxifen, which is bone protective in postmenopausal women (but has the opposite effect in premenopausal women), AIs have adverse effects on BMD and possibly also fracture risk. Interpretation of studies of the effects of AIs on bone is complicated by the use of tamoxifen as a comparator in many studies and also, in some studies, the use of tamoxifen before AI therapy. In addition, comparative data on the effects of different AIs on BMD and fracture rate are currently lacking. Nevertheless, the existing biomarker data indicate that all three AIs increase bone turnover, although some studies indicate a proportionately greater effect of exemestane on formation than resorption. Increased rates of bone loss have also been reported; for example, in the ATAC (Arimidex, Tamoxifen, Alone or in Combination) trial, median rates of bone loss at the spine and hip were 4.1% and 3.9%, respectively, over 2 years [12], and in a study of letrozole versus placebo after 5 year of tamoxifen therapy, losses in the spine and hip over 2 years were 5.4% and 3.5%, respectively [13]. Significant increases in fracture rate have been shown in several trials of postmenopausal women with breast cancer, although interpretation of the data is again complicated by the use of tamoxifen as a comparator and treatment with tamoxifen before AI therapy in the majority of studies. In the ATAC study, the incidence of fractures after 5 years was 11% in women treated with anastrozole and 7.7% in those receiving tamoxifen (p < 0.001) [14]. Comparison of letrozole and tamoxifen produced similar results, with fracture rates of 8.6% and 5.8%, respectively (p < 0.001) at a median follow-up period of 51 months [15]. However, comparison of letrozole with a placebo in women with breast cancer after completion of 5 years of tamoxifen therapy did not show any significant difference in fracture rate (3.6% in letrozole group versus 2.9% placebo; p = 0.24) [16]. Finally, in the Intergroup Exemestane Study, in which women who were disease free after 2–3 years with tamoxifen were randomized to continue tamoxifen or switched to exemestane to complete a total of 5 years of therapy, a significantly higher fracture rate was seen in the women taking exemestane (7% vs 5%, p = 0.003). [17] Collectively, these data indicate that AI therapy is associated with a significantly higher fracture risk than tamoxifen but may not increase fracture risk above that seen in untreated women. This suggests that the difference in fracture risk between women treated with AIs and tamoxifen is at least partly caused by a protective effect of the latter.

Prevention of bone loss associated with AI therapy has been demonstrated with intravenous zoledronic acid, 4 mg every 6 months, oral risedronate 35 mg once weekly and denosumab 60 mg sc every 6 months, although data on fracture reduction are lacking [18–22]. At present, it seems reasonable to advise risk assessment, including BMD measurements, in all postmenopausal women treated with AIs and to perform repeat measurements at 1- to 2-year intervals in those at moderate risk (based on age, BMD, and other clinical risk factors). In view of the uncertainty about whether AIs are associated with increased fracture risk and the absence of data on antifracture efficacy of bone protective therapy, the indications for intervention are not clearly defined. One approach is to use guidelines similar to those for postmenopausal women not receiving AIs, while others advocate treating at a higher BMD threshold in view of the rapid bone loss that occurs in some women.

THIAZOLIDINEDIONES

Thiazolidinediones (TZDs) are ligands for peroxisome proliferator-activated receptor γ (PPARγ) and are widely used in the treatment of type 2 diabetes. Activation of PPARγ increases marrow adiposity, increases insulin sensitivity, and suppresses bone formation. In transgenic mice models PPARγ deficiency is associated with high bone mass whereas activation of PPARγ induces bone loss [23, 24]. The mechanisms by which suppression of bone formation occurs have not been fully established but may include inhibition of the Wnt/ß-catenin signaling pathway, inhibition of osteoblast differentiation genes including Runx2 and osterix, and suppression of insulin-like growth factor production [25]. The effects of TZDs in humans are particularly relevant in view of the increased risk of fracture associated with type 2 diabetes. In an observational study, increased rates of bone loss in the lumbar spine, trochanter, and whole body were reported in older diabetic women treated with TZDs [26]. Grey et al. [27] in a randomized controlled trial in healthy postmenopausal women, showed that administration of the TZD rosiglitazone for 14 weeks resulted in significant bone loss in the hip (mean 1.9% vs 0.2% in the placebo group) and significant suppression of biochemical markers of bone formation. In the ADOPT study (A Diabetes Outcome Progression Trial), a randomized controlled trial in 4,360 patients with type 2 diabetes, the effects of treatment with rosiglitazone were compared with those of metformin and glibenclamide [28]. In women, there was a statistically significant increase in the incidence of fractures, mainly affecting the foot, hand, and upper arm (9.3% in the rosiglitazone group versus 1.54% and 1.29% in the metformin and glibenclamide groups, respectively). Subsequent studies, including a meta-analysis [29] have confirmed the increase in fracture risk associated with TZDs. This appears to be a drug class effect and is positively associated with duration of treatment; some, although not all, studies indicate that women and men are equally affected [30]. Adverse skeletal effects, therefore, have to be considered when weighing up the risk/benefit balance associated with TZDs, particularly in high-risk individuals. The contribution to fracture of factors other than reduced BMD in this population, including alteration in collagen cross-linking as a result of the formation of advanced glycation end products and increased risk of falling, merits further study.

ACID-SUPPRESSIVE MEDICATIONS

Increased risk of fracture has been reported in individuals treated with acid-suppressive medications. Grisso et al. [31] reported an association between use of the H_2 receptor blocker, cimetidine, and hip fracture in a case-control study of men with hip fracture (OR, 2.5; 95% CI, 1.4–4.6). Subsequently, in a case-control study from Denmark, proton pump inhibitor (PPI) use within the past year was shown to be associated with a small increase in overall fracture risk (adjusted OR, 1.18; 95% CI, 1.12–1.43), hip fracture risk (adjusted OR, 1.45; 95% CI, 1.28–1.65), and spine fracture risk (adjusted OR, 1.60; 95% CI, 1.25–2.04) [32]. Interestingly, in this study, the use of histamine H_2 receptor antagonists within the past year was associated with a significantly reduced risk of fracture (adjusted OR, 0.88; 95% CI, 0.82–0.95), although a significant increase in fracture risk was observed with antacid medications other than PPIs and H_2 receptor blockers. Analysis of data from the Study of Osteoporotic Fractures (SOF) showed that women taking a PPI or H_2 receptor blocker (grouped together) had a significantly increased risk of non-spine fracture (RH, 1.18; 95% CI, 1.01–1.39), but total hip BMD and rates of bone loss at this site were similar between users and nonusers of these acid-suppressive medications [33]. In a nested case control study from the General Practice Research Database in the U.K., a significant increase in hip fracture risk was found in PPI users, which increased significantly with dose and with duration of use [34]. However, in a subsequent study using the same database, use of PPIs was not associated with increased risk of hip fracture in individuals without major risk factors, suggesting that residual confounding or effect modification may have explained the increased risk reported in the original study [35]. Similar findings were recently reported by Corley et al. in a case control study of individuals taking either PPIs or H_2 receptor antagonists [36]. In a prospective analysis of women enrolled in the Women's Health Initiative (WHI) study, PPI use was associated with an increase in clinical spine, forearm or wrist, and all clinical fractures, but not hip fractures; a robust association between PPI and BMD was not demonstrated [37]. Overall, therefore, existing data support an association between acid-suppressive medication and fracture, although the limitations of observational studies, particularly the effects of potential but unmeasured confounding factors, have to be recognized, and the discrepant findings with respect to H_2 receptor blockers remain unexplained.

The mechanism by which PPIs and possibly other acid-suppressive medications increase fracture risk is unknown. Reduced intestinal calcium absorption resulting from increased gastric pH may contribute, although the relatively small reduction in absorption of calcium carbonate shown in one study of postmenopausal women taking omeprazole seems unlikely to be solely responsible for the development of increased fracture risk within 1 year of starting therapy [38]. Inhibition of the osteoclastic proton pump would be expected to have beneficial skeletal actions, although it is unknown whether such effects are associated with PPI use *in vivo*. Finally, the data from SOF and the WHI study indicate that the increase in fracture risk associated with PPIs may not be BMD driven. Nevertheless, the association has important clinical implications because PPI use is common in the elderly population and potentially might attenuate the effects of bone protective interventions.

ANTIEPILEPTIC DRUGS

An association between antiepileptic drugs (AEDs) and increased fracture risk has been reported in several observational studies, and reduced BMD and increased rates of bone loss have also been reported [39–43]. The underlying pathogenesis is unclear; vitamin D deficiency, trauma during seizures, increased risk of falling, and co-medications including glucocorticoids may all contribute. In a few patients with severe vitamin D deficiency, osteomalacia or rickets may be present. Currently, there are insufficient data to distinguish between the skeletal effects of specific AED regimens.

Management guidelines to prevent and treat bone disease in AED users have been proposed, although at present, these lack a robust evidence base. Routine prophylaxis of vitamin D deficiency should be considered in high-risk individuals, for example, the elderly and institutionalized (higher than normal doses may be required in patients taking some AEDs), and in such cases, calcium supplements should also be given. Routine bone densitometry in all AED users cannot be justified at present, although BMD should be measured in those who present with fracture or have other clinical risk factors. Treatment of established osteoporosis in this population has not been specifically evaluated.

SELECTIVE SEROTONIN RECEPTOR UPTAKE INHIBITORS

Selective serotonin receptor uptake inhibitors (SSRIs) are widely prescribed as antidepressants and have been associated with reduced BMD in older men [44], increased rates of hip bone loss in older women [45], and increased fracture risk in men and women 50 years of age or older [46]. While depression itself may be associated with increased risk of osteoporosis [47], the regulatory role of serotonin in bone remodeling provides a plausible biological mechanism by which SSRIs may cause adverse skeletal effects [48].

HEPARIN

Long-term heparin therapy, which nowadays is virtually restricted to prophylaxis against thromboembolism in high-risk women during pregnancy, is associated with an increased risk of osteoporosis [49]. Reduced BMD, increased rates of bone loss, and increased fracture risk have all been reported during long-term heparin administration although the mechanisms responsible for bone loss have not been established [50, 51]. The use of low molecular weight heparin and of newer antithrombotic agents such as fondaparinux may be associated with fewer adverse skeletal effects. Calcium and vitamin D supplements are often advocated but, in common with other antiresorptive regimens, have not been formally evaluated in this situation.

DRUGS THAT MAY PROTECT AGAINST OSTEOPOROSIS

Beta-blockers

In some studies, a significant protective effect of β-blocker therapy on fracture risk has been reported, although this finding has not been universal [52–55]. In a case-control study, a significant decrease in hip/femur fracture risk was associated with current β-blocker use, but was only present in patients with a history of using other antihypertensive medications and was not related to cumulative exposure, suggesting that the association may not be causal [56]. In a recent large cohort study of the effect of single therapy antihypertensive medication on fracture risk no decrease in risk of fracture was seen in people receiving β-blockers, although a significant reduction was seen in those using angiotensin receptor blockers (HR 0.76; 95% CI 0.68–0.86) and thiazide diuretics (HR 0.85; 95% CI 0.76–0.97) [57].

Thiazides

Reduced fracture risk in thiazide users has been reported in several prospective observational studies, and small increases in BMD have also been shown in randomized controlled trials [53, 57–60]. Increased renal calcium reabsorption is thought to play a role in these beneficial effects, although because this is transient, other mechanisms are likely to operate.

Statins

Statins inhibit the enzyme 3-hydroxy-3-methyl-glutaryl-coenzyme A reductase in the mevalonate pathway, thus reducing cholesterol biosynthesis but also preventing the prenylation of GTP- binding proteins and thus inhibiting osteoclast activity. Beneficial skeletal effects of statins in animals have been shown *in vitro* and *in vivo* [61], but studies in humans have produced conflicting results. Two meta-analyses have been conducted: one on BMD effects and the other on fracture [62, 63]. The former concluded that statins had small but significant benefits on BMD in the hip, whereas the latter suggested that statin use is likely to reduce hip fractures, although effects on other fracture types have not been demonstrated. However, in a double-blind randomized placebo-controlled trial, clinically relevant doses of atorvastatin that lower lipid levels had no effect on BMD or biochemical indices of bone metabolism [64].

REFERENCES

1. Greenspan SL, Coates P, Sereika SM, Nelson JB, Trump DL, Resnick NM. 2005. Bone loss after initiation of

androgen deprivation therapy in patients with prostate cancer. *J Clin Endocrinol Metab* 90: 6410–6417.
2. Shahinian VB, Kuo YF, Freeman JL, Goodwin JS. 2005. Risk of fracture after androgen deprivation therapy for prostate cancer. *N Engl J Med* 352: 154–164.
3. Smith MR, Boyce SP, Moyneur E, Duh MS, Raut MK, Brandman J. 2006. Risk of clinical fractures after gonadotrophin-releasing hormone agonist therapy for prostate cancer. *J Urol* 175: 136–139.
4. Melton LJ 3rd, Lieber MM, Atkinson EJ, Achenbach SJ, Zincke H, Therneau TM, Khosla S. 2011. Fracture risk in men with prostate cancer: A population-based study. *J Bone Miner Res* 26(8): 1808–1815.
5. Smith MR, Eastham J, Gleason DM, Shasha D, Tchekmedyian S, Zinner M. 2003. Randomised controlled trial of zoledronic acid to prevent bone loss in men receiving androgen deprivation therapy for non-metastatic prostate cancer. *J Urol* 169: 2008–2012.
6. Michaelson MD, Kaufman DS, Lee H, McGovern FJ, Kantoff PW, Fallon MA, Finkelstein JS, Smith MR. 2007. Randomised controlled trial of annual zoledronic acid to prevent gonadotropin-releasing hormone agonist-induced bone loss in men with prostate cancer. *J Clin Oncol* 25: 1038–1042.
7. Greenspan SL, Nelson JB, Trump DL, Resnick NM. 2007. Effect of once-weekly oral alendronate on bone loss in men receiving androgen deprivation therapy for prostate cancer. *Ann Intern Med* 146: 416–424.
8. Smith MR, Egerdie B, Hernández Toriz N, Feldman R, Tammela TL, Saad F, Heracek J, Szwedowski M, Ke C, Kupic A, Leder BZ, Goessl C; Denosumab HALT Prostate Cancer Study Group. 2009. Denosumab in men receiving androgen-deprivation therapy for prostate cancer. *N Engl J Med* 361: 745–755.
9. Smith MR, Morton RA, Barnette KG, Sieber PR, Malkowicz SB, Rodriguez D, Hancock ML, Steiner MS. 2010. Toremifene to reduce fracture risk in men receiving androgen deprivation therapy for prostate cancer. *J Urol* 184: 1316–1321.
10. Saylor PJ, Smith MR. 2010. Adverse effects of androgen deprivation therapy: Defining the problem and promoting health among men with prostate cancer. *J Natl Compr Canc Netw* 8: 211–223.
11. McCloskey E. 2006. Effects of third-generation aromatase inhibitors on bone. *Eur J Cancer* 42: 1044–1051.
12. Eastell R, Hannon RA, Cuzick J, Dowsett M, Clack G, Adams JE. 2006. Effect of an aromatase inhibitor on BMD and bone turnover markers: 2-year results of the anastrozole, tamoxifen, alone or in combination (ATAC) trial. *J Bone Miner Res* 21: 1215–1223.
13. Perez EA, Josse RG, Pritchard KI, Ingle JM, Martino S, Findlay HP, Shenkier TN, Tozer RG, Palmer MJ, Shepherd LE, Liu S, Tu D, Goss PE. 2006. Effect of letrozole versus placebo on bone mineral density in women with primary breast cancer completing 5 or more years of adjuvant tamoxifen: A comparison study to NCIC CTG MA.17. *J Clin Oncol* 24: 3629–3635.
14. Howell A, Cuzick J, Baum M, Buzdar A, Dowsett M, Forbes JF, Hoctin-Boes G, Houghton J, Locker GY, Tobias JS. 2005. Results of the ATAC (Arimidex, Tamoxifen. Alone or in Combination) trial after completion of 5 years' adjuvant treatment for breast cancer. *Lancet* 365: 60–62.
15. Coates AS, Keshaviah A, Thurlimann B, Mouridsen H, Mauriac I, Forbes JF, Parisdaens R, Castiglione-Gertsch M, Gelber RD, Colleoni M, Lang 1, Del Mastro L, Smith I, Chirgwin J, Nogaret JM, Pienkowski T, Wardley A, Jacobsen EH, Price KN, Goldhirsch A. 2007. Five years of letrozole compared with tamoxifen as initial adjuvant therapy for postmenopausal women with endocrine-responsive early breast cancer: Update of study BIG 1-98. *J Clin Oncol* 25: 486–492.
16. Goss PE, Ingle JN, Martino S, Robert NJ, Muss HB, Picart MJ, Castiglione M, Tu D, Shepherd LE, Pritchard KI, Livingston RB, Davidson NE, Norton L, Perez EA, Abrams JS, Therasse P, Palmer MJ, Pater JL. 2003. A randomised trial of letrozole in postmenopausal women after five years of tamoxifen therapy for early stage breast cancer. *N Engl J Med* 349: 1793–1802.
17. Coleman RE, Banks LM, Girgis SI, Kilburn LS, Vrdoljak E, Fox J, Cawthorn J, Patel A, Snowdon CF, Hall E, Bliss JM, Coombes RC. 2007. Skeletal effects of exemestane on bone mineral density, bone biomarkers, and fracture incidence in postmenopausal women with early breast cancer participating in the Intergroup Exemestane Study (IES): A randomised controlled study. *Lancet Oncol* 8: 119–127.
18. Brufsky A, Harker WG, Beck JT, Carroll R, Tan-Chiu E, Seidler C, Hohneker J, Lacerna L, Petrone S, Perez EA. 2007. Zoledronic acid inhibits adjuvant letrozole-induced bone loss in postmenopausal women with early breast cancer. *J Clin Oncol* 25: 829–836.
19. Markopoulos C, Tzoracoleftherakis E, Polychronis A, Venizelos B, Dafni U, Xepapadakis G, Papadiamantis J, Zobolas V, Misitzis J, Kalogerakos K, Sarantopoulou A, Siasos N, Koukouras D, Antonopoulou Z, Lazarou S, Gogas H. 2010. Management of anastrozole-induced bone loss in breast cancer patients with oral risedronate: Results from the ARBI prospective clinical trial. *Breast Cancer Res* 12: R24. Epub 2010 Apr 16.
20. Hines SL, Sloan JA, Atherton PJ, Perez EA, Dakhil SR, Johnson DB, Reddy PS, Dalton RJ, Mattar BI, Loprinzi CL. 2010. Zoledronic acid for treatment of osteopenia and osteoporosis in women with primary breast cancer undergoing adjuvant aromatase inhibitor therapy. *Breast* 19: 92–96.
21. Van Poznak C, Hannon RA, Mackey JR, Campone M, Apffelstaedt JP, Clack G, Barlow D, Makris A, Eastell R. 2010. Prevention of aromatase inhibitor-induced bone loss using risedronate: The SABRE trial. *J Clin Oncol* 28: 967–975.
22. Ellis GK, Bone HG, Chlebowski R, Paul D, Spadafora S, Fan M, Kim D. 2009. Effect of denosumab on bone mineral density in women receiving adjuvant aromatase inhibitors for non-metastatic breast cancer: Subgroup analyses of a phase 3 study. *Breast Cancer Res Treat* 118: 81–87
23. Cock TA, Back J, Elefteriou F, Karsenty G, Kastner P, Chan S, Auwerx J. 2004. Enhanced bone formation

23. in lipodystrophic PPARgamma (hyp/hyp) mice relocates haematopoiesis to the spleen. *EMBO Rep* 5: 1007–1012.
24. Ali AA, Weinstein RS, Stewart SA, Parfitt AM, Manolagas SC, Jilka RL. 2005. Rosiglitazone causes bone loss in mice by suppressing osteoblast differentiation and bone formation. *Endocrinology* 146: 1226–1235.
25. Lecka-Czernik B, Ackert-Bicknell C, Adamo ML, Marmolejos V, Churchill GA, Shockley KR, Reid IR, Grey A, Rosen C. 2007. Activation of peroxisome proliferator-activated receptor gamma (PPARgamma) by rosiglitazone suppresses components of the insulin-like growth factor regulatory system in vitro and in vivo. *Endocrinology* 148: 903–911.
26. Schwartz AV, Sellmeyer DE, Vittinghoff E, Palermo L, Lecka-Czernik B, Feingold KR, Strotmeyer ES, Resnick HE, Carbone L, Beamer BA, Park SW, Lane NE, Harris TB, Cummings SR. 2006. Thiazolidinedione use and bone loss in older diabetic adults. *J Clin Endocrinol Metab* 91: 3276–3278.
27. Grey A, Bolland M, Gamble G, Wattie D, Horne A, Davidson J, Reid IR. 2007. The peroxisome-proliferator-activated receptor gamma agonist rosiglitazone decreases bone formation and bone mineral density in healthy postmenopausal women: A randomized controlled trial. *J Clin Endocrinol Metab.* 92: 1305–1310.
28. Kahn SE, Haffner SM, Heise MA, Herman WH, Holman RR, Jones NP, Kravitz BG, Lachin JM, O'Neill MC, Zinman B, Viberti G. 2006. Glycaemic durability of rosiglitazone, metformin, or glyburide monotherapy. *N Engl J Med* 355: 2427–2443.
29. Loke YK, Singh S, Furberg CD. 2009. Long-term use of thiazolidinediones and fractures in type 2 diabetes: A meta-analysis. *CMAJ* 180: 32–39.
30. Lecka-Czernik B. 2010. Bone loss in diabetes: Use of anti-diabetic thiazolidinediones and secondary osteoporosis. *Curr Osteoporos Rep* 8: 178–184.
31. Grisso JA, Kelsey JL, O'Brien LA, Miles CG, Sidney S, Maislin G, LaPann K, Moritz D, Peters B. 1997. Risk factors for hip fracture in men. Hip Fracture Study Group. *Am J Epidemiol* 145: 786–793.
32. Vestergaard P, Rejnmark L, Mosekilde L. 2006. Proton pump inhibitors, histamine H2 receptor antagonists, and other antacid medications and the risk of fracture. *Calcif Tissue Int* 19: 76–83.
33. Yu EW, Blackwell T, Ensrud KE, Hillier TA, Lane NE, Orwoll E, Bauer DC. 2008. Acid-suppressive medications and risk of bone loss and fracture in older adults. *Calcif Tissue Int* 83: 251–259.
34. Yang Y-X, Lewis JD, Epstein S, Metz DC. 2006. Long-term proton pump inhibitor therapy and risk of hip fracture. *JAMA* 296: 2947–2953.
35. Kaye JA, Jick H. 2008. Proton pump inhibitor use and risk of hip fractures in patients without major risk factors. *Pharmacotherapy* 28: 951–959.
36. Corley DA, Kubo A, Zhao W, Quesenberry C. 2010. Proton pump inhibitors and histamine-2 receptor antagonists are associated with hip fractures among at-risk patients. *Gastroenterology* 139: 93–101.
37. Gray SL, LaCroix AZ, Larson J, Robbins J, Cauley JA, Manson JE, Chen Z. 2010. Proton pump inhibitor use, hip fracture, and change in bone mineral density in postmenopausal women: Results from the Women's Health Initiative. *Arch Intern Med* 170: 765–771.
38. O'Connell MB, Madden DM, Murray AM, Heaney RP, Kerzner LJ. 2005. Effects of proton pump inhibitors on calcium carbonate absorption in women: A randomised crossover trial. *Am J Med* 120: 778–781.
39. Petty SJ, O'Brien TJ, Wark JD. 2007. Anti-epileptic medication and bone health. *Osteoporos Int* 18: 129–142.
40. Ensrud KE, Walczak TS, Blackwell TL, Ensrud ER, Barrett-Connor E, Orwoll ES; Osteoporotic Fractures in Men (MrOS) Study Research Group. 2008. Antiepileptic drug use and rates of bone loss in older men: A prospective study. *Neurology* 71: 723–773.
41. Lee RH, Lyles KW, Colón-Emeric C. 2010. A review of the effect of anticonvulsant medications on bone mineral density and fracture risk. *Am J Geriatr Pharmacother* 8: 34–46.
42. Carbone LD, Johnson KC, Robbins J, Larson JC, Curb JD, Watson K, Gass M, Lacroix AZ. 2010. Antiepileptic drug use, falls, fractures, and BMD in postmenopausal women: Findings from the women's health initiative (WHI). *J Bone Miner Res* 25: 873–881.
43. Jetté N, Lix LM, Metge CJ, Prior HJ, McChesney J, Leslie WD. 2011. Association of antiepileptic drugs with non-traumatic fractures: A population-based analysis. *Arch Neurol* 68: 107–112.
44. Haney EM, Chan BK, Diem SJ, Ensrud KE, Cauley JA, Barrett-Connor E, Orwoll E. Bliziotes MM. 2007. Association of low bone mineral density with selective serotonin reuptake inhibitors in older men. *Arch Intern Med* 167: 1246–1251.
45. Diem SJ, Blackwell TL, Stone KL, Yaffe K, Haney EM, Bliziotes MM, Ensrud KE. 2007. Use of antidepressants and rates of hip bone loss in older women: The study of osteoporotic fractures. *Arch Intern Med* 167: 1240–1245.
46. Richards JB, Papaioannou A, Adachi JD, Joseph L, Whitson HE, Prior JC, Goltzman D. 2007. Effect of selective serotonin reuptake inhibitors on the risk of fracture. *Arch Intern Med* 167: 188–194.
47. Cizza G, Primma S, Coyle M, Gourgiotis L, Csako G. 2010. Depression and osteoporosis: A research synthesis with meta-analysis. *Horm Metab Res* 42: 467–482.
48. Warden SJ, Robling AG, Haney EM, Turner CH, Bliziotes MM. 2010. The emerging role of serotonin (5-hydroxytryptamine) in the skeleton and its mediation of the skeletal effects of low-density lipoprotein receptor-related protein 5 (LRP5). *Bone* 46: 4–12.
49. de Sweit M, Ward P, Fidler A, Horsman A, Katz D, Letsky E, Peacock M, Wise PH. 1983. Prolonged heparin therapy in pregnancy causes bone demineralisation. *Br J Obstet Gynaecol* 90: 1129–1134.
50. Dalhman T. 1993. Osteoporotic fractures and the recurrence of thromboembolism during pregnancy and the puerperium in 184 women undergoing thromboprophylaxis with heparin. *Am J Obstet Gynecol* 168: 1265–1270.

51. Barbour L, Kick S, Steiner J, LoVerde M, Heddleston L, Lear J, Baron A, Barton P. 1994. A prospective study of heparin-induced osteoporosis in pregnancy using bone densitometry. *Am J Obstet Gynecol* 170: 862–869.
52. Wiens M, Etminan M, Gill SS, Takkouche B. 2006. Effects of antihypertensive drug treatments on fracture outcomes: A meta-analysis of observational studies. *J Intern Med* 260: 350–362.
53. Reid IR, Gamble GD, Grey AB, Black DM, Ensrud KE, Browner WS, Bauer DC. 2005. Beta-blocker use, BMD, and fractures in the study of osteoporotic fractures. *J Bone Miner Res* 20: 613–618.
54. Bonnet N, Gadois C, McCloskey E, Lemineur G, Lespressailles E, Courteix D, Benhamou CL. 2007. Protective effect of beta blockers in postmenopausal women: Influence on fractures, bone density, micro and macro-architecture. *Bone* 40: 1209–1216.
55. de Vries F, Souverein PC, Cooper C, Leufkens HGM, van Staa TP. 2007. Use of beta-blockers and the risk of hip/femur fracture in the United Kingdom and the Netherlands. *Calcif Tissue Int* 80: 69–75.
56. Yang S, Nguyen ND, Center JR, Eisman JA, Nguyen TV. 2011. Association between beta-blocker use and fracture risk: The Dubbo Osteoporosis Epidemiology Study. *Bone* 48: 451–455.
57. Solomon DH, Mogun H, Garneau K, Fischer MA. 2011. The risk of fractures in older adults using antihypertensive medications. *J Bone Miner Res* 26: 1561–1567.
58. La Croix AZ, Wienpahl J, White LR, Wallace RB, Scherr PA, George LK, Cornoni-Huntley J, Ostfield AM. 1990. Thiazide diuretic agents and the incidence of hip fracture. *N Engl J Med*. 322: 286–290.
59. La Croix AZ, Ott SM, Ichikawa L, Scholes D, Barlow WE. 2000. Low dose hydrochlorothiazide and preservation of bone mineral density in older adults. A randomised double-blind placebo controlled trial. *Ann Intern Med* 133: 516–526.
60. Bolland MJ, Ames RW, Horne AM, Orr-Walker BJ, Gamble GD, Reid IR. 2007. The effect of treatment with a thiazide diuretic for 4 years on bone density in normal postmenopausal women. *Osteoporos Int* 18: 479–486.
61. Mundy G, Garret R, Harris S, Chan JC, Chen D, Rossini G, Boyce B, Zhao M, Gutierrez G. 1999. Stimulation of bone formation in vitro and in rodents by statins. *Science* 286: 1946–1949.
62. Uzzan B, Cohen RM, Nicolas P, Cucherat M, Perret G-Y. 2007. Effect of statins on bone mineral density: A meta-analysis of clinical studies. *Bone* 40: 1581–1587.
63. Nguyen D, Wang CY, Eisman JA, Nguyen TV. 2007. On the association between statins and fracture: A Bayesian consideration. *Bone* 40: 813–820.
64. Bone HG, Kiel DP, Lindsay RS, Lewiecki EM, Bolognese MA, Leary ET, Lowe W, McClung MR. 2007. Effects of atorvastatin on bone in postmenopausal women with dyslipidemia: a double-blind, placebo-controlled, dose-ranging trial. *J Clin Endocrinol Metab* 92: 4671–4677.

65
Orthopedic Surgical Principles of Fracture Management

Manoj Ramachandran and David G. Little

Introduction 527
Treatment Principles 527
Surgical Options 528
Common Fractures in Osteoporotic Patients 529

Intervention for Vertebral Fractures in Osteoporotic Patients 529
References 530

INTRODUCTION

A broad range of fractures and associated injuries present to the orthopedic or trauma surgeon for management. The principle that fracture immobilization supports both alignment and union of the fracture, while minimizing discomfort, has been known since antiquity. The energy of the injury, associated soft tissue injury, and displacement of the fracture usually guide intervention.

The initial management of fractures consists of realignment of the broken limb segment and immobilization of the fractured extremity once the initial assessment, evaluation, and management of any life-threatening injury are completed. The aim of fracture treatment is to obtain union of the fracture in the most anatomical position compatible with maximal functional recovery of the extremity or the spine, with minimal complications. This is accomplished by obtaining and subsequently maintaining reduction of the fracture with an immobilization technique that allows the fracture to heal and, at the same time, provides the patient with functional aftercare. Either nonoperative or operative means may be used. Any surgical technique, if chosen, should minimize additional soft tissue and bone injury, which can delay fracture healing.

In open fractures, in addition to the treatment aims outlined above, the prevention of infection is vital [1]. This is achieved by urgent wound irrigation and debridement (with serial irrigations and debridements every 24–48 hours until the wounds are clean and closed), antibiotic administration and tetanus vaccination. If soft-tissue coverage over the injury is inadequate, soft-tissue transfers or free flaps are performed when the wound is clean and the fracture is definitively treated.

TREATMENT PRINCIPLES

Fracture management can be divided into nonoperative and operative techniques. Nonoperative technique consists of a closed reduction (required if the fracture is significantly displaced or angulated), achieved by applying traction to the long axis of the injured limb and then reversing the mechanism of injury. This is followed by a period of immobilization with casting. Casts are made from fiberglass or plaster of paris. Complications of casts can include the development of pressure ulcers, thermal burns during plaster hardening, and joint stiffness. Traction (skin or skeletal) is rarely used nowadays for definitive fracture management.

If the fracture cannot be reduced (e.g., due to soft tissue interposition), surgical intervention may be required. Indications for surgery include failed nonoperative management, unstable fractures that cannot be adequately maintained in a reduced position, displaced intra-articular fractures (greater than 2mm), impending pathologic fractures, unstable or complicated open fractures, fractures in growth areas in skeletally immature individuals who

---

Primer on the Metabolic Bone Diseases and Disorders of Mineral Metabolism, Eighth Edition. Edited by Clifford J. Rosen.
© 2013 American Society for Bone and Mineral Research. Published 2013 by John Wiley & Sons, Inc.

have increased risk for growth arrest, and nonunions or malunions that have failed to respond to nonoperative treatment [2]. Contraindications to internal fixation include active infection (local or systemic), soft tissues that compromise the overlying fracture or the surgical approach (e.g., burns and previous surgical scars), medical conditions that contraindicate surgery or anesthesia (e.g., recent myocardial infarction), and cases in which amputation would be more appropriate (e.g., severe neurovascular injury).

SURGICAL OPTIONS

The treatment goals for surgical fracture management, as outlined by the Association for the Study of Internal Fixation (ASIF), are anatomic reduction of the fracture fragments (for the correction of length, angulation, and rotation, and for intra-articular fractures for the restoration of the joint surface), stable internal fixation to cope with physiological biomechanical demands, preservation of blood supply, and active, pain-free mobilization of adjacent muscles and joints [3]. The objectives of open reduction and internal fixation include adequate exposure and reduction of the fracture, followed by maintenance of the reduction by stabilization using one or a combination of the methods below.

Kirschner wires

Kirschner wires, or K-wires, placed percutaneously or through a mini-open approach, are commonly used for temporary and definitive treatment of fractures. However, they have poor resistance to torque and bending forces, and rely on friction with the bone for maintenance of reductions. Therefore, when they are used as the sole form of fixation, casting or splinting is used in conjunction. K-wires are commonly used as adjunctive fixation for screws or plates and screws that involve fractures around joints.

Plates and screws

Plates and screws are commonly used in the management of articular fractures as they provide strength and stability to neutralize the forces on the injured limb for functional postoperative aftercare. Plate designs vary, depending on the anatomic region and size of the bone the plate is used for. All plates should be applied with minimal stripping of the soft tissue. Five main plate designs exist:

1. *Buttress (antiglide) plates* counteract the compression and shear forces that commonly occur with fractures that involve the metaphysis and epiphysis. These plates are commonly used with interfragmentary screw fixation and require anatomical contouring to achieve stable fixation.
2. *Compression plates* counteract bending, shear, and torsional forces by providing compression across the fracture site via the eccentrically loaded holes in the plate. Compression plates are commonly used in the long bones, especially the fibula, radius, and ulna, and in nonunion or malunion surgery.
3. *Neutralization plates* are used in combination with interfragmentary screw fixation. The interfragmentary compression (lag) screws provide compression at the fracture site. This plate function neutralizes bending, shear, and torsional forces on the lag screw fixation, as well as increases the stability of the construct. These plates are commonly used for fractures involving the fibula, radius, ulna, and humerus.
4. *Bridge plates* are useful in the management of multifragmented diaphyseal and metaphyseal fractures. Indirect reduction techniques are preferred in bridge plating without disrupting the soft-tissue attachments to the bone fragments.
5. *The tension band plate* technique converts tension forces into compressive forces, thereby providing stability, e.g., in oblique olecranon fractures.

More recently, *locking plates* have been introduced. A locking plate acts like an internal fixator. There is no need to anatomically contour the plate onto the bone, thus reducing bone necrosis and allowing for a minimally invasive technique. Locking screws directly anchor and lock onto the plate, thereby providing angular and axial stability. These screws are incapable of toggling, sliding, or becoming dislodged, thus reducing the possibility of a secondary loss of reduction, as well as eliminating the possibility of intraoperative overtightening of the screws. The locking plate is indicated for osteoporotic fractures, for short and metaphyseal segment fractures, and for bridging comminuted areas. These plates are also appropriate for metaphyseal areas where subsidence may occur or prostheses are involved.

The technique of minimally invasive percutaneous plate osteosynthesis (MIPPO) with indirect reduction is becoming increasingly popular. This involves the use of anatomically pre-shaped plates and instrumentation to safely and effectively insert the plate percutaneously or through limited incisions [4]. Advantages of MIPPO may include faster bone healing, reduced infection rate, decreased need for bone grafting, less postoperative pain, faster rehabilitation, and more aesthetic results. Some disadvantages include difficulty with indirect reduction, increased radiation exposure due to more intraoperative radiographs being utilized, malunion, pseudoarthrosis through diastases, and delayed union with flexible fixation in simple fractures.

Intramedullary nails

These nails operate like an internal splint that shares the load with the bone and can be flexible or rigid, locked or unlocked, and reamed or unreamed. Locked intramedullary nails provide relative stability to maintain bone alignment and length and to limit rotation. Ideally, the intramedullary nail allows for compressive forces at the fracture site, which stimulates bone healing.

Intramedullary nails are commonly used for femoral and tibial diaphyseal fractures, and, occasionally, humeral diaphyseal fractures. The advantages of intramedullary nails include minimally invasive procedures, early postoperative ambulation, and early range of motion being permitted in adjacent joints. Reaming may also increase union rates possibly by providing the equivalent of a bone graft to the fractured region.

External fixation

External fixation provides fracture stabilization at a distance from the fracture site, without interfering with the soft-tissue structures that are near the fracture. This technique not only provides stability for the extremity and maintains bone length, alignment, and rotation without requiring casting, but it also allows for inspection of the soft-tissue structures that are vital for fracture healing. Indications for external fixation (temporarily or as definitive care) are as follows [5]:

1. Open fractures that have significant soft-tissue disruption (e.g., type II or III open fractures)
2. Soft-tissue injury (e.g., burns)
3. Pelvic fractures
4. Severely comminuted and unstable fractures
5. Fractures associated with bony defects
6. Limb-lengthening and bone transport procedures
7. Fractures associated with infection or nonunion

Complications of external fixation include pin tract infection, pin loosening or breakage, interference with joint motion, neurovascular damage during pin placement, malalignment caused by poor placement of the fixator, delayed union, and malunion. Modern external fixators, such as the Taylor Spatial Frame, allow for postoperative adjustment to effect more anatomical reduction of the fracture in the weeks following intervention.

COMMON FRACTURES IN OSTEOPOROTIC PATIENTS

Vertebral compression fractures, Colles' (distal radius) fractures, hip fractures and other peripheral (nonvertebral) fractures all occur in patients with osteoporosis. These fractures may indicate the need for treatment of osteoporosis for secondary fracture prevention.

Osteoporotic fractures are usually low energy injuries. Some fractures such as fatigue fractures of the pelvis have no displacement. Colles' fractures and hip fractures are usually displaced and require reduction and fixation. They are sometimes held in a cast but may require wire fixation or low-profile plating to maintain reduction. Colles' fractures nearly always heal, but significant malunion can interfere with function. Intertrochanteric hip fractures are usually fixed with sliding hip screw devices that allow compression of the fracture fragments on weight bearing. Subcapital neck of femur fractures require joint arthroplasty of some form in the majority of cases because of the high incidence of nonunion and avascular necrosis. In recent times, the development of locking plate technology has improved the surgeon's ability to internally fix fractures in osteoporotic bone, but further research is needed as in many cases optimal fixation cannot be achieved. It is a principle of management for all osteoprorotic fractures to institute load bearing and functional tasks as quickly as possible to minimize loss of function or mobility.

INTERVENTION FOR VERTEBRAL FRACTURES IN OSTEOPOROTIC PATIENTS

Acute vertebral compression fractures can be painful and lead to disability, whereas multiple "silent" compression fractures lead to kyphosis and loss of height. While in many clinically apparent fractures the pain settles in a few weeks, it has been estimated that one-third of fractures can remain chronically painful. While most vertebral compression fractures are managed nonoperatively, there has been an increase in intervention for painful acute fracture to minimize morbidity. These techniques are known as vertebroplasty and kyphoplasty.

In vertebroplasty, a percutaneous approach to the vertebral body is made under fluoroscopic control, either through or adjacent to the pedicles. Bone cement, usually polymethylmethacrylate (PMMA) is injected into the fracture under pressure via a cannula while the cement is in a fluid state. This acutely stabilizes the fracture and results in an immediate reduction of pain that is significant enough in many cases to allow immediate return to activities of daily living. Pain reduction is thought to be from the stabilization of the fracture, although heat necrosis of nerve endings from the exothermic setting of the cement has also been suggested. Vertebroplasty makes little difference to the spinal alignment as the injection of cement usually does little to change the wedging deformity of the fracture.

Kyphoplasty is designed to address these limitations. In this technique, cannulae are placed usually bilaterally to allow the introduction of balloon tamps that are expanded with radio-opaque saline. Some elevation of the endplate and thus correction of deformity can be achieved by this method. The balloon expansion creates a void into which cement is injected at a slightly more viscous state than in vertebroplasty. The literature available suggests that while correction of vertebral morphology is achievable, overall spinal balance is usually not affected as alterations in shape can be accommodated by disc spaces and deformity at other levels [6].

When one considers comparisons with conservative care, both procedures appear to be effective in relieving pain in a few days in the majority of individuals. One nonrandomized study showed decreases in pain, rapid return to function, and decreased hospital stay in vertebroplasty versus conservatively treated patients [7]. A systematic review favored a therapeutic effect and some superiority for kyphoplasty over nonoperative treatment [8]. There is a suggestion from this meta-analysis that kyphoplasty may be associated with fewer cement leakage events than vertebroplasty. Serious complications include neurological sequelae and have been reported to run at about 1%.

Recent randomized trials of vertebroplasty have placed these apparent benefits into doubt. Two separate sham-controlled vertebroplasty trials showed no benefit of active over sham treatment [9, 10]. These trials have been criticized for including patients who were not "acute"; however, a study using the data of both trials found no difference even in the acute subset [11]. Further trials are underway but as of the time of writing this chapter, these trials place the utility of these procedures in doubt.

REFERENCES

1. Gustilo RB, Merkow RL, Templeman D. 1990. The management of open fractures. *J Bone Joint Surg Am* 72(2): 299–304.
2. Canale ST. 2003. *Campbell's Operative Orthopaedics, 10th Ed.* St Louis, MO: Mosby-Year Book.
3. Ruedi TP, Buckley R, Moran C (eds.) 2007. *AO Principles of Fracture Management, 2nd Ed.* New York: Thieme Medical Publishers, Inc.
4. Krettek C, Schandelmaier P, Miclau T, Tscherne H. 1997. Minimally invasive percutaneous plate osteosynthesis (MIPPO) using the DCS in proximal and distal femoral fractures. *Injury* 28 (Suppl 1): A20–30.
5. Bucholz RW, Heckman JD, Court-Brown C, Tornetta P III, Koval KJ, Wirth MA (eds.) 2005. *Rockwood & Green's Fractures in Adults, 6th Ed.* Philadelphia, PA: Lippincott Williams & Wilkins.
6. Pradhan BB, Bae HW, Kropf MA, Patel VV, Delamarter RB. 2006. Kyphoplasty reduction of osteoporotic vertebral compression fractures: Correction of local kyphosis versus overall sagittal alignment. *Spine* 31: 435–41.
7. Diamond TH, Bryant C, Browne L, Clark WA. 2006. Clinical outcomes after acute osteoporotic vertebral fractures: A 2-year non-randomised trial comparing percutaneous vertebroplasty with conservative therapy. *Med J Aust* 184: 113–7.
8. Taylor RS, Fritzell P, Taylor RJ. 2007. Balloon kyphoplasty in the management of vertebral compression fractures: An updated systematic review and meta-analysis. *Eur Spine J* 16: 1085–100.
9. Buchbinder R, Osborne RH, Ebeling PR, Wark JD, Mitchell P, Wriedt C, Graves S, Staples MP, Murphy B. 2009. A randomized trial of vertebroplasty for painful osteoporotic vertebral fractures. *N Engl J Med* 361(6): 557–68.
10. Kallmes DF, Comstock BA, Heagerty PJ, Turner JA, Wilson DJ, Diamond TH, Edwards R, Gray LA, Stout L, Owen S, Hollingworth W, Ghdoke B, Annesley-Williams DJ, Ralston SH, Jarvik JG. 2009. A randomized trial of vertebroplasty for osteoporotic spinal fractures. *N Engl J Med* 361(6): 569–79.
11. Staples MP, Kallmes DF, Comstock BA, Jarvik JG, Osborne RH, Heagerty PJ, Buchbinder R. 2011. Effectiveness of vertebroplasty using individual patient data from two randomised placebo controlled trials: Meta-analysis. *BMJ* 343: d3952.

66

Abnormalities in Bone and Calcium Metabolism After Burns

Gordon L. Klein

Introduction 531
The Inflammatory Response 531
The Stress Response 531
Other Possible Contributing Factors 532

Treatment 532
Applicability to Other Conditions 532
References 532

INTRODUCTION

Burn injury is an example of how the body's nonspecific adaptive response can adversely affect bone. Because humans did not evolve a specific means of protection against burn injury, the responses are nonspecific and have unintended consequences. The two major adaptive responses in question are the inflammatory response and the stress response.

THE INFLAMMATORY RESPONSE

A systemic inflammatory response occurs within 24 hours of severe burn injury and includes high circulating levels of the pro-inflammatory cytokines interleukin (IL)-1β and IL-6 [1]. Both cytokines stimulate osteoblast receptor activator of nuclear factor-κB ligand (RANKL) production with subsequent marrow stem cell differentiation into osteoclasts, thus increasing bone resorption. That this occurs early on after burns can be inferred from the success of acute administration of the bisphosphonate pamidronate within the first 10 days after a burn injury in preventing both total body and lumbar spine bone loss, both acutely [2] and for up to 2 years following this acute intervention [3]. In contrast those, who did not receive pamidronate lost approximately 3% of total body bone mineral content (BMC) in the ensuing 6 months and 7% of lumbar spine bone mineral density (BMD) in the first 6 weeks postburn [2].

Moreover, children with severe burns develop acute sustained hypocalcemia and hypoparathyroidism with urinary calcium wasting [4], suggesting cytokine-mediated upregulation of the parathyroid calcium-sensing receptor (CaSR). A 50% upregulation of the CaSR has been shown to occur within 48 hours in a sheep model of burn injury [5] and suggests that the CaSR upregulation may serve to clear the excess calcium that enters the blood following cytokine-mediated bone resorption.

THE STRESS RESPONSE

The three- to eightfold increase in urine free cortisol [1, 6] within the first 24 hours of the burn injury could well act synergistically with the inflammatory cytokines during the first 2 weeks in stimulating osteoblast RANKL production and subsequent bone resorption. However, by the second week postburn there are no visible surface osteoblasts on bone biopsy [6], markedly reduced surface uptake of tetracycline label and disappearance of double labeling (Figs. 66.1 and 66.2), as well as a failure of the marrow stromal cells cultured from the bone biopsies of burned children to exhibit normal quantities of markers of osteoblast differentiation [6]. Along with the marked reduction in bone formation, urinary deoxypyridinoline,

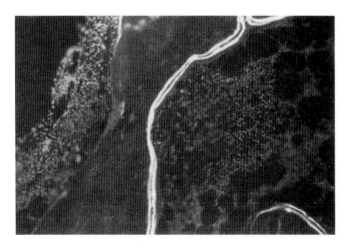

Fig. 66.1. Iliac crest bone biopsy from a normal subject showing full surface doxycycline uptake and normal double label.

Fig. 66.2. Iliac crest bone biopsy from a burned subject showing patchy surface doxycycline uptake and absent double label.

as a marker of resorption, also fell [1] creating an adynamic bone despite persistent high circulating levels of resorptive cytokines.

Thus, there would appear to be two stages of postburn bone loss: a primary resorption mediated by both inflammatory cytokines and endogenous glucocorticoids, and a secondary adynamic stage mediated primarily by endogenous glucocorticoids.

Bone remodeling resumes by 12 months postburn [3] but the lumbar spine BMD Z-scores in burned children who did not receive acute bisphosphonate therapy remained significantly lower than in those who did [3].

OTHER POSSIBLE CONTRIBUTING FACTORS

Vitamin D deficiency, beginning sometime in the first year postburn, is caused by at least two factors: lack of routine supplementation [7] and failure of the skin to convert normal quantities of 7-dehydrocholesterol to vitamin D_3 [8]. The deficiency is progressive such that at 2 years, while all serum levels of 25(OH)D are low, all levels of 1,25(OH)$_2$D are normal. By 7 years postburn, not only are all serum levels of 25(OH)D low but half of all levels of 1,25(OH)$_2$D are also low [7].

Immobilization following burn injury has not been adequately studied with regard to bone loss. Between skin grafts the patient's mobility is restricted, but the effects of immobilization are mediated by the sympathetic nervous system via beta adrenergic receptors on the osteoblast [9]. By 2 weeks postburn, in the presence of osteoblast apoptosis it is unclear what kind of effect sympathetic drive can have on bone.

TREATMENT

Current treatment consists of a 1-year course of oral oxandrolone, which increases both lean body mass followed by an increase in BMC and bone area, though not BMD [10, 11]. It acts via insulin-like growth factor-1 (IGF-1) stimulation, although the remainder of the signaling pathway is unclear at present [11]. Recombinant human growth hormone had been used in burns, and at doses of 0.5–1 mg/kg per day subcutaneously for a year works in a similar way as oxandrolone, with increase in BMC and area secondary to an increase in lean body mass [12]. However, at a dose of 0.2 mg/kg per day, direct bone resorption is observed [12]. While pamidronate has been used to prevent the postburn bone loss, it is not universally used to treat patients following burn injury.

APPLICABILITY TO OTHER CONDITIONS

Acute and chronic conditions, such as sepsis, arthritis, and inflammatory bowel disease may lead to clinically silent bone loss mediated by these adaptive responses. The clinician should be aware of these possibilities. Further studies are required to determine the extent to which patients with these and other conditions lose bone in a manner similar to burns.

REFERENCES

1. Klein GL, Herndon DN, Goodman WG, Langman CB, Phillips WA, Dickson IR, Eastell R, Naylor KE, Maloney NA, Desai M, Benjamin D, Alfrey AC. 1995. Histomorphometric and biochemical characterization of bone following acute severe burns in children. *Bone* 17: 455–460.
2. Klein GL, Wimalawansa SJ, Kulkarni G, Sherrard DJ, Sanford AP, Herndon DN. 2005. The efficacy of acute

administration of pamidronate on the conservation of bone mass following severe burn injury in children: A double-blind, randomized, controlled study. *Osteoporos Int* 16: 631–635.
3. Przkora R, Herndon DN, Sherrard DJ, Chinkes DL, Klein GL. 2007. Pamidronate preserves bone mass for at least 2 years following acute administration for pediatric burn injury. *Bone* 41: 297–302.
4. Klein GL, Nicolai M, Langman CB, Cuneo BF, Sailer DE, Herndon DN. 1997. Dysregulation of calcium homeostasis after severe burn injury in children: Possible role of magnesium depletion. *J Pediatr* 131: 246–251.
5. Murphey ED, Chattopadhyay N, Bai M, Kifor O, Harper D, Traber DL, Hawkins HK, Brown EM, Klein GL. 2000. Up-regulation of the parathyroid calcium-sensing receptor after burn injury in sheep: A potential contributory factor to post-burn hypocalcemia. *Crit Care Med* 28: 3885–3890.
6. Klein GL, Bi LX, Sherrard DJ, Beavan SR, Ireland D, Compston JE, Williams WG, Herndon DN. 2004. Evidence supporting a role of glucocorticoids in short-term bone loss in burned children. *Osteoporos Int* 15: 468–474.
7. Klein GL, Langman CB, Herndon DN. 2002. Vitamin D depletion following burn injury in children: A possible factor in post-burn osteopenia. *J Trauma* 52: 346–350.
8. Klein GL, Chen TC, Holick MF, Langman CB, Price H, Celis MM, Herndon DN. 2004. Synthesis of vitamin D in skin after burns. *Lancet* 363: 291–292.
9. Takeda S, Elefteriou F, Levasseur F, Liu X, Zhao L, Parker KL, Armstrong D, Ducy P, Karsenty G. 2002. Leptin regulates bone formation via the sympathetic nervous system *Cell* 111: 305–317.
10. Murphy KD, Thomas S, Mlcak RP, Chinkes DL, Klein GL, Herndon DN. 2004. Effects of long-term oxandrolone administration in severely burned children. *Surgery* 136: 219–224.
11. Porro LJ, Herndon DN, Rodriguez NA, Jennings K, Klein GL, Mlcak RP, Meyer WJ, Lee JO, Suman OE, Finnerty CC. 2012. Five-year outcomes after oxandrolone administration in severely burned children: A randomized clinical trial of safety and efficacy. *J Am Coll Surg* 214: 489–502.
12. Branski LK, Herndon DN, Barrow RE, Kulp GA, Klein GL, Suman OE, Przkora R, Meyer W 3rd, Huang T, Lee JO, Chinkes DL, Mlcak RP, Jeschke MG. 2009. Randomized controlled trial to determine the efficacy of long-term growth hormone treatment in severely burned children. *Ann Surg* 250: 514–523.

Section VI
Disorders of Mineral Homeostasis

Section Editors Marie Demay and Suzanne M. Jan de Beur

Chapter 67. Approach to Parathyroid Disorders   537
John P. Bilezikian

Chapter 68. Primary Hyperparathyroidism   543
Shonni J. Silverberg

Chapter 69. Familial Primary Hyperparathyroidism (Including MEN, FHH, and HPT-JT)   553
Andrew Arnold and Stephen J. Marx

Chapter 70. Non-Parathyroid Hypercalcemia   562
Mara J. Horwitz, Steven P. Hodak, and Andrew F. Stewart

Chapter 71. Hypocalcemia: Definition, Etiology, Pathogenesis, Diagnosis, and Management   572
Anne L. Schafer and Dolores Shoback

Chapter 72. Hypoparathyroidism and Pseudohypoparathyroidism   579
Mishaela R. Rubin and Michael A. Levine

Chapter 73. Pseudohypoparathyroidism   590
Harald Jüppner and Murat Bastepe

Chapter 74. Disorders of Phosphate Homeostasis   601
Mary D. Ruppe and Suzanne M. Jan de Beur

Chapter 75. Vitamin D–Related Disorders   613
Paul Lips, Natasja M. van Schoor, and Nathalie Bravenboer

Chapter 76. Vitamin D Insufficiency and Deficiency   624
J. Christopher Gallagher

Chapter 77. Pathophysiology of Chronic Kidney Disease Mineral Bone Disorder (CKD–MBD)   632
Keith A. Hruska and Michael Seifert

Chapter 78. Treatment of Chronic Kidney Disease Mineral Bone Disorder (CKD–MBD)   640
Hala M. Alshayeb and L. Darryl Quarles

Chapter 79. Disorders of Mineral Metabolism in Childhood   651
Thomas O. Carpenter

Chapter 80. Paget's Disease of Bone   659
Ethel S. Siris and G. David Roodman

Primer on the Metabolic Bone Diseases and Disorders of Mineral Metabolism, Eighth Edition. Edited by Clifford J. Rosen.
© 2013 American Society for Bone and Mineral Research. Published 2013 by John Wiley & Sons, Inc.

67

Approach to Parathyroid Disorders

John P. Bilezikian

Intrinsic Functional Abnormalities of the
Parathyroid Gland(s) 537
Hypercalcemic and Hypocalcemic Disorders Not Due to an
Intrinsic Abnormality of the Parathyroid Glands 540

Acknowledgments 541
References 541

Disorders of the parathyroid glands are an important consideration in disorders of mineral metabolism. In this section, chapters focus upon many of these disorders either due to excessive or inadequate secretion of parathyroid hormone (PTH). The section also features disorders in which the parathyroid glands overproduce or underproduce PTH as a normal physiologic adjustment to other inciting pathophysiological events that lead either to hypercalcemia or hypocalcemia. The primary parathyroid disorders, in which parathyroid glandular activity is intrinsically abnormal (e.g., primary hyperparathyroidism, hypoparathyroidism) and the secondary parathyroid disorders, in which increased or decreased parathyroid glandular activity is a normal adjustment to another pathophysiological process (e.g., vitamin D deficiency, chronic renal disease), have given us new insights into the importance of PTH not only in the regulation of the serum calcium concentration but also in terms of skeletal health. Not covered in this section, but elsewhere, is a consequence of intense interest in and investigation of PTH: namely the use of PTH as a treatment for osteoporosis. Based upon our expanded knowledge base about the parathyroids, I offer, in this chapter, an approach to the primary and secondary parathyroid disorders. More detailed information will be found in the individual chapters that follow.

INTRINSIC FUNCTIONAL ABNORMALITIES OF THE PARATHYROID GLAND(S)

Primary oversecretion of PTH: Primary hyperparathyroidism

Primary hyperparathyroidism (PHPT) is the classic endocrine disorder associated with parathyroid gland hyperfunction [1, 2]. It is most often a sporadic occurrence with only one of the four parathyroid glands involved in a benign, adenomatous process of excessive synthetic and secretory activity. Arnold provides a comprehensive discussion of the genetics of the hyperparathyroid diseases when they present in their many other manifestations such as familial hypercalciuric hypercalcemia (FHH), hyperparathyroidism-jaw tumor syndrome, and in multiple glandular syndromes such as MEN1 and MEN2 [3].

While known since the late 1920s, PHPT has undergone a rather dramatic change in its clinical phenotype from a symptomatic disorder of "bones and stones" to one that in most parts of the world is asymptomatic. It is typically discovered incidentally during the course of a calcium measurement on a routine biochemical screening test [4]. Assisted by new technologies with which the

---

Primer on the Metabolic Bone Diseases and Disorders of Mineral Metabolism, Eighth Edition. Edited by Clifford J. Rosen.
© 2013 American Society for Bone and Mineral Research. Published 2013 by John Wiley & Sons, Inc.

skeleton can be evaluated, these and other aspects of PHPT have spawned greater interest than ever before in this disease.

Hypercalcemia is the major clinical clue to the diagnosis of PHPT. Several highly useful PTH assays expedite the diagnosis and clearly distinguish this disease from non-parathyroid etiologies of hypercalcemia [5]. Even with a PTH level that is not frankly elevated, PTH levels in the mid-range of normal establishes the diagnosis. The exquisite physiological regulation of PTH by calcium indicates that readily detectable levels of PTH in the context of hypercalcemia essentially rule out most other causes of hypercalcemia. The use of lithium, thiazide diuretics, and the exceedingly rare example of true ectopic PTH secretion are exceptions.

The skeleton, one of the major target organs in PHPT, has been a rich source of knowledge about PTH action. First shown by dual energy X-ray absorptiometry (DXA) and then followed by histomorphometric analyses of bone biopsies, the trabecular skeleton is relatively well preserved [6, 7]. DXA and bone biopsies also have established that the skeletal compartment that preferentially shows deterioration is cortical. These findings suggest that the nonvertebral skeleton, composed substantially of cortical bone, should be the greatest risk for fracture in this disease. Such expectations remain speculative because prospective epidemiology studies to address this question are lacking. Nevertheless, with successful surgery, both cortical and trabecular elements improve as shown by DXA [8] and more recently by high-resolution peripheral computed tomography [9].

While these skeletal features are noteworthy, kidney stones are the most common overt complications of PHPT, with incidence figures in the neighborhood of 20% in most series [10].

With appreciation that PHPT historically had protean manifestations, it has been exceedingly difficult to identity nontraditional target organs that are specifically affected by the hyperparathyroid process in asymptomatic PHPT. For example, several groups have attempted to discern neurocognitive and cardiovascular manifestations of asymptomatic PHPT [11–14], but it is still uncertain whether, and to what extent, such observations can be directly linked to the disease and/or whether they are reversible upon successful parathyroidectomy. The most recent guidelines for surgery in PHPT do not include these putative nontraditional aspects of PHPT, but they are acknowledged for their potential importance. Rather, the guidelines are directed to more easily measurable and traditional end points, such as bone density, kidney stones, and renal function [15].

Primary hyperparathyroidism is being increasingly recognized in subjects whose serum total and ionized calcium are consistently normal. "Normocalcemic PHPT" is associated with levels of PTH that are consistently elevated [16]. Secondary causes of an elevated PTH level must be ruled out before normocalcemic PHPT can be seriously considered. An important consideration is the 25-hydroxyvitamin D level, which is the index of body stores of vitamin D. Reduced 25-hydroxyvitamin D can account for an elevated PTH. The controversial matter of how vitamin D adequacy is defined is covered in this section of the *Primer* [17, 18]. For the purposes of establishing the diagnosis of normocalcemic PHPT, many experts require a 25-hydroxyvitamin D level that is greater than 30 ng/mL. This level helps to ensure that the patient is not demonstrating a subtle form of secondary hyperparathyroidism with levels that some people might regard as normal, such as between 20 and 30 ng/mL.

The normal serum calcium level in normocalcemic PHPT might lead one to expect weaker evidence for target organ involvement than in subjects whose PHPT is accompanied by overt hypercalcemia. However, in the experience so far with normocalcemic PHPT, many subjects demonstrate reduced BMD [16]. This may be due to the fact the most of the published literature on normocalcemic PHPT has dealt with a referral population that is being evaluated for a specific reason, such as reduced bone mineral density (BMD). Screening an unselected population for normocalcemic PHPT might lead to the identification of a normocalcemic cohort whose parathyroid disease is minimal [19].

Approach to the patient. After the patient with PHPT is evaluated, parathyroidectomy may be recommended. The recommendation for surgical intervention is based usually on meeting one or more of the criteria for parathyroid surgery as set forth by the latest International Workshop [15]. However, these guidelines are not rules. Patients who meet one or more criteria may decide against surgery; patients who do not meet any criteria might opt for the operation [20]. This latter view is acceptable if there are no medical contraindications to surgery. Virtually all parathyroid surgeons now require successful preoperative localization of the abnormal parathyroid gland. Advances in imaging with high-resolution modalities such as four-dimensional (4D) computed tomography has made it possible to identify abnormal parathyroid tissue in the vast majority of patients with PHPT [21]. With preoperative localization and a highly experienced surgeon, parathyroidectomy can be performed under local anesthesia and conscious sedation with an outstanding outcome record [22]. Successful parathyroidectomy is defined by a fall in intraoperative PTH levels by more than 50%, into the normal range, after removal of the offending adenoma. In patients who do not meet or refuse the recommendation of parathyroidectomy, observation is needed with annual serum calcium measurements and regular BMD monitoring. Pharmacological approaches to these individuals include the use of bisphosphonate [23] or the calcimimetic, cinacalcet [24].

Undersecretion of PTH: Hypoparathyroidism

In contrast to primary hypersecretion of PTH, which is relatively common, the hypoparathyroid states in which PTH is undersecreted are uncommon. In fact, hypoparathyroidism is defined as an orphan disease because there are fewer than 200,000 affected individuals in the United States [25]. Hypoparathyroidism presents a contrast with its more common counterpart in another way, namely in terms of symptomatology. Asymptomatic PHPT is the most common way the hypersecretion syndrome presents. The widespread use of screening biochemical panels is responsible, in large part, for this observation. In hypoparathyroidism, on the other hand, subjects generally are not asymptomatic. They are invariably discovered only when classical signs and symptoms of hypocalcemia are present [25–28]. It remains to be seen whether there is a variant of hypoparathyroidism, similar to normocalcemic PHPT, in which asymptomatic individuals can be discovered by screening an unselected cohort with calcium and PTH levels [29].

In hypoparathyroidism, the serum calcium is typically below normal limits, and the PTH level is either undetectable or inappropriately low for the hypocalcemic state. Similar to the physiology that governs suppression of PTH in hypercalcemic states not due to PHPT, the physiology of hypocalcemia indicates that the PTH should be elevated, if not due to an intrinsic functional abnormality or absence of the parathyroids. When hypocalcemia is not associated with elevated PTH levels, the diagnosis of a hypoparathyroidism is clear. Autoimmune destruction of the parathyroids and the sequelae of neck surgery are the two most common causes of hypoparathyroidism. The genetics of the autoimmune form and other much less common heredity manifestations of hypoparathyroidism, either in isolated form or involving multiple organs systems, are reviewed in this section of the *Primer* [26].

In virtually all forms of hypoparathyroidism, the condition is permanent. The one exception is the setting of severe hypomagnesemia. In this situation, the marked magnesium deficiency impairs parathyroid secretory function, mimicking a hypoparathyroidism [30]. However, with severe magnesium deficiency, the secretory abnormality is reversible, after magnesium is administered. Thus, in the evaluation of anyone who has hypocalcemia and low PTH levels, the serum magnesium should be measured.

The skeletal manifestations of hypoparathyroidism present an opportunity to address questions related to the role of PTH in skeletal metabolism and structure. The detailed knowledge of skeletal abnormalities in PHPT provides a counterpoint and permits the delineation of some of the findings to PTH itself. PHPT is a high turnover disease; hypoparathyroidism is a low turnover disease. While the two ends of the bone turnover spectrum, high and low, are not always appreciated by the measurement of circulating or urinary bone turnover markers, such as P1NP, osteocalcin, CTX, or urinary NTX, it is clear by dynamic histomorphometry of bone biopsies that the two disorders are diametrically different from each other. By double tetracycline labeling, there is very little bone turnover in hypoparathyroidism [31] while bone turnover is generally high in PHPT [1, 2]. In hypoparathyroidism, BMD is above average, using age-specific norms (Z-scores) or young normative databases (T-score). It is not unusual for individuals with hypoparathyroidism to show BMD values that are one- to threefold higher than normal (Z- and T-scores +1 to +3). By bone biopsy, there appears to be a cortical predominance in hypoparathyroidism that contrasts with the trabecular predominance of the hyperparathyroid skeleton [32]. These insights suggest a role for PTH is helping to regulate the distribution of cortical and trabecular bone at given sites. Clearly, overall site specificity to the distribution of cortical and trabecular bone is unlikely to be governed by PTH, but within a given site, (lumbar spine, hip, or forearm) PTH might be a key modulator. Improvement of the abnormalities in these skeletal compartments after parathyroidectomy in PHPT and after PTH administration in hypoparathyroidism helps to assign this modulatory role to PTH [33].

Besides skeletal abnormalities, subjects with hypoparathyroidism are prone to deposition of calcium–phosphate complexes in the kidney, basal ganglia, and other soft tissues. Similar to PHPT, in hypoparathyroidism, there is a spectrum on nonspecific observations that are not clearly related to the disease. In many nonspecific ways, individuals with hypoparathyroidism do not feel well. They complain of lack of energy, easy fatigability, arthritic symptoms (without frank arthritis) and "brain fog." To what extent these nonspecific symptoms can be reversed by PTH administration is not yet clear.

The administration of PTH represents a logical step in the definitive treatment of a disease characterized by lack of PTH [28, 33]. In fact, hypoparathyroidism is the last endocrine deficiency disease for which the missing hormone is not yet available as an approved therapy. This situation may be changing because of promising results of studies in hypoparathyroidism with the foreshortened peptide of PTH [PTH(1-34)] and the full length PTH molecule itself [PTH(1-84)] [33–35]. With PTH administration in hypoparathyroidism, bone turnover rapidly increases within the first year and then returns to more normal baseline values. Bone structure improves. Subjects require less calcium and vitamin D [33]. Anecdotal reports, which are now being substantiated, indicate that these individuals really do feel better and that their quality of life is improved.

Approach to the patient. Hypoparathyroidism usually presents as a symptomatic condition with complaints ranging from mild paresthesias to tetany and even seizures. In the patients who have not had ever had neck surgery (note, the neck surgery may have occurred decades before), it is important to consider multiglandular autoimmune endocrine deficiency syndromes [26]. It is also important to evaluate the patient for the possibility of ectopic calcifications (basal ganglia and other brain sites), renal calcifications, and if symptomatic, joint calcifications. The mainstay of treatment is calcium and vitamin D. While most patients can be controlled with calcium and vitamin D, some patients can present a challenge because of unexplained wide swings in the level of control. In addition, many patients require very large doses of calcium and 1,25-dihydroxyvitamin D that raises additional concerns about the long term sequelae of such chronic treatment. The use of calcium and vitamin D (I prefer the judicious use of both parent vitamin D, either ergo- or cholecalciferol, and 1,25-dihydroxyvitamin D) is clearly not equivalent to replacement therapy with PTH. If PTH becomes available for the treatment of hypoparathyroidism, many patients are likely to show improved control with less need for calcium and vitamin D, thus reducing concerns of large doses of calcium and vitamin D therapy. Moreover, if quality of life is documented to improve, as appears to be the case, this would be another major benefit of treating hypoparathyroidism with PTH.

HYPERCALCEMIC AND HYPOCALCEMIC DISORDERS NOT DUE TO AN INTRINSIC ABNORMALITY OF THE PARATHYROID GLANDS

Non-parathyroid dependent hypercalcemia: Reduced PTH levels

As already noted, the differential diagnosis of hypercalcemia due to PTH or other cause is straightforward because of the PTH assay. An undetectable PTH level, the normal physiological response to hypercalcemia, argues in a compelling manner that the hypercalcemia is due to a non-PTH dependent mechanism [5]. When the hypercalcemia is associated with suppressed PTH, the next task is to establish the cause of hypercalcemia. The initial focus is usually malignancy, particularly if the patient presents with constitutional signs. If the malignancy is lung, breast, renal, or is myeloma, the diagnosis is usually readily established. If the parathyroid hormone related peptide (PTHrP) level is elevated, a squamous cell cancer becomes most suspect. Other malignancies such as pancreatic cancer or lymphoma may require extensive diagnostic testing, which is generally not recommended unless there are clinical or biochemical clues. For example, a lymphomatous or granulomatous process might be suspected if the 1,25-dihydroxyvitamin D level is elevated. If the 25-hydroxyvitamin D is elevated, on the other hand, the diagnostic possibility focuses upon exogenous ingestion of vitamin D. There are times when after an extensive search, the etiology of the non-parathyroid hormone dependent hypercalcemia is not clear. In that scenario, time usually declares the condition.

Hypocalcemia: Elevated PTH levels

The PTH assay helps to distinguish between hypocalcemia due to a hypoparathyroid state and hypocalcemic conditions associated with a normal physiological response to hypocalcemia. If the PTH level is elevated, the search begins for the cause of the hypocalcemia. Often this is readily apparent such as a malabsorption syndrome, or renal or liver failure. In these so-called secondary hyperparathyroid states, the serum calcium can be low, but it is often in the lower range of normal. The therapeutic approach depends upon adequate control of the stimulus for PTH secretion. In the secondary hyperparathyroidism of renal disease, the pathophysiology and subsequent therapeutic approaches can be complex [36, 37]. A major goal in the secondary hyperparathyroidism of renal failure is to ascertain what the PTH level is and then to control PTH so that it is unlikely to be associated with unwanted effects of the secondary hyperparathyroid state. There are various official guidelines that suggest goals for maintaining PTH levels within a range [38, 39]. All the guidelines acknowledge, however, that an acceptable PTH level in renal failure will be higher than the normal range. This point takes into account the fact that there are circulating inactive fragments of PTH that accumulate in renal failure and are detected by the commonly used intact assay for PTH.

The dictum that in primary hyperparathyroidism, the serum calcium is elevated and that in the secondary hyperparathyroidism states, such as renal failure, the serum calcium is not elevated, has to be qualified by several points. First, we are now recognizing a primary hyperparathyroidism (normocalcemic PHPT) that is characterized by normal serum calcium levels, as described above. This is the reason why secondary causes for an elevated PTH must be ruled out before considering the diagnosis of normocalcemic PHPT. Second, with prolonged stimulation of the parathyroid glands due to a renal or severe gastrointestinal disease, the serum calcium can rise to levels above normal. In this setting, the chronic stimulation of PTH secretion leads to the emergence of a semi-autonomous state due to the selection and proliferation of a clone of parathyroid cells into a single adenomatous gland. Thus, a prolonged secondary hyperparathyroidism can "morph" into a primary hyperparathyroidism. Whenever the serum calcium becomes chronically elevated in someone who has had a prolonged stimulus for PTH secretion, this possibility should be considered.

In the absence of renal disease, liver disease, and overt vitamin D deficiency, identifying the stimulus for high PTH levels can be a challenge. The phosphorus level, if elevated, can be a clue to the diagnosis of pseudohypoparathyroidism, a classic genetic disorder associated with PTH resistance [40]. If the physical signs of pseudohypoparathyroidism are present (the Type 1a variant), the diagnosis is straightforward. However, there are other forms of pseudohypoparathyroidism, in which the classical physical phenotype is not present. Such variants of pseudohypoparathyroidism can present without any physical findings.

ACKNOWLEDGMENTS

The authors acknowledge NIH grants DK 32333 and DK 069350.

REFERENCES

1. Silverberg SJ, Bilezikian JP. 2011. Primary hyperparathyroidism. In: Wass JAH, Stewart PM, (eds.) *Oxford Textbook of Endocrinology and Diabetes, 2nd Ed.* Oxford, U.K.: University Press. pp 653–664.
2. Silverberg SJ. 2012. Primary hyperparathyroidism. In: Rosen C (ed.) *Primer on the Metabolic Bone Diseases and Disorders of Mineral Metabolism, 8th Ed.* Hoboken: Wiley.
3. Arnold A, Marx SJ. 2012. Familial hyperparathyroidism. In: Rosen C (ed.) *Primer on the Metabolic Bone Diseases and Disorders of Mineral Metabolism, 8th Ed.* Hoboken: Wiley.
4. Bilezikian JP. 2012. Primary hyperparathyroidism. In: DeGroot L (ed.), *Singer F (section ed.).* Diseases of Bone and Mineral Metabolism. www.ENDOTEXT.org. MDTEXT.COM, Inc., S. Dartmouth, MA.
5. Horwitz MJ, Hodak SP, Stewart AF. 2012. Non-parathyroid hypercalcemia. In: Rosen C (ed.) *Primer on the Metabolic Bone Diseases and Disorders of Mineral Metabolism, 8th Ed.* Hoboken: Wiley.
6. Silverberg SJ, Shane E, De La Cruz L, Dempster DW, Feldman F, Seldin D, Jacobs TP, Siris ES, Cafferty M, Parisien MV, Lindsay R, Clemens TL, Bilezikian JP. 1989. Skeletal disease in primary hyperparathyroidism. *J Bone Min Res* 4: 283–291.
7. Parisien M, Dempster DW, Shane E, Bilezikian JP. 2001. Histomorphometric analysis of bone in primary hyperparathyroidism. In: Bilezikian JP, Marcus R, Levine M (eds.) *The Parathyroids, 2nd Ed.* San Diego, CA: Academic Press. pp. 423–436.
8. Silverberg SJ, Shane E, Jacobs TP, Siris E, Bilezikian JP. 1999. A 10-year prospective study of primary hyperparathyroidism with or without parathyroid surgery. *New Eng J Med* 341: 1249–1255.
9. Hansen S, Hauge EM, Rasmussen L, Jensen JE, Brixen K. 2012. Parathyroidectomy improves bone geometry and microarchitecture in female patients with primary hyperparathyroidism: A 1-year prospective controlled study using high resolution peripheral quantitative computed tomography. *J Bone Miner Res* 27(5): 1150–8.
10. Rejnmark L, Vestergaard P, Mosekilde L. 2011. Nephrolithiasis and renal calcifications in primary hyperparathyroidism. *J Clin Endocrinol Metab* 96: 2377–2385.
11. Walker, MD, McMahon DK, Inabnet WB, et al. 2009. Neuropsychological features in primary hyperparathyroidism: A prospective study. *J Clin Endocrinol Metab* 94: 1951–8.
12. Roman SA, Sosa JA, Pietrzak RH, Snyder PJ, Thomas DC, Udelsman R, Mayes L. 2011. The effects of serum calcium and parathyroid hormone changes on psychological and cognitive function in patients undergoing parathyroidectomy for primary hyperparathyroidism. *Ann Surg* 253: 131–137.
13. Walker M, Fleischer J, DiTullio MR, et al. 2010. Cardiac structure and diastolic function in mild primary hyperparathyroidism. *J Clin Endocrinol Metab* 95: 2172–2179.
14. Rubin MR, Maurer MS, McMahon DJ, Bilezikian JP, Silverberg SJ. 2005. Arterial stiffness in mild primary hyperparathyroidism. *J Clin Endocrinol Metab* 90: 3326–3330.
15. Bilezikian JP, Khan AA, Potts JT Jr; Third International Workshop on the Management of Asymptomatic Primary Hyperthyroidism. 2009. Guidelines for the management of asymptomatic primary hyperparathyroidism: Summary statement from the third international workshop. *J Clin Endocrinol Metab* 94: 335–339.
16. Lowe H, McMahon DJ, Rubin MR, Bilezikian JP, Silverberg SJ. 2007. Normocalcemic primary hyperparathyroidism: Further characterization of a new clinical phenotype. *J Clin Endocrinol Metab* 92: 3001–3005.
17. Lips P, van Schoor NM, Bravenboer N. 2012. Vitamin D-related disorders. In: Rosen C (ed.) *Primer on the Metabolic Bone Diseases and Disorders of Mineral Metabolism, 8th Ed.* Hoboken: Wiley.
18. Gallagher JC. Vitamin D insufficiency and deficiency. 2012. In: Rosen C (ed.) *Primer on the Metabolic Bone Diseases and Disorders of Mineral Metabolism, 8th Ed.* Hoboken: Wiley.
19. Cusano N, Wang P, Cremers S, Haney E, Bauer D, Orwoll E, Bilezikian J. 2011. Asymptomatic normocalcemic primary hyperparathyroidism: Characterization of a new phenotype of normocalcemic primary hyperparathyroidism. *J Bone Miner Res* 26 (Suppl 1). Available at http://www.asbmr.org/Meetings/AnnualMeeting/AbstractDetail.aspx?aid=5f84a1bd-66ae-48dc-9938-efeb9bb60968. Accessed October 18, 2011.
20. Marcocci C, Getani F. 2011. Primary hyperparathyroidism. *N Engl J Med* 365: 2389–2397.
21. Starker LF, Mahajan A, Bjorklund P, Sze G, Udelsman R, Carling T. 2011. 4D parathyroid CT as the initial localization study for patients with de novo primary hyperparathyroidism. *Ann Surg Oncol* 18: 1723–1728.
22. Udelsman R, Lin Z, Donovan P. 2011. The superiority of minimally invasive parathyroidectomy based on 1650 consecutive patients with primary hyperparathyroidism. *Ann Surg* 253: 585–591.
23. Khan AA, Bilezikian JP, Kung AWC, Ahmed MM, Dubois SJ, Ho AYY, Schussheim D, Rubin MR, Shaikh AM, Silverberg SJ, Standish TI, Syed Z, Syed ZA. 2004. Alendronate in primary hyperparathyroidism: A double-blind, randomized, placebo-controlled trial. *J Clin Endocrinol Metab* 89: 3319–3325
24. Peacock M, Bilezikian JP, Bolognese MA, et al. 2011. Cinacalcet HCl reduces hypercalcemia in primary hyperparathyroidism across a wide spectrum of disease severity. *J Clin Endocrinol Metab* 96: E9–E18.

25. Bilezikian JP, Khan A, Potts JT Jr, Brandi ML, Clarke BL, Shoback D, Juppner H, D'Amour P, Fox J, Rejnmark L, Mosekilde L, Rubin MR, Dempster D, Gafni R, Collins M, Sliney J, Sanders J. 2011. Hypoparathyroidism in the adult: Epidemiology, diagnosis, pathophysiology, target organ involvement, treatment, and challenges for future research. *J Bone Min Res* 26: 2317–2337.
26. Schafer A, Shoback D. 2012. Hypocalcemia: Definition, etiology, pathogenesis, diagnosis, and management. In: Rosen C (ed.) *Primer on the Metabolic Bone Diseases and Disorders of Mineral Metabolism, 8th Ed.* Hoboken: Wiley.
27. Shoback D. 2008. Hypoparathyroidism. *N Engl J Med* 359: 391–403.
28. Rubin M, Levine MA. 2012. Hypoparathyroidism. In: Rosen C (ed.) *Primer on the Metabolic Bone Diseases and Disorders of Mineral Metabolism, 8th Ed.* Hoboken: Wiley.
29. Cusano N, Wang P, Cremers S, Haney E, Bauer D, Orwoll E, Bilezikian J. 2011. Subclinical hypoparathyroidism: A new variant based upon a cohort from MrOS. *J Bone Miner Res* 26 (Suppl 1). Available at http://www.asbmr.org/Meetings/AnnualMeeting/AbstractDetail.aspx?aid=3b2d7e19-59db-4026-b5f8-d11f3e6d6b34. Accessed October 18, 2011.
30. Rude RK. 2012. Magnesium depletion and hypermagnesemia. In: Rosen C (ed.) *Primer on the Metabolic Bone Diseases and Disorders of Mineral Metabolism, 7th Ed.* Hoboken: Wiley.
31. Rubin MR, Dempster DW, Zhou H, Shane E, Nickolas T, Sliney J Jr, Silverberg SJ, Bilezikian JP. 2008. Dynamic and structural properties of the skeleton in hypoparathyroidism. *J Bone Miner Res* 23: 2018–2024.
32. Rubin MR, Dempster DW, Kohler T, Zhou H, Shane E, Nickolas T, Stein E, Sliney J Jr, Silverberg SJ, Bilezikian JP, Muller R. 2009. Three-dimensional cancellous bone structure in hypoparathyroidism. *Bone* 46: 190–195.
33. Rubin MR, Dempster DW, Sliney J, Zhou H, Nickolas TL, Stein EM, Dworakowski E, Dellabadia M, Ives R, McMahon DJ, Zhang C, Silverberg SJ, Shane E, Cremers S, Bilezikian JP. 2011. PTH (1-84) administration reverses abnormal bone remodeling dynamics and structure in hypoparathyroidism. *J Bone Miner Res* 26: 2727–2736.
34. Winer KK, Zhang B, Shrader JA, Peterson D, Smith M, Albert PS, Cutler GB Jr. 2012. Synthetic human parathyroid hormone 1-34 replacement therapy: A randomized crossover trial comparing pump versus injections in the treatment of chronic hypoparathyroidism. *J Clin Endocrinol Metab* 97: 391–399.
35. Sikjaer T, Rejnmark L, Thomsen JS, Tietze A, Bruel A, Andersen G, Mosekilde L. 2012. Changes in 3-dimensional bone structure indices in hypoparathyroid patients treated with PTH(1-84): A randomized controlled study. *J Bone Miner Res* 27: 781–788.
36. Hruska KA, Seifert M. 2012. The chronic kidney disease mineral bone disorder (CKD–MBD). In: Rosen C (ed.) *Primer on the Metabolic Bone Diseases and Disorders of Mineral Metabolism, 8th Ed.* Hoboken: Wiley.
37. Alshayeb HM, Quarles D. 2012. Treatment of chronic kidney disease mineral bone disorder (CKD–MBD). In: Rosen C (ed.) *Primer on the Metabolic Bone Diseases and Disorders of Mineral Metabolism, 8th Ed.* Hoboken: Wiley.
38. Uhlig K, Berns JS, Kestenbaum B, Kumar R, Leonard MB, Martin KJ, Sprague SM, Goldfarb S. 2010. KDOQI US commentary on the 2009 KDIGO Clinical Practice Guideline for the Diagnosis, Evaluation, and Treatment of CKD-Mineral and Bone Disorder (CKD-MBD). *Am J Kidney Dis* 55: 773–799.
39. Kidney Disease: Improving Global Outcomes (KDIGO) CKDMBD Work Group. 2009. KDIGO clinical practice guideline for the diagnosis, evaluation, prevention, and treatment of Chronic Kidney Disease-Mineral and Bone Disorder (CKD–MBD). *Kidney Int Suppl* 113: S1–130
40. Jüppner H, Bastepe M. Pseudohypoparathyroidism. In: Rosen C (ed.) *Primer on the Metabolic Bone Diseases and Disorders of Mineral Metabolism, 8th Ed.* Hoboken: Wiley.

68
Primary Hyperparathyroidism

Shonni J. Silverberg

Introduction 543
Signs and Symptoms 544
Clinical Forms of Primary Hyperparathyroidism 545
Evaluation and Diagnosis of Primary Hyperparathyroidism 545

Treatment of Primary Hyperparathyroidism 546
Acknowledgments 549
References 549

INTRODUCTION

Primary hyperparathyroidism is a disorder resulting from excessive secretion of PTH from one or more of the four parathyroid glands. In 90% of cases, hypercalcemia can be explained by primary hyperparathyroidism or malignancy. Other potential causes of hypercalcemia are considered after the first two are ruled out or if there is reason to believe that a different cause is likely. The differential diagnosis of hypercalcemia, as well as features of hypercalcemia of malignancy, are considered elsewhere in this *Primer*. This chapter will deal exclusively with the clinical presentation, evaluation, and therapy of primary hyperparathyroidism.

Primary hyperparathyroidism is a relatively common endocrine disease, with an incidence as high as 1 in 500 to 1 in 1,000. A four- to fivefold increase in incidence in what was once a rare disorder was noted in the early 1970s, when the widespread use of the multichannel biochemistry autoanalyzer provided physicians with serum calcium levels in patients being evaluated for unrelated complaints [1]. The incidence has leveled off or even declined slightly [2, 3]. Women are affected more often than men by a ratio of 3:1. The majority of patients are postmenopausal women, often presenting within the first decade after menopause. The disease can, however, present at all ages. When diagnosed in childhood, an unusual event, it is important to consider the possibility that the disease is a harbinger of a genetic endocrinopathy, such as multiple endocrine neoplasia type (MEN) I or II.

Primary hyperparathyroidism is caused by a benign, solitary adenoma 80–85% of the time [4]. A parathyroid adenoma is a collection of chief cells surrounded by a rim of normal tissue at the outer perimeter of the gland. In the patient with a parathyroid adenoma, the remaining three parathyroid glands are usually normal. Less commonly (approximately 10–15% of cases), the disease is caused by hyperplasia of all four parathyroid glands. Four-gland hyperplasia may occur sporadically or in association with multiple endocrine neoplasia type I or II. Multiple adenomas account for the balance of benign disease, with parathyroid carcinoma, a very rare and severe form of primary hyperparathyroidism, occurring in fewer than 1% of patients [5]. Pathological examination of the malignant tissue might show mitoses, vascular or capsular invasion, and fibrous trabeculae. However, unless gross local or distant metastases are present, the diagnosis of parathyroid cancer is difficult to make histologically. Specific genetic studies and immunohistochemical analyses (*HRPT2* gene and parafibromin staining) may help to distinguish benign from malignant parathyroid tissue when standard approaches are not clear [6, 7].

The pathophysiology of primary hyperparathyroidism relates to the loss of normal feedback control of PTH by

---

Primer on the Metabolic Bone Diseases and Disorders of Mineral Metabolism, Eighth Edition. Edited by Clifford J. Rosen.
© 2013 American Society for Bone and Mineral Research. Published 2013 by John Wiley & Sons, Inc.

extracellular calcium. Under virtually all other hypercalcemic conditions, there is feedback suppression of the parathyroid glands, and parathyroid hormone (PTH) levels are very low or undetectable. In adenomas, the parathyroid cell loses its normal sensitivity to calcium, whereas in primary hyperparathyroidism caused by hyperplasia of the parathyroid glands, the "set point" for calcium is not changed for a given parathyroid cell; instead, an increase in the number of cells gives rise to hypercalcemia.

The etiology of primary hyperparathyroidism is apparent in only a small minority of patients. These include those who received external neck irradiation in childhood (an average of 20–40 years before disease presentation), or through exposure during a radiation leak more than 20 years previously. A recent report on Chernobyl victims suggests a markedly increased risk of primary hyperparathyroisim (odds ratio 63, CI 36–113) [8]. Lithium, which decreases parathyroid sensitivity to calcium, can be associated with hypercalcemia and hyperparathyroidism in a small percentage of patients [9].

The molecular basis for primary hyperparathyroidism remains elusive in the vast majority of patients (see the chapter on familial hyperparathyroidism for more extensive discussion of genetic causes of primary hyperparathyroidism) [10–14]. The clonal origin of most parathyroid adenomas suggests a defect at the level of the gene controlling growth of the parathyroid cell or the expression of PTH. A gain of function mutation has been identified in 20–40% of sporadic parathyroid adenomas, in which the relocation of the cyclin D1 gene on chromosome 11 results in overexpression of this cell cycle regulator. Genes involved in MEN syndromes have also been investigated in sporadic disease. Inactivating mutations of the MEN1 gene have been reported in less than 10–15% of cases of sporadic parathyroid adenomas. The role of this tumor suppressor gene and menin, its gene product, in the pathogenesis of these sporadic adenomas is unknown. RET proto-oncogene gain-of-function mutations that are seen in MEN 2A or 2B have not been reported in sporadic primary hyperparathyroidism. Inactivating HRPT2 mutations have been described in familial hyperparathyroidism-jaw tumor syndrome (HPT-JT), in which they are associated with an increased risk of parathyroid cancer. While mutations of this gene have been reported in patients with sporadic parathyroid carcinoma, they are not a common feature in the absence of this familial hyperparathyroid syndrome. The possibility of mutations in the Wnt-β-catenin signaling pathway is also under investigation.

SIGNS AND SYMPTOMS

"Classical" primary hyperparathyroidism was a symptomatic disease associated with a typical skeletal disorder (osteitis fibrosa cystica), nephrolithiasis, and neuromuscular complaints. While this phenotype is still occasionally seen in the United States, today the overwhelming majority of patients lack these symptoms [15]. Osteitis fibrosa cystica, characterized by subperiosteal resorption of the distal phalanges, tapering of the distal clavicles, a "salt and pepper" appearance of the skull, bone cysts, and brown tumors of the long bones, is now seen in fewer than 5% of patients with primary hyperparathyroidism.

The incidence of kidney stones has also declined from 33% in the 1960s to 15–20% now. Nephrolithiasis, nevertheless, is still the most common overt complication of the hyperparathyroid process. Other renal features of primary hyperparathyroidism include diffuse deposition of calcium-phosphate complexes in the parenchyma (nephrocalcinosis). Hypercalciuria (daily calcium excretion of more than 250 mg for women or 300 mg for men) is seen in up to 30% of patients. In the absence of any other cause, primary hyperparathyroidism may be associated with a reduction in creatinine clearance.

The classic neuromuscular syndrome of primary hyperparathyroidism included a definable myopathy that has virtually disappeared [16]. In its place, however, is a less well-defined syndrome characterized by easy fatigue, a sense of weakness, and a feeling that the aging process is advancing faster than it should. This is sometimes accompanied by an intellectual weariness and a sense that cognitive faculties are less sharp. Some psychiatric complaints (depression and anxiety) are more common in these patients, and a recent study demonstrates some specific reversible areas of mild cognitive impairment (i.e., visual memory) [17]. However, the cumulative results of three randomized studies of surgery in patients with mild disease do not suggest that an individual patient can expect specific reversible changes in constitutional or neuropsychiatric complaints following parathyroidectomy [18–20]. Thus, the available data do not support surgical intervention for the purpose of improving neuropsychiatric symptoms at this time [21].

Gastrointestinal manifestations of primary hyperparathyroidism have classically included peptic ulcer disease and pancreatitis. Peptic ulcer disease is not likely to be linked in a pathophysiologic way to primary hyperparathyroidism unless type I multiple endocrine neoplasia is present. Pancreatitis is virtually never seen anymore as a complication of primary hyperparathyroidism because the hypercalcemia tends to be so mild.

In classical primary hyperparathyroidism, cardiovascular features included myocardial, valvular, and vascular calcification, with subsequent increased cardiovascular mortality. The limited available data suggest that cardiovascular mortality is not increased in mild disease [22]. Although overt cardiovascular involvement is not seen, there are data supporting increased vascular stiffness in mild disease, and reports of other subtle cardiovascular manifestations. Although there may be an increased incidence of hypertension in primary hyperparathyroidism, it is rarely corrected or improved after successful surgery [23]. Other potential organ systems that in the past were affected by the hyperparathyroid state are now relegated to being archival curiosities. These include gout and pseudogout, anemia, band keratopathy, and loose teeth.

CLINICAL FORMS OF PRIMARY HYPERPARATHYROIDISM

In the United States today, classical primary hyperparathyroidism is rarely seen [15]. Instead, the most common clinical presentation of primary hyperparathyroidism is characterized by asymptomatic hypercalcemia with serum calcium levels within 1 mg/dl above the upper limits of normal. Most patients do not have specific complaints and do not show evidence for any target organ complications. They have often been discovered in the course of a routine multichannel screening test. Rarely, a patient will show serum calcium levels in the life-threatening range, so-called acute primary hyperparathyroidism or parathyroid crisis [24]. These patients are invariably symptomatic due to their hypercalcemia. This unusual presentation of primary hyperparathyroidism should be considered in any patient who presents with acute hypercalcemia of unclear etiology. Other unusual clinical presentations of primary hyperparathyroidism include multiple endocrine neoplasia types I and II, familial primary hyperparathyroidism not associated with other endocrine disorders, familial cystic parathyroid adenomatosis, and neonatal primary hyperparathyroidism.

Normocalcemic primary hyperparathyroidism has recently been recognized as a specific phenotype of the disease [15, 25, 26]. These patients have normal serum calcium concentration with elevated PTH levels in the absence of an identifiable cause for secondary hyperparathyroidism. This diagnosis is often made in individuals who are undergoing evaluation for low bone mineral density (BMD) or who are receiving comprehensive screening tests for their skeletal health. In some patients this constellation of findings may represent the earliest manifestation of hypercalcemic primary hyperparathyroidism, when PTH alone is elevated and serum calcium is still normal. In order to make this diagnosis, both serum total and ionized calcium must be normal, and alternate causes for PTH elevation ruled out (i.e., renal failure, renal calcium loss, calcium malabsorption). It is particularly important to be sure that these individuals do not have vitamin D deficiency or insufficiency, a cause of secondary hyperparathyroidism that can also lower serum calcium levels in hypercalcemic primary hyperparathyroidism into the normal range. Little is known about the natural history of this phenotype. While some patients quickly progress to hypercalcemic disease, others do not. It is also not known whether current surgical criteria for hypercalcemic disease apply to patients with normocalcemic disease.

EVALUATION AND DIAGNOSIS OF PRIMARY HYPERPARATHYROIDISM

The history and the physical examination rarely give any clear indications of primary hyperparathyroidism but are helpful because they may suggest an alternate cause for hypercalcemia. The diagnosis of primary hyperparathyroidism is instead established by laboratory tests [27]. The biochemical hallmarks of primary hyperparathyroidism are hypercalcemia (except in patients with normocalcemic primary hyperparathyroidism) and elevated levels of PTH. In the presence of hypercalcemia, an elevated level of PTH virtually establishes the diagnosis. A PTH level in the mid or upper end of the normal range in the face of hypercalcemia is also consistent with the diagnosis of primary hyperparathyroidism. The standard assays for measurement of PTH are the second generation immunoradiometric (IRMA) or immunochemiluminometric (ICMA) assays that measures the "intact" molecule. This assay also measures large carboxyterminal fragments [PTH(7-84) is one such fragment] of PTH in addition to full-length PTH(1-84) and can therefore overestimate the amount of bioactive hormone in the serum. A third-generation assay specific for PTH(1-84) may offer somewhat increased diagnostic sensitivity in cases where the intact IRMA is within the normal range (albeit inappropriately), or in patients with renal insufficiency or failure. Most data suggest that either second- or third-generation PTH assays can be used in the diagnosis of the disease.

The clinical importance of PTH measurement in the differential diagnosis of hypercalcemia is a result of the fact that all non-parathyroid mediated causes of hypercalcemia (including malignancy) are associated with suppressed PTH levels. There is no cross reactivity between PTH and PTH-related peptide (PTHrP; the major causative factor in humoral hypercalcemia of malignancy) in the above assays for PTH. PTH concentration and serum calcium levels are also elevated in familial hypocalciuric hypercalcemia (FHH) [28, 29]. Some cases of reversible primary hyperparathyroidism are related to lithium or thiazide diuretic use [9, 30]. More commonly, thiazide diuretics unmask primary hyperparathyroidism by inhibiting calcium excretion. Ultimately, the only secure way to make the diagnosis of drug-related primary hyperparathyroidism is to withdraw the medication (if it is safe to do so) and confirm persistent hypercalcemia and elevated PTH levels 2–3 months later.

In addition to calcium and PTH abnormalities, serum phosphorus tends to be in the lower range of normal and is frankly low in approximately one-third of patients. Both total serum alkaline phosphatase activity and more specific markers of bone turnover (bone formation markers: bone-specific alkaline phosphatase, osteocalcin and bone resorption markers: urinary deoxypyridinoline and N-telopeptide of collagen) are elevated when there is active bone involvement but otherwise tend to be in the upper range of normal. The actions of PTH to alter acid–base handling in the kidney can lead to a small increase in the serum chloride concentration and a concomitant decrease in the serum bicarbonate concentration. Only 30% of patients are hypercalciuric. The circulating 1,25-dihydroxy-vitamin D concentration is elevated in 25% of patients while 25-hydroxyvitamin D levels tend to be in the lower end of the normal range. Although the definition of vitamin D deficiency remains controversial, most patients with primary hyperparathyroidism have levels that are below 30 ng/ml and many are below 20 ng/ml [31,32]. New guidelines suggest that 25-hydroxyvitamin D be measured in all patients with primary hyperparathyroidism, and that those with levels under 20 ng/ml be treated with vitamin D, although guidelines for repletion are lacking [21].

Although osteitis fibrosa cystica is rarely seen in the United States today, the skeleton remains an important target organ in primary hyperparathyroidism [33–35]. Plain X-rays are generally unrevealing and are not recommended. On the other hand, bone mineral densitometry has proven to be an essential component of the evaluation because of its great sensitivity in detecting early changes in bone mass. Patients with primary hyperparathyroidism tend to show a pattern of bone involvement that preferentially affects the cortical, as opposed to the cancellous, skeleton (Fig. 68.1) [33]. Typically, BMD of the distal third of the forearm, a site enriched in cortical bone, is reduced while the lumbar spine, a site enriched in cancellous bone, is relative preserved. The hip region (femoral neck) tends to show values intermediate between the distal radius and the lumbar spine because its composition is a more equal mixture of cortical and cancellous elements. The preferential reduction of bone density at the distal forearm underscores the importance of measuring BMD at that site in primary hyperparathyroidism. A small subset of patients (15%) present with an atypical BMD profile, characterized by vertebral osteopenia or osteoporosis, while occasionally patients may show reductions in BMD at all sites [34]. Bone densitometry gives an accurate assessment of the degree of skeletal involvement in primary hyperparathyroidism, information that is used to make recommendations for parathyroid surgery (see Table 68.1) [21]. As useful as bone density measurement is, it remains a proxy for the end point of fracture in primary hyperparathyroidism. Although case-control and cohort studies suggest that there may be an increase risk of fracture, questions of ascertainment bias exist, and this question remains unresolved [36–39].

TREATMENT OF PRIMARY HYPERPARATHYROIDISM

Parathyroid surgery

Surgery provides the only option for cure of primary hyperparathyroidism [4]. While surgery is indicated in all patients with classical symptoms of primary hyperparathyroidism (overt bone disease or kidney stones, or if they have survived an episode of acute primary hyperparathyroidism with life-threatening hypercalcemia), there is considerable controversy concerning the need for intervention in patients who have no clear signs or symptoms of their disease. To date, three national and international conferences (1990, 2002, 2008) have updated guidelines for surgical intervention (Table 68.1) [21, 40, 41].

Surgical guidelines (Table 68.1) [21]. Asymptomatic patients are now advised to have surgery if they have: (1) serum calcium greater than 1 mg/dl above the upper limit of normal; (2) significantly reduced creatinine clearance (less than 60 cc/min) (for the first time marked hypercalciuria is not an indication for surgery); (3) bone density more than 2.5 SD below young normal control subjects at any site (T-score: –2.5 or below) or fragility fracture; (4) younger than 50 years of age. These patients are at

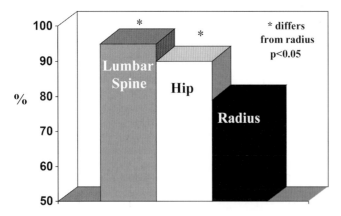

Fig. 68.1. Bone mineral density in primary hyperparathyroidism. Data are shown as percent of expected by site (adapted from Ref. 33).

Table 68.1. Guidelines for Use in Patients with Asymptomatic Primary Hyperparathyroidism

Measure	Surgical Criteria	Guidelines for Follow-Up of Nonsurgical Patients
Serum calcium	>1 mg/dl above normal	Annually
Urinary calcium	Do not measure	Do not measure
Renal function: Creatinine clearance	< 60 cc/min	Do not measure
Serum creatinine		Measure annually
BMD	T-score < –2.5 any site or fragility fracture	Every 1–2 years at spine, hip and forearm
Age	<50 years	

Guidelines describe patients for whom surgical intervention is desirable and criteria for follow-up of nonsurgical patients. It is recommended that all symptomatic patients be sent for parathyroidectomy. [Adapted from the 2008 International Workshop on Asymptomatic Primary Hyperparathyroidism (Ref. 21).]

greater risk for disease progression than older patients. Because surgery is an acceptable approach even in patients who do not meet surgical guidelines, some physicians will recommend surgery for all patients with primary hyperparathyroidism; others will not recommend surgery unless clear-cut complications of primary hyperparathyroidism are present. Similarly, some patients do not want to live with a curable disease while others may be unwilling to face the risks of surgery, although surgical indications are present. Finally, advances in surgical techniques may also shift the balance in favor of intervention in the eyes of some patients and their physicians [4].

Preoperative localization was initially used to identify the location of an ectopic parathyroid gland. Today it is used to identify candidates for minimally invasive parathyroidectomy (MIP) as well as in disease that is recurrent or persistent after surgery [42–45]. These techniques should not be used to make the diagnosis; rather, they should guide the surgeon once a diagnosis is made. The most widely used localization modalities are technetium-99m-sestamibi [with or without single photon emission computed tomography (SPECT)] or ultrasound. The former is excellent in single gland disease but is often inaccurate in multigland disease. Other localization techniques include four-dimensional (4D) CT scanning, magnetic resonance imaging (MRI), and the invasive modalities of selective venous sampling or arteriography. Radioisotopic imaging and ultrasound are best for parathyroid tissue that is located in proximity to the thyroid, whereas CT and MRI approaches are better for ectopically located parathyroid tissue. Arteriography and selective venous studies are reserved for those individuals in whom the noninvasive studies have not been successful. In patients who have undergone prior unsuccessful surgery, localization by two different modalities is suggested.

Surgery. Even without localization, an experienced parathyroid surgeon will find the abnormal parathyroid gland(s) 95% of the time in the patient who has not had previous neck surgery. The glands are notoriously variable in location, requiring the surgeon's knowledge of typical ectopic sites such as intrathyroidal, retroesophageal, the lateral neck, and the mediastinum. Four-gland exploration was long considered the gold standard surgical approach. It remains the procedure of choice in patients with no suggestive localization studies and those with hereditary disease or lithium-induced disease, in whom multigland involvement is common. Today, focused MIP is rapidly becoming the procedure of choice in patients in whom preoperative localization has localized single gland disease [4, 45]. MIP or unilateral exploration require the capability to measure intraoperative PTH levels. Taking advantage of the short half-life of PTH (3–5 minutes), an intraoperative PTH level is drawn shortly after resection [46]. If the PTH level falls by 50% and is within the normal range, the adenoma that has been removed is considered to be the only source of abnormal glandular activity, and the operation is terminated. There is some concern about the 50% decline rule, because if the PTH level falls by more than 50% but remains frankly elevated, other glandular sources of PTH may remain. In the case of multiglandular disease, the approach is to remove all tissue except for a fragment of parathyroid tissue that is left *in situ* or autotransplanted into the nondominant forearm. Potential complications of surgery include damage to the recurrent laryngeal nerve, which can lead to hoarseness and reduced voice volume, and permanent hypoparathyroidism in those who have had previous neck surgery or who undergo subtotal parathyroidectomy (for multiglandular disease).

Postoperatively, the patient may experience a brief period of transient hypocalcemia, during which time the normal but suppressed parathyroid glands regain their sensitivity to calcium. This happens within the first few days after surgery but can be prevented in most cases by providing patients with several grams of calcium on a daily basis during the first postoperative week. Prolonged postoperative symptomatic hypocalcemia as a result of rapid deposition of calcium and phosphate into bone ("hungry bone syndrome") is rare today. Such patients may require parenteral calcium for symptomatic hypocalcemia.

After successful surgery, the patient is cured. Serum biochemistries and PTH levels normalize. Long-term observational data confirm short-term randomized trial findings showing that bone density improves in the first several years after surgery [18–20, 47, 48]. The cumulative increase in bone mass at the lumbar spine and femoral neck can be as high as 12%, an increase that is sustained for years after parathyroidectomy (Fig. 68.2). It is noteworthy that substantial improvement is seen at the lumbar spine, a site where PTH seems to protect from age-related and estrogen-deficiency bone loss. Because patients who have vertebral osteopenia or osteoporosis sustain an even more impressive improvement in spine BMD after cure, they should be routinely referred for surgery, regardless of the severity of their hypercalcemia.

Nonsurgical management

Most patients who are not surgical candidates for parathyroidectomy do well when they are managed conservatively. In most such patients, biochemical indices (serum calcium, PTH, 1,25-dihydroxyvitamin D, and urinary calcium excretion) and BMD remain stable over the first decade of observation [47]. However, those patients followed for longer than that period begin to show evidence of bone loss, particularly at the more cortical sites (hip and radius; Fig. 68.2) [48]. In a 15-year observational cohort, 37% of patients with asymptomatic primary hyperparathyroidism had biochemical or bone densitometric evidence of disease progression. Those under the age of 50 years have a far higher incidence of progressive disease than do older patients (65% vs 23%), supporting the notion that younger patients should be referred for parathyroidectomy [49]. Finally, today, as in the day of classical primary hyperparathyroidism, patients with symptomatic disease do poorly

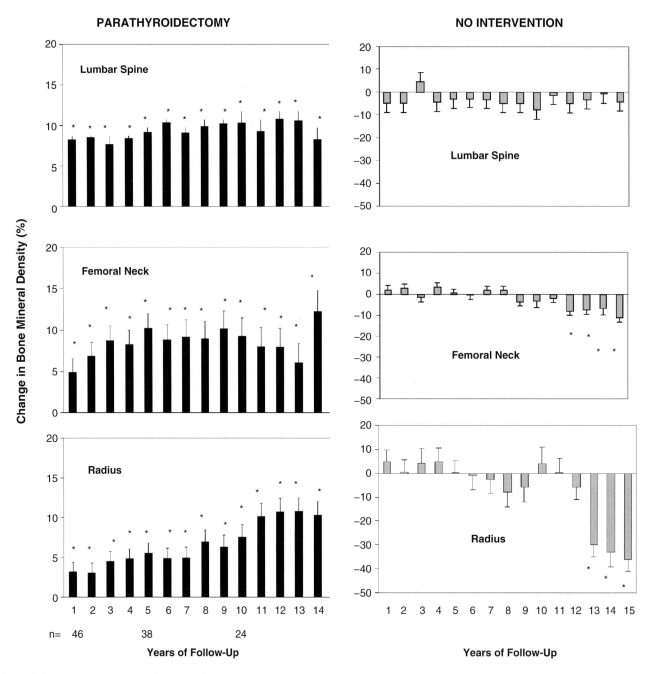

Fig. 68.2. Long-term effect of surgery (left panel) versus nonintervention (right panel) in patients with mild primary hyperparathyroidism (adapted from Ref. 48).

when observed without surgery. Thus, the data support the safety of observation without surgery only in selected patients with asymptomatic primary hyperparathyroidism, and even in those, indefinite observation is not clearly desirable.

A set of general medical guidelines is recommended for patients who do not undergo surgery (Table 68.1) [21]. Serum calcium levels should be measured once to twice yearly with annual assessment of serum creatinine and annual or biannual bone densitometry at the spine, hip, and distal one-third site of the forearm. Adequate hydration and ambulation are always encouraged. Thiazide diuretics and lithium should be avoided if possible, because they may worsen hypercalcemia. There is no good evidence that patients with primary hyperparathyroidism show significant fluctuations of their serum calcium as a function of dietary calcium intake. Dietary calcium intake should therefore be moderate, as low calcium diets could theoretically lead to further stimulation of PTH secretion. High calcium intake (more than 1 g/day) should be avoided in patients whose 1,25-dihydroxyvitamin D levels are elevated.

We still lack an effective and safe therapeutic agent approved for the medical management of primary hyperparathyroidism in most patients [50]. **Oral phosphate** will lower the serum calcium in patients with primary hyperparathyroidism by 0.5–1 mg/dl. Phosphate seems to act by three mechanisms: (1) interference with absorption of dietary calcium, (2) inhibition of bone resorption, and (3) inhibition of renal production of 1,25-dihydroxyvitamin D. Phosphate, however, is not recommended as an approach to management, because of concerns related to ectopic calcification in soft tissues as a result of increasing the calcium–phosphate product. Moreover, oral phosphate may lead to an undesirable further elevation of PTH levels. Gastrointestinal intolerance is another limiting feature of this approach. In postmenopausal women, **estrogen therapy** remains an option in those women desiring hormone replacement for treatment of symptoms of menopause [51, 52]. The rationale for estrogen use in primary hyperparathyroidism is based on the known antagonism by estrogen of PTH-mediated bone resorption. Although the serum calcium concentration does tend to decline after estrogen administration (by 0.5 mg/dl), PTH levels and the serum phosphorous concentration do not change. Estrogen replacement may improve BMD in these patients as well. Preliminary data suggest that the **selective estrogen receptor modulator**, raloxifene, may have a similar effect on serum calcium levels in postmenopausal women with primary hyperparathyroidism [53]. **Bisphosphonates** have been investigated as a possible medical approach to primary hyperparathyroidism [54, 55]. Alendronate improves vertebral BMD in patients with primary hyperparathyroidism who choose not to have surgery but does not affect the underlying disorder. While most of the agents listed above (except for phosphate) act by inhibiting bone resorption, calcimimetics target the excess PTH secretion that underlies the disease. **Calcimimetic** agents alter the function of the extracellular calcium-sensing receptor, increasing the affinity of the parathyroid cell calcium receptor for extracellular calcium, leading to increased intracellular calcium, a subsequent reduction in PTH synthesis and secretion, and ultimately a fall in the serum calcium. Clinical trials of one such agent, cinacalcet HCl, have shown normalization of serum calcium for up to 5.5 years and sustained decreases in serum calcium levels across a wide range of disease severity [56–58]. However, this agent does not provide the equivalent of a "medical parathyroidectomy" because it does not improve bone density. There are no data on the effect of calcimimetics on constitutional or neuropsychological symptoms or fracture. Cinacalcet is approved for use in parathyroid cancer and was recently FDA approved for the treatment of severe hypercalcemia in patients with primary hyperparathyroidism who are unable to undergo parathyroidectomy. However, it is not currently recommended for general use in patients who are candidates for parathyroidectomy.

Primary hyperparathyroidism in pregnancy

Although rare, primary hyperparathyroidism does occur during pregnancy. Severe hypercalcemia can have deleterious effects on both mother and fetus. The fetal parathyroid gland is suppressed, which can lead to hypocalcemia, tetany, and even fetal demise. Preconception surgery is optimal if the diagnosis is known. Otherwise, surgery is recommended during the second trimester for patients with symptoms or significant hypercalcemia. Observation through pregnancy may be appropriate in the case of asymptomatic patients with mild hypercalcemia, although in these cases it is imperative that the neonatologists be aware of possible hypocalcemia in the neonate [59, 60].

ACKNOWLEDGMENTS

This work was supported in part by NIH Grants DK32333, DK084986, and DK66329.

REFERENCES

1. Heath H, Hodgson SF, Kennedy MA. 1980. Primary hyperparathyroidism: Incidence, morbidity, and potential economic impact in a community. *N Engl J Med* 302(4): 189–193.
2. Wermers RA, Khosla S, Atkinson EJ, Hodgson SF, O'Fallon WM, Melton LJ. 1997. The rise and fall of primary hyperparathyroidism: A population based study in Rochester, Minnesota, 1965–1992. *Ann Intern Med* 126(6): 433–440.
3. Wermers RS, Khosla S, Atkinson EJ, Achenbach SJ, Oberg AN, Grant CS, Melton LJ. 2006. Incidence of primary hyperparathyroidism in Rochester, Minnesota, 1993–2001: An update on the changing epidemiology of the disease. *J Bone Miner Res* 21(1): 171–177.
4. Udelsman R, Pasieka JL, Sturgeon C, Young JEM, Clark OH. 2009. Surgery for asymptomatic primary hyperparathyroidism: Proceedings of the Third International Workshop. *J Clin Endocrinol Metab* 94(2): 366–372.
5. Marcocci C, Cetani F, Rubin MR, Silverberg SJ, Pinchera A, Bilezikian JP. 2008. Parathyroid carcinoma. *J Bone Min Res* 23(12): 1869–1880.
6. Shatuck TM, Välimäki S, Obara T, Gaz RD, Clark OH, Shoback D, Wieman ME, Tojo K, Robbins CM, Carpten JD, Famebo LO, Larsson C, Arnold A. 2003. *N Engl J Med* 349(18): 1722–1729.
7. Tan MH, Morrison C, Wang P. Yang X, Haven CJ, Zhang C, Zhao P, Tretiakova MS, Korpi-Hyovalti E, Burgess JR, Soo KC, Cheah WK, Cao B, Resau J, Morreau H, Teh BT.

2004. Loss of parafibromin immunoreactivity is a distinguishing feature of parathyroid carcinoma. *Clin Cancer Res* 10(19): 6629–6637.

8. Boehm BO, Rosinger S, Belyi D, Dietrich JW. 2011. The parathyroid as a target for radiation damage. *N Engl J Med* 365(7): 676–678.

9. Mallette LE, Eichhorn E. 1986. Effects of lithium carbonate on human calcium metabolism. *Arch Intern Med* 146(4)770–776.

10. Hendy GN, Arnold A. Molecular basis of PTH overexpression. 2008. In: Bilezikian JP, Raisz LG, Martin TJ (eds.) *Principles of Bone Biology*. San Diego, CA: Academic Press. pp. 1311–1326.

11. Westin G, Björklund P, Akerström G. 2009. Molecular genetics of parathyroid disease. *World J Surg* 33(11): 2224–2233.

12. Pausova Z, Soliman E, Amizuka N, Janicic N, Konrad EM, Arnold A, Goltzman D, Hendy GN. 1996. Role of the RET proto-oncogene in sporadic hyperparathyroidism and in hyperparathyroidism of multiple endocrine neoplasia type 2. *J Clin Endocrinol Metab* 81(7): 2711–2718.

13. Björklynd P, Lindberg D, Akerström G, Westin G. 2008. Stabilizing mutation of CTNNB1/beta-catenin and protein accumulation analyzed in a large series of parathyroid tumors of Swedish patients. *Mol Cancer* 7: 53.

14. Falchetti A, Marini F, Giusti F, Cavalli L, Cavalli T, Brandi ML. 2009. DNA-based test: When and why to apply it to primary hyperparathyroidism clinical phenotypes. *J Intern Med* 266(1): 69–83.

15. Silverberg SJ, Lewiecki EM, Mosekilde L, Peacock M, Rubin MR. 2009. Presentation of asymptomatic primary hyperparathyroidism: Proceedings of the Third International Workshop. *J Clin Endocrinol Metab* 94(2): 351–365.

16. Turken SA, Cafferty M, Silverberg SJ, de la Cruz L, Cimino C, Lange DJ, Lovelace RE, Bilezikian JP. 1989. Neuromuscular involvement in mild, asymptomatic primary hyperparathyroidism. *Am J Med* 87(5): 553–557.

17. Walker MD, McMahon DJ, Inabnet WB, Lazar RM, Brown I, Vardy S, Cosman F, Silverberg SJ. 2009. Neuropsychological features in primary hyperparathyroidism: A prospective study. *J Clin Endocrinol Metab* 94(6): 1951–1958.

18. Bollerslev J, Jannson S, Mollerup CL, Nordenström J, Lundgren E, Tørring O, Varhaug JE, Baranowski M, Aanderud S, Franco C, Freyschuss B, Isaksen GA, Ueland T, Rosen T. 2007. Medical observation, compared with parathyroidectomy, for asymptomatic primary hyperparathyroidism: A prospective, radomized trial. *J Clin Endocrinol Metab* 92(5): 1687–1692.

19. Ambrogini E, Centani F, Cianferotti L, Vignali E, Banti C, Viccia G, Oppo A, Miccoli P, Berti P, Bilezikian JP, Pinchera A, Marcocci C. 2007. Surgery or surveillance for mild asymptomatic primary hyperparathyroidism: A prospective, randomized clinical trial. *J Clin Endocrinol Metab* 92(8): 3114–3121.

20. Rao DS, Phillips ER, Divine GW, Talpos GB. 2004. Randomized controlled clinical trial of surgery versus no surgery in patients with mild asymptomatic primary hyperparathyroidism. *J Clin Endocrinol Metab* 89(11): 5415–5422.

21. Bilezikian, JP, Khan, AA, Potts JT Jr. 2009. Third International Workshop on the Management of Asymptomatic Primary, Hyperthyroidism. Guidelines for the management of asymptomatic primary hyperparathyroidism: Summary statement from the Third International Workshop. *J Clin Endocrinol Metab* 94(2): 335–339.

22. Wermers RA, Khosla S, Atkinson EJ, Grant CS, Hodgson SF, O'Fallon M, Melton LJ 3rd. 1998. Survival after the diagnosis of hyperparathyroidism: A population-based study. *Am J Med* 104(2): 115–122.

23. Bollerslev J, Rosen T, Mollerup CL, Nordenström J, Baranowski M, Franco C, Pernow Y, Isaksen GA, Godang K, Ueland T, Jansson S; SIPH Study Group. 2009. Effect of surgery on cardiovascular risk factors in mild primary hyperparathyroidism. *J Clin Endocrinol Metab* 94(7): 2255–2261.

24. Fitzpatrick LA. 2001. Acute primary hyperparathyroidism. In: Bilezikian JP (ed.) *The Parathyroids: Basic and Clinical Concepts*. New York: Academic Press. pp. 527–534.

25. Silverberg SJ, Bilezikian JP. 2003. "Incipient" primary hyperparathyroidism: A "forme fruste" of an old disease. *J Clin Endocrinol Metab* 88(11): 5348–5352.

26. Lowe H, McMahon DJ, Rubin MR, Bilezikian JP, Silverberg SJ. 2007. Normocalcemic primary hyperparathyroidism: Further characterization of a new clinical phenotype. *J Clin Endocrinol Metab* 92(8): 3001–3005.

27. Eastell, R, Arnold, A, Brandi, ML, Brown EM, D'Amour P, Hanley DA, Rao DS, Rubin MR, Goltzman D, Silverberg SJ, Marx SJ, Peacock M, Mosekilde L, Bouillon R, Lewiecki EM. 2009. Diagnosis of asymptomatic primary hyperparathyroidism: Proceedings of the Third International Workshop. *J Clin Endocrinol Metab* 94(2): 340–350.

28. Christensen SE, Nissen PH, Vestergaard P, Heickendorff L, Brixen K, Mosekilde L. 2008. Discriminative power of three indices of renal calcium excretion for the distinction between familial hypocalciuric hypercalcemia and primary hyperparathyroidism: A follow-up study on methods. *Clin Endocrinol (Oxf)* 69(5): 713–720.

29. Fuleihan Gel-H. 2002. Familial benign hypocalciuric hypercalcemia. *J Bone Miner Res* 17 Suppl 2: N51–56.

30. Wermers RA, Kearns AE, Jenkins GD, Melton LJ 3rd. 2007. Incidence and clinical spectrum of thiazide-associated hypercalcemia. *Am J Med* 120(10): 911. e9–15.

31. Silverberg SJ, Shane E, Dempster DW, Bilezikian JP. 1999. The effects of Vitamin D insufficiency in patients with primary hyperparathyroidism. *Am J Med* 107(6): 561–567.

32. Boudou P, Ibrahim F, Cormier C, Sarfati E, Souberbielle JC. 2006. A very high incidence of low 25-hydroxy-

vitamin D serum concentration in a French population of patients with primary hyperparathyroidism. *J Endocrinol Invest* 29(6): 511–515.
33. Silverberg SJ, Shane E, de la Cruz L, Dempster DW, Feldman F, Seldin D, Jacobs TP, Siris ES, Cafferty M, Parisien MV, Lindsay R, Clemens TL, Bilezikian JP. 1989. Skeletal disease in primary hyperparathyroidism. *J Bone Miner Res* 4(3): 283–291.
34. Silverberg SJ, Locker FG, Bilezikian JP. 1996. Vertebral osteopenia: A new indication for surgery in primary hyperparathyroidism. *J Clin Endocrinol Metab* 81(11): 4007–4012.
35. Dempster DW, Muller R, Zhou H, Kohler T, Shane E, Parisien M, Silverberg SJ, Bilezikian JP. 2007. Preserved three-dimensional cancellous bone structure in mild primary hyperparathyroidism. *Bone* 41(1): 19–24.
36. Dauphine RT, Riggs BL, Scholz DA. 1975. Back pain and vertebral crush fractures: An unemphasized mode of presentation for primary hyperparathyroidism. *Ann Intern Med* 83(3): 365–367.
37. Larsson K, Ljunghall S, Krusemo UB, Naessén T, Lindh E, Persson I. 1993. The risk of hip fractures in patients with primary hyperparathyroidism: A population-based cohort study with a follow-up of 19 years. *J Intern Med* 234(6): 585–593.
38. Khosla S, Melton LJ, Wermers RA, Crowson CS, O'Fallon W, Riggs B. 1999. Primary hyperparathyroidism and the risk of fracture: A population-based study. *J Bone Miner Res* 14(10): 1700–1707.
39. Vignali E, Viccica C, Diacinti D, Cetani F, Cianferotti L, Ambrogini E, Banti C, Del Fiacco R, Bilezikian JP, Pinchera A, Marcocci, C. 2009. Morphometric vertebral fractures in postmenopausal women with primary hyperparathyroidism. *J Clin Endocrinol Metab* 94(7): 2306–2312.
40. [No authors listed]. 1991. National Institutes of Health: Consensus development conference statement on primary hyperparathyroidism. *J Bone Miner Res* 6(Suppl 2): S9–S13.
41. Bilezikian JP, Potts JT Jr, Fuleihan Gel-H, Kleerekoper M, Neer R, Peacock M, Rastad J, Silverberg SJ, Udelsman R, Wells SA. 2002. Summary statement from a workshop on asymptomatic primary hyperparathyroidism: A perspective for the 21st century. *J Clin Endocrinol Metab* 87(12): 5353–5361.
42. Moure D, Larrañaga E, Dominquez-Gadea L, Luque-Ramirez M, Nattero L, Gómez-Pan A, Marazuela M. 2008. 99MTc-sestamibi as sole technique in selection of primary hyperparathyroidism patients for unilateral neck exploration. *Surgery* 144(3): 454–459.
43. Lindqvist V, Jacobsson H, Chandanos E, Bäckdahl M, Kjellman M, Wallin G. 2009. Preoperative 99Tc(m)-sestamibi scintigraphy with SPECT localizes most pathologic parathyroid glands. *Langenbecks Arch Surg* 394(5): 811–815.
44. Van Husen R, Kim LT. 2004. Accuracy of surgeon-performed ultrasound in parathyroid localization. *World J Surg* 28(11): 1122–1126.
45. Fraker DL, Harsono H, Lewis R. 2009. Minimally invasive parathyroidectomy: Benefits and requirements of localization, diagnosis, and intraoperative PTH monitoring long-term results. *World J Surg* 33(11): 2256–2265.
46. Irvin CL 3rd, Solorzano CC, Carneiro DM. 2004. Quick intraoperative parathyroid hormone assay: Surgical adjunct to allow limited parathyroidectomy, improved success rate and predict outcome. *World J Surg* 28(12): 1287–1292.
47. Silverberg SJ, Shane E, Jacobs TP, Siris E, Bilezikian JP. 1999. A 10-year prospective study of primary hyperparathyroidism with or without parathyroid surgery. *N Engl J Med* 341(17): 1249–1255.
48. Rubin MR, Bilezikian JP, McMahon DJ, Jacobs T, Shane E, Siris E, Udesky J, Silverberg SJ. 2008. The natural history of primary hyperparathyroidism with or without parathyroid surgery after 15 years. *J Clin Endocrinol Metab* 93(9): 3462–3470.
49. Silverberg SJ, Brown I, Bilezikian JP. 2002. Age as a criterion for surgery in primary hyperparathyroidism. *Am J Med* 113(8): 681–684.
50. Khan A, Grey A, Shoback D. 2009. Medical management of asymptomatic primary hyperparathyroidism: Proceedings of the Third International Workshop. *J Clin Endocrinol Metab* 94(2): 373–381.
51. Marcus R, Madvig P, Crim M, Pont A, Kosek J. 1984. Conjugated estrogens in the treatment of postmenopausal women with hyperparathyroidism. *Ann Intern Med* 100(5): 633–640.
52. Grey AB, Stapleton JP, Evans MC, Tatnell MA, Reid IR. 1996. Effect of hormone replacement therapy on BMD in post-menopausal women with primary hyperparathyroidism. A randomized controlled trial. *Ann Intern Med* 125(5): 360–368.
53. Rubin MR, Lee K, Silverberg SJ. 2003. Raloxifene lowers serum calcium and markers of bone turnover in primary hyperparathyroidism. *J Clin Endocrinol Metab* 88(3): 1174–1178.
54. Kahn AA, Bilezikian JP, Kung AW, Ahmed MM, Dubois SJ, Ho AY, Schussheim DH, Rubin MR, Shaikh AM, Silverberg SJ, Standish TI, Syed Z, Syed ZA. 2004. Alendronate in primary hyperparathyroidism: A double-blind, randomized, placebo-controlled trial. *J Clin Endocrinol Metab* 89(7): 3319–3325.
55. Chow CC, Chan WB, Li JK, Chan NN, Chan MH, Ko GT, Lo KW, Cockram CS. 2003. Oral Alendronate increases bone mineral density in postmenopausal women with primary hyperparathyroidism. *J Clin Endocrinol Metab* 88(2): 581–587.
56. Peacock M, Bilezikian JP, Klassen P, Guo MD, Turner SA, Shoback DS. 2005. Cinacalcet hydrochloride maintains long-term normocalcemia in patients with primary hyperparathyroidism. *J Clin Endocrinol Metab* 90(1): 135–141.
57. Peacock M, Bolognese MA, Borofsky M, Scumpia S, Sterling LR, Cheng S, Shoback D. 2009. Cinacalcet treatment of primary hyperparathyroidism: Biochemical and

bone densitometric outcomes in a five-year study. *J Clin Endocrinol Metab* 94(12): 4860–4867.
58. Peacock M, Bilezikian JP, Bolognese MA, Borofsky M, Scumpia S, Sterling LR, Cheng S, Shoback D. 2011. Cinacalcet HCl reduces hypercalcemia in primary hyperparathyroidism across a wide spectrum of disease severity. *J Clin Endocrinol Metab* 96(1): E9–18.
59. Schnatz PF, Curry SL. Primary hyperparathyroidism in pregnancy: Evidence-based management. 2002. *Obstet Gynecol Surv* 57(6): 365.
60. McMullen TP, Learoyd DL, Williams DC, et al. 2010. Hyperparathyroidism in pregnancy: Options for localization and surgical therapy. *World J Surg* 34: 1811.

69
Familial Primary Hyperparathyroidism (Including MEN, FHH, and HPT-JT)

Andrew Arnold and Stephen J. Marx

Introduction 553
Familial Hypocalciuric Hypercalcemia 553
Neonatal Severe Primary Hyperparathyroidism 555
Multiple Endocrine Neoplasia Type 1 556
Multiple Endocrine Neoplasia Type 2A 557
Hyperparathyroidism-Jaw Tumor Syndrome 558

Familial Isolated Hyperparathyroidism 559
Overlapping Considerations among All Forms of Familial 1°HPT 559
Acknowledgments 560
Suggested Reading 560

INTRODUCTION

Persons with familial primary hyperparathyroidism (1°HPT), defined by the combination of hypercalcemia and elevated or nonsuppressed serum PTH, are a small and important subgroup of all cases with 1°HPT (about 5%). Their familial syndromes include multiple endocrine neoplasia (MEN) [types 1, 2A, and 4], familial (benign) hypocalciuric hypercalcemia (FHH), neonatal severe primary hyperparathyroidism (NSHPT), hyperparathyroidism-jaw tumor syndrome (HPT-JT), and familial isolated primary hyperparathyroidism (FIHPT). These syndromes exhibit Mendelian inheritance patterns, and the four main genes that can cause the syndromes in most families have been identified (Table 69.1). As more knowledge accumulates on genetic contributions to complex phenotypes, contributions are being identified from additional genes, including some for less penetrant and more subtle predispositions to 1°HPT.

FAMILIAL HYPOCALCIURIC HYPERCALCEMIA

Familial benign hypocalciuric hypercalcemia (FHH) (OMIM 145980) is an autosomal dominant syndrome. The prevalence of FHH is similar to MEN1; either accounts for about 2% of 1°HPT cases.

Clinical expressions

Persons with FHH usually have mild or no symptoms. Easy fatigue, weakness, thought disturbance, or polydipsia are less common and less severe than in typical 1°HPT. Nephrolithiasis or hypercalciuria are as uncommon as in normals. Bone radiographs are usually normal. Either chondrocalcinosis or premature vascular calcification can be present but are generally silent clinically. Bone mass and susceptibility to fracture are normal. Hypercalcemia has virtually 100% penetrance at all ages. Onset of 1°HPT is otherwise rare in infancy and thus can be useful for diagnosis.

The range of serum calcium levels is similar to that in typical 1°HPT with a normal ratio of free to bound calcium in serum. Serum magnesium is typically in the high range of normal or modestly elevated, and serum phosphate is modestly depressed. Urinary excretion of calcium is normal, with hypercalcemic and unaffected family members showing a similar distribution of values. Normal urine calcium with high serum calcium explains the concept of relative hypocalciuria in FHH.

Parathyroid function, including serum parathyroid hormone (PTH) and 1,25(OH)$_2$D, is usually normal, with modest elevations of either index in 5–10% of cases. Such "normal" parathyroid function indices in the presence of lifelong hypercalcemia are inappropriate, diagnostically useful, and reflect the primary role for the parathyroids in causing this hypercalcemia.

Table 69.1. Outline of Syndromes of Familial Primary Hyperparathyroidism with Emphasis on Major Features That Distinguish among the Syndromes

Syndrome	Main Genes and Mutation Types[a]	Parathyroid Gland Aspects	Aspects Outside of the Parathyroids
FHH	CASR –[b]	Hypercalcemia begins at birth Increased secretion not growth Persist after subtotal PTX Avoid PTX	Relative hypocalciuria
NSHPT	CASR =	Hypercalcemia begins at birth Ca above 16 mg % Four very large PT glands Needs urgent total PTX	Relative hypocalciuria
MEN1	MEN1 –	Begins at average age 20 yr Asymmetric adenomas Recur 12 yr after successful subtotal PTX	Tumors among more than 20 tissues (Pituitary, pancreaticoduodenal, foregut, carcinoid, dermis, etc.)
MEN2A	RET +	Like MEN1 but later, less intense, and less symmetrical	C-cell cancer that is preventable Find and treat pheochromocytoma(s) before thyroid or parathyroid surgery
HPT-JT	HRPT2 –	Hypercalcemia can occur by age 10, often later Parathyroid cancer in 15% Benign or malignant Microcystic histology sometimes	Benign jaw tumors, renal cysts, and/or uterine tumors
FIHPT	Occult expression of MEN1-, CASR-, or HRPT2- in 30%. Other gene(s) not identified.	No specific features	None by definition. Another occult syndrome may emerge later.

FHH: familial hypocalciuric hypercalcemia; NSHPT: neonatal severe primary hyperparathyroidism; MEN: multiple endocrine neoplasia (can be type 1, 2A, or 4); HPT-JT: hyperparathyroidism jaw tumor syndrome; FIHPT: familial isolated primary hyperparathyroidism; PTX: parathyroidectomy.
[a]Mutation types in germline: – heterozygous inactivating; = homozygous inactivating; + heterozygous activating.
[b]Other genes identified for similar or identical syndromes are: FHH from AP2S1- and GA11-; MEN1 from a cyclin-dependent kinase inhibitor (CDKI) mutation (p27-, p15-, p18-, or p21-). MEN1 from p27- mutation was termed MEN4 by OMIM [Online Mendelian Inheritance in Man].

FHH cases often have mild enlargement of the parathyroid glands that may not be recognized at surgery and that may be evident only by careful measurement. Standard subtotal parathyroidectomy in FHH results in only a very transient lowering of serum calcium, followed by persistence of the hypercalcemia within a few days after surgery.

Pathogenesis/Genetics

Most cases result from heterozygous inactivating mutation of the CASR gene, which encodes a calcium-sensing receptor (CaSR). From the surface of the parathyroid cell, this CaSR "reports" the level of ionized calcium in serum. About 30% of probands and kindreds with FHH lack mutation in CASR. Some of these express mutation in AP2S1 or GA11, but others do not have mutation in an identified gene. One rare case had homozygous CASR mutation despite only a mild form of FHH; the parents and other relatives with heterozygous CASR mutation were normocalcemic, pointing to a milder part of the spectrum of heterozygous (and homozygous) mutation.

The parathyroid cell in FHH shows decreased sensitivity to elevations of extracellular calcium, due to the inactivating mutation in its CaSR. This results in impaired calcium suppression of PTH secretion, usually with little or no increase of parathyroid cell proliferation.

There also is a disturbed calcium-sensing function intrinsic to the kidneys in FHH. Normally, the CaSR functions in the kidney to maintain tubular calcium reabsorption in the direction that would correct for changes in the serum calcium. The tubular reabsorption of calcium is also increased by rises of PTH; in FHH, it is high and remains high even after an intended or unintended total parathyroidectomy. CaSRs are normally expressed in additional tissues outside of the parathyroids and kidneys, but clinical dysfunction there has not been reported in FHH or even in NSHPT.

The usual distinctions between typical 1°HPT and the 1°HPT of CASR mutation can be blurred in some cases or, rarely, in all carriers in an entire kindred. In particular, affected members in one large kindred with a germline missense mutation in the *CASR* had a typical 1°HPT syndrome, unlike the atypical 1°HPT of FHH. Several other small families with *CASR* loss-of-function mutations have contained some members with one or more features resembling typical 1°HPT. Hypocalciuric hypercalcemia can also be caused by antibodies against the CaSR and can then be associated with other autoimmune features, but this is without *CASR* mutation; this is rare and generally not familial.

Diagnosis of the family and the carrier

In the presence of hypercalcemia, a normal PTH just like a relatively low urine calcium warns about possible FHH. The family diagnosis usually is made from typical clinical features in one or more members of a family such as hypercalcemia, relative hypocalciuria, and failed parathyroidectomy. Recognition of hypercalcemia before age 10 is almost specific for the diagnosis of FHH in that kindred.

Family screening for FHH traits can be important to establish the syndrome in a proband, in an entire family, and eventually to diagnose additional relatives. Because of high penetrance for hypercalcemia in all FHH carriers, an accurate assignment for each relative at risk can usually be made from one determination of serum calcium (preferably ionized or albumin-adjusted).

Because urinary calcium excretion in a fixed interval depends heavily on glomerular filtration rate (GFR) and collection interval, total calcium excretion is not a valid index to distinguish a case of FHH from typical 1°HPT. The ratio of renal calcium clearance to creatinine clearance,

$$Ca_{Cl}/Cr_{Cl} = [Ca_u \times V/Ca_s] / [Cr_u \times V/Cr_s]$$
$$= [Ca_u \times Cr_s] / [Cr_u \times Ca_s]$$

is an empirical and useful index for this specific comparison. In hypercalcemic cases with FHH, this clearance ratio averages one-third of that in typical 1°HPT, and values below 0.01 (valid units will all cancel out) are suggestive of FHH.

CASR mutation analysis has an occasional role in the diagnosis of the syndrome, particularly with an inconclusive clinical evaluation of the family. *CASR* mutation may be undetectable if located outside the tested coding exons; this may account for much of the 30% lack of identified *CASR* mutation in typical families with FHH.

Management

Despite lifelong hypercalcemia, FHH can be compatible with survival into the ninth decade. Chronic hypercalcemia in FHH should rarely be treated, and it has been resistant to several types of drugs (diuretics, bisphosphonates, phosphates, or estrogens). Calcimimetics are a type of drug that acts like Ca^{2+} to stimulate the normal or even the mutated CaSR on the parathyroid cell and, thereby, decrease release of PTH. They might be effective (if used off-label) in rare cases of FHH for which treatment is appropriately contemplated. This potential for efficacy would be expected to depend on the domain of the mutated residues within the CaSR.

Because of their generally benign course and lack of response to subtotal parathyroidectomy, very few cases should undergo parathyroidectomy. In rare situations, such as relapsing pancreatitis, very high PTH, or very high serum calcium (persistently higher than 14 mg/dl), debulking parathyroidectomy and even total parathyroidectomy may be indicated.

Sporadic hypocalciuric hypercalcemia

Without a positive family history or *CASR* mutation, the management of sporadic hypocalciuric hypercalcemia is challenging. This should generally be managed as if it is typical FHH, unless the features of another 1°HPT syndrome become more prominent.

NEONATAL SEVERE PRIMARY HYPERPARATHYROIDISM

Clinical expressions

Neonatal severe primary hyperparathyroidism (NSHPT) (OMIM 239200) is an extremely rare neonatal state of life-threatening, severe hypercalcemia, very high PTH, rib fractures, hypotonia, respiratory distress, and massive enlargement of all parathyroid glands. The rare case to survive without early surgery is likely to show general impairments to development. The main relevance of NSHPT here is toward understanding some severe ways that *CASR* mutation can disrupt the parathyroids selectively.

Pathogenesis/Genetics

This disorder typically results from homozygous or compound heterozygous *CASR* inactivating mutation. It is uncertain if the hypercellular parathyroids are polyclonal or are overgrown by a monoclonal component.

Diagnosis

Diagnosis is usually based on the unique clinical features, often combined with parental consanguinity and/or FHH in first-degree relatives.

Management

Urgent total parathyroidectomy for severe symptoms and signs can be lifesaving. This may allow the patient to have a normal life, with treated hypoparathyroidism. Persistence of the intrinsic renal defect makes treatment of this hypoparathyroidism simpler.

MULTIPLE ENDOCRINE NEOPLASIA TYPE 1

MEN1 (OMIM 131100) is a rare and often heritable disorder with an estimated prevalence of 2–3 per 100,000 in unselected persons. Approximately 2% of all cases of 1°HPT are caused by MEN1. It is defined by consensus as tumors in two of its three main tissues (parathyroids, pituitary, and pancreaticoduodenal endocrine); affected persons are also predisposed to tumors in many other hormonal and nonhormonal tissues. By extension, familial MEN1 is defined as MEN1 with a first-degree relative showing tumor in at least one of the three main tissues.

Clinical expressions

1°HPT is the most penetrant hormonal component of MEN1 and is the initial clinical manifestation of the disorder in most cases. In MEN1, 1°HPT has several features different from the common sporadic (nonfamilial) forms of 1°HPT. The female-to-male ratio is about 1.0 in MEN1, in contrast to about threefold female predominance in sporadic 1°HPT. 1°HPT presents about 30 years earlier in MEN1, typically in the second to fourth decade of life and has been found as early as 8 years of age. Much earlier onset of primary hyperparathyroidism is likely to explain the much earlier onset of osteoporosis in MEN1.

Multiple parathyroid tumors are typical in MEN1; these tumors may vary widely in size, with an average 10:1 ratio between the largest and smallest tumor. A powerful drive to parathyroid tumorigenesis exists in MEN1, reflected by an impressively high rate of recurrent 1°HPT; this averages about 50% at 12 years after parathyroidectomy, which had been clearly successful for this patient in the earlier years postoperatively. Because of tumors in multiple parathyroid glands, an ectopic location is more likely in MEN1 than in common adenoma.

Some of the other tumors associated with MEN1 include duodenal gastrinomas, insulinoma, nonhormonal islet tumors, bronchial or thymic carcinoids, gastric enterochromaffin-like tumors, adrenocortical adenomas, lipomas, facial angiofibromas, and truncal collagenomas. In a family with few and/or mainly young affected members, tumors of MEN1 may be expressed in only the parathyroids, or in the parathyroids plus only in one additional tissue. Such families should be followed for development of other tumors of MEN1.

Pathogenesis/Genetics

Familial MEN1 shows an autosomal dominant inheritance pattern, and the main genetic basis is an inactivating germline mutation of the *MEN1* tumor suppressor gene. *MEN1* encodes menin, whose molecular pathways and detailed functions remain under study. *MEN1* mutations are distributed across the translated parts of the gene without any apparent pattern, other than the majority predicting truncation or absence of menin. Individuals with MEN1 have typically inherited one inactivated copy of the *MEN1* gene from an affected parent, but up to 10% may have a spontaneous or new germline mutation.

The outgrowth of a tumor follows from the subsequent somatic (i.e., acquired) inactivation of the normal, remaining copy of the *MEN1* gene in one cell. Such a cell, for example in the parathyroid gland, would have become devoid of *MEN1*'s tumor suppressor function, contributing to a selective growth advantage over its neighbors and leading to its clonal proliferation. This process of mutational loss of function is analogous to that for many tumor suppressor genes such as *HRPT2*, *p53*, *BRCA1*, and *APC*.

MEN1 mutation is not identified in 30% of probands and families with MEN1. Several cases among this group have MEN1 from mutation in *CDKN1B*, encoding the p27KIP1 cyclin-dependent kinase inhibitor (CDKI). This combination has been termed MEN4. Few among the remainder have mutation in *p15*, *p18*, or *p21*, three other CDKI genes. These CDKI defects suggest that menin and CDKIs share a molecular pathway, important in endocrine tumorigenesis.

Diagnosis of carriers

Direct sequencing for germline *MEN1* mutations is commercially available, albeit costly, and the indications for such testing remain under discussion. Gene analyses, typically limited to the coding region and near to it, fail to detect *MEN1* mutation in about 30% of typical MEN1 kindreds. Some of these kindreds may have unrecognized *MEN1* defects, such as large deletions or small noncoding mutations, that cannot be detected by the current testing methods. The yield of detectable *MEN1* or CDKI mutation in cases with a sporadic but true MEN1 phenotype, limited to parathyroid plus pituitary tumor, is much lower (about 7%); this suggests the existence of other predisposing gene(s) in this subgroup. Still undetermined gene(s) may contribute.

In contrast to the importance of testing for *RET* mutations for prevention or cure of a cancer, presymptomatic gene diagnosis has not been established to broadly improve morbidity or mortality in MEN1. Periodic screening with serum calcium, PTH, etc., provides a non-DNA-based alternative for carrier ascertainment. *MEN1* gene testing can be helpful for diagnosis or rarely for intervention in an MEN1-like proband, when the clinical diagnosis is inconclusive but a suspicion of MEN1 exists: for example, in a young adult with sporadic or familial isolated multigland 1°HPT, or in Zollinger Ellison Syndrome (ZES). In the latter, *MEN1* mutation occurs in one-quarter of cases, and identifying it can lead to avoidance of abdominal surgery that would otherwise be indicated.

When *MEN1* mutation is not identified in a proband or family, classical ascertainment testing with physical or biochemical traits is the back-up method. Facial angiofibromas or hypercalcemia are rather robust traits for this.

Diagnosis of other tumors

For MEN1-like probands or for established MEN1 carriers or other family members at risk for developing manifestations of MEN1, screening for tumor indices at baseline or at follow-up for the emergence of tumors is recommended, because a benefit seems likely. Tumor screening and other management of the pituitary, pancreaticoduodenal, and other MEN1 tumors outside of the parathyroids is mostly beyond the scope of this chapter.

Management

Once the biochemical diagnosis is established, the indications for referral to surgery are similar to those in sporadic 1°HPT. Osteoporosis is a surgical indication that is frequent in women with MEN1 by 35 years of age.

An age of under 50, though a criterion for referral to surgery in sporadic 1°HPT, should not be accepted as a sufficient criterion for referral to surgery in 1°HPT of MEN1. Because direct evidence is lacking, opinions differ about the optimal timing of surgery for 1°HPT in MEN1. Early and presymptomatic surgery might, on one hand, lead to better long-term bone health. On the other hand, the parathyroids may be smaller and harder to identify. Furthermore, with the high rate of recurrent 1°HPT after surgery in MEN1, a policy of early surgery might increase a person's total number of operations and, thereby, increase the risk of intraoperative morbidity.

It should be emphasized that MEN1-associated cancers in nonparathyroid tissues cause fully one-third of the deaths in MEN1 cases. For most of these cancers, no effective prevention or cure currently exists, partly because of their problematic locations. New drugs and other promising treatments, including some drugs already FDA approved for common islet tumors, still need more evaluation in similar tumors of MEN1.

Parathyroid surgery in MEN1

Preoperative imaging for tumor localization is useful before reoperation in MEN1 cases with recurrent or persistent 1°HPT. The usefulness of preoperative imaging is lower in unoperated cases due to multiplicity of parathyroid tumors in MEN1, the efforts to identify all abnormal parathyroid glands particularly at initial surgery, and the inability to image reliably all parathyroid tumors in any one case. For the same reasons, a suspected or firm preoperative diagnosis of MEN1 argues against performing minimally invasive parathyroidectomy. Even at initial surgery for multiple parathyroid tumors, intraoperative PTH measurement can be helpful.

The initial operation most frequently performed in MEN1 is 3.5 gland subtotal parathyroidectomy with transcervical near-total thymectomy. A parathyroid remnant is usually left on its native vascular pedicle in the neck and may be marked with a clip. Alternately, the remnant can be immediately autotransplanted to the forearm during an intended complete parathyroidectomy (see below). Transcervical thymectomy in MEN1 has an unproven benefit but is widely used, because it may prevent or cure thymic carcinoids (mainly in males); in addition, the thymus is a common site for parathyroid tumors in MEN1 patients with recurrent 1°HPT. Involvement of an experienced parathyroid surgical team is crucial to optimal outcome.

MULTIPLE ENDOCRINE NEOPLASIA TYPE 2A

MEN2 is subclassified into three clinical syndromes with mutations in the same gene, *RET*. These are MEN2A, MEN2B, and familial medullary thyroid cancer (FMTC). Of these three syndromes, MEN2A (OMIM 171400) is the most common and the only one that manifests 1°HPT. Even further subdivision of MEN2 according to family expression or mutated codon is likely in the future.

Clinical expressions

MEN2A is a heritable predisposition to medullary thyroid or C-cell cancer (MTC), pheochromocytoma, and 1°HPT. The frequencies of these tumors in an adult carrier of MEN2A are above 90% for MTC, 40–50% for pheochromocytoma, and 20% for 1°HPT.

This intermediate penetrance of 1°HPT in MEN2A contrasts with the higher penetrance found in every other familial 1°HPT syndrome. Similarly, the 1°HPT in MEN2A is generally milder, more asymmetrical, and later in onset than in MEN1. Unlike in MEN1, late recurrence 1°HPT is infrequent after an apparently successful parathyroidectomy in MEN2A. This is similar to the presumably excellent long-term outcome of surgery with nonfamilial multiple primary parathyroid tumors.

MTC, a potentially lethal manifestation of MEN2A, evolves from preexisting parafollicular C-cell hyperplasia, and its calcitonin release can give a useful marker for recognizing the emergence of early tumors or for monitoring a large tumor burden. Despite the pharmacologic properties of calcitonin, mineral metabolism is generally normal in the setting of impressive hypercalcitoninemia. Pheochromocytomas in MEN2A can be unilateral or bilateral.

Pathogenesis/genetics

MEN2A is inherited in an autosomal dominant pattern, with both genders affected in equal proportions; the gene defect is germline gain of function mutation of the *RET* proto-oncogene. Germline *RET* mutation is detectable in above 95% of MEN2A families. *RET* mutation at codon 634 accounts for 85% of MEN2A and is more highly associated with the expression of 1°HPT.

There are both differences and much overlap in the specific *RET* gene mutations underlying MEN2A and FMTC; in contrast, MEN2B is caused by one of two entirely distinct *RET* mutations. Why parathyroid disease fails to develop in the latter two MEN2 syndromes remains unclear. Unlike the numerous and seemingly random different inactivated codons of *MEN1* that are typical of a tumor suppressor mechanism, the mutated *RET* codons in MEN2A are limited in number, reflecting the need for highly specific changes in selected domains of the RET protein to activate this oncoprotein.

The RET protein is a plasma membrane spanning tyrosine kinase that normally transduces growth and differentiation signals in developing tissues, including around the neural crest. Knowledge of its molecular pathway and of bioassays is promoting new drugs and ongoing clinical trials, directed at MTC and other *RET*-related thyroid cancers.

Diagnosis of a kindred, of carriers, and of parathyroid tumors

For diagnosis of MEN2A, *RET* sequence testing is superior to immunoassay of basal or stimulated calcitonin, excepting, of course, in the very rare MEN2 family without detectable *RET* mutation. Gene sequencing for germline *RET* mutations is central to clinical management of MEN2A, particularly to guide thyroidectomy in management/prevention of MTC.

1°HPT in MEN2A is often asymptomatic and diagnosed and treated incidentally to thyroid surgery. Otherwise, its biochemical diagnosis, as well as indications for surgery, parallel those in sporadic 1°HPT.

Management

1°HPT is a less urgent expression of MEN2A than MTC or pheochromocytoma. Conceptually consistent with its genetics, 1°HPT in MEN2A is multiglandular, but less than four overtly enlarged glands may be present at one time. Thus, bilateral neck exploration to identify all abnormal parathyroid glands is advisable in known or suspected MEN2A; resection of enlarged parathyroid glands (up to 3.5 glands) is the most common operation. Issues of preoperative tumor localization in unoperated MEN2A patients expressing 1°HPT are similar to MEN1.

RET testing during childhood can secure a preventive or curative thyroidectomy (i.e., sufficiently early in childhood as to minimize the likelihood that extracapsular metastases of C-cell cancer will have already occurred).

If MEN2A is a possibility, evidence of pheochromocytoma should be sought before parathyroidectomy or thyroidectomy. If evident, pheochromocytoma(s) must be treated before thyroid or parathyroid surgery. Laparoscopic adrenalectomy has greatly improved the management of pheochromocytoma in MEN2A. Incompletely treated postoperative chronic hypocortisolism remains a major cause of morbidity, and even death after bilateral adrenalectomy in MEN2A.

HYPERPARATHYROIDISM-JAW TUMOR SYNDROME

Clinical expressions

Hyperparathyroidism-jaw tumor syndrome (HPT-JT) (OMIM 145001) is a rare, autosomal dominant combination of 1°HPT, ossifying or cementifying fibromas of the mandible and maxilla, renal manifestations including cysts, hamartomas, or Wilms tumors, and uterine tumors. Among adults of "classical" HPT-JT kindreds, 1°HPT is the most highly penetrant manifestation at 80%, next to ossifying fibromas of the maxilla or mandible at 30%, and a renal lesion slightly less frequently. A wide range of penetrance values occurs for uterine tumors.

1°HPT in HPT-JT may develop as early as during the first decade of life. Although all parathyroids are at risk, surgical exploration can show a solitary parathyroid tumor (even solitary atypical adenoma or solitary cancer) rather than multigland disease. Parathyroid neoplasms can be macro- or microcystic, and, whereas most tumors are classified as adenomas, parathyroid carcinoma (15–20% of 1°HPT) is markedly overrepresented in HPT-JT cases. In contrast, parathyroid cancer almost never occurs in MEN1, MEN2, or FHH. Dissemination of parathyroid cancer to the lungs can occur in the early 20s in HPT-JT. After a period of euparathyroidism, operated cases may manifest recurrent 1°HPT, and a solitary tumor asynchronously originating in a different parathyroid gland may prove responsible.

Ossifying fibromas in HPT-JT are generally benign but still may be large and destructive. More often they are small, asymptomatic, and identified incidentally on dental radiographs. They are clearly distinct from the classic, osteoclast-rich "brown tumors" of the jaws.

Pathogenesis/genetics

Germline mutation of the *HRPT2* gene (also called *CDC73*) causes HPT-JT. The yield of *HRPT2* mutation in HPT-JT kindreds is about 60–70%; the remaining kindreds may have *HRPT2* mutations that evade detection because they are not amplified by polymerase chain reaction (PCR) prior to sequencing. Mutations of *HRPT2* cause tumors, including parathyroid cancer, by inactivating or eliminating its protein product, parafibromin, consistent with a classical "two-hit" tumor suppressor mechanism. Both parafibromin's normal cellular roles and its influence (via loss of function) in tumorigenesis are under study.

Importantly, many cases with seemingly sporadic presentations of parathyroid carcinoma (OMIM 608266) also harbor a germline mutation in *HRPT2*, thus representing newly ascertained HPT-JT, occult HPT-JT, or another variant syndrome.

Diagnosis of carriers and cancers

The biochemical diagnosis of 1°HPT in HPT-JT parallels that in sporadic 1°HPT. Recognition of germline *HRPT2* mutation in classic or variant HPT-JT has opened the door to DNA-based diagnosis in probands or individuals with apparently sporadic parathyroid carcinoma, and for carrier identification in at-risk family members, aimed at preventing or curing parathyroid malignancy.

Prior to parathyroid surgery of a known or likely carrier of HPT-JT, the surgeon should be alerted to the possibility of parathyroid cancer.

Management

Management in HPT-JT centers around monitoring and surgery to address the high risk for current or future parathyroid malignancy. The finding of biochemical 1°HPT should lead promptly to surgery. All parathyroids should be identified at operation, signs of malignancy sought, and resection of abnormal glands performed. Because of the potential for malignancy, the consideration of prophylactic total parathyroidectomy (perhaps even for euparathyroid carriers) has been raised as an alternative approach. This is not favored in view of the difficulty in removing all parathyroid tissue, difficulties with lifelong hypoparathyroidism, the incomplete penetrance of parathyroid cancer in the syndrome, and the preliminary belief that close biochemical monitoring for recurrent 1°HPT will promote successful management of cancer.

Any case of apparently sporadic parathyroid cancer without identified *HRPT2* mutation should be monitored for potential HPT-JT, including consideration of screening the serum calcium periodically in first degree relatives.

FAMILIAL ISOLATED HYPERPARATHYROIDISM

Clinical expressions and diagnosis

Familial isolated primary hyperparathyroidism (FIHPT) (OMIM 145000) is clinically defined as familial 1°HPT without the extraparathyroid manifestations of another syndromal category. The diagnosis of FIHPT should therefore be changed, if features of another 1°HPT syndrome develop. Partly because this category is likely to encompass several occult or even totally unknown causes, the spectrum of its 1°HPT is broad.

Pathogenesis/Genetics

FIHPT is genetically heterogeneous and can be caused by incomplete expression from germline mutation in *MEN1*, *HRPT2*, or *CASR*; however, the majority of families do not have a detectable mutation in any of these three genes. One unidentified gene may be on the short arm of chromosome 2 (OMIM 610071); one or more other genes likely accounts for other families.

Mutation testing should be considered (e.g., when results might impact on the advisability of, or approach to, management of parathyroid and other tumors or to more gene testing in relatives). However, in FIHPT the low yield of mutations in the three known genes must be considered.

Management

Management is very similar to that in common 1°HPT. Monitoring and management must recognize that additional features of a genetically defined 1°HPT syndrome or a previously unidentified 1°HPT syndrome could become detectable. For example, the heightened risk of parathyroid carcinoma must be borne in mind in FIHPT, when occult HPT-JT is possible.

OVERLAPPING CONSIDERATIONS AMONG ALL FORMS OF FAMILIAL 1°HPT

Multifocal parathyroid gland hyperfunction

For three hereditary 1°HPT syndromes (MEN1, MEN2A, and HPT-JT), the germline mutation in the parathyroid cell causes susceptibility to postnatal and gradual overgrowth of mono- or oligoclonal parathyroid tumor. For two others (FHH and NSHPT), the phenotype is fully expressed around birth (i.e., with no postnatal delay). Stated differently, an underlying feature of these five multiorgan syndromes is that every parathyroid cell carries the same syndromal germline mutation; this is either sufficient to cause the overfunction phenotype in all parathyroid cells immediately or to put each parathyroid cell at risk to yield clonal proliferation many years later.

Detection of asymptomatic carriers

Once a 1°HPT syndrome has been diagnosed, testing for carriers among asymptomatic relatives should be considered. The concept of the carrier must include disease predisposition in a relative, even without an identifiable syndromal mutation in that family. Testing of germline DNA is often the gold standard for detection of carriers; however, carrier testing by use of traits that are expressed early and with high penetrance (such as hypercalcemia for the early carrier in FHH) or lower penetrance (such as hypercalcemia for delayed parathyroid cancer in HPT-JT) is sometimes a useful alternative.

The possible benefits of germline mutation testing include providing information to the subject, family, and physician.

Among all 1°HPT syndromes, carrier testing only for MEN2 can lead to a major intervention of almost certain efficacy in reducing mortality (from medullary thyroid carcinoma). Testing for HPT-JT can lead to management that may lessen mortality associated with parathyroid malignancy. Testing in other 1°HPT syndromes is mainly for information to the physician and patient and is less urgent. Such information about silent or even affected carriers is widely used to plan screening for tumors at baseline and during follow-up.

Monitoring for tumors

Tumor monitoring is best done with a syndrome-specific protocol; it should be followed in each carrier of a syndrome of 1°HPT. Monitoring for parathyroid and other tumors should address tumors present at the time of initial ascertainment of the carrier and the tumors that emerge during periodic follow-up. The plan for monitoring must deal with issues of cost and effectiveness.

Special approaches in surgery for familial parathyroid tumors

Many aspects of surgery for parathyroid adenoma require modification, when multiple parathyroid glands have the potential to be overactive. Efforts are made intraoperatively to know when sufficient pathologic tissue has been removed. Traditionally, the identification of all four parathyroid glands has been pursued for this reason. Rapid measurement of PTH intraoperatively to see a major drop from high values also can be useful immediately to judge if any overactive parathyroid tissue remains *in vivo*.

Efforts to identify all four parathyroid glands and remove several glands result in increased frequency of postoperative hypoparathyroidism. To minimize permanent hypoparathyroidism, in some centers, small fragments of the most normal-appearing parathyroid tumor are autografted immediately to the nondominant forearm. In other centers, because of difficulty in establishing such grafts, the surgeon leaves a small remnant of parathyroid in the neck, attached to its own vascular pedicle. In either case, one may cryopreserve fragments of the most normal-appearing tissue for possible delayed autograft for late postoperative hypoparathyroidism. However, such cryopreservation is not permitted in many centers due to concerns about legal liabilities. A normal-sized gland is not sufficiently large to give a satisfactory fresh or cryopreserved autograft; thus, tumor tissue is used. Fortunately, it can sustain euparathyroidism for many years. The potential for malignancy of any parathyroid tissue in HPT-JT argues against the autografting option there.

Beyond hypoparathyroidism, other complications are also more frequent after surgery for multigland 1°HPT, including familial 1°HPT, than after surgery for adenoma. These include complications to surrounding tissues (such as a recurrent laryngeal nerve) and postoperative persistent 1°HPT. The most obvious reason for the latter is an incomplete exploration for multigland disease.

True recurrent 1°HPT is a late complication that is much more frequent in MEN1 and other familial 1°HPT, than in common adenoma. True recurrence is defined for convenience as 1°HPT after a 3- to 6-month postoperative period of documented euparathyroidism. It could arise when a small tumorous remnant becomes overactive or when a previously normal parathyroid cell progresses into a tumor clone.

ACKNOWLEDGMENTS

This work was supported in part by the intramural program of NIDDK.

SUGGESTED READING

Agarwal SK, Mateo C, Marx SJ. 2009 Rare germline mutations in cyclin-dependent kinase inhibitor genes in MEN1 and related states. *J Clin Endocrinol Metab* 94: 1826–34.

Carling T, Szabo E, Bai M, Ridefelt P, Westin G, Gustavsson P, Trivedi S, Hellman P, Brown EM, Dahl N, Rastad J. 2000. Familial hypercalcemia and hypercalciuria caused by a novel mutation in the cytoplasmic tail of the calcium receptor. *J Clin Endocrinol Metab* 85: 2042–2047.

El-Hajj Fuleihan G, Brown EM, Heath H III. 2002. Familial benign hypocalciuric hypercalcemia and neonatal primary hyperparathyroidism. In: Bilezikian JP, Raisz LG, Rodan GA (eds.) *Principles of Bone Biology, 2nd Ed.* San Diego, CA: Academic Press. pp. 1031–1045.

Kouvaraki MA, Shapiro SE, Perrier ND, Cote GJ, Gagel RF, Hoff AO, Sherman SI, Lee JE, Evans DB. 2005. RET proto-oncogene: A review and update of genotype-phenotype correlations in hereditary medullary thyroid cancer and associated endocrine tumors. *Thyroid* 15: 531–544.

Lietman SA, Tenenbaum-Rakover Y, Jap TS, Yi-Chi W, De-Ming Y, Ding C, Kussiny N, Levine MA. 2009. A novel loss-of-function mutation, Gln459Arg, of the calcium-sensing receptor gene associated with apparent autosomal recessive inheritance of familial hypocalciuric hypercalcemia. *J Clin Endocrinol Metab* 94: 4372–9.

Marx SJ, Simonds WF, Agarwal SK, Burns AL, Weinstein LS, Cochran C, Skarulis MC, Spiegel AM, Libutti SK, Alexander HR Jr, Chen CC, Chang R, Chandrasekharappa SC, Collins FS. 2002. Hyperparathyroidism in hereditary syndromes: Special expressions and special managements. *J Bone Miner Res* 17 Suppl 2: N37–N43.

Marx SJ. 2005. Molecular genetics of multiple endocrine neoplasia types 1 and 2. *Nat Rev Cancer* 5: 367–375.

Nesbit MA, Hannan F, Howles SA, Reed AAC, Cranston T, Thakker CE, Gregory L, Rimmer AJ, Rust N, Graham U, Morrison PJ, Hunter SJ, Whyte M, McVean G, Buck D, Thakker R, Mutations in AP2S1 cause familial hypocalciuric hypercalcemia type 3. Nat Genet OnLine, 9 December 2012; doi: 10.1038/ng.2492.

Newey PJ, Bowl MR, Cranston T, Thakker RV. 2010. Cell division cycle protein 73 homolog (CDC73) mutations in the hyperparathyroidism-jaw tumor syndrome (HPT-JT) and parathyroid tumors. *Hum Mutation* 31: 295–307.

Pellegata NS, Quintanilla-Martinez L, Siggelkow H, Samson E, Bink K, Hofler H, Fend F, Graw J, Atkinson MJ. 2006. Germ-line mutations in p27Kipl cause a multiple endocrine neoplasia syndrome in rats and humans. *Proc Natl Acad Sci USA* 103: 15558–15563. Erratum in: 2006. *Proc Natl Acad Sci USA* 103: 19213.

Shattuck TM, Valimaki S, Obara T, Gaz RD, Clark OH, Shoback D, Wierman ME, Tojo K, Robbins CM, Carpten JD, Farnebo LO, Larsson C, Arnold A. 2003. Somatic and germ-line mutations of the *HRPT2* gene in sporadic parathyroid carcinoma. *N Engl J Med* 349: 1722–1729.

Simonds WF, James-Newton LA, Agarwal SK, Yang B, Skarulis MC, Hendy GN, Marx SJ. 2002. Familial isolated hyperparathyroidism: Clinical and genetic characteristics of 36 kindreds. *Medicine (Baltimore)* 81: 1–26.

Thakker RV, Newey PJ, Walls GV, Bilezikian J, Dralle H, Ebeling PR, Melmed S, Sakurai A, Tonelli F, Brandi ML. 2012. Clinical practice guidelines for multiple endocrine neoplasia type 1 (MEN1). *J Clin Endocrinol Metab* 97: 2990–3011.

70
Non-Parathyroid Hypercalcemia
Mara J. Horwitz, Steven P. Hodak, and Andrew F. Stewart

Pathophysiology of Hypercalcemia 562
Clinical Signs and Symptoms of Hypercalcemia 562
Disorders That Lead to Hypercalcemia 563
Approach to Diagnosis 568

Management 569
Acknowledgments 569
Suggested Readings 570

PATHOPHYSIOLOGY OF HYPERCALCEMIA

As discussed elsewhere in the *Primer*, the normal total serum calcium concentration of 9.5 mg/dl can be divided into three components: ionized serum calcium (approximately 4.2 mg/dl); serum calcium complexed to anions such as phosphate, sulfate, carbonate, etc., (approximately 0.3 mg/dl); and calcium bound to serum proteins, principally albumin (approximately 4.5 mg/dl). Hypercalcemia is defined as a serum calcium greater than 2 SD above the normal mean in a given laboratory, commonly 10.6 mg/dl for total serum calcium, and 1.25 mmol/L for ionized serum calcium. There is no formal grading system for defining the severity of hypercalcemia. In general, however, serum calcium concentrations less than 12 mg/dl can be considered mild, those between 12 and 14 mg/dl moderate, and those greater than 14 mg/dl severe.

The serum calcium concentration is tightly regulated by the flux of serum ionized calcium to and from four physiologic compartments: the skeleton, the intestine, the kidney, and serum binding proteins. Hypercalcemia, therefore, always results from an abnormality in calcium flux between the extracellular fluid (ECF) and one or a combination these compartments. Said another way, hypercalcemia can result only from one of four mechanisms: abnormal binding of calcium to serum proteins, or abnormal flux of calcium into extracellular fluid from the gastrointestinal (GI) tract, the skeleton, or the kidney.

Combinations of the above three mechanisms are common. Understanding hypercalcemia in this kind of mechanistic construct is critical for accurate diagnosis and is essential for effective treatment of hypercalcemia. For example, hypercalcemia from vitamin D intoxication or milk-alkali syndrome commonly arises from increased GI absorption of calcium and would not be expected to respond to antiresorptive agents such as bisphosphonates. Conversely, humoral hypercalcemia of malignancy results principally from increased skeletal resorption and renal calcium reabsorption, and therefore is not influenced by restricting dietary calcium intake.

CLINICAL SIGNS AND SYMPTOMS OF HYPERCALCEMIA

Hypercalcemia raises the electrical potential difference across cell membranes and increases the depolarization threshold. Clinically, this is manifested as a spectrum of neurological symptoms ranging from mild tiredness to obtundation to coma. There is no precise serum calcium level that leads to impaired neurologic function. Instead, the presence or absence and degree of neurologic symptoms depends on the abruptness of onset of hypercalcemia, the age and underlying neurological status of the patient, and comorbidities and medications such as narcotics and neuroleptics.

Primer on the Metabolic Bone Diseases and Disorders of Mineral Metabolism, Eighth Edition. Edited by Clifford J. Rosen.
© 2013 American Society for Bone and Mineral Research. Published 2013 by John Wiley & Sons, Inc.

Hypercalcemia acts directly at the nephron to prevent normal reabsorption of water, leading to a functional form of nephrogenic diabetes and polyuria. This may lead to thirst, prerenal azotemia, and significant dehydration, which are common clinical features of hypercalcemia. Hypercalcemia may also cause precipitation of calcium phosphate salts in the renal interstitium (nephrocalcinosis), vasculature, cardiac conduction system, the cornea (visible as so-called band keratopathy), and the gastric mucosa. Hypercalcemia may lead to renal failure from obstructive uropathy, from nephrolithiasis, from nephrocalcinosis, and from prerenal causes, including dehydration and a reversible component of hypercalcemia-induced afferent arteriolar vasoconstriction.

Hypercalcemia can also lead to electrocardiographic abnormalities, the most specific of which is a prolonged Q-Tc interval. Hypercalcemia increases the depolarization threshold of skeletal and smooth muscle, making them more refractory to neuronal activation. Resulting decreases in muscle contraction manifest clinically as skeletal muscle weakness and as constipation. Nausea, anorexia, vomiting, and flushing are also common. Finally, hypercalcemia may lead to abdominal pain and pancreatitis.

DISORDERS THAT LEAD TO HYPERCALCEMIA

The complete differential diagnosis of hypercalcemia is shown in Table 70.1. In this chapter, we consider non-parathyroid causes of hypercalcemia. The parathyroid hormone (PTH)-dependent family of hypercalcemic disorders, including primary and tertiary hyperparathyroidism, their inherited variants, and familial hypocalciuric hypercalcemia, also known as familial benign hypercalcemia, are discussed elsewhere in the *Primer*.

Cancer

Malignancy-associated hypercalcemia (MAHC) accounts for approximately 90% of hypercalcemia encountered among hospitalized patients and is a negative prognostic factor. The first patient with MAHC was reported in 1921, immediately following the development of clinical methods to measure serum calcium. MAHC can be subdivided into four mechanistic subtypes: (1) humoral hypercalcemia of malignancy (HHM); (2) local osteolytic hypercalcemia (LOH); (3) 1,25(OH)$_2$-vitamin D-induced hypercalcemia; and, (4) authentic ectopic hyperparathyroidism.

HHM

HHM is the most common form of MAHC, accounting for about 80% of subjects in large series of unselected patients with MAHC. HHM results from the secretion of parathyroid hormone-related protein (PTHrP) (dis-

Table 70.1. Differential Diagnoses of Hypercalcemia

*PTH-Dependent Hypercalcemia
Cancer
   Humoral hypercalcemia of malignancy (HHM)[1–10]
   *Local osteolytic hypercalcemia (LOH)[1–3, 6–10]
   1,25(OH)$_2$ vitamin D and lymphoma/dysgerminoma[11–12]
   Authentic ectopic PTH secretion[13–14]
   Other
Granulomatous Disorders[15–18]
Endocrine Disorders[19–26]
Immobilization[27–30]
Milk-Alkali Syndrome[31–33]
Total Parenteral Nutrition[34, 35]
Abnormal Protein Binding[36, 37]
Medications[38–51]
   Vitamin D
   Vitamin A
   Lithium
   Parathyroid hormone
   Estrogen/SERMs
   Aminophylline and theophylline
   Foscarnet
   Growth hormone
   8-chloro-cyclic AMP
Chronic and Acute Renal Failure[52, 53]
End Stage Liver Disease[54]
Manganese Intoxication[55]
Fibrin Glue[56]
Hypophosphatemia
*Pediatric Syndromes[57]

*This topic is covered in another section of the *Primer*.
Superscript numbers refer to the "Suggested Readings" section at the end of the chapter.

cussed elsewhere in the *Primer*) by HHM-associated tumors. Whereas almost any kind of tumor may cause HHM, the most common types are squamous carcinomas of any origin (lung, esophagus, skin, cervix are common sites), and breast and renal carcinomas.

HHM was first described in the 1940s and '50s in patients in whom hypercalcemia associated with cancer was corrected following successful removal of the responsible tumor. Investigators deduced the humoral nature of the syndrome from these events, but the responsible "humor" was not identified until 1987, when PTHrP was purified, sequenced, and its cDNA and gene were cloned. We now understand that continuous secretion of PTHrP by tumors leads to a dramatic uncoupling of bone resorption from formation, by activating osteoclastic bone resorption and suppressing osteoblastic bone formation (Fig. 70.1). As a result, enormous net quantities of calcium of up to 700–1,000 mg/day leave the skeleton, causing marked hypercalcemia. In addition, the anti-calciuric effects of PTHrP prevent or restrict effective renal calcium clearance. Finally, HHM is associated with reductions in circulating 1,25(OH)$_2$D levels, which in turn limit intestinal calcium absorption. Thus, in pathophysiologic

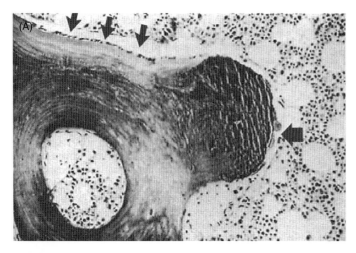

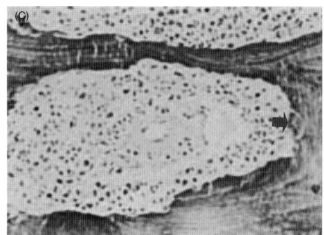

Fig. 70.1. Comparison of bone histology in a patient with hyperparathyroidism (HPT) (Panel A), humoral hypercalcemia of malignancy (HHM) (Panel B), and local osteolytic hypercalcemia due to leukemia (LOH) (Panel C). In HPT, HHM, and LOH osteoclastic activity is accelerated (large, thick arrows), although it is higher in HHM than in HPT. In HPT, osteoblastic activity (thin arrows) and osteoid are increased, but both are markedly decreased in HHM and LOH. This uncoupling of formation from resorption in HHM and LOH plays the major role in causing hypercalcemia. (From Stewart AF, Vignery A, Silvergate A, Ravin ND, LiVolsi V, Broadus AE, Baron R. 1982. Quantitative bone histomorphometry in humoral hypercalcemia of malignancy: Uncoupling of bone cell activity. *J Clin Endo Metab* 55: 219–227. Copyright 1982 The Endocrine Society. Used with permission.)

terms, HHM results from enhanced skeletal resorption coupled with an inability to clear calcium through the kidney.

HHM is also associated with a reduction in the renal phosphorus threshold, which results in phosphaturia and hypophosphatemia. The HHM syndrome is also characterized by a marked increase in cyclic adenosine monophosphate (AMP) excretion by the kidney, termed "nephrogenous cyclic AMP" or "NcAMP." The bone resorption, increased renal tubular calcium resorption, phosphaturia, and increase in NcAMP excretion that characterize HHM reflect interaction of circulating PTHrP with the common PTH/PTHrP receptor in the skeleton and the renal tubule. Surprisingly, HHM syndrome is also associated with paradoxical reductions in $1,25(OH)_2D$ and osteoblastic bone formation [Fig. 70.1(B)]. These observations stand in striking contrast to primary hyperparathyroidism (HPT), in which $1,25(OH)_2D$ is increased, and both osteoblast and osteoclast activity are increased but remain coupled [Fig. 70.1(A)]. Why PTH and PTHrP, which act through the identical receptor, should produce directionally opposite physiologic effects was unexplained until the recent demonstration by Dean et al. that PTHrP dissociates far more rapidly from the PTH1 receptor than PTH, thereby limiting its potency.

Bone scans, bone biopsies, and autopsy reveal few or no skeletal metastases in patients with HHM. This finding emphasizes the humoral nature of the syndrome and stands in contrast to patients with LOH, described below.

In an interesting turn of events, it is now also clear that systemic overproduction of PTHrP by tumors does not inevitably indicate the presence of cancer. Thus, it has been reported that in very rare instances, benign neoplastic lesions also may lead to hypercalcemia by systemic overproduction of PTHrP, a condition that has been referred to as "humoral hypercalcemia of benignancy." For example, benign uterine fibroids, benign ovarian tumors, insulinomas, pheochromocytomas, and massive mammary hyperplasia have all been documented to secrete PTHrP and thereby cause hypercalcemia that reverses on tumor excision or reduction mammoplasty.

LOH

The tumors that most commonly produce LOH are breast cancer and hematological neoplasms (myeloma, lymphoma, leukemia) in subjects with widespread skeletal involvement. By the 1940s, series of patients with hypercalcemia associated with these malignancies were reported, all of which documented extensive marrow invasion by the tumor. In the 1960s and '70s, these malignancies were shown to be associated with marked activation of osteoclasts adjacent to the sites of marrow infiltration by the malignancy. In this era, it was widely believed that hypercalcemia in the majority of patients with MAHC resulted from LOH. However, today, large series have shown that LOH accounts for only about 20% of patients with MAHC. Indeed, evidence suggests that the frequency of MAHC due to LOH may be declining with the widespread use of bisphosphonates to prevent skeletal fractures, metastases, and pain in patients with myeloma and breast cancer.

In the 1970s, several authors began to search for the osteoclast-activating factors (OAFs) responsible for LOH. Locally produced osteoclast-activating cytokines are now known to include interleukins-1 and -6, PTHrP, and macrophage inflammatory protein-1α, among others. This area has been reviewed elsewhere in the *Primer*.

Patients with LOH are characterized at bone biopsy or autopsy by extensive skeletal metastases or marrow infiltration [Fig. 70.1(C)]. Bone scintigraphic scans are generally intensely and widely positive in patients with metastatic disease from solid tumors but may be completely negative despite extensive marrow involvement in patients with multiple myeloma, reflecting a reduction on bone formation.

In mechanistic terms, LOH can be thought of as primarily a resorptive (skeletally derived) form of hypercalcemia in which massive removal of calcium from the skeleton exceeds the normal ability of the kidney to clear calcium. As the dehydration associated with such marked hypercalcemia occurs, the hypercalcemia is exacerbated by a typical decline in renal function as well.

1,25(OH)₂D-induced hypercalcemia

In the 1980s, reports began to appear describing patients with lymphomas in whom hypercalcemia occurred as a result of increased production of 1,25(OH)$_2$D. There have been about 60 patients described to date with this syndrome. There is no particular histopathological correlation: all types of lymphomas have been reported to cause this syndrome. In the past several years, the same syndrome has been reported in patients with ovarian dysgerminomas.

The primary pathophysiological abnormality in this syndrome is that the malignant cells or adjacent normal cells overexpress the enzyme 1-α-hydroxylase, which converts normal circulating quantities of the precursor 25(OH)D to abnormally elevated circulating concentrations of the active form of vitamin D, 1,25(OH)$_2$D. This pathophysiology can be viewed as the malignant counterpart of events that occur in sarcoidosis (see section on granulomatous diseases below). Since 1,25(OH)$_2$D activates intestinal calcium absorption, this syndrome is principally an absorptive form of hypercalcemia, although decreased renal clearance of calcium may also develop as a consequence of the dehydration caused by the hypercalcemia. In addition, there is emerging evidence that 1,25(OH)$_2$D may increase osteoclast mediated bone resorption by directly activating the receptor activator of nuclear factor-κB ligand (RANK-L) pathway, and therefore worsening hypercalcemia. It is of note that circulating levels of 1,25(OH)$_2$D may not always reflect the high local concentrations seen in patients with marrow tumor involvement.

Authentic ectopic hyperparathyroidism

From the 1950s through the 1980s, when HHM syndrome was finally shown to be caused by PTHrP, the hypercalcemia of HHM was widely ascribed to "ectopic secretion of PTH" by offending neoplasms. With the demonstration in the 1980s that the responsible factor in HHM was PTHrP, not PTH, authentic ectopic secretion of PTH by cancers was believed not to occur. This changed in the 1990s with the description of rare cases in which production of authentic PTH, and not PTHrP, by tumors was shown to cause MAHC. At the time of this writing, approximately 10 cases have been reported in which convincing evidence exists for hypercalcemia resulting from authentic ectopic secretion of PTH from malignant tumors. Thus, authentic ectopic secretion of PTH does exist, but it is very rare.

Other mechanisms for MAHC

The four categories described above account for more than 99% of patients with MAHC. Occasionally, however, patients who do not fit any of these categories have been described. For example, there are rare case reports in which none of the above scenarios could be invoked, and in which elevated circulating concentrations of prostaglandin E2 may have been responsible. This is thought to result from the ability of PGE2 to induce RANK-L expression via the EP4 receptor that in turn activates osteoclastic activity.

Granulomatous diseases

Almost every single disease associated with granuloma formation has been reported to cause hypercalcemia. The most common is sarcoidosis, but tuberculosis (both from *M. bovis* and *M. avium*), berylliosis, histoplasmosis, coccidoimycosis, *Pneumocystis*, inflammatory bowel disease, histocytosis X, foreign body granulomas, and granulomatous leprosy have all been associated with the syndrome. In the case of sarcoidosis, approximately 10% of patients become hypercalcemic, and 20% hypercalciuric, during the course of their disease.

The mechanism in sarcoidosis and tuberculosis is inappropriate production of 1,25(OH)$_2$D by the granulomas, as a result of increased activity of 1-α-hydroxylase, the enzyme that converts 25(OH)D to its active form, 1,25(OH)$_2$D. The reasons for this are unknown, but this results in elevated circulating concentrations of 1,25(OH)$_2$D, which in turn lead to intestinal hyperabsorption of calcium, hypercalciuria, and ultimately hypercalcemia.

The syndrome reverses with the eradication of granulomas, (e.g., by glucocorticoids or antituberculosis medications), and by oral or intravenous hydration coupled with lowering dietary intake of vitamin D and calcium. Since sunlight is a source of vitamin D, sunlight exposure should be reduced.

Other endocrine disorders

Hyperparathyroidism is the classical endocrine disorder associated with hypercalcemia, but there are four other endocrine disorders that may cause hypercalcemia as well.

Hyperthyroidism

Hyperthyroidism has been reported to cause increases in ionized or total serum calcium in up to 50% of affected patients. In general, the hypercalcemia is mild (in the 10.7–11.0 mg/dl range), but it has been reported to be as high as 13 mg/dl in rare cases. It is believed to result from increases in osteoclastic bone resorption due to thyroid hormone induced increases in RANK-L, a key regulator of osteoclast function.

Addisonian crisis

Addisonian crisis has also been reported to cause hypercalcemia with increases in both ionized as well as total calcium. In general, hypercalcemia is mild and responds to standard therapy for hypoadrenalism (fluid resuscitation and intravenous glucocorticoids). The cause is not known, but, at least in some cases, it may be due to the underlying disorder, such as tuberculosis, that led to the hypoadrenalism. In others, it is possible that the associated volume contraction may have led to factitious increases in total serum calcium through relative hyperalbuminemia. In such cases, though the measured concentration of total serum calcium is elevated, ionized serum calcium remains normal. However, at least in some cases, ionized serum calcium has been reported to have been elevated.

Pheochromocytoma

Pheochromocytoma has been associated with hypercalcemia. In some cases, the hypercalcemia is due to primary hyperparathyroidism in the setting of multiple endocrine neoplasia type 2, and hypercalcemia corrects with parathyroid surgery. In some patients, however, hypercalcemia reverses with removal of the pheochromocytoma, and some of these tumors have been shown to secrete PTHrP. In other cases, it has been suggested that catecholamine secretion by the pheochromocytoma is sufficient to activate bone resorption.

VIPoma syndrome

VIPoma syndrome is caused by vasoactive intestinal polypeptide (VIP) secretion by the pancreatic islet or other neuroendocrine tumors, and is associated with severe watery diarrhea ("pancreatic cholera"), hypokalemia, and achlorhydria (the WDHA syndrome). Interestingly, 90% of patients with this rare syndrome have been reported to be hypercalcemic, although the mechanism is unknown. VIP has been demonstrated to stimulate osteoclastic bone resorption *in vitro*, suggesting at least one potential mechanism.

Immobilization

Immobilization in association with another cause of high bone turnover (such as the high turnover associated with youth, hyperparathyroidism, myeloma or breast cancer with bone metastases, and Paget's disease) may cause hypercalcemia. The classic examples are the hypercalcemia that was prevalent in the polio epidemics and that still regularly accompanies paraplegia and quadriplegia. The age dependence of this phenomenon is evident from the observation that hypercalcemia never occurs in elderly subjects with strokes that result in complete immobilization, yet regularly occurs in children and young adults with similar degrees of immobilization from spinal cord injury, or other causes of complete immobilization.

Immobilization suppresses osteoblastic bone formation and markedly increases osteoclastic bone resorption, leading to complete uncoupling of these two normally tightly coupled processes. The result is massive loss of calcium from the skeleton, with resultant hypercalcemia and reductions in bone mineral density (BMD). It has been suggested that this process is mediated by sclerostin, which is elevated in patients who are immobilized and appears to inhibit bone formation. The process is most effectively reversed by restoration of normal weight bearing. Alternate options are bisphosphonates and measures to increase renal calcium clearance (hydration and loop diuretics).

Milk-alkali syndrome

Originally reported in 1949, this syndrome initially described patients who developed moderate or severe hypercalcemia when treated with large amounts of milk (several quarts or gallons per day) and absorbable antacids (e.g., baking soda or sodium bicarbonate) for peptide ulcer disease. Additional features of the syndrome were a metabolic alkalosis due to antacid ingestion and renal failure due to hypercalcemia.

Contemporary reports occur primarily in patients taking large amounts of calcium carbonate for peptic ulcer or esophageal reflux symptoms. Since calcium is absorbed with only moderate efficiency, normal dietary intake (800–2,000 mg/day) does not cause hypercalcemia. In contrast, daily doses in excess of 4,000 mg/day can induce hypercalciuria and hypercalcemia in normal adults. Indeed, in many case reports, the daily intake has been in the range of 10–20 gm of elemental calcium per day. Since one standard antacid tablet contains approximately 120–200 mg of elemental calcium, and each packet or roll contains some 10 tablets, it is clear that subjects must consume multiple packages of antacids per day for this syndrome to develop. The hypercalcemia reverses with hydration and correction of excessive calcium ingestion. Renal damage, however, may be permanent.

Total parenteral nutrition

Patients with short bowel syndrome or otherwise unable to eat normally by mouth who receive chronic total parenteral nutrition (TPN) have been reported to develop hypercalcemia. In some cases, it was a result of the inclusion of excessive amounts of calcium or vitamin D supplements in early TPN solutions. In other early cases, it appears to have been associated with the use of collagen lysates contaminated with aluminum. Case reports have been rare in recent years, but the etiology of recent cases is uncertain.

Abnormal protein binding

Hypercalcemia may be "artifactual" or "factitious" in some settings. In general, this refers to situations in which the total serum calcium is elevated, but the ionized serum calcium is normal. In one example, severe dehydration may lead to increases in serum albumin concentration and in the albumin-bound component of total serum calcium. This results in an increase in total, but not ionized, serum calcium. This can be suspected in the setting of volume contraction with hyperalbuminemia and confirmed by measurement of ionized serum calcium.

An analogous situation has been reported in subjects with multiple myeloma or Waldenstrom's macroglobulinemia whose monoclonal immunoglobulin specifically recognizes calcium ion. In these cases, patients have displayed severe increases in total serum calcium, in the absence of neurologic or EKG abnormalities, and in the absence of symptoms or signs of hypercalcemia. In these cases, the ionized serum calcium was found to be normal, as was the urinary calcium excretion, and the patients' immunoglobulins were shown to have an abnormal affinity for serum calcium. Treatment with agents to lower serum calcium such as mithramycin precipitated hypocalcemic seizures, despite a total serum calcium that remained in or above the normal range.

Medications

A number of medications may cause hypercalcemia. Calcium containing antacids are included above in the section on milk-alkali syndrome.

Vitamin D intoxication from standard vitamin D preparations has been reported in association with inappropriate addition by dairies or manufacturers of vitamin D to milk or to infant formula. Vitamin D intoxication may also occur with use of doses of vitamin D in excess of 50,000 units two or three times per week as a treatment for hypoparathyroidism or metabolic bone diseases such as osteoporosis. Vitamin D analogs such as calcitriol [$1,25(OH)_2D$] used in the treatment of hypoparathyroidism, chronic renal failure, and metabolic bone disease may also cause hypercalcemia. The mechanism in all of the above is a combination of increased intestinal calcium absorption and bone resorption induced by vitamin D, together with reductions in renal ability to clear calcium as a result of dehydration.

Vitamin A intoxication can cause hypercalcemia. This may occur through the excessive use of vitamin supplements, or, in Antarctic explorers, as a result of eating sled dog livers. More recently, the use of retinoic acid derivatives for the treatment of dermatologic disorders or as chemotherapy agents has also been associated with induction of hypercalcemia.

Thiazide diuretics such as hydrochlorthiazide or chlorthalidone commonly cause mild hypercalcemia. This has been ascribed to their ability to induce distal renal tubular calcium reabsorption, although it has been reported to occur in anephric subjects on dialysis, suggesting additional mechanisms.

Lithium has been reported to cause hypercalcemia in as many as 5% of patients. It has been suggested that lithium may actually induce parathyroid hyperplasia or induce parathyroid adenomas, but it is also possible that the coincidence of lithium use with hyperparathyroidism may represent the simultaneous occurrence of two common clinical syndromes. There are also well-documented patients treated with lithium whose hypercalcemia reversed with cessation of lithium therapy, and in whom hypercalcemia therefore was clearly lithium induced. Proposed mechanisms include lithium-induced activation of PTH secretion and lithium-induced stimulation of renal calcium reabsorption, both of which have been documented *in vitro* or in animals. Whether these are responsible for the hypercalcemia that occurs in humans is not certain.

Parathyroid hormone, both the PTH(1-34) and PTH(1-84) forms used for treatment of osteoporosis, are associated with hypercalcemia in a substantial minority of patients so treated. In general, it is mild and requires little or no treatment, or a reduction in the dose of PTH or supplemental calcium, but it can be severe and require discontinuation of PTH therapy.

Other medications that are known to cause hypercalcemia with no known mechanism include: estrogens and the selective estrogen receptor modifier (SERM) tamoxifen in women with breast cancer and extensive skeletal metastatic disease; aminophylline and theophylline when used in large, supratherapeutic doses in subjects with bronchospastic disease; Foscarnet, an antiviral agent used in HIV/AIDS; growth hormone treatment in subjects with severe burns and also in subjects with HIV/AIDS; and 8-chloro-cyclic AMP employed as an anticancer agent.

Acute and chronic renal failure

The recovery phase from acute renal failure caused by rhabdomyolysis has been associated with hypercalcemia. Typically, this follows an episode of severe hyperphosphatemia and hypocalcemia in the acute, oliguric phase, accompanied by severe secondary hyperparathyroidism. It has been ascribed to residual effects of PTH on bone turnover, as well as release of calcium phosphate precipitated into soft tissues such as skeletal muscle during the early hypocalcemic, hyperphosphatemic phase.

Chronic renal failure and dialysis are associated with hypercalcemia. Frequently, it is associated with the use of calcitriol or other vitamin D analogs used to prevent secondary hyperparathyroidism, or with the use of oral calcium binding agents and supplements. Hypercalcemia in this population may also result from tertiary hyperparathyroidism, as discussed elsewhere in the *Primer*. Hypercalcemia has also been observed after kidney transplant, particularly in cases of moderate to severe secondary hyperparathyroidism.

Hypophosphatemia

Severe dietary phosphate deprivation in rats causes hypophosphatemia associated with hypercalcemia. This has not been documented in humans but merits attention when considering the management of patients with hypercalcemia, since moderate to severe hypophosphatemia so commonly accompanies hyperparathyroidism, cancer hypercalcemia, and other disorders that cause hypercalcemia. It is the authors' anecdotal experience that hypercalcemia may be refractory to treatment in the presence of severe hypophosphatemia, but responds nicely to appropriate measures once the hypophosphatemia has been corrected.

Miscellaneous

Several other diseases and syndromes associated with hypercalcemia include: end-stage liver disease in patients awaiting liver transplantation; manganese intoxication in people exposed to well water contaminated with manganese derived from improperly disposed batteries; fibrin glue when used to treat refractory recurrent pneumothorax in children; idiopathic infantile hypercalcemia due to decreased 24-hydroxylase activity by a CYP24A1 mutation and other pediatric syndromes, which are covered in more detail elsewhere in the *Primer*.

APPROACH TO DIAGNOSIS

Although space precludes a detailed description of the specific approach to the differential diagnosis of each of the causes of hypercalcemia in Table 70.1, it is helpful to consider several broad guidelines. First, the causes of hypercalcemia can be divided into two broad categories: those associated with elevated PTH values (for example, primary and tertiary hyperparathyroidism, ectopic hyperparathyroidism, familial hypocalciuric hypercalcemia (FHH), and occasionally lithium treatment) and those in which PTH is low normal or frankly suppressed. This algorithm requires that the PTH samples be collected properly and performed in a state-of-the-art two site PTH-immunoassay as described elsewhere in the *Primer*.

Second, common diseases are common. Thus, the most common cause of hypercalcemia among outpatients is primary hyperparathyroidism, and the most common cause among inpatients is cancer. Thus, it is reasonable to begin a diagnostic strategy with these two disorders in mind.

Third, most patients with MAHC have large, bulky tumors that are obvious on initial screening exams and CT scans; thus, if no tumor is apparent after a careful physical exam and appropriate imaging procedures, attention should be focused on the less common items in Table 70.1. Exceptions to this guideline include small neuroendocrine tumors such as pheochromocytomas, bronchial carcinoids, and pancreatic islet tumors that may be small and difficult to find.

Fourth, although it is tempting to initially select what seems an obvious diagnosis and manage the patient with this diagnosis in mind, many of the less common items in Table 70.1 are overlooked unless specifically considered. Because so many of these are easily treatable, it is critical to consider every diagnosis in Table 70.1 in every patient. For example, a patient with breast cancer can also have primary hyperparathyroidism, which is easily treated and may change the overall prognostic perception. Similarly, patients with lung cancer may also have tuberculosis, and hypercalcemia in this setting may reverse with appropriate antitubercular treatment, again, altering the overall prognostic perception. Another common example is milk-alkali syndrome that occurs in a patient whose hypercalcemia has inappropriately been attributed to a coexisting cancer.

Fifth, carefully documenting the duration of hypercalcemia is helpful. Most hypercalcemia syndromes are unstable, and rapidly become more severe if left untreated (MAHC, immobilization, and vitamin D intoxication are examples), whereas long-term (longer than 6 months) stable hypercalcemia has a relatively short differential diagnosis that includes largely primary and tertiary hyperparathyroidism, FHH, thiazide, and lithium use, and occasional cases of sarcoid.

And sixth, it is useful to consider the principal underlying pathophysiologic mechanism responsible for hypercalcemia before completely committing to a diagnosis or therapeutic plan. Is the hypercalcemia mainly postprandial (i.e., GI in origin) as might occur in sarcoid or vitamin D intoxication? Or is it equally apparent in fasting conditions, which may suggest inability of the kidney to clear calcium (as in FHH or thiazide use) or excessive bone resorption (as in MAHC)? And between these two, is the urinary calcium/creatinine ratio very high [suggesting GI causes (such as sarcoid, milk-alkali syndrome, or vitamin D intoxication) or bone resorption (from cancer, immobilization, etc.)], or is it normal? Is the patient taking oral calcium supplements or antacids? These considerations help to narrow the diagnostic possibilities and suggest specific laboratory tests, such as PTH, PTHrP measurements, vitamin D metabolites, thyroid indices, serum angiotensin converting enzyme (ACE), serum/urine protein electrophoresis, lithium levels, bone marrow biopsy, liver biopsy, etc.

Considering these tenets while considering each of the items in Table 70.1 will facilitate and accelerate accurate diagnosis.

MANAGEMENT

Management of hypercalcemia optimally targets the underlying pathophysiology. For example, removal of a parathyroid adenoma, discontinuing or reducing the dose of an offending medication such as vitamin D or PTH, or eradication of a tumor responsible for hypercalcemia with chemotherapy or surgery would be optimal management strategies. Of course, sometimes these are not possible, or therapy must begin before a definitive diagnosis is made. In these cases, targeting the underlying pathophysiology is most appropriate. Thus, in patients whose hypercalcemia is primarily based on accelerated bone resorption (e.g., LOH, HHM, immobilization), therapy should include agents that block bone resorption, such as intravenous bisphosphonates, zoledronate, or pamidronate. In patients whose hypercalcemia is principally GI in origin [e.g., sarcoidosis, milk-alkali syndrome, vitamin D intoxication, 1,25(OH)2D-secreting lymphomas], reducing or eliminating oral calcium and vitamin D intake and sunlight exposure may be most appropriate.

For those with important renal contributions to hypercalcemia (e.g., dehydration), increasing renal calcium clearance by increasing the glomerular filtration rate (GFR) using saline infusions, and blocking renal calcium absorption using loop diuretics such as furosemide are appropriate. Of course, many patients have contributions from several sources (e.g., dehydration plus increased bone resorption in LOH, or increased GI calcium absorption plus dehydration in patients with sarcoidosis, or a combination of PTHrP-induced bone resorption plus PTHrP-induced renal calcium retention in HHM), and optimal therapy targets each of these components.

For hyperparathyroidism and its variants, specific therapy is discussed elsewhere in the *Primer*. For patients with cancer, the most effective long-term therapy is tumor eradication. If that is not possible, or while waiting for a response to chemotherapy, aggressive hydration with saline, keeping a careful watch for signs of congestive heart failure, accompanied by a loop diuretic such as furosemide are appropriate. Limiting oral calcium intake (e.g., milkshakes, ice cream) is not important in HHM and LOH, since intestinal calcium absorption is already low as a result of the low $1,25(OH)_2D$ concentrations in these patients, and because cachexia is a common feature of these patients. On the other hand, in patients with $1,25(OH)_2D$-induced hypercalcemia from lymphoma or dysgerminoma, reducing oral calcium and vitamin D intake is important. While some physicians wait to see the magnitude of the decline in serum calcium induced by hydration and diuresis, the authors' practice, when the serum calcium exceeds 12.0 mg/dl, is to institute antiresorptive therapy with an intravenous bisphosphonate such as zoledronate or pamidronate soon after the discovery of hypercalcemia. Specific doses and regimens have been reviewed in detail recently.

For the granulomatous diseases, correcting the underlying cause is critical, where possible (e.g., tuberculosis). In sarcoid, limiting calcium and vitamin D intake and sun exposure are important, together with oral or parenteral hydration. Glucocorticoid therapy may be necessary to treat the granulomas, to lower intestinal calcium absorption, and to lower $1,25(OH)_2D$ concentrations.

For immobilization-induced hypercalcemia, weight-bearing ambulation is the mainstay of therapy. Often, however, this is not possible because of spinal cord injury or pain. Here, aggressive hydration and intravenous bisphosphonates are effective and important.

For the remainder of the diagnoses in Table 70.1, correcting the underlying disorder or withdrawing or reducing the dose of the offending medication corrects the serum calcium.

ACKNOWLEDGMENTS

This work was supported by NIH grants DK51081 and DK073039.

SUGGESTED READINGS

1. Horwitz MJ, Stewart AF. 2010. Malignancy-associated hypercalcemia and medical management. In: DeGroot L, Jameson L (eds.) *Endocrinology, 6th Ed.* Philadelphia, PA: Saunders Elsevier. pp. 1198–1211.
2. Stewart AF, Horst R, Deftos LJ, Cadman EC, Lang R, Broadus AE. 1980. Biochemical evaluation of patients with cancer-associated hypercalcemia: Evidence for humoral and non-humoral groups. *N Engl J Med* 303: 1377–1383.
3. Burtis WJ, Brady TG, Orloff JJ, Ersbak JB, Warrell RP, Olson BR, Wu TL, Mitnick, MA, Broadus AE, Stewart AF. 1990. Immunochemical characterization of circulating parathyroid hormone-related protein in patients with humoral hypercalcemia of malignancy. *New Engl J Med* 322: 1106–1112.
4. Stewart AF, Vignery A, Silvergate A, Ravin ND, LiVolsi V, Broadus AE, Baron R. 1982. Quantitative bone histomorphometry in humoral hypercalcemia of malignancy: Uncoupling of bone cell activity. *J Clin Endo Metab* 55: 219–227.
5. Horwitz MJ, Tedesco MB, Sereika SK, Prebehala L, Gundberg CM, Hollis BW, Bisello A, Carneiro RM, Garcia-Ocaña A, Stewart AF. 2011. A 7-day continuous infusion of PTH or PTHrP suppresses bone formation and uncouples bone turnover. Modeling hyperparathyroidism, humoral hypercalcemia of malignancy and lactation in humans: continuous infusion of PTH or PTHrP suppresses bone formation and uncouples bone turnover. *J Bone Miner Res* 26(9):2287–2297.
6. Stewart AF. 2005. Hypercalcemia associated with cancer. *N Engl J Med* 352: 373–379.
7. Dean T, Vilardaga J-P, Potts JT, Gardella TJ. 2008. Altered selectivity of parathyroid hormone (PTH) and PTH-related protein (PTHrP) for distinct conformations of the PTH/PTHrP receptor. *Mol Endo* 22: 156–166.
8. Knecht TP, Behling CA, Burton DW, Glass CK, Deftos LJ. 1996. The humoral hypercalcemia of benignancy. A newly appreciated syndrome. *Am J Clin Pathol* 105: 487–492.
9. Khosla S, van Heerden JA, Gharib H, Jackson IT, Danks J, Hayman JA, Martin TJ. 1990. Parathyroid hormone-related protein and hypercalcemia secondary to massive mammary hyperplasia. *N Engl J Med* 322: 1157.
10. Roodman GD. 2004. Mechanisms of bone metastasis. *N Engl J Med* 350: 1655–1664.
11. Rosenthal N, Insogna KL, Godsall JW, Smaldone L, Waldron JA, Stewart AF. 1985. Elevations in circulating 1,25 dihydroxyvitamin D in three patients with lymphoma-associated hypercalcemia. *J Clin Endocrinol Metab* 60: 29–33.
12. Evans KN, Taylor H, Zehnder D, Kilby MD, Bulmer JN, Shah F, Adams JS, Hewison M. 2004. Increased expression of 25-hydroxyvitamin D-1alpha-hydroxylase in dysgerminomas: A novel form of humoral hypercalcemia of malignancy. *Am J Pathol* 165: 807–813.
13. Nussbaum SR, Gaz RD, Arnold A. 1990. Hypercalcemia and ectopic secretion of PTH by an ovarian carcinoma with rearrangement of the gene for PTH. *N Engl J Med* 323: 1324–1328.
14. VanHouten JN, Yu N, Rimm D, Dotto J, Arnold A, Wysolmerski JJ, Udelsman R. 2006. Hypercalcemia of malignancy due to ectopic transactivation of the parathyroid hormone gene. *J Clin Endocrinol Metab* 91: 580–583.
15. Barbour GL, Coburn JW, Slatopolsky E, Norman AW, Horst RL. 1981. Hypercalcemia in an anephric patient with sarcoidosis: Evidence for extrarenal generation of 1,25-dihydroxyvitamin D. *N Engl J Med* 305: 440–443.
16. Adams JS, Gacad MA. 1985. Characterization of 1 hydroxylation of vitamin D_3 sterols by cultured alveolar macrophages from patients with sarcoidosis. *J Exp Med* 161: 755–765.
17. Parker MS, Dokoh S, Woolfenden JM, Buchsbaum HW. 1984. Hypercalcemia in coccidioidomycosis. *Am J Med* 76: 341–343.
18. Gkonos PJ, London R, Hendler ED. 1984. Hypercalcemia and elevated 1,25-dihydroxyvitamin D levels in a patient with end stage renal disease and active tuberculosis. *N Engl J Med* 311: 1683–1685.
19. Ross DS, Nussbaum SR. 1989. Reciprocal changes in parathyroid hormone and thyroid function after radioiodine treatment of hyperthyroidism. *J Clin Endocrinol Metab* 68: 1216–1219.
20. Rosen HN, Moses AC, Gundberg C, Kung VT, Seyedin SM, Chen T, Holick M, Greenspan SL. 1993. Therapy with parenteral pamidronate prevents thyroid hormone-induced bone turnover in humans. *J Clin Endocrinol Metab* 77: 664–669.
21. Muls E, Bouillon R, Boelaert J, Lamberigts G, Van Imschool S, Daneels P, DeMoor P. 1982. Etiology of hypercalcemia in a patient with Addison's disease. *Calcif Tissue Int* 34: 523–526.
22. Vasikaran SD, Tallis GA, Braund WJ. 1994. Secondary hypoadrenalism presenting with hypercalcaemia. *Clin Endocrinol* 41: 261–265.
23. Verner JV, Morrison AB. 1974. Endocrine pancreatic islet disease with diarrhea. *Arch Intern Med* 133: 492–500.
24. Ghaferi AA, Chojnacki KA, Long, WD, Cameron JL, Yeo CJ. 2008. Pancreatic VIPomas: Subject review and one institutional experience. *J Gastrointest Surg* 12: 382–393.
25. Stewart AF, Hoecker J, Segre GV, Mallette LE, Amatruda T, Vignery A. 1985. Hypercalcemia in pheochromocytoma: Evidence for a novel mechanism. *Ann Intern Med* 102: 776–779.
26. Mune T, Katakami H, Kato Y, Yasuda K, Matsukura S, Miura K. 1993. Production and secretion of parathyroid hormone–related protein in pheochromocytoma: Participation of an α-adrenergic mechanism. *J Clin Endocrinol Metab* 76: 757–762.
27. Stewart AF, Adler M, Byers CM, Segre GV, Broadus AE. 1982. Calcium homeostasis in immobilization: An

example of resorptive hypercalciuria. *N Engl J Med* 306: 1136–1140.
28. Chappard D, Minaire P, Privat C, Berard E, Mendoza-Sarmiento J, Tournebise H, Basle MH, Audran W, Rebel A, Picot C. 1995. Effects of tiludronate on bone loss in paraplegic patients. *J Bone Min Res* 10: 112–118.
29. Bergstrom WH. 1978. Hypercalciuria and hypercalcemia complicating immobilization. *Am J Dis Child* 132: 553–554.
30. Gaudino A, Pennisi P, Bratengeier C, Torrisi V, Lindner B, Mangiafico RA, Pulvirenti I, Hawa G, Tringali G, Fiore C. 2010. Increased sclerostin serum levels associated with bone formation and resorption markers in patients with immobilization-induced bone loss. *J Clin Endocrinol Metab* 95: 2248–2253.
31. Orwoll ES. 1982. The milk-alkali syndrome: Current concepts. *Ann Intern Med* 97: 242–248.
32. Beall DP, Scofield RH. 1995. Milk-alkali syndrome associated with calcium carbonate consumption. *Medicine* 74: 89–96.
33. Holick MF, Shao Q, Liu WW, Chen TC. 1992. The vitamin D content of fortified milk and infant formula. *N Engl J Med* 326: 1178–1181.
34. Ott SM, Maloney NA, Klein GL, Alfrey AC, Ament ME, Lobourn JW. 1983. Aluminum is associated with low bone formation in patients receiving chronic parenteral nutrition. *Ann Intern Med* 96: 910–914.
35. Klein GL, Horst RL, Norman AW, Ament ME, Slatopolsky E, Coburn JW. 1981. Reduced serum levels of 1 alpha,25-dihydroxyvitamin D during long-term total parenteral nutrition. *Ann Intern Med* 94: 638–643.
36. Merlini G, Fitzpatrick LA, Siris ES, Bilezikian JP, Birken A, Beychok A, Osserman EF. 1984. A human myeloma immunoglobulin G binding four moles of calcium associated with asymptomatic hypercalcemia. *J Clin Immunol* 4: 185–196.
37. Elfatih A, Anderson NR, Fahie-Wilson MN, Gama R. 2007. Pseudo-pseudohypercalcaemia, apparent primary hyperparathyroidism and Waldenström's macroglobulinaemia. *J Clin Pathol* 60: 436–437.
38. Haden ST, Stoll AL, McCormick S, Scott J, Fuleihan GE. 1979. Alterations in parathyroid dynamics in lithium-treated subjects. *J Clin Endocrinol Metab* 82: 2844–2848.
39. Porter RH, Cox BG, Heaney D, Hostetter TH, Stinebaugh BJ, Suki WN. 1978. Treatment of hypoparathyroid patients with chlorthalidone. *N Engl J Med* 298: 577.
40. Wermers RA, Kearns AE, Jenkins GD, Melton LJ 3rd. 2007. Incidence and clinical spectrum of thiazide-associated hypercalcemia. *Am J Med* 120: 911.e9–911.e15.
41. McPherson ML, Prince SR, Atamer E, Maxwell DB, Ross-Clunis H, Estep H. 1986. Theophylline-induced hypercalcemia. *Ann Intern Med* 105: 52–54.
42. Saunders MP, Salisbury AJ, O'Byrne KJ, Long L, Whitehouse RM, Talbot DC, Mawer EB, Harris AL. 1997. A novel cyclic adenosine monophosphate analog induces hypercalcemia via production of 1,25-dihydroxyvitamin D in patients with solid tumors. *J Clin Endocrinol Metab* 83: 4044–4048.
43. Gayet S, Ville E, Durand JM, Mars ME, Morange S, Kaplanski G, Gallais H, Soubeyrand J. 1997. Foscarnet-induced hypercalcemia in AIDS. *AIDS* 11: 1068–1070.
44. Knox JB, Demling RH, Wilmore DW, Sarraf P, Santos AA. 1995. Hypercalcemia associated with the use of human growth hormone in an adult surgical intensive care unit. *Arch Surg* 130: 442–445.
45. Sakoulas G, Tritos NA, Lally M, Wanke C, Hartzband P. 1997. Hypercalcemia in an AIDS patient treated with growth hormone. *AIDS* 11: 1353–1356.
46. Miller PD, Bilezikian JP, Diaz-Curiel M, Chen P, Marin F, Krege JH, Wong M, Marcus R. 2007. Occurrence of hypercalciuria in patients with osteoporosis treated with teriparatide. *J Clin Endocrinol Metab* 92: 3535–3541.
47. Valente JD, Elias AN, Weinstein GD. 1983. Hypercalcemia associated with oral isotretinoin in the treatment of severe acne. *JAMA* 250: 1899.
48. Villablanca J, Khan AA, Avramis VI, Seeger RC, Matthay KC, Ramsay NK, Reynolds CP. 1995. Phase I trial of 13-*cis*-retinoic acid in children with neuroblastoma following bone marrow transplantation. *J Clin Oncol* 13: 894–901.
49. Valentin-Opran A, Eilon G, Saez S, Mundy GR. 1985. Estrogens stimulate release of bone-resorbing activity in cultured human breast cancer cells. *J Clin Invest* 72: 726–731.
50. Ellis MJ, Gao F, Dehdashti F, Jeffe DB, Marcom PK, Carey LA, Dickler MN, Silverman P, Fleming GF, Kommareddy A, Jamalabadi-Majidi S, Crowder R, Siegel BA. 2009. Lower-dose vs. high-dose oral estradiol therapy of hormone receptor-positive, aromatase inhibitor-resistant advance breast cancer. *JAMA* 302: 774–780.
51. Jacobus CH, Holick MF, Shao Q, Chen TC, Holm IA, Kolodny JM, Fuleihan GE, Seely EW. 1992. Hypervitaminosis D associated with drinking milk. *N Engl J Med* 326: 1173–1177.
52. Llach F, Felsenfeld AJ, Haussler MR. 1981. The pathophysiology of altered calcium metabolism in rhabdomyolysis-induced acute renal failure. *N Engl J Med* 305: 117–123.
53. Messa P, Cafforio C, Alfieri C. 2010. Calcium and phosphate changes after renal transplantation. *J Nephrol* 23 Suppl 16: 175–181.
54. Gerhardt A, Greenberg A, Reilly JJ, Van Thiel DH. 1987. Hypercalcema complication of advanced chronic liver disease. *Arch Intern Med* 147: 274–277.
55. Chandra SV, Shukla GS, Srivastava RS. 1981. An exploratory study of manganese exposure to welders. *Clin Toxicol* 18: 407–416.
56. Sarkar S, Hussain N, Herson V. 2003. Fibrin glue for persistent pneumothorax in neonates. *J Perinatology* 23: 82–84.
57. Schlingman KP, Kaufman M, Weber S, Irwin A, Ulrike J, Misselwitz J, Klaus G, Guran T, Hoenderop JG, Bindels RJ, Prosser DE, Jones G, Konrad M. 2011. Mutations of CYP24A1 and idiopathic infantile hypercalcemia. *N Engl J Med* 365: 410–421.

71

Hypocalcemia: Definition, Etiology, Pathogenesis, Diagnosis, and Management

Anne L. Schafer and Dolores Shoback

INTRODUCTION

Hypocalcemia, defined as an ionized calcium (Ca^{2+}) concentration that falls below the lower limit of the normal range, is a commonly encountered clinical problem with multiple causes. A normal level of ionized Ca^{2+} (usually 1.00 to 1.25 mM) is critical for many vital cellular functions including hormonal secretion, skeletal and cardiac muscle contraction, cardiac conduction, blood clotting, and neurotransmission. Approximately 50% of the total serum Ca^{2+} is in the ionized fraction, with the remainder being protein-bound (45–50%, predominantly to albumin), or complexed to circulating anions such as phosphate. Total serum Ca^{2+} is usually the only value a clinician has when making an initial determination of the state of serum Ca^{2+} homeostasis in a patient, because ionized Ca^{2+} determinations are not routine measurements in most clinical settings. Therefore, the clinician must make the first assessment based on total serum Ca^{2+} levels.

The total serum Ca^{2+} concentration is a reliable indicator of the serum ionized Ca^{2+} concentration under most, but not all, circumstances. One important common situation where total serum Ca^{2+} poorly reflects the ionized Ca^{2+} concentration is when hypoalbuminemia is present. When serum albumin is depressed, total serum Ca^{2+} often falls to subnormal levels. This can be mistaken for hypocalcemia. Many recommend that a bedside estimation of the corrected serum total Ca^{2+} be done in the hypoalbuminemic patient to determine whether there is real concern for hypocalcemia. This estimation is often done using the following formula: adjusted total Ca^{2+} = measured total Ca^{2+} + [0.8 × (4.0-measured serum albumin)]. It is far better, however, when there is any question, to establish that the ionized Ca^{2+} is truly low by making a direct measurement. Estimates of the ionized Ca^{2+} are poor surrogates for actual measurements because, in addition to albumin, disturbances in pH and other circulating substances (e.g., citrate, phosphate, paraproteins) can influence the serum total Ca^{2+}, and these confounding factors are not considered in this estimation. It is imperative that the clinician establishes that the ionized Ca^{2+} concentration is indeed reduced, before an exhaustive search for an etiology of hypocalcemia is undertaken. Full evaluation may be costly and is unjustified if there is only weak evidence that the serum ionized Ca^{2+} level is subnormal.

ETIOLOGY AND PATHOGENESIS

There are numerous etiologies of low ionized Ca^{2+}. The disorders can be classified broadly as ones in which there is inadequate parathyroid hormone (PTH) or vitamin D production, PTH or vitamin D resistance, or a miscellaneous cause (Table 71.1). The last category encompasses a large and diverse spectrum of conditions that the endocrine clinician encounters in the course of practice.

Table 71.1. Etiologies for Hypocalcemia

Inadequate PTH production
  PTH gene mutations
    Autosomal recessive (168450.0002)
    Autosomal dominant (168450.0001)
  X-linked hypoparathyroidism
  Parathyroid gland agenesis
    GCMB mutations (603716)
  Postsurgical
  Autoimmune
    Isolated
    Polyglandular failure syndrome type 1 (240300_and 607358)
    Acquired antibodies that activate the CaSR
  Postradiation therapy
  Secondary to infiltrative processes
    Iron overload: hemochromatosis, thalassemia after transfusions
    Wilson's disease
    Metastatic tumor
  Constitutively active CaSR mutations (145980)
  Magnesium excess
  Magnesium deficiency

Syndromes with component of hypoparathyroidism
  DiGeorge syndrome (188400)
  HDR (hypoparathyroidism, deafness, renal anomalies) syndrome (146255 and 256340)
  Blomstrand lethal chondrodysplasia (215045)
  Kenney-Caffey syndrome (244460)
  Sanjad-Sakati syndrome (241410)
  Kearns-Sayre syndrome (530000)

Inadequate vitamin D production
  Vitamin D deficiency
    Nutritional deficiency
    Lack of sunlight exposure
    Malabsorption
    Post-gastric bypass surgery
    End-stage liver disease and cirrhosis
    Chronic kidney disease

PTH resistance
  Pseudohypoparathyroidism
  Magnesium depletion

Vitamin D resistance
  Pseudovitamin D deficiency rickets (vitamin D-dependent rickets type 1)
  Vitamin D-resistant rickets (vitamin D-dependent rickets type 2)

Miscellaneous
  Substances interfering with the laboratory assay for total Ca^{2+}—certain gadolinium salts in contrast agents given during MRI/MRA, particularly in patients with chronic renal failure
  Hyperphosphatemia
    Phosphate retention caused by acute or chronic renal failure
    Excess phosphate absorption caused by enemas, oral supplements
    Massive phosphate release caused by tumor lysis or crush injury
  Drugs
    Intravenous bisphosphonate therapy—especially in patients with vitamin D insufficiency or deficiency
    Foscarnet
    Imatinib mesylate—lowers both calcium and phosphate
  Rapid transfusion of large volumes of citrate-containing blood
  Acute critical illness—multiple contributing etiologies
  "Hungry bone syndrome" or recalcification tetany
    Post-thyroidectomy for Grave's disease
    Post-parathyroidectomy
  Osteoblastic metastases
  Acute pancreatitis
  Rhadomyolysis

It is incumbent on the clinician to be aware of the etiologies, the pathogenic mechanisms, the intricacies of diagnostic testing including sequencing for mutations and other genetic analysis, and the best approaches to therapies in patients with hypocalcemia.

Hypoparathyroidism is a rare diagnosis. It is most commonly the sequela of parathyroid or thyroid surgery during which most or all functioning parathyroid tissues are damaged, devitalized, and/or inadvertently removed. Perhaps next in frequency is the mild hypocalcemia and hypoparathyroidism caused by constitutively activating mutations of the Ca^{2+}-sensing receptor (CaSR). These mutations lead to the inappropriate suppression of PTH secretion at subnormal serum Ca^{2+} levels. The disorder presents as autosomal dominant hypocalcemia (ADH) in families and may go unrecognized because the hypocalcemia is often mild. The biochemical hallmark of ADH is often impressive hypercalciuria, which worsens with attempts to treat the hypocalcemia with Ca^{2+} salts and vitamin D metabolites. Exacerbation of hypercalciuria with nephrocalcinosis and renal failure can result from these efforts. Such renal complications occur because the constitutively active CaSRs in the kidney misperceive prevailing serum Ca^{2+} concentrations as higher than they are, and this enhances renal excretion of Ca^{2+}. It has recently been appreciated that patients can develop antibodies that activate parathyroid and renal CaSRs. This produces an acquired form of hypocalcemia with low PTH and elevated urinary Ca^{2+} levels, thus mimicking the genetic disorder. These rare individuals often have other autoimmune disorders. Destruction of the parathyroid glands on an immune basis can occur in isolation or as part of the type 1 autoimmune polyglandular syndrome (APS1). This is an autosomal recessive disorder caused by mutations in the autoimmune regulator (AIRE-I) gene. APS1 includes mucocutaneous candidiasis and adrenal insufficiency most commonly, as well as other autoimmune manifestations, and typically presents in childhood and adolescence. The parathyroid autoantigen in approximately 50% of patients with APS1 appears to be the NACHT leucine rich-repeat protein 5 (NALP5), a putative signaling molecule, whose role as yet in parathyroid physiology is unknown. Patients with APS1 and with isolated autoimmune hypoparathyroidism also often generate antibodies against the CaSR, although it is unclear what the role, if any, the CaSR plays in the pathogenesis of the tissue destruction.

As noted above, there are multiple modes of inheritance of hypoparathyroidism, depending on the molecule involved. Autosomal recessive mutations in the gene encoding the transcription factor glial cell missing B (GCMB) are a rare cause of hypoparathyroidism. GCMB is essential for the development of the parathyroid glands. Mutations in a gene near SOX3 on the X chromosome underlie the pathogenesis of X-linked hypoparathyroidism. Another syndrome, HDR (hypoparathyroidism, deafness, renal anomalies), is caused by mutations in the transcription factor GATA3. Variable penetrance of the renal anomalies and hearing deficits has been observed.

The well-known DiGeorge syndrome, a result of multiple developmental anomalies of the third and fourth branchial pouches, includes a spectrum of hypoparathyroidism, thymic aplasia and immunodeficiency, cardiac defects, cleft palate, and abnormal facies. A variety of other very rare genetic syndromes is worth considering when a patient presents with a constellation of features that includes hypoparathyroidism (e.g., Kenney-Caffey, Kearns-Sayre, Sanjad-Sakati, and other syndromes; Table 71.1). The different forms of PTH resistance, or pseudohypoparathyroidism, are described elsewhere in the *Primer*.

In contrast to the rarity of hypoparathyroidism, vitamin D deficiency and disordered vitamin D metabolism are more common causes of hypocalcemia. Resistance to vitamin D, like resistance to PTH, remains very rare. Whereas vitamin D deficiency and insufficiency occur in multiple clinical settings (elderly patients, postmenopausal women with fractures, nursing home residents, and so forth), it is much less common to see frankly low ionized Ca^{2+} values in such patients, particularly when 25-hydroxyvitamin D [25(OH)D] levels are just mildly depressed. Generally, low ionized Ca^{2+} values result from longstanding severe vitamin D deficiency, and chronically low 25(OH)D levels and are accompanied by a significant degree of secondary hyperparathyroidism. Nevertheless, it is essential in the evaluation of patients with low ionized Ca^{2+} values that one carefully considers vitamin D inadequacy, disorders of vitamin D activation by the kidney, and reduced vitamin D-mediated signaling as possible contributors to the etiology of the hypocalcemia.

Disorders of magnesium (Mg^{2+}) homeostasis bear mentioning because both Mg^{2+} excess and deficiency can produce hypocalcemia that is generally mild and caused by functional (and reversible) hypoparathyroidism. Hypomagnesemia, often of a transient and correctable nature, accompanies a vast number of clinical situations particularly in ill and hospitalized patients (e.g., malnutrition, pancreatitis, chronic alcohol abuse, diarrhea, diuretic and antibiotic therapy, and chemotherapeutic agents such as cisplatin derivatives). Low serum Mg^{2+} levels seen in conjunction with these clinical entities require evaluation and often at least short-term therapy. Primary renal Mg^{2+} wasting states such as Gitelman's syndrome, caused by mutations in the renal thiazide-sensitive NaCl cotransporter, are more persistent and require long-term Mg^{2+} and other electrolyte replacement therapy to correct the biochemical parameters and clinical symptoms. Other rare entities that involve primary renal Mg^{2+} wasting include autosomal recessive disorders caused by mutations in the paracellin-1 gene or in the Na-K ATPase subunit (FXYDZ gene). Hypomagnesemia also interferes with PTH action at its target organs, bone and kidney, particularly the PTH receptor-mediated activation of adenylate cyclase through the stimulatory G protein alpha subunit (Gsα). Mg^{2+} is a cofactor for the adenylate cyclase enzyme complex. Hence, chronic hypomagnesemia produces a functional state of PTH resistance. More importantly, the normal physiologic response to hypocalcemia is lacking in the patient with Mg^{2+} depletion. Intact PTH levels are inappropriately low or normal in the presence of hypomagnesemia. Once the Mg^{2+} depletion is corrected, parathyroid function returns to normal. Hypermagnesemia, in contrast, activates parathyroid CaSRs, thereby suppressing PTH secretion directly. Mg^{2+} levels high enough to stimulate the CaSR tend to occur only in patients with chronic kidney disease or in the rare instance when Mg^{2+} is used for tocolytic therapy for preterm labor.

The experienced clinician will recognize that hypocalcemia occurs in the heterogeneous conditions listed in the category "Miscellaneous" (Table 71.1). In terms of frequency, pancreatitis is the most common disorder and is often associated with a low serum Ca^{2+}. This has been ascribed to the precipitation of Ca^{2+}-containing salts in the inflamed pancreatic tissue and the presence of excess free fatty acids in the circulation. In many patients, pancreatitis may progress rapidly with hemorrhage, hypotension, and sepsis as complicating features. Thus, hypocalcemia in patients with pancreatitis often correlates with illness severity.

Acute and chronic hyperphosphatemia can cause low total serum Ca^{2+}. The most common cause for chronic hyperphosphatemia is chronic kidney disease (CKD). Hypocalcemia in CKD has many contributing factors, including poor nutrition, low 1,25-dihydroxyvitamin D production, and malabsorption. Acute changes in phosphate balance can also lower serum Ca^{2+}. In any situation where large amounts of phosphate are rapidly absorbed into the intravascular compartment, there is the potential for the serum ionized Ca^{2+} to fall, even to symptomatic levels. This can be seen with phosphate-containing enemas and supplements, especially when the latter are given intravenously for the treatment of hypophosphatemia. Also, in the setting of acute tumor lysis caused by cytolytic therapy for high-grade lymphomas, sarcomas, leukemias, and solid tumors, cell breakdown with the rapid release of phosphate from intracellular nucleotides can quickly depress serum ionized Ca^{2+}.

Treating patients who are normocalcemic with intravenous aminobisphosphonates (e.g., zoledronic acid, pamidronate), which block bone resorption dramatically, has the potential to cause a low ionized Ca^{2+}. However, this is relatively infrequent unless concomitant vitamin D deficiency/insufficiency has gone unaddressed. The drug foscarnet, used to treat immunocompromised patients with refractory cytomegalovirus or herpes infections, can lower both serum Ca^{2+} and Mg^{2+} to symptomatic levels. Recent reports indicate that the tyrosine kinase inhibitor imatinib mesylate, used to treat chronic myeloid leukemia and gastrointestinal stromal tumors, can cause both hypocalcemia and hypophosphatemia, likely due to its direct skeletal effects.

The "hungry bone syndrome," or recalcification tetany, can occur after parathyroidectomy for any form of hyperparathyroidism or after thyroidectomy for hyperthyroidism. Skeletal uptake of Ca^{2+} and phosphate is intense because of the presence of a mineral-depleted bone matrix and the sudden removal by surgery of the stimulus for maintaining high rates of bone resorption (either PTH or the thyroid hormones). Depending on the severity of the bone hunger, hypocalcemia and hypophosphatemia can persist for weeks and require large doses of Ca^{2+} and vitamin D metabolites to control. If there has been no permanent damage to the parathyroid glands, intact PTH levels should eventually rise appropriately to supranormal levels. In some cases, however, the viability of the remaining parathyroid glands or the suppression of function of the remaining glands by a previously dominant adenoma may confuse the picture. Careful management and repeated mineral and PTH analyses will usually allow the diagnosis to become clear over time.

The entity of pseudohypocalcemia caused by gadolinium (Gd^{3+})-containing magnetic resonance imaging (MRI) agents has received considerable attention lately because of the frequency of performing MR angiography in general and the use of the procedure in patients with CKD. The clearance of Gd^{3+} in such patients is very prolonged. Total serum Ca^{2+} levels, as measured by standard arsenazo III reagents, will appear to be low in patients with Gd^{3+}-containing contrast agents in the circulation. This is because Gd^{3+} complexes with this Ca^{2+}-sensitive dye and blocks the colorimetric detection of Ca^{2+}. Because ionized Ca^{2+} is measured in a completely different manner, ionized Ca^{2+} levels will be normal in these individuals, and there are no symptoms of hypocalcemia.

Acute and critical illness, often in the intensive care unit setting, is frequently accompanied by hypocalcemia, including frankly low ionized Ca^{2+} values. This entity is typically multifactorial with poor nutrition, vitamin D insufficiency, renal dysfunction, acid–base disturbances, cytokines, and other factors contributing. It is prudent to follow the ionized Ca^{2+} values and treat as deemed appropriate based on clinical circumstances.

SIGNS AND SYMPTOMS

Patients with low ionized Ca^{2+} values can present with no symptoms or with significant morbidity (Table 71.2). Their presentation depends on the severity and chronicity of the disturbance. Chronic hypocalcemia, despite even very low levels of ionized Ca^{2+}, can be asymptomatic, and the only clue is the presence of the positive Chvostek's sign. Neuromuscular irritability is the most frequent cause of symptoms that include tetany, carpopedal spasms, muscle twitching and cramping, circumoral tingling, abdominal cramps, and in severe cases laryngospasm, bronchospasm, seizures, and even coma. Basal ganglia and other intracerebral calcifications can be seen on imaging studies. Ocular findings include cataracts, particularly when there are longstanding elevations in the Ca^{2+}-phosphate product, and pseudotumor cerebri may be present. Longstanding hypocalcemia can also cause cardiomyopathy and congestive heart failure,

Table 71.2. Signs and Symptoms of Hypocalcemia

Symptoms
- Paresthesias
  - Circumoral and acral tingling
- Increased neuromuscular irritability
  - Tetany
  - Muscle cramping and twitching
  - Muscle weakness
  - Abdominal cramping
- Laryngospasm
- Bronchospasm
- Altered central nervous system function
  - Seizures of all types: grand mal, petit mal, focal
  - Altered mental status and sensorium
  - Papilledema, pseudotumor cerebri
  - Choreoathetoid movements
  - Depression
  - Coma
- Generalized fatigue
- Cataracts
- Congestive heart failure

Signs
- Chvostek's sign
- Trousseau's sign
- Prolongation of the QTc interval
- Basal ganglia and other intracerebral calcifications

which reverses with management of the usually very low ionized Ca^{2+} levels. Hypocalcemia is well known for its effects on cardiac conduction; the effects are manifested on the ECG as prolongation of the QTc interval. In addition, the patient often feels a sense of generalized weakness, fatigue, and depression that often lifts as the mineral disturbances (and vitamin D deficiency, if present) are successfully treated.

DIAGNOSIS: TESTS AND INTERPRETATION

The mainstays of diagnostic testing are determination of the serum ionized Ca^{2+}, total Mg^{2+}, phosphate, intact PTH, and 25(OH)D values. Measuring the 25(OH)D level is the best way to exclude vitamin D deficiency or insufficiency. Considerable attention has been directed to the reliability of contemporary 25(OH)D assays and clinically relevant cut points for diagnosing vitamin D deficiency/insufficiency. These important clinical issues are discussed elsewhere in the *Primer*. Intact PTH measured in a reliable two-site assay will readily disclose inappropriately low, low normal, or even undetectable values in hypoparathyroid states and generally normal levels (but inappropriately so) in patients with Mg^{2+} depletion. In marked contrast, patients with vitamin D deficiency and pseudohypoparathyroidism have elevated levels of PTH or secondary hyperparathyroidism. Phosphate levels are low in vitamin D deficiency and are elevated (or at the high end of the normal range) in patients with hypoparathyroidism and pseudohypoparathyroidism, an important distinguishing measurement to make. Accurate determination of 24-hour urinary Ca^{2+} or Mg^{2+} excretion, depending on the primary disturbance (hypocalcemia versus hypomagnesemia), can be extremely helpful. Strikingly elevated urinary Ca^{2+} levels in the asymptomatic mildly hypocalcemic patient suggest ADH. Milder elevations are expected in hypoparathyroid states. In contrast, vitamin D deficiency with secondary hyperparathyroidism classically produces hypocalciuria, in an effort by the kidney, under the influence of PTH, to conserve Ca^{2+} for systemic needs. The presence of significant Mg^{2+} in the urine in a hypomagnesemic patient strongly suggests primary renal Mg^{2+} wasting and not gastrointestinal losses.

MANAGEMENT OF HYPOCALCEMIA: ACUTE AND CHRONIC

The goals of treatment are to alleviate symptoms, heal demineralized bones when osteomalacia is present, maintain an acceptable ionized Ca^{2+} or total serum Ca^{2+}, and avoid hypercalciuria. Avoiding hypercalciuria (urine Ca^{2+} >300 mg/24 hours) is critical for the prevention of renal dysfunction, stones, and nephrocalcinosis. When clinical circumstances dictate urgent treatment, intravenous Ca^{2+} salts are used. Seizures, severe tetany, laryngospasm, bronchospasm, or altered mental status are strong indicators for intravenous therapy. In contrast, if the patient is minimally symptomatic despite low numbers, the oral regimen outlined below can be used. The preferred intravenous salt is Ca gluconate (1 gm = 1 amp = 10 mL 10% solution = 4.65 mEq = 90 mg elemental Ca^{2+}). Two grams (20 mL 10% solution, or 2 ampules) may be infused slowly over 10 min to address symptoms. Generally, an infusion is begun. One infusion method is to prepare 11 g Ca gluconate in D5W to provide a final volume of 1000 mL (~1 mg elemental Ca^{2+}/mL), and to infuse it at a rate of 0.5–2.0 mg elemental Ca^{2+}/kg/hour. The goals are to control symptoms, to restore the ionized Ca^{2+} to the lower end of the normal range (~1.0 mM), and to normalize the QTc interval. Higher rates (up to 2.0 mg/kg/h) may sometimes be needed to stabilize the patient. Ionized Ca^{2+} level should be monitored closely (e.g., after 1 hour and then every 4 hours). Once the serum ionized Ca^{2+} is stabilized, a chronic oral regimen is begun. Infusion rates are tapered down as serum ionized Ca^{2+} reaches the target, symptoms resolve, and oral medications are tolerated.

Chronic management of hypocalcemia uses oral Ca^{2+} supplements, vitamin D metabolites, and sometimes thiazide diuretics. When Mg^{2+} depletion is the source of hypocalcemia, Mg^{2+} deficits are generally large and poorly reflected by the serum Mg^{2+} level, because Mg^{2+} is predominantly an intracellular cation. Supplementation with Mg^{2+} salts over an extended period of time will usually be needed to replenish total body Mg^{2+} stores. Serum Ca^{2+} and PTH secretory capacity will return to normal in nearly all cases unless there are ongoing and unaddressed losses.

Ca^{2+} supplements of all types work to treat hypocalcemia. A few general principles are worth emphasizing. It is best to divide up the supplements throughout the day and to time their administration to coincide with meals, because this will enhance absorption (especially of Ca carbonate, which requires the acidic environment of the gastrointestinal tract for full absorption). The most efficient means of supplementation is in the form of Ca carbonate or citrate salts. The former is approximately 40% and the latter approximately 21% elemental Ca^{2+} by weight. Generally, 500–1,000 mg elemental Ca^{2+} two or three times a day is a reasonable starting dose and can be escalated upward. This is done based on patient tolerance, compliance, and clinical goals. When Ca^{2+} supplements are insufficient to reach the target serum Ca^{2+}, vitamin D metabolites are prescribed. When renal function is intact and some PTH is present for vitamin D activation, ergocalciferol (vitamin D_2) or cholecalciferol (vitamin D_3) may be used. Doses in the range of 25,000–50,000 IU daily (occasionally even more) may be needed to treat hypoparathyroidism, pseudohypoparathyroidism, or vitamin D deficiency in patients with malabsorption. Care must be exercised, because these forms of vitamin D have a long tissue half-life (weeks to months) due to long-term storage in fat, and toxicity may be difficult to predict and to treat. Calcitriol (1,25-dihydroxyvitamin D; 0.25 to 1.0 mcg once or twice daily) is necessary in patients with inadequate 1-α hydroxylase activity and preferred by many clinicians because of its rapid onset and offset of action (1–3 days), and ease of titration, despite its greater costs. Other vitamin D metabolites are alternatives for treatment, including alfacalcidiol and dihydrotachysterol. In patients who develop hypercalciuria on these regimens and/or have difficulty achieving the serum Ca^{2+} goal safely, one can take advantage of the Ca^{2+}-retaining actions of thiazide diuretics. Effective doses of hydrochlorthiazide typically range from 50–100 mg/day, although lower doses can be tried. Serum Ca^{2+}, phosphorus, and creatinine levels should be monitored regularly along with 25(OH)D levels if vitamin D therapy is used, to avoid toxicity.

For the treatment of hypoparathyroidism, replacement therapy with PTH has been investigated in both adults and children. In one randomized trial of PTH(1-34) versus oral calcitriol and calcium in 27 adults with hypoparathyroidism, twice daily PTH(1-34) maintained serum calcium levels at or just below normal range over 3 years, with normal urinary calcium levels. Similar findings were reported from a 3-year randomized trial in 12 children (aged 5–14 years) with hypoparathyroidism. In an open-label study of PTH(1-84) in adults with hypoparathyroidism, PTH(1-84) reduced supplemental calcium and calcitriol requirements without changing serum and urinary calcium levels meaningfully, and other trials of PTH(1-84) are underway. Currently, no form of PTH is approved by the U.S. Food and Drug Administration for this indication.

SUGGESTED READING

Abu-Alfa AK, Younes A. 2010. Tumor lysis syndrome and acute kidney injury: Evaluation, prevention, and management. *Am J Kidney Dis* 55 (5 Suppl 3): S1–13.

Alexander RT, Hoenderop JG, Bindels RJ. 2008. Molecular determinants of magnesium homeostasis: Insights from human disease. *J Am Soc Nephrol* 19: 1451–8.

Ali A, Christie PT, Grigorieva IV, Harding B, Van Esch H, Ahmed SF, Bitner-Glindzicz M, Blind E, Bloch C, Christin P, Clayton P, Gecz J, Gilbert-Dussardier B, Guillen-Navarro E, Hackett A, Halac I, Hendy GN, Lalloo F, Mache CJ, Mughal Z, Ong AC, Rinat C, Shaw N, Smithson SF, Tolmie J, Weill J, Nesbit MA, Thakker RV. 2007. Functional characterization of GATA3 mutations causing the hypoparathyroidism-deafness-renal (HDR) dysplasia syndrome: Insight into mechanisms of DNA binding by the GATA3 transcription factor. *Hum Mol Genet* 16: 265–75.

Alimohammadi M, Bjorklund P, Hallgren A, Pontynen N, Szinnai G, Shikama N, Keller MP, Ekwall O, Kinkel SA, Husebye ES, Gustafsson J, Rorsman F, Peltonen L, Betterle C, Perheentupa J, Akerstrom C, Westin G, Scott HS, Hollander GA, Kampe O. 2008. Autoimmune polyendocrine syndrome type 1 and NALP5, a parathyroid autoantigen. *N Engl J Med* 358: 1018–28.

Berman E, Nicolaides M, Maki RG, Fleisher M, Chanel S, Scheu K, Wilson B, Heller G, Sauter N. 2006. Altered bone and mineral metabolism in patients receiving imatinib mesylate. *N Engl J Med* 354: 2006–13.

Bowl MR, Nesbit MA, Harding B, Levy E, Jefferson A, Volpi E, Rizzoti K, Lovell-Badge R, Schlessinger D, Whyte MP, Thakker RV. 2005. An interstitial deletion-insertion involving chromosomes 2p25.3 and Xq27.1, near SOX3, causes X-linked recessive hypoparathyroidism. *J Clin Invest* 115: 2822–31.

Brasier AR, Nussbaum SR. 1988. Hungry bone syndrome: Clinical and biochemical predictors of its occurrence after parathyroid surgery. *Am J Med* 84: 654–60.

Brown EM. 2009. Anti-parathyroid and anti-calcium sensing receptor antibodies in autoimmune hypoparathyroidism. *Endocrinol Metab Clin North Am* 38: 437–45.

Cooper MS, Gittoes NJ. 2008. Diagnosis and management of hypocalcaemia. *BMJ* 336: 1298–302.

Forsythe RM, Wessel CB, Billiar TR, Angus DC, Rosengart MR. 2008. Parenteral calcium for intensive care unit patients. *Cochrane Database Syst Rev* 8: CD006163.

Holick MF. 2007. Vitamin D deficiency. *N Engl J Med* 357: 266–81.

Kemp EH, Gavalas NG, Krohn KJ, Brown EM, Watson PF, Weetman AP. 2009. Activating autoantibodies against the calcium-sensing receptor detected in two patients with autoimmune polyendocrine syndrome type 1. *J Clin Endocrinol Metab* 94: 4749–56.

Kifor O, McElduff A, LeBoff MS, Moore FD Jr, Butters R, Gao P, Cantor TL, Kifor I, Brown EM. 2004. Activating antibodies to the calcium-sensing receptor in two patients with autoimmune hypoparathyroidism. *J Clin Endocrinol Metab* 89: 548–56.

Kobrynski LJ, Sullivan KE. 2007. Velocardiofacial syndrome, DiGeorge syndrome: The chromosome 22q11.2 deletion syndromes. *Lancet* 370: 1443–52.

Liamis G, Milionis HJ, Elisaf M. 2009. A review of drug-induced hypocalcemia. *J Bone Miner Metab* 27: 635–42.

Lienhardt A, Bai M, Lagarde JP, Rigaud M, Zhang Z, Jiang Y, Kottler ML, Brown EM, Garabedian M. 2001. Activating mutations of the calcium-sensing receptor: Management of hypocalcemia. *J Clin Endocrinol Metab* 86: 5313–23.

Malloy PJ, Feldman D. 2010. Genetic disorders and defects in vitamin D action. *Endocrinol Metab Clin North Am* 39: 333–46.

Prince MR, Erel HE, Lent RW, Blumenfeld J, Kent KC, Bush HL, Wang Y. 2003. Gadodiamide administration causes spurious hypocalcemia. *Radiology* 227: 639–46.

Rubin MR, Sliney J, McMahon DJ, Silverberg SJ, Bilezikian JP. 2010. Therapy of hypoparathyroidism with intact parathyroid hormone. *Osteoporos Int* 21: 1927–34.

Shoback D. 2008. Clinical practice. Hypoparathyroidism. *N Engl J Med* 359: 391–403.

Shoback D, Sellmeyer D, Bikle D. 2007. Mineral metabolism and metabolic bone disease. In: Shoback D, Gardner D (eds.) *Basic and Clinical Endocrinology*, 8th Ed. New York: Lange Medical Books/McGraw-Hill. pp. 281–345.

Thakker RV. 2004. Genetics of endocrine and metabolic disorders: Parathyroid. *Rev Endocrinol Metab Dis* 5: 37–51.

Thomee C, Schubert SW, Parma J, Le PQ, Hashemolhosseini S, Wegner M, Abramowicz MJ. 2005. GCMB mutation in familial isolated hypoparathyroidism with residual secretion of parathyroid hormone. *J Clin Endocrinol Metab* 90: 2487–92.

Vandyke K, Fitter S, Dewar AL, Hughes TP, Zannettino AC. 2010. Dysregulation of bone remodeling by imatinib mesylate. *Blood* 115: 766–74.

Vivien B, Langeron 0, Morell E, Devilliers C, Carli PA, Riou B. 2006. Early hypocalcemia in severe trauma. *Crit Care Med* 33: 1946–52.

Winer KK, Sinaii N, Reynolds J, Peterson D, Dowdy K, Cutler GB. 2010. Long-term treatment of 12 children with chronic hypoparathyroidism: A randomized trial comparing synthetic human parathyroid hormone 1-34 versus calcitriol and calcium. *J Clin Endocrinol Metab* 85: 2680–8.

Winer KK, Ko CW, Reynolds JC, Dowdy K, Keil M, Peterson D, Gerber LH, McGarvey C, Culter GB. 2003. Long–term treatment of hypoparathyroidism: A randomized controlled study comparing parathyroid hormone-(1-34) *versus* calcitriol and calcium. *J Clin Endocrinol Metab* 88: 4214–20.

72

Hypoparathyroidism and Pseudohypoparathyroidism

Mishaela R. Rubin and Michael A. Levine

Introduction 579
Clinical Manifestations of Functional Hypoparathyroidism 579
Differential Diagnosis of Functional Hypoparathyroidism 580
Diagnostic Algorithm 588
Therapy 588
Suggested Reading 589

INTRODUCTION

The term "functional hypoparathyroidism" refers to a group of metabolic disorders in which hypocalcemia and hyperphosphatemia occur either from a failure of the parathyroid glands to secrete adequate amounts of biologically active parathyroid hormone (PTH) or, less commonly, from an inability of PTH to elicit appropriate biological responses in its target tissues. Plasma concentrations of PTH are low or absent in patients with true hypoparathyroidism (HP). By contrast, plasma concentrations of PTH are elevated in patients with pseudohypoparathyroidism (PHP), and reflect the failure of target tissues to respond appropriately to the biological actions of PTH.

There are several important features that distinguish PHP from HP. First, circulating levels of PTH are elevated in patients with PHP (i.e., secondary or adaptive hyperparathyroidism) and low in patients with HP. Second, the fractional excretion of urinary calcium is elevated in HP and low in PHP. In the absence of PTH, active transport of calcium from the glomerular filtrate in the distal renal tubule is decreased. Urinary excretion of calcium is low or normal in HP patients who are hypocalcemic, but as treatment restores the serum calcium level to normal the renal filtered load of calcium increases, and urinary calcium excretion increases *pari passu*. By contrast, the distal renal tubule is responsive to PTH in patients with PHP, so that urinary calcium excretion remains low relative to the filtered load of calcium (unless PTH levels are suppressed by overtreatment). Second, skeletal responsiveness to PTH is intact in patients with PHP, and thus PTH can also induce release of calcium (and phosphate) from skeletal storage pools when circulating levels of $1,25(OH)_2D$ are normal. And third, HP is associated with low bone turnover and reduced biochemical markers of bone resorption and formation whereas bone turnover is normal or increased in PHP, particularly in patients who have elevated levels of circulating PTH. Protracted low bone turnover in HP leads to bone mass that is greater than age- and sex-matched controls, especially at the lumbar spine. Bone histomorphometry shows that subjects with HP have significantly increased cancellous bone volume, trabecular width and cortical width, as well as suppressed dynamic skeletal indices (Fig. 72.1). These changes may be associated with increased expression of the *SOST* gene encoding sclerostin in osteocytes. Sclerostin is an antagonist of the Wnt signaling pathway, and is suppressed by PTH.

CLINICAL MANIFESTATIONS OF FUNCTIONAL HYPOPARATHYROIDISM

The signs and symptoms of functional hypoparathyroidism are principally manifestations of a reduced concentration of ionized extracellular calcium. It remains

---

Primer on the Metabolic Bone Diseases and Disorders of Mineral Metabolism, Eighth Edition. Edited by Clifford J. Rosen.
© 2013 American Society for Bone and Mineral Research. Published 2013 by John Wiley & Sons, Inc.

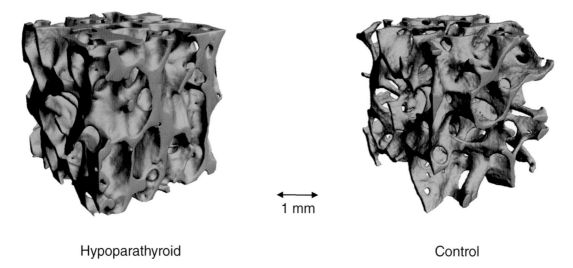

Hypoparathyroid Control

Fig. 72.1. Reconstructed μCT images of cancellous bone from a hypoparathyroid (*left*) and a control subject (*right*). Note the dense trabecular structure in hypoparathyroidism. Reproduced with permission from Rubin MR, Dempster DW, Kohler T, et al. 2010. Three-dimensional cancellous bone structure in hypoparathyroidism. *Bone* 46: 190–5.

uncertain whether PTH deficiency *per se* has any clinical characteristics. Hypocalcemia causes increased neuromuscular irritability, a condition termed "tetany." Patients may complain of paresthesias, particularly in the distal extremities and face, as well as muscle cramps. When hypocalcemia is severe, patients may experience laryngospasm, seizures, or reversible heart failure. Clinical signs of latent tetany include the Chvostek sign and Trousseau sign. Other clinical features of chronic hypocalcemia include pseudopapilledema, increased intracranial pressure, and dry, rough skin. Prolonged hypocalcemia and hyperphosphatemia, with an elevated calcium–phosphate product, will lead to posterior subcapsular cataracts and calcification of intracranial structures, which can be detected by computed tomography (CT) or magnetic resonance imaging (MRI). Rarely, calcification of the basal ganglia can cause extrapyramidal neurological dysfunction. Spondyloarthopathy has also been described. Hypocalcemia can cause prolongation of the corrected QT interval on an EKG and reversible heart failure. Patients can adapt to chronic hypocalcemia, and occasionally asymptomatic patients will be diagnosed only after a low serum calcium is detected after routine blood screening. Occasional patients with postsurgical HP report neuromuscular symptoms of hypocalcemia and reduced quality of life despite normal serum calcium levels.

DIFFERENTIAL DIAGNOSIS OF FUNCTIONAL HYPOPARATHYROIDISM

The response to exogenous PTH can distinguish between HP and the several variants of PHP (Fig. 72.2). Patients

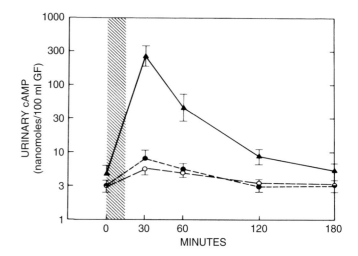

Fig. 72.2. PTH infusion. cAMP excretion in urine in response to the intravenous administration of bovine parathyroid extract (300 USP units) from 9–9:15 a.m. The peak response in normals (▲) is 50- to 100-fold times basal; patients with PHP type Ia (●) or PHP type Ib (○) show only a 2- to 5-fold response.

with HP show a robust increase in urinary excretion of nephrogenous cyclic adenosine monophosphate (cAMP) and phosphate. By contrast, patients with PHP type 1 fail to show an appropriate increase in plasma cAMP or urinary excretion of both cAMP and phosphate, while subjects with the less common type 2 form show normal increases in plasma and urinary cAMP but do not manifest a phosphaturic response (see below). The causes of PHP and HP are presented in Table 72.1.

Table 72.1. Classification of Hypoparathyroidism

Disorders of Parathyroid Gland Formation	MIM	Genetic Defect
DiGeorge sequence	188400	22q11; *TBX1* mutation 10p; intrauterine exposure to alcohol, diabetes, isotretinoin
Hypoparathyroidism, sensorineural deafness, and renal dysplasia syndrome (HDR)	146255	*GATA3*
Hypoparathyroidism-retardation-dysmorphism (HRD), Kenny-Caffey/Sanjad-Sakati Syndromes	241410, 244460	*TBCE*
Autosomal recessive/dominant hypoparathyroidism	146200	*GCM2*
X-linked hypoparathyroidism	307700	Xq27
Parathyroid Gland Destruction		
Surgery		
Radiation therapy and infiltration		
Autoimmune polyendocrinopathy-candidiasis-esctodermal dystrophy (APECED)	240300	21q22.3; *AIRE*
Reduced Parathyroid Gland Function		
Autosomal dominant hypocalcemic hypercalciuria (ADHH)	146200	3q13.3-21; *CASR*
PTH gene mutations		11p15; *PTH*
Antibodies to the CASR		
Other Causes of Hypoparathyroidism		
Mitochondrial disease (see text)		Mitochondrial tRNA
Burns		
Resistance to PTH		
Pseudohypoparathyroidism	103580, 603233, 174800	*GNAS*
Transient pseudohypoparathyroidism of the newborn		
Hypomagnesemia		

Syndromes of PTH resistance

Pseudohypoparathyroidism (PHP) type 1

The blunted nephrogenous cAMP response to PTH in subjects with PHP type 1 is caused by a deficiency of the alpha subunit of Gs (Gα_s), the signaling protein that couples PTH1R to stimulation of adenylyl cyclase. Molecular and biochemical studies have provided a basis for distinguishing between two forms of PHP type 1: patients with generalized deficiency of Gα_s, due to mutations within exons 1-13 of the *GNAS* gene, are classified as PHP type 1a (PHP 1a; OMIM 103580), whereas patients with more restricted deficiency of Gα_s, due to mutations that affect imprinting of *GNAS*, are classified as PHP type 1b (PHP 1b; OMIM 603233). In most cases, patients with PHP 1a can be distinguished from patients with PHP 1b on the basis of the pattern of hormone resistance and the presence of additional somatic features. PHP type 1c is likely a variant of PHP 1a (see below).

Pseudohypoparathyroidism 1a

PHP 1a (MIM 103580) is the most common variant and most readily recognized form of PHP. Subjects have a constellation of features termed "Albright hereditary osteodystrophy (AHO)" that include short stature, round faces, brachydactyly of hands and/or feet, mild to moderate mental retardation, and subcutaneous ossifications plus in most cases obesity (Fig. 72.3). PHP 1a results from heterozygous mutations on the maternal allele of the imprinted *GNAS* gene (20q13.2-q13.3) that reduce expression or function of the Gα_s protein (Fig. 72.4). As Gα_s is required for normal transmembrane signal transduction by many hormones and neurotransmitters, subjects with PHP 1a also have resistance to other hormones (e.g, thyroid stimulating hormone, gonadotropins, calcitonin, and growth hormone releasing hormone) whose target tissues show predominant expression of the maternal *GNAS* allele. Primary hypothyroidism, without goiter, and growth hormone deficiency are common associated endocrinopathies. Obesity is common, and based on studies of a mouse model, PHP 1a is due to a loss of Gα_s expression in imprinted regions of the hypothalamus. Obesity appears to be the result of reduced energy expenditure rather than increased calorie intake.

Responsiveness to other hormones (e.g., ACTH, vasopressin) is normal in tissues in which *GNAS* is not imprinted and both parental alleles are expressed. Subjects with paternally inherited *GNAS* mutations have phenotypic features of AHO without hormonal resistance, a condition termed "pseudopseudohypoparathyroidism (pseudoPHP)." Subjects with pseudoPHP have

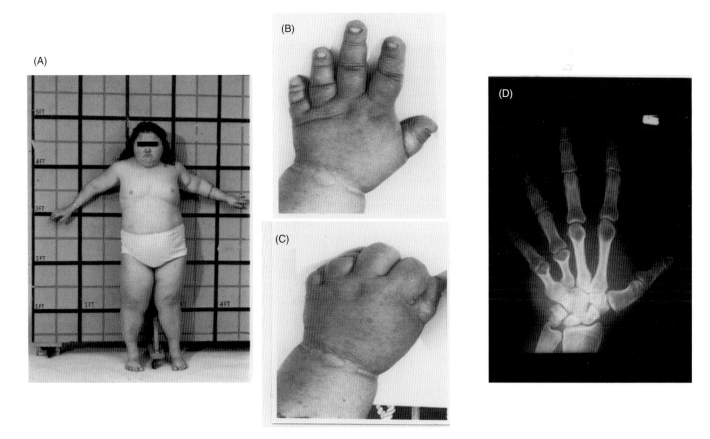

Fig. 72.3. Typical features of Albright hereditary osteodystrophy. The female in the picture demonstrates (A) short stature, sexual immaturity, and (B) brachydactyly. Note the extreme shortening of digit IV and distal phalanx of digit I (width greater than length) in (B) and (D), and the replacement of knuckles by dimples (Archibald sign) in (C).

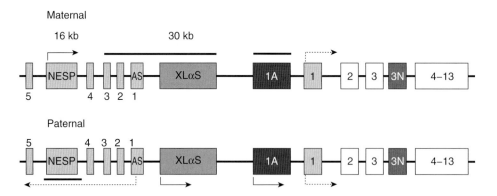

Fig. 72.4. *GNAS* gene. General organization of the *GNAS* gene complex. The *GNAS* gene complex consists of 13 exons that encode the signaling protein Gα_s. Upstream of exon 1 are three alternative first exons that are labeled exon 1A, XLαs, and Nesp55; exons 1-5 for the NESP antisense transcript (AS) are also depicted. The three alternative exons are spliced to exons 2-13 to produce unique transcripts (see text). The DMRs are denoted above the respective promoters, and arrows denote the direction of transcription. Nesp55 is transcribed exclusively from the maternal allele; XLαs and exon 1A are transcribed exclusively from the paternal allele. Nesp AS (anti-sense) and exon 1a transcripts produce noncoding RNAs. Gα_s transcripts are biallelically expressed except in a small number of tissues, such as the renal proximal tubules, thyroid, gonads, and pituitary somatotrophs, where expression is preferentially from the maternal allele.

normal renal responsiveness to PTH, which distinguishes them from occasional patients with PHP 1a who maintain normal serum calcium levels without treatment. It is not unusual to find extended families in which some members will have only AHO (i.e., pseudoPHP) while others will have hormone resistance as well (i.e., PHP 1a), based on the parental origin of the identical *GNAS* mutation.

The *GNAS* gene is a complex transcriptional unit that derives considerable plasticity through use of alternative first exons, alternative splicing of downstream exons, antisense transcripts, and reciprocal imprinting (Fig. 72.4). Gα_s is encoded by exons 1-13, and is synthesized as a 52- or 45-kDa protein based on the inclusion or exclusion of exon 3, respectively. There seems to be little difference in the signal transducing properties of the two Gα_s isoforms. Upstream of exon 1 are three alternative first exons that each splice onto exons 2-13 to create novel transcripts (Fig. 72.4). These include XL, which is expressed only from the paternal allele and which generates a transcript with overlapping open reading frames that encodes XLα_s and ALEX. The two proteins are interacting cofactors, and are specifically expressed in neuroendocrine cells. XLαs is a much larger signaling protein than Gα_s ($\approx$78 kDa vs 45–52 kDa) and can interact with receptors for PTH and a variety of other hormones *in vitro*, but the native receptors that interact with XLαs *in vivo* are presently unknown. A second alternative promoter encodes the secretory protein Nesp55, which is expressed only from the maternal allele and shares no protein homology with Gα_s. The third alternative first exon, termed exon 1A or A/B (associated first exon) is transcribed only from the paternal allele. These transcripts may be translated from an initiator codon in the second exon and encode an amino-terminal truncated protein that functions as a competitive inhibitor of Gα_s. These alternative first exons are associated with promoters that contain differentially methylated regions (DMRs) that are methylated on the non-expressed allele (Fig. 72.4). In contrast, the promoter for exon 1 is within a CpG island but is unmethylated on both alleles in all tissues. The cis-acting elements that control tissue-specific paternal imprinting of Gα_s appear to be located within the primary imprint region in exon 1A, as paternal deletion of the exon 1A DMR in mice is associated with increased Gα_s expression.

Private mutations have been found in nearly all of the AHO kindreds studied, although a 4-base deletion in exon 7 has been detected in multiple families and an unusual missense mutation in exon 13 (A366S; see below) has been identified in two unrelated young boys, suggesting that these two regions may be genetic "hot spots." Small deletions or point mutations can be identified in approximately 80% of AHO patients using polymerase chain reaction (PCR)-based techniques, and larger genomic rearrangements or uniparental disomy may account for AHO in other patients.

Post-zygotic somatic mutations in the *GNAS* gene that enhance activity of the protein are found in many autonomous endocrine tumors and affected tissues of patients with the McCune-Albright syndrome (MIM 174800). These mutations lead to constitutive activation of adenylyl cyclase, and result in proliferation and autonomous hyperfunction of hormonally responsive cells. Clinically significant effects are more likely to ensue when *GNAS* activating mutations occur on the maternally derived allele, which is preferentially expressed in imprinted tissues. The clinical significance of Gα_s activity as a determinant of hormone action is further emphasized by the description by of two unrelated males with both precocious puberty and PHP 1a. These two subjects had identical *GNAS* mutations in exon 13 (A366S) that resulted in a temperature-sensitive form of Gα_s. This Gα_s is constitutively active in the cooler environment of the testis, while being rapidly degraded in other tissues at normal body temperature. Thus, different tissues in these two individuals could show hormone resistance [to PTH and thyroid stimulating hormone (TSH)], hormone responsiveness [to adrenocorticotropic hormone (ACTH)], or hormone independent activation (to LH).

Pseudohypoparathyroidism 1b

Subjects with PHP 1b (MIM 603233) lack typical features of AHO but may have mild brachydactyly. Levels of Gα_s are normal in accessible tissues. PTH resistance is the principal manifestation of hormone resistance, but some patients have slightly elevated serum levels of TSH and normal serum concentrations of thyroid hormones as evidence of associated TSH resistance.

An epigenetic defect that results in switching of the maternal *GNAS* allele to a paternal pattern of methylation (i.e., paternal epigenotype) is a consistent finding in sporadic and familial PHP 1b. Causative mutations have been identified in most cases of familial PHP 1b, including two microdeletions in the *STX16* gene located approximately 220 kb centromeric of *GNAS* exon 1A, and deletions that remove the DMR encompassing exon NESP55 and exons 3 and 4 of the antisense transcript (Fig. 72.4). In each case, inheritance of a mutation from a female (or spontaneous mutation of a maternally derived allele) abolishes the maternal *GNAS* epigenotype. Small mutations have not been identified in sporadic PHP 1b, but some patients have uniparenal isodisomy, where both *GNAS* alleles have been inherited from the father. It is likely that the conversion of the maternal *GNAS* allele to a "paternal" epigenotype, or inheritance of two paternal alleles, leads to biallelic transcriptional silencing of the Gαs promoter in imprinted tissues, with the result that little or no Gα_s is expressed in these tissues.

Pseudohypoparathyroidism type 1c

In rare instances, patients with PHP type 1 and features of AHO show resistance to multiple hormones in the absence of a demonstrable biochemical defect in G$_s$ or G$_i$. Molecular studies suggest that these patients have *GNAS* mutations that result in functional defects of Gα$_s$ that are not apparent in conventional *in vitro* assays.

Albright hereditary osteodystrophy

Haploinsufficiency of *GNAS* is associated with an unusual constellation of skeletal anomalies that are collectively termed AHO, and which include short stature, round faces, brachydactyly, dental defects, and heterotopic ossification of the skin and subcutaneous tissues (Fig. 72.3). Additional features had been associated with AHO, such as obesity and sensory-neural abnormalities, but these defects appear limited to patients with PHP 1a who have maternal *GNAS* mutations that cause abnormal Gs$_\alpha$ signaling in the central nervous system. Patients who manifest AHO and have normal hormonal responsiveness due to inactivating mutations on the paternal *GNAS* allele are considered to have the genetically related disorder pseudoPHP (see above). Short stature and brachydactyly may be due in part to premature fusion of epiphyses in tubular and long bones, which implies a requirement of two functional copies of *GNAS* for normal growth plate maturation.

The heterotopic ossifications are the most unique feature of AHO and distinguish true AHO from a variety of clinical phenocopies. The ossifications are not calcifications, and are unrelated to the serum levels of calcium and phosphorus. Rather, the ossifications are islands of true intramembranous bone. These bone islands occur in the absence of a preexisting or associated lesion, as opposed to secondary types of cutaneous ossification that occur by metaplastic reaction to inflammatory, traumatic, and neoplastic processes. Recent studies suggest that Gs$_\alpha$ deficiency can induce ectopic expression of Cbfa1/Runx2 in mesenchymal stem cells with consequent differentiation of these cells into osteoblasts.

AHO is a unifying feature of PHP 1a and pseudoPHP, and based on the parental origin of the *GNAS* mutation, within a given kindred affected members of one generation can have only AHO (i.e., pseudoPHP) or can have hormone resistance as well (i.e., PHP 1a).

Osteoma cutis and progressive osseous heteroplasia

Osteoma cutis and progressive osseous heteroplasia (POH) represent alternative manifestations of AHO in which only heterotopic ossification occurs. In osteoma cutis ectopic ossification is limited to the superficial skin, whereas in POH heterotopic ossification involves the skin, subcutaneous tissue, muscles, tendons, and ligaments. POH can be disabling as extensive dermal ossification occurs during childhood, followed by widespread ossification of skeletal muscle and deep connective tissue. Nodules and lace-like webs of heterotopic bone extend from the skin into the subcutaneous fat and deep connective tissues, and may cross joints, thus leading to stiffness, joint locking, and permanent immobility.

Heterozygous inactivating *GNAS* mutations have been identified in most patients with osteoma cutis and POH, and in each case the defective allele was paternally inherited. Although patients with POH lack other features of AHO or PHP, maternal transmission of the defective *GNAS* allele leads to the complete PHP 1a phenotype in affected children.

Pseudohypoparathyroidism type 2

In subjects with PHP type 2, PTH resistance is characterized by a reduced phosphaturic response to administration of PTH, despite a normal increase in urinary cAMP excretion. These observations suggest that the PTH receptor–adenylyl cyclase complex functions normally to increase nephrogenous cAMP in response to PTH, but that intracellular cAMP is unable to act upon downstream targets such as the sodium phosphate transporter. PHP type 2 lacks a clear genetic or familial basis, and a similar clinical and biochemical picture occurs in patients with severe deficiency of vitamin D. Taken together, it is likely that most, if not all, cases of PHP type 2 are actually examples of unsuspected vitamin D deficiency.

Transient pseudohypoparathyroidism of the newborn

Although most newborns with late onset hypocalcemia (i.e., onset of hypocalcemia after days 5–7 of life) have low levels of PTH, about 25% of affected babies will have elevated levels of PTH. Twitchiness or seizure is most often the initial clinical sign of hypocalcemia. Hypocalcemia is associated with hyperphosphatemia, which is due to a high transport maximum of the phosphate/glomerular filtration rate despite elevated PTH levels. Serum levels of magnesium and vitamin D metabolites are typically normal. Intravenous administration of PTH(1-34) produces normal responses of plasma and/or urine cyclic AMP, but the phosphaturic response to the PTH infusion is typically impaired. Affected newborns respond to treatment with calcium and/or 1α hydroxylated metabolites of vitamin D. The condition appears to be transient, with normal serum levels of calcium, phosphorus, and PTH levels maintained without treatment by age 6 months. These features are suggestive of delayed maturation of the post-cyclic AMP signaling pathway in the proximal renal tubule.

Circulating inhibitors as a cause of PTH resistance

Past studies had noted an apparent dissociation between circulating levels of immunoreactive and bioactive PTH in patients with PHP type 1, and plasma from many of these patients diminished the biological activity of exogenous PTH in *in vitro* cytochemical bioassays. Although the basis for this inhibitory effect is unknown, one potential explanation is the accumulation of N-terminally truncated PTH fragments, such as hPTH(7-84), that can inhibit the calcemic and phosphaturic actions of hPTH(1-

34) or hPTH(1-84). Circulating levels of PTH(7-84) immunoreactivity are elevated in patients with PHP 1a and 1b, and the proportion of PTH(7-84)-like fragments to biologically active PTH(1-84) is increased. Calcitriol treatment of patients with PHP type 1 increases the phosphaturic response to PTH (1-34), and one mechanism to explain this observation may be suppression of PTH (7-84)-like fragment levels. Although it is conceivable that circulating hPTH(7-84)-like fragments contribute to PTH resistance in some patients with PHP, it is likely that these circulating antagonists arise as a consequence of sustained secondary hyperparathyroidism and do not have a significant role in the primary pathophysiology of the disorder.

Magnesium deficiency

Magnesium deficiency can impair parathyroid secretion of PTH as well as PTH action. In either case, treatment with magnesium will restore parathyroid function and/or PTH responsiveness.

Syndromes of deficient synthesis or secretion of PTH

Developmental disorders

Genetic forms of HP may derive from dysgenesis of the parathyroid glands. The parathyroid glands (PGs) derive from the pharyngeal pouches, which are transient structures that differentiate from the foregut endoderm during embryonic development. The PGs are evolutionarily homologous to pharyngeal gill slits in fish, and development of both of these organs is critically dependent upon expression of an evolutionarily conserved hierarchy of gene expression. The relevant genes for PG development include the transcription factors Hoxa3, Pax9, Eya1, GCM2, and Tbx1. In humans the PGs first appear during the fifth week of gestation; the superior glands derive from the fourth pharyngeal pouches and the thymus and inferior PGs derive from the third pharyngeal pouches.

DiGeorge sequence/CATCH-22

The DiGeorge sequence (DGS) results from dysembryogenesis of the third and fourth pharyngeal pouches, and is associated with hypoplasia of the thymus and parathyroid glands. The term "DiGeorge sequence" or "anomaly" is more appropriate than "syndrome" as the constellation of defects does not result from a single cause, but rather the failure of an embryological field to develop normally. These patients also often manifest conotruncal cardiac abnormalities, cleft palate, and dysmorphic facies. Hypoparathyroidism is present in up to 60% of patients with DGS. DGS is the leading cause of persistent hypocalcemia of the newborn, but hypoparathyroidism may resolve during childhood. Thymic defects are associated with impaired T-cell mediated immunity and frequent infections.

Molecular mapping has attributed most (70–80%) cases of DGS to hemizygous microdeletions within a critical 250-kb region of 22q11.21-q11.23. Although many genes are located within this region, the presence in some patients with DGS of point mutations that inactivate the *TBX1* gene suggests that this may be the critically important gene. *TBX1* encodes a T-box transcription factor that is widely expressed in non-neural crest cells, cranial mesenchyme, and pharyngeal pouches, and its distribution certainly corresponds to the clinical phenotype of DGS. DGS most commonly arises from *de novo* mutations, but autosomal dominant inheritance can occur.

Microdeletions of 22q11 are the most common cause of continuous gene deletion syndromes in humans, and are present in approximately 1:3,000 newborns. In addition to DGS, deletions within 22q11 can cause the conotruncal anomaly face syndrome and the velocardiofacial syndrome (VCFS). VCFS is typically diagnosed later in childhood, and hypocalcemia has been found to be present in up to 20% of cases. Because of the phenotypic variability of the various overlapping syndromes, these conditions are all included within the acronym "CATCH-22," representing a syndrome of **C**ardiac abnormality, **A**bnormal facies, **T**hymic hypoplasia, **C**left palate, and **H**ypocalcemia with deletion or chromosome **22**q11.

DGS has also been reported to arise in patients with deletions of 10p13, 17p13, and 18q21. Gestational diabetes, as well as exposure to alcohol and other toxins (e.g., retinoids) in the intrauterine stage, can also cause similar phenotypic syndromes.

Hypoparathyroidism, sensorineural deafness, and renal dysplasia syndrome

Deletions within two nonoverlapping regions of 10p have been found to contribute to a phenotype similar to DGS, namely the hypoparathyroidism, sensorineural deafness, and renal dysplasia syndrome (HDR, MIM146255). Unlike DGS/CATCH-22, individuals with HDR do not exhibit cardiac, palatal, or immunologic abnormalities. However, growth hormone insufficiency has been found to be associated with HDR syndrome and may contribute to the typical short stature. The HDR disorder is due to haploinsufficiency of the GATA binding protein-3 (*GATA3*) gene, which is located within a 200-kb critical HDR deletion region on 10p14-10pter and encodes a carboxy-terminal zinc-finger protein essential for DNA binding. It is expressed in the developing vertebrate kidney, otic vesicle, and parathyroids, as well as the central nervous system and organs of T-cell development. A novel missense *GATA3* mutation, Thr272Ile, has been recently identified and shown to result in reduced DNA binding, a partial loss of cofactor FOG2 interaction, and a decrease in gene transcription. The important role of GATA3 in parathyroid function was further highlighted by demonstration that Gata3+/- mice had decreased expression of the parathyroid-specific transcription factor Gcm2 (see below) and developed mild hypoparathyroidism.

Hypoparathyroidism-retardation-dysmorphism and Kenny-Caffey Syndromes

The hypoparathyroidism-retardation-dysmorphism syndrome (HRD, MIM241410), also known as the Sanjad-Sakati syndrome, is a rare form of autosomal recessive hypoparathyroidism associated with other developmental anomalies. In addition to parathyroid dysgenesis, affected patients have severe growth and mental retardation, microcephaly, microphthalmia, small hands and feet, and abnormal teeth. This disorder is found almost exclusively in individuals of Arab descent. The Kenny-Caffey (KS, MIM244460) Syndrome is an allelic disorder that is characterized by hypoparathyroidism, dwarfism, medullary stenosis of the long bones, and eye abnormalities. Both disorders are due to mutations in the tubulin-specific chaperone E (*TBCE*) gene on chromosome 1q42-43, which encodes a chaperone protein required for folding of α-tubulin and its heterodimerization with β-tubulin, although a second gene locus for this disorder is also probable (KS2).

Isolated hypoparathyroidism

Hypoparathyroidism may also occur as an isolated condition with either an X-linked or autosomal patterns of inheritance. The leading cause of autosomal recessive isolated hypoparathyroidism is inactivation of the *GCM2* (*GCMB*) gene at 6p23-24 (MIM146200). *GCM2* encodes a member of a small family of unique transcription factors that were originally identified in Drosophilia that lacked glial cells, hence the name *gcm* for "glial cells missing." Two mammalian homologs exist: *GCM1*, which is principally expressed in the placenta and controls placental branching and vasculogenesis, and *GCM2*, which is expressed predominantly, if not exclusively, in the developing and mature parathyroid gland. Recessive inactivating mutations or dominant inhibitor mutations of *GCM2* cause most cases of isolated hypoparathyroidism in newborns. Heterozygous mutation of *GCM2* does not appear to cause any disturbance in parathyroid development or function.

Isolated hypoparathyroidism can also be inherited as an X-linked recessive trait (MIM307700). Affected males present with infantile hypocalcemic seizures while hemizygous females are unaffected. Autopsy of an affected individual revealed complete agenesis of the parathyroid glands as the cause of hypoparathyroidism. Linkage analysis has localized the underlying mutation to a 1.5-Mb region on Xq26-q27, and recent molecular studies have identified a deletion–insertion involving chromosomes Xq27 and 2p25 as the basis for the defect. These findings have also suggested that a gene known as Sry-box 3 (*SOX3*), within a 906-kb cross-linked recessive HPT critical region, might play a role in the embryonic development of the parathyroid glands.

Parathyroid gland destruction
Surgery

The most common cause of hypoparathyroidism in adults is surgical excision of or damage to the parathyroid glands as a result of total thyroidectomy for thyroid cancer, radical neck dissection for other cancers, or repeated operations for primary hyperparathyroidism. Prolonged hypocalcemia, which may develop immediately or weeks to years after neck surgery, suggests permanent hypoparathyroidism. Postoperative hypoparathyroidism occurs in approximately 1% of thyroid and parathyroid procedures. In patients with a higher risk of developing permanent hypoparathyroidism, parathyroid tissue may be autotransplanted into the brachioradialis or sternocleidomastoid muscle at the time of parathyroidectomy or cryopreserved for subsequent transplantation as necessary. However, transplantation of parathyroid tissue has not advanced to the point where it is a reliable therapeutic option.

Radiation therapy and infiltration

Rarely, hypoparathyroidism has also been described in a small number of patients who receive extensive radiation to the neck and mediastinum. It is also reported in metal overload diseases such as hemochromatosis and Wilson's disease, and in neoplastic or granulomatous infiltration of the parathyroid glands. Hypoparathyroidism may be present in as many as 14% of patients with thalassemia who develop iron overload due to frequent blood transfusion. Hypoparathyroidism has also been observed in association with HIV disease.

Autoimmune polyendocrinopathy-candidiasis-esctodermal dystrophy (APECED)

Autoimmune destruction of the parathyroids occurs most commonly in association with the complex of immune-mediated disorders that comprises the APECED (autoimmune polyendocrinopathy-candidiasis-esctodermal dystrophy, MIM 240300) syndrome, also known as APS I (autoimmune polyglandular syndrome type I). The genetic etiology has been traced to mutations of the autoimmune regulator (*AIRE*) gene on chromosome 21q22.3, which encodes a unique protein with characteristics of a transcription factor. APS I can be either sporadic or autosomal recessive, and has been associated with more than 40 different mutations of the *AIRE* gene. Although the disorder occurs worldwide, it is most prevalent in Finns, Sardinians, and Iranian Jews, and common gene mutations in some populations suggest possible founder effects.

The syndrome's classic triad constitutes the "HAM" complex of hypoparathyroidism, adrenal insufficiency, and mucocutaneous candidiasis. The immune defect may be associated with cytotoxic antibodies that damage or destroy the parathyroid glands. Recently, antibodies that react with the NACHT leucine-rich-repeat protein

5 (NALP5), which is expressed predominantly in the cytoplasm of parathyroid chief cells, have been identified in approximately 50% of APS I patients with HP. An alternative pathophysiological mechanism has been proposed that is based on the presence of circulating antibodies bind and activate the calcium-sensing receptor (CaSR), thereby reducing PTH secretion from parathyroid cells. The temporal progression of the HAM complex is quite predictable, with the appearance of mucocutaneous candidiasis and hypoparathyroidism in the first decade of life, followed by primary adrenal insufficiency before 15 years of age. The candidiasis may affect the skin, nails, and mucous membranes of the mouth and vagina, and is often resistant to treatment. Addison's disease can mask the presence of hypoparathyroidism, or may manifest only after improvement of the hypoparathyroidism, with a reduced requirement for calcium and vitamin D. By diminishing gastrointestinal absorption of calcium and increasing renal calcium excretion, glucocorticoid therapy for the adrenal insufficiency may exacerbate the hypocalcemia and could cause complications if introduced before the hypoparathyroidism is recognized.

Some patients do not manifest all three primary elements of the HAM complex, while other individuals may develop additional endocrinopathies such as hypogonadism, insulin-dependent diabetes, hypothyroidism, and hypophysitis. Non-endocrine components of the disorder that occur frequently include malabsorption, pernicious anemia, vitiligo, alopecia, nail and dental dystrophy, autoimmune hepatitis, and biliary cirrhosis.

Reduced synthesis or secretion of PTH

Autosomal dominant hypocalcemic hypercalciuria (ADHH)

Autosomal dominant hypocalcemia (also autosomal dominant hyopcalcemic hypercalciuria, MIM 146200) most commonly occurs as a result of an activating mutation of the *CASR* gene (3q13.3-21) encoding the CaSR. Most mutations have been identified in the transmembrane and extracellular domains of the receptor and lower the set point for extracellular calcium sensing. The CaSR is widely expressed, but activating mutations have their most profound effects on calcium-induced signaling in the parathyroid gland and the kidney. The effect of the activating mutation on the parathyroid cell is to reduce PTH secretion and thereby produce a state of functional hypoparathyroidism. In the tubule cells of the thick ascending limb of the loop of Henle, activated CaSRs stimulate calciuresis and increase the fractional excretion of calcium (FeCa), thus producing relative (or absolute) hypercalciuria relative to the filtered load of calcium. Nephrocalcinosis and nephrolithiasis are common complications of vitamin D therapy. Although in most cases the degree of hypocalcemia and hypercalciuria are mild and well tolerated, in some patients severe hypocalcemia occurs. Some patients can develop a Bartter syndrome type picture (i.e., Type 5 Bartter syndrome).

PTH gene mutations

Rare mutations in the *PTH* gene (11p15.3-p15.1) have been associated with impaired synthesis and secretion of PTH. The human *PTH* gene contains three exons that encode the preproPTH hormone. Isolated hypoparathyroidism has been found in a family with a single base substitution in exon 2 of the *PTH* gene, apparently impeding conversion of preproPTH to proPTH. In another family with autosomal recessive isolated hypoparathyroidism, the entire exon 2 of the *PTH* gene was deleted, preventing generation of a mature secretory peptide. A missense mutation in exon 2 of the preproPTH gene has also been found, potentially impairing proteolytic cleavage of the preprohormone.

Anti-CaSR antibodies

Autoimmune hypoparathyroidism had been previously thought to be caused by the binding of cytotoxic autoantibodies to parathyroid cells. However, many patients with late onset primary hypoparathyroidism have circulating antibodies that activate the CaSR, and which do not produce irreversible destruction of the parathyroid glands. It appears that there might be a specific autoimmune reaction against the CaSR on parathyroid cells, although detection of antibodies against the receptor appears to be influenced by the assay system used. Recently, it was reported that 2 out of 53 patients with idiopathic hypoparathyroidism had remission after 1 year off calcium supplementation, supporting the idea of a possible temporary alteration in CaSR function as an etiology of hypoparathyroidism.

Other causes of hypoparathyroidism

Mitochondrial disease

Several syndromes due to deletions in mitochondrial DNA have been associated with hypoparathyroidism. These include the Kearns-Sayre syndrome (encephalomyopathy, ophthalmoplegia, retinitis pigmentosa, heart block), the Pearson Marrow-Pancreas syndrome (sideroblastic anemia, neutropenia, thrombocytopenia, pancreatic dysfunction), and the maternally inherited diabetes and deafness syndrome. Hypoparathyroidism has also been described in the MELAS (**M**itochondrial myopathy, **E**ncephalopathy, **L**actic Acidosis, and **S**troke-like episodes) syndrome, due to point mutations in mitochondrial tRNA. Because renal magnesium wasting is frequently seen in these conditions, a readily reversible form of hypoparathyroidism caused by hypomagnesemia should also be considered. In addition, mutations in the mitochondrial trifunctional protein (MTP), resulting in long-chain 3-hydroxy-acyl-coenzyme A dehydrogenase (LCHAD) deficiency, or combined MTP deficiency, have been associated with hypoparathyroidism in a few unrelated patients. This condition manifests as nonketotic hypoglycemia, cardiomyopathy, hepatic dysfunction, and developmental delay, and is associated with maternal fatty liver of pregnancy.

Post-burn

In individuals who have sustained severe burns, there is evidence for upregulation of the CaSR. Lower than normal concentrations of serum calcium suppress PTH secretion, thus leading to hypocalcemia and hypoparathyroidism.

Maternal hyperparathyroidism and hypomagnesemia

An infant who is exposed *in utero* to maternal primary hyperparathyroidism or hypercalcemia can have suppressed parathyroid function and hypocalcemia during the first few weeks of life, but the duration may be up to 1 year of age. Although therapy may be required acutely, the disorder is usually self-limited. Hypomagnesemia caused by defective intestinal absorption or renal tubular reabsorption of magnesium may impair secretion of PTH and cause hypoparathyroidism. Magnesium replacement will correct the hypoparathyroidism.

DIAGNOSTIC ALGORITHM

PHP should be considered in any patient with functional hypoparathyroidism (i.e., hypocalcemia and hyperphosphatemia) and an elevated plasma concentration of PTH. Hypomagnesemia and severe vitamin D deficiency can produce biochemical features of PTH resistance in some patients, and thus plasma concentrations of magnesium and 25(OH)D must be measured. Unusual initial manifestations of PHP include neonatal hypothyroidism, unexplained cardiac failure, seizures, intracerebral calcification of basal ganglia and frontal lobes, dyskinesia and other movement disorders, and spinal cord compression.

PHP or pseudoPHP may be suspected in patients who present with somatic features of AHO. However, several aspects of AHO, such as obesity, round face, brachydactyly, and mental retardation also occur in other congenital disorders (e.g., Prader-Willi syndrome, acrodysostosis, Ullrich-Turner syndrome). An interesting phenocopy of AHO occurs in subjects who have small terminal deletions of chromosome 2q37 [del(2)(q37.3)]. These patients have normal endocrine function and normal $G\alpha_s$ activity.

The classical tests for PHP, the Ellsworth-Howard test, and later modifications by Chase, Melson, and Aurbach, involved the administration of 200–300 USP units of purified bovine PTH or parathyroid extract. Although these preparations are no longer available, a form of teriparatide, the 1-34 sequence of human PTH, is available and several protocols for its use in the differential diagnosis of hypoparathyroidism have been developed. These protocols are based on intravenous infusion of the peptide, but similar results may be obtained following subcutaneous injection. The patient should be fasting except for fluids (250 ml of water hourly from 6 a.m. to noon). Two control urine specimens are collected before 9 a.m. Teripartide (0.625 µg/kg body weight to a maximum of 25 µg for intravenous use and 40 µg for subcutaneous use) is administered at 9 a.m. either by subcutaneous injection or intravenous infusion over 15 min, and experimental urine specimens are collected from 9:00 to 9:30, 9:30 to 10:00, 10:00 to 11:00, and 11:00 to 12:00. Blood samples should be obtained at 9 a.m. and 11 a.m. for measurement of serum creatinine and phosphorus concentrations. Urine samples are analyzed for cAMP, phosphorus, and creatinine concentrations, and results are expressed as nanomoles of cAMP per 100 ml GF and TmP/GFR.

Normal subjects usually show a 10- to 20-fold increase in urinary cAMP excretion and a 20–30% decrease in TmP/GFR, whereas patients with PHP type 1 (both 1a and 1b), regardless of their serum calcium concentration, will show markedly blunted responses (Fig. 72.2). Normal children and patients with HP have more robust responses. Thus, this test can distinguish patients with so-called normocalcemic PHP (i.e., patients with PTH resistance who are able to maintain normal serum calcium levels without treatment) from subjects with pseudoPHP (who will have a normal urinary cAMP response to PTH). Measurement of plasma cAMP or plasma 1,25-dihydroxyvitamin D after infusion of hPTH(1-34) may also differentiate PHP type 1 from other causes of hypoparathyroidism.

The diagnosis of PHP type 2 requires exclusion of magnesium depletion or vitamin D deficiency. Documentation of elevated serum PTH and nephrogenous cAMP is a prerequisite for a definitive diagnosis of PHP type 2. These subjects have a normal urinary cAMP response to infusion of PTH but characteristically fail to show a phosphaturic response. Unfortunately, interpretation of the phosphaturic response to PTH is often complicated by random variations in phosphate clearance, and it is sometimes not possible to classify a phosphaturic response as normal or subnormal regardless of the criteria employed.

Genetic testing can assist with the diagnosis of HP and PHP. Mutational analysis of the *CASR*, *AIRE*, *GCM2*, and *GNAS* genes is available from several approved clinical laboratories. By contrast, genetic testing for PHP 1b and most other forms of genetic hypoparathyroidism is still considered investigational.

THERAPY

Calcium and vitamin D

The goal of therapy in hypoparathyroidism is to restore serum calcium as close to normal as possible. The main pharmacologic agents available are supplemental calcium and vitamin D preparations. Phosphate binders, to lower serum phosphate, and thiazide diuretics, to decrease urinary calcium excretion, may be useful ancillary agents. The major limitation to restoration of normocalcemia is the development of hypercalciuria, with a resulting risk for nephrolithiasis. With the loss of the renal calcium-conserving effect of PTH, the enhanced calcium absorption of the gut induced by vitamin D therapy results in an increased filtered load of calcium that is readily cleared through the kidney. Consequently, urinary calcium excretion frequently increases in response to vitamin D supplementation well before serum calcium is normalized. It is thus often advisable to maintain a low normal serum calcium concentration, so as to prevent chronic hypercalciuria. Avoidance of hypercalciuria is probably most important for patients with hypercalciuric hypocalcemia caused by activating mutations of the *CaSR* gene. Fortunately, patients with PHP type 1 rarely develop hypercalciuria as long as PTH levels remain unsuppressed as the distal tubule remains responsive.

Approximately 1–2 grams (25–50 mg/kg) of elemental calcium per day are recommended to supply adequate calcium and to manage dietary phosphorus intake, and optimal results are achieved when calcium supplements are taken with meals. Patients are unable to convert parent compounds of vitamin D to fully active forms as the lack of PTH and hyperphosphatemia both inhibit the renal 1-α hydroxylase enzyme that converts 25-hydroxyvitamin D to 1,25-dihydoxyvitamin D. The 1-α hydroxylated vitamin D metabolites (e.g., calcitriol) are the preferred forms of vitamin D as these drugs circumvent the enzymatic block in vitamin D activation. Because calcitriol has a plasma half-life of only hours and body stores do not accumulate, calcitriol must be given several times each day. By contrast, alfacalcidol (1α (OH)D$_3$) has a longer half-life and may be given once daily. A variety of parent vitamin D preparations have been used in the past, including vitamin D$_3$ or D$_2$; however, typically very large concentrations are required to raise the serum calcium. In this case, body stores of vitamin D can accumulate in massive amounts, increasing the risk of severe and prolonged vitamin D toxicity. Hydrochlorothiazide therapy has been effective in reducing the vitamin D requirement, but potassium supplementation is necessary to offset the thiazide-induced hypokalemia.

PTH treatment

Treatment of HP with PTH is an attractive option because it represents replacement with the hormone that is missing. Moreover, reducing calcium and calcitriol intake in HP offers the potential advantage of reducing urinary calcium excretion. An additional possible advantage is that because of its phosphaturic properties, PTH use may reduce the risk of soft tissue deposition of calcium in the kidneys (nephrocalcinosis, nephrolithiasis) and possibly in other soft tissues.

Replacement therapy using teriparatide in adults and children with HP has demonstrated that twice-daily administration could maintain serum calcium concentrations in the low-normal or just below normal range for prolonged periods of time. Markers of bone turnover are elevated in patients taking teriparatide. Bone density as measured by dual-energy X-ray absorptiometry (DXA) remains stable in adults, but, Z-scores at the 1/3 radius site are significantly decreased in children treated with teriparatide compared to treatment with calcitriol.

Treatment of adult patients with HP with PTH(1-84) every other day has shown a significant reduction in requirements for supplemental calcium and calcitriol. The every-other-day regimen produced significant increases in bone density at the lumbar spine and modest decreases at the distal 1/3 radius. PTH(1-84) treatment was associated with increases in markers of bone turnover, and bone biopsy showed structural changes that are consistent with an increased remodeling rate in both trabecular and cortical compartments, with tunneling resorption in the former and increased porosity in the latter.

SUGGESTED READING

1. Ding C, Buckingham B, Levine MA. 2001. Familial isolated hypoparathyroidism caused by a mutation in the gene for the transcription factor GCMB. *J Clin Invest* 108: 1215–20.
2. Kelly A, Levine MA. 2009. Disorders of calcium, phosphate, parathyroid hormone and vitamin D. In: Kappy, MS, Allen, DB, and Geffner, ME (eds.) *Pediatric Practice: Endocrinology*. Springfield: Charles C. Thomas Publisher, LTD. pp. 191–256.
3. Mantovani, G. 2011. Pseudohypoparathyroidism: Diagnosis and treatment. *J Clin Endocrinol Metab* 96: 3020–30.
4. Rubin, MR, Dempster DW, Sliney J Jr, Zhou H, Nickolas TL, Stein EM, Dworakowski E, Dellabadia M, Ives R, McMahon DJ, et al. 2011. PTH(1-84) administration reverses abnormal bone-remodeling dynamics and structure in hypoparathyroidism. *J Bone Miner Res* 26: 2727–36.
5. Sikjaer T, Rejnmark L, Mosekilde L. 2011. PTH treatment in hypoparathyroidism. *Curr Drug Saf* 6: 89–99.
6. Winer KK, Sinaii N, Reynolds J, Peterson D, Dowdy K, Cutler GB Jr. 2010. Long-term treatment of 12 children with chronic hypoparathyroidism: A randomized trial comparing synthetic human parathyroid hormone 1-34 versus calcitriol and calcium. *J Clin Endocrinol Metab* 95: 2680–88.
7. Shoback D. 2008. Clinical practice. Hypoparathyroidism. *N Engl J Med* 359(4): 391–403.
8. Bilezikian JP, Khan A, Potts JT Jr, Brandi ML, Clarke BL, Shoback D, Juppner H, D'Amour P, Fox J, Rejnmark L, Mosekilde L, Rubin MR, Dempster D, Gafni R, Collins MT, Sliney J, Sanders J. 2011. Hypoparathyroidism in the adult: Epidemiology, diagnosis, pathophysiology, target-organ involvement, treatment, and challenges for future research. *J Bone Miner Res* 26(10): 2317–37.

73

Pseudohypoparathyroidism

Harald Jüppner and Murat Bastepe

PHP-Ia, PHP-Ic, PPHP, and POH: PHP Variants That Are Caused by Coding Mutations in Gsα 591	PHP-Ib: Hormone-Resistance Caused by Abnormal Gsα Imprinting 594
Gsα Transcripts Are Derived in Some Tissues Only from the Maternal Allele 592	Treatment 595
	Summary and Conclusions 596
Multiple Maternal or Paternal Transcripts Are Derived from the GNAS Locus 594	References 596

The term "pseudohypoparathyroidism (PHP)" was first used in 1942 to describe a disorder of hormone resistance in which the most prominent defect involves target organ resistance to the actions of parathyroid hormone (PTH) thus leading to hypocalcemia and hyperphosphatemia [1], which is combined with reduced serum concentrations of 1,25-dihydroxyvitamin D_3 [2, 3]. As an indication of PTH-resistance, rather than PTH-deficiency as in hypoparathyroidism, serum PTH levels are elevated and administration of exogenous biologically active PTH fails to result in an appropriate increase in urinary phosphate and cyclic adenosine monophosphate (cAMP) excretion [1, 4]. PTH-resistance occurs in the proximal renal tubule, whereas no resistance appears to exist in other PTH target tissues, such as bone [5, 6] and the thick ascending tubule [7]. It is probably due to these non-impaired PTH functions that patients can sometimes maintain normocalcemia without treatment for prolonged periods of time. However, clinical manifestations of hypocalcemia, such as increased neuromuscular excitability or seizures, usually develop unless patients are appropriately treated with oral calcium supplements and 1,25 dihydroxyvitamin D. Treatment is also recommended for asymptomatic patients with normal calcium and phosphate levels if serum PTH levels are elevated, since long-term secondary hyperparathyroidism can result in severe hyperparathyroid bone disease [8].

Since the first description of PHP by Albright and colleagues [1], different variants of this disorder have been defined (Table 73.1). Patients with PHP type I (PHP-I) exhibit both impaired nephrogenous cAMP generation and impaired phosphate excretion following exogenous PTH administration [1, 4], while patients with PHP type II (PHP-II) exhibit a dissociation between these two responses, i.e., these patients have normal nephrogenous cAMP generation but impaired phosphate excretion [9]. Recently, it was shown that acrodysostosis, which represents a form of PHP-II, is caused by a mutation in the regulatory subunit of protein kinase A (PRKAR1A) downstream of Gsα because of increased basal and PTH-stimulated urinary cAMP excretion and the lack of a PTH-induced phosphaturic response. Consistent with these molecular findings, the impaired response to PTH was associated with resistance toward multiple other hormones that mediate their actions through several G protein-coupled receptors. These findings further highlight the importance of cAMP/PKA-mediated actions of PTH and other hormones that show resistance in different PHP variants [10]. However, this chapter will largely focus on the clinical and molecular definition of PHP-I and the subtypes thereof.

Primer on the Metabolic Bone Diseases and Disorders of Mineral Metabolism, Eighth Edition. Edited by Clifford J. Rosen.
© 2013 American Society for Bone and Mineral Research. Published 2013 by John Wiley & Sons, Inc.

Table 73.1. Clinical and Molecular Features of Patients with the Different PHP-I Forms

	PTH-Resistance	Additional Hormone Resistance	Typical AHO Features	GNAS Defects
PHP-Ia/c	Yes	Yes	Yes	Gsα mutations
PPHP	No	No	Yes	Gsα mutations
POH	No	No	No	Gsα mutations
PHP-Ib	Yes	Some cases	No	STX16 or NESP55 deletions affecting GNAS imprinting; patUPD20q; sporadic cases
Acrodysostosis (PHP-II variant)	Yes	Yes	No	PRKAR1A

PHP-Ia, PHP-Ic, PPHP, AND POH: PHP VARIANTS THAT ARE CAUSED BY CODING MUTATIONS IN Gsα

Patients with PHP-Ia present with resistance to PTH and Albright's hereditary osteodystrophy (AHO), a constellation of physical features, which may include obesity, short stature, ectopic ossifications, brachydactyly, and/or mental retardation. PHP-Ia patients often also show resistance to other hormones, including thyroid-stimulating hormone (TSH), gonadotropins, calcitonin, and growth hormone-releasing hormone (GHRH) [11–16]. A common feature of these hormones is that their actions require cell surface receptors that couple to the stimulatory G protein. Accordingly, PHP-Ia is caused by heterozygous inactivating mutations located in GNAS, the gene encoding the α-subunit of the stimulatory G protein (Gsα) (Fig. 73.1); these mutations are all located on the maternal allele.

Gsα is a ubiquitous signaling protein required for agonist activated stimulation of adenylyl cyclase, which in turn generates cyclic AMP (cAMP), an intracellular second messenger involved in numerous cellular responses throughout the body [17]. Illustrating the importance of the Gsα-mediated cellular responses, homozygous ablation of Gsα leads to embryonic lethality in mice [18–20]. Numerous different GNAS mutations, including missense and nonsense amino acid changes, as well as insertions and deletions, have been identified in patients with PHP-Ia (see Online Mendelian Inheritance in Man #103580 at http://www.ncbi.nlm.nih.gov/ for a list of allelic variants). Consistent with their loss-of-function effects, these mutations are located in nearly all of the 13 exons that encode Gsα and lead to an approximately 50% reduction in Gsα mRNA/protein, which can be demonstrated in skin fibroblasts and erythrocyte membranes derived from PHP-Ia patients [12].

PHP-Ic is a variant originally used to describe patients who present with clinical findings that are identical to those found in patients with PHP-Ia but demonstrate normal Gsα bioactivity in red blood cells [21]. However, recent studies suggest that a substantial portion of PHP-Ic cases are allelic variants of PHP-Ia [22–24]. Several PHP-Ic patients have been reported to carry mutations that render Gsα seemingly functional in routinely used Gsα bioactivity tests in which the stimulation is done by guanosine triphosphate (GTP). The mutations disrupt the receptor-coupling ability of Gsα but do not impair its basal activity, thus allowing a full response to GTP stimulation. However, it remains possible that some PHP-Ic patients carry mutations in a different gene, which, when mutated, prevents elevation of cAMP. This gene could perhaps encode one of the cAMP phosphodiesterases or adenylyl cyclases.

Inactivating Gsα mutations are also found in patients who display AHO features but lack hormone resistance. This disorder has been named pseudopseudohypoparathyroidism (PPHP) [25]. Patients with PHP-Ia and those with PPHP are typically found in the same kindreds but never within the same sibship. While AHO occurs in the offspring irrespective of the gender of the parent transmitting the Gsα mutation, development of hormone resistance is subject to imprinting such that it develops only in the offspring of female obligate carriers [26, 27]. Thus, paternal inheritance of a Gsα mutation results in PPHP (AHO only), whereas maternal inheritance of the same mutation results in PHP-Ia (AHO and hormone resistance). Consistent with the imprinted mode of inheritance observed for the hormone resistance, recent investigations of the GNAS locus have revealed predominantly maternal expression of Gsα in some, but not all, tissues, including renal proximal tubules, thyroid, pituitary, and gonads (see below).

Heterozygous inactivating Gsα mutations are also found in patients with progressive osseous heteroplasia (POH), who have severe heterotopic ossifications that affect skeletal muscle and deep connective tissue [28]. Some of these Gsα mutations are identical to those found in PHP-Ia or PPHP patients, indicating that POH is an extreme manifestation of heterotopic ossifications associated with AHO [28–32]. Only few POH patients present with hormone resistance and/or AHO features [29, 33],

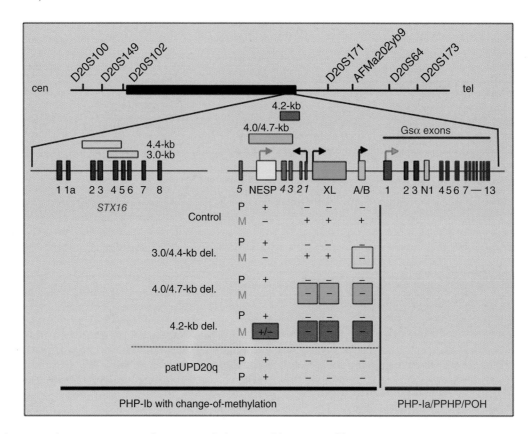

Fig. 73.1. The *GNAS* locus: parent-specific exon methylation and locations of heterozygous mutations. Gsα mutations (dark blue exons) cause PHP-Ia/PPHP/POH (Refs. 96, 97). AD-PHP-Ib is caused by *STX16* deletions located more than 200 kb upstream of *GNAS* (yellow and light blue horizontal bars) or by deletions within *GNAS* (green and purple bars); these disease variants are associated with loss of *GNAS* methylation restricted to exon A/B alone or with methylation changes at multiple exons. Sporadic PHP-Ib can be caused by paternal uniparental isodisomy for chromosome 20q (patUPD20q). Boxes, exons; connecting lines, introns. Paternal (P), maternal (M); methylated (+); unmethylated (−); paternal transcription (black arrows); maternal transcription (orange arrow); green arrow, biallelic Gsα transcription in some tissues.

and consistent with this observation, nearly all cases with POH are caused by paternally inherited mutations. The reasons for this strong parental bias in the development of POH currently remains unknown, although it is possible that deficiency of other products of the *GNAS* gene, which show imprinted expression and are also disrupted by most of these mutations, contribute to the molecular pathology.

Gsα TRANSCRIPTS ARE DERIVED IN SOME TISSUES ONLY FROM THE MATERNAL ALLELE

Inactivating Gsα mutations can lead to various different phenotypes, and genomic imprinting appears to play an important role in the development of these phenotypes. Genomic imprinting refers to differential expression of genes specifically from either maternal or paternal alleles [34–36]. This monoallelic, parental origin-specific expression correlates with allele-specific epigenetic marks within the imprinted gene (often at the promoter region), including methylation of the cytosine residues in CpG dinucleotides. Although the Gsα promoter itself lacks differential methylation [37–39], several studies in genetically manipulated animals have established that Gsα expression is predominantly maternal in a small number of tissues, i.e., Gsα transcription from the paternal *GNAS* allele is silenced [18, 40, 41]. Consistent with these findings in mutant mouse strains, kindreds with PHP-Ia and PPHP have revealed the importance of Gsα silencing in different tissues, particularly the proximal renal tubules where Gsα is predominantly or exclusively derived from the maternal allele. Thus, hormonal resistance develops in the proximal tubule if a Gsα mutation occurs on the maternal allele, while distal tubular functions, such as PTH-dependent calcium reabsorption, remain intact because of the biallelic expression of Gsα in this portion of the renal tubules. Note that such a mutation would reduce the level of Gsα by 50% in the distal tubule, but this reduction appears to be compatible with complete function (Fig. 73.2).

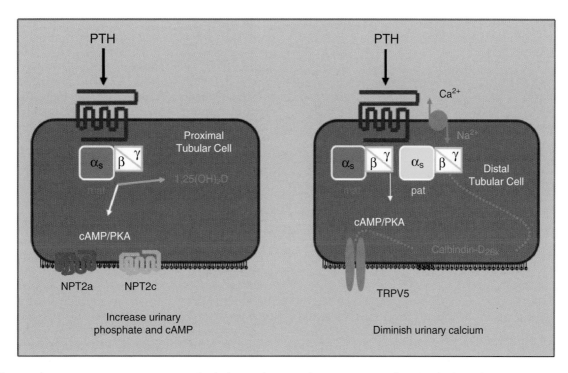

Fig. 73.2. PTH/PTHrP receptor activation in the kidney. The stimulatory G protein (Gsα), which mediates most PTH actions, is transcribed in the proximal renal tubules predominantly from the maternal allele; Gsα is biallelically transcribed in the distal tubules. PTH-stimulated cAMP production in proximal tubules and subsequent PKA activation reduces NPT2a and NPT2c (Refs. 98, 99) and increases 1,25(OH)$_2$ vitamin D production (Refs. 100, 101); cAMP is exported through unknown transporters into urine, a unique mechanism for limiting the intracellular level of this second messenger. In the distal tubules, PTH enhances expression of the calcium transporter TRPV5, at least partially through cAMP/PKA-dependent mechanisms, thereby reducing urinary calcium losses (Refs. 10, 102).

Analysis of various fetal and adult human tissues has also revealed imprinting of Gsα in some other tissues. Besides the proximal renal tubules, Gsα expression appears to be predominantly maternal in thyroid gland [42–44], gonads [42], and pituitary [45]. Conversely, Gsα expression has been shown to be biallelic in a number of different fetal tissues [46], as well as adult adrenal gland, bone, and adipose tissue [42, 47]. In addition, one study demonstrated biallelic Gsα expression in human fetal renal cortices [48]. Although apparently contradictory to the findings in mice [18], the latter finding may suggest that Gsα imprinting in this tissue is postnatal and/or exists only in a small number of renal cortical cells. Overall, however, the tissue distribution of imprinted Gsα expression appears to correlate well with the tissue distribution of hormone resistance in patients with PHP-Ia. This is consistent with the prediction that an inactivating Gsα mutation leads to a dramatic decrease in the Gsα level/activity and, thereby, hormone resistance only in tissues in which Gsα expression is predominantly maternal. Conversely, the same mutation causes no detectable change in the Gsα level/activity in the same tissues following paternal inheritance. Tissues in which Gsα is expressed biallelically are predicted to have approximately 50% reduction of Gsα level/activity. While this reduction may be sufficient for maintaining normal cellular responses in some cells, it may lead to defective function in others, i.e., haploinsufficiency. In fact, because AHO features appear to develop independently of the gender of the parent transmitting the Gsα mutation, these features are thought to result from haploinsufficiency of Gsα signaling in various tissues. This conclusion is consistent with the growth plate findings in mice chimeric for wild-type and *Gnas* exon 2-disrupted embryonic stem cells that show only minor differences in the acceleration of hypertrophic differentiation irrespective of whether the genetically manipulated cells were maternally or paternally derived [49]. Similarly, subcutaneous heterotopic ossifications indistinguishable from those observed in AHO occur with disruption of *Gnas* exon 1 on either parental allele [50]. Nonetheless, given the patient-to-patient variability in the expression and severity of individual AHO features, it remains possible that imprinting of Gsα (or other imprinted *GNAS* transcripts) also contributes to the development of certain AHO features. In fact, it has been recently revealed that obesity and cognitive impairment, two classic features of AHO, develop primarily after maternal inheritance of Gαs mutations [51, 52], and this imprinted mode of inheritance correlates well with the monoallelic expression of Gαs in certain brain regions, as demonstrated in mice [53]. The imprinting of *GNAS* could also play a role in the pathogenesis of POH, considering the predominantly paternal inheritance of POH [30–32].

MULTIPLE MATERNAL OR PATERNAL TRANSCRIPTS ARE DERIVED FROM THE *GNAS* LOCUS

Recent studies have revealed that in addition to Gsα, the *GNAS* locus gives rise to multiple coding and non-coding transcripts that show parental origin-specific expression (Fig. 73.1). Gsα is encoded by 13 exons that span about 20 kb [54]. There are four different splice variants of Gsα formed through the alternative use of exon 3, as well as an additional codon inserted alternatively at the 5′ end of exon 4 [55, 56]. In addition, the use of another alternative exon (N1) between exons 3 and 4 leads to a truncated Gsα mRNA and protein [57].

Studies of the genomic region comprising *GNAS* have led to the identification of several novel exons located upstream of exon 1 both in humans and mice. XLαs and NESP55 transcripts individually use separate upstream promoters and first exons that splice onto exons 2-13 of the Gsα transcript [38, 39]. Protein sequence identity over this region, however, exists only between XLαs and Gsα [58], since exons 2-13 in the NESP55 transcript are part of the 3′ untranslated region [59, 60]. XLαs and the neuroendocrine secretory protein-55 (NESP55) are oppositely imprinted; while XLαs is expressed from the paternal *GNAS* allele [38, 39], NESP55, which is a chromogranin-like protein abundant in neuroendocrine tissues, is expressed from the maternal *GNAS* allele [39, 60]. XLαs also shows abundant expression in neuroendocrine tissues and the central nervous system, although its mRNA is detected in multiple other tissues, including adipose tissue, pancreas, and kidney [61–63]. In addition to these protein products, the *GNAS* locus gives rise to a sense and an antisense non-coding transcript, each of which is expressed from the paternal allele. The transcript from the antisense strand (AS) has a promoter that is adjacent to the promoter of the XLαs transcript [64, 65]. The non-coding transcript from the sense strand, termed "A/B" (also referred to as 1A or 1′), comprises exons 2-13, but uses, like XLαs and NESP55 transcripts, a separate promoter and first exon [37, 66, 67]. As in other imprinted genomic loci, the promoters of these additional *GNAS* transcripts, but not the promoter of Gsα, are within differentially methylated regions (DMRs), and the non-methylated promoter drives the expression in each case [37–39, 60, 64, 65].

Currently, the biological significance of the imprinted *GNAS* transcripts is poorly understood. As mentioned above, NESP55 belongs to the family of chromogranins and is associated with constitutive secretory pathway in neurons and neuroendocrine cells [68]. Targeted disruption of the NESP55 protein in mice leads to abnormal reactivity to novel environments [69]. XLαs shares marked amino acid sequence identity with Gsα and comprises most of the domains shown to be functionally important for the latter [58]. Accordingly, XLαs can mediate basal and agonist-induced adenylyl cyclase stimulation [70, 71]. In addition, when targeted to the renal proximal tubule in mice, XLαs can mitigate the PTH-resistance phenotype caused by Gsα deficiency [72]. Targeted disruption of the XLαs protein in mice, however, has thus far failed to reveal phenotypes that are unequivocally caused by a loss of the "Gsα-like" signaling activity of this protein *in vivo* [63]. In fact, current evidence from the XLαs knockout mice suggests that XLαs may oppose Gsα actions in certain tissues, such as brown fat [63]. Mice lacking XLαs fail to adapt to feeding and show high early postnatal mortality. In addition, these mice show various defects in energy and glucose metabolism, including reduced adiposity in both brown and white fat and impaired metabolic responses to hypoglycemia [63]. The subcellular trafficking of XLαs upon activation has been shown to differ markedly from that of Gsα [73], and this difference could perhaps provide some explanation as to why XLαs appears to oppose Gsα actions in certain settings. Furthermore, recent studies have shown that an exon A/B-derived transcript results in a Gsα variant that is amino-terminally truncated and reduces the activity of full-length Gsα [74].

PHP-Ib: HORMONE-RESISTANCE CAUSED BY ABNORMAL Gsα IMPRINTING

PHP-Ib was initially defined as a disorder characterized by selective PTH-resistance, normal Gsα activity in fibroblasts and red blood cells, and lack of AHO features [12]. However, although PTH resistance remains the most prominent biochemical defect, numerous PHP-Ib patients were shown to present also with resistance toward TSH [44, 75, 76]; furthermore, recent findings have revealed usually mild, but distinct AHO features in a significant number of patients with PHP-Ib due to *GNAS* methylation changes [77–81].

Based on epigenetic changes at different *GNAS* exons/promoters and different microdeletion within or upstream of the *GNAS* locus, several different forms of PHP-Ib have been defined (see Figs. 73.1 and 73.3). As in PHP-Ia, the genetic defects in the autosomal dominant forms of PHP-Ib (AD-PHP-Ib) lead to hormone resistance only after maternal transmission. The most frequent genetic mutation leading to AD-PHP-Ib is a 3-kb deletion within the gene encoding syntaxin-16 (*STX16*) located about 220 kb centrometric of the *GNAS* locus [82–85]. This deletion is associated with a loss of methylation at *GNAS* exon A/B and biallelic expression of this transcript, which is thought to reduce Gsα expression in those tissues that normally express Gsα from the maternal allele, thus causing hormonal resistance. In a single AD-PHP-Ib kindred, another heterozygous, maternally inherited *STX16* deletion has been discovered, which is 4.4 kb in size [86]. The latter deletion overlaps with the

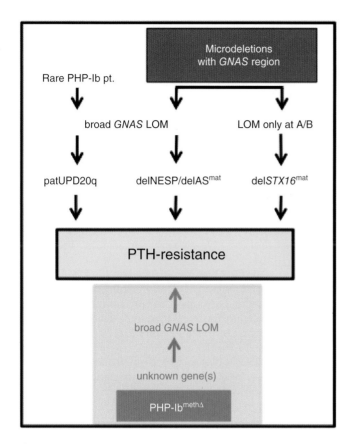

Fig. 73.3. Grouping of PHP-Ib variants according to their epigenetic and genetic findings. Autosomal dominant PHP-Ib (AD-PHP-Ib) can be caused either by *STX16* deletions with associated loss-of-methylation (LOM) at *GNAS* exon A/B alone (del*STX-16*^{mat}) (Refs. 82–86) or deletions involving NESP55 and/or antisense exons 3-4 with associated LOM at *GNAS* exons XL, AS, and A/B (delNESP/delAS^{mat}) (Refs. 88, 89). Sporadic PHP-Ib with complete LOM at *GNAS* exons XL, AS, and A/B, and complete gain-of-methylation at exon NESP55 can be due to uniparental paternal isodisomy involving chromosome 20q (patUPD20q) (Refs. 76, 91). Sporadic PHP-Ib variants with LOM at *GNAS* exons XL, AS, and A/B (often incomplete) and occasionally incomplete gain-of-methylation at exon NESP55 (PHP-Ib^{methΔ}); linkage to *GNAS* has been excluded in some of these sporadic patients and is highly unlikely in others (Refs. 90, 94).

frequently observed 3-kb deletion and removes exons 2-4 of *STX16*. Thus, both the 3-kb and 4.4-kb deletions are predicted to lead to an inactive syntaxin-16 protein on one allele. Nevertheless, given that PHP-Ib develops after maternal inheritance of the genetic defect only, which is consistent with the inheritance mode of the identified mutations in affected individuals, loss of one copy of *STX16* could cause PHP-Ib only if this gene were also imprinted. However, based on methylation and allelic expression analyses using lymphoblastoid cells derived from PHP-Ib patients and normal controls, *STX16* does not appear to be imprinted [86], and therefore it appears unlikely that syntaxin-16 is involved in the molecular pathogenesis of PHP-Ib. In fact, disruption of *Stx16* in mice does not lead to any epigenetic or laboratory abnormalities in mice [87].

In other AD-PHP-Ib forms, deletions within the *GNAS* locus that remove either exon NESP55 and antisense exons 3 and 4 [88], or both antisense exons alone [89] have been identified (see Fig. 73.1). These deletions are associated with loss of all maternal *GNAS* methylation imprints, although the biochemical abnormalities appear to be indistinguishable from those observed in patients with AD-PHP-Ib caused by *STX16* deletions [90]. Demonstrating that defective *GNAS* imprinting alone can lead to PTH-resistance, some sporadic cases of PHP-Ib are caused by paternal uniparental isodisomy of chromosome 20q and, as a consequence, a "paternal-only" imprinting profile throughout the *GNAS* locus [76, 91, 92].

Similar to the PHP-Ib caused by *GNAS* deletions or patUPD20q, most sporadic PHP-Ib cases exhibit methylation defects involving exon A/B and all or some of the other *GNAS* DMRs [75, 80, 88, 90, 93–95]. NESP55 or antisense deletions similar to those found in three AD-PHP-Ib kindreds [88, 89] or patUPD20q have been ruled out in a large number of such sporadic PHP-Ib cases; in fact, in some patients the *GNAS* locus could be excluded genetically [94]. It is therefore conceivable that sporadic PHP-Ib cases carry defects in a gene that is distinct from *GNAS*. The identification of the genetic defects in sporadic PHP-Ib cases may help identify additional genes that are involved in establishing or maintaining *GNAS* methylation or that contribute to Gsα silencing in different tissues.

TREATMENT

In treatment, the goal is to correct abnormal serum biochemistries that result from PTH and, in some cases, other hormone resistance. The distal tubular actions of PTH in PHP patients are not impaired, providing sufficient calcium reabsorption from the glomerular filtrate. This makes clinical management of hypocalcemia in PHP easier than that of hypoparathyroidism. The treatment involves oral calcium supplements and 1,25(OH)$_2$D (calcitriol) preparations. It is important to note that the active form of vitamin D is required in the treatment due to the lowered capacity of the proximal tubule to convert 25(OH)D into the biologically active 1,25(OH)$_2$D. The treatment should aim to keep the serum PTH level within or close to the normal range rather than simply avoiding symptomatic hypocalcemia, because persistent elevation of serum PTH will increase bone resorption and may eventually lead to hyperparathyroid bone disease. During the course of the treatment, urinary calcium typically does not rise owing to intact PTH actions in the distal tubule. Nevertheless, annual monitoring of blood chemistries and urinary calcium excretion is necessary in patients undergoing treatment. More frequent analyses are recommended during pubertal development and once skeletal growth is completed, as the dosing of calcium supplements and 1,25(OH)$_2$D preparations may need to be adjusted.

SUMMARY AND CONCLUSIONS

Maternally inherited mutations involving the *GNAS* exons that encode Gsα are responsible for PHP-Ia, while these mutations, when inherited paternally, lead to PPHP or POH. AD-PHP-Ib is caused by maternally inherited microdeletions within or upstream of *GNAS*, and these are associated with loss of some or all maternal *GNAS* methylation imprints; thus, impaired Gsα expression in those tissues where paternal Gsα expression is silenced. Indistinguishable *GNAS* methylation changes are also observed in PHP-Ib due to patUPD20q, while the cause of sporadic PHP-Ib and the associated *GNAS* methylation changes remains to be determined.

REFERENCES

1. Albright, F, Burnett, CH, Smith, PH, Parson W. 1942. Pseudohypoparathyroidism—An example of "Seabright-Bantam syndrome." *Endocrinology* 30: 922–932.
2. Breslau NA, Weinstock RS. 1988. Regulation of 1,25(OH)$_2$D synthesis in hypoparathyroidism and pseudohypoparathyroidism. *Am J Physiol* 255: E730–E736.
3. Drezner MK, Neelon FA, Haussler M, McPherson HT, Lebovitz HE. 1976. 1,25-dihydroxycholecalciferol deficiency: The probable cause of hypocalcemia and metabolic bone disease in pseudohypoparathyroidism. *J Clin Endocrinol Metab* 42: 621–628.
4. Chase LR, Melson GL, Aurbach GD. 1969. Pseudohypoparathyroidism: Defective excretion of 3′,5′-AMP in response to parathyroid hormone. *J Clin Invest* 48: 1832–1844.
5. Ish-Shalom S, Rao LG, Levine MA, Fraser D, Kooh SW, Josse RG, McBroom R, Wong MM, Murray TM. 1996. Normal parathyroid hormone responsiveness of bone-derived cells from a patient with pseudohypoparathyroidism. *J Bone Miner Res* 11: 8–14.
6. Murray T, Gomez Rao E, Wong MM, Waddell JP, McBroom R, Tam CS, Rosen F, Levine MA. 1993. Pseudohypoparathyroidism with osteitis fibrosa cystica: Direct demonstration of skeletal responsiveness to parathyroid hormone in cells cultured from bone. *J Bone Miner Res* 8: 83–91.
7. Stone M, Hosking D, Garcia-Himmelstine C, White D, Rosenblum D, Worth H. 1993. The renal response to exogenous parathyroid hormone in treated pseudohypoparathyroidism. *Bone* 14: 727–735.
8. Farfel Z. 1999. Pseudohypohyperparathyroidism-pseudohypoparathyroidism type Ib. *J Bone Miner Res* 14: 1016.
9. Drezner M, Neelon FA, Lebovitz HE. 1973. Pseudohypoparathyroidism type II: A possible defect in the reception of the cyclic AMP signal. *N Engl J Med* 289: 1056–1060.
10. Linglart A, Menguy C, Couvineau A, Auzan C, Gunes Y, Cancel M, Motte E, Pinto G, Chanson P, Bougneres P, et al. 2011. Recurrent PRKAR1A mutation in acrodysostosis with hormone resistance. *N Engl J Med* 364: 2218–2226.
11. Weinstein LS, Yu S, Warner DR, Liu J. 2001. Endocrine manifestations of stimulatory g protein alpha-subunit mutations and the role of genomic imprinting. *Endocr Rev* 22: 675–705.
12. Levine MA. 2002. Pseudohypoparathyroidism. In: Bilezikian JP, Raisz LG, Rodan GA (eds.) *Principles of Bone Biology*. New York: Academic Press. pp. 1137–1163.
13. Mantovani G, Maghnie M, Weber G, De Menis E, Brunelli V, Cappa M, Loli P, Beck-Peccoz P, Spada A. 2003. Growth hormone-releasing hormone resistance in pseudohypoparathyroidism type ia: New evidence for imprinting of the Gs alpha gene. *J Clin Endocrinol Metab* 88: 4070–4074.
14. Germain-Lee EL, Groman J, Crane JL, Jan de Beur SM, Levine MA. 2003. Growth hormone deficiency in pseudohypoparathyroidism type 1a: Another manifestation of multihormone resistance. *J Clin Endocrinol Metab* 88: 4059–4069.
15. Vlaeminck-Guillem V, D'Herbomez M, Pigny P, Fayard A, Bauters C, Decoulx M, Wemeau JL. 2001. Pseudohypoparathyroidism Ia and hypercalcitoninemia. *J Clin Endocrinol Metab* 86: 3091–3096.
16. Zwermann O, Piepkorn B, Engelbach M, Beyer J, Kann P. 2002. Abnormal pentagastrin response in a patient with pseudohypoparathyroidism. *Exp Clin Endocrinol Diabetes* 110: 86–91.
17. Weinstein LS, Liu J, Sakamoto A, Xie T, Chen M. 2004. Minireview: GNAS: Normal and abnormal functions. *Endocrinology* 145: 5459–5464.
18. Yu S, Yu D, Lee E, Eckhaus M, Lee R, Corria Z, Accili D, Westphal H, Weinstein LS. 1998. Variable and tissue-specific hormone resistance in heterotrimeric G$_s$ protein α-subunit (Gsα) knockout mice is due to tissue-specific imprinting of the Gsα gene. *Proc Natl Acad Sci U S A* 95: 8715–8720.
19. Chen M, Gavrilova O, Liu J, Xie T, Deng C, Nguyen AT, Nackers LM, Lorenzo J, Shen L, Weinstein LS. 2005. Alternative Gnas gene products have opposite effects on glucose and lipid metabolism. *Proc Natl Acad Sci U S A* 102: 7386–7391.
20. Germain-Lee EL, Schwindinger W, Crane JL, Zewdu R, Zweifel LS, Wand G, Huso DL, Saji M, Ringel MD, Levine MA. 2005. A mouse model of Albright hereditary osteodystrophy generated by targeted disruption of exon 1 of the gnas gene. *Endocrinology* 146: 4697–4709.
21. Farfel Z, Brothers VM, Brickman AS, Conte F, Neer R, Bourne HR. 1981. Pseudohypoparathyroidism: Inheritance of deficient receptor-cyclase coupling activity. *Proc Natl Acad Sci U S A* 78: 3098–3102.

22. Linglart A, Carel JC, Garabedian M, Le T, Mallet E, Kottler ML. 2002. GNAS1 lesions in pseudohypoparathyroidism Ia and Ic: Genotype phenotype relationship and evidence of the maternal transmission of the hormonal resistance. *J Clin Endocrinol Metab* 87: 189–197.

23. Linglart A, Mahon MJ, Kerachian MA, Berlach DM, Hendy GN, Jüppner H, Bastepe M. 2006. Coding GNAS mutations leading to hormone resistance impair in vitro agonist- and cholera toxin-induced adenosine cyclic 3′,5′-monophosphate formation mediated by human XLαs. *Endocrinology* 147: 2253–2262.

24. Thiele S, de Sanctis L, Werner R, Grotzinger J, Aydin C, Jüppner H, Bastepe M, Hiort O. 2011. Functional characterization of GNAS mutations found in patients with pseudohypoparathyroidism type Ic defines a new subgroup of pseudohypoparathyroidism affecting selectively Gsalpha-receptor interaction. *Hum Mutat* 32: 653–660.

25. Albright F, Forbes AP, Henneman PH. 1952. Pseudo-pseudohypoparathyroidism. *Trans Assoc Am Physicians* 65: 337–350.

26. Davies AJ, Hughes HE. 1993. Imprinting in Albright's hereditary osteodystrophy. *J Med Genet* 30: 101–103.

27. Wilson LC, Oude-Luttikhuis MEM, Clayton PT, Fraser WD, Trembath RC. 1994. Parental origin of Gsα gene mutations in Albright's hereditary osteodystrophy. *J Med Genet* 31: 835–839.

28. Kaplan FS, Shore EM. 2000. Progressive osseous heteroplasia. *J Bone Miner Res* 15: 2084–2094.

29. Eddy MC, De Beur SM, Yandow SM, McAlister WH, Shore EM, Kaplan FS, Whyte MP, Levine MA. 2000. Deficiency of the alpha-subunit of the stimulatory G protein and severe extraskeletal ossification. *J Bone Miner Res* 15: 2074–2083.

30. Shore EM, Ahn J, Jan de Beur S, Li M, Xu M, Gardner RJ, Zasloff MA, Whyte MP, Levine MA, Kaplan FS. 2002. Paternally inherited inactivating mutations of the GNAS1 gene in progressive osseous heteroplasia. *N Engl J Med* 346: 99–106.

31. Adegbite NS, Xu M, Kaplan FS, Shore EM, Pignolo RJ. 2008. Diagnostic and mutational spectrum of progressive osseous heteroplasia (POH) and other forms of GNAS-based heterotopic ossification. *Am J Med Genet A* 146A: 1788–1796.

32. Lebrun M, Richard N, Abeguile G, David A, Coeslier Dieux A, Journel H, Lacombe D, Pinto G, Odent S, Salles JP, et al. 2010. Progressive osseous heteroplasia: A model for the imprinting effects of GNAS inactivating mutations in humans. *J Clin Endocrinol Metab* 95: 3028–3038.

33. Ahmed SF, Barr DG, Bonthron DT. 2002. GNAS1 mutations and progressive osseous heteroplasia. *N Engl J Med* 346: 1669–1671.

34. Bartolomei MS, Tilghman SM. 1997. Genomic imprinting in mammals. *Annu Rev Genet* 31: 493–525.

35. Tilghman SM. 1999. The sins of the fathers and mothers: Genomic imprinting in mammalian development. *Cell* 96: 185–193.

36. Reik W, Walter J. 2001. Genomic imprinting: Parental influence on the genome. *Nat Rev Genet* 2: 21–32.

37. Liu J, Yu S, Litman D, Chen W, Weinstein L. 2000. Identification of a methylation imprint mark within the mouse Gnas locus. *Mol Cell Biol* 20: 5808–5817.

38. Hayward B, Kamiya M, Strain L, Moran V, Campbell R, Hayashizaki Y, Bronthon DT. 1998. The human GNAS1 gene is imprinted and encodes distinct paternally and biallelically expressed G proteins. *Proc Natl Acad Sci U S A* 95: 10038–10043.

39. Peters J, Wroe SF, Wells CA, Miller HJ, Bodle D, Beechey CV, Williamson CM, Kelsey G 1999. A cluster of oppositely imprinted transcripts at the Gnas locus in the distal imprinting region of mouse chromosome 2. *Proc Natl Acad Sci U S A* 96: 3830–3835.

40. Skinner J, Cattanach B, Peters J. 2002. The imprinted oedematous-small mutation on mouse chromosome 2 identifies new roles for Gnas and Gnasxl in development. *Genomics* 80: 373.

41. Williamson CM, Ball ST, Nottingham WT, Skinner JA, Plagge A, Turner MD, Powles N, Hough T, Papworth D, Fraser WD, et al. 2004. A cis-acting control region is required exclusively for the tissue-specific imprinting of Gnas. *Nat Genet* 36: 894–899.

42. Mantovani G, Ballare E, Giammona E, Beck-Peccoz P, Spada A. 2002. The Gsalpha gene: Predominant maternal origin of transcription in human thyroid gland and gonads. *J Clin Endocrinol Metab* 87: 4736–4740.

43. Germain-Lee EL, Ding CL, Deng Z, Crane JL, Saji M, Ringel MD, Levine MA. 2002. Paternal imprinting of Galpha(s) in the human thyroid as the basis of TSH resistance in pseudohypoparathyroidism type 1a. *Biochem Biophys Res Commun* 296: 67–72.

44. Liu J, Erlichman B, Weinstein LS. 2003. The stimulatory G protein α-subunit Gsα is imprinted in human thyroid glands: Implications for thyroid function in pseudohypoparathyroidism types 1A and 1B. *J Clin Endocrinol Metabol* 88: 4336–4341.

45. Hayward B, Barlier A, Korbonits M, Grossman A, Jacquet P, Enjalbert A, Bonthron D. 2001. Imprinting of the G(s)alpha gene GNAS1 in the pathogenesis of acromegaly. *J Clin Invest* 107: R31–36.

46. Campbell R, Gosden CM, Bonthron DT. 1994. Parental origin of transcription from the human GNAS1 gene. *J Med Genet* 31: 607–614.

47. Mantovani G, Bondioni S, Locatelli M, Pedroni C, Lania AG, Ferrante E, Filopanti M, Beck-Peccoz P, Spada A. 2004. Biallelic expression of the Gsalpha gene in human bone and adipose tissue. *J Clin Endocrinol Metab* 89: 6316–6319.

48. Zheng H, Radeva G, McCann JA, Hendy GN, Goodyer CG. 2001. Gαs transcripts are biallelically expressed in the human kidney cortex: Implications for pseudohypoparathyroidism type Ib. *J Clin Endocrinol Metab* 86: 4627–4629.

49. Bastepe M, Weinstein LS, Ogata N, Kawaguchi H, Jüppner H, Kronenberg HM, Chung UI. 2004. Stimulatory G protein directly regulates hypertrophic

50. Huso DL, Edie S, Levine MA, Schwindinger W, Wang Y, Jüppner H, Germain-Lee EL. 2011. Heterotopic ossifications in a mouse model of albright hereditary osteodystrophy. *PLoS One* 6: e21755.
51. Long DN, McGuire S, Levine MA, Weinstein LS, Germain-Lee EL. 2007. Body mass index differences in pseudohypoparathyroidism type 1a versus pseudopseudohypoparathyroidism may implicate paternal imprinting of Galpha(s) in the development of human obesity. *J Clin Endocrinol Metab* 92: 1073–1079.
52. Mouallem M, Shaharabany M, Weintrob N, Shalitin S, Nagelberg N, Shapira H, Zadik Z, Farfel Z. 2008. Cognitive impairment is prevalent in pseudohypoparathyroidism type Ia, but not in pseudopseudohypoparathyroidism: Possible cerebral imprinting of Gsalpha. *Clin Endocrinol (Oxf)* 68: 233–239.
53. Chen M, Wang J, Dickerson KE, Kelleher J, Xie T, Gupta D, Lai EW, Pacak K, Gavrilova O, Weinstein LS. 2009. Central nervous system imprinting of the G protein G(s)alpha and its role in metabolic regulation. *Cell Metab* 9: 548–555.
54. Kozasa T, Itoh H, Tsukamoto T, Kaziro Y. 1988. Isolation and characterization of the human Gsα gene. *Proc Natl Acad Sci USA* 85: 2081–2085.
55. Robishaw JD, Smigel MD, Gilman AG. 1986. Molecular basis for two forms of the G protein that stimulates adenylate cyclase. *J Biol Chem* 261: 9587–9590.
56. Bray P, Carter A, Simons C, Guo V, Puckett C, Kamholz J, Spiegel A, Nirenberg M. 1986. Human cDNA clones for four species of G alpha s signal transduction protein. *Proc Natl Acad Sci U S A* 83: 8893–8897.
57. Crawford JA, Mutchler KJ, Sullivan BE, Lanigan TM, Clark MS, Russo AF. 1993. Neural expression of a novel alternatively spliced and polyadenylated Gs alpha transcript. *J Biol Chem* 268: 9879–9885.
58. Kehlenbach RH, Matthey J, Huttner WB. 1994. XLαs is a new type of G protein. *Nature* 372: 804–809. [Erratum in *Nature* 1995 375: 253].
59. Ischia R, Lovisetti-Scamihorn P, Hogue-Angeletti R, Wolkersdorfer M, Winkler H, Fischer-Colbrie R. 1997. Molecular cloning and characterization of NESP55, a novel chromogranin-like precursor of a peptide with 5-HT1B receptor antagonist activity. *J Biol Chem* 272: 11657–11662.
60. Hayward BE, Moran V, Strain L, Bonthron DT. 1998. Bidirectional imprinting of a single gene: GNAS1 encodes maternally, paternally, and biallelically derived proteins. *Proc Natl Acad Sci U S A* 95: 15475–15480.
61. Pasolli H, Huttner W. 2001. Expression of the extra-large G protein alpha-subunit XLalphas in neuroepithelial cells and young neurons during development of the rat nervous system. *Neurosci Lett* 301: 119–122.
62. Pasolli H, Klemke M, Kehlenbach R, Wang Y, Huttner W. 2000. Characterization of the extra-large G protein alpha-subunit XLalphas. I. Tissue distribution and subcellular localization. *J Biol Chem* 275: 33622–33632.
63. Plagge A, Gordon E, Dean W, Boiani R, Cinti S, Peters J, Kelsey G. 2004. The imprinted signaling protein XLalphas is required for postnatal adaptation to feeding. *Nat Genet* 36: 818–826.
64. Hayward B, Bonthron D. 2000. An imprinted antisense transcript at the human GNAS1 locus. *Hum Mol Genet* 9: 835–841.
65. Wroe SF, Kelsey G, Skinner JA, Bodle D, Ball ST, Beechey CV, Peters J, Williamson CM. 2000. An imprinted transcript, antisense to Nesp, adds complexity to the cluster of imprinted genes at the mouse Gnas locus. *Proc Natl Acad Sci U S A* 97: 3342–3346.
66. Swaroop A, Agarwal N, Gruen JR, Bick D, Weissman SM. 1991. Differential expression of novel Gs alpha signal transduction protein cDNA species. *Nucleic Acids Res* 19: 4725–4729.
67. Ishikawa Y, Bianchi C, Nadal-Ginard B, Homcy CJ. 1990. Alternative promoter and 5′ exon generate a novel $G_s\alpha$ mRNA. *J Biol Chem* 265: 8458–8462.
68. Fischer-Colbrie R, Eder S, Lovisetti-Scamihorn P, Becker A, Laslop A. 2002. Neuroendocrine secretory protein 55: A novel marker for the constitutive secretory pathway. *Ann N Y Acad Sci* 971: 317–322.
69. Plagge A, Isles AR, Gordon E, Humby T, Dean W, Gritsch S, Fischer-Colbrie R, Wilkinson LS, Kelsey G. 2005. Imprinted Nesp55 influences behavioral reactivity to novel environments. *Mol Cell Biol* 25: 3019–3026.
70. Klemke M, Pasolli H, Kehlenbach R, Offermanns S, Schultz G, Huttner W. 2000. Characterization of the extra-large G protein alpha-subunit XLalphas. II. Signal transduction properties. *J Biol Chem* 275: 33633–33640.
71. Bastepe M, Gunes Y, Perez-Villamil B, Hunzelman J, Weinstein LS, Jüppner H. 2002. Receptor-mediated adenylyl cyclase activation through XLalpha(s), the extra-large variant of the stimulatory G protein alpha-subunit. *Mol Endocrinol* 16: 1912–1919.
72. Liu Z, Segawa H, Aydin C, Reyes M, Erben RG, Weinstein LS, Chen M, Marshansky V, Frohlich LF, Bastepe M. 2011. Transgenic overexpression of the extra-large Gsα variant XLαs enhances Gsα-mediated responses in the mouse renal proximal tubule in vivo. *Endocrinology* 152: 1222–1233.
73. Liu Z, Turan S, Wehbi VL, Vilardaga JP, Bastepe M. 2011. The extra-long Gαs variant XLαs escapes activation-induced subcellular redistribution and is able to provide sustained signaling. *J Biol Chem* 286(44): 38558–38569.
74. Puzhko S, Goodyer CG, Mohammad AK, Canaff L, Misra M, Jüppner H, Bastepe M, Hendy GN. 2011. Parathyroid hormone signaling via Gαs is selectively inhibited by an NH(2)-terminally truncated Gαs: Implications for pseudohypoparathyroidism. *J Bone Miner Res* 26: 2473–2485.
75. Bastepe M, Pincus JE, Sugimoto T, Tojo K, Kanatani M, Azuma Y, Kruse K, Rosenbloom AL, Koshiyama H, Jüppner H. 2001. Positional dissociation between the genetic mutation responsible for pseudohypoparathy-

roidism type Ib and the associated methylation defect at exon A/B: Evidence for a long-range regulatory element within the imprinted *GNAS1* locus. *Hum Mol Genet* 10: 1231–1241.

76. Bastepe M, Lane AH, Jüppner H. 2001. Paternal uniparental isodisomy of chromosome 20q (patUPD20q)—and the resulting changes in *GNAS1* methylation—as a plausible cause of pseudohypoparathyroidism. *Am J Hum Genet* 68: 1283–1289.

77. de Nanclares GP, Fernandez-Rebollo E, Santin I, Garcia-Cuartero B, Gaztambide S, Menendez E, Morales MJ, Pombo M, Bilbao JR, Barros F, et al. 2007. Epigenetic defects of GNAS in patients with pseudohypoparathyroidism and mild features of Albright's hereditary osteodystrophy. *J Clin Endocrinol Metab* 92: 2370–2373.

78. Unluturk U, Harmanci A, Babaoglu M, Yasar U, Varli K, Bastepe M, Bayraktar M. 2008. Molecular diagnosis and clinical characterization of pseudohypoparathyroidism type-Ib in a patient with mild Albright's hereditary osteodystrophy-like features, epileptic seizures, and defective renal handling of uric acid. *Am J Med Sci* 336: 84–90.

79. Mariot V, Maupetit-Mehouas S, Sinding C, Kottler ML, Linglart A. 2008. A maternal epimutation of GNAS leads to Albright osteodystrophy and parathyroid hormone resistance. *J Clin Endocrinol Metab* 93: 661–665.

80. Mantovani G, de Sanctis L, Barbieri AM, Elli FM, Bollati V, Vaira V, Labarile P, Bondioni S, Peverelli E, Lania AG, et al. 2010. Pseudohypoparathyroidism and GNAS epigenetic defects: Clinical evaluation of Albright hereditary osteodystrophy and molecular analysis in 40 patients. *J Clin Endocrinol Metab* 95: 651–658.

81. Sanchez J, Perera E, Jan de Beur S, Ding C, Dang A, Berkovitz GD, Levine MA. 2011. Madelung-like deformity in pseudohypoparathyroidism type 1b. *J Clin Endocrinol Metab* 96: E1507–E1511.

82. Bastepe M, Fröhlich LF, Hendy GN, Indridason OS, Josse RG, Koshiyama H, Körkkö J, Nakamoto JM, Rosenbloom AL, Slyper AH, et al. 2003. Autosomal dominant pseudohypoparathyroidism type Ib is associated with a heterozygous microdeletion that likely disrupts a putative imprinting control element of GNAS. *J Clin Invest* 112: 1255–1263.

83. Laspa E, Bastepe M, Jüppner H, Tsatsoulis A. 2004. Phenotypic and molecular genetic aspects of pseudohypoparathyroidism type ib in a Greek kindred: Evidence for enhanced uric acid excretion due to parathyroid hormone resistance. *J Clin Endocrinol Metab* 89: 5942–5947.

84. Liu J, Nealon JG, Weinstein LS. 2005. Distinct patterns of abnormal GNAS imprinting in familial and sporadic pseudohypoparathyroidism type IB. *Hum Mol Genet* 14: 95–102.

85. Mahmud FH, Linglart A, Bastepe M, Jüppner H, Lteif AN. 2005. Molecular diagnosis of pseudohypoparathyroidism type Ib in a family with presumed paroxysmal dyskinesia. *Pediatrics* 115: e242–e244.

86. Linglart A, Gensure RC, Olney RC, Jüppner H, Bastepe M. 2005. A novel STX16 deletion in autosomal dominant pseudohypoparathyroidism type Ib redefines the boundaries of a cis-acting imprinting control element of GNAS. *Am J Hum Genet* 76: 804–814.

87. Fröhlich LF, Bastepe M, Ozturk D, Abu-Zahra H, Jüppner H. 2007. Lack of Gnas epigenetic changes and pseudohypoparathyroidism type Ib in mice with targeted disruption of syntaxin-16. *Endocrinology* 148: 2925–2935.

88. Bastepe M, Fröhlich LF, Linglart A, Abu-zahra HS, Tojo K, Ward LM, Jüppner H 2005. Deletion of the NESP55 differentially methylated region causes loss of maternal GNAS imprints and pseudohypoparathyroidism type-Ib. *Nat Genet* 37: 25–37.

89. Chillambhi S, Turan S, Hwang DY, Chen HC, Jüppner H, Bastepe M. 2010. Deletion of the noncoding GNAS antisense transcript causes pseudohypoparathyroidism type Ib and biparental defects of GNAS methylation in cis. *J Clin Endocrinol Metab* 95: 3993–4002.

90. Linglart A, Bastepe M, Jüppner H. 2007. Similar clinical and laboratory findings in patients with symptomatic autosomal dominant and sporadic pseudohypoparathyroidism type Ib despite different epigenetic changes at the GNAS locus. *Clin Endocrinol (Oxf)* 67: 822–831.

91. Bastepe M, Altug-Teber O, Agarwal C, Oberfield SE, Bonin M, Jüppner H. 2011. Paternal uniparental isodisomy of the entire chromosome 20 as a molecular cause of pseudohypoparathyroidism type Ib (PHP-Ib). *Bone* 48(3): 659–662.

92. Fernandez-Rebollo E, Lecumberri B, Garin I, Arroyo J, Bernal-Chico A, Goni F, Orduna R, Castano L, Perez de Nanclares G. 2010. New mechanisms involved in paternal 20q disomy associated with pseudohypoparathyroidism. *Eur J Endocrinol* 163: 953–962.

93. Liu J, Litman D, Rosenberg M, Yu S, Biesecker L, Weinstein L. 2000. A GNAS1 imprinting defect in pseudohypoparathyroidism type IB. *J Clin Invest* 106: 1167–1174.

94. Fernandez-Rebollo E, Perez de Nanclares G, Lecumberri B, Turan S, Anda E, Perez-Nanclares G, Feig D, Nik-Zainal S, Bastepe M, Jüppner H. 2011. Exclusion of the GNAS locus in PHP-Ib patients with broad GNAS methylation changes: Evidence for an autosomal recessive form of PHP-Ib? *J Bone Miner Res* 26: 1854–1863.

95. Maupetit-Mehouas S, Mariot V, Reynes C, Bertrand G, Feillet F, Carel JC, Simon D, Bihan H, Gajdos V, Devouge E, et al. 2011. Quantification of the methylation at the GNAS locus identifies subtypes of sporadic pseudohypoparathyroidism type Ib. *J Med Genet* 48: 55–63.

96. Weinstein LS, Gejman PV, Friedman E, Kadowaki T, Collins RM, Gershon ES, Spiegel AM. 1990. Mutations of the Gs alpha-subunit gene in Albright hereditary osteodystrophy detected by denaturing gradient gel electrophoresis. *Proc Natl Acad Sci U S A* 87: 8287–8290.

97. Patten JL, Johns DR, Valle D, Eil C, Gruppuso PA, Steele G, Smallwood PM, Levine MA. 1990. Mutation in the gene encoding the stimulatory G protein of adenylate cyclase in Albright's hereditary osteodystrophy. *New Engl J Med* 322: 1412–1419.
98. Magagnin S, Werner A, Markovich D, Sorribas V, Stange G, Biber J, Murer H. 1993. Expression cloning of human and rat renal cortex Na/Pi cotransport. *Proc Natl Acad Sci U S A* 90: 5979–5983.
99. Segawa H, Kaneko I, Takahashi A, Kuwahata M, Ito M, Ohkido I, Tatsumi S, Miyamoto K. 2002. Growth-related renal type II Na/Pi cotransporter. *J Biol Chem* 277: 19665–19672.
100. Brenza HL, Kimmel-Jehan C, Jehan F, Shinki T, Wakino S, Anazawa H, Suda T, DeLuca HF. 1998. Parathyroid hormone activation of the 25-hydroxyvitamin D_3-1α-hydroxylase gene promoter. *Proc Natl Acad Sci U S A* 95: 1387–1391.
101. Kong XF, Zhu XH, Pei YL, Jackson DM, Holick MF. 1999. Molecular cloning, characterization, and promoter analysis of the human 25-hydroxyvitamin D3-1alpha-hydroxylase gene. *Proc Natl Acad Sci U S A* 96: 6988–6993.
102. Mensenkamp AR, Hoenderop JG, Bindels RJ. 2007. TRPV5, the gateway to Ca2+ homeostasis. *Handb Exp Pharmacol*: 207–220.

74

Disorders of Phosphate Homeostasis

Mary D. Ruppe and Suzanne M. Jan de Beur

Introduction 601
Hypophosphatemia 601
Tumor-Induced Osteomalacia 602
X-Linked Hypophosphatemic Rickets 605
Autosomal Dominant Hypophosphatemic Rickets 606
Other Disorders of Renal Phosphate Wasting 607
Hyperphosphatemia 608
Familial Tumoral Calcinosis 608
Treatment of Hyperphosphatemia 609
References 609

INTRODUCTION

Phosphorus is a critical element in skeletal development, bone mineralization, membrane composition, nucleotide structure, and cellular signaling. The physiological control of phosphate homeostasis and the major hormonal regulators of phosphate homeostasis [e.g., fibroblast growth factor 23 (FGF23), 1,25(OH)$_2$D, PTH] are discussed in detail elsewhere in the *Primer*. Serum phosphorus concentration is regulated by diet, hormones, pH, and changes in renal, skeletal, and intestinal function. The focus of this chapter is the molecular basis of human disorders of phosphate homeostasis. In recent years, advances in defining the precise molecular defects in both acquired and inherited hypo and hyperphosphatemic syndromes have catapulted our understanding of phosphate homeostatic mechanisms to a new level.

HYPOPHOSPHATEMIA

Clinical consequences

Hypophosphatemia is common and is observed in up to 5% of hospitalized patients [1, 2]. Among alcoholic patients and those with severe sepsis, up to a 30–50% prevalence has been reported.

The clinical manifestations of hypophosphatemia are dependent on the severity and chronicity of the phosphorus depletion.

Common clinical settings in which severe hypophosphatemia is observed include chronic alcoholism, nutritional repletion in at risk individuals, treatment of diabetic ketoacidosis, and in critical illness.

The symptoms of hypophosphatemia are a direct consequence of intracellular phosphorus depletion: (1) tissue hypoxia caused by reduced 2,3-diphosphoglycerate (2,3-DPG) in the erythrocyte that increases affinity of hemoglobin for oxygen and (2) diminished tissue content of ATP that compromises cellular function.

Causes of hypophosphatemia

The three major mechanisms by which hypophosphatemia can occur are (1) redistribution of phosphorus from extracellular fluid into cells, (2) increased urinary excretion, and (3) decreased intestinal absorption. The diagnosis of hypophosphatemia is often evident from the history; if, however, the diagnosis remains obscure, measurement of urinary phosphate excretion is indicated. Urinary phosphate excretion can be measured from a 24-hour collection or from a random urine specimen by calculation of fractional excretion of filtered phosphate (FEPO4): $FEPO_4 = [UPO_4 \times PCr \times 100]/[PPO_4 \times UCr]$, where U and P are urine and plasma concentrations of phosphate (PO$_4$)

and creatinine (Cr). In the setting of hypophosphatemia, fractional excretion of phosphate greater than 5% or a 24-hour urinary phosphate excretion that exceeds 100 mg/day is indicative of renal phosphate wasting.

The causes of hypophosphatemia are found in Table 74.1. Hypophosphatemia secondary to renal phosphate wasting has a wide differential diagnosis. These disorders can result from primary renal transport defects, excess PTH, FGF23, KLOTHO or other phosphaturic proteins (Fig. 74.1). The etiologies of hypophosphatemia due to renal phosphate wasting are discussed in more detail below.

TUMOR-INDUCED OSTEOMALACIA

Tumor-induced osteomalacia (TIO), or oncogenic osteomalacia, is an acquired, paraneoplastic syndrome of renal phosphate wasting that resembles genetic forms of hypophosphatemic rickets. First described in 1947 [3], clinical and experimental studies implicate the humoral factor(s) that tumors produce in the profound biochemical and skeletal alterations that characterize TIO. To date 337 cases have been reported in the literature [4].

Clinical and biochemical manifestations

Although the preponderance of TIO patients are adults (usually diagnosed in the sixth decade), this syndrome may present at any age. These patients report long-standing progressive muscle and bone pain. Children with TIO display rachitic features including gait disturbances, growth retardation, and skeletal deformities. The occult nature of TIO delays its recognition, and the average time from onset of symptoms to a correct diagnosis often exceeds 2.5 years [5]. Once the syndrome is recognized, an average of 5 years elapses from the time of diagnosis to the identification of the underlying tumor [6]. Until the underlying tumor is identified, other renal phosphate wasting syndromes must be considered. Identification of a previously normal serum phosphorus level in an adult patient supports the diagnosis of TIO, although in rare instances patients with autosomal dominant hypophosphatemic rickets (ADHR) can present in adulthood. In situations when inherited hypophosphatemic rickets must be excluded, genetic testing for mutations in the *PHEX* gene, *FGF23* gene, and *DMP1* and *ENPP1* gene is indicated.

The biochemical hallmarks of TIO are low serum concentrations of phosphorus, phosphaturia (secondary to reduced proximal renal tubular phosphate reabsorption), and frankly low or inappropriate normal levels of serum calcitriol [1,25(OH)$_2$D$_3$] that are expected to be elevated in the face of hypophosphatemia (Table 74.2). Calcium and PTH are typically normal. Bone histomorphometry shows severe osteomalacia with clear evidence of a mineralization defect with increased mineralization lag time

Table 74.1. Causes of Hypophosphatemia

Decreased intestinal absorption
  Vitamin D deficiency or resistance
    Nutritional deficiency
      Low sun exposure, low dietary intake
    Malabsorption
      Celiac disease, Crohn's disease
      Gastrectomy, bowel resection, gastric bypass
      Pancreatitis
      Chronic diarrhea
    Chronic liver disease
    Chronic renal disease
    Increased catabolism
      Anticonvulsant therapy
    Vitamin D receptor defects
      Vitamin D-dependent rickets, type 2
    Vitamin D synthetic defects
      CYP27B1 (vitamin D-dependent rickets, type 1)
      CYP27A1
  Nutritional deficiencies
    Alcoholism, anorexia, starvation
  Antacids containing aluminum or magnesium

Increased urinary losses
  Renal phosphate wasting disorders (Table 74.2)
  Primary and secondary hyperparathyroidism
  Diabetic ketoacidosis (osmotic diuresis)
  Medications
    Calcitonin, diuretics, glucocorticoids, bicarbonate
  Acute volume expansion

Intracellular shifts
  Increased insulin
    Re-feeding, treatment of DKA, insulin therapy
  Hungry bone syndrome
  Acute respiratory alkalosis
  Tumor consumption
    Leukemia blast crisis, lymphoma
  Sepsis
  Sugars
    Glucose, fructose, glycerol
  Recovery from metabolic acidosis

and excessive osteoid (Fig. 74.2). The dual defect of renal phosphate wasting in concert with impaired calcitriol [1,25(OH)$_2$D$_3$] synthesis results in poor bone mineralization and fractures [7].

The mesenchymal tumors that are associated with TIO are characteristically slow-growing, polymorphous neoplasms with the preponderance being phosphaturic mesenchymal tumor, mixed connective tissue type [8, 9] (PMTMCT; Fig. 74.2). Characterized by an admixture of spindle cells, osteoclast-like giant cells, prominent blood vessels, cartilage-like matrix, and metaplastic bone, these tumors occur equally in soft tissue and bone. Although typically benign, malignant variants of PMTMCT have been described. These mesenchymal tumors ectopically express and secrete FGF23 and other

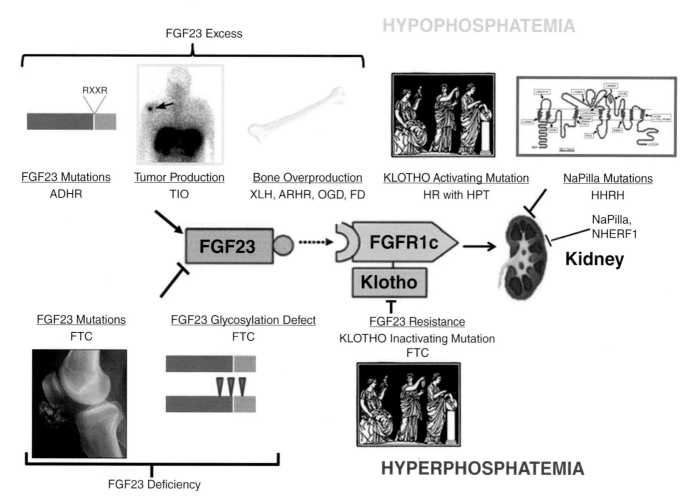

Fig. 74.1. Molecular mechanisms of disorders of phosphate homeostasis. Three major mechanisms of hypophosphatemia are (1) FGF23 excess owing to ectopic production as in TIO; (2) excess bone production seen in XLH, ARHR, ADHR, FD and OGD; and (3) mutation in the FGF23 gene that renders the protein resistant to inactivation. Hypophosphatemia may also be secondary to excess KLOTHO, the cofactor necessary for FGF23 signaling as seen in a patient with hypophosphatemic rickets with hyperparathyroidism. Finally, homozygous inactivating mutations in SLC34A3, which encodes NaPiIIc, or dominant negative mutations in SLC34A2 that encodes NaPiIIa, result in phosphate wasting due to absence of sodium–phosphate cotransporters.

Hyperphosphatemia is due to FGF23 deficiency, either through inactivating mutations in FGF23, aberrant glycosylation of FGF23 owing to GALNT3 mutations, or FGF23 resistance due to inactivating KLOTHO mutations.

phosphaturic proteins [10, 11]. FGF23, a circulating fibroblast growth factor produced by osteocytes and osteoblasts [12, 13] has two currently known physiologic functions: first, FGF23 promotes internalization of NaPiIIa and NaPiIIc from the renal brush border membrane and thus reduces reabsorption of urinary phosphorus resulting in hypophosphatemia [14, 15]. Second, it diminishes protein expression of the 25 hydroxy-l-α-hydroxylase enzyme that converts vitamin D to its active form, $1,25(OH)_2D_3$ [16] while increasing the activity of the vitamin D 24-hydroxylase enzyme, which converts vitamin D to an inactive form. This leads to disruption in the compensatory increase in $1,25(OH)_2D_3$ triggered by hypophosphatemia [17]. Circulating levels of FGF23 are elevated in most patients with TIO [18]. After surgical resection, FGF23 levels plummet. Other secreted proteins such as MEPE (matrix extracellular phosphoglycoprotein), FGF7, and sFRP4 (secreted frizzled related protein 4) are highly expressed in mesenchymal tumors associated with TIO, but the role of each of these "phosphatonins" in the disease process remains obscure.

Treatment

Detection and localization of the culprit tumor in TIO is imperative because complete surgical resection is curative. However, the mesenchymal tumors that cause this syndrome are often small, slow growing, and frequently found in a variety of anatomical locations, including the

Table 74.2. Characteristics of Renal Phosphate Wasting Disorders

Disease (OMIM)	Defect	Pathogenesis
TIO	Mesenchymal tumor	Ectopic, unregulated production of FGF23 and other phosphatonins sFRP4, MEPE, FGF7
XLH (307800)	*PHEX* mutation	Inappropriate FGF23 synthesis from bone
ADHR (193100)	*FGF23* mutation	Increased circulating intact FGF23 caused by mutations that render it resistant to cleavage
HHRH (241530)	*SLC34A3* mutation	Loss of function NaPiIIc mutations that result in renal phosphate wasting without a defect in $1,25(OH)_2D_3$ synthesis
ARHR1 (241520)	*DMP1* mutation	Loss of DMP1 causes impaired osteocyte differentiation and increased production of FGF23
ARHR2 (613312)	*ENPP1* mutation	Increased production of FGF23
HR and HPT (612089)	α-*KLOTHO* translocation	Increased KLOTHO, FGF23, and downstream FGF23 signaling
Fibrous dysplasia (139320)	*GNAS* mutation	Increased FGF23 production from the dysplastic bone
Linear nevus sebaceous syndrome	Excess FGF23 production	Increased FGF23 production from the dysplastic bone and from the nevi
OGD (166250)	*FGFR1* mutation	Increased FGF23 production from the dysplastic bone
NPHLOP1 (612286)	*SLC34A1* mutation	Renal phosphate wasting without a defect in $1,25(OH)_2D_3$ synthesis
NPHLOP2 (612287)	*SLC9A3R1* mutation	Renal phosphate wasting through potentiation of PTH-mediated cAMP production
FRTS2 (613388)	*SLC34A1* mutation	Renal phosphate wasting without a defect in $1,25(OH)_2D_3$ synthesis

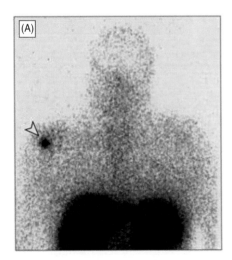
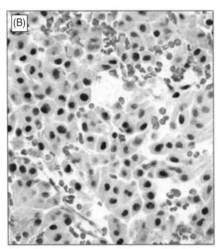
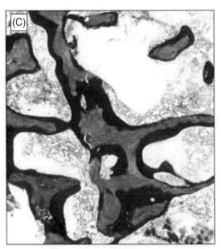

JAMA. 2005;294:1260-1267. © American Medical Association

Fig. 74.2. Radiographic and histologic features in TIO. (A) Octreotide scan showing a small mesenchymal tumor in the head of the humerus. (B) Hemiangiopericytoma with numerous pericytes and vascular channels (H&E strain). (C) Bone biopsy with Goldner stain. Excessive osteoid or unmineralized bone matrix composed mainly of collagen stains pink. Mineralized bone stains blue. This bone biopsy shows severe osteomalacia.

long bones, the distal extremities, the nasopharynx, the sinuses, and the groin. A thorough physical examination should be performed in order to assess any palpable masses as the tumors have been found in the subcutaneous tissues. The size and obscure locations make the tumors difficult to localize with conventional imaging techniques. Frequently, a combination of functional and anatomical approaches must be pursued in order to localize the tumor [4]. Because *in vitro* studies show that many mesenchymal tumors express somatostatin receptors (SSTRs)[111]In-pentetreotide scintigraphy (octreotide scan; Fig. 74.2), a scanning technique that uses a radiolabeled somatostatin analog, has been used to successfully detect and localize these tumors in some patients with TIO [19–21]. Successful tumor localization has been reported with other imaging techniques such as whole body MRI [22] and FDG-PET/CT [23]. Venous sampling for FGF23 has also been used in order to localize causative tumors in patients with TIO but seems to be more suited for confirmation that a mass seen on imaging is producing FGF23 than the *de novo* localization of the tumor [24].

The definitive treatment for TIO is complete tumor resection, which results in rapid correction of the biochemical perturbations and remineralization of bone. Radiofrequency ablation has been reported to have been beneficial in a case report [25]. However, even after the diagnosis of TIO is made, the tumor often remains obscure or incompletely resected. Therefore, medical management is frequently necessary. The current practice is to treat TIO with phosphorus supplementation in combination with calcitriol. The phosphorus supplementation serves to replace ongoing renal phosphorus loss, and the calcitriol supplements replace insufficient renal production of $1,25(OH)_2D_3$ and enhances renal and gastrointestinal phosphorus reabsorption. Generally, patients are treated with phosphorus (1–2 g/day), in divided doses 3–4 doses daily, and calcitriol (1–3 μg/day). In some cases, administration of calcitriol alone may improve the biochemical abnormalities seen in TIO and heal the osteomalacia. Therapy and dosing should be tailored to improve symptoms and normalize alkaline phosphatase. With appropriate treatment, muscle and bone pain will improve, and healing of the osteomalacia will ensue.

Monitoring for therapeutic complications is important to prevent unintended hypercalcemia, nephrocalcinosis, and nephrolithiasis. The true incidence of hyperparathyroidism with prolonged treatment with phosphorus (alone or in combination with calcitriol) is unknown. To assess safety and efficacy of therapy, monitoring of serum and urine calcium, renal function, and parathyroid status is recommended at least monthly at the initiation of treatment and every 3 months when on chronic therapy. There have been reports of the use cinacalcet in patients who did not tolerate medical therapy with phosphorus and calcitriol [26]. This therapy allowed a decrease in the phosphorus dose to one that was tolerated. Octreotide *in vitro* and *in vivo* has been shown to inhibit secretion of hormones by many neuroendocrine tumors; however, because of the limited and mixed experience with octreotide treatment in TIO [21, 27] this therapy should be reserved for the most severe cases that are refractory to current medical therapy.

X-LINKED HYPOPHOSPHATEMIC RICKETS

First described by Albright in 1939 [28], X-linked hypophosphatemic rickets (XLH) is characterized by growth retardation, rachitic and osteomalacic bone disease, and dental abscesses. It is the most common disorder of renal phosphate wasting, occurring in 3.9–5 per 100,000 live births.

Genetics

Although the X-linked inheritance was first detailed in 1958 [29], it was not until the 1990s that the genetic basis of XLH was elucidated as mutations in *PHEX* (phosphate-regulating gene with homologies to endopeptidases on the X chromosome) [30]. To date, more than 285 mutations have been described (PHEX database: www.PHEXdb.mcgill.ca). The *PHEX* gene codes for a protein of unknown function that is a member of the M13 family of membrane-bound metalloproteases that is present in osteoblasts, osteocytes, and odontoblasts but not in kidney tubules [31].

Clinical and biochemical manifestations

Before children begin to walk, clinical findings may be limited. The majority of testing that is done on infants is because of a known family history of XLH. After the child is ambulatory, progressive lower extremity bowing may become apparent with a decrease in height velocity and be associated with bone and/or joint pain. There may be dental manifestations, including abscessed noncarious teeth, enamel defects, enlarged pulp chambers, and taurodontism. Cranial abnormalities with frontal bossing and an increased anteroposterior skull length have been reported. Adults may exhibit bone and joint pain from osteomalacia, pseudofractures, and enthesopathy. Biochemically, the laboratory findings in XLH are indistinguishable from TIO (Table 74.1), with low serum concentrations of phosphorus, phosphaturia, and frankly low or inappropriate normal levels of serum $1,25(OH)_2D_3$. Calcium and parathyroid hormone (PTH) are typically normal. In infants the diagnosis may be difficult to establish as the phosphorus levels may initially be normal. This can be compounded by the fact that the infant normal range is substantial higher than that of older children and so the low phosphorus may go unrecognized. The biochemical profile suggests that regulators of phosphate other than PTH and vitamin D play a role in the development of XLH. Increasing evidence suggests that FGF23 is key in the pathogenesis of XLH. Whereas PHEX does not seem to cleave FGF23 directly, serum FGF23 levels are inappropriately normal or elevated in XLH patients, and FGF23 expression is increased in the bones of *hyp* mice, a murine model of XLH [18, 32–36]. These observations suggest that PHEX is involved in downregulation and control of FGF23; however, the precise interplay between FGF23 and PHEX is not currently understood. The diagnosis of XLH is based on a consistent medical history and physical examination, radiological evidence of rachitic disease, appropriate biochemical findings, and a family history consistent with multigenerational or sporadic occurrence of the disorder. Mutational analysis of the *PHEX* gene is available; however, studies have shown that mutations can only be found in 50–70% of affected individuals [37].

Treatment and complications

Treatment is similar for most patients and consists of oral phosphate administered three to five times daily and high-dose calcitriol. The treatment is generally started at a low dose to avoid gastrointestinal side effects. The doses are then titrated to a weight-based dose of calcitriol at 20 to 30 ng/kg/day along with phosphorus at 20 to 40 mg/kg/day administered in three to five divided doses [38]. Some clinicians favor a high-dose phase of treatment for up to a year. The high-dose phase consists of calcitriol at 50–70 ng/kg/day (up to a maximum dose of 3.0 mcg daily) along with the phosphate [39]. There has been no comparison between the two different regimens.

The treatment leads to resolution of radiographic rickets and improved, but not normal, growth. Subsequent studies have shown that age at initiation of therapy, height at initiation of therapy, and possibly the sex of the patient influence peak height attainment. Patients who do not respond to pharmacologic therapy may require surgical intervention to correct the lower extremity deformities. Because of lower bone turnover and the closure of the epiphyseal plates, the therapeutic requirements drop dramatically as children enter adulthood. In adults, the role of therapy is unclear. As adults, patients may be treated with no medications, a low dose of calcitriol, low doses of phosphorus, or with both calcitriol and phosphorus. It is unknown which patients require long-term treatment and which patients can have their treatment safely discontinued as adults. Symptomatic adults may benefit from therapy with improvement of symptoms and in bone architecture and prevention of osteomalacia. In adults, treatment is indicated in those with spontaneous insufficiency fractures, pending orthopedic procedures, biochemical evidence of osteomalacia and disabling skeletal pain [40]. There is a current clinical trial evaluating the use of a monoclonal antibody to FGF23 for the treatment of XLH in adults (http://www.clinicaltrials.gov, keyword KRN23). Although data have shown that bone mineral density (BMD) may be increased at the spine of patients with XLH, this may reflect, in part, calcific enthesopathy, and it is unclear if there is a change in long-term fracture rates. It has been shown that serum FGF23 levels increase with calcitriol and phosphorus treatment. The clinical significance of this is unknown [41]. Interestingly, in a small study of short-term treatment of adult XLH patients with subcutaneous calcitonin, there was a transient increase in serum phosphorus with a decrease in serum FGF23 levels [42]. Further studies are needed to determine if this is an efficacious treatment for XLH. Complications of treatment are similar to those in TIO.

AUTOSOMAL DOMINANT HYPOPHOSPHATEMIC RICKETS

Autosomal dominant hypophosphatemic rickets (ADHR) is a rare form of hypophosphatemic rickets with clinical characteristic similar to XLH.

Genetics

Multiple early reports documented an inheritance renal phosphate wasting syndrome with a pattern that included male-to-male transmission that distinguished it from XLH [43, 44]. Furthermore, the incomplete penetrance of disease observed within affected families was atypical of XLH. Positional mapping, cloning, and sequence analysis established that FGF23 was mutated in ADHR [45, 46]. Missense mutations in one of two arginine residues at positions 176 or 179 have been identified in affected members of ADHR families. The mutated arginine residues, located in the consensus proprotein convertase cleavage RXXR motif, prevent inactivation of FGF23 and thus result in prolonged or enhanced FGF23 action [47].

Clinical and biochemical manifestations

The clinical and biochemical findings in ADHR are similar to those observed in XLH (Table 74.1). In contrast to XLH, there are instances of delayed onset and rarely, resolution of the phosphate wasting [48]. FGF23 levels have been shown to vary with disease status in ADHR [49]. Within the same family, there can be variable presentations with two subgroups of affected individuals described. Those with childhood onset have biochemical and clinical similarities to XLH, whereas those presenting later often lack lower extremity deformities, presumably because of fusion of the growth plate before the development of hypophosphatemia. Recent studies have implicated iron-deficient states in the late manifestation of ADHR. Furthermore, low levels of iron in ADHR subjects are associated with higher FGF23 levels [50].

Treatment

Treatment is similar to that of patients with XLH, consisting of phosphate and calcitriol. As in XLH, patients who do not respond to pharmacologic therapy may require surgical intervention to straighten bowed limbs.

OTHER DISORDERS OF RENAL PHOSPHATE WASTING

Hereditary hypophosphatemic rickets with hypercalciuria

Hereditary hypophosphatemic rickets with hypercalciuria (HHRH) is a rare genetic form of hypophosphatemic rickets characterized by hypophosphatemia, renal phosphate wasting, and preserved responsiveness of $1,25(OH)_2D_3$ to hypophosphatemia (Table 74.1). This appropriate increase in calcitriol leads to increased calcium absorption from the gastrointestinal tract and thus to hypercalciuria and nephrolithiasis. The genetic defect in HHRH is loss of function mutations in the gene that encodes NaPiIIc (SLC34A3) [51, 52], one of three subtypes of the type II sodium phosphate cotransporters. The mutations can be either heterozygous or homozygous loss-of-function mutations in SLC34A3 [53]. HHRH is clinically similar to TIO, yet the distinction is easily made with biochemical testing, with HHRH exhibiting elevated levels of calcitriol and hypercalciuria. Treatment consists of phosphate supplements alone.

Autosomal recessive hypophosphatemic rickets

Mutations in two different genes have been identified in autosomal recessive hypophosphatemic rickets (ARHR). Loss-of-function mutations in dentin matrix protein 1 (DMP-1), a matrix protein related to MEPE and a member of the SIBLING (small integrin binding ligand N-linked glycoprotein) family have been described in ARHR1 [54, 55]. Interestingly, this protein seems to have two functions: it translocates into the nucleus to regulate gene transcription early in osteocyte proliferation and, likely in response to calcium fluxes, becomes phosphorylated and is exported to the extracellular matrix to facilitate mineralization by hydroxyapatite in a process that requires appropriate cleavage of the full-length protein. Loss of DMP-1 function in ARHR leads to modestly and variably increased serum FGF23, dramatically increased expression of FGF23 in bone, defects in osteocyte differentiation, and impaired skeletal mineralization. It seems that the immature osteocytes overproduce FGF23, which acts on the kidney, resulting in phosphaturia and impaired calcitriol synthesis.

Mutations in a second gene, ectonucleotide pyrophosphatase/phosphodiesterase 1 (ENPP1) have also been described in ARHR (ARHR2). ENPP1 has been previously linked to the development of generalized arterial calcification of infancy. It is a regulator of extracellular pyrophosphate and is required for bone mineralization with phosphate. Loss-of-function mutations were observed in five families with ARHR2 [56, 57]. As in XLH, treatment consists of high dose phosphate and calcitriol.

Hypophosphatemic rickets with hyperparathyroidism

Brownstein et al. reported a patient with hypophosphatemic rickets and hyperparathyroidism [58]. The patient had renal phosphate wasting, inappropriately normal $1,25(OH)_2D_3$, and hyperparathyroidism secondary to a genetic translocation resulting in increased levels of α-KLOTHO, the cofactor necessary for FGF23 to bind and activate its receptor (Table 74.1). Interestingly and somewhat unexpectedly, FGF23 serum levels are also markedly elevated in this disorder. These findings implicate α-KLOTHO in the regulation of serum phosphate, FGF23 expression, and parathyroid function.

Fibrous dysplasia

Polyostotic fibrous dysplasia is caused by activating missense mutations in *GNAS* gene, which leads to hormone-independent activation of G-protein (Gsα)-coupled signaling. In fibrous dysplasia, there is replacement of medullary bone and bone marrow with undermineralized bone and fibrotic bone marrow. Evaluation of the fibrous dysplastic bone with pyrosequencing has shown that the presence of GSα mutations in the bone can differentiate fibrous dysplasia from benign fibrous osseous lesions [59].

McCune-Albright syndrome is characterized by the triad of precocious puberty, *café au lait* lesions, and fibrous dysplasia, although other hyperfunctioning endocrine disorders and phosphate wasting may also be found in this condition. Recent evidence has implicated FGF23 as a key component of the phosphate loss in both McCune-Albright related and isolated fibrous dysplasia, the degree of fibrous dysplasia correlating with the degree of phosphate wasting [13]. There has been some debate over whether the hypophosphatemia should be treated if there is no evidence for pathologic rickets. When it is treated, treatment and complication considerations are as for XLH [60].

Linear nevus sebaceous syndrome

Linear nevus sebaceous syndrome (also known as epidermal nevus syndrome) is a rare form of hypophosphatemic rickets. Affected individuals have clinical evidence of multiple cutaneous nevi with radiological evidence of fibrous dysplasia. It is frequently associated with a severe form of hypophosphatemic rickets. Elevated levels of FGF23 are thought to contribute to the renal phosphate wasting [61]. As in XLH, treatment consists of frequently dosed phosphate and calcitriol.

Osteoglophonic dysplasia

As in fibrous dysplasia, renal phosphate wasting and lower than expected calcitriol levels have been observed in osteoglophonic dysplasia (OGD). This is a rare autosomal dominantly inherited form of dwarfism caused by activating mutations in FGF receptor 1 (FGFR1; Table 74.1) [62]. Some but not all patients display phosphate wasting, with the mechanism of the phosphate wasting presumed to be from the high burden of nonossifying bony lesions often seen in these patients. FGF23 produced by the abnormal bone is likely responsible, because the extent of bone lesions is correlated with FGF23 levels and the degree of phosphate wasting.

Hypophosphatemic nephrolithiasis/osteoporosis 1 and 2

Two patients with hypophosphatemia secondary to renal phosphate wasting and osteopenia or nephrolithiasis (NPHLOP1) were found to have heterozygous, dominant negative, mutations in the renal type IIa sodium–phosphate cotransporter gene (SLC34A1) [63]. The prominent symptoms of bone pain and muscle weakness seen in TIO were absent in the affected patients.

Hypophosphatemia associated with nephrolithiasis and low BMD (NPHLOP2) has also been found in connection with mutations in the sodium/hydrogen exchanger regulatory factor 1 (SLC9A3R1)[64].

Fanconi renotubular syndrome 2

Another hypophosphatemic syndrome associated with SLC34A1 mutations has been described [65]. Two patients with autosomal recessive renal Fanconi's syndrome and hypophosphatemic rickets (FRTS2) harbored in-frame duplication of SLC34A1. The patients presented with bone deformities, fractures, and severe short stature. Similar to HHRH, in both SLC34A1 loss of function syndromes, there are hypercalciuria and elevated calcitriol levels.

In summary, either through mutations of the sodium–phosphate transporters themselves, damage to the proximal renal tubule or through aberrant regulation of FGF23, decreased expression or function of the renal sodium–phosphate cotransporters likely represents the common pathway in renal phosphate wasting observed in these syndromes (Fig. 74.1).

HYPERPHOSPHATEMIA

Serum inorganic phosphorus levels are generally maintained 2.5–4.5 mg/dl (0.8075–1.45 mmol/L) in adults and between 6 and 7 mg/dl in children younger than 2 years old. In steady state, oral phosphate loads of up to 4,000 mg/day can be efficiently excreted by the kidneys with minimal rise in serum phosphorus through down-regulation of the sodium–phosphate cotransporter in the proximal renal tubules. Increased PTH secretion also contributes to increased renal phosphate excretion because excess phosphorus complexes with calcium, resulting in a decrease in ionized calcium, which stimulates PTH secretion. There are four general mechanisms whereby phosphate entry into the extracellular fluid can outstrip the rate of renal excretion: (1) acute exogenous phosphate loads, (2) redistribution of intracellular phosphate to the extracellular space, (3) decreased renal excretion, and (4) pseudohyperphosphatemia caused by interference with analytical detection methods.

Clinical Manifestations of hyperphosphatemia

The most common and clinically significant consequence of acute, short-term hyperphosphatemia is hypocalcemia and tetany. With rapid elevations in phosphate load, hypocalcemia and tetany can occur. Hyperphosphatemia suppresses the renal 1 α hydroxylase enzyme, reducing circulating $1,25(OH)_2D_3$, which further aggravates hypocalcemia by impairing intestinal calcium absorption.

In contrast, consequences of chronic hyperphosphatemia include soft tissue calcification and in the case of renal failure, secondary hyperparathyroidism, and renal osteodystrophy. In the setting of chronic kidney disease, the defenses against mineralization are compromised and hyperphosphatemia promotes mineral deposition in soft tissues. Hyperphosphatemia stimulates vascular cells to undergo osteogenic differentiation. Calcification of coronary arteries and heart valves is associated with hypertension, congestive heart failure, coronary artery disease, and myocardial infarction. Medial calcification of peripheral arteries associated with hyperphosphatemia may lead to calciphylaxis, a disorder with high mortality and morbidity.

Hyperphosphatemia that results from renal failure plays a major role in the development of secondary hyperparathyroidism and renal osteodystrophy. This is discussed in detail in Chapter 77 and Chapter 78.

Causes of hyperphosphatemia

The causes of hyperphosphatemia are enumerated in Table 74.3. Genetic causes of hyperphosphatemia are discussed in more detail below.

FAMILIAL TUMORAL CALCINOSIS

Hyperphosphatemic familial tumoral calcinosis (HFTC; MIM# 211900) is an inherited disorder notable for pro-

Table 74.3. Causes of Hyperphosphatemia

Mechanism	Etiology
Decreased renal excretion	Renal insufficiency/failure
	Hypoparathyroidism
	Pseudohypoparathyroidism
	Tumoral calcinosis
	Acromegaly
	Bisphosphonates
Acute phosphate load	Phosphate containing laxatives
	Fleet's phosphosoda enemas
	Intravenous phosphate
	Parental nutrition
Redistribution to the extracellular space	Tumor lysis
	Rhabdomyolysis
	Acidosis
	Hemolytic Anemia
	Severe Hyperthermia
	Fulminant Hepatitis
	Systemic infections
Pseudohyperphosphatemia	Hyperglobulinemia
	Hyperlipidemia
	Hemolysis
	Hyperbilirubinemia

gressive deposition of calcium phosphate crystals in periarticular spaces and soft tissues. There are both hyperphosphatemic and normophosphatemic forms (NFTC; MIM# 610455) of this disorder.

Genetics

To date, there have been mutations found in four different genes among the families with FTC. In the normophosphatenic form, mutations in sterile α motif domain-containing-9 protein (SAMD9; MIM# 610456) have been described. For the hyperphosphatemic form, inactivating mutations have been found in UDP-N-acetyl-α-D-galactosamine: polypeptide N-acetylgalactosaminyltransferase 3(GALNT3; MIM# 601756)[66], FGF23 [67], and KLOTHO [68] (Fig. 74.1). The mutations lead to loss of function resulting in inadequate FGF23 protein levels or FGF23 action. All patients have biallelic mutations indicating an autosomal recessive inheritance pattern.

Clinical and biochemical manifestations

Patients with FTC have heterotopic calcifications that are typically painless and slow growing; however, the masses can become painful if they infiltrate into adjacent structures. The clinical complications are related to masses near and the infiltration of skin, marrow, teeth, blood vessels, and nerves. Range of motion is generally not affected unless the masses become large. A variably present feature of the disease is an abnormality in dentition, characterized by short bulbous roots, pulp stones, and radicular dentin deposited in swirls. Biochemically, along with the hyperphosphatemia, there is increased $1,25(OH)_2D_3$ with normal calcium and alkaline phosphatase levels. Urinary phosphate excretion is frequently low. Radiographs show large aggregates of irregularly dense calcified lobules.

TREATMENT OF HYPERPHOSPHATEMIA

Treatment of hyperphosphatemia should address the underlying etiology. In the case of acute exogenous phosphorus overload, prompt discontinuation of supplemental phosphate and hydration allow for rapid renal excretion and correction of hyperphosphatemia. When transcellular shifts are the cause of hyperphosphatemia (e.g., tumor lysis, rhabdomyolysis), dietary phosphate restriction and diuresis are often successful. In diabetic ketoacidosis (DKA), treatment with insulin and treatment of acidosis reverses the hyperphosphatemia. Phosphate binders such as calcium salts, sevelamer, and lanthanum carbonate, along with dietary restriction, are indicated in renal failure. Hemodialysis may be indicated in acute hyperphosphatemia in the setting of renal dysfunction.

In familial tumoral calcinosis (FTC), medical therapy with aluminum hydroxide along with dietary phosphate and calcium deprivation has been utilized.

Use of the phosphate binder, sevelamer, and the carbonic anhydrase inhibitor, acetazolamide, have been reported to successfully decrease tumor burden. Surgical intervention is generally considered an option in patients when the masses are painful, interfere with function, or are cosmetically unacceptable.

REFERENCES

1. Halevy J, Bulvik S. 1988. Severe hypophosphatemia in hospitalized patients. *Arch Intern Med* 148: 153–5.
2. Larsson L, Rebel K, Sorbo B. 1983. Severe hypophosphatemia—A hospital survey. *Acta Medica Scandinavica* 214: 221–3.
3. McCance R. 1947. Osteomalacia with Looser's nodes (Milkman's syndrome) due to a raised resistance to vitamin D acquired about the age of 15 years. *Q J Med* 16: 33–46.
4. Chong WH, Molinolo AA, Chen CC, Collins MT. 2011. Tumor-induced osteomalacia. *Endocr Relat Cancer* 18(3): R53–77.
5. Drezner MK. 1999. Tumor-induced osteomalacia. In: Favus MJ (ed.) *Primer on Metabolic Bone Diseases*

and *Disorders of Mineral Metabolism*. Philadelphia: Lippincott-Raven. pp. 331–7.
6. Jan de Beur, SM. 2005. Tumor-induced osteomalacia. *JAMA* 294: 1260–7.
7. Kumar, R. 2000. Tumor-induced osteomalacia and the regulation of phosphate homeostasis. *Bone* 27: 333–8.
8. Folpe AL, Fanburg-Smith JC, Billings SD, Bisceglia M, Bertoni F, Cho JY, Econs MJ, Inwards CY, Jan de Beur SM, Mentzel T, Montgomery E, Michal M, Miettinen M, Mills SE, Reith JD, O'Connell JX, Rosenberg AE, Rubin BP, Sweet DE, Vinh TN, Wold LE, Wehrli BM, White KE, Zaino RJ, Weiss SW. 2004. Most osteomalacia-associated mesenchymal tumors are a single histopathologic entity: An analysis of 32 cases and a comprehensive review of the literature. *Am J Surg Pathol* 28: 1–30.
9. Weidner N, Santa Cruz D. 1987. Phosphaturic mesenchymal tumors. A polymorphous group causing osteomalacia or rickets. *Cancer* 59: 1442–54.
10. De Beur SM, Finnegan RB, Vassiliadis J, Cook B, Barberio D, Estes S, Manavalan P, Petroziello J, Madden SL, Cho JY, Kumar R, Levine MA, Schiavi SC. 2002. Tumors associated with oncogenic osteomalacia express genes important in bone and mineral metabolism. *J Bone Miner Res* 17: 1102–10.
11. Shimada T, Mizutani S, Muto T, Yoneya T, Hino R, Takeda S, Takeuchi Y, Fujita T, Fukumoto S, Yamashita T. 2001. Cloning and characterization of FGF23 as a causative factor of tumor-induced osteomalacia. *Proc Natl Acad Sci U S A* 98(11): 6500–5.
12. Sitara D, Razzaque MS, Hesse M, Yoganathan S, Taguchi T, Erben RG, Jüppner H, Lanske B. 2004. Homozygous ablation of fibroblast growth factor-23 results in hyperphosphatemia and impaired skeletogenesis, and reverses hypophosphatemia in Phex-deficient mice. *Matrix Biol* 23: 421–32.
13. Riminucci M, Collins MT, Fedarko NS, Cherman N, Corsi A, White KE, Waguespack S, Gupta A, Hannon T, Econs MJ, Bianco P, Gehron Robey P. 2003. FGF-23 in fibrous dysplasia of bone and its relationship to renal phosphate wasting. *J Clin Invest* 112: 683–92.
14. Segawa H, Yamanaka S, Ohno Y, Onitsuka A, Shiozawa K, Aranami F, Furutani J, Tomoe Y, Ito M, Kuwahata M, Imura A, Nabeshima Y, Miyamoto K. 2007. Correlation between hyperphosphatemia and type II Na-Pi cotransporter activity in klotho mice. *Am J Physiol Renal Physiol* 292: F769–79.
15. Gattineni J, Bates C, Twombley K, Dwarakanath V, Robinson ML, Goetz R, Mohammadi M, Baum M. 2009. FGF23 decreases renal NaPi-2a and NaPi-2c expression and induces hypophosphatemia in vivo predominantly via FGF receptor 1. *Am J Physiol Renal Physiol* 297: F282–91.
16. Shimada T, Hasegawa H, Yamazaki Y, Muto T, Hino R, Takeuchi Y, Fujita T, Nakahara K, Fukumoto S, Yamashita T. 2004. FGF-23 is a potent regulator of vitamin D metabolism and phosphate homeostasis. *J Bone Miner Res* 19: 429–35.
17. Strom TM, Juppner H. 2008. PHEX, FGF23, DMP1 and beyond. *Curr Opin Nephrol Hypertens* 17: 357–62.
18. Jonsson KB, Zahradnik R, Larsson T, White KE, Sugimoto T, Imanishi Y, Yamamoto T, Hampson G, Koshiyama H, Ljunggren O, Oba K, Yang IM, Miyauchi A, Econs MJ, Lavigne J, Jüppner H. 2003. Fibroblast growth factor 23 in oncogenic osteomalacia and X-linked hypophosphatemia. *N Engl J Med* 348: 1656–63.
19. Duet M, Kerkeni S, Sfar R, Bazille C, Lioté F, Orcel P. Clinical impact of somatostatin receptor scintigraphy in the management of tumor-induced osteomalacia. *Clin Nucl Med* 33: 752–6.
20. Jan de Beur SM, Streeten EA, Civelek AC, McCarthy EF, Uribe L, Marx SJ, Onobrakpeya O, Raisz LG, Watts NB, Sharon M, Levine MA. 2002. Localisation of mesenchymal tumours by somatostatin receptor imaging. *Lancet* 359: 761–3.
21. Seufert J, Ebert K, Müller J, Eulert J, Hendrich C, Werner E, Schuüze N, Schulz G, Kenn W, Richtmann H, Palitzsch KD, Jakob F. 2001. Octreotide therapy for tumor-induced osteomalacia. *N Engl J Med* 345: 1883–8.
22. Fukumoto S, Takeuchi Y, Nagano A, Fujita T. 1999. Diagnostic utility of magnetic resonance imaging skeletal survey in a patient with oncogenic osteomalacia. *Bone* 25: 375–7.
23. Dupond JL, Mahammedi H, Prié D, Collin F, Gil H, Blagosklonov O, Ricbourg B, Meaux-Ruault N, Kantelip B. 2005. Oncogenic osteomalacia: Diagnostic importance of fibroblast growth factor 23 and F-18 fluorodeoxyglucose PET/CT scan for the diagnosis and follow-up in one case. *Bone* 36: 375–8.
24. Andreopoulou P, Dumitrescu CE, Kelly MH, Brillante BA, Peck CM, Wodajo FM, Chang R, Collins MT. 2011. Selective venous catheterization for the localization of phosphaturic mesenchymal tumors. *J Bone Miner Res* 26(6): 1295–302. doi: 10.1002/jbmr.316.
25. Hesse E, Rosenthal H, Bastian L. 2007. Radiofrequency ablation of a tumor causing oncogenic osteomalacia. *N Engl J Med* 357: 422–4.
26. Geller JL, Khosravi A, Kelly MH, Riminucci M, Adams JS, Collins MT. 2007. Cinacalcet in the management of tumor-induced osteomalacia. *J Bone Miner Res* 22: 931–7.
27. Paglia F, Dionisi S, Minisola S. 2002. Octreotide for tumor-induced osteomalacia. *N Engl J Med* 346: 1748–9; author reply 1748–9.
28. Albright F, Butler A, Bloomberg E. 1939. Rickets resistant to vitamin D therapy. *Am J Dis Child* 54: 529–47.
29. Winters RW, Graham JB, Williams TF, Mcfalls VW, Burnett CH. 1958. A genetic study of familial hypophosphatemia and vitamin D resistant rickets with a review of the literature. *Medicine (Baltimore)* 37: 97–142.
30. [No authors listed]. 1995. A gene (PEX) with homologies to endopeptidases is mutated in patients with X-linked hypophosphatemic rickets. The HYP Consortium. *Nat Genet* 11: 130–6.
31. Beck L, Soumounou Y, Martel J, Krishnamurthy G, Gauthier C, Goodyer CG, Tenenhouse HS. 1997. Pex/PEX tissue distribution and evidence for a deletion in the 3′ region of the Pex gene in X-linked hypophosphatemic mice. *J Clin Invest* 99: 1200–9.

32. Weber TJ, Liu S, Indridason OS, Quarles LD. 2003. Serum FGF23 levels in normal and disordered phosphorus homeostasis. *J Bone Miner Res* 18: 1227–34.
33. Yamazaki Y, Okazaki R, Shibata M, Hasegawa Y, Satoh K, Tajima T, Takeuchi Y, Fujita T, Nakahara K, Yamashita T, Fukumoto S. 2002. Increased circulatory level of biologically active full-length FGF-23 in patients with hypophosphatemic rickets/osteomalacia. *J Clin Endocrinol Metab* 87: 4957–60.
34. Liu S, Tang W, Fang J, Ren J, Li H, Xiao Z, Quarles LD. 2009. Novel regulators of Fgf23 expression and mineralization in Hyp bone. *Mol Endocrinol* 23: 1505–18.
35. Ruppe MD, Brosnan PG, Au KS, Tran PX, Dominguez BW, Northrup H. 2011. Mutational analysis of PHEX, FGF23 and DMP1 in a cohort of patients with hypophosphatemic rickets. *Clin Endocrinol (Oxf)* 74: 312–8.
36. Carpenter TO, Imel EA, Holm IA, Jan de Beur SM, Insogna KL. A clinician's guide to X-linked hypophosphatemia. *J Bone Miner Res* 26(7): 1381–8.
37. Sabbagh Y, Tenenhouse HS, Econs MJ. 2008. The online metabolic & molecular bases of inherited disease. In: Valle D (ed.) *Mendelian Hypophosphatemias*. New York: *McGraw-Hill Companies*.
38. Carpenter TO, Imel EA, Holm IA, Jan de Beur SM, Insogna KL. A clinician's guide to X-linked hypophosphatemia. *J Bone Miner Res* 26(7): 1381–8.
39. Imel EA, DiMeglio LA, Hui SL, Carpenter TO, Econs MJ. 2010. Treatment of X-linked hypophosphatemia with calcitriol and phosphate increases circulating fibroblast growth factor 23 concentrations. *J Clin Endocrinol Metab* 95: 1846–50.
40. Liu ES, Carpenter TO, Gundberg CM, Simpson CA, Insogna KL. 2011. Calcitonin administration in X-linked hypophosphatemia. *N Engl J Med* 364: 1678–80.
41. Wilson DR, York SE, Jaworski ZF, Yendt ER. 1965. Studies in hypophosphatemic vitamin D-refractory osteomalacia in adults. *Medicine (Baltimore)* 44: 99–134.
42. Harrison HE, Harrison HC, Lifshitz F, Johnson AD. 1966. Growth disturbance in hereditary hypophosphatemia. *Am J Dis Child* 112: 290–7.
43. Econs MJ, McEnery PT, Lennon F, Speer MC. 1997. Autosomal dominant hypophosphatemic rickets is linked to chromosome 12p13. *J Clin Invest* 100: 2653–7.
44. ADHR Consortium. 2000. Autosomal dominant hypophosphataemic rickets is associated with mutations in FGF23. *Nat Genet* 26: 345–8.
45. Shimada T, Muto T, Urakawa I, Yoneya T, Yamazaki Y, Okawa K, Takeuchi Y, Fujita T, Fukumoto S, Yamashita T. 2002. Mutant FGF-23 responsible for autosomal dominant hypophosphatemic rickets is resistant to proteolytic cleavage and causes hypophosphatemia in vivo. *Endocrinology* 143: 3179–82.
46. Econs MJ, McEnery PT. 1997. Autosomal dominant hypophosphatemic rickets/osteomalacia: Clinical characterization of a novel renal phosphate-wasting disorder. *J Clin Endocrinol Metab* 82: 674–81.
47. Imel EA, Hui SL, Econs MJ. 2007. FGF23 concentrations vary with disease status in autosomal dominant hypophosphatemic rickets. *J Bone Miner Res* 22: 520–6.
48. Imel EA, Peacock M, Gray AK, Padgett LR, Hui SL, Econs MJ. 2011. Iron modifies plasma FGF23 differently in autosomal dominant hypophosphatemic rickets and healthy humans. *J Clin Endocrinol Metab* 96: 3541–9.
49. Bergwitz C, Roslin NM, Tieder M, Loredo-Osti JC, Bastepe M, Abu-Zahra H, Frappier D, Burkett K, Carpenter TO, Anderson D, Garabedian M, Sermet I, Fujiwara TM, Morgan K, Tenenhouse HS, Juppner H. 2006. SLC34A3 mutations in patients with hereditary hypophosphatemic rickets with hypercalciuria predict a key role for the sodium-phosphate cotransporter NaPi-IIc in maintaining phosphate homeostasis. *Am J Hum Genet* 78: 179–92.
50. Lorenz-Depiereux B, Benet-Pagès A, Eckstein G, Tenenbaum-Rakover Y, Wagenstaller J, Tiosano D, Gershoni-Baruch R, Albers N, Lichtner P, Schnabel D, Hochberg Z, Strom TM. 2006. Hereditary hypophosphatemic rickets with hypercalciuria is caused by mutations in the sodium-phosphate cotransporter gene SLC34A3. *Am J Hum Genet* 78: 193–201.
51. Tencza AL, Ichikawa S, Dang A, Kenagy D, McCarthy E, Econs MJ, Levine MA. 2009. Hypophosphatemic rickets with hypercalciuria due to mutation in SLC34A3/type IIc sodium-phosphate cotransporter: Presentation as hypercalciuria and nephrolithiasis. *J Clin Endocrinol Metab* 94: 4433–8.
52. Lorenz-Depiereux B, Bastepe M, Benet-Pagès A, Amyere M, Wagenstaller J, Müller-Barth U, Badenhoop K, Kaiser SM, Rittmaster RS, Shlossberg AH, Olivares JL, Loris C, Ramos FJ, Glorieux F, Vikkula M, Jüppner H, Strom TM. 2006. DMP1 mutations in autosomal recessive hypophosphatemia implicate a bone matrix protein in the regulation of phosphate homeostasis. *Nat Genet* 38: 1248–50.
53. Feng JQ, Ward LM, Liu S, Lu Y, Xie Y, Yuan B, Yu X, Rauch F, Davis SI, Zhang S, Rios H, Drezner MK, Quarles LD, Bonewald LF, White KE. 2006. Loss of DMP1 causes rickets and osteomalacia and identifies a role for osteocytes in mineral metabolism. *Nat Genet* 38: 1310–5.
54. Levy-Litan V, Hershkovitz E, Avizov L, Leventhal N, Bercovich D, Chalifa-Caspi V, Manor E, Buriakovsky S, Hadad Y, Goding J, Parvari R. 2010. Autosomal-recessive hypophosphatemic rickets is associated with an inactivation mutation in the ENPP1 gene. *Am J Hum Genet* 86: 273–78.
55. Lorenz-Depiereux B, Schnabel D, Tiosano D, Häusler G, Strom TM. 2010. Loss-of-function ENPP1 mutations cause both generalized arterial calcification of infancy and autosomal-recessive hypophosphatemic rickets. *Am J Hum Genet* 86: 267–72.
56. Brownstein CA, Adler F, Nelson-Williams C, Iijima J, Li P, Imura A, Nabeshima Y, Reyes-Mugica M, Carpenter TO, Lifton RP. 2008. A translocation causing increased alpha-klotho level results in hypophosphatemic rickets and hyperparathyroidism. *Proc Natl Acad Sci U S A* 105: 3455–60.
57. Liang Q, Wei M, Hodge L, Fanburg-Smith JC, Nelson A, Miettinen M, Foss RD, Wang G. 2011. Quantitative

analysis of activating alpha subunit of the G protein (Gsalpha) mutation by pyrosequencing in fibrous dysplasia and other bone lesions. *J Mol Diagn* 13: 137–42.
58. Dumitrescu CE Collins MT. 2008. McCune-Albright syndrome. *Orphanet J Rare Dis* 3: 12.
59. Hoffman WH, Jueppner HW, Deyoung BR, O'dorisio MS, Given KS. 2005. Elevated fibroblast growth factor-23 in hypophosphatemic linear nevus sebaceous syndrome. *Am J Med Genet A* 134: 233–6.
60. White KE, Cabral JM, Davis SI, Fishburn T, Evans WE, Ichikawa S, Fields J, Yu X, Shaw NJ, McLellan NJ, McKeown C, Fitzpatrick D, Yu K, Ornitz DM, Econs MJ. 2005. Mutations that cause osteoglophonic dysplasia define novel roles for FGFR1 in bone elongation. *Am J Hum Genet* 76: 361–7.
61. Prié D, Huart V, Bakouh N, Planelles G, Dellis O, Gérard B, Hulin P, Benqué-Blanchet F, Silve C, Grandchamp B, Friedlander G. 2002. Nephrolithiasis and osteoporosis associated with hypophosphatemia caused by mutations in the type 2a sodium-phosphate cotransporter. *N Engl J Med* 347: 983–91.
62. Karim Z, Gérard B, Bakouh N, Alili R, Leroy C, Beck L, Silve C, Planelles G, Urena-Torres P, Grandchamp B, Friedlander G, Prié D. 2008. NHERF1 mutations and responsiveness of renal parathyroid hormone. *N Engl J Med* 359: 1128–35.
63. Magen D, Berger L, Coady MJ, Ilivitzki A, Militianu D, Tieder M, Selig S, Lapointe JY, Zelikovic I, Skorecki K. 2010. A loss-of-function mutation in NaPi-IIa and renal Fanconi's syndrome. *N Engl J Med* 362: 1102–9.
64. Topaz O, Shurman DL, Bergman R, Indelman M, Ratajczak P, Mizrachi M, Khamaysi Z, Behar D, Petronius D, Friedman V, Zelikovic I, Raimer S, Metzker A, Richard G, Sprecher E. 2004. Mutations in GALNT3, encoding a protein involved in O-linked glycosylation, cause familial tumoral calcinosis. *Nat Genet* 36: 579–81.
65. Benet-Pagès A, Orlik P, Strom TM, Lorenz-Depiereux B. 2005. An FGF23 missense mutation causes familial tumoral calcinosis with hyperphosphatemia. *Hum Mol Genet* 14: 385–90.
66. Ichikawa S, Guigonis V, Imel EA, Courouble M, Heissat S, Henley JD, Sorenson AH, Petit B, Lienhardt A, Econs MJ. 2007. Novel GALNT3 mutations causing hyperostosis-hyperphosphatemia syndrome result in low intact fibroblast growth factor 23 concentrations. *J Clin Endocrinol Metab* 92: 1943–7.
67. Ichikawa S, Lyles KW, Econs MJ. 2005. A novel GALNT3 mutation in a pseudoautosomal dominant form of tumoral calcinosis: evidence that the disorder is autosomal recessive. *J Clin Endocrinol Metab* 90: 2420–3.
68. Ichikawa S, Imel EA, Kreiter ML, Yu X, Mackenzie DS, Sorenson AH, Goetz R, Mohammadi M, White KE, Econs MJ. 2007. A homozygous missense mutation in human KLOTHO causes severe tumoral calcinosis. *J Clin Invest* 117(9): 2684–91.

75

Vitamin D–Related Disorders

Paul Lips, Natasja M. van Schoor, and Nathalie Bravenboer

Introduction 613
Nutritional Rickets and Osteomalacia 613
Vitamin D Deficiency: Definition, Threshold 614
Epidemiology and Risk Groups 615
Pathogenesis 615
Clinical Picture 616
Laboratory 616
Radiology 616

Bone Histology 616
Treatment of Vitamin D Deficiency 617
Prevention of Falls and Fractures 617
Guidelines for Vitamin D Intake 618
Rickets Caused by Impaired 1α-Hydroxylation or by a Defective Vitamin D Receptor 619
References 620

INTRODUCTION

Nutritional vitamin D deficiency causes insufficient mineralization and rickets in the growing child (Fig. 75.1) or osteomalacia in the adult, when the epiphyseal lines have closed. This is the classical vitamin D–related disease [1]. The active vitamin D metabolite 1,25-dihydroxyvitamin D [$1,25(OH)_2D$] stimulates the absorption of calcium and phosphate from the gut and makes calcium and phosphate available for mineralization. In mild or moderate vitamin D deficiency, a lower serum calcium concentration causes stimulation of the parathyroid glands. The increased serum parathyroid hormone (PTH) increases the conversion of 25-hydroxyvitamin D [$25(OH)D$] to $1,25(OH)_2D$ as a compensatory mechanism. However, the rise of PTH increases bone resorption. In this way, vitamin D deficiency may also cause bone loss and contribute to the pathogenesis of osteoporosis [2].

Rickets and osteomalacia are associated with muscular weakness, and recently it has become clear that mild and moderate vitamin D deficiency may be associated with decreased physical performance and falls [3–5]. In vitro and in vivo studies have shown that $1,25(OH)_2D$ has many actions such as stimulating osteoblasts and longitudinal growth, stimulating the development of the immune system, decreasing proliferation and stimulating differentiation of many cell types, stimulating insulin release and increasing insulin sensitivity [6–8]. During the past few years, vitamin D deficiency has been associated with autoimmune diseases such as diabetes mellitus type 1 and multiple sclerosis, infectious diseases such as tuberculosis, diabetes mellitus type 2, cardiovascular diseases, several types of cancer, and depression [9–11]. In this chapter, the more classical vitamin D–related disorders will be discussed such as nutritional rickets and osteomalacia, genetic disorders of vitamin D synthesis and function. Vitamin D intoxication will be discussed.

NUTRITIONAL RICKETS AND OSTEOMALACIA

The classical picture of nutritional rickets was first described in the 17th century by Whistler and Glisson [2]. The association between rickets and low sunshine exposure was first recognized in the 19th century, and in the beginning of the 20th century rickets, it was experimentally cured by artificial UV light or sunshine [12]. The spectrum of rickets and osteomalacia includes

614 *Disorders of Mineral Homeostasis*

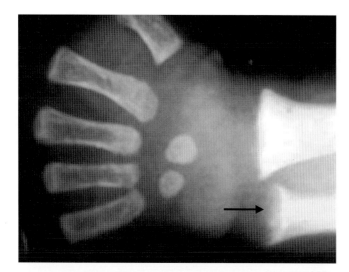

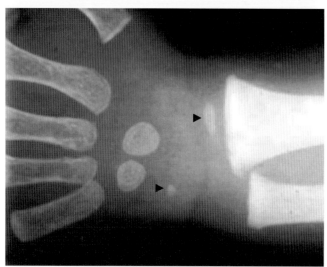

Fig. 75.1. Rickets in a child of 1.5 years. Upper panel shows florid rickets with typical unsharp concave margin of the ulna (arrow). Lower panel shows healing of rickets and appearance of new ossification centers (arrow heads).

Table 75.1. Causes of Rickets and Osteomalacia

Vitamin D–related rickets/osteomalacia
- Nutritional: low sunshine exposure, low dietary intake
- Malabsorption: celiac disease, Crohn's disease, gastrectomy, gastric bypass, bowel resection, pancreatitis
- Impaired hydroxylation in liver: severe chronic liver disease
- Impaired renal function: renal osteodystrophy/osteomalacia
- Increased renal loss: nephrotic syndrome
- Increased catabolism: anticonvulsant therapy
- Non-functioning 25-hydroxylase OMIM #600081
- Absent 1α-hydroxylase: pseudo-vitamin D deficiency rickets (vitamin D–dependent rickets type 1) OMIM #264700*
- Non-functioning VDR: hereditary vitamin D–resistant rickets (vitamin D–dependent rickets type 2), OMIM #277440

Hypophosphatemic rickets/osteomalacia: renal phosphate wasting
- X-linked hypophosphatemic rickets, OMIM #307800*
- Autosomal dominant hypophosphatemic rickets, OMIM #193100*
- Autosomal recessive hypophosphatemic rickets type 1, OMIM # 241520, type 2 OMIM #613312
- Hereditary hypophosphatemic rickets with hypercalciuria, OMIM #241530
- Oncogenic osteomalacia*
- Fanconi syndrome, metabolic acidosis

Calcium deficiency: very low calcium intake in children

Miscellaneous:
- Aluminium intoxication
- Cadmium intoxication
- Etidronate overdose (in Paget's disease)
- Hypophosphatasia, OMIM #146300

*Associated with low serum 1,25(OH)$_2$D.

vitamin D–related and other causes such as liver disease, malabsorption, drugs and renal diseases (Table 75.1).

VITAMIN D DEFICIENCY: DEFINITION, THRESHOLD

Vitamin D$_3$ is produced by the skin after sunlight exposure and is for a small part available from dietary sources. Fatty fish is the most important dietary source. Fortification of foods, e.g., milk, with vitamin D is practiced in the United States of America, Sweden, and Ireland. Following production in the skin, vitamin D is hydroxylated in the liver to 25(OH)D, which is further hydroxylated in the kidney to 1,25(OH)$_2$D, the active metabolite.

Vitamin D status is usually assessed by measuring the serum concentration of 25(OH)D [13]. There is no consensus about the required level of serum 25(OH)D. It has been defined by some as the concentration of serum 25(OH)D that maximally suppresses serum PTH. In some studies among healthy elderly, serum PTH was at its lowest point when serum 25(OH)D was around 75 nmol/l (30 ng/ml), but in other studies serum PTH still decreased when serum 25(OH)D was higher than 100 nmol/l [14]. Another approach is to define thresholds of serum 25(OH)D for different outcomes. Clinical rickets or osteomalacia usually occurs with severe vitamin D deficiency, i.e., very low levels of serum 25(OH)D, often lower than 15 nmol/l (6 ng/ml). At this concentration, serum 1,25(OH)$_2$D is decreased as is calcium absorption [15]. In a round table discussion of experts, the proposed

Table 75.2. Classification of Vitamin D Deficient Status

	25(OH)D nmol/l (ng/ml)	1,25(OH)$_2$D pmol/l (pg/ml)	PTH Increase	Bone Histology
Severe deficiency	<12.5 (<5)	(relatively) low	>30%*	incipient or overt osteomalacia
Deficiency	12.5–25 (5–10)	normal	15–30%	high turnover
Insufficiency	25–50 (10–20)	normal	5–15%	normal or high turnover
Replete	>50 (>20)	normal	–	normal

*Often accompanied by low serum calcium and phosphate, increased alkaline phosphatase, and low 24-hour urinary calcium excretion.

minimally required level of serum 25(OH)D, optimal for fracture prevention, varied between 50 and 75 nmol/l (20–30 ng/ml) [16]. According to a review based on epidemiological data, the most advantageous serum concentration of 25(OH)D for bone mineral density, lower extremity function, falls, fractures, and colorectal cancer is more than 75 nmol/l (30 ng/ml) [17]. In a Dutch cohort study, optimal serum 25(OH)D levels were observed at 40 nmol/l (16 ng/ml) for bone turnover markers, at 50 nmol/l (20 ng/l) for bone mineral density, and at 50–60 nmol/l for physical performance [14]. A consensus at the 13th Vitamin D Workshop agreed that serum 25(OH)D should be higher than 50 nmol/l (20 ng/ml) [18]. In 2010, the Institute of Medicine defined the required serum 25(OH)D level to be 50 nmol/l (20 ng/ml). The Recommended Daily Allowance, i.e., the intake level that is likely to meet the needs of about 97.5% of the population, was defined as 600 IU/day for children from 1 year to adulthood, and 800 IU/day for persons older than 70 years [19]. For the purpose of this chapter, vitamin D deficiency is defined as serum 25(OH)D lower than 25 nmol/l (10 ng/ml). Higher levels of serum 25(OH)D between 25 and 50 nmol/l (10–20 ng/ml), i.e., vitamin D insufficiency, are also associated with secondary hyperparathyroidism and decreased physical performance [14]. Both vitamin D deficiency and insufficiency should be prevented, i.e., serum 25(OH)D should be over 50 nmol/l. The staging of vitamin D status is presented in Table 75.2.

EPIDEMIOLOGY AND RISK GROUPS

Vitamin D deficiency is very common in certain risk groups, such as children (especially premature and dysmature children), pregnant women, the elderly, and non-Western immigrants. Low serum 25(OH)D levels were not only reported in children [20], but also in adolescents and young adults [21]. This may be caused by low sun exposure, reduced synthesis or intake of vitamin D_3 (e.g., being born to a vitamin D deficient mother, dark skin color), malabsorption (e.g., small-bowel disorders), or increased degradation of 25(OH)D (antiepileptics) [22]. Elderly people have decreased dermal synthesis, and especially elderly people who do not come outside frequently are at high risk. Non-Western immigrants migrating to countries at higher latitudes with limited UV-B irradiation are at high risk because of more pigmented skin, the habit to stay out of the sun, the wearing of well-covering clothes, and their diet. Non-Western pregnant women and their children are at a very high risk, and serum 25(OH)D was lower than 25 nmol/l (10 ng/ml) in 85% of pregnant women in a Dutch survey [23, 24].

Large differences in vitamin D status exist between various countries worldwide. It was estimated that less than 50% of the world population has an adequate vitamin D status (greater than 50 nmol/l or 20 ng/ml) [25]. In Europe, a north–south gradient was observed for serum 25(OH)D with higher levels in Scandinavia and lower levels in Southern and Eastern European countries [26]. This points to other determinants, e.g., nutrition, food fortification and supplement use, and cultural habits, e.g., sun-seeking or sun-avoiding behavior. Severe vitamin D deficiency has been reported in the Middle East, China, Mongolia, and India [25].

PATHOGENESIS

While 25(OH)D is the major circulating metabolite and store of vitamin D, almost all vitamin D actions are ascribed to the active metabolite 1,25(OH)$_2$D. This metabolite behaves as a steroid hormone that functions through a nuclear receptor, the vitamin D receptor (VDR). After binding to the VDR and dimerization with the retinoid X receptor, the complex binds to specific DNA regions and more than 300 genes can be activated [8]. Proteins are formed such as the calcium binding protein or calbindin thereby increasing the active calcium absorption in the gut. The enzymatic conversion of 25(OH)D into 1,25(OH)$_2$D is stimulated by PTH. The active metabolite also stimulates bone resorption and decreases PTH secretion. In osteocytes 1,25(OH)$_2$D stimulates fibroblast growth factor 23 (FGF23) expression. FGF-23 functions in the kidney as a phosphaturic hormone, and inhibits 1-alphahydroxylase, resulting in less 1,25(OH)$_2$D. This results in a balanced serum calcium concentration [27].

Studies in the vitamin D receptor null mouse have demonstrated the pleiotropic effect of 1,25(OH)$_2$D.

Studies in transgenic mice demonstrated that direct binding of 1,25(OH)$_2$D to the VDR in cells of the osteoblastic lineage results in a catabolic or anabolic effect depending on the maturation of these cells. The predominant action seems enhancement of the receptor activator of nuclear factor-κB ligand (RANKL) and reduction of osteoprotegerin, thereby stimulating osteoclast formation, but stimulation of osteoblastic genes involved in bone formation such as osteocalcin and osteopontin has also been demonstrated [28, 29].

In cases of children with vitamin D deficiency, the newly formed bone of the growth plate does not mineralize, and cartilage proliferation is prolonged. The growth plate becomes thick, wide and irregular. This results in the clinical diagnosis of rickets [1].

In adults, the newly formed bone matrix, the osteoid, does not mineralize and osteomalacia occurs. Osteomalacia may be suspected on a clinical basis, especially in case of risk factors, including malabsorption, celiac disease or severe liver disease (Table 75.1). Laboratory abnormalities may also point to the diagnosis. Vitamin D also has many nonskeletal effects. The VDR is not only observed in the classical target cells for 1,25(OH)$_2$D but also in the skin, promyelocytes, lymphocytes, colon cells, pituitary cells, and ovarian cells [11]. The function of vitamin D in these tissues has not yet been fully elucidated. The active vitamin D metabolite decreases cell proliferation, and this is applied in practice for the treatment of psoriasis [12]. It is a potent immunomodulator and may play a role in the prevention of autoimmune diseases such as multiple sclerosis and diabetes mellitus type 1 [10]. It has also been implicated in the defense against tuberculosis and other respiratory infections, insulin release and insulin sensitivity, and emotional and cognitive disorders.

CLINICAL PICTURE

The causes of rickets and osteomalacia are summarized in Table 75.1. The clinical picture of rickets is characterized by decreased longitudinal growth, widening of the epiphyseal zones, and painful swelling around these zones [30]. The bowing of the tubular bones is due to poor mineralization of the growing skeleton. Special features include the rachitic rosary caused by swelling of the cartilage of the ribs. Rachitic children do not grow well and exhibit slowed developmental milestones such as walking. The clinical features of osteomalacia include bone pain, muscular weakness, and difficulty with walking. The muscular weakness is preferential to the proximal muscles around shoulder and pelvic girdle. It is manifest with standing up from a chair or stair climbing. When serum calcium is very low in patients with rickets or osteomalacia, symptoms of low serum calcium level may prevail such as tetany and convulsions. Fractures may occur because the quantity of mineralized bone in osteomalacia is low, similar to that in osteoporosis.

LABORATORY

Vitamin D deficiency is diagnosed by measuring serum 25(OH)D. Rickets and osteomalacia are associated with a serum 25(OH)D lower than 25 nmol/l, and often lower than 12.5 nmol/l [15, 31]. Characteristic laboratory findings include a low serum calcium and phosphate concentration or overt hypophosphatemia and an increased level of alkaline phosphatase. The serum concentration of PTH is usually increased (secondary hyperparathyroidism). The serum concentration of 1,25(OH)$_2$D is not helpful as it is normal in most patients with vitamin D deficiency [13, 15]. When serum 25(OH)D is very low, the serum 1,25(OH)$_2$D level may be low due to substrate deficiency [2, 15, 32]. The urinary excretion of calcium over a period of 24 hours is low to very low in patients with vitamin D deficiency due to low calcium absorption from the gut and enhanced tubular reabsorption of calcium due to secondary hyperparathyroidism.

RADIOLOGY

Radiological pictures of patients with rickets show decreased mineralization around the epiphysis, indistinct bone margins and less contrast. The number of ossification centers is decreased. Treatment is followed by rapid improvement (Fig. 75.1).

In osteomalacia, the radiographs show less contrast and are less sharp as if the patient had moved during the X-ray [33]. The classical sign of osteomalacia is the pseudofracture, also named Looser zone. It is a radiolucent line through one cortical plate, often with sclerosis at the margins (Fig. 75.2). The pseudofractures are visible on bone scintigraphy as hot spots.

The assessment of bone mineral density (BMD) by dual X-ray absorptiometry may reveal low values compatible with osteopenia (T-score -1 to -2.5) or osteoporosis (T-score below -2.5). It indicates that the amount of mineralized bone in patients with osteomalacia is low and can be similar to that in patients with osteoporosis [33]. The amount of nonmineralized bone (osteoid tissue) may be high, leading to large and fast increases of BMD (up to 50%) when proper therapy has been instituted.

BONE HISTOLOGY

A transiliac bone biopsy after tetracycline double labeling provides a certain diagnosis of osteomalacia or can exclude it. The biopsy should be included in methylmethacrylate without prior decalcification. Sections should be stained with Goldner or von Kossa. Unstained

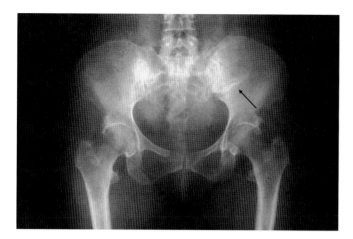

Fig. 75.2. Pseudofracture (arrow) in the left os ilium in a patient with osteomalacia.

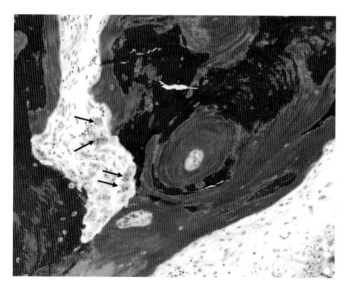

Fig. 75.3. Osteomalacia in a patient with celiac disease. Goldner stain, mineralized bone is black, osteoid tissue is gray. Besides thick osteoid seams, increased bone resorption by multinucleated osteoclasts (arrows) is visible.

sections should be made for tetracycline fluorescence. Typically for osteomalacia, the nonmineralized bone matrix (osteoid) comprises more than 5% of the bone (Fig. 75.3) (osteoid volume more than 5%, usually more than 10%). The osteoid surface is extended, more than 70%, and osteoid seams are thick, more than 15 μm or more than 4 lamellae [1]. Bone resorption is usually increased due to secondary hyperparathyroidism. Howship's (resorption) lacunae are visible with many multinucleated osteoclasts (Fig. 75.3). In frank osteomalacia, double tetracycline labels are not visible, but labeling may appear diffuse and of low intensity. The mineralization index, including osteoid thickness, osteoid volume, mineralization rate and bone formation rate, may be helpful [34].

TREATMENT OF VITAMIN D DEFICIENCY

The symptoms of rickets and osteomalacia may disappear with remarkably low doses of vitamin D_3 (e.g., 800–1,200 IU/day, 20–30 μg/day) unless intestinal absorption is impaired as in celiac disease [35]. However, a higher initial dose (vitamin D_3 2,000 IU/day, 50 μg/day) may be prescribed to bring serum 25(OH)D in the sufficient range (greater than 50 nmol/l or greater than 20 ng/ml) within a few weeks. For maintenance, infants and small children may require as little as 400 IU (10 μg) of vitamin D_3 per day. Adults require 600 to 800 IU (15–20 μg) per day as a maintenance dose. Either vitamin D_3 (cholecalciferol) or D_2 (ergocalciferol) may be prescribed, but vitamin D_2 is less effective [36]. A calcium supplement should always be added. It can be prescribed as calcium carbonate, citrate or glubionate. A dose of 1,000 mg elemental calcium per day in children with rickets and 1,500–2,000 mg per day in adults with osteomalacia is necessary. A special problem is rickets or osteomalacia in patients with celiac disease. A gluten-free diet is essential, but higher doses of vitamin D and calcium may be required, and serum 25(OH)D should be regularly assessed.

Following treatment of severe vitamin D deficiency, serum calcium and phosphate quickly rise into the normal range. The alkaline phosphatase may initially increase but then falls, within weeks to months to normal levels. At the same time, the urinary excretion of calcium increases to the normal range.

The radiological changes of rickets may heal very rapidly. Ossification centers may become visible within weeks, and healing within a few months is the rule (Fig. 75.1). Pseudofractures in adults may heal more slowly and complete disappearance may take more than one year.

Recently, the required doses of vitamin D and the appropriate level of 25(OH)D have been discussed at international meetings and in scientific papers [14, 16, 17, 37]. The required dose depends on the baseline serum 25(OH)D level, the desired level, and the dosing interval. When the baseline serum 25(OH)D is around 20 nmol/l, a dose of 400 IU/day leads to a mean serum 25(OH)D of 55 nmol/l, and a dose of 800 IU/day to 70 nmol/l after 3 months [38]. In general, 800 IU/day (20 μg/day) is a sufficient dose to obtain a serum 25(OH)D level over 50 nmol/l (20 ng/ml) in most people. The Institute of Medicine has defined the required serum 25(OH)D level at 50 nmol/l [19]. When the desired 25(OH)D level is set at 75 nmol/l much higher vitamin D doses are required [39].

PREVENTION OF FALLS AND FRACTURES

The pathway from vitamin D deficiency to falls and fractures is shown in Fig. 75.4.

In a meta-analysis of randomized controlled trials, it was found that a vitamin D dose higher than 400 IU/day

(10μg/day) with or without calcium reduced the risk of hip fracture by 18% and any nonvertebral fracture by 20% versus calcium or placebo [40]. In a Cochrane review, only the combination of vitamin D with calcium reduced hip fractures [pooled relative risk (RR) of 8 trials = 0.84, 95% CI: 0.73 to 0.96], and the authors concluded that vitamin D and calcium decrease hip fracture risk in frail institutionalized older people [41]. Another meta-analysis also showed that the addition of calcium to vitamin D is essential for its preventive effect on fractures [42].

Mild to moderate vitamin D deficiency is also associated with reduced muscle strength, impaired physical performance and falls [3–5, 43]. Vitamin D supplementation in combination with calcium may reduce the risk of falls. A meta-analysis showed that vitamin D_3 700–1,000 IU/day reduced fall risk by 19% compared with placebo (pooled RR from 7 trials = 0.81, 95% CI: 0.71–0.92) [44]. The positive effect on falls was confirmed in a recent meta-analysis [45]. While in most trials daily vitamin D doses were used, the effect of a yearly dose of 500,000 IU was compared with a placebo in an Australian study. The incidence of falls and fractures was higher in the vitamin D group than in the placebo group hazard ratio [HR 1.16 CI: 1.05–1.28 and 1.26 CI: 0.99–1.59, respectively]. This indicates that very high doses of vitamin D leading to very high serum 25(OH)D levels (greater than 120 nmol/l) may have adverse effects and that the curve may be U-shaped [46].

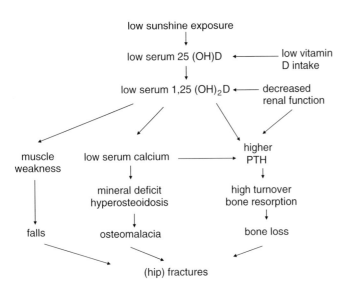

Fig. 75.4. The pathway from vitamin D deficiency to falls and fractures.

GUIDELINES FOR VITAMIN D INTAKE

In Table 75.3, the official recommendations for dietary intake in the U.S. as established by the Institute of Medicine [19] are presented and compared with other guidelines of the Endocrine Society [47], the European Commission (http://ec.europa.eu/food/fs/sc/scf/out157_en.pdf), and The Netherlands [48]. When comparing the recommendations of the Institute of Medicine with those of the Endocrine Society [19, 47], one may conclude that the required level of serum 25(OH)D and the daily requirement of vitamin D are still debated. Some experts

Table 75.3. Recommended Dietary Intake of Vitamin D in μg/day[a]

Age	IOM[b]	Endocrine Society[c]	Europe[d]	Netherlands[g,h] Light-colored skin	Netherlands[g,h] Dark skin or veiled
0–1 yrs	10[g]	10–25	10 (infants)[e]	10 (0–4 yrs)	10 (0–4 yrs)
1–18	15	15–25	0–10 (4–10 yrs) 0–15 (11–17 yrs)		
19–70 yrs	15	37.5–50	0–10 (18–64 yrs) 10 (>65 yrs)	10 (F: 50+ yrs)	10 (F: 4–50; M: 4–70) 20 (F: 50–70)
>70 yrs	20	37.5–50	10[f]	10	20
Pregnant or lactating women	15	37.5–50 (14–18 yrs: 15–25)	10	10	10

IOM = Institute of Medicine.
[a] μg (micrograms) can be converted to International Units (IU) by multiplying by 40
[b] RDA = Recommend Dietary Allowance
[c] Daily requirement
[d] Individual European countries often have more detailed recommendations
[e] Infants from 4 weeks onwards; 6–11 months 10–25 μg/day; 1–3 years 10 μg/day; Scientific Committee on Food of the European Commission (http://ec.europa.eu/food/fs/sc/scf/out157_en.pdf)
[f] The standing Committee of European Doctors now recommends 15–20 μg/day for older persons; see: www.cpme.eu (2009/179 final)
[g] Adequate intake
[h] 20 μg is also recommended for persons having osteoporosis and residents of homes for the elderly and nursing homes

believe that the required levels should be higher. In a recent nursing home study, it was shown that an intake of 600 IU (15 µg) daily reduced the percentage of persons having a serum 25(OH)D below 50 nmol/l from 100% to 10.9% within 4 months [49]. A dose of 800 IU (20 µg) should be sufficient in most people. The latter dose is supported by a meta-analysis on fracture prevention [40]. The Institute of Medicine defined the tolerable upper intake level of vitamin D as 4,000 IU (100 µg) per day for adults, including pregnant and lactating women [19]; the Endocrine Society defined the tolerable upper intake level of vitamin D of 10,000 IU (250 µg) for adults, including pregnant and lactating women aged 19 years and older [47]; the European Commission defined the tolerable upper intake as 2,000 IU (50 µg) for adults, including pregnant and lactating women (http://ec.europa.eu/food/fs/sc/scsf/out157_en.pdf). For children and adolescents, lower tolerable upper intake levels were defined.

However, long-term use of lower doses may also lead to adverse effects as was shown in the Women's Health Initiative, in which significantly more kidney stones occurred in the group receiving 1,000 mg calcium and 400 IU (10 µg) vitamin D daily as compared to a placebo for a period of 7 years (HR = 1.17; 95% CI: 1.02–1.34) [50]. No significant differences in mortality, cardiovascular disease, cancer, and gastrointestinal symptoms were observed.

As discussed above, the Australian trial comparing vitamin D 500,000 IU once per year with placebo resulted in very high serum 25(OH)D (greater than 120 nmol/l) and more falls and fractures in the first 3 months after dosing [46].

RICKETS CAUSED BY IMPAIRED 1α-HYDROXYLATION OR BY A DEFECTIVE VITAMIN D RECEPTOR

The observation that some forms of rickets could not be cured by regular doses of vitamin D led to the discovery of rare inherited abnormalities of vitamin D metabolism or the vitamin D receptor. These inborn errors of metabolism have been confirmed by several mouse knockout models [11, 51].

Pseudovitamin D deficiency rickets (vitamin D–dependent rickets type I, OMIM #264700)

After the discovery of 1,25(OH)$_2$D, it became apparent that some patients with congenital rickets had very low serum concentrations of 1,25(OH)$_2$D that did not increase after vitamin D supplementation [52]. It turned out that these patients did not have 1α-hydroxylase activity. Inactivating mutations of the 1α-hydroxylase gene were identified in affected children [53]. The disease is autosomal recessive.

Clinical features

Patients with pseudovitamin D deficiency rickets present soon after birth with rickets and signs of hypocalcemia, tetany, or convulsions. Laboratory investigations show low serum concentrations of calcium and phosphate and elevated alkaline phosphatase. Serum 25(OH)D is within normal range, but serum 1,25(OH)$_2$D is very low or undetectable [13]. Radiologic examination and bone biopsy show features indistinguishable from nutritional rickets.

Treatment

The patients should be treated with 1,25(OH)$_2$D (calcitriol 0.5–1 µg/day) or 1α-hydroxyvitamin D (alphacalcidol 0.5–1.5 µg/day) to restore serum 1,25(OH)$_2$D to normal levels. This leads to cure of rickets within a few months with normalization of serum calcium, phosphate, and alkaline phosphatase. Another marker of adequate therapy is serum PTH, which should be within reference limits. Overtreatment results in hypercalciuria, hypercalcemia, urolithiasis, and nephrocalcinosis. Serum calcium, serum creatinine, and the urinary excretion of calcium over a period of 24 hours should be checked at regular intervals: in children every month and when stable every 3 months, in adults every 3–6 months. Patients and parents should be instructed on signs and symptoms of underdosage, i.e., hypocalcemia (numbness, pins and needles, tetany) and overdosage, i.e., hypercalcemia (thirst, polyuria, nausea, headache). The requirement of calcitriol or alphacalcidol increases in pregnancy 50% to 100% and frequent monitoring is necessary [54]. Therapy should be continued for life, and accidental discontinuation will lead to hypocalcemia within a few days, later followed by signs of rickets or osteomalacia within weeks to months.

Hereditary vitamin D resistant rickets (vitamin D–dependent rickets type 2, OMIM #277440)

True resistance to 1,25(OH)$_2$D was discovered as some children with congenital rickets did not respond to treatment with calcitriol [55]. In fact, these children had a high serum 1,25(OH)$_2$D, leading to the suspicion of a (post)receptor defect. After cloning of the vitamin D receptor (VDR), mutations have been identified at the DNA binding domain, the 1,25(OH)$_2$D binding domain, and other domains [56]. Hereditary vitamin D–resistant rickets is an autosomal recessive disease.

Clinical features

Affected children born to normal heterozygote parents present early in life with rickets and signs of hypocalcemia, including tetany and convulsions. The first kindreds had alopecia, but later the disease was recognized in children without alopecia. The degree of hormone resistance, i.e., hypocalcemia, may also vary. Laboratory examination shows low serum calcium and serum phosphate and a high alkaline phosphatase level. Serum PTH is increased and serum 1,25(OH)$_2$D is elevated [13]. The latter is caused by stimulation of the renal 1α-hydroxylase by the increased PTH level.

Radiological examination shows the signs of rickets with widened epiphyseal zones and unsharp radiolucent bones. Pseudofractures may be present.

Treatment

The success of treatment is variable and depends on the degree of hormone resistance. When some VDR function is present, a pharmacologic rather than physiologic dose of calcitriol or alphacalcidol can improve calcium absorption and heal the rickets. In case of complete resistance, active vitamin D metabolites are ineffective. Such severely affected individuals can be treated with calcium infusion that overcomes the defective calcium absorption [57]. Rickets can be cured by calcium infusion, and this confirms that mineralization depends on the presence of adequate calcium and phosphate concentrations rather than on the action of $1,25(OH)_2D$. As calcium absorption from the intestine also has a passive component by diffusion—which is independent of vitamin D—very high doses of oral calcium can be effective as well.

Vitamin D intoxication

Vitamin D intoxication is caused by increased ingestion of vitamin D or one of its active metabolites, e.g., alphacalcidol or calcitriol. In case of intoxication by regular vitamin D, the pathogenesis is not completely clear. The serum 25(OH)D concentration is increased over 200 nmol/l up until 1,500 nmol/l, but the serum $1,25(OH)_2D$ level usually is normal [13]. Several mechanisms have been proposed. The high amount of 25(OH)D could act on the vitamin D receptor. More probably, the high serum 25(OH)D could compete with $1,25(OH)_2D$ for binding places of vitamin D binding globulin, thereby increasing the bioavailable (free) serum $1,25(OH)_2D$ levels. In addition, extrarenal activation of 25(OH)D by local 1α-hydroxylase in intestine and bone tissue may contribute to the hypercalcemia.

Clinical features

The patient presents with signs and symptoms of hypercalcemia, i.e., thirst, polyuria, nausea, headache. The cause may be obvious, either supplementation with high doses of vitamin D, or a somewhat high dose of active metabolite. Less obvious causes may be an over-the-counter supplement or table sugar to which vitamin D has been added [58, 59] The patient may appear dehydrated, even confused. Laboratory examination shows hypercalcemia, normal, or increased serum phosphate, increased serum creatinine or BUN, a very high 25(OH)D level, a normal $1,25(OH)_2D$ level, and suppressed serum PTH.

Treatment

The patient should be rehydrated with saline, and corticosteroids may be given to decrease the formation of $1,25(OH)_2D$. The half-life of 25(OH)D is around 25 days, so a complete cure may take some time. Calcium intake should be restricted as long as hypercalcemia is present.

Intoxication with active metabolites

Intoxication with alphacalcidol or calcitriol may appear quickly, but, when recognized, also disappears soon with appropriate measures as the half-life is short, about 7 hours. Treatment should consist of dose reduction or temporary arrest of the active metabolite, rehydration, and restriction of oral calcium intake.

Increased endogenous production of $1,25(OH)_2D$

Endogenous intoxication with calcitriol occurs in the patient with granulomatous disorders or lymphoproliferative diseases when committed macrophages contain 1α-hydroxylase, thus permitting the inappropriate conversion of 25(OH)D into $1,25(OH)_2D$. The latter has been reported from patients with sarcoidosis, tuberculosis, rheumatoid arthritis, Hodgkin's disease, and non-Hodgkin's lymphoma [60–62]. Whereas the renal formation of $1,25(OH)_2D$ is tightly regulated by feedback control, the extrarenal production in these diseases is not. The production of $1,25(OH)_2D$ depends on the amount of substrate i.e., 25(OH)D, and a positive correlation between serum 25(OH)D and $1,25(OH)_2D$ is observed in these disorders, while it is not in normal circumstances [13]. The treatment consists of management of the hypercalcemia and hypercalciuria, and treatment of the underlying disease. Glucocorticoids decrease the hydroxylation of 25(OH)D into $1,25(OH)_2D$ and may help to correct the hypercalcemia.

REFERENCES

1. Parfitt AM. 2005. Vitamin D and the pathogenesis of rickets and osteomalacia. In: Feldman D, Pike JW, Glorieux FH (eds.) *Vitamin D, 2nd Ed.* San Diego: Elsevier Academic Press. pp. 1029–48.
2. Lips P. 2001. Vitamin D deficiency and secondary hyperparathyroidism in the elderly: Consequences for bone loss and fractures and therapeutic implications. *Endocr Rev* 22: 477–501.
3. Bischoff-Ferrari HA, Dietrich T, Orav EJ, Hu FB, Zhang YQ, Karlson EW, Dawson-Hughes B. 2004. Higher 25-hydroxyvitamin D concentrations are associated with better lower-extremity function in both active and inactive persons aged >= 60 y. *Am J Clin Nutr* 80: 752–758.
4. Wicherts IS, van Schoor NM, Boeke AJ, Visser M, Deeg DJ, Smit J, Knol DL, Lips P. 2007. Vitamin D status predicts physical performance and its decline in older persons. *J Clin Endocrinol Metab* 92: 2058–2065.
5. Snijder MB, van Schoor NM, Pluijm SM, van Dam RM, Visser M, Lips P. 2006. Vitamin D status in relation to

one-year risk of recurrent falling in older men and women. *J Clin Endocrinol Metab* 91: 2980–2985.
6. Nagpal S, Na S, Rathnachalam R. 2005. Noncalcemic actions of vitamin D receptor ligands. *Endocr Rev* 26: 662–687.
7. Norman AW. 2006. Minireview: Vitamin D receptor: New assignments for an already busy receptor. *Endocrinology* 147: 5542–5548.
8. Lips P. 2006. Vitamin D physiology. *Prog Biophys Mol Biol* 92: 4–8.
9. Peterlik M, Cross HS. 2005. Vitamin D and calcium deficits predispose for multiple chronic diseases. *Eur J Clin Invest* 35: 290–304.
10. Hypponen E, Laara E, Reunanen A, Jarvelin MR, Virtanen SM. 2001. Intake of vitamin D and risk of type 1 diabetes: A birth-cohort study. *Lancet* 358: 1500–1503.
11. Bouillon R, Carmeliet G, Verlinden L, van Etten E, Verstuyf A, Luderer HF, Lieben L, Mathieu C, Demay M. 2008. Vitamin D and human health: Lessons from vitamin D receptor null mice. *Endocr Rev* 29: 726–776.
12. Holick MF. 1994. McCollum Award Lecture, 1994: Vitamin D—New horizons for the 21st century. *Am J Clin Nutr* 60: 619–630.
13. Lips P. 2007. Relative value of 25(OH)D and 1,25(OH)(2)D measurements. *J Bone Miner Res* 22: 1668–1671.
14. Kuchuk NO, Pluijm SM, van Schoor NM, Looman CW, Smit JH, Lips P. 2009. Relationships of serum 25-hydroxyvitamin D to bone mineral density and serum parathyroid hormone and markers of bone turnover in older persons. *J Clin Endocrinol Metab* 94: 1244–1250.
15. Need AG, O'Loughlin PD, Morris HA, Coates PS, Horowitz M, Nordin BE. 2008. Vitamin D metabolites and calcium absorption in severe vitamin D deficiency. *J Bone Miner Res* 23: 1859–1863.
16. Dawson-Hughes B, Heaney RP, Holick MF, Lips P, Meunier PJ, Vieth R. 2005. Estimates of optimal vitamin D status. *Osteoporos Int* 16: 713–716.
17. Bischoff-Ferrari HA, Giovannucci E, Willett WC, Dietrich T, Dawson-Hughes B. 2006. Estimation of optimal serum concentrations of 25-hydroxyvitamin D for multiple health outcomes. *Am J Clin Nutr* 84: 18–28.
18. Norman AW, Bouillon R, Whiting SJ, Vieth R, Lips P. 2007. 13th Workshop consensus for vitamin D nutritional guidelines. *J Steroid Biochem Mol Biol* 103: 204–205.
19. Ross AC, Manson JE, Abrams SA, Aloia JF, Brannon PM, Clinton SK, Durazo-Arvizu RA, Gallagher JC, Gallo RL, Jones G, Kovacs CS, Mayne ST, Rosen CJ, Shapses SA. 2011. The 2011 report on dietary reference intakes for calcium and vitamin D from the Institute of Medicine: What clinicians need to know. *J Clin Endocrinol Metab* 96: 53–58.
20. Mansbach JM, Ginde AA, Camargo CA Jr. 2009. Serum 25-hydroxyvitamin D levels among US children aged 1 to 11 years: Do children need more vitamin D? *Pediatrics* 124: 1404–1410.
21. Prentice A. 2008. Vitamin D deficiency: A global perspective. *Nutr Rev* 66: S153–S164.
22. Munns C, Zacharin MR, Rodda CP, Batch JA, Morley R, Cranswick NE, Craig ME, Cutfield WS, Hofman PL, Taylor BJ, Grover SR, Pasco JA, Burgner D, Cowell CT. 2006. Prevention and treatment of infant and childhood vitamin D deficiency in Australia and New Zealand: A consensus statement. *Med J Aust* 185: 268–272.
23. van der Meer IM, Karamali NS, Boeke AJP, Lips P, Middelkoop BJC, Verhoeven I, Wuister D 8/2006 High prevalence of vitamin D deficiency in pregnant non-Western women in The Hague, Netherlands. *Am J Clin Nutr* 84: 350–353.
24. Dijkstra SH, van Beek A, Janssen JW, de Vleeschouwer LH, Huysman WA, van den Akker EL. 2007. High prevalence of vitamin D deficiency in newborns of high-risk mothers. *Arch Dis Child Fetal Neonatal Ed* 92(9): 750–3.
25. Lips P, van Schoor NM. 2011. Worldwide Vitamin D Status. In: Feldman D, Pike JW, Adams JS (eds.) *Vitamin D, 3rd Ed., Vol. 1*. San Diego: Elsevier Academic Press. pp. 947–963.
26. Kuchuk NO, van Schoor NM, Pluijm SM, Chines A, Lips P. 2009. Vitamin D status, parathyroid function, bone turnover, and BMD in postmenopausal women with osteoporosis: Global perspective. *J Bone Miner Res* 24: 693–701.
27. Saji F, Shigematsu T, Sakaguchi T, Ohya M, Orita H, Maeda Y, Ooura M, Mima T, Negi S. 2010. Fibroblast growth factor 23 production in bone is directly regulated by 1{alpha},25-dihydroxyvitamin D, but not PTH. *Am J Physiol Renal Physiol* 299: F1212–F1217.
28. Christakos S, Dhawan P, Peng X, Obukhov AG, Nowycky MC, Benn BS, Zhong Y, Liu Y, Shen Q. 2007. New insights into the function and regulation of vitamin D target proteins. *J Steroid Biochem Mol Biol* 103: 405–410.
29. Goltzman D. 2007. Use of genetically modified mice to examine the skeletal anabolic activity of vitamin D. *J Steroid Biochem Mol Biol* 103: 587–591.
30. Pettifor JM. 2005. Vitamin D deficiency and nutritional rickets in children. In: Feldman D, Pike JW, Glorieux FH (eds.) *Vitamin D, 2nd Ed*. San Diego: Elsevier Academic Press. pp. 1065–83.
31. Lips P. Which circulating level of 25-hydroxyvitamin D is appropriate? *J Steroid Biochem Mol Biol* 89–90: 611–614.
32. Bouillon RA, Auwerx JH, Lissens WD, Pelemans WK. 1987. Vitamin D status in the elderly: Seasonal substrate deficiency causes 1,25-dihydroxycholecalciferol deficiency. *Am J Clin Nutr* 45: 755–763.
33. Rabelink NM, Westgeest HM, Bravenboer N, Jacobs MAJM, Lips P. 2011. Bone pain and extremely low bone mineral density due to severe vitamin D deficiency in celiac disease. *Arch Osteoporosis* 6(1-2): 209–213.
34. Parfitt AM, Qiu S, Rao DS. 2004. The mineralization index—A new approach to the histomorphometric appraisal of osteomalacia. *Bone* 35: 320–325.
35. Holick MF. 2007. Vitamin D deficiency. *N Engl J Med* 357: 266–281.
36. Armas LA, Hollis BW, Heaney RP. 2004. Vitamin D2 is much less effective than vitamin D3 in humans. *J Clin Endocrinol Metab* 89: 5387–5391.

37. Lips P. 2004. Which circulating level of 25-hydroxyvitamin D is appropriate? *J Steroid Biochem Mol Biol* 89–90: 611–614.
38. Lips P, Wiersinga A, van Ginkel FC, Jongen MJ, Netelenbos JC, Hackeng WH, Delmas PD, van der Vijgh WJ. 1988. The effect of vitamin D supplementation on vitamin D status and parathyroid function in elderly subjects. *J Clin Endocrinol Metab* 67: 644–650.
39. Heaney RP, Davies KM, Chen TC, Holick MF, Barger-Lux MJ. 2003. Human serum 25-hydroxycholecalciferol response to extended oral dosing with cholecalciferol. *Am J Clin Nutr* 77: 204–210.
40. Bischoff-Ferrari HA, Willett WC, Wong JB, Stuck AE, Staehelin HB, Orav EJ, Thoma A, Kiel DP, Henschkowski J. 2009. Prevention of nonvertebral fractures with oral vitamin D and dose dependency: A meta-analysis of randomized controlled trials. *Arch Intern Med* 169: 551–561.
41. Avenell A, Gillespie WJ, Gillespie LD, O'Connell DL. 2005. Vitamin D and vitamin D analogues for preventing fractures associated with involutional and postmenopausal osteoporosis. *Cochrane Database Syst Rev* (3): CD000227.
42. Boonen S, Lips P, Bouillon R, Bischoff-Ferrari HA, Vanderschueren D, Haentjens P. 2007. Need for additional calcium to reduce the risk of hip fracture with vitamin D supplementation: Evidence from a comparative metaanalysis of randomized controlled trials. *J Clin Endocrinol Metab* 92: 1415–1423.
43. Visser M, Deeg DJH, Lips P. 2003. Low vitamin D and high parathyroid hormone levels as determinants of loss of muscle strength and muscle mass (Sarcopenia): The Longitudinal Aging Study Amsterdam. *J Clin Endocrinol Metab* 88: 5766–5772.
44. Bischoff-Ferrari HA, Dawson-Hughes B, Staehelin HB, Orav JE, Stuck AE, Theiler R, Wong JB, Egli A, Kiel DP, Henschkowski J. 2009. Fall prevention with supplemental and active forms of vitamin D: A meta-analysis of randomised controlled trials. *BMJ* 339: b3692.
45. Kalyani RR, Stein B, Valiyil R, Manno R, Maynard JW, Crews DC. 2010. Vitamin D treatment for the prevention of falls in older adults: Systematic review and meta-analysis. *J Am Geriatr Soc* 58: 1299–1310.
46. Sanders KM, Stuart AL, Williamson EJ, Simpson JA, Kotowicz MA, Young D, Nicholson GC. 2010. Annual high-dose oral vitamin D and falls and fractures in older women: A randomized controlled trial. *JAMA* 303: 1815–1822.
47. Holick MF, Binkley NC, Bischoff-Ferrari HA, Gordon CM, Hanley DA, Heaney RP, Murad MH, Weaver CM. 2011. Evaluation, treatment, and prevention of vitamin D deficiency: An endocrine society clinical practice guideline. *J Clin Endocrinol Metab* 96: 1911–1930.
48. Health Council of the Netherlands. 2008. Towards an adequate intake of vitamin D. Vol. 2008/15. The Hague. http://www.gezondheidsraad.nl/en/publications/towards-adequate-intake-vitamin-d-0.
49. Chel V, Wijnhoven HA, Smit JH, Ooms M, Lips P. 2007. Efficacy of different doses and time intervals of oral vitamin D supplementation with or without calcium in elderly nursing home residents. *Osteoporos Int* 19: 663–71.
50. Jackson RD, LaCroix AZ, Gass M, Wallace RB, Robbins J, Lewis CE, Bassford T, Beresford SAA, Black HR, Blanchette P, Bonds DE, Brunner RL, Brzyski RG, Caan B, Cauley JA, Chlebowski RT, Cummings SR, Granek I, Hays J, Heiss G, Hendrix SL, Howard BV, Hsia J, Hubbell FA, Johnson KC, Judd H, Kotchen JM, Kuller LH, Langer RD, Lasser NL, Limacher MC, Ludlam S, Manson JE, Margolis KL, McGowan J, Ockene JK, O'Sullivan MJ, Phillips L, Prentice RL, Sarto GE, Stefanick ML, Van Horn L, Wactawski-Wende J, Whitlock E, Anderson GL, Assaf AR, Barad D. 2006. Calcium plus vitamin D supplementation and the risk of fractures. *N Engl J Med* 354: 669–683.
51. Panda DK, Miao D, Bolivar I, Li J, Huo R, Hendy GN, Goltzman D. 2004. Inactivation of the 25-hydroxyvitamin D 1alpha-hydroxylase and vitamin D receptor demonstrates independent and interdependent effects of calcium and vitamin D on skeletal and mineral homeostasis. *J Biol Chem* 279: 16754–16766.
52. Fraser D, Kooh SW, Kind HP, Holick MF, Tanaka Y, DeLuca HF. 1973. Pathogenesis of hereditary vitamin-D-dependent rickets. An inborn error of vitamin D metabolism involving defective conversion of 25-hydroxyvitamin D to 1 alpha,25-dihydroxyvitamin D. *N Engl J Med* 289: 817–822.
53. Kitanaka S, Takeyama K, Murayama A, Sato T, Okumura K, Nogami M, Hasegawa Y, Niimi H, Yanagisawa J, Tanaka T, Kato S. 1998. Inactivating mutations in the 25-hydroxyvitamin D-3 1 alpha-hydroxylase gene in patients with pseudovitamin D-deficiency rickets. *N Engl J Med* 338: 653–661.
54. Glorieux FH. 1990. Calcitriol treatment in vitamin D–dependent and vitamin D–resistant rickets. *Metabolism* 39: 10–12.
55. Brooks MH, Bell NH, Love L, Stern PH, Orfei E, Queener SF, Hamstra AJ, DeLuca HF. 1978. Vitamin-D-dependent rickets type II. Resistance of target organs to 1,25-dihydroxyvitamin D. *N Engl J Med* 298: 996–999.
56. Malloy PJ, Hochberg Z, Tiosano D, Pike JW, Hughes MR, Feldman D. 1990. The molecular basis of hereditary 1,25-dihydroxyvitamin D3 resistant rickets in seven related families. *J Clin Invest* 86: 2071–2079.
57. Balsan S, Garabedian M, Larchet M, Gorski AM, Cournot G, Tau C, Bourdeau A, Silve C, Ricour C. 1986. Long-term nocturnal calcium infusions can cure rickets and promote normal mineralization in hereditary resistance to 1,25-dihydroxyvitamin D. *J Clin Invest* 77: 1661–1667.
58. Koutkia P, Chen TC, Holick MF. 2001. Vitamin D intoxication associated with an over-the-counter supplement. *N Engl J Med* 345: 66–67.
59. Vieth R, Pinto TR, Reen BS, Wong MM. 2000. Vitamin D poisoning by table sugar. *Lancet* 359: 672.
60. Barbour GL, Coburn JW, Slatopolsky E, Norman AW, Horst RL. 1981. Hypercalcemia in an anephric patient

with sarcoidosis: Evidence for extrarenal generation of 1,25-dihydroxyvitamin D. *N Engl J Med* 305: 440–443.
61. Davies M, Mawer EB, Hayes ME, Lumb GA. 1985. Abnormal vitamin D metabolism in Hodgkin's lymphoma. *Lancet* 1: 1186–1188.
62. Hewison M, Kantorovich V, Liker HR, Van Herle AJ, Cohan P, Zehnder D, Adams JS. 2003 Vitamin D–mediated hypercalcemia in lymphoma: Evidence for hormone production by tumor-adjacent macrophages. *J Bone Miner Res* 18: 579–582.

76

Vitamin D Insufficiency and Deficiency

J. Christopher Gallagher

Definition 624
What Serum 25OHD Level Is Clinically Important? 624
Background 624
Pathophysiology 625
Causes 625

Fractures 626
Bone Mineral Density (BMD) 627
Summary 629
References 629

DEFINITION

Vitamin D deficiency, insufficiency, and sufficiency are categories used to describe the nutritional status of vitamin D and are defined by the level of serum 25 hydroxyvitamin D (25OHD). The term vitamin D "deficiency" applies to the most severe depletion of vitamin D in the body. It is characterized by intestinal malabsorption of calcium and phosphorus, hypocalcaemia, low 1,25 dihydroxyvitamin D, secondary hyperparathyroidism, and demineralization of bone. Unmineralized bone that accumulates as an excessive amount of osteoid tissue is described as osteomalacia in adults and rickets in children. Vitamin D "insufficiency" has less pronounced secondary hyperparathyroidism but increased bone loss and osteoporosis. Probably the changes in serum vitamin D are a continuum with insufficiency and osteoporosis at one end of the spectrum and severe deficiency and osteomalacia at the other end.

WHAT SERUM 25OHD LEVEL IS CLINICALLY IMPORTANT?

Controversy has arisen over the level of serum 25OHD that defines vitamin D insufficiency and deficiency. In 2011, the Institute of Medicine (IOM) defined insufficiency as less than 20ng/ml [1], however. In 2011 the Endocrine Society (ES) defined it as 20–29ng/ml [2]. These differences are summarized in Fig. 76.1.

One clear biochemical difference between the categories of vitamin D deficiency and insufficiency is serum 1,25 dihydroxyvitamin D because its level decreases when serum 25OHD declines to less than 10–12 ng/ml, provided renal function is normal [3].

BACKGROUND

The best-known consequences of severe vitamin D deficiency are rickets in children and osteomalacia in adults. First described in Northern Europe, it was noted that rickets occurred in as many as 60% of cases at postmortem at the end of the winter and early spring when vitamin D depletion was most extreme [4]. It was originally referred to as the "English disease." Several factors caused this problem: lack of sunlight in the northern latitudes, reduced ultraviolet light because of cloud cover, pollution from smoke due to industrial and coal fired heating in homes, and poor nutrition. In 1921 Hess was amongst the first to show that sunlight and ultraviolet light could cure rickets in children [5]. Supplementation of infant foods eradicated childhood rickets in Northern Europe. However, in certain populations such as Asian immigrants or even in the Middle Eastern coun-

---

Primer on the Metabolic Bone Diseases and Disorders of Mineral Metabolism, Eighth Edition. Edited by Clifford J. Rosen.
© 2013 American Society for Bone and Mineral Research. Published 2013 by John Wiley & Sons, Inc.

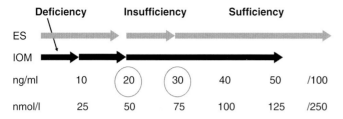

Fig. 76.1. Definition of vitamin status according to the WHO and IOM (2010) and Endocrine Society (2011).

tries that have cultural habits such as purdah and the wearing of the burkah that covers all the body and face, rickets and osteomalacia still can occur, although diet may also be a contributing factor [6, 7]. In North America rickets occurs occasionally in African-American children who only breast feed and in whom maternal levels of 25OHD during pregnancy and lactation are low [8]. Low levels of serum 25OHD are found in high-risk groups such as elderly subjects in nursing homes who have low sunlight exposure [9]. In an elderly population in England, bone biopsies of high-risk patients showed that the incidence of osteomalacia was approximately 4% [10], and in Spain low levels of serum 25OHD less than 6 ng/ml were associated with a finding of osteomalacia in bone biopsies [11]. Several other diseases that can lower serum 25OHD levels are malabsorption syndromes, bypass surgery, liver disease, chronic kidney disease, and drugs such as corticosteroids, phenytoin, ketoconazole, cholestyramine; these can result in osteomalacia [12]. Serum 25OHD levels are also influenced by certain genes. Recent genome-wide screening identified shown four genes associated with serum 25OHD levels, and a combination of three of these genes was able to predict low levels of serum 25OHD [13].

PATHOPHYSIOLOGY

Severe vitamin D deficiency causes malabsorption of calcium and phosphorus from the diet, secondary hyperparathyroidism and defective mineralization of bone leading to osteoporosis and osteomalacia (Fig. 76.2). The increase in serum parathyroid hormone (PTH) that occurs with low serum 25OHD is quite variable [14]. It is more commonly elevated when serum 25OHD levels are less than 15 ng/ml; however, in 30–50% of subjects serum PTH can be low even when serum 25OHD levels are below 15 ng/ml [15]. In elderly subjects, the age-related decrease in absorption of calcium and a low calcium intake may contribute to secondary hyperparathyroidism [16, 17]. At any stage, depending on the level of 25OHD, the sequence can be interrupted. This would occur particularly in summer when sunlight increases serum 25OHD, and mineralization of osteoid occurs partially or completely; thus osteomalacia fluctuates with season [18]. Theoretically a high calcium and phosphorus intake

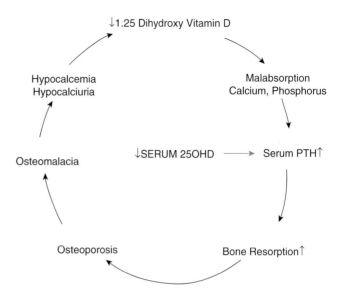

Fig. 76.2. Low serum 25 hydroxyvitamin D (25OHD) stimulates secondary hyperparathyroidism and bone resorption leading to osteoporosis. If 25OHD deficiency is severe < 10 ng/ml (25 nmol/L) it leads to both osteoporosis and osteomalacia.

can prevent demineralization and osteomalacia but in many patients with osteomalacia nutrition is also poor.

Diagnosis of osteomalacia can only be truly confirmed by bone biopsy, although in children there are characteristic radiologic changes that show irregularity of the metaphysis of bone and softening of the bone that leads to bowing of the legs, a finding that alerts parents and physician. However, the biochemical results of low ionized calcium, hypophosphatemia, and low serum 25OHD makes the diagnosis very likely [19]. In elderly people in Europe with osteoporotic hip fractures, histological evidence of osteomalacia was found in 20–30% [18, 20], though another study found a much lower incidence of 12% in the elderly, even though many had very low levels of serum 25OHD [21]. A recent study of bone biopsies in sudden death cases in Germany found that 25% of cases had evidence of osteomalacia based on the criteria of an osteoid volume greater than 2% [21]. This figure of 2% may be low, and others suggest the upper range should be 3% [22]. Ninety-nine percent of these cases occurred when serum 25OHD was 0–20 ng/ml and the other 1% when serum 25OHD was between 20 and 30 ng/ml. Of interest was the finding that in many cases with no detectable serum 25OHD in the blood, there was no evidence of osteomalacia, which suggests that other factors contribute to osteomalacia besides vitamin D.

CAUSES

The two major natural sources of vitamin D are sunlight and diet. Sunlight is the major contributor because normal diets have such a small content of vitamin D.

Table 76.1. Mean Diet and Total Vitamin D Intake and Mean Serum 25OHD for the United States, 2005–2006, (NHANES 2005–6)

Age Group (years)	Vitamin D Intake (IU/day)		Mean Serum 25OHD ng/ml (±SE)[c]
	Diet Alone[a]	Total Intake[b]	
Males			
14–18 yrs	244 ± 16	276 ± 20	24 ± 0.6
19–30	204 ± 12	264 ± 16	23 ± 0.5
31–50	216 ± 12	316 ± 12	24 ± 0.4
51–70	204 ± 12	352 ± 16	24 ± 0.5
>70	224 ± 16	428 ± 28	24 ± 0.4
Female			
14–18 yrs	152 ± 8	200 ± 20	24 ± 0.7
19–30	144 ± 12	232 ± 12	25 ± 0.8
31–50	176 ± 12	308 ± 20	23 ± 0.5
51–70	156 ± 16	404 ± 40	23 ± 0.4
>70	180 ± 8	400 ± 20	23 ± 0.4

NOTE: IU = International Units; SE = standard error. [a]Date are mean ± SE for foods only. [b]Date are mean ± SE for total intake: foods and dietary supplements. [c]Date are mean ± SE.

The amount of vitamin D generated by sunlight (UV-B) depends on latitude, season, time of day, and time of exposure. There is a marked seasonal variation in serum 25OHD. Levels increase between May to September by about 10 ng/ml in summer at a latitude of 40 degrees [23] but only 4 ng/ml at a higher latitude of 55 degrees [24]. As discussed earlier, factors that also decrease UV exposure are sunscreen and darker skin color [25, 26]. A typical diet averages between 100 and 200 IU daily (Table 76.1) but will be higher in areas where fish consumption is higher (300–400 IU/day). In North America, dairy products are fortified with vitamin D; milk has 400 IU added to a quart (approximately 1,000 ml), and cow's milk naturally has about 50 IU. Many people take supplements after age 50, and the average vitamin D intake increases to approximately 400 IU/day (Table 76.1) [1].

The **prevalence** of vitamin D deficiency, insufficiency, and sufficiency varies according to the distance from the equator and therefore the amount of sunlight exposure. In Northern Europe and Canada (latitude >50 degrees) summer seasons are shorter because they are further from the equator, and cloud cover also limits ultraviolet exposure. Recent results from the NHANES study in North America show that 8% of 25OHD values are <10 ng/ml and 24% are between 10 and 20 ng/ml [27].

Assessment of vitamin D status. Vitamin D is absorbed quickly from the diet or from skin, it has a short half-life of 24 hours, and serum levels range from 0 to 100 ng/ml. Vitamin D is transported to the liver and converted to 25OHD that has a half-life of about 3 weeks; levels vary between 10 and 50 ng/ml with a strong seasonal change. From the liver, 25OHD is transported to the kidney and converted to 1,25 dihydroxyvitamin D that has a half-life of about 4 hours and measures 25–50 pg/ml. Because of its longer half-life, serum 25OHD has become the optimal measurement for assessment of nutritional status. Today there are several techniques that are satisfactory for measurement of serum 25OHD such as radioimmunoassay, high performance liquid chromotogaphy, and tandem mass spectroscopy. All methodologies have potential errors but they are continually improving. Generally, if well performed and combined with quality controls, one can be confident of the measurement's accuracy. Although one-third of Caucasians have serum 25OHD levels less than 20 ng/ml, such low values may not have the same implications for people with darker skin color. For example, African-Americans have much lower levels of serum 25OHD than Caucasians [26], but their bone mass is higher and they have fewer fractures; so how should low vitamin D status be classified in people of color? Biological markers or clinical diseases associated with deficiency/insufficiency other than bone have yet to be identified for these groups.

FRACTURES

Most evidence-based data suggest that a serum 25OHD above 20 ng/ml is sufficient for skeletal health. This conclusion is based on five large studies totaling 6,562 subjects that show a significant increase in hip fractures in women and men with serum 25OHD below 20 ng/ml (reviewed in Ref. 28). The evidence linking clinical nonvertebral fractures to low serum 25OHD is different between men and women and ethnicity. In men, non-hip fractures are increased only in those with serum 25OHD below 20 ng/ml [29]. In women, non-hip fractures are significantly decreased in groups with higher serum 25OHD of 20–30 ng/ml and greater than 30 ng/ml compared to less than 20 ng/ml in Caucasian women. However, the opposite is found in African-American and Asian women—fracture events are more common in the groups with serum 25OHD above 20 ng/ml compared to below 20 ng/ml [30]. These are relatively small case control studies and the accuracy of reporting fractures is approximately 80%, so larger studies are needed to confirm these results. Further support for the importance of a low serum 25OHD is the finding that bone resorption markers were increased only in women with a serum 25OHD less than 20 ng/ml [14] and a study in men showing an increased rate of bone loss in those groups with serum 25OHD less than 20 ng/ml compared to higher serum 25OHD above 20 ng/ml [31].

Effect of vitamin D on fractures

The studies are difficult to interpret because none of the individual studies have a dose response design; doses vary from 300 to 1,000 IU/day, and few studies are placebo controlled. There are actually only five studies of vitamin

D alone where the primary outcome was fractures and taken together the results were not significant. Most of the trials usually compare a combination of vitamin D plus calcium compared to a calcium control group or placebo. This has led to various meta-analyses of fracture studies; most use the same data set, and the difference lies in the inclusion criteria used for that particular analysis. The most comprehensive analyses are in the AHRQ-Tufts report that analyzed 17 trials [32]. A reasonable summary of the results show that addition of vitamin and calcium compared to calcium or placebo (14 trials) reduces fractures by 10% (nonsignificant) but a comparison of vitamin D plus calcium compared only to placebo (8 trials) significantly reduces fractures by 13%. In institutionalized patients the reduction in fractures was higher at 31%. Evidence that links a serum 25OHD threshold to treatment with vitamin D in fracture studies is not clear. This is usually because in half the studies the baseline serum 25OHD level is above 20 ng/ml, and efficacy does not seem any different between studies where the baseline starts above or below 20 ng/ml. A study would have to start at a much lower serum 25OHD levels and then exceed 20 ng/ml to address this question. Another issue is that almost all studies are single dose, so a threshold dose cannot be defined. Furthermore, meta-analyses that try to define a threshold response based on serum 25OHD are challenging because it means combining serum 25OHD values from studies and assays performed over a 20-year period between 1992 and 2009 when external independent standards were not used in assays.

BONE MINERAL DENSITY (BMD)

With regard to peak bone mass in young adults, a cross-sectional study from NHANES that included all ages showed an increase in BMD and higher serum 25OHD in groups with 25OHD above 19 ng/ml with an overall gain of about 4% in Caucasians, 1.5% in Hispanics, and 1.5% in African-Americans [33]. But in a cross-sectional study of 7,000 older women with a mean age of approximately 65 years, BMD was significantly lower only in the group with serum 25OHD less than 10 ng/ml compared to groups below 20 ng/ml or above 20 ng/ml [34]. With regard to older women there was only fair evidence to support a link between low serum 25OHD and BMD or change in BMD, with nine studies showing no correlation and 10 showing a correlation [32]. Subsequently, the MrOs study showed increased bone loss in the hip only in the groups with serum 25OHD less than 20 ng/ml [31]

Bone biopsies

As discussed earlier in a study of 675 sudden death autopsies where a bone biopsy and serum 25OHD was measured, the criteria used for osteomalacia was an osteoid volume greater than 2%. Twenty-five percent of the biopsies had osteomalacia, all associated with serum 25OHD below 30 ng/ml; however, in 99% of the samples, serum 25OHD was below 20 ng/ml [21].

Nonskeletal effects

Vitamin D insufficiency and serum parathyroid hormone (PTH)

Many studies have shown that serum PTH is inversely related to serum 25OHD, and that with the increase in serum 25OHD there is a decrease in serum PTH that usually reaches a plateau. In the Endocrine Society guidelines, the statement is made that serum PTH reaches a plateau at a threshold of 30 ng/ml, but this is supported by only three references. In a recent review of 79 studies from the literature [14], the level of serum 25OHD at which serum PTH reached a plateau varied from 12 to 50 ng/ml, with most values being less than 30 ng/ml. In seven studies there was no threshold: thus results do not support a serum 25OHD threshold of 30 ng/ml (Fig. 76.3).

Vitamin D insufficiency and calcium absorption

As vitamin D levels decline, secondary hyperparathyroidism increases and stimulates 1,25 dihydroxyvitamin D production in the kidney, but after age 75 years, as renal function declines, the production of 1,25 dihydroxyvitamin D also declines [35]. There is also a separate age-related decrease in calcium absorption that

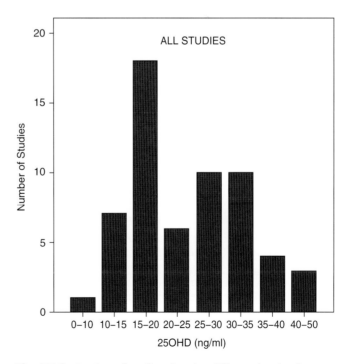

Fig. 76.3. Review of studies showing different levels of serum 25OHD when serum PTH shows a plateau.

occurs a decade earlier [16, 17], and this probably accounts for the intestinal resistance to vitamin D [36]. Renal production of 1,25(OH)$_2$D also can be impaired by serum 25OHD deficiency, and this has been said to occur at a serum 25OHD level of 12 ng/ml by limiting the supply of 25OHD substrate [3, 37]. The most convincing data on calcium absorption support a serum 25OHD threshold of 5–10 ng/ml for normal calcium absorption [37].

Physical function

There is cross-sectional data showing an association between physical function and serum 25OHD. An analysis from 4,100 subjects in the NHANES study showed a relationship between serum 25OHD and two physical performance tests: a timed walking test over 8 feet and the timed rise [38]. The major improvement in physical function occurred between the reference group (lowest quintile) with a serum 25OHD between 2 and 18 ng/ml and the second quintile with a serum 25OHD between 19 and 23 ng/ml; higher serum 25OHD was not associated with further improvements in physical performance. In a 3-year longitudinal study of 1,234 subjects in Northern Europe, physical performance was significantly worse in subjects with serum 25OHD below 20 ng/ml [39]. Prospective studies on the effect of vitamin D and calcium on physical performance tests such as quadriceps strength, grip strength, timed rise, timed walk, and timed up and go between studies have been inconsistent [40].

Frailty

A study has associated greater frailty with serum 25OHD over a 6-year follow-up in older women [41]. The odds ratio for developing frailty was 1.46 (1.19–1.42) if serum 25OHD was below 15 ng/ml; 1.24 (0.99–1.54) if values were 15–20 ng/ml, and 1.32 if serum 25OHD was above 30 ng/ml—the referent was a serum 25OHD of 20–30 ng/ml. Similarly in men, frailty was more likely to occur in men with a serum 25OHD below 20 ng/ml [42].

Falls

In a meta-regression analysis, the IOM was unable to show a significant effect of vitamin D on falls or any significant threshold effect [1].

Association studies with low serum 25OHD

Most tissues in the body have a vitamin D receptor and have the ability to synthesize 1,25(OH)2D from 25OHD in local cells. This has raised questions about the effect of low serum 25OHD on nonskeletal disease. It is very difficult to prove what level of vitamin D deficiency or insufficiency reduces extra renal synthesis. There have been a number of studies that showed a cross-sectional association with several disease such as diabetes and cancer, but due to lack of any clinical trials on the efficacy of vitamin D it is not possible yet to draw any conclusions.

Screening

The recent evidence-based IOM report suggests that a serum 25OHD below 20 ng/ml defines vitamin D insufficiency [1]. The Endocrine Society guidelines recommend that the target level should be 30 ng/ml, and they also recommend that large groups of the normal population undergo screening, including all African-Americans, all Hispanics, all pregnant women, all obese people above 30 kg/m^2 [2]. An estimate shows that if these screening guidelines were adopted, about 50 million people in the United States would need to be screened. Conservatively, the initial cost would be $15 billion dollars without taking into account physician charges and follow-up tests. It seems difficult to justify screening these groups since there are no studies that demonstrate improvement in disease outcomes in older people with osteoporosis and a serum 25OHD below 20 ng/ml.

Certain high risk groups can be identified, particularly the elderly who are institutionalized in nursing homes, those who avoid sunlight exposure, as well as medical conditions such as chronic renal disease, liver disease, malabsorption syndromes, medications known to interfere with vitamin D metabolism, anticonvulsants, corticosteroids, antifungals, and HIV drugs. For these groups screening makes sense.

Management of vitamin D deficiency and insufficiency

Because many elderly people with low vitamin D below 20 ng/ml have increased serum PTH, they also have decreased calcium absorption and a low calcium intake so treatment with a combination of vitamin D and calcium is optimal. Based on the data reviewed above, the minimum target for treatment is to exceed a serum 25OHD of 20 ng/ml; whether it should be 30 ng/ml is arguable pending better data.

It has been suggested that serum 25OHD increases by 0.7 to 1 ng/ml for every 100 IU administered [2]. However, a recent dose response study shows that this is a simplification because the dose response is not linear but quadratic, and the increase is actually four times larger per 100 IU when serum 25OHD levels are less than 20 ng/ml [43]. Vitamin D$_3$, 600–800 IU/day, will increase almost all subjects to a scrum 25OHD greater than 20 ng/ml and 1600–2000 IU/day will increase all subjects to a serum 25OHD greater than 30 ng/ml. It is advisable to add a calcium supplement if the diet calcium cannot be increased. How much calcium should be added? Probably the total calcium intake should be between 800 and 1,000 mg daily. However, a calcium intake of 600–800 mg/day may be all that is needed if the patient is on antiresorptive drugs, since net bone resorption is close to zero and there is no need for calcium supplementation unless calcium intake is less than 600 mg/day.

If the end point is serum 25OHD, then daily doses of 1,000 IU of vitamin D$_2$ (ergocalciferol) are not different from vitamin D$_3$ in the effect on serum 25OHD [44]. Monthly doses of 50,000 IU of vitamin D$_2$ produce a serum 25OHD level that is about 5 ng/ml lower than on 50,000 D_3 because D_2 has a shorter half-life; however, levels are adequate [45] and there is no evidence that a slightly higher serum level on D_3 translates to better clinical efficacy since there are no studies. There is no evidence that the initial treatment of people with vitamin D insufficiency (serum 25OHD less than 20ng/ml) with large doses of vitamin D 50,000 IU weekly for 8 weeks as proposed is any more effective than daily vitamin D. In fact, a study that compared an average of 1,500 IU/day given on a daily, weekly, or monthly dosing schedule showed no difference in final levels of serum 25OHD after 5–8 weeks [46]. In other words, there is no urgent need to increase serum 25OHD with large weekly doses.

Calcitriol (1,25-dihydroxyvitamin D) may be a better choice in the elderly with a low glomerular filtration rate (GFR) less than 40 mls/min because they have of impaired conversion of 25OHD to 1,25-dihydroxyvitamin D [34]. Lower GFR is associated with worse physical performance and more falls [47, 48].

Calcitriol has a rapid onset of action and a half-life of 4–6 hours. It is available in doses of 0.25 and 0.5 micrograms, and the usual dose is 0.25 mcg twice daily. If used, the calcium intake should be restricted to 600–800 mg daily because of a higher incidence of hypercalcemia and hypercalciuria [49].

The main natural sources of vitamin D are diet and sunlight. The natural food sources of vitamin D are fatty fish, cod, salmon, and eggs. In North America, many foods, especially dairy products and cereals, are fortified with vitamin D. Milk is fortified with 400 IU per quart (~ liter) as is orange juice and yogurt. Because all foods and liquids are labeled with the vitamin D content, it is easy to estimate the intake of vitamin D.

Sunlight

For many people, only wintertime supplementation with vitamin D is necessary because in winter the lowest levels of serum 25OHD are from January to March, and levels increase to normal by mid-summer. The mild secondary hyperparathyroidism of winter returns to normal in summer [30]. In Northern Europe, the sunny season is 2 months shorter so that the increase in serum 25OHD is less. Even short exposure to UV-B light increases serum 25OHD efficiently. In North America, serum 25OHD levels increase by about 10 ng/ml, which is equivalent to a dose of 800–1,000 IU vitamin D daily. Sunscreen will block most of the UV effect [32]. In summary, for people who are normally active and outside in the summer, low levels of serum 25OHD occur only for 4–6 months of the year and a low dose of vitamin D 600–800 IU could be used in winter months to avoid vitamin D insufficiency. For those who avoid direct sunlight in summer the dose should be increased to 800 IU.

Safety

There are no safety data on vitamin D alone. The IOM recommended 600–800 IU daily for older people. It also recommended a tolerable upper limit of 4000 IU/day; however, this was not a recommended intake but a guide to avoid toxicity. Although the IOM guidelines for people over 50 years suggest a calcium intake of 1,200 mg/day and vitamin D 800–1,000 IU/day there is no discussion of safety mainly because of the lack of data on different doses [1]. Yet vitamin D 400 IU/day and a supplement of calcium carbonate 1000 mg/day were shown to increase the incidence of kidney stones by 17% over 7 years in the Women's Health Initiative study of 36,000 women [50]. There is little other data on long-term toxicity but mortality data from the NHANES study showed an increase in all cause mortality once serum 25OHD exceeded approximately 50 ng/ml [1] so a conservative use of vitamin D and calcium is prudent pending further results.

SUMMARY

Vitamin D deficiency and insufficiency is an important clinical problem in all parts of the world. Even using a conservative definition of a serum 25OHD of below 20 ng/ml, the studies show that one-third of Caucasian have low levels especially during winter. Screening of high-risk groups for low serum 25OHD and then treatment with vitamin D 600–800 IU and calcium 500–750 mg daily can reduce fractures. At this time there are no prospective data on prevention of nonskeletal diseases potentially linked to vitamin D insufficiency; only clinical trials can provide an answer.

REFERENCES

1. Institute of Medicine. 2011. *Dietary Reference Intakes for Calcium and Vitamin D*. Washington, DC: The National Academies Press. Available from: http://www.ncbi.nlm.nih.gov/books/NBK56072
2. Holick MF, Binkley NC, Bischoff-Ferrari HA, Gordon CM, Hanley DA, Heaney RP, Murad MH, Weaver CM. 2011. Evaluation, treatment, and prevention of vitamin D deficiency: An Endocrine Society clinical practice guideline. *J Clin Endocrinol Metab* 96: 1911–30.
3. Bouillon RA, Auwerx JH, Lissens WD, Pelemans WK. 1987. Vitamin D status in the elderly: Seasonal substrate deficiency causes 1,25-dihydroxycholecalciferol deficiency. *Am J Clin Nutr* 45: 755–63.
4. Schmorl. 1909. Die Pathologische anatomie der rachitischen knochenerkrankuwa. *Ergebn Inn Med Kinderheilk* 4: 403.
5. Hess AF, Gutman P. 1921. Cure of infantile rickets by sunlight as demonstrated by a chemical alteration of the blood. *Pro Soc Exp Biol Med* XIX: 31.
6. El-Sonbaty MR, Abdul-Ghaffar NU. 1996. Vitamin D deficiency in veiled Kuwaiti women. *Eur J Clin Nutr* 50: 315–18.

7. Holvik K, Meyer HE, Haug E, Brunvand L. 2005. Prevalence and predictors of vitamin D deficiency in five immigrant groups living in Oslo, Norway: The Oslo Immigrant Health Study. *Eur J Clin Nutr* 59: 57–63.
8. Kreiter SR, Schwartz RP, Kirkman HN Jr, Charlton PA, Calikoglu AS, Davenport ML. 2000. Nutritional rickets in African American breast-fed infants. *J Pediatr* 137: 153–7.
9. McKenna MJ, Freaney R, Meade A, Muldowney FP. 1985. Hypovitaminosis D and elevated serum alkaline phosphatase in elderly Irish people. *Am J Clin Nutr* 41: 101–9.
10. Campbell GA, Kemm JR, Hosking DJ, Boyd RV. 1984. How common is osteomalacia in the elderly? *Lancet* 2(8399): 386–88.
11. Gifre L, Peris P, Monegal A, Martinez de Osaba MJ, Alvarez L, Guañabens N. 2011. Osteomalacia revisited: A report on 28 cases. *Clin Rheumatol* 30: 639–45.
12. Bhan A, Rao AD, Rao DS. 2010. Osteomalacia as a result of vitamin D deficiency. *Endocrinol Metab Clin North Am* 39: 321–31.
13. Wang TJ, Zhang F, Richards JB, Kestenbaum B, van Meurs et al. 2010. Common genetic determinants of vitamin D insufficiency: A genome-wide association study. *Lancet* 376(9736): 180–8
14. Sai AJ, Walters RW, Fang X, Gallagher JC. 2011. Relationship between vitamin D, parathyroid hormone and bone health. *J Clin Endocrinol Metab* 18: 1101–12
15. Sahota O, Mundey MK, San P, Godber IM, Lawson N, Hosking DJ. 2004. The relationship between vitamin D and parathyroid hormone: Calcium homeostasis, bone turnover, and bone mineral density in postmenopausal women with established osteoporosis. *Bone* 35: 312–9.
16. Bullamore JR, Gallagher JC, Wilkinson R, Nordin BEC. 1970. Effect of age on calcium absorption. *Lancet* 11: 535–37.
17. Gallagher JC, Riggs BL, Eisman J, Hamston A, Arnaud SB, DeLuca HF. 1979. Intestinal calcium absorption and serum vitamin D metabolites in normal and osteoporotic patients. Effect of age and dietary calcium. *J Clin Invest* 64: 729–36.
18. Aaron J, Gallagher JC, Nordin BEC. 1974. Seasonal variations of histological osteomalacia in femoral neck fracture. *Lancet* 2(7872): 84–5.
19. Francis RM, Selby PL. 1997. Osteomalacia. *Baillieres Clin Endocrinol Metab* 1: 145–63.
20. Wilton TJ, Hosking DJ, Pawley E, Stevens A, Harvey L. 1987. Osteomalacia and femoral neck fractures in the elderly patient. *J Bone Joint Surg Br* 69: 388–90.
21. Priemel M, von Domarus C, Klatte TO, Kessler S, Schlie J, Meier S, Proksch N, Pastor F, Netter C, Streichert T, Püschel K, Amling M. 2010. Bone mineralization defects and vitamin D deficiency: Histomorphometric analysis of iliac crest bone biopsies and circulating 25-hydroxyvitamin D in 675 patients. *J Bone Miner Res* 25: 305–12.
22. Recker RR, Kimmel DB, Parfitt AM, Davies KM, Keshawarz N, Hinders S. 1988. Static and tetracycline-base bone histomorphometric data from 34 normal postmenopausal females. *J Bone Miner Res* 3: 133–44.
23. Rapuri PB, Kinyamu HK, Gallagher JC, Haynatzka V. 2002. Seasonal changes in calciotropic hormones, bone markers, and bone mineral density in elderly women. *J Clin Endocrinol Metab* 87: 2024–32.
24. Macdonald HM, Mavroeidi A, Barr RJ, Black AJ, Fraser WD, Reid DM. 2008. Vitamin D status in postmenopausal women living at higher latitudes in the UK in relation to bone health, overweight, sunlight exposure and dietary vitamin D. *Bone* 5: 996–1003.
25. Holick MF, Matsuoka LY, Wortsman J. 1995. Regular use of sunscreen on vitamin D levels. *Arch Dermatol* 131: 1337–9.
26. Nesby-O'Dell S, Scanlon KS, Cogswell ME, Gillespie C, Hollis BW, Looker AC, Allen C, Doughertly C, Gunter EW, Bowman BA. 2002. Hypovitaminosis D prevalence and determinants among African American and white women of reproductive age: Third National Health and Nutrition Examination Survey, 1988–1994. *Am J Clin Nutr* 76: 187–92.
27. Looker AC, Johnson CL, Lacher DA, et al. 2011. *Vitamin D Status: United States, 2001–2006. NCHS Data Brief, No. 59*. Hyattsville, MD: National Center for Health Statistics.
28. Gallagher JC, Sai AJ. 2010. Vitamin D insufficiency, deficiency, and bone health. *J Clin Endocrinol Metab* 95: 2630–30.
29. Cauley JA, Parimi N, Ensrud KE, Bauer DC, Cawthon PM, Cummings SR, Hoffman AR, Shikany JM, Barrett-Connor E, Orwoll E. 2010. Osteoporotic fractures in men research G. Serum 25-hydroxyvitamin D and the risk of hip and nonspine fractures in older men. *J Bone Miner Res* 25: 545–53.
30. Cauley JA, Danielson ME, Boudreau R, Barbour KE, Horwitz MJ, Bauer DC, Ensrud KE, Manson JE, Wactawski-Wende J, Shikany JM, Jackson RD. 2011. Serum 25-hydroxyvitamin D and clinical fracture risk in a multiethnic cohort of women: The Women's Health Initiative (WHI). *J Bone Miner Res* 26: 2378–88.
31. Ensrud KE, Taylor BC, Paudel ML, Cauley JA, Cawthon PM, Cummings SR, Fink HA, Barrett-Connor E, Zmuda JM, Shikany JM, Orwoll ES; Osteoporotic Fractures in Men Study Group. 2009. Serum 25-hydroxyvitamin D levels and rate of hip bone loss in older men. *J Clin Endocrinol Metab* 94: 2773–80.
32. Chung M, Balk EM, Brendel M, Ip S, Lau J, Lee J, Lichtenstein A, Patel K, Raman G, Tatsioni A, Terasawa T, Trikalinos TA. 2009. *Vitamin D and Calcium: A Systematic Review of Health Outcomes. Evidence Report No. 183*. Prepared by the Tufts Evidence-based Practice Center under Contract No. HHSA 290-2007-10055-I. AHRQ Publication No. 09-E015. Rockville, MD: Agency for Healthcare Research and Quality.
33. Bischoff-Ferrari HA, Dietrich T, Orav EJ, Dawson-Hughes B. 2004. Positive association between 25-hydroxy vitamin D levels and bone mineral density: A population-based study of younger and older adults. *Am J Med*. 116: 634–9.

34. Lips P, Duong T, Oleksik A, Black D, Cummings S, Cox D, Nickelsen T. 2001. Global study of vitamin D status and parathyroid function in postmenopausal women with osteoporosis: Baseline data from the multiple outcomes of raloxifene evaluation clinical trial. *J Clin Endocrinol Metab* 86: 1212–21.
35. Kinyamu HK, Gallagher JC, Petranick KM, Ryschon KL.1996. Effect of parathyroid hormone (hPTH[1-34]) infusion on serum 1,25-dihydroxyvitamin D and parathyroid hormone in normal women. *J Bone Miner Res* 11: 1400–5.
36. Francis RM, Peacock M, Taylor GA, Storer JH, Nordin BE. 1984. Calcium malabsorption in elderly women with vertebral fractures: Evidence for resistance to the action of vitamin D metabolites on the bowel. *Clin Sci (Lond)* 66: 103–7.
37. Need AG, O'Loughlin PD, Morris HA, Coates PS, Horowitz M, Nordin BE. 2008. Vitamin D metabolites and calcium absorption in severe vitamin D deficiency. *J Bone Miner Res* 23: 1859–63.
38. Bischoff-Ferrari HA, Dietrich T, Orav EJ, Hu FB, Zhang Y, Karlson EW, Dawson-Hughes B. 2004. Higher 25-hydroxyvitamin D concentrations are associated with better lower-extremity function in both active and inactive persons aged > or =60 y. *Am J Clin Nutr* 80: 752–8.
39. Wicherts IS, van Schoor NM, Boeke AJ, Visser M, Deeg DJ, Smit J, Knol DL, Lips P. 2007. Vitamin D status predicts physical performance and its decline in older persons. *J Clin Endocrinol Metab* 92: 2058–65.
40. Stockton KA, Mengersen K, Parat JD, Kandiah D, Bennell KL. 2011. Effect of vitamin D supplementation on muscle strength: A systematic review and meta-analysis. *Osteoporos Int* 22: 859–87.
41. Ensrud KE, Ewing SK, Fredman L, Hochberg MC, Cauley JA, Hillier TA, Cummings SR, Yaffe K, Cawthon PM. 2010. Circulating 25-hydroxyvitamin D levels and frailty status in older women. *J Clin Endocrinol Metab* 95: 5266–73.
42. Ensrud KE, Blackwell TL, Cauley JA, Cummings SR, Barrett-Connor E, Dam TT, Hoffman AR, Shikany JM, Lane NE, Stefanick ML, Orwoll ES, Cawthon PM. 2011. Circulating 25-hydroxyvitamin D levels and frailty in older men: The osteoporotic fractures in men study. *J Am Geriatr Soc* 59: 101–6.
43. Gallagher JC, Sai A, Templin T 2nd, Smith L. 2012. Dose response to vitamin D supplementation in postmenopausal women: A randomized trial. *Ann Intern Med* 156(6): 425–37.
44. Holick MF, Biancuzzo RM, Chen TC, Klein EK, Young A, Bibuld D, Reitz R, Salameh W, Ameri A, Tannenbaum AD. 2008. Vitamin D2 is as effective as vitamin D3 in maintaining circulating concentrations of 25-hydroxyvitamin D. *J Clin Endocrinol Metab* 93: 677–81.
45. Heaney RP, Recker RR, Grote J, Horst RL, Armas LAG. 2011. Vitamin D3 is more potent than vitamin D2 in humans. *J Clin Endocrinol Metab* 96: E447–E452.
46. Ish-Shalom S, Segal E, Salganik T, Raz B, Bromberg IL, Vieth R. 2008. Comparison of daily, weekly, and monthly vitamin D3 in ethanol dosing protocols for two months in elderly hip fracture patients. *J Clin Endocrinol Metab* 9: 3430–5.
47. Gallagher JC, Rapuri P, Smith L. 2007. Falls are associated with decreased renal function and insufficient calcitriol production by the kidney. *J Steroid Biochem Mol Biol* 103: 610–13.
48. Dukas L, Schacht E, Runge M. 2010. Independent from muscle power and balance performance, a creatinine clearance below 65 ml/min is a significant and independent risk factor for falls and fall-related fractures in elderly men and women diagnosed with osteoporosis. *Osteoporos Int* 21: 1237–45.
49. Gallagher JC, Fowler SE, Detter JR, Sherman SS. 2001. Combination treatment with estrogen and calcitriol in the prevention of age-related bone loss. *J Clin Endocrinol Metab* 86: 3618–28.
50. Jackson RD, LaCroix AZ, Gass M, Wallace RB, Robbins J, Lewis CE, et al. 2006. Calcium plus vitamin D supplementation and the risk of fractures. *N Engl J Med* 354: 669–83.

77

Pathophysiology of Chronic Kidney Disease Mineral Bone Disorder (CKD–MBD)

Keith A. Hruska and Michael Seifert

Definition 632
Pathobiology 632
Acknowledgments 637
References 637

DEFINITION

The chronic kidney disease–mineral bone disorder (CKD–MBD), is a term coined by the organization Kidney Disease: Improving Global Outcomes (KDIGO) in 2006 [1, 2] for the syndrome associated with chronic kidney diseases (CKDs) in which disorders of mineral metabolism and renal osteodystrophy are key contributors to the excess mortality observed in CKD [3–5]. Secondly, the skeletal remodeling disorders caused by CKD contribute directly to the heterotopic mineralization, especially vascular calcification, and the disordered mineral metabolism that accompany CKD [6, 7]. Thirdly, CKD or renal injury impairs skeletal anabolism decreasing osteoblast function and bone formation rates [7, 8]. As a result of recent advances, a multiorgan system has been defined involving the kidney, skeleton, parathyroid glands, fat and the cardiovasculature that fails in CKD (see Fig. 77.1).

PATHOBIOLOGY

Abnormalities of bone in the CKD–MBD are seen after a relatively mild reduction in the glomerular filtration rate (creatinine clearances between 40 and 70 ml/min, (stage 2 CKD) [9, 10]. In addition, elevated PTH levels and elevated fibroblast growth factor-23 (FGF23) levels are observed before detectable changes in the serum phosphorus, calcitriol, or calcium [11, 12]. The early increase in FGF23 in CKD [12, 13] makes it a powerful biomarker indicating that renal injury has affected osteocytic secretion [13]. If hyperparathyroidism is prevented, the prevalence of a low turnover osteodystrophy, the adynamic bone disorder increases, further demonstrating the effects of kidney injury on the skeleton [8]. By end-stage kidney disease (ESKD), skeletal histologic pathology is found in virtually all patients [14]. The pandemic of CKD and ESKD and the role of the CKD–MBD in the associated mortality, make the syndrome a major health issue for Americans and all developed societies [5, 15, 16].

Pathogenesis

Renal injuries produce circulating signals that affect skeletal osteocytes (see Fig. 77.1) [17]. The resulting stimulation of FGF23 and sclerostin production yield an inhibition to skeletal remodeling that becomes a background as adaptive hyperparathyroidism maintains and even increases remodeling rates. In addition to FGF23, the increases in Dickkopf 1 and sclerostin powerful inhibitors of Wnt signaling in early CKD represent biomarker evidence of skeletal inhibition produced by renal injury. The loss of skeletal anabolism in CKD occurs in the presence of normal parathyroid hormone (PTH), vitamin D, Ca, and PO4 levels, but it is not usually observed as the adynamic bone disorder because disturbed homeostasis in these factors stimulate PTH secretion. The sustained increase in PTH levels produced through adaptation to CKD increase remodeling rates and eventually produce an unwanted high turnover disorder of skeletal remodeling, osteitis fibrosa, (see below).

Primer on the Metabolic Bone Diseases and Disorders of Mineral Metabolism, Eighth Edition. Edited by Clifford J. Rosen.
© 2013 American Society for Bone and Mineral Research. Published 2013 by John Wiley & Sons, Inc.

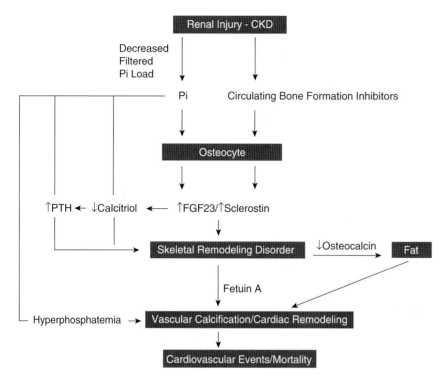

Fig. 77.1. A multiorgan system "wired" together in part by circulating communications is activated in chronic kidney disease. Renal injury increases the circulating levels of inhibitors of bone formation such as Dickkopf-1 and the soluble frizzled related protein family. Renal injury also decreases the filtered load of Pi, and this disturbance, though not sufficient to detectably change serum Pi, stimulates secretion of FGF23 by osteocytes. Through changes in FGF23, sclerostin, PTH, and calcitriol (among others), a skeletal remodeling disorder evolves, which in turn contributes to vascular calcification, cardiac remodeling and cardiovascular morbidity/mortality. Direct connections between each organ have been partially defined in the case of kidney–bone and kidney–heart communications. The contribution of the skeleton and the kidney to hyperphosphatemia, the role of hyperphosphatemia in causing vascular calcification, and the effects of vascular calcification on the heart have been established.

Pathogenetic factors in the CKD–MBD

Fibroblast growth factor 23

Fibroblast growth factor 23 (FGF23) is the original phosphatonin (phosphate excretion regulating hormone) discovered in studies of autosomal dominant hypophosphatemic rickets and oncogenic osteomalacia [18, 19]. FGF23 levels are stimulated by mild renal injury and progressively rise during the course of CKD [20, 21] due to increased secretion, loss of feedback inhibition, and decreased catabolism by the diseased kidney. FGF23 is produced by osteocytes and osteoblasts, and it represents direct bone–kidney and bone–parathyroid connections in the multiorgan system involved in the CKD–MBD (not shown in Fig. 77.1). FGF23 signaling through FGF receptors requires the coreceptor function of α-Klotho [22]. Klotho has limited tissue distribution and thus, defines the tissue targets of FGF23 as a hormone. The off-target actions of FGF23 are associated with CKD (discussed below) and perhaps with anomalous expression of Klotho (also discussed below). The principal tissue targets of FGF23 are the proximal and distal renal tubules, the parathyroid glands and the brain. FGF23 suppresses PTH secretion, and PTH gene expression through FGFR/Klotho signaling [23]. In late CKD, the very high levels of FGF23 may permit anomalous FGF receptor activation independent of Klotho and result in FGF23 stimulated toxic reactions such as cardiac myocyte hypertrophy [24].

Klotho

Klotho was originally identified as an aging suppressor, and its gene product is a single-pass transmembrane protein that can be shed through membrane-anchored proteases [25]. Secreted Klotho functions in an endocrine fashion as an enzyme or possibly as a hormone [26]. Klotho is highly expressed in the renal distal tubule, but its major actions as a FGF23 coreceptor are in the proximal tubule [27], where FGF23/Klotho regulate the activity of NaPi2a and NaPi2c, CYP27B1 (the 1α-hydroxylase) and CYP24A1 (the 24 hydroxylase), the parathyroid gland, and the brain. Recent studies suggest that Klotho is expressed in the arteries of humans and that this renders them an FGF23 target tissue [28]. Klotho deficiency in CKD is associated with vascular calcification [26, 28], and rescue by transgenic overexpression impaired phosphate stimulated vascular calcification.

Hyperphosphatemia

As renal injury decreases nephron number, phosphate excretion is maintained by reductions in the fraction of filtered phosphate reabsorbed by the remaining nephrons under the influence of FGF23 and PTH. The increase in phosphate excretion per remaining nephron maintains phosphate homeostasis at the cost of higher PTH and FGF23 levels. In stages IV and V CKD, when renal injury is severe enough that the glomerular filtration rate reaches levels of less than 30% of normal, hyperphosphatemia develops due to decreased renal excretion despite high PTH and FGF23 levels [29]. Studies demonstrate that the failure of calcium and phosphorous deposition into the skeleton or excess resorption of the skeleton also contribute to abnormal serum phosphorus and calcium levels in CKD and ESKD [7, 30]. Hyperphosphatemia decreases serum calcium through physicochemical complexation and suppresses 1α-hydroxylase activity, which results in further lowering of circulating calcitriol levels (Fig. 77.1). Moreover, a direct stimulatory effect of phosphorus on parathyroid gland cells, independent of calcium and calcitriol, produces increased secretion and nodular hyperplasia of parathyroid gland cells (Fig. 77.1) [31, 32]. A direct stimulatory effect of Pi, perhaps in the exchangeable Pi pool, regulates osteocyte FGF23 secretion. Finally, hyperphosphatemia is a signaling mechanism for induction of heterotopic mineralization of the vasculature in CKD and ESKD (Fig. 77.1) [33–35].

Calcitriol deficiency

The physiologic actions of FGF23 from the osteocyte include inhibition of proximal tubular CYP27B1, 25OH vitamin D 1α hydroxylase, and stimulation of CYP24A1, vitamin D 24-hydroxylase, decreasing calcitriol production, increasing catabolism and producing vitamin D deficiency in early CKD (Fig. 77.1). As CKD advances, the functioning nephron mass is decreased, and this, combined with an increased phosphate load in the remaining nephrons and increased FGF23 levels, results in calcitriol deficiency [36]. Calcitriol deficiency, in turn, decreases intestinal calcium absorption and leads to hypocalcemia. Calcitriol deficiency in cases of advanced kidney failure in turn diminishes tissue levels of vitamin D receptors (VDRs), in particular, the VDR of parathyroid gland cells [37]. Because the chief cell VDR suppresses the expression of pre-pro-PTH mRNA, lower circulating calcitriol levels together with a low number of VDRs in patients with ESKD result in stimulation of both synthesis and secretion of PTH [38].

Hypocalcemia

As CKD progresses, hypocalcemia develops due to decreased intestinal calcium absorption. Low blood levels of ionized calcium stimulate PTH secretion, whereas high calcium concentrations suppress it. The action of calcium on parathyroid gland chief cells is mediated through a calcium sensor, a G-protein coupled plasma membrane receptor (CASR) expressed in chief cells, kidney tubular epithelia, and widely throughout the body at lower levels [39, 40]. The short-term stimulation of PTH secretion induced by low calcium is due to exocytosis of PTH packaged in granules, and longer-term stimulation results from an increase in the number of cells that secrete PTH. More prolonged hypocalcemia induces changes in intracellular PTH degradation and mobilization of a secondary storage pool. Within days or weeks of the onset of hypocalcemia, pre-pro-PTH mRNA expression is stimulated. This effect is exerted through a negative calcium response element located in the upstream flanking region of the gene for PTH. Expression of the calcium receptor has been shown to be suppressed by calcitriol deficiency and stimulated by calcitriol administration, suggesting an additional regulatory mechanism of the active vitamin D metabolite on PTH production. The decreased number of calcium-sensing receptors with low circulating calcitriol may, at least in part, explain the relative insensitivity of parathyroid gland cells to calcium in patients undergoing dialysis.

Calcium balance is deranged in CKD as expected from the discussion above, but often it is affected by medical therapy aimed at phosphate binding or by the use of calcium supplements, surprisingly to the positive side. In CKD, there is little ability to adjust calcium excretion through the kidney for unknown reasons, and the result of increased intake and inadequately regulated absorption is positive balance. This is a major factor in the stimulation of vascular calcification highly prevalent in CKD.

Hyperparathyroidism

All of the mechanisms discussed above result in increased production of PTH and nodular hyperplasia of the parathyroid glands in CKD. The size of the parathyroid glands progressively increases during CKD and, in dialyzed patients, paralleling serum PTH levels. This increase in gland size is mainly due to diffuse cellular hyperplasia. Monoclonal chief cell growth also develops, resulting in the formation of nodules. Nodular hyperplastic glands have less vitamin D receptor and calcium-sensing receptors compared to diffusely hyperplastic glands, promoting parathyroid gland resistance to calcitriol and calcium. Sustained elevation in PTH levels, while adaptive to maintain osteoblast surfaces, produces an abnormal phenotype of osteoblast function and osteocyte stimulation with relatively less type 1 collagen and more receptor activator of nuclear factor-κB ligand (RANKL) production than anabolic osteoblasts. A key component of the RANKL stimulation is osteocytic in origin. This leads to a high turnover osteodystrophy, PTH receptor desensitization, and excess bone resorption.

Hypogonadism

Patients with ESKD have various states of gonadal dysfunction. Estrogen and testosterone deficiency significantly contribute to renal osteodystrophy (ROD) pathogenesis.

Other factors

Inflammatory mediators, acidosis, aluminum, leptin, and retained catabolites are all potentially critical factors in CKD–MBD that have not been well studied, or have been clinically eliminated (space prevents further description of their roles). Some patients with CKD are treated with glucocorticoids, which have an impact on bone metabolism. Patients maintained on chronic dialysis have retention of β2-microglobulin. Additionally, alterations in growth factors and other hormones involved in the regulation of bone remodeling may be disordered in CKD/ESRD, thus affecting bone remodeling and contributing to the development of ROD.

Table 77.1. Pathology and Diagnosis of Bone Turnover in CKD

I. Predominant hyperparathyroidism, high-turnover ROD
   a. Intact PTH >500 pg/ml
   b. Elevated alkaline phosphatase or bone-specific alkaline phosphatase
   c. Sclerostin levels elevated but may be less than in the ABD

II. Low-turnover disease
   a. Adynamic bone disorder (ABD)
      1. Intact PTH <100 pg/ml
      2. Normal alkaline phosphatase or bone-specific alkaline phosphatase
      3. Low osteocalcin
      4. Elevated sclerostin, especially early in CKD before stage V
   b. Osteomalacia
      1. Intact PTH <100 pg/ml
      2. Normal alkaline phosphatase or bone-specific alkaline phosphatase
      3. Low osteocalcin
      4. Elevated Al^{+3}

III. Mixed uremic osteodystrophy
   a. PTH >300 pg/ml
   b. Elevated Al^{+3}

IV. Unknown
   a. PTH > 100 < 500 pg/ml

ROD, renal osteodystrophy, refers to the skeletal pathology component of the CKD–MBD.

Pathology of CKD–MBD (also referred to as ROD)

Renal osteodystrophy (ROD) is a term for the skeletal pathology component of the CKD–MBD syndrome, and this specific definition replaces the previous use of the term in a broader context, which is now specified by the new syndrome term, CKD–MBD. ROD is not a uniform disorder. Depending on the relative contribution of the different pathogenic factors discussed above and their treatment, various pathologic patterns of bone remodeling are expressed in CKD and ESKD. See Table 77.1 for the pathology and diagnosis of bone turnover in CKD.

Predominant hyperparathyroid bone disease, high turnover ROD, osteitis fibrosa

Sustained excess parathyroid hormone results in increased bone turnover. Osteoclasts, osteoblasts, and osteocytes are found in abundance. Disturbed osteoblastic activity results in a disorderly production of collagen, which results in formation of woven bone. Accumulation of fibroblastic osteoprogenitors not in the osteoblastic differentiation program results in collagen deposition (fibrosis) in the peritrabecular and marrow space. The nonmineralized component of bone, osteoid, is increased, and the normal three-dimensional architecture of the osteoid is frequently lost. Osteoid seams no longer exhibit their usual birefringence under polarized light; instead, a disorderly arrangement of woven osteoid and woven bone with a typical crisscross pattern under polarized light is seen. The mineral apposition rate and number of actively mineralizing sites are increased, as documented under fluorescent light after the administration of time-spaced fluorescent (tetracycline) markers.

Low-turnover bone disease, adynamic bone disorder

Low-turnover uremic osteodystrophy is the other end of the spectrum of renal osteodystrophy. The histologic hallmark of these disorders is a profound decrease in bone turnover, due to a low number of active remodeling sites, suppression of bone formation, and resorption. Bone resorption is not as decreased as formation. The result is a low turnover osteopenic condition. The majority of trabecular bone is covered by lining cells, with few osteoclasts and osteoblasts. Bone structure is predominantly lamellar. The extent of mineralizing surfaces is markedly reduced. Usually only a few thin, single tetracycline labels are observed. Two subgroups can be identified in this type of renal osteodystrophy, depending on the cause of events leading to a decline in osteoblast activity: adynamic bone disorder (ABD) and low turnover osteomalacia from Al^{+3} intoxication, bisphosphonate administration, or other factors.

Low-turnover osteomalacia is characterized by an accumulation of unmineralized matrix in which a diminution in mineralization precedes or is more pronounced than the inhibition of collagen deposition. Unmineralized bone represents a sizable fraction of trabecular bone volume. The increased lamellar osteoid volume is due to the presence of wide osteoid seams that cover a large portion of the trabecular surface. The occasional presence of woven bone buried within the trabeculae indicates past high bone turnover. When osteoclasts are present, they are usually seen within trabecular bone or at the small fraction of trabecular surface left without osteoid coating.

Mixed uremic osteodystrophy, high turnover ROD plus a mineralization defect

Mixed uremic osteodystrophy is caused primarily by hyperparathyroidism and defective mineralization with or without increased bone formation. These features may coexist in varying degrees in different patients. Increased numbers of heterogeneous remodeling sites can be seen. The number of osteoclasts is usually increased. Because active foci with numerous cells, woven osteoid seams, and peritrabecular fibrosis coexist next to lamellar sites with a more reduced activity, greater production of lamellar or woven osteoid causes an accumulation of osteoid with normal or increased thickness of osteoid seams. Whereas active mineralizing surfaces increase in woven bone with a higher mineralization rate and diffuse labeling, mineralization surfaces may be reduced in lamellar bone with a decreased mineral apposition rate.

Associated features

Osteoporosis and osteosclerosis

With progressive loss of renal function, cancellous bone volume may be increased along with a loss of cortical bone, but this is in part due to deposition of woven immature collagen fibrils instead of lamellar fibrils. Thus, bone strength suffers despite the increase in mass detected by dual energy X-ray absorptiometry (DXA). Patients undergoing chronic dialysis might have a loss or gain in bone volume depending on bone balance. When the bone balance is positive, osteosclerosis may be observed when osteoblasts are active in depositing new bone (especially woven), thus superseding bone resorption. This is rare in the 21st century due to improved therapy of secondary hyperparathyroidism.

In the case of negative bone balance, bone loss occurs in cortical and cancellous bone and is more rapid when bone turnover is high. In those cases, bone densitometry will detect osteopenia or osteoporosis. The prevalence of osteoporosis in the population with CKD exceeds the prevalence in the general population [41–43]. Osteoporosis is observed in CKD before dialysis is required for end-stage kidney failure [44]. When bone turnover is high, as in secondary hyperparathyroidism with osteitis fibrosa, bone resorption rates are in excess of bone formation, and osteopenia progressing to osteoporosis may result. When bone turnover is low, although both bone formation rates and bone resorption may be reduced, resorption is in excess and loss of bone mass occurs. Thus, osteoporosis may be observed with either high turnover [44–47] or low turnover [48] forms of osteodystrophy. When bone resorption exceeds bone formation rates in CKD, positive phosphorus and calcium balance results in hyperphosphatemia and hypercalcemia without an increase in skeletal mineral deposition, but with a stimulation of heterotopic mineralization, especially of the vasculature. The failure of the skeleton to absorb positive phosphate balance in CKD is an important stimulus to heterotopic mineralization, and links the skeleton and osteoporosis in CKD to cardiovascular events and mortality [35]. From the above discussion, it is apparent that four forms of osteoporosis complicate the CKD–MBD: due to high turnover ROD, due to low turnover ROD, osteoporosis preexistent to renal disease, and osteoporosis due to gonadal hormone deficiency.

Bone aluminum, iron, lanthanum, bisphosphonate accumulation

These substances accumulate in bone at the mineralization front, at the cement lines, or diffusely. The extent of stainable aluminum at the mineralization front correlates with histologic abnormalities in mineralization. Aluminum deposition is most severe in cases of low-turnover osteomalacia. However, it can be observed in all histologic forms of renal osteodystrophy. In patients in whom an increased aluminum burden develops, bone mineralization and bone turnover progressively decrease. These abnormalities are reversed with the removal of the aluminum. Iron also accumulates at the mineralization front and can cause low turnover forms of ROD similar to aluminum, although much less is known about iron intoxication than aluminum. Lanthanum has recently been added as a rare earth ion administered to CKD and ESKD patients as a phosphate binder. It is poorly absorbed, and its levels in bone are much less than aluminum. It has proven not to have long-term toxic effects. Five-year data suggest that the levels of skeletal lanthanum accumulation remain below those with any biologic or toxic effects. Lanthanum disappearance from bone deposits is slow, but not as slow as bisphosphonate disappearance. Bisphosphonates are drugs used in the treatment of osteoporosis and hypercalcemia. There are increasing instances of bisphosphonate use in patients with CKD and ESKD, especially for treatment of vascular calcification and calciphylaxis. However, the nature of the bone remodeling abnormalities in CKD, especially with woven bone formation and mineralization defects, lend a high level of risk to skeletal deposition of a substance that once deposited may not be removed. Such a risk of long-term retention of an active drug inhibiting bone turnover is now being recognized with use of bisphosphonates in osteogenesis imperfecta, and the rare side effect of the drugs in osteonecrosis of the jaw and atypical femoral fractures [49]. The risk of these side effects is obviously increased in CKD, appropriately limiting their use to emergent situations.

Clinical manifestations

Patients with mild to moderate kidney insufficiency are rarely symptomatic due to ROD and its skeletal pathology. However, fracture risk is increased more than twofold above age-matched cohorts without CKD. Stimulation of vascular calcification in the setting of disordered skeletal remodeling is a life-threatening complication of the CKD–MBD. It produces vascular stiffness, which in CKD causes an increase in systolic blood pressure, a widening of the pulse pressure and an increase in pulse wave velocity. All of these lead to cardiac hypertrophy, heart failure, and cardiovascular mortality.

The pathogenesis of vascular calcification in CKD is complex, and pathologically is of two types: neointimal and arterial medial. Atherosclerotic neointimal calcification is multifactorial, but it involves activation of an osteoblastic differentiation program in cells of the neointima of atherosclerotic plaques. The skeleton contributes to the pathogenesis of vascular calcification. Signals deriving from the skeleton are direct causes of the vascular mineralization. One such signal is hyperphosphatemia [35]. Diffuse calcification of arterial tunica media is referred to as Mönckeberg's sclerosis. CKD is the most common cause of Mönckeberg's sclerosis, especially when it complicates diabetes mellitus.

Heterotopic mineralization, calciphylaxis, and tumoral calcinosis

Heterotopic tissue calcification may occur in the eyes and manifest as band keratopathy in the sclera or induce an inflammatory response known as the "red eye syndrome" in the conjunctiva. Calcium deposits are also found in the lungs and lead to restrictive lung disease. Deposits in the myocardium might cause arrhythmias, annular calcifications, valvular calcification, or myocardial dysfunction. Calcification of the kidney may contribute to progression of kidney failure. Most soft tissue calcifications are attributed to the increased calcium phosphate product contributed to by renal osteodystrophy and excess bone resorption. The syndrome of calciphylaxis is characterized by vascular calcification in the tunica media of peripheral arteries. These calcifications induce painful violaceous skin lesions that progress to ischemic necrosis. This syndrome is associated with serious complications and often death.

Tumoral calcinosis is a form of soft tissue calcification that involves the periarticular tissues. Calcium deposits may grow to enormous size and interfere with the function of adjacent joints and organs. Although this type of calcification is usually associated with high calcium phosphate products, its exact pathogenesis is poorly understood. The recent discoveries of three single gene mutations in FGF23, Klotho, and GALNT3 that cause inherited tumoral calcinosis shed light on the role of hyperphosphatemia in its pathogenesis [50–52]. The role of GALNT3 in hyperphosphatemia is associated with enzymatic degradation of FGF23.

Bone pain, fractures, and skeletal deformities

Symptoms of ROD related to the skeleton appear in patients with advanced kidney failure. Clinical manifestations are preceded, however, by an abnormal biochemical profile that should alert the physician and prompt steps to prevent more severe complications. When symptoms related to the skeleton occur, they are usually insidious, subtle, nonspecific, and slowly progressive.

Bone pain is usually vague, ill defined, and deep seated. It may be diffuse or localized in the lower part of the back, hips, knees, or legs. Weight bearing and changes in position commonly aggravate it. Bone pain may progress slowly to the degree that patients are completely incapacitated. Bone pain in patients with ESKD usually does not cause physical signs; however, local tenderness may be apparent with pressure. Occasionally, pain can occur suddenly at one joint of the lower extremities and mimic acute arthritis or periarthritis not relieved by heat or massage. A sharp chest pain may indicate rib fracture. Spontaneous fractures or fractures after minimal trauma may also occur in vertebrae (crush fractures) and in tubular bones.

Bone pain and bone fractures can be observed in all patients with ESKD independent of the underlying histologic bone disease, especially when osteoporosis is present [41]. However, low-turnover osteomalacia and aluminum-related bone disease are associated with the most severe bone pain and the highest incidence of fractures and incapacity.

Skeletal deformities can be observed in children and adults. Most children with ESKD have growth retardation, and bone deformities may develop from vitamin D deficiency (rickets) or secondary hyperparathyroidism. In rickets, bowing of the long bones is seen, especially the tibiae and femora, with typical genu valgum that becomes more severe with adolescence. Long-standing secondary hyperparathyroidism in children may be responsible for slipped epiphyses secondary to impaired transformation of growth cartilage into regular metaphyseal spongiosa. This complication most commonly affects the hips and becomes obvious in preadolescence. It causes limping but is usually painless. When the radius and ulna are involved, ulnar deviation of the hands, and local swelling may occur. In adults, skeletal deformities can be observed in cases of severe osteomalacia or osteoporosis and include lumbar scoliosis, thoracic kyphosis, and recurrent rib fractures.

The diagnosis and treatment of the CKD–MBD are covered in Chapter 78, and this discussion concludes with recall of the data leading to Fig. 77.1. Renal injury disturbs the complex physiology of bone and mineral metabolism, promotes vascular calcification, and compromises cardic function. In addition, the effects of kidney disease on the skeleton contribute through the bone–cardiovascular axes to the major effect of kidney injury to stimulate cardiovascular disease and produce the high mortality rates associated with kidney disease.

ACKNOWLEDGMENTS

The writing of this chapter was supported by NIH grants DK070790 and AR41677.

REFERENCES

1. Moe S, Drueke T, Cunningham J, Goodman W, Martin K, Olgaard K, et al. 2006. Definition, evaluation, and classification of renal osteodystrophy: A position statement from kidney disease: Improving Global Outcomes (KDIGO). *Kidney Int* 69(11): 1945–53.
2. Olgaard K (ed.) 2006. *KDIGO:Clinical Guide to Bone and Mineral Metabolism in CKD*. New York: National Kidney Foundation.
3. Stevens LA, Djurdjev O, Cardew S, Cameron EC, Levin A. 2004. Calcium, phosphate, and parathyroid hormone levels in combination and as a function of dialysis duration predict mortality: Evidence for the complexity of

the association between mineral metabolism and outcomes. *J Am Soc Nephrol* 15(3): 770–9.
4. Block GA, Hulbert-Shearon TE, Levin NW, Port FK. 1998. Association of serum phosphorus and calcium X phosphate product with mortality risk in chronic hemodialysis patients: a national study. *Am J Kidney Dis* 31(4): 607–17.
5. Slinin Y, Foley RN, Collins AJ. 2005. Calcium, phosphorus, parathyroid hormone, and cardiovascular disease in hemodialysis patients: The USRDS waves 1, 3, and 4 study. *J Am Soc Nephrol* 16(6): 1788–93
6. Mathew S, Lund R, Strebeck F, Tustison KS, Geurs T, Hruska KA. 2007. Reversal of the adynamic bone disorder and decreased vascular calcification in chronic kidney disease by sevelamer carbonate therapy. *J Am Soc Nephrol* 18(1): 122–30.
7. Davies MR, Lund RJ, Mathew S, Hruska KA. 2005. Low turnover osteodystrophy and vascular calcification are amenable to skeletal anabolism in an animal model of chronic kidney disease and the metabolic syndrome. *J Am Soc Nephro* 16(4): 917–28.
8. Lund RJ, Davies MR, Brown AJ, Hruska KA. 2004. Successful treatment of an adynamic bone disorder with bone morphogenetic protein-7 in a renal ablation model. *J Am Soc Nephrol* 15(2): 359–69.
9. Bacchetta J, Boutroy S, Vilayphiou N, Juillard L, Guebre-Egziabher F, Rognant N, et al. 2010. Early impairment of trabecular microarchitecture assessed with HR-pQCT in patients with stage II-IV chronic kidney disease. *J Bone Miner Res* 25(4): 849–57.
10. Malluche HH, Ritz E, Lange HP. 1976. Bone histology in incipient and advanced renal failure. *Kidney Int* 9(4): 355–62.
11. Al-Douahji M, Brugarolas J, Brown PA, Stehman-Breen CO, Alpers CE, Shankland SJ. 1999. The cyclin kinase inhibitor p21WAF1/CIP1 is required for glomerular hypertrophy in experimental diabetic nephropathy. *Kidney Int* 56: 1691–9.
12. Isakova T, Wahl P, Vargas GS, Gutierrez OM, Scialla J, Xie H, et al. 2011. Fibroblast growth factor 23 is elevated before parathyroid hormone and phosphate in chronic kidney disease. *Kidney Int* 79(12): 1370–8.
13. Pereira RC, Juppner H, Azucena-Serrano CE, Yadin O, Salusky IB, Wesseling-Perry K. 2009. Patterns of FGF-23, DMP1 and MEPE expression in patients with chronic kidney disease. *Bone* 45(6): 1161–8.
14. Malluche HH, Faugere MC. 1990. Renal bone disease 1990: Challenge for nephrologists. *Kidney Int* 38: 193–211.
15. Foley RN, Parfrey PS, Sarnak MJ. 1998. Clinical epidemiology of cardiovascular disease in chronic renal disease. *Am J Kidney Dis* 32(5 Suppl 3): S112–S9.
16. Shlipak MG, Sarnak MJ, Katz R, Fried LF, Seliger SL, Newman AB, et al. 2005. Cystatin C and the risk of death and cardiovascular events among elderly persons. *New Engl J Med* 352(20): 2049–60.
17. Hruska K, Mathew S, Lund R, Fang Y, Sugatani T. 2011. Cardiovascular risk factors in chronic kidney disease: Does phosphate qualify? *Kidney Int Suppl* (121): S9–13.
18. White KE, Evans WE, O'Riordan JLH, Speer MC, Econs MJ, Lorenz-Depiereux B, et al. 2000. Autosomal dominant hypophosphataemic rickets is associated with mutations in FGF23. *Nature Genetics* 26: 345–8.
19. White KE, Jonsson KB, Carn G, Hampson G, Spector TD, Mannstadt M, et al. 2001. The autosomal dominant hypophosphatemic rickets (ADHR) gene is a secreted polypeptide overexpressed by tumors that cause phosphate wasting. *J Clin Endocrinol Metab* 86(2): 497–500.
20. Larsson T, Nisbeth U, Ljunggren O, Juppner H, Jonsson KB. 2003. Circulating concentration of FGF-23 increases as renal function declines in patients with chronic kidney disease, but does not change in response to variation in phosphate intake in healthy volunteers. *Kidney Int* 64(6): 2272–9.
21. Shimada T, Muto T, Urakawa I, Yoneya T, Yamazaki Y, Okawa K, et al. 2002. Mutant FGF-23 responsible for autosomal dominant hypophosphatemic rickets is resistant to proteolytic cleavage and causes hypophosphatemia in vivo. *Endocrinology* 143(8): 3179–82.
22. Kurosu H, Ogawa Y, Miyoshi M, Yamamoto M, Nandi A, Rosenblatt KP, et al. 2006. Regulation of fibroblast growth factor-23 signaling by Klotho. *J Biol Chem* 281(10): 6120–3.
23. Ben-Dov IZ, Galitzer H, Lavi-Moshayoff V, Goetz R, Kuro-o M, Mohammadi M, et al. 2007. The parathyroid is a target organ for FGF23 in rats. *J Clin Invest* 117(12): 4003–8.
24. Faul C, Amaral AP, Oskouei B, Hu MC, Sloan A, Isakova T, et al. 2011. FGF23 induces left ventricular hypertrophy. *J Clin Invest* 121(11): 4393–408.
25. Kuro-o M, Matsumura Y, Aizawa H, Kawaguchi H, Suga T, Utsugi T, et al. 1997. Mutation of the mouse klotho gene leads to a syndrome resembling ageing. *Nature* 390: 45–51.
26. Hu MC, Shi M, Zhang J, Qui+Ýones H, Griffith C, Kuro-o M, et al. 2011. Klotho deficiency causes vascular calcification in chronic kidney disease. *J Am Soc Nephrol* 22: 124–36.
27. Hu MC, Shi M, Zhang J, Pastor J, Nakatani T, Lanske B, et al. 2010. Klotho: A novel phosphaturic substance acting as an autocrine enzyme in the renal proximal tubule. *FASEB J* 24: 3438–50.
28. Lim K, Lu T-S, Molostvov G, Lee C, Lam FT, Zehnder D, et al. 2012. Vascular Klotho deficiency potentiates the development of human artery calcification and mediates resistance to fibroblast growth factor 23/clinical perspective. *Circulation* 125(18): 2243–55.
29. Slatopolsky E, Robson AM, Elkan I, Bricker NS. 1968. Control of phosphate excretion in uremic man. *J Clin Invest* 47(8): 1865–74.
30. Kurz P, Monier-Faugere MC, Bognar B, Werner E, Roth P, Vlachojannis J, et al. 1994. Evidence for abnormal calcium homeostasis in patients with adynamic bone disease. *Kidney Int* 46(3): 855–61.

31. Moallem E, Kilav R, Silver J, Naveh-Many T. 1998. RNA-protein binding and post-transcriptional regulation of parathyroid hormone gene expression by calcium and phosphate. *J Biol Chem* 273(9): 5253–9.
32. Naveh-Many T, Rahamimov R, Livni N, Silver J. 1995. Parathyroid cell proliferation in normal and chronic renal failure rats. *J Clin Invest* 96(4): 1786–93.
33. Jono S, McKee MD, Murry CE, Shioi A, Nishizawa Y, Mori K, et al. 2000. Phosphate regulation of vascular smooth muscle cell calcification. *Circ Res* 87(7): e10–e7.
34. Li X, Yang HY, Giachelli CM. 2006. Role of the sodium-dependent phosphate cotransporter, Pit-1, in vascular smooth muscle cell calcification. *Circ Res* 98(7): 905–12.
35. Mathew S, Tustison KS, Sugatani T, Chaudhary LR, Rifas L, Hruska KA. 2008. The mechanism of phosphorus as a cardiovascular risk factor in chronic kidney disease. *J Am Soc Nephrol* 19(6): 1092–105.
36. Goodman WG, Quarles LD. 2007. Development and progression of secondary hyperparathyroidism in chronic kidney disease: Lessons from molecular genetics. *Kidney Int* 74: 276–88.
37. Naveh-Many T, Marx R, Keshet E, Pike JW, Silver J. 1990. Regulation of 1,25-dihydroxyvitamin D3 receptor gene expression by 1,25-dihydroxyvitamin D3 in the parathyroid in vivo. *J Clin Invest* 86(6): 1968–75.
38. Silver J, Russell J, Sherwood LM. 1985. Regulation by vitamin D metabolites of messenger ribonucleic acid for preproparathyroid hormone in isolated bovine parathyroid cells. *Proc Natl Acad Sci U S A* 82(12): 4270–3.
39. Brown EM, Gamba G, Riccardi D, Lombardi M, Butters R, Kifor O, et al. 1993. Cloning and characterization of an extracellular Ca2+-sensing receptor from bovine parathyroid. *Nature* 366(6455): 575–80.
40. Brown EM, Hebert SC. 1995. A cloned Ca 2+ -sensing receptor; a mediator of direct effects of extracellular Ca 2+ on renal function? *J Am Soc Nephrol* 6(6): 1530–40.
41. Alem AM, Sherrard DJ, Gillen DL, Weiss NS, Beresford SA, Heckbert SR, et al. 2000. Increased risk of hip fracture among patients with end-stage renal disease. *Kidney Int* 58(1): 396–9.
42. Cunningham J, Sprague S, Cannata-Andia J, Coco M, Cohen-Solal M, Fitzpatrick L, et al. 2004. Osteoporosis in chronic kidney disease. *Am J Kidney Dis* 43(3): 566–71.
43. Stehman-Breen C. 2004. Osteoporosis and chronic kidney disease. *Semin Nephrol* 24(1): 78–81.
44. Rix M, Andreassen H, Eskildsen P, Langdahl B, Olgaard K. 1999. Bone mineral density and biochemical markers of bone turnover in patients with predialysis chronic renal failure. *Kidney Int* 56(3): 1084–93.
45. Bonyadi M, Waldman SD, Liu D, Aubin JE, Grynpas MD, Stanford WL. 2003. Mesenchymal progenitor self-renewal deficiency leads to age-dependent osteoporosis in Sca-1/Ly-6A null mice. *Proc Natl Acad Sci U S A* 100: 5840–5.
46. Stehman-Breen C. 2001. Bone mineral density measurements in dialysis patients. *Semin Dial* 14(3): 228–9.
47. Stehman-Breen C, Sherrard D, Walker A, Sadler R, Alem A, Lindberg J. 1999. Racial differences in bone mineral density and bone loss among end-stage renal disease patients. *Am J Kidney Dis* 33(5): 941–6.
48. Coco M, Rush H. 2000. Increased incidence of hip fractures in dialysis patients with low serum parathyroid hormone. *Am J Kidney Dis* 36(6): 1115–21.
49. Khosla S, Burr D, Cauley J, Dempster DW, Ebeling PR, Felsenberg D, et al. 2007. Bisphosphonate-Associated osteonecrosis of the jaw: Report of a task force of the American Society for Bone and Mineral Research. *J Bone Miner Res.* 2007;22(10): 1479–91.
50. Ichikawa S, Imel EA, Kreiter ML, Yu X, Mackenzie DS, Sorenson AH, et al. 2007. A homozygous missense mutation in human KLOTHO causes severe tumoral calcinosis. *J Clin Invest* 117(9): 2684–91.
51. Ichikawa S, Lyles KW, Econs MJ. 2005. A novel GALNT3 mutation in a pseudoautosomal dominant form of tumoral calcinosis: Evidence that the disorder is autosomal recessive. *J Clin Endocrinol Metab* 90(4): 2420–3.
52. Benet-Pages A, Orlik P, Strom TM, Lorenz-Depiereux B. 2005. An FGF23 missense mutation causes familial tumoral calcinosis with hyperphosphatemia. *Hum Mol Genet* 14(3): 385–90.

78

Treatment Of Chronic Kidney Disease Mineral Bone Disorder (CKD–MBD)

Hala M. Alshayeb and L. Darryl Quarles

Introduction 640
Physiology 640
Pathogenesis 641
Diagnosis 642

Treatment of CKD–MBD 643
Treatment of CKD–MBD in Stage CKD 3-4 648
Parathyroidectomy 648
References 648

INTRODUCTION

Loss of glomerular filtration and other renal functions in chronic kidney disease leads to complex abnormalities of mineral homeostasis, referred to as mineral and bone disorder (CKD–MBD) [1, 2], that include biochemical abnormalities [i.e., elevations in serum fibroblast growth factor 23 (FGF23) and parathyroid hormone (PTH), reductions in 25(OH) and 1,25(OH)$_2$ vitamin D concentrations, hyperphosphatemia, hypocalcemia], metabolic bone disease, extraskeletal calcification, and increased cardiovascular and all cause mortality [3]. The discovery of the phosphaturic and vitamin D regulatory hormone FGF23 has provided new insights into the pathogenesis of secondary hyperparathyroidism [4]. A heightened interest in CKD–MBD has arisen because of compelling, but not incontrovertible, evidence that elevations of serum FGF23, PTH, and phosphorus, along with reductions in 25(OH)D and 1,25(OH)$_2$D, contribute to derangements of immune and cardiovascular functions leading to the observed increases in morbidity and mortality associated with CKD. Moreover, new therapeutic approaches and bundling of payments for management of ESRD have created an opportunity to reconsider the optimal therapeutic approaches to treat disordered mineral metabolism in CKD.

PHYSIOLOGY

Understanding the pathophysiology of abnormal mineral metabolism is a prerequisite for making informed decisions regarding therapies. Three hormones (PTH, FGF23, and 1,25(OH)$_2$D) and four organ systems [parathyroid gland (PTG), bone, small intestines, and kidney] are involved in regulation mineral homeostasis.

PTH, the key calcemic hormone, is produced and secreted by PTG and is responsible for the tight minute-to-minute regulation of serum ionized calcium levels through stimulation of calcium reabsorption in distal tubule [5], calcium efflux from bone due to increased osteoclastic bone resorption [6], and conversion of 25(OH)D to 1,25(OH)$_2$D through the simulation of Cyp27b1 activity in the proximal tubule of the kidney [7]. 1,25(OH)$_2$D stimulates calcium and phosphate absorption through the gut. The principal factor regulating PTH secretion is extracellular calcium acting through the G-protein coupled calcium-sensing receptor (CasR) located in chief cells in the parathyroid gland. 1,25(OH)$_2$D, acting through classical vitamin D receptor (VDR) nuclear pathways [8], and hyperphosphatemia through unknown mechanisms [9], respectively, inhibit and stimulate PTH production. Mouse genetic studies that compare the effects of ablation of CasR [10] and VDR in the PTG on

Primer on the Metabolic Bone Diseases and Disorders of Mineral Metabolism, Eighth Edition. Edited by Clifford J. Rosen.
© 2013 American Society for Bone and Mineral Research. Published 2013 by John Wiley & Sons, Inc.

hyperparathyroidism [11], however, suggest that hypocalcemia through CasR is the dominant over 1,25(OH)$_2$D VDR pathway in the development of hyperparathyroidism. FGF23 is purported to suppress PTH in animal studies [12], but PTH is elevated in the presence of elevated FGF23 in CKD, possibly due to downregulation of the FGFR1/Klotho coreceptor for FGF2 in the PTG in CKD [13]. Actions of PTH to increase serum calcium also results in enhanced phosphorus release from the bone and increased gastrointestinal phosphate absorption through 1,25(OH)$_2$D effects [7]. The phosphaturic actions of PTH by downregulating sodium phosphate cotransporters in the proximal tubule of the kidney increases phosphate excretion, thereby avoiding a positive phosphate balance [14].

Native vitamin D is a secosteroid that is available in the diet, cholecalciferol (vitamin D$_3$, from animal and ergocalciferol (vitamin D$_2$) from plant sources, respectively) and is produced endogenously by the conversion of 7-dehydrocholesterol by exposure UV light. Vitamin D is stored in fat and circulates bound to vitamin D binding proteins (DBPs). Vitamin D metabolism involves both systemic and intracrine pathways. In the systemic pathway, vitamin D undergoes two key hydroxylation steps to form active 1,25(OH)$_2$D. Both vitamin D$_2$ and vitamin D$_3$ are metabolized in the liver by a group of 25-hydroxylase (including CYP27A) to form 25(OH)D and 25OHD$_3$, respectively. This hydroxylation step in the liver is substrate dependent and is not thought to be regulated by hormones or alterations in calcium or phosphate. The major site of regulation is in the conversion of 25(OH)D to the active 1,25(OH)$_2$D by 1α-hydroxylase (Cyp27b1) in the kidneys, where 25(OH)D and DBPs are reabsorbed in the proximal nephron from tubular fluids by a megalin-dependent mechanism. 25(OH)D undergoes hydroxylation of the carbon atom located at position 1 of the A ring of 25-hydroxyvitamin D in the mitochondria of epithelial cells. The kidney is the major source of circulating 1,25(OH)$_2$D and its production is stimulated by PTH and decreased by phosphate and FGF23. In addition, 25(OH)D and 1,25(OH)$_2$D are catabolized by 24 α-hydroxylase (Cyp24) to inactive metabolites, a step that stimulated by FGF-23 and 1,25(OH)$_2$D itself [15].

In the intracrine pathway, 25(OH)D is also converted to 1,25(OH)$_2$D and catabolized by Cyp24 in extrarenal tissues, such as monocytes and macrophages, where it has been shown to be important regulator of the innate immune response. Antigen stimulation is required for the intracrine production of 1,25(OH)$_2$D and extrarenal tissues do not contribute to the circulating 1,25(OH)$_2$D. Other local/intracrine functions of 1,25(OH)$_2$D have been demonstrated, including effects on cardiovascular function through the regulatory effects on the renin–angiotensen system and may also have effects to prevent certain types of cancers. These extrarenal effects have been used as the biological basis for current recommendations to provide vitamin D supplementation to CKD patients.

FGF23 is a bone-derived hormone that is mainly produced by osteocytes and osteoblast. In the kidneys, FGF23 acts as phosphaturic and vitamin D–regulatory hormone by increasing renal phosphate (P) excretion through inhibition of sodium-dependent phosphate reabsorption and decreases 1,25(OH)$_2$D production through the downregulation of CYP27b1 gene and the upregulation of CYP 24 gene expression and activity [16]. Physiologically, FGF23 participates in a bone–kidney axis to prevent vitamin D toxicity, where 1,25(OH)$_2$D stimulates FGF23 production, which, in turn, targets the kidney to suppress 1,25(OH)$_2$D synthesis and increase its degradation. Another putative function of FGF23 may be to coordinate bone formation and mineralization with renal handling of phosphate. Impaired mineralization and/or bone formation leads to alteration in intrinsic matrix derived factors that stimulate FGF23 release to match bone phosphate buffering capacity with renal phosphate handling. The link between bone mineralization and FGF23 expression in bone is supported by inactivating mutations of Phex, Dmp1, and Ank, which regulate various aspects of extracellular matrix mineralization [17]. The main systemic stimulators for FGF23 secretion are increased serum 1,25(OH)$_2$D levels, dietary phosphorus intake [16] and PTH. 1,25(OH)$_2$D binds the vitamin D response element in the promoter region of the FGF23 gene in osteocytes and directly increases FGF23 gene transcription [12]. In contrast, alterations of dietary phosphate have failed to show a consistent effect to increase increased FGF23 secretion [18], or have small effects on serum FGF23 concentrations, suggesting that phosphate itself may not directly regulate FGF23. PTH effects are also variable, with some studies showing that PTH stimulates, while others show that PTH inhibits FGF23 expression in bone. These discrepancies of PTH effects on FGF23 are likely due to differential anabolic and catabolic effects of PTH on bone that are dependent on the dose and frequency of PTH administration and to interactions between these direct PTH and 1,25(OH)$_2$D on FGF23 gene transcription [4, 12].In addition to its role in regulating phosphate and vitamin D metabolism, FGF23 is associated with adverse cardiovascular events, metabolic syndrome, and mortality in both CKD and the general population through mechanisms that remain to be elucidated [19–21].

PATHOGENESIS

New insights into the regulation of mineral homeostasis derived from understanding the regulation and function of FGF23 have challenged some of the traditional views of the pathogenesis of secondary hyperparathyroidism as well as the mechanism and significance of reduced circulating 25(OH)D and 1,25(OH)$_2$D in CKD. Traditionally, secondary hyperparathyroidism in CKD has been viewed as vitamin D deficient state, where decreased glomerular filtration rate (e-GFR) and functioning nephron mass results in diminished 1-α hydroxylase enzyme (Cyp27b1) in the proximal tubule leading to

Table 78.1. Target Levels for Calcium, Phosphorus, and Parathyroid Hormone

		KDOQI	KDIGO
Calcium	CKD stage 3-5	Normal range	Normal range
	CKD 5D	Normal range but preferably 8.4–9.5 mg/dL	Normal range
Phosphorus	CKD stage 3-4	2.7–4.6 mg/dL	Normal range
	CKD stage 5	3.5–5.5 mg/dL	Normal range
	CKD stage 5D	3.5–5.5 mg/dL	Toward normal range
Intact PTH	CKD stage 3	35–70 pg/mL	Optimal level is unknown
	CKD stage 4	70–110 pg/mL	Optimal level is unknown
	CKD stage 5	200–300 pg/mL	Optimal level is unknown
	CKD stage 5D	200–300 pg/mL	2–9 times above upper limit

Ca: calcium; CKD: chronic kidney disease; GFR: glomerular filtration rate; KDOQI: Kidney/Dialysis Outcomes Quality Initiative; P: phosphorus; PTH: parathyroid hormone.
Modified from National Kidney Foundation DOQI™ Kidney Disease Outcomes Quality Initiative. 2004. *Am J Kidney Dis* 43. S1–S201; and the National Kidney Foundation and Kidney Disease: Improving Global Outcomes (KDIGO).

decreased 1,25(OH)$_2$D production and phosphate retention due to impaired phosphate clearance. The resulting reduction in 1,25(OH)$_2$D, hypocalcemia, and hyperphosphatemia results in the stimulation of PTH release from the parathyroid gland, which undergoes hypertrophy and hyperplasia, characteristic of secondary hyperparathyroidism. In the integrated view of the FGF23–PTH–vitamin D endocrine axis in CKD, initial secretion of FGF23 by the bone triggered by unknown stimuli is the inciting event leading to secondary hyperparathyroidism in CKD. Increments in serum FGF23 concentrations suppress Cyp27b1-mediated 1,25(OH)$_2$D production and enhances Cyp24-mediated 1,25(OH)$_2$D catabolism leading to reduced circulating 1,25(OH)$_2$D. In addition, FGF23 inhibits renal phosphate reabsorption to compensate for phosphate retention due to the reduced GFR in people with CKD [4]. Although FGF23 has been shown to suppress PTH production in healthy individuals, in CKD, FGFR1, and Klotho are downregulated in PTG, thus PTH is increased [4]. As CKD progresses, increased PTH stimulates bone remodeling and efflux of calcium and phosphorus from the bone and a further increase FGF23 production. In addition, PTH stimulates the production of 1, 25(OH)$_2$D production [12], offsetting the effects of FGF23 to suppress 1,25(OH)$_2$D. The interactions between FGF23–PTH–vitamin D favor net phosphate loss while maintaining calcium homeostasis [4].

DIAGNOSIS

Biochemical determinations of PTH, 25(OH)D, calcium, and phosphate are used to define and monitor CKD–MBD. While assessment of serum FGF23 may be useful, it is currently not routinely used in the assessment of CKD–MBD. In addition, while bone markers, such as bone-specific alkaline phosphatase, osteocalcin, procollagen C-terminal propeptide (PICP) and procollagen N-terminal propeptide (PINP), radiographic characterization of bone density and soft-tissue calcifications, and histological assessment of bone structure, formation, and resorption are informative, these analytical tools are not commonly being used in the diagnosis and management of CKD–MBD. The biochemical targets representing optimal management of CKD–MBD are not known, and recent efforts to establish guidelines have produced disparate recommendations.

Clinical practice guidelines for bone metabolism and disease in CKD stages 3 to 5 have been developed by the National Kidney Foundation – Kidney Disease: Outcomes Quality Initiative (KDOQI) in 2003 [22] and Kidney Disease: Improving Global Outcomes (KDIGO) in 2009 [2] (Table 78.1). The differences between the two guidelines reflect uncertainty regarding the importance of biochemical measurements in predicting outcomes. KDOQI guidelines are supported by data from observational studies that serum phosphorus concentrations less than 3.5 and above 5.5 mg/dl are associated with increased mortality, and calcium concentration greater than 9.5 mg/dl and high calcium phosphorus products greater than 55 are associated with increased vascular calcifications. PTH values between 150 and 300 pg/ml are associated with "normal" bone remodeling. The difficulty in achieving these targets in end-stage renal disease (ESRD) patients, the variability in PTH assays in predicting bone remodeling, and lack of evidence from randomized controlled trials to support the safety, efficacy, and survival advantage of attaining these biochemical ranges have lead to alternative recommendations. In this regard, KDIGO differs from KDOQI in that decision making needs to be based on a trend rather than a single laboratory value, In addition, KDIGO raises the upper limit of serum calcium to that in the general population and recommends achieving phosphate control also toward the normal range of 2.5 to 4.5 mg/dl.

PTH is a measure of PTG activity; however, it correlates poorly with indices of bone turnover. There is lack of consensus on the optimal PTH range in the dialysis patient, as reflected by different recommended PTH ranges in different guidelines. For example, the KDOQI goals for serum laboratory values for PTH in ESRD is 150–300 pg/ml, whereas in KDIGO, the upper limit of PTH in ESRD is 600 pg/ml (and upward changes in PTH levels are an indication for treatment below this limit). In contrast, the Japanese Society for Dialysis Therapy sets the PTH range to 60–180 pg/mL in ESRD patients, a upper limit roughly twofold greater than the upper normal limit in the general population. These variable recommendations are due in part to the variability in different PTH assays and differing assessments of the risks of lower PTH on bone metabolism and soft-tissue calcifications. Too liberal of an upper limit is likely to increase the risk of progressive parathyroid disease leading to tertiary hyperparathyroidism and parathyroidectomy, whereas too low of a limit might increase the risk of adynamic bone disease. The observation that serum levels of PTH greater than 600 pg/ml are associated with an increased risk of death and cardiovascular events in CKD [23] has influenced the establishment of a broader range of PTH levels.

Currently, other than bone biopsies, there is no diagnostic test that identifies the different subtypes of bone diseases in CKD, although it is not clear that such information is clinically important or can be used to direct therapy, which is typically directed at preventing progressive elevations in PTH, avoiding hypercalcemia, and minimizing the severity of hyperphosphatemia. Biomarkers of bone formation (bone-specific alkaline phosphatase, osteocalcin) and bone resorption (urinary collagen breakdown products, TRAP5b), do not predict bone histology in uremic bone disease as they do in assessing bone turnover in patients with osteoporosis.

Serum FGF23 concentrations progressively increases during CKD, reaching over 1,000-fold above normal in ESRD patients [24]. Observational studies showed that elevated FGF23 is associated with refractory hypreparathyrodism (HPTH), progression of renal failure, left ventricular hypertrophy, and vascular calcifications and increased mortality [19–21, 24]. The factors leading to increased FGF23 in CKD is not clear, but use of active vitamin D analogues results in increased FGF23 levels [16], whereas calcimemetics are associated with decreased FGF23 levels [25]. FGF23 levels are also high in patients with secondary HPTH, and parathyroidectomy results in lowering of circulating FGF23 levels. FGF23 levels are not currently used in clinical practice.

The best biochemical index of vitamin D nutrition is considered to be 25(OH)D. Low 25(OH) levels are highly prevalent in CDK. Definitions for vitamin D insufficiency (25(OH)D concentration of 20 to 30 ng/mL), and deficiency (25(OH)D level less than 20 ng/mL) are the same in CKD as the general population [26].

Skeletal X-rays are insensitive measures of bone density but can detect vascular calcifications. Measurement of bone mineral density (BMD) by dual-energy X-ray absorptiometry (DXA) scanning is limited in CKD patients, since BMD correlates poorly with osteopenia and fracture risk because of the effects of secondary HPTH to cause extraskeletal calcifications, osteosclerosis, and osteomalacia. Thus, it is an insensitive tool to evaluate bone mass in CKD patients [27].

TREATMENT OF CKD–MBD

Therapeutic goals

The goals of therapy of CKD–MBD are: (1) to attain neutral phosphorus balance without causing protein malnutrition, (2) to control PTH levels in a range that prevents the progression of secondary to tertiary HPTH while maintaining normal bone health, and (3) to provide adequate vitamin D supplementation to optimize biological effects of this hormone that are not necessarily related to mineral metabolism. Since hyperphosphatemia, vascular calcifications, and elevated FGF23 are associated with increased cardiovascular disease and mortality in CKD, a desirable outcome of management would also be to reduce the contribution of CKD–MBD to the increased morbidity and mortality. Although retrospective evaluations of large databases establish a positive association with vitamin D usage and improved survival, there are no perspective studies demonstrating that any specific treatment paradigm results in a reduction in mortality. Recent prospective trials, however, have compared protocols that use progressive titration of active vitamin D analogues to use of protocols with fixed vitamin D replacement and suppression of PTH using escalating doses of cinacalcet. Most of these trials demonstrate a greater ability of the cinacalcet-based regimens when used in combination with lower doses of active vitamin D analogues to more effectively suppress PTH without elevating serum phosphate levels [28, 29]. However, neither prospective trials that attempt to establish a salutary effect of use of activate vitamin D analogues to suppress the renin–angiotensin system and prevent left ventricular hypertrophy (LVH) nor the use of cinacalect to reduce vascular calcification and cardiovascular mortality have been positive. Consequently, at the present time there are no prospective clinical trials to define the optimal therapy for CKD–MBD. Therefore, combined pharmacological therapy that targets multiple molecular components of the disorder pathways, including phosphate binders, nutritional vitamin D and active vitamin D analogues, and calcimimetics, is the current standard of care. Treatment considerations depend on the stage of CKD, the mechanism of action of the various therapeutic agents, potential toxicities associated with various treatments, and cost/reimbursement of the various therapies.

Achieving neutral phosphate balance

CKD patients are at increased risk of positive phosphate balance and hyperphosphatemia because the amount of phosphate absorbed from dietary intake in the setting of concomitant active vitamin D therapy exceeds the amount that can be bound by phosphate binders and removed by the standard 4-hour thrice weekly dialysis regimen. Phosphate is absorbed in the small intestines by passive diffusion and by active transport mediated by the sodium-dependent phosphate transporter Napi2b, a step that is stimulated by the use of active vitamin D analogues. In addition, though not contributing to a net positive phosphate balance, increased phosphate efflux from bone under the effects of excessive PTH levels can contribute to hyperphosphatemia. Based on the potential sources of hyperphosphemia and methods to limit phosphate absorption and removal, there are multiple steps to controlling serum phosphate concentrations, including (1) dietary phosphate restriction; (2) administration of oral agents phosphate binders to prevent gastrointestinal phosphate absorption; (3) minimizing active phosphate transport in the small intestines by reducing doses of active vitamin D analogues or use of niacin; (4) reducing PTH levels to decrease bone efflux of phosphate; and (5) increasing dialysis duration and/or frequency to increase phosphate removal.

Dietary phosphate restriction

Dietary phosphate restriction to 800–1000 mg per day is difficult to achieve because of variable phosphate content of processed foods and patient compliance with complex dietary regimens. In addition, recent studies suggest that dietary phosphate restriction may be associated with poorer indices of nutritional status and adverse outcomes, and is not associated with survival benefits [30]. Avoidance of high dietary phosphate (as in dairy products, certain vegetables, many processed foods, and colas) and intake of high biologic value sources of protein (such as meat and eggs) are part of standard nutritional counseling. Dietary counseling of patients with CKD also should include the source of protein from which the phosphate is derived as there is some evidence that vegetarian protein diet leads to lower serum phosphorus levels and decreased FGF23 levels in comparison to meat protein diet [31]. In stage CKD-5D, the recommended protein intake is 1.2 (hemodialysis) and 1.3 g/kg/day (peritoneal dialysis), which provides at least 1,000 mg per day of phosphorus.

The use of phosphate binders

Given the difficulty of achieving dietary phosphate restriction, the risk of protein malnutrition and the low compliance rate associated with phosphate restriction, phosphate binders are the mainstay of therapy. Both cation- and resin-based binders act by forming poorly soluble complexes with phosphate in the intestinal lumen that is poorly absorbed and are maximally effective when they are ingested with meals. However, the low compliance rate associated with the requirement of high bill burdens, gastrointestinal side effects that are associated with some binders, and interactions between binders and other medications limit their effectiveness in preventing hyperphosphatemia. The currently available phosphate binders include the following.

Calcium-based binders

Calcium-based binders include calcium carbonate, which contains 500 mg of elemental calcium in a 1,250-mg tablet, and calcium acetate (Phoslo), which has 169 mg of elemental calcium in a 667-mg tablet. Calcium citrate should be avoided since citrate augments aluminum and calcium absorption from the gastrointestinal tract [32]. Although calcium acetate has a greater binding capacity for phosphate than calcium carbonate *in vitro*, potentially allowing a lower dose to be used, the incidence of hypercalcemia is similar with these two forms of therapy, possibly due to the greater bioavailability of calcium in the acetate preparation [33]. Calcium-based binders are inexpensive, effective phosphate binders when used in large doses in advanced cases of CKD. Calcium-based phosphate binders should be taken with meals to bind dietary phosphate. Vitamin D analogues can stimulate active calcium transport in the small intestines and the concomitant use of calcium binders can lead to increased risk of hypercalcemia and positive calcium balance [34]. The K/DOQI guidelines recommend that the total daily calcium intake be limited to 2,000 mg/day (500 mg calcium from diet, 1,500 mg calcium binders). On the basis of the binding characteristics of calcium, 1,500 mg elemental calcium in calcium acetate binds 238 mg phosphate per day; in contrast, 1,500 mg per day elemental calcium in calcium carbonate binds 166 mg of phosphorus per day.

Phosphate-binding resins

Sevelamer is a non-absorbed polymer that is available in the forms of sevelamer carbonate and sevelamer hydroclauric acid (HCL). Sevelamer selectively binds phosphorus in the gut, but its low affinity for phosphate (i.e., 800 mg of sevelamer binds 64 mg of phosphate) necessitates a large pill burden to achieve phosphate control [35]. In a prospective trial of hemodialysis patients, sevelamer achieved similar phosphate control but resulted in less hypercalcemia and progressive coronary and aortic calcification compared to treatment with calcium containing phosphate binders [36]. The most common side effects of sevelamer are gastrointestinal and its ability to bind other drugs such as furosemide, tacrolimus, L-thyroxine, and quininolones. Sevelamer HCL has the potential to provide an acid load due to the exchange of bicarbonate for chloride on the polymer, which contributes to a lowering of the serum bicarbonate levels in treated patients. Sevelamer's binding of short fatty acids in the large intestines also has a lipid lowering effects that may be beneficial [36].

Bivalent and trivalent cation binders

Lanthanum salts. The capacity of lanthanum to bind phosphate *in vitro* is similar to that of aluminium hydroxide, and it is greater than that of calcium acetate, calcium carbonate, and sevelamer [37, 38]. In general, adherence with lanthanum is better than sevalmer and calcium-containing phosphate binders because of the lower pills burden. Each 1,000 mg of elemental lanthanum binds approximately 130 mg of phosphate. Lanthanum accumulates in the bone, but to date there has been no bone toxicity demonstrated. Long-term treatment with lanthanum carbonate had no effect on bone mineralization or remodeling in a 3-year follow-up study [39].

Aluminium salts. Aluminium salts have the greatest phosphate-binding capacity per weight. However, the use of aluminium salts is associated with a risk of aluminum toxicity, which is manifested by osteomalacia, encephalopathy, and anemia. Aluminum contaminated dialysate and concomitant use of calcium citrate, which increases aluminum GI absorption [32], confound the precise assessment of the risk of low dose oral aluminum containing phosphate binders. When used, the duration of treatment is typically limited to periods of 1 to 2 months for one course only. and the doses kept as low as possible.

Iron and magnesium salts. Magnesium hydroxide and carbonate have been used phosphate binders. However, because of the risk for hypermagnesemia and adverse GI side effects, oral magnesium binders have not gained widespread clinical use as phosphate binders.

Fermagate contains magnesium and ferric iron held in an insoluble hydrotalcite structure with carbonate groups, which are exchanged for phosphate. In a phase II trial, 1 g given three times daily before meals was associated with reduced serum phosphate levels, but a higher dose (6 g per day) was associated with adverse GI events [40].

Although there are differences in these various phosphate binders with regard to efficacy, potential for calcium loading, and costs, there are no definitive controlled trials that favor the use of one phosphorus binder over the other. The risk of vascular calcifications and hypercalcemia from the use of calcium-based binders, especially with the current use of vitamin D analogues in the management of hyperparathydosim in CKD, however, has led to a greater use of non-calcium-containing phosphorus binders. However, there is no solid evidence to support this practice. In a CARE-1 study [41], calcium acetate was more effective in lowering serum phosphorus and PTH levels than sevelamer hydrochloride, but in the Treat to Goal study [42], the change in the magnitude of coronary artery and aorta calcifications was less in patients treated with sevelamer hydrochloride. Similarly, the large open–labeled Dialysis Clinical Outcomes Revisited trial (DCOR) trial [43], found a beneficial effect of sevelamer on vascular calcifications but no mortality benefit. In a recent meta-analysis of all available trials that studied various forms of phosphorus binders, there was no comparative superiority of non-calcium-binding agents over calcium-containing phosphate binders for important patient-level outcomes, including as all-cause mortality and cardiovascular end points, hospitalizations, and end-of-treatment serum calcium–phosphorus product levels [44]. Sevelamer is associated with reduced and calcium binders with an increased risk of hypercalcemia, whereas calcium binders are more effective in suppressing serum PTH [42]. The salutary effect of sevelamer to reduce vascular calcification may be due to its lipid-lowering effects. In a CARE-2 study [36], 203 prevalent hyperphosphatemic hemodialysis patients were randomized to receive calcium acetate plus atorvastatin or sevelamer without atorvastatin for 12 months. There was no difference in the progression of calcification between the two groups, providing that serum lipids were kept within the same range [36]. Because of the associated risk of hypercalcemia, non-calcium phosphate binders are typically used in combination with vitamin D analogues, whereas calcium binders offer potential advantages with cincalcet therapy, which is associated with reductions in serum calcium. Intestingly, the use of non-calcium-containing phosphate binders has not been universally associated with reduced vascular calcifications.

Minimizing active phosphate transport in the small intestines by reducing doses of vitamin D analogues or use of niacin

All vitamin D analogues have the potential to increase serum phosphorus through enhancement of the active transport of phosphorus in the small intestines by the sodium-dependent phosphate transporter Napi2b. Vitamin D analogues increase serum calcium and phosphate [23, 45], and protocols that use less active vitamin D analogues, including those using cinacalcet, improved phosphate control [28, 46]. Therefore, a reduction of active vitamin D analogues dose can result in reductions in serum phosphate concentrations.

Niacin has a direct inhibitory effect on active transport-mediated phosphorus absorption in the mammalian small intestine [47] and has been shown to decrease serum phosphorus levels [48] in patients with end stage renal disease.

Suppressing PTH to lower serum phosphate

Bone is an underappreciated source contributing to serum phosphorus concentrations in CKD patients due to increased phosphate efflux from bone caused by excessive PTH-dependent increase in bone turnover [49]. Indirect evidence for the contribution of bone to serum phosphorus levels is derived from association between elevated serum phosphorus and PTH levels, the decrease in serum phosphorus levels after parathyroidectomy, and the reductions in serum phosphate in subjects treated with cinacalcet to lower PTH [28, 29]. In addition, whereas PTH infusions lower serum phosphate in patients with normal renal function, it increases serum phosphate in patients with ESRD. The contribution of bone remodeling to serum phosphate is typically in the range of 0.5 to 1.5 mg/dl.

Frequent daily dialysis

A 4-hour standard dialysis treatment removes approximately 900 mg of phosphate [50]. Thrice 4-hour weekly dialysis removes roughly 2.7 gm of phosphorus per week [50], which is insufficient to prevent a positive phosphate balance, especially in the presence of increased gastrointestinal phosphate absorption due active vitamin D analogue therapy, unless efforts to limit phosphate absorption are initiated. Recent studies found that the total weekly phosphorus removal with nocturnal hemodialysis six times per week is more than twice that removed by thrice weekly in in-center hemodialysis [51]. Indeed, nocturnal dialysis performed six times per week results in lowering of the serum phosphate and discontinuation of phosphate binder therapy compared to patients on standard thrice-weekly hamodialysis. Although the addition of an extra dialysis session would be predicted to lower serum phosphate, there are limited data at present to support this, possibly due to the fact that the other steps to control phosphate were not initiated.

Control PTH levels within an acceptable range that prevents progressive HPT while maintaining normal bone and vascular health

The second goal for treating CKD-MBD is to reduce PTH levels. The rationale for reducing PTH are several, including prevention of parathyroid gland hyperplasia and need for parathyroidectomy, reduction in metabolic bone disease and fracture risk, better control of serum phosphate, and reduction in cardiovascular mortality. PTH levels in excess of 600 pg/ml are associated with increased mortality in patients on hemodialysis, whereas consistent maintenance of PTH in a range between 150 and 300 pg/ml is associated with reduced mortality [52]. This can be achieved by the use of either vitamin D analogues or calcimimetics. Of these approaches, targeting the CaSR in parathyroid glands with calcimimetics offers the most direct method for suppressing PTH. The use of vitamin D analogues to target the VDR is effective too, but recent data suggest that the VDR in the parathyroid gland is not essential for regulation of PTH production. The actions of vitamin D analogues on gastrointestinal calcium absorption, as well as effects on the PTH gene transcription, account for the suppression of PTH. There are two different treatment paradigms that can be used to suppress PTH: titration of active vitamin D analogues or titration of cinacalcet with fixed dose of vitamin D analogues.

Vitamin D analogues

Vitamin D analogues include 1,25(OH)$_2$D, calcitriol, prodrug alfacalcidol, and the active vitamin D analogues: paricalcitol, doxercalciferol, falecalcitriol, and 22-oxacalcitriol. All vitamin D analogues have been shown to be effective in the treatment of secondary hypreparathyrodism (HPTH) in CKD-5 and dialysis patients. On the other hand, active vitamin D analogues are associated with increased circulating FGF23 levels [16]. The significance of this is not known, but elevated FGF23 is the strongest predictor of mortality in ESRD [19–21, 24].

There is no uniform agreement or clinical evidence to support the use of one type of vitamin D analogue, the frequency of administration, or dose. In the only prospective randomized trial that has directly compared the efficacy of calcitriol versus paricalcitol, both agents were comparable in their PTH suppressive effects [53]. Because of reimbursement issues and patient compliance, intermittent intravenous administration with hemodialysis is often used in the U.S., whereas the oral route of administration is the more common in other countries. It remains to be established if differences in the route and dose of administration are associated with better clinical outcomes and less toxicities.

The current recommendations are to administer active vitamin D sterols to CKD-5 or dialysis patients who have intact PTH greater than 300 pg/ml, provided their phosphorus is less than 5.5 and corrected calcium for albumin is less than 9.5 mg/dl. The bioequivalence of intravenous calcitriol, paricalcitol, and doxercalciferol are 0.5, 2.5, and 5 mg, respectively [54].

High doses of vitamin D analogues are associated with hypercalcemia and hyperphosphatemia. Although some of active vitamin D analogues may have less calcemic and phosphatemic effects than calcitriol because of fewer gastrointestinal effects on calcium and phosphorus absorption, clinical trials also show that all analogues have the potential to increase serum calcium and phosphate when administered in high doses to achieve target PTH levels recommended by the KDOQI [55], which are associated in observational studies with increased all-cause and cardiovascular mortality in dialysis patients [56]. Active vitamin D analogues are one of the main stimulators of FGF23 production [16], which has been shown in observational studies to be a powerful, independent predictor of mortality on hemodialysis. Observational studies have shown that active vitamin D analogues are associated with improved survival in both end-stage and predialysis CKD populations [57]. In one retrospective study, the use of active vitamin D analogues in chronic hemodialysis patients is associated with 2-year survival advantage and a lower risk of cardiovascular mortality compared to patients who did not receive active vitamin D therapy [57]. However, a meta-analysis in 2007 of 76 trials of 3,667, mostly dialysis patients, provided opposing evidence that vitamin D compounds do not reduce the risk of death or vascular calcification, and calcitriol increases the risk of hypercalcemia and hyperphosphatemia, while not consistently reducing PTH levels and newer analogues increases calcium levels, in comparison to placebo [55]. Calcitriol, by binding with its intracellular VDR in multiple tissues

Calcimimetic

It is known that calcium suppresses PTH secretion by its action on CasR. In a prospective trial of 52 dialysis patients, large doses of calcium (about five times the recommended doses of elemental calcium by KDOQI) was as effective as daily oral calcitriol and intermittent intravenous calcitriol in PTH suppression with less frequent hyperphosphatemia [59]. However, concerns of the risk of increased vascular calcifications with calcium loading limits the use of calcium binders. Calcimimetics are agents that allosterically increase the sensitivity of the CasR in the PTG to calcium and suppress PTH without increasing serum calcium or providing a calcium load. Cinacalcet is the most effective agent in suppressing PTH. It targets the CaSR in parathyroid chief cells and dose dependently suppresses both PTH gene transcription and secretion, leading to reductions in serum PTH levels that are proportional to the severity of secondary hyperparathyroidism [60]. Cinacalcet is only available in oral form. The initial dose of cinacalcet is 30 mg/day, and subsequent doses are titrated every 2 to 3 weeks in 30-mg increments until a maximum dose of 180 mg/day is achieved. The most common side effects of cinacalcet use are nausea and vomiting, which are reported to be mitigated by taking the medication with meals. In addition, hypocalcemia is a prevalent side effect of cinacalcet that requires close monitoring of serum calcium. Cinacalcet should only be used if serum calcium levels are greater than 8.4 mg/dl [60]. Cinacalcet has also been shown to be effective in reducing PTG volume in hemodialysis patients, suggesting that use of this treatment may alter the natural history of PTG hyperplasia [61]. In addition, cinacalcet reduces serum FGF23 and serum phosphate concentrations [25]. A recent analysis of the ACHIEVE trial showed that treatment with Cinacalcet plus low-dose calcitriol analogues resulted in lower FGF23 levels compared with a treatment regimen using calcitriol analogues alone in ESRD [25]. Cinacalcet can be used effectively to treat secondary hyperparathyroisim with reductions in the doses of active vitamin D analogues and minimizes the risk of hypercalcemia. Ongoing clinical trials are assessing the effects of cinacalcet on mortality and other outcomes in hemodialysis patients, but *post hoc* analysis of phase III clinical trials suggest that cinacalcet may lead to reductions in the risk of parathyroidectomy, fracture, and cardiovascular hospitalization, along with improvements in self-reported physical function and diminished pain [62]. There is no evidence that any of these treatment regimens reduces preexisting vascular calcifications or improves mortality in patients with ESRD.

Two treatment paradigms: High dose vitamin D analogues versus cinacalcet and low dose vitamin D analogues for treating CKD–MBD in ESRD (CKD-5D)

The different effects of the two main medications to suppress PTH on serum calcium and phosphate have resulted in two distinct strategies to suppress PTH. One approach titrates active vitamin D analogues to suppress PTH with the potential side effects of increased serum calcium, phosphate (requiring increased doses of phosphate binders), and FGF23, but with observational data suggesting that use of vitamin D is associated with improved survival that is attributed to the importance of vitamin D in processes unrelated to its function to regulate mineral metabolism. The other approach titrates the dose on cinacalcet to suppress PTH through its actions on CasR in the PTG, and uses nutritional and/or active vitamin D analogues in low, physiological doses (e.g., 0.5 mg calcitirol) to reduce the risk of hypocalcemia and to provide the purported survival benefit of vitamin D therapy. Phosphate binders are used to control phosphate in both approaches. The differential effects of vitamin D analogues (raises) and cinacalcet (lowers) on serum phosphate in ESRD may reflect the effects of cinacalcet to suppress PTH-mediated phosphate efflux from bone and active vitamin D analogues actions to increase gastrointestinal absorption of phosphate.

These two approaches have been compared in the OPTIMA trial [29]: (OPen label, Randomized Study Using Cinacalcet To IMprove Achievement of KDOQI™ Targets in Patients With ESRD). Overall, in the cinacalcet-treated group, mean intact parathyroid hormone (iPTH) levels decreased 46%, whereas the mean PTH in the unrestricted conventional care did not change. The percentage of patients achieving the target phosphorus level was higher in the group treated with cinacalcet (63%) compared to the conventional care group (50%, P = 0.002). This study is limited due to enrollment of patients already proven to be resistant to conventional treatment with vitamin D analogues and the more limited use of vitamin D analogues due to hyperphosphatemia [29]. The ACHIEVE study [28] represents a comparative trial in ESRD patients with moderate hyperparathyroidism of escalating doses of paricalcitol compared to fixed doses of paricalcitol and escalating doses of cinacalcet. Although cinacalcet suppressed PTH to a greater degree than paricalcitol, and was associated with fewer hypercalcemic episodes, this study found no difference between paricalcitol and cinacalcet-based therapies in attaining the primary end point of suppressing PTH to a target range of 150–300 pg/ml (i.e., cinacalcet tended to over suppress PTH at the doses used).

At present there are insufficient data to support one treatment approach over another. Regardless of which paradigm is initiated, most patients will ultimately require combinations of calcium and non-calcium phosphate binders, vitamin D analogues, and cinacalcet to optimally manage the biochemical abnormalities present in ESRD. Recognizing that there are no high-quality trials showing either net beneficial or harmful effects of any of these therapies, it seems reasonable to use these agents at doses that minimizes their recognized toxicities, namely, hypercalcemia and hyperphosphatemia with vitamin D analogues, and hypocalcemia and a tendency to oversuppress PTH with calcimimetics while taking advantage of their different molecular targets and biological actions (e.g., cinacalcet targets CasR and is a more potent suppressor of PTH, whereas vitamin D analogues target the widely expressed VDR, which has potential effects on both mineral homeostasis and other biological processes such as innate immunity and cardiovascular function).

TREATMENT OF CKD–MBD IN STAGE CKD 3-4

In earlier stages of CKD, the severity of the renal dysfunction, secondary HPTH, bone disease, and vascular calcifications are less severe than in ESRD patients, which provides the opportunity to initiate therapy early to prevent the progressive complications of CKD–MBD. Since hypocalcemia and hyperphosphatemia appear late in the course of CKD, it would seem logical to initiate therapy in CKD 3 (GFR < 60 ml/min per 1.73 m^2), when FGF23 and PTH levels first begin to increase. The initial treatment of CKD–MBD in CKD 3-4 is dietary phosphorus restriction. This approach is supported by evidence in animal models and human studies that phosphate restriction, in early stages of CKD can increase endogenous production of 1,25(OH)$_2$D and delay the development of secondary hyperparathyroidism [37]. Because CKD is associated with hypocaluria [38] and there is residual renal function, the use of calcium-based binders does not have the same risk as in ESRD, and they could be used in combination with dietary phosphate restriction. In the absence of hyperphosphatemia, however, monitoring the effectiveness of phosphate binder therapy might include assessment of urinary phosphate and calcium levels and assessment of PTH (since FGF23 measurements are currently used in clinical practice). At present, phosphate binders have not been approved for treatment of patients with impaired renal function and normal phosphate levels. Assessment for nutritional vitamin D deficiency by measurement of serum 25(OH)D levels and treatment of 25(OH)D deficiency is supported by evidence of modest reductions in iPTH levels with nonactivated vitamin D therapy in CKD patients. KDOQI guidelines suggest measurement of serum 25(OH)D levels in patients with stage 3 and 4 CKD who have increased iPTH levels; optimal levels should be greater than 30 ng/mL [22]. Vitamin D deficiency has also been shown to be associated with increased mortality and morbidity in the general population and in CKD patients.

If calcium binders and phosphate restriction fail to achieve target PTH levels, then active vitamin D analogues can be used as a second line of therapy. There are no studies in CKD patients that show that the superiority of one vitamin D analogue over the other or compares the use of vitamin D analogues with treatment with calcium binders only. Cinacalcet is currently not approved for the treatment of secondary hyperparathyroidism in CKD 3-4 patients, and its use in this setting is associated with increased serum phosphate levels, requiring increase use of phosphate binders. The differences between vitamin D analogues and cinacalcet on serum phosphate levels in earlier stages of CKD may be due to their differential effects on circulating concentrations of FGF23.

PARATHYROIDECTOMY

The surgical correction remains the final, symptomatic therapy of the most severe forms of secondary hyperparathyroidism (sometimes called tertiary hyperparathyroidism), which cannot be controlled by medical management. Indications for parathyroidectomy in ESRD patients include symptomatic or severe hypercalcemia or hyperphosphatemia, symptomatic bone disease or spontaneous fracture, calciphylaxis, and persistent hyperparathyroidism for more than 1 year after transplantation. The monitoring of serum calcium, phosphorus, magnesium, and potassium levels in the immediate post parathyroidectomy course is necessary as these patients at risk of developing hungry bone syndrome characterized by symptomatic hypocalcemia, decreases in serum phosphate concentrations, and occasionally hyperkalemia. The changes in calcium and phosphate reflect increased uptake into bone due to the transition from bone resorption to formation.

REFERENCES

1. Goodman WG. 2004. The consequences of uncontrolled secondary hyperparathyroidism and its treatment in chronic kidney disease. *Semin Dial* 17(3): 209–216.
2. Kidney Disease: Improving Global Outcomes (KDIGO) CKD-MBD Work Group. 2009. KDIGO clinical practice guideline for the diagnosis, evaluation, prevention, and treatment of Chronic Kidney Disease-Mineral and Bone Disorder (CKD-MBD). *Kidney Int Suppl* (113): S1–130.
3. Goodman WG, London G, Amann K, et al. 2004. Vascular calcification in chronic kidney disease. *Am J Kidney Dis* 43(3): 572–579.
4. Quarles LD. 2011. The bone and beyond: "Dem bones" are made for more than walking. *Nat Med* 17(4): 428–430.
5. van Abel M, Hoenderop JG, van der Kemp AW, Friedlaender MM, van Leeuwen JP, Bindels RJ. 2005. Coordi-

nated control of renal Ca(2+) transport proteins by parathyroid hormone. *Kidney Int* 68(4): 1708–1721.
6. Talmage RV, Elliott JR. 1958. Removal of calcium from bone as influenced by the parathyroids. *Endocrinology* 62(6): 717–722.
7. Armbrecht HJ, Hodam TL, Boltz MA. 2003. Hormonal regulation of 25-hydroxyvitamin D3-1alpha-hydroxylase and 24-hydroxylase gene transcription in opossum kidney cells. *Arch Biochem Biophys* 409(2): 298–304.
8. Lopez-Hilker S, Galceran T, Chan YL, Rapp N, Martin KJ, Slatopolsky E. 1986. Hypocalcemia may not be essential for the development of secondary hyperparathyroidism in chronic renal failure. *J Clin Invest* 78(4): 1097–1102.
9. Laflamme GH, Jowsey J. 1972. Bone and soft tissue changes with oral phosphate supplements. *J Clin Invest* 51(11): 2834–2840.
10. Ho C, Conner DA, Pollak MR, et al. 1995. A mouse model of human familial hypocalciuric hypercalcemia and neonatal severe hyperparathyroidism. *Nat Genet* 11(4): 389–394.
11. Li YC, Amling M, Pirro AE, et al. 1998. Normalization of mineral ion homeostasis by dietary means prevents hyperparathyroidism, rickets, and osteomalacia, but not alopecia in vitamin D receptor-ablated mice. *Endocrinology* 139(10): 4391–4396.
12. Liu S, Tang W, Zhou J, et al. 2006. Fibroblast growth factor 23 is a counter-regulatory phosphaturic hormone for vitamin D. *J Am Soc Nephrol* 17(5): 1305–1315.
13. Canalejo R, Canalejo A, Martinez-Moreno JM, et al. 2010. FGF23 fails to inhibit uremic parathyroid glands. *J Am Soc Nephrol* 21(7): 1125–1135.
14. Murer H, Lotscher M, Kaissling B, Levi M, Kempson SA, Biber J. 1996. Renal brush border membrane Na/Pi-cotransport: Molecular aspects in PTH-dependent and dietary regulation. *Kidney Int* 49(6): 1769–1773.
15. Lips P. 2006. Vitamin D physiology. *Prog Biophys Mol Biol* 92(1): 4–8.
16. Shimada T, Hasegawa H, Yamazaki Y, et al. 2004. FGF-23 is a potent regulator of vitamin D metabolism and phosphate homeostasis. *J Bone Miner Res* 19(3): 429–435.
17. Feng JQ, Ward LM, Liu S, et al. 2006. Loss of DMP1 causes rickets and osteomalacia and identifies a role for osteocytes in mineral metabolism. *Nat Genet* 38(11): 1310–1315.
18. Nishida Y, Taketani Y, Yamanaka-Okumura H, et al. 2006. Acute effect of oral phosphate loading on serum fibroblast growth factor 23 levels in healthy men. *Kidney Int* 70(12): 2141–2147.
19. Parker BD, Schurgers LJ, Brandenburg VM, et al. 2010. The associations of fibroblast growth factor 23 and uncarboxylated matrix Gla protein with mortality in coronary artery disease: The Heart and Soul Study. *Ann Intern Med* 152(10): 640–648.
20. Smith K, Defilippi C, Isakova T, et al. 2013. Fibroblast growth factor 23, high-sensitivity cardiac troponin, and left ventricular hypertrophy in CKD. *Am J Kidney Dis* 61(1): 67–73.
21. Fliser D, Kollerits B, Neyer U, et al. 2007. Fibroblast growth factor 23 (FGF23) predicts progression of chronic kidney disease: The Mild to Moderate Kidney Disease (MMKD) Study. *J Am Soc Nephrol* 18(9): 2600–2608.
22. National Kidney Foundation. 2003. K/DOQI clinical practice guidelines for bone metabolism and disease in chronic kidney disease. *Am J Kidney Dis* 42(4 Suppl 3): S1–201.
23. Tentori F, Blayney MJ, Albert JM, et al. 2008. Mortality risk for dialysis patients with different levels of serum calcium, phosphorus, and PTH: The Dialysis Outcomes and Practice Patterns Study (DOPPS). *Am J Kidney Dis* 52(3): 519–530.
24. Gutierrez OM, Mannstadt M, Isakova T, et al. 2008. Fibroblast growth factor 23 and mortality among patients undergoing hemodialysis. *New Engl J Med* 359(6): 584–592.
25. Wetmore JB, Liu S, Krebill R, Menard R, Quarles LD. 2010. Effects of cinacalcet and concurrent low-dose vitamin D on FGF23 levels in ESRD. *Clin J Am Soc Nephrol* Jan 2010;5(1): 110–116.
26. Holick MF. Vitamin D deficiency. 2007. *New Engl J Med* 357(3): 266–281.
27. Ott SM. 2009. Review article: Bone density in patients with chronic kidney disease stages 4-5. *Nephrology (Carlton)* Jun 2009;14(4): 395–403.
28. Fishbane S, Shapiro WB, Corry DB, et al. 2008. Cinacalcet HCl and concurrent low-dose vitamin D improves treatment of secondary hyperparathyroidism in dialysis patients compared with vitamin D alone: the ACHIEVE study results. *Clin J Am Soc Nephrol* 3(6): 1718–1725.
29. Messa P, Macario F, Yaqoob M, et al. 2008. The OPTIMA study: Assessing a new cinacalcet (Sensipar/Mimpara) treatment algorithm for secondary hyperparathyroidism. *Clin J Am Soc Nephrol* 3(1): 36–45.
30. Lynch KE, Lynch R, Curhan GC, Brunelli SM. 2011. Prescribed dietary phosphate restriction and survival among hemodialysis patients. *Clin J Am Soc Nephrol* 6(3): 620–629.
31. Guida B, Piccoli A, Trio R, et al. 2011. Dietary phosphate restriction in dialysis patients: A new approach for the treatment of hyperphosphataemia. *Nutr Metab Cardiovasc Dis* 21(11): 879–884.
32. Molitoris BA, Froment DH, Mackenzie TA, Huffer WH, Alfrey AC. 1989. Citrate: A major factor in the toxicity of orally administered aluminum compounds. *Kidney Int* 36(6): 949–953.
33. Fournier A, Moriniere P, Ben Hamida F, et al. 1992. Use of alkaline calcium salts as phosphate binder in uremic patients. *Kidney Int Suppl* 38: S50–61.
34. Goldsmith D, Ritz E, Covic A. 2004. Vascular calcification: A stiff challenge for the nephrologist: Does preventing bone disease cause arterial disease? *Kidney Int* 66(4): 1315–1333.
35. Chertow GM, Burke SK, Dillon MA, Slatopolsky E. 2000. Long-term effects of sevelamer hydrochloride on the calcium x phosphate product and lipid profile of haemodialysis patients. *Nephrol Dial Transplan* 15(4): 559.

36. Qunibi W, Moustafa M, Muenz LR, et al. 2008. A 1-year randomized trial of calcium acetate versus sevelamer on progression of coronary artery calcification in hemodialysis patients with comparable lipid control: The Calcium Acetate Renagel Evaluation-2 (CARE-2) study. *Am J Kidney Dis* 51(6): 952–965.
37. Portale AA, Booth BE, Halloran BP, Morris RC Jr. 1984. Effect of dietary phosphorus on circulating concentrations of 1,25-dihydroxyvitamin D and immunoreactive parathyroid hormone in children with moderate renal insufficiency. *J Clin Invest* 73(6): 1580–1589.
38. Massry SG, Friedler RM, Coburn JW. 1973. Excretion of phosphate and calcium. Physiology of their renal handling and relation to clinical medicine. *Arch Intern Med* 131(6): 828–859.
39. Hutchison AJ. 1999. Calcitriol, lanthanum carbonate, and other new phosphate binders in the management of renal osteodystrophy. *Perit Dial Int* 19 Suppl 2: S408–412.
40. McIntyre CW, Pai P, Warwick G, Wilkie M, Toft AJ, Hutchison AJ. 2009. Iron-magnesium hydroxycarbonate (fermagate): A novel non-calcium-containing phosphate binder for the treatment of hyperphosphatemia in chronic hemodialysis patients. *Clin J Am Soc Nephrol* 4(2): 401–409.
41. Qunibi WY, Hootkins RE, McDowell LL, et al. 2004. Treatment of hyperphosphatemia in hemodialysis patients: The Calcium Acetate Renagel Evaluation (CARE Study). *Kidney Int* 65(5): 1914–1926.
42. Chertow GM, Burke SK, Raggi P. 2002. Sevelamer attenuates the progression of coronary and aortic calcification in hemodialysis patients. *Kidney Int* 62(1): 245–252.
43. Suki WN, Zabaneh R, Cangiano JL, et al. 2007. Effects of sevelamer and calcium-based phosphate binders on mortality in hemodialysis patients. *Kidney Int* 72(9): 1130–1137.
44. Navaneethan SD, Palmer SC, Craig JC, Elder GJ, Strippoli GF. 2009. Benefits and harms of phosphate binders in CKD: A systematic review of randomized controlled trials. *Am J Kidney Dis* 54(4): 619–637.
45. Tentori F. Mineral and bone disorder and outcomes in hemodialysis patients: Results from the DOPPS. *Semin Dial* 23(1): 10–14.
46. Chertow GM, Blumenthal S, Turner S, et al. 2006. Cinacalcet hydrochloride (Sensipar) in hemodialysis patients on active vitamin D derivatives with controlled PTH and elevated calcium x phosphate. *Clin J Am Soc Nephrol* 1(2): 305–312.
47. Eto N, Miyata Y, Ohno H, Yamashita T. 2005. Nicotinamide prevents the development of hyperphosphataemia by suppressing intestinal sodium-dependent phosphate transporter in rats with adenine-induced renal failure. *Nephrol Dial Transplan* 20(7): 1378–1384.
48. Cheng SC, Young DO, Huang Y, Delmez JA, Coyne DW. 2008. A randomized, double-blind, placebo-controlled trial of niacinamide for reduction of phosphorus in hemodialysis patients. *Clin J Am Soc Nephrol* 3(4): 1131–1138.
49. Malluche HH, Mawad HW, Monier-Faugere MC. 2011. Renal osteodystrophy in the first decade of the new millennium: Analysis of 630 bone biopsies in black and white patients. *J Bone Miner Res* 26(6): 1368–1376.
50. Indridason OS, Quarles LD. 2002. Hyperphosphatemia in end-stage renal disease. *Adv Renal Replace Th* 9(3): 184–192.
51. Kooienga L. 2007. Phosphorus balance with daily dialysis. *Semin Dial* 20(4): 342–345.
52. Danese MD, Belozeroff V, Smirnakis K, Rothman KJ. 2008. Consistent control of mineral and bone disorder in incident hemodialysis patients. *Clin J Am Soc Nephrol* 3(5): 1423–1429.
53. Caravaca F, Cubero JJ, Jimenez F, et al. 1995. Effect of the mode of calcitriol administration on PTH-ionized calcium relationship in uraemic patients with secondary hyperparathyroidism. *Nephrol Dial Transplan* 10(5): 665–670.
54. Zisman AL, Ghantous W, Schinleber P, Roberts L, Sprague SM. 2005. Inhibition of parathyroid hormone: A dose equivalency study of paricalcitol and doxercalciferol. *Am J Nephrol* 25(6): 591–595.
55. Palmer SC, McGregor DO, Macaskill P, Craig JC, Elder GJ, Strippoli GF. 2007. Meta-analysis: Vitamin D compounds in chronic kidney disease. *Ann Intern Med* 147(12): 840–853.
56. Melamed ML, Eustace JA, Plantinga L, et al. 2006. Changes in serum calcium, phosphate, and PTH and the risk of death in incident dialysis patients: A longitudinal study. *Kidney Int* 70(2): 351–357.
57. Teng M, Wolf M, Ofsthun MN, et al. 2005. Activated injectable vitamin D and hemodialysis survival: A historical cohort study. *J Am Soc Nephrol* 16(4): 1115–1125.
58. Rostand SG, Warnock DG. 2008. Introduction to Vitamin D Symposium, March 14, 2008. *Clin J Am Soc Nephrol* 3(5): 1534.
59. Indridason OS, Quarles LD. 2000. Comparison of treatments for mild secondary hyperparathyroidism in hemodialysis patients. Durham Renal Osteodystrophy Study Group. *Kidney Int* 57(1): 282–292.
60. Lindberg JS, Moe SM, Goodman WG, et al. 2003. The calcimimetic AMG 073 reduces parathyroid hormone and calcium x phosphorus in secondary hyperparathyroidism. *Kidney Int* 63(1): 248–254.
61. Komaba H, Nakanishi S, Fujimori A, et al. 2010. Cinacalcet effectively reduces parathyroid hormone secretion and gland volume regardless of pretreatment gland size in patients with secondary hyperparathyroidism. *Clin J Am Soc Nephrol* 5(12): 2305–2314.
62. Cunningham J, Danese M, Olson K, Klassen P, Chertow GM. 2005. Effects of the calcimimetic cinacalcet HCl on cardiovascular disease, fracture, and health-related quality of life in secondary hyperparathyroidism. *Kidney Int* 68(4): 1793–1800.

79

Disorders of Mineral Metabolism in Childhood

Thomas O. Carpenter

Disorders of Calcium Homeostasis 651
Disorders of Phosphate Homeostasis 654
Disorders of Magnesium 655

Skeletal Manifestations of Disorders of Calcium and
Phosphate 656
References 656

Disorders of mineral homeostasis in children may present differently than in adults. This chapter outlines disorders of mineral metabolism that occur in children emphasizing the specific features of the age group.

DISORDERS OF CALCIUM HOMEOSTASIS

Hypocalcemia

Clinical presentation

In the newborn with acute hypocalcemia, jitteriness, hyperacusis, irritability, and limb-jerking may occur, with progression to generalized or focal clonic seizures. Laryngospasm may lead to a misdiagnosis of croup. Atrioventricular heart block occurs in prematures with hypocalcemia, and electrocardiograms should be performed in newborns with significant bradycardia [1]. Apnea, tachycardia, tachypnea, cyanosis, edema, and vomiting have been reported in newborns with hypocalcemia.

Transient hypocalcemia of the newborn

Early neonatal hypocalcemia occurs during the first 3 days of life and is seen in prematures, infants of diabetic mothers, and asphyxiated infants. The premature infant has an exaggerated postnatal depression in circulating calcium, such that total calcium levels may drop below 7.0 mg/dl, but the proportional drop in ionized calcium is less. Parathyroid hormone (PTH) insufficiency may contribute to early neonatal hypocalcemia in prematures; a delay in the phosphaturic action of PTH and resultant hyperphosphatemia may further decrease serum calcium.

Late neonatal hypocalcemia presents as tetany between 5 and 10 days of life, occurs more frequently in term than in premature infants, and is usually not correlated with birth trauma or asphyxia. Late neonatal hypocalcemia is associated with maternal vitamin D insufficiency. An increased occurrence of late neonatal hypocalcemia in winter has been noted.

Hypocalcemia associated with magnesium (Mg) deficiency may present as late neonatal hypocalcemia. Severe hypomagnesemia (circulating levels of magnesium below 0.8 mg/dl) may occur in congenital defects of intestinal magnesium absorption or renal tubular reabsorption [2]. Hypocalcemia in this setting may be refractory to therapy unless magnesium levels are corrected.

Maternal hyperparathyroidism may result in neonatal hypocalcemia. Serum Pi is often greater than 8 mg/dl and symptoms may be exacerbated by high Pi intake. Maternal hypercalcemia results in increased calcium delivery to the fetus, thereby suppressing parathyroid gland responsivity. As a result, normal calcium levels are not maintained postpartum due to persistent parathyroid gland suppression.

Persistent hypocalcemia presenting in childhood
Hypoparathyroidism
Persistent hypocalcemia detected in childhood may be due to congenital hypoparathyroidism. Mutations in genes involved in parathyroid gland development, PTH processing, PTH secretion, PTH structure and PTH

Primer on the Metabolic Bone Diseases and Disorders of Mineral Metabolism, Eighth Edition. Edited by Clifford J. Rosen.
© 2013 American Society for Bone and Mineral Research. Published 2013 by John Wiley & Sons, Inc.

resistance have been identified (see Chapter 72). The most frequently identified disorder of parathyroid gland developmental is the DiGeorge anomaly (OMIM #188400), which comprises hypoparathyroidism, T-cell incompetence due to a partial or absent thymus, and conotruncal heart defects (e.g., tetralogy of Fallot, truncus arteriosus) or aortic arch abnormalities. These structures are derived from the third and fourth pharyngeal pouches and are seen in association with microdeletions of chromosome 22q11.2 [3]. Cleft palate and facial dysmorphism may occur. Some present late in childhood due to mild parathyroid defects not apparent in infancy [4]. Deletion of *TBX1*, which encodes a T-box transcription factor, is sufficient to cause the cardiac, parathyroid, thymic, facial, and vellopharyngeal features of the syndrome; however, variability in phenotype with similar genetic defects occurs [5]. One recently reported case demonstrates a child with a deletion in this region inherited from the father, who had a compensating duplication of the region on the other chromosome and was therefore asymptomatic [6]. A similar phenotype identified as DiGeorge 2 syndrome has been attributed to a deletion of distal end of chromosome 10p [7] (601362); GATA3 may be the responsible gene for the hypoparathyroidism accompanying this deletion, which likely represents variants of the HDR (hypoparathyroidism, sensorineural deafness, renal dysplasia) syndrome (146255). Other genetic defects result in disrupted parathyroid gland development (146200, 307700) [e.g., loss of *TBCE* (241410 and 244460), or *GCM2* (603716)], abnormal PTH processing and molecular structure (*PTH*, 168450), abnormal PTH secretory dynamics (*CaSR*), and resistance to PTH action (*GNAS*, 103580). Individuals with classic PTH resistance (pseudohypoparathyroidism), due to loss of function of the Gs alpha protein, often do not develop clinically evident hypocalcemia until a few years of age.

Acquired hypoparathyroidism in children is most commonly caused by autoimmune destruction of the glands [autoimmune polyendocrinopathy syndrome type 1 (APS1) (240300)]. Manifestations of APS1 include adrenal insufficiency, mucocutaneous candidiasis, and hypoparathyroidism; loss of function mutations in the *AIRES* gene, which encodes an autoimmune regulator with features of a transcription factor is often identified in such cases. Surgery for thyroid disorders may result in inadvertent removal of parathyroid tissue, resulting in acquired hypothyroidism, as can heavy metal deposition associated with thalassemia and Wilson's disease.

Vitamin D-related hypocalcemia

Vitamin D deficiency, increasingly reported in North America, is most commonly observed in African-American infants who are breastfed or have limited dietary intake of dairy products. Older age groups may be affected [8]. Rarely, inherited defects in vitamin D metabolism (mutations in *CYP1b*) or the receptor (*VDR*) cause vitamin D-related hypocalcemia. Dilated cardiomyopathy can accompany severe vitamin D-deficiency hypocalcemia in children and is reversible [9].

Other causes of hypocalcemia

Severe hypocalcemia has been induced in children when Pi enema preparations have been administered rectally or orally [10]. The Pi load can result in extreme hyperphosphatemia (up to 20 mg/dl), life-threatening hypocalcemia, and hypomagnesemia. Such preparations should never be administered to infants younger than 2 years of age. Hyperphosphatemia with rhabdomyolysis has resulted in hypocalcemia. Rotavirus infections may induce malabsorption-related hypocalcemia [11]. Ethylenediaminetetraacetic acid (EDTA) chelation therapy has resulted in acute hypocalcemia associated with fatality in a child with autism [12]. Hungry bone syndrome following treatment of severe rickets or after parathyroidectomy can result in transient, albeit severe, hypocalcemia [13].

Calcium malnutrition has been considered rare in North America; however, recent trends of decreased dietary intake of dairy products have resulted in an increasing incidence of this diagnosis in infants [14].

Infantile osteopetrosis may present with hypocalcemia due to impaired bone resorption; however, hypercalcemia may ensue with successful bone marrow transplant therapy and subsequent resorption of osteopetrotic bone [15]. Decreases in ionized calcium occur in infants who undergo exchange transfusions with citrated blood products or receive lipid infusions. Citrate and fatty acids may complex with ionized calcium, reducing the free calcium compartment in the serum and may result in symptoms of hypocalcemia. Alkalosis secondary to adjustments in ventilatory assistance may provoke a shift from ionized to protein-bound calcium. Prolonged pharmacologic inhibition of bone resorption, as with potent bisphosphonate therapy, may also precipitate hypocalcemia [16].

Treatment of hypocalcemia

In the neonate

Early neonatal hypocalcemia is usually treated when total serum calcium is below 6 mg/dl (1.25–1.50 mmol/L) (or ionized calcium below 3 mg/dl, 0.62–0.72 mmol/L) in prematures or below 7 mg/dl (1.75 mmol/L) in term infants. Therapy of acute tetany consists of intravenous (never intramuscular) calcium gluconate (10% solution) given slowly (<1 ml/min); 1 to 3 ml will usually arrest seizure activity. Doses should not exceed 20 mg of elemental Ca/kg body weight and may be repeated up to four times per 24 hours. After successful management of acute emergencies, maintenance therapy is achieved by intravenous administration of 20–50 mg of elemental Ca/kg body weight per 24 hours. Calcium glubionate is a commonly used oral supplement (most preparations provide 115 mg of elemental Ca/5 ml). Management of late neonatal tetany should include low-phosphate formula such as Similac PM 60/40, in addition to calcium supplements. Therapy can usually be discontinued after several weeks.

The place of vitamin D in the management of transient hypocalcemia is less clear. A significant portion of intestinal calcium absorption in newborns occurs by facilitated diffusion and is not vitamin D dependent. Thus, vitamin D metabolites may not be as useful for the short-term management of transient hypocalcemia as added calcium. In persistent hypoparathyroidism, calcitriol is used in the long term. Four hundred to 800 units of vitamin D daily will prevent vitamin D deficiency in premature infants. Overt vitamin D deficiency rickets should respond within 4 weeks to 1,000–2,000 units of daily oral vitamin D. It is important to monitor the clinical response to pharmacologic vitamin D therapy in infants and small children because of the risk for hypervitaminosis D [17]. Such patients should receive at least 40 mg of elemental Ca/kg body weight/day.

In the older child

Calcium and active vitamin D metabolites are usually titrated to maintain serum calcium in an asymptomatic range without incurring hypercalciuria. In hypoparathyroidism, a target serum calcium of 7.5–9.0 mg/dl is recommended, as higher serum calcium levels are more likely to result in hypercalciuria. In *autosomal dominant hypocalcemia*, due to activating mutations in the calcium-sensing receptor (CaSR), thiazide diuretics can be helpful if symptomatic hypocalcemia coexists with hypercalciuria. Hypercalciuria is not typically observed in patients with pseudohypoparathyroidism type 1a when serum calcium is maintained in the usual normal range.

Disorders of hypercalcemia

Infants are usually asymptomatic with mild to moderate hypercalcemia (11.0–12.5 mg/dL). More severe hypercalcemia may lead to failure to thrive, poor feeding, hypotonia, vomiting, seizures, lethargy, polyuria, dehydration, and hypertension. Hypercalcemia is discussed in detail elsewhere in the *Primer*. Syndromes with specific childhood features are described below.

Severe neonatal hyperparathyroidism (SNHP, 239200) presents in the first few days of life. Serum calcium levels may reach 30 mg/dL. Serum Pi is low, and serum PTH is elevated. Nephrocalcinosis may be present on ultrasonagraphic examination. SNHP is a rare autosomal recessive disorder due to homozygous CaSR loss-of-function mutations [18], occurring in families with *familial hypocalciuric hypercalcemia* (FHH, 145980); SNHP is life threatening, usually requiring emergency extirpation of the parathyroid glands.

In *Williams syndrome* (194050) hypercalcemia may occur in the neonatal period. Growth failure, a characteristic facies, cardiovascular abnormalities (usually supravalvular aortic stenosis or peripheral pulmonic stenosis), delayed psychomotor development, and selective mental deficiency may be present. A deletion of the elastin gene is often found in Williams syndrome. Hypercalcemia usually subsides spontaneously by 1 year of age but may rarely persist for longer. Treatment has traditionally consisted of a vitamin D-free, low-calcium diet, and in severe settings, corticosteroids; pamidronate can be used, with fewer potential complications than glucocorticoids [19].

Subcutaneous fat necrosis is a self-limited disorder of infancy presenting with hypercalcemia and erythematous or violacious skin. Affected areas contain mononuclear cell infiltrates, often coexistent with calcification. Pamidronate is useful when significant hypercalcemia is unresponsive to dietary calcium and vitamin D restriction.

In *vitamin D intoxication*, increased circulating 25-OHD levels occur, but 1,25(OH)$_2$D levels are usually low. *Vitamin A intoxication* may result in bone pain, hypercalcemia, headache, pseudotumor cerebri, and an exfoliative erythematous rash. Alopecia and ear discharge may be present. Hypercalcemia is mediated by increased bone resorption. In order to establish the diagnosis of vitamin A intoxication, serum retinyl ester levels should be determined.

Other conditions in children in which hypercalcemia may be manifest include *Down syndrome*, *skeletal dysplasias* (such as *Jansen's*, 156400), *hypophosphatasia* (241500, 241510), SHORT syndrome [**S**hort stature, **H**yperextensibility of joints and/or hernia (inguinal), **O**cular depression, **R**ieger anomaly, and **T**eething delay) (OMIM) [20], and *osteogenesis imperfecta* (120150 and others). Endogenous overproduction of 1,25(OH)$_2$D occurs in granulomatous diseases, such as *cat-scratch disease*. Inflammatory disorders may generate hypercalcemia via increased bone resorption as in *Crohn's disease*. Other causes include those commonly encountered in adults: immobilization, malignancy, and acquired hyperparathyroidism. Iatrogenic causes include parenteral nutrition, dialysis, and drugs (e.g., thiazides and antifungals) [21].

Idiopathic infantile hypercalcemia has been recently attributed to loss of function mutations in CYP24A1, the vitamin D 24 hydroxylase, invoking impaired catabolism of 1,25(OH)$_2$D as a mechanism for this syndrome [22, 23]. Phosphate losses or inadequate supply results in hypophosphatemia, accompanied by hypercalcemia, typical of the breastfed premature infant (and the inadequacy of breast milk phosphate to meet increased skeletal demands).

Treatment of hypercalcemia

Management of acute hypercalcemia consists of administration of intravenous saline. Furosemide (1 mg/kg) may be given intravenously at 6- to 8-hour intervals. Bisphosphonate therapy for unremitting hypercalcemia in children has become widely accepted [19]. Pamidronate has been highly successful in the management of hypercalcemia.

DISORDERS OF PHOSPHATE HOMEOSTASIS

Disorders of hypophosphatemia

Serum Pi is relatively higher in young children compared to adults. Unfortunately, lapses in diagnosis of childhood hypophosphatemia occur because this clinical difference is not always recognized (see Table 79.1).

Etiology of hypophosphatemia in children

Hypophosphatemia may result from decreased Pi supply, excessive renal losses, or intracellular/extracellular compartmental movement of Pi. "Supply" problems result from dietary deficiency or limited intestinal absorption of Pi. Reduced dietary intake occurs in breastfed premature infants, as human milk is relatively low in Pi content. Fortifiers have been developed to restore mineral content to human milk; these may result in hypercalcemia, so monitoring may be indicated when used. Rickets due to inadequate dietary Pi can be treated with 20–25 mg of elemental phosphorus/kg body weight/day, given orally in three or four divided doses.

Hypophosphatemia secondary to renal losses is encountered clinically in the setting of several primary Pi wasting disorders, of which *X-linked hypophosphatemia* (XLH, 307800) is the most common [24]. XLH may be suspected in the setting of a family history, but affected adult members may never have been correctly diagnosed. XLH typically presents in the second or third year of life, with progressive leg bowing. The progression of bowing over a 6-month period in a child over 18 months old requires further investigation. Children may be incorrectly diagnosed with other disorders (typically metaphyseal dysplasias). The delay in a correct diagnosis of XLH may result in the child missing early medical therapy, which has beneficial effects on growth and leg alignment. Circulating fibroblast growth factor 23 (FGF23) is elevated in XLH, and mediates renal Pi wasting. Additionally, FGF23 decreases 25-OHD 1-hydroxylase (*CYP27B1*) message synthesis, limiting $1,25(OH)_2D$ production, despite hypophosphatemia. Medical treatment consists of the administration of Pi in conjunction with $1,25(OH)_2D_3$ (calcitriol). Doses vary and require adjustment based on clinical monitoring. The range of doses for both Pi and calcitriol on a per weight basis are extremely wide; doses for calcitriol are usually 20–50 ng/kg/day, usually given in two divided doses, and phosphorus, 0.25–2 gms/day in three to five divided doses. A detailed compendium of treatment guidelines for XLH is available [25]. Autonomous hyperparathyroidism and/or vitamin D intoxication may complicate the therapy, so monitoring is suggested at 3- to 4-month intervals in children with measurement of serum calcium, Pi, alkaline phosphatase, and urinary calcium and creatinine excretion; circulating PTH should be measured at least twice yearly. Accurate height measurements and assessment of the bow defect should be performed at all visits. Radiographs of the epiphyses of the distal femur and proximal tibia are obtained every 2 years or more frequently if bow deformities fail to correct, or if progressive skeletal disease is grossly evident. Therapy in adulthood is variably recommended, dependent upon symptomatology. Pi should never be given as monotherapy in XLH, but always with calcitriol.

Soft-tissue calcification of the renal medullary pyramids (nephrocalcinosis) may occur secondary to the mineral load that this treatment provides and can be detected with sonograms. Most compliant patients demonstrate nephrocalcinosis within 3–4 years of beginning therapy, but significant clinical sequelae in patients with mild nephrocalcinosis are not generally observed. Calcification of the entheses are described in the third decade, but not thought to be associated with treatment.

Several other hypophosphatemic disorders are described in which elevated FGF23 occurs, including *tumor-induced osteomalacia* (due to neoplastic overproduction of FGF23) (see Chapter 74), *autosomal dominant hypophosphatemic rickets* (193100) due to mutations that disrupt proteolytic cleavage of FGF23 [26], and in *fibrous dysplasia/McCune Albright syndrome* (174800) [27]. Most recently, *autosomal recessive hypophosphatemic rickets* (ARHR, 241520), due to mutations in dentin matrix protein 1 (DMP1), or ARHR2 (613312), due to mutations in ectonucleotide pyrophosphatase/phophodiesterase 1 (ENPP1) have been described [28–30]. In that PHEX, FGF23, and DMP1 are products of the osteocyte, a central role for this cell in regulation of mineralization has been speculated. This group of disorders is discussed in detail in Chapter 79. Finally, this constellation of biochemical findings in concert with parathyroid hyperplasia have been seen in the setting of overexpression of klotho (612089) [31].

Renal Pi losses occur in *hereditary hypophosphatemic rickets with hypercalciuria* (HHRH, 241530) secondary

Table 79.1. Normative Values in mg/dL (mmol/L) for Serum Phosphate by Age

Age (y)	Mean	2.5th Percentile	97.5th Percentile
0–0.5	6.7 (2.15)	5.8 (1.88)	7.5 (2.42)
> 2	5.6 (1.81)	4.4 (1.43)	6.8 (2.20)
> 4	5.5 (1.77)	4.3 (1.38)	6.7 (2.15)
> 6	5.3 (1.72)	4.1 (1.33)	6.5 (2.11)
> 8	5.2 (1.67)	4.0 (1.29)	6.4 (2.06)
10	5.1 (1.63)	3.8 (1.24)	6.2 (2.01)
12	4.9 (1.58)	3.7 (1.19)	6.1 (1.97)
14	4.7 (1.53)	3.6 (1.15)	6.0 (1.92)
16	4.6 (1.49)	3.4 (1.10)	5.8 (1.88)
20	4.3 (1.39)	3.1 (1.01)	5.5 (1.78)
Adult	3.6 (1.15)	2.7 (0.87)	4.4 (1.41)

Source: Brodehl J, Gellissen K, Weber HP. 1982. Postnatal development of tubular phosphate reabsorption. *Clin Nephrol* 17(4): 163–71.

to mutations in the renal NaPi (2c) cotransporter (*SLC34A3*) [32]. In contrast to XLH and other FGF23-mediated disorders, circulating 1,25(OH)₂D is elevated in HHRH; hypercalciuria is common and renal stones may occur. Osteoporosis may develop. HHRH is treated with oral phosphate, without vitamin D metabolites. Generalized tubular dysfunction (*Fanconi syndrome*, 134600) may occur in conditions such as *cystinosis*, *Lowe's syndrome*, and *Wilson's disease*. Finally, hypophosphatemia may occur in *Dent's disease* (300009), an X-linked recessive disorder due to mutations in *CLCN5*, which encodes a renal tubular chloride channel.

Intracellular/extracellular shifts

Acute hypophosphatemia may occur with acute movement from the extracellular space to intracellular compartments. This is typically seen with correction of diabetic ketoacidosis when insulin-induced intracellular Pi uptake can acutely decrease serum Pi. Another such setting is in the refeeding syndrome, as with nutritional rehabilitation of anorexia nervosa. Pi levels reach a nadir within a week of refeeding; slow oral feeds minimize the severity of this phenomenon. Targeting a minimum 4-day weight gain between 0.36 kg and 0.55 kg has been recommended to reduce complications of refeeding in anorexic adolescents. [33].

Hyperphosphatemia

Increases in serum Pi above age-appropriate ranges are uncommon in children with normal renal function. A massive exogenous phosphate load may occur with use of phosphate-containing enemas resulting in hyperphosphatemia and hypernatremia. Such acute elevations in circulating Pi are often accompanied by a reciprocal decrease in serum calcium, precipitating tetany and seizures. When serum Pi and/or calcium concentrations are sufficiently elevated to result in chronic elevations in the calcium × Pi product (in children, higher than $60\,\text{mg}^2/\text{dl}^2$ is often quoted as undesirable), there is a risk of soft tissue calcification, involving blood vessels, renal parenchyma, skin, cornea, and joints. Serum Pi may increase with rapid lysis of bulky tumors (tumor lysis syndrome). Hyperphosphatemia is a biochemical hallmark of hypoparathyroidism and pseudohypoparathyroidism, accompanied by hypocalcemia. Hyperphosphatemia is observed in chronic kidney disease, where progressive loss of nephrons results in limited capacity to reabsorb Pi. A primary disorder of renal Pi excretion, *hyperphosphatemic tumoral calcinosis* (HTC, 211900), can result from loss-of-function mutations in FGF23 [34], *GALNT3*, which encodes an enzyme that initiates O-glycosylation of FGF23, an important step in its secretion and/or trafficking [35], or Klotho, a membrane protein necessary for FGF23 signaling through fibroblast growth factor receptors (FGFRs) [36]. As expected from the impaired FGF23 activity, HTC has the converse biochemical phenotype of XLH (hyperphosphatemia due to an increased tubular maximum threshold for Pi reabsorption and increased circulating 1,25(OH)₂D].

DISORDERS OF MAGNESIUM

Disorders of hypomagnesemia

Familial hypomagnesemia with secondary hypocalcemia (602014) is an autosomal recessive disease due to mutations in the TRPM6 ion channel, resulting in electrolyte abnormalities in the newborn period [37]. This syndrome presents at several weeks of age with tetany or seizures. Hypocalcemia may be refractory to therapy unless magnesium levels are corrected. Mutations in a renal tubular paracellular transport protein, paracellin, of the claudin family, and encoded by *CLDN16*, may also cause hypomagnesemia, hypocalcemia, and hypercalciuria (248250) [38]. CLDN19, another member of the claudin family, has been implicated as causal to heritable hypomagnesemia (248190) [39]. *Gitelman's syndrome* (263800) is an autosomal recessive disorder of magnesium and potassium wasting with metabolic alkalosis and hypocalciuria, due to mutations in the gene encoding a thiazide-sensitive Na–Cl cotransporter (*SLC12A3*) [2]. Recently, individuals with mutations in HNF1B, known to be important in renal development, have been shown to manifest hypomagnesemia due to renal tubular wasting [40]. Hypomagnesemia may accompany renal losses due to the tubulopathy observed with the use of drugs: chemotherapeutic agents, aminoglycosides, and cyclosporine are typical culprits in children. Refeeding syndrome, which may result in hypophosphatemia and hypokalemia, may also involve hypomagnesemia [41]. Low serum magnesium levels may also accompany hypoparathyroidism.

Treatment of hypomagnesemia

For acute symptomatic hypomagnesemia, magnesium sulfate is given intravenously using cardiac monitoring or intramuscular as a 50% solution at a dose of 0.1–0.2 ml/kg. One or two doses may treat transient hypomagnesemia; a dose may be repeated after 12 to 24 hours. Patients with primary defects in magnesium metabolism require long-term oral magnesium supplements; it is best to give these in several divided doses through the day as to avoid diarrhea. We begin oral supplementation at 5 mg of elemental Mg/kg body weight/day. A variety of salts are available for oral use, we have had limited complications with magnesium oxide.

Hypermagnesemia

Hypermagnesemia is unusual in pediatrics but may occur transiently after fetal exposure to maternal magnesium infusions used in the management of eclampsia/

preeclampsia, or with excessive use of cathartic preparations [42]. Severe hypermagnesemia can result in apnea, respiratory depression, and cardiac arrhythmias. Hypocalcemia may also result from hypermagnesemia.

SKELETAL MANIFESTATIONS OF DISORDERS OF CALCIUM AND PHOSPHATE

The typical skeletal abnormality in the growing child with a paucity of available calcium or Pi is rickets. The use of the term "rickets" in clinical settings generally refers to the growth plate cartilage abnormalities observed in the long bones. Rickets is manifest by radiographic findings such as widened metaphyses, irregular or "frayed" metaphyseal edges, and "cupped" metaphyseal deformations (Fig. 79.1).

Rickets is usually accompanied by *osteomalacia* in bone tissue. The histologic correlate of the radiographic findings at the growth plate is the expansion of the hypertrophic zone of chondrocytes [43]. Weight-bearing on the undermineralized skeleton results in characteristic bowing. A child with overt rickets may have minimal leg deformity prior to walking; however, enlarged wrists or costochondral junctions ("rachitic rosary") are typical. An osteomalacic skull (craniotabes) may be present.

Deficiency, or *nutritional rickets* refers to restricted vitamin D stores, which may result from limited intake and/or limited sunlight exposure. Dietary calcium deficiency can result in a similar clinical picture, and some children have a mixed deficiency of calcium and vitamin D [14]. Treatment is accomplished by provision of adequate vitamin D, traditionally suggested as 2,000 units daily, and providing adequate dietary calcium. Clinical and biochemical findings observed in certain inherited conditions mimic severe nutritional rickets. Mutations in encoding 1 alpha-hydroxylase (*CYP27B1*) (*1-alpha hydroxylase deficiency*, or *vitamin D-dependent rickets, type 1*) (264700), or the vitamin D receptor (VDR) (*hereditary vitamin D resistance*, or *vitamin D-dependent rickets, type 2*) (277440) are well described. Mutations in a vitamin D 25-hydroxylase enzyme have been described as well (60081) [44]. Therapy with calcitriol is generally recommended for these conditions, although hereditary vitamin D resistance may require exceptionally high dosages of this metabolite and is sometimes completely unresponsive to calcitriol. Provision of intravenous calcium has been useful in such settings, with gradual progression to enteral calcium treatment [45].

X-linked hypophosphatemia, XLH, is discussed above in the section on hypophosphatemia and in Chapter 79.

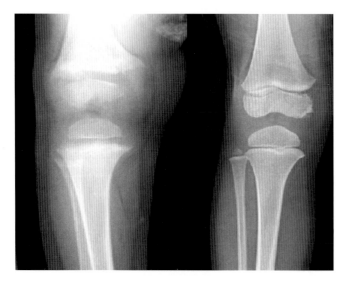

Fig. 79.1. Left: Radiograph of the right knee in an infant with vitamin D-deficiency rickets. Classic epiphyseal deformities are seen, with metaphyseal flaring and frayed edges of the metaphyseal-growth plate junction. Right: A normal knee is shown for comparison.

REFERENCES

1. Stefanaki E, Koropuli M, Stefanaki S, Tsilimigaki A. 2005. Atrioventricular block in preterm infants caused by hypocalcemia: A case report and review of the literature. *Eur J Obst Gyn Reproduct Biol* 120(1): 115–6.
2. Schlingmann KP, Konrad M, Seyberth HW. 2004. Genetics of hereditary disorders of magnesium homeostasis. *Pediatr Nephrol* 19(1): 13–25.
3. Webber SA, Hatchwell E, Barber JC, Daubeney PE, Crolla JA, Salmon AP, Keeton BR, Temple IK, Dennis NR. 1996. Importance of microdeletions of chromosomal region 22q11 as a cause of selected malformations of the ventricular outflow tracts and aortic arch: A three-year prospective study. *J Pediatr* 129(1): 26–32.
4. Sykes KS, Bachrach LK, Siegel-Bartelt J, Ipp M, Kooh SW, Cytrynbaum C. 1997. Velocardiofacial syndrome presenting as hypocalcemia in early adolescence. *Archiv Pediatr Adolescent Med* 151(7): 745–7.
5. Thakker RV. 2004. Genetics of endocrine and metabolic disorders: Parathyroid. *Revs Endocr Metabol Disorders* 5(1): 37–51.
6. Carelle-Calmels N, Saugier-Veber P, Girard-Lemaire F, Rudolf G, Doray B, Guérin E, Kuhn P, Arrivé M, Gilch C, Schmitt E, Fehrenbach S, Schnebelen A, Frébourg T, Flori E. 2009. Genetic compensation in a human genomic disorder. *N Engl J Med* 360(12): 1211–6.
7. Daw SC, Taylor C, Kraman M, Call K, Mao J, Schuffenhauer S, Meitinger T, Lipson T, Goodship J, Scambler P. 1996. A common region of 10p deleted in DiGeorge and velocardiofacial syndromes. *Nat Genet* 13(4): 458–60.
8. Gordon CM, DePeter KC, Feldman HA, Grace E, Emans SJ. 2004. Prevalence of vitamin D deficiency among healthy adolescents. *Arch Pediatr Adolescent Med* 158(6): 531–7.
9. Verma S, Khadwal A, Chopra K, Rohit M, Singhi S. 2011. Hypocalcemia nutritional rickets: A curable cause of dilated cardiomyopathy. *J Trop Pediatr* 57(2): 126–8.

10. Walton DM, Thomas DC, Aly HZ, Short BL. 2000. Morbid hypocalcemia associated with phosphate enema in a six-week-old infant. *Pediatrics* 106(3): E37.
11. Foldenauer A, Vossbeck S, Pohlandt F. 1998. Neonatal hypocalcaemia associated with rotavirus diarrhoea. *Eur J Peds* 157(10): 838–42.
12. Baxter AJ, Krenzelok EP. 2008. Pediatric fatality secondary to EDTA cheltion. *Clin Toxicol (Phila)* 46(10): 1083–4.
13. Yesilkaya E, Cinaz P, Bideci A, Camurdan O, Demirel F, Demircan S. 2009. Hungry bone syndrome after parathyroidectomy caused by an ectopic parathyroid adenoma. *J Bone Miner Metab* 27(1): 101–4.
14. DeLucia MC, Mitnick ME, Carpenter TO. 2003. Nutritional rickets with normal circulating 25-hydroxyvitamin D: A call for re-examining the role of dietary calcium intake in North American children. *J Clin Endocrinol Metab* 88(8): 3539–45.
15. Martinez C, Polgreen LE, DeFor TE, Kivisto T, Petryk A, Tolar J, Orchard PJ. 2010. Characterization and management of hypercalcemia following transplantation for osteopetrosis. *Bone Marrow Transplant* 45(5): 939–44.
16. Perman MJ, Lucky AW, Heubi JE, Azizkhan RG. 2009. Severe symptomatic hypocalcemia in a patient with RDEB treated with intravenous zoledronic acid. *Arch Dermatol* 145(1): 95–6.
17. Vanstone MB, Udelsman RD, Cheng DW, Carpenter TO. 2011. Rapid correction of bone mass after parathyroidectomy in an adolescent with primary hyperparathyroidism. *J Clin Endocrinol Metab* 96(2): E347–50.
18. Pidasheva S, D'Souza-Li L, Canaff L, Cole DE, Hendy GN. 2004. CASRdb: Calcium-sensing receptor locus-specific database for mutations causing familial (benign) hypocalciuric hypercalcemia, neonatal severe hyperparathyroidism, and autosomal dominant hypocalcemia. *Hum Mutat* 24(2): 107–11.
19. Lteif AN, Zimmerman D. 1998. Bisphosphonates for treatment of childhood hypercalcemia. *Pediatrics* 102(4 Pt 1): 990–3.
20. Reardon W, Temple IK. 2008. Nephrocalcinosis and disordered calcium metabolism in two children with SHORT syndrome. *Am J Med Genet* 146A(10): 1296–8.
21. Smith PB, Steinbach WJ, Cotten CM, Schell WA, Perfect JR, Walsh TJ, Benjamin DK Jr. 2007. Caspofungin for the treatment of azole resistant candidemia in a premature infant. *J Perinatol* 27(2): 127–9.
22. Schlingmann KP, Kaufmann M, Weber S, Irwin A, Goos C, John U, Misselwitz J, Klaus G, Kuwertz-Bröking E, Fehrenbach H, Wingen AM, Güran T, Hoenderop JG, Bindels RJ, Prosser DE, Jones G, Konrad M. 2011. Mutations in CYP24A1 and idiopathic infantile hypercalcemia. *N Engl J Med* 365(5): 410–21.
23. Dauber A, Hirschhorn JN, Abrams SA. 2011. Extremely elevated calcium absorption in idiopathic infantile hypercalcemia using calcium isotopic measurement. *Endocr Rev* 32(03 Meeting Abstracts): OR44–1
24. Holm IA, Econs MJ, Carpenter TO. 2003. Familial hypophosphatemia and related disorders. In: Glorieux FH, Pettifor JM, Jüppner H (eds.) *Pediatric Bone: Biology & Diseases, 1st Ed*. San Diego: Academic Press. pp. 603–31.
25. Carpenter TO, Imel EA, Holm IA, Jan de Beur SM, Insogna KL. 2011. A clinician's guide to X-linked hypophosphatemia. *J Bone Min Res* 26(7): 1381–8.
26. White KE, Jonsson KB, Carn G, Hampson G, Spector TD, Mannstadt M, Lorenz-Depiereux B, Miyauchi A, Yang IM, Ljunggren O, Meitinger T, Strom TM, Jüppner H, Econs MJ. 2001. The autosomal dominant hypophosphatemic rickets (ADHR) gene is a secreted polypeptide overexpressed by tumors that cause phosphate wasting. *J Clin Endocrinol Metab* 86(2): 497–500.
27. Riminucci M, Collins MT, Fedarko NS, Cherman N, Corsi A, White KE, Waguespack S, Gupta A, Hannon T, Econs MJ, Bianco P, Gehron Robey P. 2003. FGF-23 in fibrous dysplasia of bone and its relationship to renal phosphate wasting. *J Clin Invest* 112(5): 683–92.
28. Feng JQ, Ward LM, Liu S, Lu Y, Xie Y, Yuan B, Yu X, Rauch F, Davis SI, Zhang S, Rios H, Drezner MK, Quarles LD, Bonewald LF, White KE. 2006. Loss of DMP1 causes rickets and osteomalacia and identifies a role for osteocytes in mineral metabolism. *Nat Genet* 38(11): 1310–5.
29. Lorenz-Depiereux B, Schnabel D, Tiosano D, Häusler G, Strom TM. 2010. Loss-of-function ENPP1 mutations cause both generalized arterial calcification of infancy and autosomal-recessive hypophosphatemic rickets. *Am J Hum Genet* 86(2): 267–72.
30. Levy-Litan V, Hershkovitz E, Avizov L, Leventhal N, Bercovich D, Chalifa-Caspi V, Manor E, Buriakovsky S, Hadad Y, Goding J, Parvari R. 2010. Autosomal-recessive hypophosphatemic rickets is associated with an inactivation mutation in the ENPP1 gene. *Am J Hum Genet* 86(2): 273–78.
31. Brownstein CA, Adler F, Nelson-Williams C, Iijma J, Imura A, Nabehsima Y, Carpenter TO, Lifton RP. 2008. A translocation causing increased α–Klotho level results in hypophosphatemic rickets and hyperparathyroidism. *Proc Nat Acad Sci U S A* 105(9): 3455–60.
32. Bergwitz C, Roslin NM, Tieder M, Loredo-Osti JC, Bastepe M, Abu-Zahra H, Carpenter TO, Anderson D, Garabédian M, Sermet I, Fujiwara TM, Morgan KN, Tenenhouse HS, Jüppner H. 2006. SLC34A3 mutations in patients with hereditary hypophosphatemic rickets with hypercalciuria (HHRH) predict a key role for the sodium-phosphate cotransporter NaPi-IIc in maintaining phosphate homeostasis and skeletal function. *Am J Hum Genet* 78(2): 179–92.
33. Fisher M. 2006. Treatment of eating disorders in children, adolescents, and young adults. *Pediatr Rev* 27(1): 5–16.
34. Benet-Pagès A, Orlik P, Strom TM, Lorenz-Depiereux B. 2005. An FGF23 missense mutation causes familial tumoral calcinosis with hyperphosphatemia. *Hum Mol Genet* 14(3): 385–90.
35. Topaz O, Shurman DL, Bergman R, Indelman M, Ratajczak P, Mizrachi M, Khamaysi Z, Behar D, Petronius D, Friedman V, Zelikovic I, Raimer S, Metzker A, Richard G, Sprecher E. 2004. Mutations in *GALNT3*, encoding a

protein involved in O-linked glycosylation, cause familial tumoral calcinosis. *Nat Genet* 36(6): 579–81.
36. Ichikawa S, Imel EA, Kreiter ML, Yu X, Mackenzie DS, Sorenson AH, Goetz R, Mohammadi M, White KE, Econs MJ. 2007. A homozygous missense mutation in human KLOTHO causes severe tumoral calcinosis. *J Clin Invest* 117(9): 2684–91.
37. Schlingmann KP, Weber S, Peters M, Niemann Nejsum L, Vitzhum H, Klingel K, Kratz M, Haddad E, Ristoff E, Dinour D, Syrrou M, Nielsen S, Sassen M, Waldegger S, Seyberth HW, Konrad M. 2002. Hypomagnesemia with secondary hypocalcemia is caused by mutations in *TRPM6*, a new member of the TRPM gene family. *Nat Genet* 31(2): 166–70.
38. Simon DB, Lu Y, Choate KA, Velazquez H, Al-Sabban E, Praga M, Casari G, Bettinelli A, Colussi G, Rodriguez-Soriano J, McCredie D, Milford D, Sanjad S, Lifton RP. 1999. Paracellin-1, a renal tight junction protein required for paracellular Mg2+ resorption. *Science* 285(5424): 103–6.
39. Konrad M, Schaller A, Seelow D, Pandey AV, Waldegger S, Lesslauer A, Vitzthum H, Suzuki Y, Luk JM, Becker C, Schlingmann KP, Schmid M, Rodriguez-Soriano J, Ariceta G, Cano F, Enriquez R, Jüppner H, Bakkaloglu SA, Hediger MA, Gallati S, Neuhauss SC, Nurnberg P, Weber S. 2006. Mutations in the tight-junction gene claudin 19 (CLDN19) are associated with renal magnesium wasting, renal failure, and severe ocular involvement. *Am J Hum Genet* 79(5): 949–57.
40. Adalat S, Woolf AS, Johnstone KA, Wirsing A, Harries LW, Long DA, Hennekam RC, Ledermann SE, Rees L, van't Hoff W, Marks SD, Trompeter RS, Tullus K, Winyard PJ, Cansick J, Mushtaq I, Dhillon HK, Bingham C, Edghill EL, Shroff R, Stanescu H, Ryffel GU, Ellard S, Bockenhauer D. 2009. HNF1B mutations associate with hypomagnesemia and renal magnesium wasting. *J Am Soc Nephrol* 20(5): 1123–31.
41. Fuentebella J, Kerner JA. 2009. Refeeding syndrome. *Ped Clin North Am* 56(5): 1201–10.
42. Kutsal E, Aydemir C, Eldes N, Demirel F, Polat R, Taspnar O, Kulah E. 2007. Severe hypermagnesemia as a result of excessive cathartic ingestion in a child without renal failure. *Pediatr Emerg Care* 23(8): 570–2.
43. Sabbagh Y, Carpenter TO, Demay MB. 2005. Hypophosphatemia leads to rickets by impairing caspase-mediated apoptosis of hypertrophic chondrocytes. *Proc Nat Acad Sci U S A* 102(27): 9637–42.
44. Cheng JB, Levine MA, Bell NH, Mangelsdorf DJ, Russell DW. 2004. Genetic evidence that the human CYP2R1 enzyme is a key vitamin D 25-hydroxylase. *Proc Nat Acad Sci U S A* 101(20): 7711–5.
45. Balsan S, Garabédian M, Larchet M, Gorski AM, Cournot G, Tau C, Bourdeau A, Silve C, Ricour C. 1986. Long-term nocturnal calcium infusions can cure rickets and promote normal mineralization in hereditary resistance to 1,25-dihydroxyvitamin D. *J Clin Invest* 77(5): 1661–7.

80
Paget's Disease of Bone
Ethel S. Siris and G. David Roodman

Etiology 659
Pathology 660
Biochemical Indices in Paget's Disease 661
Clinical Features 662

Diagnosis 663
Treatment 664
References 666

Paget's disease of bone is a localized disorder of bone remodeling. The process is initiated by increases in osteoclast-mediated bone resorption, with subsequent compensatory increases in new bone formation, resulting in a disorganized mosaic of woven and lamellar bone at affected skeletal sites. This structural change produces bone that is expanded in size, less compact, more vascular, and more susceptible to deformity or fracture than is normal bone [1]. Clinical signs and symptoms will vary from one patient to the next depending on the number and location of affected skeletal sites, as well as on the degree and extent of the abnormal bone turnover. It is believed that most patients are asymptomatic, but a substantial minority may experience a variety of symptoms, including bone pain, secondary arthritic problems, bone deformity, excessive warmth over bone from hypervascularity, fracture and a variety of neurological complications caused in most instances by compression of neural tissues adjacent to pagetic bone.

ETIOLOGY

Although Paget's disease is the second most common bone disease after osteoporosis, the factors involved in its pathogenesis are just beginning to be clarified. Both genetic and environmental factors have been implicated in the pathophysiology of Paget's disease. Paget's disease occurs commonly in families and can be transmitted vertically in an autosomal dominant pattern. Fifteen to 30% of Paget's disease patients have positive family histories of the disorder [2–4], and familial aggregation studies in a U.S. population [5] suggest that the risk of a first-degree relative of a pagetic subject developing the condition is seven times greater than is the risk for someone who does not have an affected relative.

Multiple genetic loci have been linked to familial Paget's disease, and three genes have been identified. Recent genome-wide association studies have identified several susceptibility loci for Paget's disease. These include variants in the *CSF-1* gene, the *RANK* gene, *PML* gene, and three other genes [6, 7]. An insertion mutation in the *RANK* gene has been reported [8], but this mutation is rarely found in patients with familial Paget's disease [9].

The most frequent mutations linked to Paget's disease are in a gene on 5q35-QTER, which encodes an ubiquitin binding protein, sequestasome-1 (SQSTM1/p62) [10]. Mutations in SQSTM1 occur in 30% of patients with familial Paget's disease with the P392L mutation being the most frequent [11]. Mutations in the SQSTM1 gene have been associated with the severity of Paget's disease. Mutation carriers had an earlier age of onset and more commonly required surgery and bisphosphonate therapy [12]. Sequestasome-1 plays an important role in the nuclear factor-κB (NFκB) signaling pathway. Patients with SQSTM1 mutations can have a variable clinical phenotype, including no evidence of Paget's disease in at least one or two individuals, and no gene dose effect can

---

Primer on the Metabolic Bone Diseases and Disorders of Mineral Metabolism, Eighth Edition. Edited by Clifford J. Rosen.
© 2013 American Society for Bone and Mineral Research. Published 2013 by John Wiley & Sons, Inc.

be seen between heterozygotes and homozygotes individuals. Recent studies have reported that the P392L mutation in p62 is either a predisposing mutation for Paget's disease or can result in Paget's disease in experimental animal models [13, 14]. Human osteoclast precursors transfected with the p62^{P392L} do not form osteoclasts characteristic of Paget's disease, and transgenic mice with the p62^{P392L} mutation targeted to the osteoclast lineage develop progressive osteopenia and not Paget's disease. However, mice in which the normal p62 gene has been replaced with p62^{P392L} either do not develop Paget's disease [13] or develop pagetic-like lesions predominantly in their femurs [14].

There is a restricted geographic distribution for the occurrence of Paget's disease: It is most common in Europe, North America, Australia, and New Zealand in persons of Anglo-Saxon descent and is extremely uncommon in Asia, Africa, and Scandinavia.

Some recent studies have reported an apparent decline in the frequency and severity of Paget's disease in both Great Britain and New Zealand [15, 16]. The basis for this decline is unknown, but the changes are too rapid to be explained by genetic factors and cannot be explained by migration patterns of persons with a predisposition to Paget's disease.

For more than 30 years, studies have suggested that Paget's disease may result from a chronic paramyxoviral infection. This is based on ultrastructural studies by Rebel and coworkers [17] who demonstrated that nuclear and, less commonly, cytoplasmic inclusions that were similar to nucleocapsids from paramyxoviruses were present in osteoclasts from Paget's disease patients. Mills and Singer [18] also reported that the measles virus nucleocapsid antigen was present in osteoclasts from patients with Paget's disease, but not from patients with other bone diseases. In some specimens, both measles virus and respiratory syncytial virus nucleocapsid proteins were demonstrated by immunocytochemistry on serial sections. Gordon and colleagues [19], using *in situ* hybridization studies, found canine distemper virus nucleocapsid protein in 11 of 25 Paget's disease patients, and Mee and coworkers [20], using highly sensitive *in situ* polymerase chain reaction (PCR) techniques, found that osteoclasts from 12 of 12 English patients with Paget's disease expressed canine distemper virus nucleocapsid transcripts.

Kurihara et al. [21] provided evidence for a pathophysiologic role for the measles virus in the abnormal osteoclast activity in Paget's disease both *in vitro* and *in vivo*. Transfection of the measles virus nucleocapsid gene into normal human osteoclast precursors resulted in formation of osteoclasts that expressed many of the abnormal characteristics of pagetic osteoclasts. However, other workers have been unable to confirm the presence of measles virus or CDV in pagetic osteoclasts [22]. Kurihara et al. also targeted the measles virus nucleocapsid gene to cells in the osteoclast lineage in transgenic mice and found that 29% of these mice developed localized bone lesions that are similar to lesions seen in patients with Paget's disease [23]. More recently, these investigators reported that mice expressing the MVNP gene and the p62^{P392L} mutation develop exuberant pagetic lesions [24]. They further demonstrated that many of the effects of MVNP seen in these mice were moderated by IL-6 [24].

Among the many questions that still remain to be explained to understand the contributions of environmental and genetic factors to Paget's disease are: (1) Since paramyxoviral infections such as measles virus occur worldwide, why does Paget's disease have a very restricted geographic distribution? (2) How does the virus persist in osteoclasts in patients who are immunocompetent for such long periods of time, since measles virus infections generally occur in children rather than adults, and Paget's disease is usually diagnosed in patients over the age of 55? (3) Why does Paget's disease remain so highly localized in patients after diagnosis? (4) What is the explanation for the variable phenotypic presentation of patients with familial Paget's disease, especially that some of these patients who carry the mutated gene do not have Paget's disease even though they are over 70 years of age?

PATHOLOGY

The initiating lesion in Paget's disease is an increase in bone resorption due to an abnormality in the osteoclasts found at affected sites. Pagetic osteoclasts are more numerous than normal and contain substantially more nuclei than do normal osteoclasts, with up to 100 nuclei per cell. In response to the increase in bone resorption, numerous osteoblasts are recruited to pagetic sites where active and rapid new bone formation occurs. It is generally believed that the osteoblasts are intrinsically normal [25, 26].

In the earliest phases of Paget's disease, increased bone resorption dominates, and lytic changes are seen on radiographs. After this, there is a combination of increased resorption and relatively tightly coupled new-bone formation, produced by the large numbers of osteoblasts present at these sites. During this phase, and presumably because of the accelerated nature of the process, the new bone that is made is abnormal. Newly deposited collagen fibers are laid down in a haphazard rather than a linear fashion, creating more primitive woven bone. The end product is the so-called mosaic pattern of woven bone plus irregular sections of lamellar bone linked in a disorganized way by numerous cement lines representing the extent of previous areas of bone resorption. The bone marrow becomes infiltrated by excessive fibrous connective tissue and by an increased number of blood vessels, explaining the hypervascular state of the bone. Bone matrix is typically normally mineralized, and tetracycline labeling shows increased calcification rates. It is not unusual, however, to find areas of pagetic biopsies in which widened osteoid seams are apparent, perhaps reflecting inadequate calcium/phosphorus products in

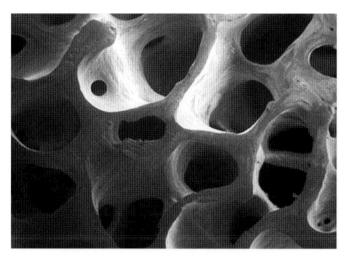 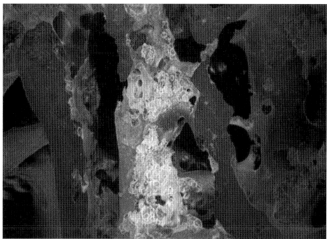

Fig. 80.1. Scanning electron micrographs with sections of normal bone (left) and pagetic bone (right). Both samples were taken from the iliac crest. The normal bone shows the trabecular plates and marrow spaces to be well preserved, whereas the pagetic bone has completely lost this architectural appearance. Extensive pitting of the pagetic bone is apparent, due to dramatically increased osteoclastic bone resorption. (Photographs courtesy of Dr. David Dempster; reproduced from Siris ES, Canfield RE. 1995. Paget's disease of bone. In: Becker KL (ed.) *Principles and Practice of Endocrinology and Metabolism, 2nd Ed.* Philadelphia: JB Lippincott. pp. 585–594. Used with permission.)

localized areas where rapid bone turnover heightens mineral demands.

In time, the hypercellularity at a locus of affected bone may diminish, leaving the end product of a sclerotic, pagetic mosaic without evidence of active bone turnover, so-called burned-out Paget's disease. Typically, all phases of the pagetic process can be seen at the same time at different sites in a particular subject. The chaotic architectural changes that occur in pagetic bone contribute to the loss of structural integrity. Figure 80.1 compares the appearances of normal and of pagetic bone by scanning electron microscopy.

BIOCHEMICAL INDICES IN PAGET'S DISEASE

Measurements of biochemical markers of bone turnover are useful clinically in the assessment of the extent and severity of disease in the untreated state and for monitoring the response to treatment [27]. Increases in serum or urine levels of biomarkers of bone resorption such as the C- and N-terminal telopeptides of collagen, CTX and NTX, reflect the increases in osteoclast mediated bone resorption. Secondary increases in osteoblastic activity are associated with elevated levels of bone formation markers including serum total alkaline phosphatase (SAP), bone specific alkaline phosphatase and procollagen type-1 N-terminal propeptide (P1NP). In untreated patients, the values of serum CTX or urine NTX and SAP rise in proportion to each other, reflecting the preserved coupling of resorption and formation. The magnitude of the increase in markers offers an estimate of the extent or severity of the abnormal bone turnover, with higher levels reflecting a more active, ongoing localized metabolic process. Active monostotic disease may have lower SAP values than polyostotic disease. Lower values (e.g., less than 3 times the upper limit of normal) may indicate fewer pagetic sites or a lesser degree of increased bone turnover at affected sites. However, mild elevations in a patient with limited and highly localized disease (e.g., the proximal tibia) may still be associated with symptoms and clear progression of disease at that site. Even a so-called normal SAP (e.g., at the upper limit of the normal range) may not truly be normal for the pagetic patient. To be confident that the SAP reflects quiescent disease, a result in the middle of the normal range is probably required.

Potent bisphosphonates are capable of normalizing the biochemical markers, an indication of a remission of the bone-remodeling abnormality, in a majority of patients and bringing the markers to near normal in most others so that monitoring the markers is helpful in assessing treatment effects. CTX or NTX may become normal in days to a few weeks after bisphosphonate therapy is initiated. It is often adequate, however, to monitor SAP alone, with a baseline measure pretreatment, a post-treatment test 1 to 3 months after treatment is completed and at 6–12-month intervals thereafter to determine duration of the effect of that treatment course.

Serum calcium is typically normal in untreated Paget's disease, but secondary hyperparathyroidism and transient decreases in serum calcium can occur in some patients who are being treated with potent bisphosphonates. This results from the early suppression of bone resorption in the setting of not yet reduced new-bone formation [28]. As restoration of coupling occurs with time, parathyroid hormone (PTH) levels fall. The problem can be largely avoided by being certain that such patients are and remain replete in both calcium and vitamin D.

CLINICAL FEATURES

Paget's disease affects both men and women, with most series describing a slight male predominance. It is rarely observed to occur in individuals younger than age 25 years and thought to develop as a clinical entity after the age of 40 in most instances. It is most commonly diagnosed in people over the age of 50. In a survey of over 800 selected patients in the U.S., 600 of whom had symptoms, the average age at diagnosis was 58 years [29]. It seems likely that many patients have the disorder for a period of time before any diagnosis is made, especially because it is often an incidental finding.

Paget's disease may be monostotic, affecting only a single bone or portion of a bone (Fig. 80.2), or may be polyostotic, involving two or more bones. Sites of disease are often asymmetric. Clinical observation suggests that in most instances, sites affected with Paget's disease when the diagnosis is made are the only ones that will show pagetic change over time. Although progression of disease within a given bone may occur, the sudden appearance of new sites of involvement years after the initial diagnosis is uncommon.

The most common sites of involvement include the pelvis, femur, spine, skull, and tibia. The humerus, clavicle, scapula, ribs, and facial bones are less commonly involved, and the hands and feet are only rarely affected. It is believed that most patients with Paget's disease are asymptomatic and that the disorder is most often diagnosed when an elevated SAP is noted on routine screening or when a radiograph taken for an unrelated problem reveals typical skeletal changes. The development of symptoms or complications of Paget's disease is influenced by the particular areas of involvement, the interrelationship between affected bone and adjacent structures, the extent of metabolic activity, and presence or absence of disease progression within an affected site.

Signs and symptoms

Bone pain from a site of pagetic involvement, experienced either at rest or with motion, is probably the most common symptom. Pagetic bone associated with a high turnover state has an increased vascularity, leading to a sensation of warmth of the skin overlying bone (e.g., skull or tibia) that some patients perceive as an unpleasant sensation. Small transverse lucencies along the expanded cortices of involved weight-bearing bones or advancing, lytic, blade-of-grass lesions sometimes cause pain.

A bowing deformity of the femur or tibia can cause clinical problems. A bowed limb is typically shortened,

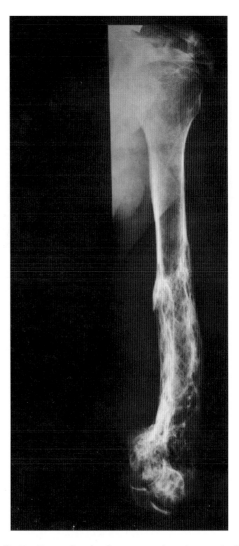

Fig. 80.2. Radiograph of a humerus showing typical pagetic change in the distal half, with cortical thickening, expansion, and mixed areas of lucency and sclerosis, contrasted with normal bone in the proximal half.

resulting in specific gait abnormalities that can lead to abnormal mechanical stresses. Clinically severe secondary arthritis can occur at joints adjacent to pagetic bone (e.g., the hip, knee, or ankle).

Back pain may result from enlarged pagetic vertebrae. Vertebral compression fractures can occur since the bone is of suboptimal quality. Lumbar spinal stenosis with neural impingement may arise, producing radicular pain and possibly motor impairment. Degenerative changes in the spine that are unrelated to Paget's disease may also contribute to a patient's symptoms. Kyphosis may occur, or there may be a forward tilt of the upper back, particularly when a compression fracture or spinal stenosis is present. Paget's disease in the thoracic spine may rarely cause direct spinal cord compression with motor and sensory changes. Several cases of apparent direct cord compression have been documented to have resulted from a vascular steal syndrome, whereby hypervascular pagetic bone "steals" blood from the neural tissue [30].

Paget's disease of the skull may be asymptomatic, but common complaints in up to one-third of patients with diffuse skull involvement may include an increase in head size with or without frontal bossing or deformity, or headache, sometimes described as a band-like tightening around the head. Hearing loss may occur as a result of isolated or combined conductive or neurosensory abnormalities; cochlear damage from pagetic involvement of the temporal bone with loss of bone density in the cochlear capsule may be an important component [31]. Cranial nerve palsies (such as in nerves II, VI, and VII) occur rarely. With extensive skull involvement, a softening of the base of the skull may produce flattening and basilar invagination, so that the odontoid process begins to extend upward as the skull sinks downward upon it. Rarely basilar invagination can produce direct brainstem compression or an obstructive hydrocephalus and increased intracranial pressure caused by blockage of cerebrospinal fluid flow. Pagetic involvement of the facial bones may cause facial deformity, dental problems, and, rarely, narrowing of the airway.

Fracture through pagetic bone can occur, particularly in long bones with active areas of advancing lytic disease; the most common sites are the femoral shaft or subtrochanteric area [32]. Increased vascularity of high turnover pagetic bone (i.e., with a moderately increased SAP) may lead to substantial blood loss in the presence of fractures due to trauma. Fractures also may occur in the presence of areas of malignant degeneration, a rare complication of Paget's disease. Far more common are the small fissure fractures along the convex surfaces of bowed lower extremities, which may be asymptomatic, stable, and persistent for years, but sometimes a more extensive transverse lucent area extends medially from the cortex, typically with symptoms of discomfort, and may lead to a clinical fracture with time. These painful lesions warrant treatment and careful radiographic follow-up over time. Fracture through pagetic bone usually heals normally, although some groups have reported as high as a 10% rate of nonunion.

Neoplastic degeneration is a relatively rare event, occurring with an incidence of less than 1%. This lesion typically presents as severe new pain at a pagetic site and has a grave prognosis. The majority of the tumors are classified as osteogenic sarcomas, although both fibrosarcomas and chondrosarcomas are also seen. The most common site of sarcomatous change appears to be the pelvis, with the femur and humerus next in frequency [33]. Typically, osteosarcomas are osteolytic, although these lesions involve cells of osteoblastic lineage [34].

Benign giant-cell tumors also may occur in bone affected by Paget's disease. These may present as localized masses at the affected site. Radiographic evaluation may disclose lytic changes. Biopsy reveals clusters of large osteoclast-like cells, which some authors believe represent reparative granulomas [35]. These tumors usually show a remarkable sensitivity to high dose glucocorticoids, and the mass will shrink or even disappear after treatment with prednisone or dexamethasone [36], although some will grow back after treatment ends. Anecdotal evidence also suggests possible shrinkage of benign giant-cell tumors in pagetic bone with thalidomide.

DIAGNOSIS

When Paget's disease is suspected, the diagnostic evaluation should include a careful medical history, including family history of the condition and symptom history, and a focused physical examination. The physical exam should note the presence or absence of warmth, tenderness, or bone deformity in the skull, spine, pelvis, and extremities, as well as evidence of loss of range of motion at major joints or leg length discrepancy.

Laboratory tests include measurement of SAP and in some cases a marker of bone resorption, as described earlier. It is also reasonable to assure normalcy of serum calcium and 25-hydroxy-vitamin D. Radiographic studies (bone scans and conventional radiographs) complete the initial evaluation. Bone biopsy is not usually indicated, as the characteristic radiographic and laboratory findings are diagnostic in most instances.

Bone scans are the most sensitive means of identifying possible pagetic sites but are nonspecific, and also can be positive in nonpagetic areas that have degenerative changes or, more ominously, metastatic disease. Plain radiographs of bones noted to be positive on the bone scan provide the most specific information, because radiographic findings are usually characteristic to the point of being pathognomonic. Enlargement or expansion of bone, cortical thickening, coarsening of trabecular markings, and typical lytic and sclerotic changes may be found. Radiographs also show the condition of the joints adjacent to involved sites, identify fissure fractures, indicate the degree to which lytic or sclerotic lesions predominate, and demonstrate the presence or absence of deformity or fracture.

Repeated scans or radiographs are usually unnecessary in observing patients over time, unless new symptoms develop or current symptoms become significantly worse. The possibility of an impending fracture or, rarely, of sarcomatous change should be borne in mind in these situations. Although imaging studies such as computed tomography (CT) or magnetic resonance imaging (MRI) scans are not usually required in routine cases, a CT scan may be helpful in the assessment of a fracture where radiographs are not sufficient, and MRI scans are useful in assessing the possibility of sarcoma, giant-cell tumor, or metastatic disease at a site of Paget's disease. Anecdotal data suggest that positron emission tomography (PET) scans of sclerotic lesions in patients with Paget's disease may help distinguish pagetic lesions from bone metastases, as the former are likely to be minimally to non-metabolic as compared with marked hypermetabolic changes seen with bone metastases [37].

The characteristic X-ray and clinical features of Paget's disease usually eliminate problems with differential diagnosis. However, an older patient may occasionally present with severe bone pain, elevations of the serum alkaline phosphatase and urinary N-telopeptide, a positive bone scan, and less-than-characteristic radiographic areas of lytic or blastic change. Here the possibility of metastatic disease to bone or some other form of metabolic bone disease (e.g., osteomalacia with secondary hyperparathyroidism) must be considered. Old radiographs and laboratory tests are very helpful in this setting, as normal studies a year earlier would make a diagnosis of Paget's disease less likely. A similar dilemma occurs when someone with known and established Paget's disease develops multiple painful new sites; here, too, the likelihood of metastatic disease must be carefully considered, and bone biopsy for a tissue diagnosis may be indicated.

TREATMENT

Specific antipagetic therapy consists of those agents capable of suppressing the activity of pagetic osteoclasts. Currently approved agents available by prescription in the United States include six bisphosphonate compounds: orally administered etidronate, tiludronate, alendronate, and risedronate, and intravenously administered pamidronate and zoledronic acid; and parenterally administered synthetic salmon calcitonin. Several of these are discussed briefly below. A more detailed review of these agents including more information on dosing regimens, clinical trial results, and side effects has been published [38].

Other symptomatic treatments for Paget's disease, including analgesics, anti-inflammatory drugs, use of orthotics or canes, and selected orthopedic and neurosurgical interventions, have important roles in management in many patients.

Two logical indications for medical treatment of Paget's disease are to relieve symptoms and to prevent future complications. It has been shown that suppression of the pagetic process by any of the available agents can effectively ameliorate certain symptoms in the majority of patients. Bone aches or pain, excessive warmth over bone, headache due to skull involvement, low-back pain secondary to pagetic vertebral changes, and some syndromes of neural compression (e.g., radiculopathy and some examples of slowly progressive brainstem or spinal cord compression) are the most likely to be relieved. Pain due to a secondary arthritis from pagetic bone involving the spine, hip, knee, ankle, or shoulder may or may not respond to antipagetic treatment. Filling in of osteolytic blade-of-grass lesions in weight-bearing bones has been reported in some treated cases with either calcitonin or bisphosphonates. On the other hand, a bowed extremity or other bone deformity will not change after treatment, and clinical experience indicates that deafness is unlikely to improve, although limited studies suggest that progression of hearing loss may be slowed [39] or even, in one case with pamidronate, reversed [40].

A second indication for treatment is to prevent the development of late complications in those patients deemed to be at risk, based on their sites of involvement and evidence of active disease, as shown by elevated levels of bone turnover markers. Admittedly, it has not been proven that suppression of pagetic bone turnover will prevent future complications. However, there is a restoration of normal patterns of new bone deposition in biopsy specimens after suppression of pagetic activity. It is also clear that active, untreated disease can continue to undergo a persistent degree of abnormal bone turnover for many years, with the possibility of severe bone deformity over time. Indeed, substantial (e.g., 50%) but incomplete suppression of elevated indices of bone turnover with older and less effective therapies has been associated with disease progression [41]; with potent bisphosphonates, however, indices become normal after treatment for extended periods in the majority of patients and approach normal in most of the rest.

Some treatment guides recommend, therefore, that the presence of asymptomatic but active disease (i.e., SAP above normal) at sites where the potential for later problems or complications exists (e.g., weight-bearing bones, areas near major joints, vertebral bodies, extensively involved skull) is an indication for treatment [38]. The need for treatment in this setting may be particularly valid in patients who are younger, for whom many years of coexistence with the disorder is likely. However, even in the elderly, one can justify treatment if a degree of bone deformity is present that might create serious problems in the next few years. Others argue that the evidence does not support such use; in the PRISM clinical trial—with a median of only 3 years of observation—disease suppression with bisphosphonates failed to reduce short-term complications or improve quality of life [42].

Although controlled studies are not available to prove efficaciousness in this situation, the use of a potent bisphosphonate before elective surgery on metabolically active pagetic bone also is recommended [43]. The goal is to reduce the hypervascularity associated with moderately active disease (e.g., a threefold or more elevation in SAP) to minimize blood loss at operation.

Recommendations for the management of Paget's disease have been published as guideline or management documents by consensus panels in the U.S. [38], U.K. [44], and Canada [45].

Bisphosphonates

Studies with etidronate [46], tiludronate [47], alendronate [48], risedronate [49], pamidronate [50] and zoledronic acid (also referred to as zoledronate) [51] have all demonstrated the efficacy of these agents in suppressing the localized bone turnover abnormality and in improving many symptoms in patients with Paget's disease, and all of them are available and approved for use in the United States.

In most instances, the drug of choice, based on the efficaciousness of the agent and patient preferences regarding an intravenous or an oral regimen, is intravenous zoledronate or oral risedronate. Generic alendronate, 40 mg per day for 6 months (with the potential to repeat after a drug-free interval) and generic pamidronate, with several possible dosing approaches based on the patient's status [38], are also available at lower cost but with less convenient dosing regimens. Etidronate (400 mg per day for 6 months with repeated 6-months-on, 6-months-off cycles as needed) and tiludronate (3 months of 400 mg per day) are much less frequently used as they are of lesser potency than the other four bisphosphonates and may typically reduce elevated bone turnover markers by about 50% rather than achieve biochemical remission in the majority of patients.

Risedronate is prescribed as a daily oral dose of 30 mg for 2 months—note that this is a different dosing regimen than that for osteoporosis. The pill is taken after an overnight fast upon arising each morning with 8 ounces of plain water. The patient must remain upright and take nothing else by mouth for 30 minutes, after which he or she should eat. A follow-up measurement of SAP 1 to 2 months after completing the course is useful; if the value is not yet normal or near normal, a third or fourth month of risedronate could be offered with a good likelihood of normalcy or near normalcy of indices thereafter. In the pivotal clinical trial 80% of the patients had achieved a normal SAP 6 months after initiation of 2 months of treatment, with a period of subsequent disease suppression of up to 18 months [49]. Periodic measurement of SAP (every 6–12 months) should be done, and re-treatment is suggested, if indicated, if and when SAP rises above normal or increases by more than 25% of the nadir value if full remission was not achieved.

Zoledronic acid at a dose of 5 mg is administered as a single 15-minute intravenous infusion. In the pivotal clinical trial comparing one 5 mg infusion of zoledronic acid with 2 months of 30 mg per day oral risedronate, a normal SAP was achieved by 89% of zoledronic acid subjects compared with 58% of risedronate subjects [51]. A period of biochemical remission after the first zoledronic acid infusion of up to 18 months following the end of the initial observation period of 6 months was also noted in this study population. In practice, if a patient has a very high SAP pretreatment that fails to come to normal or near normal by a few months after the infusion, a second infusion can be provided. For patients who enter a biochemical remission or near remission after one (or two) doses, 6- to 12-month follow-up SAP measurements are suggested. Once the SAP begins to rise above normal or more than 25% above nadir levels if remission was not achieved, and if treatment is again indicated based on symptoms or concerns about complications, another dose can be provided. Again, note that treating at variable intervals based on biochemical remission and relapse differs from the regimen that is used when zoledronic acid is given for osteoporosis.

Secondary resistance to the effects of a given bisphosphonate (failure to achieve remission or a similar reduction in turnover markers with repeated courses of treatment) has been noted anecdotally with etidronate and has been reported with pamidronate [52] in some patients. Changing treatment at that point to a different bisphosphonate appears to be effective [52].

It is important to emphasize the need for full repletion of both calcium and vitamin D prior to and during treatment with potent bisphosphonates to avoid hypocalcemia and secondary hyperparathyroidism. Calcium and vitamin D repletion should be maintained thereafter in these patients as a general principle.

Side effects with alendronate and risedronate include upper gastrointestinal symptoms consistent with esophageal irritation in a minority of individuals. Excessive doses of etidronate (e.g., more than 6 months without a drug-free interval before re-treating) can infrequently induce a transient mineralization defect and osteomalacia. The first-ever dose of either pamidronate or zoledronic acid in a patient who has not previously received a nitrogen containing bisphosphonate can be associated with a flu-like reaction for 1–2 days after treatment with fever, headache, myalgia and arthralgia, ameliorated by using acetaminophen or an nonsteroidal anti-inflammatory drug (NSAID); this reaction is unlikely to occur with subsequent doses. Finally, relatively rare cases of uveitis or iritis have been described with nitrogen containing bisphosphonates. In such patients, either etidronate or tiludronate can be given, as these compounds do not contain the nitrogen atom.

Osteonecrosis of the jaw has recently been described as a complication typically following dental extractions in patients receiving relatively high doses of potent bisphosphonates given primarily for management of bone metastases. At least seven patients with Paget's disease have also been reported to have had this complication, most of whom were given very high doses for prolonged periods of time outside the usual prescribing guidelines [53]. This topic is discussed in detail elsewhere in the *Primer*.

Calcitonin

Synthetic salmon calcitonin is available as a subcutaneous injection. It is less effective than the nitrogen containing bisphosphonates and is most useful in the rare patient who is intolerant of all bisphosphonates or if bisphosphonate therapy is contraindicated. The usual starting dose is 100 U (0.5 ml; the drug is available in a 2-ml vial), generally self-injected subcutaneously, initially on a daily basis. Symptomatic benefit may be apparent in a few weeks, and the biochemical benefit (typically about a 50% reduction from baseline in SAP) is usually seen after 3 to 6 months of treatment. After this period, many clinicians reduce the dose to 50 to 100 U every other day or three times weekly. Escape from the efficacy of salmon calcitonin may sometimes occur after a variable period of benefit. The main side effects of parenteral salmon calcitonin include, in a minority of patients, the development of nausea or queasiness, with or without flushing of the skin of the face and ears. Intranasal calcitonin is not indicated for use in Paget's disease, but anecdotal experience suggests it may relieve some symptoms and lower elevated bone turnover markers in patients with mild disease.

Other therapies

Analgesics such as acetaminophen, aspirin, and NSAIDs may be tried empirically with or without antipagetic therapy to relieve pain. In particular, pain from pagetic arthritis (i.e., osteoarthritis caused by deformed pagetic bone at a joint space) is often helped by some of these agents.

Surgery on pagetic bone [54] may be necessary in the setting of established or impending fracture. Elective joint replacement, more complex with Paget's disease than with typical osteoarthritis, is often very successful in relieving refractory pain. Rarely, osteotomy is performed to alter a bowing deformity in the tibia. Neurosurgical intervention is sometimes required in cases of spinal cord compression, spinal stenosis, or basilar invagination with neural compromise. Although medical management may be beneficial and adequate in some instances, all cases of serious neurological compromise require immediate neurological and neurosurgical consultation to allow the appropriate plan of management to be developed.

REFERENCES

1. Kanis JA. 1998. *Pathophysiology and Treatment of Paget's Disease of Bone, 2nd Ed.* London: Martin Dunitz Ltd.
2. Siris ES, Canfield RE, Jacobs TP. 1980. Paget's disease of bone. *Bull NY Acad Med* 56: 285–304.
3. Morales-Piga AA, Rey-Rey JS, Corres-Gonzalez J, Garcia-Sagredo IM, Lopez-Abente G. 1995. Frequency and characteristics of familial aggregation of Paget's disease of bone. *J Bone Miner Res* 10: 663–670.
4. McKusick VA. 1972. *Heritable Disorders of Connective Tissue, 5th Ed.* St. Louis, MO: CV Mosby. pp. 718–723.
5. Siris ES, Ottman R, Flaster E, Kelsey JL. 1991. Familial aggregation of Paget's disease of bone. *J Bone Miner Res* 6: 495–500.
6. Albagha OM, Visconti MR, Alonso N, Langston AL, Cundy T, Dargie R, Dunlop MG, Fraser WD, Hooper MJ, Isaia G, Nicholson GC, del Pino Montes J, Gonzalez-Sarmiento R, di Stefano M, Tenesa A, Walsh JP, Ralston SH. 2010. Genome-wide association study identifies variants at CSF1, OPTN and TNFRSF11A as genetic risk factors for Paget's disease of bone. *Nat Genet* 42(6): 520–524.
7. Albagha OM, Wani SE, Visconti MR, Alonso N, Goodman K, Brandi ML, Cundy T, Chung PY, Dargie R, Devogelaer JP, Falchetti A, Fraser WD, Gennari L, Gianfrancesco F, Hooper MJ, Van Hul W, Isaia G, Nicholson GC, Nuti R, Papapoulos S, Montes JD, Ratajczak T, Rea SL, Rendina D, Gonzalez-Sarmiento R, Di Stefano M, Ward LC, Walsh JP, Ralston SH; Genetic Determinants of Paget's Disease (GDPD) Consortium. 2011. Genome-wide association identifies three new susceptibility loci for Paget's disease of bone. *Nat Genet.* 43(7): 685–689.
8. Sparks AB, Peterson SN, Bell C, Loftus BJ, Hocking L, Cahill DP, Frassica FJ, Streeten EA, Levine MA, Fraser CM, Adams MD, Broder S, Venter JC, Kinzler KW, Vogelstein B, Ralston SH. 2001. Mutation screening of the TNFRSF11A gene encoding receptor activator of NF kappa B (RANK) in familial and sporadic Paget's disease of bone and osteosarcoma. *Calcif Tissue Int* 68: 151–155.
9. Hocking L, Slee F, Haslam SI, Cundy T, Nicholson G, van Hul W, Ralston SH. 2000. Familial Paget's disease of bone: Patterns of inheritance and frequency of linkage to chromosome 18q. *Bone* 26: 577–580.
10. Laurin N, Brown JP, Morissette J, Raymond V. 2002. Recurrent mutation of the gene encoding sequestosome 1 (SQSTM1/p62) in Paget disease of bone. *Am J Hum Genet* 70: 1582–1588.
11. Hocking LJ, Herbert CA, Nicholls RK, Williams F, Bennett ST, Cundy T, Nicholson GC, Wuyts W, Van Hul W, Ralston SH. 2001. Genomewide search in familial Paget disease of bone shows evidence of genetic heterogeneity with candidate loci on chromosomes 2q36, 10p13, and 5q35. *Am J Hum Genet* 69: 1055–1061.
12. Visconti MR, Langston AL, Alonso N, Goodman K, Selby PL, Fraser WD, Ralston SH. 2010. Mutations of SQSTM1 are associated with severity and clinical outcome in Paget disease of bone. *J Bone Miner Res* 25(11): 2368–2373. PMID: 20499339.
13. Kurihara N, Hiruma Y, Zhou H, Subler MA, Dempster DW, Singer FR, Reddy SV, Gruber HE, Windle JJ, Roodman GD. 2007. Mutation of the Sequestosome 1 (p62) gene increases osteoclastogenesis but does not induce Paget's disease. *J Clin Invest* 117: 133–142
14. Daroszewska A, van 't Hof RJ, Rojas JA, Layfield R, Landao-Basonga E, Rose L, Rose K, Ralston SH. 2011. A point mutation in the ubiquitin-associated domain of SQSMT1 is sufficient to cause a Paget's disease-like disorder in mice. *Hum Mol Genet* 20(14): 2734–2744.
15. Cooper C, Schafheutle K, Dennison E, Kellingray S, Guyer P, Barker D. 1999. The epidemiology of Paget's disease in Britain: Is the prevalence decreasing? *J Bone Miner Res* 14: 192–197.
16. Cundy T, McAnulty K, Wattie D, Gamble G, Rutland M, Ibbertson HK. 1997. Evidence for secular changes in Paget's disease. *Bone* 20: 69–71
17. Rebel A, Malkani K, Basle M, Bregeon C. 1997. Is Paget's disease of bone a viral infection? *Calcif Tissue Res* 22 Suppl: 283–286.

18. Mills BG, Singer FR, Weiner LP, Suffin SC, Stabile E, Holst P. 1984. Evidence for both respiratory syncytial virus and measles virus antigens in the osteoclasts of patients with Paget's disease of bone. *Clin Orthop* 183: 303–311.
19. Gordon MT, Mee AP, Sharpe PT. 1994. Paramyxoviruses in Paget's disease. *Semin Arthritis Rheum* 23: 232–234.
20. Mee AP, Dixon JA, Hoyland JA, Davies M, Selby PL, Mawer EB. 1998. Detection of canine distemper virus in 100% of Paget's disease samples by in situ-reverse transcriptase-polymerase chain reaction. *Bone* 23: 171–175.
21. Kurihara N, Reddy SV, Menaa C, Anderson D, Roodman GD. 2000. Osteoclasts expressing the measles virus nucleocapsid gene display a pagetic phenotype. *J Clin Invest* 105: 607–614.
22. Ooi CG, Walsh CA, Gallagher JA, Fraser WD. 2000. Absence of measles virus and canine distemper virus transcripts in long-term bone marrow cultures from patients with Paget's disease of bone. *Bone* 27: 417–421.
23. Kurihara N, Zhou H, Reddy SV, Garcia-Palacios V, Subler MA, Dempster DW, Windle JJ, Roodman GD. 2006. Expression of measles virus nucleocapsid protein in osteoclasts indures Paget's disease-like bone lesions in mice. *J Bone Miner Res* 21: 446–455.
24. Kurihara N, Hiruma Y, Yamana K, Michou L, Rousseau C, Morissette J, Galson DL, Teramachi J, Zhou H, Dempster DW, Windle JJ, Brown JP, Roodman GD. 2011. Contributions of the measles virus nucleocapsid gene and the SQSTM1/p62(P392L) mutation to Paget's disease. *Cell Metab* 13(1): 23–34. PMID: 21195346.
25. Rebel A, Basle M, Pouplard A, Malkani K. Filmon R, Lepatezour A. 1980. Bone tissue in Paget's disease of bone: Ultrastructure and immunocytology. *Arthritis Rheum* 23: 1104–1114.
26. Singer FR, Mills BG, Gruber HE, Windle JJ, Roodman GD. 2006. Ultrastructure of bone cells in Paget's disease of bone. *J Bone Miner Res* 21 Suppl 2: P51–P54.
27. Shankar S, Hosking DJ. 2006. Biochemical assessment of Paget's disease of bone. *J Bone Miner Res* 21 Suppl 2: P22–P27
28. Siris ES, Canfield RE. 1994. The parathyroids and Paget's disease of bone. In: Bilezikian J, Levine M, Marcus R (eds.) *The Parathyroids*. New York: Raven Press. pp. 823–828.
29. Siris ES. 1991. Indications for medical treatment of Paget's disease of bone. In: Singer FR, Wallach S (eds.) *Paget's Disease of Bone: Clinical Assessment. Present and Future Therapy*. New York: Elsevier. pp. 44–56.
30. Herzberg L, Bayliss E. 1980. Spinal cord syndrome due to non-compressive Paget's disease of bone: A spinal artery steal phenomenon reversible with calcitonin. *Lancet* 2: 13–15.
31. Monsell EM. 2004. The mechanism of hearing loss in Paget's disease of bone. *Laryngoscope* 114: 598–606.
32. Barry HC. 1980. Orthopedic aspects of Paget's disease of bone. *Arthritis Rheum* 23: 1128–1130.
33. Wick MR, Siegal GP, Unni KK, McLeod RA, Greditzer HB. 1981. Sarcomas of bone complicating osteitis deformans (Paget's disease). *Am J Surg Pathol* 5: 47–59.
34. Hansen MF, Seton M, Merchant A. 2006. Osteosarcoma in Paget's disease of bone. *J Bone Miner Res* 21 Suppl 2: P58–P63.
35. Upchurch KS, Simon LS, Schiller AL, Rosenthal DI, Campion EW, Krane SM. 1983. Giant cell reparative granulomas of Paget's disease of bone: A unique clinical entity. *Ann Intern Med* 98: 35–40.
36. Jacobs TP, Michelsen J, Polay J, D'Adamo AC, Canfield RE. 1979. Giant cell tumor in Paget's disease of bone: Familial and geographic clustering. *Cancer* 44: 742–747.
37. Sundaram M. 2006. Imaging of Paget's disease and fibrous dysplasia of bone. *J Bone Miner Res* 21 Suppl 2: P28–P30.
38. Siris ES, Lyles KW, Singer FR, Meunier PJ. 2006. Medical management of Paget's disease of bone: Indications for treatment and review of current therapies. *J Bone Miner Res* 21 Suppl 2: P94–98.
39. El-Sammaa M, Linthicum FH, House HP, House JW. 1986. Calcitonin as treatment for hearing loss in Paget's disease. *Am J Otol* 7: 241–243.
40. Murdin L, Yeoh LH. 2005. Hearing loss treated with pamidronate. *J R Soc Med* 98: 272–274.
41. Meunier PI, Vignot E. 1995. Therapeutic strategy in Paget's disease of bone. *Bone* 17: 489S–491S.
42. Langston AL, Campbell MK, Fraser WD, MacLennan GS, Selby PL, Ralston SH. 2010. Randomized trial of intensive bisphosphonate treatment versus symptomatic management in Paget's disease of bone. *J Bone Miner Res* 25: 20–31.
43. Kaplan FS. 1999. Surgical management of Paget's disease. *J Bone Miner Res* 14S2: 34–38.
44. Selby PL, Davie MWJ, Ralston SH, Stone MD. 2002. Guidelines on the management of Paget's disease of bone. *Bone* 31: 366–373.
45. Drake WM, Kendler DL, Brown JP. 2001. Consensus statement on the modern therapy of Paget's disease of bone from a Western Osteoporosis Alliance Symposium. *Clin Ther* 23: 620–626.
46. Canfield R, Rosner W, Skinner J, McWhorter J, Resnick L, Feldman F, Kammerman S, Ryan K, Kunigonis M, Bohne W. 1977. Diphosphonate therapy of Paget's disease of bone. *J Clin Endocrinol Metab* 44: 96–106.
47. McClung MR, Tou CPK, Goldstein NH, Picot C. 1995. Tiludronate therapy for Paget's disease of bone. *Bone* 17: 493S–496S.
48. Siris E, Weinstein RS, Altman R, Conte JM, Favus M, Lombardi A, Lyles K, McIlwain H, Murphy WA Jr, Reda C, Rude R, Seton M, Tiegs R, Thompson D, Tucci JR, Yates AJ, Zimering M. 1996. Comparative study of alendronate vs. etidronate for the treatment of Paget's disease of bone. *J Clin Endocrinol Metab* 81:961–967.
49. Miller PD, Adachi JD, Brown JP, Khairi RA, Lang R, Licata AA, McClung MR, Ryan WG, Singer FR, Siris ES, Tenenhouse A, Wallach S, Bekker PJ, Axelrod DW. 1997. Risedronate vs. etidronate: Durable remission

with only two months of 30 mg risedronate. *J Bone Miner Res* 12: S269

50. Harinck HI, Papapoulos SE, Blanksrna HJ, Moolenaar AJ, Vermeij P, Bijvoet OL. 1987. Paget's disease of bone: Early and late responses to three different modes of treatment with aminohydroxypropylidene bisphosphonate (APD). *Br Med J* 295: 1301–1305.

51. Reid IR, Miller P, Lyles K, Fraser W, Brown J, Saidi Y, Mesenbrink P, Su G, Pak J, Zelenakas K, Luchi M, Richardson P, Hosking D. 2005. A single infusion of zoledronic acid improves remission rates in Paget's disease: A randomized controlled comparison with risedronate. *N Engl J Med* 353: 898–908.

52. Gutteridge DH, Ward LC, Stewart GO, Retallack RW, Will RK, Prince RL, Criddle A, Bhagat CI, Stuckey BG, Price RI, Kent GN, Faulkner DL, Geelhoed E, Gan SK, Vasikaran S. 1999. Pagct's disease: Acquired resistance to one aminobisphonate with retained response to another. *J Bone Miner Res* 14S2: 79–84.

53. Khosla S, Burr D, Cauley J, et al. 2007. Bisphosphonate-associated osteonecrosis of the jaw: Report of a task force of the American Society for Bone and Mineral Research. *J Bone Miner Res* 22: 1479–1491.

54. Parvizi J, Klein GR, Sim FH. 2006. Surgical management of Paget's disease of bone. *J Bone Miner Res* 21 Suppl 2: P75–P82.

Section VII
Cancer and Bone
Section Editor Theresa A. Guise

Chapter 81. Overview of Mechanisms in Cancer Metastases to Bone 671
Gregory A. Clines

Chapter 82. Clinical and Preclinical Imaging in Osseous Metastatic Disease 677
Geertje van der Horst and Gabri van der Pluijm

Chapter 83. Metastatic Solid Tumors to Bone 686
Rachelle W. Johnson and Julie A. Sterling

Chapter 84. Hematologic Malignancies and Bone 694
Rebecca Silbermann and G. David Roodman

Chapter 85. Osteogenic Osteosarcoma 702
Jianning Tao, Yangjin Bae, Lisa L. Wang, and Brendan Lee

Chapter 86. Skeletal Complications of Breast and Prostate Cancer Therapies 711
Catherine Van Poznak and Pamela Taxel

Chapter 87. Bone Cancer and Pain 720
Patrick W. O'Donnell and Denis R. Clohisy

Chapter 88. Radiation Therapy-Induced Osteoporosis 728
Jeffrey S. Willey, Shane A.J. Lloyd, and Ted A. Bateman

Chapter 89. Skeletal Complications of Childhood Cancer 734
Ingrid A. Holm

Chapter 90. Treatment and Prevention of Bone Metastases and Myeloma Bone Disease 741
Jean-Jacques Body

Chapter 91. Radiotherapy of Skeletal Metastases 754
Edward Chow, Luluel M. Khan, and Øyvind S. Bruland

Chapter 92. Concepts and Surgical Treatment of Metastatic Bone Disease 760
Kristy Weber and Scott L. Kominsky

81

Overview of Mechanisms in Cancer Metastases to Bone

Gregory A. Clines

Introduction 671
Malignant Transformation 671
Cancer Cell Homing and Adhesion to Bone 671
Osteolytic Bone Disease 672
Osteoblastic Bone Disease 672
Conclusion 673
References 673

INTRODUCTION

The metastasis of cancer cells to the skeleton is a complication of malignancy that disrupts normal bone homeostasis and remodeling, weakens bone and results in pathological fractures [1]. New mechanisms of bone metastasis have been implicated that involve all steps of metastasis: escape of rogue cancer cells from the primary tumor, systemic embolization, adherence to bone marrow vascular endothelium, extravasation into the bone microenvironment, propagation within a hospitable environment, and interaction with osteoblasts and/or osteoclasts to adversely alter bone remodeling. This multistep program requires the cooperation of many cell types and genes, each having a unique role during metastasis. This was illustrated in a study to identify genes contributing to osteolysis in a breast cancer cell line [2]. No one gene was solely responsible, but the cooperation of the genes, each implicated in a critical step collectively contributed to the formation of bone metastasis.

MALIGNANT TRANSFORMATION

Metastasis begins with malignant transformation of individual cells with characteristics that allow survival outside the confines of the primary tumor. An early step is epithelial-to-mesenchymal transition (EMT), a neoplastic process of cellular dedifferentiation and enhanced migratory potential. The transcription factors Twist, Snail, and Slug participate in EMT with resultant downregulation of cell adhesion factors such as E-cadherin [3–6]. Also cooperating are proteases such as matrix metalloproteinases, urokinase, and cathepsins that dissolve the primary tumor basement membrane and encourage escape of malignant cells [7–9]. Cancer cells may also alter basement membrane composition and architecture to encourage further escape of malignant cells by regulating basement membrane laminins that bind integrin receptors on the cancer cells [10]. Cancer cells that successfully pass through the primary tumor basement membrane and invade surrounding vasculature are primed for systemic embolization to distant organs.

CANCER CELL HOMING AND ADHESION TO BONE

The traditional view that cancer cells home to organs that provide the right microenvironment is evolving with the proposed existence of a premetastatic niche, a concept where metastatic sites are preselected and groomed for the arrival of metastatic cancer cells. In one example, breast cancer-secreted osteopontin, acting as an endocrine factor, activates bone marrow-derived hematopoietic progenitor cells that are released into the circulation

---

Primer on the Metabolic Bone Diseases and Disorders of Mineral Metabolism, Eighth Edition. Edited by Clifford J. Rosen.
© 2013 American Society for Bone and Mineral Research. Published 2013 by John Wiley & Sons, Inc.

and deposited at distant premetastatic sites [11]. Vascular endothelial growth factor receptor 1 may serve as a marker and identify these prometastatic hematopoietic progenitor cells [12]. The premetastatic niche concept also applies to bone metastasis. Heparanase, produced by primary breast cancers, promotes shedding of tumor syndecan-1 [13]. This proteoglycan enhances osteoclastogenesis and bone remodeling, altering the bone microenvironment and preparing for the arrival of metastatic cancer cells [14].

Cancer cells entering circulation survive in part through the assistance of platelets [15]. Platelets direct fibrin deposition onto circulating cancer cells, thereby offering protection from immune surveillance and enhancing the ability to adhere to disrupted vascular endothelium [16, 17]. Cancer cells also promote platelet aggregation and the release of platelet lysophosphatidic acid (LPA) [18]. LPA receptors on cancer cells enhance cellular proliferation and stimulate the production of osteolytic factors [18].

The high prevalence of bone metastasis in advanced breast and prostate cancers suggests that they possess a unique "address" that directs circulating cancer cells to bone. The chemokine receptor CXCR4 expressed on cancer cells has an important role in delivery to bone [2, 19, 20]. Its ligand, CXCL12 or stromal-cell-derived factor 1a, is present in tissues that represent common sites of metastasis, including bone marrow and therefore cooperates with CXCR4 in cancer homing to bone [20–22]. Integrins also have important and complementary functions [23]. The integrin $\alpha v \beta 3$, which binds the RGD peptide sequence found on a variety of extracellular matrix proteins, is important in tumor cell homing [24, 25]. In animal models of bone metastasis, $\alpha v \beta 3$ antagonists prevented invasion of cancer cells into bone [26]. The integrin $\alpha 2 \beta 1$ is a high-affinity type I collagen receptor that participates in prostate cancer homing to bone [27, 28]. Tumor secretion of osteopontin and bone sialoprotein, small integrin-binding proteins, also supports metastasis by promoting tumor cell invasion, extracellular matrix degradation and survival at metastatic sites [29]. An initial interaction of metastatic tumor cells with bone marrow megakaryocytes may limit the bone metastasis potential [30].

OSTEOLYTIC BONE DISEASE

Once tumor cells arrive in bone, a "vicious cycle" of reciprocal bone–cancer cell signals regulate osteolytic metastasis. The secretion of breast cancer parathyroid hormone-related protein (PTHrP), supported by the Hedgehog signaling target GLI2, into the metastatic microenvironment stimulates osteoclastic bone resorption and release of transforming growth factor-β (TGFβ) from the mineralized bone matrix, which in turn stimulates the cancer cells to produce even more PTHrP [31–33]. Activation of osteoclast Notch signaling by expression of Jagged1 ligand in breast cancer cells serves as a complementary mechanism to increase osteoclastogenesis and the subsequent release of bone-derived TGFβ [34]. TGFβ in combination with the bone metastatic hypoxic environment regulates other prometastatic factors that enhance bone metastasis such as interleukins-6, -8, and -11, tumor necrosis factor-α, vascular endothelial growth factor, and CXCR4, and as such represents a central target for bone metastases therapy [35–38].

RUNX2, a transcription factor that specifies osteoblast fate, is also expressed in breast cancer cells and promotes breast cancer cells to function as osteoblast surrogates to support osteoclast formation [39]. Osteoclastic resorption of bone releases high concentrations of ionized calcium during dissolution of the bone. Calcium activates the calcium-sensing receptor (CaSR) expressed by breast cancer cells [40] that regulates tumor secretion of PTHrP further increasing the osteolytic response to the tumor cells [41, 42].

Multiple myeloma is far less common than breast cancer but osteolytic bone lesions are a nearly universal feature in advanced cases. Although not typically characterized as a metastatic malignancy, multiple myeloma shares many of the molecular mechanisms of pathologic bone resorption with breast cancer bone metastasis. Myeloma cells secrete a complement of osteoclast activating factors that include macrophage inflammatory protein-1α [43], receptor activator of nuclear factor-κB ligand (RANKL) [44], and interleukins-3 and -6 [45, 46]. Osteolysis is further enhanced with the secretion of inhibitors of osteoblast formation that include dickkopf homolog 1 (DKK1), secreted frizzled-related protein 2 and interleukin-3 by myeloma cells [47–49]. Other malignancies that result in osteolytic bone disease with varying frequency include non-small-cell lung cancer, melanoma, non-Hodgkin lymphoma, thyroid cancer, and acute lymphoblastic leukemia [50–54].

OSTEOBLASTIC BONE DISEASE

Endothelin 1 (ET-1) is abundantly secreted by prostate cancers and is a principal factor of the osteoblastic response to metastasis [55]. ET-1 stimulates the osteoblast, via the endothelin A receptor (ETAR), through downregulation of the Wnt signaling inhibitor DKK1 to form pathological new bone [55, 56]. The Wnt signaling pathway is a key osteoblast regulatory pathway critical for normal osteoblast differentiation and function [57]. The formation of osteoblastic bone metastasis is likely dually dependent on the downregulation of microenvironment DKK1 secretion from osteoblasts via tumor-produced ET-1 and from prostate cancer cells themselves [56, 58]. Although bone morphogenetic proteins (BMPs) cooperating with Wnt proteins have been implicated in osteoblastic disease [59], BMP-7 has been shown in some studies to prevent bone metastases. This secreted factor inhibits bone metastasis by antagonizing TGFβ signaling and reversing the epithelial-to-mesenchymal transition of cancer cells [60]. Platelet-derived growth factor may promote osteoblastic disease as well as invasive behavior of prostate epithelium [61]. A puzzling question has been the role of prostate-specific antigen (PSA) in osteoblastic metastasis. Despite convincing *in vitro* evidence that PSA has biological activity contributing to bone metastasis, *in vivo* studies demonstrating a clear causal relationship are lacking [62].

CONCLUSION

Metastatic cancer cells with an affinity to bone flourish within the bone microenvironment. These cells have the ability not only to proliferate in bone but to coax osteoblasts and osteoclasts to produce factors within the bone microenvironment that further stimulate cancer cell growth. Discovery of the molecular mechanisms that control bone metastasis has translated into therapeutic preclinical and clinical trials. Denosumab, the monoclonal antibody targeting RANKL, is now approved for preventing skeletal-related events in bone metastasis [63, 64]. Blockade of PTHrP and/or TGFβ to reduce bone metastasis has shown promise in preclinical models [31, 33, 39, 65–67]. ETAR blockade reduces the progression of osteoblastic lesions and may reduce mortality as well [68, 69].

Many unanswered questions remain regarding the pathophysiology and treatment of bone metastasis. Does accelerated bone turnover facilitate tumor growth in bone, as suggested in animal models [70, 71]? What controls tumor dormancy in bone? What is the role of angiogenesis? What is the benefit of preemptive bone targeted therapies to prevent bone metastasis and the role of combined therapies? Will inhibitors of cathepsin K be as effective as denosumab or bisphosphonates to treat metastatic bone disease? What are the most effective radiopharmaceutical therapies and pain therapies? The answer to many of these critical issues will surely lead to the identification of additional therapeutic targets. A combination of therapies, each targeting a specific metastatic step or signaling pathway, is a likely formula for a bone metastasis cure.

REFERENCES

1. Weilbaecher KN, Guise TA, McCauley LK. 2011. Cancer to bone: A fatal attraction. *Nat Rev Cancer* 11(6): 411–25.
2. Kang Y, Siegel PM, Shu W, Drobnjak M, Kakonen SM, Cordon-Cardo C, Guise TA, Massague AJ. 2003. A multigenic program mediating breast cancer metastasis to bone. *Cancer Cell* 3(6): 537–49.
3. Onder TT, Gupta PB, Mani SA, Yang J, Lander ES, Weinberg RA. 2008. Loss of E-cadherin promotes metastasis via multiple downstream transcriptional pathways. *Cancer Res* 68(10): 3645–54.
4. Yang J, Mani SA, Donaher JL, Ramaswamy S, Itzykson RA, Come C, Savagner P, Gitelman I, Richardson A, Weinberg RA. 2004. Twist, a master regulator of morphogenesis, plays an essential role in tumor metastasis. *Cell* 117(7): 927–39.
5. Batlle E, Sancho E, Franci C, Dominguez D, Monfar M, Baulida J, Garcia De Herreros A. 2000. The transcription factor snail is a repressor of E-cadherin gene expression in epithelial tumour cells. *Nat Cell Biol* 2(2): 84–9.
6. Chou TY, Chen WC, Lee AC, Hung SM, Shih NY, Chen MY. 2009. Clusterin silencing in human lung adenocarcinoma cells induces a mesenchymal-to-epithelial transition through modulating the ERK/Slug pathway. *Cell Signal* 21(5): 704–11.
7. Shiomi T, Okada Y. 2003. MT1-MMP and MMP-7 in invasion and metastasis of human cancers. *Cancer Metastasis Rev* 22(2–3): 145–52.
8. Rabbani SA, Ateeq B, Arakelian A, Valentino ML, Shaw DE, Dauffenbach LM, Kerfoot CA, Mazar AP. 2010. An anti-urokinase plasminogen activator receptor antibody (ATN-658) blocks prostate cancer invasion, migration, growth, and experimental skeletal metastasis in vitro and in vivo. *Neoplasia* 12(10): 778–88.
9. Vasiljeva O, Papazoglou A, Kruger A, Brodoefel H, Korovin M, Deussing J, Augustin N, Nielsen BS, Almholt K, Bogyo M, Peters C, Reinheckel T. 2006. Tumor cell-derived and macrophage-derived cathepsin B promotes progression and lung metastasis of mammary cancer. *Cancer Res* 66(10): 5242–50.
10. Kusuma N, Denoyer D, Eble JA, Redvers RP, Parker BS, Pelzer R, Anderson RL, Pouliot N. 2012. Integrin-dependent response to laminin-511 regulates breast tumor cell invasion and metastasis. *Int J Cancer* 130(3): 555–66.
11. McAllister SS, Gifford AM, Greiner AL, Kelleher SP, Saelzler MP, Ince TA, Reinhardt F, Harris LN, Hylander BL, Repasky EA, Weinberg RA. 2008. Systemic endocrine instigation of indolent tumor growth requires osteopontin. *Cell* 133(6): 994–1005.
12. Kaplan RN, Riba RD, Zacharoulis S, Bramley AH, Vincent L, Costa C, MacDonald DD, Jin DK, Shido K, Kerns SA, Zhu Z, Hicklin D, Wu Y, Port JL, Altorki N, Port ER, Ruggero D, Shmelkov SV, Jensen KK, Rafii S, Lyden D. 2005. VEGFR1-positive haematopoietic bone marrow progenitors initiate the pre-metastatic niche. *Nature* 438(7069): 820–7.
13. Kelly T, Suva LJ, Nicks KM, MacLeod V, Sanderson RD. 2010. Tumor-derived syndecan-1 mediates distal crosstalk with bone that enhances osteoclastogenesis. *J Bone Miner Res* 25(6): 1295–304.
14. Kelly T, Suva LJ, Huang Y, Macleod V, Miao HQ, Walker RC, Sanderson RD. 2005. Expression of heparanase by primary breast tumors promotes bone resorption in the absence of detectable bone metastases. *Cancer Res* 65(13): 5778–84.
15. Gay LJ, Felding-Habermann B. 2011. Contribution of platelets to tumour metastasis. *Nat Rev Cancer* 11(2): 123–34.

16. Palumbo JS, Talmage KE, Massari JV, La Jeunesse CM, Flick MJ, Kombrinck KW, Jirouskova M, Degen JL. 2005. Platelets and fibrin(ogen) increase metastatic potential by impeding natural killer cell-mediated elimination of tumor cells. *Blood* 105(1): 178–85.
17. Palumbo JS, Talmage KE, Massari JV, La Jeunesse CM, Flick MJ, Kombrinck KW, Hu Z, Barney KA, Degen JL. 2007. Tumor cell-associated tissue factor and circulating hemostatic factors cooperate to increase metastatic potential through natural killer cell-dependent and -independent mechanisms. *Blood* 110(1): 133–41.
18. Boucharaba A, Serre CM, Gres S, Saulnier-Blache JS, Bordet JC, Guglielmi J, Clezardin P, Peyruchaud O. 2004. Platelet-derived lysophosphatidic acid supports the progression of osteolytic bone metastases in breast cancer. *J Clin Invest* 114(12): 1714–25.
19. Smith MC, Luker KE, Garbow JR, Prior JL, Jackson E, Piwnica-Worms D, Luker GD. 2004. CXCR4 regulates growth of both primary and metastatic breast cancer. *Cancer Res* 64(23): 8604–12.
20. Sun YX, Schneider A, Jung Y, Wang J, Dai J, Wang J, Cook K, Osman NI, Koh-Paige AJ, Shim H, Pienta KJ, Keller ET, McCauley LK, Taichman RS. 2005. Skeletal localization and neutralization of the SDF-1(CXCL12)/CXCR4 axis blocks prostate cancer metastasis and growth in osseous sites in vivo. *J Bone Miner Res* 20(2): 318–29.
21. Muller A, Homey B, Soto H, Ge N, Catron D, Buchanan ME, McClanahan T, Murphy E, Yuan W, Wagner SN, Barrera JL, Mohar A, Verastegui E, Zlotnik A. 2001. Involvement of chemokine receptors in breast cancer metastasis. *Nature* 410(6824): 50–6.
22. Wang J, Loberg R, Taichman RS. 2006. The pivotal role of CXCL12 (SDF-1)/CXCR4 axis in bone metastasis. *Cancer Metastasis Rev* 25(4): 573–87.
23. Schneider JG, Amend SR, Weilbaecher KN. 2011. Integrins and bone metastasis: Integrating tumor cell and stromal cell interactions. *Bone* 48(1): 54–65.
24. Sung V, Stubbs JT 3rd, Fisher L, Aaron AD, Thompson EW. 1998. Bone sialoprotein supports breast cancer cell adhesion proliferation and migration through differential usage of the alpha(v)beta3 and alpha(v)beta5 integrins. *J Cell Physiol* 176(3): 482–94.
25. Felding-Habermann B, O'Toole TE, Smith JW, Fransvea E, Ruggeri ZM, Ginsberg MH, Hughes PE, Pampori N, Shattil SJ, Saven A, Mueller BM. 2001. Integrin activation controls metastasis in human breast cancer. *Proc Natl Acad Sci U S A* 98(4): 1853–8.
26. Clezardin P, Clement-Lacroix P. 2006. Prevention of breast cancer bone metastasis by an integrin alpha v beta 3 antagonist. *Cancer Treat Rev* 32(Suppl 3): S18.
27. Hall CL, Dai J, van Golen KL, Keller ET, Long MW. 2006. Type I collagen receptor (alpha 2 beta 1) signaling promotes the growth of human prostate cancer cells within the bone. *Cancer Res* 66(17): 8648–54.
28. Hall CL, Dubyk CW, Riesenberger TA, Shein D, Keller ET, van Golen KL. 2008. Type I collagen receptor (alpha-2beta1) signaling promotes prostate cancer invasion through RhoC GTPase. *Neoplasia* 10(8): 797–803.
29. Bellahcene A, Castronovo V, Ogbureke KU, Fisher LW, Fedarko NS. 2008. Small integrin-binding ligand N-linked glycoproteins (SIBLINGs): Multifunctional proteins in cancer. *Nat Rev Cancer* 8(3): 212–26.
30. Li X, Koh AJ, Wang Z, Soki FN, Park SI, Pienta KJ, McCauley LK. 2011. Inhibitory effects of megakaryocytic cells in prostate cancer skeletal metastasis. *J Bone Miner Res* 26(1): 125–34.
31. Guise TA, Yin JJ, Taylor SD, Kumagai Y, Dallas M, Boyce BF, Yoneda T, Mundy GR. 1996. Evidence for a causal role of parathyroid hormone-related protein in the pathogenesis of human breast cancer-mediated osteolysis. *J Clin Invest* 98(7): 1544–9.
32. Yin JJ, Selander K, Chirgwin JM, Dallas M, Grubbs BG, Wieser R, Massague J, Mundy GR, Guise TA. 1999. TGF-beta signaling blockade inhibits PTHrP secretion by breast cancer cells and bone metastases development. *J Clin Invest* 103(2): 197–206.
33. Sterling JA, Oyajobi BO, Grubbs B, Padalecki SS, Munoz SA, Gupta A, Story B, Zhao M, Mundy GR. 2006. The hedgehog signaling molecule Gli2 induces parathyroid hormone-related peptide expression and osteolysis in metastatic human breast cancer cells. *Cancer Res* 66(15): 7548–53.
34. Sethi N, Dai X, Winter CG, Kang Y. 2011. Tumor-derived JAGGED1 promotes osteolytic bone metastasis of breast cancer by engaging notch signaling in bone cells. *Cancer Cell* 19(2): 192–205.
35. Kakonen SM, Kang Y, Carreon MR, Niewolna M, Kakonen RS, Chirgwin JM, Massague J, Guise TA. 2002. Breast cancer cell lines selected from bone metastases have greater metastatic capacity and express increased vascular endothelial growth factor (VEGF), interleukin-11 (IL-11), and parathyroid hormone-related protein (PTHrP). *J Bone Miner Metab* 17(Suppl 1): M060.
36. de la Mata J, Uy HL, Guise TA, Story B, Boyce BF, Mundy GR, Roodman GD. 1995. Interleukin-6 enhances hypercalcemia and bone resorption mediated by parathyroid hormone-related protein in vivo. *J Clin Invest* 95(6): 2846–52.
37. Bendre MS, Gaddy-Kurten D, Mon-Foote T, Akel NS, Skinner RA, Nicholas RW, Suva LJ. 2002. Expression of interleukin 8 and not parathyroid hormone-related protein by human breast cancer cells correlates with bone metastasis in vivo. *Cancer Res* 62(19): 5571–9.
38. Dunn LK, Mohammad KS, Fournier PG, McKenna CR, Davis HW, Niewolna M, Peng XH, Chirgwin JM, Guise TA. 2009. Hypoxia and TGF-beta drive breast cancer bone metastases through parallel signaling pathways in tumor cells and the bone microenvironment. *PLoS One* 4(9): e6896.
39. Javed A, Barnes GL, Pratap J, Antkowiak T, Gerstenfeld LC, van Wijnen AJ, Stein JL, Lian JB, Stein GS. 2005. Impaired intranuclear trafficking of Runx2 (AML3/CBFA1) transcription factors in breast cancer cells inhibits osteolysis in vivo. *Proc Natl Acad Sci U S A* 102(5): 1454–9.
40. Yamaguchi T, Chattopadhyay N, Brown EM. 2000. G protein-coupled extracellular Ca2+ (Ca2+o)-sensing

receptor (CaR): Roles in cell signaling and control of diverse cellular functions. *Adv Pharmacol* 47: 209–53.
41. Buchs N, Manen D, Bonjour JP, Rizzoli R. 2000. Calcium stimulates parathyroid hormone-related protein production in Leydig tumor cells through a putative cation-sensing mechanism. *Eur J Endocrinol* 142(5): 500–5.
42. Sanders JL, Chattopadhyay N, Kifor O, Yamaguchi T, Butters RR, Brown EM. 2000. Extracellular calcium-sensing receptor expression and its potential role in regulating parathyroid hormone-related peptide secretion in human breast cancer cell lines. *Endocrinology* 141(12): 4357–64.
43. Choi SJ, Cruz JC, Craig F, Chung H, Devlin RD, Roodman GD, Alsina M. 2000. Macrophage inflammatory protein 1-alpha is a potential osteoclast stimulatory factor in multiple myeloma. *Blood* 96(2): 671–5.
44. Pearse RN, Sordillo EM, Yaccoby S, Wong BR, Liau DF, Colman N, Michaeli J, Epstein J, Choi Y. 2001. Multiple myeloma disrupts the TRANCE/osteoprotegerin cytokine axis to trigger bone destruction and promote tumor progression. *Proc Natl Acad Sci U S A* 98(20): 11581–6.
45. Lee JW, Chung HY, Ehrlich LA, Jelinek DF, Callander NS, Roodman GD, Choi SJ. 2004. IL-3 expression by myeloma cells increases both osteoclast formation and growth of myeloma cells. *Blood* 103(6): 2308–15.
46. Cheung WC, Van Ness B. 2002. Distinct IL-6 signal transduction leads to growth arrest and death in B cells or growth promotion and cell survival in myeloma cells. *Leukemia* 16(6): 1182–8.
47. Tian E, Zhan F, Walker R, Rasmussen E, Ma Y, Barlogie B, Shaughnessy JDJ. 2003. The role of the Wnt-signaling antagonist DKK1 in the development of osteolytic lesions in multiple myeloma. *New Engl J Med* 349(26): 2483–94.
48. Oshima T, Abe M, Asano J, Hara T, Kitazoe K, Sekimoto E, Tanaka Y, Shibata H, Hashimoto T, Ozaki S, Kido S, Inoue D, Matsumoto T. 2005. Myeloma cells suppress bone formation by secreting a soluble Wnt inhibitor, sFRP-2. *Blood* 106(9): 3160–5.
49. Ehrlich LA, Chung HY, Ghobrial I, Choi SJ, Morandi F, Colla S, Rizzoli V, Roodman GD, Giuliani N. 2005. IL-3 is a potential inhibitor of osteoblast differentiation in multiple myeloma. *Blood* 106(4): 1407–14.
50. Coleman RE. 1997. Skeletal complications of malignancy. *Cancer* 80(8 Suppl): 1588–94.
51. des Grottes JM, Dumon JC, Body JJ. 2001. Hypercalcaemia of melanoma: Incidence, pathogenesis and therapy with bisphosphonates. *Melanoma Res* 11(5): 477–82.
52. Takasaki H, Kanamori H, Takabayashi M, Yamaji S, Koharazawa H, Taguchi J, Fujimaki K, Ishigatsubo Y. 2006. Non-Hodgkin's lymphoma presenting as multiple bone lesions and hypercalcemia. *Am J Hematol* 81(6): 439–42.
53. Marcocci C, Pacini F, Elisei R, Schipani E, Ceccarelli C, Miccoli P, Arganini M, Pinchera A. 1989. Clinical and biologic behavior of bone metastases from differentiated thyroid carcinoma. *Surgery* 106(6): 960–6.
54. Ganesan P, Thulkar S, Gupta R, Bakhshi S. 2009. Childhood aleukemic leukemia with hypercalcemia and bone lesions mimicking metabolic bone disease. *J Pediatr Endocrinol Metab* 22(5): 463–7.
55. Yin JJ, Mohammad KS, Kakonen SM, Harris S, Wu-Wong JR, Wessale JL, Padley RJ, Garrett IR, Chirgwin JM, Guise TA. 2003. A causal role for endothelin-1 in the pathogenesis of osteoblastic bone metastases. *Proc Natl Acad Sci U S A* 100(19): 10954–9.
56. Clines GA, Mohammad KS, Bao Y, Stephens O, Suva LJ, Shaughnessy JD, Fox JW, Chirgwin JM, Guise TA. 2007. Dickkopf homolog 1 mediates endothelin-1-stimulated new bone formation. *Mol Endocrinol* 22: 486–98.
57. Westendorf JJ, Kahler RA, Schroeder TM. 2004. Wnt signaling in osteoblasts and bone diseases. *Gene* 341: 19–39.
58. Hall CL, Bafico A, Dai J, Aaronson SA, Keller ET. 2005. Prostate cancer cells promote osteoblastic bone metastases through Wnts. *Cancer Res* 65(17): 7554–60.
59. Dai J, Hall CL, Escara-Wilke J, Mizokami A, Keller JM, Keller ET. 2008. Prostate cancer induces bone metastasis through Wnt-induced bone morphogenetic protein-dependent and independent mechanisms. *Cancer Res* 68(14): 5785–94.
60. Buijs JT, Rentsch CA, van der Horst G, van Overveld PGM, Schwaninger R, Henriquez NV, Papapoulos SE, Pelger RCM, Vukicevic S, Cecchini MG, Lowik CWGM, van der Pluijm G. 2007. BMP7, a putative regulator of epithelial homeostasis in the human prostate, is a potent inhibitor of prostate cancer bone metastasis in vivo. *Am J Pathol* 171: 1047–57.
61. Yi B, Williams PJ, Niewolna M, Wang Y, Yoneda T. 2002. Tumor-derived platelet-derived growth factor-BB plays a critical role in osteosclerotic bone metastasis in an animal model of human breast cancer. *Cancer Res* 62(3): 917–23.
62. Williams SA, Singh P, Isaacs JT, Denmeade SR. 2007. Does PSA play a role as a promoting agent during the initiation and/or progression of prostate cancer? *Prostate* 67(3): 312–29.
63. Stopeck AT, Lipton A, Body JJ, Steger GG, Tonkin K, de Boer RH, Lichinitser M, Fujiwara Y, Yardley DA, Viniegra M, Fan M, Jiang Q, Dansey R, Jun S, Braun A. 2010. Denosumab compared with zoledronic acid for the treatment of bone metastases in patients with advanced breast cancer: a randomized, double-blind study. *J Clin Oncol* 28(35): 5132–9.
64. Henry DH, Costa L, Goldwasser F, Hirsh V, Hungria V, Prausova J, Scagliotti GV, Sleeboom H, Spencer A, Vadhan-Raj S, von Moos R, Willenbacher W, Woll PJ, Wang J, Jiang Q, Jun S, Dansey R, Yeh H. 2011. Randomized, double-blind study of denosumab versus zoledronic acid in the treatment of bone metastases in patients with advanced cancer (excluding breast and prostate cancer) or multiple myeloma. *J Clin Oncol* 29(9): 1125–32.
65. Mohammad KS, Stebbins EG, Niewolna M, Mckenna CR, Walton H, Peng XH, Li G, Murphy A, Chakravarty S, Higgins LS, Wong DH, Guise TA. 2006. TGFbeta

signaling blockade reduces osteolytic bone metastases and enhances bone mass. *Cancer Treat Rev* 32(Suppl 3): S29.

66. Javelaud D, Mohammad KS, McKenna CR, Fournier P, Luciani F, Niewolna M, Andre J, Delmas V, Larue L, Guise TA, Mauviel A. 2007. Stable overexpression of Smad7 in human melanoma cells impairs bone metastasis. *Cancer Res* 67(5): 2317–24.

67. Mohammad KS, Javelaud D, Fournier PG, Niewolna M, McKenna CR, Peng XH, Duong V, Dunn LK, Mauviel A, Guise TA. 2011. TGF-beta-RI kinase inhibitor SD-208 reduces the development and progression of melanoma bone metastases. *Cancer Res* 71(1): 175–84.

68. Nelson JB. 2005. Endothelin receptor antagonists. *World J Urol* 23(1): 19–27.

69. James ND, Caty A, Borre M, Zonnenberg BA, Beuzeboc P, Morris T, Phung D, Dawson NA. 2009. Safety and efficacy of the specific endothelin-A receptor antagonist ZD4054 in patients with hormone-resistant prostate cancer and bone metastases who were pain free or mildly symptomatic: A double-blind, placebo-controlled, randomised, phase 2 trial. *Eur Urol* 55(5): 1112–23.

70. Schneider A, Kalikin LM, Mattos AC, Keller ET, Allen MJ, Pienta KJ, McCauley LK. 2005. Bone turnover mediates preferential localization of prostate cancer in the skeleton. *Endocrinology* 146(4): 1727–36.

71. Padalecki SS, Carreon MR, Grubbs BR, Guise TA. 2003. Hypogonadism causes bone loss and increased bone metastases in a model of mixed osteolytic/oteoblastic metastases: Prevention by zoledronic acid. *J Bone Miner Res* 18(Suppl 2): S36.

82

Clinical and Preclinical Imaging in Osseous Metastatic Disease

Geertje van der Horst and Gabri van der Pluijm

Introduction 677
Cancer Imaging 677
Morphological/Anatomical Imaging Techniques 678
Molecular/Functional Imaging Techniques 680
Image-Guided Surgery 682
Multimodality Imaging and Functional Imaging 682
Summary 682
References 682

INTRODUCTION

Bone metastases are frequent complications of cancer, occurring in up to 80% of patients with advanced breast or prostate cancer and in approximately 15–30% of patients with carcinoma of the thyroid, lung, bladder, or kidney [1]. In addition, melanomas and multiple myeloma also readily metastasize to the skeleton. Bone metastases can be classified into osteolytic, osteoblastic, or a mixed phenotype. Once tumors metastasize to bone, the disease is incurable, and patients may experience several skeletal-related events such as severe bone pain, hypocalcaemia, nerve compression syndromes, and pathological fractures [2, 3]. This severely increases morbidity and diminishes the quality of life of the patients.

Imaging has become an indispensible tool in the diagnosis and treatment of skeletal metastasis both in the clinic and experimentally. Novel multimodality imaging approaches are emerging rapidly and will greatly facilitate experimental studies in preclinical models of metastatic bone disease (Fig. 82.1, Table 82.1).

CANCER IMAGING

In the clinic, imaging is traditionally used for detection and staging of the disease, monitoring of the response to therapy, and identifying the risk for pathologic fractures. However, functional and molecular imaging can also be used to study processes involved in tumor growth and metastasis, including angiogenesis and pharmacokinetics.

Preclinical imaging, using small animals, can be used to visualize and observe changes at the organ, tissue, cell, or molecular level (e.g., the location or the size of the tumor or the characteristics of the tumor or tumor-surrounding stroma).

Current imaging modalities can be classified into morphological/anatomical imaging techniques [ultrasound imaging, computed tomography (CT), and magnetic resonance imaging (MRI)] and molecular imaging techniques [positron emission tomography (PET), single photon emission computed tomography (SPECT), and optical imaging fluorescence (FLI) and bioluminescence imaging (BLI)]. For imaging metastatic bone disease in the clinic, several modalities are used, each with their own advantages and drawbacks. To detect and stage the disease, Tc^{99m} methylene diphosphonate ($Tc^{99m}MDP$) bone scintigraphy is mainly used, supported by plain film radiography or SPECT. Bone scintigraphy can be followed by MRI, CT, or even PET/SPECT. For preclinical imaging, several of the imaging modalities are adapted for small animal studies. In addition to these, optical imaging is widely used in small animals to study the pathogenesis and experimental treatment of metastatic bone disease.

Primer on the Metabolic Bone Diseases and Disorders of Mineral Metabolism, Eighth Edition. Edited by Clifford J. Rosen.
© 2013 American Society for Bone and Mineral Research. Published 2013 by John Wiley & Sons, Inc.

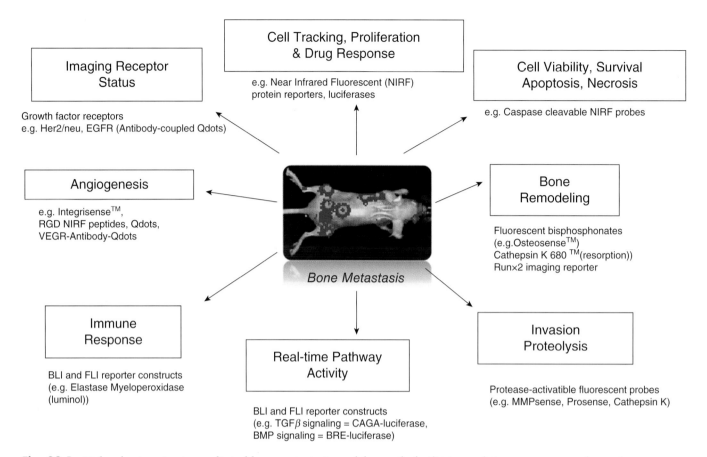

Fig. 82.1. Molecular imaging in preclinical bone metastasis models greatly facilitates real-time assessment of complex processes involved in the pathogenesis of skeletal metastasis. In the center, a representative bioluminescence image of a preclinical bone metastasis model is shown. For this, 100,000 human luciferase-expressing PC-3M-Pro4 prostate cancer cells were inoculated into the left cardiac ventricle of immuno-deficient mice to allow real-time cell tracking of skeletal metastases.

MORPHOLOGICAL/ANATOMICAL IMAGING TECHNIQUES

Plain Film Radiography measures the decline of X-rays passing through tissues. Because of its high resolution, radiography can be used to measure details such as cortical thickness, which is important to assess the risk of a pathologic fracture [4]. However, in the clinic, whole-body plain film radiography is not very often used for detection of bone metastasis. This is due to the fact that an osteolytic bone lesion must display bone resorption of at least 40–50% to be consistently detectable. Only for patients with multiple myeloma is plain film radiography still used, since osteolytic lesions of multiple myeloma have a characteristic appearance and are difficult to detect using a Tc^{99m}MDP bone scan.

Computed tomography (CT) measures the decline of X-rays passing through tissues. The X-rays decline at different rates depending on the density of the tissues. In the clinic, CT is used to obtain three-dimensional (3D) anatomical information, to obtain characterization of bone lesion size and to evaluate soft tissues. Due to the differences in shape, location, and density, CT imaging can detect many tumors. However, many small tumors or micrometastases are not detectable on a CT scan. In preclinical imaging, micro CT (μCT) is now a well-established imaging modality that can be used to image small animals and generate 3D images. Numerous experiments have been performed using μCT mainly because of the high spatial resolution of the images (50–100 μm^3) and the relative rapid data acquisition time (minutes) [5]. This technique is cost-effective and can be used to examine the architecture of bones. In addition, μCT can detect soft tissue and skeletal abnormalities as well as tumors in small animals.

Dynamic CT (DCT) is a technique that can be used to study perfusion, permeability, and blood flow. Instead of a high-resolution image, DCT acquires a series of lower-resolution CT images. A contrast agent is administered to the subject and a series of CT images are acquired imaging the wash-in and wash-out of the contrast. Using a kinetic model, quantitative maps of blood flow, tissue permeability, and perfusion can be derived [6].

However, one of the major drawbacks of both CT and μCT is the low tissue contrast (compared to MRI).

Table 82.1. Comparison of Different Imaging Modalities Including Source, Possible Applications, and the Benefits and Drawbacks of Each Technique*

Imaging Modality	Source	Possible Applications	Advantages	Drawbacks
MRI: magnetic resonance imaging	Electromagnetic waves	– Structural, anatomical information on tissue architecture – Long-term monitoring of cell viability and metastatic bone tumor burden [Transferrin, Ferritin, Tyrosine (Refs. 11, 21)]. – Monitoring of metabolism, tissue composition and detection of molecular probes (Refs. 11, 12, 21)	– High spatial resolution – Good soft tissue contrast – Quantitative – Small cell number imaging possible	– Low throughput – Highly trained personnel
CT: computed tomography	Radiation (X-rays)	– Structural, anatomical information on the bone architecture (Refs. 8, 9).	– High spatial resolution – High penetration depth	– Low sensitivity – Radiation dose
PET: positron emission tomography	Radiation (high energy γ-rays)	– Early cell localization and bone homing [e.g., 18^F-FDG (Refs. 11, 21, 25, 26)]. – Long-term monitoring of cell viability and metastatic bone tumor burden [HSV1-tk (Ref. 63)]. – Visualization of molecular and metabolic processes [e.g., whole-body pharmacokinetics or receptor expression (Refs. 24, 21, 25)]. – *Real-time signaling pathway activity [reporter constructs (Ref. 64)].*	– High sensitivity – High penetration depth	– Limited spatial resolution – Radiation – High costs – Not quantitative
SPECT: single photon emission computed tomography	Radiation (low energy γ-rays)	– Early cell localization and bone homing [e.g., 111In (Ref. 21)]. – Long-term monitoring of cell viability and metastatic bone tumor burden [NIS, somatostatib (Refs. 65, 66)]. – Visualization of molecular and metabolic processes [e.g., apoptosis or receptor expression levels (Refs. 21, 24, 25)].	– High sensitivity (lower than PET) – High penetration depth	– Limited spatial resolution – Radiation – Not quantitative
BLI: bioluminescence imaging	Bioluminescent light	– *Early cell localization and bone homing, small cell numbers (Refs. 41, 43).* – *Long-term monitoring of cell viability and metastatic bone tumor burden [functional and drug response studies (Refs. 41–43)].* – *Real-time signaling pathway activity [reporter constructs (Ref. 64)].* – *Visualization of molecular and metabolic processes [e.g., VEGFR activity (Ref. 67)].*	– High sensitivity – High throughput – Low costs – Small cell number imaging possible	– Limited spatial resolution – Limited penetration depth – Not clinically applicable
FLI: fluorescence imaging	Fluorescent light	– *Early cell localization and bone homing (Ref. 68).* – *Long-term monitoring of cell viability and metastatic bone tumor burden [functional and drug response studies NIRFs (Refs. 69–71)].* – *Real-time signaling pathway activity (reporter constructs).* – *Real-time monitoring of processes involved in bone metastasis [smart probes, e.g., MMP-sense (Ref. 55)].* – Image guided surgery (Ref. 56).	– High sensitivity – High throughput – Low costs	– Limited spatial resolution – Limited penetration depth – Auto fluorescence

*Possible applications in *italic* are only available for preclinical imaging.

Moreover, the level of radiation necessary may be high enough to induce changes in the immune response and other biological pathways, thus imitating radiotherapy [7-9]. Especially in longitudinal studies, the cumulative dose can be very high, although in response to the concerns over the mounting radiation exposure, imaging strategies have already been developed that deliver the lowest dose necessary to obtain the desired information with CT [9].

MOLECULAR/FUNCTIONAL IMAGING TECHNIQUES

Several imaging modalities can be used for molecular imaging, which is defined as: "the visualization, characterization, and measurement of biological processes at the molecular and cellular levels in humans and other living systems. Molecular imaging agents are probes used to visualize, characterize and measure biological processes in living systems. Both endogenous molecules and exogenous probes can be molecular imaging agents [10]."

Molecular imaging can be used to identify processes involved in tumor development, progression, and metastasis. Moreover, both pathway activity and pharmacokinetics can be studied using molecular imaging.

Radionuclide bone scintigraphy comprises intravenous administration of a radioactive bone-seeking agent (Tc^{99m} methylene diphosphonate or Tc^{99m} MDP), imaged using a single-photon camera. The bone scans visualize the secondary effect of bone metastases, since they bind to hydroxyapatite mineral in the bone matrix in proportion to the blood flow and are therefore markers of bone turnover and perfusion. Osteoblastic metastases, which display locally increased bone turnover, assemble more Tc^{99m} MDP than normal bone. Bone scintigraphy is a highly sensitive and cost-effective screening modality [4]; even a 5% change in bone turnover can be detected, whereas plain film radiography or CT can only detect 40-50% loss of mineral. Disadvantages of bone scintigraphy include the specificity; false positives can be seen from degenerative change, inflammation, trauma, and Paget's disease. In addition, bone scans display relatively poor spatial resolution and reduced sensitivity for tumor cells that reside in the bone marrow for very aggressive metastasis or for tumor cells that induce little healing reaction, such as multiple myeloma [4]. Therefore, SPECT imaging (see below) is often used as a complementary imaging technique to increase the sensitivity of bone scintigraphy (20-50% increase in lesion detection) and to detect and localize vertebral lesions.

Magnetic resonance imaging (MRI) allows visualization of the properties of various tissues by differences in their absorption and emission of electromagnetic energy. MRI is a very useful technique to show detailed anatomical images. However, not only structural information, but also information regarding tissue composition, perfusion, oxygenation, tissue elasticity, metabolism, detection of molecular probes, and more can be obtained (reviewed in Refs. 11 and 12). In the clinic, MRI is more sensitive than plain radiography, CT, or bone scintigraphy in detecting early lesions in bone before they provoke an osseous response [e.g., confirmed skeletal metastases (91%) versus bone scintigraphy (85%)] [13-16]. For example, in prostate cancer bone metastases, MRI can measure changes prior to the osteoblastic response in the bone marrow [17]. In preclinical imaging, one of the interesting opportunities provided by MRI imaging is its high resolution. MRI can detect early, small-sized tumors or even small amount of cells; for example, 100 embryonic stem cells were labeled by superparamagnetic iron oxide (SPIO) nanoparticles and could be imaged with MRI both *in vitro* and *in vivo*, and even a small cell cluster of 10 cells could be measured in a gel phantom [18]. These data indicate that MRI can potentially be used to image small amounts of cells (e.g., cancer stem cells) [18-20].

The disadvantages of MRI in both the preclinical and clinical setting include poor sensitivity in terms of molecular reactions, the requirement of highly trained personnel, the long acquisition time (minutes to hours), and high costs [21, 22].

Although the previously described conventional imaging methods (e.g., MRI and CT) provide structural information on the bone architecture, which is often altered during the progression of skeletal metastasis, other imaging methods are required for the imaging of other processes important for the progression of the metastases such as tumor cell biology and metabolic processes. For this, highly sensitive methods such as **positron emission tomography (PET)** and **single photon emission computed tomography (SPECT)** imaging modalities are favorable.

PET is an imaging technique that detects pairs of high energy γ-rays emitted indirectly by a positron-emitting radioactive atom (radiotracer), which was introduced into the subject. SPECT combines conventional radionuclide imaging with image reconstruction from projections and is a very promising technique with a variety of available tracers [23]. SPECT can provide true 3D information and uses low-energy γ-rays. PET and SPECT both have excellent temporal resolution and good sensitivity, although SPECT displays a lower sensitivity compared to PET. PET and SPECT imaging cannot only be used to visualize tumors, but also to visualize metabolic processes. Moreover, PET and SPECT allow functional evaluation including whole-body pharmacokinetics (reviewed in Refs. 24 and 25). One of the advantages of SPECT is that multiple molecular functions can be imaged at the same time by using different energy radioisotopes.

In the clinic, the radiotracer [18] F-fluoro-2-deoxyD-glucose (FDG) can be used for detection and staging [25]. In both the clinical and preclinical setting, **FDG-PET** measures metabolism in the tumor cells and the increased glucose transport and glycolysis associated with several tumors [25, 26]. Because FDG-PET detects metabolic activity of the tumor cells, it can detect metastases prior to the osseous response. This technique can accurately detect bone metastasis in osteolytic bone disease (e.g., multiple myeloma). However, FDG-PET is not very sensitive in detecting osteoblastic lesions (e.g., prostate carcinoma). For prostate carcinoma PET tracers such as ^{11}C-acetate, ^{11}C-choline, and ^{18}F-choline have been investigated as alternatives to FDG [27]. In addition, it has been shown that ^{18}F-NaF PET is superior to ^{99m}Tc-MDP planar scintigraphy or SPECT in detecting prostate cancer bone metastases [28]. PET using ^{18}F-NaF PET detected more lesions and showed higher contrast between tumor and normal bone. Moreover, the detection efficiency of ^{18}F-NaF PET is independent of the localization of the lesion (in contrast to bone scintigraphy) [29]. For mixed phenotypes, FDG-PET is complementary to the bone scintigraphy and allows detection of some of the lesions that are missed by the Tc^{99m} MDP bone scan [30].

Another example of clinical bone imaging using PET is ^{18F}**sodium fluoride**. This fluoride, like Tc^{99m} MDP, is incorporated into the bone as part of the bone mineralization process. One of the advantages compared to bone scintigraphy using Tc^{99m} MDP is the ability to provide quantitative information. In addition, ^{18F}sodium fluoride-PET displays higher contrast and superior spatial resolution compared to bone scintigraphy. PET ^{18F}sodium fluoride has also been used successfully in preclinical models [31].

In addition, other biological properties important for tumor growth and survival, including hypoxia, proliferation, apoptosis, and receptor expression are being investigated with PET or SPECT imaging modalities using a wide range or radiotracers [24, 25]. PET is mostly used to visualize molecules with fast kinetics whereas SPECT is used to image molecules with slow kinetics.

One of the main limitations of PET imaging alone is its relatively low spatial resolution and the lack of a clear anatomical reference. Another major drawback of PET and SPECT is the high cost of the technique. Not only a very expensive imaging modality should be purchased, but also expensive cyclotrons are needed to generate the radioisotopes. Another drawback is the possibility that the radiation influences the tumor size, as it imitates radiotherapy.

Noninvasive, whole-body **optical imaging** permits longitudinal real-time gene expression, cellular localization, and drug response studies in small laboratory animals through the use of direct-targeting probes and reporter systems. Optical imaging encompasses *bioluminescence* imaging and *fluorescence* imaging. Of these, only fluorescence imaging has potential clinical applications (e.g., image-guided surgery).

Bioluminescence imaging (BLI) is most commonly used for tracking cancer cells and studying their distribution and activity *in vivo* because it is easy to use, cost-effective, and very sensitive (reviewed in Ref. 32). BLI is ideally suited to image fundamental biological processes *in vivo* due to the high signal-to-noise ratio, low background, and short acquisition time. Bioluminescence is based on the detection of photons emitted by the enzymatic reaction in which a substrate (either D-luciferin or coelenterazin) is oxidized by a luciferase. Several different luciferases exist including *Photinis pyralis* (firefly) luciferase, red and green *Pyrophorus plagiophthalamus* (click beetle luciferase), *Gaussia* luciferase, and *Renilla* luciferase [33–35]. *Firefly* luciferase (FFLuc), which catalyzes the substrate luciferin, is the most extensively used luciferase in cell-based bioluminescence imaging [36–40]. The use of the new generation of FFLuc (the mammalian codon-optimized FFLuc2), especially, enables extremely sensitive imaging and is very useful for monitoring cell tracking and survival in small animals [41–43].

Although bioluminescence imaging has several advantages, including high sensitivity, a low background signal, a high signal-to-noise ratio, and a short acquisition time, BLI mainly provides planar imaging with limited depth information and signal localization, and the spatial resolution of BLI is relatively poor. For example, changes in depth of the signal can be confused with changes in cell survival. In addition, the technique is restricted to small animals and is very superficial in larger objects. The color of the animals is also important, since dark skin pigmentation or fur color can attenuate the bioluminescent light (the same holds true for fluorescent light).

Other markers for live-cell imaging are fluorescent proteins or fluorochromes targeted to specific cell compartments or molecules. **Fluorescence imaging** is based on the detection of emitted light subsequent to their excitation by light of a specific wavelength. Fluorescence imaging can be divided into fluorescence reflectance imaging (FRI) and fluorescence molecular tomography (FMT), the latter being able to provide 3D information [44]. Fluorescence optical imaging can be used for whole-body imaging and two-photon imaging. Two-photon imaging can be used to visualize molecular processes or cells in a surgically exposed region of the body. It provides a microscopic perspective and can be used to track real time *in vivo* effects [45]. The drawbacks of two-photon imaging are autofluorescence and the need to expose the region of interest. For deeper noninvasive imaging of small animals, the fluorescent proteins optimally need to be in the far-red or near-infrared (NIR), because then autofluorescence of the tissue is less prominent and tissue penetration is improved (reviewed in Ref. 46). The use of fluorescent proteins for imaging cancer dynamics *in vivo* at the tumor and cellular level has been reviewed by Hoffman [47], describing the use of fluorescent proteins to differentially label cancer cells in the nucleus and cytoplasm to visualize the nuclear-cytoplasmic dynamics of cancer cell trafficking in both blood vessels and lymphatic vessels in small animals. The drawbacks of fluorescence imaging include autofluorescence, limited depth information and poor spatial resolution. In addition, the technique is restricted to small animals and is very superficial in larger objects.

Several "smart probes" have been developed to study different biological processes in preclinical models. For example, the skeleton can specifically be studied using bone-specific probes including the fluorescently labeled bisphosphonate Osteosense™ (PerkinElmer) or the fluorescently labeled tetracycline derivative Bone Tag (LI-COR Biosciences) [48, 49]. These bone probes are incorporated into the bone at sites with high bone turnover, which occurs during cancer-induced bone remodeling [50, 51]. Angiosense (PerkinElmer) is a fluorescent agent that remains localized in the vasculature (Fig. 82.1) and can be used to visualize angiogenesis.

Another interesting probe is the Prosense probe, which can be used to visualize Cathepsin K activity and is highly present at osteolytic lesions and sites of active bone resorption [52–54]. Prosense is one of the smart probes, a cleavable probe that provides information about the activity of an enzyme. The substrate of the enzymes is coupled to a fluorophore, which is quenched due to the structure and location of the fluorophore. Upon cleavage by the enzyme, the fluorophore is released and can be detected.

Other smart probes were designed to visualize the matrix degradation process, which is an important process for cancer cell motility and invasiveness. For example, MMPsense (PerkinElmer) is activated by the matrix metalloproteases MMP2 and MMP9 [55]. These proteases may be produced by the cancer cells as well as the reactive stroma.

IMAGE-GUIDED SURGERY

Clinical application of optical fluorescence imaging is a relatively novel medical imaging technique. Because of the limited depth of light penetration, the imaging can only be performed for surfaces only (e.g., skin or during an endoscopic or surgical procedure). Near-infrared fluorescence imaging during surgery can aid surgeons to search for tissue that needs to be resected (e.g., tumor tissue) and tissue that needs to be spared such as blood vessels or nerves (reviewed in Ref. 56).

In preclinical models, tumor-specific, NIR fluorescent probes (e.g., probes targeting epidermal growth factor receptor) have been used to visualize oral squamous cell carcinoma and cervical lymph node metastasis [57].

MULTIMODALITY IMAGING AND FUNCTIONAL IMAGING

Taking into account that every imaging modality has its advantages and drawbacks, combining two or more imaging modalities may provide a better solution to overcome the limitations of the independent techniques (multimodality imaging).

Combining the imaging modalities will improve and expand the scope of information available. Recently, multimodality instruments have been developed, e.g., combined PET/CT or SPECT/CT scanners [58, 59]. When combining PET and CT imaging, the low resolution and high sensitivity of the PET image is compensated with the high-resolution CT image. Fusion of the functional PET data and the anatomical CT findings has been shown to increase the diagnostic performance of PET alone [60]. Other combinations of imaging modalities are also feasible. For example, the combination of µCT and optical imaging would avoid the use of radioactive tracers. However, this combination can only be used in small animal imaging because of the low light intensity and tissue penetration of fluorescent or bioluminescent light. Integration of PET and MRI imaging allows the acquisition of functional imaging data, which are projected on data sets with anatomical information [61, 62].

In the future, multimodality imaging promises to make significant contribution to both the staging and management of osseous metastatic disease.

SUMMARY

Clinically, planar bone scintigraphy using Tc^{99m} MDP are often sufficient for diagnosis and may be enhanced by SPECT. FDG-PET, CT, and MRI can offer complementary information. Multimodality imaging is, therefore, a powerful tool to acquire spatially and temporally correlated data showing, for example, the distribution of PET radiotracers or MRI contrast agents with registration of the anatomical location. Preclinical imaging has the advantage of adding the sensitive optical imaging modality to this multimodality approach.

Better understanding of the processes involved in skeletal metastasis and the development of new therapies requires noninvasive high-resolution *in vivo* imaging. The development, and clinical application, of new, more sensitive molecular imaging methods will significantly improve the detection of yet undetectable micrometastasis (minimal residual disease) and may eventually lead to better treatment and/or surgical removal of the primary tumor, affected locoregional lymph nodes and bone metastases.

REFERENCES

1. Coleman RE. 1997. Skeletal complications of malignancy. *Cancer* 80: 1588–94.
2. Mundy GR. 2002. Metastasis to bone: Causes, consequences and therapeutic opportunities. *Nat Rev Cancer* 2: 584–93.
3. Roodman GD. 2004. Mechanisms of bone metastasis. *N Engl J Med* 350: 1655–64.
4. Rosenthal DI. 1997. Radiologic diagnosis of bone metastases. *Cancer* 80: 1595–607.
5. de Kemp RA, Epstein FH, Catana C, Tsui BM, Ritman EL. 2010. Small-animal molecular imaging methods. *J Nucl Med* 51 Suppl 1: 18S–32S.
6. Thornton MM. 2004. Multi-modality imaging of musculoskeletal disease in small animals. *J Musculoskelet Neuronal Interact* 4: 364.

7. Beckmann N, Kneuer R, Gremlich HU, Karmouty-Quintana H, Ble FX, Muller M. 2007. In vivo mouse imaging and spectroscopy in drug discovery. *NMR Biomed* 20: 154–85.
8. Schambach SJ, Bag S, Schilling L, Groden C, Brockmann MA. 2010. Application of micro-CT in small animal imaging. *Methods* 50: 2–13.
9. Marin D, Nelson RC, Rubin GD, Schindera ST. 2011. Body CT: Technical advances for improving safety. *AJR Am J Roentgenol* 197: 33–41.
10. Mankoff DA. 2007. A definition of molecular imaging. *J Nucl Med* 48: 18N, 21N.
11. Koba W, Kim K, Lipton ML, Jelicks L, Das B, Herbst L, et al. 2011. Imaging devices for use in small animals. *Semin Nucl Med* 41: 151–65.
12. Handelsman H. 1994. Magnetic resonance angiography: Vascular and flow imaging. *Health Technol Assess (Rockv)* Oct(3): 1–20.
13. Taoka T, Mayr NA, Lee HJ, Yuh WT, Simonson TM, Rezai K, et al. 2001. Factors influencing visualization of vertebral metastases on MR imaging versus bone scintigraphy. *AJR Am J Roentgenol* 176: 1525–30.
14. Wu LM, Gu HY, Zheng J, Xu X, Lin LH, Deng X, et al. 2011. Diagnostic value of whole-body magnetic resonance imaging for bone metastases: A systematic review and meta-analysis. *J Magn Reson Imaging* 34: 128–35.
15. Meaney JF, Fagan A. 2010. Whole-body MR imaging in a multimodality world: Current applications, limitations, and future potential for comprehensive musculoskeletal imaging. *Semin Musculoskelet Radiol* 14: 14–21.
16. Steinborn MM, Heuck AF, Tiling R, Bruegel M, Gauger L, Reiser MF. 1999. Whole-body bone marrow MRI in patients with metastatic disease to the skeletal system. *J Comput Assist Tomogr* 23: 123–9.
17. Messiou C, Cook G, deSouza NM. 2009. Imaging metastatic bone disease from carcinoma of the prostate. *Br J Cancer* 101(8): 1225–32.
18. Stroh A, Faber C, Neuberger T, Lorenz P, Sieland K, Jakob PM, et al. 2005. In vivo detection limits of magnetically labeled embryonic stem cells in the rat brain using high-field (17.6 T) magnetic resonance imaging. *Neuroimage* 24: 635–45.
19. Hu SL, Lu PG, Zhang LJ, Li F, Chen Z, Wu N, et al. 2011. In vivo magnetic resonance imaging tracking of SPIO labeled human umbilical cord mesenchymal stem cells. *J Cell Biochem* 113(3): 1005–12.
20. Politi LS, Bacigaluppi M, Brambilla E, Cadioli M, Falini A, Comi G, et al. 2007. Magnetic-resonance-based tracking and quantification of intravenously injected neural stem cell accumulation in the brains of mice with experimental multiple sclerosis. *Stem Cells* 25: 2583–92.
21. Gross S, Piwnica-Worms D. 2006. Molecular imaging strategies for drug discovery and development. *Curr Opin Chem Biol* 10: 334–42.
22. Sandanaraj BS, Kneuer R, Beckmann N. 2010. Optical and magnetic resonance imaging as complementary modalities in drug discovery. *Future Med Chem* 2: 317–37.
23. Jaszczak RJ, Tsui BM. 1995. Single photon emission computed tomography: General principles. In: Wagner HN, Szabo Z, Buchanan JW (eds.) *Principles of Nuclear Medicine*. Philadelphia: Saunders. pp. 317–28.
24. Chatziioannou AF. 2005. Instrumentation for molecular imaging in preclinical research: Micro-PET and Micro-SPECT. *Proc Am Thorac Soc* 2: 533–11.
25. Wahl RL, Herman JM, Ford E. 2011. The promise and pitfalls of positron emission tomography and single-photon emission computed tomography molecular imaging-guided radiation therapy. *Semin Radiat Oncol* 21: 88–100.
26. Turlakow A, Larson SM, Coakley F, Akhurst T, Gonen M, Macapinlac HA, et al. 2001. Local detection of prostate cancer by positron emission tomography with 2-fluorodeoxyglucose: Comparison of filtered back projection and iterative reconstruction with segmented attenuation correction. *Q J Nucl Med* 45: 235–44.
27. Akin O, Hricak H. 2007. Imaging of prostate cancer. *Radiol Clin North Am* 45: 207–22.
28. Even-Sapir E, Metser U, Mishani E, Lievshitz G, Lerman H, Leibovitch I. 2006. The detection of bone metastases in patients with high-risk prostate cancer: 99mTc-MDP Planar bone scintigraphy, single- and multi-field-of-view SPECT, 18F-fluoride PET, and 18F-fluoride PET/CT. *J Nucl Med* 47: 287–97.
29. Schirrmeister H, Guhlmann A, Elsner K, Kotzerke J, Glatting G, Rentschler M, et al. 1999. Sensitivity in detecting osseous lesions depends on anatomic localization: Planar bone scintigraphy versus 18F PET. *J Nucl Med* 40: 1623–9.
30. Schirrmeister H, Arslandemir C, Glatting G, Mayer-Steinacker R, Bommer M, Dreinhofer K, et al. 2004. Omission of bone scanning according to staging guidelines leads to futile therapy in non-small cell lung cancer. *Eur J Nucl Med Mol Imaging* 31: 964–8.
31. Berger F, Lee YP, Loening AM, Chatziioannou A, Freedland SJ, Leahy R, et al. 2002. Whole-body skeletal imaging in mice utilizing microPET: Optimization of reproducibility and applications in animal models of bone disease. *Eur J Nucl Med Mol Imaging* 29: 1225–36.
32. O'Neill K, Lyons SK, Gallagher WM, Curran KM, Byrne AT. 2010. Bioluminescent imaging: a critical tool in pre-clinical oncology research. *J Pathol* 220: 317–27.
33. Henriquez NV, van Overveld PG, Que I, Buijs JT, Bachelier R, Kaijzel EL, et al. 2007. Advances in optical imaging and novel model systems for cancer metastasis research. *Clin Exp Metastasis* 24: 699–705.
34. Kaijzel EL, van der Pluijm G, Lowik CW. 2007. Whole-body optical imaging in animal models to assess cancer development and progression. *Clin Cancer Res* 13: 3490–7.
35. Snoeks TJ, Khmelinskii A, Lelieveldt BP, Kaijzel EL, Lowik CW. 2011. Optical advances in skeletal imaging applied to bone metastases. *Bone* 48: 106–14.
36. Buijs JT, Rentsch CA, van der Horst G, van Overveld PG, Wetterwald A, Schwaninger R, et al. 2007. BMP7, a putative regulator of epithelial homeostasis in the

human prostate, is a potent inhibitor of prostate cancer bone metastasis in vivo. *Am J Pathol* 171: 1047–57.
37. van den Hoogen C, van der Horst G, Cheung H, Buijs JT, Lippitt JM, Guzman-Ramirez N, et al. 2010. High aldehyde dehydrogenase activity identifies tumor-initiating and metastasis-initiating cells in human prostate cancer. *Cancer Res* 70: 5163–73.
38. Nakatsu T, Ichiyama S, Hiratake J, Saldanha A, Kobashi N, Sakata K, et al. 2006. Structural basis for the spectral difference in luciferase bioluminescence. *Nature* 440: 372–6.
39. Contag PR, Olomu IN, Stevenson DK, Contag CH. 1998. Bioluminescent indicators in living mammals. *Nat Med* 4: 245–7.
40. Rehemtulla A, Stegman LD, Cardozo SJ, Gupta S, Hall DE, Contag CH, et al. 2000. Rapid and quantitative assessment of cancer treatment response using in vivo bioluminescence imaging. *Neoplasia* 2: 491–5.
41. Kim JB, Urban K, Cochran E, Lee S, Ang A, Rice B, et al. 2010. Non-invasive detection of a small number of bioluminescent cancer cells in vivo. *PLoS One* 5: e9364.
42. Caysa H, Jacob R, Muther N, Branchini B, Messerle M, Soling A. 2009. A redshifted codon-optimized firefly luciferase is a sensitive reporter for bioluminescence imaging. *Photochem Photobiol Sci* 8: 52–6.
43. van der Horst G, van Asten JJ, Figdor A, van den Hoogen C, Cheung H, Bevers RF, et al. 2011. Real-time cancer cell tracking by bioluminescence in a preclinical model of human bladder cancer growth and metastasis. *Eur Urol* 60(2): 337–43.
44. Graves EE, Weissleder R, Ntziachristos V. 2004. Fluorescence molecular imaging of small animal tumor models. *Curr Mol Med* 4: 419–30.
45. Wang BG, Konig K, Halbhuber KJ. 2010. Two-photon microscopy of deep intravital tissues and its merits in clinical research. *J Microsc* 238: 1–20.
46. Hilderbrand SA, Weissleder R. 2010. Near-infrared fluorescence: application to in vivo molecular imaging. *Curr Opin Chem Biol* 14: 71–9.
47. Hoffman RM. 2009. Imaging cancer dynamics in vivo at the tumor and cellular level with fluorescent proteins. *Clin Exp Metastasis* 26: 345–55.
48. Kozloff KM, Weissleder R, Mahmood U. 2007. Noninvasive optical detection of bone mineral. *J Bone Miner Res* 22: 1208–16.
49. Snoeks TJ, Khmelinskii A, Lelieveldt BP, Kaijzel EL, Lowik CW. 2011. Optical advances in skeletal imaging applied to bone metastases. *Bone* 48: 106–14.
50. van der Pluijm G, Que I, Sijmons B, Buijs JT, Lowik CW, Wetterwald A, et al. 2005. Interference with the microenvironmental support impairs the de novo formation of bone metastases in vivo. *Cancer Res* 65: 7682–90.
51. Zaheer A, Lenkinski RE, Mahmood A, Jones AG, Cantley LC, Frangioni JV. 2001. In vivo near-infrared fluorescence imaging of osteoblastic activity. *Nat Biotechnol* 19: 1148–54.
52. Teitelbaum SL. 2000. Bone resorption by osteoclasts. *Science* 289: 1504–8.
53. Drake FH, Dodds RA, James IE, Connor JR, Debouck C, Richardson S, et al. 1996. Cathepsin K, but not cathepsins B, L, or S, is abundantly expressed in human osteoclasts. *J Biol Chem* 271: 12511–6.
54. Kozloff KM, Quinti L, Patntirapong S, Hauschka PV, Tung CH, Weissleder R, et al. 2009. Non-invasive optical detection of cathepsin K-mediated fluorescence reveals osteoclast activity in vitro and in vivo. *Bone* 44: 190–8.
55. Bremer C, Tung CH, Weissleder R. 2001. In vivo molecular target assessment of matrix metalloproteinase inhibition. *Nat Med* 7: 743–8.
56. Gioux S, Choi HS, Frangioni JV. 2010. Image-guided surgery using invisible near-infrared light: Fundamentals of clinical translation. *Mol Imaging* 9: 237–55.
57. Keereweer S, Kerrebijn JD, Mol IM, Mieog JS, Van Driel PB, Baatenburg de Jong RJ, et al. 2011. Optical imaging of oral squamous cell carcinoma and cervical lymph node metastasis. *Head Neck* 34(7): 1002–8.
58. Basu S, Kwee TC, Surti S, Akin EA, Yoo D, Alavi A. 2011. Fundamentals of PET and PET/CT imaging. *Ann N Y Acad Sci* 1228: 1–18.
59. Gnanasegaran G, Barwick T, Adamson K, Mohan H, Sharp D, Fogelman I. 2009. Multislice SPECT/CT in benign and malignant bone disease: When the ordinary turns into the extraordinary. *Semin Nucl Med* 39: 431–42.
60. Metser U, Golan O, Levine CD, Even-Sapir E. 2005. Tumor lesion detection: When is integrated positron emission tomography/computed tomography more accurate than side-by-side interpretation of positron emission tomography and computed tomography? *J Comput Assist Tomogr* 29: 554–9.
61. Zaidi H, Del GA. 2011. An outlook on future design of hybrid PET/MRI systems. *Med Phys* 38: 5667–89.
62. Wehrl HF, Judenhofer MS, Wiehr S, Pichler BJ. Preclinical PET/MR: Technological advances and new perspectives in biomedical research. *Eur J Nucl Med Mol Imaging* 36 Suppl 1: S56–68.
63. Cao F, Drukker M, Lin S, Sheikh AY, Xie X, Li Z, et al. 2007. Molecular imaging of embryonic stem cell misbehavior and suicide gene ablation. *Cloning Stem Cells* 9: 107–17.
64. Serganova I, Moroz E, Vider J, Gogiberidze G, Moroz M, Pillarsetty N, et al. 2009. Multimodality imaging of TGFbeta signaling in breast cancer metastases. *FASEB J* 23: 2662–72.
65. Schipper ML, Riese CG, Seitz S, Weber A, Behe M, Schurrat T, et al. 2007. Efficacy of 99mTc pertechnetate and 131I radioisotope therapy in sodium/iodide symporter (NIS)-expressing neuroendocrine tumors in vivo. *Eur J Nucl Med Mol Imaging* 34: 638–50.
66. Yang D, Han L, Kundra V. 2005. Exogenous gene expression in tumors: Noninvasive quantification with functional and anatomic imaging in a mouse model. *Radiology* 235: 950–8.
67. Zhang N, Fang Z, Contag PR, Purchio AF, West DB. 2004. Tracking angiogenesis induced by skin wounding

and contact hypersensitivity using a Vegfr2-luciferase transgenic mouse. *Blood* 103: 617–26.
68. Jaiswal JK, Mattoussi H, Mauro JM, Simon SM. 2003. Long-term multiple color imaging of live cells using quantum dot bioconjugates. *Nat Biotechnol* 21: 47–51.
69. Schroeder T. 2008. Imaging stem-cell-driven regeneration in mammals. *Nature* 453: 345–51.
70. Rothbauer U, Zolghadr K, Tillib S, Nowak D, Schermelleh L, Gahl A, et al. 2006. Targeting and tracing antigens in live cells with fluorescent nanobodies. *Nat Methods* 3: 887–9.
71. Shaner NC, Steinbach PA, Tsien RY. 2005. A guide to choosing fluorescent proteins. *Nat Methods* 2: 905–9.

83
Metastatic Solid Tumors to Bone
Rachelle W. Johnson and Julie A. Sterling

Importance of the Problem 686
Concept of the Vicious Cycle 686
Animal Models for Studying Bone Metastases 687
Molecular Mechanisms of Bone Metastasis (Osteolytic) 687
Molecular Mechanisms of Bone Metastasis (Osteoblastic) 688
Tumor Microenvironment Regulation of Bone Disease 688
Current Promising Clinical Targets 689
References 690

IMPORTANCE OF THE PROBLEM

Cancer metastasis to bone is a common problem in patients with metastatic tumors, such as breast, prostate, and lung cancers. For example, it is estimated that 70% of patients with metastatic breast cancer and 90% of patients with metastatic prostate cancer will develop tumors in bone [1]. Once tumors have established in bone they disrupt normal bone remodeling that can result in an excess of poorly formed bone (osteoblastic-prostate) or the destruction of bone (osteolytic-breast and lung), though many patients display a mixture of both. This disruption in bone results in an increase in fractures, bone pain, and hypercalcemia, resulting in a decrease in quality of life and poor clinical outcome. Importantly, bone metastases are incurable, and while treatments have improved, the patients will eventually die from the disease.

CONCEPT OF THE VICIOUS CYCLE

Dr. Gregory Mundy coined the term "vicious cycle" about 20 years ago to refer to the important interactions between tumor cells and the bone microenvironment. In this model, factors from the tumor stimulate osteoclast mediated bone destruction, which results in growth factors being released from the bone to stimulate further growth of the tumor cells and thus the production of more tumor-derived osteolytic factors (Fig. 83.1) [2].

While the concept is overly simplified in order to illustrate the hypothesis, the "vicious cycle" has held up well over the past 20 years with the mechanistic and clinical data collected. This is especially true in studies using osteoclast inhibitors, such as bisphosphonates. These studies have demonstrated that inhibiting the bone resorption phase also reduces the tumor burden. While there is disagreement in the field as to whether bisphosphonates directly inhibit tumor cell growth, it is likely that inhibiting bone resorption inhibits tumor cell growth [3]. Furthermore, inhibiting tumor-produced factors, such as parathyroid hormone—related protein (PTHrP) inhibits bone destruction and further tumor growth [4, 5].

Our notion of the vicious cycle continues to emerge as we learn more about the interactions between tumor cells and the bone (Fig. 83.1). It has become clear that many factors within the bone microenvironment interact with the tumor cells, and that these interactions are far more complicated than the original vicious cycle model alone can explain. For example, we now know that tumor cells can closely interact with osteoblasts [6–8], bone stroma [9], and immune cells [10, 11].

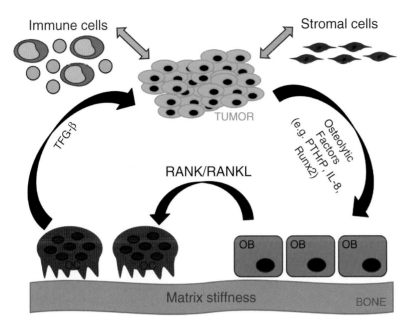

Fig. 83.1. Updated "vicious cycle" diagram. Data collected over many years suggest a complex interaction between tumor cells that metastasize to bone and many cell types that reside in bone marrow. These include immune and stromal cells as well as osteoblasts and osteoclasts. Factors produced by the tumor cells alter the growth of the cells in the microenvironment; in turn, tumor cell behavior is influenced by interactions with these cells. Furthermore, interactions with the physical microenvironment appear to also influence tumor cell behavior and gene expression.

ANIMAL MODELS FOR STUDYING BONE METASTASES

Some of the most common models of tumor-induced bone disease include tail vein [12–14], cardiac [15–17], and intratibial [18–20] inoculation of breast tumor cells. Many breast cancer models rely on cardiac injections, in which tumor cells are inoculated directly into the left cardiac ventricle. This allows for tumor cells to disperse throughout the body, but bypasses the direct blood flow to the lungs. Other tumor–bone models, such as the myeloma models, rely on tail vein injections, which for breast cancer cells result in a large tumor burden and premature death due to lung metastases. In certain cases tumors are inoculated directly into the tibia. While not a model of metastasis, this approach is extremely important for prostate models that do not readily metastasize to bone when injected intravenously. In addition, for certain breast cancer models, intratibial injections can indicate how alterations in gene expression of a tumor cell alter tumor growth in bone separately from metastasis. While these models are the primary models available for the study of human metastasis to bone, the ideal models are those that spontaneously metastasize from the mammary fat pad. Unfortunately, there are currently no human tumors that metastasize from the fat pad in mouse models. There are mouse models, such as the 4T1/BalbC model, which can metastasize to bone from the primary site [21, 22], but this model lacks the clinical significance that can only be achieved through human metastasis models.

Prostate models are particularly challenging. The only metastatic models available utilize PC—3 cells, which unlike most prostate tumors in patients are purely osteolytic. Therefore, these cells are rarely used for tumor–bone models. Instead the field mostly relies on orthotopic models developed by Leland Chung's group from LNCaP C4—2b cells derived from a lymph node metastasis [23, 24] and numerous cell lines isolated by the Vessella group from their rapid autopsy program [25]. These cells can also be directly injected into bone, where they develop into strongly osteoblastic tumors.

The most popular models of cancer metastasis to bone have traditionally focused on breast or prostate, primarily since these cancers tend to be the most prevalent in the population; however, a recent shift in the prevalence of lung cancer metastases to bone has stimulated the use of the RWGT2 model of lung carcinoma, which was generated by the Mundy group [26]. This recent interest can be largely attributed to advancements in the treatment of lung tumors at the primary site, which extends patient survival from the primary tumor but appears to facilitate the opportunity for tumor cells to metastasize to bone. The RWGT2 intracardiac model causes significant bone destruction, similar to the clinical disease.

MOLECULAR MECHANISMS OF BONE METASTASIS (OSTEOLYTIC)

Breast and lung tumors that have metastasized to bone frequently cause significant bone destruction through activation of osteoclasts. This is primarily initiated by tumor cell upregulation and secretion of factors that stimulate receptor activator of nuclear factor-κB ligand (RANKL) expression in osteoblasts and results in osteoclast activation through RANK/RANKL binding. While several factors that will be discussed later in this review have been implicated, PTHrP has been identified as a major osteolytic factor produced by tumor cells to activate bone destruction [2, 27]. We have identified the developmental transcription factor Gli2 as a key regulator of PTHrP transcription and secretion [15]. Whereas Gli2 is regulated by the Hedgehog (Hh) signaling pathway during development, we have shown that Gli2 expression in osteolytic tumor cells is regulated by a number of factors independent of Hh signaling, such as transforming growth factor-β (TGF-β) [28]. Gli2 was found to be necessary for TFG-β stimulation of PTHrP mRNA expression and is therefore an excellent target in blocking osteolysis. Gli2 blockade has indeed been successful in blocking osteolytic bone destruction in a genetic model [28], and studies are underway to inhibit Gli proteins through small molecules, such as the GANT compounds [29] and *E. agallocha* extracts from Japanese herbals [30].

PTHrP remains the most well studied of the osteolytic factors produced by bone metastatic tumor cells, but there are other molecules that contribute significantly to tumor-induced bone destruction, such as RANKL, IL—8, Runx2, and MMPs. Interleukin—8 (IL—8) has been shown to correlate with increased bone metastases in human breast cancer cells [31] and stimulate osteoclastogenesis independent of osteoprotegerin (OPG) [32] and RANKL [33]. Runx2, which is required for normal bone formation, is elevated in breast and prostate tumors and cells that metastasize to bone; inhibition of Runx2 in breast tumor cells inhibits tumor-induced bone destruction [34]. This inhibition is in part due to alterations in MMP-9 signaling, which is downstream of Runx2, and is important for tumor cell invasive properties [35]. There is also evidence to support significant involvement of MMPs in regulating tumor–bone interactions. For example, MMP-7 has been shown to enhance osteolysis through solubilization of osteoblast-derived RANKL in the PC3 prostate cancer model of bone destruction [36] and was recently found to play a similar role in breast tumor-induced osteolysis [37].

MOLECULAR MECHANISMS OF BONE METASTASIS (OSTEOBLASTIC)

Some cancers such as prostate, and less frequently breast, induce osteoblastic metastases upon establishment in bone. The mechanism driving the formation of osteoblastic lesions is unclear, but bone resorptive markers are often elevated in these patients, suggesting that osteoclast activation may still be necessary. Similar to patients with primarily osteolytic bone metastases, prostate cancer patients frequently present with mixed lesions throughout the skeleton. It is believed that at least in some cases, osteoblastic metastases begin with a heavy resorption component then switch to blastic lesions. This is further supported by the successful use of bisphosphonates, which inhibit osteoclast activity, in the treatment of prostatic bone metastases and associated bone pain [38].

The molecular mechanisms that stimulate bone formation in osteoblastic metastases remain largely undiscovered, but there is evidence to suggest that endothelin-1 (ET-1) and the Wnt signaling pathway make significant contributions. Dikkopf—1 (Dkk-1) is a downstream target and inhibitor of the Wnt signaling pathway, and has been shown to inhibit prostate tumor-induced osteoblastic metastases when overexpressed in prostatic tumor cells by inhibiting transcription factor 4 (TCF-4) activity in osteoblast precursors [39]; conversely, Dkk—1 has been identified as a mediator of tumor-derived ET-1-induced bone formation [40], suggesting an autocrine loop, whereby Dkk-1 and ET-1 upregulate one another. Differences here may be attributable to cell-line specificity, since Thudi et al. utilized ACE-1 prostate tumor cells, and Clines et al. did not test this model in tumor cells. More data are needed to determine the extent of ET-1 regulation of Wnt signaling.

TUMOR MICROENVIRONMENT REGULATION OF BONE DISEASE

Although certain tumor types rely heavily on autocrine signaling, they are not as autonomous as once thought. In recent years, increasing emphasis has been placed on tumor microenvironment influence on tumor cell behavior and the interactions between tumor cells and the stroma in particular. It has become evident that the microenvironment plays a tremendous role in both promoting and hindering tumor progression at different sites of the disease. While the host microenvironment significantly impacts tumor cell evolution, cancer cells may also influence host cells at the site of metastasis in addition to systemic alterations. This is particularly evident in the vicious cycle of bone destruction, which requires extensive interactions and complex signaling between tumor cells, bone, and the bone marrow stroma. While there are numerous interactions occurring within the bone microenvironment, some of the microenvironmental factors that have been identified as significant mediators of tumor–host interactions thus far include matrix stiffness, the reactive stroma surrounding the tumor, TGF-β, Wnt signaling, and Ephrins.

The extracellular matrix (ECM) has long been thought of as a structural scaffold on which tumor cells could rely on for support, but there is now substantial evidence to suggest that breast tumor cells modify their behavior in response to ECM signaling and the changing stiffness of the ECM at the primary site [41]. In the primary site of breast cancer, tumor progression has been connected with increased collagen cross-linking as well as integrin activation [42], and more recently we have shown that the rigidity of bone modulates tumor cell gene expression in a Rho-associated protein kinase (ROCK)-dependent mechanism [43]. The response of tumor cells to the physical microenvironment is complex and still actively being explored.

Much of the signaling that modulates tumor cell behavior in bone originates from the surrounding stroma. The contribution of the reactive stroma in tumor cell evolution has been well established, particularly in prostate cancer through studies conducted by the Cunha and Hayward groups [44, 45], but the impact of bone stroma on tumor cell behavior is becoming appreciated. Most of the published data support a significant role for bone-derived growth factors and chemokines (e.g., TGF-β, SDF-1/CXCR4) [9, 46, 47], and an emerging role for immune cells such as myeloid derived suppressor cells (MDSCs). It is well established that stromal-derived factor-1 (SDF-1/CXCL12) is involved in tumor cell homing to bone [48], where it has been shown that bone metastatic breast cancer cells express elevated levels of the SDF-1 receptor CXCR4 [49, 50], and conditioned media from primary human bone marrow contains CXCL12 [50], which presumably promotes metastasis to bone. Indeed, tumor cell migration can be blocked when the CXCR4/CXCL12 interactions are blocked by a neutralizing antibody against CXCR4 [50].

TGF-β signaling is known to regulate such cell autonomous events as proliferation, migration, apoptosis, and tumor invasion [51], and has been shown to directly promote tumor cell growth in bone following release from the mineralized bone matrix [2]. Interestingly, these same processes are controlled during development in part by the Wnt signaling pathway [52], which has been identified as an important stromal component in metastatic prostate cancer [53]. Convergence of TGF-β and Wnt signaling has been repeatedly demonstrated at the transcriptional level on the Gli proteins, which are downstream of the Hh signaling pathway. Previous reports have indicated that Wnt signaling can both regulate and be regulated by Gli proteins [54], although most publications place Hh signaling upstream of Wnt during development [55, 56]. Likewise, Hh and TGF-β signaling have been reported to crosstalk [51], and our group, as well as others, has reported that Gli2 transcription is regulated at least in part by TGF-β signaling through Smads [28, 57], suggesting the importance of this crosstalk for regulating bone destruction.

More recently, Ephrin/Eph signaling has been identified as an important mediator in tumor promotion and angiogenesis, and preclinical studies have shown the soluble EphB4 and EphB4 neutralizing antibody to be efficacious treatment in breast and prostate solid tumors [58, 59]; however, disruption of the Ephrin/Eph family signaling results in skeletal abnormalities and increased bone resorption [60–62]. Further studies are needed to determine whether blockade of Ephrin/Eph signaling is detrimental in patients with potential bone metastatic disease, but it remains possible that Ephrin/Eph signaling, similar to TGF-β, may play a biphasic role in tumor metastasis.

Immune cells such as MDSCs have been implicated in tumor promotion and metastasis to distant organs [10], and more recently these cells have been suspected to play a key role in tumor metastasis to bone [63]. For example, it has been previously demonstrated that Gr-1+ CD11b+ MDSCs are capable of differentiating into osteoclasts in the 5T model of multiple myeloma [63], but further studies are needed to determine if this same potential exists in solid tumor metastasis to bone. More recent data examining CD8+ T cells in breast tumor metastasis to bone suggest that immune cells influence tumor cell metastasis independent of tumor effects on bone, specifically showing that osteoclast activation of T cells reduces tumor burden but increases osteolysis [64]. As indicated by Fowler et al., the B-cell population may be an important factor in the development of myeloma, since RAG—2 mice that lack B cells develop 5T myeloma while nude mice do not [65]; however, it was also suggested that the elevated natural killer cell population in nude mice may contribute to this phenomenon. More studies are required to determine if these cell populations are significant in solid tumor metastases to bone, but the evidence on MDSCs and T cells is compelling.

CURRENT PROMISING CLINICAL TARGETS

Despite decades of research focusing on the treatment of bone metastases, they remain incurable. Bisphosphonates are currently the clinical standard of care for treating patients with bone metastases and have been highly effective in patients to reduce fractures and improve quality of life [66]. However, they are not a permanent cure and are not without risk. Severe side effects are rare but include severe muscle pain and osteonecrosis of the jaw and atypical subtrochanteric fractures in patients with long-term high-dose treatments [67–69]. Because of these risks, newer drugs are currently of great interest. The emerging drug for clinical use, denosumab or XGEVA™ (Amgen), which targets RANKL, has been demonstrated to reduce bone turnover and increase bone density in clinical studies [70]. Interestingly, this drug seems to have the capability of inhibiting tumor growth, which may be an indirect effect [71]. However, XGEVA™ also presents safety risks, including hypocalcemia and osteonecrosis of the jaw, though the prevalence is unclear due to limited clinical data [67]. Despite the clinical success of these drugs, they may not target the tumor, and therefore do not address the root of the disease and are not a cure for tumors growing in the bone. Other groups are investigating drugs that can target the tumor cells directly.

One such approach is by inhibiting TGF-β. TGF-β is thought to affect both the tumor cells and the microenvironment, suggesting that an inhibitor of this pathway may be able to block bone destruction through multiple targets. Thus far, TGF-β inhibitors have been very successful with several groups, demonstrating that they can inhibit metastasis to soft tissues [72, 73] and bone [74]. In addition to direct effects on the tumor, they appear to have positive effects on bone in non-tumor-bearing mice. It has been reported by two independent groups that TGF-β inhibitors 1D11 (Genzyme, Cambridge, MA) and SD-208 (Scios, Inc., Sunnyvale, CA) increase bone density and biomechanical properties as well as numerous measures of bone quality (e.g., trabecular bone architecture and the mineral:collagen ratio) [75, 76]. These studies also demonstrate an increase in osteoblast number and a decrease in osteoclast number in the bone marrow of healthy mice. Taken together, these data suggest that TGF-β inhibitors may be ideal drugs for bone metastases, since they can both inhibit tumor growth and improve bone quality. Despite their promise in preclinical models, their use may be complicated in patients due to the biphasic effects of TGF-β in tumor growth and its effects throughout the body on normal cells [77], with particular concerns in patients with residual primary tumor or multiple metastases.

It is clear that more specific drugs are needed to target tumors growing in bone. Some potential pathways are being studied by multiple groups to block metastasis and establishment in bone through inhibition of integrins [78], SDF-1/CXCR4 [79], and ROCK [43, 80]. Other groups have specifically targeted certain cell types, such as inhibiting osteoclasts using Src inhibitors [81, 82] and stimulating osteoblasts using DKK-1 inhibitors, while still others are investigating inhibitors of the tumors themselves. Some current promising approaches include the histone deacetylase inhibitor vorinostat, which inhibits tumor growth in bone but can stimulate bone loss [83], suggesting possible combination treatments to protect bone. Erlotinib, an inhibitor against epidermal growth factor receptor (EGF-R) tyrosine kinase, has been successfully tested in a preclinical model of non-small-cell lung cancer metastasis to bone and effectively inhibited release of osteolytic factors such as PTHrP and IL—8 [84]. In our studies and others', inhibiting the transcription factor GLI2 appears promising in melanoma [85] and breast cancer metastasis [15, 28]. Inhibitors to this pathway are currently under investigation in preclinical models of prostate cancer; they are promising for treatment of primary tumor burden [29]. For bone they are an attractive target since GLI2 has limited expression in adult tissues, suggesting minimal side effects.

REFERENCES

1. Boxer DI, Todd CE, Coleman R, Fogelman I. 1989. Bone secondaries in breast cancer: The solitary metastasis. *J Nucl Med* 30: 1318–20.
2. Mundy GR. 1997. Mechanisms of bone metastasis. *Cancer* 80: 1546–56.
3. Body JJ. 2011. New developments for treatment and prevention of bone metastases. *Curr Opin Oncol* 23: 338–42.
4. Gallwitz WE, Guise TA, Mundy GR. 2002. Guanosine nucleotides inhibit different syndromes of PTHrP excess caused by human cancers in vivo. *J Clin Invest* 110: 1559–72.
5. Guise TA, Yin JJ, Taylor SD, Kumagai Y, Dallas M, Boyce BF, Yoneda T, Mundy GR. 1996. Evidence for a causal role of parathyroid hormone-related protein in the pathogenesis of human breast cancer-mediated osteolysis. *J Clin Invest* 98: 1544–9.
6. Shiirevnyamba A, Takahashi T, Shan H, Ogawa H, Yano S, Kanayama H, Izumi K, Uehara H. 2011. Enhancement of osteoclastogenic activity in osteolytic prostate cancer cells by physical contact with osteoblasts. *Br J Cancer* 104: 505–13.
7. Sieh S, Lubik AA, Clements JA, Nelson CC, Hutmacher DW. 2010. Interactions between human osteoblasts and prostate cancer cells in a novel 3D in vitro model. *Organogenesis* 6: 181–8.
8. Roodman GD. 2011. Osteoblast function in myeloma. *Bone* 48: 135–40.
9. Langley RR, Fidler IJ. 2011. The seed and soil hypothesis revisited—The role of tumor-stroma interactions in metastasis to different organs. *Int J Cancer* 128: 2527–35.
10. Schmid MC, Varner JA. 2010. Myeloid cells in the tumor microenvironment: Modulation of tumor angiogenesis and tumor inflammation. *J Oncol* 2010: 201026.
11. Tadmor T, Attias D, Polliack A. 2011. Myeloid-derived suppressor cells—Their role in haemato-oncological malignancies and other cancers and possible implications for therapy. *Br J Haematol* 153: 557–67.
12. Edwards CM, Lwin ST, Fowler JA, Oyajobi BO, Zhuang J, Bates AL, Mundy GR. 2009. Myeloma cells exhibit an increase in proteasome activity and an enhanced response to proteasome inhibition in the bone marrow microenvironment in vivo. *Am J Hematol* 84: 268–72.
13. Edwards CM, Edwards JR, Lwin ST, Esparza J, Oyajobi BO, McCluskey B, Munoz S, Grubbs B, Mundy GR. 2008. Increasing Wnt signaling in the bone marrow microenvironment inhibits the development of myeloma bone disease and reduces tumor burden in bone in vivo. *Blood* 111: 2833–42.
14. Oyajobi BO, Garrett IR, Gupta A, Flores A, Esparza J, Munoz S, Zhao M, Mundy GR. 2007. Stimulation of new bone formation by the proteasome inhibitor, bortezomib: Implications for myeloma bone disease. *Br J Haematol* 139: 434–8.
15. Sterling JA, Oyajobi BO, Grubbs B, Padalecki SS, Munoz SA, Gupta A, Story B, Zhao M, Mundy GR. 2006. The hedgehog signaling molecule Gli2 induces parathyroid hormone-related peptide expression and osteolysis in metastatic human breast cancer cells. *Cancer Res* 66: 7548–53.
16. Hiraga T, Myoui A, Choi ME, Yoshikawa H, Yoneda T. 2006. Stimulation of cyclooxygenase-2 expression by bone-derived transforming growth factor-beta enhances bone metastases in breast cancer. *Cancer Res* 66: 2067–73.
17. Yin JJ, Selander K, Chirgwin JM, Dallas M, Grubbs BG, Wieser R, Massague J, Mundy GR, Guise TA. 1999. TGF-beta signaling blockade inhibits PTHrP secretion

by breast cancer cells and bone metastases development. *J Clin Invest* 103: 197–206.
18. Johnson LC, Johnson RW, Munoz SA, Mundy GR, Peterson TE, Sterling JA. 2010. Longitudinal live animal microCT allows for quantitative analysis of tumor-induced bone destruction. *Bone* 48(1): 141–51.
19. Morrissey C, Dowell A, Koreckij TD, Nguyen H, Lakely B, Fanslow WC, True LD, Corey E, Vessella RL. 2010. Inhibition of angiopoietin-2 in LuCaP 23.1 prostate cancer tumors decreases tumor growth and viability. *Prostate* 70: 1799–808.
20. Nagae M, Hiraga T, Yoneda T. 2007. Acidic microenvironment created by osteoclasts causes bone pain associated with tumor colonization. *J Bone Miner Metab* 25: 99–104.
21. Hiraga T, Hata K, Ikeda F, Kitagaki J, Fujimoto-Ouchi K, Tanaka Y, Yoneda T. 2005. Preferential inhibition of bone metastases by 5′-deoxy-5-fluorouridine and capecitabine in the 4T1/luc mouse breast cancer model. *Oncol Rep* 14: 695–9.
22. Rose AA, Pepin F, Russo C, Abou Khalil JE, Hallett M, Siegel PM. 2007. Osteoactivin promotes breast cancer metastasis to bone. *Mol Cancer Res* 5: 1001–14.
23. Thalmann GN, Anezinis PE, Chang SM, Zhau HE, Kim EE, Hopwood VL, Pathak S, von Eschenbach AC, Chung LW. 1994. Androgen-independent cancer progression and bone metastasis in the LNCaP model of human prostate cancer. *Cancer Res* 54: 2577–81.
24. Thalmann GN, Sikes RA, Wu TT, Degeorges A, Chang SM, Ozen M, Pathak S, Chung LW. 2000. LNCaP progression model of human prostate cancer: Androgen-independence and osseous metastasis. *Prostate* 44: 91–103.
25. Corey E, Quinn JE, Buhler KR, Nelson PS, Macoska JA, True LD, Vessella RL. 2003. LuCaP 35: A new model of prostate cancer progression to androgen independence. *Prostate* 55: 239–46.
26. Guise TA, Yoneda T, Yates AJ, Mundy GR. 1993. The combined effect of tumor-produced parathyroid hormone-related protein and transforming growth factor-alpha enhance hypercalcemia in vivo and bone resorption in vitro. *J Clin Endocrinol Metab* 77: 40–5.
27. Sterling JA, Edwards JR, Martin TJ, Mundy GR. 2011. Advances in the biology of bone metastasis: How the skeleton affects tumor behavior. *Bone* 48(1): 6–15.
28. Johnson RW, Nguyen MP, Padalecki SS, Grubbs BG, Merkel AR, Oyajobi BO, Matrisian LM, Mundy GR, Sterling JA. 2011. TGF-beta promotion of Gli2-induced expression of parathyroid hormone-related protein, an important osteolytic factor in bone metastasis, is independent of canonical Hedgehog signaling. *Cancer Res* 71: 822–31.
29. Lauth M, Bergstrom A, Shimokawa T, Toftgard R. 2007. Inhibition of GLI-mediated transcription and tumor cell growth by small-molecule antagonists. *Proc Natl Acad Sci U S A* 104: 8455–60.
30. Rifai Y, Arai MA, Sadhu SK, Ahmed F, Ishibashi M. 2011. New Hedgehog/GLI signaling inhibitors from Excoecaria agallocha. *Bioorg Med Chem Lett* 21: 718–22.
31. Bendre MS, Gaddy-Kurten D, Mon-Foote T, Akel NS, Skinner RA, Nicholas RW, Suva LJ. 2002. Expression of interleukin 8 and not parathyroid hormone-related protein by human breast cancer cells correlates with bone metastasis in vivo. *Cancer Res* 62: 5571–9.
32. Bendre MS, Montague DC, Peery T, Akel NS, Gaddy D, Suva LJ. 2003. Interleukin-8 stimulation of osteoclastogenesis and bone resorption is a mechanism for the increased osteolysis of metastatic bone disease. *Bone* 33: 28–37.
33. Bendre MS, Margulies AG, Walser B, Akel NS, Bhattacharrya S, Skinner RA, Swain F, Ramani V, Mohammad KS, Wessner LL, Martinez A, Guise TA, Chirgwin JM, Gaddy D, Suva LJ. 2005. Tumor-derived interleukin-8 stimulates osteolysis independent of the receptor activator of nuclear factor-kappaB ligand pathway. *Cancer Res* 65: 11001–9.
34. Pratap J, Lian JB, Javed A, Barnes GL, van Wijnen AJ, Stein JL, Stein GS. 2006. Regulatory roles of Runx2 in metastatic tumor and cancer cell interactions with bone. *Cancer Metastasis Rev* 25: 589–600.
35. Pratap J, Javed A, Languino LR, van Wijnen AJ, Stein JL, Stein GS, Lian JB. 2005. The Runx2 osteogenic transcription factor regulates matrix metalloproteinase 9 in bone metastatic cancer cells and controls cell invasion. *Mol Cell Biol* 25: 8581–91.
36. Lynch CC, Hikosaka A, Acuff HB, Martin MD, Kawai N, Singh RK, Vargo-Gogola TC, Begtrup JL, Peterson TE, Fingleton B, Shirai T, Matrisian LM, Futakuchi M. 2005. MMP-7 promotes prostate cancer-induced osteolysis via the solubilization of RANKL. *Cancer Cell* 7: 485–96.
37. Thiolloy S, Halpern J, Holt GE, Schwartz HS, Mundy GR, Matrisian LM, Lynch CC. 2009. Osteoclast-derived matrix metalloproteinase-7, but not matrix metalloproteinase-9, contributes to tumor-induced osteolysis. *Cancer Res* 69: 6747–55.
38. Saad F, Gleason DM, Murray R, Tchekmedyian S, Venner P, Lacombe L, Chin JL, Vinholes JJ, Goas JA, Chen B. 2002. A randomized, placebo-controlled trial of zoledronic acid in patients with hormone-refractory metastatic prostate carcinoma. *J Natl Cancer Inst* 94: 1458–68.
39. Thudi NK, Martin CK, Murahari S, Shu ST, Lanigan LG, Werbeck JL, Keller ET, McCauley LK, Pinzone JJ, Rosol TJ. 2011. Dickkopf-1 (DKK-1) stimulated prostate cancer growth and metastasis and inhibited bone formation in osteoblastic bone metastases. *Prostate* 71: 615–25.
40. Clines GA, Mohammad KS, Bao Y, Stephens OW, Suva LJ, Shaughnessy JD Jr, Fox JW, Chirgwin JM, Guise TA. 2007. Dickkopf homolog 1 mediates endothelin-1-stimulated new bone formation. *Mol Endocrinol* 21: 486–98.
41. Sterling JA, Guelcher SA. 2011. Bone structural components regulating sites of tumor metastasis. *Curr Osteoporos Rep* 9: 89–95.
42. Leventhal KR, Yu H, Kass L, Lakins JN, Egeblad M, Erler JT, Fong SF, Csiszar K, Giaccia A, Weninger W, Yamauchi M, Gasser DL, Weaver VM. 2009. Matrix crosslinking forces tumor progression by enhancing integrin signaling. *Cell* 139: 891–906.

43. Ruppender NS, Merkel AR, Martin TJ, Mundy GR, Sterling JA, Guelcher SA. 2010. Matrix rigidity induces osteolytic gene expression of metastatic breast cancer cells. *PLoS One* 5: e15451.
44. Hayward SW, Cunha GR. 2000. The prostate: Development and physiology. *Radiol Clin North Am* 38: 1–14.
45. Hayward SW, Cunha GR, Dahiya R. 1996. Normal development and carcinogenesis of the prostate. A unifying hypothesis. *Ann N Y Acad Sci* 784: 50–62.
46. Onishi T, Hayashi N, Theriault RL, Hortobagyi GN, Ueno NT. 2010. Future directions of bone-targeted therapy for metastatic breast cancer. *Nat Rev Clin Oncol* 7: 641–51.
47. Nannuru KC, Singh RK. 2010. Tumor-stromal interactions in bone metastasis. *Curr Osteoporos Rep* 8: 105–13.
48. Taichman RS, Cooper C, Keller ET, Pienta KJ, Taichman NS, McCauley LK. 2002. Use of the stromal cell-derived factor-1/CXCR4 pathway in prostate cancer metastasis to bone. *Cancer Res* 62: 1832–7.
49. Kang Y, Siegel PM, Shu W, Drobnjak M, Kakonen SM, Cordon-Cardo C, Guise TA, Massague J. 2003. A multigenic program mediating breast cancer metastasis to bone. *Cancer Cell* 3: 537–49.
50. Muller A, Homey B, Soto H, Ge N, Catron D, Buchanan ME, McClanahan T, Murphy E, Yuan W, Wagner SN, Barrera JL, Mohar A, Verastegui E, Zlotnik A. 2001. Involvement of chemokine receptors in breast cancer metastasis. *Nature* 410: 50–6.
51. Guo X, Wang XF. 2009. Signaling cross-talk between TGF-beta/BMP and other pathways. *Cell Res* 19: 71–88.
52. Barker N. 2008. The canonical Wnt/beta-catenin signalling pathway. *Methods Mol Biol* 468: 5–15.
53. Placencio VR, Sharif-Afshar AR, Li X, Huang H, Uwamariya C, Neilson EG, Shen MM, Matusik RJ, Hayward SW, Bhowmick NA. 2008. Stromal transforming growth factor-beta signaling mediates prostatic response to androgen ablation by paracrine Wnt activity. *Cancer Res* 68: 4709–18.
54. Mullor JL, Dahmane N, Sun T, Ruiz i Altaba A. Wnt signals are targets and mediators of Gli function. *Curr Biol* 11: 769–73.
55. Yazawa S, Umesono Y, Hayashi T, Tarui H, Agata K. 2009. Planarian Hedgehog/Patched establishes anterior–posterior polarity by regulating Wnt signaling. *Proc Natl Acad Sci U S A* 106: 22329–34.
56. Alvarez-Medina R, Le Dreau G, Ros M, Marti E. 2009. Hedgehog activation is required upstream of Wnt signalling to control neural progenitor proliferation. *Development* 136: 3301–9.
57. Dennler S, Andre J, Alexaki I, Li A, Magnaldo T, ten Dijke P, Wang XJ, Verrecchia F, Mauviel A. 2007. Induction of sonic hedgehog mediators by transforming growth factor-beta: Smad3-dependent activation of Gli2 and Gli1 expression in vitro and in vivo. *Cancer Res* 67: 6981–6.
58. Kertesz N, Krasnoperov V, Reddy R, Leshanski L, Kumar SR, Zozulya S, Gill PS. 2006. The soluble extracellular domain of EphB4 (sEphB4) antagonizes EphB4-EphrinB2 interaction, modulates angiogenesis, and inhibits tumor growth. *Blood* 107: 2330–8.
59. Krasnoperov V, Kumar SR, Ley E, Li X, Scehnet J, Liu R, Zozulya S, Gill PS. 2010. Novel EphB4 monoclonal antibodies modulate angiogenesis and inhibit tumor growth. *Am J Pathol* 176: 2029–38.
60. Compagni A, Logan M, Klein R, Adams RH. 2003. Control of skeletal patterning by ephrinB1-EphB interactions. *Dev Cell* 5: 217–30.
61. Martin TJ, Allan EH, Ho PW, Gooi JH, Quinn JM, Gillespie MT, Krasnoperov V, Sims NA. 2010. Communication between ephrinB2 and EphB4 within the osteoblast lineage. *Adv Exp Med Biol* 658: 51–60.
62. Allan EH, Hausler KD, Wei T, Gooi JH, Quinn JM, Crimeen-Irwin B, Pompolo S, Sims NA, Gillespie MT, Onyia JE, Martin TJ. 2008. EphrinB2 regulation by PTH and PTHrP revealed by molecular profiling in differentiating osteoblasts. *J Bone Miner Res* 23: 1170–81.
63. Yang L, Edwards CM, Mundy GR. 2010. Gr-1+CD11b+ myeloid-derived suppressor cells: Formidable partners in tumor metastasis. *J Bone Miner Res* 25: 1701–6.
64. Zhang K, Kim S, Cremasco V, Hirbe A, Novack D, Weilbaecher K, Faccio R. 2011. CD8+ T cells regulate bone tumor burden independent of osteoclast resorption. *Cancer Res* 71(14): 4799–808.
65. Fowler JA, Mundy GR, Lwin ST, Lynch CC, Edwards CM. 2009. A murine model of myeloma that allows genetic manipulation of the host microenvironment. *Dis Model Mech* 2: 604–11.
66. Shane E. 2010. Evolving data about subtrochanteric fractures and bisphosphonates. *N Engl J Med* 362: 1825–7.
67. Fizazi K, Carducci M, Smith M, Damiao R, Brown J, Karsh L, Milecki P, Shore N, Rader M, Wang H, Jiang Q, Tadros S, Dansey R, Goessl C. 2011. Denosumab versus zoledronic acid for treatment of bone metastases in men with castration-resistant prostate cancer: A randomised, double-blind study. *Lancet* 377: 813–22.
68. Coleman R, Woodward E, Brown J, Cameron D, Bell R, Dodwell D, Keane M, Gil M, Davies C, Burkinshaw R, Houston SJ, Grieve RJ, Barrett-Lee PJ, Thorpe H. 2011. Safety of zoledronic acid and incidence of osteonecrosis of the jaw (ONJ) during adjuvant therapy in a randomised phase III trial (AZURE: BIG 01-04) for women with stage II/III breast cancer. *Breast Cancer Res Treat* 127: 429–38.
69. Black DM, Kelly MP, Genant HK, Palermo L, Eastell R, Bucci-Rechtweg C, Cauley J, Leung PC, Boonen S, Santora A, de Papp A, Bauer DC. 2010. Bisphosphonates and fractures of the subtrochanteric or diaphyseal femur. *N Engl J Med* 362: 1761–71.
70. McClung MR. 2006. Inhibition of RANKL as a treatment for osteoporosis: Preclinical and early clinical studies. *Curr Osteoporos Rep* 4: 28–33.
71. Gonzalez-Suarez E, Jacob AP, Jones J, Miller R, Roudier-Meyer MP, Erwert R, Pinkas J, Branstetter D, Dougall WC. 2010. RANK ligand mediates progestin-induced mammary epithelial proliferation and carcinogenesis. *Nature* 468: 103–7.
72. Biswas S, Guix M, Rinehart C, Dugger TC, Chytil A, Moses HL, Freeman ML, Arteaga CL. 2007. Inhibition

of TGF-beta with neutralizing antibodies prevents radiation-induced acceleration of metastatic cancer progression. *J Clin Invest* 117: 1305–13.
73. Ganapathy V, Ge R, Grazioli A, Xie W, Banach-Petrosky W, Kang Y, Lonning S, McPherson J, Yingling JM, Biswas S, Mundy GR, Reiss M. 2010. Targeting the transforming growth factor-beta pathway inhibits human basal-like breast cancer metastasis. *Mol Cancer* 9: 122.
74. Mohammad KS, Javelaud D, Fournier PG, Niewolna M, McKenna CR, Peng XH, Duong V, Dunn LK, Mauviel A, Guise TA. 2011. TGF-beta-RI kinase inhibitor SD-208 reduces the development and progression of melanoma bone metastases. *Cancer Res* 71: 175–84.
75. Edwards JR, Nyman JS, Lwin ST, Moore MM, Esparza J, O'Quinn EC, Hart AJ, Biswas S, Patil CA, Lonning S, Mahadevan-Jansen A, Mundy GR. 2010. Inhibition of TGF-beta signaling by 1D11 antibody treatment increases bone mass and quality in vivo. *J Bone Miner Res* 25: 2419–26.
76. Mohammad KS, Chen CG, Balooch G, Stebbins E, McKenna CR, Davis H, Niewolna M, Peng XH, Nguyen DH, Ionova-Martin SS, Bracey JW, Hogue WR, Wong DH, Ritchie RO, Suva LJ, Derynck R, Guise TA, Alliston T. 2009. Pharmacologic inhibition of the TGF-beta type I receptor kinase has anabolic and anti-catabolic effects on bone. *PLoS One* 4: e5275.
77. Tan AR, Alexe G, Reiss M. 2009. Transforming growth factor-beta signaling: Emerging stem cell target in metastatic breast cancer? *Breast Cancer Res Treat* 115: 453–95.
78. Schneider JG, Amend SR, Weilbaecher KN. 2011. Integrins and bone metastasis: Integrating tumor cell and stromal cell interactions. *Bone* 48: 54–65.
79. Hirbe AC, Morgan EA, Weilbaecher KN. 2010. The CXCR4/SDF-1 chemokine axis: A potential therapeutic target for bone metastases? *Curr Pharm Des* 16: 1284–90.
80. Liu S, Goldstein RH, Scepansky EM, Rosenblatt M. 2009. Inhibition of rho-associated kinase signaling prevents breast cancer metastasis to human bone. *Cancer Res* 69: 8742–51.
81. Yang JC, Bai L, Yap S, Gao AC, Kung HJ, Evans CP. 2010. Effect of the specific Src family kinase inhibitor saracatinib on osteolytic lesions using the PC-3 bone model. *Mol Cancer Ther* 9: 1629–37.
82. Araujo J, Logothetis C. Dasatinib: A potent SRC inhibitor in clinical development for the treatment of solid tumors. *Cancer Treat Rev* 36: 492–500.
83. Pratap J, Akech J, Wixted JJ, Szabo G, Hussain S, McGee-Lawrence ME, Li X, Bedard K, Dhillon RJ, van Wijnen AJ, Stein JL, Stein GS, Westendorf JJ, Lian JB. 2010. The histone deacetylase inhibitor, vorinostat, reduces tumor growth at the metastatic bone site and associated osteolysis, but promotes normal bone loss. *Mol Cancer Ther* 9: 3210–20.
84. Furugaki K, Moriya Y, Iwai T, Yorozu K, Yanagisawa M, Kondoh K, et al. 2011. Erlotinib inhibits osteolytic bone invasion of human non-small-cell lung cancer cell line NCI-H292. *Clin Exp Metastasis* 28(7): 649–59.
85. Alexaki VI, Javelaud D, Van Kempen LC, Mohammad KS, Dennler S, Luciani F, Hoek KS, Juarez P, Goydos JS, Fournier PJ, Sibon C, Bertolotto C, Verrecchia F, Saule S, Delmas V, Ballotti R, Larue L, Saiag P, Guise TA, Mauviel A. 2010. GLI2-mediated melanoma invasion and metastasis. *J Natl Cancer Inst* 102: 1148–59.

84
Hematologic Malignancies and Bone
Rebecca Silbermann and G. David Roodman

Introduction 694
Hematologic Malignancies Involving Bone 694

Summary 698
References 698

INTRODUCTION

Hematologic malignancies can have multiple direct and indirect effects on bone including pathologic fractures, bone pain, and hypercalcemia. The frequency of bone involvement in association with these malignancies varies widely depending on the underlying diagnosis. Bone involvement in multiple myeloma occurs in nearly 80% of patients over their disease course [1], and significantly impacts quality of life, morbidity, performance status, and survival [2]. Skeletal complications in the much rarer human T-lymphotropic virus type 1(HTLV-1) associated adult T-cell leukemia and lymphoma, the second most common hematologic malignancy to affect the bone, most frequently manifest as hypercalcemia, which affects approximately 70% of patients over the course of their disease [3]. In contrast, bone involvement from Hodgkin's disease and non-Hodgkin's lymphoma is rare. Under normal conditions, interactions between bone resorbing cells [osteoclasts (OCL)] and bone forming cells [osteoblasts (OBL)] are balanced, allowing for coupled bone remodeling and normal hematopoiesis. Dysregulation of the normal bone remodeling process can occur in the setting of malignancy, resulting in either osteolytic, osteoblastic, or mixed osteolytic/osteoblastic lesions. In addition, increased production of parathyroid hormone-related protein (PTHrP) by tumor cells can increase osteoclastic bone resorption, resulting in increased renal tubular calcium resorption. This impairs glomerular filtration and can result in hypercalcemia.

This review will discuss the pathophysiology of skeletal lesions in hematologic malignancies, with a focus on the current understanding of their associated bone disease, and includes brief discussions of radiologic imaging of skeletal lesions, pharmacologic treatments, and other hematologic malignancies that affect the bone.

HEMATOLOGIC MALIGNANCIES INVOLVING BONE

Multiple myeloma

Multiple myeloma (MM) is a plasma cell malignancy characterized by monoclonal paraprotein production from terminally differentiated plasma cells and lytic bone disease [4]. Laboratory findings include an elevated monoclonal paraprotein in the serum and/or urine and decreased normal immunoglobulin (Ig) levels. Expansion of the plasma cell population in the bone marrow leads to leukopenia, anemia, and thrombocytopenia.

MM has the highest incidence of bone involvement among malignant diseases and is the second most common hematologic malignancy, accounting for approximately 15% of all hematologic malignancies. The age-adjusted incidence of myeloma in Western countries is 5.6 cases per 100,000 persons [5]. Approximately 60% of patients present with bone pain at diagnosis, 80% of patients develop bone lesions during their disease course, and up to 60% of patients develop pathologic fractures [6].

Primer on the Metabolic Bone Diseases and Disorders of Mineral Metabolism, Eighth Edition. Edited by Clifford J. Rosen.
© 2013 American Society for Bone and Mineral Research. Published 2013 by John Wiley & Sons, Inc.

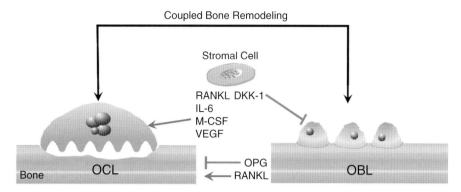

Fig. 84.1. Balanced physiologic bone remodeling. Physiologic bone remodeling is marked by balanced interactions between osteoclasts (OCL) and osteoblasts (OBL) within the bone marrow microenvironment. Locally produced cytokines and systemic hormones regulate the formation and activation of OCL. Systemic hormones (not pictured) stimulate OCL formation by inducing the expression of receptor activator of nuclear factor-κB ligand (RANKL) on marrow stromal cells and OBL. Stromal cells also produce OCL-stimulating factors including interleukin-6, macrophage colony-stimulating factor (M-CSF) and vascular endothelial growth factor (VEGF) that induce OCL formation. In addition, stromal cells produce dickkopf (DKK)-1, an OBL inhibitory factor. Coupling factors produced by OCL such as ephrins (not shown), also drive OBL differentiation while suppressing further OCL formation and activity. OBLs produce osteoprotegerin (OPG), a soluble RANKL inhibitor. Under physiologic conditions, OBL and OCL activity is balanced, in part due to the OPG/RANKL ratio. In myeloma bone disease, osteoclastogenesis is favored and osteoblastogenesis is inhibited.

While the clinical presentation of myeloma is quite variable, with approximately 20% of patients asymptomatic at presentation (disease in these patients is generally identified through routine lab studies), bone pain is the most common symptom at presentation. Bone pain, frequently centered on the chest or back and exacerbated by movement, is present in more than two-thirds of patients at diagnosis [7].

It is estimated that up to 90% of patients with myeloma have evidence of osteolysis in the form of generalized osteopenia or discrete lytic lesions [8]. In contrast to other tumors that involve bone, myeloma is rarely associated with osteosclerotic lesions except in POEMS syndrome (**P**olyneuropathy, **O**rganomegaly, **E**ndocrinopathy, **M**onoclonal Gammopathy and **S**kin changes), a multisystemic disease occurring rarely in the setting of plasma cell dyscrasias [9]. Skeletal manifestations, particularly osteolytic bone lesions, represent the most prominent source of pain and disability in MM. Bone pain occurs in 60–70% of patients and pathologic fractures in 60% of myeloma patients [10, 11]. Osteolytic lesions most often involve the axial skeleton, skull and femur. In addition, lytic lesions rarely heal, even when complete remission is attained.

Fifteen to 20% of newly diagnosed myeloma patients have hypercalcemia (defined as a corrected serum calcium level greater than 11.5 mg/dL) due to increased bone resorption, decreased bone formation, and impaired renal function, all of which are often exacerbated by immobility. Unlike other malignancies resulting in metastatic disease to the bone, PTHrP is rarely overproduced by myeloma cells. The severity of hypercalcemia in patients with myeloma is not correlated with serum PTHrP levels and instead reflects tumor burden [7]. Symptomatic hypercalcemia can result in anorexia, nausea, vomiting, confusion, fatigue, constipation, renal stones, depression, and polyuria, and is suggestive of a high tumor burden.

The bone marrow (BM) microenvironment in MM also contributes to tumor growth and the bone destructive process. It comprises cellular and extracellular elements including OBLs, OCLs, endothelial cells, immune cells, and MM cells. Interactions between MM cells and their BM microenvironment are tightly regulated. Under normal physiologic conditions, balanced interactions within the BM microenvironment result in coupled bone remodeling (Fig. 84.1). In MM, bone remodeling is uncoupled in the BM microenvironment and is characterized by generalized OCL activation and suppressed OBL function with decreased bone formation. Bone marrow biopsies from MM patients demonstrate a correlation between tumor burden, OCL number, and resorptive surface [1, 12].

MM cells produce or induce multiple osteoclastogenic factors in the bone marrow microenvironment that directly increase OCL formation and activity, and decrease production of osteoprotegerin (OPG), a soluble decoy receptor for receptor activator of nuclear factor-κB ligand (RANKL), a critical differentiation factor for OCLs produced by marrow stromal cells and OBL [13]. Myeloma cells adhere to bone marrow stromal cells via binding of surface VLA-4 ($\alpha_4\beta_1$ integrin) to VCAM-1 expressed on stromal cells, resulting in production of osteoclastogenic cytokines such as RANKL, macrophage colony-stimulating factor (M-CSF), interleukin-11, and interleukin-6 by marrow stromal cells and osteoclastogenic cytokines including macrophage inflammatory protein-1α (MIP-1α) and interleukin-3 by MM cells [14–17]. RANKL increases OCL formation and survival by

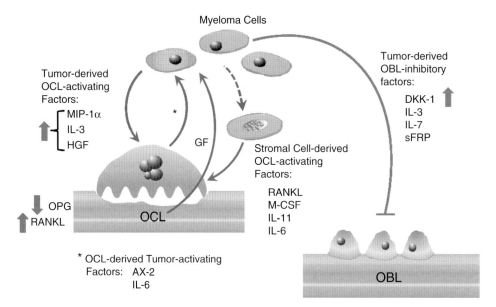

Fig. 84.2. The vicious cycle of myeloma bone disease. Myeloma cells produce factors that directly or indirectly activate osteoclasts, such as MIP-1α, IL-3, and hepatocyte growth factor (HGF). Myeloma cells also induce RANKL and IL-6 production by marrow stromal cells to enhance OCL formation. OCLs in turn produce soluble factors such as annexin II (AXII) and IL-6 that stimulate tumor growth. The bone destructive process also releases growth factors that increase the growth of myeloma cells, further exacerbating the osteolytic process, resulting in a "vicious cycle" of bone destruction. OBL differentiation is suppressed by tumor-derived OBL-inhibitory factors such as DKK-1, IL-3, IL-7, and the secreted frizzled-related proteins (sFRPs). In addition, the RANKL/OPG ratio is increased, promoting OCL development. DKK-1, MIP-1α, and RANKL levels are all increased in myeloma bone disease as compared with the physiologic state.

binding to its receptor RANK on OCL precursor cells and OCL [18], while MIP-1α acts as a chemotactic factor for OCL precursors and can induce differentiation of OCL progenitors contributing to OCL formation [19–21]. MIP-1α also directly promotes the growth, survival and migration of MM cells by inducing the activation of multiple signaling pathways crucial for MM cell growth and survival [22]. Further, the bone destructive process releases growth factors from the bone matrix that increase the proliferation of MM cells. This results in a "vicious cycle," of bone destruction, leading to increased tumor mass and further bone destruction (Fig. 84.2).

While OCLs are activated in MM, OBL activity is suppressed, with decreased bone formation and calcification despite increased bone resorption [1, 23]. As a result, serum alkaline phosphatase and osteocalcin are normal or decreased in patients with myeloma bone involvement. Co-culture experiments have demonstrated reduced myeloma cell proliferation in the presence of OBLs as compared with OCL or marrow stromal cells [24], a finding that has been confirmed in murine models of myeloma bone disease [25].

OBLs also affect myeloma cell growth indirectly via their regulation of OCLs. OBLs secrete RANKL, an OCL differentiation factor, as well as the soluble RANKL inhibitor, OPG. The balance between RANKL and OPG regulates osteoclastogenesis. In myeloma the RANKL/OPG ratio is typically increased, promoting OCL development [26, 27].

Other mechanisms of OBL suppression under investigation include downregulation of the osteogenic transcription factor runt-related transcription factor 2 (RUNX2) via direct cell-to-cell contact between MM and OB progenitor cells [28]; Wnt signaling inhibitors such as dickkopf-1 (DKK-1), soluble frizzled receptor-like protein 3 (sFRP3), and sclerostin that inhibit osteogenesis [29–31]; and modulation of OB differentiation by transforming growth factor-β (TGF-β) superfamily members including bone morphogenic protein 2 (BMP2) [32], activin A[33], and TGF-β itself [34]. DKK-1 is of particular interest as it is highly expressed in the BM of MM patients with bone lesions, appears to be involved in early bone disease, and has roles in the regulation of both OCL and OBL function [35, 36].

At this time, conventional radiography is the gold standard for evaluation of bone disease in myeloma patients. Magnetic resonance imaging (MRI) is also effective in identification of lytic lesions and is indicated for staging in patients with solitary plasmacytoma of the bone [37]. Traditional technetium bone scintigraphy scans underestimate the extent of bone disease and are thus not recommended in patients with MM [38]. The utility of F18-fluorodeoxyglucose positron emission tomography/computed tomography (FDG-PET/CT) has been evaluated in MM and has a sensitivity of approximately 85% and a specificity of approximately 90% [39]. Several studies have recently combined FDG-PET/CT and MRI of the spine and pelvis for detection of active MM. In cases where the findings from the two studies are concordant, the ability to detect sites of active disease was greater than 90% [37].

Management of MM bone disease

Bisphosphonates remain the standard of care for MM-related bone disease at this time. Bisphosphonates are potent inhibitors of osteoclast activity, and intravenous bisphosphonates given every 3 to 4 weeks are the current treatment of choice for management of pain related to bone disease and the prevention of skeletal-related events. Oral clodronate treatment has been shown to reduce the development of osteolytic lesions, fractures, hypercalcemia, and bone pain in MM [40]. The recently published Medical Research Council (MRC) IX trial demonstrated that intravenous zoledronic acid reduced the incidence of skeletal-related events (SREs) (hypercalcemia, new bone lesions, and fractures) as compared with oral clodronate in patients with newly diagnosed MM [41]. In addition, patients treated with zoledronic acid had improved disease response rates and overall survival after a median follow-up of 3.7 years, compared with patients treated with clodronate. This suggests that bisphosphonates have a direct anti-myeloma effect, a hypothesis supported by *in vitro* data [42].

Denosumab, a human monoclonal antibody that binds to RANKL with high affinity and specificity, was approved by the FDA for prevention of SREs in patients with bone metastases from solid tumors in 2010 and is currently under investigation for use in MM bone disease. A recent clinical trial has demonstrated that denosumab inhibits bone resorption and prevents SREs in patients refractory to bisphosphonate therapy [43, 44].

Bortezomib, a proteasome inhibitor active against MM, directly alters osteoblast and osteoclast activity by decreasing RANKL and DKK-1 levels in the sera of myeloma patients [45, 46]. In clinical studies of both newly diagnosed and relapsed myeloma patients, bortezomib therapy, either alone or in combination with other agents, demonstrated improvement in markers of osteoblastic activity and osteoclast inhibition [46–49]. Bortezomib's effects on osteoblast differentiation have been extensively studied. Several clinical trials showed increased bone specific alkaline phosphatase, a marker for osteoblast activation, in myeloma patients whose tumor responded to the drug [49, 50]. Some authors have interpreted these findings as evidence that bortezomib directly stimulates osteoblasts and inhibits osteoclasts. Others have suggested that biochemical markers of bone formation peak after 6 weeks of bortezomib treatment due to a direct inhibitory effect on bone resorption by osteoclasts that counteract bortezomib's initial direct osteoblast stimulatory effect [51]. Alternatively, bortezomib's direct inhibition of myeloma cells in the bone marrow microenvironment allows for normalization of osteoblast and osteoclast function, as these effects are only seen in patients whose myeloma responds to bortezomib treatment.

Other hematologic malignancies

Adult T-Cell leukemia/lymphoma

Adult T-cell leukemia/lymphoma (ATLL) is a malignancy of CD4+ T cells caused by infection with human T-lymphotropic virus type 1(HTLV-1). ATLL was initially reported in southern Japan and has been reported sporadically in areas where HTLV-1 infection is rare, including the United States. The additive lifetime risk of developing ATLL among HTLV-1 carriers varies depending on the population surveyed, but it has been reported to range between 1% and 5% in Japan and Jamaica [52]. Approximately 70% of ATLL patients develop hypercalcemia, and many of these patients also develop lytic bone lesions [3]. In contrast to the hypercalcemia that typically develops in myeloma patients, PTHrP mediates the hypercalcemia associated with ATLL. It is hypothesized that the HTLV-1 and HTLV-II tax proteins transactivate PTHrP via the cellular transcription factors activator protein (AP)-2 and AP-1, a novel mechanism of transactivation [53, 54]. However, increased transcription of PTHrP can also occur in a tax-independent manner [55]. In addition, MIP-1α has been proposed as a mediator for the hypercalcemia in ATLL, by enhancing OCL formation and inducing RANKL expression on ATLL cells in an autocrine fashion [56]. Increased RANKL expression has been noted in ATLL cells in patients with hypercalcemia [57].

ATLL cells produce chemokines that affect bone remodeling, including IL-1, IL-6, tumor necrosis factor-α (TNF-α), and MIP-1α/MIP-1β. Circulating ATLL cells infiltrate a variety of tissues, mediated by MIP-1α induction of integrin-mediated adhesion to the endothelium and subsequent transmigration [53]. As in myeloma, MIP-1α in ATLL is important for the chemotaxis of monocytes, including OCL progenitor cells, and the production of the osteoclastogenic factors IL-6, PTHrP, and RANKL by osteoblasts or stromal cells [58, 59]. In addition, IL-1 and PTHrP have also been reported to mediate bone destruction in ATLL, with elevated PTHrP levels in patients and increased concentrations of IL-1 and PTHrP in media conditioned by ATLL cells *in vitro* [60].

Bone involvement has also been infrequently reported in more classical forms of acute lymphoid leukemia and is thought to be similarly mediated by PTHrP production by malignant cells [61].

Non-Hodgkin's lymphoma

Bone involvement with non-Hodgkin's lymphoma (NHL) is rare. Less than 10% of NHL patients present with bone involvement; however, 7–25% of all patients with NHL eventually develop bone findings during the course of their disease. The most common histologic subtypes of NHL that present with bone manifestations include histiocytic, undifferentiated, and poorly differentiated NHL. In addition, lytic bone lesions are more often seen in patients with diffuse rather than nodular patterns of lymph node involvement, and frequently involve the axial skeleton [62]. As in ATLL, serum levels of PTHrP are elevated in NHL patients with hypercalcemia [63].

Hodgkin's disease

Bone involvement in Hodgkin's disease (HD) is uncommon and seldom encountered at diagnosis. Sites of involvement include the spine, pelvis, femur, humerus, ribs, sternum, scapula, and base of the skull [64, 65]; however, as with NHL, vertebral and femoral involvement is most common [66]. The most frequent presentation is that of a localized, solitary osteoblastic mass in a patient with mixed cellularity, nodular sclerosing disease [64, 65]. Bone biopsies often show fibrosis and a mixed inflammatory infiltrate with rare atypical cells. Radiologic findings include a vertebral sclerotic pattern along with a periosteal reaction and hypertrophic pulmonary osteoarthropathy [66]. As in NHL, radiographic patterns cannot predict the histologic type or the prognosis of HD and must be used with clinical staging to predict prognosis.

Bone disease in patients with HD can be lytic, blastic, or mixed. Increased new bone formation by tumor cell stimulation of osteoblast activity occurs at sites of previous osteoclastic activity. Hypercalcemia does occur in HD and is associated with excess production of $1,25(OH)_2$ vitamin D_3 or PTHrP by the lymphoma cells [66, 67].

SUMMARY

Dysregulation of physiologic bone remodeling in the setting of malignancy results in osteolytic, osteoblastic, or mixed lesions. PTHrP frequently mediates non-myelomatous bone disease. With the exception of multiple myeloma, bone involvement in hematologic malignancies is rare; however, bone lesions can significantly contribute to patient morbidity and pain. Thus, consideration of the potential consequences of skeletal manifestations of hematologic malignancies is an important component of the care of these patients.

REFERENCES

1. Taube T, Beneton MN, McCloskey EV, Rogers S, Greaves M, Kanis JA. 1992. Abnormal bone remodelling in patients with myelomatosis and normal biochemical indices of bone resorption. *Eur J Haematol* 49(4): 192–8.
2. Saad F, Lipton A, Cook R, Chen YM, Smith M, Coleman R. 2007. Pathologic fractures correlate with reduced survival in patients with malignant bone disease. *Cancer* 110(8): 1860–7.
3. Taylor GP, Matsuoka M. 2005. Natural history of adult T-cell leukemia/lymphoma and approaches to therapy. *Oncogene* 24(39): 6047–57.
4. Hideshima T, Anderson KC. 2002. Molecular mechanisms of novel therapeutic approaches for multiple myeloma. *Nat Rev Cancer* 2(12): 927–37.
5. Altekruse SF, Kosary CL, Krapcho M. 2007. *SEER Cancer Statistics Review, 1975–2007*. Bethesda, MD: National Cancer Institute. Accessed on June 21, 2011. Available from http://seer.cancer.gov/csr/1975_2007/index.html.
6. Roodman GD. 2008. Skeletal imaging and management of bone disease. In: Gewirtz AM, Muchmore EA, Burns LJ (eds.) *Hematology Am Soc Hematol Educ Program*. Washington, DC: The American Society of Hematology. pp. 313–7.
7. Roodman GD. 2009. Diagnosis and treatment of myeloma bone disease. In: Rajkumar SV, Kyle RA (eds.) *Treatment of Multiple Myeloma and Related Disorders*. New York: Cambridge University Press. pp. 64–76.
8. Roodman GD. 2004. Pathogenesis of myeloma bone disease. *Blood Cells Mol Dis* 32(2): 290–2.
9. Dispenzieri A. 2011. POEMS syndrome: 2011 update on diagnosis, risk-stratification, and management. *Am J Hematol* 86(7): 591–601.
10. Callander NS, Roodman GD. 2001. Myeloma bone disease. *Semin Hematol* 38(3): 276–85.
11. Melton LJ 3rd, Kyle RA, Achenbach SJ, Oberg AL, Rajkumar SV. 2005. Fracture risk with multiple myeloma: A population-based study. *J Bone Miner Res* 20(3): 487–93.
12. Valentin-Opran A, Charhon SA, Meunier PJ, Edouard CM, Arlot ME. 1982. Quantitative histology of myeloma-induced bone changes. *Br J Haematol* 52(4): 601–10.
13. Pearse RN, Sordillo EM, Yaccoby S, Wong BR, Liau DF, Colman N, Michaeli J, Epstein J, Choi Y. 2001. Multiple myeloma disrupts the TRANCE/osteoprotegerin cytokine axis to trigger bone destruction and promote tumor progression. *Proc Natl Acad Sci U S A* 98(20): 11581–6.
14. Gunn WG, Conley A, Deininger L, Olson SD, Prockop DJ, Gregory CA. 2006. A crosstalk between myeloma cells and marrow stromal cells stimulates production of DKK1 and interleukin-6: A potential role in the development of lytic bone disease and tumor progression in multiple myeloma. *Stem Cells* 24(4): 986–91.
15. Giuliani N, Colla S, Rizzoli V. 2004. New insight in the mechanism of osteoclast activation and formation in multiple myeloma: focus on the receptor activator of NF-kappaB ligand (RANKL). *Exp Hematol* 32(8): 685–91.
16. Choi SJ, Cruz JC, Craig F, Chung H, Devlin RD, Roodman GD, Alsina M. 2000. Macrophage inflammatory protein 1-alpha is a potential osteoclast stimulatory factor in multiple myeloma. *Blood* 96(2): 671–5.
17. Lee JW, Chung HY, Ehrlich LA, Jelinek DF, Callander NS, Roodman GD, Choi SJ. 2004. IL-3 expression by myeloma cells increases both osteoclast formation and growth of myeloma cells. *Blood* 103(6): 2308–15.
18. Ehrlich LA, Roodman GD. 2005. The role of immune cells and inflammatory cytokines in Paget's disease and multiple myeloma. *Immunol Rev* 208: 252–66.

19. Abe M, Hiura K, Wilde J, Moriyama K, Hashimoto T, Ozaki S, Wakatsuki S, Kosaka M, Kido S, Inoue D, Matsumoto T. 2002. Role for macrophage inflammatory protein (MIP)-1alpha and MIP-1beta in the development of osteolytic lesions in multiple myeloma. *Blood* 100(6): 2195–202.
20. Choi SJ, Oba Y, Gazitt Y, Alsina M, Cruz J, Anderson J, Roodman GD. 2001. Antisense inhibition of macrophage inflammatory protein 1-alpha blocks bone destruction in a model of myeloma bone disease. *J Clin Invest* 108(12): 1833–41.
21. Oyajobi BO, Franchin G, Williams PJ, Pulkrabek D, Gupta A, Munoz S, Grubbs B, Zhao M, Chen D, Sherry B, Mundy GR. 2003. Dual effects of macrophage inflammatory protein-1alpha on osteolysis and tumor burden in the murine 5TGM1 model of myeloma bone disease. *Blood* 102(1): 311–9.
22. Lentzsch S, Gries M, Janz M, Bargou R, Dorken B, Mapara MY. 2003. Macrophage inflammatory protein 1-alpha (MIP-1 alpha) triggers migration and signaling cascades mediating survival and proliferation in multiple myeloma (MM) cells. *Blood* 101(9): 3568–73.
23. Bataille R, Chappard D, Marcelli C, Dessauw P, Sany J, Baldet P, Alexandre C. 1989. Mechanisms of bone destruction in multiple myeloma: The importance of an unbalanced process in determining the severity of lytic bone disease. *J Clin Oncol* 7(12): 1909–14.
24. Yaccoby S, Wezeman MJ, Zangari M, Walker R, Cottler-Fox M, Gaddy D, Ling W, Saha R, Barlogie B, Tricot G, Epstein J. 2006. Inhibitory effects of osteoblasts and increased bone formation on myeloma in novel culture systems and a myelomatous mouse model. *Haematologica* 91(2): 192–9.
25. Edwards CM, Edwards JR, Lwin ST, Esparza J, Oyajobi BO, McCluskey B, Munoz S, Grubbs B, Mundy GR. 2008. Increasing Wnt signaling in the bone marrow microenvironment inhibits the development of myeloma bone disease and reduces tumor burden in bone in vivo. *Blood* 111(5): 2833–42.
26. Giuliani N, Bataille R, Mancini C, Lazzaretti M, Barille S. 2001. Myeloma cells induce imbalance in the osteoprotegerin/osteoprotegerin ligand system in the human bone marrow environment. *Blood* 98(13): 3527–33.
27. Qiang YW, Chen Y, Stephens O, Brown N, Chen B, Epstein J, Barlogie B, Shaughnessy JD Jr. 2008. Myeloma-derived Dickkopf-1 disrupts Wnt-regulated osteoprotegerin and RANKL production by osteoblasts: A potential mechanism underlying osteolytic bone lesions in multiple myeloma. *Blood* 112(1): 196–207.
28. Giuliani N, Colla S, Morandi F, Lazzaretti M, Sala R, Bonomini S, Grano M, Colucci S, Svaldi M, Rizzoli V. 2005. Myeloma cells block RUNX2/CBFA1 activity in human bone marrow osteoblast progenitors and inhibit osteoblast formation and differentiation. *Blood* 106(7): 2472–83.
29. Gaur T, Lengner CJ, Hovhannisyan H, Bhat RA, Bodine PV, Komm BS, Javed A, van Wijnen AJ, Stein JL, Stein GS, Lian JB. 2005. Canonical WNT signaling promotes osteogenesis by directly stimulating Runx2 gene expression. *J Biol Chem* 280(39): 33132–40.
30. Takada I, Mihara M, Suzawa M, Ohtake F, Kobayashi S, Igarashi M, Youn MY, Takeyama K, Nakamura T, Mezaki Y, Takezawa S, Yogiashi Y, Kitagawa H, Yamada G, Takada S, Minami Y, Shibuya H, Matsumoto K, Kato S. 2007. A histone lysine methyltransferase activated by non-canonical Wnt signalling suppresses PPAR-gamma transactivation. *Nat Cell Biol* 9(11): 1273–85.
31. Giuliani N, Mangoni M, Rizzoli V. 2009. Osteogenic differentiation of mesenchymal stem cells in multiple myeloma: Identification of potential therapeutic targets. *Exp Hematol* 37(8): 879–86.
32. Ryoo HM, Lee MH, Kim YJ. 2006. Critical molecular switches involved in BMP-2-induced osteogenic differentiation of mesenchymal cells. *Gene* 366(1): 51–7.
33. Vallet S, Mukherjee S, Vaghela N, Hideshima T, Fulciniti M, Pozzi S, Santo L, Cirstea D, Patel K, Sohani AR, Guimaraes A, Xie W, Chauhan D, Schoonmaker JA, Attar E, Churchill M, Weller E, Munshi N, Seehra JS, Weissleder R, Anderson KC, Scadden DT, Raje N. 2010. Activin A promotes multiple myeloma-induced osteolysis and is a promising target for myeloma bone disease. *Proc Natl Acad Sci U S A* 107(11): 5124–9.
34. Lee MH, Kwon TG, Park HS, Wozney JM, Ryoo HM. 2003. BMP-2-induced Osterix expression is mediated by Dlx5 but is independent of Runx2. *Biochem Biophys Res Commun* 309(3): 689–94.
35. Tian E, Zhan F, Walker R, Rasmussen E, Ma Y, Barlogie B, Shaughnessy JD Jr. 2003. The role of the Wnt-signaling antagonist DKK1 in the development of osteolytic lesions in multiple myeloma. *N Engl J Med* 349(26): 2483–94.
36. Raje N, Roodman GD. 2011. Advances in the biology and treatment of bone disease in multiple myeloma. *Clin Cancer Res* 17(6): 1278–86.
37. Terpos E, Moulopoulos LA, Dimopoulos MA. 2011. Advances in imaging and the management of myeloma bone disease. *J Clin Oncol* 29(14): 1907–15.
38. Dimopoulos M, Terpos E, Comenzo RL, Tosi P, Beksac M, Sezer O, Siegel D, Lokhorst H, Kumar S, Rajkumar SV, Niesvizky R, Moulopoulos LA, Durie BG. 2009. International myeloma working group consensus statement and guidelines regarding the current role of imaging techniques in the diagnosis and monitoring of multiple Myeloma. *Leukemia* 23(9): 1545–56.
39. Bredella MA, Steinbach L, Caputo G, Segall G, Hawkins R. 2005. Value of FDG PET in the assessment of patients with multiple myeloma. *AJR Am J Roentgenol* 184(4): 1199–204.
40. McCloskey EV, MacLennan IC, Drayson MT, Chapman C, Dunn J, Kanis JA. 1998. A randomized trial of the effect of clodronate on skeletal morbidity in multiple myeloma. MRC Working Party on Leukaemia in Adults. *Br J Haematol* 100(2): 317–25.
41. Morgan GJ, Davies FE, Gregory WM, Cocks K, Bell SE, Szubert AJ, Navarro-Coy N, Drayson MT, Owen RG, Feyler S, Ashcroft AJ, Ross F, Byrne J, Roddie H, Rudin C, Cook G, Jackson GH, Child JA. 2010. First-line

treatment with zoledronic acid as compared with clodronic acid in multiple myeloma (MRC Myeloma IX): A randomised controlled trial. *Lancet* 376(9757): 1989–99.

42. Aparicio A, Gardner A, Tu Y, Savage A, Berenson J, Lichtenstein A. 1998. In vitro cytoreductive effects on multiple myeloma cells induced by bisphosphonates. *Leukemia* 12(2): 220–9.

43. Fizazi K, Lipton A, Mariette X, Body JJ, Rahim Y, Gralow JR, Gao G, Wu L, Sohn W, Jun S. 2009. Randomized phase II trial of denosumab in patients with bone metastases from prostate cancer, breast cancer, or other neoplasms after intravenous bisphosphonates. *J Clin Oncol* 27(10): 1564–71.

44. Body JJ, Facon T, Coleman RE, Lipton A, Geurs F, Fan M, Holloway D, Peterson MC, Bekker PJ. 2006. A study of the biological receptor activator of nuclear factor-kappaB ligand inhibitor, denosumab, in patients with multiple myeloma or bone metastases from breast cancer. *Clin Cancer Res* 12(4): 1221–8.

45. Terpos E, Heath DJ, Rahemtulla A, Zervas K, Chantry A, Anagnostopoulos A, Pouli A, Katodritou E, Verrou E, Vervessou EC, Dimopoulos MA, Croucher PI. 2006. Bortezomib reduces serum dickkopf-1 and receptor activator of nuclear factor-kappaB ligand concentrations and normalises indices of bone remodelling in patients with relapsed multiple myeloma. *Br J Haematol* 135(5): 688–92.

46. Boissy P, Andersen TL, Lund T, Kupisiewicz K, Plesner T, Delaisse JM. 2008. Pulse treatment with the proteasome inhibitor bortezomib inhibits osteoclast resorptive activity in clinically relevant conditions. *Leuk Res* 32(11): 1661–8.

47. Giuliani N, Morandi F, Tagliaferri S, Lazzaretti M, Bonomini S, Crugnola M, Mancini C, Martella E, Ferrari L, Tabilio A, Rizzoli V. 2007. The proteasome inhibitor bortezomib affects osteoblast differentiation in vitro and in vivo in multiple myeloma patients. *Blood* 110(1): 334–8.

48. Katodritou E, Verrou E, Gastari V, Hadjiaggelidou C, Terpos E, Zervas K. 2008. Response of primary plasma cell leukemia to the combination of bortezomib and dexamethasone: Do specific cytogenetic and immunophenotypic characteristics influence treatment outcome? *Leuk Res* 32(7): 1153–6.

49. Zangari M, Esseltine D, Lee CK, Barlogie B, Elice F, Burns MJ, Kang SH, Yaccoby S, Najarian K, Richardson P, Sonneveld P, Tricot G. 2005. Response to bortezomib is associated to osteoblastic activation in patients with multiple myeloma. *Br J Haematol* 131(1): 71–3.

50. Richardson PG, Sonneveld P, Schuster MW, Irwin D, Stadtmauer EA, Facon T, Harousseau JL, Ben-Yehuda D, Lonial S, Goldschmidt H, Reece D, San-Miguel JF, Blade J, Boccadoro M, Cavenagh J, Dalton WS, Boral AL, Esseltine DL, Porter JB, Schenkein D, Anderson KC. 2005. Bortezomib or high-dose dexamethasone for relapsed multiple myeloma. *N Engl J Med* 352(24): 2487–98.

51. Lund T, Soe K, Abildgaard N, Garnero P, Pedersen PT, Ormstrup T, Delaisse JM, Plesner T. 2010. First-line treatment with bortezomib rapidly stimulates both osteoblast activity and bone matrix deposition in patients with multiple myeloma, and stimulates osteoblast proliferation and differentiation in vitro. *Eur J Haematol* 85(4): 290–9.

52. Gessain A, Mahieux R. 2005. Lymphoproliferations associated with human T-cells leukemia/lymphoma virus type I and type II infection. In: Degos L, Linch DC (eds.) *Textbook of Malignant Hematology*. Oxon: Taylor & Francis. pp. 307–41.

53. Raza S, Naik S, Kancharla VP, Tafera F, Kalavar MR. 2010. Dual-positive (CD4+/CD8+) acute adult T-cell leukemia/lymphoma associated with complex karyotype and refractory hypercalcemia: Case report and literature review. *Case Rep Oncol* 3(3): 489–94.

54. Shu ST, Martin CK, Thudi NK, Dirksen WP, Rosol TJ. 2010. Osteolytic bone resorption in adult T-cell leukemia/lymphoma. *Leuk Lymphoma* 51(4): 702–14.

55. Richard V, Lairmore MD, Green PL, Feuer G, Erbe RS, Albrecht B, D'Souza C, Keller ET, Dai J, Rosol TJ. 2001. Humoral hypercalcemia of malignancy: Severe combined immunodeficient/beige mouse model of adult T-cell lymphoma independent of human T-cell lymphotropic virus type-1 tax expression. *Am J Pathol* 158(6): 2219–28.

56. Okada Y, Tsukada J, Nakano K, Tonai S, Mine S, Tanaka Y. 2004. Macrophage inflammatory protein-1alpha induces hypercalcemia in adult T-cell leukemia. *J Bone Miner Res* 19(7): 1105–11.

57. Nosaka K, Miyamoto T, Sakai T, Mitsuya H, Suda T, Matsuoka M. 2002. Mechanism of hypercalcemia in adult T-cell leukemia: Overexpression of receptor activator of nuclear factor kappaB ligand on adult T-cell leukemia cells. *Blood* 99(2): 634–40.

58. Tanaka Y, Maruo A, Fujii K, Nomi M, Nakamura T, Eto S, Minami Y. 2000. Intercellular adhesion molecule 1 discriminates functionally different populations of human osteoblasts: Characteristic involvement of cell cycle regulators. *J Bone Miner Res* 15(10): 1912–23.

59. Han JH, Choi SJ, Kurihara N, Koide M, Oba Y, Roodman GD. 2001. Macrophage inflammatory protein-1alpha is an osteoclastogenic factor in myeloma that is independent of receptor activator of nuclear factor kappaB ligand. *Blood* 97(11): 3349–53.

60. Roodman GD. 1997. Mechanisms of bone lesions in multiple myeloma and lymphoma. *Cancer* 80(8 Suppl): 1557–63.

61. Inukai T, Hirose K, Inaba T, Kurosawa H, Hama A, Inada H, Chin M, Nagatoshi Y, Ohtsuka Y, Oda M, Goto H, Endo M, Morimoto A, Imaizumi M, Kawamura N, Miyajima Y, Ohtake M, Miyaji R, Saito M, Tawa A, Yanai F, Goi K, Nakazawa S, Sugita K. 2007. Hypercalcemia in childhood acute lymphoblastic leukemia: Frequent implication of parathyroid hormone-related peptide and E2A-HLF from translocation 17;19. *Leukemia* 21(2): 288–96.

62. Pear BL. 1974. Skeletal manifestations of the lymphomas and leukemias. *Semin Roentgenol* 9(3): 229–40.

63. Firkin F, Seymour JF, Watson AM, Grill V, Martin TJ. 1996. Parathyroid hormone-related protein in hypercal-

caemia associated with haematological malignancy. *Br J Haematol* 94(3): 486–92.
64. Ozdemirli M, Mankin HJ, Aisenberg AC, Harris NL. 1996. Hodgkin's disease presenting as a solitary bone tumor. A report of four cases and review of the literature. *Cancer* 77(1): 79–88.
65. Borg MF, Chowdhury AD, Bhoopal S, Benjamin CS. 1993. Bone involvement in Hodgkin's disease. *Australas Radiol* 37(1): 63–6.
66. Franczyk J, Samuels T, Rubenstein J, Srigley J, Morava-Protzner I. 1989. Skeletal lymphoma. *Canadian Assoc Radiol J* 40(2): 75–9.
67. Seymour JF, Gagel RF. 1993. Calcitriol: The major humoral mediator of hypercalcemia in Hodgkin's disease and non-Hodgkin's lymphomas. *Blood* 82(5): 1383–94.

85
Osteogenic Osteosarcoma

Jianning Tao, Yangjin Bae, Lisa L. Wang, and Brendan Lee

Introduction 702
Challenges in the Treatments of Osteosarcoma 703
Understanding of Genetic Factors and Pathways That Cause Human Osteosarcoma 704
Cell of Origin and Cancer Stem Cells in Osteosarcoma 706
Targeted Therapies and Application of Animal Models and Biomarkers 707
Discussion 708
Acknowledgments 708
References 708

INTRODUCTION

Osteogenic sarcoma (also known as osteosarcoma or OS), a well-defined clinical entity, is the most common primary malignant tumor of bone [1, 2]. Nevertheless, it is a rare disease, and only about 900 new cases are diagnosed annually in the United States, accounting for less than 1% of all cancers [3]. Incidence and survival rates have been recently reported based on a study of 3,482 patients with osteosarcoma from the National Cancer Institute's population-based Surveillance, Epidemiology, and End Results (SEER) Program between 1973 and 2004 [1]. Osteosarcoma affects all ages, but its incidence is bimodal, with the first peak in adolescents (8 per million at age group of 15–19 years) and the second peak in the elderly (6 per million at age group of 75–79 years) and a middle lower plateau (approximately 1–2 per million) among individuals ages 25–59 years [1, 4]. Accordingly, osteosarcoma accounts for 5% of pediatric cancers overall (about 400 new cases each year) [5]. There is a high percentage of osteosarcoma with Paget disease and osteosarcoma as a second or later cancer among the elderly [1, 4]. Males are slightly more affected than females (1.2~1.5 to 1) in all age groups. Although osteosarcoma can arise in any bone, it preferentially affects anatomic sites where rapid bone remodeling occurs such as the metaphyses of long bones (distal femur > proximal tibia > proximal humerus) [6]. In children and adolescents, these anatomical regions account for the majority of primary tumors. However, in the elderly, the distribution of anatomic sites is more variable and can include the axial skeleton and skull [1].

Biopsy is required for the diagnosis of osteosarcoma. There are several different histologic subtypes of osteosarcoma including conventional, telangiectatic, small-cell, high-grade surface, secondary, low-grade central, and periosteal and parosteal [7], but the conventional subtype, the most common in childhood and adolescence (the first peak), comprises about 85% of all cases and also has been subdivided on the basis of the predominant features of the cells (i.e., osteoblastic, chondroblastic, fibroblastic types) [8, 9]. Other subtypes comprise the remaining 15% of cases. The grades and stages of osteosarcoma [4, 9] are summarized in Table 85.1. Approximately 20% of patients will have detectable metastatic disease at the time of initial presentation, with the lungs and bones being the most common sites of metastasis. Treatment consists of surgery to remove the primary tumor and intensive chemotherapy to treat micrometastatic disease. In general, the 5-year survival rate for nonmetastatic disease is about 70% in younger patients aged younger than 25 years and approximately 45% in patients aged 60 and above. Patients with distant metastases have much poorer 5-year survival rates, on the order of 30% or less [1, 4]. Strikingly, there have been no substantial improvements in survival rates for either group of patients over the past several decades. Clearly, new treatment strategies and drugs are needed.

Primer on the Metabolic Bone Diseases and Disorders of Mineral Metabolism, Eighth Edition. Edited by Clifford J. Rosen.
© 2013 American Society for Bone and Mineral Research. Published 2013 by John Wiley & Sons, Inc.

Table 85.1. Summary of Grades and Stages of Osteosarcoma

Stage		Tumor	Metastases	Grade
I	IA	≤8 cm	No	Low
	IB	>8 cm	No	Low
II	IIA	≤8 cm	No	High
	IIB	>8 cm	No	High
III		"Skipped" to other sites in the same bone		High
IV	IVA	Any size	Only to lung	Any
	IVB	Any size	To other distant sites	Any

Notes: Grades and stages in this table are simplified according to American Joint Committee on Cancer (AJCC) Staging System. Low grade OS includes well and moderately differentiated, whereas high grade includes poorly and undifferentiated and anaplastic lesions. This system is different from a three-stage system (Stage I: low grade; II: high grade; III: metastatic disease), which is applied by the Musculoskeletal Tumor Society Staging (MSTS) System and Enneking System.

The etiology of osteosarcoma remains largely unknown. However, there has been renewed effort in understanding the molecular biology and pathogenesis of osteosarcoma over the past 2 decades [10]. Recent reviews on studies of familial syndromes, specimens, and cell lines derived from human osteosarcoma patients describe the genetic factors and signaling pathways that may be involved in several key processes in the pathogenesis, including initiation, progression, invasion, and metastasis [2, 5, 11–13]. Several genetic factors have been tested as potential diagnostic and prognostic biomarkers of disease as well as potential therapeutic targets. Based on these findings, novel agents have been tested in several molecularly targeted phase 1 clinical trials, which may lead to second-line treatment options for patients with resistant disease and/or with distant metastases [14]. Most recently, genetically engineered osteosarcoma mouse models have been generated in an attempt to recapitulate the human disease [15]. Understanding these models will broaden our knowledge of the molecular basis of osteosarcoma and will also advance preclinical studies for new therapeutic strategies [16]. This chapter updates the current understanding of osteosarcoma biology and reviews these recent findings from animal models.

CHALLENGES IN THE TREATMENTS OF OSTEOSARCOMA

Current standard treatments of osteosarcoma

Surgery and chemotherapy are two essential components of therapy for osteosarcoma. Unlike other sarcomas such as Ewing sarcoma or rhabdomyosarcoma, osteosarcoma is relatively resistant to radiation therapy. The importance of surgery was described as early as 1879 in a study of 165 cases of sarcoma of the long bones [17]. Surgery mainly consisted of amputation of the affected limb. Prior to the 1970s, before the introduction of chemotherapy for the treatment of osteosarcoma, when amputation was the sole effective treatment, the 5-year survival rates were only 10–20% [8]. Most patients developed distant metastatic disease postoperatively, most often to the lungs, despite complete removal of the tumor [9]. Thus, although only 20% of patients initially present with clinically detectable metastatic disease, as seen by modern imaging technologies, including computerized tomography, bone scintigraphy, and magnetic resonance imaging, virtually all patients already have undetectable micrometastases at the time of diagnosis [8]. After the introduction of adjuvant (postoperative) chemotherapy in the 1970s in addition to surgery, the survival rates increased dramatically, thus establishing the critical role of chemotherapy in the treatment of osteosarcoma [18, 19].

The most effective chemotherapeutic agents currently used to treat osteosarcoma include doxorubicin (or adriamycin), methotrexate, cisplatin, and ifosfamide. In the late 1970s, the concept of neoadjuvant (preoperative) chemotherapy was introduced, which offered several advantages such as early eradication of micrometastases, shrinkage of the bulk of the tumor (making surgery more feasible), and, importantly, the ability to determine the degree of tumor necrosis at the time of definitive resection [20]. The percentage of tumor necrosis (or histologic response) has been found to be a prognostic factor, with greater than 90% tumor necrosis being considered a good response and favorable prognostic factor [9]. In addition to assessment of histologic response, administration of neoadjuvant chemotherapy allows time for the orthopedic surgeon to plan for limb salvage surgery [8]. Other agents that have shown to have some response in osteosarcoma include cyclophosphamide, vincristine, melphalan, decarbazine, bleomycin, dactinomycin, and actinomycin, but their effects are more controversial since the 5-year survival rates in the past 2 decades have not significantly improved by adding them into the standard regimens [8, 21].

While chemotherapy has been important for survival in osteosarcoma, surgery is still a mainstay of therapy and is essential for survival [22]. Complete resection of tumors is difficult in some cases depending on the tumor location, for example, in the spine and pelvic bones, where complete resection is often difficult and the risk of local recurrence is high. A dismal outcome was reported for those patients with pelvic osteosarcoma where the 5-year survival rate was only about 19% [23, 24]. A major challenge is curing patients who are not eligible for metastasectomy; in those cases, less effective radiotherapy and palliative chemotherapy can be applied in the treatments [8]. Thus, despite improvements in survival from use of intensive chemotherapy and surgical regimens, there is a continued need for new therapeutic approaches.

Limitations of current standard treatments

Despite the success of current treatments, about 40% of all patients still relapse, mostly within 2 years, and half of them die in less than 5 years [2, 8]. This is attributable in large part to chemoresistance of tumor cells, which is a major challenge to current therapies. Another limitation to current treatment regimens is the toxicity associated with chemotherapy. Chemotherapeutic drugs kill not only tumor cells but normal tissues as well, causing major renal, hematologic, and cardiac toxicities [2]. Some of these toxicities occur during administration of the drugs, while others, such as cardiotoxicity from doxorubicin, can occur many years later. The recent Childhood Cancer Survivor Study (CCSS) performed on 733 long-term survivors (more than 5 years) of childhood cancer with a mean follow-up of 21.6 years, showed that 86.9% of osteosarcoma survivors experienced at least one chronic medical condition such as tinnitus and deafness. More than 50% of them experienced adverse health effects such as physical limitation, pain, and anxiety. Prospective evaluation of survivors will be important to assess both acute and long-term effects of current treatments and their impact on survivorship [25]. Another potential late effect of chemotherapy is secondary malignancy, particularly with alkylating agents such as ifosfamide. The majority of secondary malignant neoplasms such as breast, thyroid, and gastrointestinal cancers occurred around 10 years from diagnosis, and their incidence is 3–5% [25, 26].

Given the problems of chemoresistance, organ toxicity, and secondary malignancies associated with medical treatment for osteosarcoma, researchers have investigated other therapeutic avenues. Two recent pilot clinical trials have reported a modest increase of 6-year overall survival rate from 70% to 78% by the addition of the immune-enhancing drug muramyl tripeptide and an increase of 5-year overall survival rate up to 93% in treatments of localized disease with addition of the anti-osteoclast drug pamidronate to the standard 3-drug regimen [21, 27]. Another approach for the treatment of osteosarcoma in addition to novel drugs is the use of immunotherapy to eradicate tumor cells that express specific antigens recognized by the host immune system. A recent preclinical immunotherapy study showed that genetically modified T cells can recognize low levels of tumor antigen expression of Her2 and kill osteosarcoma cells [28]. However, further investigation is warranted before these new therapies can become incorporated into standard treatments.

UNDERSTANDING OF GENETIC FACTORS AND PATHWAYS THAT CAUSE HUMAN OSTEOSARCOMA

Mutations in familial syndromes

Germline mutations in *p53*, *Rb*, *RECQL4*, and *REQCL2* genes cause Li-Fraumeni, hereditary retinoblastoma, Rothmund-Thomson, and Werner syndromes, respectively, and all of these syndromes are predisposed to osteosarcoma [5]. Li-Fraumeni syndrome displays a wide spectrum of cancers, including breast, adrenocortical, brain tumors, leukemias, and osteosarcomas. P53 protein regulates cell cycle, DNA repair, and apoptosis. More than 50% of sporadic OS cases contain p53 gene mutations (point, missense, rearrangement, allelic loss) [10]. Inactivation of p53 in mice leads to osteosarcoma [29]. Along with p53, Rb protein is also a well-defined tumor suppressor in the pathogenesis of osteosarcoma [10]. Rb protein regulates the G1/S transition. About 70% of sporadic OS cases contain a genetic alteration of Rb, but few point mutations have been found. The inactivation of Rb alone in mice does not cause osteosarcoma, so it is likely that Rb is an enhancer during osteosarcomagenesis [15]. RECQL proteins have a role in maintaining genomic integrity. While less is known about the direct role of these proteins in osteosarcomagenesis, Rothmund-Thomson syndrome patients with *RECQL4* mutations have an extremely high rate of osteosarcoma as well as skeletal dysplasias [5].

Alterations of tumor suppressors in osteosarcoma

In the p53 pathway, p53 directly activates p21, a cyclin-dependent kinase (CDK) inhibitor that inhibits the activity of cyclin D-CDK4/6 or cyclin E-CDK2, thereby decreasing Rb phosphorylation, which leads to cell cycle arrest in the G1 phase [5]. P53 protein is degraded by the proteasome after binding with MDM2, an E3 ubiquitin ligase, which is negatively regulated by p14^{ARF}; therefore, p14^{ARF} acts as a tumor suppressor. In the Rb pathway, a complex of cyclin D1 and CDK4/6, which is inhibited by p16^{INK4a}, phosphorylates Rb followed by inhibition of E2F, a transcription factor that promotes DNA synthesis and cell cycle transition from G1 to S. The p16^{INK4a} protein also serves as a tumor suppressor. The p16^{INK4a} and p14^{ARF} proteins are encoded by the INK4a/ARF locus (also called CDKN2a) located on chromosome 9p21. All genes in the p53 and Rb pathways have been frequently found to be either deleted, amplified, or mutated in sporadic osteosarcoma tumors [11]. Therefore, alterations in any of the genes in these tumor suppressor pathways may contribute to osteosarcoma tumorigenesis.

Activation of oncogenes in osteosarcoma

Several oncogenes have been associated with developing osteosarcoma [5, 11]. They include *c-MYC*, *c-FOS*, *SAS*, *GLI*, *MET*, and *ERBB2*. Since MDM2, cyclin D1 and CDK4 can inhibit p53 and Rb tumor suppressor pathways, they possibly play oncogenic roles during development and progression of osteosarcoma. The nuclear transcription factor c-MYC regulates DNA replication and cell growth. JUN and c-FOS form a heterodimeric transcription complex constituting AP1 that regulates genes involved in cell growth, differentiation, transformation, and bone metabolism [30]. The c-MYC locus has been reported to be amplified up to 12% of osteosarcoma tumors. Hitherto, overexpression of either c-MYC or c-FOS in transgenic mouse models resulted in osteosarcoma development, suggesting their potential role in tumor initiation [31, 32]. However, whether any of these proteins alone can cause osteosarcoma is not yet certain.

Cytogenetic alterations

Cytogenetic abnormalities highlight the inherent complexity and instability of osteosarcomas. About 70% of OS tumors display a multitude of cytogenetic abnormalities, which vary between individuals [5]. Haploidy and up to near-hexaploidy have been found in osteosarcomas. Rearrangements can involve 1p11–p13, 1q11–q12, 1q21–q22, 11p14–p15, 14p11–p13, 15p11–p13, 17p, and 19q13. The gain of chromosome 1 and the loss of chromosomes 9, 10, 13, and 17 are most common overall. Less frequently involved chromosomal regions are 13q14 (locus of RB1), 12p12–pter (locus of KRAS), 6q11–q4, and 8p23 [30]. The most frequently detected amplifications include chromosomal regions 6p12–p21 (28%), 17p11.2 (32%), and 12q13–q14 (8%). Several other recurrent chromosomal losses (2q, 3p, 9, 10p, 12q, 13q, 14q, 15q, 16, 17p, and 18q) and chromosomal gains (Xp, Xq, 5q, 6p, 8q, 17p, and 20q) have also been identified as well as several recurrent breakpoint clusters and nonrecurrent reciprocal translocations. Identification of specific chromosomal regions perturbed in osteosarcoma will allow future detailed investigations of the affected regions for potential candidate genes that may play a role in the pathogenesis of osteosarcoma.

Altered signaling pathways contributing to the osteosarcomagenesis

Recently, several evolutionarily conserved signaling pathways have been linked with the pathogenesis of osteosarcoma. They include Wnt (wingless-type MMTV integration site), Notch, TGF/BMPs (transforming growth factor/bone morphogenetic proteins), Shh (Sonic hedgehog) and GFs (growth factors) pathways. Thus far, the Wnt and Notch pathways have been the most extensively studied. Aberrant activation of Wnt signaling is associated with many common cancers in human patients [30]. Elevated levels of cytoplasmic and/or nuclear localized β-catenin, a critical mediator of the canonical Wnt pathway, have been detected in the majority of osteosarcoma tumors, and sporadic mutations of β-catenin have also been identified among the tumors [11]. Ectopic expression of the Wnt antagonist Dkk3 suppresses invasion and motility of osteosarcoma cell lines [33]. Inactivation of Wif1, a secreted Wnt antagonist, increases β-catenin levels and accelerates development of osteosarcomas in mice [34]. Altered Notch signaling has been associated with several human cancers suggesting that Notch can act both as an oncogene and tumor suppressor gene depending on its level and temporal expression [35]. In osteosarcomas, activation of Notch signaling contributes to invasiveness and metastatic potential, and inhibition of this pathway may provide a therapeutic approach for the treatment of osteosarcomas [36–38]. In other signaling pathways such as TGFb1, TGFb2, TGFb3, BMPs, BMPRs, HGF, GLI1, FGFR2, IGF1R, and VEGF, abnormal levels of expression of the various pathway members have been demonstrated in osteosarcoma. Thus far, studies of these pathways in the pathogenesis of osteosarcoma have been limited, and further investigations are warranted.

MicroRNAs and osteosarcoma

Recent studies have shown that microRNAs (miRs) play important roles in tumorigenesis, and several miRs such as miR-34s, miR-140, miR-143, and miR-199 have been found to be dysregulated in osteosarcoma [39–42]. MiRs are non-coding RNAs of 19–25 nucleotides that mediate post-transcriptional silencing of specific target mRNAs. The miR-34 family has been identified as a direct target of p53 by p53-ChiP analysis. MiR-34 shares a similar function as p53 by inducing apoptosis, cell cycle arrest, and senescence. Ectopic expression of miR-34 partially induced cell cycle arrest and apoptosis in OS cell lines [40]. Further analysis of miR-34 genes showed the epigenetic silencing of their promoter in primary OS samples, and the miR-34 genes underwent minimal deletions in primary OS samples. These genetic alterations of miR-34 genes are associated with decreased expression of miR-34s in OS samples [40]. In the miR-140 study, expression of miR-140 was associated with chemosensitivity in osteosarcoma xenografts. Tumor cells ectopically transfected with miR-140 were more resistant to methotrexate and 5-fluorouracil (5-FU). Overexpression of miR-140 inhibited cell proliferation in osteosarcoma cell line U-2 OS (wild type p53), so miR-140 may be a candidate for developing a therapeutic strategy to overcome chemoresistance [41]. MiR-143 was found to be downregulated in osteosarcoma cell lines and primary tumor samples, and restoration of miR-143 promoted cell apoptosis and suppressed tumorigenesis by targeting Bcl-2, an antiapoptotic factor [43]. Interestingly, it was demonstrated that downregulation of miR-143 correlated with pulmonary metastasis of human osteosarcoma cells by cellular invasion, probably through elevated expression of MMP13 [42]. Also, miR-199 expression was altered in osteosarcoma cell lines and tumor tissues [39]. Ectopic expression of miR-199 precursors in osteosarcoma cell lines significantly decreased cell growth and migration by inhibition of G1/S cell cycle arrest. It has been suggested in various studies that miRNAs can be potential therapeutic targets for osteosarcoma. However, further studies are needed to dissect the molecular mechanism of miRNAs in osteosarcoma, their efficacy for treatment, and delivery of miRNAs to tumor cells.

CELL OF ORIGIN AND CANCER STEM CELLS IN OSTEOSARCOMA

Cell of origin

Mesenchymal stem cells (MSCs) reside in bone marrow and can differentiate into osteogenic, chondrogenic, adipogenic, neurogenic, or myogenic lineages [11]. During osteogenic differentiation of MSCs, the balance between proliferation and differentiation is tightly regulated in each sequential stage from preosteoblast to immature and mature osteoblasts and to terminally differentiated osteocytes [35] (Fig. 85.1). Dysregulation of this balance such as an increase in proliferation or the blocking of differentiation at early stages can lead to propensity to bone cancer [10, 11]. It remains unknown which cells along this differentiation pathway can undergo malignant transformation to osteosarcoma, and it is also unknown where this occurs at the tissue level. In a rat model, it has been shown that the transition from premalignancy to malignancy might occur in the peritrabecular regions of the metaphysis [44].

Recent studies using mouse models have suggested that the cells of origin in osteosarcoma are MSC-derived bone-forming cells. Osteosarcoma premalignant cells may arise from a MSC that acquires patterns of pathological osteoblastic differentiation during tumor progression [45]. Similarly, they may arise from either the preosteoblast or immature osteoblast that has proliferative capacity [15, 45]. In each case, these premalignant cells that initially acquire a mutation in different stages of osteoblast differentiation may transform or evolve into cancer stem cells (CSCs) or tumor initiating cells (TICs) through accumulation of additional genetic mutations during the tumor progression (Fig. 85.1). Simultaneously, the acquisition and loss of differentiation properties can be seen in the CSCs.

Cancer stem cells in osteosarcoma

The putative CSCs of osteosarcoma are able to undergo not only osteogenic but also chondrogenic and adipogenic lineage differentiation in response to appropriate environmental cues [15]. Osteosarcomas with various

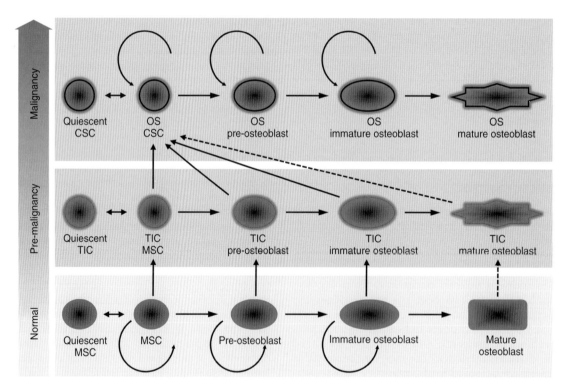

Fig. 85.1. Schematic working model for cell of origin in osteosarcoma formation. Osteogenic differentiation of the mesenchymal stem cell (MSC) occurs through multiple still poorly defined stages (preosteoblast, immature, and mature osteoblast) and culminates in the terminally differentiated osteocyte (not shown in the figure). Osteosarcoma (OS) premalignant cells, sometimes called tumor initiating cells (TICs), can arise not only from a mesenchymal stem cell, but potentially from either preosteoblast or immature osteoblast that have proliferative capacity. The potential dedifferentiation of fibroblasts into pluripotent progenitors opens the possibility that dedifferentiation of mature osteoblasts may lead to a cancer stem cell (CSC) or OS (dashed line), but this currently remains unknown. Next, the premalignant cells can transform into cancer stem cells, which may be able to undergo not only osteogenic, but also chondrogenic and adipogenic lineage differentiation in response to environmental cues.

differentiation patterns are traditionally referred as histologic subtypes. It is well recognized that many of them have mixed patterns. It is also known that the histologic subtype does not impact on chemotherapy response or outcome; therefore, patients are treated identically irrespective of subtype. This suggests that these various differentiation patterns are reflective of a single clinical disease [10].

In our model (Fig. 85.1), we propose that human osteosarcoma can originate from the various stages of osteogenic lineage differentiation from MSCs. Those premalignant cells exist at different stages of differentiation and can transform into CSCs, which can either become quiescent stem cells or differentiate pathologically into various subpopulations cells at each differentiation stage. It is thus far unclear which pools serve as the CSCs or TICs through their signature-genes clusters or surface markers. A recent study has shown that TICs may be a type of subpopulation cells that can retain dye PKH26 [46]. Thus, identifying the cell of origin and the CSCs of osteosarcoma is likely to be of critical clinical importance, especially with respect to development of molecularly targeted therapies and definition of biomarkers for diagnosis and prognosis.

TARGETED THERAPIES AND APPLICATION OF ANIMAL MODELS AND BIOMARKERS

Promise of molecularly targeted therapies in osteosarcoma patients and application of animal models

Several phase 1/2 studies have recently been conducted testing new agents in pediatric patients and have demonstrated objective responses in osteosarcoma [14]. For example, Cediranib, an orally bioavailable small molecule that potently inhibits the tyrosine kinase activity of vascular endothelial growth factor receptor 1 (VEGFR-1; Flt-1), VEGFR-2 (KDR), and VEGFR-3 (Flt-4) showed response in osteosarcoma patients with pulmonary metastases in a pediatric phase 1 trial. Another new agent is Rexin-G, a pathotropic nanoparticle incorporating a collagen matrix binding motif on its surface and bearing a dominant negative cyclin G1 construct that inhibits the cell cycle in the G1 phase. In a phase 2 study of 20 chemotherapy-resistant osteosarcoma patients, Rexin-G resulted in stable disease in 10 out of 17 patients treated [47]. There are several other currently available novel targeted agents that might potentially confer benefit for osteosarcoma patients enrolled in in targeted phase 1 or 2 trials. They include p53/MDM2 antagonists, pro-apoptotic receptor agonists, Notch inhibitors, HER2/neu antagonists, c-MET inhibitors, and IGF1R receptor antagonists [48]. Other promising novel agents include RANKL inhibitors, KIT inhibitors, mTOR inhibitors, vascular-disrupting agents, and Hsp90 inhibitors (IPI-504) [14]. Generally, all of the future targeted therapies should be able to inhibit key pathways that contribute to tumorigenesis and cell survival, and clinical trials using this approach will ultimately yield the necessary answers about efficacy. On the other hand, application of animal models [13, 15, 44, 45, 49–52] may serve as an efficient approach to identifying novel agents for phase 1 or 2 trials (Table 85.2).

Biomarkers for diagnosis and prognosis

The current challenge in application of targeted agents lies in the identification of biomarkers predictive of

Table 85.2. Summary of Osteosarcoma Animal Models

	Mouse	Rat	Dog
Approaches to generate models	Genetically engineered mice. Xenograft or allograft into immunocompromised mice	Administration of PTH 1–34 or radioactive agents	Spontaneous development
Etiology	Known	Unknown	Unknown
Penetrance	High, 60–100% depending on the Cre line used	High	Low, ~8,000 dogs per year (U.S.)
Similarity to human OS	High. OSs share common features of human OS, such as karyotypic complexity, gene signatures, histology, and metastatic potential	Only a small proportion of human OS are radiation induced	Very high. OSs share common features of human OS, high grade, metastasis to the lungs
Advantages	Amenable to genetic modeling and testing novel therapeutic approaches in xenograft models	Well-established model for studying physiology	Large size, intact immune system, amenable to modeling human treatments
Limitations	Small body size, difficult to develop devices or treatments applicable to humans	Relevance of these tumors to the human is low	Small number of cases. Mean age is about 7 years old

response/resistance and matching them with the particular histologic response of individual patients, hopefully resulting in successful translation of biology into clinical benefit [14]. Other than the presence of metastatic disease, definitive prognostic and diagnostic markers for osteosarcoma are limited. There are only a few available common laboratory tests biomarkers. For example, elevations in lactate dehydrogenase (LDH) and alkaline phosphatase (ALK) occur in 30–40% of cases and in some studies have been found to be prognostic; however, they are not used clinically for either stratification of patients or treatment decisions [9]. Recent studies have reported many potential biomarkers for diagnosis and prognosis, which include CTGF, osteocalcin, OPN, P27KIP1, LRP5, CDKN2A/p16, cyclin E1, RB1, FOS, MDM2, cyclin D1, telomerase, platelet-derived growth factor receptor, c-MYC, Her-2, S100A6, RUNX2, membrane-type matrix metalloproteinase type 1, Fas, CXCR4, CXCL6, Twist, P-glycoprotein expression, Snail2, TIMP1, CXCR4, FAS, Annexin2, and Ezrin [5, 11, 30]. However, each of these potential biomarkers was reported in small series, few have been tested prospectively, and none has yet become standard of care. Recently, a multigene classifier has been developed based on microarray technology to predict the response of osteosarcoma to preoperative chemotherapy at the time of diagnosis. Forty-five genes that could distinguish good and poor responders to primary chemotherapy were identified [8]. It is expected that, over the coming years, the new molecular markers identified through high-throughput approaches such as microarray and proteomics may be able not only to prognosticate osteosarcoma patients but also to serve as therapeutic targets and thereby further improve survival rates.

DISCUSSION

The challenges that lie ahead in osteosarcoma biology are to understand the functional interactions of the different genetic factors and pathways and how they intersect with each other during tumor initiation, progression, and metastasis. For example, mutations in tumor suppressor genes p53 and Rb are considered an initiation event in tumorigenesis. Further mutations in other tumor suppressors (e.g., the RECQ helicases) may lead to multiple genetic alterations due to loss of genomic integrity. However, p53 and Rb can also function in the maintenance of genomic stability and as such play multiple roles in limiting the process of tumorigenesis. Perhaps after those critical initiating events, other subsequent events will follow, such as alterations in oncogenes and/or signaling pathways, which may aid in tumor formation. Once established, other factors including additional oncogene overexpression and increased signaling through classical developmental pathways (such as Notch or Wnt), can then play a role in progression and metastasis [2, 5]. Dissecting the various interactions between pathways involved in cell cycle regulation, DNA metabolism, and maintenance of genomic integrity, and understanding the role of oncogenes, signaling and drug resistance pathways, will eventually provide a rational strategy for devising specific therapies that target the pathways leading to osteosarcoma.

ACKNOWLEDGMENTS

The authors are supported in part by grants from the NIH (R03-AR061565, T32-AI053831, R01-AR059063, P01-HD22657), the Cancer Prevention and Research Initiative of Texas (CPRIT grant RP101017), and the Howard Hughes Medical Institute.

REFERENCES

1. Mirabello L, Troisi RJ, Savage SA. 2009. Osteosarcoma incidence and survival rates from 1973 to 2004. *Cancer* 115(7): 1531–43.
2. Gorlick R, Khanna C. 2010. Osteosarcoma. *J Bone Miner Res* 25(4): 683–91.
3. Gurney J, Swensen A, Bulterys M. 1999. Malignant bone tumors. In: Ries L, Smith M, Gurney J, Linet M, Tamra T, Young J, Bunin G (eds.) *Cancer Incidence and Survival Among Children and Adolescents: United States SEER Program 1975–1995, National Cancer Institute, SEER Program.* NIH Pub. No. 99–4649, Bethesda, MD. pp. 99–110.
4. Jawad M, Cheung M, Clarke J, Koniaris L, Scully S. 2011. Osteosarcoma: Improvement in survival limited to high-grade patients only. *J Cancer Res Clin Oncol* 137(4): 597–607.
5. Wang LL. 2005. Biology of osteogenic sarcoma. *Cancer J* 11(4): 294–305.
6. Unni KK, Inwards CY. 2009. *Dahlin's Bone Tumors: General Aspects and Data on 10,165 Cases,* 6th Ed. Philadelphia: Lippincott Williams & Wilkins. pp. 416.
7. Yarmish G, Klein MJ, Landa J, Lefkowitz RA, Hwang S. 2010. Imaging characteristics of primary osteosarcoma: Nonconventional subtypes. *Radiographics* 30(6): 1653–72.
8. Ta HT, Dass CR, Choong PF, Dunstan DE. 2009. Osteosarcoma treatment: State of the art. *Cancer Metastasis Rev* 28(1–2): 247–63.
9. Kim HJ, Chalmers PN, Morris CD. 2010. Pediatric osteogenic sarcoma. *Curr Opin Pediatr* 22(1): 61–6.
10. Gorlick R. 2009. Current concepts on the molecular biology of osteosarcoma. *Cancer Treat Res* 152: 467–78.
11. Wagner ER, Luther G, Zhu G, Luo Q, Shi Q, Kim SH, Gao JL, Huang E, Gao Y, Yang K, Wang L, Teven C, Luo X, Liu X, Li M, Hu N, Su Y, Bi Y, He BC, Tang N, Luo J, Chen L, Zuo G, Rames R, Haydon RC, Luu HH, He TC. 2011. Defective osteogenic differentiation in the development of osteosarcoma. *Sarcoma* 2011: 325238.

12. Kansara M, Thomas DM. 2007. Molecular pathogenesis of osteosarcoma. *DNA Cell Biol* 26(1): 1–18.
13. Hock JM, Lau CC. 2009. Osteogenic osteosarcoma. In: *Primer on the Metabolic Bone Diseases and Disorders of Mineral Metabolism*. Hoboken: John Wiley & Sons, Inc. Chapter 81, pp. 382–5.
14. Subbiah V, Kurzrock R. 2011. Phase 1 clinical trials for sarcomas: The cutting edge. *Curr Opin Oncol* 23(4): 352–60.
15. Walkley CR, Qudsi R, Sankaran VG, Perry JA, Gostissa M, Roth SI, Rodda SJ, Snay E, Dunning P, Fahey FH, Alt FW, McMahon AP, Orkin SH. 2008. Conditional mouse osteosarcoma, dependent on p53 loss and potentiated by loss of Rb, mimics the human disease. *Genes Dev* 22(12): 1662–76.
16. Janeway KA, Walkley CR. 2010. Modeling human osteosarcoma in the mouse: From bedside to bench. *Bone* 47(5): 859–65.
17. Gross SWAM, MD. 1879. Sarcoma of the Long Bones; Based upon a Study of One Hundred and Sixty-five Cases. *Am J Med Sci* 78(155): 17–57; 338–77.
18. Cores EP, Holland JF, Wang JJ, Sinks LF. 1972. Doxorubicin in disseminated osteosarcoma. *JAMA* 221(10): 1132–8.
19. Jaffe N, Paed D, Farber S, Traggis D, Geiser C, Kim BS, Das L, Frauenberger G, Djerassi I, Cassady JR. 1973. Favorable response of metastatic osteogenic sarcoma to pulse high-dose methotrexate with citrovorum rescue and radiation therapy. *Cancer* 31(6): 1367–73.
20. Rosen G, Caparros B, Huvos AG, Kosloff C, Nirenberg A, Cacavio A, Marcove RC, Lane JM, Mehta B, Urban C. 1982. Preoperative chemotherapy for osteogenic sarcoma: Selection of postoperative adjuvant chemotherapy based on the response of the primary tumor to preoperative chemotherapy. *Cancer* 49(6): 1221–30.
21. Meyers PA, Schwartz CL, Krailo MD, Healey JH, Bernstein ML, Betcher D, Ferguson WS, Gebhardt MC, Goorin AM, Harris M, Kleinerman E, Link MP, Nadel H, Nieder M, Siegal GP, Weiner MA, Wells RJ, Womer RB, Grier HE. 2008. Osteosarcoma: The addition of muramyl tripeptide to chemotherapy improves overall survival—A report from the Children's Oncology Group. *J Clin Oncol* 26(4): 633–8.
22. Jaffe N, Carrasco H, Raymond K, Ayala A, Eftekhari F. 2002. Can cure in patients with osteosarcoma be achieved exclusively with chemotherapy and abrogation of surgery? *Cancer* 95(10): 2202–10.
23. Jawad MU, Haleem AA, Scully SP. 2011. Malignant sarcoma of the pelvic bones. *Cancer* 117(7): 1529–41.
24. Saab R, Rao BN, Rodriguez-Galindo C, Billups CA, Fortenberry TN, Daw NC. 2005. Osteosarcoma of the pelvis in children and young adults: The St. Jude Children's Research Hospital experience. *Cancer* 103(7): 1468–74.
25. Nagarajan R, Kamruzzaman A, Ness KK, Marchese VG, Sklar C, Mertens A, Yasui Y, Robison LL, Marina N. 2011. Twenty years of follow-up of survivors of childhood osteosarcoma. *Cancer* 117(3): 625–34.
26. Goldsby R, Burke C, Nagarajan R, Zhou T, Chen Z, Marina N, Friedman D, Neglia J, Chuba P, Bhatia S. 2008. Second solid malignancies among children, adolescents, and young adults diagnosed with malignant bone tumors after 1976. *Cancer* 113(9): 2597–604.
27. Meyers PA, Healey JH, Chou AJ, Wexler LH, Merola PR, Morris CD, Laquaglia MP, Kellick MG, Abramson SJ, Gorlick R. 2011. Addition of pamidronate to chemotherapy for the treatment of osteosarcoma. *Cancer* 117(8): 1736–44.
28. Ahmed N, Salsman VS, Yvon E, Louis CU, Perlaky L, Wels WS, Dishop MK, Kleinerman EE, Pule M, Rooney CM, Heslop HE, Gottschalk S. 2009. Immunotherapy for osteosarcoma: Genetic modification of T cells overcomes low levels of tumor antigen expression. *Mol Ther* 17(10): 1779–87.
29. Donehower LA, Harvey M, Slagle BL, McArthur MJ, Montgomery CA, Butel JS, Allan B. 1992. Mice deficient for p53 are developmentally normal but susceptible to spontaneous tumours. *Nature* 356(6366): 215–21.
30. Tang N, Song WX, Luo J, Haydon RC, He TC. 2008. Osteosarcoma development and stem cell differentiation. *Clin Orthop Relat Res* 466(9): 2114–30.
31. Jain M, Arvanitis C, Chu K, Dewey W, Leonhardt E, Trinh M, Sundberg CD, Bishop JM, Felsher DW. 2002. Sustained loss of a neoplastic phenotype by brief inactivation of MYC. *Science* 297(5578): 102–4.
32. Wang ZQ, Liang J, Schellander K, Wagner EF, Grigoriadis AE. 1995. c-fos-induced osteosarcoma formation in transgenic mice: Cooperativity with c-jun and the role of endogenous c-fos. *Cancer Res* 55(24): 6244–51.
33. Hoang BH, Kubo T, Healey JH, Yang R, Nathan SS, Kolb EA, Mazza B, Meyers PA, Gorlick R. 2004. Dickkopf 3 inhibits invasion and motility of Saos-2 osteosarcoma cells by modulating the Wnt-beta-catenin pathway. *Cancer Res* 64(8): 2734–9.
34. Kansara M, Tsang M, Kodjabachian L, Sims NA, Trivett MK, Ehrich M, Dobrovic A, Slavin J, Choong PFM, Simmons PJ, Dawid IB, Thomas DM. 2009. Wnt inhibitory factor 1 is epigenetically silenced in human osteosarcoma, and targeted disruption accelerates osteosarcomagenesis in mice. *J Clin Invest* 119(4): 837–51.
35. Tao J, Chen S, Lee B. 2010. Alteration of Notch signaling in skeletal development and disease. *Ann N Y Acad Sci* 1192: 257–68.
36. Zhang P, Yang Y, Zweidler-McKay PA, Hughes DP. 2008. Critical role of notch signaling in osteosarcoma invasion and metastasis. *Clin Cancer Res* 14(10): 2962–9.
37. Tanaka M, Setoguchi T, Hirotsu M, Gao H, Sasaki H, Matsunoshita Y, Komiya S. 2009. Inhibition of Notch pathway prevents osteosarcoma growth by cell cycle regulation. *Br J Cancer* 100(12): 1957–65.
38. Engin F, Bertin T, Ma O, Jiang MM, Wang L, Sutton RE, Donehower LA, Lee B. 2009. Notch signaling contributes to the pathogenesis of human osteosarcomas. *Hum Mol Genet* 18(8): 1464–70.
39. Duan Z, Choy E, Harmon D, Liu X, Susa M, Mankin H, Hornicek FJ. 2011. MicroRNA-199a-3p is downregulated in human osteosarcoma and regulates cell proliferation and migration. *Mol Cancer Ther* 10(8): 1337–45.

40. He C, Xiong J, Xu X, Lu W, Liu L, Xiao D, Wang D. 2009. Functional elucidation of MiR-34 in osteosarcoma cells and primary tumor samples. *Biochem Biophys Res Commun* 388(1): 35–40.
41. Song B, Wang Y, Xi Y, Kudo K, Bruheim S, Botchkina GI, Gavin E, Wan Y, Formentini A, Kornmann M, Fodstad O, Ju J. 2009. Mechanism of chemoresistance mediated by miR-140 in human osteosarcoma and colon cancer cells. *Oncogene* 28(46): 4065–74.
42. Osaki M, Takeshita F, Sugimoto Y, Kosaka N, Yamamoto Y, Yoshioka Y, Kobayashi E, Yamada T, Kawai A, Inoue T, Ito H, Oshimura M, Ochiya T. 2011. MicroRNA-143 regulates human osteosarcoma metastasis by regulating matrix metalloprotease-13 expression. *Mol Ther* 19(6): 1123–30.
43. Zhang H, Cai X, Wang Y, Tang H, Tong D, Ji F. 2010. MicroRNA-143, downregulated in osteosarcoma, promotes apoptosis and suppresses tumorigenicity by targeting Bcl-2. *Oncol Rep* 24(5): 1363–9.
44. Bensted JPM, Blackett NM, Lamerton LF. 1961. Histological and dosimetric considerations of bone tumour production with radioactive phosphorus. *Brit J Radiol* 34(399): 160–75.
45. Lin PP, Pandey MK, Jin F, Raymond AK, Akiyama H, Lozano G. 2009. Targeted mutation of p53 and Rb in mesenchymal cells of the limb bud produces sarcomas in mice. *Carcinogenesis* 30(10): 1789–95.
46. Rainusso N, Man TK, Lau CC, Hicks J, Shen JJ, Yu A, Wang LL, Rosen JM. 2011. Identification and gene expression profiling of tumor-initiating cells isolated from human osteosarcoma cell lines in an orthotopic mouse model. *Cancer Biol Ther* 12(4): 278–87.
47. Chawla SP, Chua VS, Fernandez L, Quon D, Saralou A, Blackwelder WC, Hall FL, Gordon EM. 2009. Phase I/II and phase II studies of targeted gene delivery in vivo: Intravenous Rexin-G for chemotherapy-resistant sarcoma and osteosarcoma. *Mol Ther* 17(9): 1651–7.
48. Butrynski JE, D'Adamo DR, Hornick JL, Dal Cin P, Antonescu CR, Jhanwar SC, Ladanyi M, Capelletti M, Rodig SJ, Ramaiya N, Kwak EL, Clark JW, Wilner KD, Christensen JG, Janne PA, Maki RG, Demetri GD, Shapiro GI. 2010. Crizotinib in ALK-rearranged inflammatory myofibroblastic tumor. *N Engl J Med* 363(18): 1727–33.
49. Berman SD, Calo E, Landman AS, Danielian PS, Miller ES, West JC, Fonhoue BD, Caron A, Bronson R, Bouxsein ML, Mukherjee S, Lees JA. 2008. Metastatic osteosarcoma induced by inactivation of Rb and p53 in the osteoblast lineage. *Proc Natl Acad Sci U S A* 105(33): 11851–6.
50. Lengner CJ, Steinman HA, Gagnon J, Smith TW, Henderson JE, Kream BE, Stein GS, Lian JB, Jones SN. 2006. Osteoblast differentiation and skeletal development are regulated by Mdm2-p53 signaling. *J Cell Biol* 172(6): 909–21.
51. Tashjian AH, Goltzman D. 2008. On the interpretation of rat carcinogenicity studies for human PTH(1–34) and human PTH(1–84). *J Bone Miner Res* 23(6): 803–11.
52. Selvarajah GT, Kirpensteijn J. 2010. Prognostic and predictive biomarkers of canine osteosarcoma. *Vet J* 185(1): 28–35.

86

Skeletal Complications of Breast and Prostate Cancer Therapies

Catherine Van Poznak and Pamela Taxel

Introduction 711
Breast Cancer 711
Prostate Cancer 713
Radiation and Fracture 715

Osteoclast Inhibition and Metastatic Bone Disease 715
Summary 716
References 716

INTRODUCTION

The term "cancer survivorship" represents those living following the diagnosis of cancer. It denotes a continual process beginning at the moment of diagnosis and continues lifelong, however long that may be. Anticancer therapies may be administered in an attempt to decrease the risk of distant metastases by affecting occult tumor cells or to control known tumor burden. Those with a diagnosis of cancer may be at risk for skeletal complications from the cancer and/or cancer therapy. This chapter reviews skeletal complications of cancer therapies with a focus on breast cancer and prostate cancer survivors.

BREAST CANCER

The median age at breast cancer diagnosis is 61, with 95% of cases occurring in women over the age of 40 [1]. Approximately 75% of all invasive breast cancers express the estrogen receptor (ER) and/or the progesterone receptor (PR). Blocking estrogen signaling is an important means of controlling endocrine responsive breast cancers. Adjuvant antiestrogen therapy may reduce the risk of breast cancer recurrence by approximately 30–50% in a woman with ER/PR expressing breast cancer [2]. In the metastatic setting, a first line clinical benefit rate of approximately 50% may be seen with antiestrogen therapy [3]. Hence, endocrine therapy is a standard intervention in the management of ER/PR expressing tumors. In tumors that express neither ER nor PR, chemotherapy is often used.

Endocrine therapy, breast cancer, and bone

In premenopausal and postmenopausal women with ER/PR expressing breast cancer, systemic endocrine therapies are considered as part of the cancer treatment plan. In the adjuvant setting, 5 years of endocrine therapy is a standard duration of treatment in the United States [4]. In the metastatic setting, a particular intervention is often used until there is evidence of disease progression or intolerable toxicity. The impact of endocrine therapy on bone is influenced by the menopausal status of the patient and the drugs used. Table 86.1 highlights representative studies examining adjuvant breast cancer endocrine therapy and the effect on bone-related parameters.

Tamoxifen, a selective estrogen receptor modulator (SERM), is an established adjuvant endocrine intervention in both premenopausal and postmenopausal women with ER/PR expressing breast cancer. Whether tamoxifen serves as an estrogen agonist or antagonist depends on the exposed tissue, as well as the milieu of ER coactivators and corepressors in those tissues [5]. Within the breast, tamoxifen is an estrogen antagonist regardless of menopausal status. Within bone, tamoxifen is associated with loss of bone mineral density (BMD) in premenopausal women and with gain of BMD in postmenopausal women [6].

Table 86.1. Representative Reports of the Effects of Breast Cancer Therapy on Bone

Adjuvant Endocrine Intervention	Premenopausal BMD	Premenopausal Fracture Risk	Postmenopausal BMD	Postmenopausal Fracture Risk
Chemical ovarian ablation (LHRH)	5% loss of total body bone density at 2 years (Ref. 54)	Not defined	Not applicable	
Chemical ovarian ablation (LHRH) and oral endocrine therapy (without osteoclast inhibition)	Loss of BMD at lumbar spine at 3 years Tamoxifen: 9.0% Anastrozole: 13.6% (Ref. 7).	Not defined. One study demonstrated fracture (0.2%) noted in both arms at 48 months of follow-up (Ref. 7).	Not applicable	
Oophorectomy	In a non-cancer population oophorectomy resulted in significant loss of vertebral bone mass at 12 months (Ref. 55).	A non-cancer population cohort study suggests a slight increase in risk of forearm and vertebral fractures (Ref. 56)	Non-cancer population results are mixed (Refs. 57, 58)	
Chemotherapy	Reports of chemotherapy-induced ovarian use variable definitions of amenorrhea, menopausal status, and ovarian function. Loss of lumbar spine BMD at 1 year has been reported to be 3.2–7.7% in those with ovarian dysfunction. In those who retain their menses, BMD was essentially stable (change of increase of 0.6) at 3 years (Refs. 59, 60).	Not defined	Chemotherapy and its associated supportive interventions may decrease BMD and fractures as noted in observational and epidemiologic studies (Refs. 13, 59).	
Tamoxifen	Non-cancer population results demonstrate a loss of 1.44% of lumbar spine BMD in 3 years (Ref. 6). In women who remained premenopausal after chemotherapy, tamoxifen was associated with a 4.6% loss of lumbar spine BMD at 3 years (Ref. 60).	Not defined	Non-cancer population results demonstrate an average annual increase in BMD of 1.17% in the lumbar spine (Ref. 6). In a second, non-cancer population, there was no difference in rates of fracture between tamoxifen and raloxifene (Ref. 61). Adjuvant tamoxifen is associated with an approximate 0.5–1.0% gain in bone mass per year (Ref. 62).	Current use of tamoxifen has been associated with an approximate 30% reduction in risk of osteoporotic fracture (Ref. 63).
Aromatase inhibitor	Adjuvant aromatase inhibitors are not indicated for women with preserved ovarian function		In women who have completed 5 years of adjuvant tamoxifen, continued adjuvant therapy with anastrozole decreased BMD by 5.35% at 24 months (Ref. 64). In those without prior exposure to tamoxifen, loss of lumbar spine BMD is approximately 1–2% per year of therapy. In one study, at the end of 5 years of therapy the lumbar spine BMD had decreased 6.1% (Refs. 59, 65).	During treatment with anastrozole, the annual rate of fracture was approximately 3% (Ref. 66). At 5 years of adjuvant AI therapy or more, the risk of fracture ranges from approximately 3 to 12% (Refs. 4, 67).

Decreasing ovarian estrogen production may be a component of breast cancer care in premenopausal women with ER/PR expressing tumors. Options for ovarian ablation include use of gonadotropin releasing hormone (GnRH) agonists, surgery, or radiation (not commonly used). In addition, chemotherapy may induce ovarian dysfunction. The lower circulating estrogen levels observed with ovarian ablation correlate with loss of BMD. This loss of BMD may be partially recovered upon cessation of therapy [7]. Of note, when a GnRH agonist is used with an oral endocrine therapy there does not appear to be an anticancer advantage with an aromatase inhibitor (AI) over tamoxifen, but there may be a greater loss of BMD with the AI–GnRH combination [7].

Postmenpausal women with ER/PR expressing breast cancers have the option of adjuvant treatment with tamoxifen, an aromatase inhibitor or the sequential use of tamoxifen and AI [4]. The third generation AIs (anastrozole, exemestane, letrozole) effectively block peripheral conversion of androgens to estrogens primarily in adipose tissue, as well as androstenedione to estrone in the adrenal gland by inhibiting the aromatase enzyme (cytochrome P-450, CYP 19). The AI-induced decrease in serum estradiol is associated with a decrease in risk of breast cancer recurrence and new breast cancer and is also associated with accelerated loss of BMD and fracture, regardless of the particular AI studied [4]. In spite of adjuvant studies demonstrating an increased risk of fracture with adjuvant AI use, the Phase III Mammary Prevention 3 trial (MAP.3) of exemestane versus placebo demonstrated no statistical difference between the two groups in terms of fracture [8]. The interpretation of the BMD changes associated with the AIs may be confounded in studies where tamoxifen is the comparator, given tamoxifen is a partial estrogen agonist and may have a positive impact on BMD.

Fulvestrant is a selective estrogen receptor antagonist that is FDA approved for treatment of hormone receptor positive metastatic breast cancer in postmenopausal women. In women who are gaining clinical benefit from fulvestrant, pilot data suggest that markers of bone turnover are not significantly altered over an 18-month course [9]. Concerns for loss of BMD in metastatic disease are often supplanted by the frequent presence of bone metastases and the routine use of high potency osteoclast inhibitors to decrease the risk of skeletal-related events. The impact of anticancer therapies on BMD in women with advanced breast cancer, but without osseous metastases, is not well established, although there is an increased risk of fracture in women with recurrent breast cancer who are without evidence of skeletal metastases (hazard ratio 22.7; 95% CI 9.1, 57.1; $P < 0.0001$) [10].

Chemotherapy, breast cancer, and bone

Chemotherapy regimens may also be associated with loss of BMD and fracture. Apart from chemotherapy-induced ovarian dysfunction, chemotherapy-induced bone changes are less well defined than those of endocrine therapy. Retrospective studies have suggested that adjuvant therapies have a negative impact on BMD [11] and epidemiologic data from the Women's Health Initiative Observational Study demonstrate an increased rate of fractures in women breast cancer survivors compared to women without a history of breast cancer [12]. Prospective data from a small study of postmenopausal women with breast cancer treated only with chemotherapy or observation (no endocrine therapy in either cohort) demonstrated that the majority of women treated with adjuvant chemotherapy lost 1–10% of lumbar spine BMD at 1 year [13]. The etiology of the observed loss of BMD may be due to chemotherapy itself, or the supportive therapies that may have negative impact on BMD, such as glucocorticoids.

Pharmacologic interventions to preserve BMD and prevent fractures secondary to breast cancer therapy

Randomized clinical trials investigating means to mitigate the bone loss associated with adjuvant breast cancer therapies have been reported, and additional studies are ongoing [4]. The study designs have either included established drugs, doses, and intervals used to manage postmenopausal osteopenia and ostcoporosis or the studies have included investigational and more intensive regimens of osteoclast inhibition. The vast majority of these clinical trials have demonstrated that osteoclast inhibition, with oral or intravenous bisphosphonates or denosumab, can preserve or increase BMD in the setting of cancer treatment-induced bone loss (CTIBL) [14, 15]. Sufficient fracture data does not presently exist to address the impact of osteoclast inhibition prescribed to treat CTIBL in either premenopausal or postmenopausal women with breast cancer; additional studies are needed. Given the efficacy of the FDA-approved therapies in postmenopausal women, these standard regimens are often selected in the clinic when an established indication for osteoporosis prevention or treatment is present.

PROSTATE CANCER

The median age at prostate cancer diagnosis is 67 years, and over 8% of men will develop cancer of the prostate between the ages of 50 and 70 years [16]. It is estimated that 70% of prostate cancers are androgen dependent [17] and respond to endocrine (hormonal ablation) therapy. Diminishing circulating testosterone, a growth factor for prostate cancer, is an effective anticancer therapy. Thus, androgen deprivation therapy (ADT) is a mainstay of treatment for prostate cancer and is commonly used at initial diagnosis in the management of intermediate to high risk prostate cancer.

Endocrine therapy, prostate cancer, and bone

ADT can be achieved medically with the use of GnRH agonists and/or antiandrogen medications, as well as surgically with bilateral orchidectomy (less commonly used). GnRH agonist therapy lowers testosterone and estrogen levels, both of which negatively effects skeletal health [18]. Antiandrogens such as biclutamide and flutamide are competitive inhibitors of the androgen receptor, blocking testosterone action and are typically used in conjunction with GnRH agonists. Estrogen preparations were used historically but are not commonly used now.

Approximately 33–70% of men with prostate cancer receive a GnRH agonist as primary therapy with or without antiandrogens for localized disease before (neoadjuvant) or after other (adjuvant) treatment modalities, or in men with a rising prostate-specific antigen (PSA) after primary therapy with surgery or radiotherapy [17, 19]. In patients with locally advanced disease, as well as those with high-risk localized disease (based on tumor stage, Gleason score, and PSA), neoadjuvant therapy has shown benefit in disease-specific survival, time to progression, and all-cause mortality [20]. In the adjuvant setting, ADT for more than 2 years given in conjunction with external-beam radiotherapy in men with locally advanced, high-risk prostate cancer has been demonstrated in randomized trials to improve survival in this population [20]. In metastatic disease, ADT is the first line of therapy and the duration of therapy may be measured in years. Abiraterone is a selective and irreversible inhibitor of CYP-17, reducing adrenal androgen production. Abiraterone used in combination with prednisone was FDA approved in 2011 for metastatic, castrate-resistant prostate cancer after exposure to docetaxel. The impact of abiraterone on BMD has not been reported.

Dual energy X-ray absorptiometry (DXA) hip and spine measurements of men on GnRH agonists demonstrate a loss of 2–3% per year within the first several years after the initiation of ADT therapy [19]. This loss continues, albeit more slowly, throughout therapy duration. In men who have had surgical castration with bilateral orchiectomies, hip BMD decreases by approximately 10% over 1 year, as compared with only a 1% decline in men given estrogen for medical castration [21]. In contrast, annual rates of bone loss in healthy community-dwelling men are approximately 0.5–1.0% per year [22].

Men receiving ADT have a rate of fracture as high as 20% after 5 years of treatment. In a study that utilized the Surveillance, Epidemiology and End Results (SEER) database, investigators found that 19.4% of men receiving GnRH-agonist therapy or orchiectomy had fractures compared with 12.6% of those with prostate cancer not receiving these treatments (p < .001). Longer duration of treatment was also predictive of fracture [23]. In a study of Medicare claims data, men treated with ADT for prostate cancer had an increased risk of fracture compared with a control group (hazard ratio 1.4, p < .001), and GnRH agonist treatment independently predicted fracture after controlling for age, race, comorbidity, and location [24].

Due to the typical long-standing nature and general indolence of prostate cancer in older men, many will live with the consequences of CTIBL in addition to preexisting comorbid conditions. Hyogonadism leads to decreased BMD, decreased muscle mass, increased fall risk, and impaired balance. Thus, it would stand to reason that ADT is associated with some of these adverse events and a potential for the development of the geriatric syndrome of frailty [25].

Pharmacologic interventions to preserve BMD and prevent fractures secondary to prostate cancer therapy

Bisphosphonates: Alendronate, risedronate, and zoledronic acid are FDA approved for men with osteoporosis, and only zoledronic acid is FDA approved in metastatic prostate cancer. A series of clinical trials have demonstrated efficacy of intravenous and oral bisphosphonates in preserving BMD in men on ADT [26]. In a 48-week trial, pamidronate, 60 mg intravenously every 12 weeks, maintained BMD at the spine and hip in the group receiving pamidronate as compared with controls receiving placebo; they had significant loss at these sites [19]. In a 1-year trial of a single 4-mg dose of zoledronic acid versus placebo in a similar population of men, those who received zoledronic acid had increased BMD in the spine and hip compared with loss of BMD in the placebo control group [27].

Within a double-blind, randomized, placebo-controlled trial, Greenspan et al. demonstrated that alendronate dosed at 70 mg weekly could prevent bone loss and reduce biochemical markers of bone turnover in men with prostate cancer who receive ADT [28]. After 1 year, men treated with alendronate had significant gains in BMD of 3.7% at the spine and 1.6% at the hip whereas men in the placebo group had losses of 1.4% at the spine and 0.7% at the hip.

In a smaller, double-blind, placebo-controlled, randomized trial of older men with locally advanced prostate cancer during the first 6 months of GnRH agonist therapy, 40 men completed a trial of risedronate 35 mg/week versus placebo. After 6 months, the risedronate group showed no change in femoral neck and total hip BMD while the placebo group decreased significantly by 2.0% and 2.2%, respectively. Spine BMD of the risedronate group increased significantly by 1.7% over baseline while the placebo group did not change [29]. Thus, oral and intravenous bisphosphonates can prevent loss or increase bone mass; however, long-term and larger studies are required to determine if fractures can be decreased.

Estrogens and SERMS: Historically, medical castration with estrogen, specifically diethylstilbestrol (DES) was the mainstay of treatment for metastatic prostate cancer. In a small study using transdermal estradiol patches to achieve castration, 20 men with advanced or metastatic prostate cancer showed increases in BMD at the spine and hip after 1 year of treatment [30]. Other investigators have also shown decreases in bone turnover in response to intramuscular estrogen preparation in men treated with orchiectomy [31] or DES [32] as well as in men on GnRH agonist therapy receiving 1 mg/day of oral micronized estradiol [33]. To date, long-term trials of estrogen to determine if it can improve BMD and decrease fractures have not been performed.

SERMs are not FDA approved for preserving BMD or reducing fracture risk in men with prostate cancer on ADT, but they have been evaluated. Raloxifene, a SERM, has been shown in a 12-month open label trial to have a modest effect on BMD of the spine and hip compared with men not on the medication [34]. In a study of 847 men on ADT, subjects under 80 years of age were randomized to receive either 80 mg/day of toremifene, another SERM, versus placebo for 24 months. The average time on ADT was 4 years for both groups, and the mean age of the subjects was 72 years. Vertebral fracture, a primary end point, was assessed morphometrically. One percent of the toremifene group sustained new vertebral fractures compared with 4.8% of the placebo group, a relative risk reduction of 79.5% (p < .005) and an absolute risk reduction of 3.8%. BMD increased at all sites in the toremifene group (p < .001) [35].

Denosumab: Denosumab, a human monoclonal antibody for receptor activator of nuclear factor-κB ligand (RANKL) has recently been approved for men receiving ADT at risk of fracture. In a 24-month, randomized, placebo-controlled multicenter trial, more than 770 men per group were randomized to receive denosumab, 60 mg subcutaneously or placebo every 6 months [36]. Subjects had a median duration of ADT of 20 months, with 75% over 6-months duration. BMD increased significantly compared with the placebo (6.7% difference at L-spine, 3.9% at femoral neck, 4.8% at total hip, and 5.5% at 1/3 radius, p < .001 for all comparisons). The denosumab group experienced lower rates of new vertebral fractures (morphometric) at 12, 24, and 36 months, with a decrease of 62% at 36 months (1.5% vs 3.6% in the denosumab vs placebo group, p = .006). A similar proportion of subjects in each group had adverse events. One case of hypocalcemia was reported, and there was a higher infection rate in the denosumab group; no cases of osteonecrosis of the jaws (ONJ) were documented. Denosumab is also FDA approved for the prevention of skeletal-related events in men with bone metastases from prostate cancer.

RADIATION AND FRACTURE

The mechanism of radiation therapy induced bone injury has not been fully defined and may occur secondary to alterations in the marrow microenvironment including changes to the vasculature, osteoblasts, osteoclasts, and osteocytes [37, 38]. Insufficiency fractures may occur in an irradiated field. It is essential to rule out metastasic disease and pathologic fracture to direct clinical management of fractures in patients with a history of cancer [39].

Radiation therapy and breast cancer

The clinical indications for adjuvant radiation therapy to the breast or chest wall and axillary region is directed by the surgery performed (breast conservation), surgical margin status, size of the primary tumor and number of lymph nodes involved. Adjuvant radiation therapy is associated with a small increase (less than 2%) in rib fractures, whether it be post-mastectomy, post-lumpectomy with external beam, or partial breast radiation [40, 41, 42].

Radiation therapy and prostate cancer

Men with prostate cancer may receive external beam radiation therapy (EBRT) or brachytherapy (radioactive seed implants) to the prostate as therapy for the primary tumor. In women who undergo pelvic radiation therapy for treatment of cervical, rectal, or anal cancer there is an increased risk for pelvic fractures compared to women with these types of cancers who do not receive pelvic radiation [43]. The effect of radiation therapy on pelvic fracture rates in men with prostate cancer has not been well characterized; however, there appears to be an increased risk of hip fractures associated with EBRT [44].

OSTEOCLAST INHIBITION AND METASTATIC BONE DISEASE

The incidence of bone metastases in patients with metastatic breast or prostate cancer is approximately 70%. Bone metastases are generally considered an incurable diagnosis and the goals of care are palliative. The selection of a particular osteoclast inhibitor for a patient is influenced by efficacy data, drug availability, toxicity profile, patient and clinician preferences, patient convenience, and treatment costs. Osteoclast inhibition is often initiated at the time of detecting bone metastases and may be continued indefinitely. Short-term adverse events associated with osteoclast inhibition may include hypocalcemia, bone pain, and acute phase reactions. Long-term bone specific adverse events may include osteonecrosis of the jaw (ONJ) and atypical fractures.

Osteonecrosis of the jaw (ONJ)

ONJ is an uncommon event that occurs in individuals receiving potent osteoclast inhibition. ONJ consists of necrotic, exposed bone within the oral cavity that has not healed within 8 weeks after identification by a health care provider [45]. ONJ is differentiated from osteoradionecrosis by the lack of exposure to radiation therapy to the maxillofacial region. ONJ has been reported in patients with metastatic bone disease treated with bisphosphonates and denosumab with a similar incidence of 1–2% at a median of 14 months of exposure of monthly osteoclast inhibition [46]. ONJ has been reported in patients receiving osteoclast inhibition for osteoporosis, as well as participants in adjuvant bisphosphonate clinical trials, although the incidence appears lower in these situations [45, 47]. The etiology of ONJ is presently unknown and is the subject of preclinical and clinical investigation. Risk factors for ONJ include an extended duration of osteoclast inhibition, dental extractions, and infection/inflammation [48]. Optimizing oral health and avoiding dental extractions during therapy with an osteoclast inhibitor may decrease the risk of ONJ. An observed reduction in the incidence of ONJ from 3.2% to 1.3%, when comparing pre- and post-implementation of preventive measures program, has been reported [49].

Atypical subtrochanteric femoral fractures

Atypical subtrochanteric femoral fractures have recently been reported in relation to bisphosphonate therapy [50]. These transverse femoral fractures most often occur with no or minimal trauma. Bisphosphonates may be associated with atypical subtrochanteric fractures, but causation has not been proven. It appears that these atypical fractures occur at a rate of approximately 31 per 10,000 patient years in patients treated with bisphosphonate, and at a rate of 6–13 per 10,000 patient years in untreated individuals [51].

Data on atypical fractures in patients with cancer are limited. A case report of a low trauma atypical femoral fracture of a man with renal cell carcinoma with skeletal involvement has recently been described after 18 months of intravenous zoledronic acid [52]. In a retrospective study of patients with metastatic bone disease, four patients, of a cohort of 327, were identified to have met radiographic criteria for atypical or stress fracture [53]. Three had breast cancer and one had myeloma, and they had received between 48 and 73 doses of bisphosphonates, primarily pamidronate. All experienced a prodrome of thigh pain. These investigators found no difference with regard to cumulative dose or duration of treatment of intravenous bisphosphonate between those with atypical fractures and those who did not. Thus, the incidence of these fractures appears to be low. However, prospective studies are required to determine if risk factors for atypical fractures can be elucidated.

SUMMARY

Patients with breast or prostate cancer likely exhibit a baseline risk of osteoporotic fractures due to advancing age and comorbid conditions; in addition, the management of their cancer may accelerate the risk for loss of BMD or fracture. Optimizing bone health practices through nutrition and/or supplements, weight-bearing exercises, and lifestyle modifications is essential. Clinical and radiographic assessment of fracture risk is a component of monitoring bone integrity. Pharmacologic interventions with bisphosphonates or denosumab are options to maintain BMD or reduce fracture in both the adjuvant and the metastatic setting. Participation in clinical trials is encouraged.

REFERENCES

1. American Cancer Society; http://www.cancer.org.
2. Early Breast Cancer Trialists' Collaborative Group (EBCTCG). 2005. Effects of chemotherapy and hormonal therapy for early breast cancer on recurrence and 15-year survival: An overview of the randomised trials. *Lancet* 365: 1687–717.
3. Altundag K, Ibrahim NK. 2006. Aromatase inhibitors in breast cancer: An overview. *Oncologist* 11: 553–62.
4. Burstein HJ, Prestrud AA, Seidenfeld J, Anderson H, Buchholz TA, Davidson NE, Gelmon KE, Giordano SH, Hudis CA, Malin J, Mamounas EP, Rowden D, Solky AJ, Sowers MR, Stearns V, Winer EP, Somerfield MR, Griggs JJ. 2010. American Society of Clinical Oncology clinical practice guideline: Update on adjuvant endocrine therapy for women with hormone receptor-positive breast cancer. *J Clin Oncol* 28: 3784–96.
5. Shou J, Massarweh S, Osborne CK, Wakeling AE, Ali S, Weiss H, Schiff R. 2004. Mechanisms of tamoxifen resistance: Increased estrogen receptor-HER2/neu cross-talk in ER/HER2-positive breast cancer. *J Natl Cancer Inst* 96: 926–35.
6. Powles TJ, Hickish T, Kanis JA, Tidy A, Ashley S. 1996. Effect of tamoxifen on bone mineral density measured by dual-energy X-ray absorptiometry in healthy premenopausal and postmenopausal women. *J Clin Oncol* 14(1): 78–84.
7. Gnant M, Mlineritsch B, Luschin-Ebengreuth G, Kainberger F, Kässmann H, Piswanger-Sölkner JC, Seifert M, Ploner F, Menzel C, Dubsky P, Fitzal F, Bjelic-Radisic V, Steger G, Greil R, Marth C, Kubista E, Samonigg H, Wohlmuth P, Mittlböck M, Jakesz R. 2008. Adjuvant endocrine therapy plus zoledronic acid in premenopausal women with early-stage breast cancer: 5-year follow-up of the ABCSG-12 bone-mineral density substudy. *Lancet Oncol* 9: 840–9.
8. Goss PE, Ingle JN, Alés-Martínez JE, Cheung AM, Chlebowski RT, Wactawski-Wende J, McTiernan A, Robbins J, Johnson KC, Martin LW, Winquist E, Sarto GE, Garber JE, Fabian CJ, Pujol P, Maunsell E, Farmer P, Gelmon KA, Tu D, Richardson H; NCIC CTG MAP.3 Study Investigators. 2011. Exemestane for breast-cancer prevention in postmenopausal women. *N Engl J Med* 364(25): 2381–91.

9. Agrawal A, Hannon RA, Cheung KL, Eastell R, Roberston JFR. 2009. Bone turnover markers in postmenopausal breast cancer treated with fulvestrant—A pilot study. *The Breast* 18: 3204–7.
10. Kanis JA, McCloskey EV, Powles T, Paterson AH, Ashley S, Spector T. 1999. A high incidence of vertebral fracture in women with breast cancer. *Br J Cancer* 79: 1179–81.
11. Greep NC, Giuliano AE, Hansen NM, Taketani T, Wang HJ, Singer FR. 2003. The effects of adjuvant chemotherapy on bone density in postmenopausal women with early breast cancer. *Am J Med* 114: 653–9.
12. Chen Z, Maricic M, Bassford TL, Pettinger M, Ritenbaugh C, Lopez AM, Barad DH, Gass M, Leboff MS. 2005. Fracture risk among breast cancer survivors: Results from the Women's Health Initiative Observational Study. *Arch Intern Med* 165(5): 552–8.
13. Van Poznak C, Morris PG, D'Andrea G, Schott A, Griggs J, Fornier M, Smerage J, Henry L, Collins T, Hurria A, Drullinsky P, Mills NE, Hayes DF, Hudis C. 2010. Changes in bone mineral density (BMD) of postmenopausal women who are not receiving adjuvant endocrine therapy for breast cancer. *Bone* 47(2): S308.
14. Santen RJ. 2011. Effect of endocrine therapies on bone in breast cancer patients. *J Clin Endocrinol Metab* 96: 308–19.
15. Van Poznak C. 2010. *Breast Cancer and Adjuvant Bisphosphonates*. 2010 ASCO Educational Book. pp. e62–8.
16. Howlader N, Noone AM, Krapcho M, Neyman N, Aminou R, Waldron W, Altekruse SF, Kosary CL, Ruhl J, Tatalovich Z, Cho H, Mariotto A, Eisner MP, Lewis DR, Chen HS, Feuer EJ, Cronin KA, Edwards BK (eds.) *SEER Cancer Statistics Review, 1975–2008*. National Cancer Institute: Bethesda, MD. http://seer.cancer.gov/csr/1975_2008.
17. Schally AV. 2007. Luteinizing hormone-releasing hormone analogues and hormone ablation for prostate cancer: State of the art. *BJU Int* 100 Suppl 2: 2–4.
18. Falahati-Nini A, Riggs BL, Atkinson EJ, O'Fallon WM, Eastell R, Khosla S. 2000. Relative contributions of testosterone and estrogen in regulating bone resorption and formation in normal elderly men. *J Clin Invest* 106(12): 1553–60.
19. Smith MR, McGovern FJ, Zietman AL, Fallon MA, Hayden DL, Schoenfeld DA, Kantoff PW, Finkelstein JS. 2001. Pamidronate to prevent bone loss during androgen-deprivation therapy for prostate cancer. *N Engl J Med* 345: 948–55.
20. Payne H, Mason M. 2011. Androgen deprivation therapy as adjuvant/neoadjuvant to radiotherapy for high-risk localised and locally advanced prostate cancer: Recent developments. *Br J Cancer* 105(11): 1628–34.
21. Eriksson S, Eriksson A, Stege R, Carlstrom K. 1995. Bone mineral density in patients with prostatic cancer treated with orchidectomy and with estrogens. *Calcif Tissue Int* 57: 97–9.
22. Looker AC, Wahner HW, Dunn WL, Calvo MS, Harris TB, Heyse SP, Johnston CC Jr, Lindsay R. 1998. Updated data on proximal femur bone mineral levels of US adults. *Osteoporos Int* 8: 468–89.
23. Shahinian VB, Kuo YF, Freeman JL, Orihuela E, Goodwin JS. 2005. Increasing use of gonadotropin-releasing hormone agonists for the treatment of localized prostate carcinoma. *Cancer* 103: 1615–24.
24. Smith MR, Lee WC, Brandman J, Wang Q, Botteman M, Pashos CL. 2005. Gonadotropin-releasing hormone agonists and fracture risk: A claims-based cohort study of men with nonmetastatic prostate cancer. *J Clin Oncol* 23: 7897–903.
25. Lunenfeld B, Nieschlag E. 2007. Testosterone therapy in the aging male. *Aging Male* 10(3): 139–53.
26. Adler RA. 2011. Management of osteoporosis in men on androgen deprivation therapy. *Maturitas* 68(2): 143–7.
27. Michaelson MD, Kaufman DS, Lee H, McGovern FJ, Kantoff PW, Fallon MA, Finkelstein JS, Smith MR. 2007. Randomized controlled trial of annual zoledronic acid to prevent gonadotropin-releasing hormone agonist-induced bone loss in men with prostate cancer. *J Clin Oncol* 25: 1038–42.
28. Greenspan SL, Nelson JB, Trump DL, Resnick NM. 2007. Effect of once-weekly oral alendronate on bone loss in men receiving androgen deprivation therapy for prostate cancer: A randomized trial. *Ann Intern Med* 146: 416–24.
29. Taxel P, Dowsett R, Albertsen P, Fall P, Biskup B, Raisz L. 2010. Risedronate prevents early bone loss and increased bone turnover in the first 6 months of luteinizing hormone-releasing hormone-agonist therapy for prostate cancer. *BJU Int* 106: 1473–6.
30. Ockrim JL, Lalani EN, Banks LM, Svensson WE, Blomley MJ, Patel S, Laniado ME, Carter SS, Abel PD. 2004. Transdermal estradiol improves bone density when used as single agent therapy for prostate cancer. *J Urol* 172: 2203–7.
31. Carlstrom K, Stege R, Henriksson P, Grande M, Gunnarsson PO, Pousette A. 1997. Possible bone-preserving capacity of high-dose intramuscular depot estrogen as compared to orchidectomy in the treatment of patients with prostatic carcinoma. *Prostate* 31: 193–7.
32. Scherr D, Pitts WR Jr, Vaughn ED Jr. 2002. Diethylstilbesterol revisited: Androgen deprivation, osteoporosis and prostate cancer. *J Urol* 167: 535–8.
33. Taxel P, Fall PM, Albertsen PC, Dowsett RD, Trahiotis M, Zimmerman J Ohannessian C, Raisz LG. 2002. The effect of micronized estradiol on bone turnover and calciotropic hormones in older men receiving hormonal suppression therapy for prostate cancer. *J Clin Endocrinol Metab* 87: 4907–13.
34. Smith MR, Fallon MA, Lee H, Finkelstein JS. 2004. Raloxifene to prevent gonadotropin-releasing hormone agonist-induced bone loss in men with prostate cancer: A randomized controlled trial. *J Clin Endocrinol Metab* 89(8): 3841.
35. Smith MR, Malkowicz SB, Brawer MK, Hancock ML, Morton RA, Steiner MS. 2011. Toremifene decreases vertebral fractures in men younger than 80 years receiving androgen deprivation therapy for prostate cancer. *Urol* 186(6): 2239–44.

36. Smith MR, Egerdie B, Hernández Toriz N, Feldman R, Tammela TL, Saad F, Heracek J, Szwedowski M, Ke C, Kupic A, Leder BZ, Goessl C; Denosumab HALT Prostate Cancer Study Group. 2009. Denosumab in men receiving androgen-deprivation therapy for prostate cancer. *N Engl J Med* 361: 745–55.

37. Cao X, Wu X, Frassica D, Yu B, Pang L, Xian L, Wan M, Lei W, Armour M, Tryggestad E, Wong J, Wen CY, Lu WW, Frassica FJ. 2011. Irradiation induces bone injury by damaging bone marrow microenvironment for stem cells. *Proc Natl Acad Sci U S A* 108: 1609–14.

38. Hopewell JW. 2003. Radiation-therapy effects on bone density. *Med Pediatr Oncol* 41: 208–11.

39. Moreno A, Clemente J, Crespo C, et al. 1999. Pelvic insufficiency fractures in patients with pelvic irradiation. *Int J Radiat Oncol Biol Phys* 44: 61–6.

40. Overgaard M. 1988. Spontaneous radiation-induced rib fractures in breast cancer patients treated with postmastectomy irradiation. A clinical radiobiological analysis of the influence of fraction size and dose-response relationships on late bone damage. *Acta Oncol* 27: 117–22.

41. Pierce SM, Recht A, Lingos TI, Abner A, Vicini F, Silver B, Herzog A, Harris JR. 1992. Long-term radiation complications following conservative surgery (CS) and radiation therapy (RT) in patients with early stage breast cancer. *Int J Radiat Oncol Biol Phys* 23: 915–23.

42. Brashears JH, Dragun AE, Jenrette JM. 2009. Late chest wall toxicity after MammoSite breast brachytherapy. *Brachytherapy* 8: 19–25.

43. Baxter N, Habermann E, Tepper J, Durham S, Virnig B. 2005. Risk of pelvic fractures in older women following pelvic irradiation. *J Amer Med Assoc* 294: 2587–93.

44. Elliott SP, Jarosek SL, Alanee SR Konety BR, Dusenbery KE, Virnig BA. 2011. Three-dimensional external beam radiotherapy for prostate cancer increases the risk of hip fracture. *Cancer* 17: 4557–65.

45. Khosla S, Burr D, Cauley J, Dempster D, Ebeling P, Felsenber D, Gagel R, Gilsanz V, Guise T, Koka S, McCauley L, McGowan J, McKee M, Mohla S, Pendrys D, Raisz L, Ruggiero S, Shafer D, Shum L, Silverman S, Van Poznak CH, Watts N, Woo S, Shane E. 2007. Bisphosphonate-associated osteonecrosis of the jaw: Report of a task force of the American Society for Bone and Mineral Research. *J Bone Miner Res* 22(10): 1479–91.

46. Saad F, Brown JE, Van Poznak C, Ibrahim T, Stemmer SM, Stopeck AT, Diel IJ, Takahashi S, Shore N, Henry DH, Barrios CH, Facon T, Senecal F, Fizazi K, Zhou L, Daniels A, Carrière P, Dansey R. 2012. Incidence, risk factors, and outcomes of osteonecrosis of the jaw: Integrated analysis from three blinded active-controlled phase III trials in cancer patients with bone metastases. *Ann Oncol* 23(5): 1341–7.

47. Coleman RE, Marshall H, Cameron D, Dodwell D, Burkinshaw R, Keane M, Gil M, Houston SJ, Grieve RJ, Barrett-Lee PJ, Ritchie D, Pugh J, Gaunt C, Rea U, Peterson J, Davies C, Hiley V, Gregory W, Bell R; AZURE Investigators. 2011. Breast-cancer adjuvant therapy with zoledronic acid. *N Engl J Med* 365: 1396–1405.

48. Barasch A, Cunha-Cruz J, Curro FA, Hujoel P, Sung AH, Vena D, Voinea-Griffin AE; CONDOR Collaborative Group, Beadnell S, Craig RG, DeRouen T, Desaranayake A, Gilbert A, Gilbert GH, Goldberg K, Hauley R, Hashimoto M, Holmes J, Latzke B, Leroux B, Lindblad A, Richman J, Safford M, Ship J, Thompson VP, Williams OD, Yin W. 2011. Risk factors for osteonecrosis of the jaws: A case-control study from the CONDOR dental PBRN. *J Dent Res* 90(4): 439–44.

49. Ripamonti CI, Maniezzo M, Campa T, Fagnoni E, Brunelli C, Saibene G, Bareggi C, Ascani L, Cislaghi E. 2009. Decreased occurrence of osteonecrosis of the jaw after implementation of dental preventive measures in solid tumour patients with bone metastases treated with bisphosphonates. The experience of the National Cancer Institute of Milan. *Ann Oncol* 20(1): 137–45.

50. Rizzoli R, Akesson K, Bouxsein M, Kanis JA, Napoli N, Papapoulos S, Reginster JY, Cooper C. 2011. 2011 Subtrochanteric fractures after long-term treatment with bisphosphonates: A European Society on Clinical and Economic Aspects of Osteoporosis and Osteoarthritis, and International Osteoporosis Foundation Working Group Report. *Osteoporos Int* 22(2): 373–90.

51. Abrahamsen B, Eiken P, Eastell R. 2010. Cumulative alendronate dose and the long-term absolute risk of subtrochanteric and diaphyseal femur fractures: A register-based national cohort analysis. *J Clin Endocrinol Metab* 95(12): 5258–65.

52. Bush LA, Chew FS. 2008. Subtrochanteric femoral insufficiency fracture following bisphosphonate therapy for osseous metastases. *Radiology Case Reports* 3: 232–5.

53. Puhaindran ME, Farooki A, Steensma MR, Hameed M, Healey JH, Boland PJ. 2011. Atypical subtrochanteric femoral fractures in patients with skeletal malignant involvement treated with intravenous bisphosphonates. *J Bone Joint Surg Am* 93(13): 1235–42.

54. Sverrisdottir A, Fornander T, Jacobsson H, von Schoultz E, and Rutqvist LE. 2004. Bone mineral density among premenopausal women with early breast cancer in a randomized trial of adjuvant endocrine therapy. *J Clin Oncol* 22: 3694–9.

55. Cann CE, Genant HK, Ettinger B, Gordon GS. 1980. Spinal mineral loss in oophorectomized women. *JAMA* 244(18): 2056–9.

56. Melton LJ 3rd, Crowson CS, Malkasian GD, O'Fallon WM. 1996. Fracture risk following bilateral oophorectomy. *J Clin Epidemiol* 49(10): 1111–5.

57. Antoniucci DM, Sellmeyer DE, Cauley JA, Ensrud KE, Schneider JL, Vesco KK, Cummings SR, Melton LJ 3rd; Study of Osteoporotic Fractures Research Group. 2005. Postmenopausal bilateral oophorectomy is not associated with increased fracture risk in older women. *J Bone Min Res* 20: 741–7.

58. Melton LJ 3rd, Khosla S, Malkasian GD, Achenbach SJ, Oberg AL, Riggs BL. 2003. Fracture risk after bilateral oophorectomy in elderly women. *J Bone Miner Res* 18: 900–5.

59. Body JJ. 2011. Increased fracture rate in women with breast cancer: a review of the hidden risk. *BMC Cancer* 11: 384.
60. Vehmanen L, Elomaa I, Blomqvist C, Saarto T. 2006. Tamoxifen treatment after adjuvant chemotherapy has opposite effects on bone mineral density in premenopausal patients depending on menstrual status. *J Clin Oncol* 24(4): 675–80.
61. Vogel VG, Costantino JP, Wickerham DL, Cronin WM, Cecchini RS, Atkins JN, Bevers TB, Fehrenbacher L, Pajon ER Jr, Wade JL 3rd, Robidoux A, Margolese RG, James J, Lippman SM, Runowicz CD, Ganz PA, Reis SE, McCaskill-Stevens W, Ford LG, Jordan VC, Wolmark N; National Surgical Adjuvant Breast and Bowel Project (NSABP). 2006. Effects of tamoxifen vs raloxifene on the risk of developing invasive breast cancer and other disease outcomes: The NSABP Study of Tamoxifen and Raloxifene (STAR) P-2 trial. *JAMA* 295: 2727–41.
62. Love RR, Mazess RB, Barden HS, Epstein S, Newcomb PA, Jordan VC, Carbone PP, DeMets DL. 1992. Effects of tamoxifen on bone mineral density in postmenopausal women with breast cancer. *N Engl J Med* 326: 852–6.
63. Cooke AL, Metge C, Lix L, Prior HJ, Leslie WD. 2008. Tamoxifen use and osteoporotic fracture risk: A population-based analysis. *J Clin Oncol* 26: 5227–32.
64. Goss PE, Ingle JN, Martino S, Robert NJ, Muss HB, Piccart MJ, Castiglione M, Tu D, Shepherd LE, Pritchard KI, Livingston RB, Davidson NE, Norton L, Perez EA, Abrams JS, Cameron DA, Palmer MJ, Pater JL. 2005. Randomized trial of letrozole following tamoxifen as extended adjuvant therapy in receptor-positive breast cancer: Updated findings from NCIC CTG MA.17. *J Natl Cancer Inst* 97: 1262–71.
65. Eastell R, Adams JE, Coleman RE, Howell A, Hannon RA, Cuzick J, Mackey JR, Beckmann MW, Clack G. 2008. Effect of anastrozole on bone mineral density: 5-year results from the Anastrozole, Tamoxifen, Alone or in Combination trial 18233230. *J Clin Oncol* 26: 1051–7.
66. Cuzick J, Sestak I, Baum M, Buzdar A, Howell A, Dowsett M, Forbes JF; ATAC/LATTE investigators. 2010. Effect of anastrozole and tamoxifen as adjuvant treatment for early-stage breast cancer: 10-year analysis of the ATAC trial. *Lancet Oncol* 11: 1135–41.
67. Body JJ. 2010. Prevention and treatment of side-effects of systemic treatment: Bone loss. *Ann Oncol* 21 (Suppl 7): vii180–vii185.

87

Bone Cancer and Pain

Patrick W. O'Donnell and Denis R. Clohisy

Epidemiology of Bone Cancer Pain 720
Mechanisms of Bone Cancer Pain 720
Therapeutic Strategies: Past, Present, and Future 723
References 726

EPIDEMIOLOGY OF BONE CANCER PAIN

Pain is the most common presenting symptom in patients with skeletal metastases and is directly proportional to the patient's quality of life [1, 2]. Two main types of cancer pain exist: ongoing pain, and incident or breakthrough pain. Ongoing pain is typically described as a dull and aching pain that is constant in nature and progresses according to overall disease process. Incident or breakthrough pain is most commonly associated with bone metastases and is characterized by sharp pain, intermittent in nature, exacerbated by movement. Breakthrough pain is difficult to treat, but can be found in as high as 80% of patients with advanced disease [3, 4]. Significant insight into understanding bone cancer pain and the development of new therapeutic strategies for bone cancer pain are due to the development of novel animal models and recent clinical trials.

MECHANISMS OF BONE CANCER PAIN

Rodent and canine models of bone cancer pain have been described in the literature in the past decade. Each model differs in the route of inoculation of tumor cells, type of tumor, immunocompetency of the host, and species of host [5]. Despite these differences, a wealth of information has been generated regarding the pathophysiologic mechanisms that drive bone cancer pain. Ultimately, bone cancer pain is a multifactorial process that is initiated by a complex interaction between the host cells within the affected bone and the tumor cells.

Pain generally occurs during tissue damage as the result of release of neurotransmitters, cytokines, and other factors from damaged cells, reactive or activated inflammatory cells, adjacent blood vessels, and nerve terminals. Pain is transduced at the level of the primary afferent nerve fiber that innervates peripheral tissue. Bone is densely innervated by sensory nerve fibers within the bone marrow, mineralized bone, and periosteum (Fig. 87.1) [6]. Sensory and sympathetic neurons form a mesh-like network throughout the periosteum in association with blood vessels that can detect small distortions of skeletal integrity (Fig. 87.2) [7].

The majority of metastatic skeletal malignancies are destructive in nature and produce regions of osteolysis (bone destruction). This occurs via activation, recruitment, and proliferation of osteoclasts and is characterized by an increased number and size of osteoclasts found in tumor-bearing sites [8–11]. The activation and proliferation of osteoclasts is mediated by the interaction between receptor activator for nuclear factor-κB (RANK) expressed on osteoclasts with RANK ligand (RANKL) expressed on osteoblasts. An increased expression of both RANK and RANKL has been found in tumor-bearing sites. Selective inhibition of osteoclasts using either bisphosphonates or the soluble decoy receptor for RANKL, osteoprotegerin (OPG) results in inhibition of cancer-induced osteolysis, cancer pain behaviors, and neurochemical markers of peripheral and central sensitization [12–16].

Primer on the Metabolic Bone Diseases and Disorders of Mineral Metabolism, Eighth Edition. Edited by Clifford J. Rosen.
© 2013 American Society for Bone and Mineral Research. Published 2013 by John Wiley & Sons, Inc.

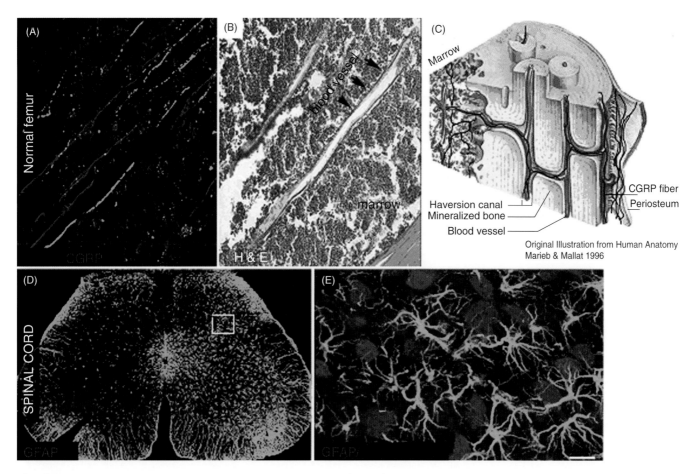

Fig. 87.1. Peripheral and central mechanisms of bone cancer. Histophotomicrographs of (A) confocal and (B) histologic serial images of (D and E) normal bone and confocal images of spinal cord of tumor-bearing mice. Note the extensive myelinated (red, NF 200) and unmyelinated (green, CGRP) nerve fibers within bone marrow that appear to course along blood vessels (arrowheads, B). (C) Schematic diagram demonstrating the innervation within periosteum, mineralized bone, and bone marrow. All three tissues may be sensitized during the various stages of bone cancer pain. (D) Confocal imaging of glial fibrillary acidic protein (GFAP) expressed by astrocytes in a spinal cord of a tumor-bearing mouse. Note increased expression only on the side ipsilateral to the tumorous limb. (E) High-power magnification of spinal cord showing hypertrophy of astrocytes (green) without changes in neuronal numbers (red, stained with neuronal marker, NeuN). NF200: neurofilament 200; CGRP: calcitonin gene-related peptide; GFAP: glial fibrillary acidic protein; NeuN: neuronal marker.

Tumor-derived cytokines, growth factors, and peptides have been shown to activate primary afferent nerve fibers that innervate bone. Prostaglandins, interleukins, protons, bradykinin, chemokines, tumor-necrosis factor-α, nerve growth factor (NGF), and endothelins are all examples of chemical mediators released from tumor cells, or the host inflammatory response, that sensitize nerve terminals resulting in cancer pain [5, 17–19]. Each mediator has a specialized receptor that converts the chemical to an electrical signal (Fig. 87.3). In bone cancer pain, chemical mediators are released that bind to respective receptors causing pain transduction. Neural sensitization occurs when constant nerve stimulation/activation leads to decreased excitation thresholds, upregulation of receptors in nerve terminals, or recruitment of previously silent pain receptors [20–23].

Central sensitization

Central sensitization is the heightened reactivity of nervous system in the face of sustained neural signals. Central sensitization can lead to allodynia, a painful condition where mechanical stimuli not normally perceived as noxious are painful. While central sensitization may occur anywhere along the central or peripheral nervous system, it is most effectively seen in the dorsal horn of the spinal cord. As verified by electrophysiologic and anatomic studies, a change in the activity and responsiveness of dorsal horn neurons occurs in response to persistent painful stimulation. Central sensitization is mediated in part by glutamate, substance P, prostaglandins, and growth factors [24].

Persistent stimulation of unmyelinated C fibers results in increased neural responsiveness of spinal neurons,

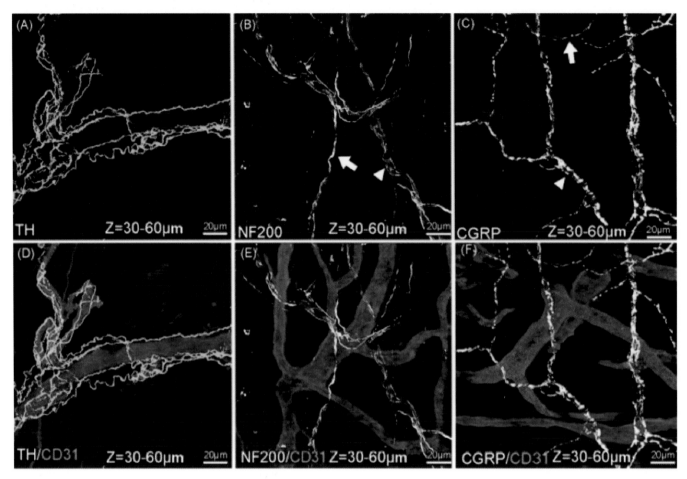

Fig. 87.2. Close association of sensory and sympathetic nerve fibers with blood vessels in the bone periosteum. High-power computed tomography scans of bone in cross section overlaid by confocal images. (A) Sympathetic nerve fibers wrapping around CD31 positive blood vessels of the periosteum (D). (B) NF200+ neurofilament positive and CGRP+ calcitonin gene-related peptide positive sensory nerve fibers (C) do not associate with CD31+ blood vessels [as seen in (E) and (F), respectively].

called "wind-up" [25]. Sensitization can also occur when persistent stimulation results in phenotypic changes in neurons that are adjacent to neurons receiving the persistent painful stimulation. Typically this adjacent sensitization occurs in A-beta fibers that normally do not transmit painful stimuli. Once sensitized, A-beta neurons are capable of transmitting both non-painful and painful information. The molecular understanding of the specific neural pathways involved in central sensitization is currently being investigated as a potential therapeutic option [26, 27].

Reorganization of peripheral and central nervous system in response to cancer pain

Several models of neural sensitization in bone cancer models exist [5, 20, 22]. In normal mice, the neurotransmitter substance P is synthesized by nociceptors and released in the spinal cord when noxious mechanical stress is applied to the femur. Substance P, in turn, binds to and activates the neurokinin-1 receptor that is expressed by a subset of spinal cord neurons, eliciting a response. In mice with bone cancer, the reorganization of nociceptive nerve fibers causes mechanical allodynia where non-painful level of mechanical stress induce the release of substance P, making the stimuli noxious [22].

Phenotypic alterations with extensive neurochemical reorganization in the innervation of tumor-bearing bones occur during the sensitization of peripheral nerves. Specific neural changes that may mediate pain include astrocyte hypertrophy and decreased expression of glutamate re-uptake transporters. The increased extracellular glutamate levels result in central nervous system excitotoxicity [28, 29]. Prolonged pain induces central sensitization, which leads to increased transmission of nociceptive information and allodynia [30].

Recently, progress has been made in understanding the pathophysiology of nociceptive nerve sprouting in prostate cancer [20]. Using a mouse model, fluorescently labeled prostate cancer cells were injected into the bone marrow of naive mice. Twenty-six days after injection,

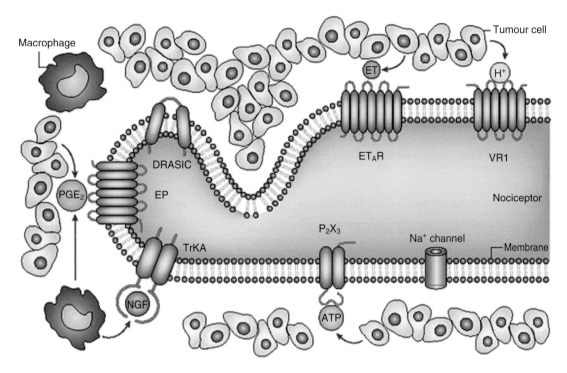

Fig. 87.3. Interaction between chemical mediators and receptors. Schematic diagram of a peripheral pain fiber expressing receptors and ion channels. Interaction between neurotransmitters and chemical mediators and their cognate receptor results in pain transduction and signaling. (H+: protons; ET: endothelin; VR1: vanilloid receptor-1; ETAR: endothelin A receptor; DRASIC: dorsal root acid sensing ion channel; EP: prostaglandin E receptor; PGE2: prostaglandin E2; TrkA: high affinity nerve growth factor tyrosine kinase receptor A; NGF: nerve growth factor; ATP: adenosine triphosphate; P2X3: purinergic ion-gated receptor; Na+: sodium). (Reprinted with permission, Mantyh PW et al. 2002. Molecular mechanisms of cancer pain. *Nat Cancer Rev* 2: 201–9. Review).

nociceptive nerve fibers showed significant new sprouting with increased fiber density and appearance, forming a network of pathological nerve fibers (Fig. 87.4). These data suggest that pathological tumoral sprouting of nociceptive nerve fibers occurs early in the metastatic prostate disease process. To further evaluate the driving force for the new nociceptive fibers, RT-PCR (reverse transcriptase polymerase chain reaction) analysis for NGF showed the surrounding tumor-associated inflammatory, immune, and stromal cells are the major source of NGF in these painful tumors [20]. Proof of concept experiments showed that local anti-NGF therapy can block ectopic sprouting and pathological reorganization of these nociceptive fibers, suggesting that prophylactic treatment may be capable of preventing bone cancer pain [21].

THERAPEUTIC STRATEGIES: PAST, PRESENT, AND FUTURE

Pain research has significantly improved our understanding of acute and chronic pain mechanisms. By highlighting key molecular mechanisms involved in pain transmission, new drugs are currently being investigated as potential novel therapies. Currently available medications such as the opioids are fraught with side effects that limit their clinical efficacy. Research is now focused on targeting pain initiation sites within the nervous system to limit systemic complications.

Therapeutic targets: Ion channels

The transient receptor potential (TRPV1) family of ion channels is located on unmyelinated C fibers and spinal nociceptive neurons that mediate pain transmission. TRPV1 channels can be activated by heat, capsaicin, and acid; mice that lack the channel are unable to develop chronic pain states. Antagonists to TRPV1 administered orally or the intrathecal space significantly decrease chronic pain [18, 31, 32]. Activation of TRPV1 initially provokes a powerful afferent nerve irritant effect, followed by desensitization and long-term analgesia. In a canine model of bone cancer, intrathecal administration of resiniferatoxin, a potent capsaicin analog, resulted in pain reduction and selective destruction of small sensory neurons [33]. As TRPV1 is only expressed on nociceptive peripheral terminals, selective blockade of TRPV1 may provide analgesia with a limited side effect profile [34]. Recent work has focused the role of TRPV1 in the acidic microenvironment of bone metastasis that mediates

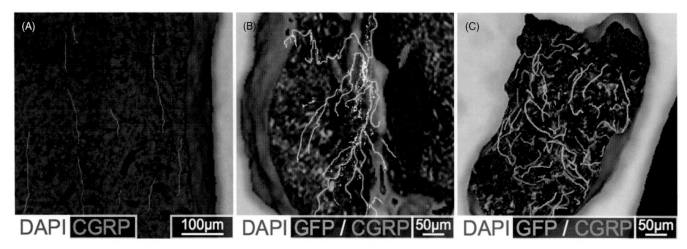

Fig. 87.4. Prostate cancer cells cause sprouting of sensory nerve fibers in bone. High-power computed tomography scans of bone in cross-section overlaid by confocal images. DAPI stained nuclei appear blue, GFP-expressing prostate cancer cells appear green, and CGRP+ sensory nerve fibers appear yellow/red. (A) Sham femur showing control level of nerve sprouting seen in characteristic linear morphology. (B) Prostate tumor-bearing femur from mice killed at the early stage of metastatic disease showing tumor colonies and marked highly branched sensory nerve sprouting. (C) Prostate tumor-bearing femur from mice killed at an advanced stage of metastatic disease with a high density of sensory nerve fibers.

pain. Specifically, acid signals received by the sensory nociceptive neurons innervating bone stimulate intracellular signaling pathways of sensory neurons. Molecular blockade of the activated intracellular transcription factors in these signaling pathways has served as a method to inhibit pain transmission [19, 35].

Therapeutic targets: Cytokines and growth factors

NGF modulates inflammatory and neuropathic pain states. In chronic pain, NGF levels are elevated in peripheral tissues and neutralizing antibodies against NGF are effective in reducing and in some cases preventing chronic pain [36]. *In vitro* studies have shown that growth and differentiation of NGF-dependent sensory nerve cell lines can be inhibited by naturally neutralizing antibodies. More recently, these same antibodies have been shown to inhibit the *in vitro* migration and metastasis of prostate cancer cells [23]. In addition, pathological sprouting of nerve fibers in a prostate cancer model is modulated in a NGF-dependent fashion (Fig. 87.5) [20]. In animal models, anti-NGF antibodies reduce continuous and breakthrough pain by blocking the nociceptive stimuli associated with the sensitization in the peripheral or central nervous system [36].

In addition to NGF, two other growth factors have been implicated in cancer-related bone pain: glial-derived growth factor (GDNF) and brain-derived growth factor (BDNF). BDNF has been implicated in the modulation of central sensitization as its expression is increased in nociceptive neurons in models of chronic neuropathy. BNDF sensitizes C fiber activity, resulting in hyperalgesia and allodynia. Inhibition of BNDF and its cognate receptor, TrkB, results in decreased C fiber firing and a reduction in pain behaviors [37, 38]. GDNF is important in the survival of sensory neurons and supporting neural cells. Neuropathic pain behaviors commonly observed in animal models of chronic pain are prevented or reversed following GDNF administration. The analgesic effects of GDNF show strong temporal and molecular regulation. Specifically, the timing of administration of GDNF directly determines whether analgesia effects are observed [38, 39].

Endothelins are a family of vasoactive peptides that are expressed by several tumors, with levels that appear to correlate with pain severity. Direct application of endothelin to peripheral nerves induces activation of primary afferent fibers and pain-specific behaviors. As such, endothelins may contribute to cancer pain by directly sensitizing nociceptors [40]. Selective blockade of endothelin receptors blocks bone cancer pain-related behaviors and spinal changes indicative of peripheral and central sensitization [41, 42]. Endothelin antagonists are a pharmacological agent that may show significant promise in the management of cancer pain [40].

Therapeutic targets: Osteoclast

Osteoclasts play an essential role in cancer-induced bone loss and contribute to the etiology of bone cancer pain. Bisphosphonates act by disrupting the activity of osteoclasts, the principle bone cell responsible for bone resorption [43]. Currently, bisphosphonates have shown clinical success in treatment of both osteoporosis and tumor-

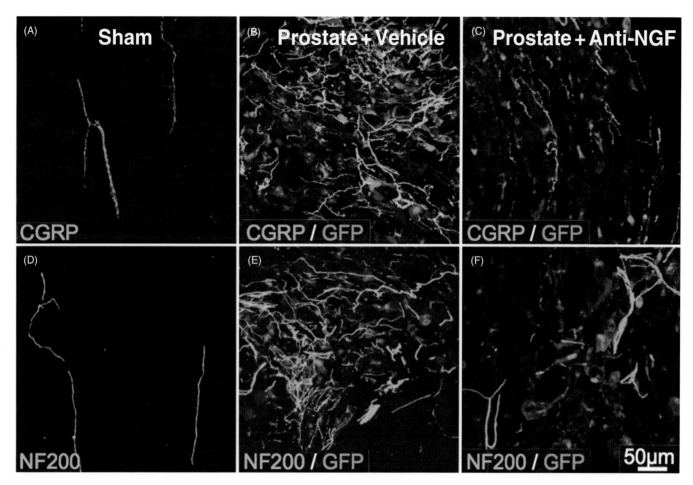

Fig. 87.5. The mesh-like network of nociceptic nerve sprouting in prostate cancer is inhibited by anti-NGF therapy. High-power computed tomography scans of bone in cross section overlaid by confocal images. CGRP+ and NF200+ nerve fibers appear orange and yellow respectively, GFP-expressing prostate cancer cells appear green. (A), (B): Sham-operated mice show regular innervation of bone by two types of nerve fibers: (A) CGRP+ and (D) NF200+. (B), (E): GFP transfected prostate cancer cells growing in bone after 26 days, with the CGRP+ and NF200+ nerve fibers. (C), (F): prevention of CGRP+ and NF200+ nerve fiber sprouting due to anti-NGF antibody therapy.

induced osteolysis. Recently, administration of bisphosphonates has shown a positive impact on overall skeletal health and quality of life in patients with breast and prostate skeletal metastasis [15, 16]. In addition, initiation of zoledronic acid thearpy prior to the development of skeletal metastasis has recently been shown to improve quality of life scores in patients with prostate cancer while decreasing measurements of clinical pain and skeletal related events such as pathological fracture [44]. The long-term beneficial effects of bisphosphonate treatment in reducing bone pain and skeletal related events (e.g. axial or appendicular pathologic fractures) and the patient reported improvement in overall quality of life are clear from clinical trials in lung, breast, and prostate cancer [44–46].

Tumor-induced osteolysis is stimulated by RANKL, and inhibited by OPG. While direct therapeutic targeting of OPG in humans has never been successful, recent work targeting RANKL has shown success in treating bone cancer pain and pathological fracture related complications. Specifically, denosumab (a human monoclonal antibody against RANKL) was recently evaluated against zoledronic acid (bisphosphonate) in a randomized clinical trial evaluating the prevention of skeletal-related events in patients with bone metastases from breast cancer. While both therapies were well tolerated and delayed or prevented skeletal-related events, denosumab trended toward superior reductions in patient reported pain and improved patient quality of life [47]. In addition to being effective in patients with breast cancer, recently, denosumab was compared to zoledronic acid in a phase III clinical trial for patients with metastatic prostate cancer. The results showed a greater decrease in skeletal-related events such as pathological fracture in patients taking denosumab than those patients taking zoledronic acid [48].

REFERENCES

1. Coleman RE. 2006. Clinical features of metastatic bone disease and risk of skeletal morbidity. *Clin Cancer Res* 12(20 Pt 2): 6243s–6249s.
2. Mantyh PW. 2006. Cancer pain and its impact on diagnosis, survival and quality of life. *Nat Rev Neurosci* 7(10): 797–809.
3. Swarm R, Abernethy AP, Anghelescu DL, et al. 2010. Adult cancer pain. *J Natl Compr Canc Netw* 8(9): 1046–1086.
4. Mercadante S, Fulfaro F. 2007. Management of painful bone metastases. *Curr Opin Oncol* 19(4): 308–314.
5. Jaggi AS, Jain V, Singh N. 2011. Animal models of neuropathic pain. *Fundam Clin Pharmacol* 25(1): 1–28.
6. Mach DB, Rogers SD, Sabino MC, et al. 2002. Origins of skeletal pain: Sensory and sympathetic innervation of the mouse femur. *Neuroscience* 113(1): 155–166.
7. Martin CD, Jimenez-Andrade JM, Ghilardi JR, Mantyh PW. 2007. Organization of a unique net-like meshwork of CGRP+ sensory fibers in the mouse periosteum: Implications for the generation and maintenance of bone fracture pain. *Neurosci Lett* 427(3): 148–152.
8. Taube T, Elomaa I, Blomqvist C, Beneton MN, Kanis JA. 1994. Histomorphometric evidence for osteoclast-mediated bone resorption in metastatic breast cancer. *Bone* 15(2): 161–166.
9. Clohisy DR, Ramnaraine ML. 1998. Osteoclasts are required for bone tumors to grow and destroy bone. *J Orthop Res* 16(6): 660–666.
10. Sterling JA, Edwards JR, Martin TJ, Mundy GR. 2011. Advances in the biology of bone metastasis: How the skeleton affects tumor behavior. *Bone* 48(1): 6–15.
11. Zhang Y, Ma B, Fan Q. 2010. Mechanisms of breast cancer bone metastasis. *Cancer Lett* 292(1): 1–7.
12. Clohisy DR, Ramnaraine ML, Scully S, et al. 2000. Osteoprotegerin inhibits tumor-induced osteoclastogenesis and bone tumor growth in osteopetrotic mice. *J Orthop Res* 18(6): 967–976.
13. Roudier MP, Bain SD, Dougall WC. 2006. Effects of the RANKL inhibitor, osteoprotegerin, on the pain and histopathology of bone cancer in rats. *Clin Exp Metastasis* 23(3–4): 167–175.
14. Lamoureux F, Moriceau G, Picarda G, Rousseau J, Trichet V, Redini F. 2010. Regulation of osteoprotegerin pro- or anti-tumoral activity by bone tumor microenvironment. *Biochim Biophys Acta* 1805(1): 17–24.
15. Diel IJ. 2007. Effectiveness of bisphosphonates on bone pain and quality of life in breast cancer patients with metastatic bone disease: A review. *Support Care Cancer* 15(11): 1243–1249.
16. Rodrigues P, Hering F, Campagnari JC. 2004. Use of bisphosphonates can dramatically improve pain in advanced hormone-refractory prostate cancer patients. *Prostate Cancer Prostatic Dis* 7(4): 350–354.
17. White FA, Jung H, Miller RJ. 2007. Chemokines and the pathophysiology of neuropathic pain. *Proc Natl Acad Sci U S A* 104(51): 20151–20158.
18. White JP, Urban L, Nagy I. 2011. TRPV1 function in health and disease. *Curr Pharm Biotechnol* 12(1): 130–144.
19. Yoneda T, Hata K, Nakanishi M, et al. 2011. Involvement of acidic microenvironment in the pathophysiology of cancer-associated bone pain. *Bone* 48(1): 100–105.
20. Jimenez-Andrade JM, Bloom AP, Stake JI, et al. 2010. Pathological sprouting of adult nociceptors in chronic prostate cancer-induced bone pain. *J Neurosci* 30(44): 14649–14656.
21. Mantyh WG, Jimenez-Andrade JM, Stake JI, et al. 2010. Blockade of nerve sprouting and neuroma formation markedly attenuates the development of late stage cancer pain. *Neuroscience* 171(2): 588–598.
22. Schmidt BL, Hamamoto DT, Simone DA, Wilcox GL. 2010. Mechanism of cancer pain. *Mol Interv* 10(3): 164–178.
23. Warrington RJ, Lewis KE. 2011. Natural antibodies against nerve growth factor inhibit in vitro prostate cancer cell metastasis. *Cancer Immunol Immunother* 60(2): 187–195.
24. Latremoliere A, Woolf CJ. 2009. Central sensitization: A generator of pain hypersensitivity by central neural plasticity. *J Pain* 10(9): 895–926.
25. Woolf CJ. 2011. Central sensitization: Implications for the diagnosis and treatment of pain. *Pain* 152(3 Suppl): S2–15.
26. Xiaoping G, Xiaofang Z, Yaguo Z, Juan Z, Junhua W, Zhengliang M. 2010. Involvement of the spinal NMDA receptor/PKCgamma signaling pathway in the development of bone cancer pain. *Brain Res* 1335: 83–90.
27. Yanagisawa Y, Furue H, Kawamata T, et al. 2010. Bone cancer induces a unique central sensitization through synaptic changes in a wide area of the spinal cord. *Mol Pain* 6: 38.
28. Schwei MJ, Honore P, Rogers SD, et al. 1999. Neurochemical and cellular reorganization of the spinal cord in a murine model of bone cancer pain. *J Neurosci* 19(24): 10886–10897.
29. Gao YJ, Ji RR. 2010. Targeting astrocyte signaling for chronic pain. *Neurotherapeutics* 7(4): 482–493.
30. Sabino MA, Mantyh PW. 2005. Pathophysiology of bone cancer pain. *J Support Oncol* 3(1): 15–24.
31. Ghilardi JR, Rohrich H, Lindsay TH, et al. 2005. Selective blockade of the capsaicin receptor TRPV1 attenuates bone cancer pain. *J Neurosci* 25(12): 3126–3131.
32. Cui M, Honore P, Zhong C, et al. 2006. TRPV1 receptors in the CNS play a key role in broad-spectrum analgesia of TRPV1 antagonists. *J Neurosci* 26(37): 9385–9393.
33. Brown DC, Iadarola MJ, Perkowski SZ, et al. 2005. Physiologic and antinociceptive effects of intrathecal resiniferatoxin in a canine bone cancer model. *Anesthesiology* 103(5): 1052–1059.
34. Premkumar LS. 2010. Targeting TRPV1 as an alternative approach to narcotic analgesics to treat chronic pain conditions. *AAPS J* 12(3): 361–370.
35. Ghilardi JR, Rohrich H, Lindsay TH, et al. 2005. Selective blockade of the capsaicin receptor TRPV1 attenuates bone cancer pain. *J Neurosci* 25(12): 3126–3131.

36. Sevcik MA, Ghilardi JR, Peters CM, et al. 2005. Anti-NGF therapy profoundly reduces bone cancer pain and the accompanying increase in markers of peripheral and central sensitization. *Pain* 115(1–2): 128–141.
37. Wright MA, Ribera AB. 2010. Brain-derived neurotrophic factor mediates non-cell-autonomous regulation of sensory neuron position and identity. *J Neurosci* 30(43): 14513–14521.
38. Hunt SP, Mantyh PW. 2001. The molecular dynamics of pain control. *Nat Rev Neurosci* 2(2): 83–91.
39. Patil SB, Brock JH, Colman DR, Huntley GW. 2011. Neuropathic pain- and glial derived neurotrophic factor-associated regulation of cadherins in spinal circuits of the dorsal horn. *Pain* 152(4): 924–935.
40. Hans G, Deseure K, Adriaensen H. 2008. Endothelin-1-induced pain and hyperalgesia: A review of pathophysiology, clinical manifestations and future therapeutic options. *Neuropeptides* 42(2): 119–132.
41. Peters CM, Lindsay TH, Pomonis JD, et al. 2004. Endothelin and the tumorigenic component of bone cancer pain. *Neuroscience* 126(4): 1043–1052.
42. Davar G. 2001. Endothelin-1 and metastatic cancer pain. *Pain Med* 2(1): 24–27.
43. Baron R, Ferrari S, Russell RG. 2011. Denosumab and bisphosphonates: Different mechanisms of action and effects. *Bone* 48(4): 677–692.
44. Saad F, Eastham J. 2010. Zoledronic acid improves clinical outcomes when administered before onset of bone pain in patients with prostate cancer. *Urology* 76(5): 1175–1181.
45. Broom R, Du H, Clemons M, et al. 2009. Switching breast cancer patients with progressive bone metastases to third-generation bisphosphonates: Measuring impact using the Functional Assessment of Cancer Therapy-Bone Pain. *J Pain Symptom Manage* 38(2): 244–257.
46. Namazi H. 2008. Zoledronic acid and survival in patients with metastatic bone disease from lung cancer and elevated markers of osteoclast activity: A novel molecular mechanism. *J Thorac Oncol* 3(8): 943–944.
47. Stopeck AT, Lipton A, Body JJ, et al. 2010. Denosumab compared with zoledronic acid for the treatment of bone metastases in patients with advanced breast cancer: A randomized, double-blind study. *J Clin Oncol* 28(35): 5132–5139.
48. Fizazi K, Carducci M, Smith M, et al. 2011. Denosumab versus zoledronic acid for treatment of bone metastases in men with castration-resistant prostate cancer: A randomised, double-blind study. *Lancet* 377(9768): 813–822.

88

Radiation Therapy-Induced Osteoporosis

Jeffrey S. Willey, Shane A.J. Lloyd, and Ted A. Bateman

Introduction 728
Fractures of Irradiated Bones after Clinical Exposure 728
Radiation and Vasculature 729
Radiation Effects on Osteoblasts 729
Radiation Effects on Osteocytes 729
Radiation Effects on Osteoclasts 730

Deterioration of Bone Quantity and Quality after Irradiation 730
Loss of Bone Strength in Animal Models after Irradiation 730
Conclusion 731
Acknowledgments 731
References 731

INTRODUCTION

Ionizing radiation is an effective cancer treatment modality. In recent decades, improvements in the diagnosis, planning, and delivery of radiotherapy (RT) have helped decrease mortality rates among some populations of cancer patients [1]. With this increase in overall long-term cancer survivorship, prevention of radiation-induced normal (noncancerous) tissue injury has received more attention. Skeletal complications, particularly fractures of the hip, represent particularly debilitating radiation late effects [2–6]. The incidental irradiation of normal skeletal tissues is not insignificant, as the dose absorbed by normal tissues in close proximity to the tumor can be substantial. For example, RT for gynecological malignancies commonly involves administration of 30 1.8 Gy fractions over 6 weeks, applying approximately a 54-Gy total dose to the tumor. Normal tissues in each hip can receive as much as half of each fraction, depending on tumor type and location. Advancements in surgical excision, concurrent chemotherapy, and modern RT procedures have helped to limit the dose absorbed by some adjacent normal skeletal elements. However, fractures that occur at irradiated skeletal sites, such as the hip and spine, remain a major source of morbidity and can have fatal consequences [7].

While radiation-induced fractures are well documented in the literature, recent animal and cellular work has helped to identify potential new mechanisms underlying these observations. Early loss of bone resulting from activation of osteoclasts likely contributes to the well-established reduction in vasculature and bone formation following irradiation [8, 9]. RT-induced osteoporosis, and the resulting increase in fracture risk, appears to result from a rapid but temporary period of bone resorption, followed by long-term suppression of bone turnover that ultimately prevents recovery of bone mass and can impair bone quality. In this review, we will outline the rationale for radiation-induced fractures as a clinical concern, as well as the current state of knowledge regarding *in vivo* and *in vitro* models of radiation-induced osteoporosis. Attention will be primarily directed toward acquired deficits in the appendicular skeleton and axial components of the trunk.

FRACTURES OF IRRADIATED BONES AFTER CLINICAL EXPOSURE

The increased incidence of pathologic fracture that occurs within the irradiated volume is largely associated with absorbed dose at or near the fracture location [10]. Patients receiving breast cancer RT exhibit rib fracture rates ranging from 1.8% [11] to 19% [12]. Similarly, patients receiving RT for various pelvic malignancies are at increased risk for fracture at pelvic skeletal locations that

---

Primer on the Metabolic Bone Diseases and Disorders of Mineral Metabolism, Eighth Edition. Edited by Clifford J. Rosen.
© 2013 American Society for Bone and Mineral Research. Published 2013 by John Wiley & Sons, Inc.

absorb dose, including structures of the "hip" (notably the femoral neck, sacrum, and acetabular rim) [2, 13–16]. Among men treated for prostate cancer, fracture incidence was shown in one study to be approximately 6.8%, with a median time to diagnosis of 20 months [17].

In women treated for gynecologic malignancies, the prevalence and timing of these fractures varies among published reports. Kwon and colleagues evaluated pelvic magnetic resonance (MR) images of 510 uterine cancer patients before and after RT: 100 patients (20%) had insufficiency fractures in the pelvis [16], and of these 100 patients, 61% had multiple fractures. The 5-year fracture prevalence rate was 45%, with the cumulative prevalence increasing from 15%, 27%, 32% and 35% at 1-, 2- 3-, and 4-years post-RT, respectively. Only 0.4% of patients experienced avascular necrosis of the femoral head. Two other studies using computed tomography (CT), rather than MR, to diagnose radiographic pelvic fractures in patients with gynecological tumors identified fracture rates of 10% and 13% [15, 18]. In all three studies, a large majority of the insufficiency fractures were near the sacroiliac joint [15, 16, 18]. In the Kwon et al. study, there was no significant difference in the fracture rate of patients who underwent concurrent chemotherapy (17%) during RT and those who did not (21%) [16].

A comprehensive study used data from the National Cancer Institute-Surveillance Epidemiology and End Results–Medicare cancer registry to examine traumatic pelvic fracture rates in nearly 6,500 women over the age of 65 with nonmalignant anal, cervical, or rectal cancers [2]. Using a proportional hazards model, they determined that irradiation increased the hazard ratios for pelvic fractures to 3.16, 1.66, and 1.65 for anal, cervical, and rectal cancers, respectively, relative to patients who did not receive external beam RT as part of treatment at any time post diagnosis. Five-years after RT, fracture rates for non-RT vs RT patients were 7.5% vs 14% (anal cancer), 5.9% vs 8.2% (cervical cancer), and 8.7% vs 11.2% (rectal cancer) [2]. Fracture rates from the arm and vertebrae, which were located outside of the radiation field, were not significantly greater following RT. Anal cancer patients, with a three-times greater risk for pelvic fracture compared to non-RT protocols, have greater doses of irradiation to the femoral head (targeted at the inguinal lymph nodes), again supporting a hypothesis that local radiation effects are a primary contributor to increased fracture risk. While systemic or non-targeted radiation effects cannot be discounted [19], the deleterious clinical response appears to be limited to the irradiated volume.

RADIATION AND VASCULATURE

Damage that occurs to bone following irradiation has historically been considered to result from physiological changes occurring within the vasculature, as well as changes to bone cell number and function [13, 20–25]. Since the turn of the 20th century, radiation-induced bone damage (previously termed "osteitis") was thought to be initiated by a reduction in vascularity throughout the bone [26]. This vascular damage is characterized by swelling and vacuolization of endothelial cells within the osteon [20, 21, 25]. In addition, sclerotic connective tissue is deposited within the marrow cavity. Late fibrosis of the subintima also occurs, along with replacement of vascular smooth muscle cells with a hyaline-like material. These changes result in eventual constriction of the vessel lumen, resulting in localized hypoxia. Bones of the skull and jaw are considered especially at risk for vascular injury due to their superficial location and relative lack of blood vessels [14]. Indeed, loss of vasculature in both the marrow cavity and Haversian system has been identified within the bone of many animal models following radiation exposure [21, 27, 28].

RADIATION EFFECTS ON OSTEOBLASTS

Damage to osteoblasts following irradiation has long been recognized as the greatest contributor to the development of reduced bone mineral density [13, 20, 25, 29]. The overall decrease in the number and activity of osteoblasts following radiation exposure has been documented, along with attenuated matrix formation [9, 28]. *In vitro* and *in vivo* data show impaired bone formation following a decrease in osteoblast proliferation and differentiation, osteoblast cell-cycle arrest, reduced collagen synthesis, and increased apoptosis [24, 30–32]. Lowered levels of runt-related transcription factor 2 (RUNX2) in osteoblast cultures stimulated with bone morphogenic protein-2 (BMP-2) have been observed following irradiation [32]. Receptor activator of nuclear factor-κB ligand (RANKL) mRNA levels in osteoblasts tend to be elevated following irradiation with photons [33]. Additionally, osteoblast precursors appear damaged following exposure [8]. Mesenchymal stem cell (MSC) colony forming ability and overall numbers are lowered after radiation exposure [28]. It may be that oxidative stress contributes to this response [8, 28]. It is important to note, however, that other groups have shown no loss of MSC viability within irradiated bone. Instead, they suggest that damage occurs as a later effect, when terminal osteoblast differentiation occurs [34].

RADIATION EFFECTS ON OSTEOCYTES

While the effects of radiation on osteocyte viability remain unclear, evidence suggests that these cells may be less radiosensitive than osteoblasts. Several studies investigating the status of osteocytes following radiation exposure indicate that the cells remain viable for several months after acute exposures in both mice and rabbits [29, 35–37]. However, other studies have identified loss of osteocytes within irradiated bone [13, 20, 21]. Osteocyte death was confirmed within the cortical lamellar and Haversian bone of monkey mandibles irradiated with 45 Gy [21]. Interestingly, osteocyte numbers were not affected within trabecular bone.

RADIATION EFFECTS ON OSTEOCLASTS

Research over the past decade has highlighted the substantial role that osteoclasts play in radiation-induced osteoporosis [8, 9]. Irradiation results in increased osteoclast number and activity within the first few days after exposure [9]. Serum tartrate-resistant acid phosphatase (TRAP5b) is increased as early as 24 hours after exposure to total-body X-rays. Osteoclast number and surface (normalized to bone surface) are both greater in rodent bones within the first 3 days after exposure [38]. This early increase in osteoclast number and activity precede a subsequent loss of bone within a week of treatment [8, 9]. However, the majority of the bone loss occurs when the number of osteoblasts and bone formation are unchanged relative to control [9]. The use of bisphosphonate antiresorptive agents (e.g., risedronate) to suppress osteoclast activity can effectively block both a radiation-induced increase in osteoclast activity and loss of bone at multiple skeletal sites [9]. The initial and early increase in osteoclast activity may then be followed by a prolonged decrease in osteoclast activity [39, 40]. A persistent reduction of both bone formation and resorption could sufficiently suppress bone remodeling as to result in damage to the material properties of skeletal tissue [41], which has been described in rodents [42]. A combination of an early yet acute increase in osteoclast activity leading to bone loss, followed by a lengthy reduction in bone formation, could compromise the structural integrity of bone as a late effect. The resulting reduction in turnover could, by lowering bone density and/or altering the mechanical or material properties of bone, lead to an increased incidence of fracture [43].

DETERIORATION OF BONE QUANTITY AND QUALITY AFTER IRRADIATION

Skeletal fractures are known to be a late effect resulting from direct irradiation. These fractures are thought to result from radiation-induced deterioration of both bone quantity and quality [2, 20, 44]. Several factors can influence the response of skeletal tissue to radiation exposure, including the total absorbed dose, radiation energy, dose per fraction, and developmental stage of the patient [12–14]. Osteopenia is often observed at 1-year after cancer RT [13, 25]. The radiation-induced deficits in bone that have been documented include demineralization, thinning, sclerosis, and loss of trabecular bone [13, 20, 25, 44]. However, trabeculae with abnormal, thickened morphology have been documented within the irradiated volumes of patients [14] and animal models [27, 45].

A more comprehensive study published nearly 20 years ago highlights substantial skeletal damage as an early response to RT. Eleven patients with cancer of the uterus/cervix underwent RT, with the third lumbar spine (L3) receiving dose as it was within the treatment field [46]. Twelve patients underwent a therapy regimen that did not include RT. CT scans were performed before treatment, at the end of RT (5 weeks), and at 3, 6, and 12 months post therapy. Quantitative computed tomography (QCT) analysis was performed to quantify L3 trabecular bone mineral content (BMC) values at each time point. Patients not receiving RT had no change in BMC over the course of the year; however, a dramatic loss of BMC was observed from the RT group. These patients experienced a loss of BMC of 32% at 5 weeks, 40% at 3 months, 47% at 6 months, and 49% at 12 months compared to the pre-RT scan [46]. During the period of RT, trabecular bone loss was exceptionally rapid (−6%/week), with significant loss continuing to 6 months without subsequent recovery. One patient experienced a compression fracture of L2 at 6 months after RT. The authors discuss radiation injury to both osteoblasts and osteoclasts, and state that "bone change due to irradiation is a type of low-turnover osteoporosis" [46]. However, the substantial loss of BMC at such early times after initiating RT supports the hypothesis that early activation of osteoclastic bone resorption has a causal role in radiation therapy-induced osteoporosis.

From animal models, acute deterioration of trabecular bone architecture has been identified following exposure to ionizing radiation. Loss of trabecular bone has been documented using microcomputed tomography (μCT) as little as 3 days after exposure to a 2-Gy dose of photons [8]. Trabecular bone microarchitecture and mass are eroded in the proximal tibia, distal femur, and throughout the fifth lumbar vertebrae of mice 1 week after exposure to a 2-Gy dose of radiation [9]. This early loss of trabecular bone appears to translate into prolonged deficits in bone quantity and quality; loss of trabecular bone persists for months after both photon and charged particle exposures [47, 48].

LOSS OF BONE STRENGTH IN ANIMAL MODELS AFTER IRRADIATION

Animal models are necessary to examine not only the mechanisms behind radiation-induced bone loss, but also to identify the cause, extent, and timing of strength reductions in bony elements. Compared with cortical bone, photon radiation appears to produce greater damage to the trabecular network [47]. However, at relatively low doses (i.e., 50 cGy), high linear energy transfer (LET) heavy ion radiation does appear to increase cortical bone porosity, cortical area, and polar moment of inertia [49]. To date, the most direct assessment of radiation-induced reduction in bone strength comes from rodent and rabbit models, although studies are limited. Reduced ultimate strength of cortical bone from rabbit tibiae has been observed at 4 and 12 months after exposure to a 50-Gy total dose [36]. Compressive testing of mouse distal femora showed a reduction of strength at 12 weeks after 5- and 12-Gy acute doses of X-rays [42]. The subsequent loss of strength as determined by compressive testing and estimated by finite element analysis (FEA) occurred despite a transient increase in bone volume and a sustained elevation of cortical bone mineral content. That is, the bone appeared to be more brittle. Changes in bone strength after irradiation may be influenced by both architectural and material properties. Additionally, a 2-Gy dose of heavy ion radiation induces a loss of vertebral stiffness when tested by compression loading and determined by FEA [50].

CONCLUSION

Clinical RT can compromise the health of normal skeletal tissue, leading to fracture. However, limited research has defined the mechanisms underlying radiation-induced bone loss. Even fewer studies have identified the magnitude and extent of bone loss in patients following RT. Based on research using animal models and cell culture studies, radiation appears to cause an early increase in active bone resorption by somehow stimulating osteoclast activity, followed by a persistent reduction in bone formation. Strength may be further compromised by lowered overall bone density, impaired bone material properties resulting from a low-turnover state, a combination of the two, or other factors. Future research should be aimed at identifying the mediators of bone damage following radiation exposure in order to develop effective therapeutic countermeasures for radiation-induced fracture.

ACKNOWLEDGMENTS

This research is supported by the National Space Biomedical Research Institute through NASA NCC 9-58 (BL01302 TAB; PF01403 JSW), NASA Cooperative Agreement NCC9-79, and the National Institutes of Health (NIAMS 1R01AR059221-01A1 TAB). Support was also provided by the National Institutes of Health (T32 CA113267 JSW). We would also like to offer thanks to Drs. Richard Loeser and Michael Robbins at Wake Forest School of Medicine, and to Dr. Henry J. Donahue (NIH R01AG13087 and R01AG015107) for partial funding support (SAJL).

REFERENCES

1. Bernier J, Hall EJ, Giaccia A. 2004. Radiation oncology: A century of achievements. *Nat Rev Cancer* 4(9): 737–747.
2. Baxter NN, Habermann EB, Tepper JE, Durham SB, Virnig BA. 2005. Risk of pelvic fractures in older women following pelvic irradiation. *JAMA* 294(20): 2587–2593.
3. Brown SA, Guise TA. 2009. Cancer treatment-related bone disease. *Crit Rev Eukaryot Gene Expr* 19(1): 47–60.
4. Guise TA. 2006. Bone loss and fracture risk associated with cancer therapy. *Oncologist* 11(10): 1121–1131.
5. Florin TA, Fryer GE, Miyoshi T, Weitzman M, Mertens AC, Hudson MM, Sklar CA, Emmons K, Hinkle A, Whitton J, Stovall M, Robison LL, Oeffinger KC. 2007. Physical inactivity in adult survivors of childhood acute lymphoblastic leukemia: A report from the childhood cancer survivor study. *Cancer Epidemiol Biomarkers Prev* 16(7): 1356–1363.
6. Oeffinger KC, Mertens AC, Sklar CA, Kawashima T, Hudson MM, Meadows AT, Friedman DL, Marina N, Hobbie W, Kadan-Lottick NS, Schwartz CL, Leisenring W, Robison LL. 2006. Chronic health conditions in adult survivors of childhood cancer. *N Engl J Med* 355(15): 1572–1582.
7. Small W Jr, Kachnic L. 2005. Postradiotherapy pelvic fractures: Cause for concern or opportunity for future research? *JAMA* 294(20): 2635–2637.
8. Kondo H, Searby ND, Mojarrab R, Phillips J, Alwood J, Yumoto K, Almeida EA, Limoli CL, Globus RK. 2009. Total-body irradiation of postpubertal mice with (137) Cs acutely compromises the microarchitecture of cancellous bone and increases osteoclasts. *Radiat Res* 171(3): 283–289.
9. Willey JS, Livingston EW, Robbins ME, Bourland JD, Tirado-Lee L, Smith-Sielicki H, Bateman TA. 2010. Risedronate prevents early radiation-induced osteoporosis in mice at multiple skeletal locations. *Bone* 46(1): 101–111.
10. Dickie CI, Parent AL, Griffin AM, Fung S, Chung PW, Catton CN, Ferguson PC, Wunder JS, Bell RS, Sharpe MB, O'Sullivan B. 2009. Bone fractures following external beam radiotherapy and limb-preservation surgery for lower extremity soft tissue sarcoma: Relationship to irradiated bone length, volume, tumor location and dose. *Int J Radiat Oncol Biol Phys* 75(4): 1119–1124.
11. Pierce SM, Recht A, Lingos TI, Abner A, Vicini F, Silver B, Herzog A, Harris JR. 1992. Long-term radiation complications following conservative surgery (CS) and radiation therapy (RT) in patients with early stage breast cancer. *Int J Radiat Oncol Biol Phys* 23(5): 915–923.
12. Overgaard M. 1988. Spontaneous radiation-induced rib fractures in breast cancer patients treated with postmastectomy irradiation. A clinical radiobiological analysis of the influence of fraction size and dose-response relationships on late bone damage. *Acta Oncol* 27(2): 117–122.
13. Mitchell MJ, Logan PM. 1998. Radiation-induced changes in bone. *Radiographics* 18(5): 1125–1136; quiz 1242–1123.
14. Williams HJ, Davies AM. 2006. The effect of X-rays on bone: A pictorial review. *Eur Radiol* 16(3): 619–633.
15. Schmeler KM, Jhingran A, Iyer RB, Sun CC, Eifel PJ, Soliman PT, Ramirez PT, Frumovitz M, Bodurka DC,

Sood AK. 2010. Pelvic fractures after radiotherapy for cervical cancer: Implications for survivors. *Cancer* 116(3): 625–630.
16. Kwon JW, Huh SJ, Yoon YC, Choi SH, Jung JY, Oh D, Choe BK. 2008. Pelvic bone complications after radiation therapy of uterine cervical cancer: Evaluation with MRI. *AJR Am J Roentgenol* 191(4): 987–994.
17. Igdem S, Alco G, Ercan T, Barlan M, Ganiyusufoglu K, Unalan B, Turkan S, Okkan S. 2010. Insufficiency fractures after pelvic radiotherapy in patients with prostate cancer. *Int J Radiat Oncol Biol Phys* 77(3): 818–823.
18. Ikushima H, Osaki K, Furutani S, Yamashita K, Kishida Y, Kudoh T, Nishitani H. 2006. Pelvic bone complications following radiation therapy of gynecologic malignancies: Clinical evaluation of radiation-induced pelvic insufficiency fractures. *Gynecol Oncol* 103(3): 1100–1104.
19. Jia D, Gaddy D, Suva LJ, Corry PM. 2011. Rapid loss of bone mass and strength in mice after abdominal irradiation. *Radiat Res* 176(5): 624–635.
20. Ergun H, Howland WJ. 1980. Postradiation atrophy of mature bone. *CRC Crit Rev Diagn Imaging* 12(3): 225–243.
21. Rohrer MD, Kim Y, Fayos JV. 1979. The effect of cobalt-60 irradiation on monkey mandibles. *Oral Surg Oral Med Oral Pathol* 48(5): 424–440.
22. Bliss P, Parsons CA, Blake PR. 1996. Incidence and possible aetiological factors in the development of pelvic insufficiency fractures following radical radiotherapy. *Br J Radiol* 69(822): 548–554.
23. Konski A, Sowers M. 1996. Pelvic fractures following irradiation for endometrial carcinoma. *Int J Radiat Oncol Biol Phys* 35(2): 361–367.
24. Gal TJ, Munoz-Antonia T, Muro-Cacho CA, Klotch DW. 2000. Radiation effects on osteoblasts in vitro: A potential role in osteoradionecrosis. *Arch Otolaryngol Head Neck Surg* 126(9): 1124–1128.
25. Hopewell JW. 2003. Radiation-therapy effects on bone density. *Med Pediatr Oncol* 41(3): 208–211.
26. Ewing J. 1926. Radiation osteitis. *Acta Radiol* 6: 399–412.
27. Furstman LL. 1972. Effect of radiation on bone. *J Dent Res* 51(2): 596–604.
28. Cao X, Wu X, Frassica D, Yu B, Pang L, Xian L, Wan M, Lei W, Armour M, Tryggestad E, Wong J, Wen CY, Lu WW, Frassica FJ. 2011. Irradiation induces bone injury by damaging bone marrow microenvironment for stem cells. *Proc Natl Acad Sci U S A* 108(4): 1609–1614.
29. Sams A. 1966. The effect of 2000 r of x-rays on the internal structure of the mouse tibia. *Int J Radiat Biol Relat Stud Phys Chem Med* 11(1): 51–68.
30. Dudziak ME, Saadeh PB, Mehrara BJ, Steinbrech DS, Greenwald JA, Gittes GK, Longaker MT. 2000. The effects of ionizing radiation on osteoblast-like cells in vitro. *Plast Reconstr Surg* 106(5): 1049–1061.
31. Szymczyk KH, Shapiro IM, Adams CS. 2004. Ionizing radiation sensitizes bone cells to apoptosis. *Bone* 34(1): 148–156.
32. Sakurai T, Sawada Y, Yoshimoto M, Kawai M, Miyakoshi J. 2007. Radiation-induced reduction of osteoblast differentiation in C2C12 cells. *J Radiat Res (Tokyo)* 48(6): 515–521.
33. Sawajiri M, Nomura Y, Bhawal UK, Nishikiori R, Okazaki M, Mizoe J, Tanimoto K. 2006. Different effects of carbon ion and gamma-irradiation on expression of receptor activator of NF-kB ligand in MC3T3-E1 osteoblast cells. *Bull Exp Biol Med* 142(5): 618–624.
34. Schonmeyr BH, Wong AK, Soares M, Fernandez J, Clavin N, Mehrara BJ. 2008. Ionizing radiation of mesenchymal stem cells results in diminution of the precursor pool and limits potential for multilineage differentiation. *Plast Reconstr Surg* 122(1): 64–76.
35. Jacobsson M, Jonsson A, Albrektsson T, Turesson I. 1985. Alterations in bone regenerative capacity after low level gamma irradiation. *Scand J Plast Reconstr Surg* 19: 231–236.
36. Sugimoto M, Takahashi S, Toguchida J, Kotoura Y, Shibamoto Y, Yamamuro T. 1991. Changes in bone after high-dose irradiation. Biomechanics and histomorphology. *J Bone Joint Surg Br* 73(3): 492–497.
37. Rabelo GD, Beletti ME, Dechichi P. 2010. Histological analysis of the alterations on cortical bone channels network after radiotherapy: A rabbit study. *Microsc Res Tech* 73(11): 1015–1018.
38. Willey JS, Lloyd SA, Robbins ME, Bourland JD, Smith-Sielicki H, Bowman LC, Norrdin RW, Bateman TA. 2008. Early increase in osteoclast number in mice after whole-body irradiation with 2 Gy X rays. *Radiat Res* 170(3): 388–392.
39. Margulies B, Morgan H, Allen M, Strauss J, Spadaro J, Damron T. 2003. Transiently increased bone density after irradiation and the radioprotectant drug amifostine in a rat model. *Am J Clin Oncol* 26(4): e106–114.
40. Sawajiri M, Mizoe J, Tanimoto K. 2003. Changes in osteoclasts after irradiation with carbon ion particles. *Radiat Environ Biophys* 42(3): 219–223.
41. Burr DB, Miller L, Grynpas M, Li J, Boyde A, Mashiba T, Hirano T, Johnston CC. 2003. Tissue mineralization is increased following 1-year treatment with high doses of bisphosphonates in dogs. *Bone* 33(6): 960–969.
42. Wernle JD, Damron TA, Allen MJ, Mann KA. 2010. Local irradiation alters bone morphology and increases bone fragility in a mouse model. *J Biomech* 43(14): 2738–2746.
43. Dhakal S, Chen J, McCance S, Rosier R, O'Keefe R, Constine LS. 2011. Bone density changes after radiation for extremity sarcomas: exploring the etiology of pathologic fractures. *Int J Radiat Oncol Biol Phys* 80(4): 1158–1163.
44. Howland W, Loeffler, RK, Starchman, DE, et. al. 1975. Post-irradiation atrophic changes of bone and related complications. *Radiology* 117: 677–685.
45. Sawajiri M, Mizoe J. 2003. Changes in bone volume after irradiation with carbon ions. *Radiat Environ Biophys* 42(2): 101–106.
46. Nishiyama K, Inaba F, Higashirara T, Kitatani K, Kozuka T. 1992. Radiation osteoporosis—An assessment using single energy quantitative computed tomography. *Eur Radiol* 2: 322–325.

47. Bandstra ER, Pecaut MJ, Anderson ER, Willey JS, De Carlo F, Stock SR, Gridley DS, Nelson GA, Levine HG, Bateman TA. 2008. Long-term dose response of trabecular bone in mice to proton radiation. *Radiat Res* 169(6): 607–614.

48. Hamilton SA, Pecaut MJ, Gridley DS, Travis ND, Bandstra ER, Willey JS, Nelson GA, Bateman TA. 2006. A murine model for bone loss from therapeutic and space-relevant sources of radiation. *J Appl Physiol* 101(3): 789–793.

49. Bandstra ER, Thompson RW, Nelson GA, Willey JS, Judex S, Cairns MA, Benton ER, Vazquez ME, Carson JA, Bateman TA. 2009. Musculoskeletal changes in mice from 20–50 cGy of simulated galactic cosmic rays. *Radiat Res* 172(1): 21–29.

50. Alwood JS, Yumoto K, Mojarrab R, Limoli CL, Almeida EA, Searby ND, Globus RK. 2010. Heavy ion irradiation and unloading effects on mouse lumbar vertebral microarchitecture, mechanical properties and tissue stresses. *Bone* 47(2): 248–255.

89
Skeletal Complications of Childhood Cancer

Ingrid A. Holm

INTRODUCTION

With improvements in therapy more and more children are surviving childhood cancer. The 5-year survival rate for children with cancer is about 82%, and the rate for acute lymphocytic leukemia (ALL), the most common childhood cancer, is about 87% [1, 2]. Survivors of childhood cancer number about 270,000, and close to 1 in 640 people between 20 and 39 years of age have had childhood cancer [3, 4]. As more and more children with cancer are surviving, the number of adults with complications from cancer is increasing, and the cumulative incidence of a chronic health condition years after the cancer diagnosis is 73% [5]. The Childhood Cancer Survivor Study (CCSS) has generated data on the long-term outcomes of pediatric cancer survivors [6]. This chapter focuses on one of the long-term sequelae: deficits in bone mineral density (BMD).

During adolescence approximately 40% of peak bone mass is accumulated. Lack of adequate bone mineral acquisition during this period may compromise peak bone mass, leading to lifelong deficits in bone mineral and predisposing to fractures. In children with cancer, nutritional deficiencies and prolonged decreases in activity may compromise bone accrual. Hormonal deficiencies secondary to cranial irradiation causing growth hormone deficiency and/or central hypogonadism, or gonadal radiation leading to secondary hypogonadism, may further compromising bone accrual [7, 8]. Finally, chemotherapeutic agents, such as glucocorticoids and methotrexate, interfere with bone mineral accretion and skeletal development [7, 8] (see Ref. 7 for a review).

ACUTE LYMPHOCYTIC LEUKEMIA (ALL)

ALL is the most common childhood malignancy and much literature on the skeletal effects of cancer focuses on this population.

At diagnosis of ALL

The ALL disease process may affect bone acquisition; osteopenia (BMD Z-score <−1.0 SD) at diagnosis is reported in 10–46% of children with ALL [9–13]. Osteopenia and/or fractures are evident on radiographs in 13% and 10% of patients, respectively [9]. Vertebral compression fractures can be seen at presentation (in 16% of ALL patients in one study [14]) and are associated with a lower BMD lumbar spine Z-score [14]. Bone formation is impaired and markers of bone formation are low [15], including osteocalcin, type I collagen carboxy-terminal propeptide (PICP), and bone-specific alkaline phosphatase (BSAP) [9, 10, 12, 13, 16–19]. Bone resorption may also be impaired [15]. Urinary N-telopeptide (NTX) is normal [10], and type I collagen carboxyl-terminal telopeptide (ICTP) is normal [12, 16] or reduced [13, 17, 18] in most, but not all [20], studies. The reduction in bone formation and resorption markers suggests a low bone turnover state is present at diagnosis [17, 21]. 1,25-dihydroxy vitamin D is reduced in some patients in most [9, 12, 16, 19], but not all [11], studies.

During treatment of ALL

Treatment of ALL is associated with additional loss of bone in some patients [10–12, 16, 19, 20]. Using phalangeal ultrasound for skeletal assessment during treatment of ALL (at diagnosis, and at 6, 12, and 24 months), a rapid deterioration of bone properties was seen after the first 6 months of therapy with subsequent progressive uncoupling of the two quantitative ultrasound (QUS) parameters, which was worse in subjects who developed skeletal complications during treatment [22].

There is not a strong relationship between changes in growth velocity and BMD during therapy. Overall growth is not significantly compromised in most children undergoing cancer treatment, which may explain why the findings of studies using areal BMD do not differ significantly from those of studies using BMAD (apparent volumetric BMD) or quantitative computed tomography (QCT) to correct for size differences [12].

Chemotherapy is associated with a decrease in BMD, and BMD is decreased in patients treated with only chemotherapy [12]. Glucocorticoids and methotrexate are known to predispose to osteopenia and fractures (see Refs. 7 and 23 for a review), and higher cumulative doses of methotrexate [16] and higher doses of glucocorticoids [10] are associated with a lower BMD. Glucocorticoids decrease bone formation, decrease intestinal absorption of calcium, and increase renal calcium excretion. Methotrexate may also inhibit new bone formation and fracture healing (see Ref. 23 for a review). Cranial irradiation is associated with a drop in BMD Z-score during treatment [11]. Lifestyle factors such as calcium intake and physical activity are not correlated with BMD during therapy [12].

Accumulation of normal bone mass may be impaired during therapy. Although induction therapy with prednisolone and methotrexate is associated with a further decrease in bone formation markers, with discontinuation of prednisolone, bone formation markers increase to normal [10, 12, 13, 15–17]; the increase is less in children who received high-dose methotrexate [17]. During induction therapy with prednisolone and methotrexate, bone resorption markers (pyridinoline, ICTP) also decrease and then increase after prednisolone is discontinued [17] to normal or elevated levels [12, 15, 16, 18], suggesting resorption is increased [21]. High doses of methotrexate are associated with an even greater increase in bone resorption markers [17, 20]. These findings suggest that osteoblast and osteoclast activity are suppressed secondary to the actions of glucocorticoids on bone, and that high doses of methotrexate continue to inhibit osteoblast proliferation but increase osteoclast activity [17]. Higher levels of bone formation markers at diagnosis (osteocalcin) and at 1 year (PICP) have been correlated with a higher BMD at 1 year [16]. The levels of 1,25-dihydroxyvitamin D have been reported to be reduced [10, 16, 24] or normal [11, 12]. In one study [12], although 1,25-dihydroxyvitamin D levels were within the normal range, the levels increased significantly in the first 32 weeks of treatment, and the increase was positively correlated with the change in lumbar spine BMD over that period of time [12]. 25-hydroxyvitamin D has been reported to be reduced [16, 24] or normal [10].

Up to 40% of patients sustain a fracture [10, 19, 20]. Fracture rates have been reported to be increased sixfold [12] and tend to occur during or shortly after discontinuation of chemotherapy [25]. Fractures are associated with osteopenia that develops during therapy [15]; children who fracture have a greater decrease in BMD Z-score than children who do not fracture [12, 16]. A decrease in BMD Z-score during treatment is a positive predictor [9, 19], and an increase in bone mineral content (BMC) a negative predictive [9], of subsequent fracture. In one study, an increase in NTX was correlated with the occurrence of a fracture [10].

There have been trials to impact on the loss of BMD with therapy. Pilot studies of oral alendronate during therapy in children with ALL [26, 27] show an increase in the BMD Z-score in most patients treated for 6 to 24 months. No placebo-controlled trials have been carried out, and many issues remain, most notably what the criteria for treatment should be. A randomized exercise intervention trial during treatment of childhood ALL showed no difference in BMD between groups that received the intervention and those who did not, although lack of compliance may have been a factor [28].

After treatment of ALL

In most studies of ALL survivors, the low BMD persists [12, 29–38] although a few studies find that BMD normalizes [39, 40]. Bone formation and resorption markers normalize [41]. In one study after remission of childhood ALL or non-Hodgkin's lymphoma, a decreased BMD was associated with dexamethasone treatment, cranial irradiation, bone marrow transplant, and total body irradiation (TBI) [38].

Most studies find some "catch-up" in the BMD after discontinuation of therapy, and the length of time since cessation of therapy is associated with a higher BMD or BMC [34, 36]. In the short term (0–3 years after discontinuation of ALL therapy), a greater increase in total body BMD is seen in patients compared to controls [41]. In a longitudinal QCT study of ALL survivors at an average of 11 years after ALL diagnosis, the mean lumbar spine BMD Z-score increased with time, although the percentage of individuals with a BMD Z-score <−1.0 SD increased over time [42], suggesting that in a subgroup of patients, catch-up growth in BMD does not occur. Thus, although there may be catch-up in the BMD after the discontinuation of therapy, peak bone mass may not be achieved.

Risk factors associated with a lower BMD in follow-up include Caucasian race [29, 35, 42]; older age at diagnosis of ALL [23, 42]; poor nutritional status [42]; alcohol consumption [42]; male gender [29, 33, 35, 42]; endocrine abnormalities including growth disturbance, growth hormone deficiency, and hypogonadism [37]; and low physical activity, including exercise capacity (as measured by peak oxygen consumption) [34], a low activity level [33, 34], and muscle strength in the extremities [43]. These physical activity findings are particularly notable as there is evidence that long-term ALL survivors are less likely to meet the Centers for Disease Control and Prevention physical activity recommendations and are more likely to report no leisure-time physical activity compared to the general population [44].

The relative contribution of chemotherapy versus cranial irradiation, with the subsequent hormonal deficiencies, on BMD is not clear. Patients who received chemotherapy in the absence of cranial irradiation have a low BMD in some [13, 25, 33], but not all [25] studies. Previous exposure to methotrexate is associated with a reduction in BMC at the spine [34], and higher doses of methotrexate have been associated with a lower BMD [33, 42]. A history of cranial irradiation has been associated with a low BMD [29, 35, 42, 45], and BMD is inversely correlated with the cranial irradiation dose [35]. Others have found that the use of cranial irradiation status did not predict BMD [36] and that the decrease in BMD seen in adults after treatment for ALL is not dependent on the severity of growth hormone deficiency secondary to cranial irradiation [30]. These findings, taken together, suggest that although chemotherapy and cranial irradiation likely play a role in the low BMD in long-term survivors, additional factors likely adversely affect BMD.

Fracture rates are increased in long-term survivors of ALL. From the time of diagnosis of ALL through the subsequent 5 years, fractures rates are reported to be double that of controls [41]. In one study, the 5-year cumulative incidence of fractures was 28% [46]. Factors associated with a higher risk of fractures include the diagnosis of ALL during adolescence [23, 46], male sex [46], and treatment with dexamethasone versus prednisone [46]. In one study, a history of cranial irradiation was not associated with fracture rate [47].

Children undergoing stem cell transplant (SCT) may be at particularly high risk for deficits in bone mass. A low BMD post transplant has been reported in 36–47% of survivors [48–50]; although in children transplanted at younger than 3 years of age, the prevalence is lower (23.5%) [51]. In children who have undergone SCT, risk factors for osteopenia include female sex [48, 49]; total body irradiation (TBI) [48], and the presence of secondary effects of TBI including delayed pubertal growth, growth hormone deficiency, hypogonadism, or chronic renal insufficiency [48, 49]. Fat mass index is positively correlated with BMD in children who have undergone bone marrow transplant [52].

Nutritional factors may play a role in the bone mineral deficits. Less than 30% of ALL survivors at least 5 years after completion of therapy met the recommended dietary intakes for vitamin D and calcium [53]. Investigators at St. Jude Children's Research Hospital have initiated a placebo-controlled double-blind randomized longitudinal study to compare the effects of nutritional counseling, 800 IU/day vitamin D, and 1,000 mg/day elemental calcium to nutritional counseling with a placebo on changes in BMD and serum and urine markers of bone metabolism in participants at least 5 years after completion of ALL therapy who have a BMD Z-score below 0 [54].

OTHER CANCERS

Survivors of solid tumors also show bone mineral deficits years after therapy. Children with brain tumors may be particularly susceptible to osteopenia due to growth hormone deficiency and hypogonadism following cranial irradiation. Children who survive with a brain tumor have a decreased BMD [55–60], and cranial irradiation may be associated with an even higher prevalence of low BMD [61]. The degree of osteopenia in brain tumor patients is correlated with a reduction in health-related quality of life [57, 61], and children with osteopenia reported more pain that significantly limited physical activity [57].

In pediatric patients with bone sarcomas, bone mineral deficits may not be present initially but may develop with time. Patients with newly diagnosed Ewing sarcoma and osteosarcoma show no deficits in lumbar spine (LS) BMD after completion of neoadjuvant chemotherapy, although in patients with a lower extremity tumor, local deficits are seen and the BMD is lower in the affected compared to the non-affected femoral neck [62]. However, at an average of over 5 years after remission, bone mineral deficits are seen in bone sarcoma survivors, including survivors of Ewing sarcoma and osteosarcomas [63–66], and in other soft tissue sarcomas (rhabdomyosarcoma and non-rhabdomyosarcoma soft tissue sarcomas) [65]. Risk factors for a low BMD included younger age at diagnosis [64, 65], male sex [64], and higher cumulative dose of cyclophosphamide [65].

In a study of children with neuroblastoma, 11.1% (3/27) presented with a BMD Z-score below −2.0, suggesting that neuroblastoma can affected BMD prior to the start of therapy [67]. Low intake of vitamin D and low physical activity, as well as treatment with methotrexate, ifosfamide, bleomycin, and cisplatin, may contribute [63].

There is evidence that Hodgkin's disease survivors have bone mineral deficits. In one series, 34% of survivors of Hodgkin's disease and non-Hodgkin's lymphoma had low ultradistal radius trabecular volumetric BMD by peripheral quantitative computed tomography (pQCT), and 45% had osteopenia (Z-score <−1.0 SD) by dual energy X-ray absorptiometry (DXA), which was negatively correlated with the cumulative dose of corticosteroid administered [68]. In another report, the proportion of Hodgkin's disease survivors with a BMD below the mean did not differ from the general population, although the proportion with a BMD Z-score <−1.5 SD was 14.7%, significantly greater than the 6.7% reported in the general population [69]. In this study, male gender was associated with an increased risk for a low BMD [69]. A third group found that among patients treated in childhood with chemotherapy alone for Hodgkin's disease, a low total body BMD and LS BMAD were found in women only, but not in men [70].

Nutritional factors may play a role in the bone mineral deficits seen in pediatric patients with a variety of solid tumors. In one study, an average of 14 months after treatment for a pediatric solid tumor (primarily cerebral tumors, Ewing sarcomas, and osteosarcomas) calcium intake was inadequate in 75% of patients, and vitamin D levels were less than 20 ng/ml in 61.5% and less than 10 ng/ml in 11.5% of patients [71]. Lack of physical activity may play a greater role in the low BMD seen in pediatric solid tumor survivors than in other cancers. In the CCSS cohort, the highest prevalence of physical performance limitations was seen in survivors of brain tumors (36.9%), bone tumors (26.6%), and Hodgkin's disease (23.3%) [72, 73].

CONCLUSIONS

An important long-term consequence of childhood cancer is deficits in bone mass accrual that often persist into adulthood, thereby increasing the risk of osteoporosis and fractures. The chemotherapies and radiation therapies used to treat these cancers are in large part responsible for the increased risk of a low BMD. However, there are a number of modifiable contributing factors. In recognition of the long-term morbidities associated with treatment of pediatric malignancies, including the bone mineral deficits the Children's Oncology Group (COG) has produced the 2006 COG long-term follow-up guidelines [7, 15, 74, 75]. Guidelines for the management of the bone morbidity that occurs after treatment of childhood cancer include recommendations directed at modifiable factors that may contribute, including exercise, adequate calcium, and vitamin D intake, and a baseline bone density scan 2 or more years after therapy has been completed [75]. In addition, early identification of growth hormone deficiency and hypogonadism may improve bone health in pediatric cancer survivors. The hope is that the early detection of bone mineral deficits and treatment of the modifiable factors will improve the prognosis in survivors of childhood cancer.

REFERENCES

1. Howlader N, Noone AM, Krapcho M, Neyman N, Aminou R, Waldron W, Altekruse SF, Kosary CL, Ruhl J, Tatalovich Z, Cho H, Mariotto A, Eisner MP, Lewis DR, Chen HS, Feuer EJ, Cronin KA, Edwards BK. 2011. *SEER Cancer Statistics Review, 1975–2008*. National Cancer Institute: Bethesda, MD. Available from: http://seer.cancer.gov/csr/1975_2008/.
2. Ellison LF, Pogany L, Mery LS. 2007. Childhood and adolescent cancer survival: A period analysis of data from the Canadian Cancer Registry. *Eur J Cancer* 43(13): 1967–75.
3. Oberfield SE. 2007. Childhood cancer cures: The ongoing consequences of successful treatments. *J Pediatr* 150(4): 332–4.
4. National Cancer Policy Board. 2003. *Childhood Cancer Survivorship: Improving Care and Quality of Life*. Washington, DC: National Academy of Sciences.
5. Oeffinger KC, Mertens AC, Sklar CA, Kawashima T, Hudson MM, Meadows AT, Friedman DL, Marina N, Hobbie W, Kadan-Lottick NS, Schwartz CL, Leisenring W, Robison LL. 2006. Chronic health conditions in adult survivors of childhood cancer. *N Engl J Med* 355(15): 1572–82.
6. Diller L, Chow EJ, Gurney JG, Hudson MM, Kadin-Lottick NS, Kawashima TI, Leisenring WM, Meacham LR, Mertens AC, Mulrooney DA, Oeffinger KC, Packer RJ, Robison LL, Sklar CA. 2009. Chronic disease in the Childhood Cancer Survivor Study cohort: A review of published findings. *J Clin Oncol* 27(14): 2339–55.
7. Wasilewski-Masker K, Kaste SC, Hudson MM, Esiashvili N, Mattano LA, Meacham LR. 2008. Bone mineral density deficits in survivors of childhood cancer: Long-term follow-up guidelines and review of the literature. *Pediatrics* 121(3): e705–13.
8. van Leeuwen BL, Kamps WA, Jansen HW, Hoekstra HJ. 2000. The effect of chemotherapy on the growing skeleton. *Cancer Treat Rev* 26(5): 363–76.
9. Halton JM, Atkinson SA, Fraher L, Webber CE, Cockshott WP, Tam C, Barr RD. 1995. Mineral homeostasis and bone mass at diagnosis in children with acute lymphoblastic leukemia. *J Pediatr* 126(4): 557–64.
10. Halton JM, Atkinson SA, Fraher L, Webber C, Gill GJ, Dawson S, Barr RD. 1996. Altered mineral metabolism and bone mass in children during treatment for acute lymphoblastic leukemia. *J Bone Miner Res* 11(11): 1774–83.
11. Henderson RC, Madsen CD, Davis C, Gold SH. 1998. Longitudinal evaluation of bone mineral density in children receiving chemotherapy. *J Pediatr Hematol Oncol* 20(4): 322–6.
12. van der Sluis IM, van den Heuvel-Eibrink MM, Hahlen K, Krenning EP, de Muinck Keizer-Schrama SM. 2002. Altered bone mineral density and body composition, and increased fracture risk in childhood acute lymphoblastic leukemia. *J Pediatr* 141(2): 204–10.
13. Boot AM, van den Heuvel-Eibrink MM, Hahlen K, Krenning EP, de Muinck Keizer-Schrama SM. 1999. Bone mineral density in children with acute lymphoblastic leukaemia. *Eur J Cancer* 35(12): 1693–7.
14. Halton J, Gaboury I, Grant R, Alos N, Cummings EA, Matzinger M, Shenouda N, Lentle B, Abish S, Atkinson S, Cairney E, Dix D, Israels S, Stephure D, Wilson B,

Hay J, Moher D, Rauch F, Siminoski K, Ward LM, Canadian SC. 2009. Advanced vertebral fracture among newly diagnosed children with acute lymphoblastic leukemia: Results of the Canadian Steroid-Associated Osteoporosis in the Pediatric Population (STOPP) research program. *J Bone Miner Res* 24(7): 1326–34.

15. Sala A, Barr RD. 2007. Osteopenia and cancer in children and adolescents: The fragility of success. *Cancer* 109(7): 1420–31.
16. Arikoski P, Komulainen J, Riikonen P, Voutilainen R, Knip M, Kroger H. 1999. Alterations in bone turnover and impaired development of bone mineral density in newly diagnosed children with cancer: A 1-year prospective study. *J Clin Endocrinol Metab* 84(9): 3174–81.
17. Crofton PM, Ahmed SF, Wade JC, Stephen R, Elmlinger MW, Ranke MB, Kelnar CJ, Wallace WH. 1998. Effects of intensive chemotherapy on bone and collagen turnover and the growth hormone axis in children with acute lymphoblastic leukemia. *J Clin Endocrinol Metab* 83(9): 3121–9.
18. Sorva R, Kivivuori SM, Turpeinen M, Marttinen E, Risteli J, Risteli L, Sorva A, Siimes MA. 1997. Very low rate of type I collagen synthesis and degradation in newly diagnosed children with acute lymphoblastic leukemia. *Bone* 20(2): 139–43.
19. Atkinson SA, Halton JM, Bradley C, Wu B, Barr RD. 1998. Bone and mineral abnormalities in childhood acute lymphoblastic leukemia: Influence of disease, drugs and nutrition. *Int J Cancer Suppl* 11: 35–9.
20. Arikoski P, Komulainen J, Riikonen P, Parviainen M, Jurvelin JS, Voutilainen R, Kroger H. 1999. Impaired development of bone mineral density during chemotherapy: A prospective analysis of 46 children newly diagnosed with cancer. *J Bone Miner Res* 14(12): 2002–9.
21. Mulder JE, Bilezikian JP. 2004. Bone density in survivors of childhood cancer. *J Clin Densitom* 7(4): 432–42.
22. Mussa A, Bertorello N, Porta F, Galletto C, Nicolosi MG, Manicone R, Corrias A, Fagioli F. 2010. Prospective bone ultrasound patterns during childhood acute lymphoblastic leukemia treatment. *Bone* 46(4): 1016–20.
23. Davies JH, Evans BA, Jenney ME, Gregory JW. 2005. Skeletal morbidity in childhood acute lymphoblastic leukaemia. *Clin Endocrinol (Oxf)* 63(1): 1–9.
24. Arikoski P, Kroger H, Riikonen P, Parviainen M, Voutilainen R, Komulainen J. 1999. Disturbance in bone turnover in children with a malignancy at completion of chemotherapy. *Med Pediatr Oncol* 33(5): 455–61.
25. van der Sluis IM, van den Heuvel-Eibrink MM, Hahlen K, Krenning EP, de Muinck Keizer-Schrama SM. 2000. Bone mineral density, body composition, and height in long-term survivors of acute lymphoblastic leukemia in childhood. *Med Pediatr Oncol* 35(4): 415–20.
26. Lethaby C, Wiernikowski J, Sala A, Naronha M, Webber C, Barr RD. 2007. Bisphosphonate therapy for reduced bone mineral density during treatment of acute lymphoblastic leukemia in childhood and adolescence: A report of preliminary experience. *J Pediatr Hematol Oncol* 29(9): 613–6.
27. Wiernikowski JT, Barr RD, Webber C, Guo CY, Wright M, Atkinson SA. 2005. Alendronate for steroid-induced osteopenia in children with acute lymphoblastic leukaemia or non-Hodgkin's lymphoma: Results of a pilot study. *J Oncol Pharm Pract* 11(2): 51–6.
28. Hartman A, te Winkel ML, van Beek RD, de Muinck Keizer-Schrama SM, Kemper HC, Hop WC, van den Heuvel-Eibrink MM, Pieters R. 2009. A randomized trial investigating an exercise program to prevent reduction of bone mineral density and impairment of motor performance during treatment for childhood acute lymphoblastic leukemia. *Pediatr Blood Cancer* 53(1): 64–71.
29. Arikoski P, Komulainen J, Voutilainen R, Riikonen P, Parviainen M, Tapanainen P, Knip M, Kroger H. 1998. Reduced bone mineral density in long-term survivors of childhood acute lymphoblastic leukemia. *J Pediatr Hematol Oncol* 20(3): 234–40.
30. Brennan BM, Rahim A, Adams JA, Eden OB, Shalet SM. 1999. Reduced bone mineral density in young adults following cure of acute lymphoblastic leukaemia in childhood. *Br J Cancer* 79(11–12): 1859–63.
31. Hoorweg-Nijman JJ, Kardos G, Roos JC, van Dijk HJ, Netelenbos C, Popp-Snijders C, de Ridder CM, Delemarre-van de Waal HA. 1999. Bone mineral density and markers of bone turnover in young adult survivors of childhood lymphoblastic leukaemia. *Clin Endocrinol (Oxf)* 50(2): 237–44.
32. Nysom K, Holm K, Michaelsen K, Hertz H, Müller J, Mølgaard C. 1998. Bone mass after treatment for acute lymphoblastic leukemia in childhood. *J Clin Oncol* 16(12): 3752–60.
33. Tillmann V, Darlington AS, Eiser C, Bishop NJ, Davies HA. 2002. Male sex and low physical activity are associated with reduced spine bone mineral density in survivors of childhood acute lymphoblastic leukemia. *J Bone Miner Res* 17(6): 1073–80.
34. Warner J, Evans W, Webb D, Bell W, Gregory J. 1999. Relative osteopenia after treatment for acute lymphoblastic leukemia. *Pediatr Res* 45(4 Pt 1): 544–51.
35. Kaste SC, Jones-Wallace D, Rose SR, Boyett JM, Lustig RH, Rivera GK, Pui CH, Hudson MM. 2001. Bone mineral decrements in survivors of childhood acute lymphoblastic leukemia: Frequency of occurrence and risk factors for their development. *Leukemia* 15(5): 728–34.
36. Alikasifoglu A, Yetgin S, Cetin M, Tuncer M, Gumruk F, Gurgey A, Yordam N. 2005. Bone mineral density and serum bone turnover markers in survivors of childhood acute lymphoblastic leukemia: Comparison of megadose methylprednisolone and conventional-dose prednisolone treatments. *Am J Hematol* 80(2): 113–8.
37. Miyoshi Y, Ohta H, Hashii Y, Tokimasa S, Namba N, Mushiake S, Hara J, Ozono K. 2008. Endocrinological analysis of 122 Japanese childhood cancer survivors in a single hospital. *Endocr J* 55(6): 1055–63.
38. Benmiloud S, Steffens M, Beauloye V, de Wandeleer A, Devogelaer JP, Brichard B, Vermylen C, Maiter D. 2010. Long-term effects on bone mineral density of different therapeutic schemes for acute lymphoblastic leukemia

38. or non-Hodgkin lymphoma during childhood. *Horm Res Paediatr* 74(4): 241–50.
39. Kadan-Lottick N, Marshall JA, Baron AE, Krebs NF, Hambidge KM, Albano E. 2001. Normal bone mineral density after treatment for childhood acute lymphoblastic leukemia diagnosed between 1991 and 1998. *J Pediatr* 138(6): 898–904.
40. Mandel K, Atkinson S, Barr RD, Pencharz P. 2004. Skeletal morbidity in childhood acute lymphoblastic leukemia. *J Clin Oncol* 22(7): 1215–21.
41. Marinovic D, Dorgeret S, Lescoeur B, Alberti C, Noel M, Czernichow P, Sebag G, Vilmer E, Leger J. 2005. Improvement in bone mineral density and body composition in survivors of childhood acute lymphoblastic leukemia: A 1-year prospective study. *Pediatrics* 116(1): e102–8.
42. Kaste SC, Rai SN, Fleming K, McCammon EA, Tylavsky FA, Danish RK, Rose SR, Sitter CD, Pui CH, Hudson MM. 2006. Changes in bone mineral density in survivors of childhood acute lymphoblastic leukemia. *Pediatr Blood Cancer* 46(1): 77–87.
43. Joyce ED, Nolan VG, Ness KK, Ferry RJ Jr, Robison LL, Pui CH, Hudson MM, Kaste SC. 2011. Association of muscle strength and bone mineral density in adult survivors of childhood acute lymphoblastic leukemia. *Arch Phys Med Rehabil* 92(6): 873–9.
44. Florin TA, Fryer GE, Miyoshi T, Weitzman M, Mertens AC, Hudson MM, Sklar CA, Emmons K, Hinkle A, Whitton J, Stovall M, Robison LL, Oeffinger KC. 2007. Physical inactivity in adult survivors of childhood acute lymphoblastic leukemia: A report from the childhood cancer survivor study. *Cancer Epidemiol Biomarkers Prev* 16(7): 1356–63.
45. Gilsanz V, Carlson ME, Roe TF, Ortega JA. 1990. Osteoporosis after cranial irradiation for acute lymphoblastic leukemia. *J Pediatr* 117(2 Pt 1): 238–44.
46. Strauss AJ, Su JT, Dalton VM, Gelber RD, Sallan SE, Silverman LB. 2001. Bony morbidity in children treated for acute lymphoblastic leukemia. *J Clin Oncol* 19(12): 3066–72.
47. Barr RD, Halton J, Willan A, Cockshott WP, Gill G, Atkinson S. 1998. Impact of age and cranial irradiation on radiographic skeletal pathology in children with acute lymphoblastic leukemia. *Med Pediatr Oncol* 30(6): 347–50.
48. Leung W, Ahn H, Rose SR, Phipps S, Smith T, Gan K, O'Connor M, Hale GA, Kasow KA, Barfield RC, Madden RM, Pui CH. 2007. A prospective cohort study of late sequelae of pediatric allogeneic hematopoietic stem cell transplantation. *Medicine (Baltimore)* 86(4): 215–24.
49. Taskinen M, Kananen K, Valimaki M, Loyttyniemi E, Hovi L, Saarinen-Pihkala U, Lipsanen-Nyman M. 2006. Risk factors for reduced areal bone mineral density in young adults with stem cell transplantation in childhood. *Pediatr Transplant* 10(1): 90–7.
50. Taskinen M, Saarinen-Pihkala UM, Hovi L, Vettenranta K, Makitie O. 2007. Bone health in children and adolescents after allogeneic stem cell transplantation: High prevalence of vertebral compression fractures. *Cancer* 110(2): 442–51.
51. Perkins JL, Kunin-Batson AS, Youngren NM, Ness KK, Ulrich KJ, Hansen MJ, Petryk A, Steinberger J, Anderson FS, Baker KS. 2007. Long-term follow-up of children who underwent hematopoeitic cell transplant (HCT) for AML or ALL at less than 3 years of age. *Pediatr Blood Cancer* 49(7): 958–63.
52. Ruble K, Hayat MJ, Stewart KJ, Chen AR. 2010. Bone mineral density after bone marrow transplantation in childhood: Measurement and associations. *Biol Blood Marrow Transplant* 16(10): 1451–7.
53. Tylavsky FA, Smith K, Surprise H, Garland S, Yan X, McCammon E, Hudson MM, Pui CH, Kaste SC. 2010. Nutritional intake of long-term survivors of childhood acute lymphoblastic leukemia: Evidence for bone health interventional opportunities. *Pediatr Blood Cancer* 55(7): 1362–9.
54. Rai SN, Hudson MM, McCammon E, Carbone L, Tylavsky F, Smith K, Surprise H, Shelso J, Pui CH, Kaste S. 2008. Implementing an intervention to improve bone mineral density in survivors of childhood acute lymphoblastic leukemia: BONEII, a prospective placebo-controlled double-blind randomized interventional longitudinal study design. *Contemp Clin Trials* 29(5): 711–9.
55. Petraroli M, D'Alessio E, Ausili E, Barini A, Caradonna P, Riccardi R, Caldarelli M, Rossodivita A. 2007. Bone mineral density in survivors of childhood brain tumours. *Childs Nerv Syst* 23(1): 59–65.
56. Pietila S, Sievanen H, Ala-Houhala M, Koivisto AM, Liisa Lenko H, Makipernaa A. 2006. Bone mineral density is reduced in brain tumour patients treated in childhood. *Acta Paediatr* 95(10): 1291–7.
57. Barr RD, Simpson T, Webber CE, Gill GJ, Hay J, Eves M, Whitton AC. 1998. Osteopenia in children surviving brain tumours. *Eur J Cancer* 34(6): 873–7.
58. Gurney JG, Kadan-Lottick NS, Packer RJ, Neglia JP, Sklar CA, Punyko JA, Stovall M, Yasui Y, Nicholson HS, Wolden S, McNeil DE, Mertens AC, Robison LL. 2003. Endocrine and cardiovascular late effects among adult survivors of childhood brain tumors: Childhood Cancer Survivor Study. *Cancer* 97(3): 663–73.
59. Hesseling PB, Hough SF, Nel ED, van Riet FA, Beneke T, Wessels G. 1998. Bone mineral density in long-term survivors of childhood cancer. *Int J Cancer Suppl* 11: 44–7.
60. Krishnamoorthy P, Freeman C, Bernstein ML, Lawrence S, Rodd C. 2004. Osteopenia in children who have undergone posterior fossa or craniospinal irradiation for brain tumors. *Arch Pediatr Adolesc Med* 158(5): 491–6.
61. Odame I, Duckworth J, Talsma D, Beaumont L, Furlong W, Webber C, Barr R. 2006. Osteopenia, physical activity and health-related quality of life in survivors of brain tumors treated in childhood. *Pediatr Blood Cancer* 46(3): 357–62.
62. Muller C, Winter CC, Rosenbaum D, Boos J, Gosheger G, Hardes J, Vieth V. 2010. Early decrements in bone

density after completion of neoadjuvant chemotherapy in pediatric bone sarcoma patients. *BMC Musculoskelet Disord* 11: 287.
63. Azcona C, Burghard E, Ruza E, Gimeno J, Sierrasesumaga L. 2003. Reduced bone mineralization in adolescent survivors of malignant bone tumors: Comparison of quantitative ultrasound and dual-energy x-ray absorptiometry. *J Pediatr Hematol Oncol* 25(4): 297–302.
64. Ruza E, Sierrasesúmaga L, Azcona C, Patiño-Garcia A. 2006. Bone mineral density and bone metabolism in children treated for bone sarcomas. *Pediatr Res* 59(6): 866–71.
65. Kaste SC, Ahn H, Liu T, Liu W, Krasin MJ, Hudson MM, Spunt SL. 2008. Bone mineral density deficits in pediatric patients treated for sarcoma. *Pediatr Blood Cancer* 50(5): 1032–8.
66. Holzer G, Krepler P, Koschat MA, Grampp S, Dominkus M, Kotz R. 2003. Bone mineral density in long-term survivors of highly malignant osteosarcoma. *J Bone Joint Surg Br* 85(2): 231–7.
67. Al-Tonbary YA, El-Ziny MA, Elsharkawy AA, El-Hawary AK, El-Ashry R, Fouda AE. 2011. Bone mineral density in newly diagnosed children with neuroblastoma. *Pediatr Blood Cancer* 56(2): 202–5.
68. Sala A, Talsma D, Webber C, Posgate S, Atkinson S, Barr R. 2007. Bone mineral status after treatment of malignant lymphoma in childhood and adolescence. *Eur J Cancer Care (Engl)* 16(4): 373–9.
69. Kaste SC, Metzger ML, Minhas A, Xiong Z, Rai SN, Ness KK, Hudson MM. 2009. Pediatric Hodgkin lymphoma survivors at negligible risk for significant bone mineral density deficits. *Pediatr Blood Cancer* 52(4): 516–21.
70. van Bcck RD, van den Heuvel-Eibrink MM, Hakvoort-Cammel FG, van den Bos C, van der Pal HJ, Krenning EP, de Rijke YB, Pieters R, de Muinck Keizer-Schrama SM. 2009. Bone mineral density, growth, and thyroid function in long-term survivors of pediatric Hodgkin's lymphoma treated with chemotherapy only. *J Clin Endocrinol Metab* 94(6): 1904–9.
71. Bilariki K, Anagnostou E, Masse V, Elie C, Grill J, Valteau-Couanet D, Kalifa C, Doz F, Sainte-Rose C, Zerah M, Mascard E, Mosser F, Ruiz JC, Souberbielle JC, Eladari D, Brugieres L, Polak M. 2010. Low bone mineral density and high incidences of fractures and vitamin D deficiency in 52 pediatric cancer survivors. *Horm Res Paediatr* 74(5): 319–27.
72. Ness KK, Hudson MM, Ginsberg JP, Nagarajan R, Kaste SC, Marina N, Whitton J, Robison LL, Gurney JG. 2009. Physical performance limitations in the Childhood Cancer Survivor Study cohort. *J Clin Oncol* 27(14): 2382–9.
73. Ness KK, Mertens AC, Hudson MM, Wall MM, Leisenring WM, Oeffinger KC, Sklar CA, Robison LL, Gurney JG. 2005. Limitations on physical performance and daily activities among long-term survivors of childhood cancer. *Ann Intern Med* 143(9): 639–47.
74. Landier W, Bhatia S, Eshelman DA, Forte KJ, Sweeney T, Hester AL, Darling J, Armstrong FD, Blatt J, Constine LS, Freeman CR, Friedman DL, Green DM, Marina N, Meadows AT, Neglia JP, Oeffinger KC, Robison LL, Ruccione KS, Sklar CA, Hudson MM. 2004. Development of risk-based guidelines for pediatric cancer survivors: The Children's Oncology Group Long-Term Follow-Up Guidelines from the Children's Oncology Group Late Effects Committee and Nursing Discipline. *J Clin Oncol* 22(24): 4979–90.
75. Blatt J, Meacham LR. 2008. *Keeping Your Bones Health after Childhood Cancer*. Children's Oncology Group. Available from: http://www.childrensoncolosgygroup.org/disc/le/pdf/BoneHealth.pdf.

90

Treatment and Prevention of Bone Metastases and Myeloma Bone Disease

Jean-Jacques Body

Clinical Aspects 741
Bisphosphonates and Denosumab in Tumor Bone Disease 742

References 750

CLINICAL ASPECTS

According to various large series, up to 80–90% of patients with advanced cancer will develop bone metastases. The skeleton is the most common site of metastatic disease. It is also the most frequent site of first distant relapse in breast and prostate cancers.

Breast cancer

Metastatic bone disease causes considerable distress to breast cancer patients. Because of the long clinical course breast cancer may follow, morbidity due to tumor bone disease also makes major demands on resources for health care provision. The term "skeletal-related events" (SREs) refers to four major objective complications of tumor bone disease: pathological fractures, need for radiotherapy, need for bone surgery, and spinal cord compression [1]. Hypercalcemia is not counted as an SRE in most trials because it is generally easily treated with a bisphosphonate, and it is often of paraneoplastic origin. Such major complications will be observed in up to one-third of the patients whose first relapse is in bone.

Bone pain can be the source of great suffering, causing the most patient concern and physician visits [2]. A recent prospective study suggests that patients with osteolytic lesions, as evaluated by a computed tomography (CT) scan of the most painful lesions, have the highest mean pain score and analgesic consumption, and the least mean scores for quality of life as compared with patients who have mixed or sclerotic lesions [3]. This distinction between osteolytic and osteosclerotic actually represents two extremes of a continuum in which dysregulation of the normal bone remodeling process occurs. Hypercalcemia classically occurs in 10–15% of the cases, spinal cord compression in 5–10%, and, when long bones are invaded, fractures will occur in 10–20% of the cases [4, 5]. Pathological fractures are a dramatic consequence of tumor bone disease. They occur with a median onset of 11 months from the initial diagnosis of bone involvement. Solitary skeletal metastases carry a better prognosis than multiple metastases and are associated with an earlier stage and a more favorable histology. Interestingly, the sternum was the most commonly involved bone in a series of 289 patients with solitary skeletal metastasis [6].

Across all tumor types, patients with breast cancer have the highest incidence of skeletal complications. Taken from data in placebo groups of randomized bisphosphonates trials, the mean skeletal morbidity rate, i.e., the mean number of SREs per year, varies between 2.2 and 4.0 [4, 5, 7–9]. Patients who have metastases only in the skeleton have a higher rate of SREs than patients who also have bone and visceral metastases, e.g., a two- to threefold increase in pathological fractures. Survival from diagnosis of bone metastases is longest for patients with only bone metastases (median survival 24 months) and least for patients with bone and liver metastases

---

Primer on the Metabolic Bone Diseases and Disorders of Mineral Metabolism, Eighth Edition. Edited by Clifford J. Rosen.
© 2013 American Society for Bone and Mineral Research. Published 2013 by John Wiley & Sons, Inc.

(5.5 months) [9]. The occurrence of SREs has a marked detrimental effect on survival in patients with bone metastases [10].

Prostate cancer

The most common sites of bone metastases are throughout the axial skeleton, whereas long bones are invaded much less often. Pain is the most common symptom. Surprisingly, there are only a few studies documenting the frequency and the nature of bone metastatic complications in patients with hormone-refractory prostate cancer [11–13]. The incidence of SREs can be best estimated by analyzing the placebo group of the large scale zoledronic acid-controlled trial [12, 13]. Inclusion criteria required that patients had at least one bone metastasis and an augmentation of prostate-specific antigen (PSA) levels while on hormonal therapy. During a follow-up period of 2 years, nearly one-half of the patients developed one or more SREs, which, in this study, also included a change in antineoplastic therapy to treat bone pain. The two most frequent complications were the need for radiation therapy and the occurrence of pathological fractures. These fractures appeared more frequently at peripheral than at vertebral sites. The median time to the first SRE was 10.5 months, whereas the mean skeletal morbidity rate per year was nearly 1.5. The median survival was 9.5 months [13].

Multiple myeloma

Bone pain is a presenting feature in three-fourths of the patients with multiple myeloma. Back pain correlates with the presence of vertebral fractures that are present in more than half the patients at diagnosis. Extensive osteolytic lesions are frequent in this aggressive bone disease, and, typically, they do not heal despite successful antineoplastic treatment. Diffuse osteoporosis can also be a presenting and misleading feature. The increased fracture rate seems to be especially high around the time of diagnosis. In a large retrospective cohort study, fracture risk was increased 16-fold in the year before diagnosis and 9-fold thereafter. Fractures of the vertebrae and ribs were the most frequent [14].

BISPHOSPHONATES AND DENOSUMAB IN TUMOR BONE DISEASE

Mode of action of bisphosphonates and denosumab: Clinical relevance

The mechanisms of action of these two classes of inhibitors of osteoclast-mediated bone resorption are reviewed in other chapters of the *Primer* and in a recent exhaustive review article [15]. Contrasting differences in the mode of action are relevant for the prevention and the treatment of bone metastases.

Bisphosphonates are widely used for the treatment and the prevention of SREs in patients with cancer and bone metastases. They concentrate in the skeleton, primarily at active remodeling sites. They are embedded in bone where they can remain inactive for years. They are released in the acidic environment of the resorption lacunae under active osteoclasts and are taken up by them. Bisphosphonates will then interrupt the "vicious cycle" of tumor-mediated osteolysis by inhibiting the activity of bone-resorbing osteoclasts and inducing their apoptosis [16].

An increased expression and secretion of receptor activator of nuclear factor-κB (RANK) ligand (RANKL) plays a key role in the pathogenesis of tumor-induced bone destruction by stimulating osteoclast formation and function [15]. Osteoclast formation and activity are regulated via RANK through the coordinated expression of RANKL and osteoprotegerin. Denosumab, a new treatment option, is a fully human monoclonal antibody against human RANKL. By binding to RANKL, denosumab affects the key signaling pathway involved in bone remodeling and disrupts the "vicious cycle" of bone destruction stimulated by the metastasized tumor cells. Denosumab inhibits the formation and the activity of osteoclasts, which probably contributes to a larger inhibitory effect on bone resorption than what is achieved with bisphosphonates (see below). On the other hand, unlike bisphosphonates, denosumab shows no evidence of sustained binding to bone surfaces and is not incorporated into the bone matrix. The inhibitory action of denosumab on bone resorption is thus more transient, and a possible rebound effect is a concern, at least theoretically, if the treatment is stopped. These considerations have evident implications for the long-term treatment of patients with bone metastases since denosumab should probably not be stopped in patients with active tumor bone disease.

Cancer hypercalcemia

Bisphosphonates were first successfully used in tumor-induced hypercalcemia. This use is reviewed in another chapter of the *Primer*.

Prevention of skeletal-Related events

Breast cancer

Bisphosphonates can relieve metastatic bone pain, decrease the frequency of SREs, and improve patient functioning and quality of life [17]. Treatment and prevention of metastatic bone pain are now part of the long-term management of metastatic bone disease with bisphosphonates. Randomized placebo-controlled trials have shown that intravenous pamidronate, clodronate, ibandronate, and zoledronic acid exert useful pain relief

[5, 18]. In a phase III trial of patients with breast cancer, pain and analgesic scores were reduced to a similar extent after 1 year with zoledronic acid and pamidronate [19]. In phase III trials of intravenous and oral ibandronate, bone pain was reduced and similarly maintained below baseline for 2 years [7, 20]. There were also significant improvements in global quality of life and physical functioning with intravenous or oral ibandronate versus placebo [20]. However, bisphosphonates have to be considered as co-analgesics, and they cannot replace opioids for the treatment of severe bone pain. In recent comparative trials, the delay in the requirement of opioids to control bone pain has actually become a measure of the relative efficacy of inhibitors of bone resorption.

Bisphosphonates constitute an effective therapeutic means for the prevention of skeletal complications secondary to bone metastases in breast cancer. Several placebo-controlled trials and fewer comparative trials against another bisphosphonate have been performed. Assessment of treatment effects has often used the first-event analyses, such as the proportion of patients with at least one SRE or time to the first event. These are quite objective and conservative end points but they do not take into account all subsequent events that occur in any given patient. From a clinical perspective, an aggregate score of symptomatic SREs is more relevant. Skeletal morbidity rate (SMR) or skeletal morbidity period rate (SMPR, the number of periods with at least 1 SRE) take into account the occurrence of multiple SREs. More recently, multiple-event analyses have been increasingly used, as they are able to model all events and the time between events. They allow the calculation of a hazard ratio that indicates the relative risk of skeletal events between two treatment groups.

Clinical trials of the bisphosphonates clodronate, in Europe, and pamidronate, in the United States and Europe, have established their effectiveness in breast cancer patients with bone metastases [8, 21–23]. Clodronate is less effective than other bisphosphonates for the prevention of SREs, as demonstrated in a limited comparative trial against pamidronate [24, 25]. Two double-blind randomized placebo-controlled trials comparing 90 mg pamidronate infusions every 4 weeks to placebo infusions for up to 2 years in addition to chemo- or hormonal therapy in large series of breast cancer patients with at least one lytic bone metastasis demonstrated that bisphosphonates can reduce SMR by more than one-third, increase the median time to the occurrence of the first SRE by almost 50%, and reduce the proportion of patients having any SRE [8, 23]. More convenient and possibly more effective aminobisphosphonates have emerged. Zoledronic acid is now widely used for patients with bone metastases from various tumors [13, 19, 26–28], and ibandronate has also been approved in many countries, but not the U.S., for the prevention of skeletal events in patients with breast cancer and bone metastases.

Three randomized, double-blind, multicenter trials assessed the efficacy of zoledronic acid in patients with breast cancer and multiple myeloma, in prostate cancer, and in lung or other solid tumors. The primary efficacy end point was the proportion of patients with at least one SRE. Secondary end points included time to first SRE, SMR, and Andersen-Gill multiple-event analysis. Patients with breast cancer or multiple myeloma (n = 1648) were randomized to a 15-minute infusion of 4 or 8 mg of zoledronic acid or a 2-hour infusion of pamidronate 90 mg every 3–4 weeks [19, 26]. The proportion of patients with at least one SRE was similar in all treatment groups and the preestablished criterion for non-inferiority of zoledronic acid to pamidronate was thus met. Zoledronic acid 8 mg was not more effective than the 4-mg dose but was associated with an increased frequency of renal adverse events, explaining why all patients in that treatment arm were switched to the lower dose of zoledronic acid during all phase III trials. Median time to the first SRE was approximately 1 year in the three treatment groups, and SMRs were also not significantly different. A preplanned multiple-event analysis, according to the Andersen-Gill model, showed, however, that 4 mg of zoledronic acid reduced the risk of developing a skeletal complication by an additional 20% over that achieved by pamidronate 90 mg in the breast cancer subgroup ($P < 0.05$) (Fig. 90.1, top panel) [26]. The short infusion time (15 minutes compared with 1 or 2 hours for pamidronate) offers a quite convenient therapy. Zoledronic acid is still viewed as the standard treatment for metastatic bone disease in breast cancer [29]. Moreover, limited data suggest that switching from either pamidronate or clodronate to second-line zoledronic acid after an SRE or progression of bone metastases can significantly improve pain control and reduce bone turnover markers [29].

In the U.S., only zoledronic acid and pamidronate, and more recently denosumab, are approved to treat patients with bone metastases. However, the efficacy of intravenous and oral ibandronate has been demonstrated in randomized double-blind, placebo-controlled studies [7, 30]. Breast cancer patients were randomized to ibandronate 6 mg or placebo infused over 1–2 hours every 3–4 weeks in the intravenous trial, whereas oral ibandronate 50 mg was given once daily, 1 hour before breakfast. The primary efficacy end point was the skeletal morbidity period rate (SMPR), defined as the number of 12-week periods with skeletal complications (vertebral fractures, nonvertebral fractures, radiotherapy to bone and surgery to bone) divided by the total observation time. Secondary end points also included a multiple-event analysis. Intravenous and oral ibandronate significantly reduced SMPR compared with placebo ($P < 0.005$ for both). Multiple-event Poisson regression analysis showed that intravenous ibandronate led to a statistically significant 40% reduction in the risk of SREs compared with placebo (Fig. 90.1, lower panel; $P < 0.005$). The effect of oral ibandronate 50 mg on the risk of SREs was similar (Fig. 90.1, lower panel; 38% reduction versus placebo, $P < 0.0001$).

Bisphosphonates may be associated with renal toxicity, which can complicate the use of these agents in

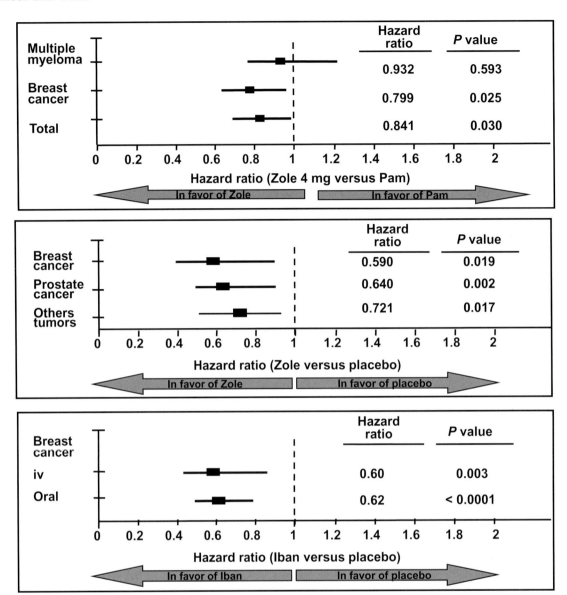

Fig. 90.1. Effects of long-term therapy with bisphosphonates on the risk of developing a skeletal complication in patients with bone metastases. Data are summarized by a multiple-event analysis (Andersen-Gill model for zoledronic acid and Poisson regression analysis for ibandronate). Hazard ratios (and 95% CI) are shown in the left part of each graph with corresponding P values indicated in the right parts. The upper graph is taken from Ref. 26; data in the middle graph are from Refs. 18, 13, and 44; data in the lower graph relate to ibandronate and are taken from Refs. 7 and 30.

patients with renal insufficiency; this is especially important in elderly patients who often have impaired renal function. Creatinine clearance should be monitored during treatment, and for zoledronic acid at least, dosing should be withheld if renal function deteriorates [31, 32]. Denosumab represents a new treatment option for the prevention of SREs in patients with bone metastases from solid tumors. Denosumab is not eliminated by the kidneys and monitoring of renal function is not necessary. Although recent American Society for Clinical Oncology (ASCO) guidelines suggest there is insufficient evidence to recommend one bone modifying agent over another in the management of metastatic bone disease in breast cancer [33], pivotal phase III trials indicate that denosumab has improved efficacy over the widely used bisphosphonate, zoledronic acid [34].

The treatment schedule of denosumab used in comparative phase III trials has been selected from two phase II studies. The first trial included women with breast cancer–related bone metastases who had not received prior intravenous bisphosphonate treatment [35, 36]. In all, 255 women were randomized to one of five denosumab regimens or open-label intravenous bisphosphonate for 24 weeks. All doses of denosumab reduced the levels of uNTx/Cr within 1 week after the first treatment dose, and this was sustained throughout the treatment period. At study week 13 (the primary end point), there was a median percentage reduction of 71% in uNTx/Cr in the pooled denosumab groups compared with a reduction of 79% with bisphosphonate therapy [35]. Similar changes in uNTx/Cr were seen at week 25 [36]. In the second phase II trial, denosumab efficacy was assessed in patients with bone metastases who had received prior intravenous bisphosphonate therapy. In this study, 111 patients with breast cancer, prostate cancer, or other solid tumors were randomized to subcutaneous denosumab 180 mg every 4 weeks or every 12 weeks, or continued intravenous bisphosphonate therapy for 25 weeks [37]. The patients had to have elevated uNTx levels at screening despite ongoing bisphosphonate therapy, and so were not considered to be responding adequately to bisphosphonates. At week 13, the proportion of patients achieving normal uNTx levels was significantly greater among individuals who received denosumab every 4 weeks (78%) and every 12 weeks (64%) than in those treated with bisphosphonates (29%). More denosumab-treated patients maintained normalized uNTx levels over 25 weeks compared with bisphosphonate-treated patients [37].

Further analysis of these two phase II studies showed that denosumab provided similar median percentage reductions in uNTx/Cr at week 25 in bisphosphonate-naïve patients (75%) and those previously treated with bisphosphonate (80%) [38]. The median time to reach uNTx levels less than 50 nmol/L BCE/mM creatinine with denosumab treatment was 9 days for both bisphosphonate-naïve and previously treated patients. This compares with 8 days for intravenous bisphosphonate therapy in naïve patients and 65 days in previously treated patients. Denosumab also provided substantially greater median percentage reductions in tartrate-resistant acid phosphatase (TRAP-5b, a surrogate marker of osteoclast number) at week 25 than intravenous bisphosphonate therapy (73% vs 11%) in patients previously treated with intravenous bisphosphonates. This suggests that functioning osteoclasts are still present in patients showing an inadequate response to bisphosphonate therapy and that switching to denosumab may help suppress their activity. This finding confirms the different mechanism of action of denosumab compared with bisphosphonates and suggests that denosumab may prove to be especially effective in patients who respond poorly to bisphosphonate therapy.

The effects of denosumab and zoledronic acid on SREs in breast cancer patients with bone metastases have been recently compared in a randomized, active-controlled, double-blind, double-dummy study including over 2,000 patients [39]. Patients were randomized equally to receive subcutaneous denosumab 120 mg and an intravenous placebo every 4 weeks, or intravenous zoledronic acid 4 mg (dose adjusted for creatinine clearance) and a subcutaneous placebo every 4 weeks. Patients were strongly recommended to take daily supplemental calcium and vitamin D. The primary end point was the time to the first on-study SRE (classically defined as pathologic fracture, radiation or surgery to bone, or spinal cord compression) assessed for treatment non-inferiority; if non-inferiority was proven, treatment superiority was tested as a secondary end point. Denosumab increased the time to first on-study SRE by 18% compared with zoledronic acid [hazard ratio (HR): 0.82; P = 0.01 for superiority; Fig. 90.2, left panel]. The median time to the first on-study SRE was 26.4 months for the zoledronic acid group and had not been reached for the denosumab treatment group. Denosumab also significantly delayed the time to the first and subsequent on-study SREs by 23% compared with zoledronic acid (multiple-event analysis, P = 0.001). The mean skeletal morbidity rate was also significantly lower with denosumab than with zoledronic acid (0.45 vs 0.58 events per patient per year; P = 0.004), which represents a reduction of 22% with denosumab. Median reductions in bone turnover markers at week 13 were greater with denosumab than with zoledronic acid. Overall survival and disease progression were similar in the two treatment groups.

It is now recommended to start bisphosphonates (or denosumab) as soon as bone metastases are diagnosed in order to delay the first SRE and to reduce the complication rate of metastatic bone disease. ASCO guidelines recommend that, once initiated, intravenous bisphosphonates should be continued until there is a substantial decline in the patient's general performance status [33]. However, criteria are lacking to determine if and how long an individual patient benefits from the administration of bisphosphonates. Promoting lifelong therapy is in contradiction with the paucity of data regarding the usefulness and the safety of treatment durations beyond 2 to 3 years. Stopping zoledronic acid therapy, at least temporarily, or reducing the frequency of the infusions (e.g., an infusion every 3 months) is often considered in patients whose bone disease is not "aggressive" and is well controlled by the antineoplastic treatment. However, no data or specific guidelines are available yet to give firm support to intermittent zoledronic acid treatments or a reduction in the frequency of infusions [40]. The pharmacokinetics of denosumab argues against intermittent treatments. Unlike bisphosphonates, denosumab is not stored in bone and interrupting its administration is probably not without risks, at least if the bone disease is not controlled by the antineoplastic treatment. A reduction in the frequency of denosumab injections, guided by the evolution of metastatic bone disease, is probably a better alternative than intermittent treatments. Long-term studies with denosumab are ongoing and should give more information on the balance between efficacy and risks of prolonged therapy.

Prostate cancer

Skeletal metastases from prostate cancer are typically osteoblastic. Therefore, it was traditionally not believed that this form of bone metastasis might respond to

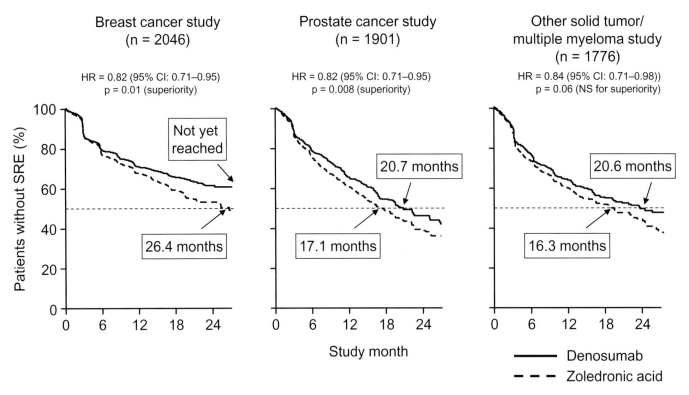

Fig. 90.2. Kaplan–Meier estimates of time to first on-study SRE with denosumab and zoledronic acid in three double-blind, double-dummy phase III clinical trials. Hazard ratios (and 95% CI) are shown above each graph with corresponding P values for superiority testing. (Adapted from Ref. 34.)

antiresorptive therapy. Meanwhile, histomorphometric analyses of bone biopsies and studies on biochemical markers of bone turnover demonstrated that enhanced bone formation in osteoblastic lesions is accompanied by a marked increase in the bone resorption rate [41].

Initial uncontrolled trials with bisphosphonates were often positive, while subsequent placebo-controlled studies were usually negative, whether for clodronate or for pamidronate [11]. The primary objective of the pivotal placebo-controlled study of zoledronic acid of 643 men with bone metastases and progressive prostate cancer after androgen deprivation therapy was to demonstrate a reduction in the frequency of objective SREs following bisphosphonate therapy. Castrate-refractory prostate cancer patients with bone metastases were randomized to receive intravenous zoledronic acid 8 mg or 4 mg or placebo every 3–4 weeks [12, 13]. As mentioned above, the group receiving 8 mg was switched during the trial to 4 mg because of renal toxicity. This trial established zoledronic acid as the first bone-modifying agent to benefit men with castrate-resistant prostate cancer and bone metastases. At the end of the core trial, there was a relative reduction of 25% in the number of patients presenting an objective bone complication. The time to the first SRE was also improved with zoledronic acid. In a multiple-event analysis, 4 mg of zoledronic acid significantly decreased the risk of developing skeletal complications by 36% compared with placebo (P < 0.005) (Fig. 90.1, middle panel). Other secondary end points, including the time to the first SRE or the percentage of patients who presented a fracture, were also significantly reduced in the 4-mg zoledronic acid group. One can speculate that part of the favorable effects of zoledronic acid, especially on the fracture rate, could be due to effective therapy of castration-induced osteoporosis. The effect on bone pain appeared to be less impressive than in breast cancer. A favorable pain response, defined as a 2-point difference on the 11-point scale of the BPI assessment, was observed in 33% of the zoledronic acid-treated patients compared with 25% for patients receiving a placebo [42].

Denosumab was recently shown to be superior to zoledronic acid in reducing the risk of SREs in a study that involved more than 1,900 men with castrate-resistant prostate cancer and bone metastases [43]. Denosumab delayed the time to the first on-study SRE by 18% compared with zoledronic acid (Fig. 90.2, middle panel; HR: 0.82; P = 0.008 for superiority). The median time to the first on-study SRE was 3.6 months longer with denosumab compared with zoledronic acid. Denosumab also significantly delayed the time to the first and subsequent on-study SREs by 18% compared with zoledronic acid (P = 0.008). As for the breast cancer study, overall survival and disease progression did not differ significantly between the treatment groups. At week 13, denosumab provided significantly greater median reductions than zoledronic acid in levels of the bone markers uNTx/Cr and serum bone alkaline phosphatase (BAP).

On the basis of these studies, zoledronic acid or denosumab should be recommended for all patients with castrate-refractory prostate cancer and bone metastases, especially when they are symptomatic. It is often tempting to start bone-modifying agents earlier in the course of the disease, especially if the bone disease causes clinically significant pain or after an SRE has occurred. One should wait, however, for the results of an ongoing prospective trial in patients with metastatic castration-sensitive prostate cancer.

Other solid tumors

In a phase III placebo-controlled study conducted in patients with lung and other solid tumors, zoledronic acid produced less impressive results than in other cancers, partly because of the short survival of lung cancer patients [44]. There was no significant effect on bone pain or quality of life. At 9 months, the primary end point (percentage of patients with an SRE) was not significantly lower with zoledronic acid 4 mg than with placebo, but a multiple-event analysis indicated a favorable effect (Fig. 90.1, middle panel). Retrospective subset analysis of patients with kidney cancer suggested a marked efficacy in that particular tumor.

Zoledronic acid was more recently compared to denosumab in 1,800 patients with bone metastases and solid tumors other than prostate or breast, or with multiple myeloma [45]. Approximately 40% of patients in that study had non-small-cell lung cancer, and 10% had multiple myeloma. Denosumab was non-inferior to zoledronic acid in delaying the time to the first on-study SRE (Fig. 90.2, right panel; HR = 0.84). However, after adjustment for multiple comparisons, the difference between the groups was not statistically significant for treatment superiority (P = 0.06). Analysis for multiple events showed a nonsignificant difference in the time to first and subsequent on-study SREs between both compounds. In tumor types other than breast, prostate, or myeloma, it is reasonable to start bisphosphonates (or denosumab) if the skeleton is one of (or the) predominant symptomatic metastatic site(s) and expected survival time is at least 3–6 months.

Multiple myeloma

A systematic review of the various therapeutic options for the management of multiple myeloma has considered the introduction of bisphosphonates as one of the two most important therapeutic advances for this disease (the other one being the use of high-dose chemotherapy) [46]. ASCO guidelines recommend starting bisphosphonates in patients with lytic disease on plain X-rays or imaging studies, or with spine compression fracture(s) from osteopenia. The panel still considers it "reasonable" to start bisphosphonates in patients with osteopenia based on plain radiographs or bone mineral density (BMD) measurement but does not recommend them in patients with solitary plasmacytoma or smoldering or indolent myeloma [47]. The recommendations of the European Myeloma Network are similar, although they firmly recommend starting bisphosphonates also upon detection of severe osteopenia or osteoporosis [48].

The Cochrane Myeloma Review Group has reported a meta-analysis based on 11 trials. This review concluded that both pamidronate and clodronate reduce the incidence of hypercalcemia, the pain index, and the number of vertebral fractures in myeloma patients [49]. In the largest placebo-controlled trial, the proportion of patients developing any SRE was significantly (P < 0.001) smaller in the pamidronate than in the placebo group, and, at the end of a second year extension of the trial, the mean number of skeletal events per year was 1.3 in the pamidronate group versus 2.2 in the placebo group [50]. The newer, more potent bisphosphonate zoledronic acid has been shown to have a comparable efficacy to pamidronate in a randomized phase III trial including myeloma patients (Fig. 90.1, upper panel) [26]. Although there is no direct comparative trial between clodronate and pamidronate or zoledronic acid, the ASCO Panel recommends only intravenous pamidronate or zoledronic acid in light of the use of the time to the first event as the primary end point and a more complete assessment of bony complications [47]. The European Myeloma Network also advises to use the intravenous route and underlies that zoledronic acid and pamidronate are equally effective in terms of reducing SREs [48]. European guidelines advise that bisphosphonates should be given for 2 years and continued if there is evidence of active myeloma bone disease [48], whereas ASCO recommends to seriously consider the discontinuation of bisphosphonates in patients with responsive or stable disease after 2 years of therapy. Bisphosphonates should then be resumed upon relapse with new-onset SREs [47]. A consensus from the Mayo Clinic advises decreasing the frequency of the infusions to every 3 months if bisphosphonates are continued after 2 years [51]. Even if this is a reasonable recommendation in order to reduce the risk of osteonecrosis of the jaw, there are no prospective trials data to support it.

Denosumab has only been studied in a limited number of myeloma patients; they comprised only 10% of patients in the phase III study [45]. There was an unfavorable trend for denosumab seen in an *ad hoc* analysis of overall survival in the subgroup of patients with myeloma. However, the denosumab and zoledronic acid arms were not well balanced for clinical subgroups of myeloma patients, so it is risky to draw conclusions from these limited data. This probably contributed to the reasoning behind why the U.S. FDA and the European Medicines Agency have so far approved denosumab in patients with solid tumors only and not in those with multiple myeloma. A randomized, double-blind study of denosumab versus zoledronic acid in newly diagnosed patients with myeloma is ongoing and should provide a clearer indication of the effects of denosumab in myeloma. Zoledronic acid has been shown to reduce the risk of SREs in patients with multiple myeloma with similar efficacy to pamidronate, and it is likely that this will also be the case for the comparison between denosumab and zoledronic acid.

Safety aspects

Although generally well tolerated, bisphosphonates and denosumab are occasionally associated with adverse events that are reviewed more extensively in other chapters of the Primer. Hypocalcemia is a side effect that may occur with all bisphosphonates, but especially in vitamin D-deficient patients. It is advisable to administer calcium and vitamin D to all patients on prolonged bisphosphonate or denosumab therapy to avoid hypocalcemia and the deleterious consequences of chronic secondary hyperparathyroidism. Hypocalcemia was reported more frequently with denosumab than with zoledronic acid in the integrated analysis of the three phase III studies (9.6% vs 5.0%) [52]. The findings are consistent with the more marked inhibition of bone resorption with denosumab suggested by the larger decrease in bone resorption markers. The majority of hypocalcemia events after zoledronic acid or denosumab administration are transient and asymptomatic, but some may be severe.

The reported incidence of renal function deterioration in clinical trials of zoledronic acid was 10.7% in patients with multiple myeloma or breast cancer, not significantly different from the pamidronate figures in that trial [19]. Rare cases of renal failure with zoledronic acid have been subsequently reported, and renal monitoring is now recommended before each infusion of zoledronic acid. The product label advocates stepwise dose reductions when baseline creatinine clearance is 30–60 ml/min [31]. Zoledronic acid is not recommended in patients with severe renal deterioration or those taking nephrotoxic medications. Prolonged use of intravenous ibandronate in patients with breast cancer has shown a low incidence of renal adverse events that is comparable to placebo [7]. The monitoring of renal function is not necessary with denosumab, as the drug is not excreted by the kidney. However, denosumab has not been studied in cancer patients with creatinine clearance less than 30 ml/min.

Osteonecrosis of the jaw (ONJ) is the most feared side effect of prolonged bisphosphonate or denosumab therapy [34, 53, 54]. Although sometimes devastating, ONJ can also present as an asymptomatic bony exposure. Its definition, early diagnosis, and follow-up have been reviewed by an American Society for Bone and Mineral Research (ASBMR) task force [54]. According to the series, its prevalence varies between 1% and 10% in patients on prolonged bisphosphonate therapy, with higher rates in myeloma than in solid tumors. It typically occurs after dental extraction but can also occur spontaneously. The prolonged inhibition of bone remodeling with lack of repair of physiologic microfractures due to constant stress from masticatory forces probably plays a key role in the development of ONJ, while the decreased intraosseous blood flow, and infections in the maxilla and/or mandible play a contributory role. The risk appears to be higher after zoledronic acid than after pamidronate, but is essentially linked to the duration of therapy. In the comparative trials between zoledronic acid and denosumab, there were numerically more cases of ONJ in the denosumab groups but the integrated analysis of the three comparative trials showed that the cumulative incidence of confirmed ONJ after 3 years was not significantly different between denosumab and zoledronic acid treatments (1.8% vs 1.3%; p = 0.13) [55]. Tooth extraction was reported for 62% of those patients. Management was mostly conservative and healing occurred in 36% of patients [55]. Patients should receive a comprehensive dental examination before treatment with bisphosphonates or denosumab to treat dental problems that may require surgical or invasive dental procedures. After therapy initiation, invasive dental procedures should be avoided if possible. Recent studies in patients with myeloma or solid tumors showed that the frequency of ONJ could be lowered by at least one-half by implementation of a preventive dental program, including a detailed assessment of dental status, regular dental care, and avoidance of invasive dental procedures during treatment with zoledronic acid [56–58].

Bisphosphonates as adjuvant therapy

Another potential major role for bisphosphonates is the prevention or at least a delay in the development of bone metastases. Bisphosphonates have the potential to reduce tumor burden in bone, whether indirectly, by decreasing bone turnover, or directly, by one or several antitumor effects [59]. Initial adjuvant studies were conducted with clodronate, but the results were contradictory. The only double-blind placebo-controlled trial involving more than 1,000 unselected breast cancer patients after surgery treated for 2 years with 1,600 mg clodronate or a placebo indicates that clodronate could reduce the incidence of bone metastases (by 31% at 5 years, P = 0.043) and may prolong survival (P = 0.048) [60]. The recently published trial of the National Surgical Adjuvant Breast Project (NSABP) B-34, conducted in 3,323 women assigned in a 1:1 ratio to either clodronate or placebo for 3 years, showed no differences in DFS or overall survival between groups. However, DFS was significantly prolonged in women age 50 years or older on study entry (61).

In early breast cancer, zoledronic acid and denosumab have been studied in the adjuvant setting for preventing aromatase inhibitor-induced bone loss (reviewed in Chapter 86 of the Primer). The phase III Austrian Breast & Colorectal Cancer Study Group (ABCSG)-12 trial investigated the adjuvant use of zoledronic acid 4 mg every 6 months in 1,800 premenopausal patients with hormone receptor-positive breast cancer receiving ovarian suppression for 3 years in combination with either tamoxifen or anastrozole. At a median follow-up of 4 years, the addition of zoledronic acid not only prevented hormone therapy-induced bone loss but also resulted in significant improvement in recurrence-free survival (HR = 0.65; P = 0.01) and disease-free survival (DFS, HR = 0.64; P = 0.01) compared with no zoledronic acid. Fewer patients who received zoledronic acid had distant disease recurrence at bone and non-bone sites, including visceral metastases, locoregional recurrences, and contralateral breast disease. A preplanned analysis showed that, in patients who were 40 years or younger at baseline, zoledronic acid did not significantly reduce the relative risk of DFS events (HR = 0.94), whereas in patients who were older than 40 years at baseline, the risk reduction (HR = 0.58) with zoledronic acid was significant. Although all patients got ovarian suppression, patients older than 40 years at baseline likely achieved more substantial estrogen deprivation. There was also a trend toward improved overall survival [61, 62]. At a median follow-up of 62 months, more than 2 years after completion of therapy, the addition of zoledronic acid to endocrine therapy produced a durable improvement in DFS overall (HR = 0.68; P = 0.009) and a trend toward improved survival (HR = 0.67; P = 0.09). Recently presented data at 68 months follow-up indicate a significant overall survival advantage. There were no reports of renal failure or ONJ [63].

The multicenter studies Zometa-Femara Adjuvant Synergy Trial, Z-FAST and ZO-FAST, are two similarly designed trials that evaluate the efficacy of zoledronic acid in preventing bone loss induced by the aromatase inhibitor letrozole in postmenopausal women with breast cancer. Patients were randomized to receive with adjuvant letrozole either upfront or delayed zoledronic acid. A combined analysis showed lower recurrence rates in the group receiving upfront zoledronic acid and the 3-year follow-up results of the ZO-FAST trial (n = 1,065) showed a significant 41% relative risk reduction for DFS events (P = 0.03) in the upfront group compared to the delayed therapy group [64]. Lastly, the Adjuvant Zoledronic acid Recurrence (AZURE) trial is a randomized open phase III trial of 3,360 patients with stage II/III breast cancer that evaluates the effect of zoledronic acid for 5 years, given at a more intensive dosing schedule than that used in ABCSG-12, in addition to standard therapy on disease-related outcomes. A subgroup analysis of that trial indicates that zoledronic acid in combination with neoadjuvant chemotherapy might confer an antitumor effect as compared to patients who did not receive zoledronic acid [65]. The final results of the AZURE trial have been recently published. For the overall study population, DFS, overall survival, and recurrence type (local, regional, or distant) did not differ between groups. The cumulative incidence of ONJ was 1.1% in the zoledronic acid group [66]. In a planned subset analysis, women who were older than 60 years of age or known to be postmenopausal for longer than 5 years had a significant improvement in DFS (HR = 0.76; P < 0.05) and overall survival (HR = 0.71; P < 0.05) with the addition of zoledronic acid. The group of postmenopausal women in the AZURE trial had very low estrogen concentrations within bone, which must also be the case for the women more than 40 years old in the ABCSG-12 or in the ZO-FAST trials [66]. This supports the hypothesis that adjuvant bisphosphonates might be most effective at reducing breast cancer recurrence in a low-estrogen, high-bone resorptive environment [67]. However, the data are not sufficiently compelling to support routine use of zoledronic acid in adjuvant therapy. Other ongoing trials are actively evaluating the role of adjuvant bisphosphonates in early breast cancer. Denosumab has also been shown to prevent aromatase inhibitor-induced bone loss and a placebo-controlled adjuvant trial in early breast cancer is ongoing.

In patients with prostate cancer, there is no evidence that bisphosphonates are useful in the adjuvant setting besides preventing bone loss induced by androgen deprivation therapy. The results of a large (n = 1,432), randomized, placebo-controlled study in patients with nonmetastatic, castration-resistant prostate cancer at high risk of developing bone metastases on the basis of PSA levels (8 ng/ml or higher) or PSA doubling time (10 months or less), have shown that denosumab (120 mg every 4 weeks) significantly improves bone metastasis-free survival by a median of 4.2 months compared with placebo; the hazard ratio was 0.85 (P = 0.028). Denosumab was also associated with increased time to first symptomatic bone metastasis. Overall survival did not differ between groups. Denosumab was associated with increased incidence of hypocalcemia and ONJ (4% at the end of year 3) [68].

Conclusions and perspectives

Bisphosphonates, especially zoledronic acid, constitute an important part of our therapeutic armamentarium to reduce the skeletal morbidity rate in patients with bone metastases and to preserve bone health in patients with early stage breast cancer [69]. Monthly infusions reduce the complications of established tumor bone disease by about 40%. Recent phase III studies have demonstrated the superiority of denosumab to zoledronic acid for delaying the time to the first SRE in patients with breast or prostate cancer and bone metastases. Non-inferiority was shown in the trial including other solid tumors and multiple myeloma. The overall burden of the disease (as assessed by an analysis of multiple events and yearly skeletal morbidity rate) was also significantly reduced in the breast and the prostate cancer studies, and in the prespecified integrated analysis that included all three studies. Denosumab is conveniently administered by subcutaneous injections and is devoid of renal toxicity. However, ONJ could be more frequent than with zoledronic acid and cost-effectiveness analyses should be performed to demonstrate that this incremental efficacy benefit is not of debatable value because of the increased expense.

Treatment with these potent inhibitors of bone resorption should be progressively "individualized.". Tailoring therapy to individual tumor and patient characteristics is increasingly accomplished in modern oncology. In metastatic breast cancer, ASCO still recommends to start bone-modifying agents as soon as metastatic bone disease is diagnosed by a radiological technique and to continue their administration until "evidence of substantial decline in a patient's general performance status" [33]. These recommendations contrast with the paucity of data beyond 2 years of treatment duration. As our understanding improves of the pathophysiology of the different steps of tumor bone destruction, and more sensitive and specific markers of cancer-induced changes in bone turnover and metabolism are developed, it is likely that future studies will enable treatment to be refined. Such studies will better define the place of intermittent treatments, according to individual disease characteristics, such as "aggressiveness" characteristics of the bone metastases, early estimation of the responsiveness to systemic therapy, and the use of biochemical parameters to monitor the influence of tumor secretory products on bone marrow, bone cells, and bone matrix. Treatments tailored to the individual patient are likely to decrease the occurrence of toxic effects, notably ONJ, and to improve the cost-effectiveness ratio of new compounds.

Tumor bone disease in breast cancer and in multiple myeloma is most often accompanied by an inhibition or disorganization of osteoblast proliferation and activity. The most promising agent to date that has been shown to increase osteoblast proliferation is a monoclonal anti-sclerostin antibody. Increased bone density and strength of non-fractured bones have been reported in animal models [70], and clinical studies have started in postmenopausal osteoporosis. Combination or sequencing therapies with denosumab and an anti-sclerostin antibody should be fruitful areas of investigation in the future. For prostate cancer, endothelin-A receptor antagonists, proto-oncogene tyrosine-protein kinase (SRC) inhibitor dasatinib, tyrosine kinase inhibitor cabozantinib, and alpha-emitting radium-223 are all under active investigation [71].

Lastly, the prevention of bone metastases is a key goal in the management of cancer patients. It appears that zoledronic acid can prevent bone metastases in estrogen-deprived patients, as suggested by the ABCSG-12 study in premenopausal women who were medically castrated and older than 40 years old, and by analysis of the subset of postmenopausal women in the adjuvant AZURE breast cancer study. These initial results are quite encouraging but bisphosphonates cannot be recommended yet in the adjuvant setting. Placebo-controlled adjuvant studies with denosumab in early breast cancer have started, and the results of a large placebo-controlled study in patients with nonmetastatic, castrate-resistant prostate cancer at high risk of developing bone metastases have shown that denosumab significantly improves bone metastasis-free survival compared with placebo. The prevention of bone metastases by "bone modifying agents" is an exciting objective for the years to come. Identifying predictors of bone metastases will allow selective use of bone-modifying agents in patients who would benefit from them most, while avoiding adverse effects and cost in patients unlikely to benefit. Recent data suggest that high serum CTX levels before initiation of adjuvant endocrine therapy are predictive for bone-only relapse [72]. Because of its importance for the growth of cancer cells in bone, the stem cell niche is another potential pharmaceutical target. Agents that stimulate normal hematopoiesis or bone formation could reduce colonization of bone by tumor stem cells and prevent metastatic progression [73].

REFERENCES

1. Body JJ, Bartl R, Burckhardt P, Delmas PD, Diel IJ, Fleisch H, Kanis JA, Kyle RA, Mundy GR, Paterson AHG, Rubens RD. 1998. Current use of bisphosphonates in oncology. International Bone and Cancer Study Group. *J Clin Oncol* 16(12): 3890–9.
2. Cleeland CS, Janjan NA, Scott CB, Seiferheld WF, Curran WJ. 2000. Cancer pain management by radiotherapists: A survey of radiation therapy oncology group physicians. *Int J Radiat Oncol Biol Phys* 47(1): 203–8.
3. Vassiliou V, Kalogeropoulou C, Giannopoulou E, Leotsinidis M, Tsota I, Kardamakis D. 2007. A novel study investigating the therapeutic outcome of patients with lytic, mixed and sclerotic bone metastases treated with combined radiotherapy and ibandronate. *Clin Exp Metastasis* 24(3): 169–78.
4. Coleman R, Rubens R. 1987. The clinical course of bone metastases from breast cancer. *Br J Cancer* 55(1): 61–6.
5. Body JJ. 2006. Breast cancer: Bisphosphonate therapy for metastatic bone disease. *Clin Cancer Res* 12(20 Pt 2): 6258s–63s.
6. Koizumi M, Yoshimoto M, Kasumi F, Ogata E. 2003. Comparison between solitary and multiple skeletal metastatic lesions of breast cancer patients. *Ann Oncol* 14(8): 1234–40.
7. Body JJ, Diel IJ, Lichinitser MR, Kreuser ED, Dornoff W, Gorbunova VA, Budde M, Bergstrom B; MF 4265 Study Group. 2003. Intravenous ibandronate reduces the incidence of skeletal complications in patients with breast cancer and bone metastases. *Ann Oncol* 14(9): 1399–405.
8. Hortobagyi GN, Theriault RL, Lipton A, Porter L, Blayney D, Sinoff C, Wheeler H, Simeone JF, Seaman JJ, Knight RD, Heffernan M, Mellars K, Reitsma DJ. 1998. Long-term prevention of skeletal complications of metastatic breast cancer with pamidronate: Protocol 19 Aredia Breast Cancer Study Group. *J Clin Oncol* 16(6): 2038–44.
9. Plunkett TA, Smith P, Rubens RD. 2000. Risk of complications from bone metastases in breast cancer. Implications for management. *Eur J Cancer* 36(4): 476–82.
10. Yong M, Jensen AO, Jacobsen JB, Norgaard M, Fryezk JP, Sorensen HT. 2011. Survival in breast cancer patients with bone metastases and skeletal-related events: A population-based cohort study in Denmark (1999–2007). *Breast Cancer Res Treat* 129(2): 495–503.
11. Small EJ, Smith MR, Seaman JJ, Petrone S, Kowalski MO. 2003. Combined analysis of two multicenter, randomized, placebo-controlled studies of pamidronate disodium for palliation of bone pain in men with metastatic prostate cancer. *J Clin Oncol* 21(23): 4277–84.

12. Saad F, Gleason DM, Murray R, Tchekmedyian S, Venner P, Lacombe L, Chin JL, Vinholes JJ, Goas JA, Chen B. 2002. A randomized, placebo-controlled trial of zoledronic acid in patients with hormone-refractory metastatic prostate carcinoma. *J Natl Cancer Inst* 94(19): 1458–68.
13. Saad F, Gleason DM, Murray R, Tchekmedyian S, Venner P, Lacombe L, Chin JL, Vinholes JJ, Goas JA, Zheng M; Zoledronic Acid Prostate Cancer Study Group. 2004. Long-term efficacy of zoledronic acid for the prevention of skeletal complications in patients with metastatic hormone-refractory prostate cancer. *J Natl Cancer Inst* 96(11): 879–82.
14. Melton LJ 3rd, Kyle RA, Achenbach SJ, Oberg AL, Rajkumar SV. 2005. Fracture risk with multiple myeloma: A population-based study. *J Bone Miner Res* 20(3): 487–93.
15. Baron R, Ferrari S, Russell RG. 2011. Denosumab and bisphosphonates: Different mechanisms of action and effects. *Bone* 48(4): 677–92.
16. Russell RG, Watts NB, Ebetino FH, Rogers MJ. 2008. Mechanisms of action of bisphosphonates: Similarities and differences and their potential influence on clinical efficacy. *Osteoporos Int* 19(6): 733–59.
17. Body JJ. 2011. New developments for treatment and prevention of bone metastases. *Curr Opin Oncol* 23(4): 338–42.
18. Kohno N, Aogi K, Minami H, Nakamura S, Asaga T, Iino Y, Watanabe T, Goessl C, Ohashi Y, Takashima S. 2005. Zoledronic acid significantly reduces skeletal complications compared with placebo in Japanese women with bone metastases from breast cancer: A randomized, placebo-controlled trial. *J Clin Oncol* 23(15): 3314–21.
19. Rosen LS, Gordon D, Kaminski M, Howell A, Belch A, Mackey J, Apffelstaedt J, Hussein M, Coleman RE, Reitsma DJ, Seaman JJ, Chen BL, Ambros Y. 2001. Zoledronic acid versus pamidronate in the treatment of skeletal metastases in patients with breast cancer or osteolytic lesions of multiple myeloma: A phase III, double-blind, comparative trial. *Cancer J* 7(5): 377–87.
20. Diel IJ, Body JJ, Lichinitser MR, Kreuser ED, Dornoff W, Gorbunova VA, Budde M, Bergstrom B; MF 4265 Study Group. 2004. Improved quality of life after long-term treatment with the bisphosphonate ibandronate in patients with metastatic bone disease due to breast cancer. *Eur J Cancer* 40(11): 1704–12.
21. Paterson AH, Powles TJ, Kanis JA, McCloskey E, Hanson J, Ashley S. 1993. Double-blind controlled trial of oral clodronate in patients with bone metastases from breast cancer. *J Clin Oncol* 11(1): 59–65.
22. Body JJ, Dumon JC, Piccart M, Ford J. 1995. Intravenous pamidronate in patients with tumor-induced osteolysis: A biochemical dose-response study. *J Bone Miner Res* 10(8): 1191–6.
23. Theriault RL, Lipton A, Hortobagyi GN, Leff R, Gluck S, Stewart JF, Costello S, Kennedy I, Simeone J, Seaman JJ, Knight RD, Mellars K, Heffernan M, Reitsma DJ. 1999. Pamidronate reduces skeletal morbidity in women with advanced breast cancer and lytic bone lesions: A randomized, placebo-controlled trial: Protocol 18 Aredia Breast Cancer Study Group. *J Clin Oncol* 17(3): 846–54.
24. Lipton A. 2003. Bisphosphonates and metastatic breast carcinoma. *Cancer* 97(3 Suppl): 848–53.
25. Jagdev SP, Purohito P, Heatley S, Herling C, Coleman RE. 2001. Comparison of the effect of intravenous pamidronate and oral clodronate on symptoms and bone resorption in patients with metastatic bone disease. *Ann Oncol* 12(10): 1433–8.
26. Rosen LS, Gordon D, Kaminski M, Howell A, Belch A, Mackey J, Apffelstaedt J, Hussein MA, Coleman RE, Reitsma DJ, Chen BL, Seaman JJ. 2003. Long-term efficacy and safety of zoledronic acid compared with pamidronate disodium in the treatment of skeletal complications in patients with advanced multiple myeloma or breast carcinoma: A randomized, double-blind, multicenter, comparative trial. *Cancer* 98(8): 1735–44.
27. Rosen LS, Gordon D, Tchekmedyian S, Yanagihara R, Hirsh V, Krzakowski M, Pawlicki M, de Souza P, Zheng M, Urbanowitz G, Reitsma D, Seaman JJ. 2003. Zoledronic acid versus placebo in the treatment of skeletal metastases in patients with lung cancer and other solid tumors: A phase III, double-blind, randomized trial—The Zoledronic Acid Lung Cancer and Other Solid Tumors Study Group. *J Clin Oncol* 21(16): 3150–7.
28. Body JJ. 2003. Zoledronic acid: An advance in tumour bone disease and a new hope for osteoporosis. *Expert Opin Pharmacother* 4(4): 567–80.
29. Clemons MJ, Dranitsaris G, Ooi WS, Yogendran G, Sukovic T, Wong BY, Verma S, Pritchard KI, Trudeau M, Cole DE. 2006. Phase II trial evaluating the palliative benefit of second-line zoledronic acid in breast cancer patients with either a skeletal-related event or progressive bone metastases despite first-line bisphosphonate therapy. *J Clin Oncol* 24(30): 4895–900.
30. Body JJ, Diel IJ, Lichinitzer M, Lazarev A, Pecherstorfer M, Bell R, Tripathy D, Bergstrom B. 2004. Oral ibandronate reduces the risk of skeletal complications in breast cancer patients with metastatic bone disease: Results from two randomised, placebo-controlled phase III studies. *Br J Cancer* 90(6): 1133–7.
31. Novartis Pharmaceuticals Corporation. Zometa (zoledronic acid): Full prescribing information. Available at: www.us.zometa.com/patient/zometa-prescribing-information.jsp
32. Novartis Pharmaceuticals Corporation. Aredia (pamidronate): Full prescribing information. Available at: www.pharma.us.novartis.com/product/pi/pdf/aredia.pdf
33. Van Poznak CH, Temin S, Yee GC, Janjan NA, Barlow WE, Biermann JS, Bosserman LD, Geoghegan C, Hillner BE, Theriault RL, Zuckerman DS, Von Roenn JH; American Society of Clinical Oncology. 2011. American Society of Clinical Oncology executive summary of the clinical practice guideline update on the role of bone-modifying agents in metastatic breast cancer. *J Clin Oncol* 29(9): 1221–7.

34. Body JJ. 2012. Denosumab for the management of bone disease in patients with solid tumors. *Expert Rev Anticancer Ther* 12(3): 307–22.
35. Lipton A, Steger GG, Figueroa J, Alvarado C, Solal-Celigny P, Body JJ, de Boer R, Berardi R, Gascon P, Tonkin KS, Coleman R, Paterson AH, Peterson MC, Fan M, Kinsey A, Jun S. 2007. Randomized active-controlled phase II study of denosumab efficacy and safety in patients with breast cancer-related bone metastases. *J Clin Oncol* 25(28): 4431–7.
36. Lipton A, Steger GG, Figueroa J, Alvarado C, Solal-Celigny P, Body JJ, de Boer R, Berardi R, Gascon P, Tonkin KS, Coleman RE, Paterson AH, Gao GM, Kinsey AC, Peterson MC, Jun S. 2008. Extended efficacy and safety of denosumab in breast cancer patients with bone metastases not receiving prior bisphosphonate therapy. *Clin Cancer Res* 14(20): 6690–6.
37. Fizazi K, Lipton A, Mariette X, Body JJ, Rahim Y, Gralow JR, Gao G, Wu L, Sohn W, Jun S. 2009. Randomized phase II trial of denosumab in patients with bone metastases from prostate cancer, breast cancer, or other neoplasms after intravenous bisphosphonates. *J Clin Oncol* 27(10): 1564–71.
38. Body JJ, Lipton A, Gralow J, Steger GG, Gao G, Yeh H, Fizazi K. 2010. Effects of denosumab in patients with bone metastases with and without previous bisphosphonate exposure. *J Bone Miner Res* 25(3): 440–6.
39. Stopeck AT, Lipton A, Body JJ, Steger GG, Tonkin K, de Boer RH, Lichinitser M, Fujiwara Y, Yardley DA, Viniegra M, Fan M, Jiang Q, Dansey R, Jun S, Braun A. 2010. Denosumab compared with zoledronic acid for the treatment of bone metastases in patients with advanced breast cancer: A randomized, double-blind study. *J Clin Oncol* 28(35): 5132–9.
40. Body JJ. 2006. Individualization of bisphosphonate therapy. In: Piccart MJ, Wood WC, Hung C-M, Solin LJ, Cardoso, F (eds.) *Breast Cancer Management and Molecular Medicine: Towards Tailored Approaches*. New-York: Springer. Chapter 27. pp. 545–64.
41. Garnero P, Buchs N, Zekri J, Rizzoli R, Coleman RE, Delmas PD. 2000. Markers of bone turnover for the management of patients with bone metastases from prostate cancer. *Br J Cancer* 82(4): 858–64.
42. Weinfurt KP, Anstrom KJ, Castel LD, Schulman KA, Saad F. 2006. Effect of zoledronic acid on pain associated with bone metastasis in patients with prostate cancer. *Ann Oncol* 17(6): 986–9.
43. Fizazi K, Carducci M, Smith M, Damião R, Brown J, Karsh L, Milecki P, Shore N, Rader M, Wang H, Jiang Q, Tadros S, Dansey R, Goessl C. 2011. Denosumab versus zoledronic acid for treatment of bone metastases in men with castration-resistant prostate cancer: A randomised, double-blind study. *Lancet* 2011 377(9768): 813–22.
44. Rosen LS, Gordon D, Tchekmedyan NS, Yanagihara R, Hirsh V, Krzakowski M, Pawlicki M, De Souza P, Zheng M, Urbanowitz G, Reitsma D, Seaman J. 2004. Long-term efficacy and safety of zoledronic acid in the treatment of skeletal metastases in patients with nonsmall cell lung carcinoma and other solid tumors: A randomized, phase III, double-blind, placebo-controlled trial. *Cancer* 100(12): 2613–21.
45. Henry DH, Costa L, Goldwasser F, Hirsh V, Hungria V, Prausova J, Scagliotti GV, Sleeboom H, Spencer A, Vadhan-Raj S, von Moos R, Willenbacher W, Woll PJ, Wang J, Jiang Q, Jun S, Dansey R, Yeh H. 2011. Randomized, double-blind study of denosumab versus zoledronic acid in the treatment of bone metastases in patients with advanced cancer (excluding breast and prostate cancer) or multiple myeloma. *J Clin Oncol* 29(9): 1125–32.
46. Kumar A, Loughran T, Alsina M, Durie BG, Djulbegovic B. 2003. Management of multiple myeloma: A systematic review and critical appraisal of published studies. *Lancet Oncol* 4(5): 293–304.
47. Kyle RA, Yee GC, Somerfield MR, Flynn PJ, Halabi S, Jagannath S, Orlowski RZ, Roodman DG, Twilde P, Anderson K; American Society of Clinical Oncology. 2007. American Society of Clinical Oncology 2007 clinical practice guideline update on the role of bisphosphonates in multiple myeloma. *J Clin Oncol* 25(17): 2464–72.
48. Terpos E, Sezer O, Croucher PI, García-Sanz R, Boccadoro M, San Miguel J, Ashcroft J, Bladé J, Cavo M, Delforge M, Dimopoulos MA, Facon T, Macro M, Waage A, Sonneveld P; European Myeloma Network. 2009. The use of bisphosphonates in multiple myeloma: Recommendations of an expert panel on behalf of the European Myeloma Network. *Ann Oncol* 20(8): 1303–17.
49. Djulbegovic B, Wheatley K, Ross J, Clark O, Bos G, Goldschmidt H, Cremer F, Alsina M, Glasmacher A. 2002. Bisphosphonates in multiple myeloma. *Cochrane Database Syst Rev* (3): CD003188.
50. Berenson JR, Lichtenstein A, Porter L, Dimopoulos MA, Bordoni R, George S, Lipton A, Keller A, Ballester O, Kovacs M, Blacklock H, Bell R, Simeone JF, Reitsma DJ, Heffernan M, Seaman J, Knight RD. 1998. Long-term treatment of advanced multiple myeloma patients reduces skeletal events. *J Clin Oncol* 16(2): 593–602.
51. Lacy MQ, Dispenzieri A, Gertz MA, Greipp PR, Gollbach KL, Hayman SR, Kumar S, Lust JA, Rajkumar SV, Russell SJ, Witzig TE, Zeldenrust SR, Dingli D, Bergsagel PL, Fonseca R, Reeder CB, Stewart AK, Roy V, Dalton RJ, Carr AB, Kademani D, Keller EE, Viozzi CF, Kyle RA. 2006. Mayo clinic consensus statement for the use of bisphosphonates in multiple myeloma. *Mayo Clin Proc* 81(8): 1047–53.
52. Lipton A, Fizazi K, Stopeck AT, Henry DH, Brown JE, Yardley DA, Richardson GE, Siena S, Maroto P, Clemens M, Bilynskyy B, Charu V, Beuzeboc P, Rader M, Viniegra M, Saad F, Ke C, Braun A, Jun S. 2012. Superiority of denosumab to zoledronic acid for prevention of skeletal-related events: a combined analysis of 3 pivotal, randomised, phase 3 trials. *Eur J Cancer* 48(16): 3082–92.
53. Woo SB, Hellstein JW, Kalmar JR. 2006. Narrative [corrected] review: Bisphosphonates and osteonecrosis of the jaws. *Ann Intern Med* 144(10): 753–61.
54. Khosla S, Burr D, Cauley J, Dempster DW, Ebeling PR, Felsenberg D, Gagel RF, Gilsanz V, Guise T, Koka S, McCauley LK, McGowan J, McKee MD, Mohla S, Pendrys DG, Raisz LG, Ruggiero SL, Shafer DM, Shum L, Silverman SL, Van Poznak CH, Watts N, Woo SB,

Shane E; American Society for Bone and Mineral Research. 2007. Bisphosphonate-associated osteonecrosis of the jaw: Report of a task force of the American Society for Bone and Mineral Research. *J Bone Miner Res* 22(10): 1479–91.

55. Saad F, Brown JE, Van Poznak C, Ibrahim T, Stemmer SM, Stopeck AT, Diel IJ, Takahashi S, Shore N, Henry DH, Barrios CH, Facon T, Senecal F, Fizazi K, Zhou L, Daniels A, Carrière P, Dansey R. 2012. Incidence, risk factors, and outcomes of osteonecrosis of the jaw: integrated analysis from three blinded active-controlled phase III trials in cancer patients with bone metastases. *Ann Oncol* 23(5): 1341–7.

56. Ripamonti CI, Maniezzo M, Campa T, Fagnoni E, Brunelli C, Saibene G, Bareggi C, Ascani L, Cislaghi E. 2009. Decreased occurrence of osteonecrosis of the jaw after implementation of dental preventive measures in solid tumour patients with bone metastases treated with bisphosphonates. The experience of the National Cancer Institute of Milan. *Ann Oncol* 20(1): 137–45.

57. Montefusco V, Gay F, Spina F, Miceli R, Maniezzo M, Teresa Ambrosini M, Farina L, Piva S, Palumbo A, Boccadoro M, Corradini P. 2008. Antibiotic prophylaxis before dental procedures may reduce the incidence of osteonecrosis of the jaw in patients with multiple myeloma treated with bisphosphonates. *Leuk Lymphoma* 49(11): 2156–62.

58. Vandone AM, Donadio M, Mozzati M, Ardine M, Polimeni MA, Beatrice S, Ciufredda L, Scoletta M. 2012. Impact of dental care in the prevention of bisphosphonate-associated osteonecrosis of the jaw: A single-center clinical experience. *Ann Oncol* 23(1): 193–200.

59. Clezardin P. 2011. Bisphosphonates' antitumor activity: An unraveled side of a multifaceted drug class. *Bone* 48(1): 71–9.

60. Powles T, Paterson S, Kanis JA, McCloskey E, Ashley S, Tidy A, Rosenqvist K, Smith I, Ottestad L, Legault S, Pajunen M, Nevantaus A, Mannisto E, Suovuori A, Atula S, Nevalainen J, Pylkkanen L. 2002. Randomized, placebo-controlled trial of clodronate in patients with primary operable breast cancer. *J Clin Oncol* 20(15): 3219–24.

61. Paterson AH, Anderson SJ, Lembersky BC, Fehrenbacher L, Falkson CI, King KM, Weir LM, Brufsky AM, Dakhil S, Lad T, Baez-Diaz L, Gralow JR, Robidoux A, Perez EA, Zheng P, Geyer CE Jr, Swain SM, Costantino JP, Mamounas EP, Wolmark N. 2012. Oral clodronate for adjuvant treatment of operable breast cancer (National Surgical Adjuvant Breast and Bowel Project protocol B-34): A multicentre, placebo-controlled, randomised trial. Lancet *Oncol* 13(7): 734–42.

62. Gnant M, Mlineritsch B, Schippinger W, Luschin-Ebengreuth G, Pöstlberger S, Menzel C, Jakesz R, Seifert M, Hubalek M, Bjelic-Radisic V, Samonigg H, Tausch C, Eidtmann H, Steger G, Kwasny W, Dubsky P, Fridrik M, Fitzal F, Stierer M, Rücklinger E, Greil R; ABCSG-12 Trial Investigators, Marth C. 2009. Endocrine therapy plus zoledronic acid in premenopausal breast cancer. *N Engl J Med* 360: 679–91.

63. Gnant M, Mlineritsch B, Stoeger H, Luschin-Ebengreuth G, Heck D, Menzel C, Jakesz R, Seifert M, Hubalek M, Pristauz G, Bauernhofer T, Eidtmann H, Eiermann W, Steger G, Kwasny W, Dubsky P, Hochreiner G, Forsthuber EP, Fesl C, Greil R; Austrian Breast and Colorectal Cancer Study Group, Vienna, Austria. 2011. Adjuvant endocrine therapy plus zoledronic acid in premenopausal women with early-stage breast cancer: 62-month follow-up from the ABCSG-12 randomised trial. *Lancet Oncol* 12(7): 631–41.

64. Eidtmann H, de Boer R, Bundred N, Llombart-Cussac A, Davidson N, Neven P, von Minckwitz G, Miller J, Schenk N, Coleman R. 2010. Efficacy of zoledronic acid in postmenopausal women with early breast cancer receiving adjuvant letrozole: 36-month results of the ZO-FAST study. *Ann Oncol* 21(11): 2188–94.

65. Coleman RE, Winter MC, Cameron D, Bell R, Dodwell D, Keane MM, Gil M, Ritchie D, Passos-Coelho JL, Wheatley D, Burkinshaw R, Marshall SJ, Thorpe H; AZURE (BIG01/04) Investigators. 2010. The effects of adding zoledronic acid to neoadjuvant chemotherapy on tumour response: Exploratory evidence for direct anti-tumour activity in breast cancer. *Br J Cancer* 102(7): 1099–105.

66. Coleman RE, Marshall H, Cameron D, Dodwell D, Burkinshaw R, Keane M, Gil M, Houston SJ, Grieve RJ, Barrett-Lee PJ, Ritchie D, Pugh J, Gaunt C, Rea U, Peterson J, Davies C, Hiley V, Gregory W, Bell R; AZURE Investigators. 2011. Breast-cancer adjuvant therapy with zoledronic acid. *N Engl J Med* 365(15): 1396–405.

67. Korde LA, Gralow JR. 2011. Can we predict who's at risk for developing bone metastases in breast cancer? *J Clin Oncol* 29(27): 3600–4.

68. Smith MR, Saad F, Coleman R, Shore N, Fizazi K, Tombal B, Miller K, Sieber P, Karsh L, Damião R, Tammela TL, Egerdie B, Van Poppel H, Chin J, Morote J, Gómez-Veiga F, Borkowski T, Ye Z, Kupic A, Dansey R, Goessl C. 2011. Denosumab and bone-metastasis-free survival in men with castration-resistant prostate cancer: Results of a phase 3, randomised, placebo-controlled trial. *Lancet* 6736(11): 61226–9.

69. Lipton A. 2011. Zoledronic acid: Multiplicity of use across the cancer continuum. *Expert Rev Anticancer Ther* 11(7): 999–1012.

70. Ominsky MS, Li C, Li X, Tan HL, Lee E, Barrero M, Asuncion FJ, Dwyer D, Han CY, Vlasseros F, Samadfam R, Jolette J, Smith SY, Stolina M, Lacey DL, Simonet WS, Paszty C, Li G, Ke HZ. 2011. Inhibition of sclerostin by monoclonal antibody enhances bone healing and improves bone density and strength of nonfractured bones. *J Bone Miner Res* 26(5): 1012–21.

71. Saylor PJ, Lee RJ, Smith MR. 2011. Emerging therapies to prevent skeletal morbidity in men with prostate cancer. *J Clin Oncol* 29(27): 3705–14.

72. Lipton A, Chapman JAW, Demers L, Shepherd LE, Han L, Wilson CF, Pritchard KI, Leitzel KE, Ali SM, Pollak M. 2011. Elevated bone turnover predicts for bone metastasis in postmenopausal breast cancer: Results of NCIC CTG MA.14. *J Clin Oncol* 29(27): 3605–10.

73. Chirgwin JM. 2012. The stem cell niche as a pharmaceutical target for prevention of skeletal metastases. *Anticancer Agents Med Chem* 12(3): 187–93.

91

Radiotherapy of Skeletal Metastases

Edward Chow, Luluel M. Khan, and Øyvind S. Bruland

Introduction 754
External Beam Radiotherapy 754
Bone-Seeking Radiopharmaceuticals 756
References 757

INTRODUCTION

Bone is the most common site of symptomatic cancer metastasis. Two-thirds to three-quarters of patients with advanced disease from breast and prostate carcinomas have skeletal metastases, and lung, thyroid, and renal carcinoma metastasize to bone in approximately 30% to 40% of cases [1]. Pain is the most common symptom [1, 2]. Additionally, clinical implications of skeletal metastases include pathological fracture, nerve entrapment and/or spinal cord compression (SCC), bone marrow insufficiency, and hypercalcaemia. Hence, bone metastases have a devastating impact on a patient's quality of life [1, 3]. SCC is of particular concern to cancer patients with a long expected survival, e.g., those with breast cancer and the diagnosis of skeletal metastasis as the first and sole metastatic event [4].

Optimal management combines medical treatment, radiation therapy, surgery, bone-targeted radiopharmaceuticals, bisphosphonates, and denosumab, depending on the biology of the disease, extent of the skeletal involvement, and the life expectancy of the patient.

EXTERNAL BEAM RADIOTHERAPY

Skeletal metastases are the single most frequent indication for palliative radiotherapy. External beam radiotherapy (EBRT) effectively relieves pain from localized sites of skeletal metastases [5, 6]. However, the lack of tumor-only selectivity limits its clinical use. Furthermore, since skeletal metastases usually are multiple and distributed throughout the axial skeleton [1, 2, 4], larger or multiple fields of irradiation are often necessary. Table 91.1 outlines factors to be considered when prescribing palliative radiotherapy for bone metastases.

Pain palliation

Solid empirical evidence has clearly documented that single-fraction (SF) EBRT provides equivalent pain relief compared with multi-fraction (MF) EBRT for uncomplicated bone metastases documented by more than 25 randomized clinical trials (RCTs) and three recent meta-analyses [6–8].

One of the first RCTs was conducted by the Radiation Therapy Oncology Group [9]. Ninety percent of patients experienced some degree of pain relief, and 54% achieved complete pain palliation. The trial initially concluded that the low-dose, short-course schedules were as effective as the high-dose protracted programs. However, this study was criticized for using physician-based pain assessment. A reanalysis of the same set of data grouped solitary and multiple bone metastases and used the end point of pain relief, taking analgesia intake into account as well as the need for re-treatment. The authors then concluded that the number of radiation fractions was significantly related to complete combined relief (absence of pain and use of narcotics) and that protracted dose-

---

Primer on the Metabolic Bone Diseases and Disorders of Mineral Metabolism, Eighth Edition. Edited by Clifford J. Rosen.
© 2013 American Society for Bone and Mineral Research. Published 2013 by John Wiley & Sons, Inc.

Table 91.1. Factors to be Considered When Prescribing Palliative Radiotherapy (EBRT) for Bone Metastases

EBRT—Single fraction (SF)	EBRT—Fractionated (MF)
Indication: "pain relief"	Indication: "local tumor control"
Short life expectancy	Expected long-term survival
Concomitant visceral metastases	Predominantly bone or bone-only metastasis
Poor performance status	Good performance status
Inflammatory pain	Neuropathic pain
Aspects of cost and inconvenience	Spinal cord compression
	Postoperative EBRT following an orthopedic procedure in selected cases
	Impending fractures where surgery is not indicated

fractionation schedules were the most effective [10]. This was contrary to the initial report and highlights that the choice of end points will influence the outcome [11].

More recently, results from several large-scale prospective RCTs have been published. The UK Bone Pain Trial Working Party randomized 765 patients with bone metastases to either an SF or MF regimen [12]. There were no significant differences in the time to first improvement in pain, time to complete pain relief, time to first increase in pain at any time up to 12 months after randomization, and no differences in the incidence of nausea, vomiting, SCC, or pathological fracture between the two groups. Re-treatment was, however, twice as common after SF than after MF radiotherapy. The study concluded that an SF of 8 Gy is as safe and effective for the palliation of metastatic bone pain for at least 12 months with greater convenience and lower cost than MF treatment.

The large Dutch Bone Metastases Study included 1,171 patients and confirmed the results mentioned above [13]. In this trial, the re-treatment rates were 25% in the single 8 Gy arm, 7% in the MF arm, and more pathological fractures were observed in the SF group but the absolute percentage was low. In a cost-utility analysis of this RCT, there was no difference in life expectancy or quality-adjusted life expectancy. The estimated cost of radiotherapy, including re-treatments and nonmedical costs, was significantly lower for the SF than for the MF schedule [14].

A Scandinavian RCT planned to recruit 1,000 patients with painful bone metastases randomized to single 8 Gy or 30 Gy (3 Gy in 10 fractions) [15]. The data monitoring committee recommended closure of the study after 376 patients had been recruited because interim analyses indicated that the treatment groups had similar outcomes. Equivalent pain relief within the first 4 months was experienced, and no differences were found for fatigue, global quality of life, and survival between the groups [15].

Two meta-analyses published in 2003 each showed no significant difference in complete and overall pain relief between SF and MF EBRT for bone metastases [6, 7]. Results were remarkably similar, with the Wu et al. paper reporting a complete response rate (absence of pain) of 33% and 32% after SF and MF EBRT, respectively, compared to 34% and 32% for Sze et al. Overall response rates from the two meta-analyses were 62% and 59% [7], compared to 60% and 59% [6], for SF and MF, respectively. When restricted to evaluable patients, overall response rates became 73% for each arm [7]. Most patients experienced pain relief in the first 2 to 4 weeks after EBRT [7]. Side effects were similar and generally consisted of nausea and vomiting.

An updated meta-analysis reviewed 16 RCTs that compared SF and MF schedules [8] involving a total of 2,513 randomizations to SF arms and 2,487 to MF arms. The overall response rate to SF EBRT was 58%, and the complete response rate was 23%, not significantly different from the 59% and 24% experienced by patients randomized to MF EBRT. No differences in acute toxicity, pathological fracture (3.2% of patients fractured after SF vs 2.8% after MF) or SCC incidence were found, thus confirming the conclusions of the 2003 systematic reviews. Recent therapeutic guidelines for the treatment of bone metastases from the American College of Radiology Appropriateness Criteria Expert Panel on Radiation Oncology also endorsed the findings [16, 17].

Palliative radiotherapy is not without side effects. Patients treated either with a single 8 Gy or multiple fractions reported a pain flare incidence of around 40% [18]. Dexamethasone as prophylaxis of radiation-induced pain flare after palliative radiotherapy for symptomatic bone metastases has been reported in a phase II study [19].

Neuropathic pain and spinal cord compression

There is some evidence that certain groups of patients would benefit from a protracted schedule. In a comparison of a single 8 Gy vs 20 Gy (4 Gy in five fractions) for 272 patients with a neuropathic pain component [20], it was found that SF was not as effective as MF; however, it was also not significantly worse. The authors recommended MF as standard radiotherapy for patients with neuropathic pain. However, in patients with short survival or poor performance status, as well as when cost/inconvenience of MF is relevant, SF could be used instead [20].

In the treatment of neoplastic SCC, one RCT has studied the outcome in patients with SCC and an estimated outcome of 6 months or less with no indication for primary surgery [21]. However, two EBRT-schedules not commonly used were compared; i.e., 16 Gy in two fractions over 1 week or a split course of 15 Gy in three fractions, followed by 4 days of rest, and the additional 15 Gy in five fractions. No significant differences were reported between the two arms [21]. More recent trials have included a single 8 Gy and concluded that SCC patients achieve palliation with minimal toxicity and inconvenience with a single fraction [22, 23]. Longer-course radiotherapy is associated with better local control of SCC [23]. Hence, patients with more favorable survival expectancy should be considered for a protracted course of radiotherapy. On the other hand, similar functional outcomes can be achieved with short-course radiotherapy. Patients with limited survival may be better served with a single fraction [24].

Until the results from a recently published RCT comparing surgery and postoperative EBRT and EBRT alone [25] were presented in favor of primary surgery, the common view was that the outcome did not differ between EBRT and surgery for patients with vertebral metastases and SCC [26].

Impending fracture and risk prediction

An impending fracture has a significant likelihood of fracture under normal physiological stresses. Although some physicians believe that all patients with proximally located femoral metastases should undergo preventive surgery, this would result in a large number of unnecessary surgical procedures [27]. Furthermore, a proportion of patients will not be candidates for an operative procedure or will refuse surgical intervention. Often, a minimum life expectancy (6–12 weeks), a reasonable performance status, manageable comorbidities, and adequate remaining bone to support the implanted hardware are required in order to justify the morbidity and mortality risk [28].

If an orthopaedic intervention is not appropriate, patients may receive EBRT alone. Although EBRT can provide pain relief and tumor control, it does not restore bone stability, and remineralization will take weeks to months [29]. Patients should be warned of the increased risk of fracture in the peri-radiation period due to an induced hyperemic response at the periphery of the tumor that temporarily weakens the adjacent bone. Pain relief may allow the patient to be more mobile and, hence, at greater risk for fracture. As such, measures to reduce anatomic forces across the lesion (crutches, a sling, or a walker) are routinely introduced during this time.

Although there is no consensus on appropriate dose fractionation, most authors recommend an MF course of EBRT in a patient with an impending or established fracture [30]. One retrospective series analyzed 27 pathologic fractures in various sites treated with doses of 40–50 Gy over 4–5 weeks. Healing with remineralization was seen in 33%, with pain relief in 67% [30]. Koswig and Budach reported in a prospective randomized trial that the recalcification showed a significant difference between patients treated with 30 Gy in 10 fractions (173%) and those treated with a single 8 Gy (120%, p < 0.0001) [31].

In practice, 20–40 Gy for established pathologic fracture is generally given over 1–3 weeks. In patients with an apparently solitary, histologically confirmed metastasis, especially after a long disease-free interval, some clinicians may wish to give even a higher dose, e.g., 40–50 Gy, under the assumption that this will provide long-term control.

Re-irradiation

Subsets of patients with metastatic disease have longer life expectancies than in the past due to advances in systemic therapy and may therefore outlive the duration of benefit provided by their initial palliative EBRT. This may require consideration of re-irradiation of previously treated sites at a later date [32].

The clinical indications, optimal dose and fractionation, and techniques for re-treatment are controversial [33] due to lack of precise quantitative data on the time course, magnitude, and tissue specificity of long-term occult radiation injury recovery [34].

Re-treatment rates after SF EBRT varied from 18% to 25% compared to 7% to 9% after MF EBRT [12, 13, 15, 35]. Sande et al. recently updated their randomized multicenter trial that patients in the single 8 Gy arm received significantly more re-irradiations as compared to the 30 Gy in 10 fractions arms (27% vs 9%, p = 0.002) [36]. The Dutch Bone Metastases Study Group recently reanalyzed their data to specifically report the efficacy of re-irradiation [37]. Of patients not responding to initial radiation, 66% who initially received a single 8 Gy responded to re-treatment, compared to 33% of patients who initially received a MF course. Re-treatment in patients after pain progression was successful in 70% of those who received SF initially, compared with 57% of those who received more than one fraction. Overall, re-irradiation was effective in 63% of all such treated patients. Jeremic et al. also noticed the efficacy of the second single 4 Gy re-irradiation for patients with painful bone metastases who had already twice received SF radiation [38].

Hence, it is important to consider re-irradiation of sites of metastatic bone pain after initial EBRT, particularly when this follows an initial period of response. There is also evidence that a proportion of initial nonresponders will respond. The preferred dose schedule, however, is at present unknown, but a large, prospective, randomized intergroup study employing common re-irradiation schedules has been launched [39].

BONE-SEEKING RADIOPHARMACEUTICALS

Treatment with intravenous (IV) injected bone-seeking radiopharmaceuticals (BSRs) is an intriguing alternative that selectively delivers ionizing radiation to targeted areas of amplified osteoblastic activity and targets multiple (symptomatic and asymptomatic) metastases simultancously. The target is Ca–OH–apatite particularly abundant in sclerotic metastases from prostate cancer but also present, although more heterogeneously distributed, in mixed sclerotic/osteolytic metastases from breast cancer. This is evident from the biodistribution image common to all BSRs—exemplified as "hot-spots" visualized on a routine diagnostic bone-scan (by ^{99m}Tc-MDP; a radiolabeled bisphosphonate). BSRs effectively relieve pain and have been thoroughly reviewed [40–44]. In the commercially available formulations, the radioisotopes involved are beta-emitters: strontium-89 dichloride (Metastron, GE Healthcare, Chalfont St. Giles, UK) and ^{153}Sm-EDTMP (Quadramet, Schering AG, Berlin, Germany, and Cytogen Co., Princeton, NJ, USA).

Because of the mm-range of the emitted electrons, the cross-irradiation of the bone marrow represents an ever-present concern. Following IV injection of a beta-emitting BSR, bone marrow is an innocent bystander and the dose-limiting organ. Furthermore, disease-associated bone-marrow suppression already present in these patients often results in delayed and unpredictable recovery. This severely limits the usefulness of beta-emitting BSRs, especially when dosages are increased to deliver potential antitumor radiation levels and/or repeated treatments are attempted. Few clinical studies to date have reported on the feasibility of combining BRSs and chemotherapy [45–48].

Due to short particle track length and potent cell killing, an alpha-emitting BSR could be an intriguing alternative [49]. In contrast to the beta emitters, the alpha-particle emitters deliver a much more energetic and localized radiation that produce densely ionizing tracks and predominantly non-reparable double DNA-strand breaks. In a phase I study of single-dosage administration of escalating amounts of the natural bone-seeker ^{223}Ra in 25 patients with bone metastases from breast and prostate cancer [50], dose-limiting hematological toxicity was not observed. Mild and reversible myelosuppression occurred; with only grade 1 toxicity for thrombocytes at the two highest doses. A phase II RCT of external beam radiation plus either saline or ^{223}Ra injections (given four times at 4-week intervals) in patients with bone metastases from castrate-resistant prostate cancer resulted in a statistically significant decrease from baseline compared with placebo both in bone alkaline phosphatase and prostate-specific antigen [51]. A favorable adverse event profile was observed with minimal bone-marrow toxicity for patients who received ^{223}Ra. Importantly, survival analyses from this phase II trial showed a significant overall survival benefit for ^{223}Ra [51]. Enrollment of 900 patients in a phase III RCT has, by January 1, 2011, been completed (see www.algeta.com).

The combination of radiotherapy or radionuclide therapy with bisphosphonates has recently been reviewed [52]. A significant response in terms of pain relief and improvement of quality of life and performance status better than either therapy alone is shown, but this needs to be vigorously tested in phase III randomized settings. There is a strong theoretical basis for synergistic activity between the two modalities [52], with enhanced healing in metastatic disease in animal studies resulting in better biochemical strength, stability, and bone microarchitecture.

REFERENCES

1. Coleman RE. 2006. Clinical features of metastatic bone disease and risk of skeletal morbidity. *Clin Cancer Res* 12(20 suppl): 6243–6249.
2. Hage WD. 2000. Incidence, location, and diagnostic evaluation of metastatic bone disease. *Orthop Clin N Am* 31: 515–528.
3. BASO: British Association of Surgical Oncology. 1999. The management of metastatic bone disease in the United Kingdom. *Eur J Surg Onc* 25: 3–23.
4. Coleman RE, Smith P, Rubens RD. 1998. Clinical course and prognostic factors following bone recurrence from breast cancer. *Br J Cancer* 77(2): 336–340.
5. Chow E, Wong R, Hruby G, Connolly R, Franssen E, Fung KW, Andersson L, Schueller T, Stefaniuk K, Szumacher E, Hayter C, Pope J, Holden L, Loblaw A, Finkelstein J, Danjoux C. 2001. Prospective patient-based assessment of effectiveness of palliative radiotherapy for bone metastases in an outpatient radiotherapy clinic. *Rad Oncol* 61: 77–82.
6. Sze WM, Shelley M, Held I, Mason M. 2003. Palliation of metastatic bone pain: Single fraction versus multi-fraction radiothcrapy—A systemic review of randomized trials. *Clin Oncol* 15: 345–352.
7. Wu JSY, Wong R, Johnston M, Bezjak A, Whelan T. 2003. Meta-analysis of dose-fractionation radiotherapy trials for the palliation of painful bone metastases. *Int J Rad Oncol Biol Phys* 55(3): 594–605.
8. Chow E, Harris K, Fan G, Tsao M, Sze WM. 2007. Palliative radiotherapy trials for bone metastases: A systemic review. *J Clin Oncol* 25(11): 1423–1436.
9. Tong D, Gillick L, Hendrickson F. 1982. The palliation of symptomatic osseous metastases: Final results of the study by the Radiation Therapy Oncology Group. *Cancer* 50: 893–899.
10. Blitzer P. 1985. Reanalysis of the RTOG study of the palliation of symptomatic osseous metastases. *Cancer* 55: 1468–1472.
11. Chow E, Wu JS, Hoskin P, Coia LR, Bentzen SM, Blitzer PH. 2002. International consensus on palliative radiotherapy endpoints for future clinical trials in bone metastases. *Radiother Oncol* 64(3): 275–280.
12. Bone Pain Trial Working Party. 1999. 8 Gy single fraction radiotherapy for the treatment of metastatic skeletal pain: Randomized comparison with multi-fraction schedule over 12 months of patient follow-up. *Radiother Oncol* 52: 111–121.
13. Steenland E, Leer J, van Houwelingen H, Post WJ, van den Hout WB, Kievit J, de Haes H, Oei B, Vonk E, van der Steen-Banasik E, Wiggenraad RGJ, Hoogenhout J, Wárlám-Rodenhuis C, van Tienhoven G, Wanders R,

Pomp J, van Reijn M, van Mierlo T, Rutten E. 1999. The effect of a single fraction compared to multiple fractions on painful bone metastases: A global analysis of the Dutch Bone Metastasis Study. *Radiother Oncol* 52: 101–109.

14. Van den Hout WB, van der Linden YM, Steenland, Wiggenraad RGJ, Kievit J, de Haes H, Leer JWH. 2003. Single- versus multiple-fraction radiotherapy in patients with painful bone metastases: Cost-utility analysis based on a randomized trial. *J Natl Cancer Inst* 95(3): 222–229.
15. Kaasa S, Brenne E, Lund J, Fayers P, Falkmer U, Holmberg M, Lagerlund M, Bruland O. 2006. Prospective randomized multicentre trial on single fraction radiotherapy (8 Gy X 1) versus multiple fractions (3 Gy X 10) in the treatment of painful bone metastases: Phase III randomized trial. *Radiother Oncol* 79(3): 278–284.
16. Janjan N, Lutz S, Bedwinek J, et al, 2009. Therapeutic guidelines for the treatment of bone metastases: A report from the American College of Radiology Appropriateness Criteria Expert Panel on Radiation Oncology. *J Palliat Med* 12(5): 417–426.
17. Janjan N, Lutz S, Bedwinek J, et al, 2009. Clinical trials and socioeconomic implication in the treatment of bone metastases: A report from the American College of Radiology Appropriateness Criteria Expert Panel on Radiation Oncology. *J Palliat Med* 12(5): 427–431.
18. Hird A, Chow E, Zhang L, et al. 2009. Determining the incidence of pain flare following palliative radiotherapy for symptomatic bone metastases: Results from three Canadian cancer centres. *Int J Radiat Oncol Phys* 75(1): 193–197.
19. Hird A, Zhang L, Holt T, et al. 2008. Dexamethasone for the prophylaxis of radiation induced pain flare after palliative radiotherapy for symptomatic bone metastases: A phase II study. *Clin Oncol* 21(4): 329–335.
20. Roos DE, Turner SL, O'Brien PC, Smith JG, Spry NA, Burmeister BH, Hoskin PJ, Ball DL. 2005. Randomized trial of 8 Gy in 1 versus 20 Gy in 5 fractions of radiotherapy for neuropathic pain due to bone metastases (Trans-Tasman Radiation Oncology Group, TROG 96.05). *Radiother Oncol* 75: 54–63.
21. Maranzano E, Bellavita R, Rossi R. 2005. Radiotherapy alone or surgery in spinal cord compression? The choice depends on accurate patient selection. *J Clin Oncol* 23(32): 8270–8272.
22. Maranzano E, Trippa F, Casale M, et al. 2009. 8 Gy single-dose radiotherapy is effective in metastatic spinal cord compression: Results of a phase III randomized multicentre Italian trial. *Radiother Oncol* 93(2): 174–179.
23. Rades D, Lange M, Veninga T, et al. 2011. Final results of a prospective study comparing the local control of short-course and long-course radiotherapy for metastatic spinal cord compression. *Int J Radiat Oncol Biol Phys* 79(2): 524–530.
24. Rades D, Abraham J. 2010. The role of radiotherapy for metastatic epidural spinal cord compression. *Nat Rev Clin Oncol* 7(10): 590–598.
25. Patchell RA, Tibbs PA, Regine WF, Payne R, Saris S, Kryscio RJ, Mohiuddin M, Young B. 2005. Direct decompressive surgical resection in the treatment of spinal cord compression caused by metastatic cancer: A randomised trial. *Lancet* 366(9486): 643–648.
26. Byrne TN. 1992. Spinal cord compression from epidural metastases. *N Engl J Med* 32: 614–619.
27. van der Linden YM, Kroon HM, Dijkstra SP, Lok JL, Noordijk EM, Leer JWH, Marijnen CAM; Dutch Bone Metastasis Study Group. 2003. Simple radiographic parameter predicts fracturing in metastatic femoral bone lesions: Results from a randomised trial. *Radiother Oncol* 69: 21–31.
28. Healey JH, Brown HK. 2000. Complications of bone metastases: Surgical management. *Cancer* 88: 2940–2951.
29. Agarawal JP, Swangsilpa T, van der Linden Y, Rades D, Jeremic B, Hoskin PJ. 2006. The role of external beam radiotherapy in the management of bone metastases. *Clin Oncol* 18: 747–760.
30. Rieden K, Kober B, Mende U. 1986. Radiotherapy of pathological fractures and skeletal lesions in danger of fractures. *Strahlenther Onkol* 162: 742–749.
31. Koswig S, Budach V. 1999. Remineralization and pain relief in bone metastases after different radiotherapy fractions (10 times 3 Gy vs. 1 time 8 Gy). A prospective study. *Strahlenther Onkol* 75 (10): 500–508.
32. Morris DE. 2000. Clinical experience with retreatment for palliation. *Sem Rad Onc* 10: 210–221.
33. Jones B, Blake PR. 1999. Retreatment of cancer after radical radiotherapy. *Br J Rad* 72: 1037–1039.
34. Nieder C, Milas L, Ang KK. 2000. Tissue tolerance to reirradiation. *Sem Rad Onc* 10: 200–209.
35. Hartsell WF, Scott CB, Bruner DW, Scarantino CW, Ivker RA, Roach M III, Suh JH, Demas WF, Movsas B, Petersen IA, Konski AA, Cleeland CS, Janjan NA, DeSilvio M. 2005. Randomized trial of short- versus long-course radiotherapy for palliation of painful bone metastases. *J Natl Cancer Inst* 97: 798–804.
36. Sande TA, Ruenes R, Lund JL, et al, 2009. Long term follow-up of cancer patients receiving radiotherapy for bone metastases: Results from a randomized multicentre trial. *Radiother Oncol* 91: 261–266.
37. van der Linden YM, Lok JJ, Steenland E, Martijn H, van Houwelingen H, Marijnen CAM, Leer JWH. 2004. Single fraction radiotherapy is efficacious: A further analysis of the Dutch Bone Metastasis Study controlling for the influence of retreatment. Dutch Bone Metastases Study Group. *Int J Radiat Oncol Biol Phys* 59: 528–537.
38. Jeremic B. Shibamoto Y, Igrutinovic I. 2002. Second single 4 Gy reirradiation for painful bone metastases. *J Pain Symptom Manage* 23(1): 26–30.
39. Chow E, Hoskin PJ, Wu J, Roos D, van der Linden Y, Hartsell W, Vieth R, Wilson C, Pater J. 2006. A phase III international randomised trial comparing single with multiple fractions for re-irradiation of painful bone metastases: National Cancer Institute of Canada Clinical Trials Group (NCIC CTG) SC 20. *Clin Oncol* 18: 125–128.

40. Lewington VJ. 2005. Bone-seeking radionuclides for therapy. *J Nucl Med* 46: 38s–47s.
41. Silberstein EB. 2000. Systemic radiopharmaceutical therapy of painful osteoblastic metastases. *Semin Radiat Oncol* 10: 240–249.
42. Finlay IG, Mason MD, Shelley M. 2005. Radioisotopes for the palliation of metastatic bone cancer: A systematic review. *Lancet Oncol* 6(6): 392–400.
43. Bauman G, Charette M, Reid R, Sathya J. 2005. Radiopharmaceuticals for the palliation of painful bone metastases—A systematic review. *Radiother Oncol* 75: 258.E1–258.E13.
44. Reisfield GM, Silberstein EB, Wilson GR. 2005. Radiopharmaceuticals for the palliation of painful bone metastases. *Am J Hosp Palliat Care* 22(1): 41–46.
45. Tu SM, Kim J, Pagliaro LC, Vakar-Lopez F, Wong FC, Wen S, General R, Podoloff DA, Lin SH, Logothetis CJ. 2005. Therapy tolerance in selected patients with androgen-independent prostate cancer following strontium-89 combined with chemotherapy. *J Clin Oncol* 23(31): 7904–7910.
46. Pagliaro LC, Delpassand ES, Williams D, Millikan RE, Tu SM, Logothetis CJ. 2003. A Phase I/II study of strontium-89 combined with gemcitabine in the treatment of patients with androgen independent prostate carcinoma and bone metastases. *Cancer* 97(12): 2988–2994.
47. Sciuto R, Festa A, Rea S, Pasqualoni R, Bergomi S, Petrilli G, Maini CL. 2002. Effects of low-dose cisplatin on 89Sr therapy for painful bone metastases from prostate cancer: A randomized clinical trial. *J Nucl Med* 43(1): 79–86.
48. Akerley W, Butera J, Wehbe T, Noto R, Stein B, Safran H, Cummings F, Sambandam S, Maynard J, Di Rienzo G, Leone L. 2002. A multiinstitutional, concurrent chemoradiation trial of strontium-89, estramustine, and vinblastine for hormone refractory prostate carcinoma involving bone. *Cancer* 94(6): 1654–1660.
49. Bruland ØS, Nilsson S, Fisher DR, Larsen RH. 2006. High-linear energy transfer irradiation targeted to skeletal metastases by the alpha-emitter 223Ra: Adjuvant or alternative to conventional modalities? *Clin Cancer Res* 12(20): 6250s–6257s.
50. Nilsson S, Balteskard L, Fosså SD, Westlin JE, Borch KW, Salberg G, Larsen RH, Bruland ØS. 2005. First clinical experiences with alpha emitter radium-223 in the treatment of skeletal metastases from breast and prostate cancer. *Clin Cancer Res* 11(12): 4451–4459.
51. Nilsson S, Franzén L, Parker C, Tyrrell C, Blom R, Tennvall J, Lennernäs B, Petersson U, Johannessen DC, Sokal M, Pigott K, Yachnin J, Garkavij M, Strang P, Harmenberg J, Bolstad B, Bruland ØS. 2007. Bone-targeted radium-223 in symptomatic, hormone refractory prostate cancer: A randomized, placebo-controlled, phase 2 study. *Lancet Oncol* 8(7): 587–594.
52. Vassiliou V, Bruland O, Janjan N, Lutz S, Kardamakis D, Hoskin P. 2009. Combining systemic bisphosphonates with palliative external beam radiotherapy or bone-targeted radionuclide therapy: Interactions and effectiveness. *Clin Oncol* 21: 665–667.

92

Concepts and Surgical Treatment of Metastatic Bone Disease

Kristy Weber and Scott L. Kominsky

Introduction 760
Biology of Metastatic Bone Lesions 760
Surgical Treatment of Metastatic Bone Disease 762
Acknowledgment 764
References 764

INTRODUCTION

Over 1.5 million people are diagnosed with cancer each year [1], and approximately 50% of those will develop bone metastasis. As treatments improve for primary and metastatic disease, patients are living longer with their disease. This often causes them to experience the morbidities of related bone disease. Although the most worrisome clinical problem is progressive disease in the skeleton, patients can also experience treatment-related osteoporosis. Additional physiologic disruptions in patients with bone metastasis include anemia and hypercalcemia. The bone lesions themselves can cause extreme pain and put the patient at risk for pathologic fractures. Patients become less mobile and may function at a lower level. Prolonged immobilization due to pain or risk of fracture creates potential problems with thromboembolic disease or decubitus ulcers. Lesions in the vertebral region can cause progressive neurologic deficits. Overall quality of life is often markedly diminished.

Comprehensive treatment of bone metastasis is beyond the scope of this chapter, but advances in chemotherapy, targeted biologic therapy, and vaccines have been variably effective for the underlying primary cancer. Different forms of radiation are used to target metastatic cancer cells within the bone to provide palliative pain relief and potentially abrogate the need for surgical intervention. Minimally invasive treatments such as radio-frequency ablation, cryoablation, and kyphoplasty are effective at relieving pain when surgery is not feasible. This chapter focuses on treatment that affects the neoplastic process as well as the bone microenvironment. A brief review of the molecular events related to metastatic bone disease will be discussed. The use of bisphosphonate therapy as well as surgical stabilization in these settings will be summarized.

BIOLOGY OF METASTATIC BONE LESIONS

Tumor–bone interactions

Upon arrival in bone, tumor cells begin a reciprocal (bilateral) interaction with the bone microenvironment. This interaction fosters tumor growth while offsetting the normally balanced process of bone remodeling toward either net bone destruction or formation, causing significant morbidity [2–7] (Fig. 92.1). The majority of research aimed at elucidating tumor–bone interactions has been performed in the field of breast cancer bone metastasis, a predominantly osteolytic disease. Studies using mouse models have provided evidence of a "vicious cycle" of tumor growth and bone destruction driven by transforming growth factor-β (TGF-β). In the course of normal bone remodeling, TGF-β is released, stimulating breast cancer cells to secrete parathyroid hormone-related protein (PTHrP) [8]. PTHrP then stimulates osteoblast precursors to increase receptor activator for nuclear factor-κB ligand (RANKL), which increases osteoclast differentiation. An increased number of active osteoclasts then destroy more

Primer on the Metabolic Bone Diseases and Disorders of Mineral Metabolism, Eighth Edition. Edited by Clifford J. Rosen.
© 2013 American Society for Bone and Mineral Research. Published 2013 by John Wiley & Sons, Inc.

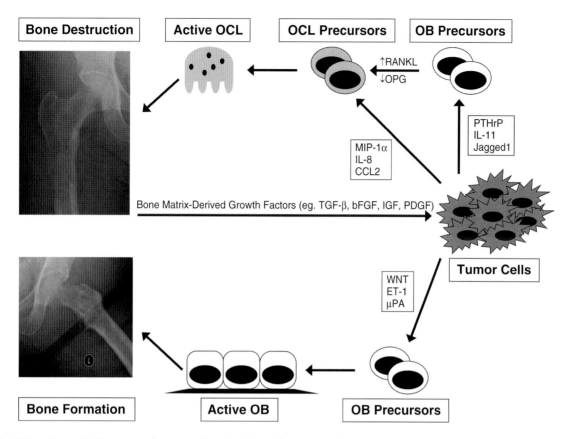

Fig. 92.1. This schematic illustrates the general cycle of bone destruction (osteolytic metastasis—lung cancer) and bone formation (osteoblastic metastasis—breast cancer).

bone. This releases numerous growth factors in addition to TGF-β, including basic fibroblast growth factor (bFGF), insulin-like growth factor (IGF), and platelet-derived growth factor (PDGF), fueling tumor growth and restarting the cycle.

Strikingly, along with PTHrP, numerous other TGF-β-regulated genes have been shown to promote osteolytic bone metastasis including IL-11, CTGF, COX-2, Jagged1, and ADAMTS1 [9–12]. Given its widespread effect on bone metastasis, the TGF-β signaling pathway has become the target of various experimental therapeutic drugs, several of which are in clinical trials [13]. In addition to TGF-β-regulated genes, studies indicate that the chemokines IL-8, CCL2, and MIP-1α promote osteolytic metastasis using mouse models of breast cancer and myeloma [14–16]. Rather than increasing osteoclast numbers via effects on osteoblasts, these factors exert direct effects on osteoclast precursors, stimulating their recruitment and differentiation into mature osteoclasts.

As opposed to other common cancer types affecting the skeleton, prostate cancer causes predominantly osteoblastic lesions in bone. Utilizing mouse models of prostate cancer, several tumor-produced paracrine factors that stimulate osteoblasts to form new bone have been identified. Endothelin-1 (ET-1) has been shown to directly stimulate the proliferation and differentiation of osteoblast precursors [17]. In addition to its direct effects, ET-1 enhances the activation of Wnt signaling in osteoblast precursors by decreasing expression of the WNT inhibitor, Dickkopf homolog 1 (Dkk1) [17]. Urokinase plasminogen activator (uPA) has also been reported to influence prostate cancer bone metastasis, wherein it induces proteases capable of performing multiple functions including the degradation of PTHrP and liberation of IGF-I from its inhibitory binding proteins, thus favoring osteoblast differentiation and activity [18, 19].

Role of bisphosphonates and a RANKL inhibitor

The use of bisphosphonates to treat patients with metastatic disease significantly decreases the incidence of skeletally related events such as pathologic fractures [20]. These compounds bind preferentially to bone matrix and are known to inhibit osteoclastic bone resorption. Bisphosphonates are separated into two classes: nitrogen-containing and non-nitrogen-containing, each having a different mechanism of action. Non-nitrogen-containing bisphosphonates cause osteoclast apoptosis following breakdown into metabolites that compete with adenosine triphosphate (ATP) during energy metabolism. Nitrogen-containing bisphosphonates target the mevalonate pathway, specifically the enzyme farnesyl diphosphate synthase (FPPS), and cause osteoclast inactivation by interfering with geranylgeranylation.

Bisphosphonates have been used with published success to treat bone pain and hypercalcemia in breast and prostate cancer and are most efficacious when used as an adjunct to systemic cancer therapies. They are a routine treatment for most patients with metastatic bone disease or multiple myeloma [21]. The most commonly used drugs in the United States are zoledronic acid (Zometa) and pamidronate (Aredia) given as intravenous injections. Based on a review of 30 randomized controlled trials of patients treated with oral or intravenous bisphosphonates for metastatic disease, these drugs are associated with a significant reduction in all skeletal morbidity end points with the exception of spinal cord compression [20]. They are given when bone metastasis are first diagnosed, as they significantly increase the time to the first skeletal-related event. Osteonecrosis of the jaw is a rare but serious complication in patients with bone metastasis on bisphosphonates and is estimated to occur with a frequency of 0.6–6.2% [22]. It is recommended that patients have a routine dental examination before starting on bisphosphonates, and risk factors include high cumulative doses, poor oral health, and a history of dental extractions.

Denosumab is a RANKL inhibitor that is approved for treatment of patients with bone metastasis and shown to be both safe and effective [23]. In a phase 3 study of patients with castration-resistant prostate cancer, denosumab was more effective than Zometa at preventing skeletally related events [24].

SURGICAL TREATMENT OF METASTATIC BONE DISEASE

Impending fractures

As patients with bone metastasis are unlikely to be surgically cured, the primary focus of orthopedic oncologists is to improve quality of life. If a bone lesion is discovered at an early stage, radiation or systemic medical treatment may prevent further destruction and allow surgery to be avoided. However, if a lesion progresses despite nonsurgical treatment or is initially discovered after there has been extensive cortical destruction and pain, surgical stabilization should be considered [25]. Patients who have prophylactic fixation of their extremity have a shorter hospitalization, quicker return to pre-morbid function, and less hardware complications [26]. Elective stabilization also allows the medical oncologist and surgeon to coordinate operative treatment and systemic chemotherapy. The difficulty lies in reliably determining which bone lesions will eventually cause a fracture. Several classifications have been proposed which involve determination of pain, cortical destruction, and/or size of the bone lesion [27, 28]. Computed tomography based structural rigidity analysis to predict vertebral fracture risk in patients with metastatic breast cancer is 100% sensitive but only 20% specific in a recent study [29].

Surgical treatment

The goals of surgical treatment of patients with bone metastasis are to improve function and decrease pain. Treatment of impending or actual fractures secondary to metastatic bone disease utilizes different principles than those used for routine traumatic fractures [25]. The underlying bone quality is often poor, and the patient may have progressive osseous destruction despite treatment.

Upper extremity bone metastases are less common than those in the lower extremity and can often be treated nonoperatively. However, if patients require their upper extremities for weight bearing (i.e., have lower extremity lesions that cause pain or inability to bear weight without assistive devices), then surgical treatment should be considered in order to improve mobility. Lesions in the scapula and clavicle are generally treated nonoperatively with radiation or minimally invasive options, as most surgical options do not improve function in these areas. Extensive bone destruction in the proximal humerus is either treated by a proximal humeral prosthetic replacement or an intramedullary device with or without supplemental methylmethacrylate if secure fixation can be achieved (Fig. 92.2). Diaphyseal humeral lesions are treated with intramedullary devices or occasionally intercalary metal spacers [30, 31]. Distal humeral lesions are less common and are stabilized with crossed intramedullary pins, dual plating, or segmental distal humeral prosthetic reconstruction. Bone metastases distal to the elbow are extremely rare and are treated on an individual basis.

Lower extremity metastases are more common than those found in the upper extremities and have a larger impact on quality of life due to the need for weight bearing. Pelvic lesions are usually treated nonoperatively or with minimally invasive techniques if the acetabulum is not affected. Acetabular lesions are treated according to specific classification schemes depending on the extent and location of bone loss [32]. Patients with severe bone loss in the acetabulum should have a reasonably long predicted lifespan and good performance status in order to make the procedure and recovery worthwhile. Metastases to the femoral neck are common and often lead to hip fractures [33]. The treatment is either a bipolar hemiarthroplasty or a total hip replacement depending on the status of the acetabulum [34] (Fig. 92.3). Internal fixation with plates and screws is not indicated, as there is a high risk of hardware failure with disease progression in this area. In the intertrochanteric and subtrochanteric regions, options for prosthetic reconstruction or intramedullary fixation are available depending on the extent and location of bone loss as well as the tumor histology. Tumors that are less responsive to systemic treatment or

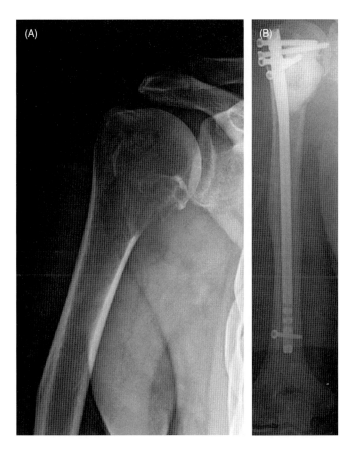

Fig. 92.2. (A) A radiograph of the right proximal humerus in a 61-year-old man with metastatic renal cell carcinoma reveals an osteolytic lesion in the metaphysis at high risk for pathologic fracture and likely to progress despite radiation and systemic treatment. There were no additional lesions in the remaining humerus. (B) Postoperative radiograph showing the stabilization of the humerus with an intramedullary nail supplemented with additional proximal screw fixation and methyl methacrylate.

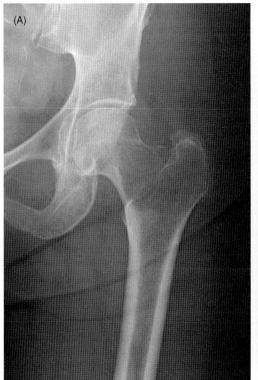
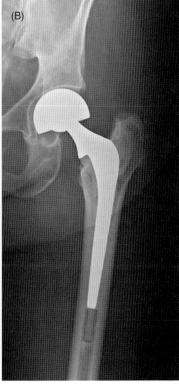

Fig. 92.3. (A) A radiograph of the left hip in a 66-year-old woman with metastatic breast cancer shows an osteolytic lesion in the intertrochanteric region at high risk for pathologic fracture. She had prior radiation to the area with progression of disease and pain. (B) Postoperative radiograph after a cemented bipolar hemiarthroplasty of the hip. There were no additional lesions in the remaining femur.

radiation (ie renal cell carcinoma) are often treated more aggressively with surgical resection. Femoral diaphyseal lesions are treated with intramedullary fixation [34]. It is important that the intramedullary device includes stabilization of the femoral neck to avoid a future hip fracture. Distal femoral lesions are treated by intramedullary fixation, plate fixation, or prosthetic reconstruction. Lesions distal to the knee are uncommon and treated on an individual basis.

The most common site of bone metastasis is the thoracic spine. If patients are neurologically intact and there are no fracture fragments impinging on the spinal cord, radiation is often the treatment of choice. If patients have intractable pain, progressive neurologic deficits or deformity progression, surgical stabilization should be considered [35, 36].

Minimally invasive options

In selected patients, minimally invasive procedures provide an alternative to surgery and can produce long-lasting pain relief. Kyphoplasty and vertebroplasty are commonly used techniques for patients who have osteolytic spine metastasis without neurologic compromise. Both techniques can be performed safely, stabilize the collapsed vertebral body, and yield quick pain relief [37, 38]. Radio-frequency ablation (RFA), cryoablation, or similar techniques are used to treat metastasis in multiple bony sites with most patients achieving some measure of pain relief [39]. Cyberknife treatment is a type of minimally invasive ouapatient radiosurgery used for spine metastasis. A comparison with external beam radiation (EBRT) showed that EBRT was more cost effective but with more acute toxicities and need for further subsequent interventions at the same vertebral level [40].

ACKNOWLEDGMENT

Neither author has any financial conflicts of interest related to the subject matter presented.

REFERENCES

1. Jemal A, Siegel R, Xu J, Ward E. 2010. Cancer statistics 2010. *CA Cancer J Clin* 60: 277–300.
2. Chirgwin JM, Mohammad KS, Guise TA. 2004. Tumor-bone cellular interactions in skeletal metastases. *J Musculoskelet Neuronal Interact* 4: 308–18.
3. Kominsky S, Doucet M, Brady K, Weber KL. 2007. TGF-β influences the development of renal cell carcinoma bone metastasis. *J Bone Miner Res* 22: 37–44.
4. Mundy GR. 2002. Metastasis to bone: Causes, consequences and therapeutic opportunities. *Nat Rev Cancer* 2: 584–93.
5. Park JI, Lee MG, Cho K, Park BJ, Chae KS, Byun DS, Ryu BK, Park YK, Chi SG. 2003. Transforming growth factor-beta1 activates interleukin-6 expression in prostate cancer cells through the synergistic collaboration of the Smad2, p38-NF-kappaB, JNK, and Ras signaling pathways. *Oncogene* 22: 4314–32.
6. Roodman GD. 1993. Role of cytokines in the regulation of bone resorption. *Calcif Tissue Int* 53 Suppl 1:S94–8.
7. Kwan Tat S, Padrines M, Théoleyre S, Heymann D, Fortun Y. 2004. IL-6, RANKL, TNF-alpha/IL-1: Interrelations in bone resorption pathophysiology. *Cytokine Growth Factor Rev* 15: 49–60.
8. Kakonen SM, Selander KS, Chirgwin JM, Yin JJ, Burns S, Rankin WA, Grubbs BG, Dallas M, Cui Y, Guise TA. 2002. Transforming growth factor-beta stimulates parathyroid hormone-related protein and osteolytic metastases via Smad and mitogen-activated protein kinase signaling pathways. *J Biol Chem* 277: 24571–8.
9. Kang Y, Siegel PM, Shu W, Drobnjak M, Kakonen SM, Cordon-Cardo C, Guise TA, Massague J. 2003. A multigenic program mediating breast cancer metastasis to bone. *Cancer Cell* 3: 537–49.
10. Singh B, Berry JA, Shoher A, Ayers GD, Wei C, Lucci A. 2007. COX-2 involvement in breast cancer metastasis to bone. *Oncogene* 26: 3789–96.
11. Sethi N, Dai X, Winter CG, Kang Y. 2011. Tumor-derived Jagged1 promotes osteolytic bone metastasis of breast cancer by engaging notch signaling in bone cells. *Cancer Cell* 19: 192–205.
12. Lu X, Wang Q, Hu G, et al. 2009. ADAMTS1 and MMP1 proteolytically engage EGF-like ligands in an osteolytic signaling cascade for bone metastasis. *Genes Dev* 23: 1882–94.
13. Buijs JT, Stayrook KR, Guise TA. 2011. TGF-β in the bone microenvironment: Role in breast cancer metastases. *Cancer Microenviron* 4(3): 261–81.
14. Bendre MS, Margulies AG, Walser B, Akel NS, Bhattacharrya S, Skinner RA, Swain F, Ramani V, Mohammad KS, Wessner LL, Martinez A, Guise TA, Chirgwin JM, Gaddy D, Suva LJ. 2005. Tumor-derived interleukin-8 stimulates osteolysis independent of the receptor activator of nuclear factor-kappa B ligand pathway. *Cancer Res* 65: 11001–9.
15. Lu X, Kang Y. 2009. Chemokine (C-C motif) ligand 2 engages CCR2$^+$ stromal cells of monocytic origin to promote breast cancer metastasis to lung and bone. *J Biol Chem* 284: 29087–96.
16. Han JH, Choi SJ, Kurihara N, Koide M, Oba Y, Roodman GD. 2001. Macrophage inflammatory protein-1 alpha is an osteoclastogenic factor in myeloma that is independent of receptor activator of nuclear factor kappa B ligand. *Blood* 97: 3349–53.
17. Clines GA, Mohammad KS, Grunda JM, Clines KL, Niewolna M, McKenna R, McKinnin CR, Yanagisawa M, Suva LJ, Chirgwin JM, Guise TA. 2011. Regulation of postnatal trabecular bone formation by the osteoblast endothelin A receptor. *J Bone Miner Res* 26(10): 2523–36.
18. Cramer SD, Chen Z, Peehl DM. 1996. Prostate specific antigen cleaves parathyroid hormone-related protein in

18. the PTH-like domain: Inactivation of PTHrP-stimulated cAMP accumulation in mouse osteoblasts. *J Urol* 156: 526–31.
19. Cohen P, Peehl DM, Graves HC, Rosenfeld RG. 1994. Biological effects of prostate specific antigen as an insulin-like growth factor binding protein-3 protease. *J Endocrinol* 142: 407–15.
20. Ross JR, Saunders Y, Edmonds PM, Patel S, Broadley KE, Johnston SRD. 2003. Systematic review of role of bisphosphonates on skeletal morbidity in metastatic cancer. *BMJ* 327: 469–75.
21. Berenson JR. 2005. Recommendations for zoledronic acid treatment of patients with bone metastases. *Oncologist* 10: 52–62.
22. Hoff AO, Toth B, Hu M, Hortobagyi GN, Gagel RF. 2011. Epidemiology and risk factors for osteoncrosis of the jaw in cancer patients. *Ann N Y Acad Sci* 1218: 47–54.
23. Lipton A, Jacobs I. 2011. Denosumab: Benefits of RANK ligand inhibition in cancer patients. *Curr Opin Support Palliat Care* 5: 258–64.
24. Fizazi K, CArducci M, Smith M, Damiao R, Brown J, Karsh L, Milecki P, Shore N, Rader M, Wang H, Jiang Q, Tadros S, Dansey R, Goessl C. 2011. Denosumab versus zoledronic acid for treatment of bone metastases in men with castration-resistant prostate cancer: A randomised, double-blind study. *Lancet* 377: 813–22.
25. Biermann JS, Holt GE, Lewis VO, Schwartz HS, Yaszemski MJ. 2010. Metastatic bone disease: diagnosis, evaluation, and treatment. *Instr Course Lect* 59: 593–606.
26. Katzcr A, Meenen NM, Grabbe F, Rueger JM. 2002. Surgery of skeletal metastases. *Arch Orthop Trauma Surg* 122(5): 251–8.
27. Beals RK, Lawton GD, Snell WE. 1971. Prophylatic internal fixation of the femur in metastatic breast cancer. *Cancer* 28: 1350–4.
28. Mirels H. 1989. Metastatic disease in long bones: A proposed scoring system for diagnosing impending pathological fractures. *Clin Orthop* 249: 256–65.
29. Snyder BD, Cordio MA, Nazarian A, Kwak SD, Chang DJ, Enterzari V, Zurakowski D, Parker LM. 2009. Noninvasive prediction of fracture risk in patients with metastatic cancer to the spine. *Clin Cancer Res* 15: 7676–83.
30. Redmond BJ, Biermann JS, Blasier RB. 1996. Interlocking intramedullary nailing of pathological fractures of the shaft of the humerus. *J Bone Joint Surg Am* 78: 891–6.
31. Damron TA, Sim FH, Shives TC, An KN, Rock MG, Pritchard DJ. 1996. Intercalary spacers in the treatment of segmentally destructive diaphyseal humeral lesions in disseminated malignancies. *Clin Orthop* 324: 233–43.
32. Marco RA, Sheth DS, Boland PJ, Wunder JS, Siegel JA, Healey JH. 2000. Functional and oncological outcome of acetabular reconstruction for the treatment of metastatic disease. *J Bone Joint Surg Am* 82: 642–51.
33. Schneiderbauer MM, Von Knoch M, Schleck CD, Harmsen WS, Sim FH, Scully SP. 2004. Patient survival after hip arthroplasty for metastatic disease of the hip. *J Bone Joint Surg Am* 86: 1684–9.
34. O'Connor M, Weber K. 2003. Indications and operative treatment for long bone metastasis with a focus on the femur. *Clinical Orthop Rel Res* 415S: 276–8.
35. Bohm P, Huber J. 2002. The surgical treatment of bony metastasis of the spine and limbs. *J Bone Joint Surg* 84B: 521–9.
36. Holman PJ, Suki D, McCutcheon I, Wolinsky JP, Rhines LD, Gokaslan ZL. 2005. Surgical management of metastatic disease of the lumbar spine: Experience with 139 patients. *J Neurosurg Spine* 2: 550–63.
37. Qian Z, Sun Z, Yang H, Gu Y, Chen K, Wu G. 2011. Kyphoplasty for the treatment of malignant vertebral compression fractures caused by metastases. *J Clin Neurosci* 18: 763–7.
38. Kassamali RH, Ganeshan A, Hoey ET, Crowe PM, Douis H, Henderson J. 2011. Pain management in spinal metastases: The role of percutaneous vertebral augmentation. *Ann Oncol* 22: 782–6.
39. Kurup AN, Callstrom MR. 2010. Ablation of skeletal metastases: Current status. *J Vasc Interv Radiol* 8(Suppl): S242–50.
40. Haley ML, Gcrszten PC, Heron DE, Chang YF, Atteberry DS, Burton SA. 2011. Efficacy and cost-effectiveness analysis of external beam and stereotactic body radiation therapy in the treatment of spine metastases: A matched-pair analysis. *J Neurosurg Spine* 14: 537–42.

Section VIII
Sclerosing and Dysplastic Bone Diseases

Section Editor Richard W. Keen

Chapter 93. Sclerosing Bone Disorders 769
Michael P. Whyte

Chapter 94. Fibrous Dysplasia 786
Michael T. Collins, Mara Riminucci, and Paolo Bianco

Chapter 95. Osteochondrodysplasias 794
Yasemin Alanay and David L. Rimoin

Chapter 96. Ischemic and Infiltrative Disorders 805
Richard W. Keen

Chapter 97. Tumoral Calcinosis—Dermatomyositis 810
Nicholas Shaw

Chapter 98. Fibrodysplasia Ossificans Progressiva 815
Frederick S. Kaplan, Robert J. Pignolo, and Eileen M. Shore

Chapter 99. Osteogenesis Imperfecta 822
Joan C. Marini

Chapter 100. Skeletal Manifestations in Marfan Syndrome and Related Disorders of the Connective Tissue 830
Emilio Arteaga-Solis and Francesco Ramirez

Chapter 101. Enzyme Defects and the Skeleton 838
Michael P. Whyte

Primer on the Metabolic Bone Diseases and Disorders of Mineral Metabolism, Eighth Edition. Edited by Clifford J. Rosen.
© 2013 American Society for Bone and Mineral Research. Published 2013 by John Wiley & Sons, Inc.

93
Sclerosing Bone Disorders
Michael P. Whyte

Introduction 769
Osteopetrosis 769
Carbonic Anhydrase II Deficiency 773
Pycnodysostosis 773
Progressive Diaphyseal Dysplasia (Camurati-Engelmann Disease) 774
Endosteal Hyperostosis 775
Osteopoikilosis 776
Osteopathia Striata 777

Melorheostosis 778
Axial Osteomalacia 778
Fibrogenesis Imperfecta Ossium 779
Pachydermoperiostosis 780
Hepatitis C-Associated Osteosclerosis 780
High Bone Mass Phenotype 781
Other Sclerosing Bone Disorders 781
References 781

INTRODUCTION

Osteosclerosis and hyperostosis refer to trabecular and cortical bone thickening, respectively. Increased skeletal mass is caused by many rare (often hereditary) osteochondrodysplasias, as well as by a variety of dietary, metabolic, endocrine, hematologic, infectious, or neoplastic disorders (Table 93.1).

OSTEOPETROSIS

Osteopetrosis (OMIM: 166600, 259700, 259710, 259720, 259730, 607634, 611490, 611497) [1, 2], sometimes called "marble bone disease," was first described in 1904 by Albers-Schönberg [3]. Traditionally, two major clinical forms are discussed [4]: the autosomal dominant adult (benign) type that is associated with relatively few symptoms [5], and the autosomal recessive infantile (malignant) type that is typically fatal (if untreated) in early childhood [6]. Additional more rare types have included an "intermediate" form that presents during childhood where the impact on life expectancy is poorly understood [7]. Osteopetrosis with renal tubular acidosis and cerebral calcification is the inborn error of metabolism, carbonic anhydrase II deficiency [4]. Neuronal storage disease with malignant osteopetrosis has been considered a distinct entity [8]. Osteopetrosis, lymphedema, anhydrotic ectodermal dysplasia, and immunodeficiency (OL-EDA-ID) is an X-linked condition that affects boys [9]. Other unusual forms of osteopetrosis have been called "lethal," "transient infantile," and "post infectious" [4]. Drug-induced osteopetrosis was first described in 2003 in a boy who received high cumulative doses of pamidronate [10]. However, revelation of the genetic defects (see below) responsible for most cases of osteopetrosis has greatly clarified this nosology while further illuminating osteoclast biology [11].

Although defects in multiple genes cause osteopetrosis [11], all true forms are due to failure of osteoclast-mediated resorption of the skeleton [4]. Consequently, primary spongiosa (calcified cartilage deposited during endochondral bone formation) persists and is the histopathological hallmark. Understandably, "osteopetrosis" has been used generically to describe radiodense skeletons yet lacking this finding. Now it is crucial to appreciate

---

Primer on the Metabolic Bone Diseases and Disorders of Mineral Metabolism, Eighth Edition. Edited by Clifford J. Rosen.
© 2013 American Society for Bone and Mineral Research. Published 2013 by John Wiley & Sons, Inc.

Table 93.1. Conditions That Cause Focal or Generalized Increases in Skeletal Mass

Dysplasias and Dysostoses
Autosomal dominant osteosclerosis
Central osteosclerosis with ectodermal dysplasia
Craniodiaphyseal dysplasia
Craniometaphyseal dysplasia
Dysosteosclerosis
Endosteal hyperostosis (van Buchem disease and sclerosteosis)
Frontometaphyseal dysplasia
Infantile cortical hyperostosis (Caffey disease)
Juvenile Paget's disease (osteoectasia with hyperphosphatasia)
Melorheostosis
Metaphyseal dysplasia (Pyle disease)
Mixed sclerosing bone dystrophy
Oculodento-osseous dysplasia
Osteodysplasia of Melnick and Needles
Osteopathia striata
Osteopetrosis
Osteopoikilosis
Progressive diaphyseal dysplasia (Camurati-Engelmann disease)
Pycnodysostosis
Trichodentoosseous dysplasia
Tubular stenosis (Kenny-Caffey syndrome)

Metabolic
Carbonic anhydrase II deficiency
Fluorosis
Heavy metal poisoning
Hepatitis C-associated osteosclerosis
Hypervitaminosis A, D
Hyper-, hypo-, and pseudohypoparathyroidism
Hypophosphatemic osteomalacia
LRP5 and 6 activation (high bone mass phenotype)
Milk-alkali syndrome
Renal osteodystrophy
X-linked hypophosphatemia

Other
Axial osteomalacia
Diffuse idiopathic skeletal hyperostosis (DISH)
Erdheim-Chester disease
Fibrogenesis imperfecta ossium
Hypertrophic osteoarthropathy
Ionizing radiation
Leukemia
Lymphomas
Mastocytosis
Multiple myeloma
Myelofibrosis
Osteomyelitis
Osteonecrosis
Paget's bone disease
Sarcoidosis
Sickle cell disease
Skeletal metastases
Tuberous sclerosis

that therapeutic approaches for genuine osteopetroses, for which this pathogenesis is documented, may be inappropriate for other sclerosing bone disorders [4].

Clinical presentation

Infantile osteopetroses manifest during the first year of life [6]. Nasal stuffiness caused by underdevelopment of the mastoid and paranasal sinuses is an early symptom. Cranial foramina do not widen, and optic, oculomotor, and facial nerves may become paralyzed. Hearing loss is common. Blindness can also be caused by retinal degeneration or raised intracranial pressure [12]. Some patients develop hydrocephalus or sleep apnea. Eruption of the dentition is delayed, and there is failure to thrive. Bones are dense but fragile. Recurrent infection and spontaneous bruising and bleeding follow myelophthisis caused by the excessive bone, abundant osteoclasts, and fibrous tissue that crowd marrow spaces. Hypersplenism and hemolysis may exacerbate the anemia. Physical findings include short stature, macrocephaly, frontal bossing, "adenoid" appearance, nystagmus, hepatosplenomegaly, and *genu valgum*. Untreated patients usually die in the first decade of life from hemorrhage, pneumonia, severe anemia, or sepsis [6].

Intermediate osteopetrosis leads to short stature, cranial nerve deficits, ankylosed teeth that predispose to osteomyelitis of the jaw, recurrent fractures, and mild or occasionally moderately severe anemia [7].

Adult osteopetrosis features radiographic abnormalities that appear during childhood. In some kindreds, "carriers" show no disturbances [5, 13]. The long bones are brittle and may fracture. Facial palsy, compromised vision or hearing, psychomotor delay, osteomyelitis of the mandible [13], carpal tunnel syndrome, slipped capital femoral epiphysis, and osteoarthritis are potential complications. Two principal types of adult osteopetrosis have been proposed [14], but so-called autosomal dominant osteopetrosis, type 1 (ADO 1) proved to be the high bone mass phenotype caused by *LRP5* gene activation (see below), whereas ADO 2 is a genuine osteopetrosis better called Albers-Schönberg disease or chloride channel 7 deficiency osteopetrosis [4].

Neuronal storage disease with osteopetrosis features especially severe skeletal manifestations with epilepsy and neurodegeneration [8]. Lethal osteopetrosis manifests *in utero* and causes stillbirth [4]. Transient infantile osteopetrosis inexplicably resolves during the first months of life [4]. Dysosteosclerosis presents during early childhood as an "osteoclast-poor" form of osteopetrosis.

Radiological features

Generalized, symmetrical increase in bone mass is the major radiographic finding [15]. Trabecular and cortical bone appear thickened. In the severe forms, all three components of skeletal development are disrupted: growth, modeling, and remodeling. Increased density is typically uniform, but alternating sclerotic and lucent bands may appear in the iliac wings and metaphyses. Metaphyses become widened and can develop a club shape or "Erlenmeyer flask" deformity (Fig. 93.1). Rarely, distal phalanges in the hands are eroded (common in pycnodysostosis). Pathological fracture of long bones is not rare. Rachitic-like changes in growth plates may occur [16] because of hypocalcemia with secondary hyperparathyroidism. The skull is usually thickened and dense, especially at the base, and the paranasal and mastoid sinuses are underpneumatized. Vertebrae can show, on lateral view, a "bone-in-bone" (endobone) configuration. Albers-Schönberg disease especially thickens the skull base with a "rugger jersey" appearance of the spine [14]. Skeletal scintigraphy helps show fractures and osteomyelitis. Magnetic resonance imaging (MRI) can assess bone marrow transplantation, because successful engraftment will enlarge medullary spaces [17]. Cranial computed tomography (CT) and MRI findings have been detailed for pediatric patients [18].

Laboratory findings

In infantile osteopetrosis, failure of bone resorption can lead to hypocalcemia because circulating calcium levels are increasingly dependent on dietary intake [19]. Secondary hyperparathyroidism with elevated serum levels of calcitriol is common. In Albers-Schönberg disease, this disturbance is mild [14]. Increased serum acid phosphatase and the brain isoenzyme of creatine kinase (BB-CK) are biomarkers for osteopetrosis [20]. Both enzymes seem to originate from the excessive or defective osteoclasts [20, 21]. Serum levels of LDH isoenzymes and sometimes AST levels are elevated in Albers-Schönberg disease [20].

Histopathological findings

The radiographic features of the osteopetroses [15] can be diagnostic; however, osteoclast failure during endochondral bone formation provides a pathognomonic histological finding because remnants of primary spongiosa persist as "islands" or "bars" of calcified cartilage within trabecular bone (Fig. 93.2). Osteoclasts may be increased, normal, or, rarely, decreased in number [22]. In infantile osteopetrosis, they are usually abundant on bone surfaces. Nuclei are especially numerous, and ruffled borders or clear zones are absent [23]. Fibrous tissue often crowds marrow spaces [23]. Adult osteopetrosis may show increased osteoid and few osteoclasts also lacking ruffled borders, or osteoclasts can be especially numerous and large [24]. Immature "woven" bone is common. Rounded, hypermultinucleated osteoclasts are transiently off of bone surfaces in bisphosphonate-induced osteopetrosis [10].

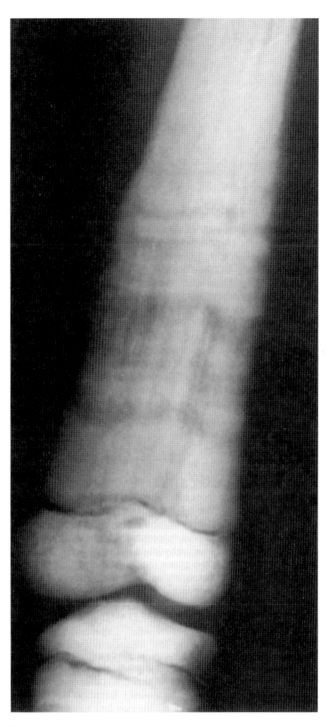

Fig. 93.1. Osteopetrosis. Anteroposterior radiograph of the distal femur of a 10-year-old boy shows a widened metadiaphysis with characteristic alternating dense and lucent bands. [Reprinted with permission from Whyte MP, Murphy WA. 1990. Osteopetrosis and other sclerosing bone disorders. In: Avioli LV, Krane SM (eds.) *Metabolic Bone Disease*, 2nd Ed. Philadelphia: Saunders. p. 618].

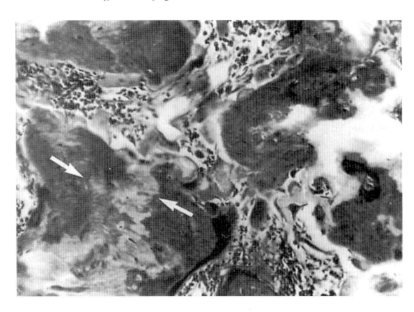

Fig. 93.2. Osteopetrosis. A characteristic area of lightly stained calcified primary spongiosa (arrows) is found within darkly stained mineralized bone.

Etiology and pathogenesis

The potential causes of osteopetrosis are many and complex [4]. Defects could involve primarily the stem cell for osteoclastogenesis or its microenvironment, mononuclear precursor cell, or mature heterokaryon. Furthermore, an osteoblast defect has been reported [25]. In theory, the bone matrix could resist resorption [4]. In osteopetrosis with neuronal storage disease (featuring accumulation of ceroid lipofuscin), lysosomes could be defective [8]. Virus-like inclusions of uncertain significance have been found in the osteoclasts of a few cases of mild osteopetrosis [26]. Synthesis of an abnormal parathyroid hormone (PTH) or defective production of interleukin (IL)-2 or superoxide have been considered [4]. In fact, leukocyte function in infantile osteoporosis may be abnormal [27]. Ultimately, impaired skeletal resorption causes bone fragility because unresorbed cartilage accumulates, collagen fibers do not interconnect osteons, woven bone remodels poorly to compact bone, and microcracks fail to heal.

The molecular basis for osteopetrosis is now known for the majority of patients [11]. Deficiency of chloride channel 7 activity caused by heterozygous mutations in *CLCN7* causes Albers-Schönberg disease [28]. Autosomal recessive infantile osteopetrosis most often involves mutations in *TCIRG1* (*ATP6I*) encoding the *a3* subunit of the vacuolar proton pump [29]. *CLCN7* defects can also cause autosomal recessive malignant or intermediate osteopetrosis [30]. Deactivation of *CA II* explains carbonic anhydrase II deficiency [4]. Accordingly, mutation of these genes that regulate acidification by osteoclasts explains osteopetrosis in most patients. Loss-of-function of the GL (grey-lethal) gene encoding "osteopetrosis associated transmembrane protein 1" (*OSTM1*) causes especially severe osteopetrosis [31]. OL-EDA-ID is caused by inactivation of a key modulator of NF-κB [9]. Especially, rare infants with osteopetrosis featuring few osteoclasts (who fail bone marrow transplantation) suffer from deactivation of the gene that encodes receptor activator of nuclear factor-κB ligand (RANKL) [22]. In others, RANK may be deactivated.

Treatment

Because the cause and outcome differ among the osteopetroses, a precise diagnosis is crucial before therapy is attempted. Diagnosis has depended on careful evaluation of the disease complications and progression, as well as investigation of the family, but now a diagnosis can be made by mutation analysis offered in commercial laboratories [11].

Bone marrow transplantation

Bone marrow transplantation (BMT) from HLA-identical donors has remarkably improved some patients with infantile osteopetrosis [32]. Nevertheless, BMT may not be beneficial for all patients [4], because the causal defect is sometimes extrinsic to the osteoclast lineage (e.g., RANKL deficiency) [22]. Hypercalcemia can occur as osteoclast function begins [33]. Severe, acute, pulmonary hypertension is a frequent complication of stem cell transplantation [34]. Patients with severely crowded medullary spaces seem less likely to engraft. Histomorphometric study of bone helps to predict the outcome of BMT, and this procedure early on seems best [32]. BMT from HLA-nonidentical donors must be improved. Administration of progenitor cells in blood from HLA-haploidentical parents has been effective [35].

Hormonal and dietary therapy

Some success has been reported with a calcium-deficient diet. Conversely, calcium supplementation may be necessary for hypocalcemia that accompanies severe osteopetrosis [16]. High doses of calcitriol to stimulate quiescent osteoclasts, while dietary calcium is limited to prevent absorptive hypercalciuria and hypercalcemia, may improve infantile osteopetrosis [36]. Nevertheless, some patients seem to become resistant to this treatment. The observation that leukocytes produce less superoxide led to recombinant human interferon γ-1b treatment for malignant osteopetrosis. High-dose glucocorticoid treatment stabilizes pancytopenia and hepatomegaly. Prednisone and a low-calcium/high-phosphate diet have been discussed as an alternative to BMT [37]. One case report describes reversal of malignant osteopetrosis after prednisone treatment [38].

Supportive

Hyperbaric oxygenation can be helpful for osteomyelitis of the jaw. Surgical decompression of the optic and facial nerves and auditory canal [39] may benefit some patients. Joint replacement is challenging but possible [40]; internal fixation may be necessary for femoral fractures. Radiographic studies occasionally detect malignant osteopetrosis late in pregnancy. Early prenatal diagnosis by sonography has generally been unsuccessful. Mutation analysis is increasingly feasible, with most severe cases caused by *TCIRG1* and *CLCN7* mutations.

CARBONIC ANHYDRASE II DEFICIENCY

In 1983, the autosomal recessive syndrome of osteopetrosis with renal tubular acidosis (RTA) and cerebral calcification was identified as carbonic anhydrase II (CA II) deficiency (OMIM: 611492) [41].

Clinical presentation

There is considerable clinical variability [42]. In infancy or early childhood, patients can suffer fractures, failure to thrive, developmental delay, short stature, optic nerve compression with blindness, and dental malocclusion. Mental subnormality is common, but not inevitable. RTA may explain the hypotonia, apathy, and muscle weakness that troubles some patients.

Periodic hypokalemic paralysis can occur. Recurrent long bone fractures, although unusual, can cause significant morbidity [42]. Life expectancy does not seem threatened, but the oldest published cases have been young adults [43]. Autopsy studies have not been reported [43].

Radiological features

CA II deficiency resembles other osteopetroses radiographically, except that cerebral calcification appears between the ages of 2 and 5 years, and the osteosclerosis and modeling defects diminish over the years. Skeletal radiographs are typically abnormal at diagnosis, although findings can be subtle at birth. The cerebral calcification resembles this finding in idiopathic hypoparathyroidism or pseudohypoparathyroidism, increases during childhood, and affects cortical and basal ganglia gray matter.

Laboratory findings

Metabolic acidosis manifests as early as the neonatal period. Both proximal and distal RTA have been described [43]; distal (type I) RTA seems better documented. Any anemia is generally mild.

Etiology and pathogenesis

CA accelerates the first step in the reaction $CO_2 + H_2O$ <-> H_2CO_3 <-> $H^+ + HCO_3^-$. CA II is present in many tissues, including brain, kidney, erythrocytes, cartilage, lung, and gastric mucosa [44]. Deactivating mutations in the gene encoding CA II cause this disorder and reveal significance for CA II in bone, kidney, and perhaps brain. [45] In heterozygous carriers, CA II levels in erythrocytes are approximately 50% of normal [41, 43]. There is a CA II knockout mouse model [45].

Treatment

Transfusion of CA II-replete erythrocytes for one patient did not improve the systemic acidosis [46]. The RTA has been treated with HCO_3^-, but the long-term impact is unknown. BMT has corrected the osteopetrosis and slowed the cerebral calcification but not altered the RTA [47].

PYCNODYSOSTOSIS

Pycnodysostosis is the autosomal recessive disorder that perhaps affected painter Henri de Toulouse-Lautrec (1864–1901) [48].

More than 100 patients have been described since 1962 [49]. Parental consanguinity is recorded in fewer than 30% of cases. Most reports are from Europe or the United States, but some come from Israel, Indonesia, India, and Africa. Pycnodysostosis seems to be especially prevalent in the Japanese [50].

Clinical presentation

Pycnodysostosis is typically diagnosed during infancy or early childhood because of disproportionate short stature and a relatively large cranium, fronto-occipital prominence, small facies and chin, obtuse mandibular angle, high-arched palate, dental malocclusion with retained deciduous teeth, proptosis, and a beaked and pointed nose [51]. The anterior fontanel and other cranial sutures are usually open. Fingers are short and clubbed from acro-osteolysis or aplasia of terminal phalanges, and the hands are small and square. The thorax is narrow and there may be *pectus excavatum*, kyphoscoliosis, and increased lumbar lordosis. Sclerae can be blue. Recurrent fractures typically involve the lower limbs and cause *genu valgum*. Rickets has been described. Adult height ranges between 4' 3" and 4' 11". Mental retardation affects fewer than 10% of cases [51]. Recurrent respiratory infections and right heart failure may occur from chronic upper airway obstruction caused by micrognathia.

Radiographic features

Pycnodysostosis resembles osteopetrosis because uniform osteosclerosis becomes apparent in childhood and increases with age, and there are recurrent fractures. The calvarium and base of the skull are sclerotic, and the orbital ridges are radiodense. However, the marked modeling defects of osteopetrosis do not occur, although long bones have narrow medullary canals. Additional findings include delayed closure of cranial sutures and fontanels (prominently the anterior), obtuse mandibular angle, wormian bones, gracile clavicles with hypoplastic ends, partial absence of the hyoid bone, and hypoplasia of the distal phalanges and ribs [52]. Endobones and radiodense striations are absent [15].

Laboratory findings

Serum levels of calcium, inorganic phosphate, and alkaline phosphatase are usually unremarkable. There is no anemia.

Electron microscopy has suggested that degradation of bone collagen might be defective [53]. In chondrocytes, inclusions have been described.In the osteoclasts of brothers, virus-like inclusions have been reported [54]. Diminished growth hormone secretion and low serum insulin-like growth factor 1 levels have been reported in five of six affected children [55].

Etiology and pathogenesis

In 1996, loss-of-function mutation within the gene that encodes cathepsin K was discovered to cause pycnodysostosis [50]. Cathepsin K, a lysosomal cysteine protease, is highly expressed in osteoclasts [56]. Impaired collagen degradation seems to be a fundamental defect [57] and compromises bone quality [58].

Treatment

There is no established medical therapy. BMT has not been reported. Fractures of long bones are typically transverse and heal at a satisfactory rate, although there can be delayed union and massive callus formation. Internal fixation of long bones or extraction of teeth is difficult because of skeletal hardness. Jaw fracture has occurred. Osteomyelitis of the mandible may require antibiotics and surgery.

PROGRESSIVE DIAPHYSEAL DYSPLASIA (CAMURATI-ENGELMANN DISEASE)

Progressive diaphyseal dysplasia (PDD; OMIM: 131300) was characterized by Cockayne in 1920 [59]. Camurati recognized the autosomal dominant inheritance. Engelmann described the severe form in 1929. In 2001, mutations were identified within a specific region of the gene that encodes transforming growth factor-β1(TGF-β1) [60].

All races are affected. Clinical severity is quite variable [61]. Hyperostosis occurs gradually on both the periosteal and endosteal surfaces of long bones. In severe cases, the axial skeleton and skull are also involved. Some carriers have no radiographic changes, but bone scintigraphy is abnormal.

Clinical presentation

PDD typically presents during childhood with limping or a broad-based and waddling gait, leg pain, muscle wasting, and decreased subcutaneous fat in the extremities mimicking a muscular dystrophy [62]. Severely affected individuals also have a characteristic body habitus that includes a large head with prominent forehead, proptosis, tall stature, and thin limbs exhibiting thickened painful bones and little muscle mass. Cranial nerve palsies may develop when the skull is involved. Puberty is sometimes delayed. Raised intracranial pressure can occur. Physical findings include palpable widened bones and skeletal tenderness. Some patients have hepatosplenomegaly, Raynaud's phenomenon, and other findings suggestive of vasculitis [63].

Although radiological studies typically show progressive skeletal disease, the clinical course is variable, and spontaneous improvement sometimes occurs during adult life [64].

Radiological features

Hyperostosis of major long bone diaphyses involves both the periosteal and endosteal surfaces [15]. The thickening is fairly symmetrical and gradually spreads to include metaphyses, yet typically spares epiphyses (Fig. 93.3). Diaphyses slowly widen and develop irregular surfaces. The tibias and femurs are most commonly affected; less frequently, the radii, ulnas, humeri, scapulae, clavicles, and pelvis, and, occasionally, short tubular bones are involved. Age-of-onset, rate of progression, and degree of

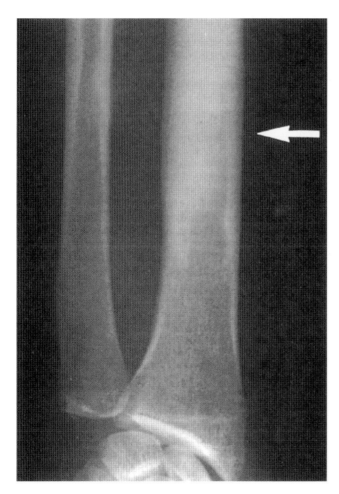

Fig. 93.3. Progressive diaphyseal dysplasia (Camurati-Engelmann disease). The distal radius of this 20-year-old woman has characteristic patchy thickening (arrow) of the periosteal and endosteal surfaces of the diaphysis.

bony change are highly variable. With relatively mild PDD, especially in adolescents or young adults, radiographic and scintigraphic abnormalities may be confined to the lower limbs. In severely affected children, regional osteopenia is possible.

Clinical, radiographic, and scintigraphic findings are generally concordant. Bone scanning typically shows focally increased radionuclide accumulation [65]. In some patients, however, advanced and metabolically quiescent disease features unremarkable bone scintigraphy [65]. Conversely, markedly increased radioisotope accumulation with minimal radiographic findings can represent early skeletal disease [65]. MRI and CT have delineated the cranial findings [66].

Laboratory findings

Serum alkaline phosphatase and urine hydroxyproline are elevated in some PDD patients. Modest hypocalcemia and significant hypocalciuria sometimes occur with severe disease, probably because of markedly positive calcium balance [64].

Other biochemical parameters of bone and mineral metabolism are typically normal. Mild anemia and leukopenia and elevated erythrocyte sedimentation rate have been reported [63]. Histopathology shows new bone formation along diaphyses with nascent woven bone undergoing centripetal maturation and then incorporation into the cortex. Electron microscopy of muscle has shown myopathic and vascular changes [62].

Etiology and pathogenesis

PDD is caused by mutation within one region of the gene that encodes transforming growth factor-β1 (TGF-β1) (*TGFβ1*). A latency-associated peptide remains bound to TGF-β1, keeping it active in skeletal matrix [67]. Mild PDD can reflect variable expressivity. PDD has been described as more severe in ensuing generations ("anticipation") [68]. Furthermore, there does seem to some locus heterogeneity [69]. The clinical and laboratory features and responsiveness to glucocorticoid treatment have suggested severe PDD is a systemic disorder (i.e., an inflammatory connective tissue disease) [63].

Treatment

PDD is somewhat unpredictable. Symptoms may remit during adolescence or adult life. Prednisone given in small doses on alternate days is effective for bone pain and can correct histological abnormalities of bone [70]. Resection of a cortical bone window has relieved localized pain. Bisphosphonate therapy may be helpful but has transiently increased symptoms [71].

ENDOSTEAL HYPEROSTOSIS

In 1955, van Buchem and colleagues described *hyperostosis corticalis generalisata* [72]. Subsequently, this and additional disorders were characterized as types of endosteal hyperostosis. The autosomal dominant relatively mild form is called Worth disease [73], and a second autosomal recessive severe form is sclerosteosis [74].

Van buchem disease

Van Buchem disease (OMIM: 239100) is a severe, autosomal recessive disorder. [72]

Clinical presentation

Progressive asymmetrical enlargement of the jaw occurs during puberty, causing marked thickening and a wide angle but without prognathism. Dental malocclusion is uncommon. Carriers of the gene defect may be symptom-free; however, recurrent facial nerve palsy, deafness, and optic atrophy from narrowing of cranial foramina are common and can begin as early as infancy. Long bones may be painful with applied pressure but are not fragile, and joint range-of-motion is unaffected. Sclerosteosis (see below) differs because of excessive height and syndactyly [74].

Radiological features

Endosteal thickening produces a dense diaphyseal cortex with narrow medullary canal [15]. The hyperostosis is selectively endosteal; long bones are properly modeled. However, osteosclerosis also affects the skull base, facial bones, vertebrae, pelvis, and ribs. The mandible enlarges.

Laboratory findings

Serum alkaline phosphatase activity may be increased because of high levels of the skeletal isoform, whereas calcium and phosphate concentrations are unremarkable. Van Buchem and colleagues suggested that the excessive bone was essentially of normal quality [72].

Etiology and pathogenesis

Van Buchem disease and sclerosteosis were predicated to be allelic disorders, with their differences reflecting modifying genes [74]. Actually, loss-of-function mutations in *SOST*, the gene encoding sclerostin, cause sclerosteosis [75], whereas van Buchem disease involves a 52-kb deletion that lessens downstream enhancement of *SOST* [76]. Sclerostin binds to LRP5/6, antagonizes canonical Wnt signaling [77], and promotes osteoblast apoptosis [78].

Treatment

There is no specific medical therapy. Decompression of narrowed foramina may help cranial nerve palsies [79]. Surgery has been used to recontour the mandible [80].

Sclerosteosis

Sclerosteosis (cortical hyperostosis with syndactyly; OMIM: 269500), like van Buchem disease, is an autosomal recessive endosteal hyperostosis that affects primarily Afrikaners or others of Dutch ancestry [74]. Initially, sclerosteosis was distinguished from van Buchem disease by excessive height and syndactyly. In fact, the genetic defects differ [75, 76].

Clinical presentation

At birth, only syndactyly may be noted. During early childhood, affected individuals become tall and heavy with skeletal overgrowth, especially involving the skull. This causes facial disfigurement. Deafness and facial palsy are prominent problems. The mandible has a square configuration. Raised intracranial pressure and headache may result from a small cranial cavity. The brainstem can be compressed. Syndactyly from either cutaneous or bony fusion of the middle and index fingers is typical, but of variable severity. Patients are resistant to fracture. Life expectancy is often shortened [81].

Radiological features

Except for syndactyly, the skeleton appears normal in early childhood. Then, progressive bone acquisition widens the skull and mandible [82]. Long bones develop thickened cortices. Vertebral pedicles, ribs, tubular bones, and pelvis may also appear dense. Auditory ossicles may fuse and the internal canals and cochlear aqueducts become narrow [83].

Histopathological findings

Dynamic histomorphometry of one affected skull showed an increased rate of bone formation with thickened trabeculae and osteoidosis, whereas resorption appeared quiescent [84].

Etiology and pathogenesis

Loss-of-function mutations in *SOST* cause sclerosteosis [75].

Enhanced osteoblast activity with failure of osteoclasts to compensate causes the dense bone of sclerosteosis [84]. No abnormality of calcium homeostasis or of pituitary gland function has been documented [85].

The pathogenesis of the neurological defects has been described [84].

Treatment

There is no established medical treatment. Surgery for syndactyly is difficult if there is bony fusion. Management of the neurological complications has been reviewed [84].

OSTEOPOIKILOSIS

Osteopoikilosis ("spotted bones") is an autosomal dominant radiographic curiosity. With accompanying connective tissue nevi, *dermatofibrosis lenticularis disseminata*, the disorder is the Buschke-Ollendorff syndrome [86]. In 2004, deactivating mutations in the *LEMD3* gene were identified [87].

Clinical presentation

Osteopoikilosis (OMIM: 166700) is usually an incidental finding. The bony lesions are asymptomatic, but if not understood can precipitate investigation for metastatic disease to the skeleton [88]. Family members at risk should be screened with a radiograph of a wrist and knee in early adult life. Joint contractions and limb length

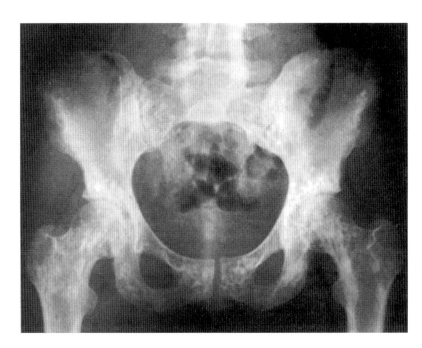

Fig. 93.4. Osteopoikilosis. The characteristic feature is the spotted appearance shown here in the pelvis and metaepiphyseal regions of the femora. [Reproduced with permission from Whyte MP. 1995 Rare disorders of skeletal formation and homeostasis. In: Becker KN (ed.) *Principles and Practice of Endocrinology and Metabolism*, 2nd Ed. Philadelphia: Lippincott-Raven Publishers. p. 598.]

inequality can occur, especially in individuals with accompanying changes of melorheostosis. The nevi usually involve the lower trunk or extremities and are small asymptomatic papules. Sometimes they are yellow or white discs or plaques, deep nodules, or streaks [86].

Radiological Features

There are numerous, small, usually round or oval, foci of osteosclerosis [15]. Commonly affected sites are the ends of the short tubular bones, metaepiphyses of long bones, and tarsal, carpal, and pelvic bones (Fig. 93.4). Lesions are unchanged for decades. Bone scanning is not abnormal [88].

Histopathological studies

Dermatofibrosis lenticularis disseminata consists of unusually broad, markedly branched, interlacing elastin fibers in the dermis; the epidermis is normal [86]. Foci of osteosclerosis are thickened trabeculae that merge with surrounding normal bone or are islands of cortical bone that include Haversian systems. Mature lesions appear to be remodeling slowly.

OSTEOPATHIA STRIATA

Osteopathia striata (OMIM: 166500) features linear striations at the ends of long bones and in the ileum [15]. Like osteopoikilosis, it is usually a radiographic curiosity, but it can also occur in a variety of important disorders including osteopathia striata with cranial sclerosis (OMIM: 300373) [89] and osteopathia striata with focal dermal hypoplasia (OMIM: 305600) [90].

Clinical presentation

Osteopathia striata alone is an autosomal dominant trait. Symptoms that may have led to the diagnosis are probably unrelated. With sclerosis of the skull, however, cranial nerve palsies are common [89], and this is due to WTX gene mutation [91]. Osteopathia striata with focal dermal hypoplasia (Goltz syndrome) is a serious condition of males, transmitted as an X-linked recessive trait, that features widespread linear areas of dermal hypoplasia through which adipose tissue can herniate and a variety of bony defects in the limbs [90].

Radiological features

Gracile linear striations are found in cancellous bone, particularly within metaepiphyses of major long bones and the periphery of the iliac bones [15]. Carpal, tarsal, and tubular bones of the hands and feet are less often and more subtly affected. The striations appear unchanged for years. Radionuclide accumulation is not increased during bone scanning [88].

Treatment

Histopathological studies of bone have not been described. Although unlikely to be misdiagnosed, radiographic screening during the young adult life of family members

MELORHEOSTOSIS

Melorheostosis (OMIM: 155950), from the Greek, refers to "flowing hyperostosis." The radiographic appearance resembles wax that has dripped down a candle. About 200 cases have been published [92] since the first description in 1922 [93]. Melorheostosis occurs sporadically, including when it accompanies osteopoikilosis.

Clinical presentation

Melorheostosis typically presents during childhood. Monomelic involvement is usual; bilateral disease is characteristically asymmetrical. Cutaneous changes may overlie the skeletal lesions and include linear scleroderma-like patches and hypertrichosis. Fibromas, fibrolipomas, capillary hemangiomas, lymphangiectasia, and arterial aneurysms can also occur [94]. Soft tissue abnormalities are often noted before the hyperostosis. Pain and stiffness are the major symptoms. Affected joints may contract. Leg length inequality can follow premature fusion of epiphyses. Bone lesions seem to advance most rapidly during childhood. In adult life, melorheostosis may or may not progress [95]. Nevertheless, pain is more frequent when there is continuing subperiosteal bone formation.

Radiological features

Dense, irregular, and eccentric hyperostosis of both periosteal and endosteal surfaces of a single bone, or several adjacent bones, is the hallmark of melorheostosis (Fig. 93.5) [15, 92]. Any bone may be affected, but the lower extremities are most commonly involved. Bone can also develop in soft tissues near skeletal lesions, particularly near joints. Melorheostotic bone is hyperemic and "hot" during bone scanning.

Laboratory findings

Serum calcium, inorganic phosphate, and alkaline phosphatase levels are normal.

Histopathological findings

Melorheostosis features endosteal thickening during growth and periosteal new bone formation during adult life [92]. Affected bones are sclerotic with thickened, irregular lamellae. Marrow fibrosis may be present [92]. In the skin, unlike in true scleroderma, the collagen of the scleroderma-like lesions appears normal and has therefore been called linear melorheostotic scleroderma [96].

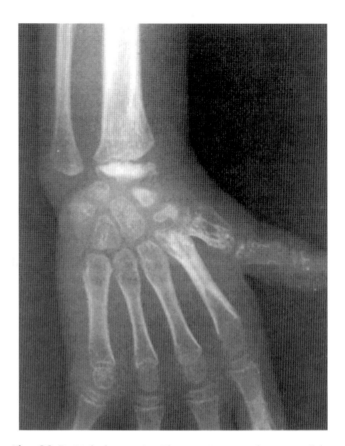

Fig. 93.5. Melorheostosis. Characteristic patchy osteosclerosis is most apparent in the radius and second metacarpal of this 8-year-old girl.

Etiology and pathogenesis

The distribution of the bone and soft tissue lesions in sclerotomes, myotomes, and dermatomes suggests a segmentary, embryonic defect [96]. Linear scleroderma may represent the primary abnormality that extends into the skeleton. In affected skin, there may be altered expression of several adhesion proteins [97]. The germline, loss-of-function mutations in *LEMD3* that cause osteopoikilosis and the Buschke-Ollendorff syndrome do not explain sporadic melorheostosis [98].

Treatment

Surgical correction of contractures can be difficult; recurrent deformity is common. Distraction techniques seem promising [99].

AXIAL OSTEOMALACIA

Axial osteomalacia (OMIM: 109130) features coarsening of trabecular bone in the axial, but not the appendicular, skeleton [100]. Fewer than 20 patients have been described. Most have been sporadic cases, but an affected mother and son have been reported [101].

Clinical presentation

Most patients have been middle-age or elderly men. Radiographic manifestations are probably detectable earlier [102]. Dull, vague, chronic, axial bone pain (often in the cervical spine) usually prompts the radiographic discovery.

Radiological features

Abnormalities are confined essentially to the spine and pelvis where the coarsened trabecular pattern resembles osteomalacia [103]. The appendicular skeleton is unremarkable. However, Looser's zones (characteristic of osteomalacia) are not reported. The cervical spine and ribs seem most severely affected. Several patients have had features of ankylosing spondylitis [102].

Laboratory studies

Serum alkaline phosphatase (bone isoform) may be increased. In a few patients, inorganic phosphate levels tended to be low [103]. For others, osteomalacia occurred despite normal levels of calcium, phosphate, 25-hydroxyvitamin D, and 1,25-dihydroxyvitamin D.

Histopathological findings

Iliac crest specimens have distinct corticomedullary junctions, but the cortices can be especially wide and porous. Total bone volume may be increased. There is excess osteoid, but collagen has a normal lamellar pattern. Tetracycline labeling shows defective skeletal mineralization [102]. Osteoblasts are flat "lining" cells, but stain intensely for alkaline phosphatase. Changes of secondary hyperparathyroidism are absent [102].

Etiology and pathogenesis

Axial osteomalacia may be caused by an osteoblast defect [104].

Treatment

Effective medical therapy has not been reported, but the natural history seems relatively benign. Methyltestosterone and stilbestrol or vitamin D_2 (as much as 20,000 U/day for 3 years) have not been helpful [104]. Slight improvement in skeletal histology, but not in symptoms, was described after calcium and vitamin D_2. Long-term follow-up of one patient showed that symptoms and radiographic findings did not change [104].

FIBROGENESIS IMPERFECTA OSSIUM

Fibrogenesis imperfecta ossium was identified in 1950. Approximately 10 cases have been reported [105]. Radiographic studies suggest generalized osteopenia, but coarse and dense appearing trabecular bone explains its designation as an osteosclerotic disorder. The clinical, biochemical, radiological, and histopathological features have been carefully contrasted with axial osteomalacia [100].

Clinical presentation

Presentation occurs during middle age or later. Both sexes are affected. Characteristically, gradual onset of intractable skeletal pain is followed by rapid deterioration. The course is debilitating with progressive immobility. Spontaneous fractures are prominent. Physical examination shows marked bony tenderness.

Radiological features

Skeletal changes affect all but the skull. Initially, there may be osteopenia only and a slightly abnormal appearance of trabecular bone [105]. Subsequently, the findings are more consistent with osteomalacia with further alterations of the trabecular bone pattern, heterogeneous bone density, and cortical thinning. Corticomedullary junctions become indistinct. Areas of the skeleton may have a mixed lytic and sclerotic appearance [105]. Remaining trabeculae appear coarse and dense in a "fish-net" pattern. Pseudofractures may develop. Some patients have a "rugger jersey" spine. Diaphyses may show periosteal reaction. The radiographic features can distinguish fibrogenesis imperfecta ossium from axial osteomalacia (generalized versus axial, respectively). The histopathological findings are also different [100].

Laboratory findings

Serum calcium and inorganic phosphate levels are normal, but alkaline phosphatase is increased. Urine hydroxyproline may be normal or elevated [105]. Typically, there is no renal tubular dysfunction. Acute agranulocytosis and macroglobulinemia have been reported.

Histopathological findings

The osseous lesion is an osteomalacia [105]. Collagen lacks birefringence where there is defective mineralization. Electron microscopy reveals thin and randomly organized collagen fibrils in a "tangled" pattern. Cortical bone in the femurs and tibias may show the least abnormality. Osteoblasts and osteoclasts can be abundant. In some regions, matrix structures of 300–500 nm diameter have been observed [106]. Unless bone is viewed with polarized-light or electron microscopy, fibrogenesis imperfecta ossium can be mistaken for osteoporosis or other forms of osteomalacia [106].

Etiology and pathogenesis

This is an acquired disorder of collagen synthesis in lamellar bone. The etiology is unknown. Genetic factors have not been implicated.

Treatment

There is no recognized therapy. Temporary improvement can occur. Treatment with vitamin D (or an active metabolite), calcium, salmon calcitonin, or sodium fluoride has not helped [105]. In fact, ectopic calcification has complicated high dose vitamin D therapy. Treatment with melphalan and prednisolone seemed to benefit one patient.

PACHYDERMOPERIOSTOSIS

Pachydermoperiostosis (hypertrophic osteoarthropathy: primary or idiopathic; OMIM: 167100) causes clubbing of the digits, hyperhidrosis and thickening of the skin especially on the face and forehead (cutis verticis gyrata), and periosteal new bone formation particularly in the distal extremities. Autosomal dominant and recessive inheritance with variable expression is established [107, 108]. In 2008, autosomal recessive pachydermoperiostosis was explained by loss-of-function mutation within the gene that encodes 15-hydroxyprostaglandin dehydrogenase [108].

Clinical presentation

Men seem to be more severely affected than women and blacks more commonly than whites. Age at presentation is variable, but usually it is during adolescence [107, 108]. All principal features (clubbing, periostitis, and pachydermia) trouble some patients; others have just one or two. Clinical manifestations emerge over a decade and can then abate [108]. Progressive enlargement of the hands and feet may cause a paw-like appearance, and there may be excessive perspiration. Acro-osteolysis can occur. Fatigue and arthralgias of the elbows, wrists, knees, and ankles are common. Stiffness and limited mobility of both the appendicular and the axial skeleton may develop. Compression of cranial or spinal nerves has been described. Cutaneous changes include coarsening, thickening, furrowing, pitting, and oiliness of the skin, especially the scalp and face. Myelophthisic anemia with extramedullary hematopoiesis may occur. Life expectancy is not compromised.

Radiological features

Severe periostitis thickens tubular bones distally: typically the radius, ulna, tibia, and fibula, and sometimes the metacarpals, tarsals/metatarsals, clavicles, pelvis, skull base, and phalanges. Clubbing is obvious, and acro-osteolysis can occur. The spine is rarely involved. Ankylosis of joints, especially in the hands and feet, may trouble older patients [15]. The major challenge in differential diagnosis is secondary hypertrophic osteoarthropathy (pulmonary or otherwise). Here, however, the radiographic features are somewhat different, featuring periosteal reaction that is typically smooth and undulating [109]. In pachydermoperiostosis, periosteal proliferation is exuberant, irregular, and often involves epiphyses. Bone scanning in either condition reveals symmetrical, diffuse, regular uptake along the cortical margins of long bones, especially in the legs, causing a "double stripe" sign.

Laboratory findings

Nascent periosteal bone roughens cortical bone surfaces and undergoes cancellous compaction so that centrally it can be difficult to distinguish histopathologically from the original cortex. There may also be osteopenia of trabecular bone from quiescent formation [15]. Mild cellular hyperplasia and thickening of blood vessels is found near synovial membranes, but synovial fluid is unremarkable [110]. Electron microscopy shows layered basement membranes.

Etiology and pathogenesis

A controversial hypothesis had suggested that some circulating factor in pachydermoperiostosis acts on vasculature to cause hyperemia, thereby altering soft tissues; later, blood flow is reduced [111].

Treatment

There is no established treatment. Painful synovial effusions may respond to nonsteroidal anti-inflammatory drugs. Colchicine reportedly helped one patient. Contractures or neurovascular compression by osteosclerotic lesions may require surgical intervention.

HEPATITIS C-ASSOCIATED OSTEOSCLEROSIS

In 1992, a new disorder featured severe, generalized osteosclerosis and hyperostosis in former intravenous drug abusers infected with the hepatitis C virus [112]. Approximately 20 cases have been reported, and hepatitis C virus infection proved common to all patients. Periosteal, endosteal, and trabecular bone thickening occurs throughout the skeleton, except the cranium. During active disease, the forearms and legs are painful. Dual energy X-ray absorptiometry (DXA) shows values 20–300% above control means. Remodeling of good quality excessive bone seems accelerated during active disease and may respond to antiresorptive therapy. Gradual, spontaneous remission with decreases in DXA values can occur. The IGF system features distinctive increases in circulating levels of IGF binding protein 2 and "big" IGF II [113].

HIGH BONE MASS PHENOTYPE

Certain activating mutations of the LRP5 gene (OMIM: 607636) encoding low-density lipoprotein receptor-related protein 5, inherited as an autosomal dominant trait, increase skeletal mass with good quality bone [114]. Enhanced Wnt signaling stimulates osteoblasts [114]. Some patients have *torus palatinus* [114], cranial nerve palsies, and oropharyngeal exostoses [115]. LRP6 gene activation can also cause high bone mass.

OTHER SCLEROSING BONE DISORDERS

Table 93.1 lists the many conditions that cause focal or generalized increases in skeletal mass. Sarcoidosis characteristically causes cysts within coarsely reticulated bone. However, sclerotic areas occasionally appear in the axial skeleton or long bones. Although multiple myeloma typically features generalized osteopenia and discrete osteolytic lesions, among indolent forms widespread osteosclerosis can occur. Lymphoma, myelofibrosis, and mastocytosis may also increase bone mass. Metastatic carcinoma, especially from the prostate, can cause dense bones. Diffuse osteosclerosis is also frequent in secondary, but not primary, hyperparathyroidism (e.g., renal disease).

REFERENCES

1. Online Mendelian Inheritance in Man 2000 OMIM. Available online at http://www.ncbi.nlm.nih.gov/oAmicm. Accessed January 25, 2013.
2. Castriota-Scanderbeg A, Dallapiccola B. 2005. *Abnormal Skeletal Phenotypes: From Simple Signs to Complex Diagnoses*. New York: Springer.
3. Albers-Schönberg H. 1904. Roentgen bilder einer seltenen Kochennerkrankung. *Munch Med Wochenschr* 51: 365.
4. Whyte MP. 2002. Osteopetrosis. In: Royce PM, Steinmann B (eds.) *Connective Tissue and Its Heritable Disorders*, 2nd Ed. New York: Wiley-Liss. pp. 789–807.
5. Johnston CC Jr, Lavy N, Lord T, Vellios F, Merritt AD, Deiss WP Jr. 1968. Osteopetrosis: A clinical, genetic, metabolic, and morphologic study of the dominantly inherited, benign form. *Medicine (Baltimore)* 47: 149–67.
6. Loría-Cortés R, Quesada-Calvo E, Cordero-Chaverri C.1977. Osteopetrosis in children: A report of 26 cases. *J Pediatr* 91: 43–7.
7. Kahler SG, Burns JA, Aylsworth AS. 1984. A mild autosomal recessive form of osteopetrosis. *Am J Med Genet* 17: 451–64.
8. Jagadha V, Halliday WC, Becker LE, Hinton D. 1988. The association of infantile osteopetrosis and neuronal storage disease in two brothers. *Acta Neuropathol (Berl)* 75: 233–40.
9. Dupuis-Girod S, Corradini N, Hadj-Rabia S, Fournet JC, Faivre L, Le Deist F, Durand P, Döffinger R, Smahi A, Israel A, Courtois G, Brousse N, Blanche S, Munnich A, Fischer A, Casanova JL, Bodemer C. 2002. Osteopetrosis, lymphedema, anhidrotic ectodermal dysplasia, and immunodeficiency in a boy and incontinentia pigmenti in his mother.. *Pediatrics* 109: e97.
10. Whyte MP, McAlister WH, Novack DV, Clements KL, Schoenecker PL, Wenkert D. 2008. Bisphosphonate-induced osteopoetrosis: Novel bone modeling defects, metaphyseal osteopenia, and osteosclerosis fractures after drug exposure ceases. *J Bone Miner Res* 23: 1698–707.
11. Balemans W, Van Wesenbeeck L, Van Hul W. 2005. A clinical and molecular overview of the human osteopetroscs. *Calcif Tissue Int* 77: 263–74.
12. Vanier V, Miller R, Carson BS. 2000. Bilateral visual improvement after unilateral optic canal decompression and cranial vault expansion in a patient with osteopetrosis, narrowed optic canals, and increased intracranial pressure. *J Neurol Neurosurg Psychiatry* 69: 405–06.
13. Waguespack SG, Hui SL, DiMeglio LA, Econs MJ. 2007. Autosomal dominant osteopetrosis: Clinical severity and natural history of 94 subjects with a chloride channel 7 gene mutation. *J Clin Endocrinol Metab* 92: 771–8.
14. Bollerslev J. 1989. Autosomal dominant osteopetrosis: Bone metabolism and epidemiological, clinical and hormonal aspects. *Endocr Rev* 10: 45–67.
15. Resnick D, Niwayama G. 2002. *Diagnosis of Bone and Joint Disorders*, 4th Ed. Philadelphia: Saunders.
16. Di Rocco M, Buoncompagni A, Loy A, Dellacqua A. 2000. Osteopetrorickets: Case report. *Eur J Paediatr Neurol* 159: 579–81.
17. Rao VM, Dalinka MK, Mitchell DG, Spritzer CE, Kaplan F, August CS, Axel L, Kressel HY. 1986. Osteopetrosis: MR characteristics at 1.5 T. *Radiology* 161: 217–20.
18. Elster AD, Theros EG, Key LL, Chen MYM. 1992. Cranial imaging in autosomal recessive osteopetrosis. Part I. Facial bones and calvarium. *Radiology* 183: 129–37; Cranial imaging in autosomal recessive osteopetrosis. Part II. Skull base and brain. *Radiology* 183: 137–44.

19. Key L, Carnes D, Cole S, Holtrop M, Bar-Shavit Z, Shapiro F, Arceci R, Steinberg J, Gundberg C, Kahn A, et al. 1984. Treatment of congenital osteopetrosis with high dose calcitriol. *N Engl J Med* 310: 409–15.
20. Whyte MP, Kempa LG, McAlister WH, Shang F, Mumm S, Wenkert D. 2010. Elevated serum lactate dehydrogenase isoenzymes and aspartate transaminase distinguish Albers-Schönberg disease (chloride channel 7 deficiency osteopetrosis) among the sclerosing bone disorders. *J Bone Miner Res* 25: 2515–26.
21. Alatalo SL, Ivaska KK, Waguespack SG, Econs MJ, Väänänen HK, Halleen JM. 2004. Osteoclast-derived serum tartrate-resistant acid phosphatase 5b in Albers-Schönberg disease (type 11 autosomal dominant osteopetrosis). *Clin Chem* 50: 883–90.
22. Sobacchi C, Frattini A, Guerrini MM, Abinun M, Pangrazio A, Susani L, Bredius R, Mancini G, Cant A, Bishop N, Grabowski P, Del Fattore A, Messina C, Errigo G, Coxon FP, Scott DI, Teti A, Rogers MJ, Vezzoni P, Villa A, Helfrich MH. 2007. Osteoclast-poor human osteopetrosis due to mutations in the gene encoding RANKL. *Nat Genet* 39: 960–62.
23. Helfrich MH, Aronson DC, Everts V, Mieremet RHP, Gerritsen EJA, Eckhardt PG, Groot CG, Scherft JP. 1991. Morphologic features of bone in human osteopetrosis. *Bone* 12: 411–19.
24. Bollerslev J, Steiniche T, Melsen F, Mosekilde L. 1986. Structural and histomorphometric studies of iliac crest trabecular and cortical bone in autosomal dominant osteopetrosis: A study of two radiological types. *Bone* 10: 19–24.
25. Lajeunesse D, Busque L, Ménard P, Brunette MG, Bonny Y. 1996. Demonstration of an osteoblast defect in two cases of human malignant osteopetrosis. Correction of the phenotype after bone marrow transplant. *Bone* 98: 1835–42.
26. Mills BG, Yabe H, Singer FR. 1988. Osteoclasts in human osteopetrosis contain viral-nucleocapsid-like nuclear inclusions. *J Bone Miner Res* 3: 101–6.
27. Beard CJ, Key L, Newburger PE, Ezekowitz RA, Arceci R, Miller R, Proto P, Ryan T, Anast C, Simons ER. 1986. Neutrophil defect associated with malignant infantile osteopetrosis. *J Lab Clin Med* 108: 498–505.
28. Cleiren E, Bénichou O, Van Hul E, Gram J, Bollerslev J, Singer FR, Beaverson K, Aledo A, Whyte MP, Yoneyama T, deVernejoul MC, Van Hul W. 2001. Albers-Schönberg disease (autosomal dominant osteopetrosis, type II) results from mutations in the CICN7 chloride channel gene. *Hum Mol Genet* 10: 2861–7.
29. Susani L, Pangrazio A. Sobacchi C, Taranta A, Mortier G, Savarirayan R. Villa A, Orchard P, Vezzoni P, Albertini A, Frattini A. Pagani F. 2004. TCIRG1-dependent recessive osteopetrosis: Mutation analysis, functional identification of the splicing defects, and in vitro rescue by U1 snRNA. *Hum Mutat* 24: 225–35.
30. Campos-Xavier AB, Saraiva JM, Ribeiro LM, Munnich A, Cormier-Daire V. 2003. Chloride channel 7 (CLCN7) gene mutations in intermediate autosomal recessive osteopetrosis. *Hum Genet* 112: 186–9.
31. Chalhoub N, Benachenhou N, Rajapurohitam V, Pata M, Ferron M, Frattini A, Villa A, Vacher J. 2003. Grey-lethal mutation induces severe malignant autosomal recessive osteopetrosis in mouse and human. *Nat Med* 9: 399–406.
32. Driessen GJ, Gerritsen EJ, Fischer A, Fasth A, Hop WC, Veys P, Porta F, Cant A, Steward CG, Vossen JM, Uckan D, Friedrich W. 2003. Long-term outcome of haematopoietic stem cell transplantation in autosomal recessive osteopetrosis: An EBMT report. *Bone Marrow Transplant* 32: 657–63.
33. Rawlinson PS, Green RH, Coggins AM, Boyle IT, Gibson BE. 1991. Malignant osteopetrosis: Hypercalcaemia after bone marrow transplantation. *Arch Dis Child* 66: 638–9.
34. Steward CG, Pellier I, Mahajan A, Ashworth MT, Stuart AG, Fasth A, Lang D, Fischer A, Friedrich W, Schulz AS; Working Party on Inborn Errors of the European Blood and Marrow Transplantation Group. 2004. Severe pulmonary hypertension: A frequent complication of stem cell transplantation for malignant infantile osteopetrosis. *Br J Haematol* 124: 63–71.
35. Tsuji Y, Ito S, Isoda T, Kajiwara M, Nagasawa M, Morio T, Mizutani S. 2005. Successful nonmyeloablative cord blood transplantation for an infant with malignant infantile osteopetrosis. *J Pediatr Hematol Oncol* 27: 495–98.
36. Key LL Jr. 1987. Osteopetrosis: A genetic window into osteoclast function. Cases Metab Bone Dis. A CPC Series. Triclinica Communications. New York, NY. 2: 1–12.
37. Dorantes LM, Mejia AM, Dorantes S. 1986. Juvenile osteopetrosis: Effects of blood and bone of prednisone and low calcium, high phosphate diet. *Arch Dis Child* 61: 666–670.
38. Iacobini M, Migliaccio S, Roggini M, Taranta A, Werner B, Panero A, Teti A. 2001. Case Report: Apparent cure of a newborn with malignant osteopetrosis using prednisone therapy. *J Bone Miner Res* 16: 2356–60.
39. Dozier TS, Duncan IM, Klein AJ, Lambert PR, Key LL Jr. 2005. Otologic manifestations of malignant osteopetrosis. *Otol Neurotol* 26: 762–6.
40. Strickland JP, Berry DJ. 2005. Total joint arthroplasty in patients with osteopetrosis: A report of 5 cases and review of the literature. *J Arthoplasty* 20: 815–20.
41. Sly WS, Hewett-Emmett D, Whyte MP, Yu YS, Tashian RE. 1983. Carbonic anhydrase II deficiency identified as the primary defect in the autosomal recessive syndrome of osteopetrosis with renal tubular acidosis and cerebral calcification. *Proc Natl Acad Sci U S A* 80: 2752–6.
42. Whyte MP. 1993. Carbonic anhydrase II deficiency. *Clin Orthop* 294: 52–3.
43. Sly WS, Shah GN. 2001. The carbonic anhydrase II deficiency syndrome: Osteopetrosis with renal tubular acidosis and cerebral calcification. In: Scriver CR, Beaudet AL, Sly WS, Valle D. Child B, Vogelstein B (eds.) *The Metabolic and Molecular Bases of Inherited Disease*, 8th Ed. New York: McGraw-Hill Book Company. pp. 5331–43.

44. Roth DE, Venta PJ, Tashian RE, Sly WS. 1992. Molecular basis of human carbonic anhydrase II deficiency. *Proc Natl Acad Sci U S A* 89: 1804–8.
45. Shah GN, Bonapace G, Hu PY, Strisciuglio P, Sly WS. 2004. Carbonic anhydrase II deficiency syndrome (osteopetrosis with renal tubular acidosis and brain calcification): Novel mutations in CA2 identified by direct sequencing expand the opportunity for genotype- phenotype correlation. *Hum Mutat* 24: 272.
46. Whyte MP, Hamm LL 3rd, Sly WS.1988. Transfusion of carbonic anhydrase-replete erythrocytes fails to correct the acidification defect in the syndrome of osteopetrosis, renal tubular acidosis, and cerebral calcification (carbonic anhydrase-II deficiency). *J Bone Miner Res* 3: 385–8.
47. McMahon C, Will A, Hu P, Shah GN, Sly WS, Smith OP. 2001. Bone marrow transplantation corrects osteopetrosis in the carbonic anhydrase II deficiency syndrome. *Blood* 97: 1947–50.
48. Maroteaux P, Lamy M. 1965. The malady of Toulouse-Lautrec. *JAMA* 191: 715–7.
49. Maroteaux P, Lamy M. 1962. [Pyknodysostosis]. [Article in French]. *Presse Med* 70: 999–1002.
50. Gelb BD, Brijmme D, Desnick RJ. 2001. Pycnodysostosis: Cathepsin K deficiency. In: Scriver CR, Beaudet AL, Sly WS, Valle D, Child B, Vogelstein B (eds.) *The Metabolic and Molecular Bases of Inherited Disease, 8th Ed.* New York: McGraw-Hill Book Company. pp. 3453–68.
51. Elmore SM. 1967. Pycnodysostosis: A review. *J Bone Joint Surg Am* 49: 153–62.
52. Soto TJ, Mautalen CA, Hoiman D, Codevilla A, PiquC J, Pangaro JA. 1969. Pycnodysostosis, metabolic and histologic studies. *Birth Defects* 5: 109–15.
53. Everts V, Aronson DC, Beertsen W. 1985. Phagocytosis of bone collagen by osteoclasts in two cases of pycnodysostosis. *Calcif Tissue Int* 37: 25–31.
54. Beneton MNC, Harris S, Kanis JA. 1987. Paramyxovirus-like inclusions in two cases of pycnodysostosis. *Bone* 8: 211–7.
55. Soliman AT, Rajab A, AlSalmi I, Darwish A, Asfour M. 1996. Defective growth hormone secretion in children with pycnodysostosis and improved linear growth after growth hormone treatment. *Arch Dis Child* 75: 242–4.
56. Fratzl-Zelman N, Valenta A, Roschger P, Nader A, Gelb BD, Fratzl P, Klaushofer K. 2004. Decreased bone turnover and deterioration of bone structure in two cases of pycnodysostosis. *J Clin Endocrinol Metab* 89: 1538–47.
57. Everts V, Hou WS, Rialland X, Tigchelaar W, Saftig P, Bromme D, Gelb BD, Beertsen W. 2003. Cathepsin K deficiency in pycnodysostosis results in accumulation of non-digested phagocytosed collagen in fibroblasts. *Calcif Tissue Int* 73: 380–6.
58. Edelson JG, Obad S, Geiger R, On A, Artul HJ. 1992. Pycnodysostosis: Orthopedic aspects, with a description of 14 new cases. *Clin Orthop* 280: 263–76.
59. Engelmann G. 1929. Ein fall von osteopathia hyperostotica (sclerotisans) multiplex infantilis. *Fortschr Geb Roentgen* 39: 1101–6.
60. Saito T, Kinoshita A, Yoshiura Kl, Makita Y, Wakui K, Honke K, Niikawa N, Taniguchi N. 2001. Domain-specific mutations of a transforming growth factor (TGF)-beta 1 latency-associated peptide cause Camurati-Engelmann disease because of the formation of a constitutively active form of TGF-beta 1. *J Biol Chem* 276: 11469–72.
61. Wallace SE, Lachman RS, Mekikian PB, Bui KK, Wilcox WR. 2004. Marked phenotypic variability in progressive diaphyseal dysplasia (Camurati-Engelmann disease): Report of a four generation pedigree, identification of a mutation in TGFB1, and review. *Am J Med Genet A* 129: 235–47.
62. Naveh Y, Ludatshcer R, Alon U, Sharf B. 1985. Muscle involvement in progressive diaphyseal dysplasia. *Pediatrics* 76: 944–9.
63. Crisp AJ, Brenton DP. 1982. Engelmann's disease of bone: A systemic disorder? *Ann Rheum Dis* 41: 183–8.
64. Smith R, Walton RJ, Corner BD, Gordon IR. 1977. Clinical and biochemical studies in Engelmann's disease (progressive diaphyseal dysplasia). *Q J Med* 46: 273–94.
65. Kumar B, Murphy WA, Whyte MP. 1981. Progressive diaphyseal dysplasia (Englemann's disease): Scintigraphic-radiologic-clinical correlations. *Radiology* 140: 87–92.
66. Applegate W, Applegate GR, Kemp SS. 1991. MR of multiple cranial neuropathies in a patient with Camurati-Engelmann disease: Case report. *AJNR Am J Neuroradiol* 557–9.
67. Janssens K, ten Dijke P, Ralston SH, Bergmann C, Van Hul W. 2003. Transforming growth factor-beta 1 mutations in Camurati-Engelmann disease lead to increased signaling by altering either activation or secretion of the mutant protein. *J Biol Chem* 278: 7718–24.
68. Saraiva JM. 2000. Anticipation in progressive diaphyseal dysplasia. *J Med Genet* 37: 394–5.
69. Hecht JT, Blanton SH, Broussard S, Scott A, Hall CR, Milunsky JM. 2001. Evidence for locus heterogeneity in the Camurati-Engelmann (DPD1) Syndrome.. *Clin Genet* 59: 198–200.
70. Naveh Y, Alon U, Kaftori JK, Berant M. 1985. Progressive diaphyseal dysplasia: Evaluation of corticosteroid therapy. *Pediatrics* 75: 321–3.
71. lnaoka T, Shuke N, Sato J, Ishikawa Y, Takahashi K, Aburano T, Makita Y. 2001. Scintigraphic evaluation of pamidronate and corticosteroid therapy in a patient with progressive diaphyseal dysplasia (Camurati-Engelmann disease). *Clin Nucl Med* 26: 680–2.
72. Van Buchem FSP, Prick JJG, Jaspar HHJ. 1976. *Hyperostosis Corticalis Generalisata Familiaris (Van Buchem's Disease)*. Amsterdam, The Netherlands: Excerpta Media.
73. Perez-Vicente JA, Rodriguez de Castro E, Lafuente J, Mateo MM, Gimenez-Roldan S. 1987. Autosomal

dominant endosteal hyperostosis. Report of a Spanish family with neurological involvement. *Clin Genet* 31: 161–9.
74. Beighton P, Barnard A, Hamersma H, van der Wouden A. 1984. The syndromic status of sclerosteosis and van Buchem disease. *Clin Genet* 25: 175–81.
75. Brunkow ME, Gardner JC, Van Ness J, Paeper BW, Kovacevich BR, Proll S. Skonier JE, Zhao L, Sabo PJ, Fu Y, Alisch RS, Gillett L, Colbert T, Tacconi P, Galas D, Hamersma H, Beighton P, Mulligan J. 2001. Bone dysplasia sclerosteosis results from loss of the SOST gene product, a novel cystine knot-containing protein. *Am J Hum Genet* 68: 577–89.
76. Loots GG, Kneissel M, Keller H, Baptist M, Chang J, Collette NM, Ovcharenko D, Plajzer-Frick I, Rubin EM. 2005. Genomic deletion of a long-range bone enhancer misregulates sclerostin in Van Buchem disease. *Genome Res* 15: 928–35.
77. Li X, Zhang Y, Kang H, Liu W, Liu P, Zhang J, Harris SE, Wu D. 2005. Sclerostin binds to LRP5/6 and antagonizes canonical Wnt signaling. *J Biol Chem* 280: 19883–7.
78. Sutherland MK. Geoghegan JC, Yu C, Turcott E, Skonier JE, Winkler DG, Latham JA. 2004. Sclerostin promotes the apoptosis of human osteoblastic cells: A novel regulation of bone formation. *Bone* 35: 828–35.
79. Ruckert EW, Caudill RJ, McCready PJ. 1985. Surgical treatment of van Buchem disease. *J Oral Maxillofac Surg* 43: 801–5.
80. Schendel SA. 1988. Van Buchem disease: Surgical treatment of the mandible. *Ann Plast Surg* 20: 462–7.
81. Hamersma H, Gardner J, Beighton P. 2003. The natural history of sclerosteosis. *Clin Genet* 63: 192–7.
82. Beighton P, Cremin BJ, Hamersma H. 1976. The radiology of sclerosteosis. *Br J Radiol* 49: 934–9.
83. Hill SC, Stein SA, Dwyer A, Altman J, Dorwart R, Doppman J. 1986. Cranial CT findings in sclerosteosis. *AJNR Am J Neuroradiol* 7: 505–11.
84. Stein SA, Witkop C, Hill S, Fallon MD, Viernstein L, Gucer G, McKeever P, Long D, Altman J, Miller NR, Teitelbaum SL, Schlesinger S. 1983. Sclerosteosis, neurogenetic and pathophysiologic analysis of an American kinship. *Neurology* 33: 267–77.
85. Epstein S, Hamersma H, Beighton P. 1979. Endocrine function in sclerosteosis. *S Afr Med J* 55: 1105–10.
86. Uitto J, Santa Cruz DJ, Starcher BC, Whyte MP, Murphy WA. 1981. Biochemical and ultrastructural demonstration of elastin accumulation in the skin of the Buschke-Ollendorff syndrome. *J Invest Dermatol* 76: 284–7.
87. Mumm S, Wenkert D, Zhang X, McAlister WH, Mier R, Whyte MP. 2007. Deactivating germline mutations in LEMD3 cause osteopoikilosis and Buschke-Ollendorff syndrome, but not sporadic melorheostosis. *J Bone Miner Res* 22: 243–50.
88. Whyte MP, Murphy WA, Seigel BA. 1978. 99m Tc-pyrophosphate bone imaging in osteopoikilosis, osteopathia striata, and melorheostosis. *Radiology* 127: 439–43.
89. Rabinow M, Unger F. 1984. Syndrome of osteopathia striata, macrocephaly, and cranial sclerosis. *Am J Dis Child* 138: 821–3.
90. Happle R, Lenz W. 1977. Striation of bones in focal dermal hypoplasia: Manifestation of functional mosaicism? *Br J Dermatol* 96: 133–8.
91. Jenkins ZA, van Kogelenberg M, Morgan T, Jeffs A, Fukuzawa R, Pearl E, Thaller C, Hing AV, Porteous ME, Garcia-Miñaur S, Bohring A, Lacombe D, Stewart F, Fiskerstrand T, Bindoff L, Berland S, Adès LC, Tchan M, David A, Wilson LC, Hennekam RC, Donnai D, Mansour S, Cormier-Daire V, Robertson SP. 2009. Germline mutations in WTX cause a sclerosing skeletal dysplasia but do not predispose to tumorigenesis. *Nat Genet* 41: 55–100.
92. Campbell CJ, Papademetriou T, Bonfiglio M. 1968. Melorheostosis: A report of the clinical, roentgenographic, and pathological findings in fourteen cases. *J Bone Joint Surg Am* 50: 1281–1304.
93. Leri A, Joanny J. 1922. Une affection non decrite des os. Hyperostose "en coulee" sur toute la longueur d'un membre ou "melorheostose." *Bull Mem Soc Med Hop Paris* 46: 1141–5.
94. Applebaum RE, Caniano DA, Sun CC, Azizkhan RA, Queral LA. 1986. Synchronous left subclavian and axillary artery aneurysms associated with melorheostosis. *Surgery* 99: 249–53.
95. Colavita N, Nicolais S, Orazi C, Falappa PG. 1987. Melorheostosis: Presentation of a case followed up for 24 years. *Arch Orthop Trauma Surg* 106: 123–5.
96. Wagers LT, Young AW Jr, Ryan SF. 1972. Linear melorheostotic scleroderma. *Br J Dermatol* 86: 297–301.
97. Kim JE, Kim EH, Han EH, Park RW, Park IH. Jun SH, Kim JC, Young MF, Kim IS. 2000. A TGF-3-inducible cell adhesion moleculae, Big-h3, is downregulated in melorheostosis and involved in oseogeneis. *J Cell Biochem* 77: 169–78.
98. Mumm S, Zhang X, McAlister WH, Wenkert D, Whyte MP. 2005. Deactivating germline mutations in LEMD3 cause osteopoikilosis and Buschke-Ollendorff syndrome, but not melorheostosis. *J Bone Miner Res* 22(2): 243–50.
99. Atar D, Lehman WB, Grant AD, Strongwater AM. 1992. The Ilizarov apparatus for treatment of melorkostosis: Case report and review of the literature. *Clin Orthop* 281: 163–7.
100. Christmann D, Wenger JJ, Dosch JC, Schraub M, Wackenheim A. 1981. L'ostéomalacie axiale: Analyse compare avec la fibrogenese imparfaite. [Axial osteomalacia. Comparative analysis with fibrogenesis imperfecta ossium (author's transl)]. [Article in French]. *J Radiol* 62: 37–41.
101. Whyte MP, Fallon MD, Murphy WA, Teitelbaum SL. 1981. Axial osteomalacia: Clinical, laboratory and genetic investigation of an affected mother and son. *Am J Med* 71: 1041–9.
102. Nelson AM, Riggs BL, Jowsey JO. 1978. Atypical axial osteomalacia. Report of four cases with two having

features of ankylosing spondylitis. *Arthritis Rheum* 21: 715–22.
103. Cortet B, Bernière L, Solau-Gervais E, Hacène A, Cotten A, Delcambre B. 2000. Axial osteomalacia with sacroiliitis and moderate phosphate diabetes: Report of a case. *Clin Exp Rheumatol* 18: 625–628.
104. Condon JR, Nassim JR. 1971. Axial osteomalacia. *Postgrad Med J* 47(554): 817–20.
105. Lang R, Vignery AM, Jenson PS. 1986. Fibrogenesis imperfecta ossium with early onset: Observations after 20 years of illness. *Bone* 7: 237–46.
106. Ralphs JR, Stamp TCB, Dopping-Hepenstal PJC, Ali SY. 1989. Ultrastructural features of the osteoid of patients with fibrogenesis imperfecta ossium. *Bone* 10: 243–9.
107. Rimoin DL. 1965. Pachydermoperiostosis (idiopathic clubbing and periostosis). Genetic and physiologic considerations. *N Engl J Med* 272: 923–31.
108. Uppal S, Diggle CP, Carr IM, et al. 2008. Mutations in 15-hydroxyprostaglandin dehydrogenase cause primary hypertrophic osteoarthropathy. *Nat Genet* 40: 789–93.
109. Ali A, Tetalman MR, Fordham EW, Turner DA, Chiles JT, Patel SL, Schmidt KD. 1980. Distribution of hypertrophic pulmonary osteoarthropathy. *AJR Am J Roentgenol* 134: 771–80.
110. Lauter SA, Vasey FB, Hüttner I, Osterland CK. 1978. Pachydermoperiostosis: Studies on the synovium. *J Rheumatol* 5: 85–95.
111. Cooper RG, Freemont AJ, Riley M, Holt PJ, Anderson DC, Jayson MI. 1992. Bone abnormalities and severe arthritis in pachydermoperiostosis. *Ann Rheum Dis* 51(3): 416–9.
112. Whyte MP, Teitelbaum SL, Reinus WR. 1996. Doubling skeletal mass during adult life: The syndrome of diffuse osteosclerosis after intravenous drug abuse. *J Bone Miner Res* 11: 554–8.
113. Khosla S, Ballard FJ, Conover CA. 2002. Use of site-specific antibodies to characterize the circulating form of big insulin-like growth factor II in patients with hepatitis C-associated osteosclerosis. *J Clin Endocrinol Metab* 87(8): 3867–70.
114. Boyden LM, Mao J, Belsky J, Mitzner L, Farhi A, Mitnick MA, Wu D, Insogna K, Lifton RP. 2002. High bone density due to a mutation in LDL-receptor-related protein 5. *N Engl J Med* 345: 1513–21.
115. Rickels MR, Zhang X, Mumm S, Whyte MP. 2005. Oropharyngeal skeletal disease accompanying high bone mass and novel LRP5 mutation. *J Bone Miner Res* 20(5): 878–85.

94

Fibrous Dysplasia

Michael T. Collins, Mara Riminucci, and Paolo Bianco

Introduction 786
Etiology and Pathogenesis 786
Clinical Features 788
Management and Treatment 789
Future Treatment 791
References 791

INTRODUCTION

Fibrous dysplasia of bone (FD) (OMIM#174800) is an uncommon skeletal disorder with a broad spectrum of clinical presentation. On one end of the spectrum, patients may present in later life with an incidentally discovered, asymptomatic radiographic finding that is of no clinical consequence. On the other end of the spectrum, patients may present early in life with a disabling disease. The disease may involve one bone (monostotic FD), multiple bones (polyostotic FD), or the entire skeleton (panostotic FD) [1–3]. FD may be associated with extraskeletal manifestations, the most common of which is areas of cutaneous hyperpigmentation commonly referred to as café au lait macules. These lesions vary widely in size but have characteristic features that include jagged, "coast of Maine" borders, some relationship with the midline, and sometimes follow the developmental lines of Blaschko [Fig. 94.1(A)–(C)]. FD can also be associated with hyperfunctioning endocrinopathies, including precocious puberty, hyperthyroidism, growth hormone (GH) excess, and Cushing syndrome. FD in combination with one or more of the extraskeletal manifestations is known as McCune-Albright syndrome (MAS) [4–7]. A renal tubulopathy, which includes renal phosphate wasting, is one of the most common extraskeletal dysfunctions associated with polyostotic disease [8]. More rarely, FD may be associated with myxomas of skeletal muscle (Mazabraud's syndrome) [9] or dysfunction of the heart, liver, pancreas, or other organs within the context of the MAS [10].

ETIOLOGY AND PATHOGENESIS

FD is caused by missense mutations of the *GNAS* complex locus on chromosome 20q13.3 [11–13]. *GNAS* encodes the alpha subunit of the stimulatory G protein ($G_s\alpha$), a unit of a trimeric protein complex involved in the cyclic adenosine monophosphate (cAMP)-dependent signaling pathway. In the majority of FD patients, the mutation replaces the arginine 201 of $G_s\alpha$ with a histidine (R201H) or a cysteine (R201C); more rarely, other amino acid substitutions are observed at the same position [14] or in other codons (Q227) [15]. In all cases, the mutation impairs the intrinsic GTPase activity of $G_s\alpha$, leading to persistent stimulation of adenylyl cyclase and aberrant production of cAMP (gain-of-function mutations) [16]. Mutations of *GNAS* associated with FD and related disorders are never inherited and could theoretically occur at any time during post-zygotic development. However, the mechanism by which the two most common amino acid replacements are established, which is aberrant methylation of the CpG dinucleotide in the R201 codon, indicates that for the majority of FD patients a restricted time window, i.e., the phase of active methylation of the genome that takes place in the inner cell mass, is the period during which the mutational event occurs [17]. This implies the involvement of a pluripotent cell as the initial target of the disease, thus explaining how the mutation can be transmitted to derivatives of all three germ layers and be broadly distributed in patients with severe forms of the disease. At the same time, differences in the size and viability of the clone

---

Primer on the Metabolic Bone Diseases and Disorders of Mineral Metabolism, Eighth Edition. Edited by Clifford J. Rosen.
© 2013 American Society for Bone and Mineral Research. Published 2013 by John Wiley & Sons, Inc.

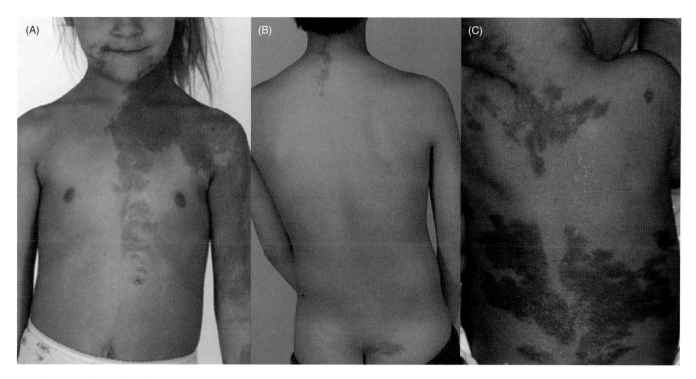

Fig. 94.1. Café au lait skin pigmentation. (A) A large pigmented macule with a classical appearance on the face, chest, and arm of a 5-year-old girl with McCune-Albright syndrome that demonstrates jagged "coast of Maine" borders, and the tendency for the lesions to respect the midline. (B) Small but typical lesions that are often found on the nape of the neck and crease of the buttocks are shown, which also show a relationship to the midline. (C) Extensive lesions on the back of an infant that follow the developmental lines of Blaschko.

arising from the original mutated pluripotent cell could account for the variability of the clinical phenotype observed in the majority of FD patients [18].

The pathology of FD (Fig. 94.2) is characterized by the development of fibro-osseous lesions that replace normal skeletal structures and impair normal skeletal functions. The fibrotic tissue is typically devoid of hematopoiesis and adipose marrow. The bony part consists of trabeculae that are abnormal in amount, distribution, and shape. Different site-specific histological patterns of the disease have been recognized (Chinese writing, sclerotic pagetoid, and sclerotic hypercellular). However, at all affected sites, recurrent subtle histopathological features (stellate-shaped osteoblasts, Sharpey fibers) can be observed as diagnostic hallmarks of the disease, distinguishing it from other skeletal fibro-osseous diseases [19]. Previous work indicated that the histology of FD results from the abnormal differentiation of postnatal multipotent skeletal stem cells and from the abnormal function of mutated osteoblasts brought about by the *GNAS* mutation [20, 21, 22]. Indeed, recent studies showed that normal human skeletal stem cells transduced with the mutated $G_s\alpha$ sequence display impaired adipogenic differentiation and deranged expression of osteogenic markers [23]. Furthermore, the selective downregulation of the mutated allele by specific RNA-interfering sequences reverts the molecular phenotype (normalization of cAMP production) and restores adipogenic differentiation [23]. Most of the histopathological features of FD emanate from the osteogenic nature of the lesional tissue. The severe osteomalacic change of the FD bone [13,24], which plays a major role in the skeletal morbidity of the disease, is related to excess production of the phosphaturic hormone, FGF23, by the expanded osteogenic tissue [25]. The inappropriate osteoclastogenesis observed within the FD tissue is dependent on osteogenic cells through secretion of interleukin-6 (IL-6) [26] and likely, as recently shown in transduced cells, through the upregulation of receptor activator of nuclear factor-κB ligand (RANKL) caused by the *GNAS* mutation [23]. Multiple determinants contribute to the establishment, evolution, and natural history of each FD lesion. The local number of mutated cells likely plays the major role in the development of the disease. However, unexpected epigenetic and nongenetic mechanisms have emerged along the way as potential adjunctive modulators of the clinical expression of the disease. The expression of the two $G_s\alpha$ alleles in clonogenic osteoprogenitor cells is asymmetric and random [27]. This implies that different skeletal sites with an equal number of mutated cells can have different levels of expression of the mutated allele and therefore a different likelihood of developing the disease. Furthermore,

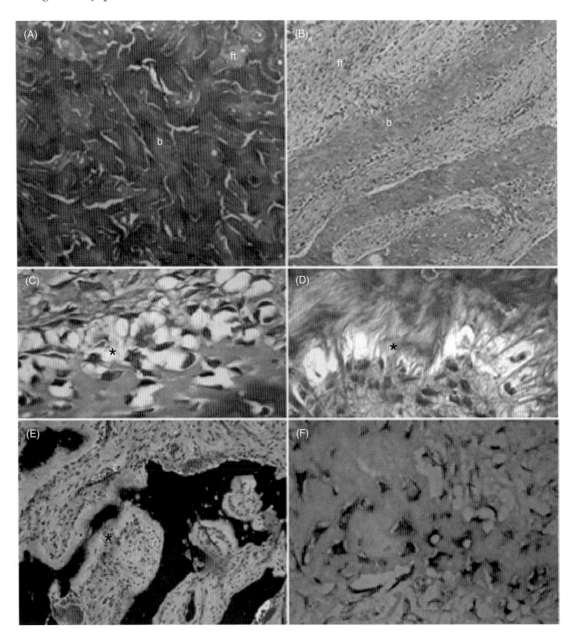

Fig. 94.2. Representative histological images of craniofacial FD. (A) Calvarial FD lesions are characterized by uninterrupted networks of bone trabeculae (b) embedded in the fibrous tissue (ft). (B) In FD, lesions from gnathic bones, newly formed bone trabeculae (b) are deposited within the fibrous tissue (ft) in a typical discontinuous and parallel pattern. (C)–(D) Recurrent histological features of FD. (C) Morphologically abnormal osteoblasts on bone surfaces (asterisk). (D) Collagen fibers perpendicularly oriented to forming bone surfaces (Sharpey fibers, asterisk). (E)–(F) Osteomalacic changes and FGF23 production in FD. (E) Excess osteoid (asterisk) and severe undermineralization of the fibrous dysplastic bone. (F) FGF23 is produced by activated FD osteogenic cells, as shown by *in situ* hybridization.

FD lesions have often been reported to improve clinically over time. Recently it has been shown that this may be dependent on the progressive consumption of mutant clonogenic progenitors followed by regeneration of normal bone and marrow by the remaining wild-type cells [28].

CLINICAL FEATURES

The sites of skeletal involvement (the "map" of affected tissues) are established early in patients with FD. Ninety percent of the craniofacial lesions are established before the age of 5, and 75% of all sites of FD are evident by the age of 15; the implication is that essentially all clinically significant disease is present very early in life, probably by the age of 5 [3]. Pathological effects of $G_s\alpha$ mutations in osteogenic cells are most pronounced and evident during the phase of rapid bone growth, and account for the fact that childhood and adolescence are the periods during which the disease most commonly presents, is the most symptomatic, and is the period of peak rate of fractures [29, 30]. Presentation in infancy is rare, and usually heralds severe, widespread disease with multiorgan involvement. The most common presenting features are a limp, pain, or a fracture. While in general children are less likely to have pain and/or complain of pain *per se*, their manifestation of pain may be to report being "tired" [31]. In adults, the complaint of pain is common and may be severe. Sites commonly affected are the ribs, long bones, and craniofacial bones, while lesions in the spine and pelvis are typically less painful [31]. Pathological fractures of weight-bearing limb bones are a major cause of morbidity. Deformity of limb bones, which is a common finding, is caused by expansion and abnormal compliance of lesional FD, fracture treatment failure, and occasionally local complications such as cyst formation [18]. Deformity of craniofacial bones is solely the result of the overgrowth of lesional bone.

Although any bone may be affected, the proximal metaphysis of the femora and the skull base are the two sites most commonly involved [31]. Femoral disease usually presents in childhood with limp, pain, fracture, and deformity that ranges from coxa vara to the classical shepherd's crook deformity [Fig. 94.3(A)]. Radiographically, the lesion may be limited to the metaphysis or extend along the diaphysis for a variable length [32]. The picture most commonly observed in children and adolescents consists of an expansile, deforming lesion that arises in the medullary space, with cortical thinning and with a "ground glass" appearance on radiograph [Fig. 94.3(A)]. The radiographic picture is significantly affected by the evolution of the lesion over time, with lesions tending to become more "sclerotic" and less homogenous with time and aging [Fig. 94.3(B)]. Sclerosis in FD lesions of the femur and other limb bones reflects less active disease.

In the skull, the skull base and the facial bones are commonly involved. The typical presentation is in childhood with facial asymmetry or a "bump" that persists, but symmetric expansion of the malar prominences and/or frontal bosses may also be seen. The disease can progress into adulthood and result in disfiguration and, as can be seen on imaging studies, encroachment on the cranial nerves and/or the otic capsule. However, vision and/or clinically significant hearing loss are uncommon but more likely to occur when FD is accompanied by GH excess in the context of MAS [33–35]. Vision loss can also occur uncommonly (fewer than 5% of the patients) in setting of a posthemorrhagic cyst, sometimes referred to as an aneurysmal bone cyst, which may arise in the highly vascular FD bone [36].

Radiographically, craniofacial FD typically has a homogenous "ground glass" appearance in children [Fig. 94.3(C)], but with age the radiographic appearance is mixed with both solid and cystic areas [Fig. 94.3(D)]. Lesions in the spine, ribs, and pelvis are common, but more difficult to appreciate on plain radiographs. These are more readily detected by bone scintigraphy, which is a more sensitive imaging technique for the detection of FD lesions [Fig. 94.3(E)–(G)]. Disease in the spine is common, frequently associated with scoliosis, may require surgery, and can be progressive into adulthood [37]. Because FD in the spine is common and can progress to a degree that can be lethal if untreated, it is important to monitor for the presence of a curve and its relative stability or progression.

Malignancy in FD is rare (less than 1%) [38]. While there is an association between the development of cancer and prior treatment with high dose external beam radiation [39, 40], it may also occur independently of prior exposure to ionizing radiation. Rapid lesion expansion and disruption of the cortex on radiographs should alert the clinician to the possibility of sarcomatous change. Osteogenic sarcoma is the most common but is not the only type of bone tumor that may complicate FD. The clinical course is usually aggressive, and surgery is the primary treatment; chemotherapeutic regimens do not appear to improve prognosis significantly.

MANAGEMENT AND TREATMENT

Diagnosis of FD must be established based on expert assessment of clinical, radiographic, and histopathological features. Markers of bone turnover are usually elevated [8]. The extent of the skeletal disease is best determined with total body bone scintigraphy, which can be used to assess the skeletal disease burden and predict functional outcome [41]. The metabolic derangements associated with FD, especially hypophosphatemia and growth hormone excess, are associated with a significantly worse clinical outcome, and therefore must be screened for and treated [24, 30, 33, 35].

Mutation analysis may be helpful in distinguishing FD from unrelated fibro-osseous lesions of the skeleton, which may mimic FD both clinically and radiographically (osteofibrous dysplasia, ossifying fibromas of jawbones; reviewed in Ref. 18). Isolated lesions of the proximal femur in adults may be improperly diagnosed and classified as distinct fibro-osseous lesions. For example, all cases of so-called liposclerosing myxofibrous tumor, in which *GNAS* mutations were sought, were found to represent monostotic fibrous dysplasia [42]. Multiple non-ossifying fibromas, skeletal angiomatosis, and Ollier's disease may sometimes enter the differential diagnosis. Distinction from these entities relies on histology and mutation analysis.

Disease of the proximal femur, in which there is fracture or impending fracture, is often best treated by

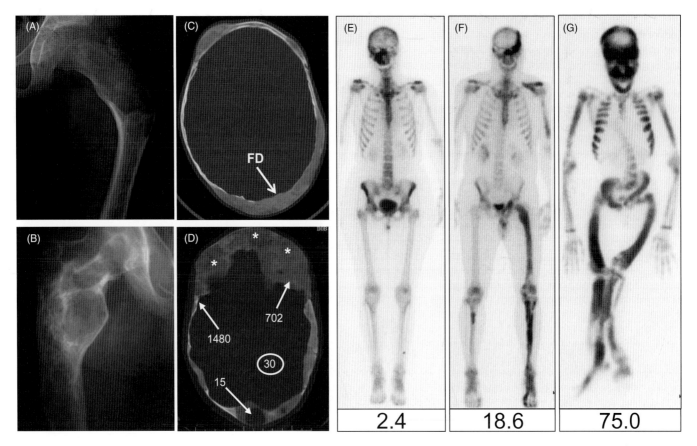

Fig. 94.3. Radiographic appearance of fibrous dysplasia. (A) A proximal femur with typical ground glass appearance and shepherd's crook deformity in a 10-year-old child is shown. (B) The appearance of FD in the femur of an untreated 40-year-old man demonstrates the tendency for FD to appear more sclerotic with time. (C) A CT image of a 10-year-old child with FD is shown. The white arrow indicates an area of FD with typical ground glass appearance that is the predominant radiographic appearance of FD in children. (D) A CT image of a 40-year-old woman demonstrates the typical appearance of craniofacial FD in an older person. Note that compared to that of the 10-year-old in (C), the FD in the frontal bone (asterisks) demonstrates a more heterogeneous pattern that is common in adults of mixed solid and "cystic" lesions. The numbers indicate the Hounsfield Unit (HU) measurements in various types of tissue/lesions that can be seen in FD. The number 1480 indicates the density of the rim of normal bone seen laterally, 702 indicates the HU of typical FD, 30 is the HU density of normal brain tissue, and 15 is the HU density of a fluid-filled lesion (true cyst) that can be rarely observed in FD. (E)–(G) Bone scintigraphy in FD. Representative ^{99}Tc-MDP bone scans that demonstrate the spectrum of disease severity and show tracer uptake at skeletal sites affected with FD. Below each image is the calculated skeletal disease burden score (see Ref. 41). (E) A 50-year-old woman with monostotic FD confined to a single focus involving contiguous bones in the craniofacial region. (F) A 42-year-old man with polyostotic FD shows the tendency for FD to be predominantly (but not exclusively) unilateral, and to involve the skull base and proximal femur. (G) A 16-year-old boy with McCune-Albright syndrome and involvement of virtually all skeletal sites is shown (panostotic FD).

insertion of intramedullary nails, in an effort to prevent serious deformity and limb length discrepancy [32, 43, 44]. In children, this may require specially designed nails or the modification of devices intended for one anatomical site for another, e.g., adult humeral rods for the femur [32, 43]. Indications for surgery of the craniofacial bones may include cosmesis, or documented vision or hearing loss. In all cases, operations should be undertaken with caution and only by experienced surgeons. Contouring or "shaving" procedures for cosmesis often invoke a rapid regrowth, especially in children, and operations for vision loss are frequently not successful at restoring vision [34, 45]. Prophylactic surgery to prevent vision loss is not advocated [34, 35]. Treatment with bisphosphonates (pamidronate, etc.) has been advocated based on observational studies with claims of reduced pain, decreased serum and urine markers of bone metabolism, and improvement in the radiographic appearance of the disease [46, 47]. However, an open label, prospective study with appropriate histological, radiographic, and clinical end points, which also demonstrated pain relief, showed no benefit radiographically or histologically [48]. Ongoing placebo controlled studies in the U.S. and Europe may help to better define the role of bisphosphonates in treating FD.

Two additional drugs that are currently approved for the treatment of other conditions, and which may prove beneficial in the treatment of FD, are tocilizumab and denosumab. Tocilizumab is an antagonist of the IL-6 receptor that is approved for the treatment of rheumatoid arthritis [49]. Based on *in vitro* data indicating a role for IL-6 excess in the pathophysiology of FD [26], treatment with tocilizumab is rational. As RANKL has also been shown to play a role in the pathophysiology of FD *in vitro* [23], treatment with the anti-RANKL antibody, denosumab, which is approved for the treatment of osteoporosis and cancer-related bone metastases [50, 51], also makes sense.

FUTURE TREATMENT

Currently available medical and surgical therapies are not satisfactory. Current efforts are directed at elucidating the effects of silencing the disease-causing gene in skeletal stem cells in *in vitro* and *in vivo* models of gene therapy [23], on devising strategies for using stem cells as either a tool or a target of therapies [23], and on identifying novel drugs that specifically target the constitutively active $G_s\alpha$ or its functional effects. As part of these approaches, it will be important to elucidate additional molecular mediators of the disease phenotype, downstream of the mutated $G_s\alpha$. Crucial to the overall goal of finding effective methods for therapy and, hopefully, curing FD, is the development of suitable animal models of the disease.

REFERENCES

1. Lichtenstein L, Jaffe HL. 1942. Fibrous dysplasia of bone: A condition affecting one, several or many bones, the graver cases of which may present abnormal pigmentation of skin, premature sexual development, hyperthyroidism or still other extraskeletal abnormalities. *Arch Pathol* 33: 777–816.
2. Collins MT. 2006. Spectrum and natural history of fibrous dysplasia of bone. *J Bone Miner Res* 21 Suppl 2: P99–P104.
3. Hart ES, Kelly MH, Brillante B, Chen CC, Ziran N, Lee JS, Feuillan P, Leet AI, Kushner H, Robey PG, Collins MT. 2007. Onset, progression, and plateau of skeletal lesions in fibrous dysplasia, and the relationship to functional outcome. *J Bone Miner Res* 22(9): 1468–74.
4. McCune DJ. 1936. Osteitis fibrosa cystica; the case of a nine year old girl who also exhibits precocious puberty, multiple pigmentation of the skin and hyperthyroidism. *Am J Dis Child* 52: 743–4.
5. Albright F, Butler AM, Hampton AO, Smith PH. 1937. Syndrome characterized by osteitis fibrosa disseminata, areas of pigmentation and endocrine dysfunction, with precocious puberty in females, report of five cases. *N Engl J Med* 216: 727–46.
6. Danon M, Crawford JD. 1987. The McCune-Albright syndrome. *Ergeb Inn Med Kinderheilkd* 55: 81–115.
7. Dumitrescu CE, Collins MT. 2008. McCune-Albright syndrome. *Orphanet J Rare Dis* 3: 12.
8. Collins MT, Chebli C, Jones J, Kushner H, Consugar M, Rinaldo P, Wientroub S, Bianco P, Robey PG. 2001. Renal phosphate wasting in fibrous dysplasia of bone is part of a generalized renal tubular dysfunction similar to that seen in tumor-induced osteomalacia. *J Bone Miner Res* 16(5): 806–13.
9. Cabral CE, Guedes P, Fonseca T, Rezende JF, Cruz Junior LC, Smith J. 1998. Polyostotic fibrous dysplasia associated with intramuscular myxomas: Mazabraud's syndrome. *Skeletal Radiol* 27(5): 278–82.
10. Shenker A, Weinstein LS, Moran A, Pescovitz OH, Charest NJ, Boney CM, Van Wyk JJ, Merino MJ, Feuillan PP, Spiegel AM. 1993. Severe endocrine and nonendocrine manifestations of the McCune-Albright syndrome associated with activating mutations of stimulatory G protein GS. *J Pediatr* 123(4): 509–18.
11. Weinstein LS, Shenker A, Gejman PV, Merino MJ, Friedman E, Spiegel AM. 1991. Activating mutations of the stimulatory G protein in the McCune-Albright syndrome. *N Engl J Med* 325(24): 1688–95.
12. Shenker A, Weinstein LS, Sweet DE, Spiegel AM. 1994. An activating Gs alpha mutation is present in fibrous dysplasia of bone in the McCune-Albright syndrome. *J Clin Endocrinol Metab* 79(3): 750–5.
13. Bianco P, Riminucci M, Majolagbe A, Kuznetsov SA, Collins MT, Mankani MH, Corsi A, Bone HG, Wientroub S, Spiegel AM, Fisher LW, Robey PG. 2000. Mutations of the GNAS1 gene, stromal cell dysfunction, and osteomalacic changes in non-McCune-Albright fibrous dysplasia of bone. *J Bone Miner Res* 15(1): 120–8.
14. Riminucci M, Fisher LW, Majolagbe A, Corsi A, Lala R, De Sanctis C, Robey PG, Bianco P. 1999. A novel GNAS1 mutation, R201G, in McCune-Albright syndrome. *J Bone Miner Res* 14(11): 1987–9.
15. Idowu BD, Al-Adnani M, O'Donnell P, Yu L, Odell E, Diss T, Gale RE, Flanagan AM. 2007. A sensitive mutation-specific screening technique for GNAS1 mutations in cases of fibrous dysplasia: The first report of a codon 227 mutation in bone. *Histopathology* 50(6): 691–704.
16. Landis CA, Masters SB, Spada A, Pace AM, Bourne HR, Vallar L. 1989. GTPase inhibiting mutations activate the alpha chain of Gs and stimulate adenylyl cyclase in human pituitary tumours. *Nature* 340(6236): 692–6.
17. Riminucci M, Saggio I, Robey PG, Bianco P. 2006. Fibrous dysplasia as a stem cell disease. *J Bone Miner Res* 21 Suppl 2: P125–31.
18. Bianco P, Gehron Robey P, Wientroub S. 2003. Fibrous dysplasia. In: Glorieux F, Pettifor JM, Juppner H (eds.) *Pediatric Bone: Biology and disease*. New York: Academic Press/Elsevier. pp. 509–39.
19. Riminucci M, Liu B, Corsi A, Shenker A, Spiegel AM, Robey PG, Bianco P. 1999. The histopathology of fibrous dysplasia of bone in patients with activating mutations

of the Gs alpha gene: Site-specific patterns and recurrent histological hallmarks. *J Pathol* 187(2): 249–58.
20. Riminucci M, Fisher LW, Shenker A, Spiegel AM, Bianco P, Gehron Robey P. 1997. Fibrous dysplasia of bone in the McCune-Albright syndrome: Abnormalities in bone formation. *Am J Pathol* 151(6): 1587–600.
21. Bianco P, Kuznetsov SA, Riminucci M, Fisher LW, Spiegel AM, Robey PG. 1998. Reproduction of human fibrous dysplasia of bone in immunocompromised mice by transplanted mosaics of normal and Gsalpha-mutated skeletal progenitor cells. *J Clin Invest* 101(8): 1737–44.
22. Robey PG, Kuznetsov S, Riminucci M, Bianco P. 2007. The role of stem cells in fibrous dysplasia of bone and the Mccune-Albright syndrome. *Pediatr Endocrinol Rev* 4 Suppl 4: 386–94.
23. Piersanti S, Remoli C, Saggio I, Funari A, Michienzi S, Sacchetti B, Robey PG, Riminucci M, Bianco P. 2010. Transfer, analysis and reversion of the fibrous dysplasia cellular phenotype in human skeletal progenitors. *J Bone Miner Res* 25(5): 1103–16.
24. Corsi A, Collins MT, Riminucci M, Howell PG, Boyde A, Robey PG, Bianco P. 2003. Osteomalacic and hyperparathyroid changes in fibrous dysplasia of bone: Core biopsy studies and clinical correlations. *J Bone Miner Res* 18(7): 1235–46.
25. Riminucci M, Collins MT, Fedarko NS, Cherman N, Corsi A, White KE, Waguespack S, Gupta A, Hannon T, Econs MJ, Bianco P, Gehron Robey P. 2003. FGF-23 in fibrous dysplasia of bone and its relationship to renal phosphate wasting. *J Clin Invest* 112(5): 683–92.
26. Riminucci M, Kuznetsov SA, Cherman N, Corsi A, Bianco P, Gehron Robey P. 2003. Osteoclastogenesis in fibrous dysplasia of bone: In situ and in vitro analysis of IL-6 expression. *Bone* 33(3): 434–42.
27. Michienzi S, Cherman N, Holmbeck K, Funari A, Collins MT, Bianco P, Robey PG, Riminucci M. 2007. GNAS transcripts in skeletal progenitors: Evidence for random asymmetric allelic expression of Gs alpha. *Hum Mol Genet* 16(16): 1921–30.
28. Kuznetsov SA, Cherman N, Riminucci M, Collins MT, Robey PG, Bianco P. 2008. Age-dependent demise of GNAS-mutated skeletal stem cells and "normalization" of fibrous dysplasia of bone. *J Bone Miner Res* 23(11): 1731–40.
29. Harris WH, Dudley HR, Barry RJ. 1962. The natural history of fibrous dysplasia. An orthopedic, pathological, and roentgenographic study. *J Bone Joint Surg Am* 44-A: 207–33.
30. Leet AI, Chebli C, Kushner H, Chen CC, Kelly MH, Brillante BA, Robey PG, Bianco P, Wientroub S, Collins MT. 2004. Fracture incidence in polyostotic fibrous dysplasia and the McCune-Albright Syndrome. *J Bone Miner Res* 19(4): 571–7.
31. Kelly MH, Brillante B, Collins MT. 2007. Pain in fibrous dysplasia of bone: Age-related changes and the anatomical distribution of skeletal lesions. *Osteoporos Int* 19(1): 57–63.
32. Ippolito E, Bray EW, Corsi A, De Maio F, Exner UG, Robey PG, Grill F, Lala R, Massobrio M, Pinggera O, Riminucci M, Snela S, Zambakidis C, Bianco P. 2003. Natural history and treatment of fibrous dysplasia of bone: A multicenter clinicopathologic study promoted by the European Pediatric Orthopaedic Society. *J Pediatr Orthop B* 12(3): 155–77.
33. Akintoye SO, Chebli C, Booher S, Feuillan P, Kushner H, Leroith D, Cherman N, Bianco P, Wientroub S, Robey PG, Collins MT. 2002. Characterization of gsp-mediated growth hormone excess in the context of McCune-Albright syndrome. *J Clin Endocrinol Metab* 87(11): 5104–12.
34. Cutler CM, Lee JS, Butman JA, FitzGibbon EJ, Kelly MH, Brillante BA, Feuillan P, Robey PG, DuFresne CR, Collins MT. 2006. Long-term outcome of optic nerve encasement and optic nerve decompression in patients with fibrous dysplasia: Risk factors for blindness and safety of observation. *Neurosurgery* 59(5): 1011–7; discussion 1017–8.
35. Lee JS, FitzGibbon E, Butman JA, Dufresne CR, Kushner H, Wientroub S, Robey PG, Collins MT. 2002. Normal vision despite narrowing of the optic canal in fibrous dysplasia. *N Engl J Med* 347(21): 1670–6.
36. Diah E, Morris DE, Lo LJ, Chen YR. 2007. Cyst degeneration in craniofacial fibrous dysplasia: Clinical presentation and management. *J Neurosurg* 107(3): 504–8.
37. Leet AI, Magur E, Lee JS, Wientroub S, Robey PG, Collins MT. 2004. Fibrous dysplasia in the spine: Prevalence of lesions and association with scoliosis. *J Bone Joint Surg Am* 86-A(3): 531–7.
38. Ruggieri P, Sim FH, Bond JR, Unni KK. 1994. Malignancies in fibrous dysplasia. *Cancer* 73(5): 1411–24.
39. Saglik Y, Atalar H, Yildiz Y, Basarir K, Erekul S. 2007. Management of fibrous dysplasia. A report on 36 cases. *Acta Orthop Belg* 73(1): 96–101.
40. Hansen MR, Moffat JC. 2003. Osteosarcoma of the skull base after radiation therapy in a patient with McCune-Albright syndrome: Case report. *Skull Base* 13(2): 79–83.
41. Collins MT, Kushner H, Reynolds JC, Chebli C, Kelly MH, Gupta A, Brillante B, Leet AI, Riminucci M, Robey PG, Bianco P, Wientroub S, Chen CC. 2005. An instrument to measure skeletal burden and predict functional outcome in fibrous dysplasia of bone. *J Bone Miner Res* 20(2): 219–26.
42. Corsi A, De Maio F, Ippolito E, Cherman N, Gehron Robey P, Riminucci M, Bianco P. 2006. Monostotic fibrous dysplasia of the proximal femur and liposclerosing myxofibrous tumor: Which one is which? *J Bone Miner Res* 21(12): 1955–8.
43. Stanton RP. 2006. Surgery for fibrous dysplasia. *J Bone Miner Res* 21 Suppl 2: P105–9.
44. Keijser LC, Van Tienen TG, Schreuder HW, Lemmens JA, Pruszczynski M, Veth RP. 2001. Fibrous dysplasia of bone: Management and outcome of 20 cases. *J Surg Oncol* 76(3): 157–66; discussion 167–8.
45. Chen YR, Chang CN, Tan YC. 2006. Craniofacial fibrous dysplasia: An update. *Chang Gung Med J* 29(6): 543–9.
46. Liens D, Delmas PD, Meunier PJ. 1994. Long-term effects of intravenous pamidronate in fibrous dysplasia of bone. *Lancet* 343(8903): 953–4.

47. Chapurlat RD, Delmas PD, Liens D, Meunier PJ. 1997. Long-term effects of intravenous pamidronate in fibrous dysplasia of bone. *J Bone Miner Res* 12(10): 1746–52.
48. Plotkin H, Rauch F, Zeitlin L, Munns C, Travers R, Glorieux FH. 2003. Effect of pamidronate treatment in children with polyostotic fibrous dysplasia of bone. *J Clin Endocrinol Metab* 88(10): 4569–75.
49. Smolen JS, Beaulieu A, Rubbert-Roth A, Ramos-Remus C, Rovensky J, Alecock E, Woodworth T, Alten R. 2008. Effect of interleukin-6 receptor inhibition with tocilizumab in patients with rheumatoid arthritis (OPTION study): A double-blind, placebo-controlled, randomised trial. *Lancet* 371(9617): 987–97.
50. Burkiewicz JS, Scarpace SL, Bruce SP. 2009. Denosumab in osteoporosis and oncology. *Ann Pharmacother* 43(9): 1445–55.
51. McClung MR, Lewiecki EM, Cohen SB, Bolognese MA, Woodson GC, Moffett AH, Peacock M, Miller PD, Lederman SN, Chesnut CH, Lain D, Kivitz AJ, Holloway DL, Zhang C, Peterson MC, Bekker PJ. 2006. Denosumab in postmenopausal women with low bone mineral density. *N Engl J Med* 354(8): 821–31.

95
Osteochondrodysplasias
*Yasemin Alanay and David L. Rimoin**

Introduction 794
Multidisciplinary Approach 795
Future Reflections of Current Research 795

References 803
Recommended Databases 804

INTRODUCTION

Genetic disorders of the skeleton are a clinically and genetically heterogeneous group of disorders of bone and/or cartilage characterized by abnormalities in growth, development, and/or homeostasis of the human skeleton [1]. They include the osteochondrodysplasias (primarily affecting bone and/or cartilage), the dysostoses (affecting a single bone or group of bones), the brachydactylies (primarily involving the hands and feet) and the lysosomal storage diseases. Although relatively rare individually, the skeletal dysplasias have an estimated birth prevalence of nearly 1/5,000 [2]. The generalized and progressive abnormalities lead to changes in the size and shape of the limbs, trunk, and/or skull, frequently resulting in disproportionate short stature [3]. In the past, most disproportionate dwarfs were referred to as having either achondroplasia [MIM #100800] (i.e., those with short limbs), or Morquio disease [MIM #253010] (i.e., those with short trunks). It is now apparent that there are over 450 distinct genetic disorders of skeleton that must be distinguished one from the other for specific genetic counseling, prognosis, and treatment. A need to develop a uniform and consistent nomenclature and classification system for these conditions led to the International Nomenclatures of Constitutional Diseases of Bone. These were initially formulated in 1972 in Paris and have since been regularly revised. The initial categorizations were purely descriptive and consisted of a mixture of the key clinical, radiographic, and pathologic features of each condition. The concept of "families" of disorders then evolved, where conditions with similar genetic backgrounds or apparent pathogenetic mechanisms were grouped together [4].

In the latest 2010 revision of the Nosology and Classification of Genetic Skeletal Disorders, there was an increase from 372 to 456 disorders in the 4 years since the classification was last revised in 2007 [5, 6]. Of these conditions, 316 are associated with one or more of 226 different genes. This increase reflects the continuing delineation of unique phenotypes among short stature conditions, which in aggregate represent about 5% of children with birth defects [7]. Some of the increase has also been driven by technological improvements in our ability to define the molecular genetic basis of these conditions, which is now known for 316 of the disorders (215 in the prior revision), with defects in 226 (140 previously) different genes. Most of the molecules and pathways recognized to be essential for normal skeletal development have been identified by characterizing the genetic basis of the skeletal dysplasias. These studies have revealed that many different mechanisms can be disrupted, leading to the short stature and deformity that characterize this group of disorders. Defects in structural molecules of the extracellular matrix, which were the first skeletal dysplasias in which the molecular basis was identified [8, 9], continue to be important in the molecular pathogenesis of these disorders [10, 11], as do defects in their post-translational modification and processing [12–15]. However, abnormalities in transcription factors, signal transduction, nucleic acid metabolism, proteoly-

*Deceased.

Primer on the Metabolic Bone Diseases and Disorders of Mineral Metabolism, Eighth Edition. Edited by Clifford J. Rosen.
© 2013 American Society for Bone and Mineral Research. Published 2013 by John Wiley & Sons, Inc.

sis, peroxisome function, osteoclast resorption, cytoskeletal structure, and cilia function, among others, have also been identified in skeletal dysplasias.

In this latest 2010 revision of the Nosology and Classification of Genetic Skeletal Disorders, 456 different conditions are listed in 40 groups (Table 95.1): Groups 1–8 are based on a common gene or pathway (collagen II group, aggrecan group). Groups 9–17 reflect radiographic changes in a specific bone structure (vertebrae, epiphyses, metaphyses, diaphysis, or combinations) or involved segment (rhizo, meso, or acro). Groups 18–20 define macroscopic criteria in addition to clinical features (bent bones, slender bones, multiple dislocations). Groups 21–25 and 28 are mineralization defects. Groups 26 and 27 include hypophosphatemic rickets and the large group of lysosomal disorders with skeletal involvement, respectively. Group 29 involves disorganization of skeletal components (enchondromas, exostoses, etc.). Group 30 (overgrowth syndromes with significant skeletal involvement) and Group 31 (genetic inflammatory/rheumatoid-like osteoarthropathies) have been added, underlining the frequent diagnostic overlap between these disorders and primary skeletal disorders. Finally, Groups 32–40 are dedicated to various dysostoses [6].

It is apparent that no single classification of these disorders will be adequate. However, using the Nosology is of benefit to the clinician, providing a list of disorders along with possible differential diagnoses.

MULTIDISCIPLINARY APPROACH

The evaluation of patients with skeletal dysplasias mandates a multidisciplinary approach involving clinical geneticists, radiologists, molecular biologists, and biochemical geneticists for diagnosis and a host of surgical specialists for management of their many complications [16]. A careful clinical examination and detailed radiographic evaluation of the skeleton are the first steps toward an accurate diagnosis. A detailed history (length at birth, growth curves, etc.) and pedigree analysis, followed by anthropometric measurements with special emphasis on body proportions are important. Serial radiographic evaluations are often necessary as some skeletal abnormalities may become evident at a later age [4, 17] (Fig. 95.1). In an adult patient, the availability of a prepubertal skeletal survey, allowing the evaluation of the epiphyses and metaphyses before epiphyseal closure, can be essential for diagnostic purposes. Comparison to normal radiographs at the particular age can be a key for diagnosis. Ossified epiphyses, which are small and/or irregular for age, suggest an epiphyseal dysplasia. Widened, flared, and/or irregular metaphyses suggest a metaphyseal dysplasia, while a diaphyseal abnormality (widening, cortical thickening or marrow space expansion) suggests an osteodysplasia. (Fig. 95.1). A combination of the aforementioned findings with platyspondyly or vertebral irregularities is helpful in recognizing chondrodysplasias coined "spondyloepiphyseal dysplasia (SED), spondyloepimetaphyseal dysplasia (SEMD), etc. [17]. Recognition of other well-known, often pathognomonic skeletal changes in the skeletal survey is helpful in narrowing the list of entities considered in the differential diagnosis [2, 17].

Morphologic examination of cartilage and bone should also be a part of the diagnostic work-up when possible [3]. Patients undergoing surgery should be considered for bone and cartilage specimen collection. Details on the collection and analysis of such specimens can be found at the International Skeletal Dysplasia Registry (www.csmc.edu/skeletaldysplasia).

FUTURE REFLECTIONS OF CURRENT RESEARCH

In the past decade, enormous progress has been made in our knowledge concerning the biochemistry and molecular genetic basis of these disorders [18, 19]. In the 1980s, elucidation of defects in the type I collagen genes in osteogenesis imperfecta paved the way to further assessment of matrix protein defects in the skeletal dysplasias [20]. A large number of matrix protein defects have now been described in the chondrodysplasias, including types I, II, IX, X, and XI collagen; matrillin 3; COMP; perlecan; and aggrecan. In addition to these defects in extracellular (matrix) structural proteins, the chondrodysplasias can also be pathogenetically grouped into disorders with defects in metabolic pathways (including enzymes, ion channels and transporters); defects in folding, processing, transport, and degradation of macromolecules; defects in hormones, growth factors, receptors and signal transduction; defects in nuclear proteins (transcription factors, homeobox genes); defects in RNA processing and metabolism; and defects in cytoskeletal proteins [21, 22].

As the underlying molecular pathogenesis of phenotypically grouped entities are unraveled, the entities are reclassified according to gene or molecular pathway. Two newly formed groups (the TRPV4 group and the aggrecan group) in the latest Nosology are perfect examples (see Table 95.1). The TRPV4 group includes disorders previously classified under different groups and provides a new prototypic spectrum ranging from mild to lethal (from autosomal dominant brachyolmia to metaphyseal dysplasia Kozlowski type to metatropic dysplasia) [19]. Identification of the molecular basis of many skeletal dysplasias has demonstrated that mutations in the same gene can produce quite distinct phenotypes, such as achondrogenesis II [MIM #200610] and Stickler syndrome [MIM #108300] with mutations in COL2A1 [21]. In contrast, similar phenotypes can be produced by mutations in different genes acting through a similar pathway. For example, Noggin and GDF5 cause multiple synostosis. As the underlying molecular basis of phenotypically grouped entities are unraveled, they are reclassified according to gene or molecular pathway (see Table 95.1).

Table 95.1. Osteochondrodysplasias [Nosology and Classification of Genetic Skeletal Disorders: 2010 Revision]

Name of Disorder	Inheritance	MIM No.	Locus	Gene	Protein
1. FGFR3 group					
Thanatophoric dysplasia type 1 (TD1)	AD	187600	4p16.3	FGFR3	FGFR3
Thanatophoric dysplasia type 2 (TD2)	AD	187601	4p16.3	FGFR3	FGFR3
SADDAN (severe achondroplasia-developmental delay- acanthosis nigricans)	AD	See 134934	4p16.3	FGFR3	FGFR3
Achondroplasia	AD	100800	4p16.3	FGFR3	FGFR3
Hypochondroplasia	AD	146000	4p16.3	FGFR3	FGFR3
Hypochondroplasia-like dysplasia	AD, SP				
Camptodactyly, tall stature, and hearing loss syndrome (CATSHL)	AD	187600	4p16.3	FGFR3	FGFR3
2. Type 2 collagen group					
Achondrogenesis type 2 (ACG2, Langer-Saldino)	AD	200610	12q13.1	COL2A1	Type 2 collagen
Platyspondylic dysplasia, Torrance type	AD	151210	12q13.1	COL2A1	Type 2 collagen
Hypochondrogenesis	AD	200610	12q13.1	COL2A1	Type 2 collagen
Spondyloepiphyseal dysplasia congenital (SEDC)	AD	183900	12q13.1	COL2A1	Type 2 collagen
Spondyloepimetaphyseal dysplasia (SEMD) Strudwick type	AD	184250	12q13.1	COL2A1	Type 2 collagen
Kniest dysplasia	AD	156550	12q13.1	COL2A1	Type 2 collagen
Spondyloperipheral dysplasia	AD	271700	12q13.1	COL2A1	Type 2 collagen
Mild SED with premature onset arthrosis	AD		12q13.1	COL2A1	Type 2 collagen
SED with metatarsal shortening (formerly Czech dysplasia)	AD	609162	12q13.1	COL2A1	Type 2 collagen
Stickler syndrome type 1	AD	108300	12q13.1	COL2A1	Type 2 collagen
Stickler-like syndrome					
3. Type 2 collagen group					
Stickler syndrome type 2	AD	604841	1p21	COL11A1	Type 2 collagen alpha-1 chain
Marshall syndrome	AD	154780	1p21	COL11A1	Type 2 collagen alpha-1 chain
Fibrochondrogenesis	AR	228520	1p21	COL11A1	Type 2 collagen alpha-1 chain
Otospondylomegaepiphyseal dysplasia (OSMED), recessive type	AR	215150	6p21.3	COL11A2	Type 2 collagen alpha-2 chain
Otospondylomegaepiphyseal dysplasia (OSMED), dominant type (Weissenbacher-Zweymüller syndrome, Stickler syndrome type 3)	AD	215150	6p21.3	COL11A2	Type 2 collagen alpha-2 chain
4. Sulphation disorders group					
Achondrogenesis type 1B (ACG1B)	AR	600972	5q32-33	DTDST	SLC26A2 sulfate transporter
Atelosteogenesis type 2 (AO2)	AR	256050	5q32-33	DTDST	SLC26A2 sulfate transporter
Diastrophic dysplasia (DTD)	AR	222600	5q32-33	DTDST	SLC26A2 sulfate transporter
MED, autosomal recessive type (rMED; EDM4)	AR	226900	5q32-33	DTDST	SLC26A2 sulfate transporter
SEMD, PAPSS2 type	AR	603005	10q23-q24	PAPSS2	PAPS-Synthetase 2

Table 95.1. (Continued)

Name of Disorder	Inheritance	MIM No.	Locus	Gene	Protein
Chondrodysplasia with congenital joint dislocations, CHST3 type (recessive Larsen syndrome)	AR	608637	10q22.1	CHST3	Carbohydrate sulfotransferase 3; chondroitin 6-sulfotransferase
Ehlers-Danlos syndrome, CHST14 type ("musculo-skeletal variant")	AR	601776	15q14	CHST14	Carbohydrate sulfotransferase 14; dermatan 4-sulfotransferase
5. Perlecan group					
Dyssegmental dysplasia, Silverman-Handmaker type	AR	224410	1q36-34	PLC (HSPG2)	Perlecan
Dyssegmental dysplasia, Rolanc-Desbuqois	AR	224400	1q36-34	PLC (HSPG2)	Perlecan
Schwartz-Jampel syndrome (myotonic chondrodystrophy)	AR	255800	1q36-34	PLC (HSPG2)	Perlecan
6. Aggrecan group					
SED, Kimberley type	AD	608361	15q26	AGC1	Aggrecan
SEMD, Aggrecan type	AR	612813	15q26	AGC1	Aggrecan
Familial osteochondritis dissecans	AD	165800	15q26	AGC1	Aggrecan
7. Filamin group and related disorders					
Frontometaphyseal dysplasia	XLD	305620	Xq28	FLNA	Filamin A
Osteodysplasty Melnick-Needles	XLD	309350	Xq28	FLNA	Filamin A
Otopalatodigital syndrome type 1 (OPD1)	XLD	311300	Xq28	FLNA	Filamin A
Otopalatodigital syndrome type 2 (OPD2)	XLD	304120	Xq28	FLNA	Filamin A
Atelosteogenesis type 1 (AO1)	AD	108720	3p14.3	FLNB	Filamin B
Atelosteogenesis type 3 (AO3)	AD	108721	3p14.3	FLNB	Filamin B
Larsen syndrome	AD	150250	3p14.3	FLNB	Filamin B
Spondylo-carpal-tarsal dysplasia	AR	272460	3p14.3	FLNB	Filamin B
Franck-ter-Haar syndrome	AR	249420	5q35.1	SH3PXD28	TKS4
Serpentine fibula-polycystic kidney syndrome	AD?	600330			
8. TRPV4 group					
Metatropic dysplasia	AD	156530	12q24.1	TRPV4	Transient receptor potential cation channel, subfamily V, member 4
Spondyloepimetaphyseal dysplasia, Maroteaux type (Pseudo-Morquio syndrome type 2)	AD	184095	12q24.1	TRPV4	Transient receptor potential cation channel, subfamily V, member 4

(Continued)

Table 95.1. (Continued)

Name of Disorder	Inheritance	MIM No.	Locus	Gene	Protein
Spondylometaphyseal dysplasia, Kozlowski type	AD	184252	12q24.1	TRPV4	Transient receptor potential cation channel, subfamily V, member 4
Brachyolmia, autosomal dominant type	AD	113500	12q24.1	TRPV4	Transient receptor potential cation channel, subfamily V, member 4
Familial digital arthropathy with brachydactyly	AD	606835	12q24.1	TRPV4	Transient receptor potential cation channel, subfamily V, member 4
9. Short-rib dysplasias (with or without polydactyly) group					
Chondroectodermal dysplasia (Ellis-van Creveld)	AR	225500	4p16	EVC1	EvC gene 1
			4p16	EVC2	EvC gene 2
SRP type 1/3 (Saldino-Noonan/Verma-Naumoff)	AR	263510	11q22.3	DYNC2H1	Dynein, cytoplasmic 2, heavy chain 1
SRP type 1/3 (Saldino-Noonan/Verma-Naumoff)	AR	263510	3q25.33	IFT80	Intraflagellar transport 80 (homolog of)
SRP type 1/3 (Saldino-Noonan/Verma-Naumoff)	AR	263510			
SRP type 2 (Majewski)	AR	263520			
SRP type 4 (Beemer)	AR	269860			
Oral-facial-digital syndrome type 4 (Mohr-Majewski)	AR	258860			
Asphyxiating thoracic dysplasia (ATD; Jeune)	AR	208500	11q22.3	DYNC2H1	Dynein, cytoplasmic 2, heavy chain 1
Asphyxiating thoracic dysplasia (ATD; Jeune)	AR	208500	3q25.33	IFT80	Intraflagellar transport 80 (homolog of)
Asphyxiating thoracic dysplasia (ATD; Jeune)	AR	208500			
Thoracolaryngopelvic dysplasia (Barnes)	AD	187760			
10. Multiple epiphyseal dysplasia and pseudoachondroplasia group					
Pseudoachondroplasia (PSACH)	AD	177170	19p12-13.1	COMP	COMP
Multiple epiphyseal dysplasia (MED) type 1 (EDM1)	AD	132400	19p13.1	COMP	COMP
Multiple epiphyseal dysplasia (MED) type 2 (EDM2)	AD	600204	1p32.2-33	COL9A2	Collagen 9 alpha-2 chain
Multiple epiphyseal dysplasia (MED) type 3 (EDM3)	AD	600969	20q13.3	COL9A3	Collagen 9 alpha-3 chain
Multiple epiphyseal dysplasia (MED) type 5 (EDM5)	AD	607078	2p23-24	MATN3	Matrilin 3
Multiple epiphyseal dysplasia (MED) type 6 (EDM6)	AD	120210	6q13	COL9A1	Collagen 9 alpha-1 chain
Multiple epiphyseal dysplasia (MED), other types					
Stickler syndrome, recessive type	AR	120210	6q13	COL9A1	Collagen 9 alpha-1 chain
Familial hip dysplasia (Beukes)	AD	142669	4q35		

Table 95.1. (Continued)

Name of Disorder	Inheritance	MIM No.	Locus	Gene	Protein
Multiple epiphyseal dysplasia with microcephaly and nystagmus (Lowry-Wood)	AR	226960			
11. Metaphyseal dysplasias					
Metaphyseal dysplasia, Schmid type (MCS)	AD	156500	6q21-22.3	COL10A1	Collagen 10 alpha-1 chain
Cartilage-hair-hypoplasia (CHH; metaphyseal dysplasia, McKusick type)	AR	250250	9p13	RMRP	RNA component of RNAse H
Metaphyseal dysplasia, Jansen type	AD	156400	3p22-21.1	PTHR1	PTH/PTHrP receptor 1
Eiken dysplasia	AR	600002	3p22-22.1	PTHR1	PTH/PTHrP receptor 1
Metaphyseal dysplasia with pancreatic insufficiency and cyclic neutropenia (Shwachman-Bodian-Diamond syndrome, SBDS)	AR	260400	7q11	SBDS	SBDS gene, function unclear
Metaphyseal anadysplasia type 1	AD, AR	309645	11q22	MMP13	Matrix metalloproteinase 13
Metaphyseal anadysplasia type 2	AR		20q13.12	MMP9	Matrix metalloproteinase 9
Metaphyseal dysplasia, Spahr type	AR	250215			
Metaphyseal acroscyphodysplasia (various types)	AD/SP	250400			
Genochondromatosis (type1/type 2)	AD/SP	137360			
Metaphyseal chondromatosis with D-2-hydroxyglutaric aciduria	AR/SP	271550			
12. Spondylometaphyseal dysplasias (SMD)					
Spondyloenchondrodysplasia (SPENCD)	AR	271550	19p13.2	ACP5	Tartrate-resistant acid phosphatase (TRAP)
Odontochondrodysplasia (ODCD)	AR	184260			
Spondylometaphyseal dysplasia Kozlowski type	AD	184252			
Spondylometaphyseal dysplasia, Sutcliffe/corner fracture type	AD	184255			
SMD with severe genu valgum	AD	184253			
SMD with cone-rod dystrophy	AR	608940			
SMD with retinal degeneration, axial type	AR	602271			
Dysspondyloenchondromatosis	SP				
Cheiro-spondyloenchondromatosis	SP				
13. Spondylo-epi(-metaphyseal dysplasias (SE(M)D)					
Dyggve-Melchior-Clausen dysplasia (DMC)	AR	223800	18q12-21.1	DYM	Dymeclin
Immuno-osseous dysplasia (Schimke)	AR	242900	2q34-36	SMARCAL1	SWI/SNF-related regulator of chromatin subfamily A-like protein 1

(Continued)

Table 95.1. (Continued)

Name of Disorder	Inheritance	MIM No.	Locus	Gene	Protein
SED Wolcott-Rallison type	AR	226980	2p12	EIF2AK3	Translation initiation factor 2-alpha kinase-3
SEMD Matrilin type	AR	608728	2p23-p24	MATN3	Matrilin 3
SEMD Missouri type	AD	602111	11q22.3	MMP13	Matrix metalloproteinase 13
Metatropic dysplasia (various forms)	AD/AR	156530			
SED tarda, X-linked (SED-XL)	XLR	313400	Xp22	SEDL	Sedlin
SPONASTRIME dysplasia	AR	271510			
SEMD short limb—abnormal calcification type	AR	271665	1q23	DDR2	Discoidin domain receptor family, member2
SEMD with joint laxity (SEMD-JL) Beighton type	AR	271640			
Spondylo-megaepiphyseal-metaphyseal dysplasia (SMMD)	AR	613330	4p16.1	NKX3	NK3 Homeobox
Spondylodysplastic Ehlers-Danlos syndrome	AR	271510	11p.11.2	SLC39A13	Zinc transporter ZIP13
SEMD with joint laxity (SEMD-JL) leptodactylic or Hall type	AD	603546			
Platyspondyly (brachyolmia) with amelogenesis imperfecta	AR	601216			
Late onset SED, autosomal recessive type	AR	609223			
Brachyolmia, Hobaek, and Toledo types	AR	271530, 271630			
14. Severe spondylodysplastic dysplasias					
Achondrogenesis type 1A (ACG1A)	AR	200600	14q32.12	TRIP11	Golgi-microtubule-associated protein, 210-kDa; GMAP210
SMD Sedaghatian type	AR	250220	7q11	SBDS	SBDS gene, function still unclear
Severe SMD Sedaghatian-like	AR				
Opsismodysplasia	AR	258480			
Schneckenbecken dysplasia	AR	269250	1p31.3	SLC35D1	Solute carrier family 35 member D1; UDP-glucuronic acid/UDP-N-acetylgalactosamine dual trasporter
15. Acromelic dysplasias					
Trichorhinophalangeal dysplasia types 1/3	AD	190350 190351	8q24	TRPS1	Zinc finger transcription factor
Trichorhinophalangeal dysplasia type 2 (Langer-Giedion)	AD	150230	8q24	TRPS1	Zinc finger transcription factor
				EXT1	Exostosin 1
Acrocapitofemoral dysplasia	AR	607778	2q33-q35	IHH	Indian hedgehog

Table 95.1. (Continued)

Name of Disorder	Inheritance	MIM No.	Locus	Gene	Protein
Cranioectodermal dysplasia (Levin-Sensenbrenner) type 1	AR	218330		WDR35	WD repeat-containing protein 35
Cranioectodermal dysplasia (Levin-Sensenbrenner) type 2	AR	613610	2p24.1		
Geleophysic dysplasia	AR	231050	9q34.2	ADAMTSL2	ADAMTS-like protein 2
Geleophysic dysplasia, other types	AR	102370			
Acromicric dysplasia	AD	101800			
Acrodysostosis	AD	105835			
Angel-shaped phalangoepiphyseal dysplasia (ASPED)	AD				
Acrolaryngeal dysplasia	AD				
Craniofacial conodysplasia	AD	606835			
Familial digital arthropathy with brachydactyly	AD	266920			
Saldino-Mainzer dysplasia	AR				
16. Acromesomelic dysplasias					
Acromesomelic dysplasia type Maroteaux	AR	602875	9p13-12	NPR2	Natriuretic peptide receptor 2
Grebe dysplasia	AR	200700	20q11.2	GDF5	Growth and differentiation factor 5
Fibular hypoplasia and complex brachydactyly (Du Pan)	AR	228900	20q11.2	GDF5	Growth and differentiation factor 5
Acromesomelic dysplasia with genital anomalies	AR	609441	4q23-24	BMPR1B	Bone morphogenetic protein receptor 1B
Acromesomelic dysplasia, Osebold-Remondini type	AD	112910			
17. Mesomelic and rhizo-mesomelic dysplasias					
Dyschondrosteosis (Leri-Weill)	Pseudo-AD	127300	Xpter-p22.32	SHOX	Short stature—homeobox gene
Langer type (homozygous dyschondrosteosis)	Pseudo-AR	249700	Xpter-p22.32	SHOX	Short stature—homeobox gene
Robinow syndrome, recessive type	AR	268310	9q22	ROR2	Receptor tyrosine kinase-like orphan receptor 2
Robinow syndrome, dominant type	AD	180700			
Mesomelic dysplasia, Korean type	AD	156232	2q24-32		
Mesomelic dysplasia, Kantaputra type	AD	163400	2q24-32		
Mesomelic dysplasia, Nievergelt type	AD				
Mesomelic dysplasia, Kozlowski-Reardon type	AR	249710			
Mesomelic dysplasia with acral synostoses (Verloes-David-Pfeiffer type)	AD	600383	8q13	SULF1 and SLCO5A1	Heparan sulfatase 6-O-endosulfatase 1 and solute carrier organic anion transporter family member 5A1

(Continued)

Table 95.1. (Continued)

Name of Disorder	Inheritance	MIM No.	Locus	Gene	Protein
Mesomelic dysplasia, Savarirayan type (Triangular Tibia-Fibular Aplasia)	SP	605274			
18. Bent bones dysplasias					
Campomelic dysplasia (CD)	AD	114290	17q24.3-25.1	SOX9	SRY-box 9
Stüve-Wiedemann dysplasia	AR	601559	5p13.1	LIFR	Leukemia inhibitory factor receptor
Cumming syndrome		211890			
Kyphomelic dysplasia, several forms		211350			
Bent bones at birth can be seen in a variety of conditions, including Antley-Bixler syndrome, cartilage-hair hypoplasia, hypophosphatasia, osteogenesis imperfecta, dyssegmental dysplasia, and others					
19. Slender bone dysplasia group					
3-M syndrome (3M1)	AR	273750	6p21.1	CUL7	Cullin 7
3-M syndrome (3M2)	AR	619921	2q35	PBSL1	Obscurin-like 1
Kenny-Caffey dysplasia type 1	AR	244460	1q42-q43	TBCE	tubulin-specific chaperone E
Kenny-Caffey dysplasia type 2	AD	127000			
Microcephalic osteodysplastic primordial dwarfism type 1/3 (MOPD1)	AR	210710	2q		
Microcephalic osteodysplastic primordial dwarfism type 2 (MOPD2; Majewski type)	AR	210720	21q	PCNT2	Pericentrin 2
Microcephalic osteodysplastic dysplasia, Saul-Wilson type	AR				
IMAGE syndrome (Intrauterine Growth Retardation, Metaphyseal Dysplasia, Adrenal Hypoplasia, and Genital Anomalies)	XL/AD	300290			
Osteocraniostenosis	SP	602361			
Hallermann-Streiff syndrome	AR	234100			
20. Dysplasias with multiple joint dislocations					
Desbuquois dysplasia (with accessory ossification center in digit 2)	AR	251450	17q25.3	CANT1	
Desbuquois dysplasia with short metacarpals and elongated phalanges)	AR	251450	17q25.3	CANT1	
Desbuquois dysplasia (other variants with or without accessory ossification center	AR				
Pseudodiastrophic dysplasia	AR	264180			

Modified and reproduced from Warman ML, Cormier-Daire V, Hall C, Krakow D, Lachman R, LeMerrer M, Mortier G, Mundlos S, Nishimura G, Rimoin DL, Robertson S, Savarirayan R, Sillence D, Spranger I, Unger S, Zabel B, Superti-Furga A. Nosology and Classification of Genetic Skeletal Disorders: 2010 Revision. Am J Med Genet 155A: 943–968. Used with permission.

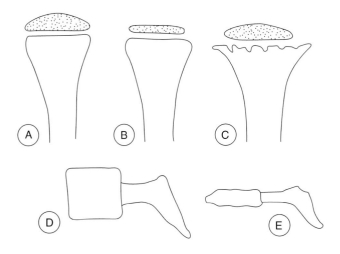

Involvement	Disease Category
A+D	Normal
B+D	Epiphyseal dysplasia
C+D	Metaphyseal dysplasia
B+E	Spondyloepiphyseal dysplasia
C+E	Spondylometaphyseal dysplasia
B+C+E	Spondyloepimetaphyseal dysplasia

Fig. 95.1. Chondrodysplasias. Classification based on radiographic involvement of long bones and vertebrae (from Ref. 3).

The technology today warrants rapid identification of novel disease-causing genes with new sequencing tools. It is now apparent that no single classification of these disorders will be adequate, necessitating the development of a multidimensional electronic classification tool, incorporating clinical, radiographic, morphologic, biochemical, molecular, and pathway data. The transformation of the Nosology into a database is therefore likely.

Accumulating knowledge on the biology of bone and cartilage and the molecular and pathogenetic defects in each of these disorders provides hope for possible therapeutic approaches in the near future for affected individuals. Finally, research in this group of monogenic disorders continues to provide genetic models for more common multifactorial complex diseases of the bones and joints, such as osteoarthritis, osteoporosis, scoliosis, and disc herniation.

REFERENCES

1. Spranger J, Brill P, Poznanski A. 2002. *Bone Dysplasias. An Atlas of Genetic Disorders of Skeletal Development*, 2nd Ed. Oxford: Oxford University Press.

2. Orioli IM, Castilla EE, Barbosa-Neto JG. 1986. The birth prevalence rates for skeletal dysplasias. *J Med Genet* 23: 328–332.

3. Unger S, Lachman RS, Rimoin DL. 2007. Chondrodysplasias. In: Rimoin DL, Connor JM, Pyeritz RE, Korf B (eds.) *Emery and Rimoin's Principles and Practice of Medical Genetics*, 5th Ed., Vol. 3. Philadelphia: Elsevier. pp. 3709–3753.

4. Spranger J. 1989. Radiologic nosology of bone dysplasias. *Am J Med Genet* 34: 96–104.

5. Superti-Furga A, Unger S. 2007. Nosology and classification of genetic skeletal disorders: 2006 revision. *Am J Med Genet A* 143: 1–18.

6. Warman ML, Cormier-Daire V, Hall C, Krakow D, Lachman R, LeMerrer M, Mortier G, Mundlos S, Nishimura G, Rimoin DL, Robertson S, Savarirayan R, Sillence D, Spranger J, Unger S, Zabel B, Superti-Furga A. 2011. Nosology and classification of genetic skeletal disorders: 2010 revision. *Am J Med Genet* 155A: 943–968.

7. Orioli IM, Castilla EE, Barbosa-Neto JG. 1986. The birth prevalence rates for the skeletal dysplasias. *J Med Genet* 23: 328–332.

8. Chu ML, Williams CJ, Pepe G, Hirsch JL, Prockop DJ, Ramirez F. 1983. Internal deletion in a collagen gene in a perinatal lethal form of osteogenesis imperfecta. *Nature* 304: 78–80.

9. Lee B, Vissing H, Ramirez F, Rogers D, Rimoin D. 1989. Identification of the molecular defect in a family with spondyloepiphyseal dysplasia. *Science* 244: 978–980.

10. Tompson SW, Merriman B, Funari VA, Fresquet M, Lachman RS, Rimoin DL, Nelson SF, Briggs MD, Cohn DH, Krakow D. 2009. A recessive skeletal dysplasia, SEMD aggrecan type, results from a missense mutation affecting the C-type lectin domain of aggrecan. *Am J Hum Genet* 84: 72–79.

11. Tompson SW, Bacino CA, Safina NP, Bober MB, Proud VK, Funari T, Wangler MF, Nevarez L, Ala-Kokko L, Wilcox WR, Eyre DR, Krakow D, Cohn DH. 2010. Fibrochondrogenesis results from mutations in the COL11A1 type XI collagen gene. *Am J Hum Genet* 87: 708–712.

12. Thiele BJ, Doller A, Kähne T, Pregla R, Hetzer R, Regitz-Zagrosek V. 2004. RNA-binding proteins heterogeneous nuclear ribonucleoprotein A1, E1, and K are involved in post-transcriptional control of collagen I and III synthesis. *Circ Res* 95: 1058–1066.

13. Morello R, Bertin TK, Chen Y, Hicks J, Tonachini L, Monticone M, Castagnola P, Rauch F, Glorieux FH, Vranka J, Bächinger HP, Pace JM, Schwarze U, Byers PH, Weis M, Fernandes RJ, Eyre DR, Yao Z, Boyce BF, Lee B. 2006. CRTAP is required for prolyl 3-hydroxylation and mutations cause recessive osteogenesis imperfecta. *Cell* 127: 291–304.

14. Alanay Y, Avaygan H, Camacho N, Utine GE, Boduroglu K, Aktas D, Alikasifoglu M, Tuncbilek E, Orhan D, Bakar FT, Zabel B, Superti-Furga A, Bruckner-Tuderman L, Curry CJ, Pyott S, Byers PH, Eyre DR, Baldridge D, Lee B, Merrill AE, Davis EC, Cohn DH, Akarsu N, Krakow D. 2010. Mutations in the gene encoding the

RER protein FKBP65 cause autosomal-recessive osteogenesis imperfecta. *Am J Hum Genet* 86: 551–559; Erratum 87: 572–573.
15. Smits P, Bolton AD, Funari V, Hong M, Boyden ED, Lu L, Manning DK, Dwyer ND, Moran JL, Prysak M, Merriman B, Nelson SF, Bonafé L, Superti-Furga A, Ikegawa S, Krakow D, Cohn DH, Kirchhausen T, Warman ML, Beier DR. 2010. Lethal skeletal dysplasia in mice and humans lacking the golgin GMAP-210. *N Engl J Med* 362: 206–216.
16. Mortier GR. 2001. The diagnosis of skeletal dysplasias: A multidisciplinary approach. *Eur J Radiol* 40: 161–167.
17. Lachman RS. 2007. *Taybi and Lachman's Radiology of Syndromes, Metabolic Disorders and Skeletal Dysplasias, 5th Ed.* Philadelphia: Elsevier.
18. Ikegawa S. 2006. Genetic analysis of skeletal dysplasia: Recent advances and perspectives in the post-genome-sequence era. *J Hum Genet* 51: 581–586.
19. Krakow D, Rimoin DL. 2010. The skeletal dysplasias. *Genet Med* 12: 327–341.
20. Rimoin DL, Cohn D, Krakow D, Wilcox W, Lachman RS, Alanay Y. 2007. The skeletal dysplasias clinical-molecular correlations. *Ann NY Acad Sci* 1117: 302–309.
21. Williams CJ, Prockop DJ. 1983. Synthesis and processing of a type I procollagen containing shortened pro-alpha 1(I) chains by fibroblasts from a patient with osteogenesis imperfecta. *J Biol Chem* 258: 5915–5921.
22. Superti-Furga A, Bonafe L, Rimoin DL. 2001. Molecular-pathogenetic classification of genetic disorders of the skeleton. *Am J Med Genet* 106: 282–293.

RECOMMENDED DATABASES

International Skeletal Dysplasia Registry: www.csmc.edu/skeletaldysplasia
International Skeletal Dysplasia Society: www.isds.ch/
Genetests: www.genetests.com

96

Ischemic and Infiltrative Disorders

Richard W. Keen

Ischemic Bone Disease (Osteonecrosis) 805
Infiltrative Disorders 807

References 808

ISCHEMIC BONE DISEASE (OSTEONECROSIS)

Introduction

Regional interruption of blood flow to the skeleton can cause ischemic (aseptic or avascular) necrosis. Ischemia, if sufficiently severe and prolonged, will kill osteoblasts and chondrocytes. Clinical problems will arise if subsequent resorption of necrotic tissue during skeletal repair compromises the bone strength sufficiently to cause fracture with subsequent deformity of bone and secondary damage to cartilage.

Epidemiology

There are little accurate data on the incidence of osteonecrosis, although it is estimated there are approximately 15,000 new cases per year in the U.S. [1]. The disease appears to occur more frequently in males than in females, with the overall male to female ratio being 8:1. The age of onset is variable, although in the majority of cases, the patient is less than 50 years of age. The average age of female cases is on average 10 years older than male cases.

Pathogenesis

Osteonecrosis is often seen in association with a number of different conditions (Table 96.1). Trauma with fracture of the femoral neck interrupts the major part of the blood supply to the head and may lead to osteonecrosis. Glucocorticoids and alcoholism are two of the main iatrogenic factors known to predispose to osteonecrosis.

An increasing number of reports have appeared in the medical literature that suggest that the use of bisphosphonates, especially intravenous bisphosphonates, is associated with osteonecrosis of the jaw (ONJ) [2]. There is no universally accepted definition of ONJ, and it may represent a distinct form of osteonecrosis, different from the other forms. Clinically, it typically appears as an area of exposed alveolar bone that can occur in the mandible or the maxilla. The majority of these reports are in patients with multiple myeloma, breast cancer, or other malignancies. In connection with these malignancies, the patients were receiving or did receive intravenous nitrogen-containing bisphosphonates in much higher doses than those used for osteoporosis and Paget's disease. The condition has also been seen in those receiving treatment with denosumab, a monoclonal antibody against receptor activator of nuclear factor-κB ligand (RANKL) [3].

A mechanical interruption to the bone's blood supply is common to most of the conditions associated with osteonecrosis. For many types of nontraumatic ischemic necrosis, the predisposed sites seem to reflect the physiological conversion of red marrow to fatty marrow with aging [4]. This process occurs from distal to proximal in the appendicular skeleton. As the transition occurs, marrow blood flow decreases. Accordingly, disorders that increase the size and/or number of adipocytes within critical areas of medullary space (e.g., alcohol abuse,

---

Primer on the Metabolic Bone Diseases and Disorders of Mineral Metabolism, Eighth Edition. Edited by Clifford J. Rosen.
© 2013 American Society for Bone and Mineral Research. Published 2013 by John Wiley & Sons, Inc.

Table 96.1. Causative Factors Associated with Osteonecrosis

Traumatic
Fracture of the femoral neck
Dislocation or fracture-dislocation of the hip
Minor fracture

Nontraumatic
Alcohol
Arteriosclerosis and other occlusive vascular disorders
Bisphosphonates
Carbon tetrachloride poisoning
Connective tissue diseases
Cushing's disease (OMIM 219090; OMIM 219080)
Denosumab (antibody to RANKL)
Diabetes mellitus (OMIM 222100, OMIM 125853)
Disordered lipid metabolism
Dysplasia
Fatty liver
Gaucher's disease (OMIM 231000)
Glucocorticoid treatment
Human immunodeficiency virus
Dysbaric conditions
Hyperuricemia and gout
Legg Calve Perthe Disease (OMIM 150600)
Osteomalacia
Pancreatitis
Pregnancy
Radiotherapy
Sickle cell anemia (OMIM 603903)
Systemic lupus erythematosus (SLE) (OMIM 152700)
Solid organ transplantation
Thrombophlebitis
Tumors

Cushing's syndrome) may ultimately compress sinusoids, thereby leading to infarction of bone. Other factors potentially involved in the pathogenesis of osteonecrosis include fat embolization, hemorrhage, and abnormalities in the quality of susceptible bone tissue. Infection and dental trauma appear important factors in the etiology of ONJ.

A genetic basis to idiopathic osteonecrosis of the femoral head is suggested by the occurrence of disease in twins and a clustering of cases in families [5, 6]. Increased incidence of osteonecrosis in specific animal models also provides further evidence of the existence of susceptibility genes [7]. In sporadic cases of osteonecrosis of the femoral neck, a number of genetic association studies have been conducted, linking specific genes to the pathogenesis of disease. The majority of the studies have, to date, focused on gene polymorphisms affecting the coagulation and fibrinolytic system.

The factor V Leiden mutation (G1691A, Arg506Gln) is a common risk factor for thrombophilia. Three out of four studies investigating the role of the mutation in osteonecrosis have reported a positive correlation of factor V Leiden with primary osteonecrosis [8–11]. Plasminogen-activating inhibitor-1 (PAI-1) was also studied in patients with osteonecrosis. Homozygosity for the 4G allele (4G/4G) has been reported to significantly increase the plasma PAI-1 level, and in two studies by the same group of investigators, the 4G/4G allele was found to be a risk factor of osteonecrosis [11, 12]. Some studies have also investigated the role of 5,10-methylenetetrahydrofolate reductase (MTHFR) gene polymorphism. The MTHFR C677T variant was overrepresented in some groups of primary studies of osteonecrosis, but not in all [11–13]. In another study, the first that did not focus on the coagulation system, the role of endothelial nitric oxide synthase (eNOS) was investigated [14]. Nitric oxide (NO) synthesized by eNOS has vasodilatory effects on vascular tone, inhibits platelet aggregation, and modulates smooth muscle proliferation. Allele 4a of a variable number tandem repeat (VNTR) polymorphism in intron 4 of the eNOS gene was found to be a risk factor for idiopathic osteonecrosis at the hip.

Genetic mutations have been identified in three families with osteonecrosis and dominant inheritance. Mutations in the type II collagen (COL2A2) gene (mapped on chromosome 12q13) proved to be the genetic cause of the disease. Type II collagen is the major structural protein in the extracellular matrix of cartilage [15].

Clinical features

The clinical presentation will be dependent on many factors, including the age of the patient, anatomical site of involvement, and the extent and severity of this involvement. The femoral head is the most common location for the development of osteonecrosis, although it may also occur at other sites including distal femur, humeral head, wrist, and foot. Patients may develop pain that can persist for weeks to months before radiographs show any change, although patients can be asymptomatic. Avascular necrosis at the hip classically will present with pain in the groin, although this can be referred to the buttock, thigh, or knee. The pain is exacerbated by weight bearing but can also be present at rest. Gait may be affected, and patients can present with a limp. Once the femoral head has begun to collapse, range of hip movement will be reduced, and leg shortening may develop.

In osteonecrosis of the hip, involvement of the contralateral hip is present in 30–70% of cases at the time of first examination. Within 3 years of diagnosis, more than 50% of cases will have progressed in the contralateral hip to such a stage where surgical intervention is required [16, 17].

Radiological features

In the earliest stages of the disease, plain radiographs will be normal. Magnetic resonance imaging (MRI) is useful to detect early pathological changes. The most characteristic image is a margin of low signal on T1- and

Ischemic and Infiltrative Disorders

Table 96.2. Staging of Osteoenecrosis

Stage	Findings	Techniques
0	All techniques normal or nondiagnostic Necrosis on biopsy	Biopsy and histology
1	Radiographs and CT normal Positive result from at least one of the additional investigations listed opposite	Radionucleotide scan MRI Biopsy and histology
2	Radiographic abnormalities without collapse (sclerosis, cysts, osteopenia)	Radionucleotide scan MRI Biopsy and histology
3	Crescent sign	Radiographs CT
4	Flattening or evident collapse	Radiographs CT
5	As for stage 4, with narrowing of joint space	Radiographs
6	As for stage 5, with destruction of joint	Radiographs

T2-weighted images, and is observed in 60–80% of cases. Radionucleotide isotope bone scans and computed tomography (CT) can also be used in cases where MRI is either contraindicated or has been inconclusive. Osteonecrosis can be staged according to the sequence of the radiological changes. These are detailed in Table 96.2.

Laboratory findings

In idiopathic osteonecrosis, laboratory investigations will generally be normal. Investigations may reveal potential contributory factors such as connective tissue disease, diabetes, hyperlipidaemia, coagulopathies, and gout.

Histopathological examination of tissue from affected bone is consistent with the pathogenesis that is suggested from the radiographic examinations. It shows that these various processes of skeletal death and repair are focal and may be occurring simultaneously [18].

Differential diagnosis

In stages 3 and 4 of the disease, the radiological features of the disease are specific. In the later stages 5 and 6, a differential diagnosis is not necessary as by this stage the bone and joint have been irreversibly damaged, and the only treatment option would be joint replacement. In the earlier stages of the disease (1 and 2), other diseases of bone, cartilage, and synovial tissue should be considered in the differential diagnosis.

Treatment

Medical treatment includes non-weight-bearing for osteonecrosis affecting load-bearing bones. This may be for between 4 and 8 weeks. Vasoactive drugs such as prostacyclin may play a role in early stages of osteonecrosis [19]. Bisphosphonates have also been shown to be effective in the treatment of osteonecrosis. Data have been observed for alendronate [20] and zoledronate [21].

Surgical treatment for osteonecrosis involves core decompression. This reduces the intramedullary pressure within the ischemic bone and has been postulated to improve circulation. The outcome from core decompression at the femoral head, with regard to resolution of radiographic changes and improvement in symptoms, varies from between 34% and 95% in the early stages of the disease [22]. These results appear better than continuing with conservative measures such as non-weight-bearing.

INFILTRATIVE DISORDERS

Systemic mastocytosis

In systemic mastocytosis (OMIM 154800) there is widespread infiltration of tissues with mast cells. These cells can be widely disributed in nearly every organ, and originate from bone marrow progenitor cells. An activating mutation in the C-KIT gene coding for c-kit (the receptor for stem cell factor) that controls mast cell development has been identified [23]. The relationship, however, between these mutations and the clinical phenotype is not fully clarified.

The clinical features of mastocytosis are produced by liberation of mast cell products. Urticaria pigmentosa has been described in between 14% and 100% of patients with systemic mastocytosis. Radiographic appearances can be variable, and the diagiosis is often confirmed on bone biopsy and histological examination. The diagnosis can also be made by the measurement of urinary excretion of mast cell mediators such as N-methyl histamine.

Treatment of systemic mastocytosis must be "tailored" in individual patients [24, 25]. Severe bone pain from advanced bone disease has been reported to respond to radiotherapy [26]. In early trials, bisphosphonates have controlled pain and improved bone density [27].

Histiocytosis-X

"Histiocytosis-X" is the term used to unify what had been regarded as three distinct entities: Letterer-Siwe disease (OMIM 246400), Hand-Schüller-Christian disease (OMIM 267700), and eosinophilic granuloma [28, 29]. An immature, clonal Langerhans cell is considered the pathognomonic and linking feature, and the condition is now called Langerhans cell histiocytosis (OMIM 604856).

Many tissues and organs can be involved, including brain, lung, oropharynx, gastrointestinal tract, skin, and bone marrow. Prognosis is age-related; infants and the elderly have poorer outcomes. The signs and symptoms of the three principal clinical forms also differ.

Letterer-Siwe disease presents between several weeks and 2 years of age with hepatosplenomegaly, lymphadenopathy, anemia, hemorrhagic tendency, fever, failure to grow, and skeletal lesions. Hand-Schüller-Christian disease is a chronic condition that begins in early childhood, although symptoms may not manifest until the third decade. The classic triad of findings consists of exophthalmos, diabetes insipidus, and bony lesions, although this is only seen in 10% of cases. Eosinophilic granuloma occurs most frequently in children between 3 and 10 years of age, and it is rare after the age of 15 years. A solitary and painful lesion in a flat bone is the most common finding.

Histiocytosis-X tends to be benign and self-limiting when there is no systemic involvement. Treatment for severe disease includes chemotherapy, radiation therapy, and immunotherapy [30].

REFERENCES

1. Steinberg ME, Steinberg DR. 1991. Avascular necrosis of the femoral head. In: Steinberg ME (ed.) *The Hip and its Disorders*. Philadelphia: WB Saunders. pp. 623–647.
2. Khosla S, Burr D, Cauley J, Dempster DW, Ebeling PR, Felsenberg D, Gagel RF, Gilsanz V, Guise T, Koka S, McCauley LK, McGowan J, McKee MD, Mohla S, Pendrys DG, Raisz LG, Ruggiero SL, Shafer DM, Shum L, Silverman SL, Van Poznak CH, Watts N, Woo SB, Shane E. 2007. Bisphosphonate-associated osteonecrosis of the jaw: Report of a task force of the American Society for Bone and Mineral Research. *J Bone Miner Res* 22: 1479–1491.
3. Saad F, Brown JE, Van Poznak C, Ibrahim T, Stemmer SM, Stopeck AT, Diel IJ, Takahashi S, Shore N, Henry DH, Barrios CH, Facon T, Senecal F, Fizazi K, Zhou L, Daniels A, Carrière P, Dansey R. 2011. Incidence, risk factors, and outcomes of osteonecrosis of the jaw: Integrated analysis from three blinded active-controlled phase III trials in cancer patients with bone metastases. *Ann Oncol* 23(5): 1341–1347.
4. Edeiken J, Dalinka M, Karasick D. 1990. *Edeiken's Roentgen Diagnosis of Diseases of Bone, 4th Ed*. Baltimore, MD: Williams and Wilkins.
5. Glueck CJ, Glueck HI, Welch M, Freiberg, Tracy T, Hamer T, Stroop D. 1994. Familial idiopathic osteonecrosis mediated by familial hypofibrinolysis with high levels of plasminogen activator inhibitor. *Thromb Haemost* 71: 195–198.
6. Nobillot R, Le Parc JM, Benoit J, Paolaggi JB. 1994. Idiopathic osteonecrosis of the hip in twins. *Ann Rheum Dis* 53: 702.
7. Boss JH, Misselevich I. 2003. Osteonecrosis of the femoral head of laboratory animals: The lessons learned from a comparative study of osteonecrosis in man and experimental animals. *Vet Pathol* 40: 345–354.
8. Zalavras CG, Vartholomatos G, Dokou E, Malizos KN. 2004. Genetic background of osteonecrosis: Associated with thrombophilic mutations? *Clin Orthop Relat Res* 422: 251–255.
9. Bjorkman A, Svensson PJ, Hillarp A, Burtscher IM, Runow A, Benoni G. 2004. Factor V Leiden and prothrombin gene mutation: Risk factors for osteonecrosis of the femoral head in adults. *Clin Orthop Relat Res* 425: 168–172.
10. Bjorkman A, Burtscher IM, Svensson PJ, Hillarp A, Besjakov J, Benoni G. 2005. Factor V Leiden and the prothrombin 20210A gene mutation and osteonecrosis of the knee. *Arch Orthop Trauma Surg* 125: 51–55.
11. Glueck CJ, Fontaine RN, Gruppo R, Stroop D, Sieve-Smith L, Tracy T, Wang P. 1999. The plasminogen activator inhibitor-1 gene, hypofibrinolysis, and osteonecrosis. *Clin Orthop Relat Res* 366: 133–146.
12. Glueck CJ, Freiberg RA, Fontaine RN, Tracy T, Wang P. 2001. Hypofibrinolysis, thrombophilia, osteonecrosis. *Clin Orthop Relat Res* 386: 19–33.
13. Zalavras CG, Malizos KN, Dokou E, Vartholomatos G. 2002. The 677C-->T mutation of the methylenetetrahydrofolate reductase gene in the pathogenesis of osteonecrosis of the femoral head. *Haematologica* 87: 111–112.
14. Koo KH, Lee JS, Lee YJ, Kim KJ, Yoo JJ, Kim HJ. 2006. Endothelial nitric oxide synthase gene polymorphisms in patients with nontraumatic femoral head osteonecrosis. *J Orthop Res* 24: 1722–1728.
15. Liu YF, Chen WM, Lin YF, Yang RC, Lin MW, Li LH, Chang YH, Jou YS, Lin PY, Su JS, Huang SF, Hsiao KJ, Fann CS, Hwang HW, Chen YT, Tsai SF. 2005. Type II collagen gene variants and inherited osteonecrosis of the femoral head. *N Engl J Med* 352: 2294–2301.
16. Jacobs B. 1978. Epidemiology of traumatic and nontraumatic osteonecrosis. *Clin Orthop Rel Res* 130: 51–67.
17. Bradway JK, Morrey BF. 1993. The natural history of the silent hip in bilateral atraumatic necrosis of the femoral head. *J Arthroplasty* 8: 383–387.
18. Plenk H Jr, Hofmann S, Eschberger J, Gstettner M, Kramer J, Schneider W, Engel A. 1997. Histomorphology and bone morphometry of the bone marrow edema syndrome of the hip. *Clin Orthop* 334: 73–84.
19. Jäger M, Tillmann FP, Thornhill TS, Mahmoudi M, Blondin D, Hetzel GR, Zilkens C, Krauspe R. 2008. Rationale for prostaglandin I2 in bone marrow oedema—From theory to application. *Arthritis Res Ther* 10: R120.
20. Lai KA, Shen WJ, Yang CY, Shao CJ, Hsu JT, Lin RM. 2005. The use of alendronate to prevent early collapse of the femoral head in patients with nontraumatic osteonecrosis. A randomized clinical study. *J Bone Joint Surg Am* 87: 2155–2159.
21. Ramachandran M, Ward K, Brown RR, Munns CF, Cowell CT, Little DG. 2007. Intravenous bisphosphonate therapy for traumatic osteonecrosis of the

femoral head in adolescents. *J Bone Joint Surg Am* 89: 1727–1734.
22. Stulberg BN, Bauer TW, Belhobek GH. 1990. Making core decompression work. *Clin Orthop* 261: 186–195.
23. Metcalfe DD, Atkin C. 2001. Matsocytosis: Molecular mechanisms and clinical disease heterogeneity. *Leuk Res* 25: 577–582.
24. Bains SN, Hsieh FH. 2010. Current approaches to the diagnosis and treatment of systemic mastocytosis. *Ann Allergy Asthma Immunol* 104: 1–10.
25. Pardanani A, Tefferi A. 2010. Systemic mastocytosis in adults: A review on prognosis and treatment based on 342 Mayo Clinic patients and current literature. *Curr Opin Hematol* 17: 125–132.
26. Johnstone PA, Mican JM, Metcalfe DD, DeLaney TF. 1994. Radiotherapy of refractory bone pain due to systemic mast cell disease. *Am J Clin Oncol* 17: 328–330.
27. Brumsen C, Hamady NAT, Papapoulos SE. 2002. Osteoporosis and bone marrow mastocytosis: Dissociation of skeletal responses and mast cell activity during long-term bisphosphonate therapy. *J Bone Miner Res* 17: 567–569.
28. Lam KY. 1997. Langerhans cell histocytosis (histiocytosis X). *Postgrad Med J* 73: 391–394.
29. Coppes-Zantinga A, Egeler RM. 2002. The Langerhans cell histiocytosis X files revealed. *Br J Haematol* 116: 3–9.
30. Abla O, Egeler RM, Weitzman S. 2010. Langerhans cell histiocytosis: Current concepts and treatments. *Cancer Treat Rev* 36: 354–359.

97

Tumoral Calcinosis—Dermatomyositis

Nicholas Shaw

Tumoral Calcinosis 810
Dermatomyositis in Children 812

References 813

TUMORAL CALCINOSIS

Tumoral calcinosis is a rare metabolic disorder characterized by the progressive deposition of calcium phosphate crystals in periarticular spaces and soft tissues. The biochemical hallmark of this condition is hyperphosphataemia due to increased renal tubular reabsorption of phosphate. This form, which is referred to as hyperphosphataemic familial tumoral calcinosis (HFTC) is an autosomal recessive disorder (OMIM #211900). However, tumoral calcinosis is also described in the absence of elevated phosphate, which is referred to as normophosphataemic familial tumoral calcinosis (NFTC) (OMIM #610455). Although the first description of this condition was in 1898, the term "tumoral calcinosis" was not used until 1943 [1].

Clinical features

Mineral deposition manifests as soft tissue masses around major joints. In one report, the order of frequency for the site of first lesions is hips, elbows, shoulders, and scapulae [2]. Onset of the lesions can vary from 22 months to adulthood, with the majority manifesting by 20 years of age. Many cases described in the Anglo-American literature have been of black ancestry with a large number of cases reported from Africa. In many reports, familial cases are described with what appears to be either an autosomal recessive or dominant pattern of inheritance. The soft tissue masses are usually painless and can progressively grow in size to that of an orange or grapefruit. Although they occur around joints they do not usually impair range of movement as they are extracapsular. They can compress adjacent neural structures such as the sciatic nerve and may also cause ulceration of overlying skin, causing a sinus tract that leaks a chalky fluid and may become infected. Some affected subjects have been reported to have features of pseudoxanthoma elasticum, i.e., skin changes, vascular calcification, and angioid streaks of the retina. A specific dental abnormality may be seen with hypoplastic teeth containing short bulbous roots and almost complete obliteration of pulp cavities with pulp stones. A related condition, which is termed "hyperostosis-hyperphosphatemia syndrome" is characterized by recurrent attacks of bone pain and swelling particularly affecting the long bones associated with an elevated level of plasma phosphate. There is now evidence that the two conditions represent a continuous spectrum of the same disease.

Radiographic findings

On plain X-ray, early and small lesions are located in regions known to be occupied by bursae and are often distributed in a para-articular fashion along the extensor surfaces of large joints [3]. These soft tissue lesions comprise multiple globular amorphous calcific components separated by radiolucent fibrous septae. Occasionally fluid levels can be seen indicating a cystic component (Fig. 97.1). An inflammatory process "diaphysitis" may also be seen on plain X-ray, computed tomography (CT),

Primer on the Metabolic Bone Diseases and Disorders of Mineral Metabolism, Eighth Edition. Edited by Clifford J. Rosen.
© 2013 American Society for Bone and Mineral Research. Published 2013 by John Wiley & Sons, Inc.

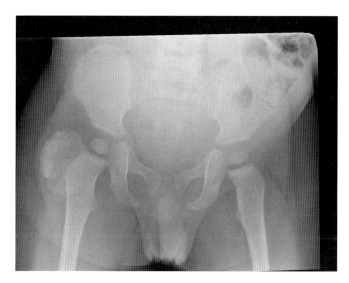

Fig. 97.1. An area of tumoral calcinosis overlying the right hip in a young child.

or magnetic resonance imaging (MRI) in some cases usually occurring in the middle third of long bones. Vascular calcification has also been reported on plain X-ray or CT. An isotope bone scan is the most reliable and simplest method for detection, localization, and assessment of extension of the calcific masses. Periarticular masses that are radiologically indistinguishable from those described in tumoral calcinosis may be seen in patients with chronic renal failure. Individuals with hyperostosis–hyperphosphatemia syndrome show radiographic changes of periosteal reaction and cortical hyperostosis.

Biochemical findings

Many subjects with tumoral calcinosis have been shown to have elevated levels of plasma phosphate and levels of serum 1,25 dihydroxyvitamin D_3 that are either elevated or inappropriately normal [4]. The tubular maximum of tubular phosphate reabsorption in relation to the glomerular filtration rate (GFR) ($TmPO_4$/GFR) is elevated, but renal function is otherwise normal. However, the condition is also described with normal levels of plasma phosphate. Plasma calcium, alkaline phosphatise, and serum parathyroid hormone are usually normal. Metabolic balance studies have demonstrated positive calcium and phosphorous balances due to increased gastrointestinal absorption and reduced renal excretion.

Histopathology

It is suggested that the early lesions are triggered by bleeding followed by aggregation of foamy histiocytes that then become transformed into cystic cavities lined by osteoclast-like giant cells and histiocytes. Movement and friction due to the periarticular location of the lesions appear to be key to the transformation. In a review of 111 cases from Zaire collected over a 30-year period histology identified exuberant cellular proliferative changes adjacent to the classical cystic form [5]. These consisted either of ill-defined reactive-like perivascular solid cell nests admixed with mononuclear and iron-loaded macrophages or well-organized fibrohistiocytic nodules of variable size embedded in a dense collagenous stroma. Mature lesions are filled with calcareous material in a viscous milky fluid.

Etiology and pathogenesis

The first identification of a genetic basis for tumoral calcinosis was in 2004 when, in a study of large Druze and African-American kindreds, the gene was mapped to 2q24-q31 with identification of biallelic mutations in the *GALNT3* gene in affected individuals [6]. The *GALNT3* gene encodes a glycosyltransferase enzyme responsible for initiating mucin-type O-glycosylation. Haplotype analysis in families with normophosphatemic familial tumoral calcinosis has excluded linkage to 2q24-q31. In a family previously felt to have autosomal dominant inheritance, individuals expressing the full phenotype were shown to have biallelic mutations in *GALNT3* while those who were heterozygous for the mutations showed incomplete expression of the condition, with increased plasma phosphate or 1,25 dihydroxyvitamin D_3 levels but no calcified deposits [7]. Thus, autosomal recessive inheritance of the condition was confirmed. Subsequently, mutations in the gene for *FGF23* were identified in affected individuals who were negative for mutations in *GALNT3* [8, 9]. In addition, elevated plasma levels of C-terminal fibroblast growth factor 23 (FGF23) have been shown in these individuals with low plasma levels of intact FGF23, suggesting failure of secretion of the intact protein from cells.

GALNT3 is now known to produce an enzyme that selectively o-glycosylates a furin-like convertase recognition sequence in FGF23 thus preventing proteolytic processing of FGF23 and allowing secretion of intact FGF23. Thus, mutations in *GALNT3* result in defective secretion of intact FGF23 resulting in hyperphosphatemia and increased synthesis of 1,25 dihydroxyvitamin D_3. Recently, a mutation in the *KLOTHO* gene was reported in a 13-year-old girl with tumoral calcinosis, who in addition to elevated plasma phosphate and 1,25 dihydroxyvitamin D_3 levels also had hypercalcemia and a high serum parathyroid hormone [10]. She had evidence of elevated levels of both C-terminal and intact FGF23 but reduced FGF23 bioactivity. KLOTHO is a cofactor required by FGF23 to enable it to bind and signal through its FGF receptors [11]. Thus, mutations in three genes, *GALNT3*, *FGF23*, and *KLOTHO* have been shown to cause the clinical and biochemical features of tumoral calcinosis. Individuals with the related condition hyperostosis-hyperphosphatemia syndrome have now been shown to have homozygous or compound heterozygous mutations in the *GALNT3* gene [12, 13] thus confirming that the two conditions represent different aspects of the continuous spectrum of the same disease. There is now evidence that normophosphatemic familiar tumoral calcinosis can be caused by mutations in the gene encoding the sterile alpha motif domain-containing-9 protein (*SAMD9*) [14].

Treatment

Surgical removal of the calcified masses may be required if they are painful, affect function, or for cosmetic reasons. Several different medical approaches to treatment have been reported, although they are usually individual case reports. Aluminium hydroxide in combination with dietary phosphate and calcium restriction has been reported to be successful [15]. Calcitonin has been used to induce phosphaturia [16], and the combination of acetazolamide with aluminium hydroxide used for a 14-year period in one patient was reported to be effective in improving the lesions [17]. Bisphosphonate therapy with alendronate was reported to alleviate symptoms within 12 weeks in one patient [18]. Another potential treatment option is the use of the phosphate binding agent sevelamer, which is primarily used in patients with end stage renal failure. It was used in conjunction with acetazolamide and a low phosphate diet in a 3-year-old girl with a large periarticular mass at the left elbow due to a homozygous mutation in the FGF23 gene [19]. This resulted in a reduction in the plasma phosphate with a significant reduction in the size of the mass.

The identification of the role of FGF23 in phosphate metabolism and tumoral calcinosis may hopefully lead to new medical approaches to treatment.

DERMATOMYOSITIS IN CHILDREN

Juvenile dermatomyositis is an idiopathic inflammatory disorder of the skin and muscle. It is characterized by progressive weakness predominantly of the proximal muscles and a rash that particularly affects the face and the extremities. It differs from adult onset dermatomyositis in that it is frequently associated with small vessel vasculitis in the skin, muscle, and gastrointestinal tract, and there is no association with malignancy. Dystrophic soft tissue calcification or "calcinosis" occurs in damaged or devitalized tissues in the presence of normal calcium/phosphorous metabolism (Fig. 97.2).

Clinical presentation

It is a rare disorder with an estimated incidence of 1.9–2.5 per million children aged under 16 years and is more common in girls than boys, with a ratio of around 2:1. In a U.K. survey, the median age of onset was 6.8 years with two peak ages of onset in the girls of 6 and 11 years [20]. Eighty-eight percent of the reported cases were Caucasian. Calcinosis is not a feature at initial presentation and is usually noted 1 to 3 years after the disease onset and is reported to occur in 20–40% of affected individuals [21]. The duration of untreated dermatomyositis is associated with pathological calcifications thus demonstrating a clear link with chronic inflammation [22]. The dystrophic calcification can cause pain, skin ulceration, limited joint mobility, contractures, and a predisposition to abscess formation. The calcification once present typically remains stable, but some spontaneous resolution is reported infrequently. The clinical course of dermatomy-

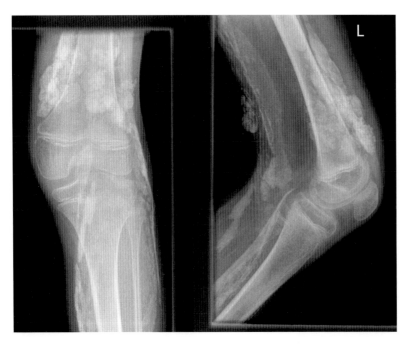

Fig. 97.2. Subcutaneous calcification in a child with dermatomyositis.

ositis in children is variable with some having long-term relapsing or persistent disease whereas others recover.

Biochemical and histological features

Levels of plasma calcium, phosphate, and alkaline phosphatase are usually normal. Urinary levels of γ-carboxyglutamic acid have been reported to be elevated particularly if there is calcinosis. The mineral present in the calcified deposits has been shown to be a poorly crystallized hydroxyapatite [21] or carbonate apatite [23]. They contain relatively more mineral than matrix with a composition more similar to enamel than bone. Bone matrix proteins such as osteopontin, sialoprotein, and osteonectin are present within the calcifications with more osteonectin than is found in human bone.

Radiographic features

Four types of dystrophic calcification can occur:

1. Superficial masses within the skin
2. Deep, discrete, subcutaneous nodular masses near joints that can impair movement (calcinosis circumscripta)
3. Deep, linear, sheet-like deposits within intramuscular fascial planes (calcinosis universalis)
4. Lacy, reticular subcutaneous deposits that encase the trunk to form a generalized "exoskeleton"

"Milk of calcium" fluid collections are a rare complication of calcinosis [24]. Although established calcification can be readily seen on plain X-rays, MRI appears to be a sensitive method for detection and localization of muscle inflammation and edema. It is also an excellent modality for monitoring progression or remission of the disease [25].

Treatment

High-dose corticosteroids soon after the onset of symptoms remains the mainstay of current treatment, reducing the potential risk of calcinosis by suppression of the inflammatory process. Additional agents that are used include methotrexate and infliximab. Several different therapies including bisphosphonates, diltiazem, and surgical extirpation have been utilized to treat the calcinosis, with individual case reports suggesting benefit [26, 27]. However, a review of the published literature over a 32-year period concluded that no treatment has convincingly prevented or reduced calcinosis with a lack of systematic study and clinical therapeutic trials [28].

REFERENCES

1. Inclan A, Leon P, Camejo MG. 1943. Tumoral calcinosis. *JAMA* 121: 490–495.
2. Slavin RE, Wen J, Kumar D, Evans EB. 1993. Familial tumoral calcinosis: A clinical, histopathologic, and ultrastructural study with an analysis of its calcifying process and pathogenesis. *Am J Surg Path* 17: 788–802.
3. Martinez S, Vogler JB, Harrelson JM, Lyles KW. 1990. Imaging of tumoral calcinosis: New observations. *Radiology* 174: 215–222.
4. Lyles KW, Halsey DL, Friedman NE, Lobaugh B. 1988. Correlations of serum concentrations of 1,25-dihydroxyvitamin D, phosphorous and parathyroid hormone in tumoral calcinosis. *J Clin Endocrinol Metab* 67: 88–92.
5. Pakasa NM, Kalengayi RM. 1997. Tumoral calcinosis: A clinicopathological study of 111 cases with emphasis on the earliest changes. *Histopathology* 31: 18–24.
6. Topaz O, Shurman DL, Bergman R, Indelman M, Ratajczak P, Mizrachi M, Khamaysi Z, Behar D, Petronius D, Friedman V, Zelikovic I, Raimer S, Metzker A, Richard G, Sprecher E. 2004. Mutations in GALNT3, encoding a protein involved in o-linked glycosylation, cause familial tumoral calcinosis. *Nature Genet* 36: 579–581.
7. Ichikawa S, Lyles KW, Econs MJ. 2005. A novel GALNT3 mutation in a pseudoautosomal dominant form of tumoral calcinosis: Evidence that the disorder is autosomal recessive. *J Clin Endocrinol Metab* 90: 2420–2423.
8. Benet-Pages A, Orlik P, Strom TM, Lorenz-Depiereux B. 2005. An FGF23 missense mutation causes familial tumoral calcinosis with hyperphosphataemia. *Human Mol Genet* 14: 385–390.
9. Larsson T, Yu X, Davis SI, Draman MS, Mooney SD, Cullen MJ, White KE. 2005. A novel recessive mutation in fibroblast growth factor-23 causes familial tumoral calcinosis. *J Clin Endocrinol Metab* 90: 2424–2427.
10. Ichikawa S, Imel EA, Kreiter ML, Yu X, Mackenzie DS, Sorenson AH, Goetz R, Mohammed M, White KE, Econs MJ. 2007. A homozygous missence mutation in human KLOTHO causes severe tumoral calcinosis. *J Clin Invest* 117: 2684–2691.
11. Urakawa I, Yamazaki Y, Shimada T, Iijima K, Hasegawa H, Okawa K, Fujita T, Fukumoto S, Yamashita T. 2006. Klotho converts canonical FGF receptor into a specific receptor for FGF23. *Nature* 444: 770–777.
12. Ichikawa S, Baujat G, Seyahi A, Garoufali AG, Imel EA, Padgett LR, Austin AM, Sorenson AH, Pejin Z, Topouchian V, Quartier P, Cormier-Daire V, Dechaux M, Malandrinou FCh, Singhellakıs PN, Le Merrer M, Econs MJ. 2010. Clinical variability of familial tumoral calcinosis caused by novel GALNT3 mutations. *Am J Med Genet A* 152A: 896–903.
13. Joseph L, Hing SN, Presneau N, O'Donnell P, Diss T, Idowu BD, Joseph S, Flanagan AM, Delaney D. 2010. Familial tumoral calcinosis and hyperostosis-hyperphosphataemia syndrome are different manifestations of the same disease: Novel missense mutations in GALNT3. *Skeletal Radiol* 39: 63–68.
14. Topaz O, Indelman M, Chefetz I, Geiger D, Metzker A, Altschuler Y, Choder M, Bercovich D, Uitto J, Bergman R, Richard G, Sprecher E. 2006. A deleterious mutation

in SAMD9 causes normophosphatemic familial tumoral calcinosis. *Am J Hum Genet* 79: 759–764.
15. Gregosiewicz A, Warda E. 1989. Tumoral calcinosis: Successful medical treatment. *J Bone Joint Surg Am* 71: 1244–1249.
16. Salvi A, Cerudelli B, Cimino A, Zuccato F, Giustina G. 1983. Phosphaturic action of calcitonin in pseudotumoral calcinosis. *Horm Metab Res* 15: 260.
17. Yamaguchi T, Sugimoto T, Imai Y, Fukase M, Fujita T, Chihara K. 1995. Successful treatment of hyperphosphatemic tumoral calcinosis with long term acetazolimide. *Bone* 16: 247S–250S.
18. Jacob JJ, Mathew K, Thomas N. 2007. Idiopathic sporadic tumoral calcinosis of the hip: Successful oral bisphosphonate therapy. *Endocr Pract* 13: 182–186.
19. Lammoglia JJ, Mericq V. 2009. Familial tumoral calcinosis caused by a novel FGF23 mutation: Response to induction of tubular renal acidosis with acetazolamide and the non-calcium phosphate binder sevelamer. *Horm Res* 71: 178–184.
20. Symmons DPM, Sills JA, Davis SM. 1995. The incidence of juvenile dermatomyositis: Results from a nation-wide study. *Br J Rheumatol* 34: 732–736.
21. Pachman LM, Veis A, Stock S, Abbott K, Vicari F, Patel P, Giczewski D, Webb C, Spevak L, Boskey A. 2006. Composition of calcifications in children with juvenile dermatomyositis. *Arthritis Rheum* 54: 3345–3350.
22. Pachman LM, Abbott K, Sinacore JM, Amoruso L, Dyer A, Lipton R, Ilowite N, Hom C, Cawkwell G, White A, Rivas-Chacon R, Kimura Y, Ray L, Ramsey-Goldman R. 2006. Duration of illness is an important variable for untreated children with untreated dermatomyositis. *J Pediatr* 148: 247–253.
23. Eidelman N, Boyde A, Bushby AJ, Howell PG, Sun J, Newbury DE, Miller FW, Robey PG, Rider LG. 2009. Microstructure and mineral composition of dystrophic calcification associated with the idiopathic inflammatory myopathies. *Arthritis Res Ther* 11(5): R159.
24. Samson C, Soulen RL, Gursel E. 2000. Milk of calcium fluid collections in juvenile dermatomyositis: MR characteristics. *Pediatr Radiol* 30: 28–29.
25. Park JH, Vital TL, Ryder NM, Hernanz-Schulman M, Leon Partain C, Price RR, Olsen NJ. 1994. Magnetic resonance imaging and P-31 magnetic spectroscopy provide unique quantitative data useful in the longitudinal management of patients with dermatomyositis. *Arthritis Rheum* 37: 736–746.
26. Mukamel M, Horev G, Mimouni M. 2001. New insights into calcinosis of juvenile dermatomyositis: A study of composition and treatment. *J Pediatr* 138: 763–766.
27. Oliveri MB, Palermo R, Mautalen C, Hubscher O. 1996. Regression of calcinosis during diltiazem treatment in juvenile dermatomyositis. *J Rheumatol* 23: 2152–2155.
28. Boulman N, Slobodin G, Rozenbaum M, Rosner I. 2005. Calcinosis in rheumatic diseases. *Semin Arthritis Rheum* 34: 805–812.

98

Fibrodysplasia Ossificans Progressiva

Frederick S. Kaplan, Robert J. Pignolo, and Eileen M. Shore

Fibrodysplasia Ossificans Progressiva 815
Progressive Osseous Heteroplasia 817

References 818

FIBRODYSPLASIA OSSIFICANS PROGRESSIVA

Fibrodysplasia ossificans progressiva (FOP: MIM #135100) is a rare heritable disorder of connective tissue characterized by congenital malformations of the great toes and progressive heterotopic endochondral ossification (HEO) in characteristic anatomic patterns [1, 2]. HEO may also occur sporadically following joint replacement, central nervous system trauma, athletic injury, war wounds, atherosclerosis, and valvular heart disease [3, 4].

FOP, first described in 1692, has more than 800 cases reported and is among the rarest of human afflictions, with an estimated incidence of 1 per 2,000,000 individuals [1, 2]. All races are affected [2]. Autosomal dominant transmission with variable expression and complete penetrance is established [5]; however, reproductive fitness is low and most cases are sporadic. Gonadal mosaicism has also been described [6].

Clinical presentation

Malformations of the great toes are present at birth in all classically affected individuals (Fig. 98.1). Typically, episodes of soft-tissue swelling (flare-ups) leading to HEO begin during the first decade of life (Fig. 98.1) [7, 8]. FOP is usually diagnosed when radiographic evidence of heterotopic ossification is noted; however, misdiagnosis is common and leads to unnecessary biopsies and invasive procedures that result in permanent harm [1, 2, 9].

The severity of FOP differs greatly among patients [5, 10]. Most individuals become immobilized and confined to a wheelchair by the third decade of life [1, 2, 7]. Wide variability in the rate of disease progression, even among identical twins, attests to the important postnatal influence of environmental factors [11].

Flare-ups appear spontaneously or may be precipitated by muscle fatigue, minor trauma, intramuscular injections or influenza-like viral illnesses [2, 12, 13]. Swellings develop rapidly during the course of several hours. Aponeuroses, fascia, tendons, ligaments, and connective tissue of voluntary muscles may be affected. Although some lesions may regress spontaneously, most mature by an endochondral pathway to form heterotopic bone with marrow elements [14]. Flare-ups are episodic and recur with unpredictable frequency. Once ossification develops, it is permanent. Disability is cumulative [7, 8].

Bony masses immobilize joints and cause contractures and deformity. Ossification around the hips, typically present by the third decade of life, often prevents ambulation [7]. Involvement of the muscles of mastication (frequently following injection of local anesthetic or overstretching of the jaw during dental procedures) leads to permanent ankylosis of the jaw, impairs nutrition, and affects quality of life [15, 16]. Ankylosis of the spine and ribs as well as chin-on-chest deformity further restrict mobility and may imperil cardiopulmonary function (Fig. 98.1) [1, 2, 7, 17, 18]. Scoliosis is common and associated with malformation of the ribs and costovertebral joints or with heterotopic bone that asymmetrically connects the ribs to the pelvis [19]. Restrictive disease of the chest wall may lead to early mortality [20, 21]. Vocal muscles,

---

Primer on the Metabolic Bone Diseases and Disorders of Mineral Metabolism, Eighth Edition. Edited by Clifford J. Rosen.
© 2013 American Society for Bone and Mineral Research. Published 2013 by John Wiley & Sons, Inc.

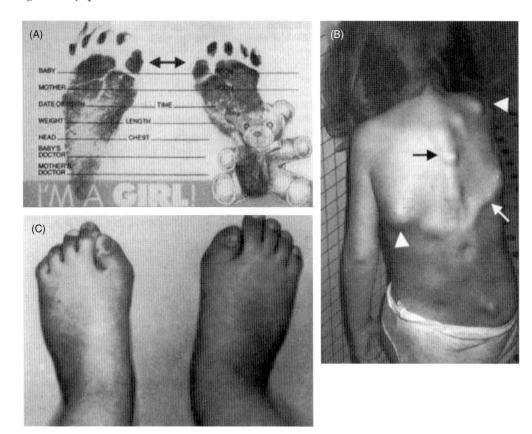

Fig. 98.1. Fibrodysplasia ossificans progressiva. Characteristic features of FOP are seen in early childhood. The presence of short malformed great toes at birth (A, arrows) heralds the later spontaneous appearance of the preosseous soft tissue lesions on the neck and back (B, arrow heads) and should provoke suspicion of FOP even before the transformation to heterotopic bone (B, arrows). An inspection of the toes (C) and/or genetic DNA sequence analysis of ACVR1 will confirm the diagnosis and may alleviate the need for a lesional biopsy (trauma) that could exacerbate the condition [from Kaplan FS and Smith RM. 1997. Clinical vignette—Fibrodysplasia ossificans progressiva (FOP). *J Bone Miner Res* 12: 855. Reproduced with permission of the American Society for Bone and Mineral Research].

smooth muscles, diaphragm, and heart are spared [1]. Hearing impairment is common [22].

Radiologic features

Skeletal anomalies and soft-tissue ossification are characteristic radiologic features of FOP [23]. Skeletal malformations involve the great toe, although other skeletal anomalies commonly occur. In some cases, the thumbs are strikingly short [1, 2]. Osteochondromas are frequent [14, 24, 25]. Progressive fusion of cervical vertebrae may be confused with Klippel-Feil syndrome [1, 26]. Malformation of the temporomandibular joints and fusion of costovertebral joints is common [15, 21]. The femoral necks may be broad and short [2, 25]. Early degenerative arthritis is common [27].

Radiographs and bone scans suggest normal modeling and remodeling of heterotopic bone [28]. Fractures are not increased and repair through normal processes in either the heterotopic or normotopic skeleton [29].

Bone scans detect abnormalities in soft tissue before heterotopic ossification can be demonstrated by conventional radiographs [28]. Computed tomographic and magnetic resonance imaging of early lesions have been described [30].

Laboratory findings

Routine biochemical studies are usually normal, although serum prostanoids, urinary basic fibroblast growth factor, and alkaline phosphatase levels may be increased during the inflammatory, fibroproliferative, and osteogenic phases of disease flare-ups, respectively [1, 2, 31–33]. Elevated numbers of circulating osteoprogenitor cells have been noted during early flare-ups [34].

Histopathology

Early preosseous FOP lesions consist of an intense aggregation of mononuclear inflammatory cells including lymphocytes, macrophages, and mast cells in the perivascular spaces of edematous muscle [35–37]. Following the catabolic phase of muscle cell death, a highly anabolic fibroproliferative phase (often mistaken for *aggressive juvenile fibromatosis*) appears and consists, in part, of mesenchymal-like stem cells that are dedifferentiated from Tie2+ cells that differentiate through an endochondral pathway into mature heterotopic bone [35, 38–41].

Etiology and pathogenesis

Similarities between FOP and the effects of the *Drosophila* decapentaplegic gene (*BMP4* homologue) mutations suggested involvement of the bone morphogenetic protein (BMP) signaling pathway in the pathogenesis of FOP [42]. The BMP signaling pathway is highly dysregulated in FOP cells [43–48]. FOP cells overexpress BMP4 and are unable to appropriately upregulate the expression of multiple BMP antagonists in response to a BMP challenge [43, 45, 46]. They exhibit a defect in BMP receptor internalization and increased activation of downstream targets, suggesting that altered BMP receptor signaling participates in ectopic bone formation in this disease [47, 48]. BMP4 transgenic mice develop an FOP-like phenotype [49].

Genome-wide linkage analysis identified linkage of FOP to 2q23-24, a locus that includes the activin A type I receptor/activin-like kinase 2 (ACVR1/ALK2) gene encoding a BMP type I receptor [50]. A recurrent heterozygous missense mutation (c.617G > A; R206H) in the glycine-serine (GS) activation domain of ACVR1/ALK2 was identified in all affected individuals with classic features of either sporadic or inherited FOP, making early molecular confirmation possible [50, 51]. Protein modeling predicted destabilization of the GS domain, consistent with dysregulated activation of ACVR1/ALK2 as the underlying cause of the ectopic chondrogenesis, osteogenesis, and joint fusions seen in FOP [50]. The GS domain is a specific binding site for FKBP1A (also known as FKBP12), a highly conserved inhibitory protein that prevents leaky activation of type I receptors in the absence of ligand. Experimental data support that the ACVR1/ALK2 (R206H) protein has reduced interaction with FKBP1A in the absence of BMP and suggests that this impaired FKBP1A-ACVR1/ALK2 interaction contributes in part to BMP-independent pathway signaling [52, 53]. Individuals with rare variants and atypical forms of FOP have been described, and all have activating mutations of ACVR1/ALK2 that, like the canonical R206H mutation, cause loss of autoinhibition of the receptor (25). Activation of the mutant receptor by a putative pH-sensitive salt bridge has been suggested [54]. Basal and ligand-stimulated dysregulation of BMP signaling are found in connective tissues progenitor cells from FOP patients, and *in vitro* and *in vivo* disease models [52, 55–60].

Treatment

There is no established medical treatment for FOP [1, 2, 61]. Medical management is currently supportive [61–62]. High-dose glucocorticoids have limited use in the management of the early inflammatory phase of flare-ups. The disorder's rarity, variability, and fluctuating clinical course pose substantial uncertainties when evaluating experimental therapies. Bone marrow transplantation is ineffective, as even a normal immune system may trigger FOP flare-ups in a genetically susceptible chimeric host [33]. Focused research based on targeted inhibition of BMP signaling [63–66], and/or inhibition of the preosseous chondrogenic anlagen of HEO offers hope for the future [67–70].

Removal of FOP lesions is often followed by significant recurrence. Surgical release of joint contractures is unsuccessful and risks new, trauma-induced HEO [1, 2]. Spinal bracing is ineffective and surgical intervention is associated with numerous complications [19]. Dental therapy should preclude mandibular blocks and stretching of the jaw [1, 2, 15, 16]. Dental techniques for focused administration of anesthetic are available. Guidelines for general anesthesia have been reported [15]. While physical therapy to maintain joint mobility may be harmful by provoking or exacerbating lesions, occupational therapy evaluations are often helpful [71]. Intramuscular injections should be avoided [12]. Prevention of falls, influenza, recurrent pulmonary infections, and complications of restrictive chest wall disease is important [20, 72].

Prognosis

Despite widespread heterotopic ossification and severe disability, some patients live productive lives into their seventh decade. Most, however, die earlier from cardiopulmonary complications of severe restrictive chest wall involvement [1, 2, 17, 21].

PROGRESSIVE OSSEOUS HETEROPLASIA

Research on FOP led to the discovery of progressive osseous heteroplasia (POH: MIM #166350), a distinct developmental disorder of heterotopic ossification [73–75]. As with FOP, POH is an autosomal dominant genetic disorder of heterotopic ossification. However, unlike in FOP, heterotopic ossification in POH commonly begins in the dermis and progresses to deeper tissues by an intramembranous, rather than an endochondral pathway [60, 75]. Identification of two patients with POH-like features who also had Albright hereditary osteodystrophy suggested the possibility of a genetic link between the two conditions [75, 76], which was confirmed in a third patient with pure POH [77]. These discoveries led to the rapid identification of paternally inherited inactivating mutations of the *GNAS* gene as the cause of POH [78], although the molecular and cellular basis for the mosaic distribution of lesions in affected individuals is unknown. There are no specific phenotype–genotype correlations that distinguish POH from the more benign forms of limited dermal ossification [79]. Reduced expression of Gs-alpha, one of several proteins encoded by *GNAS*, can induce an osteoblast-like phenotype in human mesenchymal stem cells [80]. *GNAS*-encoded G-proteins and downstream cyclic adenosine monophosphate (cAMP) signaling appear to regulate cell fate lineage decisions at an early cell commitment stage and appear to regulate osteogenesis, at least in part, through interactions with the BMP signaling pathway [81]. Heterozygous inactivation of *Gnas* by disruption of the Gs-alpha-specific exon 1 alters osteoblast differentiation in *Gnas* (+/−) mice and manifests as subcutaneous heterotopic ossification by an intramembranous process [82]. Treatment is presently supportive [75].

REFERENCES

1. Connor JM, Evans DAP. 1982. Fibrodysplasia ossificans progressiva: The clinical features and natural history of 34 patients. *J Bone Joint Surg Br* 64: 76–83.
2. Kaplan FS, Glaser DL, Shore EM, Deirmengian GK, Gupta R, Delai P, Morhart P, Smith R, Le Merrer M, Rogers JG, Connor JM, Kitterman JA. 2005. The phenotype of fibrodysplasia ossificans progressiva. *Clin Rev Bone Miner Metab* 3: 183–188.
3. Pignolo RJ, Foley KL. 2005. Nonhereditary heterotopic ossification. *Clin Rev Bone Miner Metab* 3: 261–266.
4. Mohler ER 3rd, Gannon F, Reynolds C, Zimmerman R, Keane MG, Kaplan FS. 2001. Bone formation and inflammation in cardiac valves. *Circulation* 20: 1522–1528.
5. Shore EM, Feldman GJ, Xu M, Kaplan FS. The genetics of fibrodysplasia ossificans progressiva. 2005. *Clin Rev Bone Miner Metab* 3: 201–204.
6. Janoff HB, Muenke M, Johnson LO, Rosenberg A, Shore EM, Okereke E, Zasloff M, Kaplan FS. 1996. Fibrodysplasia ossificans progressiva in two half-sisters. Evidence for maternal mosaicism. *Am J Med Genet* 61: 320–324.
7. Rocke DM, Zasloff M, Peeper J, Cohen RB, Kaplan FS. 1994. Age and joint-specific risk of initial heterotopic ossification in patients who have fibrodysplasia ossificans progressiva. *Clin Orthop* 301: 243–248.
8. Cohen RB, Hahn GV, Tabas JA, Peeper J, Levitz CL, Sando A, Sando N, Zasloff M, Kaplan FS. 1993. The natural history of heterotopic ossification in patients who have fibrodysplasia ossificans progressiva. A study of 44 patients. *J Bone Joint Surg Am* 75: 215–219.
9. Kitterman JA, Kantanie S, Rocke DM, Kaplan FS. 2005. Iatrogenic harm caused by diagnostic errors in fibrodysplasia ossificans progressiva. *Pediatrics* 116: 654–661.
10. Janoff HB, Tabas JA, Shore EM, Muenke M, Dalinka MK, Schlesinger S, Zasloff MA, Kaplan FS. 1995. Mild expression of fibrodysplasia ossificans progressiva: A report of 3 cases. *J Rheumatology* 22: 976–978.
11. Hebela N, Shore EM, Kaplan FS. 2005. Three pairs of monozygotic twins with fibrodysplasia ossificans progressiva: The role of environment in the progression of heterotopic ossification. *Clin Rev Bone Miner Metab* 3: 205–208.
12. Lanchoney TF, Cohen RB, Rocke DM, Zasloff MA, Kaplan FS. 1995. Permanent heterotopic ossification at the injection site after diphtheria-tetanus-pertussis immunizations in children who have fibrodysplasia ossificans progressiva. *J Pediatr* 126: 762–764.
13. Scarlett RF, Rocke DM, Kantanie S, Patel JB, Shore EM, Kaplan FS. 2004. Influenza-like viral illnesses and flare-ups of fibrodysplasia ossificans progressiva (FOP). *Clin Orthop Rel Res* 423: 275–279.
14. Kaplan FS, Tabas JA, Gannon FH, Finkel G, Hahn GV, Zasloff MA. 1993. The histopathology of fibrodysplasia ossificans progressiva: An endochondral process. *J Bone Joint Surg Am* 75: 220–230.
15. Luchetti W, Cohen RB, Hahn GV, Rocke DM, Helpin M, Zasloff M, Kaplan FS. 1996. Severe restriction in jaw movement after routine injection of local anesthetic in patients who have progressiva. *Oral Surg Oral Med Oral Pathol Oral Radiol Endod* 81: 21–25.
16. Janoff HB, Zasloff M, Kaplan FS. 1996. Submandibular swelling in patients with fibrodysplasia ossificans progressiva. *Otolaryngol Head Neck Surg* 114: 599–604.
17. Kussmaul WG, Esmail AN, Sagar Y, Ross J, Gregory S, Kaplan FS. 1998. Pulmonary and cardiac function in advanced fibrodysplasia ossificans progressiva. *Clin Orthop* 346: 104–109.
18. Moore RE, Dormans JP, Drummond DS, Shore EM, Kaplan FS, Auerbach JD. 2009. Chin-on-chest deformity in patients with fibrodysplasia ossificans progressiva: A case series. *J Bone Joint Surg Am* 91: 1497–1502.
19. Shah PB, Zasloff MA, Drummond D, Kaplan FS. 1994. Spinal deformity in patients who have fibrodysplasia ossificans progressiva. *J Bone Joint Surg Am* 76: 1442–1450.
20. Kaplan FS, Glaser DL. 2005. Thoracic insufficiency syndrome in patients with fibrodysplasia ossificans progressiva. *Clin Rev Bone Miner Metab* 3: 213–216.
21. Kaplan FS, Zasloff MA, Kitterman JA, Shore EM, Hong CC, Rocke DM. 2010. Early mortality and cardiorespiratory failure in patients with fibrodysplasia ossificans progressiva. *J Bone Joint Surg Am* 92: 686–691.
22. Levy CE, Lash AT, Janoff HB, Kaplan FS. 1999. Conductive hearing loss in individuals with fibrodysplasia ossificans progressiva. *Am J Audiol* 8: 29–33.
23. Mahboubi S, Glaser DL, Shore EM, Kaplan FS. 2001. Fibrodysplasia ossificans progressiva (FOP). *Pediatr Radiol* 31: 307–314.
24. Deirmengian GK, Hebela NM, O'Connell M, Glaser DL, Shore EM, Kaplan FS. 2008. Proximal tibial osteochon-

dromas in patients with fibrodysplasia ossificans progressiva. *J Bone Joint Surg Am* 90: 366–374.
25. Kaplan FS, Xu M, Seemann P, Connor JM, Glaser DL, Carroll L, Delai, P, Xu M, Seemann P, Fastnacht-Urban E, Forman SJ, Gillessen-Kaesbach G, Hoover-Fong J, Köster B, Pauli RM, Reardon W, Zaidi S-A, Zasloff M, Morhart R, Mundlos S, Groppe J, and Shore EM. 2009. Classic and atypical fibrodysplasia ossificans progressiva (FOP) phenotypes are caused by mutations in the bone morphogenetic protein (BMP) type I receptor ACVR1. *Hum Mutat* 30: 379–390.
26. Schaffer AA, Kaplan FS, Tracy MR, O'Brien ML, Dormans JP, Shore EM, Harland RM, Kusumi K. 2005. Developmental anomalies of the cervical spin in patients with fibrodysplasia ossificans progressiva are distinctly different from those in patients with Klippel Feil syndrome. *Spine* 30: 1379–1385.
27. Kaplan FS, Groppe JC, Seemann P, Pignolo RJ, Shore EM. 2010. Fibrodysplasia ossificans progressiva: Developmental implications of a novel metamorphogene. In: Bronner F, Farach-Carson MC, Roach HI (eds.) *Bone and Development*. London: Springer Verlag. Chapter 14.
28. Kaplan FS, Strear CM, Zasloff MA. 1994. Radiographic and scintigraphic features of modeling and remodeling in the heterotopic skeleton of patients who have fibrodysplasia ossificans progressiva. *Clin Orthop* 304: 238–247.
29. Einhorn TA, Kaplan FS. 1994. Traumatic fractures of heterotopic bone in patients who have fibrodysplasia ossificans progressiva. *Clin Orthop* 308: 173–177.
30. Shirkhoda A, Armin A-R, Bis KG, Makris J, Irwin RB, Shetty AN. 1995. MR imaging of myositis ossificans: Variable patterns at different stages. *J Magn Reson Imaging* 65: 287–292.
31. Lutwak L. 1964. Myositis ossificans progressiva: Mineral, metabolic, and radioactive calcium studies of the effects of hormones. *Am J Med* 37: 269–293.
32. Kaplan F, Sawyer J, Connors S, Keough K, Shore E, Gannon F, Glaser D, Rocke D, Zasloff M, Folkman J. 1998. Urinary basic fibroblast growth factor: A biochemical marker for preosseous fibroproliferative lesions in patients with FOP. *Clin Orthop* 346: 59–65.
33. Kaplan FS, Glaser DL, Shore EM, Pignolo RJ, Xu M, Zhang Y, Senitzer D, Forman SJ, Emerson SG. 2007. Hematopoietic stem-cell contribution to ectopic skeletogenesis. *J Bone Joint Surg Am* 89: 347–357.
34. Suda RK, Billings PC, Egan KP, Kim JH, McCarrick-Walmsley R, Glaser DL, Porter DL, Shore EM, Pignolo RJ. 2009. Circulating osteogenic precursor cells in heterotopic bone formation. *Stem Cells* 27: 2209–2219.
35. Gannon FH, Valentine BA, Shore EM, Zasloff MA, Kaplan FS. 1998. Acute lymphocytic infiltration in an extremely early lesion of fibrodysplasia ossificans progressiva. *Clin Orthop* 346: 19–25.
36. Gannon FH, Glaser D, Caron R, Thompson LD, Shore EM, Kaplan FS. 2001. Mast cell involvement in fibrodysplasia ossificans progressiva. *Hum Pathol* 32: 842–848.
37. Hegyi L, Gannon FH, Glaser DL, Shore EM, Kaplan, FS, Shanahan CM. 2003. Stromal cells of fibrodysplasia ossificans progressiva lesions express smooth muscle lineage markers and the osteogenic transcription factor Runx2/Cbfa-1: Clues to a vascular origin of heterotopic ossification. *J Pathol* 201: 141–148.
38. Gannon F, Kaplan FS, Olmsted E, Finkel G, Zasloff M, Shore EM. 1997. Differential immunostaining with bone morphogenetic protein (BMP) 2/4 in early fibromatous lesions of fibrodysplasia ossificans progressiva and aggressive juvenile fibromatosis. *Hum Pathol* 28: 339–343.
39. Lounev V, Ramachandran R, Wosczyna MN, Yamamoto M, Maidment AD, Shore EM, Glaser DL, Goldhamer DJ, Kaplan FS. 2009. Identification of progenitor cells that contribute to heterotopic skeletogenesis. *J Bone Joint Surg Am* 91: 652–663.
40. Medici D, Shore EM, Lounev VY, Kaplan FS, Kalluri R, Olsen BJ. 2010. Conversion of vascular endothelial cells into multipotent stem-like cells. *Nat Med* 12: 1400–1406.
41. Wosczyna MN, Biswas AA, Cogswell CA, Goldhamer DJ. 2012. Multipotent progenitors resident in skeletal muscle interstitium exhibit robust BMP-dependent osteogenic activity and mediate heterotopic ossification. *J Bone Miner Res* 27(5): 1004–1017.
42. Kaplan F, Tabas JA, Zasloff MA. 1990. Fibrodysplasia ossificans progressiva: A clue from the fly? *Calcif Tissue Int* 47: 117–125.
43. Shafritz AB, Shore EM, Gannon FH, Zasloff MA, Taub R, Muenke M, Kaplan FS. 1996. Dysregulation of bone morphogenetic protein 4 (BMP4) gene expression in fibrodysplasia ossificans progressiva. *N Engl J Med* 335: 555–561.
44. Lanchoney TF, Olmsted EA, Shore EM, Gannon FA, Rosen V, Zasloff MA, Kaplan FS. 1998. Characterization of bone morphogenetic protein 4 receptors in fibrodysplasia ossificans progressiva. *Clin Orthop* 346: 38–45.
45. Olmsted EA, Kaplan FS, Shore EM. 2003. Bone morphogenetic protein-4 regulation in fibrodysplasia ossificans progressiva. *Clin Orthop* 408: 331–343.
46. Ahn J, Serrano de La Peña L, Shore EM, Kaplan FS. 2003. Paresis of a bone morphogenetic protein antagonist response in a genetic disorder of heterotopic skeletogenesis. *J Bone Joint Surg Am* 85: 667–674.
47. Serrano de la Peña L, Billings PC, Fiori JL, Ahn J, Kaplan FS, Shore EM. 2005. Fibrodysplasia ossificans progressiva (FOP), a disorder of ectopic osteogenesis, misregulates cell surface expression and trafficking of BMPRIA. *J Bone Miner Res* 20: 1168–1176.
48. Fiori JL, Billings PC, Serrano de la Peña L, Kaplan FS, Shore EM. 2006. Dysregulation of the BMP-p38 MAPK signaling pathway in cells from patients with fibrodysplasia ossificans progressiva (FOP). *J Bone Miner Res* 21: 902–909.
49. Kan L, Hu M. Gomes WA, Kessler JA. 2004. Transgenic mice overexpressing BMP4 develop a fibrodysplasia ossificans progressiva (FOP)-like phenotype. *Am J Pathol* 165: 1107–1115.
50. Shore EM, Xu M, Feldman GJ, Fenstermacher DA, Cho T-J, Choi IH, Connor JM, Delai P, Glaser DL, Le Merrer

M, Morhart R, Rogers JG, Smith R, Triffitt JT, Urtizberea JA, Zasloff M, Brown MA, Kaplan FS. 2006. A recurrent mutation in the BMP type I receptor ACVR1 causes inherited and sporadic fibrodysplasia ossificans progressiva. *Nat Genet* 38: 525–527.
51. Kaplan FS, Xu M, Glaser DL Collins F, Connor M, Kitterman J, Sillence D, Zackai E, Ravitsky V, Zasloff M, Ganguly A, Shore EM. 2008. Early diagnosis of fibrodysplasia ossificans progressiva. *Pediatrics* 121: e1295–e1300.
52. Shen Q, Little SC, Xu M, Haupt J, Ast C, Katagiri T, Mundlos S, Seemann P, Kaplan FS, Mullins MC, Shore EM. 2009. The fibrodysplasia ossificans progressiva R206H ACVR1 mutation activates BMP-independent chondrogenesis and zebrafish embryo ventralization. *J Clin Invest* 119: 3462–3472.
53. Groppe JC, Wu J, Shore EM, Kaplan FS. 2011. In vitro analysis of the dysregulated R206H ALK2 Kinase-FKBP12 interaction associated with heterotopic ossification in FOP. *Cells Tissues Organs* 194: 291–295.
54. Groppe JC, Shore EM, Kaplan FS. 2007. Functional modeling of the ACVR1 (R206H) mutation in FOP. *Clin Orthop Rel Res* 462: 87–92.
55. Billings PC, Fiori JL, Bentwood JL, O'Connell MP, Jiao X, Nussbaum B, Caron RJ, Shore EM, Kaplan FS. 2008. Dysregulated BMP signaling and enhanced osteogenic differentiation of connective tissue progenitor cells from patients with fibrodysplasia ossificans progressiva (FOP). *J Bone Miner Res* 23: 305–313.
56. Kaplan FS, Pignolo RJ, Shore EM. 2009. The FOP metamorphogene encodes a novel type I receptor that dysregulates BMP signaling. *Cytokine Growth Factor Rev* 20: 399–407.
57. Fukuda T, Kohda M, Kanomata K, Nojima J, Nakamura A, Kamizono J, Noguchi Y, Iwakiri K, Kondo T, Kurose J, Endo KI, Awakura T, Fukushi J, Nakashima Y, Chiyonobu T, Kawara A, Nishida Y, Wada I, Akita M, Komori T, Nakayama K, Nanba A, Yoda T, Tomoda H, Yu PB, Shore EM, Kaplan FS, Miyazono K, Matsuoka M, Ikebuchi K, Ohtake A, Oda H, Jimi E, Owan I, Okazaki Y, Katagiri T. 2009. Constitutively activated ALK2 and increased SMAD1/5 cooperatively induce bone morphogenetic protein signaling in fibrodysplasia ossificans progressiva. *J Biol Chem* 284: 7149–7156.
58. van Dinther M, Visser N, de Gorter DJ, Doorn J, Goumans MJ, de Boer J, ten Dijke P. 2010. ALK2 R206H mutation linked to fibrodysplasia ossificans progressiva confers constitutive activity to the BMP type I receptor and sensitizes mesenchymal cells to BMP-induced osteoblasts differentiation and bone formation. *J Bone Miner Res* 25: 1208–1215.
59. Song GA, Kim HJ, Woo KM, Baek JH, Kim GS, Choi JY, Ryoo HM. 2010. Molecular consequences of the ACVR1 (R206H) mutation of fibrodysplasia ossificans progressiva. *J Biol Chem* 285: 22542–22553.
60. Shore EM, Kaplan FS. 2010. Inherited human diseases of heterotopic bone formation. *Nat Rev Rheumatol* 6: 518–527.
61. Glaser DL, Kaplan FS. 2005. Treatment considerations for the management of fibrodysplasia ossificans progressiva. *Clin Rev Bone Miner Metab* 3: 243–250.
62. Kaplan FS, LeMerrer M, Glaser DL, Pignolo RJ, Goldsby RE, Kitterman JA, Groppe J, Shore EM. 2008. Fibrodysplasia ossificans progressiva. *Best Pract Res Clin Rheumatol* 22: 191–205.
63. Glaser DL, Rocke DM, Kaplan FS. 1998. Catastrophic falls in patients who have fibrodysplasia ossificans progressiva. *Clin Orthop* 346: 110–116.
64. Kaplan FS, Glaser DL, Pignolo RJ, Shore EM. 2007. A new era of fibrodysplasia ossificans progressiva (FOP): A druggable target for the second skeleton. *Expert Opin Biol Ther* 7: 705–712.
65. Yu PB, Deng DY, Lai CS, Hong CC, Cuny GD, Bouxsein ML, Hong DW, McManus PM, Katagiri T, Sachidanandan C, Kamiya N, Fukuda T, Mishina Y, Peterson RT, Bloch KD. 2008. BMP type I receptor inhibition reduces heterotopic ossification. *Nat Med* 14: 1363–1369.
66. Hong CC, Yu PB. 2009. Applications of small molecule BMP inhibitors in physiology and disease. *Cytokine Growth Factor Rev* 20: 409–418.
67. Brantus J-F, Meunier PJ. 1998. Effects of intravenous etidronate and oral corticosteroids in fibrodysplasia ossificans progressiva. *Clin Orthop* 346: 117–120.
68. Zasloff MA, Rocke DM, Crofford LJ, Hahn GV, Kaplan FS. 1998. Treatment of patients who have fibrodysplasia ossificans progressiva with isotretinoin. *Clin Orthop* 346: 121–129.
69. Shimono K, Tung W-e, Macolino C, Chi AH-T, Didizian JJ, Mundy C, Chandraratna RA, Mishina Y, Enomoto-Iwamoto M, Pacifici M, Iwamoto M. 2011. Potent inhibition of heterotopic ossification by nuclear retinoic acid receptor-γ agonists. *Nat Med* 17: 454–460.
70. Kaplan FS, Shore EM. 2011. Derailing heterotopic ossification and RARing to go. *Nat Med* 17: 420–421.
71. Levy CE, Berner TF, Bendixen R. 2005. Rehabilitation for individuals with fibrodysplasia ossificans progressiva. *Clin Rev Bone Miner Metab* 3: 251–256.
72. Glaser DL, Economides AN, Wang L, Liu X, Kimble RD, Fandl JP, Wilson JM, Stahl N, Kaplan FS, Shore EM. 2003. In vivo somatic cell gene transfer or an engineered noggin mutein prevents BMP4-induced heterotopic ossification. *J Bone Joint Surg Am* 85: 2332–2342.
73. Kaplan FS, Craver R, MacEwen GD, Gannon FH, Finkel G, Hahn G, Tabas J, Gardner RJ, Zasloff MA. 1994. Progressive osseous heteroplasia: A distinct developmental disorder of heterotopic ossification. *J Bone Joint Surg Am* 76: 425–436.
74. Rosenfeld SR, Kaplan FS. 1995. Progressive osseous heteroplasia in male patients. *Clin Orthop* 317: 243–245.
75. Kaplan FS, Shore EM. 2000. Progressive osseous heteroplasia. *J Bone Miner Res* 15: 2084–2094.
76. Eddy MC, Jan De Beur SM, Yandow SM, McAlister WH, Shore EM, Kaplan FS, Whyte MP, Levine MA. 2000. Deficiency of the alpha-subunit of the stimulatory G

protein and severe extraskeletal ossification. *J Bone Miner Res* 15: 2074–2083.
77. Yeh GL, Mathur S, Wivel A, Li M, Gannon FH, Ulied A, Audi L, Olmstead EA, Kaplan FS, Shore EM. 2000. GNAS1 mutation and Cbfa1 misexpression in a child with severe congenital platelike osteoma cutis. *J Bone Miner Res* 15: 2063–2073.
78. Shore EM, Ahn J, Jan de Beur S, Li M, Xu M, Gardner RJ, Zasloff MA, Whyte MP, Levine MA, Kaplan FS. 2002. Paternally inherited inactivating mutations of the GNAS1 gene in progressive osseous heteroplasia. *N Engl J Med* 346: 99–106.
79. Adegbite NS, Xu M, Kaplan FS, Shore EM, Pignolo RJ. 2008. Diagnostic and mutational spectrum of progressive osseous heteroplasia (POH) and other forms of GNAS-based heterotopic ossification. *Am J Med Genet A* 146A: 1788–1796.
80. Leitman SA, Ding C, Cooke DW, Levine MA. 2005. Reduction in Gs-alpha induces osteogenic differentiation in human mesenchymal stem cells. *Clin Orthop Rel Res* 434: 231–238.
81. Zhang S, Kaplan FS, Shore EM. 2012. Different roles of GNAS and cAMP signaling during early and late stages of osteogenic differentiation. *Horm Metab Res* 44: 724–731.
82. Pignolo RJ, Xu M, Russell E, Richardson A, Kaplan J, Billings PC, Kaplan FS, Shore EM. 2011. Heterozygous inactivation of Gnas in adipose derived mesenchymal progenitor cells enhances osteoblast differentiation and promotes heterotopic ossification. *J Bone Miner Res* 26(11): 2647–2655.

99
Osteogenesis Imperfecta
Joan C. Marini

Introduction 822
Clinical Presentation 822
Clinical Types 823
Radiographic and Dual Energy X-ray Absorptiometry Features 825
Laboratory Findings 825
Etiology and Pathogenesis 826
Treatment 826
References 827

INTRODUCTION

Osteogenesis imperfecta (OI), also known as brittle bone disease, is a genetic disorder of connective tissue characterized by fragile bones and a susceptibility to fracture from mild trauma and even acts of daily living [1–3]. The clinical range of this condition is extremely broad, ranging from cases that are lethal in the perinatal period to cases that may be difficult to detect and can present as early osteoporosis. Individuals with OI may have varying combinations of growth deficiency, defective tooth formation (dentinogenesis imperfecta), hearing loss, macrocephaly, blue coloration of sclerae, scoliosis, barrel chest, and ligamentous laxity. Classical OI is an autosomal dominant condition caused by defects in type I collagen, the major structural component of the extracellular matrix of bone, skin, and tendon. Classical OI is generally described using the Sillence classification [4], a nomenclature based on clinical and radiographic features, which was first proposed in 1979. Subsequent biochemical and molecular studies have shown that the mild Sillence type I OI is caused by quantitative defects in type I collagen,[5] whereas the moderate and severe types are caused by structural defects in either of the two chains that form the type I collagen heterotrimer [1]. Recurrence of classical OI types in the children of unaffected parents is caused by parental mosaicism [6]. Recent exciting developments have identified the genetic causes of multiple forms of recessive OI, a relatively rare condition in which the clinical range overlaps predominantly with the lethal and severe Sillence types, although moderate forms have also been described [7]. Some cases of recessive OI are caused by defects in any of the three components of the prolyl 3-hydroxylation complex, cartilage-associated protein (CRTAP) [8–10], prolyl 3-hydroxylase 1 (P3H1) [11–13], and cyclophilin B (CyPB) [14–16]; this complex modifies the $\alpha1(I)$ chain of collagen in the endoplasmic reticulum [17]. Other cases are caused by deficiency of chaperones involved in collagen folding and intracellular transport, HSP47 [18] and FKBP10 [19–22], or a multifunctional pigment epithelium-derived factor (PEDF) which is a potent antiangiogenic factor [23, 24]. The autosomal dominant type V OI is caused by a mutation in *IFITM5*, which encodes the transmembrane protein Bril [25]. The genetic etiology of less than 5% of OI remains unknown.

CLINICAL PRESENTATION

Because the currently delineated 11 types of OI vary widely in symptoms and in the timing of their onset, the diagnosis and its differential varies with the age of the individual in question. A positive family history is usually not present, because most mutations occur *de novo*. Prenatally, severe types II, III, VII, VIII, IX, or X OI may be difficult to distinguish from thanatophoric dys-

Primer on the Metabolic Bone Diseases and Disorders of Mineral Metabolism, Eighth Edition. Edited by Clifford J. Rosen.
© 2013 American Society for Bone and Mineral Research. Published 2013 by John Wiley & Sons, Inc.

plasia, campomelic dysplasia, and achondrogenesis type I [26]. Neonatally, types III or VIII OI and infantile hypophosphatasia may have an overlapping presentation, but infantile hypophosphatasia has the radiographic distinction of spurs extending from the sides of knee and elbow joints, in addition to the biochemical distinction of a low serum alkaline phosphatase level. In childhood diagnoses of the milder forms of OI, the major distinctions are with juvenile and idiopathic osteoporosis and child abuse.

The key diagnostic element for OI is the generalized nature of the connective tissue defect, with facial features (flat midface, frontal bossing, triangular shape, bluish sclerae, yellowish or opalescent teeth), relative macrocephaly, thoracic configuration (barrel chest or pectus excavatum), joint laxity, vertebral compressions, and growth deficiency present in variable combinations in each case. The recessive forms of OI caused by defects in the collagen 3-hydroxylation complex overlap clinically with types II and III OI [8, 11], while the recessive forms caused by defects in SERPINH1, SERPINF1, and FKBP10 overlap clinically with types II, III, and IV [18–21, 23, 24]. All recessive forms have white sclerae, while dominant forms may have blue or sometimes white sclerae. When a diagnosis is still in doubt, collagen biochemical studies, DNA sequencing of type I collagen, and determination of expression levels of the recessive genes provide helpful information on the presence of a mutation.

CLINICAL TYPES

The classification proposed by Sillence in 1979 (Table 99.1) is based on clinical and radiographic criteria that distinguished four types [4]. Although both clinical and laboratory practice have subsequently evolved, the classification has continued to be useful and is still in general use in a modified form. The Sillence types have autosomal dominant inheritance; recurrence of a collagen mutation in the children of unaffected parents is almost always caused by parental mosaicism. More recently, the types of OI have been extended to include V through XI, although they are defined by different criteria than types I–IV (OMIM 166200, 166210, 259420, 66220). Type V OI (OMIM 610967) is defined by bone histology and clinical/radiographic signs [25]. Type VI OI (OMIM 610968), with autosomal recessive inheritance and a distinctive bone histology suggesting a mineralization defect, is defined both genetically, as defects in SERPINF1, and biochemically, as low to absent PEDF in serum [24]. Types VII (OMIM 610682), VIII (OMIM 610915), and IX OI (OMIM 259440) have autosomal recessive inheritance and white sclerae. Type VII was first described histologically and clinically; it was later shown to be caused by mutations in CRTAP [8, 9, 27]. Type VIII was first defined biochemically and molecularly as deficiency of P3H1 [11]. Types IX and X (OMIM 613848) were identified by consideration of PPIB and SERPINH1 as logical candidate genes. Types XI and VI were identified by homozygosity mapping in OI pedigrees without defects in the known OI genes [19, 23].

Type I OI is the mildest form of the disorder. There is postnatal onset of fractures, usually after ambulation is attained, and even beginning in early middle age when type I OI can present as early onset osteoporosis. Fractures decrease markedly after puberty. Individuals with type I OI usually have blue sclerae and often have easy bruising. They may have hearing loss (onset as early as late childhood, but usually in the 20s) or joint hyperextensibility. Growth deficiency and long bone deformity are generally mild. Type I has been divided into A and B subtypes, based on the absence or presence of dentinogenesis imperfecta.

Type II OI is usually lethal in the perinatal period, although survival for months is not uncommon, and survival to a year or more has been noted. These individuals are often born prematurely and are small for gestational

Table 99.1. Osteogenesis Imperfecta Nosology

	OI Type	Inheritance	Phenotype	Gene Defect
Classical Sillence Types	I	AD	Mild	Null COL1A1 allele
	II	AD	Lethal	COL1A1/COL1A2
	III	AD	Progressive deforming	COL1A1/COL1A2
	IV	AD	Moderate	COL1A1/COL1A2
Overactive calcification	V	AD	Distinctive histology	IFITM5 (Bril)
Mineralization Defect	VI	AR	Mineralization defect	SERPINF1 (PEDF)
3-Hydroxylation Defects	VII	AR	Severe (hypomorphic) Severe to lethal (null)	CRTAP
	VIII	AR	Severe to lethal	LEPRE1 (P3H1)
	IX	AR	Moderate to lethal	PPIB (CyPB)
Chaperone Defects	X	AR	Severe to lethal	SERPINH1 (HSP47)
	XI	AR	Progressive deforming, Bruck syndrome	FKBP10 (FKBP65)

age. Legs are usually held in the frog leg position with hips abducted and knees flexed. Radiographically, long bones are extremely osteoporotic, with *in utero* fractures and abnormal modeling (often a crumpled cylindrical shape). The skull is severely undermineralized with wide-open anterior and posterior fontanels. Scleral hue is blue-gray. The bones of these infants are composed predominantly of woven bone without haversian canals or organized lamellae. Demise is generally of pulmonary origin, especially respiratory insufficiency and pneumonias.

Type III OI is known as the progressive deforming type. Most individuals with type III OI survive childhood with severe bone dysplasia. The presentation at birth may be similar to the mild end of the type II OI spectrum. They have extremely fragile bones and, over a lifetime, will have dozens to hundreds of fractures. The long bones are soft and deform from normal muscle tension and subsequently to fractures. These individuals have extreme growth deficiency; final stature is in the range of a prepubertal child. Almost all type III cases develop scoliosis. Radiographically, metaphyseal flaring and "popcorn" formation at growth plates are seen in addition to osteoporosis. They require intensive physical rehabilitation and orthopedic care to attain assisted ambulation in childhood; many will require wheelchairs for mobility. This form is compatible with a full lifespan, although many individuals have respiratory insufficiency and *cor pulmonale* in middle age; some die in infancy and childhood from respiratory causes.

Type IV is the moderately severe Sillence form. The diagnosis may be made at birth or delayed until the toddler or school ages. Scleral hue is variable. These children often have several fractures a year and bowing of their long bones. Fractures decrease after puberty. Essentially all type IV individuals have short final stature, often in the range of pubertal children; many of these children are responsive to growth hormone for significant additional height and improved bone histology [28]. Radiographically, they have osteoporosis and mild modeling abnormalities. They may have platybasia. Many develop vertebral compressions and scoliosis. With consistent rehabilitation intervention and orthopedic management, these individuals should be able to attain independent mobility. This form is compatible with a full lifespan.

OI/EDS is a discrete subgroup of patients who have an overlap of the skeletal symptoms of OI (type IV, usually, or III) and the joint laxity of Ehlers-Danlos syndrome (EDS). Hip dysplasia occurs in some patients and early progressive scoliosis in others. Tissue is friable, requiring extra intervention for spinal fixation. These individuals have a mutation in the amino terminal region of the type I collagen chains that interferes with collagen N-propeptide processing [29].

Recently, types V–XI OI have been classified. Although these types have continued the Sillence numeration, they are based on different criteria than the Sillence types. Types V and VI were defined using bone histology distinctions and generally have a phenotype that would be included in Sillence types IV and III, respectively [25, 30]. These individuals do not have defects in type I collagen. Types VII, VIII, and IX are recessive forms, whose phenotypes overlap with Sillence types II and III [8, 11, 12, 14–16]. These forms have a deficiency of components of a collagen modification complex in the endoplasmic reticulum. Types X and XI OI have defects in collagen chaperones and overlap phenotypically with lethal and moderate/severe OI, respectively [18–21].

Type V OI is associated with a triad of findings: a radiodense metaphyseal band, hypertrophic callus at fractures or surgical sites, and calcification of the forearm interosseous membrane [25]. They have normal teeth and variable sclerae. On histology, the bone lamellae are mesh-like. Recently, a unique mutation in *IFITM5*, which adds 5 residues to the 5′ end of Bril, has been identified as causing all cases of type V OI.

Type VI OI also has distinctive bone histology [30]. The lamellae have a "fish-scale"-like appearance under the microscope, and histomorphometry suggestive of a mineralization defect. These individuals have moderate to severe skeletal disease, with normal teeth and sclerae. Alkaline phosphatase is slightly elevated. Recently, defects in *SERPINF1* have been identified as the cause of type VI OI [24].

Type VII OI is an autosomal recessive form caused by defects in *CRTAP*, cartilage-associated protein [8, 9]. The index pedigree occurs in an isolated First Nations community in northern Quebec [27]. These individuals have rhizomelia and moderate bone disease, associated with a hypomorphic mutation in *CRTAP*. Null mutations in *CRTAP* have been shown to cause a lethal form of OI, with white sclerae, rhizomelia, and a small to normal cranium [8].

Type VIII OI is an autosomal recessive form caused by defects in prolyl 3-hydroxylase 1 (P3H1, encoded by *LEPRE1*). P3H1 forms a complex in the endoplasmic reticulum (ER) with CRTAP, and there is considerable overlap in the phenotypes of OI VII and VIII. Null mutations in *LEPRE1* result in a phenotype that overlaps types II and III OI but has distinct features, including white sclerae, extreme growth deficiency, and undermineralization [11]. For example, a 15-year-old child is the length of a 3-year-old, and a 3-year-old is the length of a 3-month-old. There is a recurring *LEPRE1* mutation that occurs in contemporary West Africans and African-Americans of West African descent that is lethal in homozygous individuals [11].

Type IX OI is caused by defects in the third component of the collagen 3-hydroxylation complex, peptidyl-prolyl *cis-trans* isomerase (*PPIB*) [14–16]. Among the few cases reported, mutations causing premature stop codons or misfolded protein lead to severe lethal OI resembling types VII and VIII in collagen biochemistry and phenotype, except without rhizomelia [15, 16]. A mutation that alters the *PPIB* start codon has a moderate phenotype and normal collagen biochemistry [14].

Type X OI, a recessive form caused by defects in the ER-localized collagen chaperone HSP47, has been identified in only one infant [18]. The phenotype was severe/lethal, with dentinogenesis imperfecta and with features atypical for OI, including skin bullae and pyloric stenosis. Dermal collagen had normal post-translational modification.

Type XI OI, a recessive form caused by abnormalities in the collagen chaperone FKBP10, causes progressive deforming OI, with fractures of long bones, platyspondyly, and scoliosis, but normal sclerae and teeth [19–22]. The phenotypic spectrum of *FKBP10* mutations encompasses Bruck syndrome 1 (OI plus congenital contractures) and a predominantly congenital contracture disorder (Kuskokwim syndrome). The variability of contracture manifestations from *FKBP10* mutations was initially interpreted incorrectly as separate syndromes.

RADIOGRAPHIC AND DUAL ENERGY X-RAY ABSORPTIOMETRY FEATURES

The skeletal survey in classical OI shows generalized osteopenia. Long bones have thin cortices and a gracile appearance. In moderately to severely affected patients, long bones have bowing and modeling deformities, including a cylindrical configuration from an apparent lack of modeling, metaphyseal flaring, and "popcorn" appearance at the metaphyses [31]. Long bones of the upper extremity often seem milder than those of the lower extremity, even without weight bearing. Vertebrae often have central compressions even in mild type I OI; these often appear first at the T_{12}-L_1 level, consistent with weight-bearing stress. In moderate to severe OI, vertebrae will have central and anterior compressions and may appear compressed throughout. The compressions are generally consistent with the patient's L_1-L_4 dual energy X-ray absorptiometry (DXA) Z-score but do not correlate in a straightforward manner with scoliosis. In the lateral plane film of the spine, it is not easy to assess the asymmetry of vertebral collapse, which, along with paraspinal ligamentous laxity, is generally the cause of OI scoliosis. The skull of OI patients with a wide phenotypic range of severity has wormian bones, although this is not unique to OI. Patients with types III and IV OI may also have platybasia, which should be followed with periodic computed tomography (CT) studies for basilar impression and invagination [32].

In type V OI, radiographic features are crucial elements of the diagnostic triad of dense metaphyseal bands, ossification of the interosseus membrane of the forearm, and hypertrophic callus. The dense metaphyseal bands evolve from a rachitic appearance in the perinatal period [33]. In type VI OI, intrauterine fractures were not reported; rather, radiographic features of severe OI developed progressively but were not distinctive from those of severe OI type IV or type III OI, including vertebral compressions, long bone bowing and disorganization, protrusio acetabuli, and bulbous metaphyses [30].

The skeletal radiographs of only a few infants and children with types VII, VIII, and IX OI have been described [8, 9, 11, 14, 27]. Types VII and VIII OI have extreme osteoporosis and abnormal long bone modeling, leading to a cylindrical appearance. The bone material appears cystic and disorganized. In surviving children with type VIII, there is flaring of the metaphyses. These children also have gracile hands that appear relatively long but have shortened metacarpals.

Radiographs in the one infant reported with a mutation affecting HSP47 (OI type X) show progression over the first year of life [18]. At birth, there were multiple intrauterine fractures associated with severe undermineralization, thin ribs, bowing of long bones, and platyspondyly. By 1 year of age, the cranium was enlarged and undermineralized, the ribs were broad, and the long bones were undertubulated. Rhizomelia was apparent at birth and persisted. There have been multiple reports of radiographs in type XI OI (*FKBP10* mutations) [19–22]. These show severe osteopenia and bowing deformities of long bones, with regional lucencies but without cylindrical modeling defects. Kyphosis/scoliosis develops in childhood.

Bone densitometry by DXA (L1-L4) is useful over a wide age and severity range of OI. It aids diagnosis in milder cases and facilitates longitudinal follow-up in moderate to severe forms. There is a general correlation of Z-score and severity of OI. Type I individuals are generally in the -1 to -2 range, type IV Z-scores cluster in the -2 to -4 range, whereas type III spans -3 to -6. Children with type VIII OI have -6 to -7 Z-scores. It is important to remember that the Z-score compares the mineral quantity of the bone being studied to bone with a normal matrix structure and crystal alignment. In OI, many mutations result in irregular crystal alignment on the abnormal matrix, in addition to reduced mineral quantity. DXA does not measure bone quality, which includes bone geometry, histomorphometry, and mechanical properties.

LABORATORY FINDINGS

Serum chemistries related to bone and mineral metabolism are generally normal. Alkaline phosphatase may be elevated after a fracture and is slightly elevated in type VI [30]. PEDF has been reported to be low to absent in type VI, which may provide a useful screening test [24]. Acid phosphatase is elevated in type VIII OI and can logically be expected to be elevated in type VII. Hormones of the growth axis have normal levels [34]. Bone histomorphometry shows defects in bone modeling and in the production and thickening of trabeculae [35]. Cortical width and cancellous bone volume are decreased in all types; trabecular number and width are also decreased. Bone remodeling is increased, as are osteoblast and osteoclast surfaces. When viewed under polarized light, the lamellae of OI bone are thinner and less smooth than in controls. Mineral apposition rate is normal; crystal disorganization may contribute to bone weakness.

ETIOLOGY AND PATHOGENESIS

About 85–90% of patients who have clinical OI have abnormalities of type I collagen, the major structural protein of the bone extracellular matrix. The recessive types of OI either have a defect in a component of a complex that interacts with collagen post-translationally and 3-hydroxylates α1(I)Pro986 in the endoplasmic reticulum, and/or are deficient in a chaperone protein that assists collagen folding [7]. PEDF defects likely have a distinct mechanism, possibly related to disruption of its direct binding to collagen near the carboxyl end of the α1(I) chain, or to its antiangiogenic function [36, 37]. These recessive forms together can be estimated to make up 5–7% of OI cases. Autosomal dominant type V OI is about 5% of cases; patients share a unique mutation in *IFITM5*, which adds 5 residues to the extracellular 5′-end of the transmembrane protein Bril [25]. A small group of OI patients have unknown mutations.

The pathophysiology of OI encompasses multiple levels of dysfunction, ranging from abnormal function of bone cells and matrix, mineralization, and whole tissue. The overlapping features of dominant and recessive OI are providing important indicators of crucial pathways. Murine studies have implicated bone cell dysfunction (osteoblast ER stress, with increased expression of CHOP and GRP78, and impaired matrix production, as well as increased osteoclast numbers and activity), matrix heterogeneity resulting from heterozygosity for the mutant allele, abnormal levels of non-collagenous proteins in matrix, abnormal cell–matrix interactions, and hypermineralization of whole bone tissue demonstrated by Fourier transform infrared spectroscopy (FTIR) and bone mineral density distribution (BMDD) [38].

Cultured dermal fibroblasts are convenient cells in which to examine the collagen biochemistry of probands using gel electrophoresis. Probands with type I OI, which synthesize a reduced amount of structurally normal type I collagen because of a null *COL1A1* allele, display a relative increase in the COL3/COL1 ratio [5]. Probands with the clinically significant types II, III, and IV OI synthesize a mixture of normal collagen and collagen with a structural defect. With rare exceptions, the structural defects are either substitutions for one of the glycine residues that occur at every third position along the chain and are essential for proper helix folding (80%) or alternative splicing of an exon (20%), resulting more frequently in out-of-frame than in-frame alternative transcripts. Structural abnormalities delay helix folding, expose the constituent chains to modifying enzymes for a longer time, and result in overmodification that is detectable as slower electrophoretic migration. The biochemical test does not accurately detect abnormalities in the amino one third of the α1(I) or the amino half of the α2(I) chain [39]. Cultured fibroblasts from patients with types VII and VIII OI, and some with type IX OI, also produce collagen with overmodification of the helical regions of the chains, suggesting that deficiency of the collagen 3-hydroxylation complex delays helix folding [8, 10, 11, 15, 16].

Mutation detection by direct sequencing is more sensitive than the biochemical test, although it does not provide functional information. Collagen sequencing is available either as exon-by-exon sequencing of DNA or transcript sequencing of cDNA. Each technique will miss a small percent of unusual mutations, such as large deletions or rearrangements or low percentage splicing defects.

Genotype-phenotype modeling of the more than 800 mutations currently available has yielded different patterns for the two chains, supporting distinct roles in maintaining matrix integrity [40]. About one-third of the substitutions in α1(I) are lethal, especially those to residues with a branched or charged side chain. Two exclusively lethal regions coincide with the proposed major ligand-binding regions for the collagen monomer with integrins, matrix metalloproteinases (MMPs), fibronectin, and cartilage oligomeric matrix protein (COMP). For the α2(I) chain, only one-fifth of substitutions are lethal; these substitutions are clustered in eight regularly spaced regions along the chain, coinciding with the proteoglycan binding regions on the collagen fibril.

TREATMENT

Early and consistent rehabilitation intervention is the basis for maximizing the physical potential of individuals with OI [41, 42]. Physical therapy should begin in infancy for the severest types, promoting muscle strengthening, aerobic conditioning, and if possible, protected ambulation. Programs to ensure that children have muscle strength to lift a limb against gravity should continue between orthopedic interventions using isotonic and aerobic conditioning. Swimming should be encouraged.

Orthopedic care should be in the hands of a surgeon with experience in OI. Fractures should not be allowed to heal without reduction to prevent loss of function. The goals of orthopedic surgery are to correct the deformity for ambulation and to interrupt a cycle of fracture and refracture. The classic osteotomy procedure requires fixation with an intramedullary rod. The hardware currently in use includes telescoping rods (Bailey-Dubow [43] or Fassier-Duval [44] rods) and non-elongating rods (Rush rods). Important considerations include selection of a rod with the smallest diameter suited to the situation to avoid cortical atrophy. Children who are anticipated to have significant growth may require fewer rod revisions with either of the extensible rods. The secondary features of OI, including abnormal pulmonary functions, hearing loss, and basilar invagination, are best managed in a specialized coordinated care program. The severe growth deficiency of OI is responsive to exogenous growth hormone administration in about one-half of cases of type IV OI [28] and most type I OI [45]; some treated children can attain heights within the normal growth curves. Responders to recombinant growth hormone (rGH) also experience increased L1-L4 DXA, bone volume per total volume (BV/TV), and bone formation rate (BFR). Growth hormone remains under study for its effects on OI skeletal integrity. Four controlled trials have shown the benefits and limitations of bisphosphonate treatment for classical OI [46–49]. The trabecular bone of vertebral bodies has the most positive response. Bone mineral density (BMD) is increased, although the functional meaning of this measurement is difficult to assess because it also includes retained mineralized cartilage; the increase in Z-scores tapers after 1 to 2 years of treatment. More importantly, the vertebral ability to resist compressive forces is shown as increased vertebral area and decreased central vertebral compressions. The effect of bisphosphonate treatment on predominantly cortical long bone is more equivocal. There is a combination of increased stiffness and load bearing that is balanced by weakened bone quality [50]. There is, at best, a trend toward reduced fracture incidence or a reduced relative risk rather than a clear statistical benefit. The functional changes in ambulation, muscle strength, and bone pain reported in the uncontrolled trials have been shown to be placebo effects. The prolonged half-life and recirculation of pamidronate in children up to 8 years after treatment cessation may pose pediatric specific skeletal and reproductive risks [51]. Prolonged or high-dose administration to children can induce defective bone remodeling [52] and may lead to accumulation of bone microdamage. Delayed osteotomy healing was noted at conventional doses [53]. Our current management of bisphosphonates for classical OI is to treat for 2–3 years and then discontinue the drug but continue to follow the patient.

There is some experience in the pharmacological treatment of rare dominant and recessive types of OI. Type V OI is reported to have a response to pamidronate similar to classical OI patients [54]. Ten children with type VI OI and four children with VII OI had improved lumbar vertebral shape and BMD with pamidronate treatment, but changes in fracture rates or mobility scores were limited [55, 56].

REFERENCES

1. Byers PH, Cole WG. 2002. Osteogenesis Imperfecta. In: Royce PM, Steinmann B (eds.) *Connective Tissue and Heritable Disorders*. New York: Wiley-Liss. pp. 385–430.
2. Kuivaniemi H, Tromp G, Prockop DJ. 1997. Mutations in fibrillar collagens (types I, II, III, and XI), fibril-associated collagen (type IX), and network-forming collagen (type X) cause a spectrum of diseases of bone, cartilage, and blood vessels. *Hum Mutat* 9(4): 300–15.
3. Marini JC. 2004. Osteogenesis Imperfecta. In: Behrman RE, Kliegman RM, Jensen HB (eds.) *Nelson's Textbook of Pediatrics*. Philadelphia: Saunders. pp. 2336–38.
4. Sillence DO, Senn A, Danks DM. 1979. Genetic heterogeneity in osteogenesis imperfecta. *J Med Genet* 16: 101–16.
5. Willing MC, Pruchno CJ, Byers PH. 1993. Molecular heterogeneity in osteogenesis imperfecta type I. *Am J Med Genet* 45: 223–27.
6. Cohn DH, Starman BJ, Blumberg B, Byers PH. 1990. Recurrence of lethal osteogenesis imperfecta due to parental mosaicism for a dominant mutation in a human type I collagen gene (COL1A1). *Am J Hum Genet* 46(3): 591–601.
7. Marini JC, Cabral WA, Barnes AM, Chang W. 2007. Components of the collagen prolyl 3-hydroxylation complex are crucial for normal bone development. *Cell Cycle* 6(14): 1675–81.
8. Barnes AM, Chang W, Morello R, Cabral WA, Weis M, Eyre DR, Leikin S, Makareeva E, Kuznetsova N, Uveges TE, Ashok A, Flor AW, Mulvihill JJ, Wilson PL, Sundaram UT, Lee B, Marini JC. 2006. Deficiency of cartilage-associated protein in recessive lethal osteogenesis imperfecta. *N Engl J Med* 355(26): 2757–64.
9. Morello R, Bertin TK, Chen Y, Hicks J, Tonachini L, Monticone M, Castagnola P, Rauch F, Glorieux FH, Vranka J, Bachinger HP, Pace JM, Schwarze U, Byers PH, Weis M, Fernandes RJ, Eyre DR, Yao Z, Boyce BF, Lee B. 2006. CRTAP is required for prolyl 3-hydroxylation and mutations cause recessive osteogenesis imperfecta. *Cell* 127(2): 291–304.
10. Van Dijk FS, Nesbitt IM, Nikkels PG, Dalton A, Bongers EM, van de Kamp JM, Hilhorst-Hofstee Y, Den Hollander NS, Lachmeijer AM, Marcelis CL, Tan-Sindhunata GM, van Rijn RR, Meijers-Heijboer H, Cobben JM, Pals G. 2009. CRTAP mutations in lethal and severe osteogenesis imperfecta: The importance of combining biochemical and molecular genetic analysis. *Eur J Hum Genet* 17(12): 1560–9.
11. Cabral WA, Chang W, Barnes AM, Weis M, Scott MA, Leikin S, Makareeva E, Kuznetsova NV, Rosenbaum KN, Tifft CJ, Bulas DI, Kozma C, Smith PA, Eyre DR, Marini JC. 2007. Prolyl 3-hydroxylase 1 deficiency causes a recessive metabolic bone disorder resembling lethal/severe osteogenesis imperfecta. *Nat Genet* 39(3): 359–65.
12. Baldridge D, Schwarze U, Morello R, Lennington J, Bertin TK, Pace JM, Pepin MG, Weis M, Eyre DR, Walsh J, Lambert D, Green A, Robinson H, Michelson M, Houge G, Lindman C, Martin J, Ward J, Lemyre E, Mitchell JJ, Krakow D, Rimoin DL, Cohn DH, Byers PH, Lee B. 2008. CRTAP and LEPRE1 mutations in recessive osteogenesis imperfecta. *Hum Mutat* 29(12): 1435–42.
13. Willaert A, Malfait F, Symoens S, Gevaert K, Kayserili H, Megarbane A, Mortier G, Leroy JG, Coucke PJ, De Paepe A. 2009. Recessive osteogenesis imperfecta caused by LEPRE1 mutations: Clinical documentation and identification of the splice form responsible for prolyl 3-hydroxylation. *J Med Genet* 46(4): 233–41.
14. Barnes AM, Carter EM, Cabral WA, Weis M, Chang W, Makareeva E, Leikin S, Rotimi CN, Eyre DR, Raggio CL, Marini JC. 2010. Lack of cyclophilin B in osteogenesis imperfecta with normal collagen folding. *N Engl J Med* 362(6): 521–8.

15. Pyott SM, Schwarze U, Christiansen HE, Pepin MG, Leistritz DF, Dineen R, Harris C, Burton BK, Angle B, Kim K, Sussman MD, Weis M, Eyre DR, Russell DW, McCarthy KJ, Steiner RD, Byers PH. 2011. Mutations in PPIB (cyclophilin B) delay type I procollagen chain association and result in perinatal lethal to moderate osteogenesis imperfecta phenotypes. *Hum Mol Genet* 20(8): 1595–609.

16. van Dijk FS, Nesbitt IM, Zwikstra EH, Nikkels PG, Piersma SR, Fratantoni SA, Jimenez CR, Huizer M, Morsman AC, Cobben JM, van Roij MH, Elting MW, Verbeke JI, Wijnaendts LC, Shaw NJ, Hogler W, McKeown C, Sistermans EA, Dalton A, Meijers-Heijboer H, Pals G. 2009. PPIB mutations cause severe osteogenesis imperfecta. *Am J Hum Genet* 85(4): 521–7.

17. Vranka JA, Sakai LY, Bachinger HP. 2004. Prolyl 3-hydroxylase 1, enzyme characterization and identification of a novel family of enzymes. *J Biol Chem* 279(22): 23615–21.

18. Christiansen HE, Schwarze U, Pyott SM, AlSwaid A, Al Balwi M, Alrasheed S, Pepin MG, Weis MA, Eyre DR, Byers PH. 2010. Homozygosity for a missense mutation in SERPINH1, which encodes the collagen chaperone protein HSP47, results in severe recessive osteogenesis imperfecta. *Am J Hum Genet* 86(3): 389–98.

19. Alanay Y, Avaygan H, Camacho N, Utine GE, Boduroglu K, Aktas D, Alikasifoglu M, Tuncbilek E, Orhan D, Bakar FT, Zabel B, Superti-Furga A, Bruckner-Tuderman L, Curry CJ, Pyott S, Byers PH, Eyre DR, Baldridge D, Lee B, Merrill AE, Davis EC, Cohn DH, Akarsu N, Krakow D. 2010. Mutations in the gene encoding the RER protein FKBP65 cause autosomal-recessive osteogenesis imperfecta. *Am J Hum Genet* 86(4): 551–9.

20. Kelley BP, Malfait F, Bonafe L, Baldridge D, Homan E, Symoens S, Willaert A, Elcioglu N, Van Maldergem L, Verellen-Dumoulin C, Gillerot Y, Napierala D, Krakow D, Beighton P, Superti-Furga A, De Paepe A, Lee B. 2011. Mutations in FKBP10 cause recessive osteogenesis imperfecta and Bruck syndrome. *J Bone Miner Res* 26(3): 666–72.

21. Shaheen R, Al-Owain M, Faqeih E, Al-Hashmi N, Awaji A, Al-Zayed Z, Alkuraya FS. 2011. Mutations in FKBP10 cause both Bruck syndrome and isolated osteogenesis imperfecta in humans. *Am J Med Genet A* 155A(6): 1448–52.

22. Shaheen R, Al-Owain M, Sakati N, Alzayed ZS, Alkuraya FS. 2010. FKBP10 and Bruck syndrome: Phenotypic heterogeneity or call for reclassification? *Am J Hum Genet* 87(2): 306–7; author reply 308.

23. Becker J, Semler O, Gilissen C, Li Y, Bolz HJ, Giunta C, Bergmann C, Rohrbach M, Koerber F, Zimmermann K, de Vries P, Wirth B, Schoenau E, Wollnik B, Veltman JA, Hoischen A, Netzer C. 2011. Exome sequencing identifies truncating mutations in human SERPINF1 in autosomal-recessive osteogenesis imperfecta. *Am J Hum Genet* 88(3): 362–71.

24. Homan EP, Rauch F, Grafe I, Lietman C, Doll JA, Dawson B, Bertin T, Napierala D, Morello R, Gibbs R, White L, Miki R, Cohn DH, Crawford S, Travers R, Glorieux FH, Lee B. 2011. Mutations in SERPINF1 cause osteogenesis imperfecta type VI. *J Bone Miner Res* 26(12): 2798–803.

25. Rauch F, Moffatt P, Cheung M, Roughley P, Lalic L, Lund AM, Ramirez N, Fahiminiya S, Majewski J, Glorieux FH. 2013. Osteogenesis imperfecta type V: Marked phenotypic variability despite the presence of the IFITM5 c.-14C>T mutation in all patients. *J Med Genet* 50(1): 21–4.

26. Marini JC, Chernoff EJ. 2001. Osteogenesis Imperfecta. In: Cassidy SB, Allanson J (eds.) *Management of Genetic Syndromes*. New York: Wiley-Liss. pp. 281–300.

27. Ward LM, Rauch F, Travers R, Chabot G, Azouz EM, Lalic L, Roughley PJ, Glorieux FH. 2002. Osteogenesis imperfecta type VII: An autosomal recessive form of bone disease. *Bone* 31(1): 12–8.

28. Marini JC, Hopkins E, Glorieux FH, Chrousos GP, Reynolds JC, Gundberg CM, Reing CM. 2003. Positive linear growth and bone responses to growth hormone treatment in children with types III and IV osteogenesis imperfecta: High predictive value of the carboxyterminal propeptide of type I procollagen. *J Bone Miner Res* 18(2): 237–43.

29. Cabral WA, Makareeva E, Colige A, Letocha AD, Ty JM, Yeowell HN, Pals G, Leikin S, Marini JC. 2005. Mutations near amino end of alpha1(I) collagen cause combined osteogenesis imperfecta/Ehlers-Danlos syndrome by interference with N-propeptide processing. *J Biol Chem* 280(19): 19259–69.

30. Glorieux FH, Ward LM, Rauch F, Lalic L, Roughley PJ, Travers R. 2002. Osteogenesis imperfecta type VI: A form of brittle bone disease with a mineralization defect. *J Bone Miner Res* 17(1): 30–8.

31. Goldman AB, Davidson D, Pavlov H, Bullough PG. 1980. "Popcorn" calcifications: A prognostic sign in osteogenesis imperfecta. *Radiology* 136(2): 351–8.

32. Charnas LR, Marini JC. 1993. Communicating hydrocephalus, basilar invagination, and other neurologic features in osteogenesis imperfecta. *Neurology* 43: 2603–8.

33. Arundel P, Offiah A, Bishop NJ. 2011. Evolution of the radiographic appearance of the metaphyses over the first year of life in type V osteogenesis imperfecta: Clues to pathogenesis. *J Bone Miner Res* 26(4): 894–8.

34. Marini JC, Bordenick S, Heavner G, Rose SR, Hintz R, Rosenfeld R, Chrousos GP. 1993. The growth hormone and somatomedin axis in short children with osteogenesis imperfecta. *J Clin Endocrinol Metab* 76: 251–6.

35. Rauch F, Travers R, Parfitt AM, Glorieux FH. 2000. Static and dynamic bone histomorphometry in children with osteogenesis imperfecta. *Bone* 26(6): 581–9.

36. Hosomichi J, Yasui N, Koide T, Soma K, Morita I. 2005. Involvement of the collagen I-binding motif in the anti-angiogenic activity of pigment epithelium-derived factor. *Biochem Biophys Res Commun* 335(3): 756–61.

37. Sekiya A, Okano-Kosugi H, Yamazaki CM, Koide T. 2011. Pigment epithelium-derived factor (PEDF) shares binding sites in collagen with heparin/heparan sulfate proteoglycans. *J Biol Chem* 286(30): 26364–74.

38. Forlino A, Cabral WA, Barnes AM, Marini JC. 2011. New perspectives on osteogenesis imperfecta. *Nat Rev Endocrinol* 7(9): 540–57.
39. Cabral WA, Milgrom S, Letocha AD, Moriarty E, Marini JC. 2006. Biochemical screening of type I collagen in osteogenesis imperfecta: Detection of glycine substitutions in the amino end of the alpha chains requires supplementation by molecular analysis. *J Med Genet* 43(8): 685–90.
40. Marini JC, Forlino A, Cabral WA, Barnes AM, San Antonio JD, Milgrom S, Hyland JC, Korkko J, Prockop DJ, De Paepe A, Coucke P, Symoens S, Glorieux FH, Roughley PJ, Lund AM, Kuurila-Svahn K, Hartikka H, Cohn DH, Krakow D, Mottes M, Schwarze U, Chen D, Yang K, Kuslich C, Troendle J, Dalgleish R, Byers PH. 2007. Consortium for osteogenesis imperfecta mutations in the helical domain of type I collagen: Regions rich in lethal mutations align with collagen binding sites for integrins and proteoglycans. *Hum Mutat* 28(3): 209–21.
41. Binder H, Conway A, Hason S, Gerber LH, Marini J, Berry R, Weintrob J. 1993. Comprehensive rehabilitation of the child with osteogenesis imperfecta. *Am J Med Genet* 45(2): 265–9.
42. Gerber LH, Binder H, Weintrob JC, Grange DK, Shapiro JR, Fromherz W, Berry R, Conway A, Nason S, Marini JC 1990 Rehabilitation of children and infants with osteogenesis imperfecta: A program for ambulation. *Clin Orthop Relat Res* 251: 254–62.
43. Zionts LE, Ebramzadeh E, Stott NS. 1998. Complications in the use of the Bailey-Dubow extensible nail. *Clin Orthop Relat Res* 348: 186–95.
44. Fassier F. 2005. Experiene with the Fassier-Duval rod: Effectiveness and complications. 9th International Conference on Osteogenesis Imperfecta, June 13–16, Annapolis, MD, USA.
45. Antoniazzi F, Bertoldo F, Mottes M, Valli M, Sirpresi S, Zamboni G, Valentini R, Tato L. 1996. Growth hormone treatment in osteogenesis imperfecta with quantitative defect of type I collagen synthesis. *J Pediatr* 129(3): 432–9.
46. Gatti D, Antoniazzi F, Prizzi R, Braga V, Rossini M, Tato L, Viapiana O, Adami S. 2005. Intravenous neridronate in children with osteogenesis imperfecta: A randomized controlled study. *J Bone Miner Res* 20(5): 758–63.
47. Letocha AD, Cintas HL, Troendle JF, Reynolds JC, Cann CE, Chernoff EJ, Hill SC, Gerber LH, Marini JC. 2005. Controlled trial of pamidronate in children with types III and IV osteogenesis imperfecta confirms vertebral gains but not short-term functional improvement. *J Bone Miner Res* 20(6): 977–86.
48. Sakkers R, Kok D, Engelbert R, van Dongen A, Jansen M, Pruijs H, Verbout A, Schweitzer D, Uiterwaal C. 2004. Skeletal effects and functional outcome with olpadronate in children with osteogenesis imperfecta: A 2-year randomised placebo-controlled study. *Lancet* 363(9419): 1427–31.
49. Ward LM, Rauch F, Whyte MP, D'Astous J, Gates PE, Grogan D, Lester EL, McCall RE, Pressly TA, Sanders JO, Smith PA, Steiner RD, Sullivan E, Tyerman G, Smith-Wright DL, Verbruggen N, Heyden N, Lombardi A, Glorieux FH. 2011. Alendronate for the treatment of pediatric osteogenesis imperfecta: A randomized placebo-controlled study. *J Clin Endocrinol Metab* 96(2): 355–64.
50. Uveges TE, Kozloff KM, Ty JM, Ledgard F, Raggio CL, Gronowicz G, Goldstein SA, Marini JC. 2009. Alendronate treatment of the brtl osteogenesis imperfecta mouse improves femoral geometry and load response before fracture but decreases predicted material properties and has detrimental effects on osteoblasts and bone formation. *J Bone Miner Res* 24(5): 849–59.
51. Papapoulos SE, Cremers SC. 2007. Prolonged bisphosphonate release after treatment in children. *N Engl J Med* 356(10): 1075–6.
52. Whyte MP, Wenkert D, Clements KL, McAlister WH, Mumm S. 2003. Bisphosphonate-induced osteopetrosis. *N Engl J Med* 349(5): 457–63.
53. Munns CF, Rauch F, Zeitlin L, Fassier F, Glorieux FH. 2004. Delayed osteotomy but not fracture healing in pediatric osteogenesis imperfecta patients receiving pamidronate. *J Bone Miner Res* 19(11): 1779–86.
54. Zeitlin L, Rauch F, Travers R, Munns C, Glorieux FH. 2006. The effect of cyclical intravenous pamidronate in children and adolescents with osteogenesis imperfecta type V. *Bone* 38(1): 13–20.
55. Land C, Rauch F, Travers R, Glorieux FH. 2007. Osteogenesis imperfecta type VI in childhood and adolescence: effects of cyclical intravenous pamidronate treatment. *Bone* 40(3): 638–44.
56. Cheung MS, Glorieux FH, Rauch F. 2009. Intravenous pamidronate in osteogenesis imperfecta type VII. *Calcif Tissue Int* 84(3): 203–9.

100

Skeletal Manifestations in Marfan Syndrome and Related Disorders of the Connective Tissue

Emilio Arteaga-Solis and Francesco Ramirez

Introduction 830
Fibrillinopathies 830
Pathogenesis 832
Marfan Syndrome-Related Disorders 832
Acknowledgments 834
References 834

INTRODUCTION

Marfan syndrome (MFS; OMIM 1547000) is a dominantly inherited disorder of the connective tissue with an estimated incidence of approximately 1 per 5,000 live births and cardinal manifestations in the cardiovascular, ocular and musculoskeletal systems [1]. MFS is caused by mutations that perturb the structure or expression of the gene coding for fibrillin-1 (*FBN1*), the major structural component of extracellular microfibrils [1, 2]. Fibrillin microfibrils confer structural integrity to different tissues as components of the elastic fibers or as elastin-free assemblies; additionally, they instruct cell behavior by interacting with integrin receptors and transforming growth factor-β (TGFβ) and bone morphogenetic protein (BMP) complexes [2, 3]. MFS is considered to be part of a larger group of connective tissue disorders that similarly display morphogenetic and/or functional anomalies in different organ systems (Table 100.1); strict diagnostic criteria differentiate MFS from these phenotypically related conditions (Table 100.2) [4–6]. Some of these MFS-related disorders are occasionally associated with mutations in *FBN1*, while others are caused by dysfunctions of proteins that interact with fibrillin-1 and/or regulate TGFβ signaling [7–15]. The present chapter reviews the skeletal phenotypes of MFS and related disorders.

FIBRILLINOPATHIES

Marfan syndrome

Many of the skeletal abnormalities in MFS are either commonly found in the general population or present in related connective tissue disorders (Table 100.3). Malformations of the appendicular skeleton, such as disproportionate long limbs (dolichostenomelia) and fingers (arachnodactyly), are often the most evident manifestations in MFS. Arachnodactyly and joint hypermobility gives rise to the characteristic wrist sign (overlap of the thumb covering the entire fingernail of the fifth finger when wrapped around the contralateral wrist) and thumb sign (projection of the entire distal phalanx of the thumb beyond the ulnar border of the hand when it is folded across the palm). Hindfoot valgus with forefoot abduction and lowering of the midfoot (pes planus), medial protrusion of the femoral head (protrusio acetabuli), and reduced elbow extension are additional findings in the appendicular skeleton. Craniofacial anomalies are the least specific traits of MFS and include high-arched palate, long narrow skull (dolicocephaly), recessed lower mandible (retrognathia), malar hypoplasia, sunken eye (enophthalmos) and downward-slanting palpebral fissures.

Primer on the Metabolic Bone Diseases and Disorders of Mineral Metabolism, Eighth Edition. Edited by Clifford J. Rosen.
© 2013 American Society for Bone and Mineral Research. Published 2013 by John Wiley & Sons, Inc.

Table 100.1. Marfan Syndrome and Phenotypically Related Disorders

Disorder	Inheritance	OMIM	Locus	Mutant gene
Marfan syndrome (MFS)	AD	154700	15q21.1	*FBN1*
Acromelic dysplasia (AD)	AD	102370	15q21.1	*FBN1*
Shprintzen-Goldberg syndrome (SGS)	AD	182212	15q21.1	*FBN1*
Stiff skin syndrome (SSKS)	AD	184900	15q21.1	*FBN1*
Weill-Marchesani syndrome (WMS)	AD	608328	15q21.1	*FBN1*
Geleophysic dysplasia (GD)	AD	NA	15q21.1	*FBN1*
Congenital contractural arachnodactyly (CCA)	AD	121050	5q23-q31	*FBN2*
Geleophysic dysplasia (GD)	AR	231050	9q34.2	*ADAMTSL23*
Weill-Marchesani syndrome (WMS)	AR	277600	19p13.2	*ADAMTS103*
Tooth agenesis syndrome (STHAG)	AR	613097	11q13.1	*LTBP33*
Urban-Davis-Rifkin syndrome (URDS)	AR	613177	19q13.1-q13.2	*LTBP43*
Camurati-Engelmann disease (CED)	AD	131300	19q13.2/19q13.1	*TGFB13*
Loeys-Dietz syndrome 1A (LDS1A)	AD	609192	9q22	*TGFBR13*
Loeys-Dietz syndrome 2A (LDS2A)	AD	608967	9q22	*TGFBR13*
Loeys-Dietz syndrome 1B (LDS1B)	AD	610168	3p22	*TGFBR23*
Loeys-Dietz syndrome 2B (LDS2B)	AD	610380	3p22	*TGFBR23*

AD: autosomal dominant; AR: autosomal recessive; NA: not available.

In the axial skeleton, longitudinal rib overgrowth causes anterior or posterior displacement of the sternum (pectus carinatum or pectus excavatum, respectively) [5]; surgical repair of severe pectus excavatum is recommended to ameliorate cardiac and pulmonary function [16–19]. Spine deformities in MFS include thoracolumbar kyphosis and vertebral displacement (spondylolisthesis), and more commonly thoracic lordosis and progressive scoliosis more than 20° (generally thoracic and convex to the right). Scoliosis with a greater-than-30° curve worsens in MFS patients at a higher rate than in individuals afflicted with idiopathic scoliosis [20]. In contrast to idiopathic scoliosis, spinal bracing in MFS is usually inadequate to manage severely progressive scoliosis, and surgical repair carries a greater risk of complications [21–23]. Enlargement of the outer layer of the meningeal sac in the lumbosacral spine (dural ectasia) is another common axial skeleton manifestation. Dural ectasia (usually at the L5-S2 level) is a common manifestation that affects 63–92% of adults and 40% of children with MFS [24–30]. The abnormality is often associated with herniation of the nerve root and promotion of secondary changes in the osseous structures leading to thinner pedicles and laminae, in addition to increased risk of dural tear and failure of corrective surgery [24–30]. Moderate to severe low back pain, headache, and proximal leg pain, as well as weakness and numbness are additional symptoms associated with dural ectasia [31, 32].

Osteopenia is a controversial finding in MFS, especially in pediatric patients [33–39]. Whereas some dual X-ray absorptiometry (DXA) studies have identified reduced bone mineral density (BMD) in pre- and postmenopausal women and adolescents with MFS, other studies have been more equivocal mostly due to the inherent problem of interpreting DXA scans of significantly longer bones [33–39]. As a result, the etiology and significance of low BMD in relationship to increased long-term risk for fractures remain uncertain, and application of current therapies of bone mineral replacement in MFS patients is generally viewed as premature.

Congenital contractural arachnodactyly

Congenital contractural arachnodactyly (CCA; also known as Beals syndrome) is a rare dominantly inherited condition characterized by multiple joint contractures, arachnodactyly, dolichostenomelia, scoliosis, and osteopenia (Table 100.3); additionally, CCA displays the distinguishing feature of crumpled ears [40]. Although generally milder than MFS, severe and neonatal lethal forms of CCA have also been described [41–43]. Elbow, knee, and finger contractures are found at birth in all CCA patients and generally improve with time. Kyphoscoliosis is often present at birth or early in childhood and contributes to morbidity later in life. Differential diagnosis between MFS and CCA is sometimes difficult due to the presence of overlapping features. Cases in point are the cardiovascular abnormalities and ocular complications that have been reported in severely affected CCA patients, and the crumpled ears and finger contractures (campodactyly) that have been observed in individuals with neonatal lethal MFS [40–42, 44]. Mutations in CCA cluster within a region of fibrillin-2 that corresponds to the fibrillin-1 segment where neonatal lethal MFS mutations map [45]. The functional significance of this observation remains obscure.

Table 100.2. Revised Diagnostic Criteria of Marfan Syndrome (Ref. 5)

In the absence of family history (FH)
1. Ao (Z ≥ 2) <u>and</u> EL: MFS*
2. Ao (Z ≥ 2) <u>and</u> FBN1: MFS
3. Ao (Z ≥ 2) <u>and</u> systemic score (≥7 points): MFS*
4. EL <u>and</u> FBN1 with known Ao: MFS
   - EL with <u>or</u> without systemic score <u>and</u> no FBN1 <u>or</u> with Ao <u>and</u> FBN1 unknown: Ectopia lentis syndrome
   - Ao (Z < 2) <u>and</u> systemic score (≥5) with at least one skeletal feature without EL: MASS
   - MVP <u>and</u> Ao (Z < 2) <u>and</u> systemic score (>5) without EL: MVPS

In the presence of family history (FH)**
5. EL <u>and</u> FH of MFS: MFS
6. Systemic score (≥7 points) <u>and</u> FH of MFS: MFS*
7. Ao (Z ≥ 2 above 20 years of age or ≥3 below 20 years of age) <u>and</u> FH of MFS = MFS*

Systemic score (Score ≥7 indicates systemic involvement; maximum = 20 points)
- Wrist <u>and</u> thumb sign: **3**; wrist <u>or</u> thumb sign: **1**
- Pectus carinatum deformity: **2**; pectus excavatum or chest asymmetry: **1**
- Hindfoot deformity: **2**; plain pes planus: **1**
- Pneumothorax: **2**
- Dural ectasia: **2**
- Protrusio acetabuli: **2**
- Reduced US/LS <u>and</u> increased arm/height <u>and</u> no severe scoliosis: **1**
- Scoliosis or thoracolumbar kyphosis: **1**
- Reduced elbow extension: **1**
- Facial features (3/5): **1** (dolichocephaly, enophtalmos, downslanting palpebral fissures, malar hyoplasia, retrognathia)
- Skin striae: **1**
- Myopia >3 diopters: **1**
- Mitral valve prolapse (all types): **1**

Ao: aortic diameter at the sinuses of Valsalva; **EL:** ectopia lentis; **FBN1:** fibrillin-1 mutation **MASS:** myopia, mitral valve prolapse, aortic root dilation, skeletal findings, striae syndrome **MVPS:** mitral valve prolapse syndrome; **US/LS:** upper segment/lower segment ratio; **Z:** Z-score.
*If discriminating features of SGS, LDS, or vascular EDS are present, then TGFBR1/2 testing, collagen biochemistry, and COL3A1 testing are indicated.
**When a family member was independently diagnosed using the above criteria.

PATHOGENESIS

Studies of *Fbn1* mutant mice have correlated elevated TGFβ activity with the onset and progression of cardiovascular, muscular, and lung abnormalities [46–49]. These findings have translated into a new therapeutic strategy for aortic disease progression in MFS that blunts promiscuous TGFβ signaling through the action of the AT1 receptor blocker (ARB) losartan [50]. Mice deficient for fibrillin-1 and/or fibrillin-2 have also correlated discrete perturbations of local TGFβ and BMP signaling with distinct organ-specific manifestations, such as those affecting bone growth and homeostasis [2, 3].

Fibrillin gene expression in the forming skeleton begins well before mesenchyme cell differentiation, and thereafter fibrillin proteins accumulate at several skeletal sites where they form macroaggregates whose distinct morphologies are believed to reflect the discrete mechanical properties of cartilaginous and bony matrices [51–54]. As in other organ systems, fibrillin-2 is produced in significantly lower amounts than fibrillin-1 during the formation and remodeling/repair of skeletal tissues [51]. Even though fibrillins are structural components of the same extracellular elements and bind TGFβ and BMP complexes with equal affinity, mutations in *Fbn1* or *Fbn2* give rise to distinct skeletal phenotypes due to discrete perturbations of TGFβ and/or BMP signaling. Whereas dural attenuation associated with increased TGFβ activity is solely observed in mice that underexpress fibrillin-1 (*Fbn1$^{mgR/mgR}$* mice), impaired BMP-driven digit formation is exclusively seen in mice that lack fibrillin-2 (*Fbn2$^{-/-}$* mice) [55, 56]. Characterization of reduced BMD in *Fbn1$^{mgR/mgR}$* and *Fbn2$^{-/-}$* mice has further underscored the notion that fibrillin proteins play different roles in modulating the local bioavailability of local TGFβ and BMP signals. Osteopenia in *Fbn2$^{-/-}$* mice is impaired because enhanced latent TGFβ activation inhibits osteoblast maturation while concomitantly stimulating osteoblast-dependent osteoclastogenesis via receptor activator for nuclear factor-κB ligand (RANKL) upregulation [57, 58]. Reduced BMD in *Fbn1$^{mgR/mgR}$* mice is instead associated with elevation of both TGFβ and BMP signaling, which leads to accelerated osteoblast differentiation and RANKL-dependent augmentation of osteoblast-driven osteoclastogenesis [58, 59]. These perturbations of TGFβ and BMP signaling are thought to translate into discrete material and mechanical properties of *Fbn1$^{mgR/mgR}$* and *Fbn2$^{-/-}$* long bones, as genetic evidence has excluded a prominent role of microfibrils in providing the structural framework for mineral deposition [57, 60]. However, inhibition of promiscuous TGFβ signaling through ARB treatment of *Fbn1$^{mgR/mgR}$* mice failed to improve reduced BMD even though it ameliorated aneurysm progression [59]. These findings strongly argue for a multifaceted treatment therapy in MFS that takes into account the distinct physiological roles of microfibril-controlled TGFβ and BMP bioavailability in various tissues, at different stages of organ formation and growth, and during matrix remodeling and repair.

MARFAN SYNDROME-RELATED DISORDERS

In rare instances, *FBN1* mutations can also cause the skeletal abnormalities that characterize Shprintzen-

Table 100.3. Skeletal Features of Marfan Syndrome and Related Disorders

	MFS	AD	SGS	GD	WMS	CCA	LDS1A & B	LDS2A & B
Axial Skeletal Abnormalities								
Dural ectasia	+++	–	–	–	–	–	++	++
Pectus deformity	+++	–	+++	–	–	++	+++	+++
Scoliosis	+++	++	++	–	–	+++	+++	++
Appendicular Skeletal Abnormalities								
Arachnodactyly	+++	–	+++	–	–	+++	+++	++
Brachydactyly	–	+++	–	+++	+++	–	–	–
Camptodactyly	+	–	++	–	++	+++	++	++
Club foot deformity	–	–	+	–	–	–	+++	+
Dolichostenomelia	+++	–	+++	–	–	++	+	+
Joint laxity	++	–	++	–	–	+/–	+++	+++
Joint Contracture	+/–	+	–	+++	–	++	–	–
Pes planus	++	–	–	–	–	++	++	++
Protusio acetabuli	+++	–	–	–	–	+	+	+
Short stature	–	+++	–	+++	+++	–	–	–
Osteoporosis	+/–	–	+	NR	+/–	++	+/–	+/–
Craniofacial Abnormalities								
Bifid uvula	–	–	+	–	–	–	+++	+/–
Cleft palate	–	–	+	–	–	+/–	+++	+/–
Craniosynostotis	–	–	++	–	–	+/–	++	–
Crumpled ears	–	–	–	–	–	+++	–	–
Dolicocephaly	+++	–	–	–	–	–	+	+
Down-slant of palpebral fissures	++	–	+++	+	–	–	++	++
Early and sever myopia	+++	+	+	–	+++	–	–	–
Ectopia lentis	+++	–	–	–	++	–	–	–
Enophthalmos	++	–	–	+/–	–	–	+	+
Exophthalmos	–	–	+++	–	–	–	–	–
High arched palate	+++	–	+++	–	–	++	+++	+++
Hypertelorism	–	–	+++	++	–	+/–	+++	–
Long upper lip	–	–	–	+++	–	–	–	–
Malar hypoplasia	+++	–	++	++	+	–	+++	+++
Micrognatia	++	–	+++	+	–	+	+++	+++

–: not associated; +/–: rare and or subtle; +: occasionally observed; ++: commonly observed; +++: generally observed; NR: not reported.

Goldberg syndrome (SGS) or Weill-Marchesani syndrome (WMS) [8, 9, 61, 62]. The former condition is a rare autosomal dominant disorder in which cranial malformations and neurodevelopmental deficits are present together with MFS manifestations in other organ systems; additionally, there is also significant clinical overlap between SGS and Loeys-Dietz syndrome (LDS) (Table 100.3). Heterozygous FBN1 mutations have been identified in two unrelated SGS patients, and neither of them maps to a region of the protein devoid of MFS-causing defects [8, 61, 62]. WMS is a clinically homogeneous but genetically heterogeneous disorder that exhibits MFS antithetic skeletal manifestations, such as short stature, short and stubby hands and feet (brachydactyly), and joint stiffness (Table 100.3). An in-frame FBN1 deletion has been identified in a patient with autosomal dominant WMS, whereas nonsense and frame-shift mutations in the ADAMTS10 protease have been reported in two individuals with autosomal recessive WMS [9, 63, 64]. Interestingly, in vitro evidence suggests a role for ADAMTS10 in promoting the biogenesis rather than the turnover of fibrillin microfibrils [65]. Short stature, stiff joints, and skin thickening characterize geleophysic dysplasia (GD) and acromelic dysplasia (AD), and which in some cases are also associated with domain-specific FBN1 mutations [15]. These ADs differ from each other in the mode of inheritance and facial features, as well as for the presence of unique cardiac valvular (GD) or skeletal (AD) manifestations (Table 100.3). Homozygous nonsense and missense mutations leading to misfolding and reduced secretion of ADAMTSL2 have been identified in other GD patients [66]. Additional evidence suggests that interaction between ADAMTSL2 and fibrillin-1 is important for both microfibril biogenesis and the regulation of TGFβ bioavailability [15, 66]. It is, however, unknown how dysregulated processes of microfibril assembly and

TGFβ signaling can lead to opposite skeletal phenotypes in MFS and GD/AD.

LDSs display both MFS traits and unique features and are caused by heterozygous mutations in either the *TGFBR1* (LDSs 1A and 2A) or the *TGFBR2* (LDSs 1B and 2B) gene [14, 67]. Whereas there are no phenotypic differences between LDSs 1A and B and between LDSs 2A and B, the spectrum of craniofacial anomalies distinguishes LDSs 1A and 1B from LDSs 2A and 2B (Table 100.3). Additionally, LDSs 2A and 2B overlap substantially with the vascular form of Ehlers-Danlos syndrome (EDS; OMIM 130050), a condition typically caused by mutations in type III collagen (*COL3A1*). LDS craniofacial and skeletal manifestations include hypertelorism, uvula malformations, malar hypoplasia, arched palate, retrognathia, pectus deformity, scoliosis, joint laxity, and dural ectasia; craniosynostosis, cleft palate, and clubfoot deformity can also be present, whereas overgrowth of long bones is mild or absent (Table 100.3). Even though *TGFBR* mutations in LDS are predicted to impair receptor function, experimental findings indicate that they are paradoxically associated with increased TGFβ signaling [14, 67]. Thus, it has been argued that loss-of-function *TGFBR* mutations in LDS either activate an unproductive compensatory response or have gain-of-function properties [5]. Similar mechanisms may explain the apparent paradox that increased bone mass is associated with decreased TGFβ signaling in a rare selective tooth agenesis syndrome (STHAG), but with increased TGFβ activity in a progressive diaphyseal dysplasia (CED) [12, 68, 69]. Hence, the current view of MFS and related disorders is that phenotypic outcomes reflect the spatiotemporal responses of resident cells to dysregulated TGFβ and/or BMP bioavailability rather than simply the impact of the mutations on TGFβ and/or BMP signaling.

ACKNOWLEDGMENTS

We thank Ms. Karen Johnson for organizing the manuscript. Studies from the authors' laboratory that are described in this review were supported by grants from the National Institutes of Health (AR-42044, AR-049698) and the National Marfan Foundation.

REFERENCES

1. Judge DP, Dietz HC. 2005. Marfan's syndrome. *Lancet* 366: 1965–1976.
2. Ramirez F, Dietz HC. 2007. Marfan syndrome: From molecular pathogenesis to clinical treatment. *Curr Opin Genet Dev* 17: 252–258.
3. Ramirez F, Rifkin DB. 2009. Extracellular microfibrils: Contextual platforms for TGFβ and BMP signaling. *Curr Opin Cell Biol* 21: 616–622.
4. De Paepe A, Devereux RB, Dietz HC, Hennekam RC, Pyeritz RE. 1996. Revised diagnostic criteria for the Marfan syndrome. *Am J Med Genet* 62: 417–426.
5. Loeys BL, Dietz HC, Braverman AC, Callewaert BL, De Backer J, Devereux RB, Hilhorst-Hofstee Y, Jondeau G, Faiver L, Milewicz DM, Pyeritz RE, Sponseller PD, Wordsworth P, De Paepe A. 2010. The revised Ghent nosology for the Marfan syndrome. *J Med Genet* 47: 476–485.
6. Faivere L, Collod-Beroud G, Adès L, Arbustini E, Child A, Callewaert BL, Loeys B, Binquet C, Gautier E, Mayer K, Arslan-Kirchner M, Grasso M, Beroud C, Hamroun D, Bonithon-Kopp C, Plauchu H, Robinson PN, De Backer J, Coucke P, Francke U, Bouchot O, Wolf JE, Stheneur C, Hanna N, Detaint D, De Paepe A, Boileau C, Jondeau G. 2012. The new Ghent criteria for Marfan syndrome: What do they change? *Clin Gene* 81(5): 433–442.
7. Loeys BL, Gerber EE, Riegert-Johnson D, Iqbal S, Whiteman P, McConnell V, Chillakuri CR, Macaya D, Coucke PJ, De Paepe A, Judge DP, Wigley F, Davis EC, Mardon HJ, Handford P, Keene DR, Sakai LY, Dietz HC. 2010. Mutations in fibrillin-1 cause congenital scleroderma: Stiff skin syndrome. *Sci Transl Med* 2: 23ra20.
8. Sood S, Eldadah ZA, Krause WL, McIntosh I, Dietz HC. 1996. Mutation in fibrillin-1 and the Marfanoid-craniosynostosis (Shprintzen-Goldberg) syndrome. *Nat Genet* 12: 209–211.
9. Faivre L, Gorlin RJ, Wirtz MK, Godfrey M, Dagoneau N, Samples JR, Le Merrer M, Collod-Beroud G, Boileau C, Munnich A, Cormier-Daire V. 2003. In frame fibrillin-1 gene deletion in autosomal dominant Weill-Marchesani syndrome. *J Med Genet* 40: 34–36.
10. Putnam EA, Zhang H, Ramirez F, Milewicz DM. 1995. Fibrillin-2 (FBN2) mutations result in the Marfan-like disorder, congenital contractural arachnodactyly. *Nat Genet* 11: 456–458.
11. Noor A, Windpassinger C, Vitcu I, Orlic M, Rafiq MA, Khalid M, Malik MN, Ayub M, Alman B, Vincent JB. 2009. Oligodontia is caused by mutation in LTBP3, the gene encoding latent TGF-β binding protein 3. *Am J Hum Genet* 84: 519–523.
12. Urban Z, Hucthagowder V, Schümann N, Todorovic V, Zilberberg L, Choi J, Sens C, Brown CW, Clark RD, Holland KE, Marble M, Sakai LY, Dabovic B, Rifkin DB, Davis EC. 2009. Mutations in LTBP4 cause a syndrome of impaired pulmonary, gastrointestinal, genitourinary, musculoskeletal, and dermal development. *Am J Hum Genet* 85: 593–605.
13. Kinoshita A, Saito T, Tomita H, Makita Y, Yoshida K, Ghadami M, Yamada K, Kondo S, Ikegawa S, Nishimura G, Fukushima Y, Nakogomi T, Saito H, Sugimoto T, Kamegaya M, Hisa K, Murray JC, Taniguchi N, Nikawa N, Yoshiura K. 2000. Domain-specific mutations in TGFB1 result in Camurati-Englemann disease. *Nat Genet* 26: 19–20.
14. Loeys BL, Schwarze U, Holm T, Callewaert BL, Thomas GH, Pannu H, De Backer JF, Oswald GL, Symoens S, Manouvrier S, Roberts AE, Faravelli F, Greco MA,

Pyeritz RE, Milewicz DM, Coucke PJ, Cameron DE, Braverman AC, Byers PH, De Paepe AM, Dietz HC. 2006. Aneurysm syndromes caused by mutations in the TGF-β receptor. *N Engl J Med* 355: 788–798.

15. Le Goff C, Mahaut C, Wang LW, Allali S, Abhyankar A, Jensen S, Zylberberg L, Collod-Beroud G, Bonnet D, Alanay Y, Brady AF, Cordier M-P, Devriendt K, Genevieve D, Kiper POS, Kitoh H, Kradow D, Lynch SA, Le Merrer M, Megarbane A, Mortier G, Odent S, Polak M, Rohrbach M, Silence D, Stolte-Dijkstra I, Supreti-Furga A, Rimoin DL, Topouchian V, Unger S, Zabel B, Bole-Feysot B, Nitschke P, Handford P, Casanova J-L, Boileau C, Apte SS, Munnich A, Cormier-Daire V. 2011. Mutations in the TGFβ binding-protein-like domain 5 of FBN1 are responsible for acromicric and geleophsic dysplasias. *Am J Hum Genet* 2011 89: 7–14.
16. Lawson ML, Mellins RB, Tabangin M, Kelly RE Jr, Croitoru DP, Goretsky MJ, Nuss D. 2005. Impact of pectus excavatum on pulmonary function before and after repair with the Nuss procedure. *J Pediatr Surg* 40: 174–180.
17. Coln E, Carrasco J, Coln D. 2006. Demonstrating relief of cardiac compression with the Nuss minimally invasive repair for pectus excavatum. *J Pediatr Surg* 41: 683–686.
18. Sigalet DL, Montgomery M, Harder J, Wong V, Kravarusic D, Alassiri A. 2007. Long term cardiopulmonary effects of closed repair of pectus excavatum. *Pediatr Surg Int* 5: 493–497.
19. Redlinger RE Jr, Rushing GD, Moskowitz AD, Kelly RE Jr, Nuss D, Kuhn A, Obermeyer RJ, Goretsky MJ. 2010. Minimally invasive repair of pectus excavatum in patients with Marfan syndrome and marfanoid features. *J Pediatr Surg* 45(1): 193–199.
20. Sponseller PD, Bhimani M, Solacoff D, Dormans JP. 2000. Results of brace treatment of scoliosis in Marfan syndrome. *Spine* 25: 2350–2354.
21. Jones KB, Erkula G, Sponseller PD, Dormans JP. 2002. Spine deformity correction in Marfan syndrome. *Spine* 18: 2003–2012.
22. Di Silvestre M, Greggi T, Giacomini S, Cioni A, Bakaloudis G, Lolli F, Parisini P. 2005. Surgical treatment for scoliosis in Marfan syndrome. *Spine* 30: E597–E604.
23. Li ZC, Liu ZD, Dai LY. 2011. Surgical treatment of scoliosis associated with Marfan syndrome by using posterior-only instrumentation. *J Pediatr Orthop B* 20: 63–66.
24. Pyeritz RE, Fishman EK, Bernhardt BA, Siegelman SS. 1988. Dural ectasia is a common feature of the Marfan syndrome. *Am J Hum Genet* 43: 726–732.
25. Villeirs GM, Van Tongerloo AJ, Verstraete KL. 2010. Kunnen marfanoid features. *J Pediatr Surg* 45: 193–199.
26. Jones KB, Sponseller PD, Erkula G, Sakai L, Ramirez F, Dietz HC 3rd, Kost-Byerly S, Bridwell KH, Sandell L. 2007. Symposium on the musculoskeletal aspects of Marfan syndrome: meeting report and state of the science. *J Orthop Res* 25: 413–422.
27. Villeirs GM, Van Tongerloo AJ, Verstraete KL, Kunnen MF, De Paepe AM. 1999. Widening of the spinal canal and dural ectasia in Marfan's syndrome: Assessment by CT. *Neuroradiology* 41: 850–854.
28. Fattori R, Nienaber CA, Descovich B, Ambrosetto P, Reggiani LB, Pepe G, Kaufmann U, Negrini E, von Kodolitsch Y, Gensini GF. 1999. Importance of dural ectasia in phenotypic assessment of Marfan's syndrome. *Lancet* 354: 910–913.
29. Ahn NU, Sponseller PD, Ahn UM, Nallamshetty L, Rose PS, Buchowski JM, Garrett ES, Kuszyk BS, Fishman EK, Zinreich SJ. 2000. Dural ectasia in the Marfan syndrome: MR and CT findings and criteria. *Genet Med* 2: 173–179.
30. Knirsch W, Kurtz C, Haffner N, Binz G, Heim P, Winkler P, Baumgartner D, Freund-Unsinn K, Stern H, Kaemmerer H, Molinari L, Kececioglu D, Uhlemann F. 2006. Dural ectasia in children with Marfan syndrome: A prospective, multicenter, patient-control study. *Am J Med Genet A* 140: 775–781.
31. Ahn NU, Sponseller PD, Ahn UM, Nallamshetty L, Kuszyk BS, Zinreich SJ. 2000. Dural ectasia is associated with back pain in Marfan syndrome. *Spine* 25: 1562–1568.
32. Foran JR, Pyeritz RE, Dietz HC, Sponseller PD. 2005. Characterization of the symptoms associated with dural ectasia in the Marfan patient. *Am J Med Genet A* 134: 58–65.
33. Kohlmeier L, Gasner C, Bachrack LK, Marcus R. 1995. The bone mineral status of patients with Marfan syndrome. *J Bone Min Res* 10: 1550–1555.
34. Tobias JH Dalzell N, Child AH. 1995. Assessment of bone mineral density in women with Marfan syndrome. *Br J Rheumatol* 34: 516–519.
35. Le Parc JM, Plantin P, Jondeau G, Goldschild M, Albert M, Boileau C. 1999. Bone mineral density in sixty adult patients with Marfan syndrome. *Osteoporos Int* 10: 475–479.
36. Carter N, Duncan E, Wordsworth P. 2000. Bone mineral density in adults with Marfan syndrome. *Rheumatology* 39: 307–309.
37. Giampietro PF, Peterson M, Schneider R, Davis JG, Raggio C, Myers E, Burke SW, Boachie-Adjei O, Mueller CM. 2003. Assessment of bone mineral density in adults and children with Marfan syndrome. *Osteoporos Int* 14: 559–563.
38. Moura B, Tubach F, Sulpice M, Boileau C, Jondeau G, Muti C, Chevallier B, Ounnoughene Y, Le Parc JM; Multidisciplinary Marfan Syndrome Clinic Group. 2006. Bone mineral density in Marfan syndrome: A large case-control study. *Joint Bone Spine* 73: 733–735.
39. Giampietro PF, Peterson MG, Schneider R, Davis JG, Burke SW, Boachie-Adjei O, Mueller CM, Raggio CL. 2007. Bone mineral density determinations by dual-energy x-ray absorptiometry in the management of patients with Marfan syndrome-some factors which affect the measurement. *HSS J* 3: 89–92.
40. Tunçbilek E, Alanay Y. 2006. Congenital contractural arachnodactyly (Beals syndrome). *Orphanet J Rare Dis* 1: 20–22.
41. Wang M, Clericuzio CL, Godfrey M. 1996. Familial occurrence of typical and severe lethal congenital contractural

arachnodactyly caused by missplicing of exon 34 of fibrillin-2. *Am J Hum Genet* 59: 1027–1034.
42. Gupta PA, Wallis DD, Chin TO, Northrup H, Tran-Fadulu VT, Towbin JA, Milewicz DM. 2006. FBN2 mutation associated with manifestations of Marfan syndrome and congenital contractural arachnodactyly. *J Med Genet* 41: e56.
43. Currarino G, Friedman JM. 1986. A severe form of congenital contractural arachnodactyly in two newborn infants. *Am J Med Genet* 25: 763–773.
44. Bawle E, Quigg MH. 1992. Ectopia lentis and aortic root dilatation in congenital contractural arachnodactyly. *Am J Med Genet* 42: 19–21.
45. Gupta PA, Putnam EA, Carmical SG, Kaitila I, Steinmann B, Child A, Danesino C, Metcalfe K, Berry SA, Chen E, Delorme CV, Thong MK, Ades LC, Milewicz DM. 2002. Ten novel FBN2 mutations in congenital contractural arachnodactyly: Delineation of the molecular pathogenesis and clinical phenotype. *Hum Mut* 19: 39–48.
46. Neptune ER, Frischmeyer PA, Arking E, Myers L, Bunton TE, Gayraud B, Ramirez F, Sakai LY and Dietz HC. 2003. Dysregulation of TGF-β activation contributes to pathogenesis in Marfan syndrome. *Nat Genet* 33: 407–411.
47. Ng CM, Cheng A, Myers LA, Martinez-Murillo F, Jie C, Bedja D, Gabrielson KL, Hausladen JM, Mecham RP, Judge DP, Dietz HC. 2004. TGFβ-dependent pathogenesis of mitral valve prolapse in a mouse model of Marfan syndrome. *J Clin Invest* 114: 1586–1592.
48. Habashi JP, Judge DP, Holm TM, Cohn RD, Loeys B, Cooper TK, Myers L, Klein EC, Liu G, Calvi C, Podowski M, Neptune ER, Halushka MK, Bedja D, Gabrielson K, Rifkin DB, Carta L, Ramirez F, Huso DL, Dietz HC. 2006. Losartan, an AT1 antagonist, prevents aortic aneurysm in a mouse model of Marfan syndrome. *Science* 312: 117–121.
49. Cohn RD, van Erp C, Habashi JP, Soleimani AA, Klein EC, Lisi MT, Gamradt M, ap Rhys CM, Holm TM, Loeys BL, Ramirez F, Judge DP, Ward CW, Dietz HC. 2007. Angiotensin II type 1 receptor blockade attenuates TGF-β-induced failure of muscle regeneration in multiple myopathic states. *Nat Med* 13: 204–210.
50. Brooke BS, Habashi JP, Judge DP, Patel N, Loeys B, Dietz HC. 2008. Angiotensin II blockade and aortic-root dilation in Marfan's syndrome. *N Engl J Med* 358: 2787–2795
51. Zhang H, Hu W, Ramirez F. 1995. Developmental expression of fibrillin genes suggests heterogeneity of extracellular microfibrils. *J Cell Biol* 129: 1165–1176.
52. Gigante A, Specchia N, Nori S, Greco F. 1996. Distribution of elastic fiber types in the epiphyseal region. *J Orthop Res* 14: 810–817.
53. Keene DR, Jordan CD, Reinhardt DP, Ridgway CC, Ono RN, Corson GM, Fairhurst M, Sussman MD, Memoli VA, and Sakai LY. 1997. Fibrillin-1 in human cartilage: Developmental expression and formation of special banded fibers. *J Histochem Cytochem* 45: 1069–1082.
54. Kitahama S, Gibson MA, Hatzinikolas G, Hay S, Kuliwaba JL, Evdokiou A, Atkins GJ, Findlay DM. 2000. Expression of fibrillins and other microfibril-associated proteins in human bone and osteoblast-like cells. *Bone* 27: 61–67.
55. Jones KB, Myers L, Judge DP, Kirby PA, Dietz HC, Sponseller PD. 2005. Toward an understanding of dural ectasia: A light microscopy study in a murine model of Marfan syndrome. *Spine* 30: 291–293.
56. Arteaga-Solis E, Gayraud B, Lee SY, Shum L, Sakai L, Ramirez F. 2001. Regulation of limb patterning by extracellular microfibrils. *J Cell Biol* 154: 275–281.
57. Nistala H, Lee-Arteaga S, Smaldone S, Siciliano G, Ono R, Sengle G, Arteaga-Solis E, Levasseur R, Ducy P, Sakai LY, Karsenty G, Ramirez F. 2010. Fibrillin-1 and -2 differentially modulate endogenous TGFβ and BMP bioavailability during bone formation. *J Cell Biol* 190: 1107–1121.
58. Nistala H, Lee-Arteaga S, Smaldone S, Siciliano G, Ramirez F. 2010. Extracellular microfibrils modulate osteoblast-supported osteoclastogenesis by restricting TGFβ stimulation of RANKL production. *J Biol Chem* 285: 34126–34133.
59. Nistala H, Lee-Arteaga S, Carta L, Cook JR, Smaldone S, Siciliano G, Rifkin AN, Dietz HC, Rifkin DB, Ramirez F. 2010. Differential effects of alendronate and losartan therapy on osteopenia and aortic aneurysm in mice with severe Marfan syndrome. *Hum Mol Genet* 19: 4790–4798.
60. Arteaga-Solis E, Lee-Arteaga S, Kim, M, Schaffler MB, Jepsen KJ, Pleshko N, Ramirez F. 2011. Material and mechanical properties of bones deficient for fibrillin-1 or fibrillin-2 microfibrils. *Matrix Biol* 30: 189–194.
61. Kosaki K, Takahashi D, Udaka T, Kosaki R, Matsumoto M, Ibe S, Isobe T, Tanaka Y, Takahashi T. 2006. Molecular pathology of Shprintzen-Goldberg syndrome. *Am J Hum Genet* 140A: 104–108.
62. van Steensel MA, van Geel M, Parren LJ, Schrander-Stumpel CT, Marcus-Soekarman D. 2008. Shprintzen-Goldberg syndrome associated with a novel missense mutation in TGFBR2. *Exp Dermatol* 17: 362–365.
63. Dagoneau N, Benoist-Lasselin C, Huber C, Faivre L, Megarbane A, Alswaid A, Dollfus H, Alembik Y, Munnich A, Legeai-Mallet L, Cormier-Daire V. 2004. ADAMTS10 mutations in autosomal recessive Weill-Marchesani syndrome. *Am J Hum Genet* 75: 801–806.
64. Morales J, Al-Sharif L, Khalil DS, Shinwari JM, Bavi P, Al-Mahrouqi RA, Al-Rajhi A, Alkuraya FS, Meyer BF, Al Tassan N. 2009. Homozygous mutations in ADAMTS10 and ADAMTS17 cause lenticular myopia, ectopia lentis, glaucoma, spherophakia, and short stature. *Am J Hum Genet* 85: 558–568.
65. Kurtz WE, Wang LW, Bader HL, Majors AK, Iwata K, Traboulisi El, Sakai LY, Keene DR, Apte SS. 2011. ADAMTS10 protein interacts with fibrillin-1 and promotes its deposition in extracellular matrix of cultured fibroblasts. *J Biol Chem* 286: 17156–17167.
66. Le Goff C, Morice-Picard F, Dagoneau N, Wang LW, Perrot C, Crow YJ, Bauer F, Flori E, Prost-Squarcioni C, Krakow D, Ge G, Greenspan DS, Bonnet D, Le Merrer M, Munnich A, Apte SS, Cromier-Daire V. 2008.

67. Loeys BL, Chen J, Neptune ER, Judge DP, Podowski M, Holm T, Meyers J, Leitch CC, Katsanis N, Sharifi N, Xu FL, Myers LA, Spevak PJ, Cameron DE, De Backer J, Hellemans J, Chen Y, Davis EC, Webb CL, Kress W, Coucke P, Rifkin DB, De Paepe AM, Dietz HC. 2005. A syndrome of altered cardiovascular, craniofacial, neurocognitive and skeletal development caused by mutations in TGFBR1 or TGFBR2. *Nat Genet* 37: 275–281.
68. Saito T, Kinoshita A, Yoshiura KI, Makita Y, Wakui K, Honke K, Nilkawa N, Taniguchi N. 2001. Domain-specific mutations of a transforming growth factor (TGF)-β 1 latency-associated peptide cause Camurati-Engelmann disease because of the formation of a constitutively active form of TGF-β1. *J Biol Chem* 276: 11469–11472.
69. Walton KL, Makani Y, Chen J, Wilce MC, Chan KL, Robertson DM, Harrison CA. 2010. Two distinct regions of latency-associated peptide coordinate stability of the latent transforming growth factor-β1 complex. *J Biol Chem* 285: 17029–10737.

ADAMTSL2 mutations in geleophysic dysplasia demonstrate a role for ADAMTS-like proteins in TGF-β bioavailability regulation. *Nat Genet* 40: 1119–1123.

101
Enzyme Defects and the Skeleton
Michael P. Whyte

Introduction 838
Hypophosphatasia 838
Mucopolysaccharidoses 840
Homocystinuria 840
Alkaptonuria 841
Disorders of Copper Transport 841
Acknowledgment 841
References 841

INTRODUCTION

Inborn errors of metabolism from enzyme deficiencies can importantly affect the skeleton. Five types are reviewed here.

HYPOPHOSPHATASIA

Hypophosphatasia (HPP) is the rare heritable rickets or osteomalacia (OMIM 146300, 241500, 241510) [1] characterized biochemically by subnormal activity of the tissue-nonspecific isoenzyme of alkaline phosphatase (TNSALP) [2, 3]. Although TNSALP is normally present in all tissues, HPP disturbs predominantly the skeleton and teeth. Muscle weakness is often an important finding. Approximately 350 cases have been reported, showing a remarkable range of severity with four overlapping clinical forms described according to patient age when skeletal disease is discovered: perinatal, infantile, childhood, and adult. Odonto-HPP features dental manifestations only [4]. Generally, the earlier the skeletal problems the more severe the clinical course [2, 3].

Perinatal HPP manifests *in utero* [5]. At birth, extreme skeletal hypomineralization causes caput membraneceum and short deformed limbs [3]. Such newborns usually live briefly while suffering increasing respiratory compromise. Prolonged survival is very rare. The radiographic features are pathognomonic [2]. Sometimes, the skeleton is so poorly calcified that only the skull base is seen [3]. Alternatively, the calvarium can be ossified only centrally, segments of the spinal column may appear missing, and severe rachitic changes manifest in the limbs [3].

Infantile HPP presents before 6 months of age [3, 6]. Development appears normal until poor feeding, inadequate weight gain, hypotonia, and wide fontanels are noted. Rachitic deformities are then recognized. Vitamin B_6-dependent seizures may appear [7]. Hypercalcemia and hypercalciuria can cause recurrent vomiting, nephrocalcinosis, and sometimes renal compromise [6, 8]. Functional craniosynostosis may occur with the illusion of widely open fontanels (hypomineralized calvarium). A flail chest predisposes to pneumonia [6]. There can be spontaneous improvement or progressive skeletal deterioration [3, 6] with an estimated 50% of patients dying in infancy [2, 3, 7]. Radiographic changes are characteristic but less striking than in perinatal HPP [2]. Abrupt transition from relatively normal appearing diaphyses to hypomineralized metaphyses can suggest sudden metabolic deterioration. Progressive skeletal demineralization with fractures and thoracic deformity herald a lethal outcome [6].

Childhood HPP causes premature loss of deciduous teeth (at younger than 5 years of age), without root resorption, from hyposplasia of dental cementum [4]. Lower incisors are typically shed first, but the entire dentition can be lost. Permanent teeth fare better. Delayed walking

---

Primer on the Metabolic Bone Diseases and Disorders of Mineral Metabolism, Eighth Edition. Edited by Clifford J. Rosen.
© 2013 American Society for Bone and Mineral Research. Published 2013 by John Wiley & Sons, Inc.

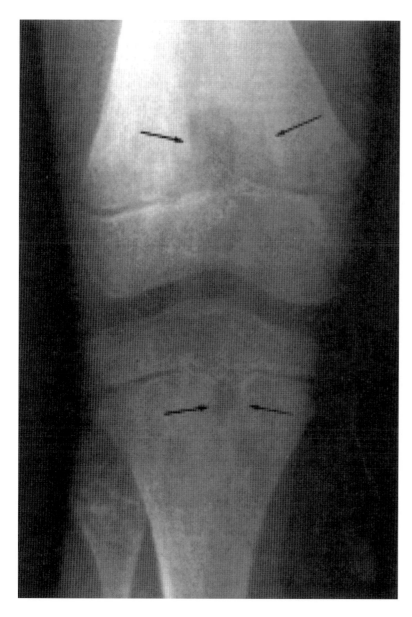

Fig. 101.1. The metaphyses at the knees of this 10-year-old boy with childhood hypophosphatasia show characteristic "tongues" of radiolucency (arrows). Note, however, that his mineralization defect (rickets) does not manifest with uniform widening of the growth plates.

with a waddling gait, short stature, and dolichocephaly are common [3]. Static myopathy is a poorly understood complication [2]. After puberty, patients seem improved, but skeletal symptoms can recur in middle age [2, 3, 9]. Radiographs show characteristic "tongues" of lucency projecting from growth plates into metaphyses (Fig. 101.1) [2]. True premature fusion of cranial sutures with craniosynostosis can cause a "beaten-copper" appearance of the skull [2].

Adult HPP usually presents during middle age, often with poorly healing, recurrent, metatarsal stress fractures [9]. Discomfort localized in the thighs or hips can reflect femoral pseudofractures [9–11]. Patients may recall rickets and/or premature loss of deciduous teeth during childhood [9]. Chondrocalcinosis, pseudogout, and pyrophosphate arthropathy from calcium pyrophosphate dihydrate crystal deposition may occur [2]. Radiographs can show osteopenia, metatarsal stress fractures, chondrocalcinosis, and subtrochanteric femoral pseudofractures [2, 9–11].

HPP rickets/osteomalacia is remarkable because serum levels of calcium and inorganic phosphate (Pi) are not reduced, and ALP activity is low, not high [3]. In fact, hypercalcemia occurs frequently in infantile HPP,[6, 8] apparently from dyssynergy between gut absorption of calcium and the defective skeletal growth and mineralization [2, 3]. In childhood and adult HPP, approximately 50% of patients are hyperphosphatemic because of enhanced renal reclamation of Pi (increased TmP/GFR) [2, 3]. Serum parathyroid hormone (PTH) and $1,25(OH)_2D$ concentrations are suppressed if there is hypercalcemia. Nondecalcified sections of HPP bone show rickets or osteomalacia without secondary hyperparathyroidism [12].

Three phosphocompounds accumulate endogenously in HPP [2, 3]: phosphoethanolamine (PEA), inorganic pyrophosphate (PPi), and pyridoxal 5′-phosphate (PLP). PPi assay is a research technique. If vitamin B_6 is not supplemented, elevated plasma PLP is an especially good marker for HPP [2, 3]. The worse the hypophosphatasemia (low serum ALP activity), the greater the plasma PLP, and the more severe the clinical manifestations [2, 3].

Perinatal and infantile HPP are autosomal recessive diseases [2, 5, 6]; carrier parents and siblings often have low or low normal serum ALP activity, and sometimes mildly elevated plasma PLP levels. Pyridoxine given orally causes exaggerated increments in plasma PLP levels in all patients and in some carriers [2, 3]. Milder forms of HPP (odonto-HPP, childhood HPP, and adult HPP) represent either autosomal dominant or autosomal recessive disorders [13, 14].

HPP is diagnosed from a consistent clinical history and physical findings, radiographic or histopathological evidence of rickets or osteomalacia, and hypophosphatasemia together with TNSALP substrate accumulation [2, 3]. Mutation analysis of *TNSALP* is available commercially. Approximately 280 mutations (approximately 80% missense) have been identified [14].

Disturbances in vitamin B_6 metabolism in HPP indicate that TNSALP functions as a cell surface enzyme [2, 3]. Extracellular accumulation of PPi, an inhibitor of hydroxyapatite crystal growth, impairs skeletal mineralization [2, 3, 15]. *TNSALP* knockout mice manifest infantile HPP [16] and have helped elucidate the role of *TNSALP* [15].

There is no established treatment for HPP. Marrow cell transplantation seemed to rescue two severely affected infants [6]. Teriparatide appeared to stimulate TNSALP biosynthesis by osteoblasts and heal fractures in a woman with HPP [10]. Bone-targeted enzyme-replacement therapy using a recombinant form of TNSALP is currently experimental and showed excellent promise in severely affected infants and young children [17]. Preliminary reports are also favorable for childhood HPP. In childhood HPP, dietary restriction and binders to correct hyperphosphatemia (to thereby reduce inhibition of TNSALP by Pi) could merit study [10].

Unless deficiencies are documented, avoiding traditional treatments for rickets or osteomalacia seems best in HPP, because circulating levels of calcium, Pi, and 25(OH)D are usually not reduced [2, 3]. In fact, supplementation could provoke or exacerbate hypercalcemia or hypercalciuria [6]. Hypercalcemia in perinatal or infantile HPP responds to restriction of dietary calcium and perhaps to salmon calcitonin and/or glucocorticoid therapy [6, 8]. Fractures can mend spontaneously, but healing may be delayed, including after osteotomy. Load-sharing intramedullary rods, rather than load-sparing plates, seem best for fractures and pseudofractures in adults [11]. Expert dental care is important. Soft foods and dentures may be necessary, even for pediatric patients.

Sonography and radiographs of fetuses have detected HPP in the second trimester [5]. Early diagnosis requires *TNSALP* mutation identification [18]. Importantly, patients with "benign prenatal" HPP manifest bowing *in utero* that corrects postnatally and resembles infantile, childhood, or odonto-HPP [5].

MUCOPOLYSACCHARIDOSES

Mucopolysaccharidoses (e.g., Hunter, Hurler, Morquio disease) are a family of disorders caused by diminished activity of the lysosomal enzymes that degrade glycosaminoglycans (acid mucopolysaccharides) [19, 20]. Accumulation of these complex carbohydrates within marrow cells somehow alters the skeleton in a pattern referred to by radiologists as "dysostosis multiplex," [20–22] featuring macrocephaly, dyscephaly, a J-shaped sella turcica, osteoporosis with coarsened trabeculae, oar-shaped ribs, widened clavicles, oval or hook-shaped vertebral bodies, dysplasia of the capital femoral epiphyses, coxa valga, epiphyseal and metaphyseal dysplasia, and proximal tapering of the second and fifth metacarpals. Joint contractures are also common [19, 20]. It may be that bone morphogenetic protein (BMP) signaling is disrupted [23]. However, the severity and precise manifestations vary according to the specific enzymopathy and the underlying gene mutations [24]. Each disorder manifests with a broad range of severity [19, 20]. Enzyme assays or genetic testing is available [1, 19, 20].

Mucopolysaccharidoses are increasingly treated by marrow cell transplantation or enzyme-replacement therapy [25, 26].

HOMOCYSTINURIA

Homocystinuria in its classic form is a rare (1:60,000–300,000) autosomal recessive disorder (OMIM 236200) caused by cystathionine β-synthase deficiency [27]. Consequently, homocystine, an intermediate in methionine metabolism, accumulates endogenously, predisposing to thrombosis and embolism and modification of connective tissue proteins including fibrillin within periosteum and perichondrium. Total homocystine is increased in plasma. Major clinical problems involve the eyes, central nervous system (CNS), vasculature, and skeleton [27]. Dislocation of the ocular lens can be the initial manifestation. Mental subnormality and thrombotic events are important complications. Patient appearance suggests Marfan syndrome; however, joint mobility is limited. Bones are elongated and overtubulated [21, 22].

There may be pectus excavatum or carinatum, arachnodactyly, and genu valgum [27]. Generalized osteoporosis occurs with "codfish" vertebrae and kyphoscoliosis [21, 22]. Mild manifestations are said to predict responsiveness (including skeletal disease) to pyridoxine (vitamin B_6) therapy, but this is controversial [27, 28]. Treatment may instead include a low methionine-cysteine diet and betaine [27].

There are other causes of hyperhomocysteinemia being studied in relationship to common forms of osteoporosis [28–31].

ALKAPTONURIA

Alkaptonuria is a rare (less than 1 in 250,000) autosomal recessive disorder (OMIM 203500) [1] caused by a deficiency of homogentisic acid oxidase from loss-of-function mutations within the *AKU* gene [32]. Consequently, phenylalanine and tyrosine degradation is blocked, leading to tissue accumulation and urine excretion of homogentisic acid. Oxidation and polymerization of homogentisic acid explains the characteristic black appearance of urine exposed to air and discoloration of connective tissues [32]. "Ochronosis" refers to the pigmentation in the sclera, skin, teeth, nails, bucchal mucosa, endocardium, intima of large vessels, hyaline cartilage of major joints, and intervertebral disks [32]. In elderly patients, pigmentation is also striking in costal, laryngeal, and tracheal cartilage, and in fibrocartilage, tendons, and ligaments. Although its pathogenesis is not well understood, severe degenerative disease from tissue fragility occurs in the spinal column where disks calcify and vertebrae fuse, and in major peripheral joints, especially the hips and knees [33]. Perhaps homogentisic acid inhibits collagen synthesis by inhibiting lysyl hydroxylase [32]. Radiographic changes in the spine are nearly pathognomonic, showing dense calcification of the remaining disk material. Calcification of ear cartilage can also occur. The shoulders and hips are most likely to develop osteoarthritis.

There is no established medical treatment, but a low-protein or other special diets seem worthwhile. Ascorbic acid may block homogentisic acid polymerization [32].

DISORDERS OF COPPER TRANSPORT

Wilson disease (OMIM 277900) and Menkes disease (OMIM 309400) [1] are genetic disorders of copper (Cu^{2+}) caused by deficiencies of Cu^{2+}-transporting ATPases in the trans-Golgi of different tissues [34].

Wilson disease affects approximately 1 in 55,000 individuals in the United States, causes impaired biliary excretion of Cu^{2+}, and leads to hepatic injury and Cu^{2+} storage in additional tissues with marked variability in the severity of clinical sequelae. Kayser-Fleischer rings in the eyes, hepatitis and cirrhosis, renal tubular dysfunction and calculi, neurological disease, and hypoparathyroidism are potential complications [34]. Skeletal disease includes osteoporosis, osteomalacia, and chondrocalcinosis with osteoarthritis and joint hypermobility. There can be hyperphosphaturia and hypercalciuria. Loss-of-function mutation disrupts the ATPase, Cu^{2+} transporting, beta-polypeptide gene, *ATP7B* [1]. Cu^{2+} chelation using penicillamine is effective for most cases [34].

In Menkes disease, inherited as an X-linked recessive trait [1], boys develop Cu^{2+} deficiency that leads to kinky, sparse hair, and CNS disease including mental retardation, seizures, and intracranial hemorrhage [34]. The skeletal sequelae include short stature, microcephaly, brachycephaly, wormian bones, metaphyseal dysplasia featuring widening and spurs, osteoporosis, and joint laxity [21, 22]. Death usually occurs by the age of 3 years. A mild form is called "occipital horn syndrome" [34]. Serum Cu^{2+} and ceruloplasmin levels are low because of mutation within the ATPase, Cu^{2+} transporting, alpha-polypeptide gene, *ATP7A*. [1]

ACKNOWLEDGMENT

The author states that he consults for and has research grant support from Alexion Pharmaceuticals, Cheshire, CT, USA.

REFERENCES

1. McKusick-Nathans Institute of Genetic Medicine, Johns Hopkins University. 2008. Online Mendelian Inheritance in Man. Available online at http://www.ncbi.nlm.nih.gov/omim/. Accessed November 24, 2011.
2. Whyte MP. 2008. Hypophosphatasia: Nature's window on alkaline phosphatase function in humans. In: Bilezikian JP, Raisz LG, Martin TJ (eds.) *Principles of Bone Biology*, 3rd Ed. San Diego: Academic Press. pp. 1573–1598.
3. Whyte, MP. 2013. Hypophosphatasia. In: Thakker RV, Whyte MP, Eisman J, Igarashi I (eds.) *Genetics of Bone Biology and Skeletal Disease*. San Diego, CA: Elsevier (Academic Press). Chapter 22, pp. 337–360.
4. van den Bos T, Handoko G, Niehof A, Ryan LM, Coburn SP, Whyte MP, Beertsen W. 2005. Cementum and dentin in hypophosphatasia. *J Dent Res* 84: 1021–1025.
5. Wenkert D, McAlister WH, Coburn SP, Zerega JA, Ryan LM, Ericson KL, Hersh JH, Mumm S, Whyte MP, 2011. Hypophosphatasia: Non-lethal disease despite skeletal presentation in utero (17 new cases and literature review). *J Bone Miner Res* 26: 2389–2398.
6. Cahill RA, Wenkert D, Perman SA, Steele A, Coburn SP, McAlister WH, Mumm S, Whyte MP. 2007. Infantile hypophosphatasia: Transplantation therapy trial using bone fragments and cultured osteoblasts. *J Clin Endocrinol Metab* 95: 2923–2930.
7. Baumgartner-Sigl SB, Haberlandt E, Mumm S, Sergi C, Ryan L, Ericson KL, Whyte MP, Hogler W. 2007. Pyridoxine-responsive seizures as the first symptom of infantile hypophosphatasia caused by two novel missense

mutations (c.677T>C, p.M226T; c.1112C>T, p.T371I) of the tissue-nonspecific alkaline phosphatase gene. *Bone* 40: 1655–1661.
8. Barcia JP, Strife CF, Langman CB. 1997. Infantile hypophosphatasia: Treatment options to control hypercalcemia, hypercalciuria, and chronic bone demineralization. *J Pediatr* 130: 825–828.
9. Khandwala HM, Mumm S, Whyte MP. 2006. Low serum alkaline phosphatase activity with pathologic fracture: Case report and brief review of adult hypophosphatasia. *Endocr Pract* 12: 676–681.
10. Whyte MP, Mumm S, Deal C. 2007. Adult hypophosphatasia treated with teriparatide. *J Clin Endocrinol Metab* 92: 1203–1208.
11. Coe JD, Murphy WA, Whyte MP. 1986. Management of femoral fractures and pseudofractures in adult hypophosphatasia. *J Bone Joint Surg Am* 68: 981–990.
12. Fallon MD, Weinstein RS, Goldfischer S, Brown DS, Whyte MP. 1984. Hypophosphatasia: Clinicopathologic comparison of the infantile, childhood, and adult forms. *Medicine (Baltimore)* 63: 12–24.
13. Henthorn PS, Raducha M, Fedde KN, Lafferty MA, Whyte MP. 1992. Different missense mutations at the tissue-nonspecific alkaline phosphatase gene locus in autosomal recessively inherited forms of mild and severe hypophosphatasia. *Proc Natl Acad Sci U S A* 89(20): 9924–9928.
14. Mornet E. 2005. Tissue nonspecific alkaline phosphatase gene mutations database. Available online at http://www.sesep.uvsq.fr/Database.html. Accessed November 24, 2011.
15. Harmey D, Hessle L, Narisawa S, Johnson KA, Terkeltaub R, Millan JL. 2004. Concerted regulation of inorganic pyrophosphate and osteopontin by akp2, enppl, and ank: An integrated model of the pathogenesis of mineralization disorders. *Am J Pathol* 164: 1199–1209.
16. Fedde KN, Blair L, Silverstein J, Coburn SP, Ryan LM, Weinstein RS, Waymire K, Narisawa S, Millan JL, MacGregor GR, Whyte MP. 1999. Alkaline phosphatase knock-out mice recapitulate the metabolic and skeletal defects of infantile hypophosphatasia. *J Bone Miner Res* 14: 2015–2026.
17. Whyte MP, Greenberg CR, Salman NJ, Bober MB, McAlister WH, Wenkert D, Van Sickle B, Simmons JH, Edgar TS, Bauer ML, Hamdan M, Bishop N, Lutz RE, McGinn M, Craig S, Moore JN, Taylor JW, Cleveland RH, Cranley WR, Lim R, Thacher ID, Mayhew JE, Downs M, Millan JL, Skrinar A, Crine P, Landy H. 2012. Enzyme replacement therapy for life-threatening hypophosphatasia. *N Engl J Med* 366(10): 904–913.
18. Henthorn PS, Whyte MP. 1995. Infantile hypophosphatasia: Successful prenatal assessment by testing for tissue-nonspecific alkaline phosphatase gene mutations. *Prenat Diagn* 15: 1001–1006.
19. Neufeld EF, Muenzer J. 2001. The mucopolysaccharidoses. In: Scriver CR, Beaudet AL, Sly WS, Valle D, Childs B, Vogelstein B (eds.) *The Metabolic and Molecular Bases of Inherited Disease, 8th Ed*. New York: McGraw-Hill. pp. 3421–3452.
20. Leroy JG, Wiesmann U. 2001. Disorders of lysosomal enzymes. In: Royce PM, Steinmann B (eds.) *Connective Tissue and Its Heritable Disorders*. New York: Wiley-Liss. pp. 8494–8499.
21. Taybi H. Lachman RS. 2006. *Radiology of Syndromes, Metabolic Disorders, and Skeletal Dysplasias, 5th Ed*. St. Louis, MO: Mosby.
22. Resnick D, Niwayama G. 2002. *Diagnosis of Bone and Joint Disorders, 4th Ed*. Philadelphia: WB Saunders.
23. Khan SA, Nelson MS, Pan C, Gaffney PM, Gupta P. 2008. Endogenous heparan sulfate and heparin modulate bone morphogenetic protein-4 signaling and activity. *Am J Physiol Cell Physiol* 294: C1387–C1397.
24. Muenzer J. 2004. The mucopolysaccharidoses: A heterogeneous group of disorders with variable pediatric presentations. *J Pediatr* 144: S27–S34.
25. Schiffmann R, Brady RO. 2002. New prospects for the treatment of lysosomal storage diseases. *Drugs* 62: 733–742.
26. Braunlin EA, Stauffer NR, Peters CH, Bass JL, Berry JM, Hopwood JJ, Krivit W. 2003. Usefulness of bone marrow transplantation in the Hurler syndrome. *Am J Cardiol* 92(7): 882–886.
27. Mudd SH, Levy HL, Kraus JP. 2001. Disorders of transsulfuration. In: Scriver CR, Beaudet AL, Sly WS, Valle D, Childs B, Vogelstein B (eds.) *The Metabolic and Molecular Bases of Inherited Disease, 8th Ed*. New York: McGraw-Hill. pp. 2007–2056.
28. Green TJ, McMahon JA, Skeaff CM, Williams SM, Whiting SJ. 2007. Lowering homocysteine with B vitamins has no effect on biomarkers of bone turnover in older persons: a 2-y randomized controlled trial. *Am J Clin Nutr* 85: 460–464.
29. Cagnacci A, Bagni B, Zini A, Cannoletta M, Generali M, Volpe A. 2008. Relation of folates, vitamin B12 and homocysteine to vertebral bone mineral density change in postmenopausal women. A five-year longitudinal evaluation. *Bone* 42: 314–320.
30. Herrmann W, Herrmann M. 2008. Is hyperhomocysteinemia a risk factor for osteoporosis? *Expert Rev Endocrinol Metab* 3: 309–313.
31. Salari P, Larijani B, Abdollahi M. 2008. Association of hyperhomocysteinemia with osteoporosis: A systematic review. *Therapy* 5: 215–222.
32. La Du BN. 2001. Alkaptonuria. In: Scriver CR, Beaudet AL, Sly WS, Valle D, Childs B, Vogelstein B (eds.) *The Metabolic and Molecular Bases of Inherited Disease, 8th Ed*. New York: McGraw-Hill. pp. 2109–2123.
33. Mannoni A, Selvi E, Lorenzini S, Giorgi M, Airó P, Cammelli D, Andreotti L, Marcolongo R, Porfirio B. 2004. Alkaptonuria, ochronosis, and ochronotic arthropathy. *Sem Arthrit Rheum* 33: 239–248.
34. Culotta VC, Gitlin JD. 2001. Disorders of copper transport. In: Scriver CR, Beaudet AL, Sly WS, Valle D, Childs B, Vogelstein B (eds.) *The Metabolic and Molecular Bases of Inherited Disease, 8th Ed*. New York: McGraw-Hill. pp. 3105–3126.

Section IX
Approach to Nephrolithiasis

Section Editor Rajesh V. Thakker

Chapter 102. Renal Tubular Physiology of Calcium Excretion 845
Peter A. Friedman and David A. Bushinsky

Chapter 103. Epidemiology of Nephrolithiasis 856
Murray J. Favus

Chapter 104. Diagnosis and Evaluation of Nephrolithiasis 860
Stephen J. Knohl and Steven J. Scheinman

Chapter 105. Kidney Stones in the Pediatric Patient 869
Amy E. Bobrowski and Craig B. Langman

Chapter 106. Treatment of Renal Stones 878
John R. Asplin

Chapter 107. Genetic Basis of Renal Stones 884
Rajesh V. Thakker

Primer on the Metabolic Bone Diseases and Disorders of Mineral Metabolism, Eighth Edition. Edited by Clifford J. Rosen.
© 2013 American Society for Bone and Mineral Research. Published 2013 by John Wiley & Sons, Inc.

102

Renal Tubular Physiology of Calcium Excretion

Peter A. Friedman and David A. Bushinsky

Introduction 845
Proximal Tubule 845
Henle's Loop 845
Distal Tubule 846

Effect of Diuretics 847
Lessons from the Genetic Hypercalciuric Stone-Forming Rats 847
References 850

INTRODUCTION

The kidneys are primarily responsible for control of extracellular calcium balance. Renal calcium absorption occurs by a series of sequential events as the urine passes through the nephron. Most of the filtered calcium is recovered by proximal tubules, with progressively smaller fractions retrieved as the insipient urine passes through consecutive tubule segments. The bulk of calcium absorption is not hormonally regulated. Hormonal and pharmacological, regulation of calcium absorption is achieved by adjusting the magnitude of calcium recovery in distal segments.

PROXIMAL TUBULE

Approximately 60% of filtered calcium is reabsorbed by proximal tubules [1, 2]. Absorption is mostly passive, driven by the favorable electrochemical gradient generated by the primary absorption of chloride and water [3–5] and proceeding through the lateral intercellular spaces that form the paracellular pathway [Fig. 102.1(A)]. A small fraction of active calcium absorption is present in rats under conditions where passive calcium movement was eliminated [6]. Although minor by comparison with the magnitude of passive paracellular calcium transport, active cellular absorption by proximal tubules amounts to some 20 μmol/min [6], or approximately twice that of the distal nephron, where calcium absorption is entirely cellular. Thus, though slight by comparison with passive proximal calcium absorption, active calcium absorption is substantial compared to that occurring in more distal nephron segments.

HENLE'S LOOP

Thin descending and ascending limbs of Henle's loop exhibit low calcium permeability [7, 8] and do not contribute meaningfully to calcium economy.

Thick ascending limbs of Henle's loop, in contrast, are major sites of calcium absorption. Approximately 20–25% of the filtered calcium is absorbed by thick ascending limbs. Calcium is recovered by both medullary and cortical portions of thick ascending limbs. Unlike proximal tubules, where calcium absorption is mostly passive and proceeds through the paracellular pathway, calcium movement in thick ascending limbs is characterized by parallel movement mediated by active transcellular movement and passive paracellular calcium transport [Fig. 102.1(B)]. Passive calcium absorption is driven by the lumen-positive transepithelial voltage, which provides the driving force for movement through calcium-permeable tight junctions. In this setting, the

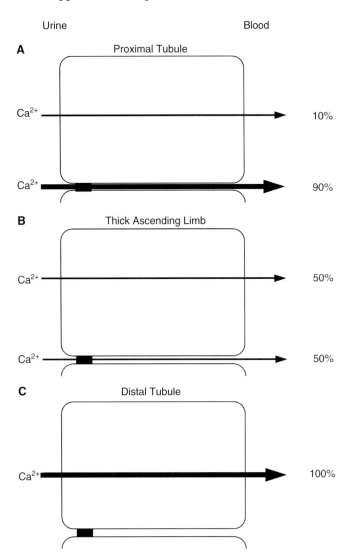

Fig. 102.1. Relative magnitude of cellular and paracellular calcium absorption in proximal tubules, thick ascending limbs, and distal tubules. In thick ascending limbs, transcellular calcium absorption stimulated by PTH approximates the extent of paracellular calcium transport. Notably, in distal tubules calcium absorption proceeds entirely through a cellular pathway.

rate and magnitude of calcium transport are parallel and proportional to those of sodium. Physiological actions or pharmacological or pathological events that enhance sodium absorption increase the voltage and thereby augment calcium absorption. Conversely, maneuvers that decrease sodium absorption reduce calcium absorption. Such interventions commonly involve diuretics such as furosemide or bumetanide that inhibit sodium absorption by thick ascending limbs and increase calcium excretion. Less commonly, mutations of the Na–K–2Cl cotransporter (Slc12a1), ROMK apical K$^+$ channel (KCNJ1), or basolateral ClC–K2 basolateral Cl$^-$ channel (CLCNKB) associated with the different forms of Bartter's syndrome, are accompanied by hypercalciuria, underscoring the parallel nature of sodium and calcium absorption by thick ascending limbs [9].

Active, transcellular calcium transport by medullary and cortical thick ascending limbs also contributes to net absorption by thick ascending limbs [7, 10].

DISTAL TUBULE

Distal convoluted tubules reabsorb 5–10% of the filtered calcium. Here, calcium absorption is entirely transcellular, proceeding against a steep electrochemical gradient [Fig. 1(C)] [11–13]. Transcellular calcium movement is a two-step process, wherein calcium enters the cell across apical membranes and following diffusion across the cell is extruded across basolateral membranes. Calcium entry is mediated by calcium ion channels [14–17], and extrusion is accomplished by the plasma membrane Ca^{2+}-ATPase and Na$^+$/Ca^{2+} exchange [14, 15, 18]. Calcium entry is generally believed to be mediated by TrpV5 (ECaC1, CAT2) [19–21]. TrpV5-null mice exhibit appreciable renal calcium wasting, thus establishing its participation in some aspects of calcium homeostasis in rodents [50]. TrpV5 is a homotetramer that is constitutively active, which is difficult to reconcile with a regulated, negative feedback process [51]. Recent studies show that by stimulating protein kinase A, parathyroid hormone (PTH) phosphorylates and activates TrpV5 [52].

Notably, calcium and sodium transport are inversely related in distal convoluted tubules and this is a hallmark of calcium absorption by this nephron segment. Evidence for the inverse relationship comes from the pharmacological effects of thiazide diuretics, which increase sodium excretion while decreasing that of calcium. Independent evidence comes from patients with Gitelman's syndrome. This inherited disorder is due to inactivating mutations of the Na–Cl cotransporter (NCC; Slc12a3) that mediates sodium absorption by distal convoluted tubules. Patients [53, 54] and experimental animals [55] with Na–Cl cotransporter mutations exhibit characteristic hypocalciuria. Gordon syndrome, which results from mutations in the lysine deficient protein kinase WNK4 causes constitutive activation of NCC. As expected, this not only increases Na absorption but also promotes hypercalciuria, underscoring the reciprocal nature of distal tubule Na and Ca movement.

It has been proposed recently that α-klotho modulates renal calcium homeostasis by acting on distal tubules [56]. Homologous recombination of α-klotho results in a phenotype that includes osteoporosis and renal calcium wasting [57, 58]. The proposed mechanism for this effect involves enhanced Ca^{2+} entry secondary to augmentation of the Na$^+$ gradient following recruitment by α-klotho of the Na$^+$,K$^+$-ATPase to basolateral cell membranes. It was suggested that this action, in cooperation with TRPV5, calbindin$_{28k}$, and the NCX1 Na$^+$/Ca^{2+} calcium exchanger, enhances calcium absorption. Although there is little doubt that α-klotho participates in renal calcium balance, we believe it unlikely that the mechanism proposed by Imura accounts for this action because it would result in parallel increases of Ca^{2+} and Na^+ transport, whereas the signature of distal tubule calcium absorption is the well-established inverse relation between Ca^{2+} and Na^+ flux. Alternatively, the finding that α-klotho activates TRPV5 by hydrolyzing extracellular sugar residues provides a more compelling mechanism [59]. α-Klotho effects on PTH-dependent renal calcium absorption in intact animals, may involve indirect actions on PTH secretion [56] in combination with rather than direct effects on distal nephron calcium movement. A comprehensive picture of the mechanism of PTH action on renal calcium homeostasis requires further examination.

The contribution of cortical collecting tubules to renal calcium conservation is modest at best and uncertain. A small absorptive flux of calcium has been noted as has an equally small secretory transport [13, 60–62]. Net calcium transport in isolated perfused rabbit cortical collecting ducts varied with the magnitude and direction of the transepithelial voltage [60]. A small net secretory calcium flux was observed under ambient conditions. The Ca^{2+} permeability was very low compared with that of thick ascending limbs and proximal tubules. It is conceivable that secretory calcium transport could play a role in regulating calcium excretion when long-term changes of transepithelial voltage occur, for instance, as an adaptive response to changes in mineralocorticoid status [63].

EFFECT OF DIURETICS

Thiazide diuretics and functionally related agents such as metolazone have the unique ability to decrease renal calcium excretion, especially upon chronic administration, while simultaneously increasing sodium excretion. Thiazide diuretics are widely used in the therapeutic management of hypertension, congestive heart failure, and idiopathic hypercalciuria.

Dissociation of sodium and calcium excretion by thiazide diuretics arises primarily from increased calcium reabsorption by distal tubules, but secondarily because of a compensatory elevation of calcium recovery by proximal tubules. The anticalciuric action of thiazide diuretics makes them uniquely useful in treating hypercalciuria. This calcium-sparing effect, which may not be evident upon initial administration of the drug, is particularly beneficial in individuals who are prone to calcium stone formation. Diuretics such as furosemide and bumetanide that act on the loop of Henle's loop cause profound increases in calcium and magnesium excretion. At one time they were used in treating hypercalcemia. However, better alternatives are now available that avoid the likely distortions of extracellular electrolyte balance that accompany treatment with loop diuretics.

LESSONS FROM THE GENETIC HYPERCALCIURIC STONE-FORMING RATS

The primary end point for successful treatment of patients with calcium-containing kidney stones is a decrease in the rate of stone recurrence [64–68]. While decreasing stone formation is an important goal, what should concern clinicians equally is maintaining and improving the patient's bone mineral density (BMD) and bone quality [69]. Stone formers have both a reduction in BMD [66, 69–81] and an increase in fracture rate compared to non-stone formers [70, 71]. While acute stone episodes often are resolved quickly, patients may live the remainder of their lives in pain and with reduced function due to the osseous complications related to fractures [82].

The majority of human kidney stone formers with calcium-containing kidney stones are hypercalciuric when compared to non-stone formers [64–68]. Patients with idiopathic hypercalciuria often excrete more calcium than they absorb, indicating a net loss of total body calcium [64, 66, 78, 83–90]. The source of this additional urine (U) calcium is almost certainly the skeleton, the largest repository of body calcium [79, 91, 92]. Idiopathic hypercalciuria has been associated with markers of increased bone turnover [81, 93, 94]. Urinary hydroxyproline is increased in unselected patients with idiopathic hypercalciuria [93], and serum osteocalcin levels are elevated in stone formers who have a defect in renal tubule calcium reabsorption but not in those with only excessive intestinal calcium absorption [94]. Studies with ^{47}Ca demonstrate increased bone formation and resorption, with the latter predominating [95]. Cytokines known to increase bone resorption have also been shown to be elevated in patients with idiopathic hypercalciuria [75, 96–98].

BMD is correlated inversely with UCa excretion in both men [99] and women [100]. This relationship was confirmed in stone formers but not in non-stone formers [92]. A number of studies demonstrate that patients with nephrolithiasis have a reduction in BMD compared to matched controls [66, 69, 71–81]. After adjusting for a large number of variables, an analysis of the Third National Health and Nutrition Examination Survey (NHANES III) demonstrated that men with a history of kidney stones have a lower femoral neck BMD than those without a history of stones [70]. Analysis of almost 6,000 older men again demonstrated an association of kidney stones with decreased BMD [101]. Stone formers have an increased risk of fractures [70, 71]. In NHANES III there was an increased risk of wrist and spine fractures in stone formers [70], and, in a retrospective analysis, stone formers had an increased incidence of vertebral fractures, but not fractures at other sites [71].

To help understand the mechanism of idiopathic hypercalciuria in humans, we developed an animal model of this disorder [22–49]. Through over 90 generations of successive inbreeding of the most hypercalciuric progeny

Table 102.1. Physiology of Rodent and Human Hypercalciuria

	Hypercalciuric Stone-Forming (IH) Humans	Genetic Hypercalciuric Stone-Forming (GHS) Rats
Urine Ca Excretion	Increased (by definition)	Increased (Refs. 22–41, 46, 67, 106)
Intestinal Ca Absorption	Increased in most patients (Refs. 64, 66, 78, 83, 85, 89, 90, 95, 107–124)	Increased (Refs. 22–41)
Renal Tubular Ca Reabsorption	Decreased in many patients (Refs. 125–127)	Decreased (Ref. 30)
Bone Resorption	Increased in most patients—as evidenced by markers of bone resorption (Refs. 75, 81, 93, 94, 96–98)	Increased (Refs. 29, 33)
Bone Mineral Density	Decreased in most patients (Refs. 66, 72–80, 128, 129)	Decreased (Ref. 46)
Serum PTH	Normal to reduced (Refs. 80, 112, 130–132) or elevated (Refs. 130, 131)	Reduced (Refs. 106)
Serum 1,25(OH)$_2$D$_3$	Normal to elevated (Refs. 79, 80, 89, 91, 111, 112, 133–136)	Normal to elevated (Refs. 28, 36, 37, 42, 103)
Vitamin D Receptor	Increased number (Ref. 104) or no increase (Ref. 137) Gene polymorphisms (Ref. 138–151)	Increased number (Ref. 28, 33, 36, 103, 152)
Ca Receptor	Changes in number not reported Activating and inactivating mutations associated with hyper- and hypocalciuria, respectively (Refs. 153, 154) Gene polymorphisms (Refs. 155, 156)	Increased number (42) Treatment with cinacalcet activates the receptor – associated with increased UCa in SD but not GHS rats (Ref. 106)
Stone formation	Consequence of hypercalciuria (Refs. 64, 66, 83, 85, 87–90, 157–162)	Present (Refs. 22–24, 29, 31, 32, 34)

of the most hypercalciuric Sprague-Dawley (SD) rats, we established a strain of rats that now consistently excrete approximately 8–10 times as much urinary calcium as SD controls (Table 102.1, Fig. 102.2). Compared to SD rats, the genetic hypercalciuric rats have a normal serum calcium and absorb far more dietary calcium [38, 44], similar to observations in man [64, 66]. The increase in intestinal calcium absorption is due to a significant increase in the mucosal-to-serosal (absorptive) calcium flux with no change in the serosal to mucosal (secretory) flux [38]. When the hypercalciuric rats are fed a diet essentially devoid of calcium, UCa excretion remains significantly elevated compared with that of similarly fed SD rats, indicating a defect in renal tubule calcium reabsorption or an increase in bone resorption, or both [37], again similar to observations in humans [79, 102]. Cultured bone from the hypercalciuric rats released more calcium than the bone of SD rats when exposed to increasing amounts of calcitriol [33], and the BMD of the hypercalciuric rats is lower than that of the SD rats [46]. Administration of a bisphosphonate, which significantly inhibits bone resorption, to hypercalciuric rats fed a low-calcium diet significantly reduces urinary calcium excretion [29]. Utilizing clearance studies, a primary defect in renal calcium reabsorption is observed [30]. Thus, these hypercalciuric rats have a systemic abnormality in calcium homeostasis: they absorb more intestinal calcium, they resorb more bone, and they do not adequately reabsorb filtered calcium (Fig. 102.3). We have found that the bone, kidney and intestine of our inbred strain of hypercalciuric rats have an increased number of vitamin D receptors (VDRs) and calcium-sensing receptors (CaSR) [33, 36, 42, 45, 103]. We have recently found that the elevated levels of VDRs are regulated by a decreased level of the transcription factor Snail [48], suggesting potential underlying mechanism(s) for the hypercalciuria. In a human study, circulating monocytes from patients with idiopathic hypercalciuria were shown to have an increased number of VDRs [104]; however, we do not know if hypercalciuric humans have altered levels of Snail.

After eating standard rat chow (1.2% Ca) for 18 weeks, virtually all of these hypercalciuric rats form kidney stones while there is no evidence for stone formation in SD rats [34]. The stones contain only calcium and phosphate (P), without oxalate (Ox), and by X-ray diffraction the stones are exclusively poorly crystalline apatite [23, 24, 31, 34]. When fed additional hydroxyproline, an amino acid that is metabolized to oxalate [105], these rats formed CaOx kidney stones [22, 26], the most common kidney stones formed by man [22, 26]. As each of the hypercalciuric rats forms renal stones, they have been termed genetic hypercalciuric stone-forming (GHS) rats [22–24, 26, 29, 31, 32, 34, 44]. The pathophysiology responsible for the hypercalciuria parallels that found in hypercalciuric humans and is thus an excellent model of hypercalciuria (Table 102.1).

We have recently demonstrated that GHS rats have a reduction in BMD and bone strength even when fed a diet with ample calcium [46] (Fig. 102.4). GHS rats had

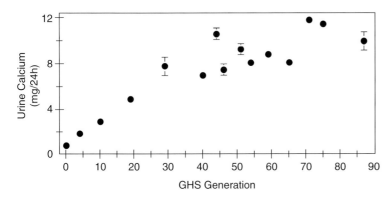

Fig. 102.2. Through successive inbreeding of the most hypercalciuric progeny of SD rats we have established a strain of genetic hypercalciuric stone-forming (GHS) rats that all excrete ~8–10 times as much urinary Ca as the parental strain. All data from published studies (Refs. 22–49).

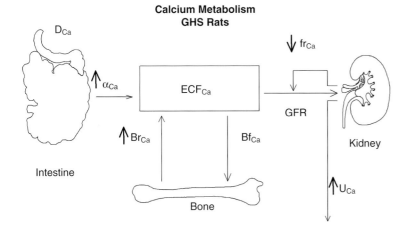

Fig. 102.3. Systemic abnormality in Ca homeostasis in GHS rats: increased intestinal Ca absorption, increased bone resorption, and decreased Ca reabsorption.

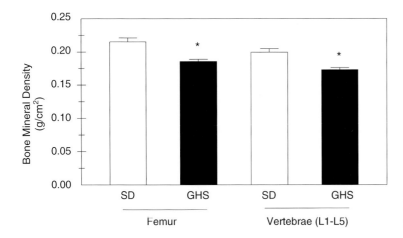

Fig. 102.4. GHS rats have a reduction in BMD in femur and vertebrae even when fed a normal Ca diet (Ref. 63).

reduced cortical (humerus) and trabecular (L1-L5 vertebrae) BMD and had a decrease in trabecular volume and thickness. GHS rats had no change in vertebral strength (failure stress), ductibility (failure strain), stiffness (modulus), or toughness, whereas in the humerus, there was reduced ductibility and toughness and an increase in modulus, indicating that the defect in mechanical properties is mainly manifested in cortical, rather than trabecular, bone. In the GHS rat, the cortical bone is more mineralized than the trabecular bone. Thus, the GHS rats, fed an ample calcium diet, have reduced BMD with reduced trabecular volume, mineralized volume, and thickness, and their bones are more brittle and fracture prone, indicating that GHS rats have an underlying defect of bone not related to a deficiency in dietary calcium.

Thiazide diuretic agents such as chlorthalidone (CTD) reduce UCa excretion in normal [163] patients with hypercalciuria [164] and rats [43, 165]. These drugs act by stimulating calcium reabsorption in the distal convoluted tubule [163] and by producing extracellular fluid volume depletion [166]. Thiazide diuretics are used to treat CaOx stone disease [65, 66]; a meta-analysis revealed that in studies of more than 2 years' duration, there was a significant reduction in stone recurrence rate [167]. A number of studies have shown that when thiazides are used to treat hypertension [168] there is a reduction of osteoporotic fractures [169, 170] and often an increase in BMD [171, 172].

We used GHS rats to test the hypothesis that CTD would have a favorable effect on BMD and bone quality. GHS rats were fed an ample calcium diet and half were also fed CTD [49]. As expected, CTD reduced UCa in GHS rats [43, 165, 173]. In the axial and appendicular skeletons, an increase in trabecular mineralization was observed with CTD compared to controls [49]. CTD also improved the architecture of trabecular bone. Using microcomputed tomography (μCT), trabecular bone volume (BV/TV), trabecular thickness and trabecular number were increased with CTD. A significant increase in trabecular thickness with CTD was confirmed by static histomorphometry. CTD also improved the connectivity of trabecular bone. Significant improvements in vertebral strength and stiffness were measured by vertebral compression. Conversely, a slight loss of bending strength was detected in the femoral diaphysis with CTD. These results obtained in hypercalciuric rats suggest that CTD can favorably influence vertebral fracture risk.

REFERENCES

1. Lassiter WE, Gottschalk CW, Mylle M. 1963. Micropuncture study of renal tubular reabsorption of calcium in normal rodents. *Am J Physiol* 204: 771–775.
2. Frick A, Rumrich G, Ullrich KJ, Lassiter WE. 1965. Microperfusion study of calcium transport in the proximal tubule of the rat kidney. *Pflugers Arch Gesamte Physiol Menschen Tiere* 286: 109–117.
3. Berry CA, Rector FC Jr. 2011. Relative sodium-to-chloride permeability in the proximal convoluted tubule. *Am J Physiol* 235: F592–F604.
4. Ng RC, Rouse D, Suki WN. 1984. Calcium transport in the rabbit superficial proximal convoluted tubule. *J Clin Invest* 74: 834–842.
5. Bomsztyk K, George JP, Wright FS. 1984. Effects of luminal fluid anions on calcium transport by proximal tubule. *Am J Physiol* 246(5 Pt 2): F600–F608.
6. Ullrich KJ, Rumrich G, Kloss S. 1976. Active Ca^{2+} reabsorption in the proximal tubule of the rat kidney. Dependence on sodium- and buffer transport. *Pfleugers Arch* 364: 223–228.
7. Rocha AS, Magaldi JB, Kokko JP. 1977. Calcium and phosphate transport in isolated segments of rabbit Henle's loop. *J Clin Invest* 59: 975–983.
8. Rouse D, Ng RCK, Suki WN. 1980. Calcium transport in the pars recta and thin decending limb of Henle of rabbit perfused in vitro. *J Clin Invest* 65: 37–42.
9. Hebert SC. 2003. Bartter syndrome. *Curr Opin Nephrol Hypertens* 12: 527–532.
10. Eknoyan G, Suki WN, Martinez-Maldonado M. 1970. Effect of diuretics on urinary excretion of phosphate, calcium, and magnesium in thyroparathyroidectomized dogs. *J Lab Clin Med* 76: 257–266.
11. Costanzo LS, Windhager EE. 1978. Calcium and sodium transport by the distal convoluted tubule of the rat. *Am J Physiol* 235: F492–F506.
12. Lau K, Bourdeau JE. 1995. Parathyroid hormone action in calcium transport in distal nephron. *Curr Opin Nephrol Hypertens* 4: 55–63.
13. Shareghi GR, Stoner LC. 1978. Calcium transport across segments of the rabbit distal nephron in vitro. *Am J Physiol* 235: F367–F375.
14. Gotch FA, Kotanko P, Thijssen S, Levin NW. 2010. The KDIGO guideline for dialysate calcium will result in an increased incidence of calcium accumulation in hemodialysis patients. *Kidney Int* 78: 343–350.
15. Poncet V, Merot J, Poujeol P. 1992. A calcium-permeable channel in the apical membrane of primary cultures of the rabbit distal bright convoluted tubule. *Pfluegers Arch* 422: 112–119.
16. Saunder JCJ, Isaacson LC. 1990. Patch clamp study of Ca channels in isolated renal tubule segments. In: Pansu D, Bronner F (eds.) *Calcium Transport and Intracellular Calcium Homeostasis*. Berlin: Springer-Verlag. pp. 27–34.
17. Lau K, Quamme G, Tan S. 1991. Patch-clamp evidence for a Ca channel in apical membrane of cortical thick ascending limb (cTAL) and distal tubule (DT) cells. *J Am Soc Nephrol* 2: 775.
18. Bacskai BJ, Friedman PA. 1990. Activation of latent Ca^{2+} channels in renal epithelial cells by parathyroid hormone. *Nature* 347: 388–391.
19. Vennekens R, Hoenderop JG, Prenen J, Stuiver M, Willems PH, Droogmans G, Nilius B, Bindels RJ. 2000. Permeation and gating properties fo the novel epithelial Ca^{2+} channel. *J Biol Chem* 275: 3963–3969.
20. Hoenderop JGJ, van der Kemp AWCM, Hartog A, Van Os CH, Willems PHGM, Bindels RJ. 1999. The epithelial calcium channel, ECaC, is activated by hyperpolarization and regulated by cytosolic calcium. *Biochem Biophys Res Commun* 261: 488–492.
21. Hoenderop JG, van der Kemp AW, Hartog A, van de Graaf SF, Van Os CH, Willems PH, Bindels RJ. 1999. Molecular identification of the apical Ca^{2+} channel in 1,25-dihydroxyvitamin D_3-responsive epithelia. *J Biol Chem* 274: 8375–8378.
22. Bushinsky DA, Asplin JR, Grynpas MD, Evan AP, Parker WR, Alexander KM, Coe FL. 2002. Calcium oxalate stone formation in genetic hypercalciuric stone-forming rats. *Kidney Int* 61: 975–987.
23. Bushinsky DA, Grynpas MD, Asplin JR. 2001. Effect of acidosis on urine supersaturation and stone formation

in genetic hypercalciuric stone forming rats. *Kidney Int* 59: 1415–1423.
24. Bushinsky DA, Parker WR, Asplin JR. 2000. Calcium phosphate supersaturation regulates stone formation in genetic hypercalciuric stone-forming rats. *Kidney Int* 57: 550–560.
25. Scheinman SJ, Cox JPD, Lloyd SE, Pearce SHS, Salenger PV, Hoopes RR, Bushinsky DA, Wrong O, Asplin J, Langman CB, Norden AG, Thakker RV. 2000. Isolated hypercalciuria with mutation in CLCN5: Relevance to idiopathic hypercalciuria. *Kidney Int* 57: 232–239.
26. Evan AP, Bledsoe SB, Smith SB, Bushinsky DA. 2004. Calcium oxalate crystal localization and osteopontin immunostaining in genetic hypercalciuric stone-forming rats. *Kidney Int* 65: 154–161.
27. Hoopes RR, Reid R, Sen S, Szpirer C, Dixon P, Pannet A, Thakker RV, Bushinsky DA, Scheinman SJ. 2003. Quantitative trait loci for hypercalciuria in a rat model of kidney stone disease. *J Am Soc Nephrol* 14: 1844–1850.
28. Yao J, Kathpalia P, Bushinsky DA, Favus MJ. 1998. Hyperresponsiveness of vitamin D receptor gene expression to 1,25-dihydroxyvitamin D_3: A new characteristic of genetic hypercalciuric stone-forming rats. *J Clin Invest* 101: 2223–2232.
29. Bushinsky DA, Neumann KJ, Asplin J, Krieger NS. 1999. Alendronate decreases urine calcium and supersaturation in genetic hypercalciuric rats. *Kidney Int* 55: 234–243.
30. Tsuruoka S, Bushinsky DA, Schwartz GJ 1997 Defective renal calcium reabsorption in genetic hypercalciuric rats. *Kidney Int* 51: 1540–1547.
31. Asplin JR, Bushinsky DA, Singharetnam W, Riordon D, Parks JH, Coe FL. 1997. Relationship between supersaturation and crystal inhibition in hypercalciuric rats. *Kidney Int* 51: 640–645.
32. Bushinsky DA, Bashir MA, Riordon DR, Nakagawa Y, Coe FL, Grynpas MD. 1999. Increased dietary oxalate does not increase urinary calcium oxalate saturation in hypercalciuric rats. *Kidney Int* 55: 602–612.
33. Krieger NS, Stathopoulos VM, Bushinsky DA. 1996. Increased sensitivity to 1,25(OH)$_2$D$_3$ in bone from genetic hypercalciuric rats. *Am J Physiol* 271: C130–C135.
34. Bushinsky DA, Grynpas MD, Nilsson EL, Nakagawa Y, Coe FL. 1995. Stone formation in genetic hypercalciuric rats. *Kidney Int* 48: 1705–1713.
35. Bushinsky DA, Kim M, Sessler NE, Nakagawa Y, Coe FL. 1994. Increased urinary saturation and kidney calcium content in genetic hypercalciuric rats. *Kidney Int* 45: 58–65.
36. Li X-Q, Tembe V, Horwitz GM, Bushinsky DA, Favus MJ. 1993. Increased intestinal vitamin D receptor in genetic hypercalciuric rats: A cause of intestinal calcium hyperabsorption. *J Clin Invest* 91: 661–667.
37. Kim M, Sessler NE, Tembe V, Favus MJ, Bushinsky DA. 1993. Response of genetic hypercalciuric rats to a low calcium diet. *Kidney Int* 43: 189–196.
38. Bushinsky DA, Favus MJ. 1988. Mechanism of hypercalciuria in genetic hypercalciuric rats: inherited defect in intestinal calcium transport. *J Clin Invest* 82: 1585–1591.
39. Bushinsky DA. 1996. Genetic hypercalciuric stone forming rats. *Semin Nephrol* 16: 448–457.
40. Bushinsky DA. 1999. Genetic hypercalciuric stone-forming rats. *Curr Opin Nephrol Hypertens* 8: 479–488.
41. Bushinsky DA. 2000. Bench to bedside: Lessons from the genetic hypercalciuric stone forming rat. *Am J Kidney Dis* 36: 61–64.
42. Yao J, Karnauskas AJ, Bushinsky DA, Favus MJ. 2005. Regulation of renal calcium-sensing receptor gene expression in response to 1,25(OH)$_2$D$_3$ in genetic hypercalciuric stone-forming rats. *J Am Soc Nephrol* 16: 1300–1308.
43. Bushinsky DA, Asplin JR. 2005. Thiazides reduce brushite, but not calcium oxalate, supersaturation and stone formation in genetic hypercalciuric stone-forming rats. *J Am Soc Nephrol* 16: 417–424.
44. Bushinsky DA, Frick KK, Nehrke K. 2006. Genetic hypercalciuric stone-forming rats. *Curr Opinion Nephrol Hypertens* 15: 403–418.
45. Hoopes RR Jr, Middleton FA, Sen S, Hueber PA, Reid R, Bushinsky DA, Scheinman SJ. 2006. Isolation and confirmation of a calcium excretion quantitative trait locus on chromosome 1 in genetic hypercalciuric stone-forming congenic rats. *J Am Soc Nephrol* 17: 1292–1304.
46. Grynpas M, Waldman S, Holmyard D, Bushinsky DA. 2009. Genetic hypercalciuric stone-forming rats have a primary decrease in bone mineral density and strength. *J Bone Miner Res* 24: 1420–1426.
47. Asplin JR, Donahue SE, Lindeman C, Michalenka A, Strutz KL, Bushinsky DA. 2009. Thiosulfate reduces calcium phosphate nephrolithiasis. *J Am Soc Nephrol* 20: 1246–1253.
48. Bai S, Wang.H, Shen J, Zhou R, Bushinsky DA, Favus MJ. 2010. Elevated vitamin D receptor levels in genetic hypercalciuric stone-forming rats are associated with downregulation of Snail. *J Bone Miner Res* 25: 830–840.
49. Bushinsky DA, Willett T, Asplin JR, Culbertson C, Che SPY, Grynpas M. 2011. Chlorthalidone improves vertebral bone quality in genetic hypercalciuric stone-forming rats. *J Bone Miner Res* 26: 1904–1912.
50. Hoenderop JG, van Leeuwen JP, van der Eerden BC, Kersten FF, Van Der Kemp A, Merillat AM, Waarsing JH, Rossier BC, Vallon V, Hummler E, Bindels RJ. 2003. Renal Ca^{2+} wasting, hyperabsorption, and reduced bone thickness in mice lacking TRPV5. *J Clin Invest* 112: 1906–1914.
51. Nilius B, Vennekens R, Prenen J, Hoenderop JG, Bindels RJ, Droogmans G. 2000. Whole-cell and single channel monovalent cation currents through the novel rabbit epithelial Ca^{2+} channel monovalent cation currents through the novel rabbit epithelial Ca^{2+} channel ECaC. *J Physiol* 527 Pt 2: 239–248.

52. de Groot T, Lee K, Langeslag M, Xi Q, Jalink K, Bindels RJM, Hoenderop JGJ. 2009. Parathyroid hormone activates TRPV5 via PKA-dependent phosphorylation. *J Am Soc Nephrol* 20: 1693–1704.
53. Gitelman HJ, Graham JB, Welt LG. 1966. A new familial disorder characterized by hypokalemia and hypomagnesemia. *Trans Assoc Am Physicians* 79: 221–235.
54. Bettinelli A, Bianchetti MG, Girardin E, Caringella A, Cecconi M, Appiani AC, Pavanello L, Gastaldi R, Isimbaldi C, Lama G, et al. 1992. Use of calcium excretion values to distinguish two forms of primary renal tubular hypokalemic alkalosis: Bartter and Gitelman syndromes. *J Pediatr* 120: 38–43.
55. Schultheis PJ, Lorenz JN, Meneton P, Nieman ML, et al. 1998. Phenotype resembling Gitelmans's syndrome in mice lacking the apical Na^+-Cl^- cotransporter of the distal convoluted tubule. *J Biol Chem* 273: 29150–29155.
56. Imura A. 2007. alpha-Klotho as a regulator of calcium homeostasis. *Science* 316: 1615–1618.
57. Kuro-o M, Matsumura Y, Aizawa H, Kawaguchi H, Suga T, Utsugi T, Ohyama Y, Kurabayashi M, Kaname T, Kume E, Iwasaki H, Iida A, Shiraki-Iida T, Nishikawa S, Nagai R, Nabeshima Y. 1997. Mutation of the mouse klotho gene leads to a syndrome resembling ageing. *Nature* 390: 45–51.
58. Tsuruoka S, Nishiki K, Ioka T, Ando H, Saito Y, Kurabayashi M, Nagai R, Fujimura A. 2006. Defect in parathyroid-hormone-induced luminal calcium absorption in connecting tubules of Klotho mice. *Nephrol Dial Transplant* 21: 2762–2767.
59. Chang Q, Hoefs S, van der Kemp AW, Topala CN, Bindels RJ, Hoenderop JG. 2005. The beta-glucuronidase klotho hydrolyzes and activates the TRPV5 channel. *Science* 310: 490–493.
60. Bourdeau JE, Hellstrom-Stein RJ. 1982. Voltage-dependent calcium movement across the cortical collecting duct. *Am J Physiol* 242: F285–F292.
61. Imai M. 1981. Effects of parathyroid hormone and N^6,O^2-dibutyryl cyclic AMP on calcium transport across the rabbit distal nephron segments perfused in vitro. *Pflugers Arch* 390: 145–151.
62. Shimizu T, Yoshitomi K, Nakamura M, Imai M. 1990. Effects of PTH, calcitonin, and cAMP on calcium transport in rabbit distal nephron segments. *Am J Physiol* 259: F408–F414.
63. O'Neil RG. 1990. Aldosterone regulation of sodium and potassium transport in the cortical collecting duct. *Semin Nephrol* 10: 365–374.
64. Monk RD, Bushinsky DA. 2010. Nephrolithiasis and nephrocalcinosis. In: Frehally J, Floege J, Johnson RJ (eds.) *Comprehensive Clinical Nephrology, 4th Ed.* St. Louis, MO: Elsevier. pp. 687–701.
65. Bushinsky DA, Coe FL, Moe OW. 2008. Nephrolithiasis. In: Brenner BM (ed.) *The Kidney, 8th Ed.* Philadelphia: W.B. Saunders. pp. 1299–1349.
66. Monk RD, Bushinsky DA. 2008. Kidney stones. In: Kronenberg HM, Melmed S, Polonsky KS, Larsen PR (eds.) *Williams Textbook of Endocrinology, 11th ed.* Philadelphia: W.B Saunders. pp. 1311–1326.
67. Bushinsky DA. 2008. Calcium nephrolithiasis. In: Rosen CJ (cd.) *Primer on the Metabolic Bone Diseases and Disorders of Mineral Metabolism, 7th Ed.* Washington, DC: American Society of Bone and Mineral Research. pp. 460–464.
68. Worcester EM, Coe FL. 2010. Calcium kidney stones. *N Engl J Med* 363: 954–963.
69. Sakhaee K, Maalouf NM, Kumar R, Pasch A, Moe OW. 2011. Nephrolithiasis-associated bone disease: Pathogenesis and treatment options. *Kidney Int* 79: 393–403.
70. Lauderdale DS, Thisted RA, Wen M, Favus M. 2001. Bone mineral density and fracture among prevalent kidney stone cases in the Third National Health and Nutrition Examination Survey. *J Bone Miner Res* 16: 1893–1898.
71. Melton LJI, Crowson CS, Khosla S, Wilson DM, Fallon WM. 1998. Fracture risk among patients with urolithiasis: A population based cohort study. *Kidney Int* 53: 459–464.
72. Pietschmann F, Breslau NA, Pak CYC. 1992. Reduced vertebral bone density in hypercalciuric nephrolithiasis. *J Bone Miner Res* 7: 1383–1388.
73. Jaeger P, Lippuner K, Casez JP, Hess B, Ackerman D, Hug C. 1994. Low bone mass in idiopathic renal stone formers: Magnitude and significance. *J Bone Miner Res* 9: 1525–1532.
74. Giannini S, Nobile M, Sartori L, Calo L, Tasca A, Dalle Carbonare L, Ciuffreda M, D'Angelo A Pagano F Crepaldi G. 1998. Bone density and skeletal metabolism are altered in idiopathic hypercalciuria. *Clin Nephrol* 50: 94–100.
75. Misael da Silva AM, dos Reis LM, Pereira RC, Futata E, Branco-Martins CT, Noronha IL, Wajchemberg BL, Jorgetti V. 2002. Bone involvement in idiopathic hypercalciuria. *Clin Nephrol* 57: 183–191.
76. Tasca A, Cacciola A, Ferrarese P, Ioverno E, Visona E, Bernardi C, Nobile M, Giannini S. 2002. Bone alterations in patients with idiopathic hypercalciuria and calcium nephrolithiasis. *Urology* 59: 865–869.
77. Heilberg IP, Martini LA, Teixeira SH, Szejnfeld VL, Carvalho AB, Lobao R, Draibe SA. 1998. Effect of etidronate treatment on bone mass of male nephrolithiasis patients with idiopathic hypercalciuria and osteopenia. *Nephron* 79: 430–437.
78. Bushinsky DA. 2002. Recurrent hypercalciuric nephrolithiasis—Does diet help? *N Engl J Med* 346: 124–125.
79. Coe FL, Favus MJ, Crockett T, Strauss AL, Parks JH, Porat A, Gantt C, Sherwood LM. 1982. Effects of low-calcium diet on urine calcium excretion, parathyroid function and serum $1,25(OH)_2D_3$ levels in patients with idiopathic hypercalciuria and in normal subjects. *Am J Med* 72: 25–32.
80. Bataille P, Achard JM, Fournier A, Boudailliez B, Westell PF, Esper NE, Bergot C, Jans I, Lalau JD, Petit J, Henon G, Jeantet MAL, Bouillon R, Sebert JL. 1991.

Diet, vitamin D and vertebral mineral density in hypercalciuric calcium stone formers. *Kidney Int* 39: 1193–1205.
81. Heilberg IP, Weisinger JR. 2006. Bone disease in idiopathic hypercalciuria. *Curr Opin Nephrol Hypertens* 15: 394–402.
82. Trombetti A, Herrmann F, Hoffmeyer P, Schurch MA, Bonjour JP, Rizzoli R. 2002. Survival and potential years of life lost after hip fracture in men and age-matched women. *Osteoporos Int* 13: 731–737.
83. Bushinsky DA. 2000. Renal lithiasis. In: Humes HD (ed.) *Kelly's Textbook of Medicine*. New York: Lippincott Williams & Wilkens. pp. 1243–1248.
84. Bushinsky DA, Parker WR, Alexander KM, Krieger NS. 2001. Metabolic, but not respiratory, acidosis increases bone PGE_2 levels and calcium release. *Am J Physiol Renal Physiol* 281: F1058–F1066.
85. Bushinsky DA. 1998. Nephrolithiasis. *J Am Soc Nephrol* 9: 917–924.
86. Consensus Conference. 1988. Prevention and treatment of kidney stones. *JAMA* 260: 977–981.
87. Pak CYC. 1992. Pathophysiology of calcium nephrolithiasis. In: Seldin DW, Giebisch G (eds.) *The Kidney: Physiology and Pathophysiology, 2nd Ed*. New York: Raven Press, Ltd. pp. 2461–2480.
88. Coe FL, Parks JH, Asplin JR. 1992. The pathogenesis and treatment of kidney stones. *N Engl J Med* 327: 1141–1152.
89. Coe FL, Favus MJ, Asplin JR. 2004. Nephrolithiasis. In: Brenner BM, Rector FC Jr. (eds.) *The Kidney, 7th Ed*. Philadelphia: W.B. Saunders Company. pp. 1819–1866.
90. Coe FL, Bushinsky DA. 1984. Pathophysiology of hypercalciuria. *Am J Physiol Renal Physiol* 247: F1–F13.
91. Monk RD, Bushinsky DA. 1996. Pathogenesis of idiopathic hypercalciuria. In: Coe F, Favus M, Pak C, Parks J, Preminger G (eds.) *Kidney Stones: Medical and Surgical Management*. Philadelphia: Lippincott-Raven. pp. 759–772.
92. Asplin JR, Bauer KA, Kinder J, Muller G, Coe BJ, Parks JH, Coe FL. 2003. Bone mineral density and urine calcium excretion among subjects with and without nephrolithiasis. *Kidney Int* 63: 662–669.
93. Sutton RAL, Walker VR. 1986. Bone resorption and hypercalciuria in calcium stone formers. *Metabolism* 35: 485–488.
94. Urivetzky M, Anna PS, Smith AD. 1988. Plasma osteocalcin levels in stone disease: A potential aid in the differential diagnosis of calcium nephrolithiasis. *J Urol* 139: 12–14.
95. Liberman UA, Sperling O, Atsmon A, Frank M, Modan M, deVries A. 1968. Metabolic and calcium kinetic studies in idiopathic hypercalciuria. *J Clin Invest* 47: 2580–2590.
96. Pacifici R, Rothstein M, Rifas L, et al. 1990. Increased monocyte interleukin-1 activity and decreased vertebral bone density in patients with fasting idiopathic and hypercalciuria. *J Clin Endocrinol Metab* 71: 138–145.
97. Weisinger JR, Alonzo E, Bellorin-Font E, et al. 1996. Possible role of cytokines on the bone mineral loss in idiopathic hypercalciuria. *Kidney Int* 49: 244–250.
98. Ghazali A, Fuentes V, Desaint C, et al. 1997. Low bone mineral density and peripheral blood monocyte activation profile in calcium stone formers with idiopathic hypercalciuria. *J Clin Endocrinol Metab* 82: 32–38.
99. Vezzoli G, Soldati L, Ardila M, et al. 2005. Urinary calcium is a determinant of bone mineral density in elderly men participating in the InCHIANTI study. *Kidney Int* 67: 2006–2014.
100. Giannini S, Nobile M, Dalle Carbonare L, et al. 2003. Hypercalciuria is a common and important finding inpostmenopausal women with osteoporosis. *Eur J Endocrinol* 149: 209–213.
101. Cauley JA, Fullman RL, Stone KL, et al. 2005. Factors associated with the lumbar spine and proximal femur bone mineral density in older men. *Osteoporos Int* 16: 1525–1537.
102. Pak CY. 1997. Nephrolithiasis. *Curr Ther Endocrinol Metab* 6: 572–576.
103. Karnauskas AJ, van Leeuwen JP, van den Bemd GJ, Kathpalia PP, DeLuca HF, Bushinsky DA, Favus MJ. 2005. Mechanism and function of high vitamin D receptor levels in genetic hypercalciuric stone-forming rats. *J Bone Miner Res* 20: 447–454.
104. Favus MJ, Karnauskas AJ, Parks JH, Coe FL. 2004. Peripheral blood monocyte vitamin D receptor levels are elevated in patients with idiopathic hypercalciuria. *J Clin Endocrinol Metab* 89: 4937–4943.
105. Hagler L, Herman RH. 1973. Oxalate metabolism. I. *Am J Clin Nutr* 26: 758–765.
106. Bushinsky DA, LaPlante K, Asplin JR. 2006. Effect of cinacalcet on urine calcium excretion and supersaturation in genetic hypercalciuric stone-forming rats. *Kidney Int* 69: 1586–1592.
107. Birge SJ, Peck WA, Berman M, Whedon GD. 1969. Study of calcium absorption in man: A kinetic analysis and physiologic model. *J Clin Invest* 48: 1705–1713.
108. Wills MR, Zisman E, Wortsman J, Evens RG, Pak CYC, Bartter FC. 1970. The measurement of intestinal calcium absorption by external radioisotope counting: Application to study of nephrolithiasis. *Clin Sci* 39: 95–106.
109. Pak CYC, East DA, Sanzenbacher LJ, Delea CS, Bartter FC. 1972. Gastrointestinal calcium absorption in nephrolithiasis. *J Clin Endocrinol Metab* 35: 261–270.
110. Pak CYC, Ohata M, Lawrence EC, Snyder W. 1974. The hypercalciurias: Causes, parathyroid functions, and diagnostic criteria. *J Clin Invest* 54: 387–400.
111. Kaplan RA, Haussler MR, Deftos LJ, Bone H, Pak CYC. 1977. The role of 1,25 dihydroxyvitamin D in the mediation of intestinal hyperabsorption of calcium in primary hyperparathyroidism and absorptive hypercalciuria. *J Clin Invest* 59: 756–760.
112. Shen FH, Baylink DJ, Nielsen RL, Sherrard DJ, Ivey JL, Haussler MR. 1977. Increased serum 1,25-dihydroxyvitamin D in idiopathic hypercalciuria. *J Lab Clin Med* 90: 955–962.

113. Frick KK, Bushinsky DA. 2003. Molecular mechanisms of primary hypercalciuria. *J Am Soc Nephrol* 14: 1082–1095.
114. Lemann J Jr. 1992. Pathogenesis of idiopathic hypercalciuria and nephrolithiasis. In: Coe FL, Favus MJ (eds.) *Disorders of Bone and Mineral Metabolism*. New York: Raven Press, Ltd. pp. 685–706.
115. Henneman PH, Benedict PH, Forbes AP, Dudley HR. 1958. Idiopathic hypercalciuria. *N Engl J Med* 259: 802–807.
116. Jackson WPU, Dancaster C. 1959. A consideration of the hypercalciuria in sarcoidosis, idiopathic hypercalciuria, and that produced by vitamin D. A new suggestion regarding calcium metabolism. *J Clin Endocrinol Metab* 19: 658–681.
117. Edwards NA, Hodgkinson A. 1965. Metabolic studies in patients with idiopathic hypercalciuria. *Clin Sci* 29: 143–157.
118. Harrison AR. 1959. Some results of metabolic investigation in cases of renal stone. *Br J Urol* 31: 398.
119. Dent CE, Harper CM, Parfitt AM. 1964. The effect of cellulose phosphate on calcium metabolism in patients with hypercalciuria. *Clin Sci* 27: 417–425.
120. Nassim JR, Higgins BA. 1965. Control of idiopathic hypercalciuria. *Br Med J* 1: 675–681.
121. Caniggia A, Gennari C, Cesari L. 1965. Intestinal absorption of ^{45}Ca in stone-forming patients. *Br Med J* 1: 427–429.
122. Ehrig U, Harrison JE, Wilson DR. 1974. Effect of long-term thiazide therapy on intestinal calcium absorption in patients with recurrent renal calculi. *Metabolism* 23: 139–149.
123. Barilla DE, Tolentino R, Kaplan RA, Pak CYC. 1978. Selective effects of thiazide on intestinal absorption of calcium in absorptive and renal hypercalciurias. *Metabolism* 27: 125–131.
124. Zerwekh JE, Pak CYC. 1980. Selective effect of thiazide therapy on serum 1, 25-dihydroxyvitamin D, and intestinal absorption in renal and absorptive hypercalciuria. *Metabolism* 29: 13–17.
125. Pak CYC, Kaplan R, Bone H. 1975. A simple test for the diagnosis of absorptive, resorptive and renal hypercalciurias. *New Engl J Med* 292: 497–500.
126. Pak CY. 1998. Kidney stones. *Lancet* 351: 1797–1801.
127. Pak CYC, Britton F, Peterson R, Ward D, Northcutt C, Breslau NA, McGuire J, Sakhaee K, Bush S, Nicar M, Norman D, Peters P. 1980. Ambulatory evaluation of nephrolithiasis: Classification, clinical presentation and diagnostic criteria. *Am J Med* 69: 19–30.
128. Barkin J, Wilson DR, Manuel MA, Bayley A, Murray T, Harrison J. 1985. Bone mineral content in idiopathic calcium nephrolithiasis. *Min Electro Metab* 11: 19–24.
129. Alhava EM, Juuti M, Karjalainen P. 1976. Bone mineral density in patients with urolithiasis. *Scan J Urol Nephrol* 10: 154–156.
130. Coe FL, Canterbury JM, Firpo JJ, Reiss E. 1973. Evidence for secondary hyperparathyroidism in idiopathic hypercalciuria. *J Clin Invest* 52: 134–142.
131. Bordier P, Ryckewart A, Gueris J, Rasmussen H. 1977. On the pathogenesis of so-called idiopathic hypercalciuria. *Am J Med* 63: 398–409.
132. Burckhardt P, Jaeger P. 1981. Secondary hyperparathyroidism in idiopathic renal hypercalciuria: Fact or theory? *J Clin Endocrinol Metab* 55: 550–555.
133. Haussler MR, Baylink DJ, Hughes MR. 1976. The assay of 1,25-dihydroxy vitamin D_3: physiologic and pathologic modulation of circulating hormone levels. *Clin Endocrinol* 5: s151–s165.
134. Gray RW, Wilz DR, Caldas AE, Lemann J Jr. 1977. The importance of phosphate in regulating plasma $1,25(OH)_2$ vitamin D levels in humans: Studies in healthy subjects, in calcium stone formers and in patients with primary hyperparathyroidism. *J Clin Endocrinol Metab* 45: 299–306.
135. Broadus AE, Insogna KL, Lang R, Ellison AF, Dreyer BE. 1984. Evidence for disordered control of 1,25-dihydroxyvitamin D production in absorptive hypercalciuria. *N Engl J Med* 311: 73–80.
136. Insogna KL, Broadus AE, Dryer BE, Ellison AF, Gertner JM. 1985. Elevated production rate of 1,25-dihydroxyvitamin D in patients with absorptive hypercalciuria. *J Clin Endocrinol Metab* 61: 490–495.
137. Zerwekh JE, Reed BY, Heller HJ, Gonzalez GB, Haussler MR, Pak CY. 1998. Normal vitamin D receptor concentration and responsiveness to 1,25-dihydroxyvitamin D_3 in skin fibroblasts from patients with absorptive hypercalciuria. *Miner Electrolyte Metab* 24: 307–313.
138. Rendina D, Mossetti G, Viceconti R, Sorrentino M, Castaldo R, Manno G, Guadagno V, Strazzullo P, Nunziata V. 2004. Association between vitamin D receptor gene polymorphisms and fasting idiopathic hypercalciuria in recurrent stone-forming patients. *Urology* 64: 833–838.
139. Vezzoli G, Soldati L, Proverbio MC, Adamo D, Rubinacci A, Bianchi G, Mora S. 2002. Polymorphism of vitamin D receptor gene start codon in patients with calcium kidney stones. *J Nephrol* 15: 158–164.
140. Bid HK, Kumar A, Kapoor R, Mittal RD. 2005. Association of vitamin D receptor gene (FokI) polymorphism with calcium oxalate nephrolithiasis. *J Endourol* 19: 111–115.
141. Bid HK, Chaudhary H, Mittal RD. 2005. Association of vitamin D and calcitonin receptor gene polymorphism in paediatric nephrolithiasis. *Pediatr Nephrol* 20: 773–776.
142. Nishijima S, Sugaya K, Naito A, Morozumi M, Hatano T, Ogawa Y. 2002. Association of vitamin D receptor gene polymorphism with urolithiasis. *J Urol* 167: 2188–2191.
143. Chen WC, Chen HY, Lu HF, Hsu CD, Tsai FJ. 2001. Association of the vitamin D receptor gene start codon Fok I polymorphism with calcium oxalate stone disease. *BJU Int* 87: 168–171.
144. Valdivielso JM, Fernandez E. 2006. Vitamin D receptor polymorphisms and diseases. *Clin Chim Acta* 371: 1–12.

145. Ozkaya O, Soylemezoglu O, Misirlioglu M, Gonen S, Buyan N, Hasanoglu E. 2003. Polymorphisms in the vitamin D receptor gene and the risk of calcium nephrolithiasis in children. *Eur Urol* 44: 150–154.

146. Scott P, Ouimet D, Valiquette L, Guay G, Proulx Y, Trouve ML, Gagnon B, Bonnardeaux A. 1999. Suggestive evidence for a susceptibility gene near the vitamin D receptor locus in idiopathic calcium stone formation. *J Am Soc Nephrol* 10: 1007–1013.

147. Jackman SV, Kibel AS, Ovuworie CA, Moore RG, Kavoussi LR, Jarrett TW. 1999. Familial calcium stone disease: TaqI polymorphism and the vitamin D receptor. *J Endourol* 13: 313–316.

148. Mossetti G, Vuotto P, Redina D, Numis FG, Viceconti R, Giordano F, Cioffi M, Scopacasa F, Nunziata V. 2003. Association between vitamin D receptor gene polymorphisms and tubular citrate handling in calcium nephrolithiasis. *J Intern Med* 253: 194–200.

149. Mossetti G, Rendina D, Viceconti R, Manno G, Guadagno V, Strazzullo P, Nunziata V. 2004. The relationship of 3' vitamin D receptor haplotypes to urinary supersaturation of calcium oxalate salts and to age at onset and familial prevalence of nephrolithiasis. *Nephrol Dial Transplant* 19: 2259–2265.

150. Ruggiero M, Pacini S, Amato M, Aterini S, Chiarugi V. 1999. Association between vitamin D receptor gene polymorphism and nephrolithiasis. *Miner Electrolyte Metab* 25: 185–190.

151. Uitterlinden AG, Fang Y, Van Meurs JB, Pols HA, van Leeuwen JP. 2004. Genetics and biology of vitamin D receptor polymorphisms. *Gene* 338: 143–156.

152. Favus MJ. 1994. Hypercalciuria: Lessons from studies of genetic hypercalciuric rats. *J Am Soc Nephrol* 5: S54–S58.

153. Gambaro G, Vezzoli G, Casari G, Rampoldi L, D'Angelo A, Borghi L. 2004. Genetics of hypercalciuria and calcium nephrolithiasis: From the rare monogenic to the common polygenic forms. *Am J Kidney Dis* 44: 963–986.

154. Chattopadhyay N, Brown EM. 2006. Role of calcium-sensing receptor in mineral ion metabolism and inherited disorders of calcium-sensing. *Mol Genet Metab* 89: 189–202.

155. Vezzoli G, Tanini A, Ferrucci L, Soldati L, Bianchin C, Franceschelli F, Malentacchi C, Porfirio B, Adamo D, Terranegra A, Falchetti A, Cusi D, Bianchi G, Brandi ML. 2002. Influence of calcium sensing receptor gene on urinary calcium excretion in stone-forming patients. *J Am Soc Nephrol* 13: 2517–2523.

156. Scillitani A, Guarnieri V, De Geronimo S, Muscarella LA, Battista C, D'Agruma L, Bertoldo F, Florio C, Minisola S, Hendy GN, Cole DEC. 2004. Blood ionized calcium is associated with clustered polymorphisms in the carboxyl-terminal tail of the calcium-sensing receptor. *J Clin Endocrinol Metab* 89: 5634–5638.

157. Parks JH, Coe FL. 1996. Pathogenesis and treatment of calcium stones. *Semin Nephrol* 16: 398–411.

158. Coe FL. 1983. Uric acid and calcium oxalate nephrolithiasis. *Kidney Int* 24: 392–403.

159. Maschio G, Tessitore N, D'Angelo A, Fabris A, Pagano F, Tasca A, Graziani G, Aroldi A, Surian M, Colussi G, Mandressi A, Trinchieri A, Rocco F, Ponticelli C, Minetti L. 1981. Prevention of calcium nephrolithiasis with low-dose thiazide, amiloride and allopurinol. *Am J Med* 71: 623–626.

160. Coe FL. 1977. Treated and untreated recurrent calcium nephrolithiasis in patients with idiopathic hypercalciuria, hyperuricosuria, or no metabolic disorder. *Ann Intern Med* 87: 404–410.

161. Coe FL, Parks JH, Nakagawa Y. 1991. Protein inhibitors of crystallization. *Semin Nephrol* 11: 98–109.

162. Coe FL, Parks JH. 1990. Familial (idiopathic) hypercalciuria. In: Coe FL, Parks JH (eds.) *Nephrolithiasis: Pathogenesis and Treatment, 2nd Ed*. Chicago: Year Book Medical Publishers, Inc. pp. 108–138.

163. Friedman PA, Bushinsky DA. 1999. Diuretic effects on calcium metabolism. *Semin Nephrol* 19: 551–556.

164. Coe FL, Parks JH, Bushinsky DA, Langman CB, Favus MJ. 1988. Chlorthalidone promotes mineral retention in patients with idiopathic hypercalciuria. *Kidney Int* 33: 1140–1146.

165. Bushinsky DA, Favus MJ, Coe FL. 1984. Mechanism of chronic hypocalciuria with chlorthalidone: Reduced calcium absorption. *Am J Physiol* 247: F746–F752.

166. Breslau NA, Moses AM, Weiner IM. 1976. The role of volume contraction in the hypocalciuric action of chlorothiazide. *Kidney Int* 10: 164–170.

167. Pearle MS, Roehrborn CG, Pak CYC. 1999. Meta-analysis of randomized trials for medical prevention of calcium oxalate nephrolithiasis. *J Endourol* 13: 679–685.

168. Ernst ME, Carter BL, Zheng S, Grimm RH. 2010. Meta-analysis of dose-response characteristics of hydrochlorothiazide and chlorthalidone: Effects on systolic blood pressure and potassium. *Am J Hypertens* 23: 440–446.

169. Renjmark L, Vestergaard P, Mosekilde L. 2005. Reduced fracture risk in users of thiazide diuretics. *Calc Tiss Int* 76: 167–175.

170. Feskanisch D, Willett WC, Stampfer MJ, Golditz GA. 1997. A prospective study of thiazide use and fractures in women. *Osteoporos Int* 7: 79–84.

171. La Croix AZ, Ott S, Ichikawa L, Scholes D, Barlow WE. 2000. The low-dose hydrochlorothiazide and preservation of bone mineral density in older adults. A randomized, double-blind, placebo-controlled trial. *Ann Intern Med* 133: 516–526.

172. Sigurdsson G, Franzson L. 2001. Increased bone mineral density in a population-based group of 70-year-old women on thiazide diuretics, independent of parathyroid hormone levels. *J Intern Med* 250: 51–56.

173. Favus MJ, Coe FL, Kathpalia SC, Porat A, Sen PK, Sherwood LM. 1982. Effects of chlorothiazide on 1,25-dihydroxyvitamin D_3, parathyroid hormone, and intestinal calcium absorption in the rat. *Am J Physiol* 242: G575–G581.

103
Epidemiology of Nephrolithiasis
Murray J. Favus

Introduction 856
Geographic Distribution of Prevalence and Incidence 856
Nutrition and Lifestyle 857
Genetics 858
Other Disorders 858
References 858

INTRODUCTION

Nephrolithiasis is a frequent disease that affects about 10% of people in Western countries and is a common cause of emergency room visits, hospitalizations, and surgical procedures. Population studies indicate that the prevalence of calcium oxalate stones has been constantly increasing during the past 50 years in industrialized countries. As the composition of stones varies depending on gender and age of patients, dietary patterns, and associated disorders, a review of the epidemiology of kidney stones provides insight into potential causes and contributing factors.

GEOGRAPHIC DISTRIBUTION OF PREVALENCE AND INCIDENCE

Prevalence (patients with a history of kidney stones) and incidence (rate of new kidney stones in the population) of kidney stones reveal its global distribution and the increasing frequency throughout the Western world. In the United States, the prevalence of kidney stones among men and women has been tracked from the periodic National Health and Nutrition Examination Surveys (NHANES)[1]. From the 2003 report, the prevalence of kidney stones increased from the years 1976 to 1980 (3.8%) to the years 1988 to 1994 (5.2%) and then stabilized (Table 103.1). Incidence data reported by the Mayo Clinic [3] show a decrease in men between 1970 and 2000 from 155 to 105 per 10,000 population. At the same time, incidence in women rose from 43.2 to 68.4 per 10,000 population.

In Germany, Hesse et al. [4] reported increased prevalence in the population between 1979 and 2001 from 4.0% to 4.7% and a greater increase in the incidence rates among those 14-years-old and above. A summary of population changes shows the stone prevalence rates and incidence per 100,000 affected subjects (Table 103.1) also increased in the population of other European countries.

In several countries in Asia, the prevalence of kidney stones is higher than in the United States. Reports from China [5] show the prevalence in men is 8.0% and 5.1% in women. In Korea [6], men and women have a kidney stone prevalence of 6.0% and 1.8%, respectively. The highest reported prevalence in Asia is from Taiwan [7] with an overall frequency of 9.6%, including 14.5% among men and 4.3% among women.

It remains unknown as to whether the increased prevalence and incidence are due to the increase in calcium oxalate and/or calcium phosphate stones, or to the less common stones composed of uric acid, struvite (infection), or cysteine. The widespread increases in prevalence may be a real increase in stones or may be due to the discovery of stones in otherwise asymptomatic subjects because of the use of sensitive techniques such as ultrasound and computed tomography (CT). Nevertheless, the global distribution of stones by gender has consistently

Primer on the Metabolic Bone Diseases and Disorders of Mineral Metabolism, Eighth Edition. Edited by Clifford J. Rosen.
© 2013 American Society for Bone and Mineral Research. Published 2013 by John Wiley & Sons, Inc.

Table 103.1. Kidney Stone Prevalence and Incidence by Year in All Populations of Several Countries*

Country	Year	Prevalence (%)	Year	Incidence**
United States	1964–1972	2.62	1971	122
	1976–1980	3.8	1977	208
	1982	5.4%	1978	164
	1988–1994	5.2	2000	116[#]
Italy	1983	1.17		
	1993–1994	1.72		
Scotland	1977	3.83		
	1987	3.5		
Spain	1979	3.0	1977	810
	1984	4.16	1980	500
	1987	2.0	1984	270
	1991	10.0		
Turkey	1989	14.8		
Germany	1979	4.0	1979	120[+]
	2001	4.7	2000	720[+]
Japan			1965	54.2
			1971	58.6
			1975	56.4
			1980	55.7
			1985	62.0
			1990	58.4
			1990	58.4
			1990	58.4
			1995	68.9
			2000	114
Sweden			1954	130
			1969	200

*Adapted from Romero et al. 2010 (Ref. 2). Used with permission.
**Incidence: Affected subjects/10,000.
[#]Ages 18 to 65 years.
[+]Over the age of 14 years.

water intake are other factors that favor excessive urine concentration of solutes. Restoring dietary balance is the first advice to prevent stone recurrence. However, the striking increase in the common calcium oxalate stones suggests that steps be taken to prevent stone formation.

Fluid intake, climate, and occupation

Observation studies [3, 8] suggest high fluid intake reduces the risk for calcium oxalate kidney stones, and the results of one randomized controlled trial of high fluid intake resulted in fewer kidney stones [9].

Water hardness is due to increased calcium concentration, and drinking hard water has been associated with increased urine calcium excretion [10] but not an increase in kidney stones.

Stone prevalence may vary by region, suggesting a contribution of climate; however, separation of climate from other regional variables such as genetic factors and dietary preferences has been difficult. In general, within a country, warmer regions have higher kidney stone prevalence. In the hot and humid climate of the southeast region of the United States, the local climate is thought to be a major influence in the high prevalence of kidney stones [1]. There is little comparable information on the relationship of climate and stone formation in other countries.

Certain workplace conditions such as being hot (steel mills) or largely outdoors (lifeguards) have long been recognized as associated with increased stone risk. Shared clinical experience suggests that occupations that limit regular fluid intake such as long-distance truck drivers, teachers, and executives confined to day-long meetings may be at increased risk for kidney stone formation.

Calcium intake

The long-standing treatment of calcium oxalate kidney stones by reducing dietary calcium intake has given way to a substantial body of evidence from both epidemiologic data [11–14] and intervention trials that show reduced calcium intake does not lower kidney stone risk. Indeed, higher calcium intake reduces stone formation rates [15–17]. A randomized clinical trial of male hypercalciuric calcium oxalate stone formers showed that those fed a high calcium diet of 1,200 mg per day with restricted salt, oxalate, and protein had a 51% lower risk of new stone formation compared to those ingesting a diet low in calcium of 400 mg per day and low in oxalate [15]. In contrast to the effects of dietary calcium, calcium supplements appear not to reduce the incidence of kidney stones. A large randomized clinical trial in which postmenopausal women maintained their usual calcium intake and were supplemented with additional 1,000 mg calcium was associated with a 17% increase in the appearance of new kidney stones [16].

favored men to women by ratios of 5:1 to 2:1. In the past 30 years, a trend for increased stones among women may be related to the concomitant rapid increased prevalence of obesity in women compared to men [8].

NUTRITION AND LIFESTYLE

Nutritional risk factors for stone disease are well known. They include excessive consumption of animal proteins, sodium chloride, and the rapidly absorbed monosaccharides, and insufficient dietary intake of fruits and potassium-rich vegetables. As a consequence, an excessive production of hydrogen ions may induce several changes in urine chemistries including low urine pH, high urine calcium and uric acid excretion, and low urine citrate excretion. Excess in caloric intake, high chocolate consumption inducing hyperoxaluria, and low

Salt

Dietary salt has increased over the past 200 years and its effects include increased urinary calcium excretion [18]. Epidemiologic data suggest a positive correlation between salt ingestion and first-time kidney stone formation in women but not men [11, 12].

Protein

Population studies of the impact of protein intake on kidney stone formation have yielded variable results. Two large follow-up cohort studies concluded that protein intake increased stone formation only in men with normal body mass index (BMI) [19], and that there was no association of animal protein intake and stone incidence in premenopausal women [14]. In idiopathic hypercalciuria [15], lower protein intake was associated with lower stone formation, but the diet associated with lower stone formation was also higher in dietary calcium, which by itself can reduce stone disease (see above). Randomized controlled trials have failed to confirm that low protein intake reduces kidney stone risk. It is possible that the source of protein may differ in the rate of stones, possibly because of differences in sulfate-rich amino acid content, urinary acid excretion, and urine pH [20, 21].

Oxalate

Prospective cohort studies in men and women reveal only a modest effect of dietary oxalate on the risk of kidney stone formation with the difference in risk increasing by about 20% when lowest and highest quartiles of dietary oxalate are compared [17]. The difference in oxalate intake across quartiles was largely due to the amount of spinach ingested. Thus, urinary oxalate excretion is an important contributor to calcium oxalate stone formation; and urinary oxalate excretion is not solely determined by dietary oxalate.

GENETICS

Heritable factors in stone disease are largely from retrospective kindred studies in which 40% of stone formers reported a first-degree relative with a history of kidney stones [22]. In a prospective study, men with a family history of stones had a risk of incident stones double those without such a family history [23]. A twin study showed that a concordance of stones in monozygotic twins approached twice the rate observed in dizygotic twins [24]. Using this data, heritability was estimated to account for 56% of stone prevalence.

OTHER DISORDERS

Obesity and diabetes

Men and women with increased BMI are at increased risk for kidney stone formation [25]. Weight gain in the adult years carries a greater risk compared to those whose weight was stable. The worldwide increase in the prevalence of obesity may be contributing to the increased stone risk among women [26]. Obesity influences the urine composition and therefore the crystal composition of stones. For example, in both men and women, the lower urine pH in obesity is associated with an increase in uric acid stones (uric acid stones accounted for 63% of stones in obese stone formers compared to 11% uric acid stone among in non-obese stone formers) [27]. A greater number of uric acid stone formers were diabetic (27.8%) compared to 6.9% of calcium stone formers [28] and may reflect the high frequency of coexistent diabetes and obesity.

Two large prospective cohort epidemiologic studies concluded that diabetes is also a risk factor for stones [29] with the relative risk of incident stones of 1.3 in diabetic versus nondiabetic older women and 1.6 in younger women. No such association was found for men.

REFERENCES

1. Stamatelou KK, Francis ME, Jones CA, et al. 2003. Time trends in reported prevalence of kidney stones in the USA: 1976–1994. *Kidney Int* 64: 1817–1823.
2. Romero V, Akpinar H, Assimos DG. 2010. Kidney stones: A global picture of prevalence, incidence, and associated risk factors. *Rev Urol* 12: e86–e96.
3. Lieske JC, Pena de la Vega LS, Slezak JM, et al. 2006. Renal stone epidemiology in Rochester, Minnesota: An update. *Kidney Int* 69: 760–764.
4. Hesse A, Brandle E, Wilbert D, et al. 2003. Study on the prevalence and incidence of urolithiasis in Germany comparing the years 1979 vs. 2000. *Eur Urol* 44: 709–713.
5. Peng J, Zhou HB, Cheng JQ, Dong SF, Shi LY, Zhang D. 2003. Study on the epidemiology and risk factors of renal calculi in special economic zone of Shenzhen city. [Article in Chinese]. *Zhonghua Liu Xing Bing Xue Za Zhi* 24: 1112–1114.
6. Kim H, Jo MK, Kwak C, et al. 2002. Prevalence and epidemiologic characteristics of urolithiasis in Seoul, Korea. *Urology* 59: 517–521.
7. Lee YH, Huang EC, Tsai JY, et al. 2002. Epidemiological studies on the prevalence of upper urinary calculi in Taiwan. *Urol Int* 68: 172–177.
8. Hedley AA, Ogden CL, Johnson CL, et al. 2004. Prevalence of overweight and obesity among US children, adolescents, and adults 1999–2002. *JAMA* 291: 2847–2850.

9. Borghi L, Meshci T, Amato F, et al. 1996. Urinary volume, water and recurrence in idiopathic calcium nephrolithiasis: A 5-year randomized prospective study. *J Urol* 155: 839–843.
10. Scheartz BF, Schenkman NS, Bruce JE, et al. 2002. Calcium nephrolithiasis: Effect of water hardness on urinary electrolytes. *Urology* 60: 23–27.
11. Curhan GC, Willett WC, Rimm EB, et al. 1993. A prospective study of dietary calcium and other nutrients and the risk of symptomatic kidney stones. *New Engl J Med* 328: 833–838.
12. Curhan GC, Willett WC, Speizer FE, et al. 1997. Comparison of dietary calcium with supplemental calcium and other nutrients as factors affecting the risk of kidney stones in women. *Ann Intern Med* 126: 497–504.
13. Taylor EN, Stampfer MJ, Curhan GC. 2004. Dietary factors and the risk of incident kidney stones in men: New insights after 14 years of follow-up. *J Am Soc Nephrol* 15: 3225–3232.
14. Curhan GC, Willett WC, Knight EL, Stampfer MJ. 2004. Dietary factors and the risk of incident kidney stones in younger women: Nurses' Health Study II. *Arch Intern Med* 164: 885–891.
15. Borghi L, Schianchi T, Meschi T, et al. 2002. Comparison of two diets for the prevention of recurrent stones in idiopathic hypercalciuria. *N Engl J Med* 346: 77–84.
16. Jackson RD, LaCroix AZ, Gass M, et al. 2006. Calcium plus vitamin D supplementation and the risk of fractures. *N Engl J Med* 354: 669–683.
17. Taylor EN, Curhan GC. 2007. Oxalate and the risk for nephrolithiasis. *Am J Soc Nephrol* 18: 2198–2204.
18. Morris RC Jr, Schmidlin O, Frassetto LA, Sebastian A. 2006. Relationship and interaction between sodium and potassium. *J Am Coll Nut* 25: 262S–270S.
19. Nguyen QV, Kalin A, Drouve U, et al. 2001. Sensitivity to meat protein intake and hyperoxaluria in idiopathic calcium stone formers. *Kidney Int* 59: 2273–2281.
20. Giannini S, Nobile M, Sartori L, et al. 1999. Acute effects of moderate dietary protein restriction in patients with idiopathic hypercalciuria and calcium nephrolithiasis. *Am J Clin Nutr* 69: 267–271.
21. Meschi T, Maggiore U, Fiaccadori E, et al. 2004. The effect of fruit and vegetables on urinary stone risk factors. *Kidney Int* 66: 2402–2410.
22. Ljunghall S, Danielson BG, Fellstrom B, et al. 1985. Family history of renal stones in recurrent stone patients. *Brit J Urol* 57: 370–374.
23. Curhan GC, Willett WC, Rimm EB, Stampfer MJ. 1997. Family history and risk of kidney stones. *J Am Soc Nephrol* 8: 1568–1573.
24. Goldfarb DS, Fischer ME, Keich Y, Goldberg J. 2005. A twin study of genetic and dietary influences on nephrolithiasis: A report from the Vietnam Era Twin (VET) Registry. *Kidney Int* 67: 1053–1061.
25. Taylor EN, Stampfer MJ, Curhan GC. 2005. Obesity, weight gain, and the risk of kidney stones. *JAMA* 293: 455–462.
26. Ogden CL, Carroll MD, Curtis LR, et al. 2006. Prevalence of overweight and obesity in the United States, 1999–2004. *JAMA* 295: 1549–1555.
27. Ekeruo WO, Tan YH, Young MD, et al. 2004. Metabolic risk factors and the impact of medical therapy on the management of nephrolithiasis in obese patients. *J Urol* 172: 159–163.
28. Daudon M, Traxer O, Conort P, et al. 2006. Type 2 diabetes increases the risk for uric acid stones. *J Am Soc Nephrol* 17: 2026–2033.
29. Taylor EN, Stampfer MJ, Curhan GC. 2005. Diabetes mellitus and the risk of nephrolithiasis. *Kidney Int* 68: 1230–1235.

104
Diagnosis and Evaluation of Nephrolithiasis

Stephen J. Knohl and Steven J. Scheinman

Introduction 860
Basic Evaluation of the Stone Patient 860
Radiographic Evaluation 861
Stone Analysis 861

Metabolic Profiling 863
Acknowledgment 866
References 867

INTRODUCTION

The physician evaluating a patient with nephrolithiasis needs to keep in mind the relative prevalence of various stone types. Calcium salts (oxalate, phosphate, or mixed) are the most common stone types, together accounting for 70–75% of nephrolithiasis; low urinary volume, hypercalciuria, hyperoxaluria, hypocitraturia, and hyperuricosuria are identified risk factors [1]. Struvite stones (magnesium ammonium phosphate) represent approximately 10–15% [1] and form in an alkaline urine with high ammonium concentration that results from infection with urease-producing bacteria [2]. Uric acid stones, accounting for 5–10%, are associated primarily with an acidic urine; hyperuricosuria and low urinary volume are additional risk factors, as is insulin resistance [1, 3]. Cystine stones account for approximately 1%. They are the only clinical consequence of hereditary cystinuria, which is the only cause of cystine stones. Stones composed of crystallized drugs represent less than 1% of all stones worldwide.

BASIC EVALUATION OF THE STONE PATIENT

Essential to every evaluation of renal colic, whether an initial or repeat event, is a detailed history, basic metabolic profile, serum uric acid, urinalysis with urine culture, non-contrast helical computed tomography (CT) of the abdomen (i.e., "stone-protocol CT kidneys"), and, if a stone has been retrieved, stone analysis [4]. The history should include stone burden (both in the past and present), the number and type of stone-related urologic procedures, previous medical therapies aimed at risk reduction, and related medical illnesses that may contribute to stone-forming risk [e.g., malabsorption syndromes, inflammatory bowel disease, bowel resection, sarcoidosis, hyperparathyroidism, gout, renal tubular acidosis (RTA), recurrent urinary tract infections (UTIs), neoplasm, immobilization, renal anomaly]. Stone formers have a higher incidence of obesity, dyslipidemia, hypertension, physical inactivity, and diabetes mellitus [5]. A family history of stone disease is associated with an increased risk of stone occurrence and recurrence [6]. A careful dietary history is useful to identify foods that contain excesses of oxalate (such as nuts, chocolate, legumes, soy, green leafy vegetables, berries, and rhubarb), uric acid (examples include shellfish, organ meats, and yeast), and animal protein. Dietary history must also include total daily sodium and calcium consumption. Finally, it is imperative to get a detailed accounting of type and amount of fluid per day. Daily excretion of large amounts of water (greater than 2 liters per day) reduces stone-forming risk. Coffee and tea have traditionally been considered to have high oxalate content. Alcoholic beverages have been purported to increase the risk of stones, but prospective studies have found that beer and wine decrease risk of stone formation [7], possibly due to inhibition of anti-diuretic hormone (ADH) release.

Primer on the Metabolic Bone Diseases and Disorders of Mineral Metabolism, Eighth Edition. Edited by Clifford J. Rosen.
© 2013 American Society for Bone and Mineral Research. Published 2013 by John Wiley & Sons, Inc.

Grapefruit juice has been associated with an increased risk of stones in epidemiologic studies.

Medications should be reviewed to identify drugs that increase stone-forming risk (acetazolamide, loop diuretics, topiramate, excess vitamin D, and calcium) as well as those that can themselves crystallize and, thus, precipitate to form stones (triamterene, indinavir, nelfinavir, amprenavir, atazanavir, acyclovir, and sulfonamide antibiotics) [8–11].

The basic metabolic profile is useful for clues to the etiology of stone disease. Hypokalemia promotes intracellular acidosis, which increases citrate reabsorption in the proximal tubule; hypocitraturia is a known risk factor for calcium stone formation. Hypobicarbonatemia with a normal anion gap may indicate the presence of a distal RTA, a risk factor for calcium phosphate crystallization. Hypercalcemia should alert one to the possibility of primary hyperparathyroidism, sarcoidosis, other granulomatous diseases, or malignancy as an underlying etiology of calcium stones. Blood urea nitrogen (BUN) and creatinine measurements (and the associated 4-model modification of diet in renal disease (MDRD) glomerular filtration rate (GFR) estimates that now often accompany basic metabolic profiles) mark the level of renal function, with acute changes indicating the need for more urgent medical attention as obstruction to urine flow may be present.

A serum uric acid is useful because hyperuricemia, when the result of overproduction or excessive intake, is associated with hyperuricosuria. The presence of hyperuricosuria has been linked with an increased risk of both uric acid and calcium oxalate stones, although the association with calcium oxalate has been called into question [12].

The urinalysis can provide an abundance of useful information. A high specific gravity indicates a concentrated urine and is useful to follow compliance with fluid intake, as a low urine volume is a risk factor for stones of all types. Dipstick urine pH, while not as accurate as a metered urine pH, is useful if measured on fresh urine. An alkaline pH on a first-morning void is consistent with distal renal tubular acidosis, a risk factor for calcium phosphate stones. A pH of 8 or higher, however, is not consistent with normal physiology and indicates infection with urease-producing bacteria, a risk factor for magnesium ammonium phosphate (struvite) stones, particularly when associated with positive tests for nitrite and leukocyte esterase. Hematuria could indicate current stone disease or, in children, the presence of hypercalciuria [13]. On urine microscopy, crystals, if present, can identify the culprit stone-type (Fig. 104.1).

RADIOGRAPHIC EVALUATION

In a patient with renal colic, radiography is essential in assessing stone burden, stone location(s), and the presence of obstruction. Studies commonly used include the plain abdominal film [kidney-ureter-bladder (KUB) film]; ultrasound; intravenous pyelogram (IVP); and helical, non-contrast CT scanning. Except for ultrasound, care should be taken in ordering these tests given the growing epidemiologic evidence linking the amount of ionizing radiation from imaging procedures and malignancy. Historically, stones were classified as radiopaque or radiolucent based on their appearance on KUB (Fig. 104.2). Struvite, cystine, and calcium-containing stones are radiopaque. Cystine stones are radiodense because of the disulfide bonds. Indinavir stones and pure uric acid stones are radiolucent on KUB; uric acid stones, but not indinavir stones, are visible on helical CT scanning. A KUB, however, has an unfavorable sensitivity and specificity profile (45–59% and 71–77%, respectively) and is not usually the best option in the initial stone evaluation [14]. Ultrasonography has high specificity (97%), provides excellent visualization of hydronephrosis, and avoids exposure to radiation, but has a lower sensitivity than KUB because of difficulty identifying calculi that are small or that are located in the ureter [14] (Fig. 104.3). For this reason, ultrasound is preferred only for patients in whom it is important to limit exposure to ionizing radiation (e.g., pregnancy). IVP, previously the gold standard, has better sensitivity (64–87%) than KUB or ultrasound and excellent specificity similar to that of ultrasonography [14]. It may also play a therapeutic role by promoting an osmotic diuresis. However, compared to non-contrast, helical CT of the kidneys, IVP has a lower sensitivity, it takes more time, and it exposes the patient to iodinated contrast and higher doses of radiation [15]. Indications for IVP include evaluation of a patient in whom indinavir stones or medullary sponge kidney are suspected.

In most circumstances the best option for imaging is non-contrast, helical CT of the kidneys. Sensitivity and specificity for this modality are both at or above 95% [14] (Fig. 104.4). While more expensive than other imaging modalities, the higher cost of helical CT is offset by speed of diagnosis, ease of use, and better sensitivity [16]. It has the added advantage of detecting other causes of abdominal pain that may present with symptoms resembling renal colic [17].

Refinements in CT imaging have led to the development of multidetector computed tomography (MDCT), which tries to determine the chemical composition of urinary tract stones based on individual crystals' Hounsfield characteristics. Because of overlap in these Hounsfield values with this technique, new dual-source CT (DSCT) machines have been developed and are able to distinguish pure uric acid, mixed uric acid, cystine, and calcium stones [18, 19]. This technology, currently offered in only a handful of medical centers, still requires additional study before it can become widely available.

STONE ANALYSIS

Stone analysis should be performed on all patients in whom a specimen is available, as it is the most direct

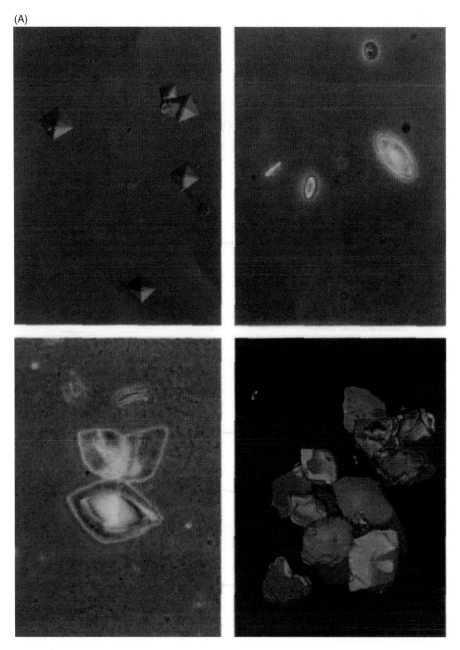

Fig. 104.1. Crystals. (A) Clockwise from top left: Typical bipyramidal calcium oxalate dihydrate crystals (interference-contrast, 640x). Ovoid monohydrate calcium oxalate crystals (phase-contrast, 640x). Rhomboid uric acid crystals (phase-contrast, 400x). Uric acid crystals under polarized light (250x). Reprinted with permission from Ref. 37.

way to identify stone type and allows for stone-specific therapy. Currently accepted methods of analysis include polarization microscopy, infrared spectroscopy, and X-ray crystallography/diffractometry [20]. Polarization microscopy, a technique that identifies crystals based on their interaction with polarized light, is inexpensive, quick, and requires very little stone material; the disadvantages include less reliability in identifying calcium phosphate, uric acid, and mixed stones [20]. Infrared spectroscopy is based on the interaction of infrared light and the molecules making up a stone (Fig. 104.5). Its advantages include ease of use, small sample requirement, and ability to identify noncrystalline material (which often makes up 5% of an individual stone) while the major disadvantage is the amount of time required to perform the test [20]. X-ray diffractometry is based on the diffraction of X-rays by the crystalline structure (Fig. 104.6). The advantage of this technique is its ability to identify any crystal type; the main drawback, however, is its inability to identify noncrystalline material [20].

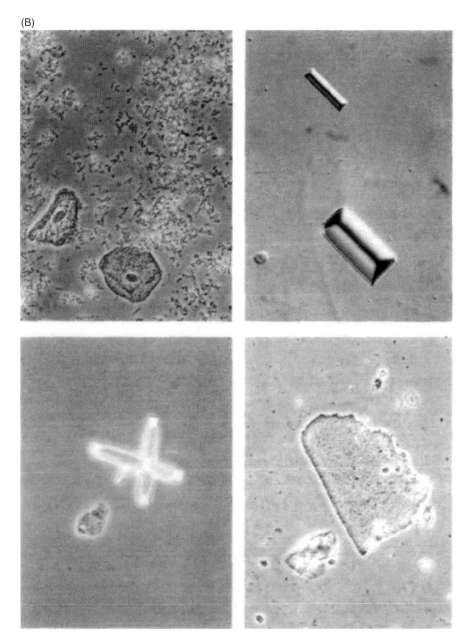

Fig. 104.1. (B) Clockwise from top left: Amorphous phosphates (phase contrast, 400x). Triple phosphate crystals (interference-contrast, 400x). Calcium phosphate plate (phase-contrast, 400x). A star-like calcium phosphate crystal (phase-contrast, 400x). Reprinted with permission from Ref. 37.

METABOLIC PROFILING

Which patients should have full metabolic profiling?

Beyond this evaluation of patients with acute renal colic, the cost-effectiveness of additional testing is not established for all adult first-time stone formers. It is clear that a child with stones, even on the first event, deserves a thorough metabolic evaluation given the increased likelihood of finding an inherited cause. Furthermore, metabolic evaluation of adults with multiple or recurrent stones has been shown to reduce overall costs [21, 22]. Other patients in whom metabolic evaluation should be considered include patients with noncalcareous stones, reduced renal function, single functioning kidney, a strong family history of kidney stones, calcium oxalate stone formers with an increased risk of enteric oxaluria (those with a history of bariatric surgery, short bowel syndrome, chronic diarrhea, or malabsorption), anatomic urinary tract abnormalities, and recurrent urinary tract

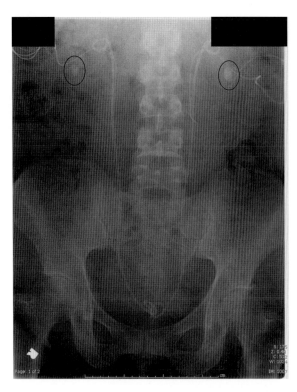

Fig. 104.2. Plain film of the abdomen (KUB). The black ovals mark the bilateral radiopaque calculi identified on this plain film. Also noted are bilateral double J-stents in this 36-year-old female with calcium phosphate stones in the face of distal renal tubular acidosis secondary to Sjogren syndrome.

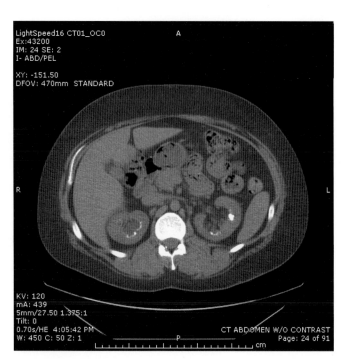

Fig. 104.4. Non-contrast, helical CT of the kidneys. A calculus is identified in the left kidney while nephrocalcinosis is present bilaterally in the same patient described in Figs. 104.2 and 104.3.

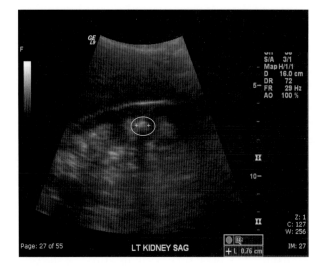

Fig. 104.3. Renal ultrasound. The white oval highlights a calculus with posterior shadowing identified in the left kidney in the same patient described in Fig. 104.2.

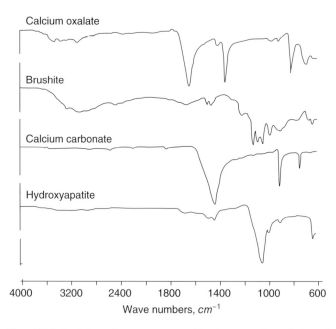

Fig. 104.5. Infrared spectroscopy. Fourier transform infrared microspectroscopy (l-FTIR) spectra. Reprinted with permission from Ref. 38.

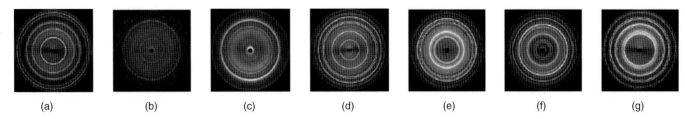

Fig. 104.6. X-ray diffractometry. Scatter patterns from X-ray diffractometry for the seven most common stone components: (a) calcium oxalate monohydrate, (b) calcium oxalate dihydrate, (c) calcium phosphate, (d) calcium phosphate dihydrate, (e) struvite, (f) uric acid, and (g) cystine. Reprinted with permission from Ref. 39.

infections [4]. The core of this additional testing is 24-hour urine profiling, which should be performed 2 to 4 weeks after an acute episode of renal colic has resolved. The components of a 24-hour urine study include measurements of volume, creatinine, calcium, oxalate, citrate, uric acid, sodium, potassium, chloride, magnesium, phosphorus, ammonium, sulfate, urine urea nitrogen, and pH. Specialized laboratories will use these data to derive the protein catabolic rate and supersaturation measures for calcium oxalate, calcium phosphate, and uric acid. If the stone type is not known and cystine is in the differential diagnosis, qualitative cystine screening should be performed; the screen involves a colorimetric test using cyanide-nitroprusside on a 24-hour urine collection. If this is positive, or if the stone is known to be cystine, quantitative testing should be performed using solid-phase assay [23]. Given the day-to-day variability in urine chemical composition, the initial metabolic evaluation should include two 24-hour urine collections [24]. Adequacy of the urine collection can only be evaluated by reviewing the creatinine excretion, which should be 20–25 mg/kg lean body weight in males and 15–20 mg/kg in females. Additional laboratory studies depend on the results of the 24-hour testing.

Supersaturation

Unique to every salt is a characteristic termed the ion activity product (AP). The solubility product (SP) is defined as the AP below which crystallization does not occur. The formation product (FP) is defined as the AP at which crystallization does occur. Supersaturation is a value obtained by the ratio AP/SP: A value less than 1 indicates that crystallization will not occur, while a value greater than or equal to 1 indicates that crystallization can (but, in contradistinction to FP, does not have to) occur [25]. FP can be classified further into homogeneous or heterogeneous. Homogenous FP is the AP at which crystallization will occur independent of other urine components. Heterogeneous FP, however, is an AP dependent on the promoters and inhibitors present in the urine. The interval between SP and $FP_{homogeneous}$ is termed the "metastable range of supersaturation"; crystallization of a salt with an AP in this range depends on the balance between promoters (e.g., calcium, oxalate, sodium, urine pH, and low urine volume) and inhibitors (e.g., citrate, pyrophosphate, potassium, magnesium, and high urine volume) of stone formation.

In human urine, calcium oxalate AP normally exists in the metastable range of supersaturation [26]. A major goal of stone prevention is to reduce supersaturation, particularly for the salt of the stone type the patient has experienced, if that is known. Increasing urinary volume will reduce the supersaturation of all salts proportionately. Therapeutic measures for altering specific solutes are discussed in Chapter 106.

Hypercalciuria

Hypercalciuria, the most commonly identified metabolic abnormality on 24-hour urine profiling, is defined as excretion of more than 200 mg/day of calcium in females and more than 250 mg/day in males [27]. The differential diagnosis of hypercalciuria is discussed in Chapter 102. If hypercalciuria is documented, laboratory evaluation should include measurement of intact parathyroid hormone (PTH), phosphorus, 1,25-dihydroxy vitamin D_3, magnesium, and thyroid-stimulating hormone (TSH). Given the growing evidence of bone demineralization in patients with hypercalciuria, a dual energy X-ray absorptiometry (DXA) scan to evaluate for osteoporosis may be considered as well [28]. Since dietary calcium restriction below a recommended daily intake of 800–1,000 mg is not of therapeutic value in patients with stones, categorization of patients into subsets of hypercalciuria based on intestinal absorption, bone resorption, or renal reabsorption, is not clinically useful.

Hyperoxaluria

Hyperoxaluria is typically defined as an oxalate excretion of more than 45 mg/day [27]. Most cases of hyperoxaluria are diet-mediated (foods rich in oxalate include dark green leafy vegetables, legumes, berries, nuts, chocolate, soy, and ripe rhubarb). Ethylene glycol (i.e., antifreeze), when ingested, is metabolized to oxalate and can cause stones.

Oxalate levels greater than 80–100 mg/day suggest that enteric oxaluria is likely contributing as well. Enteric oxaluria can result from a malabsorptive process (such as in pancreatic insufficiency, cystic fibrosis, or Crohn's disease, or following gastric bypass or bariatric surgery) in which undigested fatty acids in the colon saponify calcium, allowing more free oxalate to remain in solution and not be complexed with calcium, resulting in enhanced oxalate absorption. Non-absorbed bile salts also increase colonic permeability to oxalate. At levels above 100 mg/day, primary hyperoxaluria, a rare inherited disease that results in enhanced conversion of glyoxylate to oxalate, should be considered as an etiology. This topic is also discussed further in Chapter 102.

Hyperuricosuria

Hyperuricosuria is defined as urinary uric acid excretion exceeding 750 mg/day and 800 mg/day in women and men, respectively [27]. It is a risk factor for both calcium oxalate and uric acid nephrolithiasis (again, the former association has recently been called into question) [3, 12]. A purine-rich diet is often a factor, but it is believed that many patients may have mild metabolic overproduction of uric acid. Other less common causes include gout, leukemia, tumor lysis syndrome, Lesch-Nyhan syndrome, enhanced phosphoribosylpyrophosphate synthetase activity, xanthinuria, and 2,8-dihydroxyadeninuira. Uric acid stones are significantly more common among nephrolithiasis patients with type 2 diabetes, obesity, and/or metabolic syndrome. The principal metabolic feature responsible for this association is an overly acidic urine, which leads to the precipitation of sparingly soluble uric acid crystals in urine and subsequent development of stones [29]. It is hypothesized that insulin resistance impairs ammoniagenesis, thereby reducing the urine pH, resulting in a milieu favorable for uric acid crystallization.

Hypocitraturia

Hypocitraturia is defined as less than 500 mg/day in women and less than 350 mg/day in men [27]. Citrate, by forming a soluble salt with calcium, inhibits calcium stone formation. Acidemia reduces citrate excretion while alkalemia augments it. Most cases are related to the protein or acid content of the diet, but patients with hypocitraturia should be assessed for hypokalemia, diarrhea, infection, distal renal tubular acidosis, and for the use of acetazolamide.

Urine pH

Uric acid and cystine crystallization is favored by acidic urine, while calcium phosphate and struvite stone formation occurs in an alkaline urine. A persistently alkaline urine combined with hypocitraturia is highly suggestive of a distal renal tubular acidosis.

Other factors

Calcium excretion parallels sodium excretion; reducing sodium intake (and, thus, excretion) may reduce calciuria. It has also been suggested that excessive sodium intake is predictive of stone risk independently of its effects in increasing the excretion of calcium, uric acid, and oxalate [8]. Hypokalemia, by causing intracellular acidosis, can result in hypocitraturia, thereby increasing stone risk. Increased potassium excretion, on the other hand, decreases the risk of stone formation [30]. Magnesium in the urine has been shown to inhibit calcium oxalate crystallization, and enteral magnesium may reduce intestinal oxalate absorption [31]. Phosphate measurements are required to evaluate urinary supersaturation and usually are reflective of dietary intake. When evaluated in the context of serum levels of PTH, phosphate, and calcium, urinary phosphate excretion can yield information regarding disorders of the calcium–vitamin D–PTH axis. Ammonium levels are increased in urine that is infected with urease-producing bacteria, promoting struvite formation; ammonium excretion is physiologically increased in acid-loading states [32, 33]. Under conditions of impaired acid excretion (i.e., a distal renal tubular acidosis), ammonium excretion may be low.

Markers of protein excretion, such as urinary sulfate, urine urea nitrogen, and protein catabolic rate, serve as an indirect measure of acid load; the acid, which requires buffering from bone, can result in hypercalciuria [34, 35].

Finally, an initial stone risk profile assessment should include cystine screening. Cystinuria is suspected when urine concentrations are greater than 75 mg/L [36].

Effect of gender on normal ranges of urinary constituents

Women typically excrete less calcium and uric acid, and more citrate, than men, and the distribution curves that define the "normal" ranges in men and women reflect this. The normally higher levels of two important stone constituents, and normally lower levels of citrate, explain much, though not all, of the increased risk of stones in men. However, it should be remembered that the distribution of these solutes is continuous, and that a given level of solute has the same relevance to stone risk in a man or a woman. In this regard, the supersaturation ratios for each stone salt that are provided by specialty laboratories are particularly helpful.

ACKNOWLEDGMENT

The authors thank Dr. Anil Singh for valuable assistance in the preparation of this chapter.

REFERENCES

1. Moe OW. 2006. Kidney stones: Pathophysiology and medical management. *Lancet* 367(9507): 333–44.
2. Healy KA, Ogan K. 2007. Pathophysiology and management of infectious staghorn calculi. *Urol Clin North Am* 34(3): 363–74.
3. Shekarriz B, Stoller ML. 2002. Uric acid nephrolithiasis: Current concepts and controversies. *J Urol* 168(4 Pt 1): 1307–14.
4. Miller NL, Lingeman JE. 2007. Management of kidney stones. *BMJ* 334(7591): 468–72.
5. Ramey SL, Franke WD, Shelley MC 2nd. 2004. Relationship among risk factors for nephrolithiasis, cardiovascular disease, and ethnicity: Focus on a law enforcement cohort. *AAOHN J* 52(3): 116–21.
6. Ljunghall S, Danielson BG, Fellstrom B, Holmgren K, Johansson G, Wikstrom B. 1985. Family history of renal stones in recurrent stone patients. *Br J Urol* 57(4): 370–4.
7. Curhan GC, Willett WC, Rimm EB, Spiegelman D, Stampfer MJ. 1996. Prospective study of beverage use and the risk of kidney stones. *Am J Epidemiology* 143(3): 240–7.
8. Parmar MS. 2004. Kidney stones. *BMJ* 328(7453): 1420–4.
9. Chan-Tack KM, Truffa MM, Struble KA, Birnkrant DB. 2007. Atazanavir-associated nephrolithiasis: Cases from the US Food and Drug Administration's Adverse Event Reporting System. *AIDS* 21(9): 1215–8.
10. Feicke A, Rentsch KM, Oertle D, Strebel RT. 2008. Same patient, new stone composition: Amprenavir urinary stone. *Antivir Ther* 13(5): 733–4.
11. Engeler DS, John H, Rentsch KM, Ruef C, Oertle D, Suter S. 2002. Nelfinavir urinary stones. *J Urol* 167(3): 1384–5.
12. Curhan GC, Taylor EN. 2008. 24-h uric acid excretion and the risk of kidney stones. *Kidney Int* 73(4): 489–96.
13. Stapleton FB, Roy S 3rd, Noe HN, Jerkins G. 1984. Hypercalciuria in children with hematuria. *N Engl J Med* 310(21): 1345–8.
14. Portis AJ, Sundaram CP. 2001. Diagnosis and initial management of kidney stones. *Am Fam Physician* 63(7): 1329–38.
15. Catalano O, Nunziata A, Altei F, Siani A. 2002. Suspected ureteral colic: Primary helical CT versus selective helical CT after unenhanced radiography and sonography. *AJR Am J Roentgenol* 178(2): 379–87.
16. Chen MY, Zagoria RJ. 1999. Can noncontrast helical computed tomography replace intravenous urography for evaluation of patients with acute urinary tract colic? *J Emerg Med* 17(2): 299–303.
17. Ha M, MacDonald RD. 2004. Impact of CT scan in patients with first episode of suspected nephrolithiasis. *J Emerg Med* 27(3): 225–31.
18. Thomas C, Heuschmid M, Schilling D, Ketelsen D, Tsiflikas I, Stenzl A, Claussen CD, Schlemmer HP. 2010. Urinary calculi composed of uric acid, cystine, and mineral salts: Differentiation with dual-energy CT at a radiation dose comparable to that of intravenous pyelography. *Radiology* 257: 402–9.
19. Boll DT, Patil NA, Paulson EK, Merkle EM, Simmons WN, Pierre SA, Preminger GM. 2009. Renal stone assessment with dual-energy multidetector CT and advanced postprocessing techniques: Improved characterization of renal stone composition—Pilot study. *Radiology* 250(3): 813–20.
20. Schubert G. 2006. Stone analysis. *Urol Res* 34(2): 146–50.
21. Parks JH, Coe FL. 1996. The financial effects of kidney stone prevention. *Kidney Int* 50(5): 1706–12.
22. Robertson WG. 2006. Is prevention of stone recurrence financially worthwhile? *Urol Res* 34(2): 157–61.
23. Coe FL, Clark C, Parks JH, Asplin JR. 2001. Solid phase assay of urine cystine supersaturation in the presence of cystine binding drugs. *J Urol* 166(2): 688–93.
24. Parks JH, Goldfisher E, Asplin JR, Coe FL. 2002. A single 24-hour urine collection is inadequate for the medical evaluation of nephrolithiasis. *J Urol* 167(4): 1607–12.
25. Tiselius H. 2005. Aetiological factors in stone formation. In: Davison AMA, Cameron JS, Grunfeld J-P, Ponticelli C, Van Ypersele C, Ritz E, Winearls C (eds.) *Oxford Textbook of Clinical Nephrology, 3rd Ed.* Oxford: Oxford University Press. pp. 1199–224.
26. Mandel N. 1996. Mechanism of stone formation. *Semin Nephrol* 16(5): 364–74.
27. Coe FL, Evan A, Worcester E. 2005. Kidney stone disease. *J Clin Invest* 115(10): 2598–608.
28. Asplin JR, Donahue S, Kinder J, Coe FL. 2006. Urine calcium excretion predicts bone loss in idiopathic hypercalciuria. *Kidney Int* 70(8): 1463–7.
29. Maalouf NM. 2011. Metabolic syndrome and the genesis of uric acid stones. *J Ren Nutr* 21(1): 128–31.
30. Curhan GC, Willett WC, Rimm EB, Stampfer MJ. 1993. A prospective study of dietary calcium and other nutrients and the risk of symptomatic kidney stones. *N Engl J Med* 328(12): 833–8.
31. Massey L. 2005. Magnesium therapy for nephrolithiasis. *Magnes Res* 18(2): 123–6.
32. Griffith D. 1983. Infection induced urinary stones. In: Roth RA, Finlayson B (eds.) *Stones—Clinical Management of Urolithiasis. Volume 6, International Perspectives in Urology.* Baltimore: The Williams & Wilkins Co. pp. 210–27.
33. Parivar F, Low RK, Stoller ML. 1996. The influence of diet on urinary stone disease. *J Urol* 155(2): 432–40.
34. Martini LA, Wood RJ. 2000. Should dietary calcium and protein be restricted in patients with nephrolithiasis? *Nutr Rev* 58(4): 111–7.
35. Bingham SA. 2003. Urine nitrogen as a biomarker for the validation of dietary protein intake. *J Nutr* 133 Suppl 3: 921S–924S.
36. Finocchiaro R, D'Eufemia P, Celli M, Zaccagnini M, Viozzi L, Troiani P, Mannarino O, Giardini O. 1998. Usefulness of cyanide-nitroprusside test in detecting incomplete recessive heterozygotes for cystinuria: A standardized dilution procedure. *Urol Res* 26(6): 401–5.

37. Fogazzi GB. 1996. Crystalluria: A neglected aspect of urinary sediment analysis. *Nephrol Dial Transplant* 11(2): 379–87.
38. Evan AP, Lingeman JE, Coe FL, Shao Y, Parks JH, Bledsoe SB, Phillips CL, Bonsib S, Worcester EM, Sommer AJ, Kim SC, Tinmouth WW, Grynpas M. 2005. Crystal-associated nephropathy in patients with brushite nephrolithiasis. *Kidney Int* 67(2): 576–91.
39. Davidson MT, Batchelar DL, Velupillai S, Denstedt JD, Cunningham IA. 2005. Analysis of urinary stone components by x-ray coherent scatter: Characterizing composition beyond laboratory x-ray diffractometry. *Phys Med Biol* 50(16): 3773–86.

105
Kidney Stones in the Pediatric Patient
Amy E. Bobrowski and Craig B. Langman

Introduction 869
Predisposing Conditions 869
Clinical Evaluation 872
Surgical Management 873

Medical Management 874
Prognosis 875
References 875

INTRODUCTION

Kidney stone disease, or nephrolithiasis, results from urinary crystal aggregation in a protein matrix within the urinary tract. This condition reflects a deviation from the equipoise in normal human urine between stone promoters and inhibitors, in favor of promoters. In general, higher urinary pH (with the exception of struvite stones, which form in alkaline urine), higher urine volume and dilution, higher urinary citrate, and free flow of urine serve as natural stone inhibitors. Kidney stones occur about one-tenth less frequently in children than in adults, representing from 1 in 1,000 to 1 in 7,600 pediatric hospital admissions in previous studies [1]. More recent analysis of the United States Kids' Inpatient Database revealed diagnoses of kidney or ureteral stones in 0.2% of hospital discharges, which would suggest an increasing prevalence in this population. In this study, girls were more likely affected than boys by a 2:1 ratio, and the fraction of those with stones increased markedly with age [2]. This differs from the adult disease, where there is a male preponderance (3:1). Within subtypes of stone disease, however, male children do have a slightly higher incidence of stones related to hypercalciuria and urinary tract abnormalities [3].

Predisposing factors for nephrolithiasis can be determined in the majority of children affected. These include metabolic abnormalities in 48–86%, urinary tract infection in 14–75%, and coexisting structural urinary tract abnormalities in 10–40%.[3–5]. The recurrence rate of kidney stones in children has been reported as anywhere from 6.5% to 54% [3, 4, 6–8], and children with metabolic disorders as a cause of their stone are nearly five times more likely to have a recurrence [6]. Calcium oxalate stones are the most common found in children, with a frequency of 45–65%. These are followed by calcium phosphate (14–30%), struvite (13%), cystine (5%), uric acid (4%), and mixed (4%) stones [3, 9].

PREDISPOSING CONDITIONS

Abnormalities of the urinary tract

Anatomic abnormalities of the genitourinary tract may produce urinary stasis, allowing for crystal aggregation by heterotopic nucleation and/or infection. Most stones in North American children are located within the kidney or ureter. When recurrent bladder stones are seen, they are most often in the face of complex urologic abnormalities [10]. However, because most patients with urologic anomalies do not form stones, a full metabolic evaluation should still be performed in such children presenting with nephrolithiasis.

Infection/struvite stones

Urinary tract infections may coexist with kidney stones that arise from a metabolic origin, making it necessary

---

Primer on the Metabolic Bone Diseases and Disorders of Mineral Metabolism, Eighth Edition. Edited by Clifford J. Rosen.
© 2013 American Society for Bone and Mineral Research. Published 2013 by John Wiley & Sons, Inc.

to rule out metabolic etiologies in such cases. Infection-related stones are more common in males, and more than one-half of children with these stones have genitourinary abnormalities [3]. Patients with surgically augmented bladders, particularly those augmented with intestinal segments, are at especially high risk for developing struvite stones in the bladder [11]. Formation of struvite stones ($Mg-NH_4-PO_4$) and calcium phosphate apatite-based stones is favored by an alkaline pH, resulting from NH4 production by urea-splitting bacteria. *Proteus* species are the most common culprit, but *Pseudomonas, Klebsiella, Steptococcus, Serratia, Staphyloococcus, Candida*, and *Mycoplasma* species can produce urease, too. The stones produced in this setting tend to grow rapidly, often forming staghorn calculi, a term signifying complete filling of the urinary infundibulum with stone.

Metabolic abnormalities

Hypercalciuria

Calcium oxalate and calcium phosphate stones in children are most frequently caused by hypercalciuria, defined as a urinary calcium excretion of more than 4 mg/kg/day. Patients with hypercalciuria may present with microscopic or gross hematuria, dysuria, or urgency, even in the absence of any stones. Such children often have a positive family history of kidney stones and may have up to a 17% chance of subsequently developing urolithiasis [12, 13].

Familial idiopathic hypercalciuria (FIH) is the most common subset of hypercalciuria. Although the genetic basis of this condition is unknown, it seems to be inherited in an autosomal dominant pattern with incomplete penetrance. The pathophysiology of FIH is not yet well defined but may include any combination of the following, to varying degrees: a primary kidney tubular reduction in calcium resorption, increased dietary calcium absorption in the gastrointestinal tract secondary to excessive 1,25-dihydroxy-vitamin D action, and increased bone resorption [14]. The contribution of bone resorption to this disorder has important clinical implications, because restricting calcium intake in these patients may worsen their propensity toward significant osteoporosis [15].

Causes of hypercalciuria are listed in Table 105.1, but this list is not exhaustive. Dent's disease is an X-linked recessive condition of nephrolithiasis and subsequent kidney failure, linked to mutations in the *CLCN5* gene located on chromosome Xp 11.22 [16]. This gene is responsible for the transduction of a voltage-gated chloride channel in the kidney, the lack of which leads to hypercalciuria, low molecular weight proteinuria, nephrolithiasis, nephrocalcinosis, and varying degrees of glycosuria, aminoaciduria, and phosphaturia [17]. Approximately 15% of those with the phenotype of Dent's disease, however, have a separate defect in the *OCRL1* gene. This gene is located on the X chromosome, encodes a phosphatidylinositol 4,5-bisphosphate 5-phosphatase, and is the

Table 105.1. Causes of Hypercalciuria

Associated with Hypercalcemia	Associated With Normal or Low Serum Calcium
• Primary hyperparathyroidism • Sarcoid • Idiopathic infantile hypercalcemia • Immobilization • Bartter syndrome • Thyrotoxicosis • Bone metastases • Hypervitaminosis D • Williams syndrome	• Familial idiopathic hypercalciuria • Dent's disease • Bartter syndrome • Familial hypomagnesemia-hypercalciuria • Immobilization • Prematurity, often associated with furosemide • Distal renal tubular acidosis • Ketogenic diet • Activating mutation of the extracellular calcium-sensing gene (generally with hypocalcemia) • Medullary sponge disease • Inflammatory diseases (e.g., JRA) • Corticosteroid therapy

same gene implicated in Lowe syndrome [18]. Bartter syndrome occurs with one of a series of mutations of genes coding for transporters in the thick ascending limb of the Loop of Henle. These genes include *NKCC2*, which transduces the Na–K–2Cl transporter (type I Bartter syndrome); *ROMK*, which transduces the potassium channel (type II); and *CLCNKB*, which transduces the chloride channel (type III). There is also a type IV Bartter syndrome, or Bartter syndrome with sensorineural deafness, which is caused by a mutation in the gene for barttin, a β-subunit of the chloride channel. Type V has a similar phenotype as type IV but is caused by defects in one or both of the chloride channels that colocalize with barttin: ClC-Ka and ClC-Kb. Distal renal tubular acidosis (dRTA) is a condition of metabolic acidosis, growth retardation, hypercalciuria, and nephrocalcinosis. When associated with a mutation in the *ATP6B1* gene responsible for a vacuolar H^+-ATPase, it is associated with deafness and an autosomal dominant inheritance. Familial hypomagnesemia-hypercalciuria is associated with a mutation in the *PLCN-1* gene for the tight junction protein paracellin-1. Pseudohypoaldosteronism type II is seen with mutations in WNK kinases expressed in the distal nephron and presents with hypertension, hyperkalemia, and metabolic acidosis in addition to hypercalciuria [19].

Other causes of hypercalciuria with normocalcemia include medullary sponge kidney, systemic inflammatory diseases, and iatrogenic, resulting from medications such as loop diuretics and corticosteroids. If hypercalcemia is detected, primary hyperparathyroidism, sarcoidosis, immobilization, thyroid disease, osteolytic metastases, hypervitaminosis D, and Williams syndrome should be considered on the differential [20]. Hypocalcemia with hypercalciuria may occur with activating mutations of the calcium-sensing receptor (CaSR) in the parathyroid gland and kidney, especially when vitamin D and calcium supplementation is employed to raise serum calcium levels, leading to an even further increase in calcium excretion [21].

Hypocitraturia

Hypocitraturia is a contributory cause to nephrolithiasis, because citrate is necessary for the formation of a soluble calcium salt to prevent calcium stone crystallization. Most commonly it is seen in dRTA, but it is also present in a subset of patients with FIH. Hypocitraturia can also occur in concert with other forms of hypercalciuria, hyperuricosuria, or hyperoxaluria. Chronic diarrhea, a high-protein diet, and hypokalemia can also induce low urinary citrate levels and a predisposition to stone formation [22], because citrate absorption in the proximal tubule is stimulated by intracellular acidosis and potassium depletion [23].

Hyperoxaluria

Oxalate is a metabolic product made in the liver and excreted by the kidney, but can also be ingested and absorbed, or be formed, from dietary sources. Type 1 primary hyperoxaluria (PH1) is an autosomal recessive disease in which there is a reduction in, or absence of, alanine glyoxylate aminotransferase-1 (AGT1) activity, leading to increased conversion of glyoxylate to oxalate. Excessive urinary oxalate excretion may lead to crystallization and deposition in the urinary tract and kidney parenchyma. This in turn can result in kidney failure and systemic oxalosis, a clinical situation in which calcium oxalate precipitates in multiple organs and joints. Disease severity in PH1 varies widely. The course may be mild and fully responsive to medical therapy such as vitamin B_6 (pyridoxine) or may present aggressively in infancy with rapid kidney failure and severe systemic manifestations.

Because AGT is predominantly expressed in the liver, diagnosis has in the past relied solely on liver biopsy to assess the presence and activity of AGT. The gene encoding AGT (AGXT), located on chromosome 2q37.3, to date has over 100 mutations that have been described that either eliminate, or decrease substantially, enzyme activity [24, 25]. The relative ease in modern laboratory medicine at performance of sequence analysis, and the delineation of the molecular basis of many of the mutations behind PH1, has led to the proposal of molecular diagnostic algorithms that may obviate the need for invasive biopsy procedures. Monico et al. [24] have reported comprehensive mutation screening across the entire AGXT coding region in 55 probands with PH1, showing a 96–98% sensitivity in this population. When limited to sequencing of exons 1, 4, and 7, the sensitivity was 77%. Given the relatively small size of the gene, complete molecular analysis should not involve prohibitive expense. An algorithm beginning with limited sequencing of exons 1, 4, and 7, followed by direct sequencing of the entire gene if inconclusive, would make intuitive sense and would eliminate the need for liver biopsy in most patients.

Type 2 primary hyperoxaluria (PH2) results from a deficiency of activity in the enzyme glyoxylatereductase/hydroxypyruvatereductase (GRHPR), which is more widely distributed in the human than AGT1, with a predominance in muscle, liver, and kidney. As a group, patients with PH2 seem to have less morbidity and mortality than those with PH1, with a lower incidence of end-stage kidney disease (ESKD) and an older age at onset of symptoms [26, 27].

A third group of patients has been described who present with calcium oxalate renal stone disease in early childhood and a phenotype similar to PH1 or PH2, but with normal enzymatic activity of AGT and GRHPR, and no signs or symptoms of enteric hyperoxaluria (see below). These patients have been referred to as non-PH1/PH2 patients, and represent about 5% of all cases of PH. In a series of elegant experiments, investigators have recently identified a common genetic locus in affected families and a likely candidate gene, DHDPSL, with an expression pattern in liver and kidney cells. Experiments are ongoing to prove the assumption that this gene encodes the enzyme 4-hydroxy-2-oxoglutarate aldolase, activating mutations of which would lead to excessive production of glyoxylate, and therefore downstream enhanced oxalate as well. It is now proposed that patients with this form of PH caused by this mutation be classified as primary hyperoxaluria type 3 (PH3) [28].

Unfortunately, up to one-third of patients with PH in some case series present at end stage, when uremia develops [29, 30]. For this reason, PH should be considered in patients with recurrent calcium oxalate nephrolithiasis, unexplained nephrocalcinosis, or unexplained ESKD in which the kidneys are echodense with calcium.

Secondary (enteric) hyperoxaluria can result from increased oxalate absorption in the colon caused by small bowel malabsorption of fatty and bile acids. These substances increase colonic permeability to oxalate by binding luminal calcium, freeing unbound oxalate to be absorbed. Epithelial damage in these states also increases colonic absorption, and low dietary calcium intake can exacerbate the condition. There has been more recent interest in enteric hyperoxaluria in the population undergoing bariatric surgery, where nearly 50% of patients may develop hyperoxaluria early in the postoperative period, and stone disease occurs [31]. Depletion of Oxalobacter formigenes, an enteric oxalate-degrading bacterium, has also been postulated to contribute to enteric hyperoxaluria, but human data remain inconclusive to date. Other rare secondary causes are pyridoxine deficiency (a cofactor for AGT activity) and excessive intake of oxalate-containing foods (rhubarb gluttony) or oxalate precursors (ascorbic acid, ethyleneglycol).

Hyperuricosuria

Uric acid is the end product of purine metabolism. Hyperuricosuria may occur either in the face of uric acid overproduction or with normal serum uric acid concentrations and can predispose to both uric acid stones and calcium oxalate nephrolithiasis, acting as a heterotopic nucleation factor [32]. Lesch-Nyhan syndrome (complete deficiency of hypoxanthine-guanine phosphoribosyltransferase) and type 1 glycogen storage disease (glucose-6-phosphatase deficiency) are both inborn errors of metabolism that may present with hyperuricemia and hyperuricosuria/urolithiasis [33]. Gout caused by a partial hypoxanthine-guanine phosphoribosyltransferase deficiency can cause uric acid nephrolithiasis in older children. Myeloproliferative disorders and other causes of cell breakdown are other secondary causes of uric acid stones. Ketogenic diets, excessive protein intake, and uricosuric drugs such as high-dose aspirin, probenecid, and ascorbic acid can also cause hyperuricosuria. Normal or low serum uric acid levels may be associated with uricosuria secondary to proximal renal tubular defects. These may be caused by a single defect in the renal urate exchanger URAT1 [34] or disorders of generalized proximal tubule dysfunction. Insulin resistance, as seen in type 2 diabetes mellitus and metabolic syndrome, may also predispose to uric acid stones and an overly acidic urine by decreasing renal ammonia excretion and impairing hydrogen ion buffering [35]. In another perturbation of uric acid metabolism, xanthine stones are formed in an autosomal recessive disorder of the gene for xanthine dehydrogenase, whereby uric acid cannot be formed from xanthine precursors.

Cystinuria

Cystinuria is an autosomal recessive disease of disordered dibasic amino acid transport in kidney and may occasionally be diagnosed by the discovery of flat hexagonally shaped cystine crystals in the urine. Children with this condition have elevated urinary cystine, ornithine, arginine, and lysine levels, because all of these amino acids share common dibasic amino acid transporters. Mutations of the *SLC3A1* gene on chromosome 2 and the *SLC7A9* gene on chromosome 19 have been identified [36, 37], and patients may be either homozygous or compound or obligate heterozygotes [38]. Affected homozygous children usually excrete more than 1,000 umol/g creatinine of cystine by the age of 1 year, with a mean excretion of 4,500 umol/g creatinine, exceeding its solubility and leading to lifelong recurrent nephrolithiasis [10].

Melamine poisoning

Melamine is a synthetic chemical used in a variety of commercial products. It has a high non-protein nitrogen content, and when added to milk or animal feed, can falsely elevate their apparent protein content. Melamine nephrotoxicity was brought to international attention in 2008 after reports of nephrolithiasis among Chinese infants who drank milk-based formula contaminated with high levels of melamine, used in an effort to demonstrate a high protein content in the formula without the actual need for milk-based products. The chemical is rapidly excreted in urine and can precipitate in distal tubules. It forms an especially insoluble structure when it complexes with another common commercial chemical, cyanuric acid, in a lower urinary pH. Melamine calculi are radiolucent, best assessed by ultrasound, and may involve the kidney pelvis and/or ureters bilaterally, causing obstruction. Treatment is hydration, urinary alkalinization, pain management, and renal replacement therapy and/or surgical therapy if needed [39]. Screening of asymptomatic, possibly exposed Chinese infants now residing in other areas of the world is not universally recommended, as most stones found in such children have passed easily with hydration [40].

Other causes of kidney stones

Additional clinical situations in which patients are predisposed to forming kidney stones include patients with cystic fibrosis, who may have multiple risk factors, including an absence of the oxalate-degrading bacterium, *O. formigenes*, hyperoxaluria, hypercalciuria, and/or hypocitraturia. Patients taking protease inhibitors for HIV, especially the poorly soluble indinavir, may have urinary excretion of crystallized drug product. Patients on a ketogenic diet for seizure control are predisposed to hypercalciuria and/or hypocitraturia associated with the chronic metabolic acidosis.

CLINICAL EVALUATION

The clinical presentation of urinary tract stones in children may differ from that in adults. Less than 50% of young children with nephrolithiasis will present with abdominal, flank, or pelvic pain as seen in older children, adolescents, and adults with urinary tract stones. Gross or microscopic hematuria (occurring in 33–90% of affected children), dysuria, frequency, emesis, and urinary tract infections are additional common presenting signs in younger patients. A detailed history and physical should guide evaluation of kidney stones. Family history should focus on members with kidney stones (positive in greater than one-third of affected children), gout, arthritis, or chronic kidney disease [3]. The presence of a concomitant urinary tract infection must be sought but should not be accepted as the cause of the stone. Patients should also be advised to submit any passed stones or stone fragments for analysis by polarization microscopy or X-ray diffraction but not by simple chemical analysis.

Useful imaging studies may include plain abdominal radiology, ultrasonography, and helical computed tomography (CT). Conventional abdominal radiographs may show only radiopaque but not radiolucent stones, whereas ultrasound of the urinary tract may show both radiolucent and radiopaque stones, in addition to the presence of urinary obstruction or nephrocalcinosis. Ultrasound has largely taken the place of intravenous pyelography (IVP) as an initial study for stone presence, secondary to concerns about radiation and contrast exposure with the latter procedure. Non-contrast helical CT has been found to have high sensitivity and specificity in identifying even small stones without requiring intravenous contrast administration. It may precisely localize stones, detect obstruction and hydronephrosis, and is much more sensitive than the previously mentioned imaging modalities [41].

Because the majority of children with stones may have a metabolic problem that is discoverable and generally amenable to therapy, diagnostic urinary and blood tests for stone evaluation should be obtained while the patient is on their routine activity schedule and diet. At least two 24-hour urine collections should be performed, waiting at least 2 weeks after any acute stone event. This time frame allows for the resumption of the child's normal intake of food and fluids after recovery from pain and/or surgical intervention, which is critical for correct assignment of metabolic disturbances. These collections can assess urinary volume as a reflection of fluid intake and creatinine excretion for completeness of the 24-hour collection (at least 10–15 mg/kg/day in children older than 2 years of age; 6–9 mg/kg/day in children younger than 2 years of age) and measurement of levels of lithogenic substances such as calcium, oxalate, uric acid, and cystine. The collections can evaluate for decreased stone inhibitor levels as well, such as citrate and magnesium. Normal values of these substances are shown in Table 105.2. Serum levels of uric acid, potassium, calcium, phosphorus, creatinine, bicarbonate (total CO), and bio-intact parathyroid hormone (PTH) (if hypercalcemia is present) should be obtained as well at the end of the urinary collections.

Consultation with an expert in pediatric stone disorders is encouraged if questions arise about the results of these diagnostic studies.

Table 105.2. Normative Data for Urinary Solute Excretion*

Substance	Reference Range
Calcium	≤3.5 mg/kg/day
Citrate	>400 mg/g creatinine (spot citrate/Cr ratio >0.51 g/g)
Oxalate	≤0.5 mmol/1.73 m^2/day (<40 mg/1.73 m^2/day)
Uric Acid	Varies with age, up to 815 mg/1.73 m^2/day
Cystine	<60 mg/1.73 m^2/day

*Refs. 14, 63.

SURGICAL MANAGEMENT

The goals of the management of patients with kidney stones are to remove existing stones and prevent stone recurrence, with preservation of kidney function. Pediatric patients usually pass ureteral stones up to 5 mm in size [6]. In the absence of infection or persistent pain, such stones can be safely observed for up to 6 weeks. Larger stones and kidney-located stones, however, require the consideration of surgical intervention, with a goal of achieving and maintaining a stone-free state. The choice of surgical modality depends on stone composition, size, and location along the urinary tract. Shock wave lithotripsy (SWL) uses the generation and focusing of shock wave energy toward the stone. Pulverized fragments are subsequently passed, and multiple treatment sessions are sometimes required. One large pediatric series (n = 344) showed a 92% stone-free rate for renal pelvis stones smaller than 1 cm, a 68% rate for stones 1–2 cm, and a 50% rate for stones larger than 2 cm. Calyceal stone clearance rates were lower [42]. Overall, stone-free rates in children treated using this procedure have ranged from 67% to 99% in various studies [43], with the highest success rates appearing to be in the youngest children [44]. This procedure seems to be safe in young children and infants, with no evidence of long-term changes in glomerular filtration rate or in functional renal parenchymal scarring before and after treatment in the affected kidney [45, 46]. Minor complications such as bruising, renal colic, and hematuria may occur with SWL treatment. Small children may require the use of lung shielding to prevent pulmonary contusion, as well as reduced power settings to avoid injury. Ureteral stenting may also be required for larger stone burdens. In general, large stone burden (greater than 2 cm) and anatomic abnormalities are risk factors for unsuccessful SWL, and alternative urological approaches should be considered in these cases [43]. Struvite, calcium oxalate dehydrate, and uric acid stones are especially amenable to fragmentation with SWL, whereas cystine, brushite, and calcium oxalate monohydrate stones are all resistant to SWL treatment [47].

Percutaneous nephrolithotomy (PNL) is an alternative procedure that may be used alone or in conjunction with SWL in patients with large stone burden, significant renal obstruction, and/or staghorn calculi. It is also commonly used to remove lower pole calculi smaller than 1 cm in size. Percutaneous access to the collecting system of the kidney is achieved, and a wire is advanced to dilate the tract to accommodate a nephroscope.

Stones may be removed or pulverized under direct visualization, making this approach ideal for complex upper tract stones. A nephrostomy tube is often placed postoperatively, although a small series in adults did show a decrease in pain and recovery time, with no increase in complications, with the use of smaller or no nephrostomy tubes [48]. The development of smaller nephroscopes has made PNL available for children, with stone-free rates ranging from 83% to 98% [43].

Table 105.3. Surgical Treatment Options by Stone Size and Location

Location/Size	Shock Wave Lithotripsy	Ureteroscopy	Percutaneous Nephrolithotomy
Renal			
<1 cm	Most common	Optional	Optional
1–2 cm	Most common	Optional	Optional
>2 cm	Optional	Rare	Most common
Lower Pole			
<1 cm	Most common	Optional	Optional
>1 cm	Optional	Optional	Most common
Ureteral			
Proximal	Most common	Optional	Occasional
Distal	Optional	Most Common	Rare

From Durkee CT, Balcom A. 2006. Surgical management of urolithiasis. *Pediatr Clin N Am* 53: 465–477. Used with permission.

Ureteroscopy is most ideally used for the removal and/or fragmentation of distal ureteral stones. Whereas SWL has good efficacy for some smaller ureteral stones, stone-free rates in those with stones larger than 10 mm in size have been found to be markedly higher with ureteroscopy (93%) than with SWL (50%) [49]. Smaller rigid and flexible ureteroscopes have made this procedure an option for pediatric patients and have made the need for concomitant ureteral balloon dilation (and possible risks of stricture and vesicoureteral reflux) less frequent. Once the ureteroscope is passed, laser energy is used to fragment any visualized stones, and flexible wire baskets can be used to remove fragments. Postoperative stenting may be used to facilitate passage of residual fragments or to prevent ureteral obstruction in the face of edema caused by trauma to the ureteral wall. Stenting is not usually done in uncomplicated procedures, with easy passage of the scope [50]. A summary of surgical treatment options by stone size and location is shown in Table 105.3.

MEDICAL MANAGEMENT

Nonspecific management of urolithiasis includes an increase in fluid intake to increase urinary volume, urinary dilution, and inducing stone particle motion through the urinary tract. Other specific measures depend on the underlying predisposing diagnosis.

Hypercalciuria may be treated with a low sodium diet, thiazide diuretics, and adequate potassium intake. Thiazide therapy (e.g., hydrochlorothiazide 1 mg/kg/day, maximum of 25 mg/day) in FIH significantly decreases urinary calcium excretion and rate of stone formation in adults [51]. A decrease in urinary calcium excretion with thiazide treatment in children with FIH has also been shown [52]. In a population with hypercalciuria from immobilization, 18 of 42 children were found to be hypercalciuric, with a higher rate of fractures [53]. A 3-week course of hydrochlorothiazide and amiloride reduced the mean urinary calcium to creatinine ratio by 57.7%. Dietary calcium intake should not be limited. Citrate therapy (e.g., potassium citrate 2 mmol/kg once daily) is also appropriate in cases of documented hypocitraturia.

Another important issue for some patients is that of bone mineral density (BMD) in FIH. One study of 40 girls with FIH and their premenopausal mothers showed that BMD lumbar spine Z-scores were significantly lower in these patients compared with controls [54]. Others have shown that thiazide treatment, in addition to decreasing urinary calcium excretion, can also improve BMD scores in children. Average Z-score improved from −1.3 to −0.22 over 1 year of treatment with hydrochlorothiazide and potassium citrate in one study of 18 children [52].

The treatment of struvite stones rests on the eradication of stones, correction of any urinary obstruction, and treatment/prevention of urinary tract infections. Urinary acidification could theoretically be used to prevent crystallization, but evidence for such an approach is thus far lacking. The urease inhibitor acetohydroxaminic acid (AHA) may have some clinical use [55], but its use is limited by a high incidence of neurologic and gastrointestinal side effects.

Hyperuricosuria may be treated with dietary sodium limitation, oral bicarbonate or citrate supplementation, or addition of allopurinol or newer agents (rasburicase) if increased uric acid production and hyperuricemia are present.

Patients with suspected primary hyperoxaluria should be given a therapeutic course of vitamin B_6 (pyridoxine), and urinary oxalate levels should be used to monitor success or to suggest the need for dose escalation. Orally administered *Oxalobacter formigenes*, an anaerobic bacterium that degrades oxalate, had been found to reduce urine and plasma oxalate levels in PH patients in a small open-label study [56]. However, a randomized controlled trial of 43 subjects with PH and good kidney function was unable to demonstrate any significant difference in urinary oxalate reduction between subjects who received experimental therapy or a placebo, although the *O. formigenes* treatment was safe and well tolerated [57].

For patients with reduced kidney function, intensive dialysis followed by liver/kidney transplant can be curative, because a new liver replaces the enzymatic defect in PH-l [58, 59]. Pretransplant hemodialysis for five to six times per week, and perhaps with additional nightly peritoneal dialysis, is needed to lessen the systemic oxalate burden and prevent recurrence of disease in the transplant kidney. Prompt referral to a pediatric center with expertise in this disorder is warranted.

Secondary hyperoxaluria that results from enteric hyperoxaluria may be treated with a low-sodium/low-fat diet, high fluid intake, and a dietary calcium intake at the upper end of the daily recommended intake. Limitation of oxalate-containing foods such as chocolate, rhubarb, nuts, and spinach should be advised, as well as possible supplementation with magnesium, phosphorus, and citrate salts.

Cystinuria is treated with fluids (minimum of 3 liters/ 1.73m2/day) and provision of alkali salts, such as citrate. Low sodium intake can also decrease urinary excretion of cystine. Chelating agents such as D-penicillamine, or more recently, a-mercaptopropionylglycine (tiopronin or Thiola; Mission Pharmacal, San Antonio, TX, USA) may also be prescribed by someone skilled in pediatric stone disease. D-penicillamine may cause a severe serum sickness-like reaction, so its use is not advised, but side effects are less severe with Thiola [22]. Angiotensin-converting enzyme (ACE) inhibition with captopril therapy has been found beneficial in some patients with cystinuria (captopril-cystine complexes are 200 times more soluble than cystine alone) resistant to alkalinization and fluid therapy alone, and perhaps with less bothersome side effects [60]. Captopril, however, is not as effective as thiolcompound therapy.

PROGNOSIS

Estimates of the recurrence rate of urolithiasis in children have ranged from 16% to 67% [3, 33]. Patients with residual fragments after treatment are at risk for symptomatic growth of those fragments [61], and the existence of metabolic disorders is a strong predictor of this growth in children [62]. Additionally, prognosis for nephrolithiasis depends on the type of stones and adherence to therapy. Cystine stones have a high recurrence rate, and obstruction may impair kidney function. Primary hyperoxaluria type I is a progressive disease, often leading to progressive loss of kidney function even with optimal compliance to medical therapy, unless pyridoxine responsiveness is established. Kidney stones from hyperuricosuria may continue to occur with or without symptoms, despite treatment. Therefore, pediatric patients with stone disease should be referred to a subspecialist for appropriate diagnosis, treatment, and long-term nephrology follow-up.

REFERENCES

1. Stapleton FB. 1989. Nephrolithiasis in children. *Pediatr Rev* 11: 21–30.
2. Schaeffer AJ, Feng Z, Trock BJ, Mathews RI, Neu AM, Gearhart JP, Matlaga BR. 2011. Medical comorbidities associated with pediatric kidney stone disease. *Urology* 77(1): 195–199.
3. Milliner DS, Murphy ME. 1993. Urolithiasis in pediatric patients. *Mayo Clin Proc* 68: 241–248.
4. Diamond DA, Rickwood AM, Lee PH, Johnston JH. 1994. Infection stones in children: A twenty-seven-year review. *Urology* 43: 525–527.
5. Coward RJ, Peters CJ, Duffy PG, Corry D, Kellett MJ, ChoongS, van't Hoff WG. 2003. Epidemiology of pediatric stone disease in the UK. *Arch Dis Child* 88: 962–965.
6. Pietrow PK, Pope JC, Adams MC, Shyr Y, Brock JW 3rd. 2002. Clinical outcome of pediatric stone disease. *J Urol* 167: 670–673.
7. Choi H, Snyder HM 3rd, Duckett JW. 1987. Urolithiasis in childhood: Current management. *J Pediatr Surg* 72: 158–164.
8. Gearhart JR, Herzberg GZ, Jeffs RD. 1991. Childhood urolithiasis: Experiences and advances. *Pediatrics* 87: 445–450.
9. Stapleton FB, McKay CP, Noe HN. 1987. Urolithiasis in Children: The role of hypercalciuria. *Pediatr Ann* 16: 980–992.
10. Milliner DS 2004. Urolithiasis. In: Avner ED, Harmon WE, Niaudet P (eds.) *Pediatric Nephrology*. Philadelphia: Lippincott Williams &Wilkins. pp. 1091–1111.
11. Gillespie RS, Stapleton FB. 2004. Nephrolithiasis in children. *Pediatr Rev* 25: 131–138.
12. Stapleton FB. 1990. Idiopathic hypercalciuria: Association with isolated hematuria and risk for urolithiasis in children: The Southwest Pediatric Nephrology Study Group. *Kidney Int* 37: 807–811.
13. Garcia CD, Miller LA, Stapleton FB. 1991. Natural history of hernaturia associated with hypercalciuria in children. *Am J Dis Child* 145: 1204–1207.
14. Stapleton FB. 2002. Childhood stones. *Endocrinol Metab Clin North Am* 31: 1001–1015.
15. Langman CB, Schmeissing KJ, Sailer DM. 1994. Children with genetic hypercalciuria exhibit thiazide-response to osteopenia. *Pediatr Res* 35: 368A.
16. Lloyd SE, Pearce SHS, Fisher JE, Steinmeyer K, Schwappach B, Scheinman SJ, Harding B, Bolino A, Devoto M, Goodyer P, RigdenSP, Wrong O, Jentsch TJ, Craig IW, Thakker RV. 1996. A common molecular basis for three inherited molecular kidney stone diseases. *Nature* 379: 445–449.

17. Dent CE, Friedman M. 1964. Hypercalciuric rickets associated with renal tubular damage. *Arch Dis Child* 39: 240–249.
18. Utsch B, Bokenkamp A, Benz MR, Besbas N, Dotsch J, Franke I, Frund S, Gok F, Hoppe B, Karle S, Kuwertz-Broking E, Laube G, Beb M, Nuutinen M, Ozaltin F, Rascher W, Ring T, Tasic V, van Wijk JA, Ludwig M. 2006. Novel OCRL1 mutations in patients with the phenotype of Dent disease. *Am J Kidney Dis* 48(6): 942.
19. Thomas SE, Stapleton FB. 2000. Leave no "stone" unturned: Understanding the genetic basis of calcium-containing urinary stones in childhood. *Adv Pediatr* 47: 199–221.
20. Nicoletta JA, Lande MB. 2006. Medical evaluation and treatment of urolithiasis. *Pedatr Clin North Am* 53: 479–491.
21. Pearce SH, Williamson C, Kifor O, Bai M, Coulthard MG, Davies M, Lewis-Barned N, McCredie D, Powell H, Kendall-Taylor P, Brown EM, Thakker RV. 1996. A familial syndrome of hypocalcemia with hypercalciuria due to mutations in the calcium-sensing receptor. *N Engl J Med* 335(15): 1115–1122.
22. Bartosh SM. 2004. Medical management of pediatric stone disease. *Urol Clin North Am* 31: 575–587.
23. Reddy ST, Wang CY, Sakhaee K, Brinkley L, Pak CYC. 2002. Effect of low-carbohydrate high-protein diets on acid-base balance, stone-forming propensity, and calcium metabolism. *Am J Kidney Dis* 40: 265–274.
24. Monico CG, Rossetti S, Schwanz HA, Olson JB, Lundquist PA, Dawson DB, Harris PC, Milliner DS. 2007. Comprehensive mutation screening in 55 probands with type 1 primary hyperoxaluria shows feasibility of a gene-based diagnosis. *J Am Soc Nephrol* 18: 1905–1914.
25. Williams E, Rumsby G. 2007. Selected exonic sequencing of the AGXT gene provides a genetic diagnosis in 50% of patients with primary hyperoxaluria type 1. *Clin Chem* 53: 1216–1221.
26. Milliner D, Wilson D, Smith L. 2001. Phenotypic expression of primary hyperoxaluria: Comparative features of types I and II. *Kidney Int* 59: 31–36.
27. Milliner D, Wilson D, Smith L. 1998. Clinical expression and long-term outcomes of primary hyperoxaluria types 1 and 2. *J Nephrol* 11(Suppl): 56–59.
28. Belostotsky R, Seboun E, Idelson GH, Milliner DS, Becker-Cohen R, Rinat C, Monico CG, Feinstein S, Ben-Shalom E, Magen D, Weissman I, Charon C, Frishberg Y. 2010. Mutations in DHDPSL are responsible for primary hyperoxaluria type III. *Am J Hum Genet* 87: 392–399.
29. Hoppe B, Langman C. 2003. A United States survey on diagnosis, treatment, and outcomes of primary hyperoxaluria. *Pediatr Nephrol* 18: 986–991.
30. Jamieson N. 2007. The European PH1 Transplant Registry Report 1984–2007: Twenty-three years of combined liver and kidney transplantation for primary hyperoxaluria PH1. Presented at the 8th International Primary Hyperoxaluria Workshop, University College London, London, UK, June 29–30, 2007.
31. Duffey BG, Alanee S, Pedro RN, Hinck B, Kriedberg C, Ikramuddin S, Kellogg T, Stessman M, Moeding A, Monga M. 2010. Hyperoxaluria is a long-term consequence of Roux-en-Y Gastric bypass: A 2-year prospective longitudinal study. *J Am Coll Surg* 211(1): 8–15.
32. Pak CY, Waters O, Arnold L, Holt K, Cox C, Barilla D. 1977. Mechanism for calcium urolithiasis among patients with hyeruricosuria. *J Clin Invest* 59: 426–431.
33. Polinsky MS, Kaiser BA, Baluarte HJ. 1987. Urolithiasis in childhood. *Pediatr Clin North Am* 34: 683–710.
34. Enomoto A, Kimura H, Chairoungdua A, Shigeta Y, Jutabha P, Cha SH, Hosoyamada M, Takeda M, Sekine T, Igarashi T, Matsuo H, Kikuchi Y, Oda T, Ichida K, Hosoya T, Shimokata K, Niwa T, Kanai Y, Endou H. 2002. Molecular identification of a renal urate anion exchanger that regulates blood urate levels. *Nature* 417: 447–452.
35. Maalouf NM, Sahaee K, Parks JH, Coe FL, Adams-Huet B, Pak CY. 2004. Association of urinary pH with body weight in nephrolithiasis. *Kidney Int* 65: 1422–1425.
36. Chesney RW. 1998. Mutational analysis of patients with cystinuria detected by a genetic screening network: Powerful tools in understanding the several forms of the disorder. *Kidney Int* 54: 279–280.
37. Feliubadaló L, Font M, Purroy J, Rousaud F, Estivill X, Nunes V, Golomb E, Centola M, Aksentijevich I, Kreiss Y, Goldman B, Pras M, Kastner DL, Pras E, Gasparini P, Bisceglia L, Beccia E, Gallucci M, de Sanctis L, Ponzone A, Rizzoni GF, Zelante L, Bassi MT, George AL Jr, Manzoni M, De Grandi A, Riboni M, Endsley JK, Ballabio A, Borsani G, Reig N, Fernández E, Estévez R, Pineda M, Torrents D, Camps M, Lloberas J, Zorzano A, Palacín M; International Cystinuria Consortium. 1999. Non-type I cystinuria caused by mutations in SLC7A9, encoding a subunit of rBAT. *Nat Genet* 23(1): 52–57.
38. Goodyer P, Saadi I, Ong P, Elkas G, Rozen R. 1998. Cystinuria subtype and the risk of nephrolithiasis. *Kidney Int* 54(1): 56–61.
39. Bhalla V, Grimm PC, Chertow GM, Pao AC. 2009. Melamine nephrotoxicity: An emerging epidemic in an era of globalization. *Kidney Int* 75: 774–779.
40. Langman CB. 2009. Melamine, powdered milk, and nephrolithiasis in Chinese infants. *N Engl J Med* 360(11): 1139–1141.
41. Jackman SV, Potter SR, Regan F, Jarrett TW. 2000. Plain abdominal x-rays versus computerized tomography screening: Sensitivity for stone localization after nonenhanced spiral computerized tomography. *J Urol* 164: 308–310.
42. Muslumanoglu AY, Tefekli A, Sarilar O, Binbay M, Altunrende F, Ozkuvanci U. 2003. Extracorporeal shock wave lithotripsy as first line treatment alternative for urinary tract stones in children: A large scale retrospective analysis. *J Urol* 170: 2405–2408.
43. Desai M. 2005. Endoscopic management of stones in children. *Curr Opin Urol* 15: 107–112.
44. Aksoy Y, Ozbey I, Atmaca AF, Polat O. 2004. Extracorporeal shock wave lithotripsy in children: Experience

using a mpi-9000 lithotriptor. *World J Urol* 22: 115–119.
45. Goel MC, Baserge NS, Babu RV, Sinha S, Kapoor R. 1996. Pediatric kidney: Functional outcome after extracorporeal shock wave lithotripsy. *J Urol* 155: 2044–2046.
46. Lottmann HB. Archambaud F, Hallal B, Pageyral BM, Cendron M. 1998. 99mTechnetium-dimercapto-succinic acid renal scan in the evaluation of potential long-term renal parenchymal damage associated with extracorporeal shock wave lithotripsy in children. *J Urol* 159: 521–524.
47. Saw KC, Lingeman JE. 1999. Management of calyceal stones: Lesson 20. *AUA Update Series* 20: 154–159.
48. Desai MR, Kukreja RA, Desai MM, Mhaskar SS, Wani KA, Patel SH, Bapat SD. 2004. A prospective randomized comparison of type of nephrostomy drainage following percutaneous nephrostolithotomy: Large bore versus small bore versus tubeless. *J Urol* 172: 565–567.
49. Lam JS, Greene TD, Gupta M. 2002. Treatment of proximal ureteric calculi: Holmium:YAG laser ureterolithotrimv versus ESWL. *J Urol* 167: 1972–1976.
50. Durkee CT, Balcom A. 2006. Surgical management of urolithiasis. *Pediatr Clin North Am* 53: 465–477.
51. Ohkawa M, Tokunaga S, Nakashima T, Orito M, Hisazumi H. 1992. Thiazide treatment for calcium urolithiasis in patients with idiopathic hypercalciuria. *Br J Urol* 69: 571–576.
52. Reusz GS, Dobos M, Vásárhelyi B, Sallay P, Szabó A, Horváth C, Szabó A, Byrd DJ, Thole HH, Tulassay T. 1998. Sodium transport and bone mineral density in hypercalciuria with thiazide treatment. *Pediatr Nephrol* 12: 30–34.
53. Bentur L, Alon U, Berant M. 1987. Hypercalciuria in chronically institutionalized bedridden children: Frequency, predictive factors and response to treatment with thiazides. *Int J Pediatr Nephrol* 8: 29–34.
54. Garcia-Nieto V, Navarro JF, Monge M, Garcia-Rodriguez VE. 2003. Bone mineral density in girls and their mothers with idiopathic hypercalciuria. *Nephron Clin Pract* 94: C81–C82.
55. Griffith DP, Gleeson MJ, Lee H, Longreit R, Deman E, Earle N. 1991. Randomized, double-blind trial of lithostat (acetohydroxaminic acid) in the palliative treatment of infection-induced urinary calculi. *Eur Urol* 20: 243–247.
56. Hoppe B, Beck B, Gatter N, et al. 2006. Oxalobacter formigenes: A potential tool for the treatment of primary hyperoxaluria type 1. *Kidney Int* 70: 1305–1311.
57. Hoppe B, Groothoff JW, Hulton S, Cochat P, Niaudet P, Kemper MJ, Deschenes G, Unwin R, Milliner D. 2011. Efficacy and safety of Oxalobacter formigenes to reduce urinary oxalate in primary hyperoxaluria. *Nephol Dial Transplant* 26(11): 3609–3615.
58. Jamieson N. 2005. A 20-year experience of combined liver/kidney transplantation for primary hyperoxaluria (PH1): The European PHI Transplant Registry Experience 1984–2004. *Am J Nephrol* 25: 282–289.
59. Cibrik D, Kaplan B, Arndorfer J, Mier-Kriesche H. 2002. Renal allograft survival in patients with oxalosis. *Transplantation* 74: 707–710.
60. Coulthard MG, Richardson J, Fleetwood A. 1995. The treatment of cystinuria with captopril. *Am J Kidney Dis* 25: 661–662.
61. Streem SB, Yost A, Mascha E. 1996. Clinical implications of clinically insignificant stone fragments after extracorporeal shock wave lithotripsy. *J Urol* 155: 1186–1100.
62. Afshar K, McLorie G, Papanikolaou F, Malek R, Harvcy E, Pippi-Salle JL, Bagli DJ, Khoury AE, Farhat W. 2004. Outcome of small residual stone fragments following shock wave lithotripsy in children. *J Urol* 172: 1600–1603.
63. Santos-Victoriano M, Brouhard BH, Cunningham RJ 3rd. 1998. Renal stone disease in children. *Clin Pediatr* 37: 583–599.

106
Treatment of Renal Stones

John R. Asplin

Introduction 878
Calcium Stones 878
Uric Acid Stones 880
Cystine Stones 880
Struvite Stones 881
References 881

INTRODUCTION

Renal stones form in urine that is supersaturated with respect to the salt of which the stone is composed. Crystallization can be induced at low levels of supersaturation if there are promoters of crystallization in the urine, whereas urinary inhibitors oppose crystallization. Medical treatment of nephrolithiasis works by lowering urine saturation of the stone forming salt, reducing promoters of crystallization, and/or increasing inhibitors of crystallization. Optimal treatment is based on the type of stone and the lithogenic risk factors identified in 24-hour urine chemistries. Treatment may consist of diet, medication, or a combination of both.

Before a discussion of medical therapy, consideration must be given to the goals of treatment. The aim of treatment is to prevent new stone formation and reduce the rate of growth of any existing stones. Radiologic evaluation is a critical component of any treatment regimen. A knowledge of the number and location of all stones at the time treatment is initiated is required to monitor the success of therapy. If a patient develops renal colic and the stone passing is clearly a new stone, then medical therapy has not been successful; the cause of the failure should be studied and the treatment regimen reevaluated. Not only is the knowledge of old versus new stone important for the physician, but also for the patient. Patients who pass a stone despite following the prescribed therapy will become discouraged and stop therapy unless they understand that the stone that passed was present all along. Serial radiologic evaluation (every 1 to 2 years) is also needed to insure that new asymptomatic stones are not forming during treatment.

High fluid intake is the foundation of all medical regimens for renal stone prevention. By keeping urine dilute, urine saturation of all stone forming salts is lowered. Although high fluid intake has long been considered the standard of care, Borghi et al. have performed the only prospective randomized study of high fluid intake for stone prevention [1]. Over the 5 years of follow-up, a 55% reduction of stone recurrence in the group treated with high fluid intake compared to the control group was observed. The high fluid group had an average 24-hour urine volume of 2.5 liters, which provides a reasonable therapeutic goal for patients.

CALCIUM STONES

There are four abnormalities of urine chemistries known to cause calcium oxalate kidney stones: hypercalciuria, hyperoxaluria, hypocitraturia and hyperuricosuria. Therapy is directed against whichever abnormality the patient has.

Hypercalciuria

In the past, low-calcium diets had often been prescribed for calcium stone formers. Although dietary calcium

---

Primer on the Metabolic Bone Diseases and Disorders of Mineral Metabolism, Eighth Edition. Edited by Clifford J. Rosen.
© 2013 American Society for Bone and Mineral Research. Published 2013 by John Wiley & Sons, Inc.

restriction clearly lowers urine calcium, it has never been shown to reduce stone formation in a prospective controlled trial. In addition, the long-term safety of a low-calcium diet has been questioned because hypercalciuria has been associated with reduced bone mineral density (BMD) and increased risk of fractures [2, 3]. However, other dietary interventions can be used to lower urine calcium. Both high-sodium and high-protein diets increase urine calcium excretion, so restriction of these dietary components seems a reasonable intervention. Borghi et al. tested the efficacy of a normal calcium (1,200 mg/day), low-sodium, low-protein intake diet compared with the classic low-calcium diet (400 mg/day) in 120 men with recurrent kidney stones [4]. Stone recurrence was reduced by 48% in the group treated with the low-sodium, low-protein diet compared with the low-calcium diet. Although this study does not answer the question of whether low-calcium diets reduce stone formation, it does show there is an alternative diet therapy that is more efficacious and likely better for general health.

For patients with frequent stones or whose stone disease persists despite dietary intervention, thiazide diuretics should be used to lower urine calcium. There are three prospective, placebo-controlled, 3-year trials of thiazide type diuretics, all of which show a significant reduction in stone formation in patients with recurrent kidney stones [5–7]. There has not been a comparison between the various thiazides to determine if one has a greater degree of anti-hypercalciuric effect than the others. Diet sodium restriction should be continued when thiazides are used, because excess salt in the diet will reduce the effectiveness of a thiazide. Thiazides cause potassium wasting; therefore, serum potassium should be monitored because hypokalemia can lead to hypocitraturia, creating a new risk factor for stone formation [8]. Thiazides also have an additional benefit in this population because they improve calcium balance and prevent osteopenia in hypercalciuric stone formers [9, 10].

Hyperoxaluria

Oxalate excreted in the urine is derived from both intestinal absorption of oxalate in the diet and from endogenous production of oxalate [11]. Oxalate is an end product of human metabolism; whatever oxalate is absorbed or produced by the body must be excreted by the kidneys. Because a significant portion of urine oxalate is derived from diet, low-oxalate diets are recommended for all hyperoxaluric stone patients, although the efficacy of a low-oxalate diet in preventing stone formation has not been studied. General guidelines for low-oxalate diets are available on the Internet [12]. Other dietary advice focuses on foods that have a secondary effect on oxalate excretion. Low-calcium diets can increase urine oxalate excretion because low calcium concentration in the intestinal lumen leaves more oxalate in the unbound state to be absorbed [11]. The magnitude of this effect in clinical practice is uncertain but it seems prudent to have patients on a diet with 1,000 to 1,200 mg of calcium per day. Endogenous production of oxalate may be increased by an increased intake of oxalate precursors. A number of amino acids can be metabolized to oxalate, so it is not surprising that high-protein diets have been shown to increase urine oxalate excretion in some patients with hyperoxaluria [13]. A recent study showed a 1,200-mg calcium, low-animal-protein, low-sodium diet reduced urine oxalate excretion by 30% in patients with mild hyperoxaluria [14]. Vitamin C can also be metabolized to oxalate, so stone patients should avoid high doses of vitamin C supplements [15].

There are no FDA-approved medications for hyperoxaluria at this time. Pyridoxine has been reported to lower urine oxalate in some patients with idiopathic hyperoxaluria, but there are no controlled trials showing that pyridoxine reduces stone formation in such patients [16]. Magnesium has been proposed as a therapy for stone disease because it may lower urine oxalate by complexing oxalate in the intestine and act as a crystal inhibitor in the urine. There has been only one controlled trial of magnesium supplements in calcium stone disease, and it did not show a benefit to therapy [7]. Currently under study is the use of oxalate degrading bacteria as probiotics [17, 18]. These agents would act by lowering intestinal oxalate concentration, limiting absorption and perhaps even increasing intestinal secretion of oxalate. However, further studies need to be done to prove their efficacy in humans with hyperoxaluric stone disease [19].

Enteric hyperoxaluria and primary hyperoxaluria can lead to severe hyperoxaluria and cause kidney failure in addition to stone disease. Enteric hyperoxaluria can be seen in patients with extensive small bowel disease or bowel resection. Recently, it has been shown that modern bariatric surgery can also cause enteric hyperoxaluria [20]. To lower urine oxalate, low-oxalate diets are prescribed, and calcium supplements are provided with meals to bind oxalate in the diet [21]. Calcium citrate is preferred over calcium carbonate as many bariatric procedures alter stomach anatomy, and there may not be sufficient acid exposure to dissolve calcium carbonate. If fat malabsorption is present, a low-fat diet should be recommended; cholestyramine may also lower oxalate absorption in patients with steatorrhea. Primary hyperoxaluria (PH) is a rare genetic form of hyperoxaluria. In PH type 1 (OMIM 259900), about 30% of patients will have a significant reduction in oxalate excretion with pyridoxine therapy [22]. If oxalate excretion cannot be lowered with pyridoxine, then patients should be treated with neutral phosphate salts and/or potassium citrate as well as high fluid intake [23, 24]. PH type 2 (OMIM 260000) does not respond to pyridoxine therapy but otherwise is treated the same as PH type 1. Recently, PH type 3 (OMIM 613616) was identified and found to be caused by mutations in the gene encoding 4-hydroxy-2-oxoglutarate aldolase [25]. Although no treatment trials for this specific form of PH have been performed, that the enzyme of interest is in the pathway of hydroxyproline metabolism suggests that lowering hydroxyproline, a known precursor of oxalate, may be of benefit.

Hypocitraturia

Citrate is a normal constituent of human urine that prevents calcium stone formation by forming a soluble complex with calcium, thereby reducing the free calcium ion concentration in the urine. Hypocitraturia may be seen with acidemic states such as renal tubular acidosis, hypokalemia, or it may be idiopathic in origin [8]. Urine citrate excretion is highly dependent on systemic acid–base status, increasing with alkali loading. Reducing dietary animal protein and increasing consumption of fruits and vegetables will increase urine citrate excretion [26], but dietary intervention is usually not sufficient to correct significant hypocitraturia. Some authorities have proposed using lemonade as a treatment for hypocitraturia because of the high citric acid content of lemon juice [27]. However, significant increases in urine citrate have not been found in all studies [28].

The standard treatment for hypocitraturia is alkali supplementation. Alkali may be provided as either citrate or bicarbonate, but citrate is most commonly used. Citrate is usually provided as potassium citrate, because sodium salts can increase urine calcium excretion. If the patient cannot tolerate a potassium load, then sodium alkali can be used, although dietary sodium restriction should be encouraged to minimize the increase in urine calcium. There have been two randomized placebo-controlled trials of citrate salts in the treatment of calcium stone disease, one with potassium citrate and the other using potassium–magnesium citrate [29, 30]. Both showed significant reductions in stone recurrence rates in the group receiving the active drug. Potassium citrate also reduced stone recurrence and growth of residual stone fragments in patients who had undergone extracorporeal lithotripsy [31]. In addition to the increase in citrate, alkali salts can also cause a mild reduction in the urine calcium excretion by buffering daily metabolic acid production. In fact, long-term treatment of calcium stone formers with citrate salts has been shown to increase BMD as well, presumably from the favorable effects on urine calcium excretion and calcium balance [32]. One potential adverse consequence of alkali therapy is an increase in urine pH, which, in patients with elevated urine calcium, can increase the risk of calcium phosphate stone formation. Such patients should have urine calcium controlled by the addition of a thiazide diuretic.

Drugs that inhibit carbonic anhydrase cause renal tubular acidosis and increase the risk of kidney stones. Topiramate inhibits carbonic anhydrase and is being increasingly used as a treatment for seizures and migraine headache. Use of topiramate has been associated with nephrolithiasis, and metabolic evaluation shows the expected alkaline urine and hypocitraturia [33]. To prevent continued stone formation, replacing topiramate with an alternative therapy is recommended. If topiramate cannot be discontinued, alkali therapy is recommended, although no trials have shown that alkali reduces stone formation.

Hyperuricosuria

Uric acid is an end product of purine metabolism, produced by the oxidation of xanthine via the enzyme xanthine oxidase. Hyperuricosuria promotes calcium oxalate crystallization in the urine by a salting out phenomenon [34]. Hyperuricosuria is most commonly due to diets high in purines, although some patients may have a metabolic defect that leads to overproduction of uric acid [35]. A low purine diet is a reasonable therapy for patients with hyperuricosuria, although its effectiveness in preventing stone formation has not been tested. Allopurinol, a xanthine oxidase inhibitor, has been shown to reduce stone formation in calcium oxalate stone formers with hyperuricosuria in a randomized prospective trial [36]. It has not been shown to be effective in treatment of patients without hyperuricosuria.

URIC ACID STONES

Three factors contribute to uric acid stone formation: urine flow rate, uric acid excretion rate, and urine pH [37]. Low urine pH (less than 5.8 in a 24-hour urine collection) is the most common abnormality found in uric acid stone formers [38]. Uric acid stones seldom form without a low urine pH, unless the patient has marked overproduction of uric acid from a metabolic defect. Overly acidic urine may be seen in patients with high intake of animal protein, chronic diarrheal states, chronic kidney disease, or in metabolic syndrome [39]. There are no prospective placebo-controlled trials for prevention of uric acid stones, nor are there likely to be, because increasing urine pH with alkali therapy is universally regarded as effective. Alkali at an initial dose of 40 to 60 meq per day in two or three divided doses, will be sufficient for most patients. The desired end point is an average 24-hour urine pH of 6.0 to 6.5. Excessive alkalinization does not offer additional protection from uric acid stone formation, but it will increase the risk of calcium phosphate stone formation. Low-purine diets will help lower uric acid excretion, and by also lowering animal protein intake, will decrease metabolic acid production. Allopurinol can be used for uric acid stones if the patient also has gout, very high levels of uric acid excretion, or if the patient cannot tolerate sufficient alkali to raise urine pH above 6.0. Generally, allopurinol is considered a second line therapy for uric acid stones.

CYSTINE STONES

Cystine is an amino acid formed by the linkage of two cysteine molecules via a disulfide bond. It is poorly soluble in urine, and patients with cystinuria (OMIM 220100) can develop severe, recurrent stone disease.

Therapy to reduce stone formation is focused on lowering cystine concentration in the urine or increasing urine cystine solubility. Patients with cystinuria are instructed to increase fluid intake and maintain a high urine flow rate in order to keep urine cystine concentration below 250 mg/l; most patients need to produce at least 3 liters of urine per day. The patient should awaken at least once per night to void and drink additional water. Dietary sodium and protein restriction lower urine cystine excretion moderately and should be encouraged in all patients [40, 41].

There are two treatment options to improve cystine solubility. Solubility increases as urine pH increases, although the effect is not significant until urine pH is above 7 [42]. Since the desired pH is higher than for other forms of stone disease, it may require larger doses of alkali, although the individual response is highly variable. Potassium alkali is preferred because of the undesirable effect of high sodium intake on urine cystine excretion, although sodium salts should be used if the patient is not able to excrete a high potassium load. The other way to increase solubility of cystine is to form a new disulfide complex that is more soluble than cystine itself. Tiopronin and d-penicillamine contain thiol groups and will undergo a disulfide exchange, leading to the formation of a drug-cysteine complex that is much more soluble than cystine. These medications have been shown to reduce cystine saturation in cystine stone formers [43]. Due to the high frequency of side effects [44], thiol-containing drugs are usually reserved for patients with a very high level of cystine excretion or those who continue to form stones despite dietary and alkali therapy. Captopril, an angiotensin-converting enzyme inhibitor, contains a thiol group and has been suggested as therapy for cystinuria. However, at the doses used, it is unlikely to complex sufficient cystine to alter stone risk, and studies conflict regarding the effectiveness of captopril [45, 46].

STRUVITE STONES

Struvite (magnesium–ammonium–phosphate) stones require elevated urine pH and high ammonium concentration to form. In normal human physiology, urine ammonium excretion falls when urine pH rises above 7; the only time human urine displays high pH and high ammonium is when there is infection with a urease-containing organism such as *P. mirabilis*. Struvite forms large staghorn stones that can lead to loss of renal function. Treatment requires a combined medical and surgical approach. The best outcomes require complete removal of all stone material because the stone itself often harbors the offending bacteria, and if stone is left in the urinary system, infection and stones will recur. Extracorporeal shock wave lithotripsy and percutaneous nephrolithotomy may be used alone or in combination in order to remove all stone material [47]. Stones should be cultured to guide the choice of antibiotics. Antibiotics should be employed to sterilize the urinary system. In patients who are not surgical candidates or in whom stones continue to recur, the urease inhibitor acetohydroxamic acid (AHA) may be considered. Although AHA will not eradicate infection, it does prevent the breakdown of urea and the change in the urinary environment that leads to struvite stone formation. There are three placebo-controlled trials that prove the effectiveness of AHA, but the significant number of side effects such as headache, tremulousness, and possibly deep venous thrombosis limit its utilization [48–50].

REFERENCES

1. Borghi L, Meschi T, Amato F, Briganti A, Novarini A, Giannini A. 1996. Urinary volume, water and recurrences in idiopathic calcium nephrolithiasis: A 5-year randomized prospective study. *J Urol* 155(3): 839–4.
2. Melton LJ 3rd, Crowson CS, Khosla S, Wilson DM, O'Fallon WM. 1998. Fracture risk among patients with urolithiasis: A population-based cohort study. *Kidney Int* 53(2): 459–64.
3. Lauderdale DS, Thisted RA, Wen M, Favus MJ. 2001. Bone mineral density and fracture among prevalent kidney stone cases in the Third National Health and Nutrition Examination Survey. *J Bone Miner Res* 16(10): 1893–8.
4. Borghi L, Schianchi T, Meschi T, Guerra A, Allegri F, Maggiore U, Novarini A. 2002. Comparison of two diets for the prevention of recurrent stones in idiopathic hypercalciuria. *N Engl J Med* 346(2): 77–84.
5. Borghi L, Meschi T, Guerra A, Novarini A. 1993. Randomized prospective study of a nonthiazide diuretic, indapamide, in preventing calcium stone recurrences. *J Cardiovasc Pharmacol* 22 Suppl 6: S78–S86.
6. Laerum E, Larsen S. 1984. Thiazide prophylaxis of urolithiasis: A double-blind study in general practice. *Acta Med Scand* 215: 383–9.
7. Ettinger B, Citron JT, Livermore B, Dolman LI. 1988. Chlorthalidone reduces calcium oxalate calculous recurrence but magnesium hydroxide does not. *J Urol* 139: 679–84.
8. Hamm LL. 1990. Renal handling of citrate. *Kidney Int* 38(4): 728–35.
9. Adams J, Song C, Kantorovich V. 1999. Rapid recovery of bone mass in hypercalcuric, osteoporotic men treated with hydrochlorothiazide. *Ann Intern Med* 130(8): 658–60.
10. Coe FL, Parks JH, Bushinsky DA, Langman CB, Favus MJ. 1988. Chlorthalidone promotes mineral retention in patients with idiopathic hypercalciuria. *Kidney Int* 33(6): 1140–6.
11. Holmes RP, Goodman HO, Assimos DG. 2001. Contribution of dietary oxalate to urinary oxalate excretion. *Kidney Int* 59(1): 270–6.

12. Harvard School of Public Health Nutrition Department's file download site. Oxalate Table of Foods. Updated February 2008. Accessed September 7, 2011. Available at https://regepi.bwh.harvard.edu/health/Oxalate/files/.
13. Nguyen QV, Kalin A, Drouve U, Casez JP, Jaeger P. 2001. Sensitivity to meat protein intake and hyperoxaluria in idiopathic calcium stone formers. *Kidney Int* 59(6): 2273–81.
14. Nouvenne A, Meschi T, Guerra A, Allegri F, Prati B, Fiaccadori E, Maggiore U, Borghi L. 2009. Diet to reduce mild hyperoxaluria in patients with idiopathic calcium oxalate stone formation: A pilot study. *Urology* 73(4): 725–30, 730.e1.
15. Traxer O, Huet B, Poindexter J, Pak CY, Pearle MS. 2003. Effect of ascorbic acid consumption on urinary stone risk factors. *J Urol* 170(2 Pt 1): 397–401.
16. Edwards P, Nemat S, Rose GA. 1990. Effects of oral pyridoxine upon plasma and 24-hour urinary oxalate levels in normal subjects and stone formers with idiopathic hypercalciuria. *Urol Res* 18(6): 393–6.
17. Hatch M, Cornelius J, Allison M, Sidhu H, Peck A, Freel RW. 2006. Oxalobacter sp. reduces urinary oxalate excretion by promoting enteric oxalate secretion. *Kidney Int* 69(4): 691–8.
18. Campieri C, Campieri M, Bertuzzi V, Swennen E, Matteuzzi D, Stefoni S, Pirovano F, Centi C, Ulisse S, Famularo G, et al. 2001. Reduction of oxaluria after an oral course of lactic acid bacteria at high concentration. *Kidney Int* 60(3): 1097–105.
19. Goldfarb DS, Modersitzki F, Asplin JR. 2007. A randomized, controlled trial of lactic acid bacteria for idiopathic hyperoxaluria. *Clin J Am Soc Nephrol* 2(4): 745–9.
20. Asplin JR, Coe FL. 2007. Hyperoxaluria in kidney stone formers treated with modern bariatric surgery. *J Urol* 177(2): 565–9.
21. Worcester EM. 2002. Stones from bowel disease. *Endocrinol Metab Clin North Am* 31(4): 979–99.
22. Monico CG, Rossetti S, Olson JB, Milliner DS. 2005. Pyridoxine effect in type I primary hyperoxaluria is associated with the most common mutant allele. *Kidney Int* 67(5): 1704–9.
23. Milliner DS, Eickholt JT, Bergstralh EJ, Wilson DM, Smith LH. 1994. Results of long-term treatment with orthophosphate and pyridoxine in patients with primary hyperoxaluria. *N Engl J Med* 331(23): 1553–8.
24. Leumann E, Hoppe B, Neuhaus T. 1993. Management of primary hyperoxaluria: Efficacy of oral citrate administration. *Pediatr Nephrol* 7(2): 207–11.
25. Belostotsky R, Seboun E, Idelson GH, Milliner DS, Becker-Cohen R, Rinat C, Monico CG, Feinstein S, Ben-Shalom E, Magen D, Weissman I, Charon C, Frishberg Y. 2010. Mutations in DHDPSL are responsible for primary hyperoxaluria type III. *Am J Hum Genet* 87(3): 392–9.
26. Breslau NA, Brinkley L, Hill KD, Pak CY. 1988. Relationship of animal protein-rich diet to kidney stone formation and calcium metabolism. *J Clin Endocrinol Metab* 66(1): 140–6.
27. Seltzer MA, Low RK, McDonald M, Shami GS, Stoller ML. 1996. Dietary manipulation with lemonade to treat hypocitraturic calcium nephrolithiasis. *J Urol* 156(3): 907–9.
28. Odvina CV. 2006. Comparative value of orange juice versus lemonade in reducing stone-forming risk. *Clin J Am Soc Nephrol* 1(6): 1269–74.
29. Ettinger B, Pak CYC, Citron JT, Thomas C, Adams-Huet B, Vangessel A. 1997. Potassium-magnesium citrate is an effective prophylaxis against recurrent calcium oxalate nephrolithiasis. *J Urol* 158: 2069–73.
30. Barcelo P, Wuhl O, Servitge E, Roussaud A, Pak C. 1993. Randomized double-blind study of potassium citrate in idiopathic hypocitraturic calcium nephrolithiasis. *J Urol* 150: 1761–4.
31. Soygur T, Akbay A, Kupeli S. 2002. Effect of potassium citrate therapy on stone recurrence and residual fragments after shockwave lithotripsy in lower caliceal calcium oxalate urolithiasis: a randomized controlled trial. *J Endourol* 16(3): 149–52.
32. Pak CY, Peterson RD, Poindexter J. 2002. Prevention of spinal bone loss by potassium citrate in cases of calcium urolithiasis. *J Urol* 168(1): 31–4.
33. Welch BJ, Graybeal D, Moe OW, Maalouf NM, Sakhaee K. 2006. Biochemical and stone-risk profiles with topiramate treatment. *Am J Kidney Dis* 48(4): 555–63.
34. Grover P, Ryall R, Marshall V. 1993. Dissolved urate promotes calcium oxalate crystallization: Epitaxy is not the cause. *Clin Sci (Lond)* 85: 303–7.
35. Coe FL, Moran E, Kavalich AG. 1976. The contribution of dietary purine over-consumption to hyperpuricosuria in calcium oxalate stone formers. *J Chronic Dis* 29(12): 793–800.
36. Ettinger B, Tang A, Citron JT, Livermore B, Williams T. 1986. Randomized trial of allopurinol in the prevention of calcium oxalate calculi. *N Engl J Med* 315(22): 1386–9.
37. Asplin J. 1996. Uric acid stones. *Semin Nephrol* 16(5): 412–24.
38. Sakhaee K, Adams-Huet B, Moe OW, Pak CY. 2002. Pathophysiologic basis for normouricosuric uric acid nephrolithiasis. *Kidney Int* 62(3): 971–9.
39. Abate N, Chandalia M, Cabo-Chan AV Jr, Moe OW, Sakhaee K. 2004. The metabolic syndrome and uric acid nephrolithiasis: Novel features of renal manifestation of insulin resistance. *Kidney Int* 65(2): 386–92.
40. Rodman JS, Blackburn P, Williams JJ, Brown A, Pospischil MA, Peterson CM. 1984. The effect of dietary protein on cystine excretion in patients with cystinuria. *Clin Nephrol* 22(6): 273–8.
41. Lindell A, Denneberg T, Edholm E, Jeppsson JO. 1995. The effect of sodium intake on cystinuria with and without tiopronin treatment. *Nephron* 71(4): 407–15.
42. Nakagawa Y, Asplin JR, Goldfarb D, Parks JH, Coe FL. 2000. Clinical use of cystine supersaturation measurements. *J Urol* 164: 1481–5.
43. Dolin DJ, Asplin JR, Flagel L, Grasso M, Goldfarb DS. 2005. Effect of cystine-binding thiol drugs on urinary cystine capacity in patients with cystinuria. *J Endourol* 19(3): 429–32.

44. Pak CY, Fuller C, Sakhaee K, Zerwekh JE, Adams BV. 1986. Management of cystine nephrolithiasis with alpha-mercaptopropionylglycine. *J Urol* 136(5): 1003–8.
45. Cohen TD, Streem SB, Hall P. 1995. Clinical effect of captopril on the formation and growth of cystine calculi. *J Urol* 154(1): 164–6.
46. Michelakakis H, Delis D, Anastasiadou V, Bartsocas C. 1993. Ineffectiveness of captopril in reducing cystine excretion in cystinuric children. *J Inherit Metab Dis* 16(6): 1042–3.
47. Preminger GM, Assimos DG, Lingeman JE, Nakada SY, Pearle MS, Wolf JS. 2005. AUA guidelines on management of staghorn calculi: Diagnosis and treatment recommendations. *J Urol* 173: 1991–2000.
48. Griffith DP, Gleeson MJ, Lee H, Longuet R, Deman E, Earle N. 1991. Randomized, double-blind trial of Lithostat (acetohydroxamic acid) in the palliative treatment of infection-induced urinary calculi. *Eur Urol* 20(3): 243–7.
49. Griffith DP, Khonsari F, Skurnick JH, James KE. 1988. A randomized trial of acetohydroxamic acid for the treatment and prevention of infection-induced urinary stones in spinal cord injury patients. *J Urol* 140(2): 318–24.
50. Williams JJ, Rodman JS, Peterson CM. 1984. A randomized double-blind study of acetohydroxamic acid in struvite nephrolithiasis. *N Engl J Med* 311(12): 760–4.

107

Genetic Basis of Renal Stones

Rajesh V. Thakker

Introduction 884
Genetics 884
Monogenic Forms of Hypercalciuric Nephrolithiasis 885

Conclusions 889
References 889
Electronic Databases 891

INTRODUCTION

Renal stones affect approximately 8% of the population by the seventh decade and are usually associated with a metabolic abnormality that may include hypercalciuria, hyperphosphaturia, hyperoxaluria, hypocitraturia, hyperuricosuria, cystinuria, a low urinary volume and a defect of urinary acidification [1–3]. The etiology of these metabolic abnormalities and of renal stones is multifactorial and involves interactions between environmental and genetic determinants [4–6]. The environmental determinants include dietary intake of salt, protein, calcium and other nutrients, fluid intake, urinary tract infections, socioeconomic status of the individual, lifestyle, and climate [7, 8]. This chapter will focus on the genetics of renal stones (Table 107.1) and in particular those associated with hypercalciuric renal stone disease (nephrolithiasis) in man.

GENETICS

The greatest risk factor for nephrolithiasis, after controlling for known dietary determinants, is having an affected family member [9]. Thus, between 35% and 65% of renal stone formers will have relatives with nephrolithiasis, whereas only 5–20% of nonrenal stone formers will have relatives with nephrolithiasis [9–11]. The first degree relative risk (λ_R) among recurrent stone formers has been estimated to be in the range of 2 to 16 [9, 12, 13]. The wide range of these estimates is largely due to differences in the study designs and the methods used to ascertain the occurrence of renal stones in relatives. Moreover, genetic contributions to nephrolithiasis and hypercalciuria have been confirmed by two studies that have investigated the occurrence of these conditions in twins and shown that the heritability (h^2) of kidney stones and urinary calcium excretion are 56% and 52%, respectively [14, 15]. Both of these studies support the notion that there is a strong genetic contribution to renal stone disease and in the regulation of renal calcium excretion. Progress in the identification of the responsible genes by linkage studies utilizing affected sibling pairs or small families in genome-wide searches has been slow because of the lack of availability of appropriate patients and their relatives, and also because of difficulties in correctly ascertaining the phenotype, which may require radiological investigations and 24-hour urine collections. However, investigation of some families with monogenic forms of hypercalciuric nephrolithiasis and a genome-wide association study in families from Iceland and the Netherlands have yielded insights into the renal tubular mechanisms regulating calcium excretion [16, 17]; these will be briefly reviewed.

Primer on the Metabolic Bone Diseases and Disorders of Mineral Metabolism, Eighth Edition. Edited by Clifford J. Rosen.
© 2013 American Society for Bone and Mineral Research. Published 2013 by John Wiley & Sons, Inc.

Table 107.1. Genetics of Monogenic Forms of Nephrolithiasis

Renal Stone Disease[a]	Mode of Inheritance[b]	Gene[c]	Chromosomal Location	OMIM Number[d]
Associated with Hypercalciuria				
IH	A-d	SAC	1q23.2-q24	143870
IH	A-d	VDR	12q12-q14	601769
IH	A-d	?	9q33.2-q34.2	?
ADHH	A-d	CASR	3q21.1	601199
Hypercalcemia with hypercalciuria	A-d	CASR	3q21.1	601199
Bartter syndrome				
Type I	A-r	SLC12A1/ NKCC2	15q15-q21.1	601678
Type II	A-r	KCNJ1 /ROMK	11q24	241200
Type III	A-r	CLCNKB	1q36	607364
Type IV	A-r	BSND	1q31	602522
Type V	A-d	CASR	3q21.1	601199
Type VI	X-r	CLCN5	Xp11.22	300009
Dent's disease	X-r	CLCN5	Xp11.22	300009
Lowe's syndrome	X-r	OCRL1	Xq25	309000
HHRH	A-r	NPT2c/SLC34A3	9q34	241530
Nephrolithiasis, osteoporosis, and hypophosphatemia	A-d	NPT2a/SLC34A1	5q35	182309
Familial hypomagnesemia with hypercalciuria and nephrocalcinosis	A-r	PCLN1/CLDN16	3q28	248250
Familial hypomagnesemia with hypercalciuria and nephrocalcinosis with ocular abnormalities	A-r	CLDN19	1p34.2	248190
dRTA	A-d	SLC4A1/kAE1	17q21.31	179800
dRTA with sensorineural deafness	A-r	ATP6B1/ ATP6V1B1	2p13	267300
dRTA with preserved hearing	A-r	ATP6N1B/ ATP6V0A4	7q34	602722
Not Associated with Hypercalciuria				
Primary hyperoxaluria type 1	A-r	AGXT	2q37.3	259900
Primary hyperoxaluria type 2	A-r	GRHPR	9p13.2	260000
APRT deficiency	A-r	APRT	16q24.3	102600
Cystinuria type A	A-r	SLC3A1	2p16.3	220100
Cystinuria type B	A-r	SLC7A9	19q13.1	604144
Wilson's disease	A-r	ATP7B	13q14.3	277900

[a]IH: idiopathic hypercalciuria; ADHH: autosomal dominant hypocalcemia with hypercalciuria; HHRH: hereditary hypophosphatemic rickets with hypercalciuria; dRTA: distal renal tubular acidosis; APRT: adenine phosphoribosyltransferase.
[b]A-d: autosomal dominant; A-r: autosomal recessive; X-r: X-linked recessive.
[c]SAC: human soluble adenylyl cyclase; VDR: vitamin D receptor; CASR: calcium sensing receptor; SLC12A1: solute carrier family 12, member 1; NKCC2: sodium–potassium–chloride cotransporter 2; ROMK: renal outer-medullary potassium channel; CLCNKB: chloride channel Kb; BSND: Barttin; CLCN5: chloride channel 5; OCRL1: oculocerebrorenal syndrome of Lowe; NPT2c/a: sodium–phosphate cotransporter type 2c/a; SLC34A1/3: solute carrier family 34, member 1/3; PCLN1: paracellin; CLDN16/19: claudin 16/19; kAE1: kidney anion exchanger 1; ATP6B1: ATPase, H⁺ transporting (vacuolar proton pump), V1 subunit B1; ATP6N1B: ATPase, H⁺ transporting, lysosomal V0 subunit a4; AGXT: alanine glyoxylate aminotransferase; GRHPR: glyoxalate reductase/hydroxypyruvate reductase; APRT: adenine phosphoribosyl transferase; SLC3A1: solute carrier family 3, member 1; SLC7A9: solute carrier family 7, member 9; ATP7B: ATPase, Cu⁺⁺ transporting, beta polypeptide; ?: unknown.
[d]Online Mendelian Inheritance in Man (OMIM) reference number.

MONOGENIC FORMS OF HYPERCALCIURIC NEPHROLITHIASIS

Idiopathic hypercalciuria

Families with idiopathic hypercalciuria (IH) and recurrent calcium oxalate stones usually reveal an autosomal dominant mode of inheritance [18]. Studies of such families have established linkage between hypercalciuric nephrolithiasis and loci on chromosome 1q23.3-q24 [19], which contains the human soluble adenylyl cyclase (SAC) gene [20]; chromosome 12q12-q14, which contains the vitamin D receptor (VDR) gene; and chromosome 9q33.2-q34.2 [21], from which an appropriate candidate gene remains to be identified.

Autosomal dominant hypocalcaemic hypercalciuria (ADHH) due to CaSR mutations

The human calcium-sensing receptor (CaSR) is a 1,078 amino acid cell surface protein, which is predominantly expressed in the parathyroids and kidney and is a member of the family of G protein-coupled receptors [22, 23]. The CaSR allows regulation of parathyroid hormone (PTH) secretion and renal tubular calcium reabsorption in response to alterations in extracellular calcium concentrations. The human CaSR gene is located on chromosome 3q21.1 and loss-of-function CaSR mutations have been reported in the hypercalcaemic disorders of familial benign (hypocalciuric) hypercalcaemia (FBHH), neonatal severe primary hyperparathyroidism (NSHPT) and familial isolated hyperparathyroidism (FIHP) [22, 23]. However, gain-of-function CaSR mutations result in autosomal dominant hypocalcaemia with hypercalciuria (ADHH), and Bartter's syndrome type V (see below) [24–27].

Patients with ADHH usually have mild hypocalcaemia that is generally asymptomatic but may in some patients be associated with carpopedal spasm and seizures [24]. The serum phosphate concentrations in patients with ADHH are either elevated or in the upper-normal range, and the serum magnesium concentrations are either low or in the low-normal range. These biochemical features of hypocalcaemia, hyperphosphatemia, and hypomagnesemia are consistent with hypoparathyroidism and pseudohypoparathyroidism. However, these patients have serum PTH concentrations that are in the low normal range [24]. Thus, they are not hypoparathyroid, which would be associated with undetectable serum PTH concentrations, or pseudohypoparathyroid, which would be associated with elevated serum PTH concentrations. These patients were therefore classified as having autosomal dominant hypocalcaemia (ADH), and the association of hypercalciuria with this condition led to it being referred to as autosomal dominant hypocalcaemia with hypercalciuria [24, 28]. Treatment with active metabolites of vitamin D to correct the hypocalcemia has been reported to result in marked hypercalciuria, nephrocalcinosis, nephrolithiasis, and renal impairment, which was partially reversible after cessation of the vitamin D treatment [24]. Thus, it is important to identify and avoid vitamin D treatment in such ADHH patients and their families whose hypocalcaemia is due to a gain-of-function CaSR mutation and not hypoparathyroidism. More than 40 different CaSR mutations have been identified in ADHH patients, and over 50% of these are in the extracellular domain [23]. Almost every ADHH family has its own unique missense heterozygous CaSR mutation. Expression studies of the ADHH associated CaSR mutations have demonstrated a gain-of-function, whereby there is a leftward shift in the dose-response curve such that the extracellular calcium concentration needed to produce a half-maximal (EC_{50}) increase in the total intracellular calcium ions (or inositol trisphosphate, IP_3), is significantly lower than that required for the wild-type receptor [23, 24, 28].

Bartter syndrome

Bartter syndrome is a heterogeneous group of autosomal hereditary disorders of electrolyte homeostasis characterized by hypokalaemic alkalosis, renal salt wasting that may lead to hypotension, hyperreninemic hyperaldosteronism, increased urinary prostaglandin excretion, and hypercalciuria with nephrocalcinosis [23]. Mutations of several ion transporters and channels have been associated with Bartter syndrome, and six types (Table 107.1) are now recognized [23]. Thus, type I is due to mutations involving the bumetanide-sensitive sodium–potassium–chloride cotransporter (NKCC2 or SLC12A2); type II is due to mutations of the renal outer-medullary potassium channel (ROMK); type III is due to mutations of the voltage-gated chloride channel (CLC-Kb); type IV is due to mutations of Barttin, which is a beta subunit that is required for the trafficking of CLC-Kb and CLC-Ka (this form is also associated with deafness as Barttin, CLC-Ka, and CLC-Kb are also expressed in the marginal cells of the scala media of the inner ear that secrete potassium ion-rich endolymph); and type V is due to activating mutations of the CaSR. Patients with Bartter syndrome type V have the classical features of the syndrome, i.e., hypokalemic metabolic alkalosis, hyperreninemia, and hyperaldosteronism [25, 26]. In addition, they develop hypocalcemia, which may be symptomatic and lead to carpopedal spasm, and an elevated fractional excretion of calcium, which may be associated with nephrocalcinosis [25, 26]. Such patients have been reported to have heterozygous gain-of-function CaSR mutations, and *in vitro* functional expression of these mutations not only revealed a leftward shift in the dose-response curve for the receptor but also showed them to have a much lower EC_{50} than that found in patients with ADHH [25, 26]. This suggests that the additional features that occur in Bartter syndrome type V when compared to ADHH are due to severe gain-of-function mutations of the CaSR. Bartter syndrome type VI has been reported in one child from Turkey and was associated with a *CLCN5* mutation [27]; mutations in this gene are usually observed in Dent's disease (see below).

Dent's disease

Dent's disease is an X-linked recessive renal tubular disorder characterized by a low molecular weight proteinuria, hypercalciuria, nephrocalcinosis, nephrolithiasis, and eventual renal failure [29]. Dent's disease is also associated with the other multiple proximal tubular defects of the renal Fanconi syndrome, which include aminoaciduria, phosphaturia, glycosuria, kaliuresis, uricosuria, and impaired urinary acidification [29]. With the exception of rickets, which occurs in a minority of patients, there appear to be no extrarenal manifestations in Dent's disease [29]. The gene causing Dent's disease, *CLCN5*, encodes the chloride/proton antiporter, CLC-5 [30]. CLC family members, which are usually voltage-gated chloride channels, have important diverse functions that include the control of membrane excitability, transepithelial transport, and regulation of cell volume [31]. CLC-5, which is predominantly expressed in the kidney and in particular the proximal tubule, thick ascending limb of Henle, and the alpha intercalated cells of the collecting duct, has been reported to be critical for acidification in the endosomes that participate in solute reabsorption and membrane recycling in the proximal tubule [32, 33]. CLC-5 is also known to alter membrane trafficking via the receptor-mediated-endocytic pathway that involves megalin and cubulin [34, 35]. CLC-5 mutations associated with Dent's disease impair chloride flow and likely lead to impaired acidification of the endosomal lumen, and thereby also disrupt trafficking of endosomes back to the apical surface [34–36]. This will result in impairment of solute reabsorption by the renal tubule and in the defects observed in Dent's disease [37]. Mice that are deficient for CLC-5 develop the phenotypic abnormalities associated with Dent's disease, but some patients with Dent's disease, who do not have CLC-5 mutations, have been reported to have mutations of the gene encoding an inositol polyphosphate 5-phosphatase, which also results in the Lowe syndrome [38, 39] (see below).

Occulocerebrorenal syndrome of lowe

Oculocerebrorenal syndrome of Lowe (OCRL) is an X-linked recessive disorder that is characterized by congenital cataracts, mental retardation, muscular hypotonia, rickets, and defective proximal tubular reabsorption of bicarbonate, phosphate, and amino acids [40, 41]. Some patients may also develop hypercalciuria and renal calculi [41]. The OCRL1 gene is located on Xq25 and encodes a member of the type II family of inositol polyphosphate 5-phosphatases [42]. These enzymes hydrolyse the 5-phosphate of inositol 1, 4, 5-trisphosphate and of inositol 1, 3, 4, 5-tetrakisphosphate, phosphatidylinositol 4, 5-bisphosphate, and phosphatidylinositol 3, 4, 5-trisphosphate, thereby presumably inactivating them as second messengers in the phosphatidylinositol signaling pathway [42]. The preferred substrate of OCRL1 is phosphatidylinositol 4, 5-bisphosphate, and this lipid accumulates in renal proximal tubular cells from patients with Lowe syndrome [42]. OCRL1 has been localized to lysosomes in renal proximal tubular cells and to the trans-Golgi network in fibroblasts. OCRL1 has also been shown to interact with clathrin and indeed colocalizes with clathrin on endosomal membranes that contain transferrin and mannose 6-phosphate receptors [40]. Thus, it seems likely the OCRL1 mutations in Lowe syndrome patients result in OCRL1 protein deficiency, which leads to disruptions in lysosomal trafficking and endosomal sorting [40, 43].

Hereditary hypophosphatemic rickets with hypercalciuria

Two different heterozygous mutations (Ala48Phe and Val147Met) in NPT2a (also referred to as SLC34A1), the gene encoding a sodium-dependent phosphate transporter, have been reported in patients with urolithiasis or osteoporosis and persistent idiopathic hypophosphatemia due to decreased renal tubular phosphate reabsorption [44]. When expressed in Xenopus laevis oocytes, the mutant NPT2a showed impaired function. However, these in vitro findings were not confirmed in another study using oocytes and OK cells, raising the concern that the identified NPT2a mutation could not explain the findings in the described patients [45]. However, homozygous ablation of Npt2a in mice (Npt2a−/−) results in increased urinary phosphate excretion, hypophosphatemia, an appropriate elevation in the serum levels of 1,25-dihydroxyvitamin D, hypercalcemia, decreased serum parathyroid hormone levels, increased serum alkaline phosphatase activity and hypercalciuria [46]. Some of these biochemical features are observed in patients with hereditary hypophosphatemic rickets with hypercalciuria (HHRH), but there are important differences [47]. Thus, HHRH patients develop rickets, short stature, with an increased renal phosphate clearance, and hypercalciuria but have normal serum calcium levels, an increased gastrointestinal absorption of calcium and phosphate due to an elevated serum concentration of 1, 25-dihydroxyvitamin D, suppressed parathyroid function, and normal urinary cyclic adenosine monophosphate (AMP) excretion [47]. However, HHRH patients do not have NPT2a mutations [48], and studies have demonstrated that HHRH patients harbor homozygous or compound heterozygous mutations of SLC34A3, the gene encoding the sodium–phosphate cotransporter NPT2c [49, 50]. These findings indicate that NPT2c has a more important role in phosphate homeostasis than previously thought.

Familial hypomagnesaemia with hypercalciuria and nephrocalcinosis due to paracellin-1 (claudin 16) mutations

Familial hypomagnesaemia with hypercalciuria and nephrocalcinosis (FHHNC) is an autosomal recessive renal tubular disorder that is frequently associated with progressive kidney failure [51]. FHHNC often presents in childhood with seizures, or tetany due to hypocalcemia and hypomagnesemia. Other recurrent clinical manifestations include urinary tract infections, polyuria, polydipsia, and failure to thrive. Investigations reveal hypomagnesemia, hypocalcemia, hyperuricemia, hypermagnesuria, hypercalciuria, incomplete distal renal tubular acidosis, hypocitraturia, and renal calcification [52]. Treatment consists of high-dose enteral magnesium to restore normomagnesemia. Children with FHHNC who receive such treatment early develop normally. Linkage studies in 12 FHHNC kindreds localized the disease locus to chromosome 3q27, and positional cloning studies identified mutations in the gene encoding Paracellin-1 (PCLN-1), which is also referred to as claudin

16 (*CLDN16*) [52]. FHHNC patients were either homozygotes or compound heterozygotes for PCLN-1 mutations, consistent with the autosomal recessive inheritance of the disorder [52, 53]. The PCLN-1 mutations consisted of premature termination codons, splice-site mutations, and missense mutations [52, 53]. The PCLN-1 protein, which consists of 305 amino acids, has sequence and structural similarity to the members of the claudin family, and is therefore also referred to as CLDN16 [54, 55]. Claudins are membrane-bound proteins that form the intercellular tight junction barrier in a variety of epithelia [54,55]. They have four transmembrane domains and intracellular amino- and carboxy–termini. The two luminal loops mediate cell–cell adhesion via homo- and heterotypic interactions with claudins on a neighboring cell. In addition, claudins form paracellular ion channels, which facilitate renal tubular paracellular transport of solutes [54]. CLDN16 is exclusively expressed in the thick ascending limb of Henle's loop, where it forms the paracellular channels that are driven by an electrochemical gradient and allow reabsorption of calcium and magnesium [55]. Hence, loss of function of CLDN16 that would arise from FHHNC mutations would result in urinary calcium and magnesium loss and lead to hypocalcemia and hypomagnesemia, respectively. A CLDN16 missense mutation (Thr233Arg) has also been identified in two families with self-limiting childhood hypercalciuria [56]. The hypercalciuria decreased with age and was not associated with progressive renal failure. The Thr233Arg mutation resulted in inactivation of a PDZ-domain binding motif, and this disrupted the association with the tight junction scaffolding protein, ZO-1, with accumulation of the mutant CLDN16 protein in lysosomes and no localization to the tight junctions [56]. Thus, CLDN16 mutations may result in different abnormalities of renal tubular cell function and hence lead to differences in the clinical phenotype. A form of FHHNC with severe ocular involvement reported in one Swiss and eight Spanish/Hispanic families was recently mapped to chromosome 1p34.2 [53]. This region contains *CLDN19*, the gene that encodes claudin 19, a tight-junction protein expressed in the kidney and eye. A Gly20Asp mutation located in the first transmembrane domain of CLDN19 was identified in all but one of the Spanish/Hispanic families, and a Gln57Glu mutation in the first extracellular loop of CLDN19 was found in the Swiss family. In addition, a Leu90Pro mutation in CLDN19 was identified in a consanguineous family of Turkish origin with FHHNC and severe ocular involvement [53].

Distal renal tubular acidosis

In distal renal tubular acidosis (dRTA), the tubular secretion of hydrogen ions in the distal nephron is impaired, and this results in a metabolic acidosis that is often associated with hypokalemia due to renal potassium wasting, hypercalciuria with nephrocalcinosis, and metabolic bone disease. Distal RTA may be familial with autosomal dominant or recessive inheritance.

Autosomal dominant dRTA due to erythrocyte anion exchanger (band 3, AE1) mutations

The family of anion exchangers (AEs) are widely distributed and involved in the regulation of transcellular transport of acid and base across epithelial cells, cell volume, and intracellular pH [57]. For example, AE1, which is a major glycoprotein of the erythrocyte membrane, mediates exchange of chloride and bicarbonate [57]. AE1 is also found in the basolateral membrane of the α-intercalated cells of renal collecting ducts that are involved in acid secretion [57]. Patients with autosomal dominant dRTA, the majority of whom had hypercalciuria, renal stones, and nephrocalcinosis, and a few who had erythrocytosis, were found to have AE1 mutations [58]. These AE1 mutations resulted in several functional abnormalities that included reductions in chloride transport, and trafficking defects that lead to a cellular retention of AE1 or mistargeting of AE1 to the apical membrane. AE1 mutations may also be associated with autosomal recessive dRTA in Southeast Asian kindreds that have ovalocytosis [59].

Autosomal recessive distal renal tubular acidosis due to proton pump (H$^+$-ATPase) mutations

Proton pumps are ubiquitously expressed, and one such multi-unit H$^+$-ATPase is found in abundance on the apical (luminal) surface of the α-intercalated cells of the cortical collecting duct, which regulates urinary acidification. Failure of vectorial proton transport by these α-intercalated cells results in an inability of urinary acidification and in disorders of dRTA. The molecular basis of two types of autosomal recessive dRTA due to proton pump abnormalities have been characterized. The gene causing one type of autosomal recessive dRTA that was associated with sensorineural hearing loss was mapped to chromosome 2p13, which contained the ATP6B1gene that encodes the B1 subunit of the apical proton pump (H$^+$ATPase) [60]. Mutations, which would likely result in a functional loss of ATP6B1, were identified in over 30% of families with this form of autosomal recessive dRTA that occurred with deafness in more than 85% of families [60]. The association of dRTA and deafness is consistent with the renal and cochlear expression of ATP6B1. ATP6B1 plays a critical role in regulating the pH of the inner ear endolymph. Dysfunction of this would lead to an alkaline microenvironment in the inner ear, which has been proposed to impair hair cell function and result in progressive deafness. The gene causing autosomal recessive dRTA with normal hearing was localized to chromosome 7q33-q34, which contained the ATP6N1B gene that encodes the non-catalytic accessory subunit of the proton pump of the α-intercalated cells of the collecting duct. ATP6N1B mutations, which are predicted to result in a functional loss, were identified in more than

85% of kindreds with autosomal recessive dRTA associated with normal hearing; this is consistent with the expression of ATP6N1B in the kidney but not other organs. Approximately 15% of families with autosomal recessive dRTA were not found to have mutations in ATP6B1 or ATP6N1B mutations, and this indicates mutations in other genes are likely to be involved in the etiology of autosomal recessive dRTA.

CONCLUSIONS

Renal stone disease (nephrolithiasis) affects 5% of adults and is often associated with hypercalciuria. Hypercalciuric nephrolithiasis is a familial disorder in over 35% patients and may occur as a monogenic disorder or as a polygenic trait. Studies of monogenic forms of hypercalciuric nephrolithiasis, e.g., Bartter syndrome, Dent's disease, autosomal dominant hypocalcemic hypercalciuria, hypercalciuric nephrolithiasis with hypophosphatemia, and familial hypomagnesemia with hypercalciuria have helped to identify a number of transporters, channels, and receptors that are involved in regulating the renal tubular reabsorption of calcium. These studies have provided valuable insights into the renal tubular pathways that regulate calcium reabsorption and predispose to kidney stones and bone disease.

REFERENCES

1. Stamatelou KK, Francis ME, Jones CA, Nyberg LM, Curhan GC. 2003. Time trends in reported prevalence of kidney stones in the United States: 1976–1994. *Kidney Int* 63: 1817–1823.
2. Frick KK, Bushinsky DA. 2003. Molecular mechanisms of primary hypercalciuria. *J Am Soc Nephrol* 14: 1082–1095.
3. Scheinman SJ. 1999. Nephrolithiasis. *Semin Nephrol* 19: 381–388.
4. Robertson WG, Peacock M, Marshall RW, Speed R, Nordin BE. 1975. Seasonal variations in the composition of urine in relation to calcium stone-formation. *Clin Sci Mol Med* 49: 597–602.
5. Moe OW, Bonny O. 2005. Genetic hypercalciuria. *J Am Soc Nephrol* 16: 729–745.
6. Parry ES, Lister IS. 1975. Sunlight and hypercalciuria. *Lancet* 1: 1063–1065.
7. Curhan GC, Willett WC, Rimm EB, Stampfer MJ. 1993. A prospective study of dietary calcium and other nutrients and the risk of symptomatic kidney stones. *N Engl J Med* 328: 833–838.
8. Serio A, Fraioli A. 1999. Epidemiology of nephrolithiasis. *Nephron* 81 Suppl 1: 26–30.
9. Resnick M, Pridgen DB, Goodman HO. 1968. Genetic predisposition to formation of calcium oxalate renal calculi. *N Engl J Med* 278: 1313–1318.
10. Polito C, La Manna A, Nappi B, Villani J, Di Toro R. 2000. Idiopathic hypercalciuria and hyperuricosuria: Family prevalence of nephrolithiasis. *Pediatr Nephrol* 14: 1102–1104.
11. Curhan GC, Willett WC, Rimm EB, Stampfer MJ. 1997. Family history and risk of kidney stones. *J Am Soc Nephrol* 8: 1568–1573.
12. McGeown MG. 1960. Heredity in renal stone disease. *Clin Sci* 19: 465–471.
13. Trinchieri A, Mandressi A, Luongo P, Coppi F, Pisani E. 1988. Familial aggregation of renal calcium stone disease. *J Urol* 139: 478–481.
14. Goldfarb DS, Fischer ME, Keich Y, Goldberg J. 2005. A twin study of genetic and dietary influences on nephrolithiasis: A report from the Vietnam Era Twin (VET) Registry. *Kidney Int* 67: 1053–1061.
15. Hunter DJ, Lange M, Snieder H, MacGregor AJ, Swaminathan R, Thakker RV, Spector TD. 2002. Genetic contribution to renal function and electrolyte balance: A twin study. *Clin Sci (Lond)* 103: 259–265.
16. Stechman MJ, Loh NY, Thakker RV. 2009. Genetic Causes of hypercalciuric nephrolithiasis. *Pediatr Nephrol* 24: 2321–2332.
17. Thorleifsson G, Holm H, Edvardsson V, Walters GB, Styrkarsdottir U, Gudbjartsson DF, Sulem P, Halldorsson BV, de Vegt F, d'Ancona FC, den Heijer M, Franzson L, Christiansen C, Alexandersen P, Rafnar T, Kristjansson K, Sigurdsson G, Kiemeney LA, Bodvarsson M, Indridason OS, Palsson R, Kong A, Thorsteinsdottir U, Stefansson K. 2009. Sequence variants in the CLDN14 gene associate with kidney stones and bone mineral density. *Nat Genet* 41: 926–930.
18. Coe FL, Parks JH, Moore ES. 1979. Familial idiopathic hypercalciuria. *N Engl J Med* 300: 337–340.
19. Reed BY, Heller HJ, Gitomer WL, Pak CY. 1999. Mapping a gene defect in absorptive hypercalciuria to chromosome 1q23.3-q24. *J Clin Endocrinol Metab* 84: 3907–3913.
20. Scott P, Ouimet D, Valiquette L, Guay G, Proulx Y, Trouve ML, Gagnon B, Bonnardeaux A. 1999. Suggestive evidence for a susceptibility gene near the vitamin D receptor locus in idiopathic calcium stone formation. *J Am Soc Nephrol* 10: 1007–1013.
21. Wolf MT, Zalewski I, Martin FC, Ruf R, Muller D, Hennies HC, Schwarz S, Panther F, Attanasio M, Acosta HG, Imm A, Lucke B, Utsch B, Otto E, Nurnberg P, Nieto VG, Hildebrandt F. 2005. Mapping a new suggestive gene locus for autosomal dominant nephrolithiasis to chromosome 9q33.2-q34.2 by total genome search for linkage. *Nephrol Dial Transplant* 20: 909–914.
22. Pollak MR, Brown EM, Chou YH, Hebert SC, Marx SJ, Steinmann B, Levi T, Seidman CE, Seidman JG. 1993. Mutations in the human Ca(2+)-sensing receptor gene cause familial hypocalciuric hypercalcemia and neonatal severe hyperparathyroidism. *Cell* 75: 1297–1303.
23. Thakker RV. 2004. Diseases associated with the extracellular calcium-sensing receptor. *Cell Calcium* 35: 275–282.
24. Pearce SH, Williamson C, Kifor O, Bai M, Coulthard MG, Davies M, Lewis-Barned N, McCredie D, Powell H,

Kendall-Taylor P, Brown EM, Thakker RV. 1996. A familial syndrome of hypocalcemia with hypercalciuria due to mutations in the calcium-sensing receptor. *N Engl J Med* 335: 1115–1122.

25. Watanabe S, Fukumoto S, Chang H, Takeuchi Y, Hasegawa Y, Okazaki R, Chikatsu N, Fujita T. 2002. Association between activating mutations of calcium-sensing receptor and Bartter's syndrome. *Lancet* 360: 692–694.

26. Vargas-Poussou R, Huang C, Hulin P, Houillier P, Jeunemaitre X, Paillard M, Planelles G, Dechaux M, Miller RT, Antignac C. 2002. Functional characterization of a calcium-sensing receptor mutation in severe autosomal dominant hypocalcemia with a Bartter-like syndrome. *J Am Soc Nephrol* 13: 2259–2266.

27. Bcsbas N, Ozaltin F, Jeck N, Seyberth H, Ludwig M. 2005. CLCN5 mutation (R347X) associated with hypokalaemic metabolic alkalosis in a Turkish child: An unusual presentation of Dent's disease. *Nephrol Dial Transplant* 20: 1476–1479.

28. Pollak MR, Brown EM, Estep HL, McLaine PN, Kifor O, Park J, Hebert SC, Seidman CE, Seidman JG. 1994. Autosomal dominant hypocalcaemia caused by a Ca(2+)-sensing receptor gene mutation. *Nat Genet* 8: 303–307.

29. Wrong OM, Norden AG, Feest TG. 1994. Dent's disease; a familial proximal renal tubular syndrome with low-molecular-weight proteinuria, hypercalciuria, nephrocalcinosis, metabolic bone disease, progressive renal failure and a marked male predominance. *QJM* 87: 473–493.

30. Lloyd SE, Pearce SH, Fisher SE, Steinmeyer K, Schwappach B, Scheinman SJ, Harding B, Bolino A, Devoto M, Goodyer P, Rigden SP, Wrong O, Jentsch TJ, Craig IW, Thakker RV. 1996. A common molecular basis for three inherited kidney stone diseases. *Nature* 379: 445–449.

31. Jentsch TJ, Neagoe I, Scheel O. 2005. CLC chloride channels and transporters. *Curr Opin Neurobiol* 15: 319–325.

32. Gunther W, Luchow A, Cluzeaud F, Vandewalle A, Jentsch TJ. 1998. ClC-5, the chloride channel mutated in Dent's disease, colocalizes with the proton pump in endocytotically active kidney cells. *Proc Natl Acad Sci U S A* 95: 8075–8080.

33. Devuyst O, Christie PT, Courtoy PJ, Beauwens R, Thakker RV. 1999. Intra-renal and subcellular distribution of the human chloride channel, CLC-5, reveals a pathophysiological basis for Dent's disease. *Hum Mol Genet* 8: 247–257.

34. Piwon N, Gunther W, Schwake M, Bosl MR, Jentsch TJ. 2000. ClC-5 Cl- -channel disruption impairs endocytosis in a mouse model for Dent's disease. *Nature* 408: 369–373.

35. Reed AAC, Loh NY, Lippiat JD, Partridge CJ, Galvanovskis J, Williams SE, Terryn S, Jouret F, Wu FTF, Courtoy PJ, Nesbit MA, Devuyst O, Rorsman P, Ashcroft FM, Thakker RV. 2010. Renal albumin endocytosis involves a CLC-5 and KIF3B interaction that facilitates vesicle and microtubular trafficking. *Am J Physiol Renal Physiol* 298: F365–F380.

36. Smith AJ, Reed AA, Loh NY, Thakker RV, Lippiat JD. 2009. Characterization of Dent's disease mutations of CLC-5 reveals a correlation between functional and cell biological consequences and protein structure. *Am J Physiol Renal Physiol* 296: F390–397.

37. Gunther W, Piwon N, Jentsch TJ. 2003. The ClC-5 chloride channel knock-out mouse—an animal model for Dent's disease. *Pflugers Arch* 445: 456–462.

38. Hoopes RR Jr, Shrimpton AE, Knohl SJ, Hueber P, Hoppe B, Matyus J, Simckes A, Tasic V, Toenshoff B, Suchy SF, Nussbaum RL, Scheinman SJ. 2005. Dent disease with mutations in OCRL1. *Am J Hum Genet* 76: 260–267.

39. Shrimpton AE, Hoopes RR Jr, Knohl SJ, Hueber P, Reed AA, Christie PT, Igarashi T, Lee P, Lehman A, White C, Milford DV, Sanchez MR, Unwin R, Wrong OM, Thakker RV, Scheinman SJ. 2009. OCRL1 mutations in Dent 2 patients suggest a mechanism for phenotypic variability. *Nephron Physiology* 112: 27–36.

40. Lowe M. 2005. Structure and function of the Lowe syndrome protein OCRL1. *Traffic* 6: 711–719.

41. Sliman GA, Winters WD, Shaw DW, Avner ED. 1995. Hypercalciuria and nephrocalcinosis in the oculocerebrorenal syndrome. *J Urol* 153: 1244–1246.

42. Leahey AM, Charnas LR, Nussbaum RL. 1993. Nonsense mutations in the OCRL-1 gene in patients with the oculocerebrorenal syndrome of Lowe. *Hum Mol Genet* 2: 461–463.

43. Ungewickell AJ, Majerus PW. 1999. Increased levels of plasma lysosomal enzymes in patients with Lowe syndrome. *Proc Natl Acad Sci U S A* 96: 13342–13344.

44. Prie D, Huart V, Bakouh N, Planelles G, Dellis O, Gerard B, Hulin P, Benque-Blanchet F, Silve C, Grandchamp B, Friedlander G. 2002. Nephrolithiasis and osteoporosis associated with hypophosphatemia caused by mutations in the type 2a sodium-phosphate cotransporter. *N Engl J Med* 347: 983–991.

45. Virkki LV, Forster IC, Hernando N, Biber J, Murer H. 2003. Functional characterization of two naturally occurring mutations in the human sodium-phosphate cotransporter type IIa. *J Bone Miner Res* 18: 2135–2141.

46. Beck L, Karaplis AC, Amizuka N, Hewson AS, Ozawa H, Tenenhouse HS. 1998. Targeted inactivation of Npt2 in mice leads to severe renal phosphate wasting, hypercalciuria, and skeletal abnormalities. *Proc Natl Acad Sci U S A* 95: 5372–5377.

47. Tieder M, Modai D, Samuel R, Arie R, Halabe A, Bab I, Gabizon D, Liberman UA. 1985. Hereditary hypophosphatemic rickets with hypercalciuria. *N Engl J Med* 312: 611–617.

48. Jones A, Tzenova J, Frappier D, Crumley M, Roslin N, Kos C, Tieder M, Langman C, Proesmans W, Carpenter T, Rice A, Anderson D, Morgan K, Fujiwara T, Tenenhouse H. 2001. Hereditary hypophosphatemic rickets with hypercalciuria is not caused by mutations in the Na/Pi cotransporter NPT2 gene. *J Am Soc Nephrol* 12: 507–514.

49. Bergwitz C, Roslin NM, Tieder M, Loredo-Osti JC, Bastepe M, Abu-Zahra H, Frappier D, Burkett K, Carpenter TO, Anderson D, Garabedian M, Sermet I, Fujiwara

TM, Morgan K, Tenenhouse HS, Juppner H. 2006. SLC34A3 mutations in patients with hereditary hypophosphatemic rickets with hypercalciuria predict a key role for the sodium-phosphate cotransporter NaPi-IIc in maintaining phosphate homeostasis. *Am J Hum Genet* 78: 179–192.
50. Lorenz-Depiereux B, Benet-Pages A, Eckstein G, Tenenbaum-Rakover Y, Wagenstaller J, Tiosano D, Gershoni-Baruch R, Albers N, Lichtner P, Schnabel D, Hochberg Z, Strom TM. 2006. Hereditary hypophosphatemic rickets with hypercalciuria is caused by mutations in the sodium-phosphate cotransporter gene SLC34A3. *Am J Hum Genet* 78: 193–201.
51. Paunier L, Radde IC, Kooh SW, Conen PE, Fraser D. 1968. Primary hypomagnesemia with secondary hypocalcemia in an infant. *Pediatrics* 41: 385–402.
52. Simon DB, Lu Y, Choate KA, Velazquez H, Al-Sabban E, Praga M, Casari G, Bettinelli A, Colussi G, Rodriguez-Soriano J, McCredie D, Milford D, Sanjad S, Lifton RP. 1999. Paracellin-1, a renal tight junction protein required for paracellular Mg^{2+} resorption. *Science* 285: 103–106.
53. Konrad M, Schaller A, Seelow D, Pandey AV, Waldegger S, Lesslauer A, Vitzthum H, Suzuki Y, Luk JM, Becker C, Schlingmann KP, Schmid M, Rodriguez-Soriano J, Ariceta G, Cano F, Enriquez R, Juppner H, Bakkaloglu SA, Hediger MA, Gallati S, Neuhauss SC, Nurnberg P, Weber S. 2006. Mutations in the tight-junction gene claudin 19 (CLDN19) are associated with renal magnesium wasting, renal failure, and severe ocular involvement. *Am J Hum Genet* 79: 949–957.
54. Colegio OR, Van Itallie CM, McCrea HJ, Rahner C, Anderson JM. 2002. Claudins create charge-selective channels in the paracellular pathway between epithelial cells. *Am J Physiol Cell Physiol* 283: C142–147.
55. Konrad M, Schlingmann KP, Gudermann T. 2004. Insights into the molecular nature of magnesium homeostasis. *Am J Physiol Renal Physiol* 286: F599–605.
56. Muller D, Kausalya PJ, Claverie-Martin F, Meij IC, Eggert P, Garcia-Nieto V, Hunziker W. 2003. A novel claudin 16 mutation associated with childhood hypercalciuria abolishes binding to ZO-1 and results in lysosomal mistargeting. *Am J Hum Genet* 73: 1293–1301.
57. Wagner S, Vogel R, Lietzke R, Koob R, Drenckhahn D. 1987. Immunochemical characterization of a band 3-like anion exchanger in collecting duct of human kidney. *Am J Physiol* 253: F213–221.
58. Bruce LJ, Cope DL, Jones GK, Schofield AE, Burley M, Povey S, Unwin RJ, Wrong O, Tanner MJ. 1997. Familial distal renal tubular acidosis is associated with mutations in the red cell anion exchanger (Band 3, AE1) gene. *J Clin Invest* 100: 1693–1707.
59. Bruce LJ, Wrong O, Toye AM, Young MT, Ogle G, Ismail Z, Sinha AK, McMaster P, Hwaihwanje I, Nash GB, Hart S, Lavu E, Palmer R, Othman A, Unwin RJ, Tanner MJ. 2000. Band 3 mutations, renal tubular acidosis and South-East Asian ovalocytosis in Malaysia and Papua New Guinea: Loss of up to 95% band 3 transport in red cells. *Biochem J* 350 Pt 1: 41–51.
60. Karet FE, Finberg KE, Nelson RD, Nayir A, Mocan H, Sanjad SA, Rodriguez-Soriano J, Santos F, Cremers CW, DiPietro A, Hoffbrand BI, Winiarski J, Bakkaloglu A, Ozen S, Dusunsel R, Goodyer P, Hulton SA, Wu DK, Skvorak AB, Morton CC, Cunningham MJ, Jha V, Lifton RP. 1999. Mutations in the gene encoding B1 subunit of H+ATPase cause renal tubular acidosis with sensorineural deafness. *Nat Genet* 21: 84–90.

ELECTRONIC DATABASES

1. OMIM – http://www.ncbi.nlm.nih.gov/sites/entrez?db=omim
   Use OMIM numbers in Table 107.1 to search for details of mutations.
2. EMSEMBL – http://www.emsembl.org/index.html
   Use gene symbols in Table 107.1 to obtain DNA sequences, exon–intron structure, evolutionary conservation, and chromosomal locations.
3. PUBMED – http://www.ncbi.n/m.nih.gov/sites/entrez
   Use disease names and gene symbols in Table 107.1 to obtain published articles.

Section X
Oral and Maxillofacial Biology and Pathology

Section Editor Laurie K. McCauley

Chapter 108. Development of the Craniofacial Skeleton 895
Maiko Matsui and John Klingensmith

Chapter 109. Development and Structure of Teeth and Periodontal Tissues 904
Petros Papagerakis and Thimios Mitsiadis

Chapter 110. Craniofacial Disorders Affecting the Dentition: Genetic 914
Yong-Hee P. Chun, Paul H. Krebsbach, and James P. Simmer

Chapter 111. Pathology of the Hard Tissues of the Jaws 922
Paul C. Edwards

Chapter 112. Bisphosphonate-Associated Osteonecrosis of the Jaws 929
Hani H. Mawardi, Nathaniel S. Treister, and Sook-Bin Woo

Chapter 113. Periodontal Diseases and Oral Bone Loss 941
Mary G. Lee and Keith L. Kirkwood

Chapter 114. Oral Manifestations of Metabolic Bone Diseases 948
Roberto Civitelli and Charles Hildebolt

108
Development of the Craniofacial Skeleton

Maiko Matsui and John Klingensmith

Introduction 895
Origins of Craniofacial Skeletal Elements 895
Palatal Development 896
Orofacial Cleft 898
Mandibular Development 898

Micrognathia 899
Development of the Neurocranium 899
Craniosynostosis and Skull Base Deformities 900
Conclusion 901
References 901

INTRODUCTION

Craniofacial skeletal elements are the most developmentally complex skeletal structures in mammals. We humans have distinct, identifying facial features, the result of small variations in craniofacial development. Such differences are usually benign and result in features that make us distinct individuals. However, variable effects of gene expression or environmental factors can lead to cosmetic and functional abnormalities. Craniofacial malformations are common birth defects among humans, occurring both in isolated forms and as part of malformation syndromes. Aberrations in normal embryogenesis cause both types of defects. In this chapter, we summarize current knowledge of normal craniofacial morphogenesis and the developmental basis of its major anomalies, focusing on craniofacial skeletal elements.

ORIGINS OF CRANIOFACIAL SKELETAL ELEMENTS

Mammalian craniofacial skeletal elements consist of more than 20 small bones and cartilages that are formed precisely during development to create functional structures—the face and the head. Most facial skeletal elements, collectively the viscerocranium, are derived from cranial neural crest cells (NCCs). These pluripotent cells are collectively sometimes called "the fourth germ layer" because of the diversity of tissues they form. In early craniofacial development, presumptive cranial NCCs arise in dorsal midline ectoderm of the midbrain and the hindbrain rhombomeres, undergo an epithelial-to-mesenchymal transition, delaminate, then migrate ventrolaterally between the ectoderm and endoderm [Fig. 108.1(A)]. While the rostral cranial NCCs develop the frontonasal skeleton and the skull vault, NCCs from each rhombomere faithfully take distinct pathways to populate different pharyngeal arches (PA), numbered rostrocaudally from 1 to 6. NCCs from rhombomeres 1 and 2 migrate into PA1 and the frontonasal process. PA1 gives rise to the incus and malleus of the ears, the mandible, and the maxilla. The frontonasal process gives rise to tissues in the upper half of the face, including the forehead, nose, eyes, and philtrum (the vertical groove between the nose and the upper lip). NCCs from rhombomeres 3 and 4 migrate into PA2, which gives rise to the stapes bone of the middle ear, the styloid process of temporal bone, and a part of the hyoid bone [Fig. 108.1(A)].

The rest of the mammalian craniofacial skeletal elements that enclose and support the brain and cranial sense organs are called the neurocranium [Fig. 108.1(B)]. They comprise the skull vault and base. While the ventral craniofacial bones and anterior skull base are derived from cranial NCCs, most of the bones at the back of the head, including the parietal bones and occipital bone, and

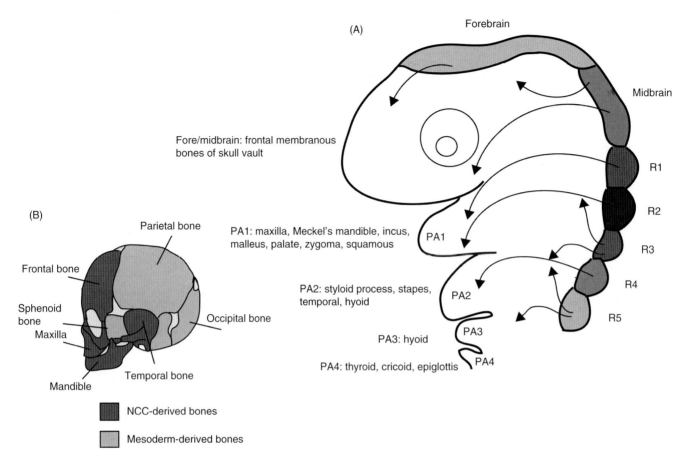

Fig. 108.1. Cranial NCC migration and NCC-derived cartilage and bones. (A) NCCs go through epithelial-mesenchymal transition and migrate ventrolaterally from rhombomeres (R) to populate pharyngeal arches (PA). NCCs in R3 and R5 merge with streams of NCCs from neighboring rhombomeres. Bones and cartilage derived from each PA are listed. (B) Facial and frontal bones are derived from NCCs. Posterior skull base and vault are mostly derived from somitic mesoderm.

the posterior part of the skull base are derived from both NCCs and paraxial mesoderm [1, 2]. The skull vault is formed through intramembranous ossification and the cranial base is formed through endochondral ossification [3].

PALATAL DEVELOPMENT

Palatogenesis is a dynamic process, with each step crucial for proper development of the palate. The palate is divided into two main parts: the primary and secondary palate. Both are mostly derived from NCCs of PA1. The primary palate develops from the intermaxillary segment, where the medial nasal processes fuse with the maxillary process, to form the most anterior part of the definitive hard palate [Fig. 108.2(A) and (B)]. Development of the secondary palate begins with downward protrusion of the two palatal shelves of the maxillary prominences on either side of the tongue, at the fifth week in humans and between embryonic day (E) 11.5 and E12.5 in mice [Figs. 108.2(B) and 108.3(A)]. Interactions between epithelial and mesenchymal cells influence the survival and continued proliferation of NCCs necessary for morphogenesis of palatal shelves [4].

The downward protrusions of the palatal shelves turn medially and elevate horizontally to meet and fuse at the midline, at about the eighth week in humans and E14.5 in mice. The tongue, meanwhile, moves downward to give enough space for the palatal shelves to elevate. The primary and secondary palates also fuse together to form the complete palate, compartmentalizing the oral and nasal cavities. Fusion of palatal shelves involves loss of medial edge epithelium (MEE) [Fig. 108.3(A)], and the cellular processes of apoptosis, epithelial migration, and epithelial-to-mesenchymal transition [5].

The ventral two-thirds of the secondary palate becomes the hard palate, containing bone, and the rest becomes the muscular soft palate. Postnatal development of the palate may be influenced by osteogenic responses to retain functional integrity against mechanical stress; for example, elevated masticatory load increased the bone density of secondary hard palate in growing rabbits [6].

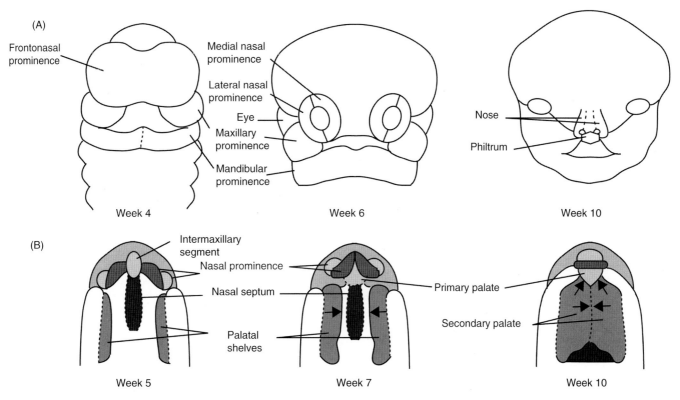

Fig. 108.2. Development of the face and the palate in humans. (A) Frontonasal, maxillary, and mandibular prominences form in the fourth week. By the sixth week, nasal prominences develop from frontonasal prominence. (B) Ventral view of palate development. Around the fifth week, palatal shelves from maxillary prominences grow downward on either side of the tongue. Growing palatal shelves start elevating medially in the seventh week. By tenth week, secondary and primary palates fuse together to form the complete palate.

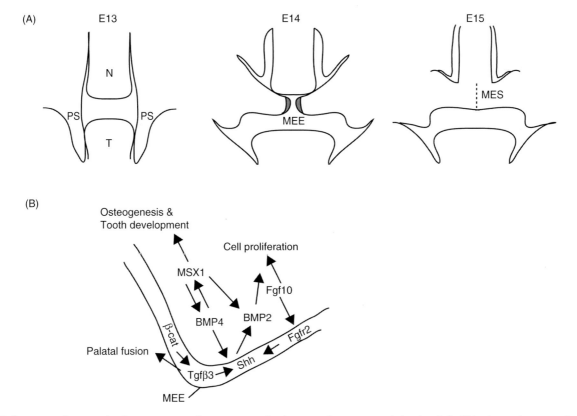

Fig. 108.3. Coronal view of palatogenesis and gene networks during palatogenesis. (A) Palatal shelf (PS), nasal septum (N), tongue (T), medial edge epithelium (MEE), medial edge seam (MES). (B) Complex gene networks in the palatal shelves. Molecular signaling events during epithelial-mesenchymal interactions of palatogenesis are depicted.

In addition to environmental risk factors, several genes and signaling pathways important for palatogenesis have been identified [Fig. 108.3(B)]. One key pathway involved in palatogenesis is bone morphogenetic protein (BMP) signaling. In early embryogenesis, BMPs influence cell migration, differentiation, proliferation, apoptosis, and condensation. Later, among other roles, BMP signaling promotes osteogenesis and chondrogenesis in skeletal formation throughout the body. For example, BMP signaling is important for mesenchymal condensation in progenitors of palatal bone [7]. Altering BMP signaling in NCCs or in the oral epithelium results in cleft palate (CP) [8, 9].

BMP2 and BMP7 expression in palatal mesenchyme is regulated by the Msx1 homeobox gene [Fig. 108.3(B)], which plays an important role in epithelial-mesenchymal interaction throughout embryogenesis. Msx1 null mice display multiple deformities in craniofacial skeletal elements, including secondary CP and abnormal tooth and mandible development [10]. Msx1 is also required for expression of Shh in the MEE, the site of palatal shelf fusion [11]. Expression of Shh and BMP2, as well as CP, in Msx1 mutant mice is rescued by transgenic expression of Bmp4, suggesting a gene network of BMP and Shh signaling pathways in development of palatal shelves [11].

The Hedgehog signaling pathway also interacts with fibroblast growth factor (Fgf) signaling in palatogenesis. Fgf10 is expressed in mesenchyme, while its receptor, Fgfr2b, is expressed in epithelium. Fgf10 is required for Shh expression in epithelium. Loss of either can cause CP due to insufficient survival or proliferation of NCCs [4].

Current models suggest that apoptosis is the major cellular mechanism for palatal fusion [12–14]. Transforming growth factor-β-3 (Tgfβ3) promotes apoptosis in juxtaposed MEE to form a single layered medial edge seam (MES) [Fig. 108.3(A)]. Tgfβ3 expression in MEE is regulated by Wnt/β-catenin signaling [15]. Table 108.1 shows the expression of some of key genes during palatogenesis.

OROFACIAL CLEFT

Orofacial clefts, among the most common congenital malformations, result from failure in the joining of bilateral facial structures during development. They are divided in three large groups, cleft lip with cleft palate (CL/P), cleft lip (CL) only, and CP only. Approximately half of the CL patients also manifest CP, considered a secondary consequence of the primary imperfection of premaxillary fusion. Thus, CP only may be etiologically distinct from CL/P. CP emanates from defective unification of the palatal shelves. Mechanisms leading to this failure include defects in palatal shelf growth, palatal elevation, and palatal shelf adhesion.

CP is subdivided into hard palate cleft, soft palate cleft, complete palate cleft, and bilateral palate cleft (Fig. 108.4). Patients often suffer from complications causing speech and feeding difficulties, frequent ear infections, dental problems as well as problems in psychological development [16]. About 70 percent of orofacial clefts arise as an isolated form where no other anomalies are associated in those patients, while the remaining cases occur in syndromes; about 400 syndromes are known to cause orofacial clefts [17]. Recent studies show that genes causing syndromic CL/P have a significant overlapping etiology with non-syndromic clefting [16]. Studying those genes and related molecular pathways will help us further understand pathogenesis of human facial clefts.

In addition to genetic factors associated with orofacial clefts, numerous nongenetic risk factors have been identified. Maternal smoking during the periconceptional period increased the incidence of CL/P and CP by twofold [18]. Smoking during pregnancy also increases the chance of causing orofacial clefts in fetuses with certain susceptibility loci [19]. In addition, the Food and Drug Administration (FDA) recently announced that certain types of antiepileptic medications increase the risk of CL/P and CP.

MANDIBULAR DEVELOPMENT

The mandible develops through unique steps. Bilaterally symmetric mandibular buds of PA1, stuffed with mesenchymal NCCs and paraxial mesoderm and enclosed by endodermal and ectodermal epithelial layers, dramatically grow and fuse to each other at the midline [20]. Within this fused mandibular process arises Meckel's cartilage (MC), a rod-shaped transient structure that functions as a template for mandibular formation. NCCs surround MC to form a membranous sheath, which later undergoes intramembranous ossification to form the mandible proper. The distal portion of MC undergoes endochondral ossification to become a part of the mandibular bone supporting the incisors, whereas the proximal portion contributes to the inner ear ossicles. The central part of MC disappears soon after birth [21]. MC

Table 108.1. Expression of Key Genes during Palatogenesis

Stage	Mesenchyme	Epithelium
E12–E13 Palatal shelf growth	Fgf10, Fgf7, Msx1	Fgfr2b, Shh
E13–E14 Palatal shelf elevation	Fgf10, Fgfr1b, Fgf2b, Msx1, TGFβ1, TGFβ2	Fgfr1, Fgfr2b, Shh, Wnt11
E14–E15 Palatal fusion	Bmp2, Bmp3, Bmp4, Fgf10, Msx1	Bmp3, Fgfr2, Shh, TGFβ3

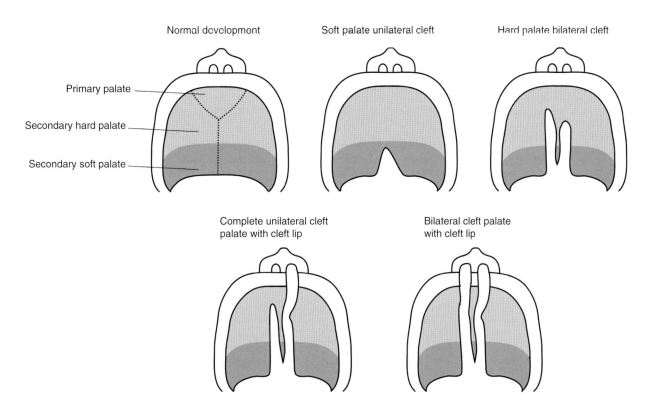

Fig. 108.4. Normal palate morphology and types of cleft palate. The palate is composed of the primary palate and secondary palate. The secondary palate is further subdivided into the hard palate and the soft palate. The degree of severity varies case by case; some only affect the soft palate while others affect both the secondary and primary palates as well as lip formation.

degradation may be induced by feed-forward secretion of interleukin-1β by infiltrating macrophages and chondrocytes [22].

Proper mandibular morphogenesis requires signaling interactions between mesenchyme and epithelial cells in PA1. In addition, signaling from pharyngeal endoderm plays a significant role in mandibular growth. The survival of mesenchyme, MC outgrowth, and osteogenesis of mandibular bones is closely regulated by stage- and region-specific roles of BMP and FGF signaling pathways [23–25]. For example, ectopic BMP4 application to mandibular explants increases apoptosis and represses Fgf8 transcription. Fgf8 signaling is considered as a source of survival signaling for mesenchyme in PA1. Mouse embryos lacking the BMP antagonists Chordin and Noggin show reduced Fgf8 in the pharyngeal ectoderm and increased apoptosis, resulting in a whole spectrum of mandibular outgrowth defects [26]. Perturbing FGF receptor 3 (FGFR3) in MC before chondrogenic and osteogenic condensations results in defects in development of the mandibular process and MC and a lack of mandibular bones [25].

MICROGNATHIA

Micrognathia, characterized by mandibular hypoplasia, is another example of a common craniofacial structural malformation. It is generally considered a defect in which too few NCCs populate PA1 due to insufficient production or migration during the fourth week of human gestation. Such NCC defects are often caused by misregulation of signaling pathways elicited by genetic and environmental factors. Patients who are affected show a wide range of malformation, from almost normal to agnathia, a complete lack of the lower jaw. While multiple syndromes include micrognathia, isolated micrognathia may cause sequential deformities. A hypoplastic mandible can displace the tongue posteriorly, which in turn prevents elevation of palatal shelves, leading to CP or CL/P as in Pierre Robin sequence [27]. Children with micrognathia often have problems with feeding and upper airway breathing as well as sleep apnea. In severe cases, surgical intervention may be required to allow the mandibular bone to elongate [28]. Although it has been believed that in some cases of mandibular hypoplasia, outgrowth may eventually attain normal proportions during childhood, recent findings suggest the underdeveloped mandible generally remains small [29, 30].

DEVELOPMENT OF THE NEUROCRANIUM

The neurocranium is composed of two parts: the skull vault and the base. The skull vault consists of multiple

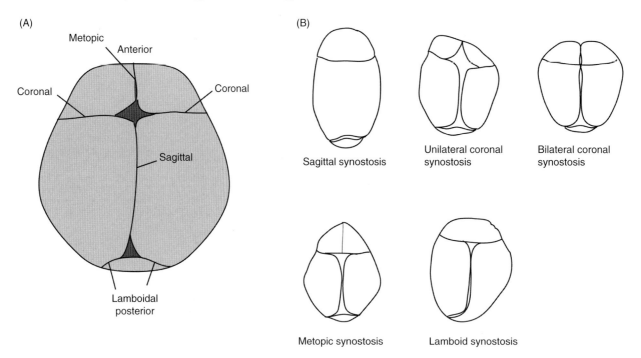

Fig. 108.5. Cranial sutures and types of craniosynostosis viewed from above. (A) Positions of the major cranial sutures. (B) Diagnostic features of craniosynostosis. Sagittal synostosis results in a long and narrow head. Right unilateral coronal synostosis shows a flattened forehead on the affected side. Bilateral coronal synostosis causes a flattened head. Metopic synostosis displays hypotelorism and a pointy forehead. Lamboid synostosis causes a flattening of the back of the head.

separate membranous bones—frontal, parietal, and a part of the occipital bones. The frontal bones are derived from NCCs, while others are mostly derived from mesoderm cells [1, 2]. Connective tissues called cranial sutures articulate these bones, resulting in formation of fontanelles at the boundaries of the cranial bones [Fig. 108.5(A)]. These sutures are the primary sites of osteogenesis during skull development. Cranial sutures are formed between neighboring membranous bones when cells at the osteogenic front proliferate. Tissue and molecular signaling interactions with underlying dura mater influence the development and maintenance of cranial sutures. Dura mater releases and takes up cytokines, mediates biochemical signaling, and contributes cells to suture mesenchyme. FGFs are expressed in dura mater just below the forming sutures. FGF signaling is critical for cranial suture biology [31, 32]. Maintenance of suture patency at birth helps the skull of a baby go through the birth canal and, until 12–18 months of age, is essential for proper brain growth and development [33].

The skull base includes midline structures such as the ethimoid, sphenoid, basioccipital bones, and parts of the temporal bones. These bones play an important role in supporting the brain. The anterior-most skull base is derived from NCCs, while the posterior skull base is derived from paraxial mesoderm [34]. The timing of growth, ossification, and maturation of the anterior and posterior skull base is not synchronous. A distinct feature of skull base bones is that, unlike other craniofacial skeletal elements, these bones develop through endochondral ossification. The skull base first forms from multiple paired cartilaginous anlagen from caudal to rostral. These grow, extend, and fuse together, resulting in formation of skull base sutures between each cartilaginous element. Eventually, they form a single perforated baseplate. Chondrogenesis in neural crest derived cartilages requires Sox9. Mesoderm-derived occipital bone is present while NCC-derived sphenoid bone is absent in mice lacking Sox9 [35]. The development of the skull base is also critical for facial growth. The size and shape of the skull base affect the position of mandible and may cause consequential craniofacial deformities such as CP [36–39].

CRANIOSYNOSTOSIS AND SKULL BASE DEFORMITIES

Premature suture fusion, craniosynostosis, occurs 1 in 2,500 human births [40] (Table 108.2). The form of craniosynostosis depends on the affected suture(s) [Fig. 108.5(B)]. The nature of the alteration in cranial vault shape depends on which sutures are fused prematurely. Different types of craniosynostosis involve distinct etiologies.

Table 108.2. Syndromes with Craniosynostosis and Implicated Genes

Gene	Syndromes
Fgfr1	Jackson-Weiss syndrome, Kallmann syndrome 2, Pfeiffer syndrome
Fgfr2	Apert syndrome, Crouzon syndrome, Jackson-Weiss syndrome, LADD syndrome, Saethre-Chotzen syndrome
Fgfr3	Crouzon syndrome, LADD syndrome, Muenke syndrome
Twist1	Saethre-Chotzen syndrome
Fbn1	Marfan syndrome, MASS syndrome
Tgfbr1 & Tgfbr2	Loeys-Dietz syndrome

Primary craniosynostosis is caused by premature ossification of the skull vault and union of one or more cranial sutures. Accordingly, an abnormal shape of the cranial vault and/or premature closure of the fontanelles result. In contrast, secondary craniosynostosis is caused by insufficient brain growth, which is usually caused in turn by the failure of cephalic neural tube closure in the first month of human embryonic development. More than 180 syndromes include craniosynostosis in humans, including Apert, Crouzon, and many others [41] (Table 108.2). Mutations in FGFs, genes involved in FGF signaling, and the osteogenic transcription factor Runx2 have been identified as the molecular lesions underlying syndromes with craniosynostosis [42].

The frequency of deformities in the skull base is less than that of defects in other parts of the craniofacial skeleton. The NCC-derived anterior skull base is more susceptible to defects than the posterior mesoderm-derived skull base [36]. Abnormalities of the skull base can affect the position of the jaws. One study suggested correlations between anterior skull base morphology and midfacial retrusion [43]. Some cases of skull base deformities are also observed in the syndromes that cause craniosynostosis [36]. Research so far has been unable to conclude if skull base abnormalities are the cause or consequences of craniosynostosis or craniofacial deformities. Defects in the skull base are often observed in skull base sutures and cartilage growth plate called synchondroses. The location and timing of premature skull base fusion and synchondroses can influence the craniofacial symmetry of a child. Premature skull base closures and abnormal growth in synchondroses are recognized in many syndromes such as Crouzon and Apert [44]. However, as Goodrich suggests [44], the development of each unit—the skull vault, facial skeleton, and the skull base—influences the other units of the head skeleton. Thus, it is critical to maintain proper regulation of gene expression, tissue growth, and patterning for development of craniofacial skeletal elements.

CONCLUSION

Among the congenital deformities found in newborns, craniofacial defects such as orofacial clefts, micrognathia, and craniosynostosis occur at a frequency second only to congenital heart defects. The severity of craniofacial deformities varies case by case. The deformities frequently occur as part of genetic syndromes. They also appear sporadically as isolated forms. In some cases, teratogenic chemicals are responsible. Such craniofacial deformities may be fatal, but more often impact quality of life because they affect normal breathing and feeding. Unfortunately, there are often negative social responses to the deformities, which take a psychological toll on the affected person. The etiology of most congenital craniofacial anomalies has just begun to be understood. It is essential to clarify the normal and pathological development of craniofacial skeletal elements and the underlying molecular and cellular mechanisms. This knowledge will enhance our ability to design strategies for future treatment and prevention of craniofacial birth defects.

REFERENCES

1. Jeong J, Mao J, Tenzen T, Kottmann AH, McMahon AP. 2004. Hedgehog signaling in the neural crest cells regulates the patterning and growth of facial primordia. *Genes Dev* 18: 937–951.
2. McBratney-Owen B, Iseki S, Bamforth SD, Olsen BR, Morriss-Kay GM. 2008. Development and tissue origins of the mammalian cranial base. *Dev Biol* 322: 121–132.
3. Sahar DE, Longaker MT, Quarto N. 2005. Sox9 neural crest determinant gene controls patterning and closure of the posterior frontal cranial suture. *Dev Biol* 280: 344–361.
4. Rice R, Spencer-Dene B, Connor EC, Gritli-Linde A, McMahon AP, Dickson C, Thesleff I, Rice DP. 2004. Disruption of Fgf10/Fgfr2b-coordinated epithelial-mesenchymal interactions causes cleft palate. *J Clin Invest* 113: 1692–1700.
5. Jin JZ, Ding J. 2006. Analysis of cell migration, transdifferentiation and apoptosis during mouse secondary palate fusion. *Development* 133: 3341–3347.
6. Menegaz RA, Sublett SV, Figueroa SD, Hoffman TJ, Ravosa MJ. 2009. Phenotypic plasticity and function of the hard palate in growing rabbits. *Anat Rec (Hoboken)* 292: 277–284.
7. Baek JA, Lan Y, Liu H, Maltby KM, Mishina Y, Jiang RL. 2011. Bmpr1a signaling plays critical roles in palatal shelf growth and palatal bone formation. *Dev Biol* 350: 520–531.
8. He F, Xiong W, Wang Y, Matsui M, Yu X, Chai Y, Klingensmith J, Chen Y. 2010. Modulation of BMP signaling by Noggin is required for the maintenance of palatal

epithelial integrity during palatogenesis. *Dev Biol* 347: 109–121.
9. Li L, Lin MK, Wang Y, Cserjesi P, Chen Z, Chen YP. 2011. BmprIa is required in mesenchymal tissue and has limited redundant function with BmprIb in tooth and palate development. *Dev Biol* 349: 451–461.
10. Satokata I, Maas R. 1994. Msx1 deficient mice exhibit cleft-palate and abnormalities of craniofacial and tooth development. *Nature Genet* 6: 348–356.
11. Zhang Z, Song Y, Zhao X, Zhang X, Fermin C, Chen Y. 2002. Rescue of cleft palate in Msx1-deficient mice by transgenic Bmp4 reveals a network of BMP and Shh signaling in the regulation of mammalian palatogenesis. *Development* 129: 4135–4146.
12. Kaartinen V, Voncken JW, Shuler C, Warburton D, Bu D, Heisterkamp N, Groffen J. 1995. Abnormal lung development and cleft palate in mice lacking TGF-beta 3 indicates defects of epithelial-mesenchymal interaction. *Nat Genet* 11: 415–421.
13. Proetzel G, Pawlowski SA, Wiles MV, Yin MY, Boivin GP, Howles PN, Ding JX, Ferguson MWJ, Doetschman T. 1995. Transforming growth factor-beta-3 is required for secondary palate fusion. *Nat Genet* 11: 409–414.
14. Taya Y, O'Kane S, Ferguson MWJ. 1999. Pathogenesis of cleft palate in TGF-beta 3 knockout mice. *Development* 126: 3869–3879.
15. He F, Xiong W, Wang Y, Li L, Liu C, Yamagami T, Taketo MM, Zhou C, Chen Y. 2011. Epithelial Wnt/beta-catenin signaling regulates palatal shelf fusion through regulation of Tgfbeta3 expression. *Dev Biol* 350: 511–519.
16. Stanier P, Moore GE. 2004. Genetics of cleft lip and palate: Syndromic genes contribute to the incidence of non-syndromic clefts. *Hum Mol Genet* 13: R73–R81.
17. Zucchero TM, Cooper ME, Maher BS, Daack-Hirsch S, Nepomuceno B, Ribeiro L, Caprau D, Christensen K, Suzuki Y, Machida J, Natsume N, Yoshiura KI, Vieira AR, Orioli IM, Castilla EE, Moreno L, Arcos-Burgos M, Lidral AC, Field LL, Liu YE, Ray A, Goldstein TH, Schultz RE, Shi M, Johnson MK, Kondo S, Schutte BC, Marazita ML, Murray JC. 2004. Interferon regulatory factor 6 (IRF6) gene variants and the risk of isolated cleft lip or palate. *N Engl J Med* 351: 769–780.
18. Honein MA, Rasmussen SA, Reefhuis J, Romitti PA, Lammer EJ, Sun LX, Correa A. 2007. Maternal smoking and environmental tobacco smoke exposure and the risk of orofacial clefts. *Epidemiology* 18: 226–233.
19. Lammer EJ, Shaw GM, Iovannisci DM, Van Waes J, Finnell RH. 2004. Maternal smoking and the risk of orofacial clefts: Susceptibility with NAT1 and NAT2 polymorphisms. *Epidemiology* 15: 150–156.
20. Chai Y, Jiang XB, Ito Y, Bringas P, Han J, Rowitch DH, Soriano P, McMahon AP, Sucov HM. 2000. Fate of the mammalian cranial neural crest during tooth and mandibular morphogenesis. *Development* 127: 1671–1679.
21. Frommer J, Margolie, MR. 1971. Contribution of Meckel's cartilage to ossification of mandible in mice. *J Dent Res* 50: 1260–1267.
22. Tsuzurahara F, Soeta S, Kawawa T, Baba K, Nakamura M. 2011. The role of macrophages in the disappearance of Meckel's cartilage during mandibular development in mice. *Acta Histochem* 113: 194–200.
23. Wilke TA, Gubbels S, Schwartz J, Richman JM. 1997. Expression of fibroblast growth factor receptors (FGFR1, FGFR2, FGFR3) in the developing head and face. *Dev Dyn* 210: 41–52.
24. Mina M, Wang YH, Ivanisevic AM, Upholt WB, Rodgers B. 2002. Region- and stage-specific effects of FGFs and BMPs in chick mandibular morphogenesis. *Dev Dyn* 223: 333–352.
25. Mina M, Havens B, Velonis DA. 2007. FGF signaling in mandibular skeletogenesis. *Orthod Craniofac Res* 10: 59–66.
26. Stottmann RW, Anderson RM, Klingensmith J. 2001. The BMP antagonists Chordin and Noggin have essential but redundant roles in mouse mandibular outgrowth. *Dev Biol* 240: 457–473.
27. Weseman CM. 1959. Congenital micrognathia. *AMA Arch Otolaryngol* 69: 31–44.
28. Figueroa AA. 2002. Long-term outcome study of bilateral mandibular distraction: A comparison of Treacher Collins and Nager syndromes to other types of micrognathia; discussion. *Plast Reconstr Surg* 109: 1826–1827.
29. Suri S, Ross RB, Tompson BD. 2010. Craniofacial morphology and adolescent facial growth in Pierre Robin sequence. *Am J Orthod Dentofacial Orthop* 137: 763–774.
30. Daskalogiannakis J, Ross RB, Tompson BD. 2001. The mandibular catch-up growth controversy in Pierre Robin sequence. *Am J Orthod Dentofacial Orthop* 120: 280–285.
31. Ogle RC, Tholpady SS, McGlynn KA, Ogle RA. 2004. Regulation of cranial suture morphogenesis. *Cells Tissues Organs* 176: 54–66.
32. De Coster PJ, Mortier G, Marks LA, Martens LC. 2007. Cranial suture biology and dental development: Genetic and clinical perspectives. *J Oral Pathol Med* 36: 447–455.
33. Richtsmeier JT, Aldridge K, DeLeon VB, Panchal J, Kane AA, Marsh JL, Yan P, Cole TM 3rd. 2006. Phenotypic integration of neurocranium and brain. *J Exp Zool B Mol Dev Evol* 306: 360–378.
34. Couly GF, Coltey PM, Ledouarin NM. 1993. The triple origin of skull in higher vertebrates: A study in quail-chick chimeras. *Development* 117: 409–429.
35. Mori-Akiyama Y, Akiyama H, Rowitch DH, de Crombrugghe B. 2003. Sox9 is required for determination of the chondrogenic cell lineage in the cranial neural crest. *Proc Natl Acad Sci U S A* 100: 9360–9365.
36. Nie XG. 2005. Cranial base in craniofacial development: Developmental features, influence on facial growth, anomaly, and molecular basis. *Acta Odontologica Scandinavica* 63: 127–135.
37. Lieberman DE, Pearson OM, Mowbray KM. 2000. Basicranial influence on overall cranial shape. *J Hum Evol* 38: 291–315.
38. Bastir M, Rosas A, Stringer C, Cuetara JM, Kruszynski R, Weber GW, Ross CF, Ravosa MJ. 2010. Effects of

brain and facial size on basicranial form in human and primate evolution. *J Hum Evol* 58: 424–431.
39. Harris EF. 1993. Size and form of the cranial base in isolated cleft lip and palate. *Cleft Palate Craniofac J* 30: 170–174.
40. Cohen MM, MacLean RE. 2000. *Craniosynostosis: Diagnosis, Evaluation, and Management, 2nd Ed.* New York: Oxford University Press.
41. Hennekam RC, Van den Boogaard MJ. 1990. Autosomal dominant craniosynostosis of the sutura metopica. *Clin Genet* 38: 374–377.
42. Kimonis V, Gold JA, Hoffman TL, Panchal J, Boyadjiev SA. 2007. Genetics of craniosynostosis. *Semin Pediatr Neurol* 14: 150–161.
43. Lozanoff S, Jureczek S, Feng T, Padwal R. 1994. Anterior cranial base morphology in mice with midfacial retrusion. *Cleft Palate Craniofac J* 31: 417–428.
44. Goodrich JT. 2005. Skull base growth in craniosynostosis. *Childs Nerv Syst* 21: 871–879.

109
Development and Structure of Teeth and Periodontal Tissues

Petros Papagerakis and Thimios Mitsiadis

Introduction 904
Stages of Tooth Morphogenesis and Its Molecular Control 904
Signals Controlling Dental Cell Differentiation 906
Dental Mineralized Tissues Formation 906
Stem Cells during Tooth Repair 911
Conclusion 911
Acknowledgments 912
References 912

INTRODUCTION

Tooth development or odontogenesis is the complex process by which dental mineralized tissues form from embryonic cells that differentiate into ameloblasts that secrete enamel, odontoblasts that produce dentin, and cementoblasts that make cementum. Enamel is of epithelial origin and covers the crown of each tooth. In contrast, dentin and cementum are of mesenchymal origin. Dentin forms the bulk of the tooth and extends within both the crown and root. It has a yellow color, in contrast to the much whiter and harder enamel. Cementum is deposited only in the root area on the recently mineralized dentin matrix. The tooth is anchored onto its socket (alveolar bone) by the periodontal ligament (PDL), a connective tissue structure that surrounds the tooth root and connects each tooth to the alveolar bone through a specialized set of collagen fibers.

STAGES OF TOOTH MORPHOGENESIS AND ITS MOLECULAR CONTROL

Mammalian teeth have distinctive crown and root morphologies, which are highly adapted to their particular masticatory function. The generation of individual teeth relies upon interactions between the oral epithelium and mesenchymal cells that are derived from the cranial neural crest cells (CNCCs). Their formation involves a precisely orchestrated series of molecular and morphogenetic events. Although many diverse types of teeth exist in different species, non-human tooth development is largely the same as in humans. Therefore, we have used here examples of mouse tooth development as a prototype model for understanding human tooth formation. When possible we tried to make connections to human pathologies.

Morphologically, tooth development begins with a thickening of the oral epithelium that forms a structure known as the dental lamina. Within the dental lamina, cells start to proliferate and to invaginate the underlying mesenchyme in precise positions to form the dental placodes (which define where the teeth will be positioned into the jaws). Tooth development proceeds through a series of morphological stages that necessitate sequential and reciprocal interactions between the oral epithelium and the underlying cranial neural crest-derived mesenchyme. In mice, the oral epithelium starts thickening at embryonic day 10.5 (E10.5) and progressively acquires the bud (E13.5), cap (E14.5), and bell (E16.5) configurations (Fig. 109.1). At the bell stage, two mesenchymal cell populations can be distinguished: the dental follicle and dental pulp. Dental pulp cells adjacent to the dental epithelium differentiate into odontoblasts, while epithe-

---

Primer on the Metabolic Bone Diseases and Disorders of Mineral Metabolism, Eighth Edition. Edited by Clifford J. Rosen.
© 2013 American Society for Bone and Mineral Research. Published 2013 by John Wiley & Sons, Inc.

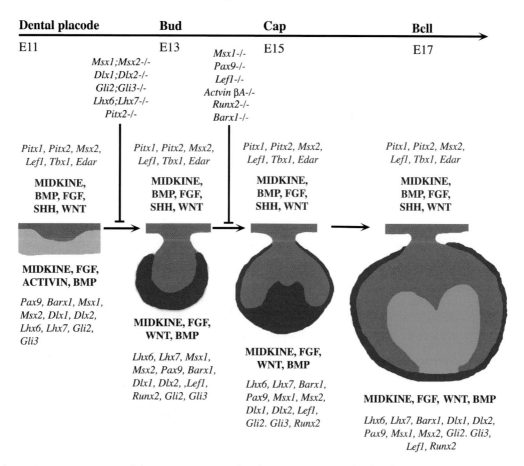

Fig. 109.1. Schematic representation of the various stages of embryonic mouse molar development. The most important signaling molecules (in bold capital letters) and transcription factors (in italic letters) that are expressed in tooth epithelium (red) and mesenchyme (variations of blue) are shown.

lial cells juxtaposing the dental pulp differentiate into ameloblasts [1, 2]. Dental follicle gives rise to specialized PDL cells.

Signaling molecules control all steps of tooth formation by coordinating cell proliferation, differentiation, apoptosis, extracellular matrix synthesis, and mineral deposition. The same molecules are repetitively used during the different stages of tooth development and are regulated according to a precise timing mechanism [3, 4]. Signals produced at a wrong time lead to abnormal cell proliferation, differentiation, apoptosis, thus affecting the overall tooth development and shape.

Numerous studies during the past 20 years have shown that bone morphogenic proteins (BMPs) regulate epithelial-mesenchymal interactions during tooth initiation [3], Wnts [4], and Sonic hedgehog (Shh) [5] regulate cell proliferation, migration, and differentiation, and fibroblast growth factors (FGFs) regulate tooth specific gene expression and cell proliferation [6] (Fig. 109.1).

The territory of the mammalian dentition is defined early, before any obvious sign of tooth development. The earliest marker that defines the oral epithelial area where teeth will grow is the transcription factor *Pitx2* [7]. Mutations of *PITX2* in humans result in Rieger's syndrome, which is characterized by eye and tooth defects, including anodontia (lack of teeth).

The mesenchyme of teeth is derived from CNCCs that form a pool of multipotent progenitors. CNCCs migrate from the dorsal part of the neural tube and subsequently generate craniofacial structures of unique morphology and function, such as teeth [8]. Malformations and syndromes that arise due to defects in neural crest cell development are collectively called neurocristopathies in humans.

Dental fields within the oral epithelium are established by epithelium-derived signals that form morphogenic gradients, providing positional information [1]. These signals determine the display and fate of the CNCCs, leading to the generation of distinct tooth shapes. As an example, ectodysplasin (Eda) signaling molecules have been shown to be involved in the determination of the size of the dental fields in the oral epithelium, and thus in the proportion of the size and number of teeth [9]. Non-syndromic hypodontia or congential absence of one or more permanent teeth is a common anomaly of dental development in humans.

Mutations in the EDA gene are related with X-linked recessive hypohidrotic ectodermal dysplasia (XLHED). XLHED is a genetic disease characterized by the defective morphogenesis of teeth, hair, and sweat glands [10].

SIGNALS CONTROLLING DENTAL CELL DIFFERENTIATION

Dental cell differentiation results in the formation of the three dental mineralized tissues (enamel, dentin, cementum) (Fig. 109.2,) which are connected through the periodontal ligament (PDL) to the alveolar bone.

The specification of the various dental cell types during dental cell differentiation (i.e., stratum intermedium, stellate reticulum, outer and inner dental epithelial cells, ameloblasts, dental pulp fibroblasts, odontoblasts, periodontal ligament fibroblasts) involves differential expression of specialized genes with restricted developmental and circadian patterns during odontogenesis. It is possible that the determination of cell fates in teeth occurs via inhibitory interactions between adjacent dental cells. These interactions seem to be mediated mainly through the Notch signaling pathway.

BMPs and FGFs have opposite effects on the expression of Notch receptors and ligands (i.e., Delta and Jagged) in dental tissues [11], indicating that cell fate choices during odontogenesis are under the concomitant control of the Notch and BMP/FGF signaling pathways. Notch-mediated lateral inhibition has a pivotal role in the establishment of the tooth morphology, as shown in *Jagged2* mutant mice where the overall development and structure of their teeth is severely affected [12].

Mutations in Notch signaling pathway members cause developmental phenotypes that affect the liver, skeleton, heart, eye, face, kidney, and vasculature in humans [13]. The tooth phenotype is still unclear in these patients. In addition, genetic findings have shown that Tbx1, a Notch signaling target, plays a significant role in the early determination of epithelial cells to adopt the ameloblast fate. Indeed, hypoplastic incisors that lack enamel are observed in mice where the *Tbx1* gene was deleted [14]. Furthermore, TBX1 mutations in humans are associated with the DiGeorge syndrome, which is characterized by abnormal cell differentiation resulting in defects of many organs including heart and tooth [15].

DENTAL MINERALIZED TISSUES FORMATION

Dentin

During tooth development, dental pulp cells start to differentiate into odontoblasts at the dental-enamel junction (DEJ) under the future cusp tip. Young odontoblasts are columnar cells that secrete an mantle matrix, called predentin, which is rich in type I collagen and matrix vesicles. After predentin deposition, the basal lamina associated with the inner enamel epithelium disintegrates [16] followed by a major upregulation in the production and secretion of enamel matrix proteins and metalloproteinase-20 (Mmp-20) [17].

Dentin is tough and elastic, and its prime feature is its penetration by odontoblast tubules, which radiate out from the dental pulp to the periphery (Fig. 109.3A, B). These, with their many side branches that remain in the

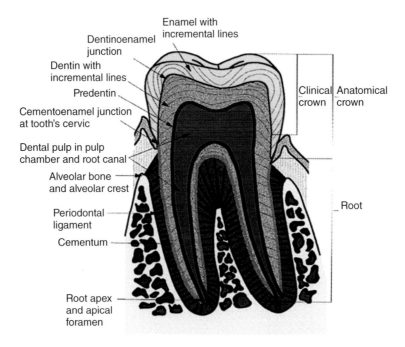

Fig. 109.2. Schematic representation of dental mineralized tissues and their structural organization.

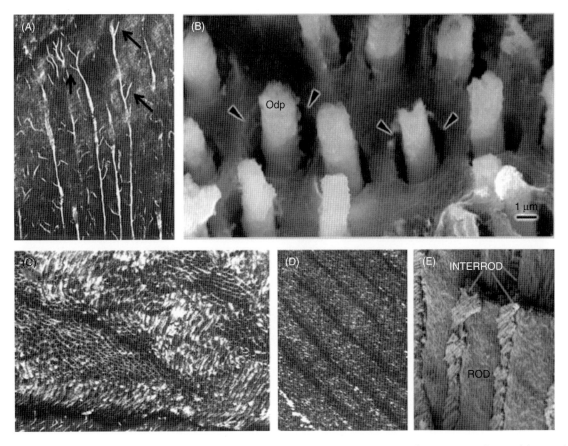

Fig. 109.3. (A) Dentin processes transverse the whole primary dentin matrix. Many ramifications are observed (arrows). (B) Close capture of odontoblast processes. Peritubular (arrowheads) and intertubular dentin can be seen. (C) Decussation (crossing in an "X" fashion) of the enamel prisms, with zones of prisms with contrasting 3D courses forming the Hunter-Schreger bands can be seen. (D) Enamel prisms. (E) Enamel rods and interrod enamel. Crystals in rod and interrod enamel are structurally similar, but have different orientations. Odp: Odontoblast process.

tubules within the dentin, are analogous to the canaliculi that house osteocyte processes in bone. Functional odontoblasts are highly polarized cells with a specialized cellular process, the odontoblast process, which transverses the heterogeneous layers of primary dentin within the dentin tubules [18].

Primary dentin forms most of the tooth prior to root eruption. Primary dentin is composed of peritubular or intratubular dentin (which creates the wall of the dentinal tubule); intertubular dentin found between the tubules; mantle dentin (which is the first predentin that forms within the tooth and is devoid of tubules); and circumpulpal dentin (which is the inner layer of dentin around the outer pulpal wall). The continuous secretion of dentin matrix (called "orthodentin") is associated with the progressive lengthening of the odontoblast cell process and retraction of the odontoblasts toward the dental pulp. In contrast, during physiopathological situations in humans, several other types of dentin are formed, i.e., sclerotic/reactionary orthodentin, where tubules may be obliterated; fibrodentin; and also osteodentin.

Odontoblasts secrete tooth-related proteins, such as dentin sialoprotein (DSP), dentin phosphoprotein (DPP), and dentin glycoprotein (DGP), which are encoded by a single gene called dentin sialophosphoprotein(DSPP) [19]. However, the majority of dentin is composed of proteins common to both dentin and bone. These proteins include type I, III, and V collagens, bone sialoprotein (BSP), osteopontin (OPN), dentin matrix protein-1 (DMP-1), osteocalcin (OC), and osteonectin (ON) [20]. Several proteins involved in calcium and phosphate handling are also synthesized jointly by osteoblasts and odontoblasts, including calbindin-D28k [21], calcium pump [22], and alkaline phosphatase [23].

Dentin continues to be slowly deposited even after tooth eruption and complete root formation. This dentin could be either the regularly deposited secondary dentin or irregular tertiary dentin, as a response of the pulp-dentin complex to attrition or disease [24]. Nerves pass from the dental pulp between odontoblasts and extend into the dentin tubules for variable distances.

The dental pulp shows signs of aging [25], which may include diffuse or local calcifications and the formation of dental stones as well as limited reparative dentin formation.

Enamel

Dental enamel is the highest mineralized tissue found in mammals. Mature enamel contains less than 1% of organic material, and it is acellular and contains no collagen. Enamel forms in an extracellular space lined by ameloblasts, which control both the ionic and the organic contents of the enamel extracellular space [26]. Enamel mineral is mainly formed by calcium hydroxyapatite, exhibiting peculiar dimensions and organization of its crystallites. Enamel crystals are only approximately 25 nm thick and 65 nm wide but are believed to extend uninterrupted from the DEJ to the surface of the tooth [27]. The secretory end of the ameloblast ends in a six-sided pyramid-like projection known as the Tomes' process. Tomes' processes organize the enamel crystallites into rods (prisms; Fig. 109.3D). The angulation of the Tomes' process is significant in the orientation of enamel rods (Fig. 109.3E). Enamel formation is divided into secretory, transition, and maturation stages [28, 29]. The stages are shown in Fig. 109.4B [30].

During the secretory stage, enamel matrix deposition is orchestrated by both pre-ameloblasts and secretory ameloblasts. Ameloblast differentiation initiates at the cusps of the tooth germ, where the first epithelial cells differentiate into pre-ameloblasts. The signals driving ameloblast differentiation derive from the recently differentiated odontoblasts [18]. Pre-ameloblasts are secretory cells that have not yet formed a Tomes' process and deposit a thin layer of aprismatic enamel on the dentin surface. Thereafter, secretory ameloblasts initiate prismatic enamel deposition. During the secretory stage, mineral is deposited rapidly on the tips of the crystals and very sowly on the crystals sides (for review, see Ref. 31). As the crystals extend, the enamel layer expands. During the secretory stage, ameloblasts secrete mostly amelogenin, enamelin, and ameloblastin [32]. These enamel specific proteins catalyze the extension of ribbon-like enamel crystallites comprising the mineral calcium hydroxyapatite. Amelogenins are the predominant enamel matrix proteins and assemble into spheres that occupy the spaces between the crystal ribbons, serving to separate and support them [33]. Ameloblastin is thought to be important for ameloblasts attachment to the mineralizing matrix [34]. Enamelin is probably responsible for initiation of mineralization at the mineralization front [35]. As ameloblasts secrete enamel proteins and extend the mineral ribbons, they retreat from the existing enamel surface, thus increasing the thickness of the enamel extracellular space (Fig. 109.4C). Unlike collagen-based mineralization processes, which are two-step processes (secretion of an organic matrix followed by matrix mineralization), enamel secretion and mineralization occur in a single step.

During the maturation stage, mineral is deposited exclusively on the sides of the crystals. Crystals grow in width and thickness until further growth is prevented by contact with adjacent crystals. Maturation stage ameloblasts oscillate between smooth-ended and ruffle-ended morphologies (Fig. 109.5C). Rapid mineralization occurs beneath the ruffle-ended ameloblasts, which is associated with a substantial drop in pH (acidity is generated by mineral deposition) [36]. Smooth-ended ameloblasts neutralize this acid by secreting bicarbonate.

Ameloblasts create the space necessary for continued increase in the volume of mineral by gradually removing of enamel proteins progressively degraded by stage specific proteases (i.e., Mmp-20 at the secretory stage and kallikrein 4 (klk4) at the maturation stage) [37]. Maturation stage ameloblasts secrete and assemble a specialized basal lamina that securely attaches them to the enamel surface to assure its mineralization [38]. Two ameloblast-specific basal lamina proteins have recently been characterized: amelotin (*AMTN*) and Apin (*ODAM*) [39]. Transition and maturation stage ameloblasts also express genes necessary for enamel mineralization such as carbonic anhydrase II (CA2), which allows the cells to secrete bicarbonate to neutralize the acid generated by hydroxyapatite formation; calcium binding protein (Calb1); calcium-sensing receptor (CaSR); anion exchanger Ae2 (SLC4a4), and alkaline phosphatase (TNAP). The expression of several growth factors, hormones, and transcription factors is also characterized during amelogenesis [40]. Of interest, genes initially found to be expressed only in ameloblasts were also detected in odontoblasts (Fig. 109.4A) [41], and, conversely, odontoblast-specific genes were found to be expressed in ameloblasts [42].

As ameloblasts move away from the dentin, they travel in groups across the surface that they make. This results in decussation (crossing in an "X" fashion) of the enamel prisms, with zones of prisms with contrasting three-dimensional (3D) courses forming the Hunter-Schreger bands (Fig. 109.3C). The unerupted enamel is protected from resorption by a layer of cells termed the "reduced enamel epithelium," generated of mature ameloblasts remnants. These cells disappear once the tooth erupts.

Cementum and periodontal ligament

The dental follicle, a sac of loose connective tissue that separates the developing tooth from its bony crypt, is essential for eruption and will become the periodontal ligament (PDL) on tooth eruption.

The cementum is an avascular and unnerved mineralized tissue with ultrastructural similarity to bone that covers the entire root surface (Fig. 109.5A). It is the interface between the dentin and the periodontal ligament and contributes to periodontal tissue repair and regeneration after damage. The organic extracellular matrix of cementum contains proteins that selectively enhance the attachment and proliferation of cell populations residing within the PDL space [43].

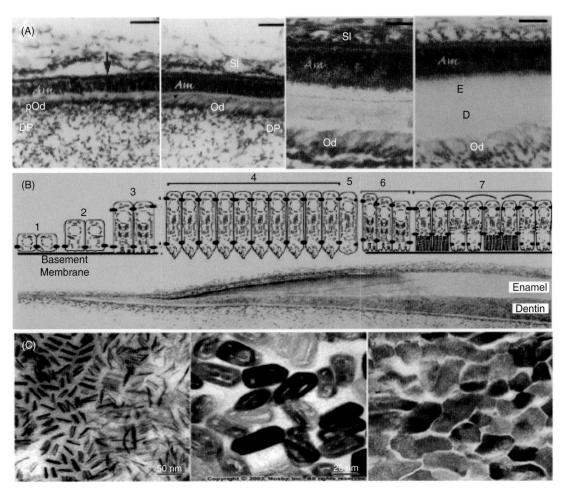

Fig. 109.4. (A) Stage-specific gene expression is a characteristic of amelogenesis. Here, as an example, amelogenin mRNA shows low expression in pre-ameloblasts and maturation stage ameloblasts. In contrast, secretory stage ameloblasts show strong expression of amelogenin mRNA using *in situ* hybridization (black spots). Am: ameloblasts; Od: odontoblasts; pOd: pre-odontoblasts; E: enamel; D: dentin; SI: stratum intermedium. (B) Ameloblast changes during enamel formation. The epithelial cells of the inner enamel epithelium (1) rest on a basement membrane. These cells increase in length as the ameloblasts differentiate above the predentin matrix (2). Presecretory ameloblasts send processes through the degenerating basement membrane as they initiate the secretion of enamel proteins on the dentin surface (3). After establishing the DEJ and mineralizing a thin layer of aprismatic enamel, secretory ameloblasts develop a secretory specialization, or Tomes' process. Along the secretory face of the Tomes' process, in place of the absent basement membrane, ameloblasts secrete proteins at a mineralization front where the enamel crystals grow in length. Amelogenin (Amel), Ameloblastin (Ambn), Enamelin (Enam) and metalloproteinase-20 (Mmp-20) are genes specifically expressed in secretory ameloblasts (4). At the end of the secretory stage, ameloblasts lose their Tomes' processes and produce a thin layer of aprismatic enamel (5). At this point the enamel has achieved its final thickness. During the transition stage of amelogenesis, ameloblasts undergo a major restructuring that diminishes their secretory activity and changes the types of proteins secreted. Carbonic anhydrase II (CA II), calbindin-D 28K (Calb1), alkaline phosphatase, as well as calcium receptor and anion exchanger Ae2 epitopes are preferentially expressed during the maturation stage of enamel formation (6). Kallikrein 4 (KLK4) is secreted in the matrix, which degrades the accumulated enamel proteins, and amelotin (AMTN) is secreted as part of the new basement membrane. During the subsequent maturation stage, ameloblasts modulate between ruffled- and smooth-ended phases (7). Their activities harden the enamel layer by promoting the deposition of mineral on the sides of enamel crystals laid down during the secretory stage. (C) Enamel crystals are very thin at the secretory stage (left panel). Enamel crystals thicken during the maturation stage (center and right panels).

Cementum is deposited initially on the newly mineralized dentin matrix of the root by cells derived from the dental follicle and/or by the epithelial-mesenchymal transition of root sheath epithelial cells [44]. Secretory proteins from the cells of the epithelial root sheath may be included in the first-formed cementum matrix. Embryologically, there are two types of cementum. These are the primary cementum (which is acellular and develops slowly as the tooth erupts) and the secondary cementum (which is formed after the tooth is in occlusion). Where cementum is deposited very rapidly, it is cellular, containing cementocytes that resemble the osteocytes of bone.

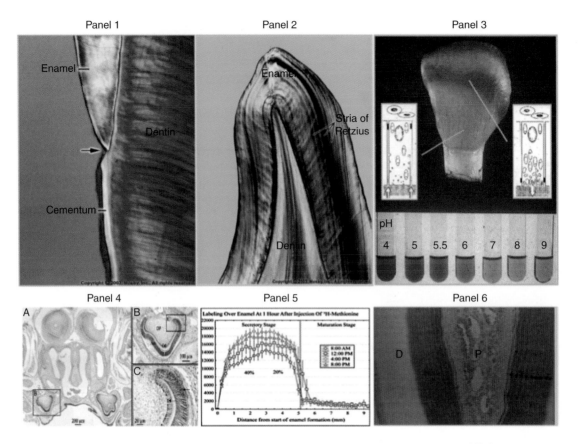

Fig. 109.5. (Panel 1) Enamel, dentin, and cementum constitute the three dental mineralized tissues. All three tissues are controlled by circadian rhythms. (Panel 2) Striae of Retzius (SR) lines also called "long-period incremental lines" are seen in enamel. About seven cross-striations are observed as vertical lines between the SR. (Panel 3) Maturation ameloblasts cycle between ruffle-ended (and smooth-ended) phases. Beneath the ruffle-ended ameloblasts the crystals rapidly mineralize and the pH drops below 6 (red). Beneath smooth-ended ameloblasts, the enamel is neutralized and rises above 7.0 (orange). (Panel 4) Immunohistochemistry of clock protein in a postnatal day 4 mouse. (A) Clock protein expression was detected in the developing first molars. (B), (C) Higher magnifications showing that the nuclei (arrows) of ameloblasts (AM) and odontoblasts (OD) have strong clock expression relative to the dental pulp (DP) cells. (Panel 5) Daily variation in ameloblast secretion of proteins containing methionine. The graph of means with a ±95% confidence interval illustrates the total amount of newly synthesized proteins released into developing enamel on rat mandibular incisors by 1 hour after a single intravenous injection of 3H-methionine administered at different times of the day. Substantially greater amounts of secretory activity for enamel proteins occur in the late afternoon (4 p.m., diamonds) compared with early morning (8 a.m., circles) throughout the secretory stage. These differences are noticeably larger (up to 40%) for inner enamel formation (distance, 0.5–3.0 mm) than for outer enamel formation (20%). (Panel 6) Similar to enamel, dentin also contains marks of short- and long-period growth lines. Lines of Owen, the equivalent of SR, can be traced over considerable distances and are deposited 6 to 10 days apart in different individuals. A radiograph of a transverse dentin section shows nine densely labeled circumpulpal bands after infusion with labeled proline for 10 days.

The main collagen of both the extrinsic and intrinsic fibers is type I. The non-collagenous proteins of cementum identified are similar to bone matrix proteins making it difficult to distinguish cementum from other calcified connective tissues. So far only the cementum attachement protein (CAP) may be specific to cementum but its specificity is controversial.

During root development, the dental follicle becomes rapidly organized into the PDL that supports the tooth, provides nutrition and mechanosensation, and allows physiological tooth movement. PDL is unique among the various ligament and tendon systems of the body, in the sense that it is the only soft tissue to span between two distinct hard tissues, namely the cementum of the dental root and the alveolar bone [45].

Through the groups of fibers of the periodontal ligament, comprising type I and type III collagen, functioning teeth are linked to each other, the gingiva, and the alveolar bone. The principal fibers of the ligament are incorporated on either side of each root within the cementum and alveolar bone. Within the ligament, there is constant adaptive remodeling of the soft tissue. The PDL collagen fibers are categorized according to their orientation and location along the tooth. The completeness and vitality of the PDL fibers are essential for the functioning of the tooth. Damage to the PDL fibers may result in ankylosis of the tooth, making the tooth lose its continuous eruption ability. Dental trauma, such as subluxation, may cause tearing of the PDL fibers and pain during mastication. The pathological damage of the periodontal tissues due to inflammatory conditions is termed "periodontitis." This disease may eventually lead to extraction or loss of the adult tooth due to lack of supporting tissue.

The alveolar bone is a part of the periodontal tissues, functioning as an anchorage of the tooth root to the alveoli and resorbing the forces generated by the function of mastication. Progenitor cells, which are responsible for alveolar bone formation, lie in the periosteal region, the PDL, or around the blood vessels. Alveolar bone marrow is considered as a useful and easily accessible source of progenitor cells, as they have similar osteogenic potential to those derived from the iliac crest [46]. The periosteum is also considered a suitable cell source for bone regeneration.

As the permanent tooth erupts, the alveolar bone is resorbed to allow its passage, its root develops, and the crown transverses the oral mucosa that contributes to a tight ring seal of epithelial cells on the enamel close to the junction of crown and root. The complex molecular signaling cascades that control eruption and root growth are unclear. At eruption, the root of the tooth is not yet fully formed, and root completion in humans takes about 18 months more in the deciduous teeth and up to 3 years in the permanent teeth. Cementum in permanent teeth sees little remodeling, but the surface of alveolar bone is continually resorbing and forming to allow the tooth to move in response to eruption, growth drift, or changing functional forces. Resorption of deciduous tooth roots begins shortly after their completion, appearing first and most extensively on the aspect adjacent to the successional tooth.

Circadian formation of dental mineralized tissues

Dental mineralized tissues form by additive modes of growth that preserve within the hard tissues short- and long-period lines of incremental growth. In dental enamel there are two regularly occurring incremental markers: daily cross-striations and long-period striae of Retzius (SR). These lines correspond to what was, at precise points in time during the secretory stage of amelogenesis, the enamel surface (Fig. 109.5, Panel 2). Daily incremental lines (called von Ebner's lines) are also observed in dentin (Fig. 109.5, Panel 6). Cross-striations in enamel and von Ebner's lines in dentin delineate the amount of mineral deposited in a single day. Circadian rhythms have been demonstrated using ^{3}H-proline tracers that label collagen in dentin formation [47]. Twice as much collagen is secreted during the daylight 12 hours as during the night time 12 hours (Fig. 109.5, Panel 6). Similarly to collagen, enamel proteins (Fig. 109.5, Panel 5) and amelogenin show circadian rhythms [48, 49]. Consistently, ameloblasts and odontoblasts strongly express clock genes [50].

Although cementum is laid down centrifugally from the cementum-dentin junction and is marked by incremental lines, close to nothing is known regarding the circadian control of cementum formation.

Elucidating the role of circadian control on dental mineralized tissue formation may help in understanding the phenotypic differences in tooth structure and form among individuals. Altered expression or polymorphisms of clock genes may also be related to dental disease predisposition, as it has shown for other pathological situations such as diabetes and cancer.

STEM CELLS DURING TOOTH REPAIR

Stem cells play a critical role in tissue homeostasis and repair. Their fate is regulated by cell intrinsic determinants and signals from a specialized microenvironment. The reparative mechanisms following dental injury involve a series of highly conserved processes that share genetic programs that occur throughout embryogenesis (reviewed in Ref. 51). In a severe injury, the dying odontoblasts are replaced by stem/progenitor cells, which differentiate into a new generation of odontoblasts that produce the reparative dentin [51]. Signaling molecules released at the injury site may attract dental pulp stem cells and thus initiate the healing process. Notch molecules and nestin are involved in the dynamic processes triggered by pulp injury [51]. Notch expression is activated in cells close to the injury site, as well as in cells located at the root apex, suggesting that these sites represent stem cell niches within the dental pulp. Activation of the Notch molecules in endothelial cells after injury may reflect another pool of stem cells. Epithelial stem cells also exist in human teeth in the root area (Athanassiou et al., unpublished data) and may be implicated in the regeneration of cementum and PDL tissues [52].

CONCLUSION

Elucidating the controls of tooth initiation, pattern, and mineralization necessitates a thorough understanding at the cellular and molecular level. Understanding when and how signaling molecules control these events will open new horizons and create new challenges. Although much has been discovered on the signaling pathways that dictate tooth morphogenesis, very little is known on the control of dental cells differentiation and subsequent matrix formation. Multiscale mathematical modeling of complex pathway interactions during tooth morphogenesis and differentiaion may result in a better understanding of development and diseases [48]. Novel scientific knowledge together with tissue engineering approaches are likely to support development of novel therapies in dentistry.

ACKNOWLEDGMENTS

We would like to thank all the members of the Laboratory of Dental Research at the University of Michigan and in particular Drs. Hu, Simmer, and Yamakoshi for scientific interactions, discussions, and material shared. The research presented here is partially supported by the NIH grant DE018878-01A1 to Petros Papagerakis.

REFERENCES

1. Mitsiadis TA, Graf D. 2009. Cell fate determination during tooth development and regeneration. *Birth Defects Res C Embryo Today* 87: 199–211.
2. Bluteau G, Luder HU, De Bari C, Mitsiadis TA. 2008. Stem cells for tooth engineering. *Eur Cell Mater* 16: 1–9.
3. Vainio S, Karavanova I, Jowett A, Thesleff I. 1993. Identification of BMP-4 as a signal mediating secondary induction between epithelial and mesenchymal tissues during early tooth development. *Cell* 75: 45–58.
4. Dassule HR, McMahon AP. 1998. Analysis of epithelial-mesenchymal interactions in the initial morphogenesis of the mammalian tooth. *Dev Biol* 202: 215–227.
5. Khan M, Seppala M, Zoupa M, Cobourne MT. 2007. Hedgehog pathway gene expression during early development of the molar tooth root in the mouse. *Gene Expr Patterns* 7: 239–243.
6. Bei M. 2009. Moleculer genetics of ameloblast cell lineage. *J Exp Zool B Mol Dev Evol* 312B: 437–444.
7. Mucchielli ML, Mitsiadis TA, Raffo S, Brunet JF, Proust JP, Goridis C. 1997. Mouse Otlx2/RIEG expression in the odontogenic epithelium precedes tooth initiation and requires mesenchyme-derived signals for its maintenance. *Dev Biol* 189(2): 275–284.
8. Trainor PA, Krumlauf R. 2000. Patterning the cranial neural crest: Hindbrain segmentation and Hox gene plasticity. *Nat Rev Neurosci* 1: 116–124.
9. Mikkola ML. 2008. TNF superfamily in skin appendage development. *Cytokine Growth Factor Rev* 19: 219–230.
10. Zhang J, Han D, Song S, Wang Y, Zhao H, Pan S, Bai B, Feng H. 2011. Correlation between the phenotypes and genotypes of X-linked hypohidrotic ectodermal dysplasia and non-syndromic hypodontia caused by ectodysplasin-A mutations. *Eur J Med Genet* 54: e377–382.
11. Mitsiadis TA, Hirsinger E, Lendahl U, Goridis C. 1998. Delta-notch signaling in odontogenesis: Correlation with cytodifferentiation and evidence for feedback regulation. *Dev Biol* 204: 420–431.
12. Mitsiadis TA, Graf D, Luder H, Gridley T, Bluteau G. 2010. BMPs and FGFs target Notch signalling via jagged 2 to regulate tooth morphogenesis and cytodifferentiation. *Development* 137: 3025–3035.
13. Penton A, Leonard L, Spinner N. 2012. Notch signaling in human development and disease. *Semin Cell Dev Biol* 23(4): 450–457.
14. Caton J, Luder HU, Zoupa M, Bradman M, Bluteau G, Tucker AS, Klein O, Mitsiadis TA. 2009. Enamel-free teeth: Tbx1 deletion affects amelogenesis in rodent incisors. *Dev Biol* 328: 493–505.
15. Toka O, Kari M, Dittrich S, Holst A. 2010. Dental aspects in patients with DiGeorge syndrome. *Quintessence Int* 41: 551–556.
16. Reith EJ. 1967. The early stage of amelogenesis as observed in molar teeth of young rats. *J Ultrastruct Res* 17: 503–526.
17. Inai T, Kukita T, Ohsaki Y, Nagata K, Kukita A, Kurisu K. 1991. Immunohistochemical demonstration of amelogenin penetration toward the dental pulp in the early stages of ameloblast development in rat molar tooth germs. *Anat Rec* 229: 259–270.
18. Sasaki T, Garant PR. 1996. Structure and organization of odontoblasts. *Anat Rec* 245: 235–249.
19. Yamakoshi Y, Hu JC, Fukae M, Zhang H, Simmer JP. 2005. Dentin glycoprotein: The protein in the middle of the dentin sialophosphoprotein chimera. *J Biol Chem* 280: 17472–17479.
20. Butler WT, Ritchie H. 1995. The nature and functional significance of dentin extracellular matrix proteins. *Int J Dev Biol* 39: 169–179.
21. Bailleul-Forestier I, Davideau JL, Papagerakis P, Noble I, Nessmann C, Peuchmaur M, Berdal A. 1996. Immunolocalization of vitamin D receptor and calbindin-D28k in human tooth germ. *Pediatr Res* 39: 636–642.
22. Borke JL, Zaki AE, Eisenmann DR, Ashrafi SM, Ashrafi SS, Penniston JT. 1993. Expression of plasma membrane Ca pump epitopes parallels the progression of mineralization in rat incisor. *J Histochem Cytochem* 41: 175–181.
23. Goseki M, Oida S, Nifuji A, Sasaki S. 1990. Properties of alkaline phosphatase of the human dental pulp. *J Dent Res* 69: 909–912.
24. Smith AJ, Cassidy N, Perry H, Begue-Kirn C, Ruch JV, Lesot H. 1995. Reactionary dentinogenesis. *Int J Dev Biol* 39: 273–280.
25. Mitsiadis TA, De Bari C, About I. 2008. Apoptosis in developmental and repair-related human tooth remodeling: A view from the inside. *Exp Cell Res* 314: 869–877.
26. Simmer JP, Fincham AG. 1995. Molecular mechanisms of dental enamel formation. *Crit Rev Oral Biol Med* 6: 84–108.
27. Daculsi G, Menanteau J, Kerebel LM, Mitre D. 1984. Length and shape of enamel crystals. *Calcif Tissue Int* 36: 550–555.
28. Nanci A. 2003. Enamel: composition, formation, and structure. In: Nanci A (ed.) *Ten Cate's Oral Histology Development, Structure, and Function*. St. Louis, MO: Mosby. pp. 145–191.
29. Smith CE, Nanci A. 1995. Overview of morphological changes in enamel organ cells associated with major events in amelogenesis. *Int J Dev Biol* 39: 153–161.
30. Hu JC, Chun YH, Al Hazzazzi T, Simmer JP. 2007. Enamel formation and amelogenesis imperfecta. *Cells Tissues Organs* 186: 78–85.

31. Simmer JP, Papagerakis P, Smith CE, Fisher DC, Rountrey AN, Zheng L, Hu JC. 2010. Regulation of dental enamel shape and hardness. *J Dent Res* 89: 1024–1038.
32. Robinson C, Brookes SJ, Shore RC, Kirkham J. 1998. The developing enamel matrix: Nature and function. *Eur J Oral Sci* 106(Suppl 1): 282–291.
33. Fincham AG, Moradian-Oldak J, Diekwisch TG, Lyaruu DM, Wright JT, Bringas P Jr, Slavkin HC. 1995. Evidence for amelogenin "nanospheres" as functional components of secretory-stage enamel matrix. *J Struc Biol* 115: 50–59.
34. Fukumoto S, Kiba T, Hall B, Iehara N, Nakamura T, Longenecker G, Krebsbach PH, Nanci A, Kulkarni AB, Yamada Y. 2004. Ameloblastin is a cell adhesion molecule required for maintaining the differentiation state of ameloblasts. *J Cell Biol* 167: 973–983.
35. Hu JC, Hu Y, Smith CE, McKee MD, Wright JT, Yamakoshi Y, Papagerakis P, Hunter GK, Feng JQ, Yamakoshi F, Simmer JP. 2008. Enamel defects and ameloblast-specific expression in enamelin knockout/LACZ knockin mice. *J Biol Chem* 283: 10858–10871.
36. Smith CE. 1989. Cellular and chemical events during enamel maturation. *Crit Rev Oral Biol Med* 9: 128–161.
37. Lu Y, Papagerakis P, Yamakoshi Y, Hu J, Bartlett J, Simmer JP. 2008. Functions of KLK4 and MMP-20 in dental enamel formation. *Biol Chem* 389: 695–700.
38. Al Kawas S, Warshawsky H. 2008. Ultrastructure and composition of basement membrane separating mature ameloblasts from enamel. *Arch Oral Biol* 53: 310–317.
39. Moffatt P, Smith CE, St-Arnaud R, Nanci A. 2008. Characterization of Apin, a secreted protein highly expressed in tooth-associated epithelia. *J Cell Biochem* 103: 941–956.
40. Davideau JL, Papagerakis P, Hotton D, Lezot F, Berdal A. 1996. In situ investigation of vitamin D receptor, alkaline phosphatase, and osteocalcine gene expression in oro-facial mineralized tissues. *Endocrinology* 137: 3577–3585.
41. Papagerakis P, MacDougall M, Bailleul-Forestier I, Oboeuf M, Berdal A. 2003. Expression of amelogenin in odontoblasts. *Bone* 32: 228–240.
42. Papagerakis P, Berdal A, Mesbah M, Peuchmaur M, Malaval L, Nydegger Y, Simmer J, MacDougall M. 2002. Investigation of osteocalcin, osteonectin, and dentin sialophosphoprotein in developing human teeth. *Bone* 30: 377–385.
43. MacNeil RL, Somerman MJ. 1993. Molecular factors regulating development and regeneration of cementum. *J Periodontal Res* 28: 550–559.
44. Huang X, Bringas P Jr, Slavkin HC, Chai Y. 2009. Fate of HERS during tooth root development. *Dev Biol* 334: 22–30.
45. McCulloch CA, Lekic P, McKee MD. 2000. Role of physical forces in regulating the form and function of the periodontal ligament. *Periodontol 2000* 24: 56–72.
46. Matsubara T, Suardita K, Ishii M, et al. 2005. Alveolar bone marrow as a cell source for regenerative medicine: Differences between alveolar and iliac bone marrow stromal cells. *J Bone Miner Res* 20: 399–409.
47. Ohtsuka M, Saeki S, Igarashi K, Shinoda H. 1998. Circadian rhythms in the incorporation and secretion of 3H-proline by odontoblasts in relation to incremental lines in rat dentin. *J Dent Res* 77: 1889–1895.
48. Athanassiou-Papaefthymiou, M, Kim, D, Harbron, L, Papagerakis, S, Schnell, S, Harada, H, Papagerakis P. 2011. Molecular and circadian controls of ameloblasts. *Eur J Oral Sci* 119 Suppl 1: 35–40.
49. Zheng L, Papagerakis S, Schnell SD, Hoogerwerf WA, Papagerakis P. 2011. Expression of clock proteins in developing tooth. *Gene Expr Patterns* 11: 202–206.
50. Zheng L, Seon YJ, Mourão MA, Schnell S, Kim D, Harada H, Papagerakis S, Papagerakis P. 2013. Circadian rhythms regulate amelogenesis. *Bone* (in press).
51. Mitsiadis TA, Rahiotis C. 2004. Parallels between tooth development and repair: Conserved molecular mechanisms following carious and dental injury. *J Dent Res* 83: 896–902.
52. Mitsiadis TA, Papagerakis P. 2011. Regenerated teeth: The future of tooth replacement? *Regen Med* 6: 135–139.

110

Craniofacial Disorders Affecting the Dentition: Genetic

Yong-Hee P. Chun, Paul H. Krebsbach, and James P. Simmer

Genetic Disorders Affecting the Dentition 914
Oral Manifestations of Metabolic Bone Diseases of Genetic Origin 918

References 919

GENETIC DISORDERS AFFECTING THE DENTITION

The skeleton contains two of the five mineralized tissues in the body: bone and calcified cartilage. The other three mineralized tissues—dentin, enamel, and cementum—are found in teeth. The mineral in each of these hard tissues is a biological apatite resembling calcium hydroxyapatite $Ca_{10}(PO_4)_6(OH)_2$ in structure, with the most common substitutions being carbonate (CO_3^{2-}) for phosphate (PO_4^{3-}), and fluoride (F^-) for hydroxyl (OH^-). Therefore, disorders involving the regulation of calcium and phosphate metabolism potentially affect multiple hard tissues. In every mineralizing tissue, biomineralization occurs in a defined extracellular space. Establishing these extracellular mineralizing environments involves the synthesis and secretion of extracellular matrix proteins, the transport of ions, and the regulated deposition of mineral. While each mineralizing tissue is in many ways unique, there are common elements that, when defective, lead to pathologies in multiple hard tissues. Changes in the dentition and its supporting oral structures may occur in response to disorders of mineral metabolism. The clinical presentations in these disorders may vary from mild asymptomatic changes to alterations that severely alter the form and function of craniofacial structures. In some cases, the oral phenotype may be the earliest or most obvious sign of a broader syndrome involving bone and mineral metabolism and lead to the original diagnosis. This chapter provides a concise overview of the dental manifestations of selected disorders of bone and mineral metabolism (Table 110.1).

Genetic diseases affecting the number of teeth: Familial tooth agenesis and supernumerary teeth

The initiation of tooth development depends on interactions between oral epithelium and mesenchyme derived from cranial neural crest cells [1]. The initiation step in tooth development is formation of a dental placode, or local thickening of the oral epithelium. Similar ectodermal placodes initiate the development of hairs and nails, as well as the mammary, salivary, sweat, and sebaceous glands [2]. The principal molecular participants in the epithelial-mesenchymal interactions that initiate formation of ectodermal organs are signaling molecules, their receptors, and transcription factors. Some of the specific regulatory molecules involved contribute to the formation of multiple ectodermal organs. Genetic diseases affecting early events in tooth formation typically are manifested as misshapen teeth and by alterations in tooth number, as in supernumerary teeth or familial tooth agenesis. These dental phenotypes may occur in isolation or be associated with other developmental anomalies.

Hypohidrotic ectodermal dysplasia (HED) is an inherited condition featuring missing teeth, thin and sparse hair, missing sweat glands, and defective nails and salivary glands. HED is caused by mutations in ectodysplasin (*EDA*, Xq12-q13.1). In mild cases, hypodontia is the

---

Primer on the Metabolic Bone Diseases and Disorders of Mineral Metabolism, Eighth Edition. Edited by Clifford J. Rosen.
© 2013 American Society for Bone and Mineral Research. Published 2013 by John Wiley & Sons, Inc.

Table 110.1. Inherited Conditions Affecting the Dentition

Familial Tooth Agenesis	
Isolated (Oligodontia)	*AXIN2* (17q24); *MSX1* (4p16.1); *PAX9* (14q12)
Syndromic	
Hypochidrotic ectodermal dysplasia (HED)	*EDA* (Xq12)
Aplasia of the lacrimal and salivary glands (ALSG)	*FGF10* (5p13)
Lacrimo-auriculo-dento-digital syndrome (LADD)	*FGFR3* (4p16.3); *FGFR3* (4p16.3)
Supernumerary Teeth	
Adenomatous polyposis of the colon (APC)	*APC* (5q21)
Cleidocranial dysplasia (CCD)	*RUNX2* (6p21)
Inherited Dentin & Enamel Defects	
Dentinogenesis imperfecta (DGI) types II and III	*DSPP* (4q21.3)
Dentin dysplasia (DD) type II	*DSPP* (4q21.3)
Amelogenesis imperfecta (AI)	
X-linked AI	*AMELX* (Xp22.3)
Autosomal-dominant AI (ADAI)	*ENAM* (4q13.3)
Autosomal-dominant hypocalcified AI (ADRCAI)	*FAM83H* (8q24.3)
Autosomal-recessive AI (ARAI)	*MMP20* (11q22.3); *KLK4* (19q13.41); *WDR72* (15q21.3)
Mucopolysaccharidosis Type IVA	*GALNS* (16q24.3)
Cherubism	*SH3BP2* (4p16.3)
Vitamin D-Dependent Rickets Type I	*CYP27B1* (12q13.3)
Hypophosphatasia	*ADHR* (12p13), *PHEX* (Xp22.2); *DMP1* (4q21)
Hypophosphatasia	*ALPL* (1p36.1)

only phenotype [3]. Several related syndromes are caused by mutations in *FGF10* (5p13) and its receptors (i.e., *FGFR3*, 4p16.3), such as aplasia of the lacrimal and salivary glands (ALSG) [4] and lacrimo-auriculo-dento-digital syndrome (LADD) [5]. LADD features aplasia or hypoplasia of the lacrimal and salivary systems associated with a variety of dental phenotypes: agenesis of the lateral maxillary incisors (hypodontia), small (microdontia) and peg-shaped laterals, mild enamel dysplasia, and delayed tooth eruption.

Especially noteworthy among the genetic disturbances affecting early tooth formation are those that disrupt the Wnt signaling system. Wnt signaling is processed by a cytosolic complex of proteins consisting of axis inhibition protein 2 (AXIN2), glycogen synthase kinase-3β (GSK-3ß) and adenomatous polyposis coli (APC), which dephosphorylate and stabilize β-catenin. β-catenin translocates into the nucleus and regulates the activity of key transcription factors. Mutations in *APC* (5q21-q22) cause adenomatous polyposis of the colon, a syndrome featuring radiopaque lesions in the jaw, that are composed of clumped toothlets (odontomas) and gastrointestinal polyps that usually undergo malignant change by the fourth decade [6]. *AXIN2* (17q24) mutations cause a severe form of familial tooth agenesis also associated with gastrointestinal polyps that turn malignant by the fourth decade [7]. *APC* and *AXIN2* phenotypes are inherited in an autosomal dominant pattern. The discovery of odontomas or familial tooth agenesis on panorex radiographs should raise concerns about gastrointestinal polyps, especially when the oral phenotype appears to have arisen spontaneously (not observed in either of the parents), as the family will not show a history of intestinal cancer.

Familial tooth agenesis occurs in the absence of other phenotypic features [8]. Msh homeobox 1 (*MSX1*, 4p16.1) and paired box 9 (*PAX9*, 14q12-q13) express interacting transcription factors that are critical for the progression of tooth development beyond the bud stage [9, 10]. Mutations in *MSX1* and *PAX9* cause similar patterns of familial tooth agenesis, with *PAX9* mutations being more likely to include second molars, while *MSX1* mutations are more likely to include maxillary first bicuspids among the teeth missing [11].

Cleidocranial dysplasia is an autosomal dominant condition featuring skeletal and dental anomalies, caused by mutations in the transcription factor gene *RUNX2* (6p21). The most notable feature is hypoplasia or aplasia of the clavicles, permitting abnormal apposing of the shoulders. Exfoliation of the primary dentition is delayed or fails, and multiple supernumerary teeth, which typically fail to erupt, are observed on dental radiographs [12].

Genetic diseases affecting dentin: Osteogenesis imperfecta, dentinogenesis imperfecta, and dentin dysplasia (DD)

Defects in either the alpha 1 chains or alpha 2 chain of type I collagen can cause osteogenesis imperfecta or OI (see Chapter 99). OI is associated with assorted dentin defects that are collectively designated as dentinogenesis imperfecta (DGI) [13]. In rare cases, the dentin defects are the only prominent phenotype in OI [14]. It has been reported that 10–50% of patients afflicted with OI also have dentinogenesis imperfecta (DGI). This assessment, however, may underestimate the true prevalence, since mild forms of DGI may require microscopic analysis for diagnosis [15]. The range of dental defects observed in OI are similar to those observed in kindreds with DGI and dentin dysplasia (DD).

The most abundant non-collagenous proteins in dentin are proteolytic cleavage products of a large chimeric protein known as dentin sialophosphoprotein (DSPP). The N-terminal cleavage product is dentin sialoprotein (DSP), a proteoglycan with both N- and O-linked glycosylations and two glycosaminoglycan attachments [16, 17]. The C-terminal cleavage product is dentin phosphoprotein (DPP), a highly phosphorylated protein with the lowest (most acidic) isoelectric point (approximately 1) of any known protein [18]. DPP is thought to participate in the nucleation of hydroxyapatite crystallites on collagen [19]. To date, 35 different *DSPP* (4q21.3) mutations have been shown to cause DD-type II, DGI-type II, and DGI-type III [20]. The dentin defects appear to result from pathology caused by the mutant protein, rather than from a loss of DSPP function [21].

The clinical classification system most often used to categorize inherited defects of dentin was established over 30 years ago, prior to recent discoveries concerning their genetic etiologies, and divided the phenotypes into two disease groups with five subtypes: dentinogenesis imperfecta (DGI, types I–III) and dentin dysplasia (DD, types I and II), with all forms showing an autosomal dominant pattern of inheritance [22]. Type I DGI is a collective designation for OI with DGI and has largely been abandoned in deference to the current OI classification system. An alternative designation for isolated inherited dentin defects is hereditary opalescent dentin [23]. Type II DGI is the most prevalent inherited dentin phenotype. Clinically, the teeth of individuals with DGI are characterized by an amber-like appearance (Fig. 110.1). The teeth are narrower at the cervical margins and thus exhibit a bulbous or bell-shaped crown. Microscopic anomalies of affected dentin include fewer and irregular dentin tubules containing vesicles and abnormally thick collagen fibers [24]. The mineral content of DGI teeth is reduced, being about 30% less than normal dentin. Intrafibrillar collagen mineralization is absent. The structurally abnormal dentin may not provide adequate support for the overlying enamel. The enamel is usually chemically and structurally normal in individuals with DGI and a lack of support from dentin leads to fracturing and severe attrition of the teeth, the distinguishing clinical feature of type II DGI. Type III DGI is a rare form that is also known as the "Brandywine isolate," after the prototype kindred identified in Brandywine, Maryland. This form features multiple pulp exposures in the deciduous teeth, which show considerable variation radiographically, ranging from shell teeth, to normal pulp chambers, to pulpal obliteration. The permanent teeth are the same as in type II DGI. In type I dentin dysplasia both the permanent and deciduous teeth appear to have a normal shape and color clinically. Dental radiographs, however, show the teeth have short roots with periapical radiolucencies in non-carious teeth (Fig. 110.2). The primary teeth show total obliteration of the pulp. Type II dentin dysplasia appears to be a mild form of type II DGI, featuring amber tooth coloration with total pulpal obliteration in the primary teeth and a thistle-tube pulp configuration with ubiquitous pulp stones and normal to near normal coloration in the permanent teeth [22, 25, 26].

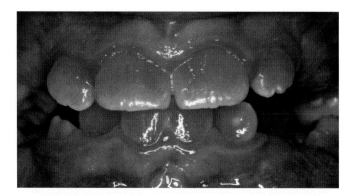

Fig. 110.1. Dental manifestations of teeth from a patient with dentinogenesis imperfecta. The permanent teeth of this patient exhibit the characteristic blue-gray or opalescent appearance associated with dentinogenesis imperfecta.

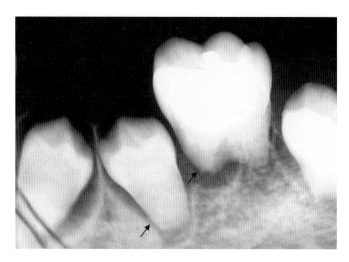

Fig. 110.2. Radiograph of teeth from a patient with dentin dysplasia type I. The roots are abnormally short or absent (arrows) and the pulp chamber is obliterated (courtesy of Dr. Sharon Brooks).

Genetic diseases affecting enamel: Amelogenesis imperfecta

Amelogenesis imperfecta (AI) is a heterogeneous group of isolated inherited defects in the enamel layer of teeth [27]. The enamel may be thin, soft, rough, and/or pigmented. When the various enamel phenotypes and the pattern of inheritance are considered, 14 subtypes are recognized [28]. Enamel formation is regulated by a toolbox of specialized extracellular matrix molecules. Amelogenin (*AMELX*, Xp22.3) mutations cause X-linked AI, with females showing vertical bands of normal and defective enamel, while males usually have little or no enamel at all [29]. Enamelin (*ENAM*, 4q13.3) mutations cause autosomal dominant AI, with distinctive horizontal bands often evident in the cervical third of the crown [30]. Truncation mutations in the last exon of *FAM83H* (family with sequence similarity 83 member H; 8q24.3) cause autosomal-dominant hypocalcified AI (ADHCAI) [31]. Mutations in WD (tryptophan-aspartate, a two-amino-acid sequence that is commonly found in the conserved domain of this family of proteins) repeat-containing protein 72 (*WDR72*; 15q21.3) and the genes that encode the proteolytic enzymes enamelysin (*MMP20*, 11q22.3) and kallikrein 4 (*KLK4*, 19q13.41) cause autosomal recessive AI, featuring relatively soft, pigmented enamel, usually of normal thickness, that may tend to chip off during function [32–34]. AI has also been used to describe enamel defects associated with inherited syndromes. There are over 70 such conditions. Recently it has been learned that recessive mutations in the metal ion transporter cyclin M4 (*CNNM4*; 2q11) cause cone-rod dystrophy and AI [35, 36] and mutations in family with sequence similarity 20 member A (*FAM20A*; 17q24.2) cause AI and gingival fibromatosis syndrome [37].

Mucopolysaccharidoses

Lysosomal storage disorders comprise over 50 inherited diseases that are caused primarily by defects in genes encoding lysosomal enzymes [38]. Among the lysosomal storage disorders are the mucopolysaccharidoses (MPS), which are characterized by the accumulation of partially degraded glycosaminoglycans (previously called mucopolysaccharides) within lysosomes, as well as in the urine. There are 10 enzymes involved in the stepwise degradation of glycosaminoglycans, and deficiencies in these activities give rise to the MPS. There are seven MPS types: I, II, III, IV, VI, VII, and IX (types V and VIII have been retired, and type IX is extremely rare). Although heterogeneous, several craniofacial characteristics are similar between the different types. The oral manifestations may include a short and broad mandible with abnormal condylar development and limited temporomandibular joint function. The teeth are often peg-shaped and exhibit increased interdental spacing, perhaps due to the frequently observed gingival hyperplasia and macroglossia. Some forms of MPS have abnormally thin enamel covering the clinical crowns or radiographic evidence of cystic lesions surrounding the molar teeth that contain excessive dermatan sulfate and collagen [39–42].

Mucopolysaccharidosis type IVA (Morquio A syndrome, MPS IVA) is an autosomal recessive disorder caused by deficiency of the lysosomal hydrolase, N-acetylgalactosamine 6-sulfatase (GALNS), encoded by a gene on human chromosome 16q24.3 [43]. Mucopolysaccharidosis type IVA is the only MPS associated with dental enamel malformations [44], although mucopolysaccharides accumulate in the developing teeth in other MPS syndromes, such as Hurler (MPS I) [45], Hunter (MPS II), and Maroteaux-Lamy (MPS VI) syndromes. In MPS IVA, enamel malformations are a consistent feature. The enamel is dull gray in color, thin, pitted, and tends to flake off from the underlying dentin. The thin enamel layer is of normal hardness and radiodensity. MPS IVA patients often show severe bone dysplasia and dwarfism. The identification of the genes involved and the invention of new molecular biology tools are leading to the development of specific enzyme replacement therapies (ERTs) for some of the mucopolysaccharidoses [46].

Cleft lip and palate

Cleft lip and/or palate are relatively common craniofacial malformations (1 in 700 births) that can have profound impact on nutritional, speech, dental, and psychological development. The causes of these disorders are complex and are known to involve both genetic and environmental factors. Most of these facial clefting birth defects are multifactoral and non-syndromic. Although genetics may play a role in the cause of cleft lip and/or palate, they are not associated with well-defined syndromes. Between 15% and 50% of cleft lips and/or palates are associated with defined syndromes. In fact, there are nearly 300 recognized syndromes that may include a facial cleft as a manifestation. However, only about 10 genes have been identified as associated with these syndromes. Common syndromes with cleft palates include Apert's, Stickler's, and Treacher Collins. Van der Woude's and Waardenberg's syndromes are associated with cleft lip with or without cleft palate. In at least one condition—Van der Woude's syndrome—a haplotype gene test may be used to identify variants in the interferon regulatory factor 6 gene and provide a correlation with an increased risk of facial clefting [47].

Cherubism

Cherubism is an autosomal dominant condition caused by mutations in *SH3BP2* (4p16.3) [48]. SHF3BP2 expresses a protein that binds to c-Abl, a tyrosine kinase involved in diverse cell signaling cascades. In patients with cherubism, multiple cystic giant cell lesions of the jaw appear typically between 2 and 5 years of age. The cysts replace bone and cause enlargement of the maxilla and mandible, which stabilizes or remits after puberty. Teeth are displaced during development, with root resorption, tooth agenesis, retention of deciduous teeth, ectopic eruption, and malocclusion observed as sequelae [49].

ORAL MANIFESTATIONS OF METABOLIC BONE DISEASES OF GENETIC ORIGIN

Metabolic diseases of bone are disorders of bone remodeling that characteristically involve the entire skeleton and are often manifest in the oral cavity, which can lead to the diagnosis of the underlying systemic disease. Numerous studies suggest that subclinical derangements in calcium homeostasis and bone metabolism may also contribute to a variety of dental abnormalities including alveolar ridge resorption and periodontal bone loss in predisposed individuals. The significance of this spectrum of diseases and their overall impact on oral health and dental management are likely to increase as the elderly segment of the population increases in the coming decades [50].

Vitamin D deficiency

In vitamin D-resistant rickets (see Chapter 75), the primary oral abnormality is similar to dentin dysplasia. Enamel is usually reported to be normal, but in some instances may be hypoplastic. Patients also suffer from delayed tooth eruption, and radiographically, teeth often display enlarged pulp chambers. Other salient radiographic findings include decreased alveolar bone density, thinning of bone trabeculae, loss of lamina dura, and retarded tooth calcification [51].

Vitamin D-dependent rickets type I is an autosomal recessive defect in vitamin D metabolism caused by mutations in the *CYP27B1* gene (12q13.3-q14) encoding 25-hydroxyvitamin D-1-alpha-hydroxylase. Decreased 1,25(OH)$_2$D results in teeth with yellow-brown color, pitted enamel, short roots, and a tendency to develop chronic periodontal disease [52].

Hypophosphatemia

Phosphorus homeostasis is incompletely understood and regulated by the actions of many hormones, including parathyroid hormone (PTH), calcitonin, and Vitamin D. Recently, however, a new set of important phosphate-regulating factors have been discovered. Fibroblast growth factor 23 (FGF23) is secreted by osteocytes into the circulation and inhibits phosphate reabsorption and 1,25(OH)$_2$D production by the kidney. Specific defects in the FGF23 gene (*ADHR*, 12p13) cause autosomal dominant hypophosphatemic rickets, although deletion of the gene encoding FGF23 in mice causes hyperphosphatemia. PHEX, an endopeptidase, and dentin matrix protein 1 (DMP1), a proteoglycan involved in bone mineralization, are synthesized by osteocytes and both inhibit FGF23 expression. Mutations in *PHEX* (Xp22.2) and *DMP1* (4q21) cause X-linked and autosomal recessive hypophosphatemia, respectively [53–55]. It is increasingly clear that autosomal dominant, autosomal recessive, and X-linked hypophosphatemia involve various aspects of a common pathophysiologic mechanism that reduces phosphate reabsorption by the renal tubuli and lowers 1,25(OH)$_2$D levels. This leads to chronic hyperphosphaturia and hypophosphatemia, and causes rickets in children and osteomalacia in adults [56].

In familial hypophosphatemia (see Chapter 74), dental findings are often the first clinically noticeable signs of the disease and resemble those seen in rickets and osteomalacia. Patients may present with abscessed primary or permanent teeth that have no signs of dental caries [57]. Although the enamel is reported to be normal, microbial infection of the pulp is thought to occur through invasion of dentinal tubules exposed by attrition of enamel or through enamel microfractures [58]. A better understanding of the mechanisms of phosphate homeostasis is providing new insights into the genetic etiologies of hypophosphatemias [59].

Hypophosphatasia

Hypophosphatasia is an inherited disorder caused by a defect in the alkaline phosphatase (*ALPL*, 1p36.1-p34) gene. Osteoblasts show the highest level of ALPL expression, and profound skeletal hypomineralization occurs in the severest forms of hypophosphatasia. The hard tissue that appears to be most sensitive to an *ALPL* defect is cementum [60]. The classic oral presentation of childhood hypophosphatasia is the premature loss of fully rooted primary teeth (Fig. 110.3). Histological examination indicates that these teeth lack cementum on their root surface, so that the attachment apparatus fails to develop properly. The periodontal ligament fibers do not connect the alveolar bone to the root and the teeth exfoliate prematurely. In the permanent teeth, large pulp spaces, late eruption and delayed apical closure are often observed. Bone loss is primarily horizontal, and in the adult form of the disease there may be widespread dental caries.

The URL for data presented herein is as follows:
Online Mendelian Inheritance in Man (OMIM), http://www.ncbi.nlm.nih.gov/Omim/

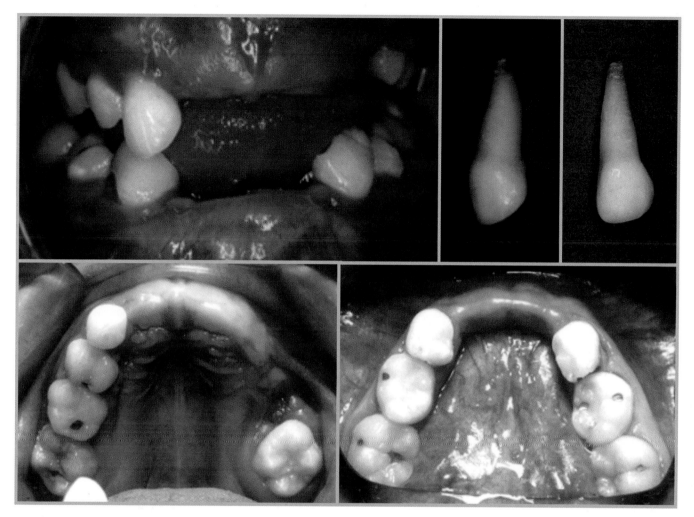

Fig. 110.3. Exfoliation of fully rooted primary teeth in hypophosphatasia. Oral photographs show the dental condition of a patient with childhood hypophosphatasia at age 6. The maxillary cuspid and incisor (upper right) were brought by the parents to the visit and had fallen out naturally about 2 years previously. This patient also showed periodontal attachment problems in her primary posterior teeth, which were mobile. Some childhood hypophosphatasia patients present with enamel hypoplasia. These dental findings are diagnostic of childhood hypophosphatasia and are often the first symptoms leading to the diagnosis (contributed by Dr. Jan C-C Hu).

REFERENCES

1. Chai Y, Jiang X, Ito Y, Bringas P Jr, Han J, Rowitch D, Soriano P, McMahon A, Sucov H. 2000. Fate of the mammalian cranial neural crest during tooth and mandibular morphogenesis. *Development* 127: 1671–9.
2. Thesleff I. 2006. The genetic basis of tooth development and dental defects. *Am J Med Genet A* 140(23): 2530–5.
3. Han D, Gong Y, Wu H, Zhang X, Yan M, Wang X, Qu H, Feng H, Song S. 2008. Novel EDA mutation resulting in X-linked non-syndromic hypodontia and the pattern of EDA-associated isolated tooth agenesis. *Eur J Med Genet* 51(6): 536–46.
4. Entesarian M, Matsson H, Klar J, Bergendal B, Olson L, Arakaki R, Hayashi Y, Ohuchi H, Falahat B, Bolstad AI, Jonsson R, Wahren-Herlenius M, Dahl N. 2005. Mutations in the gene encoding fibroblast growth factor 10 are associated with aplasia of lacrimal and salivary glands. *Nat Genet* 37(2):125–7.
5. Milunsky JM, Zhao G, Maher TA, Colby R, Everman DB. 2006. LADD syndrome is caused by FGF10 mutations. *Clin Genet* 69(4): 349–54.
6. Oner AY, Pocan S. 2006. Gardner's syndrome: A case report. *Br Dent J* 200(12): 666–7.
7. Lammi L, Arte S, Somer M, Jarvinen H, Lahermo P, Thesleff I, Pirinen S, Nieminen P. 2004. Mutations in AXIN2 cause familial tooth agenesis and predispose to colorectal cancer. *Am J Hum Genet* 74(5): 1043–50.
8. Nieminen P. 2009. Genetic basis of tooth agenesis. *J Exp Zool B Mol Dev Evol* 312B(4): 320–42.
9. Stockton DW, Das P, Goldenberg M, D'Souza RN, Patel PI. 2000. Mutation of PAX9 is associated with oligodontia. *Nat Genet* 24(1): 18–9.

10. Vastardis H, Karimbux N, Guthua SW, Seidman JG, Seidman CE. 1996. A human MSX1 homeodomain missense mutation causes selective tooth agenesis. *Nat Genet* 13(4): 417–21.
11. Kim JW, Simmer JP, Lin BP, Hu JC. 2006. Novel MSX1 frameshift causes autosomal-dominant oligodontia. *J Dent Res* 85(3): 267–71.
12. Cooper SC, Flaitz CM, Johnston DA, Lee B, Hecht JT. 2001. A natural history of cleidocranial dysplasia. *Am J Med Genet* 104(1): 1–6.
13. O'Connell AC, Marini JC. 1999. Evaluation of oral problems in an osteogenesis imperfecta population. *Oral Surg Oral Med Oral Path Oral Radiol Endod* 87(2): 189–96.
14. Pallos D, Hart PS, Cortelli JR, Vian S, Wright JT, Korkko J, Brunoni D, Hart TC. 2001. Novel COL1A1 mutation (G559C) [correction of G599C] associated with mild osteogenesis imperfecta and dentinogenesis imperfecta. *Arch Oral Biol* 46(5): 459–70.
15. Waltimo J, Ojanotko-Harri A, Lukinmaa PL. 1996. Mild forms of dentinogenesis imperfecta in association with osteogenesis imperfecta as characterized by light and transmission electron microscopy. *J Oral Pathol Med* 25(5): 256–64.
16. Yamakoshi Y, Hu JC, Fukae M, Iwata T, Kim JW, Zhang H, Simmer JP. 2005. Porcine dentin sialoprotein is a proteoglycan with glycosaminoglycan chains containing chondroitin 6-sulfate. *J Biol Chem* 280(2): 1552–60.
17. Yamakoshi Y, Nagano T, Hu JC, Yamakoshi F, Simmer JP. 2011. Porcine dentin sialoprotein glycosylation and glycosaminoglycan attachments. *BMC Biochem* 12(1): 1–6.
18. Jonsson M, Fredriksson S, Jontell M, Linde A. 1978. Isoelectric focusing of the phosphoprotein of rat-incisor dentin in ampholine and acid pH gradients. Evidence for carrier ampholyte-protein complexes. *J Chromatogr* 157: 234–42.
19. George A, Bannon L, Sabsay B, Dillon JW, Malone J, Veis A, Jenkins NA, Gilbert DJ, Copeland NG. 1996. The carboxyl-terminal domain of phosphophoryn contains unique extended triplet amino acid repeat sequences forming ordered carboxyl-phosphate interaction ridges that may be essential in the biomineralization process. *J Biol Chem* 271(51): 32869–73.
20. Nieminen P, Papagiannoulis-Lascarides L, Waltimo-Siren J, Ollila P, Karjalainen S, Arte S, Veerkamp J, Walton VT, Kustner EC, Siltanen T, Holappa H, Lukinmaa PL, Alaluusua S. 2011. Frameshift mutations in dentin phosphoprotein and dependence of dentin disease phenotype on mutation location. *J Bone Miner Res* 26(4): 873–80.
21. McKnight DA, Suzanne Hart P, Hart TC, Hartsfield JK, Wilson A, Wright JT, Fisher LW. 2008. A comprehensive analysis of normal variation and disease-causing mutations in the human DSPP gene. *Hum Mutat* 29(12): 1392–404.
22. Shields ED, Bixler D, el-Kafrawy AM. 1973. A proposed classification for heritable human dentine defects with a description of a new entity. *Arch Oral Biol* 18(4): 543–53.
23. Witkop CJ Jr. 1971. Manifestations of genetic diseases in the human pulp. *Oral Surg Oral Med Oral Pathol* 32(2): 278–316.
24. Waltimo J. 1994. Hyperfibers and vesicles in dentin matrix in dentinogenesis imperfecta (DI) associated with osteogenesis imperfecta (OI). *J Oral Pathol Med* 23(9): 389–93.
25. Giansanti JS, Allen JD. 1974. Dentin dysplasia, type II, or dentin dysplasia, coronal type. *Oral Surg Oral Med Oral Pathol* 38(6): 911–7.
26. Lukinmaa PL, Ranta H, Ranta K, Kaitila I. 1987. Dental findings in osteogenesis imperfecta: I. Occurrence and expression of type I dentinogenesis imperfecta. *J Craniofac Genet Dev Biol* 7(2): 115–25.
27. Wright JT. 2006. The molecular etiologies and associated phenotypes of amelogenesis imperfecta. *Am J Med Genet A* 140(23): 2547–55.
28. Witkop CJ Jr. 1989. Amelogenesis imperfecta, dentinogenesis imperfecta and dentin dysplasia revisited: Problems in classification. *J Oral Pathol* 17(9-10): 547–53.
29. Wright JT, Hart PS, Aldred MJ, Seow K, Crawford PJ, Hong SP, Gibson CW, Hart TC. 2003. Relationship of phenotype and genotype in X-linked amelogenesis imperfecta. *Connect Tissue Res* 44 Suppl 1: 72–8.
30. Hu JC, Yamakoshi Y. 2003. Enamelin and autosomal-dominant amelogenesis imperfecta. *Crit Rev Oral Biol Med* 14(6): 387–98.
31. Kim JW, Lee SK, Lee ZH, Park JC, Lee KE, Lee MH, Park JT, Seo BM, Hu JC, Simmer JP. 2008. FAM83H mutations in families with autosomal-dominant hypocalcified amelogenesis imperfecta. *Am J Hum Genet* 82(2): 489–94.
32. El-Sayed W, Parry DA, Shore RC, Ahmed M, Jafri H, Rashid Y, Al-Bahlani S, Al Harasi S, Kirkham J, Inglehearn CF, Mighell AJ. 2009. Mutations in the beta propeller WDR72 cause autosomal-recessive hypomaturation amelogenesis imperfecta. *Am J Hum Genet* 85(5): 699–705.
33. Hart PS, Hart TC, Michalec MD, Ryu OH, Simmons D, Hong S, Wright JT. 2004. Mutation in kallikrein 4 causes autosomal recessive hypomaturation amelogenesis imperfecta. *J Med Genet* 41(7): 545–9.
34. Kim JW, Simmer JP, Hart TC, Hart PS, Ramaswami MD, Bartlett JD, Hu JC. 2005. MMP-20 mutation in autosomal recessive pigmented hypomaturation amelogenesis imperfecta. *J Med Genet* 42(3): 271–5.
35. Parry DA, Mighell AJ, El-Sayed W, Shore RC, Jalili IK, Dollfus H, Bloch-Zupan A, Carlos R, Carr IM, Downey LM, Blain KM, Mansfield DC, Shahrabi M, Heidari M, Aref P, Abbasi M, Michaelides M, Moore AT, Kirkham J, Inglehearn CF. 2009. Mutations in CNNM4 cause Jalili syndrome, consisting of autosomal-recessive cone-rod dystrophy and amelogenesis imperfecta. *Am J Hum Genet* 84(2): 266–73.
36. Polok B, Escher P, Ambresin A, Chouery E, Bolay S, Meunier I, Nan F, Hamel C, Munier FL, Thilo B, Megarbane A, Schorderet DF. 2009. Mutations in CNNM4 cause recessive cone-rod dystrophy with amelogenesis imperfecta. *Am J Hum Genet* 84(2): 259–65.

37. O'Sullivan J, Bitu CC, Daly SB, Urquhart JE, Barron MJ, Bhaskar SS, Martelli-Junior H, Dos Santos Neto PE, Mansilla MA, Murray JC, Coletta RD, Black GC, Dixon MJ. 2011. Whole-Exome sequencing identifies FAM20A mutations as a cause of amelogenesis imperfecta and gingival hyperplasia syndrome. *Am J Hum Genet* 88(5): 616–20.
38. Schultz ML, Tecedor L, Chang M, Davidson BL. 2011. Clarifying lysosomal storage diseases. *Trends Neurosci* 34(8): 401–10.
39. Downs AT, Crisp T, Ferretti G. 1995. Hunter's syndrome and oral manifestations: A review. *Pediatr Dent* 17(2): 98–100.
40. Keith O, Scully C, Weidmann GM. 1990. Orofacial features of Scheie (Hurler-Scheie) syndrome (alpha-L-iduronidase deficiency). *Oral Surg Oral Med Oral Pathol* 70(1): 70–4.
41. Kinirons MJ, Nelson J. 1990. Dental findings in mucopolysaccharidosis type IV A (Morquio's disease type A). *Oral Surg Oral Med Oral Pathol* 70(2): 176–9.
42. Smith KS, Hallett KB, Hall RK, Wardrop RW, Firth N. 1995. Mucopolysaccharidosis: MPS VI and associated delayed tooth eruption. *Int J Oral Maxillofac Surg* 24(2): 176–80.
43. Baker E, Guo XH, Orsborn AM, Sutherland GR, Callen DF, Hopwood JJ, Morris CP. 1993. The morquio A syndrome (mucopolysaccharidosis IVA) gene maps to 16q24.3. *Am J Hum Genet* 52(1): 96–8.
44. Witkop CJ Jr, Sauk JJ Jr. 1976. Heritable defects of enamel. In: Stewart RE, Prescott GH (eds.) *Oral Facial Genetics*. St. Louis, MO: C.V. Mosby Co. pp. 151–226.
45. Gardner DG. 1971. The oral manifestations of Hurler's syndrome. *Oral Surg Oral Med Oral Pathol* 32(1): 46–57.
46. Giugliani R, Federhen A, Rojas MV, Vieira T, Artigalas O, Pinto LL, Azevedo AC, Acosta A, Bonfim C, Lourenco CM, Kim CA, Horovitz D, Bonfim D, Norato D, Marinho D, Palhares D, Santos ES, Ribeiro E, Valadares E, Guarany F, de Lucca GR, Pimentel H, de Souza IN, Correa JN, Fraga JC, Goes JE, Cabral JM, Simionato J, Llerena J Jr, Jardim L, Giuliani L, da Silva LC, Santos ML, Moreira MA, Kerstenetzky M, Ribeiro M, Ruas N, Barrios P, Aranda P, Honjo R, Boy R, Costa R, Souza C, Alcantara FF, Avilla SG, Fagondes S, Martins AM. 2010. Mucopolysaccharidosis I, II, and VI: Brief review and guidelines for treatment. *Genet Mol Biol* 33(4): 589–604.
47. Zuccati G. 1993. Implant therapy in cases of agenesis. *J Clin Orthod* 27(7): 369–73.
48. Lo B, Faiyaz-Ul-Haque M, Kennedy S, Aviv R, Tsui LC, Teebi AS. 2003. Novel mutation in the gene encoding c-Abl-binding protein SH3BP2 causes cherubism. *Am J Med Genet A* 121A(1): 37–40.
49. Pontes FS, Ferreira AC, Kato AM, Pontes HA, Almeida DS, Rodini CO, Pinto DS Jr. 2007. Aggressive case of cherubism: 17-year follow-up. *Int J Pediatr Otorhinolaryngol* 71(5): 831–5.
50. Solt DB. 1991. The pathogenesis, oral manifestations, and implications for dentistry of metabolic bone disease. *Curr Opin Dent* 1(6): 783–91.
51. Neville B, Damm D, Allen C, Bouquot J. 2001. *Oral and Maxillofacial Pathology, 1st Ed.* Philadelphia: W.B.Saunders.
52. Zambrano M, Nikitakis NG, Sanchez-Quevedo MC, Sauk JJ, Sedano H, Rivera H. 2003. Oral and dental manifestations of vitamin D-dependent rickets type I: Report of a pediatric case. *Oral Surg Oral Med Oral Pathol Oral Radiol Endod* 95(6): 705–9.
53. Feng JQ, Ward LM, Liu S, Lu Y, Xie Y, Yuan B, Yu X, Rauch F, Davis SI, Zhang S, Rios H, Drezner MK, Quarles LD, Bonewald LF, White KE. 2006. Loss of DMP1 causes rickets and osteomalacia and identifies a role for osteocytes in mineral metabolism. *Nat Genet* 38(11): 1310–5.
54. Lorenz-Depiereux B, Bastepe M, Benet-Pages A, Amyere M, Wagenstaller J, Muller-Barth U, Badenhoop K, Kaiser SM, Rittmaster RS, Shlossberg AH, Olivares JL, Loris C, Ramos FJ, Glorieux F, Vikkula M, Juppner H, Strom TM. 2006. DMP1 mutations in autosomal recessive hypophosphatemia implicate a bone matrix protein in the regulation of phosphate homeostasis. *Nat Genet* 38(11): 1248–50.
55. Roetzer KM, Varga F, Zwettler E, Nawrot-Wawrzyniak K, Haller J, Forster E, Klaushofer K. 2007. Novel PHEX mutation associated with hypophosphatemic rickets. *Nephron Physiol* 106(1): p8–12.
56. Alizadeh Naderi AS, Reilly RF. 2011. Hereditary disorders of renal phosphate wasting. *Nat Rev Neph* 6(11): 657–65.
57. Goodman JR, Gelbier MJ, Bennett JH, Winter GB. 1998. Dental problems associated with hypophosphataemic vitamin D resistant rickets. *Int J Paediatr Dent* 8(1): 19–28.
58. Hillmann G, Geurtsen W. 1996. Pathohistology of undecalcified primary teeth in vitamin D-resistant rickets: Review and report of two cases. *Oral Surg Oral Med Oral Pathol Oral Radiol Endod* 82(2): 218–24.
59. Bastepe M, Juppner H. 2008. Inherited hypophosphatemic disorders in children and the evolving mechanisms of phosphate regulation. *Rev Endocr Metab Disord* 9(2): 171–80.
60. Whyte MP. 2010. Physiological role of alkaline phosphatase explored in hypophosphatasia. *Ann N Y Acad Sci* 1192: 190–200.

111
Pathology of the Hard Tissues of the Jaws
Paul C. Edwards

Introduction 922
Tooth Demineralization and Caries 922
Cysts and Tumors of the Jaws 923
References 927

INTRODUCTION

The mandible and maxilla are unique among the bones in that they contain an overlying osseous structure—the alveolar process—that functions, in the adult, to support 32 highly specialized hard tissue organs, the teeth. The teeth, composed of unique hard tissues that are not found anywhere else in the body (enamel, dentin, and cementum), develop through a process involving sequential and reciprocal interactions between oral epithelium and ectomesenchyme. By necessity, because growth of the jaws requires the sequential formation of two sets of dentition, this process occurs over a period extending from the fetal period to the late teens.

As a result of the unique developmental processes involved in the formation of teeth and the fact that these specialized mineralized tissues are continuously exposed to the harsh oral environment, the mandible and maxilla are home to a distinctive set of pathologic entities. This chapter provides a concise review of the more common and interesting of these entities: tooth demineralization, dental caries, and odontogenic cysts and tumors.

TOOTH DEMINERALIZATION AND CARIES

Dental caries ("tooth decay") is one of the most prevalent chronic diseases affecting modern society. Once viewed as a disease primarily of children, it has become evident that as a result of the reduction in tooth loss over the past 30 years, adults today are equally as susceptible as children to developing dental caries [1, 2, 3].

The caries process represents the end result of a complex interaction between transmissible cariogenic oral microflora, primarily *Streptococcus mutans* and *Lactobacillus* spp., and fermentable dietary carbohydrates [4]. Oral microflora, vertically transmitted to the child through the mother, colonize the teeth through a process involving adhesion of bacterial surface proteins to salivary products adsorbed on the enamel surface [5]. Through the action of *Streptococcus mutans*-derived glucosyl transferase, an adherent extracellular polysaccharide matrix composed of water insoluble glucan is produced [4]. Within this dental biofilm, refined carbohydrates such as sucrose are metabolized, resulting in lactic acid production. The resultant drop in pH at the biofilm–tooth interface results in dissolution of the mineral component of the tooth. This process is a dynamic one, in which demineralization is countered by remineralization as the local pH level returns to normal through the buffering capacity of saliva. During the remineralization phase, diffusion of phosphate and calcium ions back into the hydroxyapatite mineral component of the tooth predominates. The progression of caries results when the rate of demineralization exceeds that of remineralization (Fig. 111.1).

In addition to the frequency and duration of exposure to fermentable dietary carbohydrate, other factors involved in an individual's risk of developing caries include the virulence of the specific pathogenic geno-

---

Primer on the Metabolic Bone Diseases and Disorders of Mineral Metabolism, Eighth Edition. Edited by Clifford J. Rosen.
© 2013 American Society for Bone and Mineral Research. Published 2013 by John Wiley & Sons, Inc.

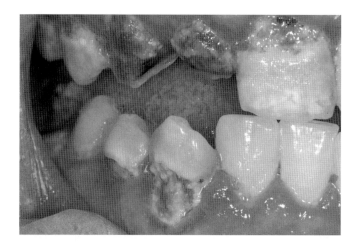

Fig. 111.1. Patient with rampant caries. This patient was a chronic methamphetamine abuser [6].

types of colonizing *S. mutans*, host immune response, and the buffering capacity and protein and mineral composition of saliva [7]. By disrupting the tooth surface biofilm, oral hygiene practices such as brushing and flossing can reduce the caries risk.

On exposure to trace quantities (approximately 1 ppm) of fluoride ion, fluoridated hydroyapatite and fluorapatite are formed. These are significantly more resistant to acid dissolution than hydroxyapatite. Studies [2] suggest that the overall magnitude of caries reduction afforded by fluoride averages 25% in both children and adults, whether delivered professionally in the form of a fluoride gel or foam, self-administered in toothpaste, or by community water fluoridation.

Although the caries process in the heavily mineralized enamel outer layer of the tooth is primarily a physicochemical process, the underlying dentin contains approximately 20% organic matrix, primarily type I collagen and a smaller component of non-collagenous proteins. Continued acid dissolution of the mineralized component of dentin results in exposure of the organic matrix to enzymatic degradation by both bacterial collagenases and host derived matrix metalloproteinases [8]. The presence of tubules within the dentin, through which odontoblast cell processes extend from the dental pulp, results in the formation of a tightly integrated "dentin-pulp complex." Inflammation within the confined space of the pulp cavity, either as a result of direct extension of bacteria into the pulp tissue, or by secondary strangulation of venous blood flow, may lead to pulpal necrosis. Inflammatory mediators resulting from degradation products of bacteria and necrotic pulp tissue lead to the formation of a mass of chronically inflamed granulation tissue and associated bone destruction at the apex of the non-vital tooth root (a "periapical granuloma").

CYSTS AND TUMORS OF THE JAWS

Origin

Jaw lesions can be categorized as being of odontogenic origin, referring to those that involve or recapitulate structures involved in the development of the teeth, or of nonodontogenic origin. By definition, odontogenic cysts and tumors are unique to the oral and maxillofacial region.

During tooth development, a thin epithelial structure from which the individual teeth ultimately form, termed the "dental lamina," arises from an ingrowth of surface epithelium into the underlying connective tissue [9]. Remnants of these structures are believed to be the source of the epithelium involved in the formation of many of the developmental odontogenic cysts. The molecular events involved in the development of the odontogenic neoplasms remain poorly understood [10, 11].

Clinical and radiographic presentation

The vast majority of odontogenic cysts and tumors originate in the tooth-bearing areas of the jaws (e.g., above the inferior alveolar nerve canal in the mandible) and are characterized by replacement of bone by soft tissue or, less commonly, a mixture of soft and hard tissue. In the absence of secondary infection or significant expansion, odontogenic cysts and tumors typically cause no symptoms and are usually identified following routine radiographic examination of the jaws (Figs. 111.2(A) and (B)).

The radiographic and clinical presentations of these lesions, though often characteristic, are not pathognomonic. As with extragnathic bone lesions, correlation of the radiographic features with the histopathologic findings is often required in order to arrive at a definitive diagnosis.

Odontogenic cysts

Odontogenic cysts can be further subclassified into inflammatory cysts and those of developmental origin.

Inflammatory odontogenic lesions
Periapical granuloma and periapical cyst

Both the periapical granuloma and the periapical (radicular) cyst are common, slow-growing lesions that develop at the apex or mid-root area of teeth exhibiting pulpal necrosis, usually representing the end result of caries extending into the pulp or as a result of previous trauma to the dental pulp. Continued bone destruction can lead to cortical bone perforation [Fig. 111.3(A)] and the formation of a mass of acute and chronically inflamed granulation tissue in the oral cavity (a "parulis"; Fig. 111.3(B)). The periapical granuloma, consisting of a localized

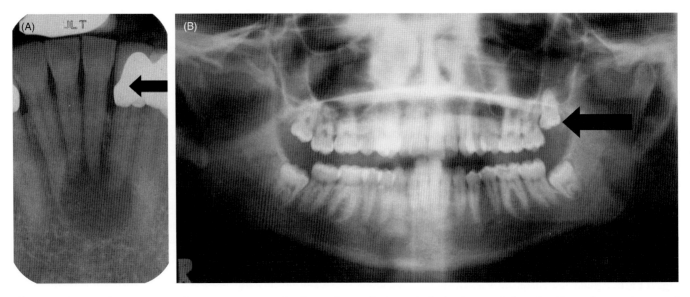

Fig. 111.2. The dentist's typical armamentarium of radiographic techniques includes intraoral dental radiographs supplemented with extraoral panoramic imaging. (A) Intraoral radiograph (periapical film) demonstrating the outline of the mandibular anterior teeth and surrounding alveolar bone. A unilocular radiolucent lesion with a well-defined corticated border is evident in the center of the radiograph at the apices of the incisors. The left lateral incisor (arrow) was non-vital. The differential diagnosis was periapical cyst versus granuloma. (B) Extraoral panoramic radiography provides a complete overview of the maxillary and mandibular bones and neighboring structures. This radiographic technique is widely used as a screening tool. This panoramic radiograph, taken on an 18-year-old patient, reveals the presence of unerupted third molars ("wisdom teeth"). A small well-defined radiolucent area with corticated borders is evident around the crown of the maxillary left third molar (arrow). As the width of the radiolucent area is less than 4 mm and the associated third molar has not completed its root development, this most likely represents a normal dental follicle surrounding the developing tooth.

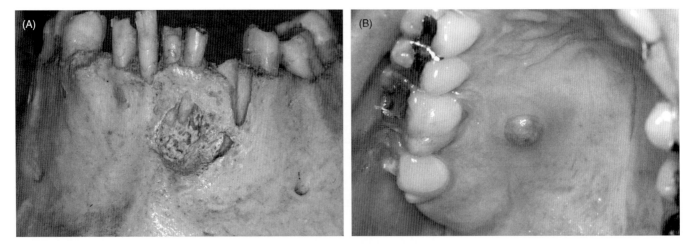

Fig. 111.3. (A) Cadaveric mandible with loss of buccal cortical plate. The crowns of the mandibular incisors are fractured, presumably resulting in pulpal necrosis and the subsequent development of a periapical granuloma or cyst. (B) 25-year-old male with a necrotic right maxillary first molar presented with a soft tissue nodule composed of acute and chronically inflamed epithelial-lined granulation tissue ("parulis") overlying the palatal root. The palate is an uncommon location for the development of a parulis, as these typically are noted on the buccal aspect of the alveolar process. In this case, the clinical differential diagnosis would also include a soft tissue neoplasm of salivary or connective tissue origin.

collection of chronically inflamed granulation tissue, is the precursor to the periapical cyst. In the presence of inflammatory mediators, cytokines, and growth factors released from mononuclear inflammatory cells and neighboring stromal cells as a result of the degradation of products from the necrotic pulp, residual epithelial cell rests involved in initial tooth formation are stimulated to proliferate [12]. Although the mechanism is poorly understood, the end result is the formation of a cystic lining (a "periapical cyst").

Non-vital teeth with periapical radiolucencies suggestive of a periapical granuloma or cyst are definitively treated by extraction of the causative tooth with conservative curettage of any lining from the cyst cavity. Failure to excise the cyst lining can lead to continued cyst expansion, termed a "residual cyst." When sufficient tooth structure remains to allow for restoration of the tooth, nonsurgical endodontic treatment, in which the degradation products are mechanically removed from the pulp chamber and canals, is usually the preferred treatment approach. If endodontic therapy is performed, the patient should be followed radiographically to confirm bone regeneration. In extremely rare cases, the epithelial lining of a periapical cyst or residual cyst can undergo malignant transformation [13].

Developmental odontogenic cysts
Dentigerous cyst

The dentigerous ("follicular") cyst, a slow-growing lesion capable of causing significant destruction of bone, is only seen in association with the crown of an unerupted ("impacted") tooth. The third molar and maxillary canine teeth are most commonly involved, since these teeth are the last to come into the mouth in the normal tooth eruption sequence, and consequently are the teeth most prone to being prevented from erupting as a result of crowding [14].

Treatment involves enucleation of the cyst lining, usually with extraction of the associated impacted tooth. Rarely, the cyst lining can undergo transformation into an odontogenic tumor, most commonly a cystic ameloblastoma [15], or, rarely, a squamous cell carcinoma [16].

Odontogenic keratocyst ("keratocystic odontogenic tumor")

The odontogenic keratocyst (OKC) demonstrates a preference for the posterior mandible and is characterized by a tendency to cause significant bone destruction if untreated and a high recurrence rate after conservative treatment [Figs. 111.4(A), (B)] [17]. OKCs range in size from small unilocular radiolucent lesions, sometimes associated with an impacted tooth, to destructive multilocular lesions involving large areas of the mandible or maxilla.

Approximately 5% of OKCs are associated with the nevoid basal cell carcinoma syndrome (NBCCS, Gorlin syndrome), an autosomal dominantly inherited condition. Additional stigmata include the development of multiple basal cell carcinomas of the skin at an early age, the presence of small pit-like developmental defects in the palms of the hands and plantar surfaces of the feet, and an increased incidence of neoplasms including medulloblastoma and meningioma [18]. NBCCS is associated with germline loss of function mutations in the *patched-1* (*PTCH1*) gene, a tumor suppressor gene that is a component of the sonic hedgehog pathway [19].

Similar *PTCH1* mutations have been documented in approximately 30% of sporadic, non-syndrome-associated OKCs [20, 21]. Recently, the World Health Organization has recommended that the OKC be renamed "keratocystic odontogenic tumor" to emphasize its aggressive, neoplastic-like behavior [22]. Additional findings supportive of a neoplastic nature include the demonstration of increased proliferative activity in the epithelial lining [23] and loss of heterozygosity at loci associated with the *p16* and *p53* tumor suppressor genes [24]. However, this proposed change in terminology has yet to be been universally accepted. Arguments offered against reclassifying the OKC as a "benign cystic neoplasm" is the documentation of *PTCH1* mutations in other developmental odontogenic cysts [25], the suggestion that molecular genetic criteria alone are not necessarily sufficient to define neoplasia, and, most significantly, that histologically, the OKC remains an epithelial-lined pathologic cavity of developmental origin.

Treatment options vary from marsupialization and decompression prior to definitive treatment, to surgical curettage with adjuvant cryotherapy or chemical fixation, to complete surgical resection [26]. Some surgeons choose to treat maxillary lesions more aggressively than mandibular lesions due to the greater risk of recurrent lesions spreading into neighboring vital structures. Recurrences have been documented up to 10 years after treatment [27].

Lateral periodontal cyst

The lateral periodontal cyst (LPC) is an uncommon lesion with limited growth potential that is often overlooked in the clinical differential diagnosis of a radiolucency occurring along the lateral root surface of an anterior tooth [28]. In contrast to the periapical cyst, there is no causal relationship to pulpal necrosis. Treatment involves conservative debridement of the lesion with preservation of the associated tooth. In order to avoid unnecessary endodontic therapy or tooth extraction, assessment of pulpal vitality of all teeth with associated radiolucent lesions is recommended.

Glandular odontogenic cyst

The recently described [29] glandular odontogenic cyst (GOC) is a rare lesion (one study [30] documented only 11 cases among 55,000+ oral cavity biopsies) with histologic features reminiscent of both the lateral periodontal cyst and mucoepidermoid carcinoma, a malignant salivary gland neoplasm rarely identified in the jaws. The

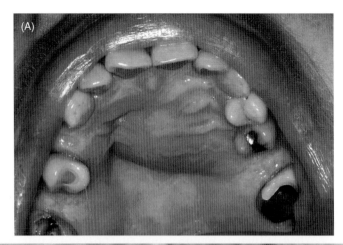
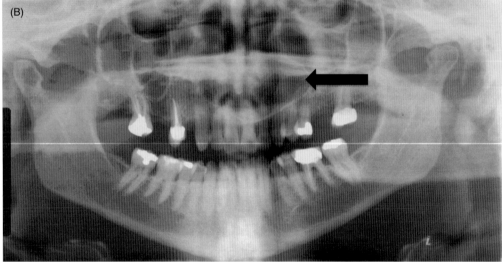

Fig. 111.4. (A) Fifty-six year old woman presents with a palatal expansion resulting from an odontogenic keratocyst ("keratocystic odontogenic tumor") that had been untreated over a 4-year period. (B) Panoramic radiography reveals a large well-defined corticated radiolucent lesion extending from the maxillary right canine to the left second premolar area [arrow; same patient as in (A)].

majority of GOCs occur in the anterior mandible, often crossing the midline [31]. The GOC exhibits a wide spectrum of clinical behavior, ranging from a benign process to a destructive lesion with features more suggestive of a malignant process.

Odontogenic tumors

Benign odontogenic tumors of the jaws

Ameloblastoma

The ameloblastoma is a locally destructive tumor with a propensity to cause significant cortical expansion, a marked predilection for the posterior mandible, and a high rate of recurrence [32]. Its incidence has been estimated at 0.3–2.3 new cases per million persons per year [33].

The unicystic ameloblastoma, seen predominantly in teenagers at an average 20 years earlier than the conventional ameloblastoma, represents a cystic version of the conventional ameloblastoma, possibly associated with a lower risk of recurrence [34]. Often associated with an impacted mandibular third molar, it is frequently mistaken radiographically for a dentigerous cyst.

Malignant odontogenic tumors

Malignant counterparts to the benign odontogenic tumors are extremely rare [35]. They commonly present as destructive lesions with irregular, poorly defined radiographic margins. Pain, paresthesia, and a tendency for early lymph node metastasis are characteristic. More commonly, malignancies involving the jaws originate by direct extension from neighboring soft tissue (e.g., oral squamous cell carcinoma), represent metastatic lesions from extraoral sites (especially breast, colon, and prostate carcinoma), or are of primary nonodontogenic origin. Rarely, maxillary sinus malignancies can mimic pain of dental origin [36].

Nonodontogenic tumors of the jaws
Central giant cell granuloma

A large number of nonodontogenic cysts, pseudocysts, and tumors also occur in the jaws. Among these is a lesion unique to the jaws: the central giant cell granuloma (CGCG). Interestingly, histologically identical lesions are seen in a number of other conditions including cherubism and hyperparathyroidism. Syndromes associated with an increased incidence of CGCG-like lesions include Noonan syndrome [37, 38] and neurofibromatosis type-1 [39, 40].

The classic CGCG is a variably aggressive nonneoplastic reactive lesion with an estimated incidence of 1.1 per million persons per year in the general population [41]. The CGCG is characterized histologically by the presence of multinucleated giant-cells (MGCs) in a background of spindle-shaped mesenchymal cells. The MGCs are usually concentrated in areas of hemorrhage and are believed to develop from the fusion of mononuclear phagocytes [42]. Although they share similarities with the osteoclast [43], phenotypic differences exist [44]. Intraosseous hemorrhage and abnormal repair of bone have been suggested as potential etiologic factors [45] in light of the observation that CGCGs are occasionally noted in association with other preexisting bone lesions such as fibrous dysplasia and ossifying fibroma [46].

Isolated lesions are commonly treated by surgical curettage. Treatment options for large or multiple lesions include intralesional corticosteroid injection [47], subcutaneous or intranasal administration of calcitonin [48], and therapy with interferon alpha-2a [49].

REFERENCES

1. Bagramian RA, Garcia-Godoy F, Volpe AR. 2009. The global increase in dental caries. A pending public health crisis. *Am J Dent* 22: 3–8.
2. Griffin SO, Regnier E, Griffin PM, Huntley V. 2007. Effectiveness of fluoride in preventing caries in Adults. *J Dent Res* 86: 410–415.
3. Edwards PC, Kanjirath P. 2010. Recognition and management of common acute conditions of the oral cavity resulting from tooth decay, periodontal disease and trauma: An update for the family physician. *J Am Board Fam Med* 23: 285–294.
4. Loesche WJ. 1986. Role of *Streptococcus mutans* in human dental decay. *Microbiol Rev* 50: 353–380.
5. Napimoga MH, Hofling JF, Klein MI, Kamiya RU, Goncalves RB. 2005. Transmission, diversity and virulence factors of *Streptococcus mutans* genotypes. *J Oral Sci* 47: 59–64.
6. Shaner JW, Kimmes N, Saini T, Edwards PC. 2006. Meth mouth: Rampant caries in methamphetamine abusers. *AIDS Patient Care and STDs* 20: 4–8.
7. Selwitz RH, Ismail AI, Pitts NB. 2007. Dental caries. *Lancet* 369: 51–59.
8. Chaussain-Miller C, Fioretti F, Goldberg M, Menashi S. 2006. The role of matrix metalloproteinases (MMPs) in human caries. *J Dent Res* 85: 22–32.
9. Cobourne MT, Sharpe PT. 2003. Tooth and jaw: Molecular mechanisms of patterning in the first branchial arch. *Arch Oral Biol* 48: 1–14.
10. Kumamoto H. 2006. Molecular pathology of odontogenic tumors. *J Oral Pathol Med* 35: 65–74.
11. Gomes CC, Duarte AP, Diniz MP, Gomez RS. 2010. Current concepts of ameloblastoma pathogenesis. *J Oral Pathol Med* 39: 585–591.
12. Lin LM, Huang GT, Rosenberg PA. 2007. Proliferation of epithelial cell rests, formation of apical cysts, and regression of apical cysts after periapical wound healing. *J Endod* 33: 908–916.
13. Whitlock RI, Jones JH. 1967. Squamous cell carcinoma of the jaw arising in a simple cyst. *Oral Surg Oral Med Oral Pathol* 24: 530–536.
14. Daley TD, Wysocki GP. 1995. The small dentigerous cyst: A diagnostic dilemma. *Oral Surg Oral Med Oral Pathol Oral Radiol Endod* 79: 77–81.
15. Holmlund HA, Anneroth G, Lundquist G, Nordenram A. 1991. Ameloblastoma originating from odontogenic cysts. *J Oral Pathol Med* 20: 318–321.
16. Bodner L, Manor E, Shear M, van der Waal I. 2011. Primary intraosseous squamous cell carcinoma arising in an odontogenic cyst: A clinicopathologic anaylsis. *J Oral Med Pathol* 40(10): 733–738.
17. Myoung H, Hong SP, Hong SD, Lee JI, Lim CY, Choung PH, Lee JH, Choi YJ, Seo BM, Kim MJ. 2001. Odontogenic keratocyst: Review of 256 cases for recurrence and clinicopathologic parameters. *Oral Surg Oral Med Oral Pathol Radiol Endod* 91: 328–333.
18. Kimonis VE, Goldstein AM, Pastakia B, Yang ML, Kase R, DiGiovanna JJ, Bale AE, Bale SJ. 1997. Clinical Manifestations in 105 persons with nevoid basal cell carcinoma syndrome. *Am J Med Genet* 69: 299–308.
19. Hahn H, Wicking C, Zaphiropoulous PG, Gailani MR, Shanley S, Chidambaram A, Vorechovsky I, Holmberg E, Unden AB, Gillies S, Negus K, Smyth I, Pressman C, Leffell DJ, Gerrard B, Goldstein AM, Dean M, Toftgard R, Chenevix-Trench G, Wainwright B, Bale AE. 1996. Mutations of the human homolog of Drosophila *patched* in the nevoid basal cell carcinoma syndrome. *Cell* 85: 841–851.
20. Gu XM, Zhao HS, Sun LS, Li TJ. 2006. *PTCH* mutations in sporadic and Gorlin-syndrome-related odontogenic keratocysts. *J Dent Res* 85: 859–863.
21. Li TJ. 2011. The odontogenic keratocyst: A cyst, or a cystic neoplasm. *J Dent Res* 90: 133–142.
22. Philipsen HP. 2005. Keratocystic odontogenic tumor. In: Barnes L, Eveson JW, Reichart P, Sidransky D (eds.) *Pathology and Genetics of Head and Neck Tumors*. Lyons, France: IARC Press. pp. 306–307.
23. Slootweg PJ. 1995. p53 protein and Ki-67 reactivity in epithelial odontogenic lesions. An immunohistochemical study. *J Oral Pathol Med* 24: 393–397.
24. Henley J, Summerlin DJ, Tomich C, Zhang S, Cheng L. 2005. Molecular evidence supporting the neoplastic

24. nature of odontogenic keratocyst: A laser capture microdissection study of 15 cases. *Histopathol* 47: 582–586.
25. Pavelic B, Levanat S, Crnic I, Kobler P, Anic I, Manojlovic S, Sutalo J. 2001. *PTCH* gene altered in dentigerous cysts. *J Oral Pathol Med* 30: 569–576.
26. Blanas N, Freund B, Schwartz M, Furst IM. 2000. Systematic review of the treatment and prognosis of the odontogenic keratocyst. *Oral Surg Oral Med Oral Pathol Radiol Endod* 90: 553–558.
27. Kolokythas A, Fernandes RP, Pazoki A, Ord RA. 2007. Odontogenic keratocyst: To decompress or not to decompress? A comparative study of decompression and enucleation versus resection/peripheral ostectomy. *J Oral Maxillofac Surg* 65: 640–644.
28. Fantasia JE. 1979. Lateral periodontal cyst. An analysis of forty-six cases. *Oral Surg Oral Med Oral Pathol Radiol Endod* 48: 237–243.
29. Gardner DG, Kessler HP, Morency R, Schaffner DL. 1988. The glandular odontogenic cyst: An apparent entity. *J Oral Pathol* 17: 359–366.
30. Jones AV, Craig GT, Franklin CD. 2006. Range and demographics of odontogenic cysts diagnosed in a UK population over a 30-year period. *J Oral Pathol Med* 35: 500–507.
31. Hussain K, Edmondson HD, Browne RM. 1995. Glandular odontogenic cysts: Diagnosis and treatment. *Oral Surg Oral Med Oral Pathol Oral Radiol Endod* 79: 593–602.
32. Reichart PA, Philipsen HP, Sonner S. 1995. Ameloblastoma: Biological profile of 3677 cases. *Eur J Cancer B Oral Oncol* 31B: 86–99.
33. Shear M, Singh S. 1978. Age-standardized incidence rates of ameloblastoma and dentigerous cyst on the Witwatersrand, South Africa. *Community Dent Oral Epidemiol* 6: 195–199.
34. Philipsen HP, Reichart PA. 1998. Unicystic ameloblastoma. A review of 193 cases from the literature. *Oral Oncology* 34: 317–325.
35. Slootweg PJ. 2002. Malignant odontogenic tumors: An overview. *Mund Kiefer Gesichtschir* 6: 295–302.
36. Edwards PC, Hess S, Saini T. 2006. Sinonasal undifferentiated carcinoma of the maxillary sinus. *J Can Dent Assoc* 72: 161–165.
37. Cohen MM Jr, Gorlin RJ. 1991. Noonan-like/multiple giant cell lesion syndrome. *Am J Med Genet* 40: 159–166.
38. Edwards PC, Fox J, Fantasia JE, Goldberg J, Kelsch RD. 2005. Bilateral central giant cell granulomas of the mandible in an 8-year-old girl with Noonan syndrome (Noonan-like/multiple giant cell lesions syndrome). *Oral Surg Oral Med Oral Pathol Oral Radiol Endod* 99: 334–340.
39. Ruggieri M, Pavone V, Polizzi A, Albanase S, Magro G, Merino M, Duray P. 1999. Unusual form of recurrent giant cell granuloma of the mandible and lower extremities in a patient with neurofibromatosis type 1. *Oral Surg Oral Med Oral Pathol Oral Radiol Endod* 87: 67–72.
40. Edwards PC, Fantasia JE, Saini T, Rosenberg T, Ruggiero S. 2006. Clinically aggressive central giant cell granulomas in two patients with neurofibromatosis 1. *Oral Surg Oral Med Oral Pathol Oral Radiol Endod* 102: 765–772.
41. de Lange J, van den Akker HP. 2005. Clinical and radiological features of central giant cell lesions of the jaw. *Oral Surg Oral Med Oral Pathol Oral Radiol Endod* 99: 464–470.
42. Abe E, Mocharla H, Yamate T, Taguchi Y, Manolagas SC. 1999. Meltrin-alpha, a fusion protein involved in multinucleated giant cell and osteoclast formation. *Calcif Tissue Int* 64: 508–515.
43. Liu B, Yu SF, Li TJ. 2003. Multinucleated giant cells in various forms of giant cell containing lesions of the jaws express features of osteoclasts. *J Oral Pathol Med* 32: 367–375.
44. Tobon-Arroyave SI, Franco-Gonzalez LM, Isaza-Guzman DM, Florez-Moreno GA, Bravo-Vasquez T, Castaneda-Pelaez DA, Vieco-Duran B. 2005. Immunohistochemical expression of RANK, GR-alpha and CTR in central giant cell granulomas of the jaws. *Oral Oncol* 41: 480–488.
45. Dorfman HD, Czerniak B. 1998 Giant cell lesions. In: Dorfman HD, Czerniak B (eds.) *Bone Tumors*. St. Louis, MO: Mosby. pp. 559–606.
46. Penfold CN, McCullagh P, Eveson JW, Ramsay A. 1993. Giant cell lesions complicating fibro-osseous conditions of the jaws. *Int J Oral Maxillofac Surg* 22: 158–162.
47. Terry BC, Jacoway JR. 1994. Management of central giant cell lesion: An alternative to surgical therapy. *Oral Maxillofac Surg Clin North Am* 6: 579–600.
48. de Lange J, Rosenberg AJ, Van den Akker HP, Koole R, Wirds JJ, VandenBerg H. 1999. Treatment of central giant cell granuloma of the jaw with calcitonin. *Int J Oral Maxillofac Surg* 28: 372–376.
49. Kaban LB, Mulliken JB, Ezekowitz RA, Ebb D, Smith PS, Folkman J. 1999. Antiangiogenic therapy of a recurrent giant cell tumor of the mandible with interferon alfa-2a. *Pediatrics* 103: 1145–1149.

112

Bisphosphonate-Associated Osteonecrosis of the Jaws

Hani H. Mawardi, Nathaniel S. Treister, and Sook-Bin Woo

Definition of Bisphosphonate-Associated Osteonecrosis of the Jaws 929
Etiology 930
Risk Factors and Prevalence 930
Clinical Presentation 931

Radiographic and Histologic Findings 932
Management 934
Osteonecrosis of the Jaw Associated with Non-Bisphosphonate Medications 935
References 935

DEFINITION OF BISPHOSPHONATE-ASSOCIATED OSTEONECROSIS OF THE JAWS

Bisphosphonate-associated osteonecrosis of the jaws (BONJ) is a clinical entity that was first reported in the literature in several case series in 2003, although the earliest case reported to the U.S. Food and Drug Administration was in 1989 [1–4]. Of the over several thousand cases that have been reported in the literature, the vast majority are in patients with multiple myeloma and metastatic cancers (from the breast, prostate, and lung) to the skeletal system who were treated with intravenous bisphosphonates; however, this condition has been identified (albeit far less frequently) in patients with osteoporosis treated only with oral formulations [4–8]. All cases have affected either the maxilla or the mandible, except for several single case reports involving other bones including the external auditory canal, thumb, and middle ear/base of skull [9–11]. The strong association between bisphosphonate use and osteonecrosis of the jaws is now well accepted, and appropriately designed prospective research is ongoing.

The American Association of Oral and Maxillofacial Surgeons (AAOMS) has published a working definition for BONJ in which the following three characteristics must be fulfilled: (1) current or previous treatment with a bisphosphonate; (2) exposed, necrotic bone in the maxillofacial region that has been present for at least 8 weeks; and (3) no history of radiation therapy to the jaws [12]. Stage 0 BONJ is defined as cases without exposed necrotic bone with nonspecific symptoms or clinical and/or radiographic abnormalities that are highly suggestive of BONJ such as pain, sinus tracts, and other signs or symptoms that cannot be attributed to an odontogenic infection [13, 14]. The AAOMS definition excludes such conditions as poorly healing dental extraction sites (e.g., alveolar osteitis, or "dry socket"), benign sequestration of the lingual plate, necrotizing periodontitis, noma, and osteoradionecrosis [15–20]. The American Society for Bone and Mineral Research has the same definition for *confirmed* cases; *suspected* cases fulfill all the criteria of a *confirmed* case except that the exposed bone has been present for fewer than 8 weeks [21]. These definitions are a critical initial step toward promoting uniform reporting for research and epidemiological reporting purposes, and may need to be revised as new diagnostic parameters come to light. Application of such criteria in well-designed prospective studies will help to move research forward [22]. At the present time, we rely on a good patient history, clinical examination, and imaging studies for the diagnosis of BONJ.

ETIOLOGY

The etiopathogenesis has been attributed primarily to suppression of bone turnover (due to bisphosphonate-induced inhibition of osteoclast activity and apoptosis) coupled with conditions that are unique to the mandible and maxilla. First, the jaw bones are separated from the oral cavity and commensal microflora by only a very thin mucosal barrier that is readily breached during normal physiologic functions (e.g., chewing). Second, the jaws are involved frequently by infection either through the periodontium (i.e., periodontal disease) or endodontically, through the dental pulp (i.e., caries extending into the pulp chamber, through the roots, and into the bone). Third, dentoalveolar surgical procedures are common (e.g., tooth extractions and periodontal surgeries) during which bone is damaged and exposed to a bacteria-rich environment. Fourth, at least in certain animal models, the rate of turnover of the jawbones is higher than for the long bones, which, if true in humans, may result in greater uptake and higher local concentrations of bisphosphonates [23, 24]. With profound osteoclast inhibition, the hypodynamic bone may be unable to respond to repair processes associated with physiologic trauma or infection, although how this mechanistically translates into bone necrosis is still unclear. The fact that tooth sockets in patients who develop BONJ may be evident years after extraction is further evidence that bone turnover and remodeling in these areas are severely compromised [25]. Microdamage accumulation may play a role in the pathogenesis although the exact mechanism is unclear [26–28].

There is evidence that bisphosphonates have antiangiogenic properties that may also contribute to poor wound healing [29–31]. Bisphosphonates have long been known to cause direct epithelial toxicity in the digestive tract, and it has recently been proposed that bisphosphonate-induced soft tissue injury may also be a contributing factor in the development of BONJ [32–38].

Several BONJ animal models have been developed, and although none of these models mimic all aspects of BONJ in humans, they may help to better elucidate the underlying driving factors that lead to necrosis [34, 39–41].

RISK FACTORS AND PREVALENCE

Several risk factors for BONJ have been identified. One of the most consistently observed is the use of nitrogen-containing bisphosphonates and in particular zoledronic acid, which is the drug of choice for treating multiple myeloma and metastatic cancers in the U.S. due to its shorter infusion time and equivalent clinical efficacy compared to pamidronate. Patients with cancer on zoledronic acid alone have a 9.1 to 15 times the risk of developing BONJ as compared to a 1.6–4 times risk when pamidronate is followed by zoledronic acid [42–44]. The cumulative dose of bisphosphonates, particularly intravenous preparations, is probably the single most important factor, with the risk increasing over time; one study showed cumulative hazard of 1% in the first year of treatment with zoledronic acid, and 15% at 4 years [45]. The median time of treatment with zoledronic acid to development of BONJ in patients with cancer ranges from 9 to 30 months (although cases have been reported after only 3 months) and appears to be significantly shorter when compared with other agents [5, 44–46]. The route of administration is also important, but this is likely a surrogate factor for the use of pamidronate and zoledronic acid, both of which are administered intravenously, have high potency, and are used more frequently in the oncologic population.

Approximately 60% of cases have a preceding history of tooth extraction or other oral surgical procedures [47–49] but only 0.5% of extractions result in BONJ [50]. It is unclear to what extent preexisting infection (e.g., periodontal and dental disease), the most common indication for dentoalveolar surgery, may be a contributing factor [5, 47, 51, 52]. In some cases, it is likely that a preexisting, evolving BONJ lesion was what caused a tooth to become painful and mobile in the first place, necessitating extraction, so that the process of tooth removal exposes necrotic bone.

Additionally, nonsurgical trauma appears also to be a risk factor, as many cases occur on the mylohyoid ridge of the lingual mandible (where the mucosa is especially thin and susceptible to masticatory forces), on tori, and in patients wearing dentures, with no associated dental infection or history of oral surgical procedure [53]. Genetic markers and risk factors are being studied [54–56]. The role played by conditions associated with poor healing such as the presence of cancer, use of other immunosuppressive agents, concomitant use of thalidomide or bortezomib in patients with myeloma, vitamin D deficiency, and diabetes have yet to be clearly elucidated [57–59].

Prevalence estimates based on retrospective chart reviews range from 2–11% in the oncology population [42, 44, 60–63]. More recently reported prospectively collected data have suggested that the true incidence is closer to 2–3% [64]. The prevalence of BONJ in those taking oral bisphosphonates has been reported to be between 1 in 2,260 to 8,470, or 0.1–0.5% [65–67]. However, some of these figures do not differentiate between those taking oral and intravenous bisphosphonates. The median time of development of BONJ in patients with osteoporosis is 3.5 years [67]. The hazard ratio for jaw surgery and inflammatory jaw conditions/osteomyelitis after exposure to intravenous (IV) bisphosphonates versus non-exposure is 3.15 and 11.48, respectively [68]. The odds ratios for patients with osteoporosis treated with oral compared with IV bisphosphonates who developed BONJ was 0.65 and 4.01, respectively, and for those requiring jaw surgery, 0.86 and 7.80, respectively [69].

CLINICAL PRESENTATION

BONJ presents with exposed, yellowish-white necrotic bone that varies in size from a few millimeters to several centimeters. The exposed bone may have rough and sharp edges, or may be smooth. The mandible is more frequently affected than the maxilla (2:1) and lesions are more common in areas with very thin mucosa overlying bony prominences such as mandibular and maxillary tori, bony exostoses, and the mylohyoid ridge which are also typically considered sites with high trauma risk (Fig. 112.1) [4, 70]. In some cases, osteonecrosis is associated with bone expansion, in particular affecting tori, putatively attributed to excess bone deposition without accompanying remodeling (Fig. 112.2) [71]. Lesions often develop in sites of previous dental extractions (Fig. 112.3); however in many cases, there is no antecedent surgical procedure (Fig. 112.4). The bone may project outward, causing trauma to the adjacent soft tissue. Loose bone fragments in the process of sequestration may also cause pain due to soft tissue irritation. Teeth that are located within the necrotic bone often become progressively mobile and symptomatic and may spontaneously exfoliate. Pathologic fractures of the mandible may occur in severe cases with extensive involvement; however, this is an infrequent complication [4, 48]. Approximately 60% of cases report some pain at initial presentation; however, many cases may remain asymptomatic for weeks, months, or years [4, 47, 48, 63]. Not surprisingly,

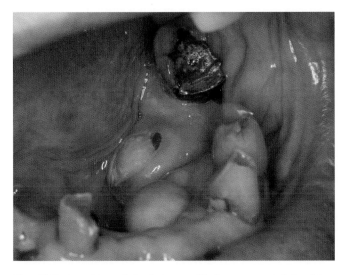

Fig. 112.1. BONJ of the left mandibular torus with healthy-appearing surrounding soft tissue.

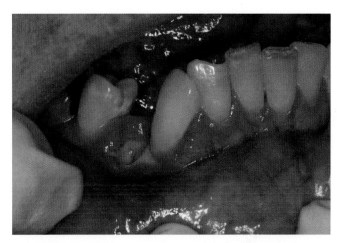

Fig. 112.3. BONJ in the area of a nonhealing mandibular premolar extraction site. The surrounding soft tissue is healthy appearing.

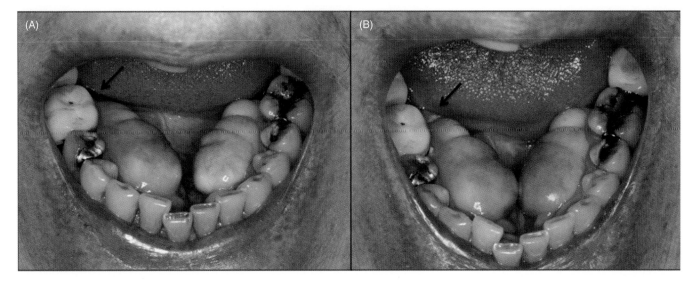

Fig. 112.2. (A) BONJ of the right mandibular torus at the time of diagnosis, and (B) 1 year later, demonstrating bony expansion bilaterally with a small area of exposed bone (arrows).

patients with BONJ have a diminished quality of life that is directly related to the severity of the condition [72].

The necrotic bone is more often than not superficially colonized by normal oral flora, similar to teeth being colonized by dental plaque, forming a biofilm [73]. The most common colonizing pathogens are Actinnomycetes and Eikenella [74, 75]. Infection of the bone leads to pain and suppuration, and osteomyelitis may ensue. The surrounding soft tissue may become erythematous and edematous, sometimes with an associated purulent discharge with intraoral or even extraoral sinus tract/fistula formation. Secondary maxillary sinusitis, with or without an oral-antral fistula, may be the initial presenting symptom in those with maxillary involvement (Fig. 112.5) [4, 70]. Symptoms of paresthesia and anesthesia may develop as the local neurovascular bundle (e.g., inferior alveolar nerve) becomes affected by inflammation or infection around the necrotic bone, in addition to nerve compression due to narrowing of canal/foramen by bone deposition [76].

RADIOGRAPHIC AND HISTOLOGIC FINDINGS

A number of radiographic signs have been identified in BONJ from both plain films and more advanced imaging studies [e.g., computed tomography (CT) and magnetic resonance imaging (MRI)] [77–82]. These include osteosclerosis, osteolysis, mottling, thickening or loss of the lamina dura, widening of the periodontal ligament space, sequestrum formation, and persistent extraction sockets (Fig. 112.6) [25, 79, 80, 83]. Radiographic changes often mimic periapical pathology, severe periodontal infection, and banal osteomyelitis. In cancer patients, radiographic changes may raise the suspicion of myeloma or metastatic bone disease [84]. As a bone biopsy may induce BONJ, this is only indicated when there is a strong clinical suspicion of metastatic disease and/or if the diagnosis will alter clinical management.

Intraoral periapical radiographs and panoramic radiographs provide high resolution and readily demonstrate evidence of BONJ, especially when disease is advanced [79, 81]; however, they fail to detect many early changes or mild disease and are limited in their ability to evaluate extent of involvement due to their two-dimensional nature. While CT scans provide more accurate three-dimensional information regarding the extent of changes, these also have not proved useful for early detection in

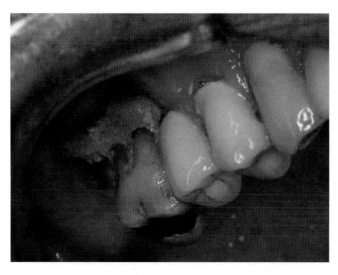

Fig. 112.4. Spontaneous area of BONJ of the right posterior maxilla arising in an area of preexisting periodontal disease.

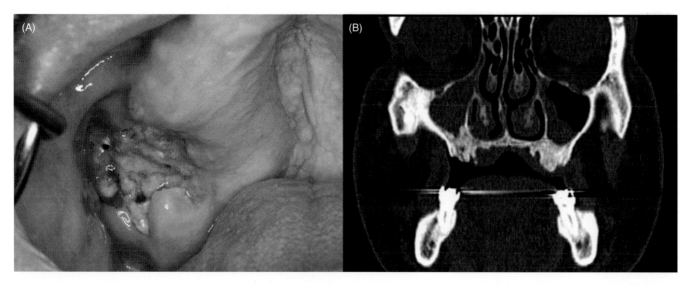

Fig. 112.5. (A) Extensive BONJ of the right posterior maxilla with associated soft tissue infection and purulence. (B) CT scan, with mottled right maxillary alveolar process and complete opacification of the right and partial opacification of the left maxillary sinuses.

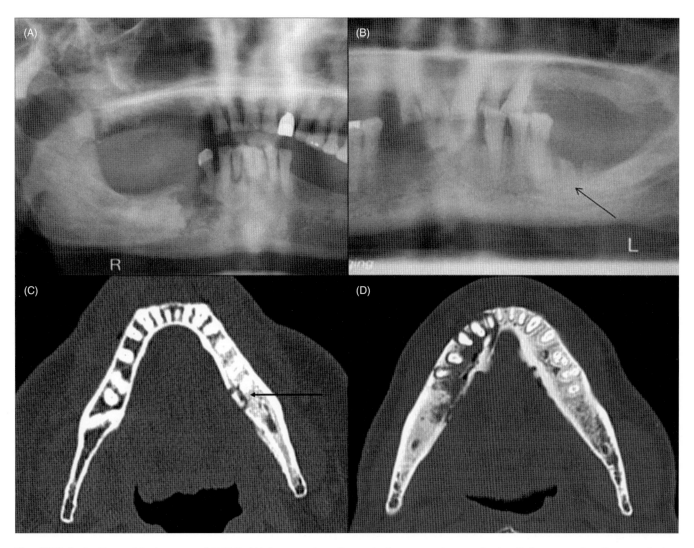

Fig. 112.6. Radiographic features of BONJ. (A) Panoramic radiograph demonstrating a mixed radiolucent and radiopaque lesion of the right mandible with a larger area of diffuse osteosclerosis. (B) Panoramic radiograph demonstrating a persistent extraction socket of the left mandible years after extraction. (C) Axial CT scan of the mandible demonstrating sclerosis with lingual sequestrum formation (arrow). (D) Axial CT scan of the mandible demonstrating extensive sclerosis and mottled changes.

asymptomatic individuals [74, 79, 82]. Head and neck cone beam CT is an imaging modality that uses approximately 10% of the radiation of conventional spiral CT and has acceptable resolution for diagnosis of BONJ [82]. MRI may be of some utility in evaluating the marrow and associated soft tissue changes but is not useful for interpreting bony pathology [79]. Nuclear medicine techniques have demonstrated that Tc-99m is not taken up in clinical areas of BONJ (although these may be positive in early or subclinical BONJ); however, ^{18}F-fluorodeoxyglucose positron emission tomography (FDG-PET) and NaF-PET scans integrated with CT may demonstrate focal uptake, indicating ongoing bone formation despite osteoclastic suppression [55, 79, 80]. The precise role for these imaging modalities in risk assessment, diagnosis, and clinical management remains to be determined.

Histopathologic examination of debrided bone fragments demonstrates necrotic bone often with extensive surface bacterial colonization, and granulation tissue, all nonspecific findings (Fig. 112.7) [85–87]. In spite of inhibition of osteoclastic activity, abnormal, giant osteoclasts are present in the vicinity of the necrotic bone [87, 88]. Bacterial cultures from the exposed surface or purulent discharge typically reveal normal oral flora [86, 89]. Actinomycetes are a normal component of the oral flora, and their identification, histopathologically or microbiologically, must be carefully interpreted. A diagnosis of actinomycosis should only be made if the culture was obtained from a sterile location (i.e., a nonexposed surface), if pain and suppuration/sinus tracts are present, and/or if sulfur granules are noted either clinically or histologically [90].

MANAGEMENT

Management of BONJ is focused primarily on preserving quality of life by controlling pain and discomfort, managing infection, avoiding procedures that may lead to the development of new areas of necrosis, and encouraging sequestration of the bone and reepithelialization of the mucosa. Published treatment algorithms/guidelines are principally based on expert opinion consensus and not prospective studies [12, 70, 83, 91]. A clinical staging system has been developed to categorize patients, guide treatment and collect data (Table 112.1) [12].

Patients on bisphosphonate therapy without BONJ

For those exposed to bisphosphonates, whether oral or IV, but without BONJ, management is directed toward optimizing dental health by educating patients regarding their risk for developing BONJ, evaluating for odontogenic infections (through clinical examination and full mouth intraoral and panoramic radiographs), and treating areas of active odontogenic infection.

For patients on IV therapy for oncologic indications, elective surgeries should be avoided if possible. When surgical procedures are necessary, patients should be followed closely postoperatively for any signs of infection or poor healing until the area is completely healed [92]. For patients on oral or IV bisphosphonates for non-oncologic indications, the AAOMS guidelines suggest no alteration in planned surgery for patients who have had less than 3 years of exposure to oral bisphosphonates, unless there has also been concomitant therapy with corticosteroids [12].

Patients on bisphosphonates therapy with BONJ

When sharp edges of necrotic bone traumatize the adjacent soft tissue, or when fragments are mobile, conservative nonsurgical sequestrectomy and/or smoothing of the involved bone provides symptomatic relief. When signs and symptoms of infection are present (i.e., inflamed surrounding soft tissue, purulence, and sinus tracts), management with both systemic antibiotics (e.g., penicillin, amoxicillin/clavulanate, clindaymycin, or metronidazole) and topical antimicrobial agents (chlorhexidine gluconate) are effective in reducing or eliminating signs and symptoms. However, long-term daily antibiotic therapy for several months may be necessary to prevent recurrence of infection in patients prone to relapse, and a subgroup of patients may require IV antibiotic therapy.

Surgical intervention has typically been reserved for those cases that are refractory to medical management with significant associated morbidity [4, 12]. Surgical management includes localized debridement and in some cases more aggressive resection of involved necrotic bone, and recent reports have demonstrated successful outcomes, with 87–100% of cases showing complete soft

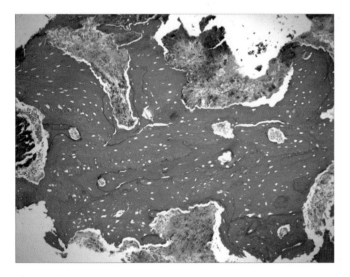

Fig. 112.7. Photomicrograph of necrotic bone with surface bacterial colonization (hematoxylin and eosin, magnification 100x).

Table 112.1. The American Association of Oral and Maxillofacial Surgeons Clinical Staging System for Bisphosphonate-Associated Osteonecrosis of the Jaw

Stage 0	Stage 1	Stage 2	Stage 3
No clinical evidence of exposed necrotic bone; presence of nonspecific clinical findings, and symptoms including sinus tact, localized periodontal pocket, and jaw pain	Exposed, necrotic bone that is asymptomatic	Exposed, necrotic bone associated with pain and infection	Exposed, necrotic bone with pain, infection, and one or more of the following: • pathologic fracture • extraoral fistula • osteolysis extending to the inferior border

tissue healing with variable follow-up periods [93–96]. One study demonstrated no statistical difference in outcomes when comparing surgical debridement alone for cases that failed antimicrobial therapy compared with antimicrobial therapy alone [97]. In addition to surgery and antimicrobial therapy, a number of other ancillary treatments have been reported primarily as case reports and small uncontrolled case series. These include hyperbaric oxygen therapy [98, 99] topically applied autologous platelet derived growth factor with surgery, low-level laser therapy, ozone therapy, pentoxifylline and tocopherol therapy, and teriparatide therapy [100–108]. Further research is necessary to better evaluate the potential benefits of these therapies.

Complete resolution of BONJ occurs in 18–53% of cases with or without intervention [5, 6, 14, 49, 109, 110]. However, it should be borne in mind that even if there is epithelialization and normalization of the visible mucosa, in most cases, the defect in the underlying bone does not change for months or years so that the term "complete resolution" should be interpreted with care.

Drug holiday and bisphosphonate dose reduction

Long-term discontinuation of IV bisphosphonates may be beneficial in stabilizing established lesions although it is unclear whether it prevents the development of new lesions because of the approximately 10-year half-life of the drug [12, 109]. Two studies showed that patients who discontinued bisphosphonate therapy demonstrated stabilization or resolution of BONJ compared to those who did not when followed for at least 6 months [5, 98, 111, 112]. Furthermore, animal studies have shown that epithelial healing is improved if bisphosphonates are discontinued for 2 weeks [35, 113]. Regardless, due to the growing recognition of BONJ as a potentially serious clinical complication of bisphosphonate therapy, a number of institutions have modified treatment protocols and reduced the dose and/or frequency of infusions according to disease signs and symptoms [114–118].

At this time, monitoring of patients consists primarily of history and periodic examination. Radiographic evaluation may be of some utility in assessing progression over time. There have been no controlled studies of serologic markers for bone turnover such as bone specific alkaline phosphatase, N-telopeptide cross-linked (NTX) and C-telopeptide cross-linked (CTX) that show a statistically significant relationship with the development of BONJ [5, 50, 119]. Furthermore, such measurements reflect remodeling activity in the entire skeletal system and not the jaws alone, and there is wide variation within the population. Similar to treatments for BONJ, prospective controlled studies that demonstrate the usefulness and cost-effectiveness of using such markers are necessary before their clinical application can be justified [12].

OSTEONECROSIS OF THE JAW ASSOCIATED WITH NON-BISPHOSPHONATE MEDICATIONS

Recent reports indicate that osteonecrosis may also occur in association with novel agents used in the treatment of osteoporosis and malignancies such as denosumab and bevacizumab. Denosumab is a human monoclonal antibody to receptor activator of nuclear factor-κB ligand (RANKL) that was FDA approved for the management of osteoporosis and metastatic cancer [120]. In a phase 3 study of denosumab compared to zoledronic acid in prostate cancer patients, the reported cumulative incidence of osteonecrosis with denusumab was 1% compared to 2% with zoledronic acid, further suggesting a key role for osteoclast suppression in the pathogenesis of BONJ [121].

Antiangiogenic agents, including bevacizumab, sunitinib, and sorafenib, which have been used in management of cancer patients, have been linked to osteonecrosis of the jaw [122, 123], especially when combined with bisphosphonates [123–127]. Further investigation of the potential role of antiangiogenic agents in the pathobiology of osteonecrosis is needed.

REFERENCES

1. Edwards BJ, Gounder M, McKoy JM, Boyd I, Farrugia M, Migliorati C, Marx R, Ruggiero S, Dimopoulos M, Raisch DW, Singhal S, Carson K, Obadina E, Trifilio S, West D, Mehta J, Bennett CL. 2008. Pharmacovigilance and reporting oversight in US FDA fast-track process: Bisphosphonates and osteonecrosis of the jaw. *Lancet Oncol* 9(12): 1166–72.
2. Migliorati CA. 2003. Bisphosphanates and oral cavity avascular bone necrosis. *J Clin Oncol* 21(22): 4253–4.
3. Marx RE. 2003. Pamidronate (Aredia) and zoledronate (Zometa) induced avascular necrosis of the jaws: A growing epidemic. *J Oral Maxillofac Surg* 61(9): 1115–7.
4. Ruggiero SL, Mehrotra B, Rosenberg TJ, Engroff SL. 2004. Osteonecrosis of the jaws associated with the use of bisphosphonates: A review of 63 cases. *J Oral Maxillofac Surg* 62(5): 527–34.
5. Marx RE, Cillo JE Jr, Ulloa JJ. 2007. Oral bisphosphonate-induced osteonecrosis: Risk factors, prediction of risk using serum CTX testing, prevention, and treatment. *J Oral Maxillofac Surg* 65(12): 2397–410.
6. Yarom N, Yahalom R, Shoshani Y, Hamed W, Regev E, Elad S. 2007. Osteonecrosis of the jaw induced by orally administered bisphosphonates: Incidence, clinical features, predisposing factors and treatment outcome. *Osteoporos Int* 18(10): 1363–70.
7. Fitzpatrick SG, Stavropoulos MF, Bowers LM, Neuman AN, Hinkson DW, Green JG, Bhattacharyya I, Cohen DM. 2012. Bisphosphonate-related osteonecrosis of jaws in 3 osteoporotic patients with history of

oral bisphosphonate use treated with single yearly zoledronic acid infusion. *J Oral Maxillofac Surg* 70(2): 325–30.
8. Lee JJ, Cheng SJ, Wang YP, Jeng JH, Chiang CP, Kok SH. 2013. Osteonecrosis of the jaws associated with the use of yearly zoledronic acid: Report of 2 cases. *Head Neck* 35(1): E6–E10.
9. Polizzotto MN, Cousins V, Schwarer AP. 2006. Bisphosphonate-associated osteonecrosis of the auditory canal. *Br J Haematol* 132(1): 114.
10. Longo R, Castellana MA, Gasparini G. 2009. Bisphosphonate-related osteonecrosis of the jaw and left thumb. *J Clin Oncol* 27(35): e242–3.
11. Froelich K, Radeloff A, Kohler C, Mlynski R, Muller J, Hagen R, Kleinsasser NH. 2011. Bisphosphonate-induced osteonecrosis of the external ear canal: A retrospective study. *Eur Arch Otorhinolaryngol* 268(8): 1219–25.
12. Ruggiero SL, Dodson TB, Assael LA, Landesberg R, Marx RE, Mehrotra B. 2009. American Association of Oral and Maxillofacial Surgeons position paper on bisphosphonate-related osteonecrosis of the jaws—2009 update. *J Oral Maxillofac Surg* 67(5 Suppl): 2–12.
13. Mawardi H, Treister N, Richardson P, Anderson K, Munshi N, Faiella RA, Woo SB. 2009. Sinus tracts—An early sign of bisphosphonate-associated osteonecrosis of the jaws? *J Oral Maxillofac Surg* 67(3): 593–601.
14. Fedele S, Porter SR, D'Aiuto F, Aljohani S, Vescovi P, Manfredi M, Arduino PG, Broccoletti R, Musciotto A, Di Fede O, Lazarovici TS, Campisi G, Yarom N. 2010. Nonexposed variant of bisphosphonate-associated osteonecrosis of the jaw: A case series. *Am J Med* 123(11): 1060–4.
15. Marx RE. 1983. Osteoradionecrosis: A new concept of its pathophysiology. *J Oral Maxillofac Surg* 41(5): 283–8.
16. Peters E, Lovas GL, Wysocki GP. 1993. Lingual mandibular sequestration and ulceration. *Oral Surg Oral Med Oral Pathol* 75(6): 739–43.
17. Novak MJ. 1999. Necrotizing ulcerative periodontitis. *Ann Periodontol* 4(1): 74–8.
18. Houston JP, McCollum J, Pietz D, Schneck D. 2002. Alveolar osteitis: A review of its etiology, prevention, and treatment modalities. *Gen Dent* 50(5): 457–63; quiz 464–5.
19. Enwonwu CO, Falkler WA Jr, Phillips RS. 2006. Noma (cancrum oris). *Lancet* 368(9530): 147–56.
20. Advisory Task Force on Bisphosphonate-Related Osteonecrosis of the Jaws, American Association of Oral and Maxillofacial Surgeons. 2007. American Association of Oral and Maxillofacial Surgeons position paper on bisphosphonate-related osteonecrosis of the jaws. *J Oral Maxillofac Surg* 65(3): 369–76.
21. Khosla S, Burr D, Cauley J, Dempster DW, Ebeling PR, Felsenberg D, Gagel RF, Gilsanz V, Guise T, Koka S, McCauley LK, McGowan J, McKee MD, Mohla S, Pendrys DG, Raisz LG, Ruggiero SL, Shafer DM, Shum L, Silverman SL, Van Poznak CH, Watts N, Woo SB, Shane E. 2007. Bisphosphonate-associated osteonecrosis of the jaw: Report of a task force of the American Society for Bone and Mineral Research. *J Bone Miner Res* 22(10): 1479–91.
22. Shane E, Goldring S, Christakos S, Drezner M, Eisman J, Silverman S, Pendrys D. 2006. Osteonecrosis of the jaw: More research needed. *J Bone Miner Res* 21(10): 1503–5.
23. Jaeger P, Jones W, Baron R, Hayslett JP. 1984. Modulation of hypercalcemia by dichloromethane diphosphonate and by calcitonin in a model of chronic primary hyperparathyroidism in rats. *Prog Clin Biol Res* 168: 375–80.
24. Huja SS, Fernandez SA, Hill KJ, Li Y. 2006. Remodeling dynamics in the alveolar process in skeletally mature dogs. *Anat Rec A Discov Mol Cell Evol Biol* 288(12): 1243–9.
25. Groetz KA, Al-Nawas B. 2006. Persisting alveolar sockets-a radiologic symptom of BP-ONJ? *J Oral Maxillofac Surg* 64(10): 1571–2.
26. Allen MR, Burr DB. 2008. Mandible matrix necrosis in beagle dogs after 3 years of daily oral bisphosphonate treatment. *J Oral Maxillofac Surg* 66(5): 987–94.
27. Mashiba T, Mori S, Burr DB, Komatsubara S, Cao Y, Manabe T, Norimatsu H. 2005. The effects of suppressed bone remodeling by bisphosphonates on microdamage accumulation and degree of mineralization in the cortical bone of dog rib. *J Bone Miner Metab* 23 Suppl: 36–42.
28. Hoefert S, Schmitz I, Tannapfel A, Eufinger H. 2010. Importance of microcracks in etiology of bisphosphonate-related osteonecrosis of the jaw: A possible pathogenetic model of symptomatic and non-symptomatic osteonecrosis of the jaw based on scanning electron microscopy findings. *Clin Oral Investig* 14(3): 271–84.
29. Wood J, Bonjean K, Ruetz S, Bellahcene A, Devy L, Foidart JM, Castronovo V, Green JR. 2002. Novel antiangiogenic effects of the bisphosphonate compound zoledronic acid. *J Pharmacol Exp Ther* 302(3): 1055–61.
30. Santini D, Vincenzi B, Dicuonzo G, Avvisati G, Massacesi C, Battistoni F, Gavasci M, Rocci L, Tirindelli MC, Altomare V, Tocchini M, Bonsignori M, Tonini G. 2003. Zoledronic acid induces significant and long-lasting modifications of circulating angiogenic factors in cancer patients. *Clin Cancer Res* 9(8): 2893–7.
31. Ziebart T, Koch F, Klein MO, Guth J, Adler J, Pabst A, Al-Nawas B, Walter C. 2011. Geranylgeraniol—A new potential therapeutic approach to bisphosphonate associated osteonecrosis of the jaw. *Oral Oncol* 47(3): 195–201.
32. de Groen PC, Lubbe DF, Hirsch LJ, Daifotis A, Stephenson W, Freedholm D, Pryor-Tillotson S, Seleznick MJ, Pinkas H, Wang KK. 1996. Esophagitis associated with the use of alendronate. *N Engl J Med* 335(14): 1016–21.
33. Reid IR, Bolland MJ, Grey AB. 2007. Is bisphosphonate-associated osteonecrosis of the jaw caused by soft tissue toxicity? *Bone* 41(3): 318–20.

34. Sonis ST, Watkins BA, Lyng GD, Lerman MA, Anderson KC. 2008. Bony changes in the jaws of rats treated with zoledronic acid and dexamethasone before dental extractions mimic bisphosphonate-related osteonecrosis in cancer patients. *Oral Oncol* 45(2): 164–72.
35. Landesberg R, Cozin M, Cremers S, Woo V, Kousteni S, Sinha S, Garrett-Sinha L, Raghavan S. 2008. Inhibition of oral mucosal cell wound healing by bisphosphonates. *J Oral Maxillofac Surg* 66(5): 839–47.
36. Scheper MA, Badros A, Chaisuparat R, Cullen KJ, Meiller TF. 2009. Effect of zoledronic acid on oral fibroblasts and epithelial cells: A potential mechanism of bisphosphonate-associated osteonecrosis. *Br J Haematol* 144(5): 667–76.
37. Walter C, Klein MO, Pabst A, Al-Nawas B, Duschner H, Ziebart T. 2009. Influence of bisphosphonates on endothelial cells, fibroblasts, and osteogenic cells. *Clin Oral Investig* 14(1): 35–41.
38. Ravosa MJ, Ning J, Liu Y, Stack MS. 2011. Bisphosphonate effects on the behaviour of oral epithelial cells and oral fibroblasts. *Arch Oral Biol* 56(5): 491–8.
39. Altundal H, Guvener O. 2004. The effect of alendronate on resorption of the alveolar bone following tooth extraction. *Int J Oral Maxillofac Surg* 33(3): 286–93.
40. Hikita H, Miyazawa K, Tabuchi M, Kimura M, Goto S. 2009. Bisphosphonate administration prior to tooth extraction delays initial healing of the extraction socket in rats. *J Bone Miner Metab* 27(6): 663–72.
41. Allen MR, Kubek DJ, Burr DB, Ruggiero SL, Chu TM. 2011. Compromised osseous healing of dental extraction sites in zoledronic acid-treated dogs. *Osteoporos Int* 22(2): 693–702.
42. Zervas K, Verrou E, Teleioudis Z, Vahtsevanos K, Banti A, Mihou D, Krikelis D, Terpos E. 2006. Incidence, risk factors and management of osteonecrosis of the jaw in patients with multiple myeloma: A single-centre experience in 303 patients. *Br J Haematol* 134(6): 620–3.
43. Corso A, Varettoni M, Zappasodi P, Klersy C, Mangiacavalli S, Pica G, Lazzarino M. 2007. A different schedule of zoledronic acid can reduce the risk of the osteonecrosis of the jaw in patients with multiple myeloma. *Leukemia* 21(7): 1545–8.
44. Hoff AO, Toth BB, Altundag K, Johnson MM, Warneke CL, Hu M, Nooka A, Sayegh G, Guarneri V, Desrouleaux K, Cui J, Adamus A, Gagel RF, Hortobagyi GN. 2008. Frequency and risk factors associated with osteonecrosis of the jaw in cancer patients treated with intravenous bisphosphonates. *J Bone Miner Res* 23(6): 826–36.
45. Dimopoulos MA, Kastritis E, Anagnostopoulos A, Melakopoulos I, Gika D, Moulopoulos LA, Bamia C, Terpos E, Tsionos K, Bamias A. 2006. Osteonecrosis of the jaw in patients with multiple myeloma treated with bisphosphonates: Evidence of increased risk after treatment with zoledronic acid. *Haematologica* 91(7): 968–71.
46. Fehm T, Beck V, Banys M, Lipp HP, Hairass M, Reinert S, Solomayer EF, Wallwiener D, Krimmel M. 2009. Bisphosphonate-induced osteonecrosis of the jaw (ONJ): Incidence and risk factors in patients with breast cancer and gynecological malignancies. *Gynecol Oncol* 112(3): 605–9.
47. Woo SB, Hande K, Richardson PG. 2005. Osteonecrosis of the jaw and bisphosphonates. *N Engl J Med* 353(1): 99–102; discussion 99–102.
48. Marx RE, Sawatari Y, Fortin M, Broumand V. 2005. Bisphosphonate-induced exposed bone (osteonecrosis/osteopetrosis) of the jaws: Risk factors, recognition, prevention, and treatment. *J Oral Maxillofac Surg* 63(11): 1567–75.
49. Van den Wyngaert T, Huizing MT, Vermorken JB. 2006. Bisphosphonates and osteonecrosis of the jaw: Cause and effect or a post hoc fallacy? *Ann Oncol* 17(8): 1197–204.
50. Kunchur R, Need A, Hughes T, Goss A. 2009. Clinical investigation of C-terminal cross-linking telopeptide test in prevention and management of bisphosphonate-associated osteonecrosis of the jaws. *J Oral Maxillofac Surg* 67(6): 1167–73.
51. Maerevoet M, Martin C, Duck L. 2005. Osteonecrosis of the jaw and bisphosphonates. *N Engl J Med* 353(1): 99–102; discussion 99–102.
52. Amarasena IU, Walters JA, Wood-Baker R, Fong K. 2008. Platinum versus non-platinum chemotherapy regimens for small cell lung cancer. *Cochrane Database Syst Rev* (4): CD006849.
53. Kyrgidis A, Vahtsevanos K, Koloutsos G, Andreadis C, Boukovinas I, Teleioudis Z, Patrikidou A, Triaridis S. 2008. Bisphosphonate-related ostenecrosis of the jaws: A case-control study of risk factors in breast cancer patients. *J Clin Oncol* 26(28): 4634–8.
54. Sarasquete ME, Garcia-Sanz R, Marin L, Alcoceba M, Chillon MC, Balanzategui A, Santamaria C, Rosinol L, de la Rubia J, Hernandez MT, Garcia-Navarro I, Lahuerta JJ, Gonzalez M, San Miguel JF. 2008. Bisphosphonate-related osteonecrosis of the jaw is associated with polymorphisms of the cytochrome P450 CYP2C8 in multiple myeloma: A genome-wide single nucleotide polymorphism analysis. *Blood* 112(7): 2709–12.
55. Raje N, Woo SB, Hande K, Yap JT, Richardson PG, Vallet S, Treister N, Hideshima T, Sheehy N, Chhetri S, Connell B, Xie W, Tai YT, Szot-Barnes A, Tian M, Schlossman RL, Weller E, Munshi NC, Van Den Abbeele AD, Anderson KC. 2008. Clinical, radiographic, and biochemical characterization of multiple myeloma patients with osteonecrosis of the jaw. *Clin Cancer Res* 14(8): 2387–95.
56. Katz J, Gong Y, Salmasinia D, Hou W, Burkley B, Ferreira P, Casanova O, Langaee TY, Moreb JS. 2011. Genetic polymorphisms and other risk factors associated with bisphosphonate induced osteonecrosis of the jaw. *Int J Oral Maxillofac Surg* 40(6): 605–11.
57. Khamaisi M, Regev E, Yarom N, Avni B, Leiterdorf E, Raz I, Elad S. 2007. Possible association between diabetes and bisphosphonate-related jaw osteonecrosis. *J Clin Endocrinol Metab* 92(3): 1172–5.

58. Badros A, Terpos E, Katodritou E, Goloubeva O, Kastritis E, Verrou E, Zervas K, Baer MR, Meiller T, Dimopoulos MA. 2008. Natural history of osteonecrosis of the jaw in patients with multiple myeloma. *J Clin Oncol* 26(36): 5904–9.
59. Hokugo A, Christensen R, Chung EM, Sung EC, Felsenfeld AL, Sayre JW, Garrett N, Adams JS, Nishimura I. 2010. Increased prevalence of bisphosphonate-related osteonecrosis of the jaw with vitamin D deficiency in rats. *J Bone Miner Res* 25(6): 1337–49.
60. Bamias A, Kastritis E, Bamia C, Moulopoulos LA, Melakopoulos I, Bozas G, Koutsoukou V, Gika D, Anagnostopoulos A, Papadimitriou C, Terpos E, Dimopoulos MA. 2005. Osteonecrosis of the jaw in cancer after treatment with bisphosphonates: Incidence and risk factors. *J Clin Oncol* 23(34): 8580–7.
61. Jadu F, Lee L, Pharoah M, Reece D, Wang L. 2007. A retrospective study assessing the incidence, risk factors and comorbidities of pamidronate-related necrosis of the jaws in multiple myeloma patients. *Ann Oncol* 18(12): 2015–9.
62. Pozzi S, Marcheselli R, Sacchi S, Baldini L, Angrilli F, Pennese E, Quarta G, Stelitano C, Caparotti G, Luminari S, Musto P, Natale D, Broglia C, Cuoghi A, Dini D, Di Tonno P, Leonardi G, Pianezze G, Pitini V, Polimeno G, Ponchio L, Masini L, Musso M, Spriano M, Pollastri G. 2007. Bisphosphonate-associated osteonecrosis of the jaw: A review of 35 cases and an evaluation of its frequency in multiple myeloma patients. *Leuk Lymphoma* 48(1): 56–64.
63. Wang EP, Kaban LB, Strewler GJ, Raie N, Troulis MJ. 2007. Incidence of osteonecrosis of the jaw in patients with multiple myeloma and breast or prostate cancer on intravenous bisphosphonate therapy. *J Oral Maxillofac Surg* 65(7): 1328–31.
64. Morgan GJ, Davies FE, Gregory WM, Cocks K, Bell SE, Szubert AJ, Navarro-Coy N, Drayson MT, Owen RG, Feyler S, Ashcroft AJ, Ross F, Byrne J, Roddie H, Rudin C, Cook G, Jackson GH, Child JA. 2010. First-line treatment with zoledronic acid as compared with clodronic acid in multiple myeloma (MRC Myeloma IX): A randomised controlled trial. *Lancet* 376(9757): 1989–99.
65. Mavrokokki T, Cheng A, Stein B, Goss A. 2007. Nature and frequency of bisphosphonate-associated osteonecrosis of the jaws in Australia. *J Oral Maxillofac Surg* 65(3): 415–23.
66. Hong JW, Nam W, Cha IH, Chung SW, Choi HS, Kim KM, Kim KJ, Rhee Y, Lim SK. 2010. Oral bisphosphonate-related osteonecrosis of the jaw: The first report in Asia. *Osteoporos Int* 21(5): 847–53.
67. Lo JC, O'Ryan FS, Gordon NP, Yang J, Hui RL, Martin D, Hutchinson M, Lathon PV, Sanchez G, Silver P, Chandra M, McCloskey CA, Staffa JA, Willy M, Selby JV, Go AS. 2010. Prevalence of osteonecrosis of the jaw in patients with oral bisphosphonate exposure. *J Oral Maxillofac Surg* 68(2): 243–53.
68. Wilkinson GS, Kuo YF, Freeman JL, Goodwin JS. 2007. Intravenous bisphosphonate therapy and inflammatory conditions or surgery of the jaw: A population-based analysis. *J Natl Cancer Inst* 99(13): 1016–24.
69. Cartsos VM, Zhu S, Zavras AI. 2008. Bisphosphonate use and the risk of adverse jaw outcomes: a medical claims study of 714,217 people. *J Am Dent Assoc* 139(1): 23–30.
70. Woo SB, Hellstein JW, Kalmar JR. 2006. Narrative [corrected] review: Bisphosphonates and osteonecrosis of the jaws. *Ann Intern Med* 144(10): 753–61.
71. Goldman ML, Denduluri N, Berman AW, Sausville R, Guadagnini JP, Kleiner DE, Brahim JS, Swain SM. 2006. A novel case of bisphosphonate-related osteonecrosis of the torus palatinus in a patient with metastatic breast cancer. *Oncology* 71(3-4): 306–8.
72. Miksad RA, Lai KC, Dodson TB, Woo SB, Treister NS, Akinyemi O, Bihrle M, Maytal G, August M, Gazelle GS, Swan JS. 2011. Quality of life implications of bisphosphonate-associated osteonecrosis of the jaw. *Oncologist* 16(1): 121–32.
73. Kumar SK, Gorur A, Schaudinn C, Shuler CF, Costerton JW, Sedghizadeh PP. 2010. The role of microbial biofilms in osteonecrosis of the jaw associated with bisphosphonate therapy. *Curr Osteoporos Rep* 8(1): 40–8.
74. Sedghizadeh PP, Kumar SK, Gorur A, Schaudinn C, Shuler CF, Costerton JW. 2008. Identification of microbial biofilms in osteonecrosis of the jaws secondary to bisphosphonate therapy. *J Oral Maxillofac Surg* 66(4): 767–75.
75. Naik NH, Russo TA. 2009. Bisphosphonate-related osteonecrosis of the jaw: The role of actinomyces. *Clin Infect Dis* 49(11): 1729–32.
76. Favia G, Pilolli GP, Maiorano E. 2009. Histologic and histomorphometric features of bisphosphonate-related osteonecrosis of the jaws: An analysis of 31 cases with confocal laser scanning microscopy. *Bone* 45(3): 406–13.
77. [No authors listed]. 2005. Alendronate (Fosamax) and risedronate (Actonel) revisited. *Med Lett Drugs Ther* 47(1207): 33–5.
78. Agarwala S, Jain D, Joshi VR, Sule A. 2005. Efficacy of alendronate, a bisphosphonate, in the treatment of AVN of the hip. A prospective open-label study. *Rheumatology (Oxford)* 44(3): 352–9.
79. Chiandussi S, Biasotto M, Dore F, Cavalli F, Cova MA, Di Lenarda R. 2006. Clinical and diagnostic imaging of bisphosphonate-associated osteonecrosis of the jaws. *Dentomaxillofac Radiol* 35(4): 236–43.
80. Catalano L, Del Vecchio S, Petruzziello F, Fonti R, Salvatore B, Martorelli C, Califano C, Caparrotti G, Segreto S, Pace L, Rotoli B. 2007. Sestamibi and FDG-PET scans to support diagnosis of jaw osteonecrosis. *Ann Hematol* 86(6): 415–23.
81. Treister N, Sheehy N, Bae EH, Friedland B, Lerman M, Woo S. 2009. Dental panoramic radiographic evaluation in bisphosphonate-associated osteonecrosis of the jaws. *Oral Dis* 15(1): 88–92.
82. Treister NS, Friedland B, Woo SB. 2010. Use of cone-beam computerized tomography for evaluation of

bisphosphonate-associated osteonecrosis of the jaws. *Oral Surg Oral Med Oral Pathol Oral Radiol Endod* 109(5): 753–64.
83. Ruggiero SL, Fantasia J, Carlson E. 2006. Bisphosphonate-related osteonecrosis of the jaw: background and guidelines for diagnosis, staging and management. *Oral Surg Oral Med Oral Pathol Oral Radiol Endod* 102(4): 433–41.
84. Bedogni A, Saia G, Ragazzo M, Bettini G, Capelli P, D'Alessandro E, Nocini PF, Lo Russo L, Lo Muzio L, Blandamura S. 2007. Bisphosphonate-associated osteonecrosis can hide jaw metastases. *Bone* 41(6): 942–5.
85. Merigo E, Manfredi M, Meleti M, Corradi D, Vescovi P. 2005. Jaw bone necrosis without previous dental extractions associated with the use of bisphosphonates (pamidronate and zoledronate): A four-case report. *J Oral Pathol Med* 34(10): 613–7.
86. Badros A, Weikel D, Salama A, Goloubeva O, Schneider A, Rapoport A, Fenton R, Gahres N, Sausville E, Ord R, Meiller T. 2006. Osteonecrosis of the jaw in multiple myeloma patients: Clinical features and risk factors. *J Clin Oncol* 24(6): 945–52.
87. Hansen T, Kunkel M, Weber A, James Kirkpatrick C. 2006. Osteonecrosis of the jaws in patients treated with bisphosphonates—Histomorphologic analysis in comparison with infected osteoradionecrosis. *J Oral Pathol Med* 35(3): 155–60.
88. Weinstein RS, Roberson PK, Manolagas SC. 2009. Giant osteoclast formation and long-term oral bisphosphonate therapy. *N Engl J Med* 360(1): 53–62.
89. Sedghizadeh PP, Kumar SK, Gorur A, Schaudinn C, Shuler CF, Costerton JW. 2009. Microbial biofilms in osteomyelitis of the jaw and osteonecrosis of the jaw secondary to bisphosphonate therapy. *J Am Dent Assoc* 140(10): 1259–65.
90. Russo T. 2005. Agents of actinomycosis. In: Mandell G, Bennett J, Dolin R (eds.) *Mandell, Douglas, and Bennett's Principles and Practice of Infectious Diseases, 6th Ed.* Philadelphia: Elsevier Churchill Livingston. pp. 2924–34.
91. Migliorati CA, Casiglia J, Epstein J, Jacobsen PL, Siegel MA, Woo SB. 2005. Managing the care of patients with bisphosphonate-associated osteonecrosis: An American Academy of Oral Medicine position paper. *J Am Dent Assoc* 136(12): 1658–68.
92. Ruggiero SL, Dodson TB, Assael LA, Landesberg R, Marx RE, Mehrotra B. 2009. American Association of Oral and Maxillofacial Surgeons position paper on bisphosphonate-related osteonecrosis of the jaw—2009 update. *Aust Endod J* 35(3): 119–30.
93. Carlson ER, Basile JD. 2009. The role of surgical resection in the management of bisphosphonate-related osteonecrosis of the jaws. *J Oral Maxillofac Surg* 67(5 Suppl): 85–95.
94. Williamson RA. 2010. Surgical management of bisphosphonate induced osteonecrosis of the jaws. *Int J Oral Maxillofac Surg* 39(3): 251–5.
95. Wilde F, Heufelder M, Winter K, Hendricks J, Frerich B, Schramm A, Hemprich A. 2011. The role of surgical therapy in the management of intravenous bisphosphonates-related osteonecrosis of the jaw. *Oral Surg Oral Med Oral Pathol Oral Radiol Endod* 111(2): 153–63.
96. Bedogni A, Saia G, Bettini G, Tronchet A, Totola A, Bedogni G, Ferronato G, Nocini PF, Blandamura S. 2011. Long-term outcomes of surgical resection of the jaws in cancer patients with bisphosphonate-related osteonecrosis. *Oral Oncol* 47(5): 420–4.
97. Scoletta M, Arduino PG, Dalmasso P, Broccoletti R, Mozzati M. 2010. Treatment outcomes in patients with bisphosphonate-related osteonecrosis of the jaws: A prospective study. *Oral Surg Oral Med Oral Pathol Oral Radiol Endod* 110(1): 46–53.
98. Magopoulos C, Karakinaris G, Telioudis Z, Vahtsevanos K, Dimitrakopoulos I, Antoniadis K, Delaroudis S. 2007. Osteonecrosis of the jaws due to bisphosphonate use. A review of 60 cases and treatment proposals. *Am J Otolaryngol* 28(3): 158–63.
99. Freiberger JJ, Padilla-Burgos R, Chhoeu AH, Kraft KH, Boneta O, Moon RE, Piantadosi CA. 2007. Hyperbaric oxygen treatment and bisphosphonate-induced osteonecrosis of the jaw: a case series. *J Oral Maxillofac Surg* 65(7): 1321–7.
100. Vescovi P, Merigo E, Meleti M, Manfredi M. 2006. Bisphosphonate-associated osteonecrosis (BON) of the jaws: A possible treatment? *J Oral Maxillofac Surg* 64(9): 1460–2.
101. Adornato MC, Morcos I, Rozanski J. 2007. The treatment of bisphosphonate-associated osteonecrosis of the jaws with bone resection and autologous platelet-derived growth factors. *J Am Dent Assoc* 138(7): 971–7.
102. Agrillo A, Ungari C, Filiaci F, Priore P, Iannetti G. 2007. Ozone therapy in the treatment of avascular bisphosphonate-related jaw osteonecrosis. *J Craniofac Surg* 18(5): 1071–5.
103. Harper RP, Fung E. 2007. Resolution of bisphosphonate-associated osteonecrosis of the mandible: Possible application for intermittent low-dose parathyroid hormone [rhPTH(1-34)]. *J Oral Maxillofac Surg* 65(3): 573–80.
104. Manfredi M, Merigo E, Guidotti R, Meleti M, Vescovi P. 2011. Bisphosphonate-related osteonecrosis of the jaws: A case series of 25 patients affected by osteoporosis. *Int J Oral Maxillofac Surg* 40(3): 277–84.
105. Lau AN, Adachi JD. 2009. Resolution of osteonecrosis of the jaw after teriparatide [recombinant human PTH-(1-34)] therapy. *J Rheumatol* 36(8): 1835–7.
106. Cheung A, Seeman E. 2010. Teriparatide therapy for alendronate-associated osteonecrosis of the jaw. *N Engl J Med* 363(25): 2473–4.
107. Bashutski JD, Eber RM, Kinney JS, Benavides E, Maitra S, Braun TM, Giannobile WV, McCauley LK. 2010. Teriparatide and osseous regeneration in the oral cavity. *N Engl J Med* 363(25): 2396–405.
108. Epstein MS, Wicknick FW, Epstein JB, Berenson JR, Gorsky M. 2010. Management of bisphosphonate-associated osteonecrosis: Pentoxifylline and tocopherol

108. in addition to antimicrobial therapy. An initial case series. *Oral Surg Oral Med Oral Pathol Oral Radiol Endod* 110(5): 593–6.
109. Van den Wyngaert T, Claeys T, Huizing MT, Vermorken JB, Fossion E. 2009. Initial experience with conservative treatment in cancer patients with osteonecrosis of the jaw (ONJ) and predictors of outcome. *Ann Oncol* 20(2): 331–6.
110. Lazarovici TS, Yahalom R, Taicher S, Elad S, Hardan I, Yarom N. 2009. Bisphosphonate-related osteonecrosis of the jaws: A single-center study of 101 patients. *J Oral Maxillofac Surg* 67(4): 850–5.
111. Kwon YD, Kim DY, Ohe JY, Yoo JY, Walter C. 2009. Correlation between serum C-terminal cross-linking telopeptide of type I collagen and staging of oral bisphosphonate-related osteonecrosis of the jaws. *J Oral Maxillofac Surg* 67(12): 2644–8.
112. Narongroeknawin P, Danila MI, Humphreys LG Jr, Barasch A, Curtis JR. 2010. Bisphosphonate-associated osteonecrosis of the jaw, with healing after teriparatide: A review of the literature and a case report. *Spec Care Dentist* 30(2): 77–82.
113. Scheper M, Chaisuparat R, Cullen K, Meiller T. 2010. A novel soft-tissue in vitro model for bisphosphonate-associated osteonecrosis. *Fibrogenesis Tissue Repair* 3: 6.
114. Lacy MQ, Dispenzieri A, Gertz MA, Greipp PR, Gollbach KL, Hayman SR, Kumar S, Lust JA, Rajkumar SV, Russell SJ, Witzig TE, Zeldenrust SR, Dingli D, Bergsagel PL, Fonseca R, Reeder CB, Stewart AK, Roy V, Dalton RJ, Carr AB, Kademani D, Keller EE, Viozzi CF, Kyle RA. 2006. Mayo clinic consensus statement for the use of bisphosphonates in multiple myeloma. *Mayo Clin Proc* 81(8): 1047–53.
115. Kyle RA, Yee GC, Somerfield MR, Flynn PJ, Halabi S, Jagannath S, Orlowski RZ, Roodman DG, Twilde P, Anderson K. 2007. American Society of Clinical Oncology 2007 clinical practice guideline update on the role of bisphosphonates in multiple myeloma. *J Clin Oncol* 25(17): 2464–72.
116. Weitzman R, Sauter N, Eriksen EF, Tarassoff PG, Lacerna LV, Dias R, Altmeyer A, Csermak-Renner K, McGrath L, Lantwicki L, Hohneker JA. 2007. Critical review: Updated recommendations for the prevention, diagnosis, and treatment of osteonecrosis of the jaw in cancer patients—May 2006. *Crit Rev Oncol Hematol* 62(2): 148–52.
117. Terpos E, Sezer O, Croucher PI, Garcia-Sanz R, Boccadoro M, San Miguel J, Ashcroft J, Blade J, Cavo M, Delforge M, Dimopoulos MA, Facon T, Macro M, Waage A, Sonneveld P. 2009. The use of bisphosphonates in multiple myeloma: Recommendations of an expert panel on behalf of the European Myeloma Network. *Ann Oncol* 20(8): 1303–17.
118. Watts NB, Diab DL. 2010. Long-term use of bisphosphonates in osteoporosis. *J Clin Endocrinol Metab* 95(4): 1555–65.
119. Lee CY, Suzuki JB. 2010. CTX biochemical marker of bone metabolism. Is it a reliable predictor of bisphosphonate-associated osteonecrosis of the jaws after surgery? Part II: A prospective clinical study. *Implant Dent* 19(1): 29–38.
120. Lipton A, Steger GG, Figueroa J, Alvarado C, Solal-Celigny P, Body JJ, de Boer R, Berardi R, Gascon P, Tonkin KS, Coleman RE, Paterson AH, Gao GM, Kinsey AC, Peterson MC, Jun S. 2008. Extended efficacy and safety of denosumab in breast cancer patients with bone metastases not receiving prior bisphosphonate therapy. *Clin Cancer Res* 14(20): 6690–6.
121. Fizazi K, Carducci M, Smith M, Damiao R, Brown J, Karsh L, Milecki P, Shore N, Rader M, Wang H, Jiang Q, Tadros S, Dansey R, Goessl C. 2011. Denosumab versus zoledronic acid for treatment of bone metastases in men with castration-resistant prostate cancer: A randomised, double-blind study. *Lancet* 377(9768): 813–22.
122. Miller K, Wang M, Gralow J, Dickler M, Cobleigh M, Perez EA, Shenkier T, Cella D, Davidson NE. 2007. Paclitaxel plus bevacizumab versus paclitaxel alone for metastatic breast cancer. *N Engl J Med* 357(26): 2666–76.
123. Christodoulou C, Pervena A, Klouvas G, Galani E, Falagas ME, Tsakalos G, Visvikis A, Nikolakopoulou A, Acholos V, Karapanagiotidis G, Batziou E, Skarlos DV. 2009. Combination of bisphosphonates and anti-angiogenic factors induces osteonecrosis of the jaw more frequently than bisphosphonates alone. *Oncology* 76(3): 209–11.
124. Estilo CL, Fornier M, Farooki A, Carlson D, Bohle G 3rd, Huryn JM. 2008. Osteonecrosis of the jaw related to bevacizumab. *J Clin Oncol* 26(24): 4037–8.
125. Greuter S, Schmid F, Ruhstaller T, Thuerlimann B. 2008. Bevacizumab-associated osteonecrosis of the jaw. *Ann Oncol* 19(12): 2091–2.
126. Aragon-Ching JB, Ning YM, Chen CC, Latham L, Guadagnini JP, Gulley JL, Arlen PM, Wright JJ, Parnes H, Figg WD, Dahut WL. 2009. Higher incidence of osteonecrosis of the jaw (ONJ) in patients with metastatic castration resistant prostate cancer treated with anti-angiogenic agents. *Cancer Invest* 27(2): 221–6.
127. Guarneri V, Miles D, Robert N, Dieras V, Glaspy J, Smith I, Thomssen C, Biganzoli L, Taran T, Conte P. 2010. Bevacizumab and osteonecrosis of the jaw: Incidence and association with bisphosphonate therapy in three large prospective trials in advanced breast cancer. *Breast Cancer Res Treat* 122(1): 181–8.

113
Periodontal Diseases and Oral Bone Loss
Mary G. Lee and Keith L. Kirkwood

Introduction 941
Etiology 941
Diagnosis and Classification of Periodontal Diseases 942
Treatment 942
Future Perspectives 946
References 946
Recommended Databases 947

INTRODUCTION

Periodontal diseases constitute a variety of inflammatory conditions affecting the health of the periodontium. The primary etiological component is bacterial-derived plaque accumulation around teeth. The two major categories of periodontal diseases are gingivitis and periodontitis, which are distinguished from one another based upon the extent of tissue loss that directly supports the teeth. Gingivitis is limited to the soft tissues surrounding the teeth where the extension has not caused any bone loss whereas periodontitis is characterized by soft tissue and alveolar bone loss resulting in decreased supporting structures for the teeth.

ETIOLOGY

Although more than 500 bacterial species have been identified within the oral cavity, plaque-associated periodontal diseases are associated with a relatively narrow subset of periodontal pathogens [1]. Susceptibility to these mixed infections is often highly variable host immune response to these pathogens. Both genetic and environmental factors can influence the immune response and modify susceptibility to infection (Fig. 113.1). Periodontal pathogens produce harmful products and enzymes (e.g., hyaluronidases, collagenases, proteases) that break down extracellular matrices such as collagen as well as host cell membranes in order to produce nutrients for their growth [2] and subsequent tissue invasion (see Fig. 113.2). Many of the microbial surface proteins and lipopolysaccharide (LPS) molecules are responsible for eliciting an immune response in the host resulting in local tissue inflammation [3]. *P. gingivalis*, *A. actinomycetemcomitans*, and other periodontal pathogens possess multiple virulence factors such as cytoplasmic membranes, peptidoglycans, outer membrane proteins, LPS, capsules, and cell surface fimbriae (with *P. gingivalis*) [4]. Once immune and inflammatory processes are initiated, various inflammatory molecules such as proteases, matrix metalloproteinases (MMPs), cytokines, prostaglandins, and host enzymes are released from leukocytes, gingival fibroblasts, osteoblasts, or other tissue-derived cells [5, 6]. Proteases can degrade the collagen, permitting leukocyte infiltration [7]. Although the production of collagenase from infiltrating neutrophils and other resident periodontal tissues is part of the natural host response to infection, in periodontal disease there is an imbalance between the level of activated tissue-destroying MMPs and their endogenous inhibitors [8, 9].

The inflammatory infiltrate from the gingival tissue can initiate tissue and alveolar bone destruction through the activation of several proinflammatory cytokines, including interleukin (IL)-1β, tumor necrosis factor-α (TNFα), and IL-6. Within the diseased periodontal tissues, activated osteoclasts are an integral component of bone destruction [10, 11]. Multiple inflammatory signals have

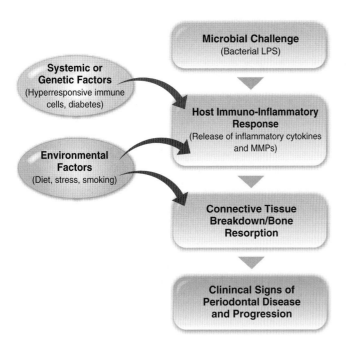

Fig. 113.1. Etiology of periodontal disease progression and factors that impact the disease progression. Periodontal diseases can occur when periodontal pathogenic bacteria is present in a susceptible host. Genetic and environmental factors modify the host immune response toward bacteria, initiating tissue and bone destruction manifested as periodontitis.

been shown to modulate the receptor activator of nuclear factor-κB ligand (RANKL), RANK, or osteoprotegerin (OPG) within periodontal tissues [12]. In the presence of periodontal pathogens (e.g., A. actinomycetemcomitans), CD4+ T cells display increased expression of RANKL, triggering the activation of osteoclasts that leads to bone loss [13]. As long as the subgingival plaque persists, and this increase in microbial density propagates, the destructive periodontal lesion will remain. As the pocket deepens, the flora becomes more anaerobic, and the host response becomes more destructive and chronic. Periodontal tissue destruction can result in tooth loss if left unabated [14, 15].

DIAGNOSIS AND CLASSIFICATION OF PERIODONTAL DISEASES

Despite our current appreciation regarding the etiology and progression of periodontal diseases, the diagnosis and classification is still made from clinical assessments [16, 17]. Information routinely collected includes medical history and history of previous/current periodontal problems, along with a thorough clinical exam. Clinical parameters of inflammation (e.g., bleeding upon probing into the periodontal sulcus), pocket depth measurements, clinical attachment levels, evidence of plaque/calculus, and tooth mobility are used to arrive at a clinical diagnosis. Radiographic assessments of the extent of bone loss are primary criteria to determine if bone loss has occurred.

Plaque-induced gingival diseases have traditionally been divided into two general categories: gingivitis and periodontitis [16]. Gingivitis is characterized by inflammation without any loss of periodontal supporting structures (i.e., connective tissue and bone), whereas the hallmark of periodontitis is pathological detachment of connective tissue leading to the resorption of alveolar bone support around teeth. In 1999, the International Workshop of Periodontal Diseases and Conditions recognized a new classification system of seven different forms of plaque induced periodontal diseases [17]. See Table 113.1 for the classification and differences between the various types of periodontal diseases. Table 113.2 shows the diagnosis of periodontal diseases.

TREATMENT

In general, periodontal therapy is directed toward the reduction and elimination of inflammation, thereby allowing the gingival tissue to heal. Both personal and professional maintenance is critical in preventing the recurrence of inflammation. Since gingivitis is defined as a reversible condition where no bone loss has been documented, therapy for gingivitis is usually limited plaque removal and personal plaque control instructions.

Therapy for periodontitis varies considerably depending upon the severity of attachment loss, anatomical location variations, along with type of periodontitis and the therapeutic objectives [18]. Since periodontitis destroys the supporting alveolar bone and surrounding connective tissues, therapy is directed toward arresting the progression of ongoing disease and resolution of inflammation along with repair or regeneration of destroyed tissues. Therapeutic approaches for periodontitis fall into two main categories: (1) nonsurgical therapy and (2) surgical therapy (see Tables 113.3 and 113.4) [18].

The most effective nonsurgical intervention is scaling and root planing combined with personal plaque control. Clear benefits have been validated from scaling and root planning, including reduction of inflammation, decreased pocket depths, gain of clinical attachment, and decreased progression of disease [18].

Adjunctive pharmacological agents have been used to manage both gingivitis periodontitis. The benefit of topical antibacterial agents to help reduce bacterial plaque and prevent gingivitis has been documented in clinical trials. There are three agents currently accepted by the American Dental Association (ADA) approved as antigingival agents (either mouth rinses or dentifrices) [19]. One product lists essential oils (thymol, menthol, eucalyptol) and methyl salicylate as the active ingredients. The other two contain either chlorhexidine gluconate or triclosan. To carry the ADA seal, these agents must show plaque and inflammation reduction in 6-month trials

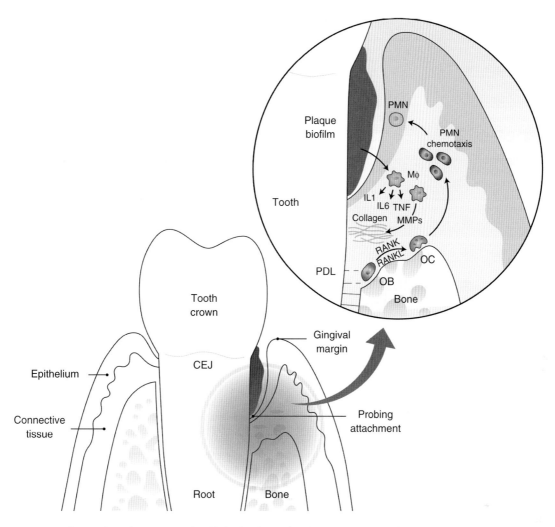

Fig. 113.2. Anatomy of periodontal tissues and cellular biology of inflammatory bone loss. The periodontal tissues surrounding the tooth include epithelium, connective tissue, periodontal ligament (PDL), and alveolar bone. The insert depicts cell responses to bacterial plaque biofilm on the root surface. Polymorphonuclear lymphocytes (PMNs) and macrophages (Mφ) secrete inflammatory cytokines and matrix metalloproteinases (MMPs) to increase the immune response and degrade the connective tissue matrix. Osteoclastogenesis is induced to these stimuli to increase bone resorption through the RANKL/RANK system.

without any adverse effects. For periodontitis, both systemic and locally delivered chemotherapeutics have been used successfully for management of periodontal diseases. Systemically administered antibiotics have been used when mechanical therapy has not been sufficient (Table 113.3) [19]. However, judicious use of antibiotics is recommended to treat acute infections or management of periodontitis in immune-compromised patients to avoid emergence of drug-resistant organisms.

Other systemic therapies focus on modulation of host-immune responses. Considerable efforts have focused on non-steroidal anti-inflammatory drugs (NSAIDS) and the collagenase inhibitor subantimicrobial dose doxycycline. Both have shown to have beneficial effects, although long-term administration with any systemic agent may have potential negative side effects (Table 113.3). Controlled local delivery within periodontal pockets can change the pathogenic microflora and improve the clinical parameters of periodontitis. The FDA has approved several systems, including vinyl acetate fibers containing tetracycline, biodegradable chips containing chlorhexidine, and a minocycline-containing polymer for use as adjuncts to scaling and root planing [19]. In addition, a doxycycline bio-absorbable polymer gel has been approved as a monotherapy for reduction of periodontal disease clinical parameters.

Surgical therapy is primarily needed to access and facilitate mechanical instrumentation of the root surface debridement. The main objectives of periodontal surgery are to decrease bacterial etiological factors (e.g., subgingival calculus), reduce the pocket depth, and regenerate lost periodontal supporting tissues. Flap access surgeries increase the surgeon's ability to instrument deep pocket depth areas and areas between roots (furcations) for more

Table 113.1. General Classification of Periodontal Diseases and Conditions*

Disease Classification	Comments
Gingivitis	Generally plaque-induced but can be modified by endocrine factors, medications, and malnutrition. No alveolar bone loss.
Chronic periodontitis	Alveolar bone loss can be localized or generalized. Not based upon age. Slow onset/progression.
Aggressive periodontitis	Alveolar bone loss can be localized or generalized. Not based upon age. Rapid onset/progression.
Periodontitis as a manifestation of systemic diseases	Can be associated with hematological (e.g., netropenia), or genetic disorders (e.g., Papillon-Lefèvre syndrome)
Necrotizing periodontal diseases	Could be gingival or periodontal forms
Abscesses of the periodontium	Could be gingival or periodontal abscess
Periodontitis associated with endodontic lesions	Associated with infections within the tooth

*Adapted from Ref. 17. Used with permission from the American Academy of Periodontology.

Table 113.2. Diagnosis of Periodontal Diseases

Diagnostic Assessments	Specific Parameter	Comments
Clinical	Probing depth	Shallow pocket depths associated with less disease progression
	Bleeding upon Probing	Lack of bleeding associated with less inflammation
	Radiographs	Absence of bone loss associated with less disease progression
Microbiological	Culture for periodontal pathogens	May be useful to detect exact pathogens in patients refractory to conventional treatment
	DNA/biochemical probes	Identify non-cultivatable organisms
Biomarkers	Inflammatory cytokines	IL-1 genotype may identify susceptible patients
	Pyridinoline cross-linked carboxyterminal telopeptide of type I collagen	Correlates well with clinical parameters and presence of periodontal pathogens

Table 113.3. Nonsurgical Treatment of Periodontitis

	Treatment Category	Treatment or Agent	Comments
Mechanical	Mechanical debridement	Scaling and root planing—manual instrumentation and ultrasonics	Decrease inflammation Decrease probing depth Increase clinical attachment levels
Chemotherapy	Mouth rinses	Chlorhexidine	ADA approved for reduction of gingivitis
	Dentifrices (toothpastes)	Triclosan Essential Oils	
	Locally applied; sustained release	Tetracyclines (doxycycline and minocycline) Chlorhexidine	Can be used as adjunctive therapies with scaling and root planing
	Systemic antibiotics	Tetracyclines Clindamycin Metronidazole/amoxicillin	May be used in more aggressive periodontitis cases
Host Response	MMP inhibitors	Low dose doxycycline	May decrease progression of attachment loss
	NSAIDs	Flurbiprophen	May decrease progression of attachment loss

Table 113.4. Surgical Treatment of Periodontitis

	Treatment Category	Clinical Procedure	Comments
Pocket Reduction	Pocket reduction surgery	Mucoperiosteal flap for access to debride root surface	Dec. pocket depth. Inc. patient access for oral hygiene. Can enhance access for restorative dentistry.
Regeneration	Guided tissue regeneration	Mucoperiosteal flap to gain access and a barrier membrane placed to facilitate new periodontal ligament, cementum, and bone on previously diseased roots	Can use resorbable or non-resorbable membranes with or without bone grafts. Dec. pocket depths. Inc. clinical attachment. Fill osseous defects.
	Bone grafts (autogenous, allogenic, xenografts)		Can use resorbable or non-resorbable membranes with or without bone grafts. Dec. pocket depths. Inc. clinical attachment. Fill osseous defects. Autogenous graft is gold standard.
	Biologicals (enamel-derived matrix proteins, BMPs, platelet-derived growth factors [PDGFs])		Fill osseous defects. Improved clinical attachment and dec. pocket depth. BMPs not FDA approved for intraoral use, PDGFs with absorbable carrier are FDA approved.
	Autologous growth factors (platelet-rich plasma, platelets rich in growth factors)	Whole blood drawn from the patient and centrifuged to separate and collect the platelet-rich layer	Enhance bone and soft tissue healing.
Replacement	Ridge augmentation (autogenous or allogenic block grafts, particulate grafts, sinus augmentations)	Mucoperiosteal flap for access to graft the insufficient alveolar ridge. Barrier membranes will be used to exclude the soft tissue and provide space maintenance for the graft	Grafting for height or width to provide adequate alveolar bone for implant placement. May be enhanced by using additional growth factors or biologicals.
	Titanium dental implants	Osteotomy prepared with irrigation and increasingly larger diameter drills into alveolar ridge	Replacement of missing teeth with surgically placed titanium dental implants followed by abutment and crown restorations.

complete calculus removal. Osseous resection has traditionally been performed to more effectively reduce pocket depths. However, the optimal goal is to restore lost periodontal tissues. Several different regenerative surgical strategies have been employed, including bone grafting (autogenous and allografts), along with guided tissue regeneration (GTR) with or without bone grafts [20]. A concentrated source of autologous growth factors can be obtained from the patient and used to enhance the regenerative procedures. In the past 10 years, several biological agents have been used to stimulate periodontal regeneration, including growth factors (e.g., platelet-derived growth factor), other proteins (e.g., enamel-derived growth factor), or synthetic peptides (e.g., a 15 amino acid type I collagen fragment) (see Table 113.4) [18, 20]. Recently, recombinant parathyroid hormone (PTH) administration was shown to improve clinical outcomes in the treatment of chronic periodontitis. Patients treated with PTH showed greater resolution of alveolar defects and accelerated osseous wound healing in the oral cavity [21].

The periodontal surgeon also addresses oral bone loss due to factors other than periodontal disease when treatment planning dental implants for missing teeth. Whether teeth were lost due to periodontal disease or decay, the remaining ridge may be inadequate for proper implant placement. Ridge augmentation to improve the maxillary or mandibular width and height of alveolar bone can be achieved with autograft, allograft blocks, or particulate bone. In addition to ridge augmentation, the maxillary sinus cavity causes insufficient height for implant placement in the posterior maxilla. A surgical procedure, termed "sinus augmentation" or "sinus lift" procedure can be performed to gain the necessary ridge height (See Table 113.4).

FUTURE PERSPECTIVES

Current basic research has focused on a newer field called "osteoimmunology," centered on the molecular mechanisms the immune system and the impact on inflammatory bone loss and the skeletal system. Clinical diagnostics that address novel salivary diagnostics measuring biomarkers that may be predictive of periodontal severity, susceptibility, or perhaps predict future disease are actively being pursued. In the clinical setting, tissue engineering and regeneration remain active areas of investigation and new biologicals undergo clinical trials to enhance predictability of grafting materials used for periodontal regeneration or dental implant therapy to replace missing teeth.

REFERENCES

1. [No Authors Listed]. 1999. The pathogenesis of periodontal diseases. *J Periodontol* 70(4): 457–70.
2. Bartold PM, Page RC. 1986. The effect of chronic inflammation on gingival connective tissue proteoglycans and hyaluronic acid. *J Oral Pathol* 15(7): 367–74.
3. Darveau RP, Tanner A, Page RC. 1997. The microbial challenge in periodontitis. *Periodontol 2000* 14: 12–32.
4. Offenbacher S. 1996. Periodontal diseases: Pathogenesis. *Ann Periodontol* 1(1): 821–78.
5. Graves DT, Jiang Y, Valente AJ. 1999. The expression of monocyte chemoattractant protein-1 and other chemokines by osteoblasts. *Front Biosci* 4: D571–80.
6. Graves, DT, Oskoui M, Volejnikova S, Naguib G, Cai S, Desta T, Kakouras A, Jiang Y. 2001. Tumor necrosis factor modulates fibroblast apoptosis, PMN recruitment, and osteoclast formation in response to P. gingivalis infection. *J Dent Res* 80(10): 1875–9.
7. Andrian E, Grenier D, Rouabhia M. 2004. In vitro models of tissue penetration and destruction by Porphyromonas gingivalis. *Infect Immun* 72(8): 4689–98.
8. Uchida M, Shima M, Shimoaka T, Fujieda A, Obara K, Suzuki H, Nagai Y, Ikeda T, Yamato H, Kawaguchi H. 2000. Regulation of matrix metalloproteinases (MMPs) and tissue inhibitors of metalloproteinases (TIMPs) by bone resorptive factors in osteoblastic cells. *J Cell Physiol* 185(2): 207–14.
9. Golub LM, Lee HM, Greenwald RA, Ryan ME, Sorsa T, Salo T, Giannobile WV. 1997. A matrix metalloproteinase inhibitor reduces bone-type collagen degradation fragments and specific collagenases in gingival crevicular fluid during adult periodontitis. *Inflamm Res* 46(8): 310–9.
10. Assuma R, Oates T, Cochran D, Amar S, Graves DT. 1998. IL-1 and TNF antagonists inhibit the inflammatory response and bone loss in experimental periodontitis. *J Immunol* 160(1): 403–9.
11. Crotti T, Smith MD, Hirsch R, Soukoulis S, Weedon H, Capone M, Ahern MJ, Haynes D. 2003. Receptor activator NF kappaB ligand (RANKL) and osteoprotegerin (OPG) protein expression in periodontitis. *J Periodontal Res* 38(4): 380–7.
12. Lerner UH. 2004. New molecules in the tumor necrosis factor ligand and receptor superfamilies with importance for physiological and pathological bone resorption. *Crit Rev Oral Biol Med* 15(2): 64–81.
13. Teng YT, Nguyen H, Gao X, Kong YY, Gorczynski RM, Singh B, Ellen RP, Penninger JM. 2000. Functional human T-cell immunity and osteoprotegerin ligand control alveolar bone destruction in periodontal infection. *J Clin Invest* 106(6): R59–67.
14. Kinane DF, Lappin DF. 2001. Clinical, pathological and immunological aspects of periodontal disease. *Acta Odontol Scand* 59(3): 154–60.
15. Lappin DF, MacLeod CP, Kerr A, Mitchell T, Kinane DF. 2001. Anti-inflammatory cytokine IL-10 and T cell cytokine profile in periodontitis granulation tissue. *Clin Exp Immunol* 123(2): 294–300.
16. Armitage GC; Research, Science and Therapy Committee of the American Academy of Periodontology. 2003. Diagnosis of peridontal diseases. *J Periodontol* 74(8): 1237–47.

17. Armitage GC. 1999. Development of a classification system for periodontal diseases and conditions. *Ann Periodontol* 4(1): 1–6.
18. Research, Science and Therapy Committee of the American Academy of Periodontology. 2001. Treatment of plaque-induced gingivitis, chronic periodontitis, and other clinical conditions. *J Periodontol* 72(12): 1790–1800.
19. Oringer RJ; Research, Science, and Therapy Committee of the American Academy of Periodontology. 2002. Modulation of host response in periodontal therapy. *J Periodontol* 73(4): 460–70.
20. Wang HL, Greenwell H, Fiorellini J, Giannobile W, Offenbacher S, Salkin L, Townsend C, Sheridan P, Genco RJ; Research, Science and Therapy Committee. 2005. Periodontal regeneration. *J Periodontol* 76(9): 1601–22.
21. Bashutski JD, Eber RM, Kinney JS, Benavides E, Maitra S, Braun TM, Giannobile WV, McCauley LK. 2010. Teriparatide and osseous regeneration in the oral cavity. *N Engl J Med* 363(25): 2396–405.

RECOMMENDED DATABASES

http://www.ada.org/prof/resources/topics/gum.asp
http://www.perio.org/resources-products/posppr2.html

114
Oral Manifestations of Metabolic Bone Diseases

Roberto Civitelli and Charles Hildebolt

Changes in Oral Bone Mass and Structure with Aging 948
Methods to Assess Oral Bone Loss 948
Relationship between Oral and Postcranial Bone Mass 950
Osteoporosis and Tooth Loss 951
Mechanisms of Oral Bone Loss 952
Assessment of Oral Bone as a Screening Tool for Osteoporosis 952
Oral Bone Loss as a Predictor of Fractures 953
Impact of Osteoporosis Therapy on Oral Bone Health 953
Paget's Bone Disease 954
Primary Hyperparathyroidism 954
Renal Osteodystrophy 954
Summary 954
References 955

CHANGES IN ORAL BONE MASS AND STRUCTURE WITH AGING

As the structure of the mandibular and maxillary bones is similar to that of the rest of the skeleton, it is reasonable to expect that with aging these skeletal segments undergo a process of bone loss, with decreased trabeculation, cortical thinning, and increased porosity, similar to what occurs in the rest of the skeleton. Based on standard bone histology, it has been known since the 1960s that there is an age-dependent decrease in mandibular and maxillary bone mass. Thus, oral bone cortical porosity increases with age, and the severity of this process depends upon the area of oral bone studied [1]. The porosity of the buccal cortex of the mandible is dependent upon the presence or absence of teeth, and trabecular bone loss is observed with tooth extraction. As a corollary to the increased cortical porosity, an increase in endosteal trabeculation also occurs with age. Studies in cadavers also showed that bone mass decreases concordantly with age in the mandible and the radius, while cortical porosity increases. Furthermore, in females, femoral porosity is inversely associated with the height of the alveolar bone [2].

METHODS TO ASSESS ORAL BONE LOSS

Dentists typically use clinical measures to assess the loss of the tooth attachment apparatus, which is in part, though not entirely, due to loss of alveolar bone. The attachment consists of the tissues that support the tooth and includes the gingiva, periodontal ligament, alveolar bone, and tooth cementum. A periodontal probe is inserted into the gingival sulcus until it encounters firm resistance from the base of the sulcus (Fig. 114.1). On the probe, the distance from the base of the sulcus to the tooth's cementoenamel junction (CEJ) is measured and is termed "attachment loss" (AL). In many cases, AL is used as a surrogate measurement for bone loss. Probing depth, another clinical parameter used in dental medicine and clinical studies, is the distance from the depth of the sulcus to the crest of the gingiva, with depths less than 3 mm usually considered clinically acceptable.

Techniques that more directly assess oral bone loss typically involve the use of radiographic measurements. The most common include measures of alveolar crestal height (ACH) and assessment of alveolar crestal height (ACH). ACH is commonly assessed with dental radiographic images by measuring (adjacent to the tooth) the

---

Primer on the Metabolic Bone Diseases and Disorders of Mineral Metabolism, Eighth Edition. Edited by Clifford J. Rosen.
© 2013 American Society for Bone and Mineral Research. Published 2013 by John Wiley & Sons, Inc.

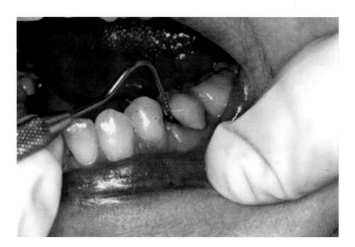

Fig. 114.1. Clinical measurement of attachment loss. A periodontal probe (with mm markings) is inserted into the gingival sulcus until it encounters firm resistance from the base of the sulcus, and the distance from the base of the sulcus to the cementoenamel junction (CEJ) is measured as attachment loss. Probing depth is the distance from the depth of the sulcus to the crest of the gingiva.

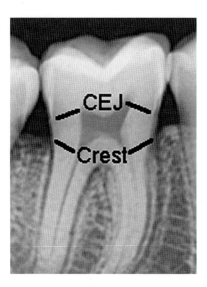

Fig. 114.2. Cementoenamel junction (CEJ) and alveolar crest as used to measure alveolar crest height (ACH) on a dental radiographic image. Logarithmic image, enhanced for display.

distance from the CEJ to the top of the alveolar bone crest (Fig. 114.2). ACH is usually measured on the mesial (toward the front of the mouth) and distal (toward the back of the mouth) sides of teeth and is reported as the average loss of bone height for all teeth measured in the mouth (mean ACH). The larger the mean ACH, the worse the bone loss surrounding the teeth. After a tooth is lost, there is loss of the bone in that region. This is called residual ridge resorption (RRR). Most often, the extent of RRR is described in edentulous subjects, but RRR can be used in dentate subjects who have lost one or more teeth (partially edentulous). BMD of oral bone has been assessed using a variety of techniques including dual-energy X-ray absorptiometry (DXA), single photon absorptiometry (SPA), dual photon absorptiometry (DPA), quantitative computed tomography (QCT), and radiographic absorptiometry (RA). For clinical research, the most widely used images for oral bone assessment are digital radiographic images (periapical, bitewing, and panoramic). This is because these images are commonly available and relatively inexpensive. These images can be obtained by digitizing plain films or images on phosphor plates, or by using images captured by charge-coupled devices (CCDs), complementary metal oxide semiconductors (CMOSs), or charge induction devices (CIDs). All imaging techniques are limited by cost, accuracy, reliability, precision, and/or practicality. QCT provides a valuable assessment of oral BMD because it allows for assessment of density in regions that with other modalities would be obstructed by teeth. However, it is expensive, and there is high exposure to radiation. Cone beam computed tomography (CBCT) has become widely available in dentistry. Although linear measurements using this technology appear to be relatively accurate and reliable, and images of oral bone can be studied in three dimensions, there is a large amount of scattered radiation, which makes CBCT determinations of BMDs problematic. SPA has been used to assess the maxilla, and DPA and DXA can be used to assess the mandible; however, positioning, reproducibility, and obstruction of teeth in dentate subjects make BMD measurement difficult.

Our group uses a vacuum-coupled, positioning device and phosphor plates to obtain images that can be used to determine changes in oral bone (Fig 114.3). Using this repositioning device, two images taken at different times are registered by minimizing trabecular noise and subtracted (Fig. 114.4). With this approach, we are able to achieve a least significant change (threshold that represents real biological change with 95% confidence) of 0.06 mm for crest height change, compared to 0.49 mm least significant changes for cementoenamel junction-alveolar crestal (CEJ-AC) distance [3]. This dramatic improvement in precision of oral bone loss determination is due in large part to the fact that this image subtraction approach overcomes the inherent limitation of standard CEJ-AC measurements, which rely on two-dimensional radiographic representation of complex three-dimensional structures that are often not clearly visible in sequential images. A limitation of our method is that it is time intensive.

A number of other measurements of oral bone have been developed for use with panoramic radiographs. The gonial index (GI) was introduced first and represents the thickness of the inferior mandibular cortex at the gonion [4]. A similar index, the antegonial index (AI) is a measurement made in the area of the antegonial notch. The panoramic mandibular index (PMI) was proposed as the ratio of mandibular cortical thickness (at the mental foramen) to the distance from the mental foramen

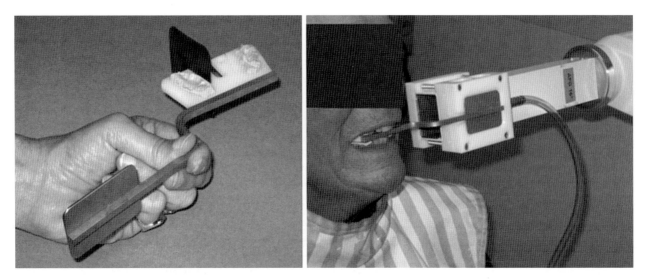

Fig. 114.3. A cross-arch, precision patient positioning bar for the molar–premolar region. The device is positioned with bite-registration material; the positioning ring is rigidly aligned with the X-ray tube by vacuum coupling. A phosphor screen, backed by two lead backscatter foils in a septic barrier envelope, is held in a vertical slot.

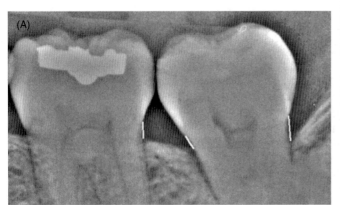
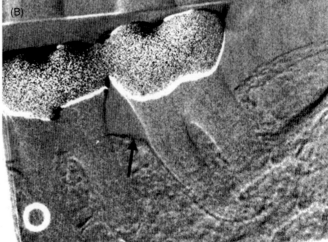

Fig. 114.4. Determination of alveolar bone loss using digital subtraction imaging. (A) High-pass-filtered image (features >40 pixels removed). Subsequent images are registered by minimizing trabecular noise. (B) Subtraction image. The arrow points to a discrepancy in crest ridge height between the two original images, showing as a dark area in this subtraction image, representing a change of −0.084 mm (adapted from Ref. 3).

(inferior or superior border) to the inferior border of the mandible [5]. Another panoramic radiograph image-based index, the mandibular cortical index (MCI), was later introduced. It is determined by visually assessing (on panoramic radiographs) the degree of erosion of the mandibular cortex posterior to the mental foramen [6]. Other indices such as the mental index (MI) or mandibular cortical width (MCW), which measure the cortical width below the mental foramen have also been proposed [6, 7]. Several of these indices are illustrated in a recent article [8]. In addition, an automated, computer-aided system was developed to measure cortical width on panoramic radiographs (see below) [9]. Limitations of panoramic radiographs are that they are not as widely used as are periapical and bitewing radiographs, and measurements vary with subject positioning.

RELATIONSHIP BETWEEN ORAL AND POSTCRANIAL BONE MASS

Earlier studies using bone mineral density (BMD) measurements from dental radiographs or absorptiometric methods established that oral BMD is significantly associated with total body calcium (determined by neutron activation analysis) and forearm and vertebral BMD, and that loss of alveolar bone or residual ridge is part of the osteoporotic syndrome [10–13]. Postmenopausal women with radiographic evidence of vertebral fractures had decreased mandibular bone mass, fewer teeth, and a lower GI (thinner gonial mandibular cortexes) than age-matched subjects with no evidence of vertebral fractures [14]. Furthermore, in women over 50 years of age who had no radiographic indications of vertebral fractures, mandibular bone mass measures were correlated with lumbar spine and forearm BMD [15].

Direct assessment of bone mineral content (BMC) of the mandible using a DXA device demonstrated that mandibular bone mass decreases 1.5%/year in old women and 0.9%/year in old men, and it is correlated with BMD at the forearm [16, 17]. Good correlations between appendicular (forearm) or central (vertebral, proximal femur) BMD measured by DXA and oral bone measurements derived from dental radiographs have been consistently reported [18, 19]. Similarly, computed tomography (CT) methods have confirmed a significant correlation between mandible and lumbar vertebrae bone mass [20], and longitudinal studies have shown that ACH and alveolar radiodensitometric parameters were correlated with changes in femoral and lumbar BMD [21–23].

Significant correlations were found between the PMI index and femoral and lumbar BMD in 355 postmenopausal women (ages 48–56) [6, 24] and with self-reported osteoporosis, though the association was weak [25]. Simple visual inspection of the mandibular cortex on panoramic radiographs produced similar results for predicting femoral and lumbar BMD.[26] Using the latest automated methods for measurement of mandibular cortical thickness, a decrease in mandibular cortical thickness of similar magnitude as for bone loss in the proximal femur was detected in women aged 42.5 years or older [27]. In a small cohort of men, measures of cortical bone in panoramic radiographs revealed higher oral bone loss in men with osteoporosis [8]. Several studies have also reported significant correlations between the MCI index and biochemical markers of bone turnover (serum total alkaline phosphatase and urinary N-teleopeptide crosslinks of type I collagen) in women and men [28, 29]. Such correlation is important in that bone turnover fluctuates widely with age, menopausal status, and within a day, yet the correlation indicates that oral bone loss is associated with active bone remodeling.

OSTEOPOROSIS AND TOOTH LOSS

Because teeth can be lost from decay and trauma and clinicians use various thresholds for determining when teeth should be extracted, it is difficult to determine whether there is an association between tooth loss and osteoporosis. Nonetheless, most published studies indicate that a lower BMD corresponds to a lower number of teeth. A cross-sectional study of mandibular bone mass determined on dental radiographs found that tooth loss and edentulism were significantly more common in osteoporotic subjects relative to a control group [15]. Subsequent studies confirmed the correlation between decreased mandibular cortical bone width and tooth loss, particularly in women older than 70 years of age [30]. Likewise, tooth loss was found to be highly correlated with low alveolar and vertebral bone mass and with vertebral fractures [31, 32].

An inverse correlation between postcranial bone mass and self-reported tooth loss was found to be significant in men, though it was not significant in women after correction for BMI, age, and smoking [33]. In another study of lumbar-spine and forearm BMD in postmenopausal women, it was concluded that systemic bone loss may contribute to tooth loss [34]. The same group of investigators later reported that women using estrogen replacement after menopause had a higher number of teeth relative to nonusers, and duration of estrogen use independently predicted the number of remaining teeth; estrogen users (1–4 years) had on average 1.1 more teeth than nonusers [35]. A 3-year, prospective Swedish study of 14,375 older men and women also found that women with the fewest teeth at baseline (lowest tertile) had a risk of hip fracture that was twice that of the women in the highest two tertiles, with the risk of fracture being threefold higher in men [36]. In a cross-sectional study of 70 dentate Caucasian women (51–78 years of age), it was found that interproximal alveolar bone loss was related to hip BMD [37]. More recently as part of the OSTEODENT collaboration, it was found that after controlling for age and smoking, tooth loss was significantly associated with osteoporosis [38].

However, in a group of Finnish women no correlation between postcranial BMD and tooth loss was found, although dental practices in Finland may have led to extractions for preventative purposes rather than as a result of underlying disease, as 25% of women had all their mandibular teeth extracted before the age of 30 [39]. Data from our own group are also not supportive of a clear relationship between spine or femur BMD and number of remaining teeth among subjects enrolled in a hormone replacement study trial [21, 40]. However, subjects were relatively young postmenopausal women who had 10 or more teeth and were generally periodontally healthy, limiting the ability to detect an association between BMD and tooth loss. Likewise, no relationship was found between BMD and tooth number in white women between the ages of 45 and 59 who were within 12 years of menopause, after adjustment for age, years since menopause, and hormone replacement [41]. Finally, a prospective study of a subgroup of participants from the Study of Osteoporotic Fractures found absolute BMD and percentage change in BMD to be similar in dentate women and edentulous at baseline examination [42]. Therefore, the issue as to whether systemic bone loss in osteoporosis contributes to tooth loss is not yet settled. The available data would, nevertheless, suggest that at least in elderly people, low BMD is often associated with a lower number of teeth, even though this may not be the case in younger people and in early postmenopausal women.

A common concern in the dental treatment of patients with osteoporosis is whether or not osseointegration and osseous regeneration is compromised. Recent reviews suggest rates (especially in the maxilla) may be lowered in patients with osteoporosis [43, 44]; however, the diagnosis of osteoporosis is not a contraindication for implant therapies. Studies of this area are complicated by the frequent lack of information and standardization of therapies (primarily, bisphosphonates) that are used by patients.

MECHANISMS OF ORAL BONE LOSS

In principle, the same systemic processes that cause age-dependent bone loss in the postcranial skeleton are active in oral bone, where the reduced alveolar bone may be particularly susceptible to destruction by periodontal disease, with concomitant reduction in alveolar crest height and loosening of the tooth attachment apparatus [37, 45]. Genetic factors that predispose a person to systemic bone loss also predispose an individual to alveolar bone loss. Likewise, certain lifestyle factors, such as cigarette smoking and suboptimal calcium and vitamin D intakes, as well as pathologic conditions that facilitate local infections (such as diabetes) may increase the risk for systemic bone loss and deterioration of the alveolar bone. In addition to these systemic factors, a key contributor to alveolar bone loss is periodontal disease, during which resorption of the alveolar bone occurs by recruitment and activation of osteoclasts, by secretion of inflammatory cytokines (during the local inflammatory process) in particular interleukin 1 (IL-1), IL-6, and tumor necrosis factor α (TNFα)—all potent osteoclastogenic factors—in response to bacterial infection [46]. The fundamental pathogenetic mechanism of alveolar bone resorption in periodontal disease is, therefore, activation of a normal biologic process (osteoclast formation and activity) by an abnormal local condition (infection driven inflammation).

ASSESSMENT OF ORAL BONE AS A SCREENING TOOL FOR OSTEOPOROSIS

The idea of using parameters of dental health to estimate the risk of systemic bone loss and/or fractures has been extensively studied in light of the correlations between oral and postcranial bone mass. However, as discussed earlier, such correlations are not always very good, and there is no consensus on which method or parameter should be used to assess oral bone in this regard. Radiometric parameters determined in panoramic radiographs, such as the MCI and PMI, and other measures of cortical bone have been proposed. Initial studies revealed poor repeatability for these measures, probably because of the need of a manual calculation of the mandibular bone parameters from standard radiographs by general practitioners [47, 48]. Specific training for dental practitioners marginally improved the predictive value of the oral measures [49]. Nonetheless, when these measures have good repeatability, sensitivities greater than 80% for predicting osteoporosis from mandibular radiometric measurements have been reported [50, 51].

A semiautomated, computer-aided system has been developed to measure MCI on panoramic radiographs. Although these methods do not seem to significantly improve the predictive value for estimating postcranial bone mass relative to manual measurements, they do offer in principle improved practical applicability to routine clinical settings [9, 52]. Using a combination of computer-aided measurements of cortical width and the MCI, it might be possible to improve specificity [9]. Data from the OSTEODENT consortium indicate that the diagnostic value of panoramic, mandibular cortical thickness measures at the mental foramen was only modestly lower in predicting osteoporosis than algorithms based upon clinical risk factors [53]. Combining the two methods increases specificity but lowers sensitivity. Different radiometric parameters have been studied in the OSTEODENT project, and overall, measurement of cortical thickness appears to perform better than MCI for screening patients who would be recommended to undergo DXA testing for osteoporosis [54]. Again, inclusion of clinical data might improve the diagnostic performance of these measurements [55].

Trabecular patterns have also been used to assess oral bone. Endosteal mandibular trabeculation increases with age [56], and trabecular coarseness on periapical radiographs was found to be highly correlated with DXA measures of the forearm [57]. With periapical radiographs, morphologic features of trabecular oral bone had an accuracy of 92% in classifying osteoporosis and control patients [58]. In an expanded study of 598 women (mean age = 77 years), trabecular features had an accuracy for predicting hip fractures that was similar to that obtained with risk assessment tools [59]. These methods were not as successful when used with panoramic radiographs. An OSTEODENT-based collaboration also found that trabecular patterns were accurate in identifying osteoporotic women, with success being slightly better with panoramic radiographs than with periapical radiographs [60]. In a subsequent study, trabecular features plus age had a sensitivity of 0.75 and a specificity of 0.78 in predicting osteoporosis status [61]. Similar high classification rates were found in other studies [62, 63].

More recently, automated determinations of cortical width were combined with other clinical risk factors in an OSTEODENT index that was tested in 339 women against the World Health Organization (WHO) fracture risk assessment tool (FRAX®) [64]. The OSTEODENT index had the same predictive value for identifying subjects who should be treated for osteoporosis as had FRAX without inclusion of DXA data [65]. It is possible that with further refinements in the OSTEODENT automated measurement of mandibular cortical parameters the predictive value may increase, thus offering a new platform for osteoporosis screening in dental clinics. Because two-thirds of the U.S. population visits dentists each year [66], and dentists regularly perform dental radiographic examinations, it may become reasonable for dentists to use dental radiographs to screen for osteoporosis and refer for further evaluation patients who appear to be at risk—just as they screen for oral cancers when they perform oral examinations [7].

ORAL BONE LOSS AS A PREDICTOR OF FRACTURES

As association between oral bone mass and low-trauma (osteoporotic) fractures has been consistently reported using different methods for assessing oral health. Earlier studies using DPA showed that women with fractures had lower forearm and mandibular BMC than did women without fractures [67]. In a small cohort of women with osteoporosis, cortical thickness at the gonial angle was determined on digitized panoramic radiographs. The best predictor of vertebral compression fractures were pixel gray-scale values, although fractal dimensions and microdensitometry measurements were also effective [68]. Furthermore, in a study of 501 subjects (91 of whom had self-reported osteoporotic fractures), the thickness of the mandibular cortex below the mental foramen (MI) and the MCI were found to be associated with osteoporotic fractures [69]. However, another study on relatively healthy subjects who were followed for 5 years did not confirm such findings, as neither panoramic cortical width nor MCI was predictive of those who had fractures [70]. A limitation of this study was that fracture events were self-reported and were not confirmed radiologically. Overall, there is evidence of an association between oral health and fracture risk, but routine screening through dental examinations cannot yet be recommended, primarily because of the limitations of the methodologies to assess oral bone health and their applicability to routine clinical settings.

Another clinically important question is whether tooth loss is correlated with fractures. As noted earlier, alveolar bone loss occurs with age and weakens the tooth attachment apparatus. Teeth are lost when there is not enough remaining alveolar bone to retain them. In a 3-year prospective study of 14,375 subjects, tooth loss and fracture risk were found to be significantly associated [36], suggesting that edentulous people or individuals with fewer teeth are more likely to be at risk of fractures than people with teeth. However, considering the multiple causes of tooth loss, and periodontal disease in particular, the predictive value of tooth loss for fracture risk is likely to be low.

IMPACT OF OSTEOPOROSIS THERAPY ON ORAL BONE HEALTH

One important, modifiable factor contributing to age-dependent bone loss is inadequate intake of vitamin D. Although no randomized clinical trial exists as of yet, a number of small-scale, noncontrolled studies point to beneficial effects of vitamin D on periodontal health (reviewed in Ref. 71). Results of a recent pilot study from our group corroborates this notion, revealing that subjects taking oral calcium (1,000 mg or more daily) and vitamin D (400 IU or less daily) supplementation had better clinical parameters of periodontal health and less ACH loss relative to individuals who did not take supplements [72]. Such differences were maintained after 1 year, although consistent periodontal care (scaling and root planing) minimizes such differences [73]. Thus, vitamin D supplementation may positively impact periodontal health.

Clinical studies have shown that estrogen deficiency decreases alveolar bone density [74], and a recent study in ovariectomized monkeys has revealed eroded endosteal surfaces and decreased cortical bone density in histologic sections of mandibular bone, associated with enlarged Haversian canals [75]. Accordingly, estrogen replacement therapy is beneficial to oral bone health. In the Leisure World Cohort and the Nurses' Health Study, estrogen use was found to reduce tooth loss [76, 77], and data from the Third United States National Health and Nutrition Examination Survey (NHANES III) demonstrated that women who had taken estrogen had less attachment loss relative to nonusers [78]. Furthermore, in a double-blind, randomized, 3-year controlled trial hormone/estrogen replacement therapy improved alveolar bone mass as well as femoral and vertebral bone density [21, 79].

Bisphosphonates, potent inhibitors of bone resorption, should in theory protect from alveolar as well as systemic bone loss, since the two processes are fundamentally similar. A few placebo-controlled, randomized trials have been conducted to test this hypothesis. Patients with periodontal disease who received periodontal maintenance therapy and were randomized to take a placebo or either oral alendronate (10 mg daily) or risedronate (5 mg daily) for 1 year experienced significantly greater improvements in AL, probing depth, and gingival bleeding relative to a placebo group [80]. However, in another randomized trial, alendronate (70 mg once weekly) did not significantly improve alveolar bone loss in subjects of both genders (71% had periodontal disease) who also received periodontal care, although there was a detectable positive effect in a subgroup of subjects with low alveolar bone mass at baseline [81]. Intriguingly, a more recent trial reported that alendronate delivered locally as a 1% gel to patients with aggressive, chronic periodontal disease significantly improved clinical parameters of periodontal health (probing depth, plaque index, and AL) relative to placebo, as an adjunct therapy to scaling and root planing [82, 83]. However, the use of bisphosphonates for alveolar bone loss is at present overshadowed by the concerns engendered by the reported association between bisphosphonate use and osteonecrosis of the jaw, a topic discussed in another chapter. Such a severe side effect seems to be linked primarily to the use of very high doses of bisphosphonates in cancer patients, while the incidence of this event in patients treated with bisphosphonates for osteoporosis is quite low. A balanced assessment of risks and benefits of this class of drugs for alveolar bone health is complicated because standardized principles for the diagnosis of osteonecrosis of the jaw are seldom applied, leading to overestimation of the risk, and the widespread use of bisphosphonates increases the likelihood of random association with osteonecrosis of the jaw. There is also concern that bisphosphonates increase the risk of implant failure; however, studies indicate that the risk appears to be low [84, 85].

PAGET'S BONE DISEASE

Up to 17% of patients with Paget's bone disease have manifestations in the oral cavity [86]. The lesions in the oral cavity are more commonly found in the maxilla than in the mandible [87]. Alveolar ridge enlargement is observed and can lead to spreading of the teeth and an abnormal occlusal pattern. Because of alveolar ridge enlargement, edentulous patients with Paget's disease may require new dentures more frequently to compensate for the enlargement. Enlargement of the middle third of the face can also be observed [88, 89]. Another complication of Paget's disease is hypercementosis (excessive deposition of the mineralized cementum structure of the tooth root), which may result in tooth ankylosis. Conversely, Paget's disease may lead to loosening of the teeth during its osteolytic phase [87]. Because the majority of affected patients have no symptoms until bone deformities emerge, the diagnosis is primarily attained by incidental biochemical testing or radiologic examination [90]. Radiographically, in the beginning stage, the bone exhibits reduced radiodensity and may resemble cemento-osseous dysplasia. At a more advanced stage, numerous irregular radiopaque areas become evident, taking on the typical "cotton wool" appearance of Paget's bone disease [86]. Histologically, there are increased numbers of osteoclasts; however, osteoblastic activity is also observed with evidence of a continuous process of abnormal deposition and resorption of bone [91]. There is little information on the effect of systemic therapy for Paget's disease with oral localization. One case report suggested that a 6-month course with alendronate in a patient with Paget's bone disease had no obvious negative effects on placement of dental implants [92].

PRIMARY HYPERPARATHYROIDISM

Several manifestations of hyperparathyroidism occur in the oral cavity. These include partial or complete loss of the lamina dura, increased periodontal ligament width, decreased alveolar bone density, and at more advanced stages, brown tumor formation [93, 94]. Loss of the lamina dura is not a pathognomonic sign of the disease because this is also seen in other diseases such as Cushing's syndrome and osteomalacia. With hyperparathyroidism, there is subperiosteal bone resorption, which is typical of parathyroid bone disorders occurring at other skeletal sites (acroosteolysis). Brown tumor is a chronic osseous lesion with abundant hemorrhage, is normally well demarcated, and can be unilocular or multilocular. Brown tumors are now rarely observed because primary hyperparathyroidism is typically diagnosed at early stages, before any long-term consequence of parathyroid hormone excess becomes manifest. A recent study found reduced cortical density in the mandible but in non-tooth-associated bone [95]. Parameters of periodontal health such as AL, probing depth, and bleeding on probing were unchanged in primary hyperparathyroidism patients, but there was a correlation between increased widening of the periodontal ligament and serum parathyroid hormone levels. Decreased cortical bone density and an increased presence of tori (which may reflect an anabolic-like action of primary hyperparathyroidism in the oral cavity) have also been reported [95].

RENAL OSTEODYSTROPHY

The oral manifestations of renal osteodystrophy are primarily the consequence of secondary hyperparathyroidism and share many features of primary hyperparathyroidism including loss of the lamina dura, "ground glass" appearance of the bone, loss of trabeculation, and brown tumor formation [96]. Probably as a consequence of the stimulatory effect of PTH on periosteal bone formation, an enlargement of the jaws may occur in renal osteodystrophy, associated with cementum resorption [97]. However, renal osteodystrophy does not lead to widening of the periodontal ligament, and indices of periodontal disease are unchanged in patients with secondary hyperparathyroidism from chronic renal failure [98].

SUMMARY

Several important questions remain regarding the correlation between systemic and oral bone mass. A better understanding of the rate of bone loss in the mandible compared with other skeletal regions with age, and the effect of menopause is still needed. Longitudinal progression of mandibular bone loss and the effects of different therapies on mandibular density compared with other skeletal sites also remain to be determined. In particular, establishing the potential benefits of calcium and vitamin D supplementation on oral bone health is an issue of high clinical relevance, considering the widespread use of these supplements in elderly populations and their relatively modest costs. Randomized clinical trials in large cohorts are needed, and a positive outcome could have a high impact on public health.

Methodologies to assess oral density and alveolar bone loss need to be further refined and improved, especially those based on radiodensitometric approaches. Application of relatively simple and reliable methods to assess oral bone status in routine clinical settings could also be helpful in identifying subjects with or at risk of osteoporosis. Additional studies on this potential application of oral bone mass assessment are of potential high impact. Metabolic bone diseases and periodontal disease are major health concerns in the United States, especially in older populations. Studies that improve our understanding of the mechanisms by which metabolic bone diseases are associated with oral bone are needed and will be increasingly important, as will quality of life issues related to these very prevalent disorders in older Americans.

REFERENCES

1. Von Wowern N. 1982. Microradiographic and histomorphometric indices of mandibles for diagnosis of osteopenia. *Scand J Dent Res* 90(1): 47–63.
2. Henrikson PA, Wallenius K. 1974. The mandible and osteoporosis. 1. A qualitative comparison between the mandible and the radius. *J Oral Rehabil* 1(1): 64–74.
3. Hildebolt C, Couture R, Garcia N, Dixon D, Milcy DD, Langenwalter E, et al. 2009. Alveolar bone measurement precision for phosphor-plate projection images. *Oral Surg Oral Med Oral Pathol Oral Radiol Endod* 108: e96–e107.
4. Bras J, van Ooij CP, Abraham-Inpijn L, Kusen GJ, Wilmink JM. 1982. Radiographic interpretation of the mandibular angular cortex: A diagnostic tool in metabolic bone loss. Part I. Normal state. *Oral Surg Oral Med Oral Pathol* 53(5): 541–5.
5. Benson BW, Prihoda TJ, Glass BJ. 1991. Variations in adult cortical bone mass as measured by a panoramic mandibular index. *Oral Surg Oral Med Oral Pathol* 71(3): 349–56.
6. Klemetti E, Kolmakov S, Kroger H. 1994. Pantomography in assessment of the osteoporosis risk group. *Scand J Dent Res* 102(1): 68–72.
7. Taguchi A. 2010. Triage screening for osteoporosis in dental clinics using panoramic radiographs. *Oral Dis* 16(4): 316–27.
8. Dagistan S, Bilge OM. 2010. Comparison of antegonial index, mental index, panoramic mandibular index and mandibular cortical index values in the panoramic radiographs of normal males and male patients with osteoporosis. *Dentomaxillofac Radiol* 39(5): 290–4.
9. Nakamoto T, Taguchi A, Ohtsuka M, Suei Y, Fujita M, Tsuda M, et al. 2008. A computer-aided diagnosis system to screen for osteoporosis using dental panoramic radiographs. *Dentomaxillofac Radiol* 37(5): 274–81.
10. Kribbs PJ, Smith DE, Chesnut CH 3rd. Oral findings in osteoporosis. Part I: Measurement of mandibular bone density. *J Prosthet Dent* 50(4): 576–9.
11. Kribbs PJ, Smith DE, Chesnut CH 3rd. 1983. Oral findings in osteoporosis. Part II: Relationship between residual ridge and alveolar bone resorption and generalized skeletal osteopenia. *J Prosthet Dent* 50(5): 719–24.
12. Kribbs PJ, Chesnut CH, 3rd, Ott SM, Kilcoyne RF. Relationships between mandibular and skeletal bone in an osteoporotic population. *J Prosthet Dent* 62(6): 703–7.
13. Kribbs PJ, Chesnut CH 3rd. 1984. Osteoporosis and dental osteopenia in the elderly. *Gerodontology* 3(2): 101–6.
14. Kribbs PJ. 1990. Comparison of mandibular bone in normal and osteoporotic women. *J Prosthet Dent* 63(2): 218–22.
15. Kribbs PJ, Chesnut CH 3rd, Ott SM, Kilcoyne RF. 1990. Relationships between mandibular and skeletal bone in a population of normal women. *J Prosthet Dent* 63(1): 86–9.
16. von Wowern N. 1988. Bone mineral content of mandibles: Normal reference values—Rate of age-related bone loss. *Calcif Tissue Int* 43(4): 193–8.
17. von Wowern N. 1985. In vivo measurement of bone mineral content of mandibles by dual-photon absorptiometry. *Scand J Dent Res* 93(2): 162–8.
18. Hildebolt CF. 1997. Osteoporosis and oral bone loss. *Dentomaxillofac Radiol* 26(1): 3–15.
19. White SC. 2002. Oral radiographic predictors of osteoporosis. *Dentomaxillofac Radiol* 31(2): 84–92.
20. Taguchi A, Tanimoto K, Suei Y, Ohama K, Wada T. 1996. Relationship between the mandibular and lumbar vertebral bone mineral density at different postmenopausal stages. *Dentomaxillofac Radiol* 25(3): 130–5.
21. Civitelli R, Pilgram TK, Dotson M, Muckerman J, Lewandowski N, Armamento-Villareal R, et al. 2002. Alveolar and postcranial bone density in postmenopausal women receiving hormone/estrogen replacement therapy: A randomized, double-blind, placebo-controlled trial. *Arch Intern Med* 162(12): 1409–15.
22. Hildebolt CF, Pilgram TK, Yokoyama-Crothers N, Vannier MW, Dotson M, Muckerman J, et al. 2002. The pattern of alveolar crest height change in healthy postmenopausal women after 3 years of hormone/estrogen replacement therapy. *J Periodontol* 73(11): 1279–84.
23. Jacobs R, Ghyselen J, Koninckx P, van Steenberghe D. 1996. Long-term bone mass evaluation of mandible and lumbar spine in a group of women receiving hormone replacement therapy. *Eur J Oral Sci* 104(1): 10–6.
24. Klemetti E, Kolmakov S, Heiskanen P, Vainio P, Lassila V. 1993. Panoramic mandibular index and bone mineral densities in postmenopausal women. *Oral Surg Oral Med Oral Pathol* 75(6): 774–9.
25. Persson RE, Hollender LG, Powell LV, MacEntee MI, Wyatt CC, Kiyak HA, et al. 2002. Assessment of periodontal conditions and systemic disease in older subjects. I. Focus on osteoporosis. *J Clin Periodontol* 29(9): 796–802.

26. Lee K, Taguchi A, Ishii K, Suei Y, Fujita M, Nakamoto T, et al. 2005. Visual assessment of the mandibular cortex on panoramic radiographs to identify postmenopausal women with low bone mineral densities. *Oral Surg Oral Med Oral Pathol Oral Radiol Endod* 100(2): 226–31.
27. Roberts M, Yuan J, Graham J, Jacobs R, Devlin H. 2011. Changes in mandibular cortical width measurements with age in men and women. *Osteoporos Int* 22(6): 1915–25.
28. Deguchi T, Yoshihara A, Hanada N, Miyazaki H. 2008. Relationship between mandibular inferior cortex and general bone metabolism in older adults. *Osteoporos Int* 19(7): 935–40.
29. Vlasiadis KZ, Damilakis J, Velegrakis GA, Skouteris CA, Fragouli I, Goumenou A, et al. 2008. Relationship between BMD, dental panoramic radiographic findings and biochemical markers of bone turnover in diagnosis of osteoporosis. *Maturitas* 59(3): 226–33.
30. Taguchi A, Tanimoto K, Suei Y, Wada T. 1995. Tooth loss and mandibular osteopenia. *Oral Surg Oral Med Oral Pathol Oral Radiol Endod* 79(1): 127–32.
31. Taguchi A, Tanimoto K, Suei Y, Otani K, Wada T. 1995. Oral signs as indicators of possible osteoporosis in elderly women. *Oral Surg Oral Med Oral Pathol Oral Radiol Endod* 80(5): 612–6.
32. Taguchi A, Suei Y, Ohtsuka M, Otani K, Tanimoto K, Hollender LG. 1999. Relationship between bone mineral density and tooth loss in elderly Japanese women. *Dentomaxillofac Radiol* 28(4): 219–23.
33. May H, Reader R, Murphy S, Khaw KT. 1995. Self-reported tooth loss and bone mineral density in older men and women. *Age Ageing* 24(3): 217–21.
34. Krall EA, Dawson-Hughes B, Papas A, Garcia RI. 1994. Tooth loss and skeletal bone density in healthy postmenopausal women. *Osteoporos Int* 4(2): 104–9.
35. Krall EA, Dawson-Hughes B, Hannan MT, Wilson PW, Kiel DP. 1997. Postmenopausal estrogen replacement and tooth retention. *Am J Med* 102(6): 536–42.
36. Astrom J, Backstrom C, Thidevall G. 1990. Tooth loss and hip fractures in the elderly. *J Bone Joint Surg Br* 72(2): 324–5.
37. Tezal M, Wactawski-Wende J, Grossi SG, Ho AW, Dunford R, Genco RJ. 2000. The relationship between bone mineral density and periodontitis in postmenopausal women. *J Periodontol* 71(9): 1492–98.
38. Nicopoulou-Karayianni K, Tzoutzoukos P, Mitsea A, Karayiannis A, Tsiklakis K, Jacobs R, et al. 2009. Tooth loss and osteoporosis: The OSTEODENT Study. *J Clin Periodontol* 36(3): 190–7.
39. Klemetti E, Vainio P. 1993. Effect of bone mineral density in skeleton and mandible on extraction of teeth and clinical alveolar height. *J Prosthet Dent* 70(1): 21–5.
40. Hildebolt CF, Pilgram TK, Dotson M, Yokoyama-Crothers N, Muckerman J, Hauser J, et al. 1997. Attachment loss with postmenopausal age and smoking. *J Periodontal Res* 32(7): 619–25.
41. Earnshaw SA, Keating N, Hosking DJ, Chilvers CE, Ravn P, McClung M, et al. 1998. Tooth counts do not predict bone mineral density in early postmenopausal Caucasian women. EPIC study group. *Int J Epidemiol* 27(3): 479–83.
42. Famili P, Cauley J, Suzuki JB, Weyant R. 2005. Longitudinal study of periodontal disease and edentulism with rates of bone loss in older women. *J Periodontol* 76(1): 11–5.
43. Alsaadi G, Quirynen M, Michiles K, Teughels W, Komarek A, van Steenberghe D. 2008. Impact of local and systemic factors on the incidence of failures up to abutment connection with modified surface oral implants. *J Clin Periodontol* 35(1): 51–7.
44. Erdogan O, Shafer DM, Taxel P, Freilich MA. 2007. A review of the association between osteoporosis and alveolar ridge augmentation. *Oral Surg Oral Med Oral Pathol Oral Radiol Endod* 104(6): 738 e1–13.
45. Wactawski-Wende J, Grossi SG, Trevisan M, Genco RJ, Tezal M, Dunford RG, et al. 1996. The role of osteopenia in oral bone loss and periodontal disease. *J Periodontol* 67(10 Suppl): 1076–84.
46. Offenbacher S. 1996. Periodontal diseases: Pathogenesis. *Ann Periodontol* 1(1): 821–78.
47. Devlin H, Horner K. 2002. Mandibular radiomorphometric indices in the diagnosis of reduced skeletal bone mineral density. *Osteoporos Int* 13(5): 373–8.
48. Devlin CV, Horner K, Devlin H. 2001. Variability in measurement of radiomorphometric indices by general dental practitioners. *Dentomaxillofac Radiol* 30(2): 120–5.
49. Sutthiprapaporn P, Taguchi A, Nakamoto T, Ohtsuka M, Mallick PC, Tsuda M, et al. 2006. Diagnostic performance of general dental practitioners after lecture in identifying post-menopausal women with low bone mineral density by panoramic radiographs. *Dentomaxillofac Radiol* 35(4): 249–52.
50. Taguchi A, Tsuda M, Ohtsuka M, Kodama I, Sanada M, Nakamoto T, et al. 2006. Use of dental panoramic radiographs in identifying younger postmenopausal women with osteoporosis. *Osteoporos Int* 17(3): 387–94.
51. Taguchi A, Asano A, Ohtsuka M, Nakamoto T, Suei Y, Tsuda M, et al. 2008. Observer performance in diagnosing osteoporosis by dental panoramic radiographs: Results from the osteoporosis screening project in dentistry (OSPD). *Bone* 43(1): 209–13.
52. Arifin AZ, Asano A, Taguchi A, Nakamoto T, Ohtsuka M, Tsuda M, et al. 2006. Computer-aided system for measuring the mandibular cortical width on dental panoramic radiographs in identifying postmenopausal women with low bone mineral density. *Osteoporos Int* 17(5): 753–9.
53. Karayianni K, Horner K, Mitsea A, Berkas L, Mastoris M, Jacobs R, et al. 2007. Accuracy in osteoporosis diagnosis of a combination of mandibular cortical width measurement on dental panoramic radiographs and a clinical risk index (OSIRIS): The OSTEODENT project. *Bone* 40(1): 223–9.
54. Horner K, Karayianni K, Mitsea A, Berkas L, Mastoris M, Jacobs R, et al. 2007. The mandibular cortex on radiographs as a tool for osteoporosis risk assessment:

The OSTEODENT Project. *J Clin Densitom* 10(2): 138–46.

55. Nackaerts O, Jacobs R, Devlin H, Pavitt S, Bleyen E, Yan B, et al. 2008. Osteoporosis detection using intraoral densitometry. *Dentomaxillofac Radiol* 37(5): 282–7.
56. Von Wowern N, Stoltze K. 1979. Age differences in cortical width of mandibles determined by histoquantitation. *Scand J Dent Res* 87(3): 225–33.
57. Jonasson G, Bankvall G, Kiliaridis S. 2001. Estimation of skeletal bone mineral density by means of the trabecular pattern of the alveolar bone, its interdental thickness, and the bone mass of the mandible. *Oral Surg Oral Med Oral Pathol Oral Radiol Endod* 92(3): 346–52.
58. White SC, Rudolph DJ. 1999. Alterations of the trabecular pattern of the jaws in patients with osteoporosis. *Oral Surg Oral Med Oral Pathol Oral Radiol Endod* 88(5): 628–35.
59. White SC, Atchison KA, Gornbein JA, Nattiv A, Paganini-Hill A, Service SK, et al. 2005. Change in mandibular trabecular pattern and hip fracture rate in elderly women. *Dentomaxillofac Radiol* 34(3): 168–74.
60. Geraets WG, Verheij JG, van der Stelt PF, Horner K, Lindh C, Nicopoulou-Karayianni K, et al. 2007. Prediction of bone mineral density with dental radiographs. *Bone* 40(5): 1217–21.
61. Verheij JG, Geraets WG, van der Stelt PF, Horner K, Lindh C, Nicopoulou-Karayianni K, et al. 2009. Prediction of osteoporosis with dental radiographs and age. *Dentomaxillofac Radiol* 38(7): 431–7.
62. Licks R, Licks V, Ourique F, Radke Bittencourt H, Fontanella V. 2010. Development of a prediction tool for low bone mass based on clinical data and periapical radiography. *Dentomaxillofac Radiol* 39(4): 224–30.
63. Lindh C, Horner K, Jonasson G, Olsson P, Rohlin M, Jacobs R, et al. 2008. The use of visual assessment of dental radiographs for identifying women at risk of having osteoporosis: the OSTEODENT project. *Oral Surg Oral Med Oral Pathol Oral Radiol Endod* 106(2): 285–93.
64. Kanis JA, Oden A, Johansson H, Borgstrom F, Strom O, McCloskey E. 2009. FRAX and its applications to clinical practice. *Bone* 44(5): 734–43.
65. Horner K, Allen P, Graham J, Jacobs R, Boonen S, Pavitt S, et al. 2010. The relationship between the OSTEODENT index and hip fracture risk assessment using FRAX. *Oral Surg Oral Med Oral Pathol Oral Radiol Endod* 110(2): 243–9.
66. NOHSS. 2008. *Dental Visits*. Atlanta, GA: National Oral Health Surveillance System: Centers for Disease Control. http://apps.nccd.cdc.gov/nohss/ListV.asp?qkey=5&DataSet=2. Accessed on January 21, 2013.
67. von Wowern N, Kollerup G. 1992. Symptomatic osteoporosis: A risk factor for residual ridge reduction of the jaws. *J Prosthet Dent* 67(5): 656–60.
68. Law AN, Bollen AM, Chen SK. 1996. Detecting osteoporosis using dental radiographs: A comparison of four methods. *J Am Dent Assoc* 127(12): 1734–42.
69. Bollen AM, Taguchi A, Hujoel PP, Hollender LG. 2000. Case-control study on self-reported osteoporotic fractures and mandibular cortical bone. *Oral Surg Oral Med Oral Pathol Oral Radiol Endod* 90(4): 518–24.
70. Okabe S, Morimoto Y, Ansai T, Yoshioka I, Tanaka T, Taguchi A, et al. 2008. Assessment of the relationship between the mandibular cortex on panoramic radiographs and the risk of bone fracture and vascular disease in 80-year-olds. *Oral Surg Oral Med Oral Pathol Oral Radiol Endod* 106(3): 433–42.
71. Hildebolt C. 2005. Effect of vitamin D and calcium on periodontitis. *J Periodontol* 76: 1576–87.
72. Miley DD, Garcia MN, Hildebolt CF, Shannon WD, Couture RA, Anderson Spearie CL, et al. 2009. Cross-sectional study of vitamin D and calcium supplementation effects on chronic periodontitis. *J Periodontol* 80(9): 1433–9.
73. Garcia MN, Hildebolt CF, Miley DD, Dixon DA, Couture RA, Spearie CL, et al. 2011. One-year effects of vitamin D and calcium supplementation on chronic periodontitis. *J Periodontol* 82(1): 25–32.
74. Payne JB, Reinhardt RA, Nummikoski PV, Patil KD. 1999. Longitudinal alveolar bone loss in postmenopausal osteoporotic/osteopenic women. *Osteoporos Int* 10(1): 34–40.
75. Tanaka M, Yamashita E, Anwar RB, Yamada K, Ohshima H, Nomura S, et al. 2011. Radiological and histologic studies of the mandibular cortex of ovariectomized monkeys. *Oral Surg Oral Med Oral Pathol Oral Radiol Endod* 111(3): 372–80.
76. Grodstein F, Colditz GA, Stampfer MJ. 1996. Postmenopausal hormone use and tooth loss: A prospective study. *J Am Dent Assoc* 127(3): 370–7, quiz 92.
77. Paganini-Hill A. 1995. The benefits of estrogen replacement therapy on oral health. The Leisure World cohort. *Arch Intern Med* 155(21): 2325–9.
78. Ronderos M, Jacobs DR, Himes JH, Pihlstrom BL. 2000. Associations of periodontal disease with femoral bone mineral density and estrogen replacement therapy: Cross-sectional evaluation of US adults from NHANES III. *J Clin Periodontol* 27(10): 778–86.
79. Hildebolt CF, Pilgram TK, Dotson M, Armamento-Villareal R, Hauser J, Cohen S, et al. 2004. Estrogen and/or calcium plus vitamin D increase mandibular bone mass. *J Periodontol* 75(6): 811–6.
80. Lane N, Armitage GC, Loomer P, Hsieh S, Majumdar S, Wang HY, et al. 2005. Bisphosphonate therapy improves the outcome of conventional periodontal treatment: Results of a 12-month, randomized, placebo-controlled study. *J Periodontol* 76(7): 1113–22.
81. Jeffcoat MK, Cizza G, Shih WJ, Genco R, Lombardi A. 2007. Efficacy of bisphosphonates for the control of alveolar bone loss in periodontitis. *J Int Acad Periodontol* 9(3): 70–6.
82. Sharma DA, Pradeep DA. 2012. Clinical efficacy of 1% alendronate gel in adjunct to mechanotherapy in the treatment of aggressive periodontitis: A randomized controlled clinical trial. *J Periodontol* 83(1): 19–26.
83. Sharma A, Pradeep AR. 2012. Clinical efficacy of 1% alendronate gel as local drug delivery system in the

treatment of chronic periodontitis—A randomized controlled clinical trial. *J Periodontol* 83(1): 11–8.
84. Bornstein MM, Cionca N, Mombelli A. 2009. Systemic conditions and treatments as risks for implant therapy. *Int J Oral Maxillofac Implants* 24 Suppl: 12–27.
85. Madrid C, Sanz M. 2009. What impact do systemically administered bisphosphonates have on oral implant therapy? A systematic review. *Clin Oral Implants Res* 20 Suppl 4: 87–95.
86. Neville B, Damm D. 2002. *Bone Pathology in Oral and Maxillofacia Pathology*, 2nd Ed. Philadelphia: Saunders.
87. Smith BJ, Eveson JW. 1981. Paget's disease of bone with particular reference to dentistry. *J Oral Pathol* 10(4): 233–47.
88. Akin RK, Barton K, Walters PJ. 1975. Paget's disease of bone. Report of a case. *Oral Surg Oral Med Oral Pathol* 39(5): 707–12.
89. Carrillo R, Morales A, Rodriguez-Peralto JL, Lizama J, Eslava JM. 1991. Benign fibro-osseous lesions in Paget's disease of the jaws. *Oral Surg Oral Med Oral Pathol* 71(5): 588–92.
90. Delmas PD, Meunier PJ. 1997. The management of Paget's disease of bone. *N Engl J Med* 336(8): 558–66.
91. Gherardi G, Lo Cascio V, Bonucci E. 1980. Fine structure of nuclei and cytoplasm of osteoclasts in Paget's disease of bone. *Histopathology* 4(1): 63–74.
92. Pirih FQ, Zablotsky M, Cordell K, McCauley LK. 2009. Case report of implant placement in a patient with Paget's disease on bisphosphonate therapy. *J Mich Dent Assoc* 91(5): 38–43.
93. Daniels JS. 2004. Primary hyperparathyroidism presenting as a palatal brown tumor. *Oral Surg Oral Med Oral Pathol Oral Radiol Endod* 98(4): 409–13.
94. Silverman S Jr, Gordan G, Grant T, Steinbach H, Eisenberg E, Manson R. 1962. The dental structures in primary hyperparathyroidism. Studies in forty-two consecutive patients. *Oral Surg Oral Med Oral Pathol* 15: 426–36.
95. Padbury AD Jr, Tozum TF, Taba M Jr, Ealba EL, West BT, Burney RE, et al. 2006. The impact of primary hyperparathyroidism on the oral cavity. *J Clin Endocrinol Metab* 91(9): 3439–45.
96. Silverman S Jr, Ware WH, Gillooly C Jr. 1968. Dental aspects of hyperparathyroidism. *Oral Surg Oral Med Oral Pathol* 26(2): 184–9.
97. Goultschin J, Eliezer K. 1982. Resorption of cementum in renal osteodystrophy. *J Oral Med* 37(3): 84–6.
98. Frankenthal S, Nakhoul F, Machtei EE, Green J, Ardekian L, Laufer D, et al. 2002. The effect of secondary hyperparathyroidism and hemodialysis therapy on alveolar bone and periodontium. *J Clin Periodontol* 29(6): 479–83.

Section XI
The Skeleton and Its Integration with Other Tissues

Section Editor Mone Zaidi

Chapter 115. Central Neuronal Control of Bone Remodeling 961
Shu Takeda and Paul Baldock

Chapter 116. The Pituitary–Bone Connection 969
Mone Zaidi, Tony Yuen, Li Sun, Terry F. Davies, Alberta Zallone, and Harry C. Blair

Chapter 117. Skeletal Muscle Effects on the Skeleton 978
William J. Evans

Chapter 118. Glucose Control and Integration by the Skeleton 986
Patricia Ducy and Gerard Karsenty

Chapter 119. Obesity and Skeletal Mass 993
Sue Shapses and Deeptha Sukumar

Chapter 120. Neuropsychiatric Disorders and the Skeleton 1002
Itai Bab and Raz Yirmiya

Chapter 121. Vascular Disease and the Skeleton 1012
Dwight A. Towler

Chapter 122. Spinal Cord Injury: Skeletal Pathophysiology and Clinical Issues 1018
William A. Bauman and Christopher P. Cardozo

Chapter 123. Hematopoiesis and Bone 1028
Benjamin J. Frisch and Laura M. Calvi

Chapter 124. Bone and Immune Cell Interactions 1036
Brendan F. Boyce

Primer on the Metabolic Bone Diseases and Disorders of Mineral Metabolism, Eighth Edition. Edited by Clifford J. Rosen.
© 2013 American Society for Bone and Mineral Research. Published 2013 by John Wiley & Sons, Inc.

115

Central Neuronal Control of Bone Remodeling

Shu Takeda and Paul Baldock

Introduction 961
Leptin and the Sympathetic Nervous System 961
Serotonin 962
Neuromedin U 963
The Neuropeptide Y System 963

Hypothalamic NPY2 Receptor Effects on Bone 963
Osteoblastic NPY1 Receptor Effects on Bone 964
The Cannabinoid Receptors 964
The Melanocortin System 964
References 965

INTRODUCTION

All homeostatic functions, including those in bone, are controlled by the brain. Indeed, clinical evidence that traumatic brain injury (TBI) accelerates the healing of fractures suggests that there is a link between the central nervous system and bone remodeling [1]. The discovery that leptin regulates bone formation through the central nervous system initiated a new research field: neuronal control of bone remodeling [2]. Since then, other neuropeptides and neurotransmitters, such as neuropeptide Y (NPY) [3], cocaine- and amphetamine-regulated transcript (CART) [4], neuromedin U [5], and, more recently, serotonin [6, 7], have been demonstrated to possess bone-regulating activities.

LEPTIN AND THE SYMPATHETIC NERVOUS SYSTEM

Leptin is a 16 kDa peptide hormone that is synthesized by adipocytes [8]. Leptin affects appetite and energy metabolism through the pro-opiomelanocortin (POMC) pathway, which increases food intake and energy expenditure, and the agouti-related protein (AgRP)/NPY pathway, which decreases food intake and energy expenditure [8]; both pathways are active in the arcuate nucleus.

Ob/ob mice that lack functional leptin are obese and sterile [8]. In spite of their hypogonadism, the most common cause of osteoporosis, *ob/ob* mice and *db/db* mice that lack a functional leptin receptor exhibit high bone mass [9]. Intracerebroventricular (ICV) infusion of leptin to *ob/ob* mice or wild-type mice at a minimal dose, without any detectable leakage into general circulation, results in reduced bone mass [9]. This observation was subsequently verified in both rats and sheep [10, 11]. Moreover, mice lacking leptin receptors specifically in the central nervous system demonstrate an identical bone phenotype to mice lacking leptin receptors in the entire body (*ob/ob* mice), whereas mice lacking leptin receptors only in osteoblasts have normal bone metabolism [12]. Thus, leptin relies on the central nervous system to regulate bone mass.

Leptin stimulates activity of the sympathetic nervous system, causing a decrease in bone mass mainly through adrenergic α2 receptor (adrb2), the most strongly expressed adrenergic receptor in osteoblasts [13]: The treatment of wild-type mice with the nonselective beta-agonist, isoproterenol, or the adrb2-selective agonist, clenbuterol or salbutamol, decreases bone mass [13, 14]. In contrast, mice with decreased sympathetic activity (*dopamine-beta-hydroxylase-/-* mice [13], *adrb2-/-* mice [4], and mice treated with nonselective beta blockers [13]) demonstrate high bone mass due to an increase in bone formation and a decrease in bone resorption. In addition, these mice are resistant to the effects of leptin in decreasing bone mass,

Primer on the Metabolic Bone Diseases and Disorders of Mineral Metabolism, Eighth Edition. Edited by Clifford J. Rosen.
© 2013 American Society for Bone and Mineral Research. Published 2013 by John Wiley & Sons, Inc.

demonstrating that the main downstream pathway of leptin in bone metabolism is the sympathetic nervous system. Moreover, osteoblast-specific *adrb2-/-* mice recapitulate the bone abnormality of *adrb2-/-* mice [15], indicating that adrb2 in osteoblasts, not in other tissues, is responsible for the effect of the sympathetic nervous system on bone.

Sympathetic signaling in osteoblasts involves two different pathways to control bone formation and bone resorption. The former is via cyclic adenosine monophosphate (cAMP) response element binding protein (CREB) and v-Myc myelocytomatosis viral oncogene homolog (c-myc) transcription factors [15, 16]. Upon stimulation of the sympathetic nervous system–adrb2 pathway in osteoblasts, CREB phosphorylation is inhibited by an unidentified mechanism[15] that results in the inhibition of further downstream effectors, i.e., molecular clocks, such as Per, Cry, and AP-1 transcription factors [16]. Molecular experiments revealed that Per1 and Per2 negatively regulate the expression of c-myc and G1 cyclins, and thus osteoblast proliferation, and that the absence of Per1 and Per2 or Cry1 and Cry2 favors bone mass accrual [16]. In contrast, AP-1 stimulates the expression of c-myc and G1 cyclins [16]. As a combined effect of these antagonistic proteins, the sympathetic nervous system (SNS) decreases bone formation [16]. The latter pathway, which regulates bone resorption, is also mediated through osteoblasts [4]. Upon stimulation of the SNS–adrb2 pathway, osteoblastic ATF4 is phosphorylated, inducing *Rankl* expression [4]. As the bone resorption abnormality in *adrb2-/-* mice cannot be corrected by leptin ICV infusion, it has been suggested that leptin signaling is dependent on the SNS to regulate bone resorption; however, given that osteoclasts do exist in *adrb2-/-* mice, SNS signaling is not essential for osteoclast differentiation.

Various mouse models of osteoporosis, such as ovariectomy-induced [13], unloading-induced [17], or depression-induced bone loss [18] are all ameliorated by the concomitant treatment with beta-blockers, though there are some conflicting reports. These discrepancies may be related to the amount of beta-blocker used in the study: a low dose of propranolol that does not affect any cardiovascular functions was sufficient to increase bone formation parameters, and increasing the doses of propranolol progressively decreases its beneficial effect on bone [19].

Recently, it was shown that the norepinephrine content in bone was decreased by TBI via cannabinoid receptor 1 signaling and that the TBI-induced stimulation of osteogenesis was restrained by a beta agonist [20]. Many epidemiological studies also confirmed the effect of a beta-blocker on bone mass or fracture [21, 22]. Though there are some conflicting results showing either beneficial or indifferent effects of beta-blockers in the prevention of osteoporotic fractures, a meta-analysis of eight studies demonstrated that the use of beta-blockers is associated with the reduction of hip fracture risk and the risk of any fracture [22]. Considering the widespread usage of beta-blockers in clinical medicine, beta-blockers can also be easily applied to the treatment of osteoporosis. However, because most of the studies addressing the relationship of beta-blockers and osteoporotic fracture are observational studies, randomized clinical trials are strongly needed.

Other adrenergic receptors and muscarinic receptor are also involved in bone remodeling. *M3 muscarinic receptor-/-* mice and neuron-specific *M3 muscarinic receptor-/-* mice, both of which show increases in sympathetic nervous activity, demonstrate a low bone mass phenotype due to a decrease in bone formation and an increase in bone resorption [23], whereas osteoblast-specific *M3 muscarinic receptor-/-* mice have no bone abnormality [23]. These results demonstrate that the parasympathetic nervous system affects bone mass by targeting neurons and that the balance between the autonomic nervous systems defines bone mass. Moreover, *α2A/α2C adrenergic receptor-/-* mice demonstrate high bone mass despite an increase in sympathetic nervous system activity and selective α2AR agonists increase osteoclast formation [24], indicating that α-adrenergic receptors are also involved in bone remodeling. Whether these observations may be applied to human beings remains unknown.

SEROTONIN

Serotonin is a monoamine compound that is produced exclusively by the action of Tph2 in the central nervous system [25]. *Tph2-/-* mice develop an osteoporotic phenotype due to an increase in SNS activity, and *ob/ob* mice demonstrate an increase in the concentration of serotonin in the brain [6]. The antiosteogenic and anorexigenic actions of leptin were hampered in mice lacking leptin receptors exclusively in serotonergic neurons located in the brainstem [6], whereas mice lacking leptin receptors specifically in the arcuate nucleus or ventromedial hypothalamus, which are known to be indispensable for the anorexigenic action of leptin [8], have normal bone mass [6], demonstrating a pivotal role of serotonin for the action of leptin. Thus, brainstem-derived serotonin favors bone mass accrual by its binding to the 5-HT2c receptor [6]. However, contradictory observations using identical mouse models is also reported [26]: serotonergic neurons do not express leptin receptors and hence serotonergic neuron-specific leptin receptor-/- mice have normal bone mass and normal body weight. Currently, the cause of discrepancy is not known, although it has been proposed that serotonergic neuron-specific leptin receptor-/- mice by Yadav et al. might have been globally leptin receptor-/- mice. Moreover, selective serotonin reuptake inhibitors (SSRIs), which inhibit serotonin uptake and are believed to stimulate the serotonin signaling pathway, increase the risk of fracture [27]. Thus, further study is needed to thoroughly address the role of serotonin in the brain for the control of bone metabolism.

In the periphery, serotonin is mostly produced by the action of Tph1 by enterochromaffin cells in the gastrointestinal tract [25]. *Tph1-/-* mice whose serum concentration of serotonin is markedly decreased present high bone mass [7]. Peripheral serotonin acts on HTr1b and inhibits CREB phosphorylation and osteoblast proliferation [7]. Unexpectedly, gut-derived serotonin is shown to be a downstream mediator of the skeletal effect of LRP5: the gut-specific inactivation of LRP5 or the activation of LRP5 signaling only in the gut fully recapitulates the low bone mass in *LRP5-/-* mice or high bone mass in mice carrying an LRP5-activating mutation in the entire body [7]. Moreover, a low tryptophan diet normalizes the high serum serotonin concentration and low bone mass in *LRP5-/-* mice.[7]

A negative association between bone mineral density and serum serotonin levels is also reported in human [28], and, importantly, an inhibitor of Tph1, which does not affect Tph2, favors bone mass accrual in rodents to a similar extent as parathyroid hormone (PTH) injection [29]. Thus, serotonin and Tph1 is an attractive target for a novel bone anabolic therapy for osteoporosis. However, contradictory observations using identical mouse models are reported [30]: Osteocyte-specifc *LRP5-/-* mice have a decrease in bone mass that is identical to what was observed in global *LRP5-/-* mice, whereas gut-specific *LRP5-/-* mice and *Tph1-/-* mice have a normal bone mass. Moreover, LRP5 does not affect *Tph1* expression or serum serotonin concentration in mice, treatment with a different Tph1 inhibitor does not affect bone mass in mice, and hence LRP5 regulates bone metabolism exclusively acting in bone. Currently, the cause of discrepancy is unknown and requires further corroboration.

NEUROMEDIN U

Neuromedin U (NMU) is a neuropeptide produced in the gastrointestinal tract and in the brain, and NMU inhibits food intake by a leptin-independent mechanism [31]. *Nmu-/-* mice present a high bone mass phenotype with an isolated increase in bone formation [5], similar to *ob/ob* mice. This phenotype is not cell autonomous, as *Nmu-/-* osteoblasts are indistinguishable from wild-type osteoblasts *in vitro* [5]. In contrast, ICV infusion of NMU to *Nmu-/-* mice and wild-type mice decreases their bone formation and bone mass [5]. Importantly, leptin ICV infusion or isoproterenol treatment does not decrease bone mass in *Nmu-/-* mice, demonstrating that NMU mediates the action of leptin and SNS in the regulation of bone formation [5]. Further analysis reveals that NMU in the hypothalamus affects only the negative regulator of osteoblast proliferation, namely, the molecular clock [5, 16].

THE NEUROPEPTIDE Y SYSTEM

The neuropeptide Y (NPY) system comprises three ligands; NPY, peptide YY (PYY), and pancreatic polypeptide (PP), mediating their actions through five Y receptor subtypes: NPY1R, NPY2R, NPY4R, NPY5R, and Npy6R [32, 33]. NPY is predominantly neural, produced by central and peripheral neurons and is often co-secreted with noradrenaline [34]. NPY-ergic neurons are abundant in the brain, with high levels in several hypothalamic nuclei (the arcuate nucleus and ventromedial hypothalamus) [35–37]. Early studies identified NPY-immunoreactive fibers in bone, associated with blood vessels [38–41], but also cells in the periosteum and bone lining cells [38, 39]. Central NPY treatment was associated with a reduction in bone mass [9], while NPY treatment in osteoblastic cell lines inhibited the cAMP response to PTH and norepinephrine [42, 43], suggesting the presence of functional Y receptors and a possible regulatory role for NPY in bone forming cells.

A recent publication has confirmed the role of NPY in skeletal metabolism. NPY null mice demonstrated a generalized bone anabolic phenotype [44], without significant changes in body weight. Despite early reports of no effect [45], the negative relationship between hypothalamic NPY and bone formation is consistent with previous reports of reduced bone formation following overexpression of NPY in hypothalamic neurons [46] or cerebrospinal fluid [9] of wild-type mice, as well increased bone mass following loss of NPY receptors (discussed below). Interestingly, central NPY overexpression represents a model of forced central starvation, similar to that evident in leptin deficient *ob/ob* mice [47]. Importantly, elevation of central NPY (mimicking the conditions encountered in the hypothalamus during starvation [48]), decreased bone mass despite marked increases in body weight, as evident in *ob/ob* [47]. In this manner, weight may be matched to bone mass: Calorie restriction reduces body weight and increases central NPY [49], which inhibits bone formation, as a component of the whole body energy-conservative response. Conversely, excessive calorie intake increases body weight but reduces NPY expression, which stimulates bone formation, thereby matching bone mass to increases in body mass. Thus, the central perception of body weight, i.e., as evident by alterations in central NPY may act to pair bone mass to changes in body weight [44]; a process occurring in addition to the well-described mechanical responses to altered body weight. Interestingly, NPY is also expressed in osteoblasts and osteocytes, and expression is reduced *in vitro* by mechanical loading [44, 50]. Thus, NPY signaling may regulate multiple processes as part of a system to coordinate bone and energy homeostasis.

HYPOTHALAMIC NPY2 RECEPTOR EFFECTS ON BONE

Two NPY receptors have been connected with skeletal homeostasis: NPY1R and NPY2R. Both receptors are expressed in the hypothalamus as well as in peripheral nerves [51–53]. Analysis of the distal femur of germline

NPY2r-/- mice revealed a greater cancellous bone volume associated with a greater rate of bone formation, due to increased osteoblast activity, without an increase in mineralizing surface [3, 54]. Parameters of bone resorption were unchanged. Critically, the bone phenotype of germline *NPY2r-/-* mice was recapitulated in adult mice following selective deletion of NPY2R solely from the hypothalamus, demonstrating a role for central NPY2R in this pathway. Moreover, the skeletal changes observed in germline and conditional *NPY2r-/-* mice occurred in the absence of measurable changes in bone active endocrine factors. Thus, these findings indicated that the anabolism resulting from NPY2R deletion was mediated through a neural mechanism originating within the hypothalamus.

Importantly, a recent study demonstrated that ablation of hypothalamic NPY2R specifically from NPY-ergic neurons produced only moderate increases in cancellous bone volume, and no effect on cortical bone mass [55]. This result indicates that within the hypothalamus the NPY2R-mediated regulation of bone mass is mediated through neuronal populations other than NPY neurons. Preliminary indications suggest that sympathetic neurons emanating from the paraventricular nucleus, the target region for arcuate NPY, may be responsible for the efferent pathway (unpublished observation).

OSTEOBLASTIC NPY1 RECEPTOR EFFECTS ON BONE

The NPY1 receptor has recently been confirmed as a second Y receptor active in the regulation of bone. Similar to NPY2R-deficient mice, loss of NPY1R expression resulted in a generalized anabolic phenotype, with greater bone mass and formation [56] although with an additional increase in bone resorption. The bone phenotype, however, differed from *NPY2r-/-* mice in several critical aspects. Most importantly, conditional deletion of hypothalamic NPY1R receptors had no effect on bone homeostasis, indicating a noncentral mechanism for NPY1R action in bone. The existence of a direct NPY1R-mediated effect on anabolism was suggested following identification of NPY1R expression in osteoblastic cells *in vivo* [56]. Deletion of NPY1R from osteoblastic cells *in vitro* recapitulated the bone anabolic changes evident in germline *NPY1r-/-* mice, although bone resorption was not different from the wild type [57]. Moreover, treatment of wild-type osteoblast-like cultures with NPY resulted in a decrease in cell number, a response that was completely absent in *NPY1r-/-* cultures, which indicates functional osteoblastic NPY1R.

Moreover, this osteoblastic NPY1R expression may be directly involved in the *NPY2r-/-* phenotype. *NPY1r-/-* /*NPY2r-/-* mice do not display an additive phenotype in bone, and NPY1R expression is substantially reduced in osteoblast-like cultures from *NPY2r-/-* mice. Although their role in the control of bone homeostasis is yet to be fully elucidated, these studies indicate that NPY1R signaling may be a critical downstream component of the neural regulation of bone mass.

THE CANNABINOID RECEPTORS

The endocannabinoid system mediates its actions via two cannabinoid receptors, CB1-R and CB2-R, which couple to inhibitory G proteins [58]. CB1-R is primarily found within the central nervous system [59], while CB2-R is predominantly expressed in peripheral tissues [20]. Cannabinoid receptors are also expressed in osteoblasts and osteoclasts, and play a role in the control of bone homeostasis by a centrally mediated and direct mechanism. Figure 115.1 is a schematic diagram of the central neuronal control of bone remodeling.

The CB1 receptor plays a significant role in regulating bone mineral density (BMD) [60]. It has been demonstrated that mice with inactivation of CB1-R, have increased BMD and, additionally, are protected against ovariectomy-induced bone loss [60]. Furthermore, the synthetic cannabinoid receptor antagonists inhibit osteoclast formation and bone resorption *in vitro* and protect against ovariectomy-induced bone loss *in vivo* [60].

There is limited evidence concerning CB2-R action in bone mass in humans. Karsak et al. provide evidence that the CNR2 gene, encoding the CB2-R, returned a significant association of a single polymorphism and haplotypes encompassing the CNR2 gene on human chromosome 1p36, associated with low BMD [61].

CB2-R-deficient mice display accelerated age-related cancellous bone loss and cortical expansion, albeit with unaltered cortical thickness [62]. Despite the loss of bone, *CB2r-/-* mice exhibit increased mineral appositional rate and bone formation rate. This low bone mass, associated with high bone turnover, is another phenotypic parallel with postmenopausal osteoporosis [62]. Functional CB2-R has been demonstrated in both the osteoblast and osteoclast lineages [62]. Combined, these studies indicate that CB2-R signaling contributes to the maintenance of bone mass by two mechanisms: (1) stimulating stromal cells/osteoblasts directly, and (2) inhibiting monocytes/osteoclasts, both directly and inhibiting osteoblast/stromal cell receptor activator of nuclear factor-κB ligand (RANKL) expression. Jointly, these data suggest that the cannabinoid system plays an important role in the regulation and maintenance of bone mass through the signaling of both the CB1-R and CB2-R.

THE MELANOCORTIN SYSTEM

Melanocortins are a complex family comprising a number of endogenous agonists that are all derived from a single precursor, pro-opiomelanocortin (POMC), of which α-, β- and γ-MSH (melanocyte-stimulating hormone) and

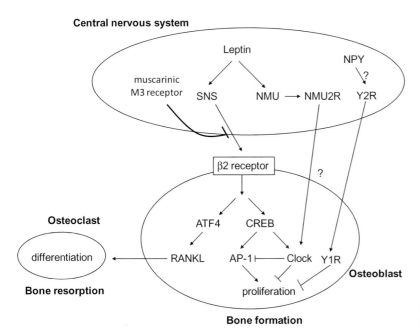

Fig. 115.1. A schematic diagram of central neuronal control of bone remodeling.

adrenocorticotropic hormone (ACTH) elicit their action by interacting with five melanocortin receptors (MCRs), identified as G-protein coupled receptors MCR1–5 [63, 64]. In addition to the melanocortin agonists, agouti-related protein (AgRP) has been identified as a high-affinity antagonist [65].

The regulation of bone homeostasis by this system centers around the action of the melanocortin 4 receptor (MC4R) expressed in hypothalamic neurons. Patients deficient in MC4R are known to exhibit high BMD, resulting from a decrease in bone resorption [66]. Importantly, the greater BMD is still evident following correction of the obesity that is characteristic of MC4R deficiency [66]. Mechanistic studies in mice have enabled dissection of this pathway to bone and interestingly, have implicated another hypothalamic neuropeptide, cocaine and amphetamine-regulated transcript (CART). Hypothalamic CART expression is increased in *MC4R-/-* mice, which display a high bone mass phenotype due to decreased osteoclast number and function [4, 67], as evident in human studies. Additionally, MC4R mutant mice lacking one or two copies of Cart exhibited a significantly lower bone mass [4, 67], demonstrating increased CART signaling, is critical to the low-bone-resorption/high-bone mass phenotype observed in MC4R-deficient mice.

REFERENCES

1. Perkins R, Skirving AP. 1987. Callus formation and the rate of healing of femoral fractures in patients with head injuries. *J Bone Joint Surg Br* 69: 521–4.
2. Takeda S, Karsenty G. 2008 Molecular bases of the sympathetic regulation of bone mass. *Bone* 42: 837–40.
3. Baldock PA, Sainsbury A, Couzens M, Enriquez RF, Thomas GP, Gardiner EM, Herzog H. 2002. Hypothalamic Y2 receptors regulate bone formation. *J Clin Invest* 109: 915–21.
4. Elefteriou F, Ahn JD, Takeda S, Starbuck M, Yang X, Liu X, Kondo H, Richards WG, Bannon TW, Noda M, Clement K, Vaisse C, Karsenty G. 2005. Leptin regulation of bone resorption by the sympathetic nervous system and CART. *Nature* 434: 514–20.
5. Sato S, Hanada R, Kimura A, Abe T, Matsumoto T, Iwasaki M, Inose H, Ida T, Mieda M, Takeuchi Y, Fukumoto S, Fujita T, Kato S, Kangawa K, Kojima M, Shinomiya K, Takeda S. 2007. Central control of bone remodeling by neuromedin U. *Nat Med* 13: 1234–40.
6. Yadav VK, Oury F, Suda N, Liu ZW, Gao XB, Confavreux C, Klemenhagen KC, Tanaka KF, Gingrich JA, Guo XE, Tecott LH, Mann JJ, Hen R, Horvath TL, Karsenty G. 2009. A serotonin-dependent mechanism explains the leptin regulation of bone mass, appetite, and energy expenditure. *Cell* 138: 976–89.
7. Yadav VK, Ryu JH, Suda N, Tanaka KF, Gingrich JA, Schütz G, Glorieux FH, Chiang CY, Zajac JD, Insogna KL, Mann JJ, Hen R, Ducy P, Karsenty G. 2008. Lrp5 controls bone formation by inhibiting serotonin synthesis in the duodenum. *Cell* 135: 825–37.
8. Gautron L, Elmquist JK. 2011. Sixteen years and counting: An update on leptin in energy balance. *J Clin Invest* 121: 2087–93.
9. Ducy P, Amling M, Takeda S, Priemel M, Schilling AF, Beil FT, Shen J, Vinson C, Rueger JM, Karsenty G. 2000. Leptin inhibits bone formation through a hypothalamic relay: a central control of bone mass. *Cell* 100: 197–207.
10. Pogoda P, Egermann M, Schnell JC, Priemel M, Schilling AF, Alini M, Schinke T, Rueger JM, Schneider E, Clarke I, Amling M. 2006. Leptin inhibits bone formation not

11. Guidobono F, Pagani F, Sibilia V, Netti C, Lattuada N, Rapetti D, Mrak E, Villa I, Cavani F, Bertoni L, Palumbo C, Ferretti M, Marotti G, Rubinacci A. 2006. Different skeletal regional response to continuous brain infusion of leptin in the rat. *Peptides* 27: 1426–33.
12. Shi Y, Yadav VK, Suda N, Liu XS, Guo XE, Myers MG Jr, Karsenty G. 2008. Dissociation of the neuronal regulation of bone mass and energy metabolism by leptin in vivo. *Proc Natl Acad Sci U S A* 105: 20529–33.
13. Takeda S, Elefteriou F, Levasseur R, Liu X, Zhao L, Parker KL, Armstrong D, Ducy P, Karsenty G. 2002. Leptin regulates bone formation via the sympathetic nervous system. *Cell* 111: 305–17.
14. Bonnet N, Brunet-Imbault B, Arlettaz A, Horcajada MN, Collomp K, Benhamou CL, Courteix D. 2005. Alteration of trabecular bone under chronic beta2 agonists treatment. *Med Sci Sports Exerc* 37: 1493–501.
15. Kajimura D, Hinoi E, Ferron M, Kode A, Riley KJ, Zhou B, Guo XE, Karsenty G. 2011. Genetic determination of the cellular basis of the sympathetic regulation of bone mass accrual. *J Exp Med* 208: 841–51.
16. Fu L, Patel MS, Bradley A, Wagner EF, Karsenty G. 2005. The molecular clock mediates leptin-regulated bone formation. *Cell* 122: 803–15.
17. Kondo H, Nifuji A, Takeda S, Ezura Y, Rittling SR, Denhardt DT, Nakashima K, Karsenty G, Noda M. 2005. Unloading induces osteoblastic cell suppression and osteoclastic cell activation to lead to bone loss via sympathetic nervous system. *J Biol Chem* 280: 30192–200.
18. Yirmiya R, Goshen I, Bajayo A, Kreisel T, Feldman S, Tam J, Trembovler V, Csernus V, Shohami E, Bab I. 2006. Depression induces bone loss through stimulation of the sympathetic nervous system. *Proc Natl Acad Sci U S A* 103: 16876–81.
19. Bonnet N, Laroche N, Vico L, Dolleans E, Benhamou CL, Courteix D. 2006. Dose effects of propranolol on cancellous and cortical bone in ovariectomized adult rats. *J Pharmacol Exp Ther* 318: 1118–27.
20. Tam J, Trembovler V, Di Marzo V, Petrosino S, Leo G, Alexandrovich A, Regev E, Casap N, Shteyer A, Ledent C, Karsak M, Zimmer A, Mechoulam R, Yirmiya R, Shohami E, Bab I. 2008. The cannabinoid CB1 receptor regulates bone formation by modulating adrenergic signaling. *FASEB J* 22: 285–94.
21. Pasco JA, Henry MJ, Sanders KM, Kotowicz MA, Seeman E, Nicholson GC; Geelong Osteoporosis Study. 2004. Beta-adrenergic blockers reduce the risk of fracture partly by increasing bone mineral density: Geelong Osteoporosis Study. *J Bone Miner Res* 19: 19–24.
22. Wiens M, Etminan M, Gill SS, Takkouche B. 2006. Effects of antihypertensive drug treatments on fracture outcomes: a meta-analysis of observational studies. *J Intern Med* 260: 350–62.
23. Shi Y, Oury F, Yadav VK, Wess J, Liu XS, Guo XE, Murshed M, Karsenty G. 2010. Signaling through the M(3) muscarinic receptor favors bone mass accrual by decreasing sympathetic activity. *Cell Metab* 11: 231–8.
24. Fonseca TL, Jorgetti V, Costa CC, Capelo LP, Covarrubias AE, Moulatlet AC, Teixeira MB, Hesse E, Morethson P, Beber EH, Freitas FR, Wang CC, Nonaka KO, Oliveira R, Casarini DE, Zorn TM, Brum PC, Gouveia CH. 2011. Double disruption of alpha2A- and alpha2C-adrenoceptors results in sympathetic hyperactivity and high-bone-mass phenotype. *J Bone Miner Res* 26: 591–603.
25. Ducy P, Karsenty G. 2010. The two faces of serotonin in bone biology. *J Cell Biol* 191: 7–13.
26. Lam DD, Leinninger GM, Louis GW, Garfield AS, Marston OJ, Leshan RL, Scheller EL, Christensen L, Donato J Jr, Xia J, Evans ML, Elias C, Dalley JW, Burdakov DI, Myers MG Jr, Heisler LK. 2011. Leptin does not directly affect CNS serotonin neurons to influence appetite. *Cell Metab* 13: 584–91.
27. Wu Q, Bencaz AF, Hentz JG, Crowell MD. 2012. Selective serotonin reuptake inhibitor treatment and risk of fractures: A meta-analysis of cohort and case-control studies. *Osteoporos Int* 23(1): 365–75.
28. Mödder UI, Achenbach SJ, Amin S, Riggs BL, Melton LJ 3rd, Khosla S. 2010. Relation of serum serotonin levels to bone density and structural parameters in women. *J Bone Miner Res* 25: 415–22.
29. Yadav VK, Balaji S, Suresh PS, Liu XS, Lu X, Li Z, Guo XE, Mann JJ, Balapure AK, Gershon MD, Medhamurthy R, Vidal M, Karsenty G, Ducy P. 2010. Pharmacological inhibition of gut-derived serotonin synthesis is a potential bone anabolic treatment for osteoporosis. *Nat Med* 16: 308–12.
30. Cui Y, Niziolek PJ, MacDonald BT, Zylstra CR, Alenina N, Robinson DR, Zhong Z, Matthes S, Jacobsen CM, Conlon RA, Brommage R, Liu Q, Mseeh F, Powell DR, Yang QM, Zambrowicz B, Gerrits H, Gossen JA, He X, Bader M, Williams BO, Warman ML, Robling AG. 2011. Lrp5 functions in bone to regulate bone mass. *Nat Med* 17: 684–91.
31. Brighton PJ, Szekeres PG, Willars, GB. 2004. Neuromedin U and its receptors: Structure, function, and physiological roles. *Pharmacol Rev* 56: 231–48.
32. Blomqvist AG, Herzog H. 1997. Y-receptor subtypes—How many more? *Trends Neurosci* 20: 294–8.
33. Lin S, Boey D, Couzens M, Lee N, Sainsbury A, Herzog H. 2005. Compensatory changes in [125I]-PYY binding in Y receptor knockout mice suggest the potential existence of further Y receptor(s). *Neuropeptides* 39: 21–8.
34. Grundemar L, Hakanson R. 1993. Multiple neuropeptide Y receptors are involved in cardiovascular regulation. Peripheral and central mechanisms. *Gen Pharmacol* 24: 785–96.
35. Chronwall BM, DiMaggio DA, Massari VJ, Pickel VM, Ruggiero DA, O'Donohue TL. 1985. The anatomy of neuropeptide-Y-containing neurons in rat brain. *Neuroscience* 15: 1159–81.
36. Hökfelt T, Broberger C, Zhang X, Diez M, Kopp J, Xu Z, Landry M, Bao L, Schalling M, Koistinaho J, DeArmond

SJ, Prusiner S, Gong J, Walsh JH. 1998. Neuropeptide Y: some viewpoints on a multifaceted peptide in the normal and diseased nervous system. *Brain Res Brain Res Rev* 26: 154–66.

37. Lindefors N, Brené S, Herrera-Marschitz M, Persson H. 1990. Regulation of neuropeptide Y gene expression in rat brain. *Ann N Y Acad Sci* 611: 175–85.

38. Ahmed M, Bjurholm A, Kreicbergs A, Schultzberg M. 1993. Neuropeptide Y, tyrosine hydroxylase and vasoactive intestinal polypeptide-immunoreactive nerve fibers in the vertebral bodies, discs, dura mater, and spinal ligaments of the rat lumbar spine. *Spine (Phila Pa 1976)* 18: 268–73.

39. Hill EL, Turner R, Elde R. 1991. Effects of neonatal sympathectomy and capsaicin treatment on bone remodeling in rats. *Neuroscience* 44: 747–55.

40. Lindblad BE, Nielsen LB, Jespersen SM, Bjurholm A, Bünger C, Hansen ES. 1994. Vasoconstrictive action of neuropeptide Y in bone. The porcine tibia perfused in vivo. *Acta Orthop Scand* 65: 629–34.

41. Sisask G, Bjurholm A, Ahmed M, Kreicbergs A. 1996. The development of autonomic innervation in bone and joints of the rat. *J Auton Nerv Syst* 59: 27–33.

42. Bjurholm, A. 1991. Neuroendocrine peptides in bone. *Int Orthop* 15: 325–9.

43. Bjurholm A, Kreicbergs A, Schultzberg M, Lerner UH. 1992. Neuroendocrine regulation of cyclic AMP formation in osteoblastic cell lines (UMR-106-01, ROS 17/2.8, MC3T3-E1, and Saos-2) and primary bone cells. *J Bone Miner Res* 7: 1011–9.

44. Baldock PA, Lee NJ, Driessler F, Lin S, Allison S, Stehrer B, Lin EJ, Zhang L, Enriquez RF, Wong IP, McDonald MM, During M, Pierroz DD, Slack K, Shi YC, Yulyaningsih E, Aljanova A, Little DG, Ferrari SL, Sainsbury A, Eisman JA, Herzog H. 2009. Neuropeptide Y knockout mice reveal a central role of NPY in the coordination of bone mass to body weight. *PLoS One* 4: e8415.

45. Elefteriou F, Takeda S, Liu X, Armstrong D, Karsenty G. 2003. Monosodium glutamate-sensitive hypothalamic neurons contribute to the control of bone mass. *Endocrinology* 144: 3842–7.

46. Baldock PA, Sainsbury A, Allison S, Lin EJ, Couzens M, Boey D, Enriquez R, During M, Herzog H, Gardiner EM. 2005. Hypothalamic control of bone formation: Distinct actions of leptin and y2 receptor pathways. *J Bone Miner Res* 20: 1851–7.

47. Sainsbury A, Schwarzer C, Couzens M, Herzog H. 2002. Y2 receptor deletion attenuates the type 2 diabetic syndrome of ob/ob mice. *Diabetes* 51: 3420–7.

48. de Rijke CE, Hillebrand JJ, Verhagen LA, Roeling TA, Adan RA. 2005. Hypothalamic neuropeptide expression following chronic food restriction in sedentary and wheel-running rats. *J Mol Endocrinol* 35: 381–90.

49. Lauzurica N, Garcia-Garcia L, Pinto S, Fuentes JA, Delgado M. 2010. Changes in NPY and POMC, but not serotonin transporter, following a restricted feeding/repletion protocol in rats. *Brain Res* 1313: 103–12.

50. Igwe JC, Jiang X, Paic F, Ma L, Adams DJ, Baldock PA, Pilbeam CC, Kalajzic I. 2009. Neuropeptide Y is expressed by osteocytes and can inhibit osteoblastic activity. *J Cell Biochem* 108(3): 621–30.

51. Kishi T, Elmquist JK. 2005. Body weight is regulated by the brain: A link between feeding and emotion. *Mol Psychiatry* 10: 132–46.

52. Kopp J, Xu ZQ, Zhang X, Pedrazzini T, Herzog H, Kresse A, Wong H, Walsh JH, Hökfelt T. 2002. Expression of the neuropeptide Y Y1 receptor in the CNS of rat and of wild-type and Y1 receptor knock-out mice. Focus on immunohistochemical localization. *Neuroscience* 111: 443–532.

53. Naveilhan P, Neveu I, Arenas E, Ernfors P. 1998. Complementary and overlapping expression of Y1, Y2 and Y5 receptors in the developing and adult mouse nervous system. *Neuroscience* 87: 289–302.

54. Baldock PA, Allison S, McDonald MM, Sainsbury A, Enriquez RF, Little DG, Eisman JA, Gardiner EM, Herzog H. 2006. Hypothalamic regulation of cortical bone mass: Opposing activity of Y2 receptor and leptin pathways. *J Bone Miner Res* 21: 1600–7.

55. Shi YC, Lin S, Wong IP, Baldock PA, Aljanova A, Enriquez RF, Castillo L, Mitchell NF, Ye JM, Zhang L, Macia L, Yulyaningsih E, Nguyen AD, Riepler SJ, Herzog H, Sainsbury A. 2010. NPY neuron-specific Y2 receptors regulate adipose tissue and trabecular bone but not cortical bone homeostasis in mice. *PLoS One* 5: e11361.

56. Baldock PA, Allison SJ, Lundberg P, Lee NJ, Slack K, Lin EJ, Enriquez RF, McDonald MM, Zhang L, During MJ, Little DG, Eisman JA, Gardiner EM, Yulyaningsih E, Lin S, Sainsbury A, Herzog H. 2007. Novel role of Y1 receptors in the coordinated regulation of bone and energy homeostasis. *J Biol Chem* 282: 19092–102.

57. Lee NJ, Nguyen AD, Enriquez RF, Doyle KL, Sainsbury A, Baldock PA, Herzog H. 2010. Osteoblast specific Y1 receptor deletion enhances bone mass. *Bone* 48: 461–7.

58. Howlett AC, Barth F, Bonner TI, Cabral G, Casellas P, Devane WA, Felder CC, Herkenham M, Mackie K, Martin BR, Mechoulam R, Pertwee RG. 2002. International Union of Pharmacology. XXVII. Classification of cannabinoid receptors. *Pharmacol Rev* 54: 161–202.

59. Mackie, K. 2008. Signaling via CNS cannabinoid receptors. *Mol Cell Endocrinol* 286(1-2 Suppl 1): S60–5.

60. Idris AI, van 't Hof RJ, Greig IR, Ridge SA, Baker D, Ross RA, Ralston SH. 2005. Regulation of bone mass, bone loss and osteoclast activity by cannabinoid receptors. *Nat Med* 11: 774–9.

61. Karsak M, Cohen-Solal M, Freudenberg J, Ostertag A, Morieux C, Kornak U, Essig J, Erxlebe E, Bab I, Kubisch C, de Vernejoul MC, Zimmer A. 2005. Cannabinoid receptor type 2 gene is associated with human osteoporosis. *Hum Mol Genet* 14: 3389–96.

62. Ofek O, Karsak M, Leclerc N, Fogel M, Frenkel B, Wright K, Tam J, Attar-Namdar M, Kram V, Shohami E, Mechoulam R, Zimmer A, Bab I. 2006. Peripheral cannabinoid receptor, CB2, regulates bone mass. *Proc Natl Acad Sci U S A* 103: 696–701.

63. Beltramo M, Campanella M, Tarozzo G, Fredduzzi S, Corradini L, Forlani A, Bertorelli R, Reggiani A. 2003. Gene expression profiling of melanocortin system in neuropathic rats supports a role in nociception. *Brain Res Mol Brain Res* 118: 111–8.
64. Nijenhuis WA, Oosterom J, Adan RA. 2001. AgRP(83-132) acts as an inverse agonist on the human-melanocortin-4 receptor. *Mol Endocrinol* 15: 164–71.
65. Emmerson PJ, Fisher MJ, Yan LZ, Mayer JP. 2007. Melanocortin-4 receptor agonists for the treatment of obesity. *Curr Top Med Chem* 7: 1121–30.
66. Farooqi IS, Yeo GS, Keogh JM, Aminian S, Jebb SA, Butler G, Cheetham T, O'Rahilly S. 2000. Dominant and recessive inheritance of morbid obesity associated with melanocortin 4 receptor deficiency. *J Clin Invest* 106: 271–9.
67. Ahn JD, Dubern B, Lubrano-Berthelier C, Clement K, Karsenty G. 2006. Cart overexpression is the only identifiable cause of high bone mass in melanocortin 4 receptor deficiency. *Endocrinology* 147: 3196–202.

116

The Pituitary–Bone Connection

Mone Zaidi, Tony Yuen, Li Sun, Terry F. Davies, Alberta Zallone, and Harry C. Blair

---

Introduction 969
Pituitary Hormones as Diverse and Evolutionarily Conserved Signals 969
Growth Hormone Action on Bone Occurs Primarily *via* IGF-1 970
FSH Contributes to Bone Loss 970
TSH Uncouples Bone Remodeling 971

ACTH Regulates Vascular Survival and Growth in Bone 972
Contrasting Age-Related Effects of Prolactin on Bone 972
Oxytocin Is a Bone Anabolic Hormone 973
Conclusion 973
Acknowledgments 973
Disclosures 973
References 974

INTRODUCTION

Traditionally, a specific, limited function has been ascribed to each anterior and posterior pituitary hormone. However, the recent use of mouse genetics has led to the realization that these hormones and their receptors have more ubiquitous functions in integrative physiology, particularly in organs such as the skeleton that are regulated by, and respond to, both local factors and systemic signals related to central metabolism and reproduction. While the skeleton expresses steroid-family receptors, which play major roles in regulation, the major pituitary hormones also have critical direct actions in skeletal homeostasis.

The skeletal expression of pituitary glycoprotein receptors further reflects that the function of these signals in endocrine control is evolutionarily more recent [1]. Thus, growth hormone (GH), follicle stimulating hormone (FSH), thyroid stimulating hormone (TSH), adrenocorticotrophic hormone (ACTH), and oxytocin all affect bone, and in mice, the haploinsufficiency of either the ligand and/or receptor often yields a skeletal phenotype with the primary target organ remaining untouched. Recognition and in-depth analysis of the mechanism of action of each pituitary hormone have improved our understanding of bone pathophysiology and opened new avenues for therapy. In this chapter, we discuss the interaction of each pituitary hormone with bone and the potential it holds in understanding and treating osteoporosis.

PITUITARY HORMONES AS DIVERSE AND EVOLUTIONARILY CONSERVED SIGNALS

ACTH is the clearest example of a pituitary hormone being part of a widely distributed G-protein coupled receptor (GPCR) system that is known to participate in local cell differentiation in several contexts. Yet, this distributed function is overshadowed by its pituitary–adrenal signaling function. There are five melanocortin receptors, including the ACTH receptor (MC2R), which regulate various physiological activities such as pigment production, appetite, and sexual function. All are controlled by ligands processed from a single large pro-hormone, pro-opiomelanocorticin (POMC). Hormone production occurs by tissue-specific regulated proteolysis, with ACTH being the predominant product in the anterior pituitary. At other sites, POMC, three melanotropins, and β-endorphin are synthesized from the same precursor. There are reports of ACTH production by human macrophage/monocyte cells [2], making it possible that MC2Rs in bone may be activated by local, instead of pituitary-derived ACTH.

Primer on the Metabolic Bone Diseases and Disorders of Mineral Metabolism, Eighth Edition. Edited by Clifford J. Rosen.
© 2013 American Society for Bone and Mineral Research. Published 2013 by John Wiley & Sons, Inc.

Such decentralized control is also exemplified by corticotropin releasing factor (CRF), which, in the adult, stimulates pituitary ACTH production [3]. In the fetus, it stimulates cortisol synthesis directly [3]. This fetal system shows that evolution for centralization of ACTH as the CRF second messenger has not yet completely supplanted an ancestral regulatory system.

TSH and FSH are two of a group of hormones, along with chorionic gonadotropin (hCG) and luteinizing hormone (LH), which are heterodimeric proteins that share a common α-chain. Their specificity, however, depends on their differing β-chains. These hormones are particularly interesting in that simpler phyla have distributed functions. In coelenterates, which have a primitive nervous system but no endocrine glands, a TSHR family gene is readily identifiable, widely expressed, and shows the intron–exon structure found in mammals [4]. In lower vertebrates, such as in bony fish, the TSHR is abundant in the thyroid, but is also detectable in ovaries, heart, muscle, and brain [5]. In fish, the gonadal expression of the receptors, LH receptor (LHR) and FSH receptor (FSHR), is established, and all higher orders retain this. In fact, multiple differently processed forms of the FSHR occur in fish [6]; this may reflect isoforms with differing functions (below). Further, in fish the FSHR binds both FSH and LH, whereas the LHR recognizes only LH [7]. While high-level FSHR expression is restricted to gonads, low-level expression is seen in the spleen [8], quite similarly to findings in human cells (below).

Low-level TSH production by bone marrow cells has likewise been reported [9]; in this case, a splice variant activates the TSHR, and may do so locally in bone. Lymphocytes also express TSH [10, 11], but such production is unlikely to affect circulating levels. There is no evidence, however, for bone or marrow cell production of FSH, although coproduction of TSHβ and FSH is noted in CD11β cells from mouse thyroid [12]. Overall, therefore, the presence of GPCRs in tissues other than traditional endocrine targets, such as the skeleton, and in cases, coexistence of their ligands, comes as no surprise. What does, however, come as a surprise, is that the skeleton appears to be more sensitive to GPCR stimulation than the primary target organs, at least in mouse genetic and limited human studies.

GROWTH HORMONE ACTION ON BONE OCCURS PRIMARILY VIA IGF-1

Growth hormone (GH), a single-chain polypeptide, plays a vital role in skeletal homeostasis. It directly affects bone through a GPCR, but its primary action occurs via its release of insulin-like growth factors (IGFs). The predominant IGF, IGF-1, is synthesized mainly in the liver and approximately 80% that circulates is bound to IGF binding protein-3 (IGFBP3) and the acid labile subunit (ALS).

The importance of IGF-1 over GH in skeletal homeostasis is borne out by the demonstration that growth retardation and osteoporosis in GHR-deficient mice are arrested by the overexpression of IGF-1 [13]. Furthermore, despite elevated GH levels, mice lacking both liver IGF-1 (LID) and ALS, with depleted serum IGF-1, show reduced bone growth and bone strength [14]. These results suggest that the skeletal effects of GH require IGF-1. In fact, the induction of osteoclastic activity by GH also appears to require IGF-1 made from bone marrow stromal cells, which then activates bone resorption by acting on osteoclastic receptors, as well as by altering receptor activator of nuclear factor-κB ligand (RANKL) expression [15–17]. However, there is limited evidence to suggest that GH can act independently of IGF. For example, GH replacement reverses the increased adiposity in hypophysectomized rats, while IGF-1 replacement does not [18]. Furthermore, in ovariectomized LID mice, GH reverses osteopenia [19]. While these findings point to a direct action of GH on bone and other tissues, selective deletion of this GPCR in osteoblasts and other cells should provide further clarity.

FSH CONTRIBUTES TO BONE LOSS

We discovered that FSH directly stimulates bone resorption by osteoclasts [20]. Several studies have now confirmed direct effects of FSH on the skeleton in rodents and humans. Amenorrheic women with a higher mean serum FSH (approximately 35 IU/L) have greater bone loss than those with lower levels (approximately 8 IU/L) in the face of near-equal estrogen levels [21]. Likewise, in a recent study, patients with functional hypothalamic amenorrhea, in whom both FSH and estrogen were low, showed slight to moderate skeletal defects [22]. Importantly, women harboring an activating FSHR polymorphism, rs6166, have lower bone mass and high resorption markers [23]; this attests to a role for FSHRs in human physiology. Consistent with these human studies, exogenous administration of FSH to rats augments ovariectomy-induced bone loss, and a FSH antagonist reduces bone loss after ovariectomy or FSH injection [24, 25].

Clinical correlations between bone loss and serum FSH levels have been documented extensively. Most impressive is the Study of Women's Health across the Nations (SWAN), a longitudinal cohort of 2,375 perimenopausal women. Not only was there a strong correlation between serum FSH levels and markers of bone resorption, a change in FSH levels over 4 years predicted decrements in bone mass [26]. Analyses of data from Chinese women showed similar trends: a significant association between bone loss and high serum FSH [27, 28]. In a group of southern Chinese women aged between 45 and 55 years, those in the highest quartile of serum FSH lost bone at a 1.3- to 2.3-fold higher rate than those in the lowest quartile [29]. Likewise, examination of a National Health and Nutrition Evaluation Survey III (NHANES III) cohort of women between the ages of 42 and 60 years showed a strong correlation between serum FSH and femoral neck bone mineral density (BMD) [30]. A recent cross-sectional analysis of 92 postmenopausal women found that serum osteocalcin and C-telopeptide cross-linked (CTX) were both positively correlated with FSH, but not with estradiol [31]. The Bone Turnover Range of Normality (BONTURNO) Study group likewise showed that women with serum FSH levels of greater than 30 IU/mL had significantly higher serum bone turnover markers than age-matched women, despite having normal menses [32]. In contrast, Gourlay et al. failed to show a strong relationship between bone mass and FSH or, indeed, estrogen [33]. Interestingly, however, the same authors showed an independent correlation between FSH and lean mass [34]. This latter association makes biological sense inasmuch as FSHRs are present on mesenchymal stem cells [20], known to have the propensity for adipocytic differentiation. However, studies have yet to determine whether FSH inhibits adipogenesis. Nonetheless, the weight of the evidence prompts the use of FSH at least as a serum marker for identifying "fast bone losers" during the early phases of the menopausal transition [35].

Mechanistically, FSH increases osteoclast formation, function, and survival through a distinct FSHR isoform [20, 36–38]. Wu et al. further showed that the osteoclastogenic response to FSH was abolished in mice lacking immunoreceptor tyrosine-based activation motif (ITAM) adapter signaling molecules [38]. This suggests an interaction between FSH and immune receptor complexes, although the significance of this remains unclear. In a separate study, FSHR activation was shown to enhance RANK receptor expression [39]. In addition, FSH indirectly stimulates osteoclast formation by releasing osteoclastogenic cytokines, namely IL-1β, tumor necrosis factor-α (TNF-α), and IL-6 in proportion to the surface expression of FSHRs [40, 41]. In a study of 36 women between the ages of 20 and 50, serum FSH concentrations correlated with circulating cytokine concentrations [41, 42].

Two groups have, however, failed to identify FSHRs on osteoclasts, having likely used primers targeted to the ovarian isoform [43, 44]. We very consistently find FSHR in human CD14$^+$ cells and osteoclasts using nested primers and sequencing to verify the specificity of the reaction, and amplifying regions that contain an intron to avoid the pitfall of genomic DNA contamination [36].

It has been difficult to tease out the action of FSH from that of estrogen *in vivo*, as FSH releases estrogen, and the actions of FSH and estrogen on the osteoclast are opposed. The injection of FSH into mice with intact ovaries [44], or its transgenic overexpression [43], even in *hpg* mice, is unlikely to reveal pro-resorptive actions of FSH. This is because direct effects of FSH on the osteoclast will invariably be masked by the antiresorptive and anabolic actions of the ovarian estrogen so released in response to FSH.

Whether lowering FSH in a hypogonadal state to prevent bone loss can be leveraged therapeutically is unclear.. There is evidence that women with low FSH levels undergo less bone loss [21] and that the effectiveness of estrogen therapy is related to the degree of FSH suppression [45]. With that said, patients with pituitary hypogonadism lose bone. Luperide treatment, and hence the lowering of FSH, has not been shown to prevent hypogonadal hyper-resorption [46]. While this proves that low estrogen is a *cause of* acute hypogonadal bone loss, it does not exclude a role for FSH in human skeletal homeostasis [46]. Rather than blocking FSH in acute hypogonadism, where the effect of low estrogen is likely to be overwhelming, FSH inhibition during the late perimenopause, particularly when estrogen levels are normal and FSH is high, could potentially be of therapeutic significance. A highly selective approach, such as the use of a blocking antibody, is thus envisaged [47].

TSH UNCOUPLES BONE REMODELING

TSH is a direct inhibitor of osteoclasts [48]. The haploinsufficiency of TSHRs in heterozygotic TSHR$^{+/-}$ mice results in osteoporosis, while thyroid hormone levels remain unaffected [48]. Furthermore, TSHR$^{-/-}$ mice are osteoporotic, a phenotype that cannot be explained by the known pro-osteoclastic action of thyroid hormones, particularly as TSHR$^{-/-}$ mice are hypothyroid [49]. Furthermore, skeletal runting, but not the osteoporotic phenotype, is reversed upon rendering TSHR$^{-/-}$ mice euthyroid by thyroid hormone replacement [48]. Thus, TSH acts on bone independently of thyroid hormones, and the osteoporosis of hyperthyroidism may, in part, be due to low TSH [50, 51].

The osteoporosis of TSHR deficiency is of the high-turnover variety. TSHR$^{-/-}$ mice show evidence of increased osteoclastic activity, similarly to hyt/hyt mice that have defective TSHR signaling [52, 53]. Studies show that recombinant TSH attenuates the genesis, function, and survival of osteoclasts *in vitro* in bone marrow [48] and murine ES cell cultures [54]. The latter suggests a role early in bone development [54]. In contrast, the overexpression of constitutively activated TSHR in osteoclast precursor cells [55], or transgenically, in mouse precursors [53], inhibits osteoclastogenesis. In postmenopausal women, a single subcutaneous injection of TSH drastically lowers serum C-telopeptide to premenopausal levels within 2 days, with recovery at day 7 [56]. In none of the studies with TSH replacement did thyroid hormones increase, exemplifying again that the pituitary–bone axis is more primitive than the pituitary–thyroid axis.

This anti-osteoclastogenic action of TSH is mediated by reduced NF-κB and Janus N-terminus kinase (JNK) signaling, and TNF-α production [48, 55]. The effect of TSH on TNF-α synthesis is mediated transcriptionally by binding of two high mobility group box proteins, HMGB1 and HMGB2, to a TNF-α gene promoter [57]. TNF-α production is expectedly upregulated in osteoporotic TSHR$^{-/-}$ mice [48], and the genetic deletion of TNF-α in these mice reverses the osteoporosis [55], proving that the TSHR⁻/⁻ phenotype is mediated by TNF-α, at least in part.

The role of TSH is osteoblast regulation is less defined. While it inhibits osteoblastogenesis in bone marrow-derived cell cultures, it stimulates differentiation and mineralization in murine cell cultures through a Wnt5a-dependent mechanism [58]. Likewise, in vivo, intermittently administered TSH is anabolic in both rats and mice [53, 59]. In rats, TSH, injected up to once every 2 weeks, inhibits ovariectomy-induced bone loss, 28 weeks following ovariectomy [53]. Calcein-labeling studies are consistent with a direct anabolic action of intermittent TSH [59]. Furthermore, in humans, Martini et al. [60] show an increase in N-terminal propeptide of human procollagen type I (PINP), a marker of bone formation, validating the conclusion that a bolus dose of TSH is anabolic.

Epidemiologically, a 4.5-fold increase in the risk of vertebral fractures and 3.2-fold increase in the risk of nonvertebral fractures is seen at TSH levels less than 0.1 IU/L [61]. There is also a strong negative correlation between low serum TSH and high C-telopeptide levels, without an association with thyroid hormone [62]. In patients on L-thyroxine, a significantly greater bone loss has been noted in those with a suppressed TSH than those without suppression [63–65]. The Tromso study supports this: participants with serum TSH below 2 SD had a significantly lower BMD, those with TSH above 2 SD had a significantly increased BMD, whereas there was no association between TSH and BMD at normal TSH levels [66]. In patients taking suppressive doses of thyroxine for thyroid cancer, the serum level of cathepsin K, a surrogate but yet unvalidated resorption marker, was elevated [67], and the HUNT 2 study found a positive correlation between TSH and BMD at the distal forearm [68].

Analysis of data from NHANES data has shown that the odds ratio for correlations between TSH and bone mass ranged between 2 and 3.4 [69]. Furthermore, euthyroid women with serum TSH in the lower tertile of normal have a higher incidence of vertebral fractures, independent of age, BMD, and thyroid hormones [70]. Importantly, patients harboring the TSHR-D727E polymorphism have high bone mass [71]; similar allelic associations have been reported from the United Kingdom and in the Rotterdam study [72, 73]. Another polymorphism, T+140974TC, seen in the Korean population, is also associated with increased BMD, moreso in patients with an elevated TSH, again substantiating the role of TSH in protecting against bone loss [72].

Physiologically, therefore, TSH uncouples bone remodeling by inhibiting osteoclastic bone resorption and stimulating osteoblastic bone formation, particularly when given intermittently. Furthermore, absent TSH, signaling stimulates bone remodeling directly, and through TNF-α production, causes net bone loss. Low TSH levels may thus contribute to the pathophysiology of osteoporosis of hyperthyroidism, which has traditionally been attributed to high thyroid hormone levels alone.

ACTH REGULATES VASCULAR SURVIVAL AND GROWTH IN BONE

Glucocorticoids, under natural regulation mainly by ACTH, are important coregulators of many processes including vascular tone, central metabolism, and immune response. At higher, pharmacological levels, they become anti-inflammatory and immunosuppressant drugs, with incident complications including diabetes, osteoporosis, and osteonecrosis. Osteonecrosis, in particular, is a painful debilitating condition that affects metabolically active bone, typically the femoral head [74], and invariably requires surgical treatment. The underlying mechanisms of glucocorticoid-induced osteonecrosis are poorly understood, although a key finding is that osteonecrosis occurs prior to macroscopic vascular changes [75].

Isales et al. [76] discovered that bone-forming units strongly express melacortin receptor 2 (MC2R). We showed that as with the adrenal cortex, ACTH induces vascular endothelial growth factor (VEGF) production in osteoblasts through its action on MC2Rs [77]. This likely translates into the protection by ACTH of glucocorticoid-induced osteonecrosis in a rabbit model [77]. An independent report with consistent findings was recently published [78]. We speculate that VEGF suppression secondary to ACTH suppression may contribute to bone damage with long-term glucocorticoid therapy. Much needs to be done to validate this idea to a therapeutic advantage, considering that ACTH analogs are already approved for human use.

CONTRASTING AGE-RELATED EFFECTS OF PROLACTIN ON BONE

Prolactin (PRL), a peptide hormone, is secreted by the anterior pituitary. It primarily acts to induce and maintain lactation and prevent another pregnancy by suppressing folliculogenesis and libido. During pregnancy, it increases the calcium bioavailability for milk production and fetal skeletalogenesis by promoting intestinal calcium absorption and skeletal mobilization [79]. Accelerated bone turnover and bone loss is noted in hyperprolactinemic adults [80]. Antagonism of PRL by bromocriptine, a dopamine agonist, reverses the bone loss [81].

This osteoclastic action of PRL is traditionally thought to arise from the accompanying hypoestrogenemia [82]. However, it has been shown that osteoblasts express PRL receptors (PRLRs) [83], suggesting a direct interaction between PRL and the osteoblast. In fact, the pattern of bone loss is distinct in PRL-exposed and ovariectomized rats [84].

Ex vivo, PRL decreases osteoblast differentiation markers [84], in part, through the PI3K signaling pathway [85]. *In vivo*, PRL accelerates bone resorption in adult mice [84] through an indirect action on osteoclasts, notably by increasing the RANK/OPG (osteoprotegerin) ratio [84]. Osteoclasts, themselves, do not possess PRLRs [83]. In contrast, in infant rats, PRL causes net bone gain [86] and increased osteocalcin expression. Likewise, in human fetal osteoblast cells, PRL decreases the RANKL/OPG ratio [85]. It appears therefore that the net effect of PLR on bone depends on the biological maturity of the organism. In the fetal stage, it promotes bone growth and mineralization, while accelerating bone resorption in the mother to make nutrients available. Further, insight is needed to clarify the role of PRL in bone metabolism and determine the cellular pathways.

OXYTOCIN IS A BONE ANABOLIC HORMONE

Oxytocin (OT) is a nanopeptide synthesized in the hypothalamus and released into circulation via the posterior pituitary. Its primary function is to mediate the milk ejection reflex in nursing mammals. It also stimulates uterine contraction during parturition; however, OT is not a requirement for this function. Thus, OT-null mice can deliver normally but are unable to nurse. Subcutaneous OT injection completely rescues the milk ejection phenotype, attesting to this being a peripheral, as opposed to a central, action [87]. Central actions of OT include the regulation of social behavior, including sexual and maternal behavior, affiliation, social memory, as well as penile erection and ejaculation [88–91]. It also controls food, predominantly carbohydrate, intake centrally [92]. Thus, the social amnesia, aggressive behavior, and overfeeding observed in $OT^{-/-}$ and $OTR^{-/-}$ mice are reversed on intracerebroventricular OT injection [93].

OT acts on a GPCR, present in abundance on osteoblasts [94], osteoclasts, and their precursors [95]. In line with the ubiquitous distribution of OT receptors (OTRs), cells of bone marrow also synthesize OT, suggesting the existence of autocrine and paracrine interactions [96]. In vitro, OT stimulates osteoblast differentiation and bone formation. Thus, $OT^{-/-}$ and $OTR^{-/-}$ mice, including the haploinsufficient heterozygotes with normal lactation, display severe osteoporosis due to a bone-forming defect [97]. This not only indicates that the osteoblast is the target for OT, but also that bone is more sensitive to OT than the breast, hitherto considered its primary target. Once again, the finding emphasizes a relatively primitive pituitary–bone axis. Effects of OT on bone resorption in vivo appear minimal, as OT stimulates osteoclastogenesis but inhibits the activity of mature osteoclasts, with a net zero effect on resorption.

In vivo gain-of-function studies document a direct effect of OT on bone. Intraperitoneal OT injections result in increased BMD and ex vivo osteoblast formation [97]. In contrast, short-term intracerebroventricular OT does not affect bone turnover markers. OT injections in wild-type rats alter the RANKL/OPG ratio in favor on bone formation, again attesting to an anabolic action [98].

Although unproven, OT may have a critical role in bone anabolism during pregnancy and lactation. Both are characterized by excessive bone resorption in favor of fetal and postpartum bone growth, respectively [99]. This bone loss is, however, completely reversed upon weaning by a yet unidentified mechanism [100]. OT peaks in blood during late pregnancy and lactation, and while its pro-osteoclastogeneic action may contribute to intergenerational calcium transfer, its anabolic action could enable the restoration of the maternal skeleton. $OT^{-/-}$ pups show hypomineralized skeletons, and $OT^{-/-}$ moms display reduced bone formation markers. The question whether estrogen, via its positive regulation of osteoblastic OT production, can synergize this action through a local feed-forward loop, remains to be determined. These studies nonetheless pave the way for a greater understanding of pregnancy and lactation-associated osteoporosis, as well as new potential therapeutic options.

CONCLUSION

The discovery of glycoprotein hormones and their direct regulation of bone helps explain some of the inconsistencies of older models that assumed that pituitary signaling was mediated entirely via endocrine organs through steroid-family signals. Important direct responses include actions of TSH, FSH, ACTH, and OT in bone. It is important, in evaluating these new signaling mechanisms, to consider that the skeletal responses may or may not have similar mechanisms to the responses of the traditional endocrine targets, and that the signals may vary in importance due to secondary endocrine and paracrine control. The discovery of direct skeletal responses of pituitary hormones nevertheless offers a new set of therapeutic opportunities.

ACKNOWLEDGMENTS

Mone Zaidi, Li Sun, Terry F. Davies, and Harry C. Blair are supported by grants from the National Institutes of Health. Alberta Zallone[2] is supported by the Ministry of Education, Italy.

DISCLOSURES

Mone Zaidi consults for Merck and Takeda and is a named inventor of a pending patent application related to osteoclastic bone resorption filed by the Mount Sinai School of Medicine (MSSM). In the event the pending or issued patent is licensed, he would be entitled to a share of any proceeds MSSM receives from the licensee. All other authors have nothing to disclose.

REFERENCES

1. Blair HC, Robinson LJ, Sun L, Isales C, Davies TF, Zaidi M. 2011. Skeletal receptors for steroid-family regulating glycoprotein hormones: A multilevel, integrated physiological control system. *Ann N Y Acad Sci* 1240: 26–31.
2. Pallinger E, Csaba G. 2008. A hormone map of human immune cells showing the presence of adrenocorticotropic hormone, triiodothyronine and endorphin in immunophenotyped white blood cells. *Immunology* 123: 584–589.
3. Sirianni R, Rehman KS, Carr BR, Parker CR Jr, Rainey WE. 2005. Corticotropin-releasing hormone directly stimulates cortisol and the cortisol biosynthetic pathway in human fetal adrenal cells. *J Clin Endocrinol Metab* 90: 279–285.
4. Vibede N, Hauser F, Williamson M, Grimmelikhuijzen CJ. 1998. Genomic organization of a receptor from sea anemones, structurally and evolutionarily related to glycoprotein hormone receptors from mammals. *Biochem Biophys Res Commun* 252: 497–501.
5. Kumar RS, Ijiri S, Kight K, Swanson P, Dittman A, Alok D, Zohar Y, Trant JM. 2000. Cloning and functional expression of a thyrotropin receptor from the gonads of a vertebrate (bony fish): Potential thyroid-independent role for thyrotropin in reproduction. *Mol Cell Endocrinol* 167: 1–9.
6. Kobayashi T, Andersen O. 2008. The gonadotropin receptors FSH-R and LH-R of Atlantic halibut (Hippoglossus hippoglossus), 1: Isolation of multiple transcripts encoding full-length and truncated variants of FSH-R. *Gen Comp Endocrinol* 156: 584–594.
7. Bogerd J, Granneman JC, Schulz RW, Vischer HF. 2005. Fish FSH receptors bind LH: How to make the human FSH receptor to be more fishy? *Gen Comp Endocrinol* 142: 34–43.
8. Kumar RS, Ijiri S, Trant JM. 2001. Molecular biology of the channel catfish gonadotropin receptors: 2. Complementary DNA cloning, functional expression, and seasonal gene expression of the follicle-stimulating hormone receptor. *Biol Reprod* 65: 710–717.
9. Vincent BH, Montufar-Solis D, Teng BB, Amendt BA, Schaefer J, Klein JR. 2009. Bone marrow cells produce a novel TSHbeta splice variant that is upregulated in the thyroid following systemic virus infection. *Genes Immun* 10: 18–26.
10. Smith EM, Phan M, Kruger TE, Coppenhaver DH, Blalock JE. 1983. Human lymphocyte production of immunoreactive thyrotropin. *Proc Natl Acad Sci U S A* 80: 6010–6013.
11. Harbour DV, Kruger TE, Coppenhaver D, Smith EM, Meyer WJ 3rd. 1989. Differential expression and regulation of thyrotropin (TSH) in T cell lines. *Mol Cell Endocrino* 64: 229–241.
12. Klein JR, Wang HC. 2004. Characterization of a novel set of resident intrathyroidal bone marrow-derived hematopoietic cells: Potential for immune-endocrine interactions in thyroid homeostasis. *J Exp Biol* 207: 55–65.
13. De Jesus K, Wang X, Liu JL. 2009. A general IGF-I overexpression effectively rescued somatic growth and bone deficiency in mice caused by growth hormone receptor knockout. *Growth Factors* 27: 438–447.
14. Yakar S, Rosen CJ, Beamer WG, Ackert-Bicknell CL, Wu Y, Liu JL, Ooi GT, Setser J, Frystyk J, Boisclair YR, LeRoith D. 2002. Circulating levels of IGF-1 directly regulate bone growth and density. *J Clin Invest* 110: 771–781.
15. Guicheux J, Heymann D, Rousselle AV, Gouin F, Pilet P, Yamada S, Daculsi G. 1998. Growth hormone stimulatory effects on osteoclastic resorption are partly mediated by insulin-like growth factor I: An in vitro study. *Bone* 22: 25–31.
16. Hou P, Sato T, Hofstetter W, Foged NT. 1997. Identification and characterization of the insulin-like growth factor I receptor in mature rabbit osteoclasts. *J Bone Miner Res* 12: 534–540.
17. Rubin J, Ackert-Bicknell CL, Zhu L, Fan X, Murphy TC, Nanes MS, Marcus R, Holloway L, Beamer WG, Rosen CJ. 2002. IGF-I regulates osteoprotegerin (OPG) and receptor activator of nuclear factor-kappaB ligand in vitro and OPG in vivo. *J Clin Endocrinol Metab* 87: 4273–4279.
18. Menagh PJ, Turner RT, Jump DB, Wong CP, Lowry MB, Yakar S, Rosen CJ, Iwaniec UT. 2010. Growth hormone regulates the balance between bone formation and bone marrow adiposity. *J Bone Miner Res* 25: 757–768.
19. Fritton JC, Emerton KB, Sun H, Kawashima Y, Mejia W, Wu Y, Rosen CJ, Panus D, Bouxsein M, Majeska RJ, Schaffler MB, Yakar S. 2010. Growth hormone protects against ovariectomy-induced bone loss in states of low circulating insulin-like growth factor (IGF-1). *J Bone Miner Res* 25: 235–246.
20. Sun L, Peng Y, Sharrow AC, Iqbal J, Zhang Z, Papachristou DJ, Zaidi S, Zhu LL, Yaroslavskiy BB, Zhou H, Zallone A, Sairam MR, Kumar TR, Bo W, Braun J, Cardoso-Landa L, Schaffler MB, Moonga BS, Blair HC, Zaidi M. 2006. FSH directly regulates bone mass. *Cell* 125: 247–260.
21. Devleta B, Adem B, Senada S. 2004. Hypergonadotropic amenorrhea and bone density: New approach to an old problem. *J Bone Miner Metab* 22: 360–364.
22. Podfigurna-Stopa A, Pludowski P, Jaworski M, Lorenc R, Genazzani AR, Meczekalski B. 2012. Skeletal status and body composition in young women with functional hypothalamic amenorrhea. *Gynecol Endocrinol* 28: 299–304.
23. Rendina D, Gianfrancesco F, De Filippo G, Merlotti D, Esposito T, Mingione A, Nuti R, Strazzullo P, Mossetti G, Gennari L. 2010. FSHR gene polymorphisms influence bone mineral density and bone turnover in postmenopausal women. *Eur J Endocrinol* 163: 165–172.
24. Liu S, Cheng Y, Fan M, Chen D, Bian Z. 2010. FSH aggravates periodontitis-related bone loss in ovariectomized rats. *J Dent Res* 89: 366–371.

25. Liu S, Cheng Y, Xu W, Bian Z. 2010. Protective effects of follicle-stimulating hormone inhibitor on alveolar bone loss resulting from experimental periapical lesions in ovariectomized rats. *J Endod* 36: 658–663.
26. Sowers MR, Greendale GA, Bondarenko I, Finkelstein JS, Cauley JA, Neer RM, Ettinger B. 2003. Endogenous hormones and bone turnover markers in pre- and perimenopausal women: SWAN. *Osteoporos Int* 14: 191–197.
27. Xu ZR, Wang AH, Wu XP, Zhang H, Sheng ZF, Wu XY, Xie H, Luo XH, Liao EY. 2009. Relationship of age-related concentrations of serum FSH and LH with bone mineral density, prevalence of osteoporosis in native Chinese women. *Clin Chim Acta* 400: 8–13.
28. Wu XY, Wu XP, Xie H, Zhang H, Peng YQ, Yuan LQ, Su X, Luo XH, Liao EY. 2010. Age-related changes in biochemical markers of bone turnover and gonadotropin levels and their relationship among Chinese adult women. *Osteoporos Int* 21: 275–285.
29. Cheung E, Tsang S, Bow C, Soong C, Yeung S, Loong C, Cheung CL, Kan A, Lo S, Tam S, Tang G, Kung A. 2011. Bone loss during menopausal transition among southern Chinese women. *Maturitas* 69: 50–56.
30. Gallagher CM, Moonga BS, Kovach JS. 2010. Cadmium, follicle-stimulating hormone, and effects on bone in women age 42–60 years, NHANES III. *Environ Res* 110: 105–111.
31. Garcia-Martin A, Reyes-Garcia R, Garcia-Castro JM, Rozas-Moreno P, Escobar-Jimenez F, Munoz-Torres M. 2012. Role of serum FSH measurement on bone resorption in postmenopausal women. *Endocrine* 41: 302–308.
32. Adami S, Bianchi G, Brandi ML, Giannini S, Ortolani S, DiMunno O, Frediani B, Rossini M. 2008. Determinants of bone turnover markers in healthy premenopausal women. *Calcif Tissue Int* 82: 341–347.
33. Gourlay ML, Preisser JS, Hammett-Stabler CA, Renner JB, Rubin J. 2011. Follicle-stimulating hormone and bioavailable estradiol are less important than weight and race in determining bone density in younger postmenopausal women. *Osteoporos Int* 22: 2699–2708.
34. Gourlay ML, Specker BL, Li C, Hammett-Stabler CA, Renner JB, Rubin JE. 2012. Follicle-stimulating hormone is independently associated with lean mass but not BMD in younger postmenopausal women. *Bone* 50: 311–316.
35. Zaidi M, Turner CH, Canalis E, Pacifici R, Sun L, Iqbal J, Guo XE, Silverman S, Epstein S, Rosen CJ. 2009. Bone loss or lost bone: Rationale and recommendations for the diagnosis and treatment of early postmenopausal bone loss. *Curr Osteoporos Rep* 7: 118–126.
36. Robinson LJ, Tourkova I, Wang Y, Sharrow AC, Landau MS, Yaroslavskiy BB, Sun L, Zaidi M, Blair HC. 2010. FSH-receptor isoforms and FSH-dependent gene transcription in human monocytes and osteoclasts. *Biochem Biophys Res Commun* 394: 12–17.
37. Sun L, Zhang Z, Zhu LL, Peng Y, Liu X, Li J, Agrawal M, Robinson LJ, Iqbal J, Blair HC, Zaidi M. 2010. Further evidence for direct pro-resorptive actions of FSH. *Biochem Biophys Res Commun* 394: 6–11.
38. Wu Y, Torchia J, Yao W, Lane NE, Lanier LL, Nakamura MC, Humphrey MB. 2007. Bone microenvironment specific roles of ITAM adapter signaling during bone remodeling induced by acute estrogen-deficiency. *PLoS One* 2: e586.
39. Cannon JG, Kraj B, Sloan G. 2011. Follicle-stimulating hormone promotes RANK expression on human monocytes. *Cytokine* 53: 141–144.
40. Iqbal J, Sun L, Kumar TR, Blair HC, Zaidi M. 2006. Follicle-stimulating hormone stimulates TNF production from immune cells to enhance osteoblast and osteoclast formation. *Proc Natl Acad Sci U S A* 103: 14925–14930.
41. Cannon JG, Cortez-Cooper M, Meaders E, Stallings J, Haddow S, Kraj B, Sloan G, Mulloy A. 2010. Follicle-stimulating hormone, interleukin-1, and bone density in adult women. *Am J Physiol Regul Integr Comp Physiol* 298: R790–798.
42. Gertz ER, Silverman NE, Wise KS, Hanson KB, Alekel DL, Stewart JW, Perry CD, Bhupathiraju SN, Kohut ML, Van Loan MD. 2010. Contribution of serum inflammatory markers to changes in bone mineral content and density in postmenopausal women: A 1-year investigation. *J Clin Densitom* 13: 277–282.
43. Allan CM, Kalak R, Dunstan CR, McTavish KJ, Zhou H, Handelsman DJ, Seibel MJ. 2010. Follicle-stimulating hormone increases bone mass in female mice. *Proc Natl Acad Sci U S A* 107: 22629–22634.
44. Ritter V, Thuering B, Saint Mezard P, Luong-Nguyen NH, Seltenmeyer Y, Junker U, Fournier B, Susa M, Morvan F. 2008. Follicle-stimulating hormone does not impact male bone mass in vivo or human male osteoclasts in vitro. *Calcif Tissue Int* 82: 383–391.
45. Kawai H, Furuhashi M, Suganuma N. 2004. Serum follicle-stimulating hormone level is a predictor of bone mineral density in patients with hormone replacement therapy. *Arch Gynecol Obstet* 269: 192–195.
46. Drake MT, McCready LK, Hoey KA, Atkinson EJ, Khosla S. 2010. Effects of suppression of follicle-stimulating hormone secretion on bone resorption markers in postmenopausal women. *J Clin Endocrinol Metab* 95: 5063–5068.
47. Zhu LL, Blair H, Cao J, Yuen T, Latif R, Guo L, Turkova IL, Li J, Davies TF, Sun L, Bian Z, Rosen C, Zallone A, New MI, Zaidi M. 2012. Blocking antibody to the β-subunit of FSH prevents bone loss by inhibiting bone resorption and stimulating bone synthesis. *Proc Natl Acad Sci U S A* 109: 14574–14579.
48. Abe E, Marians RC, Yu W, Wu XB, Ando T, Li Y, Iqbal J, Eldeiry L, Rajendren G, Blair HC, Davies TF, Zaidi M. 2003. TSH is a negative regulator of skeletal remodeling. *Cell* 115: 151–162.
49. Novack DV. 2003. TSH, the bone suppressing hormone. *Cell* 115: 129–130.
50. Zaidi M, Sun L, Davies TF, Abe E. 2006. Low TSH triggers bone loss: Fact or fiction? *Thyroid* 16: 1075–1076.

51. Baliram R, Sun L, Cao J, Li J, Latif R, Huber AK, Yuen T, Blair HC, Zaidi M, Davies TF. 2012. Hyperthyroid-associated osteoporosis is exacerbated by the loss of TSH signaling. *J Clin Invest* 122: 3731–3741.
52. Britto JM, Fenton AJ, Holloway WR, Nicholson GC. 1994. Osteoblasts mediate thyroid hormone stimulation of osteoclastic bone resorption. *Endocrinology* 134: 169–176.
53. Sun L, Vukicevic S, Baliram R, Yang G, Sendak R, McPherson J, Zhu LL, Iqbal J, Latif R, Natrajan A, Arabi A, Yamoah K, Moonga BS, Gabet Y, Davies TF, Bab I, Abe E, Sampath K, Zaidi M. 2008. Intermittent recombinant TSH injections prevent ovariectomy-induced bone loss. *Proc Natl Acad Sci U S A* 105: 4289–4294.
54. Ma R, Latif R, Davies TF. 2009. Thyrotropin-independent induction of thyroid endoderm from embryonic stem cells by activin A. *Endocrinology* 150: 1970–1975.
55. Hase H, Ando T, Eldeiry L, Brebene A, Peng Y, Liu L, Amano H, Davies TF, Sun L, Zaidi M, Abe E. 2006. TNFalpha mediates the skeletal effects of thyroid-stimulating hormone. *Proc Natl Acad Sci U S A* 103: 12849–12854.
56. Mazziotti G, Sorvillo F, Piscopo M, Cioffi M, Pilla P, Biondi B, Iorio S, Giustina A, Amato G, Carella C. 2005. Recombinant human TSH modulates in vivo C-telopeptides of type-1 collagen and bone alkaline phosphatase, but not osteoprotegerin production in postmenopausal women monitored for differentiated thyroid carcinoma. *J Bone Miner Res* 20: 480–486.
57. Yamoah K, Brebene A, Baliram R, Inagaki K, Dolios G, Arabi A, Majeed R, Amano H, Wang R, Yanagisawa R, Abe E. 2008. High-mobility group box proteins modulate tumor necrosis factor-alpha expression in osteoclastogenesis via a novel deoxyribonucleic acid sequence. *Mol Endocrinol* 22: 1141–1153.
58. Baliram R, Latif R, Berkowitz J, Frid S, Colaianni G, Sun L, Zaidi M, Davies TF. 2011. Thyroid-stimulating hormone induces a Wnt-dependent, feed-forward loop for osteoblastogenesis in embryonic stem cell cultures. *Proc Natl Acad Sci U S A* 108: 16277–16282.
59. Sampath TK, Simic P, Sendak R, Draca N, Bowe AE, O'Brien S, Schiavi SC, McPherson JM, Vukicevic S. 2007. Thyroid-stimulating hormone restores bone volume, microarchitecture, and strength in aged ovariectomized rats. *J Bone Miner Res* 22: 849–859.
60. Martini G, Gennari L, De Paola V, Pilli T, Salvadori S, Merlotti D, Valleggi F, Campagna S, Franci B, Avanzati A, Nuti R, Pacini F. 2008. The effects of recombinant TSH on bone turnover markers and serum osteoprotegerin and RANKL levels. *Thyroid* 18: 455–460.
61. Bauer DC, Ettinger B, Nevitt MC, Stone KL. 2001. Risk for fracture in women with low serum levels of thyroid-stimulating hormone. *Ann Intern Med* 134: 561–568.
62. Zofkova I, Hill M. 2008. Biochemical markers of bone remodeling correlate negatively with circulating TSH in postmenopausal women. *Endocr Regul* 42: 121–127.
63. La Vignera S, Vicari E, Tumino S, Ciotta L, Condorelli R, Vicari LO, Calogero AE. 2008. L-thyroxin treatment and post-menopausal osteoporosis: Relevance of the risk profile present in clinical history. *Minerva Ginecol* 60: 475–484.
64. Flynn RW, Bonellie SR, Jung RT, MacDonald TM, Morris AD, Leese GP. 2010. Serum thyroid-stimulating hormone concentration and morbidity from cardiovascular disease and fractures in patients on long-term thyroxine therapy. *J Clin Endocrinol Metab* 95: 186–193.
65. Baqi L, Payer J, Killinger Z, Susienkova K, Jackuliak P, Cierny D, Langer P. 2010. The level of TSH appeared favourable in maintaining bone mineral density in postmenopausal women. *Endocr Regul* 44: 9–15.
66. Grimnes G, Emaus N, Joakimsen RM, Figenschau Y, Jorde R. 2008. The relationship between serum TSH and bone mineral density in men and postmenopausal women: The Tromso study. *Thyroid* 18: 1147–1155.
67. Mikosch P, Kerschan-Schindl K, Woloszczuk W, Stettner H, Kudlacek S, Kresnik E, Gallowitsch HJ, Lind P, Pietschmann P. 2008. High cathepsin K levels in men with differentiated thyroid cancer on suppressive L-thyroxine therapy. *Thyroid* 18: 27–33.
68. Svare A, Nilsen TI, Bjoro T, Forsmo S, Schei B, Langhammer A. 2009. Hyperthyroid levels of TSH correlate with low bone mineral density: The HUNT 2 study. *Eur J Endocrinol* 161: 779–786.
69. Morris MS. 2007. The association between serum thyroid-stimulating hormone in its reference range and bone status in postmenopausal American women. *Bone* 40: 1128–1134.
70. Mazziotti G, Porcelli T, Patelli I, Vescovi PP, Giustina A. 2010. Serum TSH values and risk of vertebral fractures in euthyroid post-menopausal women with low bone mineral density. *Bone* 46: 747–751.
71. Heemstra KA, van der Deure WM, Peeters RP, Hamdy NA, Stokkel MP, Corssmit EP, Romijn JA, Visser TJ, Smit JW. 2008. Thyroid hormone independent associations between serum TSH levels and indicators of bone turnover in cured patients with differentiated thyroid carcinoma. *Eur J Endocrinol* 159: 69–76.
72. van der Deure WM, Uitterlinden AG, Hofman A, Rivadeneira F, Pols HA, Peeters RP, Visser TJ. 2008. Effects of serum TSH and FT4 levels and the TSHR-Asp727Glu polymorphism on bone: the Rotterdam Study. *Clin Endocrinol (Oxf)* 68: 175–181.
73. Albagha OME, Natarajan R, Reid DM, Ralston SH. 2005. The D727E polymorphism of the human thyroid stimulating hormone receptor is associated with bone mineral density and bone loss in women from the UK. *J Bone Miner Res* 20 (Suppl 1): S341.
74. Mankin HJ. 1992. Nontraumatic necrosis of bone (osteonecrosis). *N Engl J Med* 326: 1473–1479.
75. Eberhardt AW, Yeager-Jones A, Blair HC. 2001. Regional trabecular bone matrix degeneration and osteocyte death in femora of glucocorticoid- treated rabbits. *Endocrinology* 142: 1333–1340.

76. Isales CM, Zaidi M, Blair HC. 2010. ACTH is a novel regulator of bone mass. *Ann N Y Acad Sci* 1192: 110–116.
77. Zaidi M, Sun L, Robinson LJ, Tourkova IL, Liu L, Wang Y, Zhu LL, Liu X, Li J, Peng Y, Yang G, Shi X, Levine A, Iqbal J, Yaroslavskiy BB, Isales C, Blair HC. 2010. ACTH protects against glucocorticoid-induced osteonecrosis of bone. *Proc Natl Acad Sci U S A* 107: 8782–8787.
78. Wang G, Zhang CQ, Sun Y, Feng Y, Chen SB, Cheng XG, Zeng BF. 2010. Changes in femoral head blood supply and vascular endothelial growth factor in rabbits with steroid-induced osteonecrosis. *J Int Med Res* 38: 1060–1069.
79. Lotinun S, Limlomwongse L, Krishnamra N. 1998. The study of a physiological significance of prolactin in the regulation of calcium metabolism during pregnancy and lactation in rats. *Can J Physiol Pharmacol* 76: 218–228.
80. Naylor KE, Iqbal P, Fledelius C, Fraser RB, Eastell R. 2000. The effect of pregnancy on bone density and bone turnover. *J Bone Miner Res* 15: 129–137.
81. Lotinun S, Limlomwongse L, Sirikulchayanonta V, Krishnamra N. 2003. Bone calcium turnover, formation, and resorption in bromocriptine- and prolactin-treated lactating rats. *Endocrine* 20: 163–170.
82. Meaney AM, Smith S, Howes OD, O'Brien M, Murray RM, O'Keane V. 2004. Effects of long-term prolactin-raising antipsychotic medication on bone mineral density in patients with schizophrenia. *Br J Psychiatry* 184: 503–508.
83. Coss D, Yang L, Kuo CB, Xu X, Luben RA, Walker AM. 2000. Effects of prolactin on osteoblast alkaline phosphatase and bone formation in the developing rat. *Am J Physiol Endocrinol Metab* 279: E1216–1225.
84. Seriwatanachai D, Thongchote K, Charoenphandhu N, Pandaranandaka J, Tudpor K, Teerapornpuntakit J, Suthiphongchai T, Krishnamra N. 2008. Prolactin directly enhances bone turnover by raising osteoblast-expressed receptor activator of nuclear factor kappaB ligand/osteoprotegerin ratio. *Bone* 42: 535–546.
85. Seriwatanachai D, Charoenphandhu N, Suthiphongchai T, Krishnamra N. 2008. Prolactin decreases the expression ratio of receptor activator of nuclear factor kappaB ligand/osteoprotegerin in human fetal osteoblast cells. *Cell Biol Int* 32: 1126–1135.
86. Krishnamra N, Seemoung J. 1996. Effects of acute and long-term administration of prolactin on bone 45Ca uptake, calcium deposit, and calcium resorption in weaned, young, and mature rats. *Can J Physiol Pharmacol* 74: 1157–1165.
87. Nishimori K, Young LJ, Guo Q, Wang Z, Insel TR, Matzuk MM. 1996. Oxytocin is required for nursing but is not essential for parturition or reproductive behavior. *Proc Natl Acad Sci U S A* 93: 11699–11704.
88. Young WS 3rd, Shepard E, DeVries AC, Zimmer A, LaMarca ME, Ginns EI, Amico J, Nelson RJ, Hennighausen L, Wagner KU. 1998. Targeted reduction of oxytocin expression provides insights into its physiological roles. *Adv Exp Med Biol* 449: 231–240.
89. Insel TR, Harbaugh CR. 1989. Lesions of the hypothalamic paraventricular nucleus disrupt the initiation of maternal behavior. *Physiol Behav* 45: 1033–1041.
90. Mantella RC, Vollmer RR, Li X, Amico JA. 2003. Female oxytocin-deficient mice display enhanced anxiety-related behavior. *Endocrinology* 144: 2291–2296.
91. Argiolas A, Collu M, Gessa GL, Melis MR, Serra G. 1988. The oxytocin antagonist d(CH2)5Tyr(Me)-Orn8-vasotocin inhibits male copulatory behaviour in rats. *Eur J Pharmacol* 149: 389–392.
92. Sclafani A, Rinaman L, Vollmer RR, Amico JA. 2007. Oxytocin knockout mice demonstrate enhanced intake of sweet and nonsweet carbohydrate solutions. *Am J Physiol Regul Integr Comp Physiol* 292: R1828–1833.
93. Ferguson JN, Young LJ, Hearn EF, Matzuk MM, Insel TR, Winslow JT. 2000. Social amnesia in mice lacking the oxytocin gene. *Nat Genet* 25: 284–288.
94. Copland JA, Ives KL, Simmons DJ, Soloff MS. 1999. Functional oxytocin receptors discovered in human osteoblasts. *Endocrinology* 140: 4371–4374.
95. Colucci S, Colaianni G, Mori G, Grano M, Zallone A. 2002. Human osteoclasts express oxytocin receptor. *Biochem Biophys Res Commun* 297: 442–445.
96. Colaianni G, Di Benedetto A, Zhu LL, Tamma R, Li J, Greco G, Peng Y, Dell'Endice S, Zhu G, Cuscito C, Grano M, Colucci S, Iqbal J, Yuen T, Sun L, Zaidi M, Zallone A. 2011. Regulated production of the pituitary hormone oxytocin from murine and human osteoblasts. *Biochem Biophys Res Commun* 411: 512–515.
97. Tamma R, Colaianni G, Zhu LL, DiBenedetto A, Greco G, Montemurro G, Patano N, Strippoli M, Vergari R, Mancini L, Colucci S, Grano M, Faccio R, Liu X, Li J, Usmani S, Bachar M, Bab I, Nishimori K, Young LJ, Buettner C, Iqbal J, Sun L, Zaidi M, Zallone A. 2009. Oxytocin is an anabolic bone hormone. *Proc Natl Acad Sci U S A* 106: 7149–7154.
98. Elabd SK, Sabry I, Hassan WB, Nour H, Zaky K. 2007. Possible neuroendocrine role for oxytocin in bone remodeling. *Endocr Regul* 41: 131–141.
99. Wysolmerski JJ. 2002. The evolutionary origins of maternal calcium and bone metabolism during lactation. *J Mammary Gland Biol Neoplasia* 7: 267–276.
100. Sowers M, Eyre D, Hollis BW, Randolph JF, Shapiro B, Jannausch ML, Crutchfield M. 1995. Biochemical markers of bone turnover in lactating and nonlactating postpartum women. *J Clin Endocrinol Metab* 80: 2210–2216.

117
Skeletal Muscle Effects on the Skeleton
William J. Evans

Introduction 978
Sarcopenia and Bone 979
Cachexia and Bone 982

Summary 983
References 983

INTRODUCTION

A reduction in lean body mass and an increase in fat mass is one of the most striking and consistent changes associated with advancing age. Skeletal muscle [1] and bone mass are the principal (if not exclusive) components of lean body mass to decline with age. These changes in body composition occur throughout life and have important functional and metabolic consequences. The term "sarcopenia" refers to this age-associated loss of skeletal muscle mass and strength. It is a term was first used by Irv Rosenberg [2] and originally described by Evans and Campbell [3] and further defined [4]. The loss of skeletal muscle with advancing age is universal and contributes to a number of conditions that are common in elderly people. Much as bone density is associated with the risk of a bone fracture, it was thought that sarcopenia would predict the risk of disability and loss of independence in elderly people. Skeletal muscle has an important role in the development of a number of geriatric syndromes besides its vital role in movement and in the etiology of mobility disability. It is the primary site of glucose disposal, and muscle insulin resistance is a key feature in the etiology of type 2 diabetes. Sarcopenia is the primary reason for the age-associated decrease in basal metabolic rate and a declining energy requirement [1]. A number of authors have defined sarcopenia more specifically as a subgroup of older persons with muscle mass depletion, usually defined as being 2 standard deviations below the mean muscle mass of younger persons (usually age 35 years of age) [5]. More recently, there has emerged a consensus that skeletal muscle mass alone is not a strong predictor of outcomes related to risk of disability, and a definition of who is sarcopenic must include a functional measurement (habitual gait speed, for example) along with a measure of skeletal muscle mass. Sarcopenia has become recognized as an important geriatric condition and a key precursor to the development of frailty [6, 7]. Much like osteopenia (bone density) predicts risk of a bone fracture, sarcopenia is a powerful predictor of late-life disability.

Sarcopenia very likely begins in early adulthood [8] with atrophy and loss of type II muscle fibers [9, 10], and continues throughout life as a result of the complex interaction of environmental and genetic causes. Longitudinal studies have shown a clear decline in muscle mass, strength, and power beginning at approximately 35 years of age [11]. Strength and power decline to a greater extent than does muscle mass [12]. In addition to sarcopenia, intramusculater lipid, termed "myosteatosis" [13], increases with age and increasing body fatness. Janssen et al. [14] estimated that sarcopenia results in an excess cost to the health-care system of the United States of $18.4 billion a year (year 2001), due to associated disability. In the New Mexico Study, the prevalence of sarcopenia was originally determined to be over 50% in persons older than 80 years of age [5]. Subsequent studies in this population using more direct estimates found a prevalence of 12% for persons 60 to 70 years old and nearly 30% for persons over 80 [5, 15]. Janssen et al. [16], using the large cohort from National Health and Nutri-

Primer on the Metabolic Bone Diseases and Disorders of Mineral Metabolism, Eighth Edition. Edited by Clifford J. Rosen.
© 2013 American Society for Bone and Mineral Research. Published 2013 by John Wiley & Sons, Inc.

Table 117.1. Prevalence of Sarcopenia Estimated from Different Cohort by Differing Methods[†]

Author	Method	Sex	n	Age (years)	Prevalence
Baumgartner et al. 1998 (Ref. 5)	Anthropomorphic	m/f	883	61–70	13%
				71–80	24%
				80+	50%
Morley et al. 2001 (Ref. 15)	DXA	m/f	199	<70	12%
				80+	30%
Janssen et al. 2002 (Ref. 16)	Bioelectrical impedance	m/f	4,504	60+	7%
		f	4,321		
Tanko et al. 2002 (Ref. 20)	DXA	f	754	70+	12.3%
Melton et al. 2000 (Ref. 21)	DXA	m	50	70–79	14%
			50	≥80	42%
		f	51	70–79	4.7%
			48	≥80	16.3%
Gillette-Guyonnet et al. 2003 (Ref. 18)	DXA	f	1,321	76–80	8.9%
				86–95	10.9%
Castillo et al. 2004 (Ref. 17)	Bioelectrical impedance	m	694	55–98	Overall 6%
		f	1,006	70–75	4% men; 3% women
				85+	13% women; 16% men
Ianuzzi-Sacich et al. 2002 (Ref. 19)	DXA	m	142	64–92	26.8%
		f	195		22.6%
Newman et al. 2003 (Ref. 64)	DXA	m	1,435	74.5±2.8	12–30%[*]
		f	1,549	73.8±2.8	12–30%
Schaap et al. (Ref. 65)	DXA	m/f	328	74.0±8	15%[#]

[†]DXA = Dual X-ray absorptiometry; f = female; m = male.
[*]BMI < 30.
[#]Longitudinal analysis with sarcopenia defined as a loss of appendicular muscle mass of >3% in 3 years.

tion Evaluation Survey III (NHANES III), found the prevalence of sarcopenia (–2 SD) in persons 60 years of age and older was 7% to 10%. Women were more likely to be sarcopenic than men in this study, but others have reported the opposite [17–21]. Table 117.1 compares a number of different studies on the prevalence of sarcopenia. A common finding of all of these studies is that sarcopenia is highly prevalent in older people and that it increases with advancing age.

SARCOPENIA AND BONE

The loss of skeletal muscle and bone mass with advancing age is closely related [22]. Doyle et al. [23] examined the ash weight of the third lumbar vertebrate and compared it to the weight of the left psoas muscle in male and female cadavers and reported a strong relationship ($r = 0.722$, $p < 0.001$, Fig. 117.1). They proposed that the weight of the psoas was a reflection of the forces the muscle exerts on the bone, resulting in increased or decreased bone. Using neutron activation to determine total body Ca (TBCa) and measurement of total body ^{40}K (TBK) as an indication of active cell mass (skeletal muscle), Cohn and coworkers [24] measured 79 men and

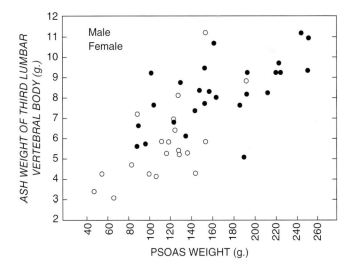

Fig. 117.1. Relationship ($r = 0.722$, $p < 0.001$) of ash weight of L3 vertebral body and psoas weight in 26 male and 20 female cadavers (age range 26–83). From Doyle et al. (Ref. 23).

women between the ages of 30 and 90. They found that at each decade of life, the TBK/TBCa ratio remained constant with advancing age, concluding that "the mechanism responsible for loss of bone with age, whether nutritional deficiency or decreased gonadal function and physical activity may also be responsible for the loss of muscle mass with age." Ellis and Cohn [25] observed a TBK/TBCa ratio of 0.122 ± 0.008 for men and 0.100 ± 0.007 for women, indicating greater muscle mass per unit skeletal mass that remained constant with advancing age. Capozza and coworkers [26] evaluated the relationship between bone mineral content and lean body mass (BMC/LM) in a cohort of 3,205 pre- and postmenopausal women. In addition, they examined pre- and postmenopausal women who had experienced a recent fracture. They found no differences in this ratio between premenopausal women with and without fractures. However, in postmenopausal women, both LM and BMC showed similar changes with age, and women experiencing a fracture had significantly lower BMC/LM, particularly in women with a hip fracture.

Because skeletal muscle mass is an important determiner of bone density and functional status, loss of skeletal muscle also results in loss of bone density. Disuse is discussed below. Weight loss through dietary restriction results in a loss of body fat, lean mass, and bone density. Langlois et al. [27] examined the association between weight change and risk of hip fracture in a cohort of 2,413 community-dwelling white men aged 67 years or older during an 8-year follow-up period. Extreme weight loss (10% or more) beginning at age 50 years was associated in a proportional hazards model with increased risk of hip fracture (relative risk, 1.8; 95% confidence interval, 1.04–3.3). Weight loss of 10% or more was associated with several indicators of poor health, including physical disability, low mental status score, and low physical activity ($P < 0.05$). Weight gain of 10% or more beginning at age 50 years provided borderline protection against the risk of hip fracture (relative risk, 0.4; 95% confidence interval, 0.1–1.00). Despite differences between older men and women in the incidence of, and risk factors for, hip fracture, weight history is also an important determinant of the risk of hip fracture among older men. Weight loss of 10% or more beginning at age 50 years increases the risk of hip fracture in older white men; weight gain of 10% or more decreases the risk of hip fracture. The relationship between extreme weight loss and poor health suggests that weight loss is a marker of frailty that may increase the risk of hip fracture in older men. Physicians should include weight history in their assessment of the risk of hip fracture among older men. Ensrud et al. [28] examined effects of changes in body weight on bone density in a group of 6,785 women over 65 years over a period of 5.7 years. They demonstrated that the greatest loss of bone occurred in women who lost weight (compared to those who were weight stable or gained weight) and this weight loss doubled the risk for a hip fracture irrespective of current weight or intention to lose weight. They concluded, "These findings indicate that even voluntary weight loss in overweight elderly women increases hip fracture risk." Villareal et al. [29] also demonstrated that weight loss is associated with a substantial loss of bone density. A 1-year program of calorie restriction and weight loss resulted in approximately 2% loss of bone mineral density (BMD) from the total hip and intertrochanter. Women who participated in a regular exercise program and lost a similar amount of weight and demonstrated no change in BMD.

There are a number of factors associated with the risk of a bone fracture; bone density and skeletal strength are the primary ones. Skeletal muscle strength and functional capacity are strongly associated with risk of a fall in elderly people [30]. The skeleton adapts to stress and mechanical loads and skeletal muscle exerts powerful loading forces on bone. The increased muscle mass/unit bone demonstrated by the greater TBK/TBCa ratio in men compared to women, observed by Ellis and Cohn, helps to explain the greater bone density and lower fracture risk in men. Using the measurement of TBK, Aloia et al. [31] reported that muscle mass declines during menopause and TBCa and bone density were associated with this loss. They also found that measures of TBK, bone density, and TBCa were positively related to body mass index (BMI) but not body fatness. Indeed, a high percentage of older women who experience a fracture of the hip are sarcopenic [32]. In this group of 313 women with a hip fracture, dual energy X-ray absorptiometry (DXA) assessment of lean mass and bone density showed a significant relationship between the prevalence of sarcopenia and osteopenia ($p = 0.026$) (Table 117.2).

However important bone density is in risk for a fracture, risk of a fall is equally as important, as at least 90% of hip fractures result from a fall. Gunter et al. [33] found that functional capacity was powerful predictor of those who were classified as fallers or non-fallers. They concluded that "given that the 'get up and go' discriminates between fallers and non-fallers and is associated with lower extremity strength and power, fall prevention

Table 117.2. Count and Expected Count of Women with Sarcopenia and Osteoporsis*

	With osteoporosis	Without osteoporosis	Total
With sarcopenia	141	39	180
(expected count)	132.3	47.7	
Without sarcopenia	89	44	133
(expected count)	97.7	35.3	
Total	230	83	313

*From Di Monaco et al. (Ref. 32). In 313 older women who have experienced a hip fracture, the actual number of women with sarcopenia (defined as LBM > 2 SD less than young adult) and osteoporosis, from DXA estimates. This population has a higher prevalence of both conditions than age-matched Caucasian women with no history of fracture.

strategies should focus on improving both functional mobility and lower extremity strength and power." In very old (89 ± 7 years old) and frail residents of a long-term care facility, Greenspan et al. [34] demonstrated that a fall to the side, low femoral bone density, and impaired mobility were independent risk factors for a hip fracture. In this population of frail elderly men and women, strategies for improving muscle function may significantly decrease the risk of an injurious fall and should be considered as a first-line prevention of an initial fall or recurrent fall. This may be particularly important as prevention or treatment of very low bone density in this population may take a long time to be effective or may not improve bone strength sufficiently to prevent a fracture. The responses of very old and frail long-term care residents to a progressive resistance training program are both substantial and rapid [35, 36]. The positive effects of this exercise training includes greatly improved strength, increased stair climbing power, improved balance, and increased level of spontaneous activity, each of which may decrease the risk of an injurious fall.

Orwell et al. [37] examined 2,587 community-based men between the ages of 65 and 99 years of age. Incident falls were ascertained every 4 months during 4 years of follow-up. Men in the lowest quartile for testosterone had a 40% greater risk for falling than men in the highest quartile. The men with the low testosterone also had a lower score for physical function but had a higher risk of falling even after controlling for functional capacity. These data suggest that muscle mass may play a strong role in preventing falls and subsequent bone fracture.

Disuse and physical activity

The positive effects of physical activity on bone density are well described. Regular, weight-bearing, aerobic exercise such as walking slows the age-associated rate of loss of bone density [38, 39]. In 96 healthy, Caucasian women, a positive relationship was seen between measures of muscle strength, physical activity, and bone density [40]. While regular physical activity helps to preserve LM [41], progressive resistance training (PRT) has also been shown to have a dramatic effect on both muscle mass, strength, and bone density. Nelson et al. [42] examined the effects of a twice-a-week, year-long PRT (80% of one repetition maximum) program in postmenopausal women who were randomly assigned PRT or to a sedentary control group. Femoral neck BMD and lumbar spine BMD increased by $0.005 \pm 0.039\,g/cm^2$ ($0.9\% \pm 4.5\%$) (mean ± SD) and $0.009 \pm 0.033\,g/cm^2$ ($1.0\% \pm 3.6\%$), respectively, in the strength-trained women, and decreased by $-0.022 \pm 0.035\,g/cm^2$ ($-2.5\% \pm 3.8\%$) and $-0.019 \pm 0.035\,g/cm^2$ ($-1.8\% \pm 3.5\%$), respectively, in the controls ($P = 0.02$ and 0.04). In addition, the women in the exercise group also increased muscle mass, strength, and balance when compared to the control group. These results demonstrate that a twice-a-week exercise program positively affects not just bone density, but other important causes of a fall and fracture in older women.

Walsh et al. [43] examined the relationship between sarcopenia and osteopenia and found that in premenopausal osteopenic women the prevalence of sarcopenia was 12.5%. In postmenopausal women it was 25% for those with osteopenia and 50% for those with osteoporosis. Physical activity was independently related to skeletal muscle mass (beta = 0.222, p = 0.0001), but diet and hormone replacement therapy (HRT) were not. They found that after adjusting for physical activity, muscle mass was not significantly related to BMD. These data strongly suggest that, while muscle mass may be linked to bone health, it is the forces that muscle exerts on the skeleton that may be the most important in maintaining bone density with advancing age.

Muscle strength is a predictor of a number of outcomes including risk of disability and bone fracture. Muscle strength is generally a reflection of skeletal muscle mass; however, strength shows a greater decrease with advancing age than does muscle mass. In 89 frail, very old nursing home residents, muscle mass was significantly associated with leg strength [44]. However, after controlling for gender, $r^2 = 0.0625$, indicating that that muscle mass explained only a very small amount of the variability in muscle strength. Reduction in muscle strength may result from a number of factors, including decreased level of physical activity, while in subjects who undertake progressive resistance exercise training, increased strength is seen before any changes in muscle size. One important factor affecting muscle strength in elderly people is intramyocellular fat. Goodpaster and coworkers [45] have demonstrated that aging is associated with increased lipid infiltration of muscle, which results in decreased force production even after controlling for muscle cross-sectional area. Fat infiltration increases the risk of a bone fracture in older people. Schafer et al. [46] examined 2,762 older subjects in longitudinal health and reported that increased intramyocellular fat was associated with a 19% increased risk of a fracture. They also showed that when compared with subjects with normal glucose tolerance, those with type 2 diabetes had increased rate of fracture.

Inactivity has a powerful effect on skeletal muscle strength, size, and bone density. A large number of studies using bed rest as a model for exposure to a microgravity environment have demonstrated losses of muscle, exercise capacity, cardiovascular function, and bone density. However, elderly people with reduced exercise reserve, low muscle mass, and low bone density are most often hospitalized and placed on complete bed rest for long periods of time as therapy. The consequences of prolonged bed rest on body composition and functional capacity in elderly people has not received a great deal of attention. Covinsky et al. [47] showed that activities of daily living (ADL) function were often significantly reduced as a result of hospitalization. Kortebein et al. [48] examined the effects of 10 days of bed rest in healthy elderly men and women. These subjects experienced a reduction in LM of 3.2% (p = 0.004) and a decrease in leg lean mass of 0.95 kg (−6.3%, p = 0.003). This reduction of leg muscle mass was associated with a 30% decrease in the fractional synthetic rate of muscle protein. They also experienced a decrease in physical activity, strength (−15.6%), and VO_{2peak} (−15.0%). The reduction in VO_{2peak} was equivalent to the effects of 15 years of aging [49, 50]. Younger men and women undergoing 28 days bed rest as a model for microgravity have been shown to lose about 400 g of muscle tissue [51]. These data demonstrated that healthy older people lose more than twice the amount of muscle compared to young subjects in one-third the amount of time during periods of bed rest.

There are few studies that have examined the recovery from bed-rest-induced changes in muscle and bone. Manske et al. examined the role of muscle force production in triggering bone adaptations in mice. By using a Botox muscle injection to immobilize muscle, they discovered that, as expected, muscle and bone losses were similar during immobilization. However, contrary to their hypothesis that recovery of muscle size and strength would precede increased bone density, they found that muscle and bone properties recovered at similar rates.

Muscle–bone interactions

The skeleton adapts to changes in physical activity, and these adaptations are manifested through loading forces produced by skeletal muscle. This may be one important reason why strengthening exercises that produce high amount of force for a short period of time appear to have a more powerful effect than lower intensity endurance exercise. Nelson et al. [42] examined the effects of a 1-year-long resistance exercise training program on bone density, muscle mass and strength, and functional status. After 1 year of a twice-weekly exercise program, the strength-trained women experienced an average 0.9% ± 4.5% increase in lumbar spine BMD and a 1.0% ± 3.6% increase in femoral neck BMD, while the non-exercise control group showed a −2.5% ± 3.8% and −1.8% ± 3.5% decrease, respectively. In addition, the women in the exercise group (and not in the control group) increased muscle mass, strength, balance, and overall level of physical activity, each of which have been demonstrated to be risk factors for an injurious fall. Increased mechanical loading of bone [52] by skeletal muscle may be an important reason that resistance exercise, during which muscles generate a far greater amount of force than during endurance-type exercise, has a more powerful effect on bone density than does lower intensity exercise. Bassey and coworkers [53–55] have examined the effects of high impact exercises. In one study [53], premenopausal women performed 50 heel-drop exercises (standing on toes with hip and knee extended and dropping on to the heel) each day for 6 months. The women performing the exercises showed a 3–4% increase in trochanteric density compared to controls. However, a similar study in postmenopausal women [54] failed to show an effect of a similar exercise program. In a direct comparison of high-impact, jumping exercise in pre- and postmenopausal women [55], positive effects on bone density were observed in the younger compared to older subjects.

Skeletal muscle expresses growth factors that may also exert an anabolic effect on the skeleton. Indeed, Hamrick et al. [56] demonstrated that bone is rich in receptor proteins for growth factors secreted by skeletal muscle, particularly in the muscle–bone interface. Muscle growth and hypertrophy are associated with secretion of insulin-like growth factor-1 (IGF-1) and mechano-growth factor (MGF), which stimulates bone formation by oseoprogenitor cells in the periosteum that express IGF-1 receptor (IGF-1R). Muscle injury promotes release of fibroblast growth factor-2 (FGF-2), which induces bone formation and stimulates fracture healing by perioseal osteoprogenitor cells expressing FGF-R2. Hamrick [57] points out that a number of osteogenic factors are secreted by skeletal muscle including IGF-1, insulin-like growth factor binding protein, basic fibroblast growth factor, osteonectin, transforming growth factor-α1, matrix metalloproteinase 2, and leukemia inhibitory factor. These hormone-like "myokines" are increased during times of muscle hypertrophy, such as that seen with resistance exercise training. They also point out that myostatin, a negative regulator of muscle growth, has a direct effect on bone and periods of muscle atrophy due to disuse also result in loss of bone.

CACHEXIA AND BONE

Cachexia [58] has been defined as:

> a complex metabolic syndrome associated with underlying illness and characterized by loss of muscle with or without loss of fat mass. The prominent clinical feature of cachexia is weight loss in adults (corrected for fluid retention) or growth failure in children (excluding endocrine disorders). Anorexia, inflammation, insulin resistance and increased muscle protein breakdown are frequently associated with cachexia. Cachexia is distinct from starvation, age-related loss of muscle mass, primary depression, malabsorption and hyperthyroidism and is associated with increased morbidity.

While cachexia is defined as a muscle-wasting syndrome, little attention has been paid to the concomitant loss of bone density. Prostate and breast cancer have been associated with low bone density and increased fracture risk, which has been described as the consequences of adjuvant therapy [59]. Heart failure is also associated with low bone density [60]. Men with HIV infection have a high prevalence of sarcopenia and low bone density [61]. Patients with chronic obstructive pulmonary disease (COPD) have been demonstrated have low bone density. They also have poor functional capacity, suffer from cachexia (muscle wasting) and frequently use corticosteroids. However, an additional factor that may cause an accelerated loss of muscle and bone in these patients is inflammation. Bone et al. [62] examined 40 COPD patients awaiting lung transplantation. They demonstrated that 95% of these patients had low bone density and also showed that biomarkers of bone formation and resorption were related to tumor necrosis factor-α (TNF-α) and interleukin 2. A strong relationship was seen between airway obstruction and BMD in patients with emphysema [63].

SUMMARY

The strong relationship between skeletal muscle mass and bone density has been described in a number of studies. This relationship is seen in children, adults, men, and women. Indeed, conditions that result in loss (weight loss, disuse, cachexia) or gain (resistance exercise, androgen therapy) in skeletal muscle also cause a gain in bone density. This coordinated loss of muscle and bone should be considered and treated, particularly in elderly people. When elderly people undergo hospitalization, they are often discharged with considerably lower muscle mass and bone density as well as orthostatic intolerance and poor functional capacity. Although increased body fat can increase the risk of type 2 diabetes and disability in older people, traditional weight loss regimens using energy restriction result in losses of bone and muscle and an increase in risk of a fracture. Use of exercise can be of great benefit. However, pharmacological tools to preserve skeletal muscle and the skeleton should be strongly considered. As new muscle anabolic therapies such as selective androgen receptor modulators or anti-myostatin compounds are discovered, the promise of a single medication that enhances muscle mass and quality as well as improves bone health is great.

REFERENCES

1. Tzankoff SP, Norris AH. 1978. Longitudinal changes in basal metabolic rate in man. *J Appl Physiol* 33: 536–9.
2. Rosenberg IH. 1989. Summary comments. *Am J Clin Nutr* 50: 1231–3.
3. Evans WJ, Campbell WW. 1993. Sarcopenia and age-related changes in body composition and functional capacity. *J Nutr* 123: 465–8.
4. Evans W. 1995. What is sarcopenia? *J Gerontol A Biol Sci Med Sci* 50 Spec No: 5–8.
5. Baumgartner RN, Koehler KM, Gallagher D, Romero L, Heymsfield SB, Ross RR, Garry PJ, Lindeman RD. 1998. Epidemiology of sarcopenia among the elderly in New Mexico. *Am J Epidemiol* 147(8): 755–63.
6. Morley JE. 2001. Anorexia, sarcopenia, and aging. *Nutrition* 17(7–8): 660–3.
7. Morley JE, Kim MJ, Haren MT, Kevorkian R, Banks WA. 2005. Frailty and the aging male. *Aging Male* 8(3–4): 135–40.
8. Lexell J, Henriksson-Larsen K, Wimblod B, Sjostrom M. 1983. Distribution of different fiber types in human skeletal muscles: Effects of aging studied in whole muscle cross sections. *Muscle Nerve* 6: 588–95.
9. Larsson L. 1978. Morphological and functional characteristics of the aging skeletal muscle in man. *Acta Physiol Scand Suppl* 457 (Suppl.): 1–36.
10. Larsson L. 1983. Histochemical characteristics of human skeletal muscle during aging. *Acta Physiol Scand* 117: 469–71.
11. Frontera WR, Hughes VA, Fielding RA, Fiatarone MA, Evans WJ, Roubenoff R. 2000. Aging of skeletal muscle: A 12-yr longitudinal study. *J Appl Physiol* 88(4): 1321–6.
12. Ferrucci L, Guralnik JM, Buchner D, Kasper J, Lamb SE, Simonsick EM, Corti MC, Bandeen-Roche K, Fried LP. 1997. Departures from linearity in the relationship between measures of muscular strength and physical performance of the lower extremities: The Women's Health and Aging Study. *J Gerontol A Biol Sci Med Sci* 52(5): M275–85.
13. Delmonico MJ, Harris TB, Lee JS, Visser M, Nevitt M, Kritchevsky SB, Tylavsky FA, Newman AB. 2007. Alternative definitions of sarcopenia, lower extremity performance, and functional impairment with aging in older men and women. *J Am Geriatr Soc* 55(5): 769–74.
14. Janssen I, Shepard DS, Katzmarzyk PT, Roubenoff R. 2004. The healthcare costs of sarcopenia in the United States. *J Am Geriatr Soc* 52(1): 80–5.
15. Morley JE, Baumgartner RN, Roubenoff R, Mayer J, Nair KS. Sarcopenia. 2001. *J Lab Clin Med* 137(4): 231–43.
16. Janssen I, Heymsfield SB, Ross R. 2002. Low relative skeletal muscle mass (sarcopenia) in older persons is associated with functional impairment and physical disability. *J Am Geriatr Soc* 50(5): 889–96.
17. Castillo EM, Goodman-Gruen D, Kritz-Silverstein D, Morton DJ, Wingard DL, Barrett-Connor E. 2003. Sarcopenia in elderly men and women: The Rancho Bernardo study. *Am J Prev Med* 25(3): 226–31.
18. Gillette-Guyonnet S, Nourhashemi F, Andrieu S, Cantet C, Albarede JL, Vellas B, Grandjean H. 2003. Body composition in French women 75+ years of age: The EPIDOS study. *Mech Ageing Dev* 124(3): 311–6.
19. Iannuzzi-Sucich M, Prestwood KM, Kenny AM. 2002. Prevalence of sarcopenia and predictors of skeletal muscle mass in healthy, older men and women. *J Gerontol A Biol Sci Med Sci* 57(12): M772–7.
20. Tanko LB, Movsesyan L, Mouritzen U, Christiansen C, Svendsen OL. 2002. Appendicular lean tissue mass and the prevalence of sarcopenia among healthy women. *Metabolism* 51(1): 69–74.
21. Melton LJ 3rd, Khosla S, Crowson CS, O'Connor MK, O'Fallon WM, Riggs BL. 2000. Epidemiology of sarcopenia. *J Am Geriatr Soc* 48(6): 625–30.
22. Crepaldi G, Maggi S. 2005. Sarcopenia and osteoporosis: A hazardous duet. *J Endocrinol Invest* 28(10 Suppl): 66–8.
23. Doyle F, Brown J, Lachance C. 1970. Relation between bone mass and muscle weight. *Lancet* 21: 391–3.

24. Cohn SH, Vaswani A, Zanzi I, Aloia JF, Roginsky MS, Ellis KJ. 1976. Changes in body chemical composition with age measured by total-body neutron activation. *Metabolism* 25(1): 85–96.
25. Ellis K, Cohn S. 1975. Correlation between skeletal calcium mass and muscle mass in man. *J Appl Physiol* 38(3): 455–60.
26. Capozza RF, Cure-Cure C, Cointry GR, Meta M, Cure P, Rittweger J, Ferretti JL. 2008. Association between low lean body mass and osteoporotic fractures after menopause. *Menopause* 15(5): 905–13.
27. Langlois JA, Visser M, Davidovic LS, Maggi S, Li G, Harris TB. 1998. Hip fracture risk in older white men is associated with change in body weight from age 50 years to old age. *Arch Intern Med* 158(9): 990–6.
28. Ensrud KE, Ewing SK, Stone KL, Cauley JA, Bowman PJ, Cummings SR. 2003. Intentional and unintentional weight loss increase bone loss and hip fracture risk in older women. *J Am Geriatr Soc* 51(12): 1740–7.
29. Villareal DT, Fontana L, Weiss EP, Racette SB, Steger-May K, Schechtman KB, Klein S, Holloszy JO. 2006. Bone mineral density response to caloric restriction-induced weight loss or exercise-induced weight loss: A randomized controlled trial. *Arch Intern Med* 166(22): 2502–10.
30. de Rekeneire N, Visser M, Peila R, Nevitt MC, Cauley JA, Tylavsky FA, Simonsick EM, Harris TB. 2003. Is a fall just a fall: Correlates of falling in healthy older persons. The Health, Aging and Body Composition Study. *J Am Geriatr Soc* 51(6): 841–6.
31. Aloia JF, McGowan DM, Vaswani AN, Ross P, Cohn SH. 1991. Relationship of menopause to skeletal and muscle health. *Am J Clin Nutr* 53: 1378–83.
32. Di Monaco M, Vallero F, Di Monaco R, Tappero R. 2010. Prevalence of sarcopenia and its association with osteoporosis in 313 older women following a hip fracture. *Arch Gerontol Geriatr* 52(1): 71–4.
33. Gunter KB, White KN, Hayes WC, Snow CM. 2000. Functional mobility discriminates nonfallers from one-time and frequent fallers. *J Gerontol A Biol Sci Med Sci* 55(11): M672–6.
34. Greenspan SL, Myers ER, Kiel DP, Parker RA, Hayes WC, Resnick NM. 1998. Fall direction, bone mineral density, and function: risk factors for hip fracture in frail nursing home elderly. *Am J Med* 104(6): 539–45.
35. Fiatarone MA, Marks EC, Ryan ND, Meredith CN, Lipsitz LA, Evans WJ. 1990. High-intensity strength training in nonagenarians. Effects on skeletal muscle. *JAMA* 263: 3029–34.
36. Fiatarone MA, O'Neill EF, Ryan ND, Clements KM, Solares GR, Nelson ME, Roberts SB, Kehayias JJ, Lipsitz LA, Evans WJ. 1994. Exercise training and nutritional supplementation for physical frailty in very elderly people. *N Engl J Med* 330(25): 1769–75.
37. Orwoll E, Lambert LC, Marshall LM, Blank J, Barrett-Connor E, Cauley J, Ensrud K, Cummings SR. 2006. Endogenous testosterone levels, physical performance, and fall risk in older men. *Arch Intern Med* 166(19): 2124–31.
38. Nelson ME, Dilmanian FA, Dallal GE, Evans WJ. 1991. A one-year walking program and increased dietary calcium in postmenopausal women: Effects on bone. *Am J Clin Nutr* 53: 1304–11.
39. Coupland CA, Cliffe SJ, Bassey EJ, Grainge MJ, Hosking DJ, Chilvers CE. 1999. Habitual physical activity and bone mineral density in postmenopausal women in England. *Int J Epidemiol* 28(2): 241–6.
40. Sinaki M, Fitzpatrick LA, Ritchie CK, Montesano A, Wahner HW. 1998. Site-specificity of bone mineral density and muscle strength in women: Job-related physical activity. *Am J Phys Med Rehab* 77(6): 470–6.
41. Hughes VA, Frontera WR, Roubenoff R, Evans WJ, Singh MA. 2002. Longitudinal changes in body composition in older men and women: Role of body weight change and physical activity. *Am J Clin Nutr* 76(2): 473–81.
42. Nelson ME, Fiatarone MA, Morganti CM, Trice I, Greenberg RA, Evans WJ. 1994. Effects of high-intensity strength training on multiple risk factors for osteoporotic fractures. A randomized controlled trial. *JAMA* 272(24): 1909–14.
43. Walsh MC, Hunter GR, Livingstone MB. 2006. Sarcopenia in premenopausal and postmenopausal women with osteopenia, osteoporosis and normal bone mineral density. *Osteoporos Int* 17(1): 61–7.
44. Fiatarone MA, O'Neill EF, Ryan ND, Clements KM, Solares GR, Nelson ME, Roberts SB, Kehayias JJ, Lipsitz LA, Evans WJ. 1994. Exercise training and nutritional supplementation for physical frailty in very elderly people. *N Engl J Med* 330(25): 1769–75.
45. Goodpaster BH, Carlson CL, Visser M, Kelley DE, Scherzinger A, Harris TB, Stamm E, Newman AB. 2001. Attenuation of skeletal muscle and strength in the elderly: The Health ABC Study. *J Appl Physiol* 90(6): 2157–65.
46. Schafer AL, Vittinghoff E, Lang TF, Sellmeyer DE, Harris TB, Kanaya AM, Strotmeyer ES, Cawthon PM, Cummings SR, Tylavsky FA, Scherzinger AL, Schwartz AV. 2010. Fat infiltration of muscle, diabetes, and clinical fracture risk in older adults. *J Clin Endocrinol Metab* 95(11): E368–72.
47. Convinsky KE, Palmer RM, Fortinsky RH, Counsel SR, Stewart AL, Kresevic D, Burant CJ, Landefeld CS. 2003. Loss of independence in activities of daily living in older adults hospitalized with medicial illnesses: Increased vulnerability with age. *J Am Geriatr Soc* 51: 451–8.
48. Kortebein P, Ferrando A, Lombeida J, Wolfe R, Evans WJ. 2007. Effect of 10 days of bed rest on skeletal muscle in healthy older adults. *JAMA* 297(16): 1772–4.
49. Fleg JL, Lakatta EG. 1988. Role of muscle loss in the age-associated reduction in VO2 max. *J Appl Physiol* 65(3): 1147–51.
50. Fleg JL, Morrell CH, Bos AG, Brant LJ, Talbot LA, Wright JG, Lakatta EG. 2005. Accelerated longitudinal decline of aerobic capacity in healthy older adults. *Circulation* 112(5): 674–82.
51. Ferrando AA, Tipton KD, Bamman MM, Wolfe RR. 1997. Resistance exercise maintains skeletal muscle

protein synthesis during bed rest. *J Appl Physiol* 82(3): 807–10.
52. Rubin CT, Lanyon LE. 1985. Regulation of bone mass by mechanical strain magnitude. *Calcif Tissue Int* 37: 411–7.
53. Bassey EJ, Ramsdale SJ. 1994. Increase in femoral bone density in young women following high-impact exercise. *Osteoporos Int* 4(2): 72–5.
54. Bassey EJ, Ramsdale SJ. 1995. Weight-bearing exercise and ground reaction forces: A 12-month randomized controlled trial of effects on bone mineral density in healthy postmenopausal women. *Bone* 16(4): 469–76.
55. Bassey EJ, Rothwell MC, Littlewood JJ, Pye DW. 1998. Pre- and postmenopausal women have different bone mineral density responses to the same high-impact exercise. *J Bone Miner Res* 13(12): 1805–13.
56. Hamrick MW, McNeil PL, Patterson SL. 2010. Role of muscle-derived growth factors in bone formation. *J Musculoskelet Neuronal Interact* 10(1): 64–70.
57. Hamrick MW. 2011. A role for myokines in muscle-bone interactions. *Exerc Sport Sci Rev* 39(1): 43–7.
58. Evans WJ, Morley JE, Argiles J, Bales C, Baracos V, Guttridge D, Jatoi A, Kalantar-Zadeh K, Lochs H, Mantovani G, Marks D, Mitch WE, Muscaritoli M, Najand A, Ponikowski P, Rossi Fanelli F, Schambelan M, Schols A, Schuster M, Thomas D, Wolfe R, Anker SD. 2008. Cachexia: A new definition. *Clin Nutr* 27(6): 793–9.
59. Daniell HW, Dunn SR, Ferguson DW, Lomas G, Niazi Z, Stratte PT. 2000. Progressive osteoporosis during androgen deprivation therapy for prostate cancer. *J Urol* 163(1): 181–6.
60. Anker SD, Sharma R. 2002. The syndrome of cardiac cachexia. *Int J Cardiol* 85(1): 51–66.
61. Buehring B, Kirchner E, Sun Z, Calabrese L. 2012. The frequency of low muscle mass and its overlap with low bone mineral density and lipodystrophy in individuals with HIV–a pilot study using DXA total body composition analysis. *J Clin Densitom* 15(2): 224–32.
62. Bon JM, Zhang Y, Duncan SR, Pilewski JM, Zaldonis D, Zeevi A, McCurry KR, Greenspan SL, Sciurba FC. 2010. Plasma inflammatory mediators associated with bone metabolism in COPD. *COPD* 7(3): 186–91.
63. Bon J, Fuhrman CR, Weissfeld JL, Duncan SR, Branch RA, Chang CC, Zhang Y, Leader JK, Gur D, Greenspan SL, Sciurba FC. 2010. Radiographic emphysema predicts low bone mineral density in a tobacco-exposed cohort. *Am J Respir Crit Care Med* 183(7): 885–90.
64. Newman AB, Kupelian V, Visser M, Simonsick E, Goodpaster B, Nevitt M, Kritchevsky SB, Tylavsky FA, Rubin SM, Harris TB. 2003. Sarcopenia: Alternative definitions and associations with lower extremity function. *J Am Geriatr Soc* 51(11): 1602–9.
65. Schaap LA, Pluijm SM, Deeg DJ, Visser M. 2006. Inflammatory markers and loss of muscle mass (sarcopenia) and strength. *Am J Med* 119(6): 526 e9–17.

# 118

Glucose Control and Integration by the Skeleton

Patricia Ducy and Gerard Karsenty

Introduction 986
Osteocalcin-Mediated Bone Regulation of Glucose Metabolism 986
Bone as An Insulin Target Tissue 988
When Bone Remodeling Meets Glucose Homeostasis 989
Osteocalcin and Energy Metabolism in Humans 989
References 990

INTRODUCTION

The cross-talk between the regulation of bone mass accrual and energy metabolism has two major tenets. The first one is represented by the various influences exerted by multiple organs involved in the control of energy intake and metabolism on bone mass accrual. This regulation includes, but is not limited to, an influence of the gastrointestinal tract via the inhibitory effect of gut-derived serotonin on bone formation and the role of stomach acidity on bone extracellular matrix mineralization, and the opposite influences on bone formation and bone resorption of leptin, and through it of brain-derived serotonin and the sympathetic nervous system on bone mass accrual [1–7]. The second tenet arose from a question raised by the existence of this broad, complex, and multifaceted regulation of bone mass by energy metabolism: could bone, as an endocrine organ, exert a feedback regulation on some aspects of energy metabolism? It turned out that this is indeed the case, and this chapter will review the main aspects of this recently discovered regulation.

OSTEOCALCIN-MEDIATED BONE REGULATION OF GLUCOSE METABOLISM

As is not so infrequent in biological research, the stepping-stone of this novel endocrine function of bone came out of an unrelated study that appeared to be a failure. *Osteocalcin* is an osteoblast-specific gene that encodes a secreted protein present in the bone extracellular matrix (ECM) and the general circulation [8, 9]. Upon synthesis by osteoblasts, osteocalcin is carboxylated on three glutamic residues, which are then called GLA residues (hence the other name given to osteocalcin: bone GLA protein). This post translational modification confers to proteins high affinity to mineral ions and thereby allows gamma-carboxylated osteocalcin (GLA-Ocn) to bind to the hydroxyapatite mineral present in mineralized bone. Not surprisingly, this mineral-binding property, along with the great abundance of osteocalcin in the bone ECM, led to the very logical assumption that osteocalcin may be involved in controlling the mineralization of the bone ECM. The reality, however, turned out to be different. Indeed, neither loss- nor gain-of-function mutations of *Osteocalcin* nor its ectopic expression in other ECMs had any consequences on ECM mineralization in bone or in other tissues [10, 11]. This failure contrasted strikingly with the fact that a related GLA protein present in the cartilaginous ECM, matrix gla protein, is indeed a powerful inhibitor of ECM mineralization [12].

This negative result left unanswered the question that had motivated the original study: What is/are the functions of osteocalcin and how does it achieve them? A peculiarity of the *Osteocalcin*-deficient mice became informative in that regard; although this did not affect their overall body weight, these mice were abnormally fatty. This phenotype, along with the fact that osteocalcin is present in the general circulation, raised the

Primer on the Metabolic Bone Diseases and Disorders of Mineral Metabolism, Eighth Edition. Edited by Clifford J. Rosen.
© 2013 American Society for Bone and Mineral Research. Published 2013 by John Wiley & Sons, Inc.

hypothesis that it may act as a hormone regulating some aspects of energy metabolism. That it was the case was first established through co-culture assays. Indeed, co-culture of wild-type osteoblasts with pancreatic islets or adipocytes increased the expression of *insulin* in these islets and of *Adiponectin* in adipocytes [13]. In contrast, *Osteocalcin*-null osteoblasts failed to increase gene expression in either case. These *in vitro* data found support in the analysis of the *Osteocalcin*-deficient mice [13]. Both insulin expression and secretion were decreased in *Osteocalcin*–/– mice, as was the proliferation of the insulin-producing β cells of the pancreas. Likewise, adiponectin expression and secretion were also decreased in absence of *Osteocalcin*, although expression of other markers of the adipocyte function, such as leptin and resistin, were not changed. There were two other phenotypes of interest in the *osteocalcin*–/– mice. The first one was a decrease in insulin sensitivity in muscle, liver, and white adipose tissue (WAT). This phenotype was established molecularly by studying gene expression in liver, muscle, and WAT, and physiologically by performing insulin tolerance tests and hyperinsulinemic euglycemic clamps [13]. The second additional phenotype observed in the *Osteocalcin*–/– mice was a decrease in energy expenditure [13]. As a result of all these abnormalities, *Osteocalcin*–/– mice were glucose intolerant, abnormally fat, and insulin resistant, all features of metabolic syndrome. These observations indicate that osteocalcin is a hormone that, in the sphere of energy metabolism, has four main functions: it favors β-cell proliferation, insulin expression and secretion, insulin sensitivity, and energy expenditure (Fig. 118.1). In contrast, osteocalcin does not affect appetite, a trait consistent with the normal level of leptin observed in the *Osteocalcin*-deficient mice [13].

The next question was to determine which form of osteocalcin is active: the carboxylated one, the uncarboxylated one, or both? This question was first addressed in cell culture systems. *In vitro* assays provided a quite clear answer by showing that it is the uncarboxylated form of osteocalcin that could enhance *Insulin* and *Adiponectin* expression in isolated islets or cultured β cells and in adipocytes, respectively, while the carboxylated form could not [13]. Likewise, recombinant uncarboxylated osteocalcin could induce the secretion of insulin in islets perifusion assays [14]. Using a transgenic approach to confirm this finding *in vivo* was not a valid option since the gamma-carboxylase, the enzyme responsible for osteocalcin carboxylation, is ubiquitously expressed and therefore overexpression or ectopic expression of an *Osteocalcin* transgene would mostly lead to increased levels of carboxylated, i.e., inactive, osteocalcin. Pharmacologic strategies, though, could demonstrate that uncarboxylated osteocalcin is metabolically active. Indeed, both long-term subcutaneous infusion and once-daily injections of recombinant uncarboxylated osteocalcin were shown to increase glucose tolerance and insulin sensitivity in wild-type mice [15, 16]. Furthermore, these treatments could partially restore insulin sensitivity and glucose tolerance in mice fed a high-fat diet as well as decrease weight gain by stimulating energy expenditure. Altogether, these studies established that uncarboxylated osteocalcin is a metabolically active form or this hormone.

That osteoblasts regulate glucose metabolism is surprising, but what is even more surprising is that there is a second gene expressed in osteoblasts that exerts an influence exactly opposite to the one of osteocalcin on this physiological process. This gene, *Esp*, encodes a little-known protein tyrosine phosphatase (ESP, also termed "OST-PTP") that is present in only three cell types: the embryonic stem cell, the Sertoli cell of the testes, and the osteoblast [17, 18]. This restricted pattern of expression justified the study, in the context of the entire animal, of the function of ESP; this was done in two different laboratories and in two different ways. One group generated mice lacking *Esp* in all cells while a second group generated mice lacking it only in osteoblasts [13, 17]. Remarkably, both animal models develop the same phenotype indicating that ESP's main, if not only, function occurs in osteoblasts [13].

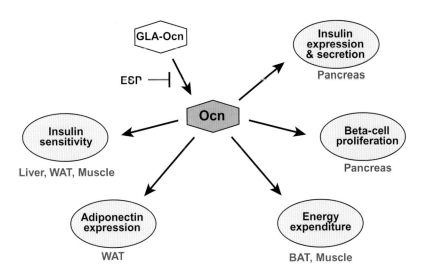

Fig. 118.1. Target tissues of the metabolic function of osteocalcin. Osteocalcin regulates glucose and energy homeostasis by acting on beta cell proliferation, insulin production, insulin sensitivity, and energy expenditure. To achieve these functions, it acts on multiple organs such as the pancreas, liver, white adipose tissue (WAT), brown adipose tissue (BAT), and muscle. Reducing the level of ESP, a tyrosine phosphatase expressed in osteoblasts, negatively regulates these effects.

The phenotype of the *Esp*-deficient mice was in all points the mirror image of the one seen in *Osteocalcin*-deficient mice. Indeed, *Esp*–/– mice are hypoglycemic because they are hyperinsulinemic; they are lean, hypersensitive to insulin, and display an increase in energy expenditure [13]. Consequently, when put on a high-fat diet, *Esp*–/– mice do not develop type 2 diabetes and become less obese than wild-type mice. That *Osteocalcin*–/– and *Esp*–/– mice have opposite phenotypes suggested that they act in the same genetic, if not biochemical, pathway and that, in fact, the *Esp*–/– mice are a model of a gain-of-function of osteocalcin. This assumption was verified genetically through the generation of *Esp*–/– mice lacking one allele of *Osteocalcin*. These latter mutant mice had normal insulin secretion, insulin sensitivity, and normal energy expenditure, thus establishing that *Esp* is indeed genetically upstream of *Osteocalcin* [13] (Fig. 118.1). Biochemically, however, the link was less obvious as ESP is a tyrosine phosphatase and osteocalcin is not a phosphorylated molecule, nor is gamma-carboxylase, the enzyme responsible for its carboxylation [19]. The solution of this puzzle came by approaching it through another angle.

BONE AS AN INSULIN TARGET TISSUE

The insulin receptor is a receptor tyrosine kinase; it is, therefore, conceivable that it could serve as a substrate for the phosphatase moiety of ESP and, through such functional interaction, would impact osteocalcin biology. That *Insulin* expression is regulated by osteocalcin is also a reason to study whether, in a feedback or feedforward loop, insulin was affecting *Osteocalcin* expression, secretion, or activity. A third reason to study this question stems from an important aspect of insulin biology. Traditionally, liver, muscle, and WAT are seen as the main target tissues of insulin signaling, yet inactivation of the insulin receptor in myoblasts or in white adipocytes does not result in glucose intolerance [20, 21]. This observation implies that insulin should signal in additional cell types and tissues to achieve glucose homeostasis. The regulation of its secretion by osteocalcin made it conceivable that this tissue was bone. This hypothesis was tested by two laboratories that essentially reached the same conclusions.

A first line of experiments demonstrated through a classical substrate trapping assay that the insulin receptor is indeed a substrate for the tyrosine phosphatase moiety of ESP in osteoblasts [19]. This observation provided a biochemical basis to the notion that insulin signaling in osteoblasts could be a regulator of osteocalcin biology. As mentioned above, the specific deletion of the insulin receptor in osteoblasts in the mouse was engineered by two laboratories [19, 22], each using a different mouse Cre driver active in osteoblasts only [23, 24]. Both mouse models, thereafter named $InsR_{osb}$–/– mice, were hyperglycemic, and glucose tolerance tests (GTTs) showed that they were glucose intolerant. Glucose-stimulated insulin secretion tests showed that insulin secretion is also hampered in the absence of insulin signaling in osteoblasts. Lastly, insulin sensitivity in peripheral tissues was decreased in either $InsR_{osb}$–/– model. Hence, the ablation of insulin signaling in osteoblasts could do what its ablation in myoblasts could not: It created a glucose intolerance phenotype resembling type 2 diabetes [19, 22]. Two aspects of these experiments further highlight their biological relevance. First, it is important to note that those results were obtained in mice fed a normal diet, i.e., challenged only by the genetic manipulation they harbor. This indicates that the role of insulin signaling in osteoblasts is important in normal physiology. Second, since those two $InsR_{osb}$–/– mouse strains were cell-specific gene inactivation models in which deletion of the gene encoding the insulin receptor in osteoblasts was, in both cases, only partial, they are only hypomorphic models of its absence. That decreasing only its potency could lead to a significant metabolic phenotype highlights how important this regulatory pathway is for whole body glucose homeostasis.

The phenotype of the $InsR_{osb}$–/– mice is the mirror image of the one displayed by *Esp*–/– mice, an expected result given that the insulin receptor is a substrate of ESP and that its dephosphorylation hampers its activity [25]. A more remarkable finding was that there was more carboxylated osteocalcin and less active, undercarboxylated, osteocalcin in the serum of the $InsR_{osb}$–/– mice than in wild-type mice as quantified by a sensitive enzyme-linked immunosorbent assay (ELISA) specifically developed to detect these different forms of osteocalcin in the mouse [19, 26]. Although these findings provided an explanation for the fact that the phenotype of the $InsR_{osb}$–/– mice was a phenocopy of the one of *Osteocalcin*–/– mice, they raised another question: How could insulin signaling in osteoblasts be regulating osteocalcin levels and carboxylation, i.e., bioactivity?

Analysis of the bone phenotype of the $InsR_{osb}$–/– mice provided two answers to this question. First, gene expression studies revealed that expression the *Twist 1* and *Twist 2* genes, which encode nuclear factors inhibiting the transcriptional activity the osteoblast-specific factor Runx2 [27, 28], is enhanced in these mutant mice [22]. As a result, expression of the *Osteocalcin* genes is hampered, and less osteocalcin is produced (Fig. 118.2). In agreement with these data, infusing recombinant uncarboxylated osteocalcin in the $InsR_{osb}$–/– can rescue their metabolic phenotype [22]. Second, there is an increase in bone resorption parameters in the $InsR_{osb}$–/– mice, which is due to a decrease in expression of *Osteoprotegerin* (*Opg*) in osteoblasts [19]. One regulator of *Opg* expression is FoxO1, and insulin signaling in osteoblasts inactivates FoxO1 through phosphorylation just as it does in other cell types (Fig. 118.2) [19, 29]. This observation is consistent with the fact that inactivation of FoxO1 specifically in osteoblasts improves glucose tolerance and insulin sensitivity, at least in part, by promoting osteocalcin decarboxylation [30, 31]. A further analysis showed that two key genes are overexpressed in the osteoclasts of $InsR_{osb}$–/– mice: *Tcirg1*, a component of the proton pump

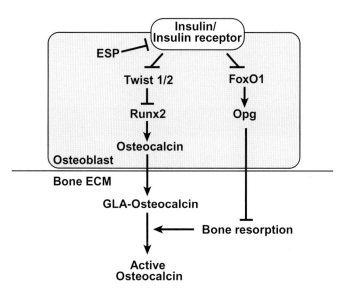

Fig. 118.2. Downstream effects of insulin/insulin receptor signaling in osteoblasts. On the one hand, insulin receptor signaling inhibits the expression of the Twist genes, releasing their negative effect on the transcriptional activity of Runx2, thereby promoting osteocalcin production. On the other hand, insulin receptor signaling inhibits the FoxO1-mediated induction of Opg expression, leading to an increase in bone resorption. This in turn promotes the release from the bone matrix and activation (via Glu13 decarboxylation) of osteocalcin. Through its ability to dephosphorylate, the insulin receptor ESP inhibits this process.

that is necessary for acidification of the resorption lacuna, and *CathepsinK*, which encodes a protein found in lysosomes that has an optimal enzymatic activity in acidic conditions [19, 32, 33]. In other words, through its negative regulation of *Opg* expression, insulin signaling in osteoblasts indirectly favors acidification within the resorption lacuna, a process that is mandatory for bone resorption to occur [32]. This aspect of insulin biology, which apparently has no relation to its regulation of glucose homeostasis, is, in fact, intimately intertwined with it.

WHEN BONE REMODELING MEETS GLUCOSE HOMEOSTASIS

It has been known for more than 20 years that the only means to decarboxylate a protein outside the cell is through an acid pH [34]. Accordingly, fully carboxylated osteocalcin incubated at pH 4.5 for a week becomes undercarboxylated on residue Glu 13 and is then able to enhance insulin secretion from isolated β cells. In contrast, when it is incubated at pH 7.4 for the same amount of time, carboxylated osteocalcin remains fully carboxylated and is unable to stimulate insulin secretion by β cells [19]. This set of observations implies that bone resorption, through the acid pH it requires, could enhance osteocalcin undercarboxylation, i.e., activation. The demonstration that *oc/oc* mice, which harbor a loss-of-function mutation in *Tcirg1* and therefore cannot properly generate the acidic milieu necessary for bone resorption, have less active osteocalcin and display glucose intolerance, insulin resistance, and a decrease in insulin secretion is in full agreement with this contention [19].

That insulin signaling in osteoblasts favors bone resorption and more specifically acidification of the bone ECM suggests that insulin uses the interplay between osteoblasts and osteoclasts to favor its own secretion via activation of osteocalcin. This hypothesis was demonstrated using several mouse models. Analyses of wild-type mice fed a high-fat diet and treated with receptor activator of nuclear factor-κB ligand (RANKL), of *Esp*−/− mice (a gain-of-function model of insulin signaling in osteoblasts) treated with alendronate, and of heterozygous compound $oc/+; InsR_{osb}+/-$ mice all concur to demonstrate that insulin signaling in osteoblasts activates osteocalcin through its positive regulation of bone resorption and thereby, in a feed-forward manner, favors insulin secretion by pancreatic β cells [19].

OSTEOCALCIN AND ENERGY METABOLISM IN HUMANS

There is no example yet of a molecule identified as a hormone in mice that has abruptly lost this characteristic in humans. Nevertheless, as with all findings primarily made using animal models that investigate whether osteocalcin plays the same role in humans and mice, this has been an important question.

Following the original paper describing osteocalcin's endocrine function, a continuous flow of clinical studies have shown that the blood level of this new hormone is a reliable marker of glucose intolerance and more generally of a metabolic dysregulation in humans. For instance, levels of circulating osteocalcin have been inversely correlated with body mass index (BMI), fat mass, plasma glucose, fasting insulin, and insulin resistance in adult men and women, diabetic or not [35–44]. More generally, lower osteocalcin levels were observed in obese or type 2 diabetic subjects compared to nonobese or nondiabetic control individuals [35, 36, 45]. Likewise, decreased levels of osteocalcin have been associated with premature myocardial infarction in young patients and coronary heart disease in older individuals, in which it could be associated with the progression of atherosclerosis [43, 46, 47]. On the other hand, a lower odds ratio of developing metabolic syndrome was associated with higher levels of osteocalcin in black as well as white non-Hispanic individuals [38, 48]. Lastly, two recent interventional studies have correlated osteocalcin levels with glucose homeostasis even more directly. In the first one, weight loss and a decrease in homeostatic model assessment (HOMA) was associated with an increase in osteocalcin levels in obese children [45]. In the second one, an increase in blood glucose levels was observed in patients upon resection of an osteoid osteoma, a begnin osteoblastic tumor that produces high levels of osteocalcin [49]. Altogether, this now quite large body of clinical studies has established that osteocalcin, in humans as in mice, is associated with the regulation of glucose homeostasis and energy metabolism.

A second point that needed to be addressed is whether osteocalcin's osteoclast-dependent mode of activation also occurs in humans. This question was more so important because ESP is a pseudogene in humans [50]. A combination of molecular biology and human genetic efforts showed that PTP-1B, a well-characterized protein tyrosine phosphatase that has been previously shown to dephosphorylate the insulin receptor in other cell types [51, 52], could play the same role instead of ESP in regulating osteocalcin activation in humans. Indeed, PTP-1B is highly expressed in human osteoblasts, can interact with the insulin receptor in these cells, and can regulate expression of *OPG* [19]. At the genetic levels it was shown that patients with osteopetrosis due to a defect in ECM acidification have abnormally high circulating levels of the fully carboxylated, i.e., inactive, form of osteocalcin and are hypoinsulinemic [19]. These results provide strong evidence that, as one would expect, osteocalcin is activated through the same mechanisms in mice and in humans.

Altogether, these clinical findings establish the importance of proper levels of osteocalcin for the good of energy homeostasis. They also identify a need for targeted investigations that will specifically evaluate the impact of the bone–osteocalcin–metabolism connection in the pathogenesis of human diseases and their treatments. Among such studies, one question to address is testing whether antiresorptive treatments, which are the main therapeutic approach to osteoporosis at the present time, might, in some rare circumstances, have a worsening effect on an already compromised glucose tolerance or insulin sensitivity. We do not believe, however, that this will be a major concern for the general population. Likewise, anticoagulating agents that affect the function of vitamin K, and therefore enhance osteocalcin gamma-carboxylation, could potentially have a mild and so far unrecognized effect on metabolism. It appears less likely, however, that increasing vitamin K levels would have such an effect since osteocalcin is already almost fully carboxylated in normal conditions. On the other hand, conditions and drugs usually associated with the control of glucose metabolism should also be reassessed for a potential effect on bone remodeling. Moreover, it could be possible that some glucose-lowering agents act, at least partly, by modifying the production or activation of osteocalcin.

In our view, a most important point is that the role of osteocalcin as a sole marker of bone formation needs to be reevaluated. Indeed, variations in its serum levels could be caused by variations in insulin levels or by a need for a transient metabolic adaptation rather than by changes in bone turnover. For instance, a negative association could be found between osteocalcin levels and prediabetes in overweight children but not with bone mineral content [53]. This observation is consistent with the growing number of clinical studies suggesting that osteocalcin, in addition of being a marker of bone formation, is a marker of insulin sensitivity.

REFERENCES

1. Schinke T, Schilling AF, Baranowsky A, Seitz S, Marshall RP, Linn T, Blaeker M, Huebner AK, Schulz A, Simon R, Gebauer M, Priemel M, Kornak U, Perkovic S, Barvencik F, Beil FT, Del Fattore A, Frattini A, Streichert T, Pueschel K, Villa A, Debatin KM, Rueger JM, Teti A, Zustin J, Sauter G, Amling M. 2009. Impaired gastric acidification negatively affects calcium homeostasis and bone mass. *Nat Med* 15: 674–681.
2. Yadav VK, Oury F, Suda N, Liu ZW, Gao XB, Confavreux C, Klemenhagen KC, Tanaka KF, Gingrich JA, Guo XE, Tecott LH, Mann JJ, Hen R, Horvath TL, Karsenty G. 2009. A serotonin-dependent mechanism explains the leptin regulation of bone mass, appetite, and energy expenditure. *Cell* 138: 976–989.
3. Yadav VK, Balaji S, Suresh PS, Liu XS, Lu X, Li Z, Guo XE, Mann JJ, Balapure AK, Gershon MD, Medhamurthy R, Vidal M, Karsenty G, Ducy P. 2010. Pharmacological inhibition of gut-derived serotonin synthesis is a potential bone anabolic treatment for osteoporosis. *Nat Med* 16: 308–312.
4. Yadav VK, Ryu JH, Suda N, Tanaka KF, Gingrich JA, Schutz G, Glorieux FH, Chiang CY, Zajac JD, Insogna KL, Mann JJ, Hen R, Ducy P, Karsenty G. 2008. Lrp5 controls bone formation by inhibiting serotonin synthesis in the duodenum. *Cell* 135: 825–837.
5. Ducy P, Amling M, Takeda S, Priemel M, Schilling AF, Beil FT, Shen J, Vinson C, Rueger JM, Karsenty G. 2000. Leptin inhibits bone formation through a hypothalamic relay: A central control of bone mass. *Cell* 100: 197–207.
6. Takeda S, Elefteriou F, Levasseur R, Liu X, Zhao L, Parker KL, Armstrong D, Ducy P, Karsenty G. 2002. Leptin regulates bone formation via the sympathetic nervous system. *Cell* 111: 305–317.
7. Elefteriou F, Ahn JD, Takeda S, Starbuck M, Yang X, Liu X, Kondo H, Richards WG, Bannon TW, Noda M, Clement K, Vaisse C, Karsenty G. 2005. Leptin regulation of bone resorption by the sympathetic nervous system and CART. *Nature* 434: 514–520.
8. Hauschka PV, Lian JB, Cole DE, Gundberg CM. 1989. Osteocalcin and matrix Gla protein: Vitamin K-dependent proteins in bone. *Physiol Rev* 69: 990–1047.
9. Price PA. 1989. Gla-containing proteins of bone. *Connect Tissue Res* 21: 51–57; discussion 57–60.
10. Ducy P, Desbois C, Boyce B, Pinero G, Story B, Dunstan C, Smith E, Bonadio J, Goldstein S, Gundberg C, Bradley A, Karsenty G. 1996. Increased bone formation in osteocalcin-deficient mice. *Nature* 382: 448–452.
11. Murshed M, Schinke T, McKee MD, Karsenty G. 2004. Extracellular matrix mineralization is regulated locally;

different roles of two gla-containing proteins. *J Cell Biol* 165: 625–630.
12. Luo G, Ducy P, McKee MD, Pinero GJ, Loyer E, Behringer RR, Karsenty G. 1997. Spontaneous calcification of arteries and cartilage in mice lacking matrix GLA protein. *Nature* 386: 78–81.
13. Lee NK, Sowa H, Hinoi E, Ferron M, Ahn JD, Confavreux C, Dacquin R, Mee PJ, McKee MD, Jung DY, Zhang Z, Kim JK, Mauvais-Jarvis F, Ducy P, Karsenty G. 2007. Endocrine regulation of energy metabolism by the skeleton. *Cell* 130: 456–469.
14. Hinoi E, Gao N, Jung DY, Yadav V, Yoshizawa T, Myers MG Jr, Chua SC Jr, Kim JK, Kaestner KH, Karsenty G. 2008. The sympathetic tone mediates leptin's inhibition of insulin secretion by modulating osteocalcin bioactivity. *J Cell Biol* 183: 1235–1242.
15. Ferron M, Hinoi E, Karsenty G, Ducy P. 2008. Osteocalcin differentially regulates beta cell and adipocyte gene expression and affects the development of metabolic diseases in wild-type mice. *Proc Natl Acad Sci U S A* 105: 5266–5270.
16. Ferron M, McKee MD, Levine RL, Ducy P, Karsenty G. 2012. Intermittent injections of osteocalcin improve glucose metabolism and prevent type 2 diabetes in mice. *Bone* 50(2): 568–575.
17. Dacquin R, Mee PJ, Kawaguchi J, Olmsted-Davis EA, Gallagher JA, Nichols J, Lee K, Karsenty G, Smith A. 2004. Knock-in of nuclear localised beta-galactosidase reveals that the tyrosine phosphatase Ptprv is specifically expressed in cells of the bone collar. *Dev Dyn* 229: 826–834.
18. Mauro LJ, Olmsted EA, Skrobacz BM, Mourey RJ, Davis AR, Dixon JE. 1994. Identification of a hormonally regulated protein tyrosine phosphatase associated with bone and testicular differentiation. *J Biol Chem* 269: 30659–30667.
19. Ferron M, Wei J, Yoshizawa T, Del Fattore A, DePinho RA, Teti A, Ducy P, Karsenty G. 2010. Insulin signaling in osteoblasts integrates bone remodeling and energy metabolism. *Cell* 142: 296–308.
20. Bluher M, Michael MD, Peroni OD, Ueki K, Carter N, Kahn BB, Kahn CR. 2002. Adipose tissue selective insulin receptor knockout protects against obesity and obesity-related glucose intolerance. *Dev Cell* 3: 25–38.
21. Bruning JC, Michael MD, Winnay JN, Hayashi T, Horsch D, Accili D, Goodyear LJ, Kahn CR. 1998. A muscle-specific insulin receptor knockout exhibits features of the metabolic syndrome of NIDDM without altering glucose tolerance. *Mol Cell* 2: 559–569.
22. Fulzele K, Riddle RC, DiGirolamo DJ, Cao X, Wan C, Chen D, Faugere MC, Aja S, Hussain MA, Bruning JC, Clemens TL. 2010. Insulin receptor signaling in osteoblasts regulates postnatal bone acquisition and body composition. *Cell* 142: 309–319.
23. Dacquin R, Starbuck M, Schinke T, Karsenty G. 2002. Mouse alpha1(I)-collagen promoter is the best known promoter to drive efficient Cre recombinase expression in osteoblast. *Dev Dyn* 224: 245–251.
24. Zhang M, Xuan S, Bouxsein ML, von Stechow D, Akeno N, Faugere MC, Malluche H, Zhao G, Rosen CJ, Efstratiadis A, Clemens TL. 2002. Osteoblast-specific knockout of the insulin-like growth factor (IGF) receptor gene reveals an essential role of IGF signaling in bone matrix mineralization. *J Biol Chem* 277: 44005–44012.
25. Schlessinger. J. 2000. Cell signaling by receptor tyrosine kinases. *Cell* 103: 211–225.
26. Ferron M, Wei J, Yoshizawa T, Ducy P, Karsenty G. 2010. An ELISA-based method to quantify osteocalcin carboxylation in mice. *Biochem Biophys Res Commun* 397: 691–696.
27. Bialek P, Kern B, Yang X, Schrock M, Sosic D, Hong N, Wu H, Yu K, Ornitz DM, Olson EN, Justice MJ, Karsenty G. 2004. A twist code determines the onset of osteoblast differentiation. *Dev Cell* 6: 423–435.
28. Ducy P, Zhang R, Geoffroy V, Ridall AL, Karsenty G 1997 Osf2/Cbfa1: A transcriptional activator of osteoblast differentiation. *Cell* 89: 747–754.
29. Puigserver P, Rhee J, Donovan J, Walkey CJ, Yoon JC, Oriente F, Kitamura Y, Altomonte J, Dong H, Accili D, Spiegelman BM. 2003. Insulin-regulated hepatic gluconeogenesis through FOXO1-PGC-1alpha interaction. *Nature* 423: 550–555.
30. Rached MT, Kode A, Silva BC, Jung DY, Gray S, Ong H, Paik JH, DePinho RA, Kim JK, Karsenty G, Kousteni S. 2010. FoxO1 expression in osteoblasts regulates glucose homeostasis through regulation of osteocalcin in mice. *J Clin Invest.* 120(1): 357–368.
31. Kode A, Mosialou I, Silva BC, Joshi S, Ferron M, Rached MT, Kousteni S. 2012. FoxO1 protein cooperates with ATF4 protein in osteoblasts to control glucose homeostasis. *J Biol Chem.* 287(12): 8757–8768.
32. Teitelbaum SL. 2000. Bone resorption by osteoclasts. *Science* 289: 1504–1508.
33. Teitelbaum SL, Ross FP. 2003. Genetic regulation of osteoclast development and function. *Nat Rev Genet* 4: 638–649.
34. Engelke JA, Hale JE, Suttie JW, Price PA. 1991. Vitamin K-dependent carboxylase: Utilization of decarboxylated bone Gla protein and matrix Gla protein as substrates. *Biochim Biophys Acta* 1078: 31–34.
35. Kindblom JM, Ohlsson C, Ljunggren O, Karlsson MK, Tivesten A, Smith U, Mellstrom D. 2009. Plasma osteocalcin is inversely related to fat mass and plasma glucose in elderly Swedish men. *J Bone Miner Res* 24: 785–791.
36. Lee YJ, Lee H, Jee SH, Lee SS, Kim SR, Kim SM, Lee MW, Lee CB, Oh S. 2010. Serum osteocalcin is inversely associated with adipocyte-specific fatty acid-binding protein in the Korean metabolic syndrome research initiatives. *Diabetes Care* 33: e90.
37. Pittas AG, Harris SS, Eliades M, Stark P, Dawson-Hughes B. 2009. Association between serum osteocalcin and markers of metabolic phenotype. *J Clin Endocrinol Metab* 94: 827–832.
38. Saleem U, Mosley TH Jr, Kullo IJ. 2010. Serum osteocalcin is associated with measures of insulin resistance,

adipokine levels, and the presence of metabolic syndrome. *Arterioscler Thromb Vasc Biol* 30: 1474–1478.
39. Hwang YC, Jeong IK, Ahn KJ, Chung HY. 2009. The uncarboxylated form of osteocalcin is associated with improved glucose tolerance and enhanced beta-cell function in middle-aged male subjects. *Diabetes Metab Res Rev* 25: 768–772.
40. Kanazawa I, Yamaguchi T, Yamamoto M, Yamauchi M, Kurioka S, Yano S, Sugimoto T. 2009. Serum osteocalcin level is associated with glucose metabolism and atherosclerosis parameters in type 2 diabetes mellitus. *J Clin Endocrinol Metab* 94: 45–49.
41. Im JA, Yu BP, Jeon JY, Kim SH. 2008. Relationship between osteocalcin and glucose metabolism in postmenopausal women. *Clin Chim Acta* 396: 66–69.
42. Bao YQ, Zhou M, Zhou J, Lu W, Gao YC, Pan XP, Tang JL, Lu HJ, Jia WP. 2011. Relationship between serum osteocalcin and glycemic variability in type 2 diabetes. *Clin Exp Pharmacol Physiol* 38(1): 50–54.
43. Zhou M, Ma X, Li H, Pan X, Tang J, Gao YC, Hou X, Lu H, Bao Y, Jia W. 2009. Serum osteocalcin concentrations in relation to glucose and lipid metabolism in Chinese individuals. *Eur J Endocrinol* 161(5): 723–729.
44. Kanazawa I, Yamaguchi T, Yamauchi M, Yamamoto M, Kurioka S, Yano S, Sugimoto T. 2011. Serum undercarboxylated osteocalcin was inversely associated with plasma glucose level and fat mass in type 2 diabetes mellitus. *Osteoporos Int* 22: 187–194.
45. Reinehr T, Roth CL. 2010. A new link between skeleton, obesity and insulin resistance: Relationships between osteocalcin, leptin and insulin resistance in obese children before and after weight loss. *Int J Obes (Lond)* 34: 852–858.
46. Kanazawa I, Yamaguchi T, Sugimoto T. 2011. Relationship between bone biochemical markers versus glucose/lipid metabolism and atherosclerosis; a longitudinal study in type 2 diabetes mellitus. *Diabetes Res Clin Pract* 92: 393–399.
47. Goliasch G, Blessberger H, Azar D, Heinze G, Wojta J, Bieglmayer C, Wagner O, Schillinger M, Huber K, Maurer G, Haas M, Wiesbauer F. 2011. Markers of bone metabolism in premature myocardial infarction (</=40years of age). *Bone* 48(3): 622–626.
48. Yeap BB, Chubb SA, Flicker L, McCaul KA, Ebeling PR, Beilby JP, Norman PE. 2010. Reduced serum total osteocalcin is associated with metabolic syndrome in older men via waist circumference, hyperglycemia, and triglyceride levels. *Eur J Endocrinol* 163: 265–272.
49. Confavreux CB, Borel O, Lee F, Vaz G, Guyard M, Fadat C, Carlier MC, Chapurlat R, Karsenty G. 2012. Osteoid osteoma is an osteocalcinoma affecting glucose metabolism. *Osteoporos Int* 23(5): 1645–1650.
50. Cousin W, Courseaux A, Ladoux A, Dani C, Peraldi P. 2004. Cloning of hOST-PTP: The only example of a protein-tyrosine-phosphatase the function of which has been lost between rodent and human. *Biochem Biophys Res Commun* 321: 259–265.
51. Delibegovic M, Bence KK, Mody N, Hong EG, Ko HJ, Kim JK, Kahn BB, Neel BG. 2007. Improved glucose homeostasis in mice with muscle-specific deletion of protein-tyrosine phosphatase 1B. *Mol Cell Biol* 27: 7727–7734.
52. Delibegovic M, Zimmer D, Kauffman C, Rak K, Hong EG, Cho YR, Kim JK, Kahn BB, Neel BG, Bence KK. 2009. Liver-specific deletion of protein-tyrosine phosphatase 1B (PTP1B) improves metabolic syndrome and attenuates diet-induced endoplasmic reticulum stress. *Diabetes* 58: 590–599.
53. Pollock NK, Bernard PJ, Wenger K, Misra S, Gower BA, Allison JD, Zhu H, Davis CL. 2010. Lower bone mass in prepubertal overweight children with prediabetes. *J Bone Miner Res* 25: 2484–2493.

119
Obesity and Skeletal Mass

Sue Shapses and Deeptha Sukumar

Introduction 993
Stromal Cell Adipocyte/Osteoblast 993
Obesity and Bone Marrow Fat 994
Bone Mineral Density in Obesity 994
Fracture Risk 995
Bone and Body Composition 996

Adipokines and Endocrine Profile in Obesity:
Actions on Bone 996
Weight Reduction and Bone 997
Summary 997
References 997

INTRODUCTION

The number of persons who are overweight or obese in the United States has reached epidemic proportions at 69% [1]. Obesity is a national health problem and its prevalence is increasing in developing countries. Although obesity is associated with higher bone mineral density (BMD) and bone mineral content (BMC), it has been shown that bone quality may be compromised in adults [2] and in children [3–5] and that fracture risk is higher at certain bone sites. Also, for a given amount of bone in obesity, fracture risk is higher [6]. The etiology for a relationship between bone and adiposity is complex due to the multiple factors that are involved. Both osteoblasts and adipocytes are derived from a common mesenchymal stem cell (MSC), and factors that stimulate adipogenesis typically inhibit osteoblast differentiation [7, 8]. In addition, the hormonal profile and adipokines in obesity may contribute to bone quality. Furthermore, mechanical forces in obesity and during a fall may have specific effects on bone quality and fracture risk at certain bone sites compared to individuals with a lower body weight. Factors related to lifestyle in obesity such as poor dietary habits, sedentary activity, or repeated attempts to lose weight would also potentially influence fracture risk.

STROMAL CELL ADIPOCYTE/OSTEOBLAST

The pluripotent stromal cell is influenced by different factors that lead to the differentiation into their mature cell types that eventually constitute the balance between bone and adipose tissue, as well as cartilage and muscle mass. It was classically assumed that reduced bone mass and osteoblastogenesis that occurs with aging are associated with an increase in marrow adipogenesis [9]. However, the relationship between bone and fat formation within the bone marrow microenvironment is complex. There are a number of transgenic or gene-deficient (knockout) mice with phenotypes reflecting altered bone marrow adipogenesis and/or osteogenesis [7]. While Runt-related transcription factor 2 (RUNX2), which is also known as core-binding factor subunit alpha-1 (CBF-alpha-1) (Runx2/Cbfa1), and osterix are bone-specific transcription factors required for osteoblast differentiation [8], the peroxisome proliferator activated receptor δ (PPARδ) plays a central role in initiating adipogenesis and inhibiting osteoblastogenesis. Ligands for PPARδ include the thiazolidinedione class of antidiabetic drugs (i.e., rosiglitazone) that increase insulin sensitivity [10], has also been shown to increase adiposity, reduce bone mass, and increase fracture risk [11]. This

---

Primer on the Metabolic Bone Diseases and Disorders of Mineral Metabolism, Eighth Edition. Edited by Clifford J. Rosen.
© 2013 American Society for Bone and Mineral Research. Published 2013 by John Wiley & Sons, Inc.

may occur due to regulation through increased PPARγ and fibroblast growth factor 21 (FGF21). For example, pharmacological doses of FGF21 induce severe bone loss by favoring the differentiation of bone marrow mesenchymal cells into adipocytes instead of osteoblasts in mice [12]. Glucocorticoids are another example of drugs that increase adiposity, especially visceral fat, at the expense of osteoblast differentiation. Both osteoporosis and aging-related bone loss increase marrow adiposity and suggest a switch in differentiation of stromal cells from the osteoblastic to the adipocytic lineage. It is possible that low-density lipoprotein oxidation products promote osteoporotic bone loss by directing progenitor marrow stromal cells to undergo adipogenic instead of osteogenic differentiation [13]. Also, skeletal unloading increases adipocyte differentiation and inhibits osteoblast differentiation, and both processes can be reversed by transforming growth factor–β or β2 [14]. Overall, the plasticity between adipose and osteoblast cells can help explain the relationship between different components of body composition. Continued study in this area could potentially result in drugs that could inhibit marrow adipogenesis and further our understanding of obesity and osteoporosis related to aging and to disease states.

OBESITY AND BONE MARROW FAT

Although the interaction between obesity and BMD has traditionally been considered a positive one, the relationship between fat and bone within the bone marrow differs. Previous studies have shown that an increase in bone marrow fat [also known as bone marrow adipose tissue (BMAT)] is associated with lower BMD [15, 16], which may have racial differences [17]. Bone marrow fat is also elevated after loss of BMD associated with caloric restriction (CR) [18]. Therefore, in contrast to the loss of subcutaneous or visceral adipose tissue depots, there is an increase in bone marrow fat after weight loss. It is possible that reduced concentrations of leptin, estrogen, and insulin-like growth factor-1 (IGF-1), and increased cortisol, which are known to be altered by CR, contribute to BMAT deposition. FGF21 also increases with CR [12, 19] and results in greater adipogenesis, growth hormone resistance, and lowers IGF-1 [20]. All of these actions could promote marrow adipocytes and BMAT [18, 21]. In obesity, the amount of BMAT may be influenced by hormones and adipokines related to visceral adiposity, since one study showed that vertebral bone marrow fat is positively associated with visceral fat [22]. This same study in obese women showed that BMAT is inversely associated with IGF-1 and BMD [22] but shows no association with body mass index (BMI), subcutaneous, or total abdominal adipose tissue. Hence, greater visceral fat associated with an altered metabolic profile and/or metabolic syndrome in obesity may be a marker of greater bone marrow fat and lower BMD. The specific role of marrow adipose tissue and its relationship with BMD in obesity remains unclear.

BONE MINERAL DENSITY IN OBESITY

It is well established that a higher body weight is associated with a higher BMD and indeed the fracture risk assessment tool (FRAX) model includes BMI along with other clinical risk factors to assess fracture risk [23]. More recently, it has become clear that while total BMD is higher in the obese, trabecular and cortical volumetric BMD (vBMD) show a different pattern compared to normal-weight individuals. In children, a higher body weight is associated with a higher trabecular bone, but not cortical vBMD, and may decrease bone strength [3, 5]. A lower bone strength in the forearm of overweight children may be attributed to the greater fat-to-muscle ratio as compared to normal-weight children [24]. In adults, obesity is also associated with higher trabecular and lower cortical vBMD [2]. Even when bone parameters were controlled for confounders such as lean mass, physical activity level, and macronutrient intake, the positive effect of higher body weight on the trabecular parameters still remained, but a higher BMI did not have a positive effect on cortical parameters such as BMC, area and thickness and strength indices, and it remained negatively associated with cortical vBMD [2]. Obesity is associated with an altered hormonal profile, and in particular the chronically elevated serum parathyroid hormone (PTH) [2, 25] may be affecting bone quality due to its known catabolic actions on trabecular and cortical bone [26]. Thus, a greater understanding of the etiology and implications for differences in the trabecular and cortical compartments of the bone in obesity is needed. Evidence points in the direction that bone quality is compromised, and it may explain newer findings [6, 27] of increased fracture risk in the obese.

Potential measurement errors in obesity

Excess adiposity may artificially decrease BMD, especially in axial sites [28]. Long-term precision of BMD using dual energy X-ray absorptiometry (DXA) with weight change may also be a concern, with lower precision in heavier persons [29, 30]. Also, the practicality of positioning an obese person within the scan area on the DXA bed can be challenging. Despite these concerns, the trends observed in clinical trials for higher BMD in obesity are consistent with findings in rodent studies [31]. Peripheral bone measurements may have an advantage since there is less surrounding fat thickness compared to axial sites. In addition, quantitative computed tomography (QCT) can be more accurate since it measures vBMD rather than two-dimensional areal BMD (g/cm^2) as measured by DXA. Furthermore, obesity in older patients increases the risk of spinal osteoarthritis, which can overestimate bone mass [32]. Careful examination of the lumbar spine measurement for vertebral exclusion is important in the interpretation of BMD [33] and is a special consideration for the obese. Thus, clinical exams and research examining BMD in the obese individual requires additional effort to ensure accuracy and the T-score interpretation.

FRACTURE RISK

It is well established that lower weight individuals are at greatest risk of fracture compared to overweight and obese individuals. A low body weight or BMI (less than $20\,\text{kg/m}^2$) is associated with low BMD and higher fracture risk (approximately six times higher) compared to a BMI of approximately $30\,\text{kg/m}^2$ [34]. Higher fracture risk is exacerbated in the absence of physical activity [34]. On the other hand, De Laet et al. [35] found that risk of fracture increases again when BMI is greater than $35\,\text{kg/m}^2$. This is consistent with a reduction in the effective modulus in obese compared to nonobese rats [36]. The etiology of higher fracture risk in these two extremes in body weight may be related to differences in bone quality, biomechanical disadvantages, hormonal influences, or factors related to a poor diet and/or reduced physical activity. One or more of these factors may explain the higher risk of low trauma fractures in obese postmenopausal women who have a normal BMD [6]. Similarly, obese men have a higher risk of fracture, adjusted for BMD, compared to normal weight men [27]. The greater risk of falling and impact upon falling may increase fracture risk at certain sites in the obese. In a population-based longitudinal study in 10,902 women, those with high BMI had a significantly higher risk of proximal humerus and ankle fractures while they had a lower risk of forearm, vertebral, and hip fractures [37]. The higher risk of humerus fracture in obese women has also been found by others [38]. This finding of higher risk of nonvertebral fracture is consistent with higher forearm fracture risk in children with high body weight [39]. Similar to findings in normal-weight individuals, it can be expected that racial or ethnic differences will also affect fracture risk in the obese. There are a few reasons why fracture risk will differ at specific bone sites in the obese compared to normal-weight individuals, such as a different risk of falling due to altered gait [40], a different biomechanical impact upon falling, and/or greater padding due to excess soft tissue at certain sites, such as surrounding the axial sites. Overall, obesity is associated with lower hip fracture but a higher risk of proximal humerus fracture and possibly ankle and vertebral fracture in women (Fig. 119.1). Importantly, it has also been shown that once a fracture occurs in an obese individual, there are more complications along with a longer recovery time [41].

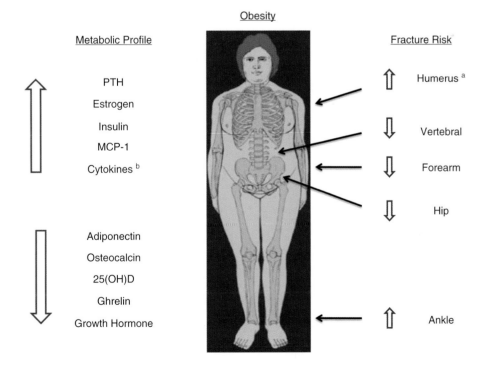

Fig. 119.1. Obesity: Metabolic profile and risk of fracture. The direction of fracture risk due to obesity at each site is similar in women and men, except the humerus (a), which is higher and lower with increasing BMI in women and men, respectively. (b) Pro-inflammatory cytokines.

BONE AND BODY COMPOSITION

Fat mass and fat-free soft tissue contribute to BMC and BMD [42, 43]. Fat-free soft tissue and strength are significant predictors of bone, and this positive effect may be due to exercise, lifestyle factors, estrogen sufficiency, or a combination of these factors. Low muscle mass is a risk factor for low BMD, whereas increased adiposity in adults is protective only when it is associated with substantial muscle mass. Total fat mass has shown to have both a positive and negative effect on BMD: the positive effect due to adipose-derived synthesis of estrogen and other hormones [42, 43] and the negative effect possibly due to release of other hormones and/or cytokines (Fig. 119.1). Fat mass accounts for approximately 15–30% of total body weight in normal-weight men and women, with much greater amounts in the obese. A fat threshold may exist whereby the higher adiposity may be detrimental due to increased release of adipokines and inflammatory cytokines, or reduced fat-free soft tissue may permit the threshold to be attained more readily. In addition, the distribution of adipose tissue in the body may be an important in the relationship with bone. In postmenopausal obese women it was found that the waist-to-hip ratio was a better predictor of bone at the radius than total fat mass [44]. In addition, abdominal fat has been found to be the most important positive predictor of BMD in women [45, 46], and low abdominal fat is associated with higher fracture risk [47], but the effect was not independent of body weight. In contrast, a greater visceral adiposity has been shown to be associated with reduced BMD in men [48] or young women [49, 50]. Others found an inverse relationship between BMD and visceral and subcutaneous fat corrected for lean tissue mass in adults across a wide age range [51]. Hence, studies do not show a consistent inverse relationship between visceral fat and bone, although these different relationships may depend on the specific bone site. Therefore, the site in which fat accumulates and the specific population of interest are important in understanding the link between adiposity and bone mass.

ADIPOKINES AND ENDOCRINE PROFILE IN OBESITY: ACTIONS ON BONE

Obesity may induce changes in bone due to several mechanisms, including an altered hormonal milieu because of factors secreted by adipose tissue, known as adipokines, which may be important mediators in the bone–fat relationship [8, 42, 43]. Adipose tissue is a highly metabolically active tissue containing a vast variety of cell types, and the more abundant cell types present are adipocytes, preadipocytes, immune cells, and endothelial cells. The adipocyte-derived factors have an effect on many organs in the body, including the bone. For example, excess adiposity has been shown to stimulate the osteoblast through increased production of the hormones insulin, estrogen, and leptin. Adipose tissue is a site of peripheral synthesis of estrogen from androstenedione resulting in higher serum levels of estrogen in obesity. Therefore, excess body fat is expected to be physiologically beneficial to bone health. In contrast, excess adipose tissue may hinder bone growth by enhancing the role of oxidized lipids in accelerating atherogenesis, thus calcifying vascular cells and inhibiting osteoblastic differentiation [52]. Other adipose-derived hormones include leptin, resistin, and adiponectin. Leptin is best known for its effect on suppressing appetite and increasing energy expenditure, but it has both a direct and central mediated effect on bone remodeling. The centrally mediated effect on bone occurs through sympathetic tone that inhibits bone formation and enhances bone resorption [53] yet *in vitro* has been shown to have direct effects on osteoblast differentiation [54]. Clinical trials have reported both positive and negative effects of leptin on bone [42, 43, 55, 56]. Serum levels of resistin are higher in obesity, and resistin is expressed in osteoblasts and osteoclasts. Resistin likely has a role in bone remodeling, but its overall effect is not established, and clinical trials do not consistently show an association with BMD [57]. The low levels of adiponectin in obesity may have a negative effect on bone since there are receptors expressed in human osteoblasts [58] but there are mixed findings in clinical studies [42, 59]. In addition, gut peptides [i.e., ghrelin, incretins, cholecystokinin (CCK), peptide YY–PYY, pancreatic polypeptide (PPY)] that regulate satiety are often altered in obesity and also regulate bone. For example, ghrelin is an appetite stimulant that is lower in obese subjects; a 1-year double-blind trial showed a positive effect of ghrelin treatment on bone [60]. However, a recent systematic review examining 59 studies concluded that there is a lack of convincing data to support an association between resistin, visfatin, or ghrelin with BMD or fractures [57].

Bone-regulating hormones that are altered in obesity include lower 25-hydroxyvitamin D [25(OH)D] [61] and higher levels of PTH [25]. The lower 25(OH)D is attributed to greater deposition into adipose tissue, or lower exposure in heavier individuals. Higher BMI and body fat are also associated with lower active 1,25(OH)D [62, 63]. Lower circulating 25(OH)D may explain the higher serum PTH in obesity; however, it is also possible that PTH has an independent effect to maintain higher adiposity. Greater adiposity due to hyperparathyroidism is associated with a decrease in fat oxidation [64]. However, it has been shown that correction of excess PTH following parathyroidectomy does not reduce body weight [65]. Thus, the cause and effect of the PTH and adiposity relationship is not entirely clear. Higher chronic levels of PTH reduce cortical bone but results in similar or higher trabecular bone [26, 66]. The overall effect of higher PTH on bone in obesity is not clear but has been shown to be associated with higher lumbar spine BMD and similar or lower bone mass at the femoral neck [67] and more cortical rich sites [2].

Obesity is also characterized by abnormal growth hormone (GH) secretion and insufficiency. Obese patients with no evidence of pituitary disease generally have normal or increased IGF-1 [68] that could be lower in the presence of central adiposity [69]. Hence, the effect of IGF-1 on bone in obesity is less well understood, yet the lower GH would be expected to have a negative effect on bone. Higher cortisol levels are associated with excess visceral adiposity, and excess endogenous cortisol may also lead to bone loss [70]. Osteocalcin, a bone formation marker secreted by the osteoblast, might also function as an important regulator of energy homeostasis [53] by regulating insulin sensitivity. Interestingly, serum osteocalcin levels are lower in obese individuals [71], and whether or not this has a direct effect on bone in this population is not known. Finally, higher circulating concentrations of inflammatory cytokines, including interleukin-6, monocyte chemoattractant protein-1 (MCP-1), and C-reactive protein, are found in obesity, and which, in general, have a negative effect on bone, but their specific effect on bone in obesity is not as clear [42]. For example, PTH upregulates serum MCP-1 levels in both normal-weight and obese subjects [72], and MCP-1 may mediate the anabolic effects associated with intermittent PTH on the osteoblast. Overall, the altered hormonal milieu and change in cytokines in obesity theoretically supports both positive and negative effects of higher body weight on BMD and the risk of fracture.

WEIGHT REDUCTION AND BONE

Loss of body weight in older men or women, whether voluntary or involuntary, is associated with reduced BMD and increased hip fracture risk [73, 74]. In addition, others have shown that only a 5% weight loss will increase fracture risk by 33% at the distal forearm [75]. Initial body weight is a strong predictor of the bone loss in response to weight reduction [43]. There may only be partial bone recovery with weight regain [76] so risk of fracture may be greater in individuals with a history of weight cycling. In overweight and obese older women who lose about 10% of their body weight, there is a 1–2% loss of bone at the trochanter, 1/3 radius, and lumbar spine [43, 77, 78]. The loss of bone can be attributed to a few factors including a decrease in micronutrient intake, calcium absorption, sex steroids, and/or mechanical loading [43]. Loss of bone due to moderate weight reduction is often not reported in younger women or men (younger than 50 years of age) [43, 79, 80], and this may be attributed to the well-maintained muscle mass and performance in men and young women. Weight reduction generally results in loss of highly trabecular rich regions [31, 78], but extreme loss of body weight (approximately 30% weight loss), proportionally increases bone loss [81], primarily from cortical regions [67, 82]. This suggests that moderate, compared to extreme, weight loss may differentially influence bone quality. Dietary interventions can attenuate bone loss such as with higher intakes of calcium and protein [78, 83]. Exercise may also attenuate bone loss with weight reduction [79, 84], and may also attenuate the risk of falling in older obese individuals due to an increased level of physical function [85]. There is strong evidence for bone loss and increased fracture risk due to weight reduction; the extent of loss is related to age, nutrient intake, exercise, as well as the amount and rate of weight loss.

SUMMARY

Obesity has a positive relationship with bone, but the quality of the bone differs compared to normal-weight individuals. Obesity is associated with lower cortical vBMD and higher risk of fracture for a given BMD or T-score. Hence, while the low- or normal-weight population remains at highest risk for fracture, new evidence shows that obesity also raises fracture risk. Many obese persons have a history of dieting, which may increase fracture risk. Also, adiposity-induced metabolic alterations, poor dietary intake, sarcopenia, and/or biomechanical disadvantages in obesity may contribute to the risk of fracture in obese children and adults. The benefits of weight reduction in the obese are multiple, and importantly, a higher protein intake, a multivitamin/mineral supplementation, and exercise should be encouraged to increase calcium and other micronutrients that are reduced during dieting to minimize bone loss. Due to the emerging data that obese patients are also at risk of fracture, this population should be included in osteoporosis screening and preventive measures should be encouraged.

REFERENCES

1. Flegal KM, Carroll MD, Kit BK, Ogden CL. 2012. Prevalence of obesity and trends in the distribution of body mass index among US adults, 1999–2010. *JAMA* 307(5): 491–7.
2. Sukumar D, Schlussel Y, Riedt CS, Gordon C, Stahl T, Shapses SA. 2011. Obesity alters cortical and trabecular bone density and geometry in women. *Osteoporos Int* 22(2): 635–45.
3. Pollock NK, Laing EM, Baile CA, Hamrick MW, Hall DB, Lewis RD. 2007. Is adiposity advantageous for bone strength? A peripheral quantitative computed tomography study in late adolescent females. *Am J Clin Nutr* 86(5): 1530–8.
4. Wetzsteon RJ, Petit MA, Macdonald HM, Hughes JM, Beck TJ, McKay HA. 2008. Bone structure and volumetric BMD in overweight children: A longitudinal study. *J Bone Miner Res* 23(12): 1946–53.
5. Cole ZA, Harvey NC, Kim M, Ntani G, Robinson SM, Inskip HM, Godfrey KM, Cooper C, Dennison EM. 2012. Increased fat mass is associated with increased bone size

but reduced volumetric density in pre pubertal children. *Bone* 50(2): 562–7.
6. Premaor MO, Pilbrow L, Tonkin C, Parker RA, Compston J. 2010. Obesity and fractures in postmenopausal women. *J Bone Miner Res* 25(2): 292–7.
7. Gimble JM, Zvonic S, Floyd ZE, Kassem M, Nuttall ME. 2006. Playing with bone and fat. *J Cell Biochem* 98(2): 251–66.
8. Rosen CJ, Bouxsein ML. 2006. Mechanisms of disease: Is osteoporosis the obesity of bone? *Nat Clin Pract Rheumatol* 2(1): 35–43.
9. Meunier P, Aaron J, Edouard C, Vignon G. 1971. Osteoporosis and the replacement of cell populations of the marrow by adipose tissue. A quantitative study of 84 iliac bone biopsies. *Clin Orthop Relat Res* 80: 147–54.
10. Tontonoz P, Spiegelman BM. 2008. Fat and beyond: The diverse biology of PPARgamma. *Annu Rev Biochem* 77: 289–312.
11. Habib ZA, Havstad SL, Wells K, Divine G, Pladevall M, Williams LK. 2010. Thiazolidinedione use and the longitudinal risk of fractures in patients with type 2 diabetes mellitus. *J Clin Endocrinol Metab* 95(2): 592–600.
12. Wei W, Dutchak PA, Wang X, Ding X, Wang X, Bookout AL, Goetz R, Mohammadi M, Gerard RD, Dechow PC, et al. 2012. Fibroblast growth factor 21 promotes bone loss by potentiating the effects of peroxisome proliferator-activated receptor gamma. *Proc Natl Acad Sci U S A* 109(8): 3143–8.
13. Parhami F, Jackson SM, Tintut Y, Le V, Balucan JP, Territo M, Demer LL. 1999. Atherogenic diet and minimally oxidized low density lipoprotein inhibit osteogenic and promote adipogenic differentiation of marrow stromal cells. *J Bone Miner Res* 14(12): 2067–78.
14. Ahdjoudj S, Lasmoles F, Holy X, Zerath E, Marie PJ. 2002. Transforming growth factor beta2 inhibits adipocyte differentiation induced by skeletal unloading in rat bone marrow stroma. *J Bone Miner Res* 17(4): 668–77.
15. Shen W, Chen J, Punyanitya M, Shapses S, Heshka S, Heymsfield SB. 2007. MRI-measured bone marrow adipose tissue is inversely related to DXA-measured bone mineral in Caucasian women. *Osteoporos Int* 18(5): 641–7.
16. Bredella MA, Fazeli PK, Miller KK, Misra M, Torriani M, Thomas BJ, Ghomi RH, Rosen CJ, Klibanski A. 2009. Increased bone marrow fat in anorexia nervosa. *J Clin Endocrinol Metab* 94(6): 2129–36.
17. Shen W, Chen J, Gantz M, Punyanitya M, Heymsfield SB, Gallagher D, Albu J, Engelson E, Kotler D, Pi-Sunyer X, et al. 2012. Ethnic and sex differences in bone marrow adipose tissue and bone mineral density relationship. *Osteoporos Int* 23(9): 2293–301.
18. Devlin MJ, Cloutier AM, Thomas NA, Panus DA, Lotinun S, Pinz I, Baron R, Rosen CJ, Bouxsein ML. 2010. Caloric restriction leads to high marrow adiposity and low bone mass in growing mice. *J Bone Miner Res* 25(9): 2078–88.
19. Kubicky RA, Wu S, Kharitonenkov A, De LF. 2012. Role of fibroblast growth factor 21 (FGF21) in undernutrition-related attenuation of growth in mice. *Endocrinology* 153(5): 2287–95.
20. Fazeli PK, Bredella MA, Misra M, Meenaghan E, Rosen CJ, Clemmons DR, Breggia A, Miller KK, Klibanski A. 2010. Preadipocyte factor-1 is associated with marrow adiposity and bone mineral density in women with anorexia nervosa. *J Clin Endocrinol Metab* 95(1): 407–13.
21. Devlin MJ. 2011. Why does starvation make bones fat? *Am J Hum Biol* 23(5): 577–85.
22. Bredella MA, Torriani M, Ghomi RH, Thomas BJ, Brick DJ, Gerweck AV, Rosen CJ, Klibanski A, Miller KK. 2011. Vertebral bone marrow fat is positively associated with visceral fat and inversely associated with IGF-1 in obese women. *Obesity (Silver Spring)* 19(1): 49–53.
23. Kanis JA, Johnell O, Oden A, Johansson H, McCloskey E. 2008. FRAX and the assessment of fracture probability in men and women from the UK. *Osteoporos Int* 19(4): 385–97.
24. Ducher G, Bass SL, Naughton GA, Eser P, Telford RD, Daly RM. 2009. Overweight children have a greater proportion of fat mass relative to muscle mass in the upper limbs than in the lower limbs: Implications for bone strength at the distal forearm. *Am J Clin Nutr* 90(4): 1104–11.
25. Bolland MJ, Grey AB, Ames RW, Horne AM, Gamble GD, Reid IR. 2006. Fat mass is an important predictor of parathyroid hormone levels in postmenopausal women. *Bone* 38(3): 317–21.
26. Charopoulos I, Tournis S, Trovas G, Raptou P, Kaldrymides P, Skarandavos G, Katsalira K, Lyritis GP. 2006. Effect of primary hyperparathyroidism on volumetric bone mineral density and bone geometry assessed by peripheral quantitative computed tomography in postmenopausal women. *J Clin Endocrinol Metab* 91(5): 1748–53.
27. Nielson CM, Marshall LM, Adams AL, Leblanc ES, Cawthon PM, Ensrud K, Stefanick ML, Barrett-Connor E, Orwoll ES. 2011. BMI and fracture risk in older men: The osteoporotic fractures in men study (MrOS). *J Bone Miner Res* 26(3): 496–502.
28. Bolotin HH. 1998. A new perspective on the causal influence of soft tissue composition on DXA-measured in vivo bone mineral density. *J Bone Miner Res* 13(11): 1739–46.
29. Tothill P, Hannan WJ, Cowen S, Freeman CP. 1997. Anomalies in the measurement of changes in total-body bone mineral by dual-energy X-ray absorptiometry during weight change. *J Bone Miner Res* 12(11): 1908–21.
30. Rajamanohara R, Robinson J, Rymer J, Patel R, Fogelman I, Blake GM. 2011. The effect of weight and weight change on the long-term precision of spine and hip DXA measurements. *Osteoporos Int* 22(5): 1503–12.
31. Hawkins J, Cifuentes M, Pleshko NL, Ambia-Sobhan H, Shapses SA. 2010. Energy restriction is associated with lower bone mineral density of the tibia and femur in lean but not obese female rats. *J Nutr* 140(1): 31–7.
32. Liu G, Peacock M, Eilam O, Dorulla G, Braunstein E, Johnston CC. 1997. Effect of osteoarthritis in the lumbar spine and hip on bone mineral density and diagnosis of

osteoporosis in elderly men and women. *Osteoporos Int* 7(6): 564–9.
33. Lewiecki EM, Gordon CM, Baim S, Leonard MB, Bishop NJ, Bianchi ML, Kalkwarf HJ, Langman CB, Plotkin H, Rauch F, et al. 2008. International Society for Clinical Densitometry 2007 Adult and Pediatric Official Positions. *Bone* 43(6): 1115–21.
34. Armstrong ME, Spencer EA, Cairns BJ, Banks E, Pirie K, Green J, Wright FL, Reeves GK, Beral V. 2011. Body mass index and physical activity in relation to the incidence of hip fracture in postmenopausal women. *J Bone Miner Res* 26(6): 1330–8.
35. De Laet C, Kanis JA, Oden A, Johanson H, Johnell O, Delmas P, Eisman JA, Kroger H, Fujiwara S, Garnero P, et al. 2005. Body mass index as a predictor of fracture risk: A meta-analysis. *Osteoporos Int* 16(11): 1330–8.
36. Woo DG, Lee BY, Lim D, Kim HS. 2009. Relationship between nutrition factors and osteopenia: Effects of experimental diets on immature bone quality. *J Biomech* 42(8): 1102–7.
37. Holmberg AH, Johnell O, Nilsson PM, Nilsson J, Berglund G, Akesson K. 2006. Risk factors for fragility fracture in middle age. A prospective population-based study of 33,000 men and women. *Osteoporos Int* 17(7): 1065–77.
38. Gnudi S, Sitta E, Lisi L. 2009. Relationship of body mass index with main limb fragility fractures in postmenopausal women. *J Bone Miner Metab* 27(4): 479–84.
39. Goulding A, Grant AM, Williams SM. 2005. Bone and body composition of children and adolescents with repeated forearm fractures. *J Bone Miner Res* 20(12): 2090–6.
40. Ko S, Stenholm S, Ferrucci L. 2010. Characteristic gait patterns in older adults with obesity—Results from the Baltimore Longitudinal Study of Aging. *J Biomech* 43(6): 1104–10.
41. Leet AI, Pichard CP, Ain MC. 2005. Surgical treatment of femoral fractures in obese children: Does excessive body weight increase the rate of complications? *J Bone Joint Surg Am* 87(12): 2609–13.
42. Zhao LJ, Jiang H, Papasian CJ, Maulik D, Drees B, Hamilton J, Deng HW. 2008. Correlation of obesity and osteoporosis: Effect of fat mass on the determination of osteoporosis. *J Bone Miner Res* 23(1): 17–29.
43. Shapses SA, Riedt CS. 2006. Bone, body weight, and weight reduction: What are the concerns? *J Nutr* 136(6): 1453–6.
44. Tarquini B, Navari N, Perfetto F, Piluso A, Romano S, Tarquini R. 1997. Evidence for bone mass and body fat distribution relationship in postmenopausal obese women. *Arch Gerontol Geriatr* 24(1): 15–21.
45. Kuwahata A, Kawamura Y, Yonehara Y, Matsuo T, Iwamoto I, Douchi T. 2008. Non-weight-bearing effect of trunk and peripheral fat mass on bone mineral density in pre- and post-menopausal women. *Maturitas* 60(3-4): 244–7.
46. Warming L, Ravn P, Christiansen C. 2003. Visceral fat is more important than peripheral fat for endometrial thickness and bone mass in healthy postmenopausal women. *Am J Obstet Gynecol* 188(2): 349–53.
47. Nguyen ND, Pongchaiyakul C, Center JR, Eisman JA, Nguyen TV. 2005. Abdominal fat and hip fracture risk in the elderly: The Dubbo Osteoporosis Epidemiology Study. *BMC Musculoskelet Disord* 6: 11.
48. Jankowska EA, Rogucka E, Medras M. 2001. Are general obesity and visceral adiposity in men linked to reduced bone mineral content resulting from normal ageing? A population-based study. *Andrologia* 33(6): 384–9.
49. Gilsanz V, Chalfant J, Mo AO, Lee DC, Dorey FJ, Mittelman SD. 2009. Reciprocal relations of subcutaneous and visceral fat to bone structure and strength. *J Clin Endocrinol Metab* 94(9): 3387–93.
50. Russell M, Mendes N, Miller KK, Rosen CJ, Lee H, Klibanski A, Misra M. 2010. Visceral fat is a negative predictor of bone density measures in obese adolescent girls. *J Clin Endocrinol Metab* 95(3): 1247–55.
51. Katzmarzyk PT, Barreira TV, Harrington DM, Staiano AE, Heymsfield SB, Gimble JM. 2012. Relationship between abdominal fat and bone mineral density in white and African American adults. *Bone* 50(2): 576–9.
52. Parhami F, Morrow AD, Balucan J, Leitinger N, Watson AD, Tintut Y, Berliner JA, Demer LL. 1997. Lipid oxidation products have opposite effects on calcifying vascular cell and bone cell differentiation. A possible explanation for the paradox of arterial calcification in osteoporotic patients. *Arterioscler Thromb Vasc Biol* 17(4): 680–7.
53. Karsenty G, Oury F. 2010. The central regulation of bone mass, the first link between bone remodeling and energy metabolism. *J Clin Endocrinol Metab* 95(11): 4795–801.
54. Thomas T, Gori F, Khosla S, Jensen MD, Burguera B, Riggs BL. 1999. Leptin acts on human marrow stromal cells to enhance differentiation to osteoblasts and to inhibit differentiation to adipocytes. *Endocrinology* 140(4): 1630–8.
55. Rosen CJ, Bouxsein ML. 2006. Mechanisms of disease: Is osteoporosis the obesity of bone? *Nat Clin Pract Rheumatol* 2(1): 35–43.
56. Yamauchi M, Sugimoto T, Yamaguchi T, Nakaoka D, Kanzawa M, Yano S, Ozuru R, Sugishita T, Chihara K. 2001. Plasma leptin concentrations are associated with bone mineral density and the presence of vertebral fractures in postmenopausal women. *Clin Endocrinol (Oxf)* 55(3): 341–7.
57. Biver E, Salliot C, Combescure C, Gossec L, Hardouin P, Legroux-Gerot I, Cortet B. 2011. Influence of adipokines and ghrelin on bone mineral density and fracture risk: A systematic review and meta-analysis. *J Clin Endocrinol Metab* 96(9): 2703–13.
58. Berner HS, Lyngstadaas SP, Spahr A, Monjo M, Thommesen L, Drevon CA, Syversen U, Reseland JE. 2004. Adiponectin and its receptors are expressed in bone-forming cells. *Bone* 35(4): 842–9.
59. Barbour KE, Zmuda JM, Boudreau R, Strotmeyer ES, Horwitz MJ, Evans RW, Kanaya AM, Harris TB, Bauer

DC, Cauley JA. 2011. Adipokines and the risk of fracture in older adults. *J Bone Miner Res* 26(7): 1568–76.
60. Nass R, Pezzoli SS, Oliveri MC, Patrie JT, Harrell FE Jr, Clasey JL, Heymsfield SB, Bach MA, Vance ML, Thorner MO. 2008. Effects of an oral ghrelin mimetic on body composition and clinical outcomes in healthy older adults: A randomized trial. *Ann Intern Med* 149(9): 601–11.
61. Wortsman J, Matsuoka LY, Chen TC, Lu Z, Holick MF. 2000. Decreased bioavailability of vitamin D in obesity. *Am J Clin Nutr* 72(3): 690–3.
62. Konradsen S, Ag H, Lindberg F, Hexeberg S, Jorde R. 2008. Serum 1,25-dihydroxy vitamin D is inversely associated with body mass index. *Eur J Nutr* 47(2): 87–91.
63. Parikh SJ, Edelman M, Uwaifo GI, Freedman RJ, Semega-Janneh M, Reynolds J, Yanovski JA. 2004. The relationship between obesity and serum 1,25-dihydroxy vitamin D concentrations in healthy adults. *J Clin Endocrinol Metab* 89(3): 1196–9.
64. Gunther CW, Lyle RM, Legowski PA, James JM, McCabe LD, McCabe GP, Peacock M, Teegarden D. 2005. Fat oxidation and its relation to serum parathyroid hormone in young women enrolled in a 1-y dairy calcium intervention. *Am J Clin Nutr* 82(6): 1228–34.
65. Bollerslev J, Rosen T, Mollerup CL, Nordenstrom J, Baranowski M, Franco C, Pernow Y, Isaksen GA, Godang K, Ueland T, et al. 2009. Effect of surgery on cardiovascular risk factors in mild primary hyperparathyroidism. *J Clin Endocrinol Metab* 94(7): 2255–61.
66. Bilezikian JP, Silverberg SJ, Shane E, Parisien M, Dempster DW. 1991. Characterization and evaluation of asymptomatic primary hyperparathyroidism. *J Bone Miner Res* 6 Suppl 2: S85–9.
67. Goode LR, Brolin RE, Chowdhury HA, Shapses SA. 2004. Bone and gastric bypass surgery: Effects of dietary calcium and vitamin D. *Obes Res* 12(1): 40–7.
68. Nam SY, Lee EJ, Kim KR, Cha BS, Song YD, Lim SK, Lee HC, Huh KB. 1997. Effect of obesity on total and free insulin-like growth factor (IGF)-1, and their relationship to IGF-binding protein (BP)-1, IGFBP-2, IGFBP-3, insulin, and growth hormone. *Int J Obes Relat Metab Disord* 21(5): 355–9.
69. Gram IT, Norat T, Rinaldi S, Dossus L, Lukanova A, Tehard B, Clavel-Chapelon F, van Gils CH, van Noord PA, Peeters PH, et al. 2006. Body mass index, waist circumference and waist-hip ratio and serum levels of IGF-I and IGFBP-3 in European women. *Int J Obes (Lond)* 30(11): 1623–31.
70. Tauchmanova L, Rossi R, Nuzzo V, del PA, Esposito-del PA, Pizzi C, Fonderico F, Lupoli G, Lombardi G. 2001. Bone loss determined by quantitative ultrasonometry correlates inversely with disease activity in patients with endogenous glucocorticoid excess due to adrenal mass. *Eur J Endocrinol* 145(3): 241–7.
71. Cifuentes M, Johnson MA, Lewis RD, Heymsfield SB, Chowdhury HA, Modlesky CM, Shapses SA. 2003. Bone turnover and body weight relationships differ in normal-weight compared with heavier postmenopausal women. *Osteoporos Int* 14(2): 116–22.
72. Sukumar D, Partridge NC, Wang X, Shapses SA. 2011. The high serum monocyte chemoattractant protein-1 in obesity is influenced by high parathyroid hormone and not adiposity. *J Clin Endocrinol Metab* 96(6): 1852–8.
73. Ensrud KE, Ewing SK, Stone KL, Cauley JA, Bowman PJ, Cummings SR. 2003. Intentional and unintentional weight loss increase bone loss and hip fracture risk in older women. *J Am Geriatr Soc* 51(12): 1740–7.
74. Meyer HE, Sogaard AJ, Falch JA, Jorgensen L, Emaus N. 2008. Weight change over three decades and the risk of osteoporosis in men: The Norwegian Epidemiological Osteoporosis Studies (NOREPOS). *Am J Epidemiol* 168(4): 454–60.
75. Omsland TK, Schei B, Gronskag AB, Langhammer A, Forsen L, Gjesdal CG, Meyer HE. 2009. Weight loss and distal forearm fractures in postmenopausal women: The Nord-Trondelag health study, Norway. *Osteoporos Int* 20(12): 2009–16.
76. Fogelholm GM, Sievanen HT, Kukkonen-Harjula TK, Pasanen ME. 2001. Bone mineral density during reduction, maintenance and regain of body weight in premenopausal, obese women. *Osteoporos Int* 12(3): 199–206.
77. Gozansky WS, Van Pelt RE, Jankowski CM, Schwartz RS, Kohrt WM. 2005. Protection of bone mass by estrogens and raloxifene during exercise-induced weight Loss. *J Clin Endocrinol Metab* 90(1): 52–9.
78. Sukumar D, Ambia-Sobhan H, Zurfluh R, Schlussel Y, Stahl TJ, Gordon CL, Shapses SA. 2011. Areal and volumetric bone mineral density and geometry at two levels of protein intake during caloric restriction: A randomized, controlled trial. *J Bone Miner Res* 26(6): 1339–48.
79. Redman LM, Rood J, Anton SD, Champagne C, Smith SR, Ravussin E. 2008. Calorie restriction and bone health in young, overweight individuals. *Arch Intern Med* 168(17): 1859–66.
80. Uusi-Rasi K, Sievanen H, Kannus P, Pasanen M, Kukkonen-Harjula K, Fogelholm M. 2009. Influence of weight reduction on muscle performance and bone mass, structure and metabolism in obese premenopausal women. *J Musculoskelet Neuronal Interact* 9(2): 72–80.
81. Fleischer J, Stein EM, Bessler M, Della BM, Restuccia N, Olivero-Rivera L, McMahon DJ, Silverberg SJ. 2008. The decline in hip bone density after gastric bypass surgery is associated with extent of weight loss. *J Clin Endocrinol Metab* 93(10): 3735–40.
82. Hamrick MW, Ding KH, Ponnala S, Ferrari SL, Isales CM. 2008. Caloric restriction decreases cortical bone mass but spares trabecular bone in the mouse skeleton: Implications for the regulation of bone mass by body weight. *J Bone Miner Res* 23(6): 870–8.
83. Riedt CS, Cifuentes M, Stahl T, Chowdhury HA, Schlussel Y, Shapses SA. 2005. Overweight postmenopausal women lose bone with moderate weight reduction and 1 g/day calcium intake. *J Bone Miner Res* 20(3): 455–63.

84. Silverman NE, Nicklas BJ, Ryan AS. 2009. Addition of aerobic exercise to a weight loss program increases BMD, with an associated reduction in inflammation in overweight postmenopausal women. *Calcif Tissue Int* 84(4): 257–65.

85. Villareal DT, Chode S, Parimi N, Sinacore DR, Hilton T, Armamento-Villareal R, Napoli N, Qualls C, Shah K. 2011. Weight loss, exercise, or both and physical function in obese older adults. *N Engl J Med* 364(13): 1218–29.

120

Neuropsychiatric Disorders and the Skeleton

Itai Bab and Raz Yirmiya

Introduction 1002
Major Depressive Disorder 1002
Anorexia Nervosa 1005
Schizophrenia 1005
Traumatic Brain Injury 1006

Alzheimer's Disease 1006
Neuropsychiatric Medications 1007
Summary 1008
References 1008

INTRODUCTION

Studies published in the past decade report previously unexpected evidence on the control of bone metabolism by the central nervous system (CNS) and by neurotransmitters released in the skeleton by bone cells and their microenvironment. Bone is densely innervated by autonomic and sensory nerve fibers [1]. Bone cells, mainly osteoblasts, express receptors for neurotramitters and neuropeptides such as acetylcholine [2], norepinephrin (NE) [3], endocannabinoids (ECs) [4], neuropeptide Y [5], calcitonin gene related peptide, and substance P [1]. Central interleukin-1 (IL-1) signaling has been also implicated in the regulation of bone metabolism [6]. To date, the best experimentally characterized brain-to-bone pathway is the sympathetic nervous system (SNS) that mediates the skeletal effects of leptin and serotonin via a hypothalamic serotonergic relay [7]. Sympathetic nerve terminals form a synaptic-like junction with osteoblasts, which express β2-adrenergic receptors (β2AR). These receptors are activated by NE released from the sympathetic terminals, thus tonically restraining bone formation [8]. Activation of β2AR in osteoblasts also leads to stimulation of receptor activator of nuclear factor-κB ligand (RANKL) expression and their stromal cell progenitors, which in turn results in increased osteoclast number and activity [9]. The skeletal sympathetic tone is downregulated by brain-stem-derived serotonin and M3 acetylcholine muscarinic receptors expressed in the brain [7, 10]. NE discharge by skeletal sympathetic terminals is attenuated by the endocannabinoid 2-arachidonoylglycerole (2-AG) released into the sympatho-osteoblast junction by osteoblasts and activates CB1 cannabinoid receptors present in the prejunctional membrane.[3]

Neuropsychiatric diseases are mental disorders attributable to organic nervous system pathology, mainly brain pathology. The number of psychiatric diseases included in this category is still growing, as their organic etiology is being progressively established. In addition, clinical conditions involving cognition and/or behavior caused by brain injury or disease are sometimes included. Table 120.1 lists some of the major neuropsychiatric disorders and examples of related neural processes. It is now well established that some of these diseases, mainly major depressive disorder (MDD), anorexia nervosa (AN), and traumatic brain injury (TBI) are associated with changes in bone mass. Also, initial indications have been published suggesting that schizophrenia and Alzheimer's disease (AD) are accompanied by changes in bone density. Importantly, the skeletal status in these conditions is often complicated by psychiatric medication that affects bone directly and indirectly.

MAJOR DEPRESSIVE DISORDER

MDD is a mental disorder characterized by an all-encompassing low mood accompanied by loss of interest or pleasure in normally enjoyable activities, as well as

---

Primer on the Metabolic Bone Diseases and Disorders of Mineral Metabolism, Eighth Edition. Edited by Clifford J. Rosen.
© 2013 American Society for Bone and Mineral Research. Published 2013 by John Wiley & Sons, Inc.

Table 120.1. Some Neuropsychiatric Conditions and Related Neural Processes	
Major depressive disorder	Prefrontal-subcortical (limbic) dysregulation of monoamines, neurotrophins, hormonal systems, and neurogenesis
Bipolar disorder	Prefrontal cortex, anterior cingulate, hippocampus, and amygdala
Eating disorder	Atypical serotonergic system, right frontal, and temporal lobe damage
Schizophrenia	Structural alterations in frontal and temporal lobes, abnormal mesolimbic and mesocortical dopaminergic circuitry, changes in glutamatergic neurotransmission
Visual hallucination	Retino-geniculo-calcarine tract, ascending brainstem modulatory structures
Auditory hallucination	Frontotemporal functional connectivity
Obsessive compulsive disorder	Structural and functional alterations in the basal ganglia, particularly right caudate activity, changes in serotonergic neurotransmission
Post-traumatic stress disorder	Dysregulation of amygdala by ventromedial prefrontal cortex
Autism spectrum disorders	Genetic variations and mutations in single or multiple genes
Altzheimer's disease/dementia	Formation of plaques and tangles, neuronal death, neuroinflammation
Huntington's disease	Death of medium-sized neurons in striatum and cortex
Traumatic brain injury	Focal ischemia, neuroinflammation

alterations in appetite, body weight, sleep patterns, psychomotor activity, fatigue, worthlessness and low self-esteem, cognitive impairments, and suicidal ideation [11]. The term "depression" is ambiguous as it often refers to other mood disorders or to lower mood states lacking clinical significance. MDD is a disabling condition that adversely affects a person's family, work or school life, sleeping and eating habits, and general health. More than half of the individuals who commit suicide have MDD or another mood disorder [12]. It is generally believed that MDD is the consequence of a physiological vulnerability to stressful events. Although several central systems and processes have been implicated in depression, including alterations in serotonergic and adrenergic transmission, hormonal dysregulation, reduced neutrophines, and neurogenesis, as well as neuroinflammation and IL-1 secretion, the mechanisms involved are not well understood [13, 14].

Like osteoporosis, MDD is a prevalent disease, considered the second-leading cause of years of life lived with disability [15]. Both MDD and osteoporosis are substantially more common in women than men [11, 16]. Recently, data from the second Nord-Trondelag Health Study have demonstrated a cross-sectional association between depression and reduced bone mineral density (BMD) [17]. The relationship between MDD and low bone density has been established by several meta-analyses (reviewed in Ref. 18). The most comprehensive of these analyses includes 23 studies comparing a total of 2,327 depressed with 21,141 nondepressed subjects [19]. It shows that individuals with MDD have lower BMD and higher bone resorption markers than nondepressed subjects. The association of MDD with low BMD is significant in the spine, femoral neck, and distal radius, suggesting that the MDD-associated low BMD involves a multiplicity of trabecular bone sites throughout the skeleton. Consistent with the Nord-Trondelag Health Study [17], the overall effect size of depression on BMD for the entire population, comprising adult males and females, is rather small, with a moderate effect on serum and urine bone resorption markers.[19]

Women are significantly more vulnerable to depression-associated low BMD [19, 20]. This gender difference may be related to the greater sensitivity of females to stress in general, and to the greater responsiveness of depressed women to various stressors in particular (reviewed in Ref. 19). Still, the effect of MDD in men is more robust, i.e., it is not influenced by various moderating variables. In contrast, the association between MDD and BMD in women is heterogeneous, as it is significantly moderated by the menopausal status. Premenopausal women display a greater MDD-associated decrease in BMD compared to postmenopausal subjects [19], consistent with a decreased rate of bone mass accrual in teenager females with MDD [21], portraying reduced peak bone mass as one determinant leading to decreased BMD in depressed premenopausal women. The greater MDD-associated decrease in BMD in pre- as oppsed to postmenopausal subjects does not necessarily mean that MDD is not associated with low bone mass after menopause. However, in postmenopausal women this association may be masked by the multiplicity of factors contributing to the development of low bone mass, such as estrogen depletion, reduced physical activity, nutritional disturbances, and drug treatments (reviewed in Ref. 10).

Several studies report that the contribution of potential moderator variables to MDD-associated low BMD such as body weight and height, number of previous depressive episodes, total duration of disease, history of estrogen treatment, smoking, and race do not modulate the association between MDD and bone density. Low levels of physical activity, characteristic of depressed patients, were also suggested to be associated with low BMD (reviewed in Ref. 19). However, studies that assessed the levels of exercise found no evidence that it modulates the association between MDD and BMD [19]. In addition, in

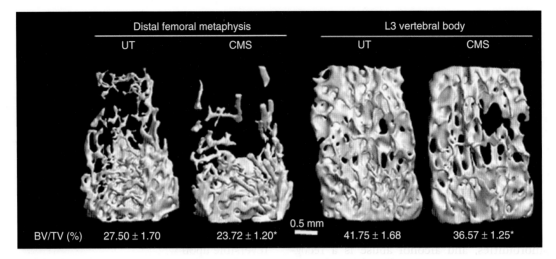

Fig. 120.1. Depression-induced structural impairment of the skeleton in mice exposed to chronic mild stress (CMS) for 4 weeks or left untreated (UT). μCT analysis. National Academy of Sciences, USA, copyright 2006 (Ref. 26).

several studies demonstrating depression-associated low BMD, the levels of endocrine factors believed to affect BMD do not differ between depressed and nondepressed subjects. These factors include serum 25-hydroxyvitamin D, parathyroid hormone, free T3, insulin-like growth factor I, and thyroid stimulating hormone (reviewed in Ref. 19). It is therefore unlikely that these variables are involved in the depression-associated low BMD.

Antidepressant therapy, especially by selective serotonin reuptake inhibitors (SSRIs), could be also a confounding variable affecting the above meta-analyses (discussed below). However, studies using antidepressant therapy as a covariate found no evidence to its effect on the association between MDD and BMD. It thus appears that the association between MDD and BMD is independent of the deleterious effects of antidepressants on the skeleton.

Taken together, these findings imply that all MDD patients are at risk for developing osteoporosis, with depressed women, particularly those who are premenopausal, showing the highest risk. These patients should be periodically evaluated for progression of bone loss and imbalances in bone remodeling and treated to prevent the progression of skeletal deterioration.

Causal relationship between depression and low bone mass

In the early 1980s, osteoporosis researchers suggested that depression was one of the major negative consequences of bone loss and osteoporotic fractures. They believed that osteoporosis occurred first, leading to a reactive depression. A similar, but distinct psychiatry literature reported that low BMD appeared to be an undesirable consequence of depression (reviewed in Ref. 22). The perspective that osteoporosis causes depression argues that the latter results from the pain and discomfort associated with osteoporotic fractures. The other approach, that depression is the causal process, claims that most studies demonstrate an association of depression with low BMD rather than with increased fracture rate. That depression is the causal attribute has been further proposed based on the well-established depression-induced increases in glucocorticoids and NE [23], agents also known to suppress bone formation and bone mass [24, 25]. In support of the latter concept, loss of bone mass and architecture has been demonstrated in mice with chronic mild stress (CMS), an established rodent model for depression (Fig. 120.1) [26]. The bone loss in this model results mainly from decreased bone formation. Both the reduced bone formation and the trabecular bone loss were apparently generalized as they were recorded in both the distal femoral metaphysis and lumber vertebral bodies. These skeletal deficits as well as the depressive symptoms (reduced sucrose preference and social exploration) could be prevented by the antidepressant drug imipramine [26]. The depressive-like state was associated with increased NE levels in bone and elevated serum corticosterone. Furthermore, the CMS-induced bone loss, but not the depressive-like state, could be prevented by the β-adrenergic antagonist, propranolol, portraying bone sympathetic innervation as a brain-to-bone pathway communicating depressive signals to the skeleton. In spite of the elevated serum corticosterone in mice subjected to CMS, the role of the hypothalamic–pituitary–adrenal (HPA) axis in CMS-induced bone loss is still unraveled, since it is unclear whether this elevation is sufficient to cause a negative bone remodeling balance and bone loss. Interestingly, although leptin has been implicated in both depression and the regulation of bone remodeling [27, 28], no relationship between leptin serum levels and CMS-induced bone loss could be established [26].

The adrenergic system and HPA axis are the most studied pathways mediating depressive signals from the CNS to the periphery. However, several other systems, implicated in both depression and low bone mass, could be involved in this process, such as the endocannabinoid system and inflammatory cytokines such as IL-1, IL-6, and tumor necrosis factor-α (TNF-α) (reviewed in Ref. 18). In addition, dietary and behavioral patterns commonly observed in psychiatric patients may also contribute to the pathogenesis of low bone mass. Another contributing factor could be cigarette smoking, which is more common within psychiatric populations. It increases the risk for the onset of MDD and has repeatedly been shown to negatively influence bone mass in cross-sectional studies of both men and women. Likewise, depression and excessive alcohol consumption are common comorbidities, and alcohol abuse is a recognized risk factor for osteoporosis. Finally, although with a less well-defined cause–effect relationship, changes in food consumption are typically associated with depression, and certain nutrients reported to be deficient in patients with MDD are required for the maintenance of good bone health (reviewed in Ref. 18).

ANOREXIA NERVOSA

Anorexia nervosa (AN) is an eating disorder characterized by refusal to maintain a healthy body weight and an obsessive fear of gaining weight. It is often coupled with a distorted self-image and associated cognitive biases that modify the body self-perception [29, 30]. The average caloric intake in AN patients is 600–800 calories per day, but complete self-starvation is also known [31]. AN patients exhibit a high incidence of a spectrum of comorbidities such as prolonged amenorrhea, hair loss and/or thinning, and depression [11]. AN presents the highest mortality rate of any psychiatric disorder. It usually begins in adolescence with the number of female patients being tenfold higher compared to males [32]. AN can occur at any age. Noticeably, 0.2–4% of the affected individuals are adolescent girls and college-aged women [33].

AN is highly heritable with multiple polymorphisms in more than 40 genes involved in eating behavior, motivation and reward, personality traits, and emotion. Some of these polymorphisms have been identified in genes related to skeletal remodeling such as brain-derived neurotrophic factor and norepinephrine transporter. In addition, an abnormal brain response to the anorexic effects of estrogen and serotonin dysregulation has been reported [34, 35]. Also, autoantibodies for melanocortin, a neuropeptide implicated in personality traits associated with eating disorders and the control of bone mass, have been demonstrated in AN [36]. However, a causal relationship between these genetic and molecular changes and AN has not been established.

Low bone density is a serious complication in nearly all AN patients, with almost half of them presenting osteoporosis. It results from decelerated bone mass accrual as well as loss of established bone mass and is accompanied by structural changes such as reduced trabecular and cortical thicknesses, decreased trabecular number and greater trabecular separation (reviewed in Ref. 37). Measurement of serum bone remodeling markers suggests that lower rate of bone mass accrual occurs consequent to a low bone turnover, whereas adult AN patients present an absolute decrease in bone formation with increased bone resorption [38, 39]. Studies carried out so far do not point to impaired neuronal or neuroendocrine activity as the primary cause for the poor skeletal status in AN. Rather, their low BMD is a consequence of reduced muscle mass, hypogonadism, low insulin-like growth factor-1 (IGF-1), hypercortisolemia and profound alterations in energy metabolism. It is only partially reversible upon weight regain [37]. In the face of the vast neuroendocrine complications, the only remedy that can be currently offered to improve the skeletal status in AN is rescue of weight loss and menstrual function.

SCHIZOPHRENIA

Schizophrenia is a mental disorder characterized by disintegration of thinking and emotional responsiveness. It is commonly manifested as auditory hallucinations, paranoid or bizarre delusions, and/or disorganized speech and thinking, and is associated with significant social and occupational dysfunction. The onset of symptoms typically occurs in adolescence or young adulthood, with a global lifetime prevalence of about 0.5% [40]. Although impaired neurodevelopment and several neurodegenerative processes have been implicated in the etiology and pathogenesis of schizophrenia, its causes are poorly understood.

Most of the studies assessing the possible association of schizophrenia and osteoporosis report a positive relationship between schizophrenia, impaired bone metabolism, low BMD, and increased prevalence of osteoporotic fractures [41, 42]. In the absence of definitive organic pathogenic processes underlying schizophrenia, this relationship has been attributed to schizophrenia-associated factors such as hypogonadism, undernutrition, smoking, and polydipsia [41]. Of these factors, hyperprolactinemia, induced by neuroleptic medications (see below) and the resultant hypogonadism have received much attention. However, this approach has been recently challenged by studies from the Psychiatric Department of Yuli Veterans Hospital in Taiwan. These studies compared 965 adult schizophrenia patients and 405 nonschizophrenic community members living in the same district. They report that young male and female schizophrenia patients already have lower bone density than nonschizophrenics at the age of 20, consistent with impaired bone mass accrual prior to that age. However, the schizophrenia patients did not exhibit age-related bone loss, with female patients even showing some gain. The bone density in schizophrenics older than 60 years was higher compared to the normal sample [43]. Interestingly, such a "protection" against age-related bone loss may result from mutations at the Dgcr8 gene involved in the 22q11.2 microdeletions shown in schizophrenia patients [44], as silencing of Dgcr8 leads to decreased osteoclastogenesis and a mild increase in bone mass [45]. Analyzing subsamples of schizophrenia patients with a low bone density (T-score equal to or lower than −2.5) and gender-matched patients with normal density, the Yuli study further reports no significant association between reduced bone density and serum prolactin or determinants involved in bone metabolism such as calcium, phosphate, osteocalcin, collagen type I N-terminal crosslinks (NTX), thyroid-stimulating hormone, and parathyroid hormone levels. In addition, there was no association between the bone density measurements and the type of antipsychotics or the menstruation status in the female patients [46]. Taken together, it appears that the changes in skeletal status reported in schizophrenia patients may be linked to comorbidities and behavioral patterns as well as schizophrenia-induced impairment of brain-to-bone communication.

TRAUMATIC BRAIN INJURY

Intracranial traumatic brain injury (TBI) occurs when an external force traumatically injures the brain. The trauma is caused by a direct impact or by acceleration alone and can involve damage to structures other than the brain, such as the scalp and skull. In addition to the damage caused at the time of injury, a sequence of events take place in the minutes-to-weeks following the injury. These pathological processes, which include alterations in cerebral blood flow and increased intracranial pressure, contribute substantially to the damage from the initial injury and often lead to an array of physical, cognitive, social, emotional, and behavioral effects. The outcome can range from complete recovery to permanent disability or death.

Many TBI patients show increased bone formation manifested mainly as heterotopic ossification (HO) and accelerated fracture healing. The reported incidence of TBI-induced HO is 15–40%, and it appears that the occurrence of symptomatic HO is correlated with the severity of TBI [47]. TBI-induced HO can cause pain and reduced range of motion of joints, which in 8–10% of cases results in severe functional limitation and even complete bony ankylosis of joints in 5% of cases [48].

There is clinical and experimental circumstantial evidence that the TBI-induced systemic increase in bone formation is associated with the release of osteogenic factors from the brain into the cerebrospinal fluid and blood circulation. These factors include fibroblast growth factor, growth hormone, IGF-1, parathyroid hormone, and bone morphogenetic proteins (reviewed in Ref. [47]). However, so far a causal relationship between the serum levels of any of these factors and TBI stimulated bone formation has not been demonstrated.

There is also evidence that the TBI-generated central signals are communicated to the skeleton by the SNS, thus stimulating bone formation. Studies in mice have shown a decrease in the inhibitory sympathetic tone of bone formation within hours after TBI. It has been further demonstrated that the TBI-induced decrease of this sympathetic tone is mediated by increased bone ECs levels, leading to super activation of CB1 cannabinoid receptors expressed in the skeletal sympathetic nerve terminals, which in turn inhibits NE release [3]. Therefore, cannabinoid therapy could be effective for the prevention of TBI-induced HO.

ALZHEIMER'S DISEASE

Dementia syndromes affect 10% of the population aged over 65. Alzheimer's disease (AD) is the commonest cause of dementia. It accounts for 65% of the dementia cases either alone or in combination with vascular dementia [49]. Although the etiology of AD is essentially unknown, the disease presents progressive neuronal loss associated with a number of molecular changes in the brain, including reduced acetylcholine levels [50], mutations in the amyloid beta (Aβ) precursor protein (APP) gene, and the consequent accumulation of Aβ deposits as well as soluble Aβ oligomers. Also, hyperphosphorylation of tau protein leads to the formation of intracellular neurofibrillary tangles inside nerve cell bodies [51]. In addition, autophagy due to valosin-containing protein (VCP) gene mutations and interference to proteosomal degradation has been suggested to cause neuronal apoptosis leading to frontotemporal dementia, a major feature of AD [52].

Like other forms of dementia, AD is usually diagnosed in patients older than 65. Although osteoporosis can be observed by the age of 50, all its consequences become apparent from the seventh decade and onward [49]. Expectedly, this circumstantial relationship results in a significant correlation between the two conditions. This correlation may be also enhanced by common risk factors shared by the two diseases, such as low body mass, nutritional deficiencies (mainly vitamin K and vitamin D deficiencies), less exposure to sunlight, and reduced physical activity. That this relationship may not be merely circumstantial is suggested by (1) the higher incidence of osteoporosis in AD patients compared to subjects in the same age group without memory impairment (reviewed in Ref. 50) and (2) a similar incidence of osteoporotic fractures in women and men with AD as compared to several-fold higher incidence of fractures in females in the general population [53]. Importantly, the treatment and rehabilitation in postfracture AD patients are more difficult than in non-AD patients, with less than half of the AD patient regaining pre-fracture functional status. Also, patients with dementia fall more frequently, thus displaying an increased incidence of femoral neck fractures [54].

Although only 5% of patients with AD have familial Alzheimer's disease (FAD) [6], the discovery of genes for FAD facilitated the generation of transgenic mice that express these genes, mainly mutated human APP. These transgenic mouse models reproduce many critical aspects of AD, mainly $A\beta_{1-40}$ and $A\beta_{1-42}$ plaque formation in the brain, including cerebrovascular $A\beta$ deposition (reviewed in Refs. 55 and 56). Accumulating evidence from these mice suggests that intraneuronal $A\beta_{1-42}$ triggers early neuronal loss and synaptic deficits [57]. Alterations in synaptic integrity is one of the earliest events in AD pathogenesis, which precede neuronal loss.[58]

A couple of recent reports in mice expressing *APPswe*, the Swedish APP mutation, demonstrate a possible causal relationship between AD and bone loss. One of these studies demonstrated a biphasic effect on osteoclastogenesis with an increase in mice younger than 4 months old and a decrease in older animals. The increase in the young mice appears to be mediated by $A\beta$ oligomers and receptors for advanced glycation end products (RAGE) in bone marrow macrophages (BMMs), whereas the decrease of osteoclast formation and activity in the older mice may be due to an increase in soluble RAGE, an inhibitor of RANKL-induced osteoclastogenesis [59]. The second study used animals coexpressing APPswe and PS1 (encoding a human presenilin 1 mutation). These mice showed an accelerated age-related bone loss [60]. While either study suggests that $A\beta$ can stimulate bone loss, it is still unknown whether $A\beta$ is produced in bone in AD patients (or otherwise) or dysregulates central mechanisms that control bone remodeling.

NEUROPSYCHIATRIC MEDICATIONS

A number of first-line agents used in the treatment of neuropsychiatric disorders are associated with low BMD and increased fracture risk, potentially amplifying the negative impact of these disorders on the skeleton. These drugs include mainly the SSRI family of antidepressants and tranquilizing psychiatric (or neuroleptic) medications.

SSRIs

Serotonin [5-hydroxytryptamine (5-HT)] is a biogenic monoamine neurotransmitter. Like many of these amines, its synaptic and extracellular concentrations are regulated by a transporter (5-HTT) acting by high-affinity 5-HT reuptake from the extracellular to the intracellular milieu. The transporter 5-HTT is a major target for a family of antidepressants that inhibit its activity, thereby potentiating the effect of 5-HT. The skeletal action of these drugs, called SSRIs, has attracted substantial interest because of its potential impact on osteoporosis and resultant fractures.

Serotonin (5-HT) is best known for its role in the CNS, gastrointestinal (GI) tract, and cardiovascular (CV) system. In the CNS, it is produced by presynaptic neurons and is released into the synaptic gap, leading to activation of pre- and postsynaptic 5-HT receptors, thus influencing a myriad of behavioral, physiological, and cognitive functions [61, 62]. In the GI tract, 5-HT is synthesized and secreted by enterochromaffin cells and diffuses to enteric nerve endings to stimulate peristalsis [63, 64]. In both the CNS and GI tract, the duration and intensity of the serotonergic activity is enhanced by the sodium chloride-dependent 5-HTT [65, 66]. In the CV system, 5-HT is taken up primarily by platelets via 5-HTT and stored in dense granules [67]. It is released by activated platelets and induces blood vessel constriction or dilation,[68] and smooth muscle cell hypertrophy and hyperplasia.[69]

Osteoblasts, osteocytes and osteoclasts express functional 5-HT receptors and 5-HTT (reviewed in Ref. 70). In osteoblasts, 5-HT receptor agonists influence cell proliferation, potentiate the parathyroid-hormone-induced increase in AP-1 activity and modulate the cellular response to mechanical stimulation. In osteocytes, 5-HT increases whole-cell cyclic adenosine monophosphate (cAMP) and PGE2 levels, which are also involved in the transduction of mechanical stimuli [71]. In osteoclasts, 5-HT and 5-HTT have been shown to affect differentiation but not activity [70].

What is the source of 5-HT in bone tissue? The CNS does not appear to be a likely source of 5-HT available to bone cells, as the blood–brain barrier is impermeable to 5-HT and serotonergic innervation has not yet been demonstrated in the skeleton. As in the case of other neurotransmitters, such as ECs [3], 5-HT could be synthesized and released by bone cells and act in an autocrine/paracrine manner. Indeed, mRNA transcripts for tryptophan hydroxylase-1 (Tph1), a rate-limiting enzyme in 5-HT synthesis, have been detected in osteoblast and osteocyte cell lines [72]. Most of an organism's 5-HT is produced in the GI tract and stored in dense granules in platelets. Because 5-HT from this source is released only upon platelet activation [67], it is an unlikely activator of bone cell 5-HT receptors. However, a small fraction of the GI-derived 5-HT remains in the serum and serum 5-HT has been suggested as a negative regulator of osteoblast proliferation, bone formation and bone mass.[73, 74]

What is the physiologic role of the skeletal serotonergic system? The diversity of actions of 5-HT results from the occurrence of multiple 5-HTRs, which are divided into seven classes based on their signaling pathways [75]. Of these, only $5\text{-HT}_{1A}R$, $5\text{-HT}_{2A}R$, $5\text{-HT}_{1B}R$, and $5\text{-HT}_{2B}R$ are expressed in osteoblasts, and only the expression of $5\text{-HT}_{2B}R$ is increased during osteoblast differentiation. Mice deficient in $5\text{-HT}_{2B}R$ have accelerated age-related, low turnover bone loss, secondary to impaired osteoblast recruitment and proliferation [76]. In line with these findings, rats treated with 5-HT have increased BMD [77]. In contrast, mice deficient in osteoblastic $5\text{-HT}_{1B}R$

have a high bone mass phenotype, secondary to increases in osteoblast number and bone formation [74]. Disruption of the 5-HTT gene or pharmacological inhibition of 5-HTT by SSRIs leads to a low bone mass phenotype in growing mice [70]. These findings suggest that 5-HT has different age-dependent effects that inhibit peak bone mass accrual in the growing skeleton and keep bone remodeling and bone mass in balance in the adult skeleton. However, this explanation is challenged by the deleterious effects of SSRIs, both on trabecular and cortical bone, in adult mice [78]. These apparently paradoxical data may be explained by the osteoblast stimulatory effect of central 5-HT signaling, which by activating hypothalamic $HT_{2C}R$ alleviates the inhibitory sympathetic tone of bone formation [7].

In humans, SSRIs have emerged as popular drugs in the treatment of mood disorders. Most clinical studies report that antidepressants in general, but mainly SSRIs, are associated with low BMD and a dose-dependent increase in the risk of fractures and low bone mass in children (reviewed in Ref. 79). The reason for these deleterious effects may be linked to dysregulation of the balance between the skeletal, gastrointestinal, and central serotonergic systems, especially after prolonged administration [80].

Neuroleptics

Neuroleptic medications are classified as *typical antipsychotics* that were discovered in the 1950s, and the more recently developed second generation known as *atypical antipsychotics*. Either class of neuroleptics, but mainly the typical antipsychotics, includes members that adversely induce hyperprolactinemia and, consequently, bone loss.

Prolactin is produced by lactotroph cells in the anterior pituitary. Prolactin production is controlled by the dopaminergic and serotonergic systems. The predominant factor is dopamine that tonically inhibits prolactin production by acting on D_2 receptors on lactotrophs, thus restraining prolactin gene transcription. In contrast, 5-HT stimulates prolactin secretion by activating $5HT_{1A}R$ and $5HT_2R$ [81]. Both typical and atypical neuroleptics tend to antagonize receptors in the brain dopamine pathways, including differential inhibition of the D_2 and 5-HT receptors, and cause a dose-dependent increase in serum prolactin levels (reviewed in Ref. 41).

While the major function of prolactin is to promote the development of mammary epithelium and lactation, its clinical significance is related to the reproductive system and the skeleton, as sustained hyperprolactinemia leads to suppression of gonadotropin-releasing hormone (GnRH) secretion, inhibition of LH and FSH, and decreased secretion of estradiol and testosterone. Indeed, hyperprolactinemia-induced bone loss, frequently occurring during pregnancy and lactation, has been mainly attributed to sex steroid deficiency. The severity of bone loss is related to the level and duration of hyperprolactinemia, which impacts on the skeleton of female and male adults and adolescents (reviewed in Ref. 82). In addition, prolactin may affect the skeleton directly through prolactin receptors expressed in osteoblasts and their precursors. *In vitro* studies using differentiating and mature osteoblastic cells suggest that the osteoblastic processes affected depend on the level of hyperprolactinemia. The prolactin concentration typically recorded during lactation (100 ng/ml) inhibits preosteoblast proliferation, osteoblast number, and production of a mineralized extracellular matrix. This prolactin concentration stimulates RUNX2 and alkaline phosphatase expression in early stage osteoblast differentiation and inhibits the same genes in late stages [83]. Higher concentrations (up to 500 ng/ml), such as those often induced by neuroleptics, increase the osteoblastic expression of osteoclastogenic factors such as RANKL, MCP-1, Cox-2, TNF-α, IL-1, and ephrin-B1 [84]. Hence, multiple pathways may communicate the negative impact of neuroleptic medications and hyperprolactinemia on the skeleton.

SUMMARY

The association between neuropsychiatric and skeletal disorders is emerging as an important theme in bone and psychiatric pathophysiology. Comorbidities shared by the two disciplines appear to impact millions of patients of both sexes and all ages, and present a wealth of unanswered basic mechanistic questions and unresolved clinical issues. Perhaps the most urgent of these issues is the conflict between the beneficial effects of psychiatric medications, particularly in mood disorders and schizophrenia, and the deleterious effects of the very same medications on the skeleton.

REFERENCES

1. Imai S, Matsusue Y. 2002. Neuronal regulation of bone metabolism and anabolism: Calcitonin gene-related peptide-, substance P-, and tyrosine hydroxylase-containing nerves and the bone. *Microsc Res Tech* 58: 61–69.
2. Sato T, Abe T, Chida D, Nakamoto N, Hori N, Kokabu S, Sakata Y, Tomaru Y, Iwata T, Usui M, Aiko K, Yoda T. 2010. Functional role of acetylcholine and the expression of cholinergic receptors and components in osteoblasts. *FEBS Lett* 584: 817–824.
3. Tam J, Trembovler V, Di Marzo V, Petrosino S, Leo G, Alexandrovich A, Regev E, Casap N, Shteyer A, Ledent C, Karsak M, Zimmer A, Mechoulam R, Yirmiya R, Shohami E, Bab I. 2008. The cannabinoid CB1 receptor regulates bone formation by modulating adrenergic signaling. *FASEB J* 22: 285–294.
4. Bab I, Ofek O, Tam J, Rehnelt J, Zimmer A. 2008. Endocannabinoids and the regulation of bone metabolism. *J Neuroendocrinol* 20 Suppl 1: 69–74.

5. Franquinho F, Liz MA, Nunes AF, Neto E, Lamghari M, Sousa MM. 2010. Neuropeptide Y and osteoblast differentiation—The balance between the neuro-osteogenic network and local control. *FEBS J* 277: 3664–3674.
6. Bajayo A, Goshen I, Feldman S, Csernus V, Iverfeldt K, Shohami E, Yirmiya R, Bab I. 2005. Central IL-1 receptor signaling regulates bone growth and mass. *Proc Natl Acad Sci U S A* 102: 12956–12961.
7. Yadav VK, Oury F, Suda N, Liu ZW, Gao XB, Confavreux C, Klemenhagen KC, Tanaka KF, Gingrich JA, Guo XE, Tecott LH, Mann JJ, Hen R, Horvath TL, Karsenty G. 2009. A serotonin-dependent mechanism explains the leptin regulation of bone mass, appetite, and energy expenditure. *Cell* 138: 976–989.
8. Takeda S, Elefteriou F, Levasseur R, Liu X, Zhao L, Parker KL, Armstrong D, Ducy P, Karsenty G. 2002. Leptin regulates bone formation via the sympathetic nervous system. *Cell* 111: 305–317.
9. Elefteriou F, Ahn JD, Takeda S, Starbuck M, Yang X, Liu X, Kondo H, Richards WG, Bannon TW, Noda M, Clement K, Vaisse C, Karsenty G. 2005. Leptin regulation of bone resorption by the sympathetic nervous system and CART. *Nature* 434: 514–520.
10. Shi Y, Oury F, Yadav VK, Wess J, Liu XS, Guo XE, Murshed M, Karsenty G. 2010. Signaling through the M(3) muscarinic receptor favors bone mass accrual by decreasing sympathetic activity. *Cell Metab* 11: 231–238.
11. American Psychiatric Association. 2000. *Diagnostic and Statistical Manual of Mental Disorders, Fourth Edition, Text Revision: DSM-IV-TR.* Washington, DC: American Psychiatric Publishing, Inc.
12. Barlow DH. 2005. *Abnormal Psychology: An Integrative Approach (5th Ed.).* Belmont, CA: Thomson Wadsworth.
13. Levinson DF. 2006. The genetics of depression: A review. *Biol Psychiatry* 60: 84–92.
14. Goshen I, Kreisel T, Ben-Menachem-Zidon O, Licht T, Weidenfeld J, Ben-Hur T, Yirmiya R. 2008. Brain interleukin-1 mediates chronic stress-induced depression in mice via adrenocortical activation and hippocampal neurogenesis suppression. *Mol Psychiatry* 13: 717–728.
15. Murray CJ, Lopez AD. 1997. Alternative projections of mortality and disability by cause 1990–2020: Global Burden of Disease Study. *Lancet* 349: 1498–1504.
16. Riggs BL, Khosla S, Melton LJ 3rd. 2002, Sex steroids and the construction and conservation of the adult skeleton. *Endocr Rev* 23: 279–302.
17. Williams LJ, Bjerkeset O, Langhammer A, Berk M, Pasco JA, Henry MJ, Schei B, Forsmo S. 2011. The association between depressive and anxiety symptoms and bone mineral density in the general population: The HUNT Study. *J Affect Disord* 131:164–171.
18. Bab I, Yirmiya R. 2010. Depression, selective serotonin reuptake inhibitors, and osteoporosis. *Curr Osteoporos Rep* 8: 185–191.
19. Yirmiya R, Bab I. 2009. Major depression is a risk factor for low bone mineral density: A meta-analysis. *Biol Psychiatry* 66: 423–432.
20. Wu Q, Magnus JH, Liu J, Bencaz AF, Hentz JG. 2009. Depression and low bone mineral density: A meta analysis of epidemiologic studies. *Osteoporos Int* 20: 1309–1320.
21. Dorn LD, Susman EJ, Pabst S, Huang B, Kalkwarf H, Grimes S. 2008. Association of depressive symptoms and anxiety with bone mass and density in ever-smoking and never-smoking adolescent girls. *Arch Pediatr Adolesc Med* 162: 1181–1188.
22. Gold DT, Solimeo S. 2006. Osteoporosis and depression: A historical perspective. *Curr Osteoporos Rep* 4: 134–139.
23. Ilias I, Alesci S, Gold PW, Chrousos GP. 2006. Depression and osteoporosis in men: Association or causal link? *Hormones (Athens)* 5: 9–16.
24. Weinstein RS, Jilka RL, Parfitt AM, Manolagas SC. 1998. Inhibition of osteoblastogenesis and promotion of apoptosis of osteoblasts and osteocytes by glucocorticoids. Potential mechanisms of their deleterious effects on bone. *J Clin Invest* 102: 274–282.
25. Elefteriou F. 2008. Regulation of bone remodeling by the central and peripheral nervous system. *Arch Biochem Biophys* 473: 231–236.
26. Yirmiya R, Goshen I, Bajayo A, Kreisel T, Feldman S, Tam J, Trembovler V, Csernus V, Shohami E, Bab I. 2006. Depression induces bone loss through stimulation of the sympathetic nervous system. *Proc Natl Acad Sci U S A* 103: 16876–16881.
27. Lu XY. 2007. The leptin hypothesis of depression: A potential link between mood disorders and obesity? *Curr Opin Pharmacol* 7: 648–652.
28. Karsenty G. 2006. Convergence between bone and energy homeostases: Leptin regulation of bone mass. *Cell Metab* 4: 341–348.
29. Cooper MJ. 2005. Cognitive theory in anorexia nervosa and bulimia nervosa: Progress, development and future directions. *Clin Psychol Rev* 25: 511–531.
30. Brooks S, Prince A, Stahl D, Campell IC, Treasure J. 2010. A systematic review and meta-analysis of cognitive bias to food stimuli in people with disordered eating behaviour. *Clin Psychol* 31: 37–51.
31. Frude, N. 1998. *Understanding Abnormal Psychology.* Oxford: Blackwell Publishing.
32. Attia E. 2010. Anorexia nervosa: Current status and future directions. *Ann Rev Med* 61: 425–435.
33. Lucas AR, Beard CM, O'Fallon WM, Kurland LT. 1991. 50-year trends in the incidence of anorexia nervosa in Rochester, Minn.: A population-based study. *Am J Psychiatry* 148: 917–922.
34. Rask-Andersen M, Olszewski PK, Levine AS, Schiöth HB. 2009. Molecular mechanisms underlying anorexia nervosa: Focus on human gene association studies and systems controlling food intake. *Brain Res Rev* 62: 147–164.
35. Yamashiro T, Fukunaga T, Yamashita K, Kobashi N, Takano-Yamamoto T. 2001. Gene and protein expression of brain-derived neurotrophic factor and TrkB in bone and cartilage. *Bone* 28: 404–409.
36. Ahn JD, Dubern B, Lubrano-Berthelier C, Clement K, Karsenty G. 2006. Cart overexpression is the only

identifiable cause of high bone mass in melanocortin 4 receptor deficiency. *Endocrinology* 147: 3196–3202.
37. Misra M, Klibanski A. 2011. The neuroendocrine basis of anorexia nervosa and its impact on bone metabolism. *Neuroendocrinology* 93: 65–73.
38. Grinspoon S, Baum H, Lee K, Anderson E, Herzog D, Klibanski A. 1996. Effects of short-term recombinant human insulin-like growth factor I administration on bone turnover in osteopenic women with anorexia nervosa. *J Clin Endocrinol Metab* 81: 3864–3870.
39. Soyka L, Misra M, Frenchman A, Miller K, Grinspoon S, Schoenfeld D, Klibanski A. 2002. Abnormal bone mineral accrual in adolescent girls with anorexia nervosa. *J Clin Endocrinol Metab* 87: 4177–4185.
40. van Os J, Kapur S. 2009. Schizophrenia. *Lancet* 374: 635–645.
41. Misra M, Papakostas GI, Klibanski A. 2004. Effects of psychiatric disorders and psychotropic medications on prolactin and bone metabolism. *J Clin Psychiatry* 65: 1607–1618.
42. Abraham G, Friedman RH, Verghese C, de Leon J. 1995. Osteoporosis and schizophrenia: Can we limit known risk factors? *Biol Psychiatry* 38: 131–132.
43. Renn JH, Yang NP, Chueh CM, Lin CY, Lan TH, Chou P. 2009. Bone mass in schizophrenia and normal populations across different decades of life. *BMC Musculoskelet Disord* 10: 1.
44. Stark KL, Xu B, Bagchi A, Lai WS, Liu H, Hsu R, Wan X, Pavlidis P, Mills AA, Karayiorgou M, Gogos JA. 2008. Altered brain microRNA biogenesis contributes to phenotypic deficits in a 22q11-deletion mouse model. *Nat Genet* 40: 751–760.
45. Sugatani T, Hruska KA. 2009. Impaired micro-RNA pathways diminish osteoclast differentiation and function. *J Biol Chem* 284: 4667–4678.
46. Renn JH, Yang NP, Chou P. 2010. Effects of plasma magnesium and prolactin on quantitative ultrasound measurements of heel bone among schizophrenic patients. *BMC Musculoskelet Disord* 11: 35.
47. Toffoli AM, Gautschi OP, Frey SP, Filgueira L, Zellweger R. 2008. From brain to bone: Evidence for the release of osteogenic humoral factors after traumatic brain injury. *Brain Injury* 22: 511–518.
48. Subbarao JV, Garrison SJ. 1999. Heterotopic ossification: Diagnosis and management, current concepts and controversies. *J Spinal Cord Med* 22: 273–283.
49. Tysiewicz-Dudek M, Pietraszkiewicz F, Drozdzowska B. 2008. Alzheimer's disease and osteoporosis: Common risk factors or one condition predisposing to the other? [Article in English, Polish] *Ortop Traumatol Rehabil* 10: 315–332.
50. Francis PT, Palmer AM, Snape M, Wilcock GK. 1999. The cholinergic hypothesis of Alzheimer's disease: A review of progress. *J Neurol Neurosurg Psychiatr* 66: 137–147.
51. Polvikoski T, Sulkava R, Haltia M, Kainulainen K, Vuorio A, Verkkoniemi A, Niinistö L, Halonen P, Kontula K. 1995. Apolipoprotein E, dementia, and cortical deposition of beta-amyloid protein. *N Engl J Med* 333: 1242–1247.
52. Watts GD, Thomasova D, Ramdeen SK, Fulchiero EC, Mehta SG, Drachman DA, Weihl CC, Jamrozik Z, Kwiecinski H, Kaminska A, Kimonis VE. 2007. Novel VCP mutations in inclusion body myopathy associated with Paget disease of bone and frontotemporal dementia. *Clin Genet* 72: 420–426.
53. Cumming RG, Nevitt MC, Cumming RR. 1997. Epidemiology of hip fractures. *Epidemiol Rev* 19: 244–257.
54. Myers AH, Young Y, Langlois JA. 1996. Prevention of falls in the elderly. *Bone* 18: 87S–101S.
55. Philipson O, Lord A, Gumucio A, O'Callaghan P, Lannfelt L, Nilsson LN. 2010. Animal models of amyloid-beta-related pathologies in Alzheimer's disease. *FEBS J* 277: 1389–1409.
56. Schaeffer DL, Figueiro M, Gattaz WF. 2011. Insights into Alzheimer disease pathogenesis from studies in transgenic animal models. *Clinics* 66: 45–54.
57. Oddo S, Caccamo A, Shepherd JD, Murphy MP, Golde TE, Kayed R, Metherate R, Mattson MP, Akbari Y, LaFerla FM. 2003. Triple-transgenic model of Alzheimer's disease with plaques and tangles: Intracellular Abeta and synaptic dysfunction. *Neuron* 39: 409–421
58. Rutten BP, Van der Kolk NM, Schafer S, van Zandvoort MA, Bayer TA, Steinbusch HW, Schmitz C. 2005. Age-related loss of synaptophysin immunoreactive presynaptic boutons within the hippocampus of APP751SL, PS1M146L, and APP751SL/PS1M146L transgenic mice. *Am J Pathol* 167: 161–173.
59. Cui S, Xiong F, Hong Y, Jung JU, Li XS, Liu JZ, Yan R, Mei L, Feng X, Xiong WC. 2011. APPswe/Aβ regulation of osteoclast activation and RAGE expression in an age-dependent manner. *J Bone Min Res* 26: 1084–1098.
60. Yang MW, Wang TH, Yan PP, Chu LW, Yu J, Gao ZD, Li YZ, Guo BL. 2011. Curcumin improves bone microarchitecture and enhances mineral density in APP/PS1 transgenic mice. *Phytomedicine* 18: 205–213.
61. Kroeze WK, Kristiansen K, Roth BL. 2002. Molecular biology of serotonin receptors structure and function at the molecular level. *Curr Top Med Chem* 2: 507–528.
62. Raymond JR, Mukhin YV, Gelasco A, Turner J, Collinsworth G, Gettys TW, Grewal JS, Garnovskaya MN. 2001. Multiplicity of mechanisms of serotonin receptor signal transduction. *Pharmacol Ther* 92: 179–212.
63. Gershon MD. 2005. Nerves, reflexes, and the enteric nervous system: Pathogenesis of the irritable bowel syndrome. *J Clin Gastroenterol* 39(5 Suppl 3): S184–S193.
64. Talley NJ. 2001. Serotoninergic neuroenteric modulators. *Lancet* 358: 2061–2068.
65. Murphy DL, Lerner A, Rudnick G, Lesch KP. 2004. Serotonin transporter: Gene, genetic disorders, and pharmacogenetics. *Mol Interv* 4: 109–112.
66. Wade PR, Chen J, Jaffe B, Kassem IS, Blakely RD, Gershon MD. 1996. Localization and function of a 5-HT transporter in crypt epithelia of the gastrointestinal tract. *J Neurosci* 16: 2352–2364.

67. McNicol A, Israels SJ. 1999. Platelet dense granules: Structure, function and implications for haemostasis. *Thromb Res* 95: 1–18.
68. Egermayer P, Town GI, Peacock AJ. 1999. Role of serotonin in the pathogenesis of acute and chronic pulmonary hypertension. *Thorax* 54: 161–168.
69. Lee SL, Wang WW, Lanzillo JJ, Fanburg BL. 1994. Serotonin produces both hyperplasia and hypertrophy of bovine pulmonary artery smooth muscle cells in culture. *Am J Physiol* 266(1 Pt 1): L46–L52.
70. Warden SJ, Bliziotes MM, Wiren KM, Eshleman AJ, Turner CH. 2005. Neural regulation of bone and the skeletal effects of serotonin (5-hydroxytryptamine). *Mol Cell Endocrinol* 242: 1–9.
71. Cherian PP, Cheng B, Gu S, Sprague E, Bonewald LF, Jiang JX. 2003. Effects of mechanical strain on the function of Gap junctions in osteocytes are mediated through the prostaglandin EP2 receptor. *J Biol Chem* 278: 43146–43156.
72. Bliziotes M, Eshleman A, Burt-Pichat B, Zhang XW, Hashimoto J, Wiren K, Chenu C. 2006. Serotonin transporter and receptor expression in osteocytic MLO-Y4 cells. *Bone* 39: 1313–1321.
73. Rand M, Reid G. 1951. Source of "serotonin" in serum. *Nature* 168: 385.
74. Yadav VK, Ryu JH, Suda N, Tanaka KF, Gingrich JA, Schütz G, Glorieux FH, Chiang CY, Zajac JD, Insogna KL, Mann JJ, Hen R, Ducy P, Karsenty G. 2008. Lrp5 controls bone formation by inhibiting serotonin synthesis in the duodenum. *Cell* 135: 825–837.
75. Hoyer D, Hannon JP, Martin GR. 2002. Molecular, pharmacological and functional diversity of 5-HT receptors. *Pharmacol Biochem Behav* 71: 533–554.
76. Collet C, Schiltz C, Geoffroy V, Maroteaux L, Launay JM, de Vernejoul MC. 2008. The serotonin 5-HT2B receptor controls bone mass via osteoblast recruitment and proliferation. *FASEB J* 22: 418–427.
77. Gustafsson BI, Westbroek I, Waarsing JH, Waldum H, Solligård E, Brunsvik A, Dimmen S, van Leeuwen JP, Weinans H, Syversen U. 2006. Long-term serotonin administration leads to higher bone mineral density, affects bone architecture, and leads to higher femoral bone stiffness in rats. *J Cell Biochem* 97: 1283–1291.
78. Bonnet N, Bernard P, Beaupied H, Bizot JC, Trovero F, Courteix D, Benhamou CL. 2007. Various effects of antidepressant drugs on bone microarchitectecture, mechanical properties and bone remodeling. *Toxicol Appl Pharmacol* 221: 111–118.
79. Williams LJ, Pasco JA, Jacka FN, Henry MJ, Dodd S, Berk M. 2008. Depression and bone metabolism. A review. *Psychother Psychosom* 78: 16–25.
80. Ziere G, Dieleman JP, van der Cammen TJ, Hofman A, Pols HA, Stricker BH. 2008. Selective serotonin reuptake inhibiting antidepressants are associated with an increased risk of nonvertebral fractures. *J Clin Psychopharmacol* 28: 411–417.
81. Durham RA, Johnson JD, Eaton MJ, Moore KE, Lookingland KJ. 1998. Opposing roles for dopamine D1 and D2 receptors in the regulation of hypothalamic tuberoinfundibular dopamine neurons. *Eur J Pharmacol* 355: 141–147.
82. Shibli-Rahhal A, Schlechte J. 2009. The effects of hyperprolactinemia on bone and fat. *Pituitary* 12: 96–104.
83. Seriwatanachai D, Krishnamra N, van Leeuwen JPTM. 2009. Evidence for direct effects of prolactin on human osteoblasts: Inhibition of cell growth and mineralization. *J Cellular Biochem* 107: 677–685.
84. Wongdee K, Tulalamba W, Thongbunchoo J, Krishnamra N, Charoenphandhu N. 2011. Prolactin alters the mRNA expression of osteoblast-derived osteoclastogenic factors in osteoblast-like UMR106 cells. *Mol Cell Biochem* 349: 195–204.

121
Vascular Disease and the Skeleton

Dwight A. Towler

Introduction 1012
Avascular Necrosis 1012
Atherosclerosis, Arteriosclerosis, and Skeletal Health 1014
Conclusions and Future Directions 1015
Acknowledgments 1015
References 1015

INTRODUCTION

Bone never forms without vascular interactions [1, 2]. This obvious statement of biological fact does not do justice to the physiological, clinical, and pharmacologic implications of the relationship. The vasculature provides the conduit for osteoprogenitor invasion of avascular cartilaginous templates during endochondral bone formation [2]. Moreover, the vasculature is the conduit for mineral exchange between bone and the rest of the body to acutely meet systemic demands on calcium phosphate homeostasis with feast and famine [3, 4]. The vasculature provides the sustentacular niche for the birth of osteoprogenitors [5]. Mature osteoblasts, in turn, establish the hematopoietic stem cell niche and control the egress of circulating myeloid and endothelial progenitor cells (ePCs) in concert with skeletal CD169+ macrophages and autonomic nerves [6]. Furthermore, vascular endothelial cells elaborate osteogenic morphogens such as BMP2 and Wnt7 that promote osteogenic differentiation of neighboring mesenchymal progenitors [7]. Osteoblasts reciprocally elaborate the expression of vascular endothelial growth factor (VEGF), the prototypic stimulus for angiogenesis and vascular and canalicular permeability in bone [8].

In this chapter, we very briefly relate the vital interplay between vascular biology and bone physiology as relevant to metabolic bone disease. We review data that highlight how prevalent vascular disease processes impair skeletal homeostatic mechanisms.

AVASCULAR NECROSIS

Avascular necrosis (AVN), or osteonecrosis, represents the best-recognized clinical disorder of vascular disease and the skeleton. A large number of risk factors and causes have been identified (Table 121.1) [9, 10]; these can be usefully viewed as reflecting bone ischemia from (a) traumatic, anatomic, or thromboembolic compromise of extrinsic macrovascular blood supply to bone or a bone segment; (b) inflammatory, metabolic, prothrombotic, infectious, or hyperviscosity states that compromise intrinsic microvascular function and bone perfusion; and (c) marrow infiltrative processes or venous thrombosis that increase intraosseous "back-pressure," thus decreasing the pressure gradient supporting marrow perfusion [10]. To be sure, in most clinical settings pathobiology is multifactorial. For example, glucocorticoid excess not only causes fatty marrow expansion but also reduces skeletal VEGF production [11]. Bone microvascular occlusion with sickle cell crisis engenders pain, marrow edema, and congestion that can also alter intraosseous perfusion [12]. AVN arising with rheumatoid arthritis or inflammatory bowel disease often, but not always, occurs in concert with glucocorticoid treatment. Like VEGF antagonists, amino-bisphosphonates exhibit potent angiostatic properties and suppress VEGF expression [13]. Moreover, osteoclast activity is important for skeletal angiogenic responses [14]. However, osteonecrosis of the jaw arising with amino-bisphosphonate use most often occurs in the setting of malignancy with attendant chemotherapeutics and/or radiotherapy [15].

Primer on the Metabolic Bone Diseases and Disorders of Mineral Metabolism, Eighth Edition. Edited by Clifford J. Rosen.
© 2013 American Society for Bone and Mineral Research. Published 2013 by John Wiley & Sons, Inc.

Table 121.1. Risk Factors for Avascular Necrosis/Osteonecrosis

Glucocorticoid therapy, Cushing's disease	Fracture (e.g., Garden type IV fracture of the hip)
Arterial thromboembolic disease	Joint dislocation with artery impingement
Sickle cell disease	Decompression sickness (caisson disease, the bends)
Excessive alcohol use, fatty marrow expansion	Gaucher disease
Radiation and chemotherapy	Bisphosphonate use in osteonecrosis of the jaw
Hyperviscosity syndromes (e.g., CML with leukostasis, Waldenström's macroglobulinemia)	Anti-VEGF signaling therapy (e.g., bevacizumab, sunitinib)
HAART* for HIV/AIDS	Hyperlipidemia (cholesterol, triglycerides) with or without HIV/AIDS
Hyperhomocysteinemia	Type 2 diabetes
Chronic renal insufficiency, transplantation	Idiopathic but anatomic (e.g., Legg–Calvé–Perthes disease of the femoral head; Kienböck's disease of the lunate)
Systemic lupus, rheumatoid arthritis, juvenile dermatomyositis	Inflammatory bowel disease, Behçet's disease, pancreatitis
Chronic graft vs host disease	Osteomyelitis

*Highly active antiretroviral therapy.

Importantly, certain anatomic venues are exquisitely susceptible to impaired arteriovenous perfusion. Examples include the femoral head—primarily supplied by the lone foveal nutrient artery [16]—and those individuals with a single palmar artery branch to the lunate [17]. The relative paucity of collateral vascular supply to the femoral head makes perfusion tenuous [16, 18]. Moreover, the posterior superior retinacular arteries that provide this collateral supply can readily become "kinked" with hip fracture. Thus, increased marrow adiposity, traumatic misalignment, and edema arising in any number of settings will critically impair vascular perfusion in these specific anatomic venues. In addition to impaired vascularity, the attendant colony forming unit-fibroblast (CFU-F) population that gives rise to bone-forming osteoblasts may also be reduced [19].

Diagnosis of AVN is made on the basis of clinical setting and radiographic findings [20]. In AVN of the femoral head or distal femur, pain (e.g., a deep dull ache) is often the only initial symptom often mimicking stress ("hairline") fracture [20]. Magnetic resonance imaging (MRI) with or without technetium-99m medronate bone scintigraphy readily distinguishes between the two. When disease is due to trauma or is severe or long standing, plain radiographs are revealing. Flattening of the femoral head (Fig. 121.1), joint space narrowing with sclerosis, subchondral fracture (crescent sign), and cortical collapse occur in the hip. Sclerotic changes are also seen in other venues such as the wrist, shoulder, vertebrae, and ribs [21]. Disability is related to reduced mobility secondary to pain and osteoarthritic joint malfunction [20]; thus, therapy is usually surgical with decompression or total joint arthroplasty, with recent studies evaluating potential roles for mesenchymal stem cell autografts [22]. In Kienbock's disease, revascularization with bone graft, radial shortening, or complete resection of the lunate may be required [17].

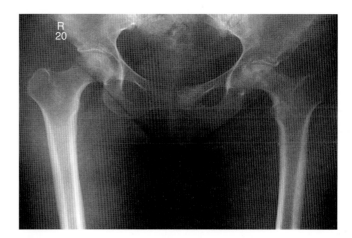

Fig. 121.1. Bilateral avascular necrosis of the femoral head. Pelvic radiograph demonstrating bilateral avascular necrosis of the femoral head in a bone marrow transplant patient with chronic graft versus host disease and steroid-induced osteoporosis. Note the flattening of the femoral head that distorts the normal spherical contour. Disease is worse in the left hip.

Medical approaches to AVN are currently very limited. Several anecdotal reports of treating bisphosphonate-associated osteonecrosis of the jaw (ONJ) with PTH(1-34) (teriparatide) have been recently published [23]. While not rigorously established, this is certainly biologically plausible since parathyroid hormone (PTH) signaling upregulates VEGF expression and reorients skeletal microvascular supply with sites of bone formation at the basic multicellular unit (BMU) [24]. Perhaps paradoxically, amino-bisphosphonates have been used to preserve femoral head structure and function in preclinical rabbit models of hip AVN as one strategy to preserve femoral head shape and stave off osteoarthritic responses [25]. One very small study suggests potential benefit in humans as well [26]. At present, medical approaches to AVN emphasize mitigation of the primary insults causing disease and pain control.

ATHEROSCLEROSIS, ARTERIOSCLEROSIS, AND SKELETAL HEALTH

As McCarthy recently reviewed, blood flow demands in the skeletal vary dependent upon the histoanatomic envelope [27]. Microsphere perfusion assays in preclinical models indicate flow values ranging between 20 mL/min/100 gm tissue in cancellous bone and 5 ml/min/100 gm in periosteal and cortical bone. The endocortical marrow compartment experiences the lowest flow rates (approximately 1 ml/min/100 gm) that potentially relate to the intricacies and needs of the bone remodeling compartment (BRC) [28] and endosteal hematopoietic niche [29]. In healthy young bone, perfusion is largely centrifugal [30]; principal nutrient arteries transverse cortical bone via the nutrient foramen and enter marrow space and bifurcate to form ascending and descending medullary arteries. Radial branches penetrate the endosteal cortex and ramify to form the Haversian capillaries that supply cortical bone and ultimately drain into the periosteal venous plexus. Marrow capillaries form sinusoidal structures that ultimate drain into a central venous sinus; this route of flow exits bone via the nutrient canal as one or two nutrient veins. Importantly, in preclinical models, pharmacological strategies that enhance nutrient artery-mediated marrow blood flow during exercise augment trabecular bone accrual [31]. Arterioles of the periosteum also supply vessels to perforating cortical canals similar to Volkmann's canals that interconnect Haversian systems. With age, this periosteal vascular supply becomes increasingly prominent, such that a centripetal flow vector, i.e., periosteum to cortex to marrow cavity, predominates [30]. The reasons for this are not entirely clear, but may relate to changes in marrow pressure, fat content, and vascular tone. Brinker and colleagues first highlighted that the principal nutrient arteries are more responsive to vasoconstrictors and relatively insensitive to vasodilators [32].

With aging, diabetes, dyslipidemia, hypertension, and uremia, conduit vessels throughout the body undergo arteriosclerosis, i.e., mechanical stiffening [33]. This impairs Windkessel physiology, the rubbery elasticity of conduit arteries necessary for smooth distal tissue perfusion throughout systole and diastole [34]. With each cardiac cycle, a portion of the kinetic energy elaborated during ventricular contraction of systole is stored as potential energy throughout the vascular tree; this energy is released during diastole and helps maintain uniformity of flow in distal capillary beds during the cardiac cycle [33]. Indeed, in preclinical models of murine arteriosclerosis, blood flow to femur as measured by microsphere perfusion is significantly decreased, even though blood flow to other organs such as the kidney is relatively preserved [35]. Thus, age- and metabolism-related changes in macrovascular stiffness and concomitant endothelial dysfunction in nutrient arteries interact to reduce and alter the pattern of skeletal blood flow. Consistent with this Windkessel model, Brenneise and Squier demonstrated that diet-induced carotid atherosclerosis reduces osseous blood flow by 80% to rhesus monkey maxillary and mandibular bone—even in the absence of altered nutrient artery diameter, osseous lumen area, intraosseous vessel wall thickness, or vessel lumen area/tissue area [36]. Of note, arterial stiffness [37] and carotid atherosclerosis [38] are increased in postmenopausal women with osteoporosis or low bone mineral density (BMD).

What are the clinical implications of these findings? Vogt et al. demonstrate that reduced blood flow to the lower extremities was associated with greater rates of bone loss at the hip and calcaneous [39]. As Meunier and colleagues went on to establish, skeletal blood flow is a critical determinant of bone formation and BMU activity in humans [40]. Subsequent noninvasive studies have demonstrated strong correlations between MRI measures of hip bone marrow perfusion and BMD in women, with concomitant increases in marrow fat [41]. Collins and colleagues identified that arteriosclerotic disease was an independent risk factor for low BMD in the hip in men [42]. They went on to show that hip bone loss and non-spine fracture risk in the MrOS study was significantly increased in men with peripheral arterial disease (PAD) as assessed by ankle-brachial indices (ABI) [42]. Of note, both low ABI values (index of conduit vessel atherosclerotic obstruction) and high ABI values (greater than 1.3; index of conduit vessel arteriosclerotic stiffening via medial artery calcification [43, 44]) were associated with increased rates of hip bone loss [42].

Diabetes, dyslipidemia, and renal insufficiency all induce macrovascular arteriosclerotic responses and endothelial dysfunction in addition to compromising bone mass accrual and conveying enhanced fracture risk. The mechanisms are only beginning to be understood, but relate to inflammation, phosphate metabolism, oxidative stress and oxylipid signaling [45]. Demer, Tintut, and colleagues recently demonstrated that multiple oxylipids derived from LDL—a key contributor to atherosclerosis and arteriosclerosis—suppresses osteoblast-mediated bone formation, enhances T-cell receptor activator of nuclear factor-κB ligand (RANKL) production, and impairs skeletal anabolic responses to PTH [46, 47]. Importantly, PTH/PTHrP (PTH-related protein) receptor signaling simultaneously suppresses arteriosclerotic conduit vessel stiffening while supporting skeletal bone mass accrual [48, 49]. Thus, the metabolic milieu that impairs arterial conduit vessel function necessary for healthy bone physiology also favors the "uncoupling" of bone formation and bone resorption.

Fracture, of course, is not the only musculoskeletal manifestation of vascular disease. In diabetes and PAD, the lower extremities bear the brunt of the burden, with amputation and arthropathy also confronting patient and physicians [50]. Multiple investigators have now demonstrated that arteriosclerotic calcification conveys risk for lower extremity amputation [51–55]. Iliofemoral medial artery calcification in type 2 diabetes increases amputation risk threefold [53]. In bone diabetic and nondiabetic patients, computed tomography (CT) scanning and Agatston scoring to quantify tibial artery calcification outperforms ABI assessment in predicting lower extremity amputation risk [52]. The increasing prevalence of type 2 diabetes, metabolic syndrome, and renal insufficiency will significantly contribute to the arteriosclerotic disease burden in our society and convey risk for fracture, impaired oral bone health, and lower extremity amputation. Whether medical strategies that aggressively target the initiation and progression of vascular disease simultaneously reduce fracture risk and amputation risk remains to be examined.

CONCLUSIONS AND FUTURE DIRECTIONS

Preclinical and clinical studies performed over the past two decades have converged to reinforce and highlight the critical important of bone–vascular interactions in skeletal health. Arterial compliance, perfusion, and remodeling are vital to normal bone anabolic responses, and are regulated by prototypic osteoanabolic hormones like PTH [24, 48]. Signaling is bidirectional, with osteoblast hormones such as fibroblast growth factor-23 (FGF23) [56] and PTH-regulated, marrow-derived ePCs [57] coordinating systemic responses that maintain normal vascular health and perfusion. The perspectives of experts in endocrinology, cardiology, developmental biology, orthopedics, biochemistry, genetics, pathology, engineering and hematology often emphasize different features of the relationship between the vasculature and bone. This integrated picture provides a "toolbox" for devising novel strategies to address unmet needs in skeletal health. Moreover, consideration of the bone–vascular relationship helps explain why, in the setting of malignancy, many of our best therapeutic strategies oftentimes have mechanism-based skeletal toxicities; these are mediated in great part via perturbation of the bone–vascular relationship, tilting the risk-benefit ratio of current antiresorptive and angiostatic regimens toward risk with respect to oral and systemic skeletal health in some patients. As our understanding of bone–vascular interactions improves, so too will our capacity to meet the clinical needs of our patients afflicted with musculoskeletal and vascular diseases.

ACKNOWLEDGMENTS

Supported by NIH grants HL69299, HL81138, and HL88651 to D.A. Towler and by the Barnes-Jewish Hospital Foundation.

REFERENCES

1. Zelzer E, McLean W, Ng YS, Fukai N, Reginato AM, Lovejoy S, D'Amore PA, Olsen BR. 2002. Skeletal defects in VEGF(120/120) mice reveal multiple roles for VEGF in skeletogenesis. *Development* 129(8): 1893–1904.
2. Maes C, Kobayashi T, Selig MK, Torrekens S, Roth SI, Mackem S, Carmeliet G, Kronenberg HM. 2010. Osteoblast precursors, but not mature osteoblasts, move into developing and fractured bones along with invading blood vessels. *Dev Cell* 19(2): 329–344.
3. Mailhot G, Petit JL, Dion N, Deschenes C, Ste-Marie LG, Gascon-Barre M. 2007. Endocrine and bone consequences of cyclic nutritional changes in the calcium, phosphate and vitamin D status in the rat: An in vivo depletion-repletion-redepletion study. *Bone* 41(3): 422–436.
4. Namgung R, Tsang RC. 2003. Bone in the pregnant mother and newborn at birth. *Clin Chim Acta* 333(1): 1–11.
5. Bianco P. 2011. Bone and the hematopoietic niche: A tale of two stem cells. *Blood* 117(20): 5281–5288.
6. Chow A, Lucas D, Hidalgo A, Mendez-Ferrer S, Hashimoto D, Scheiermann C, Battista M, Leboeuf M, Prophete C, van Rooijen N, Tanaka M, Merad M, Frenette PS. 2011. Bone marrow CD169+ macrophages promote the retention of hematopoietic stem and progenitor cells in the mesenchymal stem cell niche. *J Exp Med* 208(2): 261–271.
7. Bostrom KI, Rajamannan NM, Towler DA. 2011. The regulation of valvular and vascular sclerosis by osteogenic morphogens. *Circ Res* 109(5): 564–577.
8. Riddle RC, Khatri R, Schipani E, Clemens TL. 2009. Role of hypoxia-inducible factor-1alpha in angiogenic-osteogenic coupling. *J Mol Med (Berl)* 87(6): 583–590.
9. Lafforgue P. 2006. Pathophysiology and natural history of avascular necrosis of bone. *Joint Bone Spine* 73(5): 500–507.
10. Boss JH, Misselevich I. 2003. Osteonecrosis of the femoral head of laboratory animals: The lessons learned from a comparative study of osteonecrosis in man and experimental animals. *Vet Pathol* 40(4): 345–354.
11. Harada S, Nagy JA, Sullivan KA, Thomas KA, Endo N, Rodan GA, Rodan SB. 1994. Induction of vascular endothelial growth factor expression by prostaglandin E2 and E1 in osteoblasts. *J Clin Invest* 93(6): 2490–2496.
12. Aguilar C, Vichinsky E, Neumayr L. 2005. Bone and joint disease in sickle cell disease. *Hematol Oncol Clin North Am* 19(5): 929–941, viii.
13. Wood J, Bonjean K, Ruetz S, Bellahcene A, Devy L, Foidart JM, Castronovo V, Green JR. 2002. Novel antiangiogenic effects of the bisphosphonate compound zoledronic acid. *J Pharmacol Exp Ther* 302(3): 1055–1061.
14. Cackowski FC, Anderson JL, Patrene KD, Choksi RJ, Shapiro SD, Windle JJ, Blair HC, Roodman GD. 2010. Osteoclasts are important for bone angiogenesis. *Blood* 115(1): 140–149.

15. Khosla S, Burr D, Cauley J, Dempster DW, Ebeling PR, Felsenberg D, Gagel RF, Gilsanz V, Guise T, Koka S, McCauley LK, McGowan J, McKee MD, Mohla S, Pendrys DG, Raisz LG, Ruggiero SL, Shafer DM, Shum L, Silverman SL, Van Poznak CH, Watts N, Woo SB, Shane E. 2007. Bisphosphonate-associated osteonecrosis of the jaw: Report of a task force of the American Society for Bone and Mineral Research. *J Bone Miner Res* 22(10): 1479–1491.
16. Moon ES, Mehlman CT. 2006. Risk factors for avascular necrosis after femoral neck fractures in children: 25 Cincinnati cases and meta-analysis of 360 cases. *J Orthop Trauma* 20(5): 323–329.
17. Schuind F, Eslami S, Ledoux P. 2008. Kienbock's disease. *J Bone Joint Surg Br* 90(2): 133–139.
18. Giannoudis PV, Kontakis G, Christoforakis Z, Akula M, Tosounidis T, Koutras C. 2009. Management, complications and clinical results of femoral head fractures. *Injury* 40(12): 1245–1251.
19. Tauchmanova L, De Rosa G, Serio B, Fazioli F, Mainolfi C, Lombardi G, Colao A, Salvatore M, Rotoli B, Selleri C. 2003. Avascular necrosis in long-term survivors after allogeneic or autologous stem cell transplantation: A single center experience and a review. *Cancer* 97(10): 2453–2461.
20. Mont MA, Marker DR, Zywiel MG, Carrino JA. 2011. Osteonecrosis of the knee and related conditions. *J Am Acad Orthop Surg* 19(8): 482–494.
21. Ejindu VC, Hine AL, Mashayekhi M, Shorvon PJ, Misra RR. 2007. Musculoskeletal manifestations of sickle cell disease. *Radiographics* 27(4): 1005–1021.
22. Jones KB, Seshadri T, Krantz R, Keating A, Ferguson PC. 2008. Cell-based therapies for osteonecrosis of the femoral head. *Biol Blood Marrow Transplant* 14(10): 1081–1087.
23. Cheung A, Seeman E. 2010. Teriparatide therapy for alendronate-associated osteonecrosis of the jaw. *N Engl J Med* 363(25): 2473–2474.
24. Prisby R, Guignandon A, Vanden-Bossche A, Mac-Way F, Linossier MT, Thomas M, Laroche N, Malaval L, Langer M, Peter ZA, Peyrin F, Vico L, Lafage-Proust MH. 2011. Intermittent PTH(1-84) is osteoanabolic but not osteoangiogenic and relocates bone marrow blood vessels closer to bone-forming sites. *J Bone Miner Res* 26(11): 2583–2596.
25. Hofstaetter JG, Wang J, Yan J, Glimcher MJ. 2009. The effects of alendronate in the treatment of experimental osteonecrosis of the hip in adult rabbits. *Osteoarthritis Cartilage* 17(3): 362–370.
26. Agarwala S, Jain D, Joshi VR, Sule A. 2005. Efficacy of alendronate, a bisphosphonate, in the treatment of AVN of the hip. A prospective open-label study. *Rheumatology (Oxford)* 44(3): 352–359.
27. McCarthy I. 2006. The physiology of bone blood flow: A review. *J Bone Joint Surg Am* 88 Suppl 3: 4–9.
28. Eriksen EF. 2010. Cellular mechanisms of bone remodeling. *Rev Endocr Metab Disord* 11(4): 219–227.
29. Nakamura Y, Arai F, Iwasaki H, Hosokawa K, Kobayashi I, Gomei Y, Matsumoto Y, Yoshihara H, Suda T. 2010. Isolation and characterization of endosteal niche cell populations that regulate hematopoietic stem cells. *Blood* 116(9): 1422–1432.
30. Bridgeman G, Brookes M. 1996. Blood supply to the human femoral diaphysis in youth and senescence. *J Anat* 188 (Pt 3): 611–621.
31. Dominguez JM 2nd, Prisby RD, Muller-Delp JM, Allen MR, Delp MD. 2010. Increased nitric oxide-mediated vasodilation of bone resistance arteries is associated with increased trabecular bone volume after endurance training in rats. *Bone* 46(3): 813–819.
32. Brinker MR, Lippton HL, Cook SD, Hyman AL. 1990. Pharmacological regulation of the circulation of bone. *J Bone Joint Surg Am* 72(7): 964–975.
33. Safar ME, Boudier HS. 2005. Vascular development, pulse pressure, and the mechanisms of hypertension. *Hypertension* 46(1): 205–209.
34. Westerhof N, Lankhaar JW, Westerhof BE. 2009. The arterial Windkessel. *Med Biol Eng Comput* 47(2): 131–141.
35. Shao JS, Sierra OL, Cohen R, Mecham RP, Kovacs A, Wang J, Distelhorst K, Behrmann A, Halstead LR, Towler DA. 2011. Vascular calcification and aortic fibrosis: A bifunctional role for osteopontin in diabetic arteriosclerosis. *Arterioscler Thromb Vasc Biol* 31(8): 1821–1833.
36. Brenneise CV, Squier CA. 1985. Blood flow in maxilla and mandible of normal and atherosclerotic rhesus monkeys. *J Oral Pathol* 14(10): 800–808.
37. Sumino H, Ichikawa S, Kasama S, Takahashi T, Kumakura H, Takayama Y, Kanda T, Sakamaki T, Kurabayashi M. 2006. Elevated arterial stiffness in postmenopausal women with osteoporosis. *Maturitas* 55(3): 212–218.
38. Uyama O, Yoshimoto Y, Yamamoto Y, Kawai A. 1997. Bone changes and carotid atherosclerosis in postmenopausal women. *Stroke* 28(9): 1730–1732.
39. Vogt MT, Cauley JA, Kuller LH, Nevitt MC. 1997. Bone mineral density and blood flow to the lower extremities: The study of osteoporotic fractures. *J Bone Miner Res* 12(2): 283–289.
40. Reeve J, Arlot M, Wootton R, Edouard C, Tellez M, Hesp R, Green JR, Meunier PJ. 1988. Skeletal blood flow, iliac histomorphometry, and strontium kinetics in osteoporosis: A relationship between blood flow and corrected apposition rate. *J Clin Endocrinol Metab* 66(6): 1124–1131.
41. Griffith JF, Yeung DK, Tsang PH, Choi KC, Kwok TC, Ahuja AT, Leung KS, Leung PC. 2008. Compromised bone marrow perfusion in osteoporosis. *J Bone Miner Res* 23(7): 1068–1075.
42. Collins TC, Ewing SK, Diem SJ, Taylor BC, Orwoll ES, Cummings SR, Strotmeyer ES, Ensrud KE. 2009. Peripheral arterial disease is associated with higher rates of hip bone loss and increased fracture risk in older men. *Circulation* 119(17): 2305–2312.
43. Aboyans V, Ho E, Denenberg JO, Ho LA, Natarajan L, Criqui MH. 2008. The association between elevated ankle systolic pressures and peripheral occlusive arterial disease in diabetic and nondiabetic subjects. *J Vasc Surg* 48(5): 1197–1203.

44. Brooks B, Dean R, Patel S, Wu B, Molyneaux L, Yue DK. 2001. TBI or not TBI: That is the question. Is it better to measure toe pressure than ankle pressure in diabetic patients? *Diabet Med* 18(7): 528–532.
45. Shao JS, Cheng SL, Sadhu J, Towler DA. 2010. Inflammation and the osteogenic regulation of vascular calcification: A review and perspective. *Hypertension* 55(3): 579–592.
46. Demer L, Tintut Y. 2011. The roles of lipid oxidation products and receptor activator of nuclear factor-kappaB signaling in atherosclerotic calcification. *Circ Res* 108(12): 1482–1493.
47. Sage AP, Lu J, Atti E, Tetradis S, Ascenzi MG, Adams DJ, Demer LL, Tintut Y. 2011. Hyperlipidemia induces resistance to PTH bone anabolism in mice via oxidized lipids. *J Bone Miner Res* 26(6): 1197–1206.
48. Cheng SL, Shao JS, Halstead LR, Distelhorst K, Sierra O, Towler DA. 2010. Activation of vascular smooth muscle parathyroid hormone receptor inhibits Wnt/beta-catenin signaling and aortic fibrosis in diabetic arteriosclerosis. *Circ Res* 107(2): 271–282.
49. Shao JS, Cheng SL, Charlton-Kachigian N, Loewy AP, Towler DA. 2003. Teriparatide (human parathyroid hormone (1-34)) inhibits osteogenic vascular calcification in diabetic low density lipoprotein receptor-deficient mice. *J Biol Chem* 278(50): 50195–50202.
50. Boulton AJ, Vileikyte L, Ragnarson-Tennvall G, Apelqvist J. 2005. The global burden of diabetic foot disease. *Lancet* 366(9498): 1719–1724.
51. Everhart JE, Pettitt DJ, Knowler WC, Rose FA, Bennett PH. 1988. Medial arterial calcification and its association with mortality and complications of diabetes. *Diabetologia* 31(1): 16–23.
52. Guzman RJ, Brinkley DM, Schumacher PM, Donahue RM, Beavers H, Qin X. 2008. Tibial artery calcification as a marker of amputation risk in patients with peripheral arterial disease. *J Am Coll Cardiol* 51(20): 1967–1974.
53. Lehto S, Niskanen L, Suhonen M, Ronnemaa T, Laakso M. 1996. Medial artery calcification. A neglected harbinger of cardiovascular complications in non-insulin-dependent diabetes mellitus. *Arterioscler Thromb Vasc Biol* 16(8): 978–983.
54. Nelson RG, Gohdes DM, Everhart JE, Hartner JA, Zwemer FL, Pettitt DJ, Knowler WC. 1988. Lower-extremity amputations in NIDDM. 12-yr follow-up study in Pima Indians. *Diabetes Care* 11(1): 8–16.
55. Wang CL, Wang M, Lin MC, Chien KL, Huang YC, Lee YT. 2000. Foot complications in people with diabetes: A community-based study in Taiwan. *J Formos Med Assoc* 99(1): 5–10.
56. Stubbs JR, Liu S, Tang W, Zhou J, Wang Y, Yao X, Quarles LD. 2007. Role of hyperphosphatemia and 1,25-dihydroxyvitamin D in vascular calcification and mortality in fibroblastic growth factor 23 null mice. *J Am Soc Nephrol* 18(7): 2116–2124.
57. Napoli C, William-Ignarro S, Byrns R, Balestrieri ML, Crimi E, Farzati B, Mancini FP, de Nigris F, Matarazzo A, D'Amora M, Abbondanza C, Fiorito C, Giovane A, Florio A, Varricchio E, Palagiano A, Minucci PB, Tecce MF, Giordano A, Pavan A, Ignarro LJ. 2008. Therapeutic targeting of the stem cell niche in experimental hindlimb ischemia. *Nat Clin Pract Cardiovasc Med* 5(9): 571–579.

122

Spinal Cord Injury: Skeletal Pathophysiology and Clinical Issues

William A. Bauman and Christopher P. Cardozo

Introduction 1018
Bone Loss after Acute and Chronic SCI 1019
Fracture after SCI 1019
Bone Metabolism after SCI 1019
Heterotopic Ossification 1020
Evaluation of Skeletal Mineral Density by Dual-Energy Absorptiometry 1020
Vitamin D and Calcium 1021
Potential Pharmacological Therapies to Prevent or Reduce Bone Loss 1021
Effects of Mechanical Loading and Electromagnetic Fields on Bone 1022
Combined Pharmacological and Mechanical Therapy 1023
Grant Sources 1024
References 1024

INTRODUCTION

The National Spinal Cord Injury Statistical Center estimates the incidence of acute spinal cord injury (SCI) in the United States to be about 40 cases per million population each year, or 12,000 individuals annually, with the number of new cases rising with the increasing size of the general U.S. population at risk [1]. It has been estimated that the prevalence of SCI in the U.S. is 721 cases per million population; using assumptions of incidence and life expectancy, it is estimated that there will be about 300,000 persons living with SCI in the United States by 2014 [2]. SCI results in varying degrees of partial or complete loss of sensation, motor function, and autonomic nervous system regulation to tissues and organs below the neurological level of injury. Sympathetic outflow is almost invariably interrupted to tissues and organs below the level of injury, and both sympathetic and parasympathetic efferents to the lower extremities may be disrupted [3]. Approximately 80% of individuals with SCI are male, and tetraplegia occurs in just over half of the cases of new SCI. Complete loss of motor function and sensation below the level of injury occurs in slightly less than half of individuals. With advances in care, except for those with the highest, most complete injuries, patients with SCI are now living for decades after their injury, with life spans approaching those in the general population.

This review will discuss several aspects of the pathophysiology, general clinical considerations, and relevant experimental findings in individuals with acute and chronic SCI. Animal studies are kept at a minimum but are used to clarify the cellular events that occur and appear to be responsible for the clinical changes observed after acute injury. The unique sublesional skeletal resorption that occurs after SCI compared to other conditions associated with generalized, rapid, and substantial bone loss will be described, as well as the far greater magnitude of the sublesional skeletal changes with SCI. In addition, the heightened risk of fracture, as well as its SCI population-specific risk factors, the occurrence of and contributing factors to heterotopic ossification (HO), and the prevalence of vitamin D deficiency in those with SCI, and the reasons for its high prevalence, will be discussed. In closing, the pharmacological and/or mechanical approaches attempted to date in persons with acute and chronic SCI will be presented. However, none has proven to be of convincing efficacy in persons with complete motor SCI who lack the ability to weight bear.

Primer on the Metabolic Bone Diseases and Disorders of Mineral Metabolism, Eighth Edition. Edited by Clifford J. Rosen.
© 2013 American Society for Bone and Mineral Research. Published 2013 by John Wiley & Sons, Inc.

BONE LOSS AFTER ACUTE AND CHRONIC SCI

Loss of bone is an important secondary consequence of acute SCI and is due to a state of heightened skeletal resorption [4, 5]. This becomes a clinical issue during the initial period after SCI when calcium is rapidly resorbed from bone, producing hypercalciuria, which may be complicated by urinary tract stones; individuals with high bone turnover states and/or impaired renal function are also at increased risk for hypercalcemia. In contrast to the more global bone loss associated with postmenopausal or nonparalyzed immobilized individuals, bone loss due to SCI is restricted to that below the neurological level of injury (i.e., sublesional sites).

Bone loss after neurologically more complete forms of SCI is unique for its rate, distribution, and resistance to currently available treatments. In individuals with neurologically motor complete SCI, bone loss may occur at rates approaching 1% of bone mineral density (BMD) per week for the first 6 to 12 months after injury [6–8]. The rate of bone loss after SCI is substantially greater than that observed with microgravity (0.25% per week), bed rest (0.1% per week), or in postmenopausal women who are not taking antiresorptive medications (3–5% per year) [9–11]. The increased rate of bone loss appears to continue beyond the first 12 months of injury and continues at least over the next 3 to 7 years, albeit at a slower rate than initially. Why bone is lost at such an accelerated rate after SCI as compared to immobilization or spaceflight is unknown. This dramatic phenomenon may be related to any number of factors including, but not limited to: the loss of anabolic factors, such as circulating testosterone and/or growth hormone; factors in the local bone milieu; the presence of catabolic factors, such as the administration of methylprednisolone at extremely high doses within hours of the acute event or the production of inflammatory mediators/cytokines locally; and/or the loss of positive central and peripheral neural influences on bone. Bone is richly innervated by sensory and sympathetic nerves, and the former has been suggested to have anabolic influences on bone [12]. There is evidence from a study of monozygotic twins discordant for SCI to indicate that the difference in sublesional BMD between the twins increased with time after SCI for several decades, suggesting that loss of bone may continue at an accelerated rate in those with SCI for an extended period of time [13].

The areas of greatest bone loss are the distal femur and proximal tibia, and fractures are most common in these locations. In a cross-sectional study of eight SCI patients, a change of 35.3% in BMD of the tibial trabecular bone was noted within the first 2 years after SCI, whereas there was only a 12.9% reduction in tibial cortical bone [14]. In a cross-sectional study of 31 patients with SCI for greater than 1 year, Dauty et al. [15] demonstrated a demineralization of –52% for the distal femur and –70% for the proximal tibia, in agreement with the findings of several other investigators [16–19]. In a cross-sectional study of men with motor complete SCI, bone loss in the epiphysis, which is predominantly trabecular bone, was exponential with time [19]. On average, loss of epiphyseal bone was 50% in the femur and 60% in the tibia [19]. In the diaphyses, which is predominantly cortical bone, losses were 35% in the femur and 25% in the tibia and involved erosion of the thickness of cortical bone by 0.25 mm/year over the initial 5 to 7 years after SCI [19]. This level of depletion places the bones, particularly at the knee, below clinically accepted thresholds for fracture.

FRACTURE AFTER SCI

Peripheral fracture is a significant complication of osteoporosis after SCI. In a prospective study of a cohort of persons with SCI, Morse et al. reported that fractures most often involved the tibia or fibula, followed by the distal femur, with upper extremity fractures being much less common, albeit occurring more frequently in higher cord lesions [20]. Falls from a wheelchair and transfers were the most common causes of fracture, although fractures can also result from merely performing range of motion activities. These findings agree with a recent cross-sectional study that demonstrated 15 out of 98 individuals with SCI who were followed longitudinally sustained 39 fragility fractures of the legs over 1,010 years of combined observation; the mean time to first fracture was about 9 years, with a 1% fracture rate within the first 12 months and 4.6%/year fracture rate if more than 20 years post occurrence of SCI [21]. From these reports, increasing risk for fracture was associated with motor complete SCI, lower level of lesion (e.g., paraplegia versus tetraplegia because of their greater ability to more actively participate in exercise regimens and other forms of physical activity), longer duration of injury, and greater alcohol consumption [20, 21]. The risk of fractures, as expected, has been found to be closely related to the BMD of epiphyseal trabecular bone [22]. Because persons with SCI may lack pain sensation due to the disruption of sublesional somatosensory, they often lack awareness of the acute occurrence of fracture, and they may seek medical attention only for resultant symptoms, which may include swelling, fever, increased muscle spasticity, and/or autonomic dysreflexia.

BONE METABOLISM AFTER SCI

Skeletal unloading as a consequence of SCI, or other severe immobilizing conditions, results in the eventual uncoupling of the osteoblast–osteoclast relationship. However, immediately after injury, there appears to be an increase in both osteoclast and osteoblast function [23]. Over the ensuing months, histomorphometry has revealed an increase in trabecular osteoclastic resorptive surfaces and an early depression of osteoblastic bone formation [24]. The major osseous effect of SCI in humans is to increase bone resorption, reflected by hypercalciuria [24–26] and elevated markers of bone resorption, specifically serum or urine N-terminal telopeptide (NTX), C-terminal telopeptide (CTX), and pyridinoline and deoxypyridinoline cross-links [26, 27]. Consistent with these changes in metabolic markers of bone resorption, cultures of bone marrow from the iliac crest, a region below the level of the SCI, produced greater numbers of osteoclasts [28]. In persons with long-standing SCI, the bone turnover rate is assumed to be depressed, somewhat analogous to senile osteoporosis.

HETEROTOPIC OSSIFICATION

Ectopic bone formation, or heterotopic ossification (HO), occurs in up to half of patients after acute SCI, depending on the sensitivity of the diagnostic technique employed. The mechanisms for its occurrence have not been elucidated, but its etiology has been postulated to be due to trauma, hemorrhage, deep venous thrombosis, and/or immobility [29, 30]. HO often develops at the major joints, including the hip, which accounts for about 90% of cases, and the knee, distal femur, elbow, and shoulder joints. Clinically significant HO that reduces the range of motion to interfere with function occurs in about 10–20% of cases. The complications of HO may include the inability to sit (due to reduced range of motion), chronic pain, pressure ulcers, deep venous thrombosis, and increased muscle spasticity.

Local swelling may initially occur with the development of HO. Serum alkaline phosphatase (AP) activity is an early (i.e., 2 to 3 weeks after it first appears), nonspecific marker for HO; circulating AP levels do not correlate with the degree of bone activity, and, as such, AP should not be relied on to determine the severity or resolution of the condition. Some additional nonspecific biomarkers include C-reactive protein, erythrocyte sedimentation rate, creatine phosphokinase, hydroxyproline, and 24-hour urinary prostaglandin E_2. The three-phase technetium-99 labeled diphosphonate bone scan remains the earliest and most sensitive indicator to make the diagnosis of HO. Initial radiographic manifestations of HO lag behind its appearance on technetium scans by 3 to 4 weeks, with periarticular bone formation not appearing for 2 to 3 months on radiographs.

Nonsteroidal anti-inflammatory drugs, if administered early after acute SCI, have been shown to have the greatest efficacy in the prevention of HO [31–33]; however, this approach has not been routinely implemented in clinical practice. Treatment has included range of motion exercises with gentle stretching after the acute inflammatory period has resolved, usually after 1 to 2 weeks. Careful mobilization of the affected joints is recommended to prevent further loss of joint mobility. Sodium etidronate administration has been reported to be efficacious in reducing the severity of the condition by reducing the rate of new bone formation once HO has been identified to be present; this agent has no effect on ectopic bone *per se*. In one study, 20 of 27 patients had a decrease in swelling over the initial 2 days of intravenous etidronate therapy (prescribed intravenously for 3 to 5 days), which was followed by 6 months of oral therapy [34]. Longer therapy with etidronate may be more efficacious in preventing progression of HO but would also be anticipated to be associated with defective mineralization. Unfortunately, there is only one report in the literature of five patients using a newer generation bisphosphonate, pamidronate, to treat HO [35]. Radiation therapy and surgical resection of the ectopic calcified tissue have been performed for more severe cases to prevent joint ankylosis [33, 35, 36].

EVALUATION OF SKELETAL MINERAL DENSITY BY DUAL-ENERGY ABSORPTIOMETRY

With the wide availability of dual-energy X-ray absorptiometry (DXA), skeletal integrity in persons after SCI may now be routinely evaluated. Use of DXA methodology has been limited by clinicians because of a lack of accessible automated procedures for analyzing the most relevant regions of interest, specifically the distal femur and the proximal tibia; specialized software applications for the analysis of the knee are only now becoming available through the manufacturers of these imaging devices. Despite the well-appreciated prevalence and severity of bone loss after SCI, DXA scanning was performed in only a fraction of patients admitted with fractures after SCI [20]. As will be discussed below, effective treatments to reduce bone loss after complete motor SCI are not currently available. As such, any screening method for osteoporosis has limited relevance in the most neurologically compromised individuals because it does not provide information that could be acted upon with regard to a therapeutic approach. There may still be value in providing health care professionals information on bone mass in the regions of interest, which would permit them to inform those individuals whose BMD has decreased well below fracture thresholds and to counsel such patients to avoid high risk activities that may result in fracture.

Posterior–anterior (PA) DXA scanning has been shown to provide spurious results when assessing the spine because this technique cannot distinguish vertebral bone from other calcified entities, such as HO, or osteophytes and other degenerative changes that are common after SCI. As may have been expected, SCI subjects with moderate degenerative joint disease (DJD) had significantly higher T-scores by PA DXA than those without, or with mild, DJD [37]. Thus, when assessment of vertebral spine for bone loss is deemed necessary, lateral DXA methodology or computerized tomography should provide more reliable information than that of standard PA DXA imaging [37, 38].

VITAMIN D AND CALCIUM

After acute SCI, and because of rapid bone resorption, serum ionized calcium levels are elevated, as reflected by a markedly increased renal clearance of calcium [4]. In the presence of high ionized serum calcium levels, parathyroid hormone (PTH) is suppressed, reducing PTH-mediated tubular absorption of calcium and further aggravating urinary calcium losses. Gut absorption of calcium would be expected to be reduced due to a reduction in renal 1-α hydroxylase activity. As such, modifying calcium intake should have little effect on urinary calcium excretion, as has been shown in patients after subacute SCI [4].

Individuals with severe chronic conditions, including disabilities, have long been recognized to be at high risk for the development of vitamin D deficiency because of reduced sunlight exposure due to lifestyle changes and/or institutionalization and medications [39, 40]. Those with disabilities tend to have a greatly reduced sunlight exposure, and conversion of vitamin D precursors to the active form requires ultraviolet exposure. Several of the drugs prescribed in persons with SCI, especially anticonvulsants and psychotropic agents, may accelerate hydroxylation of vitamin D and increase its renal clearance [40]. Persons with SCI often have a lower calcium intake than the general population, including vitamin D-fortified milk, which is the major dietary source of vitamin D, excluding supplements [41].

Vitamin D deficiency has been reported to be extremely prevalent in the SCI population, as well as in those with other disabilities [42, 43]. Bauman et al. demonstrated that approximately one-third of a veteran SCI population had an absolute deficiency of vitamin D [42]. Oleson and Wuermser recently reported that in subjects with chronic SCI, 81% had 25(OH)D levels below 32 ng/mL in the summer, and the percentage of those deficient in the winter increased to 96%; 54% of these subjects had 25(OH)D levels below 13 ng/dL during the winter months [43]. In one retrospective inpatient study of 100 SCI patients who were consecutively admitted to an acute inpatient rehabilitation facility, the prevalence of 25 hydroxyvitamin [25(OH)D] relative or absolute deficiency was 93% with a mean 25(OH)D value of 16.3 ± 7.7 ng/mL [44]. It should be appreciated that if an efficacious antiresorptive agent becomes available to suppress the unchecked bone resorption of acute SCI, then one must ensure that 25(OH)D levels are sufficient to facilitate gut absorption of calcium to prevent hypocalcemia. In an effort to restore serum vitamin D levels to a level that would serve to optimize gastrointestinal absorption of calcium [45], cholecalciferol, or vitamin D_3, 2,000 IU/day was administered to individuals with SCI in conjunction with daily oral calcium supplementation of 1.3 g elemental calcium/day; in vitamin D-deficient subjects, vitamin D supplementation for 3 months raised the level of vitamin D into the normal range in 85% of subjects, and the mean level of vitamin D for the total group was well above the lower limit of the normal range [25(OH)D greater than 30 ng/mL] [46].

POTENTIAL PHARMACOLOGICAL THERAPIES TO PREVENT OR REDUCE BONE LOSS

Bisphosphonates have a strong affinity for bone and inhibit osteoclast bone resorption, and this class of agents has been used in an attempt to prevent or reduce bone resorption associated with acute SCI. Bisphosphonates act by inhibiting farnesyl pyrophosphate synthase, which interferes with isoprenylation of guanosine triphophatases (GTPases) at the ruffled border of osteoclasts, thereby preventing attachment of osteoclasts to the bone surface, halting resorption, and initiating cell death [47]. However, our experience has raised questions concerning the efficacy of this class of medications in persons with SCI who are non-weight-bearing (i.e., complete motor SCI).

In various animal models of disuse osteoporosis, bisphosphonate administration was not effective at preventing bone loss or maintaining cortical bone strength, findings that differentiate disuse osteoporosis from other models of osteoporosis in which bisphosphonates have been shown to be of greater efficacy [48, 49]. In a human bed rest study in non-SCI individuals, bisphosphonates reduced osteoclast number in cancellous bone, but not in cortical bone [50]. Two relatively small, non-randomized case series have demonstrated the benefit of bisphosphonates in persons with incomplete motor SCI who are able to bear weight and ambulate [51, 52]. This is an important insight for consideration in the use of bisphosphonates in the SCI population. It may be difficult to determine the efficacy of this class of therapeutic agents unless the authors prospectively and specifically address in their experimental design the varying degrees of motor impairment, the ability to bear weight, and/or ambulate. In the few frequently referenced reports that have addressed the effect of bisphosphonates on BMD after acute paralysis, the results are expressed as a composite of patients with varying degrees of completeness of motor lesions and ability to bear weight [53, 54]. In addition, as is often the case after acute SCI [55], high-dose glucocorticoids were administered in an attempt to preserve neurological function, but the potential effects of glucocorticoids were not controlled for in the experimental design in at least one of these studies, confounding interpretation of the results [54]. Zoledronic acid has been administered to patients with acute motor complete SCI with apparent benefit at the hip at 6 months (e.g., end points determined were BMD, cross-sectional area, and measures of bone strength), but these salutary effects were almost totally lost by 12 months [56]. Despite the appropriate interest in BMD of the hip and parameters of bone strength in the general population, it should be appreciated that patients with chronic SCI tend to fracture the distal femur and proximal tibia. In an attempt to preserve BMD in patients with acute complete motor SCI, sequential doses of pamidronate preserved BMD of the total leg for the initial several months after acute injury but measurements at multiple sublesional skeletal regions 2 years after SCI were not different from the control group [57]. A recent review on this topic concluded:

> Data were insufficient to recommend routine use of bisphosphonates for fracture prevention in these patients. Current studies are limited by heterogeneity of patient populations and outcome measures. Uniform bone density measurement sites with rigorous quality control and compliance monitoring are needed to improve reliability of outcomes. Future studies should address specific populations (acute or chronic SCI) and should assess fracture outcomes [58].

The major regulators of bone remodeling are the receptor activator of nuclear factor-κB ligand (RANKL), which serves to positively drive osteoclastogenesis and function, and osteoprotegerin (OPG), which serves as a "decoy" receptor for RANKL to downregulate osteoclastic differentiation and activity. Denosamub (XGEVA®, Amgen, Thousand Oaks, CA), a human monoclonal antibody to RANKL, represents an immunological/pharmacological approach to the treatment of osteoporosis that has been recently approved by the FDA. Thus, the mechanism of action of denosamub, while inhibiting the osteoclast, is distinctly different from that of bisphosphonates. Histomorphometric, bone turnover, and bone density findings suggest that the effects of denosumab on bone remodeling are more potent than those of bisphosphonates [59, 60]. The dramatic reduction in eroded bone surfaces with denosumab may be especially relevant to preventing or reducing bone loss after acute SCI because of the robust osteoclastosis/bone resorption that develops shortly after paralysis.

Animal models of immobilization have shown that paralysis, casting of limbs, prolonged bed rest, or weightlessness cause substantial bone loss due to an initial phase of accelerated bone resorption and a prolonged phase of decreased formation [61, 62]. Manipulation of the OPG/RANKL system was highly effective at maintaining cortical bone mass in several disuse models, including spaceflight, tail suspension, and sciatic nerve injury [63]. Our group has preliminary evidence in a rodent model of SCI to suggest that there is a several-fold increase in RANKL expression after acute SCI, while the expression of OPG is reduced, making the ratio of RANKL to OPG highly unfavorable acutely after SCI (unpublished observation). In *ex vivo* cell cultures from our animal model of acute SCI, there was also an almost twofold increase in osteoclast differentiation markers; in contrast, osteoblast differentiation markers were markedly depressed. Thus, in acute SCI there appears to be the dual challenge to maintain bone integrity in the presence of increased osteoclastic activity and reduced osteoblastic function. Because we have found excessive RANKL expression after acute SCI in a rodent model, it may be postulated that the drug of choice may be an agent that directly antagonizes RANKL to markedly reduce function of the osteoclast. However, denosumab has yet to be investigated in the "high bone turnover" state of acute/subacute SCI.

EFFECTS OF MECHANICAL LOADING AND ELECTROMAGNETIC FIELDS ON BONE

Insights into the possibility that reloading may preserve or rebuild bone after prolonged unloading come from studies of bone mass in individuals returning to Earth after prolonged periods in space, in whom substantial bone loss had occurred during zero-gravity conditions. After 4–6 months of unweighting during spaceflight, 1 year of loading from the Earth's gravity led to substantial increases in BMD [64]. These findings provide insight, as well as hope, that it may be possible to increase bone mass and strength after the extensive sublesional skeletal loss that occurs after SCI.

Static mechanical loading of bone has been ineffective in slowing bone loss after SCI [65, 66]. Cyclical reloading provided by partial body-weight-supported treadmill training has also been ineffective in slowing bone loss or increasing bone mass in persons with SCI [67, 68]. To date, the mechanical intervention that has been shown to be beneficial to bone after SCI is cyclical muscle contraction elicited by functional electrical stimulation (FES). During FES, electrical stimulation of nerves is employed to stimulate muscular contraction, usually with surface electrodes. One study that was initiated within several months of acute SCI evaluated the effects of isometric contraction of the soleus muscle of one leg for 4 to 6 years of treatment using the tibia of the contralateral leg for comparison. Whereas the untreated leg lost BMD progressively over time, FES partially preserved bone along the posterior aspect of the tibia; and BMD of this region was more than double that for the untrained leg and only 25% lower than that in able-bodied controls [69]. Trabecular BMD of the central core of the tibia cross-section was 41% higher in the FES-trained leg than in the untrained leg. There is preliminary evidence to suggest that combining FES with standing may have a greater beneficial effect on BMD of the hip and the knee in a subset of subjects with SCI than FES alone, whereas the standing alone subset had a decrease in BMD of these regions [70]. FES was less effective in increasing bone mass in individuals with SCI who had long-established bone loss [71] that was likely associated with greater loss of trabecular architecture; however, the finding that modest increases in bone mass occurred with FES training using a cycle ergometer [72] or knee extension against resistance [73] suggests that even years after SCI it may be possible to stimulate net accrual of bone if bone integrity is still somewhat preserved. In the latter study, the effect of FES-mediated extension of the left knee against resistance for 1 hour a day, 5 days each week for 24 weeks in a group of 14 individuals with SCI applied for greater than 1 year was compared to the right leg, which was untreated; this approach of FES-resistance training resulted in a 30% increase in BMD at the distal femur and proximal tibia [73]. Thus FES, if administered at the time of acute injury, prevents bone loss at the site applied, or, if administered sometime thereafter, improves bone mass at the site that is receiving the load. Translation to clinical care has been fraught with considerable difficulty due to the labor-intensive nature of the present approaches and the awareness that any effect on bone is rapidly lost when FES is discontinued.

Low-intensity, high-frequency vibration is an intervention delivered by an oscillating platform that has been found to reduce bone loss in children with disuse osteoporosis due to neurological impairments and in postmenopausal women [74, 75]. A study has recently demonstrated the feasibility of delivering low-intensity mechanical signals through the lower appendicular and axial skeleton while supine in subjects with SCI; transmission of signal increased with increasingly upright body positions [76]. This observation supports the feasibility for the use of low intensity vibration as a possible mechanical intervention to be studied in the prevention or reduction of bone loss in patients with SCI. Future studies are needed to investigate the effectiveness of long-term exposure to low-intensity vibration on preservation of BMD in persons with acute and subacute SCI. Animal and clinical trials should be considered to define the effects of frequency, intensity, and duration of low-intensity vibration on regional BMD and bone microarchitecture.

Findings from studies of electromagnetic stimulation in tissue culture and animal models have shown a remarkable effect to modulate human mesenchymal stem cell osteogenesis, to favorably alter cytokine profiles during disuse osteoporosis, to promote bone formation and repair, to regulate proteoglycan and collagen synthesis, and to increase bone formation in models of endochondral ossification [77–80]. In the able-bodied population, electromagnetic field stimulation is of possible value in successful spine fusion. In 201 patients who underwent spine fusion surgery and who were evaluated at 9 months, 64% of those administered electromagnetic treatment postoperatively healed compared with 43% of those treated with placebo devices [81]. In a review comprising four studies of 125 patients who were diagnosed with union or nonunion of the long bones (predominantly nonunion of the tibia) who received either electromagnetic fields (three of the studies) or capacitive coupled electric fields (one study), electromagnetic field stimulation appeared to offer some benefit in the treatment of delayed union and nonunion long bone fractures; overall, though, these findings were inconclusive [82]. The single report to date in individuals with SCI addressed the effect of pulsed electromagnetic fields applied unilaterally at the knee in subjects at least 2 years after acute SCI [83]. This study revealed increased BMD of the stimulated knee and decreased BMD of the contralateral knee at 3 months, a return to baseline levels of BMD at 6 months, and a fall in BMD of both knees by 12 months, with generally larger magnitude of changes closer to the site of stimulation; these findings suggested a complex effect that included both local and systemic effects of pulsed electromagnetic field stimulation on BMD above and below the knee [83].

COMBINED PHARMACOLOGICAL AND MECHANICAL THERAPY

The possibility of using a combined mechanical and pharmacological approach to reduce or prevent bone loss after acute SCI may also be entertained. Because of the obvious difficulty in instituting a mechanical intervention immediately after acute SCI, it may be necessary to begin therapy with drug interventions. It is conceivable that early after SCI a pharmacological approach to target and inhibit RANKL in an effort to suppress rampant osteoclastosis may be prove efficacious, as well as practical. Another consideration is that patients who are acutely or subacutely paralyzed are commonly hypogonadal [84], a hormonal deficiency state that may occur precipitously and, as such, may be assumed to worsen the bone loss of acute immobilization. Because high-dose methylprednisolone is frequently prescribed at the time of acute paralysis [55], and there are recent animal studies supporting the role of androgens to antagonize the effects of glucocorticoids on muscle [85, 86] and bone [87], this may be yet another reason to consider testosterone replacement therapy if a patient is diagnosed as hypogonadal shortly after injury.

When the patient is sufficiently stabilized after the acute traumatic event, which may be 2 to 6 months after acute SCI, a combination approach with mechanical intervention with at least one pharmacological therapy may be envisioned. Albeit purely conjecture at this time, it may be postulated that the sooner a physical intervention can be integrated into the care plan in the form of practical and cost-effective mechanical stimulation to restore forces that had been suddenly removed from the skeleton at time of injury, the greater the likelihood of preserving bone mass. The initiation of a mechanical approach may be reinforced by innovative, and temporally appropriate, pharmacological maneuvers. Thus, it may be hypothesized that along with a mechanical stimulation, a pharmacological anabolic therapy, such as teriparatide (FORTEO®, Eli Lilly and Company, Indianapolis, IN), an anabolic steroid, or another drug to stimulate depressed function of the osteoblast, may prove of added clinical value.

GRANT SOURCES

Veterans Affairs Rehabilitation Research and Development Service (B4162C) and the James J. Peters VA Medical Center.

REFERENCES

1. National Spinal Cord Injury Statistical Center. 2010. Spinal cord injury facts and figures at a glance. *J Spinal Cord Med* 33(4): 39–40.
2. DeVivo MJ, Chen Y. 2011. Trends in new injuries, prevalent cases, and aging with spinal cord injury. *Arch Phys Med Rehabil* 92(3): 332–8.
3. Krassioukov AV, Karlsson AK, Wecht JM, Wuermser LA, Mathias CJ, Marino RJ. 2007. Assessment of autonomic dysfunction following spinal cord injury: Rationale for additions to International Standards for Neurological Assessment. *J Rehabil Res Dev* 44(1): 103–12.
4. Stewart AF, Adler M, Byers CM, Segre GV, Broadus AE. 1982. Calcium homeostasis in immobilization: An example of resorptive hypercalciuria. *N Engl J Med* 306(19): 1136–40.
5. Naftchi NE, Viau AT, Sell GH, Lowman EW. 1980. Mineral metabolism in spinal cord injury. *Arch Phys Med Rehabil* 61(3): 139–42.
6. Szollar SM, Martin EM, Sartoris DJ, Parthemore JG, Deftos LJ. 1998. Bone mineral density and indexes of bone metabolism in spinal cord injury. *Am J Phys Med Rehabil* 77(1): 28–35.
7. Garland DE, Adkins RH, Kushwaha V, Stewart C. 2004. Risk factors for osteoporosis at the knee in the spinal cord injury population. *J Spinal Cord Med* 27(3): 202–6.
8. Warden SJ, Bennell KL, Matthews B, Brown DJ, McMeeken JM, Wark JD. 2002. Quantitative ultrasound assessment of acute bone loss following spinal cord injury: A longitudinal pilot study. *Osteoporos Int* 13(7): 586–92.
9. Vico L, Collet P, Guignandon A, Lafage-Proust MH, Thomas T, Rehaillia M, Alexandre C. 2000. Effects of long-term microgravity exposure on cancellous and cortical weight-bearing bones of cosmonauts. *Lancet* 355(9215): 1607–11.
10. Leblanc AD, Schneider VS, Evans HJ, Engelbretson DA, Krebs JM. 1990. Bone mineral loss and recovery after 17 weeks of bed rest. *J Bone Miner Res* 5(8): 843–50.
11. Recker R, Lappe J, Davies K, Heaney R. 2000. Characterization of perimenopausal bone loss: A prospective study. *Journal of bone and mineral research: The official journal of the American Society for Bone and Mineral Research* 15(10): 1965–73.
12. Qin W, Bauman WA, Cardozo CP. 2010. Evolving concepts in neurogenic osteoporosis. *Curr Osteoporos Rep* 8(4): 212–8.
13. Bauman WA, Spungen AM, Wang J, Pierson RN Jr, Schwartz E. 1999. Continuous loss of bone during chronic immobilization: A monozygotic twin study. *Osteoporos Int* 10(2): 123–7.
14. de Bruin ED, Dietz V, Dambacher MA, Stussi E. 2000. Longitudinal changes in bone in men with spinal cord injury. *Clin Rehabil* 14(2): 145–52.
15. Dauty M, Perrouin Verbe B, Maugars Y, Dubois C, Mathe JF. 2000. Supralesional and sublesional bone mineral density in spinal cord-injured patients. *Bone* 27(2): 305–9.
16. Biering-Sorensen F, Bohr HH, Schaadt OP. 1990. Longitudinal study of bone mineral content in the lumbar spine, the forearm and the lower extremities after spinal cord injury. *Eur J Clin Invest* 20(3): 330–5.
17. Finsen V, Indredavik B, Fougner KJ. 1992. Bone mineral and hormone status in paraplegics. *Paraplegia* 30(5): 343–7.
18. Frey-Rindova P, de Bruin ED, Stussi E, Dambacher MA, Dietz V. 2000. Bone mineral density in upper and lower extremities during 12 months after spinal cord injury measured by peripheral quantitative computed tomography. *Spinal Cord* 38(1): 26–32.
19. Eser P, Frotzler A, Zehnder Y, Wick L, Knecht H, Denoth J, Schiessl H. 2004. Relationship between the duration of paralysis and bone structure: A pQCT study of spinal cord injured individuals. *Bone* 34(5): 869–80.
20. Morse LR, Battaglino RA, Stolzmann KL, Hallett LD, Waddimba A, Gagnon D, Lazzari AA, Garshick E. 2009. Osteoporotic fractures and hospitalization risk in chronic spinal cord injury. *Osteoporos Int* 20(3): 385–92.
21. Zehnder Y, Luthi M, Michel D, Knecht H, Perrelet R, Neto I, Kraenzlin M, Zach G, Lippuner K. 2004. Long-term changes in bone metabolism, bone mineral density, quantitative ultrasound parameters, and fracture incidence after spinal cord injury: A cross-sectional observational study in 100 paraplegic men. *Osteoporos Int* 15(3): 180–9.
22. Eser P, Frotzler A, Zehnder Y, Denoth J. 2005. Fracture threshold in the femur and tibia of people with spinal cord injury as determined by peripheral quantitative computed tomography. *Arch Phys Med Rehabil* 86(3): 498–504.
23. Chantraine A, Nusgens B, Lapiere CM. 1986. Bone remodeling during the development of osteoporosis in paraplegia. *Calcif Tissue Int* 38(6): 323–7.
24. Minaire P, Neunier P, Edouard C, Bernard J, Courpron P, Bourret J. 1974. Quantitative histological data on disuse osteoporosis: Comparison with biological data. *Calcif Tissue Res* 17(1): 57–73.
25. Bergmann P, Heilporn A, Schoutens A, Paternot J, Tricot A. 1977. Longitudinal study of calcium and bone metabolism in paraplegic patients. *Paraplegia* 15(2): 147–59.
26. Roberts D, Lee W, Cuneo RC, Wittmann J, Ward G, Flatman R, McWhinney B, Hickman PE. 1998. Longitudinal study of bone turnover after acute spinal cord injury. *J Clin Endocrinol Metab* 83(2): 415–22.

27. Reiter AL, Volk A, Vollmar J, Fromm B, Gerner HJ. 2007. Changes of basic bone turnover parameters in short-term and long-term patients with spinal cord injury. *Eur Spine J* 16(6): 771–6.
28. Demulder A, Guns M, Ismail A, Wilmet E, Fondu P, Bergmann P. 1998. Increased osteoclast-like cells formation in long-term bone marrow cultures from patients with a spinal cord injury. *Calcif Tissue Int* 63(5): 396–400.
29. Perkash A, Sullivan G, Toth L, Bradleigh LH, Linder SH, Perkash I. 1993. Persistent hypercoagulation associated with heterotopic ossification in patients with spinal cord injury long after injury has occurred. *Paraplegia* 31(10): 653–9.
30. Chantraine A, Minaire P. 1981. Para-osteo-arthropathies. A new theory and mode of treatment. *Scand J Rehabil Med* 13(1): 31–7.
31. Banovac K, Williams JM, Patrick LD, Levi A. 2004. Prevention of heterotopic ossification after spinal cord injury with COX-2 selective inhibitor (rofecoxib). *Spinal Cord* 42(12): 707–10.
32. Banovac K, Williams JM, Patrick LD, Haniff YM. 2001. Prevention of heterotopic ossification after spinal cord injury with indomethacin. *Spinal Cord* 39(7): 370–4.
33. Teasell RW, Mehta S, Aubut JL, Ashe MC, Sequeira K, Macaluso S, Tu L. 2010. A systematic review of the therapeutic interventions for heterotopic ossification after spinal cord injury. *Spinal Cord* 48(7): 512–21.
34. Banovac K, Gonzalez F, Wade N, Bowker JJ. 1993. Intravenous disodium etidronate therapy in spinal cord injury patients with heterotopic ossification. *Paraplegia* 31(10): 660–6.
35. Schuetz P, Mueller B, Christ-Crain M, Dick W, Haas H. 2005. Amino-bisphosphonates in heterotopic ossification: First experience in five consecutive cases. *Spinal Cord* 43(10): 604–10.
36. Meiners T, Abel R, Bohm V, Gerner HJ. 1997. Resection of heterotopic ossification of the hip in spinal cord injured patients. *Spinal Cord* 35(7): 443–5.
37. Bauman WA, Kirshblum S, Cirnigliaro C, Forrest GF, Spungen AM. 2010. Underestimation of bone loss of the spine with posterior-anterior dual-energy X-ray absorptiometry in patients with spinal cord injury. *J Spinal Cord Med* 33(3): 214–20.
38. Bauman WA, Schwartz E, Song IS, Kirshblum S, Cirnigliaro C, Morrison N, Spungen AM. 2009. Dual-energy X-ray absorptiometry overestimates bone mineral density of the lumbar spine in persons with spinal cord injury. *Spinal Cord* 47(8): 628–33.
39. Lifshitz F, Maclaren NK. 1973. Vitamin D-dependent rickets in institutionalized, mentally retarded children receiving long-term anticonvulsant therapy. I. A survey of 288 patients. *J Pediatr* 83(4): 612–20.
40. Hahn TJ, Hendin BA, Scharp CR, Haddad JG Jr. 1972. Effect of chronic anticonvulsant therapy on serum 25-hydroxycalciferol levels in adults. *N Engl J Med* 287(18): 900–4.
41. Walters JL, Buchholz AC, Martin Ginis KA. 2009. Evidence of dietary inadequacy in adults with chronic spinal cord injury. *Spinal Cord* 47(4): 318–22.
42. Bauman WA, Zhong YG, Schwartz E. 1995. Vitamin D deficiency in veterans with chronic spinal cord injury. *Metabolism* 44(12): 1612–6.
43. Oleson CV, Patel PH, Wuermser LA. 2010. Influence of season, ethnicity, and chronicity on vitamin D deficiency in traumatic spinal cord injury. *J Spinal Cord Med* 33(3): 202–13.
44. Nemunaitis GA, Mejia M, Nagy JA, Johnson T, Chae J, Roach MJ. 2010. A descriptive study on vitamin D levels in individuals with spinal cord injury in an acute inpatient rehabilitation setting. *PM R* 2(3): 202–8; quiz 28.
45. Dawson-Hughes B, Heaney RP, Holick MF, Lips P, Meunier PJ, Vieth R. 2005. Estimates of optimal vitamin D status. *Osteoporos Int* 16(7): 713–6.
46. Bauman WA, Emmons RR, Cirnigliaro CM, Kirshblum SC, Spungen AM. 2011. An effective oral vitamin D replacement therapy in persons with spinal cord injury. *J Spinal Cord Med* 34(5): 455–60.
47. Thompson DD, Seedor JG, Weinreb M, Rosini S, Rodan GA. 1990. Aminohydroxybutane bisphosphonate inhibits bone loss due to immobilization in rats. *J Bone Miner Res* 5(3): 279–86.
48. Kodama Y, Nakayama K, Fuse H, Fukumoto S, Kawahara H, Takahashi H, Kurokawa T, Sekiguchi C, Nakamura T, Matsumoto T. 1997. Inhibition of bone resorption by pamidronate cannot restore normal gain in cortical bone mass and strength in tail-suspended rapidly growing rats. *J Bone Miner Res* 12(7): 1058–67.
49. Li CY, Price C, Delisser K, Nasser P, Laudier D, Clement M, Jepsen KJ, Schaffler MB. 2005. Long-term disuse osteoporosis seems less sensitive to bisphosphonate treatment than other osteoporosis. *J Bone Miner Res* 20(1): 117–24.
50. Chappard D, Petitjean M, Alexandre C, Vico L, Minaire P, Riffat G. 1991. Cortical osteoclasts are less sensitive to etidronate than trabecular osteoclasts. *J Bone Miner Res* 6(7): 673–80.
51. Pearson EG, Nance PW, Leslie WD, Ludwig S. 1997. Cyclical etidronate: Its effect on bone density in patients with acute spinal cord injury. *Arch Phys Med Rehabil* 78(3): 269–72.
52. Nance PW, Schryvers O, Leslie W, Ludwig S, Krahn J, Uebelhart D. 1999. Intravenous pamidronate attenuates bone density loss after acute spinal cord injury. *Arch Phys Med Rehabil* 80(3): 243–51.
53. Gilchrist NL, Frampton CM, Acland RH, Nicholls MG, March RL, Maguire P, Heard A, Reilly P, Marshall K. 2007. Alendronate prevents bone loss in patients with acute spinal cord injury: A randomized, double-blind, placebo-controlled study. *J Clin Endocrinol Metab* 92(4): 1385–90.
54. Bubbear JS, Gall A, Middleton FR, Ferguson-Pell M, Swaminathan R, Keen RW. 2011. Early treatment with zoledronic acid prevents bone loss at the hip following acute spinal cord injury. *Osteoporos Int* 22(1): 271–9.

55. Bracken MB, Holford TR. 2002. Neurological and functional status 1 year after acute spinal cord injury: Estimates of functional recovery in National Acute Spinal Cord Injury Study II from results modeled in National Acute Spinal Cord Injury Study III. *J Neurosurg* 96(3 Suppl): 259–66.
56. Shapiro J, Smith B, Beck T, Ballard P, Dapthary M, BrintzenhofeSzoc K, Caminis J. 2007. Treatment with zoledronic acid ameliorates negative geometric changes in the proximal femur following acute spinal cord injury. *Calcif Tissue Int* 80(5): 316–22.
57. Bauman WA, Wecht JM, Kirshblum S, Spungen AM, Morrison N, Cirnigliaro C, Schwartz E. 2005. Effect of pamidronate administration on bone in patients with acute spinal cord injury. *J Rehabil Res Dev* 42(3): 305–13.
58. Bryson JE, Gourlay ML. 2009. Bisphosphonate use in acute and chronic spinal cord injury: A systematic review. *J Spinal Cord Med* 32(3): 215–25.
59. Reid IR, Miller PD, Brown JP, Kendler DL, Fahrleitner-Pammer A, Valter I, Maasalu K, Bolognese MA, Woodson G, Bone H, Ding B, Wagman RB, San Martin J, Ominsky MS, Dempster DW. 2010. Effects of denosumab on bone histomorphometry: The FREEDOM and STAND studies. *J Bone Miner Res* 25(10): 2256–65.
60. Kendler DL, Roux C, Benhamou CL, Brown JP, Lillestol M, Siddhanti S, Man HS, San Martin J, Bone HG. 2010. Effects of denosumab on bone mineral density and bone turnover in postmenopausal women transitioning from alendronate therapy. *J Bone Miner Res* 25(1): 72–81.
61. Weinreb M, Rodan GA, Thompson DD. 1989. Osteopenia in the immobilized rat hind limb is associated with increased bone resorption and decreased bone formation. *Bone* 10(3): 187–94.
62. Thompson DD, Rodan GA. 1988. Indomethacin inhibition of tenotomy-induced bone resorption in rats. *J Bone Miner Res* 3(4): 409–14.
63. Kearns AE, Khosla S, Kostenuik PJ. 2008. Receptor activator of nuclear factor kappaB ligand and osteoprotegerin regulation of bone remodeling in health and disease. *Endocr Rev* 29(2): 155–92.
64. Lang TF, Leblanc AD, Evans HJ, Lu Y. 2006. Adaptation of the proximal femur to skeletal reloading after long-duration spaceflight. *J Bone Miner Res* 21(8): 1224–30.
65. Dudley-Javoroski S, Shields RK. 2008. Muscle and bone plasticity after spinal cord injury: Review of adaptations to disuse and to electrical muscle stimulation. *J Rehabil Res Dev* 45(2): 283–96.
66. Qin W, Bauman W, Cardozo C. 2010. Bone and muscle loss after spinal cord injury: Organ interactions. *Ann N Y Acad Sci* 1211: 66–84.
67. Coupaud S, Jack LP, Hunt KJ, Allan DB. 2009. Muscle and bone adaptations after treadmill training in incomplete spinal cord injury: A case study using peripheral quantitative computed tomography. *J Musculoskelet Neuronal Interact* 9(4): 288–97.
68. Giangregorio LM, Webber CE, Phillips SM, Hicks AL, Craven BC, Bugaresti JM, McCartney N. 2006. Can body weight supported treadmill training increase bone mass and reverse muscle atrophy in individuals with chronic incomplete spinal cord injury? *Appl Physiol Nutr Metab* 31(3): 283–91.
69. Dudley-Javoroski S, Shields RK. 2008. Asymmetric bone adaptations to soleus mechanical loading after spinal cord injury. *J Musculoskelet Neuronal Interact* 8(3): 227–38.
70. Forrest GF, Harkema SJ, Angeli CA, Faghri PD, Kirshblum SC, Cirnigliaro CM, Garbarini E, Bauman WA. Preliminary results on the differential effect on bone of applying multi-muscle electrical stimulation to the leg while supine or standing in patients with SCI: The importance of combining a mechanical intervention with gravitational loading. *J Spinal Cord Med*. Accepted for publication.
71. Shields RK, Dudley-Javoroski S. 2007. Musculoskeletal adaptations in chronic spinal cord injury: Effects of long-term soleus electrical stimulation training. *Neurorehabil Neural Repair* 21(2): 169–79.
72. Frotzler A, Coupaud S, Perret C, Kakebeeke TH, Hunt KJ, Donaldson Nde N, Eser P. 2008. High-volume FES-cycling partially reverses bone loss in people with chronic spinal cord injury. *Bone* 43(1): 169–76.
73. Belanger M, Stein RB, Wheeler GD, Gordon T, Leduc B. 2000. Electrical stimulation: Can it increase muscle strength and reverse osteopenia in spinal cord injured individuals? *Arch Phys Med Rehabil* 81(8): 1090–8.
74. Ward K, Alsop C, Caulton J, Rubin C, Adams J, Mughal Z. 2004. Low magnitude mechanical loading is osteogenic in children with disabling conditions. *J Bone Miner Res* 19(3): 360–9.
75. Rubin C, Recker R, Cullen D, Ryaby J, McCabe J, McLeod K. 2004. Prevention of postmenopausal bone loss by a low-magnitude, high-frequency mechanical stimuli: A clinical trial assessing compliance, efficacy, and safety. *J Bone Miner Res* 19(3): 343–51.
76. Asselin P, Spungen AM, Muir JW, Rubin CT, Bauman WA. 2011. Transmission of low-intensity vibration through the axial skeleton of persons with spinal cord injury as a potential intervention for preservation of bone quantity and quality. *J Spinal Cord Med* 34(1): 52–9.
77. Tsai MT, Li WJ, Tuan RS, Chang WH. 2009. Modulation of osteogenesis in human mesenchymal stem cells by specific pulsed electromagnetic field stimulation. *J Orthop Res* 27(9): 1169–74.
78. Shen WW, Zhao JH. 2010. Pulsed electromagnetic fields stimulation affects BMD and local factor production of rats with disuse osteoporosis. *Bioelectromagnetics* 31(2): 113–9.
79. Jansen JH, van der Jagt OP, Punt BJ, Verhaar JA, van Leeuwen JP, Weinans H, Jahr H. 2010. Stimulation of osteogenic differentiation in human osteoprogenitor cells by pulsed electromagnetic fields: An in vitro study. *BMC Musculoskelet Disord* 11: 188.
80. Aaron RK, Ciombor DM, Wang S, Simon B. 2006. Clinical biophysics: The promotion of skeletal repair by physical forces. *Ann N Y Acad Sci* 1068: 513–31.

81. Linovitz RJ, Pathria M, Bernhardt M, Green D, Law MD, McGuire RA, Montesano PX, Rechtine G, Salib RM, Ryaby JT, Faden JS, Ponder R, Muenz LR, Magee FP, Garfin SA. 2002. Combined magnetic fields accelerate and increase spine fusion: A double-blind, randomized, placebo controlled study. *Spine (Phila Pa 1976)* 27(13): 1383–9; discussion 1389.
82. Griffin XL, Costa ML, Parsons N, Smith N. 2011. Electromagnetic field stimulation for treating delayed union or non-union of long bone fractures in adults. *Cochrane Database Syst Rev* (4): CD008471.
83. Garland DE, Adkins RH, Matsuno NN, Stewart CA. 1999. The effect of pulsed electromagnetic fields on osteoporosis at the knee in individuals with spinal cord injury. *J Spinal Cord Med* 22(4): 239–45.
84. Clark MJ, Schopp LH, Mazurek MO, Zaniletti I, Lammy AB, Martin TA, Thomas FP, Acuff ME. 2008. Testosterone levels among men with spinal cord injury: Relationship between time since injury and laboratory values. *Am J Phys Med Rehabil* 87(9): 758–67.
85. Qin W, Pan J, Wu Y, Bauman WA, Cardozo C. 2010. Protection against dexamethasone-induced muscle atrophy is related to modulation by testosterone of FOXO1 and PGC-1alpha. *Biochem Biophys Res Commun* 403(3–4): 473–8.
86. Wu Y, Zhao W, Zhao J, Zhang Y, Qin W, Pan J, Bauman WA, Blitzer RD, Cardozo C. 2010. REDD1 is a major target of testosterone action in preventing dexamethasone-induced muscle loss. *Endocrinology* 151(3): 1050–9.
87. Crawford BA, Liu PY, Kean MT, Bleasel JF, Handelsman DJ. 2003. Randomized placebo-controlled trial of androgen effects on muscle and bone in men requiring long-term systemic glucocorticoid treatment. *J Clin Endocrinol Metab* 88(7): 3167–76.

123
Hematopoiesis and Bone

Benjamin J. Frisch and Laura M. Calvi

Introduction 1028
The HSC Niche 1028
A Niche for B Lymphopoiesis 1030
Malignancy and the HSC Niche 1030
Conclusions 1031
References 1031

INTRODUCTION

One of the critical and unique functions of the skeleton in terrestrial vertebrates is to provide the anatomical spaces for production and storage of hematopoietic progenitors and precursors. The most immature cells of the hematopoietic system, hematopoietic stem cells (HSCs), are also found in bone marrow. A true HSC is defined as a single cell capable of generating the entire hematopoietic system throughout the lifetime of an adult [1]. To perform this task HSCs must constantly balance self-renewal and differentiation. While cell autonomous processes play a role in HSC cell fate decisions, *in vitro* studies have continuously demonstrated a loss of long-term engraftment capacity when HSCs are grown in culture without a supportive stromal cell layer [2]. In addition, several studies have shown that disruptions in the microenvironment can lead to aberrant hematopoiesis and even hematopoietic malignancies in mice [3–5]. This has led to the hypothesis that a specific microenvironment or niche is necessary for the maintenance and regulation of normal adult hematopoiesis and HSCs. Due to the singular homing of HSC to bone marrow spaces, it has long been suspected that non-hematopoietic cells contained in the skeleton, as well as their secreted products, including the matrix, may provide cellular and molecular components that are critical for the regulation of hematopoiesis and HSCs. This chapter will review the principal cell types and skeletal signals that have been implicated as regulatory cellular components of the HSC niche. These components not only illustrate the extraordinary complexity of skeletal tissue, but also provide critical clues to novel therapeutic targets for HSC expansion in conditions of myeloablative injury.

THE HSC NICHE

Osteoblastic cells

Schofield first proposed the concept that HSCs are regulated by microenvironmental interactions in 1978 [6]. It would take nearly 20 years before the first studies elucidating the cellular components of the HSC niche would be performed, and osteoblastic cells were identified early on as supportive of HSCs. *In vitro*, osteoblastic cells can maintain the stem cell activity of bone marrow [7–9]. Additional murine studies have demonstrated that transplanted HSCs preferentially engraft at the endosteal surface and in close contact with osteoblastic cells [10–13]. They also demonstrated that, *in vivo*, HSCs with high proliferative and long-term engrafting potential tightly adhered to the endosteal surface [14]. Increased osteoblastic numbers correlate with increased HSCs [9, 12], and the converse experiments have shown that a loss of the osteoblastic lineage of cells results in disrupted hematopoiesis [15, 16].

While osteoblasts produce many soluble factors that are known to be important for hematopoiesis [7, 9, 17], direct cell contact is necessary for osteoblastic cell

Primer on the Metabolic Bone Diseases and Disorders of Mineral Metabolism, Eighth Edition. Edited by Clifford J. Rosen.
© 2013 American Society for Bone and Mineral Research. Published 2013 by John Wiley & Sons, Inc.

support of HSC self-renewal *in vitro* [9]. Treatment with parathyroid hormone (PTH) is capable of stimulating self-renewal of HSCs and increasing the HSC population despite the fact that HSCs do not express the PTH receptor (PTH1R), indicating a microenvironmentally mediated effect [9]. In addition to increasing HSCs, studies utilizing either PTH treatment or the constitutive activation of PTH1R on osteoblastic cells revealed that PTH signaling increases bone volume and the expression of the Notch ligand Jagged1 on osteoblastic cells [9, 18]. Notch signaling is known to support stem cell self-renewal in other stem cell systems and has been implicated in the support of HSCs [19–22]. Further, the inhibition of Notch signaling ablates the PTH-dependent increase in HSC number [9, 18]. Additional studies have further suggested that Notch signaling between osteoblasts and HSCs plays a role in maintaining repopulating potential [23, 24]. A separate study questioned the importance of canonical Notch signaling in the HSC niche, and work remains to be done elucidating the precise role Notch signaling plays in microenvironmental regulation of HSCs [25].

Annexin 2 (Anxa2) is expressed by both osteoblasts and endothelial cells in the bone marrow, and treatment of mice with Anxa2 inhibitors impairs HSC homing and engraftment [26]. Angiopoietin-1 (Ang-1) is expressed by osteoblasts and promotes stronger adhesion and quiescence in HSCs through interactions with the receptor tyrosine kinase Tie2 [27]. Genetic deletion of both Tie1 and Tie2 in mice results in a loss of maintenance of HSCs in the marrow, again suggesting a role of Ang-1 [28]. Osteoblastic cells also express thrombopoietin, a regulator of HSC quiescence [29, 30]. Moreover, osteolineage cells produce CXCL12, a critical regulator of HSC homing and maintenance [31].

Short–lived secreted signals are also potential osteoblastic-dependent HSC regulators. Prostaglandin E_2 (PGE$_2$) is an arachidonic acid derivative that is produced by multiple cell types within the bone marrow, including osteoblastic cells. Treatment of mice with PGE$_2$ increases short-term repopulating HSCs [32] and *ex vivo* treatment of HSCs with dimethyl-PGE$_2$ increases HSC repopulating potential both in mouse bone marrow and in human cord blood samples [33–35], one of the first illustrations of how identification of HSC regulatory signals can be used pharmacologically for therapeutic purposes.

Osteoblastic cells are also critical regulators of osteoclastic activity, and recent data suggest that osteoclastic cells are necessary for the formation of the osteoblastic HSC niche [36]. Osteoclastic cells were first implicated as playing a direct role in HSC regulation when data demonstrated that lack of the calcium sensor impaired the inability of HSCs to home to the skeleton [37]. High concentrations of calcium would be expected to colocalize with actively resorbing functional osteoclasts. Moreover, it was noted that treatment with strontium, which increases the osteoblastic pool but inhibits osteoclasts, did not expand HSCs [38]. In addition, increased osteoclastic bone resorption reduces the levels of local CXCL12 and stimulates mobilization of HSCs from the bone marrow [39, 40]. Together, these data have suggested a supportive role for osteoclasts in the HSC niche, and inhibition of osteoclasts has recently been associated with reduction of HSC numbers *in vivo* [38]. However, recent data using genetic murine models have suggested that osteoclasts are dispensable for HSC maintenance and mobilization [41].

While numerous studies, using both *in vivo* and *in vitro* experimental models, firmly implicate cells of the osteoblastic lineage in HSC maintenance and regulation, the differentiation stage of the osteolineage cell that supports HSCs has begun to emerge, and it has supported the hypothesis that immature cells in the osteoblastic lineage are critical for HSC regulation. Self-renewing osteoprogenitor cells in the marrow can form supportive HSC niches [42]. Most recently, Nestin positive putative mesenchymal stem cells (MSCs) have been demonstrated to express high levels of genes implicated in HSC maintenance. Additionally, these Nestin positive cells are spatially colocated with HSCs, and depletion of the Nestin positive population results in HSC mobilization [43]. A recent study defines the population of proliferative osteoblastic progenitors as MX-1 positive bone marrow stromal cells. This cell population is functionally restricted to an osteoblastic fate *in vivo*, and further investigation into this population's possible role in the HSC niche is warranted [44]. Conversely, even activation of the PTH1R in terminally mature osteocytes, in spite of expansion in the osteoblastic pool, did not increase microenvironmental support for HSCs [45]. The precise characteristics of the osteolineage cells responsible for HSC support remain of great interest since they could predict whether therapeutic strategies that stimulate osteoblastic cells could also achieve beneficial HSC effects.

Endothelial cells

The endothelium gives rise to the first definitive HSCs in the developing embryo [46, 47], and there is mounting evidence that endothelial cells also play an important role in the maintenance of HSCs in the adult marrow. HSCs and hematopoietic progenitor cells localize to endothelial structures near the endosteal surface [48, 49]. Angiocrine factors produced by endothelial cells support HSCs following myeloablation [50, 51] and expand immature hematopoietic cells *ex vivo* [52]. Additionally, genetic models have demonstrated that endothelial cells and a heterogeneous population of perivascular cells express high levels of CXCL12. Further, the conditional deletion of CXCL12 from endothelial cells resulted in the loss of HSC retention in the marrow [53]. *In vivo* imaging shows HSC localization to areas of close spatial relationship between vascular endothelial cells and endosteal osteoblasts, suggesting that both structures may be necessary components of a singular niche [11, 54].

CXCL12 abundant reticular cells

C-X-C motif ligand 12 (CXCL12) also known as stromal cell-derived factor 1 (SDF1) is a chemokine produced by osteoblasts and endothelial cells [31]. CXCL12 signals through its receptor C-X-C motif receptor 4 (CXCR4) on HSCs to induce migration to and retention in the bone marrow [40, 55, 56]. HSCs were shown to be closely associated with CXCL12 abundant reticular (CAR) cells that are in turn closely associated with endothelial cells [57]. The identity of these cells remains to be further characterized.

Sympathetic nervous system neuronal cells and glia

Treatment with granulocyte colony stimulating factor (G-CSF) results in migration of HSCs from the marrow into the bloodstream through inhibition of osteoblastic cells and downregulation of the C-X-C motif ligand 12 (CXCL12) [58–61]. Convincing data have suggested that this phenomenon is at least in part regulated by sympathetic nervous system (SNS) neurons [62]. SNS neurons coordinate the circadian oscillation of HSC numbers in the marrow, and ablation of SNS neurons results in a loss of circadian controlled release of HSCs into the periphery. In addition, sympathectomy of one tibia in a mouse results in altered expression of CXCL12 while the sham-operated contralateral tibia is unaffected [63]. Transforming growth factor β (TGF-β) also regulates HSC dormancy *ex vivo* [64]. Glial cells are a major source of activated TGF-β in the bone marrow, are closely associated with HSCs, and produce numerous factors previously identified as playing roles in the HSC niche [65]. Ablation of this population in the bone marrow results in a loss of HSC dormancy and ultimately of HSC numbers [65]. Therefore, neuronal cells and glia have been recently implicated in HSC regulation in the bone marrow.

Adipocytes

Adipocytic cells make up a large portion of the adult marrow. One adypocytic product, the adipokine adiponectin, has been demonstrated to increase proliferation of HSCs while retaining their repopulating potential [66]. Adiponectin, however, is not solely produced by adipocytes in the marrow, but is also expressed by osteoblastic cells [67]. In contrast to these pro-HSC effects of adipocytes, adipocyte numbers have more recently been described as inversely related to the number of HSCs in the marrow by comparing anatomically distinct regions of the skeleton that display varying levels of adiposity [68]. Additional studies demonstrated that loss of adipocytes in the marrow either genetically or pharmacologically resulted in enhanced engraftment of HSCs and improved hematopoietic recovery following myeloablative injury [68]. Therefore, a dominant inhibitory effect of adipocytes on HSC has recently been elucidated.

Macrophages and monocytes

Macrophages and monocytes are recent additions to the HSC niche's cellular milieu. They are involved in the G-CSF dependent mobilization of HSCs out of the marrow, and depletion of the macrophage population results in loss of osteoblastic cells and increased HSC mobilization into the periphery [69, 70]. In addition, reduction of monocytic phagocytes from the marrow reduces the expression of genes associated with HSC retention in Nestin positive cells, and specific ablation of macrophage populations resulted in the egress of HSCs from the marrow [71]. It is unclear whether macrophages and monocytes specifically interact with HSCs or function only in support of other niche components; however, it has been recently established that they play a pivotal role in the regulation of the HSC niche.

A NICHE FOR B LYMPHOPOIESIS

In addition to HSCs, there is emerging evidence that B lymphopoiesis is regulated by microenvironmental factors in the bone marrow. B-cell precursors are in direct contact with stromal cells in the marrow space that express CXCL12 [72]. Osteoblastic cells can support the development of B lymphocytes from HSCs *in vitro* [73], and ablation of osteoblasts results in a loss of B lymphocytes prior to the loss of HSCs [16, 73]. Osteoblastic support for B lymphopoiesis appears to be mediated by $G_s\alpha$ signaling as the loss of $G_s\alpha$ in osteolineage cells results in a decrease in B-cell precursors in the bone marrow [74] (Fig. 123.1).

MALIGNANCY AND THE HSC NICHE

While cancer metastasis is described elsewhere in the *Primer*, it is important to note that recent data have suggested that metastatic cancers home to benign HSC niches [75]. Therefore, as therapeutic targets are identified for HSC expansion through manipulation of niche components, the effect of their stimulation on malignant cells should be monitored, particularly in the setting of a prior cancer diagnosis.

An area of current interest is the role of the benign niche in hematopoietic malignancies. Data in xenograft models have suggested that malignant leukemic cells disrupt normal interactions between HSCs and their microenvironment [76] and induce changes in the marrow that decrease normal HSC niches and increase malignant niches. Recent data also demonstrate an inhibitory effect of leukemic cells on osteoblastic cells in a murine model

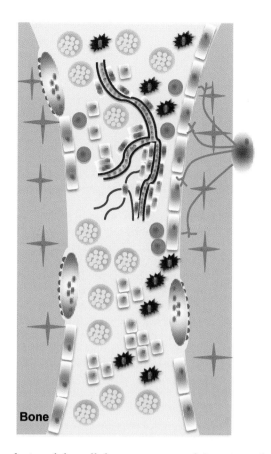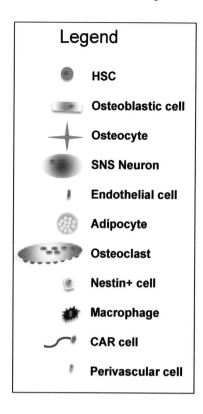

Fig. 123.1. Complexity of the cellular components of the HSC niche. Schematic representation of the cellular components of the HSC niche within the bone marrow space.

of acute myelogenous leukemia (AML), indicating a likely effect on the HSC niche [77]. Additional data have suggested that the marrow microenvironment can protect malignant stem cells as immature AML cells have been shown to reside near the endosteal surface of bone, and these endosteal associated AML cells were resistant to chemotherapy treatment in xenograft models [78, 79]. Further, migration of AML cells out of the marrow by disrupting the CXCL12/CXCR4 axis can sensitize malignant cells to chemotherapeutics [80, 81]. Future discoveries of the cellular and molecular components of normal and malignant niches are therefore needed in order to safely manipulate the microenvironment in the context of malignancy.

CONCLUSIONS

The importance of the skeletal microenvironment in hematopoietic regulation is beginning to be elucidated and holds great promise for translational therapies aimed at hematopoietic recovery and treatment of hematopoietic malignancies as well as metastatic disease. While it is tempting to postulate there are as many HSC niches in the marrow as cell types that have been demonstrated as supportive of HSCs, the reality is almost certainly much more complex. Many cell types likely coordinate to provide one regulatory niche, and equally as likely there are separate niches providing support for distinct and separate populations of HSCs, and/or different potential HSC fates. Further study in this area has benefited greatly from interactions of the field of hematopoiesis with bone biology and will likely continue to rapidly progress to increase our understanding and therapeutic use of the complex cellular relationships within the bone marrow.

REFERENCES

1. Baum CM, Weissman IL, Tsukamoto AS, Buckle AM, Peault B. 1992. Isolation of a candidate human hematopoietic stem-cell population. *Proc Natl Acad Sci U S A* 89(7): 2804–8.
2. Lawal RA, Calvi LM. The niche as a target for hematopoietic manipulation and regeneration. 2011. *Tissue Eng Part B Rev* 17(6): 415–22.
3. Raaijmakers MH, Mukherjee S, Guo S, Zhang S, Kobayashi T, Schoonmaker JA, Ebert BL, Al-Shahrour F,

Hasserjian RP, Scadden EO, Aung Z, Matza M, Merkenschlager M, Lin C, Rommens JM, Scadden DT. 2010. Bone progenitor dysfunction induces myelodysplasia and secondary leukaemia. *Nature* 464(7290): 852–7.

4. Walkley CR, Olsen GH, Dworkin S, Fabb SA, Swann J, McArthur GA, Westmoreland SV, Chambon P, Scadden DT, Purton LE. 2007. A microenvironment-induced myeloproliferative syndrome caused by retinoic acid receptor gamma deficiency. *Cell* 129(6): 1097–110.

5. Walkley CR, Shea JM, Sims NA, Purton LE, Orkin SH. 2007. Rb regulates interactions between hematopoietic stem cells and their bone marrow microenvironment. *Cell* 129(6): 1081–95.

6. Schofield R. The relationship between the spleen colony-forming cell and the haemopoietic stem cell. 1978. *Blood Cells* 4(1–2): 7–25.

7. Taichman RS, Emerson SG. 1994. Human osteoblasts support hematopoiesis through the production of granulocyte colony-stimulating factor. *J Exp Med* 179(5): 1677–82.

8. Taichman RS, Reilly MJ, Emerson SG. 1996. Human osteoblasts support human hematopoietic progenitor cells in vitro bone marrow cultures. *Blood* 87(2): 518–24.

9. Calvi LM, Adams GB, Weibrecht KW, Weber JM, Olson DP, Knight MC, Martin RP, Schipani E, Divieti P, Bringhurst FR, Milner LA, Kronenberg HM, Scadden DT. 2003. Osteoblastic cells regulate the haematopoietic stem cell niche. *Nature* 425(6960): 841–6.

10. Nilsson SK, Johnston HM, Coverdale JA. 2001. Spatial localization of transplanted hemopoietic stem cells: Inferences for the localization of stem cell niches. *Blood* 97(8): 2293–9.

11. Lo Celso C, Fleming HE, Wu JW, Zhao CX, Miake-Lye S, Fujisaki J, Cote D, Rowe DW, Lin CP, Scadden DT. 2009. Live-animal tracking of individual haematopoietic stem/progenitor cells in their niche. *Nature* 457(7225): 92–6.

12. Zhang J, Niu C, Ye L, Huang H, He X, Tong WG, Ross J, Haug J, Johnson T, Feng JQ, Harris S, Wiedemann LM, Mishina Y, Li L. 2003. Identification of the haematopoietic stem cell niche and control of the niche size. *Nature* 425(6960): 836–41.

13. Xie Y, Yin T, Wiegraebe W, He XC, Miller D, Stark D, Perko K, Alexander R, Schwartz J, Grindley JC, Park J, Haug JS, Wunderlich JP, Li H, Zhang S, Johnson T, Feldman RA, Li L. 2009. Detection of functional haematopoietic stem cell niche using real-time imaging. *Nature* 457(7225): 97–101.

14. Haylock DN, Williams B, Johnston HM, Liu MC, Rutherford KE, Whitty GA, Simmons PJ, Bertoncello I, Nilsson SK. 2007. Hemopoietic stem cells with higher hemopoietic potential reside at the bone marrow endosteum. *Stem Cells* 25(4): 1062–9.

15. Visnjic D, Kalajzic I, Gronowicz G, Aguila HL, Clark SH, Lichtler AC, Rowe DW. 2001. Conditional ablation of the osteoblast lineage in Col2.3deltatk transgenic mice. *J Bone Miner Res* 16(12): 2222–31.

16. Visnjic D, Kalajzic Z, Rowe DW, Katavic V, Lorenzo J, Aguila HL. 2004. Hematopoiesis is severely altered in mice with an induced osteoblast deficiency. *Blood* 103(9): 3258–64.

17. Marusic A, Kalinowski JF, Jastrzebski S, Lorenzo JA. 1993. Production of leukemia inhibitory factor mRNA and protein by malignant and immortalized bone cells. *J Bone Miner Res* 8(5): 617–24.

18. Weber JM, Forsythe SR, Christianson CA, Frisch BJ, Gigliotti BJ, Jordan CT, Milner LA, Guzman ML, Calvi LM. 2006. Parathyroid hormone stimulates expression of the Notch ligand Jagged1 in osteoblastic cells. *Bone* 39(3): 485–93.

19. Milner LA, Bigas A. 1999. Notch as a mediator of cell fate determination in hematopoiesis: Evidence and speculation. *Blood* 93(8): 2431–48.

20. Karanu FN, Murdoch B, Gallacher L, Wu DM, Koremoto M, Sakano S, Bhatia M. 2000. The notch ligand jagged-1 represents a novel growth factor of human hematopoietic stem cells. *J Exp Med* 192(9): 1365–72.

21. Karanu FN, Murdoch B, Miyabayashi T, Ohno M, Koremoto M, Gallacher L, Wu D, Itoh A, Sakano S, Bhatia M. 2001. Human homologues of Delta-1 and Delta-4 function as mitogenic regulators of primitive human hematopoietic cells. *Blood* 97(7): 1960–7.

22. Karanu FN, Yuefei L, Gallacher L, Sakano S, Bhatia M. 2003. Differential response of primitive human CD34– and CD34+ hematopoietic cells to the Notch ligand Jagged-1. *Leukemia* 17(7): 1366–74.

23. Chitteti BR, Cheng YH, Poteat B, Rodriguez-Rodriguez S, Goebel WS, Carlesso N, Kacena MA, Srour EF. 2010. Impact of interactions of cellular components of the bone marrow microenvironment on hematopoietic stem and progenitor cell function. *Blood* 115(16): 3239–48.

24. Weber JM, Calvi LM. 2010. Notch signaling and the bone marrow hematopoietic stem cell niche. *Bone* 46(2): 281–5.

25. Maillard I, Koch U, Dumortier A, Shestova O, Xu L, Sai H, Pross SE, Aster JC, Bhandoola A, Radtke F, Pear WS. 2008. Canonical notch signaling is dispensable for the maintenance of adult hematopoietic stem cells. *Cell Stem Cell* 2(4): 356–66.

26. Jung Y, Wang J, Song J, Shiozawa Y, Havens A, Wang Z, Sun YX, Emerson SG, Krebsbach PH, Taichman RS. 2007. Annexin II expressed by osteoblasts and endothelial cells regulates stem cell adhesion, homing, and engraftment following transplantation. *Blood* 110(1): 82–90.

27. Arai F, Hirao A, Ohmura M, Sato H, Matsuoka S, Takubo K, Ito K, Koh GY, Suda T. 2004. Tie2/angiopoietin-1 signaling regulates hematopoietic stem cell quiescence in the bone marrow niche. *Cell* 118(2): 149–61.

28. Puri MC, Bernstein A. 2003. Requirement for the TIE family of receptor tyrosine kinases in adult but not fetal hematopoiesis. *Proc Natl Acad Sci U S A* 100(22): 12753–8.

29. Qian H, Buza-Vidas N, Hyland CD, Jensen CT, Antonchuk J, Mansson R, Thoren LA, Ekblom M, Alexander WS, Jacobsen SE. 2007. Critical role of thrombopoietin

in maintaining adult quiescent hematopoietic stem cells. *Cell Stem Cell* 1(6): 671–84.
30. Yoshihara H, Arai F, Hosokawa K, Hagiwara T, Takubo K, Nakamura Y, Gomei Y, Iwasaki H, Matsuoka S, Miyamoto K, Miyazaki H, Takahashi T, Suda T. 2007. Thrombopoietin/MPL signaling regulates hematopoietic stem cell quiescence and interaction with the osteoblastic niche. *Cell Stem Cell* 1(6): 685–97.
31. Ponomaryov T, Peled A, Petit I, Taichman RS, Habler L, Sandbank J, Arenzana-Seisdedos F, Magerus A, Caruz A, Fujii N, Nagler A, Lahav M, Szyper-Kravitz M, Zipori D, Lapidot T. 2000. Induction of the chemokine stromal-derived factor-1 following DNA damage improves human stem cell function. *J Clin Invest* 106(11): 1331–9.
32. Frisch BJ, Porter RL, Gigliotti BJ, Olm-Shipman AJ, Weber JM, O'Keefe RJ, Jordan CT, Calvi LM. 2009. In vivo prostaglandin E(2) treatment alters the bone marrow microenvironment and preferentially expands short-term hematopoietic stem cells. *Blood* 114(19): 4054–63.
33. North TE, Goessling W, Walkley CR, Lengerke C, Kopani KR, Lord AM, Weber GJ, Bowman TV, Jang IH, Grosser T, Fitzgerald GA, Daley GQ, Orkin SH, Zon LI. 2007. Prostaglandin E2 regulates vertebrate haematopoietic stem cell homeostasis. *Nature* 447(7147): 1007–11.
34. Goessling W, Allen RS, Guan X, Jin P, Uchida N, Dovey M, Harris JM, Metzger ME, Bonifacino AC, Stroncek D, Stegner J, Armant M, Schlaeger T, Tisdale JF, Zon LI, Donahue RE, North TE. 2011. Prostaglandin E2 enhances human cord blood stem cell xenotransplants and shows long-term safety in preclinical nonhuman primate transplant models. *Cell Stem Cell* 8(4): 445–58.
35. Hoggatt J, Singh P, Sampath J, Pelus LM. 2009. Prostaglandin E2 enhances hematopoietic stem cell homing, survival, and proliferation. *Blood* 113(22): 5444–55.
36. Mansour A, Abou-Ezzi G, Sitnicka E, SE WJ, Wakkach A, Blin-Wakkach C. 2012. Osteoclasts promote the formation of hematopoietic stem cell niches in the bone marrow. *J Exp Med* 209(3): 537–49.
37. Adams GB, Chabner KT, Alley IR, Olson DP, Szczepiorkowski ZM, Poznansky MC, Kos CH, Pollak MR, Brown EM, Scadden DT. 2006. Stem cell engraftment at the endosteal niche is specified by the calcium-sensing receptor. *Nature* 439(7076): 599–603.
38. Lymperi S, Horwood N, Marley S, Gordon MY, Cope AP, Dazzi F. 2008. Strontium can increase some osteoblasts without increasing hematopoietic stem cells. *Blood* 111(3): 1173–81.
39. Cho KA, Joo SY, Han HS, Ryu KH, Woo SY. 2010. Osteoclast activation by receptor activator of NF-kappaB ligand enhances the mobilization of hematopoietic progenitor cells from the bone marrow in acute injury. *Int J Mol Med* 26(4): 557–63.
40. Kollet O, Dar A, Shivtiel S, Kalinkovich A, Lapid K, Sztainberg Y, Tesio M, Samstein RM, Goichberg P, Spiegel A, Elson A, Lapidot T. 2006. Osteoclasts degrade endosteal components and promote mobilization of hematopoietic progenitor cells. *Nat Med* 12(6): 657–64.
41. Miyamoto K, Yoshida S, Kawasumi M, Hashimoto K, Kimura T, Sato Y, Kobayashi T, Miyauchi Y, Hoshi H, Iwasaki R, Miyamoto H, Hao W, Morioka H, Chiba K, Yasuda H, Penninger JM, Toyama Y, Suda T, Miyamoto T. 2011. Osteoclasts are dispensable for hematopoietic stem cell maintenance and mobilization. *J Exp Med* 208(11): 2175–81.
42. Sacchetti B, Funari A, Michienzi S, Di Cesare S, Piersanti S, Saggio I, Tagliafico E, Ferrari S, Robey PG, Riminucci M, Bianco P. 2007. Self-renewing osteoprogenitors in bone marrow sinusoids can organize a hematopoietic microenvironment. *Cell* 131(2): 324–36.
43. Mendez-Ferrer S, Michurina TV, Ferraro F, Mazloom AR, Macarthur BD, Lira SA, Scadden DT, Ma'ayan A, Enikolopov GN, Frenette PS. 2010. Mesenchymal and haematopoietic stem cells form a unique bone marrow niche. *Nature* 466(7308): 829–34.
44. Park D, Spencer JA, Koh BI, Kobayashi T, Fujisaki J, Clemens TL, Lin CP, Kronenberg HM, Scadden DT. 2012. Endogenous bone marrow MSCs are dynamic, fate-restricted participants in bone maintenance and regeneration. *Cell Stem Cell* 10: 259–72.
45. Calvi LM, Bromberg O, Rhee Y, Lee R, Weber JM, Smith JN, Basil M, Frisch BJ, Bellido T. 2012. Osteoblastic expansion induced by parathyroid hormone receptor signaling in murine osteocytes is not sufficient to increase hematopoietic stem cells. *Blood* 119(11): 2489–99.
46. Chen MJ, Li Y, De Obaldia ME, Yang Q, Yzaguirre AD, Yamada-Inagawa T, Vink CS, Bhandoola A, Dzierzak E, Speck NA. 2011. Erythroid/myeloid progenitors and hematopoietic stem cells originate from distinct populations of endothelial cells. *Cell Stem Cell* 9(6): 541–52.
47. Chen AT, Zon LI. 2009. Zebrafish blood stem cells. *J Cell Biochem* 108(1): 35–42.
48. Dahlberg A, Delaney C, Bernstein ID. 2011. Ex vivo expansion of human hematopoietic stem and progenitor cells. *Blood* 117(23): 6083–90.
49. Kiel MJ, Yilmaz OH, Iwashita T, Terhorst C, Morrison SJ. 2005. SLAM family receptors distinguish hematopoietic stem and progenitor cells and reveal endothelial niches for stem cells. *Cell* 121(7): 1109–21.
50. Butler JM, Nolan DJ, Vertes EL, Varnum-Finney B, Kobayashi H, Hooper AT, Seandel M, Shido K, White IA, Kobayashi M, Witte L, May C, Shawber C, Kimura Y, Kitajewski J, Rosenwaks Z, Bernstein ID, Rafii S. 2010. Endothelial cells are essential for the self-renewal and repopulation of Notch-dependent hematopoietic stem cells. *Cell Stem Cell* 6(3): 251–64.
51. Kobayashi H, Butler JM, O'Donnell R, Kobayashi M, Ding BS, Bonner B, Chiu VK, Nolan DJ, Shido K, Benjamin L, Rafii S. 2010. Angiocrine factors from Akt-activated endothelial cells balance self-renewal and differentiation of haematopoietic stem cells. *Nat Cell Biol* 12(11): 1046–56.
52. Chute JP, Muramoto GG, Dressman HK, Wolfe G, Chao NJ, Lin S. 2006. Molecular profile and partial functional analysis of novel endothelial cell-derived growth factors that regulate hematopoiesis. *Stem Cells* 24(5): 1315–27.

53. Ding L, Saunders TL, Enikolopov G, Morrison SJ. 2012. Endothelial and perivascular cells maintain haematopoietic stem cells. *Nature* 481(7382): 457–62.
54. Ellis SL, Grassinger J, Jones A, Borg J, Camenisch T, Haylock D, Bertoncello I, Nilsson SK. 2011. The relationship between bone, hemopoietic stem cells, and vasculature. *Blood* 118(6): 1516–24.
55. Broxmeyer HE, Orschell CM, Clapp DW, Hangoc G, Cooper S, Plett PA, Liles WC, Li X, Graham-Evans B, Campbell TB, Calandra G, Bridger G, Dale DC, Srour EF. 2005. Rapid mobilization of murine and human hematopoietic stem and progenitor cells with AMD3100, a CXCR4 antagonist. *J Exp Med* 201(8): 1307–18.
56. Peled A, Petit I, Kollet O, Magid M, Ponomaryov T, Byk T, Nagler A, Ben-Hur H, Many A, Shultz L, Lider O, Alon R, Zipori D, Lapidot T. 1999. Dependence of human stem cell engraftment and repopulation of NOD/SCID mice on CXCR4. *Science* 283(5403): 845–8.
57. Sugiyama T, Kohara H, Noda M, Nagasawa T. 2006. Maintenance of the hematopoietic stem cell pool by CXCL12-CXCR4 chemokine signaling in bone marrow stromal cell niches. *Immunity* 25(6): 977–88.
58. Petit I, Szyper-Kravitz M, Nagler A, Lahav M, Peled A, Habler L, Ponomaryov T, Taichman RS, Arenzana-Seisdedos F, Fujii N, Sandbank J, Zipori D, Lapidot T. 2002. G-CSF induces stem cell mobilization by decreasing bone marrow SDF-1 and up-regulating CXCR4. *Nat Immunol* 3(7): 687–94.
59. Levesque JP, Hendy J, Takamatsu Y, Simmons PJ, Bendall LJ. 2003. Disruption of the CXCR4/CXCL12 chemotactic interaction during hematopoietic stem cell mobilization induced by GCSF or cyclophosphamide. *J Clin Invest* 111(2): 187–96.
60. Semerad CL, Christopher MJ, Liu F, Short B, Simmons PJ, Winkler I, Levesque JP, Chappel J, Ross FP, Link DC. 2005. G-CSF potently inhibits osteoblast activity and CXCL12 mRNA expression in the bone marrow. *Blood* 106(9): 3020–7.
61. Christopher MJ, Liu F, Hilton MJ, Long F, Link DC. 2009. Suppression of CXCL12 production by bone marrow osteoblasts is a common and critical pathway for cytokine-induced mobilization. *Blood* 114(7): 1331–9.
62. Katayama Y, Battista M, Kao WM, Hidalgo A, Peired AJ, Thomas SA, Frenette PS. 2006. Signals from the sympathetic nervous system regulate hematopoietic stem cell egress from bone marrow. *Cell* 124(2): 407–21.
63. Mendez-Ferrer S, Lucas D, Battista M, Frenette PS. 2008. Haematopoietic stem cell release is regulated by circadian oscillations. *Nature* 452(7186): 442–7.
64. Yamazaki S, Iwama A, Takayanagi S, Eto K, Ema H, Nakauchi H. 2009. TGF-beta as a candidate bone marrow niche signal to induce hematopoietic stem cell hibernation. *Blood* 113(6): 1250–6.
65. Yamazaki S, Ema H, Karlsson G, Yamaguchi T, Miyoshi H, Shioda S, Taketo MM, Karlsson S, Iwama A, Nakauchi H. 2011. Nonmyelinating schwann cells maintain hematopoietic stem cell hibernation in the bone marrow niche. *Cell* 147(5): 1146–58.
66. DiMascio L, Voermans C, Uqoezwa M, Duncan A, Lu D, Wu J, Sankar U, Reya T. 2007. Identification of adiponectin as a novel hemopoietic stem cell growth factor. *J Immunol* 178(6): 3511–20.
67. Berner HS, Lyngstadaas SP, Spahr A, Monjo M, Thommesen L, Drevon CA, Syversen U, Reseland JE. 2004. Adiponectin and its receptors are expressed in bone-forming cells. *Bone* 35(4): 842–9.
68. Naveiras O, Nardi V, Wenzel PL, Hauschka PV, Fahey F, Daley GQ. 2009. Bone-marrow adipocytes as negative regulators of the haematopoietic microenvironment. *Nature* 460(7252): 259–63.
69. Christopher MJ, Rao M, Liu F, Woloszynek JR, Link DC. 2011. Expression of the G-CSF receptor in monocytic cells is sufficient to mediate hematopoietic progenitor mobilization by G-CSF in mice. *J Exp Med* 208(2): 251–60.
70. Winkler IG, Sims NA, Pettit AR, Barbier V, Nowlan B, Helwani F, Poulton IJ, van Rooijen N, Alexander KA, Raggatt LJ, Levesque JP. 2010. Bone marrow macrophages maintain hematopoietic stem cell (HSC) niches and their depletion mobilizes HSCs. *Blood* 116(23): 4815–28.
71. Chow A, Lucas D, Hidalgo A, Mendez-Ferrer S, Hashimoto D, Scheiermann C, Battista M, Leboeuf M, Prophete C, van Rooijen N, Tanaka M, Merad M, Frenette PS. 2011. Bone marrow CD169+ macrophages promote the retention of hematopoietic stem and progenitor cells in the mesenchymal stem cell niche. *J Exp Med* 208(2): 261–71.
72. Tokoyoda K, Egawa T, Sugiyama T, Choi BI, Nagasawa T. 2004. Cellular niches controlling B lymphocyte behavior within bone marrow during development. *Immunity* 20(6): 707–18.
73. Zhu J, Garrett R, Jung Y, Zhang Y, Kim N, Wang J, Joe GJ, Hexner E, Choi Y, Taichman RS, Emerson SG. 2007. Osteoblasts support B-lymphocyte commitment and differentiation from hematopoietic stem cells. *Blood* 109(9): 3706–12.
74. Wu JY, Purton LE, Rodda SJ, Chen M, Weinstein LS, McMahon AP, Scadden DT, Kronenberg HM. 2008. Osteoblastic regulation of B lymphopoiesis is mediated by Gs(alpha)-dependent signaling pathways. *Proc Natl Acad Sci U S A* 105(44): 16976–81.
75. Shiozawa Y, Pedersen EA, Havens AM, Jung Y, Mishra A, Joseph J, Kim JK, Patel LR, Ying C, Ziegler AM, Pienta MJ, Song J, Wang J, Loberg RD, Krebsbach PH, Pienta KJ, Taichman RS. 2011. Human prostate cancer metastases target the hematopoietic stem cell niche to establish footholds in mouse bone marrow. *J Clin Invest* 121(4): 1298–312.
76. Colmone A, Amorim M, Pontier AL, Wang S, Jablonski E, Sipkins DA. 2008. Leukemic cells create bone marrow niches that disrupt the behavior of normal hematopoietic progenitor cells. *Science* 322(5909): 1861–5.
77. Frisch BJ, Ashton JM, Xing L, Becker MW, Jordan CT, Calvi LM. 2012. Functional inhibition of osteoblastic cells in an in vivo mouse model of myeloid leukemia. *Blood* 119(2): 540–50.

78. Lane SW, Wang YJ, Lo Celso C, Ragu C, Bullinger L, Sykes SM, Ferraro F, Shterental S, Lin CP, Gilliland DG, Scadden DT, Armstrong SA, Williams DA. 2011. Differential niche and Wnt requirements during acute myeloid leukemia progression. *Blood* 118(10): 2849–56.
79. Ishikawa F, Yoshida S, Saito Y, Hijikata A, Kitamura H, Tanaka S, Nakamura R, Tanaka T, Tomiyama H, Saito N, Fukata M, Miyamoto T, Lyons B, Ohshima K, Uchida N, Taniguchi S, Ohara O, Akashi K, Harada M, Shultz LD. 2007. Chemotherapy-resistant human AML stem cells home to and engraft within the bone-marrow endosteal region. *Nat Biotechnol* 25(11): 1315–21.
80. Nervi B, Ramirez P, Rettig MP, Uy GL, Holt MS, Ritchey JK, Prior JL, Piwnica-Worms D, Bridger G, Ley TJ, Dipersio JF. 2008. Chemosensitization of AML following mobilization by the CXCR4 antagonist AMD3100. *Blood* 113(24): 6206–14.
81. Zeng Z, Shi YX, Samudio IJ, Wang RY, Ling X, Frolova O, Levis M, Rubin JB, Negrin RR, Estey EH, Konoplev S, Andreeff M, Konopleva M. 2009. Targeting the leukemia microenvironment by CXCR4 inhibition overcomes resistance to kinase inhibitors and chemotherapy in AML. *Blood* 113(24): 6215–24.

124

Bone and Immune Cell Interactions

Brendan F. Boyce

Introduction 1036
Roles for RANKL and NF-κB in Osteoclast Formation and Immune Responses 1036

References 1040

INTRODUCTION

For many years before the identification, in the mid-1990s, of receptor activator of nuclear factor-κB ligand (RANKL) as the critical osteoclastogenic factor expressed by osteoblastic cells, it was known that osteoblastic cells regulated osteoclast precursor (OCP) differentiation and osteoclast (OC) activation [1]. It was also recognized that inflammatory and infectious diseases affecting the skeleton induced localized or generalized osteolysis through the effects of pro-inflammatory cytokines and other factors. However, the molecular mechanisms mediating bone loss and how immune cells interacted with bone cells in pathologic processes were poorly understood. In the mid-to-late 1980s, nuclear factor-kappa B (NF-κB) was identified as a set of transcription factors that regulate immune cell differentiation and immune responses [2], and later was shown to play essential roles in osteoclastogenesis downstream of RANK, the receptor for RANKL [3, 4]. Calcitonin and a few inflammatory mediators, including PGE2 and interferon γ, were known to inhibit OC-mediated bone resorption, but there were no data to suggest that OCs or OCPs might positively or negatively regulate their own functions or those of osteoblastic and other cell types.

Since the discovery of RANKL [5], understanding of the molecular mechanisms that regulate osteoclastogenesis and how bone and immune cells interact with one another has increased significantly. In addition, numerous other factors and mechanisms have been shown to regulate interactions among mesenchymal stem cells (MSCs), hematopoietic stem cells (HSCs), and osteoblastic, hematopoietic, and immune cells in normal and pathologic states [6, 7, 8]. These interactions maintain skeletal and hematopoietic homeostasis, and they are readily disturbed in a number of pathologic conditions affecting the skeleton. RANKL and NF-κB mediate many normal and pathologic processes; they play a central role linking bone and immune cell functions and immune responses, and discovery of their roles in these processes has spawned the new field of osteoimmunology. This chapter reviews current understanding of how these interactions affect the skeleton.

ROLES FOR RANKL AND NF-κB IN OSTEOCLAST FORMATION AND IMMUNE RESPONSES

RANKL

Bone cells consist of osteoblasts (OBs), osteocytes, chondroblasts, chondrocytes, and their progenitors, which are all derived from mesenchymal stem cells (MSCs) [8, 9]. Osteoclasts, which degrade bone, are hematopoietic cells and like immune cells, such as B cells, T cells, and macrophages and their various progenitors, they are derived from hematopoietic stem cells (HSCs) [8, 9]. Bone and hematopoietic cells are in close contact with one another

---

Primer on the Metabolic Bone Diseases and Disorders of Mineral Metabolism, Eighth Edition. Edited by Clifford J. Rosen.
© 2013 American Society for Bone and Mineral Research. Published 2013 by John Wiley & Sons, Inc.

within the marrow cavity. Osteoblastic/stromal cells from the marrow have long been considered to regulate OC formation from precursors in the myeloid lineage by a combination of direct cell–cell contact and expression of a variety of factors and cytokines [9], including macrophage-colony stimulating factor (M-CSF) and RANKL [5, 9]. M-CSF induces expression of RANK on the surface of OC precursors [10], thus priming them for completion of their differentiation into mature OCs. Recent studies reported the surprising finding that during adult bone remodeling in mice, osteocytes within bone, rather than marrow stromal cells, appear to regulate OC formation by expressing RANKL [11, 12]. Interestingly, osteocytes do not fulfill this function in developing long bones during skeletogenesis. Other studies suggest that during long bone growth, hypertrophic chondrocytes in growth plates may be a major source of RANKL, which they express in response to vitamin D_3 [13] and locally expressed bone morphogenetic protein 2 (BMP2) [14]. RANKL secreted at the interface between the cartilage and newly formed bone could attract OCPs circulating in the blood [15, 16] from the prominent vascular channels seen at this site. The precise mechanism whereby RANKL expressed by osteocytes regulates bone remodeling remains to be identified, but presumably the cells secrete RANKL locally at sites destined for resorption to attract OCPs.

NF-κB

RANKL interaction with RANK activates NF-κB, a family of transcription factors that has essential functions in OCPs for their differentiation into OCs [5]. Importantly, in the context of immune responses and bone, NF-κB also regulates the differentiation and activation of immune cells, including B and T cells, maintenance of immune responses, and the development of lymph nodes [3, 4]. B and T cells also express RANKL and thus can cooperate with osteoblastic cells to enhance OC formation at sites of inflammation in bone where expression of pro-inflammatory cytokines, such as tumor necrosis factor (TNF), interleukin-1 (IL-1), and IL-6, typically is increased [17]. TNF and IL-1, like RANKL, also induce NF-κB expression in OCPs and other immune cells [18]; NF-κB in turn upregulates the expression of TNF, IL-1, and IL-6 [19]. Thus, inflammatory reactions tend to induce autocrine and paracrine auto-amplifying cycles within immune cells that maintain and upregulate immune responses, and thus can lead to more aggressive bone destruction [19].

TNF, which is secreted by macrophages in inflamed joints, promotes OCP egress from the bone marrow by inhibiting production of stromal cell-derived factor-1 (SDF-1) by marrow stromal cells and increases osteoclast precursor mobilization from bone marrow to the peripheral blood [20]. It also appears to attract OCPs to affected joints by increasing local expression of SDF-1 [20]. TNF not only induces expression of RANKL by accessory cells, but it also induces OC formation directly [21].

T and B lymphocytes

Numerous inflammatory processes affect the skeleton and typically result in localized or generalized bone resorption. These include common conditions, such as acute and chronic bacterial osteomyelitis and rheumatoid arthritis (RA). In patients with RA, important roles have been identified for subsets of T cells [17], particularly helper T (Th) and regulatory T (Treg) cells, B cells, and macrophages, which are abundant in the synovium of affected joints [22–24]. Inflamed synoviocytes and activated T cells secrete RANKL, which can attract circulating OCPs to inflamed joints and are now considered to be major players in inflammation-mediated joint destruction in RA [22–24]. In addition to the activated T cells that induce bone resorption, there are other T cells that produce cytokines, which potently inhibit osteoclastogenesis. For example, Tregs overall appear to inhibit OC formation mainly by direct contact with OCPs through cytotoxic T lymphocyte antigen 4 [25] and production of IL-4 and IL-10, which can potently inhibit osteoclastogenesis [17, 22].

B cells and plasma cells are found commonly in chronic inflammatory processes in bone.

Recent studies have implicated T cells in the bone loss that occurs as a result of sex steroid deficiency following ovariectomy (Ovx) in mice, which is associated with increased production of pro-inflammatory cytokines by T cells and macrophages. For example, Pacifici and colleagues have reported that T-cell-deficient nude mice, wild-type mice depleted of T cells, and mice lacking the T-cell co-stimulatory molecule, CD40 ligand, do not lose bone after Ovx [26, 27]. These results have been confirmed by some, but not other, investigators, leaving the findings somewhat controversial [26]. However, they are supported by other studies showing that estrogen inhibits differentiation of the Th17 subset of $CD4^+$ helper cells, which express high levels of IL-17, RANKL, and TNF, and low levels of the osteoclastogenesis inhibitor, interferon gamma (IFNγ) [26]. Although the roles of RANKL and TNF are well established in Ovx-induced bone loss, the role of IL-17 is controversial [26]. The role of T cells in Ovx-induced bone loss is also supported by reports that transgenic mice overexpressing Tregs are protected against Ovx-induced bone loss; that estrogen increases Treg numbers and regulates TNF production by T cells by inhibiting production of the osteoclastogenic cytokine, IL-7; and that estrogen deficiency expands the pool of TNF-producing T cells [26].

Ovx is also associated with increased production of reactive oxygen species (ROS) in the bone marrow by numerous cell types, including T cells and macrophages [28]. ROS increase the production of mature dendritic cells (DCs) expressing the co-stimulatory molecule, CD80, increase DC-mediated antigen presentation, and increase osteoclast formation [26, 28]. Upregulation of expression of major histocompatability molecules (MHCs) and co-stimulatory molecules, such as CD40 and CD80, by mature DCs is necessary for antigen presentation by them and for their activation of T cells during immune responses. Importantly, like DCs, OCs also express the Fc receptor common γ subunit (FcRγ), MHC, CD40, and CD80, and function as antigen-presenting cells to activate T cells [29]. Thus, OCs could potentially amplify Ovx-induced T-cell proliferation and activation. However, OCs also can inhibit T-cell proliferation and suppress T-cell production of TNFα and IFNγ [30], and thereby potentially could also limit the role of T cells in Ovx-induced bone loss. Full discussion of the mechanisms whereby immune cells, including OCs, regulate sex steroid-related bone loss is beyond the scope of this review, but there clearly is a lot more to be learned about the interactions between bone and immune cells in this setting.

Plasma cells develop from B cells and secrete antibodies in chronic inflammation, but they can also express autoantibodies in autoimmune diseases. These include rheumatoid factor and anti-cyclic citrullinated peptide antibody in patients with RA where increased serum levels correlate with the level of disease activity [31]. B cells express numerous cytokines, including RANKL and OPG, but conflicting data that reported increased production of both RANKL [31] and OPG [32] have resulted in their roles in osteoclastogenesis being controversial and requiring further investigation. Full coverage of all aspects of the important field of osteoimmunology is beyond the scope of this review, but they have been covered in a number of recent reviews [17, 22, 33].

Other roles for RANKL

Numerous other cells types express RANKL in various pathologic settings, in addition to osteoblastic, synovial, and immune cells. For example, RANKL is expressed by intestinal cells in inflammatory bowel disease where it may contribute to generalized bone loss in affected patients [34]; it centrally controls body temperature and fever in females [35]; it is expressed in atheromatous plaques in arteries where immune cells have a pathogenetic role, and aberrant expression of RANKL and OPG has been implicated in hypertension, cardiovascular disease [36], and diabetes; and it regulates the proliferation and metastasis of breast cancer cells to bone [37, 38].

Osteoclast-rich lesions outside bone

Given the widespread expression of RANKL, it is perhaps not surprising that osteoclasts are also found in inflammatory and neoplastic lesions outside of bone. For example, they are present in the nodular lesions around joints in pigmented villonodular synovitis (PVNS) [39]; in giant cell tumor of tendon sheath, an inflammatory lesion similar to PVNS; and in some malignant tumors, including leiomyosarcoma, and carcinomas of pancreas, breast, and bladder [40]. Stromal cells in PVNS and giant cell tumor of tendon sheath express M-CSF, and the cells derived from them form OCs in response to RANKL [39].

Although it is likely that malignant tumors with OCs present within them express RANKL, to date there have been no reports describing this, let alone what the functions of OCs in these pathologic settings might be. However, as will be seen below, numerous positive and negative regulatory roles have been identified for OCs and OCPs in the past few years, and it is possible that they have supportive roles in some pathologic processes.

Co-stimulatory signaling in osteoclastogenesis

In addition to RANKL expressed by various cell types during inflammatory processes, so-called co-stimulatory signaling can also directly induce osteoclast formation [22, 41] through immunoglobulin-like receptors, including triggering receptor expressed in myeloid cells-2 (TREM-2) and osteoclast-associated receptor (OSCAR) [38]. Adaptor molecules, including DNAX-activating protein 12 (DAP12) and FcRγ, associate with these receptors. Through a series of phosphorylation events involving immunoreceptor tyrosine-based activation motifs (ITAMs) within these adaptor proteins [22] and phospholipase Cγ (PLCγ), calcium fluxes increase within OCPs and nuclear factor of activated T cells (NFATc1), the so-called master regulator of osteoclastogenesis, is activated by its dephosphorylation by calcineurin [22]. The ligands for co-stimulatory signaling receptors remain largely unidentified; however, OSCAR appears to be activated in OCPs by parts of collagen fibers [42], which become exposed in resorbing bone surfaces. These recent findings suggest that OSCAR can induce OC formation during normal bone resorption and presumably also at sites of inflammation in bone.

RANK and co-stimulatory signaling actually appear to be linked directly within OCPs through scaffold proteins to increase PLCγ/calcineurin-induced activation of NFATc1 and thus OC formation [43]. For example, in response to RANKL, Bruton (Btk) and Tec tyrosine kinases bind to RANK, and they also phosphorylate PLCγ. [22] They associate with B-cell linker protein (BLNK), which also binds to proteins in the co-stimulatory signaling pathway [43]. Thus, RANK and co-stimulatory signaling together activate PLCγ through BLNK and likely augment OC formation to mediate enhanced OC formation in inflammatory diseases affecting the skeleton. Calcium fluxes induced by RANK and co-stimulatory signaling also involve activation of cyclic adenosine monophosphate responsive-element-binding protein (CREB) by calcium/calmodulin-dependent protein kinase IV (CaMKIV) [44]. This results in activation of c-Fos, another transcription factor downstream from NF-κB whose expression is also necessary for OC formation by directing myeloid precursors otherwise destined to be macrophages down the OC differentiation pathway [18, 45]. Although these are complex intracellular signaling systems, they play crucial roles in basal and inflammation-mediated osteoclastogenesis; fuller understanding of the molecular mechanisms involved could lead to development of novel therapies to prevent not only bone resorption in common inflammatory disease, but also the inflammation that accompanies it.

Acute and chronic osteomyelitis

Acute and chronic bone infections (osteomyelitis) typically result in localized lytic lesions at affected sites where increased osteoclastogenesis is induced by a number of mechanisms. Osteomyelitis is typically blood-borne from a distant site of acute infection, particularly in children. However, it can also result from direct inoculation of bacteria, fungi, or other organisms. This can occur in compound fractures that result from severe force and cause the bone to protrude through the skin or through infected gangrenous soft tissues of patients with severe peripheral vascular disease. However, most cases of post-traumatic osteomyelitis are caused by hospital-acquired bacteria [46]. Osteomyelitis can also occur in healthy patients undergoing elective arthroplasty for degenerative joint disease where methicillin-resistant *Staphylococcus aureus* (MRSA) infections have reemerged as a significant, predominantly hematogenous complication. In addition, new infections caused by organisms, such as *Acinetobacter baumannii*, have emerged at unprecedented levels in soldiers wounded in recent military conflicts in the Middle East.

Acute bacterial infections quickly activate complements in the innate immune response, leading to local vasodilatation, migration of polymorphonuclear leukocytes (PMNs), bacterial coating with ospsonins, which facilitate their intracellular uptake, and release of cytokines, including IL-1, IL-6, and TNF. These cytokines attract and activate PMNs and macrophages, which release reactive oxygen species, and, like cytokines, they stimulate OC formation and activity. The acquired immune response is activated days later and involves cytotoxic or CD8+ T cells, which lyse bacteria-infected host cells, and B lymphocytes, which produce antibacterial antibodies that clear persistent bacteria and prevent recurrence of the same infection. Osteoblastic cells also appear to be a significant line of defense against colonization by bacteria and bacterial formation of biofilms, which protect the organisms from phagocytosis in infections in bone. They express Toll-like receptors (TLRs) 2, 4, and 9 [47], which enable them to respond to bacterial cell surface lipopolysaccharides and DNA (CpG oligonucleotides). Following activation of these receptors by bacterial ligands, osteoblasts produce a variety of antimicrobial peptides to limit the effects of infection in bone. These include beta defensin-3 [48], the chemokines, CCL2 and −5, CXCL8 and −10, inflammatory cytokines (IL-6), co-stimulatory molecules (CD40), and MHC II, which are expressed typically by lymphocytes and antigen-presenting phagocytic immune cells in soft tissues and other organs during infections. Thus, these findings suggest that osteoblasts play an unexpected important role in the cellular immune response to bacterial infections. Osteoblasts also appear to internalize bacteria *in vitro* [49] and *in vivo* as well as the *S. aureus* sigma B regulon, a key mediator of bacterial infection, which could promote infection by inducing osteoblast apoptosis.

Interactions between osteoblastic and osteoclastic cells that influence bone mass in inflammatory responses

Bone formation typically is inhibited within sites of acute and chronic infections in bone where bone resorption is stimulated, but is seen around the periphery of affected sites as part of the inflammatory/repair process. Resolution of the infection typically is accompanied by osteoblastic new bone formation, which eventually fills in the lytic lesions. The molecular mechanisms that initially inhibit and later stimulate new bone formation in this setting have not been studied in detail, but high concentrations of pro-inflammatory cytokines, such as TNF, can inhibit OB formation and survival. Interestingly, TNF can also stimulate MSC differentiation into OBs at lower concentrations, which appear to be present in the early phases of fracture repair [50]. Thus, it is possible that as cytokine concentrations fall in the resolving phase of inflammation, they stimulate new bone formation. Another possibility is that secretion of OB inhibitors, such as dikkoff-1 by immune cells [51], decreases as inflammation is resolving.

Recent studies suggest that bone mass may be affected by direct interactions between osteoclastic and osteoblastic cells in normal and disease states. For example, osteoblastic cells appear to regulate OC formation negatively by direct cell–cell interaction through so-called reverse signaling via the ligand, ephrin B2, expressed by OCPs during its interaction with its receptor, Eph4, on the surface of osteoblast precursors. The reverse signaling downregulates expression in OCPs of c-Fos and NFATc1 and thus inhibits osteoclastogenesis. In contrast, forward signaling through Eph4 inactivates RhoA and thus stimulates OB differentiation from precursors [52]. This could provide an additional mechanism to mediate new bone formation in resolving inflammation. Eph4B expression by OBs and osteocytes is increased in samples of subchondral bone from patients with osteoarthritis (OA), which typically is sclerotic, and in OBs from bone samples from some OA subjects [53]. However, Eph4B activation in these cells is inhibited by IL-1, IL-6, and matrix metalloproteases (MMPs), which mediate joint destruction in OA [53].

Ephrin/Eph signaling regulates numerous other cellular functions, such as those in which cell processes extend over relatively large distances. These include neuronal axon pathfinding and arteriovenous linkup during embryonic development, interactions between osteoblastic cells where their signaling is increased by PTH or PTHrP, migration of endothelial cells, and T lymphocytes during immune responses, tissue development, and angiogenesis [54, 55]. Ephrin/Eph signaling can also be disturbed in pathologic processes involving immune cells. For example, myeloma cells, like OCPs, appear to downregulate expression of Eph4 in osteoprogenitors; this could account in part for the inhibited bone formation seen typically in lytic myeloma lesions [56]. Osteoclastic cells can also negatively regulate OB formation by means of a secretory mechanism mediated by Atp6v0d2, a subunit of the proton pump, which secretes H⁺ to combine with Cl⁻ and generate HCl to dissolve the mineral component of bone. Atp6v0d2 also positively regulates OCP fusion, a role identified unexpectedly when Atp6v0d2–/– mice were generated and found to be osteopetrotic due to a combination of defective OCP fusion and increased new bone formation [57]. How or if these recently identified regulatory mechanisms are functional in inflammatory bone lesions and exactly how they regulate basal bone remodeling remains to be determined.

REFERENCES

1. Rodan GA, Martin TJ. 1982. Role of osteoblasts in hormonal control of bone resorption—A hypothesis. *Calcif Tissue Int* 34(3): 311.
2. Vallabhapurapu S, Karin M. 2009. Regulation and function of NF-kappaB transcription factors in the immune system. *Annu Rev Immunol* 27: 693–733.
3. Franzoso G, Carlson L, Xing L, Poljak L, Shores EW, Brown KD, Leonardi A, Tran T, Boyce BF, Siebenlist U. 1997. Requirement for NF-kappaB in osteoclast and B-cell development. *Genes Dev* 11(24): 3482–96.
4. Iotsova V, Caamaño J, Loy J, Yang Y, Lewin A, Bravo R. 1997. Osteopetrosis in mice lacking NF-kappaB1 and NF-kappaB2. *Nat Med* 3(11): 1285–9.
5. Boyle WJ, Simonet WS, Lacey DL. 2003. Osteoclast differentiation and activation. *Nature* 423(6937): 337–42.
6. Kollet O, Dar A, Lapidot T. 2007. The multiple roles of osteoclasts in host defense: Bone remodeling and hematopoietic stem cell mobilization. *Annu Rev Immunol* 25: 51–69.
7. Battiwalla M, Hematti P. 2009. Mesenchymal stem cells in hematopoietic stem cell transplantation. *Cytotherapy* 11(5): 503–15.
8. Ratajczak MZ, Zuba-Surma EK, Wojakowski W, Ratajczak J, Kucia M. 2008. Bone marrow—Home of versatile stem cells. *Transfus Med Hemother* 35(3): 248–59.
9. Karsenty G, Wagner EF. 2002. Reaching a genetic and molecular understanding of skeletal development. *Dev Cell* 2(4): 389–406.
10. Arai F, Miyamoto T, Ohneda O, Inada T, Sudo T, Brasel K, Miyata T, Anderson DM, Suda T. 1999. Commitment and differentiation of osteoclast precursor cells by the sequential expression of c-Fms and receptor activator of nuclear factor kappaB (RANK) receptors. *J Exp Med* 190(12): 1741–54.
11. Nakashima T, Hayashi M, Fukunaga T, Kurata K, Oh-Hora M, Feng JQ, Bonewald LF, Kodama T, Wutz A, Wagner EF, Penninger JM, Takayanagi H. 2011. Evidence for osteocyte regulation of bone homeostasis through RANKL expression. *Nat Med* 17(10): 1231–4.
12. Xiong J, Onal M, Jilka RL, Weinstein RS, Manolagas SC, O'Brien CA. 2011. Matrix-embedded cells control osteoclast formation. *Nat Med* 17(10): 1235–41.
13. Masuyama R, Stockmans I, Torrekens S, Van Looveren R, Maes C, Carmeliet P, Bouillon R, Carmeliet G. 2006. Vitamin D receptor in chondrocytes promotes osteoclastogenesis and regulates FGF23 production in osteoblasts. *J Clin Invest* 116(12): 3150–9.
14. Usui M, Xing L, Drissi H, Zuscik M, O'Keefe R, Chen D, Boyce BF. Murine and chicken chondrocytes regulate osteoclastogenesis by producing RANKL in response to BMP2. *J Bone Miner Res* 23(3): 314–25.
15. Henriksen K, Karsdal M, Delaisse JM, Engsig MT. 2003. RANKL and vascular endothelial growth factor (VEGF) induce osteoclast chemotaxis through an ERK1/2-dependent mechanism. *J Biol Chem* 278(49): 48745–53.
16. Muto A, Mizoguchi T, Udagawa N, Ito S, Kawahara I, Abiko Y, Arai A, Harada S, Kobayashi Y, Nakamichi Y, Penninger JM, Noguchi T, Takahashi N. 2011. Lineage-committed osteoclast precursors circulate in blood and settle down into bone. *J Bone Miner Res* 26(12): 2978–90.
17. Takayanagi H. 2007. Osteoimmunology: Shared mechanisms and crosstalk between the immune and bone systems. *Nat Rev Immunol* 7(4): 292–304.
18. Boyce BF, Xing L. 2008. Functions of RANKL/RANK/OPG in bone modeling and remodeling. *Arch Biochem Biophys* 473(2): 139–46.
19. Boyce BF, Yao Z, Xing L. 2010. Functions of nuclear factor kappaB in bone. *Ann N Y Acad Sci* 1192: 367–75.
20. Zhang Q, Badell IR, Schwarz EM, Boulukos KE, Yao Z, Boyce BF, Xing L. 2005. Tumor necrosis factor prevents alendronate-induced osteoclast apoptosis in vivo by stimulating Bcl-xL expression through Ets-2. *Arthritis Rheum* 52(9): 2708–18.
21. Yao Z, Xing L, Boyce BF. 2009. NF-kappaB p100 limits TNF-induced bone resorption in mice by a TRAF3-dependent mechanism. *J Clin Invest* 119(10): 3024–34.
22. Okamoto K, Takayanagi H. 2011. Regulation of bone by the adaptive immune system in arthritis. *Arthritis Res Ther* 13(3): 219.
23. Sato K, Suematsu A, Okamoto K, Yamaguchi A, Morishita Y, Kadono Y, Tanaka S, Kodama T, Akira S, Iwakura Y, Cua DJ, Takayanagi H. 2006. Th17 functions as an osteoclastogenic helper T cell subset that links T cell activation and bone destruction. *J Exp Med* 203(12): 2673–82.
24. Kong YY, Feige U, Sarosi I, Bolon B, Tafuri A, Morony S, Capparelli C, Li J, Elliott R, McCabe S, Wong T, Campagnuolo G, Moran E, Bogoch ER, Van G, Nguyen LT, Ohashi PS, Lacey DL, Fish E, Boyle WJ, Penninger JM. 1999. Activated T cells regulate bone loss and joint destruction in adjuvant arthritis through osteoprotegerin ligand. *Nature* 402(6759): 304–9.

25. Zaiss MM, Axmann R, Zwerina J, Polzer K, Gückel E, Skapenko A, Schulze-Koops H, Horwood N, Cope A, Schett G. 2007. Treg cells suppress osteoclast formation: A new link between the immune system and bone. *Arthritis Rheum* 56(12): 4104–12.
26. Pacifici R. 2012. Role of T cells in ovariectomy induced bone loss-revisited. *J Bone Miner Res* 27(2): 231–9.
27. Pacifici R. 2010. The immune system and bone. *Arch Biochem Biophys* 503(1): 41–53.
28. Manolagas SC. 2010. From estrogen-centric to aging and oxidative stress: A revised perspective of the pathogenesis of osteoporosis. *Endocr Rev* 31(3): 266–300.
29. Li H, Hong S, Qian J, Zheng Y, Yang J, Yi Q. 2010. Cross talk between the bone and immune systems: Osteoclasts function as antigen-presenting cells and activate CD4+ and CD8+ T cells. *Blood* 116(2): 210–7.
30. Grassi F, Manferdini C, Cattini L, Piacentini A, Gabusi E, Facchini A, Lisignoli G. 2011. T cell suppression by osteoclasts in vitro. *J Cell Physiol* 226(4): 982–90.
31. Townsend MJ, Monroe JG, Chan AC. 2010. B-cell targeted therapies in human autoimmune diseases: An updated perspective. *Immunol Rev* 237(1): 264–83.
32. Manabe N, Kawaguchi H, Chikuda H, Miyaura C, Inada M, Nagai R, Nabeshima Y, Nakamura K, Sinclair AM, Scheuermann RH, Kuro-o M. 2001. Connection between B lymphocyte and osteoclast differentiation pathways. *J Immunol* 167(5): 2625–31.
33. Schett G, David JP. 2010. The multiple faces of autoimmune-mediated bone loss. *Nat Rev Endocrinol* 6(12): 698–706.
34. Tilg H, Moschen AR, Kaser A, Pines A, Dotan I. 2008. Gut, inflammation and osteoporosis: Basic and clinical concepts. *Gut* 57(5): 684–94.
35. Hanada R, Leibbrandt A, Hanada T, Kitaoka S, Furuyashiki T, Fujihara H, Trichereau J, Paolino M, Qadri F, Plehm R, Klaere S, Komnenovic V, Mimata H, Yoshimatsu H, Takahashi N, von Haeseler A, Bader M, Kilic SS, Ueta Y, Pifl C, Narumiya S, Penninger JM. 2009. Central control of fever and female body temperature by RANKL/RANK. *Nature* 462(7272): 505–9.
36. Lieb W, Gona P, Larson MG, Massaro JM, Lipinska I, Keaney JF Jr, Rong J, Corey D, Hoffmann U, Fox CS, Vasan RS, Benjamin EJ, O'Donnell CJ, Kathiresan S. 2010. Biomarkers of the osteoprotegerin pathway: Clinical correlates, subclinical disease, incident cardiovascular disease, and mortality. *Arterioscler Thromb Vasc Biol* 30(9): 1849–54.
37. Jones DH, Nakashima T, Sanchez OH, Kozieradzki I, Komarova SV, Sarosi I, Morony S, Rubin E, Sarao R, Hojilla CV, Komnenovic V, Kong YY, Schreiber M, Dixon SJ, Sims SM, Khokha R, Wada T, Penninger JM. 2006. Regulation of cancer cell migration and bone metastasis by RANKL. *Nature* 440(7084): 692–6.
38. Chen G, Sircar K, Aprikian A, Potti A, Goltzman D, Rabbani SA. 2006. Expression of RANKL/RANK/OPG in primary and metastatic human prostate cancer as markers of disease stage and functional regulation. *Cancer* 107(2): 289–98.
39. Taylor R, Kashima TG, Knowles H, Gibbons CL, Whitwell D, Athanasou NA. 2011. Osteoclast formation and function in pigmented villonodular synovitis. *J Pathol* 225(1): 151–6.
40. Shishido-Hara Y, Kurata A, Fujiwara M, Itoh H, Imoto S, Kamma H. 2010. Two cases of breast carcinoma with osteoclastic giant cells: Are the osteoclastic giant cells pro-tumoural differentiation of macrophages? *Diagn Pathol* 5: 55.
41. Koga T, Inui M, Inoue K, Kim S, Suematsu A, Kobayashi E, Iwata T, Ohnishi H, Matozaki T, Kodama T, Taniguchi T, Takayanagi H, Takai T. 2004. Costimulatory signals mediated by the ITAM motif cooperate with RANKL for bone homeostasis. *Nature* 428(6984): 758–63.
42. Barrow AD, Raynal N, Andersen TL, Slatter DA, Bihan D, Pugh N, Cella M, Kim T, Rho J, Negishi-Koga T, Delaisse JM, Takayanagi H, Lorenzo J, Colonna M, Farndale RW, Choi Y, Trowsdale J. 2011. OSCAR is a collagen receptor that costimulates osteoclastogenesis in DAP12-deficient humans and mice. *J Clin Invest* 121(9): 3505–16.
43. Shinohara M, Koga T, Okamoto K, Sakaguchi S, Arai K, Yasuda H, Takai T, Kodama T, Morio T, Geha RS, Kitamura D, Kurosaki T, Ellmeier W, Takayanagi H. 2008. Tyrosine kinases Btk and Tec regulate osteoclast differentiation by linking RANK and ITAM signals. *Cell* 132(5): 794–806.
44. Sato K, Suematsu A, Nakashima T, Takemoto-Kimura S, Aoki K, Morishita Y, Asahara H, Ohya K, Yamaguchi A, Takai T, Kodama T, Chatila TA, Bito H, Takayanagi H. 2006. Regulation of osteoclast differentiation and function by the CaMK-CREB pathway. *Nat Med* 12(12): 1410–6.
45. Wang ZQ, Ovitt C, Grigoriadis AE, Möhle-Steinlein U, Rüther U, Wagner EF. 1992. Bone and haematopoietic defects in mice lacking c-fos. *Nature* 360(6406): 741–5.
46. Tsukayama DT. 1999. Pathophysiology of posttraumatic osteomyelitis. *Clin Orthop* 360: 22–9.
47. Bar-Shavit Z. 2008. Taking a toll on the bones: Regulation of bone metabolism by innate immune regulators. *Autoimmunity* 41(3): 195–203.
48. Varoga D, Wruck CJ, Tohidnezhad M, Brandenburg L, Paulsen F, Mentlein R, Seekamp A, Besch L, Pufe T. 2009. Osteoblasts participate in the innate immunity of the bone by producing human beta defensin-3. *Histochem Cell Biol* 131(2): 207–18.
49. Jevon M, Guo C, Ma B, Mordan N, Nair SP, Harris M, Henderson B, Bentley G, Meghji S. 1999. Mechanisms of internalization of Staphylococcus aureus by cultured human osteoblasts. *Infect Immun* 67(5): 2677–81.
50. Glass GE, Chan JK, Freidin A, Feldmann M, Horwood NJ, Nanchahal J. 2011. TNF-alpha promotes fracture repair by augmenting the recruitment and differentiation of muscle-derived stromal cells. *Proc Natl Acad Sci U S A* 108(4): 1585–90.
51. Diarra D, Stolina M, Polzer K, Zwerina J, Ominsky MS, Dwyer D, Korb A, Smolen J, Hoffmann M, Scheinecker C, van der Heide D, Landewe R, Lacey D, Richards WG,

Schett G. 2007. Dickkopf-1 is a master regulator of joint remodeling. *Nat Med* 13(2): 156–63.
52. Zhao C, Irie N, Takada Y, Shimoda K, Miyamoto T, Nishiwaki T, Suda T, Matsuo K. 2006. Bidirectional ephrinB2-EphB4 signaling controls bone homeostasis. *Cell Metab* 4(2): 111–21.
53. Kwan Tat S, Pelletier JP, Amiable N, Boileau C, Lajeunesse D, Duval N, Martel-Pelletier J. 2008. Activation of the receptor EphB4 by its specific ligand ephrin B2 in human osteoarthritic subchondral bone osteoblasts. *Arthritis Rheum* 58(12): 3820–30.
54. Suzuki K, Kumanogoh A, Kikutani H. 2008. Semaphorins and their receptors in immune cell interactions. *Nat Immunol* 9(1): 17–23.
55. Takamatsu H, Takegahara N, Nakagawa Y, Tomura M, Taniguchi M, Friedel RH, Rayburn H, Tessier-Lavigne M, Yoshida Y, Okuno T, Mizui M, Kang S, Nojima S, Tsujimura T, Nakatsuji Y, Katayama I, Toyofuku T, Kikutani H, Kumanogoh A. 2010. Semaphorins guide the entry of dendritic cells into the lymphatics by activating myosin II. *Nat Immunol* 11(7): 594–600.
56. Pennisi A, Ling W, Li X, Khan S, Shaughnessy JD Jr, Barlogie B, Yaccoby S. 2009. The ephrinB2/EphB4 axis is dysregulated in osteoprogenitors from myeloma patients and its activation affects myeloma bone disease and tumor growth. *Blood* 114(9): 1803–12.
57. Lee SH, Rho J, Jeong D, Sul JY, Kim T, Kim N, Kang JS, Miyamoto T, Suda T, Lee SK, Pignolo RJ, Koczon-Jaremko B, Lorenzo J, Choi Y. 2006. v-ATPase V0 subunit d2-deficient mice exhibit impaired osteoclast fusion and increased bone formation. *Nat Med* 12(12): 1403–9.

Index

Page references followed by f denote figures. Page references followed by t denote tables.

ABD (adynamic bone disorder), 635, 635t
Acetohydroxaminic acid (AHA), 874, 881
Achondroplasia, genetic testing for, 340
Acid-suppressive medications, skeletal effects of, 522
ACTH (adrenocorticotrophic hormone), 969–970, 972
Activity product, 865
Activity reporters, 73
Acute lymphocytic leukemia (ALL), 734–736
Acute myelogenous leukemia (AML), 1031
Adaptive immunity, vitamin D and, 241–242
Addisonian crisis, hypercalcemia and, 566
Adenomatous polyposis coli (APC), 915, 915t
ADHH (autosomal dominant hypocalcemic hypercalciuria), 587, 886
ADHR. *See* Autosomal dominant hypophosphatemic rickets
Adipocyte, 82–83, 993–994, 1030
Adipokines, 996, 1030
Adiponectin, 996, 1030
Adolescent pregnancy and lactation, 162
Adrenocorticotrophic hormone (ACTH), 969–970, 972
Adult T-cell leukemia/lymphoma, 697
Adynamic bone disorder (ABD), 635, 635t
Aging
  bone mass and structure changes with, 367–369, 368f–369f
  effects on skeleton in men, 508–509
  non-sex steroid hormone changes with, 372
  oral bone mass and structure cahnges with, 948
  sarcopenia, 978–981, 979t
  sex steroid loss and, 200–201, 200f, 202f, 369–372
    bone loss, 369–372, 371t, 372f
    in men, 370–372, 371t, 372f
    in women, 369–370
AGT1 (alanine glyoxylate aminotransferase-1), 871
AHA (acetohydroxaminic acid), 874
AHO (Albright hereditary osteodystrophy), 581, 582f, 583–584, 591, 593, 818
AIRE gene, 586
Alanine glyoxylate aminotransferase-1 (AGT1), 871
Albers-Schönberg disease, 771, 772
Albright hereditary osteodystrophy (AHO), 581, 582f, 583–584, 591, 593, 818
Albumin, 50–51, 51t

Alendronate
  in acute lymphocytic leukemia (ALL), 735
  after organ transplantation, 498, 499t
  for BMD preservation in prostate cancer therapy, 714
  for glucocorticoid-induced osteoporosis, 476, 477t
  for osteoporosis, 413f–414f, 414–415
  for Paget's disease, 664–665
  for primary hyperparathyroidism, 549
  for tumoral calcinosis, 812
Alkaline phosphatase
  in bone composition, 51, 53t
  hypophosphatasia and, 838, 918
  in osteogenesis imperfecta, 825
  in Paget's disease, 661
Alkali supplementation
  for cystine stones, 880
  for hypocitraturia, 880
  for uric acid stones, 880
Alkaptonuria, 841
ALL (acute lymphocytic leukemia), 734–736
Allopurinol, for uric acid stones, 880
Alphacalcidol
  for hereditary vitamin D resistant rickets, 620
  for pseudovitamin D deficiency rickets, 619
αKlotho, 191–192
  action on distal tubules, 846–847
  PTH secretion regulation, 211
  vitamin D and, 238–239, 239f
ALSG (aplasia of the lacrimal and salivary glands), 915, 915t
Aluminum accumulation in bone, 636
Aluminum hydroxide, for tumoral calcinosis, 812
Aluminum salts, as phosphate binders, 645
Alveolar bone formation, 911
Alzheimer's disease, 1006–1007
Ameloblastin, 908, 909f
Ameloblastoma, 926
Ameloblasts, 908, 909f
Amelogenesis, 908, 909f
Amelogenesis imperfecta, 915t, 917
Amelogenin, 917
AML (acute myelogenous leukemia), 1031
Amniotic fluid, fetal calcium metabolism and, 181

Primer on the Metabolic Bone Diseases and Disorders of Mineral Metabolism, Eighth Edition. Edited by Clifford J. Rosen.
© 2013 American Society for Bone and Mineral Research. Published 2013 by John Wiley & Sons, Inc.

Anabolic therapy for osteoporosis
  antiresorptive therapy after, 446
  combination anabolic and antiresorptive, 444–446
  concurrent antiresorptive therapy, 445–446
  future therapies, 463–464
    inhibition of dickkopf 1 (DKK-1) action, 463
    inhibition of GSK-3, 464
    inhibition of sclerostin, 464
    safety and specificity, 464
    Wnt signaling targets, 463–464
Anastrozole, 521
Androgen deprivation therapy, 520–521, 713–714
Androgens, 195–204
  deficiency and osteoporosis development, 201–203, 203f
  hormone biosynthesis, 195–196, 196f
  loss and aging, 200–201, 200f, 202f
  overview, 195
  receptors and molecular mechanisms of action, 196–197
  selective androgen receptor modulators (SARMs), 204
  skeletal growth, effects on, 197–198
    linear growth, 197
    periosteal expansion, 197–198
  skeletal maintenance, effects on, 198–200
Animal models
  allelic determinants for BMD, 76–79
    future directions, 79
    phenotypes, 76–77
    themes of existing data, 77–79
  genetic manipulation, 69–73
Ankylosing spondylitis, 484–485
Ankylosis, in fibrodysplasia ossificans progressiva, 815
Annexin 2, 1029
Anorexia nervosa, 1005
Antiepileptic drugs, skeletal effects of, 522–523
Antihormonal drugs, skeletal effects of, 520–521
Antipsychotics, 1008
Antiresorptives
  bisphosphonates (see Bisphosphonates)
  future osteoporosis therapies, 461–463
    cathepsin K inhibitors, 462–463
    inhibition of osteoclast action, 462–463
    inhibition of osteoclast formation, 461–462
    monoclonal antibody inhibiting RANKL action, 461–462
APC (adenomatous polyposis coli), 915, 915t
Aplasia of the lacrimal and salivary glands (ALSG), 915, 915t
Apoptosis, osteocyte, 37
APPswe mutation, 1007
APS1 (autoimmune polygrandular syndrome type 1), 574
Arachnodactyly, in Marfan syndrome, 830, 833t
ARHR. See Autosomal recessive hypophosphatemic rickets
Aristotle, 195
Aromatase inhibitor, 521, 712t, 713, 749
Arteriosclerosis, 1014–1015
Arzoxifene, 409
ATF4, 84
Atherosclerosis, 1014
Attachment loss, 948, 949f
Atypical subtrochanteric femoral fractures, 716
Authentic ectopic hyperparathyroidism, 565
Autoimmune polygrandular syndrome type 1 (APS1), 574
Autophagy, 37
Autosomal dominant hypocalcemic hypercalciuria (ADHH), 587, 886
Autosomal dominant hypophosphatemic rickets (ADHR), 606–607
  clinical and biochemical manifestations, 607
  fibroblast growth factor-23 (FGR23) association, 190, 654
  genetics of, 607
  genetic testing for, 339
  treatment, 607

Autosomal recessive hypophosphatemic rickets (ARHR), 607
  fibroblast growth factor-23 (FGR23) association, 191, 654
  genetic testing for, 339
Avascular necrosis, 1012–1014, 1013f, 1013t
Avp3, 27–28
Axial osteomalacia, 778–779
  clinical presentation, 779
  etiology and pathogenesis, 779
  histopathological findings, 779
  laboratory findings, 779
  radiological features, 779
  treatment, 779
Axial skeletal patterning, 3–4, 5f
Axis inhibition protein 2 (AXIN2), 915

Back pain, in Paget's disease, 662
Bacterial infection in osteomyelitis, 1039
Balanced structure variants, 338
Bartter syndrome, 870, 886
Base-pair substitutions, 336–337
  missense variants, 336–337
  non-synonymous variants, 336–337
  splicing variants, 337
  synonymous variants, 336
BDNF (brain-derived growth factor), 724
Beals syndrome, 42, 831
Bed rest, effect on skeletal muscle and bone, 981–982
Beta2 adrenergic receptors, bone mass and, 84–85
Beta-blockers
  effect on fracture risk, 84
  skeletal effects of, 523
β-catenin
  genome-wide association studies of Wnt-β-catenin signaling pathway, 380
  preventing of phosphorylation of, 464
Bevacizumab, 935
Bicarbonate, for hypocitraturia, 880
Biglycan gene, 51, 52t
Biochemical markers of bone turnover in osteoporosis, 297–302
  analytical and preanalytical variability, 297–299, 298t
  bone turnover rate
    bone loss and, 299
    fracture risk and, 299–300
    monitoring osteoporosis treatment, 300–302
  in men, 302
  reference values, 299
  table of, 298t
Bioluminescence imaging (BLI), 679t, 681
Biomarkers for osteosarcoma, 707–708
Biopsy. See Bone biopsy
Bisphosphonates, 283
  accumulation in bone, 636
  after organ transplantation, 498, 499t–500t, 501–503, 503f
  antifracture efficacy of, 413–415, 414f
  atypical subtrochanteric femoral fracture association, 716
  for BMD preservation in cancer therapy
    breast cancer, 713
    prostate cancer, 714
  for bone pain reduction, 724–725
  for fibrous dysplasia, 790
  for hypercalcemia in children, 653
  hypocalcemia with, 575, 748
  intermittent administration of, 414
  long-term effects on bone fragility, 415
  for metastatic bone disease, 688, 761–762
  mode of action, 742
  for multiple myeloma related bone disease, 697
  oral bone health, impact on, 953–954
  for osteogenesis imperfecta, 827

osteonecrosis of the jaw from, 748, 762, 929–935
   clinical presentation, 931–932, 931f–932f
   defined, 929
   discontinuation of bisphosphonates, 935
   etiology, 930
   histologic findings, 933, 934f
   management, 934–935, 934t
   monitoring of patients, 935
   prevalence, 930
   radiographic findings, 932–933, 933f
   risk factors, 930
   staging system, 934, 934t
osteopetrosis induced by, 771
for osteoporosis, 412–417
   adverse effects, 416–417
   atypical femur fractures, 416, 416f
   cost-effectiveness, 416
   excessive suppression of bone remodeling, 415–416
   glucocorticoid-induced, 476, 477t
   jaw osteonecrosis, 416
   in men, 511–512
   premenopausal, 517
for Paget's disease, 664–665
pharmacology, 412–413
for prevention of skeletal-related events in metastatic cancer, 742–750, 744f, 746f
   as adjuvant therapy, 748–749
   breast cancer, 742–745, 746f, 748–749
   multiple myeloma, 746f, 747–748
   prostate cancer, 746–747, 746f
for primary hyperparathyroidism, 549
safety, 748
in spinal cord injury, 1021–1022
structure of, 412, 413f
for tumoral calcinosis, 812
variable response to, 382
B lymphocytes, 689, 1030, 1037–1038
BMAT (bone marrow adipose tissue), 994
BMC. *See* Bone mineral content
BMD. *See* Bone mineral density
BMP-2, as therapy for bone healing, 94
Bone
   aging effects, 367–369, 368f–369f
   calcium sensing receptor (CaSR) role in maintaining calcium homeostasis, 229–231
   macroarchitecture, 359
   microarchitecture, 358–359, 358f
   skeletal healing, 90–95, 91f
Bone acquisition, ethnic differences in, 135–139, 137t
Bone anabolism, oxytocin and, 973
Bone biopsy, 307–314
   example photomicrographs, 308f
   histomorphometric findings in metabolic bone disease, 310–313, 311t
      gastrointestinal bone disease, 312
      glucocorticoid-induced osteoporosis, 312
      hypogonadism, 312
      hypophosphatemic osteopathy, 312
      hypovitaminosis osteopathy, 312
      menopausal osteoporosis, 312
      primary hyperparathyroidism, 312
      renal osteodystrophy, 312–313
   histomorphometric variables, 309
   in idiopathic juvenile osteoporosis, 470
   indications for, 314, 314t
   interpretation of findings, 309–310
      abnormal osteoid morphology, 310
      accumulation of unmineralized osteoid, 310
      altered bone remodeling, 310
      cortical bone deficit, 310
      reference data, 309–310
      replacement of normal marrow elements, 310
   microscopy, 313–314
   organization and function of bone cells, 307–308, 308f
      bone remodeling process, 308
      intermediary organization of the skeleton, 307
      osteoblasts, 308, 308f
      osteoclasts, 307, 308f
      osteocytes, 308
   in osteomalacia, 625, 627
   in osteosarcoma, 702
   postburn, 531, 532f
   procedure, 313
   remodeling process, 308
   specimen, procedures for obtaining, 313
   specimen processing and analysis, 313–314
   in vitamin D deficiency, 616–617, 617f, 625, 627
Bone cement, 529–530
Bone composition, 49–56
   cell types, 49
   collagen, 50, 50t
   lipids, 56
   mineral, 49–50
   minor components, 56
   noncollagenous proteins, 50–56, 51t–55t
      gla-containing proteins, 55–56, 55t
      glycosylated proteins, 51, 53, 53t
      proteoglycans, 51, 52t
      RGD-containing glycoproteins, 54t, 55
      serum derived proteins, 50–51, 51t
      small integrin-binding ligand N-glycosylated (SIBLING) protein, 53, 54t, 55
   water, 56
Bone density. *See* Bone mineral density (BMD)
Bone development, human fetal and neonatal, 121–124
   extrinsic influences, 122–123
   inherited defects, 123–124
   physiology, 121–122
Bone fragility. *See also* Fracture; Osteoporosis
   bisphosphonates, long-term effects of, 415
   components of, 358t
Bone loss. *See also* Osteopenia; Osteoporosis
   bone turnover rate and, 299
   follicle stimulating hormone (FSH) contribution to, 970–971
   during immobilization, 982
   inactivity and, 981–982
   inflammation-induced in rheumatic disease, 482–485
      ankylosing spondylitis, 484–485
      overview, 482–483
      rheumatoid arthritis, 483–484
      systemic lupus erythematosus, 485
   menopause and, 165–166, 167f
   oral (*see* Oral bone loss)
   prolactin and, 972–973
   sex steroids and, 369–372, 371t, 372f
      in men, 370–372, 371t, 372f
      in women, 369–370
   spinal cord injury, 1019
   weight reduction and, 997
Bone marrow
   cell-cell interactions in, 30–31, 30f
   fat, 994
   in multiple myeloma, 695–696
   replacement of normal elements in bone biopsy findings, 310
   transplantation, 502–503
      bone loss and, 496
      experiments on circulating osteogenic cells, 112
      for osteopetrosis, 772

Bone marrow adipose tissue (BMAT), 994
Bone mass
  adipocyte-driven central control of, 82–83
  aging and, 367–369, 368f–369f
  decrease and sarcopenia, 978–981, 979f
  diet role in maintaining, 362–363
  ethnic differences, 138
  osteoblast-osteoclast interaction in inflammation, 1039–1040
  peak (see Peak bone mass)
  persistence of childhood bone adaptation, 152–153, 152f
  sclerosing bone disorders, 769–781, 770t
Bone metabolism
  after spinal cord injury, 1019–1020
  ethnic differences in, 135–136
Bone mineral content (BMC)
  accrual of, 149–150, 150f
  after irradiation, 730
  persistence of childhood bone adaptation, 152–153, 152f
  postburn injury, 531–532
  seasonal influence on newborn, 122–123
Bone mineral density (BMD)
  after spinal cord injury, 1019
  after stem cell transplantation, 736
  allelic determinants of, 76–79
  areal BMD (BMD$_a$), 251–260
  calcium impact on, 403–404
  chemotherapy and, 735, 736
  childhood nutrition and, 142–145
  decrease
    with breast cancer therapies, 711, 713
    in cachexia, 982–983
    with chemotherapy, 713
    with tamoxifen therapy, 711
  depression and, 1003–1004
  evaluation with dual-energy X-ray absorptiometry (DXA), 251–260, 1020–1021
  follicle stimulating hormone (FSH) levels and, 970–971
  FRAX®, 289–295, 291t, 294f
  in hypogonadism, 510
  increase
    with denosumab, 422, 424
    with strontium ranelate, 439, 440
  magnetic resonance imaging and, 279–280
  in obesity, 994–995
  oral, 949–952
  for osteoporosis evaluation in men, 511–512
  oxytocin effects on, 973
  persistence of childhood bone adaptation, 152–153, 152f
  phenotypes, 76–77
  physical activity and, 981–982
  potential measurement errors in obesity, 994–995
  premenopausal women with low, 514–515
  preservation during cancer therapy
    breast cancer, 713
    prostate cancer, 714–715
  quantitative computed tomography (QCT) for diagnosis of BMD status in
    adults, 268
    children, 267–268
  skeletal complications of childhood cancer, 734–737
  urinary calcium excretion and, 847
  vitamin D impact on, 403–404, 627
Bone morphogenetic protein (BMP), 44
  in ankylosing spondylitis, 485
  as chemoattractant for circulating osteogenic cells, 115
  described, 6–8, 7f
  in fracture healing, 93
  signaling, 16–17, 17f
  in fibrodysplasia ossificans progressiva, 817
  in mandibular development, 899
  in palatogenesis, 898
  in tooth development, 906
Bone pain
  in cancer, 720–725, 721f–725f
    breast cancer, 741
    central sensitization in, 721–722
    epidemiology, 720
    mechanisms, 720–723, 721f–723f
    multiple myeloma, 694–695, 742
    nervous system reorganization in response to, 722–723
    nociceptive nerve sprouting, 722–723, 724f–725f
    prostate, 742
    therapeutic targets, 723–725
  in fibrous dysplasia, 789
  in Paget's disease, 662
  radiotherapy for palliation, 754–755
  in renal osteodystrophy, 637
  "wind-up," 722
Bone remodeling, 128–130, 128f–129f
  balanced physiologic, 695, 695f
  bone biopsy findings, 310
  bone's external size, shape and internal architecture, 128–130, 128f–129f
  central neuronal control, 961–965, 965f
    cannabinoid receptors, 964
    leptin and, 961–962
    melanocortin system, 964–965
    neuromedin U, 963
    neuropeptide Y, 963
    NPY1 receptor, 964
    NPY2 receptor, 963–964
    serotonin, 962–963
  excessive suppression by bisphosphonates, 415–416
  glucose homeostasis and, 989
  mediators of, 176
  in multiple myeloma, 695
  neuronal regulation of, 82–86
    adipocyte-driven central control, 82–83
    CART, 85
    leptin and, 82–85
    parasympathetic nervous system, 85
    sensory nerves, role of, 86
    serotonin and, 83
    skeletogenic neuropeptides, 85–86
    sympathetic nervous system, 83–85, 84f
  postburn injury, 532
  process described, 308
  in rheumatoid arthritis, 484
  thyroid stimulating hormone (TSH) and, 971–972
Bone resorption
  in acute lymphocytic leukemia (ALL), 734–736
  calcium release, 176
  immobilization and, 566
  insulin in osteoblasts, 989
  osteoclast biology and, 25–31, 26f–27f
  osteolytic bone disease in cancer metastasis, 672
  in Paget's disease, 660–661, 661f
  perimenopausal, 165–166, 167f
  prolactin acceleration and, 972–973
Bone scan. See Radionuclide scintigraphy
Bone-seeking radiopharmaceuticals (BSRs), 756–757
Bone sialoprotein, 54t, 55
Bone strength, 151–152, 151f
  architecture of bone and, 358–359, 358f
  components of, 358t

loss after irradiation, 730–731
magnetic resonance imaging-derived measures, 279–280
Bone turnover
in acute lymphocytic leukemia (ALL), 734–736
chronic kidney disease-mineral bone disorder (CKD-MBD), 635–636, 635t
Bone turnover markers, 297–302
analytical and preanalytical variability, 297–299, 298t
bone turnover rate
bone loss and, 299
fracture risk and, 299–300
monitoring osteoporosis treatment, 300–302
in men, 302
reference values, 299
table of, 298t
Bone water quantification measurements, magnetic resonance imaging for, 280
Bortezomib, for multiple myeloma related bone disease, 697
Bowed limbs, in Paget's disease, 662
Brain-derived growth factor (BDNF), 724
Brain injury, traumatic, 1006
Brain tumors, osteopenia and, 736
Breast, calcium sensing receptor (CaSR) role in, 231
Breast cancer, 672
metastasis to bone, 687–688, 760, 761f, 763f
bisphosphonates as adjuvant therapy, 748–749
clinical aspects, 741–742
prevention and treatment, 742–745, 746f
treatment, 711–713, 712t
chemotherapy, 712t, 713
endocrine therapy, 711, 712t, 713
of metastasis to bone, 742–745, 746f
to preserve BMD and prevent fractures, 713
radiation therapy, 715
Breastfeeding, 144
Bridge plates, 528
BSRs (bone-seeking radiopharmaceuticals), 756–757
Burns, 531–532
inflammatory response, 531
PTH secretion suppression after, 588
stress response, 531–532, 532f
treatment, 532
Buschke-Ollendorff syndrome, 778
Buttress (antiglide) plates, 528

Cachexia, 982–983
Café au lait skin pigmentation, 786, 787f
Calcification
in dermatomyositis, 812–813, 812f
in renal osteodystrophy, 637
Calcimimetic agents, for primary hyperparathyroidism, 549
Calcinosis
dermatomyositis, 812–813, 812f
tumoral calcinosis, 608–609, 810–812
biochemical findings, 811
clinical features, 810
etiology and pathogenesis, 811–812
fibroblast growth factor-23 (FGR23) association, 191–192, 811–812
genetics, 609
histopathology, 811
radiographic findings, 810–811, 811f
in renal osteodystrophy, 637
treatment, 812
Calciotropic hormones. See also specific hormones
bone remodeling and, 176
fetal calcium metabolism, 180–181

in lactation, 159–160
in pregnancy, 156–157
Calcitonin, 409–410
in fetal and neonatal bone development, 122
for Paget's disease, 665–666
Calcitonin gene-related peptide (cGRP), 86
Calcitriol
after organ transplantation, 498, 499t–500t, 501
deficiency and chronic kidney disease-mineral bone disorder (CKD-MBD), 634
endogenous intoxication with, 620
for hereditary vitamin D resistant rickets, 620
for pseudohypoparathyroidism, 595–596
for pseudovitamin D deficiency rickets, 619
for vitamin D deficiency, 629
for X-linked hypophosphatemic rickets (XLH), 606
Calcium, 403–406
absorption, effect of vitamin D deficiency on, 627–628
binding by calcium sensing receptor (CaSR), 226, 226f
bone mineral density, impact on, 403–404
calcium-based phosphate binders, 644
disorders in children, 651–653
distribution, 173–174
blood levels, 173–174, 184–185
cell levels, 173
fetal, 184–185
total body, 173
ethnic differences in dietary calcium and bone mass, 138
familial hypocalciuric hypercalcemia (FHH), 553–555, 554t
familial hypomagnesemia with secondary hypocalcemia, 655
fracture rates, impact on, 404
during growth, 142–143
homeostasis (see Calcium homeostasis)
hypercalcemia (see Hypercalcemia)
hypercalciuria, 865, 870–871, 870t, 874, 878–879, 885–889, 885t
in hyperparathyroidism, 537–538
hypocalcemia (see Hypocalcemia)
intake
nephrolithiasis and, 857
requirements, 405
intestinal calcium transport, 174–175, 175t
in lactation
average daily calcium loss, 159
homeostasis, 157f
intestinal absorption, 160
low intake of calcium, 162
renal excretion, 160
skeletal metabolism of calcium, 160–161, 161f
levels after spinal cord injury, 1021
malnutrition, 652
metabolism (see Calcium metabolism)
parathyroid hormone (PTH) regulation by extracellular calcium, 210–211, 210f
pharmacotherapy role of, 404–405
placental transport, 181–182, 185
postburn metabolism, 531
in pregnancy
homeostasis, 157f
intestinal absorption, 157
low intake of calcium, 159
renal excretion, 157
serum levels, 156, 157f
skeletal metabolism of calcium, 157–158
renal calcium handling, 175–176, 176t

1048 *Index*

Calcium (contd.)
  renal excretion, 845–850
    distal tubule, 846–847, 846f
    diuretic effects, 847
    genetic hypercalciuric stone-forming rat studies, 847–850, 848t, 849f
    Henle's loop, 845–846, 846f
    proximal tubule, 845, 846f
  safety, 405–406
  supplementation
    after organ transplantation, 498, 499t–500t
    for hypocalcemia, 576–577
    for hypocalcemia in children, 652–653
    for pseudohypoparathyroidism, 595–596
  target levels for, 642t
  transport across placenta, 121–122
  water concentration of, 857
Calcium homeostasis, 173–174, 174f
  calcium sensing receptor (CaSR) roles in tissues maintaining, 228t, 229–231
    bone and cartilage, 230–231
    breast and placenta, 231
    C cells, 229
    intestine, 230
    kidneys, 229–230
    parathyroid, 229
  fetal, 184–185, 185f
    blood calcium regulation, 184–185
    placental calcium transfer, 185
    skeletal mineralization, 185
    sources of calcium, 184, 184f
  regulation of hormone production and actions on, 176
Calcium metabolism
  disorders in children, 651–653
    hypercalcemia, 653
    hypocalcemia, 651–653
  distribution, 173–174
  fetal, 180–185, 184f–185f
    amniotic fluid, 181
    calcium sensing receptor (CaSR), 181
    integrated calcium homeostasis, 184–185, 185f
    kidney actions, 181
    mineral ions and calciotropic hormones, 180–181
    overview, 180
    placental transport, 181–182
    response to maternal hyperparathyroidism, 182–183
    response to maternal hypoparathyroidism, 183
    response to vitamin D deficiency, 183–184
    skeleton, 182, 185
    sources of calcium, 184, 184f
  fetal skeleton, 182, 185
  homeostasis, 173–174, 174f
    calcium sensing receptor (CaSR) roles in tissues maintaining, 228t, 229–231
    fetal, 184–185, 185f
    regulation of hormone production and actions on, 176
  menopause effects, 166
  postburn, 531
  PTH and 1,25(OH)$_2$D$_3$ actions on target tissues, 174–176
    bone remodeling, 176
    bone resorption and calcium release, 176
    intestinal calcium transport, 174–175, 175t
    renal calcium handling, 175–176, 176t
  regulation of, 173–176
  vitamin D and, 238–239, 239f
Calcium sensing receptor (CaSR), 224–231
  agonists, 226–227, 227f
  anti-CaSR antibodies, 587
  autosomal dominant hypocalcaemic hypercalciuria (ADHH) and, 886
  binding of Ca^{2+}, 226, 226f
  biogenesis of, 225–226
  in breast cancer, 672
  fetal calcium metabolism, 181
  function, 225–228, 228t
  gene, 225
  hypoparathyroidism and, 574
  overview, 224
  parathyroid hormone regulation, 210–211
  postburn upregulation of, 531
  properties of, 225
  roles in tissues maintaining calcium homeostasis, 228t, 229–231
    bone and cartilage, 230–231
    breast and placenta, 231
    C cells, 229
    intestine, 230
    kidneys, 229–230
    parathyroid, 229
  signaling, 225–228
    activation of, 226
    agonists and, 226–227, 227f
    intracellular, 227–228, 228t
    as magnesium sensor, 227
  structure of, 224–226, 225f
    binding of Ca^{2+}, 226, 226f
    biogenesis and, 225–226
    gene, 225
Calcium sensing receptor gene, 123
Calcium stones, treatment of, 878–880
Calibration, 66
cAMP (cyclic adenosine monophosphate), 580
Camurati-Engelmann disease. *See* Progressive diaphyscal dysplasia
Cancer. *See also specific cancer types*
  childhood, 734–737
  hematologic malignancies, 694–698
  hematopoietic malignancies, 1030–1031
  imaging
    bioluminescence imaging (BLI), 679t, 681
    comparison of modalities, 679t
    computed tomography, 678, 680
    FDG-PET, 680–681
    fluorescence imaging, 679t, 681–682
    magnetic resonance imaging (MRI), 679t, 680
    molecular/functional techniques, 680–682
    molecular imaging in preclinical bone metastasis, 677, 678f
    morphological/anatomical techniques, 678, 680
    multimodality, 682
    optical imaging, 681
    positron emission tomography (PET), 679t, 680–682
    radiography, 678
    radionuclide bone scintigraphy, 680
    single photon emission computed tomography, 679t, 680–681
    surgery, image-guided, 682
  importance of problem, 686
  malignancy-associated hypercalcemia (MAHC), 563–565
  metastasis to bone, 671–673, 686–690, 741–750
    animal models, 687
    biology of metastatic lesions, 760–761, 761f
    breast cancer, 741–745, 746f
    cancer cell homing and adhesion to bone, 671–672
    clinical aspects, 741–742
    clinical targets for treatment, 689–690
    imaging, 677–682, 678f, 679t
    malignant transformation, 671
    molecular mechanisms of, 687

multiple myeloma, 742, 746f, 747–748
osteoblastic bone disease, 672–673
osteoclast inhibition and, 715–716
osteogenic osteosarcoma, 702
osteolytic bone disease, 672
overview, 671
prevention, 741–750
prostate cancer, 742, 745–747, 746f
radionuclide scintigraphy, 283
radiotherapy, 754–757, 757t
tumor microenvironment regulation of bone disease, 688–689
"vicious cycle" concept, 686, 687f, 760
osteogenic osteosarcoma, 702–708, 703t, 706f, 707t
pain, 720–725, 721f–725f
radiotherapy, 754–757, 757t
bone-seeking radiopharmaceuticals, 756–757
external beam, 754–756, 755t
radiation therapy-induced osteoporosis, 728–731
treatment
bisphosphonates, 761–762
denosumab, 762
of metastasis to bone, 741–750, 760–764
minimally invasive options, 764
surgery, 762–764, 763f
Cancer stem cells in osteosarcoma, 706–707
Cancer treatment-induced bone loss (CTIBL), 713, 714. *See also* Bone loss
Cannabinoid receptors, 86, 964
Captopril, 881
Carbonated beverage consumption, 145
Carbonic anhydrase II deficiency, 773
clinical presentation, 773
etiology and pathogenesis, 773
laboratory findings, 773
radiological features, 773
treatment, 773
Cardiac transplantation, 501
Cardiovascular system, PTHrP (parathyroid hormone-related protein) effects on, 219
Caries, 922–923, 923f
Cart (cocaine and amphetamine-regulated transcript), 85
Cartilage
calcium sensing receptor (CaSR) role in maintaining calcium homeostasis, 229–231
chondrocyte proliferation and differentiation in, 7–9, 7f
embryonic formation, 6–7, 7f
knockout and overexpression studies in, 72
promoters specific for, 69–70
CaSR. *See* Calcium sensing receptor (CaSR)
Cataract surgery, fall and fracture reduction with, 391
CATCH-22, 585
Cathepsin K
inhibitors, 462–463
pycnodysostosis and, 774
Cation binders, 644–645
Cat-scratch disease, 653
CCA (congenital contractural arachnodactyly), 42, 831, 831t, 833t
CCD (cleidocranial dysplasia), 15, 124, 915, 915t
C cells, 229
CD11b promoter, 70
CDK (cyclin-dependent kinase) inhibitor, 704
Cediranib, 707
Cell-cell interactions in bone marrow, 30–31, 30f
Cement, bone, 529–530
Cementum, 908–911
Central giant cell granuloma (CGCG), 927

Central nervous system
central neuronal control of bone remodeling, 961–965, 965f
PTHrP (parathyroid hormone-related protein) effects, 220
Central sensitization, in cancer bone pain, 721–722
CGCG (central giant cell granuloma), 927
CGH (comparative genomic hybridization) arrays, 337–338
CGRP (calcitonin gene-related peptide), 86
Chelation therapy, hypocalcemia and, 652
Chemotherapy. *See also specific chemotherapeutic agents*
bone mineral density (BMD) and, 713, 735–736
for breast cancer, 712t, 713
for osteosarcoma, 703–704
Cherubism, 915t, 917–918
Children
bone formation in
fetal and neonatal bone development, 121–124
mechanical loading, importance of, 149–153
persistence of childhood bone adaptation, 152–153, 152f
window of opportunity, 149–150
cancer, skeletal complications of, 734–737
dermatomyositis, 812–813, 812f
dual-energy X-ray absorptiometry (DXA) in, 258–260, 259f
early life influences on adult fragility fracture, 354
effect of illnesses on bone morphology as maturational stage-specific, 131
fetal and neonatal bone development, 121–124
extrinsic factors in, 122–123
environmental, 122
mechanical, 122
nutritional, 122
inherited defects, 123–124
in bone matrix production, 123
in cranial suture closure and osteogenesis, 124
in mineral deposition, 123
in mineral homeostasis, 123
in osteoclast function, 124
physiology of, 121–122
fetal calcium metabolism, 180–185, 184f–185f
fractures in, 130–131, 354
growth of metaphyses in, 130–131, 131f
infantile osteopetroses, 770
kidney stones, 869–875
clinical evaluation, 872, 873t
incidence of, 869
medical management, 874–875
predisposing conditions, 869–872, 870t
prognosis, 875
surgical management, 873–874, 874t
low bone mass in, 354
maternal hyperparathyroidism, effect of, 588, 651
maternal hypomagnesemia, effect of, 588
melorheostosis in, 778, 778f
mineral metabolism disorders, 651–656
calcium, 651–653
magnesium, 654–656, 654t
phosphate, 654–655, 654t
rickets, 656, 656f
nutrition during growth, 142–145
osteoporosis in, 468–471, 469f, 470t–471t
progressive diaphyseal dysplasia in, 774
pycnodysostosis in, 773–774
quantitative computed tomography (QCT) for diagnosis of BMD status, 267–268
sex differences in bone morphology, 130
skeletal growth and peak bone strength, 127–131
transient pseudohypoparathyroidism of the newborn, 584
CHL (crown-heel length), 127
Chloride/proton antiporter CLC-5, 886–887

Chondrocytes
  embryonic development, 6–7, 7f
  knockout and overexpression studies in, 72
  overexpression of target genes, 69
  proliferation and differentiation in, 7–9, 7f
  PTHrP actions on, 217, 217f
  survival, regulation of, 9–10
Chondrodysplasias. *See* Osteochondrodysplasias
Chronic kidney disease (CKD), 495–496
  fibroblast growth factor-23 (FGR23) association, 192, 633
  mineral bone disorders (*see* Chronic kidney disease-mineral bone disorder (CKD-MBD))
Chronic kidney disease-mineral bone disorder (CKD-MBD), 632–637
  clinical manifestations, 636–637
    bone pain, 637
    calcifications, 637
    fractures, 637
    heterotropic mineralization, 637
    skeletal deformities, 637
    tumoral calcinosis, 637
  definition, 632
  diagnosis, 642–643
  osteoporosis and osteosclerosis associated with, 636
  pathogenesis, 632, 633f
  pathogenetic factors in, 633–634
    calcitriol deficiency, 634
    fibroblast growth factor 23 (FGF23), 633
    hyperparathyroidism, 634
    hyperphosphatemia, 633–634
    hypocalcemia, 634
    hypogonadism, 634
    Klotho, 633
  pathology of, 635–636, 635t
    bone turnover, 635–636, 635t
    low-turnover bone disease, adynamic bone disorder, 635, 635t
    mixed uremic osteodystrophy, 635–636, 635t
    predominant hyperparathyroid bone disease, high turnover renal osteodystrophy, osteitis fibrosa, 635, 635t
  treatment, 640–649
    achieving neutral phosphate balance, 643–646
    calcimimetic, 647
    dietary phosphate restriction, 644
    in end-stage renal disease, 647
    goals for, 643
    overview, 640
    parathyroidectomy, 648
    pathogenesis, 641–642
    phosphate binders, 644–645
    physiology, 640–641
    PTH level control, 646–647
    PTH suppression, 645
    reducing use of niacin, 645
    in stage 3-4 chronic kidney disease, 648
    vitamin D analogues, 645, 646–647
Chronic obstructive pulmonary disease (COPD), 490, 982–983
Chvostek sign, 575, 580
Cigarette smoking, impaired fracture healing with, 94
Cimetidine, 522
Cinacalcet HCl, for primary hyperparathyroidism, 549
Circadian formation of dental mineralized tissues, 910f, 911
Circulating osteogenic precursor (COP) cells, 111–116, 115f
  bone marrow transplantation experiments, 112
  ectopic bone formation experiments, 112
  homing of, 114–116, 115f
  in human diseases, 112
  isolation approaches, 112–114
    from hematopoietic lineage cells, 113–114
    plastic adherence, 112–113
    plastic nonadherence cells, 113
    from vascular lineage cells, 113
  MSC mobilization studies, 112
  overview of, 114–116, 115f
  parabiosis experiments, 111–112
  physiologic and pathologic functions of, 114
Citrate in hypocitraturia, 866, 871, 880
CKD. *See* Chronic kidney disease
CKD-MBD. *See* Chronic kidney disease-mineral bone disorder
Claudin16 *(CLDN16)* gene/protein, 887–888
CLCN7 gene, 772, 773
Cleft lip, 898, 917
Cleft palate, 898, 917
Cleidocranial dysplasia (CCD), 15, 124, 915, 915t
Climate, effect on kidney stone formation, 857
Clodronate, 413f, 414f, 415
  for metastatic bone disease
    as adjuvant therapy, 748
    breast cancer, 742–743
    multiple myeloma, 747
  for multiple myeloma related bone disease, 697
Clustering, 78
c-MYC, 704–705
CNCCs (cranial neural crest cells), 904, 905
Cocaine and amphetamine-regulated transcript (CART), 965
COL2A2 gene, 806
Col11a2 promoter, 69
Col1a1 proximal promoter, 70
COLIA1 and COLIA2 genes/proteins, 339
Collagen
  bone composition, 50, 50t
  genetic diseases affecting dentin, 915–916
  genetic testing for osteogenesis imperfecta, 339–340
  in osteogenesis imperfecta, 339–340, 822–826
  periodontal ligament, 910
Colles' fractures, 529
Comparative genomic hybridization (CGH) arrays, 337–338
Complex traits, 76
Compliance and persistence with osteoporosis medications, 448–452
  complexity of issue, 451
  healthy adherer effect, 451–452
  improving, 452
  measurement of, 450
    claims data, 450
    medication possession ratio (MPR), 450
    prospective studies, 450
  poor nature of, 448–449
  terminology issues, 449
Compression plates, 528
Computed tomography (CT)
  for assessment of fracture healing, 101
  dynamic (DCT), 678
  in fibrous dysplasia, 790f
  in hypoparathyroidism, 580, 580f
  micro (µCT), 678
  in nephrolithiasis, 861, 864f, 872–873
  for oral cavity assessment, 949
  in osseous metastatic disease, 678, 679t, 680
  in osteonecrosis, 807, 807t
  in osteonecrosis of the jaw, 932–933, 932f–933f
  in osteopetrosis, 771
  in Paget's disease, 663
  quantitative computed tomography (QCT), 264–272, 328
  for vertebral fracture assessment, 268, 327–328, 328f

Congenital contractural arachnodactyly (CCA), 42, 831, 831t, 833t
Congestive heart failure, 496
Connective tissue
　disorders of, 830–834, 831t–833t
　pathways that regulate growth factors, 42–46, 46f
COP cells. See Circulating osteogenic precursor (COP) cells
COPD (chronic obstructive pulmonary disease), 490, 982–983
Copper transport disorders, 841
Copy number variants, 337–338
Cortical bone deficit, in bone biopsy findings, 310
Corticosteroids, for dermatomyositis, 813
Corticotropin releasing factor (CRF), 970
Cost-effectiveness
　comparator, choice of, 455–456
　incremental cost-effectiveness ratio (ICER), 455
　methods for analysis of, 455–457
　model-based analyses, 456
　of osteoporosis treatment, 455–458, 456t
　quality-adjusted life years (QALY), 456–457, 458f
Covariation, 77
Cranial irradiation, skeletal complications of, 735, 736
Cranial neural crest cells (CNCCs), 904, 905
Cranial suture closure, defects in, 124
Craniofacial development, 895–901
　anomalies in Marfan syndrome, 830, 833t
　craniosynostosis, 900–901, 900f, 901t
　mandibular development, 898–899
　micrognathia, 899
　neurocranium development, 899–900, 900f
　origins of skeletal elements, 895–896, 896f
　orofacial cleft, 898
　palatal development, 896–898, 897f, 898t, 899f
　skull base deformities, 901
Craniofacial patterning, 3
Craniosynostosis, 124, 900–901, 900f, 901t
CREB (cyclic adenosine monophosphate responsive-element-binding protein), 962, 1038
Cre-loxP system, 71–73
CRF (corticotropin releasing factor), 970
Crohn's disease, 490
Crown-heel length (CHL), 127
CRTAP, 822–824
Cryoablation, 764
Crystals in urine, 861, 862f–863f. See also Nephrolithiasis
CT. See Computed tomography
CTIBL. See Cancer treatment-induced bone loss
CTX-I, 297–302
CXCL12, 689
CXCL12 abundant reticular cells, 1030
CXCR4 receptor, 115–116, 672, 689
Cyberknife treatments, 764
Cyclic adenosine monophosphate (cAMP), 580
Cyclic adenosine monophosphate responsive-element-binding protein (CREB), 962, 1038
Cyclin-dependent kinase (CDK) inhibitor, 704
Cyclooxygenase activity, in bone repair, 93
Cyclosporine, 497
CYP27B1 gene, 918
Cystathionine β-synthase deficiency, 840
Cystine stones, treatment of, 880–881
Cystinosis, 655
Cystinuria, 866, 872, 875
Cysts
　dentigerous, 925
　odonotogenic, 923–926, 924f
　periapical, 923, 925
Cytogenetic abnormalities, in osteosarcoma, 705

Cytokines
　cancer bone pain and, 724
　in obesity, 99
　postburn, 99, 531
Cytoskeleton, mechanical signals and, 398

1,25(OH)$_2$D$_3$. See Vitamin D
Degenerative joint disease (DJD), 1020
Deletions, 337
Demeclocycline for fluorochrome labeling, 313
Dementia, 1006–1007
Dendritic cells, 1037–1038
Denosumab
　for BMD preservation in cancer therapy
　　breast cancer, 713
　　prostate cancer, 715
　for bone metastatic tumors, 689
　for cancer-related bone diseases, 424, 744–750, 746f
　clinical concerns, 423–424
　　immune compromise, 423
　　oversuppression of bone turnover, 423
　　vascular effects, 423–424
　early studies, 420–421, 421f
　for fibrous dysplasia, 791
　for glucocorticoid-induced osteoporosis, 477–478, 477t
　hypocalcemia with, 748
　longer-term responses, 422
　for metastatic bone disease, 762
　mode of action, 742
　for multiple myeloma related bone disease, 697
　osteonecrosis link to, 748, 935
　for osteoporosis, 420–424, 477–478, 477t
　for Paget's disease, 424
　pivotal trials, 421–422, 422f
　for prevention of skeletal-related events in metastatic cancer
　　breast cancer, 744–745, 746f
　　multiple myeloma, 746f, 747–748
　　prostate cancer, 746–747, 746f
　response to discontinuing, 422–423
　for rheumatoid arthritis, 424
　safety and tolerability, 423, 748
　in spinal cord injury patients, 1022
Dentigerous cyst, 925
Dentin
　development of, 906–908, 907f
　genetic diseases affecting, 915–916, 915t, 916f
Dentin dysplasia, 915t, 916, 916f
Dentin matrix protein-1 (DMP1), 191, 607, 918
Dentinogenesis imperfecta, 915t, 916, 916f
Dentin sialophosphoprotein (DSPP), 916
Dentition. See also Teeth
　development and structure of teeth and periodontal tissues, 904–911, 905f–907f, 909f–910f
　genetic disorders affecting, 914–919, 915t
　　amelogenesis imperfecta, 915t, 917
　　cherubism, 915t, 917–918
　　cleft lip and palate, 917
　　dentin, 915–916, 915t, 916f
　　dentin dysplasia, 915t, 916, 916f
　　dentinogenesis imperfecta, 915t, 916, 916f
　　enamel defects, 915t, 917
　　familial tooth agenesis, 914–915, 915t
　　hypohidrotic ectodermal dysplasia (HED), 914–915, 915t
　　mucopolysaccharidoses, 915t, 917
　　number of teeth, 914–915
　　osteogenesis imperfecta, 915–916
　　supernumerary teeth, 914–915, 915t

Dentition. *See also* Teeth (contd.)
  metabolic bone diseases affecting, 918, 919f
    hypophosphatasia, 918, 919f
    hypophosphatemia, 918
    vitamin D deficiency, 915t, 918
Dent's disease, 655, 870, 886–887
Depression
  low bone mass and, 1004–1005
  major depressive disorder (MDD), 1002–1005, 1004f
Dermatofibrosis lenticularis disseminata, 776–777
Dermatomyositis, 812–813, 812f
  biochemical and histological features, 813
  clinical presentation, 812–813
  radiographic features, 812f, 813
  treatment, 813
DES (diethylstilbestrol), 714
Developmental odontogenic cysts, 925–926
DFGF signaling, 906
Dgcr8 gene, 1006
Diabetes mellitus
  arteriosclerosis and skeletal health, 1014–1015
  impaired fracture healing with, 94
  as nephrolithiasis risk factor, 858
  osteoporosis associated with, 490–491
Dialysis, for chronic kidney disease-mineral bone disorder (CKD-MBD), 645
Diaphysitis, 810
Dickkopf 1. *See* DKK-1
Diet. *See also* Nutrition
  building peak bone mass, 361–362, 362f
  dietary bioactive constituents, 364
  dietary patterns, 363
  maintaining bone mass, 362–363
  osteoporosis and, 361–364
  in pregnancy, 143–144
  therapy for osteopetrosis, 772–773
Diethylstilbestrol (DES), 714
DiGeorge syndrome, 574, 585
Dimensions, genetics of bone, 77
Distal convoluted tubules, in calcium excretion physiology, 846–847, 846f
Distal renal tubular acidosis, 870, 888–889
Distraction osteogenesis, 102
Disuse, effect on skeletal muscle and bone, 981–982
Diuretics, effect on renal calcium excretion, 847
DJD (degenerative joint disease), 1020
DKK-1, 463, 483, 672, 688, 690, 696, 761
DMP1, 191, 607, 918
Dolichostenomelia, in Marfan syndrome, 830, 833t
Down syndrome, 653
Doxorubicin
  cardiotoxicity, 704
  for osteosarcoma, 703
DPA (dual photon absorptiometry), oral cavity assessment, 949
D-penicillamine, for cystine stones, 881
Drugs. *See also specific drugs*
  hypercalcemia and, 567–568
  with skeletal effects, 520–523, 521t
    acid-suppressive medications, 522
    androgen deprivation therapy, 520–521
    antiepileptic drugs, 522–523
    antihormonal drugs, 520–521
    aromatase inhibitors, 521
    beta-blockers, 523
    heparin, 523
    selective serotonin reuptake inhibitors (SSRIs), 523
    statins, 523
    thiazides, 523
    thiazolidinediones, 521–522
DSPP (dentin sialophosphoprotein), 916
Dual-energy X-ray absorptiometry (DXA), 251–260
  advantages and limitations of central DXA, 252t
  after spinal cord injury, 1020–1021
  artifacts, 256, 256t
  BMD phenotypes and, 107
  bone strength measurement, 152
  in children, 258–260, 259f
  for ethnic differences comparisons, 137
  fracture prediction, 253–256, 255f
  in hepatitis C-associated osteosclerosis, 781
  manual (MXA), 325, 326
  measurement errors in obesity, 994
  monitoring BMD_a change, 258
  in osteogenesis imperfecta, 825
  overview, 251
  peripheral (pDXA), 253
  precision errors, 258
  printout, 253f–254f
  quantitative computed tomography (QCT) compared, 267, 268
  scan interpretation
    treatment criteria from National Osteoporosis Foundation Guidelines, 258t
    using FRAX®, 257–258, 257f, 257t
    using T- and Z-scores, 256–257
  scan sites, 252–253
  technical aspects, 251–252
  vertebral fracture assessment (VFA), 258, 325–327, 326f–327f, 330, 331t
  in vitamin D deficiency, 616
Dual photon absorptiometry (DPA), oral cavity assessment, 949
Dural ectasia, in Marfan syndrome, 831, 833t
DXA. *See* Dual-energy X-ray absorptiometry
Dysosteosclerosis, 770

EBRT. *See* External beam radiotherapy
ECM. *See* Extracellular matrix
Ectopic bone formation, 112, 1020
Ehlers-Danlos syndrome, 824
Electromagnetic stimulation, in spinal cord injury patients, 1022–1023
Ellsworth-Howard test, 588
Embryonic development, 3–10, 4f–5f, 7f, 10f
Embryonic stem (ES) cells, gene targeting in, 70–71
Enamel
  development of, 908, 909f
  genetic diseases affecting, 915t, 917
Enamelin, 917
Endocrine disorders, hypercalcemia and, 566
Endocrine profile in obesity, 996–997
Endocrine therapy
  for breast cancer, 711, 712t, 713
  for prostate cancer, 714
Endosteal hyperostosis, 775–776
  sclerosteosis, 776
  Van Buchem disease, 775–776
Endothelial cells, 1029
Endothelial nitric oxide (eNOS), 806
Endothelin-1 (ET-1), 672
  in osteoblastic metastases, 688
  osteoblast stimulation by, 761
Endothelin A receptor (ETAR), 672, 673
Endothelins and cancer bone pain, 724
End-stage renal disease, histomorphometry findings in, 312–313
Energy metabolism, osteocalcin and, 989–990, 997
ENOS (endothelial nitric oxide), 806

ENPP1 gene/protein, 191
Environmental influences, on fetal/neonatal bone development, 122
Enzyme defects
 alkaptonuria, 841
 copper transport disorders, 841
 homocystinuria, 840–841
 hypophosphatasia (HPP), 838–840, 839f
 mucopolysaccharidoses, 840
Eosinophilic granuloma, 807–808
Ephrin/Eph signaling, 689, 1039–1040
Epidermal nevus syndrome, 191, 607–608
Epidermis, differentiation and vitamin D (1,25(OH)$_2$D$_3$), 242
Epigenetics
 epigenetic variants, 337
 pseudohypoparathyroidism (PHP) type 1b, 583, 594–595, 595f
Epiphican gene, 51
Epithelial-to-mesenchymal transition, 671, 673
Ergocalciferol, 628–629
Erlenmeyer flask deformity, 771, 771f
Erlotinib, 690
Esp gene/protein, 987–988, 990
ESR1 gene, 381
Estrogen, 195–204
 agonist/antagonists, 09
 for BMD preservation in prostate cancer therapy, 714–715
 bone loss in women, 369–370
 deficiency and chronic kidney disease-mineral bone disorder (CKD-MBD), 634
 follicle stimulating hormone (FSH) and, 971
 genome-wide association studies of estrogen pathway, 381
 hormone biosynthesis, 195–196, 196f
 hypercalcemia from, 568
 loss and aging, 200–201, 200f, 202f
 osteoporosis
  deficiency and osteoporosis development, 201–203, 203f
  pathogenesis, 369–370
  treatment, 203–204, 408–409
 overview, 195, 408–409
 perimenopausal effects
  on bone, 165–166
  on calcium metabolism, 166
 for primary hyperparathyroidism, 549
 receptors and molecular mechanisms of action, 196–197
 replacement therapy and oral bone health, 953
 skeletal growth, effects on, 197–198
  linear growth, 197
  periosteal expansion, 197–198
 skeletal maintenance, effects on, 198–200
  effects on osteoblasts, 199
  effects on osteoclasts, 198–199
  effects on osteocytes, 199–200
ET-1. *See* Endothelin-1 (ET-1)
ETAR (endothelin A receptor), 672, 673
Ethnic differences
 in bone acquisition, mass, and geometry, 136–137, 137t
 in bone and mineral metabolism, 135–136
 in dietary calcium and bone mass, 138
 South Africa case study, 138
Etidronate, 413f, 664–665
Ewing sarcoma, 736–737
Exemestane, 521, 713
Exercise
 for fall prevention, 391
 influence on peak bone mass, 149–150, 150f
 for osteoporosis prevention, 396–400, 397f
  mechanical factors regulating bone cell response, 396–397
  mechanically response bone cells, 397–398
  transducing mechanical signals into cellular response, 398, 399f
  translating mechanical signals to the clinic, 398–400
External beam radiotherapy (EBRT), 754–756, 755t, 764
 fractures and, 756
 for neuropathic pain, 755
 for pain palliation, 754–755
 re-irradiation, 756
 for spinal cord compression, 756–757
External fixation, 529
Extracellular matrix (ECM)
 bone composition, 49–56
 proteins, 42
 signaling in bone metastatic tumors, 688

Factor V Leiden mutation, 806
Falls
 costs of, 389
 definition, 389–390
 epidemiology of, 389
 inclusion of fall risk in fracture risk prediction, 390
 mechanics and fracture risk, 390
 nonskeletal fall-fracture construct, 390, 390f
 prevention, 389–392
 risk factors for, 390–391
 skeletal muscle strength and, 980–981
 strategies for prevention, 391–392
  tai chi, 391
  weight-bearing exercise programs, 391
 testosterone level and, 981
 vitamin D deficiency and, 617–618, 618f
Familial benign (hypocalciuric) hypercalcaemia (FBHH), 886
Familial hypocalciuric hypercalcemia (FHH), 553–555, 554t
 clinical expressions, 553–554
 diagnosis of family and the carrier, 555
 management, 555
 pathogenesis/genetics, 554–555
 in pregnancy, 159
 sporadic, 555
Familial hypomagnesaemia with hypercalciuria and nephrocalcinosis (FHHNC), 887–888
Familial hypomagnesemia-hypercalciuria, 870
Familial hypomagnesemia with secondary hypocalcemia, 655
Familial hypophosphatemic rickets (FHR), genetic testing for, 339
Familial idiopathic hypercalciuria (FIH), 870, 874, 885, 885t
Familial isolated primary hyperparathyroidism (FIHPT), 554t, 559, 886
 clinical expressions, 559
 diagnosis, 559
 management, 559
 pathogenesis/genetics, 559
Familial primary hyperparathyroidism, 553–560, 554t
 detection of asymptomatic carriers, 560
 familial hypocalciuric hypercalcemia (FHH), 553–555, 554t
  clinical expressions, 553–554
  diagnosis of family and the carrier, 555
  management, 555
  pathogenesis/genetics, 554–555
  sporadic, 555
 familial isolated primary hyperparathyroidism (FIHPT), 554t, 559
  clinical expressions, 559
  diagnosis, 559
  management, 559
  pathogenesis/genetics, 559

Familial primary hyperparathyroidism (contd.)
  hyperparathyroidism-jaw tumor syndrome (HPT-JT), 554t, 558–559
    clinical expressions, 558
    diagnosis of carriers and cancers, 559
    management, 559
    pathogenesis/genetics, 559
  monitoring for tumors, 560
  multifocal parathyroid gland hyperfunction, 559–560
  multiple endocrine neoplasia type 1 (MEN1), 554t, 556–557
    clinical expressions, 556
    diagnosis of carriers, 556–557
    diagnosis of other tumors, 557
    management, 557
    parathyroid surgery for, 557
    pathogenesis/genetics, 556
  multiple endocrine neoplasia type 2A (MEN2A), 554t, 557–558
    clinical expressions, 557–558
    diagnosis of kindred, of carriers, and parathyroid tumors, 558
    management, 558
    pathogenesis/genetics, 558
  neonatal severe primary hyperparathyroidism (NSHPT), 554t, 555–556
    clinical expressions, 555
    diagnosis, 555
    management, 555–556
    pathogenesis/genetics, 555
  surgery for familial parathyroid tumors, 560
Familial tooth agenesis, 914–915, 915t
Familial tumoral calcinosis (FTC), 608–609. See also Tumoral calcinosis
  fibroblast growth factor-23 (FGR23) association, 191–192, 811–812
  genetics, 609
Fanconi renotubular syndrome 2, 608
Fat infiltration, 981
Fat mass, 996
Fat necrosis, subcutaneous, 653
FBHH (familial benign (hypocalciuric) hypercalcaemia), 886
FDG (^{18}F-fluorodeoxyglucose) PET, 285, 680–681, 933
FDG (^{18}F-fluorodeoxyglucose) PET/computed tomography (FDG-PET/CT), in multiple myeloma, 696–697
Femoral fractures
  atypical subtrochanteric, 716
  bisphosphonates and atypical, 416, 416f
Femoral head, avascular necrosis of the, 1013, 1013f
FES (functional electrical stimulation), 1022–1023
Fetal calcium metabolism, 180–185, 184f–185f
  amniotic fluid, 181
  calcium sensing receptor (CaSR), 181
  homeostasis, 184–185, 185f
    blood calcium regulation, 184–185
    placental calcium transfer, 185
    skeletal mineralization, 185
    sources of calcium, 184, 184f
  kidney actions, 181
  mineral ions and calciotropic hormones, 180–181
  overview, 180
  placental transport, 181–182
  response to maternal hyperparathyroidism, 182–183
  response to maternal hypoparathyroidism, 183
  response to vitamin D deficiency, 183–184
  skeleton, 182
  sources of calcium, 184, 184f
Fetal development, craniofacial, 895–901
^{18}F-fluorodeoxyglucose positron emission tomography (FDG-PET). See FDG (^{18}F-fluorodeoxyglucose) PET
FGF2 (fibroblast growth factor-2), 982

FGF21 (fibroblast growth factor-21), 994
FGF23. See Fibroblast growth factor-23 (FGF23)
FGFR3-related disorders, genetic testing for, 340
FHH. See Familial hypocalciuric hypercalcemia
FHR (familial hypophosphatemic rickets), genetic testing for, 339
Fibrillin, 54t
  described, 43
  molecular mechanisms orchestrated on scaffold, 44–45
  regulation of growth factors by, 43–44
Fibrillin gene, 830–833, 831t
Fibrillinopathies, 42–43, 830–831. See also specific disorders
Fibroblast growth factor-2 (FGF2), 982
Fibroblast growth factor-21 (FGF21), 994
Fibroblast growth factor-23 (FGF23), 188–192
  activity, 188–189
  chronic kidney disease-mineral bone disorder (CKD-MBD) and, 633
  in fibrous dysplasia, 787
  gene and protein, 188
  gene defects and hypophosphatemia, 918
  overview, 188
  PTH secretion regulation, 211
  receptors, 189
  regulation in vivo, 189
  serum assays, 189
  syndromes associated with, 189–192, 190t
    autosomal dominant hypophosphatemic rickets (ADHR), 190, 654
    autosomal recessive hypophosphatemic rickets (ARHR), 191, 654
    chronic kidney disease (CKD), 192
    epidermal nevus syndrome (ENS), 191
    familial tumoral calcinosis, 191–192, 811–812
    McClune Albright syndrome, 191, 654
    opsismodysplasia, 191
    osteoglophonic dysplasia, 191
    tumor-induced osteomalcia (TIO), 190, 654
    X-linked hypophosphatemic rickets (XLH), 190, 654
  vitamin D and, 238–239, 239f, 615
Fibroblast growth factors (FGFs), 19–20
  in fracture healing, 93
  signaling, 4, 5f, 7–8, 7f, 8, 19–20, 898–899, 901, 906
    in craniosynostosis, 901t
    in mandibular development, 899
    in palatogenesis, 898
    tooth development, 906
Fibrocytes, circulating osteogenic cells isolated from, 113–114
Fibrodysplasia ossificans progressiva (FOP), 815–817
  clinical presentation, 815–816, 816f
  etiology and pathogenesis, 817
  histopathology, 817
  laboratory findings, 816
  overview, 815
  prognosis, 817
  radiologic features, 816
  treatment, 817
Fibrogenesis imperfecta ossium, 779–780
  clinical presentation, 779
  etiology and pathogenesis, 780
  histopathological findings, 779–780
  laboratory findings, 779
  radiological features, 779
  treatment, 780
Fibronectin, 54t
Fibrous dysplasia, 607, 786–791
  café au lait skin pigmentation, 786, 787f
  clinical features, 788–789

etiology and pathogenesis, 786–788
histology of, 787, 788f
management and treatment, 789–791
overview, 786
radiographic appearance of, 789, 790f
FIH (familial idiopathic hypercalciuria), 870, 874
FIHP. *See* Familial isolated primary hyperparathyroidism
FIHPT. *See* Familial isolated primary hyperparathyroidism
Fine element analysis (FEA)
   bone strength loss after irradiation, 731
   in vertebral fracture assessment, 327–328
Finite element modeling, QCT-based, 270–271, 271f
FKBP10 gene, 823t, 825
Flavonoids, 364
Fluid intake
   effect on kidney stone formation, 857
   for renal stone prevention, 878
Fluorescence imaging, 679t, 681–682
Fluorochrome labeling, 313
Follicle stimulating hormone (FSH), 969–971
Follicular cyst, 925
FOP. *See* Fibrodysplasia ossificans progressiva
Formation product, 865
FoxO1, 988, 989f
Fracture(s). *See also specific causes; specific fracture locations*
   in acute lymphocytic leukemia (ALL), 735–736
   assessment with radionuclide scintigraphy, 283–284, 284f
   atypical subtrochanteric femoral fractures, 716
   biomechanics of healing, 99–102
      assessment, 99–102, 100f
      mechanobiology of healing, 102
      models of, 103f
      stages of healing, 100–101, 101f
   in children, 130–131, 354
   epidemiology
      clustering of fractures in individuals, 350
      early life influences on adult fragility fracture, 354
      geography, 352
      hip fracture, 349, 349f, 349t, 352f, 353
      impact of fractures, 349t
      incidence and prevalence, 348–350, 349f, 351f–352f
      low bone mass in children, 354
      in men, 509, 509f
      morbidity, 353
      mortality, 353
      osteoporotic, 348–354
      time trends and future projections, 350–352
      vertebral fracture, 349–350, 349f, 349t, 351f, 353
      wrist (distal forearm) fracture, 349f, 349t, 350, 353
   falls and, 389–390
   in metastatic bone disease, 762–764, 763f
   oral bone loss as predictor of, 953
   in osteogenesis imperfecta, 822–827
   in Paget's disease, 662, 663
   pharmacologic interventions to prevent in cancer therapy
      breast cancer, 713
      prostate cancer, 714–715
   prediction with dual-energy X-ray absorptiometry (DXA), 253–256, 255f
   prevention
      bisphosphonates for, 413–415, 414f
      cost-effectiveness of, 457
      strontium ranelate for, 437–441
   PTH for acceleration of repair, 433
   in pycnodysostosis, 774
   radiation-induced, 715, 728–729
   radiotherapy and, 756
   in renal osteodystrophy, 637

   risk
      bone turnover rate and, 299–300
      in obesity, 995, 995f
      relative risk, 253–256, 255f
      skeletal muscle strength and, 980–981
      weight reduction and, 997
   risk assessment
      fall risk inclusion in, 390
      FRAX®, 257–258, 257f, 257t, 289–295
      magnetic resonance imaging-derived measures, 279–280
      quantitative computed tomography (QCT), 268–270
         clinical interpretation, 269–270
         hip fractures, 268–269
         mixed fracture groups, 269
         vertebral fractures, 268
   skeletal healing, 90–95, 91f
   spinal cord injury and, 1019
   vitamin D deficiency and, 617–618, 618f, 626–627
Fracture Risk Assessment Tool (FRAX®), 257–258, 257f, 257t, 289–295
   applications, 294
      vertebral fracture assessment, 318
   guidelines
      in North America, 294
      in United Kingdom, 293–294, 294f
   input and output, 289–290
   interventions and assessment thresholds, 292–293, 293f
   limitations, 292
   management algorithm, 293, 293f
   performance characteristics, 290–292
Frailty
   vitamin D deficiency and, 628
   weight loss as marker of, 980
FRAX®. *See* Fracture Risk Assessment Tool
Fruit intake during growth, 143
FSH (follicle stimulating hormone), 370, 969–971
FTC. *See* Familial tumoral calcinosis
Fulvestrant, 713
Functional electrical stimulation (FES), 1022–1023
Functional genomics, 73
Furosemide, for hypercalcemia in children, 653

Gadolinium-containing MRI agents, 575
GALNT3 gene/protein, 191, 811–812
Gα_s, 581, 582f, 583–584. *See also* Gsα
GAL4/UAS system, 70
Gap junctions, role in osteocyte communication, 38–39
Gastrointestinal bone disease, histomorphometry findings in, 312
GATA3 transcription factor, 574, 585–586
GCMB (glial cell missing B), 574
GCM2 gene, 586
G-CSF (granulocyte colony stimulating factor), 1030
GDNF (glial-derived growth factor), 724
Geleophysic dysplasia, 831t, 833–834, 833t
Gene expression
   expression QTLs, 77
   profile during fracture healing, 92
Gene knockout, 71–73
Genes, overexpression of target, 69–70
Gene targeting, 70–71
Genetic Factors for Osteoporosis (GEFOs) Consortium, 107, 108f, 109
Genetic manipulation
   animal models, 69–73
   functional genomics, 73
   gene targeting, 70–71
   lineage tracing and activity reporters, 73

Genetic manipulation (contd.)
  overexpression of target genes
    advantages/disadvantages of approach, 70
    chondrocytes, 69
    osteoblasts, 70
    osteoclasts, 70
    tendons and ligaments, 70
  tissue-specific and inducible knockout and overexpression, 71–73
    cartilage, 72
    considerations for inducible knockout use, 72–73
    mesenchyme, 72
    osteoblasts, 72
    osteoclasts, 72
    overview, 71–72
Genetic Markers for Osteoporosis (GENOMOS), 107
Genetics
  allelic determinants of bone mineral density, 76–79
  autosomal dominant hypophosphatemic rickets, 607
  bone phenotypes, 377–378
  connective tissue disorders, 830, 831t, 832–834
  craniofacial disorders affecting the dentition, 914–918, 915t
  craniosynostosis, 901
  epigenetics and pseudohypoparathyroidism (PHP) type 1b, 583, 594–595, 595f
  familial hypocalciuric hypercalcemia (FHH), 554–555
  familial isolated primary hyperparathyroidism (FIHPT), 559
  genomic imprinting in pseudohyoparathyroidism, 592–595
  hyperparathyroidism-jaw tumor syndrome (HPT-JT), 559
  hyperphosphatemic familial tumoral calcinosis, 609
  hypoparathyroidism, 574, 581–587, 581t, 582f
  multiple endocrine neoplasia type 1 (MEN1), 556
  multiple endocrine neoplasia type 2A (MEN2A), 558
  neonatal severe primary hyperparathyroidism (NSHPT), 555
  nephrolithiasis, 858
  osteoporosis, 376–383
    bone phenotypes, 377–378
    candidate gene studies, 378, 379t
    clinical application: individualized prognosis, 381–382
    clinical application: pharmacogenetics, 382
    genome-wide studies, 378–381, 379t, 381t
    linkage analysis, 378–380, 379t
    overview, 376–377
  palatogenesis, 898, 898t
  pharmacogenetics, 382
  pseudohypoparathyroidism, 581–584, 581t, 582f, 590–596, 591t, 592f, 595f
  QTLs, 76–79
  tooth development, 905
  X-linked hypophosphatemic rickets, 605
Genetic testing, 336–341
  described, 336
  evolving approaches of DNA-based testing, 338
  methods for detecting small scale changes in DNA, 337
  for skeletal disorders, 338–340
    achondroplasia, 340
    familial hypophosphatemic rickets (FHR), 339
    FGFR3-related disorders, 340
    hypophosphatemia, 339
    metabolic bone disease, 339
    multiple epiphyseal dysplasia (MED), 340
    osteogenesis imperfecta, 339–340
    skeletal dysplasias, 339–340
    vitamin D-related disorders, 339
  types available, 336–338
    balanced structure variants, 338
    base-pair substitutions, 336–337
    copy number variants, 337–338
    epigenetic variants, 337
    indels, 337
    large-scale variants, 337
    loss of heterozygosity (LOH), 338
    repeat expansions, 337
    small-scale variants, 336–337
  when to order, 340–341
Genome-wide association (GWA) studies, human, 106–109, 108f
Genome-wide studies, of osteoporosis genetics, 376, 378–381, 379t, 381t
  association studies, 376, 380–381, 381t
  linkage analysis, 376, 378–380, 379t
Genomic imprinting, in pseudohyoparathyroidism, 592–595
Genomics, functional, 73
GENOMOS (Genetic Markers for Osteoporosis), 107
Ghrelin, 996
GIO. See Glucocorticoid-induced osteoporosis
Gitelman's syndrome, 574, 655
Gla-containing proteins, 55–56, 55t
Glandular odontogenic cyst, 925–926
Gli2, 688–690
Glial cell missing B (GCMB), 574
Glial-derived growth factor (GDNF), 724
Glucocorticoid-induced osteoporosis, 473–478, 972
  clinical features, 473–474
  histomorphometry findings in, 312, 474–475
  management
    bisphosphonates, 476, 477t
    denosumab, 477–478, 477t
    general measures, 475–476
    in men, 512
    pharmaceutical agents, 476–478, 477t
    PTH (teriparatide), 433, 476–477, 477t
    responsible glucocorticoid therapy, 475
    risk factors, 474, 474t
    science of, 474
  osteonecrosis, 475
  overview, 473
  premenopausal, 517
Glucocorticoids
  ACTH regulation, 972
  adiposity increase with, 994
  bone mineral density decrease with, 735
  in fibrodysplasia ossificans progressiva, 817
  transplantation osteoporosis and, 496–497
Glucose, skeleton control and integration of, 986–990
  bone remodeling, 989
  insulin effects on bone tissue, 988–989, 989f
  osteocalcin and energy metabolism, 989–990
  osteocalcin-mediated regulation, 986–988, 987f
Glycogen storage disease, 872
Glycosylated proteins, in bone composition, 51, 53, 53t
Glyoxylatereductase/hydroxypyruvatereductase (GRHPR), 870
GNAS gene, 581, 582f, 583–584, 591–596, 591t, 592f, 595f, 607
  in fibrous dysplasia, 786–787
  in progressive osseous heteroplasia (POH), 818
GnRH. See Gonadotropin releasing hormone
Gonadal steroids, 195–204. See also specific hormones
  deficiency and chronic kidney disease-mineral bone disorder (CKD-MBD), 634
  hormone biosysnthesis, 195–196, 196f
  loss and aging, 200–201, 200f, 202f
  osteoporosis
    deficiency and osteoporosis development, 201–203, 203f
    treatment, 203–204
  overview, 195
  receptors and molecular mechanisms of action, 196–197

skeletal growth, effects on, 197–198
  linear growth, 197
  periosteal expansion, 197–198
skeletal maintenance, effects on, 198–200
  effects on osteoblasts, 199
  effects on osteoclasts, 198–199
  effects on osteocytes, 199–200
Gonadotropin releasing hormone (GnRH)
  agonists
    for breast cancer, 713
    for prostate cancer, 714–715
  analogs, 520
Gout, 872
GPCR (G-protein coupled receptor) system, 969–970
G-protein coupled receptor (GPCR) system, 969–970
GRAII algorithm, 107
Granulocyte colony stimulating factor (G-CSF), 1030
Granuloma
  central giant cell, 927
  periapical, 923, 925
Granulomatous diseases, hypercalcemia and, 565–566
Great toes, malformations of, 815, 816f
GRHPR (glyoxylatereductase/hydroxypyruvatereductase), 870
Growth factors
  cancer bone pain and, 724
  connective tissue pathways that regulate, 42–46, 46f
Growth hormone
  action on bone, 970
  in obesity, 99
  for osteogenesis imperfecta, 826–827
Gsα, 590–594, 591t, 592f–593f, 607, 786–787. See also Gα$_s$
GSK-3, inhibition of, 464
GTPases
  osteoclast regulation by, 28f
  Rac, 28
  Rho, 28
GWA (genome-wide association) studies, human, 106–109, 108f

Hair follicle cycling and vitamin D, 242
Hand-Schüller-Christian disease, 807–808
HCG (human chorionic gonadotropin), 970
HDR (hypoparathyroidism, deafness, renal anomalies) syndrome, 574
Healthy adherer effect, 451–452
Heart transplantation, 501
HED (hypohidrotic ectodermal dysplasia), 914–915, 915t
Hedgehog signaling, 19
  in fracture healing, 93
  in palatogenesis, 898
  in skeletal patterning, 4, 5f, 6
Hematologic malignancies, 694–698
  adult T-cell leukemia/lymphoma, 697
  Hodgkin's lymphoma, 698
  non-Hodgkin's lymphoma, 697–698
  overview, 694
Hematopoiesis, 1028–1031, 1031f
  adipocytes, 1030
  B lymphopoiesis, 1030
  CXCL12 abundant reticular cells, 1030
  endothelial cells, 1029
  macrophages, 1030
  malignancy, 1030–1031
  monocytes, 1030
  osteoblastic cells, 1028–1029
Hematopoietic lineage, circulating osteogenic cells isolated from, 113–114
Hematopoietic malignancies, 1030–1031

Hematopoietic stem cells, 1028–1031, 1031f
Hemichannels, role in osteocyte communication, 38–39
Henle's loop, in calcium excretion physiology, 845–846, 846f
HEO (heterotopic endochondral ossification), 815–817
Heparin, skeletal effects of, 523
Hepatitis C-associated osteosclerosis, 780–781
Hereditary hypophosphatemic rickets with hypercalciuria (HHRH), 607, 654–655, 887
Hereditary vitamin D resistant rickets, 619–620
Heritability, 77
Heterotopic endochondral ossification (HEO), 815–817
Heterotopic ossification, 1020
Heterotropic mineralization, in renal osteodystrophy, 637
HFTC (hyperphosphatemic familial tumoral calcinosis), 810
HHM (humoral hypercalcemia of malignancy), 563–565
High bone mass phenotype, 781
Hip dysplasia, 824
Hip fracture
  ethnic differences in, 135
  gradient of risk, 291t
  incidence by age, 291f
  osteoporotic
    incidence, 349, 349f, 349t, 352f
    morbidity, 353
    mortality, 353
    orthopedic surgery, 529
  probability by age and glucocorticoid dose, 292t
  quantitative computed tomography (QCT) fracture risk assessment, 268–269
  skeletal muscle strength and, 980–981
Histiocytosis-X, 807–808
Histology/histopathology
  in axial osteomalacia, 779
  in fibrodysplasia ossificans progressiva, 817
  in fibrogenesis imperfecta ossium, 779–780
  in fibrous dysplasia, 787, 788f
  in hyperparathyroidism, 564f
  in melorheostosis, 778
  in osteonecrosis of the jaw, 933, 934f
  in osteopetrosis, 771
  in osteopoikilosis, 777
  in sclerosteosis, 776
  in tumoral calcinosis, 811
  in vitamin D deficiency, 616–617, 617f
Histomorphometry, 307–314
  findings in metabolic bone disease, 310–313, 311t
    gastrointestinal bone disease, 312
    glucocorticoid-induced osteoporosis, 312, 474–475
    hypogonadism, 312
    hypophosphatemic osteopathy, 312
    hypovitaminosis osteopathy, 312
    menopausal osteoporosis, 312
    primary hyperparathyroidism, 312
    renal osteodystrophy, 312–313
  in glucocorticoid-induced osteoporosis, 312, 474–475
  indications for, 314, 314t
  interpretation of findings, 309–310
    abnormal osteoid morphology, 310
    accumulation of unmineralized osteoid, 310
    altered bone remodeling, 310
    cortical bone deficit, 310
    reference data, 309–310
    replacement of normal marrow elements, 310
  microscopy, 313–314
  organization and function of bone cells, 307–308, 308f
  specimen, procedures for obtaining, 313
  specimen processing and analysis, 313–314

Histomorphometry (contd.)
    variables, 309
        kinetic features, 309
        structural features, 309
Hodgkin's disease, 698, 736–737
Homcystinuria, 840–841
Homogentisic acid oxidase, 841
Hormone response elements (HREs), 196
Hormones. See also specific hormones
    obesity and, 996–997
    pituitary, 969–973
Hormone therapy. See also specific conditions; specific hormones
    for primary hyperparathyroidism, 549
    PTH combined with, 432
HPP. See Hypophosphatasia (HPP)
HPT -JT. See Hyperparathyroidism-jaw tumor syndrome
H_2 receptor blockers, skeletal effects of, 522
HREs (hormone response elements), 196
HRPT2 gene, 543–544
Human chorionic gonadotropin (hCG), 970
Humoral hypercalccmia of malignancy (HHM), 563–565
Hungry bone syndrome, 575, 652
HVO (hypovitaminosis osteopathy), 312
Hydroxylapatite, 49
Hyperbaric oxygenation for osteomyelitis of the jaw, 773
Hypercalcemia
    in breast cancer metastasis, 741
    in children
        idiopathic infantile hypercalcemia, 653
        severe neonatal hyperparathyroidism (SNHP), 653
        subcutaneous fat necrosis, 653
        treatment, 653
        vitamin D intoxication, 653
        Williams syndrome, 653
    clinical signs and symptoms, 562–563
    diagnosis approach, 568–569
    differential diagnosis, 563–568, 563t
        abnormal protein binding, 567
        Addisonian crisis, 566
        cancer, 563–565
        endocrine disorders, 566
        granulomatous diseases, 565–566
        hyperthyroidism, 566
        hypophosphatemia, 568
        immobilization, 566
        medications, 567–568
        milk-alkali syndrome, 566–567
        pheochromocytoma, 566
        renal failure, acute and chronic, 568
        total parenteral nutrition (TPN), 567
        VIPoma syndrome, 566
    familial hypocalciuric hypercalcemia (FHH), 553–555, 554t
    in Hodgkin's disease, 698
    in hyperparathyroidism, 537–538
    malignancy-associated hypercalcemia (MAHC), 563–565
        authentic ectopic hyperparathyroidism, 565
        humoral hypercalcemia of malignancy (HHM), 563–565
        local osteolytic hypercalcemia (LOH), 565
        vitamin D-induced hypercalcemia, 565
    management, 569
    non-parathyroid, 540, 562–569
    parathyroid hormone levels in, 540
    pathophysiology of, 562
    treatment in children, 653
Hypercalciuria, 865
    autosomal dominant hypocalcemic hypercalciuria (ADHH), 587
    causes of, 870–871, 870t, 885–889

genetic hypercalciuric stone-forming rat studies, 847–850, 848t, 849f
    genetics of, 885–889, 885t
    hereditary hypophosphatemic rickets with hypercalciuria (HHRH), 607
    nephrolithiasis and, 870–871
    treatment, 874, 878–879
Hypermagnesemia in children, 655–656
Hyperostosis-hyperphosphatemia syndrome, 810
Hyperoxaluria, 865–866, 871, 874–875, 879
Hyperparathyroidism, 537–538, 543–549
    approach to patient, 538
    causes, 543, 544
    chronic kidney disease-mineral bone disorder (CKD-MBD) and, 634
    clinical forms of, 545
    evaluation and diagnosis of, 545–546, 546f
    familial primary, 553–560, 554t
        familial hypocalciuric hypercalcemia (FHH), 553–555, 554t
        familial isolated primary hyperparathyroidism (FIHPT), 554t, 559
        hyperparathyroidism-jaw tumor syndrome (HPT-JT), 554t, 558–559
        multiple endocrine neoplasia type 1 (MEN1), 554t, 556–557
        multiple endocrine neoplasia type 2A (MEN2A), 554t, 557–558
        neonatal severe primary hyperparathyroidism (NSHPT), 554t, 555–556
    fetal response to maternal, 182–183
    histomorphometry findings in, 312
    hypophosphatemic rickets with hyperparathyroidism, 607
    incidence of, 543
    maternal, 588, 651
    oral manifestations, 954
    pathophysiology of, 543–544
    in pregnancy, 549
    primary in pregnancy, 158–159
    radionuclide scintigraphy, 285–286
    severe neonatal hyperparathyroidism (SNHP), 653
    signs and symptoms, 544
    treatment of, 546–549, 546t, 548f
        nonsurgical management, 546t, 547–549, 548f
        parathyroid surgery, 546–547, 546t, 548f
Hyperparathyroidism-jaw tumor syndrome (HPT-JT), 554t, 558–559
    clinical expressions, 558
    diagnosis of carriers and cancers, 559
    management, 559
    pathogenesis/genetics, 559
Hyperphosphatemia, 608–609
    causes of, 608, 609t
    in children, 655
    chronic kidney disease-mineral bone disorder (CKD-MBD) pathogenesis and, 633–634
    clinical and biochemical manifestations, 609
    clinical manifestations, 608
    familial tumoral calcinosis (FTC), 608–609
    genetics, 609
    hypocalcemia and, 575
    treatment, 609
Hyperphosphatemic familial tumoral calcinosis (HFTC), 810
Hyperthyroidism, hypercalcemia and, 566
Hypertrophic osteoarthropathy. See Pachydermoperiostosis
Hyperuricemia, in nephrolithiasis, 861
Hyperuricosuria, 866, 872
    in nephrolithiasis, 861
    treatment, 874, 880

Hypocalcemia, 572–577
  autosomal dominant hypocalcemic hypercalciuria (ADHH), 587
  in children, 651–653
    calcium malnutrition, 652
    chelation therapy, 652
    clinical presentation, 651
    hungry bone syndrome, 652
    hypoparathyroidism, 651–652
    infantile osteopetrosis, 652
    phosphate enemas, 652
    transient hypocalcemia of the newborn, 651
    treatment, 652–653
    vitamin D-related, 652
  chronic kidney disease-mineral bone disorder (CKD-MBD) and, 634
  clinical features of, 579–580
  definition of, 572
  diagnostic tests and interpretation, 576
  etiology, 572, 573t, 574
  familial hypomagnesemia with secondary hypocalcemia, 655
  in hypoparathyroidism, 538–540
  management, 576–577
  parathyroid hormone levels in, 540
  pathogenesis, 574–575
  pseudohypocalcemia, 575
  signs and symptoms, 575–576, 576t
  treatment
    in children, 652–653
    as side effect of bisphosphonates and denosumab, 748
Hypocitraturia, 866, 871, 880
Hypodentia, 914–915
Hypogonadism
  chronic kidney disease-mineral bone disorder (CKD-MBD) and, 634
  histomorphometry findings in, 312
  osteoporosis in men, 510, 512
Hypohidrotic ectodermal dysplasia (HED), 914–915, 915t
Hypomagnesemia
  causes, 655
  in children, 655
  treatment, 655
Hypoparathyroidism, 538–540. See also Pseudohypoparathyroidism
  approach to patient, 539–540
  in children, 651–652
  classification, 581t
  diagnostic algorithm, 588
  fetal response to maternal, 183
  functional, 579–589
    clinical manifestations, 579–580
    defined, 579
    differential diagnosis, 580–588
    therapy, 588–589
  genetics, 574, 581–587, 581t, 582f
  "HAM" complex, 586–587
  hypocalcemia and, 574
  in lactation, 161
  in pregnancy, 159
  PTH resistance and
    circulating inhibitors as cause of, 584–585
    magnesium deficiency and, 585
    pseudohypoparathyroidism (PHP) type 1, 581–584
    pseudohypoparathyroidism (PHP) type 2, 584
    syndromes of, 581–585
    transient pseudohypoparathyroidism of the newborn, 584
  syndromes of deficient PTH synthesis or secretion, 585–588
    anti-CaSR antibodies, 587
    autoimmune polyendocrinopathy-candidiasis-exctodermal dystrophy (APECED), 586–587
    autosomal dominant hypocalcemic hypercalciuria (ADHH), 587
    developmental disorders, 585–586
    DiGeorge sequence/CATCH-22, 585
    hypoparathyroidism, sensorineural deafness, and renal dysplasia syndrome (HDR), 585–586
    hypoparathyroidism-retardation-dysmorphism syndrome (HRD), 586
    isolated hypoparathyroidism, 586
    Kenny-Caffey syndrome, 586
    maternal hyperparathyroidism and hypomagnesemia, 588
    mitochondrial disease, 587–588
    parathyroid gland destruction, 586–587
    post-burn, 586
    PTH gene mutations, 587
    radiation therapy, 586
    surgery, 586
  therapy, 588–589
Hypoparathyroidism, sensorineural deafness, and renal dysplasia syndrome, 585–586
Hypoparathyroidism-retardation-dysmorphism syndrome, 586
Hypophosphatasia (HPP), 653, 838–840, 839f, 915t, 918, 919f
Hypophosphatemia, 601–602, 915t, 918
  causes of, 601–602, 602t, 654–655
  in children, 654–655
  clinical consequences, 601
  genetic testing for, 339
  hypercalcemia and, 568
Hypophosphatemic nephrolithiasis/osteoporosis 1 and 2, 608
Hypophosphatemic osteopathy, histomorphometry findings in, 312
Hypophosphatemic rickets with hyperparathyroidism, 607
Hypovitaminosis osteopathy (HVO), 312

Ibandronate, 413f–414f, 415–416
  after organ transplantation, 498, 499t–500t
  for metastatic bone disease in breast cancer, 742–743, 744f
ICER (incremental cost-effectiveness ratio), 455
Idiopathic hypercalciuria, 847–850, 848t, 849f
Idiopathic infantile hypercalcemia, 653
Idiopathic juvenile osteoporosis (IJO), 468–471, 469f, 470t–471t
  biochemical findings, 470
  bone biopsy, 470
  clinical features, 469, 469f
  differential diagnosis, 470, 470t–471t
  overview, 468–469
  pathophysiology, 469
  prevalence of, 468–469
  radiographic features, 469–470, 469f
Ifosfamide, 704
IGF-1. See Insulin-like growth factor-1
Ihh. See Indian hedgehog
IJO. See Idiopathic juvenile osteoporosis
IL-1 (interleukin), osteoclast regulation and, 28–29
Illnesses, effect on bone morphology, 131
Image resolution, voxel size and, 63–64, 64f
Imaging of bone. See also specific modalities
  issues associated with, 63–66, 64f–66f
    calibration, 66
    image resolution, 63–64
    segmentation, 64–65, 65f
    skeletal site and volume of interest, 65–66, 66f
    voxel size, 63–64, 64f

Imaging of bone. *See also specific modalities (contd.)*
  in osseous metastatic disease, 677–682, 678f, 679t
  in rodents, 59–66
IMKC (International Mouse Knockout Consortium), 73
Immobilization
  bone loss during, 982
  hypercalcemia and, 566
Immune cells
  bone interactions, 1036–1040
  in bone metastatic tumors, 689
Immunity, vitamin D and
  adaptive immunity, 241–242
  innate immunity, 241
Immunosuppressive drugs, skeletal effects of, 496–497
  calcineurin inhibitors, 497
  glucocorticoids, 496–497
Immunotherapy, for osteosarcoma, 704
IMO (infantile malignant osteopetrosis), 123
Implant osseointegration, with strontium ranelate, 440, 441f
Inactivity, effect on skeletal muscle and bone, 981–982
Incremental cost-effectiveness ratio (ICER), 455
Indels, 337
Indian hedgehog (Ihh)
  in fracture healing, 93
  PTHrP actions on, 217, 217f
  signaling, 6, 7f, 8, 93
Infantile malignant osteopetrosis (IMO), 123
Infantile osteopetrosis, 652
Infection
  osteomyelitis, 1039
  periodontal disease, 941–943
  urinary tract, 869–870
Infiltrative disorders, 807–808
  histiocytosis-X, 807–808
  systemic mastocytosis, 807
Inflammation
  bone loss in rheumatic disease, 482–485
    ankylosing spondylitis, 484–485
    overview, 482–483
    rheumatoid arthritis, 483–484
    systemic lupus erythematosus, 485
  burn injury, 531
  dermatomyositis, 812–813, 812f
  lymphocytes, 1037–1038
  odontogenic lesions, 923, 924f, 925
  osteoblast-osteoclast interaction and, 1039–1040
  in periodontal disease, 941–942, 943f
  secondary osteoporosis-associated
    chronic obstructive pulmonary disease (CPOD), 490
    inflammatory arthritis, 483–484, 489
    inflammatory bowel diseases, 490
Inflammatory bowel diseases, 490
Infliximab, for dermatomyositis, 813
Inherited fetal/neonatal bone disorders, 123–124
  in bone matrix production, 123
  in mineral deposition, 123
  in mineral homeostasis, 123
  in osteoclastic function, 123
Innate immunity, vitamin D and, 241
Insertions, 337
Insulin, bone as target tissue, 988–989, 989f
Insulin-like growth factor-1 (IGF-1), 19, 372
  bone marrow adipose tissue and, 994
  in fracture healing, 93
  growth hormone action on bone, 970
  muscle growth stimulation of, 982
  in obesity, 99
Insulin-like growth factor-2 (IGF-2), 372

Integrins, 27–28, 672, 690
Interferon γ (IFNγ), osteoclast regulation and, 29
Interfragmentary strain theory, 102
Interleukin (IL-1), osteoclast regulation and, 29
Interleukins, postburn, 531
International Mouse Knockout Consortium (IMKC), 73
Intestine
  absorption of calcium
    in lactation, 160
    in pregnancy, 157
  calcium sensing receptor (CaSR) role in maintaining calcium homeostasis, 230
  inflammatory bowel diseases, 490
  transplantation, 496, 502
  vitamin D effects, 240
Intracellular signaling pathways, 31
Intramedullary nails, 528–529
Ion channels, as cancer pain therapeutic targets, 723–724
Iron accumulation in bone, 636
Iron salts, as phosphate binders, 645
Ischaemic bone disease, 805–807. *See also* Osteonecrosis
  causative factors, 806t
  clinical features, 806
  differential diagnosis, 807
  epidemiology, 805
  laboratory findings, 807
  pathogenesis, 805–806
  radiological features, 806–807
  staging, 807t
  treatment, 807

Jagged1, 672
Jagged2, 906
Jaw
  cysts and tumors, 923–927
    clinical and radiographic presentation, 923, 924f
    nonodontogenic tumors, 927
    odontogenic cysts, 923–926, 925f–926f
    odontogenic tumors, 926
    origin, 923
  hyperbaric oxygenation for osteomyelitis of, 773
  hyperparathyroidism-jaw tumor syndrome (HPT-JT), 554t, 558–559
  osteonecrosis (*see* Osteonecrosis of the jaw)
  pathology of hard tissues, 922–927
  tooth demineralization and caries, 922–923
  van Buchem disease, 775

Kenny-Caffey syndrome, 586
Keratocystic odontogenic tumor, 925, 926f
Kidney. *See also* Chronic kidney disease-mineral bone disorder
  calcium sensing receptor (CaSR) role in maintaining calcium homeostasis, 229–230
  chronic kidney disease, 495–496
    CKD-mineral bone disorder (CKD-MBD), 632–637
    fibroblast growth factor-23 (FGR23) association, 192, 633
  disorders of renal phosphate wasting, 607–608
    autosomal recessive hypophosphatemic rickets (ARHR), 607
    Fanconi renotubular syndrome 2, 608
    fibrous dysplasia, 607
    hereditary hypophosphatemic rickets with hypercalciuria (HHRH), 607
    hypophosphatemic nephrolithiasis/osteoporosis 1 and 2, 608
    hypophosphatemic rickets with hyperparathyroidism, 607
    linear nevus sebaceous syndrome, 607–608
    osteoglophonic dysplasia (OGD), 608
    table of characteristics, 604t
  fetal calcium metabolism, 181

renal failure, hypercalcemia and, 568
renal handling of calcium
  in lactation, 160
  in pregnancy, 157
renal osteodystrophy
  bone pain, 637
  calcifications, 637
  clinical manifestations, 636–637
  fractures, 637
  heterotropic mineralization, 637
  high turnover, 635, 635t
  high turnover plus mineralization defect, 635–636, 635t
  histomorphometry findings in, 312–313
  low turnover, 635, 635t
  oral manifestations, 954
  radionuclide scintigraphy, 286–287, 286f
  skeletal deformities, 637
  transplantation osteoporosis, 495–496, 498
  tumoral calcinosis, 637
renal tubular acidosis in carbonic anhydrase II deficiency, 773
renal tubular physiology of calcium excretion, 845–850, 848t, 849f
transplantation, 498, 501
vitamin D effects, 240
Kidney-pancreas transplantation, 501
Kidney stones. See Nephrolithiasis; Renal stones
Kienbock's disease, 1013
Kirschner wires, 528
Klotho, 191–192
  action on distal tubules, 846–847
  chronic kidney disease-mineral bone disorder (CKD-MBD) and, 633
  PTH secretion regulation, 211
  vitamin D and, 238–239, 239f
Knockout studies, 71–73
K wires, 528
Kyphoplasty, 529–530, 764
Kyphosis, 529, 662

Lactation, 159–162
  adolescent, 162
  breastfeeding, 144
  calciotropic hormone levels, 159–160
  calcium
    average daily calcium loss, 159
    homeostasis, 157f
    intestinal absorption, 160
    low intake of calcium, 162
    renal excretion, 160
    skeletal metabolism of calcium, 160–161, 161f
  hypoparathyroidism, 161
  osteoporosis, 161
  pseudohypoparathyroidism, 161–162
  vitamin D deficiency and insufficiency, 162
Lacunocanalicular network, 34, 35f
LADD (larimo-auriculo-dento-digital syndrome), 915, 915t
Langerhans cell histiocytosis, 807–808
Lanthanum accumulation in bone, 636
Lanthanum salts, as phosphate binders, 644–645
Larimo-auriculo-dento-digital syndrome (LADD), 915, 915t
Lateral periodontal cyst, 925
LDS. See Loeys-Dietz syndrome
LEMD3 gene, 778
LEPRE1 gene, 823t, 824
Leptin
  bone mass and, 82–85, 84f
  obesity and, 83, 996

osteocalcin and, 987
sympathetic nervous system and, 83–85, 84f, 961–962
Lesch-Nyhan syndrome, 872
Letrozole, 521, 749
Letterer-Siwe disease, 807–808
Leukemia
  acute lymphocytic leukemia (ALL), 734–736
  acute myelogenous leukemia (AML), 1031
  adult T-cell leukemia/lymphoma, 697
LH (luteinizing hormone), 970
Li-Fraumeni syndrome, 704
Limb patterning, 4–6, 5f
Lineage tracing, 73
Linear nevus sebaceous syndrome, 607–608
Linkage analysis, 106
  in fibrodysplasia ossificans progressiva, 817
  of osteoporosis genetics, 376, 378–380, 379t
Lipids, in bone, 56
Lithium, hypercalcemia from, 567
Liver disease, end-stage, 496
Liver transplantation, 501–502
Loading
  characteristics of an effective loading prescription, 150–151, 151f
  importance of, 149–153
  persistence of childhood bone adaptation, 152–153, 152f
  in spinal cord injury patients, 1022–1023
Local osteolytic hypercalcemia (LOH), 565
Locking plates, 528
Loeys-Dietz syndrome (LDS), 44, 831t, 832–834, 833t
LOH (local osteolytic hypercalcemia), 565
Loss of heterozygosity, 338
Lowe's syndrome, 655
LRP5 gene, 380, 781
LTBPs (latent TGFβ binding proteins), 43
Luciferases, 681
Lung cancer
  metastasis to bone, 687
  osteolytic metastasis, 761f
  RWGT2 model of, 687
Lung transplantation, 501
Luteinizing hormone (LH), 970
Lymphocytes, 1037–1038
Lymphoma
  adult T-cell leukemia/lymphoma, 697
  Hodgkin's, 698
  non-Hodgkin's, 697–698, 736
Lysophosphatidic acid, platelet, 672
Lysosomal storage disorders, 917

Macrophage-colony stimulating factor (M-CSF), 25, 26f, 28–30, 92–93, 1037–1038
Macrophages, 1030
Magnesium
  calcium sensing receptor (CaSR) as Mg^{2+} sensor, 227
  cellular content of, 177
  deficiency and parathyroid hormone (PTH) resistance, 585
  distribution, 177
  homeostasis, 177
  hypermagnesemia, 655–656
  hypocalcemia and, 574–575
  hypomagnesemia, 655
  intestinal absorption, 177
  maternal hyperparathyroidism and hypomagnesemia, 588
  metabolism disorders in children, 654–655, 654t
  renal handling, 177
  supplementation, 655

Magnesium-ammonium-phosphate stones, 881. *See also* Struvite stones
Magnesium salts, as phosphate binders, 645
Magnetic resonance imaging (MRI), 277–280
    applications using cortical bone measures, 280
    for assessment of bone mass and microarchitecture in rodents, 61–62, 61f
    of cortical bone, 279, 280
    in multiple myeloma, 696–697
    in osseous metastatic disease, 679t, 680
    in osteonecrosis, 806–807, 807t
    in osteopetrosis, 771
    overview of, 277
    in Paget's disease, 663
    pseudohypocalcemia from gadolinium-containing agents, 575
    of trabecular bone, 277–280, 278f
    for vertebral fracture assessment, 328–330, 329f
MAHC. *See* Malignancy-associated hypercalcemia
Major depressive disorder (MDD), 1002–1005
Major histocompatibility complex (MHC), 1037, 1039
Malignancies. *See* Cancer
Malignancy-associated hypercalcemia (MAHC), 563–565
    authentic ectopic hyperparathyroidism, 565
    humoral hypercalcemia of malignancy (HHM), 563–565
    local osteolytic hypercalcemia (LOH), 565
    vitamin D-induced hypercalcemia, 565
Malignant odontogenic tumors, 926
Malignant transformation, 671
Mammary gland, PTHrP (parathyroid hormone-related protein) effects on, 218–219, 218f, 219f
Mandibular development, 898–899
Marble bone disease. *See* Osteopetrosis
Marfan syndrome, 42
    diagnostic criteria, 832t
    genetics of, 830, 831t
    pathogenesis, 832
    skeletal manifestations, 830–831, 833t
Marfan syndrome-related disorders, 831t, 832–834, 833t
Marrow stromal stem cells (MSCs). *See* Mesenchymal stem cells (MSCs)
Mastocytosis, 491, 807
Matrix gla protein, 55, 55t
Matrix metalloproteinase (MMP), 688
Mazabraud's syndrome, 786
McCune-Albright syndrome, 191, 583, 607, 786
MCP-1 (monocyte chemoattractant protein-1), 997
MC2R (melacortin receptor 2), 972
M-CSF. *See* Macrophage-colony stimulating factor
MDCT (multidetector computed tomography), 327
MDD (major depressive disorder), 1002–1005
MDP (methylene diphosphonate), 283
MDSCs (myeloid derived suppressor cells), 689
Mechanical performance, genetic determinants of, 77
Mechanical signals
    exercise and osteoporosis prevention, 396–400, 397f
        mechanical factors regulating bone cell response, 396–397
        mechanically response bone cells, 397–398
        transducing mechanical signals into cellular response, 398, 399f
        translating mechanical signals to the clinic, 398–400
    influences on fetal/neonatal bone development, 122
    loading (*see* Loading)
Mechano-growth factor (MGF), 982
Mechanoreceptors, 398
Mechanosensation, 38
Mechanotransduction, 38

Meckel's cartilage, 898
MED (multiple epiphyseal dysplasia), genetic testing for, 340
Medication possession ratio (MPR), 450
Medications, neuropsychiatric, 1007–1008
Melacortin receptor 2 (MC2R), 972
Melamine poisoning, 872
Melanocortins, 964–965
MELAS syndrome, 587
Melorheostosis, 778
    clinical presentation, 778
    etiology and pathogenesis, 778
    histopathological findings, 778
    laboratory findings, 778
    radiographic features, 778, 778f
    treatment, 778
Men
    aging effects on skeleton, 508–509
    bone turnover markers in, 302
    fracture epidemiology, 509, 509f
    osteoporosis in, 201–203, 203f, 508–512
        aged-related, 511–512
        BMD measurements, 510–511
        causes, 509–510, 510t
        clinical evaluation, 511
        evaluation, 510–511, 511t
        glucocorticoid-induced, 512
        hypogonadism, 510, 512
        idiopathic osteoporosis, 510
        laboratory testing, 511, 511t
        prevention, 511
        treatment, 432, 511–512
    sex steroids and bone loss, 370–372, 371t, 372f
    skeletal development in, 508
    testosterone level and risk of falling, 981
    urinary constituents, normal range of, 866
MEN-1. *See* Multiple endocrine neoplasia type 1
MEN2A. *See* Multiple endocrine neoplasia type 2A
Menkes disease, 841
Menopause, 165–167
    bone effects, 165–166, 167f
    calcium metabolism effects, 166
    described, 165
    osteoporosis, histomorphometry findings in, 312
Mesenchymal stem cells (MSCs)
    adipocytes from, 993
    circulating osteogenic precursor (COP) cells, 111–113, 115f
    mechanical signals and, 398
    osteoblast differentiation, 15, 16f, 993
    osteosarcoma differentiation, 706, 706f
Mesenchyme, knockout and overexpression studies in, 72
Metabolic bone disease
    genetic testing for, 339
    histomorphometry findings in, 310–313, 311t
    oral manifestations of, 948–955
    radionuclide scintigraphy, 283–287
Metabolism disorders in children, 651–656
Metaphyses, growth in childhood, 130–131, 131f
Metastasis. *See* Cancer, metastasis to bone
Methotrexate
    bone mineral density decrease with, 735
    for dermatomyositis, 813
Methylene diphosphonate (MDP), 283
Methylenetetrahydrofolate reductase (MTHFR) gene, 806
Metolazone, effect on renal calcium excretion, 847
MGF (mechano-growth factor), 982
MHC (major histocompatibility complex), 1037, 1039
Microcomputed tomography (μCT), for assessment of bone mass and microarchitecture in rodents, 62–63, 62f–63f

Micrognathia, 899
microRNAs and osteosarcoma, 705
Microscopy. *See* Bone biopsy; Histology/histopathology
Milk
   avoidance, 145
   effect on bone mass, 362f, 363
Milk-alkali syndrome, 566–567
"Milk of calcium," 813
Mineral bone disorder. *See* Chronic kidney disease-mineral bone disorder (CKD-MBD)
Mineralization. *See also* Calcification
   fetal skeleton, 182, 185
   heterotrophic, 637
Minerals. *See also specific mineral disorders; specific minerals*
   composition of bone, 49–50
   ethnic differences in metabolism, 135–136
   in fetal/neonatal bone development
     defects in deposition, 123
     defects in homeostasis, 123
   lactation levels of, 159–160
   metabolism disorders in children, 651–656
     calcium, 651–653
     magnesium, 654–656, 654t
     phosphate, 654–655, 654t
     rickets, 656, 656f
   placental transport, 181–182
   pregnancy levels of, 156–157
   transport across placenta, 121–122
Missense variants, 336–337
Mitochondrial disease, 587–588
MLPA (multiplex ligation dependent probe amplification), 337
MMP (matrix metalloproteinase), 688
MNHPT. *See* Neonatal severe primary hyperparathyroidism
Molecular control of fracture healing, 92–93
Molecular therapies to enhance bone healing, 94–95
Monoclonal antibody inhibiting RANKL action, 461–462
Monocyte chemoattractant protein-1 (MCP-1), 997
Monocytes, 113–114, 1030
Mouse Genome Informatics, 73
MPR (medication possession ratio), 450
MRI. *See* Magnetic resonance imaging
MSCs. *See* Mesenchymal stem cells
Msx1 homeobox gene, 897f, 898
MTHFR (methylenetetrahydrofolate reductase) gene, 806
Mucopolysaccharidoses, 840, 915t, 917
Multidetector computed tomography (MDCT), 327
Multinucleated giant cells (MGCs), 927
Multiple endocrine neoplasia type 1 (MEN1), 554t, 556–557
   clinical expressions, 556
   diagnosis of carriers, 556–557
   diagnosis of other tumors, 557
   management, 557
   parathyroid surgery for, 557
   pathogenesis/genetics, 556
Multiple endocrine neoplasia type 2A (MEN2A), 554t, 557–558
   clinical expressions, 557–558
   diagnosis of kindred, of carriers, and parathyroid tumors, 558
   management, 558
   pathogenesis/genetics, 558
Multiple epiphyseal dysplasia (MED), genetic testing for, 340
Multiple myeloma, 694–697
   bone remodeling in, 695
   clinical presentation, 695
   incidence, 694
   metastasis to bone
     clinical aspects, 742
     treatment and prevention, 746f, 747–748
   sclerosis bone disorder, 781
   vicious cycle of, 696, 696f
Multiplex ligation dependent probe amplification (MLPA), 337
Muramyl tripeptide, 704
Muscle mass, low, 996
Musculoskeletal system, vitamin D deficiency impact on, 243
Mutant Mouse Regional Resource Centers, 73
Myeloid derived suppressor cells (MDSCs), 689
Myokines, 982
Myosteatosis, 978
Myxomas of skeletal muscle, 786

Nanocomputed tomography (nCT), for assessment of bone mass and microarchitecture in rodents, 63, 64f
National Osteoporosis Foundation (NOF), 258, 258t, 457
National Osteoporosis Foundation Guidelines, 258t
National Osteoporosis Guidelines Group (NOGG), 258, 293, 294f, 318
NCCs (neural crest cells), 3, 4f
Neonatal severe primary hyperparathyroidism (NSHPT), 123, 554t, 555–556, 886
   clinical expressions, 555
   diagnosis, 555
   management, 555–556
   pathogenesis/genetics, 555
Neoplastic vertebral fracture, differentiating osteoporotic from, 329–330, 330t
Nephrolithiasis
   diagnosis and evaluation of, 860–866
     basic patient evaluation, 860–861
     metabolic profiling, 861, 863–866
     in pediatric patients, 872–873
     radiographic evaluation, 861, 864f, 872–873
     stone analysis, 861–863, 864f–865f
     urinalysis, 861, 862f–863f
   epidemiology of, 856–858
     diabetes as risk factor, 858
     genetics, 858
     geography of, 856
     lifestyle risk factors, 858
     nutritional risk factors, 857–858
     obesity as risk factor, 858
     prevalence and incidence, 856–857
   genetic basis of renal stones, 884–889, 885t
   hypophosphatemic nephrolithiasis/osteoporosis 1 and 2, 608
   metabolic profiling, 861, 863–866
     hypercalciuria, 865
     hypercitraturia, 866
     hyperoxaluria, 865–866
     hyperuricosuria, 866
     patient selection for, 863, 865
     supersaturation, 865
     urine pH, 866
   nutritional risk factors
     calcium intake, 857
     fluid intake, 857
     oxalate, 858
     protein intake, 858
     salt intake, 858
   in pediatric patient, 869–875
     clinical evaluation, 872, 873t
     incidence of, 869
     medical management, 874–875
     predisposing conditions, 869–872, 870t
     prognosis, 875
     surgical management, 873–874, 874t

Nephrolithiasis (contd.)
  predisposing conditions, 869–875, 870t
    cytinuria, 872
    hypercalciuria, 870–871, 870t
    hypercitraturia, 871
    hyperoxaluria, 871
    hyperuricosuria, 872
    infection, 869–870
    melamine poisoning, 872
    metabolic abnormalities, 870
    struvite stones, 870
    urinary tract abnormalities, 869
  treatment, 878–881
    calcium stones, 878–880
    cystine stones, 880–881
    struvite stones, 881
    uric acid stones, 880
Nerve growth factor (NGF), 721, 723–724, 723f
Nervous system
  central
    central neuronal control of bone remodeling, 961–965, 965f
    PTHrP (parathyroid hormone-related protein) effects, 220
    parasympathetic in bone remodeling, 85
    PTHrP (parathyroid hormone-related protein) effects, 220
  sympathetic
    hematopoiesis and, 1030
    leptin and, 83–85, 84f, 961–962
Neural crest cells (NCCs), 3, 4f, 895–896, 896f, 898–901
Neuroblastoma, 736
Neurocranium development, 899–900, 900f
Neuroleptic medications, 1008
Neuromedin U (NMU), 963
Neuronal form of NO synthase (nNOS), 86
Neuronal storage disease with osteopetrosis, 770
Neuropathic pain, radiotherapy for, 755
Neuropeptides, skeletogenic, 85–86
Neuropeptide U (NPU), 86
Neuropeptide Y (NPY), 85, 963
Neuropsychiatric disorders, 1002–1008
  Alzheimer's disease, 1006–1007
  anorexia nervosa, 1005
  low bone mass and depression, 1004–1005
  major depressive disorder (MDD), 1002–1005, 1004f
  medications for, 1007–1008
    neuroleptics, 1008
    SSRIs, 1007–1008
  schizophrenia, 1005–1006
  table of, 1003t
  traumatic brain injury, 1006
Neutralization plates, 528
Next generation sequencing, 338
NF-κB. *See* Nuclear factor-kappa B
NFTC (normophosphatemic familial tumoral calcinosis), 810
NGF (nerve growth factor), 721, 723–724, 723f
Niacin, 645
NMU (neuromedin U), 963
nNOS (neuronal form of NO synthase), 86
Nociceptive nerve sprouting, 722–723, 724f–725f
NOF (National Osteoporosis Foundation), 258, 258t, 457
NOGG (National Osteoporosis Guidelines Group), 258, 293, 294f, 318
Non-Hodgkin's lymphoma, 697–698, 736
Nonsense mutations, 337
Non-steroidal anti-inflammatory drugs (NSAIDs), negative impact on bone repair, 93
Non-synonymous variants, 336–337
Normophosphatemic familial tumoral calcinosis (NFTC), 810
NO synthase, neuronal (nNOS), 86

Notch signaling, 4, 20
  osteosarcomagenesis, 705
  tooth development, 906
NPU (neuropeptide U), 86
NPY (neuropeptide Y), 85, 963
NPY1 receptor, 964
NSAIDs (non-steroidal anti-inflammatory drugs), negative impact on bone repair, 93
Nuclear factor-kappa B (NF-κB)
  discovery of, 1036
  roles in osteoclast formation, 1037
Nutrition. *See also* Diet
  after acute lymphocytic leukemia treatment, 736
  childhood tumors and, 736–737
  fetal/neonatal bone development, 122
  during growth, 142–145
    breastfeeding, 144
    calcium, 142–143
    fruits and vegetables, 143
    milk avoidance, 145
    pregnancy, 143–144
    salt, 144–145
    soft drinks, 145
    vitamin D, 143
  nephrolithiasis risk factors, 857–858
    calcium intake, 857
    fluid intake, 857
    oxalate, 858
    protein intake, 858
    salt intake, 858
  osteoporosis and, 361–364
    building peak bone mass, 361–362, 362f
    dietary bioactive constituents, 364
    dietary patterns, 363
    maintaining bone mass, 362–363
    overview, 361

Obesity
  bone and body composition, 996
  bone marrow fat, 994
  bone mineral density (BMD) in, 994–995
  endocrine profile in, 996–997
  fracture risk, 995, 995f
  as nephrolithiasis risk factor, 858
  prevalence of, 993
  as state of leptin resistance, 83
  stromal cell adipocyte/osteoblast differentiation, 993–994
Ob/ob mice, 82–83, 85, 961, 963
Oculocerebrorenal syndrome of Lowe (OCRL), 887
Odanacatib, 462–463
Odontoblasts, 906–907, 907f
Odontogenesis. *See* Teeth, development
Odontogenic cysts
  dentigenous cysts, 925
  developmental, 925–926
  glandular, 925–926
  inflammatory lesions, 923, 924f, 925
  lateral periodontal cysts, 925
  odontogenic keratocyst, 925, 926f
  periapical cyst, 923, 925
Odontogenic keratocyst (OKC), 925, 926f
Odontogenic tumors, 926
OKC (odontogenic keratocyst), 925, 926f
Oncogene activation in osteosarcoma, 704–705
1,000 Genomes Project, 107
OPG. *See* Osteoprotegerin
Opsismodysplasia, 191

Opsonins, 1039
Optical imaging, 681
Oral bone loss, 948–954
   aging and, 948
   assessment
      methods, 948–950, 949f–950f
      as screening tool for osteoporosis, 952–953
   as fracture predictor, 953
   mechanisms of, 952
   osteoporosis and tooth loss, 951–952
   osteoporosis therapy impact on, 953
   periodontal diseases and, 941–946
      diagnosis and classification, 942, 944t
      etiology, 941–942, 942f–943f
      treatment, 942–946, 944t–945t
   primary hyperparathyroidism, 954
   relationship to postcranial bone mass, 950–951
   renal osteodystrophy, 954
Oral cavity
   aging effects, 948
   hyperparathyroidism and, 954
   metabolic bone disease manifestations, 948–954
   osteonecrosis of the jaw, 929–935
   osteoporosis
      impact of therapy, 953–954
      screening, 952–953
      tooth loss, 951–952
   Paget's bone disease and, 954
   periodontal disease, 941–943
   renal osteodystrophy and, 954
Orofacial cleft, 898
Orthopedic surgery, 527–530
   in osteoporotic patients
      common fractures, 529
      vertebral fracture intervention, 529–530
   plates and screws, 528
      bridge plates, 528
      buttress (antiglide) plates, 528
      compression plates, 528
      locking plates, 528
      neutralization plates, 528
      tension band plate, 528
   surgical options, 528–529
      external fixation, 529
      intramedullary nails, 528–529
      Kirschner wires, 528
      plates and screws, 528
   treatment principles, 527–528
OSCAR (osteoclast-associated receptor), 1038
Ossification
   endochondral, 815–817
   in fibrodysplasia ossificans progressiva, 815–817
   heterotopic, 815–817, 1020
   physiology of fetal and neonatal bone development, 121–122
   in progressive osseous heteroplasia (POH), 817–818
Osteitis fibrosa, 635
Osteoblastic bone disease, 672–673
Osteoblasts
   appearance of, 308, 308f
   differentiation, 993–994
   embryonic development, 6–7
   estrogen effects, 165–166, 199
   glucose metabolism regulation, 986–987
   hematopoiesis and, 1028–1029
   insulin signaling, 988–989, 989f
   knockout and overexpression studies in, 72
   markers for, 34
   mechanical stimulation of, 398
   in multiple myeloma, 695–696, 696f
   NPY1 receptor, 964
   osteoclast interaction in inflammation, 1039–1040
   overexpression of target genes, 70
   prolactin effects on, 972–973
   radiation effects on, 729
   signal transduction cascades controlling differentiation, 15–21
   suppression by immobilization, 566
   sympathetic nervous system and, 84, 962
   thyroid stimulating hormone (TSH) regulation of, 972
Osteocalcin, 55, 55t
   energy metabolism and, 989–990, 997
   insulin signaling, 988, 989f
   osteocalcin-mediated bone regulation of glucose metabolism, 986–988, 987f
   target tissues of, 987f
Osteochondrodysplasias, 794–803, 796t–802t, 803f
   classification, 795, 796t–802t, 803
   multidisciplinary approach, 795
   overview, 794–795
Osteoclast-associated receptor (OSCAR), 1038
Osteoclasts
   appearance of, 307, 308f
   cancer bone pain and, 724–725
   cell biology of, 25–28, 26f–27f
   co-stimulatory signaling in osteogenesis, 1038–1039
   defects in function, 123
   estrogen effects on, 166, 198–199
   factors regulating formation and/or function, 28–30
   follicle stimulating hormone (FSH) stimulation of, 971
   future osteoporosis therapies
      inhibiting osteoclast action, 462–463
      inhibiting osteoclast formation, 461–462
   inhibition in metastatic bone disease, 715–716
   knockout and overexpression studies in, 72
   in multiple myeloma, 695–696, 696f
   nuclear factor-kappa B (NF-κB) and, 1037
   osteoblast interaction in inflammation, 1039–1040
   osteoclast-rich lesions outside bone, 1038
   overexpression of target genes, 70
   prolactin effects on, 972–973
   radiation effects on, 730
   RANKL and, 1036–1037
   signaling pathways, 27f
   stimulation by immobilization, 566
   sympathetic nervous system control of, 84
   thyroid stimulating hormone (TSH) stimulation of, 971–972
Osteocyte halo, 38
Osteocytes, 34–39
   appearance of, 308
   cell death and apoptosis, 37
   estrogen effects on, 199–200
   lacunocanalicular network, 34, 35f
   markers for, 34, 36, 36t
   mechanical responsiveness of, 397–398
   mechanosensation and transduction, 38
   modification of their microenvironment by, 37–38
   ontogeny, 34–36
   as orchestrators of bone (re)modeling, 36, 37f
   as percentage of bone cells, 34
   radiation effects on, 729–730
   role in bone disease, 39
   role of gap junctions and hemichannels in osteocyte communication, 38–39
Osteodystrophy. *See specific conditions*
Osteogenesis, defects in, 124

Osteogenesis imperfecta, 123, 653
　clinical presentation, 822–823
　clinical types, 823–825, 823t
　dual energy X-ray absorptiometry of, 825
　etiology and pathogenesis, 826
　genetic testing for, 339–340
　laboratory findings, 825
　overview, 822
　radiographic features, 825
　treatment, 826–827
Osteogenic cells, circulating, 111–116, 115f
Osteogenic osteosarcoma, 702–708
　animal models, 707
　biomarkers, 707–708
　cancer stem cells, 706–707
　cell of origin, 706, 706f
　diagnosis, 702, 708
　5-year survival rate, 702
　genetic factors, 704–705
　　alterations of tumor suppressors, 704
　　cytogenetic alterations, 705
　　microRNAs, 705
　　mutations in familial syndromes, 704
　　oncogene activation, 704–705
　　signaling pathway alterations, 705
　grades and stages, 702, 703t
　incidence, 702
　metastasis, 702
　overview of, 702–703
　treatment, 703–704, 707
Osteoglophonic dysplasia, 191, 608
Osteoid
　abnormal morphology, 310
　accumulation of unmineralized, 310
Osteolysis
　in multiple myeloma, 695
　osteolytic, 37
　tumor-derived, 724–725
Osteomalacia. *See also* Rickets
　axial, 778–779
　bone biopsy, 625, 627
　in children, 656
　diagnosis, 625
　in fibrous dysplasia, 787
　hypophosphatasia (HPP), 838–840, 839f
　low-turnover, 635
　nutritional rickets and, 613–614, 614f, 614t
　radiologic findings, 616, 617f
　radionuclide scintigraphy, 287
　table of causes, 614t
　treatment, 617
Osteomyelitis, 1039
Osteonecrosis, 805–807
　causative factors, 806t
　clinical features, 806
　differential diagnosis, 807
　epidemiology, 805
　glucocorticoid-induced, 475, 972
　jaw (*see* Osteonecrosis of the jaw)
　laboratory findings, 807
　pathogenesis, 805–806
　radiological features, 806–807
　staging, 807t
　treatment, 807
Osteonecrosis of the jaw
　bisphosphonate-associated, 762, 929–935
　　clinical presentation, 931–932, 931f–932f
　　defined, 929
　　discontinuation of bisphosphonates, 935
　　etiology, 930
　　histologic findings, 933, 934f
　　management, 934–935, 934t
　　monitoring of patients, 935
　　prevalence, 930
　　radiographic findings, 932–933, 933f
　　risk factors, 930
　　staging system, 934, 934t
　non-bisphosphonate medications and, 935
　osteoclast inhibition and, 715–716
Osteonectin, 53, 53t
Osteopathia striata, 777–778
　clinical presentation, 777
　radiological features, 777
　treatment, 777–778
Osteopathy. *See specific conditions*
Osteopenia, 396
　after irradiation, 730
　in brain tumor patients, 736
　in Marfan syndrome, 831
Osteopetrosis, 769–773
　clinical presentation, 770
　etiology and pathogenesis, 772
　histopathological findings, 771, 772f
　laboratory findings, 771
　mineral crystals in, 50
　radiological features, 770–771, 771f
　treatment, 772–773
Osteopoikilosis, 776–777
　clinical presentation, 776–777
　histopathological studies, 777
　radiological features, 777, 777f
Osteopontin, 54t, 55
Osteoporosis
　biochemical markers of bone turnover in, 297–302
　　analytical and preanalytical variability, 297–299, 298t
　　bone turnover rate and bone loss, 299
　　bone turover rate and fracture risk, 299–300
　　bone turover rate and monitoriing, 300–302
　　in men, 302
　　reference values, 299
　　table of, 298t
　chronic kidney disease-mineral bone disorder (CKD-MBD) and, 636
　estrogen(s)
　　deficiency and osteoporosis development, 201–203, 203f
　　treatment, 203–204
　exercise and prevention, 396–400, 397f
　　mechanical factors regulating bone cell response, 396–397
　　mechanically response bone cells, 397–398
　　transducing mechanical signals into cellular response, 398, 399f
　　translating mechanical signals to the clinic, 398–400
　fall prevention, 389–392
　fracture epidemiology, 348–354
　　clustering of fractures in individuals, 350
　　early life influences on adult fragility fracture, 354
　　geography, 352
　　impact of fractures, 349t
　　incidence and prevalence, 348–350, 349f, 351f, 352f
　　low bone mass in children, 354
　　morbidity, 353
　　mortality, 353
　　time trends and future projections, 350–352
　genetics of, 376–383
　　bone phenotypes, 377–378
　　candidate gene studies, 378, 379t

clinical application: individualized prognosis, 381–382
clinical application: pharmacogenetics, 382
genome-wide studies, 378–381, 379t, 381t
linkage analysis, 378–380, 379t
overview, 376–377
glucocorticoid-induced, 473–478
   histomorphometry findings in, 312
   in men, 512
   premenopausal, 517
   PTH treatment, 433
hip fracture
   incidence, 349, 349f, 349t, 352f
   morbidity, 353
   mortality, 353
histomorphometric findings in
   glucocorticoid-induced, 312
   menopausal, 312
idiopathic juvenile (IJO), 468–471, 469f, 470t–471t
   biochemical findings, 470
   bone biopsy, 470
   clinical features, 469, 469f
   differential diagnosis, 470, 470t–471t
   overview, 468–469
   pathophysiology, 469
   prevalence of, 468–469
   radiographic features, 469–470, 469f
of lactation, 161
magnetic resonance imaging-derived measures, 279–280
in men, 201–203, 203f, 508–512
   aged-related, 511–512
   BMD measurements, 510–511
   causes, 509–510, 510t
   clinical evaluation, 511
   evaluation, 510–511, 511t
   glucocorticoid-induced, 512
   hypogonadism, 510, 512
   idiopathic osteoporosis, 510
   laboratory testing, 511, 511t
   prevention, 511
   treatment, 432, 511–512
mineral crystals in, 50
nutrition and, 361–364
   building peak bone mass, 361–362, 362f
   dietary bioactive constituents, 364
   dietary patterns, 363
   maintaining bone mass, 362–363
   overview, 361
oral bone as screening tool for, 952–953
orthopedic surgery
   common fractures, 529
   surgical options, 528–529
   treatment principles, 527–528
   vertebral fracture intervention, 529–530
overview, 345–347, 346f
pathogenesis
   overview, 357–359, 358f, 358t
   sex steroids and, 367–372
in pregnancy, 158
prematurity and small gestational age associated with, 123
premenopausal, 514–518
   bisphosphonates for, 517
   BMD interpretation, 515
   evaluation, 516
   glucocorticoid-induced, 517
   history of low-trauma fracture, 514
   low bone mineral density (BMD), 514–515
   management, 516–518
   secondary causes, 515–516, 515t
   teriparatide (PTH) for, 517–518
radionuclide scintigraphy, 283–284, 284f
sarcopenia and, 980, 980t, 981
secondary, 489–492
   causes associated with increased fracture risk, 291t
   chronic obstructive pulmonary disease (CPOD), 490
   diabetes mellitus, 490–491
   inflammatory arthritis, 483–484, 489
   inflammatory bowel diseases, 490
   mastocytosis, 491
   overview, 489
   screening patients for, 491
   table of causes, 358t
thyroid stimulating hormone (TSH) receptor deficiency, 971
tooth loss, 951–952
transplantation, 495–504
   candidates for bone marrow transplantation, 496
   candidates for intestinal transplantation, 496
   diagnostic strategies, 497
   management of, 497–503, 499t–500t
   overview, 495
   preexisting bone disease, 495–496
   risk factors, 497, 498t
   skeletal effects of immunosuppressive drugs, 496–497
treatment
   anabolic agents, 463–464
   antiresorptives, 461–463
   bisphosphonates, 412–417
   calcitonin, 409–410
   combination anabolic and antiresorptive, 444–446
   compliance and persistence with medications, 448–452
   cost-effectiveness of, 455–458
   denosumab, 420–424
   estrogen, 203–204, 408–409
   future, 461–464
   orthopedic surgery, 527–530
   parathyroid hormone, 428–434
   selective estrogen receptor modulators (SERMs), 204, 409
   strontium ranelate, 437–440
vertebral fracture
   differentiating from neoplastic fracture, 329–330, 329f, 330t
   incidence, 349–350, 349f, 349t, 351f
   morbidity, 353
   mortality, 353
   nonunion, 328–329
   orthopedic surgery, 529–530
wrist (distal forearm) fracture
   incidence, 349f, 349t, 350
   morbidity, 353
Osteoprotegerin (OPG), 25, 26f, 28–30
   B lymphocyte expression of, 1038
   estrogen effect on, 370
   FoxO1 regulation of, 988
   genome-wide association studies of RANK-RANKL-OPG pathway, 380
   in skeletal healing, 92, 93
   tumor-derived osteolysis and, 725
   vitamin D effects on, 616
Osteosarcoma, 702–708, 703t, 706f, 707t, 736
Osteosclerosis
   chronic kidney disease-mineral bone disorder (CKD-MBD) and, 636
   hepatitis C-associated, 780–781
Osterix transcription factor, 16
Ovarian ablation, 712t, 713
Overexpression of target genes, 69–70

Oxalate
    dietary intake and kidney stone risk, 858
    hyperoxaluria, 865–866, 871, 879
Oxidative stress, 200–201, 200f, 202f
Oxytocin, 973

Pachydermoperiostosis, 780
    clinical presentation, 780
    etiology and pathogenesis, 780
    laboratory findings, 780
    radiological features, 780
    treatment, 780
PAD (peripheral arterial disease), 1014
Paget's disease, 659–666, 954
    biochemical indices in, 661–662
    clinical features, 662–663, 662f
    diagnosis, 663–664
    etiology, 659–660
    overview, 659
    pathology, 660–661, 661f
    radiography, 662f, 663–664
    radionuclide scintigraphy, 284–285, 285f
    signs and symptoms, 662–663
    treatment, 664–666
        analgesia, 666
        bisphosphonates, 664–665
        calcitonin, 665–666
        indications for, 664
        surgery, 666
PAI-1 (plasminogen activating inhibitor-1), 806
Pain. See Bone pain
Palatal development, 896–898, 897f, 898t, 899f
Pamidronate, 413f, 414f
    after organ transplantation, 498, 499t–500t, 502–503
    for BMD preservation in prostate cancer therapy, 714
    for fibrous dysplasia, 790
    for hypercalcemia in children, 653
    for metastatic bone disease
        breast cancer, 742–743
        multiple myeloma, 747–748
        prostate cancer, 746
    for osteogenesis imperfecta, 827
    for Paget's disease, 664–665
    in spinal cord injury patients, 1022
Pancreas
    kidney-pancreas transplantation, 501
    PTHrP (parathyroid hormone-related protein) effects, 219–220
    vitamin D effects, 241
Pancreatic polypeptide (PP), 963
Parabiosis experiments on circulating osteogenic cells, 111–112
Paracellin-1 (PCLN-1) gene/protein, 887–888
Paramyxovirus infection, 660
Parasympathetic nervous system (PNS), in bone remodeling, 85
Parathyroidectomy, 648
Parathyroid glands
    calcium sensing receptor (CaSR) role in maintaining calcium homeostasis, 229
    disorders, 537–540
        hyperparathyroidism, 537–538, 543–549
        hypocalcemia, 540
        hypoparathyroidism, 538–540
        intrinsic functional abnormalities of, 537–540
        non-parathyroid dependent hypercalcemia, 540
        overview, 537
        primary hyperparathyroidism, 537–538
    evolution of, 208
    fetal calcium metabolism, 181
    response to maternal hyperparathyroidism, 182–183
    response to maternal hypoparathyroidism, 183
    surgery, 546–547, 546t, 548f, 555, 557, 559–560, 648
    vitamin D effects, 240–241
Parathyroid hormone (PTH), 19
    autosomal dominant hypocalcaemic hypercalciuria (ADHH) and, 886
    bone remodeling, 176
    bone resorption and calcium release, 176
    combination anabolic and antiresorptive therapy for osteoporosis, 444–446
        antiresorptive after anabolic, 445–446
        antiresorptive followed by anabolic, 444–445
        concurrent use, 445–446
    in fetal and neonatal bone development, 122
    fetal calcium metabolism, 181, 184–185, 185f
    fetal response to maternal hyperparathyroidism, 182–183
    fetal response to maternal hypoparathyroidism, 183
    genetics of, 208–209, 209f
    hypercalcemia from, 568
    intestinal calcium transport, 174–175, 175t
    lactation levels of, 160
    levels
        after parathyroid surgery, 547
        in hypercalcemia, 540
        in hyperparathyroidism, 537–538, 543–549
        in hypocalcemia, 540
        in hypoparathyroidism, 538–540, 539
        in pseudohypoparathyroidism, 539
        serum vitamin D levels, effect on, 627, 627f
        target, 642t
    mechanism of action, 212
    in obesity, 996
    for osteoporosis, 428–434
        to accelerate fracture repair, 433
        candidates for therapy, 428–429
        combination anabolic and antiresorptive, 444–446
        cost-effectiveness, 433
        in glucocorticoid-treated patients, 433, 476–477, 477t
        in men, 432, 512
        persistence of effect, 433
        postmenopausal, 429–432
        premenopausal, 517
        PTH and antiresorptive combination/sequential therapy, 430–432
        PTH and bisphosphonates, 430–431
        PTH and hormone therapy, 432
        PTH(1-84) as monotherapy, 430
        rechallenge with PTH, 433
        teriparatide (TPTD) and raloxifene, 431
        teriparatide (TPTD) as monotherapy, 429–430, 476–477, 477t, 517–518
    pregnancy levels of, 156
    regulation
        by α-Klotho, 211
        by extracellular calcium, 210–211, 210f
        by FGF23, 211
        by plasma phosphate, 211
        of production and actions on calcium homeostasis, 176
        by vitamin D, 211
    renal calcium handling, 175–176, 176t
    resistance
        circulating inhibitors as cause of, 584–585
        magnesium deficiency and, 585
        pseudohypoparathyroidism (PHP) type 1, 581–584
        pseudohypoparathyroidism (PHP) type 2, 584

syndromes of, 581–585
  transient pseudohypoparathyroidism of the newborn, 584
response to exogenous for distinguishing hypoparathyroidism
    from pseudohypoparathyroidism, 580, 580f
signaling, 19
suppression for treating CKD-MBD, 645
syndromes of deficient synthesis or secretion, 585–588
  anti-CaSR antibiodies, 587
  autoimmune polyendocrinopathy-candidiasis-exctodermal
      dystrophy (APECED), 586–587
  autosomal dominant hypocalcemic hypercalciuria (ADHH),
      587
  developmental disorders, 585–586
  DiGeorge sequence/CATCH-22, 585
  hypoparathyroidism, sensorineural deafness, and renal
      dysplasia syndrome (HDR), 585–586
  hypoparathyroidism-retardation-dysmorphism syndrome
      (HRD), 586
  isolated hypoparathyroidism, 586
  Kenny-Caffey syndrome, 586
  maternal hyperparathyroidism and hypomagnesemia, 588
  mitochondrial disease, 587–588
  parathyroid gland destruction, 586–587
  post-burn, 586
  PTH gene mutations, 587
  radiation therapy, 586
  surgery, 586
synthesis and secretion, 209–210
as therapy for bone healing, 94–95
treatment for fracture healing, 100, 101f
Parathyroid hormone related peptide (PTHrP), 8, 215–220
  in ATLL, 697
  bone metastatic tumors and, 688
  discovery of, 215
  in fetal and neonatal bone development, 122
  fetal calcium metabolism, 181, 184–185, 185f
  genetics of, 208–209, 209f, 215–216
  in Hodgkin's disease, 698
  lactation levels of, 160
  in multiple myeloma, 695
  in non-Hodgkin's disease, 698
  nuclear, 216–217
  osteoclast stimulation by, 760
  overview of, 215
  physiological functions of, 217–220, 217f–219f
    cardiovascular system, 219
    mammary gland, 218–219, 218f–219f
    nervous system, 220
    pancreatic islets, 219–220
    placenta, 219
    skeleton, 217–218, 218f
    smooth muscle, 219
    teeth, 219
  as polyhormone, 216
  pregnancy levels of, 156–157
  receptors, 216
  sequence of, 208–209, 209f
Patched-1 (PTCH1) gene, 925
Pattern formation
  axial, 3–4, 5f
  craniofacial, 3
  defined, 3
  limb, 4–6, 5f
PBM. *See* Peak bone mass
PCLN-1 (paracellin-1) gene/protein, 887–888
PCR (polymerase chain reaction), in genetic testing, 337
PDD. *See* Progressive diaphyseal dysplasia
PDXA (peripheral dual-energy X-ray absorptiometry), 60–61

Peak bone mass (PBM)
  peak bone strength, 151–152, 151f
  role of diet in building, 361–362, 362f
  timing of, 149
Peak bone strength, 151–152, 151f
Peak height velocity (PHV), 149
PEDF (pigment epithelium-derived factor), 822–823, 826
Peptide YY (PYY), 963
Percutaneous nephrolithotomy (PNL), 873, 874t, 881
Periapical cyst, 923, 925
Periapical granuloma, 923, 925
Periodontal disease
  classification, 942, 944t
  diagnosis, 942, 944t
  etiology, 941–942, 942f–943f
  oral bone loss and, 941–946
  treatment, 644t–645t, 942–946
    nonsurgical, 942–943, 944t
    surgical, 943, 944t, 945
Periodontal ligament, development of, 908–911
Periostin, 53, 53t, 55, 55t
Peripheral arterial disease (PAD), 1014
Peripheral dual-energy X-ray absorptiometry (pDXA), 60–61
Peripheral nervous system, PTHrP effects, 220
Peripheral quantitative computed tomography (pQCT), 61
Persistence
  compliance and persistence with osteoporosis medications,
      448–452
  measurement of, 450
PET. *See* Positron emission tomography
p53 gene/protein, 704
Pharmacogenetics, 382
Phenotypes
  composite, 77
  of interest to bone biologists, 76–77
PhenX project, 106
Pheochromocytoma, 566
PHEX gene/protein, 190, 339, 605–606, 918
Phosphate
  achieving neutral phosphate balance, 643–646
  autosomal dominant hypophosphatemic rickets (ADHR),
      606–607
    clinical and biochemical manifestations, 607
    fibroblast growth factor-23 (FGR23) association, 190
    genetics, 607
    genetic testing for, 339
    treatment, 607
  dietary restriction, 644
  disorders of homeostasis, 601–609, 603f
  disorders of renal phosphate wasting, 607–608
    autosomal recessive hypophosphatemic rickets (ARHR), 607
    Fanconi renotubular syndrome 2, 608
    fibrous dysplasia, 607
    hereditary hypophosphatemic rickets with hypercalciuria
        (HHRH), 607
    hypophosphatemic nephrolithiasis/osteoporosis 1 and 2, 608
    hypophosphatemic rickets with hyperparathyroidism, 607
    linear nevus sebaceous syndrome, 607–608
    osteoglophonic dysplasia, 608
    table of characteristics, 604t
  fibroblast growth factor-23 (FGR23) and, 188–192
    activity, 188–189
    gene and protein, 188
    overview, 188
    receptors, 189
    regulation *in vivo*, 189
    serum assays, 189
    syndromes associated with, 189–192, 190t

Phosphate (contd.)
  hyperostosis-hyperphosphatemia syndrome, 810
  hyperphosphatemia, 608–609
    biochemical manifestations, 609
    causes of, 608, 609t
    in children, 655
    chronic kidney disease-mineral bone disorder (CKD-MBD) pathogenesis and, 633–634
    clinical manifestations, 608, 609
    familial tumoral calcinosis (FTC), 608–609
    genetics, 609
    hypocalcemia and, 575
    treatment, 609
  hyperphosphatemic familial tumoral calcinosis (HFTC), 810
  hypophosphatasia, 915t, 918, 919f
  hypophosphatemia, 601–602, 915t, 918
    causes of, 601–602, 602t
    in children, 654–655
    clinical consequences, 601
    genetic testing for, 339
    hypercalcemia and, 568
    hypophosphatemic osteopathy, histomorphometry findings in, 312
  metabolism
    disorders in children, 654–655, 654t
    vitamin D and, 238–239, 239f
  normative values by age, 654t
  oral for primary hyperparathyroidism, 549
  PTH secretion regulation, 211
  tumor-induced osteomalacia (TIO), 602–605, 604t
    clinical and biochemical manifestations, 602–603, 604f
    treatment, 603, 605
  X-linked hypophosphatemic rickets (XLH), 605–606
    clinical and biochemical manifestations, 605–606
    fibroblast growth factor-23 (FGR23) association, 190
    genetics of, 605
    genetic testing for, 339
    treatment and complications, 605
Phosphate binders, 609, 644–645
  aluminum salts, 645
  bivalent and trivalent cation binders, 644–645
  calcium-based, 644
  iron and magnesium salts, 645
  lanthanum salts, 644–645
  resins, 644
  sevelamer, 644
Phosphate enemas, 652
Phosphoinositides, 30
Phosphorus, target levels for, 642t
PHP. See Pseudohypoparathyroidism
PHV (peak height velocity), 149
Physical activity, positive effects of, 981–982
Physical function, vitamin D deficiency and, 628
Physical therapy, for osteogenesis imperfecta, 826
Physiology
  of fetal and neonatal bone development, 121–122
  pregnancy and lactation, 156–162
Pigmented villonodular synovitis (PVNS), 1038
Pigment epitheium-derived factor (PEDF), 822–823, 826
PINP, 297, 299, 301–302
Pituitary hormones
  ACTH, 972
  as diverse and evolutionary conserved signals, 969–970
  FSH, 970–971
  growth hormone, 970
  overview, 969
  oxytocin, 973
  prolactin, 972–973
  TSH, 971–972

PITX2 gene, 905
Placenta
  calcium sensing receptor (CaSR) role in maintaining calcium homeostasis, 231
  calcium transport, 181–182, 185
  PTHrP (parathyroid hormone-related protein) effects, 219
Plasma cells, 1037, 1038
Plasminogen activating inhibitor-1 (PAI-1), 806
Plates and screws, 528
  bridge plates, 528
  buttress (antiglide) plates, 528
  compression plates, 528
  locking plates, 528
  neutralization plates, 528
  tension band plate, 528
Pleiotropy, 78
PMNs (polymorphonuclear leukocytes), 1039
PNL (percutaneous nephrolithotomy), 873, 874t
PNS (parasympathetic nervous system), in bone remodeling, 85
POEMS syndrome (Polyneuropathy, Organomegaly, Endocrinopathy, Monoclonal Gammopathy and Skin changes), 695
POH (progressive osseous heteroplasia), 591–592, 591t, 592f, 594, 817–818
Polymerase chain reaction (PCR), in genetic testing, 337
Polymorphonuclear leukocytes (PMNs), 1039
POMC (pro-opiomelanocortin), 969
Positron emission tomography (PET)
  FDG (^{18}F-fluorodeoxyglucose) PET, 285, 680–681, 933
  in osseous metastatic disease, 679t, 680–682
  in Paget's disease, 663
Postmenopausal osteoporosis, parathyroid hormone (PTH) for, 429–432
PP (pancreatic polypeptide), 963
PPHP (pseudopseudohypoparathyroidism), 581, 583, 591, 591t, 592f
PPIB gene, 823–824, 823t
Prebiotics, 364
Preeclampsia, 159
Pregnancy, 156–159
  adolescent, 162
  calciotropic hormone levels, 156–157
  calcium
    homeostasis, 157f
    intestinal absorption, 157
    low intake of calcium, 159
    renal excretion, 157
    serum levels, 156, 157f
    skeletal metabolism of calcium, 157–158
  diet during, 143–144
  familial hypocalciuric hypercalemia, 159
  hypoparathyroidism, 159
  osteoporosis, 158
  primary hyperparathyroidism, 158–159, 549
  pseudohypoparathyroidism, 159
  vitamin D deficiency and insufficiency, 159
Premenopausal osteoporosis, 514–518
  bisphosphonates for, 517
  BMD interpretation, 515
  evaluation, 516
  glucocorticoid-induced, 517
  history of low-trauma fracture, 514
  low bone mineral density (BMD), 514–515
  management, 516–518
  secondary causes, 515–516, 515t
  teriparatide (PTH) for, 517–518
Presomitic mesoderm, 4, 5f

Primary hyperparathyroidism, 158–159. *See also* Familial primary hyperparathyroidism; Hyperparathyroidism
Principal component analysis, 77
Progressive diaphyseal dysplasia (PDD), 774–775
   clinical presentation, 774
   etiology and pathogenesis, 775
   laboratory findings, 775
   radiological features, 774–775, 775f
   treatment, 775
Progressive osseous heteroplasia (POH), 591–592, 591t, 592f, 594, 817–818
Progressive resistance training, benefits of, 981
Prolactin, 972–973, 1008
Promoters, 69–70
Pro-opiomelanocorticin (POMC), 969
Prostaglandin E_2, 1029
Prostate cancer
   animal models, 687
   metastasis to bone, 761
     clinical aspects, 742
     treatment and prevention, 745–747, 746f
   nociceptive nerve sprouting in, 722–723, 724f–725f
   treatment, 713–715
     endocrine therapy, 714
     to preserve BMD and prevent fractures, 714–715
     radiation therapy and, 715
Prostate-specific antigen (PSA), 673
Protein
   binding, hypercalcemia and abnormal, 567
   intake, kidney stone formation and, 858
   markers of urinary excretion, 866
Protein S, 55, 55t
Proteoglycans, in bone composition, 51, 52t
Proton pump inhibitors, skeletal effects of, 522
Proximal tubules, calcium excretion physiology and, 845, 846f
Prx1 promoter, 69
PSA (prostate-specific antigen), 673
Pseudohypoaldosteronism, 870
Pseudohypocalcemia, 575
Pseudohypoparathyroidism (PHP), 579–596
   clinical and molecular features of forms, 591t
   diagnostic algorithm, 588
   genetics of, 581–584, 581t, 582f, 590–596, 591t, 592f, 595f
   hypoparathyroidism distinguished from, 579
   in lactation, 161–162
   origin of term, 590
   overview of, 590
   in pregnancy, 159
   progressive osseous heteroplasia (POH), 591–592, 591t, 592f, 594
   pseudohypoparathyroidism (PHP) type 1, 581–584
     Albright hereditary osteodystrophy (AHO), 581, 582f, 583–584, 591, 593
     PHP 1a, 581, 583, 591, 591t, 592f, 593
     PHP 1b, 583, 591, 591t, 592f, 594–595, 595f
     PHP 1c, 583–584, 591
   pseudohypoparathyroidism (PHP) type 2, 584
   pseudopseudohypoparathyroidism (PPHP), 581, 583, 591, 591t
   syndromes of PTH resistance, 581–584, 582f
   therapy, 588–589
   transient pseudohypoparathyroidism of the newborn, 584
Pseudopseudohypoparathyroidism (PPHP), 581, 583, 591, 591t, 592f
Pseudovitamin D deficiency rickets, 619
PTH. *See* Parathyroid hormone
PTHrP (parathyroid hormone-related protein). *See* Parathyroid hormone related peptide

Puberty, 130
PVNS (pigmented villonodular synovitis), 1038
Pycnodysostosis, 773–774
   clinical presentation, 773–774
   etiology and pathogenesis, 774
   laboratory findings, 774
   radiological features, 774
   treatment, 774
Pyridoxine, for hyperoxaluria, 879
PYY (peptide YY), 963

Quality-adjusted life year (QALY), 456–457, 458f
Quantitative computed tomography (QCT), 264–272
   clinical application, 267–270
     diagnosis of BMD status in adults, 268
     diagnosis of BMD status in children, 267–268
     fracture risk assessment, 268–270
     monitoring changes in bone status, 270
     reference data, 267, 267f
   dual-energy X-ray absorptiometry (DXA) compared, 267–268
   for ethnic difference comparisons, 136–137
   finite element modeling, QCT-based, 270–271, 271f
   fracture risk assessment, 268–270
     clinical interpretation, 269–270
     hip fractures, 268–269
     mixed fracture groups, 269
     vertebral fractures, 268, 328
   methodology, 264–267
     hip QCT, 265
     multislice high-resolution pQCT (HRpQCT), 266, 266f
     peripheral QCT (pQT), 265–266
     radiation exposure, 266–267
     single-energy QCT, 264–265, 265f
     spinal QCT, 265
Quantitative trait loci (QTLs)
   bone mineral density and, 76–79
   clustering, 78
   concordance of rodent and human data, 78–79
   defined, 76
   expression, 77
   intersite discordance, 78
   sex limitation, 78

Rac GTPases, 28
Radiation therapy
   bone-seeking radiopharmaceuticals, 756–757
   for breast cancer, 715
   external beam radiotherapy (EBRT), 754–756, 755t
     fractures and, 756
     for neuropathic pain, 755
     pain palliation, 754–755
     re-irradiation, 756
     for spinal cord compression, 756–757
   fractures and, 715
   hypoparathyroidism after, 586
   for osteosarcoma, 703
   for prostate cancer, 715
   of skeletal metastasis, 754–755, 757t
Radiation therapy-induced osteoporosis, 728–731
   bone quality/quantity deterioration, 730
   bone strength loss in animal models, 730–731
   fractures, 728–729
   osteoblasts, effects on, 729
   osteoclasts, effects on, 730
   osteocytes, effects on, 729–730
   overview, 728
   vasculature, 729
Radio-frequency ablation (RFA), 764

Radiography
  for assessment of bone mass and microarchitecture in rodents, 59–60, 60f
  in axial osteomalacia, 779
  in carbonic anhydrase II deficiency, 773
  dental, 924f
  in dermatomyositis, 812f, 813
  in fibrodysplasia ossificans progressiva, 816
  in fibrogenesis imperfecta ossium, 779
  in fibrous dysplasia, 789, 790f
  in idiopathic juvenile osteoporosis, 469–470, 469f
  in melorheostosis, 778, 778f
  in nephrolithiasis, 861, 864f, 872–873
  oral cavity, 949–950, 949f–950f
  in osseous metastatic disease, 678
  in osteonecrosis, 806–807
  in osteopathia striata, 777
  in osteopetrosis, 770–771, 771f
  in osteopoikilosis, 777
  in pachydermoperiostosis, 780
  in Paget's disease, 662f, 663–664
  in progressive diaphyseal dysplasia, 774–775, 775f
  in pycnodysostosis, 774
  in sclerosteosis, 776
  in tumoral calcinosis, 810–811, 811f
  in van Buchem disease, 776
  in vitamin D deficiency, 614f, 616, 617f
Radionuclide scintigraphy
  in fibrodysplasia ossificans progressiva, 816
  metabolic bone disease, 283–287
    bone metastases, 283
    hyperparathyroidism, 285–286
    osteomalacia, 287
    osteoporosis, 283–284, 284f
    Paget's disease, 284–285, 285f
    renal osteodystrophy, 286–287, 286f
  in osseous metastatic disease, 680
  in osteonecrosis, 807, 807t
  overview, 283
  in Paget's disease, 663–664
Radiopharmaceuticals, bone-seeking, 756–757
Radiotherapy. See Radiation therapy
RAGE (receptors for advanced glycation end products), 1007
Raloxifene, 409
  for primary hyperparathyroidism, 549
  with PTH, 431–432
RANK. See Receptor activator of nuclear factor-κB
RANKL. See Receptor activator of nuclear factor-κB ligand
Reactive oxygen species (ROS), 200–201, 200f, 202f
Recalcification tetany, 575
Receptor activator of nuclear factor-κB (RANK), 1038
  cancer bone pain and, 720
  genome-wide association studies of RANK-RANKL-OPG pathway, 380
Receptor activator of nuclear factor-κB ligand (RANKL), 25, 26f, 28–30, 84
  antibody against (see Denosumab)
  in ATLL, 697
  cancer bone pain and, 720
  cancer metastasis to bone and, 687–688
  estrogen effect on, 370
  gene, 109
  genome-wide association studies of RANK-RANKL-OPG pathway, 380
  identification of, 1037
  inhibition by denosumab, 420–421
  mechanical response, 397–398
  monoclonal antibody inhibiting RANKL action, 461–462
  in multiple myeloma, 695–696
  oxidative stress and, 200–201, 200f
  postburn stimulation of, 531
  in rheumatoid arthritis, 483–484
  roles of, 1038
  in skeletal healing, 92–93
  in spinal cord injury patients, 1022
  tumor-derived osteolysis and, 725
  vitamin D effects on, 616
Receptors for advanced glycation end products (RAGE), 1007
RECQL proteins, 704
Remodeling. See Bone remodeling
Renal failure, hypercalcemia and, 568
Renal handling of calcium
  in lactation, 160
  in pregnancy, 157
Renal osteodystrophy
  clinical manifestations, 636–637
    bone pain, 637
    calcifications, 637
    fractures, 637
    heterotropic mineralization, 637
    skeletal deformities, 637
    tumoral calcinosis, 637
  histomorphometry findings in, 312–313
  oral manifestations, 954
  pathology of
    high turnover, 635, 635t
    high turnover plus mineralization defect, 635–636, 635t
    low turnover, 635, 635t
  radionuclide scintigraphy, 286–287, 286f
  transplantation osteoporosis, 495–496, 498
Renal stones
  genetics of, 884–889, 885t
    autosomal dominant hypocalcaemic hypercalciuria (ADHH), 886
    Bartter syndrome, 886
    Dent's disease, 886–887
    distal renal tubular acidosis, 888–889
    familial hypomagnesaemia with hypercalciuria and nephrocalcinosis (FHHNC), 887–888
    hereditary hypophosphatemic rickets with hypercalciuria (HHRH), 887
    idiopathic hypercalciuria, 885
    oculocerebrorenal syndrome of Lowe (OCRL), 887
  prevalence of, 884
  treatment, 878–881
    calcium stones, 878–880
    cystine stones, 880–881
    struvite stones, 881
    uric acid stones, 880
Renal tubular acidosis in carbonic anhydrase II deficiency, 773
Renal tubular physiology of calcium excretion, 845–850, 848t, 849f
Repeat expansions, 337
Reporters, 73
Residual cyst, 925
Residual ridge resorption, 949
Resins, phosphate-binding, 644
Resistance exercise, 982
Resistin, 996
Resorption. See Bone resorption
Respiratory failure, chronic, 496
Retinoblastoma, 704
Retinoic acid signaling, 4
Revascularization, in bone repair, 93
Rexin-G, 707
RFA (radio-frequency ablation), 764

RGD-containing glycoproteins, 54t, 55
Rhabdomyosarcoma, 736
Rheumatoid arthritis, 483–484, 1037
Rho-associated protein kinase (ROCK), 688, 690
Rho family of GTPases, 28
Rickets
　autosomal dominant hypophosphatemic rickets (ADHR), 606–607
　　clinical and biochemical manifestations, 607
　　fibroblast growth factor-23 (FGR23) association, 190, 654
　　genetics, 607
　　genetic testing for, 339
　　treatment, 607
　in children, 656, 656f
　familial hypophosphatemic rickets (FHR), 339
　hereditary hypophosphatemic rickets with hypercalciuria (HHRH), 607, 654–655, 887
　hereditary vitamin D resistant rickets, 619–620
　hypophosphatasia (HPP), 838–840, 839f
　hypophosphatemic rickets with hyperparathyroidism, 607
　nutritional rickets and osteomalacia, 613–614, 614f, 614t, 656
　prematurity and small gestational age associated with, 123
　pseudovitamin D deficiency rickets, 619
　table of causes, 614t
　treatment, 617
　vitamin D-dependent rickets type I (VDDR-I), 339, 619
　vitamin D-dependent rickets type II (VDDR-II), 339, 619–620
　X-linked hypophosphatemic rickets (XLH), 605–606
　　clinical and biochemical manifestations, 605–606
　　fibroblast growth factor-23 (FGR23) association, 190, 654
　　genetics, 605
　　genetic testing for, 339
　　treatment and complications, 605
Risedronate, 413f–414f, 414–415
　after organ transplantation, 503
　for glucocorticoid-induced osteoporosis, 476, 477t
　for Paget's disease, 664–665
ROCK (Rho-associated protein kinase), 688, 690
Rodents, assessment of bone mass and microarchitecture in, 59–66
　imaging considerations, 63–66, 64f–66f
　magnetic resonance imaging, 61–62, 61f
　microcomputed tomography, 62–63, 62f–63f
　nanocomputed tomography, 63, 64f
　peripheral dual-energy X-ray absorptiometry, 60–61
　peripheral quantitative computed tomography, 61
　radiographs, 59–60, 60f
ROS (reactive oxygen species), 200–201, 200f, 202f
Rosiglitazone, 522
Rotavirus infection, 652
Rothmund-Thomson syndrome, 704
Runx2, 6, 9, 15–16, 20f, 672
　bone metastatic tumors and, 688
　insulin signaling and, 988, 989f
　in skeletal healing, 92

Salt
　intake during growth, 144–145
　kidney stone formation and, 858
Sanjad-Sakati syndrome, 586
Sarcoidosis, 781
Sarcopenia, 978–981, 979t, 980t
SARMs (selective androgen receptor modulators), 204
Schizophrenia, 1005–1006
Scintigraphy. *See* Radionuclide scintigraphy
Scleraxis (Scx) gene, 70

Sclerosing bone disorders, 769–781, 770t
　axial osteomalacia, 778–779
　　clinical presentation, 779
　　etiology and pathogenesis, 779
　　histopathological findings, 779
　　laboratory findings, 779
　　radiological features, 779
　　treatment, 779
　carbonic anhydrase II deficiency, 773
　　clinical presentation, 773
　　etiology and pathogenesis, 773
　　laboratory findings, 773
　　radiological features, 773
　　treatment, 773
　endosteal hyperostosis, 775–776
　　sclerosteosis, 776
　　Van Buchem disease, 775–776
　fibrogenesis imperfecta ossium, 779–780
　　clinical presentation, 779
　　etiology and pathogenesis, 780
　　histopathological findings, 779–780
　　laboratory findings, 779
　　radiological features, 779
　　treatment, 780
　hepatitis C-associated osteosclerosis, 780–781
　high bone mass phenotype, 781
　melorheostosis, 778
　　clinical presentation, 778
　　etiology and pathogenesis, 778
　　histopathological findings, 778
　　laboratory findings, 778
　　radiographic features, 778, 778f
　　treatment, 778
　osteopathia striata, 777–778
　　clinical presentation, 777
　　radiological features, 777
　　treatment, 777–778
　osteopetrosis, 769–773
　　clinical presentation, 770
　　etiology and pathogenesis, 772
　　histopathological findings, 771, 772f
　　laboratory findings, 771
　　radiological features, 770–771, 771f
　　treatment, 772–773
　osteopoikilosis, 776–777
　　clinical presentation, 776–777
　　histopathological studies, 777
　　radiological features, 777, 777f
　pachydermoperiostosis, 780
　　clinical presentation, 780
　　etiology and pathogenesis, 780
　　laboratory findings, 780
　　radiological features, 780
　　treatment, 780
　progressive diaphyseal dysplasia (PDD), 774–775
　　clinical presentation, 774
　　etiology and pathogenesis, 775
　　laboratory findings, 775
　　radiological features, 774–775, 775f
　　treatment, 775
　pycnodysostosis, 773–774
　　clinical presentation, 773–774
　　etiology and pathogenesis, 774
　　laboratory findings, 774
　　radiological features, 774
　　treatment, 774
Sclerosteosis, 776
Sclerostin, 464, 579

Scoliosis
  in fibrodysplasia ossificans progressiva, 815
  in Marfan syndrome, 831, 833t
Secondary malignant neoplasms, 704
Secreted phosphoprotein 24, 53, 53t
Segmentation, image, 64–65, 65f
Selective androgen receptor modulators (SARMs), 204
Selective estrogen receptor modulators (SERMs), 204, 409
  for BMD preservation in prostate cancer therapy, 714–715
  for breast cancer, 711, 712t, 713
  hypercalcemia from, 568
  for primary hyperparathyroidism, 549
Selective serotonin reuptake inhibitors (SSRIs)
  for neuropsychiatric disorders, 1007–1008
  skeletal effects of, 523
Selective tooth agenesis syndrome (STHAG), 831t, 834
Sensory nerves, role in bone remodeling, 86
SERMs. See Selective estrogen receptor modulators
Serotonin, 962–963
  actions of, 1007–1008
  control of bone formation, 83
SERPINH1 gene, 823–824, 823t
Serum-derived proteins, in bone composition, 50–51, 51t
Severe neonatal hyperparathyroidism (SNHP), 653
Sex differences in bone morphology, 130
Sex hormone binding globulin (SHBG), 371
Sex steroids. See also Gonadal steroids
  chronic kidney disease-mineral bone disorder (CKD-MBD) and, 634
  in hypogonadism, 510, 512, 634
  osteoporosis pathogenesis, 367–372
    bone loss in men, 370–372, 371t, 372f, 510, 512
    bone loss in women, 369–370
SGS (Shprintzen-Goldberg syndrome), 831t, 832–833, 833t
SHF3BP2 gene/protein, 917–918
Shh (sonic hedgehog). See Hedgehog
Shock wave lithotripsy, 873, 874t, 881
SHORT syndrome, 653
Shprintzen-Goldberg syndrome (SGS), 831t, 832–833, 833t
SIBLING. See Small integrin-binding ligand, N-linked glycoprotein (SIBLING) family
Signaling. See also specific signaling pathways
  in bone metastatic tumors, 688–689
  bone morphogenic protein (BMP), 6–8, 7f, 16–17, 17f, 44–45, 46f, 817, 898–899, 906
  in cancer metastasis, 672–673
  in chondrocyte proliferation and differentiation, 8
  co-stimulatory signaling in osteogenesis, 1038–1039
  for dental cell differentiation, 906
  in embryonic cartilage and bone formation, 6–7, 7f
  ephrin/eph, 689, 1039–1040
  fibroblast growth factors (FGFs), 4, 5f, 7–8, 7f, 19–20, 93, 898–899, 901, 906
    in craniosynostosis, 901t
    in mandibular development, 899
    in palatogenesis, 898
    tooth development, 906
  in fibrodysplasia ossificans progressiva, 817
  in fracture healing, 93
  Hedgehog, 4, 5f, 6, 19, 93, 898
  IGF-1, 19
  Indian hedgehog (Ihh), 6, 7f, 8, 93
  insulin in osteoblasts, 988–989, 989f
  integrin, 28
  intracellular pathways, 31
  metastatic bone lesions, 760–761
  Notch, 4, 20, 705, 906
  osteoclast, 27f
  in osteosarcomagenesis, 705
  Osterix transcription factor, 16
  PTH, 19
  retinoic acid, 54
  Rho family of GTPases, 28
  Runx2 transcription factor, 15–16, 20f
    in skeletal patterning, 3–6, 5f
    tooth development, 905, 906
  transforming growth factor-β (TGF-β), 17, 43–45, 46f, 672–673, 689
  Wnt (see Wnt signaling)
Single nucleotide polymorphism (SNP), 106–109, 108f
Single photon absorptiometry (SPA), for oral cavity assessment, 949
Single photon emission tomography (SPECT), 283, 286, 680–681
Site-specific recombination, 71–73
Skeletal calcium metabolism
  in lactation, 160–161, 161f
  in pregnancy, 157–158
Skeletal development
  craniofacial, 895–901
  in men, 508
Skeletal disorders
  genetic testing for, 338–340
  in Marfan syndrome, 830–831, 833t
  in renal osteodystrophy, 637
Skeletal dysplasias
  genetic testing for, 339–340
  hypercalcemia in children with, 653
Skeletal healing, 90–95, 91f
  conditions that impair fracture healing and therapeutic modalities, 93–94
    aging, 93–94
    cigarette smoking, 94
    diabetes, 94
  molecular therapies to enhance, 94–95
  process of, 90–93, 91f
    cellular contribution, 90–92
    gene expression profile during, 92
    molecular control of, 92–93
  tissue morphogenesis during repair, 91f
Skeletal mass, obesity and, 993–997
Skeletal mineralization, fetal, 182, 185
Skeletal morphogenesis, 3–10, 4f–5f, 7f, 10f
Skeletal muscle, 978–983
  cachexia, 982–983
  fat infiltration, 981
  physical activity and, 981–982
  sarcopenia, 978–981, 979t, 980t
Skeletal patterning
  axial, 3–4, 5f
  craniofacial, 3
  limb, 4–6, 5f
Skeletogenic neuropeptides, 85–86
Skeleton
  aging effects in men, 508–509
  complications of childhood cancer, 734–737
  drug effects on, 520–523, 521t
    acid-suppressive medications, 522
    androgen deprivation therapy, 520–521
    antiepileptic drugs, 522–523
    antihormonal drugs, 520–521
    aromatase inhibitors, 521
    beta-blockers, 523
    heparin, 523
    selective serotonin reuptake inhibitors (SSRIs), 523
    statins, 523

thiazides, 523
thiazolidinediones, 521–522
fetal, calcium metabolism in, 182, 185
glucose control and integration, 986–990
intermediary organization of, 307
PTHrP (parathyroid hormone-related protein) effects, 217–218, 218f
skeletal muscle effects on, 978–983
Skull base deformities, 901
SLE (systemic lupus erythematosus), 485
SLRPs (small leucine-rich proteoglycans), 51, 52t
Small bowel transplantation, 502
Small integrin-binding ligand, N-linked glycoprotein (SIBLING) family, 53, 54t, 55, 191
Small leucine-rich proteoglycans (SLRPs), 51, 52t
Smart probes, 681–682
^{153}Sm-EDTMP, 757
Smooth muscle, PTHrP (parathyroid hormone-related protein) effects on, 219
SNHP (severe neonatal hyperparathyroidism), 653
SNP (single nucleotide polymorphism), 106–109, 108f
^{18}F sodium fluoride, 681
Sodium intake during growth, 144–145
Soft drinks, 145
Solubility product, 865
Sonic hedgehog (Shh). *See* Hedgehog
Sorafenib, 935
SOST gene/protein, 380, 464, 579, 776
Sox9 transcription factor, 6, 9
SPA (single photon absorptiometry), for oral cavity assessment, 949
SPECT (single photon emission tomography), 283, 286, 680–681
Spinal cord compression, radiotherapy for, 756–757
Spinal cord injury, 1018–1023
bone loss, 1019
bone metabolism after, 1019–1020
DXA evaluation after, 1020–1021
fractures after, 1019
heterotopic ossification, 1020
overview, 1018
therapy
bisphosphonates, 1021–1022
combined pharmacological and mechanical therapy, 1023
functional electrical stimulation (FES), 1022–1023
mechanical loading effects, 1022–1023
Spine deformities, in Marfan syndrome, 831, 833t
Splicing variants, 337
Spondylitis, ankylosing, 484–485
Spondyloarthropathy, 580
Spotted bones. *See* Osteopoikilosis
SSRIs. *See* Selective serotonin reuptake inhibitors
Statins, skeletal effects of, 523
Stem cells
cancer, 706–707
hematopoietic, 1028–1031, 1031f
during tooth repair, 911
Stem cell transplantation, 736
STHAG (selective tooth agenesis syndrome), 831t, 834
Stone analysis, 861–863, 864f–865f
Streptococcus mutans, 922–923
Stress response in burn injury, 531–532, 532f
Stromal cell-derived factor-1 (SDF-1), 115–116
Strontium-89 dichloride, 757
Strontium ranelate
fracture healing, 440
implant osseointegration, 440, 441f
for osteoporotic fracture prevention, 437–440, 438f, 441f
antifracture efficacy, 437–438, 438f
cost-effectiveness, 439
mechanisms of anti-fracture efficacy, 440
preclinical studies, 440
quality of life and, 439
recent developments, 440
reversibility of bone mineral density increase, 439
safety, 439
subgroup analysis, 438–439
overview of, 437
safety, 439
Struvite stones, 870, 874, 881
Subcutaneous fat necrosis, 653
Substance P, 86, 722
Sunitinib, 935
Sunlight, vitamin D and, 243–244, 629
Supernumerary teeth, 914–915, 915t
Supersaturation, 865
Surgery
orthopedic, 527–530
for osteosarcoma, 703
Sympathetic nervous system
hematopoiesis and, 1030
leptin and, 83–85, 84f, 961–962
Synonymous variants, 336
Systemic lupus erythematosus (SLE), 485
Systemic mastocytosis, 807

Tacrolimus, 497
Tai chi, 391
Tamoxifen, 409
for breast cancer, 711, 712t, 713
hypercalcemia from, 568
Tartrate-resistant acid phosphatase (TRAP), 128, 307
TBCE gene, 586
Tbx1 gene, 906
TCIRG1 gene, 772–773
Teeth. *See also* Dentition
caries, 922–923, 923f
demineralization, 922–923
development
aveolar bone formation, 911
cementum formation, 908–911
circadian formation of dental mineralized tissues, 910f, 911
dentin formation, 906–908, 907f
enamel formation, 908, 909f
molecular control, 904–906, 905f
periodontal ligament formation, 908–911
signals controlling dental cell differentiation, 906
stages of morphogenesis, 904–906, 905f
stem cells, 911
genetics of disorders, 914–919, 915t
of dentin, 915–916, 916f
of enamel, 917
metabolic bone diseases, 918, 919f
number of teeth, 914–915
loss with osteoporosis, 951–952
odontogenic cysts, 923–926, 924f, 926f
odontogenic tumors, 926
PTHrP (parathyroid hormone-related protein) effects, 219
stem cells, 911
structure, 906f
Tenascin, 53, 53t
Tendon and ligament, gene expression and, 70
Tension band plate, 528
Teriparatide, 429–430, 476–477, 477t, 517

1076 Index

Testosterone. *See also* Sex steroids
  in aging men, 370–372
  biosynthesis, 196, 196f
  deficiency and chronic kidney disease-mineral bone disorder (CKD-MBD), 634
  for hypogonadal osteoporosis in men, 512
  level and risk of falling, 981
Tetany, 580
Tetracycline antibiotics for fluorochrome labeling, 313
Tetranectin, 53, 53t
TGF-β. *See* Transforming growth factor-β
Thiazide diuretics
  effect on renal calcium excretion, 847
  hypercalcemia from, 567
  for hypercalciuria, 879
  skeletal effects of, 523
Thiazolidinediones, skeletal effects of, 521–522
Thrombospondins, 53, 54t
Thyroid stimulating hormone (TSH), 969–970
Tiludronate, 413f, 664–665
TIO. *See* Tumor-induced osteomalacia
Tiopronin, for cystine stones, 881
Tissue-nonspecific iso-enzyme of alkaline phosphatase (TNSALP), 123
TLRs (Toll-like receptors), 1039
T lymphocytes, 689, 1037–1038
TNF. *See* Tumor necrosis factor
TNSALP (tissue-nonspecific iso-enzyme of alkaline phosphatase), 123
Tocilizumab, for fibrous dysplasia, 791
Toll-like receptors (TLRs), 1039
Torque-twist curves, 100f
Total parenteral nutrition (TPN), hypercalcemia and, 567
Trabecular bone structure
  after irradiation, 730
  allelic determinants of phenotype, 77
TRACP promoter, 70
Transactivator, 70
Transcription factors, involved in osteoblast differentiation, 16
Transforming growth factor-β (TGF-β), 672–673
  in connective tissue disorders, 830, 831t, 832–834
  in fracture healing, 93
  hematopoiesis and, 1030
  inhibitors, 689–690
  in metastatic bone disease, 760–761
  progressive diaphyseal dysplasia and, 775
  signaling, 17, 43–45, 46f, 672–673, 689
Transient hypocalcemia of the newborn, 651
Transient pseudohypoparathyroidism of the newborn, 584
Transient receptor potential (TRPV1) family of ion channels, 723–724
Translational genetics. *See* Genetics
Transplantation osteoporosis, 495–504
  bone marrow transplantation, 502–503
  candidates for bone marrow transplantation, 496
  candidates for intestinal transplantation, 496
  cardiac transplantation, 501
  diagnostic strategies, 497
  kidney-pancreas transplantation, 501
  kidney transplantation, 498, 501
  liver transplantation, 501–502
  lung transplantation, 501
  management of, 497–503, 499t–500t
  overview, 495
  preexisting bone disease, 495–496
    chronic kidney disease, 495–496
    chronic respiratory failure, 496
    congestive heart failure, 496
    end-stage liver disease, 496
  risk factors, 497, 498t
  skeletal effects of immunosuppressive drugs, 496–497
    calcineurin inhibitors, 497
    glucocorticoids, 496–497
  small bowel transplantation, 502
TRAP (tartrate-resistant acid phosphatase), 128, 307
Traumatic brain injury, 1006
Trousswau sign, 580
TRPV1 (transient receptor potential) family of ion channels, 723–724
Tryptophan hydroxylase 2, 83
T-scores, 252–253, 253f, 256–257, 268, 292
TSH (thyroid stimulating hormone), 969–970
Tumoral calcinosis, 608–609, 810–812
  biochemical findings, 811
  clinical features, 810
  etiology and pathogenesis, 811–812
  fibroblast growth factor-23 (FGR23) association, 191–192, 811–812
  genetics, 609
  histopathology, 811
  radiographic findings, 810–811, 811f
  in renal osteodystrophy, 637
  treatment, 812
Tumor-derived osteolysis, 724–725
Tumor-induced osteomalacia (TIO), 602–605, 604t
  clinical and biochemical manifestations, 602–603, 604f
  fibroblast growth factor-23 (FGR23) association, 190
  treatment, 603, 605
Tumor necrosis factor (TNF)
  osteoclast formation and, 1037
  osteoclast regulation and, 29
Tumor suppressors in osteosarcoma, 704
Twist 1 and *Twist 2* genes, 988, 989f

Ultrasonography, in nephrolithiasis, 861, 864f, 872–873
Ureteroscopy, 874, 874t
Uric acid
  hyperuricosuria, 866, 872, 874, 880
  stones, 880
Urinalysis
  effect of gender on normal ranges of urinary constituents, 866
  nephrolithiasis, 861, 862f–863f
Urinary solute excretion, normative data for, 873t
Urinary tract
  abnormalities of, 869
  infection, 869–870
Urine calcium excretion
  in lactation, 160
  in pregnancy, 157
Urine pH, 861, 866

Van Buchem disease, 775–776
Van der Woude's syndrome, 917
Vascular disease, 1012–1015
  arteriosclerosis, 1014–1015
  atherosclerosis, 1014
  avascular necrosis, 1012–1014, 1013f, 1013t
  overview, 1012
Vascular endothelial growth factor (VEGF), 92–93, 972, 1012–1013
Vascular lineage cells, circulating osteogenic cells isolated from, 113
Vasculature, radiation and, 729
Vasoactive intestinal polypeptide (VIP), 566

VDR (vitamin D receptor) gene, 382
Vegetable intake during growth, 143
VEGF (vascular endothelial growth factor), 92–93, 972, 1012–1013
Velocardiofacial syndrome (VCFS), 585
Ventromedial hypothalamic neurons, 83
Vertebral fracture
- assessment
  - algorithm-based qualitative (ABQ) method, 325
  - dual-energy X-ray absorptiometry (DXA), 258, 325–327, 326f–327f, 330, 331t
  - magnetic resonance imaging (MRI), 328–330, 329f
  - multidetector computed tomography (MDCT), 327
  - quantitative computed tomography (QCT) fracture risk assessment, 268, 328
  - quantitative morphometry, 323–325, 326f
  - radionuclide scintigraphy, 283–284, 284f
  - semi-quantitative analysis, 322–323, 324f–325f
- defining a fracture, 322
- detection of, 319–331
  - clinical detection, 319
  - clinical recommendations for spine imaging for, 330, 331t
  - spinal radiography, 319–322, 320f–322f, 325f
- diagnosis and classification, 317–331
- differentiating osteoporotic from neoplastic fracture, 329–330, 329f, 330t
- nonunion, 328–329
- osteoporotic
  - differentiating from neoplastic fracture, 329–330, 329f, 330t
  - incidence, 349–350, 349f, 349t, 351f
  - morbidity, 353
  - mortality, 353
  - nonunion, 328–329
  - orthopedic surgery, 529–530
- in Paget's disease, 662
- pathophysiology of, 318–319, 318f
- significance of, 317–318

Vertebroplasty, 529–530, 764
VIP (vasoactive intestinal polypeptide), 566
VIPoma syndrome, 566
Vitamin D, 235–244, 403–406
- balance, impact on, 404
- bone mineral density (BMD), impact on, 403–404, 627
- bone remodeling, 176
- in calcium and phosphate metabolism, 238–239, 239f
- deficiency, 242–243, 624–629
  - bone histology, 616–617, 617f
  - bone mineral density effects, 627
  - calcium absorption, effect on, 627–628
  - causes, 625–626
  - classification of status, 615t
  - clinical findings, 616
  - definition, 614–615, 615f, 624, 625f
  - dentition abnormalities in, 915t, 918
  - epidemiology of, 615
  - falls and fractures, 617–618, 618f, 626–627
  - fetal response to, 183–184
  - frailty and, 628
  - hypocalcemia in children and, 652
  - hypocalcemia with, 574
  - impact beyond musculoskeletal system, 243
  - impact on musculoskeletal system, 243
  - laboratory findings, 616
  - management, 617, 628–629
  - nonskeletal effects, 627–628, 627f
  - nutritional rickets and osteomalacia, 613–614, 614f, 614t, 656
  - overview, 624–625
  - parathyroid hormone levels and, 627, 627f
  - pathogenesis, 615–616, 625, 625f
  - physical function and, 628
  - postburn, 532
  - prevalence of, 626
  - radiologic findings, 614f, 616, 617f
  - risk groups, 615
  - screening, 628
  - in spinal cord injury, 1021
  - sunlight for, 629
  - threshold, 614–615
  - treatment, 617, 628–629
- deficiency and insufficiency
  - in lactation, 162
  - in pregnancy, 159
- disorders related to, 613–620
  - hereditary vitamin D resistant rickets, 619–620
  - intoxication, 620
  - nutritional rickets and osteomalacia, 613–614, 614f, 614t
  - pseudovitamin D deficiency rickets, 619
- fall risk, impact on, 404, 617–618, 618f
- in fetal and neonatal bone development, 122
- fetal levels of, 183
- fortification of foods, 614
- fracture rates, impact on, 404, 626–627
- genetic testing for vitamin D-related disorders, 339
- during growth, 143
- hypercalcemia induced by, 565
- hypovitaminosis osteopathy (HVO)
  - histomorphometry findings in, 312
- immunobiology of, 241–242
  - adaptive immunity, 241–242
  - innate immunity, 241
- insufficiency (see also Vitamin D, deficiency)
  - defined, 624, 625f
  - management, 628–629
  - screening, 628
- intake
  - guidelines, 618–619, 618t
  - requirements, 405
  - statistics, 626, 626t
- internalization of metabolites, 237–238
- intestinal calcium transport, 175, 175t
- intoxication
  - with active metabolites, 620
  - in children, 653
  - clinical features, 620
  - endogenous, 620
  - hypercalcemia from, 567, 620, 653
  - pathogenesis, 620
  - treatment, 620
- keratinocyte function in epidermis and hair follicles, 242
- metabolism, 235–237, 236f
- molecular mechanism of action, 238
- muscle strength, impact on, 404
- nutritional considerations, 242–243
- in obesity, 996
- osteoclast regulation and, 29
- pharmacotherapy role of, 404–405
- production, 176, 235
- PTH secretion regulation by, 211
- receptor, 615–616
- renal calcium handling, 176
- safety, 405–406
- serum level, importance of, 624
- status assessment, 626
- structure of, 236f
- sufficiency, 242–243, 624, 625f

Vitamin D (contd.)
  supplementation, 244
    after organ transplantation, 497–498, 499t–500t, 501, 503f
    after spinal cord injury, 1021
    for deficiency/insufficiency, 628–629
    fall and fracture reduction with, 391–392
    guidelines for intake, 618–619, 618t
    for hypocalcemia in children, 653
    for nutritional rickets and osteomalacia, 617
    for pseudohypoparathyroidism, 595–596
    for pseudovitamin D deficiency rickets, 619
    safety of, 629
  target tissues, 239–241
    bone, 239
    intestine, 240
    kidney, 240
    pancreas, 241
    parathyroid glands, 240–241
  transport in blood, 237
  treatment strategies, 243–244
Vitamin D analogues, 645, 646–647
Vitamin D-dependent rickets type I (VDDR-I), 339, 619
Vitamin D-dependent rickets type II (VDDR-II), 339, 619–620
Vitronectin, 54t
Volume of interest, 65–66, 66f
Volumetric bone mineral density (vBMD)
  in children, 127–131
  in obesity, 994
Voxel size and image resolution, 63–64, 64f

Water
  in bone, 56
  hard, 857
Weight loss, as marker of frailty, 980
Weight reduction, effect on bone, 997
Weill-Marchesani syndrome, 42–43
Werner syndrome, 704
White adipose tissue, 987
Williams syndrome, 653
Wilson's disease, 655, 841
Wnt signaling, 4, 5f, 6–7, 7f, 10f, 17–19, 18f, 84
  in bone metastatic tumors, 689
  in cancer metastasis, 672
  in fracture healing, 93
  genome-wide association studies of Wnt-β-catenin signaling pathway, 380
  in high bone mass phenotype, 781
  in osteoblastic metastases, 688
  in osteosarcomagenesis, 705
  targets for future osteoporosis therapies, 463–464
    inhibition of dickkopf 1 (DKK-1) action, 463
    inhibition of GSK-3, 464
    inhibition of sclerostin, 464
    safety and specificity, 464
  targets for osteoporosis therapies, 463–464
  tooth formation and, 915
Women
  lactation (see Lactation)
  menopause (see Menopause)
  pregnancy (see Pregnancy)
  premenopausal osteoporosis, 514–518
  sex steroids and bone loss, 369–370
  urinary constituents, normal range of, 866
Wrist (distal forearm) fracture, osteoporotic
  incidence, 349f, 349t, 350
  morbidity, 353
  orthopedic surgery, 529
WTX gene, 777

X-linked hypophosphatemic rickets (XLH), 605–606
  in children, 654
  clinical and biochemical manifestations, 605–606
  fibroblast growth factor-23 (FGR23) association, 190, 654
  genetics, 605
  genetic testing for, 339
  treatment and complications, 605
X-ray diffractometry of kidney stones, 862, 864f

ZBTB40 gene, 381
Zoledronate (zoledronic acid)
  as adjuvant therapy, 748–749
  after organ transplantation, 498, 499t–500t, 502–503
  atypical subtrochanteric femoral fracture association, 716
  for BMD preservation in prostate cancer therapy, 714
  for metastatic bone disease, 762
    breast cancer, 742–743, 744f, 745, 746f
    multiple myeloma, 746f, 747–748
    prostate cancer, 746–747, 746f
  for multiple myeloma related bone disease, 697
  osteonecrosis link to, 935
  for osteoporosis, 413f, 415–416
    glucocorticoid-induced, 476, 477t
    in men, 511–512
  for Paget's disease, 664–665
  with PTH, 430–431
  in spinal cord injury patients, 1021–1022
Z-scores, 252–257, 253f–255f, 267